BIOCHEMISTRY

Contributing Authors

DANIEL E. ATKINSON
RAYMOND L. BLAKLEY
JAMES W. BODLEY
FREDERICK J. BOLLUM
RONALD BRESLOW
ANN BAKER BURGESS
RICHARD R. BURGESS
STANLEY N. COHEN
JAMES P. FERRIS
PERRY A. FREY
IRVING GEIS
ALFRED L. GOLDBERG
MAX GOTTESMAN
SUSAN GOTTESMAN
EDWARD E. HARRIS
LLOYD L. INGRAHAM
GARY R. JACOBSON
JULIUS MARMUR
RICHARD PALMITER
WILLIAM W. PARSON
R. C. PETERSON
MILTON H. SAIER, JR.
F. RAYMOND SALEMME
JACK STROMINGER
H. EDWIN UMBARGER
DAVID A. USHER
DENNIS E. VANCE
GEOFFREY ZUBAY

BIOCHEMISTRY

SECOND EDITION

Coordinating Author
GEOFFREY ZUBAY
COLUMBIA UNIVERSITY

MACMILLAN PUBLISHING COMPANY
New York

COLLIER MACMILLAN PUBLISHERS
London

Earlier edition copyright © 1983 by Addison-Wesley Publishing Company, Inc.

Portions of this book are reprinted with the permission of The Benjamin/Cummings Publishing Company, Inc., from their text titled *Genetics* by Geoffrey Zubay.

Macmillan Publishing Company
866 Third Avenue, New York, New York 10022

Collier Macmillan Canada, Inc.

Library of Congress Cataloging-in-Publication Data

Zubay, Geoffrey L.
  Biochemistry.

  Includes index.
  1. Biochemistry.   I. Title.
QP514.2.Z83   1988   574.19′2   87-28201
ISBN 0-02-432080-3

Printing: 1 2 3 4 5 6 7 8   Year: 8 9 0 1 2 3 4 5 6 7

This text is
dedicated to the memory of
FRITZ LIPMANN

PREFACE

To set forth in one volume the main reactions of biochemistry, as well as the principles that govern those reactions, is clearly a major undertaking. Biochemistry encompasses a great variety of molecules that are related, it often seems, only by the fact that they are all made for the purpose of creating and sustaining living forms. And, although biochemical reactions are subject to the familiar laws of physical chemistry and organic chemistry, they are also subject to certain new principles that relate specifically to biochemistry. These complications and principles are all addressed in this volume in as clear a manner as possible and in a form that is most suitable for classroom use.

OUR GENERAL STRATEGY

This is the second edition of *Biochemistry*; it has been updated and substantially revised. In the five years that separated the publication of the first and second editions we solicited a great number of reviews and received much unsolicited feedback. The collective response to the first edition convinced us that we were on the right track. There was general approval of the depth of coverage, the topics treated, and the multiauthor approach. But the feedback we received from users also convinced us that a more desirable text would result from a general reorganization of topics and from changes in emphasis in specific areas, as well as from a rewriting of numerous sections to improve the pedagogy. We have carefully considered all of the comments we received, and have made modifications that we felt would bring us closer to satisfying the needs of teachers and students of biochemistry. The result, we believe, is an exciting text that presents important biochemical information while emphasizing the basic principles.

Many challenges confront the writer of a biochemistry textbook. The first challenge is to accommodate users from different backgrounds. To understand the material, students of biochemistry must have a certain minimum amount of science background if the subject is to be presented in a serious way. Some training in general chemistry and organic chemistry is absolutely essential. Some understanding of biological principles is also helpful, because it provides the student with a better perspective. Nevertheless, we recognize that some chemistry majors take biochemistry with little or no biology background, and the text is written so that they will not be greatly handicapped. The introductory chapter that precedes Chapter 1 is the great equalizer in this text. It is intended to acquaint all students, regardless of their backgrounds, with the basic principles in chemistry and biology that relate to biochemistry.

Two challenges virtually unique to the writing of a biochemistry book are the enormous scope of the subject and the rapid rate at which new information is constantly being generated. The expertise required to generate a truly satisfactory text far exceeds that of one or a few individuals. By some means, the expertise of many biochemists working in different areas of biochemical research must be pooled into a collective effort. This text represents the combined efforts of twenty-eight scholars of biochemistry. The collective intellect of this group gives us overlapping strength in all major areas of biochemistry.

The efforts of this group have been carefully coordinated to assure a highly integrated text. Coordination has resulted from strong interactions between the contributors. Prior to writing, each contributor received a detailed description of the overall organization of the book. During the writing there has been ongoing communication between individual authors and many reviews of individual chapters and parts of the text by expert biochemists and teachers. Finally, and of the utmost importance, the coordinating author and a staff of expert writers and analysts have assessed all the reviews and considered the best way of presenting the material and adapting the illustrations to achieve maximum clarity. The contributing authors have responded to criticisms and agreed to modifications, even where they amounted to concessions, in the interest of the overall book.

CHANGES IN THE SECOND EDITION

Major changes made in the second edition include the following:

1. We have reorganized the material so that topics are presented in a more conventional order, making it easier to use in both one-term and two-term biochemistry courses.
2. We have done much rewriting and rephrasing to improve readability. As a further reading aid, we have added periodic summaries within the chapters and summaries at the ends of chapters.
3. We have altered many figures to improve their clarity and the integration of figures with text. In addition, many figure captions have been expanded.
4. We have made numerous additions and alterations to the text, both to bring it up to date and to make it more authoritative.

ORGANIZATION OF THE TEXT

The text is arranged so that the first half can be used in a one-term course or as the first term of a two-term course. The material is divided into six parts, and each part is divided into several chapters. An abbreviated table of contents and an expanded table of contents indicate the topics that are covered. Each part starts with a brief description of the topics that are covered in that part.

The first sixteen chapters would probably be a reasonable amount to cover in a one-term course. These chapters are tightly organized, and it is strongly recommended that they be studied in the order written. Once the knowledge of the first sixteen chapters has been mastered, the student should have a good basic understanding of the methods of biochemistry. This foundation makes it possible to do more picking and choosing in the second semester, according to the goals of the instructor.

The following is a brief description of the chapters.

Part I, comprising the first seven chapters, describes the structures and to a lesser extent the functions of all major components of the cell. We have adopted this approach in the second edition because we feel that students find it easier to cope with structure in one cluster of chapters before going into metabolism. When they get to the comparable sections on metabolism they can, when necessary, refer to the structure chapters that they have already studied.

Chapter 1 is concerned with properties of amino acids, peptides, and polypeptides. In Chapter 2, the three-dimensional structures of proteins are considered. Chapter 3 deals with methods for characterization and purification of proteins and includes a section describing the purification of two quite different proteins. This last section of Chapter 3 gives the student an appreciation of how different separation techniques are actually used in the laboratory.

Chapter 4 describes carbohydrates and their derivatives, including a section on glycoproteins that brings the discussion of carbohydrates up to date. Chapter 5 discusses the structures of lipids. This discussion was dispersed through several chapters in the first edition. Chapter 6 deals with the structure and assembly of biological membranes, and Chapter 7 deals with the structures of nucleotides and nucleic acids. When they have finished Part I, the students will be familiar with most of the molecular structures that they will encounter in the remainder of the text.

Part II is concerned with the problem of enzyme catalysis. It begins with Chapter 8, dealing exclusively with enzyme kinetics. Chapter 9 discusses the mechanisms of catalysis for different enzymes. Each enzyme that is discussed is considered from various points of view to provide a complete picture. Regulatory proteins are discussed separately in Chapter 10, beginning with a detailed account of the allosteric binding of oxygen by hemoglobin. Although hemoglobin is not an enzyme, the regulatory behavior displayed by hemoglobin in the binding of oxygen is a good model for teaching purposes. Finally, Chapter 11 concludes Part II with a discussion of the structure and function of coenzymes. The material on vitamin B_{12} has been considerably updated, and at the end of this chapter there is new material, on a completely unique set of coenzymes found only in archaebacteria.

Part III is concerned with catabolism. Carbohydrates, lipids, and amino acids are discussed under this heading. As a foundation, Chapter 12 deals with thermodynamic considerations that relate directly to metabolism. Other aspects of thermodynamics are discussed in subsections of chapters where they are most relevant. Chapter 13, a new chapter called "Overview of Metabolism," deals with the ways in which metabolism is organized and the ways in which it is regulated. This chapter is intended to give the student sufficient theoretical background to appreciate specific aspects of metabolism that are discussed later in the text.

Chapters 14 through 16 are devoted to carbohydrate metabolism in relation to the generation of ATP. This approach is designed to help the student appreciate the complex arrangement of the many reactions of carbohydrate metabolism from both an energetic and a functional point of view. Unlike the rest of Part III, Chapter 14 deals with anabolic as well as catabolic processes—namely, the synthesis of simple sugars and simple sugar polymers. This is because it is impossible to comprehend how the metabolism of sugars is regulated unless both their synthesis (anabolism) and their breakdown (catabolism) are discussed.

Chapter 17, which deals with photosynthesis, is placed immediately following the chapter on electron transport and the production of ATP (Chapter 16) because of the parallels between the electron transport process and also because of the involvement of both processes in energy-generating metabolism. Considerable attention has been devoted to explaining these complex light-catalyzed processes so that a minimum background in physics and chemistry is needed.

To conclude Part III, Chapter 18 deals with the catabolism of fatty acids and Chapter 19 with the catabolism of amino acids. In the present edition greater emphasis has been given to mammalian metabolism in the treatment of both of these topics.

In most schools a one-semester treatment of biochemistry will probably end with Chapter 19 or perhaps even as early as Chapter 16. As we indicated earlier, the first half of the text is tightly structured, leaving few options for alternative ways of presenting the material or skipping around. The same is not true of the second half of the text. The chapters in Parts IV, V, and VI can be pursued in virtually any order.

Part IV considers selected aspects of anabolism. Biosynthesis is actually the main subject of both Parts IV and Part V, but Part IV focuses on the biosynthesis of intermediary metabolites such as carbohydrates, lipids, amino acids, and nucleotides, whereas Part V is concerned with informational macromolecules such as nucleic acids and proteins.

Chapter 20 considers carbohydrates. The biosynthesis of simple sugars and simple sugar polymers was already discussed in Part III. For this reason Chapter 20 does not consider those topics but goes directly into more complex subjects, such as the synthesis of modified sugars of different types and the synthesis of complex oligosaccharides, polysaccharides, and glycoproteins. The discussion and the accompanying illustrations give perspective on the significance of the synthesis of glycoproteins in relationship to various parts of the cell.

Chapters 21, 22, and 23 deal with the metabolism of fatty acids and lipids. The material in these chapters has been revised to achieve greater clarity and to bring the discussion up to date.

Chapter 24 deals with amino acids and amino-acid-derived products. As with its sister chapter (Chapter 19), considerable modification has resulted in greater clarity, a more comprehensive treatment without increase in the length, and more emphasis on mammalian metabolism.

Part IV ends with a critical discussion of nucleotide metabolism in Chapter 25. This chapter has been substantially reorganized and updated, especially with respect to its treatment of nucleotide analogs as antimetabolites.

Part V, comprising eight chapters, deals with the metabolism of nucleic acids and proteins. In order to keep the second edition from getting too large, the separate chapters on bacteriophage λ and animal viruses have been eliminated from the second edition, and elements of these chapters have been incorporated in various chapters of Part V. For similar reasons, the two chapters on protein synthesis in the first edition have been combined into one chapter. On the other hand, the limited treatment on recombinant DNA methodology in the first edition has been expanded into a full chapter in the second edition. This change reflects the increasing importance of recombinant DNA techniques in both biochemistry and genetics. The chapter on the origins of life has been shortened, focusing on those aspects of the origins of life that deal with informational macromolecules. All of the chapters in this section have been modified to improve their clarity and update the material.

Part VI, which deals with membrane-associated reactions, contains three chapters: Chapter 34 on membrane transport, Chapter 35 on neurotransmission, and Chapter 36 on vision. Updating is most noticeable in Chapter 36 because of the tremendous progress that has been made in our understanding of the biochemistry of vision. The topics in Part VI could be studied at any time after studying Part I of this text.

TEACHING AIDS

The problems at the end of each chapter are an indispensable aid to understanding the material. Most of the problems were contributed by the authors. Drs. Jon Noble and John Oberdick reviewed all of the problems that were submitted. In many cases they modified those problems, and they devised new ones where additional problems seemed to be needed. Noble and Oberdick also reviewed the solutions if they were supplied and derived solutions if they were not. Answers to selected problems are found at the end of the text. A *Student's Solution Guide* containing worked out solutions for all problems is available from the publisher. It should be very useful to both instructors and students.

To facilitate classroom lecturing, a select set of approximately two hundred transparency masters is available. This set includes many of the illustrations that are too complex to copy on the blackboard during the course of a lecture.

Every attempt has been made to explain new terms when they are introduced in the text. Nevertheless, in such a large volume, it is not always easy to remember these definitions or where they have appeared. To aid in the understanding of new terms, a glossary has been added to the back matter.

APPRECIATION

My publisher, Macmillan, and my editor, Bob Rogers, have given me tremendous support in this undertaking. Bob Rogers consulted with the contributing authors and reviewers when his assistance was needed. He also was instrumental in recruiting the staff involved in the development and production of this book. Several members of the production staff at Macmillan deserve special mention: Dora Rizzuto, Production Supervisor; Laura Ierardi, Art and Design Director; and Ann Berlin, Associate Director of Manufacturing. Special acknowledgement is due to Emily Arulpragasam, who copyedited the manuscript. The term copy editor is totally inadequate to describe Emily's contribution to improving the phraseology and accuracy of this text. She has done a truly magnificent job.

Throughout the review process and the rewriting my constant advisor was James Funston. James's knowledge of biochemistry and good pedagogy were a major asset to this project. He played a critical role in almost every improvement in the second edition.

The molecular art presented in the text was greatly enhanced by many drawings from Irving Geis and the computer graphic art from Gary Quigley and Will Gilbert of the Massachusetts Institute of Technology. Jane Richardson generously contributed a large number of her unique ribbon drawings of protein structure.

G. Z.

LIST OF CONTRIBUTORS

DANIEL E. ATKINSON
CHAPTERS 13, 14, 15
Department of Chemistry and Biochemistry, University of California, Los Angeles, California 90024

RAYMOND L. BLAKLEY
CHAPTER 25
Division of Biochemical and Clinical Pharmacology, St. Jude Children's Research Hospital, Memphis, Tennessee 38101

JAMES W. BODLEY
CHAPTER 28
Biochemistry Department, University of Minnesota, Minneapolis, Minnesota 55455

FREDERICK J. BOLLUM
CHAPTER 26
Department of Biochemistry, Uniformed Service University of the Health Sciences, Bethesda, Maryland 20814-4799

RONALD BRESLOW
CHAPTERS 8, 9, 10
Department of Chemistry, Columbia University, New York, New York 10027

ANN BAKER BURGESS
CHAPTER 27
McArdle Laboratory for Cancer Research, University of Wisconsin, Madison, Wisconsin 53706

RICHARD R. BURGESS
CHAPTER 27
McArdle Laboratory for Cancer Research, University of Wisconsin, Madison, Wisconsin 53706

STANLEY N. COHEN
CHAPTER 32
Department of Genetics, Stanford University Medical Center, Stanford, California 94305

JAMES P. FERRIS
CHAPTER 33
Department of Chemistry, Rensselaer Polytechnic Institute, Troy, New York 12181

PERRY A. FREY
CHAPTER 11
Institute for Enzyme Research, University of Wisconsin, Madison, Wisconsin 53706

IRVING GEIS
CHAPTER 2
4700 Broadway, New York, New York 10040

ALFRED L. GOLDBERG
CHAPTER 28
Department of Physiology and Biophysics, Harvard Medical School, Boston, Massachusetts 02115

MAX GOTTESMAN
CHAPTER 29
Director, Institute for Cancer Research, Columbia University, New York, New York 10027

SUSAN GOTTESMAN
CHAPTER 29
Laboratory of Molecular Biology, National Cancer Institute, NIH, Bethesda, Maryland 20014

EDWARD E. HARRIS
CHAPTER 11
Department of Biochemistry and Biophysics, Texas A & M University, College Station, Texas 77843

LLOYD L. INGRAHAM
CHAPTER 12
Department of Biochemistry and Biophysics, University of California at Davis, Davis, California 95616

GARY R. JACOBSON
CHAPTERS 5, 6, 34, 35
Department of Biology, Boston University, Boston, Massachusetts 02215

JULIUS MARMUR
CHAPTERS 7, 26, 30
Department of Biochemistry, Albert Einstein College of Medicine, Bronx, New York 10461

RICHARD PALMITER
CHAPTER 31
Department of Biochemistry, University of Washington, Seattle, Washington 98195

WILLIAM W. PARSON
CHAPTERS 16, 17, 36
Department of Biochemistry, University of Washington, Seattle, Washington 98195

R. C. PETERSON
CHAPTER 26
Department of Biochemistry, Uniformed Service University of Health Sciences, Bethesda, Maryland 20014

MILTON H. SAIER, JR.
CHAPTERS 6, 34, 35
Department of Biology, University of California at San Diego, LaJolla, California 92093

F. RAYMOND SALEMME
CHAPTERS 1, 2, 3
Central Research and Development Department, Experimental Station, E.I. de Nemours & Co., Wilmington, Delaware 19898

JACK STROMINGER
CHAPTERS 4, 20
Department of Biochemistry and Molecular Biology, Harvard University, Cambridge, Massachusetts 02138

H. EDWIN UMBARGER
CHAPTERS 19, 24
Department of Biological Sciences, Purdue University, West Lafayette, Indiana 47907

DAVID A. USHER
CHAPTER 33
Department of Chemistry, Cornell University, Ithaca, New York 14853

DENNIS E. VANCE
CHAPTERS 5, 18, 21, 22, 23
Lipid and Microprotein Group, University of Alberta, Edmunton, Alberta, Canada T6G 2C2

GEOFFREY ZUBAY
CHAPTERS 1, 2, 3, 4, 7, 8, 9, 10, 19, 20, 24, 26, 29, 30, 32
Biosciences Department, Columbia University, New York, New York 10027

BRIEF CONTENTS

Introduction   The Essence of Living Things   1
GEOFFREY ZUBAY

=PART I=

MAJOR COMPONENTS OF THE CELL   29

Chapter 1   The Building Blocks of Proteins: Amino Acids, Peptides, and Polypeptides   31
GEOFFREY ZUBAY and F. RAYMOND SALEMME

Chapter 2   The Three-dimensional Structures of Proteins   53
GEOFFREY ZUBAY, F. RAYMOND SALEMME, and IRVING GEIS

Chapter 3   Methods for Characterization and Purification of Proteins   100
GEOFFREY ZUBAY and F. RAYMOND SALEMME

Chapter 4   Carbohydrates and Their Derivatives   131
GEOFFREY ZUBAY and JACK STROMINGER

Chapter 5   Lipids   154
DENNIS E. VANCE and GARY R. JACOBSON

Chapter 6   Structure and Assembly of Biological Membranes   176
GARY R. JACOBSON and MILTON H. SAIER, JR.

Chapter 7   Nucleotides and Nucleic Acids   211
GEOFFREY ZUBAY and JULIUS MARMUR

=PART II=

CATALYSIS   257

Chapter 8   Enzyme Kinetics   259
GEOFFREY ZUBAY and RONALD BRESLOW

Chapter 9   Mechanisms of Enzyme Catalysis   284
GEOFFREY ZUBAY and RONALD BRESLOW

Chapter 10   Regulatory Proteins and Enzymes   324
GEOFFREY ZUBAY and RONALD BRESLOW

Chapter 11   Structure and Function of Coenzymes   347
PERRY A. FREY (with assistance in section on metals by EDWARD E. HARRIS)

=PART III=

CATABOLISM   389

Chapter 12   Thermodynamics and Metabolism   392
LLOYD L. INGRAHAM

Chapter 13   Overview of Metabolism   415
DANIEL E. ATKINSON

Chapter 14   Glycolysis, Gluconeogenesis, and the Pentose Phosphate Pathway   436
DANIEL E. ATKINSON

Chapter 15   The Tricarboxylic Acid Cycle   481
DANIEL E. ATKINSON

Chapter 16   Electron Transport and the Production of ATP   512
WILLIAM W. PARSON

Chapter 17   Photosynthesis and Other Reactions Involving Light   564
WILLIAM W. PARSON

Chapter 18   Catabolism of Fatty Acids   598
DENNIS E. VANCE

Chapter 19   Catabolism of Amino Acids   615
H. EDWIN UMBARGER and GEOFFREY ZUBAY

=PART IV=

ANABOLISM   645

Chapter 20   Carbohydrates   647
GEOFFREY ZUBAY and JACK STROMINGER

xiv

Chapter 21   Fatty Acid Biosynthesis   669
DENNIS E. VANCE

Chapter 22   Complex Lipids   689
DENNIS E. VANCE

Chapter 23   Cholesterol and Related Derivatives   725
DENNIS E. VANCE

Chapter 24   Amino Acids and Amino-Acid-Derived Products   749
H. EDWIN UMBARGER and GEOFFREY ZUBAY

Chapter 25   Nucleotides   806
RAYMOND L. BLAKLEY

Chapter 29   Regulation of Gene Expression in Prokaryotes   974
GEOFFREY ZUBAY, MAX GOTTESMAN, and SUSAN GOTTESMAN

Chapter 30   Regulation of Gene Expression in Eukaryotes 1011
GEOFFREY ZUBAY and JULIUS MARMUR

Chapter 31   Hormone Action   1045
RICHARD PALMITER

Chapter 32   Recombinant DNA Methodology   1088
STANLEY N. COHEN and GEOFFREY ZUBAY

Chapter 33   Origins of Life   1120
JAMES P. FERRIS and DAVID A. USHER

= PART V =

NUCLEIC ACID AND PROTEIN METABOLISM   845

Chapter 26   DNA Metabolism   847
FREDERICK J. BOLLUM, R. C. PETERSON, JULIUS MARMUR and GEOFFREY ZUBAY

Chapter 27   RNA Metabolism   892
RICHARD R. BURGESS and ANN BAKER BURGESS

Chapter 28   Protein Metabolism   928
JAMES W. BODLEY (with assistance in sections on protein degradation by ALFRED L. GOLDBERG)

= PART VI =

MEMBRANE-ASSOCIATED REACTIONS   1153

Chapter 34   Membrane Transport   1155
GARY R. JACOBSON and MILTON H. SAIER, JR.

Chapter 35   The Nervous System: Transmission   1193
GARY R. JACOBSON and MILTON H. SAIER, JR.

Chapter 36   Vision   1213
WILLIAM W. PARSON

Glossary   1228

Answers to Selected Problems   1241

Index   1246

DETAILED CONTENTS

INTRODUCTION   THE ESSENCE OF LIVING THINGS
1

On Earth all living systems are biochemically related
        through a common evolution   1
The cell is the fundamental unit of life   3
The cell is composed of small molecules,
        macromolecules, and organelles   4
Biochemical reactions are a subset of ordinary chemical
        reactions   12
Secondary valence forces result in complex structures   17
Conditions for biochemical reactions differ from those for
        ordinary chemical reactions   18
Many biochemical reactions require energy   22
Organisms are biochemically dependent on one another
        24
Biochemical reactions are localized in the cell   25
Biochemical reactions are organized into pathways   25
Biochemical reactions are regulated in cells   26
Overview of the text   27

═ PART I ═

MAJOR COMPONENTS OF THE
CELL   29

CHAPTER 1   THE BUILDING BLOCKS OF PROTEINS:
AMINO ACIDS, PEPTIDES, AND POLYPEPTIDES   31

Amino acids   32
    Amino acids have both acid and base properties   34
    All amino acids except glycine show asymmetry   36
Peptides and polypeptides   38
        BOX 1–A   Structure of the peptide bond   39
Determination of amino acid composition of proteins   40
Determination of amino acid sequence of proteins   42
        BOX 1–B   The dansyl chloride method for N-terminal
        amino acid determination   46
Chemical synthesis of peptides and polypeptides   49

CHAPTER 2   THE THREE-DIMENSIONAL
STRUCTURES OF PROTEINS   53

The information for folding is contained in the primary
        structure   54
Forces that determine folding   56
    Water aids in determining protein conformation   56
        BOX 2–A   Visualizing molecular structures   60
    Secondary valence forces are the glue that holds
        polypeptide chains together   62
        Electrostatic Forces   62
        Van der Waals Forces   63
        Hydrogen Bond Forces   63
        Hydrophobic Forces   64
    The Ramachandran plot predicts sterically permissible
        conformations   65
Protein folding reveals a hierarchy of structural
        organization   68
    Three secondary structures are found in most proteins
        70
        The α helix   70
        β sheets   71
        β bends   71
    Relationships between amino acid sequence and
        secondary structure   74
    Fibrous proteins have repeating secondary structures
        75
    The secondary and tertiary structures of globular
        proteins are extremely varied   78
        Chiral properties influence conformation   79
        Effects of packing between secondary structures   83
        Domains are functional units of tertiary structure   86
    Quaternary structure involves subunit interaction   87
        Hemoglobin: a tetramer containing two different
        subunits   88
        Actin and myosin: a supermolecular aggregate   89
        Antibodies: tetramers containing two classes of
        subunits   90
Functional diversification of proteins   90
    Proteins with prosthetic groups   92

Evolutionary diversification of proteins 93
Molecular evolution and gene splicing 96
Protein diversification in action: antibodies 96

CHAPTER 3   METHODS FOR CHARACTERIZATION
AND PURIFICATION OF PROTEINS 100

Methods of protein characterization 100
Solubility reflects a balance of protein–solvent
interaction 101
A minimum in solubility occurs at the isolectric
point 101
Salting in and salting out 101
BOX 3–A *The meaning of ionic strength and its
effect on charge–charge interactions in proteins*
102
Several methods are available for determination of
gross size and shape 102
Osmotic pressure 102
Light scattering 103
Sedimentation analysis 104
Gel-exclusion chromatography 107
Electrophoretic methods are the best way to analyze
mixtures 108
Isoelectric focusing 109
Sodium dodecyl sulfate (SDS) gel electrophoresis
109
Radiation techniques probe fine structure and
conformation 111
UV absorption spectroscopy 111
Optical rotatory dispersion (ORD) and circular
dichroism (CD) 113
BOX 3–B *Polarized light and polarimetry* 113
X-ray diffraction and crystallography 115
Methods of protein purification 119
Differential centrifugation divides a sample into two
fractions 120
Differential precipitation is based on solubility
differences 120
Column procedures are the most versatile and
productive purification methods 121
Preparative gel-exclusion chromatography 121
Column chromatography with protein binding 121
High-performance liquid chromatography 122
Electrophoretic methods are used for preparation and
analysis 123
Purification of specific proteins involves combinations
of different procedures 123
The purification of UMP synthase from mammalian
tumor cells 123
The purification of lactose carrier protein from
Escherichia coli bacteria 126

CHAPTER 4   CARBOHYDRATES AND THEIR
DERIVATIVES 131

Monosaccharides and related compounds 131
Families of monosaccharides are structurally related
131
Monosaccharides can form intramolecular hemiacetals
134

Monosaccharides and the glycosidic bond 137
Disaccharides 138
Polysaccharides 138
BOX 4–A *Detection of sugars with aldehydic groups*
139
Cellulose is a homopolymer of glucose 140
Starch and glycogen are homopolymers of glucose 141
Configurations of glycogen and cellulose dictate their
roles 142
Chitin: a homopolymer with a different building block
144
Heteropolysaccharides: polymers with more than one
building block 144
Glycosaminoglycans contain repeating disaccharide
subunits 145
Glycosaminoglycans form proteoglycans 148
The bacterial cell wall: a peptidoglycan 149
O-antigen: a lipopolysaccharide 151
BOX 4–B *The antigenic behavior of bacterial
lipopolysaccharides* 152

CHAPTER 5   LIPIDS 154

Fatty acids 154
Fatty acids are stored as an energy reserve in
triacylglycerols 158
Phospholipids 160
Phosphoglycerides have a glycerol-3-phosphate
backbone 160
Phospholipids are amphipathic molecules 162
Sphingolipids contain a long-chain hydroxylated
secondary amine 163
Cholesterol and related steroids 166
Cholesterol is a derivative of a tetracyclic hydrocarbon
166
Steroid hormones are biosynthesized from cholesterol
168
Bile acids are the major degradation products of
cholesterol 168
Other biological lipids 170
Many vitamins are lipid-soluble 170
Terpenes are a diverse group made from isoprene
precursors 172
Polyketides are complex molecules derived from
acetate 174

CHAPTER 6   STRUCTURE AND ASSEMBLY OF
BIOLOGICAL MEMBRANES 176

Constituents of biological membranes 178
Membranes from eukaryotic cells have been isolated
and analyzed 178
Bacterial cell envelopes contain one or two membranes
180
Membrane lipids are a complex mixture 182
Membrane lipids spontaneously form ordered
structures 185
Membrane proteins are peripheral or integral 187
General properties and isolation 187
Arrangement of integral proteins within the
membrane 191

The structure of biological membranes 195
  Membrane structure is dynamic 196
  Temperature, composition, and extramembrane
    interactions affect dynamic properties 199
  Biological membranes are asymmetric 202
Biosynthesis and Assembly of Biological Membranes 205
  Topography and coordination of membrane assembly
    205
  The problem of membrane protein biogenesis 207

CHAPTER 7   NUCLEOTIDES AND NUCLEIC ACIDS
211

The genetic significance of nucleic acids 211
  Transformation is DNA-mediated 211
  Studies on viruses confirm the genetic nature of
    nucleic acid 213
Structural properties of DNA 216
  Nucleotides are the building blocks of nucleic acids
    217
  Nucleotides can form diphosphates and triphosphates
    221
  The polynucleotide chain contains mononucleotides
    linked by phosphodiester bonds 221
  Most DNAs form a double helix (duplex) structure 222
    BOX 7–A *X-ray diffraction of DNA* 224
    Hydrogen bonds and stacking forces stabilize the
      double helix 226
    Conformational variants of the double-helix structure
      226
  Duplex structures can be supercoiled 232
    BOX 7–B *Equilibrium density-gradient centrifugation*
      235
  DNA denaturation involves separation of the two
    strands 236
  DNA renaturation involves duplex formation from
    single strands 238
    Renaturation rate measures sequence complexity
      239
    BOX 7–C *Techniques using nucleic acid
      renaturation* 239
  Chemical methods are used to determine nucleotide
    sequence 242
    BOX 7–D *Applications of gel electrophoresis for
      chromosome analysis* 245
  Chemical synthesis of DNA 247
Conformational behavior of RNA 249
  Transfer RNA forms a cloverleaf structure 249
  Ribosomal RNA forms a complex folded structure 252
Nucleoproteins 252

=PART II=
CATALYSIS   257

CHAPTER 8   ENZYME KINETICS 259

Basic aspects of chemical kinetics 259
Influence of thermodynamics and kinetics on reaction
  rates 262

    BOX 8–A *Dependence of reaction rate on activation
      energy* 263
General properties of enzyme catalysts 264
Kinetics of enzyme-catalyzed reactions 266
  The Henri-Michaelis-Menten treatment assumes the
    enzyme-substrate complex is in equilibrium with
    free enzyme and substrate 267
  Steady-state kinetic analysis assumes a fixed level of
    enzyme-substrate complex 267
    Significance of the Michaelis constant, K_m 269
    Significance of the turnover number, k_{cat} 271
    Significance of the specificity constant, k_{cat}/K_m 271
Enzyme inhibition 272
  Reversible inhibitors frequently resemble substrate 272
    BOX 8–B *Derivation of the velocity expression when
      there is competitive inhibition* 274
  Irreversible inhibitors usually form covalent linkages
    275
    Examples of reversible and irreversible inhibition
      276
Effects of temperature on enzymatic activity 277
Effects of pH on enzymatic activity 277
Kinetic analysis for reactions involving two substrates
  278

CHAPTER 9   MECHANISMS OF ENZYME CATALYSIS
284

Enzyme catalysts and other chemical catalysts 284
Unique features of enzyme catalysts 286
The trypsin family of enzymes 287
The chymotrypsin catalytic mechanism 288
  Kinetic studies on peptide and ester substrates reveal
    different rate-limiting steps 292
  Steady-state kinetic analysis gives the K_m for both
    reactions 294
  Pre-steady-state measurements provide additional
    information on the kinetic parameters 295
    BOX 9–A *Description of stopped-flow apparatus*
      296
  Chymotrypsin activity is very sensitive to pH 297
Carboxypeptidase A: a zinc metalloenzyme 298
  Substrate specificity 298
  Structure of the enzyme-substrate complex 299
  Mechanism of the reaction 301
    BOX 9–B *Induced fit and multisubstrate enzyme
      specificity* 302
Pancreatic RNAse A: an example of concerted acid-base
  catalysis 303
  Chemical studies on the active site 304
  Crystal structure studies 306
  Mechanism of the reaction 307
Lysozyme 308
  Bound substrate is strained at the active site 309
  Mechanism of the reaction 313
Lactate dehydrogenase: a bisubstrate enzyme 315
  Binding of coenzyme occurs before binding of sugar
    316
  Kinetic studies reveal intermediates and slow step in
    the reaction 319
  Reaction results from concerted catalysis 319

Isoenzymes of the enzyme serve different functions 321

CHAPTER 10   REGULATORY PROTEINS AND ENZYMES   324

Hemoglobin: an allosteric oxygen-binding protein   325
  The binding of certain factors to hemoglobin influences the oxygen binding in a negative way   327
    BOX 10–A   Theory of multiple binding and the Hill plot   328
  X-ray diffraction studies indicate two conformations   329
  Changes in conformation are initiated by oxygen binding   332
Alternative theories on how hemoglobin and other allosteric proteins work   336
Aspartate carbamoyltransferase: an allosterically regulated enzyme   336
  Substrate binding to ACTase follows an ordered pathway   337
  Chemical and kinetic evidence demonstrates allosteric behavior   339
  X-ray diffraction studies confirm the allostery and reveal the groups at the active site   340
The advantages of positive cooperativity   343
Negative cooperativity   344

CHAPTER 11   STRUCTURE AND FUNCTION OF COENZYMES   347

Thiamine pyrophosphate and C—C and C—X bond cleavage   348
Pyridoxal-5′-phosphate and reactions involving α-amino acids   351
Nicotinamide coenzymes and reactions involving hydride transfers   356
Flavins and reactions involving one or two electron transfers   361
Phosphopantetheine coenzymes and reactions requiring active sulfhydryl groups   366
α-Lipoic acid and acyl group transfers   369
Biotin: a mediator of carboxylations   371
Folate coenzymes: an enzyme-dissociable agent for C_1 transfer   373
Vitamin B_{12} coenzymes and rearrangements on adjacent carbon atoms   377
Iron-containing coenzymes and redox reactions   381
Coenzymes involved in methanogenesis   383

=PART III=

CATABOLISM   389

CHAPTER 12   THERMODYNAMICS AND METABOLISM   392

Thermodynamic quantities   393
  Internal energy is conserved in a chemical reaction   393

  BOX 12–A   Some constants useful in thermodynamic calculations   395
  Internal energy and enthalpy are similar for most biochemical reactions   395
  Entropy always increases in a spontaneous reaction   396
  Free energy provides the most useful criterion for spontaneity   400
Applications of the free energy function   401
  Values of free energy are known for many compounds   401
  The standard free energy is directly related to the equilibrium constant   402
    BOX 12–B   Calculation of the equilibrium constant K_{eq} when the numbers of reactants and products are not equal   403
  Free energy is the maximum energy available for useful work   404
  Biological systems perform various kinds of work   405
  Unfavorable reactions can be driven by favorable reactions   405
    BOX 12–C   Free energy and electrical work   406
ATP, the main carrier of free energy   408
  Hydrolysis of ATP yields considerable free energy   408
    BOX 12–D   Factors affecting the equilibrium constant for ATP hydrolysis   410
Thermodynamics in open systems   411

CHAPTER 13   OVERVIEW OF METABOLISM   415

Organisms differ in sources of starting materials and energy   416
Reactions are organized into sequences   417
  Sequences are either anabolic or catabolic   418
    BOX 13–A   Functional interrelationships between pathways   422
  Enzymes are organized according to function   423
  Pathways are coupled to the ATP-ADP system   423
  Coupling in pathways assures conversions in both directions   424
Regulation of pathways   427
  Enzyme levels are regulated   427
  Enzyme activity is regulated by interaction with regulatory factors   428
  Regulatory enzymes occupy key positions in metabolic pathways   428
  Regulatory enzymes often show cooperative behavior   429
  Both anabolic and catabolic pathways are regulated by the energy charge   430
    BOX 13–B   Cooperative binding of substrate and the partitioning of substrate at branchpoints   431
  Regulation of pathways involves the interplay of kinetic and thermodynamic factors   432

CHAPTER 14   GLYCOLYSIS, GLUCONEOGENESIS, AND THE PENTOSE PHOSPHATE PATHWAY   436

Glycolysis   437
  Three hexoses play a central role in glycolysis   438

Phosphorylase converts storage carbohydrates to hexose phosphates 439

Hexokinase and glucokinase convert free sugars to hexose phosphates 440

Phosphoglucomutase interconverts glucose-1-phosphate and glucose-6-phosphate 441

Phosphohexoisomerase interconverts glucose-6-phosphate and fructose-6-phosphate 443

Formation of fructose-1,6-bisphosphate is a commitment to glycolysis 443

Fructose-1,6-bisphosphate and the two triose phosphates constitute the second metabolic pool in glycolysis 445

Aldolase cleaves fructose-1,6-bisphosphate 445

Triose phosphate isomerase interconverts the two trioses 446

The conversion of triose phosphates to phosphoglycerates occurs in two steps 447

The three-carbon phosphorylated acids constitute the third metabolic pool 448

Conversion of phosphoenolpyruvate to pyruvate generates ATP 449

Organization of glycolysis 452

Gluconeogenesis 452

Gluconeogenesis consumes ATP 453

Conversion of pyruvate to phosphoenolpyruvate requires ATP and GTP 455

Other routes from pyruvate to phosphoenolpyruvate 457

Conversion of phosphoenolpyruvate to fructose-1,6-bisphosphate uses the same enzymes as glycolysis 457

Fructose-bisphosphate phosphatase converts fructose-1,6-bisphosphate to fructose-6-phosphate 459

Conversion of hexose phosphates to polysaccharides 459

BOX 14–A   *Metabolite transfer via enzyme-enzyme complexes* 460

Other hexose derivatives 462

Excess fructose may cause metabolic complications 462

Abnormal individuals cannot metabolize galactose 463

Energetics and regulation of glycolysis and gluconeogenesis 464

Hormones and other factors regulate glycogen metabolism 467

The interconversion of fructose-6-phosphate and fructose-1,6-bisphosphate is regulated in both directions 470

Phosphofructokinase is regulated by energy charge and citrate 471

Fructose bisphosphate phosphatase is regulated by the AMP level 472

Fructose-2,6-bisphosphate regulates the flux between fructose-6-phosphate and fructose-1,6-bisphosphate 472

The pentose phosphate pathway 473

NADPH is generated by oxidation of glucose-6-phosphate 473

Transaldolase and transketolase play a variety of roles 474

Transaldolase 474

Transketolase 474

Production of ribose-5-phosphate and xylulose-5-phosphate 476

Regulation of the pentose phosphate pathway 478

CHAPTER 15   THE TRICARBOXYLIC ACID CYCLE 481

Overview of aerobic catabolism 481

Discovery of the TCA cycle 482

Steps in the tricarboxylic acid cycle 484

The oxidative decarboxylation of pyruvate leads to acetyl-CoA 485

The nature of the pyruvate dehydrogenase complex 488

Citrate synthase is the gateway to the TCA cycle 489

Aconitase catalyzes the isomerization of citrate to isocitrate 489

Isocitrate dehydrogenase catalyzes the first oxidation in the TCA cycle 490

α-Ketoglutarate dehydrogenase catalyzes the decarboxylation of α-ketoglutarate to succinyl-CoA 490

Succinate thiokinase couples the conversion of succinyl-CoA to succinate with the synthesis of GTP 490

Succinate dehydrogenase catalyzes the oxidation of succinate to fumarate 492

Fumarase catalyzes the addition of water to fumarate to form malate 492

Malate dehydrogenase catalyzes the oxidation of malate to oxaloacetate 493

Stereochemical aspects of TCA cycle reactions 493

BOX 15–A   *Potter's Demonstration of Prochirality* 495

ATP stoichiometry of the TCA cycle 495

Organization of the reactions of the TCA cycle 496

The amphibolic nature of the TCA cycle 496

BOX 15–B   *Generation of Biosynthetic Intermediates Under Anaerobic Conditions* 497

The glyoxylate cycle: an alternative fate for acetyl-coenzyme A 498

Oxidation of other substrates by the TCA cycle 501

Regulation of TCA cycle activity 502

Pyruvate dehydrogenase is regulated by small-molecule effectors and covalent modification 503

Citrate synthase is negatively regulated by NADH and the energy charge 503

Isocitrate dehydrogenase is very sensitive to the NADH/NAD$^+$ ratio 505

α-Ketoglutarate dehydrogenase is negatively regulated by NADH 508

The pyruvate branchpoint partitions pyruvate between acetyl-CoA and oxaloacetate 508

CHAPTER 16   ELECTRON TRANSPORT AND THE PRODUCTION OF ATP 512

Mitochondria: the site of aerobic ATP production 513

Electron transport 514

A chain of cytochromes transfers electrons to O_2 514
Other electron carriers work with the cytochromes 519
  Flavins 519
  Iron-sulfur centers 520
  Ubiquinone 520
Redox potentials are a measure of electron reduction
    power 521
  $E°$ values of electron carriers 525
The sequence of carriers can be determined with
    spectral analysis and inhibitors 526
  Difference spectra 526
  Inhibitors 528
Most of the electron carriers occur in large complexes
    529
  Complex I: the NADH dehydrogenase complex 530
  Complex II: the succinate dehydrogenase complex
    531
  Complex III: the cytochrome bc_1 complex 532
  The path of electrons through complex III: the Q
    cycle 532
  Complex IV: cytochrome oxidase 535
  Electron transfer from one complex to the next 535
Oxidative phosphorylation 537
  The P/O ratio measures the efficiency of coupling 538
  Inhibitors of phosphorylation act directly or as
    uncouplers 540
  Oxidative phosphorylation can be reversed 542
  The chemiosmotic theory proposes that
    phosphorylation is driven by a proton gradient
    543
  Electron transport creates an electrochemical potential
    gradient for protons 545
  How do the electron-transfer reactions drive protons
    across the membrane? 548
  The movement of protons back into the matrix drives
    the formation of ATP 549
  The proton-conducting ATP-synthase: F_1 and F_o 551
  Mechanism of action of the ATP-synthase 553
Transport of substrates, P_i, ADP, and ATP into and out of
    mitochondria 554
  BOX 16–A Natural uncouplers: thermogenesis in
    brown fat and skunk cabbage 555
  Uptake of P_i and oxidizable substrates is coupled to
    release of other compounds 555
  ATP export is coupled to ADP uptake 556
  Reducing equivalents from cytosplasmic NADH are
    imported by shuttle systems 557
  Complete oxidation of glucose yields 36 to 38 ATPs
    558
Electron transport and ATP synthesis in bacteria 559

CHAPTER 17   PHOTOSYNTHESIS AND OTHER
REACTIONS INVOLVING LIGHT   564

Photosynthesis 564
  The photochemical reactions of photosynthesis takes
    place in membranes 565
  Photosynthesis depends on the photochemical
    reactivity of chlorophyll 567
    A physical definition of light 569
    How light interacts with molecules 570

Light causes an electron-transfer reaction 571
  Photooxidation of chlorophyll generates a cationic free
    radical 572
  The reactive chlorophyll is in pigment-protein
    complexes called reaction centers 573
  In bacterial reaction centers, electrons move from P870
    to bacteriopheophytin and then to quinones 575
  A cyclic electron-transport chain drives the formation
    of ATP 576
  An antenna system transfers energy to the reaction
    centers 577
  Chloroplasts have two photosystems linked in series
    579
    BOX 17–A Carotenoids protect the cell against
    damage by O_2 580
    System I reduces NADP by way of iron-sulfur
    proteins 584
    O_2 evolution requires the accumulation of four
    oxidizing equivalents in system II 584
  The flow of electrons from H_2O to NADP is linked to
    proton transport 586
  Cyclic electron transport also occurs in chloroplasts
    587
  Carbon fixation results from the reductive pentose
    cycle 588
Photorespiration and the C-4 cycle 590
Other biochemical processes involving light 592
  The synchronization of circadian and seasonal rhythms
    in plants utilizes phytochrome 592
  Bioluminescence is biochemically produced light 593

CHAPTER 18   CATABOLISM OF FATTY ACIDS   598

Major sources of fatty acids for oxidation 598
Mobilization and transfer of fatty acids from adipose
    tissue 600
Acyl-CoA ligases: the enzymes that activate fatty acids
    602
Transport of fatty acids into the mitochondria 602
Oxidation of fatty acids 603
  The oxidation of fatty acids occurs in mitochondria
    604
  Unsaturated fatty acids are also oxidized in
    mitochondria 606
  Fatty acids with an odd number of carbons are
    oxidized to propionyl-CoA 607
  A defect in α oxidation causes Refsum's disease 608
  Fatty acids can also be oxidized at the methyl group
    610
  Ketone bodies are formed in the liver and used for
    energy in other tissues 610
  β oxidation can also occur in peroxisomes 610
  β oxidation in the heart is regulated by the supply of
    fatty acids and acetyl-CoA 612

CHAPTER 19   CATABOLISM OF AMINO ACIDS   615

Protein degradation 615
Nitrogen removal from amino acids 616
  Transamination is the most widespread form of
    nitrogen removal 616

Oxidative deamination is required for net deamination 617

The fate of nitrogen derived from amino acid breakdown 618

Urea formation in the animal liver detoxifies NH$_3$ 619

The urea cycle and the TCA cycle are linked 621

Different carriers transport ammonia to the liver 621

Amino acids as a source of carbon and energy 622

Five amino acids degrade to acetyl-CoA by way of pyruvate 623

Alanine 623

Threonine 623

Glycine and serine 624

Cysteine 625

Five amino acids degrade to acetyl-CoA by way of acetoacetyl-CoA 625

Lysine and leucine 625

Tyrosine 625

Phenylalanine 628

Tryptophan 631

Five amino acids degrade to α-ketoglatarate 631

Glutamine and glutamate 631

Proline 631

Arginine 632

Histidine 634

Catabolism of methionine, valine, and isoleucine leads to succinyl-CoA 635

Methionine 637

Aspartate and asparagine are deaminated to oxaloacetate 639

Inborn errors in catabolism of amino acids in the human 639

Regulation of amino acid breakdown 640

PART IV

ANABOLISM   645

CHAPTER 20   CARBOHYDRATES   647

Biosynthesis of hexoses 647

The hexose monophosphate pool includes glucose, fructose, and mannose 647

Galactose is not a member of the pool 649

Hexose modifications involve alterations or additions 650

Disaccharide synthesis 652

Polysaccharide synthesis 653

Energy-storage polysaccharides are simple homopolymers 653

Structural polysaccharides include homopolymers and heteropolymers 655

Oligosaccharide synthesis in higher animals 655

All glycoproteins are synthesized in the endoplasmic reticulum 656

Biosynthesis of N-linked oligosaccharides 658

Biosynthesis of O-linked oligosaccharides 662

Bacterial cell wall synthesis 663

Synthesis of the UDP-N-acetylmuramyl-pentapeptide monomer occurs in the cytoplasm 664

Formation of linear polymers of the peptidoglycan is membrane-associated 665

Cross-linking of the peptidoglycan strands occurs on the noncytoplasmic side of the plasma membrane 666

Penicillin inhibits the transpeptidation reaction 667

CHAPTER 21   FATTY ACID BIOSYNTHESIS 669

Biosynthesis of saturated fatty acids 669

Acetyl-CoA and NADPH are the main substrates for fatty acid synthesis 670

Fatty acid synthesis in *E. coli* and plants is catalyzed by separate enzymes 673

The enzymes for fatty acid synthesis in animals are organized in a complex 674

Acetyl-CoA carboxylase consists of three protein components 674

BOX 21–A *Diversity in the structure of fatty acid synthases* 675

Seven reactions are catalyzed by the fatty acid synthase 677

Binding of acetyl and malonyl-CoA 677

The condensation reaction 677

BOX 21–B *Specificity in electron transfer* 679

The reduction reactions 679

Continuation reactions 679

BOX 21–C *Assaying the activity of fatty acid synthase* 680

Biosynthesis of monounsaturated fatty acids 680

Biosynthesis of polyunsaturated fatty acids 682

Regulation of fatty acid metabolism 683

Acetyl-CoA carboxylase activity and the availability of acetyl-CoA are the controlling factors 683

Malonyl-CoA and hormones inhibit simultaneous synthesis and breakdown of fatty acids 686

Chain length and ratio of unsaturated to saturated fatty acids are regulated in *E. coli* 686

CHAPTER 22   COMPLEX LIPIDS 689

Glycerolipids 689

Phospholipids are amphipathic substances 689

Phospholipid synthesis in *E. coli* leads to phosphatidylethanolamine, phosphatidylglycerol, and cardiolipin 690

BOX 22–A *Mechanism of action of acyltransferases* 691

Phospholipid synthesis in *E. coli* is regulated at an early stage in fatty acid synthesis 691

Phospholipid synthesis in eukaryotes is more complex 694

Biosynthesis of phosphatidylcholine and phosphatidylethanolamine 696

BOX 22–B *The mechanism of methylation by S-adenosylmethionine* 698

Biosynthesis of phosphatidylserine, phosphatidylglycerol, and the inositol phospholipids 698

Biosynthesis of alkyl and alkenyl ethers 701

BOX 22–C *The phosphatidylinositol cycle* 702

Biosynthesis of triacylglycerols 704
Regulation of glycerolipid synthesis in the liver favors structural lipids over energy-storage lipids 704
Mechanisms for final distribution of lipids between membranes are still unclear 706
Phospholipases degrade phospholipids 707
Sphingolipids 709
Sphingolipids are synthesized on membranes 709
Defects in sphingolipid catabolism are associated with metabolic diseases 712
Glycosphingolipids function as structural components and as specific cell receptors 714
The eicosanoids: prostaglandins, thromboxanes, leukotrienes, and hydroxy-eicosaenoic acids 715
BOX 22–D Analysis of prostaglandins and thromboxanes 716
All eicosanoids are related to C_{20} polyunsaturated acids 717
Most eicosanoids are synthesized from arachidonic acid 718
Eicosanoids are hormones that exert their action locally 720

CHAPTER 23 CHOLESTEROL AND RELATED DERIVATIVES 725

Biosynthesis of cholesterol 725
Mevalonate is a key intermediate in cholesterol biosynthesis 726
The rate of mevalonate synthesis determines the rate of cholesterol biosynthesis 727
It takes six mevalonates and ten steps to make lanosterol, the first tetracyclic intermediate 729
BOX 23–A Stereochemistry of the conversion of mevalonate to squalene 731
From lanosterol to cholesterol takes another twenty steps 732
Lipoprotein metabolism 733
There are five types of lipoproteins in the human plasma 733
Lipoproteins are made in the endoplasmic reticulum of the liver and intestine 735
Chylomicrons and very-low-density lipoproteins (VLDL) transport cholesterol and triacylglycerol to other tissues 736
Low-density lipoproteins (LDL) are removed from the plasma by the liver, adrenals, and adipose tissue 737
Serious diseases result from cholesterol deposits 739
High-density lipoproteins (HDL) may reduce cholesterol deposits 740
Bile acid metabolism 741
Metabolism of steroid hormones 741
Overview of mammalian cholesterol metabolism 745

CHAPTER 24 AMINO ACIDS AND AMINO-ACID-DERIVED PRODUCTS 749

Amino acids and nutritional needs 750
Sources of nitrogen for amino acid biosynthesis: the nitrogen cycle 752

Nitrogen fixation involves an enzyme complex called nitrogenase 753
Nitrate and nitrite reduction play two physiological roles 755
The advantages of microorganisms for the study of amino acid biosynthesis 755
The biosynthesis of amino acids of the glutamate family: L-glutamate, L-glutamine, L-proline, L-lysine, and L-arginine 756
The direct amination of α-ketoglutarate leads to glutamate 757
Amidation of glutamate to glutamine is a highly regulated process 759
The amination of α-ketoglutarate by the amide group of L-glutamine also leads to glutamate 760
Three enzymes convert glutamate to proline 761
Arginine biosynthesis uses some reactions seen in urea cycle 761
Only in fungi does lysine biosynthesis start from α-ketoglutarate 761
The biosynthesis of amino acids of the serine family (L-serine, glycine, and L-cysteine) and the fixation of sulfur 764
Three enzymes convert 3-phospho-D-glycerate (glycerate-3-phosphate) to serine 764
Two more enzymes convert L-serine to glycine 765
Cysteine biosynthesis involves sulfhydryl transfer to activated serine 765
The biosynthesis of amino acids of the aspartate family: L-aspartate, L-asparagine, L-methionine, L-threonine, and L-isoleucine 768
Aspartate is formed from oxaloacetate in a transamination reaction 768
Asparagine biosynthesis requires ATP and glutamine 768
L-Aspartic-β-semialdehyde is a common intermediate in L-lysine, L-methionine, and L-threonine synthesis 769
The pathway to L-lysine in bacteria and plants is different from that in fungi 769
Methionine is both an end product and a precursor of S-adenosylmethionine 771
Two enzymes convert 2-homoserine into threonine 773
Isoleucine biosynthesis and valine biosynthesis are related 773
The carbon flow in the aspartate family is regulated at the aspartokinase step 773
The biosynthesis of amino acids of the pyruvate family: L-alanine, L-valine, and L-leucine 775
L-Alanine is formed from pyruvate in a transamination reaction 775
Isoleucine biosynthesis and valine biosynthesis share four enzymes 776
L-Leucine is formed from α-keto-isovalerate in four steps 779
The biosynthesis of the aromatic family of amino acids: L-trytophan, L-phenylalanine, and L-tyrosine 779
Chorismate is a key intermediate 779
Prephenate is a common intermediate in phenylalanine and tyrosine synthesis 779

Tryptophan is synthesized in five steps from chorismate   780

Carbon flow in the biosynthesis of aromatic amino acids is regulated at branchpoints   785

The biosynthesis of histidine   785

Additional aspects of the regulation of amino acid biosynthesis   788

End-product inhibition takes on various forms   788

Sometimes enzymes are activated   790

Protein–protein interactions can regulate enzyme activity   791

Nonprotein amino acids   792

A wide variety of D-amino acids are found in microbes   792

Racemases are often involved in D-amino acid formation   792

There are hundreds of natural amino acid analogs   793

The conversion of amino acids to other amino acids and to other metabolites   793

Porphyrin biosynthesis starts with the condensation of glycine and succinyl-CoA   793

BOX 24–A   *Mechanism of polymerization in linear tetrapyrrole*   796

The conversion of amino acids to biologically active amines entails decarboxylation   796

Glutathionine is γ-glutamylcysteinylglycine   797

Gramicidin is a cyclic decapeptide synthesized on a protein template   800

Phosphocreatine is an important energy reservoir in skeletal muscle   801

CHAPTER 25   NUCLEOTIDES   806

Synthesis of purine ribonucleotides *de novo*   808

Inosine monophosphate (IMP) is the first purine nucleotide formed   809

IMP is converted into AMP and GMP   811

Synthesis of pyrimidine ribonucleotides *de novo*   811

UMP is a precursor of all pyrimidine nucleotides   812

CTP is formed from UTP   814

Biosynthesis of deoxyribonucleotides   814

Thymidylate is formed from dUMP   816

Formation of nucleoside monophosphates from bases (salvage pathways)   817

Purine phosphoribosyltransferases convert purines to nucleotides   818

Conversion of pyrimidines to mononucleotides goes through nucleoside intermediates   820

Conversion of nucleoside monophosphates to triphosphates goes through diphosphates   820

Nucleoside monophosphate kinases   821

Nucleoside diphosphate kinases   821

Inhibitors of nucleotide synthesis   821

Catabolism of nucleotides   828

Intracellular catabolism of nucleotides is highly regulated   829

Purines are catabolized to uric acid and then to other products   829

Pyrimidines are catabolized to β-alanine, NH_3, and CO_2 831

Regulation of nucleotide metabolism   832

Purine biosynthesis is regulated at two levels   833

Pyrimidine biosynthesis is regulated at the level of carbamoyl aspartate formation   834

Ribonucleotide reduction is regulated by both activators and inhibitors   834

Metabolites are channeled along the nucleotide biosynthesis pathways   835

Intracellular nucleotide pools vary with the stage in the cell cycle   836

T4 bacteriophage infection stimulates the nucleotide metabolism   838

Biosynthesis of nucleotide coenzymes   838

= PART V =

NUCLEIC ACID AND PROTEIN METABOLISM   845

CHAPTER 26   DNA METABOLISM   847

DNA replicates semiconservatively   847

DNA synthesis in prokaryotes   851

Bacterial DNA replication is bidirectional   851

Growth is discontinuous on at least one strand   854

Many proteins are required for DNA replication   856

A polymerase with the expected characteristics is discovered   859

E. coli has three DNA polymerases   861

Polynucleotide ligase links chains together   862

Topoisomerases catalyze supercoiling   863

Helicases catalyze DNA unwinding   864

Single-strand binding protein stabilizes single strands   864

Replication of a viral chromosome   865

Replication of the E. coli chromosome   868

DNA synthesis and chromosomal replication in eukaryotes   869

Eukaryotic cells have several DNA polymerases   870

Eukaryotic chromosomes contain multiple origins of replication   870

Mitochondrial DNA replicates continuously on both strands   872

Some polymerases function without a DNA template   873

Control of DNA replication   874

Specific inhibitors of DNA replication   875

Some inhibitors interact directly with the DNA   875

Some inhibitors interact with the enzymes involved in DNA synthesis   876

BOX 26–A   *The Sanger method for sequencing DNA*   877

DNA degradation   878

Nucleases attack phosphodiester bonds   878

Glycohydrolases attack *N*-glycosidic bonds   881

Postreplicative modification protects DNA from certain nucleases   881

DNA repair   882

DNA recombination   886

CHAPTER 27   RNA METABOLISM   892

Different classes of RNA   893
  Messenger RNA carries the information for polypeptide
    synthesis   893
  Ribosomal RNA is an integral part of the ribosome
    894
  Transfer RNA carries amino acids to the ribosome   894
  Other cellular RNAs are involved in DNA synthesis
    and processing   895
DNA-dependent synthesis of RNA   895
  RNA polymerase is composed of several protein
    subunits   896
    Subunit structure and function   896
    Genetics of the subunits   898
  Transcription involves initiation, elongation, and
    termination   899
    Transcription units and promoter signals   900
    Binding at promoters   901
    Initiation at promoters   903
    Elongation of RNA   904
    Termination of transcription   904
    Other factors regulating transcription   905
Eukaryotic transcription   905
  There are three nuclear RNA polymerases   906
  Some organelles have their own RNA polymerases   907
  Important differences exist between eukaryotic and
    prokaryotic transcription   908
Other RNA synthesis   908
  DNA primase makes primer RNA for DNA synthesis
    908
  Many viruses make their own RNA polymerases   909
    RNA-dependent RNA polymerases of RNA viruses
      910
  3′-end addition enzymes add nucleotides to tRNA   910
    BOX 27–A   *Synthesis of viral DNA from the viral
      RNA genome*   911
  Polynucleotide phosphorylase makes polynucleotides
    from ribodiphosphates   912
  RNA ligase of bacteriophage T4 links RNAs together
    912
Posttranscriptional modification and processing of RNA
  913
  Processing and modification of tRNA requires several
    enzymes   913
  Processing of ribosomal RNA precursor leads to three
    RNAs   916
  Modification and processing of messenger RNA is of
    major importance in eukaryotes   916
    Capping the 5′ end   916
    Polyadenylation of the 3′ end   918
    Removal of intervening sequences   918
  Some RNAs are self-splicing   921
Degradation of RNA by ribonucleases   923
Inhibitors of RNA metabolism   924
  Some inhibitors act by binding to DNA   924
  Some inhibitors of transcription bind to DNA gyrase
    924
  Some inhibitors bind to RNA polymerase   924

CHAPTER 28   PROTEIN METABOLISM   928

The cellular machinery of protein synthesis   928
  Transfer RNA transports amino acids to the template
    929
  Ribosomes are the site of protein synthesis   930
    BOX 28–A   *Ribosome structure and assembly*   932
  Messenger RNA is the template for protein synthesis
    934
The steps in translation   937
  Amino acids are activated before being linked to
    transfer RNA   937
    BOX 28–B   *Specificity considerations in aminoacyl-
      tRNA formation*   939
  Polypeptide synthesis is initiated on the ribosome   939
  Elongation reactions involve peptide bond formation
    and translocation   941
  Termination of translation requires special termination
    codons   942
    BOX 28–C   *The puromycin reaction*   943
    BOX 28–D   *An in vitro assay for release factors*   946
    BOX 28–E   *The role of GTP in ribosomal reactions*
      946
Deciphering the genetic code   947
  Synthetic polynucleotides can serve as templates for
    polypeptide synthesis   948
  Nucleotide triplets stimulate ribosome binding of
    aminoacyl-tRNA   949
  Repeating-sequence synthetic polynucleotides are
    mRNAs for simple repeating polypeptides   950
  Start and stop triplets were identified by *in vitro*
    studies   951
Code word assignments   952
  The 5′ base in the anticodon can frequently pair with
    more than one base in the codon   952
  The genetic code is not completely universal   953
    BOX 28–F   *Dimensions of wobble pairs*   954
Inhibitors of protein synthesis   956
  The tetracyclines prevent anticodon binding   957
  Streptomycin leads to mistranslation   957
  Chloramphenicol inhibits peptidyl transferase   957
  Diphtheria toxin catalyzes the ADP-ribosylation of
    elongation factor EF-2   958
  Specific nucleases can inactivate ribosomes   959
Posttranslational modification of proteins   959
  Amino acids in proteins sometimes undergo covalent
    modification   959
  Proteolytic processing of proteins is a maturation
    reaction for some proteins   961
    BOX 28–G   *Observations relating to the signal
      hypothesis*   962
Intracellular protein degradation   963
  The rate of breakdown of a protein helps determine its
    intracellular level   964
  Abnormal proteins are selectively degraded   965
  The soluble pathway for protein breakdown requires
    ATP   966
  The ATP-dependent pathway in eukaryotic cells
    requires ubiquitin   968

Many proteases and acid hydrolases are localized in the lysosome 969

Lysosomes function in degradation of both endocytosed and cellular proteins 969

Overall protein breakdown in microorganisms is regulated 970

CHAPTER 29 REGULATION OF GENE EXPRESSION IN PROKARYOTES 974

Regulation of gene expression in *E. coli* 974

Initiation point is a major site for regulating gene expression 975

The *lac* operon is a cluster of three genes regulated as a unit 976

Enzyme induction 977

Discovery of the repressor gene 979

BOX 29–A *Genetic concepts and genetic notation* 979

Discovery of operator mutations 980

The operon hypothesis 981

Isolation and action of the *i* gene repressor 982

BOX 29–B *Measuring the binding of inducer to repressor* 982

Catabolite repression and the catabolite activator protein CAP 984

Gene regulatory proteins usually bind first to the DNA 986

Many genes important in catabolism are stimulated by CAP 987

BOX 29–C *Interaction between DNA and specific regulatory binding proteins* 988

Amino acid biosynthesis is regulated at the level of transcription 990

The attenuator: a means for controlling transcription after initiation 991

Genes for RNA polymerase and ribosomes are coordinately regulated 995

Control of rRNA and tRNA synthesis by the *rel* gene 995

Translational control of ribosomal protein synthesis 997

Regulation of RNA polymerase synthesis is coordinated with ribosomal protein synthesis 998

Overview of gene regulation in *E. coli* 999

Regulation of gene expression in bacterial viruses 1000

Bacteriophage λ can develop along two different pathways 1000

The establishment of lysogeny requires rapid synthesis of cI protein 1004

The maintenance of the lysogenic state is favored by cI synthesis 1004

The lytic cycle is favored by destruction of the cI repressor 1004

BOX 29–D *Control of the λ immunity regions by cI and cro repressor proteins* 1006

CHAPTER 30 REGULATION OF GENE EXPRESSION IN EUKARYOTES 1011

Gene regulation in unicellular eukaryotes 1011

Galactose metabolism exemplifies catabolite gene regulation in yeast 1012

Specific controls affect expression of the *GAL* gene 1013

Catabolite gene products are regulated by repression and inactivation 1014

Positive and negative controls of transcription are common in yeast 1015

Mating type is determined by transposable elements 1016

Gene regulation in multicellular eukaryotes 1018

Nuclear differentiation occurs during early development 1019

Chromosome structure varies with gene activity 1020

Giant chromosomes permit direct visualization of active genes 1020

In some cases entire chromosomes are inactive 1022

Histones are the main proteins found in chromatin 1023

Histones are often modified in active chromatin 1025

Active chromatin is most susceptible to DNAse degradation 1026

Active chromatin is associated with high-mobility group (HMG) proteins 1026

Enhancer sequences are general stimulators of transcription 1027

Regulatory phenomena associated with development 1029

During embryonic development in the amphibian, specific gene products are required in large amounts 1029

Ribosomal RNA in frog eggs is elevated by DNA amplification 1029

5S RNA synthesis in frog eggs requires regulatory protein 1030

Demands for histones are met by gene amplification and storage 1033

Specific genes regulate development in fruit flies 1033

Maternal effect genes 1035

Segmentation genes 1035

Homeotic genes 1037

Gene changes and cell amplification lead to specific antibody synthesis in vertebrates 1039

B cells cooperate with T cells to make antibodies 1039

DNA splicing has been demonstrated in antibody-forming cells 1040

CHAPTER 31 HORMONE ACTION 1045

Hormone synthesis 1046

Polypeptide hormones are synthesized from larger precursors 1046

Thyroid hormones and epinephrine are amino acid derivatives 1051

Steroid hormones are derivatives of cholesterol 1053

Regulation of circulating hormone concentration 1057

Hormone action is mediated by receptors 1058

Hormone binding to receptors changes their functional activity 1059

Hormone receptors are detected by their ability to bind hormone 1060

Most membrane receptors generate a diffusible intracellular signal 1062

The adenylate cyclase pathway 1062

The guanylate cyclase pathway 1066

Calcium and the inositol triphosphate pathway 1066

Steroid receptors modulate the rate of gene transcription 1068

Regulation of hormonal responses 1073

Endocrinopathies 1077

Overproduction of hormones is commonly caused by tumor formation 1078

Underproduction of hormones has multiple causes 1078

Target-cell insensitivity results from a lack of functional receptors 1079

Growth factors 1079

Interferon 1080

Plant hormones 1082

CHAPTER 32   RECOMBINANT DNA METHODOLOGY 1088

Historical perspectives and basic methodology 1089

Separate DNA fragments can be joined *in vitro* 1089

Addition of complementary homopolymeric termini to DNA 1091

Restriction enzymes cleave DNA to fragments ideal for cloning 1092

Small plasmids were the first vectors to be used for cloning 1094

The initial DNA cloning experiments involved antibiotic resistance genes 1095

DNA cloning materials and procedures in E. coli 1098

The choice of cloning vector depends on the research objective 1098

Plasmids 1098

Bacteriophage λ vectors 1099

Cosmids 1100

Bacteriophage M13 vectors 1101

Shuttle vectors 1101

The ideal clone bank (library) contains all the sequences with a minimum of repeats 1101

BOX 32–A *Construction of a cosmid* 1102

Preparation of a genomic DNA library 1103

Preparation of a cDNA library 1103

There are several approaches for picking the correct clone from a library 1107

Cloning in other systems 1109

Yeast is the most popular eukaryotic cell for cloning 1110

Cloning in mammalian cells employs tissue culture cells 1111

Introducing DNA into mammalian cells 1111

Selection of genetic markers introduced into mammalian cells 1111

Vectors used for cloning in mammalian cells 1114

Cloning in plants has been accomplished with a bacterial plasmid 1115

Uses of cloned DNA 1115

BOX 32–B *Construction of a T1 plasmid derivative suitable for cloning* 1116

DNA sequence analysis 1117

Site-directed mutagenesis 1117

Gene mapping 1117

Studies of gene transcription *in vivo* and *in vitro* 1117

Expression vectors for making large amounts of proteins 1117

Possible medical applications 1117

CHAPTER 33   ORIGINS OF LIFE 1120

What is life? 1121

Underlying assumptions about the origins of life 1121

Possible stages in chemical evolution 1122

Formation of our galaxy and solar system 1122

Cosmochemistry 1123

The atmosphere of the primitive Earth 1125

Ancient forms of life 1126

Synthesis of biomonomers 1127

Amino acids are formed in experiments using electric discharges 1127

Amino acids are formed by hydrolysis of HCN polymers 1128

Amino acids are present in carbonaceous meteorites 1128

Nucleic acid bases can be made from hydrogen cyanide and cyanoacetylene 1129

Purines 1129

Pyrimidines 1130

Ribose may have been formed from formaldehyde 1131

Nucleosides and nucleotides can be made by dry-phase heating of precursors 1131

Condensing agents may have driven prebiotic reactions 1132

Cyanamide and related structures 1132

Cyanogen, cyanoformamide, and cyanate 1133

Diaminomaleonitrile 1134

Trimetaphosphate and polyphosphates 1134

Polymerization of biomonomers 1134

Formation of polynucleotides by nontemplate reactions requires dry heat or condensing agents 1136

Polynucleotide templates favor polynucleotide synthesis 1138

Stability favors the 3′,5′ linkage over the 2′,5′ linkage 1140

RNA probably preceded DNA 1141

Formation of polypeptides 1142

Coevolution of polynucleotides and polypeptides—the origin of translation 1143

Origin of the genetic code 1144

Translation 1145

Peptide bond formation 1146
Selection of α-amino acids in a template reaction
1148
Membranes and compartmentation 1148

$=$ PART VI $=$

MEMBRANE-ASSOCIATED REACTIONS
1153

CHAPTER 34 MEMBRANE TRANSPORT 1155

The theory and thermodynamics of biological transport
1156
Most solutes are transported by specific carriers 1156
Transport against a concentration gradient requires
energy 1158
Active transport and group translocation are used to
concentrate solutes 1159
Energy-coupling mechanisms in biological transport
1160
Isotope analogs and specially prepared vesicles are
used to study transport 1160
Examples of facilitated diffusion 1163
The NA^+-K^+ ATPase is an active transport system
1164
Other ion-translocating ATPases exist for H^+, Na^+, K^+,
and Ca^{2+} 1165
Some transport systems are driven by electron transport
or light 1168
Some bacterial systems use specific binding proteins
1169
Secondary active transport can result when the uptake
of one solute is coupled to the uptake of another
solute 1171
The mitochondrial ATP/ADP exchanger expels ATP
because of a membrane potential 1174
The bacterial PEP:sugar phosphotransferase system
exemplifies group translocation 1175
Energy interconversion and active transport in bacteria
are regulated by growth conditions 1177
Lactose permease function is regulated by glucose
1178
Molecular mechanisms of biological transport 1179
Mobile carriers or pores could explain transport 1179
Most biological transport systems use pores 1181
Examples of molecular mechanisms in membrane
transport 1184
Reconstitution of purified transport proteins 1187

CHAPTER 35 THE NERVOUS SYSTEM:
NEUROTRANSMISSION 1193

Nerve-impulse propagation 1194
An unequal distribution of ionic species results in a
resting transmembrane potential 1194
An action potential is the change in membrane
potential occurring during nerve stimulation 1195
Nerve cell membranes have separate channels for K^+
and Na^+ 1198
Gated pores regulate Na^+ ion transmembrane flux .
1201
The synapse is a chemical connection for
communication between nerve cell and target cell
1203
A number of compounds serve as neurotransmitters in
addition to acetylcholine 1205
The acetylcholine receptor is the best-understood
neurotransmitter receptor 1208

CHAPTER 36 VISION 1213

The visual pigments are found in membranes in rod and
cone cells 1213
Rhodopsin consists of 11-*cis*-retinal bound to a protein,
opsin 1214
Light isomerizes the retinal of rhodopsin to all-*trans*
1216
Transformations of rhodopsin can be detected by
changes in its absorption spectrum 1216
Isomerization of the retinal causes other groups in the
protein to move 1218
Absorption of a photon causes a change in the cation
conductivity of the cytoplasmic membrane 1220
The effect of light is mediated by guanine nucleotides
1221
Rhodopsin can move around in the disk membrane
1223
Bacteriorhodopsin: a bacterial pigment-protein complex
that resembles rhodopsin 1225

Glossary 1228

Answers to Selected Problems 1241

Index 1246

INTRODUCTION

THE ESSENCE OF LIVING SYSTEMS

Biochemistry has its roots in two major scientific disciplines, biology and chemistry. As students approaching the subject for the first time, you already know a good deal about biochemistry from your previous courses. You should be familiar with the basic rules of organic chemistry and most likely you have also learned some of the basic rules governing biological systems. The object of this introduction is to remind you of the material you have already studied that is most relevant to biochemistry, and to provide an overview of the subject.

ON EARTH ALL LIVING SYSTEMS ARE BIOCHEMICALLY RELATED THROUGH A COMMON EVOLUTION

Since biochemistry is a study involving living systems, our first concern is to see to what extent life can be defined in biochemical terms. Despite the advances in knowledge made in recent years, for many people the notion of a vital force of some sort still survives. The view taken here is that such a notion is not needed for an understanding of living systems—that all actions and reactions that occur in living systems can be explained in terms of their chemistry.

A space traveler landing on a planet in a distant galaxy might have a difficult problem if given the task of determining whether there were living things on that planet. The appearance and chemistry of living forms could be totally foreign to anything encountered on earth. Failing to detect objects resembling life forms on earth, the careful observer might feel compelled to formulate a more general definition of living forms to use as a criterion to investigate further. In the process, the space traveler would probably come to the realization that a precise definition of a living form is most difficult to make. Fortunately, students of biochemistry need to consider only the commonalities between different forms of life on earth. Thus for the sake of this text we can avoid the more fundamental question of what is the essence of life and deal with the description of the various forms of life on our planet.

An appropriate beginning point is to consider the evolutionary data, which suggest that all organisms are derived from a common ancestor. Evolutionary information is conveniently summarized in the form of a tree (Figure I–1). It is customary to depict the ancestral prokaryote in this tree as the

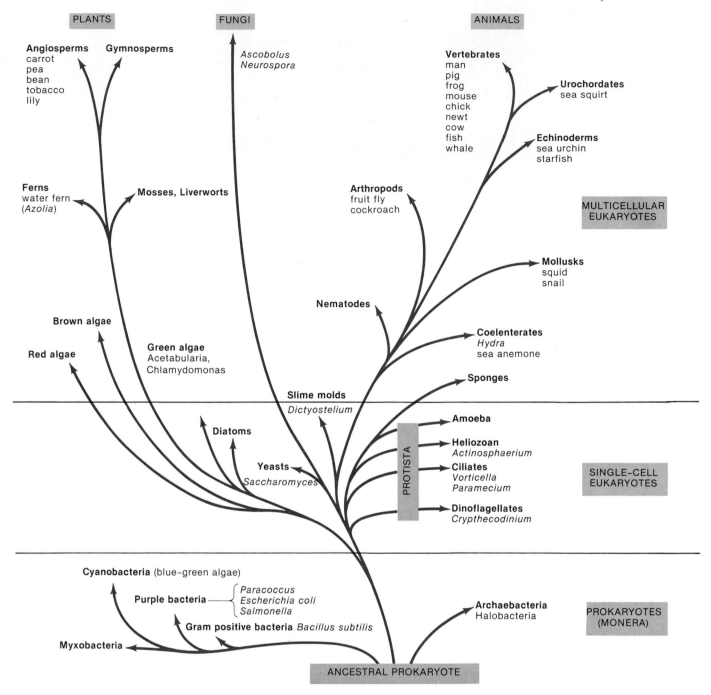

Figure I-1
Evolutionary tree. All living forms have a common origin, believed to be the ancestral prokaryote. Through a process of evolution some of these prokaryotes changed into other organisms with different characteristics. The evolutionary tree indicates the main pathways of evolution.

first living form. Through a process of evolution the ancestral prokaryote gave rise to different organisms, and these in turn underwent further evolutionary change, so that the tree became a highly branched structure. The original evidence for the evolutionary tree came from morphological comparisons. More and more, this type of data is being supplemented by comparisons of the fine structure of comparable molecules in different organisms. The molecular data have led to confirmation and refinement of previous notions of evolutionary relatedness.

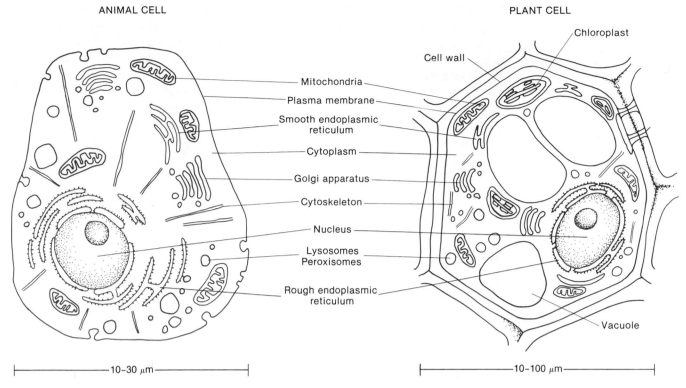

ANIMAL CELL

PLANT CELL

Chloroplast

Cell wall

Mitochondria

Plasma membrane

Smooth endoplasmic reticulum

Cytoplasm

Golgi apparatus

Cytoskeleton

Nucleus

Lysosomes
Peroxisomes

Rough endoplasmic reticulum

Vacuole

10–30 μm

10–100 μm

(a) Prototypical animal and plant cells

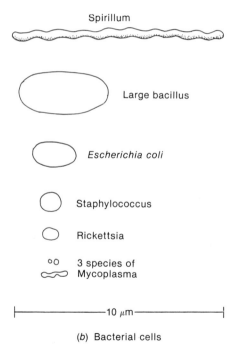

Spirillum

Large bacillus

Escherichia coli

Staphylococcus

Rickettsia

3 species of
Mycoplasma

10 μm

(b) Bacterial cells

Figure I–2

Cells are the fundamental units in all living systems, and they vary tremendously in size and shape. All cells are functionally separated from their environment by the plasma membrane that encloses the cytoplasm. (a) Generalized representations of the internal structures of animal and plant cells (eukaryotic cells). Plant cells have two structures not found in animal cells: a cellulose cell wall, exterior to the plasma membrane, and chloroplasts. (b) The many different types of bacteria (prokaryotes) are all smaller than most plant and animal cells. Bacteria, like plant cells, have an exterior cell wall, but it differs greatly in chemical composition and structure from the cell wall in plants. Like all other cells, bacteria have a plasma membrane that functionally separates them from their environment. Some bacteria also have a second membrane, the outer membrane, exterior to the cell wall.

THE CELL IS THE FUNDAMENTAL UNIT OF LIFE

Microscopic examination of any of the organisms indicated in Figure I–1 reveals that they all are composed of membrane-enclosed objects called cells. The enclosing membrane is called the _cell membrane_ or the _plasma membrane_. Cells vary enormously in size and shape. In Figure I–2 we show prototypical animal and plant cells and a variety of the shapes and sizes of some well-known bacteria. The cells in bacteria frequently exist as single cells, but sometimes they are connected into long chains. In multicellular organisms the cells associate to form specialized tissues.

The plasma membrane is a delicate, semipermeable, sheetlike structure enclosing the entire cell. By forming an enclosure it prevents gross loss of the intracellular contents; its semipermeable character permits the selective absorption of nutrients and the selective removal of metabolic waste products. In many plant and bacterial (but not animal) cells, a _cell wall_ encompasses the plasma membrane. The cell wall is a more porous structure than the plasma membrane, but it is mechanically stronger because it is made of a covalently cross-linked, three-dimensional network; this structure helps keep the cell from deforming or disrupting under stress.

The contents enclosed by the plasma membrane is called the *cytoplasm*. The purely liquid portion of the cytoplasm is called the *cytosol*. Within the cytoplasm there are a number of macromolecules and larger structures, many of which can be seen by high-power light microscopy or by electron microscopy. Some of the structures are membranous and are called *organelles*. Organelles commonly found in plant and animal cells include the *nucleus*, the *mitochondria*, the *endoplasmic reticulum*, the *golgi apparatus*, the *lysosomes*, and the *peroxisomes* (see Figure I–2). *Chloroplasts* are an important class of organelles found in many plant cells but never in animal cells. Each type of organelle represents a special biochemical factory in which certain biochemical products are synthesized. All of the organelles will be discussed in detail in later chapters. Some of them will be briefly discussed later in this introductory chapter. In addition to organelles, animal and plant cells contain filamentous structures termed *cytoskeleton*, which are important in maintaining the three-dimensional integrity of the cell.

The organelles and cytoskeleton that are found in plant, fungus, and animal cells are not present in bacteria. The biochemical functions associated with organelles are frequently present in bacteria, but they are organized in a different way.

If we refer back to Figure I–1, we see that the evolutionary tree is bisected into a lower, or prokaryotic, domain and an upper, or eukaryotic, domain. The terms *prokaryote* and *eukaryote* refer to the most basic division between cell types. The fundamental difference is that in eukaryotes the cell contains a nucleus, whereas in prokaryotes it does not. The cells of prokaryotes usually lack most of the other membrane-bounded organelles as well. Plants, fungi, and animals are examples of eukaryotes, and bacteria are examples of prokaryotes.

Cells are organized in a variety of ways in different living forms. Prokaryotes of a given type produce cells that are very similar in appearance. The most common types of prokaryotes are bacteria, which exist as single cells. Each bacterial cell replicates by a process of binary fission, in which two identical daughter cells arise from an identical parent cell. If we proceed up the evolutionary tree, we find simple eukaryotes that also exist as single nonassociating cells. Further up the tree we find eukaryotes of increasing complexity, containing many cells. Often, in multicellular organisms the same organism possesses cells of many different types, with specialized structures and functions. In humans, which contain about 10^{14} cells, we have an example of a high degree of complexity. Humans contain more than a hundred different types of cells. There are specialized cells associated with skin, connective tissue, nervous tissue, muscle, blood, sensory functions, and reproductive organs (Figure I–3). In such a complex organism, the capacity of different cells for replication is limited. When a skin cell or a muscle cell precursor replicates, it makes more cells of the same tissue type. The only cells capable of reproducing an entire organism are the germ cells, that is, the sperm and the egg.

THE CELL IS COMPOSED OF SMALL MOLECULES, MACROMOLECULES, AND ORGANELLES

In order to appreciate the functions of the various organelles and the biochemical reactions that occur in the organelles or in the cytosol, we must begin to think in terms of molecules. The water molecule is the major component of the cell, accounting for about 70 per cent of its weight (Table I–1). As a result, most of the other components exist in an aqueous environment. Water is a highly polar solvent, because of the partial separation of charge between the oxygen and the hydrogen of the water molecule (Figure I–4).

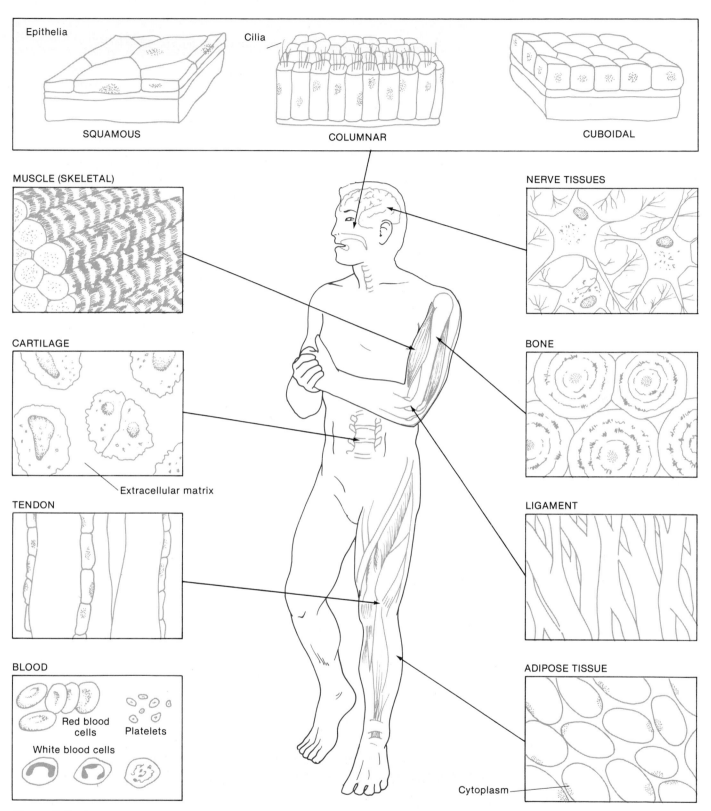

Figure I–3
Specialized cell types found in the human. Although all cells in a multicellular organism have common constituents and functions, specialized cell types have unique chemical compositions, structures, and biochemical reactions that establish and maintain their specialized functions. Such cells arise during embryonic development by the complex processes of cell proliferation and cell differentiation. Except for the sex cells, all cell types contain the same genetic information, which is faithfully replicated and partitioned to daughter cells. Cell differentiation is the process whereby some of this genetic information is activated in some cells, resulting in the synthesis of certain proteins and not other proteins. Thus specialized cells come to have different complements of enzymes and metabolic capacities.

TABLE I–1
The Approximate Chemical Composition of a Bacterial Cell

	Percent of Total Cell Weight	Number of Types of Each Molecule
Water	70	1
Inorganic ions	1	20
Sugars and precursors	3	200
Amino acids and precursors	0.4	100
Nucleotides and precursors	0.4	200
Lipids and precursors	2	50
Other small molecules	0.2	~200
Macromolecules (proteins, nucleic acids, and polysaccharides)	22	~5000

Source: Adapted from S. E. Luria, S. J. Gould and S. Singer, *A View of Life,* Benjamin/Cummings. Menlo Park, Calif., 1981.

Figure I–4

The structure of the water molecule and the secondary structures formed by multiple water molecules. (*a*) Water is a highly polar molecule. The hydrogens carry a partial positive charge (δ^+), the oxygens a partial negative charge (δ^-). (*b*) Two water molecules interact to form a hydrogen bond between the oxygen of one molecule and one hydrogen of another molecule. (*c*) In bulk water a network of hydrogen bonds is formed. (*d*) Water interacts with polar molecules by forming hydrogen bonds. (*e*) Ions also interact with water. The cations interact with the oxygens, and the anions interact with the hydrogens. (*f*) Apolar molecules such as hydrocarbons do not interact electrostatically with water molecules.

THE STRUCTURE OF WATER AND THE INTERACTION OF WATER WITH OTHER WATER MOLECULES

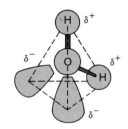

(*a*) Single water molecule

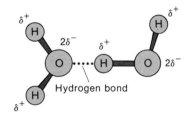

(*b*) Two interacting water molecules

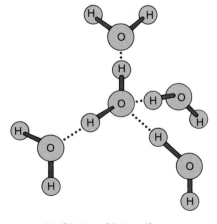

(*c*) Cluster of interacting water molecules

THE INTERACTION OF WATER WITH OTHER MOLECULES

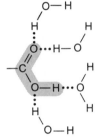

(*d*) Interaction between water and a polar side chain

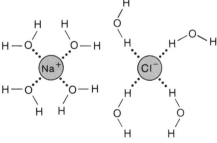

(*e*) Interaction between water and sodium chloride

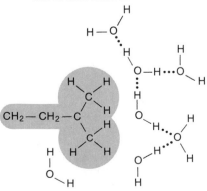

(*f*) Interaction between water and an apolar side chain

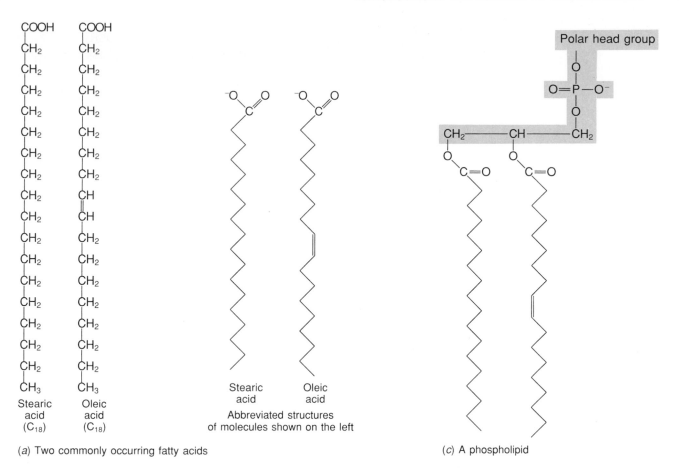

(a) Two commonly occurring fatty acids

(c) A phospholipid

(b) Triacylglycerol

Figure I–5

The structures of common lipids. (a) The structures of saturated and unsaturated fatty acids, represented here by stearic acid and oleic acid. (b) Three fatty acids covalently linked to glycerol by ester bonds form a triacylglycerol. (c) The general structure for a phospholipid consists of two fatty acids esterified to glycerol, which is linked through phosphate to a polar head group. The polar head group may be any one of several different compounds—for example, choline, serine, or ethanolamine.

Other polar molecules tend to be _hydrophilic;_ that is, they have a high affinity for water, and thus are very soluble in water. By contrast, apolar substances tend to be _hydrophobic;_ that is, they have a low affinity for water, and thus are poorly soluble in water.

Except for water, most of the molecules found in the cell are macromolecules, which can be classified into four different categories: _lipids, carbohydrates, proteins,_ and _nucleic acids._ Each type of macromolecule possesses distinct chemical properties that suit it for the functions it serves in the cell.

Lipids are primarily hydrocarbon structures (Figure I–5). They tend to be poorly soluble in water, and are therefore particularly well suited to serving as a major component of the various membrane structures found in the cells. Lipids also serve as a convenient, compact way to store chemical energy.

Carbohydrates, like lipids, are primarily hydrocarbon structures, but they also contain many polar hydroxyl (—OH) groups and are therefore very

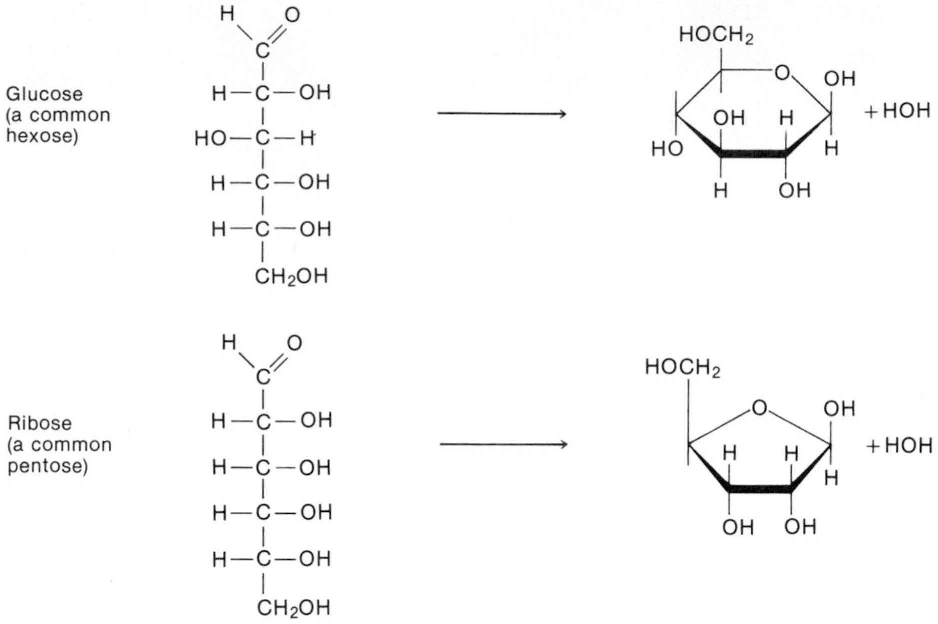

(a) Two common monosaccharides that circularize in aqueous solution

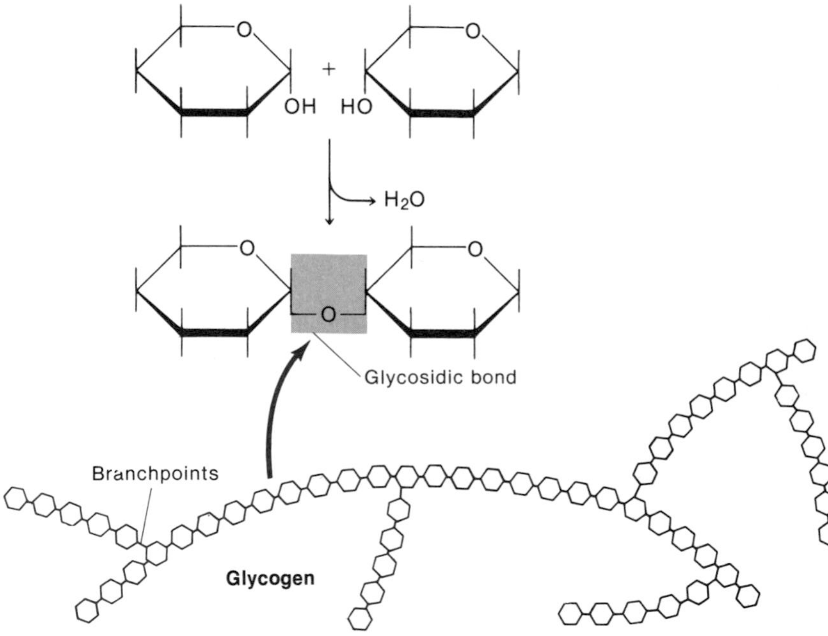

(b) Polysaccharides composed of covalently linked monosaccharides

Figure I–6
Monomers and polymers of carbohydrates. (a) The most common carbohydrates are the simple six-carbon (hexose) and five-carbon (pentose) sugars. In aqueous solution, these sugar monomers circularize to form ring structures. (b) Polysaccharides are usually composed of hexose monosaccharides covalently linked together by glycosidic bonds to form long straight-chain or branched-chain structures.

soluble in water. Large carbohydrate molecules called _polysaccharides_ consist of many small, ringlike sugar molecules, the sugar monomers, attached to one another by _glycosidic bonds_ in a linear or branched array to form the sugar polymer (Figure I–6). In the cell, such polysaccharides often form storage granules that may be readily broken down into their component sugars. With further chemical breakdown these sugars release chemical energy and may also provide the carbon skeletons for the synthesis of a variety of other molecules. Important structural functions are also served by polysaccharides. Linear polysaccharides form a major component of plant cell walls. Bacterial cell walls are composed of cross-linked polysaccharides, but are very different from plant cell walls.

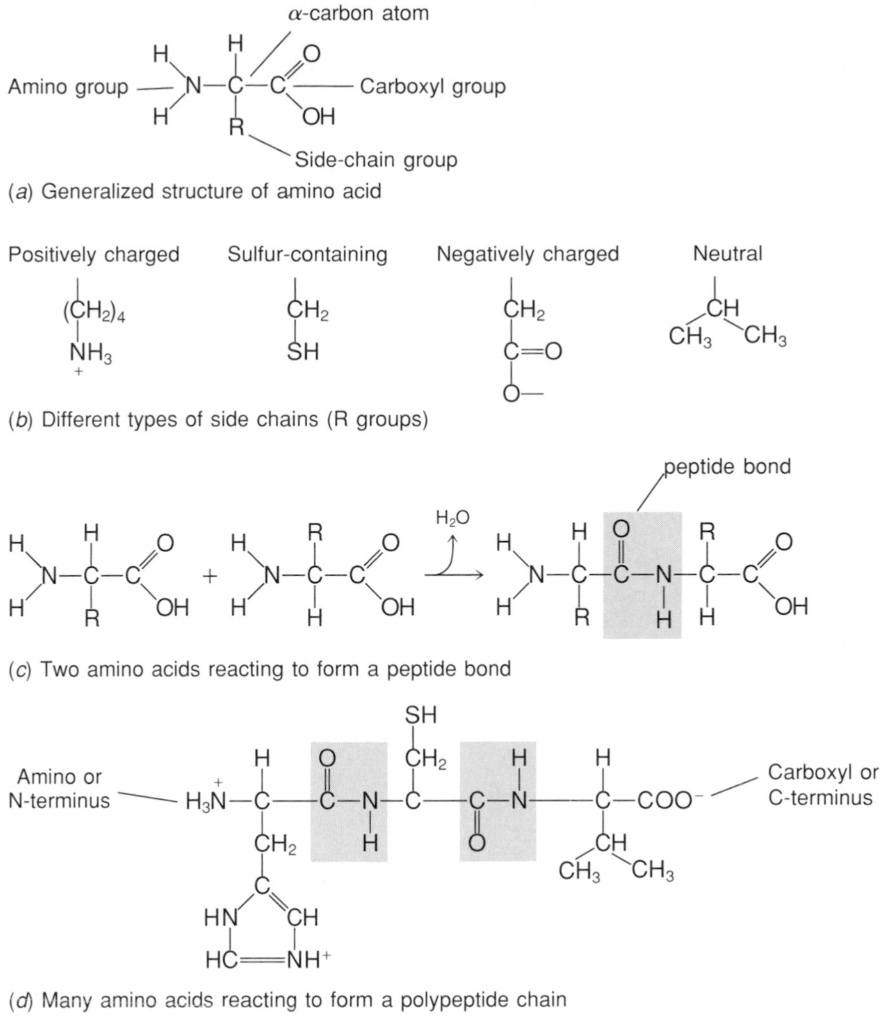

(a) Generalized structure of amino acid

(b) Different types of side chains (R groups)

(c) Two amino acids reacting to form a peptide bond

(d) Many amino acids reacting to form a polypeptide chain

Figure I–7
Amino acids and the structure of the polypeptide chain. Polypeptides are composed of L-amino acids covalently linked together in a sequential manner to form linear chains. (a) The generalized structure of the amino acid. (b) Structures of some of the R groups found for different amino acids. (c) Two amino acids become covalently linked by a peptide bond, and water is lost. (d) Repeated peptide bond formation generates a polypeptide chain, which is the major component of all proteins.

Proteins are the most complex macromolecules found in the cell. They are composed of linear polymers called _polypeptides_, which contain amino acid monomers connected by _peptide bonds_ (Figure I–7). Each amino acid contains a central carbon atom attached to four substituents: (1) a carboxyl group, (2) an amino group, (3) a hydrogen atom, and (4) an R group. The R group is what gives each amino acid its unique characteristics. There are twenty different amino acids in proteins. Some R groups are charged, some are neutral but still polar, and some are apolar.

The properties of a given protein depend on the sequences of amino acids found in its polypeptide chains. After synthesis, the linear polypeptide molecules fold in a highly specific way to form a unique three-dimensional structure. Many proteins are composed of two or more polypeptides held together by noncovalent forces. In such a structure the individual polypeptide chains are referred to as _subunits_. Certain proteins function in structural roles. Some structural proteins interact with lipids in membrane structures. Others aggregate to form part of the cytoskeleton that gives the cell its shape. Still others are the chief components of muscle or connective tissue. The substances known as _enzymes_ form a major class of proteins, which function as catalysts that direct and accelerate biochemical reactions. Each enzyme functions in a highly specific manner, usually catalyzing only one type of reaction. Even the simplest cells probably contain more than a thousand different types of enzymes.

Nucleic acids are the largest macromolecules in the cell. These exist as very long linear polymers, called *polynucleotides*, which are composed of *nucleotide* monomers. A nucleotide contains (1) a five-carbon sugar molecule, (2) one or more phosphate groups, and (3) a nitrogenous base. It is the nitrogenous base that gives the nucleotide a distinct character (Figure I–8). Of special interest are the nucleic acids known as *deoxyribonucleic acid (DNA)*. DNA contains the genetic information that is inherited by each

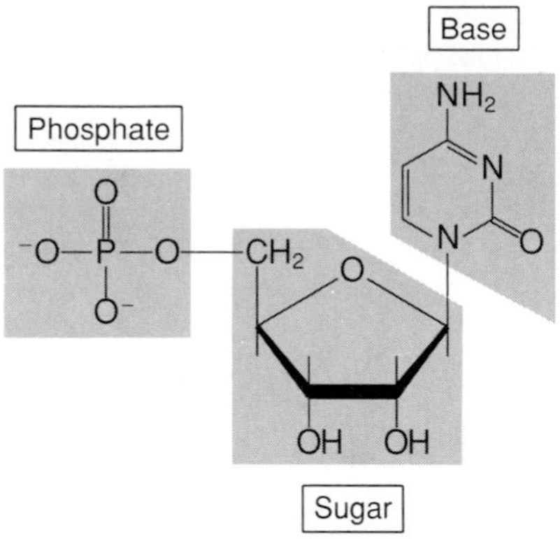

(a) Generalized structure of a nucleotide

Figure I–8
The structural components of nucleic acids. Nucleic acids are long linear polymers of nucleotides, called polynucleotides. *(a)* The nucleotide consists of a five-carbon sugar (ribose in RNA or deoxyribose in DNA) covalently linked at the 5′ carbon to a phosphate, and at the 1′ carbon to a nitrogenous base. *(b)* Nucleotides are distinguished by the types of bases they contain. These are either of the two-ring purine type or of the one-ring pyrimidine type. *(c)* When two nucleotides become linked they form a dinucleotide, which contains one phosphodiester bond. Repetition of phosphodiester bond formation leads to a polynucleotide.

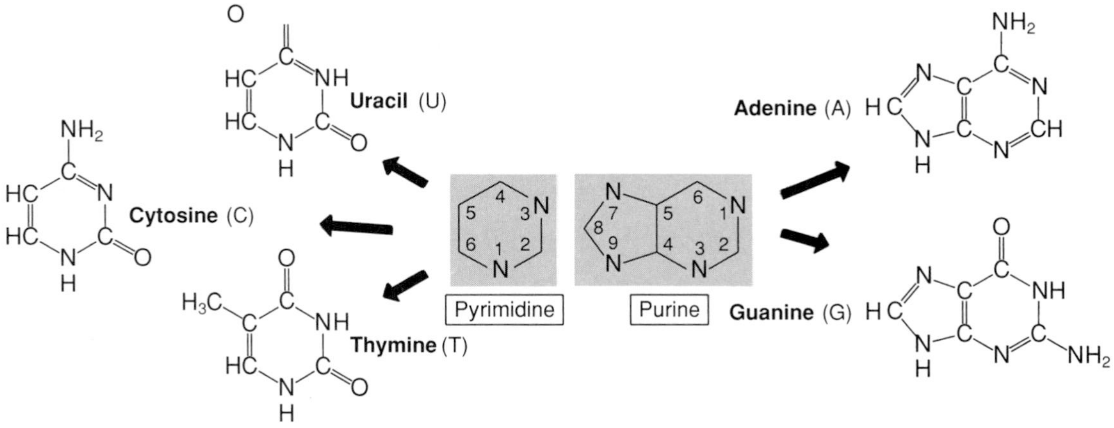

(b) Different bases found in nucleotides

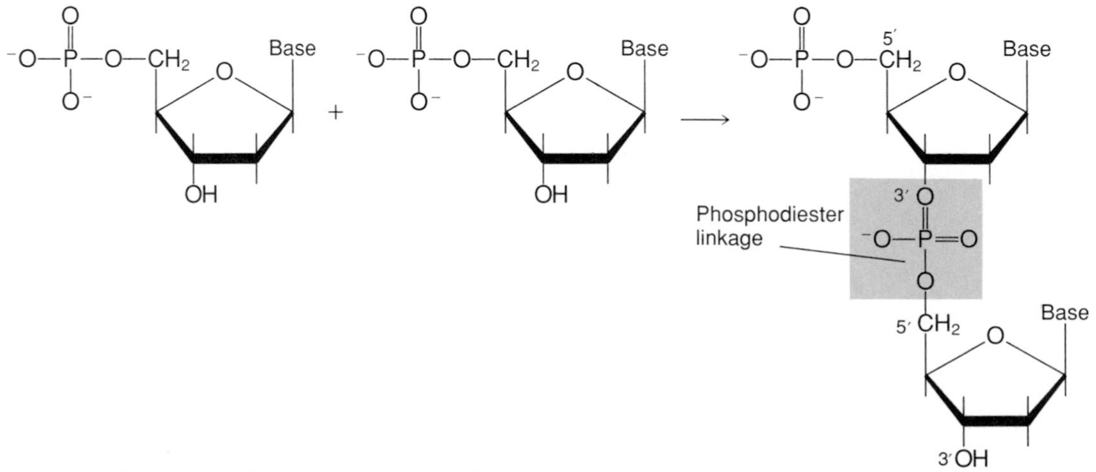

(c) Two nucleotides reacting to form a dinucleotide

daughter cell during growth when cells divide. The DNA usually exists as nucleoprotein (DNA-protein) complexes called *chromosomes*. A prokaryotic cell contains a single chromosome except just prior to cell division, when it contains two identical chromosomes. Thus chromosomal DNA replicates prior to cell division and segregates so that an identical complement of DNA goes to each of two newly formed daughter cells.

Eukaryotic cells are more complex than the cells of prokaryotes; they usually contain more DNA, and this DNA is partitioned into two, three, or many chromosomes. In both prokaryotes and eukaryotes, all cells of the same organism contain the same number of chromosomes. In eukaryotes most of the chromosomes are localized in the nucleus. Thus the DNA is isolated from the main body of the cytoplasm—a unique feature of eukaryotes and the primary distinction between prokaryotes and eukaryotes. Some organelles, notably the mitochondria and the chloroplasts, also contain a single circular chromosome.

Eukaryotic chromosomes are detectable by microscopic techniques in certain cell types only, and then only at the stage just prior to cell duplication. At this stage, called _mitosis_, chromosomes appear as elongated refractile structures that can be seen to segregate in equal numbers and types to each of the daughter cells before cell division (Figure I–9). Each chromosome carries specific hereditary (genetic) information necessary for the synthesis of specific compounds essential for cell maintenance, growth, and replication. Each chromosome contains a single DNA molecule composed of 10^6 or more nucleotides in a specific arrangement. The sequence of bases in the chromosomal DNA determines the sequence of amino acids in the protein polypeptide chains of the organism. The relationship between base sequences and resultant amino acids is known as the _genetic code_. Each grouping of three bases—called a triplet—represents a specific amino acid and is called a _codon_. The genetic code ensures that the organism's characteristics

Figure I–9

Mitosis and cell division in eukaryotes. After DNA duplication has occurred, mitosis is the process by which quantitatively and qualitatively identical DNA is delivered to daughter cells formed by cell division. Mitosis is traditionally divided into a series of stages characterized by the appearance and movement of the DNA-bearing structures, the chromosomes.

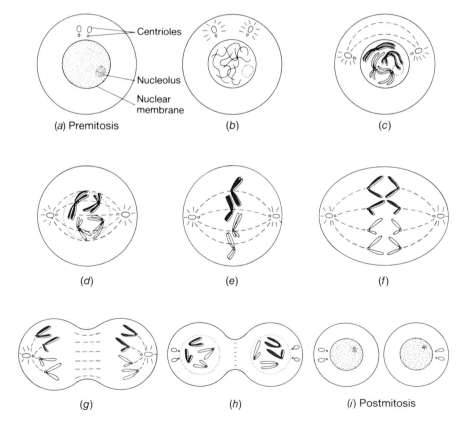

will be reflected by the sequence of nucleotides in its DNA. When chromosomes replicate, the DNA replicates precisely, so that the same nucleotide sequence is passed along to each of the daughter cells resulting from mitosis and cell division.

In the cytoplasm we find another type of nucleic acid known as _ribonucleic acid (RNA)_. RNAs are smaller than DNAs (usually 10^2–10^4 nucleotides in length), and differ in minor chemical respects from their DNA cousins. Each RNA contains a nucleotide sequence that reflects the nucleotide sequence in a specific region of the DNA. Different types of RNA molecules serve different functions. One important type of RNA molecule is referred to as _messenger RNA (mRNA)_. The mRNA transmits the genetic information from the DNA to the cytoplasm, where the information specifies the sequences of amino acids during protein synthesis. The overall process of information transfer from DNA to mRNA and from mRNA to protein is depicted in Figure I–10. Another major class of RNA is called _ribosomal RNA (rRNA)_. This is found in ribonucleoproteins called _ribosomes_, which exist in the cytosol or attached to the endoplasmic reticulum (see Figure I–2). The ribosome is the site where protein synthesis takes place. A third major class of RNA is called _transfer RNA (tRNA)_. There are transfer RNAs for each amino acid. These molecules transport the amino acids to the ribosomes, where they become linked into polypeptide chains.

In addition to water and the macromolecules and organelles, the cytosol contains a large variety of small molecules that differ greatly in both structure and function. These never make up more than a small fraction of the total cell mass despite their great variety (see Table I–1). One class of small molecules consists of the monomer precursors of the different types of macromolecules. These monomers are derived by chemical modification from the nutrients absorbed through the cell membrane. Rarely are the nutrients themselves the actual monomers used by the cell. As a rule each nutrient must undergo a series of enzymatically catalyzed alterations before it is suitable for incorporation into one of the biopolymers. The intermediate molecules between nutrients and monomers are also present in small concentrations in the cytosol. Another varied class of molecules found in the cytosol includes molecules formed as side products in important synthetic reactions and as breakdown products of the macromolecules. Finally, small bioorganic molecules known as _coenzymes_ are also present in the cytoplasm. These compounds act in union with the enzymes in a highly specific manner to catalyze a wide variety of reactions.

BIOCHEMICAL REACTIONS ARE A SUBSET OF ORDINARY CHEMICAL REACTIONS

Even though the total number of biochemical reactions is very large, it is still much smaller than the potential number of reactions that occur in ordinary chemical systems. This simplification results partly from the fact that only a limited number of elements account for the vast majority of substances found in living cells. The elements of major importance, in order of decreasing numerical abundance, are hydrogen (H), carbon (C), oxygen (O), nitrogen (N), phosphorus (P), and sulfur (S). Certain metal ions are also important, although they do not become covalently attached to other biomolecules; these include Na^+, K^+, Mg^{2+}, Ca^{2+}, Zn^{2+}, and Fe^{2+} or Fe^{3+}. Other metals and elements that are needed in very small amounts are iodine, cobalt, molybdenum, selenium, vanadium, nickel, chromium, tin, fluorine, silicon, and arsenic. In some cases we don't know the biological roles of these "trace elements" but only that they are needed by some organisms for normal growth or development.

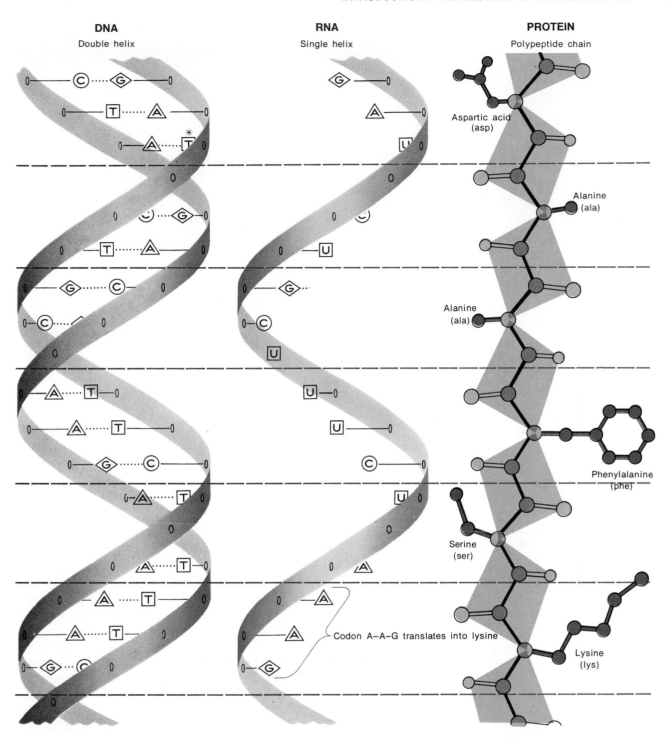

DNA
Double helix

RNA
Single helix

PROTEIN
Polypeptide chain

Aspartic acid
(asp)

Alanine
(ala)

Alanine
(ala)

Phenylalanine
(phe)

Serine
(ser)

Codon A–A–G translates into lysine

Lysine
(lys)

Thymine T in DNA becomes uracil U in RNA.

Figure I–10

Transfer of information from DNA to protein. The nucleotide sequence in DNA specifies the sequence of amino acids in a polypeptide. DNA usually exists as a two-chain structure. The information contained in the nucleotide sequence of only one of the DNA chains is used to specify the nucleotide sequence of the messenger RNA molecule (mRNA). This sequence information is used in polypeptide synthesis. A three-nucleotide sequence in the messenger RNA molecule codes for a specific amino acid in the polypeptide chain.

The types of covalent linkages most commonly found in biomolecules are also quite limited (Table I–2). Only sixteen different types of linkages account for more than 95 per cent of the linkages found in biomolecules. All the elements form single or double bonds, except for hydrogen, which can

TABLE I–2
Types of Covalent Linkages Most Commonly Found in Biomolecules

	H	C	O	N	P	S
H						
C	>C—H	>C—C< ; >C=C<				
O	—O—H	>C—O— ; >C=O				
N	>N—H	>C—N= ; >C=N—	—			
P	—	—	>P—O— ; >P=O	—		
S	—S—H	>C—S—	>S—O⁻ ; >S=O	—	—	—S—S—

TABLE I–3
Most Common Valencies Observed by Atoms in Covalent Linkages

Element	Valence
H	−1
C	+4
O	−2
P	+5
N	−3
S	+6, −2

form only single bonds; all the elements exist primarily in a single valence state except for sulfur, which is found either in a $+6$ or a -2 valence state (Table I–3). Despite this overall simplicity, many other valence states can be found in unusual cases, and some of these are very important. For example, the biochemistry of nitrogen involves consideration of all the valence states of nitrogen from $+5$ to 0 to -3. A major source of nitrogen available to biosystems is gaseous nitrogen found in the atmosphere (valence state 0). Biochemical reactions exist for the conversion of gaseous nitrogen into other forms of nitrogen. These reactions are very complex and occur only in a select group of microorganisms that possess the necessary enzyme systems.

Biochemical reactions involving the different classes of substances use a limited number of functional groups, which are illustrated in Figure I–11. Most of the reactive groups in biomolecules contain one or more of these functional groups or closely related ones. Many cellular reactions involving these functional groups are closely related to reactions that take place outside the cell under different conditions and are studied in organic chemistry.

Alcohols, which contain a hydroxyl functional group, can undergo dehydration reactions to form *esters* with either *carboxylic* or *phosphoric* acids.

| Alcohol | Carboxylic acid | | Ester | Water |

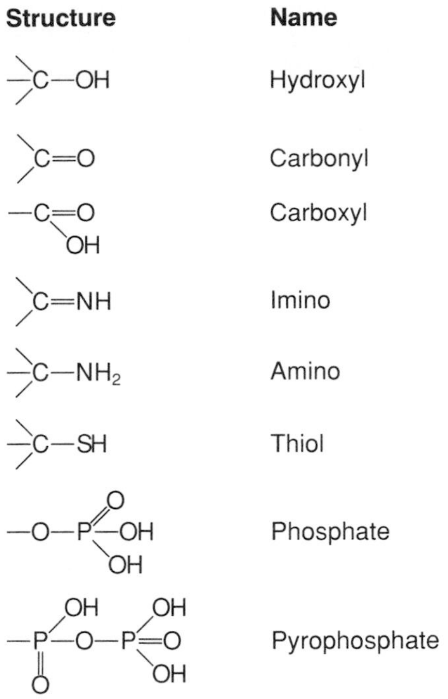

Structure	Name
$\searrow$C—OH	Hydroxyl
$\searrow$C=O	Carbonyl
—C=O with OH	Carboxyl
$\searrow$C=NH	Imino
$\searrow$C—NH$_2$	Amino
$\searrow$C—SH	Thiol
—O—P(=O)—OH with OH	Phosphate
—P(OH)—O—P(=O)(OH) with O	Pyrophosphate

Figure I–11
Different functional groups found in biomolecules. This figure includes the major functional groups. Other functional groups are found in minor amounts.

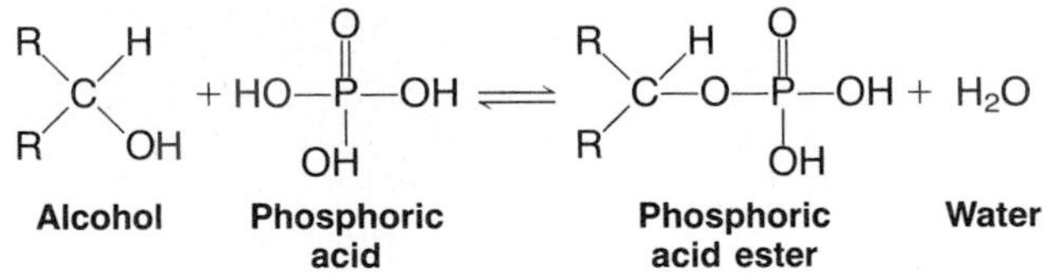

Thiols, containing, *sulfhydryl* groups (—SH), can substitute for alcohols in some reactions, leading to the formation of *thiol esters*.

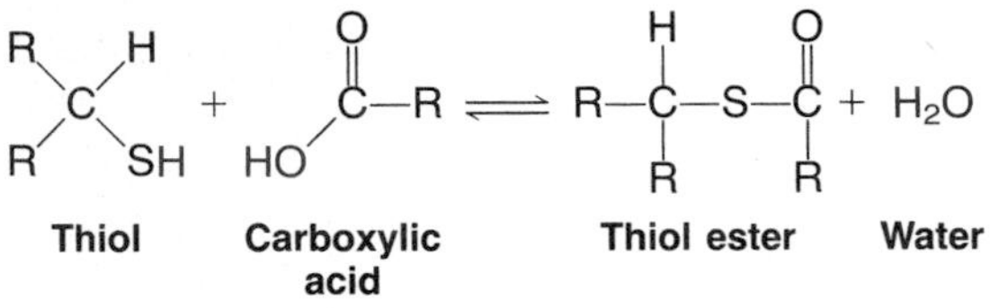

Two alcohols can react with one another to form an *ether*.

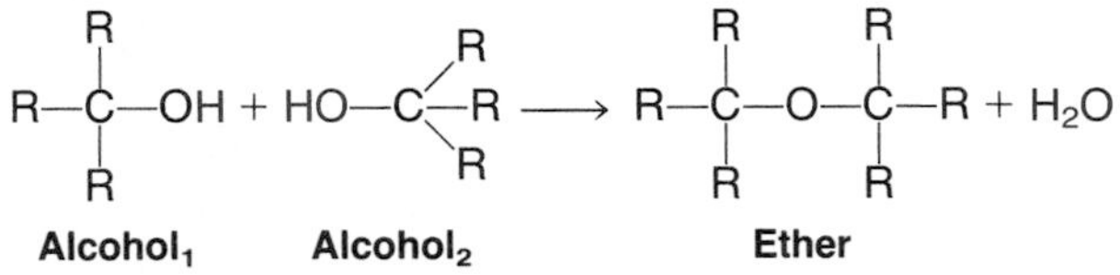

Alcohols can also undergo a dehydrogenation reaction to form a carbonyl derivative (aldehyde or ketone).

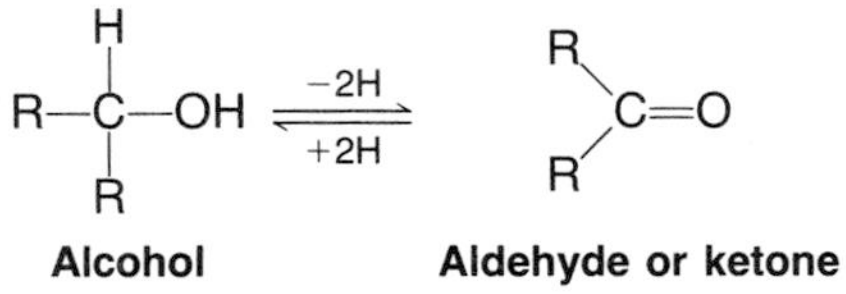

Amines undergo reactions with carboxylic acids comparable to the formation of esters from alcohols. The product is known as an *amide*.

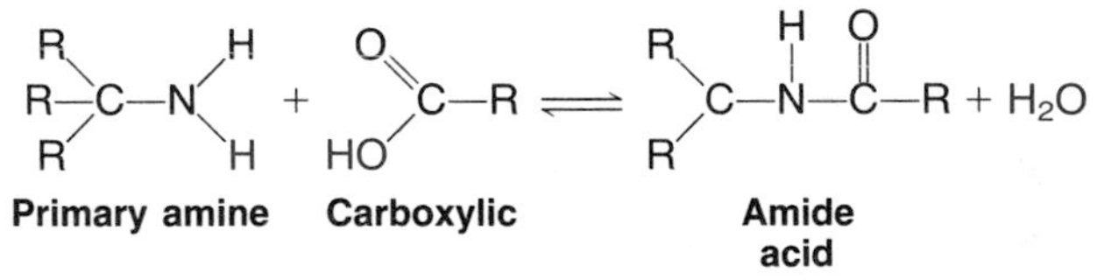

Amines can also undergo dehydrogenation reactions leading to the formation of *imines*, which are frequently unstable in water and hydrolyze to *ketones* or, in cases where one of the R groups is an H, to aldehydes:

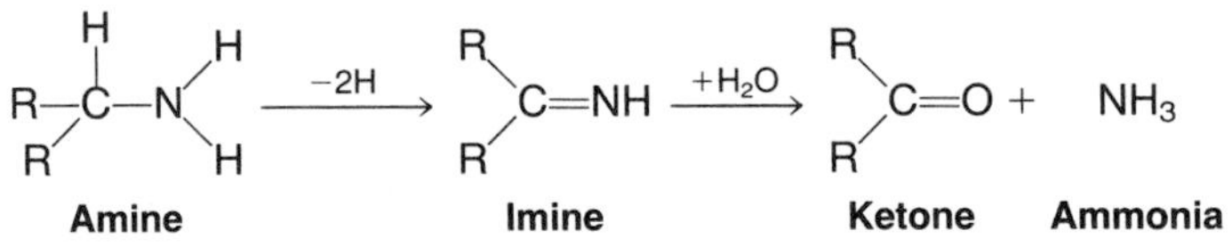

Aldehydes and ketones both may be reduced to alcohols by *hydrogenation* (see the alcohol dehydrogenation reaction). Aldehydes may react with either water or alcohol to form *aldehyde hydrates* or *hemiacetals*, respectively. Reaction of an aldehyde with two molecules of alcohol leads to *acetal* formation.

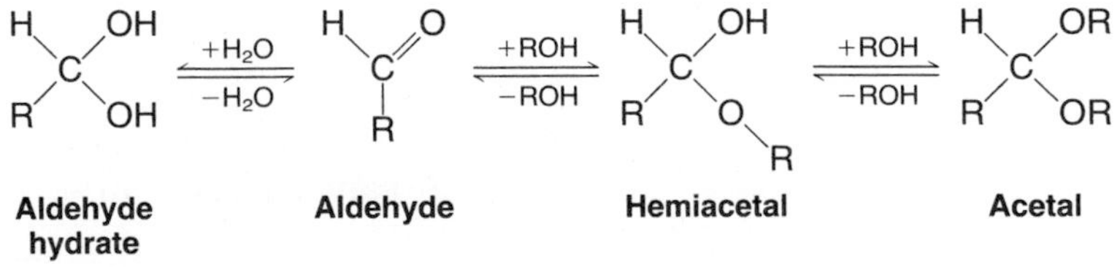

Dehydrogenation of an aldehyde hydrate leads to carboxylic acid formation.

Aldehyde hydrate → **Carboxylic acid**

Aldehydes and ketones may also *isomerize* to the *enol* form as long as the adjacent carbon atom contains at least one H atom. The reaction involves the migration of a hydrogen and the shift of the double bond.

Keto form   **Enol form**

Pyrophosphates may hydrolyze to inorganic phosphoric acid (phosphate) and an organophosphoric acid.

Organopyrophosphate   **Organophosphoric acid**   **Phosphoric acid**

Hydrolysis reactions of this sort yield considerable energy, which can be utilized in biosynthesis.

All of the functional groups that we have described are electrostatically neutral in organic solvents. However, in water many of these functional groups either lose or gain protons to become charged species. Such *ionization reactions* are very important in biochemical systems because they frequently influence solubility and reactivity.

Carboxylic and phosphoric acids lose one or more protons in water to become negatively charged. The ionized forms are stabilized by resonance as shown:

Amines usually add a proton to become positively charged.

Near neutrality (10^{-7} M H$^+$), where most biochemical systems function, the carboxyl group exists mainly in the negatively charged form, phosphoric acid exists mainly in the diionized form, and amino groups exist mainly in the positively charged form. This fact has interesting consequences for amino acids, since they contain one amino group and one carboxyl group. The amino acids are usually neutral overall, even though they contain two charged groups, one resulting from the deprotonation of the carboxyl group and the other resulting from the protonation of the amino group. Amino acids existing as dipolar ions are called *zwitterions*.

Uncharged	**Zwitterion**

These are some of the more important reactions involving covalent bond breakage or formation in biochemistry. By now two things should be apparent about biochemical reactions: (1) as stated at the outset, the number of reactions in biochemistry is much more limited than in ordinary chemistry; (2) as far as the reactants and products are concerned, biochemical reactions can be understood in the same terms as ordinary chemical reactions.

SECONDARY VALENCE FORCES RESULT IN COMPLEX STRUCTURES

In a number of important biochemical reactions that occur within or between biomolecules, *secondary valence forces* are involved. All of these interactions are tempered by the highly polar aqueous environment in which they occur. Thus polar groups in biomolecules prefer associations with other polar groups, especially water. On the other hand, apolar groups in biomolecules prefer contacts with other apolar groups and avoid contacts with water. The result is that a biopolymer adopts a three-dimensional structure in which the polar groups are oriented on the surface, where they can associate with water, whereas the apolar groups are usually buried in the interior of the structure, where they can associate with other apolar groups.

An exception to this rule is observed with polar groups that form *hydrogen bonds*. Hydrogen bonds form between the hydrogens of amino or hydroxyl groups and other nitrogen or oxygen groups that lack covalently bonded hydrogens.

Hydrogen bonds are weak electrostatic attractions. Amino and hydroxyl groups are highly polar, with a partial negative charge (δ^-) on the N and the O atoms in N—H or —O—H functional groups. When the bonding electrons in these functional groups are attracted away from the covalently linked hydrogen, there are no other electrons to shield the positive charge of the bare hydrogen nucleus, a proton. Consequently there is a very small, highly localized positive charge (δ^+) on the hydrogen that seeks to form electrostatic bonds with other N and O atoms carrying a partial negative charge. Intramolecular and intermolecular hydrogen bonds are frequently formed by complementary polar groups in biomolecules, and often these bonds play a decisive role in determining the three-dimensional structure.

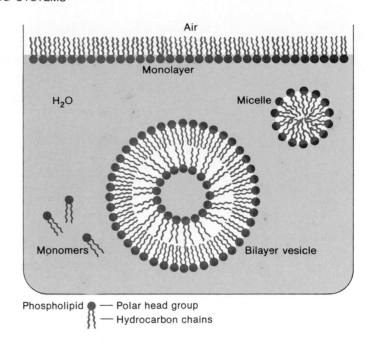

Figure I–12

Structures formed by phospholipids in aqueous solution. Phospholipids may form a monomolecular layer at the air–water interface, or they may form spherical aggregations surrounded by water. A vesicle consists of a double molecular layer of phospholipids surrounding an internal compartment of water. A micelle is a smaller aggregation of phospholipids that does not contain an internal aqueous compartment. In all these structures, the nonpolar hydrocarbon chains associate with one another and not with water. However, the polar head groups of the phospholipids do interact with water and are found on the surfaces of these structures exposed to water. Very few single phospholipid molecules will be found in the water phase.

Some structures illustrating these principles of macromolecular interaction are shown in Figures I–12, I–13, I–14. _Phospholipids_ (Figure I–5c), with a hydrophilic polar group on one end and long hydrophobic side chains attached to it, form multimolecular structures in an aqueous environment (Figure I–12). These phospholipid aggregates take the form of monomolecular layers at the air–water interface, or of micelles or bilayer vesicles within the water. In all of these structures, the polar head groups of the lipid are in contact with water, whereas the apolar side chains are excluded from the solvent structure.

As another example of polarity effects on macromolecular structure, consider polypeptide chains, which usually contain a mixture of amino acids with hydrophilic and hydrophobic side chains. Enzymes fold into complex three-dimensional globular structures with hydrophobic residues located on the inside of the structure and hydrophilic residues located on the surface, where they can interact with water (Figure I–13).

DNA forms a complementary structure involving two helically oriented polynucleotide chains (Figure I–14). In this structure the polar sugar and phosphate groups are located on the surface, where they can interact with water; the nitrogenous bases from the two chains form intermolecular hydrogen bonds in the core of the structure.

CONDITIONS FOR BIOCHEMICAL REACTIONS DIFFER FROM THOSE FOR ORDINARY CHEMICAL REACTIONS

Thus far we have stressed the ways in which biochemical reactions resemble ordinary chemical reactions. For the remainder of this introduction we will focus on some of the ways in which they differ.

Chemical reactions are frequently carried out in nonaqueous solvents, using elevated temperatures and pressures, acids or bases, or other harsh reagents. Most of the conditions used to carry out reactions in a chemistry laboratory would lead to destruction of the functional organization of a living cell. Biochemical reactions usually take place under very mild conditions in aqueous solution. Many chemical reactions will not proceed at reasonable rates under such conditions. However, despite these mild conditions, _biochemical reactions proceed at substantial rates because of_

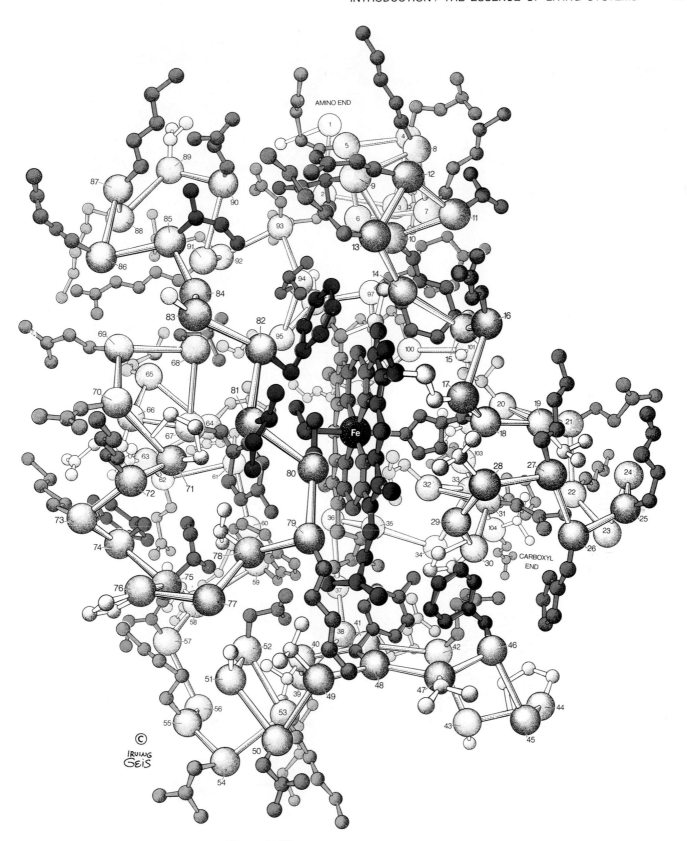

Figure I–13

A graphic representation of a three-dimensional model of the protein cytochrome *c*. Amino acids with nonpolar, hydrophobic side chains (color) are found in the interior of the molecule, where they interact with one another. Polar, hydrophilic amino acid side chains (gray) are on the exterior of the molecule, where they interact with the polar aqueous solvent.

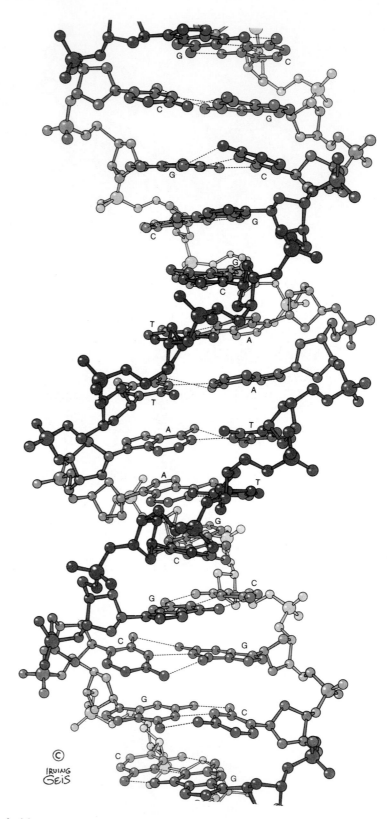

Figure I–14
The right-handed helical structure of DNA. DNA normally exists as a two-chain structure held together by hydrogen bonds (···) formed between the bases in the two chains. Along the chain the planar surfaces of these bases interact and, together with the hydrogen bonds, contribute to the stability of the two-chain structure. The negatively charged phosphate groups are on the outside of the structure, where they interact with water, ions, or charged molecules.

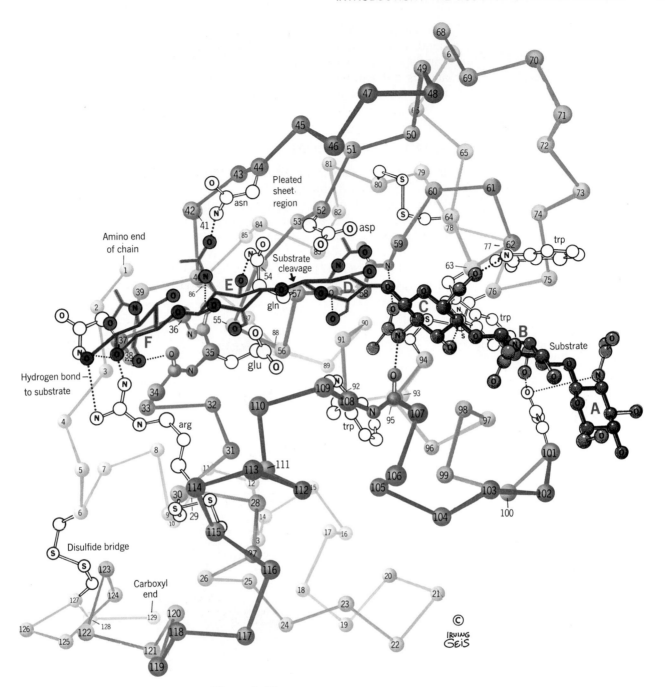

Figure I–15

The structure of the complex formed between the enzyme lysozyme and its substrate. The crevice that forms the site for substrate binding (the active site) runs horizontally across the enzyme molecule. The individual hexose sugars of the hexasaccharide substrate are shown in a darker color and labeled A–F. (Reprinted from R. E. Dickerson and I. Geis, *The Structure and Action of Proteins*, Benjamin/Cummings, Menlo Park, Calif., 1969; Coordinates courtesy of D. C. Phillips, Oxford.)

the very special nature of the enzyme catalysts used to accelerate these reactions.

Enzymes are structurally complex, highly specific catalysts; each enzyme usually catalyzes only one type of reaction. The enzyme surface is designed to bind the interacting molecules, or <u>substrates</u>, so that they are favorably disposed to react with one another (Figure I–15). The specificity of enzyme catalysis also has a selective effect, so that only one of several poten-

NH₂
|
R—C—COOH ⟶   Lipid
| Carbohydrate
H   Protein
               Nucleic acid
               Waste products

Figure I–16
The different fates of an amino acid.
Depending on which enzymes are
present and active and on the needs of
the organism, an amino acid can be
metabolized in different ways. Each of
these conversions involves one or
more steps, and usually each step
requires a specific enzyme.

tial reactions will take place. For example, a simple amino acid can be utilized in the synthesis of any of the four major classes of macromolecules or
can be simply secreted as waste product (Figure I–16). The fate of the amino
acid is determined as much by the presence of specific enzymes as by its
reactive functional groups.

MANY BIOCHEMICAL REACTIONS REQUIRE ENERGY

If we were to sum over all the reactions occurring in the living cell, we
would find that a considerable amount of energy is needed to build a cell.
Even maintaining a cell in a steady nongrowing state requires energy input.
Chemical energy is needed to drive many biochemical reactions, to do mechanical work, and for transport of substances across the plasma membrane.
The ultimate source of energy that drives a cell's reactions is derived from
sunlight (Figure I–17). Light energy is converted into chemical energy in the
chloroplasts of plant cells or in the photosynthetic grana of certain microor-

Figure I–17
Flow of energy in the biosphere. The
sun's rays are the ultimate source of
energy. These rays are absorbed and
converted into chemical energy (ATP)
in the chloroplasts. The chemical
energy is used to make carbohydrates
from carbon dioxide and water. The
energy stored in the carbohydrates is
then used, directly or indirectly, to
drive all the energy-requiring
processes in the biosphere.

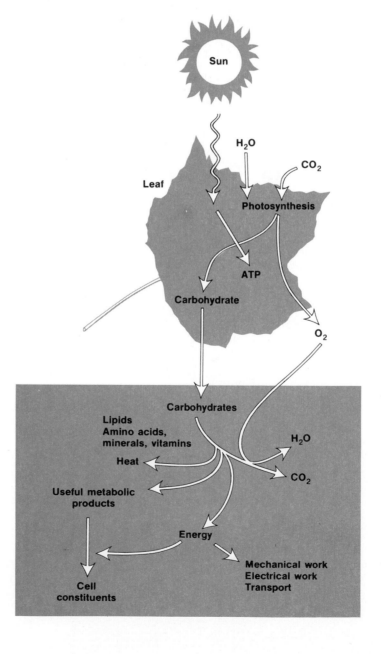

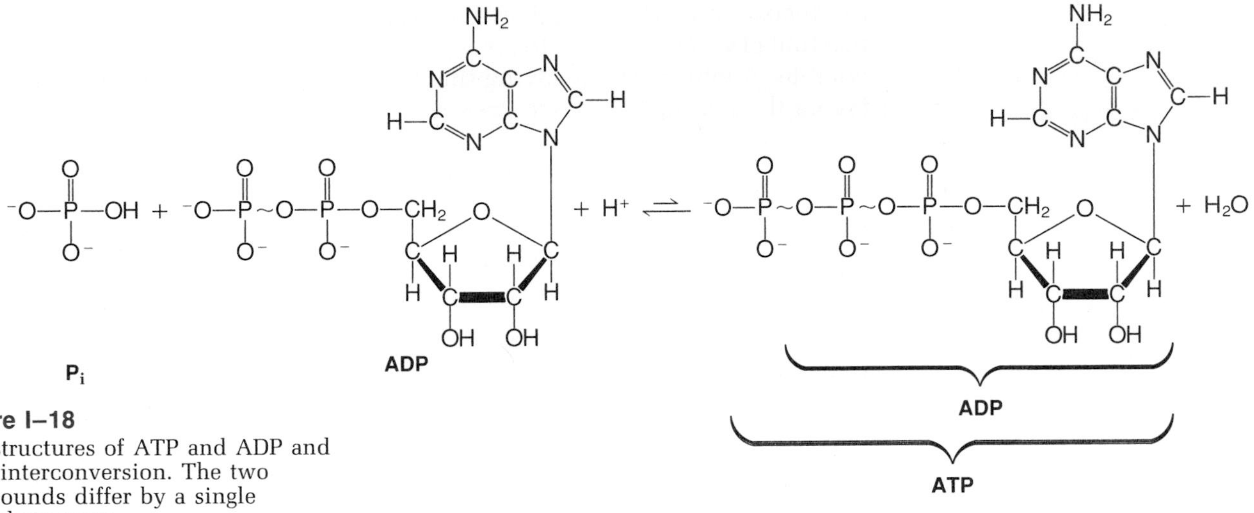

Figure I–18
The structures of ATP and ADP and their interconversion. The two compounds differ by a single phosphate group.

ganisms. The main form of chemical energy produced in the chloroplast is a nucleotide containing three phosphoric acid groups attached in sequence, called *adenosine triphosphate* or *ATP* (Figure I–18). Organisms that cannot harness the light rays of the sun themselves to make ATP are able to make ATP from the breakdown of organic nutrients originating from plants or other organisms.

Many nutrients consist of partly degraded macromolecules or various other small molecules that after absorption must be converted into a form suitable for the production of ATP. One of the simplest and yet most effective substances useful in ATP synthesis is the six-carbon sugar glucose. Degradation of a molecule of glucose can produce 38 molecules of ATP by the following overall reaction, which involves several enzymes:

$$C_6H_{12}O_6 + 6\ O_2 + 38\ ADP + 38\ P_i^* \longrightarrow 6\ CO_2 + 38\ ATP + 44\ H_2O$$

Two characteristics of this equation should be noted. First, the glucose is degraded by oxidation to CO_2 and H_2O. A substantial fraction of the energy released by this complete oxidation of glucose is used in the production of ATP. Second, ATP is being synthesized not from small molecular precursors but simply by the addition of a single phosphate (P_i) to adenosine diphosphate (ADP). Considerable energy is released when ATP is hydrolyzed to ADP to P_i, and the ADP can be reutilized many hundreds of times. These are two of the reasons that ATP is so effective as an energy source. A third reason is that ATP is quite stable in water; it does not lose its terminal phosphate readily except in enzyme-catalyzed reactions.

Most biochemical reactions fall into one of two classes: degradative or synthetic. Degradative, or *catabolic*, reactions result in the breakdown of organic compounds to simpler substances. Synthetic, or *anabolic*, reactions lead to the assembly of biomolecules from simpler molecules. Most anabolic processes require energy to drive them. This energy is usually supplied by coupling the energy-requiring biosynthetic reactions to energy-releasing catabolic reactions. Most frequently the energy-releasing reaction involves ATP hydrolysis, either directly or indirectly.

As an example of coupling the hydrolysis of ATP to an energy-requiring reaction, let us examine the use of glucose in an anabolic reaction, the synthesis of a polysaccharide containing glucose as the monomer. The first step

$$^*P_i = HO-\overset{\displaystyle O}{\underset{\displaystyle O^-}{\overset{\|}{P}}}-O^-$$

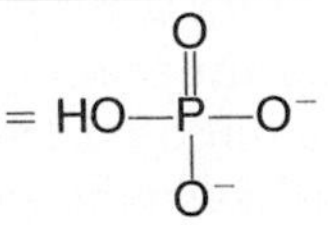

in glucose utilization involves its conversion to glucose-6-phosphate. The reaction of glucose with P_i to form glucose-6-phosphate is energetically unfavorable. Another way of saying this is that the equilibrium for the reaction favors the reactants, not the products.

$$\text{Glucose} + P_i \rightleftharpoons \text{glucose-6-phosphate}$$

Thus, even in the presence of a catalyst for this reaction, very litle glucose-6-phosphate would be formed. In order to make glucose-6-phosphate efficiently, its formation from glucose is coupled to the energetically favorable reaction of ATP breakdown:

$$\text{ATP} \rightleftharpoons \text{ADP} + P_i$$

When these two reactions are biochemically coupled, the net reaction is

$$\text{Glucose} + \text{ATP} \rightleftharpoons \text{glucose-6-phosphate} + \text{ADP}$$

The net reaction now favors the formation of glucose-6-phosphate, because more energy is released by ATP hydrolysis than is required for the phosphorylation of glucose. In the cell this is what happens when glucose-6-phosphate is synthesized. The unfavorable (energy-requiring) reaction is coupled to the favorable (energy-releasing) reaction to give an overall favorable reaction.

ORGANISMS ARE BIOCHEMICALLY DEPENDENT ON ONE ANOTHER

About 3.8 billion years ago the first organisms appeared on earth; they had to have the capacity for extracting needed nutrients from the chemical compounds that existed in prebiotic times. We have some general notions about what types of substances were present at that time. One of the most important substances that was not present at that time in significant amounts was molecular oxygen, O_2. Currently this form of oxygen is required by all forms of life visible to the naked eye.

The O_2 that is used by most organisms is ultimately converted by them into CO_2. Oxygen is utilized at a rapid rate and it would soon disappear if it were not for special classes of photosynthetic organisms that are constantly producing more O_2 by the oxidation of water.

The oxygen story is an example of the dependence of one class of organisms on another for certain chemicals. A similar situation exists with the elements carbon and nitrogen, which must be converted from gaseous forms, CO_2 and N_2, to organic forms utilizable by most organisms. Reduced carbon compounds are constantly being lost by oxidation to gaseous CO_2. The supply of organic carbon compounds required by all forms of life is replenished by photosynthetic organisms; these include most plants and certain microorganisms. Similarly, nitrogen in organic molecules is constantly being lost to the atmosphere in the form of gaseous nitrogen. The reactions required for the conversion of nitrogen to a reduced form more usable to the majority of organisms occurs in only a limited number of microorganisms; yet without these nitrogen-fixing organisms life as we know it would soon vanish.

As we ascend the evolutionary tree, we find increasingly complex multicellular forms. Such organisms generally require more complex nutrients, which must ultimately be supplied to them by simpler living forms. Bacteria like _Escherichia coli_ can make all of their own amino acids from a reduced form of nitrogen, such as NH_3, and a reduced form of carbon, such as glucose. Humans, on the other hand, must receive most of their amino acids as nutrients. Humans and other complex organisms have gained new biochemical capacities, which permit them to make the components associated with highly specialized differentiated tissues. At the same time, they have lost many of the biochemical systems required to survive on simpler nutrients.

Many biochemical reactions of great importance take place in only a limited number of organisms. This fact increases the complexity of the study of biochemistry. We must learn many reactions; we must also be aware of the biochemical potentials of different organisms. This is the only way we can understand the biochemical interdependency of organisms.

BIOCHEMICAL REACTIONS ARE LOCALIZED IN THE CELL

Biochemical reactions are organized so that different reactions occur in different parts of the cell. This organization is most apparent in eukaryotes, where membrane-bounded structures give us visible proof for the localization of different biochemical processes. For example, the synthesis of DNA and RNA takes place in the nucleus of a eukaryotic cell. The RNA is subsequently transported across the nuclear membrane to the cytoplasm, where it takes part in protein synthesis. Proteins made in the cytoplasm are used in all parts of the cell. A limited amount of protein synthesis also occurs in chloroplasts and mitochondria. Proteins made in these organelles are used exclusively in organelle-related functions. Most ATP synthesis occurs in chloroplasts and mitochondria. A host of reactions involving the transport of nutrients and metabolites occur in the plasma membrane and the membranes of various intracellular organelles. The localization of functionally related reactions to different parts of the cell permits the concentration of reactants and products at sites where they can be most efficiently utilized.

BIOCHEMICAL REACTIONS ARE ORGANIZED INTO PATHWAYS

Most biochemical reactions are integrated into multistep pathways involving several enzymes. For example, the breakdown of glucose into CO_2 and H_2O involves a series of reactions that begins in the cytosol and continues to completion in the mitochondrion. A complex series of reactions like this is referred to as a _biochemical pathway_ (Figure I–19). Synthetic reactions,

Figure I–19

Summary diagram of the breakdown of glucose to carbon dioxide and water in a eukaryotic cell. As depicted here, the process starts with the absorption of glucose at the plasma membrane and its conversion into glucose-6-phosphate. In the cytosol, this six-carbon compound is then broken down by a sequence of enzyme-catalyzed reactions into two molecules of the three-carbon compound pyruvate. After absorption by the mitochondrion, pyruvate is broken down to carbon dioxide and water by a sequence of reactions that requires molecular oxygen.

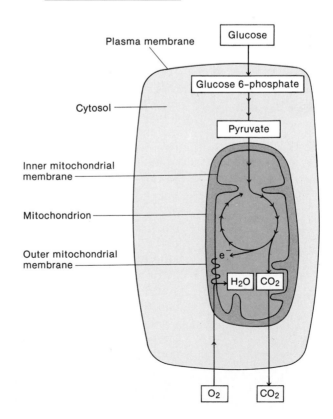

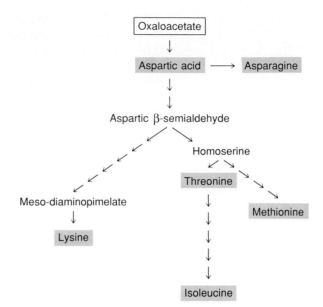

Figure I–20

Synthesis of various amino acids from oxaloacetate. Each arrow represents a discrete biochemical step requiring a unique enzyme. Thus aspartic acid is produced in one step from oxaloacetate, whereas isoleucine is produced in five steps from threonine.

such as the biosynthesis of amino acids in *E. coli,* are similarly organized into pathways (Figure I–20). Frequently pathways have branchpoints. For example, the synthesis of the amino acids threonine and lysine starts with the carbohydrate oxaloacetate. After three steps, a branchpoint is reached in the formation of the organic compound aspartic-β-semialdehyde. One branch of this pathway leads to the synthesis of the amino acid lysine, and another branch leads to the synthesis of the amino acids methionine, threonine, and isoleucine.

An understanding of the role of each biochemical reaction requires an appreciation of the pathway in which the reaction is involved and of the relationship of that pathway with other functionally related pathways.

BIOCHEMICAL REACTIONS ARE REGULATED IN CELLS

There are hundreds of biochemical reactions taking place even in the cells of relatively simple microorganisms. Living systems have evolved a sophisticated hierarchy of controls that permits them to maintain a stable intracellular environment. These controls insure that substances required for maintenance and growth are produced in adequate amounts but without huge excesses. Biochemical controls have developed in such a way that the cell can make adjustments in response to a changing external environment. Adjustments are needed because the temperature, ionic strength, acid concentration, and concentration of nutrients present in the external environment vary over much wider limits than could be tolerated inside the cell.

The rate of intracellular reactions is a function of the availability of substrates and enzymes. Enzyme activity is controlled at two different levels. First and foremost, the rate of a catalyzed reaction is regulated by the amount of the catalyzing enzyme that is present in the cell. Control of enzyme amounts is usually accomplished by regulating the rate of enzyme synthesis; in some cases the rate of enzyme degradation is also regulated. We can think of controls that regulate the total amount of enzyme present as coarse controls. They define the limits of possible enzyme activity as being anywhere from 0 to 100 percent of the full activity of the enzyme. Fine controls that act directly on enzymes are also present. Only certain special enzymes, called *regulatory enzymes*, are susceptible to this second type of regulation. Regulatory enzymes usually occupy key points in biochemical pathways, and their state of activity frequently is decisive in determining the utilization of the pathway.

A simple example will serve to illustrate how these two types of controls work. We have already mentioned that *E. coli* can synthesize all of the amino acids required for protein synthesis. Histidine is one of these amino acids; its synthesis starts from the sugar phosphate compound phosphoribosylpyrophosphate (PRPP) and requires ten enzymes. Each enzyme catalyzes one reaction in a ten-step pathway. The synthesis of all ten enzymes is regulated by the end product of the pathway, histidine. If there is sufficient histidine present for protein synthesis, then the cell ceases to make the enzymes for this pathway. The shutdown is triggered by controls that sense the level of histidine in the cell and as a result turn off synthesis of the messenger RNA required to make the enzymes. The histidine pathway also provides an example of a regulatory enzyme. The activity of the first enzyme in the pathway is directly inhibited by histidine. Thus, when there is sufficient histidine present in the cell, the activity of the first enzyme in the pathway is inhibited, and no more material or energy is funneled into the synthesis of unneeded histidine. Finally, if there is abundant histidine available from the external environment, the extracellular histidine is absorbed into the cell and the synthesis of the enzymes of the histidine pathway is brought to a halt. Bacterial cells grown in a histidine-rich growth medium for several generations contain only trace amounts of these enzymes.

The underlying principle in regulation is to maintain a favorable intracellular environment in the most economical manner. The cell makes products in the amounts that are needed. Each pathway is regulated in a somewhat different way, assuring that biochemical energy and substrates will be efficiently utilized.

OVERVIEW OF THE TEXT

The contents of this text reflect the issues raised in this introduction. It is necessary to divide up a subject for presentational purposes, but you should always remember that such division is artificial, and that by and large it does not exist in the cell.

In Part I we will consider the structures and some of the functions of the four main classes of molecules found in cells: lipids, carbohydrates, proteins, and nucleic acids. In Part II we will examine the types of reactions carried out by enzymes and the mechanisms of enzyme action. Part III is concerned with catabolism and chemical energy production. Part IV presents the biosynthetic processes involved in lipid and carbohydrate metabolism, as well as in the synthesis of the building blocks of nucleic acids (nucleotides) and the building blocks of proteins (the amino acids). In Part V we continue to study biosynthetic processes, but more specifically those processes involved in information transfer from DNA to protein. Part VI deals with reactions occurring at the membrane surface; most of these involve the transport of substances across membranes.

I

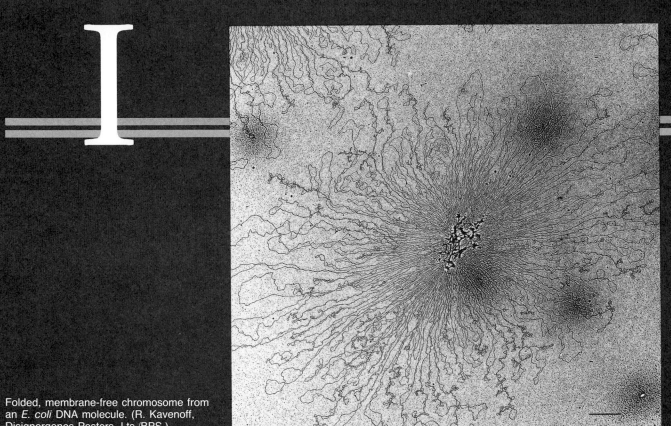

Folded, membrane-free chromosome from an *E. coli* DNA molecule. (R. Kavenoff, Disignergenes Posters, Lts./BPS.)

MAJOR COMPONENTS
OF THE CELL

PART I

In the seven chapters of Part I, we discuss the molecular structures, properties, and biological functions of the four fundamental classes of molecules from which all living things on the earth are constructed: amino acids, carbohydrates, lipids, and nucleotides.

The first three chapters deal with amino acids and their polymers, polypeptides and proteins. In Chapter 1 we describe the properties of the twenty individual amino acids found in proteins and how these amino acids are linked together to form peptides and polypeptides. The methods used to determine the amino acid composition and sequence of polypeptides are presented here, as are the requirements for their laboratory synthesis. In Chapter 2 we turn our attention to the geometry and attractive forces that determine the three-dimensional structures of protein molecules. These properties are the basis for the extraordinary variety and functional specificity of proteins in living systems. We examine both representative and unique properties of some specific proteins, namely, collagen, hemoglobin, actin, myosin, and immunoglobulin.

Chapter 3 is something of an interlude. It explores the different laboratory techniques used to isolate and characterize proteins. Many of these techniques are used in similar or modified form for other macromolecules and will be referred to often in the chapters that follow. Since what we know is dependent on and limited by the methods we use, this chapter has increasing importance as we progress further in the book.

In Chapter 4 we focus on the molecular structures and properties of the carbohydrates, a group of compounds with fundamental importance for life on earth. Carbohydrates serve as an energy source for all living things, and for some organisms they serve as the sole source of carbon, from which all other classes of molecules are synthesized. The discussion here includes the properties of carbohydrate polymers in the cell walls of plants and bacteria and in the outer surfaces of plasma membranes. In these circumstances and others, carbohydrates form covalent complexes with peptides, proteins, and lipids.

Chapters 5 and 6 deal with the lipids and the ubiquitous biological structure they form, the membrane. Chapter 5 focuses on the chemistry, molecular geometry, and biological significance of fatty acids, phospholipids, cholesterol, and related lipids. Chapter 6 discusses the composition of biological membranes and the way lipids interact in aqueous environments to form the fundamental three-dimensional structure characteristic of all biological membranes. Although the structure and the biological roles of membranes are referred to often in the chapters that follow, a detailed discussion of membrane transport is deferred until Part VI.

In the final chapter of Part I, we describe the nucleotides and the polymers they form. The extraordinary biological roles of the nucleic acids, DNA and RNA, can only be properly appreciated after their three-dimensional structures and stabilizing forces are presented in this chapter. In the cell, nucleic acids do not exist as isolated entities, and the different types of nucleic acid–protein complexes are discussed here. This chapter presents the fundamental background necessary to appreciate the references to nucleic acids that occur in Parts II, III, and IV, as well as the comprehensive discussion in Part V of the functions of DNA and RNAs in cellular information transfer, regulation, and the origin of life.

1

THE BUILDING BLOCKS OF PROTEINS: AMINO ACIDS, PEPTIDES, AND POLYPEPTIDES

In the middle of the nineteenth century, the Dutch chemist Gerardus Mulder extracted a substance common to animal tissues and the juices of plants, which he believed to be "without doubt the most important of all substances of the organic kingdom, and without it life on our planet would probably not exist." At the suggestion of the famous Swedish chemist Berzelius, Mulder named this substance _protein_ (from the Greek _proteios_, meaning "of first importance"), and assigned to it a specific chemical formula ($C_{40}H_{62}N_{10}O_{12}$). Although he was wrong about the chemistry of proteins, he was right about their being indispensable to living organisms. The term "protein" endures.

Proteins are the most abundant of cellular components. They include enzymes, antibodies, hormones, transport molecules, and even components for the cytoskeleton of the cell itself. Proteins are also informational macromolecules, the ultimate heirs of the genetic information encoded in the sequence of nucleotide bases within the chromosomes. Structurally and functionally, they are the most diverse and dynamic of molecules and play key roles in nearly every biological process. Proteins are complex macromolecules with exquisite specificity; each is a specialized player in the orchestrated activity of the cell. Together they tear down and build up molecules, extract energy, repel invaders, act as delivery systems, and even synthesize the genetic apparatus itself.

In the first three chapters of Part I we will discuss the basic structural and chemical properties of proteins. In this chapter we will concentrate on the basic structural and chemical properties of amino acids, peptides, and polypeptides—the building blocks of proteins. In Chapter 2 we will discuss the complex three-dimensional structures of proteins; and in Chapter 3 we will describe methods used for isolation and characterization of proteins.

AMINO ACIDS

Every protein molecule can be considered as a polymer of amino acids. There are twenty common amino acids. Figure 1–1a shows the structure of a single amino acid. At the center is a tetrahedral carbon atom called the α carbon (C_α). It is covalently bonded on one side to an amino group (NH_2) and on the other side to a carboxyl group (COOH). A third bond is always hydrogen, and the fourth bond is to a variable side chain (R). In neutral solution (pH 7), the carboxyl group loses a proton and the amino group gains one. Thus an amino acid in solution, while neutral overall, is a double charged species called a _zwitterion_ (Figure 1–1b).

The structures of the twenty amino acids commonly found in proteins are listed in Table 1–1. All of these amino acids except proline have an α ammonium ion ($-N^+H_3$) attached to the α carbon. In proline one of the N—H linkages is replaced by a N—C linkage forming part of a cyclic structure. Various ways of classifying amino acids according to their R groups have been proposed. In Table 1–1 we have divided the amino acids into three categories. The first category contains eight amino acids with relatively apolar R groups; the second category contains seven amino acids with uncharged polar R groups; and the third category contains five amino acids with R groups that normally exist in the charged state.

Amino acids are often abbreviated by three-letter symbols; occasionally when this proves to be too cumbersome a one-letter symbol is used. Both of these designations are given in Table 1–1, together with the molecular weight (M_r) of each amino acid.

In addition to the twenty commonly occurring α-amino acids, a variety of other amino acids are found in minor amounts in proteins and in nonprotein compounds. The unusual amino acids found in proteins result from modification of the common amino acids (see Chapter 24). The unusual amino acids found in nonprotein compounds are extremely varied in type and are formed by a number of different metabolic pathways (see Chapter 24).

Figure 1–1
Amino acid anatomy. (a) Uncharged amino acid. (b) Doubly charged zwitterion.

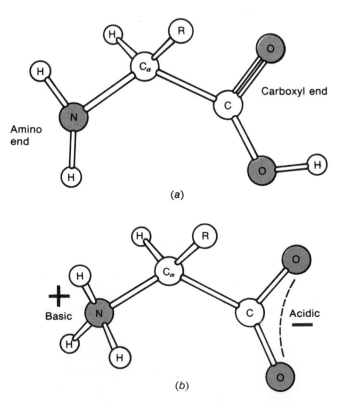

TABLE 1–1
Structure of the Twenty Amino Acids Found in Proteins

Group I. Amino Acids with Apolar R Groups

R groups

Alanine
Ala
A
M_r 89

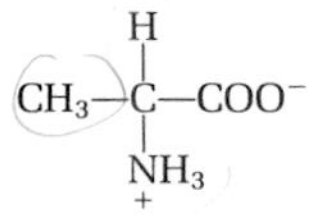

Valine
Val
V
M_r 117

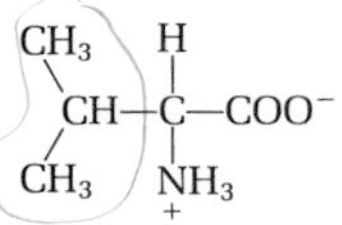

Leucine
Leu
L
M_r 131

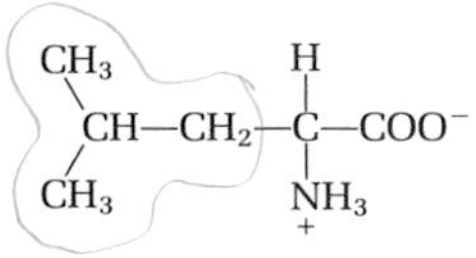

Isoleucine
Ile
I
M_r 131

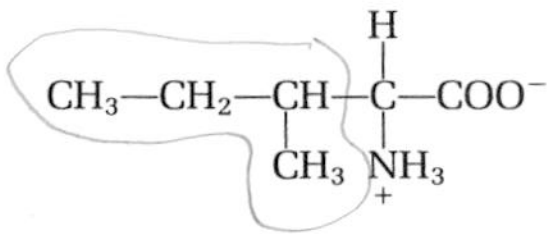

Proline
Pro
P
M_r 115

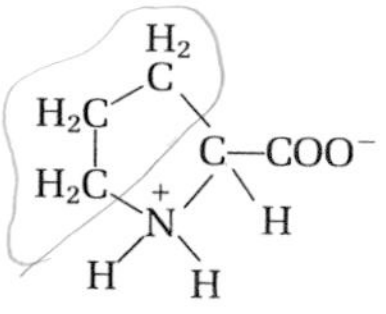

Phenylalanine
Phe
F
M_r 165

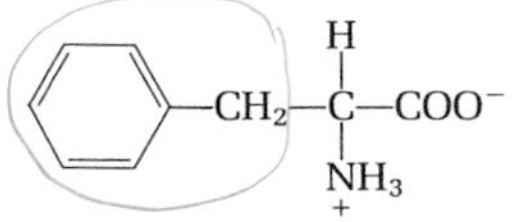

Tryptophan
Trp
W
M_r 204

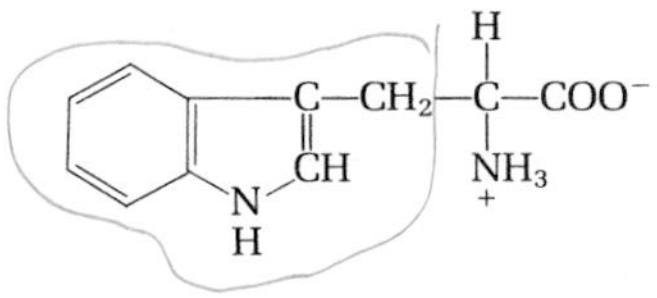

Methionine
Met
M
M_r 149

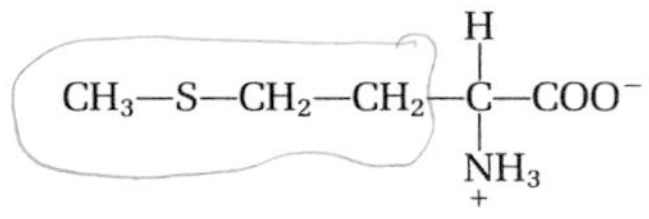

Group II. Amino Acids with Uncharged Polar R Groups

R groups

Glycine
Gly
G
M_r 75

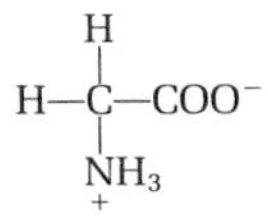

Serine
Ser
S
M_r 105

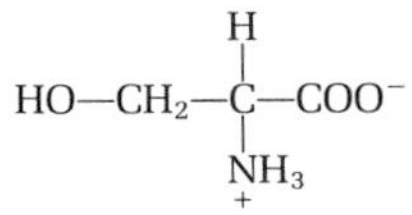

Threonine
Thr
T
M_r 119

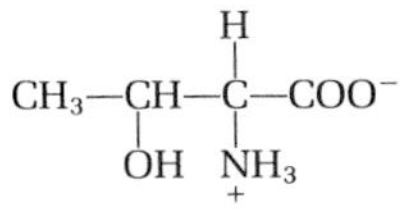

Cysteine
Cys
C
M_r 121

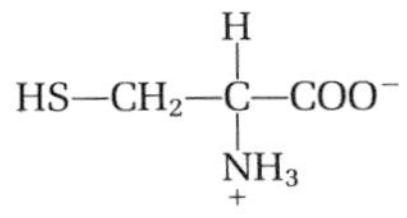

Tyrosine
Tyr
Y
M_r 181

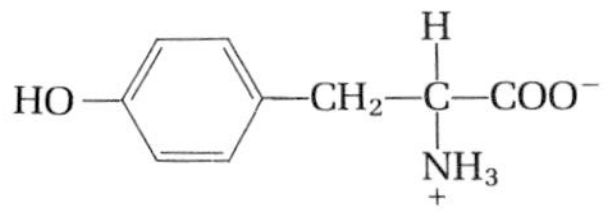

Asparagine
Asn
N
M_r 132

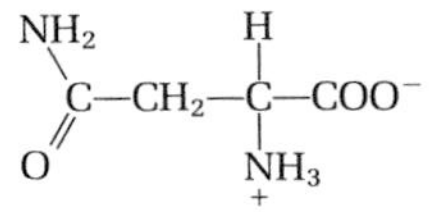

Glutamine
Gln
Q
M_r 146

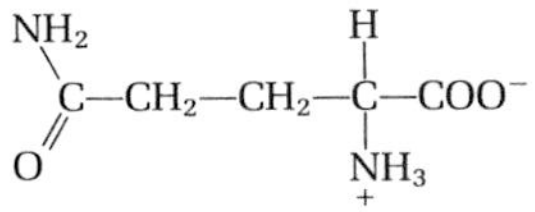

Group III. Amino Acids with Charged R Groups

R groups

Aspartic acid
Asp
D
M_r 133

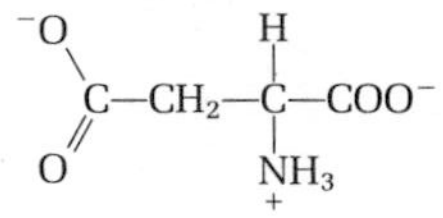

Glutamic acid
Glu
E
M_r 147

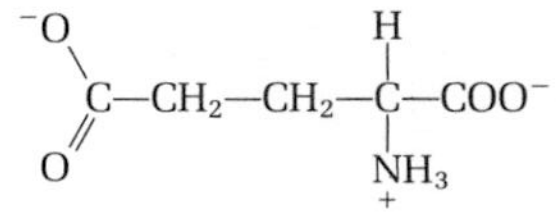

Lysine
Lys
K
M_r 146

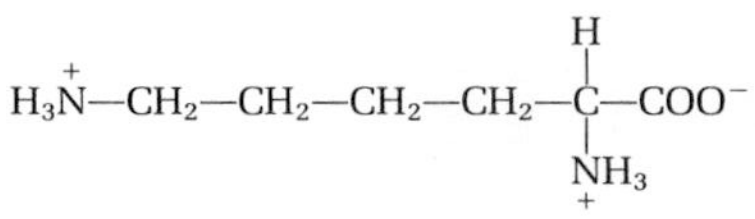

Arginine
Arg
R
M_r 174

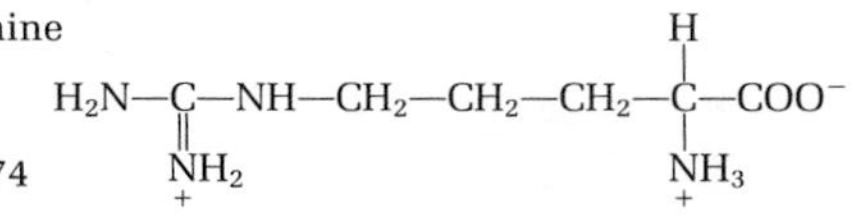

Histidine (at pH 6.0)
His
H
M_r 155

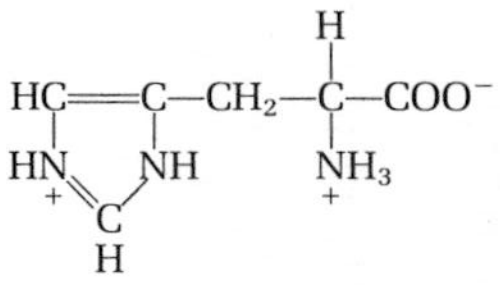

*Molecular weights in this text are expressed in units of grams per mole.

Amino Acids Have Both Acid and Base Properties

The charge properties of amino acids are very important in determining the reactivity of certain amino acid side chains and in the properties they confer on proteins. The charge properties of amino acids in aqueous solution may best be considered under the general treatment of acid–base ionization theory. We will find this treatment useful at other points in the text as well.

Recall that water can be considered a weak acid (or a weak base) because it dissociates into a proton and a hydroxide ion, according to the equilibrium

$$H_2O \rightleftharpoons H^+ + OH^- \tag{1}$$

The equilibrium expression for this reaction is

$$K_{eq} = \frac{[H^+][OH^-]}{[H_2O]} \tag{2}$$

Because water dissociates to such a small extent, the concentration of undissociated water is high and does not vary significantly for chemical reactions in aqueous solution. Therefore, the denominator in this equation is effectively constant and for simplicity is given a value of unity. The constant K_w for the dissociation of water is redefined by the expression

$$K_w = [H^+][OH^-] = 10^{-14} \ (mole/liter)^2 \tag{3}$$

at 25°C.

In pure water we expect equal amounts of H^+ ("hydrogen ion") and OH^- ("hydroxide ion"). From the above equation we can calculate the concentration of H^+ or OH^- in pure water to be 10^{-7} M. Therefore, a solution with a H^+ concentration of 10^{-7} M is defined as neutral. A H^+ concentration greater than 10^{-7} M indicates an acidic solution; a H^+ concentration less than 10^{-7} M indicates a basic solution. Rather than deal with exponentials, it is convenient to express the H^+ concentration on a pH scale, the term pH being defined by the equation:

$$pH = \log (1/[H^+]) = -\log[H^+] \tag{4}$$

According to this definition a neutral solution has a pH of 7. Other values of pH and corresponding H^+ and OH^- concentrations are given in Table 1–2.

The most common equilibria that biochemists encounter are those of acids and bases. The dissociation of an acid may be written as

$$HA \rightleftharpoons H^+ + A^- \tag{5}$$

The equilibrium constant for this reaction is called the _acid dissociation constant_, K_a, written as

$$K_a = \frac{[H^+][A^-]}{[HA]} \tag{6}$$

Strong acids in aqueous solution dissociate completely into anions and protons. The concentration of hydrogen ion $[H^+]$ is therefore equal to the total concentration C_{HA} of the acid HA that is added to the solution. Thus the pH of the solution of a strong acid is simply $-\log C_{HA}$.

The pH of the solution of a weak acid is a function of both the C_{HA} and the acid dissociation constant. The dissociation constant of a weak acid may be written in terms of the species present in the equation for the acid dissociation constant.

First solving Equation (6) for $[H^+]$ gives

$$[H^+] = \frac{K_a[HA]}{[A^-]} \tag{7}$$

TABLE 1–2
The pH Scale

pH	$[H^+]$	$[OH^-]$
0	10^0	10^{-14}
1	10^{-1}	10^{-13}
2	10^{-2}	10^{-12}
3	10^{-3}	10^{-11}
4	10^{-4}	10^{-10}
5	10^{-5}	10^{-9}
6	10^{-6}	10^{-8}
7	10^{-7}	10^{-7}
8	10^{-8}	10^{-6}
9	10^{-9}	10^{-5}
10	10^{-10}	10^{-4}
11	10^{-11}	10^{-3}
12	10^{-12}	10^{-2}
13	10^{-13}	10^{-1}
14	10^{-14}	10^0

Taking the logarithm of both sides and changing signs gives us

$$-\log[H^+] = -\log K_a + \log\frac{[A^-]}{[HA]} \qquad (8)$$

Substituting pH for $-\log[H^+]$ and pK_a for $-\log K_a$ in Equation (8), we obtain the *Henderson-Hasselbach equation:*

$$pH = pK_a + \log\left[\frac{A^-}{HA}\right] = pK_a + \log\left[\frac{base}{acid}\right] \qquad (9)$$

The Henderson-Hasselbach equation is useful for calculating the molar ratio of base (proton acceptor) to acid (proton donor) for a given pH and pK or for calculating the pK given the ratio of base (proton acceptor) to acid (proton donor). It can be seen that *when the concentration of anion or base is equal to the concentration of undissociated acid* (i.e., when the acid is half neutralized), *the pH of the solution is equal to the pK of the acid.*

The values of pK for a particular molecule are determined by titration. A typical pH dependence curve for the titration of a weak acid by a strong base is shown in Figure 1–2. The concentration of the anion equals the concentration of the acid when the acid is exactly half neutralized. Note that at this point on the curve, the pH is least sensitive to the quantity of added base (or acid). The solution is said to be *buffered* under these conditions. Biochemical reactions are typically highly dependent on the pH of the solution. Therefore, it is frequently advantageous to study reactions in buffered solutions. The ideal buffer is one that has a pK numerically equivalent to the working pH.

A simple amino acid with a nonionizable R group gives a complex titration curve with two inflection points. For example, such is the case in the titration of alanine, shown in Figure 1–3. At very low pH, alanine carries a single positive charge on the α amino group. The first inflection point occurs at a pH of 2.3. This is the pK for titration of the carboxyl group, pK_1 ($-COOH \rightarrow -COO^-$). At a pH of 6.0, alanine has an equal amount of positive and negative charge. This value is referred to as the *isoelectric point* or the *isoelectric pH*. As the titration continues, a second inflection point is

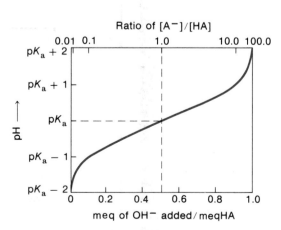

Figure 1–2
The dependence of pH on the equivalents of base added to a typical weak acid. Note that at the pKa, $[A^-] = [HA]$.

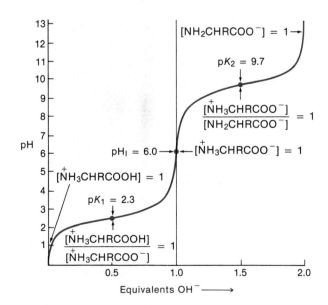

Figure 1–3
Titration curve of alanine. The predominant ionic species at each cardinal point in the titration is indicated.

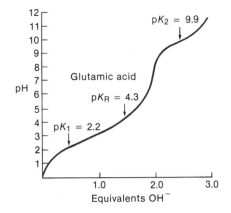

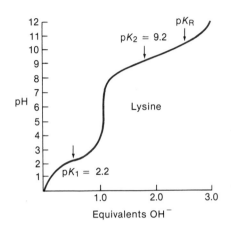

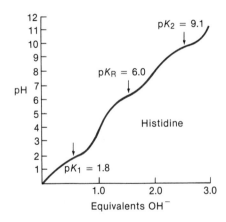

Figure 1–4
Titration curves of glutamic acid, lysine, and histidine. In each case, the pK of the R group is designated pK_R.

reached at a pH of 9.7. The pK at this point, pK_2, represents the equilibrium for the titration of the proton from the amino group ($-N^+H_3 \rightarrow NH_2$).

Amino acids with an ionizable R group show even more complex titration curves, indicative of three ionizable groups (Figure 1–4). The pK for the ionizable side chain, pK_R, is usually readily distinguishable from the pK values for the ionizable α carboxyl and α amino groups, pK_1 and pK_2, respectively, as the latter have numerical values close to the comparable pK values of alanine (see Figure 1–3 and Table 1–3). Note that the only ionizable R group with a pK_R in the vicinity of 7, where most biological systems function, is that for histidine. This means that although other ionizable groups are usually fully charged under biological conditions, the side chain of histidine can be fully charged, uncharged, or partially charged, depending on the precise situation. This variability has major implications for the way the histidine side chain functions in enzyme catalysis. The side chain can serve as either a proton donor or a proton acceptor (see discussion in Chapter 9).

An additional point should be noted from Table 1–3. Whereas the amino acid side chains (R groups) that are normally charged at physiological pH are restricted to five amino acids (aspartic acid, glutamic acid, lysine, arginine, and sometimes histidine), a number of potentially ionizable R groups are part of other amino acids. These include cysteine, serine, threonine, and tyrosine. The ionization reactions for all of the potentially ionizable side chains are indicated in Figure 1–5.

All Amino Acids Except Glycine Show Asymmetry

One of the most striking and significant properties of amino acids is their *chirality* or handedness. The word "chiral" is related to the Greek word meaning hand. Just as the right hand is related to the left hand by a mirror image, so, in general, a naturally occurring amino acid is related to a

TABLE 1–3
Values of pK for the Ionizable Groups of the Twenty Amino Acids Commonly Found in Proteins

Amino acid	pK_1 (α —COOH)	pK_2 (α —NH$_3^+$)	pK_R (R group)
Alanine	2.3	9.9	—
Arginine	1.8	9.0	12.5
Asparagine	2.0	8.8	—
Aspartic acid	2.0	10.0	3.9
Cysteine	1.8	10.8	8.3
Glutamic acid	2.2	9.9	4.3
Glutamine	2.2	9.1	—
Glycine	2.3	9.8	—
Histidine	1.8	9.1	6.0
Isoleucine	2.3	9.7	—
Leucine	2.3	9.7	—
Lysine	2.2	9.2	10.8
Methionine	2.2	9.3	—
Phenylalanine	2.6	9.2	—
Proline	1.9	10.6	—
Serine	2.2	9.4	~13
Threonine	2.1	9.1	~13
Tryptophan	2.4	9.4	—
Tyrosine	2.2	9.1	10.1
Valine	2.3	9.7	—

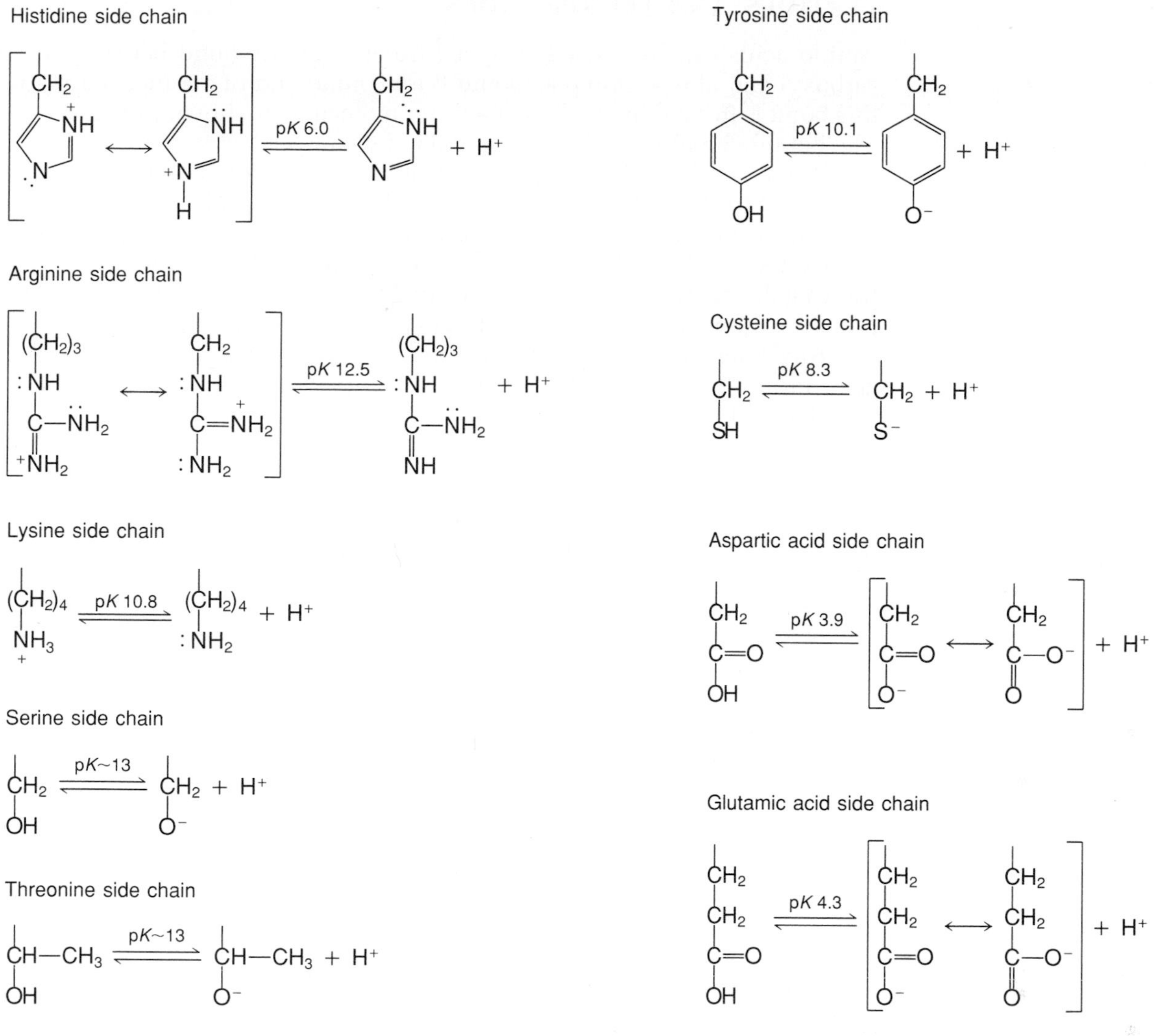

Histidine side chain

Tyrosine side chain

Arginine side chain

Cysteine side chain

Lysine side chain

Aspartic acid side chain

Serine side chain

Glutamic acid side chain

Threonine side chain

Figure 1–5

Equilibrium between charged and uncharged forms of amino acid side chains.

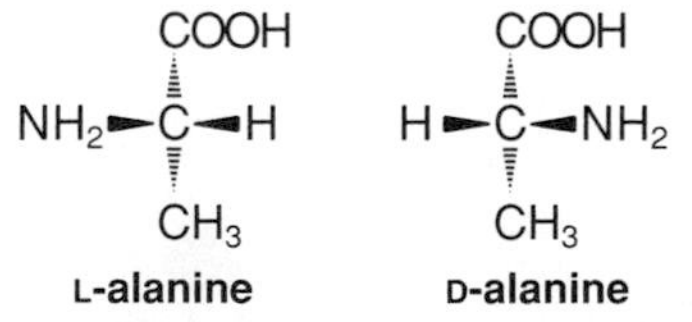

Figure 1–6

The covalent structure of alanine, showing the three-dimensional structure of the L and D sterioisomeric forms. Those bonds that project out of the plane of the paper and toward the reader are drawn as solid triangles; those bonds that lie behind the plane of the paper and away from the reader are shown as a series of dashes.

stereoisomer by its mirror image. This observation is true of nineteen out of the twenty amino acids; the one exception is glycine.

The chirality of amino acids stems from the chiral or asymmetric center, the α-carbon atom. The α-carbon atom is a chiral center if it is connected to four different substituents. Thus, glycine has no chiral center. Two of the amino acids, isoleucine and threonine, possess additional chiral centers because each has one additional asymmetric carbon. You should be able to locate these carbons by simple inspection.

Two structures that constitute a stereoisomeric pair are referred to as *enantiomers*. The two enantiomers for alanine are illustrated in Figure 1–6. These two isomers are called L-alanine and D-alanine, according to the way in which the substituents are arranged about the asymmetric carbon atom. The naming by L and D refers to a convention established by Emil Fischer many years ago. According to this convention all amino acids found in proteins are of the L form. Some D-amino acids are found in bacterial cell walls and certain antibiotics.

Another convention for referring to configurations is called the R, S convention. As the R, S convention is not as popular for amino acids or sugars as it is for other types of biomolecules, such as lipids, we will not discuss this notation in Chapter 1.

PEPTIDES AND POLYPEPTIDES

Amino acids can link together by a covalent peptide bond between the α carboxyl end of one amino acid and the α amino end of another. Formally, this bond is formed by the loss of a water molecule, as shown in Figure 1–7. The peptide bond has partial double-bond character owing to resonance effects; as a result, the C—N peptide linkage and all of the atoms directly connected to C and N lie in a planar configuration called the *amide plane*. (The effects of resonance on the peptide bond are discussed in greater depth in Box 1A.) In the following chapter we will see that this amide plane, by limiting the number of orientations available to the polypeptide chain, plays a major role in determining the three-dimensional structures of proteins.

Any number of amino acids can be joined by successive peptide linkages, forming a polypeptide chain. The polypeptide chain, like the dipeptide, has a directional sense. At one end, called the N-terminal or amino-terminal end, it has a free α amino group, whereas the other end, the C-terminal or carboxy-terminal end, has a free α carboxyl group. The sequence of main-chain atoms from the N-terminal end to the C-terminal end is C_α—C—N—C_α, etc., and in the opposite direction it is C_α—N—C—C_α, etc. Short polypeptide chains, up to a length of about 20 amino acids, are called *peptides* or *oligopeptides* if they are fragments of whole polypeptide chains. A small protein molecule may contain a polypeptide chain of only 50 amino acids; a large protein may contain chains of 3000 amino acids or more. One of the larger single polypeptide chains is that of the muscle protein myosin, which consists of approximately 1750 amino acid residues. Figure 1–8 shows a section of a polypeptide chain as a linear array with α carbons and planar amides alternating as repeating units of the main chain. Different side chains are attached to each α carbon.

Figure 1–7
Formation of a dipeptide from two amino acids. *(a)* Two amino acids. *(b)* A peptide bond (CO—NH) links amino acids by joining the α carboxyl group of one with the α amino group of another. A water molecule is lost in the reaction. It is conventional to draw dipeptides and polypeptides so that their free amino terminal is to the left and their free carboxyl terminal is to the right. The amide plane refers to six atoms that lie in the same plane.

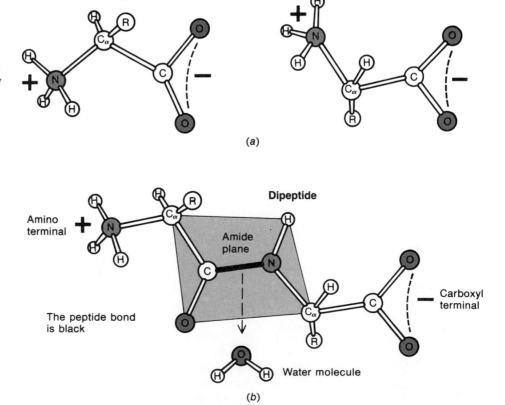

BOX 1–A

Structure of the Peptide Bond

As was first pointed out by Linus Pauling, resonance plays a major role in determining the structure of the peptide bond. This fact is immensely important in the effects it has on protein structure (see Chapter 2). The two major resonating hybrids contributing to the structure of the peptide are

In structure 1 the C—N bond is a single bond with no overlap between the nitrogen lone electron pair and the carbonyl carbon. In structure 2 there is a double bond between the amide nitrogen and the carbonyl carbon; moreover, in structure 2 the nitrogen atom bears a charge of +1 and the carbonyl oxygen bears a charge of -1.

In structure 1 the carboxyl carbon is sp^2-hybridized and is therefore planar, while the nitrogen is sp^3-hybridized and pyramidal. In structure 2 both the carboxyl carbon and the amide nitrogen are sp^2-hybridized, both are planar, and all six atoms lie in the same plane. The resulting structure is shown in the illustration. The C—N bond length is 1.325 Å, which is significantly shorter than the length of a single C—N bond, 1.47 Å. The actual structure of the peptide bonds is a compromise between structures 1 and 2.

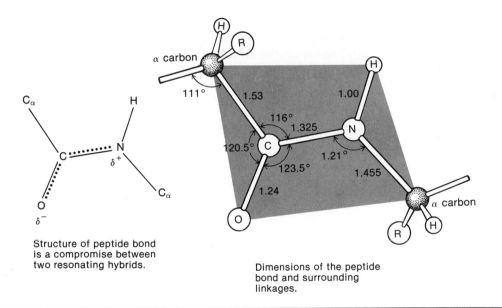

Structure of peptide bond is a compromise between two resonating hybrids.

Dimensions of the peptide bond and surrounding linkages.

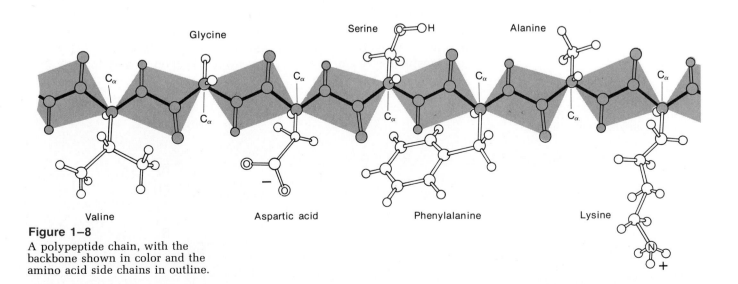

Figure 1–8

A polypeptide chain, with the backbone shown in color and the amino acid side chains in outline.

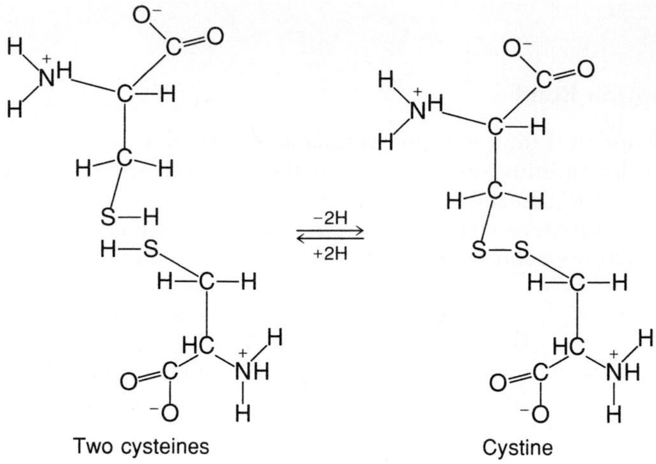

Figure 1–9

Disulfide bonds can form between two cysteines. The cysteines can exist in the cytosol as free amino acids (as shown), in which case they give rise to cysteine, or they can be on polypeptide chains. In the latter instance, they can be on the same polypeptide chains or different polypeptide chains. In either case the formation of covalent disulfide bonds stabilizes structural relationships.

In addition to the covalent peptide bonds formed between adjacent amino acids within a polypeptide chain, covalent disulfide bonds can be formed within the same polypeptide chain or between different polypeptide chains (Figure 1–9). Such disulfide linkages have an important stabilizing influence on the structures formed by many proteins (see Chapter 2).

DETERMINATION OF AMINO ACID COMPOSITION OF PROTEINS

Each protein is uniquely characterized by its amino acid composition and sequence. A protein's amino acid composition is defined simply as the number of each type of amino acid comprising the polypeptide chain. In order to determine a protein's amino acid composition it is necessary to (1) break down the polypeptide chain into its constituent amino acids, (2) separate the resulting free amino acids according to type, and (3) measure the quantities of each amino acid.

Cleavage of the peptide bonds is usually achieved by boiling the protein in 6 N HCl, which causes hydrolysis of the peptide bonds and the consequent release of free amino acids (Figure 1–10). Although acid hydrolysis is the most frequently used means of breaking a protein into its constituent amino acids, it results in the partial destruction of the indole ring of tryptophan. Consequently, the amount of tryptophan in the protein must be estimated by an alternative method (e.g., spectroscopic absorption) when using acid hydrolysis. In addition, acid hydrolysis results in the loss of ammonia from the side-chain amide groups of glutamine and asparagine, with the consequent production of glutamic and aspartic acids (Figure 1–11). Consequently, estimates of amino acid composition based on acid hydrolysis show

Figure 1–10

Acid hydrolysis of a protein or polypeptide to yield amino acids.

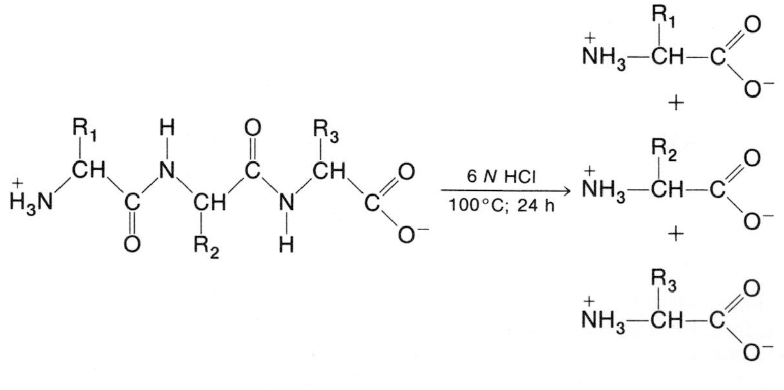

Figure 1–11

Acid hydrolysis of protein converts glutamine to glutamic acid. A similar reaction occurs for asparagine, which possesses an identical side-chain functional group.

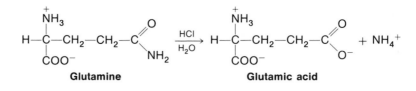

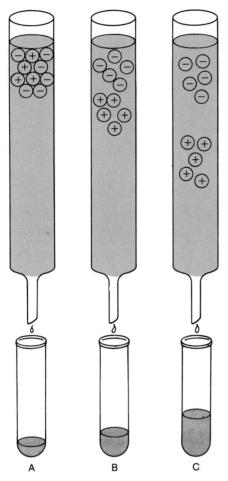

Figure 1–12

Migration of aspartic acid ⊖ and lysine ⊕ through a column with a higher affinity for aspartic acid. Views A, B, and C show the column at successively increasing time intervals after starting the elution.

glutamine and glutamic acid combined and measured as glutamic acid. Similarly, asparagine and aspartic acid are combined and measured as aspartic acid.

Separation of amino acids for quantitative analytical purposes is usually achieved by *ion-exchange chromatography*. The general efficacy of chromatographic techniques is based on a difference in affinity between each compound to be separated and an immobile phase or resin. The resin consists of some relatively chemically inert polymer that has weakly basic side-chain constituents that are positively charged at pH 7. If we were to add some of this resin to a solution containing free aspartic acid and lysine at pH 7, the negatively charged aspartic acid would have a higher affinity for the resin than would the positively charged lysine. If we were to pump a solution of these two amino acids through a column containing such a positively charged resin, the progress of the aspartic acid through the column would be retarded relative to the lysine, owing to the greater affinity of the aspartic acid for the resin (Figure 1–12).

Ion-exchange resins have been developed that have differential binding affinities for all the naturally occurring amino acids. Such resins are effective in separating a solution of amino acids into its components. It should be emphasized that the details of the forces responsible for the differential binding of amino acids to an ion-exchange resin are quite complicated and depend additionally on side-chain polarity, on subtle differences in the pK values of α-amino and α-carboxyl groups, on solvation effects, and on other factors. To enhance the separation properties of the column, such separation techniques frequently exploit changes in the pH of the solution buffer (eluting buffer) used to remove the compounds of interest. For example, a column might initially be run with the eluting buffer at a pH that results in some amino acids being so strongly bound to the resin that they are essentially immobile. However, after the separation and elution of the less strongly bound amino acids, the pH of the eluting buffer can be appropriately shifted to lessen the charge difference between the resin and the strongly bound amino acids. These amino acids can then be eluted and separated according to the newly established pattern of resin-binding affinities.

Quantitative determination of the separated amino acids is achieved by their reaction with ninhydrin to produce a colored reaction product. This product is measured spectrophotometrically. As shown in Figure 1–13, the ninhydrin reaction abstracts an amino group from each amino acid, so that the amount of colored product formed is proportional to the amount of amino acid initially present.

Figure 1–13

Reaction of ninhydrin with an amino acid yields a colored complex. The ninhydrin reaction permits qualitative location of amino acids in chromatography and quantitative assay of separated amino acids.

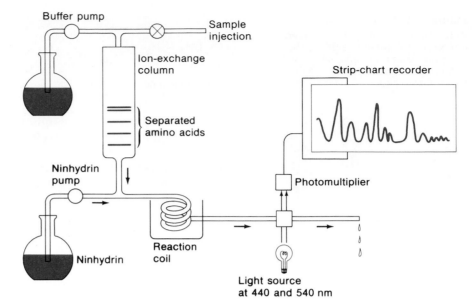

Figure 1—14
Schematic diagram of an amino acid analyzer. The amino acids are passed through an ion-exchange column and thereby separated. Eluted fractions are mixed and reacted with ninhydrin. The intensity of the resulting colored product is measured in a spectrophotometer and the results are displayed on a recording chart.

Quantitative measurements of amino acid composition are usually carried out on an *amino acid analyzer*, a device that automates the previously described operations. As illustrated in Figure 1—14, the amino acid analyzer consists of an ion-exchange column through which the appropriate eluting buffer is pumped after the amino acids are introduced at the top of the column. As the separated amino acids emerge, they are mixed with ninhydrin solution and passed through a heated coil of tubing to allow the formation of the colored ninhydrin reaction product. The separated ninhydrin reaction products then pass through a cell that measures their optical absorbance at 570 and 440 nm and plots the results on a strip-chart recorder. The absorbance is measured at two wavelengths because proline, which is substituted at its amino group, forms a different ninhydrin reaction product, with an absorption maximum that is correspondingly different from that of the remaining amino acids.

Usually the amino acid analyzer is first standardized by running through it a sample containing known quantities of amino acids, in order to account for any differences in their ninhydrin reaction properties. In this way it is possible to directly relate the amount of amino acid present to the amount of colored product formed, as measured by the area under the "peak" produced on the strip-chart recorder (see Figure 1—14). Similarly, the amino acid hydrolysate of a protein of unknown composition can be run through the analyzer, and the relative peak areas can be used to estimate the ratios of the different amino acids present.

Conversion of the relative ratios of amino acids into an estimate of actual composition requires some additional information concerning the protein's molecular weight; e.g., an analysis giving relative ratios of Ala (1.0), Gly (0.5), and Lys (2.0) could correspond to composition Ala_2-Gly-Lys_4 or any multiple thereof. The required information is usually available, and in any case, an estimation of composition based on a minimum molecular weight of the protein is always possible. Results for three proteins are shown in Table 1—4.

DETERMINATION OF AMINO ACID SEQUENCE OF PROTEINS

The most important properties of a protein are determined by the sequence of amino acids in the polypeptide chain. This sequence is called the primary structure of the protein. We know the sequences for hundreds of peptides

TABLE 1–4
Amino Acid Content of Proteins (in per cent)

Constituent	Insulin (Bovine)	Ribonuclease (Bovine)	Cytochrome (Equine)
Alanine	4.6	7.7	3.5
Amide NH_3	1.7	2.1	1.1
Arginine	3.1	4.9	2.7
Aspartic acid	6.7	15.0	7.6
Cysteine	0	0	1.7
Cystine	12.2	7.0	0
Glutamic acid	17.9	12.4	13.0
Glycine	5.2	1.6	5.6
Histidine	5.4	4.2	3.4
Isoleucine	2.3	2.7	5.4
Leucine	13.5	2.0	5.6
Lysine	2.6	10.5	19.7
Methionine	0	4.0	2.1
Phenylalanine	8.6	3.5	4.5
Proline	2.1	3.9	3.3
Serine	5.3	11.4	0
Threonine	2.0	8.9	8.4
Tryptophan	0	0	1.5
Tyrosine	12.6	7.6	4.9
Valine	9.7	7.5	2.4

and proteins, largely through the use of methods developed in Fred Sanger's laboratory and first used to determine the sequence of the peptide hormone insulin in 1953. Knowledge of the amino acid sequence is extremely useful in a number of ways: (1) it permits comparisons to be made between normal and mutant proteins (see Chapter 3); (2) it permits comparisons to be made between comparable proteins in different species and thereby has been instrumental in positioning different organisms on the evolutionary tree (see Figure 1–1, in Introduction); (3) finally and most important, it is a vital piece of information for determining the three-dimensional structure of the protein.

Determining the order of amino acids involves the sequential removal and identification of successive amino acid residues from one or the other free terminal of the polypeptide chain. However, in practice it is extremely difficult to get the required specific cleavage reaction of the desired products to proceed with 100 per cent yield. This obstacle becomes significant when sequencing long polypeptides, because the fraction of the total material of minimum polypeptide chain length becomes constantly smaller as the successive removal of terminal residues continues. Conversely, the amino acid released from the polypeptide chain becomes increasingly contaminated with amino acids released from previously unreacted chains.

Because of this fundamental chemical limitation, the polypeptide chain must be broken down into sequences short enough for the chemistry to produce reliable results. The short sequences are then reassembled to obtain the overall sequence. The steps actually involved in protein sequencing (Figure 1–15) are (1) purification of the protein; (2) cleavage of all disulfide bonds; (3) determination of the terminal amino acid residues; (4) specific cleavage of the polypeptide chain into small fragments in at least two different ways; (5) independent separation and sequence determination of peptides produced by the different cleavage methods; and (6) reassembly of the individual peptides with appropriate overlaps to determine the overall sequence.

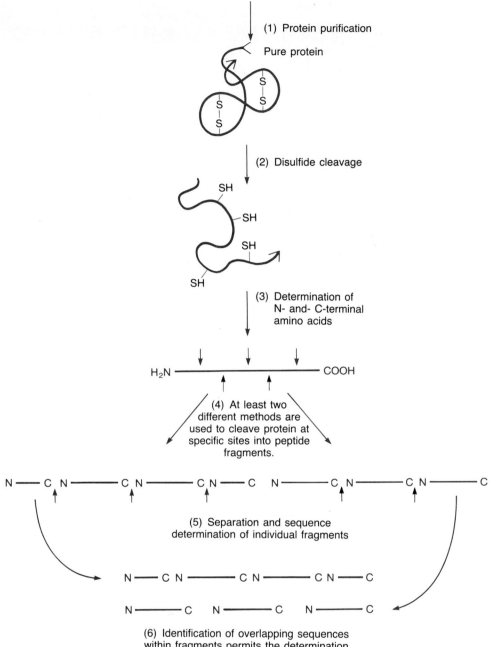

Figure 1–15
Steps involved in sequencing a protein.

The first step, protein purification, is a process that will be discussed in Chapter 3. Once a supply of pure protein is available, sequence analysis can begin, with cleavage of the disulfide bonds. Cleavage can be achieved either by oxidation of the disulfide linkage with performic acid or by reduction of the bond, followed by subsequent reaction to prevent re-formation of the disulfide linkage (Figure 1–16).

The third step is to determine the polypeptide-chain end groups. The amino-terminal amino acid can be identified by reaction with fluorodinitrobenzene (FDNB) (Figure 1–17). Subsequent acid hydrolysis releases the FDNB-labeled amino-terminal amino acid, which is highly colored and can be identified by its characteristic migration behavior on thin-

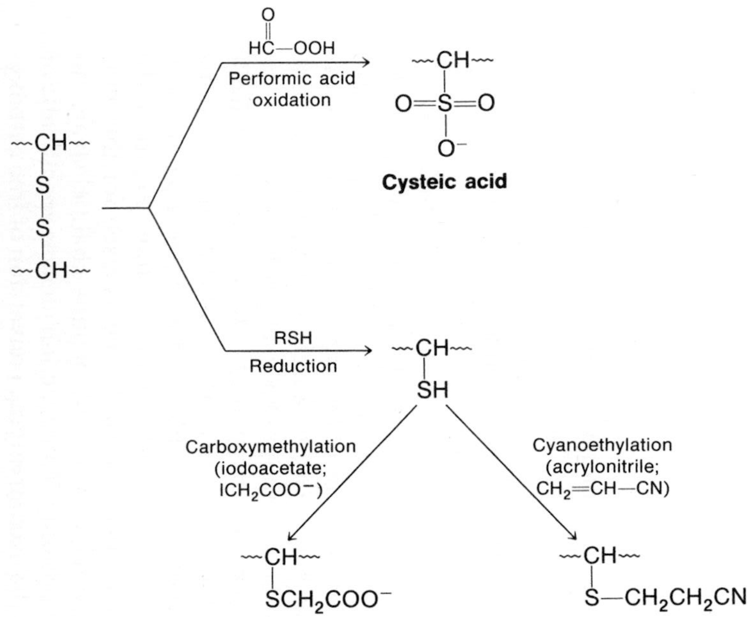

Figure 1-16

Disulfide cleavage reactions. Prior to sequence analysis interchain and intrachain disulfide linkages are irreversibly cleaved by one of the two procedures shown.

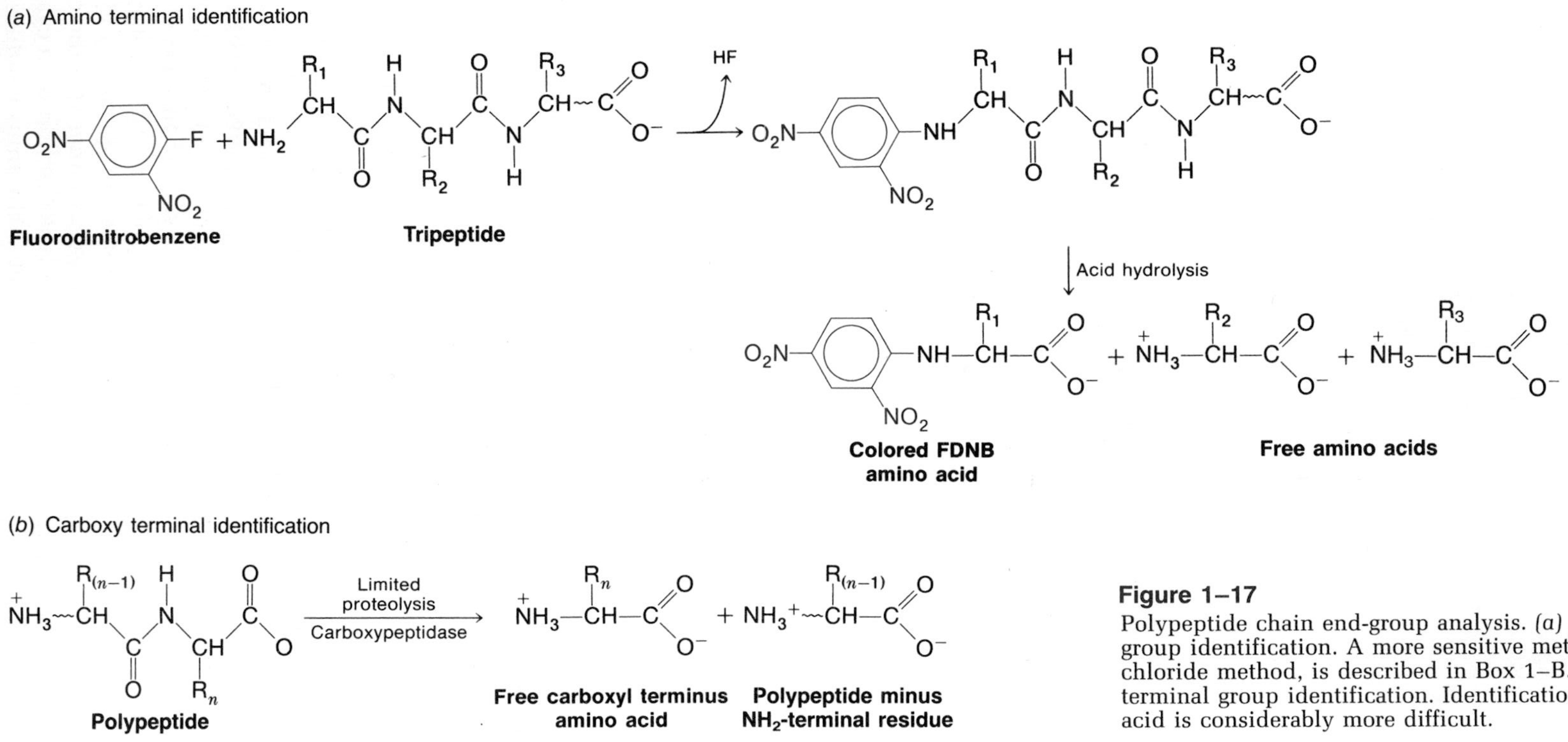

(a) Amino terminal identification

(b) Carboxy terminal identification

Figure 1-17

Polypeptide chain end-group analysis. (*a*) Amino-terminal group identification. A more sensitive method, the dansyl chloride method, is described in Box 1–B. (*b*) Carboxy-terminal group identification. Identification of this amino acid is considerably more difficult.

BOX 1–B

The Dansyl Chloride Method for N-terminal Amino Acid Determination

The dansyl chloride method provides an alternative to the Sanger method for N-terminal amino acid determination. Because it is considerably more sensitive than the Sanger method, it has become the method of choice. The reaction is diagrammed in the illustration. A polypeptide is treated with dansyl chloride to give an N-dansyl peptide derivative. This derivative is hydrolyzed to yield a highly fluorescent N-dansyl-amino acid, which is detected chromatographically.

layer chromatography or paper electrophoresis. A more sensitive method of end-group determination involves the use of dansyl chloride (see Box 1B).

Chemical methods for carboxyl end-group determination are considerably less satisfactory. Treatment of the peptide with anhydrous hydrazine at 100°C results in conversion of all the amino acid residues to amino acid hydrazides except for the carboxy-terminal residue, which remains as the free amino acid and can be isolated and identified chromatographically. Alternatively, the polypeptide can be subjected to limited breakdown (proteolysis) with the enzyme carboxypeptidase. This results in release of the carboxy-terminal amino acid as the major free amino acid reaction product. The amino acid type can then be identified chromatographically.

Step 4 involves breaking down the polypeptide chain into shorter, well-defined fragments for subsequent sequence analysis. Fragmentation can be achieved by the use of *endopeptidases*, which are enzymes that catalyze polypeptide-chain cleavage at specific sites in the protein. Figure 1–18 shows the specificity of four endopeptidases commonly used for this purpose. Another specific chemical method for polypeptide-chain cleavage involves reaction with cyanogen bromide. This reaction cleaves specifically at the methionine residues, with the accompanying conversion of free carboxy-

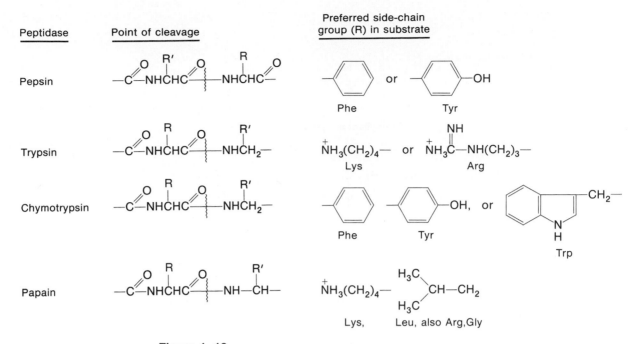

Figure 1–18
Site of action of some endopeptidases used for polypeptide chain cleavage prior to sequence analysis. Of the four different enzymes used, trypsin is used most frequently because of its high specificity.

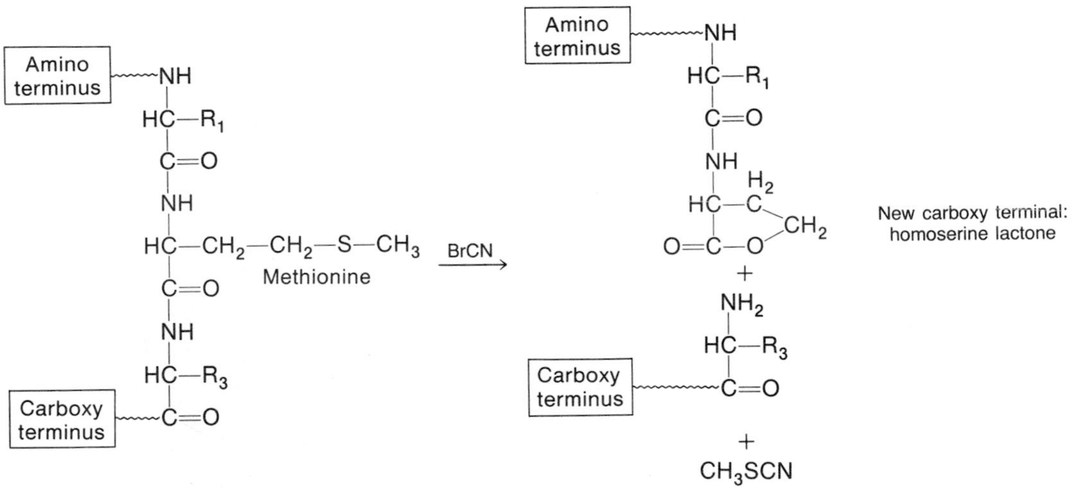

Figure 1–19
The cleavage of polypeptide chains at methionine residues by cyanogen bromide. The cleavage reaction is accompanied by the conversion of the free carboxy-terminal methionine to homoserine lactone.

terminal methionine to homoserine lactone (Figure 1–19). Although this methionine reaction product differs from the twenty naturally occurring amino acids, it is nevertheless readily identified by subsequent conversion to homoserine.

Peptides resulting from cleavage of the intact protein are generally separated by ion-exchange chromatographic methods. The isolated peptides may then be analyzed (step 5) to determine both their amino acid composition and their sequence. Sequence determination involves the stepwise removal and identification of successive amino acids from the polypeptide amino

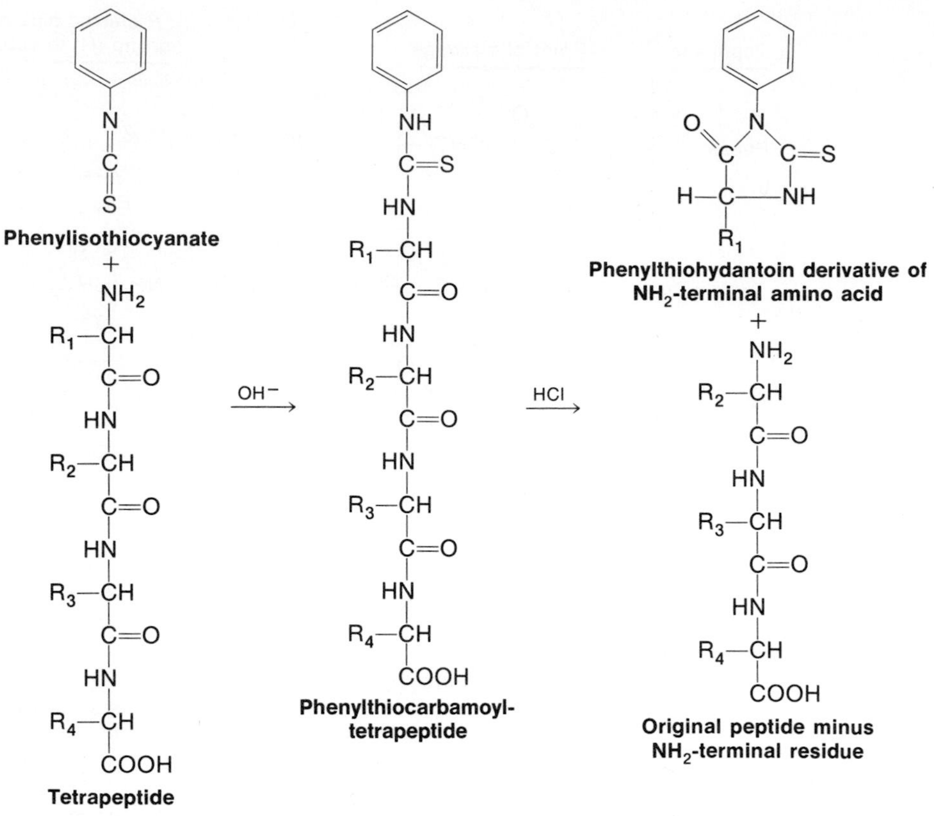

Figure 1–20

The Edman degradation method for polypeptide sequence determination. The sequence is determined one amino acid at a time, starting from the amino-terminal end of the polypeptide. First the polypeptide is reacted with phenylisothiocyanate to form a polypeptidyl phenylthiocarbamoyl derivative. Gentle hydrolysis releases the amino-terminal amino acid as a phenylthiohydantoin (PTH), which can be separated and detected spectrophotometrically. The remaining intact polypeptide, shortened by one amino acid, is then ready for further cycles of this procedure. A more sensitive reagent, dimethylaminoazobenzene isothiocyanate, can be used in place of phenylisothiocyanate. The chemistry is the same.

terminal by means of the _Edman degradation_ (Figure 1–20). This process is carried out by reacting the free amino-terminal group with phenylisothiocyanate to form a peptidyl phenylthiocarbamoyl derivative. Gentle hydrolysis with hydrochloric acid releases the amino-terminal amino acid as a phenylthiohydantoin (PTH) derivative. The remaining intact peptide, shortened by one amino acid, is then ready for further cycles of this procedure. The PTH-amino acid can be identified by its properties on thin-layer chromatography (Figure 1–21).

Devices called sequenators are available that automate the Edman degradation procedure. The success of these devices depends in large part on the technical innovation of covalently linking the peptide to be sequenced to glass beads. Attachment of the peptide through its carboxy-terminal group to this immobile phase facilitates the complete removal of potentially contaminating reaction products during successive stages of the degradation.

Finally, having established the sequences of the individual peptides, it is necessary only to establish how they are connected together in the intact protein (step 6). It is at this stage that we see why the preceding sequence analysis was performed on peptides obtained by two different specific cleavage methods. This approach makes it possible to piece together the overall sequence, because the two sets of results produce _overlapping sequences_. That is, the free amino and carboxy residues of peptides originally interconnected in the intact protein and liberated by one specific cleavage method

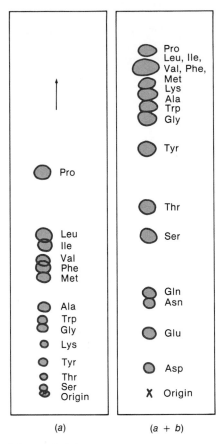

Figure 1–21
Thin-layer chromatography of amino acid–phenylthiohydantoin derivatives on silica gel plates. (a) Separation is done in a 98:2 mixture of chloroform and ethanol. (b) This is followed by further separation using an 88:2:10 mixture of chloroform, ethanol, and methanol. More sophisticated procedures using column chromatography, give superior resolution and improved sensitivity. Automated sequencers always use such procedures. A general description of the use of columns is given in Chapter 3.

will recur in the internal sequences of the peptides liberated by a second specific method.

Once the protein's primary sequence has been determined, the location of disulfide bonds in the intact protein can be established by repeating a specific enzymatic cleavage on another sample of the same protein in which the disulfide bonds have not previously been cleaved. Separation of the resulting peptides will show the appearance of one new peptide and the disappearance of two other peptides, when compared with the enzymatic digestion product of the material whose disulfide bonds have first been chemically cleaved. In fact, these difference techniques are generally useful in the detection of sites of mutations in protein molecules of previously known sequence, since a single substitution will generally affect the chromatographic properties of only a single peptide released during proteolytic digestion.

Great progress has been made in recent years in devising procedures for sequencing the DNA that encodes for proteins (Chapter 7). Knowing the sequence of coding triplets in DNA allows us to read off the amino acid sequence of the corresponding protein. Nevertheless, such studies have produced the remarkable observation that some eukaryotic DNA sequences coding for proteins are not continuous, but instead contain untranslated intervening DNA sequences. Although these results have profound implications for protein evolution, they obviously confound the general applicability of DNA-sequencing methods for the purposes of protein primary-structure determination. In cases like this, the usual solution has been to isolate the mRNA for the protein and use this to make a DNA carrying the same sequence. This procedure circumvents the intervening-sequence problem because the mRNA carries only the coding sequences (see Chapter 32).

CHEMICAL SYNTHESIS OF PEPTIDES AND POLYPEPTIDES

Knowledge about the structure-function interrelationships in proteins and peptides has encouraged biochemists to develop techniques for synthesizing peptides and proteins with predetermined sequences. To synthesize a peptide in the laboratory, we must overcome several problems related to preventing undesired groups from reacting. The amino and carboxyl groups that are to remain unlinked must be blocked; so must all reactive side chains. Some protecting groups for carboxyl and amino groups are shown in Figures 1–22 and 1–23, respectively.

After blocking those groups to be protected, we generally activate the carboxyl group; two methods for doing this are shown in Figure 1–24. It is of interest that carboxyl-group activation is also employed in natural biosynthesis in the cell (see Chapter 28). After peptide synthesis, the protecting groups must be removed by a mild method. The overall process—comprising protection, activation, coupling, and unblocking—is shown in Figure 1–25.

Figure 1–22
Carboxyl protecting groups used in peptide synthesis. The symbol, ▲, in the amino acid structure on the left, stands for one of the protecting groups (middle) leading to the named compound indicated on the right. The protecting group prevents the carboxyl group from participating in subsequent reactions involved in peptide synthesis.

Amino acid	Carboxy terminal protecting group ▲	Esters formed
	CH_3CH_2O-	Ethyl
	$O_2N-\langle\text{benzene}\rangle-CH_2-O-$	Nitrobenzyl
	$H_3C-\underset{CH_3}{\overset{CH_3}{C}}-O^-$	t-Butyl

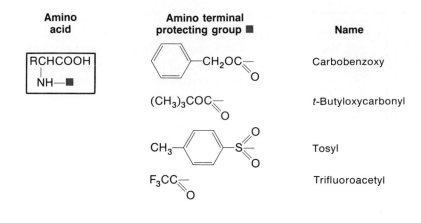

Figure 1–23
Amino protecting groups used in peptide synthesis.

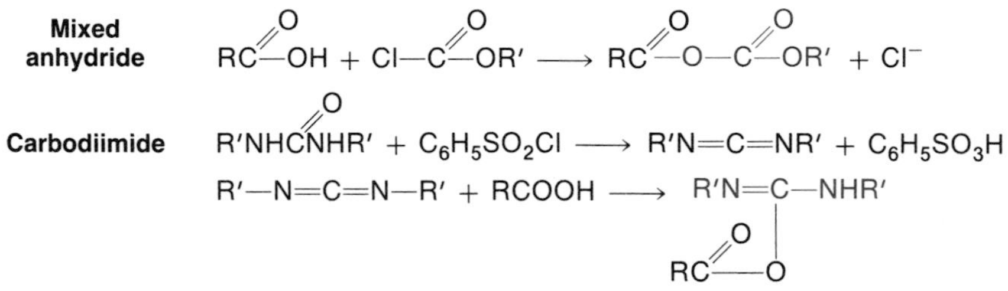

Figure 1–24
Different ways of activating the carboxyl group for peptide synthesis. Activated amino acids will react spontaneously with most α-amino acids as illustrated in Figure 1–25.

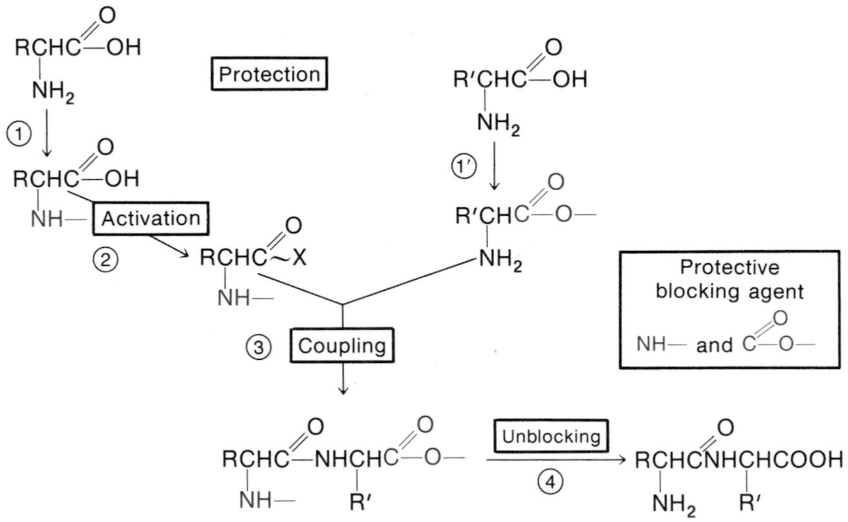

Figure 1–25
Schematic diagram illustrating the chemical method for peptide synthesis. First the amino acids to be linked are selected. The carboxyl group and the amino group that are to be excluded from peptide synthesis are protected (steps 1 and 1'). Next the amino acid containing the unprotected carboxyl group is carboxyl-activated (step 2). This amino acid is mixed and reacted with the other amino acid (steps). Protecting groups are removed from the product (step 4).

An important variation of the usual methods of peptide synthesis involves attaching a protected (*t*-butoxycarbonyl group) amino acid to a solid polystyrene resin; removal of the amino protecting group; condensation with a second protected amino acid; and so on. In the last step, the finished peptide is cleaved from the resin. This method (outlined in Figure 1–26) has the advantage that cumbersome purification between steps, often resulting in serious losses, is replaced by mere washing of the insoluble resin. Since each reaction is essentially quantitative, very long peptides, and even proteins, can be synthesized by this method. Indeed, Li synthesized the 39-amino-acid peptide ACTH by this method, and Merrifield synthesized bovine pancreatic ribonuclease, which contains 129 amino acids in a single polypeptide chain. A number of variants of ribonuclease that contain one or more changes in amino acid sequence also have been made by this method. The importance of the Merrifield process was underscored by the awarding of a Nobel Prize to Merrifield in 1984.

Figure 1–26

Merrifield procedure for solid-state dipeptide synthesis. (1) Polymer is activated. (2) Amino acid containing BOC protecting group is carboxy-linked to polymer. This amino acid will be the carboxy-terminal amino acid in the final peptide. (3) The BOC protecting group is removed from the polymer-linked amino acid. (4) A second amino acid, containing a BOC on its α-amino group and a dicyclohexylcarbodiimide (DCC) activated group, is reacted with the column-bound amino acid to form a dipeptide. (5) The dipeptide is released from the polymer and the BOC protecting group by adding hydrogen bromide (HBr) in trifluoroacetic acid.

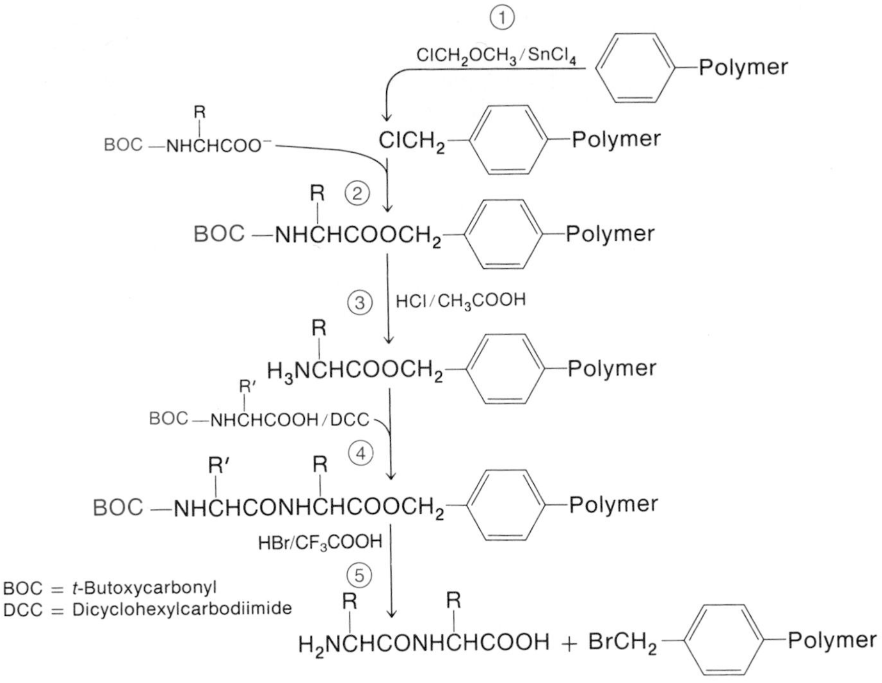

BOC = *t*-Butoxycarbonyl
DCC = Dicyclohexylcarbodiimide

SUMMARY

Proteins are polymers of twenty amino acids. Each amino acid contains a tetrahedral carbon atom, the α carbon (C_α). This atom is covalently bound to an amino group, a carboxyl group, a hydrogen, and a variable side chain (R). It is the R group that gives each of the amino acids its distinctive characteristics. The R groups are uncharged apolar (eight), uncharged polar (seven), or charged polar (five). Those amino acids (fifteen) that carry no net charge in their side chains are neutral overall. Nevertheless, at neutral pH (pH 7) they carry two charges, one on the α amino group ($-NH_3^+$) and one on the α carboxyl group ($-COO^-$).

The charged state of an amino acid is best characterized by its pH titration curve. This is obtained by titrating an aqueous solution of the amino acid with acid or base. The acid dissociation constant of the protonating group, K_a, usually expressed as pK_a ($-\log K_a$) is that region of the titration curve showing a minimum pH change as a function of added acid or base. All amino acids have two pK_a values, one for the $\alpha -NH_2$ group and one for the $\alpha -COOH$ group. The charged amino acids have an additional pK_a for the side-chain group.

All amino acids except glycine are chiral (handed) because they have four different substituents attached to their C_α atoms. Two enantiomers exist for each amino acid; these are given the designation D or L according to the arrangement of the four substituents about C_α. All amino acids found in proteins have a related L configuration.

Proteins are linear chains of amino acids that are linked by a peptide bond formed between the α carboxyl end of one amino acid and the α amino end of the adjacent amino acid. Each protein has a characteristic amino acid composition and sequence. To determine the amino acid composition the protein is usually subjected to acid hydrolysis, which degrades it into amino acids. Then the amino acids are separated by ion-exchange chromatography and the amount of each amino acid is determined with the help of a colorimetric reagent.

The sequence of amino acids in a protein is called the primary structure of the protein. Sequence analysis usually involves selective degradation of the polypeptide chains into smaller units. This is done with specific proteases that recognize specific linkages. The smaller polypeptides are analyzed by the stepwise removal of single amino acids from the amino end of the polypeptide chain. Each amino acid is treated with ninhydrin to facilitate its detection by spectrophotometry.

Chemical methods also exist for the synthesis of polypeptide chains. The most popular method, called the Merrifield process, involves first attaching the C-terminal amino acid residue to a resin. Subsequent stepwise addition of the desired amino acids is done with amino acids containing chemically activated carboxyl groups. It has been possible to synthesize small proteins by this means.

SELECTED READINGS

Eisenberg, D., and D. Crothers, *Physical Chemistry and Its Applications to the Life Sciences.* Menlo Park, Calif.: Benjamin/Cummings, 1979.

Tristram, G. R., and R. H. Smith, Amino acid composition of proteins, *Advances in Protein Chemistry* 18:227, 1963.

PROBLEMS

1. The accompanying figure is the titration curve of an unidentified amino acid.
   a. Estimate the pK_a values of the ionizable groups in this amino acid.
   b. Comparing these values to the pK_a values listed in Table 1–3, identify the amino acid.
   c. Describe the form, charged or uncharged, of each ionizable group at points A through D on the curve, and estimate the isoelectric point.
   d. Estimate the amount of amino acid in solution.
   e. Using the pK_a values given in Table 1–3, draw a titration curve for cysteine.

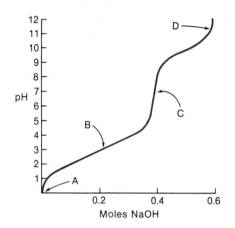

2. You are given a 10 mM solution of aspartate adjusted to pH 9.5. Assuming the pK_a values for Asp given in Table 1–3, calculate the concentrations of the principal ionized forms present. If 1 ml of 1 M NaOH is added to a liter of this solution, what will be the pH of the resulting solution?

3. Amino acids are sometimes used as buffers. A buffer is a solution whose pH changes very little when acid or base is added. Over what range(s) of pH would you expect the unidentified amino acid in Problem 1 to be an effective buffer? Using pK_a values given in Table 1–3, suggest an amino that would be appropriate as a buffer at each of these pH values: 4.5, 6.5, 8.5, and 10.5.

4. Which of the naturally occurring amino acid side chains are charged at pH 7, 2, and 13?

5. For the peptide shown below, the numbers in parentheses are the pK_a values of the ionizable groups.

(8.0)   O   O   (3.0)

NH$_3$—CH—C—NH—CH—C—NH—CH—COO$\ominus$
$\oplus$
    CH$_2$ (CH$_2$)$_4$ CH$_2$

    COO$\ominus$ NH$_3$ (10.0)   SH
                    $\oplus$

   (4.0)   (9.0)

   a. Estimate the net charge at pH 1 and at pH 14.

   b. Estimate the isoelectric pH.
   c. The values of pK_a given for the α-amino and α-carboxyl groups of this peptide differ from those described in Table 1–3. Suggest an explanation for this discrepancy.

6. Which amino acid side chains can function as hydrogen-bond donors, acceptors, or both?

7. Polyhistidine is insoluble in water at pH 7.8 but is water-soluble at pH 5.5. Suggest an explanation. [Hint: The solubility of polyhistidine in aqueous solution is a function of the polarity of the histidine side chain.]

8. An enzyme exhibits a maximum in catalytic rate at pH 6.5. Amino acid analysis of the native protein shows that it contains two histidine residues. A sample of the protein, inactivated by chemical modification, gives two products separable by chromatography. Both products, when analyzed, show the loss of one histidine. What conclusions do you reach concerning the active site of this enzyme?

9. An organism of unknown origin produces a potent inhibitor of nerve conduction that you wish to sequence. Amino acid analysis shows the peptide's composition to be Ala$_5$-Lys$_1$-Phe$_1$. Reaction of the intact peptide with FDNB followed by acid hydrolysis liberates free FDNB alanine. Trypsin cleavage gives a tripeptide and a tetrapeptide, with compositions Ala$_3$-Phe$_1$ and Lys$_1$-Ala$_2$. Reaction of the intact peptide with chymotrypsin yields a hexapeptide plus free alanine. What is the inhibitor's sequence?

10. You have isolated an octapeptide from a rare fungus that prevents baldness, and you are interested in its commercial possibilities. Analysis gives the composition Lys$_2$-Asp$_1$-Tyr$_1$-Phe$_1$-Gly$_1$-Ser$_1$-Ala$_1$. Reaction of the intact peptide with FDNB followed by acid hydrolysis liberates FDNB alanine. Cleavage with trypsin gives peptides with compositions Lys$_1$-Ala$_1$-Ser$_1$ and Gly$_1$-Phe$_1$-Lys$_1$, plus a dipeptide. Reaction with chymotrypsin liberates free aspartic acid, a tetrapeptide with composition Lys$_1$-Ser$_1$-Phe$_1$-Ala$_1$, and a tripeptide that liberates FDNB-glycine on reaction with FDNB followed by acid hydrolysis. What is the sequence?

11. A South American beetle produces a substance that extracts gold from seawater, and you desire to know its structure in order to carry out its large-scale synthesis. Analysis gives a composition Lys$_1$-Pro$_1$-Arg$_1$-Phe$_1$-Ala$_1$-Tyr$_1$-Ser$_1$. Reaction with FDNB gives no products, unless the material is first reacted with chymotrypsin, in which case both DNP serine and DNP lysine are formed from peptides with compositions Ala$_1$-Tyr$_1$-Ser$_1$ and Pro$_1$-Phe$_1$-Lys$_1$-Arg$_1$. Reaction with trypsin also produces two peptides, with compositions Arg-Pro and Phe-Tyr-Lys-Ser-Ala. What is the structure of this peptide?

2

THE THREE-DIMENSIONAL STRUCTURES OF PROTEINS

The enormous structural diversity of proteins begins with the amino acid sequences of polypeptide chains. Each protein consists of one or more unique polypeptide chains, and each of these polypeptide chains is folded into a complex three-dimensional structure. The final folded arrangement of the polypeptide chain in the protein is referred to as its _conformation_. Most proteins exist in unique conformations exquisitely suited to the highly specific function of each protein. It is the availability of a wide variety of conformations that permits proteins as a group to perform a broader range of functions than any other class of biomolecules.

Before the first x-ray diffraction results were understood, it was imagined that protein structures were relatively simple geometric arrangements of polypeptide chains, such as geometric cages, repeating zigzags, or uniform arrays of parallel rods. Indeed, the first structures determined for proteins were of this type. These structures were deduced by Pauling and Corey, using information from various sources: (1) They knew a little about the structures of peptides from small-molecule crystallography, which indicated that the peptide bond was planar and gave accurate bond lengths and angles. (2) They were already aware of the importance of hydrogen bonds in determining the orientation of amino acids, peptides, and even water in simple crystals. (3) They made shrewd guesses about the interpretation of a few spacings in the diffraction patterns of certain fibrous proteins. (4) Putting all of this information together, they experimented with molecular models until they could produce structures in reasonable agreement with all the available facts. This was a historic achievement.

It is not surprising that repeating structures with long-range order were the first protein structures to be understood. The demands on the available technology were minimal. Much more sophisticated technology was required to interpret the diffraction patterns of most proteins, which have less long-range repetition. Even Kendrew and Perutz, who led their research teams to a solution of the first such structures to be determined, myoglobin and hemoglobin, were shocked when they realized how chaotic the arrangements in such structures seemed to be. Kendrew was once introduced at Harvard as the man who proved that proteins were ugly, and Perutz, addressing his initial disappointment, said, "Could the search for ultimate truth really have revealed so hideous and visceral-looking an object?" Whether Kendrew and Perutz were truly as disappointed as they appeared to be, or whether they were actually delighted that their efforts of more than a quarter century had led them to structures that were so complex that no person could have predicted them is for science historians to determine. Suffice it to say that this crowning achievement of protein structure determination by x-ray diffraction has been repeated many times since with less ado.

Enormous advances have been made in protein chemistry and in computer technology as well. These advances have systematized the necessary research and greatly reduced the amount of work and time required to determine a protein structure. Accurate structure determinations have now been made on over 200 different proteins. It appears that detailed three-dimensional structures of proteins also may be determined either by a combination of electron diffraction and low-dose imaging in the electron microscope or by high-resolution, two-dimensional nuclear magnetic resonance spectroscopy. Nevertheless, essentially all the information summarized here comes from the results of x-ray crystallography. From this wealth of data, patterns of structure are becoming apparent that suggest, among other things, that _the final folding arrangements of proteins may some day be predictable from the amino acid sequences of the polypeptide chains._

In this chapter we will discuss the nature of the forces that cause proteins to fold. We will also see how these forces, in concert with what are basically geometric properties of the polypeptide chain, combine to produce the highly organized structures that are typical of functionally active proteins. To start, we will look at the basic properties of the structural material of proteins, to see what sorts of local structural arrangements are possible for polypeptide chains. Later we will describe the way in which these locally organized structural units can be most efficiently assembled into progressively larger and more complicated arrangements. We will also consider the relationships among a protein's sequential, functional, and structural properties.

THE INFORMATION FOR FOLDING IS CONTAINED IN THE PRIMARY STRUCTURE

The conformation of a native or highly organized protein reflects a delicate balance among a variety of interaction forces, both within the folded protein's interior and with surrounding solvent. If the protein's solvent environment is perturbed, the protein's native conformation can be disrupted, with a resulting loss of function and the production of a partially unfolded, or _denatured_, protein. Denaturation may be reversible or irreversible, and it may be partial or complete.

Proteins vary tremendously in their susceptibility to denaturation. Some proteins can be exposed to strong mineral acid without suffering irreversible loss of enzymatic activity, whereas most proteins would be irreversibly denatured by such conditions. The magnitude of the structural perturbation required for loss of function may also vary appreciably. There is a spectrum of intermediates between two extreme forms—the native and the so-called random-coil denatured state. The conditions giving rise to partial denaturation and consequent loss of function may be subtle, such as a small change in pH, temperature, ionic strength, or dielectric constant of the medium. Conversely, prolonged boiling or exposure to thiol-containing compounds and detergents, such as sodium dodecyl sulfate, or to hydrogen bond-breaking reagents, such as urea or guanidine hydrochloride, may be required for complete denaturation.

It occurred to biochemists many years ago that the native conformation of proteins might be solely determined by the amino acid sequence. This was demonstrated to be the case by a series of classic experiments performed in the early 1960s by F. White and C. Anfinsen. They chose to study bovine pancreatic ribonuclease, an enzyme containing 124 amino acid residues with four disulfide bridges. The experiment is described schematically in Figure 2–1.

First, the enzyme was denatured in a solution containing 8 M urea and β-mercaptoethanol, a thiol reagent that reduces disulfides to sulfhydryls, thus cleaving the covalent cross-links. These conditions have been used since as a general means of denaturing proteins, by completely disrupting the conformation without giving rise to coagulation or precipitation. The reduced, denatured ribonuclease is biochemically inactive because its native structure has been destroyed.

Next, the protein was allowed to slowly air-oxidize. The result was the formation of a variety of different intermediates with randomly distributed disulfide bonds. These trapped intermediates remained inactive after removal of the denaturant urea by dialysis through a semipermeable membrane.

To recover the native structures, these inactive intermediates were exposed to a trace amount of the reducing agent mercaptoethanol. This step served to catalyze the rearrangement of the disulfide cross-links, resulting finally in the spontaneous generation of an essentially fully active product. Analysis by several methods showed that the renatured product was indistinguishable from native ribonuclease and that all the correct disulfide pairings had been reestablished. (Renaturation can be carried out in a less cum-

Figure 2–1
Schematic representation of an experiment to demonstrate that the information for folding into a biologically active conformation is contained in the protein's amino acid sequence.

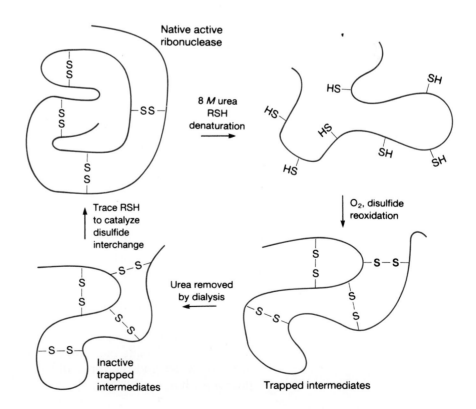

bersome manner by removing the urea before air oxidation; in this case, the native structure is a major product.)

Thus it appears that the information for folding to the native conformation is present in the amino acid sequence, for of the many possible disulfide-paired ribonuclease isomers that are possible, only the one correctly paired product was formed in major yield. Further studies have indicated that a similar result is obtained with many other proteins.

FORCES THAT DETERMINE FOLDING

We have just seen that after strenuous treatment leading to denaturation or unfolding, a structurally complex protein such as ribonuclease can be induced to renature or refold to its native conformation. This observation leads to two very important conclusions about protein structure: (1) *the conformation adopted by a protein is determined by its primary amino acid sequence, and* (2) *the spontaneity of the process indicates that the folding operation is energetically favorable.* That is, ΔG, the free energy change in going from the unfolded to the folded state, must be negative. Recall that ΔG is a composite of two very different terms that dictate the spontaneity with which a transition between two states can occur:

$$\Delta G = \Delta H - T \Delta S$$

In this equation ΔH is the enthalpy, which is approximately equal to the difference in bond energies, and ΔS is the entropy, which is a measure of the order in the system. For a given transition, a negative ΔH indicates that the system is going from more weakly to more strongly bonded interactions. A positive ΔS indicates that the system is going from a more ordered to a less ordered state.

In discussing polypeptide-chain folding it is often easy to forget the effects of water. The protein folding of concern is taking place in an aqueous environment, not in some other solvent or a vacuum. In evaluating the forces involved in protein folding, we must take account of what is happening to the water as well as to the polypeptide chain.

Other factors in aqueous solution, such as small anion and cation interactions, also can play a significant role in determining structural stability. We know, too, that there is a difference between the environment of an isolated protein in solution or in a protein crystal and the environment of a protein inside the cell. However, except in relation to membrane proteins, this potentially important consideration is not discussed in this text, because it is too difficult to assess. The omission simply reflects a gap in our current knowledge. The difference is probably least important for proteins that function as noninteracting proteins and most serious for proteins that function in aggregates, such as membrane proteins or enzymes that exist in multienzyme complexes.

Water Aids in Determining Protein Conformation

The amino acid side chains differ not only in size and shape but also in the charge they carry, their general affinity for water (*hydrophilicity*), or their general aversion to water (*hydrophobicity*). The *native conformation of proteins is a strong function of the interactions that take place within and between polypeptide chains; it is also highly dependent on the interactions that take place with water*, since proteins exist in an aqueous environment. In order to appreciate the central role of water, let us first consider the properties of water and the types of interactions that occur between water and polypeptides.

Ice is a highly organized structure in which individual molecules are held together by hydrogen bonds (H bonds) in a regular three-dimensional lattice (Figure 2–2). Water is also a highly H-bonded structure with a somewhat less regular and (judging by its greater density) more condensed structure. Most of the H bonds present in ice are also present in water. Hence water is a highly H-bonded structure, not too different from ice, but a structure in which the individual molecules have much more mobility and, consequently, a much higher entropy. Polypeptides also contain groupings that can form H bonds, and under certain conditions polypeptides form H bonds with water.

An individual water molecule has a significant *dipole*. This arises because of the greater electronegativity of the oxygen atom over the hydrogen atoms (see Figure 2–3). The dipolar properties of water molecules are responsible for the dissolution of salts, such as sodium chloride, in water (see Figure 2–3). The kinds of ion–dipole interactions that take place between water and simple ions such as Na^+ and Cl^- are also an important factor in the interactions between the charged amino acid side chains of proteins and water.

Those amino acid side chains that contain charged groups, H-bond-forming groups, or dipoles also could be classified as hydrophilic. In the

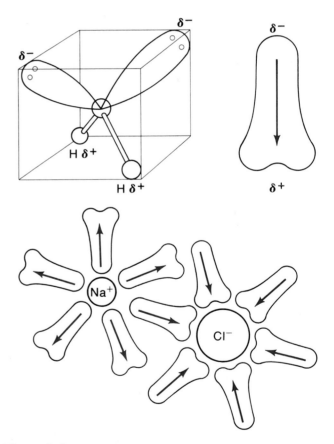

Figure 2–3
The water molecule is composed of two hydrogen atoms covalently bonded to an oxygen atom with tetrahedral (sp^3) electron orbital hybridization. As a result, two lobes of the oxygen sp^3 orbital contain pairs of unshared electrons, giving rise to a dipole in the molecule as a whole. The presence of an electric dipole in the water molecule allows it to solvate charged ions because the water dipoles can orient to form energetically favorable electrostatic interactions with charged ions.

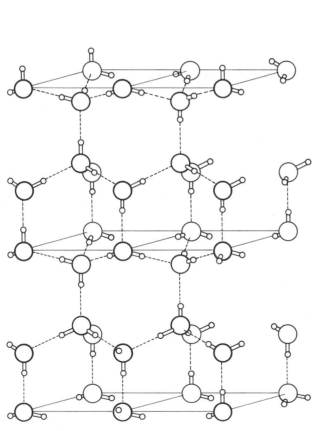

Figure 2–2
The arrangement of molecules in an ice crystal. The orientation of the water molecules, as represented in the drawing, is arbitrary; one proton along each oxygen–oxygen axis is closer to one or the other of the two oxygen atoms.

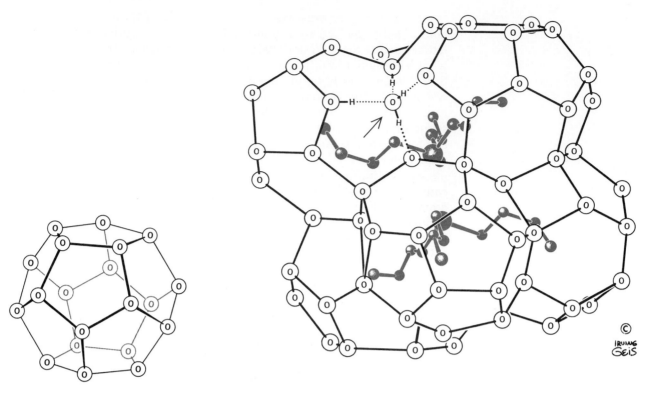

Figure 2–4
Clathrate structures are ordered cages of water molecules around hydrocarbon chains. The dodecahedral cage (left) of water molecules is a common building block in clathrates. To the right is a portion of the cage structure of $(nC_4H_9)_9S^-F^- \cdot 23H_2O$. The trialkyl sulfur ion nests within the hydrogen-bonded framework of water molecules. In the intact framework, each oxygen is tetrahedrally coordinated to four others. One such oxygen atom and its associated hydrogens are shown by the arrow. (Reprinted with permission from R. E. Dickerson and I. Geis, *The Structure and Action of Proteins*, Benjamin/Cummings, Menlo Park, Calif., 1969.)

form of small molecules, such groups tend to be highly soluble in water. By contrast, neutral organic side chains, which are hydrocarbon in character and do not contain significant dipoles or the capacity for forming H bonds, have little to gain by interacting with water. As a result, they are poorly soluble in water. When such hydrophobic molecules are present in water, the water forms a rigid so-called clathrate structure around them, which is believed to be even more ordered than the structure of ice (see Figure 2–4).

Most proteins contain both hydrophobic and hydrophilic amino acids. The hydrophilic residues are usually exposed on the outer surface of the protein, where they can interact with water. The hydrophobic residues are usually buried within the protein. Figure 2–5 shows the positions of all 104 side chains for horse heart cytochrome c.* This is one of a chain of molecules that transports electrons in the mitochondria (see Chapter 16). Hydrophobic side chains (colored) pack inside the molecule, and hydrophilic side chains (grey) are distributed over the surface of the molecule.

Whereas interaction with water is a major factor in determining protein conformation, it is not the only one. To fully understand this subject we must consider other noncovalent interactions and interactions involving polypeptide chains; we must also consider the steric limitations imposed by the covalent structure of the polypeptide chain.

*Cytochrome c in this illustration is represented by a ball-and-stick model. There are several ways in which protein structures can be illustrated; they are reviewed in Box 2–A.

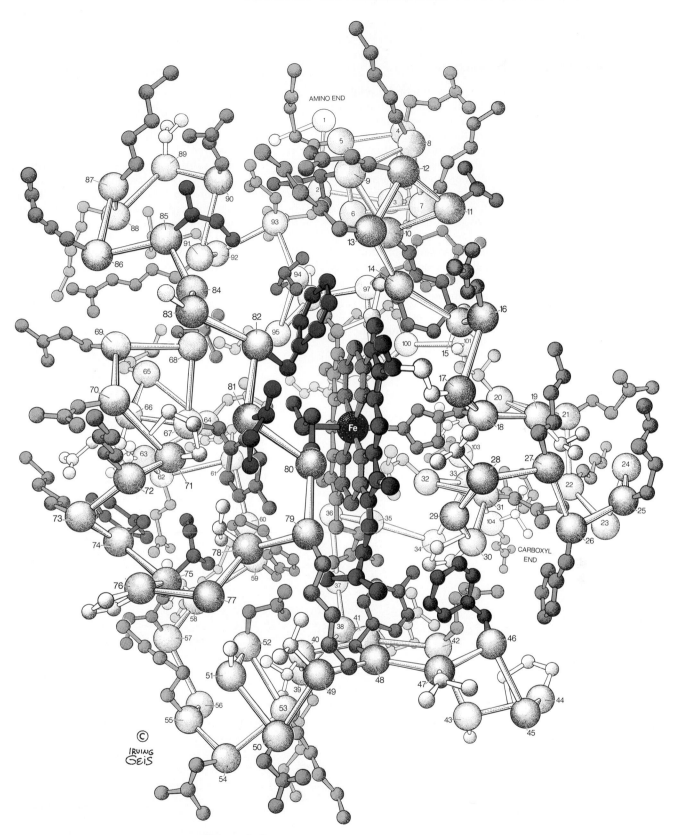

Figure 2–5
Full side-chain model of cytochrome *c*. Hydrophilic residues (in gray) are situated on the outer surface, where they can interact with water. Hydrophobic residues (in color) are situated in the interior of the protein, where they interact with one another. Cytochrome *c* contains a nonpeptidyl moiety, an iron-porphyrin. Porphyrins are discussed in the last section of this chapter.

BOX 2–A

Visualizing Molecular Structures

The primary data of protein crystallography yield a three-dimensional electron-density map, which must be interpreted in terms of a complex three-dimensional model of all atom positions in the protein. Such modeling is now generally done by computer graphics.

The difficulty with complete models is that even for relatively small proteins, their complexity is almost impossible to comprehend. Therefore, it is more common to show only selected parts of a molecule (such as only the polypeptide backbone) and to highlight particular features of interest.

Here, we show four different presentations for ribonuclease, in which each model is seen from the same perspective. Each method of presentation has its advantages. The space-filling model (Illustration 1) is excellent for displaying the volume occupied by molecular constituents and the shape of the outer surface. Illustration 2 shows a stereo pair of a space-filling model from exactly the same perspective as Illustration 1. When the pair is viewed with stereo glasses, the three-dimensional illusion is striking. Illustration 3 shows the polypeptide chain, with N and O atoms labeled and with dotted lines representing hydrogen bonds. In addition, all α-carbon positions are numbered. Illustration 4 shows an abstraction of the polypeptide chain in which the β strands are characterized as flat arrows and the α helices as spiral ribbons. This simplified style has proved useful in classifying and comparing proteins according to their secondary- and tertiary-structure folding patterns.

Illustration 1

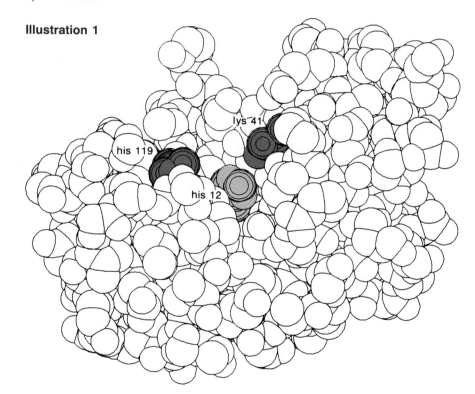

Illustration 2

The 3-D illusion can be seen without stereo glasses by the following method. With your eyes about 10 inches from the page, stare at the drawings below as if you were looking straight ahead at a far-away object. A double image will form and the central pair should drift together and fuse. Then the illusion becomes apparent. Adjust the page, if necessary, so that a horizontal line is perfectly parallel with the eyes. As an aid to seeing one image with each eye, use two cardboard tubes from toilet-paper rolls. Close the right eye and focus the left eye on the left image. Do the same for the other eye. Now with both eyes together, the 3-D illusion should appear.

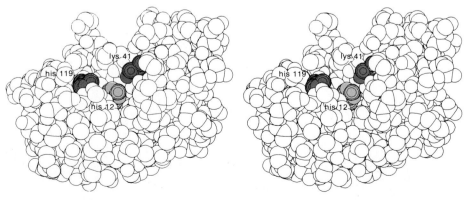

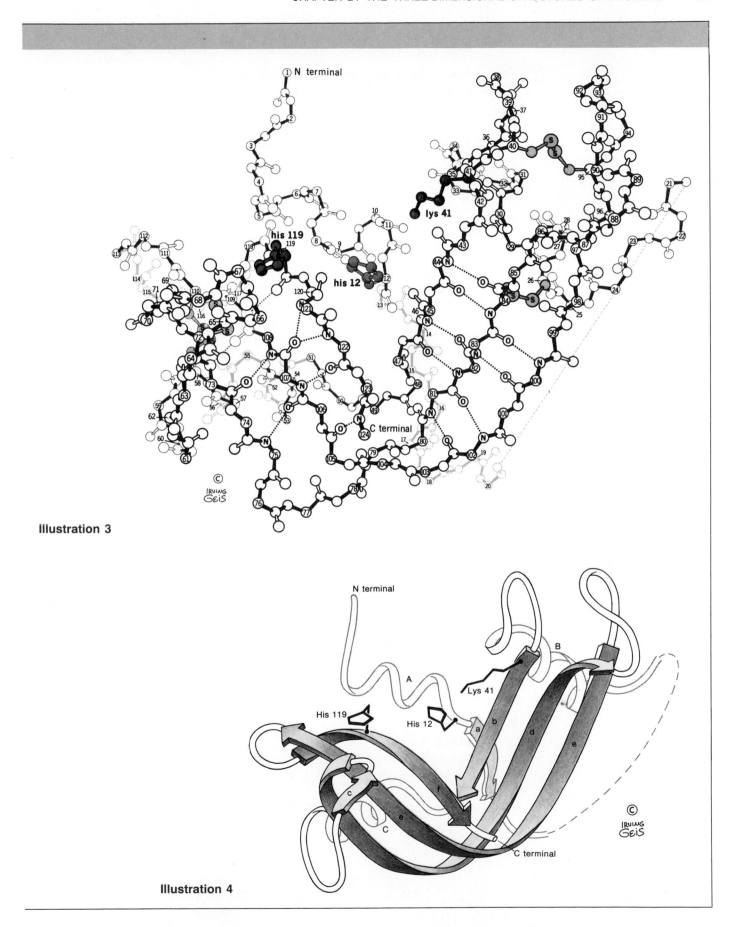

Illustration 3

Illustration 4

TABLE 2–1
Bond Energies between Some Atoms of Biological Interest

Energy Values for Single Bonds (kcal/mole)

C—C	82	C—H	99	S—H	81
O—O	34	N—H	94	C—N	70
S—S	51	O—H	110	C—O	84

Energy Values for Multiple Bonds (kcal/mole)

C=C	147	C=N	147	C=S	108
O=O	96	C=O	164	N≡N	226

Secondary Valence Forces Are the Glue that Holds Polypeptide Chains Together

Covalent bond energies range in value from 30 to 230 kcal/mole between most atoms of biological interest (Table 2–1). By contrast, intermolecular bond energies between noncovalently linked atoms range in value from 0.1 to 6 kcal/mole. Nevertheless, except for the disulfide linkages formed between two cysteines, *protein folding is mainly directed by the relatively weak intermolecular forces between noncovalently bonded atoms*. Even in the case of the disulfide linkages, the "correct" linkages form because the appropriate cysteines are brought close together by the weaker secondary forces. This effect was exemplified in the renaturation study of ribonuclease discussed earlier. The intermolecular forces that are of interest in protein structure may be classified into the following four categories.

Electrostatic Forces. In this category we may consider charge–charge interactions, charge–dipole interactions, and dipole–dipole interactions. Each of these three types of interactions shows a different functional dependence on distance between the interacting groups, as summarized in Table 2–2. For example, the energy of interaction between two charges Q_1 and Q_2 is proportional to the product of the charges and inversely proportional to the distance R between them:

$$\text{energy of interaction} \propto \frac{Q_1 Q_2}{R}$$

In solution this interaction is reduced by the dielectric constant of the surrounding medium:

$$\text{energy of interaction} \propto \frac{Q_1 Q_2}{\epsilon R}$$

TABLE 2–2
Range of Some Intermolecular Interactions Expressed as the Power of the Intermolecular Separation

Range of Interaction	Type of Interaction
$1/R$	Charge–charge
$1/R^2$	Charge–dipole
$1/R^3$	Dipole–dipole
$1/R^6$	Van der Waals (dipole–induced dipole) attractive forces
$1/R^{12}$	Van der Waals repulsive forces

If the two charges in question are relatively buried within a protein, their interaction energy can be substantially increased because the dielectric constant in the regions inaccessible to water is much lower than the dielectric constant of water.

A number of amino acid side chains are charged at physiological pH values (lysine, arginine, aspartate, glutamate, and sometimes histidine). However, favorable charge–charge interactions between oppositely charged amino acid side chains are rarely as significant in determining protein folding as are the ion–dipole interactions between the charged groups and water. As mentioned earlier, water molecules have a dipolar character that leads to favorable electrostatic interactions with the charged groups of the polypeptide. From such considerations it seems likely that charged amino acid side chains would increase protein structural stability if they were present on the surface of a protein, where they could interact with the water. As a general rule, this is the case.

Van der Waals Forces.   The term _van der Waals forces_ refers to two types of interactions, one attractive and one repulsive. The attractive forces are due to favorable interactions among the induced instantaneous dipole moments that arise from fluctuations in the electron charge densities of neighboring nonbonded atoms. Such forces tend to be small, yielding energies of 0.1 to 0.2 kcal/mole, but they can add up as the number of interactions between two molecules increases. That is why a typical hydrophobic molecule like hexane is primarily a liquid rather than a gas at room temperature. In this example, we are considering the transition between liquid and noninteracting molecules in the gaseous state. In aqueous solution, van der Waals attractive forces between the hydrophobic groups are far less important in influencing the protein structure. In fact, the interaction of so-called hydrophobic molecules such as hexane with water has a favorable enthalpy over the interaction of hexane with itself. Other factors (entropic) therefore determine why liquid hexane and water are virtually immiscible (see "Hydrophobic Forces," later in this discussion).

Repulsive van der Waals forces arise when noncovalently bonded atoms or molecules come too close together. An electron–electron repulsion arises when the charge clouds between two molecules begin to overlap. If two molecules are held together exclusively by van der Waals forces, their average separation will be governed by a balance between the van der Waals attractive and repulsive forces. The distance is known as the _van der Waals separation_. Some van der Waals radii for biologically important atoms are given in Table 2–3. The van der Waals separation between two nonbonded atoms is given by the sum of their respective van der Waals radii.

Hydrogen Bond Forces.   The strength of the hydrogen bond is due to the partially unshielded nature of the single proton that makes up the hydrogen nucleus. An attractive interaction exists between the lone pair of electrons of either a nitrogen atom or an oxygen atom and the hydrogen atom in either an N—H or an O—H chemical bond. The attraction is usually directed along the lone-pair orbital axis of the H-bond acceptor group. Evidence for a significant H bond comes from the observation of a decreased distance between the donor and acceptor groups making up the H bond. Thus from the van der Waals radii given in Table 2–3, we can calculate that the distances between nonbonded H and O atoms and between nonbonded H and N atoms are 2.6 and 2.7 Å, respectively. When an H bond is present, this distance is usually reduced to 1.8 to 1.9 Å. Some of the more important H-bond donors and acceptors are shown in Figure 2–6. As a rule the angle between the N or O acceptor and the N—H or O—H donor is close to 180°.

TABLE 2–3
Radii for Covalently Bonded and Nonbonded Atoms

Element	Single Bond	Double Bond	Triple Bond
Covalent bond radii (in Å)			
Hydrogen	0.30		
Carbon	0.77	0.67	0.60
Nitrogen	0.70		
Oxygen	0.66		
Phosphorus	1.10		
Sulfur	1.04		
Van der Waals radii (in Å)			
Hydrogen	1.2		
Carbon	2.0		
Nitrogen	1.5		
Oxygen	1.4		
Phosphorus	1.9		
Sulfur	1.8		

Hydrogen bond donors

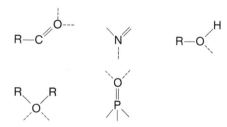

Hydrogen bond acceptors

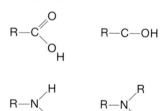

Hydrogen bond complex

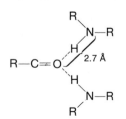

Figure 2–6
Major hydrogen-bond donor and acceptor groups found in proteins. Note that the angle between the O acceptor and the N—H donor is 180° in the hydrogen bond complex.

Polypeptides carry a number of H-bond donor and acceptor groups, both in their backbone structure and in their side chains. Water also contains a hydroxyl donor group and an oxygen acceptor group for making H bonds (Figure 2–2). Formation of the maximum number of H bonds between a polypeptide chain and water would obviously require the complete unfolding of the polypeptide chain. However, it is not obvious that such an unfolding would result in a net energy gain. The reason is that water is a highly H-bonded structure, and for every H bond formed between water and protein, an H bond within the water structure itself must be broken. *The strategy followed by most proteins is to maximize the number of intramolecular H bonds between the backbone peptide groups, but to keep most of the potential H-bond-forming side chains of the protein surfaces for interaction with water.* It seems likely that such side-chain–water interactions will occur both because of the H bonds and dipole–dipole interactions.

Hydrophobic Forces. Hydrophobic forces are both the most poorly understood and the hardest to appreciate type of interaction. The term is not a very good one either, since it is somewhat misleading. Whereas the previous forces discussed are due primarily to enthalpic factors, hydrophobic forces relate primarily to entropic factors. Furthermore, the entropic factors mainly concern the solvent, not the solute.

It was stated earlier that hexane has a small but favorable enthalpy for solution in water, but that the entropic factor was sufficiently unfavorable to make hexane highly insoluble in water. What is this mysterious factor? Owing to the weak enthalpic interactions between a hydrophobic molecule such as hexane and water, the water tends to withdraw to some extent in the region of the apolar hydrophobic molecule and form a relatively rigid hydrogen-bonded network with itself (for example, see the clathrate structures illustrated in Figure 2–4). This network effectively restricts the number of possible orientations of the water molecules forming the water–hydrophobic-group interface. Now, virtually all chemical systems tend spontaneously toward a state of maximum disorder. When hydrophobic groups are exposed to water, however, some extent of ordering is introduced in the surrounding water. This energetically unfavorable entropic effect is greater than the small, favorable enthalpic effect. The point can be difficult to grasp because one observes hydrophobic groups clustering together in aqueous solution, and it is tempting to attribute this effect mainly to van der Waals attractive forces between the hydrophobic groups. However, thermodynamic measurements do not support this explanation.

Proteins contain a number of amino acids with predominantly apolar side chains (e.g., alanine, valine, isoleucine, leucine, and phenylalanine). From what has been said, it might be anticipated that most of these side chains would be located internally in the native protein structure, as we have seen for cytochrome *c* (see Figure 2–5).

In concluding this section on the intermolecular forces involved in folding, we must refer to the fact stated at the outset—that the energetically favorable state for a protein (at physiological pH values and temperatures) is the folded state. Because of experimental difficulties, the thermodynamic parameters ΔH and ΔS for the transition between unfolded to folded states have rarely been measured. However, in the limited number of cases where data are available, it appears that the overall entropy of folding is slightly negative (unfavorable) and the overall enthalpy is also slightly negative (favorable). Thus from thermodynamics the conclusion is reached that folding is opposed by the entropy but favored by the enthalpy change and occurs because the latter factor outweighs the former. More thermodynamic data on a number of proteins could alter this view as to which thermodynamic parameter dictates a negative free energy for folding in specific instances.

The main difficulty in making predictions from a detailed consideration of the different intermolecular forces involved is a quantitative one. For example, it is known that the entropy effect owing to interaction between apolar side chains and water will encourage folding, but it is not known how big this factor is by comparison with the ordering effect in the polypeptide chain that should discourage folding. Again, we know that intramolecular H bonds formed by the polypeptide chain constitute an enthalpic factor that encourages folding, but it is not known if the net gain here (which would have to take into account the breaking of H bonds between the polypeptide backbone and water) is enough to overcome the negative overall entropy effect on folding.

Finally, it should be noted that *the native folded state of a protein reflects a delicate balance between opposing energetic contributions of enormously large magnitude.* Specifically, the total free energy of stabilization of a globular protein in the folded state relative to its unfolded form is typically equivalent to the free energy of formation of less than ten hydrogen bonds or, alternatively, a fraction of a single covalent C—H bond. This delicate energy balance has important manifestations that relate to both the functional properties and the stability of proteins in their native states.

The Ramachandran Plot Predicts Sterically Permissible Structures

Figure 2–7 shows a ball-and-stick model of a short section of a polypeptide chain. Many geometric features of this structure are fixed as a result of bonded interactions between adjacent atoms. The bond lengths and bond angles are nearly constant for all proteins. Additionally, the backbone peptide bond has substantial double-bond character owing to electron delocalization over a π orbital system involving the carbonyl oxygen, carbonyl carbon, and amide nitrogen atoms of the backbone peptide bond. As a result, all the atoms of the peptide bond, together with the connected α-carbon atoms (conventionally labeled C_α), lie in a common plane with the carbonyl oxygen and amide hydrogen in the *trans* configuration. Consequently, the only adjustable geometric features of the polypeptide-chain backbone involve rotations about the single covalent bonds that connect each residue's C_α to the adjacent planar peptide groups.

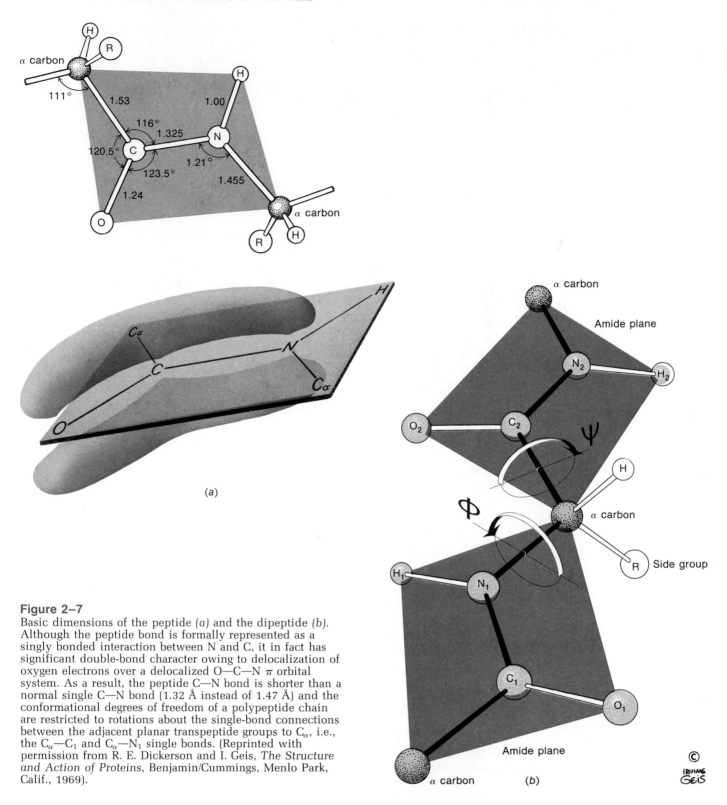

Figure 2–7
Basic dimensions of the peptide (a) and the dipeptide (b). Although the peptide bond is formally represented as a singly bonded interaction between N and C, it in fact has significant double-bond character owing to delocalization of oxygen electrons over a delocalized O—C—N π orbital system. As a result, the peptide C—N bond is shorter than a normal single C—N bond (1.32 Å instead of 1.47 Å) and the conformational degrees of freedom of a polypeptide chain are restricted to rotations about the single-bond connections between the adjacent planar transpeptide groups to C_α, i.e., the C_α—C_1 and C_α—N_1 single bonds. (Reprinted with permission from R. E. Dickerson and I. Geis, *The Structure and Action of Proteins*, Benjamin/Cummings, Menlo Park, Calif., 1969).

Rotations about the C_α—N bond are labeled with the Greek letter ϕ (phi), and rotations about the C_α-carbonyl carbon bond are labeled ψ (psi). In principle, both ϕ and ψ can have any value between $-180°$ and $+180°$, so that all possible conformations of the polypeptide chain can be described in terms of their ϕ, ψ conformational angles, a description that automatically takes account of the fixed geometric features of the polypeptide backbone. Thus any polypeptide conformation can be represented as a point on a plot of ϕ versus

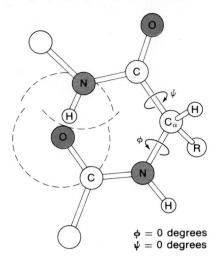

Figure 2–8
The conformation corresponding to $\phi = 0°$, $\psi = 0°$. This conformation is disallowed by the steric overlap between the H and O atoms of adjacent peptide planes. Rotation of both ϕ and ψ by 180° gives the fully extended conformation seen in Figure 2–7b. Curved arrows for ϕ and ψ indicate positive variations in angle.

$\phi = 0$ degrees
$\psi = 0$ degrees

ψ, where ϕ and ψ have values that range from $-180°$ to $+180°$. By convention, the conformation corresponding to $\phi = 0°$, $\psi = 0°$ is one in which both peptide planes that are connected to a common C_α atom lie in the same plane, as shown in Figure 2–8. Positive variations in ϕ correspond to clockwise rotations of the preceding peptide about the C_α—N_1 bond when viewed from C_α toward N_1 (see Figure 2–7b). Positive variations in ψ correspond to clockwise rotations of the succeeding peptide about the C_α—C_2 bond when viewed from C_α toward C_2 (see Figure 2–7b).

Experiments with models that approximate the polypeptide atoms as hard spheres, with appropriate van der Waals radii, quickly reveal that many ϕ, ψ angular combinations are impossible because of steric collisions between atoms along the backbone or between backbone atoms and the side-chain R group. For example, it is clear that the $\phi = 0°$, $\psi = 0°$ conformation shown in Figure 2–8 is impossible. The reason is that this conformation results in noncovalently bonded interatomic contacts that are considerably less than the sum of the van der Waals radii of the atoms involved. In fact, of all the possible ϕ, ψ combinations, only a relatively restricted number of conformations are sterically allowed. The Ramachandran plot (Figure 2–9) shows explicitly how the accessible regions of ϕ, ψ space are limited by steric interactions among the polypeptide backbone and side-chain groups, assuming that the atomic groups behave as rigid spheres having appropriate van der Waals radii. In reality, the atoms in molecules do not behave as rigid

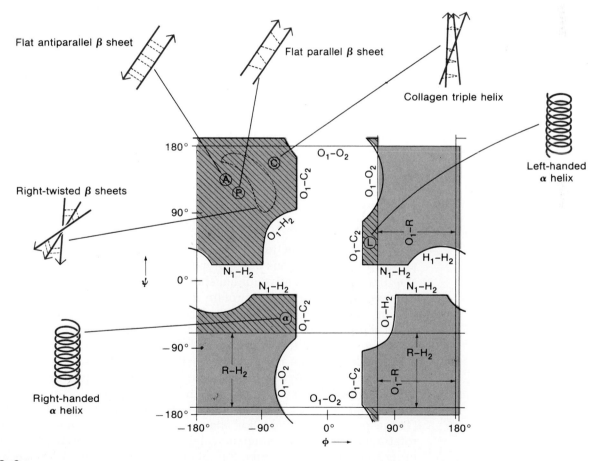

Figure 2–9
Ramachandran plot, showing which atomic collisions (using a hard-sphere approximation) produce the restrictions of the main-chain angles ϕ and ψ. The cross-hatched regions are allowed for all residues, and each boundary of a prohibited region is labeled with the atoms that collide in that conformation. Additional shaded regions are for glycine residues only. The numbering scheme for amide atoms used in the derivation diagram is given in Figure 2–7b. Each boundary of a prohibited region is labeled with the atoms that collide in that conformation. For an explanation of the various labeled structures, see the text.

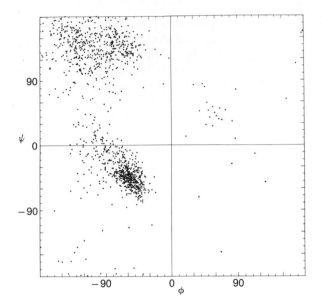

Figure 2–10
Ramachandran plot of main-chain angles ϕ and ψ, experimentally determined for approximately 1000 nonglycine residues in eight proteins whose structures have been refined at high resolution (chosen to be representative of all categories of tertiary structure).

spheres, so real proteins span a slightly greater range of values than is suggested by this plot (Figure 2–10).

Figure 2–10 shows the distribution of some observed ϕ, ψ conformational values for proteins whose three-dimensional structures are known from crystallography. The great majority of these lie within the bounds defined by allowable steric interactions. The exceptional residues are usually glycines. Glycine frequently can assume conformations that are sterically hindered in other amino acids because its R group, a hydrogen atom, is considerably smaller than the CH_2 or CH_3 groups connected to C_α in all other amino acids.

In summary, it can be seen that *owing to the basic geometric properties of the polypeptide chain, its sterically allowed conformations are severely restricted* by the occurrence of unfavorable steric interactions between various atomic groups.

PROTEIN FOLDING REVEALS A HIERARCHY OF STRUCTURAL ORGANIZATION

In the preceding section we described the energetics of polypeptide–polypeptide interactions and polypeptide–water interactions, which favor stabilization of the native folded state of a protein. Although this treatment gave a general idea of why protein folding takes place, it neglected the important question of how it takes place.

To appreciate the magnitude of this question, consider a hypothetical example involving the spontaneous folding of a polypeptide of 100 amino acid residues. In the native folded state of such a protein, each residue is spatially fixed relative to the others to produce a unique three-dimensional structure. For the sake of simplicity, the spatial orientation of each residue in the folded structure can be described in terms of three geometric parameters. (These parameters might, for example, correspond to the values of some rotational angles about single covalent bonds in the polypeptide chain.) In order to completely describe the folded protein, it is therefore necessary to uniquely specify 300 internal geometric parameters. Even if each of the parameters could assume only two possible values, the total number of potentially possible geometric arrangements of the structure would be $2^{300} = 2 \times 10^{90}$. If folding the protein involved random sampling of all these possible arrangements at a rate corresponding to typically observed frequencies of

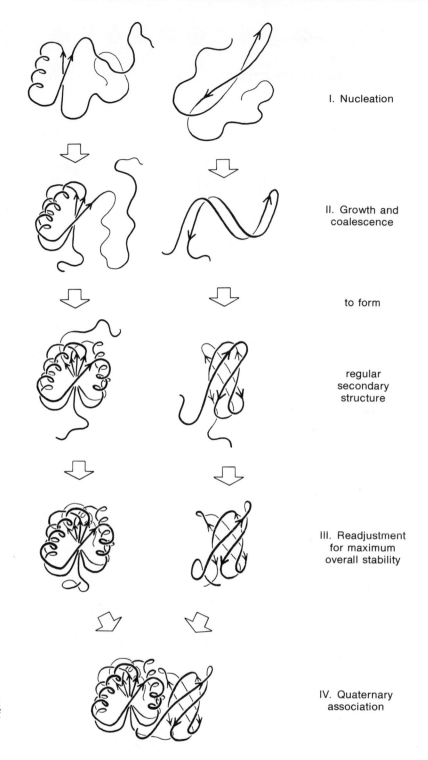

I. Nucleation

II. Growth and
coalescence

to form

regular
secondary
structure

III. Readjustment
for maximum
overall stability

IV. Quaternary
association

Figure 2–11
Possible successive steps in the
protein-folding process as they might
apply to a typical example of each of
the four major categories of structure
(see text for fuller explanation).

rotational rearrangements about single covalent bonds (about 10^{-13} sec), the estimated time required to fold the protein into the correct arrangement would exceed the age of the earth.

In actuality, newly synthesized polypeptide chains typically fold in seconds. This means that *protein folding must be a highly directed and cooperative process*. Although much remains to be learned about the details of the process, its speed and its facility suggest the existence of a sequential set of folding intermediates, each being more highly organized than the one before it (for example, Figure 2–11). In what follows we will see how the forces that

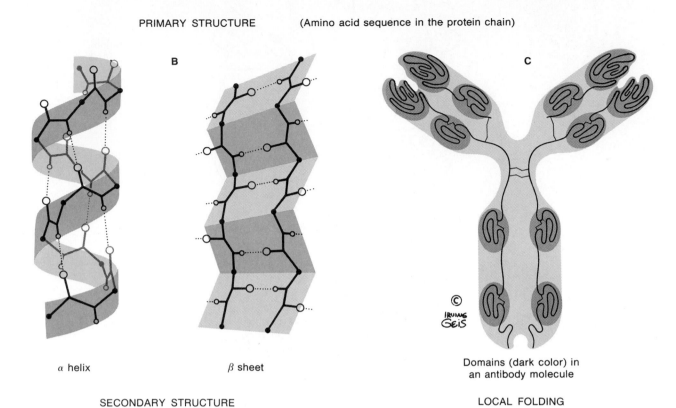

A

PRIMARY STRUCTURE   (Amino acid sequence in the protein chain)

B

α helix   β sheet

C

Domains (dark color) in
an antibody molecule

SECONDARY STRUCTURE   LOCAL FOLDING

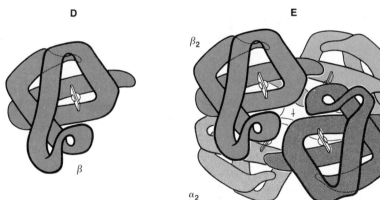

D

TERTIARY
STRUCTURE

One complete protein chain
(β chain of hemoglobin)

E

QUATERNARY STRUCTURE

The four separate chains
of hemoglobin assembled
into an oligomeric protein

F

QUATERNARY STRUCTURE

α (white) and β (color)
tubulin molecules in a
microtubule.

Figure 2–12
Hierarchies of protein structure.

stabilize proteins act in concert with related energetic and geometric factors to yield successively larger and more complex protein structural arrangements (Figure 2–12).

To begin with, we will examine the geometric properties of the polypeptide chain; as we will see, steric interactions restrict its accessible conformations and reflect features of the protein's amino acid sequence, or *primary*

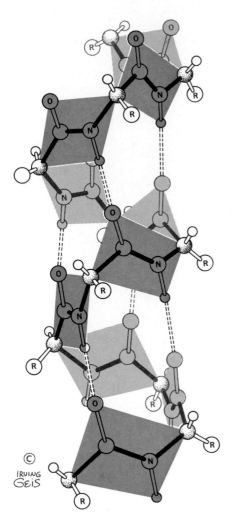

Figure 2–13
The α helix with 3.6 residues per turn. Approximately two turnings of the helix are shown. Hydrogen bonds between substituents of the rigid amide planes are shown as dashed lines. These hydrogen bonds stabilize the helical form in space. Also shown are the amino acid side chains (R groups) of the helix.

structure. Next we will show how the requirements for hydrogen-bond preservation in the folded structure result in the cooperative formation of regular structural regions in proteins. This situation arises principally because of the regularly repeating geometry of the hydrogen-bonding groups of the polypeptide backbone and leads to the formation of regular hydrogen-bonded _secondary structures_. Association between elements of secondary structure in turn results in the formation of _structural domains_, whose properties are determined both by chiral properties of the polypeptide chain and packing requirements that effectively minimize the molecule's hydrophobic surface area. Further association of domains results in the formation of the protein's _tertiary structure_, or overall spatial arrangement of the polypeptide chain in three dimensions. Likewise, fully folded protein subunits can pack together to form _quaternary structures_, which can either serve a structural role or provide a structural basis for modification of the protein's functional properties.

Three Secondary Structures Are Found in Most Proteins

A major driving force in protein folding is the necessity to minimize the extent of exposure by the hydrophobic group to solvent. This consideration involves a sacrifice of the favorable hydrogen-bonded interactions between the unfolded polypeptide backbone and water. To preserve a favorable energy balance on folding, the backbone polypeptide groups must take part in alternative hydrogen-bonded interactions between themselves in the protein's folded state.

The α Helix.   One of the most commonly observed protein secondary structures is the α helix. In an α helix, the polypeptide backbone follows the path of a right-handed helical spring to form an arrangement in which each residue's carbonyl group forms a hydrogen bond with the amide NH group of the residue four amino acids farther along the polypeptide chain (Figure 2–13). All residues in an α helix have nearly identical conformations, averaging $\psi = -45°$ to $-50°$ and $\phi = -60°$, so they lead to a regular structure in which each 360° of helical turn incorporates approximately 3.6 amino acid residues and rises 5.6 Å along the helix axis direction. The advance per amino acid residue along the helix axis is 1.5 Å. Although alternative helical arrangements having different hydrogen-bonding patterns and different geometries are also conformationally possible, the α helix is by far the most commonly observed helical arrangement found in proteins. The particular stability of the α helix appears to be related not only to the formation of stable hydrogen bonds between all the backbone carbonyl and NH groups, but also to the tight packing achieved in folding the chain to form the structures. This packing is most apparent when the helix is represented by space-filling van der Waals models (Figure 2–14). Note the close-packed arrangement of most of the atoms in the structure. And, in the tube view (Figure 2–14b), note the absence of any free space down the core of the helix. Alternative arrangements, in contrast, either have inferior hydrogen bonds or are not as tightly packed as the α helix.

Figure 2–9 shows that there is a possible but small and shallow energy minimum at the left-handed α-helical (L_α) position for nonglycine residues and that only 1 to 2 per cent of the nonglycine residues are L_α. However, for the symmetrical glycine, whose R group is the same as its C_α H and therefore has no hand at C_α, left-handed conformations are exactly equivalent to right-handed ones, and in fact, about half the glycines have positive ϕ values. Extended L_α helices have not been observed since they are difficult for the usual sequence of side chains, but isolated L_α residues are fairly common

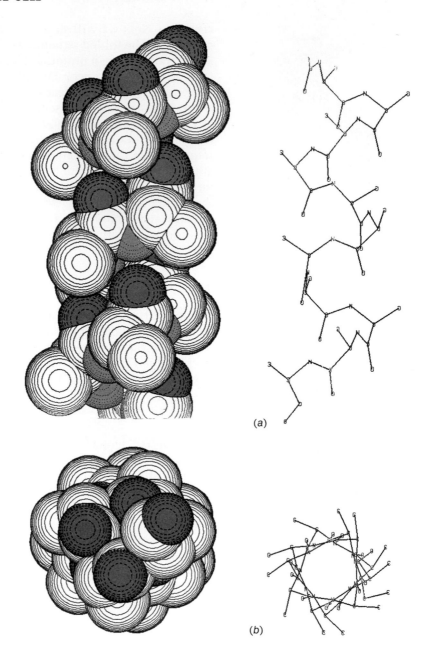

(a)

(b)

Figure 2–14
Another way of displaying the helix structure. (a) Side view. (b) Tube view. Hydrogens not shown. These models are computer-generated. The models on the left use van der Waals radii to give a picture of the space-filling characteristics of the molecule. The wire model on the right shows the connectivities.

and are quite important in producing a greater diversity of backbone conformations. They occur frequently in β bends, to be described shortly, as well as at other places where the backbone changes direction, such as at the C terminals of helices. Many of these L_α residues are glycines.

An important property stemming from the conformational regularity of the α helix, which applies to other secondary structures as well, is _cooperativity_ in folding. For example, once a single turn of α helix has been formed, addition of successive residues becomes much more likely and faster because the first turn of the helix forms a template upon which to erect successive helical residues. Owing to steric restrictions, the torsional angle ϕ is approximately correct for each additional residue. Each addition mainly involves sampling various conformations of ψ until the residue is "captured" in the correct conformation by the formation of a hydrogen bond to a group that is already fixed in the helical conformation.

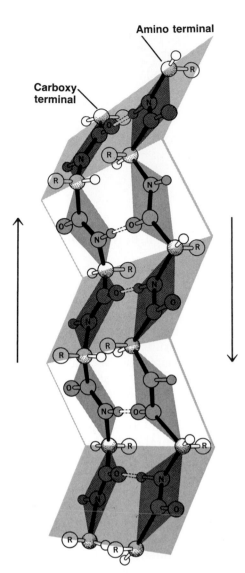

Amino terminal

Carboxy terminal

Figure 2–15
The antiparallel β sheet. This structure is composed of two or more polypeptide chains in the fully extended form, with hydrogen bonds formed between the chains. Hydrogen bonds are shown as dashed lines.

β Sheets.   A second type of commonly occurring protein secondary structure is the β-pleated sheet (Figure 2–15). Sheets are formed when two or more almost fully extended polypeptide chains are brought together side by side so that regular hydrogen bonds can form between the peptide backbone amide NH and carbonyl oxygen groups of adjacent chains. Notice that since each backbone peptide group has its NH and carbonyl groups in a *trans* orientation, it is possible to extend a β sheet into a multistranded structure simply by adding successive chains to the sheet. β sheets can occur in two different arrangements. In the first of these, the chains are arranged with the same N-to-C polypeptide sense to produce a *parallel β sheet*. Alternatively, the chains can be aligned with opposite N-to-C sense to produce an *antiparallel β sheet*. As illustrated (Figure 2–16), parallel and antiparallel β sheets are both composed of polypeptide chains that have conformations pointing alternate R groups to opposite sides of the sheet, but have their peptide planes nearly in the sheet plane to allow good interchain hydrogen bonding. Nevertheless, the chain conformation that produces the best interchain hydrogen bonding in parallel sheets is slightly less extended than that for the antiparallel arrangement. As a result, the parallel sheet has both a shorter repeat period (6.5 Å per residue pair, versus 7.0 Å for antiparallel) and more pronounced pleats than the antiparallel sheet.

β Bends.   Thus far we have described the geometry of protein secondary structures that resemble long rods or flat sheets. Obviously, to fold a polypeptide chain to a compact globular form, there must be some way to change the direction of the polypeptide chain. Such folding might, for example, be required to connect adjacent ends of the polypeptide chains in an antiparallel β sheet. A commonly observed and particularly efficient way to do this is by formation of a tight loop in which a residue's carbonyl group forms a hydrogen bond with the amide NH group of the residue three positions farther along the polypeptide chain. The resulting so-called *β bend* reverses the direction of the polypeptide chain.

Several conformational variations of the β bend have been observed that are a function of the amino acid sequence in the bend (Figure 2–17). In particular, it has been observed that the amino acids glycine and proline occur frequently in β bends. Because of its small size, glycine is conformationally more flexible than other amino acids. It can therefore serve as a flexible hinge between regions of polypeptide chains whose steric interac-

(a) Antiparallel   14.0 Å

(b) Parallel   13.0 Å

Figure 2–16
Two forms of the β-sheet structure: *(a)* the antiparallel and *(b)* the parallel β sheet. The advance per two amino acid residues is indicated for each structure.

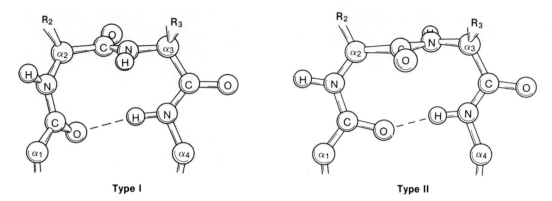

Type I   **Type II**

Figure 2–17
The two major types of tight turn of β bends (I and II). In type II, R₃ is generally glycine.

tions would otherwise keep them in more extended conformations. Proline, in contrast, is more conformationally restricted than other amino acids, since its cyclically bonded structure fixes its ϕ conformational degree of freedom. In a sense, then, part of the geometry that results in bend formation is preformed in proline-containing sequences. It appears probable that either situation might promote the formation of a bend during initial stages of protein folding and so cause structures such as antiparallel β sheets to assemble cooperatively, in a manner resembling the closure of a zipper.

The secondary structures described in this section—the α helix, the β sheet, and the β bend—are the most common structural elements found in most real proteins. Before discussing specific protein structures, we will consider the roles played by the different amino acid residues in building these structural elements.

Relationships Between Amino Acid Sequence and Secondary Structure. Examination of a large number of known three-dimensional protein structures has shown that the various amino acids have tendencies to form different secondary structures. As shown in Figure 2–18, glutamic acid, methionine, and alanine appear to be the strongest α-helix formers and valine, isoleucine, and tyrosine the most probable β-sheet formers, while proline, glycine, asparagine, aspartic acid, and serine occur most frequently in β-bend conformations.

This information is becoming of increasing value in the prediction of secondary structural regions of proteins from their amino acid sequences. The observed frequencies of occurrence of each amino acid in a given conformation can be equated with the probabilities that the same amino acid will behave similarly in a sequence whose actual secondary structure is unknown. In order to predict the secondary structure from the sequence, it is consequently only necessary to sequentially plot the individual amino acid probabilities or, better, a local average over a few adjacent residues to account for the cooperative nature of secondary-structure formation. In such a scheme, sequences such as Gly-Pro-Ser and Ala-His-Ala-Glu-Ala give high joint probabilities for being, respectively, in β-bend and α-helical conformations. However, comparisons of predicted versus directly observed polypeptide conformations give mixed results. This situation is a consequence of two facts: that several amino acids are somewhat ambiguous in their secondary-structure-forming tendencies, and that strong β-bend formers do occasionally turn up in the middle of α helices.

Nevertheless, it is clear that the structural arrangement of a protein in its

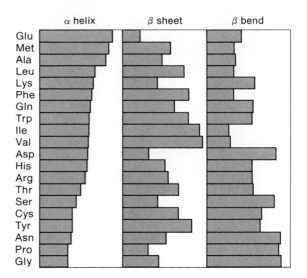

Figure 2–18
Relative probabilities that any given amino acid will occur in the α-helical, β-sheet, or β-hairpin-bend secondary structural conformations.

folded state must depend ultimately on its sequence. The information given in Figure 2–18—that given sequences have general tendencies toward forming particular sorts of structures—is a basic point of departure. The attainment of a particular folding arrangement depends on details of both the short- and long-range interactions (e.g., charge pairing and close packing) that uniquely characterize and stabilize each protein's structure. At present it is difficult to evaluate these effects in the detail required to actually produce a protein's structure from its sequence.

Fibrous Proteins Have Repeating Secondary Structures

Early in this century attempts were made to characterize proteins by their physicochemical properties, particularly their solubilities. *Globular proteins* were those proteins that were water-soluble; in addition they had an approximately round shape. *Fibrous proteins* were usually water-insoluble and highly elongated rodlike molecules. It is still convenient to use this classification scheme when discussing protein structure, because these properties have a structural basis.

Fibrous proteins can be assembled in two ways: (1) extended or α-helical polypeptide chains can twist together or bundle into fibers, or (2) globular subunits can be helically arranged to make fibers (as in the case of tubulin; see Figure 2–6e). In this section the focus will be on proteins of the first type, which have predominantly regular secondary structures.

Keratins form one of the most diverse classes of fibrous proteins. They are major constituents of such biological tissues as hair, scales, horns, hooves, wool, beaks, nails, and claws. The most prevalent forms of keratin are composed largely of polypeptide chains in the α-helical conformation (α-keratins). Alternative forms exist in bird feathers, which appear to be composed of stacked and folded β sheets (β-keratins).

Figure 2–19 illustrates the arrangement of the α helices in typical α-keratins such as hair or wool. The materials are composed of extremely long α-helical polypeptide chains that are arranged side by side to form long cables. The α helices in α-keratins are spirally twisted so that the resulting cable has an overall left-handed twist. The formation of a cable with twisted α helices appears to be the result of optimization of packing among the amino acid side-chain residues between the helices. Side-chain residues of an α helix are arranged in a spiral fashion so that residues falling on the same

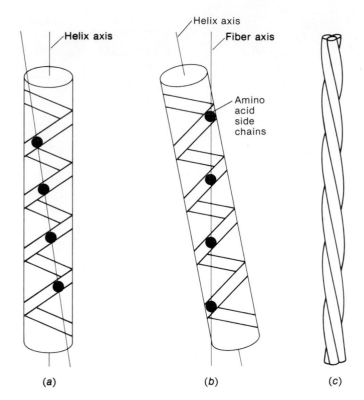

Figure 2–19
Coiling of α helices in α-keratins. Residues on the same side of an α helix form rows that are tilted relative to the helix axis. Packing helices together in fibers is optimized when the individual helices wrap around each other so that rows of residues pack together along the fiber axis.

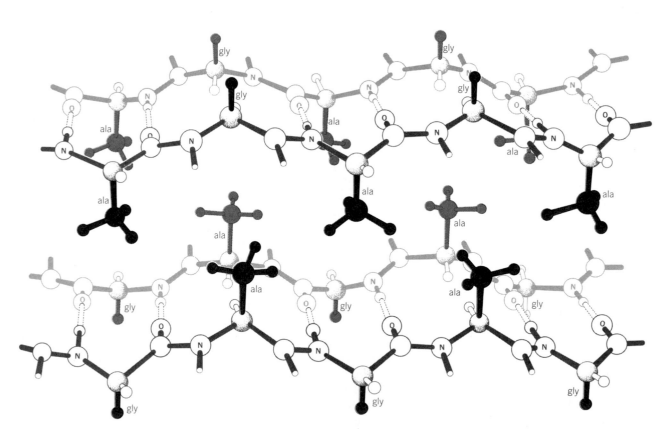

Figure 2–20
The three-dimensional architecture of silk. The side chains of one sheet nestle quite efficiently between those of neighboring sheets. The cut bonds extend to neighboring chains in the same sheet. (Reprinted with permission from R. E. Dickerson and I. Geis, *The Structure and Action of Proteins*, Benjamin/Cummings, Menlo Park, Calif., 1969.)

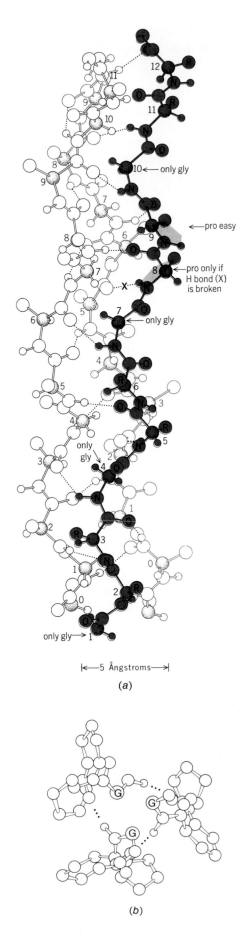

only gly→ 1

|←——5 Ångstroms——→|

(a)

(b)

side of a helix generally do not lie along a line parallel to the helix axis (Figure 2–19a). The packing together of helices is consequently optimized when the helices interact at an angle of about 18°. Obviously, if the α helices involved in such a packing interaction are straight, they will eventually become separated. However, their packing interaction can be preserved if the helices are slightly twisted around each other, with the resultant formation of the left-twisted cable structure characteristic of the α-keratins (Figure 2–19b and c). The coiled character of the α helices in such fibers consequently represents a tradeoff between some local deformations that coil the α helix and the optimization of extended side-chain packing interactions in the cable as a whole.

The springiness of hair and wool fibers results from the tendency of the α-helical cables to untwist when stretched and spring back when the external force is removed. In many forms of α-keratin, the individual α helices or fibers are covalently linked by disulfide bonds formed between cysteine residues of adjacent polypeptide chains. The pattern of these covalent interactions serves to both influence and fix the extent of curliness in the hair fiber as a whole. Chemical reactions and mechanical processes involving reductive cleavage, reorganization, and reoxidation of these interhelix disulfide bonds form the basis of the "permanent wave."

Silk is a variety of fibrous protein produced by certain insects. Silks are structurally composed of aligned and stacked antiparallel β sheets (Figure 2–20). Sequence analysis of silk proteins show them to be largely composed of glycine, serine, and alanine, where every alternate residue is glycine. Since the side-chain groups of a flat antiparallel β sheet point alternately upward and downward from the plane of the sheet, all the glycine residues are arranged on one surface of each sheet and all the substituted amino acids on the other. Two or more such sheets can consequently be intimately packed together to form an arrangement of stacked sheets in which two adjacent glycine-substituted or alanine-substituted sheet surfaces interlock with each other. Owing to both the extended conformations of the polypeptide chains in the β sheets and the interlocking of the side chains between sheets, silk is a mechanically rigid material that tends to resist stretching.

Collagen is a particularly rigid and inextensible protein that serves as a major constituent of tendons and many connective tissues. Analysis of collagen amino acid sequences shows them to be characterized by a repetitious tripeptide sequence, Gly-X-Pro or Gly-X-Hydroxyproline. In these tripeptides X can be any amino acid, and hydroxyproline is a hydroxylated derivative of proline. Owing to the repeating proline residue, collagen polypeptide chains cannot adopt either an α-helical or a β-sheet conformation. Instead, individual collagen polypeptide chains tend to assume a left-handed helical conformation, in which successive side-chain groups point toward the corners of an equilateral triangle when viewed down the polypeptide chain axis (Figure 2–21). The glycine at every three residues is strictly required because there is no room for any other amino acid inside the triple helix, where the glycine R groups are located. The collagen helix is already very extended, so it cannot easily stretch further like the α helix. Moreover, in contrast to the α helix, formation of collagen helix is not accompanied by the formation of hydrogen bonds among residues within each polypeptide chain. Instead, three collagen chains associate in a three-stranded cable, with hydrogen

Figure 2–21
The basic coiled-coil structure of collagen. Three left-handed single-chain helices wrap around one another with a right-handed twist. (a) Ball-and-stick single-collagen chain. (b) View from top of helix axis. Note that glycines are all on the inside. In this structure the C=O and N—H groups of glycine protrude approximately perpendicularly to the helix axis so as to form interchain hydrogen bonds.

bonds between each chain and its neighbors. This produces a highly inter-locked fibrous structure that is admirably suited to its biological role of pro-viding rigid connections between muscles and bones, as well as structural reinforcement for skin and connective tissues.

Although there exist in living organisms additional types of proteins, as well as polysaccharide-based structural materials, we have focused here on three arrangements whose structural properties are currently best under-stood. Two of these, the α-keratins and the silks, incorporate polypeptide secondary structures that also commonly occur in globular proteins. Col-lagen, in contrast, is a protein that evolution has developed to play a more specialized role.

The Secondary and Tertiary Structures of Globular Proteins Are Extremely Varied

Globular proteins, as their name implies, differ from fibrous proteins in that they generally have a more or less spherical shape. Nevertheless, three-dimensional structural studies of _globular proteins_ and enzymes show that they _incorporate many of the secondary structural features that typify the fibrous proteins_. Figure 2–22, for example, illustrates the first enzyme whose three-dimensional structure was determined, the 120-residue protein lysozyme. This protein has local regions of ordered α-helical and antiparal-lel β-sheet secondary structures, but it has, in addition, several extended regions incorporating β bends and extended polypeptide chains with a less regular conformation.

Among the approximately 250 proteins whose structures have been de-termined, there is a rich variety of alternative structural arrangements. Each different polypeptide sequence is associated with the formation of a unique tertiary structure. Nevertheless, careful comparisons have shown that many proteins share some fundamental structural similarities. Further, the simi-larities recur among proteins that show little similarity in sequence or func-tion. This fact suggests that the recurring features have common physical origins. They appear, in fact, to arise from two different sorts of physical effects.

The first of these is a chiral effect, or tendency for extended structural arrangements in proteins to be "handed." This tendency reflects the fact that their constituent polypeptide chains are composed of chiral L-amino acids.

Figure 2–22
Lysozome. In this and succeeding figures the polypeptide backbone is represented as a ribbon to allow the polypeptide-chain course to be followed easily.

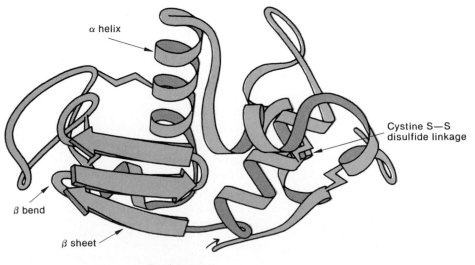

Egg lysozyme

Chiral effects manifest themselves both in the manner in which regions of secondary structure are interconnected in globular proteins and in the geometric properties of globular protein sheets. The second effect of importance in tertiary structural organization relates to how secondary structural regions, such as α helices and β sheets, can most efficiently pack together so as to minimize the protein's solvent-accessible surface area. Both of these effects are considered in the following discussion.

Chiral Properties Influence Conformation.   In the preceding discussion we noted that many potential polypeptide conformations are ruled out by unfavorable steric interactions. However, the frequent occurrence of structures such as α helices suggests that certain of these arrangements not only are allowed, but are particularly stable. As in the case of the α helix, the relative stability of a particular conformation is governed by the details of the interaction forces among the atoms comprising the polypeptide chain. Given the fact that proteins are composed primarily of chiral L-amino acids, it is not surprising that the most stable conformations of extended polypeptide chains are not straight. Instead, detailed conformational energy calculations have shown that *extended polypeptide chains prefer to be slightly twisted in a right-handed sense when viewed down the polypeptide chain axis.* Since the residues of straight, extended polypeptide chains alternate in position by 180° (e.g., see Figure 2–15), *the cumulative effect of this tendency toward right twisting is to produce extended structures that are coiled in a right-handed direction* (Figure 2–23).

The effects of this tendency for extended-chain structures to form right-handed coiled or twisted structures are revealed in two related but different structural features common to virtually all known proteins. The first is the kind of connection that occurs between the ends of parallel polypeptide strands that form β sheets in globular proteins. The connection from one β strand to the next can occur at the same end of the sheet in a simple hairpin turnaround only in antiparallel β sheets. In parallel β sheets a "crossover" connection to the other end of the sheet is required; this can be either right-handed or left-handed (Figure 2–24). *All crossover connections observed in*

Figure 2–23
The natural tendency for the polypeptide chain to twist in the right-hand direction produces structures with an overall right-handed connectivity. The structure represents a single fully extended polypeptide chain.

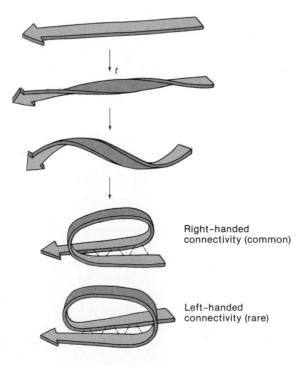

Right-handed connectivity (common)

Left-handed connectivity (rare)

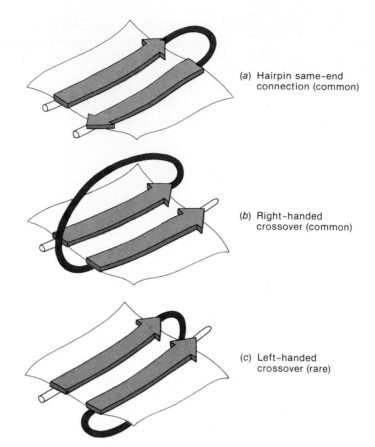

(a) Hairpin same–end connection (common)

(b) Right–handed crossover (common)

(c) Left–handed crossover (rare)

Figure 2–24
Three ways of making connections between β strands. *(a) A hairpin same-end connection is commonly found for β strands in the antiparallel orientation. (b) A right-handed crossover connection is commonly found for β strands in the parallel orientation. (c) A left-handed crossover connection is rarely found.*

protein structures are right-handed, irrespective of whether or not the strands they join are adjacent. This invariant pattern is most likely a result of the energetically favored nature of the right-handed crossover.

The second feature that reveals the influence of right-handed twisting is the geometry of globular protein parallel β sheets. The β sheets of globular proteins are always twisted in a right-handed sense when viewed along the polypeptide chain direction. The twisting behavior of β sheets is an important feature in protein structural architecture, since twisted β sheets frequently constitute the backbone of protein structures. Figure 2–25 shows the polypeptide-chain folding of four proteins that incorporate twisted parallel or mixed β sheets. These include the exoprotease carboxypeptidase A, the electron transport protein flavodoxin, and the glycolytic enzyme triose phosphate isomerase. Although these proteins all have right-twisted β sheets, it is clear that their overall geometries differ. That is, the β sheets in carboxypeptidase and flavodoxin are smoothly twisted to form saddle-shaped surfaces, while the β sheets in triose phosphate isomerase take the form of a cylinder or β barrel.

Within each overall type of parallel β sheet organization, the detailed hydrogen-bond pattern can be understood in terms of the forces acting within and between the polypeptide chains (Figure 2–26). In the case of the roughly rectangular sheets in carboxypeptidase and flavodoxin, the observed geometry reflects a competition between the tendency of the individual chains to twist and the tendency of the interchain hydrogen bonds to remain firm. Basically, the interchain hydrogen bonds tend to stretch when the sheet is twisted and so resist introduction of twist into the sheet. The observed saddle-shaped geometry reflects the uniform distribution of these conflicting forces throughout the sheet, as shown in Figure 2–26a.

β BARREL SHAPE

SADDLE SHAPE

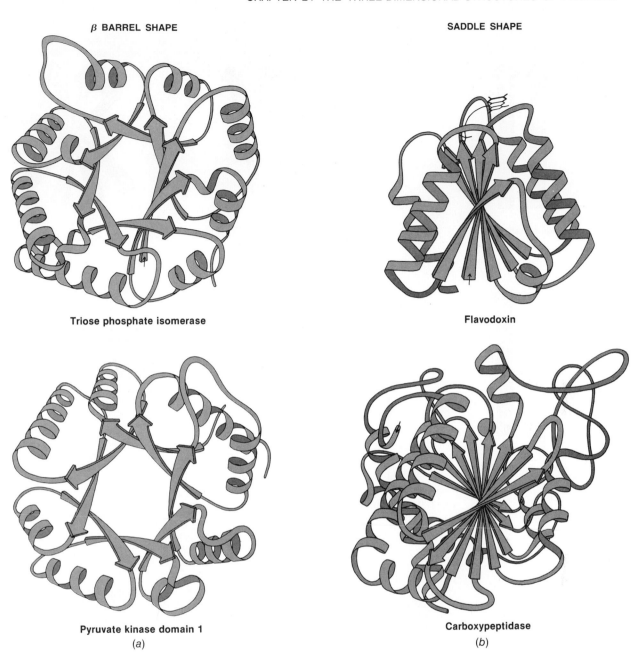

Triose phosphate isomerase

Flavodoxin

Pyruvate kinase domain 1

(a)

Carboxypeptidase

(b)

Figure 2–25
A comparison of parallel β-sheet structures forming the backbone structures in different enzymes (or parts of enzymes): (a) β-barrel arrangement, (b) saddle shape.

The β sheet that forms the barrel in triose phosphate isomerase has an hourglass-shaped surface with cylindrical curvature. Twisted β strands with a staggered hydrogen-bond pattern (Figure 2–26b) automatically produce a cylindrical curvature. Conversely, twisted strands on a cylindrical surface necessitate a staggered hydrogen-bond pattern. Again, a compromise occurs between twisting and hydrogen bonding, leading to approximately straight chains with somewhat stretched hydrogen bonds at the top and bottom, which produce the hourglass shape. The differences in the geometries of rectangular and staggered plane sheets therefore result from differences in how adjacent sheet strands are hydrogen-bonded together. In either case, the operative forces are similar, and the final result reflects a compromise between chain twisting and preservation of good interchain hydrogen bonds.

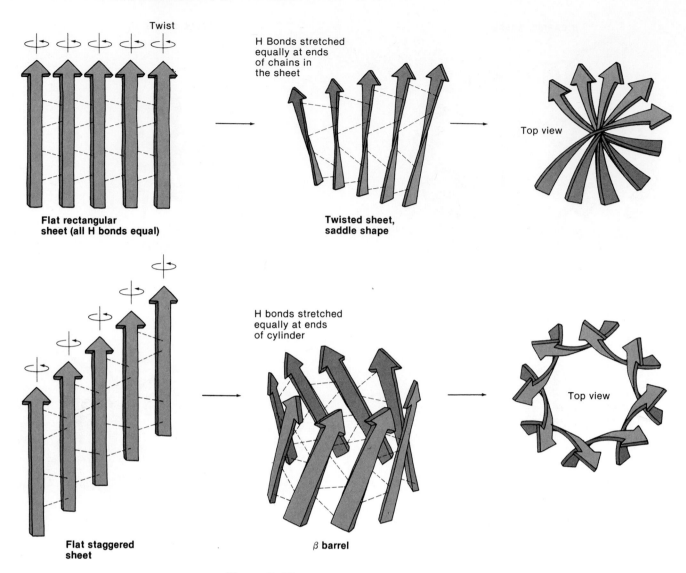

Figure 2–26

Origin of β-barrel and β-sheet conformations. The observed geometry in β-sheet structures represents a competition between the tendency of the individual chains to twist in a right-handed way and the tendency of the interchain hydrogen bonds to be preserved. Rectangularly arranged sheets give rise to the saddle shape. Staggered sheets give rise to the β-barrel shape.

Chiral preferences affect the connectivity as well as the sheet geometry in parallel β proteins. The right-handed crossover in parallel β barrels cannot go down the center, which is only large enough to accommodate the hydrophobic side chains. As a rule, the polypeptide backbone winds in a simple right-handed spiral around the barrel, moving over by one β strand at a time and packing helices or loops around the outside. Thus although these structures tend to be large, their organization (Figure 2–27a) is very simple.

The saddle-shaped parallel β sheets, such as those in carboxypeptidase or flavodoxin (see Figure 2–25), have a layer of helices and loops on each side. In order to accomplish this with right-handed crossover connections, the polypeptide chain must sometimes move along the sheet in one direction and sometimes in the other direction. The most common organizational pattern found in known protein structures starts in the middle of the sheet and winds toward one edge, with right-handed crossovers packing a layer of helices on one side of the sheet. Then the polypeptide chain returns to the

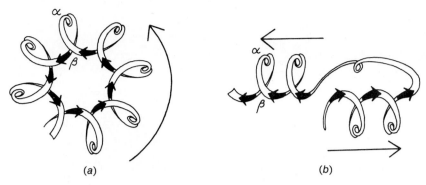

(a)   (b)

Figure 2–27
Highly simplified sketches of (a) a singly wound parallel β barrel, (b) a doubly wound β sheet. Thin arrows next to the diagrams show the direction in which the chain is progressing from strand to strand in the sheet. The α and β labels in the figure refer to regions of α helix and β sheet.

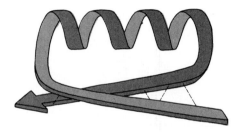

Figure 2–28
A βαβ loop. This arrangement forms the basis of many of the more extended structural arrangements, such as those shown in Figure 2–25.

middle of the sheet and winds out to the opposite edge, packing helices against the other side of the sheet (Figure 2–27b). This pattern is often known as a nucleotide-binding domain, since most of these proteins bind a mononucleotide or dinucleotide cofactor in the middle of the C-terminal end of the β sheet.

Effects of Packing Between Secondary Structures. In the preceding section we described how energetically preferred chiral properties of the polypeptide chain manifest themselves in the pattern of strand connectivity and the overall geometry of globular-protein β sheets. An additional important factor in protein structural organization is the efficient packing together of secondary structural elements to form larger units. Figure 2–28 shows one of the most commonly observed arrangements, called a βαβ loop. This is a special case of the right-handed crossover connection described in the previous section. A βαβ loop is composed of a pair of adjacent hydrogen-bonded parallel β strands that are right-connected, with a stretch of α helix that packs tightly on the surface of the sheet. Generally, the β-sheet strands in a local βαβ loop are part of a much larger β sheet. Structures such as triose phosphate isomerase (Figure 2–25) can be viewed as a series of overlapping βαβ structural units. The overlapping produces a final structural arrangement in which an inner barrel, composed of β-sheet strands, packs tightly within an outer barrel composed of α helices.

Although tertiary structures composed of βαβ loops are perhaps the most commonly observed pattern in known protein structures, other arrangements are found in proteins having either predominantly antiparallel β-sheet or predominantly α-helical conformations. The most common structural organization in antiparallel β proteins has two layers of β sheets (Figure 2–29). Other antiparallel β proteins have a single twisted β sheet covered on only one side by a layer of helices and loops (Figure 2–30).

No protein is stable as a single-layer structure, since it requires at least two layers to bury the hydrophobic core. Thus antiparallel β proteins are typically two-layer structures; antiparallel sheets are apparently quite stable when one side is exposed to solvent. Parallel sheets require at least one additional layer besides the sheet in order to make the crossover connections between the parallel β strands. In contrast to antiparallel β sheets, they apparently cannot tolerate solvent exposure on even one side and are always found as a structural "backbone" in protein interiors, with other layers of structure on both sides. Therefore, proteins with βαβ loops generally have

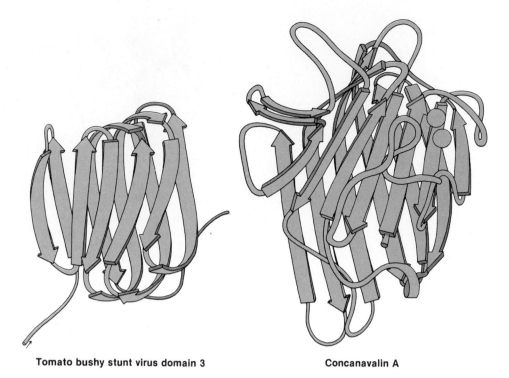

Tomato bushy stunt virus domain 3

Figure 2–29
Examples of proteins containing
β-sheet domains.

Tomato bushy stunt virus domain 3

Concanavalin A

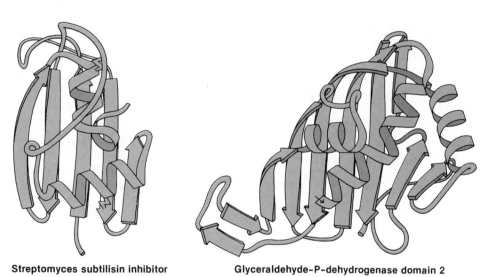

Figure 2–30
Examples of antiparallel β proteins
that are covered on only one side by
larger helices and loops.

Streptomyces subtilisin inhibitor

Glyceraldehyde–P–dehydrogenase domain 2

either three layers, as in the nucleotide-binding domains, or four layers, as in
the parallel β barrels (see Figure 2–25). Usually the outer layers are formed
of α helices, which must pack against one another and also against the surface formed by the β-sheet side chains.

The structural geometry of proteins that have only an α-helical secondary structure is also largely determined by requirements for efficient packing
between the helices. One particularly stable interhelical packing arrangement has already been encountered in the discussion of fibrous proteins (see
Figure 2–19). In the α-keratins, adjacent helices pack together with an interaction angle between helices of about 18°. This extended-helix interaction
pattern forms the basis for a protein structural motif frequently seen in globular proteins, the 4-α-helical bundle. In this arrangement, four α helices,
sequentially connected to their nearest neighbors, pack together to form an

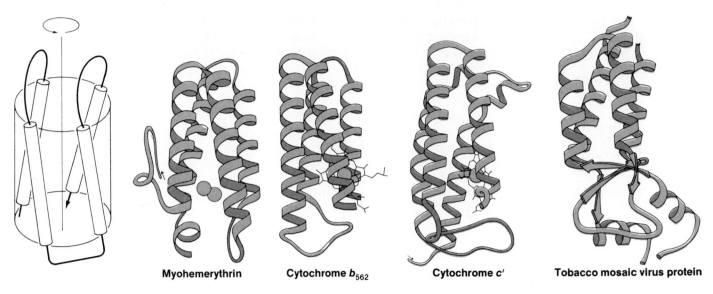

Myohemerythrin   **Cytochrome b_{562}**   **Cytochrome c'**   **Tobacco mosaic virus protein**

Figure 2–31
Examples of some proteins that share a common structural motif of four α helices.

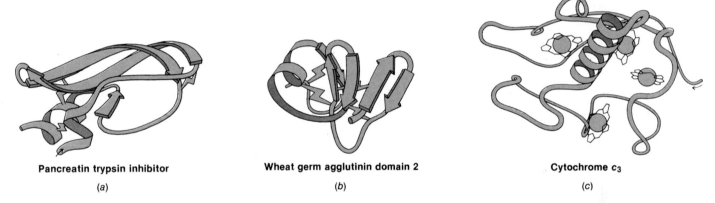

Pancreatin trypsin inhibitor   **Wheat germ agglutinin domain 2**   **Cytochrome c_3**

(a)   (b)   (c)

Figure 2–32
Examples of some small proteins or domains in which disulfide bonds (a, b) or a porphyrin group (c) are a dominant factor holding the structure together. Porphyrins are depicted in Figure 2–41.

array with a roughly square cross section. Since each helix interacts with its neighbors at an angle of about 18°, the overall bundle has a left-handed twist (Figure 2–31). This commonly observed folding domain clearly represents a minimum accessible surface area arrangement for four sequentially connected α helices of approximately equal length. Many α-helical proteins that lack these features have more complex and irregular geometries. However, even in these cases it appears that the relative orientations of adjacently packed helices reflect geometric restrictions that accompany close packing between helices.

In addition to the packing of elements of protein secondary structure, which is a dominant feature in most proteins, there are some cases, especially among the smallest structures, where the geometry and packing of disulfide bonds or nonpeptidyl groups are a dominant factor. Figure 2–32 shows examples of this sort, in which the secondary structures are short and irregular and cannot assume their native structures if the disulfides are broken (a, b), or if the nonpeptidyl groups are missing (c).

In summary, examination of a large number of protein tertiary structures

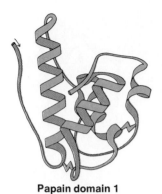

Papain domain 1

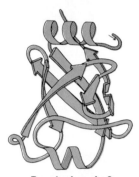

Papain domain 2

Figure 2–33
Papain, a protein in which the domains are very different from one another.

Figure 2–34
Rhodanese domains 1 and 2 as an example of a protein with two domains that resemble each other extremely closely. Rhodanase is a liver enzyme that detoxifies cyanide by catalyzing the formation of thiocyanate from thiosulfate and cyanide.

Figure 2–35
Schematic backbone drawing of the elastase molecule, showing the similar β-barrel structures of the two domains. The outside surfaces of the β barrels are stippled.

has shown that they incorporate several different sorts of recurring structural arrangements. In general, these arrangements owe their origins to physical effects, some of which predispose the most stable conformations of the polypeptide chain, and some of which govern the formation of intimately packed tertiary structures.

Domains Are Functional Units of Tertiary Structure.   The patterns of tertiary structure described in this and previous sections frequently constitute the entire protein. However, within a single folded chain or subunit, contiguous portions of the polypeptide chain often fold into compact local units called _domains_, each of which might consist, for example, of a four-helix cluster or a barrel or an antiparallel β sheet. Sometimes the domains within a protein are very different from one another, as within the protease papain (Figure 2–33), but often they resemble each other very closely, as in rhodanase (Figure 2–34).

The separateness of two domains within a subunit varies all the way from independent globular domains joined only by a flexible length of polypeptide chain, to domains with tight and extensive contact and a smooth globular surface for the outside of the entire subunit, as in the proteolytic enzyme elastase (Figure 2–35). An intermediate level of domain separateness, characterized by a definite neck or cleft between the domains, is found in phosphoglycerate kinase (Figure 2–36).

Domains as well as subunits can serve as modular bricks to aid in efficient assembly of the native conformation. Undoubtedly the existence of separate domains is important in simplifying the protein-folding process into separable, smaller steps, especially for very large proteins. There is no strict upper limit on folding size. Indeed, known domains vary in size all the way from about 40 residues to over 400.

Rhodanese domain 1   **Rhodanese domain 2**

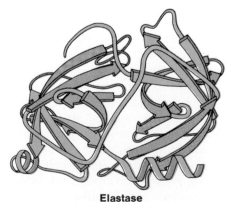

Elastase

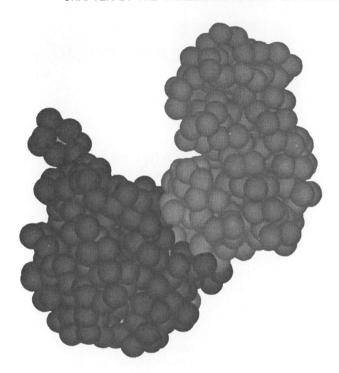

Figure 2–36
The dumbbell domain organization of phosphoglycerate kinase, with a relatively narrow neck between two well-separated domains.

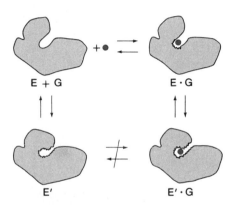

Figure 2–37
Schematic representation of the change in conformation of the hexokinase enzyme on binding substrate. E and E′ are the inactive and active conformations of the enzyme, respectively. G is the sugar substrate. Regions of protein or substrate surface excluded from contact with solvent are indicated by a crinkled line. (Adapted from W. S. Bennett and T. A. Steitz, Glucose-induced conformational change in yeast hexokinase, *Proc. Natl. Acad. Sci. USA* 75:4848, 1978.)

Another important function of domains is to allow for movement. Completely flexible hinges would be impossible between subunits because they would simply fall apart. However, flexible hinges can exist between covalently linked domains. Limited flexibility between domains is often crucial to substrate binding, allosteric control (discussed in Chapter 10), or assembly of large structures. In hexokinase, the two domains within the individual subunits hinge toward each other upon binding of the substrate glucose, enclosing it almost completely (Figure 2–37). In this manner glucose can be bound in an environment that excludes water as a competing substrate (see Chapter 14 for further details on the hexokinase reaction).

Quaternary Structure Involves Subunit Interaction

Although many globular proteins function as monomers, biological systems abound with examples of complex protein assemblies (Table 2–4). This higher-order organization of several globular subunits to form a functional aggregate is referred to as the quaternary structure of the protein. Protein quaternary structures can be classified into two types. One type involves the assembly of different kinds of subunits. Examples range from dimeric molecules that contain different molecular subunits to complex assemblies such as ribosomes (which also contain ribonucleic acid as a structural component). The organization of these sorts of quaternary structures depends on the specific nature of each interaction occurring between the different molecular subunits and their neighbors. Each intermolecular interaction generally occurs only once within a given aggregate arrangement, so that the overall complex structure has a highly irregular geometry.

A second observed pattern of quaternary structure is typified in molecular aggregates composed of multiple copies of one or more different kinds of subunits. Owing to the recurrence of specific structural interactions between the subunits, such aggregates typically form regular geometric arrangements. Structures of this type are found most frequently in the coats that surround viruses. Satellite tobacco necrosis virus and tobacco mosaic virus are examples of these (see Table 2–4). The protein tubulin (see Figure 2–12e) is an-

TABLE 2-4
Molecular Weight and Subunit Composition of Selected Proteins

Protein	Molecular Weight	Number of Subunits	Function
Glucagon	3,300	1	Hormone
Insulin	11,466	2	Hormone
Cytochrome c	13,000	1	Electron transport
Ribonuclease A (pancreas)	13,700	1	Enzyme
Lysozyme (egg white)	13,900	1	Enzyme
Myoglobin	16,900	1	Oxygen storage
Chymotrypsin	21,600	1	Enzyme
Carbonic anhydrase	30,000	1	Enzyme
Rhodanese	33,000	1	Enzyme
Peroxidase (horseradish)	40,000	1	Enzyme
Hemoglobin	64,500	4	Oxygen Transport
Concanavalin A	102,000	4	Unknown
Hexokinase (yeast)	102,000	2	Enzyme
Lactate dehydrogenase	140,000	4	Enzyme
Bacteriochlorophyll protein	150,000	3	Enzyme
Ceruloplasmin	151,000	8	Copper transport
Glycogen phosphorylase	194,000	2	Enzyme
Pyruvate dehydrogenase (E. coli)	260,000	4	Enzyme
Aspartate carbamoyltransferase	310,000	12	Enzyme
Phosphofructokinase (muscle)	340,000	4	Enzyme
Ferritin	440,000	24	Iron storage
Glutamine synthase (E. coli)	600,000	12	Enzyme
Satellite tobacco necrosis virus	1,300,000	60	Virus coat
Tobacco mosaic virus	40,000,000	2,130	Virus coat

other example. In these kinds of subunit interactions each subunit makes the same contacts with other subunits.

The forces that hold quaternary structures together are a mixture of the types of noncovalent forces that are involved in forming secondary and tertiary structures. Sometimes it is possible to disrupt quaternary structures without significantly disrupting the secondary or tertiary structures. For example, a low concentration of urea might bring about the dissociation of a multisubunit protein into its individual subunits without disrupting the secondary or tertiary structures. In other cases selective disruption of the quaternary structures has not been possible.

The formation of a subunit aggregate is usually intimately related to its function. A few examples should illustrate this point.

Hemoglobin: A Tetramer Containing Two Different Subunits. The function of hemoglobin is to deliver oxygen from the lungs to the other tissues. It does this by absorbing oxygen in the lung capillaries, where the oxygen pressure is high, and discharging the oxygen in other tissues, where the oxygen pressure tends to be lower. The quaternary structure of hemoglobin is ideally suited to this task. Hemoglobin consists of two α subunits, each with 141 amino acids, and two β subunits, each with 146 amino acids. The subunits are held together by a multitude of salt bridges and hydrogen bonds. Each subunit contains one oxygen binding site.

Two conformations for the hemoglobin tetramer are known; these are called the oxy and the deoxy forms. The transformation from one form to the other involves changes in both the tertiary and the quaternary structures. The oxy form predominates in the lungs, where hemoglobin gathers its oxygen for delivery. The deoxy form is favored at cell sites, where hemoglobin releases its bound oxygen. This two-structure motif of hemoglobin facilitates delivery of oxygen in the cells where it is needed, because the oxy form binds oxygen much more strongly than the deoxy form. If hemoglobin is dissociated into its monomeric subunits, it binds oxygen more strongly than it does in the tetrameric forms. Therefore, free subunits of hemoglobin do not release oxygen readily. This fact would have disturbing consequences if the monomeric form were to occur *in vivo*. The detailed mechanism of hemoglobin oxygen binding is considered in Chapter 10.

Actin and Myosin: A Supermolecular Aggregate.   Vertebrate muscle represents a remarkable example of a supermolecular aggregate of two proteins, actin and myosin, that are capable of undergoing a reversible reorganization. Muscle fiber is composed of many myofibrils, which when viewed in the light microscope present a striated and banded appearance. As shown in Figure 2–38, each myofibril exhibits a longitudinally repeating structure called the sarcomere. This repeating unit, 23,000 Å long, is characterized by the appearance of several distinct bands, the less optically dense band being referred to as the I band and the more dense one as the A band. Furthermore, in the center of each I band there appears a dense line called the Z line, and in the center of the A band there appears a dense narrow band, somewhat similar in appearance, called the M line. Adjacent to the M line are regions of the A band that appear less dense than the remainder and are termed the H zone.

Transverse sections of the sarcomere reveal that these band patterns result from the interdigitation of two sets of filaments (Figure 2–39). For example, when a sarcomere is sectioned through the I band, a somewhat disordered arrangement of thin filaments (about 70 Å in diameter) is seen. In contrast, when it is sectioned through the H zone, a hexagonal array of thick filaments (about 150 Å in diameter) is apparent. Finally, a transverse section in the dense region of the A band shows a regularly packed array of interdigitating thick and thin filaments. These observations led Hugh Huxley to propose that in muscle contraction the thick and thin filaments slide past each other as shown in Figure 2–39.

Figure 2–38
Schematic view of a striated muscle sarcomere, showing the appearance of filamentous structures when cross-sectioned at the locations illustrated below.

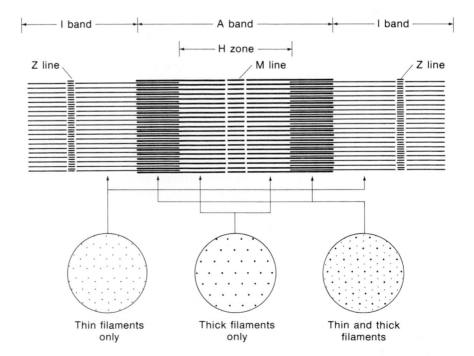

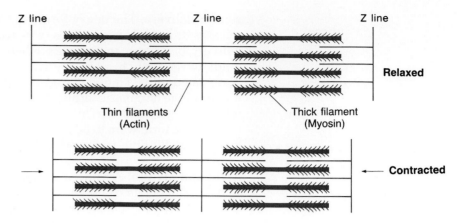

Figure 2–39
The sliding-filament model of muscle contraction. During contraction, the thick and thin filaments slide past each other so that the overall length of the sarcomere becomes shorter.

Closer examination of the subunit structure of muscle reveals regularly spaced cross-bridges between the actin and the myosin filaments. In muscular contraction the number of these cross-bridges is increased in an energy-driven process. Thus the quaternary structure formed between the actin and myosin filaments is a dynamic one that is directly responsible for muscular contraction.

Antibodies: Tetramers Containing Two Classes of Subunits. One of the major defense systems of vertebrates against invading foreign substances, such as viruses, bacteria, and cells from other organisms, is the immune system. This system produces proteins called *antibodies* that are released in response to the introduction of the foreign substances (usually proteins themselves), which are called *antigens*. Each antibody is highly specific for the particular antigen in question and combines with it to form an antibody-antigen complex in which the invading substance is rendered inactive and later removed from the organism.

Although antibody molecules differ in structure and have widely differing functional specificity, they are all variations of a common structural theme. Individual differences are restricted to a region of the molecule called the antigen binding site. Figure 2–40 gives a simplified overview of the "gross anatomy" of the major classes of immunoglobulins. The antibody molecule is composed of two light and two heavy chains (light and heavy refer to relative size or molecular weight) shaped in the form of a Y. The antigen binding sites are at the ends of the Y. The specificity of the antigen binding site is a function of both the light and heavy subunits that make up the complex. Different combinations of heavy and light chains result in antigen binding sites with specificities for different antigens. Thus the interaction of different protein subunits helps to account for the diversity of antigen binding sites. This diversity is augmented by mechanisms discussed in the next section.

FUNCTIONAL DIVERSIFICATION OF PROTEINS

The function diversification of proteins has arisen by a long evolutionary process. In addition to proteins that are structurally and functionally related, this process has resulted in proteins with similar structures and dissimilar functions, as well as proteins with dissimilar structures and similar functions.

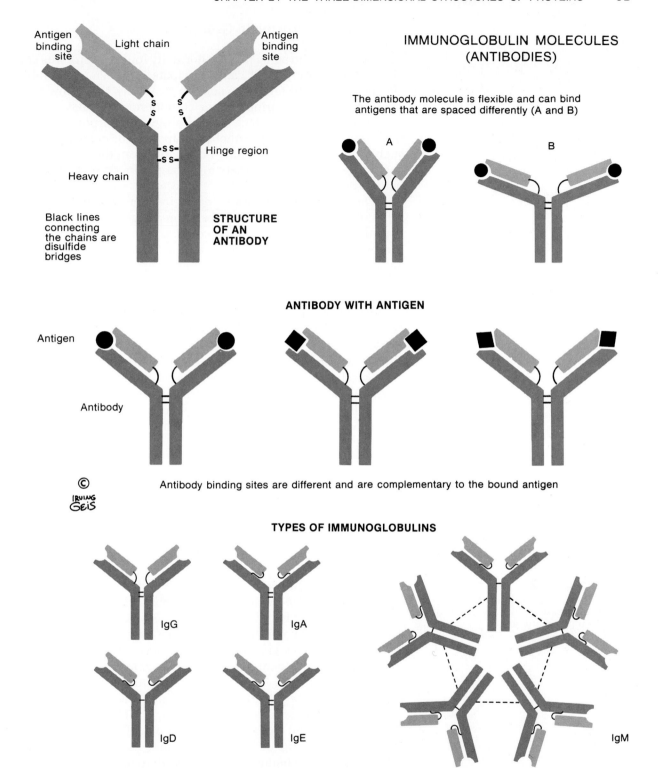

IMMUNOGLOBULIN MOLECULES (ANTIBODIES)

The antibody molecule is flexible and can bind antigens that are spaced differently (A and B)

ANTIBODY WITH ANTIGEN

Antibody binding sites are different and are complementary to the bound antigen

TYPES OF IMMUNOGLOBULINS

IgG molecules are serum proteins; IgA is found in secretions such as tears.

IgE is involved in allergies and IgM (a pentamer) responds most quickly to antigen.

IgD appears before IgM as a development marker that is not found in the mature β lephocyte.

Figure 2–40
Immunoglobulin molecules (antibodies).

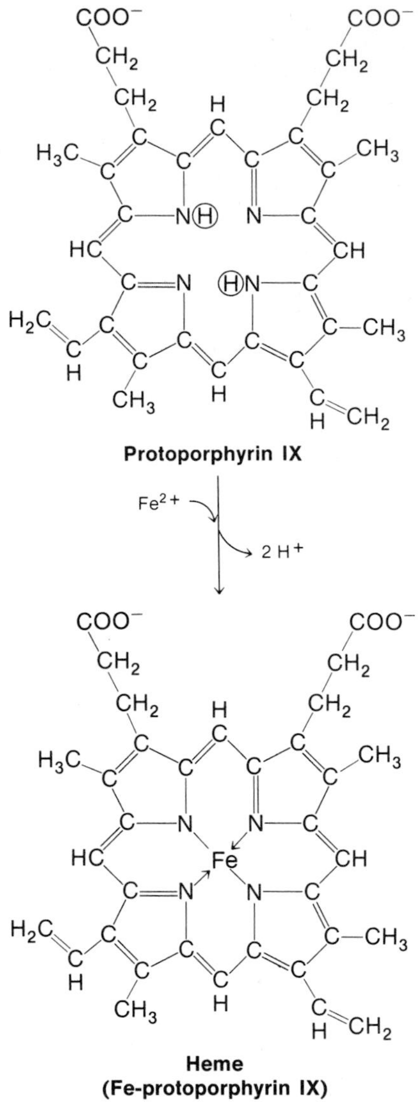

Protoporphyrin IX

Fe^{2+}

$2 H^+$

Heme
(Fe-protoporphyrin IX)

Figure 2–41
Ligation of protoporphyrin IX by an iron atom results in the formation of the heme prosthetic group.

Proteins with Prosthetic Groups

The various amino acids that are normally found in proteins supply a variety of functional groups that can participate in enzymatic catalytic processes. However, some proteins also contain structures that allow them to incorporate metal ions or various larger compounds that have potentially useful chemical properties. In fact, several proteins incorporate nonpeptidyl moieties, frequently called prosthetic groups. Of the relatively small number of such groups, some are directly obtainable from the organism's environment (e.g., metal ions), while others must be produced by their own biosynthetic pathways.

The functional properties of a given prosthetic group depend on the types of interaction that occur between the group and its surrounding polypeptide chain in a molecular complex. The differences in structure and function that arise when the same prosthetic group is wrapped up in different polypeptide chains can be seen by comparing several heme-containing proteins.

Figure 2–41 shows the structure of protoporphyrin IX, one of the most commonly occurring of a number of porphyrin structural variants synthesized by living organisms. This is a macrocyclic ring system of four pyrrole rings, with two vinyl and two propionic acid side chains attached to its periphery. Chelation of the dianionic form of protoporphyrin IX with an iron atom produces protoheme IX. Since the iron atom can potentially have different oxidation states, the net charge on the heme (neglecting the propionic acid side chains) reflects the difference between the porphyrin dianionic charge and the iron atom charge. When the porphyrin dianion is combined with Fe^{2+}, a neutral species results.

Three representative examples of heme-containing proteins of known structure are illustrated in Figure 2–42. Shown first is the structure of myoglobin. The ability of the heme iron in myoglobin to reversibly bind oxygen without itself becoming oxidized is a property conferred by interactions with residues of the surrounding polypeptide chain. On occasion, when the iron in myoglobin does become oxidized, a special enzyme system reduces it back to the Fe^{2+} state.

The second heme protein shown in Figure 2–42 is mitochondrial cytochrome c. This molecule has 104 amino acid residues and contains regions of both α-helical and extended polypeptide chain conformation. In contrast to myoglobin, the heme group in cytochrome c is covalently bound to the polypeptide backbone through thioether linkages, formed by condensation of the protoheme IX vinyl groups with two cysteine side chains. In addition, the heme iron of cytochrome c forms two coordinate bonds with polypeptide side-chain groups—a histidine imidazole nitrogen and a methionine sulfur atom. Cytochrome c functions principally as an electron carrier in the mitochrondrial electron-transport chain, during which time its heme iron is reversibly oxidized and reduced between the Fe^{2+} and Fe^{3+} oxidation states. In this molecule, the covalent interactions that occur between the polypeptide chain and the heme group serve both to regulate the iron atom's electromotive potential and to orient the heme so that electron transfer can occur through the surface-exposed edge of the delocalized porphyrin orbital system.

The third illustration in Figure 2–42 shows the structure of a dimeric cytochrome c' from a photosynthetic bacterium. Each subunit is composed primarily of four α helices. The subunit packing makes both hemes accessible from the same side of the molecule. This gives the molecule the potential to simultaneously transfer two electrons (one from each heme) to a physiological acceptor molecule.

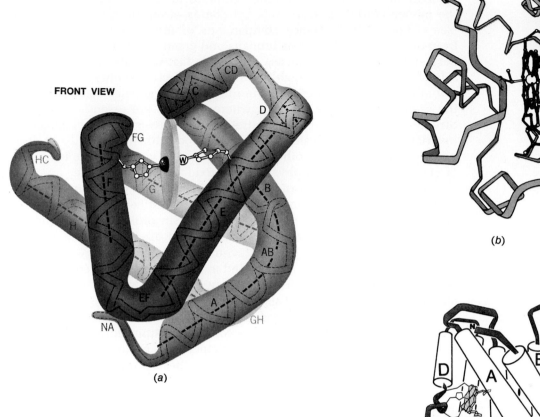

Figure 2–42
Some representative examples of structurally and functionally different heme proteins: *(a)* myoglobin, *(b)* mitochondrial cytochrome *c*, and *(c)* a dimeric cytochrome *c'* from photosynthetic bacteria.

The examples described here are but a few of the many functionally different sorts of protein molecules that incorporate heme prosthetic groups. Although all these molecules share a common prosthetic group, the specific pattern of interactions between the heme and the polypeptide chain confers different functional properties on the molecules as a whole.

Evolutionary Diversification of Proteins

An alternative way of viewing the diversification of protein function is to examine the way in which proteins change as a result of modifications in their encoding DNA. The question is, How do selective evolutionary pressures reveal themselves in the diversification of protein sequence, structure, and function?

Proteins that serve similar or identical biological functions in different living organisms are typically very similar in both their amino acid sequences and their tertiary structures. Such related families of molecules are generally assumed to result from processes of divergent evolution. That is, they are thought to have evolved through gradual and successive selective modification of one ancestral molecule, owing to the fixation of mutations in the protein's encoding DNA.

One of the most extensively studied protein families in this regard is the cytochrome *c* family. Cytochrome *c* proteins function as electron carriers in the mitochondrial electron-transport chains of all multicellular organisms (see Chapter 16). Sequence comparisons of mitochondrial cytochrome *c* from organisms as diverse as humans and green plants reveal an extraordinary extent of sequence conservation. This fact suggests that the functional role of the molecule was highly refined by selective evolutionary pressures prior to the emergence of the first multicellular organisms. Some positions in the sequence are quite variable, whereas others are essentially invariant. Structural and chemical modification studies have shown that some of the invariant amino acid residues are associated with functionally important heme interactions, while others are important in governing the interactions of cytochrome *c* with its physiological oxidase and reductase.

If the sequence and structure of mitochondrial cytochrome *c* were indeed highly refined prior to the emergence of multicellular organisms, then its evolutionary precursors should still exist in prokaryotic organisms. In fact, in virtually all prokaryotic organisms that use oxidative or photosynthetic electron-transport chains to synthesize the high-energy intermediate adenosine triphosphate (ATP), we find molecules that are exceedingly similar to mitochondrial cytochrome *c*. However, as might be expected, cytochrome *c* proteins from prokaryotes exhibit much more sequence diversity than the proteins typically found in higher organisms. In particular, the prokaryotic cytochrome *c* proteins often contain multiple amino acid insertions or deletions relative to mitochondrial cytochrome *c*. Nevertheless, from tertiary-structure determination of several of these prokaryotic proteins, we find that these molecules are all variations on a basic structural theme (Figure 2–43). Further, the prokaryotic molecules all show a strong conservation of those amino acid residues that interact in functionally important ways with the protein's heme group. The slight variations of the basic structural theme of cytochrome *c* that are observed appear to optimize the molecule's function in different organisms.

In addition to the cytochrome *c* proteins, several other families of proteins have been found to share similarities in amino acid sequence and tertiary structure. Again, the observed differences in sequence and structure among individual members reflect evolutionary pressures that modified a basic structural arrangement in order to diversify the functional properties of the molecules. Examples of such structurally and functionally related families include the oxygen-binding globins, the dehydrogenases, and the serine protease enzyme families. Generally, related members within a given enzyme family catalyze chemically similar reactions but exhibit varying specificities for structurally different substrate molecules. For example, while all serine proteases catalyze the hydrolytic cleavage of peptide bonds, different members of this molecule family cleave polypeptides at different locations, which are determined by the nature of the amino acid side chains adjacent to the cleavage site (see Chapter 9).

Although divergent evolutionary processes generally appear to result in gradual changes in protein function, in many situations mutations occur that radically alter protein function. In fact, such mutations frequently result in the synthesis of functionally defective molecules and so constitute one cause of inheritable disease. Other examples exist where amino acid substitutions in related proteins result in the generation of new functions. The enzyme lysozyme, which binds and subsequently cleaves polysaccharide chains, and the protein α-lactalbumin, which transports sugars, are very similar in both sequence and structure. In this case, it appears that relatively slight modifications of a common ancestral precursor have resulted in selection for molecules with quite different functions.

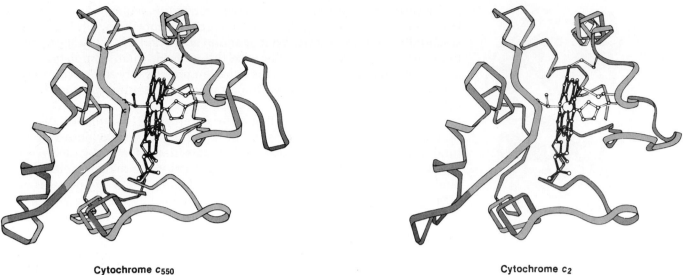

Cytochrome c_{550}

Cytochrome c_2

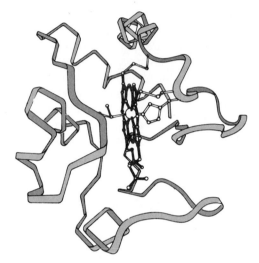

Cytochrome c

Figure 2–43
Examples of structural diversification in prokaryotic cytochromes c. Three cytochromes are shown; cytochrome C_{550} from the denitrifying bacterium *P. denitrifican*; cytochrome c from mitochondria; cytochrome c_3 from the photosynthetic bacterium *P. rubrum*. The prokaryotic cytochromes contain more residues in their polypeptide chains (shaded) than does cytochrome c.

Although many functionally related families of proteins appear to have arisen from a process of divergent evolution from a common ancestor, in other cases proteins that have extensive functional or structural similarities appear to have arisen independently. An outstanding example of two molecules that are functionally similar, but are radically different in sequence and structure, occurs in the serine proteases (Figure 2–44). Many members

Figure 2–44
Schematic illustration of two serine proteases, elastase and subtilisin. Their molecules differ totally in sequence and tertiary structure but have catalytic sites that are nearly identical. The configuration of the active site for elastase is described in Chapter 9.

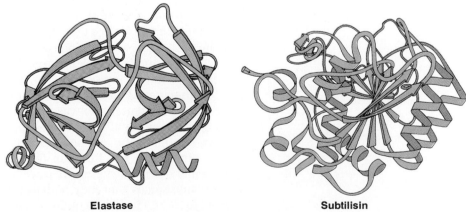

Elastase

Subtilisin

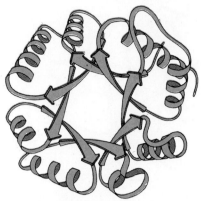

Pyruvate kinase domain 1

Pyruvate kinase domain 2

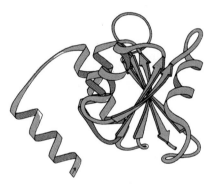

Pyruvate kinase domain 3

Figure 2–45
Pyruvate kinase domains 1, 2, and 3 as an example of a protein whose domains show no structural resemblance whatsoever.

of the serine protease family are closely related in sequence and structure. However, in *Bacillus subtilis* the serine protease subtilisin, while having an essentially identical arrangement of amino acid residues at the active site as the other serine proteases, otherwise differs from them completely in sequence and tertiary structure. This situation presumably reflects convergent evolution on a particular active site arrangement required for the protein's catalytic function.

More frequently, proteins that differ completely in sequence and function have quite similar tertiary structures. In these cases, the observed structural similarities most probably reflect selection of a particularly stable structural arrangement. Examples of such structurally related molecules include those with similarly twisted β sheets (Figure 2–25) and proteins organized as a bundle of four closely packed α helices (see Figure 2–31).

Molecular Evolution and Gene Splicing

In the preceding descriptions of protein diversification, we looked at evolutionary change that occurs as a consequence of the continuing selection of individual point mutations in the protein's encoding DNA. However, _many proteins exhibit structural characteristics that suggest that they have resulted from processes of gene splicing_. In particular, a surprisingly large fraction of known protein structures incorporate multiple copies of structurally similar domains. In many cases it appears that these molecules have arisen by the splicing together of duplicate or multiple copies of a gene coding for a given structural domain, followed by the essentially independent fixation of mutations throughout the spliced genome. The eventual result is a protein composed of sequentially different but structurally similar repeating domains.

Additional evidence for the role of gene splicing in protein evolution comes from the observation that some large proteins are composed of several different structural domains, each of which may individually structurally resemble either parts or the entirety of other known proteins. A good example is the glycolytic enzyme pyruvate kinase (Figure 2–45). This large protein is organized as three structural domains, two of which show convincing structural similarities to, respectively, the β barrel of triose phosphate isomerase (domain 1) and a twisted β-sheet domain common to many dehydrogenases (domain 3). The third domain of pyruvate kinase (domain 2) also has a convincing similarity to a common structural type, the antiparallel β barrel.

Protein Diversification in Action: Antibodies

In the preceding section, we saw that protein functional evolution can involve both the fixation of individual point mutations and larger-scale rearrangements of a protein's encoding DNA. Further, it should be apparent that the rate of fixation of mutations can differ widely for different regions within a protein's sequences. For example, even in the case of "highly evolved" molecules such as cytochrome *c*, some sequence positions show much greater interspecies variability than others. However, in all these cases the fixation of new mutations was pictured as a relatively infrequent event, resulting in the gradual evolution of proteins such as cytochrome *c*.

By contrast, one of the important biological defense mechanisms of higher organisms, the immune response, essentially depends on the very rapid generation of structurally novel molecules that can recognize and bind foreign substances that may be harmful to the organism. The molecules responsible for the initial recognition and binding of foreign substances are the

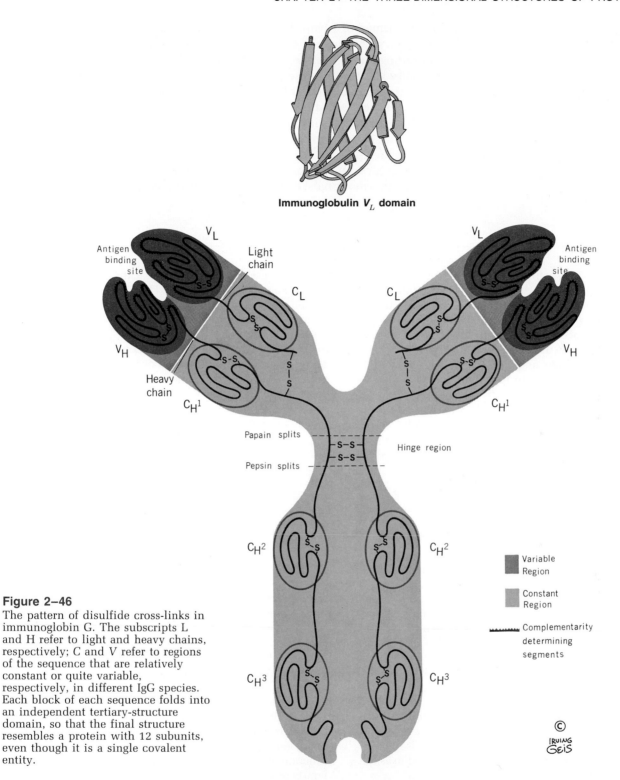

Immunoglobulin V_L domain

Figure 2–46
The pattern of disulfide cross-links in immunoglobin G. The subscripts L and H refer to light and heavy chains, respectively; C and V refer to regions of the sequence that are relatively constant or quite variable, respectively, in different IgG species. Each block of each sequence folds into an independent tertiary-structure domain, so that the final structure resembles a protein with 12 subunits, even though it is a single covalent entity.

immunoglobins. These molecules are composed of two pairs of polypeptide chains of different length that are interconnected by covalent cysteine disulfide linkages (Figure 2–46; also see Figure 2–40). Sequence studies of various immunoglobins have shown that both the heavy and light polypeptide chains contain repeating homologous sequences that are about 110 residues in length. Structural studies of the immunoglobin molecule show that the sequentially homologous regions fold individually into similar structural domains, arranged as a bilayer of antiparallel β sheets. The molecule in its entirety is formed of 12 similar structural domains, of which 8 are formed by the two heavy chains and 4 by the two light chains.

Immunoglobins that are specific for binding to different foreign substances vary greatly in the sequences found in the amino-terminal domains of both the heavy and light chains. It is these variable regions that form the binding sites between the immunoglobin molecule and the foreign substances that trigger the immune response. The remarkable property of this system is that it can rapidly diversify the sequence of variable regions by mutation, gene splicing, and RNA splicing. The net result is that the organism can produce an enormous variety of antibodies from a quite limited amount of informational DNA originating in the germ-line tissue.

SUMMARY

Proteins that have been completely denatured will renature to the original native structure with return of full functional capacity. Thus it seems clear that the information to direct the folding process is contained in the primary structure.

Since folding occurs spontaneously, a negative free energy must accompany the folding process. Free energy is a function of the enthalpy and the entropy. In evaluating changes in enthalpy and entropy, both the polypeptide chain and the solvent water must be considered.

Water is an extensively hydrogen-bonded structure, and individual water molecules are highly polar. Polar molecules interact much more strongly with water than apolar molecules. Proteins contain both types of groups. Native proteins have their polar hydrophilic amino acid side chains on the surface of the protein structure, where they can interact strongly with water, and have their apolar hydrophobic amino acid side chains buried within the protein, where they interact with each other.

The forces that maintain a protein's three-dimensional structure are mainly noncovalent. They include electrostatic forces, van der Waals forces, hydrogen bond forces, and hydrophobic forces.

Protein folding reveals a hierarchy of structural organization. Secondary structures are regular conformations maintained by hydrogen bonding between elements of the peptide backbone. Structural domains involve more complex interactions between elements of secondary structures and their side chains. The tertiary structure is the totally folded polypeptide chain, and it may include one or more domains.

Fully folded native polypeptide chains frequently interact with other polypeptide chains to form larger structures. Each polypeptide chain in such a structure is referred to as a subunit, and the overall assembly is referred to as the quaternary structure of the protein.

The two most common forms of secondary structure are the α helix and the β sheet. The α helix is a right-helical structure in which the peptide backbone forms regular intramolecular hydrogen bonds. The β sheet is an array of extended polypeptide chains in which hydrogen bonds are formed between adjacent chains. Both of these structures are commonly found in fibrous and globular proteins. A third type of regular repeating structure, involving a three-stranded helix with hydrogen bonds between the chains, is found only in fibrous proteins of the collagen family. Fibrous protein structures are generally simpler than globular protein structures. They are usually composed of regularly repeating structures.

Globular proteins, as the name implies, are roughly spherical in shape. They usually contain elements of β sheet and α helix in a complex folded array. The tremendous variety of globular protein structures reflects the complexity of the folding possible. Nevertheless, there are two properties of polypeptide chains that have a major impact on the way in which globular protein chains fold. The first property is the tendency of an extended polypeptide chain to be twisted in a right-hand sense. This tendency affects both secondary structures and tertiary structures. The second property relates to how regions of secondary structure pack together. A great deal of the complex domain and tertiary structure can be understood by a set of elementary rules relating to these two properties.

The enormous diversity of protein structures reflects the equally diverse functional roles of proteins in living organisms. And yet, many common structural motifs recur among proteins that lack any obvious relationship. In general, the recurrence of these features, such as right-twisted β sheets and preferred packing arrangements among α helices, reflects common physical features that are relatively independent of specific sequence requirements. However, many functionally similar families of proteins, such as the trypsin family of proteases, show corresponding structural similarities selected for during the course of their evolution.

SELECTED READINGS

Blake, C. C. T., and L. N. Johnson, Protein structure. *Trends in Biochem. Sci.* 9:147–151, 1984.

Cantor, C. R., and P. R. Schimmel, *Biophysical Chemistry*, Parts 1, 2, and 3. New York: W. H. Freeman and Co.,

1980. Includes several chapters (2, 5, 13, 17, 20 and 21) on the principles of protein folding and conformation.

Chothia, C., Principles that determine the structure of proteins. *Ann. Rev. Biochem.* 53:537–572, 1984.

Chothia, C., and A. Leak, Helix movements in proteins. *Trends in Biochem. Sci.* 10:116–118, 1985.

Creighton, Thomas E., *Proteins, Structures and Molecular Principles.* New York: W. H. Freeman and Co., 1984.

Very readable and reasonably comprehensive and up-to-date.

Pauling, Linus, *The Nature of the Chemical Bond*, 3rd ed. Ithaca: Cornell University Press, 1960. The classic on molecular structure.

Rose, C. D., A. R. Geselowitz, G. J. Lesser, R. H. Lee, and M. H. Zehfus, Hydrophobicity of amino acid residues in globular proteins. *Science* 229:834–838, 1985.

PROBLEMS

1. What is the principal driving force for protein folding?

2. Outline the hierarchy of protein structural organization.

3. Many proteins are anchored to membranes by sections of polypeptide chain at their amino terminals. What would one expect the probable structure to be for the sequence Met-Ala-Leu-Phe-Leu-Leu-Met-Ala-Ala-Leu-Gly-Pro-Asn-Ala-Met-Leu-Phe-Leu-Leu-Ala-Ala-Met? Why would this be a likely sequence to insert in a membrane?

4. One of the proteins shown in Figure 2–44, illustrating the handedness of crossover connections in β sheets, has a left-handed connection. Which one?

5. You have isolated several multimeric proteins with the following subunit compositions:
   a. $\alpha\beta\gamma$ c. $\alpha_3\beta_5$ e. $\alpha_{24}\beta_{24}$
   b. $\alpha_2\beta_2$ d. $\alpha_6\beta_6$ f. α_{60}.
   Which of these might you expect to have regular polyhedral quaternary structures? Draw an example of each, showing possible structural arrangements of subunits relative to each other.

6. Write a short essay outlining the relationship between protein structure, function, and evolution.

7. Which is more stable, a right-handed or a left-handed α helix of polyglycine? Why?

8. The right-handed π helix has 4.4 amino acid residues per turn with acceptable ψ and ϕ values and good intra-chain hydrogen bonds. However, the π helix has a 1-Å hole down the core. The hole makes this structure considerably less stable than the α helix. Why?

9. Molecular weight studies of an unknown protein yield the following data.

Solvent	Molecular Weight
Water	200,000
6 M guanidine hydrochloride (GH)	100,000
6 M GH + 0.1 M β-mercaptoethanol	50,000

In addition, amino terminal analysis indicates that there are two moles of alanine and two moles of serine per 200,000 grams of protein. What can you deduce about the quaternary structure of the unknown protein?

10. See the Ramachandran plot in Figure 2–9. Can you say why the region from $\phi = -40°$ to $-160°$ is particularly favorable?

11. The following table* is a comparison matrix of human inter-subunit and inter-species amino acid sequences of hemoglobin. The numbers indicate numbers of differences between two molecular species and hence evolutionary relatedness. Can you come up with a possible evolutionary tree for globin, including whale Mb, human α, β, γ, and δ Hb, and horse α and β, that is consistent with all the data?

Differences in Amino Acid Sequences

	Horse α	Human α	Horse β	Human β	Human δ	Human γ	Whale Mb
Total Residues:	141	141	146	146	146	146	153
Horse α	0	18	84	86	87	87	118
Human α	18	0	87	84	85	89	115
Horse β	84	87	0	25	26	39	119
Human β	86	84	25	0	10	39	117
Human δ	87	85	26	10	0	41	118
Human γ	87	89	39	39	41	0	121
Whale Mb	118	115	119	117	118	121	0

*From R. E. Dickerson and I. Geis, *The Structure and Action of Proteins* (Benjamin/Cummings, Menlo Park, Calif., 1969).

3

METHODS FOR CHARACTERIZATION AND PURIFICATION OF PROTEINS

In Chapter 1 we discussed protein primary structure and examined some methods for investigating protein structure at this level. We saw that all proteins are composed of one or more polypeptide chains, each with a unique linear sequence of peptide-linked amino acids. In Chapter 2 we described higher levels of order in protein structure. We saw that the specific folding patterns adopted by proteins are predetermined by the sequences of amino acids in the protein polypeptide chains. Some proteins contain only one subunit; others contain two or more subunits. Similar principles govern interactions between polypeptide chain segments, whether they originate from the same or different chains.

In this chapter we describe methods used to investigate higher levels of protein structure. The same procedures that are used to characterize proteins frequently form the basis for the techniques used for protein purification. Therefore it is appropriate that methods used to isolate specific proteins from the whole organism should also be described in this chapter. Many of the methods described in this chapter will prove to be useful in studies on other biomolecules.

METHODS OF PROTEIN CHARACTERIZATION

Many more techniques for protein analysis exist than could be covered in a single chapter. The emphasis here is on some basic biophysical and biochemical techniques and the principles involved in their use. These techniques have enabled us to describe the overall size and shape of proteins as well as to determine the atomic coordinates of the individual amino acid residues within the proteins.

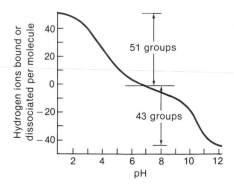

Figure 3–1

Titration curve of β-lactoglobulin. At very low pH's (<2) all ionizable groups are protonated. At a pH of about 7.2 (indicated by horizontal bar) 51 groups (mostly the glutamic and aspartic amino acids and some of the histidines) have lost their protons. At pH 12 most of the remaining ionizable groups (mostly lysine and arginine amino acids and some histidines) have lost their protons as well. (Adapted from R. H. Haschenmeyer and A. E. V. Haschenmeyer, *Proteins, A Guide to Study by Physical and Chemical Methods*, Wiley, New York, 1973, p. 204.)

Solubility Reflects a Balance of Protein–Solvent Interactions

The solubility of a protein reflects a delicate balance between different energetic interactions, both internally within the protein and between the protein and the surrounding solvent (see Chapter 2). Consequently, the protein's solvent or thermal environment could affect both its solubility and structure. As we have seen, extreme changes can lead to denaturation (see Chapter 2). In this chapter we will, for the most part, be concerned with conditions under which the native structure is maintained. Changes in protein solubility that do not destroy the molecule's structural integrity can occur in several ways.

A Minimum in Solubility Occurs at the Isoelectric Point. Proteins typically have on their surfaces charged amino acid chains that undergo energetically favorable polar interactions with the surrounding water. The total charge on the protein is the sum of the side-chain charges. However, the actual charge on the weakly acidic and basic side-chain groups also depends on the solution pH. In fact, the acidic and basic groups within the protein can be titrated just like free amino acids (Chapter 1) to determine their number and their pK_a values.

A titration curve for β-lactoglobulin is shown in Figure 3–1. This protein contains 94 potentially ionizable groups. The protein is positively charged at low pH and negatively charged at high pH. At intermediate pH values a point is found where the sum of the positive side-chain charges exactly equals the sum of the negative charges, so that the net charge on the protein is zero. This value is called the *isoelectric point* (pI) of the protein; for β-lactoglobulin the pI is about 5.2. The isoelectric point is not an invariant quantity. The binding of charged species present in the solution could raise or lower the pI, depending on their charge.

Proteins tend to show a minimum solubility at their isoelectric pH. This is illustrated for β-lactoglobulin in Figure 3–2. Here the solubility of the protein is measured as a function of both pH and salt (NaCl) concentration. The decrease in solubility at the isoelectric pH reflects the fact that the individual protein molecules, which would all have similar charges at pH values away from their isoelectric points, cease to repel each other. Instead, they coalesce into insoluble aggregates.

Salting In and Salting Out. Proteins also show a variation in solubility that depends on the concentration of salts in the solution. These frequently complex effects may involve specific interactions between charged side chains and solution ions or, particularly at high salt concentrations, may reflect more comprehensive changes in the solvent properties.

Figure 3–2 illustrates the effect of salt concentration on the solubility of β-lactoglobulin. Most globulins are sparingly soluble in pure water. The effect of salts such as sodium chloride on increasing the solubility of globulins is often referred to as *salting in*. The salting-in effect is related to the nonspecific effect the salt has on increasing the ionic strength of the solution. The higher the ionic strength, the smaller are the interactions between charged groups on the same or different proteins (see Box 3–A).

The effects of various salts on the solubility of hemoglobin at pH 7 are illustrated in Figure 3–3. All of these salts produce the salting-in effect with this protein; two of them, sodium sulfate and ammonium sulfate, also produce a greatly decreased solubility of the protein at high salt concentrations. This result is called *salting out* and occurs with salts that effectively compete with the protein for available water molecules. In this case, the protein

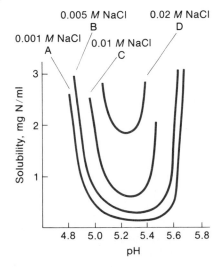

Figure 3–2

Solubility of β-lactoglobulin as a function of pH and ionic strength. The isoelectric pH (pI) for this protein is about 5.2. This corresponds to the point of minimum solubility.

BOX 3–A

The Meaning of Ionic Strength and Its Effect on Charge–Charge Interactions in Proteins

The ionic strength μ is a measure of the effective salt strength. It is defined by the equation

$$\mu = \frac{1}{2}\sum_i M_i Z_i^2$$

where M is the molarity and Z is the charge of the ion. For a univalent salt at low concentration, μ is approximately equal to the molarity. Increasing the ionic strength decreases the "sphere of influence" of each charged site on the protein. The effective sphere of influence of a charge in solution is approximated by the Debye length $1/b$, where b is calculated from the expression

$$b^2 = \frac{4\pi e^2}{\epsilon kT}\sum_i M_i Z_i^2$$

In this equation e is the charge of the electron, k is the Boltzmann constant, T is the absolute temperature, and ϵ is the dielectric constant of the medium (a measure of the charge-shielding effect of the medium).

The last term in the expression for b^2,

$$\sum_i M_i Z_i^2$$

is twice the ionic strength μ. Thus the Debye length $1/b$ is proportional to $1/\sqrt{\mu}$. In a 1 M aqueous solution of sodium chloride at 25°C, $1/b = 3.1$ Å. In a 0.01 M solution, $1/b = 31$ Å. As described earlier, many globular proteins precipitate at very low ionic strengths or in pure water. This happens because oppositely charged sites on different proteins are able to interact favorably, leading to an electrostatic complex. When this complex formation is extended between many protein molecules, it can lead to protein precipitation. Increasing the ionic strength tends to break up this complex because it results in a decrease of the Debye length.

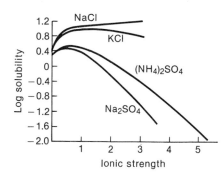

Figure 3–3
Solubility of horse carbon monoxide hemoglobin in different salt solutions. The addition of a moderate amount of salt (salting in) is required to solubilize this protein. At high concentrations, certain salts compete more favorably for solvent, decreasing the solubility of the protein and thus leading to its precipitation (salting out). (Adapted from E. J. Cohn and J. T. Edsall, *Proteins, Amino Acids, and Peptides as Ions and Dipolar Ions*, Reinhold, New York, 1942.)

molecules tend to associate with each other because at high salt concentrations, protein–protein interactions become energetically more favorable than protein–solvent interactions. Each protein has a characteristic salting-out point, and we can exploit this fact to make protein separations in crude extracts.

In the next section we will consider procedures for characterizing the size and shape of the protein in solution. For this purpose it is important to choose conditions that preserve the normal native state of the protein without undue aggregation.

Several Methods Are Available for Determination of Gross Size and Shape

Several methods are available for determining the size and shape of protein molecules in solution. Most of these give molecular weights that are accurate to a few per cent. The linear dimensions of the protein are never directly given by solution measurements; rather, we obtain a parameter, the frictional coefficient, that measures the effective size. From this we may calculate the dimensions of the protein, assuming that it has a particular shape—say, that of a rod, a random coil, or a sphere. Many fibrous proteins are rodlike; globular proteins are often approximately spherical; and denatured proteins often have the structure of a random coil.

Osmotic Pressure. For proteins with molecular weights in the range of 10,000 to 100,000, osmotic-pressure measurements give a reasonable molecular-weight estimate. This method is of both experimental and theoretical interest. The general technique consists of putting a solution containing the purified protein inside a semipermeable membrane. The membrane is permeable to the water and any salts or buffer present, but not to the protein. After equilibration, a greater pressure exists within the membrane; the pressure of the protein solution may be measured by the height of the solution in

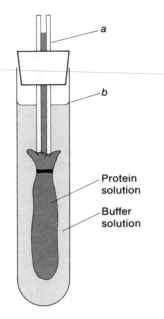

Figure 3–4
Apparatus for measuring osmotic pressure. Protein sample is placed inside a semipermeable membrane with buffer on both sides of the membrane. Osmotic pressure generated by the sample is proportional to the difference in height of the protein solution, a, and the buffer solution, b. From a knowledge of the concentration, C, and the osmotic pressure, π, the molecular weight can be estimated.

a capillary tube (Figure 3–4). This pressure, called the *osmotic pressure* and symbolized by π, is related to the concentration c of protein (in grams per liter) by van't Hoff's law:

$$\lim_{c \to 0} \frac{\pi}{RTc} = \frac{1}{M_r} \tag{1}$$

where R is the gas constant, T is the absolute temperature, and M_r is the molecular weight of the protein. In practice, the osmotic pressure is measured over a range of concentrations; the values for π/c are plotted against c, and the value of π/c, extrapolated to zero concentration, is used to calculate the molecular weight.

Light Scattering. If we illuminate a protein solution with a beam of light that is far from any absorption band in the solvent water or the protein, absorption will not occur. However, the sinusoidally varying electromagnetic field of the incoming radiation will produce a sinusoidal oscillation of the electrons within the solvent and the protein. As a result, some of the incoming radiation will scatter in directions other than the direction of the incident radiation. For a protein in solution the amount of scattering is a function of the molecular weight. In practice, a solution is placed in a scattering cell (Figure 3–5) and the scattered light is measured over a range of scattering angles and at different concentrations. The equation relating the fraction of scattered light, R_θ, to the molecular weight M_r is

$$\left. \frac{Kc}{R_\theta} \right]_{\theta=0} = \frac{1}{M_r} + 2Bc \tag{2}$$

In this equation K is a constant depending on the wavelength of the incident light, c is the concentration of the protein, M_r is the molecular weight of the protein, R_θ is the fraction of scattered light at the angle θ, and B is the second virial coefficient. B is a measure of the interaction with solvent water; it increases in proportion to the strength of the interactions between solvent and solute protein. Usually the observed data are graphed on a Zimm plot (Figure 3–6), and M_r and B are found from the intercept and the slope.

Light scattering has certain unique advantages over other procedures for molecular-weight determination. In particular, the measurement can be made very rapidly and therefore can be used to study rapid changes in molecular weight such as might accompany the association or dissociation of the subunits of a multimeric protein.

Figure 3–5
A light-scattering photometer. Collimated monochromatic light is focused on a glass cell containing a protein solution. The scattered light is measured at various angles of scattering (θ) with a photomultiplier mounted on a turntable. These measurements are used to calculate the molecular weight.

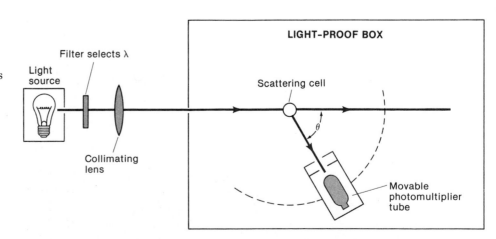

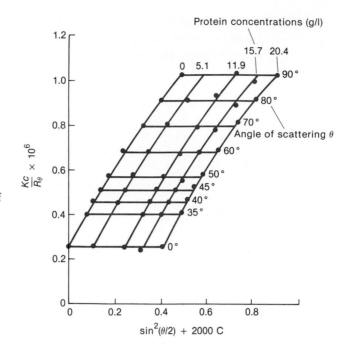

Figure 3–6

Zimm plot of light-scattering data. Scattering is determined at four different concentrations. At each concentration the scattering is determined at eight different angles of scattering between 35° and 90°. The concentration data are extrapolated to zero concentration for each angle of scattering, and the scattering data at different angles are extrapolated to 0° for each concentration. The intercept on the ordinate for the extrapolated data is equal to the reciprocal of the molecular weight. (Adapted from P. Doty and B. Bunce, *J. Am. Chem. Soc.* 74:5029, 1952.)

Sedimentation Analysis. Information concerning the molecular weight of a protein can be obtained by observing its behavior in an intense centrifugal field. To get a qualitative understanding of how this method works, we must first recognize that protein molecules are generally slightly denser than water. However, the molecules in a protein solution seldom settle out in the earth's gravitational field ($1 \times g$) because they are constantly being stirred up by collisions with surrounding solvent molecules. Nevertheless, protein molecules in solution can be made to settle if they are subjected to very high centrifugal force fields ($\sim$100,000 $\times$ g), such as can be attained in an ultracentrifuge (Figure 3–7).

The protein molecules slowly migrate toward the bottom of the centrifuge tube at a rate that is proportional to their molecular weight. The rate of sedimentation may be recorded by optical methods (see Figure 3–7) that do not interfere with the operation of the centrifuge. From this rate, the *sedimentation constant,* or *sedimentation coefficient,* is determined. This con-

Figure 3–7

Apparatus for analytical ultracentrifugation. *(a)* The centrifuge rotor and method of making optical measurements. *(b)* The optical recordings as a function of centrifugation time. As the light-absorbing molecule sediments, the solution becomes transparent.

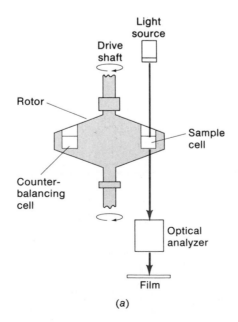

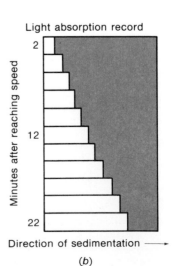

stant equals the rate at which a molecule sediments, divided by the gravitational field (angular acceleration in a spinning rotor), and is defined by the equation

$$S = \frac{dx/dt}{\omega^2 x} \tag{3}$$

where dx/dt is the rate at which the particle travels at distance x from the center of rotation, ω is the angular velocity of the rotor in radians per second (hence $\omega^2 x$ is the angular acceleration), and t is the time of centrifugation in seconds. The sedimentation constant is usually given in Svedberg units (S) where one $S = 10^{-13}$ sec.

The molecular weight cannot be calculated directly from the sedimentation constant without further information; this is because the sedimentation constant is proportional to the molecular weight and inversely proportional to the _frictional coefficient_ (f) of the protein. The coefficient f is bigger for large proteins, and, for proteins of the same molecular weight, it is larger for elongated, rodlike molecules. Another complication in calculating the molecular weight from the sedimentation constant is that the sedimentation rate is reduced by the buoyancy factor, $1 - \bar{v}_p\rho_s$, which takes into account the density difference between solvent (ρ_s) and the volume of water displaced per gram of protein ($\bar{v}_p$).* The equation that relates S, M, and f is

$$S = \frac{M(1 - \bar{v}_p\rho_s)}{Nf} \tag{4}$$

where N is Avogadro's number. Thus in order to estimate the molecular weight from the sedimentation constant we must have a means of determining $\bar{v}_p$ and f.

For most proteins $\bar{v}_p$ is about 0.75 cc/g, so its value does not present much of a problem. If sufficient protein is available it is an easy constant to measure. Even if the constant cannot be directly measured, not more than 15 per cent error can result from using the average value.

The frictional coefficient varies over a wide range, and it must usually be determined. It can be calculated from the diffusion constant, D, with the help of the equation

$$D = \frac{RT}{Nf} \qquad \text{or} \qquad f = \frac{RT}{ND} \tag{5}$$

where R is the gas constant and T is the absolute temperature. Substituting this value for f in the expression for the sedimentation constant leads us to the equation

$$S = \frac{M_r \cdot D(1 - \bar{v}_p\rho_s)}{RT} \tag{6}$$

Transposing, we get

$$M_r = \frac{RTS}{D(1 - \bar{v}_p\rho_s)} \tag{7}$$

The diffusion constant needed to use Equation (7) for molecular-weight calculations is usually determined with the help of Fick's first law of diffusion:

$$\frac{dn}{dt} = -DA\left(\frac{dc}{dx}\right)_t \tag{8}$$

This equation states that the amount dn of a substance crossing a given area A in time dt is proportional to the concentration gradient dc/dx across that

*The symbol $\bar{v}_p$ is often called the partial specific volume of the protein; it is usually given in units of cc/g.

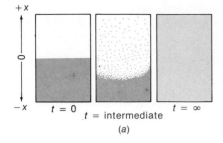

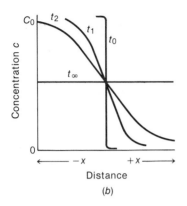

Figure 3–8
Measurement of diffusion. (a) A cell with a removable partition is filled on one side with pure solvent and on the other side with solvent plus solute (protein). The partition is removed and the rate at which the solute moves across the boundary defined by the partition and the changing concentration of solute, c, is measured as a function of time and location in the cell. (b) Graph of data from such a procedure. (Adapted from C. Tanford, *Physical Chemistry of Macromolecules*, New York, Wiley, 1961.)

area. The diffusion constant D, which is related to molecular weight and shape, is the proportionality constant. It can be measured by observing the spread of an initially sharp boundary between the protein solution and a solvent as the protein diffuses into the solvent layer (Figure 3–8). Once the diffusion constant is determined, the information is combined with the sedimentation data, and the molecular weight of the protein is calculated from Equation (7). Some data obtained in this way for various proteins are presented in Table 3–1.

Sometimes the technique of *equilibrium sedimentation* is used to measure molecular weight. In this case neither the sedimentation constant nor the diffusion constant is directly measured. Instead, the ultracentrifuge is maintained at a relatively low constant speed until the distribution of the protein molecules becomes constant. At this time, the downward movement due to sedimentation is exactly counterbalanced by the upward movement due to diffusion. The equation for calculating the molecular weight from sedimentation equilibrium is

$$M_r = \frac{2\,RT\,\ln c}{\omega^2 x^2 (1 - \overline{v}_p \rho_s)} \qquad (9)$$

Equilibrium sedimentation is a most rigorous method for molecular-weight determination, precisely because it is an equilibrium method. But the technique is more difficult and time-consuming than those previously described, and also it is not effective when other macromolecules are present. Thus its use is usually restricted to measurements on highly purified proteins in which a species of only one molecular weight is present.

Frequently sedimentation analysis is carried out in linear gradients of sucrose or glycerol. First, a linear gradient is created in the centrifuge tube, with the greatest density at the bottom of the tube. A small volume of the protein solution is layered on top of the tube, and the tube is spun for the desired length of time. The protein travels down the tube at a rate roughly proportional to its sedimentation constant. This arrangement for measuring the sedimentation constant has the advantage that the gradient stabilizes the solution so that the centrifuge may be stopped at any time without disturbing the protein distribution. The solution can be carefully removed from the tube by punching a hole in the bottom and collecting fractions in separate tubes. This technique is used both for sedimentation-constant estimation and for preparative purposes, when the aim is to purify the desired protein from a mixture.

TABLE 3–1
Physical Constants of Some Proteins

Protein	Molecular Weight	Diffusion Constant $(D \times 10^7)$	Sedimentation Constant (S)	pI (Isoelectric)
Cytochrome c (bovine heart)	13,370	11.4	1.17	10.6
Myoglobin (horse heart)	16,900	11.3	2.04	7.0
Chymotrypsinogen (bovine pancreas)	23,240	9.5	2.54	9.5
β-Lactoglobulin (goat milk)	37,100	7.5	2.9	5.2
Serum albumin (human)	68,500	6.1	4.6	4.9
Hemoglobin (human)	64,500	6.9	4.5	6.9
Catalase (horse liver)	247,500	4.1	11.3	5.6
Urease (jack bean)	482,700	3.46	18.6	5.1
Fibrinogen (human)	339,700	1.98	7.6	5.5
Myosin (cod)	524,800	1.10	6.4	—
Tobacco mosaic virus	40,590,000	0.46	198	—

When using gradients, the sedimentation constant is measured only approximately, and the molecular weight is estimated using a protein "marker" of known sedimentation constant in the same tube. For a large number of protein molecules, the sedimentation constant is roughly proportional to the two-thirds power of the molecular weight, giving the relation

$$\frac{S(\text{of unknown})}{S(\text{of standard})} = \left[\frac{M_r(\text{of unknown})}{M_r(\text{of standard})} \right]^{2/3} \tag{10}$$

This is not a rigorous relationship, and the approximation is best for spherical molecules. Most globular proteins give a reasonable fit; fibrous proteins, which are often highly asymmetrical in shape, are likely to give a poor fit.

Gel-exclusion Chromatography.   *Gel-exclusion chromatography* is used for size estimations as well as protein purifications. This popular technique exploits the availability of both natural polysaccharides and synthetic polymers (polydextrans) that can be formed into beads with varying pore sizes, depending on the extent of cross-linking between polymer chains. For example, suppose you have some polysaccharide beads with average maximum pore sizes of 30 Å and you add these to a mixed solution of proteins of molecular weights 10,000 and 50,000. The final volume of the mixture equals the sum of the volume required to hydrate and fill the beads, plus the remaining excess solution volume that fills the spaces between the beads. That is, the total solution volume $v_T = v_i + v_o$, where v_i and v_o represent, respectively, the solution volumes inside and outside the beads. For fractionation purposes, the critical point is that a low-molecular-weight protein is sufficiently small that it can readily penetrate into the beads; as a result, it is uniformly distributed over the total volume v_T. A large protein, by contrast, cannot penetrate the beads and so is concentrated in the solution volume v_o.

By simply filtering out the beads we can achieve some separation between the molecules based on their molecular weight. By repeating this process many times, eventually the concentration of the smaller protein in v_o is vanishingly small. However, it is simpler and more efficient to construct a column of the beads and pass the protein solution through it in a continuous fashion. In this case, the smaller molecule penetrates the entire column volume v_T, while the larger molecule, which is restricted to v_o, passes through the column at a more rapid rate, thus effecting a separation on the basis of molecular weight (Figure 3–9).

Figure 3–9

Polydextran column showing separation of small and large molecules. The column is immersed in solvent, which penetrates the gel particles. A separation is initiated by layering a small sample containing different size proteins on the top of the column. This sample is pushed through the column by opening the stopcock at the bottom and adding further solvent at the top to keep up with the flow. As shown, the small protein molecules can penetrate the gel particles but the big ones cannot. Therefore the big proteins move through the column much more rapidly, and a separation of the two proteins results.

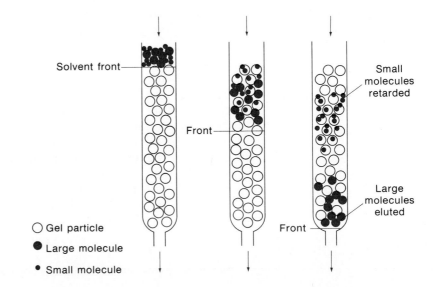

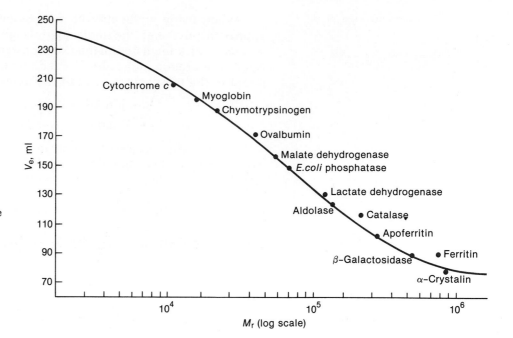

Figure 3–10
Plot of elution volume (V_e) versus the logarithm of the protein molecular weight. A cross-linked dextran (Sephadex G-2000 at pH 7.5) was used. The further apart two proteins are on this plot, the easier it is to separate them by gel-exclusion chromatography. (Adapted from P. Andrews, *Biochemical J.* 96:595, 1965.)

Although this process has been described in terms of molecules either penetrating or not penetrating the beads, by careful selection of the bead pore size, we can create a situation in which the various molecules penetrate the beads to varying extents and, consequently, migrate through the column at varying rates. By observing the elution pattern of a mixture of proteins of known molecular weights, the behavior of the column can be standardized so that the molecular weight of an unknown protein can be estimated (Figure 3–10).

Electrophoretic Methods Are the Best Way to Analyze Mixtures

We turn now from methods in which the primary object is to determine the size and shape to electrophoretic analysis, where this goal is often secondary to determining the number of components in a mixture. The transport of particles by an electrical field is termed *electrophoresis*. Electrophoresis is one of the most commonly used techniques in biochemistry. The reason is, in part, that the apparatus required is quite simple and serves multiple purposes. Electrophoresis can be used to assess the isoelectric point and the molecular weight of a protein. It is also the favored approach for assessing the number of proteins in a mixture, because of the high resolving power it offers.

Electrophoresis is very much like sedimentation, since in both cases a force gradient leads to protein transport in the direction of the force. In the case of sedimentation the force is gravity, so the rate of migration depends on the effective mass of the particle, i.e., $M_r(1 - \overline{v}_p\rho_s)$. In electrophoresis the force is the electrical potential, E, so the rate of migration depends on the net charge on the molecule, rather than its mass.

The characteristic quantity measured in electrophoresis is called the *electrophoretic mobility* (μ). This is defined in terms of the ratio of the velocity of the particle (V) to the imposed electrical field (E):

$$\mu = \frac{V}{E} = \frac{Z^*}{f} \tag{11}$$

*V is usually given in cm/sec and E is usually given in volts/cm.

Pattern before electrophoresis

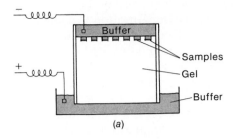

Pattern after electrophoresis

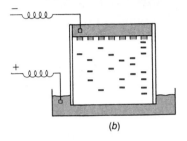

Figure 3–11
Apparatus for slab-gel electrophoresis.
(a) Samples are layered in little slots
cut in the top of the gel slab. Buffer is
carefully layered over the samples,
and a voltage is applied to the gel for
a period of usually 1 to 4 hr. (b) After
this time the proteins have moved into
the gel at a distance proportional to
their electrophoretic mobility. The
pattern shown indicates that different
samples were layered in each slot.
Some samples contained single
proteins, some as many as seven.

That is, electrophoretic mobility is equal to the net charge on the molecule
(Z) divided by its frictional coefficient (f). The frictional coefficient in this
equation is closely related to the frictional coefficient that appeared in the
equations for the sedimentation constant. The two quantities may, however,
have quite different values in high electric fields, where the molecules tend
to orient in the direction of the electric field.

Although electrophoresis can be done in pure aqueous solution, it is
almost always done in an aqueous solution supported by a gel system. The
gel is a loosely cross-linked network that functions to stabilize the protein
boundaries between the protein and the solvent, both during and after electrophoresis, so that they may be stained or otherwise manipulated. By judicious manipulation (to be explained shortly) the gel may also be used to
purify a protein in a mixture.

Electrophoretic separations are typically carried out on gels composed
of the polysaccharide agarose or polyacrylamide (Figure 3–11). The percentage of gel used is gauged according to the size of the proteins being separated. Increasing the gel concentration lowers the mobility of the protein in a
given electric field. For the finest separations, gradient gels are made with a
continuous increase in gel percentage along the length of the slab. This approach leads to optimum separation of components in a mixture and the
sharpest protein solvent boundaries. At the completion of a run, a dye that
stains the proteins can be added to the gel. The dye establishes the locations
of the protein bands.

A widely used variation of polyacrylamide gel electrophoresis involves
the use of a stacking gel. The basic idea here is to increase the resolution in
the gel by first compressing the protein mixture into a very narrow starting
zone. This aim is achieved by forming the gel in two layers. Generally the
upper, or "stacking," gel is less extensively polymerized than the lower, or
"resolving," gel, so that there is a smaller pore size in the resolving gel. More
important, the pH of the solution in the stacking gel is adjusted so that the
mobilities of proteins to be separated are higher in the stacking gel than in
the resolving gel. As a consequence of this gel arrangement, a relatively
dilute protein mixture moves rapidly and easily through the stacking gel, but
accumulates at the interface with the resolving gel, because of the difference
in pH and the smaller pore size.

Isoelectric Focusing. Another electrophoretic method frequently used for
characterizing proteins is based on differences in their isoelectric points;
this method is called *isoelectric focusing*. The apparatus usually consists of
a narrow tube containing a gel and a mixture of *ampholytes*, which are small
molecules with positive and negative charges. The ampholytes have a wide
range of isoelectric points, and are allowed to distribute in the column under
the influence of an electric field. This step creates a pH gradient from one
end of the gel to the other, as each particular ampholyte comes to rest at a
position coincident with its isoelectric point. At this stage, a solution of
proteins is introduced into the gel. The proteins migrate in the electric field
until each reaches a point at which the pH resulting from the ampholyte
gradient exactly equals its own isoelectric point. Isoelectric focusing provides a way of both accurately determining a protein's isoelectric point (pI)
and effecting separations among proteins whose isoelectric points may differ
by as little as a few hundredths of a pH unit. Some protein pIs are reported in
Table 3–1.

Sodium Dodecyl Sulfate (SDS) Gel Electrophoresis. A popular method for
protein molecular-weight estimation is called SDS gel electrophoresis. This
procedure uses the same types of polyacrylamide gels as the procedures

described earlier. The mixture of proteins to be characterized is first completely denatured by the addition of sodium dodecyl sulfate (a detergent) and mercaptoethanol and by a brief heating step. Denaturation is caused by the association of the apolar tails of the SDS molecules with protein hydrophobic groups. Any cystine disulfide bonds are cleaved by a disulfide interchange reaction with mercaptoethanol (Figure 1–16).

The resulting unfolded polypeptide chains have relatively large numbers of SDS molecules bound to them. The success of the technique for molecular-weight estimation depends on two facts: (1) that each bound SDS molecule contributes two negative charges to the denatured protein complex, so that the charge of the protein in its native state is effectively masked by the more numerous charged groups of the associated detergent molecules; and (2) that the total number of detergent molecules bound is proportional to the polypeptide-chain length or, equivalently, the protein's molecular weight. As a result, the SDS-denatured protein molecules acquire net negative charges that are approximately proportional to their molecular weights.

Typically, the behavior of an SDS gel electrophoresis system is calibrated by concurrently running standards of known molecular weight and then comparing the migration behavior of the unknowns with the standards. Although SDS gel electrophoresis gives only an approximate measure of protein molecular weight, with its accuracy depending on experimental conditions, the procedure offers both experimental simplicity and high resolving power. These advantages have led to its widespread use for the characterization of a variety of protein mixtures. An SDS gel electrophoresis pattern for a mixture of known proteins is illustrated in Figure 3–12.

A fancy extension of the electrophoretic method of separation involves combining isoelectric focusing with SDS gel electrophoresis to produce a two-dimensional electrophoretogram. This technique is most valuable for the analysis of very complex mixtures. First, the sample is run in a one-dimensional pH gradient gel (isoelectric focusing). The resulting narrow

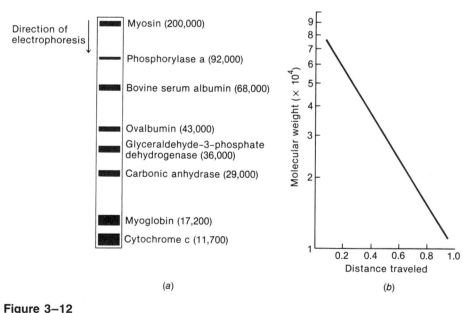

(a)   (b)

Figure 3–12
Electrophoresis behavior of proteins on an SDS-polyacrylamide gel. (a) A protein mixture was layered at the top of the gel in phosphate buffer, pH 7.2, containing 0.2% SDS. After electrophoresis the gel was removed from the apparatus and stained with Coomassie blue. The protein and its molecular weight are indicated next to each of the stained bands. (b) The logarithm of the molecular weight against the mobility (distance traveled) shows an approximately linear relationship. (Data of K. Weber and M. Osborn.)

Point of initial application of protein mixture

Isoelectric focusing ———→

×

SDS—gel electrophoresis ———→

Figure 3–13
Two-dimensional SDS-isoelectric focusing gel electrophoresis. First the sample is run in a one-dimensional pH gradient, partially separating the sample along a strip of gel. Then the strip of gel containing the sample is placed alongside an SDS gel, and the proteins are permitted to further separate by moving in the second dimension, at right angles to the first separation. Sample shown is total *E. coli* protein; individual proteins are detected by autogradiography. (Photograph provided by Patrick O'Farrell. See O'Farrell, 1975. *J. Biol. Chem.* 250, 4007.)

strip of gel, containing the partially separated mixture of proteins, is placed alongside a square slab of SDS gel. An electrical field is imposed so that the sample moves at right angles to its motion in the first gel. Figure 3–13 shows the separation of total *E. coli* protein into more than 1000 different components.

Radiation Techniques Probe Fine Structure and Conformation

Although radiation is rarely used as a tool for protein separations, many methods are available in which the interaction of radiation with proteins is used to characterize specific molecular properties of the proteins. These methods may be classified into three main categories according to the measured property of the radiation that emerges from the exposed sample:

1. The fraction of the incident radiation that is absorbed or scattered by the sample (Table 3–2, a–h).
2. The radiation emitted by a protein at wavelengths other than those used for excitation (Table 3–2, i–j).
3. The polarization of emerging radiation as altered by protein in solution (Table 3–2, k–m).

The types of information obtained by the use of the different procedures is very extensive and covers virtually every aspect of protein structure, with considerable overlap between available techniques (see Table 3–2). We have already discussed the use of light scattering. The remaining discussion of radiation techniques will be limited to a description of ultraviolet (UV) absorption spectroscopy, optical rotatory dispersion, circular dichroism, and x-ray diffraction. Excellent references explaining the use of the other radiation techniques are given at the end of the chapter.

UV Absorption Spectroscopy. The aromatic amino acids phenylalanine, tyrosine, and tryptophan all possess absorption maxima in the near-ultraviolet range (Figure 3–14). These absorption bands arise from the interaction of radiation with electrons in the aromatic rings. The near-ultraviolet absorp-

TABLE 3–2
Radiation Techniques for Examining Protein Structure

Technique	Information Obtained
(a) X-ray diffraction	Detailed atomic structure
(b) Nuclear magnetic resonance spectrometry (NMR)	Structure of specific sites; ionization state of individual residues
(c) Electron paramagnetic resonance	Structure of specific sites; proximity between specific sites
(d) Spectrometry (EPR)	
(e) Optical absorption spectroscopy	Measurement of concentrations or rate of reactions
(f) Infrared absorption spectroscopy	Type and extent of secondary structure
(g) Light scattering	Molecular weight and size
(h) Ultraviolet absorption spectroscopy	Concentration and conformation
(i) Fluorescence	Proximity between specific sites
(j) Raman scattering	Structure of specific sites
(k) Optical rotatory dispersion (ORD)	Type and extent of secondary structure
(l) Circular dichroism (CD)	Type and extent of secondary structure
(m) Fluorescence polarization	Molecular weight, shape, flexibility and orientation of secondary structure units

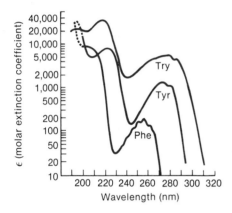

Figure 3–14
Ultraviolet absorption spectra of tryptophan (Try) tyrosine (Tyr), and phenylalanine (Phe) at pH 6. The molar absorptivity is reflected in the extinction coefficient, with the concentration of the absorbing species expressed in moles per liter. (Adapted from D. B. Wetlaufer, *Adv. Prot. Chem.* 17:303–390, 1962.)

tion properties of proteins are determined solely by their content of these three aromatic amino acids.

In dilute solutions, UV absorption can be quantified with the help of a conventional spectrophotometer and used as a measure of the concentration of the protein. The general quantitative relationship that governs all absorption processes is called the Beer-Lambert law:

$$I = I_0 10^{-\epsilon cd} \qquad (12)$$

where I_0 is the intensity of the incident radiation, I is the intensity of the radiation transmitted through a cell of thickness d (in centimeters) that contains a solution of concentration c (expressed either in moles per liter or in grams per 100 ml), and ϵ is the *extinction coefficient*, a characteristic of the substance being investigated (Figure 3–15). The spectrophotometer usually is capable of directly recording the absorbance A, which is related to I and I_0 by the equation

$$A = \log_{10}(I_0/I) \qquad (13)$$

Hence $A = \epsilon cd$, and A is a direct measure of concentration. We can see from Figure 3–14 that the ϵ values are largest for tryptophan and smallest for phenylalanine. Since protein absorption maxima in the near-ultraviolet are

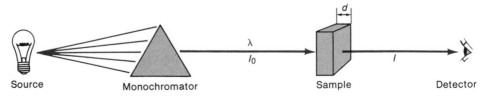

Figure 3–15
Schematic diagram of a spectrophotometer for measuring light absorption. Laboratory instruments for making measurements are much more complex than this, but they all contain the same basic components: a light source, a monochromator, a sample, and a detector. λ is the wavelength of the light, I_0 and I are the incident light intensity and the transmitted light intensity, respectively, and D is the thickness of the absorbing solution.

determined by the content of the aromatic amino acids and their respective ϵ values, most proteins have absorption maxima in the 280-nm region.

Whereas the ultraviolet absorption of a protein in the near-ultraviolet (240–300 nm) is dominated by its aromatic amino acids, the absorption in the far-ultraviolet (around 190 nm) is shown by all polypeptides regardless of their aromatic amino acid content. The reason is that the main absorption in this region is due primarily to the peptide linkage. Figure 3–16 shows ultraviolet absorption curves for poly-L-lysine, in three different conformations.

Optical Rotatory Dispersion (ORD) and Circular Dichroism (CD).   Any molecule in aqueous solution possessing one or more centers of asymmetry (*chiral centers*) is optically active. The extent of optical activity is quantitatively assessed in a *polarimeter* (see Box 3–B), which measures the rotation of plane-polarized light on passing through a measured amount of solution

---

BOX 3–B

Polarized Light and Polarimetry

Light is a form of electromagnetic radiation that oscillates sinusoidally in space and time. The oscillating electric and magnetic fields of light are perpendicular to each other and are both in a plane perpendicular to the direction of the light ray. In an unpolarized beam, there are equally strong fields with all different orientations in the plane (Illustration 1). A light beam is said to be polarized if the orientations of the fields in the plane are fixed. In *linearly polarized light* the orientation of the fields does not vary along the direction of the beam. In *circularly*

polarized light the orientation of the fields gradually changes in either a right-handed or left-handed manner along the direction of the beam.

A polarimeter is an instrument for studying the interaction of polarized light with optically active substances (Illustration 2). A cylindrical tube is filled with a solution containing the protein of interest. A monochromatic beam of polarized light is passed through the solution, and the effects on the polarized beam after passing through the solution are measured.

Illustration 1
Unpolarized and polarized light.

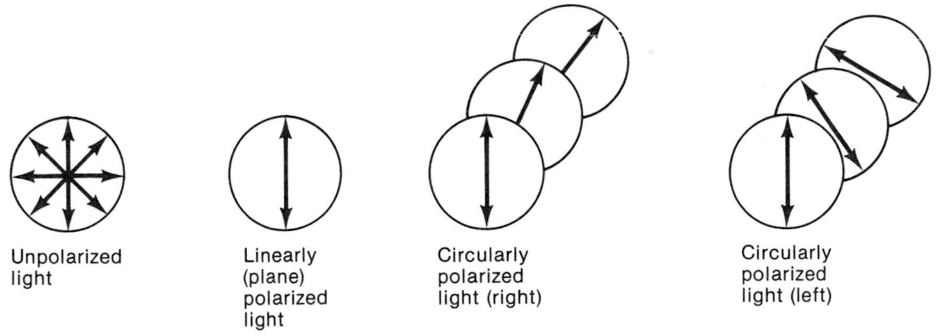

| Unpolarized light | Linearly (plane) polarized light | Circularly polarized light (right) | Circularly polarized light (left) |

Illustration 2
Simple polarimeter for measuring rotation of linearly polarized light.

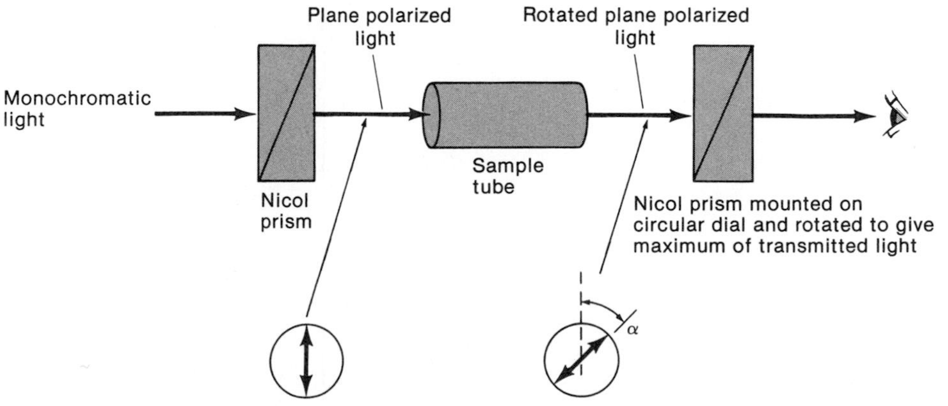

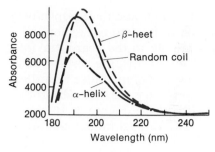

Figure 3–16

Ultraviolet absorption spectra of poly-L-lysine in different conformations: random coil, pH 6.0, 24°C; α helix, pH 10.8, 225°C; β sheet, pH 10.8, 52°C. This curve demonstrates the sensitivity of the peptide absorption band to the polypeptide chain confirmation. (Adapted from K. Rosenheck and P. Doty, *Proc. Natl. Acad. Sci. U.S.A.* 47:1775, 1961.)

containing the optically active molecule. The molecule is described as dextrorotatory or levorotatory according to the direction (right or left, respectively) in which it rotates plane-polarized light.

A polypeptide chain is optically active because of the asymmetric nature of its individual amino acid residues. In a helical conformation a polypeptide chain has an additional asymmetric component resulting from the chiral nature of the helical conformation. Optical rotation effects resulting from the helical conformation have a maximum in the 200-nm region, which, it may be recalled, is the region of optimum absorption by the peptide linkage (see Figure 3–16).

It is customary to measure optical rotation effects over a spread of wavelengths surrounding the asymmetrically oriented group that is responsible for absorption. When optical rotation is measured as a function of the wavelength of linearly polarized light, the characteristic is referred to as the optical rotatory dispersion (ORD). Circularly polarized light is also used to measure optical rotation effects. In this case the characteristic being measured is called circular dichroism (CD); it is the difference in absorption of left and right circularly polarized light. In an idealized situation, where there is a single absorption band arising from an asymmetrically oriented component, the peak in the CD curve coincides with the absorption peak of the optically active band. The ORD curve shows a value of zero at this point (Figure 3–17). ORD and CD have the same cause and are closely related. Either characteristic may be used to study conformational properties of proteins in

Figure 3–17

Unique spectral properties for a chiral molecule in solution as a function of wavelength, as shown by curves on the left. (a) The absorption band produces a symmetric curve with a maximum at the wavelength of maximum absorbance. (b) The optical rotatory dispersion curve shows a minimum at the λ_{max} of the optically active band. (c) The circular dichroism curve shows a curve similar in shape to the absorption curve. Curve (a) would be observed for any molecule that absorbs light, regardless of its configuration. Curves (b) and (c) are observed only if the absorbing species is optically active. For the enantiomer, the curve would be the same in (a), but would be opposite in sign in (b) and (c), as shown on the right.

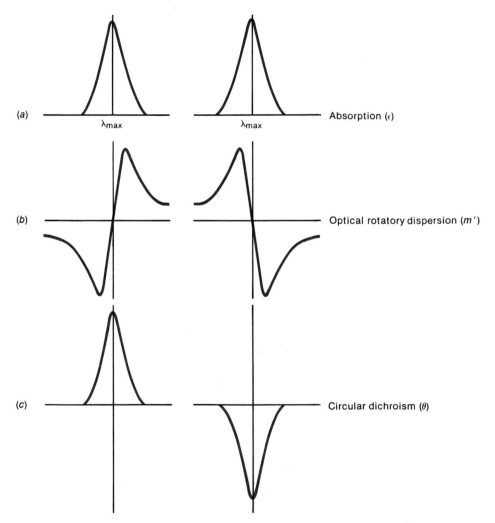

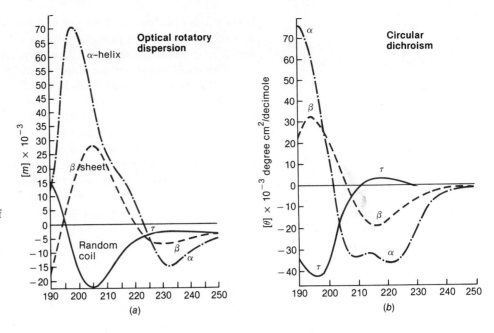

Figure 3–18
(a) Optical rotatory dispersion and (b) circular dichroism spectra for poly-L-lysine in the α helix, β sheet, and random coil conformations. These curves are more complex than those shown in Figure 3–17 because more than one type of optically active component is present in the polypeptide chain. It is clear that both of these curves are a sensitive function of the polypeptide chain conformation. The precise quantitative interpretation for each curve is not known. (Adapted from A. J. Adler, W. J. Greenfield, and G. D. Fasman, 1973. *In Methods in Enzymology*, vol. 27, ed. by C. H. W. Hirs and S. N. Timasheff, New York: Academic Press.)

solution, but circular dichroism has become more popular because it shows discrete spectral bands that may be either positive or negative.

The ORD and CD curves for poly-L-lysine are illustrated in Figure 3–18. It can be seen that conformation has a major effect on the ORD and CD curves in the region of the peptide absorption band. Effects of this sort have led to the application of these polarimetric techniques to studies of protein conformation in solution. For example, helical contents of proteins in solution can be estimated and denaturation temperatures can be determined.

Protein Tertiary-Structure Determination by X-ray Diffraction and Crystallography. Much of the fundamental understanding gained in recent years concerning the catalytic and functional properties of proteins has derived from a detailed knowledge of their tertiary structures. This knowledge has been gained in large part through the techniques of x-ray diffraction and x-ray crystallography.

The quality of the information potentially available through x-ray diffraction depends on the degree or extent of order of the protein molecules in a given sample. A sample under investigation usually consists of a hydrated purified protein. If the individual protein molecules are packed in a random order relative to one another, the distances between atoms or groups of atoms can be derived, but we cannot tell how these spacings are ordered in three dimensions.

If the protein is fibrous, it is often possible to learn its two-dimensional order. For example, when a fibrous sample consisting of long α helices is stretched, the helix axes become oriented in the direction of stretching. The resulting fibrous bundle gives a characteristic x-ray diffraction pattern indicating an ordered molecular arrangement along the helix axis. The pitch of the α helix (5.4 Å) and the advance per residue along the helix axis (1.5 Å) were detected by this technique. But the orientation of specific atoms in the helix cannot be determined from diffraction patterns of stretched fibers because of the lack of three-dimensional order. To deduce the correct three-dimensional structure for the α helix (and the β sheet), it was necessary to work with molecular models (currently computer programs are available for such purposes). By trial and error, model structures were built that were consistent with the steric limitations of the polypeptide chain that had the helix pitch and the advance per residue indicated by the diffraction pattern of stretched fibers. A comprehensive explanation of fiber diffraction methodology is given in Chapter 7, where we discuss its application to DNA.

The most information about a protein's structure is obtained from ordered three-dimensional protein crystals; this is the main interest of x-ray crystallographers. The goal in x-ray crystallography is to obtain a three-dimensional image of a protein molecule in its native state at a sufficient level of detail to locate its individual constituent atoms. The way this is done can most easily be appreciated by considering the more familiar problem of how we obtain a magnified image of an object in a conventional light microscope. In a visible-light microscope, light from a point source is projected on the object we wish to examine. When the light waves hit the object, they are scattered so that each small part of the object essentially serves as a new source of light waves. The important point is that the light waves scattered from the object contain information about its structure. The scattered waves are collected and recombined by a lens to produce a magnified image of the object (Figure 3–19).

Given this picture, we might ask what prevents us from simply putting a protein molecule in place of our object and viewing its magnified image. The basic problem here is one of resolution. The resolution, or extent of detail, that can be recovered from any imaging system depends on the wavelength of light incident upon the object. Specifically, the best resolution obtainable equals $\lambda/2$, or one-half the wavelength of the incident light. Since λ lies in the range of 4000–7000 Å for visible light, a visible-light microscope clearly does not have the resolving power to distinguish the atomic structural detail of molecules. What we need is a form of incident radiation with a wavelength comparable to interatomic distances. X-rays emitted from excited metal atoms, with wavelengths in the range of one to a few angstroms, would be most suitable.

However, simply replacing a visible-light source with an x-ray source does not solve all the problems. For example, to get a three-dimensional view of a protein, some provision must be made for looking at it from all possible angles, an obvious impossibility when dealing with a single molecule. Furthermore, when x-rays interact with proteins, only a small fraction of the rays are scattered. Most x-rays pass through the protein, but a relatively large number of them interact destructively with the protein, so that a

Figure 3–19

Schematic diagram of the procedures followed for image reconstruction in light microscopy (top) and x-ray crystallography (bottom).

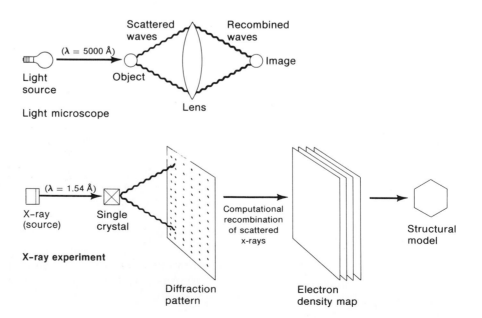

single molecule would be destroyed before scattering enough x-rays to form a useful image. Both these problems are overcome by replacing a single protein molecule with an ordered three-dimensional array of many molecules, which scatters x-rays essentially as if it were one molecule. This ordered array of protein molecules forms a single crystal, so the general technique is called protein x-ray crystallography.

The problems do not end here, because although the protein crystal readily scatters incident x-rays, there are no lens materials available that can recombine the scattered x-rays to produce an image. Instead, the best that can be done is to directly collect the scattered x-rays in the form of a diffraction pattern. Although recording the diffraction pattern results in loss of some important information, experimental techniques have been developed for recovering the lost information. Eventually the scattered waves can be mathematically recombined in a computational analog of a lens. By collecting the diffraction pattern of the crystal in many orientations, it is then possible to construct a three-dimensional image of the protein molecule.

Crystals suitable for protein x-ray studies may be grown by a variety of techniques, which generally depend on solvent perturbation methods for rendering proteins insoluble in a structurally intact state. The trick is to induce the molecules to associate with each other in a specific fashion to produce a three-dimensionally ordered array. A typical protein crystal useful for diffraction work is about 0.5 mm on a side and contains about 10^{12} protein molecules (an array 10^4 molecules long along each crystal edge). It is important to recognize that protein crystals are from 20 to 70 per cent solvent by volume. Consequently, crystalline protein is in an environment that is not substantially different from free solution.

The x-ray radiation usually employed for protein crystallographic studies is derived from the bombardment of a copper target with high-voltage ($\sim$50 kV) electrons, producing characteristic copper x-rays with $\lambda = 1.54$ Å. A schematic illustration of an x-ray diffraction pattern from a protein crystal is shown in Figure 3–20. Several features about this pattern bear explanation. First, it is apparent that the diffraction pattern consists of a regular lattice of spots of different intensities. *Diffraction from a crystal produces a series of spots that are due to destructive interference of waves scattered from the repeating unit of the crystal.* For the crystal whose diffraction pat-

Figure 3–20

Schematic view of an x-ray diffraction pattern. The spacing of the spots is reciprocally related to the dimensions of the repeating unit cell of the crystal. The symmetry of the spots (e.g., the mirror planes in the sample shown) and the pattern of missing spots (alternating spots along the mirror axes) give information on how molecules are arranged in the unit cell. Information concerning the structure of the molecule is contained in the intensities of the spots. Spots closest to the center of the film arise from large-scale or low-resolution structural features of the molecule, while those farther out correspond to progressively more detailed features. Circles show 5-Å and 3-Å regions of resolution. Mirror axes are labeled m. Spacing of vertically oriented spots, b^*, and horizontally oriented spots, a^*, are reciprocally related to b and a, the dimensions of the unit cell.

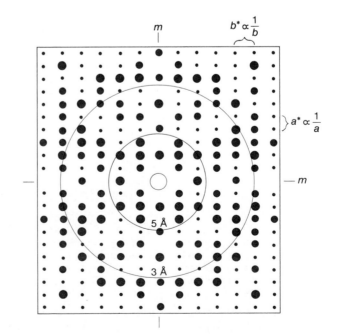

tern is shown, the repeating unit (or crystal unit cell) contains four symmetrically arranged protein molecules. Corresponding symmetrical features appear in the spot-intensity pattern. Further, *the lattice spacing of the diffraction spots is inversely proportional to the actual dimensions of the crystal's repeating unit or unit cell.* Consequently, both the crystal's unit-cell dimensions and general molecule packing arrangement can be derived from inspection of the crystal's diffraction pattern.

Information concerning the detailed structural features of the protein is contained in the intensities of the diffraction spots. A three-dimensional structural determination requires that all these spots be measured, either by scanning x-ray films with a densitometer or by measuring the diffraction spots individually with a scintillation counter. In this connection, it is important to realize that all the atoms in the protein structure make individual contributions to the intensities of each diffraction spot. Conversely, all the intensity data have to be collected to reconstruct the protein's structure.

Initial studies of a protein's tertiary structure are generally carried out at low resolution, that is, using intensity data near the origin (center) of the diffraction pattern. Diffraction data near the origin reflect large-scale structural features of the molecule, while those nearer the edge correspond to progressively more detailed features. Figure 3–21 shows some examples of electron-density maps calculated at different resolutions, illustrating how various levels of structural detail appear at different degrees of resolution.

A powerful aspect of protein crystallography is that once the native structure is known, various cofactors or enzyme substrate analogs can be bound to the molecule in the crystal. By simply remeasuring the diffraction

Figure 3–21

Views of crystallographic electron-density maps, showing how the structural detail revealed depends on the resolution of the data used to compute the maps. The actual molecular structure is inserted in its true position in the electron-density maps.

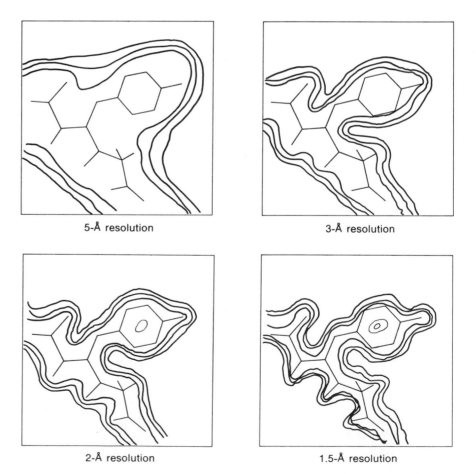

5-Å resolution

3-Å resolution

2-Å resolution

1.5-Å resolution

intensities, we can compute a new map that allows direct and explicit examination of the structural interactions between the native protein and its substrate or cofactor molecules. Detailed analysis of these interactions has provided much of the foundation for our current understanding of many protein catalytic and functional properties.

METHODS OF PROTEIN PURIFICATION

Before we can fully characterize a protein we must purify it from a natural source. Once the decision has been made to purify a particular protein, several factors must be weighed. For example, how much material is needed? What level of purity is required? The starting material should be readily available and should contain the desired protein in relative abundance. If the protein is part of a larger structure, such as the nucleus, the mitochondria, or the ribosome, then it is advisable to isolate the large structure first from a crude cell extract.

Purification must usually be performed in a series of steps, using different techniques at each step. Some purification techniques are more useful when handling large amounts of material, whereas others work best on small amounts. A purification procedure is arranged so that the techniques that are best for work with large amounts are used during early steps in the overall purification. The suitability of each purification step is evaluated in terms of the amount of purification achieved by that step and the per cent recovery of the desired protein.

Combining techniques introduces new considerations and new problems. If two purification techniques each give tenfold enrichment for the desired protein when executed independently on a crude extract, this does not mean they will give 100-fold enrichment when combined. In general, they will give somewhat less. As a rule, purification techniques that combine most effectively usually are based on different properties of the protein. For example, a technique based on size fractionation is more effectively combined with a technique based on negative charge than with another technique based on size fractionation.

Throughout the purification we must have a convenient means of assaying for the desired protein, so we can know the extent to which it is being enriched relative to the other proteins in the starting material. In addition, a major concern in protein purification is stability. Once the protein is removed from its normal habitat, it becomes susceptible to a variety of denaturation and degradation reactions. Specific inhibitors are sometimes added to minimize attack by proteases on the desired protein. During purification it is usual to carry out all operations at 5°C or below. This temperature control minimizes protease degradation problems and decreases the chances of denaturation.

In their natural habitat, proteins are usually surrounded by other proteins and organic factors. When these are removed or diluted as during purification, the protein becomes surrounded by water on all sides. Proteins react differently to a pure aqueous environment; many are destabilized and rapidly denatured. A common remedial measure is to add 5 to 20 per cent glycerol to the purification buffer. The organic surface of the glycerol is believed to simulate the environment of the protein in the intact cell. Two other ingredients that are most frequently added to purification buffers are mercaptoethanol and ethylenediamine tetraacetate (EDTA). The mercaptoethanol inhibits the oxidation of protein —SH groups, and the EDTA chelates divalent cations. The latter, even in trace amounts, can lead to aggregation problems or activate degradative enzymes.

In the remainder of this chapter we discuss the merits of different types of purification. We conclude by describing the purification of two different proteins.

Differential Centrifugation Divides a Sample into Two Fractions

A typical crude broken cell preparation contains broken cell membranes, cellular organelles, and a large number of soluble proteins, all dispersed in an aqueous buffered solution. The membranes and the organelles can usually be separated from one another and from the soluble proteins by differential centrifugation. Differential centrifugation divides a sample into two fractions: the pelleted fraction, or sediment, and the supernatant fraction, that is, the fraction that does not sediment. The two fractions may then be separated by decantation.

According to its purpose, differential centrifugation requires different speeds and different times of centrifugation (Table 3–3). For example, if the protein of interest were in the mitochondrial fraction, the crude lysate should be centrifuged first at 4000 × g for 10 min to remove cell membranes, nuclei, and (in the case of plant material) chloroplasts. The supernatant from this step contains, among other elements, the mitochondria, and would be decanted and recentrifuged at 15,000 × g for 20 min to obtain a sediment primarily containing mitochrondria. If ribosomes instead of mitochondria were the goal, then the crude lysate should be centrifuged at 30,000 × g for 30 min and the resulting supernatant decanted and centrifuged at 100,000 × g for 180 min to obtain a ribosomal sediment. If the soluble protein fraction were the goal, then the entire lysate would be centrifuged at 100,000 × g for 180 min and the resulting supernatant, containing the soluble protein, would be carefully decanted for further processing.

Differential Precipitation Is Based on Solubility Differences

We mentioned earlier that every protein has a characteristic salting-out point, which is reached by altering the concentration of salts such as ammonium sulfate, $(NH_4)_2SO_4$. We can use this fact to carry out a differential precipitation: Salting out is a relatively crude procedure, usually resulting in no more than a two- to threefold purification, but it is easy to do on any volume of material, and even a two- to threefold enrichment substantially decreases the bulk of an initial extract, making it more manageable in later steps.

Typically, the desired protein precipitates over a range of salt concentrations. If it precipitates in the range of 20–30 per cent by weight, then we would first add ammonium sulfate to a concentration slightly below 20 per cent and then centrifuge to remove by sedimentation any proteins that precipitate in the 0–20 per cent range. To the supernatant from this centrifugation we would then add more ammonium sulfate, to 30 per cent. Centrifuga-

TABLE 3–3
Sedimentation Conditions for Different Cellular Fractions

Fraction Sedimented	Centrifugal Force (× g)	Time (min)
Cells (eukaryotic)	1,000	5
Chloroplasts; cell membranes; nuclei	4,000	10
Mitochondria; bacteria cells	15,000	20
Lysosomes; bacterial membranes	30,000	30
Ribosomes	100,000	180

TABLE 3–4
Fractionation Range of Different Cross-linked Dextran Beads

Dextran Type	Fractionation Range (M_r)
G-10	0–700
G-15	0–1,500
G-25	100–5,000
G-50	1,500–30,000
G-75	3,000–80,000
G-100	4,000–150,000
G-100	5,000–300,000
G-200	5,000–600,000

tion at this point brings down the desired protein, as well as other proteins that precipitate in this range of salt concentrations. The supernatant is discarded and the sediment saved for further purification.

Column Procedures Are the Most Versatile and Productive Purification Methods

With two bulk steps behind us (differential centrifugation and ammonium sulfate precipitation), we are generally ready for a more sophisticated column step. Column procedures are the most important and the most varied of purification steps. Usually a glass cylinder with a large opening at the top and a capillary opening at the bottom is used to support the column material. The cylinder is packed with the hydrated column material to be used for the fractionation; this material could be a cross-linked polydextran (see Figure 3–9), a resin, or finely divided cellulose fibers. The column material usually contains functional groups with an affinity for specific proteins.

Preparative Gel-exclusion Chromatography. In gel-exclusion chromatography, a cross-linked dextran without any special attached functional groups is used for the column substrate. Large molecules flow more rapidly through this type of column than small ones, for the reasons explained earlier (see "Gel-exclusion Chromatography"). The dextrans have different degrees of cross-linking, making them effective over different size ranges (see Table 3–4). The dextran is first rinsed with a buffer. Then the protein sample, in a volume of less than 1/20 the column volume, is applied to the top of the column. Once all of the protein solution is in contact with the column, further buffer is passed through the column. The eluant appearing at the lower end of the column is collected, usually with the help of a fraction collector. The collector is equipped with an automatic device that collects the same amount of eluant in each of a series of test tubes (Figure 3–22). The various fractions are analyzed for both total protein concentration and the desired protein. Fractions containing an appreciable amount of the desired protein are combined (pooled) for further purification.

Column Chromatography with Protein Binding. In most column procedures, unlike gel-exclusion chromatography, the protein is first bound to the column material; a change in the elution buffer then leads to a differential elution of the bound protein.

Ion-exchange Chromatography. As described previously (Chapter 1), amino acids can be separated according to their varying affinities for ion-exchange resins immobilized in a column. Ion-exchange chromatography also can be used to separate proteins, but significant differences arise as a result of the larger size of proteins. First, cross-linked resins are rarely used for protein separations because proteins are too large to penetrate the resin beads. Instead, finely divided celluloses containing either positively or negatively charged groups are most commonly used to make such columns (Table 3–5). Second, at a given salt concentration, protein binding tends to be an all-or-nothing phenomenon, rather than an equilibrium phenomenon as it is with amino acids or resins. Consequently, the only way to achieve separations of proteins on charged cellulose columns is by changing the salt concentration. This is done in either a continuous manner (gradient elution) or a discontinuous manner (step elution).

Proteins differ enormously in their affinity for positively or negatively charged columns. This affinity is proportional to the salt concentration required to release the protein from the column material so that it will start

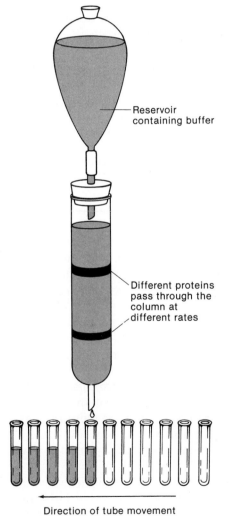

Reservoir containing buffer

Different proteins pass through the column at different rates

Direction of tube movement

Figure 3–22
Collecting fractions during column chromatography. Column material and elution procedure are chosen to effect optimal separation of the desired protein.

TABLE 3–5
Some Column Materials for Ion-exchange Chromatography of Proteins

Matrix	Functional Groups on Column
Phosphocellulose (PC)	$-PO_3^-$
Carboxymethyl cellulose (CMC)	$-CH_3-COO^-$
Diethylaminoethyl cellulose (DEAE)	$-(CH_2)_2-N\begin{array}{l} CH_2-CH_3 \\ CH_2-CH_3 \end{array}$

flowing down the column. The affinity between column and protein is strongly influenced by pH. For this reason the pH of the buffers must be carefully controlled during a column step. Typically, a column is loaded with protein solution at a low ionic strength so that most of the protein binds to the column. Once the sample is loaded and rinsed, elution is initiated by gradually increasing the salt concentration of the elution buffer. Proteins are eluted in the order of increasing affinity. Those with the greatest affinity come off last. Collections, assays, and pooling procedures are similar to those used in gel-exclusion chromatography. In addition to the proteins present in various fractions, the salt concentration in each fraction is recorded.

Additional Adsorbants Used for Column Chromatography. Finely divided celluloses may also be used in the column in conjunction with attached hydrophobic groups such as octyl alcohol. In this case, proteins with exposed hydrophobic centers bind to the column with varying affinities. These proteins may be eluted in order of decreasing affinities for the column by increasing the level of free octyl alcohol in the eluting buffer.

A finely divided calcium phosphate gel known as hydroxyl apatite is used for a wide variety of separations. The eluting buffer usually consists of increasing concentrations of phosphate buffer. Many other types of adsorbants are used on occasion in column chromatography.

Affinity Chromatography. In addition to the column techniques already described for the isolation and characterization of proteins, several other methods have been developed that exploit specific binding properties of a given protein. Methods of this sort are generically referred to as affinity chromatography.

For example, many enzymes reversibly bind organic cofactor molecules, such as adenosine triphosphate or pyridine nucleotides, in order to catalyze the chemical reactions of their substrates. Often, the separation of such enzymes from other proteins can be readily achieved by preparing a chromatographic column that is first chemically reacted with a suitable cofactor derivative. As the mixture passes through the column, those protein molecules having specific binding affinities for the cofactor bind to the column, while other proteins pass through. The cofactor binding protein can subsequently be eluted with a solution containing the same soluble cofactor that is covalently attached to the column.

The power of affinity chromatography techniques lies in the great variety of specific interactions that characterize the functional properties of protein molecules. The enrichments obtainable in single steps by affinity chromatography sometimes exceed 1000-fold, a result testifying to the high selectivity of the method.

High-performance Liquid Chromatography. Thus far we have described three types of column chromatography for purification and characterization of proteins: gel-exclusion chromatography, ion-exchange chromatography, and affinity chromatography. High-performance liquid chromatography (HPLC) is not so much a new type of chromatography as a new way of looking at old chromatographic techniques. The same principles are involved, but the column support materials usually consist of more finely divided particles made of physically stronger materials, which can withstand high pressures (5000–10,000 psi) without changing their structure. The column apparatus itself also must be designed to withstand high pressures. Finely divided column materials lead to slower flow rates, but this factor can be more than compensated for by applying high hydrostatic pressures. As a rule, much better separations are achieved in a much shorter time with the proper applications of HPLC.

Electrophoretic Methods Are Used for Preparation and Analysis

Electrophoresis is used extensively to monitor protein purity during purification procedures. Typically, SDS gel electrophoresis and isoelectric focusing are used to resolve the desired protein and the contaminating proteins into discrete bands on a stained gel. Both of these procedures, which were explained in a previous section, can also be used in themselves as purification steps. Isoelectric focusing is especially useful in this regard, because it involves a principle that is unique among purification methods. SDS gel electrophoresis is used as a purification step only if there is an easy way of renaturing the denatured protein extracted from the gel. Both procedures are limited to the processing of small amounts. For this reason, they are used at or near the final stages of purification, when the amounts of material being processed are relatively small.

We will now see how various purification steps are combined in actual protein purifications.

Purification of Specific Proteins Involves Combinations of Different Procedures

The following two examples of purification show how various techniques can be effectively combined to produce purified proteins with a minimum of effort and loss of activity.

The Purification of UMP Synthase from Mammalian Tumor Cells. The last two steps in the biosynthesis of the mononucleotide uridine 5′-monophosphate (UMP) are catalyzed by (1) orotate phosphoribosyltransferase (OPRTase) and (2) orotate 5′-monophosphate (OMP) decarboxylase.

Mary Ellen Jones and her colleagues set out to purify the enzyme or enzymes involved in these two reactions. Their main goal was to determine whether the two reactions are carried out by one protein or more than one. Their findings indicated that the two reactions were both catalyzed by the same enzyme, consisting of a single polypeptide chain. To demonstrate this fact, it was necessary to monitor both enzyme activities at each step in the purification and show that both activities *copurified*. For this purpose, Jones used specific enzyme assays for both enzyme activities. All fractions were assayed for both enzymatic activities at each stage of the purification.

The main data associated with the purification are summarized in Table 3–6. This table indicates the total protein obtained in each step, the number of enzyme units* for each enzyme, and the ratio of enzyme units to total protein, called the *specific activity*. In the absence of enzyme inactivation, the specific activity should be directly proportional to the enrichment. The

---

*Enzyme units are proportional to the amount of enzyme activity. The relationship between enzyme units and absolute amount of enzyme need not concern us here.

TABLE 3–6
Outline of Purification of UMP Synthase from Ehrlich Ascites Carcinoma

Fraction	Volume (ml)	Protein (mg)	OMPDase[a] Units[b]	Sp. act.[c]	Per cent recovery	OPRTase[a] Units[b]	Sp. act.[c]	Per cent recovery	OMPDase / OPRTase
1. Streptomycin fraction	1040	11,700	40.4	0.0034		20.5	0.0018		2.0
2. Dialyzed $(NH_4)_2SO_4$ fraction	144	311	24.3	0.0078	60	8.7	0.0028	42	2.8
3. Affinity column eluate (concentrated)	0.475	0.51	4.0	7.8[d]	10	0.35	0.69	3.3	11.4

[a]OMPDase = OMP decarboxylase; OPRTase = orotate PRTase.

[b]Units refer to total amount of enzyme activity.

[c]Specific activity refers to the units of enzyme activity divided by the total protein.

[d]This value represents a 2300-fold enrichment from fraction 1.

per cent recovery refers to the amount of enzyme activity in the indicated fraction, as compared with the amount present in fraction 1. This number is usually less than 100 per cent. The apparent losses may reflect actual losses of enzyme during purification, or they may reflect *inactivation* (usually due to unknown causes) of the enzyme during purification.

The nine steps involved in the purification of UMP synthase from starting tissue are summarized in Figure 3–23. All steps were carried out at 0–5°C. About 200 g of Ehrlich ascites cells, a mammalian tumor rich in the desired enzymes, was suspended in buffer and processed in a tissue homogenizer, which mechanically breaks down the tissue and the cell membranes (step 1). Then EDTA and an —SH reagent were added to this total cell lysate. Solid streptomycin sulfate was also added with stirring (step 2). Streptomycin sulfate aggregates nucleic acids so that they may be more easily removed by centrifugation. The resulting slurry was subjected to high-speed centrifugation, and the resulting supernatant (Table 3–6, fraction 1) was carefully decanted for further processing (step 3).

Preliminary experiments had shown that the desired enzymes were in the 18.5–28 per cent $(NH_4)_2SO_4$ fraction. This knowledge served as the basis for the next three steps. First 239 g of solid $(NH_4)_2SO_4$ was added to 1040 ml of supernatant (step 4). The resulting precipitate was removed by centrifugation (step 5). Then an additional 120 g of $(NH_4)_2SO_4$ was added to the supernatant (step 6). The resulting slurry was centrifuged. This time the supernatant was discarded after centrifugation, leaving the sediment containing the enzyme activity for further processing (step 7). The sediment was resuspended in a dilute buffer for column chromatography (Table 3–6, fraction 2). Inspection of Table 3–6 indicates only about a twofold increase in specific activity between fractions 1 and 2.

The main purification was achieved by two column steps, carried out in series. The first column was an affinity column containing an analog of UMP, 6-azauridine 5'-monophosphate (azaUMP), covalently attached to an agarose column support system. In dilute buffer, greater than 99 per cent of the protein in fraction 2 is retained on this column. After thorough rinsing of the column with dilute buffer, 5×10^{-5} M azaUMP was added to the buffer. This addition resulted in the elution of the UMP synthase (step 8). Then the column eluant carrying the two enzyme activities associated with UMP synthase was resuspended in pure buffer and passed over a phosphocellulose column (step 9). Phosphocellulose was chosen because the negatively charged phosphate groups result in the retention of proteins by electrostatic

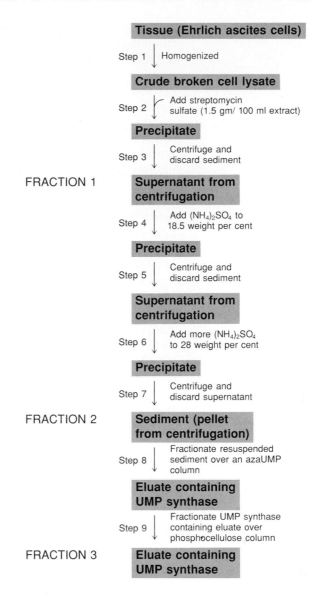

Figure 3–23
Outline of purification scheme for
UMP synthase from Ehrlich ascites
tumor cells of mice.

attraction alone, but they also resemble the phosphate groups in the naturally occurring enzyme substrate, OMP. Thus the phosphocellulose column may be thought of as an ion-exchange column and an affinity column combined. Recall that in ordinary ion-exchange chromatography, the protein, after column loading, is eluted by increasing the ionic strength with a simple inorganic salt. In this example, Jones used a more specific method to elute the enzyme, which involved adding 10^{-5} M azaUMP and 2×10^{-5} M OMP to the original loading buffer. The addition does not substantially increase the ionic strength of the buffer. Therefore the only phosphocellulose-bound proteins that are likely to be eluted by this treatment are those with an especially high affinity for either of these nucleotides (azaUMP and OMP). This fact should greatly favor selective elution of those enzymes that carry specific sites for binding these nucleotides. The two column steps together resulted in an enzyme preparation (Table 3–6, fraction 3) that was approximately 2300-fold purified over the starting material, as measured by the increase in specific activity.

The final product (Table 3–6, fraction 3) was examined by SDS gel electrophoresis and found to contain a single band with an estimated molecular weight of 51,000. The denaturing conditions of an SDS gel would be expected to dissociate a multisubunit protein. Hence the SDS gel result indicated that the enzyme contains one type of polypeptide chain, but it does not tell us whether the enzyme contains one or more of these chains. Sedimentation analysis on a sucrose density gradient in a nondenaturing buffer indicated a single band with an estimated molecular weight of about 50,000. These two results taken together demonstrate that the enzyme in its native state contains a single polypeptide chain. The purity of the enzyme was also confirmed by isoelectric focusing and two-dimensional electrophoresis, with isoelectric focusing in the first direction followed by SDS gel electrophoresis in the second direction.

The conclusion drawn from these results was that both of the enzyme activities associated with UMP synthase—OMPDase and OPRTase—are contained in a single protein. The basis for the conclusion was that both enzyme activities are always present in the same fractions throughout the multistep purification. However, inspection of Table 3–6 indicates a possible objection to this interpretation. In the columns showing per cent recovery, it can be seen that substantial amounts of enzyme activity are lost for both enzymes during purification, but that considerably more activity is lost for the OPRTase. These losses could be due to actual loss of enzymes during purification or to some sort of inactivation of the enzyme sites. The preferential loss of OPRTase activity is emphasized by the last column in Table 3–6, which gives the ratio of the two enzyme activities in the different fractions. Considered alone, these data could indicate that a separate catalytic unit of orotate PRTase is lost during purification. Jones thinks this is unlikely for two reasons: (1) the activity appears in no fractions other than with OMP decarboxylase during purification, and (2) the orotate PRTase activity is notably unstable. It is concluded that both enzyme activities exist at distinct sites on a single protein. The greater loss in activity of one enzyme activity over the other is attributed to a greater sensitivity of one reaction site over the other on the enzyme surface.

Purification of Lactose Carrier Protein from *Escherichia coli* Bacteria. The second purification procedure we will look at illustrates an unusual approach to the purification of a membrane-bound protein. The lactose carrier protein of *E. coli* is normally tightly bound to the plasma membrane. This protein is involved in the active transport of the dissaccharide lactose across the cytoplasmic membrane. When lactose carrier protein is present, the intracellular concentration of lactose can achieve levels 1000-fold higher than those found in the external medium. Ron Kaback devised a simple yet elegant procedure for the purification of this protein.

Purification of the membrane-bound lactose carrier protein is a very different problem from the purification of the soluble OMP synthase. Both the approach to purification and the assays for the protein during purification are quite novel. The assay involves reconstituting a transport system with membranes that are free of lactose carrier protein, then adding the partially purified carrier protein and radioactively labeled lactose. The activity in this assay system is proportional to the transport of radioactive lactose across the membrane in the cell-free reconstituted system.

The results of the purification steps are tabulated in Table 3–7, and the purification procedure is outlined in Figure 3–24. In this procedure advantage was taken of the fact that the carrier protein in its native state is firmly bound to the cytoplasmic membrane. Thus the first step consisted of isolating these membranes from the rest of the cell constituents. Starting from the membrane fraction only 35-fold purification was required to achieve pure carrier protein. This rapid result was possible because a special strain of *E. coli*, containing about 100 times the normal carrier protein, was used as

TABLE 3–7
Purification of the Lactose Carrier Protein

Fraction	Protein (mg)	Per cent recovery (total protein)	Per cent recovery (carrier protein)	Purification factor
1. Membrane fraction	12.5	100	100	1.0
2. Urea-extracted membrane	5.6	45	76	1.7
3. Urea/cholate-extracted membrane	2.6	21	61	2.9
4. Octylglucoside extract	0.4	3.2	38	12
5. DEAE column peak	0.056	0.4	14	35

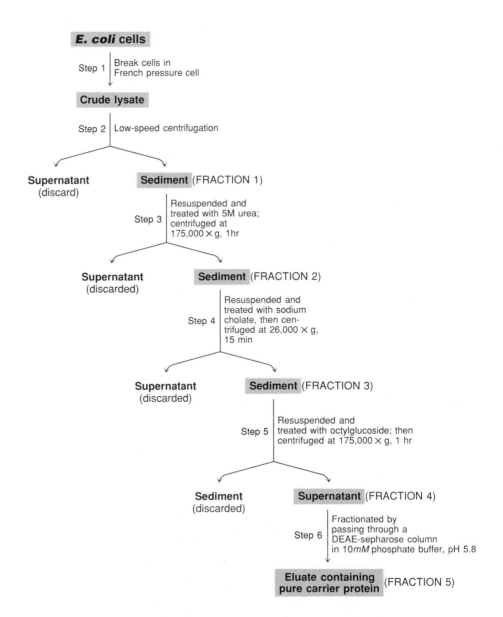

Figure 3–24
Outline of purification procedure for
lactose carrier protein from *E. coli.*

starting material for the purification. The high initial content was engineered
by putting the carrier protein gene on a multicopy plasmid, which was then
inserted into the cell—a procedure described in Chapter 32.

Bacterial cells are much tougher than mammalian cells, requiring a
more stringent procedure for cell disruption. In this case, the cells were
placed in a so-called French pressure cell and bled through an orifice from

very high pressures (~10,000 psi) to atmospheric pressure (step 1). Under these conditions the cells literally explode, fragmenting their membranes and releasing the cytoplasmic contents. The membranes were pelleted by a brief centrifugation, leaving a supernatant containing DNA, ribosomes, and cytoplasmic protein, which was removed by decantation (step 2). The pelleted membrane fraction was next resuspended and extracted with 5 M urea and then re-extracted with 6 per cent sodium cholate. The pellet obtained by high-speed centrifugation from these two extractions still contained most (61 per cent) of the carrier protein in the more rapidly sedimenting membrane fraction (fraction 3), although 79 per cent of the total membrane-bound protein was released by these treatments (steps 3 and 4).

At this point, the carrier protein was released from a suspension of the membranes by addition of the hydrophobic reagent octylglucoside in the presence of E. coli phospholipid (step 5). It is believed that the octyl part of octylglucoside competes effectively with the membrane for binding to hydrophobic centers on the carrier protein. The E. coli phospholipid facilitates dissociation of the carrier protein from the membrane fraction. The solubilized octylglucoside-containing extract was fourfold enriched in carrier protein after a high-speed centrifugation to remove residual membrane and membrane-bound proteins.

Finally, the octylglucoside-containing extract was passed over a positively charged diethylaminoethyl (DEAE) sepharose column (sepharose is a form of cross-linked polydextran) in a buffer containing 10 mM potassium phosphate, 20 mM lactose, and 0.25 mg of washed E. coli lipid per milliliter. The carrier protein passed through this column as a symmetrical peak of protein (step 6). Most of the remaining protein in the extract adsorbed to the positively charged column. The fractions containing the bulk of the carrier protein activity were judged to be pure by SDS gel electrophoretic analysis. The purified protein contained a single polypeptide chain with an estimated molecular weight of 33,000.

The two purifications described here are as different as the two proteins involved. No two purifications are exactly alike, but the principles of purification, as stated at the outset, are quite similar. Fortunately, an almost endless variety of purification techniques exist. This variety is both helpful and challenging, as a great deal of knowledge and creativity are required to exploit it. In addition to professional expertise, the two most important things required to make a purification possible are an unambiguous assay for the protein in question and a means of stabilizing the protein during purification.

SUMMARY

This chapter discusses methods of protein characterization and protein purification. There is considerable overlap between the methods used for the two goals, but insofar as they are separable, methods of protein characterization are discussed first.

Protein solubility is not a fixed quantity for a given protein. Rather it is a function of many variables. Two of these are pH and salt concentration. Proteins show a minimum solubility at their isoelectric point. Frequently, proteins require the addition of a small amount of salt to become soluble (salting in), but excessive amounts of salt lead to protein precipitation (salting out).

There are several methods for determination of molecular weight. These include osmotic pressure, light scattering, sedimentation analysis, and gel-exclusion chromatography. Sedimentation analysis may be used in two different ways: (1) by independently determining the sedimentation and diffusion rates and combining this information to calculate a molecular weight, and (2) by equilibrium ultracentrifugation. Gel-exclusion chromatography uses cross-linked polydextrans, and relates molecular weight to the rate of migration through a column.

Electrophoretic methods are used in various ways to characterize protein mixtures and purified proteins. The high resolution attainable by electrophoresis makes it ideal for determining the number of proteins in a mixture as well as their approximate size. Isoelectric focusing, a powerful electrophoretic technique for analyzing mixtures, also permits determination of the isoelectric pH.

Many techniques for protein characterizations involve the use of radiation. Every aspect of protein three-dimensional structure can be determined by one or more of the available techniques. The most productive technique has been x-ray diffraction analysis, which leads from detailed descriptions of the three-dimensional conformations down to atomic dimensions.

Methods of protein purification include differential centrifugation, differential precipitation with $(NH_4)_2SO_4$, gel-exclusion chromatography, differential electrophoretic mobility, and differential affinities for column matrices containing different functional groups. Column procedures are particularly versatile because of the large number of functional groups that can be used to bind proteins in different ways and because of the variety of conditions for differential column elution.

The chapter concludes with a description of the purification of two proteins, UMP synthase from eukaryotic cells and lactose carrier protein from *E. coli*. Each of these purifications requires the intelligent combination of several purification procedures performed in series. Throughout purification, the presence and amount of the desired proteins must be monitored by a specific protein assay. A successful purification requires careful selection of starting material and working under conditions that minimize loss of functional protein through denaturation.

SELECTED READINGS

Cantor, C. R., and P. R. Schimmel, *Biophysical Chemistry.* New York: W. H. Freeman and Co., 1980. Especially see volume 2, entitled *Techniques for Study of Biophysical Structure and Function.*

Eisenberg, D., and D. Crothers, *Physical Chemistry and Its Application to the Life Sciences.* Menlo Park, Calif.: Benjamin/Cummings, 1979.

Methods in Enzymology. New York: Academic Press. A continuing series of over 131 volumes that discusses most methods at a level suitable for laboratory application.

PROBLEMS

1. The table below summarizes a method for the purification of the enzyme 6-phosphogluconate from *E. coli*. Complete the table by calculating the specific activity, the per cent yield, and the purification factor for each step. Which step results in the greatest purification, and which step gives the best yield?

Procedure	Total Protein (mg)	Total Activity (units)	Specific Activity	Per Cent Yield	Purification Factor
Crude extract	70,000	2370			
Ammonium sulfate	25,380	2200			
Heat treatment	16,500	1980			
DEAE-cellulose	391	1683			
Phosphocellulose	47	1345			
Bio-Gel A—0.5 *M*	35	1120			

2. The following table summarizes the purification data for bovine Factor VII, a serine protease necessary for blood-coagulation.* To quantify the purification, a trace of highly purified FVII (^{3}H-FVII) was added at the outset. Calculate the per cent yield and the purification factor in terms of both ^{3}H cpm and units of enzyme activity, and fill in the table accordingly. What is odd about the per cent yield data in terms of enzyme activity? How might you account for this? [Note: Barium citrate elution is a step analogous to $(NH_4)_2SO_4$ precipitation.]

*Data are from Bach et al., *Blood* 63:393–398.

	Protein (mg)	Total Tritium (cpm × 10^{-6})	Tritium (% Yield)	Tritium Purification (-fold)	Total Activity (Factor VII U × 10^{-3})	Activity (% Yield)	Purification Factor
Plasma	6.88×10^5	6.07			4.6		
Barium citrate eluate	3.76×10^3	3.63			8.3		
DEAE pool	186	2.94			5.76		
IgG-Affi-Gel pool	5.33	1.97			0.90		
S-200 pool	1.03	1.08			0.86		

3. Using Equation (7) of the text and Table 3–1, calculate the molecular weight of β-lactoglobulin. Assume that water has a density of 0.998 and a specific volume of 0.75 (all at 20°C). Check your answer in the same table.

4. A mutant form of alkaline phosphatase is run in an iso-electric focusing experiment and found to differ from the native material by an amount corresponding to one additional positive charge relative to the native material. Assuming that the mutant's properties reflect the consequences of a single amino acid substitution, what are the possibilities?

5. Osmotic-pressure experiments on a protein in the presence of the membrane-permeable cation Ca^{2+} give plots of π/c versus c whose extrapolated intercepts on the π/c axis differ by a factor of 6. What do you conclude about this protein?

6. You are working in a laboratory that is investigating the functional properties of mutant forms of ribonuclease. An SDS gel electrophoresis run, using previously purified ribonucleases as a marker, shows an initial salt-precipitated fraction to be contaminated with two additional proteins. One contaminant has an M_r of about 13,000 (similar to ribonuclease) and a pI that is 4 pH units more acidic than that of ribonuclease. The second contaminant is a large protein of M_r 89,000. Describe an efficient purification protocol for your mutant ribonuclease.

7. You have a mixture of proteins with the following properties:

   Protein 1: $M_r = 12,000$, pI = 10
   Protein 2: $M_r = 62,000$, pI = 4
   Protein 3: $M_r = 28,000$, pI = 7

   Protein 4: $M_r = 9,000$, pI = 5

   Other factors aside, what order of emergence would you expect from these proteins when run on the following?
   a. An anion-exchange resin such as DEAE cellulose, with linear salt gradient elution.
   b. A Sephadex G-50 gel-exclusion column.

8. You wish to effect the rapid, high purification of an enzyme that binds ATP, starting from a crude extract containing several contaminating proteins. Since you are interested in large-scale purification, it is worthwhile to consider some sophisticated strategies. Which would you consider?

9. You have isolated a potent toxin from a root found in a remote area of Arizona. In an initial SDS gel electrophoresis run, the protein runs midway between the marker proteins myoglobin ($M_r = 16,900$) and β-lactoglobulin ($M_r = 37,000$). However, material treated first with mercaptoethanol and iodoacetate gives a single broad band on SDS gel electrophoresis that runs close to a cytochrome c marker ($M_r = 13,370$). Further experiments show that reaction of the native material with FDNB (see Chapter 1) followed by acid hydrolysis liberates DNP-glycine and DNP-tyrosine. What conclusions can you derive about the structure of this protein?

10. In carrying out some experiments on DNA replication, you have isolated a protein complex that sediments in the ultracentrifuge similarly to a hemoglobin marker. However, when the same complex is run in the presence of 2 M NaCl, other conditions being equivalent, the sedimentation behavior resembles that of a myoglobin marker. What conclusions do you reach concerning the properties of this complex?

4

CARBOHYDRATES AND THEIR DERIVATIVES

chemical energy → glucose
fructose
Storage energy → Starch
glycogen.

Carbohydrates make up the bulk of organic substances on the earth. They are an important source of carbon for the synthesis of other compounds and of chemical energy for immediate release, as with glucose or fructose, or for storage in the form of polysaccharides such as starch or glycogen. Many carbohydrates, such as cellulose and chitin, serve a structural function, providing support and protection. Structural polysaccharides are frequently found in combination with proteins (glycoproteins and proteoglycans) or lipids (glycolipids and lipopolysaccharides). They are also found in combination with peptides in peptidoglycans. For classification purposes, carbohydrates may be divided into monosaccharides, oligosaccharides* (i.e., disaccharides and trisaccharides, etc.), and high-molecular-weight polysaccharides. Small carbohydrates, including monosaccharides and low-molecular-weight oligosaccharides, are commonly referred to as sugars. In this chapter we discuss some of the basic structural and chemical characteristics of naturally occurring carbohydrates.

MONOSACCHARIDES AND RELATED COMPOUNDS

Monosaccharides are either polyhydroxy aldehydes (aldoses) or polyhydroxyketones (ketoses). They may be thought of as being derived from polyalcohols (polyols) by oxidation of one carbinol group to a carbonyl group. For example, the simple three-carbon triol glycerol can be converted either to the aldotriose, glyceraldehyde, or the ketotriose, dihydroxyacetone, by the loss of two hydrogens (Figure 4–1).

Since the middle carbon of glyceraldehyde is connected to four different substituents, it is a *chiral* center. This single chiral center leads to two possible forms of glyceraldehyde, known as the D and L forms. D-Glyceraldehyde

---

* Oligo is the Greek prefix for "few."

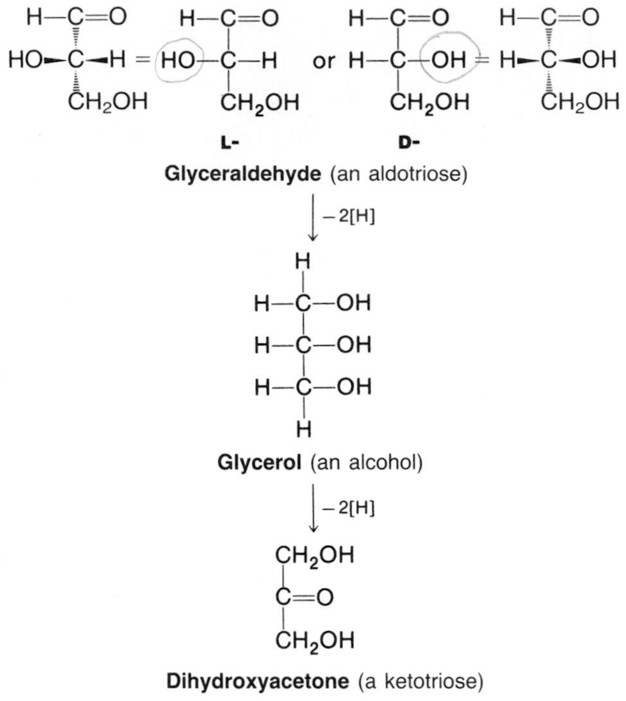

$$\begin{array}{cccc} \text{H–C}\!=\!\text{O} & \text{H–C}\!=\!\text{O} & \text{H–C}\!=\!\text{O} & \text{H–C}\!=\!\text{O} \\ \text{HO–C–H} = \text{HO–C–H} & or & \text{H–C–OH} = \text{H–C–OH} \\ \text{CH}_2\text{OH} & \text{CH}_2\text{OH} & \text{CH}_2\text{OH} & \text{CH}_2\text{OH} \end{array}$$

L-   **D-**

Glyceraldehyde (an aldotriose)

$$\downarrow -2[\text{H}]$$

$$\begin{array}{c} \text{H} \\ \text{H–C–OH} \\ \text{H–C–OH} \\ \text{H–C–OH} \\ \text{H} \end{array}$$

Glycerol (an alcohol)

$$\downarrow -2[\text{H}]$$

$$\begin{array}{c} \text{CH}_2\text{OH} \\ \text{C}\!=\!\text{O} \\ \text{CH}_2\text{OH} \end{array}$$

Dihydroxyacetone (a ketotriose)

Figure 4–1
Loss of two hydrogens by glycerol leads to the formation of glyceraldehyde or dihydroxyacetone, depending on whether the two hydrogens are lost from the end or middle position, respectively.

is illustrated in Figure 4–1 in the so-called *Fischer projection* formula, in which the —OH group attached to the central carbon atom points to the right. If the central carbon were in the plane of the paper with tetrahedrally arranged substituents, the H and OH connected to it would project above the plane of the paper and the other two substituents would project below the plane of the paper. For larger sugars, Fischer projections are written with the most highly oxidized carbon, C-1, at the top.

A molecule such as glyceraldehyde, having one center of asymmetry (chiral center), is optically active, and the two forms of the molecule can be described as the *dextrorotatory* and *levorotatory* forms, according to the way in which they rotate plane-polarized light. The symbol *d* or (+) refers to dextrorotatory rotation, and *l* or (−) refers to levorotatory rotation.

Rotation is testable by making a solution of an optically active compound and measuring the rotation of plane-polarized light through it (see Box 3–B). A mixture containing equal amounts of the two forms is optically inactive and is referred to as a *racemic* mixture. The specific rotation is defined as $[\alpha] = (100 \times A)/(c \times l)$, where A is the observed rotation in degrees, c is the concentration of the optically active substance in grams per 100 ml of solution, and l is the path length in decimeters of the solution through which the rotation is observed.

The stereochemistry of sugars is based on configurational rather than optical properties. The actual sign of rotation may still be indicated by the italic letters *d* and *l*, but the absolute configuration of the four different substituents around the asymmetric carbon atom is designated by the prefix symbols D and L. For the common sugars, the prefixes D and L refer to that center of asymmetry most remote from the aldehyde or ketone end of the molecule. By convention, all optically active centers are related to the asymmetric carbon of glyceraldehyde, the arbitrarily chosen standard of reference for stereoisomers. Isomers stereochemically related to D-glyceraldehyde are designated D, and those related to L-glyceraldehyde are designated L. We may visualize the C_4, C_5, and C_6 sugars as arising from the trioses through the stepwise condensation of formaldehyde to either glyceraldehyde or dihydroxyacetone. The actual synthesis of the sugars occurs by other means.

Families of Monosaccharides Are Structurally Related

A phosphorylated derivative of the D-glyceraldehyde is an intermediate in the degradation of carbohydrates. This aldose is reversibly converted to its ketose isomer, dihydroxyacetone (Figure 4–1), by an enzyme. The tetroses, pentoses, and hexoses related to D-glyceraldehyde are shown in Figure 4–2. A similar series exists for L-glyceraldehyde. The ketoses (e.g., fructose) are similarly related to dihydroxyacetone (Figure 4–3). Some of the better-known sugars are indicated by asterisks in the figures; we shall be referring to a number of these sugars.

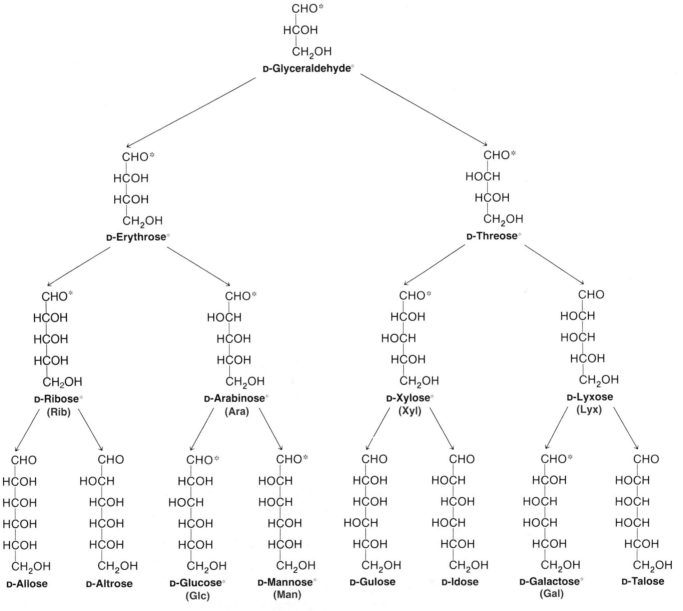

Figure 4–2
Configurational relationships of the D-aldoses. The most important sugars are starred. Note that in the two D-series the configuration about the chiral center farthest from the carbonyl is the same.

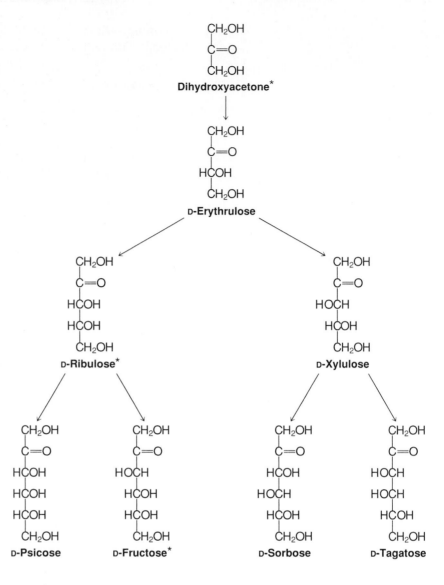

Figure 4–3
Configurational relationships of the
D-ketoses. The most important sugars
are starred.

Monosaccharides Can Form Intramolecular Hemiacetals

Aldehydes can add hydroxyl compounds to the carbonyl group. If a molecule of water is added, the product is an aldehyde hydrate, as shown in Figure 4–4. If a molecule of alcohol is added, the product is a _hemiacetal_. Addition of a second alcohol produces an _acetal_. Sugars often form _intramolecular hemiacetals_; they do so readily in water in cases where the resulting compound has a five- or six-membered ring.

The fact that sugars can exist in more than one configurational state as a result of hemiacetal formation first became apparent from optical studies of D-glucose. The optical rotation caused by a freshly dissolved sample of D-glucose is observed to change over time. This unusual phenomenon is due to the existence of two different forms that are convertible in solution; they are referred to as α-D-glucose, where $[\alpha] = 112°$, and β-D-glucose, where $[\alpha] = 19°$. A freshly prepared solution of either compound will eventually approach an intermediate value. The exact value depends on the equilibrium between the two forms. Formulas for the two hemiacetals are indicated in Figure 4–5 (top).

The convention for numbering hexoses is shown in the central structure of Figure 4–5. The α designation for the D series indicates that the aldehyde

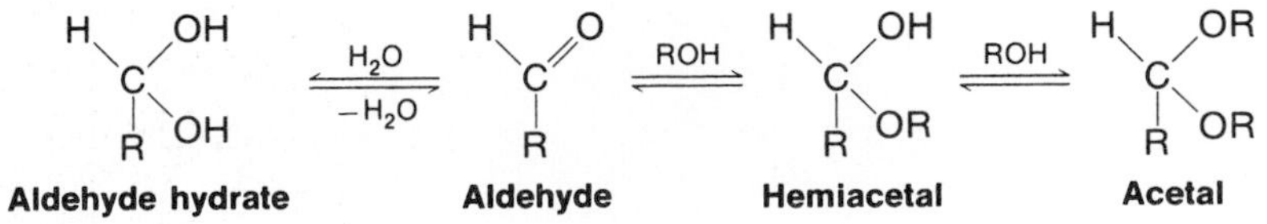

Figure 4–4
Aldehydes can add H_2O to form hydrates or can add alcohols to form hemiacetals and acetals.

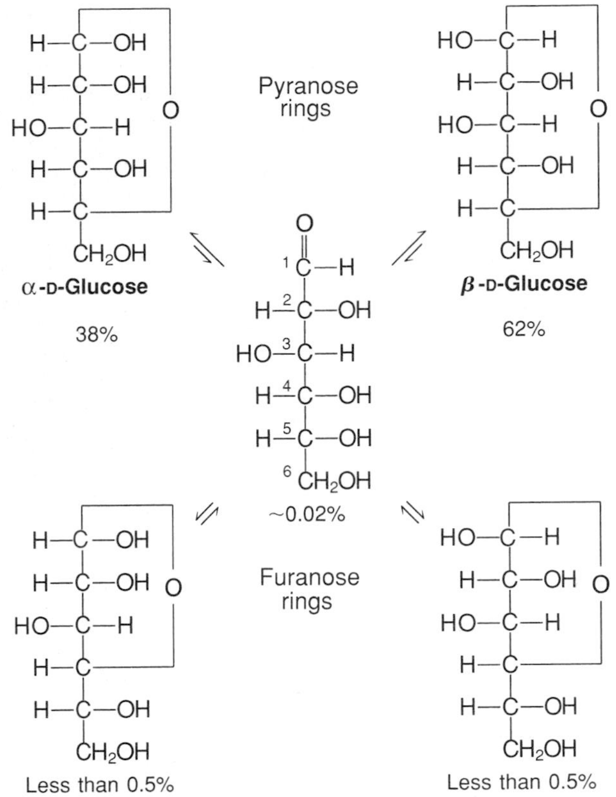

Aldehyde hydrate   Aldehyde   Hemiacetal   Acetal

Figure 4–5
Different forms of glucose that result from dissolving glucose in water. At 25°C in water, glucose reaches an equilibrium containing about 0.02% free aldehyde, 38% α-pyranose form ($[\alpha]_D = 120]$), 62% β-pyranose form ($[\alpha_D] = 19$), and less than 0.5% of the furanose forms. The anomeric carbon is shown in color.

or C-1 hydroxyl group is on the same side of the structure as the ring oxygen in the Fischer projection, and the β designation indicates the reverse situation. For the L series the opposite is the case. When the sugar is dissolved in water, the hemiacetal is in equilibrium with the straight-chain hydrated form. The straight-chain form can produce either hemiacetal, α or β. Conversion of one stereoisomer to another in solution is referred to as _mutarotation_. The two different forms are referred to as _anomers_, and the _anomeric carbon_ is the carbon that contains the reactive carbonyl. In the case of glucose this is the C-1 carbon. An equilibrium situation is reached without added catalyst in a few hours at room temperature. The open-chain form usually represents only a small fraction of the total (see Figure 4–5 for actual percentages).

Hemiacetals with five-membered rings are called _furanoses_, and hemiacetals with six-membered rings are called _pyranoses_. In cases where either five- or six-membered rings are possible, the six-membered ring usually predominates. For example, for glucose less than 0.5 per cent of the furanose forms exist at equilibrium (see Figure 4–5, bottom). The reasons for the general preponderance of the pyranose is not known. Both furanoses and pyranoses are more realistically represented by pentagons or hexagons in the _Haworth_ convention, as shown in Figure 4–6. The lower edge of the ring (heavy line) is seen as projecting toward the viewer and the opposite edge as projecting away, behind the plane of the paper.

Haworth structures are unambiguous in depicting configurations, but even they do not show correctly the spatial relationship of groups attached

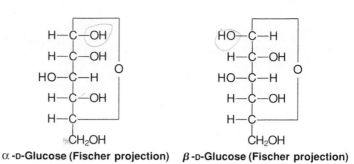

α-D-Glucose (Fischer projection)   β-D-Glucose (Fischer projection)

Figure 4–6
Comparison of the Fischer and Haworth projections for α- and β-D-glucose. The Haworth projection is closer to reality but it is more cumbersome to draw, so it is not always used.

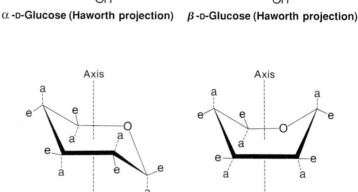

α-D-Glucose (Haworth projection)   β-D-Glucose (Haworth projection)

a = axial bond
e = equatorial bond

Figure 4–7
Chair and boat forms for a generalized pyranose ring structure. Structures of this type are more realistic than the Haworth structure, as the carbon–carbon bond angle is correct. The chair form is usually favored over the boat form. The substituents are labeled "a" (axial) and "e" (equatorial). The axis of symmetry is labeled for both forms. Axial bonds are parallel to this axis of symmetry. Equatorial bonds are parallel to the nonadjacent sides of the rings. A large substituent generally prefers to be in an equatorial location.

to rings. The normal valence angle of saturated carbon (109°) presents a stable planar arrangement for cyclohexane or the related pyranose molecule. The two most likely conformations* are the so-called *chair* and *boat* forms (Figure 4–7). Usually the chair form is considerably more stable than the boat form. The twelve substituent atomic groups of the ring carbons fall into two classes, those that are approximately perpendicular to the plane of the ring, i.e., *axial*, and those that are parallel to the plane of the ring, i.e., *equatorial*. As a rule, a substituent is at a lower energy state in the equatorial position because there is less chance of steric hindrance with other substituents. This fact becomes more important with larger substituents. The two

---

*When discussing sugars we make frequent use of the terms *configuration* and *conformation*. These terms have different meanings. Two conformations of the same molecule are interconvertible without breakage of any chemical bonds; two configurations of the same molecule are not. For example, the chair and boat forms of α-D-glucose represent two different conformations of the same molecule. But α-D-glucose and β-D-glucose represent two different configurations of D-glucose.

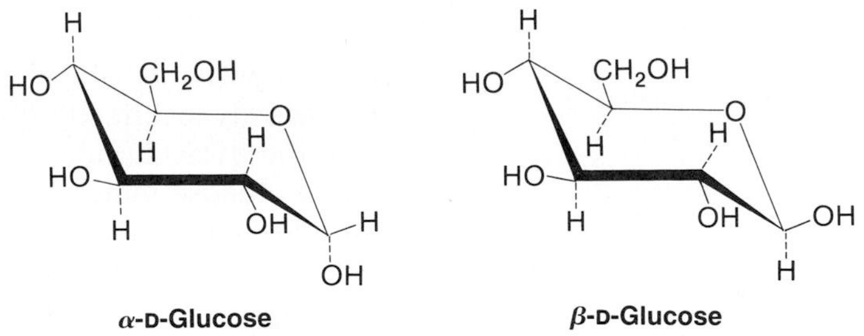

Figure 4–8
Chair configurations for the two anomers of D-glucose. Note that the largest substituent, —CH$_2$OH, is in an equatorial location in both structures. The differences between the two anomers are shown in color.

anomers for the favored chair-form conformation of D-glucose are shown in Figure 4–8. In sugar chemistry, formulating the conformations of aldohexoses is important in interpreting reactivity of hydroxyl groups and other sizable substituents. The most stable conformation for a particular aldohexose is the chair form, which places the maximum number of substituents larger than hydrogen in the equatorial position.

The furanose ring is nonplanar also and can exist in more than one conformation. The conformations for D-ribose (β-D-ribofuranose) and D-2-deoxyribose (β-D-2-deoxyribofuranose), the two pentoses found in all nucleic acids, will be discussed in Chapter 7.

MONOSACCHARIDES AND THE GLYCOSIDIC BOND

Warming glucose in methanol and acid produces a mixture of two new substances, α- and β-methylglucosides (Figure 4–9). Whereas these derivatives of glucose are referred to as _glucosides_, the comparable derivatives of galactose are referred to as galactosides, and so on. Generally the bond between a sugar and an alcohol is referred to as a _glycosidic bond_, and the compound by the generic name _glycoside_. Thus a glucoside is also a glycoside. But glycosides also includes glucosides, galactosides and many other chemically similar sugar derivatives. The formation of a glycoside from a sugar and methanol by acid catalysis is identical to the formation of an acetal from an aldehyde and an alcohol (see Figure 4–2). While the two forms of glucose in solution are in equilibrium through mutarotation, the corresponding glucosides are locked into one configuration. This is understandable, because mutarotation requires that, in the intermediate, the anomeric carbon adopt a carbonyl structure, which is not possible in a glycoside. A glycoside can be formed with aliphatic alcohols, phenols, and hydroxy carboxylic acids, as well as with another sugar.

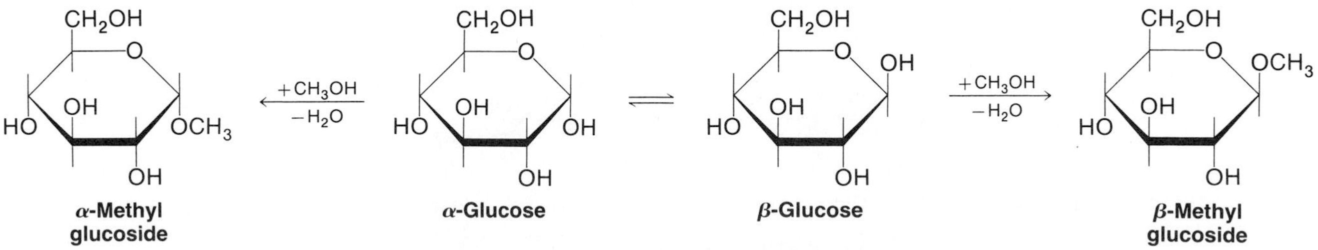

Figure 4–9
Formation of methyl glucosides. Glucosides (or glycosides) are quite stable in alkali but they hydrolyze readily in dilute acid.

DISACCHARIDES

The most important glycosides are those formed with other sugars. Monosaccharides are glycosidically linked to form disaccharides (Figure 4–10). For instance, the disaccharide maltose contains a glycosidic bond between the C-1 of one glucose molecule and the C-4 of another glucose molecule. The compound is said to have an $\alpha(1,4)$ glycosidic linkage because the anomeric C-1 carbon of one sugar is connected to the C-4 of another sugar and the configuration about the anomeric carbon is α. Maltose possesses one potentially free aldehyde group. Its presence is determined by the chemical means described in Box 4–A. The basic assay for an aldehydic function is the reduction of another chemical grouping by the aldehyde. For this reason, sugars possessing aldehydic functions are referred to as reducing sugars. The configuration about the hemiacetal hydroxyl group has not been specified in Figure 4–10 because it can undergo mutarotation. Maltose arises as a degradation product of starch.

The disaccharide cellobiose is identical with maltose except for having a $\beta(1,4)$ glycosidic linkage. Cellobiose is a degradation product of cellulose.

Lactose is a disaccharide found exclusively in the milk of mammals. Lactose contains a $\beta(1,4)$ glycosidic linkage between galactose and glucose.

Sucrose is found in abundance in sugar beets and sugar cane. On acid hydrolysis it yields equivalent amounts of D-glucose and D-fructose. Sucrose is a nonreducing sugar. Since the reducing groups are found on the C-1 and C-2 carbon atoms, respectively, of glucose and fructose, the glycosidic bond of the disaccharide must be between these two carbon atoms.

POLYSACCHARIDES

Most carbohydrates in nature exist as high-molecular-weight polymers called polysaccharides. Polysaccharides are composed of simple sugars connected by glycosidic bonds. The most common building block used in polysaccharides is D-glucose. Other sugars that are used include D-mannose, D- and L-galactose, D-xylose, L-arabinose, D-glucuronic acid, D-galacturonic acid, D-mannuronic acid, D-glucosamine, D-galactosamine, and neuraminic acid. Some of these sugar building blocks used in polysaccharides are illustrated in Figure 4–2; others are illustrated in Figure 4–11. Polymers com-

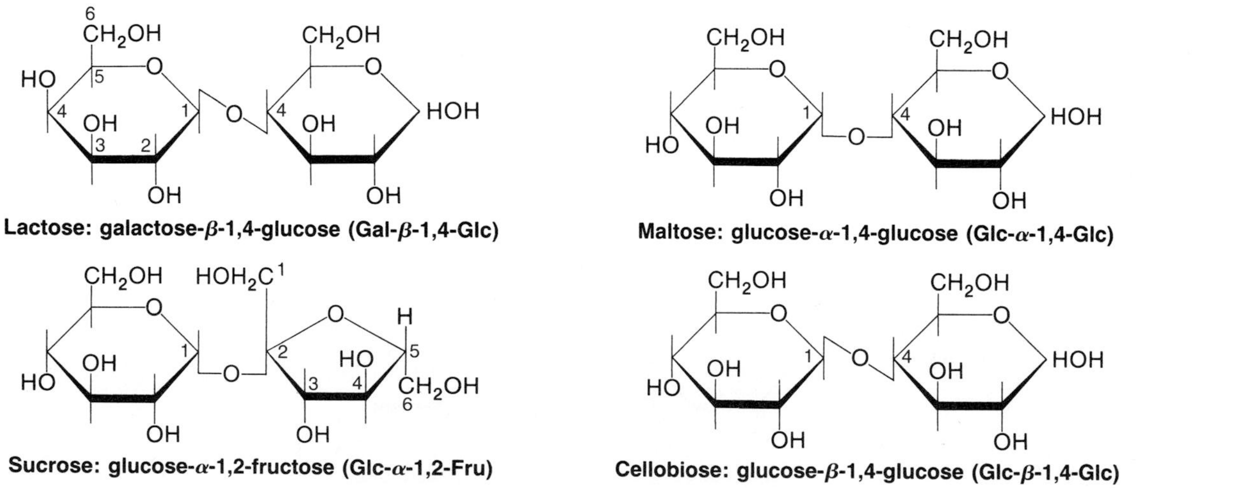

Lactose: galactose-β-1,4-glucose (Gal-β-1,4-Glc)

Maltose: glucose-α-1,4-glucose (Glc-α-1,4-Glc)

Sucrose: glucose-α-1,2-fructose (Glc-α-1,2-Fru)

Cellobiose: glucose-β-1,4-glucose (Glc-β-1,4-Glc)

Figure 4–10
Four commonly occurring disaccharides. The configuration about the hemiacetal group has not been specified for lactose, maltose, or cellobiose because both anomers exist in equilibrium.

Detection of Sugars with Aldehyde Groups

The usual test for an aldose is the same as for any organic moiety containing an aldehydic group. The test is based on the unique reducing capacity of the aldehyde among organic functional groups. Hence a sugar containing an aldehydic group is called a reducing sugar.

Several tests are available; two of them are described here:

1. Reduction of the silver ammonium complex, $Ag(NH_3)_2^+$, known as Tollen's reagent. The reaction is

$$RCHO + 2\ Ag(NH_3)_2^+ + 2\ OH^- \longrightarrow$$
$$RCO_2^- + 2\ Ag + 3\ NH_3 + NH_4^+$$

Deposition of silver metal is a positive qualitative test.

2. Fehling's solution. Cupric sulfate and sodium carbonate with a tartrate buffer are used. The reaction is

$$RCHO + 2\ Cu^{2+} + 5\ OH^- \longrightarrow$$
$$RCO_2^- + Cu_2O + 3\ H_2O$$

Deposition of brick-red Cu_2O precipitate is a positive qualitative test.

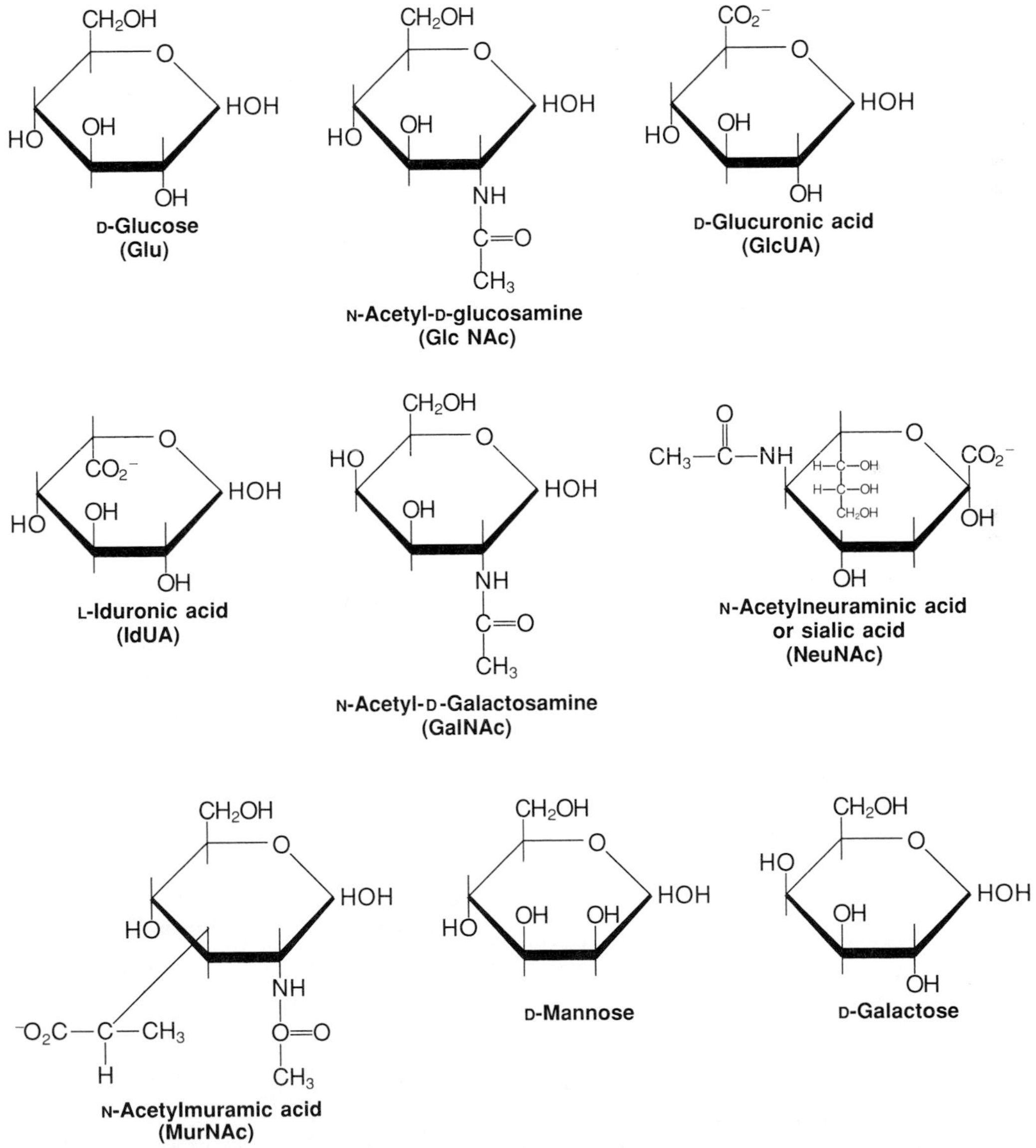

Figure 4–11
Some of the sugar building blocks found in polysaccharides.

posed of one type of building block are called *homopolymers*; those composed of more than one are called *heteropolymers*.

Polysaccharides function in two quite distinct roles: some serve as a means for storage of chemical energy and others serve a structural function.

Cellulose Is a Homopolymer of Glucose

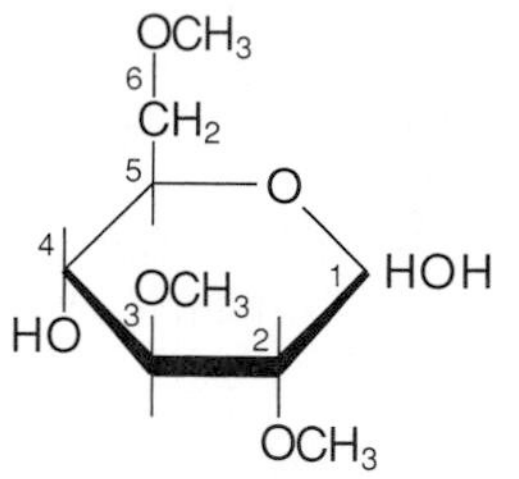

Cellulose is a structural polysaccharide found as the major component of cell walls in plants. It is the most abundant of organic compounds, constituting approximately 50 per cent of all the carbon found in plants. Upon acid hydrolysis, cellulose yields the monomer glucose and some dimer cellobiose, the latter due to incomplete hydrolysis. The reaction of polysaccharides with dimethylsulfate is often useful in structural analysis. Treatment of a saccharide with dimethylsulfate in alkali results in the conversion of all free hydroxyl groups to O-methyl ethers. Fully methylated cellulose gives 2,3,6-tri-O-methylglucose on acid hydrolysis (Figure 4–12). The absence of a methyl on the anomeric C-1 says nothing about the structure of cellulose, since such methyl derivatives are susceptible to mild acid hydrolysis. However, the absence of a methyl group at the C-4 position indicates that this position is inaccessible in the cellulose structure. This suggests that in cellulose the glycosidic linkages are of the 1,4 type. Other measurements show that the 1,4 linkages are about the anomeric carbon. The repeating unit of cellulose is indicated in Figure 4–13.

Cellulose is insoluble in water because of the high affinity of the polymer chains for one another. Individual polymeric chains have a molecular weight of 50,000 or greater. The molecular chains of cellulose interact in parallel bundles of about 2000 chains, a bundle having a diameter of 100–250 Å. Each bundle of 2000 comprises a single microfibril. Many microfibrils arranged in parallel comprise a macrofibril, which can be seen under the light microscope. Figure 4–14 shows the inner secondary walls of the plant *Valonia*; the fibrils in the secondary wall are almost pure cellulose.

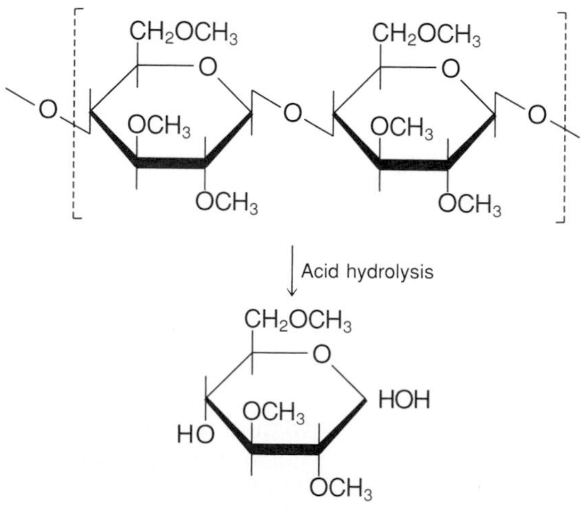

Figure 4–12
Structure of 2,3,6-tri-O-methylglucose. This is the main product resulting from exhaustive methylation of cellulose followed by acid hydrolysis.

Figure 4–13
Structure of the repeating unit of cellulose (top) and its analysis by reaction with dimethylsulfate. As can be seen, the residues in the native structure are connected by $\beta(1,4)$ linkages.

Figure 4–14
Fibril arrangements in the cell wall of *Valonia* (12,000×). (Electron micrograph from A. Frey-Wyssling and K. Mühlethaler, *Ultrastructural Plant Cytology*, Elsevier, Amsterdam, 1965, p. 298. Reprinted with permission.)

Starch and Glycogen Are Homopolymers of Glucose

Although glucose is the most important sugar involved in energy metabolism in most cells and tissues, it is not present in the cell to any large extent as the free monosaccharide. Cells store glucose for future use in the form of simple homopolymers, and thereby reduce the osmotic pressure of the stored sugar. A polysaccharide consisting of 1000 glucose units exerts an osmotic pressure that is only 1/1000 of the pressure that would result if the glucose units were all present as separate molecules. In polymeric form, glucose can be stored compactly until needed.

The two major polysaccharides used for energy storage are starch in plant cells and glycogen in animal cells. Both are $\alpha(1,4)$ homopolymers with occasional $\alpha(1,6)$ linkages at branchpoints (Figure 4–15). The two polysac-

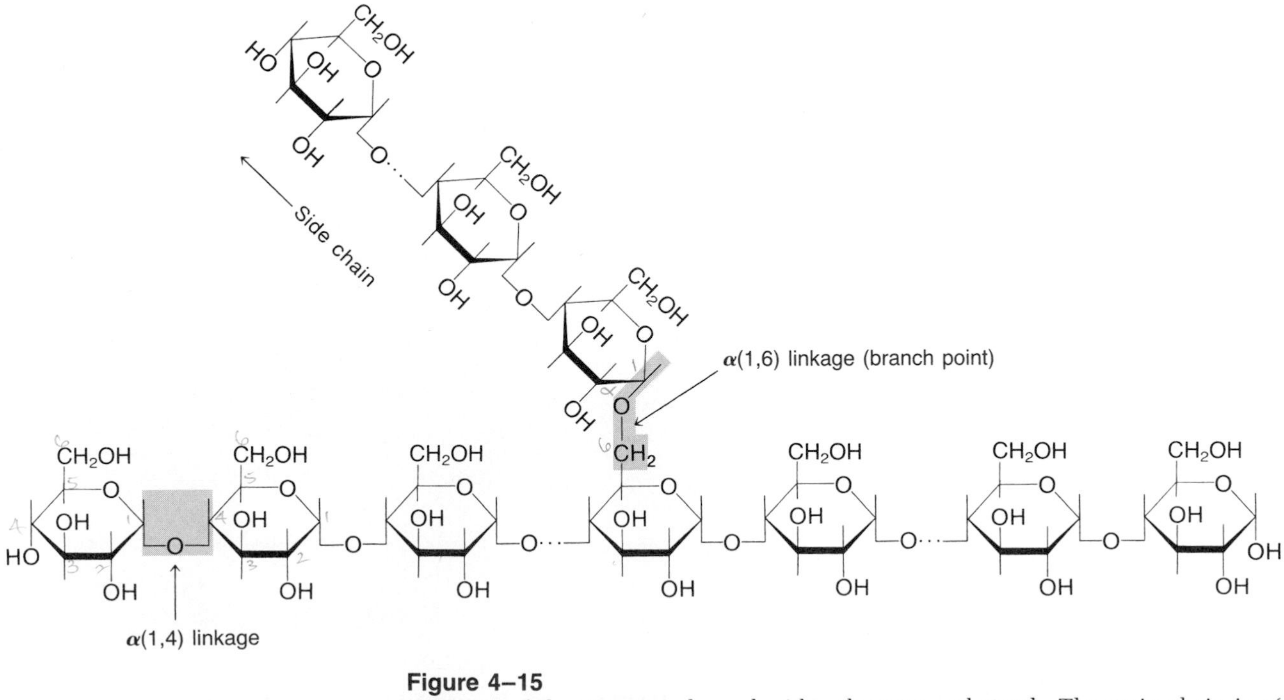

Side chain

$\alpha(1,6)$ linkage (branch point)

$\alpha(1,4)$ linkage

Figure 4–15
Structure of the storage polysaccharides glycogen and starch. The main chain is $\alpha(1,4)$ linked. Side chains are connected to the main chain by $\alpha(1,6)$ linkages.

charides, starch and glycogen, differ primarily in their chain lengths and branching patterns. Glycogen is highly branched, with an $\alpha(1,6)$ linkage occurring every 8 to 10 glucose units along the backbone, giving rise in each case to short side chains of about 8 to 12 glucose units each. Starch occurs both as unbranched *amylose* and as branched *amylopectin*. Like glycogen, amylopectin has $\alpha(1,6)$ branches, but these occur less frequently along the molecule (once every 12 to 25 glucose residues) and give rise to longer side chains (lengths of 20 to 25 glucose units are common). Starch deposits are usually about 10–30 per cent amylose and 70–90 per cent amylopectin.

Configurations of Glycogen and Cellulose Dictate Their Roles

It is a remarkable fact that the main energy-storage polysaccharides and the main structural polysaccharides found in nature both have a primary structure of 1,4-linked polyglucose. Why should two such closely related compounds be used in totally different roles? A closer look at the stereochemistry of the α and β glycosidic linkage for polyglucose suggests why this is so.

Recall that D-glucose exists in the chair form of a pyranose ring (see Figure 4–8). The ring has a rigid character to it. We can think of it as a structural building block in a polysaccharide chain, just as we think of the rigid planar peptide grouping as a structural building block in a polypeptide chain. It is also possible to specify two torsional angles ϕ and ψ for rotation about the glycosidic C—O linkage (see Figure 4–16). These angles are used

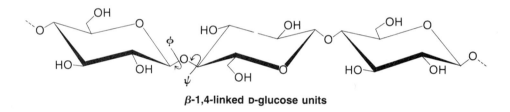

β-1,4-linked D-glucose units

Figure 4–16
Energetically favored conformations of $\beta(1,4)$-linked D-glucose (a) and $\alpha(1,4)$-linked D-glucose (b). Note that in the $\beta(1,4)$ configuration in (a), alternating residues are flipped 180° relative to one another so that long straight chains result. In the $\alpha(1,4)$ configuration (b), the chain has a natural curvature.

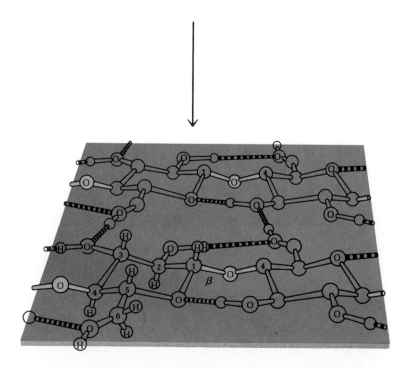

(a)   **Conformations favored by $\beta(1,4)$-linked polyglucose** extensively to discuss polypeptide configuration in proteins, about which a great deal of information is available (see Chapter 2). They have limited value in discussing polysaccharide structures, because much less information is available about such structures. Nevertheless, it is clear that _only the β(1,4)-linked polyglucose has the capacity to form straight chains_ (Figure 4–16a). A straight chain can be created by flipping each glucose unit by 180° relative to the previous one. This process should result in an almost fully extended polysaccharide chain, which is known to be characteristic of the structure of cellulose. Evidently this conformation is energetically favored.

Figure 4–16 (Continued)

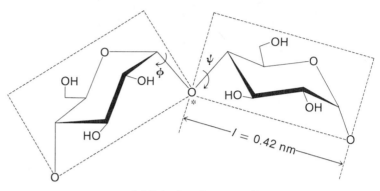

α-1,4-linked D-glucose units

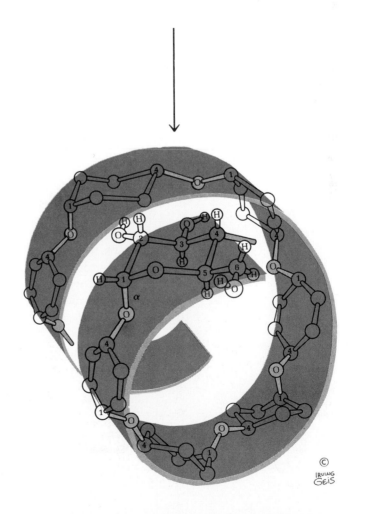

(b)   **Conformations favored by** α(1,4)-**linked polyglucose**

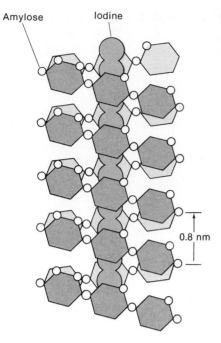

Amylose   Iodine

0.8 nm

Figure 4–17
Structure of the helical complex of amylose with iodine (I_2). The amylose forms a left-handed helix with six glucosyl residues per turn and a pitch of 0.8 nm. The iodine molecules (I_2) fit inside the helix parallel to its long axis.

By contrast, *α(1,4)-linked units in a polyglucose cause a natural turning of the chain* (Figure 4–16b). Consistent with this fact is the observation that amylose adopts a coiled helical configuration. Indeed, one of the first helical structures to be discovered (1943) was the left-handed helix of amylose wound around molecules of iodine (Figure 4–17). This structure is responsible for the characteristic blue color on the amylose-iodine complex.

The extended-chain form of polyglucose has been exploited in nature for structural purposes, leaving by default the coiled form as an energy-storage macromolecule. Correlated with this functional difference is the omnipresence of degrading enzymes for glycogen and starch and the very limited phylogenetic distribution of comparable enzymes for cellulose. Cellulose is degraded in the gastrointestinal tract of herbivores, such as the cow, or in insects, such as termites, by a protozoan that synthesizes the enzyme cellulase. Humans do not possess this enzyme and hence cannot degrade cellulose.

Chitin Contains a Different Building Block

Many polysaccharides consist of different sugars than glucose or different combinations of sugars. In some cases the new sugar is merely a derivative of glucose. Chitin is an example of a structural polysaccharide that uses a modified derivative of glucose. Chitin is found in the shells of crustaceans and insects and in the cell walls of fungi; it is a linear β(1,4) polymer of N-acetyl-D-glucosamine (Figure 4–18). A major rigid component of bacterial cell walls, the peptidoglycan, could be regarded as a substituted chitin.

HETEROPOLYSACCHARIDES: POLYMERS WITH MORE THAN ONE BUILDING BLOCK

A large variety of carbohydrates contain either modified sugars, like the N-acetylglucosamine found in chitin, or two or more different sugars in straight-chain or branched-chain linkages. Many heteropolysaccharides are linked to peptides or proteins. Peptide-linked heteropolysaccharides are known as *peptidoglycans*, whereas protein-linked heteropolysaccharides are called *proteoglycans* or *glycoproteins*.

Glycoproteins are mostly protein; the carbohydrates in glycoproteins are short, frequently branched sugar chains of 15 residues or less. By contrast, in proteoglycans the protein is the minor component. Typical proteoglycans are much larger than glycoproteins, and they usually contain 90–95 per cent carbohydrate by weight, in the form of long unbranched chains. Carbohydrates are also found linked to lipids in compounds known as *glycolipids* or *lipopolysaccharides*.

Figure 4–18
Structure of the repeating unit of chitin: a β(1,4) homopolymer of N-acetyl-D-glucosamine.

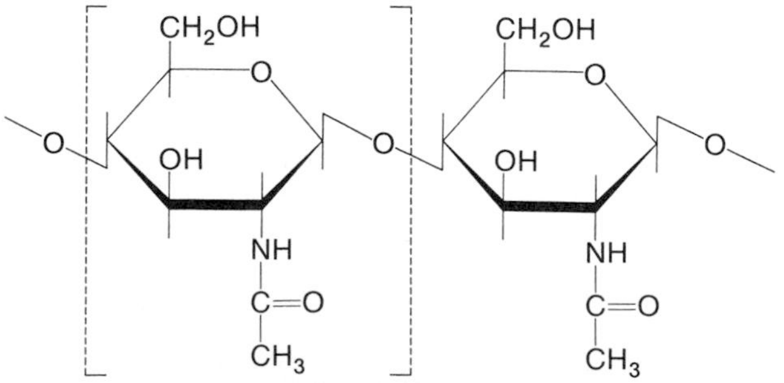

β-1,4-linked *N*-acetyl-D-glucosamine units

Glycosaminoglycans Contain Repeating Disaccharide Subunits

Structurally, the simplest and best known of the heteropolysaccharides are the _glycosaminoglycans_. These are long, unbranched polysaccharide chains composed of repeating disaccharide subunits in which one of the two sugars is either N-acetylglucosamine or N-acetylgalactosamine (Table 4–1). Glycosaminoglycans are highly negatively charged because of the presence of carboxyl or sulfate groups on many of the sugar residues. The high negative

TABLE 4–1
Structure of Glycosaminoglycans

Polysaccharide	Monosaccharide Units[a]		Substituents	Possible Repeating Unit
	A	B		
Hyaluronate	β-D-GlcUA (COO⁻, OH, OH)	β-D-GlcN (CH₂OH, HO, HNR)	$R = -C(=O)CH_3$	(COO⁻ ... CH₂OH, HNR)ₙ
Chondroitin sulfates Dermatan sulfate	β-D-GlcUA (COO⁻, OH, OH); α-L-IdUA (COO⁻, OH, OR′)	β-D-GalN (CH₂OR′, R′O, HNR)	$R = -C(=O)CH_3$; $R' = -H$ or $-SO_3^-$	(COO⁻, R′O ... CH₂OR′, HNR)ₙ
Heparan sulfate and Heparin	β-D-GlcUA (COO⁻, OH, OH); α-L-IdUA (COO⁻, OH, OR′)	α-D-GlcN (CH₂OR′, OR′, HNR)	$R = -C(=O)CH_3$ or $-SO_3^-$; $R' = -H$ or $-SO_3$	(COO⁻, OH ... CH₂OR′, OR′, HNR)ₙ
Keratan sulfate	β-D-Gal (HO, CH₂OH, OH)	β-D-GlcN (CH₂OR′, OH, HNR)	$R = -C(=O)CH_3$; $R' = -H$ or $-SO_3$	(CH₂OH, OH ... CH₂OR′, HNR)ₙ

[a]The polysaccharides are depicted as linear polymers of alternating A and B monosaccharide units.
Abbreviations: GlcUA, glucuronic acid; IdUA, iduronic acid; GlcN, galactosamine; Gal, galactose.

charge causes the polymeric chains to adopt a stretched or extended configuration. Their extended structure gives a high viscosity to the surrounding region, even in a dilute solution of the polysaccharide. Glycosaminoglycans are usually found in extracellular space in multicellular organisms, where, even at low concentrations, they produce a viscous extracellular matrix that resists compression. Such an environment can be beneficial to the organism in various ways: it can provide a passageway for cell migration, supply lubrication between joints, or help maintain certain structural shapes such as the ball of the eye.

Hyaluronic acid is a copolymer of D-glucuronic acid and _N_-acetyl-D-glucosamine (Figure 4–19). It is much larger than other glycosaminoglycans, reaching molecular weights in excess of 10^6.

Most glycosaminoglycans are linked to a core protein as lateral extensions, forming a proteoglycan with a highly extended, brushlike structure (Figure 4–20a). The linkage between the glycosaminoglycans and the core protein within the proteoglycan is mediated by a specific trisaccharide unit that is linked on one side to the repeating disaccharide unit of the glycosaminoglycan and on the other side to a serine hydroxyl group of the core

Figure 4–19

Structure of the repeating unit of hyaluronic acid: GlcUA β(1,3)-GlcNAc β(1,4).

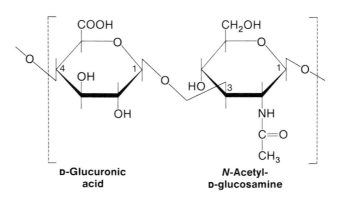

D-Glucuronic acid

_N_-Acetyl-D-glucosamine

Figure 4–20

(a) Structure of a typical proteoglycan. (b) The attachment site between a serine in the core protein and the glycosaminoglycan.

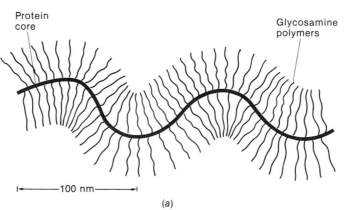

Protein core

Glycosamine polymers

—100 nm—

(a)

Protein

Serine

Trisaccharide linkage unit

Glycosamine glycan

(b)

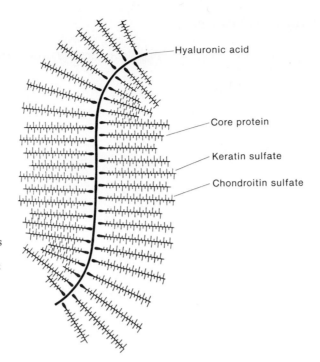

Figure 4–21

A hyaluronic acid–proteoglycan complex. The individual proteoglycans aggregate in some ill-defined way along the length of a single hyaluronic acid molecule. The individual proteoglycans contain a core protein to which are linked various glycosaminoglycans as shown in Figure 4–20. (Adapted from D. E. Metzler, *Biochemistry*, Academic Press, New York, 1977, p. 87.)

protein (Figure 4–20*b*). In the extracellular matrix hyaluronic acid and the other proteoglycans form aggregates with each other as well as with other macromolecular components, such as serum glycoproteins, collagen, elastin, or the outer plasma membrane itself. Hyaluronic acid, when complexed with other proteoglycans through their core proteins, produces very large complexes (Figure 4–21).

Thus far we have seen that glycosaminoglycans and their proteoglycans form highly extended aggregates that impart a rigid, gel-like structure to the extracellular matrix. It seems likely that many more specific reactions occur involving these compounds. However, it should be emphasized that relatively little is known about the organization or reactions of these molecules in the extracellular matrix. One exception is the case of the glycosaminoglycan *heparin*. We know that this compound functions as a highly specific anticoagulant. First it forms a complex with the plasma protein antithrombin III. This complex in turn inhibits the serine proteases of the blood clotting system.

In typical glycoproteins, the protein component predominates. Short straight-chain or branched-chain oligosaccharide groupings are frequently attached to the protein. The sugar chains are attached in *O*-glycosidic linkage to the hydroxyl groups of serine, threonine, or hydroxylysine or in *N*-glycosidic linkage to the amide nitrogen of asparagine (Figure 4–22). The

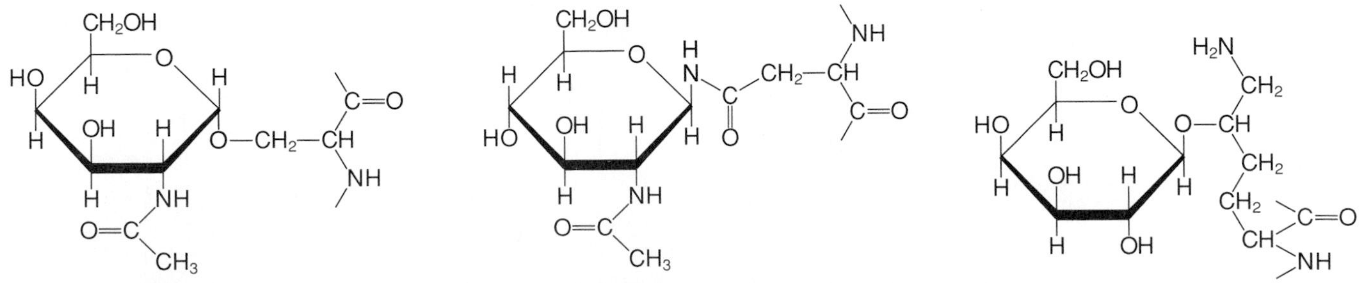

(a) The *N*-acetylgalactosamine-serine linkage

(b) The *N*-acetylglucosamine-asparagine linkage

(c) The galactose-hydroxylysine linkage

Figure 4–22

Three types of linkages found between oligosaccharides and proteins in glycoproteins.

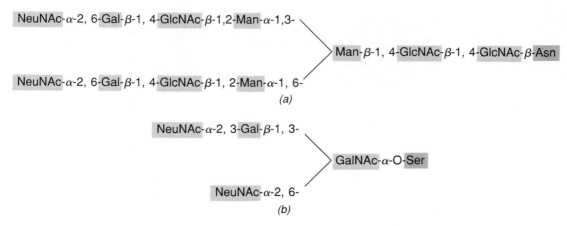

NeuNAc-α-2, 6-Gal-β-1, 4-GlcNAc-β-1,2-Man-α-1,3-

Man-β-1, 4-GlcNAc-β-1, 4-GlcNAc-β-Asn

NeuNAc-α-2, 6-Gal-β-1, 4-GlcNAc-β-1, 2-Man-α-1, 6-

(a)

NeuNAc-α-2, 3-Gal-β-1, 3-

GalNAc-α-O-Ser

NeuNAc-α-2, 6-

(b)

Figure 4–23
Representative oligosaccharide structures found in glycoproteins. *(a)* An N-glycosidically linked oligosaccharide found in human transferrin. *(b)* An O-glycosidically linked oligosaccharide found in human glycophorin, an erythrocyte membrane protein. The molecular structure of the linkage formed between the protein and the oligosaccharides in *(a)* and *(b)* here are illustrated in figures 4–22b and 4–22a, respectively.

glycoproteins involved include enzymes, hormones, and structural proteins; they are especially common among proteins attached to cell surfaces and in secreted proteins. In most cases the function of the carbohydrate moiety is completely unclear.* Typical oligosaccharides found in glycoproteins are complex and highly branched (Figure 4–23).

Glycosaminoglycans Form Proteoglycans

Most cells contain oligo- or polysaccharides on their surfaces in the forms of glycoproteins or glycolipids (Figure 4–24). The carbohydrate in the membranes is specifically confined to the noncytoplasmic surface. The principal sugars found in eukaryotic membrane glycoproteins and glycolipids are galactose, mannose, fucose, galactosamine, glucosamine, glucose, and sialic acid (see Figures 4–3 and 4–11). Sialic acid residues are frequently found at the ends of the carbohydrate side chains, where they confer a net negative charge to the membrane surface. The carbohydrates in membranes may have several possible functions. For example, they may help to orient and anchor a protein or a lipid to its proper location on the membrane. It also seems

*In the case of some glycoproteins the carbohydrate targets the protein so that it becomes absorbed by a lysosome or the plasma membrane.

Figure 4–24
Diagram of a eukaryotic cell membrane. Oligosaccharides are found on membrane-bound proteins, absorbed glycoproteins and glycolipids. The carbohydrate is always on the outside of the membrane; it never faces the cytoplasm.

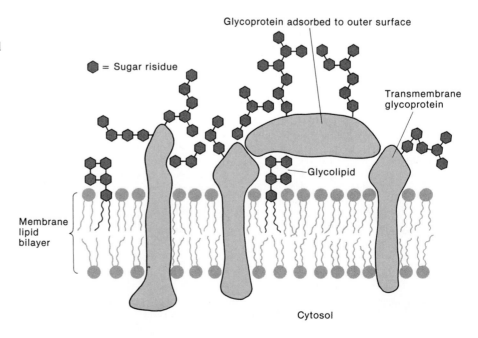

highly likely, because of their conspicuous location, that they play a major role in the specific reactions that occur on cell membrane surfaces, whether these involve cell-to-cell interactions or cell-to-protein interactions. In this regard it should be noted that the carbohydrate moiety of a membrane-bound protein is always located on the noncytoplasmic side of the membrane.

The Bacterial Cell Wall: A Peptidoglycan

All cells are surrounded by a semipermeable membrane composed of lipid and protein. Bacterial cells all have a cell wall that is a loosely cross-linked network composed of a linear heteropolysaccharide cross-linked by peptide. The cell wall gives mechanical strength and is the main reason that bacterial cells are difficult to disrupt when subject to moderate degrees of shear or when subject to osmotic shock. Some bacterial cells (Gram negative cells) also possess an outer membrane composed of lipids, proteins, and polysaccharides. The main structural features of the bacterial cell envelope, a composite of the two membranes and the cell wall, is illustrated in Figure 4–25.

Of the three structures that make up the cell envelope, the most is known about the cell wall. Here we discuss some of the structural aspects of cell walls, leaving a description of the biosynthesis to Chapter 20.

The peptidoglycan that constitutes the cell wall is a polymeric structure of a heteropolysaccharide composed of amino sugars in one dimension and

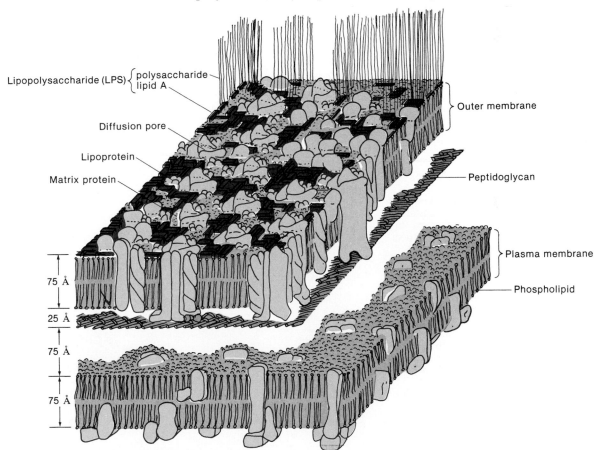

Figure 4–25
Diagram of a Gram negative cell envelope. The trimers of matrix protein of the outer membrane are associated with lipoprotein and with lipopolysaccharide (of variable polysaccharide length), and lipoprotein is covalently bound to peptidoglycan. Diagram also illustrates some general properties of membranes. Phospholipid molecules are illustrated with a circle for the polar groups and a line for each fatty acid acyl moiety. (Courtesy M. Inouye.)

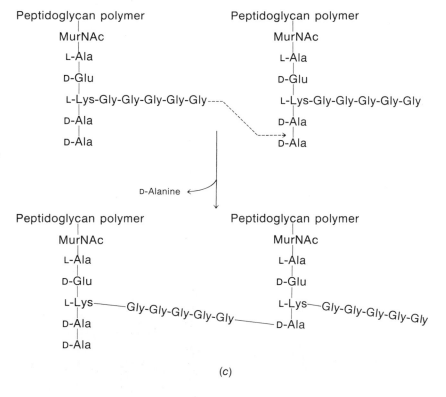

Figure 4–26
Structure of the peptidoglycan of the cell wall of *Staphylococcus aureus*. (a) In this representation, X (acetylglucosamine) and Y (acetylmuramic acid) are the two sugars in the peptidoglycan. Open circles represent the four amino acids of the tetrapeptide L-alanyl-D-glutamyl-L-lysyl-D-alanine. Solid circles are pentaglycine bridges that interconnect peptidoglycan strands. The nascent peptidoglycan units bearing open pentaglycine chains are shown at the left of each strand. TA—P is the teichoic acid antigen of the organism, which is attached to the polysaccharide through a phosphodiester linkage. Teichoic acids are discussed in Chapter 6. (b) The structure of X (N-acetylglucosamine) and Y (N-acetylmuramic acid), which are linked by β(1,4) linkages and alternate in the glycan strand. (c) The structure of a segment of the peptidoglycan before and after the final cross-linking reaction.

cross-linked through branched polypeptides in the other (Figure 4–26a). The amino sugars are N-acetylglucosamine and N-acetylmuramic acid (which is the 3-O-D-lactic acid ether of N-acetylglucosamine). These amino sugars strictly alternate in the polymer, forming the glycan strands (Figure 4–26b). The carboxyl group of the lactic acid moiety of the acetylmuramic acid is substituted by a tetrapeptide, which in the Gram positive bacterium

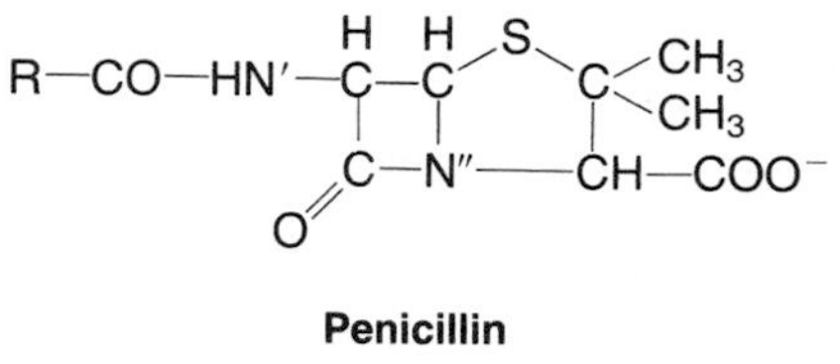

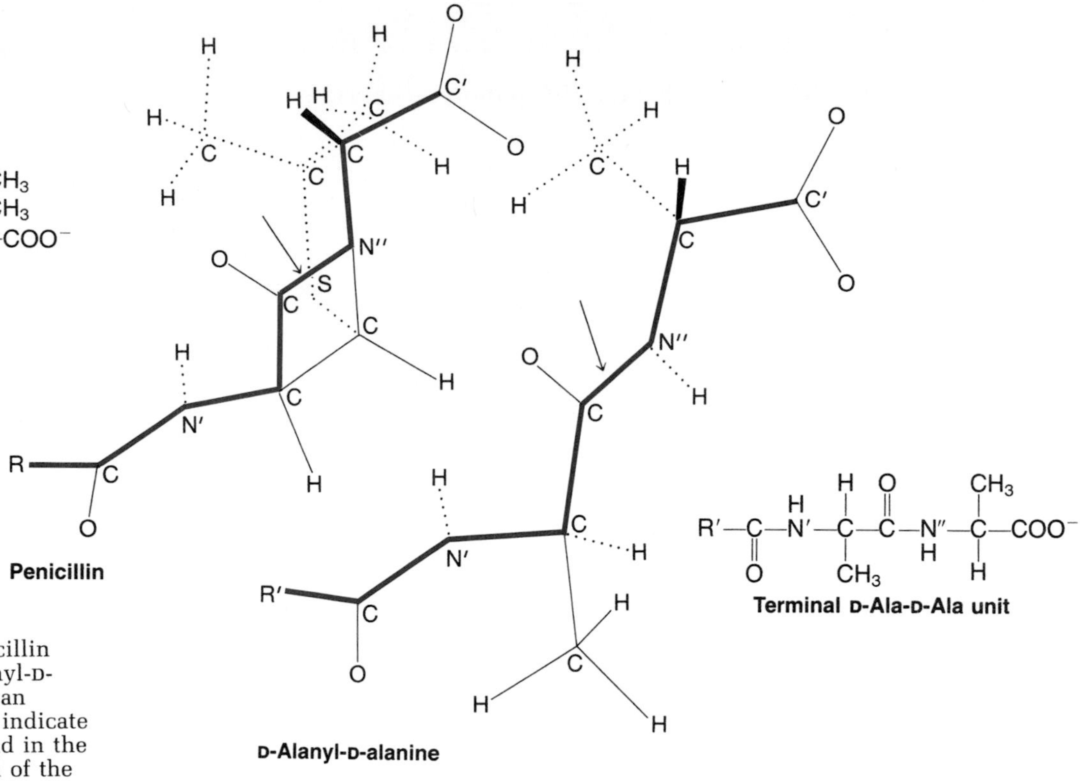

Figure 4–27

Dreiding stereomodels of penicillin (middle, left) and of the D-alanyl-D-alanine end of the peptidoglycan strand (middle, right). Arrows indicate the position of the CO—N bond in the β-lactam ring of penicillin and of the CO—N bond in D-alanyl-D-alanine at the end of the peptidoglycan strand.

Staphylococcus aureus has the sequence L-alanyl-D-γ-glutamyl-L-lysyl-D-alanine. In this tetrapeptide the glutamyl residue is attached through its γ-COOH rather than its α-COOH. All the muramic acids are substituted in this way to form peptidoglycan strands. Some close variant of this structure is characteristic of all bacterial species. The peptidoglycan strands are further linked to each other by means of an interpeptide bridge. In S. aureus, this bridge is a pentaglycine chain that extends from the terminal carboxyl group of the D-alanine residue of one tetrapeptide to the ε-NH₂ group of the third amino acid, L-lysine, in another tetrapeptide (Figure 4–26c). The third dimension is probably built up by bridges extending in different planes. This gigantic macromolecule has the mechanical stability required for the cell wall.

The biosynthesis of two-dimensional peptidoglycans in extracellular space presents some interesting and unique problems in biosynthesis, which are considered in Chapter 20. Certain enzymes involved in cell wall synthesis represent targets for the antibacterial action of *penicillins* and several other related antibiotics that kill bacteria. Penicillin is known to inhibit the enzymes that carry out the final cross-linking reaction that is depicted in Figure 4–26c. In this connection it is of great interest that the structure of penicillin is very similar to the structure of the terminal D-alanyl-D-alanine moiety that is cleaved in the cross-linking reaction (Figure 4–27). It is considered highly likely that the effectiveness of penicillin is due to its binding at the active site of the cross-linking enzyme. We will see (Chapter 8) that many enzyme inhibitors work in this way. That is, they resemble the substrate of the enzyme to the extent that they can bind at the active site of the enzyme and block its action.

O Antigen: A Lipopolysaccharide

The outer membranes of certain bacteria are composed of phospholipids as well as various proteins and lipopolysaccharides (see Figure 4–25). The lip-

---

BOX 4–B

The Antigenic Behavior of Bacterial Lipopolysaccharides

Because of their structure and location the lipopolysaccharide fibers of bacteria are highly antigenic. They are known as O antigens. Their presence stimulates the immune system to make specific antibody proteins that react with the polysaccharides. Infection of a vertebrate host by a particular strain of bacteria leads to an immune response to that organism.

The immune response consists of a complex series of reactions whereby a specific O antigen stimulates the infected organism to synthesize large amounts of a specific antibody that bonds to the O antigen. The bonding results in aggregation and elimination of the infecting bacteria. The struggle for survival does not end here, however, because bacteria have learned to switch from making one type of O antigen to another. Sometimes they can do this fast enough to cause further damage to the infected organism before antibody to the new O antigen is made. The mechanism of antigen switching in bacteria is discussed in genetics texts.

Figure 4–28
Repeat unit in O antigen of *S. typhimurium*. (Abe = abequose, Man = mannose, Rha = Rhamnose, Gal = galactose.)

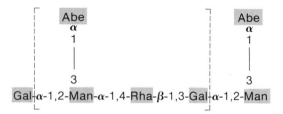

opolysaccharides contain repeating oligosaccharide units attached covalently to a basal core polysaccharide, which, in turn, is attached to a complex lipid conglomerate known as lipid A. The repeating oligosaccharide units attached to the core polysaccharide protrude as fibers of indefinite length from the outer membrane surface. These polysaccharide fibers vary widely in composition in different bacterial strains, and even the same bacterial strain has the capacity for making different polysaccharides (see Box 4–B).

The structure of an O antigen of a particular strain of *S. typhimurium*, an organism closely related to one that causes typhoid fever, is shown in Figure 4–28. The polymer contains a tetrameric repeat for four sugars—abequose, mannose, rhamnose, and galactose. The abequose is not part of the main chain, but rather forms a branch, being attached to the main chain mannose residues by an $\alpha(1,4)$ glycosidic linkage. The length of different O-specific chains using the same repeating unit varies considerably, containing as few as three repeats and as many as forty repeat units per chain.

SUMMARY

Carbohydrates are the most abundant organic substances; they are an important source of carbon compounds for all biomolecules and they also serve important energy and structural needs. Carbohydrates may be divided into monosaccharides, oligosaccharides, and polysaccharides.

The monosaccharides are either polyhydroxyl aldehydes or polyhydroxyketones. All monosaccharides are optically active because of the presence of one or more asymmetrical carbon atoms (chiral centers). Straight-chain sugars in aqueous solution tend to form ring structures known as intramolecular hemiacetals, especially when the resultant is a five-membered (furanose) ring or a six-membered (pyranose) ring. Depending on which way the ring forms about the anomeric carbon, the structure is called an α or β hemiacetal. For a given hemiacetal several conformations are possible. Usually the chair form is favored over the boat form because of the lower steric repulsion produced by side chains in the chair form. Polymer synthesis locks in a particular hemiacetal configuration.

Monosaccharides are linked by glycosidic bonds to form oligosaccharides and polysaccharides. Polysaccharides function in two quite distinct roles: some serve as a means for storage of chemical energy and others serve a structural function. Polymers that use one type of building block are called homopolymers, and those that use more than one are called heteropolymers.

Cellulose is the best known and most abundant structural polysaccharide. It is a homopolymer of glucose with $\beta(1,4)$ linkages between adjacent monomeric residues. Starch and glycogen are energy storage polysaccharides. They also are homopolymers of glucose but with $\alpha(1,4)$ linkages between adjacent residues. In addition they contain branches with $\alpha(1,6)$ linkages.

Many other polysaccharides use different sugars as well as different combinations of sugars or modified sugars. Most of these function in a structural capacity. Many heteropolysaccharides are linked to peptides (peptidoglycans) or proteins (proteoglycans or glycoproteins). Glycoproteins are mostly protein, with short, highly branched carbohydrate chains. By contrast, in proteoglycans the protein is the minor component. Carbohydrates are also found linked to lipids in compounds known as glycolipids and lipopolysaccharides.

The main component in bacterial cell walls is a peptidoglycan. The peptidoglycan is a polymeric structure made of a glycosaminoglycan extending in one dimension, cross-linked through oligopeptides extending in the other dimension. This cross-linked network completely surrounds the bacterial cell and provides it with the mechanical strength needed to resist osmotic shock and other mechanical stresses.

SELECTED READINGS

Ginsberg, V., and P. Robbins, eds., *Biology of Carbohydrates.* New York: Wiley, 1984.

Hassell, J. R., J. H. Kimina, and L. Cantly, Proteoglycan core protein families. *Ann. Rev. Biochem.* 55:539–568, 1986.

Lennarz, W. J., ed., *The Biochemistry of Glycoproteins and Proteoglycans.* New York, London: Plenum, 1980.

McNeil, M., A. G. Darvill, S. C. Fry, and P. Albersheim, Structure and function of the primary cell walls of plants. *Ann. Rev. Biochem.* 53:625–664, 1984.

Quiocho, F. A., Carbohydrate-binding proteins: tertiary structure and protein–sugar interactions. *Ann. Rev. Biochem.* 55:287–316, 1986.

Also see Selected Readings at the end of Chapter 20.

PROBLEMS

1. a. Draw a Fischer projection formula for D-glucose.
   b. Indicate the chiral carbons.
   c. On the basis of (b), how many stereoisomers of the aldohexoses should there be?
   d. Considering your answer in (c), speculate whether polysaccharides could produce the amount of diversity necessary to catalyze the manifold reactions of a living cell, as compared with the like potential of proteins.

2. The usual form of glucose found in nature has a specific rotation of $[\alpha]_D^{20} = +52.7$. Given that the specific rotation for pure β-D-glucose is $+18.7$ and for α-D-glucose is $+112.2$, calculate the fraction of each in natural glucose.

3. 5 g of glycogen was exhaustively methylated and subsequently acid hydrolyzed. Upon analysis, 4 mmol of 2,3-dimethylglucose was obtained (see text and Figures 4–12, 4–13, and 4–15).
   a. What percentage of the glucose residues occurs at 1,6 branch points?
   b. What is the average number of glucose residues per branch?
   c. How much 2,3,6-trimethylglucose was obtained?
   d. If the glycogen has a molecular weight of 2.8×10^6, how many glucose residues does it contain?

4. Both glycogen and cellulose are polyglucose.
   a. Compare and contrast the functions of these related compounds.
   b. What are the molecular reasons for this functional difference?
   c. How do you suppose a cell can produce these two separate molecules without making errors such as glycogen-cellulose hybrid molecules?

5. Penicillin kills only growing cells. Explain.

6. Penicillin has been helpful in elucidating cell wall synthesis but not in obtaining mutants in enzymes involved in cell wall synthesis. Explain.

5

LIPIDS

Lipids are broadly defined as biological molecules that are soluble in organic solvents. Although the lipids encompass a large and diverse group of compounds, they have only four major biological functions: (1) in all cells, the major structural elements of the membranes are composed of lipids; (2) certain lipids (the *triacylglycerols*) serve as efficient reserves for the storage of energy; (3) many of the vitamins and hormones found in animals are lipids or derivatives of lipids, and (4) the bile acids help to solubilize the other lipid classes during digestion.

In this chapter we will consider the structures and, briefly, the functions of the major classes of biological lipids. The roles of certain lipids in membrane structure will be considered in the following chapter, while the oxidation of fatty acids, the major energy-storing units of the triacylglycerols, will be covered in Chapter 18.

FATTY ACIDS

Compounds with the structural formula $CH_3(CH_2)_nCOOH$ that contain no carbon–carbon double bonds are known as *saturated fatty acids*. The two most abundant saturated fatty acids are *palmitic* and *stearic* acids (Table 5–1). Some other saturated fatty acids present in smaller quantities in mammalian tissues also are shown in Table 5–1. The *sphingolipids*, which will be considered later, contain longer-chain fatty acids ($n = 20-24$), as well as palmitic and stearic acids. In some tissues, short-chain fatty acids also are found, such as decanoic acid (10:0) in milk.

Fatty acids with double bonds in the aliphatic chain are called unsaturated fatty acids. *Monounsaturated* fatty acids have one double bond, and *polyunsaturated fatty acids* contain more than one double bond. The double bonds in virtually all naturally occurring fatty acids are *cis*. Fatty acids are often abbreviated as shown in Table 5–1. The number to the left of the colon indicates the number of carbon atoms of the fatty acid and the number to the right indicates the number of double bonds. The numbering begins from the carboxyl group. The position of the double bond is shown by a superscript Δ followed by the number of carbons between the double bond and the carboxyl group. Sometimes unsaturated fatty acids are numbered from the terminal methyl group. In this instance, the numbering is preceded by a lower-

TABLE 5–1
Fatty Acids

Common Name	Systematic Name	Structure	Abbreviation
Saturated fatty acids			
Myristic acid	n-Tetradecanoic acid	$CH_3(CH_2)_{12}COOH$	14:0
Palmitic acid	n-Hexadecanoic acid	$CH_3(CH_2)_{12}CH_2CH_2COOH$	16:0
Stearic acid	n-Octadecanoic acid	$CH_3(CH_2)_{12}CH_2CH_2CH_2CH_2COOH$	18:0
Arachidic acid	n-Eicosanoic acid	$CH_3(CH_2)_{12}CH_2CH_2CH_2CH_2CH_2CH_2COOH$	20:0
Behenic acid	n-Docosanoic acid	$CH_3(CH_2)_{12}CH_2CH_2CH_2CH_2CH_2CH_2CH_2CH_2COOH$	22:0
Lignoceric acid	n-Tetracosanoic acid	$CH_3(CH_2)_{12}CH_2CH_2CH_2CH_2CH_2CH_2CH_2CH_2CH_2CH_2COOH$	24:0
Cerotic acid	n-Hexacosanoic acid	$CH_3(CH_2)_{12}CH_2CH_2CH_2CH_2CH_2CH_2CH_2CH_2CH_2CH_2CH_2CH_2COOH$	26:0
Unsaturated fatty acids			
Palmitoleic acid	cis-9-Hexadecenoic acid	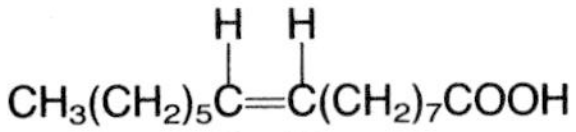 $CH_3(CH_2)_5C=C(CH_2)_7COOH$	$16:1^{\Delta 9}$
Oleic acid	cis-9-Octadecenoic acid	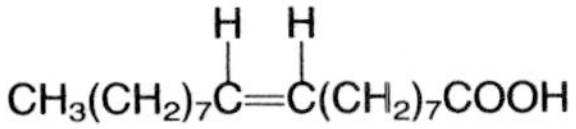 $CH_3(CH_2)_7C=C(CH_2)_7COOH$	$18:1^{\Delta 9}$
Vaccenic acid	cis-11-Octadecenoic acid	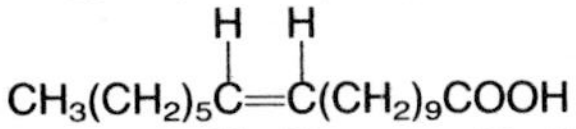 $CH_3(CH_2)_5C=C(CH_2)_9COOH$	$18:1^{\Delta 11}$
Linoleic acid	cis,cis-9,12-Octadecadienoic acid	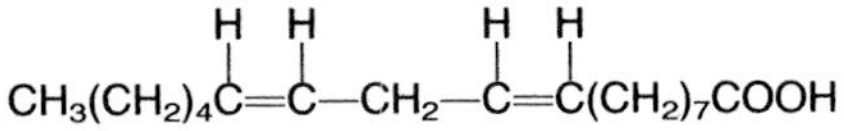 $CH_3(CH_2)_4C=C-CH_2-C=C(CH_2)_7COOH$	$18:2^{\Delta 9,12}$
α-Linolenic acid	All-cis-9,12,15-Octadecatrienoic acid	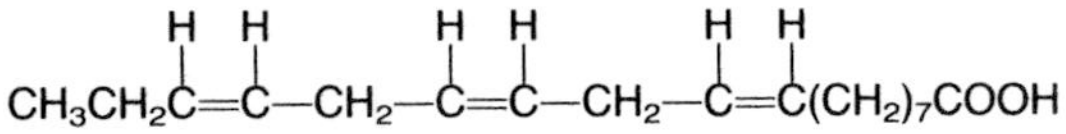 $CH_3CH_2C=C-CH_2-C=C-CH_2-C=C(CH_2)_7COOH$	$18:3^{\Delta 9,12,15}$
Arachidonic acid	All-cis-5,8,11,14-Eicosatetraenoic acid	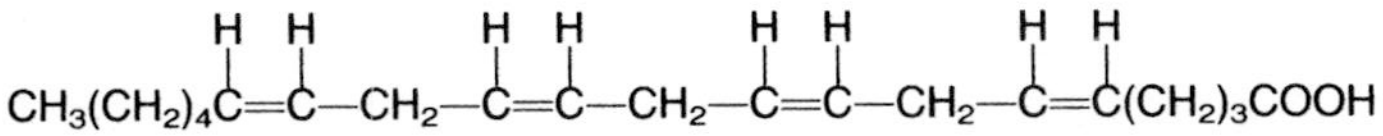 $CH_3(CH_2)_4C=C-CH_2-C=C-CH_2-C=C-CH_2-C=C(CH_2)_3COOH$	$20:4^{\Delta 5,8,11,14}$
	All-cis-4,7,10,13,16,19-Docosahexaenoic acid	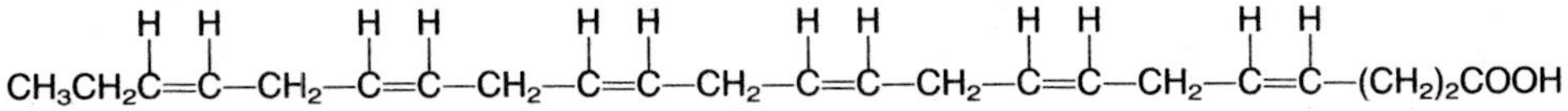 $CH_3CH_2C=C-CH_2-C=C-CH_2-C=C-CH_2-C=C-CH_2-C=C-CH_2-C=C-(CH_2)_2COOH$	$22:6^{\Delta 4,7,10,13,16,19}$
Some unusual fatty acids			
	2,4,6,8-Tetramethyl decanoic acid	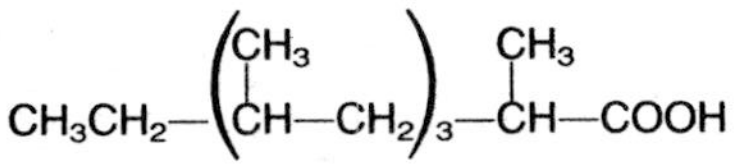 $CH_3CH_2-\left(\overset{CH_3}{\underset{}{CH}}-CH_2\right)_3-\overset{CH_3}{\underset{}{CH}}-COOH$	
Lactobacillic acid		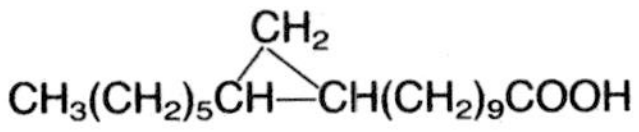 $CH_3(CH_2)_5CH-CH(CH_2)_9COOH$ with CH_2 bridge	
An α-mycolic acid		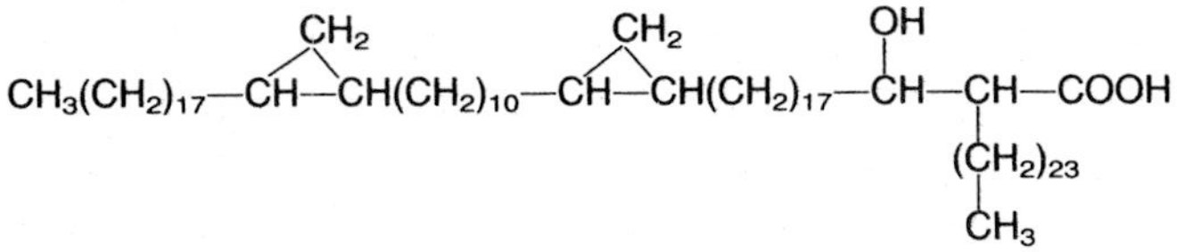 $CH_3(CH_2)_{17}CH-CH-CH(CH_2)_{10}-CH-CH(CH_2)_{17}-CH-\overset{OH}{\underset{}{CH}}-\overset{(CH_2)_{23}}{\underset{CH_3}{CH}}-COOH$ with CH_2 bridges	

155

case omega (ω). Thus linoleic acid might be called Δ9,12-octadecadienoic acid or ω-6,9-octadecadienoic acid.

The double double bonds in polyunsaturated fatty acids are always separated by one methylene group. Mammalian tissues contain all the unsaturated fatty acids listed in Table 5–1 with the exception of vaccenic acid, which is present in *Escherichia coli* and other bacteria. *E. coli* does not contain polyunsaturated fatty acids. Oleic acid is the most common monounsaturated fatty acid in mammals. Two unsaturated fatty acids, *linoleic* and *linolenic* acids, are not synthesized by mammals and are therefore important dietary requirements. Like vitamins, these two fatty acids are required for growth and good health. Hence they are called essential fatty acids. Plants are able to synthesize linoleic and linolenic acids and are the original source of these fatty acids in our diet.

Often, unsaturated fatty acids are classified according to their relationships to other fatty acids. In this case the (n-) nomenclature is used to tell how many carbons away from the methyl end the first double bond occurs. Hence linoleic acid ($18:2^{\Delta9,12}$) is often referred to as 18:2(n-6), since the first double bond is six carbons from the methyl end. This convention is useful for identifying fatty acids derived from the same precursor. As we shall see in Chapter 22, arachidonic acid [20:4(n-6)] can be biosynthesized from linoleic acid [18:2(n-6)].

In addition to the commonly occurring fatty acids, many structural variations have evolved. There are well over 100 other fatty acids found in various creatures and organisms, often associated with specialized functions. For instance, branched-chain fatty acids are found in many different tissues. The uropygial gland of the duck produces such a fatty acid (2,4,6,8-tetramethyldecanoic acid). The duck uses the fatty acids secreted by this gland to preen its feathers and thereby ensure that water continues to "run off its back." In the bacterial genus *Bacillus*, the monoenoic fatty acids are replaced by branched-chain fatty acids in which a methyl group is adjacent to either the terminal methyl group (the *iso* series) or the terminal ethyl group (the *ante-iso* series). Another example is fatty acids with a *cyclopropane ring* in the alkyl chain, found in many bacteria. The bacterium that causes the disease tuberculosis, *Mycobacterium tuberculosis*, produces a family of complex fatty acids known as *mycolic acids*, which contain cyclopropane rings. One class of these is the α-mycolic acids (an example is given in Table 5–1). Many structurally related α-mycolic acids are found in the mycobacteria and other related bacteria (nocardiae and corynebacteria). These mycolic acids appear to have a structural function in the outer part of the bacterial cell wall. There is much evidence to suggest that a major drug used in the treatment of tuberculosis (Isoniazid) functions by the inhibition of an early reaction of α-mycolic acid biosynthesis.

$$CH_3-\overset{\overset{\displaystyle CH_3}{|}}{CH}-$$
Iso series

$$CH_3CH_2-\overset{\overset{\displaystyle CH_3}{|}}{CH}-$$
Ante-Iso series

Fatty acids are usually found as components of complex lipids, and hence only a very small per cent exists as unesterified (free) fatty acids. Nevertheless, it is worth noting that the pK_a for dissociation of the acid proton is around 4.7. Therefore, at pH 7.0, the fatty acid exists primarily in the dissociated form (RCOO$^-$):

$$CH_3(CH_2)_nCOOH \rightleftharpoons CH_3(CH_2)_nCOO^- + H^+$$

Because it exists as a salt at neutral pH, it is not easily extracted from an aqueous medium by organic solvents such as hexane. However, if the pH is lowered by the addition of HCl or another strong acid, the fatty acid becomes protonated and is easily extracted by organic solvents.

Another property of fatty acids that should be noted is their physical form at room temperature. If n equals 8 or less, the fatty acid is a liquid, whereas if n equals 10 or more, the fatty acid is a solid. If a fatty acid has a double bond, it has a lower melting point than the saturated fatty acid with the same number of carbons. Unsaturated fatty acids are more condensed in length than the corresponding saturated fatty acids (Figure 5–1). The double bonds prevent the tight packing within membranes that occurs with saturated fatty acids (see Chapter 6).

When fatty acids were first discovered and chemists were involved in structural determinations, a major problem was separation of the various fatty acids into pure compounds. Consider the difficulties encountered in the separation of stearic acid from palmitic acid by the techniques available in the early 1900s (organic extractions and crystallization). However, with the advent of gas-liquid chromatography in the 1950s, it became relatively easy to analyze and purify fatty acids from complex mixtures. Today fatty acids are commonly analyzed by gas chromatography of the methyl esters. These are formed by esterification of the fatty acids with methanol [R represents hydrogen (H) or any group to which the fatty acid is esterified]:

$$CH_3(CH_2)_n\overset{O}{\underset{}{C}}-OR + CH_3OH \xrightarrow{HCl \text{ or } BF_3} CH_3(CH_2)_n\overset{O}{\underset{}{C}}OCH_3 + ROH$$

Methyl ester

An analysis of the fatty acids from rat liver is shown in Figure 5–2.

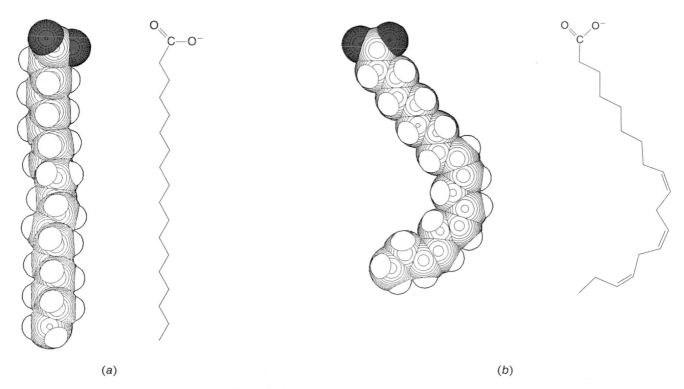

(a)　　　　　　　　　　　　　　　　　　(b)

Figure 5–1
Space-filling and conformational models of (a) stearic and (b) linolenic acids. Each of these fatty acids has 18 carbon atoms, but the three double bonds in linolenic acid create a more rigid, curved molecule that interferes with tight packing in membrane structures.

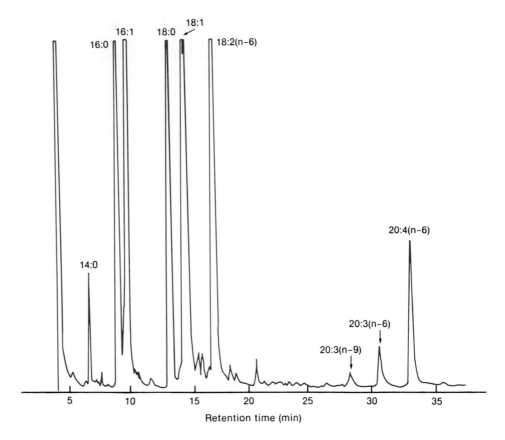

Figure 5–2
Gas chromatography of methyl esters of fatty acids derived from serum phospholipids. Analysis was at 180°C on a 50-mm capillary column coated with Silar 9 CP. (Courtesy of Dr. Harold Cook, Dalhousie University, Halifax, Nova Scotia).

Fatty Acids Are Stored as an Energy Reserve in Triacylglycerols

Fatty acids are major components of the *triacylglycerols* (Table 5–2 and Figure 5–3) and most of the complex lipids present in membranes. *It is in the triacylglycerols that fatty acids are stored as an energy reserve.* Triacylglycerols are the major uncharged glycerol derivatives found in animals. The monoacylglycerols and diacylglycerols are metabolites of triacylglycerols and of phospholipids, as discussed in Chapters 21 and 22, and are normally present in cells in very small quantities.

TABLE 5-2
Neutral Glycerides

Common Name	Systematic Name	Structure
Triglyceride	1,2,3-Triacyl-*sn*-glycerol	
Diglyceride	1,2-Diacyl-*sn*-glycerol	
Monoglyceride	1-Monoacyl-*sn*-glycerol	

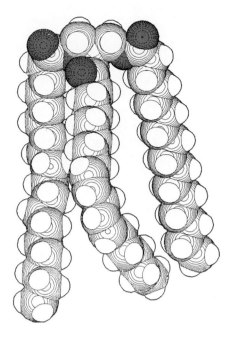

Because the substituents esterified to the first and third carbons of the glycerol derivative are usually different, the second carbon is asymmetric. In naming and numbering these glycerol derivatives, a special convention has been adopted: The prefix *sn-* (for stereospecifically numbered) immediately precedes "glycerol" and differentiates the naming of the compound from other approaches, such as the RS system described in Chapter 11. The glycerol derivative is drawn in a Fischer projection with the secondary hydroxyl to the *left* of the central carbon, and the carbons are numbered 1, 2, and 3 from the top to the bottom. The prefix *rac-* (for *racemo*) precedes the name if the compound is an equal mixture of antipodes. If the configuration is unknown or not specified, *x-* precedes the name.

Although triacylglycerols are found in the liver and intestine, they are primarily found in adipose tissue (fat), which functions as a storage depot for this lipid. The specialized cell in this tissue is called the *adipocyte*. The cytoplasm of the cell is full of lipid vacuoles that are almost exclusively triacylglycerols (Figure 5–4) and serve as an energy reserve for mammals. At times when the diet or glycogen reserves are insufficient to supply the body's need for energy, the fuel stored as fatty acyl components of the triacylglycerols is mobilized and transported to other tissues in the body. A second important function of adipose tissue is insulation of the body from cold. This function is most obvious in such cold-water mammals as the arctic whales (Beluga whales), which have vast stores of fat (blubber).

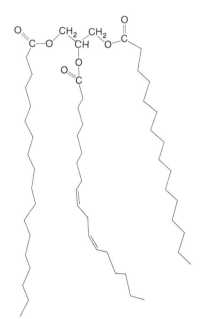

Figure 5–3
Space-filling and conformational models of triacyglycerol. Note the uncharged nature of this molecule, which serves as a means of storing fatty acids for future energy needs.

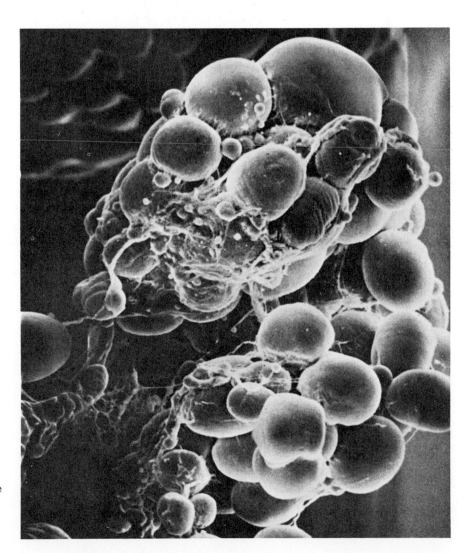

Figure 5–4
Scanning electron micrograph of white adipocytes from rat adipose tissue (600×). (Courtesy Dr. A. Angel and Dr. M. J. Hollenberg of the University of Toronto.)

PHOSPHOLIPIDS

By definition, phospholipids are lipids that contain phosphate. The *phosphoglycerides* are quantitatively the most important structural group in this lipid class and contain, in addition to phosphate, a glycerol backbone and esterified fatty acids and alcohols. Phosphoglycerides are the major structural lipids of all biological membranes (Chapter 6). The importance of phospholipids to living organisms is underscored by the nearly complete lack of genetic defects in the metabolism of these lipids in humans. Presumably, any such defects are lethal at early stages of development and, therefore, are never observed.

Phosphoglycerides Have a Glycerol-3-phosphate Backbone

All phosphoglycerides have a *glycerol-3-phosphate* backbone, as shown in Figure 5–5, and the *sn* nomenclature discussed earlier is also used for the systematic naming of the various phospholipids. The hydroxyls on carbons 1 and 2 are usually acylated with fatty acids. In most phospholipids, the fatty acid substituent at carbon 1 is saturated and at carbon 2 is unsaturated. In some instances, the substituent on carbon 1 is an *alkyl ether* or an *alkenyl ether*, as shown in Figure 5–6. The alkenyl ether phospholipids are also known as *plasmalogens*.

Phosphoglycerides are classified according to the substituent (X) on the phosphate group (Table 5–3). If X is a hydrogen, the compound is called *3-sn-phosphatidic acid*. If X—OH is choline, the lipid is called *3-sn-phosphatidylcholine* (Figure 5–7), which is often referred to by its original name, *lecithin*, and is the most abundant phospholipid in animal tissues. The name "lecithin" is derived from the Greek name for egg yolk, the source from

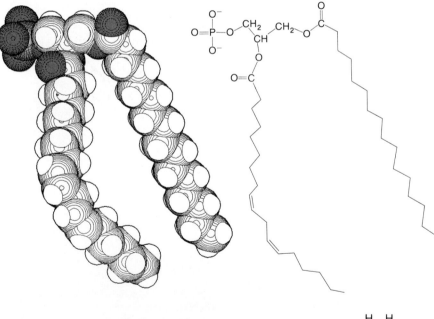

Figure 5–5
The structure of phosphatidic acid, a phosphoglyceride. The cluster of polar and charged oxygens gives phosphatidic acid its amphipathic properties. The fatty acids attached to carbons 1 and 2 are saturated and unsaturated, respectively.

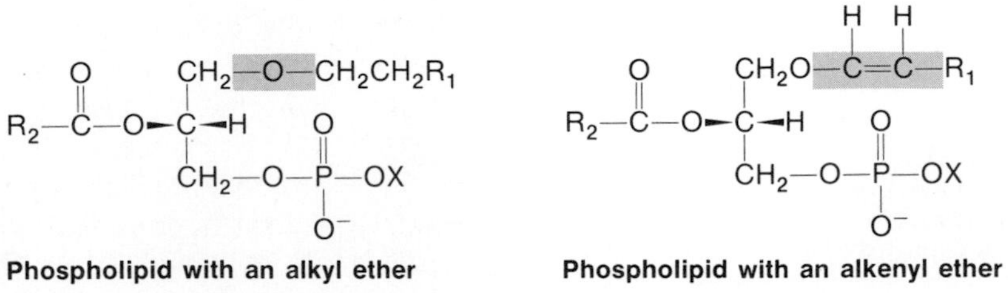

Figure 5–6
Phospholipids with alkyl or alkenyl ether substituents.

Phospholipid with an alkyl ether

Phospholipid with an alkenyl ether

TABLE 5-3
Major Classes of Phospholipids

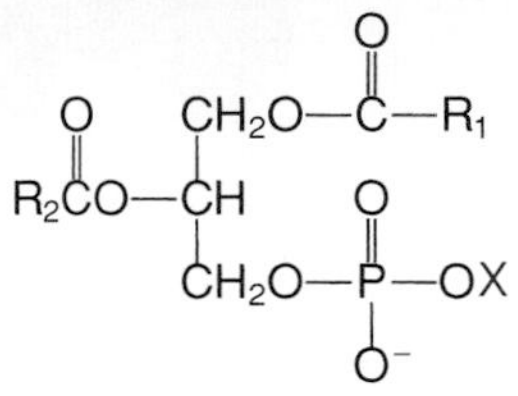

	X Substituent	
Name of X—OH	Formula of X	Name of Phospholipid
Water	—H	Phosphatidic acid
Choline	—CH$_2$CH$_2$N$^+$(CH$_3$)$_3$	Phosphatidylcholine (lecithin)
Ethanolamine	—CH$_2$CH$_2$N$^+$H$_3$	Phosphatidylethanolamine
Serine	—CH$_2$—CH (N$^+$H$_3$) (COO$^-$)	Phosphatidylserine
Glycerol	—CH$_2$CH(OH)CH$_2$OH	Phosphatidylglycerol
Phosphatidyl- glycerol	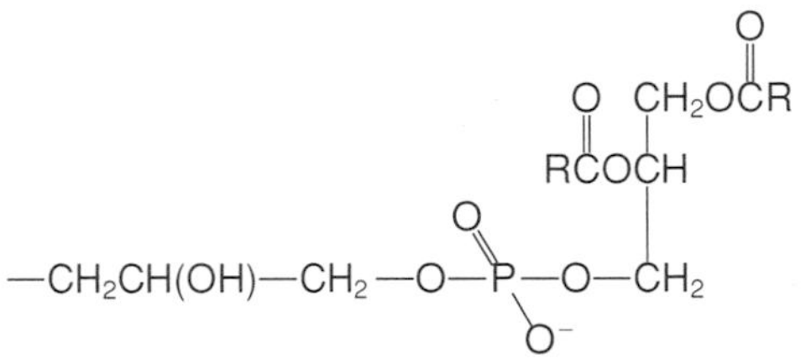 —CH$_2$CH(OH)—CH$_2$—O—P(O)(O$^-$)—O—CH$_2$ ... RCOCH, CH$_2$OCR	Diphosphatidylglycerol (cardiolipin)
myo-Inositol	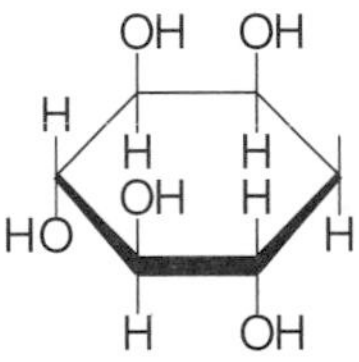 (inositol ring structure)	Phosphatidylinositol

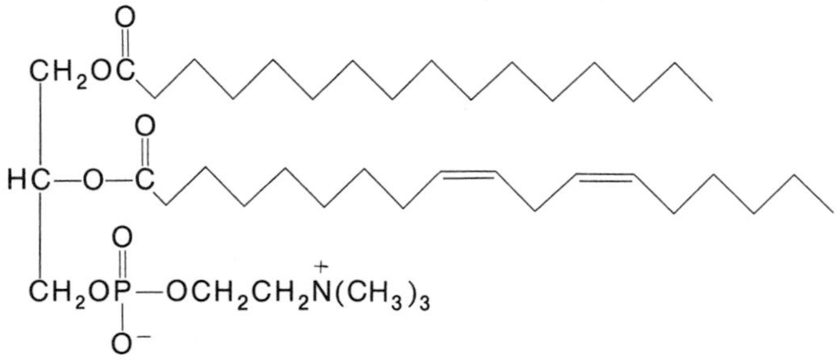

Figure 5-7
Structure of phosphatidylcholine
(lecithin).

which the lipid was first isolated. In addition to its role in membrane structure, phosphatidylcholine is an important structural component of the plasma lipoproteins and bile. The other major phospholipids are listed in Table 5–3.

Another important class of phosphoglycerides is the _lysophospholipids_, in which one of the acyl substituents (usually from position 2) is missing, as shown in Figure 5–8. If the acyl substituent at carbon 1 were removed, the

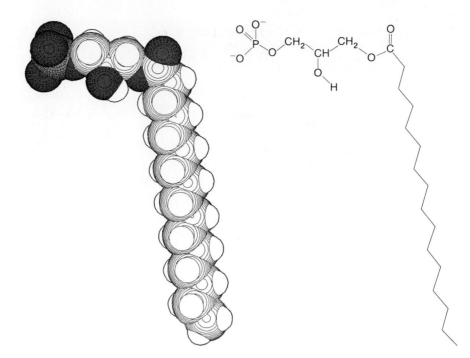

Figure 5–8
The basic structure for lysophospholipids. These phospholipids usually lack a substituent at position 2.

acyl group from position 2 would spontaneously migrate to position 1. We can differentiate between deacylation at position 1 or position 2 by analysis of the fatty acids derived from the lysophospholipid. If the fatty acid on the lysophospholipid is saturated, it is likely that the fatty acid from position 2 of the phospholipid has been cleaved. However, if the fatty acid on the lysophospholipid is mostly unsaturated, the fatty acid from position 1 has probably been cleaved and migration of the fatty acid from position 2 has occurred. The lysophospholipids are named by simply adding the prefix *lyso-* to the name of the original phospholipid (e.g., *lysophosphatidyl-choline*). The lysophospholipids account for only 1 or 2 per cent of the total phospholipids in animal cells.

Phospholipids Are Amphipathic Molecules

The phospholipids as a group are *amphipathic* ("dual sympathy") molecules because they have both polar and nonpolar portions. The *polar head groups* (e.g., the phosphate group plus the X-substituent in Table 5–3) prefer an aqueous environment (are hydrophilic), whereas the nonpolar (hydrophobic) acyl substituents ("tails") are excluded from the aqueous environment. It is this amphipathic property that causes most phospholipids to arrange spontaneously into a *bilayer* when suspended in an aqueous environment. As discussed in the next chapter, phospholipid bilayers are the major structural feature comprising most biological membranes.

Although the phospholipids are often referred to in the singular (e.g., phosphatidylcholine), we are actually describing a complex mixture of phospholipids with the same head group (choline) but with a variety of different fatty acid substituents. Thus human red blood cells have 21 different molecular species of phosphatidylcholine that differ in fatty acid substituents at either position 1 or position 2, or both. This is true in a similar fashion for the other red blood cell phospholipids. Therefore, in the membranes of red blood cells, there are only six major classes of phospholipids, but over 100 different molecular species. What initially appears to be a simple phospholipid composition for the red blood cell is actually a very com-

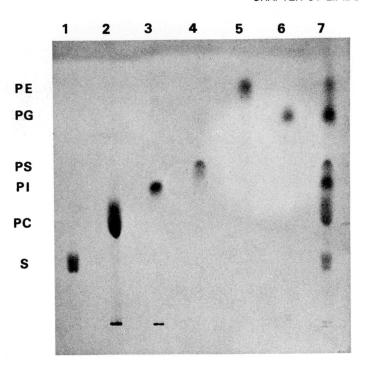

Figure 5–9
Thin-layer chromatography of the major phospholipids. Lanes 1 to 6 contain 0.1 mg of each lipid. Lane 7 is a mixture of the six phospholipids. PE = phosphatidylethanolamine, PG = phosphatidylglycerol, PS = phosphatidylserine, PI = phosphatidylinositol, PC = phosphatidylcholine, S = sphingomyelin. Each compound was spotted on a silica gel G60 thin-layer plate that had been activated at 100°C an hour before the analysis. The plate was developed in a solvent that contained $CHCl_3:CH_2OH:CH_3COOH:H_2O$ (50:25:8:4 by volume). The plate was sprayed with dilute sulfuric acid and heated in an oven to oxidize the hydrocarbon, producing a black deposit that can be visualized.

plex mixture. The advantage of such complexity is not readily apparent. However, it is clear that the fatty acid composition of the phospholipids is very important for the normal operation of membrane-associated functions such as active transport. This subject will be discussed in more detail in Chapter 34.

Lipids are extracted from cells or tissues with organic solvents (for example, $CHCl_3$). Phospholipids are resolved from uncharged lipids (neutral lipids) by adsorption column chromatography (usually the adsorbant is silicic acid) or by adsorption thin-layer chromatography. The various classes of phospholipids are usually separated on a small scale by thin-layer chromatography, as shown in Figure 5–9. Large-scale preparations usually involve column chromatography or high-pressure liquid chromatography.

Sphingolipids Contain a Long-chain, Hydroxylated Secondary Amine

The common structural feature of sphingolipids is a long-chain, hydroxylated secondary amine. There are three major long-chain bases (as they are called) (Table 5–4) that contain 18 carbons and a number of other bases that differ in chain length, number of double bonds, or branching of the alkyl chain. *Sphingosine (4-sphingenine)* is quantitatively the most important long-chain base (usually 90 per cent or more) in animal cells, whereas *phytosphingosine (4-hydroxysphinganine)* is characteristically found in plant tissues. Most bacteria, including *E. coli*, do not contain sphingolipids, whereas yeast cells do.

The sphingolipid bases occur as components of more complex lipids. The bases themselves are toxic to cells and therefore are present only in trace quantities. The sphingolipid base is acylated on the amine with a fatty acid to give *ceramide* (Figure 5–10), which is common to all the sphingolipids. The fatty acid substituents are mainly C_{16}, C_{18}, C_{22}, or C_{24}, saturated or monounsaturated. In many tissues, the acyl group also may contain an α-hydroxy residue.

$$CH_3(CH_2)_{12}\overset{\underset{\displaystyle H}{|}}{C}=\overset{\underset{\displaystyle H}{|}}{C}-\overset{\overset{\displaystyle H}{|}}{\underset{\underset{\displaystyle OH}{|}}{C}}-\overset{\overset{\displaystyle H}{|}}{\underset{\underset{\displaystyle NH}{|}}{C}}-CH_2OH$$

Figure 5–10
Structure of ceramide with sphingosine as the long-chain base. Sphingosine (shown in color) is very toxic to cells and is usually found only in trace amounts.

TABLE 5-4
Three Important Sphingolipid Bases

Structure	Systematic Name	Common Name
$CH_3(CH_2)_{12}C=C-C-C-CH_2OH$ (with H, OH, NH_3^+ substituents)	4-Sphingenine	Sphingosine
$CH_3(CH_2)_{12}CH_2CH_2-C-C-CH_2OH$ (with OH, NH_3^+ substituents)	Sphinganine	Dihydrosphingosine
$CH_3(CH_2)_{12}CH_2C-C-C-CH_2OH$ (with OH, OH, NH_3^+ substituents)	4-Hydroxy-sphinganine	Phytosphingosine

Ceramide occurs as part of more complex lipids that have on the primary hydroxyl group either phosphocholine to give *sphingomyelin* (Figure 5–11) or carbohydrate to give the class called *glycosphingolipids*. The carbohydrates most often associated with the glycosphingolipids are glucose, galactose, N-acetylglucosamine, and N-acetylgalactosamine. There is a subdivision of the glycosphingolipids called *gangliosides*, and these also contain one or more molecules of *N-acetylneuraminic acid (sialic acid)* (Figure 5–12) in addition to other carbohydrates. As the name implies, the gangliosides were first isolated from nerve tissue; subsequently, however, they were

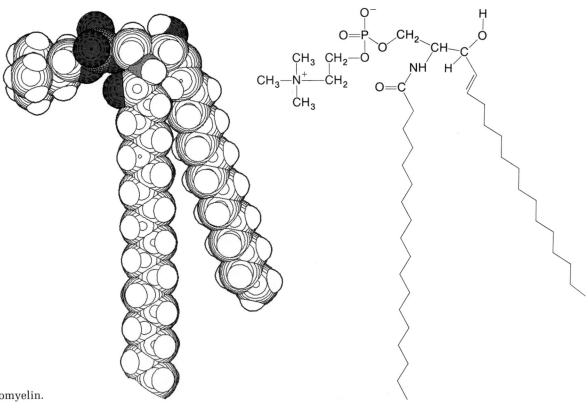

Figure 5–11
Structure of sphingomyelin.

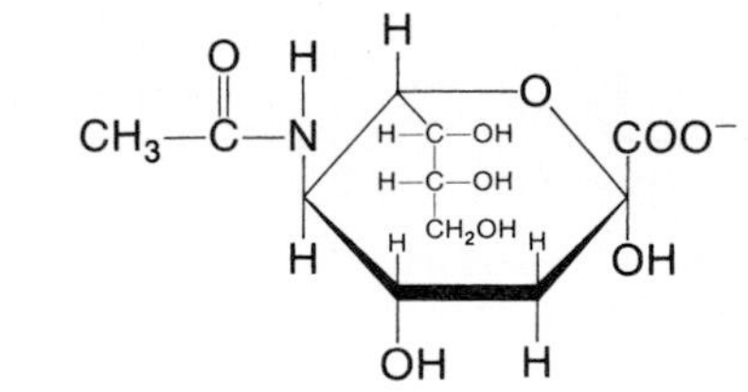

Figure 5–12
Structure of α-N-acetylneuraminic acid (sialic acid). Gangliosides are a subclass of glycosphingolipids that always contain one or more molecules of N-acetylneuraminic acid. This compound is also commonly found as a terminal residue of glycoproteins (see Chapter 4).

found in most other animal tissues. Some of the major glycosphingolipids are listed in Table 5–5. The structures of two important glycosphingolipids, globoside and GM$_2$, are shown in Figures 5–13 and 5–14. Over 50 separate classes of glycosphingolipids have been identified that differ in the structure of the oligosaccharide.

Sphingolipids are important components of the myelin sheath, a multi-layered membranous structure that protects and insulates nerve fibers. The lipids in human myelin contain 5 per cent sphingomyelin (the original source of this lipid, as the name implies) and 15 per cent *galactosylceramide (galactocerebroside)* (Figure 5–15). In addition, a sulfate derivative, *3'-sulfate-galactosylceramide*, makes up 5 per cent of the myelin lipid. The sphingolipids are found in plasma as components of *lipoproteins*, primarily *low-density lipoproteins*, which are discussed later in this chapter and in Chapter 23.

TABLE 5–5
Some Major Glycosphingolipids

Structure	Common Name	Abbreviated Name
Gal-β-1,1-ceramide	Galactosylceramide	—
Glc-β-1,1-ceramide	Glucosylceramide	—
Gal-β-1,4-Glc-β-1,1-ceramide	Lactosylceramide	—
Gal-α-1,4-Gal-β-1,4-Glc-β-1,1-ceramide	Trihexosylceramide	—
GalNAc-β-1,3-Gal-α-1,4-Gal-β-1,4-Glc-β-1,1-ceramide	Globoside	—
NeuAc-α-2,3-Gal-β-1,4-Glc-β-1,1-ceramide	Hematoside	GM3
GalNAc-β-1,4-Gal-β-1,4-Glc-β-1,1-ceramide $\quad\quad\quad\quad\quad$ | $\quad\quad\quad\quad\quad$ 3 $\quad\quad\quad\quad\quad$ | $\quad\quad\quad\quad\quad$ α-2 $\quad\quad\quad\quad\quad$ NeuAc	Tay Sachs ganglioside	GM2
Gal-β-1,3-GalNAc-β-1,4-Gal-β-1,4-Glc-β-1,1-ceramide $\quad\quad\quad\quad\quad$ | $\quad\quad\quad\quad\quad$ 3 $\quad\quad\quad\quad\quad$ | $\quad\quad\quad\quad\quad$ α-2 $\quad\quad\quad\quad\quad$ NeuAc	—	GM1

Note: Gal = galactose; Glc = glucose; GalNAc = N-acetylgalactosamine; NeuAc = N-acetylneuraminic acid.

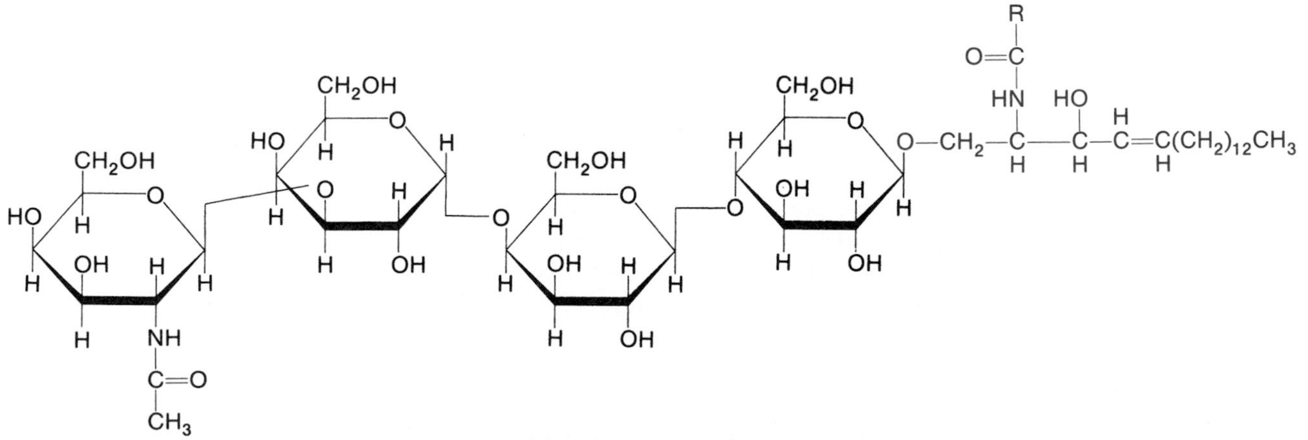

GalNAc-β-1,3-Gal-α-1,4-Gal-β-1,4-Glc-β-1,1-ceramide

Figure 5–13
Structure of globoside, a glycosphingolipid.

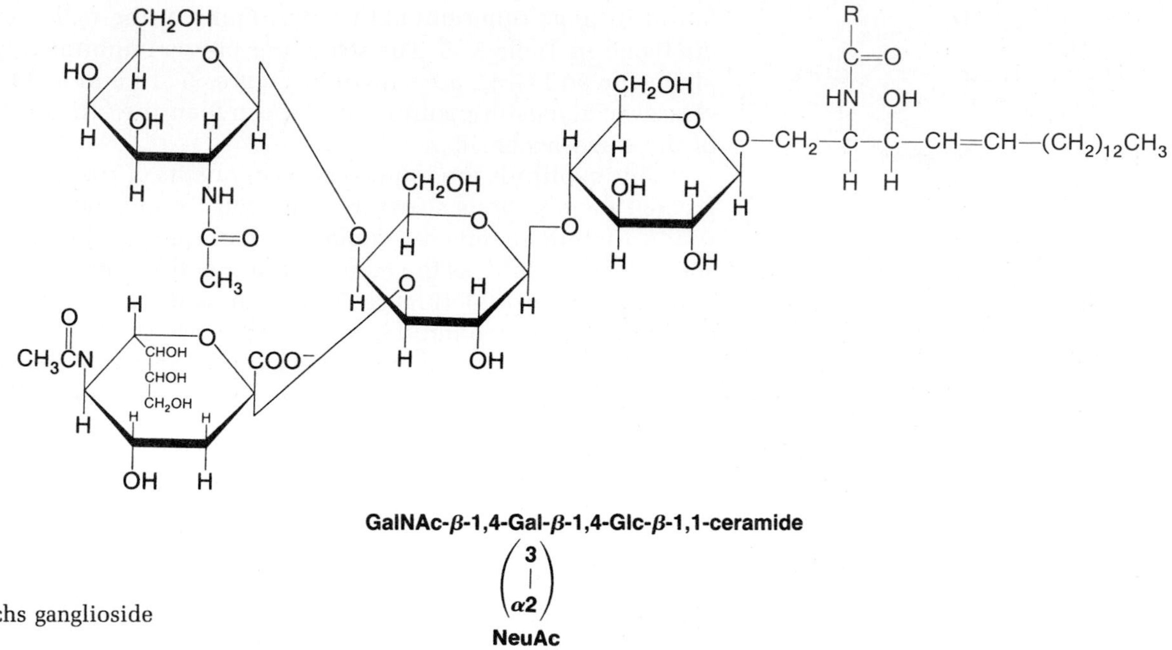

GalNAc-β-1,4-Gal-β-1,4-Glc-β-1,1-ceramide

$$\left(\begin{array}{c} 3 \\ | \\ \alpha2 \end{array}\right)$$

NeuAc

Figure 5–14
Structure of Tay-Sachs ganglioside (GM$_2$).

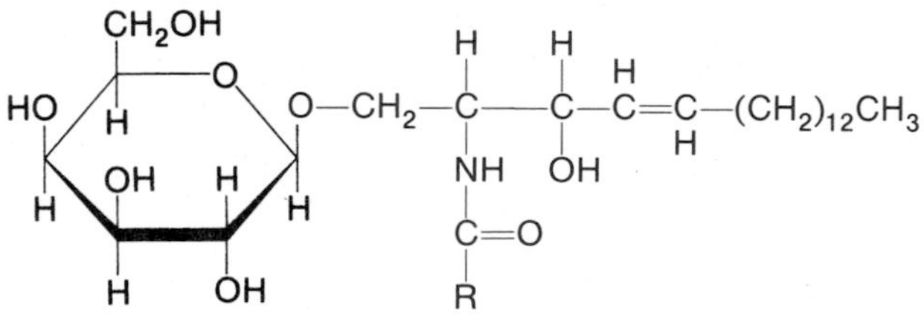

Figure 5–15
Structure of galactosylceramide. This lipid comprises 15 per cent of human myelin lipid.

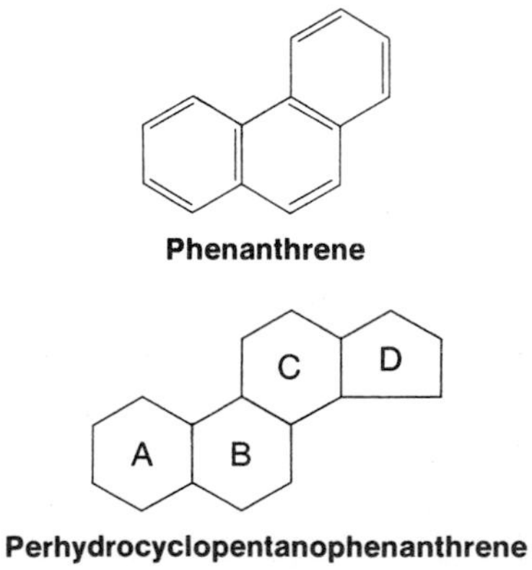

CHOLESTEROL AND RELATED STEROIDS

Cholesterol is the most prominent member of the steroid family and is infamous among the public because of the relationship between an elevated level of serum cholesterol and an increased risk of cardiovascular disease. Approximately 1100 mg of this sterol is excreted each day from a normal human adult. In a typical American diet, this is replaced by 250 mg in the diet and 850 mg by means of biosynthesis. Cholesterol is also the precursor for other important steroids in animals: _bile acids_ and _steroid hormones_. In addition, cholesterol is an important structural component of some eukaryotic membranes, but is generally absent from bacterial membranes.

Cholesterol Is a Derivative of a Tetracyclic Hydrocarbon

The name "cholesterol" is derived from the Greek words _chole_, for bile, and _stereos_, for solid. Highly purified cholesterol is a white powder at room temperature. Cholesterol, like other _steroids_, is a derivative of the tetracyclic hydrocarbon _perhydrocyclopentanophenanthrene_ (Figure 5–16). The four rings are identified by the first four letters of the alphabet and the carbons are numbered in the sequence shown in Figure 5–17. In addition to the basic ring structure, cholesterol contains a hydroxyl group at C-3, an aliphatic chain at C-17, methyl groups at C-10 and C-13, and a Δ^5 double bond.

The cyclohexane rings of steroids can adopt either the chair or the boat conformation. The chair conformation is more stable and is the preferred conformation of steroids. The conformations of _cholestanol_ and _coprostanol_,

Phenanthrene

Perhydrocyclopentanophenanthrene

Figure 5–16
Structures of phenanthrene and perhydrocyclopentanophenanthrene.

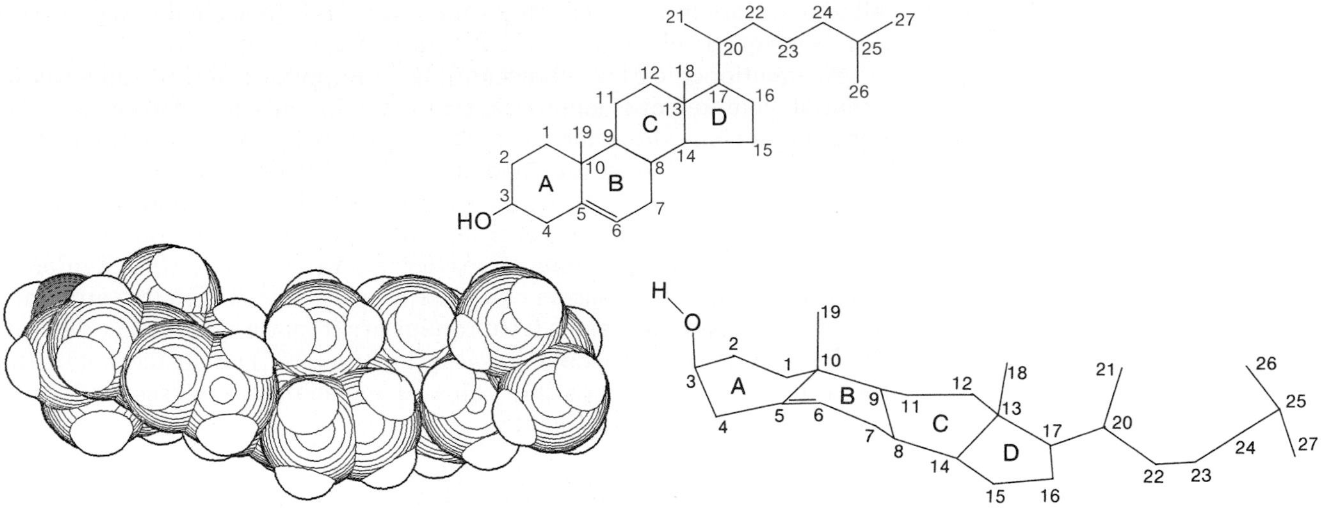

Figure 5–17

Structure of cholesterol, in three different views. The conventional projection is shown at the top center. The more realistic space-filling and conformational models are shown at the lower left and the lower right, respectively.

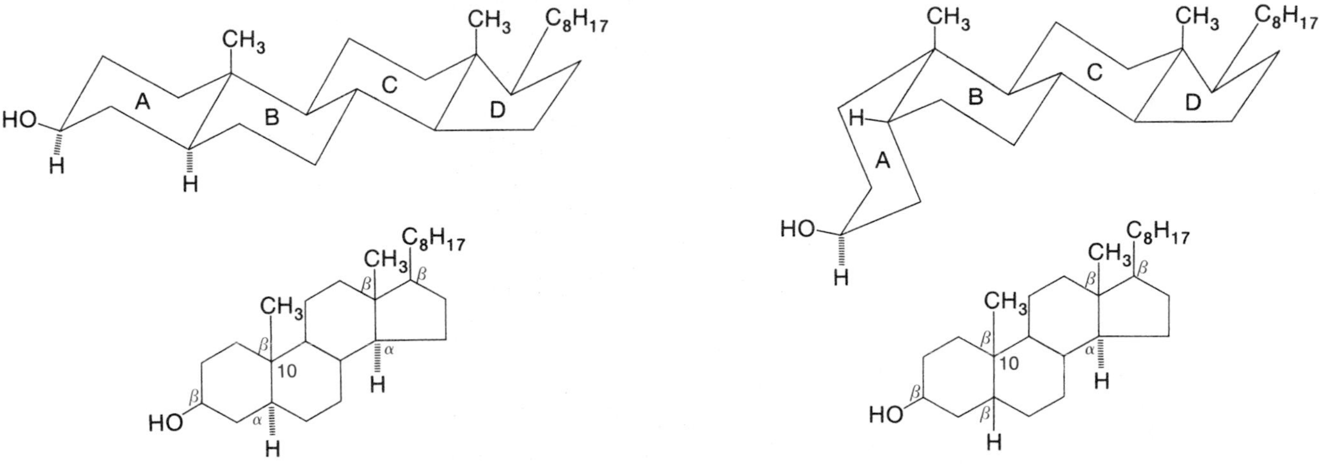

Cholesterol

Coprostanol

Figure 5–18

Conformational and conventional structures of cholestanol and coprostanol. The A and B rings are joined in a *trans* configuration in cholestanol and in a *cis* configuration in coprostanol. The methyl at position 10 is located above the plane of the rings and is said to be in the β orientation. Other substituents are labeled β or α, depending on whether they are above or below the planes of the rings.

the two saturated derivatives of cholesterol, are shown in Figure 5–18. The A and B rings can be joined in a *trans* configuration, as in cholestanol, or in a *cis* configuration, as in coprostanol. As you can see, the spatial orientation of the A and B rings of these two stereoisomers is very different.

Although it is important that we recognize the three-dimensional structures of steroids, such structural representations are too cumbersome for most uses in biochemistry. Hence a configurational convention for steroids has been adopted in which structural formulas are more easily drawn and recognized. The substituents of the steroid rings are related to the CH_3 group at position 10, which by definition projects above the plane of the rings. This methyl, which is said to be a β substituent, is indicated in structural formulas by a solid line (—). Similarly, other groups that are above the plane of the rings are referred to as β. Those substituents below the plane of the rings are called α and are indicated in structural formulas by a dashed line (---). Examples of α and β substituents are shown in Figure 5–18.

As mentioned earlier, cholesterol is an important, and in some cases an essential, component of some eukaryotic membranes (see also Chapter 6). In higher organisms, cholesterol is synthesized predominantly in the liver and is then transported to various body tissues by _lipoproteins_ in blood plasma. Plasma lipoproteins are complexes containing specific proteins, phospholipids, cholesterol, triacylglycerol, and/or cholesterol ester, and they are subdivided into several different classes based on size and density. Evidence is accumulating that both serum cholesterol levels and the relative amounts of the various classes of plasma lipoproteins are important factors in determining the risk of cardiovascular disease in humans. The structures of these lipoproteins and their possible roles in metabolic processes will be discussed in more detail in Chapter 23.

Steroid Hormones Are Biosynthesized from Cholesterol

Communication among cells of a multicellular organism is achieved by chemical messengers that pass between cells. These chemicals, known as _hormones_, serve to coordinate the metabolic activity of the various tissues, allow the organism to adapt to a changing environment, and prepare it for reproduction. Steroids constitute a major class of hormones that function in various capacities. They are synthesized in specialized glands and carried by the circulatory system to other parts of the organism, where they evoke specific responses.

Steroid hormones are biosynthesized from cholesterol (see Chapter 23). The adrenal cortex and the gonads are the major organs involved in the biosynthesis of these steroids. In pregnant females, the placenta also manufactures steroid hormones. The structures and functions of five major hormones are given in Table 5–6. The biosynthesis and the action of steroid hormones is described in Chapter 23; the way in which hormones work is considered in Chapter 31.

Bile Acids Are the Major Degradation Products of Cholesterol

Bile acids, the major degradation products of cholesterol, are dihydroxylated and trihydroxylated steroids with 24 carbons. All hydroxyl groups of the bile acids are α in orientation, and the two methyl groups are β (Figure 5–19); thus these molecules have a polar and a nonpolar face. Another distinguishing feature is the C-24 carboxylic acid group. In addition, the major bile acids have the A and B rings joined in a _cis_ configuration (see Figures 5–18 and 5–19). The most important bile acids in humans are _cholic acid_ and _chenodeoxycholic acid_ (Figure 5–19). _Deoxycholic acid_ is found in bile of other mammals and is also an important reagent used in the laboratory for solubilization of membrane proteins and enzymes. Cholate is also an important solubilization reagent. The bile acids are present in bile mostly as amide conjugates of _taurine_ or _glycine_ (Figure 5–20), and these derivatives are referred to as _bile salts_. The structures for _taurocholate_ and _glycocholate_ are given in Figure 5–20.

Bile consists of a mixture of organic and inorganic compounds. As shown in Table 5–7, the bile salts and phosphatidylcholine are quantitatively the most important constituents of bile. The bile, made in the liver, is stored in the _gall bladder_ and passes along the _common bile duct_ into the duodenum when food is present. _The bile salts act as detergents in the small intestine and thereby aid the solubilization of lipids, which are more easily degraded by intestinal lipases._ The ability of the bile salts to solubilize lipids is due to the amphipathic nature of these compounds. The polar groups,

TABLE 5-6
Major Steroid Hormones

Steroid	Function
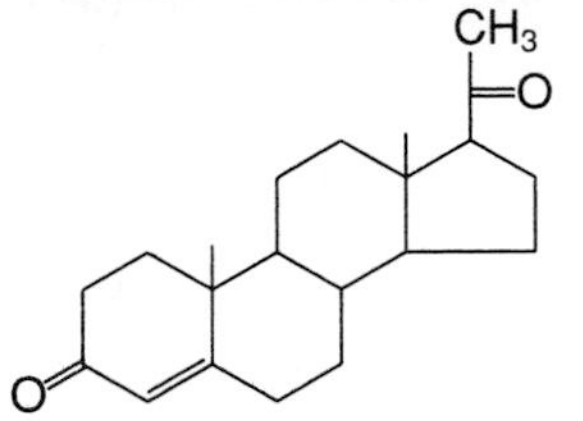 **Progesterone**	Precursor of other steroids; prepares uterus for implantation of an egg; prevents ovulation during pregnancy
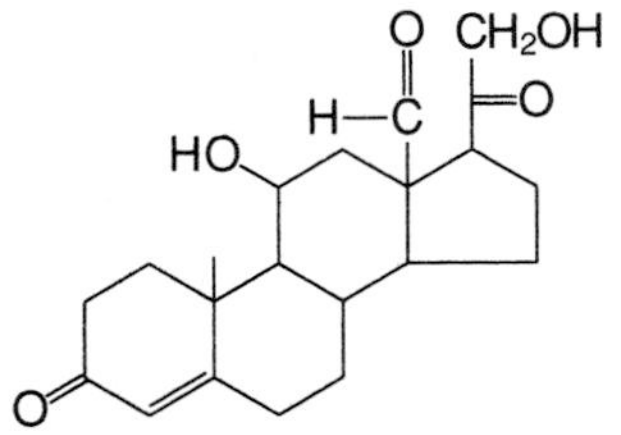 **Aldosterone** (a mineralocorticoid)	Increases retention of sodium ions by the renal tubules
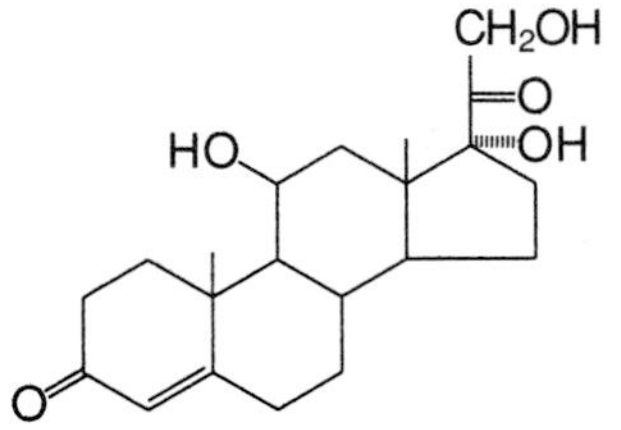 **Cortisol** (a glucocorticoid)	Promotes gluconeogenesis; suppresses inflammatory reactions
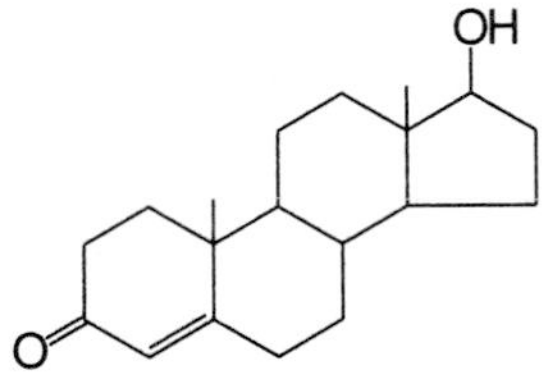 **Testosterone** (an androgen)	Promotes male sexual development; promotes and maintains male sex characteristics
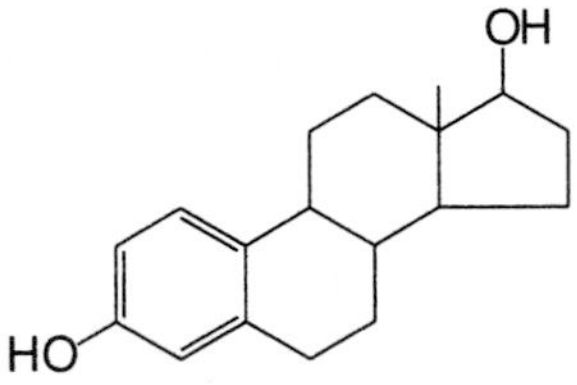 **Estradiol** (an estrogen)	Responsible for sexual development in the female; promotes and maintains female sex characteristics

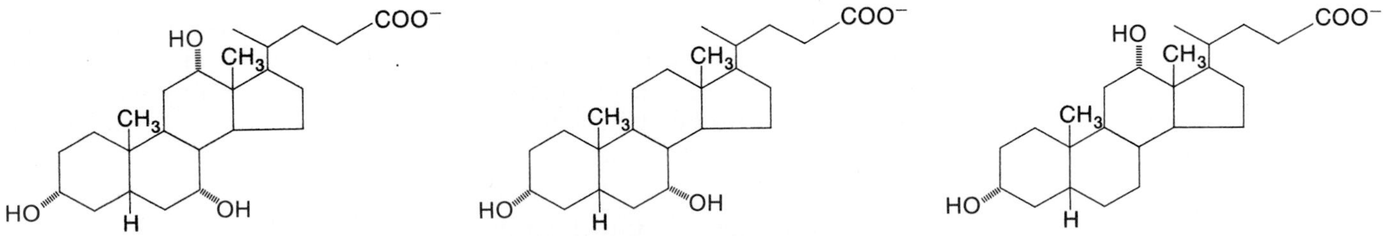

Cholic acid   **Chenodeoxycholic acid**   **Deoxycholic acid**

Figure 5–19
Structures of three bile acids. Cholic acid and chenodeoxycholic acid are important bile acids in humans. Deoxycholic acid is important in other mammals and is also used as a laboratory reagent to solubilize membrane proteins.

Figure 5–20
Structures of the taurine and glycine conjugates of cholic acid. These conjugates are referred to as bile salts.

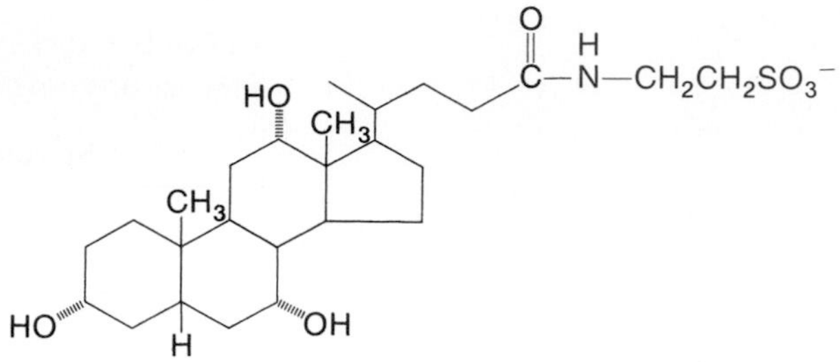

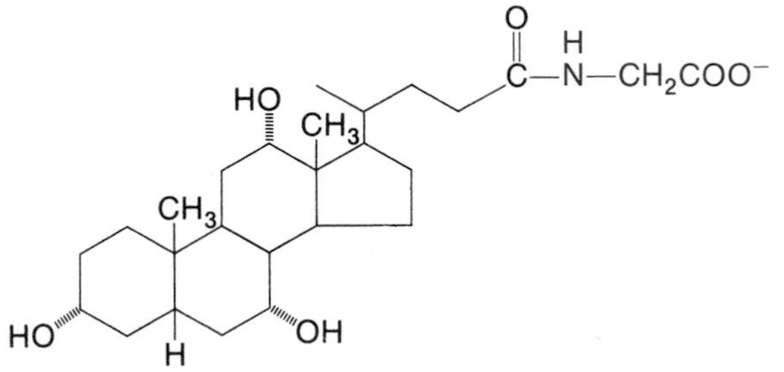

TABLE 5–7
Organic Components in Human Bile

Compound	Concentration (g/l)
Bile salts	1.2–18.0
Phosphatidyl-choline	1.4–8.0
Cholesterol	1.0–3.2
Bilirubin	0.1–0.7
Protein	0.3–3.0

which favor an aqueous environment, all lie on the α side of the steroid nucleus. The nonpolar side interacts with hydrophobic lipids. As a result of these properties and interactions, the nonpolar dietary lipids are suspended in a uniform dispersion in the aqueous medium.

Digested lipids are absorbed primarily in the upper part of the intestine, whereas most of the bile salts are resorbed in the lower intestine and returned to the liver by means of the portal blood. The bile salts are resecreted into bile and recirculate through the intestine. This cycle is called the entero-hepatic circulation of bile salts. In humans, approximately 20–30 g of bile salts is secreted by the liver, but only 0.5 g appears in the feces each day. Approximately half the cholesterol in bile is excreted in the feces either as cholesterol or, after its conversion by intestinal bacteria, as coprostanol. This represents a major mechanism for elimination of cholesterol from the body.

OTHER BIOLOGICAL LIPIDS

In addition to the major biological lipids we have described, there exist many other diverse groups of biologically important lipids. Although many of these are present in relatively minute amounts in many organisms, they nonetheless often perform vital functions. In this section, we consider the structures and functions of a number of these lipids, including lipid-soluble vitamins, terpenes, and polyketides. The eicosanoids, lipids with a hormone-like function that are active in extremely small amounts, will be discussed in Chapter 22.

Many Vitamins Are Lipid-soluble

Vitamins are essential factors that cannot be synthesized by the organism. A number of the water-soluble vitamins and their related coenzymes are discussed in Chapter 11. Here we will discuss the structures and functions of a number of the lipid-soluble vitamins.

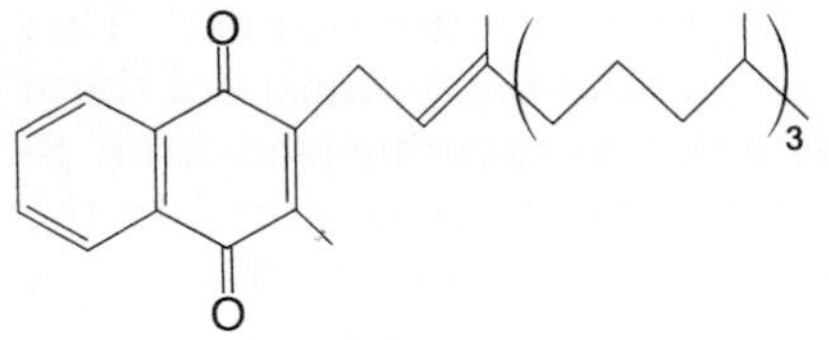

Vitamin K₁
(phylloquinone)

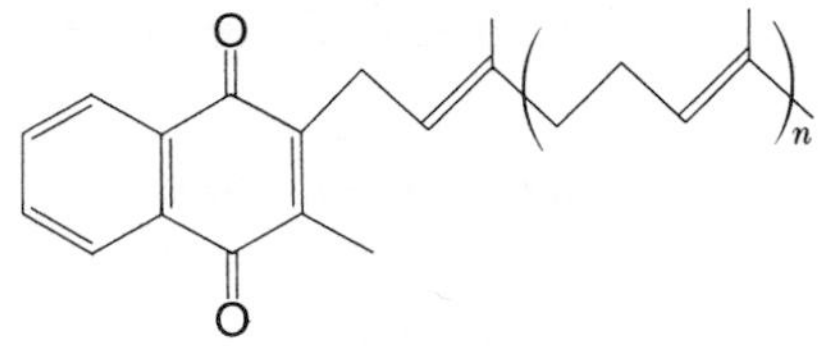

Vitamin K₂
(menaquinone)

Figure 5–21
Structures of vitamins K_1 and $K_2 \cdot K_1$ is found in plants, K_2 in animals and bacteria.

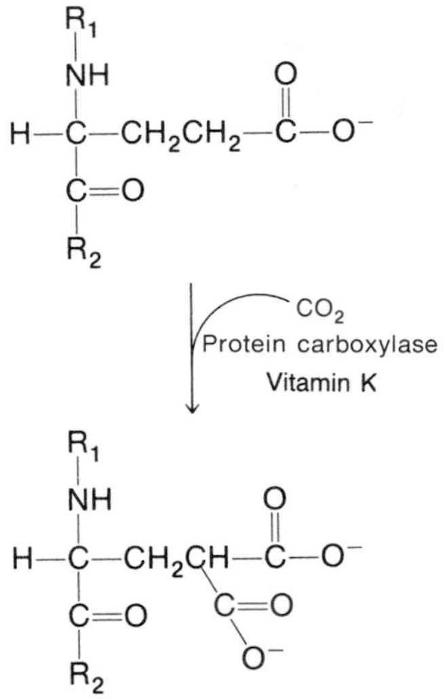

**γ-Carboxyglutamic acid
in a protein**

Figure 5–22
Vitamin-K-dependent carboxylation of a glutamic acid residue in a protein. This reaction is essential to blood clotting.

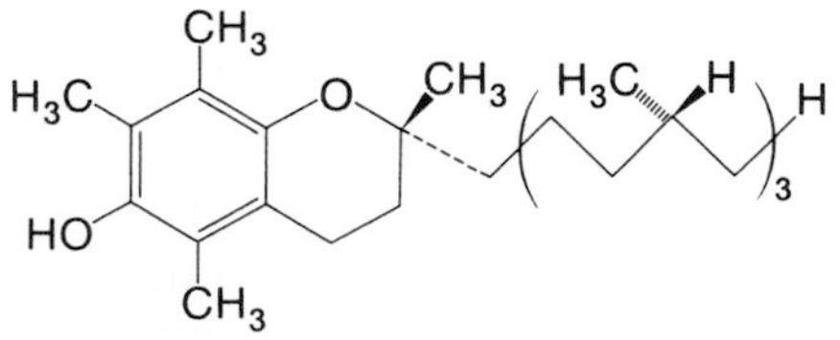

Figure 5–23
Structure of vitamin E (α-tocopherol).

Vitamin D₃ (cholecalciferol) can be made in the skin from 7-dehydrocholesterol in the presence of ultraviolet light (Chapter 31). Vitamin D_3 is formed by the cleavage of ring B of 7-dehydrocholesterol. Vitamin D_3 made in skin or absorbed from the small intestine is transported to the liver and hydroxylated at C-25 by a microsomal mixed-function oxidase. *25-Hydroxyvitamin D₃* appears to be biologically inactive until it is hydroxylated at C-1 by a mixed-function oxidase in kidney mitochondria. The *1α,25-dihydroxyvitamin D₃* is delivered to target tissues for the regulation of calcium and phosphate metabolism. The mode of action of 1,25-dihydroxyvitamin D_3 is analogous to that of the steroid hormones (see Chapter 31).

Vitamin K was discovered by Henrik Dam in Denmark in the 1920s as a fat-soluble factor important in blood coagulation (K is for *koagulation*). The structures of vitamins K_1 and K_2 (Figure 5–21) were elucidated by Edward Doisy. Vitamin K_1 is found in plants; vitamin K_2, in animals and bacteria. How this vitamin functions in blood coagulation eluded scientists until 1974, when a requirement for vitamin K was shown for the formation of *γ-carboxyglutamic acid* (Figure 5–22) in certain proteins. γ-Carboxyglutamic acid specifically binds calcium, which is important for blood coagulation. Such modified glutamic acid residues appear to be important in many other processes involving calcium transport and calcium-regulated metabolic sequences.

Vitamin E (α-tocopherol) (Figure 5–23) was recognized in 1926 as an organic-soluble compound that prevented sterility in rats. The function of this vitamin still has not been clearly established. A favorite theory is that it is an antioxidant that prevents peroxidation of polyunsaturated fatty acids. Tocopherol certainly prevents peroxidation *in vitro*, and it can be replaced by other antioxidants. However, other antioxidants will not relieve all the symptoms of vitamin E deficiency.

Vitamin A (trans-retinol) is called an *isoprenoid alcohol* (Figure 5–24) because it consists, in part, of units of a single five-carbon compound called *isoprene*:

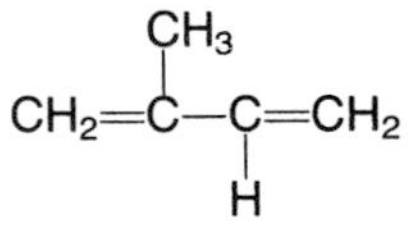

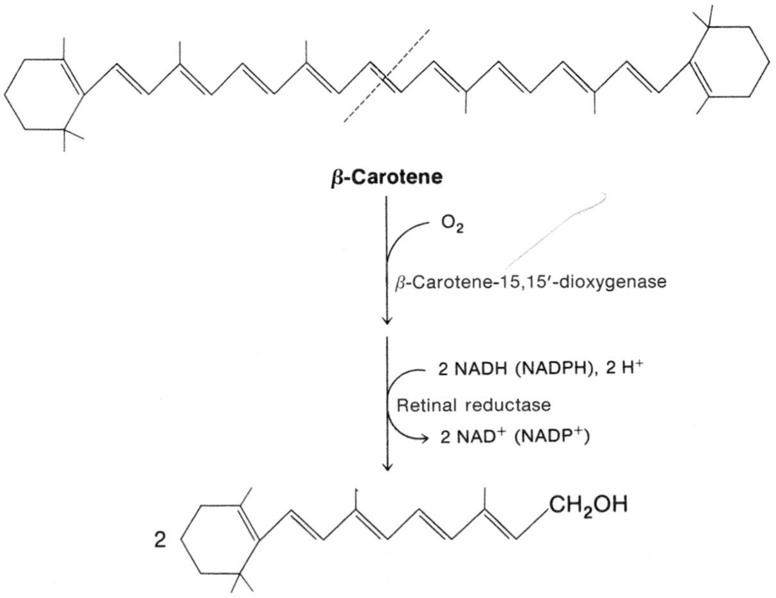

β-Carotene

Figure 5–24
Biosynthesis of vitamin A from β-carotene.

Vitamin A
(*trans*-retinol)

Isoprene is also a precursor of steroids and terpenes (discussed next). This relationship will become clear when we examine the biosynthesis of these compounds (see Chapter 23). Vitamin A is either biosynthesized from β-carotene (Figure 5–24) or absorbed in the diet. Vitamin A is stored in the liver predominantly as an ester of palmitic acid. For many decades, it has been known to be important for vision and for animal growth and reproduction. The form of vitamin A active in the visual process is *11-cis-retinal*, which combines with the protein *opsin* to form *rhodopsin*. Rhodopsin is the primary light-gathering pigment in the vertebrate retina (see Chapter 36).

Terpenes Are a Diverse Group Made from Isoprene Precursors

Terpenes form an extraordinarily diverse group of compounds that are biosynthesized from isoprene precursors. The terpenes, which have a wide range of functions, include primary metabolites and secondary metabolites. Examples of primary metabolites are steroids, certain hormones (juvenile hormone in insects), and precursors to hormones (β-carotene). The major diversity found among terpenes occurs in *secondary metabolites* (compounds that serve no obvious function in the life of the organism that produces them), which are made by plants and microorganisms.

Examples of several kinds of terpenes are shown in Figure 5–25. *Monoterpenes* (C_{10}) are found in all higher plants. One example is *limonene*, which is largely responsible for the characteristic odor of lemons. A bicyclic monoterpene, *pinene*, is the major constituent of turpentine, the solvent obtained from pine resins. *Sesquiterpenes* (C_{15}) also are widely distributed in the plant kingdom; several thousand individual sesquiterpenes have been identified. For example, *santonin* has been used in India for the treatment of intestinal worms, and *cedrol* is a constituent of cedarwood oil. The *diterpenes* are C_{20} terpenes. *Abietic acid* from pine resin is one of the most prevalent terpenes in this group. Another diterpene, *phytol*, is a component of the chlorophyll molecule. Also in this group are the *gibberellins*, which are important plant growth hormones. One of these hormones, *gibberellic acid*, is shown in Figure 5–25. *Triterpenes* are C_{30} terpenes, of which *lanosterol* is an important example discussed in Chapter 23. There are a large number of triterpenes in plants, most of which are pentacyclic. *Friedelin* (Figure 5–25) is one example of a triterpene extracted from cork. Of the *tetraterpenes* (C_{40}), the *carotenes* predominate. β-Carotene (Figure 5–25), from carrots, has been mentioned in connection with vitamin A biosynthesis. *Lycopene* (Figure 5–25), a pigment from tomatoes, is similar to β-carotene, but lacks the cyclohexene rings.

Among the various *polyprenols* (polyisoprenoid alcohols), the *dolichols* are currently under intensive investigation. Dolichols are long-chain polyprenols that consist of 16 to 22 isoprene units. They function in the form of *dolichyl phosphate* as carriers of carbohydrates in the biosynthesis of *glycoproteins* in animals. Bacteria use a similar lipid carrier for carbohydrates, *undecaprenol*, in which the number of isoprene units is 11 (see Chapter 6).

One sesquiterpene that has attracted much attention in the past two decades is *juvenile hormone III*, along with two closely related juvenile hormones, I and II (Figure 5–26). These hormones maintain insects in the larval form (see Chapter 31). During normal metamorphosis, the blood (hemolymph) levels of juvenile hormone fall and the insect matures. Consequently, the addition of this hormone inhibits the transformation of these larva into the pupa and adult forms of the insect. The hormone appears again in the adult female and plays a role in maturation of the reproductive system.

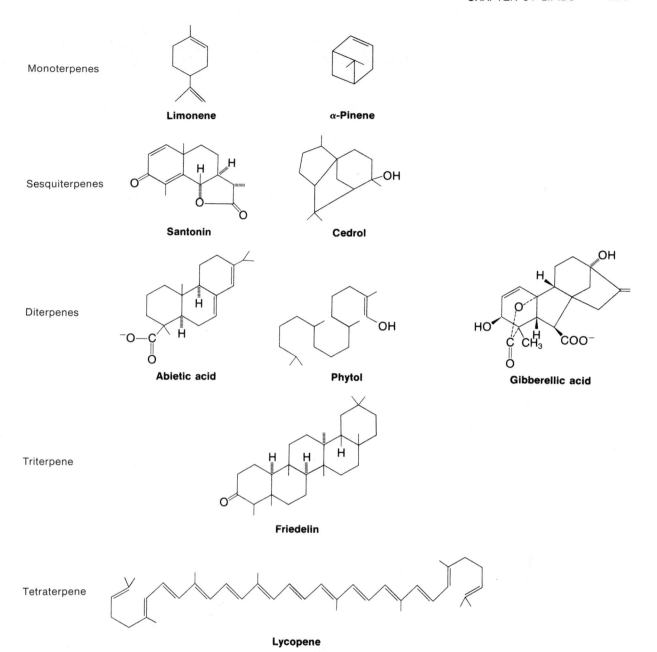

Figure 5–25
Structures of selected terpenes. These compounds are found mainly in plants.

Juvenile hormone I

Juvenile hormone II

Juvenile hormone III

Figure 5–26
Structures of juvenile hormones. These compounds are sesquiterpenes and are found in insects.

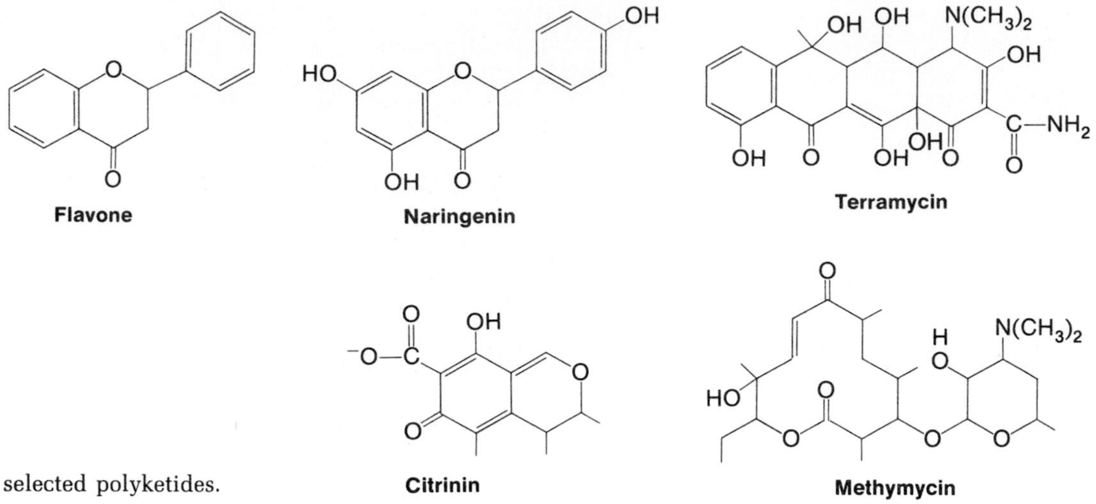

Figure 5–27
Structures of selected polyketides.

Polyketides Are Complex Molecules Derived from Acetate

Polyketides (acetogenins) are complex molecules derived from condensation of three or more units of acetate. The polyketides are responsible for many of the brilliant colors that abound in nature. They account for most colors of flowers, autumn leaves, rhubarb, sea urchins, lichens, molds, and fungi. The other two major groups of pigments that are not polyketides are the tetrapyrroles (e.g., heme and chlorophyll) and the carotenes. The *polyphenols* account for over three-quarters of the more than 1000 polyketides that have been identified.

The largest subgroup of polyketides is the *flavonoids*, which are plant, flower, and fruit pigments with the skeleton of *flavone* (Figure 5–27). One example is *naringenin* (Figure 5–27). The mold metabolite *citrinin* is related to this class of polyketides. The *tetracyclines*, which are important drugs for the treatment of bacterial infection, belong to another group of polyketides. *Terramycin* (Figure 5–27) was the first compound in this group for which the structure was determined. Another group of polyketides of great importance in medicine is the *macrolide antibiotics*, one example of which is *methymycin*.

As with all of the biological molecules described in this portion of the text, lipids must interact with other types of biomolecules to aid the growth and reproduction of cells and organisms. One example of such interaction is provided in the following chapter, in which we consider the structure and assembly of biological membranes, which consist of, or interact with, all the major types of biomolecules that we have discussed. As will be seen, the organization of biological membranes defines the boundaries of cells, and also provides matrices within which a large number of vital functions are carried out.

SUMMARY

Lipids are biological molecules that are soluble in organic solvents. They comprise a wide diversity of structures and they serve four major biological functions: (1) some serve as major structural elements of membranes; (2) some are efficient reserves for the storage of energy; (3) many vitamins and hormones are lipids or derivatives of lipids; and (4) the bile acids help to solubilize the other lipid classes during digestion.

Fatty acids are the major structural components of membrane lipids and are the biological molecules in which energy can be stored most efficiently. Fatty acids exhibit a large number of structural variations, primarily in chain length and in the number and location of double bonds. Triacylglycerols are the molecules in which fatty acids are stored for energy. They are found mostly in adipose tissue.

Phospholipids contain a polar head group and a nonpolar tail portion composed of long-chain acyl or alkyl groups. The amphipathic nature of the phospholipids allows them to form lipid bilayers spontaneously in an aqueous medium. Hence phospholipids play a major role in all biological membranes.

The sphingolipids also are important structural lipids found in eukaryotic membranes. The acylation of sphingosine produces ceramide, which reacts with phosphatidylcholine to give sphingomyelin.

Steroids are derivatives of the tetracyclic hydrocarbon perhydrocyclopentanophenanthrene. Cholesterol is the most prominent member of the steroid family of lipids. Cholesterol is an important structural component of some eukaryotic membranes, but is generally absent from bacterial membranes.

Bile acids are C_{24} carboxylic acids that are biosynthetically derived from cholesterol. The taurine or glycine conjugates of the bile acids are secreted into the small intestine and aid the solubilization and digestion of lipids.

Steroid hormones are biosynthesized from cholesterol in the adrenal cortex, gonads, and placenta. These steroids are important hormones for many specific physiological processes.

Another important class of lipids consists of the lipid-soluble vitamins, which are dietary components essential for vision, fertility, blood clotting, and many other biological functions.

The terpenes are biosynthesized from isoprene and include such metabolites as cholesterol, dolichol, carotene, and juvenile hormone. Many terpenes are secondary metabolites with no obvious function in the cell.

SELECTED READINGS

Deuel, H. J., *The Lipids: Biochemistry*, Vol. 3. New York: Interscience, 1957. A comprehensive and classical treatise on the biochemistry of lipids until the mid-1950s.

Hawthorne, J. N., and G. B. Ansell, eds., *New Comprehensive Biochemistry*, Vol. 4: *Phospholipids*. Amsterdam: Elsevier, 1982. Detailed collection of review articles on the structures, functions, and biosynthesis of the phospholipids.

Nes, W. R., and M. L. McKean, *Biochemistry of Steroids and Other Isopentenoids*. Baltimore: University Park Press,

1977. A complete and authoritative text on the structure and metabolism of steroids.

Razin, S., and S. Rottem, eds., *Current Topics in Membranes and Transport*, Vol. 17: *Membrane Lipids of Prokaryotes*. New York: Academic Press, 1982. A comprehensive summary of lipid structure and function in bacteria.

Stumpf, P. K., *The Biochemistry of Plants*, Vol. 4: *Lipids: Structure and Function*. New York: Academic Press, 1980. A general summary on the biochemistry of plant lipids.

PROBLEMS

1. If you eat only meat, eggs, and dairy products, will you eventually develop essential fatty acid deficiency? Explain.

2. Draw the structure of the major class of phospholipids found in animal tissues.

3. The term phosphatidylcholine (PC) actually describes a class of molecules. In what way do the members of this class differ?

4. Draw the structure of sphingomyelin. Indicate the aspects of its structure common to all sphingolipids.

5. Although the chemical structures of phosphoglycerides and sphingolipids are considerably different, they nonetheless have very similar roles in the structures of biological membranes, as we shall see in Chapter 6. From the space-filling models of Figures 5–5 and 5–11, explain why this similarity in role might be predicted.

6. The bile acids play an important role in the digestion of lipids. In what way are the structures of these molecules suited to this role?

7. Distinguish between the structures, properties, and roles of the major classes of steroids.

8. Which of the lipids discussed in this chapter can be considered polyisoprenoids? Do all of the polyisoprenoids have similar biological functions?

9. As we shall see later, lipids are involved in very important interactions with proteins in cells. From what you know about the chemistry and structures of lipids and proteins, predict what sorts of interactions might be expected to occur between these types of biomolecules.

6

STRUCTURE AND ASSEMBLY OF BIOLOGICAL MEMBRANES

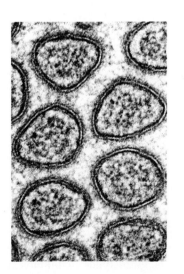

Figure 6–1
Cross section of microvilli of cat intestinal epithelial cells, showing the trilaminar structure of the cytoplasmic membranes (165,000×) (Courtesy Dr. S. Ito.)

All cells possess a cytoplasmic membrane, often referred to as a _plasma membrane_. This membrane encapsulates the cytoplasm and creates internal compartments in which essential functions are carried out. In addition to its role as a physical barrier that maintains the integrity of the cell, the plasma membrane provides functions necessary for the survival of the cell, including exclusion of harmful substances, acquisition of nutrients and energy sources, disposal of unusable and toxic materials, reproduction, locomotion, and interaction with components in the environment. All these functions require coordination both for short-range activities, such as sensation, and for long-range processes, such as growth and differentiation. The appearance of a typical plasma membrane in the electron microscope (Figure 6–1) reveals few clues as to how these processes might be carried out, and it has been largely left to biochemists and biophysicists to investigate the molecular architecture of this complex—and vital—macromolecular structure.

Most _eukaryotic_ (nucleated) cells contain numerous _organelles_ of widely differing structure and function, each of which is specialized in its function: digestion (lysosomes), respiration (mitochondria), photosynthesis (chloroplasts), secretion (endoplasmic reticulum and Golgi apparatus), or nucleic acid biosynthesis (nucleus). Each organelle is bounded by its own specialized membrane system, which has evolved to participate in these specialized functions (Figure 6–2). In contrast, _prokaryotic_ cells (bacteria) typically have all these functions integrated into the plasma membrane and lack specialized intracellular organelles. These differences in cell structure do not indicate different biochemical mechanisms, but merely the presence or absence of compartments specifically designed to fulfill essential functions. In the eukaryotic cell, each process is performed in a spatially isolated domain, whereas these processes operate largely within a single compartment in the prokaryotic cell. In either case, the membrane biochemist's task is a formidable one: to deduce how individual processes take place within a water-insoluble matrix.

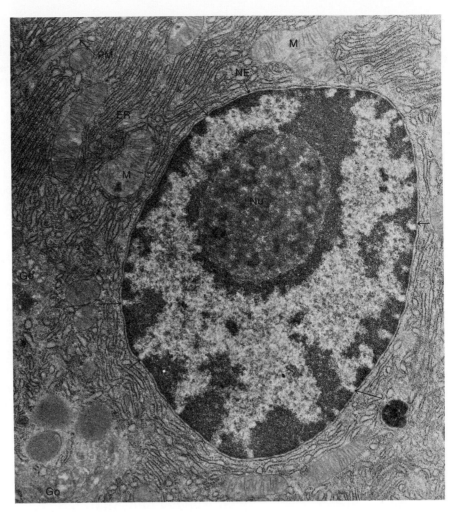

Figure 6–2
Electron micrograph of a cell from the rat pancreas, showing several different intracellular organelles. (PM = plasma membrane; NE = nuclear envelope; Nu = nucleolus; M = mitochondrion; ER = endoplasmic reticulum; Go = Golgi apparatus; arrows show pore complexes in the nuclear envelope; 24,000×.) (From S. L. Wolfe, *Biology of the Cell,* 2d ed., Wadsworth, Belmont, Calif., 1981. Reprinted with permission.)

In this chapter we examine the structures and functions of biological membranes. Since most of our current understanding of membrane structure and function stems from research done only over the last few decades, this field continues to be one of the most rapidly expanding in biochemistry. Nevertheless, *a number of fundamental principles have emerged that appear to apply to most membrane systems that have been studied*. For this reason, we shall examine aspects of membrane structure and function in systems as seemingly divergent as bacteria and mammalian mitochondria. In Chapters 16 and 34 we will find that in both cases, the biosynthesis of ATP is a membrane-associated process that occurs by a similar mechanism. To take another example, membranes of *Escherichia coli* contain a protein that behaves in artificial membrane systems much like nerve cell membrane channels, which are partly responsible for propagation of the nerve impulse in vertebrates (Chapters 34 and 35). Thus many mechanisms responsible for complex membrane phenomena are undoubtedly used repeatedly throughout the living kingdom. We therefore consider the structures and functions of the E. *coli* cell envelope to be of equal importance with those of the human erythrocyte (red blood cell) membrane.

TABLE 6–1
Chemical Compositions of Some Cell Membranes

Membrane	Protein (%)	Lipid (%)	Carbohydrate (%)
Myelin	18	79	3
Human erythrocyte plasma membrane	49	43	8
Amoeba plasma membrane	54	42	4
Mycoplasma cell membrane	58	37	1.5
Halobacterium purple membrane	75	25	0

Source: Adapted from G. Guidotti, Membrane proteins. *Ann. Rev. Biochem.* 41:731, 1972.

CONSTITUENTS OF BIOLOGICAL MEMBRANES

Typically, a biological membrane contains lipid, protein, and carbohydrate in ratios varying with the source of the membrane (Table 6–1). Nearly always, the carbohydrate is covalently associated with protein (*glycoproteins*) or with lipid (*glycolipids* and *lipopolysaccharides*). Thus the membrane can be thought of as a lipid-protein matrix in which specific functions are carried out by proteins, while the permeability barrier and the structural integrity of the membrane are provided by membrane lipids. As we shall see, current evidence favors a model of membrane structure that agrees with this overall interpretation of the roles of protein and lipid in biological membranes.

Membranes from Eukaryotic Cells Have Been Isolated and Analyzed

In order to study the structure of biological membranes, it is first necessary to isolate them in a more or less intact form from the cell. In eukaryotic cells, this problem is complicated by the existence of several different membrane systems in addition to the plasma membrane, each surrounding a specific organelle. Separation of membrane fractions initially requires the disruption of the plasma membrane under conditions that leave subcellular organelles intact. One common procedure involves <u>mild homogenization</u> in a slightly hypotonic solution,* while another method, <u>nitrogen cavitation</u>, involves forcing nitrogen gas into the cells under pressure and then rapidly releasing the pressure to "explode" the cell membrane.

Organelles can be isolated from disrupted cells by *differential centrifugation*, which separates them on the basis of size (Figure 6–3). Ruptured plasma membrane fragments can be purified from the same mixture by equilibrium-density-gradient (*isopycnic*) centrifugation because of their low density (high lipid content) relative to intact organelles (Table 6–2, column 4). This technique relies on centrifuging the sample into a preformed gradient of a solute, such as sucrose. When equilibrium is reached, each type of membrane or organelle is found in the region of the gradient corresponding to its own density (Figure 6–3). Gradients of synthetic sucrose polymers (Ficoll) or colloidal silica particles (Percoll) also are used in these separations because of their inertness, ability to form more stable gradients, and impermeability to biological membranes.

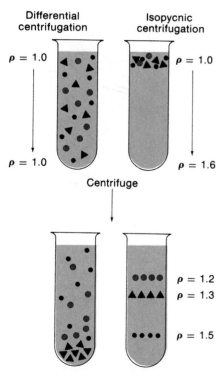

Differential centrifugation

$\rho = 1.0$

$\rho = 1.0$

Isopycnic centrifugation

$\rho = 1.0$

$\rho = 1.6$

Centrifuge

$\rho = 1.2$
$\rho = 1.3$

$\rho = 1.5$

Figure 6–3
Comparison of differential centrifugation (left), which separates on the basis of size, and isopycnic centrifugation (right), which separates on the basis of density. ρ is the density in g/milliliter.

*In a hypotonic solution the osmolarity is below the normal physiologic value.

TABLE 6–2
Properties of Rat Liver Organelles

Organelle	Per Cent of Cell Protein	Diameter (μm)	Equilibrium Density in Sucrose (g/ml)	Organelle-Specific Enzyme Marker
Liver cell	100	20	1.20	—
Nuclei	15	5–10	1.32	DNA polymerase
Golgi apparatus	2	2	1.10	Glycosyl transferases
Mitochondria	25	1	1.20	Monoamine oxidase (outer membrane); cytochrome c (inner membrane)
Lysosomes	2	0.5	1.20	Acid phosphatase
Endoplasmic reticular vesicles	20	0.1	1.15	Cytochrome b_5 reductase and cytochrome b_5; glucose-6-phosphatase
Cytoplasmic membrane	2	—	1.15	Na$^+$-K$^+$ ATPase; viral receptors
Soluble protein	30	<0.01	—	—

Source: M. H. Saier, Jr., and C. D. Stiles, *Molecular Dynamics in Biological Membranes,*
Heidelberg Science Library, Vol. 22, Springer Verlag, New York, 1975. Reprinted with
permission.

The properties of isolated rat liver organelles are summarized in Table 6–2. The entries in column 2 provide some idea of the relative proportions of these organelles in the mammalian liver. For example, mitochondria represent 25 per cent of the total cell protein, while lysosomes comprise about 2 per cent. Interestingly, the plasma membrane, which completely surrounds the cell, also represents only 2 per cent of the total protein. Although soluble proteins comprise only 30 per cent of the total, the isolated organelles also contain soluble protein, so that somewhat less than 50 per cent of the total cell protein is probably membrane-associated.

In column 3 of Table 6–2, the relative sizes of the organelles are indicated. It should be noted that the sedimentation behavior of an organelle in a sucrose density gradient (column 4) does *not* correlate with size, but rather with its density, which is determined by its chemical composition (Figure 6–3). Nucleic acid ($\rho \sim 1.7$) is more dense than protein ($\rho \sim 1.25$), and protein is more dense than lipid ($\rho = 0.9$–1.1). These facts account for the relatively high density of nuclei and the low density of the Golgi apparatus, which has a relatively high lipid content.

Since each organelle has a specific function, it also must possess a unique complement of enzymes. This prediction is verified by the subcellular localization of numerous enzymes (Table 6–2, column 5), and these specific associations have greatly facilitated the assay and isolation of organelles from eukaryotic cells.

Once the subcellular organelles have been separated, their membranes can be isolated. For those organelles enclosed by a single membrane, treatment in hypotonic buffer (*osmotic shock*) followed by centrifugal separation of the membrane fragments from the intraorganellar soluble proteins allows one to study membrane composition. Nuclei and mitochondria, however, possess two membranes, and these must be separated before their chemical and physical properties can be studied. In these cases, selective solubilization of the outer membrane can be obtained by treatment with appropriate detergents, allowing purification of intact inner membranes. Procedures such as these have made it possible to analyze in detail the lipid and protein contents of organellar membranes (Table 6–3) and have provided experimental systems in which to study the structures and functions of each different membrane system.

TABLE 6-3
Protein and Lipid Content of Organellar Membranes

Membrane	Approximate Protein/Lipid Ratio (wt/wt)	Approximate Cholesterol/ Other Lipids (Molar Ratio)
Golgi apparatus	0.7	0.08
Liver plasma membrane	1.0	0.40
Endoplasmic reticulum	1.0	0.06
Mitochondrial outer membrane	1.0	0.05
Mitochondrial inner membrane	3.0	0.03
Nuclear membrane	3.0	0.11
Lysosomal membrane	3.0	0.16

Source: Adapted from M. H. Saier, Jr., and C. D. Stiles, *Molecular Dynamics in Biological Membranes*, Heidelberg Science Library, Vol. 22, Springer Verlag, New York, 1975.

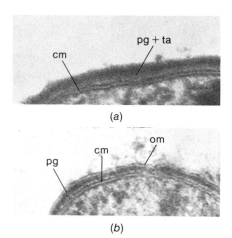

Figure 6-4
Electron micrographs of sections through the surface layers of (a) a Gram positive and (b) a Gram negative bacterium. (cm = cytoplasmic membrane; om = outer membrane; pg = peptidoglycan; ta = teichoic acid.) Note the thick cell wall in (a), compared with the distinct inner and outer trilaminar membranes separated by a thin peptidoglycan layer in (b). (Courtesy J. Stolz; 150,000×.)

Bacterial Cell Envelopes Contain One or Two Membranes

In contrast to animal cells, most prokaryotic cells are surrounded by a rather complex and rigid *cell wall*, which allows bacteria to live in a hypotonic environment without bursting and confers upon these cells their characteristic shape (rod, sphere, or spiral). In 1884, Christian Gram discovered that bacteria could be divided into those which retained a crystal violet–iodine dye complex after washing with alcohol *(Gram positive)* and those that did not *(Gram negative)*. Even today, the Gram stain reaction is a useful tool in classifying bacteria, and this difference in staining has been found to correlate with a fundamental difference in cell wall structure between Gram positive and Gram negative cells (Figure 6–4). Gram positive cells are surrounded by a cytoplasmic membrane and a thick cell wall consisting of a sugar–amino acid heteropolymer, or *peptidoglycan* (Figure 4–26), and polyol phosphate polymers called *teichoic acids* (Figure 6–5a). Gram negative bacteria have a much thinner cell wall consisting entirely of peptidoyglycan and associated proteins, and this cell wall is surrounded by a second, outer membrane comprised of lipid, *lipopolysaccharide* (Figure 6–5b), and protein. The biosyntheses of the peptidoglycan and the outer membrane lipopolysaccharide are discussed in Chapter 20. The space between the inner and outer membranes, or *periplasmic space*, also contains proteins that have a variety of functions (see the following discussion).

In order to examine the compositions and functions of the two cell layers of Gram negative bacteria, it is necessary to first separate these layers. This has been accomplished by treatment of the cells with *lysozyme* (which hydrolyzes peptidoglycan) and EDTA (which destabilizes the outer membrane) in isoosmotic* sucrose solutions (Figure 6–6). *Periplasmic proteins* are released by this first step and can be separated by sedimenting the resulting *spheroplasts*, which have lost any nonspherical shape characteristic of the original cell because their peptidoglycan cell wall has been digested. Subsequent treatment of spheroplasts with high-frequency sound *(sonication)* ruptures both outer and inner membranes, which quickly reseal into smaller spherical, closed *vesicles* (Figure 6–6). Because of their higher carbohydrate content, vesicles derived from the outer membrane have a higher density than those derived from the inner membrane and thus can be sepa-

*An isoosmotic sucrose solution has the same osmolarity as is found under normal physiological conditions.

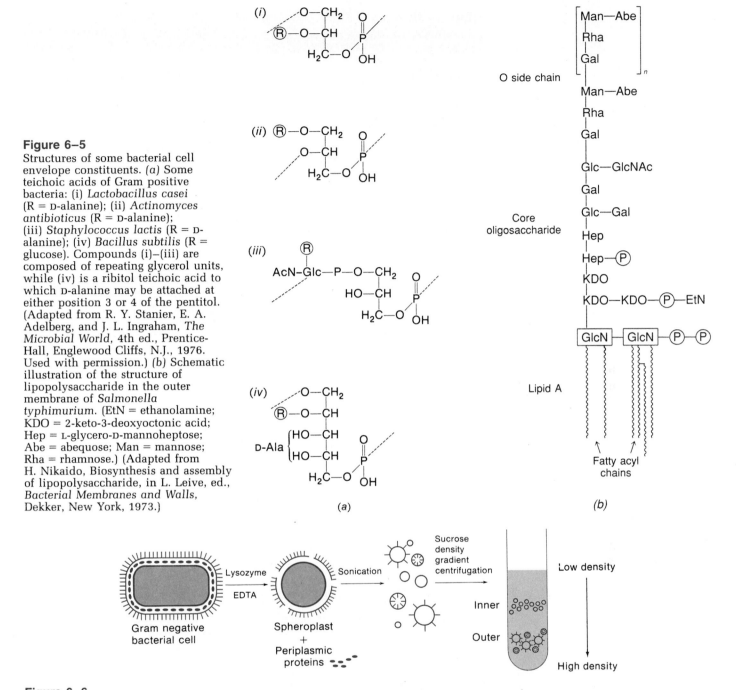

Figure 6–5
Structures of some bacterial cell envelope constituents. *(a)* Some teichoic acids of Gram positive bacteria: (i) *Lactobacillus casei* (R = D-alanine); (ii) *Actinomyces antibioticus* (R = D-alanine); (iii) *Staphylococcus lactis* (R = D-alanine); (iv) *Bacillus subtilis* (R = glucose). Compounds (i)–(iii) are composed of repeating glycerol units, while (iv) is a ribitol teichoic acid to which D-alanine may be attached at either position 3 or 4 of the pentitol. (Adapted from R. Y. Stanier, E. A. Adelberg, and J. L. Ingraham, *The Microbial World*, 4th ed., Prentice-Hall, Englewood Cliffs, N.J., 1976. Used with permission.) *(b)* Schematic illustration of the structure of lipopolysaccharide in the outer membrane of *Salmonella typhimurium*. (EtN = ethanolamine; KDO = 2-keto-3-deoxyoctonic acid; Hep = L-glycero-D-mannoheptose; Abe = abequose; Man = mannose; Rha = rhamnose.) (Adapted from H. Nikaido, Biosynthesis and assembly of lipopolysaccharide, in L. Leive, ed., *Bacterial Membranes and Walls*, Dekker, New York, 1973.)

Figure 6–6
Separation of the periplasmic proteins and inner and outer membranes of a Gram negative bacterium. The periplasmic proteins are those proteins that are concentrated in the space between the inner and outer membranes. By treating intact cells with lysozyme and EDTA in an isotonic solution, the periplasmic proteins are released, leaving the spheroplasts. The proteins are removed by sedimentation; then the spheroplasts are suspended in a buffered saline solution and sonicated. This treatment ruptures the outer and inner membranes into smaller spherical closed vesicles. These may be separated by centrifugation on a sucrose density gradient. The outer membrane vesicles sediment to a lower point in the centrifuge tube because of their higher density.

rated from them in a sucrose density gradient. By these techniques, electron transport chains, ATP synthesizing enzymes, many transport proteins, and other enzymes have been localized to the cytoplasmic membrane of Gram negative bacteria, while the outer membrane has been shown to harbor receptors for bacteriophage and bacteriocins, certain other transport proteins,

TABLE 6–4
Subcellular Location of Proteins and Nucleic Acids in Gram Negative Bacteria

Extracellular

Proteases
Lipases
Carbohydrases
Nucleases

Outer Lipopolysaccharide Membrane

Receptor proteins for viruses and bacteriocidal agents
Phospholipase A and lysophospholipase
Specific and nonspecific pore proteins

Periplasm

Solute-binding proteins involved in transmembrane transport and chemotaxis
Phosphatases and esterases

Inner Cytoplasmic Membrane

Electron-transfer chain
Proton-translocating ATPase
Some transport proteins
Lipid and cell envelope biosynthetic enzymes

Cytoplasm

Enzymes that catalyze the synthesis and degradation of soluble substrates
DNA
RNA
Ribosomes
Enzymes involved in DNA replication, transcription, translation, etc.

Source: M. H. Saier, Jr., and C. D. Stiles, *Molecular Dynamics in Biological Membranes*, Heidelberg Science Library, Vol. 22, Springer Verlag, New York, 1975. Reprinted with permission.

and various phospholipases. The periplasmic space has been shown to contain hydrolytic enzymes as well as nutrient-binding proteins involved in transmembrane transport and chemotaxis (Chapter 34). These observations on protein localization in Gram negative bacteria are summarized in Table 6–4. The organization of proteins in Gram positive bacteria is usually much simpler because these cells are surrounded by a single membrane. Thus soluble and cytoplasmic membrane proteins carry out functions similar to those of Gram negative bacteria, while macromolecular hydrolase enzymes are exported into the extracellular medium, where they function to scavenge nutrients from the environment.

Membrane Lipids Are a Complex Mixture

Once isolated by any of the procedures just described, membrane fractions can be analyzed biochemically for their content of lipids and proteins. Lipids are usually extracted from proteins by treatment of membranes with organic solvents, such as chloroform and methanol. Numerous studies have shown that the most common lipids of biological membranes are the phospholipids, including the phosphoglycerides and the sphingolipids (Chapter 5). Most biological membranes contain predominantly the phosphoglycerides listed in Table 5–3 and/or sphingomyelin (Figure 5–11). Certain membrane systems, however, may be enriched in special types of phospholipids. For example, the membranes of nerve and muscle cells, in contrast to most other types of cells, are especially enriched in plasmalogens and glycosphingolipids (also see the discussion that follows).

Because a variety of fatty acids (Chapter 5) may be found in any of the phosphoglycerides or sphingomyelin, one usually isolates a mixture of compounds when purifying a particular phospholipid class from a biological membrane. Thus, in theory at least, a sample of purified lecithin could consist of a mixture of 1-palmitoyl-2-stearoyl-phosphatidylcholine and 1-oleoyl-2-linoleoyl-phosphatidylcholine (Figure 6–7). In practice, however,

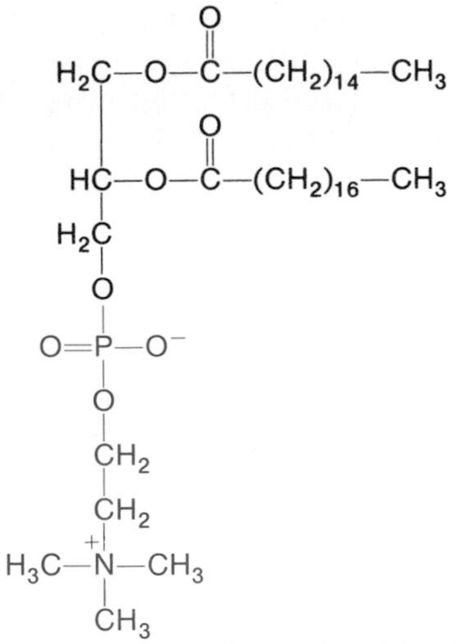

1-Palmitoyl-2-stearoyl-phosphatidylcholine

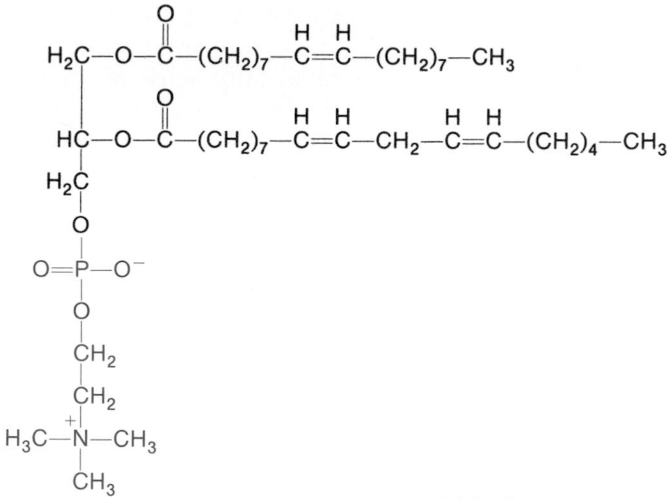

1-Oleoyl-2-linoleoyl-phosphatidylcholine

Figure 6-7
Two molecules of phosphatidylcholine (lecithin) with different fatty acyl groups. Phosphatidylcholine is really a heterogeneous class of compounds. Also see Figure 6-8. The phosphocholine parts are shown in color.

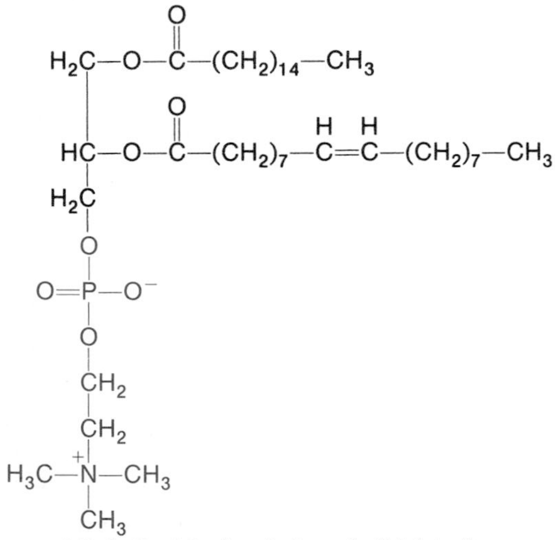

1-Palmitoyl-2-oleoyl-phosphatidylcholine

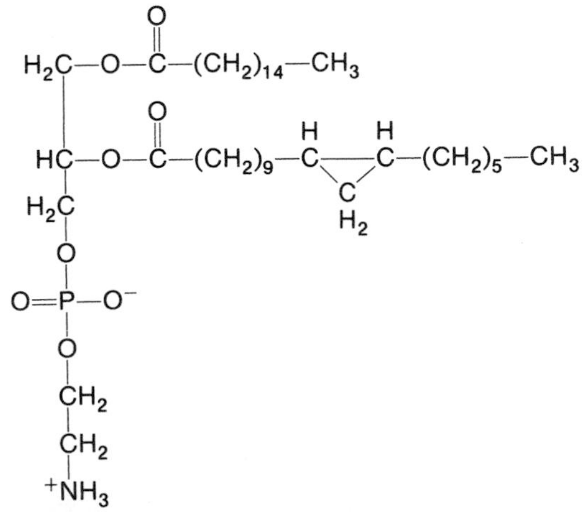

1-Palmitoyl-2-lactobacilloyl-phosphatidylethanolamine

Figure 6-8
A common constituent of bovine phosphatidylcholine. The phosphocholine part is shown in color.

Figure 6-9
A bacterial phospholipid containing the cyclopropane fatty acid lactobacillic acid at the C-2 position. The cyclopropane fatty acids are among the few types of fatty acids that contain an odd number of carbon atoms.

phosphoglycerides usually contain one saturated fatty acid at position 1 and an unsaturated fatty acid or cyclopropane fatty acid (in bacteria) in position 2 of the glycerol moiety (Figure 6-8). A common constituent of bovine lecithin is found to be 1-palmitoyl-2-oleoylphosphatidylcholine, while a typical bacterial phosphatidylethanolamine might have the structure shown in Figure 6-9.

An examination of the fatty acid compositions that have been determined for various phospholipids from biological sources indicates the variety of compounds that a given class of phospholipids can encompass because of variable fatty acid compositions (see Chapter 5). Nearly all fatty acids found in phospholipids from biological sources have an *even* number (14-24) of carbon atoms (except for cyclopropane fatty acids) (Figure 6-9).

An exception to this rule is found in some marine organisms, which contain significant amounts of fatty acids with odd numbers of carbons. The unsaturated fatty acids of phospholipids may have one, two, three, or even four double bonds. In most bacteria, however, unsaturated fatty acids containing more than one double bond are lacking. Furthermore, the majority of unsaturated fatty acids found in naturally occurring phospholipids are of the *cis* configuration. The importance of this fact will become evident when we discuss the physical properties of membranes.

Lipopolysaccharide, discussed earlier as a component of the outer membrane of Gram negative bacteria, appears to be limited to this class of organisms. Although it generally has a higher content of carbohydrate than lipid by weight, it also is often classified as a membrane lipid because its fatty acyl chains constitute an integral part of the structure of the Gram negative outer membrane. Among other functions, lipopolysaccharide confers upon a bacterium its characteristic antigenicity because the terminal carbohydrate chains (O antigens) extend to the surface of the cell (Figure 6–5). This subject was discussed in the last section of Chapter 4. Lipopolysaccharide is also the major component of *endotoxins*, which are released upon lysis of Gram negative bacteria. Endotoxins are potent pyrogens (fever producers) and contribute to the pathogenicity of a number of these organisms.

The final class of lipids commonly found in biological membranes are the *steroids* and their biosynthetic precursors, the *isoprenoids* or *terpenes*. The most common membrane steroid is cholesterol (Figure 5–17), which is found in animal cell membranes and a few microorganisms. Some of the more important terpenes found in biological membranes were considered in Chapter 5. They include *11-cis-retinal*, a derivative of vitamin A (Figure 5–24) that is found in the vertebrate retina (Chapter 36); *phytol* (Figure 5–25), a component of the chlorophylls; and *undecaprenol* (or *bactoprenol*, Figure 6–10), an important factor in cell wall biosynthesis in bacteria (Chapter 20). *Quinones*, one class of polyketides, are important lipids in the inner mitochondrial membrane, where they function as hydrogen carriers (Chapter 16).

Table 6–5 compares the lipid compositions of membranes from a number of biological sources. Several generalizations can be made from the data in this table. First, phosphatidylcholine is the major phospholipid found in membranes of animal cells, while phosphatidylethanolamine predominates in bacteria. Second, in addition to cholesterol, both sphingomyelin and glycolipids (except for lipopolysaccharides) are usually absent from prokaryotic membranes. Most of the membranes represented in the table, however, have a characteristic trilaminar appearance in the electron microscope (Figure 6–1) and yet have widely varying lipid compositions. Thus either membrane lipids have little to do with determining membrane structure, or they all have in common certain properties that allow them to contribute in a similar manner to the characteristic architecture of biological membranes. As we shall see in the following sections, the latter explanation is undoubtedly the correct one.

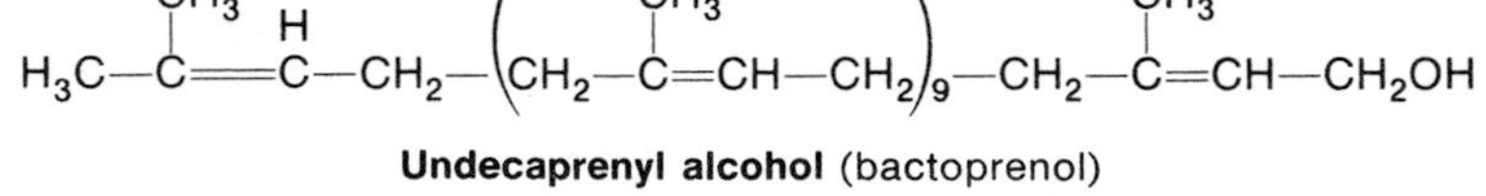

$$H_3C-\underset{\underset{CH_3}{|}}{C}=\underset{\underset{H}{|}}{C}-CH_2-\left(CH_2-\underset{\underset{CH_3}{|}}{C}=CH-CH_2\right)_9-CH_2-\underset{\underset{CH_3}{|}}{C}=CH-CH_2OH$$

Undecaprenyl alcohol (bactoprenol)

Figure 6–10
Undecaprenol (bactoprenol) is an isoprenoid lipid found in bacteria, where it functions in cell wall synthesis and lipopolysaccharide synthesis. The role in cell wall synthesis is discussed in Chapter 20.

TABLE 6–5
Lipid Compositions of Membrane Preparations

Source	Lipid Composition (lipid per cent)[a]									
	Choles-terol	PC	SM	PE	PI	PS	PG	DPG	PA	Glyco-lipids
Rat liver										
Cytoplasmic membrane	30.0	18	14.0	11	4.0	9.0	—	—	1	—
Endoplasmic reticulum (rough)	6.0	55	3.0	16	8.0	3.0	—	—	—	—
Endoplasmic reticulum (smooth)	10.0	55	12.0	21	6.7	—	—	1.9	—	—
Mitochondria (inner)	3.0	45	2.5	25	6.0	1.0	2.0	18.0	0.7	—
Mitochondria (outer)	5.0	50	5.0	23	13.0	2.0	2.5	3.5	1.3	—
Nuclear membrane	10.0	55	3.0	20	7.0	3.0	—	—	1.0	—
Golgi	7.5	40	10.0	15	6.0	3.5	—	—	—	—
Lysosomes	14.0	25	24.0	13	7.0	—	—	5.0	—	—
Rat brain										
Myelin	22.0	11	6.0	14	—	7.0	—	—	—	21
Synaptosome	20.0	24	3.5	20	2.0	8.0	—	—	1.0	—
Rat erythrocyte	24.0	31	8.5	15	2.2	7.0	—	—	0.1	3
Rat rod cell (outer segment)	3.0	41	—	37	2.0	13.0	—	—	—	—
E. coli cytoplasmic membrane	0	0	—	80	—	—	15.0	5.0	—	—
Bacillus megaterium cytoplasmic membrane	0	0	—	69	—	—	30.0	1.0	—	Trace

[a]PC = phosphatidylcholine; SM = sphingomyelin; PE = phosphatidylethanolamine;
PI = phosphatidylinositol; PS = phosphatidylserine; PG = phosphatidylglycerol;
DPG = diphosphatidylglycerol (cardiolipin); PA = phosphatidic acid.

Source: Adapted from M. K. Jain and R. C. Wagner, *Introduction to Biological Membranes*, Wiley, New York, 1980; the data for *B. megaterium* are from J. E. Rothman and E. P. Kennedy, Asymmetrical distribution of phospholipids in the membrane of *Bacillus megaterium. J. Mol. Biol.* 110:603, 1977.

Membrane Lipids Spontaneously Form Ordered Structures

One characteristic common to all of the membrane lipids we have discussed is the fact that they are amphipathic, as described in Chapter 5. Phospholipids consist of a pair of long hydrophobic tails, which are the fatty acyl chains, and a compact polar head group composed of the phosphate group and the esterified alcohol. At neutral pH, the polar head group may have a net negative charge (phosphatidic acid, phosphatidylglycerol, diphosphatidylglycerol, phosphatidylinositol, phosphatidylserine), positive charge (*O-lysylphosphatidylglycerol*), or no net charge (phosphatidylcholine, phosphatidylethanolamine). Sphingomyelin, although differing considerably in chemical structure from the phosphoglycerides, nevertheless appears very similar to the other phospholipids in space-filling models (see Chapter 5). Similarly, glycolipids and bacterial lipopolysaccharide have the hydrocarbon tails and polar head groups (in this case, sugar residues). Even cholesterol has hydrophobic and hydrophilic "sides," although differing altogether in chemical structure from the fatty acid–containing lipids.

Because of their amphipathic nature, membrane lipids exhibit only a limited solubility in aqueous solution. Thus, when phospholipid is added to water, very few lipid molecules exist freely in solution as monomers because of the large hydrophobic surface area of the molecule. Instead, a "film" of phospholipid tends to form on the water–air interface. Physical studies have shown that this film is a monolayer of phospholipid arranged such that the polar head groups are in contact with water, while the hydrocarbon tails

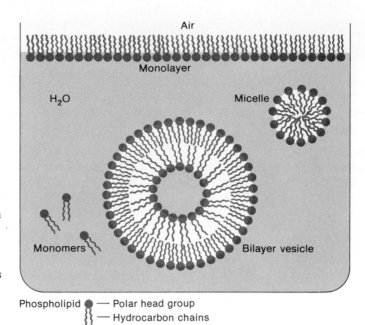

Figure 6–11
Structures formed by phospholipids in aqueous solution. Each molecule is depicted schematically as a polar headgroup (●) attached to two fatty acyl hydrocarbon chains. The monolayer at the air–water interface is the first to form. When this interface has become saturated, further phospholipid forms micelles and bilayer vesicles.

extend up into the air phase (Figure 6–11). When more phospholipid is added to the solution, saturating the air–water interface, other assemblages of phospholipids are formed, including _micelles_ and _bilayers_ (see Figure 6–11). Both these structures maximize hydrophobic and van der Waals interactions between the fatty acyl chains, effectively excluding water from their vicinity, and allow the polar head groups to interact with water molecules. Monolayers, micelles, and bilayers are the favored forms of phospholipid in aqueous solution because their formation results in an _increase_ in entropy, resulting from the fact that water molecules need not order themselves around the hydrophobic hydrocarbon tails of the phospholipid monomer.

As shown in Figure 6–11, phospholipid bilayers in aqueous solution are actually spherical "bubbles" or vesicles with water inside and out. This structure is favored over a planar bilayer because exposed hydrocarbon tails, which would occur at the periphery of a planar sheet of phospholipid, are not present. In vesicular structures, no hydrophobic groups need to be exposed to water molecules. In addition, most naturally occurring phospholipids prefer to form vesicular bilayers instead of micelles in water solution, because more efficient packing of the molecules can take place in the bilayer vesicle. In contrast, _lysophospholipids_ (which lack one fatty acyl chain), free fatty acids, and detergents (see next section) form micelles more readily than bilayers because of their geometry, which includes a smaller hydrophobic surface area relative to the phospholipids.

The implications of these properties of phospholipids and glycolipids with respect to biological membrane structure are as follows.

1. In aqueous solution, these molecules _spontaneously_ form bilayer structures; i.e., the ΔG for this process is negative owing to an increase in entropy, ΔS (recall that $\Delta G = \Delta H - T\,\Delta S$).
2. A cell is essentially a plasma membrane–encapsulated vesicle, although it is usually not a perfect sphere owing to other structural contributions to its shape.
3. The arrangement of phospholipid polar head groups in bilayers suggests an explanation for the trilaminar appearance of biological membranes in the electron microscope (Figure 6–1). Indeed, electron micrographs of

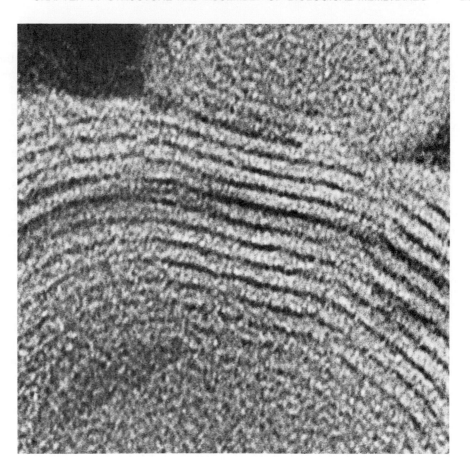

Figure 6–12
Multilayered vesicles formed from sonically dispersed lecithin containing 10% dicetylphosphate in 2% potassium phosphotungstate. Note the trilaminar structure consisting of two dark, electron-dense layers separated by a lighter layer that corresponds to the electron-sparse hydrophobic interior of each bilayer. (A. D. Bangham, M. M. Standish, and J. C. Watkins, Diffusion of univalent ions across the lamellae of swollen phospholipids, *J. Mol. Biol.* 13:238, 1965. Reprinted with permission.)

purified phospholipid dispersed in water have a similar appearance (Figure 6–12).

4. Phospholipid bilayers are relatively impermeable to most hydrophilic substances because of their hydrophobic interiors. This is also a property of biological membranes, unless a <u>transport system</u> recognizes the hydrophilic molecule (Chapter 34).

We are left with the conclusion that *biological membranes must themselves consist at least partially of lipid bilayers,* and that lipids must have a very important role in determining membrane structure. As we shall see shortly, both these conclusions are overwhelmingly confirmed by physical and chemical studies of biological membrane structure.

Membrane Proteins Are Integral or Peripheral

Although some characteristics of biological membranes can be explained by the properties of membrane lipids in aqueous solution, other characteristics, especially specific functions such as transport and enzymatic activities, must rely on membrane-associated proteins. Therefore, in any model of membrane structure, we also must consider the properties of proteins found in biological membranes.

General Properties and Isolation. Two general types of such proteins have been found in membranes: *peripheral* membrane proteins and *integral* membrane proteins. Peripheral membrane proteins are dissociable from the isolated membrane by agents that tend to disrupt ionic or hydrogen bonds—for

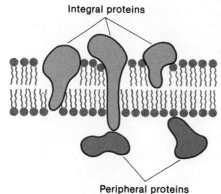

Integral proteins

Peripheral proteins

Figure 6–13
Schematic illustration of integral and peripheral membrane proteins. Peripheral membrane proteins may be removed by mild reagents without disruption of the membrane structure. Integral membrane proteins can be released only if the membranes are disrupted.

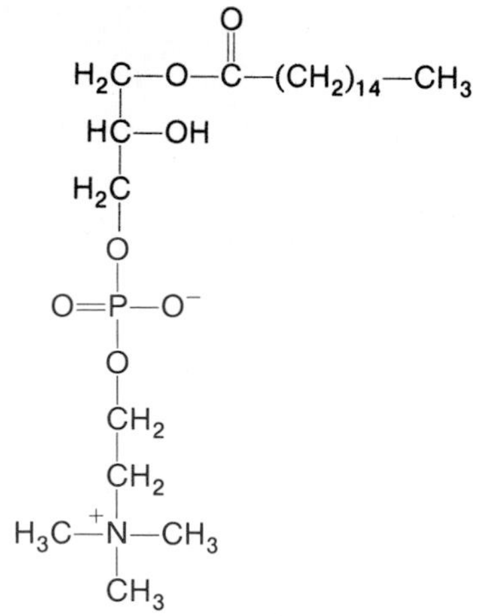

example, by high salt, by EDTA, which chelates Ca^{2+} and Mg^{2+} ions, or by urea. Integral membrane proteins, however, can be released from the membrane only by disrupting the hydrophobic interactions of membrane lipids by organic solvents or detergents. Peripheral membrane proteins probably are bound to the membrane as a result of specific interactions with exposed, hydrophilic portions of integral membrane proteins. In contrast, integral membrane proteins appear to be deeply embedded in the membrane. Significant hydrophobic interactions with membrane lipids and proteins probably are responsible for the interaction properties of integral membrane proteins. Figure 6–13 illustrates these differences, assuming that the fundamental structure of the biological membrane is, indeed, a lipid bilayer.

In order to understand the physical and chemical properties of integral membrane proteins, it is necessary to purify them. First, the proteins must be dissociated from the membrane matrix. As was just mentioned, integral membrane proteins often can be brought into solution using organic solvents or *detergents*. Detergents are naturally occurring or synthetic amphipathic molecules that disrupt membranes by intercalation into the membrane matrix and solubilization of the component lipids and proteins. Examples of detergents that are natural products include *lysolecithin* (Figure 6–14) and *sodium deoxycholate*, a steroid bile salt (Figure 5–19). Both are *ionic* detergents. Synthetic detergents include *Triton X-100* and *octylglucoside* (Figure 6–15), two *nonionic* detergents, and the ionic compounds *cetyl trimethylammonium bromide* and *sodium dodecyl sulfate* (SDS) (Figure 6–16), which is also an extremely effective protein denaturant.

Many detergents, including the nonionic ones and lysolecithin, dissolve membranes by forming detergent-lipid and detergent-lipid-protein *mixed micelles* (Figure 6–17). These detergents prefer to form micelles rather than

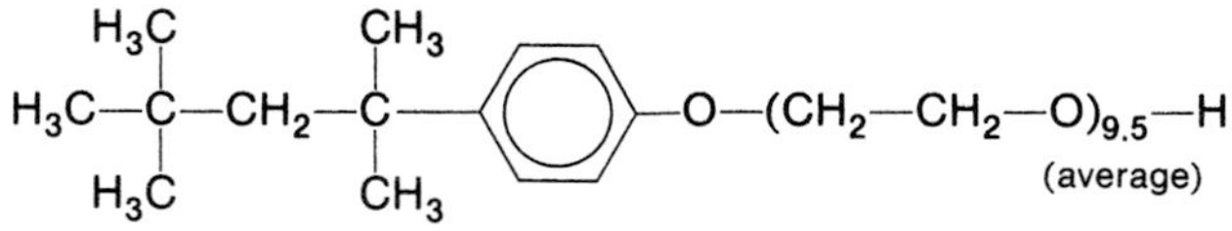

A lysolecithin

Figure 6–14
A naturally occurring lipid detergent. In lysolecithin the fatty acyl group is always at the C-1 position. The phosphocholine part is shown in color.

Triton X-100
[polyoxyethylene(9.5)*p-t*-octylphenol]

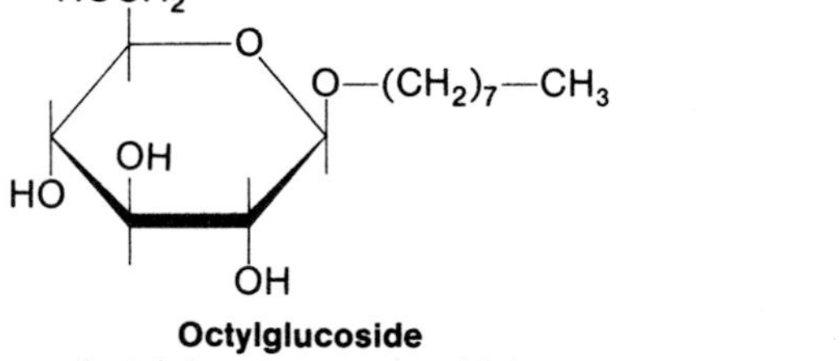

Octylglucoside
(octyl-β-D-glucopyranoside)

Figure 6–15
Two synthetically derived nonionic detergents. These detergents are useful for dissolving membranes.

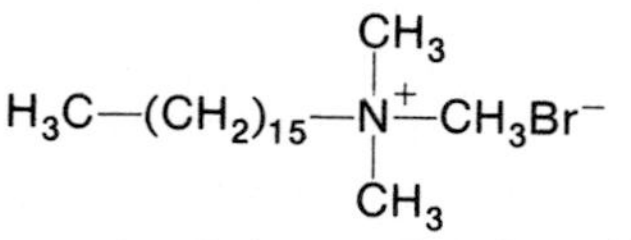

Cetyl trimethylammonium bromide

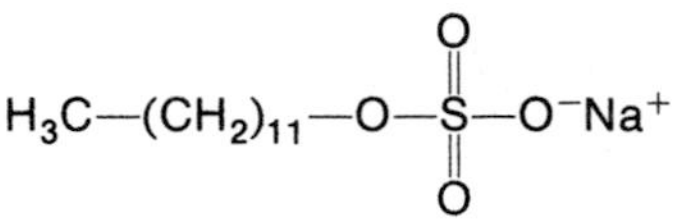

Sodium dodecyl sulfate (SDS)

Figure 6–16
Two additional synthetically derived ionic detergents that can dissolve membranes. Sodium dodecylsulfate is also an extremely effective protein denaturant and is commonly used in protein gel electrophoresis (see Chapter 3).

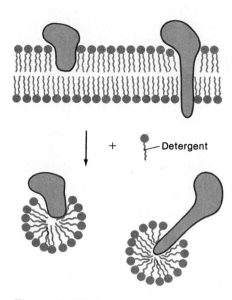

Figure 6–17
Detergent solubilization of biological membranes yields detergent-lipid-protein mixed micelles. The detergent disrupts the membrane and makes a complex with the hydrophobic portion of the integral membrane protein and the membrane lipid.

bilayers because of their particular geometries. Thus an excess of these compounds, when added to a membrane suspension, will tend to shift the equilibrium of all amphipathic molecules in the mixture from bilayer to mixed micelle (Figure 6–17).

Each detergent has a characteristic *critical micellar concentration* (CMC), above which it exists in aqueous solution almost entirely in a micellar form, and a characteristic *hydrophilic-lipophilic balance* (HLB), which is defined as the ratio of the molecular weight of the hydrophilic portion of the molecule to that of the hydrophobic portion. Both these properties appear to be important in determining the effectiveness of a particular detergent in dissolving membranes, although different membrane systems often respond differently to the same detergent.

Some of the properties of detergents commonly used in membrane biochemistry are listed in Table 6–6. Because ionic detergents are more likely to alter the conformation of hydrophilic portions of membrane proteins, which are often responsible for catalytic activity, they are more likely to destroy biological function. This observation is in agreement with the fact that only ionic detergents bind to and denature soluble proteins. Little interaction is usually observed between typical hydrophilic proteins and nonionic detergents.

Once dissociated from the membrane with detergent, specific membrane proteins can be isolated by a variety of separation techniques, provided an assay is available for the protein of interest. Gel filtration and other procedures that separate proteins on the basis of size and shape (Chapter 3) are often not particularly useful in this regard because inclusion of integral membrane proteins in detergent-lipid micelles, which can have molecular weights of 10^5 or greater, tends to mask individual size differences. Only with detergents possessing a high CMC, such as deoxycholate and octylglucoside, are separations based on size generally possible. Ion-exchange chromatography (Chapter 3) can be useful in separating solubilized membrane proteins if the detergent used is nonionic, such as Triton X-100. These difficulties have led to the search for techniques that might be especially suited to membrane protein isolation.

One tool found to be particularly useful in this regard is *hydrophobic interaction chromatography*. This technique separates proteins on the basis of their relative hydrophobicities and uses insoluble supports, such as agarose or polyacrylamide, to which hydrophobic alkyl or aryl groups have been covalently attached. Liquid chromatography columns of such *hydro-*

TABLE 6–6
Properties of Some Detergents

Detergent	Monomer M_r	CMC (mM)[a]	Micellar M_r (average)[a]
Nonionic			
Triton X-100	625	0.24	90,000
Octyl-β-D-glucoside	292	25	—
Ionic			
Deoxycholate (anion)	392	4	800
Lysolecithin (egg; mixture)	500–600	0.02–0.2	95,000
SDS	288	8	17,000

[a]It should be noted that both CMC and micellar size are dependent on a number of factors, including ionic strength and pH. These numbers should therefore be used only for general comparisons, since each was determined under slightly different conditions.

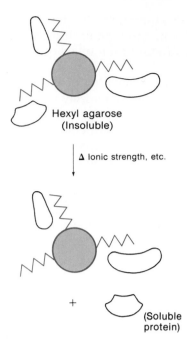

Figure 6–18

Hydrophobic interaction
chromatography. Proteins bound by
hydrophobic interactions to insoluble
alkyl agarose beads can be
differentially removed by changes in
ionic strength, hydrophobicity, pH, or
temperature. General chromatographic
procedures are discussed in Chapter 3.

phobic resins have been used in the purification of a number of integral
membrane proteins. Proteins bound to the resin by means of hydrophobic
interactions can then often be eluted sequentially, depending on the strength
of this interaction, by gradual changes in hydrophobicity, ionic strength, pH,
or temperature of the eluting buffer (Figure 6–18).

Affinity chromatography, a technique that is very useful in purifying
soluble enzymes (Chapter 3), also has been applied successfully to a number
of integral membrane proteins, such as transport proteins (Chapter 34),
which interact with specific metabolites. Ammonium sulfate and isoelectric
precipitation, isoelectric focusing, and preparative gel electrophoresis also
have been used in the purification of integral membrane proteins in the
presence of suitable solubilizing agents (see Chapter 3).

The molecular weights of membrane proteins and the complexity of a
specific membrane can be estimated by the technique of polyacrylamide gel
electrophoresis, after the membrane has been completely solubilized in the
ionic detergent SDS (Chapter 3). An example of this technique for examining
proteins of the human erythrocyte membrane is shown in Figure 6–19. In
Table 6–7, some physical and chemical properties of a number of integral
membrane proteins that have been purified to apparent homogeneity are
compared. Some of these proteins, such as bacteriorhodopsin and the lactose
permease from _E. coli_, are highly enriched in hydrophobic amino acids. This
property may reflect the proportion of the polypeptide chain embedded in
the hydrophobic portion of the membrane (see the following discussion).
One property, however, that seems to be common to many isolated integral
membrane proteins is that they usually _require a hydrophobic environment_
for maintenance of their biologically active structures. Techniques that remove detergent and residual phospholipid from these proteins often lead to

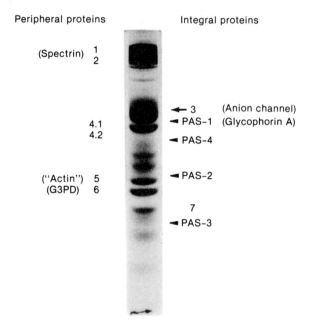

Figure 6–19

Proteins of the human erythrocyte membrane resolved by polyacrylamide gel
electrophoresis in the presence of sodium dodecyl sulfate. Protein bands were stained
with the dye Coomassie brilliant blue. The bands so stained are numbered 1 to 7.
(After G. Fairbanks, T. L. Steck, and D. F. H. Wallach, Electrophoretic analysis of the
major polypeptides of the human erythrocyte membrane, _Biochemistry_ 10:2606, 1971).
PAS-1 to PAS-4 are sialoglycoproteins, which stain heavily with a reagent that detects
carbohydrate. On this electropherogram, the anion channel (band 3) and glycophorin
A (PAS-1) are not completely resolved. (From V. T. Marchesi, H. Furthmayr, and
M. Tomita, The red cell membrane, _Ann. Rev. Biochem._ 45:667, 1976. Reprinted with
permission.)

TABLE 6-7
Properties of Some Purified Membrane Proteins

Protein	Source	Monomer M_r	Subunit Structure	Per Cent Hydrophobic Amino Acids[a]	Covalent Carbohydrate
Cytochrome b_5	Liver endoplasmic reticulum	16,000	Dimer (0.4% deoxycholate)	40	−
Cytochrome b_5 reductase	Liver endoplasmic reticulum	43,000	Monomer (0.4% deoxycholate)	48	−
Anion transport (band 3) protein	Human erythrocytes	95,000	Dimer	48	+
Glycophorin	Human erythrocytes	31,000	Dimer	38	++
Bacteriorhodopsin	*Halobacterium halobium*	27,000	Trimer	57	−
Lactose permease	*E. coli* plasma membrane	46,000[b]	Monomer or Dimer	59[b]	−
Mannitol permease	*E. coli* plasma membrane	60,000	Dimer	46	−
Porin	*E. coli* outer membrane	36,000	Trimer	34	−

[a]Mole per cent of Pro, Ala, ½ Cys, Val, Met, Ile, Leu, Phe, and Trp.

[b]Deduced from the DNA sequence of the *lacY* gene. (D. E. Büchel, B. Gronenborn, and B. Müller-Hill, Sequence of the lactose permease gene. *Nature* 283:541, 1980.)

inactivation of any enzymatic or other biological activity they possess. This inactivation is often accompanied by aggregation or precipitation.

Arrangement of Integral Proteins Within the Membrane. In order to understand how proteins contribute to the structure of biological membranes, it is also necessary to examine their three-dimensional structures with respect to the membrane system in which they are found. This has been done for a number of proteins that are major components of particular biological membrane systems. Techniques for examining membrane protein topography include the use of proteases and other membrane-impermeable modifying reagents, coupled with amino acid sequence analysis to determine which portions of the polypeptide chain are exposed on the outside or inside of the membrane and which are buried within the lipid portion of the membrane structure. In a few cases in which a membrane contains a single type of protein, it has even been possible to use electron microscopic techniques to examine the disposition of the polypeptide with respect to the membrane.

These kinds of experiments have shown that *two basic types* of integral membrane proteins exist: those with a large proportion of their mass extending beyond the hydrophobic interior of the membrane into the aqueous medium, and those in which most of the polypeptide is embedded in the membrane and is thus inaccessible to hydrophilic labeling reagents. Examples of the former include two major proteins found in the red blood cell membrane: a major sialoglycoprotein, *glycophorin*, and the so-called *band 3*, or *anion-transport*, protein (Figure 6–19).

Glycophorin has been isolated from the red blood cell membrane and has a molecular weight of about 31,000. It is 40 per cent protein and 60 per cent carbohydrate on a weight basis and bears the ABO- and MN-blood group antigenic specificities of the cell as well as serving as the receptor for influenza virus. The probable structure for this protein is shown in Figure 6–20. It consists of a single polypeptide chain on which short carbohydrate chains are covalently attached to amino acid residues comprising the N-terminal region of the protein. This portion of the molecule is polar, because of the presence of sugar and hydrophilic amino acid residues. The carboxyl end of the polypeptide is also rich in polar amino acids, but the central region of the protein contains an extremely hydrophobic stretch of about 20 amino acid residues. Labeling experiments have shown that the carbohydrate-rich N terminus is localized on the external surface of the red blood cell and that this region of the protein comprises over 80 per cent of the mass of glycophorin. Because the carboxyl terminus of the protein has been shown to be exposed to the cytoplasm of the red blood cell, *each molecule of glycophorin must span the membrane.*

Other studies have shown that the hydrophobic portion of glycophorin has a high affinity for phospholipids and cholesterol, the two principal lipid constituents of the red blood cell membrane. Moreover, this portion of the molecule forms a very stable α helix in a hydrophobic environment. The length of this helix is about 40 Å, slightly less than the known width of biological membranes. These observations provide strong experimental evidence for the structural model proposed in Figure 6–20.

A second well-characterized integral membrane protein is the anion-channel, or band 3, protein (Figure 6–21) of the human erythrocyte membrane. Labeling studies using intact cells and unsealed membrane ghosts

Figure 6–20
Topography of glycophorin in the mammalian erythrocyte membrane. Carbohydrate residues (○○○○) are all in the N-terminal domain on the outside of the cell and are attached mainly to the hydroxyl groups of serine and threonine residues of the protein. (Adapted from M. H. Saier, Jr., and C. D. Stiles, *Molecular Dynamics in Biological Membranes,* Heidelberg Science Library, Vol. 22, Springer-Verlag, New York, 1975. Used with permission.)

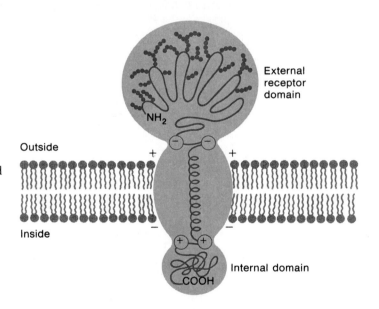

Figure 6–21
Possible disposition of the anion-channel (band 3) protein in the human erythrocyte membrane. Each identical subunit consists of 929 amino acid residues, and the N-terminal methionine has been shown to be acetylated (NAc). Glycosyl residues are linked almost exclusively on an extracellular asparagine residue. The figure also shows the positions of cleavage by chymotrypsin and trypsin, giving rise to the fragments listed in Table 6–8. (Adapted from D. Jay and L. Cantley, Structural aspects of the red cell anion exchange protein, *Am. Rev. Biochem.* 55:511, 1986.)

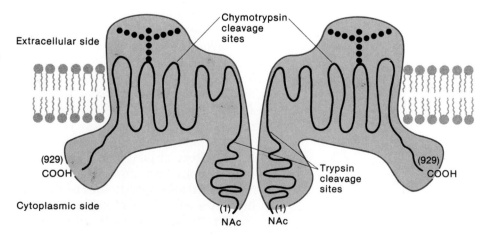

TABLE 6–8
Domains of the Anion Channel (Band 3) Protein

Domain	M_r	Isolation	Location	Carbohydrate	Per Cent Hydrophobic Amino Acids[a]
C-terminal	38,000	Chymotryptic digestion from outside the cell	Exterior, intramembrane, and cytoplasmic	++	52
N-terminal	41,000	Trypsin digestion on the cytoplasmic side of the membrane	Cytoplasm	−	44
Transmembrane	17,000	Chymotryptic digestion from outside and tryptic digestion from inside	Intramembrane	+	50

[a]Defined as in Table 6–7.

Source: Adapted from T. L. Steck, The band 3 protein of the human red cell membrane: A review. *J. Supramol. Struct.* 8:311, 1978.

have shown that this 95,000-M_r glycoprotein also spans the membrane with the carbohydrate residues on the outside of the cell (Figure 6–21). In contrast to glycophorin, however, nearly half the mass of the polypeptide is exposed on the *inside* of the cell, including the N terminus of this protein, and the polypeptide exists as a dimer within the red blood cell membrane. The C-terminal third of the molecule, including most of the amino acids to which sugar residues are attached, is at least partially available to labels from the outside of the cell, while a 17,000-M_r segment has been shown to completely penetrate the membrane. Both these domains are relatively rich in hydrophobic amino acid residues compared with the completely water-soluble cytoplasmic domain (Table 6–8). Recent evidence suggests that within these segments of the molecule, *the polypeptide chain of the anion channel traverses the membrane at least eight times*.

In a few unusual instances it has been possible to isolate a biological membrane that contains only a single kind of protein. The so-called *purple membrane* from the halophile (salt lover) *Halobacterium halobium* consists of lipid and only one protein, *bacteriorhodopsin*. This protein functions as a proton pump in response to light and is extremely important to this organism for ATP synthesis under anaerobic conditions (Chapter 34). Because bacteriorhodopsin forms a highly ordered hexagonal lattice within the plane of the purple membrane, it has been possible to deduce its structure from electron micrographs and diffraction patterns to a resolution of 7 Å. These results show that the 247-residue polypeptide chain comprising the molecule is organized as a bundle of seven α helices whose long axes are roughly perpendicular to the membrane surfaces (Figure 6–22). Estimates of the number of residues required to make seven α helices long enough to span the membrane (about 40 Å) suggest that virtually all the polypeptide chain is required to form the α helices, so that the interconnections between them must be quite short. This result has motivated model-building studies aimed at fitting the known amino acid sequence of the molecule to the observed pattern of packed α helices. Although the resolution of the current structure determination is not yet sufficiently detailed to allow structure examination at a level where individual residues can be distinguished, these studies have led to several important suggestions concerning the structure of membrane proteins. First is the observation that the *regions of structure within the membrane appear to be primarily α-helical*. As outlined in Chapter 2, α helices are secondary structures whose backbones are completely hydrogen-bonded, except at their ends. Consequently, they might be readily inserted into membranes, since there are no unsatisfied hydrogen-bonded groups

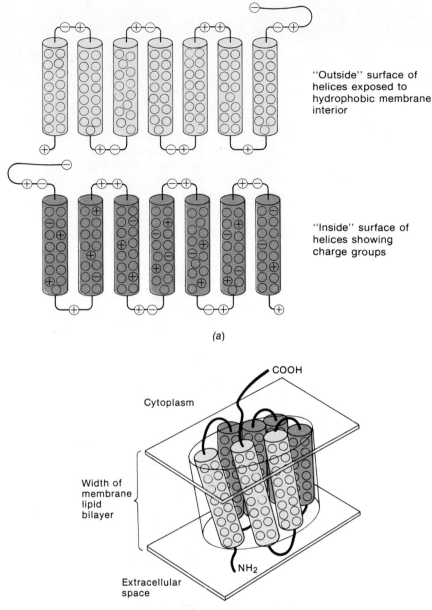

"Outside" surface of helices exposed to hydrophobic membrane interior

"Inside" surface of helices showing charge groups

(a)

COOH

Cytoplasm

Width of membrane lipid bilayer

Extracellular space

NH₂

Folded structure in the membrane with charge α groups interacting on the inside of the protein and hydrophobic surface exposed to membrane lipid side chains.

(b)

Figure 6–22
A model for the structure of bacteriorhodopsin. (a) Appearance of the membrane-exposed and membrane-removed helical sides; the hydrophobic residues are concentrated on the membrane-exposed side.
(b) The way in which the helical units are believed to be clustered.

whose stabilization would require the formation of hydrogen-bonded inter-actions with water. Second, it has been shown that by fitting the known sequence to the observed pattern of helices it is possible to find an arrange-ment that *exposes only hydrophobic side chains to the hydrophobic envi-ronment of the membrane interior* (Figures 6–22 and 6–23). This orientation agrees with the results from neutron-diffraction experiments that show that the valines are located on the outer surface. Such an arrangement would necessitate that the charged amino acid side chains that do occur in the α helices interact in complementary fashion with each other. Put simply, the membrane protein bacteriorhodopsin appears to be inside-out relative to globular proteins that exist in a polar water environment.

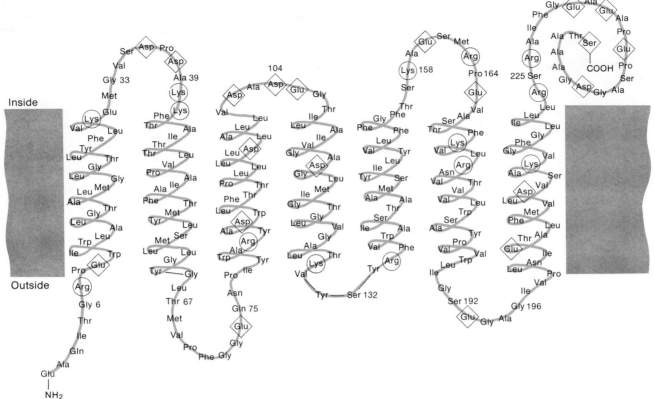

Figure 6–23
The amino acid sequence of bacteriorhodopsin, arranged as α helices situated in the bacterial membrane. The colored line indicates the course of the polypeptide chain backbone. (Adapted from D. M. Engelman and G. Zaccai, Bacteriorhodopsin is an inside-out protein, *Proc. Natl. Acad. Sci. USA* 77:5897, 1980.)

Finally, some integral membrane proteins appear not to span the membrane at all. Thus current evidence on the topography of the electron-transport chain in the inner mitochondrial membrane indicates that some integral membrane proteins are localized at the inner face and some at the outer face, while still others appear to penetrate the entire structure. This arrangement is consistent with the currently accepted model of electron transport and proton pumping during respiration (Chapter 15).

Thus the topographies of proteins found in biological membranes are considerably more variable than those of membrane lipids. The polypeptide may or may not span the membrane; considerable portions of the protein may project outside or inside the cell, and the polypeptide may have a relatively hydrophobic amino acid composition; and one protein may have its N terminus outside the cell and its C terminus projecting into the cytoplasm, while the reverse may be true for a different integral polypeptide. The only generalizations that appear valid are

1. *A significant, if small, proportion of the polypeptide must interact strongly with membrane lipids through hydrophobic interactions.*
2. *α helices are common intramembrane structures of polypeptides.*
3. *The carbohydrate portions of plasma membrane glycoproteins are always found on the outside of the cell.*

THE STRUCTURE OF BIOLOGICAL MEMBRANES

As we have seen, the most probable arrangement of lipids in biological membranes is a bilayer, based on structures formed by pure lipids in aqueous solution. The idea of a lipid bilayer was first proposed by the Dutch investigators E. Gorter and F. Grendel in 1925. By measuring the total surface area of a red blood cell and the area occupied by a monolayer of lipids isolated from the same cell, they came to the conclusion that the membrane must, in fact, be made up of a lipid bilayer. Subsequent models of membrane structure by J. Danielli and H. Davson in 1935 and by J. D. Robertson in 1959 took this fact into account but placed membrane proteins exclusively on the exterior surfaces of the lipid bilayer. However, none of these investigators had the information we now possess on the topographies of membrane lipids and proteins.

Membrane Structure Is Dynamic

The currently accepted conception of biological membrane structure, proposed by J. S. Singer and G. L. Nicolson in 1972, is shown in Figure 6–24. This *fluid mosaic model* suggests that the essential structural repeating unit is the phospholipid molecule, in a bilayer arrangement with a thickness of about 50 Å. Integral membrane proteins are "dissolved" in the bilayer in a seemingly random fashion. Some proteins (such as certain mitochondrial cytochromes) are localized at one or the other of the two surfaces of the lipid bilayer; other proteins (such as glycophorin and the anion channel) may pass from one side of the membrane to the other; and still others (such as bacteriorhodopsin) may be largely embedded in the hydrophobic matrix. Although most of the membrane phospholipids are in the bilayer array, some may be specifically associated with integral membrane proteins and essential for their biological activities. Also part of the fluid mosaic model is the idea that the entire structure is *dynamic* rather than static, with most components capable of relatively rapid lateral diffusion and of rotational motion about an axis perpendicular to the plane of the bilayer. Rotation of lipids and proteins through the plane of the bilayer, however, is proposed to be a rare event. Figure 6–24 depicts a hypothetical biological membrane at one point in space and time.

The fluid mosaic model incorporates a number of well-known features of biological membranes. The bilayer nature of membrane lipids has been well established by physical techniques. For example, the hydrocarbon tails of phospholipids and glycolipids have a lower electron density than the polar head groups. Experiments that measure the diffraction by electrons of x-rays beamed at low incident angles on biological membranes have shown a "trough" of electron density at the center of the membrane bounded by two peaks of diffraction at the two peripheries (Figure 6–25). The dimensions of these electron-rich and electron-poor areas are similar to those of the bilayer structures that pure phospholipids form in aqueous solutions. Other physicochemical measurements are fully consistent with the view that most, if not all, biological membranes contain a lipid bilayer as an essential structural feature.

It is also well-established that the lipid bilayer acts as a *permeability barrier* for hydrophilic solutes, such that only those water-soluble molecules

Figure 6–24
The fluid mosaic membrane as envisioned by Singer and Nicolson. The proteins within the bilayer have lateral mobility. Some rotational movement is also possible. (Adapted from S. J. Singer and G. L. Nicolson, The fluid mosaic model of the structure of cell membranes, *Science* 175:720, 1972.)

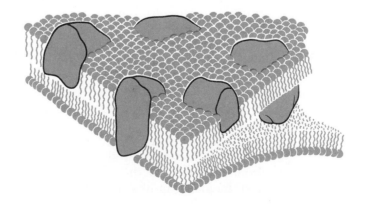

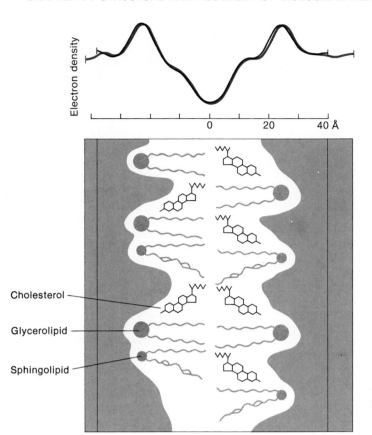

Figure 6–25
(Top) Relative electron densities of rabbit optic nerve (black line) and sciatic nerve (color) myelin membranes as a function of distance from the center, measured by x-ray diffraction. The density profile is due mainly to lipids because of the high lipid/protein ratio in myelin membranes (see Table 6–1). (Bottom) Structural interpretation of the density profile using the approximate known ratios of lipids in mammalian nerve myelins (6 cholesterols, 5 glycerolipids, and 4 sphingolipids.) The side chains in the bilayer do not appear to interdigitate to any great extent. (Adapted from D. L. D. Caspar and D. A. Kirschner, Myelin membrane at 10Å resolution, *Nature* 231:46, 1971.)

recognized by specific integral transport proteins can easily permeate the membrane (Chapter 34). In turn, these membrane proteins only function efficiently in a hydrophobic environment provided by their interactions with the nonpolar portions of membrane lipids in the interior of the bilayer. Unidirectional membrane processes, such as proton pumping out of the mitochondrion (Chapter 16), suggest an *asymmetric* organization of proton carriers across the inner mitochondrial membrane. The fact that proteins such as glycophorin and the anion channel *are always oriented in a single direction* with respect to the inside and outside of the cell shows that such asymmetry can exist. Such an arrangement can be maintained only if rotation of integral membrane proteins around an axis parallel to the lipid bilayer (flip-flopping) does not occur.

In contrast, the abilities of membrane proteins and lipids to diffuse laterally, parallel to the plane of the bilayer, have been amply demonstrated by a variety of techniques. One of the more straightforward of these is the *fluorescence photobleach recovery technique*. In this method, fluorescently labeled membrane components are bleached (their fluorescence is destroyed) in a small area of the membrane by a short, intense flash of light. By monitoring this region as a function of time, using a fluorescence microscope, one observes a reappearance of fluorescence (recovery), which is due to unbleached molecules diffusing into the treated area. If a specific membrane component is so labeled, its diffusion coefficient in the plane of the membrane can be calculated from the time course of fluorescent recovery. By using this technique, lateral movements of both proteins and lipids have been demonstrated in a number of biological membranes. For example, rhodopsin, a light-receptor protein of vertebrate retinal membranes (Chapter 36), has been shown to be highly mobile within the lipid bilayer by a photobleach recovery method utilizing its intrinsic absorbance, while other

2,2,6,6-Tetramethylpiperidine-1-oxyl (TEMPO)

(a)

Spin-label analog of phosphatidylcholine

(b)

Figure 6–26
Hydrophobic molecules containing unpaired electrons that are used to determine mobilities of lipid molecules in membranes by electron paramagnetic resonance (EPR) spectroscopy. Paramagnetic nitroxide groups are shown in color.

Egg lecithin–cholesterol

(4:1)

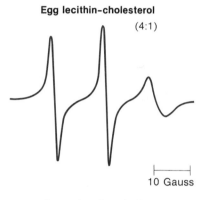

10 Gauss

Sarcoplasmic reticulum

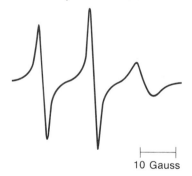

10 Gauss

Figure 6–27
EPR spectra of the compound shown in Figure 6–26b in egg lecithin-cholesterol (4:1) and in muscle sarcoplasmic reticular membranes. The nitrogen nucleus splits the spectrum due to energy absorption by the unpaired electron into 3 peaks. Relatively sharp peaks, such as those shown here, are typical of a low-viscosity, fluid-like environment. (Adapted from C. J. Scandella, P. Devaux, and H. M. McConnell, Rapid lateral diffusion of phospholipids in sarcoplasmic reticulum, *Proc. Natl. Acad. Sci. USA* 69:2056, 1972.)

proteins vary enormously in their intramembrane diffusion coefficients (see the following discussion).

In an elegant experiment also designed to test protein mobility in biological membranes, L. Frye and M. Edidin fused a mouse cell and a human cell with the aid of a virus, called Sendai virus, that facilitates such fusion. Membrane proteins of the mouse cell were labeled with specific antibodies that fluoresced green, while similar labeling of the human cells was with red fluorescent antibody. If membrane proteins failed to diffuse, then the _heterokaryon_ resulting from the fusion of these two cells should remain "half red" and "half green," even after long periods of incubation at physiological temperatures (37°C). In this experiment, however, significant intermixing of red and green labels was observed in the light microscope in less than 30 min, and complete _mosaics_ of human and mouse proteins were seen in all heterokaryons within 1 hr. These observations, therefore, also provided direct evidence for the lateral mobility of integral membrane proteins and indirect evidence for the movement of the lipid components of the bilayer itself.

Evidence for the diffusion of lipid molecules within the plane of the membrane also has been obtained by a variety of other spectroscopic techniques. One such method, _electron paramagnetic resonance_ (EPR spectroscopy), measures the energy absorbed by an unpaired electron of a free radical as a function of the magnitude of an externally applied magnetic field. Since biological membranes do not normally contain such paramagnetic groups, it is necessary to introduce into the system being examined a molecule containing a stable free radical. _Spin labels_ containing _nitroxide_ groups are often used for this purpose and may be either hydrophobic themselves (Figure 6–26a) or attached to a normal lipid component of the membrane, such as phosphatidylcholine, as shown in Figure 6–26b. Energy absorption by the unpaired electron in nitroxides is influenced by the nitrogen nucleus, which splits the spectrum into three peaks, and environmental factors, such as viscosity and polarity of the surrounding medium, which affect the shape and relative positions of these peaks (Figure 6–27). EPR spectroscopy has therefore been useful in examining the mobility of components of biological membranes.

Using the spin label TEMPO (Figure 6–26a), which is spontaneously incorporated into some biological membranes, W. Hubbell and H. McConnell showed in 1968 that the hydrophobic regions of nerve and muscle cell membranes have a low-viscosity, fluidlike nature. In parallel experiments, they also observed that vesicles of purified soybean phospholipid affected the EPR spectrum of TEMPO in a like manner. Similarly, nitroxide derivatives of phospholipids, such as that shown in Figure 6–26b, have been used to show that the diffusion rates of these paramagnetic probes in the plane of typical biological membranes are on the order of *several micrometers per second* at 37°C. Thus it is possible for a phospholipid molecule to travel from one side of an average-sized animal cell to the other in a few minutes, while the same process could take less than a second in a typical bacterium.

The types of experiments just described provide unequivocal evidence for the lateral motion of proteins and lipids in biological membranes, a major feature of the fluid mosaic model. Likewise, other physical measurements, such as [^{13}C] nuclear magnetic resonance spectroscopy, have shown that membrane lipids and proteins are free to rotate about an axis perpendicular to the plane of the bilayer. However, the specific orientations maintained by integral membrane proteins with respect to the bilayer suggest that rotation of these molecules through the plane of the bilayer (*flip-flop* or *transverse motion*) does not occur. Since most integral membrane proteins have at least some hydrophilic surface area exposed at one or both membrane faces, a transverse rotation would be an energetically unfavorable event, requiring transient interaction of these polar groups with the hydrophobic interior of the bilayer. Indeed, physical and chemical measurements have failed to detect such motions of proteins in biological membranes. Comparable measurements for phospholipids have revealed that flip-flop of these molecules can occur, but that half-times for this process can be as long as days at physiological temperatures. These rates are many orders of magnitude smaller than those of lateral diffusion and again can be explained by the high energies required for these processes. As we will see, *the asymmetric topography of biological membranes is essential to many membrane functions. This topography is maintained by the amphipathic nature of the membrane's lipid and protein constituents.*

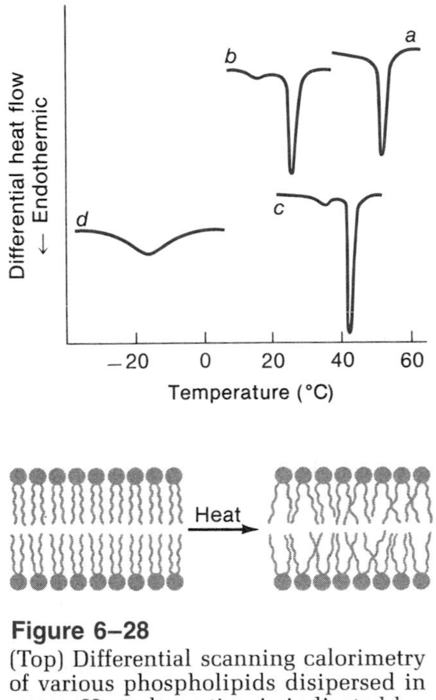

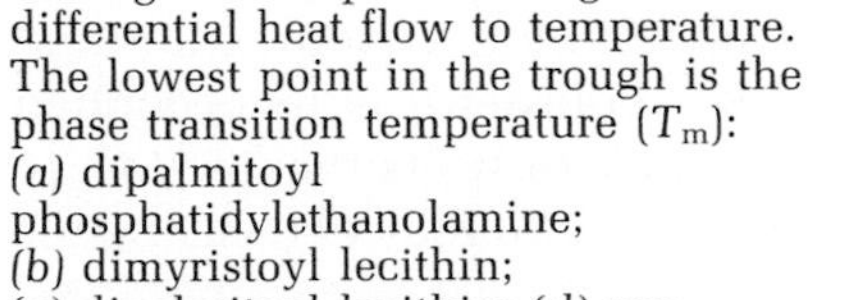

Temperature, Composition, and Extramembrane Interactions Affect Dynamic Properties

As emphasized earlier, biological membranes are fluid, dynamic structures, and these properties are undoubtedly important for a number of processes carried out on and within them (Chapter 34). It is therefore important that we examine those factors which influence or change these properties if we are to understand relationships of structure to function in membrane biochemistry.

Diffusion rates in lipid bilayers are a function both of temperature and of the composition of the membrane being examined. Bilayers consisting of a single type of phospholipid typically show an abrupt change in physical properties over a characteristic and narrow temperature range (T_m). These temperature-dependent phase transitions are due to an organizational change in the fatty acyl side chains and can be detected by a variety of physical techniques, including x-ray diffraction, EPR spectroscopy, and *differential scanning calorimetry*. This last procedure measures energy (heat) absorption as a function of temperature and shows that over the temperature range of the phase transition, a relatively large amount of heat is absorbed per degree of temperature change compared with temperature ranges well above or below the transition (Figure 6–28). It is thought that the membranes pass from a state in which the fatty acyl side chains are highly ordered (*gel*

Figure 6–28

(Top) Differential scanning calorimetry of various phospholipids dispersed in water. Heat absorption is indicated by a trough in the plot relating differential heat flow to temperature. The lowest point in the trough is the phase transition temperature (T_m): (a) dipalmitoyl phosphatidylethanolamine; (b) dimyristoyl lecithin; (c) dipalmitoyl lecithin; (d) egg lecithin (plus ethylene glycol to prevent freezing). (Adapted from D. L. Melchior and J. M. Steim, Thermotropic transitions in biomembranes, *Ann. Rev. Biophys. Bioeng.* 5:205, 1976. Used with permission.) (Bottom) Molecular interpretation of the heat-absorbing reaction during the phase transition.

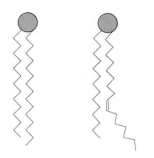

Figure 6–29
Introduction of a *cis* double bond into a fatty acyl chain results in an inflexible kink in the phospholipid tail (right); compare with a phosopholipid containing only saturated fatty acids (left). Such kinks reduce the effectiveness of hydrophobic interactions and stabilization. The result is a lowering of the temperature for phase transitions (ordered → disordered) for membranes containing unsaturated fatty acyl chains.

TABLE 6–9
Midtransition Temperatures for Aqueous Suspensions of Phospholipids

Phospholipid[a]	T_m (°C)
Di-14:0 PC	24
Di-16:0 PC	41
Di-18:0 PC	58
Di-22:0 PC	75
Di-18:1 PC	−22
1-18:0, 2-18:1 PC	3
Di-14:0 PE	51
Di-16:0 PE	63
Di-14:0 PG	23
Di-16:0 PG	41
Di-16:0 PA	67

[a]Phospholipid abbreviations are as in Table 6–5; additionally, Di-14:0, for example, refers to dimyristoyl (14 carbons, 0 double bonds).

Source: Adapted from M. K. Jain and R. C. Wagner, *Introduction to Biological Membranes*, Wiley, New York, 1980.

phase) to one in which they are far more mobile (*liquid crystalline phase*). This process is illustrated in Figure 6–28. It is accompanied by increased rotational motion about the carbon–carbon bonds of the hydrocarbon chains of the phospholipids, allowing them to assume more random, disordered conformations.

In contrast to pure phospholipid bilayers, membranes isolated from cells usually undergo such phase transitions over a much broader ($\geq 10°C$) temperature range. In some instances, distinct transitions cannot even be distinguished, because of the heterogeneity of lipids found in most biological membranes and because integral membrane proteins may decrease the mobility of lipids in their immediate vicinity. In general, lipids bearing *short* or *unsaturated* fatty acyl chains undergo phase transitions at *lower temperatures* than those containing long-chain saturated fatty acids. This is because short hydrocarbon chains have a smaller surface area with which to undergo hydrophobic interactions that stabilize the gel state and because *cis* unsaturation introduces a "kink" in the fatty acyl chain that also leads to more disorder in the bilayer (Figure 6–29 and Table 6–9). Therefore, both the length of fatty acyl groups present and the proportion of unsaturated fatty acids (as well as the position of the double bond along the hydrocarbon chain) affect the fluidity (viscosity) of a biological membrane at a given temperature. Since different lipids "melt" at different temperatures, *broad phase transitions are a general characteristic of cellular membranes*.

It is apparent from the data in Table 6–9 that the midtransition temperature of a pure phospholipid suspension also depends on the nature of the polar head group. Thus the dipalmitoyl esters of phosphatidic acid and phosphatidylethanolamine "melt" at temperatures about 20°C higher than the same derivatives of phosphatidylcholine and phosphatidylglycerol. *Divalent cations*, such as Ca^{2+} and Mg^{2+}, also affect membrane fluidity, presumably by forming ionic bonds with neighboring phosphoryl head groups, tending to "tie" phospholipid molecules together and decrease their mobility. This is undoubtedly one of the reasons that divalent cations are well-known stabilizers of biological membranes and their removal often facilitates lysis of cells as well as dissociation of peripheral membrane proteins.

Finally, cholesterol is a well-known modulator of membrane structure in eukaryotes and in one type of bacteria, the *Mycoplasmas*. In these cells cholesterol intercalates among the fatty acyl chains, with its polar hydroxyl group interacting with the polar head groups of membrane lipids. At low concentrations of cholesterol in phospholipid bilayers, separate *domains*, or patches, of cholesterol plus phospholipid and pure phospholipid appear to exist, with a resultant broadening of the phase transition profile compared with phospholipid alone. Its effects on various physical parameters of membranes depends on its proportion to other lipid components, as well as on the membrane system examined. However, a general property of membranes containing a high concentration of cholesterol appears to be an *inhibition* of processes dependent on a fluid environment. For example, the permeability of vesicles made from purified egg phosphatidylcholine to both water and glucose decreases when greater than 20 mol % cholesterol is incorporated into the vesicles. Intercalated cholesterol apparently restricts the freedom of motion of the phospholipid hydrocarbon side chains, thereby decreasing the mobilities of membrane constituents.

Most cells maintain a lipid composition that allows for relatively rapid lateral diffusion of many membrane components at the growth temperature. Membrane-associated processes, such as the vectorial reactions catalyzed by some transmembrane transport systems (Chapter 34) and endo- and exocytosis rely on a semifluid environment for their operation. Consequently organisms have evolved intricate mechanisms to maintain this environment

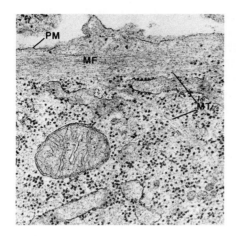

Figure 6–30
Membrane-associated cytoskeletal components of cultured mouse cells. (PM = plasma membrane, MF = microfilaments, MT = microtubules; 54,000×.) (From G. L. Nicolson, Transmembrane control of the receptors on normal and tumor cells. I. Cytoplasmic influence over cell surface components, *Biochim. Biophys. Acta* 457:57, 1976. Reprinted with permission.)

under a variety of conditions. One of the most remarkable of these is the ability of a variety of plant, animal, and bacterial cells to increase the proportion of membrane unsaturated fatty acids in response to a decrease in temperature. This ensures proper functioning of the membrane at the lower temperature. In bacteria such as *E. coli* and *Bacillus megaterium*, this modulation appears to be the result of temperature effects on the activities and/or on the induction of synthesis of enzymes involved in the biosynthesis of phospholipids containing unsaturated fatty acyl chains. In plants and yeast, increased solubility of O_2 at low temperatures apparently increases the proportion of unsaturated fatty acids because O_2 is a substrate of the desaturase enzyme that leads to their biosynthesis. Other factors that affect unsaturated fatty acid biosynthesis, and thus membrane fluidity, in various systems include light (in plants), nutrition, developmental stage, and aging.

Despite the large body of evidence in favor of the lateral mobility of many membrane constituents at physiological temperatures, it would be an oversimplification to view a biological membrane only as a random "sea" of lipids with proteins floating aimlessly about in them. In recent years it has become clear that nearly all eukaryotic cells contain within their cytoplasm a *cytoskeleton* made up of *microtubules* and *microfilaments* consisting primarily of the proteins *tubulin* and *actin*, respectively. Microfilaments, which are structurally similar to actin filaments of muscle cells, have been shown to form bundles just beneath the plasma membrane of many cells (Figure 6–30). These bundles are believed to have an important role in such processes as locomotion and phagocytosis, which involve local or general changes in the shape of the cell surface and thus in the plasma membrane. In a few cases, direct association of microfilaments with the plasma membrane has been demonstrated. For example, *fibronectin*, a peripheral, cell-surface glycoprotein of many animal cells, is believed to have a role in cell–cell and cell–substratum adhesion. A *transmembrane* association of fibronectin with cytoskeletal microfilaments has been deduced from immunofluorescent microscopy (Figure 6–31), presumably by means of one or more integral membrane proteins. As a consequence, lateral diffusion of fibronectin has

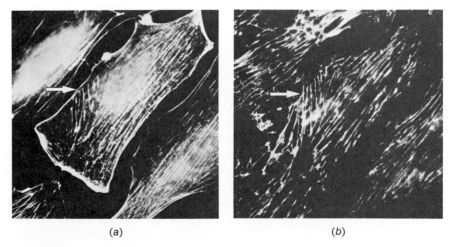

(a)   (b)

Figure 6–31
Colinearity of actin and fibronectin fibrils in cultured hamster fibroblast cells as seen by immunofluorescence in the light microscope. (a) The fluorescence is due to actin antibodies bound to intracellular microfilaments: These antibodies were produced in rabbits. Their location of binding is made visible by staining with fluorescent goat antirabbit antibodies. (b) The fluorescence is that of the fluorescent antifibronectin antibodies bound to extracellular fibronectin fibrils. The correspondence between the arrangement of actin and fibronectin filaments strongly suggests a transmembrane association between the two proteins. [From R. O. Hynes and A. T. Destree, Relationships between fibronectin (LETS protein) and actin, *Cell* 15:875, 1978.]

been shown to be at least _5000 times slower_ than freely diffusible membrane lipids and proteins. It is very likely that other peripheral and integral membrane proteins also interact with the cytoskeleton, and these interactions may similarly reduce their mobilities, thereby affecting processes that must be localized at specific points on the cytoplasmic membrane.

We also should recognize that membrane lipids may not always be freely diffusible in the plane of a biological membrane, even above the phase transition temperature. Many purified integral membrane proteins have been shown to retain bound phospholipid molecules after solubilization. For example, the enzyme _cytochrome oxidase_, isolated from beef heart mitochondria, spontaneously forms vesicular structures in aqueous suspension owing to associated lipid. When a nitroxide derivative of stearic acid is added to such a suspension, EPR spectroscopy reveals two mobility classes of the spin label—one highly diffusible and one less mobile. These observations have been interpreted as evidence for associated lipid molecules surrounding the enzyme. These "lipids of solvation" are believed to be essential for maintaining the structural and functional integrity of integral membrane proteins such as cytochrome oxidase. It should be emphasized, however, that this situation is probably not a static one. Rather, it is likely that protein-associated lipids are more or less freely exchangeable with the bulk of the lipid molecules in the fluid bilayer array. Furthermore, "patches" of membrane lipids differing in overall composition from the membrane as a whole also have been shown to exist in some membrane systems. An example was given earlier: at low concentration, cholesterol induces phospholipid-cholesterol "patches" in an otherwise pure phospholipid model membrane system. Within such areas, phospholipid molecules would be expected to be less mobile than the bulk of the phospholipid in the membrane. Indeed, this interpretation may provide the explanation for the broad phase transition profiles of such membrane systems. _Thus temperature, ionic environments, and composition can affect the general physical state of a biological membrane, while local mobilities of membrane components can be influenced by protein–protein, lipid–protein, and lipid–lipid interactions._

Biological Membranes Are Asymmetric

Chemical probes of membrane protein structure have provided ample evidence for the unidirectional, asymmetric orientation of proteins with respect to the lipid bilayer, as described earlier. This asymmetry also has been more directly observed by the technique of _freeze-fracture electron microscopy_. In this procedure, whole cells or membranes are rapidly frozen, and the specimen is then struck with a sharp knife called a microtome. Very often, the fracture plane actually passes between the outer and inner monolayers (_leaflets_) of the membrane lipid bilayer because the relatively weak hydrophobic interactions between the fatty acyl chains of the two leaflets offer a "path of least resistance" (Figure 6–32a). The inner surfaces of the two leaflets can then be viewed after heavy-metal shadowing in the electron microscope. This technique is especially useful in examining membrane morphology, because it avoids the potentially destructive fixing, embedding, and staining steps of more conventional sample-preparation procedures. Many biological membranes, when split in half by freeze-fracture, can be seen to have quite different morphologies of the inner and outer leaflet surfaces owing to preferential adherence of some membrane proteins to one of the two monolayers (Figure 6–32b). This presumably reflects the asymmetric and unidirectional disposition of these polypeptides across the lipid bilayer.

Proteins are not the only membrane constituents to show asymmetric orientations. Recent evidence favors an _unequal distribution of certain lip-_

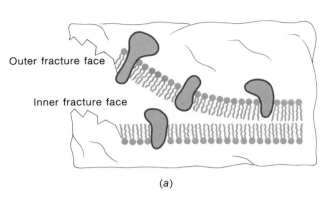

Outer fracture face

Inner fracture face

(a)

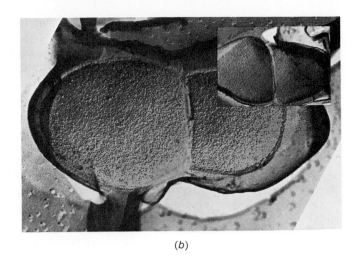

(b)

Figure 6–32
Freeze-fracture electron microscopy. (a) When struck with a sharp knife, membranes embedded in ice usually fracture between the monolayer leaflets of the lipid bilayer.
(b) Freeze-fracture electron micrograph of the plasma membrane of *Streptococcus faecalis*, showing a large number of protrusions (presumably proteins) on the outer fracture face and the relative lack of such particles on the inner fracture face (inset). (From H. C. Tsien and M. L. Higgins, Effect of temperature on the distribution of membrane particles in *Streptococcus faecalis* as seen by the freeze-fracture technique. *J. Bacteriol.* 118:725, 1974. Reprinted with permission.)

ids *between the inner and outer leaflets of the bilayer* in many biological membranes. For example, in 1972, M. S. Bretscher showed that most of the phosphatidylethanolamine and all the phosphatidylserine of human erythrocyte membranes are *inaccessible to chemical modification from outside* the cell. Both phosphoglycerides could be modified, however, in *erythrocyte ghosts*, in which both inner and outer leaflets were exposed to the chemical reagent. These results suggested that lipid asymmetry also could exist in a biological membrane. Subsequent experiments using enzymes that degrade phospholipids (*phospholipases* and *sphingomyelinase*) showed that phosphatidylcholine and sphingomyelin appear to be preferentially found in the *outer leaflet* of human red blood cell membranes (Table 6–10). Thus, although total membrane lipid is equally distributed between outer and inner

TABLE 6–10
Lipid Asymmetry in Biological Membranes

Membrane	Preferential Outside	Preferential Inside	Equal
Various erythrocytes	PC, SM, glycolipids, cholesterol	PE, PS	—
Rabbit sarcoplasmic reticulum	PE	PS	PC, lyso PC
Mouse LM cell plasma membrane	SM	PE	PC
E. coli outer membrane	—	PE	—
Bacillus megaterium	—	PE	—
Micrococcus lysodeikticus	PG	PI	DPG
PC/PE artificial vesicles	PC	PE	—

Note: Abbreviations are as in Table 6–5.

Source: Adapted from J. A. F. Op den Kamp, Lipid asymmetry in membranes. *Ann. Rev. Biochem.* 48:47, 1979.

monolayers in the erythrocyte membrane, the *lipid composition of each leaf-let appears to be different*.

Lipid asymmetry has now been found in membranes from a variety of biological sources, and some examples are given in Table 6–10. A favorite probe for examining the asymmetry of phospholipids containing a primary amino group (phosphatidylethanolamine and phosphatidylserine) has been *2,4,6-trinitrobenzene sulfonic acid* (TNBS) (Figure 6–33). Because of its polarity, TNBS does not penetrate most biological membranes rapidly at low temperature, and the fact that TNBS reacts rapidly with primary amino groups of lipids in intact cells or sealed vesicles reveals that these molecules are in the outer leaflet of the bilayer, as shown in Figure 6–33. Separation and quantitation of both modified and unmodified phosphatidylserine and phosphatidylethanolamine after reaction of cells with TNBS can lead to an estimation of their relative proportions in each half of the membrane. The distribution of other membrane lipids can be determined enzymatically, as described earlier, or by a variety of physical techniques, including NMR and EPR spectroscopy and x-ray diffraction analysis. Such experiments strongly suggest that the *asymmetric partitioning of lipids between the bilayer leaflets may be a general property of biological membranes*.

Given the asymmetric arrangements of proteins in membranes along with the nearly unidirectional nature of many membrane-associated processes (Chapter 34), it is perhaps not surprising that certain lipids might be found predominantly on one side or the other of the bilayer. This could be a consequence of the fact that some integral membrane proteins may associate with specific lipids. Thus the orientation of the protein could influence the orientation of bound lipid with respect to the bilayer. Furthermore, the solutions that contact each monolayer are usually quite different (e.g., extracellular milieu versus cytoplasm), so that lipid asymmetry also could be induced by compositional and/or biological activity differences between the inside and outside environments. It is therefore reasonably clear, from energetic arguments, how membrane asymmetry can be maintained and, from functional requirements, why it exists. An equally intriguing question, however, is: *How are asymmetric membrane structures formed in the first place?* To answer this, we must consider how biological membranes are synthesized and assembled by cells.

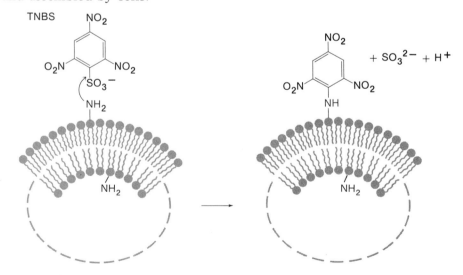

Figure 6–33
TNBS (trinitrobenzene sulfonate) reacts rapidly at low temperatures only with free primary amino groups exposed on the outside of a biological membrane. It can therefore be used to determine the distribution of lipids containing free amino groups between outer and inner leaflets of the bilayer.

BIOSYNTHESIS AND ASSEMBLY OF BIOLOGICAL MEMBRANES

In terms of its overall composition, the plasma membrane is one of the most complex cellular structures. For example, a typical bacterial cytoplasmic membrane contains over 100 different proteins and several classes of phospholipids, each of which may be subdivided into species of different fatty acid compositions. Furthermore, as we have just seen, these molecules are not arranged symmetrically in the membrane, but rather are found in specific orientations with respect to the inside and outside of the cell. Obviously, then, the process of membrane assembly, or *biogenesis*, must have programmed within it not only methods for ensuring insertion of proteins and lipids destined to become membrane components, but also mechanisms for attaining the proper orientations of these molecules in the bilayer.

A second problem related to membrane biogenesis concerns the mechanism of biosynthesis of macromolecules destined to be *exported* through the membrane of a cell or organelle. For example, how do components of the outer membrane, the peptidoglycan and periplasmic proteins, reach their ultimate destinations in Gram-negative bacteria, and how do hydrophilic, water-soluble proteins excreted by secretory organelles, such as the endoplasmic reticulum of eukaryotic cells, traverse the hydrophobic lipid bilayer?

In this section, we examine what is known about the biosynthesis and assembly of biological membranes in a number of systems, especially as they relate to the structural features we have already pointed out. Our discussion is confined mainly to the cytoplasmic and outer membranes of Gram-negative bacteria and to the endoplasmic reticulum in eukaryotic cells (the precursor of the plasma membrane), because most information currently available is about these systems. It is likely, however, that mechanisms of biogenesis of other membrane systems will turn out to be related to those found in the better-characterized systems.

Topography and Coordination of Membrane Assembly

All growing cells need to synthesize new plasma membrane components as they enlarge and eventually divide into daughter cells. Because the membrane must maintain its permeability barrier function throughout this process, it is clear that newly synthesized membrane proteins and lipids must be inserted into the preexisting bilayer without disrupting its structural continuity. Thus membrane biosynthesis involves *expansion* as well as self-assembly. Further, the biosynthetic machinery must be so organized as to ensure efficient insertion of membrane lipids and the proper orientations of integral membrane proteins with respect to the bilayer.

Not surprisingly, most of the biosynthetic enzymes for membrane lipids are themselves integral membrane proteins in both prokaryotic and eukaryotic cells. In many bacteria, nearly all the membrane lipid is phospholipid (Table 6–5), and fatty acylation of glycerol-3-phosphate and all subsequent steps in phospholipid biosynthesis (Chapter 22) occur on the cytoplasmic membrane. In eukaryotic cells, cytoplasmic membrane assembly is believed to occur by expansion of the endoplasmic reticulum followed by the migration of vesicles derived from this organelle to the cytoplasmic membrane, with which they fuse. Supporting this conclusion is the observation that lipid biosynthetic enzymes have been shown to be tightly associated with the endoplasmic reticular membrane in animal cells. In both eukaryotic and prokaryotic systems, the active sites of these enzymes have been shown to be on the *cytoplasmic face* of the membrane. This result was to be expected, since the precursors of membrane lipids, many of which are hydrophilic, are synthesized in the cytoplasm. Thus lipid is apparently inserted into biological membranes passively, as a direct result of the site of its biosynthesis.

An interesting question is posed by this topographic organization of membrane lipid biosynthesis. *Because flip-flop of preexisting membrane lipids is generally a slow process, how do newly synthesized molecules become inserted into the outer leaflet of the bilayer?* This question was explored by J. Rothman and E. Kennedy in the Gram-positive bacterium *Bacillus megaterium*. Using the membrane-impermeable probe TNBS, they demonstrated that 33 percent of the total membrane phosphatidylethanolamine was in the outer leaflet of the cytoplasmic membrane, while the rest was in the inner leaflet. By labeling newly synthesized phosphatidylethanolamine with a short pulse of $[^{32}P]$phosphate and then immediately treating with TNBS, they showed that none of the radioactive phospholipid was initially found in the outer leaflet. After 30 min at 24°C, however, about one-third of the $[^{32}P]$phosphatidylethanolamine was labeled with TNBS, the same ratio as for the preexisting, unlabeled molecules. Their experiments lead to two important conclusions about membrane lipid biosynthesis in *B. megaterium*: (1) *newly synthesized phosphatidylethanolamine is incorporated initially only in the inner leaflet, and (2) its appearance in the outer leaflet occurs nearly 10^5 times faster than known flip-flop rates for phospholipids in model bilayers.* Recent experiments with endoplasmic reticular membranes suggest that the same conclusions may hold true in eukaryotic cells. This leads to the possibility that transverse rotation of lipids through the bilayer may be catalyzed by one or more proteins in biological membranes. Indeed, recent evidence shows that phospholipid flip-flop in biological membranes requires ATP, and likely involves one or more enzymes. If such proteins exist, then their activities could play a role in establishment of asymmetric lipid distributions between the two monolayers.

In contrast to membrane phospholipids, *synthesis and insertion of integral membrane proteins appear to take place by at least two pathways* (see Chapter 28). Some are synthesized by polyribosomes in the cytoplasm, after which they are incorporated into the membrane (*posttranslational insertion*). Other membrane proteins are synthesized by polyribosomes that are bound to the membrane, and these polypeptides appear to insert as they are being translated (*cotranslational insertion*). Membrane-bound polysomes have been isolated from bacteria and from endoplasmic reticulum and have been shown to synthesize membrane and secreted proteins almost exclusively. The converse is not true, however, because isolated free (cytoplasmic) polysomes direct the synthesis of both cytoplasmic and membrane proteins. *Whether a membrane protein is synthesized on membrane-associated or free polyribosomes seems to be a function solely of the amino acid sequence of the protein itself,* as will be discussed in Chapter 28.

Because both membrane lipids and some membrane proteins are synthesized at the membrane itself, and because normal membrane biogenesis during cell growth involves insertion of both, it is reasonable to ask whether coordinate synthesis of lipid and protein is necessary for membrane expansion. It is clear that the protein composition in cell membranes can change considerably depending on age and environmental factors. For example, within 5 days after birth, levels of the enzyme glucose-6-phosphatase in rat liver endoplasmic reticular membranes increase about 30-fold. Over the same time period, NADH: cytochrome *c* reductase and cytochrome b_5 are also induced, but the increase in amount of these two electron carriers is neither coordinate with nor equivalent to that of glucose-6-phosphatase. In bacteria, rates of synthesis of many membrane proteins can be altered by changing the medium in which the cells are grown. For example, specific transport proteins for many sugars and amino acids (Chapter 34) appear in the membrane only if induced by the appropriate substrate. Thus *the rates of synthesis and insertion of integral membrane proteins are not generally coordinate with one another.*

To investigate whether ongoing membrane lipid synthesis is necessary for synthesis and insertion of membrane proteins, mutants of *E. coli* defective in phospholipid biosynthesis under some conditions have been especially useful. These mutants depend on exogenous glycerol in the medium for the synthesis of phosphoglycerides, and when glycerol is removed, net phospholipid synthesis ceases immediately. Upon glycerol starvation, these cells continue to grow, however, for about one generation, during which time the proportion of protein to lipid in both inner and outer membranes *increases about 50 percent* as a result of continued protein synthesis. These results demonstrate that *coordinate synthesis of membrane lipid is not necessary for synthesis and insertion of membrane protein*, at least in bacterial cells. Conversely, if total protein synthesis is blocked by the antibiotic *chloramphenicol* in the presence of glycerol, phospholipids continue to be synthesized and inserted into the cytoplasmic membrane without the concomitant incorporation of protein. Hence *membrane lipid synthesis can occur independently of the biosynthesis of membrane protein.*

These observations on the assembly of biological membranes seem to rule out a mechanism in which nonmembranous intracellular protein-lipid complexes are first formed, followed by their insertion into the growing membrane. They are consistent with the fact that biosynthesis of many membrane components takes place on the membrane itself. Nevertheless, *regulation of the differential rates* of membrane protein and lipid synthesis must occur to achieve relatively stable ratios of these molecules in the bilayer during cell growth. Thus *E. coli* phospholipid biosynthetic mutants that have been starved for glycerol and therefore have abnormally protein-rich membranes show a higher rate of phospholipid synthesis upon readdition of glycerol to the medium than unstarved cells. In this case, one or more of the membrane-bound phospholipid biosynthetic enzyme activities is presumably controlled by the protein-lipid ratio in the membrane.

Finally, it should be emphasized that the assembly and maintenance of membrane structures in cells is a dynamic process. Not only are components synthesized and inserted into a growing membrane, but they also are being continuously degraded, albeit at a slower rate. This *turnover* process varies with each individual type of molecule. In general, phospholipids have a shorter half-life in the membrane (higher turnover) than membrane proteins, which themselves vary enormously in "life expectancy" depending on the specific protein examined. This constant turnover allows cells to rapidly adjust membrane composition in response to changes in the environment (temperature, nutrition, and so forth).

The Problem of Membrane Protein Biogenesis

Phospholipids are relatively small molecules, with molecular weights that are usually less than 1000. It is not difficult to imagine their spontaneous incorporation into the bilayer after their synthesis by membrane-bound enzymes. Most integral membrane proteins, however, have molecular weights that range between 10^4 and 10^5 and can contain a reasonably high proportion of hydrophilic amino acids (Table 6–7). Furthermore, their insertion into the membrane also must result in an asymmetric, unidirectional orientation with respect to the bilayer. It is not surprising, therefore, that the problem of membrane protein biogenesis is a good deal more complex than that of many membrane lipids. Only recently have enough clues been obtained to the mechanism of this process to enable the formulation of models.

Related to this problem is the mechanism of biosynthesis and assembly of macromolecules and supramolecular structures whose ultimate location is _exterior to the plasma membrane_. Included in this category are the outer membrane, cell wall, and periplasmic proteins of Gram-negative bacteria as well as proteins secreted from some eukaryotic cells, including hormones, hydrolytic enzymes, and antibodies. In both cases, the question is similar: _How do large, often hydrophilic molecules partially or completely traverse the hydrophobic barrier of the membrane bilayer from their sites of synthesis within the cell?_ In the case of the bacterial peptidoglycan envelope, it seems likely that a C_{55} isoprenoid lipid carrier, bactoprenol (Figure 6–11) is involved in carrying the peptidyl-disaccharide "building block" across the plasma membrane to be added to the growing cell wall. Similarly, the terminal carbohydrate portions of the lipopolysaccharide of the Gram-negative outer membrane (O antigens) are attached to bactoprenol before being added to the lipid A core polysaccharide unit on the outside of the cytoplasmic membrane (Chapter 20). A compound related in structure to bactoprenol, _dolichol_, similarly participates in the transfer of carbohydrate residues to membrane glycoproteins in eukaryotic cells.

Although these types of mechanisms can explain bacterial cell wall biosynthesis and glycoprotein carbohydrate translocation in eukaryotic cells, they are not applicable to the problem of integral membrane protein insertion or of the secretion of some proteins by both types of cells. Mechanisms for these latter processes will be considered in Chapter 28 (see Box 28G).

SUMMARY

Biological membranes are composed of lipids, proteins, and carbohydrates in various combinations. Fractionation of lysed cells followed by purification of different kinds of membranes by various procedures has permitted workers to investigate their structures and functions in some detail.

The most abundant lipids of most biological membranes are the phospholipids, including the phosphoglycerides and sphingomyelin. Glycolipids and cholesterol are also important membrane components of many eukaryotic cells, while various isoprenoids or terpenes are found in membrane systems specialized for such functions as light reception, photosynthesis, and electron transport. Phospholipids and glycolipids spontaneously form bilayer structures in aqueous solution because of their amphipathic properties, and the lipid bilayer is fundamental to biological membranes in determining their structural and functional characteristics.

Membrane proteins can be subdivided into two classes: peripheral and integral. Peripheral proteins may be dissociated from the membrane in a water-soluble form by relatively mild procedures that disrupt hydrogen and ionic bonds. Integral proteins can be brought into solution only by disrupting the bilayer structure with detergents, to produce mixed micelles of detergent, lipid, and protein. Physical and chemical analyses of purified integral membrane proteins show that they vary widely in structure, but that most of them require a hydrophobic environment for maintenance of their biologically active structures.

The topographies of a number of integral proteins with respect to the membrane have been studied using membrane-impermeable modifying reagents and proteolytic enzymes, among other techniques. Two basic classes of such proteins exist: those with most of their mass buried within the lipid bilayer, such as bacteriorhodopsin, and those with a large proportion of the polypeptide extending into the cytosol or outside the cell, such as the anion-channel (band 3) protein and glycophorin of red blood cells. Many integral membrane proteins also have been shown to span the membrane, and if carbohydrates are bound to amino acid residues in the protein, they are always found on the external surface of the cell.

These properties of membrane lipids and proteins led to the formation of the fluid mosaic model for membrane structure by Singer and Nicholson in 1972. In this model, the essential structural unit is a relatively impermeable lipid bilayer in which protein molecules are embedded. Both protein and lipid molecules are capable of rapid lateral diffusion and of rotational motion about an axis perpendicular to the plane of the bilayer at physiological temperatures. Rotation of these molecules through the plane of the bilayer (transverse motion), however, is usually a very rare event, allowing the membrane to maintain its asymmetric structure with respect to both protein orientation and the distributions of specific lipids between the inner and outer leaflets. A number of biophysical techniques, including x-ray diffraction, fluorescence spectroscopy, and electron paramagnetic resonance spectroscopy, have confirmed these essential features of the fluid mosaic model.

Many factors have been shown to affect the general mobility of membrane components, including temperature, the fatty acid compositions of phospholipids and glycolipids, and the presence or absence of cholesterol. Likewise, local mobilities within a given membrane system are a function of protein–protein, lipid–protein, and lipid–lipid interactions. In response to environmental changes, many cells can regulate the composition of their membranes to maintain the overall semifluid environment necessary for many membrane-associated functions.

The asymmetric orientation of integral membrane proteins with respect to the bilayer is a well-established fact and is necessary for their proper functioning. Recently, evidence has accumulated in favor of lipid asymmetry in many membranes as well. It is likely that the unequal partitioning of some membrane lipids between the two leaflets of the bilayer also is necessary for the functions of some membrane systems and may be influenced by both protein asymmetry and the different environments to which the inner and outer leaflets are exposed.

The complexity and asymmetry of biological membranes pose the problem of how they are assembled in the cell. Membrane lipids are synthesized by membrane-bound enzymes, and appear to be passively inserted into the inner leaflet of the bilayer. Those lipids destined for the outer leaflet appear to be translocated there in an ATP-requiring process. Although membrane lipid and protein biosyntheses are not necessarily coordinate, these processes are regulated in most cells to achieve a stable lipid/protein ratio. The complex process of membrane protein insertion will be considered in Chapter 28.

SELECTED READINGS

Bennett, V., The membrane skeleton of human erythrocytes and its implications for more complex cells. *Ann. Rev. Biochem.* 54:273, 1985. A review of the organization of the red blood cell surface, including a discussion of the possible roles of the cytoskeleton.

Eisenberg, D., Three-dimensional structure of membrane and surface proteins. *Ann. Rev. Biochem.* 53:595–623, 1984.

Fleischer, S., and L. Packer, eds., *Methods in Enzymology*, Vols. 31 and 32. New York: Academic Press, 1974. Volume 31 of this invaluable series focuses on subcellular fractionation and membrane isolation techniques, including isopycnic and differential centrifugation. Volume 32 deals with the composition and characterization of various membranes and membrane components, including articles on model membrane systems.

Helenius, A., and K. Simons, Solubilization of membranes by detergents. *Biochim. Biophys. Acta* 415:29, 1975. Lengthy review of the properties of detergents and their use for membrane solubilization.

Houslay, M. D., and K. K. Stanley, *Dynamics of Biological Membranes*. New York: Wiley, 1982. Emphasis on organization, mobility, and structure of biological membranes.

Jain, M. K., and R. C. Wagner, *Introduction to Biological Membranes*. New York: Wiley, 1980. Chapters 1–8. Includes surveys of the electron microscopy of biological membranes, isolation and characterization of membrane components, properties of lipid bilayers, and membrane biogenesis.

Jay, D., and L. Cantley, Structural aspects of the red cell anion exchange protein. *Ann. Rev. Biochem.* 55:511, 1986. Recent review of techniques that have been used to determine the intramembrane structure of the band 3 (anion channel) protein of erythrocytes.

Rydstrom, J., et al., eds., *Membrane Proteins: Structure, Function and Assembly*, Nobel Symposium No. 66 (Chemica Scripta, Stockholm), 1987.

Singer, S. J., The molecular organization of membranes. *Ann. Rev. Biochem.* 43:805, 1974. Excellent review of membrane structure through 1973.

Voelker, B. R., Lipid assembly into membranes. In D. Vance and J. E. Vance, eds., *Biochemistry of Lipids and Membranes*. Menlo Park, Calif.: Benjamin/Cummings, 1985. An authoritative and unique review of a complex topic that is under active investigation.

PROBLEMS

1. Considerable evidence, much of it circumstantial, suggests that mitochondria may have evolved from endosymbiotic aerobic bacteria within ancestral eukaryotic cells (and likewise, that chloroplasts evolved from a photosynthetic prokaryote).
   a. From what you know about membrane structure, defend this hypothesis.
   b. Assuming that the hypothesis is correct, explain the differences in envelope structure that are observed between mitochondria and present-day bacteria.

2. Predict the effects of the following on the phase transition temperature (T_m) and phospholipid mobility in pure dipalmitoylphosphatidylcholine vesicles ($T_m = 41°C$).
   a. Raising the temperature from 30 to 50°C.
   b. Introducing dipalmitoleoylphosphatidylcholine into the vesicles.
   c. Adding dimyristoylphosphatidylcholine to the vesicles.
   d. Introducing a high concentration of cholesterol into the vesicles.

e. Incorporating integral membrane proteins into the vesicles.

3. a. Speculate on the basis for the differences in T_m between the dipalmitoyl esters of phosphatidic acid, phosphatidylethanolamine, and phosphatidylcholine (see Table 6–9).

   b. Why do you think that phosphatidylethanolamine partitions preferentially into the inner leaflet in mixed PC/PE artificial vesicles (see Table 6–10)?

4. Lidocaine and phenethylalcohol are both local anesthetics that have been shown to interact with nerve cell and other membranes. Although the precise mechanisms of their anesthetic action are not known, it seems reasonable to conclude that they involve changes in membrane structure and physical properties. Both these compounds have been shown to inhibit processing (cleavage) of signal sequences and in some cases also to affect final localization of certain membrane and periplasmic proteins of *E. coli*. From the chemical structures of these compounds given below, postulate reasonable mechanisms for their inhibition of membrane protein processing and assembly. (See Box 28G in Chapter 28.)

Phenethylalcohol   **Lidocaine**

5. Triton X-100 ($M_r = 625$, CMC = 0.24 mM) and sodium deoxycholate ($M_r = 414$; CMC = 4 mM) are both often used to solubilize crude membrane fractions before purification of constituent proteins is attempted.

   a. Which of these detergents would be most easily removed by dialysis (diffusion through a membrane permeable to molecules with molecular weights less than about 10,000) from a 0.1% (weight/volume) aqueous solution?

   b. In which of these detergents would ion-exchange chromatography probably be more successful for protein purification? How might you exchange the less favorable for the preferred detergent if this were necessary?

   c. In which of these detergents would gel filtration chromatography be likely to give a closer approximation to the molecular weight of an integral membrane protein in a 0.1% solution? Why?

6. In 1925, Gorter and Grendel measured the total area occupied by a monolayer of red blood cell lipids using a device known as a Langmuir trough. In one experiment, they found that a monolayer of lipids was achieved in an area of 0.89 m² starting with lipids isolated from 4.74×10^9 human red blood cells. The surface area they measured for one cell was 99.4 μm^2. Using this information, show how they concluded that the membrane covering a human red blood cell was two lipid molecules thick.

7. Suppose you treated a hypothetical cell that was grown at 30°C with TNBS at 3°C and found that about 75 per cent of the total membrane phosphatidylethanolamine (PE) could be labeled. At 15°C, however, you could label nearly all the PE in several hours, but 75 per cent reacted with a half-time of 2 min, while the other 25 per cent had a half-time reaction of 30 min.

   a. Assuming that no new phospholipid could be synthesized at either 3°C or 15°C, how would you interpret the preceding results?

   b. Suppose that at 30°C in growing cells, 75 per cent of the newly synthesized PE flip-flops with a half-time of 5 min from the inside monolayer to the outside one, while the rate of the reverse process is negligible. What would the distribution of PE synthesized at "time zero" be 5 min later under these conditions?

   c. Can you think of any reason why 25 per cent of the newly synthesized PE does not flip-flop at a high rate?

   d. Give an explanation, other than specific catalysis by a protein, for the high rates of transverse diffusion seen in some systems for newly synthesized membrane phospholipids.

7

NUCLEOTIDES AND NUCLEIC ACIDS

Among the four major classes of biomolecules, nucleic acids occupy a unique position. Not only are they involved in many biochemical reactions, but they carry the genetic information for the synthesis of proteins. The important reactions that nucleic acids are involved in include energy transfer reactions and nucleic acid and protein synthesis. The genetic information is carried by a special class of large nucleic acids that are centrally located in the nucleus in eukaryotic cells and in the cytoplasm in prokaryotic cells. These molecules contain all the information for duplication of their own kind as well as for the formation of other cellular nucleic acids and proteins. In this chapter we will discuss the historically significant experiments that pinpointed the unique genetic role of the nucleic acids and then we will consider their structural properties.

THE GENETIC SIGNIFICANCE OF NUCLEIC ACIDS

After the discovery, around the turn of the century, that genes are carried by chromosomes, a great deal of effort went into characterizing the sizes and shapes of chromosomes. But it was not until much later that significant progress was made in elucidating the chemical nature of the gene. Biologists were aware, as early as 1900, that chromosomes are composed of both nucleic acids and proteins. However, the seemingly simple chemical composition of nucleic acids misled early investigators into believing that nucleic acids were a purely structural component of the chromosome. It was, they thought, the arrangement of specific proteins along the chromosome that accounted for gene specificity. This notion was dispelled in 1944 when Oswald T. Avery and his colleagues at Rockefeller University demonstrated that purified deoxyribonucleic acid (DNA) contains the genetic determinants of the bacterium *Diplococcus pneumoniae*. In this section we discuss Avery's results, as well as some other historically important observations on the genetic significance of nucleic acids.

Transformation Is DNA-mediated

In 1928 Fred Griffith was experimenting with two different strains of pneumococcus. Type S bacteria (S for smooth, from the appearance of bacterial colonies on agar plates) are encapsulated by polysaccharide. The capsules protect them from the host immune system, making the S bacteria pathogenic; when S bacteria are injected into mice, death results. By contrast, R bacteria (R for rough colonies) are nonencapsulated; they are readily attacked by the mouse's immune system and consequently are nonpathogenic. Although heat-killed S bacteria by themselves are nonvirulent, Griffith found that when heat-killed S bacteria were mixed with live R bacteria and introduced into a susceptible laboratory mouse, death of the animal frequently occurred (Figure 7–1a). S bacteria could then be detected in the

Figure 7–1

In vivo and *in vitro* evidence that DNA causes transformation of *D. pneumoniae*. (a) Transformation experiment by Griffith. R bacteria are nonvirulent. S bacteria are virulent. A mixture of R bacteria and heat-killed S bacteria is also virulent if transformation has occurred. (b) Transformation experiment by Avery and coworkers. When bacteria from a liquid culture are spread on a semisolid medium, each cell adheres to the medium at random. As time passes, the cells and their offspring grow and divide, leading to visible clones or colonies, each of which arose from a single cell. R and S cells each have a distinct clonal morphology. While R cells produce small rough clones, S colonies are smooth. R cells exposed to DNA from S cells produce a mixture of both types of clones: untransformed R colonies and transformed S colonies.

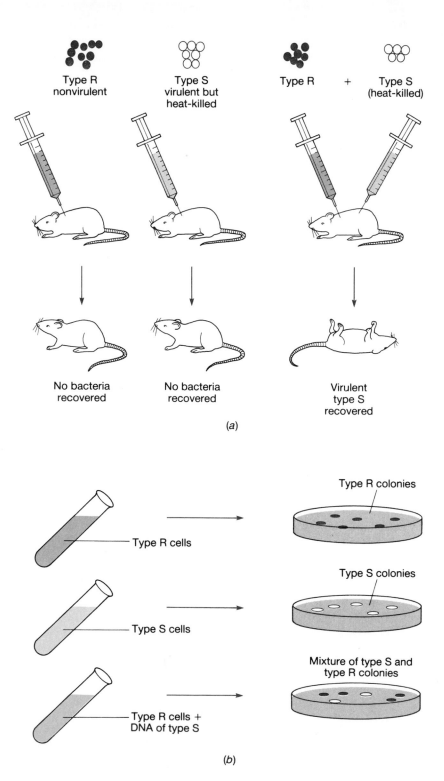

blood. Apparently the genetic factor required for encapsulation was transferred from killed S cells to live R cells.

Griffith's result was duplicated *in vitro* (i.e., outside of the animal) by Dawson and Sia in 1930. Both types of bacteria were grown in a liquid growth medium and distinguished by the distinctive appearance of their colonies on plates (see Figure 7–1b). In parallel with the results obtained in the mouse, heat-killed extracts of S cells transformed R cells into S cells. Alloway, in Avery's laboratory, succeeded in isolating a stable, alcohol-precipitable transforming agent from killed smooth cultures. The *in vitro* results represented an important advance in experimental techniques. The transforming potency of S cell extracts could be quantitatively correlated with the number of colonies of transformed R cells. The ability to fractionate the active component (or "transforming principle," as it was then called) from S cells allowed investigators to use a quantitative approach similar to that used by biochemists to purify an enzyme.

More than ten years later, Avery provided convincing proof that the active agent in the S cell extracts was the cellular DNA. Avery and his co-workers did this by purifying the DNA and showing that it was extremely active in transforming R cells into S cells *in vitro* (Figure 7–1b). The transforming activity in the purified extract was destroyed if the extract was first incubated with the enzyme DNAse, which specifically degrades DNA. The activity was not affected by RNAse, proteases, or enzymes that degrade capsular polysaccharides. It was subsequently shown that DNA could be used to transfer many other genetic traits between the appropriate pairs of donor and recipient bacterial strains. For example, resistance to the antibiotics streptomycin or penicillin could be transferred with the DNA of resistant cells to sensitive cells. The transformation studies with different traits conferred by the donor DNA showed that the active genetic material being transferred faithfully reflected the mutational history and hence the genetic patterns of the donor strains. The procedure of altering the genetic composition of one cell strain by exposing it to DNA of another strain is termed *transformation*.

Transformation occurs naturally in only a small number of bacterial species other than *Diplococcus* (now called *Streptococcus*) *pneumoniae*. In organisms that are not transformed naturally, laboratory procedures are available to partially permeabilize the cell envelope (e.g., of *E. coli*) and permit uptake of DNA. Transformation is an essential step in the cloning of genes; usually it involves special procedures of treating the recipient cells to make them competent to take up DNA. This treatment has greatly expanded the number of organisms that can be transformed and thus greatly increased the usefulness of the transformation technique (see Chapter 32).

Studies on Viruses Confirm the Genetic Nature of Nucleic Acid

At a time when doubts lingered about whether DNA, protein, or possibly capsular polysaccharide was the transforming factor, other investigators performed elegant experiments to show that nucleic acid was the genetic substance of bacterial viruses (bacteriophages). In 1952, A. D. Hershey and Martha Chase demonstrated the independent functions of viral protein and viral nucleic acid in the *E. coli* bacteriophage T2. This bacterial virus is composed of about equal weights of DNA and protein. The linear, duplex DNA is encompassed by a protein shell in a polyhedral "head" or capsid, which is connected to a protein tail that facilitates infection of cells. When a single bacteriophage particle infects a bacterium, it causes cell death and cell disruption (lysis) accompanied by the release of several hundred viruses within 30 min after infection.

The fate of the DNA and the protein components of the infecting viruses

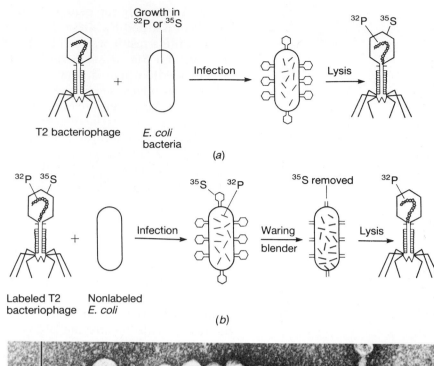

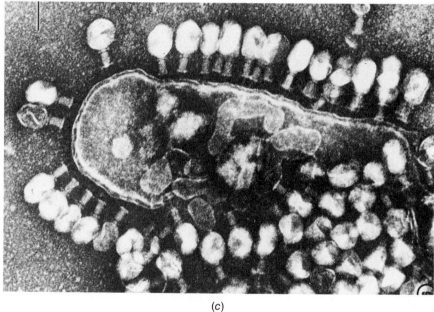

Figure 7–2
Hershey-Chase experiment
demonstrating that the DNA but not
the protein of the T2 bacteriophage is
passed from parent to progeny phage
particles. *(a)* Viruses with ^{32}P-labeled
DNA or ^{35}S-labeled protein were
prepared by growing the viruses on
bacteria in growth medium containing
^{32}P-PO$_4^{3-}$ or ^{35}S-SO$_4^{2-}$, respectively.
(b) When labeled phage particles
infect an unlabeled *E. coli* cell, only
the ^{32}P label appears in the progeny
phage particles, showing that the DNA
structure but not the protein structure
is preserved during phage replication.
(c) T4 phage (closely related to phage
T2) lysing an *E. coli* cell. (Courtesy of
Lee D. Simon, Wakesman Institute for
Microbiology, Rutgers University.)

can each be followed by labeling with an appropriate radioactive isotope
(Figure 7–2). In an initial round of infection, inorganic ^{32}P-labeled phosphate was used to label the DNA component of the virus and ^{35}S-labeled
inorganic sulfate was used to label the protein component. The phosphorus
becomes incorporated into the phosphoryl groups of the viral DNA and the
sulfate into the sulfur-containing amino acids of the viral protein. In a second round of infection, the radioactively labeled bacteriophages were added
to cells in an unlabeled medium. Shortly after the bacterial host was infected, the phage-bacterium complex was subjected to vigorous agitation in a
Waring blender. This agitation sheared the phage from the attachment sites
on the bacterial cell wall. The bacteria were then sedimented by centrifugation, leaving the DNA-free viral protein in the supernatant. Analysis for radioactivity showed that most of the ^{32}P-labeled viral DNA sedimented with the
bacteria, whereas most of the ^{35}S-labeled viral protein remained in the su-

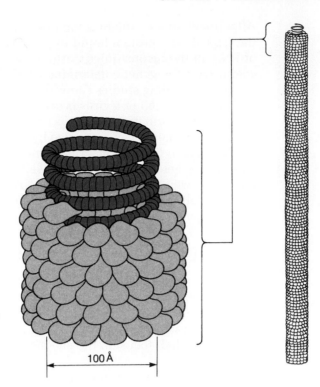

Figure 7–3
Model of tobacco mosaic virus (TMV), showing the helical array of identical proteins forming a protective coat around a single-stranded RNA molecule. (After A. Klug and D. L. D. Caspar, *Adv. Virus Res.* 7:260, 1960.)

100 Å

pernatant. That result demonstrated that most of the viral DNA entered the cells (see Figure 7–2) during infection. Quantitative isotopic analysis of the progeny phage particles indicated that they retained less than 1 per cent of the original infecting viral protein but about two-thirds of the original infecting DNA. Thus, most of the original DNA that enters the cell on infection also becomes an integral part of the progeny phage, but the protein does not. Since the genetic material should survive replication, the results strongly implicated DNA as the genetic component of bacteriophage T2. Long after these experiments were performed, it was shown that purified, deproteinized DNA of the phage could itself produce phage after being taken up by host cells. Infection of bacteria with phage-derived DNA is referred to as *transfection.*

Other early experiments on plant viruses provided even more convincing evidence identifying nucleic acid as the genetic substance of a virus. This work was done on tobacco mosaic virus (TMV), the first virus to be crystallized. TMV is a plant virus with a molecular weight (M_r) of 40×10^6. The virion contains one molecule of single-stranded RNA with a M_r of 2×10^6 and 2130 identical proteins (subunits), each with a M_r of 18,000 (Figure 7–3). In 1956 Gierer and Schramm showed that the deproteinized viral RNA produced lesions on tobacco plant leaves similar in appearance to those produced by the intact virus.

Shortly thereafter (1957) Fraenkel-Conrat and Singer obtained similar evidence that RNA was the genetic substance of TMV. Using two different strains of the virus, HR and TMV, each with readily identifiable protein components they made reconstituted viruses *in vitro* in all possible combinations:

HR-protein + HR-RNA
HR-protein + TMV-RNA
TMV-protein + HR-RNA
TMV-protein + TMV-RNA

The reconstituted virus made from purified RNA and protein components had infectious activities comparable to those of the normal viruses.

After infection the protein components were examined. For all combinations the type of viral protein found in progeny virus was determined by the type of RNA in the reconstituted particles, a very elegant proof that the nucleic acid carries the genetic determinants of the viral protein.

The preceding studies showed that the genetic information required for replication of a cell or a virus is carried by nucleic acids. In the case of a cell the genetic information is always carried by DNA. In the case of viruses the genetic information can be carried by either single- or double-stranded DNA or RNA, depending on the virus. Sometimes the viral DNAs are circular. The only case where circular RNAs have been encountered is in _viroids_, low-molecular-weight, single-stranded RNAs that cause several important diseases in cultivated plants but are not encapsidated. Viroids are the smallest known agents that cause infectious disease.

STRUCTURAL PROPERTIES OF DNA

The early work equating the genetic material with certain nucleic acids plunged genetics into an entirely new vocabulary of chemical terms. The genetic consequences of these early studies could be understood, but the chemistry was new and almost totally a mystery. In the remainder of this chapter we shall focus on the chemical and structural properties of nucleic acids.

As we have seen, genes can be composed of either DNA or RNA. However, RNA is encountered as the genetic component only in some viruses. Where DNA provides the genetic information, it is transcribed into RNA, which, in most cases, is subsequently used to direct protein synthesis. The different functions of the two cellular nucleic acids are reflected by their different locations in the cell: DNA is found almost exclusively in the cell nucleus, whereas the mature forms of RNA are located mainly in the cytoplasm, where protein synthesis takes place.

The amount of DNA per cell differs widely among different organisms (Figure 7–4). Mammalian cells contain about 1000-fold more DNA than bac-

Figure 7–4

Genome size in different cells, viruses, and plasmids. Plasmids are small circular DNA molecules that replicate autonomously in cells harboring them. Unlike viruses, they do not form any complex nucleoprotein structures. In the case of plasmids, most viruses, and bacteria, the genome size is equivalent to the size of the chromosomal DNA because there is only one chromosome. For all of the remaining eukaryotic organisms that are listed, the genome is subdivided into two or more chromosomes. Some organisms contain over one hundred chromosomes. All chromosomes are believed to contain a single DNA molecule.

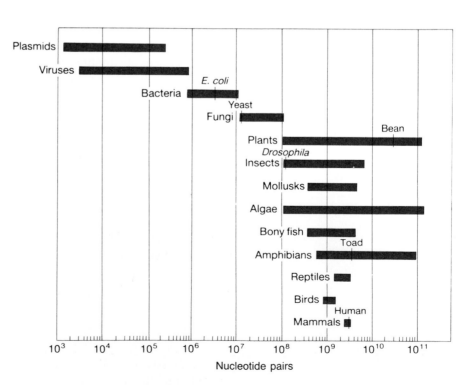

terial cells. Bacterial viruses such as the T type (T1 through T7) bacterio-phages (or phages) that infect *E. coli* contain about tenfold less DNA than the bacterial chromosome. The DNA of the smallest viruses is about one-tenth the size of the smallest T phage DNA, containing barely enough genetic material to accommodate about three to ten genes. This finding is consistent with the fact that viruses do not contain sufficient genetic information for independent growth, but can only grow parasitically in the host cells they infect. On the other hand, the amount of DNA per cell is not always directly proportional to the amount of genetic information that an organism carries. This is because complex eukaryotes contain in their chromosomes a great deal of noninformational DNA, whose function is frequently unclear.

In addition to the main DNA associated with the cell or the virus, there is informational DNA found in special organelles such as chloroplasts and mitochondria. This DNA carries genes for some of the RNAs and proteins associated with the respective organelles. The simplest type of system using informational DNA is the plasmid. They vary in size and exist as naked DNA molecules that carry anywhere from 5 to 100 genes. Plasmids are very wide-spread in bacteria. They have also been found in eukaryotes, where they are usually confined to the nucleus. Plasmids in bacteria have been useful for genetic transfers for a long time. More recently plasmids have been ex-tremely useful for DNA manipulation (see Chapter 32).

Nucleotides are the Building Blocks of Nucleic Acids

Both DNA and RNA are linear polymers composed of nucleotide building blocks. Each nucleotide is composed of three parts: (1) a heterocyclic nitrog-enous base, termed a *purine* or a *pyrimidine*; (2) a sugar, *ribose* in RNA, *2′ deoxyribose* in DNA; and (3) a *phosphoryl* group (Figure 7–5). The phos-phoryl group is in an ester linkage to the C-5′ (or C-3′) position of the sugar. (Carbon atoms in the sugar are given a prime designation to distinguish them from the numbered atoms in the purines or pyrimidines.) If the nucleotide lacks a phosphoryl group, it is called a *nucleoside*. Nucleotides differ solely by the base that is covalently bound to the C-1′ position of the sugar. This

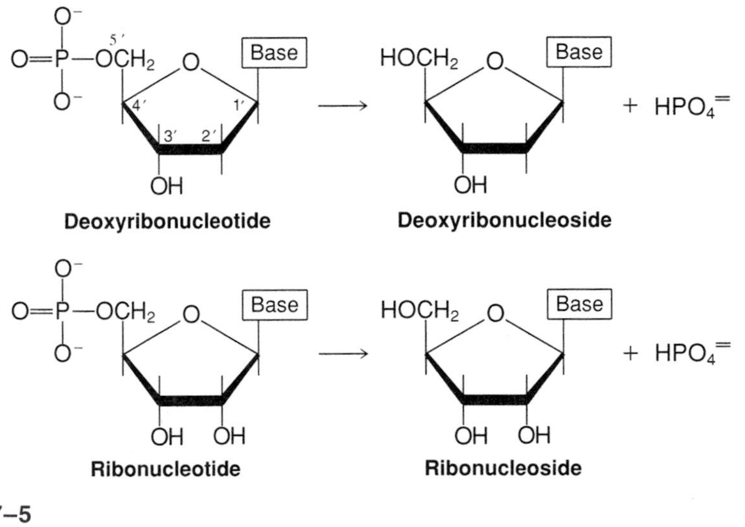

Figure 7–5

Structure of a nucleotide. A nucleotide has three components: a phosphoryl group, a pentose, and a nitrogenous base—either a purine (Pu) or a pyrimidine (Py). Deoxyribonucleotides lack a hydroxyl group at C-2′ of the pentose. The nitrogenous base is attached at C-1 of the pentose in the β configuration. The phosphoryl group may be attached at C-5′ or C-3′ or both locations; it is shown at the C-5′ location. Removal of the phosphoryl group converts the nucleotide into a nucleoside.

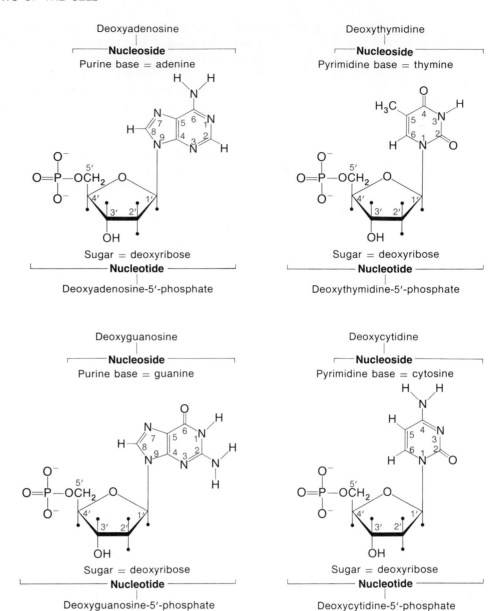

Figure 7–6
Structure of the four common deoxyribonucleotides found in DNA. Note the numbering system for the carbon and nitrogen atoms of the purine and pyrimidine bases. The carbon atoms of the sugars are usually given prime designations. Note that the nitrogen bases are *cis* relative to the C-5′ and *trans* relative to the C-3′-OH.

linkage between the base and the sugar is called an N-*glycosidic bond* (the configuration is always β).

The structures of the four commonly occurring deoxyribonucleotides found in DNA are shown in Figure 7–6. The bases in DNA are the purines, *adenine* (A) and *guanine* (G), and the pyrimidines, *thymine* (T) and *cytosine* (C). Three of the same bases are found in RNA, but thymine is replaced by *uracil* (U), which has an H atom instead of a CH_3 group on the C-5 position of the pyrimidine. The nomenclature for different nucleosides and nucleotides is given in Table 7–1.

All of the commonly occurring bases in nucleotides are capable of existing in two tautomeric forms, which differ by the placement of a proton and some electrons. For example, guanosine (G) can undergo a change from a keto form to an enol form as shown in Figure 7–7. The keto form is so strongly favored that it is difficult to detect even trace amounts of the enol form at equilibrium.

Similarly, the keto forms of thymidine (T) or uridine (U) are strongly preferred. Adenosine (A) and cytidine (C) can isomerize to imino forms, but

TABLE 7–1
Names of the Most Common Bases Found in DNA and RNA and Corresponding Names of Nucleosides and Nucleotides Containing These Bases

Type of Base	Base	Nucleoside[a]	Nucleotide[a]
Purine	Adenine (DNA or RNA)	Adenosine	Adenylate (or adenylic acid)
Purine	Guanine (DNA or RNA)	Guanosine	Guanylate (or guanylic acid)
Pyrimidine	Thymine (DNA)	Thymidine	Thymidylate (or thymidylic acid)
Pyrimidine	Uracil (RNA)	Uridine	Uridylate (or uridylic acid)
Pyrimidine	Cytosine (DNA or RNA)	Cytidine	Cytidylate (or cytidylic acid)

[a]The prefix *deoxy-* or *deoxyribo-* is added to nucleosides or nucleotides of the deoxyribose type. The prefix *ribo-* is added to nucleosides or nucleotides of the ribose type.

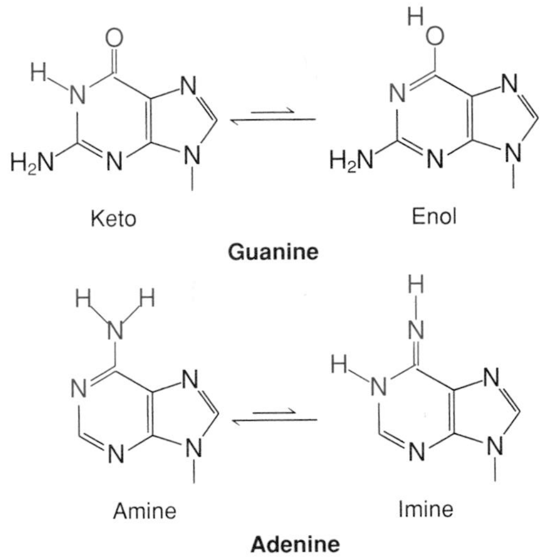

Keto   Enol
Guanine

Amine   Imine
Adenine

Figure 7–7
Tautomeric equilibrium of guanine and adenine. The keto and imino forms are strongly favored for guanine and adenine, respectively. Comparable structures could be drawn for thymine (or uracil) and cytosine.

TABLE 7–2
Ionization Constants of the Ribonucleotides (Presented as pK Values)

	Base	Secondary Phosphate	Primary Phosphate
Adenosine-5′-phosphate (5′-AMP)[a]	3.8	6.1	0.9
Uridine-5′-phosphate (5′-UMP)	9.5	6.4	1.0
Cytidine-5′-phosphate (5′-CMP)	4.5	6.3	0.8
Guanine-5′-phosphate (5′-GMP)	2.4, 9.4	6.1	0.7

[a]5′-AMP (or 5′-rAMP) refers to the ribonucleotide.

the amino forms are strongly preferred (see Figure 7–7). Even though the unusual tautomers are present in very small amounts, it is conceivable that when present in DNA they contribute to the mutation process.

Some nucleotides undergo protonation in acid and some undergo deprotonation in base; the relevant pK values are listed in Table 7–2. At neutrality there is no charge on any of the bases. Three of the bases, A, C and G, undergo protonation as the pH is lowered. The adenine moiety in adenylic acid protonates on the N-1 position of the purine rather than on the amino group (Figure 7–8). The charged form is stabilized by the resonance hybrids shown. In cytidylic acid the proton adds to the comparable N-3 ring nitrogen. In guanine a proton adds to the N-7 rather than the amino group (Figure 7–9), again indicating the unusually low basicity of the amino groups on the nucleotides compared with primary aliphatic amines. On the basic side of neutrality both 5′-UMP and 5′-GMP lose a proton from the imino nitrogens at positions 3 and 1, respectively. As might be expected, the ionization constants for the primary and secondary dissociations of the phosphate group do not differ appreciably for the various nucleotides (see Table 7–2).

Owing to the large number of conjugated double bonds in their nitrogen bases, all nucleotides show absorption maxima in the near-ultraviolet range (Figure 7–10). The spectrum is pH-dependent since protonation or deprotonation changes the electronic distribution in the base rings. The ultraviolet absorption of the nucleotides has been useful in many ways for the study of mononucleotides and polynucleotides. Indeed, it has been most useful in studying nucleic acid conformation, as we will soon see.

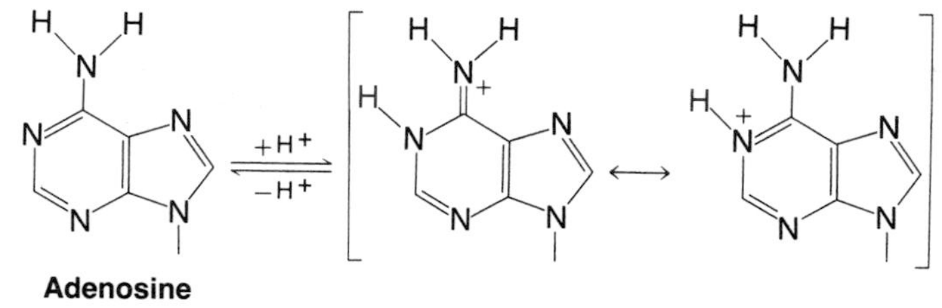

Adenosine

Figure 7–8
Uncharged and protonated forms of adenosine. The charged base resonates between the two structures shown on the right. Cytosine protonates in a similar way.

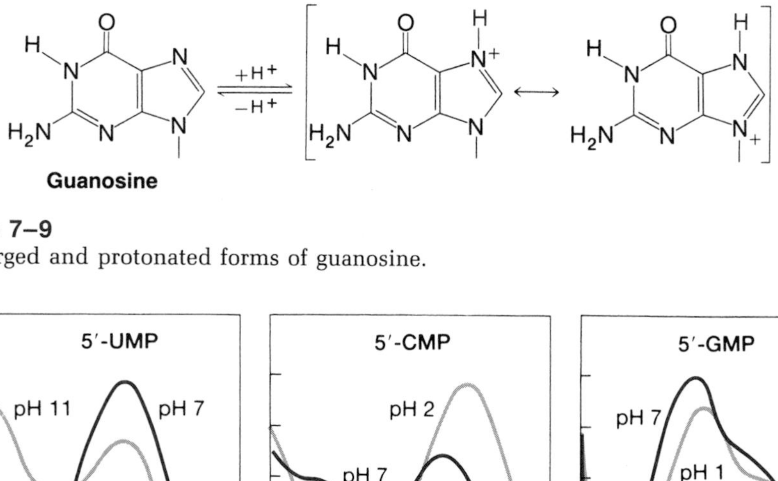

Guanosine

Figure 7–9
Uncharged and protonated forms of guanosine.

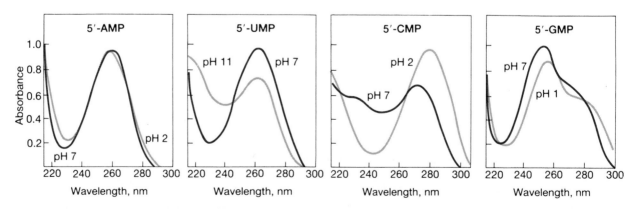

Figure 7–10
Curves showing the molar absorbances ($\times 10^{-3}$) of the 5′-ribonucleotides at several pH values. The comparable 5′-deoxyribonucleotides show comparable absorbance curves. DNA and RNA (not shown) have an absorbance maximum at 260 nm. The molar absorbance of the nucleotides in native DNA is about 60–70 per cent lower than for an equivalent mixture of the free nucleotides, and about 40 per cent lower than that for denatured DNA.

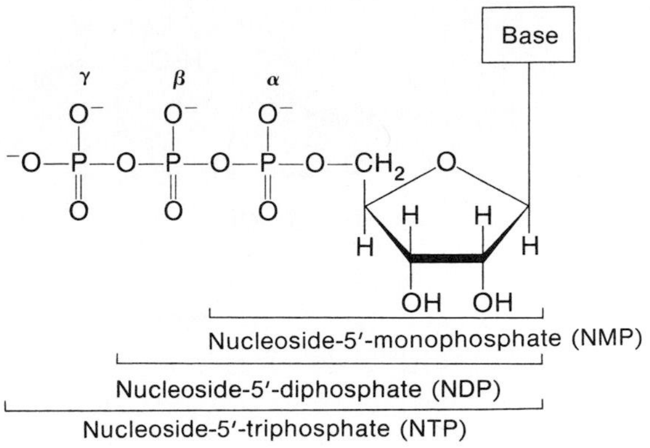

Figure 7–11
The general structure of a nucleoside monophosphate, diphosphate, and triphosphate.

Nucleotides Can Form Diphosphates and Triphosphates

All the common 5′-nucleoside monophosphates also occur in cells as 5′-diphosphates and 5′-triphosphates (Figure 7–11). The nucleoside diphosphates (NDPs) and the nucleoside triphosphates (NTPs) dissociate three and four protons, respectively, from their phosphate groups. In the triphosphates, the phosphate immediately attached in ester linkage to the 5′-carbon is designated α; the middle phosphate, in pyrophosphate linkage, is called β; and the terminal phosphate, also in pyrophosphate linkage, is called γ. The NDPs and NTPs can form complexes with Mg^{2+} or Ca^{2+} and probably exist in these complexes in the cell. The NDPs and NTPs have a number of important metabolic functions in the cell: they serve as energy-carrying enzyme cofactors (e.g., see Chapters 12–15) and as substrates for polymeric nucleic acids (see Chapters 26 and 27).

The Polynucleotide Chain Contains Mononucleotides Linked by Phosphodiester Bonds

When a 5′-phosphomononucleotide is joined by a phosphodiester bond to the 3′-OH group of another mononucleotide, a dinucleotide is formed. Repetition of this linkage leads to the formation of polydeoxyribonucleotides in DNA or polyribonucleotides in RNA; the structure of a short polydeoxyribonucleotide is shown in Figure 7–12. The polymeric structure consists of a sugar phosphate diester backbone with bases attached as distinctive side chains to the sugars. The polynucleotide chain has a directional sense with a 5′ and a 3′ end. Either of these ends may contain a free hydroxyl group or a phosphorylated hydroxyl group. The structure shown in Figure 7–12 contains a phosphate group on the 5′ end but none on the 3′ end. By convention, one writes a nucleic acid sequence from the 5′ to the 3′ end so that a comparable structure is written pTpApCpG. With no phosphate on the 5′ end, the structure is designated TpApCpG; alternatively, if the terminal phosphate is on the 3′ end rather than the 5′ end the structure is written TpApCpGp. When the phosphates are not indicated, the oligonucleotide is written TACG. If the oligonucleotide is part of a larger polynucleotide, this fact is indicated by dashes on either end: –TACG–. The letters d or r sometimes precede the capital letter of the nucleotide to indicate a deoxyribo- or a ribo-derivative.

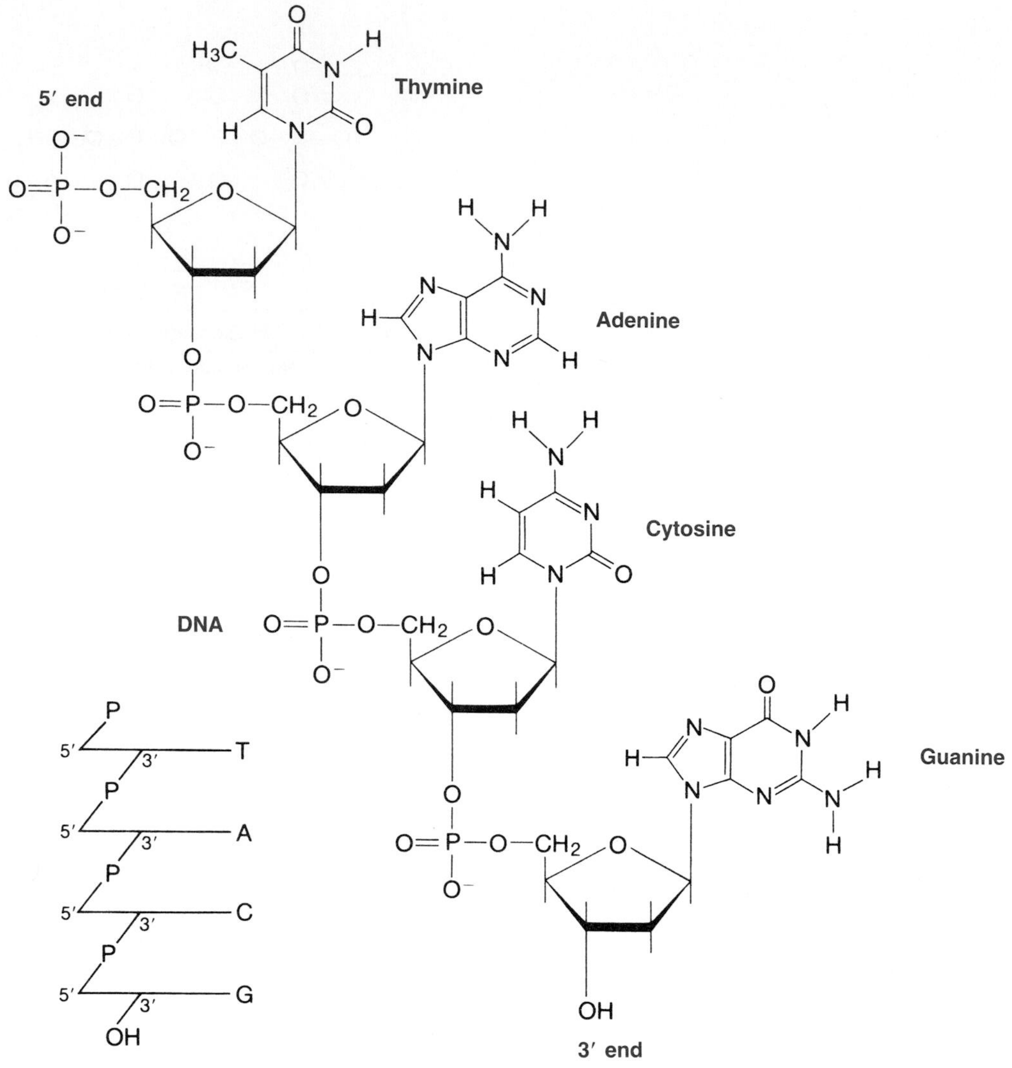

Figure 7–12
The structure of a deoxyriboligonucleotide. Drawn in abbreviated form at lower left. The illustrated structure is written pTpApCpG.

Most DNAs Form a Double Helix (Duplex) Structure

Like most other types of biological macromolecules, nucleic acids adopt highly organized three-dimensional structures. The dominant factors that determine nucleic acid conformations are the limitations imposed by the stereochemistry of the polynucleotide chains, the high negative charge resulting from the regularly repeating phosphate groups, and the noncovalent affinities between purine and pyrimidine bases.

A body of chemical information that proved vital to understanding DNA structure came from Erwin Chargaff's analyses of the nucleotide composition of duplex DNAs from various sources (Table 7–3). Although the base compositions varied over a wide range, Chargaff found that within the DNA of each source that he examined, the amount of A was very nearly equal to the amount of T, and the amount of G was very nearly equal to the amount of C. (Actually, the C is present as both unmodified C and, to a lesser extent, 5-methylcytosine, which results from postreplicative modification.) These two equalities were the first indication that regular complexes occur between A and T and between G and C in DNA.

While searching for the meaning of these equalities, James Watson discovered, with the help of molecular models, that between A and T and between G and C it was possible to form hydrogen-bonded base-paired structures that have the same overall dimensions (Figure 7–13). Two hydrogen

TABLE 7–3
Base Composition of DNAs from Different Sources

	(A) Adenine	(G) Guanine	(C) Cytosine	(5-MC) 5-Methyl-cytosine	(T) Thymine	$\dfrac{A + T}{G + C + 5\text{-MC}}$
Human	30.4	19.6	19.9	0.7	30.1	1.53
Sheep	29.3	21.1	20.9	1.0	28.7	1.38
Ox	29.0	21.2	21.2	1.3	28.7	1.36
Rat	28.6	21.4	20.4	1.1	28.4	1.33
Hen	28.0	22.0	21.6		28.4	1.29
Turtle	28.7	22.0	21.3		27.9	1.31
Trout	29.7	22.2	20.5		27.5	1.34
Salmon	28.9	22.4	21.6		27.1	1.27
Locust	29.3	20.5	20.7	0.2	29.3	1.41
Sea urchin	28.4	19.5	19.3		32.8	1.58
Carrot	26.7	23.1	17.3	5.9	26.9	1.16
Clover	29.9	21.0	15.6	4.8	28.6	1.41
Neurospora crassa	23.0	27.1	26.6		23.3	0.86
Escherichia coli	24.7	26.0	25.7		23.6	0.93
T4 Bacteriophage	32.3	17.6		16.7[a]	33.4	1.91

[a] In T bacteriophage all of the cytosine exists in the 5-hydroxymethyl form 5-HMC.

Figure 7–13
Dimensions and hydrogen bonding of (a) thymine to adenine and (b) cytosine to guanine. Note that there are two hydrogen bonds formed in the A-T base pair and three in the G-C base pair. The overall dimensions of the base pairs are the same. Consequently they will fit at any position in an otherwise regular polymeric structure. (Adapted from M. H. F. Wilkins and S. Arnott, *J. Mol. Biol.* 11:391, 1965.)

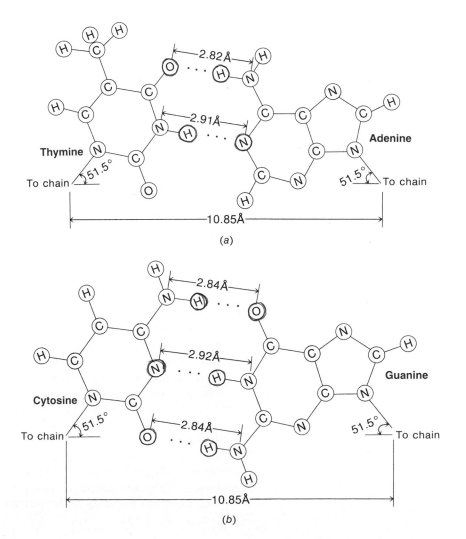

BOX 7–A

X-ray Diffraction of DNA

An x-ray pattern of DNA is obtained by holding a stretched fiber containing many DNA molecules in a vertical direction and exposing it to a collimated monochromatic beam of x-rays. Only a small percentage of the x-ray beam is diffracted. Most of the beam travels through the specimen with no change in direction. A photographic film is held in back of the specimen; a hole in the center of the film allows the incident undiffracted beam to pass through. Coherent diffraction occurs only in certain directions, specified by Bragg's law: $2d \sin \theta = n\lambda$. Here d is the distance between identical repeating struc-

tural elements, θ is the angle between the incident beam and the regularly spaced diffracting planes, λ is the wavelength of x-rays used, and n is the order of diffraction, which may equal any integer but is usually strongest for $n = 1$. The most important point is that $\sin \theta \approx \theta$ and $d \approx 1/\theta$, so that a spot far out on the photographic film is indicative of a repeating element of small dimension and vice versa.

Watson and Crick were the first to appreciate the significance of strong 3.4-Å and 34-Å spacings and the central crosslike pattern, which reflects a helix structure in the x-ray diffraction pattern of DNA (Illustration 1). They interpreted this as arising from the hydrogen-bonded antiparallel double-helix structure shown in Illustration 2.

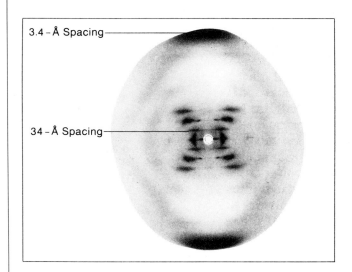

3.4 – Å Spacing

34 – Å Spacing

Illustration 1

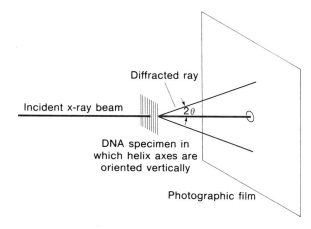

Diffracted ray

Incident x-ray beam

2θ

DNA specimen in which helix axes are oriented vertically

Photographic film

Illustration 2

bonds are formed in the A-T base-paired structure and three in the G-C base-paired structure. Watson brought this information to the attention of his crystallographer colleague Francis Crick, who perceived that the x-ray diffraction pattern produced by DNA could be interpreted in terms of a helix (see Box 7–A) composed of two polynucleotide strands with hydrogen-bonded pairs formed between the bases of opposing strands (Figure 7–14). In this complementary structure the hydrogen-bonded base pairs can form when the directional senses of the two interacting chains are opposite or antiparallel, as indicated in Figure 7–15. The planes of the base pairs are nearly perpendicular to the helix axis, and the distance between adjacent pairs along the helix axis is 3.4 Å, bringing them into close contact. The structure repeats itself after about 10 residues, or once every 34 Å along the helix axis; the repeating distance is referred to as the _pitch length_ or just the _pitch_. An average size gene is about 1000 base pairs (bp) in length, equivalent to 100 helical turns. The complementary structure of duplex DNA explains how the genetic material is faithfully replicated as well as how it is expressed.

An important feature of the helical structure is the grooved nature of the surface resulting from the helical twist. Alternating wide (major) and narrow (minor) grooves are displayed in a side view of the helix structure (see Figure 7–14). Different sections of the purine and pyrimidine bases are exposed in

Figure 7–14
The most common form of DNA. (Top) A view perpendicular to the helix axis. A single turning of the DNA duplex is shown. Note that the helix has a wide groove and a narrow groove. Different parts of the base pairs are exposed in the wide and narrow grooves as indicated below. (Bottom) A view looking down the helix axis at two adjacent base pairs is shown on the left. On the right is indicated the orientation of the groupings in the base pairs with respect to the wide and narrow grooves.

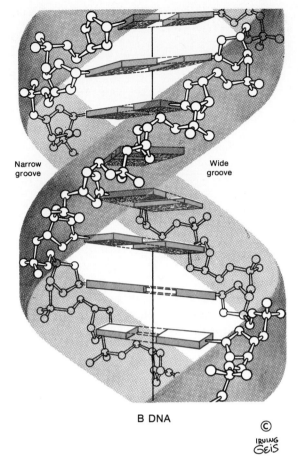

Narrow groove   Wide groove

B DNA

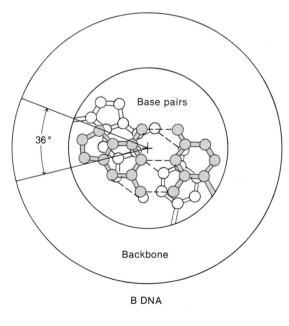

36°

Base pairs

Backbone

B DNA

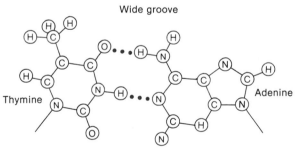

Wide groove

Thymine   Adenine

Narrow groove

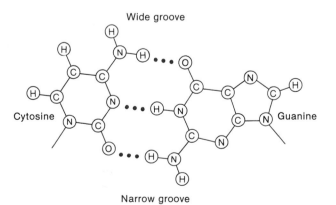

Wide groove

Cytosine   Guanine

Narrow groove

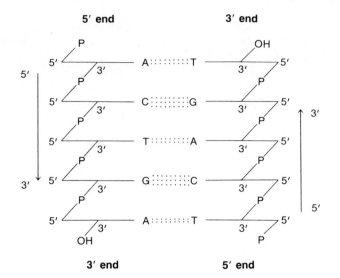

Figure 7–15
Segment of DNA duplex emphasizing the antiparallel orientation of the complementary chains.

these two grooves as indicated in Figure 7–14 (bottom). Many different proteins interact with DNA; most of the interactions occur with the phosphoryl groups on the outer surface of the structure and with the purine and pyrimidine bases in the wide groove because of its greater accessibility. Specific instances of DNA-protein interactions will be considered in later chapters (Chapters 29, 30, and 31).

Hydrogen Bonds and Stacking Forces Stabilize the Double Helix. Several factors account for the stability of the double-helix structure. The negatively charged phosphoryl groups are all located on the outer surface, where they have a minimum effect on one another. That is, repulsive electrostatic interactions generated by these charged groups are often partly neutralized by interaction with positively charged salts such as Mg^{2+} ions, basic polyamines (such as putrescine and spermidine), and the positively charged side chains of chromosomal proteins. The core of the helix is composed of the base pairs, held together by the specific hydrogen bonds and also by favorable stacking interactions between the planes of adjacent base pairs. These stacking interactions are complex, involving dipole–dipole interactions and van der Waals forces, and result in a *stacking energy* comparable in magnitude to the stabilizing energy generated by the hydrogen bonds between the base pairs. The result is that stacking is maximized in most nucleic acid structures. In this connection it is noteworthy that two fully extended polynucleotide strands can form a hydrogen-bonded base-paired complex, leading to a stepladderlike structure (Figure 7–16a). In this structure the chains do not form a helix but, lie straight, with a distance of 6.8 Å between identical residues in the direction of the long axis. This 6.8-Å gap between adjacent base pairs would presumably have to be filled by water. Such a conformation does not result in an energetically favorable structure, however, since the planes of the bases would prefer to be in close contact with each other rather than with water. The stepladder structure can be converted into the helix structure by a simple right-handed twist (Figure 7–16b). When this is done, the distance between base pairs decreases until they are in close contact, with a spacing of 3.4 Å.

Conformational Variants of the Double-helix Structure. The same base-pairing arrangement is found in all naturally occurring double-helix structures. However, *the inherent flexibility in the furanose ring of the sugar and the degrees of freedom generated by several rotatable single bonds per resi-*

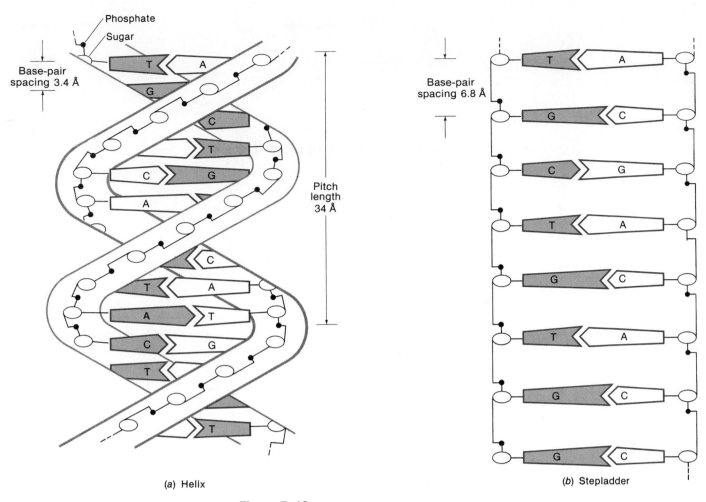

(a) Helix (b) Stepladder

Figure 7–16

Different conformations of base-paired DNA: (a) helix, (b) stepladder. The stepladder is an unstable structure that is converted into a helix by a right-handed twist. This permits the planes of the base pairs to come into close contact.

Figure 7–17

A segment of polynucleotide chain with rotatable bonds indicated by curved arrows. The presence of so many rotatable bonds permits the backbone of a polynucleotide chain to adopt many different conformations. (Adapted from W. K. Olson and P. J. Flory, *Biopolymers* 11:1, 1972.)

due, six in the sugar phosphate backbone and one in the C-1'-N-glycosidic linkage (Figure 7–17), lead to considerable variation in the conformations adopted by double-helix structures. Four puckered conformations for the sugar, with small differences in stability, are shown in Figure 7–18. In the double-helix structure shown in Figure 7–14, the furanose rings are in the C-2' endo conformation. This is believed to be the major structure adopted

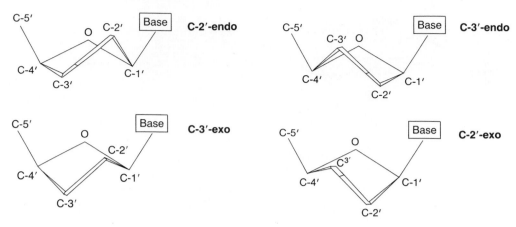

Figure 7–18
Four pucker conformations of the furanose rings of ribose and deoxyribose that are deemed energetically feasible. The flexibility of these furanose rings contributes to the possible variations in the conformation of the double helix.

by DNA, both when free in aqueous solution and in chromosomes. It is known as the B form of DNA. When some of the water is removed from the hydrated DNA fibers, the double helices push even closer together and the structure changes to the so-called A form, which has about 11 bases per turn and base pairs that are tilted about 20° with respect to the helix axis (Figure 7–19b). In the A form, the furanose rings have changed their pucker to the C-3'-endo conformation. The furanose rings in RNA have a stronger preference for the C-3'-endo conformation, with the result that RNA double helices adopt a structure similar to the DNA A form, even at high degrees of hydration.

The most striking conformational variant observed for a DNA double helix with Watson-Crick base pairing is referred to as the Z form. This structure was first detected by Alex Rich and his coworkers for the deoxyoligonucleotide d(CpGpCpGpCpG), which crystallizes into an antiparallel double helix with a left-handed rather than a right-handed twist (see Figure 7–20). The Z form is a considerably slimmer helix than the B form and contains 12 base pairs per turn rather than 10. In the Z form, the planes of the base pairs are rotated approximately 180° with respect to the helix axis from their orientation in the B form (Figure 7–21). The flipping of the base pairs involves different conformational changes in the G and C residues in the alternating GC structure. In the case of the G residues the base is rotated by 180° about the glycosidic bond, resulting in a transition from the so-called _anti_ conformation found in B DNA to the _syn_ conformation (Figure 7–22). Model-building studies indicate that the anti conformation of the nucleotide (found in B DNA) has less steric crowding than the syn conformation but that it is far easier for a purine nucleotide to adapt the syn conformation than for a pyrimidine nucleotide. In fact, cytidine remains in the _anti_ conformation in Z DNA. The flipping of the cytidine base in going from the B to Z conformation involves rotation of the entire cytidine residue while maintaining the anti conformation. The effect is to make the sugar-phosphate backbone follow a zigzag course (Figure 7–20a). Thus the name Z DNA is an appropriate descriptive designation for this structure. A number of the structural parameters associated with the A, B, and Z helices are summarized in Table 7–4.

Because of the different orientations of the G and C residues in Z DNA, this DNA conformation requires that the sequence of purine and pyrimidine bases be strictly alternating. Many other arrangements that involve alternat-

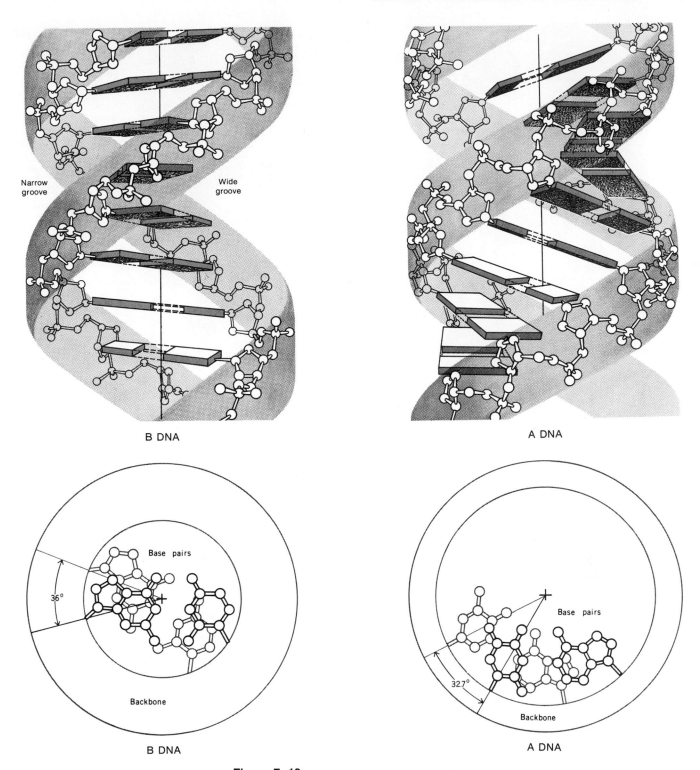

Figure 7–19
The A and B forms of DNA. (Top) A view perpendicular to the helix axis. (Bottom) A view of two adjacent base pairs, looking down the helix axis. A single turning is shown for each duplex. DNA usually is found in the B form. The A form has been observed in microcrystals of DNA from which a significant amount of the water has been removed. Dimensions of the two structures are compared in Table 7–4.

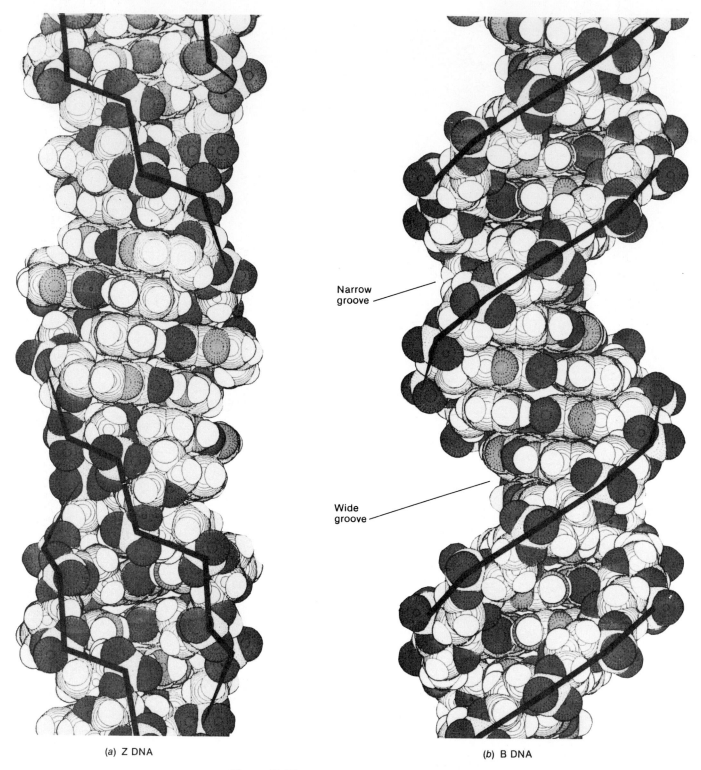

(a) Z DNA   (b) B DNA

Figure 7–20
Space-filling models of (a) Z DNA and (b) B DNA. The irregularity of the Z DNA backbone is illustrated by the heavy lines that go from phosphate to phosphate residues along the chain. In contrast, B DNA has a smooth line that connects the phosphate groups and the two grooves, neither one of which extends into the helix axis of the molecule. The space-filling model is excellent for displaying the volume occupied by molecular constituents and the shape of the outer surface. By contrast, the framework representations of DNA (Figure 7–19) show connectivities and allow one to look inside.

Figure 7–21
The change in topological relationship if a four-base-pair segment of B DNA is converted into Z DNA. Such a conversion could be accomplished by rotation of the bases relative to those in B DNA. This rotation is shown diagrammatically by coloring one surface of the bases. All of the colored areas are at the bottom in B DNA. In the segment of Z DNA, however, four of them are turned upward. The turning is indicated by the curved arrows.

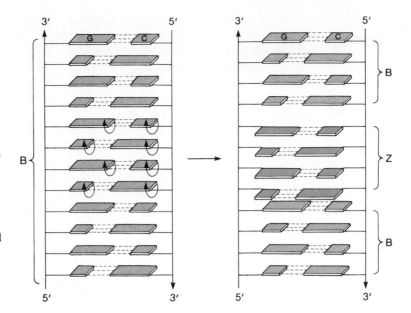

Figure 7–22
The syn and anti conformations of deoxyadenosine and deoxycytidine. Purine nucleosides can readily adapt to either conformation by a simple rotation about the C-1—N-9 glycosidic bond. Pyrimidine nucleosides are considerably less stable in the syn conformation because of steric hinderance between the sugar and the C-2 carbonyl group.

TABLE 7–4
Some Structural Parameters of A, B, and Z DNA

	A DNA	B DNA	Z DNA
Helix sense	Right-handed	Right-handed	Left-handed
Residues per turn	11	10	12 (6 dimers)
Rise per residue	2.55 Å	3.4 Å	3.7 Å
Helix pitch	28 Å	34 Å	45 Å
Base pair tilt	20 degrees	6 degrees	7 degrees
Rotation per residue	33 degrees	36 degrees	−60 degrees (per dimer)
Glycosidic conformation			
Deoxycytidine	Anti	Anti	Anti
Deoxyguanosine	Anti	Anti	Syn
Sugar pucker			
Deoxycytidine	C-3′-endo	C-2′-endo	C-2′-endo
Deoxyguanosine	C-3′-endo	C-2′-endo	C-3′-endo

ing purine-pyrimidine residues can adopt the Z conformation. For example, the duplex containing alternating T and G residues in one strand and the complementary A and C residues in the other strand can adopt a Z conformation. An alternating A-T DNA sequence has never been observed in a Z conformation. The reason is believed to be the way that water molecules orient around an A-T base pair in the Z helix. The arrangement is not so satisfactory as that observed for the G-C base pair, creating a less stable situation. Other factors that favor the stability of Z DNA are methylation of the 5 position of C residues and negative supercoiling of the DNA. Negative supercoiling is discussed in the next section.

The biological significance of Z DNA is currently unclear. However, several cellular proteins that bind specifically to Z DNA have been isolated from the nuclei of *Drosophila* fruit flies. The mere existence of such proteins suggests that they may function in some specific role when they encounter stretches of DNA that can adopt a Z conformation. A mutation-genetic analysis of such Z-DNA-binding proteins would be helpful to ascertain their role(s). Current speculation on the biological significance of the Z conformation centers on the general notion that the conformation plays a regulatory role in gene expression and possibly in recombination.

In addition to the A, B, and Z helices, many other conformations have been observed for DNA. All, however, preserve Watson-Crick hydrogen bonding. Synthetic polyribonucleotide complexes have been made that simulate the normal duplex structure. For example, poly(A) and poly(U) in 1:1 base pair ratios form a duplex resembling the A form of DNA. In 1:2 ratios of one A chain to two U chains they form a triple helix that displays additional kinds of hydrogen bonding. Two chains of A make a double helix and four chains of G can form a quadruple helix. A great variety of structures and hydrogen-bonding patterns is possible with polymers that have special sequences. None of the unusual hydrogen-bonded structures has yet been detected in nature. Still the possibility remains that some of them will be found in limited regions of the genome where the appropriate base sequences occur. To this, one should add that unusual pairings do occur in transfer RNA (described later).

Duplex Structures Can Be Supercoiled

The detection of different conformations of DNA underscores the inherent flexibility built into the DNA duplex. All the conformations discussed thus far involve regular linear duplexes. Energetically favorable interactions with other molecules, particularly proteins, can induce additional conformations that do not result in major changes in either base pairing or stacking. Several conformations are believed to play important roles in different situations. Bends are known to be important structures formed by chromosomes (discussed later in this chapter). *Cruciforms*, in which a single chain folds back on itself into a hairpinlike duplex, are important as intermediates in DNA and RNA synthesis (see Chapters 26 and 27). *Supercoiled* DNA is a very common type of tertiary structure observed in DNAs that involves the twisting of double-helix segments around each other to form supertwists. Supercoiled DNA is topologically constrained by being covalently closed and circular or by being complexed to proteins so that the ends of the DNA cannot rotate freely.

DNA can form right-handed (negatively supercoiled) or left-handed (positively supercoiled) supercoils (Figure 7–23). *Negative supercoiling imparts a torsional stress to the DNA that favors unwinding, whereas positive supercoiling favors tighter winding of the double helix.* Supercoiling imparts a more compact structure to a circular duplex, which makes it sedi-

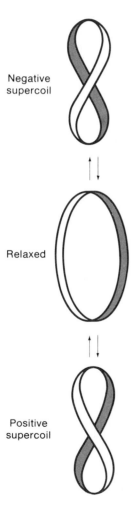

Negative
supercoil

Relaxed

Positive
supercoil

Figure 7–23
Topology of negative and positive supercoil.

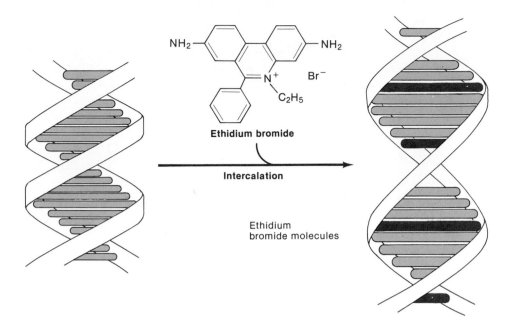

Figure 7–24
Ethidium bromide and DNA intercalated with ethidium bromide. Molecules such as ethidium bromide can intercalate DNA because they are flat rings of the same thickness as DNA base pairs. In order to accommodate the ethidium bromide molecule the duplex must untwist in the region of intercalation. This increases the separation between the planes of the base pairs. We saw the effect of complete untwisting in Figure 7–16.

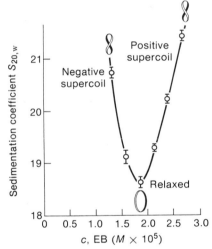

Figure 7–25
The sedimentation coefficients of closed circular phage PM2 DNA in 2.85 M CsCl containing varying amounts of ethidium bromide. The more compact the structure, the faster it will sediment. Supercoiled DNA is more compact than relaxed circular DNA. Thus when ethidium bromide is added to negatively supercoiled DNA the sedimentation constant decreases, indicating that the DNA is becoming more relaxed. A minimum in sedimentation constant is reached when the fully relaxed structure is obtained. Addition of ethidium bromide beyond this point results in an increase in the sedimentation constant again, as the DNA becomes positively supercoiled. (Adapted from J. Vinograd, *Nature (New Biol.)* 229:10, 1971.)

ment more rapidly in a centrifuge. Thus, either a positively or a negatively supercoiled DNA will sediment more rapidly than a circular duplex with no supercoiling. Jerome Vinograd demonstrated this by adding ethidium bromide (Figure 7–24) to circular DNA of the small phage PM2. This DNA has about 40 negative supercoiled turns when it is isolated from cells. Addition of ethidium bromide to duplex DNA leads to the binding of ethidium between adjacent base pairs, a type of binding known as <u>intercalation</u> (Figure 7–24). Normally the base pairs are nearly closely packed, so such binding would not be possible without an alteration in the DNA structure. In order to accommodate the ethidium, the duplex unwinds by about $-27°$ per base pair. This unwinding reduces the negative supercoiling in naturally occurring circular duplexes such as PM2 DNA. As more ethidium is added, enough unwinding takes place to eliminate all the supercoiling. At this point the DNA has its slowest sedimentation rate (Figure 7–25). Addition of further ethidium causes the DNA to adopt a positively supercoiled form that again increases its sedimentation rate.

Supercoiling of circular duplex DNA is quantitatively considered in terms of the <u>linking number</u> (L), an integer that specifies the number of complete turns made by one strand around the other. *The linking number can change only if a covalent linkage in the DNA backbone is broken*. If a molecule of DNA is projected onto a two-dimensional surface, the linking number is defined as *the excess of right-handed over left-handed crossings of one strand over the other*. Linear duplex DNA with free ends adopts a conformation in solution close to the B form, with about 10 base pairs per turn. Therefore, a closed circular duplex with this extent of twist is presumed to be under no torsional strain and is said to be *relaxed*. Because B DNA is a right-handed helix, the linking number is normally positive by the sign convention we have adopted. The linking number of relaxed DNA, $L°$, will be distributed over a narrow range of integral values centered around 1 per 10

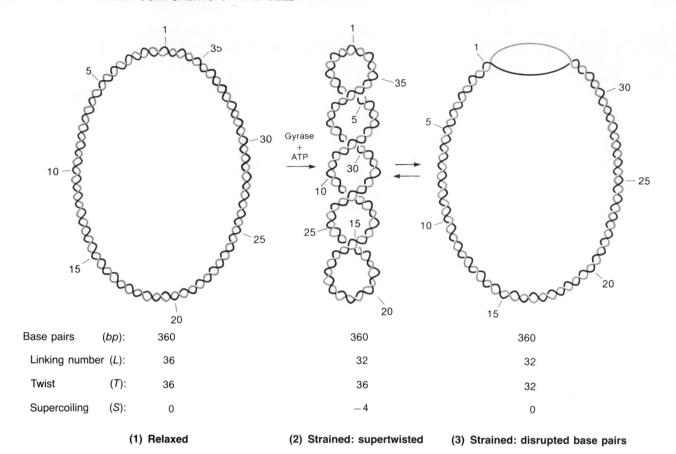

Base pairs	(bp):	360	360	360
Linking number	(L):	36	32	32
Twist	(T):	36	36	32
Supercoiling	(S):	0	−4	0

(1) Relaxed   **(2) Strained: supertwisted**   **(3) Strained: disrupted base pairs**

Figure 7–26

A circular duplex molecule in different topological states (adapted from a diagram supplied by M. Gellert). The linking number can be changed only by breakage and re-formation of the phosphodiester linkages, as shown in the conversion of the relaxed circular form (1) to the strained negatively supercoiled form (2). The strain in the negatively supercoiled form can be partitioned in different ways between twist (T) and supercoiling (S) as shown in the interconversion between (2) and (3). No phosphodiester linkages are broken in making this interconversion and consequently there is no change in the linking number (L) as indicated.

base pairs. DNA with a mean linking number smaller than this is termed _negatively supercoiled_, or _underwound_; DNA with a larger linking number is termed _positively supercoiled_, or _overwound_. The deviation of the linking number from its relaxed value, $\Delta L = L - L°$, can be partitioned between _twist_ (altered double helix coiling) and _supercoiling_:

$$L = \text{twist } (T) + \text{supercoiling } (S)^*$$

At the present time it is not known precisely how L will partition between twist and supercoiling for helices under torsional stress. For example, consider the hypothetical situation illustrated in Figure 7–26. A 360-base-pair structure in the circular relaxed form ($L = +36$, $T = +36$, $S = 0$) is indicated to the left. Exposure to the bacterial enzyme _DNA gyrase_ would introduce negative supercoils into such a structure in a reaction that requires ATP (see Chapter 26). If four negative supertwists are introduced, L will be reduced to $+32$. Barring other changes, T remains fixed and S becomes -4. In fact, the torsional strain introduced by the four negative supertwists will tend to reduce T, causing a partial unwinding of the duplex. At one extreme, this effect could lead to the unwinding of four helical turns, in which case the supercoiling would disappear entirely ($L = +32$, $T = +32$, $S = 0$). The actual situation would probably lead to a reduction in the negative value of S and a concomitant reduction in T that would be spread over the entire duplex without any localized total unwinding of the double helix as pictured. Note that _the linkage number L changes only when covalent bonds are broken, as in the case of gyrase treatment._

---

*It is common to use writhe (W) instead of supercoiling (S) in this equation. Writhe includes conformations other than supercoiled that are topologically equivalent as far as the linking number is concerned.

BOX 7–B

Equilibrium-density-gradient Centrifugation

In equilibrium-density-gradient centrifugation, macro-molecules such as nucleic acids or proteins are dissolved in a concentration of CsCl whose density is similar to that of the macromolecule. At high speeds, a nearly linear gradient of the salt is established. The macromolecules will form a narrow band (whose width is inversely related to their molecular weight) at a position where their density is equal to the density of the solution. If a mixture of DNA, RNA, and protein is centrifuged, the protein will band at a buoyant density of about 1.25 g/cc; DNA (depending on the per cent G + C) will band at 1.710 g/cc; and RNA will settle to the bottom of the gradient, since its density is unusually high, about 1.9 g/cc. The greater the G + C content of DNA, the higher its buoyant density; the density in CsCl can in fact be used to determine the G + C content of native DNA.

CsCl gradients can also be used to distinguish ^{15}N-labeled DNA from ^{14}N- or ^{14}N/^{15}N-labeled DNA, or DNA containing 5-bromouracil from unsubstituted DNA. This feature has proved useful to establish the semiconservative replication of DNA and to separate replicated from unreplicated DNA.

Finally, density-gradient centrifugation is very useful in the preparative separation of covalently closed circular (e.g., plasmid) DNA from linear (e.g., fragmented genomic) DNA in the presence of intercalating agents such as ethidium bromide. Because of its topological constraints, covalently closed circular DNA binds less of the dye. Since binding makes DNA lighter, its density is greater than that of linear (or nicked circular DNA) bound to ethidium bromide.

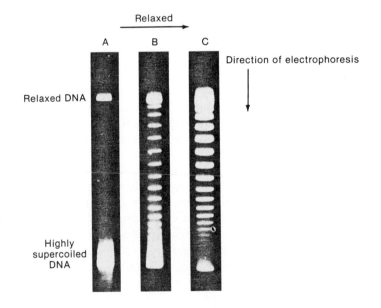

Figure 7–27
Electrophoretic patterns of highly supercoiled or partially supercoiled DNA. Strip A represents a sample of circular duplex DNA obtained by deproteinization of the animal virus SV40. In strips B and C the DNA has been exposed for increasing times to an enzyme (topoisomerase) that catalyzes relaxation. Adjacent bands differ by 1 in linking number. (From W. Keller, *Proc. Nat. Acad. Sci.* 72:2553, 1975).

For DNA with a molecular weight of less than 10^7, agarose gel electrophoresis is a most effective method for assessing the extent of supercoiling. (For analysis of higher-molecular-weight DNAs, see Box 7–B). DNA isomers differing by 1 in linking number form separate bands in the gel (Figure 7–27). The more highly supercoiled molecule will migrate more rapidly through the gel as a result of its more compact structure.

Enzymes called *topoisomerases*, which relax positively or negatively supercoiled DNA, as well as topoisomerases such as DNA gyrase (mentioned earlier), which generate negatively supercoiled DNA from relaxed DNA, are discussed in Chapter 26. Gyrases that catalyze the formation of negatively supercoiled DNA have been found only in bacteria. Consistent with this, all double-helix DNA found in bacteria that is topologically constrained by not having free ends, such as circular DNA, is negatively supercoiled. Circular DNA isolated from virus-infected eukaryotic cells (for example, simian virus 40, or SV40, DNA) is frequently found to be negatively supercoiled, but only after deproteinization. In such instances, the DNA is not supercoiled in cells. Instead, the supercoiling results from removal of the proteins that normally are bound to the DNA and cause it to be underwound in its native state.

The biological importance of supercoiling has only been clearly established for DNA in bacteria. Drugs, such as novobiocin, that specifically inhibit DNA gyrase have a lethal effect except in strains that have a novobiocin-resistant DNA gyrase. On the other hand, mutants containing an altered topoisomerase Ω, which relaxes negatively supercoiled DNA, have been isolated from E. coli. Careful analysis of such mutants has shown that they are double mutants; that is, they have an altered DNA gyrase that is less active, in addition to the defective Ω enzyme. This observation suggests that in a normal bacterial cell the extent of negative supercoiling is carefully adjusted by the relative activities of the gyrase and Ω enzymes. Negative supercoiling imparts a torsional stress or tension to the DNA structure that can be relieved to some extent by unwinding of the double helix. Both the initiation of DNA synthesis (Chapter 26) and the transcription of certain DNA sequences or genes (Chapter 27) have been found to be strongly dependent on negative supercoiling, probably because of the necessity to unwind at least one turn of the double helix during the initiation process. This same torsional stress explains why Z helix formation is encouraged by negative supercoiling; since in Z DNA the twists are left-handed, the transition from the B to Z form will tend to relax negatively supercoiled DNA.

DNA Denaturation Involves Separation of the Two Strands

The process of separating the polynucleotide strands of duplex nucleic acid structures is called denaturation. In denaturation the secondary binding forces that hold the strands together are disrupted, but covalent linkages within the polynucleotide strands are not. Recall that the secondary binding forces include the edge-to-edge hydrogen bonds between the base pairs of opposing strands and the face-to-face stacking forces between the planes of adjacent base pairs. Individually, these secondary forces are weak, but when they act cooperatively they give rise to a DNA duplex that is highly stable in aqueous solution. The conditions required to denature DNA provide us with a measure of the strength of these interactions.

One of the simplest ways to denature DNA is by heating. The extent of denaturation at any temperature can be readily measured by the change in ultraviolet absorbance of a solution of DNA. A substantial rise in absorbance (hyperchromic shift) accompanies the transition from the native to the denatured state. When a solution of native DNA is slowly heated, the absorption at the ultraviolet maximum (at 260 nm) remains constant until an elevated temperature is reached, at which point the absorbance increases by about 40 per cent over a narrow temperature range. The rise in absorbance coincides with the disruption of the regular base-paired structure, and the two polynucleotide strands are able to separate from one another. The sharpness of the disruption of the regularly hydrogen-bonded base-paired native structure may be likened to the melting of a pure organic compound. It is customary to refer to this ultraviolet absorption temperature profile as a melting curve (curve a in Figure 7–28).

The melting temperature, T_m, of DNA is defined as the temperature at the midpoint of the absorption increase. This is about 85°C for the example shown in Figure 7–28. Rapid cooling of the denatured DNA solution leads to re-formation of intrastrand hydrogen bonds, but in a nonspecific, irregular manner. The absorbance decreases, but only by about three-fourths of the total original increase, and the decrease occurs over a much broader range of temperatures, as shown by curve b in Figure 7–28. On subsequent heating

Figure 7–28

Effect of temperature on the relative absorbance of native, renatured, and denatured DNA. When native DNA is heated in aqueous solution its absorbance does not change until a temperature of about 80°C is reached, after which the absorbance rises sharply, by about 40 per cent (curve a). On cooling the absorbance falls, but along a different curve, and it does not return to its original value (curve b). Renatured DNA, in which the two strands have been brought back into perfect register, shows a sharp melting curve similar to native DNA (curve c). Renatured DNA is prepared from denatured DNA by holding the temperature at about 25°C below the denaturation temperature for an extended time. This subject is discussed in detail later in the text. The temperature at which the native DNA is half denatured is labeled T_m.

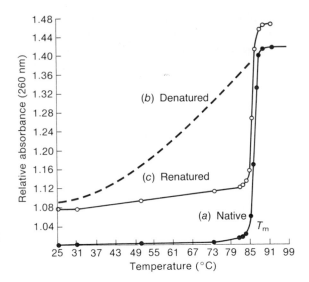

and cooling, the absorbance follows this cooling curve in the appropriate direction, indicating that denaturation results in an irreversible change.

Some DNAs do not occur naturally in the double-helix form. The melting curve is an excellent tool for detecting DNAs in the single-stranded conformation. For example, in certain bacteriophages that infect *E. coli*, ϕX174 and fd, the DNA exists as a single, circular strand. The ultraviolet absorption temperature curve for the DNA of these phages is similar in shape to that observed for denatured DNA (curve b in Figure 7–28). Broad melting curves also are characteristic of most RNAs, which rarely have regions of regular base pairing that extend for more than 10 or 20 residues.

Melting curves also have provided evidence that the stability of the double-helix structure is a function of its base composition. The midpoint of denaturation (T_m) of naturally occurring DNAs is precisely correlated with the average base composition of the DNA; the higher the mole per cent G-C base pairs, the higher the T_m (Figure 7–29). This seems reasonable, since the G-C base pair contains three hydrogen bonds, whereas the A-T base pair contains only two (see Figure 7–13); thus DNA with a greater G-C content would be expected to be more stable. As indicated above, base stacking is also believed to contribute to the stability of the duplex structure. In general, the interaction energy gained by stacking between adjacent G-C base pairs is greater than that gained by interaction between A-T base pairs. An excellent source of natural DNAs that differ in their content of G-C base pairs is bacteria. Members of the genus *Clostridium* have only 25 per cent G-C in their DNA, whereas some *Micrococcus* species have DNA with 72 per cent G-C. *E. coli* DNA is about 50 per cent G-C. Whereas the base composition of the DNA of bacteria along the genome is uniform, reflected in a narrow, symmetrical peak in a CsCl density gradient, the DNA of higher eukaryotic cells is not so uniform, with a minor fraction of DNA displaying a G-C content different from the major fraction of DNA. This minor fraction can be detected in a CsCl density gradient as a small band separate from the main band. The smaller band is termed *satellite* DNA.

Other factors present in aqueous solution can affect the stability of the double-helix structure in a positive or a negative way. For example, salt has a stabilizing effect, which is mainly due to the repulsive electrostatic interactions between the negatively charged phosphate groups. Salt shields this charge interaction and therefore stabilizes the duplex structure. Thus DNA in 0.15 M NaCl denatures at a T_m about 20°C higher than DNA in 0.01 M

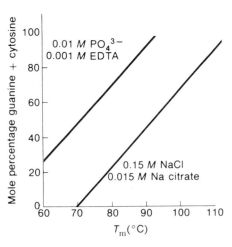

Figure 7–29

Dependence of the temperature midpoint (T_m) of DNA on the content of guanine and cytosine. As the percentage of G + C increases, the T_m increases. Two curves are shown to illustrate the point that the denaturation temperature is shifted to lower values when the ionic strength is lowered.

phosphate (see Figure 7–29). In pure water (no salt present) DNA denatures at room temperature. Extremes of pH also have a destabilizing effect on the double-helix structure. When the pH is above 11.5 or below 2.3, there is extensive deprotonization or protonization, respectively, of the hydrogen-bonding groups of the bases, which in turn disrupts the hydrogen-bonded structure. Alkali is an excellent DNA denaturant because it permits rapid separation of the strands without any degradation. Alkali both denatures and degrades RNA to 2′ (3′) mononucleotides.

Many solutes that can form hydrogen bonds also lower the melting temperature (decrease the stability) of double-helix structures. The organic compounds formamide and urea are frequently used to lower the denaturation temperature as well as to prevent reaggregation of denatured strands on subsequent cooling. This is important in DNA manipulations, where nonspecific aggregation is undesirable. Reagents that increase the solubility of the DNA bases (e.g., methanol) or disrupt the water shell around them (e.g., trifluoracetate) reduce the hydrophobic interactions between the bases and lower the T_m. Most proteins that bind to DNA inhibit denaturation. However, some DNA-binding proteins destabilize the native state. Proteins of this class usually bind preferentially to single-stranded DNA, thereby favoring separation of the double strands. For example, gene 32 of bacteriophage T4 encodes a DNA-binding protein that is essential for bacteriophage T4 DNA replication. This protein induces the local unwinding of the DNA duplex, a process that facilitates its replication (Chapter 26). Addition of appropriate amounts of the gene-32-encoded protein leads to complete denaturation of individual DNA molecules. This complete denaturation occurs because the gene 32 protein exhibits _cooperative binding_; that is, the binding of one gene 32 protein molecule to DNA facilitates binding of a second gene-32-encoded molecule at a neighboring site and continued binding at adjacent sites. A large number of DNA-binding proteins similar to the gene 32 protein have been found in both prokaryotic and eukaryotic cells. They are believed to play an important role in DNA replication.

DNA Renaturation Involves Duplex Formation from Single Strands

We have said that when a solution of heat-denatured DNA, in which the two strands of the duplex are separated, is allowed to cool rapidly, the regularly hydrogen-bonded structure does not re-form. However, reassembly of the two separated polynucleotide strands into the native structure, called _renaturation_, is possible under certain specialized conditions.

The first indication that renaturation was possible came from studies of Julius Marmur and Paul Doty, using transforming DNA in a biological assay. The basis of the assay was that denatured _Streptococcus pneumoniae_ DNA is inactive in DNA-mediated transformation, whereas native DNA is active in transferring genetic traits, such as streptomycin resistance, from a donor to a recipient strain. When transforming DNA was heated and rapidly cooled it was biologically inactive; however, when denatured DNA was slowly cooled, a large fraction of the initial transforming activity was recovered.

Further experiments showed that the optimum temperature for renaturation is about 25° below the T_m. As is the case with denaturation, the renaturation process can be followed spectrophotometrically. If the temperature of a denatured DNA solution is maintained at $T_m - 25°C$ for a long period of time, the absorbance of the solution gradually decreases until it approaches a value close to that of native DNA (see curve c in Figure 7–28). The optimum temperature for renaturation is frequently referred to as the _annealing temperature_. At such a temperature, irregularly hydrogen-bonded structures are unstable but regularly hydrogen-bonded structures are stable.

Consequently, prolonged exposure of denatured DNA at the annealing temperature allows the DNA bases to explore various configurations until complementary regions of pairing are formed between otherwise separated strands.

Spectrophotometric methods or biological assays are not the only means of monitoring renaturation. Other methods are sometimes used, especially when it is desirable to separate the denatured and renatured fractions or when small, radioactive amounts of nucleic acids are used. One of these methods involves digestion of the nucleic acid with enzymes that preferentially degrade single-stranded DNAs. For this purpose, the S1 nuclease derived from the mold *Aspergillus* is commonly used. The amount of DNA resistant to digestion by this nuclease gives an accurate measure of the amount of duplex structure in a sample of partially renatured DNA. A more commonly used method involves binding to a column made from a calcium phosphate gel known as hydroxyapatite. For reasons that are unclear, duplex DNA binds to hydroxyapatite at salt concentrations where single-stranded DNA does not bind. This method must be used with caution as a quantitative measure of renaturation, because a DNA molecule that is partially duplex and partially single-stranded will also bind to hydroxyapatite. Hydroxyapatite chromatography is also useful as a preparative technique for separating rapidly reassociating fragments from slowly reassociating fragments. The significance of rate of reassociation will be explained shortly.

Renaturation Rate Measures Sequence Complexity. When Marmur and Doty and their coworkers discovered the technique of renaturation, there was considerable interest in the experiments. Although the initial discovery represented a reaffirmation of the stability of the double-helix structure, a great many applications have grown out of the technique of annealing, and it is a useful one to molecular geneticists (see Box 7–C).

BOX 7–C

Techniques Using Nucleic Acid Renaturation

Nucleic acid hybridization takes place between any two complementary single-stranded nucleic acids that are annealed at $T_m - 25°C$ for prolonged periods of time. Since the rate of duplex formation depends on the concentration of the interacting complementary strands, the method can be used to measure the abundance of a specific nucleic acid in a mixture. Thus the number of *Drosophila* ribosomal RNA genes can be estimated by titrating a fixed amount of denatured genomic DNA with increasing amounts of rRNA. Specific DNA-RNA hybrids are detected by first immobilizing denatured DNA onto nitrocellulose filters, then by adding labeled RNA to serve as a probe. The labeled RNA will be retained by the nitrocellulose only if it is hybridized by complementarity to the DNA. The method can be modified for use in isolating specific mRNA sequences; RNA hybridized to immobilized, denatured DNA can be eluted following denaturation of the hybrids.

Hybridization can be used to detect specific sequences in the nucleus or chromosomes using a labeled probe and cells fixed to a microscope slide. Such *in situ* experiments serve to locate genes on chromosomes and were useful in demonstrating that highly repeated sequences in eukaryotic DNA are found in hetrochromatin. In the *Southern blotting* technique, individual DNA fragments generated by restriction endonuclease digestion are first separated by gel electrophoresis; they are denatured and then transferred to a nitrocellulose support. The DNA fragments can then be probed for specific sequences using a suitable labeled DNA or RNA probe. In *northern blot hybridization*, immobilized RNA separated by agarose gel electrophoresis under denaturing conditions is detected by a labeled, single-strand probe. These blotting techniques can be used to determine the size and abundance of the immobilized nucleic acid. The blotting and filter hybridization techniques are very useful in studying the level of expression of various genes and the effect of inducers, of repression, and of mutations on the level of expression.

Hybridization has also been exploited to detect specific sequences in recombinant clones. A labeled probe is applied to bacterial colonies that have been immobilized and the DNA denatured. Hybridization allows one to detect transformants harboring recombinant vectors that carry the sequence of interest.

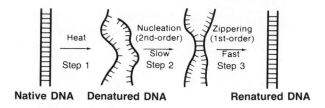

Native DNA   Denatured DNA   Renatured DNA

Figure 7–30

Steps in denaturation and renaturation of a DNA duplex. In step 1 the temperature is raised to the point where the two strands of the duplex separate. If denatured DNA is slowly cooled, the events depicted as steps 2 and 3 follow. In step 2 there is a second-order reaction in which two complementary strands of DNA must collide and form interstrand hydrogen bonds over a limited region. Step 3 is a first-order reaction in which additional hydrogen bonds form between the complementary strands that are partially hydrogen-bonded (zippering). Once complementary strands are partially bonded, the zippering reaction will occur rapidly. In the overall process, step 2 is rate-limiting.

The renaturation rate of DNA has been used extensively to compare the base sequence complexity of DNAs from different sources. For a given weight concentration of DNA, the more homogeneous in sequence a DNA sample, the more rapidly it will renature. Thus a 0.01% solution of denatured T4 bacteriophage DNA renatures much more rapidly than does a 0.1% solution of *E. coli* DNA. In turn, bacterial DNA renatures much more rapidly than mammalian DNA. The steps involved in denaturation and renaturation are indicated in Figure 7–30. Renaturation may be thought of as a two-step process. In the first step, called _nucleation_, contact is made between two complementary regions of DNA. Nucleation is followed by a relatively rapid _zippering_ up of adjoining base residues into a duplex structure. Thus nucleation is the rate-limiting step, and it is this step that must be examined to determine the time course of renaturation. Because nucleation involves interaction between two molecules, it should occur at a rate proportional to the square of the concentration of single strands.

If c is the concentration of single-stranded DNA at time t, then the second-order rate equation for loss of single-stranded DNA is

$$-\frac{dc}{dt} = k_2 c^2$$

where k_2 is the second-order rate constant. (Rate equations will be presented in detail in Chapter 8.) Starting with a concentration c_0 of completely denatured DNA, the amount of single-stranded DNA left after renaturation for time t is given by

$$\frac{c}{c_0} = \frac{1}{1 + k_2 c_0 t}$$

At time $t_{1/2}$, when half of the DNA is renatured, $c/c_0 = 0.5$ and $t = t_{1/2}$, from which it follows that

$$c_0 t_{1/2} = \frac{1}{k_2}$$

As a rule, c/c_0 is plotted as a function of $c_0 t$. The resulting curve is referred to as a "cot" curve (Figure 7–31).

Data are shown for DNAs and RNAs with varying _nucleotide complexity_ N, which is defined as the number of nucleotides in a nonrepeating sequence. If there are no repeating sequences in the cellular DNA, then N is equal to the number of nucleotides in the genome. It can be seen that $c_0 t_{1/2}$ is proportional to N for these samples. This family of curves has been used to calibrate more complex situations.

Figure 7–31

Reassociation of double-stranded nucleic acids from various sources. The genome size is indicated by arrows near the upper nomographic scale. Over a factor of 10^9, this value is proportional to the cot required for half-reaction. All DNAs were sheared so that they have approximately the same fragment size (about 400 nucleotides, single-stranded). Correction has been made to give the rate that would be observed at 0.18 M sodium ion concentration. No correction for temperature has been applied, since it was approximately optimum in all cases. The labels for the different DNAs should not concern the average reader. MS2 is the RNA sequence obtained from a bacterial virus. Mouse satellite and calf (nonrepetitive fraction) are fractions of the genome obtained from the indicated animals. (Adapted from R. J. Britten and D. E. Kohn, *Science* 161:529, 1968.)

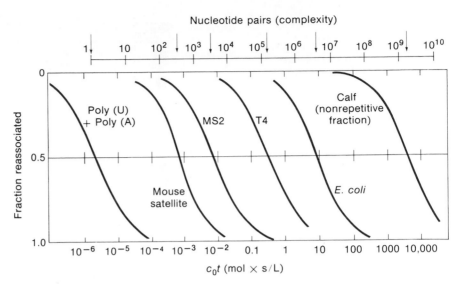

For a given nucleotide complexity, the renaturation rate is also a function of the length L of the nucleic acid, i.e., the actual number of bases per single-stranded nucleic acid molecule present in a renaturation mixture. It can be shown that

$$K_2 = \frac{L^{0.5}}{N}$$

and therefore $c_0 t_{1/2}$ should be proportional to $N/L^{0.5}$. When DNA samples of different initial length are compared, the effect of length is usually eliminated by shearing the DNAs of the different samples to a more or less uniform length, so that $c_0 t_{1/2}$ can be used directly as a measure of N. This is usually the parameter of interest in a renaturation rate experiment.

Simple monophasic cot curves with one inflection point are obtained when most prokaryotic DNAs are examined (Figure 7–31). In such instances, N is directly proportional to the amount of DNA in the chromosome. However, when the total DNA from a complex eukaryotic organism such as a human is analyzed, the curves are more complex, containing more than one inflection point. A precise interpretation of the cot curve for human DNA, shown in Figure 7–32, is not possible, but the following explanation is considered most likely. About 2 per cent of the DNA (foldback) renatures very rapidly. This is due mostly to single-stranded regions of DNA that fold back on themselves, thereby forming hairpin duplex (cruciform) structures. Structures giving rise to cruciforms must have possessed inverted repeating sequences called *palindromes*. The next class of sequences to renature (designated fast) accounts for about 5 per cent of the total DNA of higher eukaryotes and has a $c_0 t_{1/2}$ value of about 10^{-2}. The low amount of DNA, together with the rapid reassociation kinetics, suggests that about 5 per cent of the nuclear DNA (fast) is present in a very large number of copies per genome (*highly repetitive DNA*). In the same figure, a region labeled intermediate can be seen that accounts for an additional 20 per cent of the DNA (*middle repetitive DNA*). Finally, about 70 per cent of the DNA has reassociation kinetics in the range expected for *single-copy or unique DNA*, i.e., DNA whose sequence is present as a single copy per haploid genome.

This type of renaturation kinetic analysis has been used for the gross characterization of DNA in many cell types as shown in Figure 7–33. These data show that bacterial DNA (*E. coli*) contains very little repetitive DNA (0.3 per cent). This is mainly accounted for by the eight genes for *E. coli* riboso-

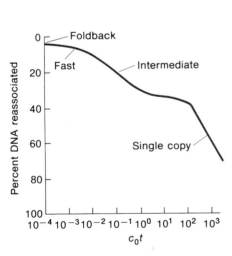

Figure 7–32

The cot curves for total human nuclear DNA. The total human DNA (and the DNA from other complex eukaryotic organisms) does not give a simple monophasic plot, but shows more than a single inflection point. Labels indicate the type of DNA that renatures at different cot values. (Adapted from unpublished data of A. R. Mitchell.)

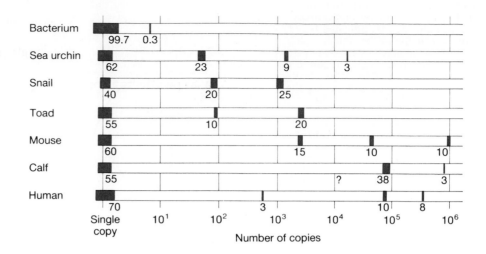

Figure 7–33

Distribution of single-copy DNA and repetitive-sequence DNA in various organisms. The width of bands and the number below the bands indicate the fraction of total cellular DNA in each class. For example, in calf about 55 per cent of the DNA sequences are single-copy, about 38 per cent are present in somewhat less than 10 copies, and about 3 per cent are present in somewhat less than 10 copies. (From R. J. Britten and D. E. Kohne, Repeated segments of DNA *Sci. Amer.* 222(4):24–31, 1970. Copyright 1970 by Scientific American, Inc. All rights reserved).

mal RNA that have nearly identical sequences. In eukaryotes, single-copy DNA (i.e., nonrepetitive DNA) accounts for 40 to 70 per cent of the DNA, most of the remainder being roughly divided between middle repetitive ($<10^4$ copies per genome) and highly repetitive ($>5 \times 10^4$ copies per genome). Further analyses of eukaryotic gene structure by a variety of other techniques have shown that most genes encode proteins that belong to the unique (single copy) class. The middle repetitive class includes transfer RNA genes and ribosomal RNA genes that are involved in protein synthesis (Chapter 27) as well as the genes encoding the nuclear proteins called histones (Chapter 30). Some highly repetitive DNA sequences appear as spacer DNA between genes; other repetitive sequences occur in tandem; and still other repetitive DNA elements are distributed at random throughout the genome. It has also been argued that some families of repetitive DNA, referred to as "selfish DNA," represent "parasitic" sequences that replicate together with the genome without conferring any positive or negative characteristics to the host cells that harbor these sequences. Satellite DNA detected in C_5Cl density gradient centrifugation is usually enriched in highly repetitive DNA.

Chemical Methods Are Used to Determine Nucleotide Sequence

Since all cellular genetic information is stored in the DNA in the form of specific sequences of the four commonly occurring bases, it would be of great value to have a convenient method that could be used for nucleic acid sequence determination. The importance of DNA sequencing was fully appreciated in the 1950s, but it was not until the 1970s that the technical means was developed for determining the base sequences in DNA.

Then two methods appeared that were almost equally effective in sequence analysis. One was developed by Gilbert and Maxam and the other in Sanger's laboratory. The Sanger method will be discussed in Chapter 26, as it requires an understanding of how DNA polymerase works before it can be understood. Here we will concentrate on the chemical methods.

As in protein sequence analysis, we must start with a highly purified molecule and then have some way of obtaining well-defined fragments that are small enough to be directly sequenced. The first DNAs to be sequenced were obtained from small viruses because they are relatively easy to purify. More recently, the techniques of recombinant DNA methodology (see Chapter 32) have permitted the isolation of purified DNA segments from virtually any source. Sequencing depends on separation of a continuous set of DNA fragments that differ in length by only one nucleotide; the chemical nature of the terminal nucleotide is identified by simple inspection of acrylamide gel electrophoretograms of four separate reaction mixtures that somehow identify each of the four bases. Either single- or double-stranded molecules are used.

First, a specific fragment of DNA is ^{32}P-labeled at its 5′ end with polynucleotide kinase. This enzyme transfers the ^{32}PO$_4$ group from [γ^{32}P]ATP to the 5′-hydroxyl end of ribonucleotide and deoxyribonucleotide chains of any size. If double-stranded DNA is used, both 5′ ends will become labeled, unless a phosphate already occupies either of the 5′ positions, so such a molecule cannot be sequenced directly. Instead, the complementary strands or the ends of a double labeled DNA must first be separated. This can be done in two ways: (1) denaturation followed by gel electrophoretic fractionation of the individual chains, or (2) treatment of the DNA with an enzyme (usually a restriction enzyme—see Chapter 26) that cleaves the molecule asymmetrically into two segments that can be separated by electrophoresis. The second method does not require separation of the individual strands, as only one of them will be labeled on the 5′ end.

Next, the DNA is treated with a chemical reagent that specifically reacts with one of the four bases. This reaction is carried out for a limited period of time, so that on the average only a few residues in the polynucleotide chain react with the reagent. The modified base introduces a linkage amenable to backbone cleavage by subsequent chemical treatment (Figure 7–34a). To a first approximation, all bases of a given type in a chain are equally susceptible to modification. The net result is a family of products labeled at the 5′ end with ^{32}P and terminating at the point of cleavage (see Figure 7–34b).

Four chemical reactions are used that cleave DNA preferentially at guanines (G > A), adenines (A > G), cytosine alone (C), and cytosines and thymines equally (C + T). When the products of the four reactions are re-

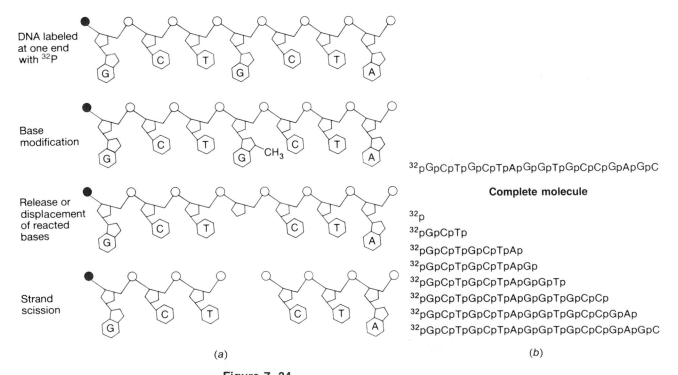

^{32}pGpCpTpGpCpTpApGpGpTpGpCpCpGpApGpC

Complete molecule

^{32}p

^{32}pGpCpTp

^{32}pGpCpTpGpCpTpAp

^{32}pGpCpTpGpCpTpApGp

^{32}pGpCpTpGpCpTpApGpGpTp

^{32}pGpCpTpGpCpTpApGpGpTpGpCpCp

^{32}pGpCpTpGpCpTpApGpGpTpGpCpCpGpAp

^{32}pGpCpTpGpCpTpApGpGpTpGpCpCpGpApGpC

(a)   (b)

Figure 7–34

Sequencing end-labeled DNA by limited, base-specific chemical cleavage. (a) A sequence of three reactions leading to strand cleavage at a guanine. If the entire DNA is subjected to this reaction sequence for a limited time, the result is (b), a family of labeled fragments that will be cleaved at different guanine sites.

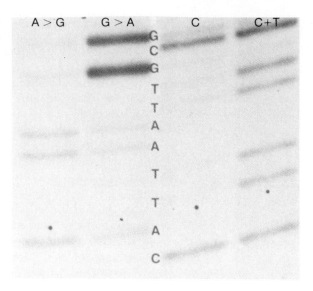

Figure 7–35

Autoradiogram of a sequencing gel according to Maxam and Gilbert. Only a portion of the autoradiogram is shown. To obtain this autoradiogram four different reaction mixtures were used. Each reaction mixture started with the same labeled DNA, then was subjected to a different set of limited digestion conditions. In column 1, the digestion conditions led to cleavage at both adenines and guanines, but more rapidly at the adenines. In column 2, the conditions were adjusted so that fragmentation was more likely at the guanines than the adenines. In column 3, digestion conditions that resulted in cleavage at cytosines were used, and finally, in column 4, digestion conditions that resulted in approximately equal extents of cleavage at the cytosines and thymines were used. The different digestion products were electrophoresed (from top to bottom in the figure). The smaller the chain size, the further the product migrated in the gel. It is possible to "read" the sequence by tracing the individual bands one step at a time, starting at the bottom of the gel. The sequence of the DNA for this gel pattern is given in the center of the autoradiogram.

solved according to size by gel electrophoresis, the DNA sequence can be read from the pattern of radioactive bands as an autoradiogram (Figure 7–35). The technique of obtaining an autoradiogram involves placing a sheet of film over the gel for a suitable exposure time. Further applications of gel electrophoresis for DNA and chromosome analysis are examined in Box 7–D.

For the purine-specific reaction, an aliquot of the DNA is treated with dimethylsulfate, which methylates the guanines in DNA at the N-7 position and the adenines at the N-3 position (Figure 7–36). The glycosidic bond of a methylated purine is cleaved on heating at neutral pH, leaving the sugar free. Alkali at 90°C will then cleave the sugar from the neighboring phosphate groups. When the resulting end-labeled fragments are resolved on a gel, the autoradiogram contains a pattern of dark and light bands (G > A lane). An adenine-enhanced cleavage can be obtained by treating the methylated DNA with acid, which releases adenine preferentially (A > G lane).

Other aliquots are reacted with hydrazine, which cleaves cytosine and thymine (Figure 7–37). After a partial reaction in aqueous hydrazine, the phosphate backbone of the DNA is cleaved with 0.5 M piperidine (Figure 7–38). The final gel pattern contains bands of similar intensity owing to the cleavages at cytosines and thymines (C + T lane). However, if 2 M NaCl is included in the hydrazine reaction, the reaction of the thymine residues is suppressed. Then the piperidine breakage produces bands only from cytosines (C lane).

DNA sequencing methods are extensively used, and the information thereby obtained can lead to a detailed understanding of the structure and

BOX 7–D

Application of Gel Electrophoresis for Chromosome Analysis

A variety of methods, both absolute (light scattering) and empirical (sedimentation, viscosity, gel electrophoresis), exist to determine the size of nucleic acids. The most accurate method is to determine the base sequence of the nucleic acid. Then, sequenced restriction fragments can be used as standards for comparison to samples whose molecular weight is being studied.

To analyze heterogeneous DNA populations, gel electrophoresis offers the best approach. However, very large DNA molecules—in the range of 150 kb or greater—tend to migrate with size-independent mobilities. (The abbreviation kb stands for "kilobase pair," or 1000 base pairs.) A greatly improved, new technique can resolve large DNA molecules, the size of the *E. coli* genome and intact, individual yeast chromosomal DNA, on agarose gels. The gently isolated DNA samples are subjected alternately to two approximately orthogonal electric fields. First, one field is applied and turned off. Then the other field is applied and turned off. This process is repeated with a pulse time of many seconds over a period of several hours. The precise pulse time depends on the size of the molecules being separated, being longer for larger molecules.

When large DNA molecules enter a gel in response to an electric field the molecule must elongate parallel to the field. When the field is shut off and a new field is applied, perpendicular to the long axis of the DNA, the molecule must reorient. The reorientation time should be quite sensitive to the size or molecular weight of the DNA. Since the length of the stretched-out DNA molecule is generally larger than the pores in an agarose gel, this reorientation is absolutely essential if the DNA is to undergo any net migration in response to the new field. Therefore, the smaller molecules have a mobility advantage over the larger molecules, which results in a size-based separation. (See Illustration 1).

Another recent development is the electrophoretic analysis of partially denatured DNA, a technique that is very useful in the detection of single base substitutions of cloned DNA. Fragments of DNA that are wild type or that contain a single base mismatch migrate into a polyacrylamide gel containing an ascending denaturing gradient of urea and formamide in the same direction as the electrical field. At a critical depth the mobility of partially denatured DNA slows down abruptly. The distance moved is a measure of the stability of the sequence, rather than the overall base composition or size of the remaining unmelted duplex. Differences in the partial denaturation of wild type and mutant DNA molecules allow one to separate and detect them. This method, developed by L. Lerman, provides a quick method to detect differences, even a single base-pair change, in cloned homologous fragments from wild-type and mutant sources.

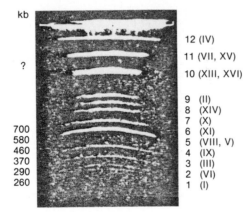

Illustration 1

Separation of intact DNA from the yeast chromosome. The chromosome assignments are on the right; size markers, in kb, are on the left. The DNA bands were detected by staining with ethidium bromide. (From G. F. Carle and M. V. Olson, *Proc. Natl. Acad. Sci.* 82, 3756–3760, 1985.)

expression of genes. To date, the complete sequences of many virus chromosomes are known and plans are being made for the total sequencing of the *E. coli* bacterial genome. About 3 per cent of the human genome has thus far been sequenced; once sequencing has been automated, a massive effort could possibly yield the sequence of the entire human genome.

DNA sequencing methods are so effective that techniques for direct sequencing of RNAs have not been extensively used. Rather, the RNA of interest in conjunction with the enzyme *reverse transcriptase*, which produces DNA from an RNA template (see Chapters 26 and 27), is used to sequence the RNA using the Sanger method described in Chapter 26. Reverse transcriptase, however, has a higher level of base misincorporation than DNA polymerase and can lead to some errors in base sequence determination.

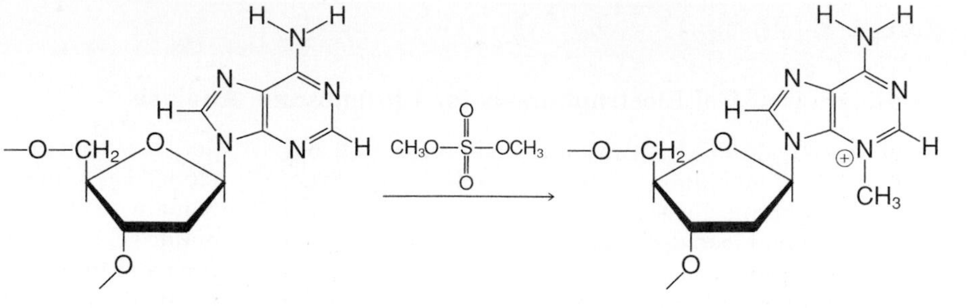

Figure 7–36
Treatment of DNA with dimethyl sulfate, resulting in methylation at N-3 in adenine (a) and at N-7 in guanine (b). Guanine is about five times more reactive than adenine. After reaction for a suitable period of time with the dimethyl sulfate, the pH is adjusted to pH 7.0 to permit removal of the modified base and subsequent cleavage of the polynucleotide chain.

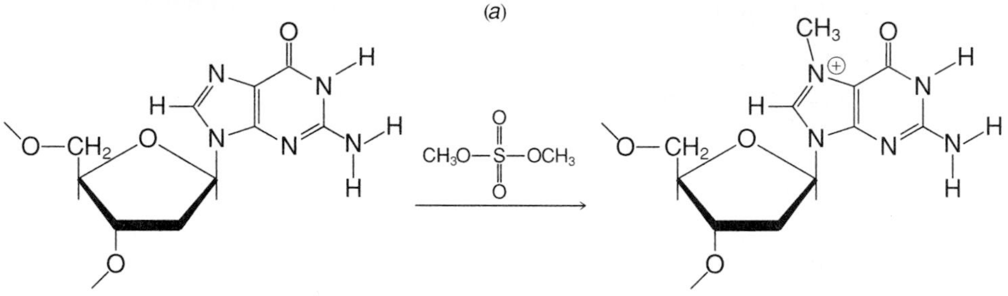

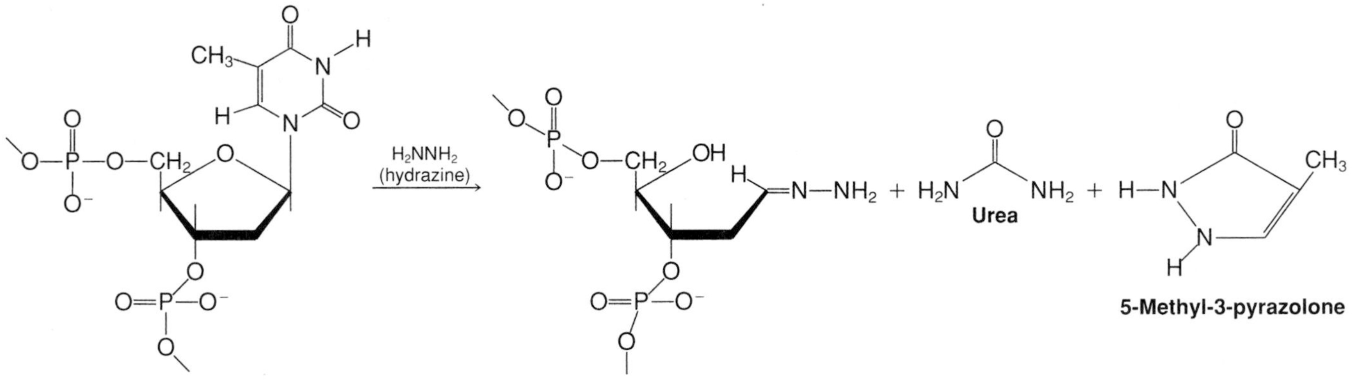

Figure 7–37
Reaction of thymine and cytosine bases with hydrazine. The products of the reaction are identical. Preferential reaction of cytosine occurs in high salt. In low salt the two bases react at approximately equal rates.

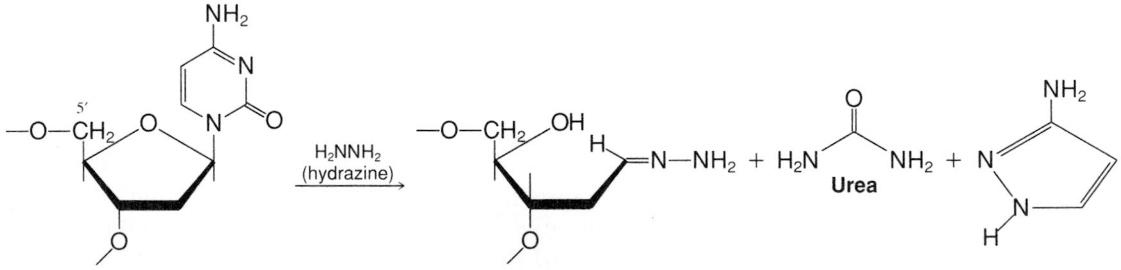

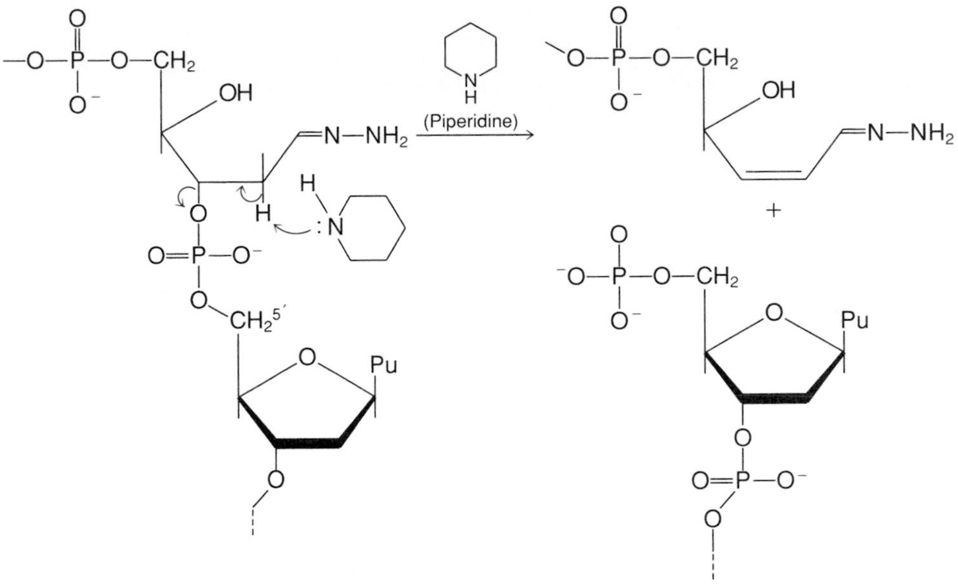

Figure 7–38
Cleavage of the hydrazine derivative of the cytosine and the thymine in the presence of piperidine.

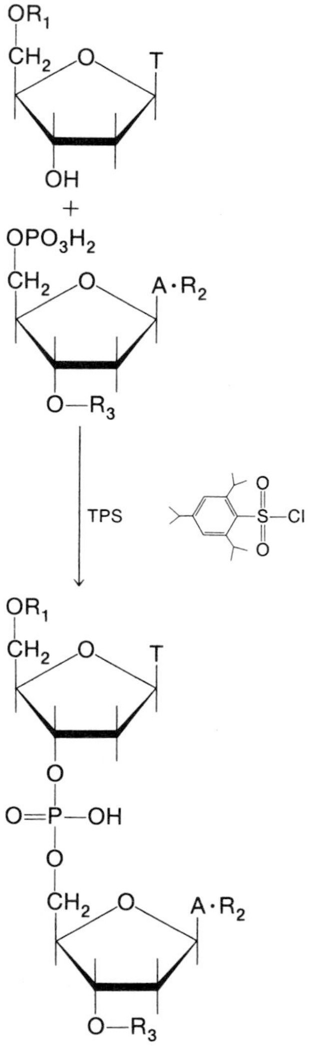

Figure 7–39
Chemical synthesis of a dinucleotide. The synthesis of TpA dinucleotide is sketched. Protecting groups are represented by R_1, R_2, and R_3. The 5′-phosphate on the suitably protected A residue is reacted with triisopropylbenzenesulfonyl chloride (TPS) and condensed with the T residue to synthesize the dinucleotide. The process can be repeated by removal of the R_3 protecting group to make a trinucleotide, and so on.

CHEMICAL SYNTHESIS OF DNA

Most of the chemical methods for making synthetic DNA originated in the laboratory of H. G. Khorana. Khorana's original goal was to chemically synthesize an entire gene and then to see if it would function properly when reinserted into the organism. He wisely chose a small gene for this purpose, the one for transfer RNA for alanine (this 76-nucleotide RNA is used in protein synthesis—see Chapter 27). His success in the total synthesis of a gene was a historic achievement and led others to refine the methods and apply them in other areas. With the new apparatus now available, one can synthesize entire genes that encode polypeptide hormones or enzymes (e.g., ribonuclease). In more routine experiments, oligonucleotides are being synthesized to be used as probes to detect complementary DNA sequences, in site-directed mutagenesis, as linkers in cloning experiments, and to make primers in DNA sequence analyses by the Sanger method, as well as in the primer extension method to determine the site of initiation of transcription.

Formation of a phosphodiester bond between two nucleotides requires activation of a monoester (5′-nucleotide) and reaction of the activated molecule with an alcoholic group (3′-OH) on another nucleotide. Chemical activation of phosphate groups produces species that will react with several of the functional groups present on the acceptor nucleotide, so protecting groups must be used to reduce undesired reactions. Several activating reagents have been investigated. One of the most useful that has been found is triisopropylbenzenesulfonyl chloride (TPS). Amino groups can be protected by acetylation, anisoylation, or isobutyrylation. Hydroxyl groups in the 3′ position are usually protected by acetylation. Monomethoxytrityl residues are used to protect the 5′-OH. Dinucleotides can be produced by the reaction shown in Figure 7–39, proceeding from blocked 5′-deoxynucleotides. Repetition of the condensation using blocked dinucleotides leads to tri-, tetra-, and eventually long polymers of nucleotides defined chemical sequence.

Another approach to the synthesis of oligodeoxynucleotides of defined sequence employing triesters proceeds from 3′-deoxynucleotides. The reac-

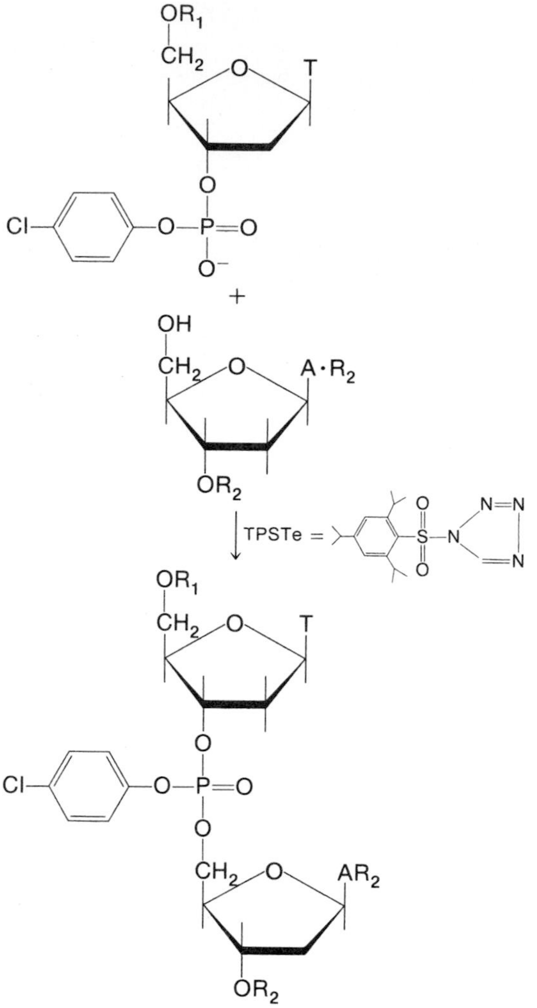

Figure 7–40
Triester method for chemical synthesis of a dinucleotide. This reaction starts from 3'-deoxynucleotides. The reaction is carried out by condensation of blocked diesters in the presence of triisopropylbenzene sulfonyl tetrazolide (TPSTe) to form products that are fully blocked triesters. The aromatic triesters for the intermediate products reduce side reactions and improve solubility in organic solvents. The triester oligodeoxynucleotides are deblocked to diesters before use in biological systems.

tions are carried out by condensation of blocked diesters in the presence of triisopropylbenzenesulfonyl tetrazolide (TPSTe) to form products that are fully blocked triesters. The use of aromatic triesters for the intermediate products reduces side reactions at the phosphate, improves solubility in organic solvents, and enhances possibilities for rapid separation. Before use in biological systems, the triester oligodeoxynucleotides are completely deblocked to diesters. The synthetic reaction employing the triester method is shown in Figure 7–40. As with the phosphodiester method, it is possible to use the procedure repetitiously to produce oligonucleotides of the desired sequence and size. Drawing on the technology developed by Merrifield for the solid-phase synthesis of polypeptides (see Chapter 1), it is possible to couple the 5'-OH group of the initial nucleotide to a resin support and proceed with the addition of nucleotides. All mixing and washing steps are thereby greatly simplified and the procedure has become amenable to automation; it is possible to produce oligonucleotides of the desired sequence without difficulty.

Having carried out the pioneering work on the chemical synthesis of defined oligodeoxynucleotide sequences, Khorana and coworkers attacked the problem of the combined chemical and enzymatic synthesis for alanine transfer RNA (tRNA[Ala]). To accomplish this, 15 oligodeoxynucleotides, each containing 6 to 20 nucleotides of strategically designed sequences comprising the complete tRNA gene, were chemically synthesized. These complementary oligonucleotides had overlaps of four to five nucleotides to allow

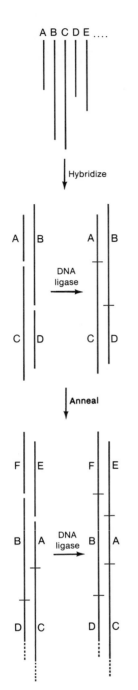

Figure 7–41

Schematic diagram showing how the tRNA gene for alanine tRNA was synthesized. First, small oligonucleotides representing different fragments of the gene were synthesized. Then these were annealed and covalently linked with the help of the enzyme DNA ligase. The process of annealing was repeated until the full gene was assembled.

annealing to adjacent sequences. Appropriate double-stranded molecules were then formed by sequential annealing (Figure 7–41) and covalently linked together by the enzyme DNA ligase (see Chapter 26). Eventually, the complete double-stranded molecule was prepared, inserted into a vector by recombinant DNA technology, and shown to be functional in protein synthesis.

CONFORMATIONAL BEHAVIOR OF RNA

The gross differences in primary structure between DNA and RNA were discussed earlier. RNA, like DNA, is a polymer of four different nucleotides. In this sense it has the same potential for carrying genetic information as DNA. Nevertheless, cellular genomes are invariably composed of DNA, and it has been argued that this is due to the greater chemical stability of a chain lacking the C-2′-OH group on the furanose residue. Despite this, the genomes of a number of viruses in both prokaryotes and eukaryotes are composed of RNA. In *E. coli* all known RNA phages exist as single strands in their mature form. Double-stranded RNA phages have been isolated from other bacteria, such as *Pseudomonas*. In plants and animals, RNA viruses of the single-strand or duplex types have been found. The double-stranded RNA genomes of viruses and phages are always segmented, the fragments each carrying different information.

Single-stranded DNA viruses form partial duplex intermediates during replication. DNA-RNA hybrid duplexes are also well known; they are intermediates in retrovirus DNA synthesis (Chapters 26 and 27) and they are presumed to be intermediates (containing about 10 base pairs) in all reactions involving DNA-directed RNA synthesis (see Chapter 27). Furthermore, DNA-RNA hybrids are made extensively in the laboratory for analytical purposes. Watson-Crick base pairing is universal in all natural duplex structures involving RNA; a conformation of the A type is favored, as stated earlier, because of the preference of ribose for the C-3′-endo conformation (see Figure 7–18).

There exists in the cell a wide range of nongenetic RNA that functions in messenger RNA processing and protein synthesis. Many aspects of RNA function are taken up in Chapters 27 and 28. In this section some of the structural properties of transfer RNA (tRNA) and ribosomal RNA will be discussed.

Transfer RNA Forms a Cloverleaf Structure

Transfer RNA carries activated amino acids to the messenger RNA template, where the amino acids are ordered and linked to form polypeptide chains. The complete three-dimensional structure of the yeast transfer RNA for phenylalanine has been determined by x-ray crystallography. The primary, secondary, and tertiary structures of phenylalanine tRNA, tRNA^{Phe}, are shown in Figure 7–42. This molecule contains 76 nucleotides in a single chain with four loops.

Most of the bases are involved in a complex secondary and tertiary structure. The four loops in the molecule (Figure 7–42) are referred to as the TψC loop, the anticodon loop, the D loop, and the variable loop. The variable loop is so named because it contains a different number of bases in different tRNAs. The remainder of the molecule (with the exception of the D loop in some instances) contains the same number of residues in all tRNAs, even though there is considerable sequence variability. The tRNA^{Phe} molecule contains 20 base pairs that are hydrogen-bonded in Watson-Crick fashion. It contains an additional 40 or so hydrogen bonds, most of which are not of the

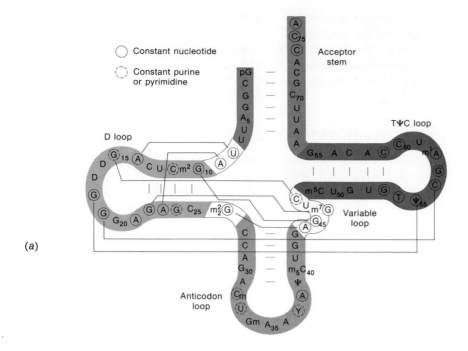

(a)

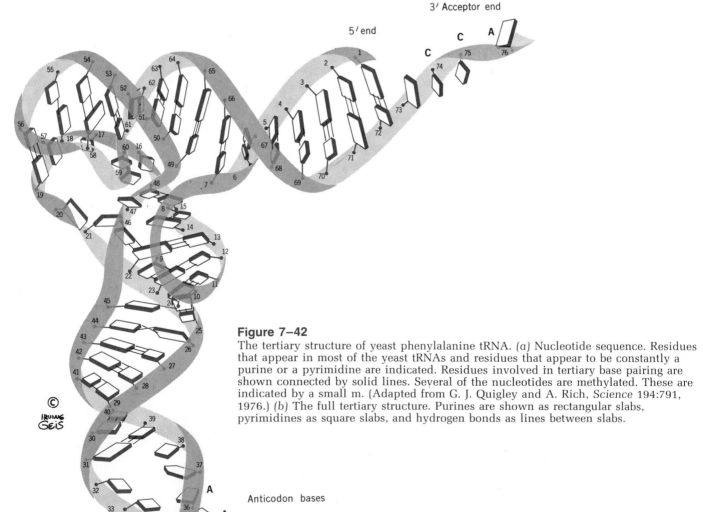

(b)

Figure 7–42

The tertiary structure of yeast phenylalanine tRNA. (a) Nucleotide sequence. Residues that appear in most of the yeast tRNAs and residues that appear to be constantly a purine or a pyrimidine are indicated. Residues involved in tertiary base pairing are shown connected by solid lines. Several of the nucleotides are methylated. These are indicated by a small m. (Adapted from G. J. Quigley and A. Rich, *Science* 194:791, 1976.) (b) The full tertiary structure. Purines are shown as rectangular slabs, pyrimidines as square slabs, and hydrogen bonds as lines between slabs.

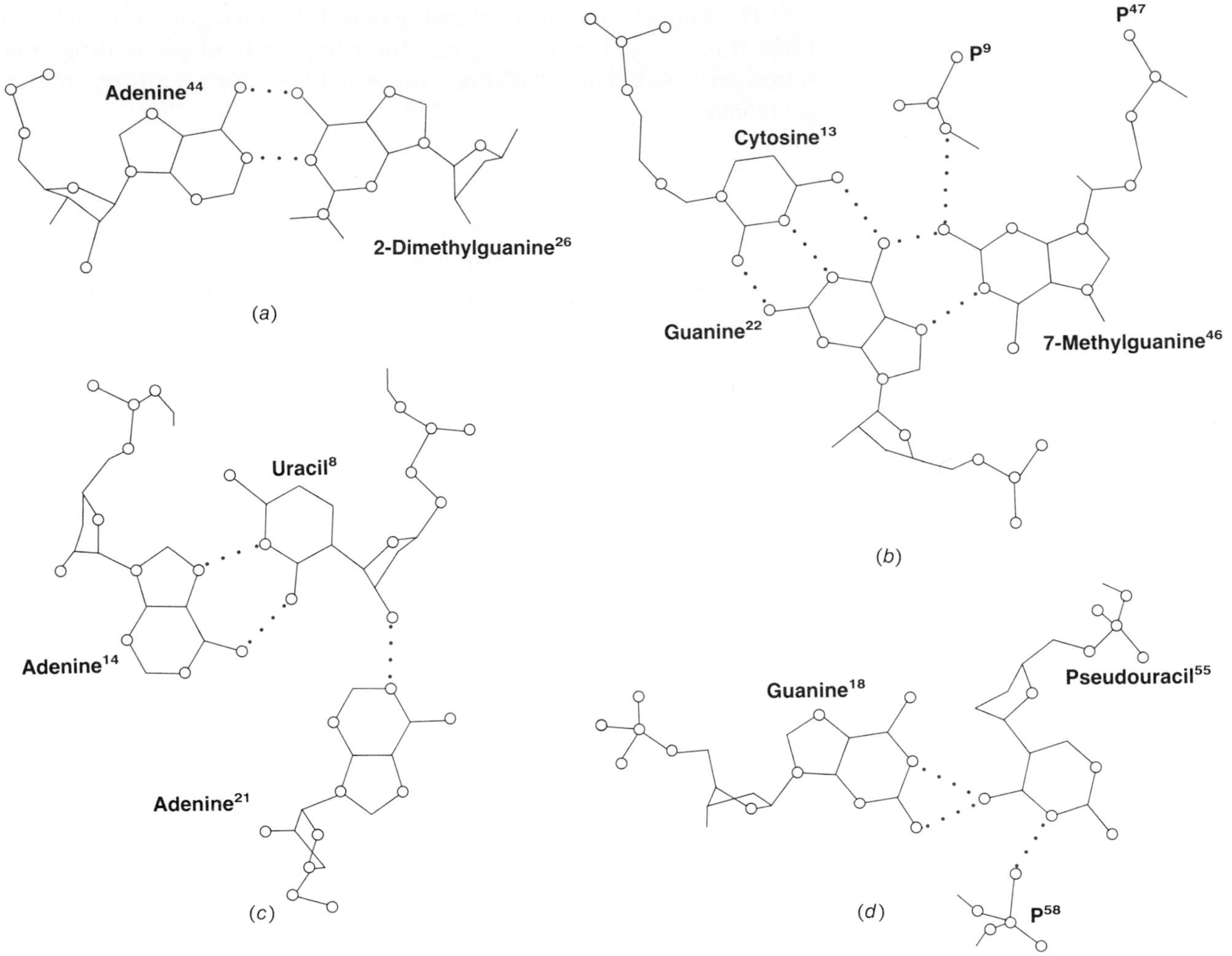

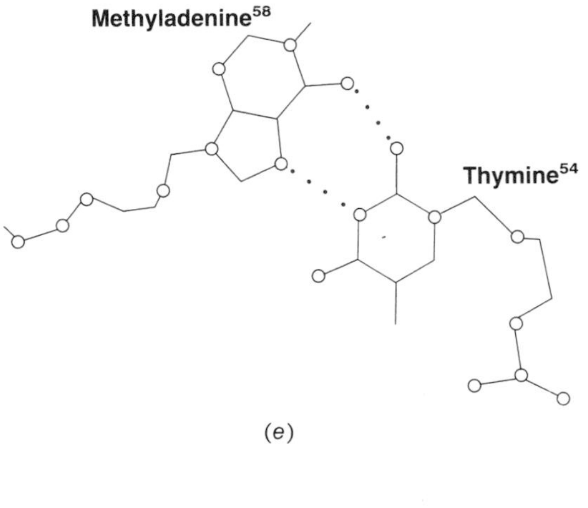

Figure 7–43
Some of the tertiary hydrogen-bonded interactions found in yeast phenylalanine tRNA. Superscripts refer to base number in tRNA molecule. It can be seen that there are many hydrogen-bonding arrangements in a tRNA structure that are not found in a duplex DNA structure. (Adapted from G. J. Quigley and A. Rich, *Science* 194:791, 1976.)

Watson-Crick type. These additional hydrogen bonds and the accompanying base stacking stabilize the tRNA in the complex folded structure shown in Figure 7–42b. Some of the unusual hydrogen-bonded interactions involved in stabilizing the complex tertiary structure are shown in Figure 7–43. These structures involve two to four residues from different regions of the RNA. Not only are the hydrogen-bonding patterns unusual, but in some cases phosphate and the C-2′-OH group participate in the hydrogen bonding. Tertiary structures of most tRNAs are very similar except for the variable loop, even though tRNAs show considerable sequence variability.

The importance of base stacking in stabilizing regular nucleic acid structures was stressed earlier. Despite the irregularity of the hydrogen bonding found in tRNA, base stacking, assessed by a close scrutiny of the three-dimensional structure, is a prominent feature of the tRNA structure. Only four of the 76 bases in the molecule (D^{16}, D^{17}, U^{47}, and G^{20}) do not participate in stacking. The D^{16} and D^{17} dihydrouracil residues cannot stack under any conditions, because these residues no longer have a planar configuration. It is believed that stacking and hydrogen bonding contribute about equally to the conformational stability of the tRNA.

Another significant feature of the tRNA structure is the high percentage of ribonucleotide bases that differ from the usual four. We will see (Chapter 28, Box 28F) that tRNA is initially synthesized from the four normal bases, which are then enzymatically modified by a large variety of enzymes. Some of the more than 30 such unusual bases so formed are shown in Figure 7–44. Of the 76 nucleotides in phenylalanine tRNA, twelve are of the modified type. The function of the modification is not known. Modification of bases in other types of RNA has also been observed but not to the extent that it occurs in tRNA.

The detailed structural investigations of tRNA greatly broadened our expectations in considering the less-well-understood structures formed by other nuclear and cytoplasmic RNAs. This single example illustrates the fact that nucleic acids with a properly adjusted primary sequence can adopt complex secondary and tertiary structures. Little is known in detail about the tertiary structures of most other RNAs.

Ribosomal RNA Forms a Complex Folded Structure

Protein synthesis occurs on very large RNA-protein complexes known as ribosomes. In *E. coli* all of the ribosomes have the same composition and structure, with 56 proteins of varying sizes and three RNA molecules: a 16S RNA of 1524 bases, a 23S RNA of 2904 bases, and a 5S RNA of 120 bases (the S refers to sedimentation coefficient, see Chapter 3). All of these RNAs have been sequenced, and attempts have been made to build, by trial and error, models in which base pairing is maximized. In so doing the G-U base pair has been considered an acceptable base pair along with the standard Watson-Crick base pairs, G-C and A-U, as in tRNA. This is because G-U can form a pair of hydrogen bonds with a slight adjustment in the sugar-phosphate backbone (see Chapter 27). The overall secondary structure of the 16S RNA obtained in this way is shown in Figure 7–45a. However, there is no direct confirmation that this structure is correct. A phylogenetic comparison of the secondary structures of 16S-like ribosomal RNAs from an archaebacterium (*Halobacterium volcanii*) and a eukaryote (*Saccharomyces cerevisiae*) suggests quite similar secondary structures for these molecules (compare Figures 7–45a, 7–45b, and 7–45c). Thus 16S-like ribosomal RNAs from widely different organisms appear to prove substantially the same secondary structures. The information that is available on the tertiary structures of these RNAs suggests that they fold into similar three-dimensional structures as well.

NUCLEOPROTEINS

All types of nucleic acids interact with proteins. Chromosomal DNA forms stable nonspecific complexes with structural proteins that stabilize their tertiary structures. They also form transient complexes with enzymes and regulatory proteins that are involved in DNA metabolism and RNA metabolism. RNAs also form protein complexes with enzymes and structural proteins.

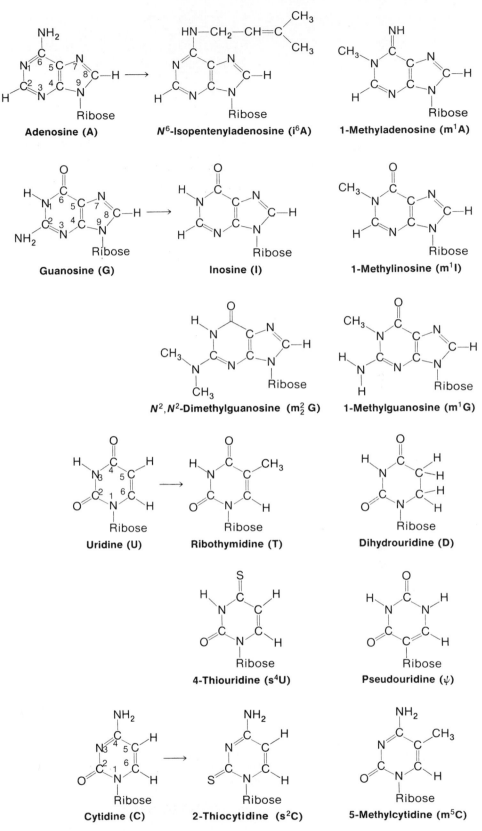

Figure 7–44
The structures of some modified bases found in tRNA. The parent ribonucleosides are shown with the numbering of the atoms in the purine and pyrimidine rings.

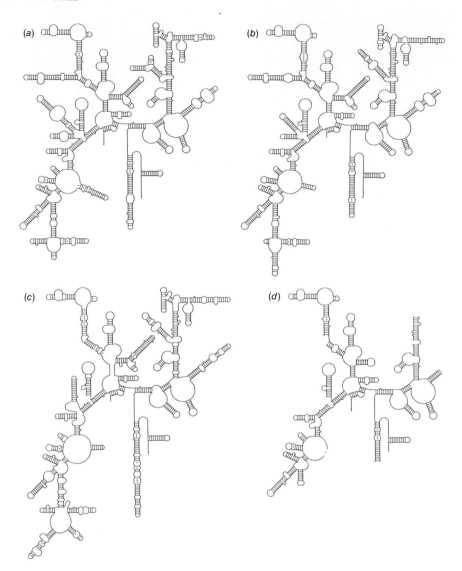

Figure 7–45
Comparison of secondary structures of 16S-like rRNAs from (a) eubacteria (E. coli), (b) archaebacteria (H. volcanii), (c) eukaryote (S. cerevisiae), and (d) a structure showing features common to all sequenced 16S-like rRNAs, including mitochondrial types. (From R. R. Gutell, B. Weiser, C. R. Woese, and H. F. Noller, *Prog. Nucleic Acid Res. Mol. Biol.* 32:155–216, 1985.)

The most studied ribonucleoprotein complexes are those formed in ribosomes.

The subject of nucleoprotein complexes is taken up at various points in this text where it can be most directly related to metabolism.

SUMMARY

The genetic material of cells and viruses consists of DNA or RNA. That DNA bears genetic information was first shown when the heritable transfer of various traits from one bacterial strain to another was found to be mediated by purified DNA.

All nucleic acids consist of covalently linked nucleotides. Each nucleotide has three characteristic components: (1) a purine or pyrimidine base; (2) a pentose; and (3) a phosphate group. The purine or pyrimidine bases are linked to the C-1' carbon of a deoxyribose sugar in DNA or a ribose sugar in RNA. The phosphate groups are linked to the sugar at the C-5' position. The purine bases in both DNA and RNA

are always adenine (A) and guanine (G). The pyrimidine bases in DNA are thymine (T) and cytosine (C); in RNA they are uracil (U) and cytosine. The bases may be postreplicatively or posttranscriptionally modified by methylation or other reactions in certain circumstances.

DNA exists most typically as a double-stranded molecule, but in rare instances it exists (in some phages and viruses) in a single-stranded form. The continuity of the strands (except in some RNA viruses, such as influenza and REO virus, that have segmented genomes) is maintained by repeating 3',5'-phosphodiester linkages formed between the sugar and the phosphate groups; they constitute the covalent backbone of the macromolecule. The side chains of the covalent backbone consist of the purine or pyrimidine bases. In double-stranded, or duplex, DNA the two chains are held together in an antiparallel arrangement. The base composition of DNA varies characteristically from one species to another in the range of 25 to 75 per cent guanine plus cytosine. Specific pairing occurs between bases on one strand and bases on the other strand. The complementary base pairs are either A and T, which can form two hydrogen bonds, or G and C, which can form three hydrogen bonds. The duplex is stabilized by the edge-to-edge hydrogen bonds formed between these planar base pairs and face-to-face interactions (stacking) between adjacent base pairs. Twisting of the duplex structure into a helix makes stacking interactions possible.

The right-handed helical structure, known as B DNA, is the most commonly occurring conformation of linear duplex DNA in nature. In this structure, the distance between stacked base pairs is ~3.4 Å, with approximately 10 base pairs per helical turn. The inherent flexibility of the structure, however, makes a variety of conformations possible under different conditions. In some instances, nucleotide sequence and degree of hydration dictate which conformations will be favored. DNA interacts with a variety of proteins inside the cell, and these proteins can also have a significant influence on its secondary and tertiary structure.

Circular DNA molecules, which are topologically confined so that their ends are not free to rotate, can form supercoils that are either right-handed (negative) or left-handed (positive). Negative supercoiling exerts a torsional tension favoring the untwisting of the primary right-handed double helix, whereas positive supercoiling has the opposite effect. Negatively supercoiled DNAs are most commonly observed in prokaryotes, which contain an enzyme that generates the supercoiled structure.

When duplex DNA (or RNA) is heated, it dissociates (denatures) into single strands. The temperature at which this dissociation occurs (the melting temperature) is a measure of the stability of the duplex and is a function of the G-C content of the DNA. A preparation of denatured DNA may be renatured to the native duplex structure by maintaining the temperature about 25°C below the melting temperature. The rate of renaturation is a measure of the sequence complexity of the DNA. In prokaryotes, which consist predominantly of unique sequences, the complexity (the number of base pairs) is approximately equal to the genome size. However, complex eukaryotic cells contain DNAs of varying sequence complexity that renature at quite different rates. The fastest-renaturing fractions are present in many copies per nucleus (usually detected as satellites in CsCl density gradients), whereas the slowest renaturing fractions are present in single copies. Analyses by other techniques have shown that some of the repetitive DNA sequences exist as tandemly repeated structures, while other types of repetitive sequences are dispersed throughout the genome.

Chemical methods exist for determining the nucleotide sequence in DNA chains. The Maxam-Gilbert method for sequence analysis involves radioactive labeling of the ends of specific segments with ^{32}P, limited degradation of each segment to produce a continuous series of fragments differing by one nucleotide, and finally analysis of the mixture by gel electrophoresis and autoradiography. Chemical methods also exist for synthesis of DNA chains of any desired sequence. Such segments have been useful in a variety of ways.

RNA can exist as single- or double-stranded linear structures, and rarely (in viroids) as single-stranded circles. Transfer RNA is the best-understood RNA structure. It is a single-stranded molecule with a complex folded conformation that contains hydrogen-bonded duplex regions formed between different segments of the single polynucleotide strand. Complex folded structures appear to be common among RNAs of different types.

SELECTED READINGS

Burlingame, R. W., W. E. Love, Bi-Chen Wang, R. Hamlin, N-H. Xuong, and E. N. Mondrianskis, Crystallographic structure of the octameric histone core of the nucleosome at a resolution of 3.3 Å. *Science* 228:546–553, 1985.

Felsenfeld, G., DNA. *Scientific American* 253(4):58–66, 1985.

Kim, S. H. Three-dimensional structure of transfer RNA. *Prog. Nuc. Acid Res. Mol. Biol.* 17:181–216, 1973.

Lerman, L. S., S. G. Fischer, I. Hurley, K. Silverstein, and N. Lumelsky, Sequence-determined DNA separations. *Ann. Rev. Biophys. Bioeng.* 13:399–423, 1983.

Noller, H. F., Structure of ribosomal RNA. *Ann. Rev. Biochem.* 53:119–162, 1984.

Pabo, C. O. and R. T. Sauer, Protein-DNA recognition. *Ann. Rev. Biochem.* 53:293–322, 1984.

Rich, A., A. Nordheim, and A. H.-J. Wang. The chemistry and biology of left-handed Z DNA. *Ann. Rev. Biochem.* 53:791–846, 1984.

Saenger, W., *Principles of Nucleic Acid Structure.* Springer-Verlag, 1983.

Schwartz, D. C., and C. R. Cantor, Separation of yeast chromosome-sized DNAs by pulsed field gradient gel electrophoresis. *Cell,* 37:67–75, 1984.

PROBLEMS

1. Why does linear duplex DNA separate into its constituent strands when the DNA is put into pure water?

2. Which of the following lowers the T_m of duplex DNA?

Explain their mode of action: T4 gene-32-encoded protein, NaCl, formamide, alkali.

3. Although the predominant structure adopted by DNA in

solution is of the B form, other conformational variants of the DNA double helix do exist, among them A and Z forms. What structural features allow for this flexibility of form in each case? DNA-RNA double helices adopt an essentially A-form conformation in solution. Suggest an explanation for this behavior.

4. a. The alternating polymer poly(dG-dC).poly(dG-dC) undergoes the transition from B to Z form only at high salt concentrations. How would you explain this salt dependence? [Hint: See Figure 7–20.]

   b. The diagram below describes the pattern obtained on gel electrophoresis of two partially relaxed plasmids under different salt conditions. What do you think is happening to the structure of plasmid A at the higher salt concentration (see Figure 7–27)? What can you infer about the sequences of plasmids A and B?

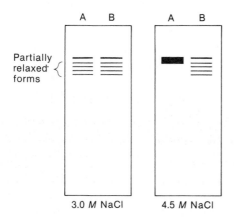

5. RNA is hydrolyzed by treatment with alkali but DNA is not. Explain why you think this may be so.

6. Linear duplex DNA can bind more ethidium bromide than covalently closed circular DNA of the same molecular weight. Suggest an explanation. How might you take advantage of this behavior to separate the two types of DNA on a CsCl gradient? (Hint: The ethidium cation has a density less than water.)

7. You are given a mixture of the following nucleic acid species, all in the same container, and all approximately of the same molecular weight. How would you separate the species from one another (assuming that your separation scheme is a very efficient one, with a logical basis) and specifically identify each species? Assume that each nucleic acid species is derived from a phage or virus and is homogeneous in size and composition.

   a. Circular phage, ϕX174 DNA (isolated from the phage).

b. Double-stranded, linear phage DNA.

c. Supercoiled DNA.

d. Double-stranded, linear DNA with hairpin loops at either end.

e. Single-stranded, linear RNA.

f. Double-stranded, linear RNA.

Use a flow diagram to illustrate your separation scheme and state the basis for the fractionation.

8. What is the structure of the nucleic acid called A, given the clues listed below?

   a. A sediments as a single species in a neutral sucrose density gradient.

   b. In an alkaline sucrose gradient, one-half (structure B) of the mass of A sediments down the tube, the other half remains at the miniscus.

   c. The buoyant density of A is much higher than that of duplex DNA of the same G + C content.

   d. Whereas A gives a sharp thermal transition profile, B's profile is broad.

   e. When A is heated and fast-cooled it gives rise to B and C, whose buoyant densities are quite different, that of C being heavier than that of B. B and C, when added together, will renature when exposed to a temperature 20°C below the T_m of A.

   What does each clue signify? Diagram the structure of A, B, and C. (Hint: Single-stranded DNA is more dense than duplex DNA, and RNA is more dense than DNA.)

9. Size is a critical parameter when separating DNAs by sucrose gradient sedimentation but not when separating DNAs by CsCl gradient sedimentation. Explain. (Hint: CsCl gradient sedimentation is an equilibrium technique, but sucrose gradient sedimentation is not.)

10. DNA-DNA duplexes are less stable than DNA-RNA duplexes. How might you take advantage of this fact in the preferential formation of DNA-RNA hybrids in a mixture containing complementary single strands of DNA together with RNA that is complementary to one of the DNA strands?

11. Renaturation of randomly sheared denatured DNA can be measured by hypochromic shift in the ultraviolet absorbance, S1 nuclease resistance, or retention on hydroxyapatite. Which method is likely to give an overestimate for the extent of renaturation?

12. Give the relative times for 50 per cent renaturation of the following pairs of denatured DNAs, starting with the same initial DNA concentrations.

    a. Bacteriophage T4 DNA and *E. coli* DNA, each sheared to an average single-strand length of 400 nucleotides.

    b. Unsheared T4 DNA and T4 DNA sheared as above.

II

CATALYSIS

PART II

Biomolecules were considered from a structural point of view in Part I. The synthesis and degradation of biomolecules is catalyzed by protein enzymes. In Part II we shall introduce the different classes of enzymes and describe the mechanisms that make enzymes extremely effective catalysts.

In Chapter 8 we review some basic aspects of chemical kinetics before discussing some of the general properties of enzyme catalysts. The enzyme-substrate complex is the focal point for the kinetic analysis of enzyme-catalyzed reactions. Different ways of measuring the rate constants and affinity constants are described. Reactions involving one or two substrates are considered. For one-substrate reactions the effects of various types of enzyme inhibitors are evaluated.

In Chapter 9 we examine the mechanisms of catalysis for a select group of enzymes whose mechanisms of action are reasonably well understood. This analysis reveals several reasons why enzymes are so effective as catalysts.

A few enzymes are designated as regulatory proteins because they have regulatory sites as well as active sites. Although regulatory proteins are in the minority, they are an extremely important means for controlling the general metabolism. The way in which regulatory enzymes work is considered in Chapter 10.

Many enzymes must incorporate other, small organic molecules to be active. These molecules, called coenzymes, are usually an integral part of the catalytic site. Indeed, they frequently are the active site. The major coenzymes and their mechanisms of action are described in Chapter 11.

8

ENZYME KINETICS

Enzymes are proteins that accelerate biochemical reactions. Experimentation on enzymes began in 1897, when Eduard Buchner demonstrated enzyme activity in a crude cell-free extract. The source of the extract was yeast and the reactions that were catalyzed were those involved in alcoholic fermentation. In 1926 Sumner demonstrated that enzymes were proteins by isolating a single protein from a crude extract and showing that it possessed a specific enzyme activity. The enzyme was named urease because it catalyzed the hydrolysis of urea to CO_2 and NH_3. In the 1930s Northrop carried out a classic series of studies that led to the isolation and characterization of several enzymes from the digestive juices. His work generalized the conclusion arrived at earlier by Sumner—that enzymes were proteins. In the next half century there was an ever increasing effort to isolate and characterize new enzymes. This effort has led to the characterization of thousands of enzymes.

In this chapter we will focus on the kinetic characterization of enzymes. We will see that many factors affect the rates of enzyme-catalyzed reactions. From a detailed kinetic analysis we gain a broad picture of the properties of the enzyme.

BASIC ASPECTS OF CHEMICAL KINETICS

Chemical kinetics is the study of the rates of chemical reactions. Before delving into the complexities of enzyme-catalyzed reactions, let us review some of the basic concepts and terminology of chemical kinetics that most of us were exposed to in general chemistry.

The _rate of a reaction_ is usually expressed in terms of the rate of disappearance of the reactants or the rate of appearance of the products. In the case where there is one reactant, A, and one product, B, the reaction can be represented as

$$A \longrightarrow B \tag{1}$$

and the rate of the reaction is given by

$$\text{Rate} = \frac{-d[A]}{dt} = \frac{+d[B]}{dt} \tag{2}$$

where [A] and [B] are the concentrations of A and B, respectively.

The mathematical relationship between the rate of reaction and the concentrations is called the _rate law_. For a simple reaction of the type described above, in which a single atom or molecule of A is converted into a product B, the rate law is given by the expression

$$\text{Rate} = k[A] \tag{3}$$

where k is the _rate constant_. The rate constant does not change with concentration but it can change with temperature or pressure or when a catalyst is present.

Radioactive decay of ^{32}P is an example of a reaction of this type. In this case

$$\text{Rate of decay} = \frac{-d[^{32}P]}{dt} = k[^{32}P]$$

where $[^{32}P]$ is the concentration of radioactive phosphorus isotope and $-d[^{32}P]/dt$ is its rate of disappearance. An intramolecular isomerization would obey a similar rate law.

The rate law indicates the _kinetic order_ of the reaction. The rate law given by Equation (3) indicates a _first-order reaction_ because the right-hand side of the equation contains the concentration of a single substance raised to the first power. In a more complex reaction the rate would involve a higher power of the concentration. For example, the reaction of monomeric A leading to dimeric A

$$A + A \longrightarrow A_2 \tag{4}$$

is given by the rate equation

$$\text{Rate} = \frac{d[A_2]}{dt} = k[A][A] = k[A]^2 \tag{5}$$

Such a reaction is said to be of _second order_.

Where two reactants are involved, e.g.,

$$A + B \longrightarrow \text{products} \tag{6}$$

the rate may be proportional to the first power of each reactant:

$$\text{Rate} = \frac{d[\text{products}]}{dt} = k[A][B] \tag{7}$$

The reaction is of first order with respect to either [A] or [B], but overall the reaction is of second order, since the right-hand side of the equation contains the product of two concentrations. In the more general situation

$$mA + nB \longrightarrow \text{products} \tag{8}$$

the rate is given by

$$\text{Rate} = k[A]^m[B]^n \tag{9}$$

The reaction is of mth order with respect to A and of nth order with respect to B, and the overall order is $m + n$.

In all of the reactions we have considered thus far, we have indicated a reaction arrow in one direction only. The reverse reaction, of products going to reactants, has not been considered. The reverse reaction can be ignored under two sets of circumstances: (1) when the rate constant for the reverse reaction is very small or (2) when the concentration of products is very small compared with the concentrations of reactants. In either of these cases the rate of the reverse reaction will be negligible compared with the rate of the forward reaction.

We sometimes have occasion to refer to the *molecularity* of a reaction, particularly when considering the mechanism. The molecularity of a reaction is the number of molecules involved in a specific step. In the case of ^{32}P decay, the reaction is *unimolecular*. In the case of A_2 dimer formation, the reaction is *bimolecular*. Notice that in both of these cases the order of the reaction and the molecularity are numerically equal. This is because both reactions are single-step reactions. In multistep reactions we can speak of the overall order of the reaction, but when speaking of the molecularity we must refer to individual steps. The kinetic order of any single step of a reaction is the same as the molecularity of that step. The vast majority of chemical reactions are either monomolecular or bimolecular. That is because the probability of a three-body collision, which would be required for a trimolecular reaction, is very low.

Frequently the order of a reaction and its molecularity are not revealed by the stoichiometry. This is most common when the reaction involves two or more steps, as in the reaction of nitrous oxide with hydrogen to form nitrogen and water. In this case the stoichiometric equation is

$$2\ NO + 2\ H_2 \longrightarrow N_2 + 2\ H_2O \tag{10}$$

and the experimentally determined rate law is

$$\text{Rate} = \frac{-d[NO]}{dt} = k[NO]^2[H_2] \tag{11}$$

We see that this reaction is of first order with respect to the hydrogen concentration even though two H_2 molecules appear in the overall reaction leading from reactants to products. A detailed kinetic analysis of this reaction has shown that it takes place in three steps and that each step has a characteristic rate constant associated with it. The slowest step involves the reaction of a dimer of the NO molecule with one H_2 molecule. Other steps in the reaction take place much more rapidly. The three steps in the reaction are

(a) $2\ NO \underset{k_{-1}}{\overset{k_1}{\rightleftharpoons}} N_2O_2$ Rapid equilibrium

$$\tag{12}$$

(b) $N_2O_2 + H_2 \overset{k_2}{\longrightarrow} N_2O + H_2O$ Slow, rate-limiting step $\tag{13}$

(c) $N_2O + H_2 \overset{k_3}{\longrightarrow} N_2 + H_2O$ Rapid step $\tag{14}$

When one of the steps in a multistep reaction is much slower than the others, the rate law and the order of the overall reaction are determined by that step. In this reaction, then, step (b) is rate-limiting. Accordingly the rate and order of the overall reaction are reflected by the rate and order of step (b).

$$\text{Rate (overall)} \cong \text{rate (step b)} = \frac{-d[NO]}{dt} = k_2[N_2O_2][H_2] \tag{15}$$

N_2O_2 is an intermediate in the overall reaction, whose concentration may be expressed in terms of the rate constants k_1 and k_{-1} and the concentration of the initial reactant NO. Since step (a) indicates that NO and N_2O_2 are in equilibrium and since an equilibrium constant is the ratio of forward and reverse rate constants,

$$N_2O_2 = \frac{k_1}{k_{-1}}[NO]^2 \tag{16}$$

Substituting this in the Equation (15), we get

$$\text{Rate (overall)} = \frac{k_1}{k_{-1}} k_2 [NO]^2[H_2] \tag{17}$$

If we compare this equation with Equation (11), we see that the rate constant for the overall reaction is a composite of three rate constants,

$$k = \frac{k_2 k_1}{k_{-1}} \tag{18}$$

From this analysis we have learned two things of general importance about kinetic analysis: (1) the rate law tells us more about the mechanism of a reaction than the stoichiometry can tell us, and (2) the analysis of multistep reactions may be greatly simplified in cases where one of the steps is rate-limiting.

INFLUENCE OF THERMODYNAMICS AND KINETICS ON REACTION RATES

The *equilibrium constant* for a reaction is strictly a function of thermodynamic factors. The most useful thermodynamic quantity for evaluating the equilibrium constant is the *free energy*, designated by the symbol ΔG. Each molecule has a fixed free energy at a given temperature and pressure. If the free energy of the reactants for a given reaction is greater than the free energy of the products, then the equilibrium constant will favor the formation of products. The factors that determine the free energy of the reactants and products and the quantitative relationship between the free energy differences and the equilibrium constant will be discussed in Chapter 12. For the time being it is adequate to have this qualitative picture in mind.

Some reactions that are highly energetically favorable occur very rapidly; others do not. The reaction of hydrogen ions with hydroxide ions to make water is an example of a highly favorable reaction that occurs rapidly. Thus when we mix equimolar amounts of a strong acid and a strong base in aqueous solution, neutralization occurs as fast as the mixing process and we are left with very small amounts of free hydrogen or hydroxide ions. By contrast when we mix H_2 and O_2 at room temperature, very little reaction occurs even though the equilibrium constant is very much in favor of the formation of water. However, if the temperature is raised, the formation of water occurs with explosive violence. This example shows that thermodynamic factors alone do not permit us to predict whether or not a reaction will take place. *Thermodynamic calculations indicate only that a certain reaction is possible; they do not assure that it will take place.*

Kinetic factors determine the rate at which a reaction will occur for a given set of circumstances. As we have already noted, the equilibrium constant is equal to the ratio of the rate constants in both directions. But either or both of these may be so small that equilibrium may take a very long time to attain even if the reaction is quite favorable in one direction. When such is the case we commonly refer to a *kinetic barrier*.

The gaseous reaction between N_2O_2 and H_2 to form N_2O and H_2O is an example of a slow reaction. It is a bimolecular reaction that requires that a molecule of N_2O_2 collide with a molecule of H_2. If all of these collisions were effective in forming product, then N_2O and H_2O would be formed very rapidly. In fact this reaction is rather slow, indicating that only a small fraction of the collisions between the reacting species are effective. Two factors must be scrutinized to determine why this is the rate-limiting step. First, how many collisions are these two molecules likely to make in a given time?

BOX 8-A

Dependence of Reaction Rate on Activation Energy

The highest free-energy state of a reacting species is referred to as the transition state, or the activated complex. Kinetic theory states that the fraction of molecules f that have enough kinetic energy to attain this state is a function of the temperature and the activation energy according to the following equation:

$$f = e^{-\Delta G^*/RT}$$

where e is the base of the natural logarithm, R is the molar gas constant, and T is the absolute temperature. The rate constant for the reaction is related to this quantity by the following equation:

$$k = Ae^{-\Delta G^*/RT} \quad \text{or} \quad \ln k = \ln A - \frac{\Delta G^*}{RT}$$

where A is a constant related to the collision frequency. A also depends on the temperature, because the frequency of collisions increases with temperature. However, this dependence is minimal and can be ignored in most treatments of interest to biochemists, because they work over a narrow range of temperatures.

From the first equation above, a rate constant k is ten times faster at 25°C if the activation energy ΔG^* is lowered by 1.38 kcal/mole. Thus for an enzyme to speed up a reaction by a factor of 10^{10} (ten billion), not an unusual acceleration, the enzyme must lower the activation energy by 13.8 kcal/mole.

And second, what percentage of the collisions will be effective in leading to products?

The frequency of collisions is a function of the concentration of the two reactants and the speed of the molecules. The probability that a given collision will be effective is a function of the _free energy of activation_ ΔG^* (see Box 8–A). This quantity is equal to the difference between the free energy of the reactants and the highest free energy state that the reactants will have to achieve before going to products. This highest free energy state is referred to as the _transition state_ or the _activated complex_ (see Figure 8–1). The higher the free energy of activation, the lower is the probability that a given collision between reactants will be effective.

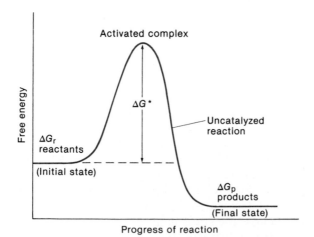
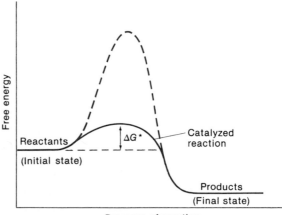

Figure 8–1

Free energy diagram for an uncatalyzed and a catalyzed reaction. The comparison applies both to purely chemical and to biochemical reactions. Note that the catalyst does not change the free energy of the reactants or the products, ΔG_r or ΔG_p, but only the free energy of the activated complex, ΔG^*.

For a given concentration of reactants, there are two ways in which a slow reaction can be accelerated. One is by raising the temperature. This causes the molecules to move faster, which increases both the number of molecular collisions and the fraction of effective collisions. The other is by introducing a factor that lowers the activation free energy. In ordinary chemical reactions both methods are used in the laboratory and in industry. In relatively delicate biological systems it is not practical to raise the temperature very much, so the usual means of accelerating a reaction is by introducing a catalyst. The catalyst is a "third party" that lowers the activation energy without itself being consumed. The catalyst does not change the free energy of the reactants or the products, only the activation energy required to achieve the state of the activated complex (see Figure 8–1).

GENERAL PROPERTIES OF ENZYME CATALYSTS

Enzymes are proteins that catalyze biochemical reactions. Each enzyme is highly specific for the type of reaction it catalyzes. The catalytic site on the enzyme contains the groupings that are directly involved in the catalysis. The catalytic site is usually only a small localized region on the enzyme surface. You may wonder why enzymes are much larger than their catalytic sites. There are several reasons. First, the folded structure of a fairly large polypeptide chain is required to define and stabilize the catalytic site. Second, the enzyme also must contain specific binding sites for reactants, or substrates, as the reactants of enzymes are usually called. The substrate binding site and the catalytic site are usually near one another and frequently overlapping. Together they are referred to as the active site. Finally, many enzymes contain additional surface features that are essential for function. Some enzymes interact with other cellular constituents, such as other enzymes in multienzyme complexes or lipoproteins in the case of membrane-bound enzymes. Isolated enzymes, as they might be studied in a laboratory, cannot of course interact with other cellular structures; this is one reason why we must be cautious in extrapolating from the properties observed for enzymes in vitro to those of the enzymes in vivo.

Frequently enzymes require additional small-molecule factors called cofactors for their activity. The cofactors may be as simple as metallic cations or as elaborate as complex organic molecules known as coenzymes. When cofactors are required, the enzyme must include additional binding sites. An enzyme lacking an essential coenzyme is referred to as an apoenzyme. The intact active enzyme is called the holoenzyme.

For most enzymes there is a common name that is used most frequently and a systematic name determined by the Enzyme Commission. The common name is often obtained by adding the suffix -ase to the name of the substrate on which the enzyme acts. In many cases function is also reflected by the common name. Thus glucose oxidase is an enzyme that catalyzes the oxidation of glucose, whereas glucose-6-phosphatase is an enzyme that hydrolyses phosphate from glucose-6-phosphate.

There are thousands of known enzymes and many more catalyzed reactions for which the enzymes are still to be discovered. A systematic scheme of classification was adopted by the International Union of Biochemistry in 1961. Each enzyme is designated by four numbers separated by periods: the main class, the subclass, the sub-subclass, and the serial number. There are six main classes: oxidoreductases, transferases, hydrolases, lyases, isomerases, and ligases or synthases. Oxidoreductases catalyze oxidation-reduction reactions. Transferases catalyze the transfer of a molecular group from one molecule to another. Hydrolases catalyze bond cleavage by the introduction of water. Lyases catalyze reactions involving removal of a group to form a double bond or addition of a group to a double bond. Isomerases catalyze reactions involving intramolecular rearrangements. Finally, ligases or synthases catalyze reactions joining together two molecules. A sampling of some of the divisions of main classes into further subdivisions with specific examples is given in Table 8–1. This list is intended to give some order to the complex array of enzyme-catalyzed reactions. Examples of most of these reactions will be discussed in this text.

TABLE 8–1
Classification of Enzymes

Number	Systematic Name	Common Name	Reaction
1	*Oxidoreductases* (oxidation-reduction reactions)		
1.1	Acting on CH—OH group of donors		
1.1.1	With NAD or NADP as acceptor		
1.1.1.1	Alcohol:NAD oxidoreductase	Alcohol dehydrogenase	Alcohol + NAD^+ $\rightleftharpoons$ aldehyde or ketone + NADH + H^+
1.1.3	With O_2 as acceptor		
1.1.3.4	β-D-glucose:oxygen oxidoreductase	Glucose oxidase	β-D-glucose + O_2 $\rightleftharpoons$ D-glucono-δ-lactose + H_2O_2
1.2	Acting on the $>$C=O group of donors		
1.2.3	With O_2 as acceptor		
1.2.3.2	Xanthine:oxygen oxidoreductase	Xanthine oxidase	Xanthine + H_2O + O_2 $\rightleftharpoons$ urate + H_2O_2
1.3	Acting on the CH—CH group of donors		
1.3.1	With NAD or NADP as acceptor		
1.3.1.1	4,5-Dihydrouracil:NAD oxidoreductase	Dihydrouracil dehydrogenase	4,5-Dihydrouracil + NAD^+ $\rightleftharpoons$ uracil + NADH
2	*Transferases* (transfer of functional groups)		
2.1	Transferring C-1 groups		
2.1.1	Methyltransferases		
2.1.1.2	S-Adenosylmethionine: guanidinoacetate N-methyltransferase	Guanidinoacetate methyltransferase	S-Adenosylmethionine + guanidinoacetate $\rightleftharpoons$ S-adenosylhomocysteine + creatine
2.1.2	Hydroxymethyltransferases and formyltransferases		
2.1.2.1	L-Serine:tetrahydrofolate 5,10-hydroxymethyltransferase	Serine hydroxymethyl-transferase	L-Serine + tetrahydrofolate $\rightleftharpoons$ glycine + 5,10-methylenetetra-hydrofolate
2.1.3	Carboxyltransferases and carbamoyltransferase		
2.2	Transferring aldehydic or ketonic residues		
2.3	Acyltransferases		
2.4	Glycosyltransferases		
2.6	Transferring N-containing groups		
2.6.1	Aminotransferases		
2.6.1.1	L-Aspartate:2-oxoglutarate	Aspartate aminotransferase	L-Aspartate + 2-oxoglutarate $\rightleftharpoons$ oxaloacetate + L-glutamate
2.7	Transferring P-containing groups		

TABLE 8–1 (*Continued*)

Number	Systematic Name	Common Name	Reaction
3	*Hydrolases* (hydrolysis reactions)		
3.1	Cleaving ester linkage		
3.1.1	Carboxylic ester hydrolases		
3.1.1.7	Acetylcholine acetylhydrolase	Acetylcholinesterase	Acetylcholine + H_2O $\rightleftharpoons$ choline + acetic acid
3.1.3	Phosphoric monoester hydrolases		
3.1.3.9	D-Glucose-6-phosphate phosphohydrolase	Glucose-6-phosphatase	D-Glucose-6-phosphate + H_2O $\rightleftharpoons$ D-glucose + H_3PO_4
3.1.4	Phosphoric diester hydrolases		
3.1.4.1	Orthophosphoric diester phosphohydrolase	Phosphodiesterase	A phosphoric diester + H_2O $\rightleftharpoons$ a phosphoric monoester + an alcohol
4	*Lyases* (addition to double bonds)		
4.1	C=C lyases		
4.1.1	Carboxy lyases		
4.1.1.1	2-Oxo-acid carboxy-lyase	Pyruvate decarboxylase	A 2-oxo-acid $\rightleftharpoons$ an aldehyde + CO_2
4.1.2	Aldehyde lyase		
4.1.2.7	Ketose-1-phosphate aldehydelyase	Aldolase	A ketose-1-phosphate $\rightleftharpoons$ dihydroxyacetone phosphate + an aldehyde
4.2	C=O lyases		
4.2.1	Hydrolases		
4.3	C=N lyases		
4.3.1	Ammonia-lyases		
4.3.1.3	L-Histidine ammonia-lyase	Histidine ammonia-lyase	L-Histidine $\rightleftharpoons$ urocanate + NH_3
5	*Isomerases* (isomerization reactions)		
5.1	Racemases and epimerases		
5.1.3	Acting on carbohydrates		
5.1.3.1	D-Ribulose-5-phosphate 3-epimerase	Ribulose phosphate epimerase	D-Ribulose-5-phosphate $\rightleftharpoons$ D-xylulose-5-phosphate
5.2	Cis-trans isomerases		
6	*Ligases* (formation of bonds with ATP cleavage)		
6.1	Forming C—O bonds		
6.1.1	Amino acid-RNA ligases		
6.1.1.1	L-Tyrosine:tRNA ligase (AMP)	Tyrosyl-tRNA synthase	ATP + L-tyrosine + tRNA $\rightleftharpoons$ AMP + pyrophosphate + L-tyrosyl-tRNA
6.2	Forming C—S bonds		
6.3	Forming C—N bonds		
6.4	Forming C—C bonds		
6.4.1	Carboxylases		
6.4.1.2	Acetyl-CoA:carbon dioxide ligase (ADP)	Acetyl-CoA carboxylase	ATP + acetyl-CoA + CO_2 + H_2O $\rightleftharpoons$ ADP + orthophosphate + malonyl-CoA

KINETICS OF ENZYME-CATALYZED REACTIONS

Kinetic analysis of enzymes was used for characterization of enzyme-catalyzed reactions even before enzymes had been isolated in pure form. As a rule kinetic analysis is done on purified or partially purified enzymes *in vitro*. But the properties so determined must always be referred back to the *in vivo* situation in some way to be sure they are physiologically relevant. We will see that a detailed kinetic analysis of an enzyme is indispensable to a comprehensive picture of an enzyme. Kinetic analysis is also a major tool used to investigate the mechanism of enzyme action (see Chapter 9).

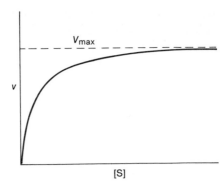

Figure 8–2

The reaction velocity, v, as a function of the substrate concentration [S] for an enzyme-catalyzed reaction. At high substrate concentrations the reaction velocity reaches a limiting value V_{max}. This curve usually has a hyperbolic shape. The velocity v is an initial velocity, measured in the brief period following mixing of enzyme and substrate, before the enzyme has had any effect on the substrate concentration.

The Henri-Michaelis-Menten Treatment Assumes the Enzyme-Substrate Complex Is in Equilibrium with Free Enzyme and Substrate

One of the first things that is measured in kinetic analysis is the variation in rate of reaction with substrate concentration. For this purpose a fixed low concentration of enzyme is used in a series of parallel experiments in which only the substrate concentration is varied. Under these conditions the initial reaction velocity increases until it reaches a substrate-independent maximum velocity at high substrate concentrations (Figure 8–2). The saturation effect is believed to reflect the fact that all the enzyme binding sites are occupied with substrate. This interpretation of the substrate saturation curve led Henri and Michaelis and Menten in the early 1900s to develop a general treatment for kinetic analysis of enzyme-catalyzed reactions. In their treatment they assumed that the enzyme-substrate complex is in equilibrium with free enzyme and substrate and that the formation of products (P) can proceed only through the enzyme-substrate complex

$$\text{Enzyme (E) + substrate (S)} \rightleftharpoons \text{enzyme-substrate (ES)} \rightleftharpoons$$
$$\text{E + products (P)} \quad (19)$$

Their objective was to develop an expression that would relate the reaction velocity to observable quantities and useful molecular parameters.

Since the slow step in the reaction described by Equation (19) is the formation of E and P from ES, the velocity of the reaction should be

$$v = k_2[\text{ES}] \quad (20)$$

The maximum velocity should be obtained when all of the enzyme, $[E_t]$, is in the form of the enzyme-substrate complex, [ES]. Thus

$$V_{max} = k_2[E_t] = k_2[E + ES] \quad (21)$$

since the total enzyme is equal to the sum of the free enzyme and the enzyme in the enzyme-substrate complex.

Dividing Equation (20) by Equation (21) gives

$$\frac{v}{V_{max}} = \frac{[\text{ES}]}{[\text{E + ES}]} \quad (22)$$

If K_S is the dissociation constant of the ES complex in Equation (19) then

$$K_S = \frac{[\text{E}][\text{S}]}{[\text{ES}]} \quad \text{and} \quad [\text{ES}] = \frac{[\text{E}][\text{S}]}{K_S} \quad (23)$$

Substituting this value of ES into Equation (22) and rearranging gives

$$\boxed{v = \frac{V_{max}[\text{S}]}{K_S + [\text{S}]}} \quad (24)$$

This is the _Henri-Michaelis-Menten equation_. It relates the initial reaction velocity to the maximum reaction velocity, the substrate concentration, and the dissociation constant for the enzyme-substrate complex.

Steady-state Kinetic Analysis Assumes a Fixed Level of Enzyme-substrate Complex

We could proceed to consider the implications of the Henri-Michaelis-Menten equation. However, it will be more useful to develop a kinetic treatment that has somewhat broader usage in that it eliminates the requirement

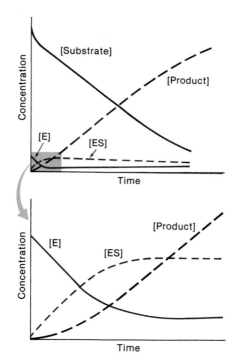

Figure 8–3
Concentration of various factors over the time course of a reaction. The shaded portion in the top curve is shown in expanded form in the bottom curve. After a very brief initial period (usually a few seconds) the concentration of enzyme-substrate complex, [ES], remains approximately constant for an extended period. This is the time when the steady-state approximation is applicable.

that the enzyme-substrate complex be in equilibrium with free enzyme and free substrate. In order to do this we introduce the concept of the *steady state*, first proposed by Briggs and Haldane in 1925. The steady state occurs shortly after mixing enzyme and substrate and constitutes the time interval when the rate of reaction is constant with time.† The concentration of the ES complex and other intermediates involved in the reaction is also nearly a constant during this time period (Figure 8–3). Usually substrate is present in much higher molar concentrations than enzyme, and since the initial period of the reaction is being examined, the free substrate concentration is approximately equal to the total substrate added to the reaction mixture.

$$E + S \rightleftharpoons ES \rightleftharpoons E + P \qquad (19')$$

In general there are four constants associated with these reactions:

$$E + S \underset{k_{-1}}{\overset{k_1}{\rightleftharpoons}} ES \underset{k_{-2}}{\overset{k_2}{\rightleftharpoons}} E + P \qquad (19'')$$

If we confine ourselves to measuring the initial rate of substrate reaction, the rate constant k_{-2} may be ignored, because not enough product will be present to make the reverse reaction proceed at a significant rate. For the rate of formation v_f of ES, we may write

$$V_f = k_1[E][S] = k_1([E_t] - [ES])[S] \qquad (25)$$

where $[E_t]$ and $[ES]$ have the same meaning as before and $[E_t] - [ES]$ represents the concentration of free or uncombined enzyme [E] as before.

The rate of disappearance V_d of ES is

$$V_d = k_{-1}[ES] + k_2[ES] \qquad (26)$$

since ES can disappear to give the initial reactants (k_{-1}) or to form products (k_2).

When the rates of formation and disappearance of ES are equal, i.e., during that phase of a reaction when the concentration of ES is virtually a constant, V_f equals V_d. Equations (25) and (26) may be combined to describe this steady-state situation:

$$k_1([E_t] - [ES])[S] = k_{-1}[ES] + k_2[ES] \qquad (27)$$

Rearrangement of Equation (27) gives

$$\frac{[S]([E_t] - [ES])}{[ES]} = \frac{k_{-1} + k_2}{k_1} = K_m \qquad (28)$$

The constant K_m is called the *Michaelis constant*. We shall see that it is a particularly useful parameter characteristic of the enzyme-substrate complex under conditions of the steady state. It is a simple matter to proceed from this point to an expression comparable to the Henri-Michaelis-Menten equation [Equation (24)] that uses K_m in place of K_S.

From Equation (28) it can be shown that

$$[ES] = \frac{[E][S]}{K_m} \qquad (29)$$

With this expression for ES we can follow the same procedure that resulted in Equation (24), this time arriving at the *Briggs-Haldane equation* for the reaction velocity

---

†The steady state is no stranger to the living cell. Most reactions in the living cell are in the steady state most of the time. Thus the steady state used originally as a convenience by kineticists is in fact a most appropriate way to analyze enzyme-catalyzed reactions.

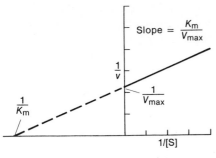

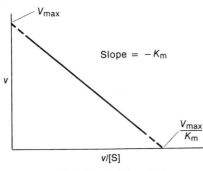

Figure 8–4

Two approaches to plotting enzyme kinetic data for the purpose of evaluating the kinetic parameters K_m and V_{max}. (a) The Lineweaver-Burk plot is the classical way of plotting enzyme kinetic data. (b) The Eadie-Hofstee plot has the advantage that data over the whole concentration range are more evenly weighted.

$$v = \frac{V_{max}[S]}{K_m + [S]} \quad \text{or} \quad K_m = [S]\left(\frac{V_{max}}{v} - 1\right) \tag{30}$$

$$\text{(a)} \qquad\qquad\qquad \text{(b)}$$

Equation (30a) is identical to the Henri-Michaelis-Menten equation [Equation (24)] except that K_S has been replaced by the more complex constant K_m. The second form of the Briggs-Haldane equation [Equation (30b)] is obtained by simple algebraic manipulation.

For the purpose of graphical representation of experimental data it is convenient to rearrange Equation (30a). Taking the reciprocal of Equation (30a) and rearranging, we get the *Lineweaver-Burk equation*

$$\frac{1}{v} = \frac{K_m}{V_{max}}\frac{1}{[S]} + \frac{1}{V_{max}} \tag{31}$$

From the Lineweaver-Burk equation we can see that when $1/v$ and $1/[S]$ are plotted on a graph, the equation should give a straight line. The slope is K_m/V_{max}, the intercept on the ordinate is $1/V_{max}$, and the intercept on the abscissa is $-1/K_m$ (Figure 8–4).

The Lineweaver-Burk equation may be rearranged to give

$$v = V_{max} - \frac{K_m v}{[S]} \tag{32}$$

A plot of v against $v/[S]$ gives a straight line with a slope of $-K_m$. The intercepts on the ordinate and the abscissa are V_{max} and V_{max}/K_m, respectively (see Figure 8–4). This *Eadie-Hofstee plot* has the advantage over the Lineweaver-Burk plot that data over the whole concentration range are more evenly weighted.

Significance of the Michaelis Constant, K_m. The Michaelis constant, K_m, is of great practical value in enzyme analysis. One of the most important relationships is that between K_m and the substrate concentration, $[S]$, at half maximum velocity. At this point these two quantities are equal. We can see this by inserting $V_{max}/2$ for v in Equation (30b). Approximate values of K_m can be readily obtained by determining the initial reaction velocity at a series of substrate concentrations and estimating the half maximum velocity (see Figure 8–2), or more reliably from a Lineweaver-Burk plot (see Figure 8–4). A number of K_m values for different enzymes are given in Table 8–2. From this table we can see that the value of K_m varies over a very wide range for different enzymes. For most enzyme-substrate pairs the value of K_m lies between 10^{-1} M and 10^{-6} M. From Table 8–2 it is also apparent that the K_m is a function of the substrate for those enzymes, like chymotrypsin, that act on more than one substrate. The K_m is also a function of other variables such as temperature, pH, and ionic strength. The K_m establishes an approximate value for the intracellular level of the substrate. It is unlikely that the substrate level would be much greater than the K_m, for this would result in an excessive amount of substrate. It is also unlikely that the substrate level would be much lower than the K_m, as this would result in a very inefficiently catalyzed reaction.

The K_m is often thought of as equivalent to the dissociation constant of the enzyme-substrate complex. In fact this is only true in certain cases. Thus according to Equation (28)

$$K_m = \frac{k_{-1} + k_2}{k_1} \tag{28'}$$

TABLE 8–2
The Michaelis Constant for Some Enzymes

Enzyme and Substrate	K_m (M)
Catalase	
H_2O_2	1.1
Hexokinase	
Glucose	1.5×10^{-4}
Fructose	1.5×10^{-3}
Chymotrypsin	
N-Benzoyltyrosinamide	2.5×10^{-3}
N-Formyltyrosinamide	1.2×10^{-2}
N-Acetyltyrosinamide	3.2×10^{-2}
Glycyltyrosinamide	1.2×10^{-1}
Aspartate Aminotransferase	
Aspartate	9.0×10^{-4}
α-Ketoglutarate	1.0×10^{-4}
Fumarase	
Fumarate	5.0×10^{-6}
Malate	2.5×10^{-5}

whereas the dissociation constant is

$$K_S = \frac{k_{-1}}{k_1}$$

Comparing these equations we can see that K_m will always be larger than K_S. However, if $k_{-1} \gg k_2$, then $K_m \cong k_{-1}/k_1$ and $K_m \cong K_S$.† But when $k_2 \gg k_{-1}$, then $K_m \cong k_2/k_1$ and $K_m \gg K_S$. If k_{-1} and k_2 are of the same order of magnitude, then all three reaction constants, k_1, k_{-1}, and k_2, are important contributors to the value of K_m. In practice all three situations have been observed with different enzyme-substrate complexes.

The relationship between K_m and K_S can be considerably more complex for situations where there are additional intermediates in the reaction pathway. Such situations are not uncommon. Imagine the following case, in which enzyme and substrate form an initial complex that undergoes two transitions before being converted to product:

$$E + S \underset{K_S}{\rightleftharpoons} ES \underset{K}{\rightleftharpoons} ES' \underset{K'}{\rightleftharpoons} ES'' \overset{k_4}{\rightleftharpoons} E + P \qquad (33)$$

Assume that k_4 is the slow step in this multistep reaction and that all the intermediates are in equilibrium. Then

$$[ES'] = K[ES] \qquad \text{and} \qquad [ES''] = K'[ES'] \qquad (34)$$

and

$$K_m = \frac{K_S}{1 + K + K'} \qquad (35)$$

One must always be wary of such possibilities, since steady-state kinetic analysis cannot discriminate between relatively simple situations involving one enzyme-substrate complex and more involved situations where additional intermediates are involved. That can be achieved only with the help of additional kinetic approaches, and ideally with analytical methods that permit detection of the various intermediates. We will consider some examples in the next chapter.

---

†In fact, this is the limiting case assumed by Michaelis and Menten, for when $k_1 \gg k_2$, then the enzyme-substrate complex is in equilibrium with the free enzyme and free substrate.

TABLE 8–3
Values of k_{cat} for Some Enzymes

Enzyme	k_{cat} (sec^{-1})
Catalase	40,000,000
Carbonic anhydrase	1,000,000
Acetylcholinesterase	14,000
Penicillinase	2,000
Lactate dehydrogenase	1,000
Chymotrypsin	100
DNA polymerase I	15
Lysozyme	0.5

Significance of the Turnover Number, k_{cat}. The *turnover number* of an enzyme, k_{cat}, is the maximum number of molecules of substrate converted to product per active site per unit time. For the simple situation in which there is only one enzyme-substrate complex and the conversion of enzyme-substrate complex to enzyme-product complex is rate-limiting, k_{cat} is equal to the kinetic constant k_2. From Equation (21), that is, $V_{max} = k_2[E_t]$, we can see that the turnover number of an enzyme is equal to the number of substrate molecules converted into product per second per mole of enzyme present when the enzyme is fully complexed with substrate. Thus in this simple situation the turnover number is a direct measure of the *catalytic efficiency of the active site*.

For more complicated reactions, k_{cat} can be a function of all the first-order rate constants, and thus it cannot be assigned to any particular process except when simplifying features occur. For example, if dissociation of product from the enzyme-product complex is slow, the rate constant for dissociation contributes to k_{cat}. In the limiting case, where the enzyme-product dissociation is much slower than the chemical steps, k_{cat} is equal to the enzyme-product dissociation rate constant.

A group of representative turnover numbers is given in Table 8–3. Some of the values are enormous, such as that observed for catalase. At the other end of the spectrum we have the sluggish turnover number for lysozyme.

Significance of the Specificity Constant, k_{cat}/K_m. We have just seen that k_{cat} gives a measure of the catalytic efficiency of the enzyme when the enzyme is saturated with substrate. However, most enzymes do not operate at saturating substrate concentrations *in vivo*. Instead, the $[S]/K_m$ ratio is more commonly in the range of 0.01 to 1.0. As long as the substrate concentration is less than saturating, the enzymatic rate will be lower than that indicated by the catalytic constant. Under these more commonly observed conditions it would be useful to have a constant that would serve as a direct measure of the enzymatic rate. For this purpose the ratio k_{cat}/K_m is quite useful. Its value can be appreciated from the equations that have already been developed.

Substituting the expression for V_{max} from Equation (21) into the Briggs-Haldane equation, Equation (30), we get

$$v = \frac{k_{cat}[E_t][S]}{K_m + S} \tag{36}$$

At low [S], $K_m \gg [S]$, and $[E_t]$ is approximately equal to the concentration of free enzyme [E]. Under these conditions Equation (36) reduces to

$$v = \frac{k_{cat}}{K_m}[E][S] \tag{37}$$

In this equation we see that k_{cat}/K_m is an apparent second-order rate constant for the reaction of E and S to form product. Thus Equation (37) gives us a direct measure of the *catalytic efficiency* of different enzymes at substrate concentrations that are significantly below saturating levels. In Table 8–4 we see a range of values of k_{cat}/K_m for a number of very efficient enzymes. The values in the 10^8 range, which are observed for a number of these enzymes, are very close to the maximum rates that could be expected from calculations of the diffusion rates of molecules.

The usefulness of k_{cat}/K_m does not end here. This ratio is sometimes called the *specificity constant*, since the discrimination between the competing substances is determined by the ratios of k_{cat}/K_m rather than by either quantity alone. This becomes clear if we consider the relative velocities of reaction of two substrates, S_1 and S_2.

TABLE 8–4
Enzymes for Which k_{cat}/K_m is Close to the Diffusion-controlled Association Rate

Enzyme	Substrate	k_{cat} (sec^{-1})	K_m (M)	k_{cat}/K_m (sec^{-1} M^{-1})
Acetylcholin-esterase	Acetylcholine	1.4×10^4	9×10^{-5}	1.6×10^8
Carbonic anhydrase	CO_2	1×10^6	0.012	8.3×10^7
	HCO_3^-	4×10^5	0.026	1.5×10^7
Catalase	H_2O_2	4×10^7	1.1	4×10^7
Crotonase	Crotonyl-CoA	5.7×10^3	2×10^{-5}	2.8×10^8
Fumarase	Fumarate	800	5×10^{-6}	1.6×10^8
	Malate	900	2.5×10^{-5}	3.6×10^7
Triosephosphate isomerase	Glyceraldehyde 3-phosphate	4.3×10^3	4.7×10^{-4}	2.4×10^8
β-Lactamase	Benzylpenicillin	2.0×10^3	2×10^{-5}	1×10^8

Source: A. Fersht, *Enzyme Structure and Mechanism*, 2d ed., Freeman, New York, 1985, p. 152. Reprinted with permission.

Given

$$\frac{-d[S_1]}{dt} = V_{S_1} = \left(\frac{k_{cat}}{K_m}\right)_{S_1} [E][S_1] \tag{38}$$

and

$$\frac{-d[S_2]}{dt} = V_{S_2} = \left(\frac{k_{cat}}{K_m}\right)_{S_2} [E][S_2] \tag{39}$$

upon dividing Equation (38) by Equation (39) we get

$$\frac{V_{S_1}}{V_{S_2}} = \frac{(k_{cat}/K_m)_{S_1} [S_1]}{(k_{cat}/K_m)_{S_2} [S_2]} \tag{40}$$

Thus at equal molar concentrations of substrate the relative rates of reaction of two substrates by the same enzyme are proportional to the ratio of the respective k_{cat}/K_m values. Values of k_{cat}/K_m are more useful than values of k_{cat} when attempting to get a realistic idea of enzyme efficiency under physiological conditions. A perusal of the values of the two quantities in Table 8–4 shows a wide range, reflecting the broad distribution of values for K_m and k_{cat}.

ENZYME INHIBITION

Enzyme inhibition (like enzyme activation) is a phenomenon of great practical importance, as well as a useful probe for determining the active site on the enzyme and other factors that control enzyme activity. In this section we will consider some general aspects of enzyme inhibition, leaving more sophisticated aspects of enzyme activity regulation to Chapter 10.

The active site of an enzyme usually represents a small part of the whole enzyme molecule. Strong evidence for the localization of active centers comes from studies with specific inhibitors of enzymes. Inhibition may be reversible or irreversible with different inhibitors.

Reversible Inhibitors Frequently Resemble Substrate

Reversible inhibitors work by a variety of mechanisms that can frequently be distinguished by kinetic analysis (Figure 8–5). In *competitive inhibition*, the binding of inhibitor and substrate to the enzyme is mutually exclusive. Usu-

REACTIONS   MODEL   LINEWEAVER-BURK EQUATION   LINEWEAVER-BURK PLOT

(a) Competitive

$$E + S \underset{K_S}{\rightleftharpoons} ES \xrightarrow{K_P} E + P$$
$$+$$
$$I$$
$$K_I \updownarrow$$
$$EI$$

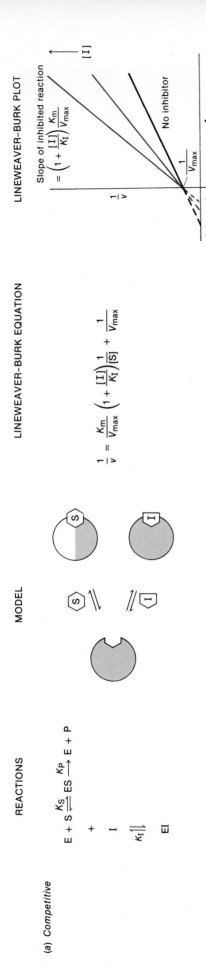

$$\frac{1}{v} = \frac{K_m}{V_{max}}\left(1 + \frac{[I]}{K_I}\right)\frac{1}{[S]} + \frac{1}{V_{max}}$$

$$\text{Slope of inhibited reaction} = \left(1 + \frac{[I]}{K_I}\right)\frac{K_m}{V_{max}}$$

No inhibitor

$\frac{1}{v}$, $\frac{1}{[S]}$, $\frac{1}{V_{max}}$

$\text{Intercepts} = \dfrac{-1}{K_m\left(1 + \frac{[I]}{K_I}\right)}$

(b) Noncompetitive

$$E + S \underset{K_S}{\rightleftharpoons} ES \xrightarrow{K_P} E + P$$
$$+ \qquad +$$
$$I \qquad I$$
$$K_I \updownarrow \qquad \updownarrow K_I$$
$$EI + S \underset{K_S}{\rightleftharpoons} ESI$$

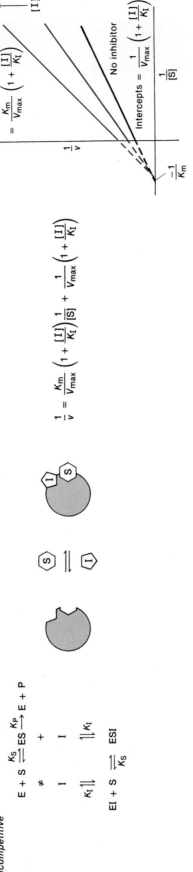

$$\frac{1}{v} = \frac{K_m}{V_{max}}\left(1 + \frac{[I]}{K_I}\right)\frac{1}{[S]} + \frac{1}{V_{max}}\left(1 + \frac{[I]}{K_I}\right)$$

$$\text{Slope of inhibited reaction} = \frac{K_m}{V_{max}}\left(1 + \frac{[I]}{K_I}\right)$$

No inhibitor

$\text{Intercepts} = \dfrac{1}{V_{max}}\left(1 + \frac{[I]}{K_I}\right)$

$\frac{1}{v}$, $\frac{1}{[S]}$

$-\dfrac{1}{K_m}$

(c) Uncompetitive

$$E + S \underset{K_S}{\rightleftharpoons} ES \xrightarrow{K_P} E + P$$
$$+$$
$$I$$
$$\updownarrow K_I$$
$$ESI$$

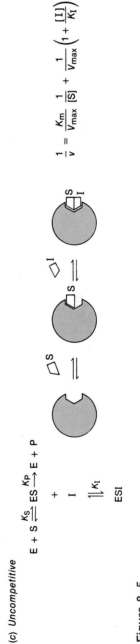

$$\frac{1}{v} = \frac{K_m}{V_{max}}\frac{1}{[S]} + \frac{1}{V_{max}}\left(1 + \frac{[I]}{K_I}\right)$$

$$\text{Slope of inhibited reaction} = \frac{K_m}{V_{max}}$$

No inhibitor

$\text{Intercepts} = \dfrac{1}{V_{max}}\left(1 + \frac{[I]}{K_I}\right)$

$\frac{1}{v}$, $\frac{1}{[S]}$

$\text{Intercepts} = -\dfrac{1 + \frac{[I]}{K_I}}{K_m}$

Figure 8–5

Three types of reversible inhibition: competitive, noncompetitive, and uncompetitive. For each type of inhibition the reaction, the mechanism (model) of inhibition, the Lineweaver-Burk equation, and the Lineweaver-Burk plot are given.

BOX 8–B

Derivation of the Velocity Expression When There Is Competitive Inhibition

The expression relating initial reaction velocity, v, to V_{max}, [S], K_S (or K_m), [I], and K_I can be derived from either the rapid equilibrium assumption (Henri-Michaelis-Menten) or the steady-state assumption (Briggs and Haldane).

When there is a competitive inhibitor present, the total enzyme exists in three forms:

$$[E_t] = [E] + [ES] + [EI] \qquad (A)$$

From Equation (20),

$$v = k_2[ES]$$

Dividing Equation (20) by Equation (A),

$$\frac{v}{[E_t]} = \frac{k_2[ES]}{[E] + [ES] + [EI]} \qquad (B)$$

By transposing Equation (21), we get

$$[E_t] = \frac{V_{max}}{k_2} \qquad (C)$$

Also, from Equations (19) and (41) we know that in the rapid equilibrium situation

$$\frac{[E][S]}{[ES]} = K_S \quad \text{or} \quad [ES] = \frac{[E][S]}{K_S} \qquad (D)$$

and

$$\frac{[E][I]}{[EI]} = K_I \quad \text{or} \quad [EI] = \frac{[E][I]}{K_I} \qquad (E)$$

After substituting the expressions for $[E_t]$, [ES], and [EI] from Equations (C), (D), and (E) into Equation (B), we get

$$v = \frac{V_{max}[S]}{K_S\left(1 + \dfrac{I}{K_I}\right) + [S]} \qquad (F)$$

In the steady-state situation K_m replaces K_S in this expression so that

$$v = \frac{V_{max}[S]}{K_m\left(1 + \dfrac{I}{K_I}\right) + [S]} \qquad (G)$$

We can see that this expression reduces to the Briggs-Haldane equation, Equation (30), when the inhibitor concentration is zero.

---

ally, but not always, competitive inhibition occurs because the inhibitor structurally resembles the substrate and binds at the active center. Most cases of inhibition that we will consider are of this type.

According to the Henri-Michaelis-Menten treatment, the sequence of reactions leading from substrate to products is

$$E + S \underset{}{\overset{K_S}{\rightleftharpoons}} ES \longrightarrow E + P$$

The equation for the formation of an enzyme-inhibitor complex is like that for formation of an enzyme-substrate complex

$$E + I \overset{K_I}{\rightleftharpoons} EI \qquad (41)$$

where I is the inhibitor and EI is the enzyme-inhibitor complex.

In the presence of a competitive inhibitor, the rate of the catalyzed reaction is dependent on the relative concentrations of substrate and inhibitor. The derivation of the relevant equations is similar to the derivation of the original Henri-Michaelis-Menten equation. The major difference is that the $[E_t]$ includes an additional form of the enzyme, EI. The dissociation constant K_I of the EI complex must also be considered (see Box 8–B).

The equation in which both competitive inhibitor and substrate binding must be considered should be compared with that in which only substrate is involved:

$$\frac{1}{v} = \frac{K_m}{V_{max}} \frac{1}{[S]} + \frac{1}{V_{max}} \qquad (32')$$

$$\frac{1}{v} = \frac{K_m}{V_{max}} \frac{1}{[S]}\left(1 + \frac{[I]}{K_I}\right) + \frac{1}{V_{max}} \qquad (42)$$

where [I] is the concentration of inhibitor and K_I is the dissociation constant of the enzyme-inhibitor complex.

High concentrations of substrate overcome the negative effects caused by the inhibitor because both substrate and inhibitor molecules are competing for binding to the same site. A reciprocal plot of $1/v$ against $1/[S]$ in the presence of this type of inhibitor will result in two nonparallel straight lines that intercept on the y axis at infinitely high substrate concentration, i.e., when $1/[S] = 0$, as shown in Figure 8–5a. The apparent K_m is altered to K_m', where $K_m' = K_m(1 + [I]/K_I)$ according to Equation (42). In this relationship K_m' is constant only if the inhibitor concentration is a constant. The higher the inhibitor concentration, the larger will be the observed value for K_m'.

Another type of inhibitor can bind to free enzyme or to the enzyme-substrate complex. Such substances are known as _noncompetitive inhibitors_. They do not prevent substrate binding but they slow or stop the enzymatic reaction by binding to the enzyme-substrate complex, forming an ESI complex (see Figure 8–5b). In the simplest cases of noncompetitive inhibition, the K_m is not altered. The Lineweaver-Burk plot for such a situation is shown in Figure 8–5b. Both the slope and the intercept on the y axis are increased. In noncompetitive inhibition no amount of substrate can relieve the inhibition. This follows from the fact that substrate and inhibitor bind to different, nonoverlapping sites.

Another form of inhibition, called _uncompetitive_, occurs when inhibitor (I) binds to the enzyme-substrate complex (ES) but not to the substrate-free enzyme (E) (Figure 8–5c); this is quite rare for enzymes with only one substrate, but more common for multisubstrate systems. With an uncompetitive inhibitor the plot of Figure 8–5c shows a line parallel to the line for no inhibitor, but higher (v is smaller, so $1/v$ is larger).

Irreversible Inhibitors Usually Form Covalent Linkages

Competitive, noncompetitive, and uncompetitive inhibition are all reversible. If the inhibitor concentration is lowered, the enzymatic activity rises. If the inhibitor is removed, the enzyme activity returns to its normal value. In the case of irreversible inhibition the inhibitor binds so strongly to the enzyme (usually by forming a covalent bond; see Table 8–5) that it will not dissociate when the inhibitor concentration is lowered. Irreversible inhibition can occur at the active site on the enzyme or elsewhere. The site of binding of the irreversible inhibitor is best determined by direct chemical methods. Irreversible inhibitors often provide valuable information about the active site on the enzyme. For example, certain amino acids that are part of the active site can be irreversibly reacted with specific reagents. This leads to an irreversible loss of enzyme activity. Some of the specific reagents that serve as irreversible inhibitors are listed in Table 8–5; others are discussed in the text.

TABLE 8–5
Some Inhibitors of Enzymes that Form Covalent Linkages with Functional Groups on the Enzyme

Inhibitor	Enzyme Group that Combines with Inhibitor
Cyanide	Fe, Cu, Zn, other transition metals
p-Mercuribenzoate	Sulfhydryl
Diisopropylfluorophosphate	Serine hydroxyl
Iodoacetate	Sulfhydryl, imidazole, carboxyl, thioether

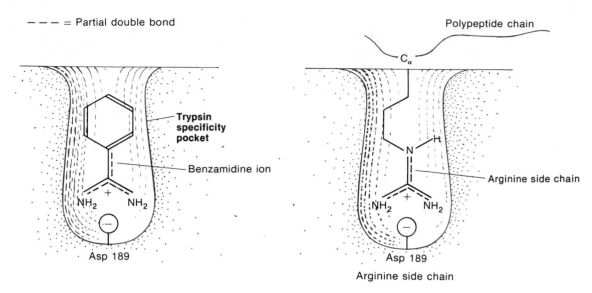

Figure 8–6
The specificity pocket of trypsin can accommodate an arginine side chain of a polypeptide substrate or a benzamidine ion, which acts as a competitive inhibitor.

Examples of Reversible and Irreversible Inhibition of Trypsin. As you may recall from Chapter 3, trypsin is a protease that cleaves polypeptide chains at the peptide linkage adjacent to a basic amino acid residue. We have seen that competitive inhibitors frequently resemble part of the structure of the substrate. This should not be surprising, since they most frequently compete with the substrate for binding at the active site. The benzamidine ion is one such competitive inhibitor for trypsin. When protonated, the amidine end of the molecule has the same flat, delocalized electronic structure as is found in protonated arginine side chains. To the specificity pocket of trypsin, a benzamidine ion looks like the outer half of an arginine side chain and is accepted and bound as such (Figure 8–6). An enzyme inhibited in this way is reversibly inactivated. Thus the enzyme is inhibited in the presence of benzamidine, but full activity is restored when inhibitor is removed.

Diisopropylfluorophosphate (DFP) is an irreversible inhibitor of trypsin and other, trypsinlike enzymes that contain an essential serine at the active site. DFP binds covalently and irreversibly to serine 195 of trypsin to form diisopropyl-serine trypsin (Figure 8–7). DFP also reacts with the comparable

Ser 195
|
CH_2
|
OH

$$\begin{matrix} CH_3 & & F & & CH_3 \\ | & & | & & | \\ CH-O-P-O-CH \\ | & & \| & & | \\ CH_3 & & O & & CH_3 \end{matrix}$$

Diisopropylfluorophosphate (DFP)

$\longrightarrow$

Ser 195
|
CH_2
|
O

$$\begin{matrix} CH_3 & & & & CH_3 \\ | & & | & & | \\ CH-O-P-O-CH \\ | & & \| & & | \\ CH_3 & & O & & CH_3 \end{matrix}$$ + HF

Diisopropyl enzyme (DIP enzyme)

Figure 8–7
Diisopropylfluorophosphate (DFP) is an irreversible inhibitor with trypsin, chymotrypsin, and elastase. It forms a covalent bond with the Ser 195 side chain of trypsin, thereby preventing subsequent catalysis. The DFP-treated enzyme is permanently inhibited.

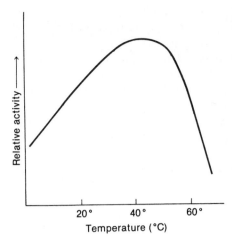

Figure 8–8
Typical reactivity curve showing temperature dependence. The rate keeps rising with temperature until a point is reached where the enzyme is no longer stable.

serine groups in the proteases chymotrypsin and elastase, thereby inactivating them. The correlation between loss of enzyme activity and reaction with this serine group shows that this serine is essential for enzyme activity.

EFFECTS OF TEMPERATURE ON ENZYMATIC ACTIVITY

For most enzymes the rate of reaction increases with temperature until a temperature is reached where the enzyme is no longer stable (Figure 8–8). At this point and above, there is a precipitous drop in activity that is frequently irreversible. At low and moderate temperatures, where the enzyme is stable, the increase in enzyme activity with temperature is a sensitive function of the activation energy for the reaction. For many enzymes the rate increases by a factor of about two for a 10°C rise in temperature. However, there are some enzymes with much higher temperature dependencies because of the higher activation energies for the reactions they catalyze.

EFFECTS OF pH ON ENZYMATIC ACTIVITY

The effect of pH on activity for any particular enzyme depends on the pK of the ionizing groups on the enzyme and the substrate that participates in the reaction. The optimum pH, unlike the K_m, is usually a characteristic more of the enzyme than of the particular substrate. Thus for the enzyme pepsin, the optimum is around 2, for trypsin around 8, and for cholinesterase 7 or greater, while for papain pH values between 4 and 8 are optimal (Figure 8–9). Whereas the pH sensitivity of the enzyme is most often an indication of an ionizable group at the site of enzyme reactivity, it can also be a measure of some change in the tertiary structure of the enzyme that affects the active site. For this reason, the sensitivity of enzyme activity to pH must be interpreted with caution. Specific cases will be discussed in Chapter 9.

Figure 8–9
Enzyme activity as a function of pH for four different enzymes. The optimum pH is usually a characteristic of the enzyme and not the substrate it interacts with. Often the pH sensitivity is an indication of an ionizable group at the active site. It can also be a measure of some change in the tertiary structure of the enzyme.

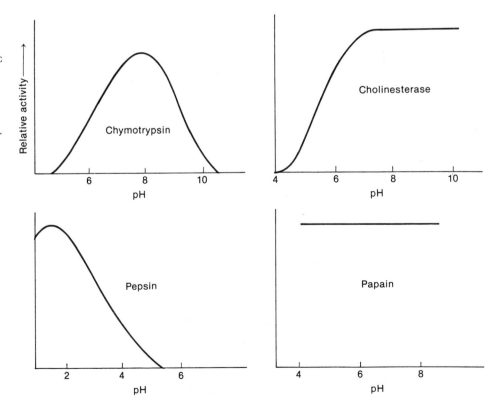

KINETIC ANALYSIS FOR REACTIONS INVOLVING TWO SUBSTRATES

Multisubstrate enzymes work in a wide variety of ways. Enzymes are known that catalyze the reaction of two, three, or even more different substrate molecules to one or more products. In some cases all the substrate molecules must be bound to the enzyme simultaneously. In other cases the first bound substrate is modified and then reacted with a subsequently bound substrate. We will confine our discussion in this section to enzymes that interact with two substrates.

There are several models for representing enzyme-catalyzed pathways involving two substrates. In the *random-order pathway*, the order of binding of the substrates S_1 and S_2 is random and reaction occurs after the two substrates become bound to the enzyme (Figure 8–10a). In the *simple ordered pathway* (Figure 8–10b) one substrate, say S_1, must be bound before the second substrate, S_2, can bind, whereupon reaction occurs as in the random pathway. In the *reactive intermediate pathway* (Figure 8–10c) one substrate becomes bound and undergoes a reaction with the enzyme before the second substrate becomes bound, leading to additional reaction(s). In the *reactive intermediate pathway that leads to an altered enzyme* (Figure 8–10d), we have another variation of the ordered pathway. In this case S_1 binds first and then reacts to form an altered enzyme E′ and a product P_1 that dissociates, making way for the altered enzyme to bind S_2 and convert it into product, P_2. Finally, P_2 dissociates from the enzyme and the original enzyme is regenerated. This pattern of reaction is referred to as the *Ping-Pong mechanism* to emphasize the alteration of the enzyme between two states, E and E′.

Each of these four possibilities can be analyzed by a steady-state kinetic equation that gives rise to a characteristic Lineweaver-Burk plot. Frequently the situations are complex, and additional information is necessary to reach an unambiguous interpretation of the steady-state kinetic data. We shall review the equations and methods of plotting for two situations, the simple ordered pathway and the Ping-Pong mechanism.

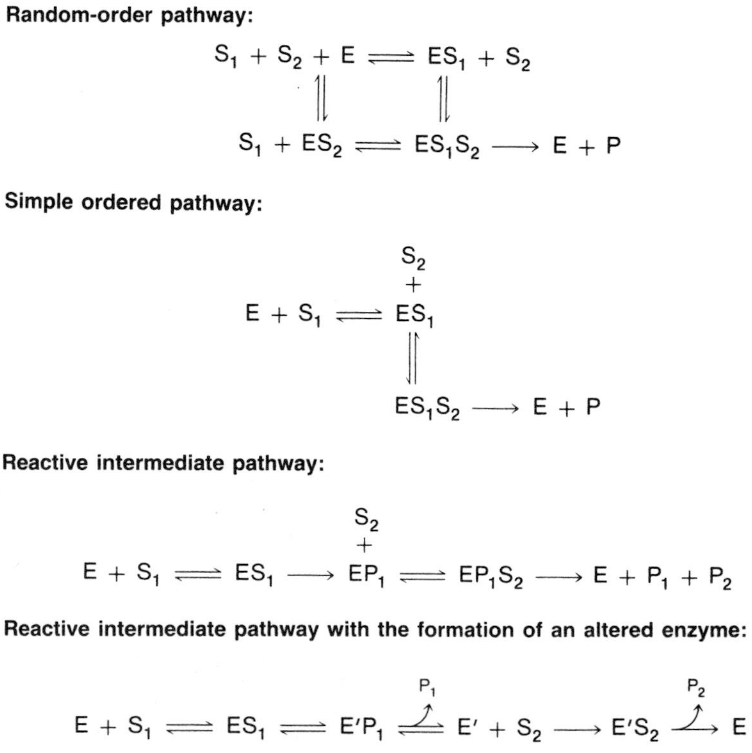

Figure 8–10
Different types of pathways seen for enzymes that use two substrates.

Random-order pathway:

$$S_1 + S_2 + E \rightleftharpoons ES_1 + S_2$$

$$S_1 + ES_2 \rightleftharpoons ES_1S_2 \longrightarrow E + P$$

Simple ordered pathway:

$$E + S_1 \rightleftharpoons ES_1 \quad \overset{S_2}{\underset{+}{}}$$

$$ES_1S_2 \longrightarrow E + P$$

Reactive intermediate pathway:

$$E + S_1 \rightleftharpoons ES_1 \longrightarrow EP_1 \rightleftharpoons EP_1S_2 \longrightarrow E + P_1 + P_2 \quad \overset{S_2}{\underset{+}{}}$$

Reactive intermediate pathway with the formation of an altered enzyme:

$$E + S_1 \rightleftharpoons ES_1 \rightleftharpoons E'P_1 \overset{P_1}{\rightleftharpoons} E' + S_2 \longrightarrow E'S_2 \overset{P_2}{\longrightarrow} E$$

In the simple ordered pathway the characteristic kinetic equation for the initial reaction velocity is given by the general expression

$$\frac{1}{v} = \frac{K_{mS_1}}{V_{max}}\frac{1}{S_1} + \frac{K_{mS_2}}{V_{max}}\left(\frac{1}{S_2}\right)\left(1 + \frac{K_{eqS_1}}{S_1}\right) + \frac{1}{V_{max}} \tag{43}$$

where K_{mS_1} and K_{mS_2} are the Michaelis constants for S_1 and S_2, respectively, and K_{eqS_1} is the equilibrium dissociation constant for the ES_1 complex. It should be noted that the expression differs from the one for the single-substrate reaction by the presence of the middle term on the right [compare with Equation (31)].

For purposes of plotting, it is convenient to rearrange Equation (43) to

$$\frac{1}{v} = \frac{1}{V_{max}}\left(1 + \frac{K_{mS_1}}{[S_1]} + \frac{K_{mS_2}}{[S_2]} + \frac{K_{eqS_1}K_{mS_2}}{[S_1][S_2]}\right) \tag{44}$$

The kinetic parameters we would like to evaluate are the maximum velocity, V_{max}, the two Michaelis constants K_{mS_1} and K_{mS_2}, and the equilibrium constant for the reversible dissociation of the complex ES_1, K_{eqS_1}.

Equation (44) produces linear plots only if one of the substrate concentrations is held constant. The usual procedure is to do two series of experiments; one in which S_1 is held constant and S_2 is varied, and the other in which S_1 is varied and S_2 is held constant. In Figure 8–11a we see the case for a family of plots in which S_2 is held constant and S_1 is varied. The kinetic parameters are evaluated from two secondary plots. In the first secondary plot (b), the intercepts from graph (a) are plotted against $1/S_2$. This permits evaluation of $1/V_{max}$ and K_{mS_2} as shown. In the other secondary plot the slopes from plot (a) are graphed against $1/S_2$. From this curve the remaining parameters, K_{mS_1} and K_{eqS_1}, are evaluated. Examples of enzymes that follow the simple ordered pathway are discussed in Chapters 9 and 10 (see lactate dehydrogenase in Chapter 9 and aspartate carbamoyltransferase in Chapter 10).

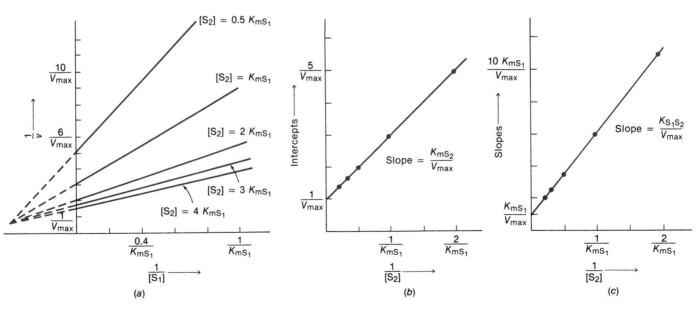

Figure 8–11
Reciprocal plots used to analyze kinetics of two-substrate enzymes. (a) Plot of $1/v$ against $1/[S_1]$ for a series of different concentrations of the second substrate, S_2. (b) A secondary plot in which the intercepts from part (a) are plotted against $1/[S_2]$. (c) A secondary plot in which the slopes from part (a) have been plotted against $1/[S_2]$. The figures have been drawn for the case that $K_{mS_1} = 10^{-3}$ M, $K_{mS_2} = 2K_{mS_1}$, and $K_{S_1S_2} = K_{eqS_1}K_{mS_2}$. Concentrations are expressed in moles per liter.

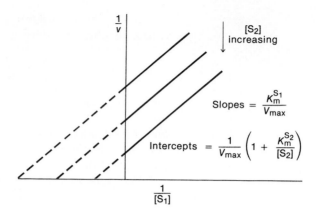

Figure 8–12
The characteristic parallel-line reciprocal plots of Ping-Pong kinetics. As the concentration of the second substrate in the sequence increases, V_{max} increases, as does the K_m for the first substrate. V_{max}/K_m, the reciprocal of the slope of the plot, remains constant.

The Ping-Pong mechanism is characterized by the kinetic equation

$$\frac{1}{v} = \frac{K_{mS_1}}{V_{max}} \frac{1}{[S_1]} + \frac{K_{mS_2}}{V_{max}[S_2]} + \frac{1}{V_{max}} \tag{45}$$

If we plot $1/v$ versus $1/[S_1]$ at various fixed concentrations of S_2, we observe a series of parallel lines with slope K_{mS_1}/V_{max}, separated by increments contributed by the second term of the equation (Figure 8–12). In this way the Ping-Pong mechanism is usually easily recognized. A full evaluation of the parameters requires secondary plots, as in the analysis of the simple ordered pathways.

Enzymes that use coenzymes frequently operate by the Ping-Pong mechanism. Reactions involving the coenzymes pyridoxal or flavins proceed through such an alternating cycle. Examples are discussed in Chapter 11. Certain membrane-transport proteins are also believed to function by the reactive intermediate Ping-Pong mechanism (see Chapter 34).

In the following chapter we will examine the mechanism of specific enzymes; we will have an opportunity to make use of some of the kinetic methods described in this chapter, as well as other sources of information.

SUMMARY

Enzyme kinetic analysis is the most broadly used tool for characterizing enzyme-catalyzed reactions. This chapter begins by reviewing certain basic principles of chemical kinetics. Every chemical reaction has a characteristic rate, which is characterized by the mathematical relationship between the rate of reaction and the concentrations of the reactants. The overall order of the reaction is the sum of the exponentials for all the molecular species given by the rate law. The molecularity is the number of interacting molecules in any particular step of a multistep reaction. The rate law for multistep reactions is frequently the rate of the slowest step.

Both thermodynamics and kinetics influence reactions in different ways. Thermodynamics leads to a prediction of the equilibrium constant of the reaction. Kinetic factors determine the rates of the forward and reverse reactions that are required to reach equilibrium.

Reaction rates are a function of the collision frequency between reacting species and the fraction of these collisions that is effective in producing products. The former is determined by the concentration and the temperature; the later is primarily a function of the activation energy.

A catalyst increases the reaction rate by lowering the activation energy. Catalysts in biochemical systems are usually proteins called enzymes. Enzymes are highly specific for the reactions they catalyze. They contain highly localized catalytic sites, substrate binding sites, and frequently a third class of sites that lead to specific interactions with other cellular components.

All known enzymes have been classified into six groups

according to the types of reactions they catalyze: oxidore-ductases, transferases, hydrolases, lyases, isomerases, and ligases.

Enzyme kinetics is usually studied on purified or partially purified enzymes *in vitro*. The most popular treatment for analyzing kinetic data was introduced in the first quarter of this century by Henri, by Michaelis and Menten, and by Briggs and Haldane. The treatment starts with a model that says the first step in an enzyme-catalyzed reaction involves complex formation with substrate. All subsequent reactions occur on the enzyme surface until the products are released. The speed of the overall reaction is proportional to the concentration of the enzyme-substrate complex. Shortly after a reaction is initiated, the concentration of the enzyme-substrate complex becomes approximately fixed for a significant duration of time, known as the steady state. Through a study of the reaction rate, as measured by the disappearance of reactants and the appearance of products, it is possible to deduce the concentration of the enzyme-substrate complex as well as its dissociation constant and the rate at which it is converted into products.

Enzymes differ widely in their affinity for substrate(s) and the efficiency of their catalytic sites. This leads to widely varying rates for different reactions.

There are a number of different ways of inhibiting enzyme reactions, which may be characterized according to the way in which they influence the enzyme kinetics. Some inhibitors compete with substrate for binding; other inhibitors do not compete with substrate for binding but inhibit the reaction of bound substrate. Irreversible inhibitors usually react covalently with specific sites on the enzyme.

Most of the analyses in this chapter are for enzymes involving single substrate molecules. There is a brief discussion of reactions that involve two substrates. Two substrates can react in an enormous number of ways; most of our discussion is limited to two important pathways—the simple ordered pathway, in which one substrate must bind before the other substrate binds, and the Ping-Pong mechanism, which is also an ordered pathway, but one in which binding of the first substrate is followed by alteration of the enzyme structure. The Henri-Michaelis-Menten analysis of these two types of pathways is described.

SELECTED READINGS

Advances in Enzymology. New York: Academic Press. An ongoing annually published volume containing monographs on selected topics.

The Enzymes. Boyer, P. D., ed., New York: Academic Press. A continuing series with more than 16 volumes of monographs written by experts on selected topics.

Fersht, A., *Enzyme Structure and Mechanism*. New York: W.H. Freeman and Co., 1985. Very readable and up-to-date.

Hammes, G. G., *Enzyme Catalysis and Regulation*. New York: Academic Press, 1982. Good for some modern approaches.

Kaiser, T., D. S. Lawrence, and S. Z. Rokita, The chemical modification of enzymatic specificity. *Ann. Rev. Biochem.* 54:597–630, 1985.

Segel, Irwin H., *Enzyme Kinetics*. New York: Wiley, 1975. A thorough treatment of kinetic analysis.

PROBLEMS

1. What is meant by the order of a reaction?

2. Define K_m, K_S, and k_{cat}.

3. K_m is all too frequently equated with K_S. In fact, in most reactions there is an appreciable disparity between the values for K_m and K_S. For the reaction A → B, define conditions under which $K_m = K_S$. Describe three conditions under which this is not true.

4. What is the difference between the activated complex and the enzyme-substrate complex?

5. What is the steady-state approximation, and under what conditions is it valid?

6. If the activation free energy E_a of an enzyme-catalyzed reaction is 10 kcal/mole, how much difference in the initial reaction velocity would you expect between 20 and 30°C? Between 30 and 40°C? Make the same calculation for an E_a of 20 kcal/mole. If an enzyme lowers the E_a from 20 to 2 kcal/mole, what is the rate enhancement factor achieved by using the enzyme? (See Box 8–A.)

7. Assume that an enzyme-catalyzed reaction follows Henri-Michalis-Menten kinetics with a K_m of $1 \times$ 10^{-6} M. If the initial reaction rate is 0.1 μmol/min at 0.1 M, what would it be at 0.01 M, 10^{-3} M, and 10^{-6} M?

8. If the K_m of an enzyme for its substrate is 10^{-5} M and the K_I of the enzyme for a competitive inhibitor is 10^{-6} M, what concentration of inhibitor will be necessary to lower the rate of the reaction by a factor of 10 when the substrate concentration is 0.1 M? 0.01 M?

9. Assume that an enzyme-catalyzed reaction has the following mechanism:

$$E + S \underset{k_2}{\overset{k_1}{\rightleftharpoons}} ES \underset{k_4}{\overset{k_3}{\rightleftharpoons}} P + E$$

and the following rate constants:

$$k_1 = 10^9 \text{ M sec}^{-1}; \quad k_2 = 10^5 \text{ sec}^{-1};$$
$$k_3 = 10^2 \text{ sec}^{-1}; \quad k_4 = 10^7 \text{ M sec}^{-1}.$$

When $K_{eq} = 0.1$, $E_0 = 10^{-10}$ M, and $S_0 = 2 \times 10^{-5}$ M, calculate the following:

a. K_m

b. V_{max}

c. Turnover number.

d. Initial velocity.

10. You have purified to apparent homogeneity the enzyme X-ase, which catalyzes the reaction X → Y following Henri-Michaelis-Menten kinetics. On SDS gel electrophoresis it resolves as a single band of molecular weight 20,000. You do a series of Henri-Michaelis-Menten plots at various enzyme concentrations. From the combined data you plot V_{max} versus $[E_t]$ (total enzyme concentration) and find that they are directly proportional as they should be (see plot below). However, the K_m increases with increasing enzyme concentration, and the catalytic efficiency increases with decreasing enzyme (see plot of k_{cat}/K_m versus [X-ase]).

a. What is probably happening here?
b. How might you test your hypothesis?
c. What is the actual K_m of X-ase?

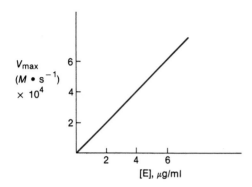

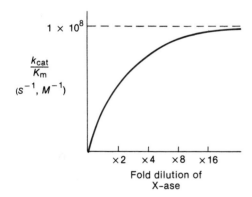

11. Lactate dehydrogenase is an example of a bisubstrate enzyme that follows a simple ordered pathway for binding of substrates. Explain how this works. If you had an enzyme that uses two substrates and wanted to determine whether it followed a random-order pathway or an ordered pathway for binding of substrate, what measurements would you make?

12. An enzyme inhibitor can act in a reversible or an irreversible manner. How would you distinguish between an irreversible inhibitor and a reversible noncompeti-

tive inhibitor? Assume you have an abundant supply of pure enzyme to work with.

13. Your mentor has just handed you a highly purified enzyme that catalyzes the reaction A → B and has asked you to make the necessary measurements for a Lineweaver-Burk plot. Explain what you would do and what the results should tell you about the reaction.

14. Factor VII (FVII) is the first in a cascade of serine proteases that are essential for blood clotting. $FVII_a$ is the proteolytically activated form of the zymogen FVII (see Chapter 9). It is thought that in order to initiate blood coagulation FVII, unlike other zymogens, must have some inherent activity. Since $FVII_a$ is much more active than FVII it is difficult to study the enzymatic properties of FVII, because during purification unknown quantities of $FVII_a$ always contaminate zymogen preparations. That is, a tiny amount of $FVII_a$ can produce most of the activity in an FVII preparation.

DFP, an irreversible inhibitor of serine proteases, has a different time course of action on FVII relative to $FVII_a$. The accompanying graph shows the data for DFP inhibition of FVII.[†] In the inhibition plot the decay of activity is best fit by a double exponential

$$y = C_1 e^{-k_1 t} + C_2 e^{-k_2 t}$$

where y is the measurable activity at time t, C_1, and C_2 are the coefficients describing the fractional contribution to overall activity of FVII and $FVII_a$, respectively, and k_1 (0.032 min^{-1}) and k_2 (0.130 min^{-1}) are the rate of constants for incorporation of DFP into FVII and $FVII_a$, respectively.

FVII can be converted to $FVII_a$ by treatment with FX_a, another enzyme in the cascade, and hence the activatability of FVII can be determined by using a standard clotting assay.

a. From the data shown, use the following equation to calculate the theoretical maximum fold activatability of FVII. The preparation under study was experimentally found to be only 45-fold activatable.

$$y = (b - C_2)/C_1$$

where y is the theoretical maximum activatability, b is the measured activatability, and C_1, and C_2 are as defined above.

b. Calculate the mole fraction of $FVII_a$ in the FVII preparation by using the following equation:

$$C_2/\{(123 \times C_1) + C_2\} = \text{mole fraction } FVII_a$$

c. The experiment in Table II makes use of the different rates of inhibition of FVII and $FVII_a$. How does it confirm your answer to part (a)?

[†]All data are from Zur et al., Initiation and inhibition of coagulation by Factor VII, *J. Biol. Chem.* 257:5623–5631, 1982.

TABLE I

The Computed Fractional Contributions to Coagulant Activity by Factor VII and Factor VII$_a$ in a Factor VII Preparation Inhibited by DFP

Experiment Number	Relative Contribution % of total	
	Factor VII	Factor VII$_a$
1	37.6 ± 2.0	62.4 ± 3.1
2	39.8 ± 2.1	59.9 ± 3.9
3	32.4 ± 1.4	67.5 ± 3.3
4	34.5 ± 2.1	65.5 ± 3.2
5	36.9 ± 2.3	63.1 ± 3.5
Mean	36.2 ± 1.28 S.E.	63.7 ± 1.31 S.E.

TABLE II

The Degree of Activation of Factor VII in Aliquots Removed from a Reaction of the Zymogen Preparation with 2 mM DFP

Time of Removal of Aliquot (min)	Clotting assay (units/ml)		Fold Activation	Calulated Residual [VII$_a$] (% of total)
	Before Activation	After Activation		
0	2,640	98,000	37	2.750
10	1,320	64,000	48	.749
20	637	37,000	58	.204
30	370	29,000	78	.055
40	185	21,000	114	.015
50	132	14,000	106	.004
60	95	9,800	103	.001
70	66	8,500	129	<.001
80	47	5,700	122	<.001

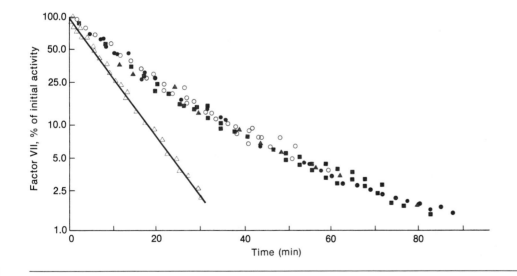

The rates of inhibition of the coagulant activity of Factor VII and of Factor VII$_a$ by DFP. The reactions were performed in the presence of 1 mg/ml of protein and 2 mM DFP, in NaCl/Tris buffer pH 7.5, at 25°C. Two preparations of Factor VII$_a$ (△) were employed, and five sets of data are included for Factor VII (○, ●, ▲, ■ (2 sets)).

9

MECHANISMS OF ENZYME CATALYSIS

In Chapter 8 we saw that enzymes can increase the rate of a reaction by many orders of magnitude over its spontaneous rate. We also saw that enzyme-catalyzed reactions are highly specific, not only in the types of reactions that they catalyze but in the particular substrates that they act on. Finally, we saw that catalysis was achieved under mild conditions of temperature, pressure, and pH. In this chapter we will explore the mechanisms of enzyme-catalyzed reactions. We will see that the effectiveness of enzymes stems from the fact that they contain highly elaborate active sites, where the substrate or substrates are bound so that they are favorably disposed for reactions with the catalytic site and the formation of products.

ENZYME CATALYSTS AND OTHER CHEMICAL CATALYSTS

Enzymes and small-molecule solution catalysts have many properties in common; both types of catalysts obey many of the same basic rules.

All chemical bonds are formed by electrons, and the rearrangement or breakage of these bonds starts with the migration of electrons. In the most general terms, reactive groups can be said to function either as electrophiles or nucleophiles. The former are electron-deficient substances that attack electron-rich substances. The latter are electron-rich substances that attack electron-deficient substances. Frequently the main job of the catalyst is to make a potentially reactive center more attractive for reaction, i.e., to potentiate the electrophilic or nucleophilic character of the reactive center of the reacting species. The range of functional groups that serve as catalysts for ordinary chemical reactions is much broader than that found in enzymes, but they serve a similar purpose.

In aqueous solution, protons (actually present as a hydronium ion) or hydroxide ions are the most common catalysts for nonenzymatic reactions. General acids or bases, defined as compounds capable of yielding protons or hydroxide ions themselves or through interaction with water, are also common forms of solution catalysts. In enzymes, general acids and bases are

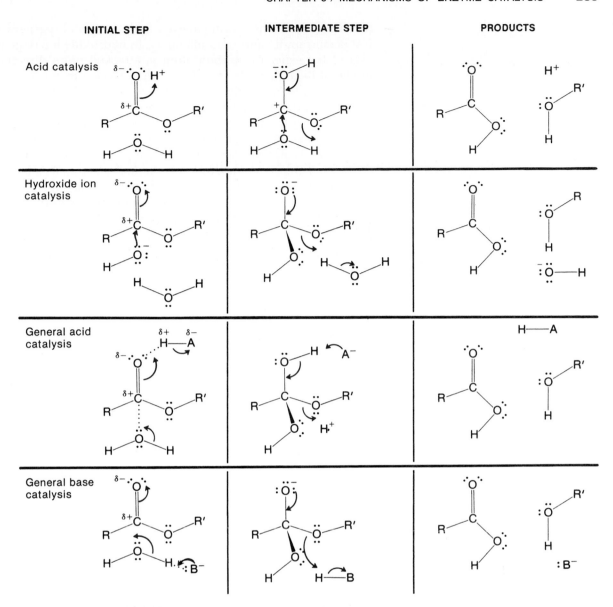

INITIAL STEP	INTERMEDIATE STEP	PRODUCTS

Acid catalysis

Hydroxide ion catalysis

General acid catalysis

General base catalysis

Figure 9–1

Different ways in which ester hydrolysis can be catalyzed. The curved (colored) arrow represents the flow of an electron pair from an electron donor to an electron acceptor. The arrowhead indicates the position of the newly formed bond at the electron acceptor. General acid and general base catalysis are more relevant to enzyme catalysis. In this case the general acid or general base groups could usually be represented by amino acid side chains.

always used in preference to free acids and bases because of the delicacy of biological systems.

The way in which catalysts work is illustrated in Figure 9–1 for the hydrolysis of an ester linkage. As a result of the electronegativity of the oxygen atom in the ester C=O group, the oxygen has a fractional negative charge δ^- and the carbon has a fractional positive charge δ^+. Hydrolysis of the ester can be accelerated by either acid or base catalysis. In acid catalysis, a proton or a generalized acid (HA) acting as an electrophile is attracted to the oxygen. This leads to an intermediate that accentuates the positive charge on the carbon atom, making it more attractive to a nucleophile, in this case water. Water is a poor nucleophile and would attack the carbon very slowly without such an inducement. This is the key step in the catalysis. The remaining reactions leading to ester hydrolysis and regeneration of the catalyst occur rapidly and spontaneously. In hydroxide ion or general base catalysis of the ester, the carbon atom is attacked more directly by a stronger nucleophile, either OH⁻ itself or a water molecule converted into an attacking hydroxide ion by the presence of the generalized base (B :). Again, after hydrolysis, the catalyst is regenerated.

An interesting feature of the general acid and general base mechanisms shown is that there is catalysis of both steps. For example, in the first step of general acid catalysis, HA adds a proton, and thus is acting as an acid, but in the second step A⁻ removes a proton and is acting as a base. In the general base catalysis mechanism the reverse sequence is followed. Such sequential catalysis of steps by a general acid or general base group is much more common in enzymatic reactions than ordinary chemical reactions.

UNIQUE FEATURES OF ENZYME CATALYSTS

Thus far we have considered only the aspects of catalysis that are common to ordinary chemical reactions and biochemical reactions. In this section we will survey the unique aspects of enzyme catalysis.

1. Enzyme catalysis is mediated by a limited number of functional groups found in amino acid side chains, coenzymes, and metal cations. Notably, the amino acid side chains of histidine, serine, tyrosine, cysteine, lysine, glutamate, and aspartate are frequently directly involved in the catalytic process. Coenzymes as well as metal cations work in conjunction with enzymes to enrich the variety of functional groups.
2. Enzyme-catalyzed reactions are sensitive to extremes of pH and frequently have narrow optimum pH ranges for the reactions they catalyze. Extremes of pH cause denaturation and consequent loss of enzyme activity, often in an irreversible manner. The narrow optimum pH range results from the involvement of general acids or bases that usually function as catalysts in only one ionization state.
3. The _catalytic site_ and the _substrate binding_ site on the enzyme are localized and highly specific, both for the type of reaction being catalyzed and for the type of substrate. In some cases the sites are rigid in structure; in other cases they are flexible and change their structure upon binding to substrates or regulatory factors.
4. Frequently the catalytic site contains more than one catalytic group so that two reactions can occur simultaneously or in very rapid sequence. This phenomenon is known as _concerted catalysis_. Concerted catalysis has two advantages: first, it eliminates the need for high-energy intermediates that might slow a reaction, and second, it prevents the loss of reactive intermediates in nonproductive side reactions.
5. The substrate of an enzyme is invariably bound near the catalytic site and is usually oriented in a manner that is highly favorable for reaction with the catalytic site on the enzyme or with the reactive groups of other substrates in a multisubstrate enzyme.
6. Just as the enzyme may sometimes change its conformation on binding, so may the substrate change its structure on binding to the enzyme. In the case of substrate binding, a current belief is that many substrates are bound in such a way that a bond strain is created in the substrate that favors the formation of the transition-state complex.

These advantages of enzyme catalysis are best appreciated by considering the following examples.

THE TRYPSIN FAMILY OF ENZYMES

Trypsin, chymotrypsin, and *elastase* represent a group of closely related digestive enzymes whose role is to hydrolyze polypeptide chains. They are synthesized in the pancreas as inactive zymogens, or *preenzymes*, and then are secreted into the digestive tract and activated. The three enzymes work as a team. Each cleaves a protein chain at an internal peptide linkage next to a different type of amino acid side group (Figure 9–2). Trypsin cuts a chain just past the carbonyl group of a basic amino acid, either lysine or arginine. Chymotrypsin preferentially cuts a polypeptide chain next to an aromatic amino acid. Elastase is less discriminating in its choice of cleavage point, but it tends to cut preferentially adjacent to small, uncharged side chains.

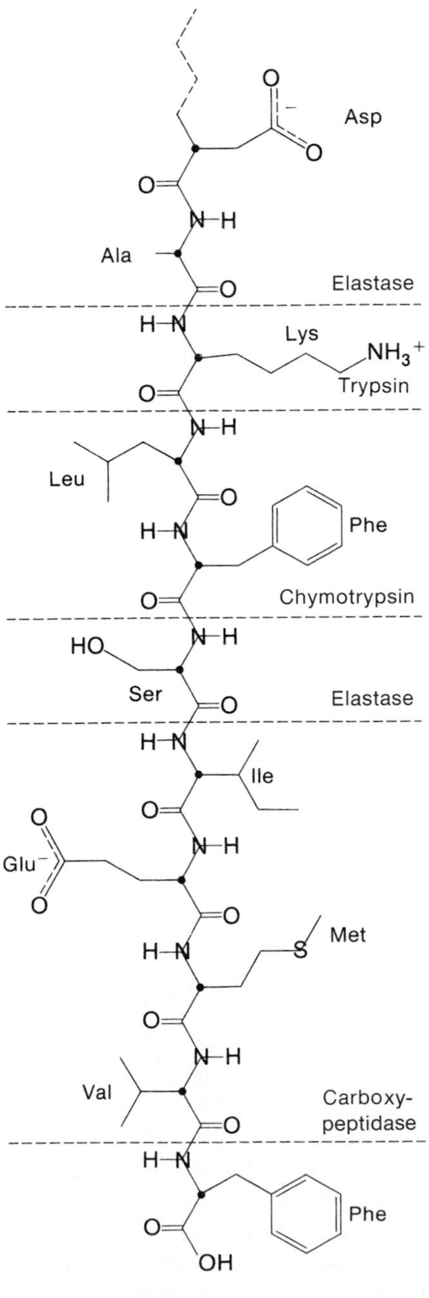

Figure 9–2
Chymotrypsin, trypsin, and elastase each cut a polypeptide chain adjacent to a different type of amino acid side chain. Chymotrypsin prefers aromatic rings; trypsin favors positively charged groups; and elastase cuts best next to small, nonpolar side chains. Carboxypeptidase cleaves one amino acid at a time from the carboxy-terminal end of the chain.

The trypsin enzymes were among the first to be comprehensively investigated because they are small (approximately 250 amino acids), easily obtainable in quantity, and relatively stable. Chymotrypsin has been the subject of more chemical studies than the others because it is less likely to digest itself in the purified state. Protein-digesting enzymes, being proteins themselves, have a tendency toward self-destruction. Chymotrypsin, however, does not tend to digest itself because the large aromatic groups that it favors are usually buried inside the molecule, whereas the groups that trypsin or elastase prefer are more likely to be exposed on the enzyme surface.

The fact that the trypsin family operates together during digestion suggests that these enzymes might be related in some way. Their amino acid sequences show striking similarities, as illustrated in Figure 9–3. They have identical amino acids at 62 of the 257 positions. Although the preenzymes are different, the native active enzymes in all three cases begin a polypeptide chain at the same place (residue 16). The four disulfide bridges that connect distant parts of the chain in elastase also are present in the other two enzymes, with chymotrypsin having one more disulfide bridge and trypsin having two. *All three enzymes have histidine at position 57, aspartic acid at 102, and serine at 195, which are the main groups involved in the catalytic mechanism.*

X-ray crystal structure analyses have revealed that these three proteolytic enzymes are folded the same way in three dimensions. The backbone skeletons for trypsin and chymotrypsin are shown in Figure 9–4. The only amino acid side chains illustrated are His 57, Asp 102, and Ser 195 at the active site and the various disulfide bridges. Because the chains are folded in the same way, positions corresponding to disulfide bridges in one enzyme also are close together even in an enzyme that does not have that particular bridge. The three-dimensional structure reveals another important feature that was not apparent from the amino acid sequences alone. The chain is folded back on itself in such a way that the three catalytic side chains (57, 102, and 195) are brought close together at a depression on the surface of the molecule. This is the active site of the enzyme, where the substrate that will be cut during digestion is bound.

Very little α helix is present in these three proteins. A short helix is found near residues 169–170 at the bottom of the molecule, and the chain ends with a helix at residues 230–245 at the left rear. A more important structural feature is a silklike twisted β sheet of nearly parallel extended chains. One of these can be seen at the upper left of trypsin in the chains that contain residues 60, 89, 105, and 53. Another twisted sheet, in the lower right of the molecule, is harder to see in this view. These two cores of twisted β sheets define the shape of the molecule, and the active site sits in a groove between them. The interiors of these enzymes are packed with hydrophobic residues, and charged side chains lie on the outside.

THE CHYMOTRYPSIN CATALYTIC MECHANISM

Trypsin, chymotrypsin, and elastase not only share a common structure, they also share a common catalytic mechanism. The mechanism has been determined from chemical and x-ray studies. The polypeptide substrate binds to the molecule with one portion hydrogen-bonded to residues 215–219 in an antiparallel manner, like adjacent chains in silk (see Chapter 2). This bonding helps to hold the substrate in place. At the bend in the substrate chain, the NH—CO bond that will be cut is brought close to His 57 and Ser 195. There is considerable distortion in this bond in the enzyme-substrate complex, favoring formation of the activated complex. The side group of the substrate just prior to this bond is inserted into a pocket in the

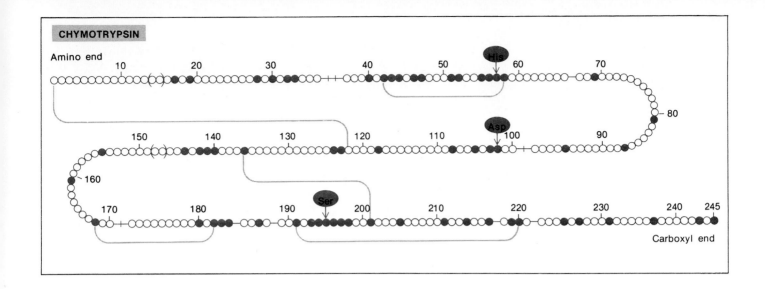

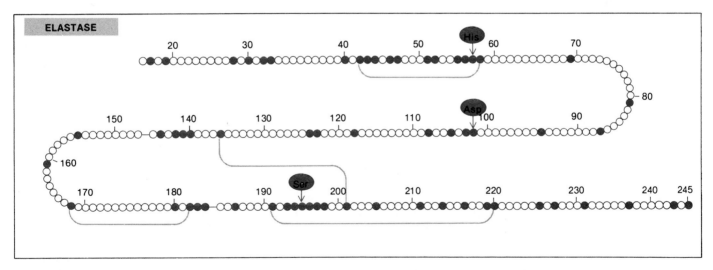

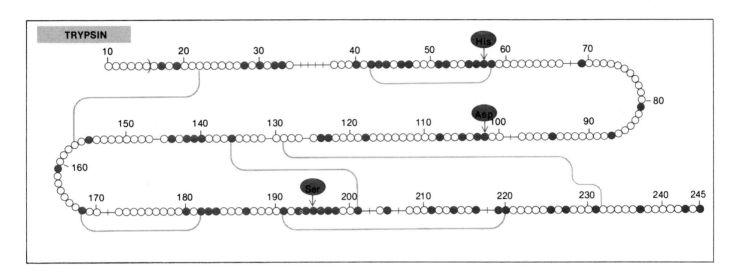

Figure 9–3

Schematic diagram of the amino acid sequence of chymotrypsin, trypsin, and elastase. Each circle represents one amino acid. Amino acid positions that are identical in all three proteins are in solid color. Long connections between nonadjacent amino acids represent disulfide bonds.

Locations of the three catalytically important amino acid side chains (histidine, aspartate, and serine) are marked. Location of sites that are cleaved to transform the preenzyme to the active enzyme are indicated by parentheses, either double ( ) or single).

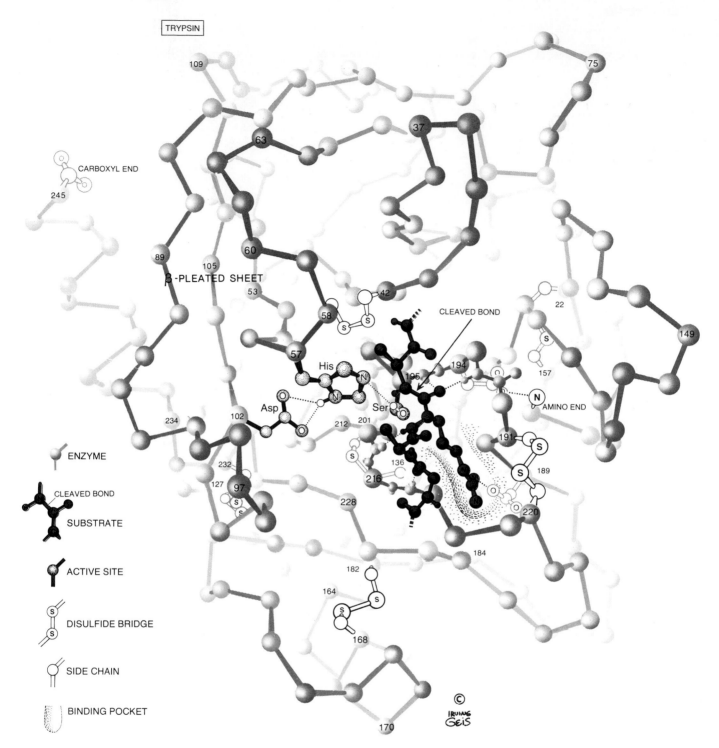

TRYPSIN

109

75

CARBOXYL END

245

63

37

89

105

β-PLEATED SHEET

60

53

42

58

57

His

CLEAVED BOND

22

194

Asp

102

234

212

201

195

Ser

157

149

191

ENZYME

232

136

216

189

CLEAVED BOND

127

97

228

220

SUBSTRATE

184

ACTIVE SITE

182

164

DISULFIDE BRIDGE

168

SIDE CHAIN

BINDING POCKET

170

IRVING GEIS

Figure 9–4

(Above) Main-chain skeleton of the trypsin molecule. The α-carbon atoms are shown by shaded spheres, with certain of them given residue numbers for identification. The —CO—NH— amide groups connecting the α-carbons are represented by straight lines. Disulfide bridges are shown in outline. A portion of the polypeptide chain substrate appears in dark color. The specificity pocket is sketched in shading, with a lysine side chain from the substrate molecule inserted. The catalytically important aspartate, histidine, and serine are poised for cleavage of the peptide bond marked by an arrow. Activation of trypsinogen involves cleaving six residues from the amino terminal end of the chain. *(Facing page)* Folding of the main chain in chymotrypsin. Notice the cut chain ends where residues 147–148 and 14–15 have been enzymatically removed in the process of activating chymotrypsin from its precursor, chymotrypsinogen.

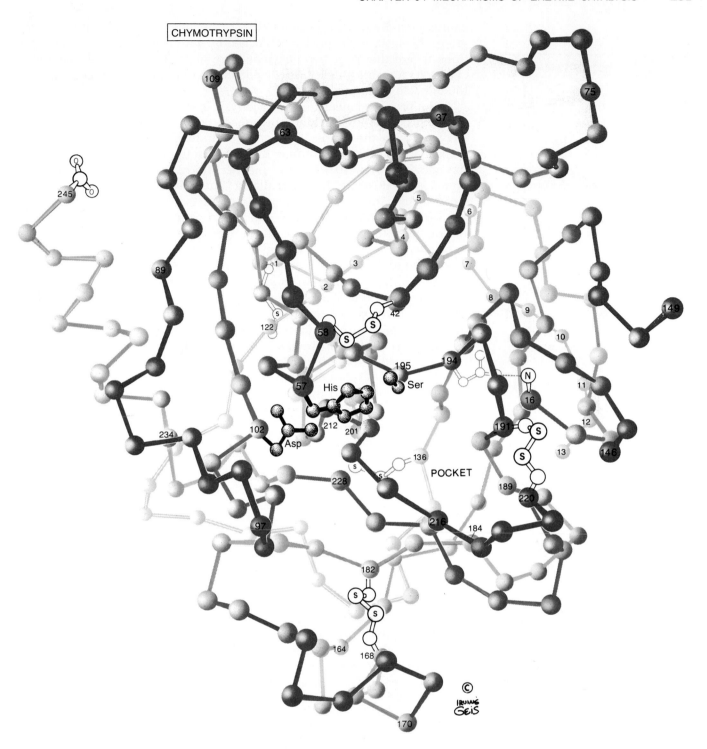

surface of the enzyme molecule. The pocket, bordered by residues 214–230, a disulfide bridge, and residues 191–195, can be seen in Figure 9–5. The rim of the pocket from residues 215 to 219 is the binding site for the substrate. *The pocket provides an explanation for the different specificities of the three enzymes for the bonds that they will cut.* In trypsin, the specificity pocket is deep and has a negative charge from asparatic acid 189 at the bottom (see Figure 9–5). The pocket is designed to accept a long, positively charged basic side chain: lysine or arginine. In chymotrypsin, the corresponding pocket is wider and completely lined with hydrophobic side

Trypsin

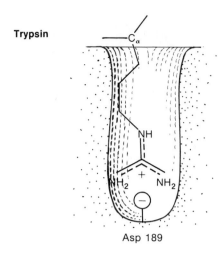

Asp 189

Chymotrypsin

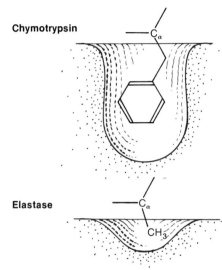

Elastase

Figure 9–5
Specificity pockets of trypsin, chymotrypsin, and elastase. The size of each pocket and the nature of the side chains lining it determine what kind of amino acid chain will be held best. This factor in turn determines at which position along a substrate chain cleavage will occur.

chains, thereby providing an efficient receptacle for a large, bulky aromatic group. In elastase, the pocket is blocked by valine and threonine at the positions where the other two enzymes have only glycine, which has no side chain. As a result, in elastase no side chain of appreciable size can bind to the enzyme surface.

The proposed mechanism of catalysis is shown in Figure 9–6. In steps 1 and 2, a polypeptide chain approaches and binds to the active site of the enzyme, with the proper type of side chain inserted into the specificity pocket. After substrate binding, the three catalytically important groups on the enzyme—Asp 102, His 57, and Ser 195—are connected by hydrogen bonds in what David Blow has called a "charge-relay system." In steps 3 and 4 of the mechanism, the histidine nitrogen acts first as a general base, pulling the serine proton toward itself, and then as a general acid, donating the proton to the lone electron pair on the nitrogen atom of the polypeptide bond to be cleaved. The aspartic acid group is pictured as helping the histidine to attract the proton by making an electrostatic linkage or hydrogen bond with the other proton on the imidazole ring. The importance of aspartate is shown by the fact that almost all serine proteases have an aspartate in this position. The hydrogen bond between an aspartate carboxylate group and the imidazole also helps aim the imidazole at the serine OH group. With only flexible single bonds holding the imidazole ring to the polypeptide backbone, a histidine needs such extra hydrogen bonding to fix it in place.

As the serine H—O bond is broken in step 3, a bond is formed between the serine oxygen and the carbonyl carbon on the polypeptide chain. The carbon becomes tetrahedrally bonded, and the effect of the negative charge on Asp 102 has in a sense been relayed to the carbonyl oxygen atom of the substrate. This negative oxygen in the tetrahedral intermediate is stabilized by hydrogen bonds to N—H groups on the enzyme backbone. The transition state of step 3 is short-lived and cannot be isolated. The enzyme passes quickly through the tetrahedral intermediate to step 4. As the polypeptide N accepts the proton from histidine, the N—C bond is broken. One half of the polypeptide chain falls away as a free amine, R—NH_2. The other half remains bound covalently to the enzyme as an acylated intermediate. This acyl-enzyme complex is stable enough to be isolated and studied in special cases where further reaction is blocked.

The steps to deacylate the enzyme and restore it to its original state (steps 5 to 8) are like the first four steps run in reverse, with H_2O playing the role of the missing half-chain. A water molecule attacks the carbonyl carbon of the acyl group in step 5 and donates one proton to histidine 57 to form another tetrahedral intermediate (step 6). This intermediate breaks down when the proton is passed on from histidine to serine (step 7). The second half of the polypeptide chain falls away (step 8), aided by charge repulsion between the newly generated carboxylate anion and the negative charge on aspartic acid 102. The enzyme is restored to its original state.

Kinetic Studies on Peptide and Ester Substrates Reveal Different Rate-limiting Steps

Chemical experiments have been done to find out how choosy chymotrypsin is among its substrates. It has been known for a long time that chymotrypsin hydrolyzes suitably substituted esters at comparable rates to peptide linkages. Many experiments using model ester substrates have been done to explore the question of substrate specificity. Small organic esters (Figure 9–7 and Table 9–1) have the general structure $R_1CHR_2COOR_3$. Results with these model substrates are reported in terms of the turnover number (or catalytic

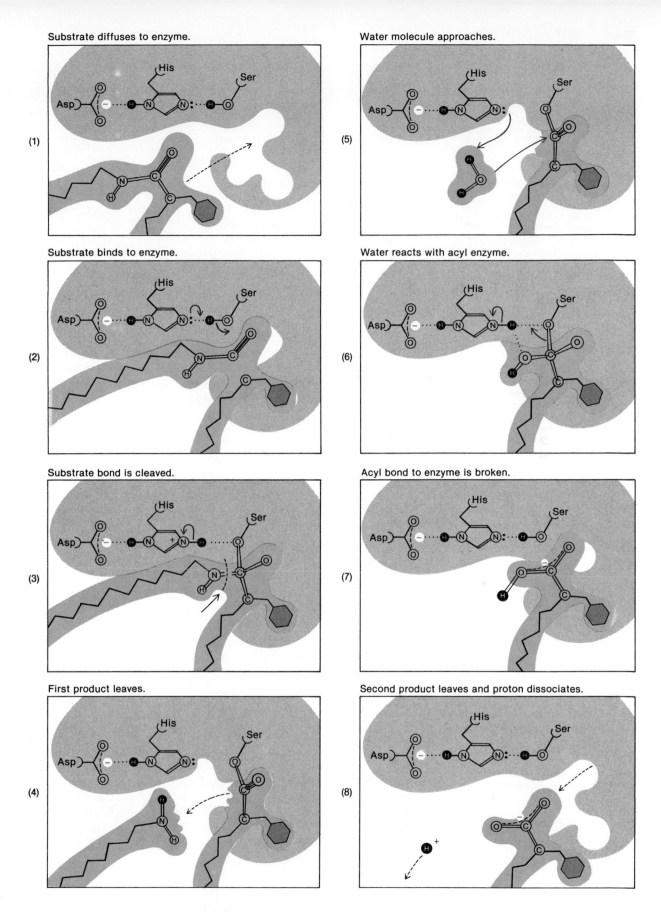

Figure 9–6
Proposed steps in chymotrypsin-catalyzed hydrolysis, showing the system of charge relay operating between serine, histidine, and aspartate.

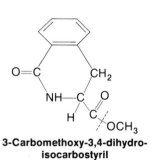

Acetylphenylalanine methyl ester

Formylphenylalanine methyl ester

Benzoylalanine methyl ester

3-Carbomethoxy-3,4-dihydro-isocarbostyril

Figure 9–7
Structures of some model organic ester compounds used to study the effects of various substituents and configurational factors on the catalytic rate. Dashed lines indicate the point of enzymatic cleavage.

TABLE 9–1
Rates of Hydrolyses of the D and L Forms of Various Esters by Chymotrypsin

Substrate	Turnover Number, k_{cat} (sec^{-1})
Acetylphenylalanine methyl ester	
L	63
D	Very low
Formylphenylalanine methyl ester	
L	Very high
D	0.0034
Benzoylalanine methyl ester	
L	0.26
D	0.011
3-Carbomethoxy-3,4-dihydroisocarbostyril	
L	0.12
D	22.7

rate constant) k_{cat}, where k_{cat} is $V_{max}/[E_t]$ (see Chapter 8). Two main points are established by the results presented in Table 9–1. First, k_{cat} is very sensitive to the side-group composition of the ester, and second, enantiomers react at very different rates, testifying to the stereospecificity of the reaction.

Steady-state Analysis Gives the K_m for Both Reactions

Although hydrolysis studies of model ester compounds have been useful in arriving at the foregoing conclusions, their use also initially led to some confusion when steady-state analysis was attempted. To understand the source of this confusion we must recall that the mechanism of hydrolysis proposed earlier can be divided conceptually into three steps: (1) formation of the enzyme-substrate complex, (2) acylation, and (3) deacylation. A careful kinetic analysis has shown that for esters, deacylation is the rate-limiting step, whereas for peptides, acylation is the rate-limiting step. Consider the following set of reactions for hydrolysis of acetylphenylalanine methyl ester by chymotrypsin.

$$E + S \underset{k_{-1}}{\overset{k_1}{\rightleftharpoons}} ES, \qquad K_S = \frac{k_{-1}}{k_1} \tag{1}$$

$$ES \underset{k_2}{\overset{k_2}{\longrightarrow}} \overset{P_1}{\nearrow} EP_2 \overset{k_3}{\longrightarrow} E + P_2 \tag{2}$$

where K_S is the dissociation constant of the enzyme-substrate complex, S is acetylphenylalanine methyl ester, P_1 is methanol, P_2 is acetylphenylalanine, and ES is the reversibly formed enzyme-substrate complex. From these equations we can see that there are two enzyme-bound compounds that must be considered, the enzyme-substrate complex ES and the covalent bound intermediate EP_2. If the steady-state assumption is applied to $[EP_2]$, it may be shown that

$$v = [E_t][S]\left\{\frac{k_2k_3/(k_2 + k_3)}{K_Sk_3/(k_2 + k_3) + [S]}\right\} \tag{3}$$

This is the Briggs-Haldane equation, in which

$$k_{cat} = \frac{V_{max}}{[E_t]} = \frac{k_2 k_3}{k_2 + k_3} \qquad (4)$$

$$K_m = K_S\left(\frac{k_3}{k_2 + k_3}\right) \qquad (5)$$

In the hydrolysis of amides or peptides, the acylation rate constant k_2 is much smaller than the deacylation rate constant k_3; therefore

$$k_{cat} = k_2 \qquad (6)$$

$$K_m = K_S \qquad (7)$$

By contrast, in the hydrolysis of esters, the acylation rate constant k_2 is much greater than the deacylation rate constant k_3. When this is the case, the preceding equations reduce to the more complex expressions

$$k_{cat} = k_3 \qquad (8)$$

$$K_m = K_S(k_3/k_2) \qquad (9)$$

Here K_m is the Michaelis constant, determined by the usual steady-state kinetic analysis and the Lineweaver-Burk plot described in Chapter 8. It is clear from the preceding considerations that the Michaelis constant K_m for ester hydrolysis is going to be much smaller than the dissociation constant K_S. Physically, the bound substrate is partly bound in the reversible unmodified form and partly as an intermediate by covalent linkage. By contrast, for amides, most of the bound substrate is bound in the reversible form so that K_m is approximately equal to K_S. For peptides, no further kinetic analysis is required to evaluate the kinetic parameters. For ester substrates, further analysis is necessary.

Pre-steady-state Measurements Provide Additional Information on the Kinetic Parameters

Analysis of reactions like the chymotrypsin-catalyzed ester hydrolysis, where additional intermediates become rate-limiting, requires more information than can be supplied by steady-state analysis alone. Frequently the required additional information can be obtained by a kinetic analysis before the steady state is reached, i.e., by pre-steady-state measurements.

In the hydrolysis of the ester p-nitrophenyl acetate, the rate of liberation of p-nitrophenol is biphasic, with a rapid initial burst of p-nitrophenol followed by a slower steady rate (Figure 9–8). There is a rapid liberation of p-nitrophenol in a concentration roughly equivalent to the concentration of enzyme, and a slow, steady-state liberation of acetate resulting from the hydrolysis of the acyl-enzyme (Figure 9–8). This result shows that the acylation rate for ester hydrolysis is quite rapid, but that the enzyme very quickly becomes blocked for further reaction because of the slow rate of deacylation. In the steady-state situation that develops after about 2 min (Figure 9–8), the reaction cannot proceed further until deacylation occurs. In the cyclic process depicted in Figure 9–9 the deacylation reaction very quickly becomes the rate-limiting step. This is the case for most esters that can be hydrolyzed by chymotrypsin. Very little of the enzyme is available for reversible binding to substrate.

The key to analyzing this situation according to Equations (1) to (3) is to show that the phenol is released in a fast step characterized by the rate constant k_2 and that the acid is released in a slower step. Information on the fast step was obtained by measurements made before the system reached a

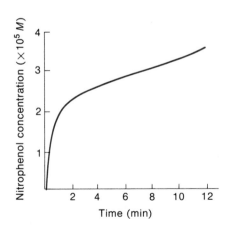

Figure 9–8
p-Nitrophenol formation as a function of time in the chymotrypsin-catalyzed hydrolysis of p-nitrophenyl acetate. A rapid initial rate of hydrolysis is followed by a much slower steady-state rate of hydrolysis. This result indicates that the formation of p-nitrophenol is not the rate-limiting step in the reaction. (Data of Hartley and Kilby.)

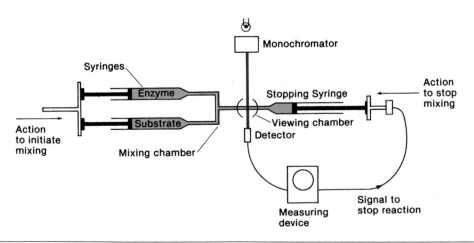

Figure 9-9
Steps in the chymotrypsin-catalyzed hydrolysis of *p*-nitrophenyl acetate to form *p*-nitrophenol and acetic acid. A rapid liberation of *p*-nitrophenol is followed by a slow rate of deacylation of the enzyme. Once the enzyme has been deacylated it is ready to digest more substrate.

steady state. This required the use of very rapid measuring techniques, using a so-called stopped-flow apparatus (see Box 9–A). Advantage was taken of the fact that absorbancy changes in the enzyme near 290 nm occur on binding of substrate. It is possible to witness four phases in the ester hydrolysis reaction. First, there is an initial rapid increase in absorbance that is complete in less than 2 ms. This phase of the reaction is considered to result from the reversible formation of enzyme-substrate complex. Second, a slower increase in absorbance leads to the formation of a steady-state intermediate (after about 20 sec), which is considered to be EP_2 in equation (2). Third is a period of time during which the absorbancy does not change and product is

---

BOX 9–A

Description of Stopped-flow Apparatus

The stopped-flow apparatus is a device for observing a reaction very soon after mixing of substrate with enzyme. The apparatus has three syringes. One is used to drive enzyme into the mixing chamber and one is used to drive substrate into the mixing chamber. The mixing chamber is directly connected to the viewing chamber, which is part of a spectrophotometer for measuring the course of the reaction. The third syringe is used to stop the flow from the two driving syringes. The reaction is initiated by a forward pressure on the two driving syringes. The flow initiated by the two driving syringes is halted by a forward pressure on the stopping syringe. The course of the reaction is observed spectrophotometrically in the viewing chamber. In this type of apparatus it is possible to make kinetic measurements as soon as 0.5 ms after mixing of enzyme and substrate.

TABLE 9–2
Rate and Equilibrium Constants Pertaining to Chymotrypsin-Catalyzed Hydrolysis of Ester Substrates at pH 5.0 and 25°C

Substrate	k_2 (sec^{-1})	k_3 (sec^{-1})	K_S (mM)	K_m (mM)
N-Acetyl-L-Trp-ethylester[a]	35	0.84	2.1	0.08
N-Acetyl-L-Phe-ethylester[b]	13	2.2	7.3	1.3

Note: k_2 and K_S were measured directly by rapid kinetic methods; k_{cat} and K_m were determined under conditions of steady state. For the reactions shown it can be seen that $k_3 \ll k_2$ and therefore $k_{cat} \approx k_3$.

[a]Data from K. G. Brandt, A. Himoe, and G. P. Hess.

[b]Data from A. Himoe, K. G. Brandt, R. J. DeSa, and G. P. Hess.

produced at a constant rate. The time interval corresponds to the steady-state phase of the reaction. Fourth, a slow decrease in absorbancy occurs as the substrate supply becomes exhausted. This phase may be identified with the decay of the steady-state intermediate to give free enzyme and product.

From this information, all the kinetic parameters for Equations (1) to (3) have been evaluated. Data for the two different esters shown in Table 9–2 indicate that the rate constant for the formation of the steady-state intermediate k_2 is larger than the rate constant k_3 for the decomposition of this intermediate. The values of K_S, the dissociation constant of the enzyme-substrate complex, obtained by rapid kinetic measurements, are much larger than the K_m value obtained from steady-state measurements.

The chymotrypsin-catalyzed hydrolysis of amides also follows the pathway described by Equations (1) to (3), but in this case the rate of formation of the acyl-enzyme is rate-limiting. Thus the acylation rate constant k_2 is smaller than the deacylation rate constant k_3 for the pathway shown by Equations (1) to (3).

In addition, the steady-state kinetic parameter K_m is approximately equal to K_S, the enzyme-substrate dissociation constant. In this case the value of K_m determined from steady-state measurements is similar to the value of K_S determined from equilibrium measurements.

Chymotrypsin Activity Is Very Sensitive to pH

Chymotrypsin-catalyzed reactions have bell-shaped pH profiles, such as the one shown in Figure 9–10 for the hydrolysis of the neutral substrate acetyl-L-tryptophanamide. Bell-shaped pH-rate profiles with neutral substrates often result when two ionizing groups of the enzyme are important in the catalytic reaction. The midpoint of the left-hand side of the pH-rate profile indicates that the rate of the catalytic reaction increases as the deprotonated form of an ionizing group with an apparent pK (pK_{app}) of about 7 is formed. The midpoint of the right side of the curve indicates that the rate of the reaction decreases as an amino acid residue with a pK_{app} of about 8.5 is deprotonated. The pK around 7 could be identified with an imidazole group of histidine that would be unprotonated in the catalytically active form of the enzyme.

The only amino acid side chain with a pK normally around 8.5 is cysteine. However, the α amino-terminal group Ile 16 has a pK with this value, and of course, in the neutral range this group would be charged. When the preenzyme chymotrypsinogen is activated by trypsin hydrolysis (see Figures 9–3 and 9–4), the Arg 15–Ile 16 linkage is cleaved to produce this amino-

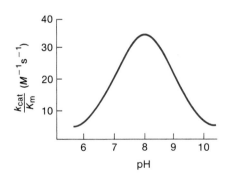

Figure 9–10
The variation of the specificity constant with pH. As indicated in Chapter 8, the specificity constant gives us an estimate of the catalytic efficiency at substrate concentrations below saturating levels. This curve was drawn from steady-state kinetic data on the α-chymotrypsin-catalyzed hydrolysis of N-acetyl-L-tryptophan amide at 25°C. Continuous measurements of released ammonia were made at each pH level over a range of substrate concentrations. This type of bell-shaped curve indicates that at least two ionizable groups are important for enzyme activity.

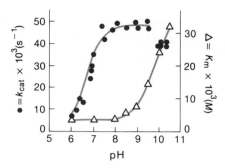

Figure 9–11
The turnover number and the Michaelis constant as a function of pH for the hydrolysis N-acetyl-L-tryptophanamide at 25°C. This analysis explains some aspects of the curve in Figure 9–10. The increase in enzyme rate between pH 6 and pH 8 is caused by an increase in the turnover number, which measures the catalytic efficiency of the active site at saturating substrate concentrations. This effect is most likely to be due to the loss of a proton at the catalytic site. The decrease in the enzyme rate between pH 8 and pH 10 is caused by an increase in the Michaelis constant. This change is most likely to be due to a decreased affinity for substrate.

terminal linkage. Crystal structure analysis indicates that Asp 194 and His 40 form an ion pair in chymotrypsinogen. In this ion pair Asp 194 occupies part of the substrate binding site. Hydrolysis of the Arg 15–Ile 16 linkage creates a positively charged terminal amino group on Ile 16. This group displaces His 40 in the salt linkage with Asp 194. In so doing, the movement in the polypeptide chain establishes a substrate binding site. The importance of the Asp 194–Ile 16 linkage is underscored by the fact that chemical modification of either of these groups to prevent ionization results in a severe lowering of enzyme activity. Hence the pH inactivation effect on the basic side appears to be due to the removal of the positive charge from the amino-terminal isoleucine, thereby interfering with the substrate binding site. Further measurements indicate that when this happens, the substrate binding site returns to the conformation it had in the preenzyme.

This interpretation is supported by a kinetic analysis of the values of k_{cat} and K_m for the hydrolysis of N-acetyl-L-tryptophanamide as a function of pH. Between pH 8 and 10.5, k_{cat} is almost constant, whereas K_m increases severalfold (Figure 9–11). Thus the substantial decrease in enzyme reaction rate in this pH range appears to be purely a function of decreased affinity for substrate and not related to a decreased effectiveness of the catalytic site.

It also can be shown that the decreased binding of substrate directly parallels a conformational change in the protein. This was done by measuring the circular dichroism of the protein and the substrate affinity over the same pH range. A sizable change in circular dichroism indicates an appreciable conformational change in the protein (see Chapter 3). The two changes do indeed parallel one another.

CARBOXYPEPTIDASE A: A ZINC METALLOENZYME

In the preceding discussion we described a group of closely related endopeptidases secreted by the pancreas—trypsin, chymotrypsin, and elastase. These digestive enzymes are accompanied in the secretions by certain exopeptidases, e.g., carboxypeptidases A and B. Carboxypeptidase A (CPA) and carboxypeptidase B (CPB) finish the job started by the endopeptidases; they hydrolyze the oligopeptides one at a time from the C-terminal end of the polypeptide chain. CPA has a preference for aromatic residues and CPB functions best with basic residues, a comparison reminiscent of the chymotrypsin-trypsin complementarity.

All these digestive enzymes are synthesized as inactive precursors, and are converted to the active enzymes just before use. In the case of CPA, the inactive precursor is a complex assembly of three subunits that is split by trypsin through a complex series of reactions in which only one of the subunits eventually becomes trimmed down to a single uninterrupted chain with 307 residues. The amino acid sequence of CPA is known, and its three-dimensional structure, both with and without bound substrate, has been determined. In addition, a great deal of work has gone into studying the mechanism of hydrolysis, making CPA one of the better-understood enzymes. The enzyme differs in two major respects from the other enzymes we have been discussing. *CPA is a zinc metalloenzyme, and it undergoes a large conformational change upon binding of the substrate* that serves the purpose of bringing together the components of the active site. We will see that the zinc plays a key role in the catalytic process.

Substrate Specificity

The peptide bond that is to be hydrolyzed must be adjacent to a terminal free carboxy group, as shown in Figure 9–12. As was stated, CPA preferentially hydrolyzes peptides when the terminal residue is hydrophobic, either aro-

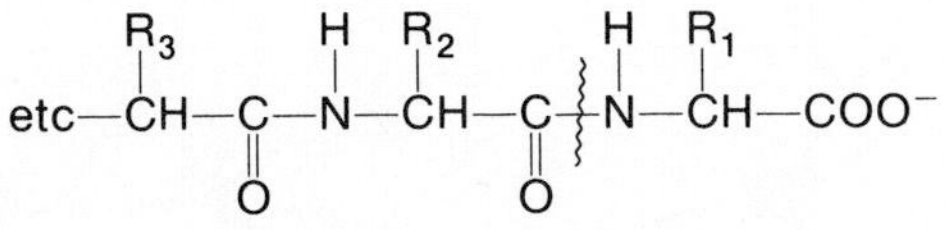

$$\text{etc—CH—C—N—CH—C}\{\text{N—CH—COO}^-$$

Figure 9–12
Substrate for carboxypeptidase. The peptide bond to be hydrolyzed is indicated by the wavy line. This bond must be adjacent to a terminal free carboxy group. The R groups represent amino acid side chains. Carboxypeptidase A has a preference for aromatic residues and carboxypeptidase B has a preference for basic residues.

matic or branched aliphatic groups make favorable substituents. The binding is also stereospecific, as that grouping must be in the L configuration. Integrity of the second peptide bond is also important for rapid hydrolysis. Thus dipeptides having a free amino group are hydrolyzed slowly, but if this group is blocked by N-acetylation, the hydrolysis is rapid. CPA also possesses esterase activity. As in the case of peptide hydrolysis, the L configuration of the C-terminal residue and an aromatic C-terminal side chain are important for rapid ester hydrolysis. Interesting differences between the mechanisms of peptide hydrolysis and ester hydrolysis have been instrumental to an understanding of the reaction, and these differences are discussed below.

Structure of the Enzyme-substrate Complex

The three-dimensional structure of the enzyme has been determined in the free state and when complexed with the dipeptide inhibitor glycyl-L-tyrosine. Although this dipeptide is a poor substrate, we can at least hope that it complexes with the enzyme in a way that closely resembles complexing with a good substrate. Figure 9–13 shows a view of the enzyme structure with inhibitor carbobenzoxy-Ala-Ala-Tyr in place. Positions of Arg 135, Tyr 248, and Glu 270 before as well as after binding of substrate are shown. The location of the left half of the inhibitor is predicted from model-building studies. The aromatic C-terminal side group of the substrate fits into a pocket in the interior of the molecule whose rim is composed of the chain of residues 245 to 251.

A good deal of movement in the protein occurs on binding substrate. Arg 145 moves 2 Å closer to interact with the substrate's terminal carboxyl group. Tyr 248 swings down to place its hydroxyl near the nitrogen of the bond to be split, and the neighboring carboxyl oxygen becomes complexed with the Zn^{2+}. The Zn^{2+} is hexacoordinated with bonds to both oxygens of the Glu 72 side chain, as well as with the amino nitrogen and the carboxyl of the dipeptide, as shown in Figure 9–14. In the absence of substrate, the Zn^{2+} is pentacoordinated, having two linkages to His 69 and His 196 as shown, two linkages to Glu 72, and one linkage to a water molecule that is expelled when substrate is present. Thus the Zn^{2+} changes its coordination number from 5 in the native state to 6 in the complex with substrate.

The overall effect of the binding of substrate to CPA is the conversion of the enzyme cavity from a water-filled to a hydrophobic region. At least four water molecules must be expelled when a substrate C-terminal side chain such as Tyr is inserted into the pocket, and one water molecule is displaced from the Zn^{2+} when the carboxyl group of the substrate is bound. In addition, the charge of Arg 145 is neutralized by its electrostatic interaction with the terminal carboxylate ion of the substrate.

Finally, Tyr 248, in making its conformational change, closes the enzyme cavity. It seems likely that the displacement of water upon binding the substrate and the resultant conversion of the active center of the enzyme to a hydrophobic area provide a major part of the driving force for substrate bind-

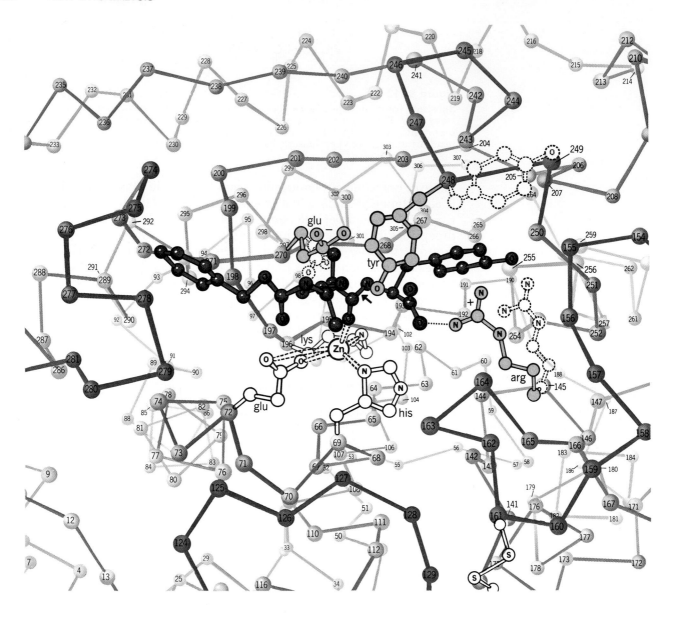

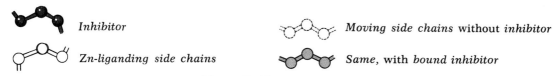

Inhibitor

Zn-liganding side chains

Moving side chains without *inhibitor*

Same, with *bound inhibitor*

Figure 9–13
Carboxypeptidase A complexed with glycyl-L-tyrosine bound to the active site. Although this dipeptide is a poor substrate, researchers hope that it complexes with the enzyme in the same way as a good substrate. Arg 145, Tyr 248, and Gly 270 are shown before substrate binding (dotted line) and after (solid line). Note the pentacoordination of the zinc cation (Zn) to His 69, Lys 196, and the carboxylate oxygens of the glutamate. (Reprinted with permission from R. E. Dickerson and I. Geis, *The Structure and Action of Proteins*, Benjamin/Cummings, Menlo Park, Calif., 1969.)

ing. Such a change in the enzyme structure on binding substrate has been called <u>*induced fit*</u> by Koshland. The enzyme in this case is like a Venus fly trap, closing up around the substrate as the result of substrate binding. Other general advantages of induced fit are discussed in Box 9–B.

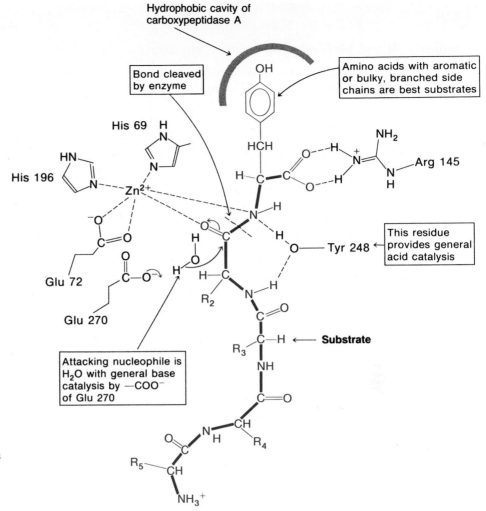

Figure 9–14
Expanded schematic view of the active site showing a hexacoordinated Zn^{2+}. In the absence of substrate, the Zn^{2+} is pentacoordinated, having two linkages to His 69 and His 196, two linkages to Glu 72, and one linkage to a water molecule that is expelled when substrate is present. In the presence of substrate, Zn^{2+} changes its coordination number from 5 to 6. The electron shifts involved in the initial cleavage reaction are indicated by the curved (colored) arrows.

The presence of a dead-end pocket, in addition to a groove, provides an explanation for the observation that CPA is an exopeptidase and not an endopeptidase. Internal peptide units in a polypeptide chain could never be effectively fitted into or around this binding site with the proper juxtaposition of the peptide bond to the active center. They also lack the negatively charged carboxylate for making the electrostatic linkage to Arg 145. Model-building studies show that it is difficult to fit the R group from a C terminal D residue into the pocket, which explains the stereospecific preference of the enzyme for substrates with the L configuration.

Mechanism of the Reaction

Schematic views of the bound substrate, derived from structural studies, and of other groups important at the active site are shown in Figure 9–14. The structure of the complex immediately suggests that the Zn^{2+} plays a key role as an electrophilic catalyst, and this view has been supported by chemical studies. A mechanism proposed by R. Breslow and D. L. Wernick is shown in Figure 9–15. The peptide substrate binds so as to displace water from the Zn^{2+}. This makes the carbon of the carbonyl more attractive for nucleophilic attack, as it results in a greater partial positive charge on this carbon. Then a water molecule is delivered by the glutamate carboxylate (Glu 270). After delivery of the hydroxyl to the carbonyl and pickup of the first proton by Glu

BOX 9–B

Induced Fit and Multisubstrate Enzyme Specificity

Koshland's idea of the induced fit of an enzyme to a substrate helps explain some of the tight ES complexes seen by x-ray, and is useful in a model for the allosteric effect (see Chapter 10). Another aspect of enzyme behavior that is explained by induced fit is the low incidence of enzymatic mistakes.

Consider a kinase enzyme that uses ATP to put a phosphate group on a substrate hydroxyl. If the substrate is not present, its place in the enzyme should presumably be taken by H_2O. But the phosphorylation of H_2O (hydrolysis of ATP) would be an unacceptable waste. And in reality, kinases do not hydrolyze ATP; instead, they wait for the other substrate to be present. This behavior makes sense if the enzyme, with only ATP bound, does not have the correct shape to catalyze any reaction. If binding of the second substrate is needed to fold the enzyme into its catalytically active form, by bringing the catalytic groups together into the active site, then we can understand the substrate specificity of the kinases.

As another example, the enzyme serine hydroxymethylase (see Chapter 24) catalyzes the loss of formaldehyde from serine, forming glycine. The formaldehyde is captured by a bound coenzyme, tetrahydrofolic acid (THF). If the reaction were to proceed even in the absence of THF, the formaldehyde would be lost; it is also rather toxic. To prevent this, the enzyme doesn't actually catalyze the first reaction, formaldehyde removal, until the second substrate, THF, is bound. Its binding folds the enzyme so the first step can occur, even though the THF isn't yet involved in the chemistry. Good evidence has been obtained for the induced fit mechanism in this case: the enzyme won't convert serine to formaldehyde and glycine in the absence of THF, but it can be fooled into doing so by some compounds that look like THF, so they bind and induce the formation of glycine.

Many other examples of such effects are known. They indicate how important conformational change in enzymes is to catalysis itself.

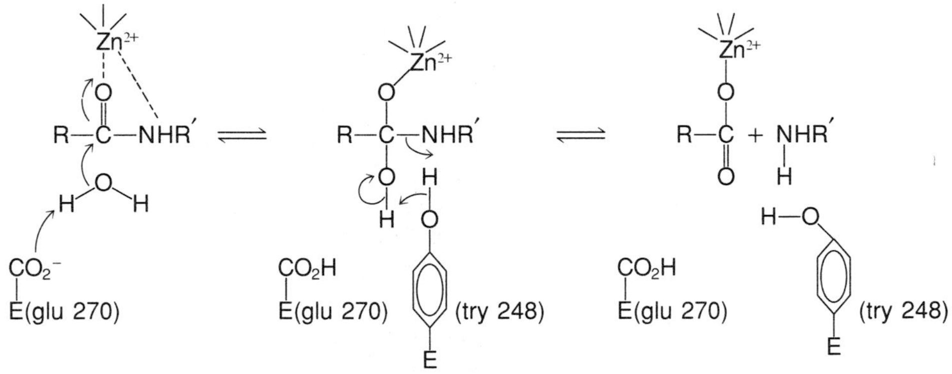

Figure 9–15
Proposed mechanism for carboxypeptidase-A-catalyzed peptide hydrolysis. The mechanism involves glutamate 270 and tyrosine 248 although there is strong evidence that tyrosine 248 is not indispensable. The zinc, by complexing with the peptide carbonyl, makes the carbon of the carbonyl a more attractive target for nucleophilic attack by water.

270, there must be an additional proton transfer to permit cleavage, so both protons of H_2O eventually are released. In the mechanism shown in Figure 9–15 this second proton transfer is mediated through the hydroxyl group of Tyr 248, which functions as a general acid catalyst.

The importance of tyrosine for peptidase activity is controversial. On the one hand it is supported by the finding that acetylation of the tyrosine destroys the peptidase activity. On the other hand it has been shown that Tyr 248 can be replaced by another amino acid without loss of activity. If Tyr 248 is not required, then the proton needed on the leaving amino group* would have to come from elsewhere. A good possibility is the CO_2H group formed

---

*In a displacement reaction where one group displaces another group, the displaced group is called the leaving group.

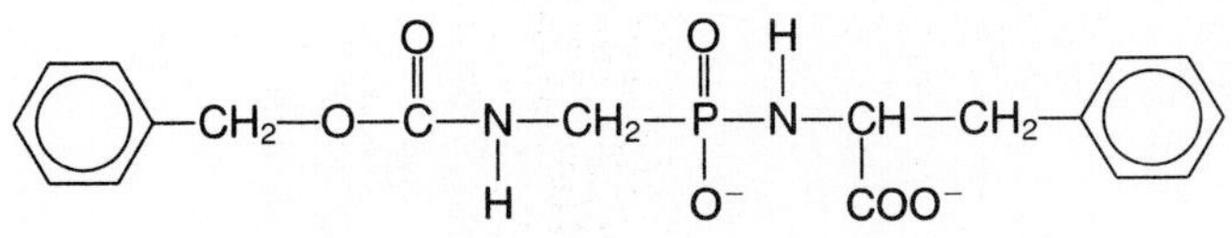

Figure 9–16
Structure of a strong competitive inhibitor of the carboxypeptidase A enzyme. Strong competitive inhibitors frequently resemble the structure of the transition state of the substrate. The phosphorus atom, with its attached oxygens and nitrogen, resembles the tetrahedral carbon atoms in the two intermediates of Figure 9–15, acidic histidine 12 to form the pyrimidine ribose 3′-phosphate product.

in the first step of the reaction. By this mechanism the carboxylate of Glu 270 would be acting first as a general base, then as a general acid, just as His 57 did in chymotrypsin.

Additional evidence to support the general mechanism shown in Figure 9–15 comes from the study of _transition-state analogs_. These are compounds that are believed to look like the substrate in its transition state. The strategy in the use of transition-state analogs is to construct a compound that closely resembles the hypothesized transition state and then to see how these compounds react with the enzyme. To be effective the transition-state analog must not be susceptible to reaction by the enzyme. Good transition-state analogs should act as competitive inhibitors of the normal substrate. If, as is often the case, the substrate binding site favors the transition state, then the analog should make a very strong competitive inhibitor. Thus the use of transition-state analogs can be used both as support for the mechanism of the reaction and as an indication of the nature of the substrate binding site. Another extremely valuable use of transition analogs is for the production of cocrystals with the enzyme that can be structurally analyzed by x-ray diffraction. We cannot get a photograph of the transition state, but we can build a molecule that looks like the enzyme–transition-state complex.

Paul Bartlett constructed the phosphorus molecule shown in Figure 9–16. This compound proved to be a very powerful competitive inhibitor of carboxypeptidase A, a result suggesting that it may bind like a transition-state analog. The phosphorus atom, with its attached oxygens and nitrogen, resembles the tetrahedral carbon atoms in the two intermediates of Figure 9–15, and the transition states of all three steps. Thus this result of Bartlett's supports the idea that the transition state involves a tetrahedral intermediate.

Carboxypeptidase is an example of a metalloprotein in which the metal ion plays a key role in the catalysis. The zinc ion in this enzyme acts as a generalized acid, stabilizing the developing O^- as water attacks the carbonyl. This is one role that zinc can play in enzymes. In carbonic anhydrase zinc plays a somewhat different role. The metal binds H_2O and makes it acidic enough to lose a proton and form a Zn—OH group. Then the zinc-bound hydroxide is delivered as a nucleophile to the substrate. The ability of zinc to act either as an electrophile or as the source of a nucleophilic group to participate in catalysis has resulted in its use in many different kinds of enzymes.

PANCREATIC RNASE A: AN EXAMPLE OF CONCERTED ACID-BASE CATALYSIS

Ribonuclease is a hydrolytic enzyme cleaving RNA at the 3′ P—O bond on the far side of pyrimidine nucleotides (Figure 9–17). The reaction is believed to occur in two discrete chemical steps with a cyclic 2′,3′-phosphate intermediate.

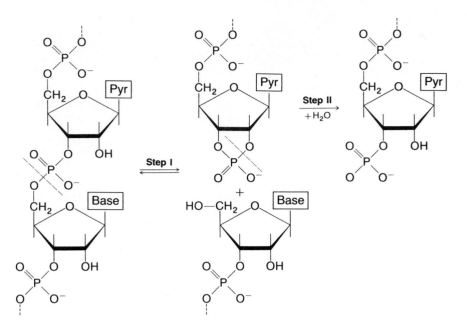

Figure 9–17
RNA chain indicating points of cleavage by pancreatic RNAse. Pyr refers to pyrimidine and Base refers to either purine or pyrimidine. The reaction is believed to be divided into two steps as shown.

Sanford Moore and William Stein, who had earlier developed ion-exchange methods for the analysis of amino acids and peptides, determined the amino acid sequence of the 124-residue enzyme from the bovine pancreas (Figure 9–18). This was only the second protein and the first enzyme to be sequenced. For their efforts Moore and Stein, together with Christian Anfinsen, were awarded the Nobel Prize in chemistry in 1972.

Chemical Studies on the Active Site

Although RNAse A was available in large quantities and could be crystallized, the early chemical investigations of this enzyme preceded the technological developments in x-ray crystallography that were necessary for a three-dimensional structural analysis. A great deal was deduced about the catalytic mechanism from chemical and kinetic studies, and this was confirmed by later developments that led to a description of the three-dimensional structure (Box 2–A, Chapter 2).

A discovery useful for correlating structure with function was the specific cleavage of ribonuclease between residues 20 and 21 by the bacterial proteolytic enzyme *subtilisin*. The resultant two peptides could be separated chromatographically in denaturing solvents. Separated peptides were catalytically inactive even after removal of the denaturant. On reincubation of the two peptide fragments in the absence of denaturant, a specific but noncovalent recombination took place to yield a fully active ribonuclease called RNAse S. This research showed that there are strong noncovalent binding interactions in the protein that can hold the chains together even when one peptide link is cut. It also presented a method for specific chemical modifications of amino acid residues on each of the two polypeptide chains independently. Thus it has become possible to study the effect of such individual modifications on the activity of the reconstituted active enzyme. From systematic chemical modifications of the S peptide, researchers have estimated the contribution of each residue to the peptide–peptide interaction and to the activity. It appears that His 12 is required for enzyme activity and that several other residues in the S peptide are important for the binding that allows reconstitution to form the RNAse S protein.

Chemical probes that react with specific side chains have been most useful in predicting groups important in the catalytic action of the enzyme.

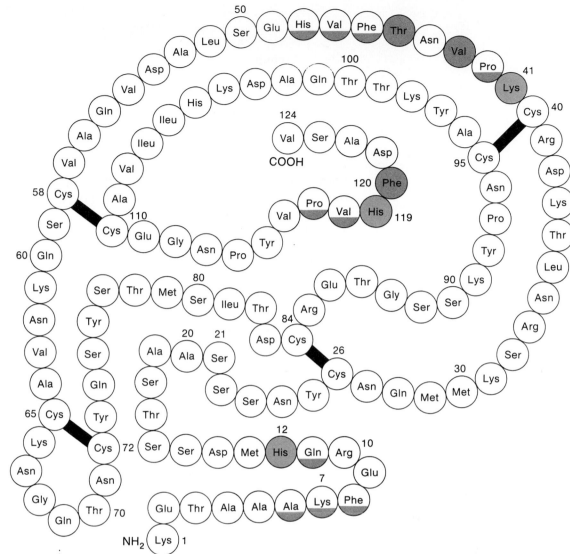

Figure 9–18
Amino acid sequence of bovine ribonuclease. Ribonuclease was the first protein to be completely synthesized from its amino acids. The final product is enzymatically indistinguishable from native ribonuclease. Four disulfides are indicated by black bars. Three residues, histidine 12, histidine 119, and lysine 41 (shown in color) are important in the active site. Other groups (shown in gray) are involved in binding the substrate to the active site: threonine 45, valine 43, and phenylalanine 120.

1-CM His 119

3-CM His 12

Figure 9–19
When RNAse is treated with equimolar amounts of iodoacetate (ICH_2COO^-), two major products obtained are the carboxymethylated derivatives of histidine 119 and histidine 12. One or the other of these derivatives has lost the majority of its catalytic activity, a result indicating the importance of both histidines in the active site.

On treatment of native RNAse with iodoacetate in equimolar quantities, carboxymethylated His 119 is the major product and carboxymethylated His 12 is a minor product (Figure 9–19). The other histidines in RNAse are much less reactive toward this reagent. Either of these derivatives has lost the majority of its catalytic activity, indicating that both histidines are important in the active site. Support for this idea comes from the observation that the alkylation is inhibited by small molecules such as cytidine-3'-phosphate, which is believed to bind to the active site. The Lys 41 residue was similarly implicated in the active site through the observation that the enzyme is inactivated under conditions in which the reagent fluorodinitrobenzene reacts selectively with the ϵ amino group of this lysine (Figure 9–20).

An important indication as to the mechanism of action of ribonuclease was provided by detection of cyclic 2',3'-phosphates in RNAse digestion of RNA. Simple nucleoside cyclic 2',3'-phosphates can serve as substrates of

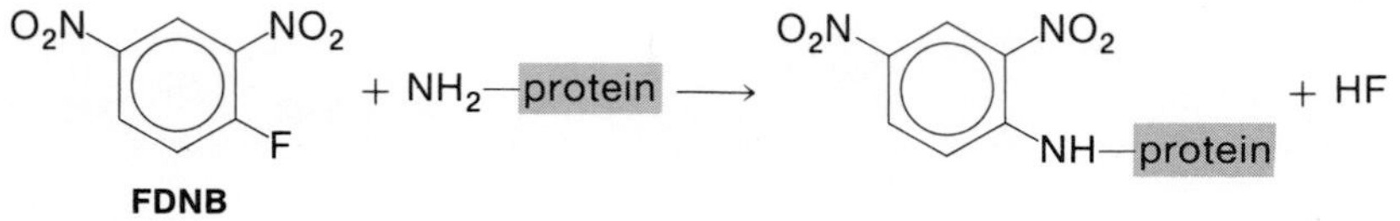

$$O_2N \diagdown \bigcirc \diagup NO_2 \diagdown_F + NH_2-\boxed{protein} \longrightarrow O_2N \diagdown \bigcirc \diagup NO_2 \diagdown_{NH-\boxed{protein}} + HF$$

FDNB

Figure 9–20
Reaction of fluorodinitrobenzene (FDNB) with the ϵ amino group of lysine. The lysine 41 residue was implicated in the active site by the observation that the enzyme is inactivated by FDNB.

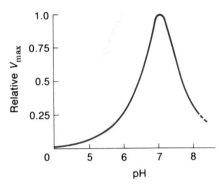

Figure 9–21
The dependence of V_{max} on pH in the ribonuclease-catalyzed hydrolysis of nucleoside cyclic 2′,3′-phosphate. The bell-shaped pH profile is consistent with the involvement of both histidines in the catalysis, one in the charged form and one in the neutral form.

RNAse. It should be remembered that the first step in the digestion of RNA by the enzyme leads to the formation of a cyclic intermediate, as shown in step 1 of Figure 9–17. The second step results in the conversion of the cyclic 2′,3′-phosphate to a 3′-phosphate derivative.

The pH dependence of the hydrolysis of these synthetic 2′,3′-cyclic nucleotide substrates, as well as of RNA, is shown in Figure 9–21. The bell-shaped profile is suggestive of a reaction requiring both acidic and basic catalytic groups, and the apparent pK_2 values are compatible with the involvement of two histidine residues, one functioning in the charged form as a general acid, the other functioning in the neutral form as a general base. The evidence cited earlier—that histidine residues of the RNAse molecule are essential for catalysis—strengthens this supposition. The pK_2 values of His 12 and His 119 are 5.8 and 6.2, respectively. These values were obtained by observing the changes in nuclear magnetic resonance spectra of bands characteristic of these amino acids as a function of pH.

Crystal Structure Studies

Despite the enormous number of chemical studies done in advance, the crystal-structure studies of RNAse and of RNAse complexed with specific inhibitors have been indispensable in understanding how the enzyme works. There really is no substitute for this type of precise structural information, and it is clear that the increasing ease of determining structures by x-ray diffraction will greatly aid future investigations.

The three-dimensional structure of the enzyme confirms the closeness of His 119, His 12, and Lys 41, which the chemical data predicted were a critical part of the active center (Figure 9–22). These functional groups are distributed around a depression that occurs near the middle of the molecule. The structure of the cocrystal made from RNAse S and an inhibitor of RNAse shows the probable binding site of substrate (Figure 9–22). The inhibitor is a modified UpA, with a methylene carbon atom replacing the oxygen between the phosphorus and the 5′-CH_2 of the ribose of adenosine (A); it is designated UpCH$_2$A. This inhibitor should resemble the substrate, but be nonhydrolyzable, because of the replacement of —O— by —CH_2—. His 12 and His 119 are in close juxtaposition to the phosphate ester linkage that is hydrolyzed. The pyrimidine and purine rings fit into specific regions on the enzyme surface. Hydrogen bonds are formed between the pyrimidine ring and both the side chain —OH and the peptide —NH— of the Thr 45. Phe 120 is located on one side of the pyrimidine ring and Val 43 on the other. Together they form a groove for pyrimidine binding. Fred Richards found that when the pyrimidine is replaced by a purine, the compound still binds to the active site, but the distance between the His 12 and the C-2′ OH of the ribose is increased by about 1.5 Å. This probably explains why the enzyme is specific for a pyrimidine group, either C or U in this position.

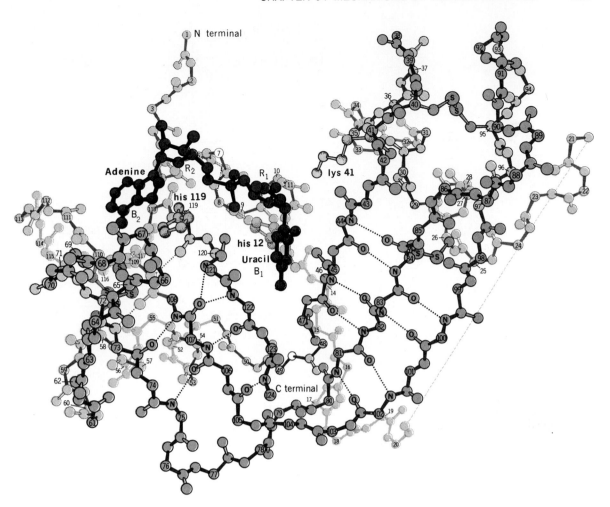

Figure 9-22
A schematic model of the binding of the dinucleotide phosphate $UpCH_2A$ (full colored) to RNAse S. This dinucleotide is a substrate analog. It cannot be hydrolyzed because an O group has been replaced by a CH_2 group. The two histidines and the one lysine crucial to the active site are indicated in grey. RNAse S is an active enzyme, although it is different from the native enzyme because of a peptide bond cleavage between residues 20 and 21 (dashed line).

Mechanism of the Reaction

His 12 and His 119 act by general acid-base catalysis (Figure 9-23). In the first step, which leads to a cyclic phosphate, His 12 serves as a base, assisting in the removal of a proton from the ribose C-2' OH that couples to the phosphate to form the cyclic 2',3'-phosphate. His 119 acts as an acid catalyst, protonating the leaving group to give a ribose C-5' OH. In the second step (opening of the cyclic phosphate ring), these roles could be reversed, with His 12 acting as an acid and His 119 as a base. It has been suggested that the function of the positively charged group on Lys 41 is to stabilize a transient pentacovalent phosphorus formed by an OH (from the C-2' OH in the cyclization step) or from water (in the ring-opening step). Only the two main steps in the overall reaction are depicted in Figure 9-23.

Note that ribonuclease needs two different imidazoles to perform functions similar to those done by one imidazole (His 57) in chymotrypsin. In both cases a proton must be removed from an attacking OH group, and a proton must be put onto the leaving group. In chymotrypsin the transient

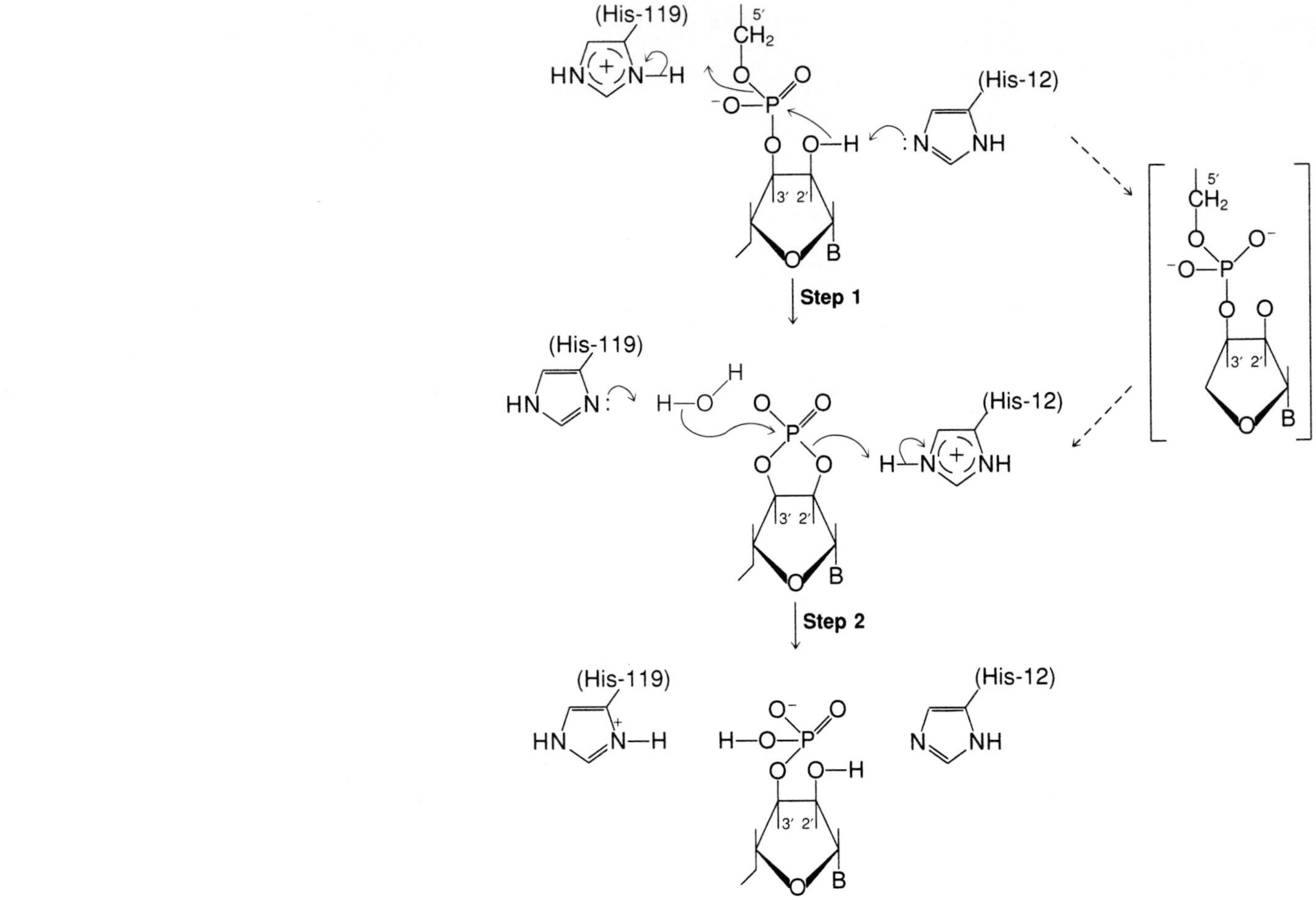

Figure 9–23

Mechanism of RNAse-catalyzed RNA hydrolysis. The concerted action of histidine 12, acting as a base to accept a proton from the ribose-2' OH, and histidine 119, acting as an acid to form a hydrogen bond with an oxygen atom of the phosphate, leads to a transition-state complex with a pentacoordinated phosphorus. The formation of the cyclic 2',3'-phosphoribose intermediate is accompanied by loss of a proton from histidine 119 and the uptake of a proton by histidine 112. Water then enters the site, donating a proton to histidine 119 and an —OH to the phosphate to form the trigonal pyramid structure, which rearranges with the aid of the acidic His 12 to form the pyrimidine ribose-3'-phosphate.

tetrahedral intermediate had the attacking and leaving groups at neighboring positions (any two positions on a tetrahedron are neighbors), so the imidazole could move between them. In the case of RNAse we are dealing with a reaction where the central ligand is phosphorus rather than carbon. Phosphate groups can undergo hydrolytic reactions by more than one mechanism. In an _in-line mechanism_ the attacking nucleophile enters opposite the leaving group (Figure 9–24a). In an _adjacent mechanism_ the attacking nucleophile enters on the same side as the leaving group (Figure 9–24b). The in-line mechanism is used by RNAse, so the attacking and leaving groups must be on opposite sides of the phosphorus. The same imidazole cannot move between groups so far apart, so two different imidazoles are used in a concerted manner.

LYSOZYME

Lysozyme, like all of the previous enzymes discussed in this chapter, is also a hydrolytic enzyme; in this case, the substrate is a polysaccharide chain. The bacteriolytic properties of hen eggwhite lysozyme were first reported by

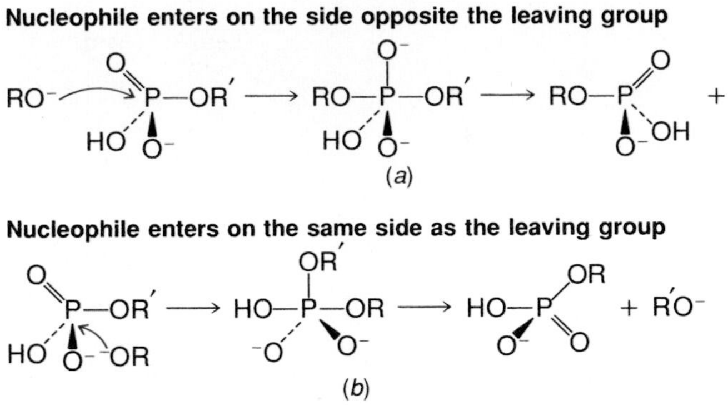

Nucleophile enters on the side opposite the leaving group

$$RO^- \longrightarrow P\text{—}OR' \longrightarrow RO\text{—}P\text{—}OR' \longrightarrow RO\text{—}P + R'O^-$$

(a)

Nucleophile enters on the same side as the leaving group

$$P\text{—}OR' \longrightarrow HO\text{—}P\text{—}OR \longrightarrow HO\text{—}P + R'O^-$$

(b)

Figure 9–24

Two routes leading to phosphate transfer. In both cases the phosphate is attacked by a nucleophile, forming a pentacovalent intermediate. (a) The in-line mechanism involves entering and leaving groups on opposite sides. (b) In the adjacent mechanism, the nucleophile enters on the same side as the leaving group.

P. Laschtchenko in 1909, long before A. Fleming published his observations that similar bacteriolytic agents are widespread in biological tissues and secretions, such as tears. It is still not clear if the function of the enzyme is to kill bacteria or to aid in their breakdown once they have been killed by other agents, or possibly both. The bacteriolytic action of lysozyme is accounted for by the ability of the enzyme to hydrolyze glycosidic linkages in bacterial cell walls. Some aspects of the structure of lysozyme have already been considered in Chapter 2 (see Figure 2–22). Other properties of lysozyme as an agent for cell-wall destruction are considered in Chapters 4 and 20. Here we will focus on the mechanism of action of the enzyme. Most of this work has been done on hen eggwhite lysozyme because eggwhite provides a bountiful source of the enzyme that is easy to purify.

Eggwhite lysozyme is a globular protein of molecular weight 14,000, with 129 amino acids in a single polypeptide chain and four disulfide linkages (Figure 9–25). Lysozyme attacks many bacteria by lysing the mucopolysaccharide structure of the cell wall (Chapter 4). Susceptible bacterial cell walls are built from β-1,4-polymer derived from β-hydroxyl monomers. The monomers are arranged in a strict alternating sequence of N-acetylglucosamine (GlcNAc) and N-acetylmuramic acid (MurNAc) residues. These chains are cross-linked by short polypeptides that are attached to the —OR side chains of MurNAc by peptide bonds. Lysozyme cuts the polysaccharide chain on the far side of a linking oxygen atom that is attached to the C-4 of a GlcNAc residue. For an alternating GlcNAc-MurNAc-GlcNAc-MurNAc polymer, the cut is between the C-1 of MurNAc and the chain-linking oxygen, as shown in Figure 9–26.

Bound Substrate Is Strained at the Active Site

The GlcNAc-GlcNAc-GlcNAc trimer was found to form a stable inhibited complex, and the structure of the enzyme was determined by x-ray diffraction with and without the trimer present. Longer GlcNAc polymers were substrates, being cleaved with increasing rapidity up to the hexasaccharide. Although it has not been possible to observe the enzyme crystallographically with more than the trimer in place, it has been possible by model building to work out the probable mode of binding of the longer polysaccharides and from this to suggest a catalytic mechanism. A crevice runs horizontally across the molecule, as shown in Figure 9–27. It was deduced from x-ray diffraction that the tri-GlcNAc inhibitor binds in the top half of the crevice of the molecule, as shown in Figure 9–27. This fact suggests that the crevice

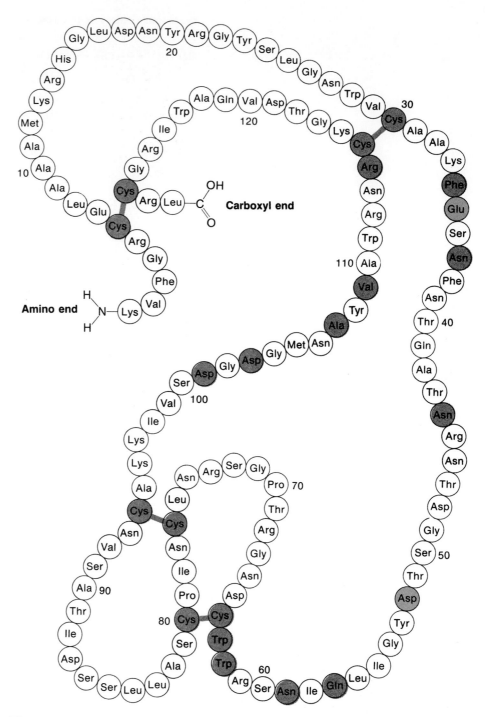

Figure 9–25
The primary sequence of amino acid residues in eggwhite lysozyme. Functional
groups of the active site are indicated in color. Some residues important in substrate
binding are shaded in gray. Cystine disulfide linkages are also shown in gray.

contains the active site. Most of the hydrophobic groups that are on the
outside of the enzyme molecule appear as lining for the crevice. This leads to
an energetically favorable interaction with the organic substrate molecule.
There are small changes in the amino acid side-chain positions when the
trimer inhibitor binds. Particularly noticeable is the movement of Trp 62 by
about 0.75 Å toward the trimer B ring; the entire side of the crevice closes
down slightly on the substrate. As stated earlier, it is possible to build three

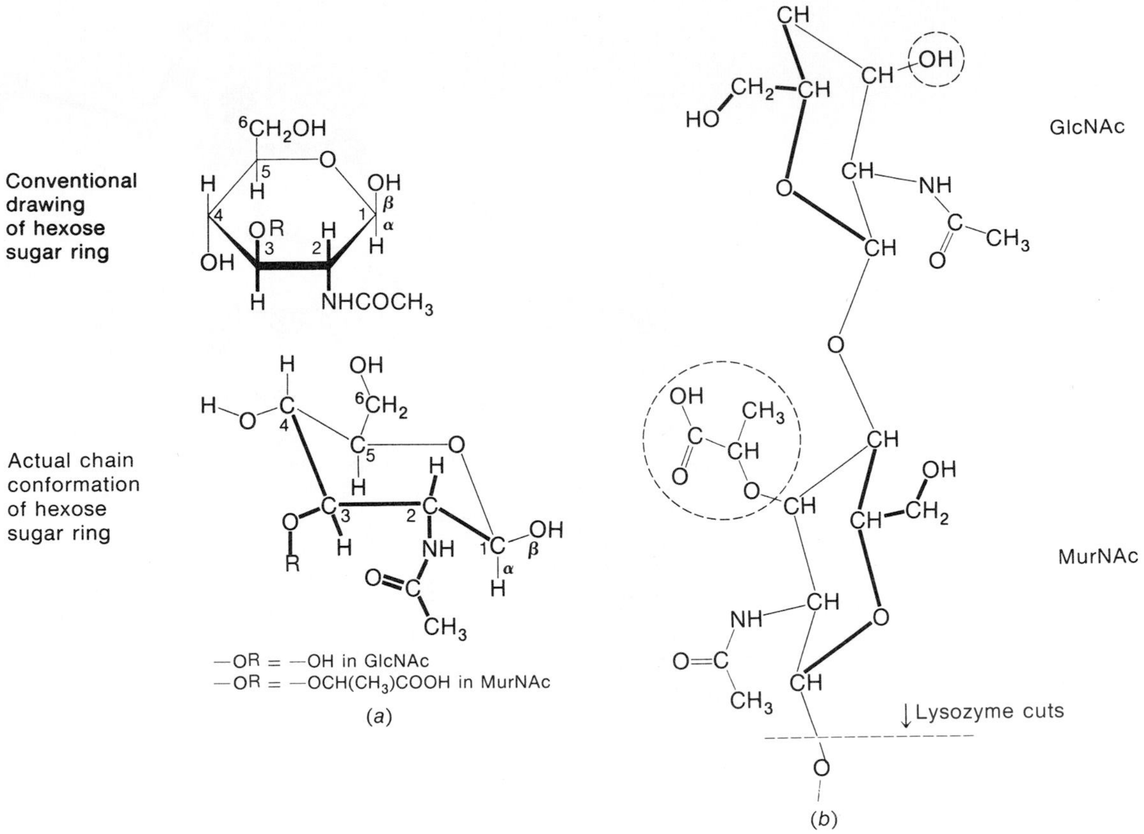

Conventional
drawing
of hexose
sugar ring

Actual chain
conformation
of hexose
sugar ring

—OR = —OH in GlcNAc
—OR = —OCH(CH₃)COOH in MurNAc

(a)

GlcNAc

MurNAc

↓ Lysozyme cuts

(b)

Figure 9–26

The conformations of N-acetylglucosamine amine (GlcNAc) and N-acetylmuramic acid
(MurNAc) as monomers (a) and when linked (b). These substituted sugar derivatives
are found in bacterial cell walls and are the substrate for lysozyme. Position of
cleavage by lysozyme is indicated in (b). (Adapted from R. E. Dickerson and I. Geis,
The Structure and Action of Proteins, Benjamin/Cummings, Menlo Park, Calif., 1969.)

more sugar rings into the left half of the crevice to form a hypothetical complex for the hexasaccharide substrate.

A schematic view of the hexamer and its interactions with enzyme is
shown in Figure 9–28; also see the more precise presentation in Figure 9–27.
Six subsites, A through F on the enzyme, bind the sugar residues. The nature
and number of these contacts are indicated in Table 9–3. Alternate sites
interact with the acetamide side chains (labeled "a" in Figure 9–27) of the
GlcNAc residues. These sites are unable to accommodate MurNAc residues
because of their bulky lactyl side chains (P). This structure is consistent with
the fact that $\beta(1,4)$ linked hexoses prefer the almost fully extended state,
with adjacent hexose units flipped over 180° relative to one another (e.g., see
Figure 4–16). Site D cannot bind a sugar residue without distortion, and the
glycosidic linkage that is cleaved binds between sites D and E as shown by
the arrow. Therein is the crux of the mechanism. Binding to the D site is
energetically unfavorable and results in a distortion of the normal chair form
of the sugar. The precise nature of this distortion of the sugar is unclear but it
seems likely that the distortion will favor the transition state relative to the
unstrained conformation of the substrate. The other sugar residues bind favorably to their sites and provide enough energy to compensate for the unfavorable binding to the D site.

The idea that binding of the substrate at the D site is unfavorable and
causes some distortion of the substrate has been supported by binding of

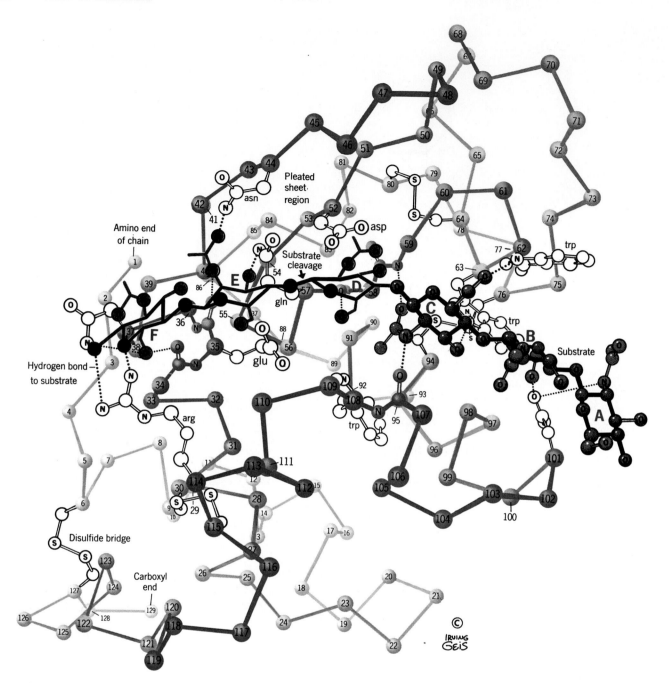

Figure 9–27

The complex formed between lysozyme and its substrate. The crevice that forms the active site runs horizontally across the molecule. The hexasaccharide substrate is shown in darker color. Rings A, B, and C to the right come from the observed trimer binding site. Trimer binds strongly but is not digested by the enzyme. It is an excellent competitive inhibitor. Rings D, E, and F are inferred from model building. The side chains, which are believed to interact with the substrate, are shown in line. Isoleucine 98 is so bulky that it helps to prevent MurNAc, with its large side group, from binding at ring position C and thus establishes the arrangements on the molecule of the alternating GlcNAc-MurNAc copolymer and points out the locus of cleavage in the active site. (Coordinates courtesy of Dr. D. C. Phillip, Oxford.) (Reprinted with permission from R. E. Dickerson and I. Geis, *The Structure and Action of Proteins*, Benjamin/Cummings, Menlo Park, Calif., 1969.)

normal and modified oligosaccharides. The strength of binding of compounds with two to six residues has been measured and the contributions to the binding have been estimated for the six sites A to F. It was found that occupancy of the D site made a positive (destabilizing) free energy contribution while occupancy of any of the others made a negative (stabilizing) con-

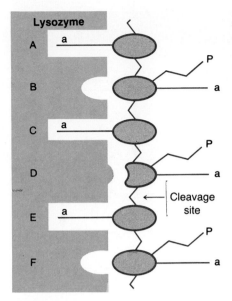

Lysozyme

A

B

C

D

← Cleavage site

E

F

Figure 9–28
Schematic diagram showing the specificity of hen eggwhite lysozyme for hexasaccharide substrates. Six subsites (A–F, same meaning as in Figure 9–27) on the enzyme bind the sugar residues. Alternate sites interact with the acetamide side chains (a) and these sites are thus unable to accommodate MurNAc residues with their lactyl side chains (P). Site D cannot bind a sugar residue without distortion and the glycosidic linkage that is cleaved binds between sites D and E, as shown by the arrow. (Adapted from drawing of T. Ionoto, L. N. Johnson, A. C. T. North, D. C. Phillips, and J. A. Ripley, Vertebrate enzymes, in P. D. Boyer, ed., *The Enzymes*, Vol. 3, p. 7113, Academic Press, New York, 1971.)

tribution to the binding energy. With all six sites occupied, the total negative free energy of binding forces occupancy of the D site. This is the case with normal substrates. However, it was also possible to synthesize a tetramer, occupying sites A to D, in which the fourth group is a sugar lactone that is flat at carbon 1. This flat lactone resembles the presumed flat cationic intermediate formed in the postulated mechanism (to be discussed next), and is thus a transition-state analog. The lactone proves to be a powerful inhibitor of lysozyme, and now occupancy of the D site is stabilizing. Once again we see that a transition-state analog supports the mechanistic ideas derived from other work.

Table 9–4 presents the results of a study of the rates of hydrolysis of a homologous series of compounds containing from three to six residues of GlcNAc and the normal GlcNAc-MurNAc hexamer. It can be seen that the catalytic rate does not become significant until the pentamer. The lower oligomers bind to the enzyme, but they do not bind across the D site because this is energetically unfavorable. Therefore they are not properly oriented about the active site with a strained substrate favorably disposed for hydrolysis.

Mechanism of the Reaction

It can be seen (Table 9–4) that the hexamer hydrolysis occurs at the highest rate and exclusively between the fourth and fifth residues. This position in the enzyme substrate complex must be located adjacent to the active site. Model building suggests that the susceptible C—O bond between saccharide units D and E falls between Asp 52 and Glu 35. The favored proposed mechanism, illustrated in Figure 9–29, takes into account the closeness of these two acidic amino acid side chains, the strained configuration of the sugar, and the fact that the hydrolyzed tetramer product retains the same configuration after hydrolysis. Thus the leaving group (—OR) and the adding group (—OH) have to approach the sugar from the same side. After binding, the

TABLE 9–3
The Main Interatomic Contacts between the Hexasaccharide Substrate and the Lysozyme Molecule

Site	Polar Contacts	Total Number of van der Waals Contacts < 4 A
A	NH-Asp 101	7
B	O_6-Asp 101	11
C	O_6-Trp 62	30
	O_3-Trp 63	
	NH-CO 107	
	CO-NH 59	
D	O_6-CO 57	35
	O_1-Glu 35	
E	O_3-Gln 57	45
	NH-CO 35	
	CO-Asn 44	
F	O_6-CO 34	13
	O_6-Asn 37	
	O_6-Arg 114	
	O_1-Arg 114	

Source: Data from J. A. Rupley, *Proc. Roy. Soc.* B167:416, 1967.

TABLE 9–4
Rates of Reaction and Cleavage Patterns Shown by Different Substrates

Compound	Turnover Number k_{cat} (sec^{-1})	Cleavage Pattern
$(GlcNAc)_3$	8.3×10^{-6}	$X_1\text{—}X_2\text{—}X_3$
$(GlcNAc)_4$	6.6×10^{-5}	$X_1\text{—}X_2\text{—}X_3\text{—}X_4$
$(GlcNAc)_5$	0.033	$X_1\text{—}X_2\text{—}X_3\text{—}X_4\text{—}X_5$
$(GlcNAc)_6$	0.25	$X_1\text{—}X_2\text{—}X_3\text{—}X_4\text{—}X_5\text{—}X_6$
$(GlcNAc\text{—}MurNAc)_3$	0.50	$X_1\text{—}X_2\text{—}X_3\text{—}X_4\text{—}X_5\text{—}X_6$

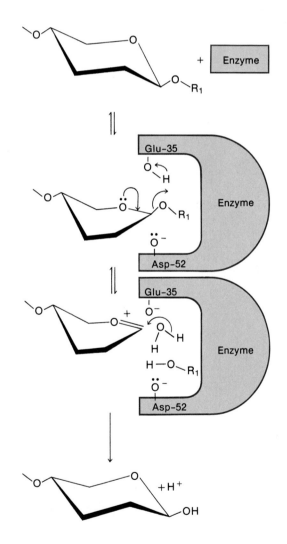

Figure 9–29
Proposed mechanism for lysozyme-catalyzed hydrolysis. Binding of the hexasaccharide to the enzyme distorts the reacting sugar (bound at the D site in Figure 9–28). This conformation of the substrate is presumed to approach that of the positively charged intermediate. The final product produced by the addition of an —OH group from water restores the chair form in the product, as well as preserving the original stereochemistry in the glycosidic bond.

substrate undergoes a bond rearrangement to yield a carbonium ion at a rate enhanced by at least three factors:

1. A ring conformation distorted toward that of the transition state.
2. A glutamic acid 35 that acts as a general acid catalyst by donating a proton to the glycosidic oxygen.
3. A negatively charged aspartate that stabilizes the positively charged carbonium ion.

The resulting aglycone (HOR_1) diffuses away, and the process is completed by reaction with water or another acceptor.

None of the factors involved in lysozyme catalysis seems able alone to account for the observed rate enhancement, but concerted attack involving all three contributions mentioned earlier appears to be necessary. Nevertheless, lysozyme is an efficient enzyme as far as its turnover number (k_{cat}) is concerned. The k_{cat} value of lysozyme (0.5 sec^{-1}) is very much smaller than that of carboxypeptidase A (100 sec^{-1}) or lactate dehydrogenase (1000 sec^{-1}), to be discussed in the next section. If the main job of lysozyme is to rupture bacterial cell walls, it may not have to break many linkages to do so. Thus the "physiological efficiency" of lysozyme may be much higher than its efficiency in terms of turnover number.

Two other aspects of lysozyme action are deserving of comment. First, the bond-strain mechanism used by lysozyme guarantees the safety of disaccharides, trisaccharides, and tetrasaccharides in its presence. It is not clear whether this feature serves a useful purpose or is merely an incidental property of the enzyme. Second, the pH optimum for lysozyme action is 5, and activity falls off rapidly on either side of that value. The rapid fall-off on the high side is consistent with Glu 35 functioning as a generalized acid catalyst. For this purpose, it must possess a proton. Various measurements indicate that the pK of Glu 35 is near 6 in the free protein and is increased somewhat by saccharide binding. On the low side, the fall-off of reaction rate could be due to protonation of Asp 52 as well as other carboxylates (e.g., Asp 101 and Asp 66) in the protein. Lysozyme is active in secretions that tend to have pH values on the acid side of neutrality. Thus if it is going to employ an amino acid for general acid catalysis, glutamic acid is a far better choice than histidine (which is used in RNAse), since the higher pK of the latter amino acid would substantially raise the pH optimum.

LACTATE DEHYDROGENASE: A BISUBSTRATE ENZYME

Lactate dehydrogenase (LDH) is a tetramer of M_r 140,000. The enzyme is widely distributed in tissues and plays a key role in energy metabolism. Lactate dehydrogenase is a bisubstrate enzyme; the substrates consist of either lactate or pyruvate and a derivative of the hydrogen-carrying coenzyme, nicotinamide adenine dinucleotide (Figure 9–30).

Many enzymes use nicotinamide adenine dinucleotide (NAD) in oxidation reactions in which NAD^+ reacts with another substrate to accept a hydride ion (two electrons and one proton) in one or more steps. The resulting NADH (reduced form of the coenzyme) must be reoxidized so that it can be used again. In glycolysis (Chapter 14) many organisms accomplish the reoxidation with the help of L-lactate dehydrogenase, which catalyzes the reduction of pyruvate to lactate according to the reaction

$$
\underset{\text{Pyruvate}}{\overset{\displaystyle\begin{array}{c}CH_3\\|\\C{=}O\\|\\COOH\end{array}}{}} + NADH + H^+ \rightleftharpoons \underset{\text{L-Lactate}}{\overset{\displaystyle\begin{array}{c}CH_3\\|\\HCOH\\|\\COOH\end{array}}{}} + NAD^+ \tag{10}
$$

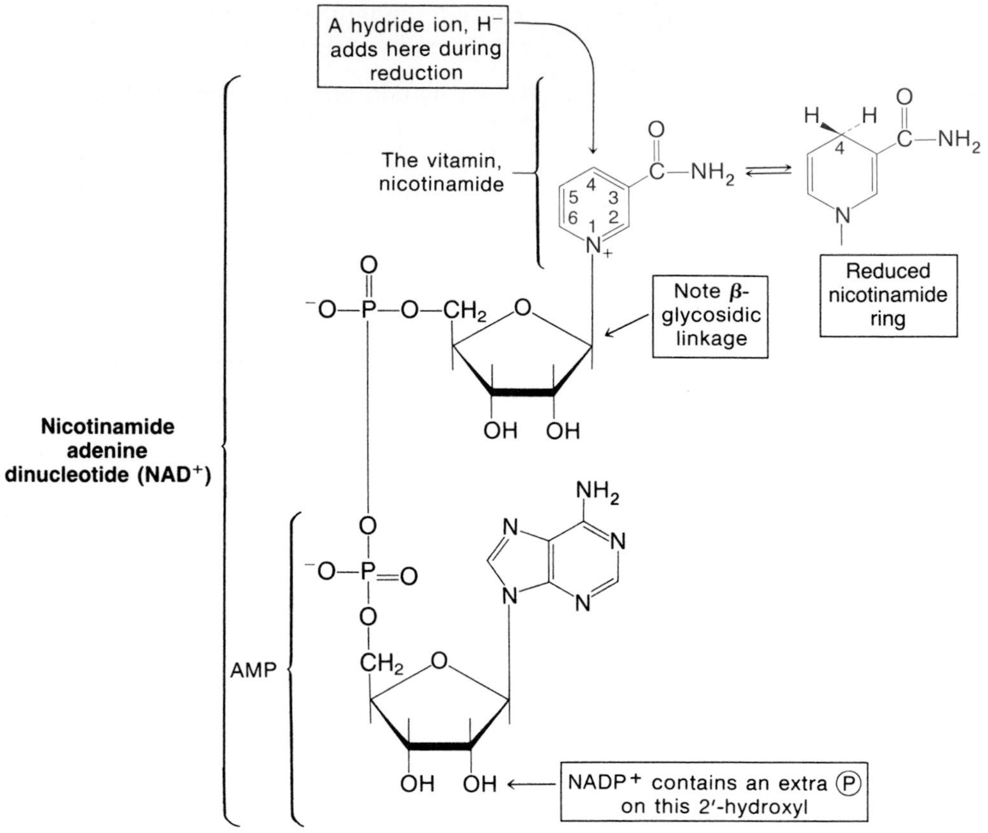

Figure 9–30
The hydrogen-carrying coenzymes NAD⁺ and NADP⁺. Note use of the abbreviations NAD⁺ and NADP⁺ even though the net charge on the entire molecule at pH 7 is negative.

The equilibrium for this reaction is strongly favored to the right at neutral pH. In spite of this, the initial rate of the reaction can be studied in the leftward direction by addition of the appropriate substrates at elevated values of pH. Physiologically it also appears, as we will see, that LDH can play useful roles in both directions, although the conversion of pyruvate to lactate under anaerobic conditions is certainly the best understood function. Lactate produced in tissues functioning anaerobically is secreted and absorbed by other tissues functioning aerobically. LDH can then convert the lactate back to pyruvate for further utilization in the Krebs cycle (see Chapter 15).

Binding of Coenzyme Occurs Before Binding of Sugar

The strategy of all NAD-dependent dehydrogenases is to orient the coenzyme and the substrate on the enzyme surface so that the C-4 atom on the nicotinamide is pointed toward the reactive carbon of the substrate. Different dehydrogenases have remarkably similar protein domains for binding NAD⁺, but dissimilar sites for binding the cosubstrate, the latter being dependent on the structure of the substrate. The coenzyme binding domains are compared for lactate dehydrogenase and two other dehydrogenases in Figure 9–31.

In LDH, binding of NADH is about 400 times stronger than the binding of NAD⁺ and about 50 times stronger than that of NADPH (the same mole-

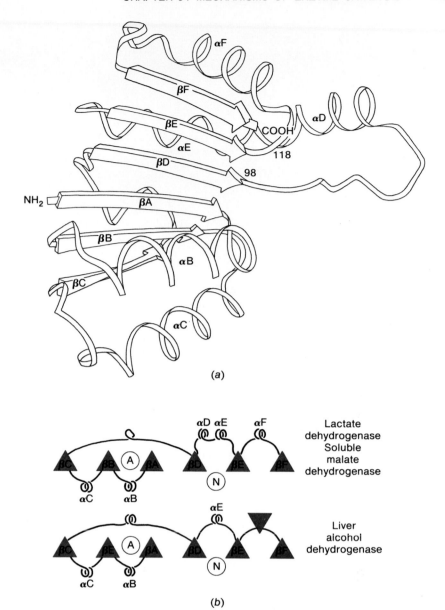

Figure 9–31
Nucleotide-binding protein domains and related tertiary structures in a series of proteins. *(a)* Schematic tertiary structure of part of lactate dehydrogenase. Arrows show strands of β sheets drawn from the amino to the carboxyl end. *(b)* Cartoons of similar tertiary structure patterns in lactate dehydrogenase and other proteins. Triangles indicate β sheets viewed with the N terminal closest to the observer. Coils indicate α helices. Circled A indicates known binding sites for the adenine moiety of NAD⁺ or ATP, and circled N indicates binding sites for the nicotinamide moiety of NAD⁺.

cule with an additional phosphate, see Figure 9–30). Some enzymes can use either NAD⁺ or NADP as coenzymes, but for LDH the binding data show that there is a strong preference for the former. A conformational change takes place in the enzyme when it interacts with the coenzyme NAD⁺ (or NADH) and a suitable substrate, usually lactate (or pyruvate). A minor movement of Asp 53 and its associated polypeptide chain occurs on binding the adenosine part of the coenzyme. There is a larger movement of the loop connecting D and E, including the helix αD (residues 98–120, see Figure 9–31), involving the main-chain displacement of up to 11 Å. In ternary complexes this flexible loop drops down and encloses the coenzyme and substrate. In the free enzyme it extends into the solvent. Arg 101 forms an ion pair with the pyrophosphate group of the coenzyme in the ternary complex (Figure 9–32). The formation of this ion pair may be the driving force for the conformational change of the loop and the subsequent rearrangements in the subunit. The guanidinium group of Arg 109 moves 14 Å and changes from being completely exposed to the solvent to having a close interaction with groups

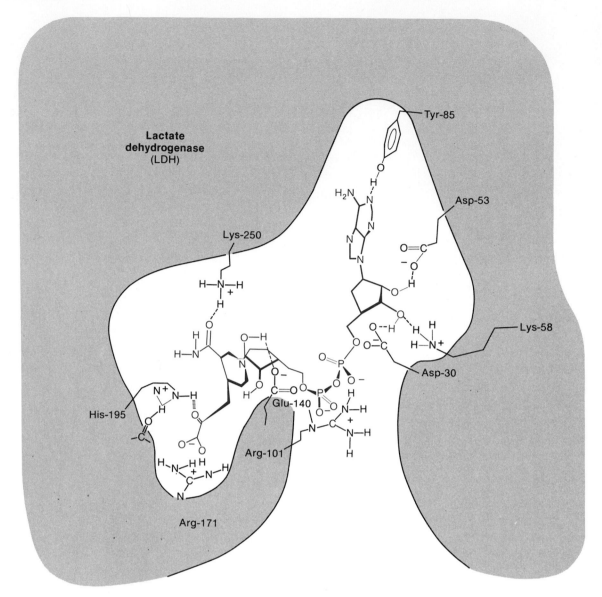

Figure 9–32
Diagrammatic representation of binding of a covalent adduct formed between pyruvate and NAD$^+$ to lactate dehydrogenase. This compound (shown in color) is a competitive inhibitor and therefore it is presumed to bind similarly to the substrates.

around the substrate site. The two connected helixes αD and αE are more angled to each other in the ternary complex.

The movement of the N-terminal part of helix αD is associated with a movement of the neighboring C-terminal part of helix αH (located in the substrate-binding domain, not shown). The overall effect of these conformational changes in the protein involves movement toward the substrate and a contraction of the subunit on binding coenzyme and substrate. The enzyme folds up around the substrates, so that the fit of enzyme and substrates has been induced by substrate binding.

In the native state, lactic dehydrogenase exists as a tetramer containing four identical subunits. With all the conformational changes taking place in going from a free enzyme to a ternary complex it is natural to suspect that there might be some effects between different binding sites. However, careful equilibrium binding studies of the interaction between enzyme and NAD^+ or NADH have led to the conclusion that the intact enzyme contains four independent noninteracting binding sites. Existence of the enzyme as a tetramer must have other advantages. Sometimes dimers or tetramers are preferable simply because they are more stable and therefore less susceptible to denaturation or degradation.

Kinetic Studies Reveal Intermediates and Slow Step in the Reaction

The forward reaction for the oxidation of lactate by LDH and NAD^+ fits the general form of the steady-state kinetic expression for the ordered pathway described in Chapter 8 [see Figures 8–10 and 8–11 and Equation (43)]. Equilibrium binding experiments have been used to demonstrate the order of binding. The coenzyme NAD^+ will bind to the enzyme even without lactate, so NAD^+ binds first; this by itself is believed to produce an alteration in the enzyme structure that creates a binding site for the lactate. Lactate won't bind to the enzyme unless NAD^+ is present and bound. The same is true for the reverse reaction; i.e., the NADH binds before pyruvate. In the release of products the sugar is always released before the coenzyme.

Optical properties of the protein and the coenzyme have been instrumental in making pre-steady-state kinetic measurements, which have been most important in determining the steps that take place after binding. NADH fluoresces weakly at 470 nm when its absorption band at 340 nm is excited. Its fluorescence is greatly increased on combination with the enzyme, so fluorescence may be used to measure coenzyme binding. The enzyme fluoresces at about 350 nm when excited by radiation at 270–305 nm. This fluorescence is due to the tryptophan residues in the enzyme. In LDH there is a substantial drop in intrinsic fluorescence on binding NADH, which is probably due to the transfer of some of the tryptophan residues to a more hydrophilic environment. Despite the fact that the exact cause of the drop in fluorescence is unclear, the change may be used to determine whether or not the protein is bound to coenzyme and substrate.

In addition, the release of H^+, a product of the reaction, can be measured optically by combination of H^+ with a dye molecule. All these optical changes have been measured as a function of time immediately after mixing the enzyme with saturating amounts of NAD^+ and lactate. Such measurements suggest at least three phases in the approach to the steady state. In the first phase, which takes less than 1 ms, a small amount of NADH can be detected with no release of proton. In the second phase, which follows a first-order rate with respect to lactate concentration, NADH and H^+ are produced in equimolar amounts, and only the proton is liberated. The third phase, which follows zero-order kinetics, begins after about 40 ms. During this steady-state phase of the reaction NADH and H^+ are liberated at the same rate. The proposed phases in the reaction after binding substrate and coenzyme are indicated in Figure 9–33. One of the important conclusions from this experiment is that the rate-limiting step after reaching the steady state is the release of the NADH product from the enzyme.

Reaction Results from Concerted Catalysis

An expanded view of the active site in the ternary complex, as inferred from the x-ray-determined structure, is shown in Figure 9–34. The substrates, lactate or pyruvate, are oriented so that they are favorably disposed for a

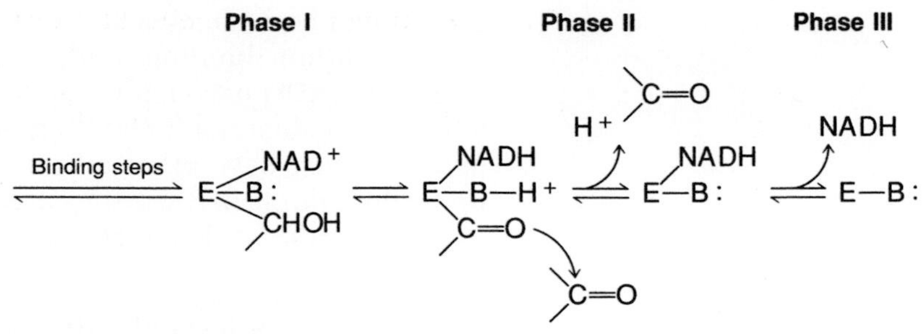

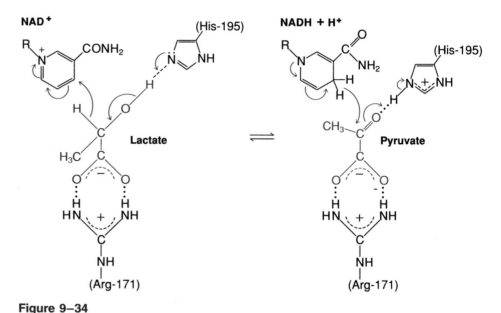

Figure 9–33
Proposed phases in the reaction of NAD⁺ and lactate after enzyme binding. Here
\>CHOH represents lactate and \>C=O represents pyruvate.

Figure 9–34
The mechanism of catalysis of lactate dehydrogenase. The binding to the amino acid
side chains of Arg 171 and His 195 and the coenzyme (NAD⁺ or NADH) is shown.
The reaction arrows leading to the interconversion are also shown. The nicotinamide
accepts a hydride from lactate in the oxidation reaction and donates a hydride to the
pyruvate in the reduction reaction.

concerted reaction with the nicotinamide ring of the coenzyme and the imidazole ring of His 194. In both forward and reverse reactions, Arg 171 helps to
anchor the substrate through H bonds and electrostatic bonds formed with
the nonreacting substrate carboxylate group. In both reactions a substrate-
orienting H bond is also formed with the imidazole group of His 195. His 195
acts as an acid-base catalyst, removing the proton from lactate during oxidation (rightward reaction in Figure 9–34) or donating it to pyruvate in the
reverse reduction reaction. The nicotinamide accepts a hydride from lactate
in the oxidation reaction and donates a hydride to the pyruvate in the reduction reaction.

Chiral aspects associated with NAD-dependent oxidation-reduction reactions are discussed in Chapter 11. Here we simply note that the hydride
adds to one face of the pyruvate group, forming L-lactic acid in a stereospecific reaction, and that the two hydrogens on NADH are not equivalent (one
forward, one back) so they do not react with equal probability.

Isoenzymes of Lactate Dehydrogenase Serve Different Functions

Most vertebrates possess at least two genes for lactate dehydrogenase, which make similar but nonidentical polypeptides called M and H. In embryonic tissue, both genes are equally active, resulting in equimolar amounts of the two gene products and a statistical array of tetramers (M_4, M_3H_1, M_2H_2, H_3M_1, and H_4 in the ratios of $1:4:6:4:1$). These forms are called *isoenzymes*, or isozymes. They can usually be detected by differing electrophoretic mobilities. As embryonic tissue multiplies and differentiates, the relative amounts of the M and H forms change. In pure heart tissue, which is considered aerobic, the H_4 tetramer predominates. In skeletal muscle, which functions anaerobically under stress, the M_4 isozyme predominates.

Although we can only speculate at this point, it seems likely that the M and H forms have evolved to serve different functions. A clue to these functions is revealed by the inhibiting effect of pyruvate on the dehydrogenase: Pyruvate can form a covalent complex with NAD^+ at the active site of the enzyme according to the following reaction:

$$\text{ADPR}-\overset{+}{\text{N}}\langle\text{ring}\rangle_{\text{CONH}_2} + CH_3COCOO^- \xrightarrow[\Delta]{H^+} \text{ADPR}-\text{N}\langle\text{ring}\rangle_{\text{CONH}_2}^{\;\;H,\;CH_2COCOO^-}$$

Indeed, this is the compound whose binding was the basis for Figure 9–32. The H_4 tetramer shows a much greater inhibition by this compound than the M_4 tetramer.

Active muscle tissue is anaerobic and produces a good deal of pyruvate. Inhibition of lactate dehydrogenases under anaerobic conditions would shrink the supply of NAD^+ and shut down the glycolytic pathway (see Chapter 14), with disastrous consequences. In fact, active muscle tissue has augmented levels of pyruvate but converts this readily to lactate. The reason is possibly that the predominant form of lactate dehydrogenase in muscle is M_4, which is only poorly inhibited by excess pyruvate.

The function of LDH in aerobic tissue is less clear. Heart muscle is aerobic tissue, and consequently, most of its pyruvate is funneled into the Krebs cycle for greater energy production (see Chapter 15). The lactate dehydrogenase of heart muscle, H_4, might be inhibited by pyruvate to prevent waste of this potential high-energy carbon source or excessive buildup of pyruvate resulting from the conversion of incoming lactate to pyruvate. Possibly the H_4 enzyme is used in such tissues to convert absorbed lactate into pyruvate.

The control of enzyme activity by isozyme type is very beneficial in certain cases. A far more widespread form of control of enzyme activity involves modulation in their activity through binding by regulatory small molecules. This topic will be the main subject of Chapter 10.

SUMMARY

Enzyme catalysts resemble ordinary chemical catalysts in several respects. Both types of catalysts function by direct association with the reacting species, making the reactive centers more attractive for reaction. In many respects enzyme catalysts are more sophisticated than ordinary chemical catalysts. Their binding sites for substrates are highly specific, resulting in optimum orientation of the substrate for reaction. Frequently the enzyme contains more than one catalytic site so that two steps in a reaction can be carried out simultaneously or in rapid succession, thereby avoiding the problem of high-energy intermediates.

The merits of enzyme catalysis are illustrated by examples of five enzymes whose catalytic mechanisms are discussed in detail. The enzymes are chymotrypsin, carboxypeptidase A, pancreatic RNAse A, hen eggwhite lysozyme, and L-lactate dehydrogenase.

Chymotrypsin is a member of the trypsin family of proteases, which includes trypsin, chymotrypsin, and elastase. These enzymes are very similar in structure, but they differ in the nature of their substrate binding site. As a result, each enzyme has a strong preference for attacking peptide linkages adjacent to different types of amino acid side chains. The active sites in all three enzymes are identical; each contains three key amino acids—serine, histidine, and aspartate. The amino acid side chains of these amino acids are oriented so that the serine hydroxyl becomes a strong nucleophilic reagent for attacking the peptide carbonyl carbon. Kinetic investigations of naturally occurring peptide substrates and synthetic ester substrates have revealed different rate-limiting reactions in the two cases. In the case of ester hydrolysis, it was necessary to analyze both the steady-state kinetics and the pre-steady-state kinetics to evaluate the important kinetic parameters.

Carboxypeptidase A is a zinc-containing exopeptidase. The zinc plays a major role in the reaction, serving as an electrophilic catalyst. Substrate binding involves an appreciable alteration in the structure of the enzyme, a fact testifying to the flexibility of the enzyme structure. A compound resembling the hypothesized transition state of the substrate has been found to make a potent competitive inhibitor.

Pancreatic RNAse A hydrolyzes RNA chains on the 3′ side of the pyrimidine group. The reaction occurs in steps, with a stable 2′3′-cyclic intermediate being formed in the first step. Each step involves the concerted action of two histidines. In the first step, one histidine functions as an acid catalyst and the other as a base catalyst. In step 2 the roles of the two histidines are reversed.

Lysozyme hydrolyzes polysaccharides found in cell walls. Because of the way in which it binds substrate, only oligosaccharides containing five or more residues are readily susceptible to hydrolysis. The susceptible linkage at the active site is bound with considerable strain, which probably increase its reactivity.

The fifth and last enzyme to be discussed is L-lactate dehydrogenase. This enzyme requires two substrates, NAD$^+$ and L-lactate, which are converted to NADH and pyruvate. The reaction is somewhat reversible, especially if the pH is altered. In either direction the nucleotide coenzymes must bind before the sugar can bind. It is believed that binding of the coenzyme alters the enzyme-coenzyme complex in a manner that creates a site for binding the sugar. In the ternary complex the sugar is aligned for reaction with the active sites on the coenzyme and the enzyme.

SELECTED READINGS

See the list at the end of Chapter 8.

PROBLEMS

1. In what ways are chymotrypsin, trypsin, and elastase similar as catalysts? In what ways do they differ? What factors in the enzyme structure are responsible for these differences?

2. For many enzymes, V_{max} is dependent on pH. At what pH would you expect the V_{max} of RNAse to be optimal? Explain.

3. If histidines were substituted for Glu 35 and Asp 52 at the active site in lysozyme, would you expect the modified enzyme to be functional? If it were functional, how might the optimum pH be altered, provided the enzyme was still pH-dependent?

4. Why should elevated pH values make it easier to observe the conversion of lactate to pyruvate by L-lactate dehydrogenase? [You may wish to refer to Equation (10).]

5. RNAse can be completely denatured by boiling or by treatment with SDS, yet it can renature to its fully active form. By contrast, the enzymes in the trypsin family and carboxypeptidase A cannot regain activity after total denaturation. What aspect of their structure precludes renaturation?

6. In this figure are shown the amino acids in the active site of papain. Based on what you know about the reaction mechanism of chymotrypsin, come up with a feasible reaction mechanism for the protease papain. Indicate analogous reactive groups and any likely intermediates.

7. Many enzymes force their substrate into a state or conformation that the substrate would not normally adopt. Their mechanisms for doing so include change of solvent, charge–charge interactions, and geometric strain placed on bond lengths and angles.
   a. In the case of lysozyme, which of these mechanisms seems to be most important?
   b. How is this mechanism relevant to catalysis?
   c. How does the enzyme "pay" for the cost of this destabilization?
   d. Another example of destabilization involves the displacement of water by substrate in carboxypeptidase A. How might this aid in the catalytic mechanism?

8. One manner in which enzymes might act as specific catalysts is by providing an environment that stabilizes a rate-limiting transition state of a given reaction. Two reports (Tramontano et al. and Pollack et al., both in *Sci-*

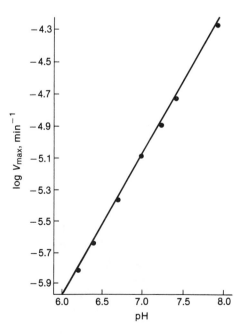

ence 234:1566, 1986) support such a view. In the latter article a monoclonal antibody (MOPC 167) specific for p-nitrophenylphosphorylcholine (NPPC) is found to have catalytic activity. Specifically it is found to hydrolyze p-nitrophenyl N-trimethyl ammonioethyl carbonate chloride 1 (carbonate 1). The transition state for such a cleavage is thought to be tetrahedral, much like the phosphate group in NPPC (see below). As expected, NPPC is a powerful inhibitor of carbonate 1 catalysis by the antibody.

a. The kinetics of catalysis of carbonate 1 by MOPC 167 is standard Henri-Michaelis-Menten:

$$v = k_{cat}[Ig][1]/\{K_m + [1]\}$$

where [Ig] is antibody concentration and [1] is substrate concentration (carbonate 1).

$$[Ig] + [1] \rightleftharpoons [Ig.1] \xrightarrow{k_{cat}} [Ig] + \text{products}$$

Recall from Chapter 8 the use of an Eadie-Hofstee plot. In the Eadie-Hofstee plot shown here, k_{obs} is the rate of hydrolysis in the presence of 11.6 μM MOPC 167, and k_{un} is the rate in the absence of catalyst. $(k_{obs} - k_{un})$ simply accounts for hydrolysis not due to MOPC 167. Calculate K_m, V_{max}, and k_{cat} as accurately as possible.

b. Consider the plot of log V_{max} versus pH. Perhaps the role of the antibody combining site is to stabilize the transition state formed by attack of an external nucleophile (OH⁻). Explain how this plot does or does not support such a model.

c. If it were possible to isolate them, what sort of transition-state analogs should you look for, such that you could make an antibody that mimics (i) trypsin? (ii) carboxypeptidase A? (iii) RNAse A? (iv) lysozyme?

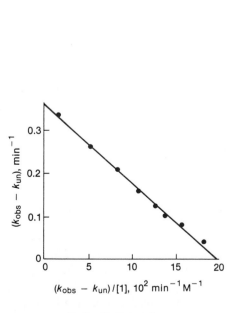

Eadie-Hofstee plot

Plot of log V_{max} as a function of pH

10

REGULATORY PROTEINS AND ENZYMES

One of the main ways that metabolic activities are controlled is by regulating enzyme activity. In Chapter 9 we saw several examples of enzyme regulation. In the pancreatic zymogens, regulation is accomplished by maintaining the enzyme in an inactive form until it is needed. Lysozyme may be said to be pH-regulated in the sense that it has a low activity around pH 7, which is the physiological pH, and maximum activity around pH 5, which is common in extracellular secretions. In the case of lactate dehydrogenase, the M_4 and the H_4 forms differ in their sensitivity to inhibition by pyruvate. The difference most likely serves an important regulatory function for the enzyme in different tissues. These three situations exemplify specialized forms of regulation in which the enzyme is predesigned to act in a certain way to a given set of environmental conditions.

Another form of regulation that is far more common is enzyme activation or inhibition that results from interaction with substrates or inhibitors. All enzymes are activated by increasing the substrate concentration, as long as the concentration is below the saturating level (see Chapter 8). Similarly, most enzymes are inhibited by small molecules (substrate analogs) that bind to the active site of the enzyme. In such cases, where we see a direct action on the active site, the explanation for the effect on enzyme activity is readily apparent from our understanding of classical enzyme kinetics.

A more elaborate form of regulation is observed for a select group of enzymes called _regulatory enzymes_. These enzymes have a built-in device that permits their activity to be reversibly altered by fine increments over wide limits.

Regulatory enzymes occupy key positions in metabolic pathways. An understanding of a particular multienzyme pathway usually suggests which enzymes might make attractive candidates for regulation. For example, many biosynthetic pathways involve a long chain of single chemical steps, each catalyzed by a discrete enzyme. The enzyme involved in the first step of a multistep reaction is typically inhibited by the product of the final step. This type of inhibition is called _feedback inhibition_. It is most economical for the first enzyme of a biosynthetic pathway to be inhibited once sufficient final product is present. This prevents the wasteful and sometimes harmful accumulation of useless intermediates or unneeded final product. Thus in a biosynthetic pathway the first enzyme in the pathway is usually regulatory but the other enzymes in the pathway are usually not.

Considerable attention is devoted to the patterns of control of biosynthetic and degradative pathways in Parts III and IV of this text. In this chapter we will focus on the structural properties of regulatory proteins. We have seen that in ordinary enzymes the active site resembles the substrate. In regulatory enzymes the inhibitor or activator need not resemble the substrate and frequently does not. Such flexibility greatly increases the possible modes for regulation so that they can be tailored to suit a given metabolic need.

For example, consider the case of histidine biosynthesis, which requires ten enzymes that act in sequential fashion to produce histidine. The first step in histidine biosynthesis involves the reaction between phosphoribosyl pyrophosphate and adenosine triphosphate. Neither of these compounds even vaguely resembles the structure of histidine, the end product of the pathway, and yet the enzyme that catalyzes the first step in histidine biosynthesis is strongly inhibited by histidine. Of the ten enzymes in the histidine pathway only the first enzyme in the pathway shows this exquisite sensitivity to histidine. To accomplish its purpose the regulatory enzyme is carefully designed. First, it contains a binding site for histidine that is distinct from the active site. Second, binding of histidine to this regulatory site causes a structural change in the enzyme that is transmitted to the active site. The most important effect of histidine binding is that it alters the structure of the active site, making it less effective. The alteration in the active site could involve the catalytic site, the substrate binding site, or both.

Let us rephrase this description of the dynamic structural organization of a regulatory enzyme in more general terms. Most commonly, regulatory enzymes contain regulatory sites in addition to their active sites. A small-molecule effector influences the enzyme activity by binding to the regulatory site. This binding causes a structural alteration in the enzyme that is transmitted to the active site. If the small-molecule effector is an activator, an increase in enzyme activity results. If the small-molecule effector is an inhibitor, a decrease in enzyme activity results. Regulatory enzymes that function in this way are called *allosteric* (Greek, *allos* = other, *stereos* = solid or space) and the small molecules that cause the change in enzyme activity are called *allosteric effectors*.

Our understanding of how regulatory proteins work is most complete for hemoglobin, which is not an enzyme at all but rather a transport protein whose main function is to deliver O_2 from the lungs to various tissues. Nevertheless, the same principles are believed to apply to hemoglobin oxygen binding as might apply to the binding of substrate to an allosteric enzyme. We will introduce the subject of allosteric proteins by a detailed consideration of how hemoglobin interacts with O_2 and then turn to an allosteric enzyme proper, aspartate carbamoyltransferase.

HEMOGLOBIN: AN ALLOSTERIC OXYGEN-BINDING PROTEIN

Hemoglobin consists of two α subunits, each with 141 amino acids, and two β subunits, each with 146 amino acids. Each subunit is capable of binding a single molecule of oxygen. The main task of hemoglobin is to transport oxygen from the lungs to the capillaries of the tissues in order to meet the oxygen needs of cellular metabolism. In muscle cells, a reserve oxygen store is provided by the myoglobin molecule, which is similar in structure to hemo-

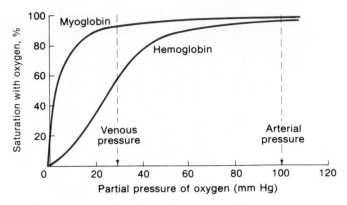

Figure 10–1

Equilibrium curves measure the affinity for oxygen of hemoglobin and of the simpler myoglobin molecule. Myoglobin, a protein of muscle, has just one polypeptide chain and resembles a single subunit of hemoglobin. The vertical axis gives the amount of oxygen bound to one of these proteins, expressed as a percentage of the total amount that can be bound. The horizontal axis measures the partial pressure of oxygen in a mixture of gases with which the solution is allowed to reach equilibrium. For myoglobin, the equilibrium curve is hyperbolic. Myoglobin absorbs oxygen readily, but becomes saturated at a low pressure. The hemoglobin curve is sigmoidal. Initially hemoglobin is reluctant to take up oxygen, but its affinity increases with oxygen uptake. At arterial oxygen pressure, both molecules are nearly saturated, but at venous pressure, myoglobin would give up only about 10 per cent of its oxygen, whereas hemoglobin releases roughly half. At any partial pressure, myoglobin has a higher affinity than hemoglobin, which allows oxygen to be transferred from blood to muscle.

globin except that it exists as a monomer. While the components of myoglobin and hemoglobin are remarkably similar, their physiological responses are very different. On a weight basis, each molecule binds about the same amount of oxygen at high oxygen tensions (pressures). At low oxygen tensions, however, hemoglobin gives up its oxygen much more readily. These differences are reflected in the oxygen-binding curves of the purified proteins in aqueous solution (Figure 10–1). The oxygen-binding curve for myoglobin is hyperbolic in shape, as would be expected for simple one-to-one association of myoglobin and oxygen:

$$Mb + O_2 \rightleftharpoons MbO_2$$

$$K_f = \frac{[MbO_2]}{[Mb][O_2]} = \text{equilibrium formation constant}$$

If y is the fraction of myoglobin molecules saturated, and if the oxygen concentration is expressed in terms of the partial pressure of oxygen $[O_2]$, then

$$K_f = \frac{y}{[1 - y][O_2]} \quad \text{and} \quad y = \frac{K_f[O_2]}{1 + K_f[O_2]}$$

This is the equation of a hyperbola, as shown in Figure 10–1.

Hemoglobin behaves differently. Its sigmoidal binding curve can be fitted by an association-constant expression with a greater-than-first-power dependence on the oxygen concentration:

$$K_f = \frac{[HbO_2]}{[Hb][O_2]^n} \quad \text{and} \quad y = \frac{K_f O_2^n}{1 + K_f O_2^n}$$

Under physiological conditions the value of n is around 2.8, indicating that the binding of oxygen molecules to the four hemes in hemoglobin is not independent and that binding to any one heme is affected by the state of the other three.* The first oxygen attaches itself with the lowest affinity, and

*An expanded development of the equations for multiple binding and the procedure for determining n from experimental data are presented in Box 10–A.

each successive oxygen is bound with a higher affinity. The exact value of n for hemoglobin is a function of the extent of oxygen binding as well as the presence of other factors discussed below. In general, a value of $n > 1$ indicates <u>cooperative binding</u> (or positive cooperativity) between small-molecule ligands, a value of $n < 1$ indicates <u>anticooperative binding</u> (or negative cooperativity) and a value of $n = 1$ indicates no cooperativity.

The cooperative binding of oxygen by hemoglobin is ideally suited to the conditions involved in oxygen transport; in the lung, where the oxygen tension is relatively high, hemoglobin can become nearly saturated with oxygen, whereas in the tissues, where the oxygen tension is relatively low, hemoglobin can release about half its oxygen (see Figure 10–1). If myoglobin were used as the oxygen transporter, less than 10 per cent of the oxygen would be released under similar conditions. The positive cooperativity associated with oxygen binding to hemoglobin is a special case of allostery in which the binding of "substrate" to one site stimulates the binding of "substrate" to another site on the same multisubunit protein. Although this is a special case of allostery, it is quite commonly observed in regulatory proteins and we will see it again when we discuss aspartate carbamoyltransferase later in this chapter.

The Binding of Certain Factors to Hemoglobin Influences Oxygen Binding in a Negative Way

Lowering the pH (i.e., increasing the H^+ concentration) decreases the affinity of hemoglobin for oxygen. For example, at pH 7.6 and 40 mm Hg of oxygen tension, hemoglobin retains more than 80 per cent of its oxygen; if the pH is lowered to 6.8, only 45 per cent of the oxygen is retained. Likewise, glycerate-2,3-bisphosphate (Figure 10–2) causes a decreased affinity between oxygen and hemoglobin (Figure 10–3). Neither H^+ nor glycerate bisphosphate at the usually observed concentrations *in vivo* would interfere with the near-saturation uptake of oxygen that takes place in the lung tissue. However, at the lower oxygen tension that exists in oxygen-consuming tissues,

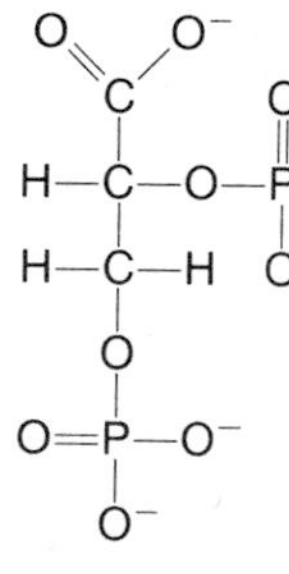

Figure 10–2

The structure of glycerate-2,3-bisphosphate, an allosteric effector for hemoglobin oxygen release.

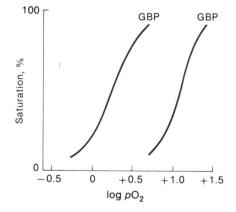

Figure 10–3

Oxygen binding curve for hemoglobin as a function of the partial pressure of oxygen. Two curves are shown, one in the absence and one in the presence of glycerate-2,3-bisphosphate (GBP). GBP decreases the affinity between oxygen and hemoglobin, as shown by the displacement of the binding curve to high oxygen concentrations in its presence.

BOX 10–A

Theory of Multiple Binding and the Hill Plot

Binding studies require a measurement of the free ligand, B, in the presence of the macromolecule, A. From this we may determine the average binding number, called y. The value of y is equal to the ratio of the total number of B molecules bound to the total number of binding sites. If there is only one binding site per A molecule, the expression for y is rather simple:

$$y = \frac{[AB]}{[A] + [AB]} \qquad (A)$$

The value of $[A] + [AB]$ is known from the total amount of A added to the solution. The value of $[AB]$ is equal to the total amount of B added, minus the experimentally observed concentration of free B after A is added:

$$y = \frac{\text{total B} - \text{free B}}{\text{total A}} \qquad (B)$$

Let us first consider the simple binding of one ligand B to a protein molecule A with a formation constant K_f:

$$A + B \xrightleftharpoons{K_f} AB; \qquad K_f = \frac{[AB]}{[A][B]} \qquad (C)$$

From Equations (B) and (C) we can express y in terms of K_f, the formation constant, or K_d, the dissociation constant:

$$y = \frac{K_f[B]}{1 + K_f[B]} \qquad (D)$$

or, since $K_f = 1/K_d$,

$$y = \frac{[B]}{[B] + K_d} \qquad (E)$$

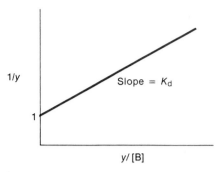

Illustration 1
A binding plot to determine the dissociation constant K_d for the simple situation where there is one ligand binding site. y is the average binding number and $[B]$ is the concentration of ligand.

Taking the reciprocal of both sides, we obtain

$$\frac{1}{y} = 1 + K_d\left(\frac{1}{[B]}\right) \qquad (F)$$

By plotting the experimentally determined value of $1/y$ against $1/[B]$, we obtain a straight line with a slope equal to the dissociation constant K_d. The intercept on the ordinate should be 1 (Illustration 1).

When there is more than one site on A for binding B, the equations become more complex and in general the plots are not linear. If we consider the average binding number for n sites on a molecule, the total average binding number is the sum of the binding number for each of these sites.

---

increasing concentrations of H^+ or glycerate bisphosphate cause a greater release of hemoglobin-bound oxygen. The net effect is that shifts in pH (7.6 to 6.8)* to low concentrations of glycerate bisphosphate (5 mM) make hemoglobin a more efficient transporter of oxygen. Both of these effects are physiologically significant. Thus proton concentrations tend to rise under the oxygen-demanding conditions of vigorous metabolic activity, and glycerate-2, 3-bisphosphate, found normally in red cells, is present in higher concentrations in organisms living at high altitudes, where the oxygen supply is lower.

Carbon dioxide also promotes the release of oxygen from hemoglobin. Since higher levels of CO_2 are found in oxygen-consuming tissues, the effect of CO_2, like that of proton, favors the delivery of oxygen to tissues where it is most needed. Together the CO_2 and H^+ effects on oxygen release are known as the Bohr effect.

The cooperative binding of oxygen to hemoglobin, as well as the inhibition of oxygen binding by protons, CO_2, or glycerate-2,3-bisphosphate, can

*This may seem like a small shift in pH to a chemist but it is a large one for the órganism.

$$y = \sum_{i=1}^{n} \frac{K_{fi}[B]}{1 + K_{fi}[B]} \tag{G}$$

There are two situations in which the data for multiple binding may be treated rather simply. First, when all binding sites bind B with the same energy, all terms are equal and the solution to the sum is simply n times each term:

$$y = \frac{nK_f[B]}{1 + K_f[B]} \tag{H}$$

Again, since $K_f = 1/K_d$,

$$y = \frac{n[B]}{[B] + K_d} \tag{I}$$

and taking the reciprocal of both sides,

$$\frac{1}{y} = \frac{1}{n} + \frac{1}{n}K_d\left(\frac{1}{[B]}\right) \tag{J}$$

If $1/y$ is plotted against $1/[B]$, the slope is equal to K_d/n and the intercept (when $1/[B]$ approaches 0) is $1/n$. Alternatively, the data may be plotted as y versus $y/[B]$ (Illustration 2), in which case the slope is $-K_d$ and the intercept at $y/[B] = 0$ is n, since

$$y = n - \frac{yK_d}{[B]} \tag{K}$$

Second, in one other situation, multiple binding can be treated rather simply. If the sites interact so strongly that only the fully saturated product AB_n is formed, the data may be treated in the following way. The formation of only one major product means that the binding of the ligand molecule to the macromolecule greatly enhances

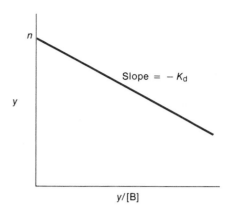

Illustration 2
A binding plot to determine the dissociation constant K_d and the number of binding sites n for the situation where there are n ligand binding sites per macromolecule with identical binding affinities.

the further binding of additional ligands such that at equilibrium, only three species exist in significant concentrations: A, B, and AB_n:

$$A + nB \rightleftharpoons AB_n$$

$$y = \frac{n[AB_n]}{[AB_n] + A} \tag{L}$$

$$y = \frac{nK_f[B]^n}{1 + K_f[B]^n} \tag{M}$$

$$\frac{1}{y} = \frac{1}{n} + \frac{1}{nK_f[B]^n} \tag{N}$$

be understood in terms of the effects these small-molecule ligands have on the hemoglobin structure. All of these ligands, i.e., oxygen, protons, and glycerate-2,3-bisphosphate, bind at different, nonoverlapping sites on the protein, meaning that the influence of one ligand binding on another must be transmitted through the protein polypeptide chains. Our understanding of these effects has resulted largely from crystallographic studies by x-ray diffraction.

X-ray Diffraction Studies Indicate Two Conformations for Hemoglobin

X-ray diffraction studies on fully oxygenated hemoglobin and deoxygenated hemoglobin have shown that the molecule is capable of existing in two states, with significant differences in tertiary and quaternary structures (Figure 10–4). Further studies on partially oxygenated hemoglobin may reveal further intermediate structures between these two extremes. Until these can be characterized in structural terms, the two-state model serves as a useful

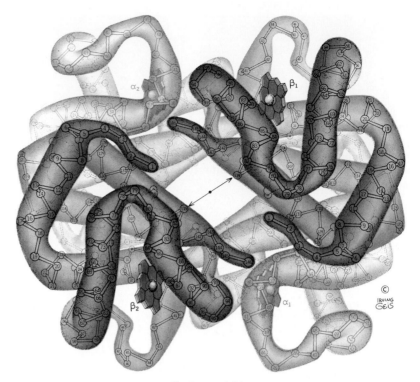

Oxyhemoglobin

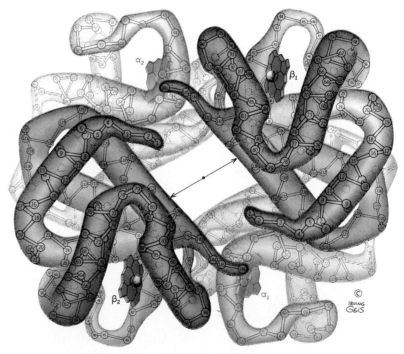

Deoxyhemoglobin

Figure 10–4

Three-dimensional structure of oxy and deoxy hemoglobin as determined by x-ray crystallography. This is a view down the dyad (two fold) axis, with the β chains on top. In the oxy-deoxy transformation (quaternary motion) $\alpha_1\beta_1$ and $\alpha_2\beta_2$ dimers move as units relative to each other. This allows glycerate-2,3-bisphosphate to bind to the larger central cavity in the deoxy conformation. A close-up of the binding site is shown in Figure 10–7.

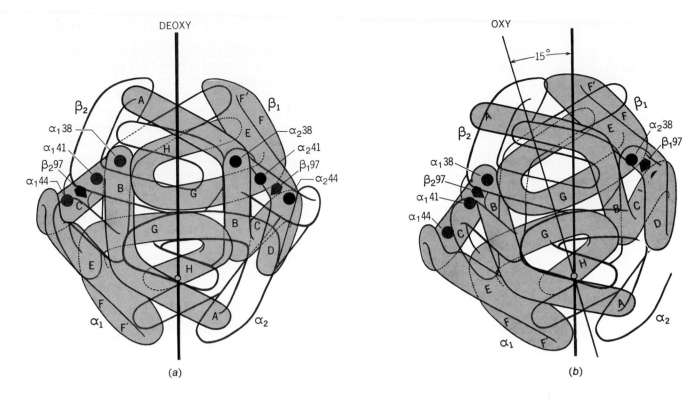

Figure 10–5

The deoxy-to-oxy shift upon binding oxygen in one hemoglobin molecule. The projection shown in this figure is approximately perpendicular to the one shown in Figure 10–4. The $\alpha_1\beta_1$ dimer moves as a unit relative to the $\alpha_2\beta_2$ dimer. The interface between the two dimers is crucial to the cooperativity effect in hemoglobin. The interface is not visible in this figure (see Figure 10–6).

conceptual framework for explaining the allosteric mechanism of the hemoglobin system.

The hemoglobin tetramer is composed of two identical halves (dimers), with the $\alpha_1\beta_1$ subunits in one dimer and the $\alpha_2\beta_2$ subunits in the other. The subunits within the dimers are tightly held together; the dimers themselves are capable of motion with respect to one another (Figure 10–5). *The interface between the movable dimers contains a network of salt bridges and hydrogen bonds when hemoglobin is in the deoxy conformation* (Figure 10–6). *The quaternary transformation that takes place on binding of oxygen causes the breakage of these bonds.*

The effects of H^+, CO_2, and glycerate-2,3-bisphosphate on oxygen binding can be understood in terms of their stabilizing effect on the deoxy conformation. The decreased oxygen binding as the pH is lowered from 7.6 to 6.8 suggests the involvement of histidine side chains, because these are the only side-chain groups in proteins that have a pK in this pH range. Certain histidines in the charged form make salt linkages that contribute to the stability of the deoxy form (see Figure 10–6b). As the pH is lowered, these histidines tend to become charged, which increases the stability of the deoxy form. Such a change should inhibit a structural transition to the oxy form and thereby lower the affinity of the protein for oxygen. Similarly, glycerate-2,3-bisphosphate binds most strongly to the deoxy form (Figure 10–7) and thereby discourages the transition to the oxy form, which lowers the affinity for oxygen. Carbon dioxide binds as bicarbonate to the α amino group in hemoglobin; this binding also favors the deoxy conformation.

Changes in Conformation Are Initiated by Oxygen Binding

*The oxygen binding at the heme group itself initiates the tertiary- and qua-
ternary-structure changes that are responsible for the cooperative effect seen
on oxygen binding.* The heme group contains an Fe^{2+} ion located near the
center of a porphyrin ring. This Fe^{2+} makes four single bonds to the nitrogens in the heme ring, and a fifth bond to a histidine side chain of the F
helix, F8 histidine (Figure 10–8). When oxygen is present it binds at the
sixth coordination position of the iron on the other side of the heme. Movement of the iron upon oxygen binding (or release) pulls the F8 histidine and
the F helix to which it is covalently attached (Figure 10–9). *The tertiary-
structure change in the F helix induces a strain in the rest of the protein that
facilitates the conversion of the deoxy to the oxy structure. This favors the
binding of further oxygen at other unoccupied sites on the tetramer.*

The Fe^{2+} ion is well suited to its job in hemoglobin, not only because it
has a natural affinity for oxygen, but also because it changes its electronic

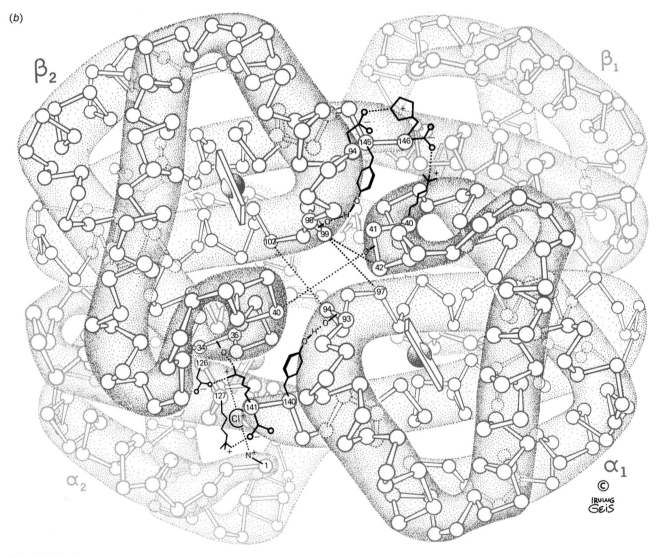

Figure 10–6

(a) The $\alpha_1\beta_2$ (and $\alpha_2\beta_1$) interface is shown schematically at
the upper left and in detail below. This is the regulatory
zone of the hemoglobin molecule, which contains crucial
hydrogen bonds and salt bridges. (b) All the hydrogen bonds
and salt bridges shown here (dotted lines) exist only in the
deoxy state, with the exception of $\alpha_1 41 – \beta_2 40$ and $\alpha_1 94 –
\beta_2 102$, which exist only in the oxy state. Only the α carbons
are shown in the backbone structure of the hemoglobin,
except in the region of the interface.

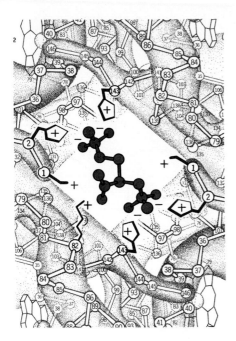

Figure 10–7
The binding of glycerate-2,3-bisphosphate in the central cavity of deoxyhemoglobin between β chains. The surrounding positively charged residues are the amino terminal, His 2, Lys 82, and His 143.

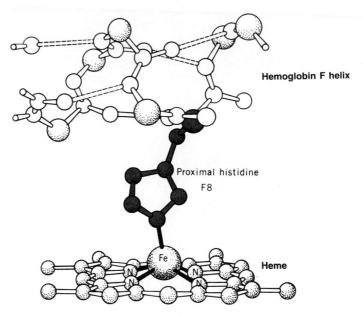

Figure 10–8
A close-up view of the iron-porphyrin complex with the F helix in deoxyhemoglobin. Note that the iron atom is displaced slightly above the plane of the porphyrin.

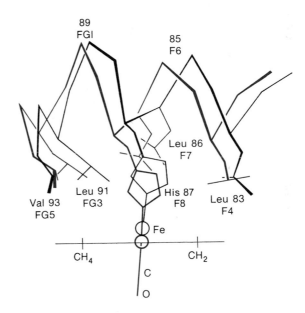

Figure 10–9
Downward movement of the iron atom and the complexed polypeptide chain on binding oxygen. The structure is shown before (black) and after (colored) binding oxygen. Movement of His F8 is transmitted to FG5 valine, straining and breaking the hydrogen bond to the penultimate tyrosine. Only the α chain is shown here.

structure in a highly significant way in so doing. Fe^{2+} is normally paramagnetic as a result of four unpaired electrons in its outer d electronic orbitals. As such, it is too large to sit precisely in the plane of a porphyrin, as studies with model compounds have shown. Fe^{2+} is also paramagnetic when it is pentacoordinated in deoxyhemoglobin, and as expected, it is displaced from the plane of the porphyrin by a few tenths of an angstrom unit. When O_2 binds, the Fe^{2+} becomes hexacoordinated and diamagnetic (no unpaired electrons). This change results in a major reorganization of its outer d orbit-

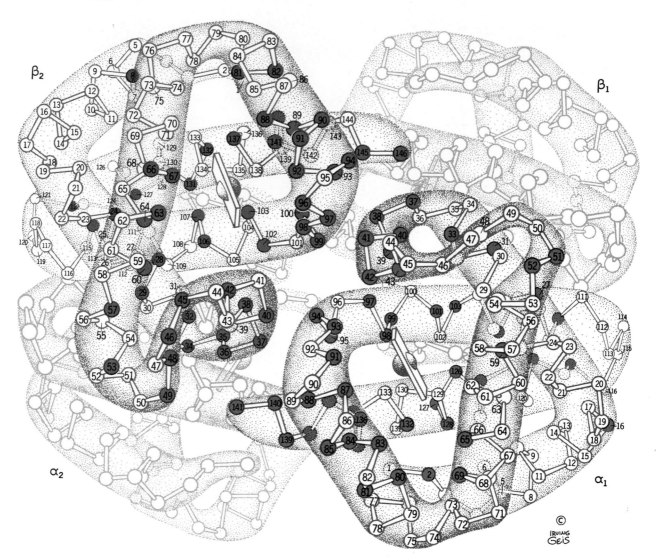

Figure 10–10
Invariant residues in the α and β chains of mammalian hemoglobin. The colored dots, indicating the positions of invariant residues, line the heme pockets as well as the crucial $\alpha_1\beta_2$ interface. The invariant residues have been found in about 60 species. There are 43 invariant positions in the hemoglobin molecule.

als, which decreases the radius of the Fe^{2+} so that it can move to an energetically more favorable position in the center of the porphyrin (Figure 10–9).

The structural arguments advanced here to explain oxygen binding by hemoglobin are supported by amino acid sequences of α and β chains from a large number of vertebrate as well as invertebrate hemoglobins. Data from 60 species of α chains and 66 species of β chains reveal 43 invariant positions in the hemoglobin molecule. These _invariants_ are plotted on a map of the hemoglobin structure in Figure 10–10. Here the invariant positions are shown by colored dots. In a sense, the positions of these dots provide a diagram of the working machinery of hemoglobin, for they line the heme pockets where oxen is bound and the crucial $\alpha_1\beta_2$ interface, which changes its orientation when oxygen binds. Electrostatic forces and hydrogen bonds stitch this interface together in the deoxy conformation when the molecule gives up its oxygen to the tissues. If there are changes, resulting from mutations, at any of these positions, then trouble is likely to develop. This is just

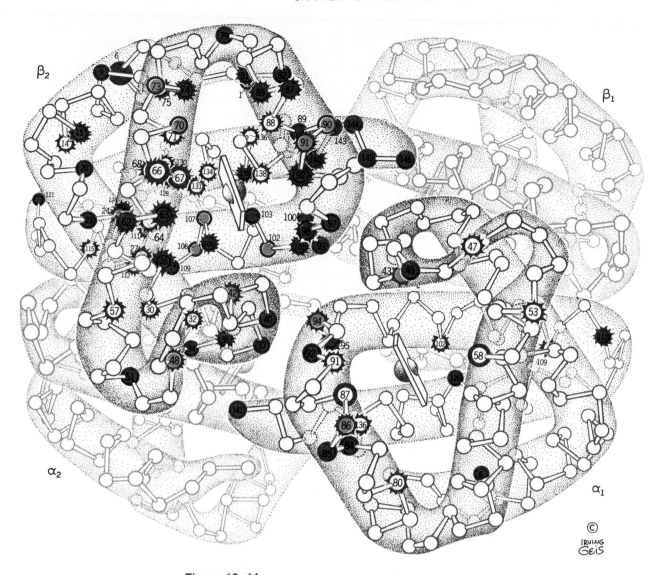

Figure 10–11

Positions of mutations in hemoglobin that produce a pathological condition.
Comparison with Figure 10–10 shows that these mutations, in general, show the same
pattern as the distribution of the invariant positions. (Dark circles indicate positions of
abnormal residues, solid black dot indicates the valine β6 mutation in sickle-cell
anemia, heavy circles indicate M (Met) hemoglobin, and jagged perimeter indicates
unstable hemoglobin. Dark color indicates increased oxygen affinity; light color
indicates decreased oxygen affinity.

what happens, as can be seen in Figure 10–11, which shows the positions of
pathological mutations in hemoglobin. Where there are changes in the heme
pockets or in the $\alpha_1\beta_2$ interface, hemoglobin abnormalities occur; many of
these are associated with serious diseases.

As a rule, invariants are the only critical loci for hemoglobin function.
One striking exception occurs at position 6 of the β chain. A hydrophobic
valine residue is substituted for glutamic acid (a charged side chain) with
disastrous results. The specific consequence of the β6 alteration is to cause
hemoglobin tetramers to aggregate when they are in the deoxy state. The
aggregates form long fibers that stiffen the normally flexible red blood cell.
The resulting distortion of the red cells leads to capillary occlusion, which
prevents proper delivery of oxygen to the tissues. This pathological condition is known as sickle-cell anemia.

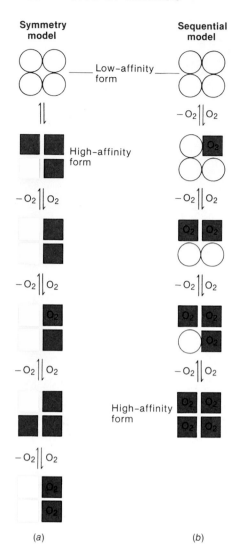

Symmetry model   Sequential model

Low-affinity form

High-affinity form

High-affinity form

(a)   (b)

Figure 10–12
Alternate models for hemoglobin allostery. (a) In the symmetry model hemoglobin can exist in only two states. (b) In the sequential model hemoglobin can exist in a number of different states. Only the subunit binding oxygen must be in the high-affinity form.

ALTERNATIVE THEORIES ON HOW HEMOGLOBINS AND OTHER ALLOSTERIC PROTEINS WORK

We have spent a great deal of time developing the hemoglobin story because it is the best-understood regulatory protein and provides us with a model system for understanding how other allosteric proteins work. Hemoglobin was first recognized as an allosteric protein by the sigmoidal shape of its oxygen binding curve (see Figure 10–1). Allosteric proteins are usually composed of two or more subunits. Different ligands may bind to quite different sites. In the case of hemoglobin four of the ligands are the same. Namely O_2, and the oxygen-binding sites also are quite similar.

Two quite different models were proposed about twenty years ago to explain the unusual nature of the hemoglobin oxygen-binding curve. These models could also be used to explain other allosteric proteins (Figure 10–12); they are introduced here because more is known about ligand binding and allosteric transitions for hemoglobin than for any other allosteric protein. The first model, introduced by Monod, Wyman, and Changeux in 1965, is called the _symmetry model_. In this model hemoglobin can exist in only two conformations, one with all four of the subunits within a given tetramer in the low-affinity form and one with all four subunits in the high-affinity form (Figure 10–12a). Also, in this model, the hemoglobin molecule is always symmetrical, i.e., all the subunits are either in one state or the other, and all of the binding sites have identical affinities. The binding of oxygen to one of the subunits favors the transition to the high-affinity form. The greater the number of oxygens that bind to the tetramer, the more likely it is that the transition from the low-affinity form to the high-affinity form will occur.

The second model, proposed by Koshland, Nemetby, and Filmer in 1966, is referred to as the _sequential model_ (Figure 10–12b). In this model the binding of an oxygen molecule to a given subunit causes that subunit to change its conformation to the high-affinity form. Because of its molecular contacts with its neighbors, the change increases the probability that another subunit in the same molecule will switch to the high-affinity form and bind a second oxygen more readily. The binding of the second oxygen has the same type of enhancing effect on the remaining unoccupied oxygen binding sites.

Either of these models (or something in between) could account for the sigmoidal oxygenated curve of hemoglobin, and either is consistent with the fact that deoxygenated and fully oxygenated hemoglobin have different conformations. The only way to rigorously discriminate between these two models is to obtain structural information on partially oxygenated hemoglobin. This information is still lacking, so no final judgment can yet be made. Even when the situation is fully resolved for hemoglobin, there is no assurance that other allosteric proteins must work in the same way.

ASPARTATE CARBAMOYLTRANSFERASE: AN ALLOSTERICALLY REGULATED ENZYME

Aspartate carbamoyltransferase (ACTase) of E. coli is one of the better-understood allosterically regulated enzymes, and we shall discuss it as an example of this class of regulatory enzymes. ACTase catalyzes the first committed reaction in the biosynthetic pathway for pyrimidines (Figure 10–13), which involves the formation of carbamoyl aspartate from carbamoyl phosphate and aspartate (the entire pathway is described in Chapter 25). The carbamoyl group appears to be transferred directly from the carbamoyl phosphate to L-aspartate without the formation of any intermediate products or a carbamoyl enzyme. The first enzyme of the pathway is inhibited by the final product of the multistep pathway, namely cytidine triphosphate.

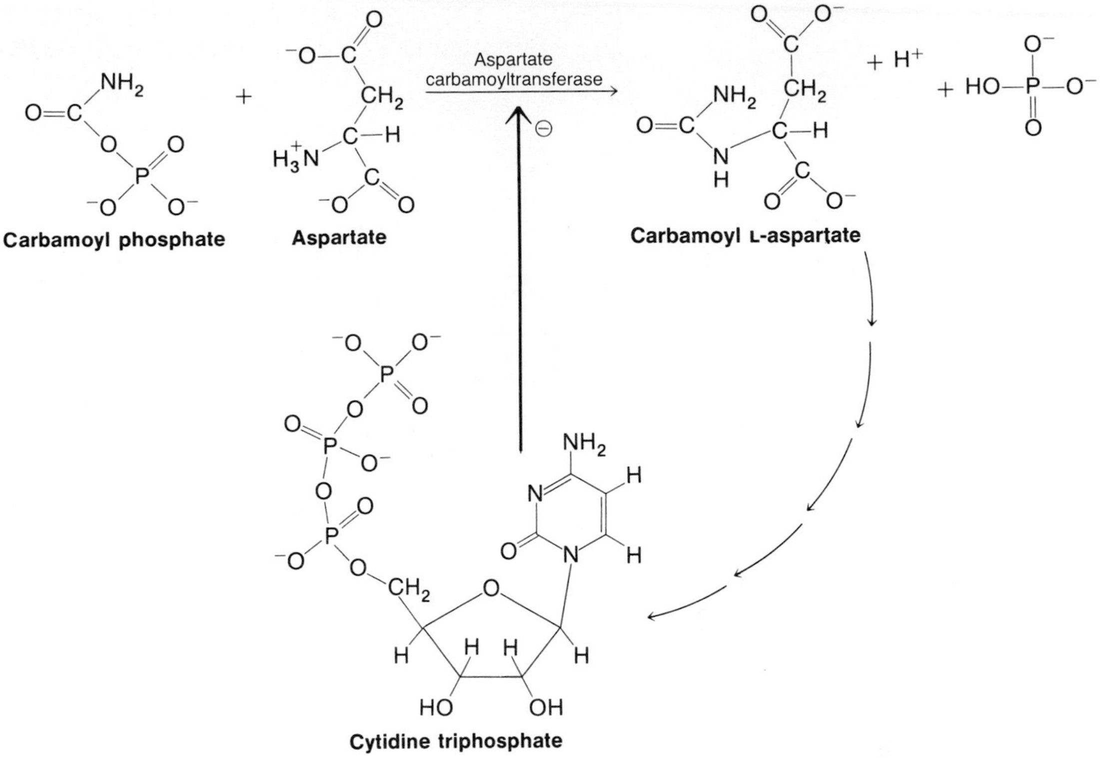

Figure 10–13
Reaction catalyzed by aspartate carbamoyltransferase and the feedback inhibition of this enzyme in *E. coli* by the pathway end product, CTP. The sequence of small arrows represents additional reaction steps in the pathway to CTP. The red arrow with the negative sign indicates the site of inhibition.

Substrate Binding to Aspartate Carbamoyltransferase Follows an Ordered Pathway

As with many other bisubstrate reactions, two pathways are possible for the order of binding of the two reactants to the enzyme. But the upper pathway shown in Figure 10–14, in which carbamoyl phosphate binds first, appears to predominate. Evidence favoring this order of binding comes from interaction studies of the two substrates in the presence of the transition-state (or bisubstrate) analog N-phosphonacetyl-L-aspartate (PALA), whose structure is shown in Figure 10–15. This compound is a strong competitive inhibitor

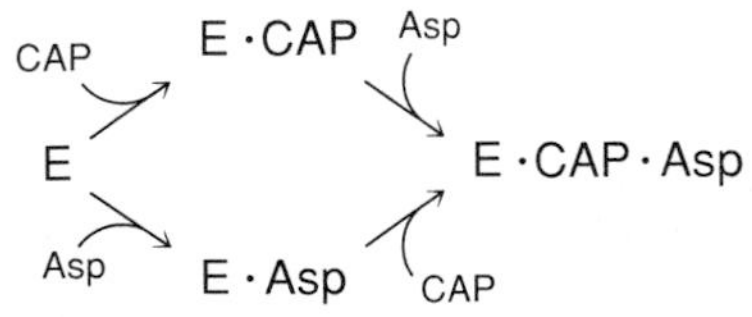

Figure 10–14
Possible pathways leading to an enzyme-carbamoyl-phosphate-L-aspartate complex. A random mechanism would have significant contributions from either pathway, whereas an ordered mechanism would go by one route or the other. In fact, the ordered mechanism shown by the pathway above predominates. CAP stands for carbamoyl phosphate and ASP stands for aspartate.

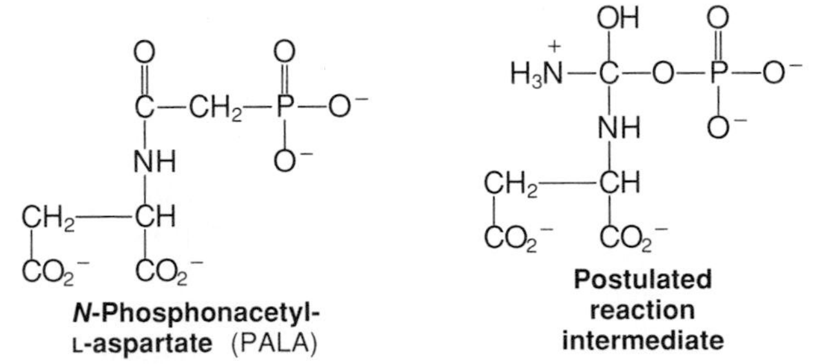

Figure 10–15
The structure of N-phosphonacetyl-L-aspartate (PALA), a transition-state analog for the aspartate carbamoyltransferase-catalyzed reaction. PALA is a competitive inhibitor for carbamoyl phosphate and a noncompetitive inhibitor for aspartate. The structure of the postulated reaction intermediate is shown for comparison. It can be seen that there is a great structural similarity between the two compounds.

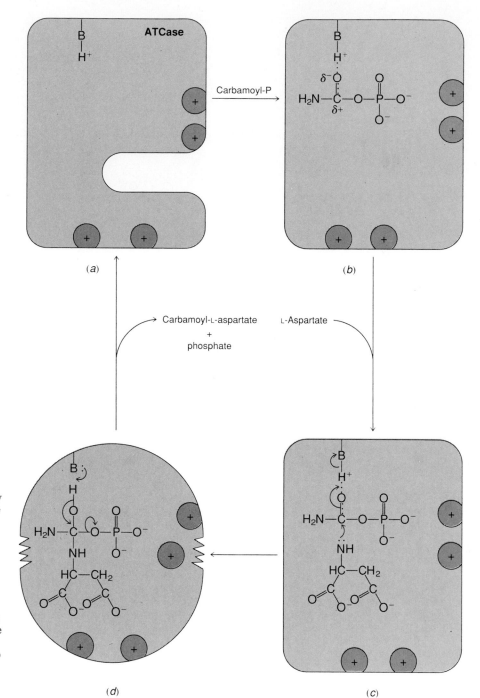

Figure 10–16
Schematic representation of the
compression mechanism for ACTase.
(a) Unliganded enzyme with positively
charged sites for substrate binding, the
acid BH$^+$, and a steric constraint on
the binding site for L-aspartate
(represented by the deep impression
in the form of the enzyme).
(b) Carbamoyl phosphate binds and
interacts with BH$^+$ through its
carbonyl oxygen. The steric constraint
on L-aspartate binding is relieved.
(c) L-Aspartate binds, and the α amino
group is in a position to react with the
bound and activated carbamoyl
phosphate. (d) Compression of the two
substrates together by a
conformational change of the enzyme
aids reaction.

for carbamoyl phosphate binding and a strong noncompetitive inhibitor for
L-aspartate binding. The interpretation of the first result is that the inhibitor
and carbamoyl phosphate compete for the same site on the enzyme. Thus
when one compound is present in a high concentration, it should be able to
displace the other compound according to the law of mass action. However,
although PALA strongly inhibits L-aspartate binding, the effect cannot be
overcome by high levels of L-aspartate. If carbamoyl phosphate is required
for aspartate binding, and if PALA prevents binding of the former, no
amount of aspartate will displace PALA from the active site.

A schematic representation of the binding site and the partial mechanism of the reaction are shown in Figure 10–16. At the active site, the free
enzyme contains positively charged sites for substrate binding, the general
acid BH$^+$, and a steric constraint on the binding site for L-aspartate. Carbam-

oyl phosphate binds and interacts with BH^+ through its carbamoyl oxygen. The steric constraint on L-aspartate binding is relieved, so that L-aspartate can bind. Its α amino group is in a position to react with the bound and activated carbamoyl phosphate. The two substrates are pushed toward one another, aided by a conformational change of the enzyme. Transition from the hypothesized tetrahedral intermediate (Figure 10–16d) to products could invoke additional base catalysis. Further information is required to verify this mechanism.

Chemical and Kinetic Evidence Demonstrates Allosteric Behavior

There is abundant evidence that aspartate carbamoyltransferase is allosterically regulated. This evidence comes first and foremost from studies of the enzyme activity as a function of substrate concentration. The explanation for this concentration dependence runs parallel to that previously given for hemoglobin (also see Box 10–A).

First let's review the theory of multisubstrate binding. If you have an enzyme with n binding sites, all of which have an equal affinity for the small molecule or ligand in question, then the binding curve may be expressed by the equation:

$$y = \frac{K_f[B]}{1 + K_f[B]}$$

where y is the average binding number, i.e., the ratio of the number of binding sites occupied to the total number of binding sites; K_f is the affinity constant, or formation constant, for the binding to a single site, and $[B]$ is the free molar concentration of the small molecule that binds. This is the equation for a hyperbola, as shown for the case of oxygen binding to myoglobin (see Figure 10–1), where $n = 1$. Even if n is not equal to 1, as in the case of lactate dehydrogenase, where n is 4, a hyperbolic binding curve is obtained. As long as the binding sites are equivalent, and binding at one site does not affect binding at another site on the same protein molecule, a hyperbolic binding curve is observed.

At the other extreme we can imagine a protein with more than one binding site, where binding at one site influences the binding at other sites on the same protein so strongly that either all the binding sites are occupied or none of them are occupied. This is the extreme case of positive cooperativity between binding sites, and the relevant equation to describe the situation is

$$y = \frac{nK_f[B]^n}{1 + K_f[B]^n}$$

When y is plotted against $[B]$, a sigmoidal dependence on binding is observed (e.g., see the curve for oxygen binding to hemoglobin, Figure 10–1). Hemoglobin has four nearly identical binding sites for oxygen, so that the largest possible value of n for oxygen binding would be 4. The experimentally determined value of n is 2.8. This value indicates some positive cooperativity between binding sites. It is not clear exactly how to translate the result into a molecular picture. Although we know the average amount of O_2 bound per hemoglobin, we do not know the distribution of oxygen binding among the different hemoglobin molecules. This is a recurrent problem in analyzing binding curves that show some positive cooperativity. At the present time, we must content ourselves with a qualitative understanding of such situations; a hyperbolic binding curve indicates little or no cooperativity between binding sites, and a sigmoidal curve indicates positive cooperativity. The numerical value of n gives us a qualitative idea of the extent of

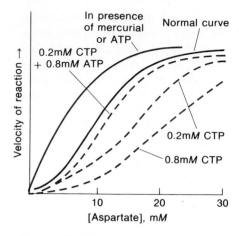

Figure 10–17

Effect of CTP, ATP, and mercurials on the affinity of aspartate carbamoyltransferase for aspartate (actually, reaction velocity is measured). Positive cooperativity on aspartate binding is reflected by the pronounced sigmoidal character of the reaction curve in the absence of other additives. ATP eliminates the positive cooperativity and CTP augments it. ATP reverses the CTP effect. Organic mercurials convert the curve from sigmoidal to hyperbolic. Other experiments show that the mercurials convert the enzyme into two trimers of the catalytic subunits and three dimers of the regulatory subunits.

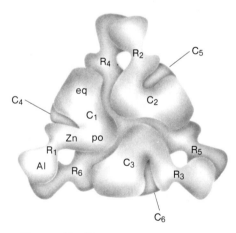

Figure 10–18

Subunit and domain structure of aspartate carbamoyltransferase. The threefold axis is in the center of the molecule as projected and extends as a vertical line above and below the plane of the page. A threefold axis of symmetry means that if the molecule is rotated by one third of a full circle about the axis, an identical projection of the structure results. Domains of the catalytic chains are labeled equatorial (eq) and polar (po), while those of the regulatory chains are labeled zinc (Zn) and allosteric (Al). (Adapted from K. L. Krause, K. W. Volz, and W. N. Lipscomb, *Proc. Natl. Acad. Sci.* 82:1643–1647, 1985.)

cooperativity. Values of n of less than 1 indicate negative cooperativity between binding sites.

For situations involving an enzyme, the velocity of reaction is plotted instead of the binding because this is easier to measure. It is generally valid to do this, as the velocity of the reaction is usually proportional to substrate bound.

In the case of aspartate carbamoyltransferase, positive cooperativity on aspartate binding is reflected by the pronounced sigmoidal character of the reaction curve (Figure 10–17). ATP eliminates the positive cooperativity and CTP augments it. Furthermore, ATP reverses the CTP effect, as would be expected if the two nucleotides compete for the same binding site on the enzyme. Both of the nucleotide effects make good physiological sense. An abundance of CTP would decrease the need for further pyrimidine synthesis, whereas an abundance of ATP might be a good signal for favoring more pyrimidine synthesis, which, in turn, should favor more biosynthesis (see Chapter 13).

The inhibiting effect of CTP increases with the concentration of CTP. However, even the effects of elevated levels of CTP (0.5 mM) are overcome by increasing the concentration of aspartate, a result showing that the inhibitory effect is a competitive one. It is clear from the results presented below that the CTP binding site in the enzyme is far removed from the active site. Therefore, competitive inhibition need not result from binding of the inhibitor at a site coincident with the substrate binding site. Allosteric enzymes in different instances show either competitive or noncompetitive behavior toward inhibitors.

Direct evidence for the allosteric nature of aspartate carbamoyltransferase began with the observation that the regulatory behavior of the enzyme disappears when it is treated with organic mercurials such as p-hydroxymercuribenzoate (see Figure 10–17). After such treatment, ATP and CTP at discrete concentrations no longer influence the enzyme activity, and the binding of aspartate no longer shows any positive cooperativity. Other measurements show that exposure to mercurials results in the dissociation of the enzyme into two fragments. In the native enzyme, each molecule contains six regulatory R-protein subunits ($M_r = 17,000$) and six catalytic C-protein subunits ($M_r = 34,000$). Treatment of the enzyme with mercurials leads to a dissociation into two trimers of the C subunit and three dimers of the R subunit. These can be observed as separate peaks by sedimentation velocity analysis in the ultracentrifuge. The separated C_3 protomers show enzyme activity, while the separated R_2 protomers show binding affinity for the usual allosteric effectors ATP and CTP, but no enzyme activity. Sensitivity of enzyme activity to the allosteric effectors ATP or CTP requires reconstitution of the enzyme from the R- and C-containing portions in the presence of Zn^{2+}, which is a constituent of the native enzyme.

X-ray Diffraction Studies Confirm the Allostery and Reveal the Groups at the Active Site

Our understanding of ACTase has been greatly augmented by the x-ray diffraction studies from Bill Lipscomb's laboratory. High-resolution structures (around 2.8 Å) have been determined for the free enzyme, the enzyme complexed to the negative effector CTP, and the enzyme complexed to the bisubstrate analog PALA (see Figure 10–15).

In the free enzyme the C subunits are centrally located and related to each other within each C_3 trimer by a threefold axis of symmetry (Figure 10–18). In the catalytic chain a polar domain shows interactions between adjacent pairs of C chains that form each C_3 trimer, while an equatorial domain shows intramolecular C_3–C_3 interactions. Each catalytic chain is also in contact with two regulatory monomers. Each regulatory monomer borders on one other regulatory chain and two catalytic chains.

At a higher level of resolution it can be seen that there is a good deal of β-sheet structure and α-helix structure present in both subunits (Figure 10–19). In the regulatory chain there is a CTP-binding domain that interacts with an adjacent regulatory subunit and a zinc-binding domain that interacts with the catalytic subunit.

The CTP binding sites are located a considerable distance from the active sites. Therefore, the inhibitory effect of CTP must result from an alteration in the protein conformation that is transmitted through the polypeptide chains to the active site. For the most part, the regulatory and catalytic chains of the free enzyme and the CTP-complexed enzyme are in similar conformations. Large conformational differences between the two forms exist, however, in the region of CTP binding to the regulatory chain. In addition, a segment of ten amino acid residues, which includes Lys 83 and Lys 84 of the catalytic chain, is disordered in the CTP-complexed enzyme. In the free enzyme these same amino acids assume fixed positions. The inhibitory effect of CTP is probably due to the movement of this portion of the catalytic chain. As we will see, Lys 84 is almost certainly involved in the binding of substrate and reaction intermediate at the active site.

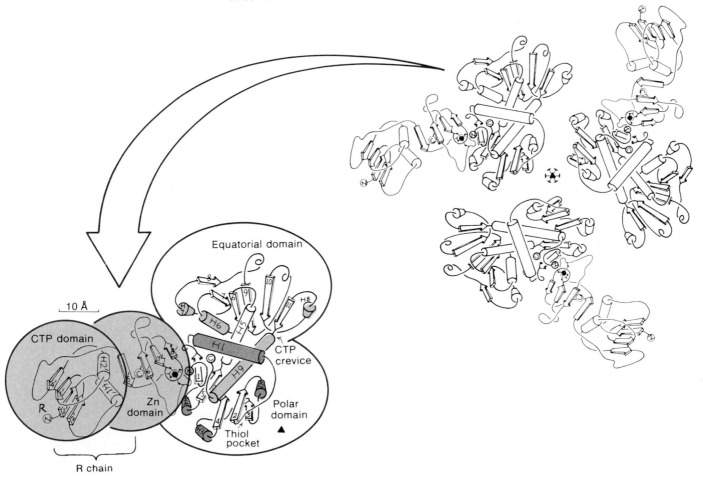

Figure 10–19

(Right) The "upper half" C_3R_3 of the molecule of aspartate carbamoyltransferase. The angle of the projection is the same as that in Figure 10–18, down the threefold axis. (Left) A single CR unit. The CTP domain of the R chain is furthest to the left, while the Zn domain is adjacent and connected. Within the CR unit the Zn domain makes contact with the polar domain in the catalytic monomer. The active site is near the CTP crevice at a boundary between adjacent C monomers in the full structure. Helices and strands of β sheet have been numbered in order from NH₂ terminal (N) of the C chain and similarly, with primes added, for the R chain. (Adapted from H. L. Monaco, J. L. Crawford, and W. N. Lipscomb, Three-dimensional structures of aspartate carbamoyltransferase from *Escherichia coli* and of its complex with cytidine triphosphate, *Proc. Natl. Acad. Sci.* 75:5277, 1978.)

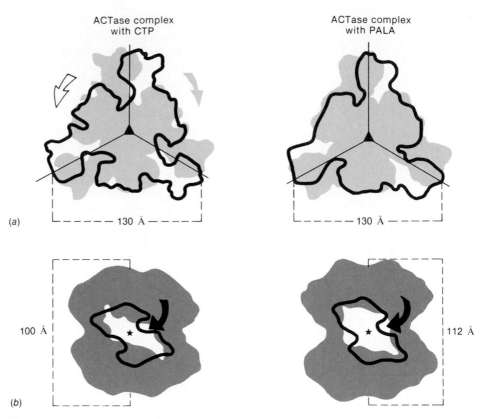

Figure 10–20

Changes in relative positions of the ACTase subunits. The CTP-enzyme complex is on the left, the PALA-enzyme complex is on the right. *(a)* View down the molecular threefold axis. The lower C_3 and the three lower R's are represented in light color. The upper C_3 and the three upper R's are represented by the heavily outlined area. The directions of C_3 rotation that accompany conversion from the one state to the other are indicated by the open arrow for the upper C_3 and the shaded arrow for the lower C_3. *(b)* View down the molecular twofold axis. The four C chains closest to the viewer are represented by the dark color. One R_2 dimer is represented by the heavily outlined area and is below the four C chains. The open space is the central cavity. The access to the cavity, which is unobstructed from this viewpoint, is marked with an arrow. The access to the central cavity is greater in the PALA-enzyme complex than in the CTP-enzyme complex. (Adapted from J. E. Ladner et al., *Proc. Natl. Acad. Sci.* 74:3126, 1982.)

Large quaternary structural changes occur when the bisubstrate analog PALA binds to the enzyme (Figure 10–20). The two catalytic trimers (C_3) move apart by 12 Å and mutually reorient by 10°. The three regulatory dimers (R_2) reorient, each about its twofold axis, by about 15°. In this transition, new polar interactions develop between equatorial domains of C chains and the Zn domains of the R chains. Within the C chain the two domains, one binding the phosphonate moiety and the other binding the aspartate moiety of PALA, move closer together. A loop including Lys 84 makes a large relocation so that this residue and Ser 80 bind to the PALA of an adjacent catalytic chain within the same C unit. A very large change in tertiary structure brings the 230–245 loop nearer the active site, allowing Arg 229 and Gln 231 to bind to the PALA.

The detailed structure of PALA at the catalytic site is shown in Figure 10–21. Since the structure of PALA is close to that of the postulated reaction intermediate (see Figure 10–15), the structure of this complex is highly relevant to the reaction mechanism. As shown, PALA interacts with four arginines and one lysine and with a number of hydrogen bond donors. Most of these interactions occur within the catalytic chain to which PALA binds; but because of a conformational change (referred to earlier) that occurs on bind-

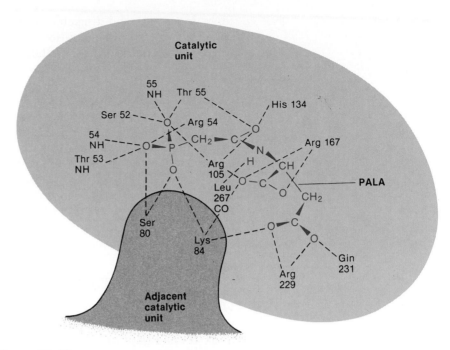

Figure 10–21

Binding of PALA to the catalytic unit of the ACTase holoenzyme. Dashed lines are drawn between various groups on the PALA and the amino acid side chains that are in close contact. Note that two of these residues, Ser 80 and Lys 84, originate from an adjacent catalytic chain. (Adapted from K. W. Volz, K. L. Krause, and W. N. Lipscomb, *Biochem. Biophys. Res. Comm.* 136:828, 1986.)

ing PALA, some residues from an adjacent catalytic chain also bind the PALA molecule. These additional contacts resulting from the conformational change help to explain the allosteric advantage of the protein-PALA complex.

The phosphonate moiety of PALA makes salt links with Arg 54 and Arg 105 and hydrogen bonds to Ser 52, Thr 55, and the peptide NH groups of residues 53, 54, and 55. Lys 84 and Ser 80 from an adjacent C chain also bind to this portion of the PALA molecule.

The α-carboxylate group is bound to Lys 84 (adjacent chain), Arg 105, and Arg 167. The β-carboxylate is also bound to Lys 84 (from the adjacent chain) and to Arg 167, Gln 231, and Arg 229.

The NH of the peptide bond of PALA forms a hydrogen bond to the carbonyl group of Leu 267, while the CO of the PALA peptide bond is hydrogen-bonded to His 134, Thr 55, and Arg 105. From its location at the active site, His 134 is the most likely candidate for the general acid-base catalyst in the postulated reaction mechanism (see Figure 10–16).

The structural information derived from the x-ray diffraction studies clearly supports the main conclusions from the chemical studies that ACTase is an allosteric enzyme. It also gives a detailed description of the catalytic site, which will undoubtedly be of great value in further studies aimed at a detailed understanding of the catalytic mechanism.

THE ADVANTAGES OF POSITIVE COOPERATIVITY

In hemoglobin and aspartate carbamoyltransferase we see two classic examples of allostery that involve positive cooperativity in substrate binding (here we are speaking loosely of oxygen as a substrate). In both cases the binding of one oxygen or one aspartate molecule encourages the binding of a second and so on. If the communication between subunits is interrupted, as by dissociation of the subunits, the binding at low levels of substrate is increased and the binding changes from a sigmoidal to a hyperbolic dependence on substrate concentration. From this result it seems unlikely that the function of the allostery in these examples of positive cooperativity is merely to increase the binding. *The unique advantage of the positive cooperativity resulting from substrate binding is that it encourages the binding to occur over a narrow range of concentrations of small molecule.* This makes the system extremely sensitive to small changes in concentration of small molecule. In the case of hemoglobin we have seen that this sensitivity improves the net transport of oxygen from the lungs (where the oxygen tension is comparatively high) to other tissues (where the oxygen tension is comparatively low). In the case of aspartate carbamoyltransferase the enzyme will tend towards low activity unless the aspartate concentration is quite high; then it will rise rapidly. The advantage of sensitivity to substrate concentration is that it permits more constant control of metabolic processes. We will elaborate on the merits of this feature in Chapter 13.

NEGATIVE COOPERATIVITY

We have seen that aspartate carbamoyltransferase shows positive cooperativity for the binding of aspartate, just as hemoglobin shows positive cooperativity for the binding of oxygen. The ACTase enzyme also shows negative cooperativity in the binding of CTP to the R subunits. The advantage of this negative cooperativity is that it limits the production of CTP according to need by inhibiting the first reaction in the pathway when CTP is present in excess. Quite remarkably, ACTase also shows another type of negative cooperativity in the binding of carbamoyl phosphate to the C subunits. Careful pre-steady-state measurements on purified enzyme show that only three of the six subunits participate readily in the binding of either of these molecules. Apparently the binding of the small molecule to one subunit causes a structural transition so that a nearby subunit cannot participate similarly in the binding process.

A number of other enzymes show the same type of negative cooperativity, which is known as *half-of-the-sites reactivity*. They include acetoacetate decarboxylase, aldolase, alkaline phosphatase, some aminoacyl-tRNA synthetases, and many other enzymes. There is not yet a convincing argument for the physiological usefulness of this widespread phenomenon, but Paul Boyer has made an interesting proposal. Ideally, after an enzyme has catalyzed a reaction it should open up to release the products rapidly. If two subunits are coupled so that closing one will open the other one, like a see-saw, then the two can alternate as catalysts. One binds the substrate and closes, while the other opens up to release products, then the roles reverse. It has also been suggested that half-of-the-sites reactivity might assist the catalytic steps, not just the product release steps.

SUMMARY

Enzymes whose activity can be reversibly altered play a major role in regulating metabolism. In the most common situation, a regulatory enzyme has, in addition to a catalytic site, a separate regulatory site for binding a small regulatory molecule. Binding of the regulatory molecule induces a structural alteration in the protein that affects the catalytic site in a positive way (positive cooperativity) or a negative way (negative cooperativity). Regulatory proteins that behave in this manner are called allosteric proteins. They are usually multisubunit proteins. The small regulatory molecules are called allosteric effectors.

In this chapter we have restricted the discussion to two allosteric proteins, hemoglobin and aspartate carbamoyltransferase. Hemoglobin is an oxygen-transporting protein rather than an enzyme. It is a tetrameric molecule with one binding site for oxygen in each monomer. The binding of oxygen shows a sigmoidal dependence on oxygen concentration, demonstrating a positive cooperativity between subunits in the binding of oxygen. The binding of oxygen to one subunit produces a conformational change in that subunit and favors the same conformational change in the remaining subunits of the same tetramer. Other factors, notably glycerate-2,3-bisphosphate, CO_2, and protons, bind favorably to deoxyhemoglobin and thereby inhibit the binding of oxygen.

In the case of the enzyme aspartate carbamoyltransferase, the first indication of allosteric behavior was a sigmoidal dependence of enzyme activity on aspartate substrate concentration. By analogy with hemoglobin it was inferred that this is the result of a conformational change initiated by aspartate binding at one site, which stimulates the binding of aspartate at other sites on the enzyme. The enzyme has 12 subunits, 6 with catalytic sites (C subunits) and 6 with regulatory sites (R subunits). Exposure to mercurials leads to dissociation of the holoenzyme into two C_3 trimers and three R_2 dimers. The C_3 trimers are catalytically active but do not behave like the normal allosteric enzyme. Thus their velocity as a function of substrate concentration is described by a hyperbolic curve rather than a sigmoidal curve, and CTP has no effect on the reaction velocity. The regulatory subunits contain binding sites for cytidine triphosphate, which inhibits the enzyme.

X-ray diffraction studies have been done on free ACTase enzyme, CTP-complexed enzyme, and enzyme complexed to a bisubstrate analog known as PALA. The conformation of the enzyme is distinct in all three cases. The conformation favored when PALA is present brings additional amino acid side chains into the region of the catalytic site, which should augment the binding of substrate and reaction intermediate.

Positive cooperativity, as seen in the examples of hemoglobin and ACTase, increases the sensitivity of the system to small fluctuations in substrate concentration. The advantages of this for oxygen transport are explained in this chapter; the advantages to metabolic processes in general are explained in Chapter 13.

Only a small percentage of enzymes show regulatory properties. It appears that as a result of natural selection processes, regulatory proteins are found only in key locations in metabolic pathways.

SELECTED READINGS

Dickerson, R. E., and I. Geis, *Hemoglobin*. Menlo Park, Calif.: Benjamin/Cummings, 1983. Comprehensive, reasonably up-to-date, and extremely well illustrated.

Monod, J., J. P. Changeux, and J. Jacob, Allosteric proteins and cellular control systems. *J. Mol. Biol.* 6:306, 1963. The grand classic.

Volz, K. W., K. L. Krause, and W. N. Lipscomb, The Binding of N-(Phosphonacetyl)-L-Aspartate to Aspartate Carbamoyltransferase of Escherichia coli. *Biochem. Biophys. Res. Comm.* 136:822, 1986. See this journal article for the latest information on structural studies of the PALA complex with aspartate carbamoyltransferase.

Also see references at the end of Chapter 8.

PROBLEMS

1. Assume that this flow diagram relates to an amino acid biosynthetic pathway in which J, G, and H are all amino acid end products and A is their common precursor. The numbers over the arrows refer to enzymes that catalyze discrete steps.

$$A \xrightarrow{1} B \xrightarrow{2} C \xrightarrow{3} D \xrightarrow{4} E \xrightarrow{5} F$$

with $D \xrightarrow{7} I \xrightarrow{8} J$, $F \xrightarrow{6} G$, and $F \xrightarrow{9} H$

Suggest a plausible scheme in which certain enzymes are regulated by end-product inhibition. Indicate which enzymes these are and by what end products they are inhibited. If this were a degradative pathway rather than a synthetic pathway, would you expect to see the same pattern of regulation?

2. Aspartate carbamoyltransferase is an allosteric enzyme in which the active site and the binding site for the allosteric effector are on different subunits. Is it possible to imagine an allosteric enzyme in which the two sites are on the same subunit? Explain.

3. Some allosteric enzymes are inhibited competitively and some are inhibited noncompetitively by the binding of an allosteric effector at a nonoverlapping site. Explain, in qualitative terms, the sets of conditions that could produce the two different results.

4. Explain what is meant by positive and negative cooperativity.

5. Mutations in hemoglobin that decrease the positive cooperativity of hemoglobin–oxygen binding are commonly found at sites other than the oxygen binding sites. Explain. Is it likely that variants of regulatory enzymes will show similar behavior?

6. If you separated hemoglobin into its respective α and β dimer subunits, would you expect it to bind more or less oxygen at low oxygen tensions? Explain your answer. What effect would you expect glycerate bisphosphate to have on the oxygen binding to the separated subunits?

7. a. What is the primary significance to the cell of cooperativity?

   b. How does the effect of CTP on ACTase illustrate this?

   c. Explain how the effect of ATP on ACTase makes good sense for the cell (refer to Chapter 13 if necessary).

8. Compare the effect of glycerate bisphosphate on the O_2 binding of hemoglobin (see Figure 10–3) with the effect of CTP on the affinity of ACTase for aspartate (see Figure 10–17).

a. What do you notice about the shapes of the curves in the two cases?

b. How might you account for this in terms of the symmetry model proposed by Monod, Wyman, and Changeux? (See Figure 10–12 and also the solution to Problem 4.)

c. For ACTase, what conditions give an analogous plot to that observed during O_2 binding studies with myoglobin or with single hemoglobin protomers?

11

STRUCTURE AND FUNCTION OF COENZYMES

Many metabolic reactions involve chemical changes that could not be brought about by the structures of the amino acid side-chain functional groups in enzymes. The amino acid side chains are limited in type and do not offer a wide range of chemical properties for catalyzing many of the known metabolic reactions. Enzymes, in catalyzing such reactions, do not operate outside the limits of chemical feasibility, however. Instead, they act in cooperation with other smaller molecules, called _coenzymes_ that have the required chemical reactivities.

In a few cases, molecules that have come to be recognized as coenzymes do not exhibit specialized chemical properties unrepresented by the functional groups of proteins. For example, the catalytically functional groups in 4-phosphopantetheine coenzymes and α-lipoic acid are sulfhydryl and disulfide groups, also found in cysteine and cystine, respectively. In these cases, the structures of the coenzymes give them group translocation capabilities that are not shared by cysteine or cystine and that are essential aspects of catalysis in certain critically important metabolic reactions involving multienzyme complexes.

A coenzyme may be defined as a molecule that possesses physicochemical properties not found in the polypeptide chain of the enzyme and that acts together with the enzyme to catalyze a biochemical reaction.* In this chapter we discuss the structures and functions of coenzymes, which will be seen to represent a wide diversity of chemical and structural properties. The compounds considered in this chapter do not exhaust the list of molecules that have been called coenzymes. Many, such as ATP, are now more properly classified as ordinary metabolites; at earlier stages in the development of biochemistry they appeared to be coenzymes because they activated enzymatic processes. As these processes became better understood, the true roles of such molecules became clear.

*Tightly bound coenzymes are sometimes referred to as prosthetic groups.

TABLE 11–1
Water-soluble Vitamins and Associated Nutritional Diseases

Vitamin	Associated Coenzyme	Deficiency Disease
Thiamine (B_1)	Thiamine pyrophosphate	Human beriberi
Riboflavin (B_2)	Flavin mononucleotide, flavin adenine dinucleotide	Rat growth factor
Pyridoxal (B_6)	Pyridoxal phosphate	Rat pellagra
Nicotinamide	Nicotinamide adenine dinucleotide, nicotinamide adenine dinucleotide phosphate	Human pellagra
Pantothenic acid	Coenzyme A	Chick dermatitis
Biotin	Biocytin	Human dermatitis
Folic acid	Tetrahydrofolic acid	Tropical macrocytic anemia
Vitamin B_{12}	Deoxyadenosyl cobalamin	Pernicious anemia

Coenzymes were originally discovered as _vitamins_ or growth factors in nutritional and medical studies. Most coenzymes are modified forms of vitamins, including thiamine pyrophosphate, pyridoxal phosphate, nicotinamide coenzymes, phosphopantetheine coenzymes, tetrahydrofolate, flavin coenzymes, and vitamin B_{12} coenzymes. α-Lipoic acid is a growth factor for certain bacteria, but it is not a vitamin because it is not a required constituent of the human diet. Vitamins were originally discovered in nutritional studies, in which they were purified from foodstuffs and shown to cure various disorders in animals maintained on deficient diets. Table 11–1 lists the water-soluble vitamins discussed in this chapter and the nutritional diseases they cure.

Vitamin deficiencies in humans and animals most often result from dietary deficiencies. However, in certain cases, the deficiency may arise from a genetic fault so that the animal fails to absorb the vitamin from a normal diet. A case in point is vitamin B_{12} deficiency, which usually results from a failure to absorb the vitamin. Intrinsic factor is an intestinal protein that binds the vitamin as a part of the absorption process. When intrinsic factor is absent or defective, the vitamin is not absorbed, leading to a deficiency.

This chapter emphasizes the _principles_ underlying the mechanisms of action of coenzymes and so does not delineate the detailed mechanism of all the various reactions in which coenzymes play essential roles. To gain proficiency in the writing of reaction mechanisms, and thereby achieve an improved understanding, you should work the exercises at the end of the chapter.

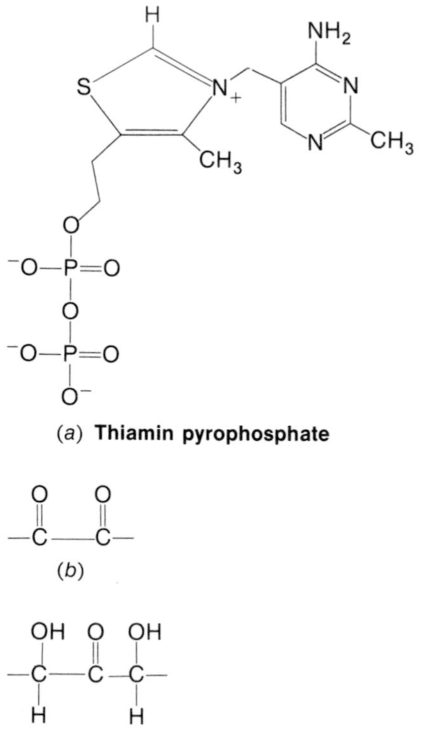

THIAMINE PYROPHOSPHATE (TPP) AND C—X BOND CLEAVAGE

The structure of thiamine pyrophosphate is given in Figure 11–1. The vitamin, thiamine or vitamin B_1, lacks the pyrophosphoryl group. _Thiamine pyrophosphate is the essential coenzyme involved in the actions of enzymes that catalyze cleavages of the bonds indicated in color in Figure 11–1._ The bond scission in Figure 11–1b is representative of those in many α-keto acid decarboxylations, nearly all of which require the action of TPP. The phosphoketolase reaction involves both cleavages shown in Figure 11–1c, while the transketolase reactions (Chapter 14) involve the cleavage of the carbon–carbon bond but not the elimination of —OH. Acetoin and acetolactate arise by the formation of the carbon–carbon bond in Figure 11–1c.

The catalytic pathways in these reactions could be understood if the

Figure 11–1
(a) Structure of thiamine pyrophosphate and (b, c) the bonds it cleaves or forms. Reactive part of the coenzyme and the bonds subject to cleavage in (b) and (c) are indicated in color.

carbon–carbon bond cleavages in Figure 11–1b and c proceeded by heterolytic scission to produce acyl anionic species as reaction intermediates. These could react in subsequent steps with protons or other electrophiles or undergo elimination reactions to form products. However, such acyl anions are so high in energy that they cannot be produced at significant rates under physiological conditions. In fact the function of thiamine pyrophosphate is to promote the bond cleavages indicated in Figure 11–1b and c. This was first understood when it was shown by Ronald Breslow, in one of the earliest applications of nuclear magnetic resonance to biochemical mechanisms, that the proton bonded to C-2 in the thiazolium ring is readily exchangeable with the protons of H_2O and deuterons of D_2O in the base-catalyzed reaction of Equation (1).

$$(1)$$

The ylid-like* intermediate undergoes nucleophilic addition to the π bond of polar carbonyl groups in substrates to produce intermediates, such as those in Equation (2), that possess the chemical reactivities required for cleaving the bonds indicated in Figure 11–1b and c.

$$(2)$$

This reaction is illustrated in Figure 11–2, which shows that the electron pair resulting from the carbon–carbon bond scission in Figure 11–1b, i.e., decarboxylation of pyruvate, is stabilized by the thiazolium ring. The immediate decarboxylation product (Figure 11–2a) is a resonance hybrid in which the electron pair is stabilized by delocalization into the thiazolium ring. The intermediate is common to many enzymatic reactions involving the decarboxylation of pyruvate. As shown in Figure 11–2d, it may be protonated to α-hydroxyethyl thiamine pyrophosphate; it may react with other electrophiles, such as the carbonyl groups of acetaldehyde or pyruvate, to form the species in Figure 11–2b and c, respectively; or it may be oxidized to acetylthiamine pyrophosphate (Figure 11–2e), depending on the reaction specificity of the enzyme.

The α,β-dihydroxy intermediate analogous to that in Figure 10–2a is shown in Figure 11–2f, and it results from carbon–carbon bond cleavage in reactions of thiamine pyrophosphate with ketose sugars, such as fructose-6-phosphate. This intermediate can undergo all the reactions cited earlier for the intermediate in Figure 11–2a, but in addition, it can eliminate the β-hydroxyl group by using the stabilized electron pair as the driving force to produce acetylthiamine pyrophosphate (Figure 11–2e), initially as its enol form. Acetylthiamine pyrophosphate can react with nucleophiles to produce, for example, acetate or acetyl phosphate, as illustrated in Figure 11–2, depending on the reaction specificity of the enzyme.

*An ylid is a compound with opposite charges on adjacent covalently bound atoms that both have an electronic octet.

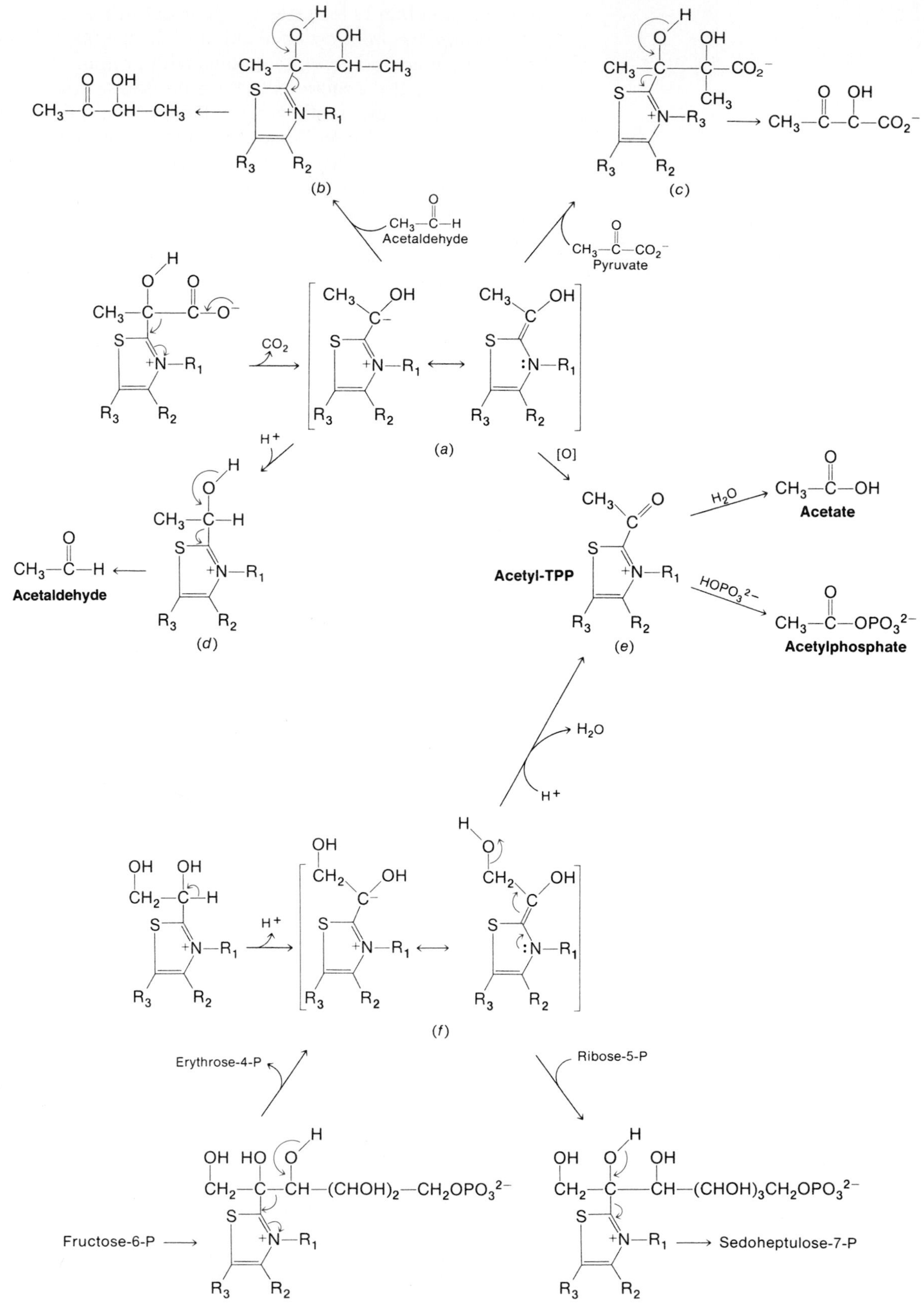

(a)
(b)
(c)
(d)
(e)
(f)
Acetaldehyde
Pyruvate
Acetate
Acetyl-TPP
Acetylphosphate
Erythrose-4-P
Ribose-5-P
Fructose-6-P
Sedoheptulose-7-P

Figure 11–2 *(facing page)*

Mechanisms of thiamine pyrophosphate action. Intermediate *(a)* is represented as a resonance-stabilized species. It arises from the decarboxylation of the pyruvate-thiamine pyrophosphate addition compound shown at left of *(a)* and in Equation (2). It can react as a carbanion with acetaldehyde, pyruvate, or H$^+$ to form *(b)*, *(c)*, or *(d)*, depending on the specificity of the enzyme. It can also be oxidized to acetyl-thiamine pyrophosphate *(e)* by other enzymes, such as pyruvate oxidase. The intermediates *(b)* through *(e)* are further transformed to the products shown by the actions of specific enzymes. Intermediate *(f)* is the resonance-stabilized species analogous to *(a)*, but derived from the cleavage of addition compounds formed between thiamine pyrophosphate and ketose sugars, such as fructose-6-phosphate. Transketolases catalyze the reaction of *(f)* with aldose phosphates to form other ketose phosphates, such as sedoheptulose-7-phosphate, as shown. Phosphoketolase catalyzes the dehydration of *(f)* to *(e)* and its further reaction with P$_i$ to acetyl-phosphate. The addition and removal of protons indicated in the interconversion of intermediates and their conversions to products are catalyzed by general acids and bases in the enzymes that catalyze the reactions.

The essential aspects of the mechanisms of the best-known biological reactions involving the action of thiamine pyrophosphate are illustrated in Figure 11–2. The experimental support for these mechanisms is, in addition to Equations (1) and (2), that hydroxyethyl-TPP and α,β-dihydroxyethyl-TPP have been synthesized and shown to be utilized as precursors of enzymatic products and of the central intermediates shown in Figure 11–2a and *f*.

The essence of the coenzymatic function of thiamine pyrophosphate is outlined in Equations (1) and (2) and Figures 11–1 and 11–2. Dissociation of the C-2 proton from thiamine pyrophosphate in Equation (1) generates the ylid-like zwitterion. This ion undergoes nucleophilic addition to the carbonyl group of a substrate in Equation (2) to produce the intermediate shown. Carbon–carbon bond cleavages in these intermediates produce the central intermediates in Figure 11–2a and *f*, in which the electron pairs resulting from the carbon–carbon bond cleavages are stabilized by delocalization into the electron-deficient thiazolium ring while being kept available for subsequent reactions. In its capability to function in this way, thiamine pyrophosphate is unique in the biosphere.

PYRIDOXAL-5′-PHOSPHATE AND REACTIONS INVOLVING α-AMINO ACIDS

Pyridoxal-5′-phosphate is the coenzyme form of vitamin B$_6$ and has the structure shown in Figure 11–3. Vitamin B$_6$ refers to any of a group of related compounds lacking the phosphoryl group, including pyridoxal, pyridoxamine, and pyridoxine (pyridoxol).

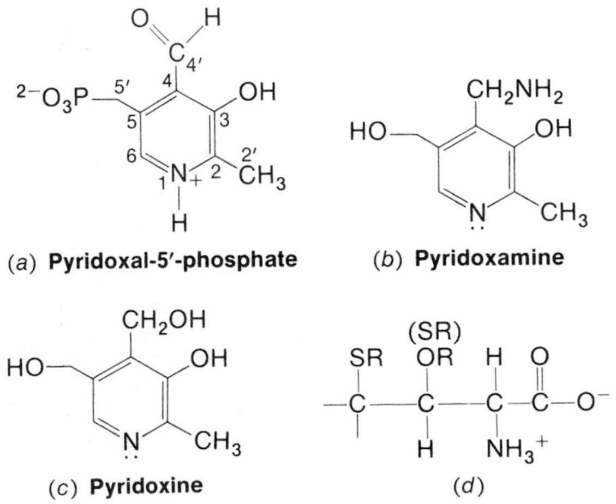

Figure 11–3

Structures of vitamin B$_6$ derivatives and the bonds cleaved or formed by the action of pyridoxal phosphate *(a)*. The reactive part of the coenzyme is shown in color in *(a)*. The bonds shown in color in *(d)* are the types of bonds in substrates that are subject to cleavage.

Pyridoxal-5′-phosphate is involved in a variety of reactions in the metabolism of α-amino acids, including transaminations, α-decarboxylations, racemizations, α,β eliminations, β,γ eliminations, aldolizations, and the β decarboxylation of aspartic acid. Equations (3) through (9) illustrate the variety of reactions in which pyridoxal-5′-phosphate exerts its essential coenzymatic functions.

$$R_1-\overset{\overset{\displaystyle H}{|}}{\underset{\underset{\displaystyle NH_3{}^+}{|}}{C}}-CO_2{}^- + R_2-\overset{\overset{\displaystyle O}{\|}}{C}-CO_2{}^- \;\xrightarrow{\text{Transaminase}}\;$$

$$R_1-\overset{\overset{\displaystyle O}{\|}}{C}-CO_2{}^- + R_2-\overset{\overset{\displaystyle H}{|}}{\underset{\underset{\displaystyle NH_3{}^+}{|}}{C}}-CO_2{}^- \quad (3)$$

$$R-\overset{\overset{\displaystyle H}{|}}{\underset{\underset{\displaystyle NH_3{}^+}{|}}{C}}-CO_2{}^- + H^+ \;\xrightarrow{\text{Decarboxylase}}\; CO_2 + R-CH_2-NH_3{}^+ \quad (4)$$

$$R-\overset{\overset{\displaystyle H}{|}}{\underset{\underset{\displaystyle NH_3{}^+}{|}}{C}}-CO_2{}^- \;\xrightarrow{\text{Racemase}}\; R-\overset{\overset{\displaystyle NH_3{}^+}{|}}{\underset{\underset{\displaystyle H}{|}}{C}}-CO_2{}^- \quad (5)$$

$$R-\overset{\overset{\displaystyle OH}{|}}{CH}-\overset{\overset{\displaystyle H}{|}}{\underset{\underset{\displaystyle NH_3{}^+}{|}}{C}}-CO_2{}^- \;\xrightarrow{\text{Dehydratase}}\; R-CH_2-\overset{\overset{\displaystyle O}{\|}}{C}-CO_2{}^- + NH_4{}^+ \quad (6)$$

$$R-\overset{\overset{\displaystyle OH}{|}}{CH}-\overset{\overset{\displaystyle H}{|}}{\underset{\underset{\displaystyle NH_3{}^+}{|}}{C}}-CO_2{}^- \;\xrightarrow{\text{Aldolase}}\; R-CHO + \underset{\underset{\displaystyle NH_3{}^+}{|}}{CH_2}-CO_2{}^- \quad (7)$$

$$H_2O + \overset{\overset{\displaystyle CH_2-\overset{\overset{\displaystyle NH_3{}^+}{|}}{CH}-CO_2{}^-}{|}}{S}-CH_2-CH_2-\overset{\overset{\displaystyle}{}}{\underset{\underset{\displaystyle NH_3{}^+}{|}}{CH}}-CO_2{}^- \;\xrightarrow{\text{Cystathionase}}\; \overset{\overset{\displaystyle SH}{|}}{CH_2}-\overset{\overset{\displaystyle}{}}{\underset{\underset{\displaystyle NH_3{}^+}{|}}{CH}}-CO_2{}^- +$$

$$CH_3CH_2-\overset{\overset{\displaystyle O}{\|}}{C}-CO_2{}^- + NH_4{}^+ \quad (8)$$

$$^-O_2C-CH_2-\overset{\overset{\displaystyle H}{|}}{\underset{\underset{\displaystyle NH_3{}^+}{|}}{C}}-CO_2{}^- + H^+ \;\xrightarrow{\text{Aspartate-}\beta\text{-decarboxylase}}\;$$

$$CO_2 + CH_3-\overset{\overset{\displaystyle H}{|}}{\underset{\underset{\displaystyle NH_3{}^+}{|}}{C}}-CO_2{}^- \quad (9)$$

Equations (3) through (9) show that pyridoxal-5′-phosphate is involved in the cleavages of the bonds shown in color in Figure 11–3d. This represents a versatility not matched by most other coenzymes. Yet these reactions can all be understood in chemical terms on the basis that pyridoxal-5′-phosphate can promote the heterolytic bond cleavages in Figure 11–3d by stabilizing the resulting electron pairs at the α- or β-carbon atoms of α-amino acids. This can be seen to be well within the chemical capabilities of pyridoxal-5′-phosphate once it is realized that the aldehyde group of the coen-

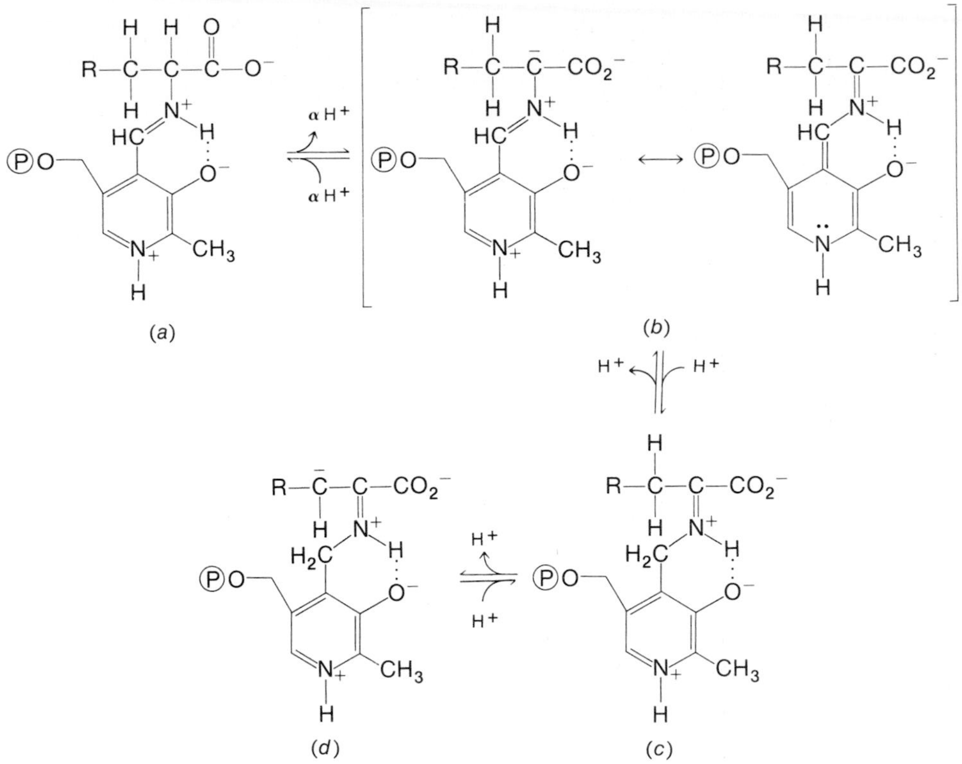

Figure 11–4

Structures of catalytic intermediates in pyridoxal-phosphate–dependent reactions. The initial aldimine intermediate resulting from Schiff's base formation between the coenzyme and the α-amino group of an amino acid (a). This aldimine is converted to the resonance-stabilized intermediate (b) by loss of a proton at the alpha carbon. Further enzyme-catalyzed proton transfers to intermediates (c) and (d) may occur, depending upon the specificity of a given enzyme. The enzymes use their general acids and bases to catalyze these proton transfers.

zyme can react with the α amino group of α-amino acids to produce an aldimine (Figure 11–4a) or Schiff's base, which is internally stabilized by H bonding. Loss of the α hydrogen as H^+ produces a resonance-stabilized species (Figure 11–4b) in which the electron pair is stabilized by delocalization into the pyridinium system. This species may undergo further reactions at the α carbon to form products determined by the reaction specificity of the enzyme. If, for example, the enzyme is a racemase, the species resulting from the loss of the proton from the α carbon may accept a proton from the opposite side to produce ultimately the enantiomer of the amino acid.

When the substrate is substituted at the β carbon with a potential leaving group, such as —OH, —SH, —OPO_3^{2-} (Figure 11–3d), the corresponding α-carbanion intermediate (Figure 11–4b) can eliminate this group, and this is an essential step in α,β eliminations, such as the serine dehydratase and threonine deaminase reactions. Upon hydrolysis, the elimination intermediate produces pyridoxal-5'-phosphate and the substrate-derived enamine, which spontaneously hydrolyzes to ammonia and an α-keto acid [Equation (6)].

In transamination, the intermediate (Figure 11–4b) is protonated at the aldimine carbon-4' of pyridoxal-5'-phosphate, producing the intermediate in Figure 11–4c. Upon hydrolysis, this forms an α-keto acid and pyridoxamine-5'-phosphate. As shown in Figure 11–5, the microscopic reverse of this sequence with a second α-keto acid accounts for the transamination reaction.

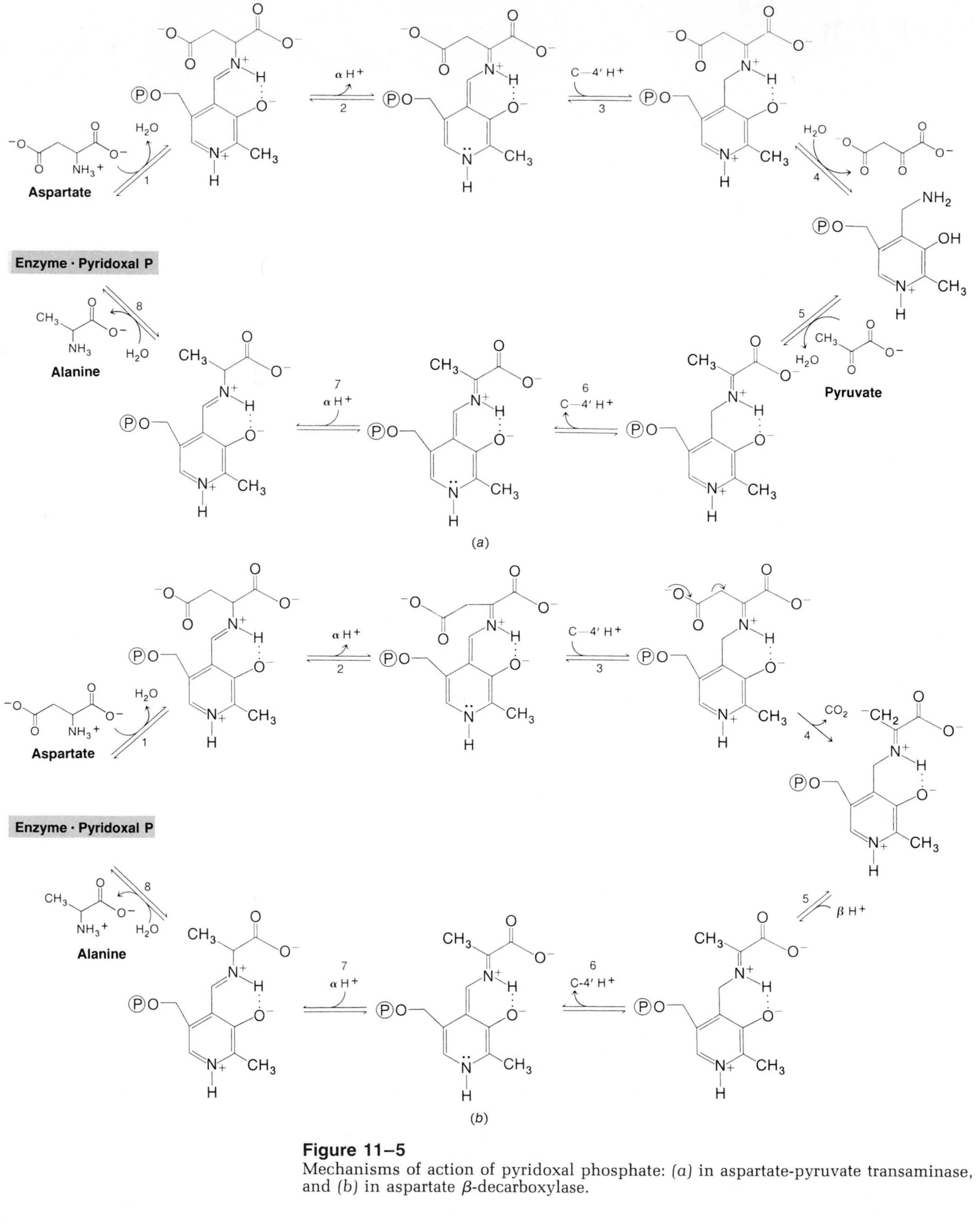

Figure 11–5
Mechanisms of action of pyridoxal phosphate: (*a*) in aspartate-pyruvate transaminase, and (*b*) in aspartate *β*-decarboxylase.

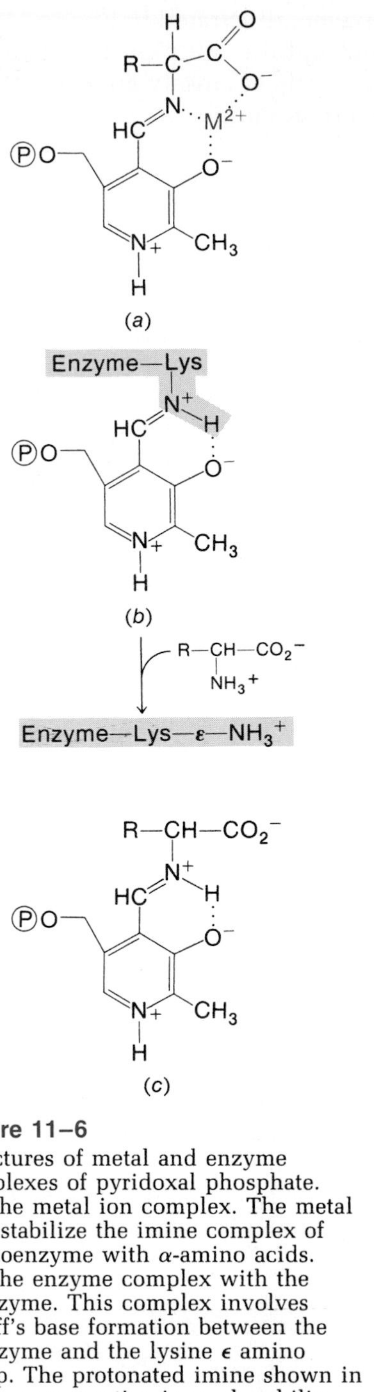

(a)

(b)

(c)

Figure 11–6
Structures of metal and enzyme complexes of pyridoxal phosphate. (a) The metal ion complex. The metal ions stabilize the imine complex of the coenzyme with α-amino acids. (b) The enzyme complex with the coenzyme. This complex involves Schiff's base formation between the coenzyme and the lysine ε amino group. The protonated imine shown in (b) is more reactive in nucleophilic addition reactions than is the carbonyl group of the unmodified coenzyme. The transimination process frees the lysine ε amino group in the process of forming the amino acid-pyridoxal-5'-phosphate aldimine as shown in (c).

An intermediate analogous to that in Figure 11–4b but generated from glycine and so lacking the β and γ carbons can react as a carbanion with an aldehyde to produce ultimately a β-hydroxy-α-amino acid. These reactions are catalyzed by aldolases such as threonine aldolase and serine transhydroxymethylase.

α-Decarboxylases [Equation (4)] generate intermediates analogous to that in Figure 11–4b by catalyzing the elimination of CO_2 instead of H^+ from the intermediate in Figure 11–4a. Protonation of the α-carbanionic intermediates by protons from H_2O followed by hydrolysis of the resulting imines produces the amines corresponding to the replacement of the carboxylate group in the substrate by a proton.

The stability of the resonance hybrid (Figure 11–4b) accounts for the catalytic action of pyridoxal-5'-phosphate in the reactions in Equations (3) through (7). The question of whether the carbanionic or the quinonoid form of the hybrid is the more important contributor to this structure is beyond the scope of this chapter. In certain cases, spectral data support the quinonoid structure as the main contributor.

β decarboxylation of aspartate and β,γ elimination by cystathionase can by understood on the basis of the stabilization of amino acid β-carbanions by pyridoxal-5'-phosphate. Returning to the intermediate in Figure 11–4c, it can be seen that elimination of a β proton produces a β-carbanion (Figure 11–4d) that is stabilized by resonance with the neighboring protonated imine. This carbanion can eliminate a good leaving group from the γ carbon, exemplified by the γ-cystathionase-catalyzed elimination of cysteine from cystathionine in Equation (8). The β decarboxylation of aspartate [Equation (9)] proceeds by elimination of a β-carbanionic intermediate like that in Figure 11–4d from the ketimine, analogous to the intermediate produced by loss of the α proton from the aldimine of aspartate with pyridoxal-5'-phosphate. This mechanism is outlined in greater detail in Figure 11–5.

Most of the enzymatic reactions that involve pyridoxal-5'-phosphate are catalyzed by the coenzyme itself in the absence of any enzyme, albeit slowly, at high temperatures, and in the presence of metal ions. The metal ions stabilize the imine complexes of the coenzyme with α-amino acids by chelation of the α-carboxylate, the imine nitrogen, and the phenolic oxygen, which is illustrated in Figure 11–6a. These reactions have been studied as models of the enzymatic reactions to gain insight into the mechanism of action of the coenzyme.

In view of the fact that the coenzyme alone catalyzes these reactions, it is pertinent to discuss the role of the enzymes in reactions involving the action of pyridoxal-5'-phosphate. The enzymes provide specificity and rate acceleration as in other enzymatic reactions. Substrate specificity is determined by binding specificity for α-amino acids, and reaction specificity is determined in substantial part by the orientations of general acids and bases in the active sites of the enzymes. Comparisons of their mechanisms show that these reactions are often distinguished by the nature and orientation of the proton-transfer processes. These are catalyzed by general acids and bases present in the active site, the positioning and states of protonation of which can determine the reaction that is catalyzed, e.g., transamination, racemization, dealdolization, etc.

The enzymes also provide rate acceleration by mechanisms other than general acid-base catalysis. Two are of special importance in pyridoxal-5'-phosphate–dependent reactions. The enzymes use their binding properties to stabilize the aldimine complexes of α-amino acids and pyridoxal-5'-phosphate. This is accomplished without the involvement of metal ions. The enzymes also promote the formation of the aldimines by binding pyridoxal-5'-phosphate as a preformed imine between the ε amino group of a lysine and the carboxaldehyde group of the coenzyme, illustrated in Figure 11–6*b*. Imine formation is known to involve nucleophilic addition of the amino group to the carbonyl group, as in Equation (10). A closely analogous reaction is Equation (11), between an imine and an amine.

$$R{-}NH_2 + \underset{\displaystyle \underset{}{}}{\overset{O}{\underset{}{C}}} \rightleftharpoons \underset{\displaystyle RNH}{\overset{OH}{\underset{}{-C-}}} \rightleftharpoons \underset{\displaystyle RN}{C} + H_2O \tag{10}$$

$$R_1{-}NH_2 + \underset{\displaystyle }{\overset{R_2NH^+}{\underset{}{C}}} \rightleftharpoons \underset{\displaystyle R_1NH}{\overset{R_2NH_2^+}{\underset{}{-C-}}} \rightleftharpoons \underset{\displaystyle R_1NH^+}{C} + R_2NH_2 \tag{11}$$

The protonated imine in Equation (11) is more reactive in nucleophilic addition reactions than is the carbonyl group. Thus the enzymes catalyze "transimination" reactions analogous to Equation (11) in forming the amino acid–pyridoxal-5'-phosphate aldimine complexes.

The transimination process frees the lysine ϵ amino group in the process of forming the amino acid–pyridoxal-5'-phosphate aldimine. The ϵ amino group then functions as the general base for the proton transfer steps. The lysine ϵ amino group therefore plays dual roles, binding pyridoxal-5'-phosphate as an imine, thereby promoting transimination, and catalyzing the proton transfer steps.

The most fundamental biochemical function of pyridoxal-5'-phosphate is the formation of aldimines with α-amino acids that stabilize the development of carbanionic character at the α and β carbons of α-amino acids in intermediates such as those in Figure 11–4*b*. Enzymes acting alone cannot stabilize these carbanions and so cannot, by themselves, catalyze reactions involving their formation as intermediates.

Pyridoxal-5'-phosphate is not unique in this function, however. There are a few enzymes that catalyze reactions similar to pyridoxal-5'-phosphate–dependent reactions but which do not involve this coenzyme. In each case, the enzyme contains a carbonyl or carbonyl-like functional group not normally found in proteins. For example, histidine decarboxylase from *Lactobacillus* and S-adenosylmethionine decarboxylase from *E. coli* and mammalian tissue contain α-ketoacyl groups that act in place of pyridoxal phosphate. Histidine and phenylalanine ammonia lyases catalyze the reactions shown in Equations (12) and (13). These enzymes contain functional groups related to dehydroalanyl groups whose precise structures and functions remain to be determined.

$$\tag{12}$$

$$\tag{13}$$

NICOTINAMIDE COENZYMES AND REACTIONS INVOLVING HYDRIDE TRANSFERS

Nicotinamide adenine dinucleotide (NAD^+), also known as diphosphopyridine nucleotide (DPN^+), is one of the two coenzymatic forms of nicotinamide (Figure 11–7). The other is nicotinamide adenine dinucleotide phos-

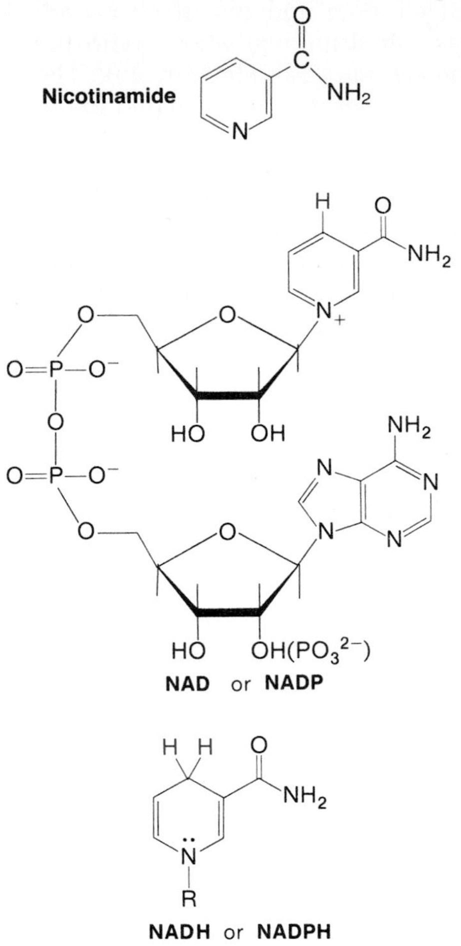

Nicotinamide

NAD or **NADP**

NADH or **NADPH**

Figure 11–7
Structures of nicotinamide and nicotinamide coenzymes. The reactive sites of the coenzymes are shown in color.

phate ($NADP^+$), also known as triphosphopyridine nucleotide (TPN^+), which differs from NAD by the presence of a phosphate group at C-2′ of the adenosyl moiety.

The nicotinamide coenzymes are biological carriers of reducing equivalents, i.e., electrons, which in most of their reactions function as cosubstrates rather than as true coenzymes. The most common function of NAD^+ is to accept two electrons and a proton (H^- equivalent) from a substrate undergoing metabolic oxidation to produce NADH, the reduced form of the coenzyme. This then diffuses or is transported to the terminal-electron transfer sites of the cell and reoxidized by terminal-electron acceptors, O_2 in aerobic organisms, with the concomitant formation of ATP (Chapter 16). Equations (14), (15), and (16) are typical reactions in which NAD^+ acts as such an acceptor.

$$NAD^+ + CH_3CH_2OH \xrightleftharpoons[]{\text{Alcohol dehydrogenase}} CH_3-\overset{O}{\overset{\|}{C}}H + NADH + H^+ \qquad (14)$$

$$^-O_2C(CH_2)_2\overset{NH_3^+}{\overset{|}{C}}HCO_2^- + NAD^+ + H_2O \xrightleftharpoons[]{\text{Glutamate dehydrogenase}}$$

$$^-O_2C(CH_2)_2\overset{O}{\overset{\|}{C}}CO_2^- + NADH + NH_4^+ + H^+ \quad (15)$$

$$HPO_4^- + {}^{2-}O_3POCH_2\overset{OH}{\overset{|}{C}}H-\overset{O}{\overset{\|}{C}}H + NAD^+ \xrightleftharpoons[]{\text{Glyceraldehyde-3P dehydrogenase}}$$

$$^{2-}O_3POCH_2\overset{OH}{\overset{|}{C}}H\overset{O}{\overset{\|}{C}}OPO_3^{2-} + NADH + H^+ \quad (16)$$

The chemical mechanisms by which NAD^+ is reduced to NADH in Equations (14) through (16) are probably similar, as represented in generalized forms in Equation (17).

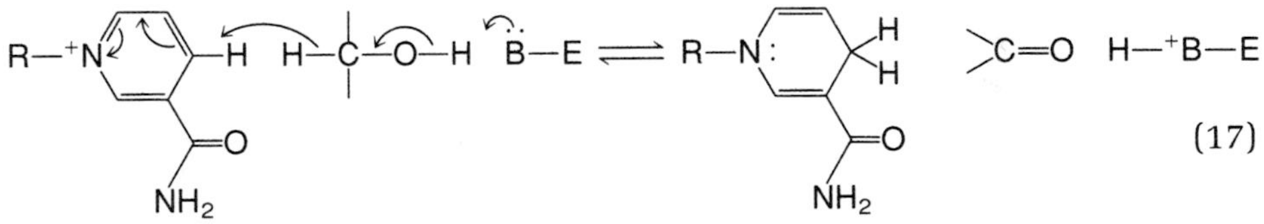

$$(17)$$

According to this formulation, the immediate oxidation product in Equation (15), where $-NH_2$ replaces $-OH$ in Equation (17), would be the imine of α-ketoglutarate, which would quickly undergo hydrolysis to α-ketoglutarate and ammonia in aqueous solution. The oxidation of an aldehyde group catalyzed by glyceraldehyde-3-P dehydrogenase [Equation (16)] also can be understood on the basis of this formulation once it is realized that there is an essential $-SH$ group at the active site that is transiently acylated during the course of the reaction. The $-SH$ group reacts with the aldehyde group of glyceraldehyde-3-P according to Equation (18), forming a thiohemiacetal that is the species oxidized. The resulting acyl-enzyme then reacts with phosphate to produce glycerate-1,3-bisphosphate.

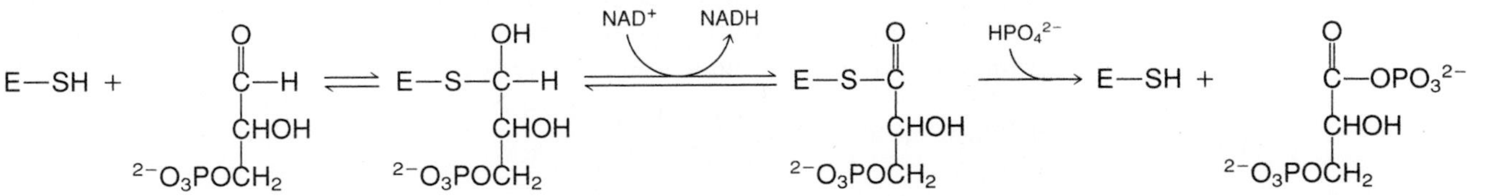

$$E-SH + \overset{O}{\underset{CHOH}{\overset{\|}{C}}}-H \rightleftharpoons E-S-\overset{OH}{\underset{CHOH}{\overset{|}{C}}}-H \xrightleftharpoons[NADH]{NAD^+} E-S-\overset{O}{\underset{CHOH}{\overset{\|}{C}}} \xrightarrow{HPO_4^{2-}} E-SH + \overset{O}{\underset{CHOH}{\overset{\|}{C}}}-OPO_3^{2-} \quad (18)$$

$${}^{2-}O_3POCH_2 \qquad {}^{2-}O_3POCH_2 \qquad {}^{2-}O_3POCH_2 \qquad {}^{2-}O_3POCH_2$$

Equation (17) implies that the hydrogen atom and two electrons are transferred in a concerted process, i.e., as a hydride equivalent, with the quaternary nitrogen in the pyridinium ring serving as an electron sink. The hydrogen atom is certainly transferred directly, and this is discussed in the following section.

In a series of classic experiments, Westheimer, Vennesland, and coworkers demonstrated that hydrogen transfer between NAD$^+$ and substrates such as those in Equations (14) through (16) are direct hydrogen transfers and occur with stereospecificity. The experiments with alcohol dehydrogenase were the first to establish these points, and they were the earliest to define the remarkable stereospecificity of enzymatic action at prochiral* centers involving chemically equivalent hydrogen atoms. Once the regiospecificity of hydrogen transfer to NAD$^+$ and the absolute configurations of the molecules were established, it became possible to formulate these processes as set forth in Equations (19) through (21) for alcohol dehydrogenase.

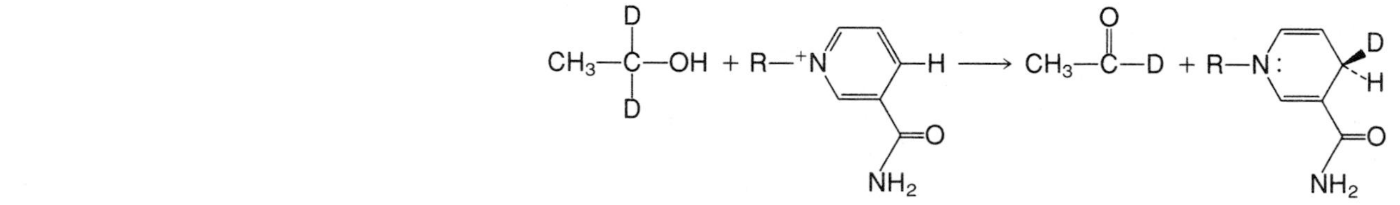

(19)

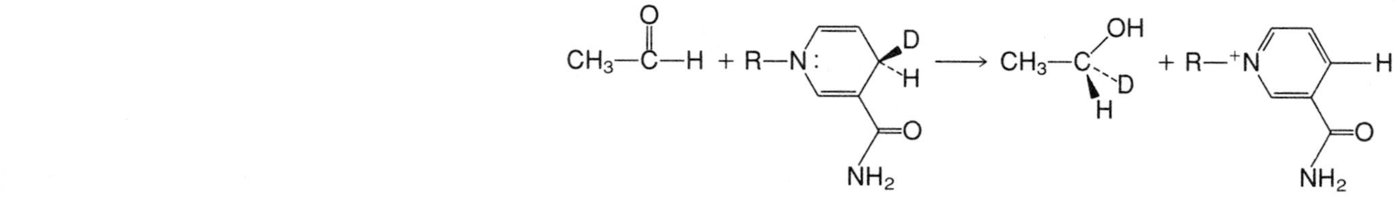

(20)

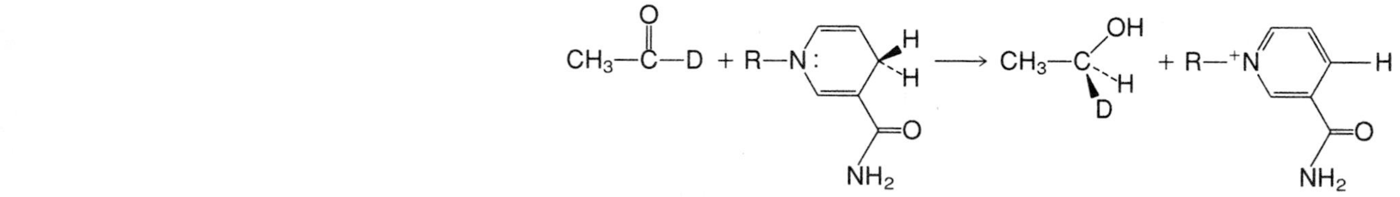

(21)

The tracing of deuterium through these transformations established quite clearly that enzymes are stereospecific in abstracting chemically equivalent hydrogens from prochiral centers and in transferring hydrogens specifically to one face of planar molecules, even in molecules as small as acetaldehyde. This specificity is thought to be a natural consequence of the fact that enzymes are asymmetrical molecules that form highly stereoselective complexes with their substrates, even those that have planes of symmetry or are themselves planar molecules. This is illustrated schematically in Figure 11–8, which shows how specific binding interactions can lead to stereospecific hydrogen transfer between acetaldehyde and NADH.

In addition to acting as a cellular electron carrier, NAD$^+$ also acts as a true coenzyme with certain enzymes. Enzymes are sometimes confronted with the problem of catalyzing such reactions as epimerizations, aldolizations, and eliminations on substrates lacking the intrinsic chemical reactivities required for these reactions to occur at significant rates. Sometimes such reactivities can be introduced into the substrate by oxidizing an appropriate alcohol group to a carbonyl group, and the enzyme is then found to contain NAD$^+$ as a tightly bound coenzyme. NAD$^+$ functions coenzymatically by transiently oxidizing the key alcohol group to the carbonyl level, producing

*See footnote under Figure 11–8.

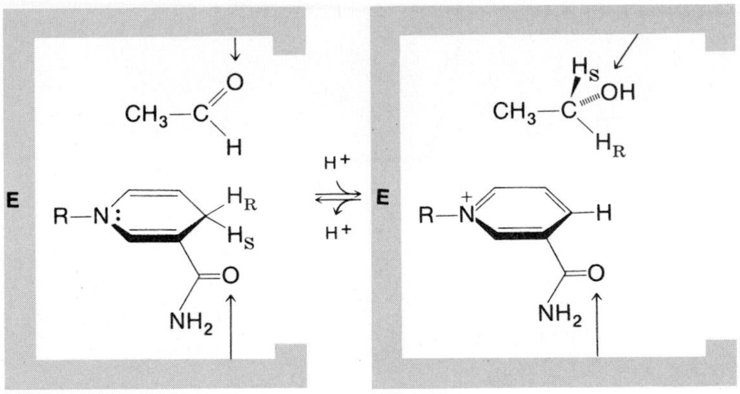

Figure 11–8

Stereospecificity of hydrogen transfer in nicotinamide coenzymes. This shows that a highly stereoselective complex with substrates can result in a stereospecific reaction even when the substrate has a plane of symmetry. The arrows represent hypothetical enzyme-binding interactions. Because of the way in which the coenzyme and the substrate are bound and oriented toward one another, one hydrogen is predisposed for transfer, the other is not. The subscripted stereochemical symbols R and S are explained in the footnote.*

Chiral and *prochiral* are derived from the Greek word $\chi\epsilon\iota\rho$, meaning "hand," and they refer to "handedness" in stereoisomeric or potentially stereoisomeric centers. The stereochemical symbols frequently used for designating two stereoisomers are R and S, which are assigned as follows. For a tetrahedral carbon (or other tetrahedral atom), the four different substituents are assigned relative priorities by the application of rules that generally accord higher priority to groups having the larger summation of atomic weights. The article by G. Popják listed in Selected Readings should be consulted for these priority rules. Once the priorities are assigned, the atom is viewed from the side opposite the lowest priority group, and the symbol R, for *rectus*, is assigned if the remaining groups appear in clockwise order from highest to lowest priority. The symbol S, for *sinister*, is assigned if they appear in counterclockwise order. For atoms whose substituent groups are $a < b < c < d$ in order of increasing priority, the following structures are those of the R and S isomers:

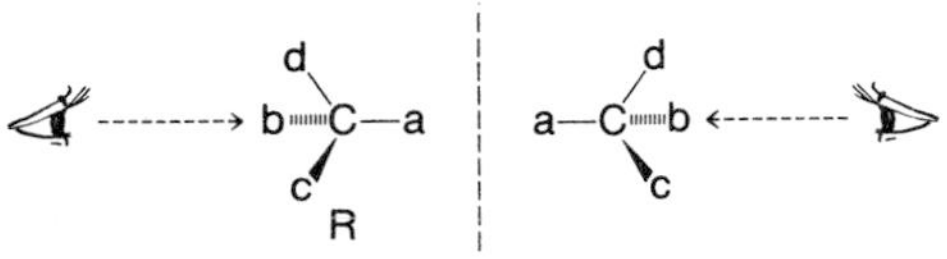

Atoms in which two of the groups are identical are called *prochiral*, since the elevation of one of the identical groups to a higher priority would lead to a chiral center. Carbon-1 in ethanol and nicotinamide C-4 in NADH are prochiral centers, since they each have two hydrogens and two other substituents. In Figure 11–8, these two hydrogens are subscripted R or S for the configurational designations that would result from according higher priority to one or the other. One means of *granting* such priority is by the use of isotopes of hydrogen. Thus deuterium has a higher priority than protium, so that the configuration of deuterioethanol in Equation (20) is R, while in Equation (21) it is S.

an oxidatively activated intermediate whose further transformation is catalyzed by the enzyme. In the last step, the carbonyl group is reduced back to the hydroxyl group by the transiently formed NADH.

Typical examples are the enzymes UDPgalactose-4-epimerase and S-adenosylhomocysteinase, both of which contain tightly bound NAD^+ (Figure 11–9). The NAD^+ in S-adenosylhomocysteinase oxidizes carbon atom 3 of the ribose ring in the substrate. The resulting carbonyl group renders the C-4 proton enolizable and promotes the elimination of homocysteine. Addition of water and reduction of the ketone by NADH completes the process. The mechanisms of many of these reactions, generalized in Figure 11–10, are based on the stabilization of a carbanionic intermediate by the carbonyl group produced by oxidative activation.

In NADH, the two hydrogens bonded to nicotinamide carbon atom 4 are chemically equivalent, so that from a purely chemical standpoint either could be transferred to an aldehyde or ketone. It is their topographic in-

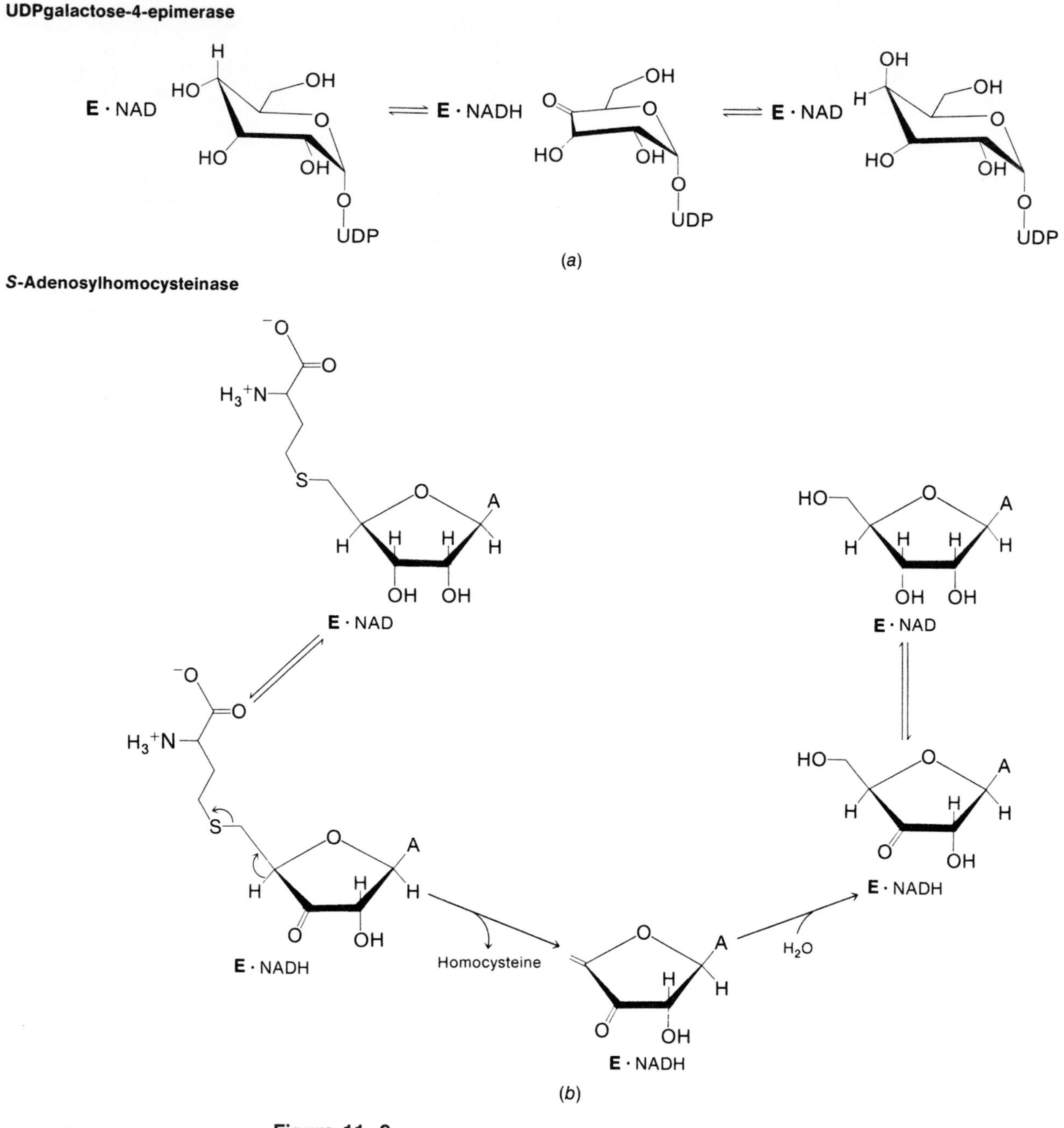

UDPgalactose-4-epimerase

(a)

S-Adenosylhomocysteinase

(b)

Figure 11–9
Mechanisms of NAD$^+$ action in (a) UDPgalactose-4-epimerase and (b) S-adenosylhomocysteinase. In the former reaction there is no net oxidation or reduction. Only the intermediate is oxidized. Note that in the latter reaction the catalytic pathway involves the intermediates analogous to (a), (c), and (d) in Figure 11–10.

equivalence that leads to stereospecificity in enzymatic reactions; however, the enzymes do not all exhibit the same stereospecificity for catalyzing hydrogen transfer from this center or in forming this center from NAD$^+$. Those enzymes such as alcohol dehydrogenase which catalyze transfer of the pro-R hydrogen in NADH (see Figure 11–8) are known as R-side-specific enzymes, and those transferring the other hydrogen, the pro-S hydrogen, are known as S-side-specific enzymes. Table 11–2 lists the RS specificities of some oxidoreductases.

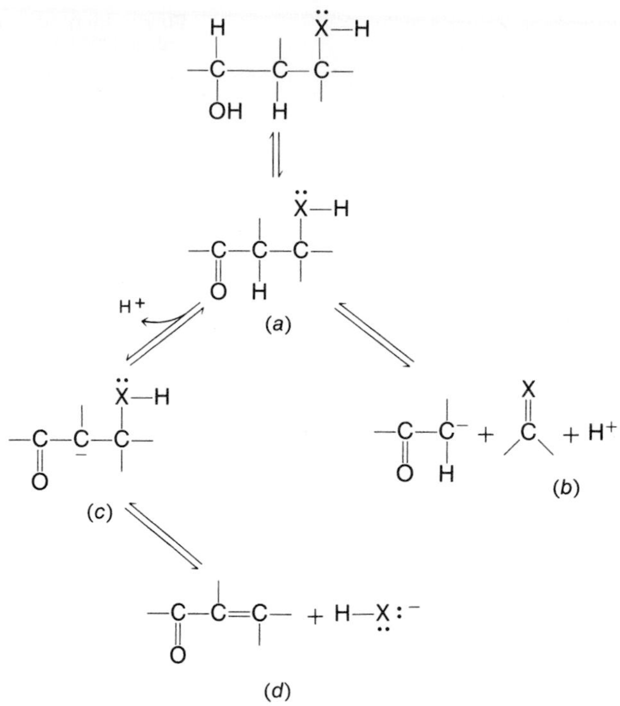

Figure 11–10
Bond cleavages and mechanisms involving oxidative activation by NAD^+. Bonds susceptible to cleavage are shown in color.

FLAVINS AND REACTIONS INVOLVING ONE OR TWO ELECTRON TRANSFERS

Flavin adenine dinucleotide (FAD) (Figure 11–11) and flavin mononucleotide (FMN) are the coenzymatically active forms of vitamin B_2, riboflavin. Riboflavin is the N^{10}-ribityl isoalloxazine portion of FAD, which is enzymatically converted into its coenzymatic forms first by phosphorylation of the ribityl C-5′ hydroxy group to FMN and then adenylylation to FAD. FMN and FAD appear to be functionally equivalent coenzymes, and which is involved with a given enzyme appears to be a matter of enzymatic binding specificity.

The catalytically functional portion of the coenzymes is the isoalloxazine ring, specifically N-5 and C-4a, which are thought to be the immediate locus of catalytic function, although the entire chromophoric system extending over N-5, C-4a, C-10a, N-1, and C-2 should be regarded as an

TABLE 11–2
Nicotinamide Side-Specificities of Dehydrogenases

Enzyme	Reaction	RS Specificity
Alcohol dehydrogenase	Ethanol + NAD^+ $\rightleftharpoons$ acetaldehyde + NADH + H^+	R
Lactate dehydrogenase	S-Lactate + NAD^+ $\rightleftharpoons$ pyruvate + NADH + H^+	R
Malate dehydrogenase	S-Malate + NAD^+ $\rightleftharpoons$ oxalacetate + NADH + H^+	R
Aldehyde dehydrogenase	Acetaldehyde + NAD^+ $\rightleftharpoons$ acetate + NADH + $2H^+$	R
Triose-P dehydrogenase	R-Glyceraldehyde-3-P + NAD^+ + HPO_4^{2-} $\rightleftharpoons$ R-glycerate 1,3-bisphosphate + NADH + H^+	S
Glutamate dehydrogenase	S-Glutamate + NAD^+ $\rightleftharpoons$ α-ketoglutarate + NADH + H^+ + NH_4^+	S
Glucose-6-P dehydrogenase	D-Glucose-6-P + $NADP^+$ $\rightleftharpoons$ 6-phosphogluconolactone + NADPH + H^+	S

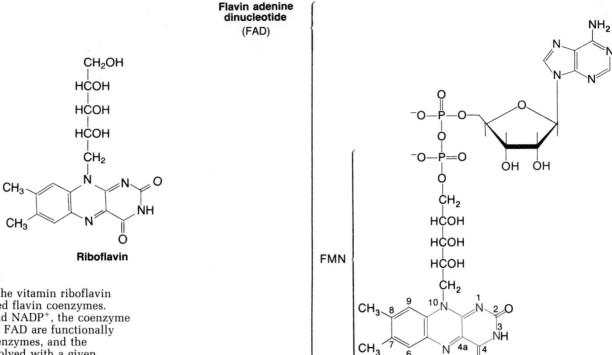

Figure 11–11
Structures of the vitamin riboflavin and the derived flavin coenzymes. Like NAD^+ and $NADP^+$, the coenzyme pair FMN and FAD are functionally equivalent coenzymes, and the coenzyme involved with a given enzyme appears to be a matter of enzymatic binding specificity. The catalytically functional portion of the coenzymes is shown in color.

indivisible catalytic entity, as are the nicotinamide, pyridinium, and thiazolium rings of NAD^+, pyridoxal-P, and thiamine pyrophosphate.

Flavin-dependent enzymes are called flavoproteins and, when purified, normally contain their full complements of FAD or FMN as tightly bound coenzymes. Flavoproteins are the bright yellow color of isoalloxazine chromophore in its oxidized form. In a few flavoproteins, the coenzyme is known to be covalently bonded to the protein by means of an enzymatic sulfhydryl or imidazole group at the C-8a methyl group and in at least one case at C-6. In most flavoproteins, the coenzymes are tightly but noncovalently bound, and many can be resolved into apoenzymes that can be reconstituted to holoenzymes by readdition of FAD or FMN.

Flavin coenzymes exist in three spectrally distinguishable oxidation states that account in part for their catalytic functions; the yellow oxidized form, the red or blue one-electron-reduced form, and the colorless two-electron-reduced form. Their structures are depicted in Figure 11–12. These and other less well-defined forms often have been detected spectrally as intermediates in flavoprotein catalysis.

Flavins are very versatile redox coenzymes. Flavoproteins are dehydrogenases, oxidases, and oxygenases that catalyze a variety of reactions on an equal variety of substrate types. Since these classes of enzymes do not consist exclusively of flavoproteins, it is difficult to define catalytic specificity for flavins. *Biological electron acceptors and donors in flavin-mediated reactions can be two-electron acceptors, such as NAD^+ or $NADP^+$, or a variety of one-electron acceptor systems, such as cytochromes (Fe^{2+}/Fe^{3+}) and quinones, and molecular oxygen is an electron acceptor for flavoprotein oxidases as well as the source of oxygen for oxygenases.* The only obviously common aspect of flavin-dependent reactions is that all are redox reactions.

Typical reactions catalyzed by flavoproteins are listed in Table 11–3, which groups flavoproteins into those that do not utilize molecular oxygen as a substrate and those that do. The significance of this is best appreciated when one realizes that $FADH_2$, a likely intermediate in many flavoprotein

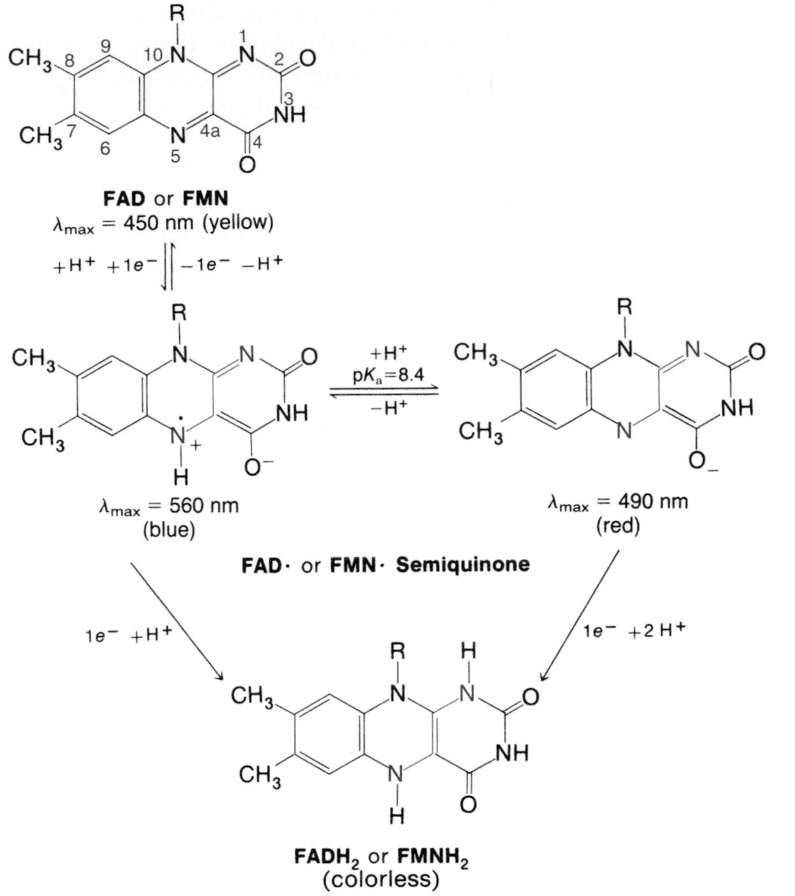

Figure 11–12
Oxidation states of flavin coenzymes. The flavin coenzymes exist in three spectrally distinguishable oxidation states that account in part for their catalytic functions. They are the yellow oxidized form, the red or blue one-electron-reduced form, and the colorless two-electron-reduced form.

FAD or **FMN**
λ_{max} = 450 nm (yellow)

$+H^+$ $+1e^-$ $\Vert$ $-1e^-$ $-H^+$

λ_{max} = 560 nm (blue)

$+H^+$
$pK_a = 8.4$
$-H^+$

λ_{max} = 490 nm (red)

FAD· or **FMN· Semiquinone**

$1e^- + H^+$ $1e^- + 2H^+$

FADH$_2$ or **FMNH$_2$**
(colorless)

TABLE 11–3
Reactions Catalyzed by Flavoproteins

Flavoprotein	Reaction
Dehydrogenases	
Glutathione reductase	$H^+ + GSSG + NADPH \rightleftharpoons 2\,GSH + NADP^+$
Acyl-CoA dehydrogenases	$RCH_2CH_2COSCoA + NAD^+ \rightleftharpoons RCH{=}CHCOSCoA + NADH + H^+$
Succinate dehydrogenase	$^-O_2CCH_2CH_2CO_2^- + E \cdot FAD \rightleftharpoons {}^-O_2CCH{=}CHCO_2^- + E \cdot FADH_2$
D-Lactate dehydrogenase	$CH_3{-}CHOH{-}CO_2^- + E \cdot FAD \rightleftharpoons CH_3{-}{-}CO{-}CO_2^- + E \cdot FADH_2$
Oxidases	
Amino acid oxidases	$R{-}\overset{\overset{\displaystyle NH_3^+}{\mid}}{C}H{-}CO_2^- + O_2 \longrightarrow R{-}CO{-}CO_2^- + H_2O_2 + NH_4^+$
Glucose oxidase	D-Glucose $+ O_2 \longrightarrow$ D-gluconolactone $+ H_2O_2$
Monoamine oxidase	$R{-}CH_2NH_2 + O_2 + H_2O \longrightarrow R{-}CHO + H_2O_2 + NH_3$
Monooxygenases	
Lactate oxidase	$CH_3{-}CHOH{-}CO_2^- + O_2 \longrightarrow CH_3{-}CO_2^- + CO_2 + H_2O$
Salicylate hydroxylase	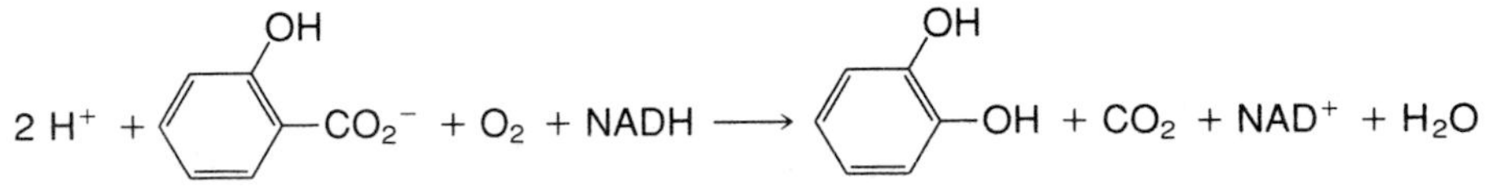 $2\,H^+ + $ (salicylate) $CO_2^- + O_2 + NADH \longrightarrow$ (catechol) $OH + CO_2 + NAD^+ + H_2O$
Ketone monooxygenase	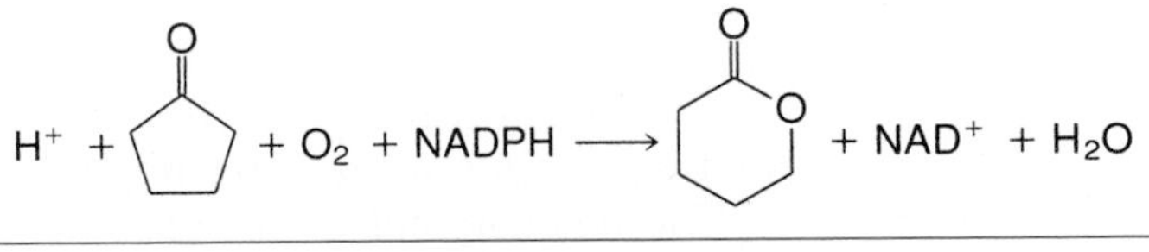 $H^+ + $ (cyclopentanone) $+ O_2 + NADPH \longrightarrow$ (lactone) $+ NAD^+ + H_2O$

reactions, spontaneously reacts with O_2 to produce H_2O_2. In the case of the dehydrogenases, therefore, either $FADH_2$ is not an intermediate or it is somehow prevented from reacting with O_2. Among the dehydrogenases are several that utilize the two-electron acceptor substrates NAD^+ or $NADP^+$, and it is reasonable to suppose that the two-electron reduction of NAD^+ by an intermediate E · $FADH_2$ might be involved. Also listed in Table 11–3 are other dehydrogenases for which the electron acceptors from E · $FADH_2$ are not given. These enzymes are membrane-bound and transfer electrons directly to membrane-bound acceptors, mainly one-electron acceptors such as quinones and cytochromes (Fe^{2+}/Fe^{3+}). The stability of the flavin semiquinone, FAD · and FMN · in Figure 11–12, gives flavins the capability to interact with one-electron acceptors in electron-transport systems.

The other classes of flavoproteins in Table 11–3 interact with molecular oxygen either as the electron-acceptor substrate in redox reactions catalyzed by oxidases or as the substrate source of oxygen atoms for oxygenases. Molecular oxygen also serves as an electron acceptor and source of oxygen for metalloflavoproteins and dioxygenases, which are not listed in Table 11–3. These enzymes catalyze more complex reactions involving catalytic redox components such as metal ions and metal-sulfur clusters in addition to flavin coenzymes.

The mechanisms of action of flavin coenzymes are currently under active investigation. A recurrent theme appears to be the probable involvement of $FADH_2$ or $FMNH_2$ as transient intermediates in a variety of flavoprotein reactions. Figure 11–13 illustrates reasonable catalytic pathways for three of the enzymes listed in Table 11–3, one from each of the three classes, and shows the probable involvement of E · $FADH_2$ in each case.

Several points should be understood about Figure 11–13. One is that the detailed mechanisms by which E · $FADH_2$ arises need not be the same in all the reactions. Note, for example, that in the case of glutathione reductase, the mechanisms by which E · FAD is reduced to E · $FADH_2$ by NADPH in the forward direction and by glutathione in the reverse direction are undoubtedly different. The mechanism shown is one recently proposed based on the nonenzymatic reaction of 3-methyl riboflavin with dithiothreitol. The reduction of FAD by NADPH might be expected to occur by concerted transfer of a C-4 proton and two electrons, a hydride equivalent, from NADH to N-5 of FAD. Transfer could not be detected directly using [4-^{3}H]NADPH because upon transfer to N-5 in $FADH_2$, the ^{3}H would quickly exchange with protons in water. Such a direct transfer can be observed, however, using a 5-deaza analog of the coenzymes so that the transferred ^{3}H in 5-deaza-$FADH_2$ is bonded to carbon and nonexchangeable.

In other cases, E · $FADH_2$ may result from a concerted two-electron reduction of FAD or by two successive one-electron reductions involving the free-radical semiquinone form of the coenzyme as an intermediate. The detailed mechanism by which oxygen reacts with $FADH_2$ to produce H_2O_2 and FAD is not known with certainty, although there is good reason based on nonenzymatic model reactions to expect C-4a flavin hydroperoxide addition compounds to be involved.

It is pertinent to consider the relationship between the amino acid oxidases and lactate oxidases. They differ essentially in that H_2O_2 and the α-keto acid dissociate quickly as products from the amino acid oxidases but not from lactate oxidase, which catalyzes further reaction of H_2O_2 with the enzyme-bound pyruvate to produce acetate and CO_2. Labeling experiments with $^{18}O_2$ are in accord with the proposed mechanism. However, they cannot distinguish it from closely related mechanisms involving flavin hydroperoxides acting in the capacity shown for H_2O_2 in Figure 11–13.

The biochemical importance of flavin coenzymes appears to be their

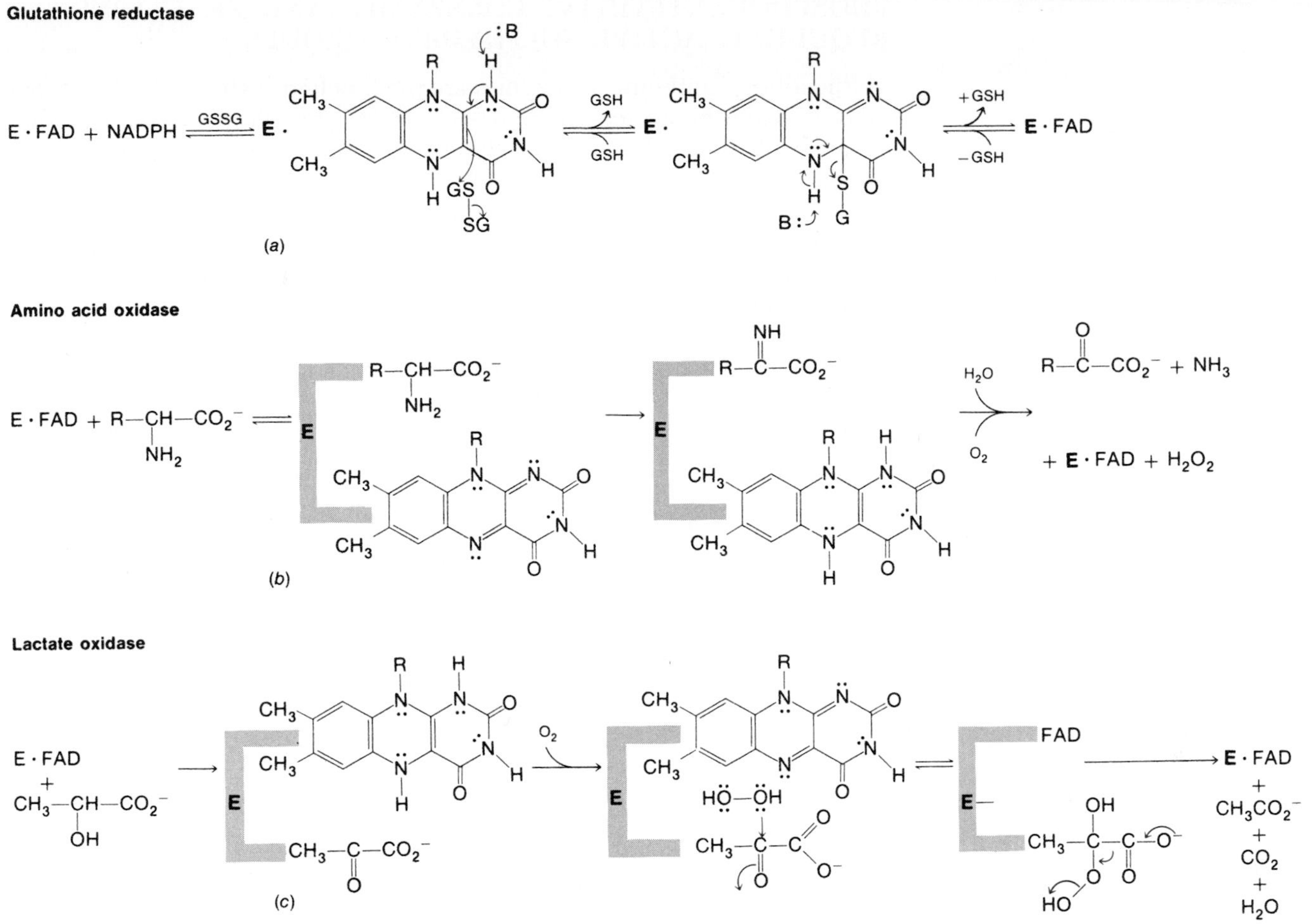

Glutathione reductase

(a)

Amino acid oxidase

(b)

Lactate oxidase

(c)

Figure 11–13

Mechanisms of flavin-dependent reactions. (a) In the glutathione reductase reaction, the first steps, not shown, involve the reduction of FAD to $FADH_2$ by NADPH and the binding of glutathione (glutathione is a sulfhydryl compound; see Figure 24–36). The mechanism by which oxidized glutathione is reduced by the $E \cdot FADH_2$ complex is shown. (b) In the amino acid oxidase reaction, the first step shown is the binding of the substrate by the $E \cdot FAD$ complex. In the second step, the amino acid is oxidized to an imino acid and FAD is reduced to $FADH_2$. In the third step, the imino acid is released from the enzyme and hydrolyzed by water to an α-ketoacid, and $FADH_2$ is oxidized by O_2 to FAD and H_2O_2. (c) In the lactate oxidase reaction, the first step shown involves the binding and oxidation of lactate, producing a complex containing $FADH_2$ and pyruvate. In the second step, O_2 oxidizes $FADH_2$ and is thereby reduced to H_2O_2. The hydrogen peroxide so produced then reacts with pyruvate at the active site, resulting in its oxidative decarboxylation to acetate and CO_2.

versatility in mediating a variety of redox processes, including electron transfer and the activation of molecular oxygen for oxygenation reactions. The detailed mechanisms of oxygen activation are not well understood. An especially important manifestation of their redox versatility is their ability to serve as the switch point from the two-electron transfer processes, which predominate in cytosolic carbon metabolism, to the one-electron transfer processes, which predominate in membrane-associated terminal electron-transfer pathways. In mammalian cells, for example, the end products of the aerobic metabolism of glucose are CO_2 and NADH (Chapter 15). The terminal electron-transfer pathway is a membrane-bound system of cytochromes, nonheme iron proteins, and copper-heme proteins—all one-electron acceptors that transfer electrons ultimately to O_2 to produce H_2O and NAD with concomitant production of ATP from ADP and P_i. The interaction of NADH with this pathway is mediated by NADH dehydrogenase, a flavoprotein that couples the two-electron oxidation of NADH with the one-electron reductive processes of the membrane.

PHOSPHOPANTETHEINE COENZYMES AND REACTIONS REQUIRING ACTIVE SULFHYDRYL GROUPS

4'-Phosphopantetheine coenzymes are the biochemically active forms of the vitamin pantothenic acid. In Figure 11–14, 4'-phosphopantetheine is shown as covalently linked to an adenylyl group in coenzyme A; or it can also be linked to a protein such as a serine hydroxyl group in acyl carrier protein (ACP). It is also found bonded to proteins that catalyze the activation and polymerization of amino acids to polypeptide antibiotics. Coenzyme A was discovered, purified, and structurally characterized by Fritz Lipmann and colleagues in work for which Lipmann was awarded the Nobel Prize in 1953.

The sulfhydryl group of the β-mercaptoethylamine (or cysteamine) moiety of phosphopantetheine coenzymes is the functional group that is directly involved in the enzymatic reactions for which they serve as coenzymes. From the standpoint of the chemical mechanism of catalysis, it is the essential functional group, although it is now recognized that phosphopantetheine coenzymes have other functions as well. Many reactions in metabolism involve acyl-group transfer or enolization of carboxylic acids that exist as unactivated carboxylate anions at physiological pH. The predominant means by which these acids are activated for acyl transfer and enolization is esterification with the sulfhydryl group of pantetheine coenzymes.

The mechanistic importance of activation is exemplified by the condensation of two molecules of acetyl-coenzyme A to acetoacetyl-coenzyme A catalyzed by β-ketothiolase:

$$CH_3-\overset{O}{\underset{\|}{C}}-SCoA + CH_3-\overset{O}{\underset{\|}{C}}-SCoA \rightleftharpoons$$

$$CH_3-\overset{O}{\underset{\|}{C}}-CH_2-\overset{O}{\underset{\|}{C}}-SCoA + CoASH \quad (22)$$

The two important steps of the reaction depend on both acetyl groups being activated, one for enolization and the other for acyl-group transfer. In the

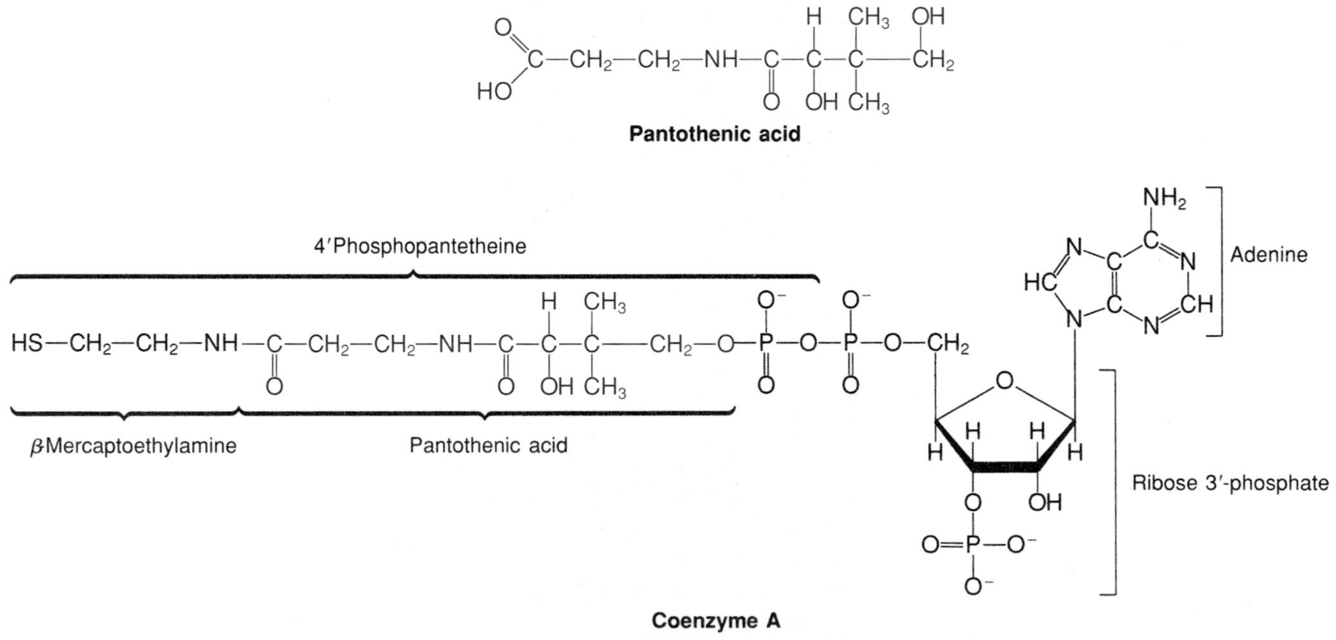

Pantothenic acid

Coenzyme A

Figure 11–14
Structures of the vitamin pantothenic acid and coenzyme A. The terminal —SH is the reactive group in coenzyme A (CoASH).

first step, one of the molecules must be enolized by the intervention of a base to remove an α proton, forming an enolate:

$$B\!:\!\curvearrowright\!\overset{}{H}\!-\!\widehat{CH_2}\!-\!\overset{O\curvearrowright}{\underset{\parallel}{C}}\!-\!SCoA \;\rightleftharpoons\; [\overset{\delta^+}{B}\!\cdots\!H\!\cdots\!\overset{\delta^-}{CH_2}\!\cdots\!\overset{\overset{\textstyle\ddot{O}}{\parallel}}{C}\!-\!SCoA] \;\rightleftharpoons\; \overset{+}{B}\!-\!H \;+\; CH_2\!\cdots\!\overset{\overset{\textstyle\ddot{O}^{-}}{\parallel}}{C}\!-\!SCoA \quad (23)$$

The enolate is stabilized by delocalization of its negative charge between the α carbon and the acyl oxygen atom, making it thermodynamically accessible as an intermediate. Moreover, this developing charge is also stabilized in the transition state preceding the enolate, so it is also kinetically accessible; that is, it is rapidly formed. Consideration of the same enolization reaction with acetate anion reveals a starkly contrasting picture, for enolization of the acetate anion would result in the generation of a second negative charge in the enolate, an energetically and kinetically unfavorable process.

The second stage of the condensation is the reaction of the enolate anion with the acyl group of the second molecule of acetyl-CoA:

$$\begin{array}{c}\overset{\overset{\textstyle O\nearrow}{\parallel}}{CH_3C}\!-\!SCoA \\[4pt] \underset{\displaystyle CH_2\!\cdots\!\overset{\overset{\textstyle\ddot{O}^{-}}{\parallel}}{C}\!-\!SCoA}{\searrow} \end{array} \rightleftharpoons \left[\begin{array}{c}\overset{\overset{\textstyle\ddot{O}^{-}}{\searrow}}{CH_3}\!-\!\overset{\displaystyle C}{\underset{\displaystyle CH_2\!-\!\overset{\overset{\textstyle O}{\parallel}}{C}\!-\!SCoA}{}}\!\!\!\nearrow\!SCoA \end{array}\right] \overset{H^+}{\rightleftharpoons} \begin{array}{c}CH_3\!-\!\overset{\overset{\textstyle O}{\parallel}}{C} \\[4pt] \underset{\displaystyle CH_2\!-\!\overset{\overset{\textstyle O}{\parallel}}{C}\!-\!SCoA}{\big|}\end{array} \;+\; CoASH \quad (24)$$

Nucleophilic addition to the neutral activated acyl group is a favored process, and coenzyme A is a good leaving group from the tetrahedral intermediate. Consideration of the occurrence of this process with the acetate anion, i.e., acetate reacting with an enolate anion, again provides a sharp contrast with the process of Equation (24), for it would entail the nucleophilic addition of an anion to an anionic center generating a dianionic transition state, an unfavorable process from both thermodynamic and kinetic standpoints. Moreover, the resulting intermediate would not have a very good leaving group other than the enolate anion itself, so the transition-state energy for acetoacetate formation would be high. Finally, the K_{eq} for the condensation of 2 moles of acetate to 1 mole of acetoacetate is not favorable in aqueous media, whereas the condensation of 2 moles of acetyl-CoA to produce acetoacetyl-CoA and coenzyme A is thermodynamically spontaneous.

The maintenance of metabolic carboxylic acids involved in enolization and acyl-group transfer reactions as coenzyme A esters provides the ideal lift over the kinetic and thermodynamic barriers to these reactions. The foregoing discussion, in emphasizing the purely electrostatic energy barriers, does not address the question of whether there is an activation advantage in thiol esters relative to oxygen esters. Why thiol esters in preference to oxygen esters? *Thiol esters are more readily enolized than oxygen esters.* They are more "ketone-like" because of their electronic structures, in which the degree of resonance-electron delocalization from the sulfur atom to the acyl group resulting from overlapping of the occupied p orbitals of sulfur with the acyl-π bond is less than that of oxygen esters.

$$\left[\,R_1\!-\!\overset{\overset{\textstyle O}{\parallel}}{C}\!-\!\ddot{S}\!-\!R_2 \;\longleftrightarrow\; R_1\!-\!\overset{\overset{\textstyle O^{-}}{\parallel}}{C}\!=\!\overset{+}{S}\!-\!R_2\,\right] \qquad \left[\,R_1\!-\!\overset{\overset{\textstyle O}{\parallel}}{C}\!-\!\ddot{O}\!-\!R_2 \;\longleftrightarrow\; R_1\!-\!\overset{\overset{\textstyle O^{-}}{\parallel}}{C}\!=\!\overset{+}{\ddot{O}}\!-\!R\,\right]$$

Another statement of this is that the charge-separated resonance form is a smaller contributor to the electronic structure in thiol esters than in oxygen esters. The reasons for this are not fully understood, but it is thought that one factor may be the larger size of sulfur relative to carbon and oxygen. This may result in a poorer energy match for the overlapping orbitals in thiol esters relative to oxygen esters.

While the pantetheine sulfhydryl group has the appropriate chemical properties for activating acyl groups, this is not unique to pantetheine coenzymes in the biosphere. Both glutathione and cysteine, as well as cysteamine, would serve, so the chemistry does not itself explain the importance of these coenzymes. Coenzyme A has many binding determinants in its large structure, especially in the nucleotide moiety, so it may serve a specificity function in the binding of coenzyme A esters by enzymes. *It also may serve as a binding "handle" in cases in which the acyl group must have some mobility in the catalytic site*, i.e., if it must enolize at one subsite and then diffuse a short distance to undergo an addition reaction to a ketonic group of a second substrate.

One system in which pantetheine almost certainly performs such a transport role is the fatty acid synthase from *E. coli* in which 4'-phosphopantetheine is a component of the acyl carrier protein (see also Chapter 21). The process of fatty acid chain elongation is represented by Equations (25) through (32).

$$CH_3-\overset{O}{\underset{\|}{C}}-SCoA + ACP-SH \underset{\text{ACP-acyltransferase}}{\rightleftharpoons} CH_3-\overset{O}{\underset{\|}{C}}-SACP + CoA \tag{25}$$

$$\beta\text{-ketoacyl-ACP synthase}\underset{SH}{|} + CH_3-\overset{O}{\underset{\|}{C}}-SACP \rightleftharpoons \beta\text{-ketoacyl-ACP synthase}\underset{\underset{O=C-CH_3}{|}}{\overset{S}{|}} + ACP-SH \tag{26}$$

$$CH_3-\overset{O}{\underset{\|}{C}}-SCoA + HOCO_2^- + ATP \xrightarrow{\text{Acetyl-CoA carboxylase}} \overset{CO_2^-}{\underset{\|}{CH_2}}-\overset{O}{\underset{\|}{C}}-SCoA + ADP + P_i \tag{27}$$

$$\overset{CO_2^-}{\underset{\|}{CH_2}}-\overset{O}{\underset{\|}{C}}-SCoA + ACP-SH \xrightarrow{\text{ACP-acyltransferase}} \overset{CO_2^-}{\underset{\|}{CH_2}}-\overset{O}{\underset{\|}{C}}-SACP + CoA \tag{28}$$

$$\beta\text{-ketoacyl-ACP synthase}\underset{\underset{O=C-CH_3}{|}}{\overset{S}{|}} + ACPS-\overset{}{\underset{\underset{O}{\|}}{C}}-CH_2-CO_2^- \xrightarrow[\text{transferase}]{\text{ACP-malonyl}}$$

$$\beta\text{-ketoacyl-ACP synthase}\underset{SH}{|} + ACP-S-\overset{O}{\underset{\|}{C}}-CH_2-\overset{O}{\underset{\|}{C}}-CH_3 + CO_2 \tag{29}$$

$$ACP-S-\overset{O}{\underset{\|}{C}}-CH_2-\overset{O}{\underset{\|}{C}}-CH_3 + NADPH + H^+ \xrightarrow[\text{reductase}]{\beta\text{-Ketoacyl-ACP}} ACP-S-\overset{O}{\underset{\|}{C}}-CH_2-\overset{OH}{\underset{|}{CH}}-CH_3 + NADP^+ \tag{30}$$

$$ACP-S-\overset{O}{\underset{\|}{C}}-CH_2-\overset{OH}{\underset{|}{CH}}-CH_3 \xrightarrow{\text{Enoyl-ACP hydrase}} ACP-S-\overset{O}{\underset{\|}{C}}-CH=CH-CH_3 + H_2O \tag{31}$$

$$ACP-S-\overset{O}{\underset{\|}{C}}-CH=CH-CH_3 + NADPH + H^+ \xrightarrow[\text{reductase}]{\text{Enoyl-ACP}} ACP-S-\overset{O}{\underset{\|}{C}}-CH_2-CH_2-CH_3 + NADP^+ \tag{32}$$

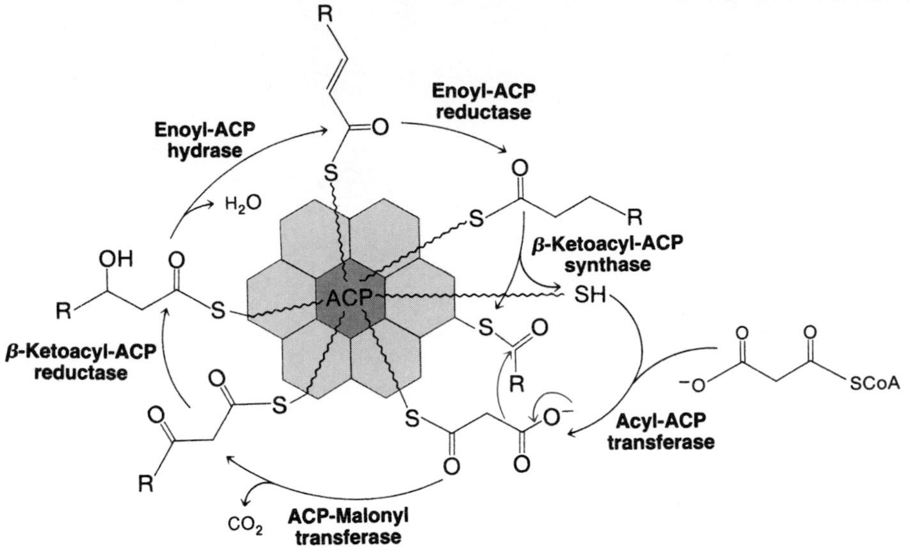

Figure 11–15
Interactions of 4′-phosphopantetheine (shown as a long wavy line) with the six enzymes of *E. coli* fatty acid synthase. In the hypothesized model shown, the central hexagon represents acyl carrier protein (ACP) and the light-colored hexagons represent the other enzymes. The single 4′-phosphopantetheine in ACP is illustrated in various orientations and acylated states interacting with the peripherally situated enzymes.

This sequence involves the interaction of the 4′-phosphopantetheinyl moiety of ACP with each of the six enzymes in the complex, either as an acyl-group donor/acceptor or as a carrier of acyl groups to the active sites of the enzymes that catalyze the reduction of β-ketoacyl-ACP to fatty acyl-ACP. This involves the physical transport of acyl groups from one active site to another. A possible organization of the fatty acid synthase is shown in Figure 11–15. *4′-Phosphopantetheine is structurally suited to this role because it consists of a 2.0 nM (20 Å) chain of atoms with torsional freedom about at least nine single bonds. This enables it to sweep out a large volume of space, encompassing the active sites of all six enzymes.*

α-LIPOIC ACID AND ACYL GROUP TRANSFERS

α-Lipoic acid is the internal disulfide of 6,8-dithioctanoic acid, whose structural formula is given in Figure 11–16. It is the coupler of electron and group transfers catalyzed by α-keto acid dehydrogenase multienzyme complexes. The pyruvate and α-ketoglutarate dehydrogenase complexes are centrally involved in the metabolism of carbohydrates by the glycolytic pathway (Chapter 14) and the tricarboxylic acid cycle (Chapter 15). They catalyze two of the three decarboxylation steps in the complete oxidation of glucose, and they produce NADH and activated acyl compounds from the oxidation of the resulting ketoacids.

$$R-\overset{O}{\overset{\|}{C}}-CO_2^- + NAD^+ + CoASH \longrightarrow CO_2 + NADH + R-\overset{O}{\overset{\|}{C}}-SCoA \qquad (33)$$

The chemical aspect of the coenzymatic action of α-lipoic acid is to mediate the transfer of electrons and activated acyl groups resulting from the decarboxylation and oxidation of α-keto acids within the complexes. In this process, lipoic acid is itself transiently reduced to dihydrolipoic acid, and this reduced form is the acceptor of the activated acyl groups. This dual role of electron and acyl-group acceptor enables lipoic acid to couple the two processes.

The interactions of α-lipoic acid in the *E. coli* pyruvate dehydrogenase complex exemplify its coenzymatic functions. The complex consists of three proteins: a pyruvate decarboxylase, which is thiamine pyrophosphate-dependent and designated E$_1$-TPP; dihydrolipoyl transacetylase, designated E$_2$-lipoyl-S$_2$, which contains α-lipoic acid covalently bonded through amide

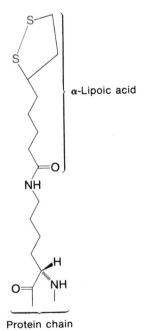

α-Lipoic acid

Protein chain

Figure 11–16
Structure of the α-lipoyl enzyme, showing the reactive disulfide in color. The lipoic acid is commonly covalently bonded through amide linkage with the ε amino group of a lysine residue as shown.

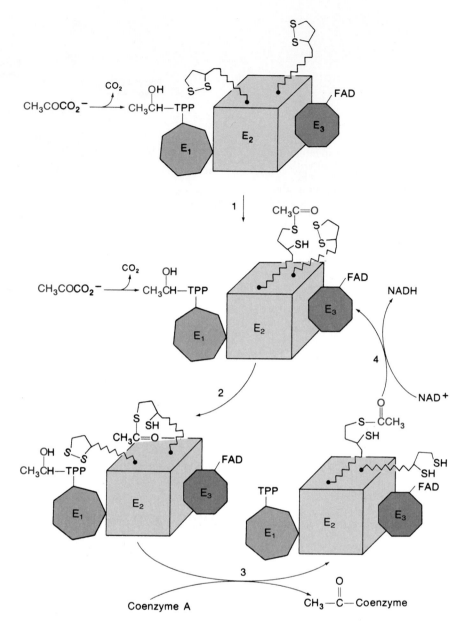

Figure 11–17
Interactions of α-lipoyl groups in the pyruvate dehydrogenase complex. The cubic structure represents the 24 subunits of dihydrolipoyl transacetylase, which constitutes the core of the complex. Two of the 48 lipoyl groups in the core are shown interacting with one of the 24 pyruvate decarboxylases ($E_1 \cdot$ TPP) and one of the 12 dihydrolipoyl dehydrogenase ($E_2 \cdot$ FAD) subunits. Note the interaction of the lipoyl groups in relaying electrons over the long distance between TPP and FAD. The lipoic acid must interact at active sites over distances too long to be covered by a single fully extended coenzyme. This problem is solved by use of a shuttle system involving two α-lipoyl groups for each transfer.

linkage with the ϵ amino group of a lysine residue (see Figure 11–17); and dihydrolipoyl dehydrogenase, a flavoprotein designated $E_3 \cdot$ FAD. A single particle of the complex consists of at least 24 chains of each of the first two enzymes and 12 of the flavoprotein. The core is composed of 24 subunits of E_2-lipoyl-S_2 arranged in cubic symmetry and containing 48 α-lipoyl groups, and associated with this core are 12 dimers of $E_1 \cdot$ TPP and 6 dimers of $E_3 \cdot$ FAD. Equations (34) through (38) represent the reaction sequence by which the complex catalyzes the reaction of pyruvate with NAD^+ and CoA.

$$H^+ + CH_3COCO_2^- + E_1 \cdot TPP \longrightarrow CO_2 + E_1 \cdot \text{hydroxyethyl-TPP} \quad (34)$$

$$E_1 \cdot \text{hydroxyethyl-TPP} + E_2\text{-lipoyl-S}_2 \longrightarrow$$

$$E_1 \cdot TPP + E_2\text{-lipoyl(SH)}{-}S{-}\overset{\overset{\displaystyle O}{\|}}{C}{-}CH_3 \quad (35)$$

$$E_2\text{-lipoyl(SH)}{-}S{-}\overset{\overset{\displaystyle O}{\|}}{C}{-}CH_3 + CoA \rightleftharpoons$$

$$E_2\text{-lipoyl(SH)}_2 + CH_3{-}\overset{\overset{\displaystyle O}{\|}}{C}{-}SCoA \quad (36)$$

$$E_2\text{-lipoyl(SH)}_2 + E_3 \cdot FAD \rightleftharpoons E_2\text{-lipoyl-S}_2 + \text{dihydro-}E_3 \cdot FAD \quad (37)$$

$$\text{Dihydro-}E_3 \cdot FAD + NAD^+ \rightleftharpoons E_3 \cdot FAD + NADH + H^+ \quad (38)$$

These equations show how α-lipoic acid couples the electron and group transfer processes involved. Note further, however, that lipoic acid must interact at active sites on all three enzymes. α-Lipoic acid bonded to E_2 is as shown in Figure 11–17, bonded through an amide linkage to a lysyl-ϵ-NH$_2$ group that places the reactive disulfide at the end of a flexible chain of atoms with rotational freedom about as many as 10 single bonds. When fully extended, this chain is 1.4 nm long, giving α-lipoic acid the potential capacity to sweep out a space having a spherical diameter of 2.8 nm. This distance turns out to be inadequate to account fully for the transport of electrons in this complex because the average distance between TPP on E_1 and FAD on E_3 has been estimated at between 4.5 and 6.0 nm by fluorescence energy-transfer measurements. The problem of long-distance interactions is overcome in the complex by the fact that each E_2 subunit contains two lipoyl groups that interact with each other and with α-lipoyl groups on other subunits according to Equation (39). This interaction facilitates the transport of electrons and acetyl groups through a network of α-lipoyl groups encompassing the entire core of E_2. *By relaying electrons and acetyl groups through two or more α-lipoyl groups, the coenzyme is able to span the distances among sites.* The relay process is illustrated schematically in Figure 11–17, which shows how two or more S-acetyldihydrolipoyl groups can interact to transport electrons over the large distances separating the sites for TPP on pyruvate dehydrogenase and FAD on dihydrolipoyl dehydrogenase.

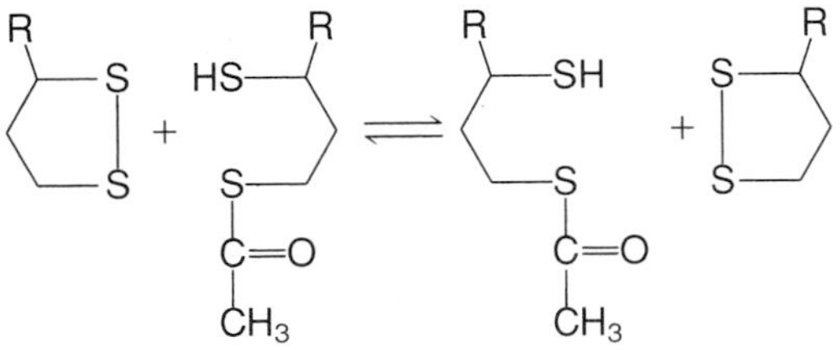

$$(39)$$

The coenzymatic capabilities of α-lipoyl groups result from a fusion of its chemical and physical properties, the ability to act simultaneously as both electron and acyl-group acceptor, the ability to span long distances to interact with sites separated by up to 2.8 nm, and the ability to act cooperatively with other α-lipoyl groups by disulfide interchange to relay electrons and acyl groups through distances that exceed its reach.

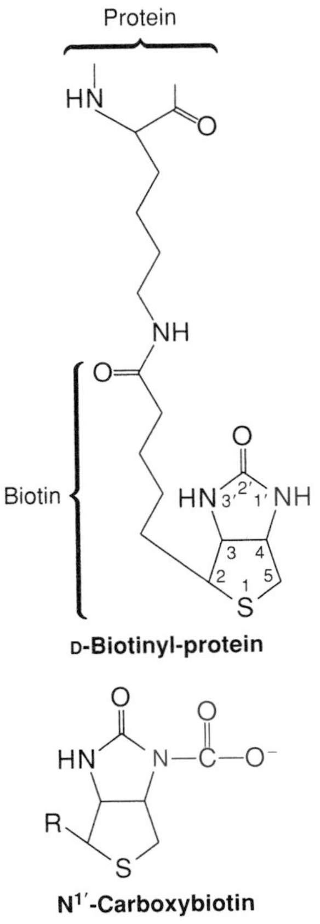

BIOTIN: A MEDIATOR OF CARBOXYLATIONS

The biotin structure shown in Figure 11–18 is an imidazolone ring that is *cis* fused to a tetrahydrothiophene ring substituted at position 2 by valeric acid. In carboxylase enzymes, biotin is covalently bonded to the proteins by an

Figure 11–18
Structures of biotinyl enzyme and N^1-carboxybiotin. The reactive portions of the coenzyme and the active intermediate are shown in color. In carboxylase enzymes, biotin is covalently bonded to the proteins by an amide linkage between its carboxyl group and a lysyl-ϵ-NH$_2$ group in the polypeptide chain.

amide linkage between its carboxyl group and a lysyl-ε-NH₂ group in the polypeptide chain. This places the imidazolone ring at the end of a long, flexible chain of atoms extending a maximum of about 1.4 nm from the α carbon of lysine.

Biotin is the essential coenzyme for carboxylation reactions involving bicarbonate as the carboxylating agent. Five such reactions have been described in which ATP-dependent carboxylation occurs at carbon atoms activated for enolization by ketonic or activated acyl groups, while in one other a nitrogen atom of urea is carboxylated. One other reaction, the transcarboxylase reaction of oxalacetate with propionyl-CoA to produce pyruvate and methylmalonyl-CoA, differs from the others in that it does not require ATP.

A general formulation of the ATP-dependent carboxylation of an α carbon by oxygen-18–enriched bicarbonate is given by Equation (40).

$$RCH_2-\overset{\overset{\textstyle O}{\|}}{C}-SCoA + ATP + HC^{18}O_3^- \xrightarrow{\text{Biotinyl carboxylase}}$$

$$R-\underset{\underset{\textstyle C^{18}O_2^-}{|}}{CH}-\overset{\overset{\textstyle O}{\|}}{C}-SCoA + ADP + HP^{18}O_4^{2-} \quad (40)$$

The appearance of ^{18}O in inorganic phosphate verifies that the function of ATP in the reaction is essentially the "dehydration" of bicarbonate. This also explains the lack of an ATP requirement by the transcarboxylase, since in that case the carboxyl-group donor is a carboxylic acid in which the carboxyl group is not hydrated as it is in bicarbonate. Thus the transfer of the carboxyl group from methylmalonyl-CoA to C-3 of pyruvate in Equation (41) is nearly isoenergetic and does not require ATP.

$$CH_3-\underset{\underset{\textstyle CO_2^-}{|}}{CH}-\overset{\overset{\textstyle O}{\|}}{C}-SCoA + CH_3-\overset{\overset{\textstyle O}{\|}}{C}-CO_2^- \rightleftharpoons$$

$$CH_3-CH_2-\overset{\overset{\textstyle O}{\|}}{C}-SCoA + {}^-O_2C-CH_2-\overset{\overset{\textstyle O}{\|}}{C}-CO_2^- \quad (41)$$

Biotin-dependent carboxylation reactions proceed in two stages, the carboxylation of imidazolone-N¹ in biotin to form N¹-carboxybiotin followed by the carboxylation of the substrate by N¹-carboxybiotin. These processes can be separated and studied as individual steps defined by Equations (42) and (43) for acetyl-CoA carboxylase.

$$ATP + HOCO_2^- + \text{biotinyl-E} \rightleftharpoons ADP + P_i + N^1\text{-carboxybiotinyl-E} \quad (42)$$

$$N^1\text{-carboxybiotinyl-E} + CH_3-\overset{\overset{\textstyle O}{\|}}{C}-SCoA \rightleftharpoons$$

$$\text{biotinyl-E} + {}^-O_2C-CH_2-\overset{\overset{\textstyle O}{\|}}{C}-SCoA \quad (43)$$

The first step is the ATP-dependent carboxylation of biotin by bicarbonate. This is believed to involve the transient formation of carbonic-phosphoric anhydride, or "carboxyphosphate," as an active carboxylation intermediate according to Equation (44):

$$HO-\overset{\overset{\textstyle O}{\|}}{C}-O^- \xrightarrow[\text{ADP}]{ATP} {}^-O-\underset{\underset{\textstyle OH}{|}}{\overset{\overset{\textstyle O}{\|}}{P}}-O-\overset{\overset{\textstyle O}{\|}}{C}-O^- \xrightarrow[P_i]{\text{Biotinyl-E}} N^1\text{-carboxybiotinyl-E} \quad (44)$$

N^1-Carboxybiotinyl enzymes have been isolated using [^{14}C]bicarbonate as the carboxylation substrate, and the ^{14}C has been shown to be bonded to N^1 of biotin. The [^{14}C]N^1-carboxybiotinyl enzymes have been shown to transfer their [^{14}C]carboxyl groups to the appropriate substrates, forming the corresponding [^{14}C]carboxylation products. The identification of N^1 as the carboxylation site in biotin was first achieved by Fyodor Lynen and associates, who found that β-methylcrotonyl-CoA carboxylase would catalyze the ATP-dependent carboxylation of *free* biotin by [^{14}C]bicarbonate. The product of this reaction was methylated with diazomethane to stabilize the [^{14}C]carboxyl group, and the methylation product was identified as the methyl ester of N^1-[^{14}C]carboxybiotin by comparison with an authentic synthetic sample. [^{14}C]Carboxybiotinyl enzymes have subsequently been degraded to N^1 carboxylated products. Synthetic N^1-carboxybiotin derivatives have since been shown to transfer carboxyl groups to acceptor substrates in the presence of the appropriate enzymes.

On the evidence of the foregoing experiments, *the coenzymatic function of biotin appears to be to mediate the carboxylation of substrates by accepting the ATP-activated carboxyl group and transferring it to the carboxyl acceptor substrate.* There is good reason to believe that the enzymatic sites of ATP-dependent carboxylation of biotin are physically separated from the sites at which N^1-carboxybiotin transfers the carboxyl group to acceptor substrates, i.e., the transcarboxylase sites. In fact, in the case of the acetyl-CoA carboxylase from *E. coli*, these two sites reside on two different subunits, while the biotinyl group is bonded to a third, a small subunit designated biotin carboxyl carrier protein. Transcarboxylase is also a multisubunit protein, one subunit being a small biotinyl protein.

Biotin appears to have just the right chemical and structural properties to mediate carboxylation. It readily accepts activated carboxyl groups at N^1 and maintains them in an acceptably stable yet reactive form for transfer to acceptor substrates. Since biotin is bonded to a lysyl group, the N^1-carboxyl group is at the end of a 1.6-nm chain with bond rotational freedom about nine single bonds, giving it the capability to transport activated carboxyl groups through space from the carboxyl activation sites to the carboxylation sites.

FOLATE COENZYMES: AN ENZYME-DISSOCIABLE AGENT FOR ONE CARBON TRANSFER

Tetrahydrofolate and its derivatives 5,10-methylenetetrahydrofolate, 5,10-methenyltetrahydrofolate, 10-formyltetrahydrofolate, and 5-methyltetrahydrofolate are the biologically active forms of folic acid, a four-electron-oxidized form of tetrahydrofolate. The structural formulas are given in Figure 11–19, which also shows how they arise from tetrahydrofolate. The structures are shown glutamylated on the carboxyl group of the p-aminobenzoyl group; the most active forms contain oligo- or polyglutamyl groups.

The tetrahydrofolates are the only coenzymes discussed in this chapter which are not known to function as tightly enzyme-bound coenzymes. They are specialized cosubstrates for a variety of enzymes involved in one-carbon metabolism. 10-Formyltetrahydrofolate and 5,10-methenyltetrahydrofolate are *formyl-group donor substrates for transformylases*. In living cells, 10-formyltetrahydrofolate is produced enzymatically from tetrahydrofolate and formate in an ATP-linked process in which formate is activated by phosphorylation to formyl phosphate; the formyl group of formyl phosphate is then transferred to N^{10} of tetrahydrofolate. 10-Formyltetrahydrofolate is a formyl donor substrate for some enzymes and is interconvertible with 5,10-

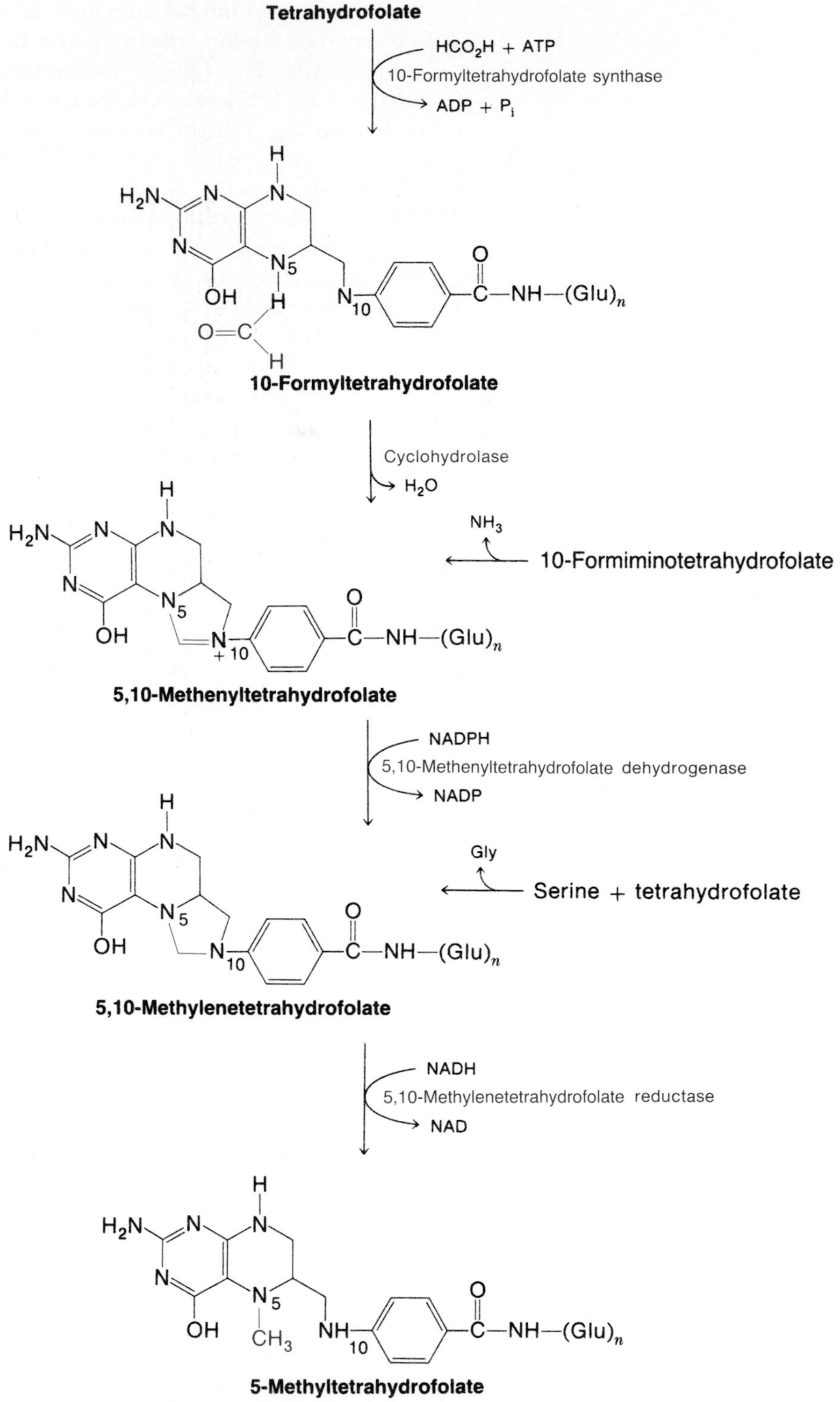

Figure 11–19
Structures and enzymatic interconversions of folate coenzymes. The reactive centers of the coenzymes are shown in color. The most active forms of the coenzyme contain oligo- or polyglutamyl groups.

methenyltetrahydrofolate by the action of cyclohydrolase. The methenyl derivative is also a formyl-donor substrate for other transformylases. 10-Formyltetrahydrofolate and 5,10-methenyltetrahydrofolate also can be synthesized in nonenzymatic reactions of tetrahydrofolate with free formic acid.

5,10-Methylenetetrahydrofolate is a _hydroxymethyl-group donor substrate_ for several enzymes and a _methyl-group donor substrate_ for thymidylate synthase (Figure 11–20). It arises in living cells from the reduction of

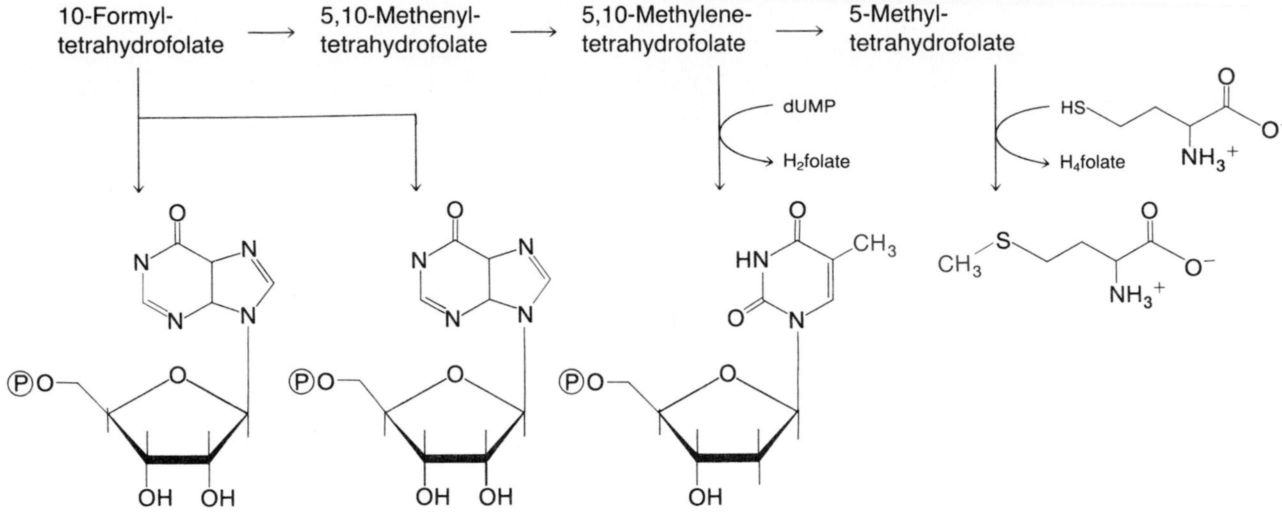

Figure 11–20
Involvement of folate coenzymes in one-carbon metabolism. Shown in color are the one-carbon units of the end products that originate with the reactive one-carbon units of the folate coenzymes.

5,10-methenyltetrahydrofolate by NADPH and also by the serine *trans*-hydroxymethylase-catalyzed reaction of serine with tetrahydrofolate. It also can be synthesized nonenzymatically by direct reaction of tetrahydrofolate with formaldehyde.

5-Methyltetrahydrofolate is the methyl-group donor substrate for methionine synthase, which catalyzes the transfer of the 5-methyl group to the sulfhydryl group of homocysteine. This and selected reactions of the other folate derivatives are outlined in Figure 11–20, which emphasizes the important role tetrahydrofolate plays in nucleic acid biosynthesis by serving as the immediate source of one-carbon units in purine and pyrimidine biosynthesis.

Note that in the thymidylate synthase reaction, 5,10-methylenetetrahydrofolate is the methyl-group donor to dUMP. The source of reducing equivalents to reduce the methylene group to the methyl level is tetrahydrofolate itself, so that the folate product of the reaction is dihydrofolate. The mechanism of the thymidylate synthase reaction is outlined in Figure 11–21.

The mechanism of action by thymidylate synthase exemplifies the reactions of 5,10-methylenetetrahydrofolate (Figure 11–21). It is not known in precisely what sequence some of the steps occur. The reaction begins with the nucleophilic addition of an enzymatic sulfhydryl group to C-6 of the uracil ring, forming an enolate species that is potentially carbanionic at C-5. Meanwhile 5,10-methylenetetrahydrofolate, upon protonation, isomerizes to a positively charged imine, which reacts with the enolate of uracil to produce a methylene-bridged intermediate. The bridged intermediate eliminates the enzymatic sulfhydryl group with concomitant loss of the C-5 proton, and either simultaneously or in a separate step, the bridging methylene group is reduced to a methyl group by the tetrahydropyrazine ring by a mechanism involving the conservation of the C-7 hydrogen, which ultimately appears as one of the three hydrogens in the methyl group.

The reduction of dihydrofolate produced by thymidylate synthase back to tetrahydrofolate by dihydrofolate reductase and NADH is an exceedingly important reaction because it maintains folate in the tetrahydro form and thereby facilitates the maintenance of all the tetrahydrofolates in Figure 11–19 at required levels. The importance of the tetrahydrofolates to cellular

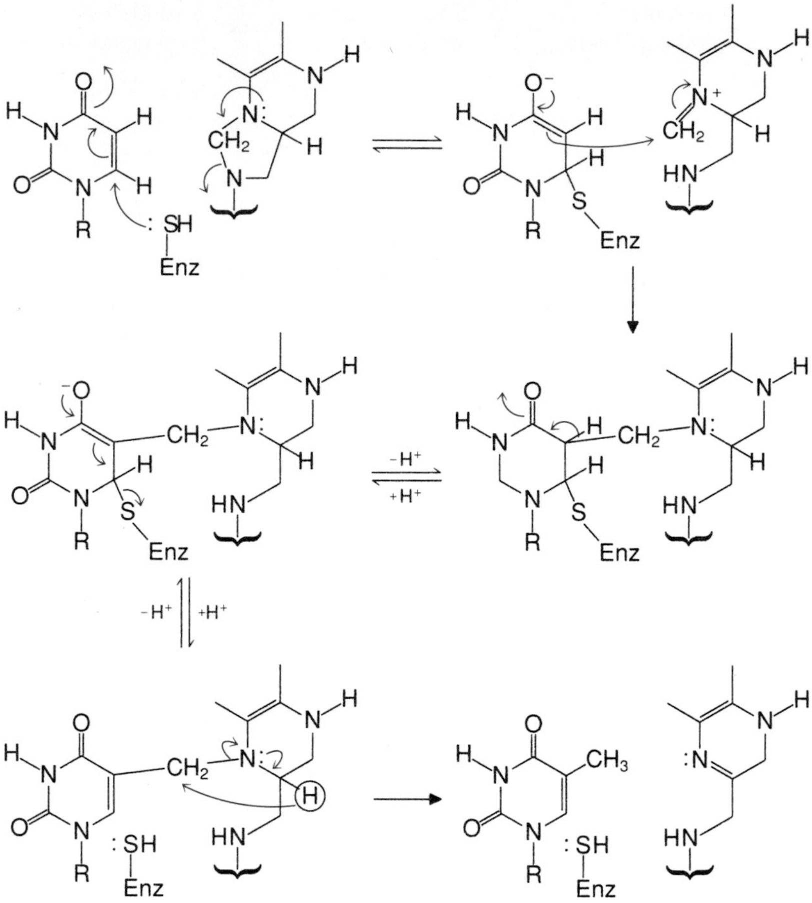

Figure 11–21
The mechanism of the thymdylate synthase reaction. It is not known in precisely what sequence some of the steps occur. The reaction begins with the nucleophilic addition of an enzymatic sulfhydryl group to C-6 of the uracil ring, forming an enolate species that is potentially carbanionic at C-5. Meanwhile 5,10-methylenetetrahydrofolate, upon protonation, isomerizes to a positively charged imine, which reacts with the enolate of uracil to produce a methylene-bridged intermediate. The bridged intermediate eliminates the enzymatic sulfhydryl group with concomitant loss of the C-5 proton, and, either simultaneously or in a separate step, the bridging methylene group is reduced to a methyl group by the tetrahydropyrazine ring by a mechanism involving the conservation of the C-7 hydrogen, which ultimately appears as one of the three hydrogens in the methyl group.

proliferation is highlighted by the fact that a potent inhibitor of dihydrofolate reductase, methotrexate (Figure 11–22), is one of the most effective and widely used drugs for cancer chemotherapy (see Chapter 25). 5-Fluorouracil (Figure 11–22) is also widely used in cancer chemotherapy, and it has been found that 5-fluoro-2'-dUMP is an exceedingly potent inhibitor of thymidylate synthase. 5-Fluoro-2'-dUMP reacts as if it were a substrate for thymidylate synthase through the first two steps of the mechanism outlined in Figure 11–21. However, the methylene-bridged intermediate cannot undergo the next step of the reaction, since it contains fluorine at C-5 instead of hydrogen and the next step involves the loss of the C-5 hydrogen as H^+. Fluorine *cannot* be eliminated as F^+. Thus the reaction is frozen at that step with the active site of the enzyme blocked.

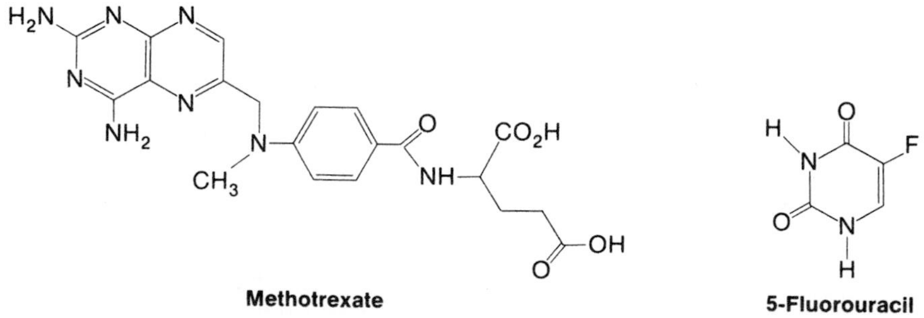

Methotrexate   **5-Fluorouracil**

Figure 11–22
Structures of methotrexate and 5-fluorouracil. Methotrexate is a potent inhibitor of dihydrofolate reductase. 5-Fluorouracil is a potent inhibitor of thymidylate synthase.

Formaldehyde is a toxic substance that reacts spontaneously with amino groups of proteins and nucleic acids, hydroxymethylating them and forming methylene-bridged cross-links between them. Free formaldehyde, therefore, wreaks havoc in living cells and could not serve as a useful hydroxymethylating agent. In the form of 5,10-methylenetetrahydrofolate, however, its chemical reactivity is attenuated but retained in a potentially available form where needed by specific enzymatic action. Formate, however, is quite unreactive under physiological conditions and must be activated to serve as an efficient formylating agent. As 10-formyltetrahydrofolate and 5,10-methenyltetrahydrofolate it is in a reactive state suitable for transfer to appropriate substrates. *The fundamental biochemical importance of tetrahydrofolate is to maintain formaldehyde and formate in chemically poised states, not so reactive as to pose toxic threats to the cell but available for essential processes by specific enzymatic action.*

VITAMIN B_{12} COENZYMES AND REARRANGEMENTS ON ADJACENT CARBON ATOMS

The principal coenzymatic form of vitamin B_{12} is 5′-deoxyadenosylcobalamin, whose structural formula is given in Figure 11–23. The structure involves a cobalt–carbon bond between the 5′ carbon of the 5′-deoxyadenosyl moiety and the cobalt atom of cobalamin. Vitamin B_{12} itself is cyanocobalamin in which the cyano group is bonded to cobalt in place of the 5′-deoxyadenosyl moiety. Other forms of the vitamin have water (aquocobalamin) or the hydroxyl group (hydroxycobalamin) bonded to cobalt.

The vitamin was discovered in liver as the antipernicious anemia factor in 1926, but the determination of its complete structure had to await its purification, chemical characterization, and crystallization, which required more than 20 years. Even then the determination of such a complex structure

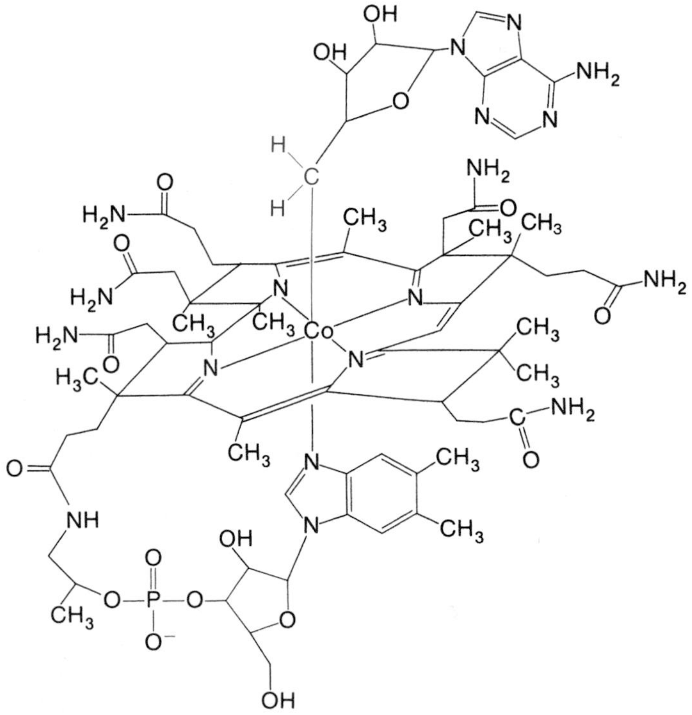

Figure 11–23
Structure of 5′-deoxyadenosylcobalamin coenzyme (vitamin B_{12}). The reactive groups are shown in color.

proved to be an elusive goal by conventional approaches of that day and had to await the elegant x-ray crystallographic study of Lenhert and Hodgkin in 1961, for which Dorothy Hodgkin was awarded the Nobel Prize in 1964.

5′-Deoxyadenosylcobalamin is the first substance to be discovered to contain a stable cobalt–carbon bond. The coenzyme was discovered by H. A. Barker and coworkers as the activating factor for glutamate mutase in the bacterium *Clostridium*. The direct bonding between cobalt and the 5′-deoxyadenosyl group was established by x-ray crystallography.

All but one of the 5′-deoxyadenosylcobalamin–dependent enzymatic reactions are rearrangements that follow the pattern of Equation (45), in which a hydrogen atom and another group (designated X) bonded to an adjacent carbon atom exchange positions, with the group X migrating from C_α to C_β.

$$a-\underset{X}{\overset{b}{C_\alpha}}-\underset{H}{\overset{c}{C_\beta}}-d \rightleftharpoons a-\underset{H}{\overset{b}{C_\alpha}}-\underset{X}{\overset{c}{C_\beta}}-d \qquad (45)$$

Three specific examples of rearrangement reactions are given in Equations (46) through (48). It is interesting and significant that the migrating groups —OH, —COSCoA, and —CH(NH$_2$)CO$_2$ have little in common and that the hydrogen atoms migrating in the opposite direction are often chemically unreactive.

$$^-O_2C-CH_2CH_2-\underset{NH_3^+}{CH}-CO_2^- \underset{mutase}{\overset{Glutamate}{\rightleftharpoons}} {}^-O_2C-\underset{NH_3^+}{\overset{CH_3}{CH}}-CH-CO_2^- \qquad (46)$$

$$^-O_2C-CH_2CH_2-\overset{O}{\underset{}{C}}-SCoA \underset{mutase}{\overset{Methylmalonyl\text{-}CoA}{\rightleftharpoons}} {}^-O_2C-\overset{CH_3}{\underset{}{CH}}-\overset{O}{\underset{}{C}}-SCoA \qquad (47)$$

$$CH_3\underset{OH}{CH}CH_2OH \xrightarrow{Dioldehydrase} CH_3CH_2\underset{OH}{CH}-OH \xrightarrow{Dioldehydrase}$$

$$CH_3CH_2CHO + H_2O \quad (48)$$

The hydrogen migrations in all the B$_{12}$ coenzyme-dependent rearrangements proceed without exchange with the protons of water; i.e., isotopic hydrogen in the substrates is conserved in the products even though, as discussed later, these migrations are not simply intramolecular 1,2 shifts. The hydrogen migrations in Equations (46) through (48) are stereospecific, as are the migrations of the —X groups. The migrations of the —CHNH$_2$CO$_2$H and —OH groups catalyzed by glutamate mutase and dioldehydrase occur with overall inversion of configuration at the terminals of migration, while that of the —COSCoA group catalyzed by methylmalonyl-CoA mutase proceeds with retention of configuration. The significance of these differences is not understood in mechanistic terms, but any general concept of the mechanisms of these rearrangements must be consistent with the stereochemistry.

The work of Robert Abeles and coworkers studying the mechanism of action of dioldehydrase shed the first light on the role of 5′-deoxyadenosylcobalamin in the rearrangement reactions. Moreover, their findings have been confirmed in other rearrangement reactions.

The first direct evidence of the role played by 5′-deoxyadenosylcobalamin in mediating hydrogen transfer was obtained in experiments showing that the conversion of [1-^{3}H$_2$]1,2-propanediol to [1,2-^{3}H]propionaldehyde led to the incorporation of tritium into *both* C-5′ hydrogen positions of the 5′-deoxyadenosyl moiety in the coenzyme. It also was shown that

[5'-^{3}H]5'-deoxyadenosylcobalamin, either isolated from the foregoing enzymatic experiments or prepared by chemical synthesis, could transfer all of its tritium into propionaldehyde when used as the coenzyme with unlabeled 1,2-propanediol and the dehydrase. These results accounted for the finding that hydrogen transfer catalyzed by this enzyme is not compulsorily intramolecular. It was subsequently shown that hydrogen transfer in a given turnover could be either intermolecular or intramolecular and that the rates of tritium transfer into the coenzyme from substrate and from the coenzyme into product account for the rate of appearance of tritium in the product.

The involvement of adenosyl-C-5' in hydrogen transfer and the fact that hydrogen transfer can be either intermolecular or intramolecular in a given turnover strongly imply that the cobalt-carbon bond in 5'-deoxyadenosylcobalamin is transiently cleaved and that the C-5' carbon transiently becomes a methyl group in the catalytic process.

Cobalamins lacking the 5'-deoxyadenosyl moiety, i.e., the vitamin itself or hydroxycobalamin, are known to exist in three oxidation states: the Co(III) state, known as B_{12a} or B_{12b} for hydroxycobalamin or aquocobalamin; the one-electron reduced form or Co(II) state, known as B_{12r}; and the two-electron reduced Co(I) state, known as B_{12s}. The three forms are distinguishable by their visible and ultraviolet absorption spectra (B_{12a} is a red compound, B_{12r} is yellow, and B_{12s} is described as grey-green in color) and by the fact that B_{12r} is paramagnetic, exhibiting a strong electron-spin resonance signal.

Spectroscopic data have implicated the Co(II) form of cobalamin as a catalytic intermediate in several 5'-deoxyadenosylcobalamin-dependent reactions, including the dioldehydrase reaction. The visible absorption spectrum of the enzyme-coenzyme complex is very similar to that of 5'-deoxyadenosylcobalamin itself, but upon addition of substrate, it quickly shifts to a spectrum characteristic of B_{12r} and then reverts almost to its initial state upon complete conversion of substrate to product. In similar experiments monitored by electron-spin resonance spectroscopy, a signal corresponding to that of B_{12r} is observed upon adding substrate to the enzyme-coenzyme complex, and signals corresponding to other free-radical species are also detected. The rates of appearance of these signals are on the catalytic time scale, indicating that the B_{12r} and other free-radical species are catalytic intermediates.

The mechanisms of the 5'-deoxyadenosylcobalamin-dependent rearrangements are not fully understood, but from the preceding hydrogen-transfer and spectroscopic experiments, the outline of a mechanism is emerging. This is given in Figure 11-24, in which the substrate and product are shown as the generalized forms of Equation (45) and the structure of the coenzyme is abbreviated to emphasize the importance of the cobalt–carbon bond.

The reaction begins by homolytic cleavage of the Co–C bond, (Figure 11-24) generating Co II (B_{12r}) and 5'-deoxyadenosyl free radical (step 1). This radical abstracts a hydrogen atom from the substrate, the migrating hydrogen in Equation (45), generating 5'-deoxyadenosine and a substrate-derived free radical as intermediates (step 2). The substrate-radical undergoes rearrangement to a product-derived free radical, which abstracts a hydrogen atom to form the final product and regenerate the coenzyme (steps 3–6). Much evidence supports the involvement of the intermediates shown in Figure 11-24; however, quantitative aspects of available data suggest the possible existence of additional species and a more complex hydrogen transfer process.

The most fundamental property of 5'-deoxyadenosyl-cobalamin leading to its unique action as a coenzyme is the weakness of the Co–C bond. *This bond has a low dissociation energy, of less than 30 kcal/mole, strong enough to be essentially stable in free solution but weak enough to be broken as a*

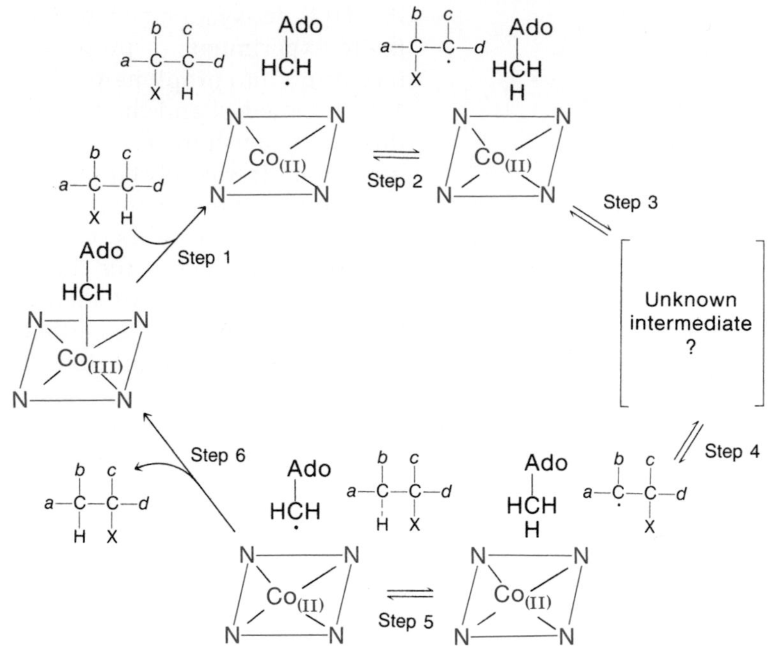

Figure 11–24
Hypothetical partial mechanism of vitamin B_{12}-dependent rearrangements. The designations Co(III) and Co(II) refer to species that are spectrally and magnetically similar to Co^{3+} and Co^{2+}, respectively. Co(III) is diamagnetic and red, while Co(II) is paramagnetic (unpaired electron) and yellow. The metal does not undergo a change in electrostatic charge when the cobalt–carbon bond breaks homolytically (i.e., without charge separation), since one electron remains with the metal and the other with 5'-deoxyadenosine. One or more unknown intermediates, symbolized by the brackets, may be involved in the rearrangement.

result of strain induced by multiple binding interactions between the enzymic binding sites and the adenosyl and cobalamin portions of the coenzyme. The free radicals resulting from cleavage of this bond and abstraction of hydrogen from substrates undergo the rearrangements characteristic of B_{12}-dependent reactions.

The one known 5'-deoxyadenosylcobalamin-dependent enzyme that does not catalyze a rearrangement is the ribonucleotide reductase from *Lactobacillus leichmanii*, which catalyzes the reduction of ribonucleoside triphosphates to deoxynucleoside triphosphates (Equation 49).

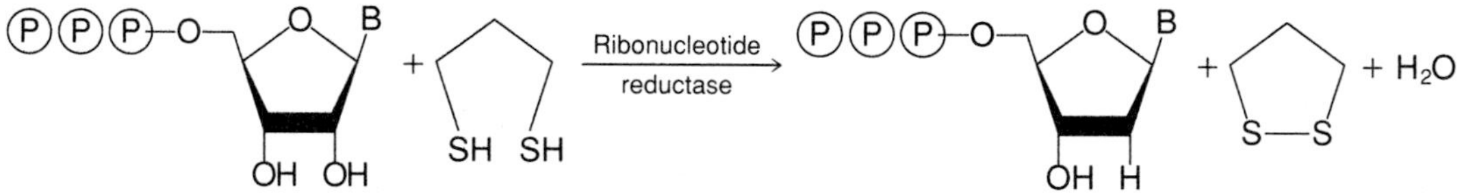

$$(49)$$

The reducing agent may be a vicinal dithiol, such as dihydrolipoic acid, or a vicinal dithiol protein, such as thioredoxin (see Chapter 25). In $[^3H]H_2O$, the enzyme catalyzes the incorporation of ^3H into the 2'-position of the 2'-deoxynucleoside triphosphate and also into deoxyadenosyl-C-5' of the coenzyme. Although this is not a rearrangement reaction, the mechanism is closely related to the other B_{12}-dependent reactions. It is known that the ribosyl-C-3' hydrogen is transiently removed and then returned to the same position in the deoxyribosyl moiety of the product. Available information is consistent with the mechanism outlined in Equations (50) and (51).

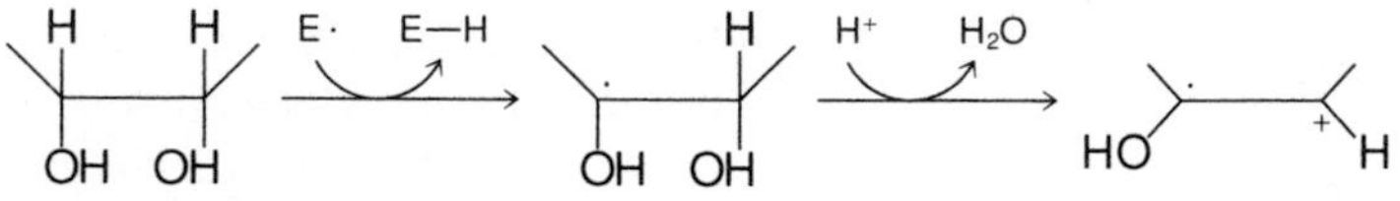

$$(50)$$

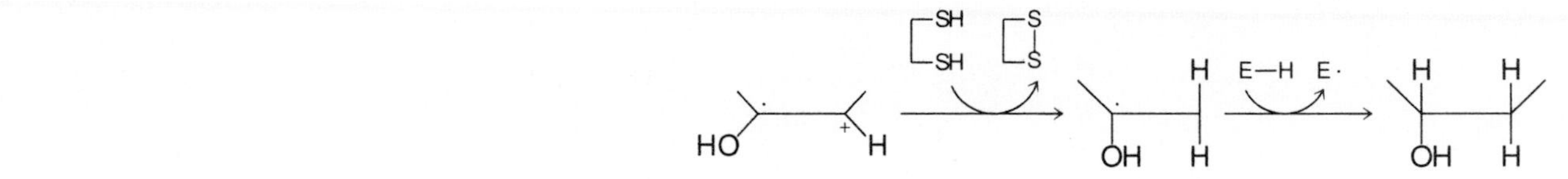

$$\tag{51}$$

The coenzyme is thought to facilitate the generation of a free radical on the enzyme that abstracts the ribosyl-C-3' hydrogen from the substrate. The resulting free radical then undergoes an enzymatic, general acid catalyzed dehydration to a radical-cation [Equation (50)]. Reduction of the radical-cation at C-2' by the dithiol leads to the 3'-free radical of the product, which reabstracts from the enzyme the same hydrogen that had originally been removed [Equation (51)].

The radical-cation in Equations (50) and (51) is a resonance-stabilized species that derives additional stabilization by the bonded hydroxyl group [see Equation (52)]. A similar intermediate may be involved in the diol-dehydrase reaction [Equation (48)]. Equation (52) outlines a rearrangement mechanism via such a resonance-stabilized species.

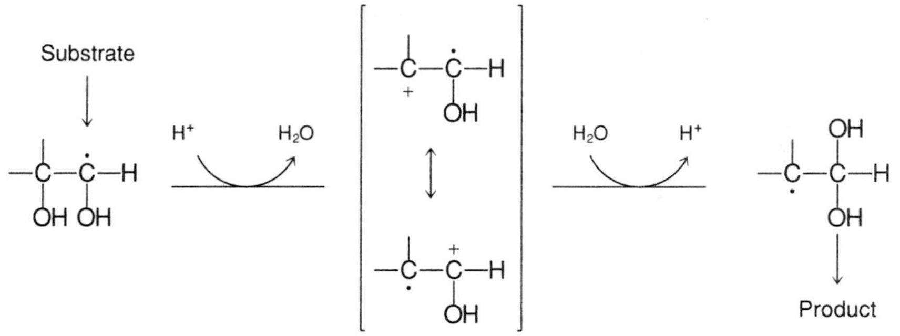

$$\tag{52}$$

The free radical generated by coenzymatic abstraction of a hydrogen from C-1 of the substrate (Figure 11−24) undergoes dehydration to a radical-cation similar to that in Equations (50) and (51). This is a species in which the positive charge and unpaired electron are delocalized, so that H_2O can add back to either carbon. When H_2O adds to C-1 the resulting aldehyde hydrate C-2-free radical abstracts a hydrogen from 5'-deoxyadenosine to form the aldehyde hydrate as product and regenerate the coenzyme.

Another coenzymatic form of vitamin B_{12} is methylcobalamin, which has a methyl group in place of the 5'-deoxyadenosyl moiety. Methylcobalamin arises as an intermediate in the methionine synthase reaction (Figure 11−20). This enzyme is a B_{12} protein that must be maintained in a reduced state to be active. The cobalamin mediates methyl transfer, presumably via the intermediate formation of a methylcobalamin.

IRON-CONTAINING COENZYMES AND REDOX REACTIONS

Iron as a cofactor in catalysis is receiving increasing attention. The metal exists in two oxidation states: Fe^{2+} and Fe^{3+}. Iron complexes are nearly all octahedral, and practically all are paramagnetic (caused by unpaired electrons in the $3d$ orbital). The most common form of iron in biological systems is heme. Heme groups (Fe^{2+}) and hematin (Fe^{3+}) most frequently involve a complex with protoporphyrin IX (Figure 11−25). They are the coenzymes (prosthetic groups) for a number of redox enzymes, including catalase, which catalyzes dismutation of hydrogen peroxide [Equation (53)], and per-

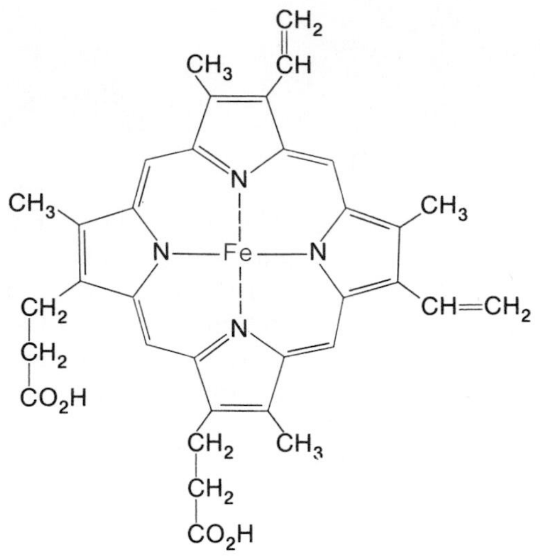

Figure 11–25
Structure of protoporphyrin IX. This coenzyme acts in conjunction with a number of different enzymes involved in oxidation and reduction reactions.

oxidases, which catalyze the reduction of alkyl hydroperoxides by such reducing agents as phenols, hydroquinones, and dihydroascorbate [represented as AH_2 in Equation (54)].

$$2 \ H_2O_2 \rightleftharpoons 2 \ H_2O + O_2 \tag{53}$$

$$R-O-O-H + AH_2 \longrightarrow A + R-O-H + H_2O \tag{54}$$

Heme proteins exhibit characteristic visible absorption spectra as a result of protoporphyrin IX; their spectra differ depending on the identities of the lower axial ligand donated by the protein and the oxidation state of the iron as well as the identities of the upper axial ligands donated by the substrates. Spectral data show clearly that the heme coenzymes participate directly in catalysis; however, the mechanisms of action of hemes are not so well understood as those of other coenzymes.

Many redox enzymes contain iron-sulfur clusters that mediate one-electron transfer reactions. The clusters consist of two or four irons and an equal number of inorganic sulfide ions clustered together with the iron, which is also liganded to cysteinyl-sulfhydryl groups of the protein (see Figure 11–26). The enzyme nitrogenase, which catalyzes the reduction of N_2 to $2 \ NH_3$, contains such clusters in which some of the iron has been replaced by molybdenum (Chapter 24). Electron-transferring proteins involved in one-electron transfer processes often contain iron-sulfur clusters. These proteins include the mitochondrial membrane enzymes NADH dehydrogenase and succinate dehydrogenase (Chapter 15), which are flavoproteins, and the small-molecular-weight proteins ferredoxin, rubredoxin, adrenodoxin, and putidaredoxin (Chapters 16, 17, and 23).

Heme coenzymes, iron-sulfur clusters, flavin coenzymes, and nicotinamide coenzymes cooperate in multienzyme systems to catalyze the chemically remarkable hydroxylations of hydrocarbons such as steroids (Chapter

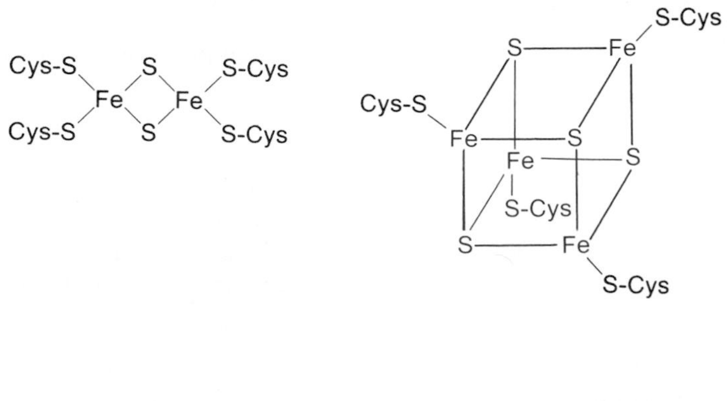

Figure 11–26
Structures of iron-sulfur clusters. Many redox enzymes contain iron-sulfur clusters that mediate one-electron transfer reactions.

23). In these hydroxylation systems, the heme proteins are known as cyto-chrome P-450, named for the wavelength corresponding to the most intense absorption band of the carbon monoxide-liganded heme, an inhibited form. The reactions catalyzed by these systems are represented in generalized form by Equation (55).

$$H^+ + R\!-\!CH_2\!-\!R' + O_2 + NADPH \longrightarrow R\!-\!\overset{\overset{\displaystyle OH}{|}}{C}H\!-\!R' + NADP^+ + H_2O \quad (55)$$

The enzymes involved usually include a cytochrome P-450, an iron-sulfur cluster–containing protein such as adrenodoxin or putidaredoxin, and a flavoprotein reductase. The detailed mechanisms by which these proteins co-operate in catalyzing hydroxylations are not understood. Cytochrome P-450 interacts directly with O_2 and receives electrons from the iron-sulfur protein, which is in turn reduced by NADPH by the action of the flavoprotein reduc-tase. An oxygenating form of cytochrome P-450 is thereby produced, but its chemical nature and the mechanism by which it hydroxylates substrates are not known.

The cytochrome P-450 hydroxylases include those in adrenal cells which hydroxylate sterols and steroids in the production of steroid hor-mones, such as aldosterone and hydrocortisone (Chapter 23). The liver mi-crosomal systems detoxify amines and polycyclic hydrocarbons by hydrox-ylating them in preparation for further transformations and eventual elimination. The hydroxylation of a secondary amine results in the dealkyla-tion of the amine according to Equation (56). The immediate hydroxylation product, an α-hydroxyamine, is an addition compound of a primary amine and an aldehyde, and it spontaneously dissociates.

$$H^+ + R\!-\!NH\!-\!CH_2\!-\!R' + O_2 + NADPH \longrightarrow$$

$$R\!-\!NH\!-\!\overset{\overset{\displaystyle OH}{|}}{\underset{\downarrow}{C}}H\!-\!R' + NADP^+ + H_2O \quad (56)$$

$$R\!-\!NH_2 + R'\!-\!CHO$$

Bacterial hydroxylation systems such as those found in *Pseudomonas* cata-lyze the hydroxylation of hydrocarbons, the first step in the oxidative degra-dation of hydrocarbons to produce energy for the organism. The hydroxy-lated hydrocarbons are further oxidized and degraded by more conventional oxidative processes involving nicotinamide and flavin coenzymes, the re-duced forms of which are reoxidized by terminal electron acceptors in reac-tions coupled to the production of ATP.

COENZYMES INVOLVED IN METHANOGENESIS

Methanogens are a primitive type of bacteria believed to reflect one of the earliest living forms to evolve on this planet. Methanogens are so named because they have the unique capability to use methyl groups as terminal electron acceptors, and to produce methane as an end product of their me-tabolism. For the reactions involved, methanogens have evolved a number of coenzymes that are believed to be confined to their species. It seems likely that a serious study of the biochemistry of methanogens will be of great evolutionary significance.

Most methanogens utilize CO_2 as an electron acceptor, reducing it all the way to methane. In these organisms CO_2 fulfills the role played by O_2 in aerobic organisms. The overall pathway by which CO_2 is reduced to CH_4 in methanogens has recently been elucidated and shown to involve six new coenzymes. The structural formulas for four of the coenzymes are shown in Figures 11–27 through 11–30. The process begins by reaction of CO_2 with

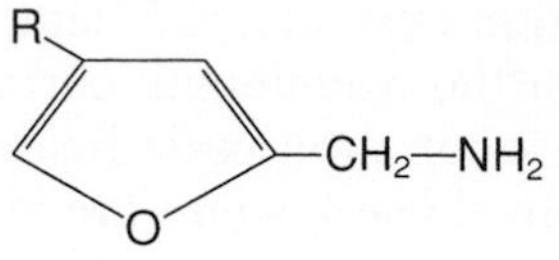

Methanofuran

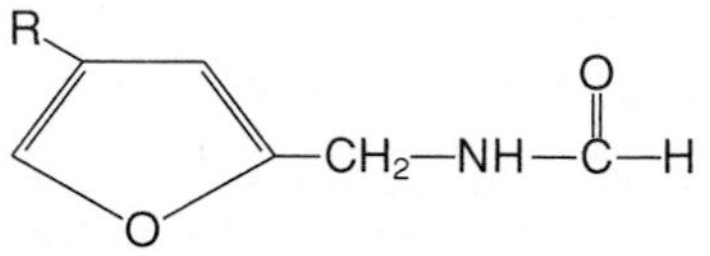

Formyl-methanofuran

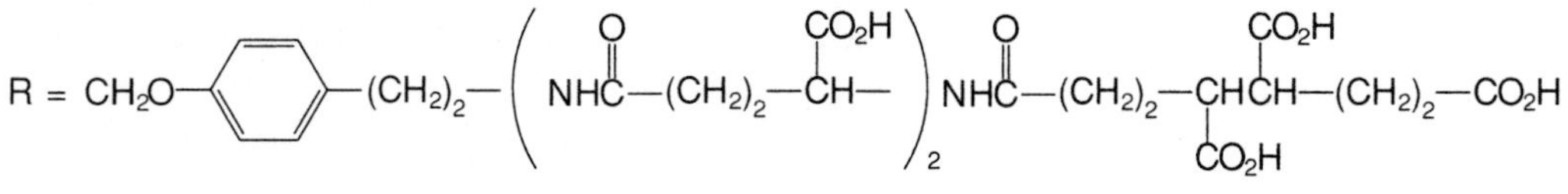

$R = CH_2O-\bigcirc-(CH_2)_2-\left(NHC-(CH_2)_2-CH-\right)_2 NHC-(CH_2)_2-CHCH-(CH_2)_2-CO_2H$

Figure 11–27
Structures of methanofuran and formyl-methanofuran, coenzymes involved in the first step of methanogenesis.

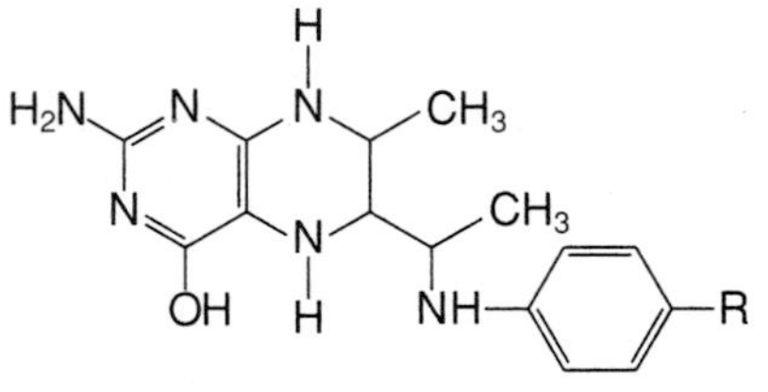

Tetrahydromethanopterin

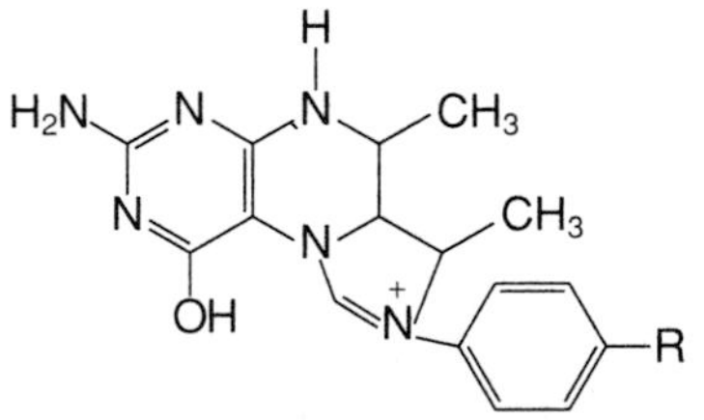

Methenyl-tetrahydromethanopterin

$R = CH_2-CH-CH-CH-CH_2$ ...

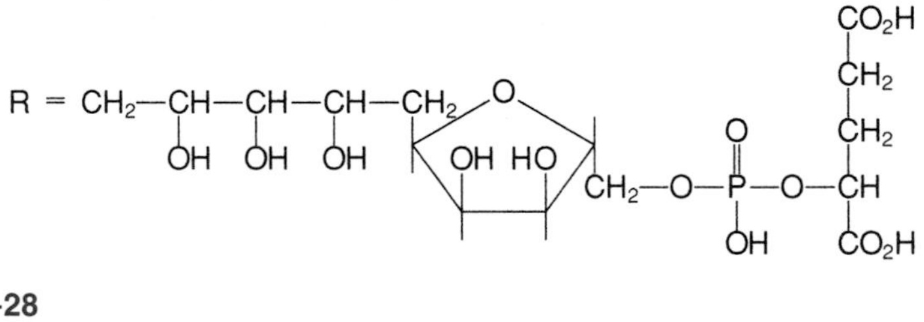

Figure 11–28
Structures of tetrahydromethanopterin and methenyl-tetrahydromethanopterin. These coenzymes are structurally related to the tetrahydrofolates. In methanogenesis methenyl-tetrahydromethanopterin is reduced first to methylene-tetrahydromethanopterin, then to methyl-tetrahydromethanopterin.

methanofuran and a reducing system to produce formyl-methanofuran (Figure 11–27), a process by which CO_2 is reduced to the level of formate. In the next step the formyl group is transferred from formyl-methanofuran to tetrahydromethanopterin (H_4MPT), forming methenyl-tetrahydromethanopterin (methenyl-H_4MPT). The structures of H_4MPT and methenyl-H_4MPT (Figure 11–28) are similar to those of the corresponding tetrahydrofolates. Methenyl-H_4MPT is then reduced in two stages to methylene-H_4MPT and methyl-H_4MPT, just as 5,10-methenyltetrahydrofolate is reduced to 5,10-methylene- and then to 5-methyltetrahydrofolate in Figure 11–20. In the

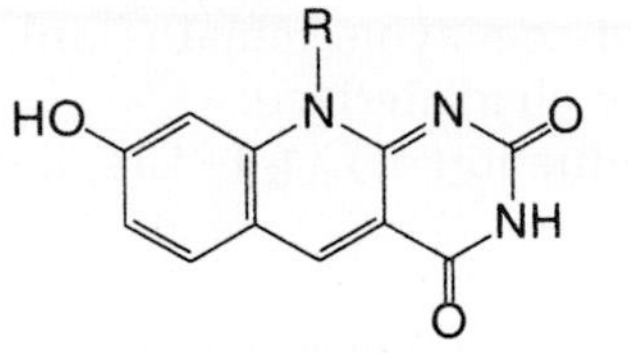

Oxidized F$_{420}$

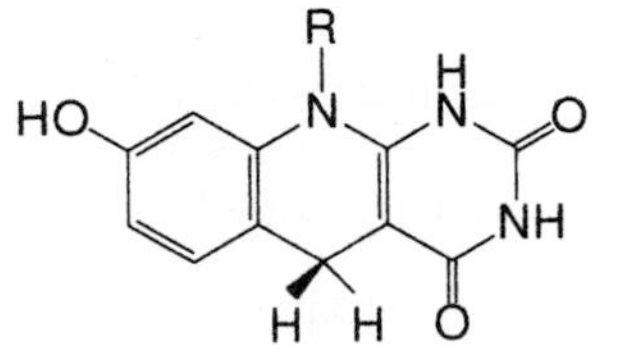

Reduced F$_{420}$

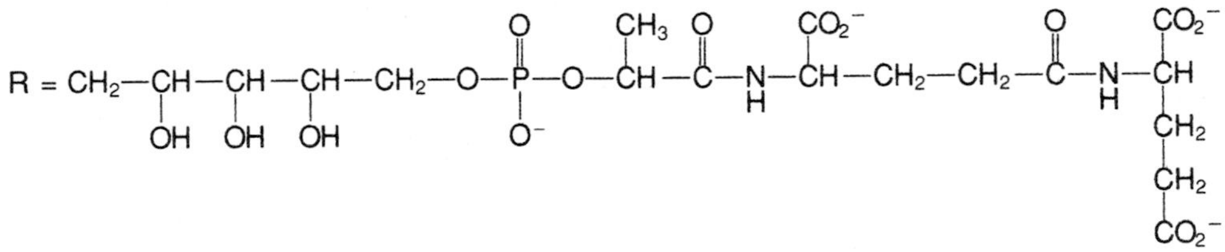

$$R = CH_2-CH-CH-CH-CH_2-O-\overset{\overset{O}{\|}}{P}-O-CH-\overset{\overset{CH_3 \; O}{\|}}{C}-N-CH-CH_2-CH_2-\overset{\overset{O}{\|}}{C}-N-CH$$

with OH OH OH on the first chain, O^- on phosphorus, CO_2^- groups, and CH_2, CH_2, CO_2^- chain.

Figure 11–29
Structure of F$_{420}$, a cofactor of the hydrogenase component of the methyl reductase system in methanogenesis.

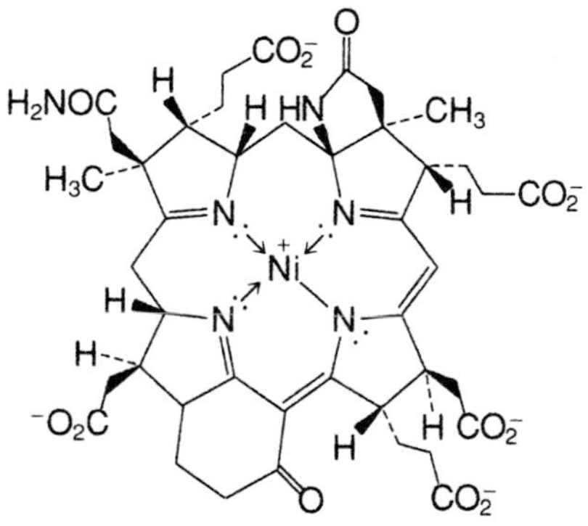

F$_{430}$

Figure 11–30
Structure of F$_{430}$, a cofactor of the methyl reductase system in methanogenesis.

next step the methyl group of methyl-H$_4$MPT is transferred to the —SH group of 2-mercaptoethanesulfonate, also known as coenzyme M or CoM. The structures of CoM and methyl-CoM are shown below.

$$HS-CH_2-CH_2-SO_3^- \qquad\qquad CH_3-S-CH_2-CH_2-SO_3^-$$
CoM   **Me-CoM**

Methyl-CoM is the precursor of methane, which is produced by the action of a complex H$_2$-dependent reductase system consisting of several proteins and three coenzymes in addition to FAD. Two of the coenzymes are F$_{420}$, a 5-deazaflavin derivative (Figure 11–29), and F$_{430}$, a nickel tetrapyrroline structurally related to vitamin B$_{12}$ (Figure 11–30). These factors are designated F$_{420}$ and F$_{430}$ for the wavelengths of maximum absorbance in their visible spectra, 420 nm and 430 nm, respectively. Both are prosthetic groups firmly bound to protein components of the reductase system. The structure of a third coenzyme remains unknown, and most of the proteins are not as yet fully characterized.

The overall reduction of CO_2 to CH_4 is summarized in Equations (57) to (62).

$$CO_2 + \text{methanofuran} \xrightarrow{[H]} \text{formyl-methanofuran} \tag{57}$$

$$\text{Formyl-methanofuran} + H_4MPT \longrightarrow \text{methenyl-}H_4MPT + \text{methanofuran} \tag{58}$$

$$\text{Methenyl-}H_4MPT \xrightarrow{[H]} \text{methylene-}H_4MPT \tag{59}$$

$$\text{Methylene-}H_4MPT \xrightarrow{[H]} \text{methyl-}H_4MPT \tag{60}$$

$$\text{Methyl-}H_4MPT + CoM \longrightarrow H_4MPT + \text{methyl-CoM} \tag{61}$$

$$\text{Methyl-CoM} + H_2 \xrightarrow[\text{proteins}]{\substack{F_{420}, F_{430}, \text{FAD,} \\ \text{new coenzyme}}} CH_4 + CoM \tag{62}$$

The enzymes catalyzing the overall reduction have not as yet been characterized, and the primary reductants for most steps are still unknown, although NADH or NADPH is likely to be involved. The reductant in the last step [Equation (62)] is H_2, which reduces the F_{420}-dependent hydrogenase component of the methyl reductase system. F_{430} and the unidentified coenzyme appear to mediate electron transfer from the hydrogenase to methyl-CoM, producing CH_4 and CoM.

The detailed chemical mechanisms of the reactions involved in methanogenesis are unknown. The most mysterious reactions are the first and last steps, Equations (57) and (62). The intervening transformylation, reductions, and transmethylation are likely to be similar to the corresponding reactions of tetrahydrofolates.

SUMMARY

Coenzymes are molecules that act in cooperation with enzymes to catalyze biochemical processes, performing functions that the enzymes are otherwise chemically or physically ill-equipped to carry out. Most coenzymes are derivatives of the water-soluble vitamins, but a few, such as hemes, lipoic acid, and iron-sulfur clusters, are biosynthesized in the body. Each coenzyme plays a unique chemical or physical role in the enzymatic processes of living cells, and the chemical structure and reactivity of each are suited to its biochemical function.

Thiamine pyrophosphate promotes the decarboxylation of α-keto acids and the cleavage of α-hydroxyl ketones by reacting with the ketone groups of substrates to produce intermediates in which electron pairs resulting from carbon–carbon bond cleavages can be stabilized by the thiazolium ring of the coenzyme. Pyridoxal-5′-phosphate promotes decarboxylations, racemizations, transaminations, aldol cleavages, and α,β and β,γ eliminations in amino acids by forming imine intermediates with amino acid substrates in which electron pairs resulting from cleavages of covalent bonds are stabilized by the pyridine ring of the coenzyme. Nicotinamide coenzymes act as intracellular electron carriers in transporting reducing equivalents from metabolic intermediates to the terminal electron-transport systems in cellular and mitochondrial membranes. Nicotinamide coenzymes also act as cocatalysts with enzymes by transiently oxidizing substrates to form reactive intermediates.

Flavin coenzymes are oxidation-reduction coenzymes that act as cocatalysts with enzymes in a large variety of biochemical redox reactions, most of which involve O_2. Phosphopantetheine coenzymes facilitate the enolization and acyl-group transfer reactions of acyl groups to which they are bonded by thioester linkages. Lipoic acid in α-keto acid dehydrogenase complexes mediates electron transfer and acyl-group transfer among the active sites in the complex by undergoing transient reduction and acylation.

Biotin mediates the carboxylation of substrates by undergoing transient carboxylation to N^1-carboxybiotin. Phosphopantetheine, lipoic acid, and biotin, by virtue of their flexible, chainlike structures, facilitate the physical translocation of chemically reactive species among catalytic sites or subunits. Vitamin B_{12} coenzymes mediate enzymatic 1,2 arrangements in which a hydrogen bonded to a given carbon exchanges places with another group bonded to an adjacent carbon.

Heme coenzymes play essential roles in a variety of

enzymatic reactions involving reductions of peroxides. Iron-sulfur clusters, composed of Fe and S^{2-} in equal numbers together with cysteinyl side chains of proteins, mediate electron-transfer processes in a variety of enzymatic reactions, including the reduction of N_2 to $2NH_3$. Nicotinamide, flavin, and heme coenzymes act cooperatively with iron-sulfur clusters in multienzyme systems to catalyze hydroxylations of hydrocarbons.

Recent investigations of primitive bacteria known as methanogens have led to the discovery of several new coenzymes involved in the conversion of CO_2 to CH_4.

SELECTED READINGS

Bruice, T. C., and S. J. Benkovic, *Bioorganic Chemistry*, Vols. 1 and 2. Menlo Park, Calif.: Benjamin, 1966. A detailed discussion of the mechanisms of bioorganic reactions, including those involving coenzymes.

Jencks, W. P., *Catalysis in Chemistry and Enzymology*. New York: McGraw-Hill, 1969. A detailed analysis of mechanisms of enzymatic and nonenzymatic reactions, including those involving coenzymes.

Phipps, D. A., *Metals and Metabolism*, Oxford Chemistry Series. Oxford: Clarendon Press, 1976. Examines the importance of metal ions in metabolic processes.

Popjak, G., Stereospecificity of enzymic reactions. In P. D. Boyer, ed., *The Enzymes*, Vol. 2. New York: Academic Press, 1970. P. 115.

Walsh, C. T., *Enzymatic Reaction Mechanisms*. San Francisco: Freeman, 1977. Provides an up-to-date discussion of the mechanisms of enzymatic reactions. An in-depth treatment of coenzymes.

PROBLEMS

1. The following reactions are catalyzed by pyridoxal-5′-phosphate–dependent enzymes. Write reaction mechanisms for them, showing how pyridoxal-5′-phosphate is involved in catalyzing them.

a. $CH_3-\underset{\underset{NH_2}{|}}{\overset{\overset{H}{|}}{C}}-CO_2H \rightleftharpoons CH_3-\underset{\underset{H}{|}}{\overset{\overset{NH_2}{|}}{C}}-CO_2H$

b. $H_2N-(CH_2)_4-\underset{\underset{NH_2}{|}}{CH}-CO_2H \longrightarrow$

$CO_2 + H_2N-(CH_2)_5-NH_2$

c. $HO_2C-CH_2-\underset{\underset{NH_2}{|}}{CH}-CO_2H \longrightarrow$

$CO_2 + CH_3-\underset{\underset{NH_2}{|}}{CH}-CO_2H$

2. Thiamine pyrophosphate-dependent enzymes catalyze the following reactions. Write mechanisms for them, showing the catalytic role of the coenzyme.

a. $CH_3-\overset{\overset{O}{\|}}{C}-CO_2H \longrightarrow CO_2 + CH_3-\overset{\overset{O}{\|}}{C}-H$

b. $ⓅOCH_2-(CHOH)_2-\underset{\underset{OH}{|}}{CH}-\overset{\overset{O}{\|}}{C}-CH_2OH + HOPO_3^{2-} \longrightarrow$

$ⓅO-CH_2-(CHOH)_2-CHO + CH_3-\overset{\overset{O}{\|}}{C}-OPO_3^{2-} + H_2O$

3. What do biotin, lipoic acid, and phosphopantetheine coenzymes have in common?

4. How is NAD involved in the following enzymatic reactions? Write the mechanisms.

a.

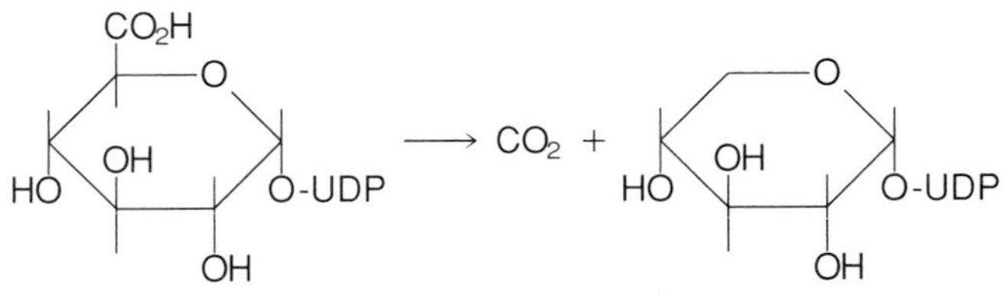

$\longrightarrow CO_2 +$

b. $\underset{\underset{NH_2}{|}}{CH_2}-CH_2-CH_2-\underset{\underset{NH_2}{|}}{CH}-CO_2H \longrightarrow$ $CO_2 + NH_3$

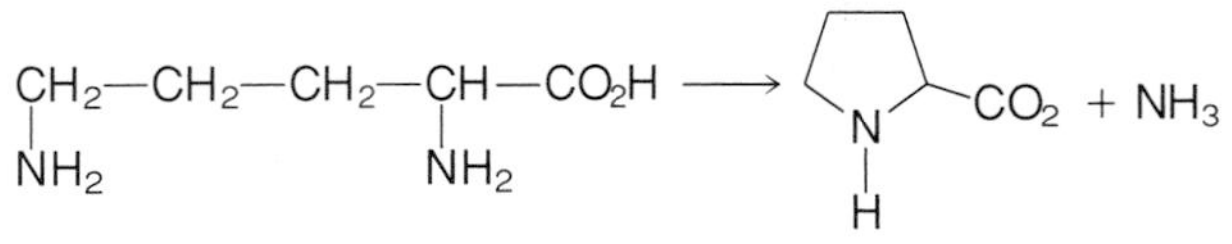

5. How is biotin involved in the following enzymatic reactions? Write the mechanisms.

a. $CH_3-\overset{\overset{O}{\|}}{C}-SCoA + HCO_3^- + ATP \longrightarrow$

$HO_2C-CH_2-\overset{\overset{O}{\|}}{C}-SCoA + ADP + HOPO_3^{2-}$

b. $CH_3-CH_2-\overset{\overset{O}{\|}}{C}-SCoA + HO_2C-CH_2-\overset{\overset{O}{\|}}{C}-CO_2H \rightleftharpoons$

$HO_2C-\underset{\underset{CH_3}{|}}{CH}-\overset{\overset{O}{\|}}{C}-SCoA + CH_3-\overset{\overset{O}{\|}}{C}-CO_2H$

6. Explain the importance of phosphopantetheine coenzymes for activating carboxylic acids toward enolization.

7. How is flavin adenine dinucleotide involved in the following reaction? Show the mechanism.

$$\alpha\text{-D-glucose} + O_2 \longrightarrow \alpha\text{-D-gluconolactone} + H_2O_2$$

8. For each of the following enzymatic reactions, identify the coenzymes involved.

a. 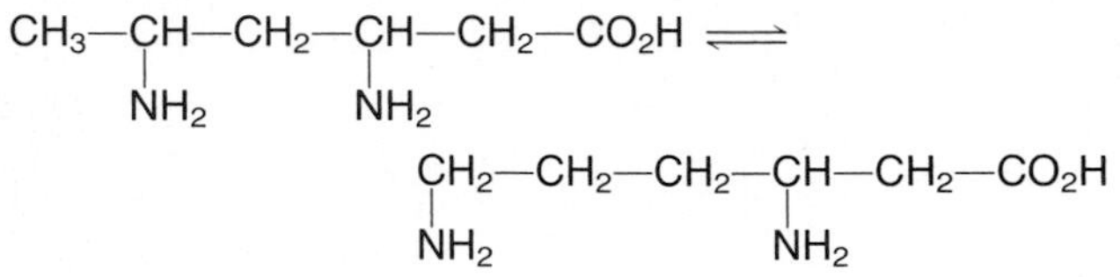 CH₃—CH—CH₂—CH—CH₂—CO₂H ⇌
          | |
          NH₂   NH₂

   CH₂—CH₂—CH₂—CH—CH₂—CO₂H
   | |
   NH₂   NH₂

b. 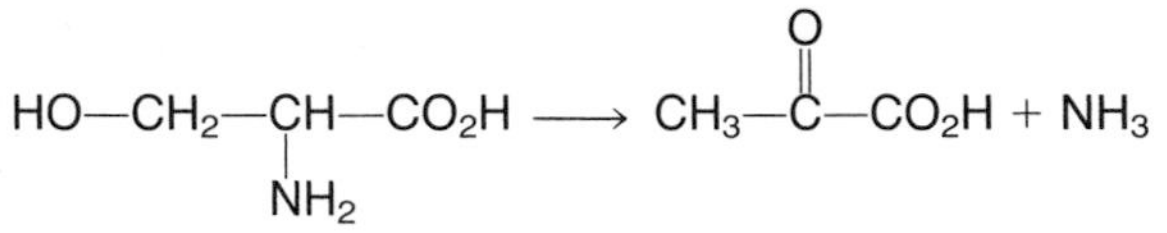 HO—CH₂—CH—CO₂H ⟶ CH₃—C(=O)—CO₂H + NH₃
            |
            NH₂

c. H₂N—(CH₂)₄—CH—CO₂H + O₂ ⟶
              |
              NH₂

   H₂N—(CH₂)₄—C(=O)—NH₂ + CO₂ + H₂O

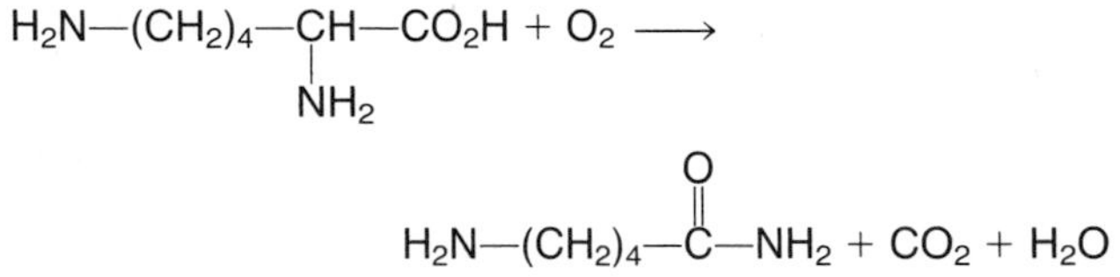

d.

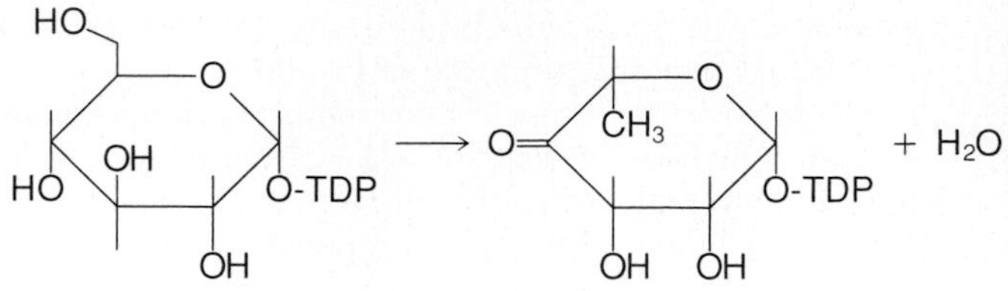

e. 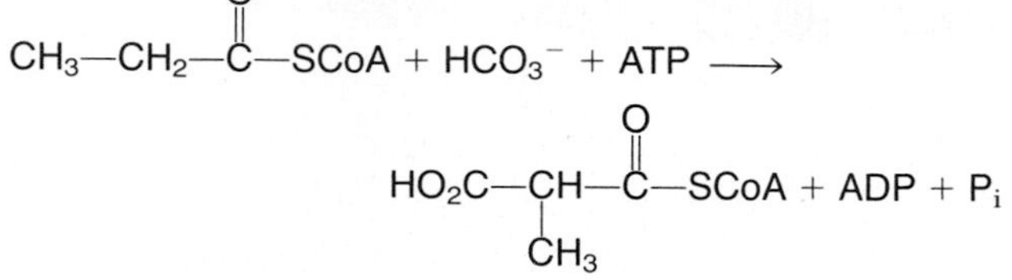 CH₃—CH₂—C(=O)—SCoA + HCO₃⁻ + ATP ⟶

   HO₂C—CH—C(=O)—SCoA + ADP + Pᵢ
          |
          CH₃

f.

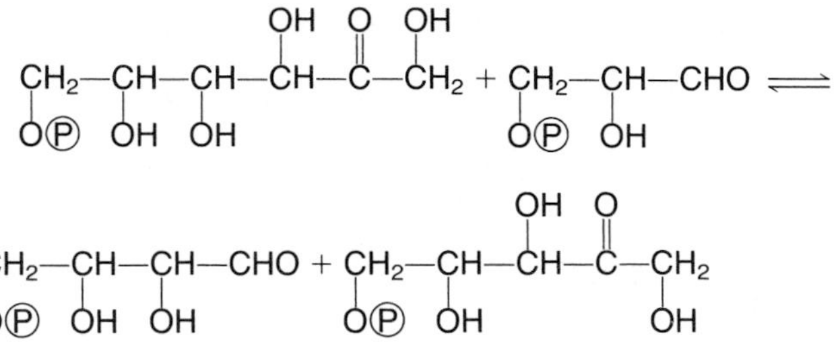

Bundle sheath cells in Zea mays (corn). The conspicuous dark granules are situated in chloroplasts and represent the product of photosynthesis. Magnification 16,700×. E. H. Newcomb and S. E. Frederick, Univ. of Wisconsin, Madison/ BPS.

CATABOLISM

PART III

Part III is the first of three parts concerned with metabolism. In Part III the emphasis will be on catabolism and its relationship to energy-generating metabolism. Biochemical energy in most living systems is generated from the breakdown (catabolism) of organic substances. Whereas the production of biochemically useful energy may be considered the main function of catabolism, it is not the only one. Catabolism often generates intermediates that are useful in biosynthesis (anabolism) of different and more complex molecules. However, sometimes the function of catabolism is merely to eliminate waste products.

Part III begins with two chapters that encompass all metabolism. Chapter 12 ("Thermodynamics and Metabolism") provides an elementary presentation of the most important thermodynamic quantities—enthalpy, entropy, and free energy. The interrelationships among these quantities are essential to our understanding of metabolism. The conditions that determine whether or not a reaction can proceed are discussed, and the relationship between free energy and the chemical equilibrium of a reaction is presented. The role of ATP as the main distributor or carrier of free energy is also described.

Chapter 13 ("Overview of Metabolism") discusses the general pattern of chemical reactions in living cells. The emphasis is on how biochemically cata-lyzed reactions differ from ordinary chemical reactions and the various stratagems that are used in biochemistry. A unique aspect of biochemical reactions is that they are functionally interrelated. Catabolic sequences generate both the ATP and the reducing power needed to drive anabolic sequences. Reaction sequences and enzymes have evolved so that the reactions that occur, as well as their rates, are optimal to the needs of the organism under certain conditions, but can be altered to meet other conditions the organism may encounter.

Chapter 14 is the first of four chapters dealing with carbohydrate catabolism and energy-generating metabolism. Carbohydrates are the main source of chemical energy. Other molecules, such as fats or amino acids, can produce energy by catabolism also, but they must first be converted to molecules that can enter the pathways of carbohydrate metabolism.

In Chapter 14 the discussion centers on glucose and other hexoses, because glucose is at the center of carbohydrate metabolism for nearly all organisms. When energy is required, glucose is catabolized to smaller compounds, thereby releasing energy that was trapped in ATP synthesized from ADP and inorganic phosphate. When glucose is not needed immediately for energy production, it is converted to polysaccharides and stored to meet later energy needs. Thus glucose can be obtained from catabolism of storage polysaccharides. However, it can also be synthesized from smaller carbon compounds by a series of reactions called gluconeogenesis.

In the absence of oxygen, glucose can be catabolized only by the glycolytic pathway to pyruvate or other compounds containing three carbon atoms. The reactions of the glycolytic pathway release only a small amount of the total energy present in glucose. When oxygen is present, glucose is oxidized completely to CO_2 and H_2O, and much more energy is released. This further oxidation requires an additional pathway known as the tricarboxylic acid (TCA) cycle (Chapter 15). The TCA cycle itself does not result in additional ATP production; rather, it leads to the production of reduced forms of the coenzymes FAD and NAD. These molecules transfer electrons to molecular oxygen, and in the process ATP is generated from ADP. This complex process, in which ATP is generated by electron transfer, is discussed in Chapter 16.

In Chapter 17 we discuss photosynthesis—the utilization of light to make sugars from CO_2 and H_2O. This process is confined to plants and certain microorganisms; it starts with the absorption of light by chlorophyll pigments, and continues with the passage of electrons from the pigments to an electron-transport chain. ATP is generated from this electron transport by a process that is similar to that used by mitochondria in oxidative phosphorylation. The ATP and the reduced NADPH that result from photosynthesis are used to make sugars from CO_2 by a pathway known as the Calvin cycle.

Chapter 18 ("Catabolism of Fatty Acids") focuses on the mobilization of fatty acids from lipid storage deposits. These storage deposits serve as the source of biochemical energy when the more immediate energy storage forms have been exhausted. The regulation of fatty acid breakdown is discussed in Part IV, where breakdown and synthesis are discussed together.

Chapter 19 deals with the complex subject of amino acid catabolism. The subject is complex because each amino acid is broken down by a specific pathway. One simplifying feature is the fact that in many cases similar or identical degradative enzymes are used. Amino acid breakdown serves multiple purposes. It eliminates amino acid excesses. It supplies the precursors for many different biosynthetic pathways. And finally, it supplies biochemical energy for utilization in energy-requiring processes.

The importance of orderly degradation processes is underscored by the existence of many genetically inherited diseases that result from a deficiency in one or more of the catabolic enzymes associated with amino acid breakdown.

12

THERMODYNAMICS AND METABOLISM

The primary usefulness of thermodynamics to the biochemist lies in the prediction of whether a process could or could not occur. We say "could" because thermodynamics can tell us only whether it is possible for a process to occur, not that it will occur. Whether or not a process will occur depends on the existence of a good chemical pathway for the process. For a biochemical pathway this means that suitable enzymes must be available. Thermodynamics is probably even more important for the biochemist than it is for the chemist, because if a chemical reaction will not proceed under the original conditions, the chemist can change the pressure or temperature or increase the concentration of reactants. An organism is under considerably more rigid constraints; it must function at a fixed temperature and pressure and within a limited range of concentrations of reactants. Since organisms live on chemical energy, it is important to know which reactions are capable of producing this type of energy.

A simple use of thermodynamics would be to predict what compounds could possibly be used as energy sources for an organism. It is well known that most oxidations by molecular oxygen produce energy. For example, wood and coal burn with a large output of energy. Similarly, organisms can consume carbohydrates, fats, and proteins to produce energy by aerobic oxidations. Some organisms oxidize hydrocarbons, some oxidize reduced forms of sulfur, and others oxidize iron. No organisms oxidize molecular nitrogen for energy, and the explanation lies in the thermodynamics. The reaction simply does not produce energy. This example illustrates the importance of thermodynamics in controlling all life. For an organism living on chemical energy, thermodynamics is not an esoteric subject. It is a matter of life or death.

In this chapter we will elaborate on the properties of certain thermodynamic quantities that we introduced in Chapter 2. We will then expand the discussion to show how the concept of free energy is used in predicting biochemical pathways, and we will explore the central role of ATP in providing energy for various reactions.

THERMODYNAMIC QUANTITIES

Thermodynamic quantities are properties of the _state_ of a substance and do not depend on how a substance was made or how it reached a certain state. In a chemical reaction it is the difference between the initial and the final state that is emphasized; the pathway that was taken to get from the initial state to the final state is not relevant in thermodynamic calculations (Figure 12–1). Expressed in chemical terms, thermodynamics is concerned, not with the mechanism of a reaction, but rather with the difference in energy between reactants and products.

The properties of a substance may be classified as _intensive_ or _extensive_. Intensive properties do not depend on the amount of material. Examples of these are density, pressure, temperature, and concentration. Extensive properties depend on the amount of material. Volume, weight, and energy are all extensive properties. Most of the thermodynamic properties we will be discussing are extensive properties. In this section we will discuss _internal energy (E), enthalpy (H), entropy (S), and free energy (G)._

Internal Energy Is Conserved in a Chemical Reaction

Of the thermodynamic quantities we have just listed, _internal energy_, designated by E, is perhaps the easiest to understand in terms of molecular forces. The internal energy of a molecule includes the translational, rotational, and vibrational energy. It also includes the electronic energy of the molecule. Electronic energies involve electron–electron interactions, electron–nucleus interactions, and nucleus–nucleus interactions.

In biochemistry we are concerned not so much with absolute energies as with changes in energies in the course of a reaction. This fact makes it easier to calculate the internal energy of a reaction, since the internal nuclear terms, most electron–nucleus terms, and many electron–electron terms do not change during the course of a chemical reaction. Electronic energies are much larger than the translational, rotational, and vibrational energies. For this reason, the electronic terms contribute the major share to the internal energy of a chemical reaction, even though many electronic terms do not change on reaction.

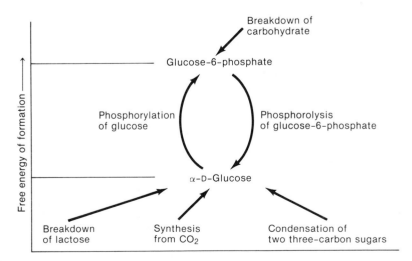

Figure 12–1
Thermodynamics considers only the difference between the initial and final states; it gives us no information about the pathway. The diagram indicates the different free energies of formation of glucose-6-phosphate and α-D-glucose. The values do not depend on how these compounds were produced.

The *first law* of thermodynamics states that *the total amount of internal energy in the universe is constant*. Although the total energy remains constant, energy may undergo transformations from one form to another. Thus the chemical energy of a molecule may be transformed into thermal, electrical, or mechanical energy.

To make useful applications of the first law we must account for all the energy changes in a reaction. This means taking into account any changes in the *system* of interest as well as any changes in the *surroundings*. The system of interest may be a reaction occurring in a test tube or a living cell. The surroundings include the rest of the universe. In practical terms they need only include the immediate surroundings of the system that are likely to be influenced by the reaction occurring in the system. According to the first law, the overall internal energy remains constant even though energy in some form may flow from the system to the surroundings or from the surroundings to the system.

The first law of thermodynamics asserts the principle of conservation of energy, as we have been discussing. It also asserts *the equivalence of heat and work*, the two means by which any specific system may gain or lose energy to its surroundings. The system may gain or lose heat to the surroundings or do work on or have work done on it by the surroundings. A common form of the first law of thermodynamics states that the change in internal energy must be equivalent to the difference between the heat absorbed by the system and the work done by the system.

$$\Delta E = q - w \qquad (1)$$

where ΔE is the change in internal energy of the system, q is the heat flow, and w is the work done. Both q and w depend on the path, but their difference, ΔE, is independent of path and therefore defines a state function. In this expression, the convention is adopted that heat absorbed by a system and work done by a system are positive quantities (Figure 12–2). For a small differential change.

$$dE = dq - dw \qquad (2)$$

If the only kind of work done involves change in the volume of the system against a fixed external pressure (known as $P\,dV$ work), then

$$dE = dq - P\,dV \qquad (3)$$

If the volume is held constant, and no other work is done,

$$dE = dq \qquad \text{or} \qquad \Delta E = q_V \qquad (4)$$

That is, the internal energy of a reaction (the change in the internal energy of the system) is equal to the heat absorbed by the system at constant volume.

The internal energy for the formation of a compound can be measured in a bomb calorimeter, in which the compound is thoroughly combusted in an enclosed container and the resulting heat is measured. Internal energies (and enthalpies, discussed next) are usually given in units of kilocalories per mole or kilojoules per mole.

It is useful to record internal energies of molecules with reference to a standard state. The standard internal energy, $\Delta E°$, reported for any given molecule is equal to the internal energy change resulting from the formation of that molecule from the constituent elements in their standard state. We will use Δ's to express all of our state functions for similar reasons. Good estimates of the energy change resulting from a chemical reaction can usually be obtained from a computation of the difference between the bond energies of the reactants and the products (see Table 2–1).

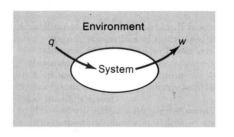

Figure 12–2
The system and the surroundings. Heat flow into the system is designated as a positive quantity q. Work done on the surroundings is designated as a positive quantity w.

BOX 12–A

Some Constants Useful in Thermodynamic Calculations

Joule: $1\,J = 1\,kg\,m^2\,s^{-1}$
          $= 1\,N\,m$ (newton meter)
          $= 1\,W\,s$ (watt second)
          $= 1\,C\,V$ (coulomb volt)

Thermochemical calorie: $1\,cal = 4.184\,J$

Large calorie: $1\,Cal = 1\,kcal = 4.184\,kJ$

Work required to raise 1 kg 1 m on earth (at sea level): 9.807 J

Work required to concentrate 1 mole of a substance 1000-fold, e.g., from 10^{-6} to $10^{-3}\,M$: $4.09\,kcal = 17.1\,kJ$

Avogadro's number, the number of particles in a mole: $N = 6.0220 \times 10^{23}$

Faraday: $1\,F = 96,485\,C\,mol^{-1}$ (coulombs per mole)

Coulomb: $1\,C = 1\,A\,s$ (ampere second)
          $= 6.241 \times 10^{18}$ electronic charges

The Boltzmann constant: $k_B = 1.3807 \times 10^{-23}\,J\,deg^{-1}$

The gas constant, $R = N\,k_B = 8.3144\,J\,deg^{-1}\,mol^{-1}$
          $= 1.9872\,cal\,deg^{-1}\,mol^{-1}$
          $= 0.082061\,atm\,deg^{-1}\,mol^{-1}$

and at 25° C, $RT = 2.479\,kJ\,mol^{-1}$

The unit of temperature is °K (or simply K); 0°C = 273.16°K

$\log X = 2.3026 \ln X$

Source: Adapted from D. E. Metzler, *Biochemistry* (Academic Press, 1977).

Internal Energy and Enthalpy Are Similar for Most Biochemical Reactions

Chemical and biochemical reactions are more often conducted at constant pressure than at constant volume. For this reason chemists use another function, the enthalpy, more frequently than the internal energy function. The change in enthalpy in a reaction is designated by the symbol ΔH, and the relationship between enthalpy and internal energy changes is given by the equation

$$\Delta H = \Delta E + \Delta PV \qquad (5)$$

For small changes in enthalpy,

$$dH = d(E + PV) = dE + P\,dV + V\,dP$$
$$= dq - dw + P\,dV + V\,dP \qquad (6)$$

For a system doing only $P\,dV$-type work, $dw = P\,dV$, and

$$dH = dq + V\,dP \qquad (7)$$

For processes occurring at constant pressure, Equation (7) becomes

$$dH = dq \qquad \text{or} \qquad \Delta H = q_P \qquad (8)$$

That is, the <u>enthalpy of a reaction is measured as the heat absorbed by the system at constant pressure</u>. In most biochemical reactions there is little change in either pressure or volume, so the difference between ΔH and ΔE can be regarded as insignificant. Keeping this in mind, we will use ΔH for the heat absorbed or released in a reaction.

A reaction system that absorbs heat causes the surroundings to cool, and the change in the enthalpy of the system is positive. When heat is evolved from the system, the ΔH of the reaction is negative. Under these conditions, energy is lost from the system to the surroundings. Most reactions that proceed spontaneously evolve heat.

Whereas enthalpy changes are useful for calculating heat changes during the course of a reaction, they do not provide us with a reliable criterion for the spontaneity of a reaction. For example, LiCl and $(NH_4)_2SO_4$ both dissolve readily in water. The former reaction releases heat, whereas the latter absorbs heat. The latter reaction is driven by the entropy effect.

Entropy Always Increases in a Spontaneous Reaction

The *second law* of thermodynamics states that *systems tend to proceed from a state of low probability* (ordered) *to a state of high probability* (disordered). This effect is measured by a term called *entropy* (denoted by the symbol S). A reaction that increases in entropy (positive ΔS) is favored to proceed over one that decreases in entropy (negative ΔS). If the system and the surroundings both are considered, the second law asserts that the entropy change in a reaction is always greater than or equal to zero. If W is the number of ways of arranging a system without changing the internal energy, or enthalpy, we may define the absolute entropy per molecule in the system by the equation

$$S = k \ln W \tag{9}$$

where k is Boltzman's constant; $k \simeq 3.4 \times 10^{-24}$ cal/°C. For a mole of substance,

$$S = Nk \ln W = R \ln W \tag{10}$$

Here N is the number of molecules in a mole (6×10^{23}) and R is the gas constant [$R \simeq 2$ cal/(°C · mole)]. Quantitative values for entropies are usually given in entropy units, 1 eu = 1 cal/°C.

The absolute entropy of a compound may be calculated from statistical mechanical methods if all the translational, vibrational, and rotational parameters are known. These include force constants, moments of inertia, and rotational barriers. Statistical mechanical calculations of entropy are possible only for relatively small molecules and certainly could not be used for most biological reactants. The entropy contributions to propane that have been calculated by means of statistical mechanics are listed in Table 12–1. These values show large translational and rotational contributions to the entropy, a small vibrational contribution, and no electronic contribution. This quantitative distribution of entropy is also typical for larger molecules found in biochemical systems. Thus *when we think of entropy we should associate it primarily with translation and rotation*. This situation is very different from enthalpy, in which electronic terms are all-important and the other terms are insignificant by comparison.

There are several molecular properties that are important in determining the entropy of a compound. If we consider some of these properties and how they affect the entropy, we will have a better qualitative appreciation of entropy.

From statistical mechanics we find that translational entropy, which is still the most important entropy term even for large molecules, depends on $\frac{3}{2}R \ln M_r$ (plus other terms), where R is the gas constant and M_r is the molecular weight. We can appreciate the importance of the translational energy by considering the effects of a simple dimerization reaction. Intuitively we expect dimerization to decrease the entropy because two molecules can no longer move independently. Quantitatively we can calculate the effect as a function of the molecular weight.

Let S be a monomer that undergoes a dimerization reaction

$$2\,S \longrightarrow S_2$$

If M_r is the molecular weight of the monomer, then the approximate translational entropy change in going from the monomeric state to the dimeric state is

TABLE 12–1
Entropy of Propane at 231 K

	kcal/(°C · mole)
Translational	36.04
Rotational	23.38
Vibrational	1.05
Electronic	0.0
Total	60.47

$$\Delta S = 2(\tfrac{3}{2}R \ln M_r) - \tfrac{3}{2}(R \ln 2M_r) \tag{11}$$

$$= \tfrac{3}{2}R \ln M_r - \tfrac{3}{2}R \ln 2$$

$$= \tfrac{3}{2}R \ln (2M_r) - 2(\tfrac{3}{2}R \ln M_r)$$

$$= \tfrac{3}{2}R \ln 2 - \tfrac{3}{2}R \ln M_r$$

$$= -\tfrac{3}{2}R \ln (M_r/2)$$

This calculation shows that the decrease in translational entropy in dimerization is a logarithmic function of the molecular weight.

Structural features that make molecules more rigid reduce rotational and vibrational contributions to entropy. Thus double bond formation or ring formation substantially decreases the entropy even when there is no effect on the molecular weight. Considerations of this sort are also important in the formation of comparatively rigid macromolecular structures from flexible polypeptide and polynucleotide chains (see Chapter 2).

The physical state is also a very important factor in the entropy of a compound. A gas has much more translational and rotational freedom than a liquid, and a liquid has much more freedom than a solid. As a result, the relative entropy increases from solid to liquid to gas.

It can be shown that the increase in entropy in an isothermal reversible process is given by

$$\Delta S = \Delta H/T = q_P/T \tag{12}$$

where T is the absolute temperature in degrees Kelvin. An isothermal process is one that occurs at constant temperature. A reversible process is one that proceeds infinitely slowly through a series of intermediate states in which the system is always at equilibrium. For any real process occurring at a finite rate, the system is not strictly in equilibrium, and ΔS is larger than the value given by Equation (12). We shall return to this point shortly.

From Equation (12), the entropy increase on vaporization and melting can be determined simply from the heat of vaporization divided by the boiling point and from the heat of fusion divided by the melting point, respectively. The entropy increase on vaporization of water is 26 eu/mole and that on melting of ice is 5.3 eu/mole. These values are consistent with our intuition that the increase in translational and rotational freedom should be much greater in going from a liquid to a gas than in going from a solid to a liquid.

The entropy of solutions is markedly affected by mixing two solvents, by solvation of solutes, and by hydrogen bonding and other associations in the solvent or between solute and solvent. The probability of two miscible liquids being mixed is greater than each staying separate in the same container. The entropy change on going from the pure liquid state to the mixed state is

$$\Delta S = n_aR \ln \frac{1}{X_a} + n_bR \ln \frac{1}{X_b} \tag{13}$$

$$= -n_aR \ln X_a - n_bR \ln X_b$$

when n_a and n_b are the number of moles of a and b that are mixed and X_a and X_b are the corresponding mole fractions. Since X_a and X_b are always less than 1, $\ln X_a$ and $\ln X_b$ will be negative. This means that the dilution of each component resulting from mixing will make a positive contribution to the entropy. This equation applies to the mixing of _ideal_ solutions, in which there is no interaction between the molecules. Any intermolecular interaction will result in a decrease in the entropy of mixing because it will restrict the translational and rotational freedom of the individual molecules.

Solvation, i.e., the interaction between solute and solvent molecules,

Solvent (H₂O)

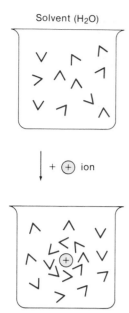

|
↓ + ⊕ ion

Figure 12–3
The entropy decrease resulting from solvation (hydration). When a small solute molecule is dissolved in water, the entropy of the salt molecules increases because of the increased possibilities for translation and rotation. But at the same time the movement of the water molecules becomes restricted in the vicinity of the solute molecules. The net effect is frequently a decrease in the entropy on solvation. The net entropy effect on solvation can, however, be either positive or negative for different solutes.

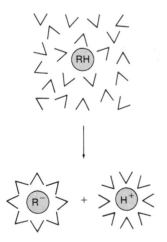

Figure 12–4
The number of free particles decreases upon ionization, instead of increasing as one might at first imagine. This effect is due to the solvation of water molecules around the ions.

makes an important negative contribution to the entropy of a compound in solution. When a solute is added to a solvent and solvation occurs, the entropy decreases because of the restricted movement of the solute as well as because of the restricted movement of the solvent in the vicinity of the solute (Figure 12–3).

Small ions are more highly solvated or hydrated than large ones with the same charge, and anions are more hydrated than cations. In general, the entropy of hydration becomes a larger negative number with increasing charge or decreasing radius. In addition to ions, molecules that have dipoles or hydrogen-bond donor or acceptor groups tend to interact strongly with water, and consequently they restrict the mobility of water. Similar decreases in entropy usually result from the hydration of such molecules.

It is noteworthy that although enthalpy and entropy depend on entirely different terms, hydration affects both values because it involves both electronic interactions and changes in translational and rotational motion. Enthalpies and entropies of solvation most frequently tend to negate each other. For charged species, the more negative the enthalpy of solvation (favorable), the more negative (unfavorable) the entropy of solvation.

An entirely different type of solvation effect occurs when an apolar molecule is added to water. The result is a decrease in entropy, but for very different reasons. The water orients on the surface of the apolar molecule to form a relatively rigid cage held together by hydrogen bonds (see Figure 2–4). This kind of solvation effect is a very important factor for the large apolar areas of proteins and other biological macromolecules (see Chapter 2).

From the preceding, you might expect that the ionization of a proton from an acid, which produces more particles, would lead to an increase in entropy. However, this effect is more than counterbalanced by hydration effects. The production of the charged anion and proton in water "freezes out" a large amount of water in the form of water of hydration, which results in a large decrease in entropy (Figure 12–4). Thus when a weak acid ionizes, there is a decrease in the total number of mobile molecules instead of an increase as you might expect. The entropy of ionization of a typical weak acid in water is usually about −22 eu/mole. The ionization of an amino acid (Figure 12–5) shows a much smaller decrease in entropy than this because the zwitterion is highly hydrated before the proton is lost. The entropy is −9.2 eu/mole.

The entropy changes on binding or absorption of molecules on the surface of a macromolecule are important to a biochemist because these are commonly occurring phenomena in the cell. Of particular interest is binding of a substrate or an inhibitor to an enzyme. When gas molecules are absorbed on a solid catalytic surface, there is a large decrease in entropy (Figure 12–6). The large translational entropy of the gas molecules disappears when they are held fixed on a surface, and the thermodynamics of the absorbed gas is much more like that of a solid. The negative change in entropy forms a barrier to the industrial use of solid catalysts acting on gases. To make a reaction go, this barrier must be overcome by a strongly negative enthalpy for the absorption of the gas by the catalyst or by increased pressure. Therefore, the binding energy of the solid catalyst for the gas must be strong enough to overcome the improbable situation of gas molecules sitting on the surface.

$$\underset{\underset{NH_3^+}{|}}{CH_2-COO^-} \longrightarrow \underset{\underset{NH_2}{|}}{CH_2-COO^-} + H^+ \qquad \Delta S° = -9.2 \text{ eu/mole}$$

Figure 12–5
The entropy change on the ionization of glycine. The entropy decrease for an amino acid on ionization is less than usual because of the double-charged nature of the neutral amino acid.

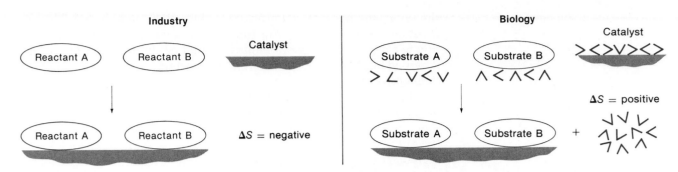

Figure 12–6
The entropy of binding to an industrial catalyst as compared with a biological catalyst. The entropy change on binding reactants to a solid industrial catalyst is negative because of the restricted movement of the reactants. By contrast, the entropy change on binding of substrate(s) to an enzyme surface is frequently positive. This is true because the restricted movement of substrate on binding is more than compensated for by the release of bound water, both from the substrate surface and from the enzyme surface.

By analogy, one might expect a highly negative entropy when a substrate is absorbed on an enzyme surface. However, these processes frequently have positive entropies, the reason being that a large number of water molecules are displaced when a substrate is bound to the surface of an enzyme. This large release of absorbed water causes an increase in entropy (see Figure 12–6). Binding between pepsin and an ester substrate has an increase of 20.6 eu/mole, and urea combining with urease has an increase of 13.3 eu/mole. Because of the favorable entropy, the binding enthalpy of an enzyme for substrate need not be nearly so large as that of industrial catalysts working on gases. Industrial and biological catalysts may seem superficially to be similar, but there are qualitative differences in the thermodynamics of each.

Another most important entropy effect when an enzyme and substrate combine could be called the _chelation effect_. This effect is best understood by discussing a relatively simple case of metal chelation. Cadmium ion tends to be quadrivalent, so if there are two amino groups in one molecule, as in ethylenediamine, the cadmium will combine with two molecules. If there is only one amino group in a molecule, as in methylamine, the cadmium ion requires four molecules, as shown in Table 12–2. Notice that the entropy is much more favorable for the combination with ethylenediamine than with methylamine, so that there is much stronger binding to the ethylenediamine than to the methylamine. Water molecules are released from cadmium ions when the ligands are added, but the increase in entropy from this factor is

TABLE 12–2
Enthalpy and Entropy Changes on Forming Complexes between Cadmium Ions and Methylamine or Ethylenediamine

	Values for Association			
	ΔH kcal/mole	ΔS eu/mole	$T\Delta S$ kcal/mole	ΔG kcal/mole
$Cd^{2+} + 2CH_3NH_2 \longrightarrow Cd(CH_3NH_2)_2{}^{2+}$	−7.03	−1.58	−.47	−6.56
$Cd^{2+} + en^a \longrightarrow Cd(en)^{2+}$	−7.02	+3.15	+.94	−7.96
$Cd^{2+} + 4\,CH_3NH_2 \longrightarrow Cd(CH_3NH_2)_4{}^{2+}$	−13.7	−16.0	−4.77	−8.93
$Cd^{2+} + 2\,en \longrightarrow Cd(en)_2{}^{2+}$	−13.5	+3.29	+.98	−14.48

[a]en = ethylenediamine.

Source: Data from Spike and Parry, Thermodynamics of chelation I. The statistical factor in chelate ring formation, _J. Am. Chem. Soc._ 75:2726, 1953.

the same whether methylamine or ethylenediamine is added. The much less favorable entropy factor resulting from association of the methylamine is due to the larger number of molecules that combine with the cadmium ions when methylamine is the complexing agent and to the correspondingly larger loss of entropy.

In general, the chelation effect means that if a molecule has n points of attachment to another molecule (a metal or a protein), then it will bind much more strongly than n molecules with one point of attachment, even though the enthalpy of each point of attachment is the same. The chelation effect is a big factor in substrate binding to an enzyme when there are several points of attachment. Several weak binding interactions can produce an overall tight binding because of the additive effect of many small enthalpy contributions and the lack of a proportional decrease in entropy. This effect is even greater in binding between two proteins or a protein and a nucleic acid (e.g., see the case of *lac* repressor binding to DNA in Chapter 29), because there are many interactions between such large molecules.

The final column in Table 12–2 indicates the free energy of formation between cadmium ion and the two amine compounds. The free energy is a function of the enthalpy and the entropy; it provides us with another criterion as to whether a reaction could proceed spontaneously, as explained in the next section.

Free Energy Provides the Most Useful Criterion for Spontaneity

As we have seen in the previous discussion, a system tends toward the lowest enthalpy and the highest entropy. The entropy always increases in a spontaneous process, but its evaluation is often difficult and requires that we observe the entropy change in both the system of immediate interest and the surrounding environment (unless, of course, the system is isolated from its surroundings).

A more convenient function for predicting the extent of reaction was discovered by Josiah Gibbs. Gibbs was the first to appreciate that *in reactions occurring at equilibrium and constant temperature*, such as the phase change of ice going to water at 0°C, *the change in entropy is numerically equal to the change in enthalpy divided by the absolute temperature*. This relationship is the one already presented in Equation (12). The equation can be transposed to

$$\Delta H - T\,\Delta S = 0 \qquad (14)$$

In search of a criterion for spontaneity, Gibbs proposed a new function called the *free energy*, defined by the equation

$$\Delta G = \Delta H - T\,\Delta S \qquad (15)$$

For a reaction occurring at equilibrium, such as the melting of ice at 0°C, the change in free energy would be zero. For the same reaction occurring in a *nonequilibrium situation*, such as for ice melting at 10°C, the ΔS is larger and the free energy change is negative. The melting occurs spontaneously. In the reverse reaction, the conversion of water to ice at 10°C, there would be a positive free energy change and the reaction would not occur.

Free energy is a valuable concept, not only because it allows us to determine whether a reaction could proceed, but also because it allows us to calculate other useful parameters of the reaction, such as the equilibrium constant, which tells us the quantitative extent to which a reaction can proceed. Free energy, like internal energy, enthalpy, and entropy, is a state function and an extensive property of a system. Provided a good pathway exists and the free energy change is favorable (negative), the reaction will occur. If the free energy change is favorable and no pathway is available for the conversion, a catalyst may be added that provides an acceptable pathway. However, if the free energy change is unfavorable (positive), no catalyst can ever make the reaction proceed.

Free energy is more useful than entropy for predicting whether or not a reaction will occur, because it applies to systems that are not isolated from their surroundings. Furthermore, as we shall see in the next section, it is relatively simple to obtain quantitative estimates of free energy.

APPLICATIONS OF THE FREE ENERGY FUNCTION

The free energy function dominates thermodynamic thinking in biochemistry. This is because it provides the most convenient criterion for determining the spontaneity of a reaction and because it is relatively easy to measure. In this section we will elaborate on these advantages of the free energy function and describe other advantages. Despite all its advantages, however, many people find the free energy function the most difficult thermodynamic function to grasp intuitively. The reason, perhaps, is that free energy is a composite of enthalpic and entropic terms, which often make opposing contributions. We have already seen examples of this confusion in arguments about stabilizing factors in protein conformation (see Chapter 2).

Values of Free Energy Are Known for Many Compounds

The _standard free energy_ of formation is known for thousands of compounds. The standard free energy of formation, ΔG_f°, is the free energy difference between a compound in its standard state (usually the pure compound) and the standard state of the elements of which the compound is composed. The standard states usually chosen for gases such as oxygen and nitrogen are O_2 and N_2, respectively. For liquids, the standard state is usually the pure liquid; for solids, the pure solid, and for solutes, the concentration of 1 M. All the standard states are defined as the form of the substance at 25°C and 1 atmosphere pressure.

The free energies of formation of many compounds can be found in tabular form in books on thermodynamics. They are usually given in units of kilocalories per mole or kilojoules per mole: 1 kilocalorie (kcal) is the amount of heat required to raise the temperature of 1 kilogram of water from 14.5 to 15.5°C. One kilojoule (kJ) is the amount of energy needed to apply 1 newton of force over a distance of 1 kilometer; 1 kcal is equivalent to 4.184 kJ. By subtracting the sum of the free energies of formation of the reactants from the sum of the free energies of formation of the products, it is possible to determine the free energy change of any reaction for which the free energies of formation of all reactants and products are known.

Some standard free energies of formation are given in Table 12–3. From the values listed in the table we can calculate the standard free energy change for the reaction

$$\text{Oxaloacetate}^{2-} + \text{H}^+ (10^{-7}\ M) \longrightarrow CO_2(g) + \text{pyruvate}^-$$

as

$$-113.44 - 94.45 + 9.87 + 190.62 = -7.4\ \text{kcal/mole.}$$

(Note that in biochemistry the standard state for hydrogen in solution is usually defined for a $10^{-7}\ M$ solution because this is close to the concentration in most systems of interest to biochemists.) Thus the free energy change of this reaction for 1 M oxaloacetate anion, $10^{-7}\ M$ hydrogen ions, carbon dioxide at 1 atm, and 1 M pyruvate anions is -7.4 kcal. From this it would appear that the reaction will proceed at these concentrations. However, the

TABLE 12-3
Standard Free Energies of Formation of Some Compounds of Biological Interest

Substance	ΔG_f° (kcal/mole)	ΔG_f° (kJ/mole)
Lactate ions	−123.76	−516
Pyruvate ions	−113.44	−474
Succinate dianions	−164.97	−690
Glycerol (1 M)	−116.76	−488
Water	−56.69	−280
Acetate anions	−88.99	−369
Oxaloacetate dianions	−190.62	−797
Hydrogen ions (10^{-7} M)	−9.87[a]	−41
Carbon dioxide (gas)	−94.45	−394
Bicarbonate ions	−140.49	−587

[a]This is the value for hydrogen ions at a concentration of 10^{-7} M. The free energy of formation at unit activity (1 M) is 0.

concentrations are not all expressed in a realistic manner. At pH values where the dianion of oxaloacetate would be formed, the carbon dioxide would not be a gas but would be bicarbonate. We must make some further calculations before the free energy we have found is of any real use to us. One of these would be to add the free energy of the reaction of carbon dioxide gas to give bicarbonate anion:

$$CO_2 + H_2O \longrightarrow HCO_3^- + H^+ \tag{16}$$

This calculation would yield a correction of −140.49 − 9.87 − (−56.69) − (−94.45) = 0.8 kcal. The free energy change for the reaction of oxaloacetate to form pyruvate and 1 M bicarbonate ions instead of carbon dioxide is −7.4 + 0.8 = −6.6 kcal/mole.

The standard free energy for a reaction may be determined by adding or subtracting the standard free energies of two other reactions that will combine to give the desired reaction if the free energies of the two reactions are known. An example is the calculation of the free energy of hydrolysis of ATP at pH 7 from the free energies of formation of glucose-6-phosphate and the hydrolysis of glucose-6-phosphate at pH 7, as shown in Table 12–4.

The Standard Free Energy Is Directly Related to the Equilibrium Constant

In biochemistry we are most concerned with applications of thermodynamics to reactions occurring in aqueous solution. Suppose we have a chemical reaction with the following stoichiometry:

$$aA + bB \rightleftharpoons cC + dD \tag{17}$$

TABLE 12-4
Determination of the Standard Free Energy of ATP Hydrolysis ($\Delta G^{\circ\prime}$) by Adding the Standard Free Energies of Two Other Reactions

	$\Delta G^{\circ\prime}$(kcal/mole)
Glucose + ATP $\rightleftharpoons$ Glucose-6-P + ADP	−4.7
Glucose-6-P + H_2O $\rightleftharpoons$ Glucose + P	−3.1
ATP + H_2O $\rightleftharpoons$ ADP + P	−7.8

where a, b, c, and d refer to the moles of A, B, C, and D, respectively, at concentrations c_A, c_B, c_C, and c_D. The free energy change for the reaction is

$$\Delta G = G_{\text{final state}} - G_{\text{initial state}} \tag{18}$$

If the reaction occurs at constant pressure and temperature, Equation (18) can be expressed as the difference of standard free energies of the products and reactants, $\Delta G°$, plus a correction for the concentrations:

$$\Delta G = \Delta G° + RT \ln \frac{c_C{}^c \, c_D{}^d}{c_A{}^a \, c_B{}^b} \tag{19}$$

The last term in Equation (19) is the correction for concentration and as such is an entropic contribution to the free energy. Note the similarity of this term to Equation (13). In a rigorous treatment we should use activities instead of concentrations in this formula, but for simplicity we will ignore the difference, although we must keep in mind that it may be substantial in certain cases.

At equilibrium, $\Delta G = 0$ and

$$\frac{c_C{}^c \, c_D{}^d}{c_A{}^a \, c_B{}^b} = K_{eq} \tag{20}$$

where K_{eq} is the equilibrium constant for the reaction. Therefore

$$\Delta G° = -RT \ln K_{eq} \tag{21}$$

Thus the equilibrium constant for a reaction can be used to calculate the free energy of a reaction. Conversely, free energies are used to calculate equilibrium constants. Indeed, Equation (21) is used for predicting the direction and extent of a chemical reaction. Because of the logarithmic relationship, the equilibrium constant has a very high dependence on the free energy. A reaction that proceeds to 99 per cent completion is, for most practical purposes, a quantitative reaction. It requires an equilibrium constant of 100 but a standard free energy difference of only -2.7 kcal, which is little more than half the free energy for the formation of a hydrogen bond (see Table 12–5).

Equations (19) and (21) are two of the most important thermodynamic equations for biochemists to remember. When the concentrations of reactants and products are at their equilibrium values, then there is no change in

TABLE 12–5
Relationship Between $\Delta G°$ and K_{eq} (at 25°C)

$\Delta G°$ (kcal/mole)	$K_{eq}{}^a$
−6.82	10^5
−5.46	10^4
−4.09	10^3
−2.73	10^2
−1.36	10
0	1
1.36	10^{-1}
2.73	10^{-2}
4.09	10^{-3}
5.46	10^{-4}
6.82	10^{-5}

$^a\Delta G°$ values at 25°C are calculated from the equation

$\Delta G° = -RT \ln K_{eq}$

$\quad = -1.98 \times 298 \times 2.3 \log K_{eq}$

$\quad = -1364 \log K_{eq}$

BOX 12–B

Calculation of the Equilibrium Constant, K_{eq}, When the Numbers of Reactants and Products Are Not Equal

In the expression for the equilibrium constant each concentration term actually represents a ratio of the concentration to the concentration in the standard state. Thus the thermodynamic equilibrium constant is a dimensionless quantity even when the numbers of reactants and products are not equal. Since it is customary to express concentrations in units of molarity and the standard state is usually 1 molar, the equilibrium constant is usually expressed in terms of the actual concentrations. It is also customary to indicate a unit such as M^1 or M^{-1} when appropriate in the numerical expression for the equilibrium constant, since this value depends on the unit used to express concentration even though it is a dimensionless quantity. The unit appears only when the numbers of products and of reactants differ. Thus there is no need to use the unit M^0 when stating the numerical value for the equilibrium constant.

Another potential complication in evaluating the equilibrium constant is how to treat water when it appears in the equilibrium expression. The problem seldom arises, since water is assumed to be in its standard state (as far as concentration is concerned) at all times. Thus it need not appear in the equilibrium expression. But clearly, we can imagine cases where water would not be in its standard state—for example, reactions occurring in the relatively anhydrous environment of a lipoprotein membrane. However, we will not treat this type of complication here.

free energy for the reactions going in either direction. In biological systems concentrations of certain compounds are often maintained at values far from their equilibrium values, so that considerable free energy is generated by their reactions. We will amplify on this strategy in the next chapter.

Free Energy Is the Maximum Energy Available for Useful Work

The free energy change of a system gives a quantitative measure of the maximum amount of *useful work* that can be obtained from a reaction. This can be shown by combining information from the first and second laws. Thus the second law says that for any spontaneous process

$$\Delta S_{system} + \Delta S_{surroundings} \geq 0 \tag{22}$$

with the equals sign applying only when the process is at equilibrium.

If the system gains an amount of heat q (that is, if this amount of heat is transferred from the surroundings to the system), then

$$\Delta S_{surroundings} = -q/T \tag{23}$$

From the first law we know also that

$$q = \Delta E_{system} + w_{system} \tag{24}$$

where ΔE_{system} and w_{system} are the change in the energy of the system and the work done by the system on the surroundings. Combining Equations (22) through (24) gives

$$\Delta S_{system} - (\Delta E_{system} + w_{system})/T \geq 0 \tag{25}$$

Multiplying by T, rearranging terms, and dropping the subscripts gives the following equation for the system:

$$w \leq -(\Delta E - T\,\Delta S) \tag{26}$$

Equation (26) says that the amount of work that the system does must be less than or equal to the quantity $-(\Delta E - T\,\Delta S)$. The amount of work will actually be equal to $-(\Delta E - T\,\Delta S)$ only if the process is reversible, in the thermodynamic sense that the process proceeds infinitely slowly, through a series of intermediate states that are all at equilibrium. Otherwise, the amount of work will be less.

If the process occurs at constant pressure, it may involve some work of volume expansion or contraction, $P\,\Delta V$. This will generally be a useless sort of work as far as biochemists are concerned. If we discount pressure-volume work, the maximum amount of *useful* work that could be obtained from the process (w_{useful}) is $w - P\,\Delta V$, which we can express as follows by the use of Equation (26):

$$w_{useful} \leq -(\Delta E + P\,\Delta V - T\,\Delta S) \tag{27}$$

or

$$w_{useful} \leq -(\Delta H \cdot T\,\Delta S) = -\Delta G \tag{27a}$$

The maximum amount of useful work that could be obtained is thus equal to $-\Delta G$. If ΔG is zero, we conclude that we could not obtain any useful work from the process, which must mean that the system is at equilibrium. If ΔG is less than zero, the process might yield useful work as the system proceeds spontaneously toward equilibrium. If ΔG is greater than zero, the process is headed away from equilibrium, and we would have to perform work *on* the system in order to drive it in this direction. The larger the amount of work that we might obtain from a reaction, the farther the reactants are from equilibrium. However, we must keep in mind that Equation (27) expresses an inequality, not an equality; $-\Delta G$ gives only the *maximum* amount of work. Work and heat, unlike ΔG, ΔE, and ΔS, are not functions of state. The amount of work that actually is obtained depends on the path that the process takes, and it can be zero even if $-\Delta G$ is large.

Biological Systems Perform Various Kinds of Work

To produce the changes necessary to sustain and propagate life requires that various types of work be done. We can ask what sorts of changes living systems must effect or what kinds of work they must do. The answer comes in terms of three major categories of cellular activities that involve change. These will be discussed in more detail elsewhere in the text, but they can be intimated here briefly. In terms of changes effected, living systems require energy for the following purposes:

1. *Mechanical work: Changes in location or position of cellular structures.* Mechanical work involves a change in the location or orientation of an organism, a cell, or a part thereof. As examples, consider the contracting muscles of a jogger, the motion of a flagellated protozoan in a pond, the migration of chromosomes toward the opposite poles of the mitotic spindle, or the movement of a ribosome along a strand of messenger RNA.

2. *Osmotic and electrical work: Changes in concentration.* Concentration work always involves the movement of chemical compounds or ions against a prevailing concentration gradient, thereby establishing the localized concentrations of specific molecules and ions upon which most essential life processes depend. Concentration work is sometimes also referred to, although less satisfactorily, as osmotic work. Examples include the uptake of glucose from the blood by the cells of your body, the pumping of sodium out of a marine microorganism, and the movement of nitrate from the soil into the cells of a plant root. When the species being concentrated is a charged ion, the resulting concentration gradient is also a potential gradient, so that electrical work becomes a subset of concentration work for charged species. Although the most dramatic example of this is the large potential developed by the electric eel, electrical work is a common phenomenon and is in fact the mechanism of membrane excitation and of conduction of nerve impulses.

3. *Synthetic work: Changes in bonds.* Synthetic work is that involved in the formation of the chemical bonds necessary to fabricate and assemble the complex organic molecules of which cells are composed. These are in general molecules of greater complexity and energy content than the simple molecules available to organisms from their environment, so that energy must clearly be expended in their synthesis. Although synthetic work is most obvious during periods of growth of an organism, it also occurs in nongrowing mature organisms, in which existing structures must be continuously repaired and replaced. The continuous expenditure of energy to elaborate and maintain ordered structures out of less-ordered raw materials is, in fact, one of the most characteristic properties of living systems.

Unfavorable Reactions Can Be Driven by Favorable Reactions

We will consider here the uses of free energy to perform chemical work, leaving applications to other types of work for subsequent chapters. One way of accomplishing chemical work is to harness the high negative free energy of one reaction (an *exergonic reaction*) to drive another reaction, which has a low negative free energy or even a positive free energy of reaction (an *endergonic reaction*). This strategy is common in living cells and stems from the

BOX 12–C

Free Energy and Electrical Work

Since the free energy change determines the maximum work other than $P\,dV$ work that a reaction can produce, we can use it to calculate energy available to carry out electrical work, i.e., the ability of a system to transfer electrons. For this purpose we introduce a term called the *electromotive force* (or *electromotive potential*). The electromotive force of a molecule is a measure of the tendency of the molecule to release or to accept electrons.

The standard free energy of a reaction involving a transfer of electrons is proportional to the negative of the number of electrons transferred times the electromotive force. The equation that expresses this is

$$\Delta G^{\circ} = -nFE^{\circ}$$

where n is the number of electrons and F is the proportionality constant between electron volts (nE°) and calories. The proportionality constant F, called the *Faraday constant*, is equal to 23,060 cal/eV. The symbol E° is for the standard potential of the reaction. This is the potential observed in a cell where all reactants and products are in their standard states.

The electromotive force is comparable to pressure in a water pipe. It is independent of the number of electrons, just as the pressure is independent of the amount of water. However, the work the water can do depends on the amount of flow times the pressure, just as the free energy depends on the number of electrons times the electromotive force. This is the reason that we must know the number of electrons when free energy is converted to electromotive force or electromotive force is converted to free energy. The use of standard potentials to calculate standard free energies for oxidation-reduction reactions is described in Chapter 16.

Another useful expression is one showing the relationship between the equilibrium constant and the standard electromotive potential:

$$E^{\circ} = \frac{RT}{nF} \ln K_{eq} = 2.3\frac{RT}{nF} \log K_{eq}$$

This relationship is easy to derive from previously described relationships between the equilibrium constant and the free energy and between the standard potential and the free energy. The logarithmic relationship means that a small change in potential will correspond to a large change in equilibrium constant. A reaction that goes to 99 per cent completion ($K_{eq} = 100$) requires a driving potential of only 118 mV.

The electromotive potential at concentrations of reactants and products other than $1\,M$ is given by the *Nernst equation*:

$$E = E^{\circ} - \frac{RT}{nF} \ln \frac{c_D{}^d c_C{}^c}{c_A{}^a c_B{}^b}$$

for the reaction $aA + bB \rightleftharpoons cC + dD$.

Electric potentials are used to drive a number of reactions at membrane surfaces. The lipid membranes that exist at cell surfaces or organelle surfaces (e.g., in mitochondria and chloroplasts) do not permit free ion flow between the inside and the outside of membrane-bound structures. This creates a situation where it is possible to achieve stable transmembrane concentration gradients and electric potentials. In order to do this, selective pumps that translocate specific ions across membranes are used. The transmembrane potentials, in turn, can be used to provide energy for driving reactions that occur at the membrane surfaces. Numerous examples relating to the usefulness of transmembrane potentials are given in Chapters 16, 17, 34, and 35.

At constant temperature and pressure, the general relationship for the difference in free energy, ΔG_j, of an ion species j across the membrane is given by the equation

$$G_j = Z_j F\,\Delta\psi + RT \ln \frac{C_j^{outside}}{C_j^{inside}}$$

where $C_j^{outside}$ and C_j^{inside} are the concentrations of ion species j on either side of the membrane-bound structure, Z_j is the valency (with sign), and $\Delta\psi$ is the transmembrane potential. If a proton is the ion species in question, this equation becomes

$$\Delta G_{H^+} = F\,\Delta\psi - 2.3RT\,\Delta pH$$

since $Z = +1$ and $pH = -\log(H^+)$. These two equations have been used as the starting point for explaining the energetics of electrochemical potentials, how they are used in active transport, and how they are believed to be coupled to endothermic reactions such as the phosphorylation of ADP to ATP at the membrane surface (see Chapters 16 and 34).

fact that reactions in living systems are usually not isolated, but are part of long complicated sequences in which the product of one reaction serves as the reactant for another reaction. In such instances the crucial thermodynamics is not that of individual reactions but that of the initial and final states of the sequence. An important consequence of this is that an individual reaction with an unfavorable equilibrium may proceed if the total sequence has a favorable equilibrium constant. In biochemical pathways, *if the cell needs a reaction to proceed that will not proceed in isolation ($+\Delta G$), the reaction is coupled to another favorable reaction ($-\Delta G$).*

The advantages of such *coupling* can be appreciated by considering the equilibrium constants for each reaction of a coupled series. For example, let us consider the following sequence:

$$A \rightleftharpoons B \rightleftharpoons C \tag{28}$$

We shall assume that the equilibrium constant between A and B is $K_1 = 0.1$ and the equilibrium constant between B and C is $K_2 = 100.0$. Because of the low equilibrium constant, only small quantities of A would be consumed if the conversion of A to B were not followed or driven by the subsequent conversion of B to C.

At equilibrium,

$$K_1 = \frac{B}{A} = 0.1 \quad \text{and} \quad K_2 = \frac{C}{B} = 100 \tag{29}$$

The equilibrium constant for the overall process is the product of the equilibrium constants for the individual reactions:

$$K_1 K_2 = \frac{C}{A} = 10 \tag{30}$$

Thus when the reaction A to B is pulled along by a subsequent favorable reaction B to C, we find that most of A reacts.

This type of coupling is a common occurrence in biochemical sequences. Consider, for example, the condensation of acetyl-CoA with oxaloacetate to form citric acid. The hydrolysis of the coenzyme A off the citric acid provides the driving force for the reaction (Figure 12–7). The first reaction in this sequence has a standard free energy change of only -0.05 kcal/mole. However, the second reaction in the sequence, involving the hydrolysis of citryl-CoA at pH 7, has a standard free energy change of -8.4 kcal/mole, which causes the overall reaction to proceed with a high negative free energy of -8.4 kcal/mole.

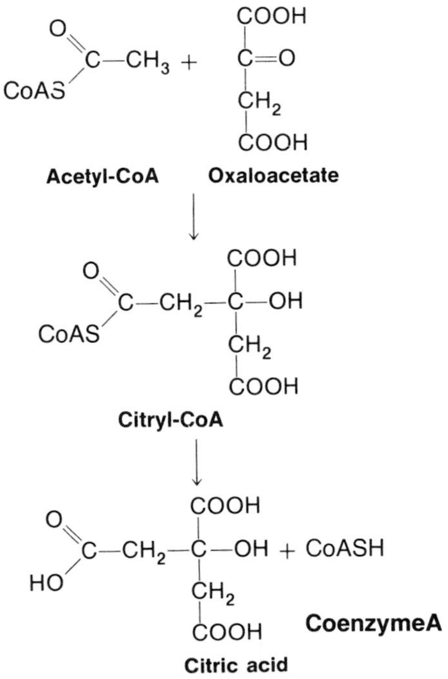

Another and perhaps more common type of coupling reaction is that in which an enzyme catalyzes the two reactions together. For example, assume the reaction $A \rightleftharpoons B$ has an equilibrium constant again of only 0.1 and the reaction of $C \rightleftharpoons D$ has an equilibrium constant of 100. If the enzyme catalyzes the overall process $A + C \rightleftharpoons B + D$, so that the two individual reactions are linked, the reaction of A to B will be driven by the C to D reaction:

$$A + C \overset{K_3}{\rightleftharpoons} B + D \tag{31}$$

$$K_1 = \frac{B}{A}, \quad K_2 = \frac{D}{C}, \quad K_3 = K_1 K_2 = 10 = \frac{[B][D]}{[A][C]} \tag{32}$$

An example of this type of reaction is the phosphorylation of glucose to ATP to produce ADP and glucose-6-phosphate. The reaction of glucose with phosphoric acid to form glucose-6-phosphate has an unfavorable equilibrium constant of 0.0062 M^{-1}. The hydrolysis of ATP has a quite favorable equilibrium constant equal to 9.2×10^5 M. When these two reactions are simultaneously coupled, the overall reaction is favorable, with an equilibrium constant equal to 5.7×10^3 M.

$$\text{Glucose} + \text{ATP} \rightleftharpoons \text{glucose-6-phosphate} + \text{ADP} \tag{33}$$

Notice from the two previous examples that when free energy is being calculated we add the free energies for the individual reactions, but that

Figure 12–7
The condensation of acetyl-CoA with oxaloacetate, followed by the hydrolysis of the citryl-CoA ester. The hydrolysis of the coenzyme A from the complex with citric acid provides the driving force for this reaction.

when the equilibrium constant is being calculated we multiply the equilibrium constants for the individual reactions.

ATP, THE MAIN CARRIER OF FREE ENERGY

By far the most important compound for energy utilization in the cell is adenosine triphosphate (ATP). ATP serves as a common medium of exchange of energy between energy-producing and energy-consuming systems. Energy-producing systems make ATP and energy-requiring systems consume ATP in coupled reactions. In humans about 5 lb of ATP is consumed every day to satisfy normal energy needs.

Hydrolysis of ATP Yields Considerable Free Energy

Adenosine triphosphate can hydrolyze by two different reactions (Figure 12–8). The α,β linkage can be hydrolyzed to form adenosinemonophosphate (AMP) and pyrophosphate ion ($HP_2O_7^{2-}$) or the β,γ linkage can be hydrolyzed to form adenosine diphosphate and phosphate ion. The free energies for these reactions at pH 7 are about equal. In the first reaction the pyrophosphate initially produced can subsequently be hydrolyzed to ensure the irreversibility of the reaction. The first reaction is used primarily in situations where any reversibility would be intolerable, such as in nucleic acid synthesis. The second reaction is most frequently used in processes where a limited amount of reversibility is tolerable. It has the advantage that less free energy is required to recharge the ADP product than would be required to recharge AMP by successive additions of phosphate.

The standard free energy values for these hydrolyses and those of a number of other organophosphate compounds are given in Table 12–6. It can

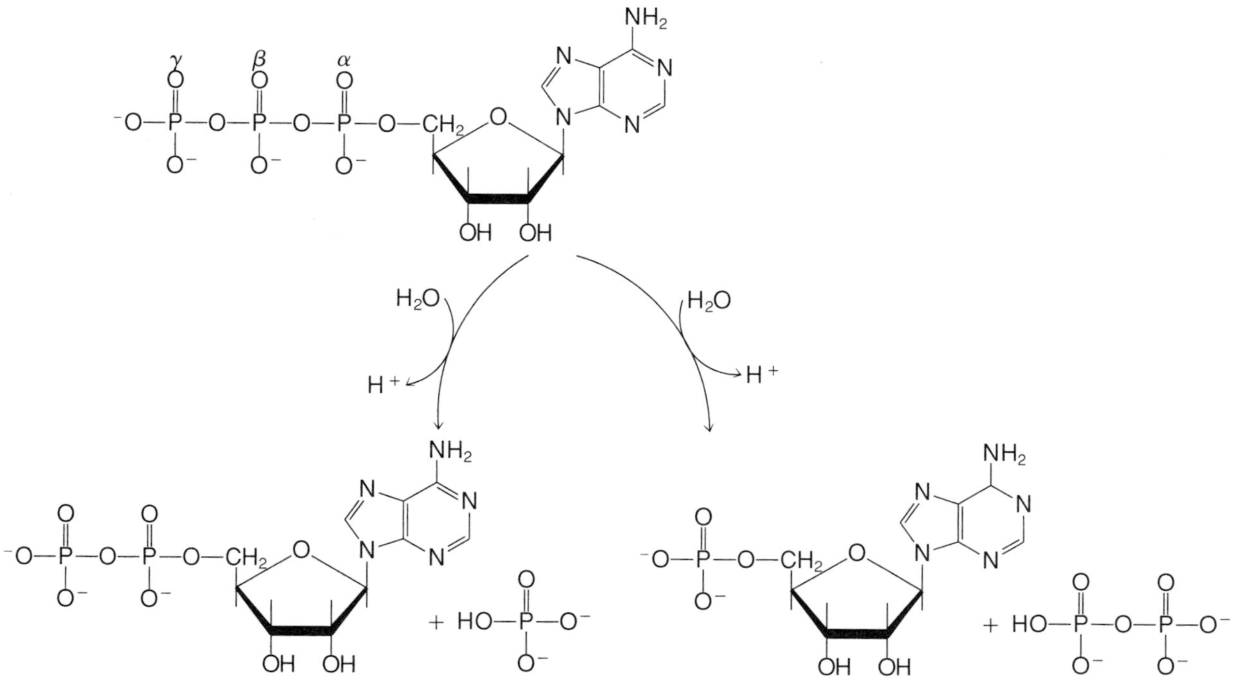

Figure 12–8
Alternative routes of ATP hydrolysis. The charged groups shown are the main ones that exist at physiological pH and ionic strength. The phosphate groups are referred to as α, β, and γ as shown. The standard free energy values for these hydrolyses are given in Table 12–5.

TABLE 12–6
Approximate Standard Free Energies of Hydrolysis for ATP ($\Delta G°'$) and Some Other High-Energy Phosphate Compounds of Biological Interest

Principal Reacting Species	Hydrolysis Products	$\Delta G°'$ (kcal/mole)
ATP^{4-}	$ADP^{3-} + HPO_4^{2-} + H^+$	-8.2
ATP^{4-}	$AMP^{2+} + H_2P_2O_7^{3-} + H^+$	-8.4
ADP^{3-}	$AMP^{2-} + HPO_4^{2-} + H^+$	-8.6
AMP^{2-}	Adenosine $+ HPO_4^{2-}$	-2.2
$HP_2O_7^{3-}$	$2\ HPO_4^{2-} + H^+$	-7.9
Acetyl phosphate	Acetate$^- + HPO_4^{2-} + H^+$	-11.3
Phosphoenolpyruvate	Pyruvate$^- + HPO_4^{2-}$	-14.7
Phosphocreatine	Creatine$^+ + HPO_4^{2-}$	-10.2
Phosphoarginine	Arginine$^+ + HPO_4^{2-}$	-9.1

be seen that considerable free energy is released when the phosphate group is hydrolyzed from either ATP or ADP but not from AMP. For this reason the linkages between the α,β and the β,γ phosphates (see Figure 12–8) are often referred to as high-energy or energy-rich linkages. In fact, the reasons for the relatively large negative free energy for the first two hydrolyses must be discussed from two standpoints—the factors leading to the relative instability of the reactant, ATP, and the factors leading to the relative stability of the products. Comparison of these reactions in Table 12–6 suggests two things. The reacting species that yield the higher free energies on hydrolysis are more highly negatively charged and release a proton on hydrolysis. Resonance also stabilizes the products more than the reactants, which increases the negative free energy change on hydrolysis. The relative contributions of these different factors to the free energy of hydrolysis are still unclear.

The main event taking place when ATP hydrolyzes to ADP and phosphate at neutrality is described by the following equation:

$$ATP^{4-} + H_2O \rightleftharpoons ADP^{3-} + HPO_4^{2-} + H^+ \qquad (34)$$

The equilibrium constant K for this reaction is equal to 0.63:

$$K = 0.63 = \frac{[ADP^{3-}][HPO_4^{2-}][H^+]}{[ATP^{4-}]} \qquad (35)$$

Biochemists commonly study reactions at or close to neutrality. Instead of continually correcting from 1 M acid or base to neutrality, the values of free energy and the equilibrium constant are commonly reported for solutions at neutrality, i.e., pH 7. These values are designated by a prime and written as $\Delta G°'$, $\Delta G'$, and K'. For the equilibrium constant K' we redefine the standard state of proton as 10^{-7} M; it should be noted that for a reaction releasing one proton, the relationship between K and K' is simply $K' = K/[H^+]$. We must be very careful, in evaluating the free energies $\Delta G°$ or $\Delta G°'$, to use the equilibrium constants K or K', respectively, since these are very different quantities for reactions such as ATP hydrolysis that involve the uptake or release of protons. Further aspects involved in evaluating the equilibrium constant for ATP hydrolysis are considered in Box 12–D.

Crucial to an understanding of the key role that ATP plays as an intermediate in energy metabolism is an appreciation of the position ATP occupies relative to other phosphate-containing compounds with which it interacts. Figure 12–9 summarizes the free energies of hydrolysis for several phosphorylated intermediates common to energy metabolism. In general the relative positions of two compounds on the scale dictate the ease with which

BOX 12–D

Factors Affecting the Equilibrium Constant for ATP Hydrolysis

Complications in evaluating the equilibrium constant for ATP hydrolysis arise because ATP^{4-}, ADP^{3-}, and HPO_4^{2-} are not the only species present at pH 7. Each of these phosphate-containing moieties is partially protonated. The most useful equilibrium constant at pH 7 would be for the reaction that takes the concentration of all species of ATP, ADP, and P into account. This equilibrium constant is designated by the symbol K^+ and represented by the equation

$$K^+ = \frac{[ADP, \text{ all forms}][P, \text{ all forms}]}{[ATP, \text{ all forms}]}$$

The values of K^+ may be calculated from K by the following expression, where $fATP$, $fADP$, and fP are the ratios of ATP^{4-} to the total ATP, of ADP^{3-} to the total ADP, and of HPO_4^{2-} to the total inorganic phosphate, respectively:

$$K^+ = \frac{K}{[H^+]} \frac{fATP}{fADP \cdot fP}$$

To evaluate the factors $fATP$, $fADP$, and fP requires a knowledge of the ionization constants of ATP and other species:

$$[ATP] \simeq [ATP^{4-}] + [ATP^{3-}] + [ATP^{2-}]$$

$$[ATP] \simeq [ATP^{4-}]\left[1 + \frac{(H^+)}{(K_1ATP)} + \frac{(H^+)^2}{(K_1ATP)(K_2ATP)}\right]$$

In this expression, K_1ATP and K_2ATP are the dissociation constants for $HATP^{3-}$ and H_2ATP^{2-}, respectively. The values of these dissociation constants are given in Table D–1. The factor $fATP$ for any given pH can be determined from these dissociation constants and the pH:

$$fATP = \frac{1}{1 + \frac{(H^+)}{(K_1ATP)} + \frac{(H^+)^2}{(K_1ATP)(K_2ATP)}}$$

Similar expressions may be derived for $fADP$ and fP. Actually, at pH 7, the differences between K' and K^+ amount to a difference of only about 2 per cent in the free energy of hydrolysis of ATP. At lower pH values, the difference is greater, whereas at higher pH values the discrepancy is less.

Thus far we have considered only the association of protons with the phosphates involved in ATP hydrolysis. Almost all reactions involving ATP hydrolysis are carried out in the presence of magnesium ions or other divalent cations that have significant associations with both ATP and ADP. Terms similar to those used for protons may be added to $fATP$, $fADP$, and fP for the magnesium chelate species and used in the equation for K^+ to determine the K^+ at any pH in the presence of magnesium or any other ion. For example, the total concentration of ATP in terms of ATP^{4-} in the presence of Mg^{2+} ions has two extra terms:

$$[ATP] = [ATP^{4-}]\left[1 + \frac{Mg^{2+}}{(KMgATP)} + \frac{(H^+)}{(K_1ATP)}\right.$$
$$\left. + \frac{Mg^{2+}}{(KMgHATP)} + \frac{(H^+)^2}{(K_1ATP)(K_2ATP)}\right]$$

The constants $KMgATP$ and $KMgHATP$ are the equilibrium constants for the dissociation of magnesium ions from MgATP and from MgHATP, respectively:

$$MgATP^{2-} \longrightarrow Mg^{2+} + ATP^{4-}$$

$$MgHATP^- \rightleftharpoons Mg^{2+} + HATP^{3-}$$

phosphate can be transferred from one compound to the other. The higher a compound is on this scale, the higher is its _phosphate group transfer potential_. Unless concentration conditions result in ΔG values that actually invert the rankings shown in Figure 12–9, it is generally true that any compound in the figure can be exergonically phosphorylated at the expense of any compound that lies above it (i.e., has a more negative $\Delta G°'$ value), but not by a compound that lies below it. Thus ADP can be phosphorylated to ATP by transfer of phosphate from glycerate-1,3-bisphosphate but not from glucose-6-phosphate. Similarly, ATP can be used to phosphorylate glycerol, but not to generate phosphoenolpyruvate from pyruvate.

It should be clear from Figure 12–9 that phosphorylated compounds actually display a broad range of $\Delta G°'$ values, with no sharp dividing line. More important than any arbitrary classification is the crucial intermediate position that ATP occupies on the scale. The whole role of the ATP-ADP system is to serve as an acceptor and donor of phosphate groups. Essential to that function is an intermediate position between the higher-energy compounds from which ATP accepts phosphate and the lower-energy compounds to which it donates phosphate.

ATP is the immediate source of energy for most energy-requiring reac-

The effect of magnesium ions on the free energy of hydrolysis of ATP is complex, since that effect depends on the pH and the ionic strength. When all the relevant calculations are made, we find that magnesium ions cause slightly less negative values for the free energy of hydrolysis of ATP to ADP and phosphate. A value of -8.4 kcal/mole for the hydrolysis of ATP to ADP at pH 7, 0.2 ionic strength, 25°C, and in the absence of magnesium becomes -7.7 kcal/mole in the presence of 0.001 M magnesium ions and decreases further to -7.5 kcal/mole in the presence of 0.01 M magnesium ions. The -7.5 value will be used for most calculations later in this text, but it should be appreciated that the true value may be considerably higher than this (especially see Table D–2).

Concentrations of the various reactants and products in the ATP hydrolysis can have a much larger effect on the computed free energy of hydrolysis. Assuming that all phosphate-containing molecules (ATP, ADP, and P) are present at the same concentration (a gross assumption), the free energies of hydrolysis at concentrations

other than the standard state may be calculated from Equation (19) (see Table D–2). The value of -8.4 kcal/mole at 1 M becomes -12.5 kcal/mole at 1 mM. In fact, the concentrations of the ATP, ADP, and P in most cells are probably somewhere between 2 and 10 mM. Also shown in Table D–2 are the variations in the free energy of hydrolysis over a range from pH 6 to 8.

TABLE D–1
Acid Dissociation Constants at 25°C and 0.2 Ionic Strength

$HATP^{3-}$	$\rightleftharpoons H^+ + ATP^{4-}$	1.12×10^{-7}
H_2ATP^{2-}	$\rightleftharpoons H^+ + HATP^{3-}$	8.71×10^{-5}
$HADP^{2-}$	$\rightleftharpoons H^+ + ADP^{3-}$	1.32×10^{-7}
H_2ADP^-	$\rightleftharpoons H^+ + HADP^{2-}$	1.18×10^{-4}
$H_2PO_4^-$	$\rightleftharpoons H^+ + HPO_4^{2-}$	1.66×10^{-7}

TABLE D–2
Free Energy of Hydrolysis for ATP as a Function of pH and Concentration

pH	$-\Delta G^+$			
	1 M	0.1 M	0.01 M	0.001 M
6.0	7.89			
6.2	7.94			
6.4	8.01			
6.6	8.12			
6.8	8.26			
7.0	8.40[a]	9.78	11.15	12.52
7.2	8.59			
7.4	8.81			
7.6	9.06			
7.8	9.32			
8.0	9.56			

[a]Calculated from R. W. Guynn and R. L. Veech, The equilibrium constants of the adenosine triphosphate hydrolysis and the adenosine triphosphate citrate lyase reactions, *J. Biol. Chem.* 248:6966, 1973.

tions in most cells. Moreover, since energy is needed continuously by all living cells, it follows that every cell must make provision for the continual generation of ATP and for careful adjustment of ATP formation to actual energy needs. The importance of continuous ATP synthesis and the immediacy with which ATP utilization is linked to its formation can be seen in the very rapid rate of ATP turnover. The time for one-half of all the ATP molecules present in the cell to be used in driving endergonic reactions and then regenerated in exergonic reactions is often only a few seconds. Even in a far more sluggish cell, half the ATP molecules can turn over in a matter of minutes. Obviously, then, cells would live for only a few seconds or minutes if suddenly deprived of further ATP-generating ability. The turnover is awesome, and the balance is fine.

THERMODYNAMICS IN OPEN SYSTEMS

Before moving on to consider the general design of metabolism in living systems, it is essential to add one disclaimer. The principles of classical thermodynamics that we have applied in this chapter were designed to be applied to equilibrium situations and in closed systems that do not exchange

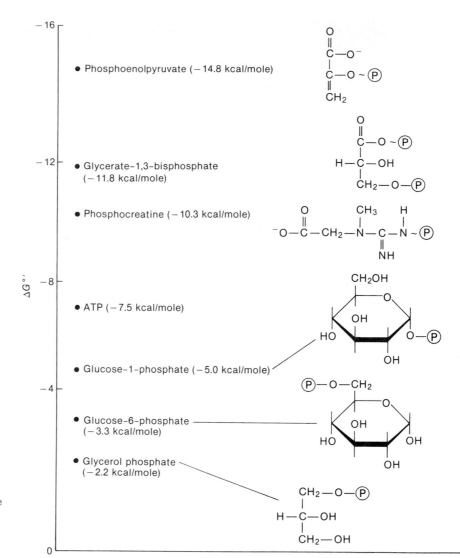

Figure 12–9
Standard free energies of hydrolysis for some common phosphorylated compounds. The $\Delta G^{\circ\prime}$ values are for the energy released during a hydrolysis reaction releasing a single phosphate. Where there is more than one phosphate, the $\Delta G^{\circ\prime}$ value refers to the hydrolysis of the phosphate bonded by a wavy line $\sim\!\!\sim$. This symbol stands for a large free energy released on hydrolysis, and does not represent anything structurally unique about the specific bond. It is important to realize that there is a large range of values for hydrolysis of esterified phosphates.

matter with their surroundings. Living systems are not at equilibrium, nor are they closed systems. They are open systems in which the flow of materials is highly significant, and they are in a nonequilibrium steady state in which the concentrations of many components, such as ATP, remain constant despite a high turnover rate.

There is a branch of thermodynamics called _nonequilibrium_ or _irreversible thermodynamics_ which has been applied to open systems. Some applications have been made to living systems. The main difficulty has been that, in its currently developed form, irreversible thermodynamics applies only to situations near equilibrium. But in living cells we deal with many situations which are far from equilibrium. Even the ratio of ATP to ADP is much higher than it would be at equilibrium, and for good reason. If it were near equilibrium, very little energy would be supplied by ATP hydrolysis.

The gap between thermodynamic theory and biologic reality poses a very real dilemma. We must be cautious in interpreting the conclusions from thermodynamic reasoning. Thermodynamics certainly gives us a good idea of what reactions are possible, as stated at the outset of this chapter, but we must be ready for surprises that cannot be explained readily by existing theories. Biochemistry, after all, is primarily an empirical science in which we are continuously discovering the unexpected. Great discoveries often lead to new theories.

SUMMARY

Thermodynamics is useful to biochemists for predicting whether or not a reaction could take place, but not whether or not it actually will take place. Thermodynamic quantities are properties of the state of a system, and thermodynamic calculations are concerned with the initial and the final state of a system, not with the mechanism whereby the system goes from one state to the other.

The most important thermodynamic quantities are the internal energy, ΔE, the enthalpy, ΔH, the entropy, ΔS, and the free energy ΔG. The internal energy of a molecule includes translational, rotational, and vibrational energy, as well as electronic energy and nuclear energy. In much of thermodynamics, including bioenergetics, we are not concerned with absolute values but rather with differences in energy between initial and final states. For this reason thermodynamic quantities are usually preceded by a delta (Δ). Electronic terms are usually the major contributor to the change in internal energy resulting from a chemical reaction. The first law of thermodynamics states that energy cannot be created or destroyed in a chemical reaction. Thus if energy is added to a system the environment must lose the equivalent amount of energy. The change in enthalpy of a compound, ΔH, is the change in internal energy of the compound plus the change in the product PV. For most biochemical reactions ΔE and ΔH are nearly equal.

Most but not all reactions that proceed spontaneously show a negative change in enthalpy; if the reaction does no work it leads to a release of heat. Those reactions that absorb heat in the absence of work have a positive change in enthalpy. Such reactions invariably show an increase in entropy.

Entropy is a measure of the order of a system. Systems that are highly ordered have a low entropy. For chemical reactions the most important factors to be evaluated to determine the entropy change are the translational and rotational terms. The second law of thermodynamics states that isolated systems proceed from a state of low entropy or low probability (ordered) to a state of high entropy or high probability (disordered).

Free energy, ΔG, is defined in terms of enthalpy and entropy and the absolute temperature by the equation $\Delta G = \Delta H - T \Delta S$. Free energy is the most suitable quantity for thermodynamic calculations because it is directly related to the equilibrium constant and the useful work that can be performed by a system. Reactions that proceed spontaneously have a negative ΔG. Like the other thermodynamic quantities, ΔE, ΔH, and ΔS, the ΔG for a reaction is simply equal to the difference in free energy between the reactants and products at equilibrium. For a reaction in solution, the standard free energy change, $\Delta G°$, is simply related to the concentrations of the reactants and products. Once their values are known, the free energy change for a reaction away from equilibrium concentrations can be calculated, as long as the concentrations are known.

ATP is the main carrier of free energy in living cells. The free energy provided by ATP hydrolysis is used to drive most energy-requiring reactions.

SELECTED READINGS

Alberty, R. A., and F. Daniels, *Physical Chemistry*, 5th ed. New York: Wiley, 1975. A well used classic textbook.

Cantor, C. R., and P. R. Schimmel, *Biophysical Chemistry*. San Francisco: Freeman, 1980. A three-volume, up-to-date treatment covering a broad range of topics in an authoritative manner.

Dickerson, R. E., *Molecular Thermodynamics*. New York: Benjamin, 1969. One of the best of the "newer" books on thermodynamics.

Ingraham, L. L., and A. B. Pardee, Free Energy and Entropy in Metabolism. In D. M. Greenberg (ed.), *Metabolic Pathways*, Vol. 1. New York: Academic Press, 1967. An

excellent review of thermodynamics stressing empirical relationships useful to biochemists.

Latimer, W. M., *Oxidation Potentials*. Englewood Cliffs, N.J.: Prentice-Hall, 1952. Although this is an old book, it does contain much useful information on oxidation-reduction potentials and free energies, enthalpies, and entropies of various compounds.

Tanford, C., *The Hydrophobic Effect*, 2d ed. New York: Wiley, 1980. A quite useful book but on a more specialized topic.

Van Holde, K. E., *Physical Biochemistry*. Englewood Cliffs, N.J.: Prentice-Hall, 1971. A lot of information presented quite precisely and mathematically.

PROBLEMS

1. In some respects thermodynamic considerations are more important to a biochemist than to a chemist. Why is this so?

2. Name three extensive and three intensive properties that relate to thermodynamic quantities. What is the basic difference between the two types of properties?

3. As a rule, what are the most significant parameters to be considered in evaluating the enthalpy and the entropy of a biochemical reaction?

4. What is meant by a state function? Why is enthalpy a state function? In what way is it useful to know that you are dealing with a state function?

5. For most biochemical reactions, internal energy and enthalpy may be equated. Why?

6. In thermodynamics, it is important to distinguish between the total system and the system being studied. Why?

7. Why do enthalpies and entropies of solvation tend to negate one another?

8. Why does the entropy of a weak acid decrease on ionization?

9. Proteins that serve as gene repressors frequently have two identical binding sites, which interact with comlementary sites on the DNA. If a repressor protein is cut in half without damaging either of its DNA binding sites, how will this affect the binding to DNA? How will the binding be affected if one of the two binding sites on the DNA is eliminated by changing the nucleotide sequence? Discuss the enthalpy, entropy, and free-energy effects.

10. Cite some of the factors that make ATP an ideal source of biochemical energy.

11. Table D–2 (in Box 12D) gives the free energy for hydrolysis of ATP at pH 7.0 for standard conditions (1 M) and for more dilute solutions. Explain how the latter values are obtained.

12. Tom and Mary both ate a complete basal diet containing all nutrients, but they decided to gain weight by eating an extra 3000 K calories a day. Tom chose to get his 3000 K calories in glycine and Mary took her 3000 K calories in glucose. Recall that nutritional calories are measured by heats of combustion to form nitrogen instead of urea as is done in the body and that the body gains weight on free energy, not enthalpy. Therefore, the caloric values of their diets are not equivalent. Who gained the most weight?

Data:

$$\text{Glucose} \quad \Delta H_{combustion} = -675.4 \text{ kcal/mol}$$
$$\Delta G_{combustion} = -686.5 \text{ kcal/mol}$$
$$\text{Glycine} \quad \Delta H_{combustion} = -232.6 \text{ kcal/mol}$$
$$\Delta G_{combustion} = -241.0 \text{ kcal/mol}$$
$$\text{Urea} \quad \Delta G_{combustion} = -158.9 \text{ kcal/mol}$$

13. Calculate the free-energy change involved in transporting 1 mole of H^+ from a region with a concentration of 10^{-7} M to a region with a concentration of 1 M.

14. Explain why a closed system cannot be in a steady state. Remember that a steady state refers to a nonequilibrium situation that prevails because of a balance between reactions that supply and remove substances.

15. In Chapter 15 the final reaction in the tricarboxylic acid cycle involves regeneration of oxaloacetate from malate. This reaction has a $\Delta G^{\circ\prime}$ of +7.1 kcal/mole. Suggest reasons why this reaction goes in the forward direction in the cell.

16. Hemoglobin has four essentially identical sites for binding O_2. Explain how you would determine whether the binding of O_2 at one site influenced the binding at another site. What are the relevant equations?

17. The oxidation of one mole of palmitate to CO_2 and water forms 129 moles of ATP from ADP and inorganic phosphate at pH 7. The free energy ($\Delta G^{\circ\prime}$) is $-2,340$ kcal/mole for the oxidation of palmitate by molecular oxygen.
   a. What is the free energy of the overall process to form ATP?
   b. What is the efficiency of the storage of energy as ATP?

18. Could an aerobic organism obtain energy by the oxidation of ammonia at 0.01 atm to form atmospheric nitrogen? If so, how much? Data for ΔG° of formation:

$$NH_3(g) \quad -3.94 \text{ kcal/mole}$$
$$H_2O(l) \quad -55.69 \text{ kcal/mole}$$

19. Pyridine binds to an enzyme with a free energy of -16 kcal/mole and benzaldehyde binds to the same enzyme with a free energy of -20 kcal/mole. The substrate for this enzyme is compound A.

A

Estimate the range of the free energy of binding for this substrate to the enzyme. Assume no interaction between the pyridine and benzaldehyde on the enzyme.

20. Calculate the pH of the following solution of acetic acid (pK = 5).
   a. 0.01 M acetic acid.
   b. 10^{-4} M acetic acid.

21. Calculate a dissociation constant for magnesium ion from ATP^{4-} from the following data. Give an estimate of the enthalpy of the reaction.

(ATP) × 10^5	Total (Mg) × 10^5	Free (Mg) × 10^5
75	20	2
41	20	4
28.6	20	6
21.4	20	8
16.4	20	10

22. How much sodium hydroxide should be added to 0.1 M acetic acid (pK = 5.0) to make a buffer of pH = 4.0?

13

OVERVIEW OF METABOLISM

In Chapter 12 we considered thermodynamic limitations on chemical change. Biochemical reactions, like all chemical reactions, can proceed only in the direction in which free energy change is negative. Yet living cells are chemically complex, and their free energies, expressed per carbon atom, are much greater than for the carbon dioxide from which all of the component compounds of living organisms are ultimately derived. That situation is possible only because organisms have evolved intricate systems of coupling between metabolic pathways. Before discussing specific pathways in later chapters, we will look here at general aspects of the functional interrelationships among metabolic sequences and the thermodynamic and kinetic principles that underlie them.

$\Delta G = -ve$

When the chemical activities of a cell or organism are compared with those that occur in a similar volume of nonliving material, four aspects of metabolic chemistry are particularly impressive:

1. The diversity of chemical activity—typically several hundred to a few thousand reactions proceed simultaneously in the tiny volume of a living cell.
2. The very high rates of the reactions, especially in view of the relatively low temperatures and the low concentrations of reactants.
3. The strict chemical specificity of the reactions.
4. The functional correlation between the reactions, that is, the interdependencies of the chemical activities of the cell.

Of these, the first three differ only in degree from conditions in the inorganic world (although the differences are very large). But the fourth is fundamentally different in kind. Functional correlations occur nowhere in the world except in living things or in objects made by living things, and the relationships among the chemical activities of a living organism are the most elaborate and complex functional interactions that are known to exist in the universe. In considering metabolism we must take the unusual number, the high velocities, and the strict specificities of reactions into account. They are striking and essential features of life. However, it is the functional aspects—the organization of reactions into functional sequences, the functional regulation of reaction rates, and the correlations between sequences—that primarily distinguish biochemistry from the chemistry of the nonliving world.

All of the activities of cells or organisms require energy, and most of them also consume specific compounds, such as biosynthetic starting materials. Large amounts of both energy and starting materials are consumed in synthesis of the many macromolecules (for example, proteins, nucleic acids, complex lipids, and polysaccharides) of which the cell is composed. Cells also must move molecules across membranes, often against concentration gradients. Much energy is devoted to such movements, which are essential for maintaining the composition and integrity of the organism. Many organisms use energy in mechanical work, such as swimming, running, or flying. We might call these activities, from cell division to running, the high-level functions of the organism.

At the higher levels of an organism's functions, rates of activity can change frequently and over wide ranges. In a growing organism or cell culture, for example, each cell will make DNA only during a specific part of the cell cycle. The rate at which materials need to be pumped across membranes will vary with conditions, and of course the rate at which energy is expended in motion varies over a wide range—in mammals, for example, from sleep to hard running. In order to meet such varying needs, energy must be constantly available and the needed small molecules must be available at essentially constant usable concentrations.

Intermediary metabolism—the synthesis and degradation of small molecules—*has the functions of (1) supplying the energy that is needed* for synthesis of macromolecules, for pumping materials across membranes, and for movement, *and (2) keeping these high-level functions supplied with starting materials or building blocks*—*amino acids for protein synthesis and nucleoside triphosphates for synthesis of nucleic acids, for example.*

Because the rates at which the high-level functions require energy and materials vary widely, the rates of the reaction sequences in intermediary metabolism must be regulated so that energy and intermediates will always be available to meet those needs. Thus *the rates of reactions* in those central or intermediary pathways *must vary over wide ranges.* But *the concentrations of certain key items* that those pathways supply for the cell *are remarkably stable.* This quality of stability is referred to as biochemical or physiological homeostasis. The fact that it is achieved under nonequilibrium conditions has led to the term *steady state** to describe the normal physiologic condition of the organism. In the steady state, constancy of key intermediary metabolites is achieved by maintaining a rate of synthesis for each intermediate that is equivalent to its rate of utilization.

In this chapter we will consider in a general way the interactions among pathways and the mechanisms by which metabolic *fluxes* (rates of flow of materials along metabolic sequences) are regulated to meet functional needs.

ORGANISMS DIFFER IN SOURCES OF STARTING MATERIALS AND ENERGY

A cell resembles a very complex economic system. Resources must be obtained and partitioned among various uses so as to meet the needs of the cell's economy. Typical aerobic heterotrophic cells (those that require oxygen and preformed organic compounds), must oxidize a large part of the

---

*The steady-state treatment was used for kinetic analysis of enzyme action in Chapter 8. Here we extend the use of the term *steady state* to a global view of the living cell.

TABLE 13–1
Classification of Organisms According to Starting Metabolites and Sources of Energy

	Starting Metabolites	Energy Sources
Autotrophs		
Chemoautotrophs (lithotrophs)	CO_2	Oxidation of inorganic molecules
Photoautotrophs	CO_2	Light
Heterotrophs		
Photoheterotrophs	Organic molecules	Light
Typical heterotrophs	Organic molecules	Oxidation of organic molecules

foods taken in to obtain energy to power the cell's many activities. There is a complicated division of labor between metabolic sequences. Some sequences supply energy or other basic needs of the cell, and others use those secondary resources in carrying out the many chemical and physical processes necessary for maintenance, growth, reproduction, and other activities of the cell or organism.

The means by which the cell obtains its fundamental needs—starting materials and energy—provide the basis for a metabolic classification of organisms (Table 13–1). The primary distinction is between *autotrophs*, self-feeding organisms that can make all of their component compounds and materials from CO_2 and other inorganic precursors, and *heterotrophs*, which must obtain nourishment by consuming other organisms or products of their metabolism. Green plants are typical autotrophs and mammals are typical heterotrophs.

As shown in the table, there are additional subdivisions within both the heterotrophs and the autotrophs. Autotrophs can produce all of the components of their cells from carbon dioxide. *Chemoautotrophs* (sometimes called lithoautotrophs) obtain energy from oxidation of inorganic materials such as hydrogen, ammonia, nitrite ion, sulfur, or sulfide ion. *Photoautotrophs*, such as green plants, use the energy of sunlight and obtain electrons for the reduction of CO_2 by the photochemical oxidation of water to oxygen.

There are also two types of heterotrophs. Typical heterotrophs, such as mammals, must obtain energy and starting materials from preformed organic compounds. Various types of pigmented bacteria, termed *photoheterotrophs*, can obtain energy from absorption of light but require organic compounds as sources of starting materials. These bacteria can use organic material more efficiently than other heterotrophs, because they do not need to degrade a major portion of the foodstuffs that they consume in order to obtain energy.

REACTIONS ARE ORGANIZED INTO SEQUENCES

In Part II of this text we took a close look at the mechanisms of action of a few enzymes. It is well to keep in mind that most of the enzyme-catalyzed reactions in a living cell are organized into sequences with specific func-

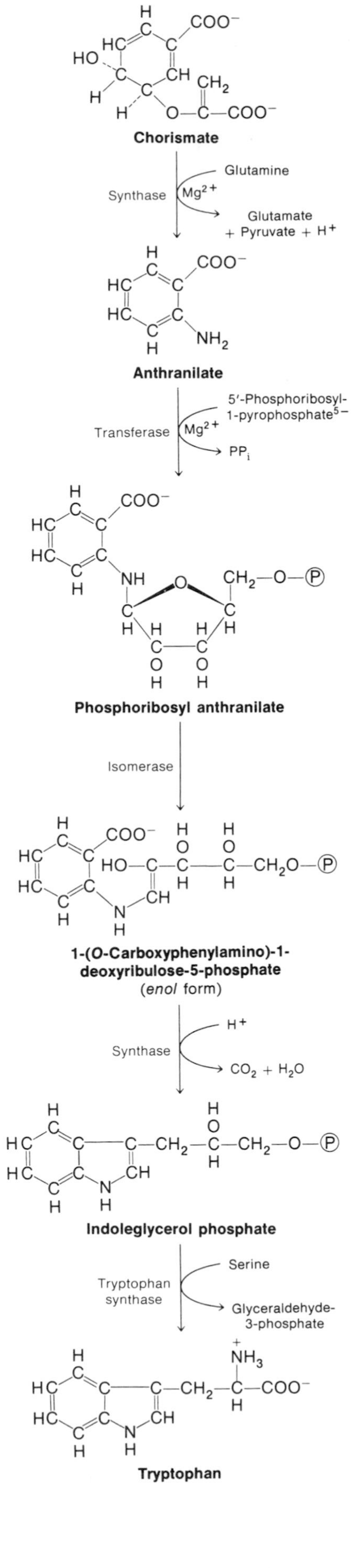

Chorismate

Synthase (Mg^{2+})
Glutamine
Glutamate + Pyruvate + H$^+$

Anthranilate

Transferase (Mg^{2+})
5'-Phosphoribosyl-1-pyrophosphate^{5-}
PP$_i$

Phosphoribosyl anthranilate

Isomerase

1-(O-Carboxyphenylamino)-1-deoxyribulose-5-phosphate (*enol* form)

Synthase
H$^+$
CO$_2$ + H$_2$O

Indoleglycerol phosphate

Tryptophan synthase
Serine
Glyceraldehyde-3-phosphate

Tryptophan

tions. *It is the sequence that serves a function.* An individual reaction within the sequence usually serves no function by itself, but only when it is an integral part of the sequence. For example, the compound chorismate is a precursor of the amino acid tryptophan. The conversion of chorismate to tryptophan occurs in five discrete steps, each requiring a specific enzymatic activity (Figure 13–1). The intermediates between chorismate and tryptophan serve no function except as precursors of tryptophan. The amino acid tryptophan has many uses, including its role as a building block in protein synthesis.

The breakdown of the six-carbon sugar glucose to the three-carbon compound pyruvate serves quite a different purpose. It supplies energy in the form of ATP. This complex process requires ten discrete enzymatically catalyzed steps (Figure 13–2). In the first six steps of the process ATP is actually consumed. It is only in the last four steps that the starting amount of ATP is recovered and further ATP is synthesized. Hence the function of this sequence will not be served until the entire sequence has been completed.

Sequences Are Either Anabolic or Catabolic

The synthesis of tryptophan and the breakdown of glucose exemplify the two main types of metabolic pathways. The biosynthetic pathway is called _anabolic_ (from the Greek *ana* = up) while the degradative pathway is called _catabolic_ (from the Greek *kata* = down). Anabolic pathways usually involve an increase in atomic order (decrease in entropy), they are often reductive in nature, and they are almost always energy-requiring (_endergonic_). Other examples of anabolic pathways are the synthetic reactions leading to proteins and polysaccharides. By contrast, catabolic pathways are usually energy-liberating (_exergonic_) usually oxidative, and usually involve a decrease in atomic order (increase in entropy).

Catabolic and anabolic pathways sometimes share a common partial sequence, which functions in one direction for synthesis and in the opposite direction for degradation. However, the catabolic route is never exactly the reverse of the anabolic pathway, for reasons that will be discussed later in this section.

An overview of some of the main metabolic pathways is depicted in Figure 13–3. This figure does not show the individual reaction steps as did the previous two figures. Only some of the main routes of anabolism and catabolism are indicated. The pathways are organized so that the arrows for the anabolic pathways are in the up direction and the arrows for the catabolic pathways are directed downwards. Pathways that either consume or produce ATP are also indicated. In all cases ATP-producing pathways are catabolic and ATP-consuming pathways are anabolic. Starting from the top, compounds are depicted as _polymers_, then _building blocks_ that are components of the polymers. At the next level down there is _activated acetate_, which is the substrate for many building blocks (in the upward direction) as well as a precursor for total oxidation by the tricarboxylic acid cycle (in the downward direction). Finally , at the bottom level we find the ultimate _end products_ of catabolism—NH$_3$, H$_2$O, and CO$_2$.

Another important design element of metabolic pathways, one that is only partially portrayed by Figure 13–3, is the connectivity between path-

Figure 13–1
The biosynthesis of tryptophan. Tryptophan is synthesized in five steps from chorismate. Each step requires a specific enzyme activity. The four intermediates between chorismate and tryptophan serve no function other than as precursors of tryptophan.

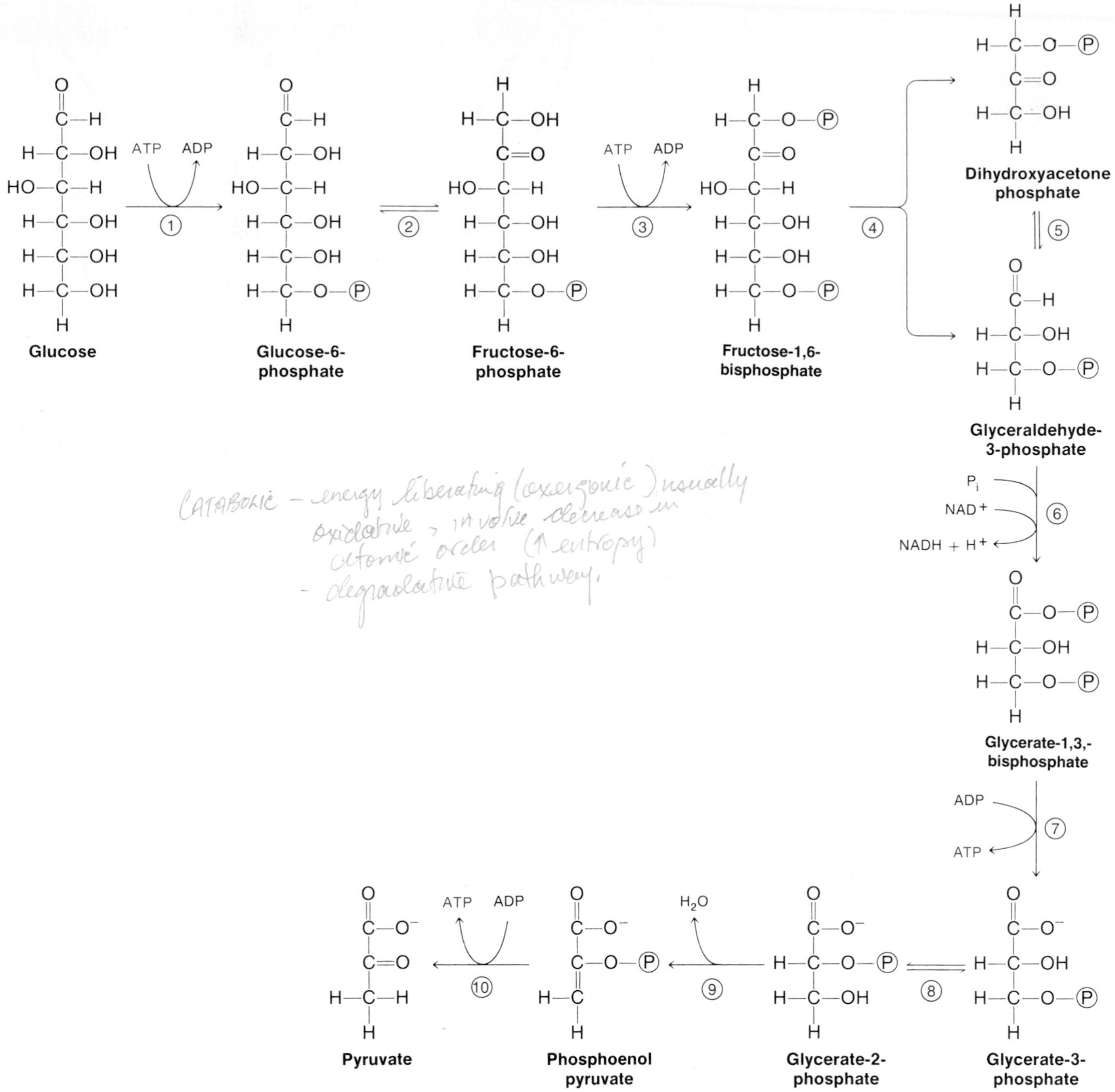

Figure 13–2

The breakdown of glucose to pyruvate. The conversion of glucose to pyruvate requires ten steps and one enzyme for each step. In steps 1 and 3, ATP is consumed. In steps 7 and 10, ATP is produced. The net production of ATP is 2 moles for each mole of glucose consumed. Many of the intermediates between glucose and pyruvate serve as starting metabolites for biosynthesis.

ways. Actual pathways are directly connected by branchpoints. In Figure 13–3 acetyl-CoA is a branchpoint, where many of the degradation products of proteins, polysaccharides, and lipids are seen to converge into a single molecule. Conversely, acetyl-CoA can be used by a series of different pathways as a starting point for the synthesis of amino acids, sugars, or fats. In general the patterns of branching observed for catabolic pathways or for anabolic pathways are strikingly different (Figure 13–4). The *catabolic pathways show a converging pattern*, starting with a broad variety of substrates

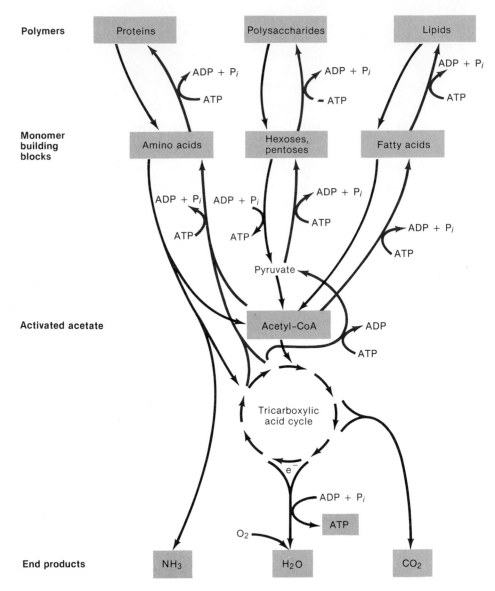

Figure 13–3
Overview of metabolism. Metabolic pathways are either anabolic (biosynthetic) or catabolic (degradative). Anabolic pathways consume ATP and catabolic pathways produce ATP. The ATP produced by the catabolic pathways is used to drive anabolism.

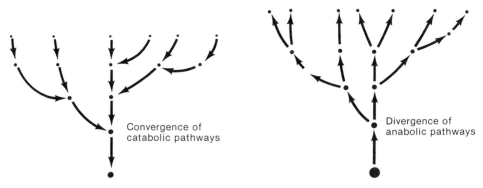

Figure 13–4
The convergence and divergence of metabolic pathways. Catabolic pathways start out from quite different substrates and converge, leading to a few common degradation products. By contrast, anabolic pathways start from common intermediates that are used to make a rich variety of products.

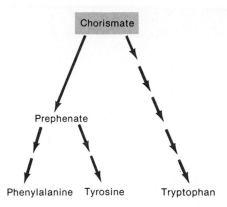

Figure 13–5

The three different end products derived from chorismate. Chorismate is a branchpoint metabolite that leads to three different amino acids. It shows a diverging pattern characteristic of anabolic pathways.

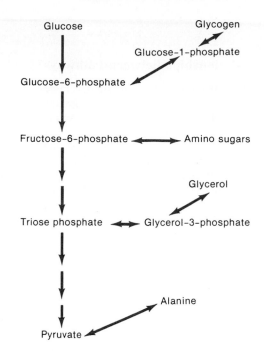

Figure 13–6

The glycolytic pathway. The breakdown of glucose to pyruvate contains numerous branchpoints. In the catabolic direction a number of substances converge into the glycolytic pathway, which ultimately leads to pyruvate. In the anabolic direction a number of the intermediates between glucose and pyruvate, including pyruvate, serve as starting metabolites for anabolic pathways.

that funnel down to common intermediates. On the other hand, the _anabolic pathways show a diverging pattern_, starting from common intermediates and ending with quite different end products.

We will see this contrast between the diverging and converging aspects of the two major types of metabolic pathways in detail in later chapters, when we consider individual pathways. Right now, however, note that chorismate (mentioned with respect to Figure 13–1) is at a branchpoint. It is not only a precursor of tryptophan, as already indicated, but it also serves as a precursor of two other amino acids, phenylalanine and tyrosine (Figure 13–5). You can see that there is even one more branchpoint in the common pathway at prephenate, which diverges to either of the aromatic amino acids, phenylalanine and tyrosine. This example shows the typical diverging pattern of a group of anabolic sequences.

In connection with Figure 13–2 we discussed the multistep conversion of glucose into pyruvate, a catabolic process. Figure 13–6 shows that several other sugars or sugar derivatives, such as glycerol and glycogen, and some amino acids funnel into this catabolic sequence. The overall pattern shows convergence.

There is one other highly significant feature of branchpoints that should be mentioned. Frequently a branchpoint is shared by an anabolic sequence on one branch and a catabolic sequence on the other branch (see Figure 13–7). This situation arises when the partial breakdown product reaches a point where it can either be broken down further or be used as a starting metabolite for an anabolic sequence. Consider the pyruvate example in Figure 13–6. Pyruvate can be further catabolized to acetyl-CoA and CO_2 and H_2O, or it can be used as a starting metabolite for the synthesis of certain amino acids such as alanine. At a branchpoint of this type a choice must be made: Is it more important for the intermediate to continue down the catabolic pathway to supply energy, or to be diverted to the anabolic pathway

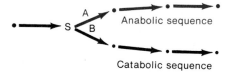

Figure 13–7

Branchpoints in metabolic pathways occur at locations where an intermediary, S, can follow more than one route. In this case S can proceed along a catabolic sequence or an anabolic sequence. The first reactions after the branchpoint are catalyzed by enzymes B and A, respectively. Once the first step after the branchpoint has been taken, the metabolic intermediate is irreversibly committed to follow that pathway.

BOX 13–A

Functional Interrelationships Between Pathways

In typical aerobic heterotrophic organisms, such as mammals, most other animals, and many bacteria and fungi, the three basic requirements of all cells—energy, reducing power, and starting materials—are supplied by the catabolic sequences at the expense of oxidative degradation of food materials such as carbohydrates, fats, and proteins. These functional relationships are shown schematically in the accompanying diagram.

In catabolic sequences carbohydrates, fats, proteins, and other foodstuffs are oxidized to carbon dioxide. In the process, electrons are made available for the reduction of $NADP^+$ to NADPH and for the conversion of ADP to ATP by oxidative phosphorylation (Chapter 15).

The catabolic sequences also supply the starting materials from which the constituents of the cell are manufactured. As indicated in the diagram, these are surprisingly few in number. All of them are intermediates in major sequences of carbohydrate metabolism. That fact is an example of the close relationships between metabolic functions: the same sequences by which foodstuffs are oxidized to supply electrons for reductions and for oxidative phosphorylation also supply the starting materials for the biosynthetic activities of the cell.

Quantitatively the most important sequences in catabolism are glycolysis and the tricarboxylic acid cycle, which are discussed in the next two chapters. These are often thought of as pathways of carbohydrate catabolism. However, they should more appropriately be considered the central pathways of the cell's entire metabolism. Fats and the amino acids that result from hydrolysis of proteins are degraded by individual pathways that lead into glycolysis or the tricarboxylic acid cycle, which thus serve a more general catabolic role than only degradation of carbohydrates. Similarly, nearly all of the components of the cell are manufactured from intermediates of these pathways. (The pentose phosphate pathway, which parallels part of glycolysis, supplies two of the starting materials seen in the diagram: five-carbon and four-carbon sugar phosphates.) *Thus glycolysis and the tricarboxylic acid cycle are not merely pathways by which ATP can be obtained from the oxidation of sugars. They participate in the catabolic degradation of nearly every compound that the cell is able to degrade, and they supply reducing power and starting materials, as well as ATP, for all of the cell's biosynthetic sequences, as well as energy for other activities.*

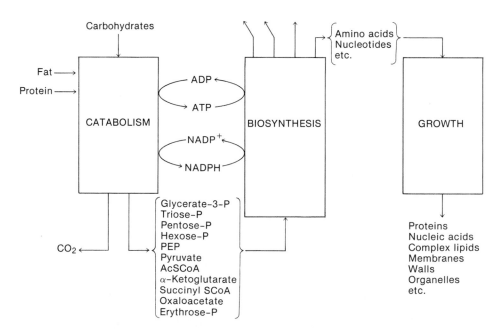

Schematic block diagram of the metabolism of a typical aerobic heterotroph. The block labeled CATABOLISM represents glycolysis, the tricarboxylic acid cycle, and the pentose phosphate pathway (which will be discussed in the following two chapters), as well as individual pathways by which fats and amino acids are converted to intermediates of those central sequences. The block labeled BIOSYNTHESIS represents the syntheses of low- to medium-molecular-weight components of the cell, such as cofactors, and of the amino acid and nucleotide building blocks from which proteins and nucleic acids are made. The block labeled GROWTH includes the synthesis of proteins, nucleic acids, and other macromolecules as well as the assembly of such compounds into the membranes, organelles, etc., of the cell.

The figure illustrates the roles of ATP/ADP and $NADP^+$/NADPH as coupling systems interrelating catabolic and anabolic sequences, and the role of catabolism as supplier of energy, reducing power, and synthetic starting materials for biosynthesis.

Although not shown in the figure, the reactions in the GROWTH block also require energy transduction from CATABOLISM by way of the ATP/ADP couple.

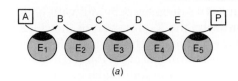

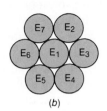

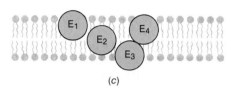

Figure 13–8
The organization of functionally related enzymes. Functionally related enzymes are organized in three ways: (a) as unlinked proteins soluble in the aqueous milieu of the same cellular compartment; (b) as components in a multiprotein complex; or (c) as components on a membrane.

and thereby serve as the starting metabolite for a biosynthetic product? We will come back to this point.

Enzymes Are Organized According to Function

We have seen that reactions are organized into sequences. It would seem logical for enzymes that function in different steps to be clustered together in the cell. In this way an intermediate resulting from the first reaction in a pathway could be directly shunted to the second enzyme in the pathway, and so on. Such an arrangement might be expected to accelerate synthesis of end product as well as to minimize the loss of intermediates.

Three types of situations can be found (Figure 13–8). In the simplest situation, all the enzymes for a particular pathway are on different proteins that exist as independent soluble proteins in the same cellular compartment. In such cases the intermediates would have to get from one enzyme to another either by free diffusion through the cytoplasm or by transfer after contact between two sequentially related enzymes, one carrying the reactive intermediate and the other ready to receive it. Some sequentially related enzymes that are isolated as independent molecules probably do actually exist *in vivo* as complexes held together by weak secondary forces. However, our knowledge on this point is still uncertain.

The second type of arrangement observed for sequentially related enzymes is exemplified by the *E. coli* fatty acid synthase (see Figure 11–15). This is probably a complex of all the enzymes involved in fatty acid synthesis. The intermediates in this case are bound to the enzyme complex until synthesis is complete.

In the third type of arrangement, functionally related enzymes are membrane-bound. This is the case for a number of the enzymes involved in the electron-transport processes associated with oxidative phosphorylation (see Chapter 16). Many of these enzymes are membrane-bound proteins in the mitochondrion.

Pathways Are Coupled to the ATP-ADP System

The stoichiometric relationships that we deal with in nonbiological chemistry are of two types. The first is simple *conservation of mass—there must be as many atoms of each type on the right-hand side of an equation as on the left-hand side*. The second deals with oxidation-reduction relationships—*if some reactants are oxidized, others must undergo an equivalent amount of reduction*. If we apply these familiar relationships to the oxidation of glucose, for example, we see that six moles of CO_2 must be formed from each mole of glucose, and that six moles of O_2 must be reduced to 12 moles of water in accepting the 24 equivalents of electrons that are lost by carbon in the process. Such stoichiometric relationships are fixed by the chemistry of a reaction and cannot be changed by catalysts, including enzymes; they apply to biological reactions exactly as they do to any others.

A third kind of stoichiometry, unique to biochemistry, is of fundamental importance to biological function. This is the *evolved coupling stoichiometry* that underlies the interlinking of metabolic sequences, especially by the ATP-ADP system. The whole functional organization of metabolism depends on the fact that all metabolic sequences are linked with the ATP-ADP system. Catabolic sequences also supply starting materials and electrons for biosynthetic reductions, but under most conditions in many types of cells the quantitatively most important consequence of catabolism is the phosphorylation of ADP.

The overall K_{eq} for oxidation of glucose is decreased by obligate coupling to the phosphorylation of ADP. Similarly, the K_{eq} for each of the cell's energy-requiring chemical, physical, and mechanical activities is made more favorable by obligate coupling to some definite number of ATP-to-ADP conversions (see Chapter 12). *The K_{eq} for any conversion may have any value, from very large to very small, depending on the number of ATP-to-ADP or ADP-to-ATP conversions that are built into it.* Catabolic sequences are coupled to ADP-to-ATP conversions, but the stoichiometric factors (the number of ADP phosphorylations that are coupled to a given catabolic sequence) are low enough so that the overall K_{eq} for the sequence remains large in the catabolic or degradative direction. Biosynthetic sequences, nearly all of which would be thermodynamically unfavorable in the absence of ATP coupling, are coupled to enough ATP-to-ADP conversions to cause their K_{eq} values to be large in the direction of synthesis.

Coupling in Pathways Assures Conversions in Both Directions

It is a fundamental generalization of chemistry that the *directions* of chemical reactions are controlled by thermodynamic factors; specifically, reactions can go only in the direction toward equilibrium. By contrast, *rates* of reactions are determined by kinetic factors. Such statements apply to enzyme-catalyzed reactions as well as to any others.

Since modifiers or other signals cannot affect the thermodynamic characteristics of reactions or reaction sequences, it might appear that metabolic signals could affect only rates of biochemical conversions and not their directions. And yet, directions of conversions in living cells *must* change in response to metabolic need.

Our livers, for example, convert glucose to glycogen, a polysaccharide storage material, after meals, and convert glycogen back to glucose to supply energy on demand between meals. If the directions of these conversions were determined only by thermodynamic considerations, the concentration of glucose would necessarily be high when glycogen was being stored. In order for stored glycogen to be used, the concentration of glucose would need to fall so far that equilibrium would favor breakdown of glycogen. An organism that operated in this way would be under severe disadvantages. The reactions of energy storage and mobilization would almost certainly be sluggish, since the conversions would always be occurring near equilibrium. The level of blood glucose would be low whenever stored reserves were being used, which is likely to be when energy is most needed—for example, to allow the organism to capture food or to escape a predator, or merely to return a deep lob in tennis or finish a mile race.

A less obvious but probably equally important disadvantage of wide fluctuations in blood glucose concentration is that they would disrupt interrelationships between organs, as well as the metabolic activities of many organs and tissues. The more complex a technological device and the more interrelationships involved in its functioning, the more important is stabilization of key potentials and other operating parameters. A living organism is extremely complex, and its proper functioning depends on an enormous number of precise interactions. In the evolution of mammals, tissues and organs have become highly interdependent, and stabilization of various systemic (organism-wide) parameters has become essential. Functioning of the central nervous system is highly disturbed, for example, if the concentration of glucose in blood rises or falls outside its narrow normal range. Stabilization within that range would not be possible if the directions of the relevant metabolic sequences were determined by simple thermodynamic considerations only.

On a longer time scale, most animals in the wild in climates with wide seasonal changes store relatively large amounts of fat in summer and fall and use this energy reserve as necessary during the winter, when food is less available. Again, if storage of fat and its breakdown to supply energy were controlled by simple thermodynamic factors, the concentrations of energy-producing intermediates would be low in winter and the animal would be severely handicapped by diminished availability of energy, just when life is most difficult and rapidly utilizable energy is essential. Similar considerations apply to many other _metabolic conversions_; they _need to go in either direction according to metabolic need, and this must be possible without wide swings in concentrations of metabolic intermediates_ if it is not to seriously upset the organism.

In all of these cases and many others, the directions of the conversions, as well as their rates, are actually regulated by metabolic signals such as concentrations of specific local metabolites, the available ATP, and hormones or other signals from other parts of the body. The means of achieving this apparently impossible result—control of _directions_ of conversions by the _kinetic_ effects of metabolic signals—may well be considered the most specifically biological feature of the chemistry of living organisms.

Any conversion could be made thermodynamically favorable by coupling it to a sufficient number of ATP-to-ADP conversions. That fact underlies the control of the directions of metabolic conversions. Metabolic sequences commonly occur in pairs. The two sequences of a pair connect the same compounds, which we will call A and Z in discussing the general case. The reactions of the two sequences are catalyzed by different enzymes, and usually most or all of the intermediates are different. The significant difference between the sequences is that, because of the specificities of their component enzymes, they are coupled to different numbers of ATP-to-ADP conversions. Remember that coupling an ATP-to-ADP conversion into a metabolic sequence changes the equilibrium ratio of products by a very large factor, so that any conversion can be made favorable by adjustment of the number of ATP-to-ADP conversions that are associated with it. In each pair of unidirectional sequences, one of the sequences involves an appropriate number of ATP-to-ADP conversions to cause the equilibrium constant for the conversion of A to Z to be so large that this direction of conversion will be favored under any conditions that could arise in a living cell. For the other sequence, the ATP-ADP stoichiometry is such that the equilibrium constant for the conversion of Z to A is large, and this direction of conversion will always be favorable _in vivo_. The general situation is illustrated schematically in Figure 13–9. It is the difference between x and y that allows the conversions in both directions to be favorable at all times.

In analyzing the control of metabolism, we must distinguish clearly between a _sequence_ and a _conversion_. A sequence, as the name applies, is a group of reactions, arranged sequentially, that carry out a biological conversion. In other words, a conversion is the metabolism of starting material A to end product Z or of Z to A; a sequence is the specific set of reactions by which the conversion is carried out. Any given conversion could be carried out by many different sequences of reactions, which might be coupled to nearly any number of ATP-to-ADP or ADP-to-ATP interconversions, depending on the specificities of the enzymes that catalyze the component reactions. The value of the overall free energy change and the overall equilibrium constant is determined by the stoichiometry of this coupling—that is, by the number of moles of ATP converted to ADP (or of ADP converted to ATP) per mole of Z that is produced or consumed. Since the equilibrium constant can be either greater than or less than 1, the ATP coupling determines which direction of conversion (A to Z, or Z to A) is thermodynamically favorable.

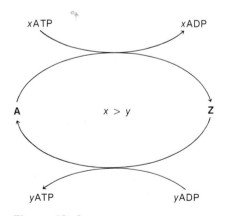

Figure 13–9
Schematic illustration of the organization of metabolic sequences into oppositely directed pairs. The two sequences result in opposite conversions. The values for the overall equilibrium constant are a function of the conversion and the number of ATP-to-ADP conversions to which each sequence is coupled.

Such pairs of oppositely directed sequences look like metabolic cycles. Indeed, if both sequences were simultaneously active to a significant degree, they would form a true cycle. But since the end products of one sequence are the starting materials for the other, such a cycle would have no metabolic consequences except that ATP would be wasted (that is, more ATP would be used in one sequence than is regenerated in the other). Because its cyclic operation would have no value, such oppositely directed paired sequences are often called _futile cycles_. The term _pseudocycle_ also is used to describe this type of cycle, because it almost never functions as a true cycle, owing to kinetic constraints.

When discussing the regulation of metabolism we will find it useful to think in terms of pseudocycles and those factors that prevent pseudocycles from operating in both directions. The oppositely directed sequences within a pseudocycle may be quite long or may be as short as one reaction in each sequence. Perhaps the most fundamental of all such pairs occurs in autotrophic organisms. The oxidation of sugars to carbon dioxide is the major source of energy in plants in the dark and in nongreen cells of plants at all times, just as it is in most aerobic heterotrophs. In the autotrophic pseudocycle, one sequence is the oxidation of sugars to CO_2, and the other is the photosynthetic conversion of CO_2 to carbohydrates.

In most metabolic conversions, all of the reactions of each sequence differ from those of the other. A major exception to this generalization is the glycolysis-gluconeogenesis pair. These two sequences catalyze the conversion of glucose to pyruvate and the conversion of pyruvate to glucose. They are, in terms of the amounts of conversion that occur, the most important pair of oppositely directed sequences in mammals. We will discuss them in the following chapter, where we will see that they differ from typical pairs of sequences in that most of the enzymes and intermediates are shared by the two sequences. Instead of one long pseudocycle, there are three small pseudocycles imbedded in the glycolysis-gluconeogenesis system. One of these, the interconversion of fructose-6-phosphate and fructose-1,6-bisphosphate, provides an especially instructive example of the importance of ATP coupling, because each "sequence" of the pseudocycle consists of only one reaction, and the "sequences" differ by one ATP-to-ADP conversion. The reactions are shown by two equations:

$$\text{Fructose-6-phosphate} + \text{ATP} \longrightarrow \text{fructose-1,6-bisphosphate} + \text{ADP}$$

$$\text{Fructose-1,6-bisphosphate} + H_2O \longrightarrow \text{fructose-6-phosphate} + P_i$$

Figure 13–10 shows these reactions in the form of a pseudocycle. The equilibrium constant for the hydrolysis of fructose-1-6-bisphosphate to fructose-6-phosphate is about 10^4. Thus this reaction will always be favorable at any feasible physiological ratio of the two compounds. The coupling of an

Figure 13–10
The interconversion of fructose-6-phosphate and fructose-1,6-bisphosphate. This reaction exemplifies the type of interconversion shown in Figure 13–9, where x = 1, y = 0, A = fructose-6-phosphate, and Z = fructose-1,6-bisphosphate.

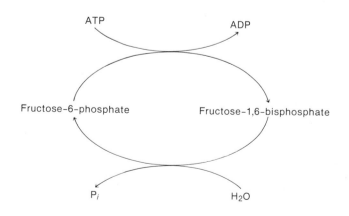

ATP-to-ADP interconversion changes the ratio of product to substrate under physiological conditions by a factor of approximately 10^8. Thus the ratio of fructose-6-phosphate to the bisphosphate could be about 10^4 if the phosphatase reaction were at equilibrium under physiological conditions, while the ratio of fructose-1-6-bisphosphate to fructose-6-phosphate could be about 10^4 if the phosphofructokinase reaction were at equilibrium. At any ratio of concentrations that can reasonably be expected in a metabolizing cell, both reactions are far from equilibrium and in the direction that is highly favorable for reaction.

To recapitulate: For nearly every metabolic conversion, two sequences with different ATP stoichiometries have evolved. In one sequence the equilibrium constant for the conversion of A to Z is large, and in the other the equilibrium constant for conversion of Z to A is large. In the next section we will see how pathways are regulated so that the rates of metabolic conversions respond to need.

REGULATION OF PATHWAYS

The two most important principles governing the regulation of pathways are *economy* and *flexibility*. Organisms must regulate their metabolic activities economically to avoid significant deficiencies or excesses of metabolic products. When there is a significant change in the environment, such as a shift in the concentration or kinds of nutrients, the organism must be flexible and alter its metabolism to suit the new growth conditions.

A number of different strategies are used to regulate metabolism. Indeed, a good deal of this text deals with problems of metabolic regulation at different levels. In the remainder of this chapter we will survey the basic modes of pathway regulation in general terms.

Enzyme Levels Are Regulated

The amount of any particular enzyme is regulated at the level of RNA and protein synthesis. The way in which this is done is consistent with the principles of flexibility and economy. Regulatory mechanisms control the supply of enzymes so that they are synthesized most when they are needed. For example, in *E. coli* the catabolic enzyme β-galactosidase is synthesized at high levels when the disaccharide lactose is present in the growth medium. The sole function of this enzyme is to degrade lactose to its constituent monosaccharides. The concentration of β-galactosidase is 1000 times higher when lactose is present in the growth medium than when it is absent. This enormous difference in enzyme level results from regulation of the synthesis of the enzyme by a mechanism that is discussed in Chapter 29. Enzyme level is also controlled in certain cases by regulation of the rate of degradation (see Chapter 28).

Not all enzymes are regulated at the level of synthesis or degradation. Enzymes that are produced at moderate levels all the time are said to be *constitutively* expressed. Enzymes that are constitutively expressed are usually involved in metabolic activities common to all cells. Examples include the enzymes involved in coenzyme synthesis and in RNA processing or modification.

Before leaving this topic we must say something about the patterns of regulation. The regulation of synthesis of the enzymes involved in the five-step conversion of chorismate to tryptophan presents a lucid example of how the level of functionally related enzymes is regulated for an anabolic pathway in *E. coli*. When tryptophan is abundant in the bacterial growth medium, the synthesis of all five enzymes uniquely associated with tryptophan synthesis is turned off. When tryptophan is lacking from the growth medium then the five enzymes are all synthesized at about the same high rate. This is a typical pattern in that the levels of all enzymes that are functionally related to the same pathway or process are coordinately regulated.

Enzyme Activity Is Regulated by Interaction with Regulatory Factors

The most common way of regulating metabolic activity is by direct control of enzyme activity. Enzyme activities are usually regulated by noncovalent interaction with small-molecule regulatory factors (see Chapter 10) or by a reversible covalent modification such as that brought about by phosphorylation or adenylation (Chapters 10 and 31) of an amino acid side chain. The effect of the regulatory factor in specific instances may be to increase or decrease the activity of the enzyme.

Regulatory Enzymes Occupy Key Positions in Metabolic Pathways

Enzymes that are susceptible to direct regulation occupy key positions in metabolic pathways. They are usually called regulatory enzymes to emphasize the fact that they have specific sites for binding regulatory factors, which are separate and distinct from their substrate binding sites. In a multienzyme pathway the first enzyme in the pathway is usually regulated, while the others are not (Figure 13–11a). In the case of CTP synthesis, discussed in Chapter 10, we saw that the enzyme aspartate carbamoyltransferase is negatively regulated by CTP. That is, excess CTP binds to the regulatory sites on this multiprotein enzyme, thereby inhibiting its activity. Aspartate carbamoyltransferase catalyzes the first step in a multistep pathway leading to CTP. Since CTP is the end product of the pathway, this type of regulation is referred to as _end-product inhibition_. End-product inhibition is very common for anabolic pathways.

The mechanism is clearly flexible and economic. If there is sufficient end product, there is no point in processing substrates down the pathway. To do so would be a waste of both energy and materials. Furthermore, it is most effective to block the first enzyme in the pathway; this by itself will rapidly reduce activity of the entire pathway. It would be redundant to block any other enzymes in the pathway, since no intermediates will be available

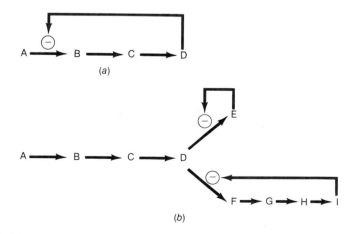

Figure 13–11
Two patterns for end-product inhibition. In end-product inhibition the first reaction in the pathway (a) or the first reactions after a branchpoint (b) are inhibited by the specific end products.

to these enzymes if the first step is blocked. Blocking only a middle enzyme in the pathway, instead of the first enzyme, would also be quite wasteful. The result would be accumulation of an intermediate that would serve no useful function. Indeed, such intermediates in excess frequently have harmful effects.

In a branched-chain pathway, end-product inhibition results in inhibition of the first enzyme after the branchpoint (Figure 13–11b). This arrangement leads to control of either pathway after the branchpoint. If the supply of the branchpoint substrate (D in Figure 13–11b) is limiting, the inhibition of one pathway after the branchpoint could increase the metabolic flow of the other pathway. This effect is often highly significant. A propos of this, anabolic sequences frequently arise as branchpoints from catabolic pathways (Figure 13–7). In such instances the control factors affect whether a catabolic intermediate will be further degraded or will serve as a substrate for biosynthesis.

Regulatory Enzymes Often Show Cooperative Behavior

It is clear from what was said in the previous section that control of the rates of metabolic sequences is most effective when it is exerted at branchpoints. As you might expect, the enzymes that catalyze reactions at branchpoints have evolved features to make the partitioning very sensitively responsive to metabolic signals. A striking feature of a typical enzyme at a branchpoint (the first enzyme in a biosynthetic sequence, for example) is that the reaction that it catalyzes is of high kinetic order with respect to its substrate (see Chapter 10). Whereas other enzymes along the pathway typically exhibit normal Michaelis kinetic responses to the concentrations of their substrates (this is often termed first-order or hyperbolic kinetics), branchpoint enzymes usually catalyze reactions of higher order (termed cooperative or sigmoid kinetics). A fourth-order response is quite common. The two types of behavior are shown in Figure 13–12. Any enzyme-catalyzed reaction must level off (to a kinetic order of zero) at high concentration because of saturation of the catalytic sites, but for the "normal Michaelis" enzyme the kinetic order at very low concentrations of substrate is one, whereas for the cooperative enzyme it has a higher value, frequently four.

A major advantage of this higher order is that it makes the system very sensitive to changes in substrate concentration. If the first reaction in the sequence responds to the fourth power of the concentration of its substrate

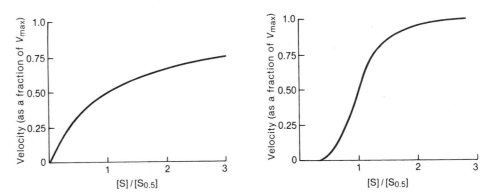

Figure 13–12
Reaction velocity as a function of substrate concentration for a first-order enzymic reaction (left) and for a cooperative enzyme with fourth-order kinetics (right). Substrate concentration is expressed as $[S]/[S_{0.5}]$ and velocity as a fraction of maximum velocity. At a $[S]/[S_{0.5}]$ value of 1 the substrate concentration is equal to $[S_{0.5}]$ and the reaction velocity is half of the maximum velocity.

(the starting material for the sequence), the flow of material into the sequence will be much more sensitive to small changes in the concentration of the starting material than if the reaction were first order with respect to substrate, and intermediates in catabolic sequences will not be diverted to biosynthetic use unless they are in good supply.

Perhaps an even more important advantage of high kinetic order is that the regulatory responses of the enzyme are greatly strengthened. We have discussed kinetic regulation of enzymes at metabolic branchpoints by negative feedback, in which the end product of the sequence binds to a regulatory site and decreases the affinity of the catalytic site for the substrate (the branchpoint metabolite). When the enzyme is constructed so as to have cooperative interactions between sites, the sensitivity of negative feedback control is greatly sharpened. This matter is considered further in Box 13–B.

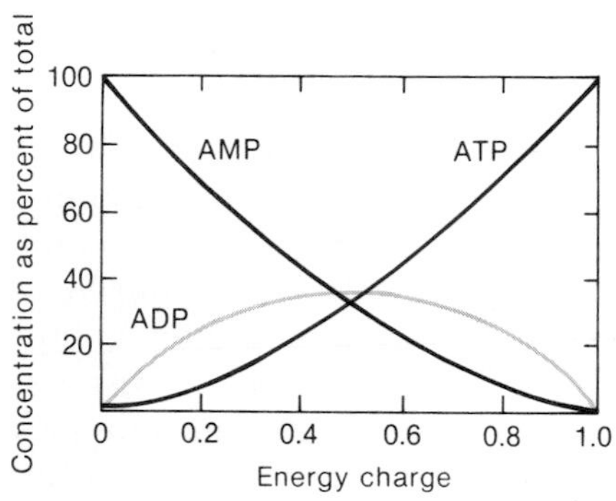

Figure 13–13
Relative concentrations of ATP, ADP, and AMP as a function of the adenylate energy charge. The adenylate kinase reaction was assumed to be at equilibrium, and a value of 1.2 was used for its effective equilibrium constant in the direction shown in the equation in the text.

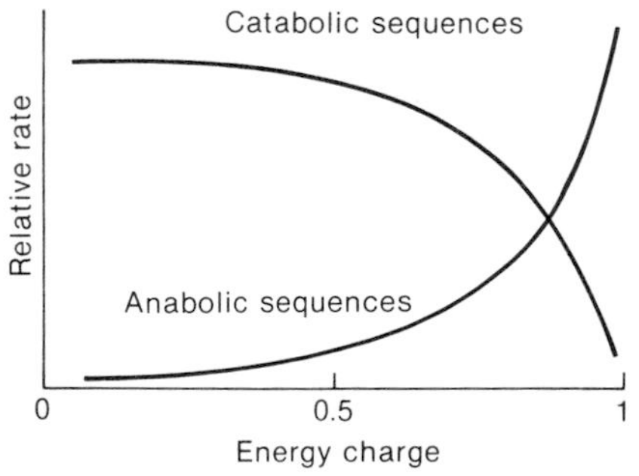

Figure 13–14
Variation in reaction rates as a function of the energy charge. As the energy charge increases, the rate of catabolic reactions decreases. Meanwhile, the rate of anabolic reactions increases. The combined effect is to stabilize the energy charge at a value around 0.9.

Both Anabolic and Catabolic Pathways Are Regulated by the Energy Charge

It is not always an advantage for the concentration of an amino acid or other end product to be the only factor that determines the rate at which that end product is synthesized. If a cell is starved for energy or catabolic intermediates, it may not be able to afford to synthesize amino acids, even if their concentrations are quite low. Continued biosynthesis under such circumstances might deplete the already scarce supply of energy to the point where essential functions, such as maintenance of concentration gradients across membranes, would be impaired. It would clearly be advantageous for the rate of biosynthesis to be regulated by the general energy status of the cell as well as the need for the specific end product.

Since it is the adenine nucleotide system ATP-ADP (and occasionally ATP-AMP) that couples energy into biosynthetic sequences, we might expect that this system would also supply the necessary signal indicating the energy status of the cell, to be sensed by the kinetic control mechanisms of biosynthesis. It is helpful to have a term that quantitatively expresses the energy status of the cell. The term used is the *energy charge*. This is defined as the effective mole fraction of ATP in the $\overline{\text{ATP/ADP/AMP}}$ pool:

$$\text{Energy charge} = (\text{ATP} + 0.5\ \text{ADP})/(\text{ATP} + \text{ADP} + \text{AMP})$$

In this equation the 0.5 in the numerator takes into account the fact that ADP is about half as effective as ATP at carrying chemical energy. Thus two ADPs can be converted into one ATP and one AMP by a reaction with an equilibrium constant of about 1.

$$\text{ATP} + \text{AMP} \rightleftharpoons 2\ \text{ADP}, \quad K_{eq} \approx 1$$

The values for the energy charge could conceivably vary from 0 to 1, as illustrated in Figure 13–13. In fact, however, the values for real cells are usually held within very narrow limits. The result is a general stabilizing effect (homeostatic effect) on the cellular metabolism. In much the same way, voltage regulation is necessary for the effective operation of many electronic devices. Now let us see how the limits are maintained.

Several reactions in anabolic and catabolic pathways have been found to respond to variation in the value of the energy charge. As might be expected, the enzymes in catabolic pathways respond in a direction opposite to that of enzymes in anabolic pathways. The two responses are compared in Figure 13–14. It is clear that if catabolic sequences, which lead to the generation of ATP, respond as shown by the upper curve, and biosynthetic sequences,

BOX 13-B

Cooperative Binding of Substrate and the Partitioning of Substrate at Branchpoints

The accompanying illustration compares competition at a metabolic branchpoint between two first-order enzymes with that between two cooperative enzymes. For simplicity, we will use A to designate both the first enzyme in the branch pathway and the reaction catalyzed by that enzyme. Similarly, we will use B for the enzyme that catalyzes the next reaction in the catabolic sequence after the branchpoint and also for the reaction that it catalyzes. In this model we will assume that the responses of the enzymes in the catabolic pathway are not changed. We will also assume for the present that V_{max} values are the same for both enzymes, although this will usually not be true. If the value of $[S_{0.5}]$* for the first enzyme in a biosynthetic pathway (enzyme A) is initially twice as high as that of the enzyme in the catabolic pathway with which it competes (enzyme B) the rate of reaction A will be smaller than that of reaction B, whether the reactions are of first order or cooperative. The figures show, however, that the difference between rates A and B is much larger in the case of the cooperative enzymes, especially at relatively low concentrations of substrate.

*$[S_{0.5}]$ is the substrate concentration that results in half-maximal velocity. It is equal to the Michaelis constant for enzymes that follow simple Michaelis kinetics.

Now suppose that the concentration of the end product of the sequence (an amino acid) falls because it is being used more rapidly, as a result of an increase in the rate of protein synthesis. Because binding of the end product to regulatory sites on enzyme A is an equilibrium process, the amount bound will decrease, and the affinity of the enzyme for its substrate will increase (the value of $[S_{0.5}]$ will decrease). For the sake of this comparison we will assume that $[S_{0.5}]$ changes by a factor of 4, so that $[S_{0.5}]$ for enzyme A is now half that of enzyme B. Now the rate of reaction A (curve A′) is larger than that of reaction B in both cases, but again the ratio of rates is much higher when the reactions are cooperative than when they are of first order. The ratios are greatest in the lower half of the figure, which is usually the region of metabolic interest. Because the rates of most if not all metabolic sequences can increase severalfold in response to metabolic need, it seems probable that under normal or unstressed conditions most enzymes *in vivo* operate at rates between about 10% and 30% of their maximal rates. Some rates are much lower than that, as shown by the fact that the rates of some metabolic conversions can increase by factors of over 100 under stress. Evidently in such cases the normal rate must be less than 1% V_{max}.

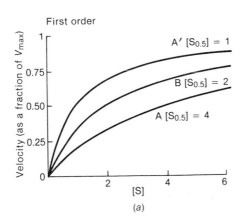

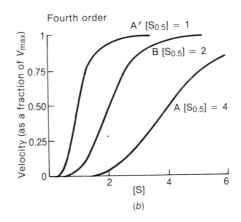

Schematic illustration of the effect of a fourfold change in the $[S_{0.5}]$ value for one of a competing pair of enzymes. Left graph, first-order enzymes; right graph, fourth-order enzymes. Velocity is expressed as a fraction of maximal velocity. The substrate concentration is in arbitrary units, and the values of $[S_{0.5}]$ are: curve A, 4; curve B, 2; curve A′, 1.

which use ATP, respond as shown by the lower curve, then the charge will be strongly stabilized at a value of about 0.9, where the two curves intersect. A tendency for the charge to fall will be resisted by the resulting general increase in the rate of catabolic sequences and the general decrease in the rate of biosynthetic sequences. A tendency for the charge to rise will be resisted by the opposite effects.

Regulation of Pathways Involves the Interplay of Kinetic and Thermodynamic Factors

The general properties of paired unidirectional sequences and their advantages to the organism are in one sense very simple. But on further consideration, we see that they depend on a rather complex interplay between kinetic and thermodynamic factors and effects.

In any kind of negative feedback system, cause and effect relationships are circular. It would be meaningless to ask whether the air temperature in a house, as sensed by a thermostat, determines the rate at which fuel is fed to the furnace or whether the rate at which fuel is supplied determines the temperature. Of course, each contributes to determining the other. Such circularity is evident at many levels in almost all aspects of the chemical activities of a living cell or organism. Homeostatic stabilization of such properties as concentrations, temperature, pH, etc., in the face of changes in external conditions or metabolic fluxes has long been recognized to be a fundamental characteristic of life, and the concept of negative feedback is fundamental to our understanding of such stabilization. We see the circularity of negative feedback at a simple level in the response of a biosynthetic sequence to the concentration of its end product. The relationships are strictly analogous to those that apply to a thermostatic system. The concentration of the end product is a major factor controlling the rate of synthesis, and the rate of synthesis is a major factor controlling the concentration of the end product.

For pairs of oppositely directed metabolic sequences, greater conceptual sophistication is needed than for simple feedback control of a synthetic pathway. In our discussion of pseudocycles thus far, we have tacitly assumed that the ATP/ADP ratio remains at a value very far from equilibrium. But of course, in the absence of highly effective controls, the ATP/ADP ratio would not remain far from equilibrium; it would rapidly approach its equilibrium value. What are the controls that keep this from happening? The answer is the enzymes that catalyze first steps in pseudocycles—the same controls that we have already discussed.

As illustrated in Figure 13–14, rates of reactions that control sequences in which there is net conversion of ATP to ADP (such as A to Z in our generalized pseudocycle) are high when the energy charge (or the ATP/ADP ratio) is high, and they decrease sharply with a decrease in those parameters. Rates of regulatory enzymes in sequences in which ATP is regenerated (see Figure 13–14) are high when the energy charge or ATP/ADP ratio is low, and decrease sharply as the charge increases. The curves, as we saw earlier, intersect at an energy charge value of about 0.9 (which corresponds to an ATP/ADP ratio of 5). The kinetic effects illustrated by Figure 13–14 play the major role in stabilizing the energy charge and the ATP/ADP ratio. From that consideration we see that the kinetic responses of these enzymes are responsible for the thermodynamic properties of the ATP-ADP system *in vivo*. So kinetic control of rates of conversion depends on the value of the ATP-ADP ratio being far from equilibrium (that is, on thermodynamic factors); but at the same time the ATP/ADP ratio itself depends on reciprocal regulation of oppositely directed sequences (that is, on kinetic factors).

As we have seen, the characteristic and essential feature of paired, oppositely directed sequences is that they differ in their ATP stoichiometries. But that difference is meaningful only because the ATP/ADP ratio is far from equilibrium in the cell. If ATP, ADP, and P_i were at equilibrium, it would not matter how many ATP-to-ADP conversions were coupled to any metabolic sequence. For any system at equilibrium ΔG is zero, and coupling to another reaction or sequence would have no effect on the free energy change or the position of equilibrium of the reaction or sequence. So the differential ATP stoichiometry would mean nothing if it were not for the kinetic controls that hold the ATP/ADP ratio at a steady-state value far from equilibrium. But kinetic regulation could not control directions of conversions and thus regulate the balance between ATP utilization and regeneration if it were not for the thermodynamic difference between the sequences of each pair, which depends on the ATP/ADP ratio. Neither the thermodynamic nor the kinetic features can be said to be more fundamental than the other. Metabolic correlation and control, and hence life, depend on an intricate interplay between thermodynamic and kinetic factors.

The concept of a steady state is very important in biochemistry. Biological steady states are of two types. At a given flux through a pathway, the concentration of each intermediate is determined by the amounts and properties of the enzyme that produces it and especially of the enzyme that catalyzes its further reaction. This *passive steady state* is similar to those found in nonbiological systems. The interplay of thermodynamic and kinetic regulatory factors that we have discussed leads to a different and more specifically biological type of steady state—one in which there are stable concentrations of the end products of biosynthetic sequences and stable ratios of coupling factors such as ATP/ADP. Such *active steady states* are in a sense more fundamental than the passive steady states seen within each sequence. Thus if the rate of protein synthesis should increase, negative feedback mechanisms will cause the rates of synthesis of amino acids to increase so as to hold the concentrations of the amino acids nearly constant. The concentrations of the intermediates of each biosynthetic sequence will, however, adjust to a new steady state that depends on the flux through the sequence.

In a textbook it is necessary to consider reaction sequences (or types of reaction sequences) individually. For example, we treat amino acid synthesis and amino acid degradation in separate chapters. But for each amino acid there is a pair of oppositely directed sequences of the type discussed here, with reciprocal controls. Even in this overview chapter, where the main goal is to emphasize interrelationships and correlations, we have had to discuss various aspects of metabolic functional regulation individually. But in the cell, different sequences and a wide variety of regulations and correlations occur continuously and simultaneously, and they would have no meaning or significance in isolation. It is obviously impossible to discuss everything at the same time, but in metabolism everything is related to nearly everything else. The direction and rate at which a metabolic conversion proceeds at any given time depends on the directions and rates of other conversions, and in turn may affect the directions and rates of some of those conversions, as well as others. In reading later chapters in this book, and whenever you think about any aspect of biochemistry or physiology, it will be very useful to ask yourself frequently how that structure or activity interacts with other aspects of the activity of the cell, tissue, or organism, and how it benefits the organism.

SUMMARY

All cells need energy and starting materials for synthesis. Ultimately these are supplied by autotrophic organisms, especially green plants; in plants the starting materials are made from CO_2, and the supply of chemical energy and reducing power is dependent on the absorption of light energy. In a heterotrophic organism, the role of catabolism is to supply those basic needs from conversions (usually oxidation) of foodstuffs.

Metabolic chemistry is characterized by functionality. Each reaction is important because of its participation in a sequence of reactions, and each sequence interacts functionally with other sequences.

Sequences may be broadly classified into two main types: biosynthetic, or anabolic, and degradative, or catabolic. Anabolic sequences are usually energy-requiring (endergonic) and catabolic sequences are usually energy-producing (exergonic).

Metabolic regulatory mechanisms have evolved so as to stabilize concentrations of key metabolites over a broad range of conditions. In the course of those stabilizations, the rates of metabolic conversions are varied over wide ranges.

Energy is coupled from catabolic sequences to the energy-requiring activities of a cell by the ATP-ADP system, which uniquely interacts with all sequences in the cell.

The stoichiometry of coupling to ATP-to-ADP conversions determines the overall equilibrium constant of a sequence and therefore the direction of conversion that will be thermodynamically favorable. Any conversion can be made favorable by coupling to an appropriate number of ATP-to-ADP conversions.

Metabolic sequences occur in pairs. Because of different ATP stoichiometries, each sequence in a pair is unidirectional, and the two are oppositely directed. It is through reciprocal kinetic control of the reaction rates for the sequences in a pair that the direction of conversion is determined.

Regulatory enzymes respond to signals in such a way that rates of biosyntheses are controlled by the need for the product, as indicated by negative feedback, and by the availability of energy and starting materials.

Metabolism is regulated primarily by adjustment of the ratios by which intermediates are partitioned at metabolic branchpoints. A high kinetic order is advantageous for sensitivity of partitioning, and enzymes at branchpoints usually catalyze reactions with high orders.

SELECTED READINGS

Atkinson, D. E. *Cellular Energy Metabolism and Its Regulation.* New York: Academic Press, 1977. A general discussion, covering some topics in this and the two following chapters in somewhat greater depth than the treatment in this book.

Cohen, P. *Control of Enzyme Activity,* 2nd Ed. London and New York: Chapman and Hall, 1983. Brief discussion of some types of regulation of activity of metabolic enzymes, emphasizing regulation by covalent modification of the enzymes.

Herman, R. H., Cohn, R. M., and McNamara, P. D. *Principles of Metabolic Control in Mammalian Systems.* New York and London: Plenum Press, 1980. Discusses various aspects of metabolic control in mammals, mainly at the intracellular level. Many references.

Hochachka, P. W., and Somera, G. N. *Biochemical Adaptation.* Princeton: Princeton U. Press, 1984. An excellent and extensive discussion of how biochemical processes, including many discussed in this book, are adapted by various types of organisms in fitting themselves for survival under specific and often difficult conditions.

PROBLEMS

1. Which are more fundamental in regards to all life, thermodynamic or kinetic considerations? Explain.

2. True or false:
   a. Most enzymes *in vivo* operate at or around V_{max}.
   b. End-product inhibition usually occurs at the first enzyme before a branchpoint.
   c. Catabolic pathways tend to converge, while anabolic ones diverge.

3. a. What is the primary significance to the cell of cooperativity?
   b. Read about aspartate carbamoyltransferase in Chapter 10. How does the effect of CTP on ACTase illustrate your answer to part (a)?
   c. Explain how the effect of ATP on ACTase makes good sense for the cell.

4. This chapter deals with concepts. Since trying to explain something is the best way to learn it, think how you might explain the following points to a young cousin who is a whiz in freshman chemistry but has had little exposure to biology. Putting the concepts into your own words, in a way that would be clear to someone else, should clarify and reinforce them in your own mind.

   a. The organization and correlation of metabolism can be discussed in terms of steering. Enzymes steer reactions along reaction pathways with lower activation energy than the corresponding noncatalyzed reaction; enzymes steer extended sequences along specific pathways, selecting at each step between several possible reactions; and metabolism is regulated by steering different fractions of the flux through a branchpoint into the alternative branches.
   b. Although catalysts can only increase the rate at which a system approaches equilibrium and cannot determine the direction of a reaction, modulation of the kinetic properties of enzymes can determine the directions of metabolic conversions.
   c. A chemical system that contains many reactive compounds and catalysts for reactions between those compounds would be expected to move rapidly toward equilibrium. The chemical composition of the mixture

would, of course, change in the process. Yet in a living cell, which contains several hundred compounds and several hundred exceedingly active catalysts, the chemical composition tends to be very nearly constant. Change is resisted even when the environment changes.

d. The feedback interactions by which most biosynthetic pathways are regulated are in principle similar to the system that regulates the level of water in the tank that supplies water for flushing a toilet.

e. When two enzymes compete for the same branchpoint metabolite, it is easily seen that a feedback metabolite will be more effective if the enzyme binds it cooperatively. However, can you explain why the system becomes much more sensitive to the effects of the feedback modifier if the enzyme binds the substrate of the reaction (the branchpoint metabolite) cooperatively?

f. In most systems that are discussed in freshman chemistry, if the concentrations of component compounds are constant the system is probably at equilibrium, and no net chemical change is occurring. In a living cell, the rates of chemical change may be very high even though the concentrations of nearly all of the components of the cell are virtually constant.

g. If a sequence of nonenzymatic reactions

$$A \longrightarrow B \longrightarrow C \longrightarrow D \longrightarrow P,$$

were at equilibrium and P began to be removed, the concentrations of the intermediates B, C, and D would decrease. In a biosynthetic pathway, if the rate of utilization of the product increases, then the concentrations of the intermediates will rise. Can you make the reasons for this difference clear to your cousin?

14

GLYCOLYSIS, GLUCONEOGENESIS, AND THE PENTOSE PHOSPHATE PATHWAY

In the last two chapters we considered the importance of thermodynamic and kinetic factors in metabolism and the strategies that have evolved to support metabolism. We saw that ATP is the immediate source of energy for most conversions that would otherwise be thermodynamically unfavorable. We saw also that one of the primary functions of catabolism is to regenerate ATP from ADP. Carbohydrates play a central role in energy metabolism in most cells, and polymers of glucose are important energy storage compounds that can be used for ATP production when needed. In the next four chapters our central concern will be with the relationship between carbohydrate metabolism and energy production.

In this chapter we will discuss the interconversions among glucose and other hexoses, the conversion of glucose to glycogen and starch, and especially the breakdown of hexoses to pyruvate and lactate and the synthesis of glucose and other hexoses from those three-carbon acids. Those conversions are of primary importance in the generation, use, and storage of metabolically useful energy. They also supply carbon skeletons for the synthesis of most other metabolites in the cell, from small cofactors to structural macromolecules.

The breakdown of glucose to pyruvate does not require oxygen, and it occurs equally readily in anaerobic and aerobic organisms. In the next chapter we will see how pyruvate is oxidized to CO_2 in the presence of oxygen. This oxidation results in the direct synthesis of further ATP and also in the reduction of nicotinamide adenine dinucleotide (NAD^+) and flavine adenine dinucleotide (FAD). In Chapter 16 we will see how the reduced forms of those electron carriers (NADH and $FADH_2$) donate electrons for additional synthesis of ATP. In Chapter 17 we will turn to a quite different topic, the utilization of light energy to drive the regeneration of ATP.

GLYCOLYSIS

Whether aerobic or anaerobic, the process of glucose catabolism involves a reaction sequence that is without doubt the single most ubiquitous pathway in all energy metabolism. _Glycolysis_ occurs in almost every living cell. It is generally regarded as a primitive process. Thus it occurs in the cytosol rather than being compartmentalized into a specific organelle within the eukaryotic cell. Furthermore, it is thought to have arisen early in biological history, before the advent of eukaryotic organelles and perhaps even before oxygen was a prominent component in the atmosphere.

The glycolytic pathway was the first major metabolic sequence to be elucidated. Most of the decisive work was done in the 1930s by the German biochemists G. Embden, O. Meyerhof, and O. Warburg, two of whom gave the sequence its alternative name, the _Embden-Meyerhof pathway_.

The glycolytic pathway appears in detail in Figure 14–1 (also see Table 14–1) and is shown in the context of overall chemotrophic energy metabolism in Figure 15–1. The essence of the process is suggested by the name, since glycolysis comes from the Greek roots _glykos_, meaning sweet, and _lysis_, meaning loosing. Literally, then, glycolysis is the loosing or splitting of something sweet, which is, of course, the starting sugar. From Figure 14–1 it is clear that the actual splitting occurs at the aldolase step. It is at this point that a six-carbon sugar is cleaved to yield two three-carbon compounds, one of which, glyceraldehyde-3-phosphate, is the only oxidizable molecule in the whole pathway. Subsequent to the cleavage, two successive ATP-generating steps occur. These represent the energy payoff of the process, since they are the only ATP-yielding reactions of the pathway under anaerobic conditions.

TABLE 14–1
Reactions, Enzymes, and Free Energies for Steps in the Glycolytic Pathway

Step	Reaction	Enzyme	$\Delta G^{\circ\prime}$
1	Glucose + ATP $\longrightarrow$ glucose-6-phosphate + ADP + H^+	Hexokinase	−4.0
2	Glucose-6-phosphate $\rightleftharpoons$ fructose-6-phosphate	Phosphohexose isomerase	+0.4
3	Fructose-6-phosphate + ATP $\longrightarrow$ fructose-1,6-bisphosphate + ADP + H^+	Phosphofructokinase	−3.4
4	Fructose-1,6-bisphosphate $\rightleftharpoons$ dihydroxyacetone phosphate + glyceraldehyde-3-phosphate	Aldolase	+5.7
5	Dihydroxyacetone phosphate $\rightleftharpoons$ glyceraldehyde-3-phosphate	Triose phosphate isomerase	+1.8
6	Glyceraldehyde-3-phosphate + P_i + NAD^+ $\rightleftharpoons$ glycerate-1,3-bisphosphate + NADH + H^+	Phosphoglyceraldehyde dehydrogenase	+1.5
7	Glycerate-1,3-bisphosphate + ADP $\rightleftharpoons$ glycerate-3-phosphate + ATP	3-Phosphoglycerate kinase	−4.5
8	Glycerate-3-phosphate $\rightleftharpoons$ glycerate-2-phosphate	Phosphoglyceromutase	+1.1
9	Glycerate-2-phosphate $\rightleftharpoons$ phosphoenolpyruvate + H_2O	Enolase	+0.4
10	Phosphoenolpyruvate + ADP + H^+ $\longrightarrow$ pyruvate + ATP	Pyruvate kinase	−7.5

Net reaction:

$$C_6H_{12}O_6 + 2\ NAD^+ + 2\ ADP + 2\ P_i \longrightarrow 2\ C_3H_4O_3 + 2\ NADH + 2\ H^+ + 2\ ATP + 2\ H_2O$$

Note: Fructose 1,6 bisphosphate = Fructose 1,6 diphosphate
glycerate 1,3 bisphosphate = 1,3 diphosphoglycerate
glycerate 3 phosphate = 3 phosphoglycerate
glycerate 2 phosphate = 2 phosphoglycerate.

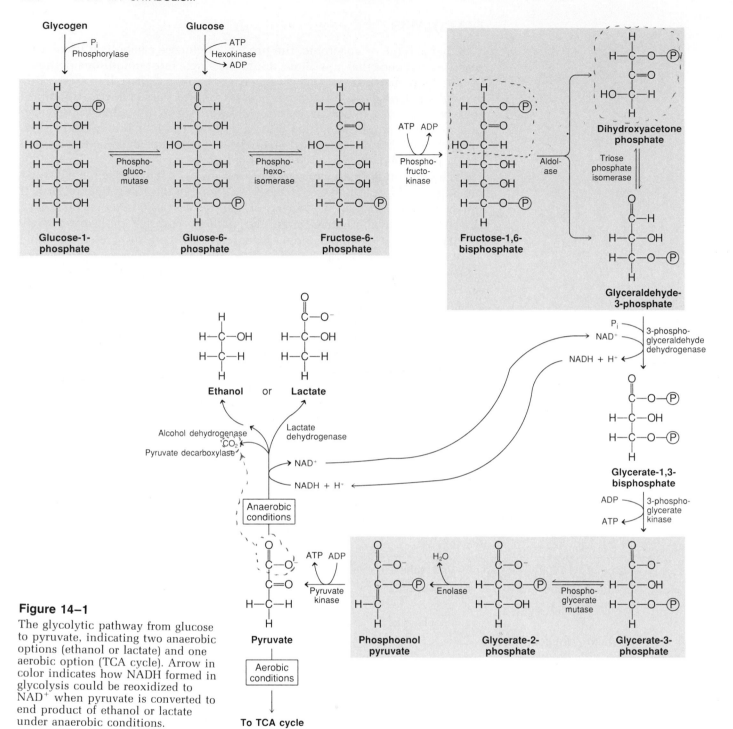

Figure 14–1

The glycolytic pathway from glucose to pyruvate, indicating two anaerobic options (ethanol or lactate) and one aerobic option (TCA cycle). Arrow in color indicates how NADH formed in glycolysis could be reoxidized to NAD^+ when pyruvate is converted to end product of ethanol or lactate under anaerobic conditions.

Three Hexoses Play a Central Role in Glycolysis

The three hexose phosphates shown in Figures 14–1 and 14–2, *glucose-1-phosphate, glucose-6-phosphate*, and *fructose-6-phosphate*, are readily interconvertible and thus constitute a single *metabolic pool*. The pool can be replenished by generation of any of its three components, depending on conditions. Glucose-1-phosphate is the first product in the utilization of storage polysaccharides; glucose-6-phosphate is the first hexose phosphate formed when free glucose is metabolized; and fructose-6-phosphate is the first hexose phosphate formed when carbohydrate is made *de novo* (from noncarbohydrate precursors).

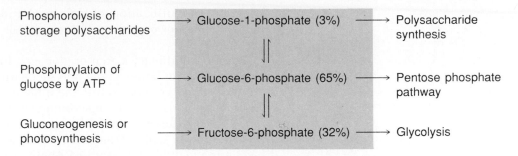

Phosphorolysis of storage polysaccharides ⟶ Glucose-1-phosphate (3%) ⟶ Polysaccharide synthesis

Phosphorylation of glucose by ATP ⟶ Glucose-6-phosphate (65%) ⟶ Pentose phosphate pathway

Gluconeogenesis or photosynthesis ⟶ Fructose-6-phosphate (32%) ⟶ Glycolysis

Figure 14–2
The hexose monophosphate pool. The equilibrium percentages of the three hexose monophosphates are indicated. Horizontal flow arrows indicate major routes of replenishment and utilization of the three hexose monophosphates.

Production of carbohydrates *de novo* is of two types. Gluconeogenesis, the synthesis of sugars from such precursors as the three-carbon compound lactate, is a process that can be carried out by many kinds of cells, heterotrophic as well as autotrophic, including the cells of several organs of our bodies. Only autotrophic organisms are capable of producing sugars *de novo* from CO_2 (see Chapter 17). In both cases fructose-6-phosphate is the first member of the hexose monophosphate pool to be formed. We will now see how these three members of the hexose monophosphate pool are formed and interconverted.

Phosphorylase Converts Storage Carbohydrates to Hexose Phosphates

The first step in the cell's use of starch or glycogen is the removal of a terminal glucose residue, a reaction catalyzed by the enzyme *phosphorylase*. The reaction involves a displacement at C-1 of the terminal residue, with an incoming phosphate group replacing the remainder of the polysaccharide molecule (Figure 14–3). The product generated is therefore glucose-1-phosphate. Phosphorylase is part of an elaborate control system by which the

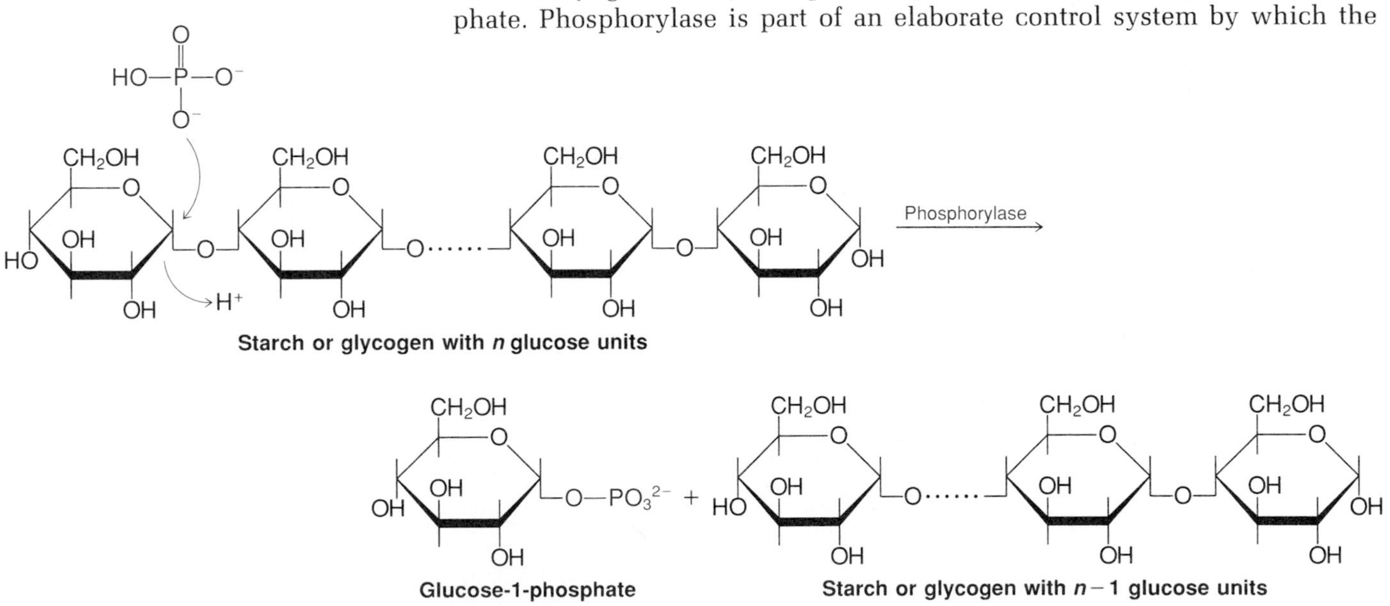

Starch or glycogen with *n* glucose units

Glucose-1-phosphate

Starch or glycogen with *n* − 1 glucose units

Figure 14–3
The reaction catalyzed by phosphorylase. An oxygen of a phosphate ion attacks C-1 of the terminal glucosyl unit of starch or glycogen, displacing the macromolecule and generating glucose-1-phosphate. The free proton shown originates from water. The catalytic role of the enzyme is not indicated.

production and utilization of polysaccharides are regulated to meet the needs of the cell or organism. Those controls will be discussed later in this chapter.

Digestion of dietary starch or glycogen in the intestine follows a different course. Interior bonds of the polysaccharide are hydrolyzed by a number of enzymes that are secreted into the intestine by the pancreas and the intestinal mucosal cells, and the macromolecule is gradually split into progressively smaller fragments. Individual residues are not removed from the polysaccharide or its major fragments; the degradation proceeds through the disaccharide maltose, glucosyl α(1,4)-glucoside. The final step in the intestine is the hydrolysis of maltose to free glucose.

The difference between the modes of hydrolysis inside cells and in the intestine reflects biological needs and functions. Phosphorylated sugars, like other polar compounds, cross biological membranes much less readily than free sugars. In the cell, phosphorylated hexoses are desirable because they will not be lost by diffusion out of the cell. In contrast, the very reason for hydrolysis of polysaccharides in the intestine is to make their component residues available for absorption into the body, so unphosphorylated glucose is preferable.

Hexokinase and Glucokinase Convert Free Sugars to Hexose Phosphates

Free glucose—for example, that obtained by mammals from the diet through intestinal hydrolysis of lactose, sucrose, glycogen, or starch—is brought into the hexose phosphate pool through the action of *hexokinase*. This enzyme catalyzes phosphorylation at the oxygen attached to C-6 of glucose (Figure 14–4). As usual in metabolism, the source of the phosphate group is ATP:

$$\text{Glucose} + \text{ATP} \longrightarrow \text{glucose-6-phosphate} + \text{ADP} \qquad (1)$$

The standard free energy change is about −5 kcal/mole, and the equilibrium constant is about 4000. Hexokinase is thus able to catalyze the conversion of

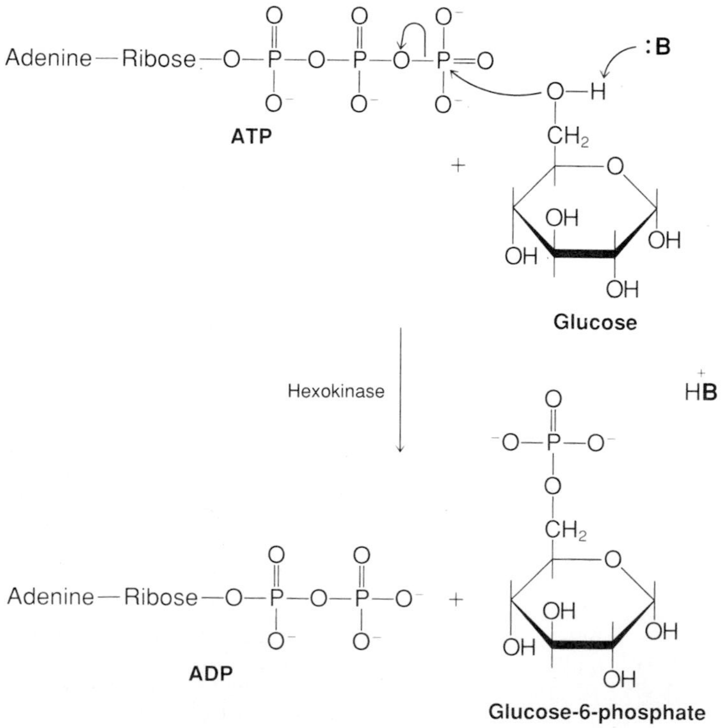

Figure 14–4

The reaction catalyzed by hexokinase. Attack on the terminal phosphorus atom of ATP is probably facilitated by proton removal by a negatively charged group in the catalytic site of the enzyme (B: in the figure). It is also facilitated by the fact that the terminal phosphate of ATP is an excellent leaving group.

glucose to glucose-6-phosphate even when the concentration of glucose is very low.

Of course, as we saw in Chapter 12, equilibrium considerations indicate only the potential for a reaction to occur; the potential can be converted to reality only by appropriate kinetic properties. That is, the favorable thermodynamic properties of the hexokinase reaction would be of limited use to the cell if the Michaelis constant of the enzyme were large, since the thermodynamic potential to use glucose at low concentrations could not be translated into actual use if the enzyme were unable to bind glucose at those low concentrations. Hexokinase has the necessary high affinity for glucose. The Michaelis constants for hexokinases from various sources range from 10 to 20 μM. Thus, by the expenditure of ATP, hexokinase can convert glucose in the micromolar range to glucose-6-phosphate in the millimolar range for utilization in metabolism.

Some kinds of bacteria, including *E. coli*, contain complex enzymic and transport systems that phosphorylate glucose in the process of bringing it into the cell (Chapter 34). These systems use phosphoenolpyruvate rather than ATP as the source of the phosphoryl group. They can thus bring glucose into the cell, protect it by phosphorylation against loss by diffusion, and produce a relatively high concentration of glucose phosphate inside the cell even when the external concentration of glucose is very low.

The liver contains another enzyme, named *glucokinase*, that catalyzes the same reaction as hexokinase. Its Michaelis constant (about 10 mM) is 1000 times as large as that of hexokinase; thus glucokinase can function only when the concentration of glucose is relatively high. Probably this enzyme is active only when blood glucose is high and the liver is taking up glucose for conversion to glycogen.

Phosphoglucomutase Interconverts Glucose-1-phosphate and Glucose-6-phosphate

The enzyme *phosphoglucomutase* catalyzes the interconversion of glucose-1-phosphate and glucose-6-phosphate (Figure 14–5). Hemiacetal phos-

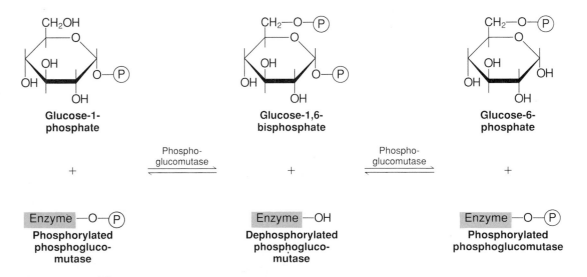

Figure 14–5
The reaction catalyzed by phosphoglucomutase. The enzyme apparently can bind glucose phosphates in two ways, allowing it to transfer phosphoryl groups to and from either the oxygen atom at C-1 or the oxygen at C-6. The direction of reaction will be driven by mass action, depending on the relative concentrations of glucose-1-phosphate and glucose-6-phosphate.

phates, such as glucose-1-phosphate, are thermodynamically less stable than ordinary phosphate esters, such as glucose-6-phosphate. The difference is small, however, and the equilibrium constant for the conversion of glucose-1-phosphate to glucose-6-phosphate is only about 19.

Phosphoglucomutase does not catalyze the direct migration of a phosphoryl group between the two ends of the glucose molecule. The transfer is indirect, by way of a phosphorylated enzyme and glucose-1,6-bisphosphate. The reactions are

$$\text{Glucose-1-phosphate} + \text{Enz-P} \rightleftharpoons$$
$$\text{glucose-1,6-bisphosphate} + \text{Enz} \quad (2)$$

$$\text{Glucose-1,6-bisphosphate} + \text{Enz} \rightleftharpoons$$
$$\text{glucose-6-phosphate} + \text{Enz-P} \quad (3)$$

Sum: $\text{Glucose-1-phosphate} \rightleftharpoons \text{glucose-6-phosphate}$ $\quad (4)$

where Enz represents phosphoglucomutase, and Enz-P represents phosphoglucomutase phosphorylated at a specific serine residue in the active site. The enzyme cycles between the free and the phosphorylated states at each conversion of glucose-1-phosphate to glucose-6-phosphate (or of glucose-6-phosphate to glucose-1-phosphate).

Glucose-1,6-bisphosphate is a catalyst in the sense that its concentration does not change as a result of the reaction. However, it is an unusual catalyst in that it is also an intermediate in the sequence, since each molecule of substrate follows the pathway

$$\text{Glucose-1-phosphate} \rightleftharpoons \text{glucose-1,6-bisphosphate}$$
$$\rightleftharpoons \text{glucose-6-phosphate} \quad (5)$$

The net synthesis of typical pathway intermediates presents no problem; their concentrations are related to the properties of the enzymes with which they interact, and can rise or fall depending on the flux through the sequence and on other factors. If the concentration of a typical metabolite should happen to fall to zero, the intermediate could be brought back to its normal concentration by the reactions of the sequence itself. From Equations (2) and (3), it is clear that the situation is different for glucose-1,6-bisphosphate. The sum of the concentrations of the bisphosphate and the phosphorylated enzyme is constant, and cannot increase as a result of the reactions of the sequence. Thus a reaction, separate from the reactions of the sequence, is needed by which the bisphosphate can be generated to meet the needs of growth and to compensate for hydrolytic and other losses of glucose-bisphosphate itself or of the phosphorylated enzyme. That accessory reaction is catalyzed by the enzyme *phosphoglucose kinase*:

$$\text{Glucose-1-phosphate} + \text{ATP} \longrightarrow \text{glucose-1,6-bisphosphate} + \text{ADP} \quad (6)$$

The special role of glucose-1,6-bisphosphate is an example of the general fact that in science definitions often become fuzzy when applied to actual situations. That should cause no concern. Definitions, by helping us to relate phenomena, are useful tools to aid in understanding the systems that we study. They become impediments to understanding only if we allow ourselves to think that the definition is important in its own right, and to try to fit phenomena to definitions rather than definitions to phenomena. In this case, glucose-1,6-bisphosphate is undeniably an intermediate, but it is also a catalyst or cofactor. Because of its special nature, it is not usually listed as a component of the hexose phosphate pool.

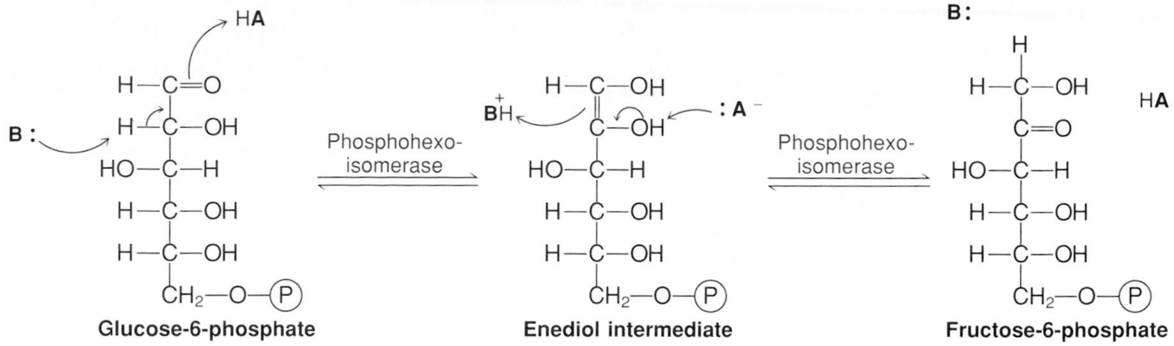

Figure 14–6
Mechanism of the interconversion of glucose-6-phosphate and fructose-6-phosphate. Loss of a proton from the oxygen attached to C-2 of the intermediate enediol leads to fructose-6-phosphate. A and B represent catalytic groups on the enzyme. It is not always known what specific groups are involved in a catalysis. In this case the HA group originates from a glutamate on the enzyme.

Phosphohexoisomerase Interconverts Glucose-6-phosphate and Fructose-6-phosphate

Glucose-6-phosphate and fructose-6-phosphate interconvert readily in weakly alkaline solution. Presumably this is because the intermediate enediol is stabilized by ionization in alkaline medium (Figure 14–6). Glucose-6-phosphate and fructose-6-phosphate are interconverted specifically by the action of the enzyme *phosphohexoisomerase*. The reaction presumably resembles the nonenzymic conversion in going by way of the enediol intermediate.

The free energies of formation of aldoses and ketoses are very similar, so at equilibrium the concentration ratio of glucose-6-phosphate to fructose-6-phosphate is about 2. From this value and the equilibrium constant of about 19 for the phosphoglucomutase reaction, it can be calculated that at equilibrium the hexose monophosphate pool consists of approximately 3% glucose-1-phosphate, 65% glucose-6-phosphate, and 32% fructose-6-phosphate. This pool, as noted earlier (see Figure 14–2), can be replenished by phosphorolysis of storage polysaccharide (forming glucose-1-phosphate), by phosphorylation of glucose (forming glucose-6-phosphate), or by gluconeogenesis or photosynthesis (forming fructose-6-phosphate). The pool is depleted in two ways. One is by catabolic sequences such as glycolysis, which uses fructose-6-phosphate as the starting point, and the pentose phosphate pathway, which uses glucose-6-phosphate. The other is by storage as polysaccharide or by use in synthesis of disaccharides or complex carbohydrates and other macromolecules that contain hexose derivatives. Glucose-1-phosphate is the starting point for these storage and biosynthetic uses. Thus *each component of the hexose monophosphate pool serves both as an entry point and as an exit*. In vertebrate liver, glucose-6-phosphate can also be hydrolyzed to form free glucose for export to maintain the normal blood glucose level. These sequences that replenish or deplete the hexose-phosphate pool will be discussed in this chapter, and some of them in more depth in later chapters. We turn our attention now to reactions leading from the hexoses to pyruvate.

Formation of Fructose-1,6-bisphosphate Is a Commitment to Glycolysis

The cleavage of hexose phosphate to yield triose phosphate is the basis for the name glycolysis ("sugar splitting"), by which the whole metabolic sequence between glucose and pyruvate, lactate, or ethanol is commonly

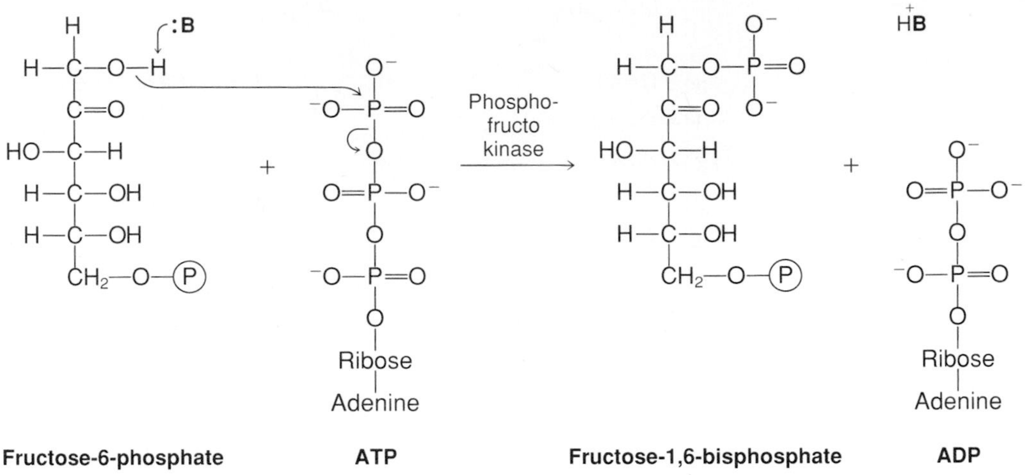

Figure 14–7

The reaction catalyzed by phosphofructokinase. The mechanism of this reaction is very similar to the hexokinase reaction shown in Figure 14–4. B ∶ is a proton acceptor at the active site.

known. Fructose-6-phosphate is converted to triose phosphates in two steps. A second phosphoryl group is added to fructose-6-phosphate before the actual cleavage occurs.

Fructose-6-phosphate is converted to fructose-1,6-bisphosphate by transfer of a phosphoryl group from ATP in the reaction catalyzed by _phosphofructokinase_ (Figure 14–7). Since net regeneration of ATP is a major function of the catabolism of carbohydrates, it may seem strange that the early steps of glucose breakdown consume two molecules of ATP for each molecule of glucose. (Only one molecule of ATP is used for each glucosyl residue when the starting material is storage glycogen or starch.) As we have noted, phosphorylation of substrates helps to protect against loss by diffusion out of the cell, but the hexose phosphates are already protected. Why should the second phosphoryl group be added? One possible answer is again related to prevention of loss of intermediates by diffusion. If fructose-6-phosphate were the substrate for aldolase, which cleaves the six-carbon sugar to two trioses, only one of the products would be phosphorylated. The unphosphorylated product might need to be phosphorylated immediately for protection against loss by diffusion, and phosphorylation before cleavage is even more effective.

But there is another advantage to phosphorylation before cleavage that may be more important. The rate at which hexose phosphates are used in glycolysis must be regulated so that the needs of the cell or organism will be met without wasteful use of more than the necessary amount of substrate. The reaction by which material is removed from the hexose phosphate pool is the point at which such control must be exerted for maximal effectiveness. Metabolic regulation is possible only at steps for which the physiological ratio is far from the equilibrium ratio, so that kinetic control mechanisms can cause increases and decreases in the rates of reactions without thermodynamic constraints.

Under physiological conditions the concentration ratio of fructose bisphosphate to fructose-6-phosphate varies considerably, depending on metabolic conditions, but is probably usually between 5 and 0.2. At equilibrium of the phosphofructokinase reaction the ratio would be about 10^4. When the concentration ratio is so far from equilibrium, there are no thermodynamic limitations on the reaction. By leading to a very large value of the equilibrium constant, the use of ATP thus allows the phosphofructokinase reaction to be controlled by kinetic factors that affect the behavior of the enzyme. Phosphofructokinase serves as the major control point in regulation of the rate at which hexose phosphates are used. We will see that both of the phosphoryl groups of fructose-1,6-bisphosphate are recovered as ATP in a later step of glycolysis, which is catalyzed by the enzyme pyruvate kinase. Thus the decrease in free energy of a later step is in effect fed back by way of ATP to build up the concentrations of hexose phosphates (the hexokinase reaction) and to make control of the early steps in the sequence possible (the phosphofructokinase reaction).

Like other kinases, phosphofructokinase catalyzes a simple attack by a specific oxygen atom of the substrate (in this case, the O attached to C-1) on the terminal phosphorus atom of ATP, with displacement of ADP (Figure 14–7), and like other kinases it requires Mg^{2+} ion, which probably participates in a chelated transition state, not shown in the figure. However, as we will discuss later in this chapter, phosphofructokinase is far more than a simple kinase. It plays a major role in the regulation of the rate of glycolysis, and responds in a complex way to a number of metabolic signals. It may be because of this regulatory complexity that phosphofructokinase is quite large, in contrast to the relatively small size of many kinases. The enzyme from muscle, for example, is a tetramer with a total molecular weight of about 360,000.

Fructose-1,6-bisphosphate and the Two Triose Phosphates Constitute the Second Metabolic Pool in Glycolysis

The metabolic pool that consists of fructose-1,6-bisphosphate and the two triose phosphates, glyceraldehyde-3-phosphate and dihydroxyacetone phosphate (DHAP), is somewhat different from the other two pools of intermediates in glycolysis because of the nature of the chemical relationships between these compounds. In the other pools the relative concentrations of the component compounds at equilibrium are independent of absolute concentrations. Because of the cleavage of one substrate into two products, the relative concentrations of fructose-1,6-bisphosphate and the triose phosphates are functions of the actual concentrations. For such reactions, the relative concentrations of the split products must increase with dilution. (For the reaction $A \rightleftharpoons B + C$, the equilibrium constant is equal to $[B][C]/[A]$. If the concentration of A decreases, for example, by a factor of 4, equilibrium is restored when the concentrations of B and C decrease by a factor of only 2.) But except for that difference, interconversions of the compounds in this pool are similar to those in the other pools in that reactions within the pool are close to equilibrium and can go in either direction, depending on which components are added to the pool and which are removed.

Aldolase Cleaves Fructose-1,6-bisphosphate

Fructose-1,6-bisphosphate is cleaved to two molecules of triose phosphate by _aldolase_ in a reversal of an aldol condensation (Figure 14–8). For some aldolases, the cleavage reaction is preceded by a condensation reaction between the carbonyl of the substrate and the amino group of a specific lysine residue at the catalytic site of the enzyme. Other aldolases (as shown in Figure 14–8) merely bind the substrate noncovalently. Most aldolases are highly specific for the "upper" end of the substrate molecule, requiring phosphorylation at C-1, a carbonyl at C-2, and a specific one of the four possible steric configurations at C-3 and C-4. The nature of the remainder of the molecule is relatively unimportant.

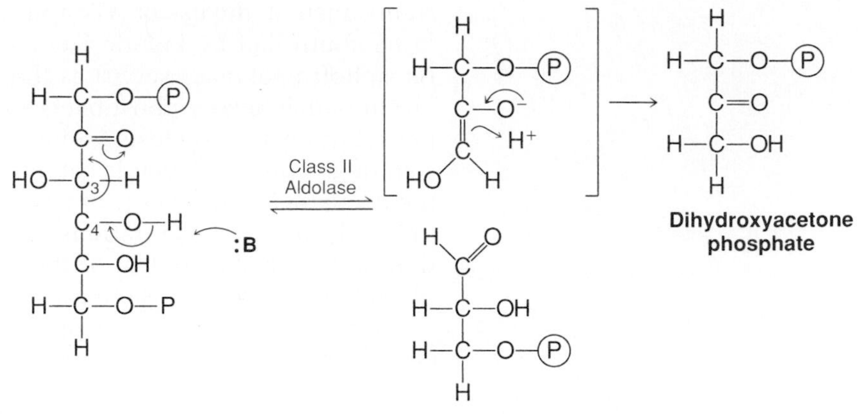

Figure 14–8

Cleavage of fructose-1,6-bisphosphate, an aldolase-catalyzed reaction. The aldolase reaction entails a reversal of the familiar aldol condensation. The first step involves abstraction of the hydrogen of the C-3 hydroxyl group, followed by elimination of an enolate anion.

Whatever the substrate, the reaction is always the same cleavage between C-3 and C-4 (Figure 14–8), and carbons 1, 2, and 3 of the substrate are always converted to dihydroxyacetone phosphate (DHAP). The identity of the other product naturally depends on the nature of the substrate. The specificity is necessarily similar for the reverse reaction, so ketose phosphates containing five to eight carbon atoms can be produced by condensations between dihydroxyacetone phosphate and appropriate aldoses. When fructose-1,6-bisphosphate is the substrate, the products of aldolase-catalyzed cleavage are the two isomeric triose phosphates, dihydroxyacetone phosphate and glyceraldehyde-3-phosphate (Figure 14–8).

The equilibrium constant of the aldolase reaction is about 10^{-4} when the concentrations are expressed in the usual units of molarity. From that value, it appears that this cleavage reaction might pose thermodynamic difficulties. The problem is more apparent than real. This is because the relative concentrations of the products are increased with dilution. The steady-state concentration of the reactant fructose-1,6-bisphosphate in a cell may be around 1 mM. With an equilibrium constant of 10^{-4}, this gives 0.32 mM for the concentration of each of the triose phosphate products. At lower concentrations, the ratio of hexose to triose would be even smaller. Thus cleavage of fructose bisphosphate presents no problem as far as thermodynamics is concerned.

Triose Phosphate Isomerase Interconverts the Two Trioses

The isomeric triose phosphates, glyceraldehyde-3-phosphate and dihydroxyacetone phosphate, bear the same relationship to each other as do glucose-6-phosphate and fructose-6-phosphate. Their interconversion, catalyzed by *triose phosphate isomerase*, is equally facile (see Figure 14–1). Dihydroxyacetone phosphate is a starting material for the synthesis of the glycerol moiety of fats (Chapter 21), but only glyceraldehyde-3-phosphate is used in glycolysis. Thus under ordinary circumstances nearly all of the dihydroxyacetone phosphate that is formed in the cleavage of fructose bisphosphate is converted to glyceraldehyde-3-phosphate by triose phosphate isomerase. All six carbons of the hexoses are thereby made available for the later steps of carbohydrate catabolism.

At equilibrium, the concentration ratio of dihydroxyacetone phosphate to glyceraldehyde-3-phosphate is about 22. That ratio, together with the K_{eq} of about 10^{-4} for the aldolase reaction, allows us to calculate the composition of this pool at equilibrium. As noted above, the composition varies with absolute concentrations, but over the likely physiological range of concentrations the pool contains enough of each component to allow free conversion in either direction. If the concentration of fructose-1,6-bisphosphate were 0.5 mM, the pool would contain 31% fructose-1,6-bisphosphate, 66% dihydroxyacetone phosphate, and 3% glyceraldehyde-3-phosphate at equilibrium. At 5 mM fructose-1,6-bisphosphate, the values would be 58%, 40%, and 2%, respectively.

The Conversion of Triose Phosphates to Phosphoglycerates Occurs in Two Steps

The oxidation of glyceraldehyde-3-phosphate to glycerate-3-phosphate is the first energy-yielding (ATP-producing) reaction in the glycolytic pathway. The mechanism by which an inorganic phosphate ion is taken up and transferred to ADP during the conversion of glyceraldehyde-3-phosphate to glycerate-3-phosphate is indirect and rather elaborate, requiring two separate enzyme-catalyzed reactions.

The first reaction is catalyzed by the enzyme *3-phosphoglyceraldehyde dehydrogenase*, sometimes called triose-phosphate dehydrogenase.

$$\text{Glyceraldehyde-3-phosphate} + \text{NAD}^+ + \text{P}_i \longrightarrow$$
$$\text{glycerate-1,3-bisphosphate} + \text{NADH} + \text{H}^+ \quad (7)$$

The reaction is initiated by a condensation of the —SH group of a specific cysteine residue at the catalytic site of the enzyme with the aldehyde carbonyl, forming a sulfhydryl adduct or thiohemiacetal (Figure 14–9, step 1). A pair of electrons, along with a proton (thus in effect a hydride ion) are then donated to a NAD$^+$ molecule that is tightly bound nearby, converting the tetrahedral hemiacetal into a thioester (step 2). Thioesters provide considerably more free energy on hydrolysis than ordinary oxygen esters. That facilitates the next step (step 3), an attack by an oxygen of a phosphate ion on the carbonyl carbon of the thioester. The sulfur atom, and thus the enzyme molecule, is displaced from covalent linkage to the reaction intermediate, and the product, glycerate-1,3-bisphosphate, is released. The hydride ion is passed on to an NAD$^+$ molecule in solution that binds to the enzyme and dissociates after reduction. Thus an NADH molecule is generated in solution and the tightly-bound NAD$^+$ is again in the oxidized form, ready to participate in another catalytic cycle.

In the second step leading to glycerate-3-phosphate, a phosphate group is transferred from glycerate-1,3-bisphosphate to ADP. This reaction is catalyzed by *3-phosphoglycerate kinase*.

The glycerate-1,3-bisphosphate is a mixed anhydride between a carboxylic acid and phosphoric acid. Like related compounds such as acyl halides, which are often used as acylating agents in organic syntheses, acyl phosphates are good acylating or phosphorylating agents. The standard free energy of hydrolysis of glycerate-1,3-bisphosphate to glycerate-3-phosphate is about −12 kcal/mole. The transfer of a phosphate group to ADP, forming ATP, thus goes with a significant decrease in standard free energy and a favorable equilibrium constant of about 2000.

$$\text{Glycerate-1,3-bisphosphate} + \text{ADP} \longrightarrow \text{glycerate-3-phosphate} + \text{ATP} \quad (8)$$

In the two steps catalyzed by 3-phosphoglyceraldehyde dehydrogenase and 3-phosphoglycerate kinase, the oxidation of glyceraldehyde-3-phosphate to glycerate-3-phosphate is coupled to the regeneration of ATP. The equilibrium constant for the overall conversion is about 160.

Since two molecules of glyceraldehyde-3-phosphate are produced from each molecule of hexose, the oxidation of the aldehyde to glycerate-3-phosphate leads to the production of two molecules of ATP per molecule of glucose or other hexose consumed. At this point the energy (ATP) gained is

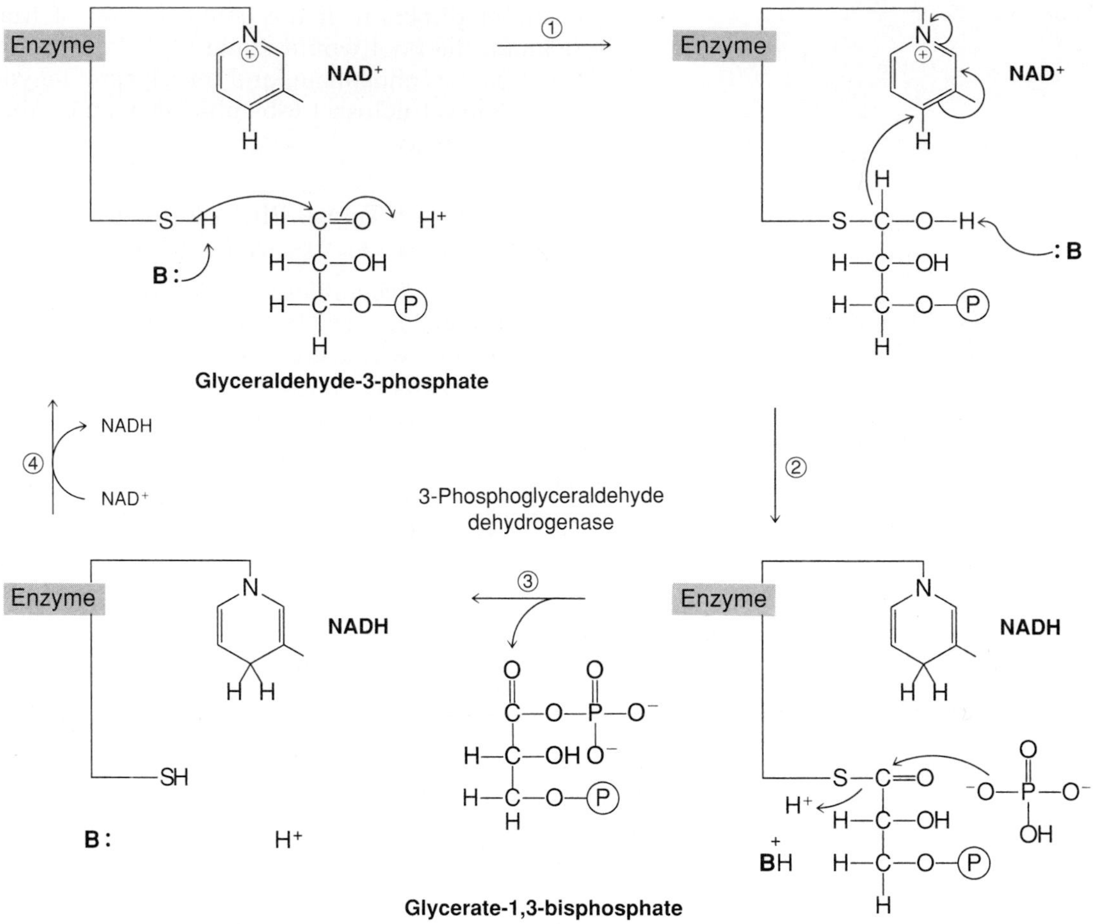

Figure 14–9
The reaction catalyzed by glyceraldehyde-3-phosphate dehydrogenase (3-phosphoglyceraldehyde dehydrogenase). This interesting and complex reaction consists of several steps. The enzyme first catalyzes a reaction of the substrate with a sulfhydryl group of a cysteine residue of the enzyme itself. The substrate is then oxidized from the aldehyde level of oxidation to the carboxylic acid level while still attached covalently to the enzyme. Displacement of the enzyme by inorganic phosphate ion liberates the product, glycerate-1,3-bisphosphate. The bound NADH of the enzyme, which became reduced when the substrate was oxidized, then transfers a pair of electrons to an unbound NAD^+, and the enzyme is ready for another catalytic cycle.

equal to the energy invested if the starting material was free hexose. The two phosphate groups that were introduced in the hexokinase and phosphofructokinase reactions are still contained in glycerate-3-phosphate. If storage glycogen or starch is the starting material, only one phosphate group is supplied by ATP, and one of the two molecules of ATP regenerated in the phosphoglycerate kinase step represents a net gain.

The Three-carbon Phosphorylated Acids Constitute the Third Metabolic Pool

Glycerate-3-phosphate, glycerate-2-phosphate, and phosphoenolpyruvate (PEP) make up another equilibrium group of metabolites that, like the hexose phosphates or the triose phosphates, may be considered to be in effect a single metabolic pool (Figure 14–1).

The first interconversion is the apparent migration of the phosphoryl group from C-3 to C-2 of the phosphoglycerate molecule (Figure 14–10). The

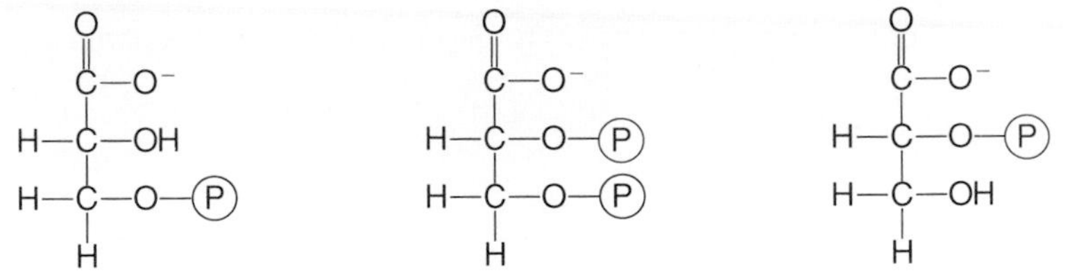

Figure 14–10
Interconversion of glycerate-3-phosphate and glycerate-2-phosphate, catalyzed by phosphoglyceromutase. The reaction closely resembles that catalyzed by phosphoglucomutase (see Figure 14–5) except that the phosphate binds to a histidine side chain instead of a serine side chain. The enzyme can transfer a covalently bound phosphoryl group either to the oxygen on C-2 of glycerate-3-phosphate or to the oxygen on C-3 of glycerate-2-phosphate. The resulting glycerate-2,3-bisphosphate, in turn, can donate either of its phosphoryl groups to the enzyme. These catalytic capabilities provide for interconversion of the two monophosphoglycerates in either direction.

actual reaction is not intramolecular, however; the reaction path is similar to that of the reaction catalyzed by phosphoglucomutase. The enzyme is reversibly phosphorylated in the course of the reaction, and glycerate-2,3-bisphosphate is a necessary cofactor. The value of $\Delta G^{\circ\prime}$ for the conversion of glycerate-3-phosphate to glycerate-2-phosphate is about 1 kcal/mole, and at equilibrium the ratio of glycerate-3-phosphate to glycerate-2-phosphate is about 6.

The dehydration of glycerate-2-phosphate, catalyzed by the enzyme _enolase_, leads to the production of phosphoenolpyruvate (see Figure 14–1). Enolase is a dimer of identical subunits (subunit $M_r = 44,000$). Magnesium ion (or Mn^{2+}) is required for the reaction. This step is one of several points at which fluoride ion inhibits metabolism. Magnesium fluorophosphate binds at the active site of enolase, preventing its catalytic action.

The standard free energy change in going from glycerate-2-phosphate to phosphoenolpyruvate is about 0.4 kcal/mole, and the equilibrium constant is about 0.5. From this value and the equilibrium constant for phosphoglycerate mutase, it can be calculated that the equilibrium composition of this metabolic pool is approximately 80% glycerate-3-phosphate, 13% glycerate-2-phosphate, and 7% phosphoenolpyruvate.

Conversion of Phosphoenolpyruvate to Pyruvate Generates ATP

The enzyme _pyruvate kinase_ was named for the reverse of the reaction it normally catalyzes in glycolysis, the phosphorylation of pyruvate. Because of the thermodynamics of the system, that reaction could never occur in a living cell. The standard free energy of hydrolysis of phosphoenolpyruvate is the most negative among the phosphorylated intermediates of the central pathways, about −14 kcal/mole. Thus even when the phosphoryl group is transferred to ADP instead of water, the standard free energy change is still about −6.5 kcal/mole (the difference between the standard free energies of hydrolysis of phosphoenolpyruvate and ATP), and the equilibrium constant is of the order of 10^5.

In this reaction two molecules of ATP are regenerated for each molecule of hexose phosphate consumed, bringing the net yield of ATP to two molecules for each molecule of glucose (two molecules of ATP were regenerated in the phosphoglycerate kinase step and two in this step, and two were consumed in the hexokinase and phosphofructokinase steps).

$$\text{Glucose} + 2\,NAD^+ + 2\,ADP + 2\,P_i \longrightarrow$$

$$2\,\text{pyruvate} + 2\,NADH + 2\,H^+ + 2\,ATP + 2\,H_2O \quad (9)$$

The net yield is three ATPs for each glucosyl residue when the substrate is storage glycogen or starch. In the absence of oxygen as an electron acceptor,

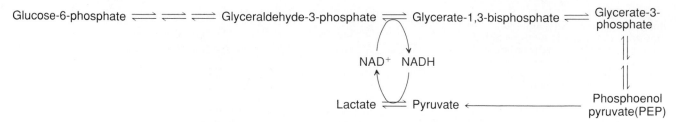

Figure 14–11
Regeneration of NAD^+ by reduction of pyruvate to lactate. Since NAD^+ is a necessary participant in the oxidation of glyceraldehyde-3-phosphate to glycerate-1,3-bisphosphate, glycolysis is possible only if there is a way by which NADH can be reoxidized.

these are the total yields of ATP that will be obtained from catabolism of sugars. Many organisms (anaerobes) are capable only of anaerobic metabolism, and others, including such familiar forms as the enteric bacterium *Escherichia coli* and many kinds of yeasts, are called *facultative anaerobes* because they can live either anaerobically or aerobically. Organisms of either type are able, in the absence of oxygen, to satisfy all their metabolic needs for growth and reproduction from the conversion of sugars to pyruvate (or of related anaerobic sequences).

When oxygen is not available as an ultimate or terminal electron acceptor, pyruvate is, in terms of energetics, the end product of the glycolytic sequence. No more energy is available from its further metabolism under those conditions. But pyruvate still serves an essential function: that of contributing to oxidation-reduction balance (Figure 14–11). In the reaction catalyzed by *3-phosphoglyceraldehyde dehydrogenase*, a molecule of NAD^+ was reduced for each molecule of phosphoglyceraldehyde that was oxidized. The NADH that was produced must be oxidized to regenerate NAD^+ if glycolysis is to continue.

Since equimolar amounts of NADH and pyruvate are produced in glycolysis, the simplest way to oxidize the NADH would be by transfer of electrons to pyruvate, forming lactate. That is the solution that is employed by many kinds of organisms. Several types of anaerobic bacteria—for example, some species of the genus *Lactobacillus* that usually live in decaying plant material—can satisfy all their needs for energy and starting materials for biosynthesis from the conversion of glucose to lactate. The reduction of pyruvate is catalyzed by the enzyme *lactate dehydrogenase*.

$$\text{Pyruvate} + \text{NADH} + H^+ \longrightarrow \text{lactate} + NAD^+ \tag{10}$$

The equilibrium constant of the reaction is about 2×10^4. This large value reflects the ease with which aldehydes and ketones are reduced to alcohols.

The conversion of a mole of glucose to two moles of lactate involves no net oxidation or reduction. The standard free energy change for this conversion alone is about -47 kcal/mole of glucose; as it actually occurs in metabolism, coupled with the conversion of two moles of ADP to ATP per mole of glucose, the standard free energy change is about $-47 + (2 \times 7.5) = -32$ kcal/mole, which corresponds to an equilibrium constant for the sequence of nearly 10^{23}. A simple calculation shows that, assuming the ratio [ATP]/[ADP] to have its normal physiological value of about 5 and the concentration of P_i to be about 10 mM, one molecule of glucose would be in equilibrium with nearly one gram of lactic acid. Such unimaginably high values of equilibrium constants are common in metabolism.

This same conversion is used also in mammalian muscles when the rate at which oxygen can be supplied becomes a limiting factor in the regenera-

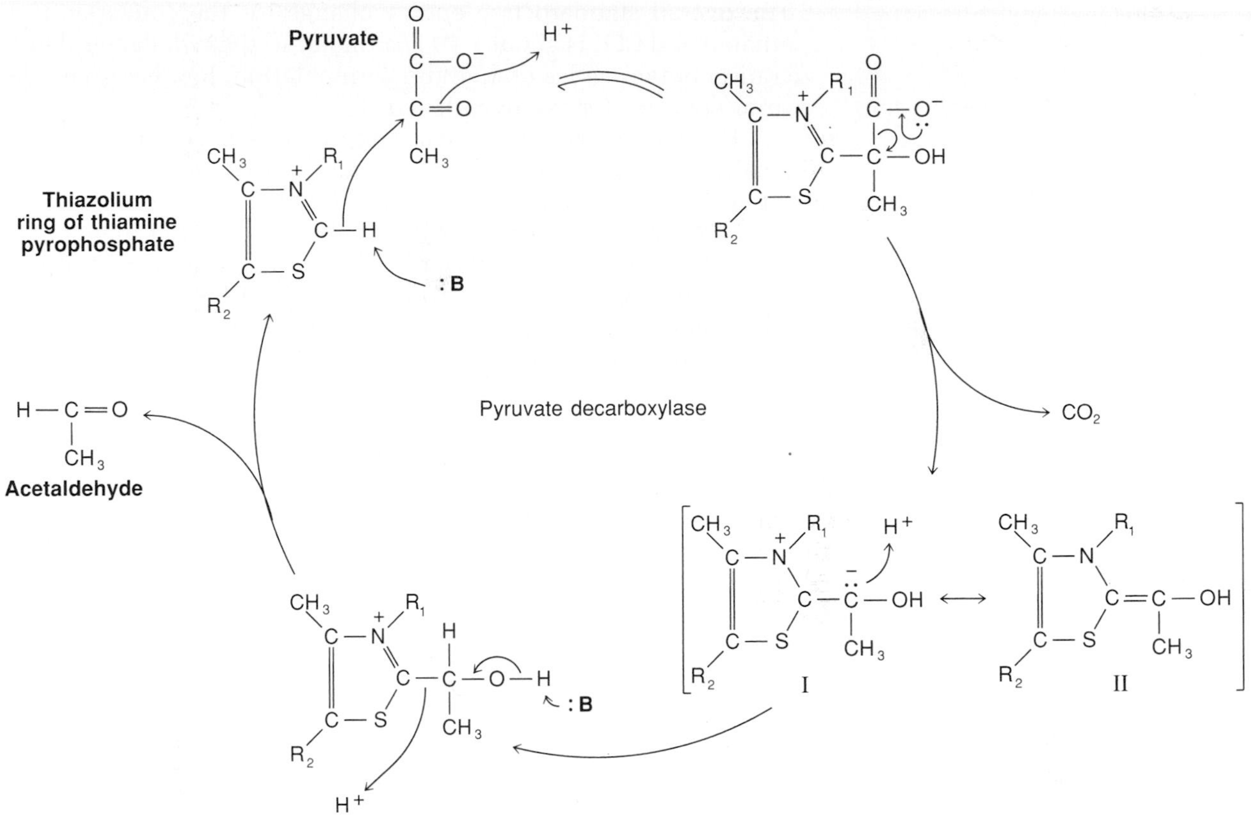

Figure 14–12

The conversion of pyruvate to acetaldehyde and carbon dioxide, catalyzed by pyruvate decarboxylase with thiamin pyrophosphate as bound cofactor. Decarboxylation is facilitated by stabilization of the product of the decarboxylation step, the anion I, by the strong electron-withdrawing power of the thiazolium ring. Alternatively, intermediate I can be said to be stabilized by the existence of the resonance form II without charge separation.

tion of ATP by oxidative phosphorylation (Chapters 15 and 16). Although only two moles of ATP are produced per mole of glucose used, the rate of glucose utilization can be very rapid, so that intense muscular activity can be supported for a short time by production of lactic acid. Alternatively, the pyruvate may be transaminated to alanine (see Chapter 19).

Yeasts and some other kinds of organisms produce ethanol and CO_2, rather than lactic acid, from the anaerobic catabolism of sugars:

$$\text{Pyruvate} + \text{NADH} + \text{H}^+ \longrightarrow \text{ethanol} + CO_2 + \text{NAD}^+ \qquad (11)$$

In these species pyruvate is decarboxylated to acetaldehyde through the action of *pyruvate decarboxylase*, which requires thiamin pyrophosphate as a cofactor (Figure 14–12). α-Keto acids, such as pyruvic acid, do not lose CO_2 readily because of the high energy of the carbanion of the keto carbon that would be a potential intermediate. As in all of the cases where thiamin pyrophosphate is used, its function is to add to the carbonyl carbon, leading to a reaction intermediate in which a negative charge on that carbon is stabilized (Figure 14–12; also see Figure 11–2).

In organisms that carry out the ethanol fermentation, NADH from the phosphoglyceraldehyde dehydrogenase step is reoxidized to NAD$^+$ by reaction with acetaldehyde, catalyzed by *ethanol dehydrogenase*:

$$\text{CH}_3\text{—CHO} + \text{NADH} + \text{H}^+ \rightleftharpoons \text{CH}_3\text{—CH}_2\text{OH} + \text{NAD}^+ \qquad (12)$$

The overall standard free energy change for the conversion of glucose to ethanol and CO_2 is about -40 kcal/mole, so that when coupled to the regeneration of two moles of ATP this fermentation, too, has a very large equilibrium constant for the overall sequence.

Unlike true facultative anaerobes such as *E. coli*, which can live anaerobically for an indefinite period, yeasts can live only for a few generations in the total absence of oxygen because they require molecular oxygen for synthesis of membrane components (Chapter 22). Rather than a true alternative life style, for yeasts alcohol fermentation appears to be part of a very effective competitive strategy. When fruits ripen and fall from the tree or bush, yeasts, which are widespread in nature, are likely to be among the first invaders. In the anaerobic interior of the fruit, yeasts rapidly convert sugars to ethanol, which they excrete. As ethanol accumulates, the growth of most other microorganisms is discouraged, but yeasts can continue to grow until the concentration of ethanol reaches about 12%. When this level is reached, and when the softened fruit breaks open, there is a thorough-going change in the basic energy metabolism of the yeast cells. They begin to take up the ethanol that they had previously excreted, and oxidize it to CO_2, with a much larger production of ATP than in the anaerobic stage (about 18 moles of ATP per mole of ethanol, as compared to two moles of ATP per mole of glucose in the anaerobic stage.) By rapidly converting the available sugars into a compound that cannot be metabolized for energy by most other organisms and in fact is toxic to them, yeasts gain an advantage over competing microorganisms, and also gain energy in the process. They then can exploit the waste product of the first stage of growth as the main nutrient for the later aerobic stage. This example illustrates that the organization of even the most central metabolic pathways can, like other properties of an organism, be shaped by competition.

ORGANIZATION OF GLYCOLYSIS

Energetically, the glycolytic pathway resembles a series of lakes connected by short rivers. This pattern is reflected in the ways that functional metabolic relationships have evolved. All of the branchpoint compounds that serve as starting materials for biosynthesis are found in the "lakes," or metabolic pools. Reactions in which ATP and ADP are involved are in the interconnecting reactions, or "rivers." That clearly is where they are to be expected; an ATP-linked reaction within a metabolic pool would make no more sense than a hydroelectric power plant in the middle of a lake. Regulatory enzymes are also found in the segments of the pathways that connect the pools. Again it is obvious that that is where they must be to be effective. Kinetic controls cannot be applied to reactions that are at or close to equilibrium, as is the case for reactions within the pools.

Further discussion of the kinetics and regulation of glycolysis will be deferred until after we have considered the reverse of glycolysis, that is, the synthesis of glucose from pyruvate.

GLUCONEOGENESIS

The term *gluconeogenesis* is used with two somewhat different meanings. In the more general sense, it refers to the production of sugars from nonsugar precursors such as lactate or amino acids. More specifically, it may be used to refer to the production of free glucose by vertebrate livers for export to the blood. It is important to distinguish between the two meanings, because they are metabolically quite distinct.

A microbiologist or biochemist who studies the metabolism of microorganisms is likely to think of gluconeogenesis as indicating nutritional abundance. That is because, when they are available beyond the needs of the cell, many kinds of nutrients, such as amino acids, can be converted to hexose phosphates for incorporation into storage polysaccharide. An animal physiologist or a biochemist who studies mammalian biochemistry may, on the other hand, consider gluconeogenesis to be a process that often indicates energy deficiency. If the blood glucose level should fall slightly, the liver will attempt to counter the decrease by synthesizing glucose from any available source, including amino acids derived from hydrolysis of proteins of the organism itself. The glucose is supplied to the blood as soon as it is produced. Even in the liver, however, gluconeogenesis does not occur only under conditions of energy stress. Lactate is produced by muscle tissue, especially during heavy exercise, and its conversion back to hexose phosphate and then on to glycogen or to free glucose is an ongoing function of the liver.

Gluconeogenesis occurs under different conditions and serves different functions in different species. It may even have different functions in the same cells under different conditions. Therefore, the systems that regulate the rate of gluconeogenesis are likely to be complex and to differ markedly between species.

Glycolysis and gluconeogenesis make up a pseudocycle, or pair of oppositely directed reaction sequences, of the type that were discussed in the preceding chapter (Figure 14–13). In vertebrates and many other kinds of heterotrophs, the fluxes (rates of flow of intermediates) through these sequences are much greater than those through any other pair of oppositely directed sequences. Perhaps for that reason, the relationship between glycolysis and gluconeogenesis is different from that of any other pseudocycle.

In other pseudocycles, the two sequences are often entirely separate. In this case, however, the intermediates in both pathways are the same (except for two added intermediates, oxaloacetate and UDP-glucose, in the conversion of three-carbon compounds to glycogen). Most of the enzymes—those that catalyze interconversions within the metabolic pools—function both in glycolysis and in gluconeogenesis. Only at three points, all outside the pools, are reactions of glycolysis and gluconeogenesis distinct and catalyzed by different enzymes (Figure 14–13).

Gluconeogenesis Consumes ATP

It is the organization of glycolysis and gluconeogenesis as a series of metabolic pools that allows most of the same enzymes to function in both of the oppositely directed sequences. The intermediates in a metabolic pool are at near-equilibrium concentrations and can proceed in either direction as a result of small changes in concentrations. The reactions that connect the metabolic pools are those that involve large changes in standard free energy, are coupled to the ATP-ADP system, and are the sites for regulatory interactions that determine the net direction of flux and the rates of conversion. Thus the properties of the energy-dependent and kinetically regulated segments between the pools will determine which components of the pools will be consumed and which will be replenished under a particular set of conditions.

Flux within a pool is determined by simple thermodynamic and mass-action considerations. Thus when glycerate-3-phosphate is being produced in the phosphoglycerate kinase reaction and phosphoenolpyruvate is being removed by action of pyruvate kinase, the flux in the three-carbon carboxylate pool will be in the glycolytic direction, from glycerate-3-phosphate toward phosphoenolpyruvate. When the pattern of activation of the regulatory

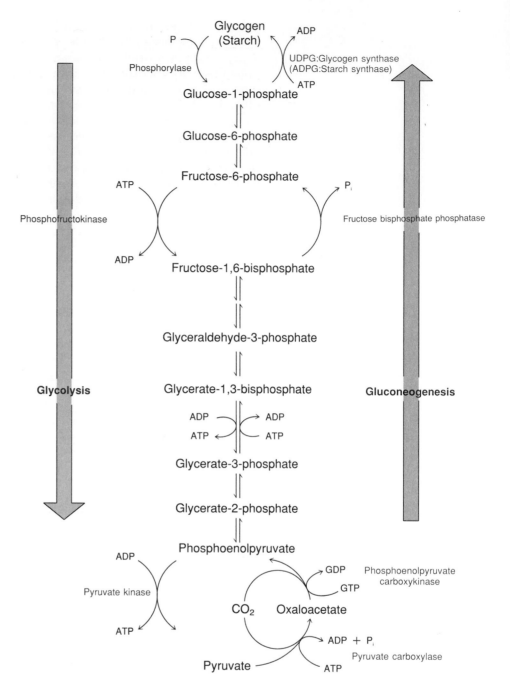

Figure 14–13
The glycolytic and gluconeogenic pathways. Only enzymes that are unique to either pathway are indicated.

enzymes is different, so that phosphoenolpyruvate is being produced and glycerate-3-phosphate is being consumed, the flux in the same pool, catalyzed by the same enzymes, will flow in the gluconeogenic direction, from phosphoenolpyruvate to glycerate-3-phosphate. Similarly for the other pools, the direction of net flux through the pool will depend on which components are flowing into the pool and which are being removed; those rates, of course, depend in turn on the regulatory properties of the enzymes that catalyze the reactions that connect the pools.

In the direction of glycolysis there is a rather large drop in the standard free energy of formation of the component compounds in going from one metabolic pool to the one below it (see Figures 14–13 and 14–1). In two cases (glyceraldehyde-3-phosphate to glycerate-3-phosphate and phosphoenolpyruvate to pyruvate), one molecule of ADP is phosphorylated to ATP for each molecule of intermediate that flows from one pool to the next.

In our lake-and-river analogy, there is a hydroelectric generating plant on the river, and useful energy is obtained from the flow between lakes. Unlike a hydroelectric plant, the ATP-generating machinery is quantized. Even if there is a large enough free energy drop to make 1.5 molecules of ATP, such fractional stoichiometry is not possible, and only one molecule can actually be made. In such a case, the equilibrium constant for the reaction is therefore large, and the reaction is effectively nonreversible. We saw that the equilibrium constant for the pyruvate kinase reaction, for example, is about 10^5. It is thus obvious that it would be impossible for that reaction ever to go in the reverse direction in a living cell. The conversion can be reversed only if it is coupled to two ATP-to-ADP conversions. This is a specialized restatement of the principle that oppositely directed sequences have different ATP stoichiometries (see Chapter 13).

Conversion of Pyruvate to Phosphoenolpyruvate Requires ATP and GTP

In gluconeogenesis it is necessary to form phosphoenolpyruvate from pyruvate that is produced by oxidation of lactate or by catabolism of alanine or other amino acids. Here the principle that any conversion can be made thermodynamically favorable by coupling in ATP-to-ADP conversions is applicable. Reversal of the pyruvate kinase reaction, which would use one ATP for each phosphoenolpyruvate produced, is not thermodynamically favorable, so it is necessary that the conversion be coupled to at least two ATP-to-ADP conversions. The means by which that is done is shown in Figure 14–14.

The first step in the conversion, the carboxylation of pyruvate to form oxaloacetate, is catalyzed by the enzyme *pyruvate carboxylase*. As in many other enzymic carboxylation reactions, the immediate CO_2 donor is a carboxylated derivative of the cofactor biotin (see Chapter 11). ATP is used in the carboxylation of biotin; thus the stoichiometry for the production of oxaloacetate is

$$\text{Pyruvate} + \text{ATP} + CO_2 \longrightarrow \text{oxaloacetate} + \text{ADP} + P_i \qquad (13)$$

Because it is intimately related to the tricarboxylic acid (TCA) cycle, we will discuss pyruvate carboxylase and its regulation in the next chapter.

The carboxylation of pyruvate supplies thermodynamic activation for the next step in the sequence. The standard free energy change for decarboxylation of β-keto carboxylic acids is negative and rather large. Thus the decarboxylation will make the formation of phosphoenolpyruvate more thermodynamically favorable than it would otherwise be. You will see another important example of this interesting strategy of activation by carboxylation when you study fatty acid synthesis, in which carboxylation of acetyl coenzyme A serves a similar function (Chapter 21).

Oxaloacetate is converted to phosphoenolpyruvate in a reaction catalyzed by *phosphoenolpyruvate carboxykinase*. In many species, including mammals, this reaction involves a GTP-to-GDP conversion, rather than ATP-to-ADP.

$$\text{Oxaloacetate} + \text{GTP} \longrightarrow \text{phosphoenolpyruvate} + \text{GDP} + CO_2 \quad (14)$$

The use of GTP, UTP, or CTP in a metabolic reaction or sequence is energetically equivalent to the use of ATP, because the nucleoside diphosphate that is produced is rephosphorylated at the expense of ATP in the reaction catalyzed by *nucleoside diphosphate kinase*:

$$\text{ATP} + \text{NDP} \rightleftharpoons \text{ADP} + \text{NTP} \qquad (15)$$

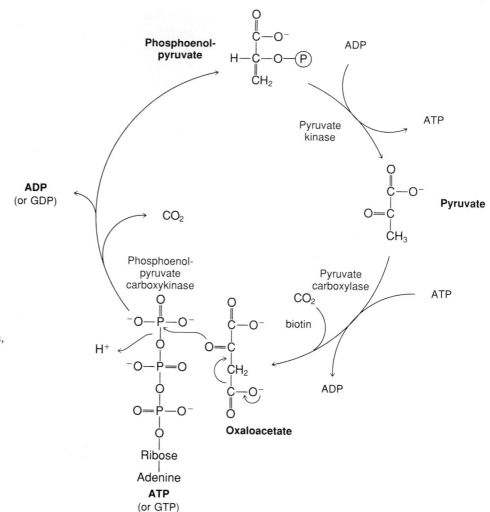

Figure 14–14
The pyruvate-phosphoenolpyruvate pseudocycle. One molecule of ATP is regenerated in the conversion of phosphoenolpyruvate to pyruvate, catalyzed by pyruvate kinase, which occurs in glycolysis. Conversion of pyruvate to phosphoenolpyruvate, which is necessary in gluconeogenesis, must be coupled to two ATP-to-ADP conversions if it is to be thermodynamically feasible. The first ATP is used indirectly, in the production of the active carboxyl transfer agent, biotin-COO⁻ (Chapter 11). Transfer of a carboxyl group to pyruvate produces oxaloacetate. Decarboxylation of oxaloacetate aids in the attack by the carbonyl oxygen on the terminal phosphorus atom of ATP or GTP, producing phosphoenolpyruvate and ADP or GDP.

NDP may be CDP, GDP, or UDP, and NTP the corresponding triphosphate. The enzyme is unusual in its lack of specificity for the acceptor nucleoside diphosphate.

Carboxylation is mechanistically, as well as energetically, an appropriate means of activating pyruvate for conversion to phosphoenolpyruvate, as seen in Figure 14–14. When a β-keto acid loses CO_2, electrons move into the α—β bond, displacing a pair of electrons from the carbonyl double bond. The usual result is that a proton is picked up from the surroundings, yielding as the initial product the enol form of the ketone. The enol form will tautomerize almost quantitatively to the keto form. However, in the case of the phosphoenolpyruvate carboxykinase reaction, the enzyme positions one GTP molecule appropriately so that the carbonyl oxygen atom of oxaloacetate, on becoming negatively charged, attacks the terminal phosphorus atom of the cofactor, displacing GDP. Thus the product is locked in the high-energy enolic form. The negative free energy changes that accompany β decarboxylation and the hydrolysis of GTP are sufficiently large to allow the conversion of oxaloacetate to phosphoenolpyruvate to proceed with a standard free energy change of about -4 kcal/mole, and an equilibrium constant of about 800.

If we add the equations for the reactions catalyzed by pyruvate carboxylase, phosphoenolpyruvate carboxykinase, and nucleoside diphosphate kinase, we obtain the overall reaction for conversion of pyruvate to phosphoenolpyruvate:

$$\text{Pyruvate} + 2\ \text{ATP} \longrightarrow \text{phosphoenolpyruvate} + 2\ \text{ADP} + P_i \quad (16)$$

Thus two molecules of ATP are used to reverse the conversion that yields one molecule of ATP in the glycolytic direction, and both directions of conversion are thermodynamically favorable.

Other Routes from Pyruvate to Phosphoenolpyruvate

Different pathways are used by bacteria and green plants for the conversion of pyruvate to phosphoenolpyruvate. Many bacterial species contain an enzyme called _phosphoenolpyruvate synthase_, which catalyzes the direct conversion of pyruvate to phosphoenolpyruvate, with the production of AMP and inorganic phosphate ion from ATP. Since a second molecule of ATP must be used to convert the product AMP to ADP, the conversion is actually coupled to two ATP-to-ADP conversions, and the overall equilibrium constant is identical to that of the conversion by way of oxaloacetate:

$$\text{Pyruvate} + \text{ATP} \longrightarrow \text{phosphoenolpyruvate} + \text{AMP} + P_i \quad (17)$$

$$\text{ATP} + \text{AMP} \rightleftharpoons 2\ \text{ADP} \quad (18)$$

Sum: $\text{Pyruvate} + 2\ \text{ATP} \longrightarrow \text{phosphoenolpyruvate} + 2\ \text{ADP} + P_i \quad (16')$

Some species of green plants (those that carry out "C-4" photosynthesis) fix CO_2 into oxaloacetate as part of a transport mechanism that increases the concentration of CO_2 at the photosynthetic centers (Chapter 17). Since phosphoenolpyruvate is the CO_2 acceptor, these plants must carry out a massive conversion of pyruvate to phosphoenolpyruvate. This conversion is effected by an enzyme called _pyruvate phosphate dikinase_, which catalyzes the transfer of one phosphate group to pyruvate and another to inorganic phosphate ion. Hydrolysis of the resulting pyrophosphate and conversion of AMP to ADP by the action of adenylate kinase results again in the same stoichiometry for conversion of pyruvate to phosphoenolpyruvate as in the other cases:

$$\text{Pyruvate} + P_i + \text{ATP} \longrightarrow \text{phosphoenolpyruvate} + \text{AMP} + PP_i \quad (19)$$

$$PP_i + H_2O \longrightarrow 2\ P_i \quad (20)$$

$$\text{ATP} + \text{AMP} \rightleftharpoons 2\ \text{ADP} \quad (18')$$

Sum: $\text{Pyruvate} + 2\ \text{ATP} \longrightarrow \text{phosphoenolpyruvate} + 2\ \text{ADP} + P_i \quad (16')$

Conversion of Phosphoenolpyruvate to Fructose-1,6-bisphosphate Uses the Same Enzymes as Glycolysis

The interconversion of the phosphorylated three-carbon acids phosphoenolpyruvate, glycerate-2-phosphate, and glycerate-3-phosphate has already been discussed in this chapter. This pool is linked to the fructose-1,6-bisphosphate/triose phosphate pool by the reactions that convert glycerate-3-phosphate to glyceraldehyde-3-phosphate. The linkage here, however, is organized differently from the reactions that link other pools. In this case the same reactions, catalyzed by the same enzymes, function in both directions (Figure 14–13). That is possible only because the free energy difference between the two pools is approximately equal to the free energy of hydrolysis of ATP, so that material can move from one pool to the other in either direction at reasonable physiological concentrations of the components of the two pools. That is, the one molecule of ATP per molecule of substrate that is generated in the pathway from the fructose-1,6-bisphosphate/triose phosphate pool to the phosphorylated acid pool is enough to make the reverse conversion thermodynamically feasible. In our water-flow analogy, the amount of electrical energy generated when water flows downhill (from fructose-1,6-bisphosphate/triose phosphate to phosphorylated acids) is equal to the amount of energy needed to pump water uphill in the reverse direction, so that no additional energy input is necessary.

From Figure 14–13 it can be seen that there is no pair of oppositely directed reactions with different ATP stoichiometries (no pseudocycle) in the whole segment of glycolysis and gluconeogenesis between phosphoenolpyruvate and fructose-1,6-bisphosphate. Thus the stoichiometry for conversions in the two directions must be the same. Both directions of conversion do, in fact, occur (from fructose-1,6-bisphosphate to phosphoenolpyruvate in glycolysis and from phosphoenolpyruvate to fructose-1,6-bisphosphate in gluconeogenesis). It follows that the components of both pools, as well as glycerate-1,3-bisphosphate, which lies between them, must be so close to equilibrium in the cell that the direction of flux will depend on mass action.

On first thought, it seems unlikely that material could flow freely between these two pools without the pumping effect of a pseudocycle. The standard free energy change for the phosphate-glycerate kinase reaction is large and negative, and we would not expect that this reaction, which leads to the formation of glycerate-3-phosphate in the glycolytic direction, could be reversed under physiological conditions. The equilibrium constants of neighboring reactions, however, are such as to bring the equilibrium constant for the overall interconversion between fructose-1,6-bisphosphate and phosphoenolpyruvate into a range where reversal is feasible. During glycolysis the value of ΔG for the conversion of fructose-1,6-bisphosphate to phosphoenolpyruvate is negative. A rather small shift in the concentrations of those compounds can cause the value of ΔG to change sign, so that the conversion of phosphoenolpyruvate to fructose-1,6-bisphosphate becomes thermodynamically favorable. This change may be enhanced by a change in the $NAD^+/NADH$ ratio.

Approximate relative concentrations of the intermediates between fructose-1,6-bisphosphate and phosphoenolpyruvate at equilibrium are shown in Figure 14–15. These concentrations are based on the values that we have used for the equilibrium constants of the individual reactions and on the ATP/ADP and $NAD^+/NADH$ ratios and the phosphate ion concentration indicated in the figure legend. According to those calculations, fructose-1,6-bisphosphate at 1 mM should be in equilibrium with phosphoenolpyruvate

Figure 14–15
Approximate equilibrium concentrations of the metabolites in glycolysis and gluconeogenesis between fructose-1,6-bisphosphate and phosphoenolpyruvate when the concentration of fructose-1,6-bisphosphate is 1 mM. Although the equilibrium concentrations of some intermediates are considerably higher or lower, the concentrations of the two compounds at the ends of the segment, fructose-1,6-bisphosphate and phosphoenolpyruvate, differ by less than a factor of ten. The conversion can proceed in either direction physiologically. Since the stoichiometry is the same in both directions, the direction of the reaction depends only on the concentrations of fructose-1,6-phosphate and phosphoenolpyruvate.

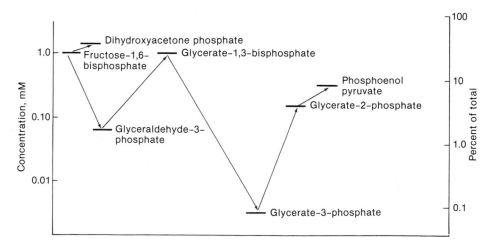

at 90 μM. At lower concentrations of fructose-bisphosphate, the ratio of fructose bisphosphate to phosphoenolpyruvate would be smaller. For example, when the concentration of fructose-1,6-bisphosphate is 100 μM, which may be closer to the normal physiological value, the expected equilibrium concentration of phosphoenolpyruvate is 28 μM.

It is important to recognize that calculations such as these cannot be precise for a number of reasons. First, the cumulative uncertainty in the values of the equilibrium constants is probably greater than the ratio of about 10, shown in the figure, between the concentrations of fructose-1,6-bisphosphate and phosphoenolpyruvate. Thus for this reason alone the two concentrations may actually be equal at equilibrium (or may differ by a greater ratio than is indicated). In addition, the effective concentration of phosphate ion may vary, and in any case is not known precisely; the effective values of the ATP/ADP ratio, and especially of the NAD^+/NADH ratio, are not known precisely; and the effects of ionic strength, etc., are ignored. Thus the values in the figure must not be considered to be quantitatively accurate. The purpose of the figure is to show that even in a reversible segment of a pathway the concentrations of intermediates may differ widely. The large $\Delta G^{\circ\prime}$ of the phosphoglycerate kinase reaction is not a bar to the participation of this reaction in both directions in metabolism. Since the concentrations in the figure were calculated for equilibrium, the value of ΔG for each step would of course be zero under those conditions.

In spite of our uncertainties as to exact concentrations, we can say with certainty that this sequence of reactions actually does proceed in both directions in a metabolizing cell (fructose-1,6-bisphosphate to phosphoenolpyruvate during glycolysis; phosphoenolpyruvate to fructose-1,6-bisphosphate during gluconeogenesis), and that the profile of relative concentrations of intermediate concentrations shown in the figure is approximately correct if all intermediates are free in solution (but see Box 14–A).

Fructose-bisphosphate Phosphatase Converts Fructose-1,6-bisphosphate to Fructose-6-Phosphate

Fructose-1,6-bisphosphate is converted to fructose-6-phosphate by hydrolysis of the phosphoryl ester bond at C-1, a reaction catalyzed by _fructose-bisphosphate phosphatase_. This reaction and that catalyzed by phosphofructokinase thus make up a metabolic pseudocycle. The standard free energy change for the hydrolysis is about -4 kcal/mole, corresponding to an equilibrium constant of nearly 1000. Thus, as in any such pseudocycle, the two conversions—the phosphorylation of fructose-6-phosphate to form fructose 1,6-bisphosphate with ATP as the phosphate donor, and the hydrolysis of fructose-bisphosphate to form fructose-6-phosphate—are both highly favorable thermodynamically under any conditions that will arise in a living cell.

CONVERSION OF HEXOSE PHOSPHATES TO POLYSACCHARIDES

The interconversion of fructose-6-phosphate, glucose-6-phosphate, and glucose-1-phosphate was discussed earlier. When fructose-6-phosphate is generated by photosynthesis (see Chapter 17) or gluconeogenesis, an equimolar amount of glucose-1-phosphate will usually be removed from the hexose monophosphate pool by conversion to storage polysaccharide (glycogen in animals and many kinds of microorganisms; starch in green plants).

The first step in the conversion of glucose-1-phosphate to starch or glycogen is activation at the expense of a nucleoside triphosphate. In green

---

BOX 14–A

Metabolite Transfer via Enzyme-Enzyme Complexes

The concentrations shown in Figure 14–15 are calculated on the assumption that metabolic intermediates exist freely in solution, and that the product of each enzymic reaction dissociates from the site at which it is produced and reaches the catalytic site of the next enzyme in the sequence by simple diffusion. This assumption may not be true in all cases.

For example, nuclear magnetic resonance studies indicate that the rate at which glycerate-1,3-bisphosphate dissociates from phosphoglycerate kinase is slower than the rate at which gluconeogenesis can occur, and much slower than the rate at which glycerate-1,3-bisphosphate can be converted to glyceraldehyde-3-phosphate *in vitro* by a combination of 3-phosphoglycerate kinase and 3-phosphoglyceraldehyde dehydrogenase. The simplest explanation of that paradox appears to be that glycerate-1,3-bisphosphate can be transferred directly from the catalytic site of phosphoglycerate kinase to the catalytic site of the dehydrogenase much more rapidly than it can be released into solution. When a preparation of the kinase bearing glycerate-1,3-bisphosphate at its catalytic sites was used as substrate for 3-phosphoglyceraldehyde kinase, researchers found that the reaction was rapid, which further supports the possibility of direct transfer. Both the kinase and the dehydrogenase are present in cells at concentrations greater than either the measured or calculated concentration of glycerate-1,3-bisphosphate, so it is likely that nearly all of the glycerate-1,3-bisphosphate is bound to catalytic sites. If glycerate-1,3-bisphosphate is actually transferred directly between the enzymes during glycolysis and gluconeogenesis, its calculated equilibrium concentration in solution, as given in Figure 14–15, is not relevant. Further evidence is needed, but at present direct transfer seems quite likely.

This and other examples of possible direct transfer are discussed by D. K. Srivastava and S. A. Bernhard (see *Science* 234:1084–1086, 1986). They also present evidence suggesting that NAD$^+$ and NADH may be transferred directly between pairs of dehydrogenases. These observations have an interesting stereochemical feature. The nicotinamide ring of NAD or NADP is chiral, and an enzyme would be expected to bind it specifically from one side or the other. In fact, dehydrogenases are known to transfer the hydrogen from C-4 of the pyridine ring stereospecifically (Chapter 11). Some enzymes transfer hydrogen only from the A side of the ring and others only from the B side. Bernhard and colleagues found evidence for direct transfer when an A-specific dehydrogenase and a B-specific dehydrogenase were present in the same solution, but not when two A-specific or two B-specific enzymes were mixed.

At present we don't know how widespread the direct transfer of intermediates and cofactors may be. For substrates that are transferred directly *in vivo*, the Michaelis constants that are determined by kinetic experiments *in vitro* with only one enzyme present are probably not relevant. We do know that in many cases the values of Michaelis constants, as estimated *in vitro*, are in functional agreement with intermediate concentrations *in vivo*. Furthermore, in the case of regulatory enzymes, modifiers that are known to affect metabolic rates *in vivo* can be shown to modulate affinities of enzymes for their substrates *in vitro* in the appropriate direction and extent. These facts suggest that such enzymes probably bind their substrates directly from solution in the cell. However, regulatory enzymes are usually those that catalyze the first reactions in their sequences, and this is not where direct transfer would be expected anyway. Thus a lack of direct transfer to regulatory enzymes does not mean that no direct transfer takes place between some—or even many or most—of the enzymes that catalyze later reactions in metabolic sequences.

---

plants the activated form is ADP-glucose, formed by transfer of an AMP moiety from ATP to the phosphate of glucose-1-phosphate by the enzyme ADP-glucose synthase. An oxygen atom of the phosphoryl group of glucose-1-phosphate attacks the α phosphorus atom of ATP, displacing pyrophosphate (Figure 14–16). Since the bond broken and the bond formed are both pyrophosphate bonds, the change in standard free energy for this reaction is close to zero and the equilibrium constant is near 1. However, if the pyrophosphate is hydrolyzed by the enzyme *pyrophosphatase*, as probably occurs under most conditions in living cells, the standard free energy change for the two reactions combined is large and negative, about −7.5 kcal/mole.

In the reaction catalyzed by *starch synthase*, the oxygen on C-4 of the terminal glucosyl residue of a starch chain attacks C-1 of a molecule of ADP-glucose, displacing ADP and adding the glucosyl unit to the polymeric starch molecule (see Figure 14–16). The standard free energy change for this reaction is about −3 kcal/mole.

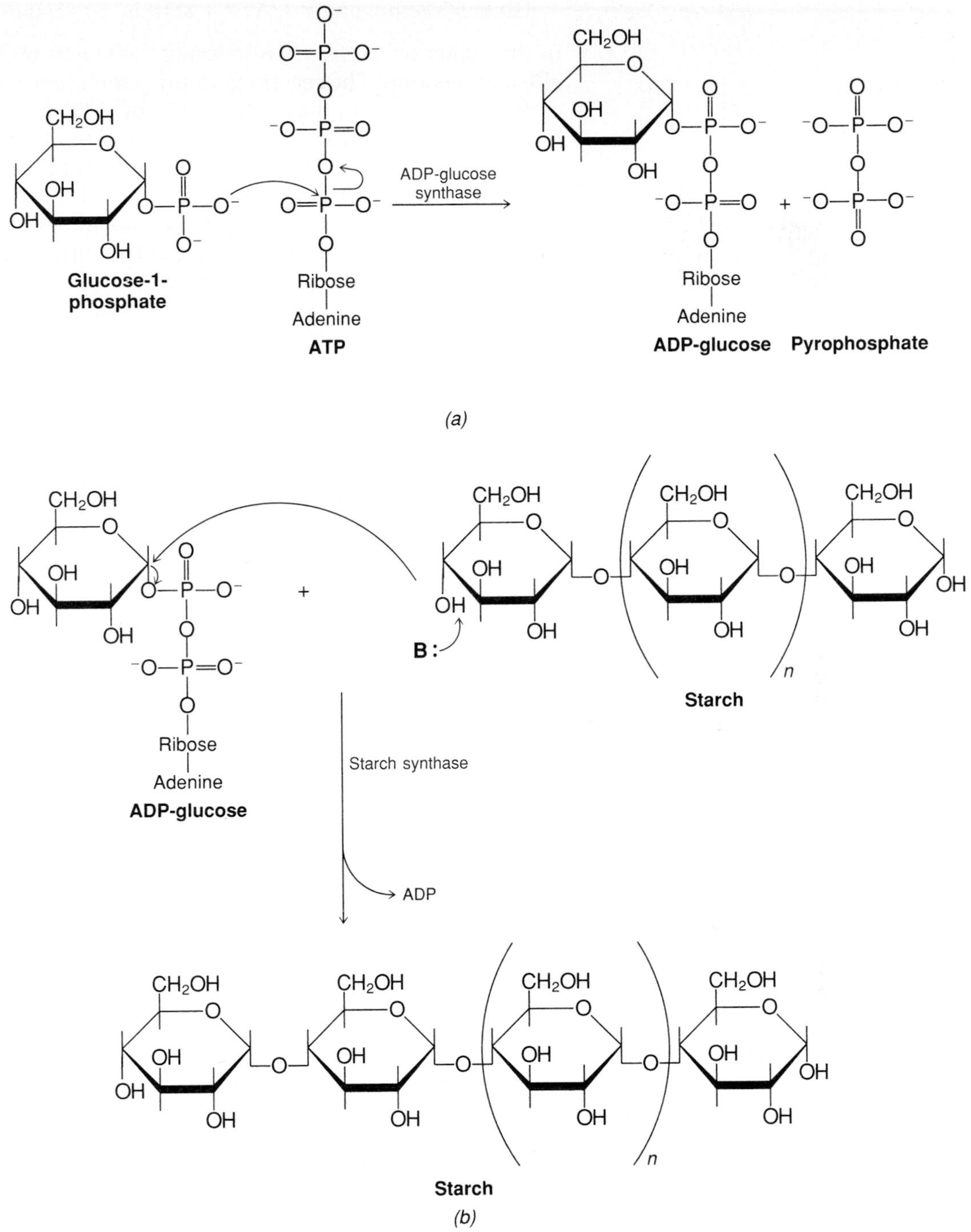

(a)

(b)

Figure 14–16

(a) The synthesis of ADP-glucose and (b) its incorporation into starch. The equilibrium constant for the synthesis of ATP-glucose is near one because the bonds that are formed and broken are equivalent. However, the reaction is favorable because the concentration of the other product, pyrophosphate ion, is made very low by hydrolysis of the pyrophosphate to phosphate, catalyzed by a pyrophosphatase (not shown). In (b), an oxygen at C-4 of the terminal glycosyl residue of a starch chain then displaces ADP from the ADP-glucose. The overall stoichiometry thus shows that one ATP is converted to ADP for each glucosyl residue incorporated into starch.

The equation for the incorporation of the glucosyl residue of glucose-1-phosphate into starch, assuming that the pyrophosphate formed is hydrolyzed, is the reverse of the equation for the phosphorolysis of a polysaccharide, except for the conversion of ATP to ADP and phosphate ion:

$$\text{Glucose-1-phosphate} + \text{ATP} + \text{starch} \longrightarrow (\text{starch}^+) + \text{ADP} + 2\,P_i \quad (21)$$

In this equation, (starch$^+$) represents the starch molecule after addition of a glucosyl residue. The reactions in this conversion, which includes cleavage of both of the pyrophosphate bonds of ATP and the formation of a new pyrophosphate bond, are a bit more complex than in the case of a simple kinase, but the thermodynamic effect is merely that an ATP-to-ADP conversion has been added in the direction of polysaccharide synthesis. Thus the pseudocycle that connects glucose-1-phosphate and starch is energetically equivalent to any other in which the two oppositely directed conversions differ by one ATP-to-ADP conversion.

In most animals, many bacterial species, and yeasts and other fungi, glycogen serves the same function as starch in plants. Glycogen resembles starch in consisting of glucose residues linked primarily by $\alpha(1,4)$ acetal bonds, but it contains a larger number of 1,6 branches (Chapter 4). Bacteria use ADP-glucose as the glucosyl donor in glycogen synthesis, but vertebrates use UDP-glucose. Energetically this makes no difference. Since UTP is regenerated from UDP at the expense of ATP, the sum of the reactions is the same as when ADP-glucose is used:

$$\text{Glucose-1-phosphate} + \text{UTP} \rightleftharpoons \text{UDP-glucose} + \text{PP}_i \quad (22)$$

$$\text{PP}_i \longrightarrow 2\,P_i \quad (20')$$

$$\text{UDP-glucose} + \text{glycogen} \longrightarrow (\text{glycogen}^+) + \text{UDP} \quad (23)$$

$$\text{UDP} + \text{ATP} \rightleftharpoons \text{UTP} + \text{ADP} \quad (24)$$

Sum:
$$\text{Glucose-1-phosphate} + \text{ATP} + \text{glycogen} \longrightarrow$$
$$(\text{glycogen}^+) + \text{ADP} + 2\,P_i \quad (25)$$

In these equations, (glycogen$^+$) represents the glycogen molecule after addition of a glucosyl residue.

Because of its roles in the synthesis of glycogen, in isomerization of hexose phosphates, and as a precursor for a number of biosynthetic intermediates, UDP-glucose might be considered the most central hexose derivative in mammalian metabolism, aside from glycolysis and gluconeogenesis. In bacteria and plants, those roles are divided between ADP-glucose (production of storage polysaccharide) and UDP-glucose (sugar interconversions and biosynthesis).

OTHER HEXOSE DERIVATIVES

The two disaccharides lactose and sucrose are of great interest because they are major sources of carbohydrates in human diets. Lactose breaks down to the monosaccharides glucose and galactose, and sucrose breaks down to glucose and fructose. Under certain conditions the metabolism may be adversely affected by either of these disaccharides.

Excess Fructose May Cause Metabolic Complications

Because of its use as a transport sugar in plants, sucrose is an ingredient of the diet of many kinds of animals, including mammals, and it has become a major food item in many human diets. Sucrose is a disaccharide containing glucose and fructose (Chapter 4). It is hydrolyzed in the small intestine, and free glucose and fructose are taken up into the blood and carried to the tissues, especially the liver. Free fructose can be phosphorylated by hexokinase (which is not highly specific for glucose). In that case, the product is fructose-6-phosphate, which equilibrates into the hexose-phosphate pool that we discussed earlier. In liver, however, much of the fructose is phosphorylated by a special fructokinase, which attaches a phosphoryl group to carbon 1. The resulting fructose-1-phosphate is split by a special aldolase, forming dihydroxyacetone phosphate and free glyceraldehyde. A special kinase converts glyceraldehyde to glyceraldehyde-3-phosphate, and the two triose phosphates than continue along the glycolytic pathway in the usual way.

The metabolism of fructose-1-phosphate bypasses the reaction catalyzed by phosphofructokinase, and so is subject to the potential disadvantages that were suggested earlier for a path in which cleavage of a hexose preceded the second phosphorylation. In particular, because it is not constrained by the normal regulation that occurs at the phosphofructokinase step, metabolism of fructose-1-phosphate may continue at an unnecessarily high rate when the need for energy or intermediates is low. Such unregulated catabolism might lead to higher than usual concentrations of later intermediates, such as acetyl-coenzyme A, which supplies all of the carbon for synthesis of fats and cholesterol (Chapter 21). Since concentrations of starting materials are among the factors that determine the flux into biosynthetic pathways, it seems possible that excessive fructose in the diet might lead to overproduction of fats and cholesterol, with increased danger of strokes and cardiovascular disease. The degree of this danger is not known, and there may be parallel regulatory factors not yet recognized that reduce it, but both the high sucrose content of typical diets and the increasing use of free fructose in prepared foods and drinks (because fructose is sweeter than other hexoses) may be matters of concern.

Abnormal Individuals Cannot Metabolize Galactose

Although its derivatives are common in structural macromolecules, free galactose does not occur in nature. Galactose is metabolically important only because the main sugar of milk, lactose, is a disaccharide containing glucose and galactose. The disaccharide is hydrolyzed in the small intestine and the free hexoses are absorbed. The major pathway by which galactose is brought into the glycolytic mainstream is:

$$\text{Galactose} + \text{ATP} \longrightarrow \text{galactose-1-phosphate} + \text{ADP} \qquad (26)$$

$$\text{Galactose-1-phosphate} + \text{UDP-glucose} \rightleftharpoons$$
$$\text{UDP-galactose} + \text{glucose-1-phosphate} \quad (27)$$

$$\text{UDP-galactose} \rightleftharpoons \text{UDP-glucose} \qquad (28)$$

Sum: $\quad \text{Galactose} + \text{ATP} \longrightarrow \text{glucose-1-phosphate} + \text{ADP} \qquad (29)$

The first reaction is catalyzed by _galactokinase_. In the second reaction, catalyzed by an enzyme called _galactose phosphate uridyltransferase_, an oxygen of the phosphoryl group of galactose-1-phosphate attacks the α phosphorus atom of UDP-glucose, displacing glucose-1-phosphate. The third reaction is catalyzed by an _epimerase_.

Galactosemia, a rare genetic disease, is caused by the absence of uridyltransferase activity. Because infants with that condition cannot metabolize the galactose derived from milk, it accumulates in their blood. Perhaps surprisingly, in view of the fact that galactose is an ordinary hexose, this accumulation has very serious consequences, and if untreated can be fatal. Symptoms include intestinal disturbances, severe damage to the liver and the central nervous system, and cataracts in the lens of the eye. However, if the condition is recognized in time and a galactose-free diet supplied, galactosemic infants can develop normally.

Since lactose is a major component of milk, young mammals necessarily secrete into the intestines *lactase*, an enzyme that hydrolyzes the disaccharide. In nature, lactose is not ingested after weaning. As part of the extensive changes in enzyme synthesis that are programmed into development, the young of each species of mammal stop production of lactase soon after the usual age of weaning. If lactose is not digested in the small intestine, it will pass on into the colon and be metabolized by bacteria, with production of acidic and gaseous products. Therefore most humans, like all other mammals, are subject to severe gastrointestinal disturbances, which may be fatal to young children, if they consume fresh milk products extensively. Yoghurt and other fermented preparations cause no difficulty because the lactose is consumed by bacteria during the fermentation. There is one known exception to this susceptibility to lactose. In human populations whose ancestors have kept domestic cattle and used milk and milk products (most western Europeans, many Middle Eastern peoples, and some African tribes), the synthesis of lactase does not turn off at the usual time but continues throughout life. Since the domestication of cattle occurred only a few thousand years ago, this is a striking illustration of the speed with which a trait can be established by selection. In milk-consuming populations, a mutation that caused continued production of lactase would increase the chance of survival through childhood, and so it spread through the population.

ENERGETICS AND REGULATION OF GLYCOLYSIS AND GLUCONEOGENESIS

The advantage that the cell or organism gains from the organization of metabolism into pairs of oppositely directed sequences is the ability to regulate the directions of metabolic conversions in response to metabolic needs, as these are reflected in the concentrations of specific metabolites that serve as signals (see Chapter 13). We may consider the whole glycolysis-gluconeogenesis system as a large pseudocycle, or as several small pseudocycles that are embedded in these sequences. When we consider overall stoichiometries and energetics, the broad view is appropriate; when we focus on mechanism and regulation, we must consider the small pseudocycles individually.

There are three small pseudocycles in the paired sequences between glycogen and pyruvate, and also three between free glucose and pyruvate (Figure 14–17). The difference of stoichiometries in each of these small pseudocycles is one ATP. Because the pyruvate-phosphoenolpyruvate conversion must occur twice for each molecule of hexose that is used or produced, that pseudocycle contributes a difference of two ATPs for each hexose, so that the stoichiometric difference between glycolysis and gluconeogenesis is four ATPs per hexose.

Several features of the glycolysis-gluconeogenesis system indicate that regulation of these sequences must be particularly important to the organism. In mammals, and in many other kinds of organisms as well, *these sequences*, together with the TCA cycle into which pyruvate flows in the presence of oxygen (see Chapter 15), *have the highest fluxes of all metabolic sequences. Not only are the fluxes high, but they change direction more frequently than the fluxes through most other pathways.* In mammalian liver, for example, there is massive glycogen production after a meal. If the meal was high in protein, much of the glucose that is incorporated into glycogen is produced from pyruvate and oxaloacetate. These metabolites are products of the catabolism of several amino acids (Chapter 19), and oxaloacetate is also produced indirectly from amino acids that are degraded to α-ketoglutarate or succinate (Chapter 19). During a period of fasting, much or all of the glycogen, depending on the length of time until the next meal, is

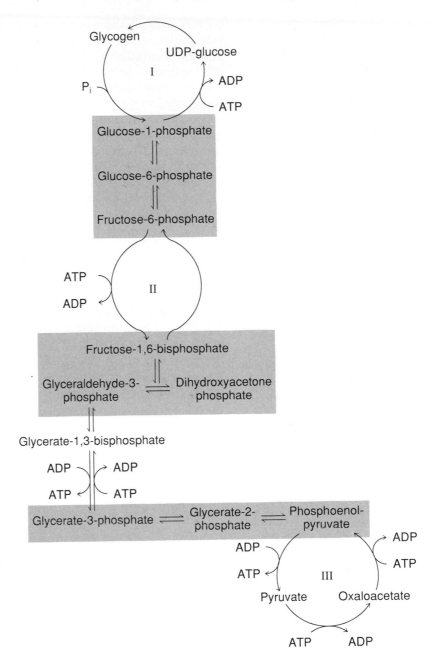

Figure 14–17
Energetic relationships in glycolysis and gluconeogenesis. Points at which ATP is produced or consumed are indicated. Compounds in the same metabolic pool are indicated by a colored box. There are three small pseudocycles (I, II, III) in the paired sequences between glycogen and pyruvate.

converted to glucose-1-phosphate and metabolized to pyruvate. In addition, this pathway and the tricarboxylic acid cycle provide most of the starting materials for the synthetic activities of the cell.

Biochemists are only beginning to understand the regulatory interactions that are related to the control of these sequences so as to balance the availability of substrates with the many demands on glycolysis and the partitioning of resources among those demands. Here we will focus on what seem to be the most important, or most general, aspects of our present incomplete knowledge of the regulation of glycolysis and gluconeogenesis.

In considering regulation of these sequences, it is important to keep in mind the distinction between liver and more typical tissues or cells. Because of its large size, easy accessibility, and high levels of a wide range of enzyme activities, the liver has been the most common source of enzymes for study by biochemists. But the resulting familiarity with liver enzymes, together with the metabolic importance of the liver, may lead us to consider liver metabolism to be representative of metabolism generally. That is not true, and the regulation of the glycolysis-gluconeogenesis system is a particularly good example of that fact.

In a unicellular organism, glycolysis and gluconeogenesis will ordinarily be regulated individually. That is, when nutrients are in short supply the cell may draw on its stored reserves of glycogen, and convert the resulting glucose-1-phosphate to lactate or ethanol and CO_2 under anaerobic conditions, or to CO_2 in the presence of oxygen. If lactate, amino acids, or protein is available in excess, the same cells will reverse the whole sequence all the way from pyruvate and oxaloacetate to glycogen. The cell lives by and for itself, and those are the obviously appropriate responses.

The situation in the liver is quite different. Although it must, of course, keep its own cells functional, *the liver is part of a complex organism, and its main function is regulation of the composition of the blood* (except for inorganic blood components, which are regulated primarily by the kidney). Therefore *the liver's massive storage of glycogen is not for its own later use, but for use in supplying glucose to the blood* as necessary to maintain a nearly constant blood glucose level of about 5 mM.

After a high-protein meal, the proteins are hydrolyzed in the intestine by enzymes secreted by the pancreas and the cells lining the intestine, and the resulting amino acids are absorbed by the intestinal mucosa and passed on to the blood, from which they are taken up by several tissues and organs, notably the liver. Much of the carbon of the amino acids is incorporated into storage glycogen. Thus far, the chain of events is in principle similar to that in a unicellular organism. It is some time after the meal, when the blood glucose level tends to fall because of uptake by the tissues, that the differences between the liver and cells that are energetically self-sufficient are most evident. Now the liver must supply glucose to the blood, and any process that generates glucose should be activated. Glycogen is broken down, but the production of hexose phosphates from pyruvate and oxaloacetate would also be useful. We will see that regulatory interactions in the liver respond to that need.

When the blood glucose level is low, material may flow into the hexose phosphate pool from both directions—from glycogen and from pyruvate and oxaloacetate—simultaneously. Glucose-6-phosphate is hydrolyzed, and free glucose is supplied to the blood. The replenishment of the hexose-phosphate pool, both from breakdown of glycogen and from gluconeogenesis simultaneously, is appropriate only because of the liver's special role in stabilizing blood concentrations. It would never be useful to a cell or tissue that metabolized for its own sake. Thus *gluconeogenesis occurs in many kinds of cells and unicellular organisms under conditions of nutrient abundance, when storage is appropriate, but in the liver it may increase not only when nutrients are abundant but also under the opposite condition of energy deficiency, when the level of blood glucose is low*. Indeed, gluconeogenesis becomes especially important in the liver when nutrients are limited for an extended period. Under those conditions, much protein from the liver itself and from other tissues, including skeletal muscle, is degraded to supply amino acids for use by the liver in generating glucose for the blood.

These considerations illustrate the important principle that *function must always be taken into account in considerations of regulatory interactions*. Many patterns, such as end-product negative feedback and response to energy charge, are similar in most types of cells but, especially in higher animals, with their extensive specialization of function between different tissues and organs, important differences have evolved to meet functional needs.

Hormones and Other Factors Regulate Glycogen Metabolism

Storage compounds are useful only if they can be made available when needed. Therefore the production and utilization of storage polysaccharide must proceed rapidly in response to metabolic conditions, especially the demand for energy and its availability.

Glycogen phosphorylase, which catalyzes the phosphorolytic cleavage of the bond between the terminal glucosyl residue and the next residue in the polyglucose chain, was mentioned briefly earlier in this chapter. It _is part of a complex regulatory system,_ and study of its properties has led to an extraordinary range of findings that contribute to our understanding of mechanisms and relationships in metabolic regulation of several kinds.

Investigations have revealed that the activity of glycogen phosphorylase is stimulated by AMP. This was the first indication of modulation of enzyme behavior as a consequence of noncovalent binding of a modifier molecule. It has been discovered also that the major form of the enzyme under most conditions, _phosphorylase b,_ can be converted by phosphorylation to a form, termed _phosphorylase a,_ with different properties. This was the first indication of modulation of enzyme behavior by covalent modification of the enzyme itself. Modulations of this type are very important in metabolic regulation. Both of those findings were made by Carl and Gerty Cori.

As a rule an increase in the concentration of AMP indicates a decrease in ATP concentration and in the value of the energy charge. Activation of glycogen phosphorylase is an appropriate response to an increased level of AMP, since an increase in the rate of production of glucose-1-phosphate will lead to an increase in the rate at which ATP can be regenerated.

However, since the liver's metabolic functions are not dictated by local needs but are directed toward the control of blood composition, additional regulatory inputs are needed. _As in many other cases in which the activities of different tissues and organs are correlated for the good of the whole organism, hormones are involved in this additional level of control_ (see Chapter 31).

When the level of glucose in the blood falls slightly, small clusters of cells in the pancreas (A or α cells of the islets of Langerhans) sense the decrease. They respond by releasing peptide hormone into the blood. On reaching the liver, the hormone _glucagon_ causes the conversion of phosphorylase b to phosphorylase a. The activity of the enzyme increases and it becomes independent of AMP control. Thus the hormone, reflecting a systemic (organism-wide) need, overrides the AMP effect that provides control in response to local needs. Such override of local control to satisfy systemic needs is a common feature of hormone action. _Epinephrine,_ a small hormone derived from tyrosine (Chapter 31), affects phosphorylase in the same way as does glucagon. Epinephrine is made in the adrenal glands and is liberated into the blood in response to signals from the central nervous system. Thus the two hormones have a similar effect, but they originate from different sources, and, as we shall see, they affect different cell types.

In the 1950s Earl Sutherland demonstrated activation of phosphorylase by epinephrine and glucagon in cell-free systems. He found that when he centrifuged a broken cell extract, the membrane-containing fraction, if supplied with ATP and either epinephrine or glucagon, produced a low-molecular-weight factor that could be isolated and would stimulate the activity of glycogen phosphorylase in the supernatant fraction. This was part of the proof that these hormones cause a membrane-bound enzyme in the particulate fraction to convert ATP to cyclic AMP (Figure 14–18; also see Figure 3–13). The cell membrane contains receptor molecules for glucagon and epinephrine on its outer surface. When either hormone binds to its receptor,

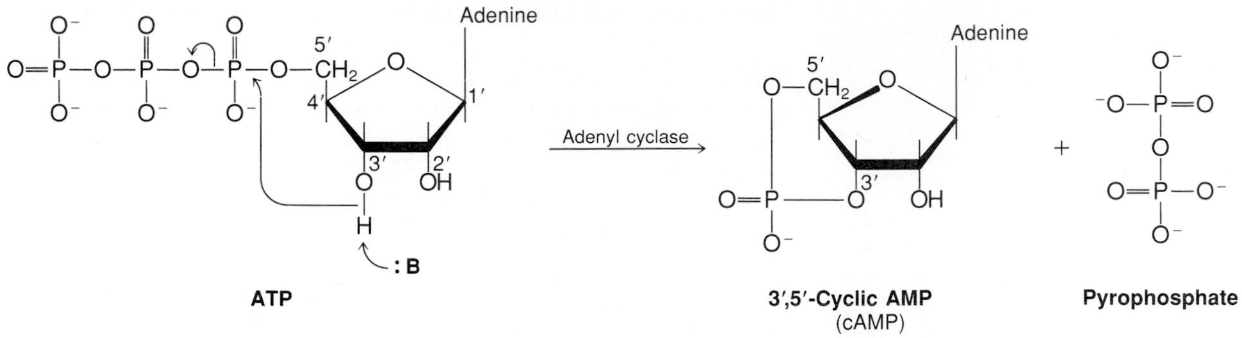

Figure 14–18
Synthesis of cyclic AMP. A catalytic site on adenyl cyclase (B :) removes a proton from the C-3 oxygen, which then attacks the α phosphate and displaces the pyrophosphate group.

the enzyme _adenylate cyclase,_ on the inner surface of the membrane, is stimulated to catalyze the production of cyclic AMP. As is discussed in Chapter 31, this activation of the cyclase involves GTP and a GTP-binding protein called the G protein.

In liver cells and many other kinds of cells, cyclic AMP activates a _protein kinase._ This enzyme exists in an inactive form in which two catalytic subunits (C) are associated with two regulatory subunits (R), which block their activity. Cyclic AMP binds to the regulatory subunits and causes a conformational change that results in the dissociation of the R and C subunits (Figure 14–19). The free C dimers are active, and catalyze the transfer of a phosphoryl group from ATP to specific sites on a number of other enzymes, causing changes in their catalytic behavior. In many cases the specificity of hormones for certain types of cells and for certain effects in those cells results from the presence of appropriate receptors on the surfaces of the target cells, and from the presence of appropriate regulatory responses when specific sites on specific enzymes in the target cells are phosphorylated by protein kinases.

In liver cells, one of the proteins that are substrates for the cyclic-AMP-dependent protein kinase is _phosphorylase kinase_, which by being phosphorylated is converted from an inactive to an active form. Phosphorylase kinase then catalyzes the conversion of phosphorylase _b_ to phosphorylase _a_ by phosphorylation of a specific serine residue. As noted above, this conversion eliminates the dependence of glycogen phosphorylase on AMP. The entire cascade of effects is shown schematically in Figure 14–20. The phosphoryl donor for all of these phosphorylations is ATP.

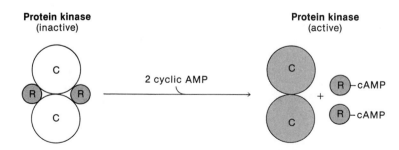

Figure 14–19
Activation of protein kinase by cyclic AMP. Two regulatory subunits (M_r = 82,000) bind tightly to two catalytic units (M_r = 49,000), inactivating them. When cyclic AMP binds to the regulatory subunits, a conformational change occurs that results in the dissociation of the subunits and the simultaneous activation of the catalytic subunits.

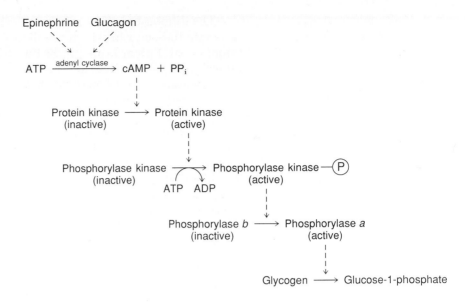

Figure 14–20

Molecular basis for the stimulation of glycogen breakdown by the hormones epinephrine and glucagon. The epinephrine or the glucagon activates adenyl cyclase, which activates protein kinase, which activates phosphorylase kinase. Phosphorylase kinase in turn activates phosphorylase, which cleaves glucose from glycogen. This cascade of regulatory interactions results in a very large amplification, so that a small amount of hormone can have a large effect on the rate of breakdown of glycogen.

Several such regulatory cascades are known. Their apparent advantage is amplification of a regulatory signal. In this case, it can be seen that the binding of a single hormone molecule to a receptor could lead to the production of many molecules of cyclic AMP. Further amplification occurs at each enzymic step in the sequence, so that extremely low concentrations of hormones or other regulatory signals can have large effects on metabolism.

Both glucagon and epinephrine lead to the conversion of phosphorylase *b* to phosphorylase *a*, and both act in the same way, through the production of cyclic AMP as a second messenger. However, the physiological significance of the two hormones is quite different. *Glucagon is part of the system that is responsible for stabilization of the concentration of glucose in the blood.* The counteracting hormone, *insulin*, is another polypeptide hormone. It is synthesized and released by other clusters of cells in the endocrine islets of the pancreas (the B or β cells) in response to an increase in the blood glucose level. Insulin stimulates the uptake of glucose by various tissues and its incorporation into glycogen in liver, and thus tends to cause a decrease in the blood glucose level.

Since the function of glucagon is to stimulate the rate at which the liver supplies glucose to the blood, it is appropriate for it to release phosphorylase from local control. The result is that glycogen will be broken down faster than is necessary to meet the needs of the liver, so that glucose can be supplied to the blood. When blood glucose is low, synthesis of glucose from lactate or amino acids is also appropriate, and glucagon also stimulates gluconeogenesis.

In contrast to the homeostatic function of glucagon (its support of the status quo), *epinephrine is an alarm signal that leads to deviation from the homeostatic status quo.* Its release is controlled by the central nervous system; specifically, it is released in response to fear or anger. Its function is to prepare the body for a sudden extreme energy output if that becomes necessary; it is often termed the "fight or flight" hormone. In addition to many physiological effects, such as dilation of the air passages leading to the lungs and redirection of the circulatory system to favor skeletal muscles, epinephrine has several biochemical effects, including conversion of phosphorylase *b* to *a* in liver and skeletal muscle. The result is to build up the levels of glycolytic intermediates, and presumably to eliminate a time lag in energy mobilization when and if the organism begins either to fight or to flee.

The differential effects of glucagon and epinephrine on liver and muscle illustrate the ways in which regulatory responses have evolved to fit functional needs. Epinephrine causes the conversion of phosphorylase *b* to phosphorylase *a* both in liver and in muscle, since it is desirable to increase the concentrations of intermediates both locally (in the muscle) and in reserve (in the liver). In contrast, glucagon, which has the same effect on liver phosphorylase as epinephrine, does not affect muscle phosphorylase. This is because muscle cells lack glucagon receptors. Muscle is a consumer tissue and does not contribute to stabilization of the level of glucose in blood; thus it should not, and does not, break down its glycogen reserves in response to low blood glucose.

In its normal or base state, *glycogen synthase* is relatively active. The same cyclic-AMP-dependent protein kinase that phosphorylates phosphorylase kinase also catalyzes the phosphorylation of glycogen synthase. *Whereas phosphorylation of phosphorylase kinase leads to increased activity of glycogen phosphorylase, the phosphorylation of glycogen synthase decreases its activity.* The phosphorylated enzyme is much less active. The result is that when glycogen breakdown is stimulated in response to epinephrine or glucagon, the synthesis of glycogen is prevented. In this way the simultaneous operation of both arms of the glycogen-hexose-monophosphate pseudocycle is prevented.

Regulatory responses must necessarily be reversible if regulation is to be sensitive to changing metabolic needs. All of the effects we have seen are promptly reversed when the need for them has ended. Cyclic AMP is hydrolyzed to ordinary 5'-AMP by an enzyme called *cyclic AMP phosphodiesterase,* and there are *protein phosphatases* that catalyze the hydrolytic removal of the phosphoryl groups that had been attached to proteins by protein kinases.

The Interconversion of Fructose-6-phosphate and Fructose-1,6-bisphosphate Is Regulated in Both Directions

The phosphorylation of fructose-6-phosphate is the first step in the glycolytic breakdown of hexose phosphates. As the enzyme catalyzing this first committed step in glycolysis, phosphofructokinase is appropriately placed to play a major role in regulation of the rate of glycolysis. Effective control can be exerted only on reactions that are thermodynamically unidirectional (those with a large decrease in standard free energy). And, as we suggested when discussing the individual enzymes of glycolysis, a plausible reason for the use of ATP at the first step in an ATP-regenerating sequence is to introduce a unidirectional step and thus facilitate kinetic regulation at this strategic point.

Kinases usually occur in synthetic or storage pathways, where it is appropriate for the rate of the reaction to increase as the energy charge (or the ATP/ADP ratio) increases. Such control can be easily produced by competition between ATP and ADP at the catalytic site, and the regulatory effect can be fine-tuned to metabolic needs by evolutionary adjustment of the relative affinities of the catalytic site for ATP and ADP. The regulatory problem at the phosphofructokinase reaction is entirely different. Although the enzyme is a kinase, catalyzing transfer of a phosphoryl group like any other kinase, it catalyzes the first step in a sequence in which ATP is regenerated. Thus *it is necessary for phosphofructokinase to respond to fluctuations in the value of the energy charge* (or the ATP/ADP ratio) *in the direction opposite from the responses of typical kinases.* A separate regulatory site with these properties has evolved.

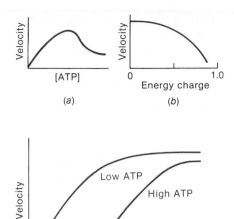

Figure 14–21

Some aspects of the behavior of fructokinase, as observed in experiments *in vitro*. (*a*) When the concentration of fructose-6-phosphate is constant and in the absence of ADP, the rate of reaction increases as the concentration of the second substrate, ATP, increases. However, at high concentrations, ATP inhibits because of increased binding at the regulatory site. (*b*) When assayed at different values of the adenylate energy charge, phosphofructokinase responds like a typical enzyme of a catabolic sequence. (*c*) The primary regulatory effect of ATP when bound at the regulatory site is to decrease the affinity of the enzyme for the other substrate, fructose-6-phosphate. The kinetic order of the reaction is also increased from about first order at low concentrations of ATP to about second order at high concentrations.

Phosphofructokinase Is Regulated by Energy Charge and Citrate

The activity of phosphofructokinase as a function of ATP, energy charge, and fructose-6-phosphate has been studied *in vitro*. The results are instructive in resolving effects at the catalytic and regulatory sites. When the reaction rate at a constant concentration of fructose-6-phosphate is plotted as a function of the concentration of ATP, the resulting curve rises to a peak and then declines as the concentration of ATP is increased further (Figure 14–21*a*). This *in vitro* arrangement is unrealistic since in a metabolizing cell ATP, ADP, and AMP are all present, and the sum of their concentrations is virtually constant. When the rate of the reaction catalyzed by phosphofructokinase at a constant concentration of fructose-6-phosphate is measured *in vitro* as a function of the energy charge, the response obtained is of the type typical of, and appropriate for, a sequence that regenerates ATP (Figure 14–21*b*).

The increase in rate with ATP concentration at low ATP concentrations in Figure 14–21*a* results from an increase in the fraction of catalytic sites at which ATP is bound. Because of the virtually constant total concentration of adenine nucleotides and the high affinity of the catalytic site for ATP, the curve for low ATP concentrations has no metabolic significance; the catalytic site *in vivo* will almost always be nearly saturated with ATP. However, the same curve at high ATP concentrations reflects the modulation resulting from ATP binding at the regulatory site. This is seen in a more functionally relevant way in Figure 14–21*b*, where the activity is plotted against the energy charge.

The actual consequence of ATP binding at the regulatory site is to increase the K_m, that is, to lower the affinity for the substrate fructose-6-phosphate at the catalytic site (Figure 14–21*c*). This decrease in affinity causes a decrease in rate of the reaction if the concentration of fructose-6-phosphate remains constant. *In vitro* it has been observed that the inhibitory effect of ATP on the rat liver phosphofructokinase is decreased by ADP, so that the regulatory response depends on the ATP/ADP ratio. For the yeast enzyme, the response is to the ATP/AMP ratio. In both cases, the response can be expressed in terms of the energy charge, which aids in the comparison of results obtained *in vitro* with conditions *in vivo*.

Phosphofructokinase is an unusual kinase in another respect, its ability to use ITP, GTP, CTP, or UTP as phosphoryl donor. However, the regulatory site is specific for ATP and either ADP or AMP. This higher specificity at the regulatory site probably has evolved because of the fundamental importance of the adenine nucleotides in the stoichiometric coupling between sequences, in energy transduction, and in regulation and correlation of metabolic sequences.

The catalytic behavior of phosphofructokinase is also modulated by the concentration of citrate, which intensifies the effect of ATP, at least in part by increasing the affinity of the regulatory site for ATP. At high concentrations, citrate also favors dissociation of phosphofructokinase from an active tetramer of identical subunits to an inactive dimer. It seems likely that the concentration of citrate in the cytosol increases with an increase in the energy charge inside the mitochondrion (Chapter 15). A normal level of citrate also indicates that the supply of biosynthetic starting materials (most of which are intermediates of glycolysis or the TCA cycle) is adequate. Thus it is appropriate that a decrease in the concentration of citrate should increase the activity of phosphofructokinase and thereby increase the flux through the glycolytic sequence, with consequent increase in the rate of regeneration of ATP and the rate of replenishment of the pools of starting materials. The concentration of citrate in the cytosol is an important regulatory input into other sequences, notably the synthesis of fatty acids. This subject will be considered in Chapters 15 and 21.

Fructose-bisphosphate Phosphatase Is Regulated by the AMP Level

On the basis of present knowledge, the regulation of the conversion of fructose-1,6-bisphosphate to fructose-6-phosphate, catalyzed by fructose-bisphosphate phosphatase, seems to be less complex than that of phosphofructokinase. This enzyme responds to slight changes in the value of the energy charge by sensing the concentration of AMP. It differs in that respect from the majority of regulatory enzymes, which respond to either the ATP/ADP or the ATP/AMP ratio. An increase in the concentration of AMP signifies a decrease in the energy charge. Thus the observed effect of AMP as a negative modifier is in the appropriate direction: gluconeogenesis is desirable when the energy charge is high and the concentration of AMP is correspondingly low.

Fructose-2,6-bisphosphate Regulates the Flux Between Fructose-6-phosphate and Fructose-1,6-bisphosphate

It is, of course, the net conversion of fructose-6-phosphate to fructose-1,6-bisphosphate or of fructose-1,6-bisphosphate to fructose-6-phosphate that matters, so regulatory responses of the two enzymes must be considered together. The regulatory responses to energy charge and citrate that we have discussed are generally typical of the enzymes from mammals, yeast, and bacteria. They participate in adjusting the rate of phosphorylation of fructose-6-phosphate to the momentary metabolic needs of the individual cell. However, as in the case of phosphorylase, such local controls can be overridden in mammalian cells, especially in the liver, by hormonal signals that reflect systemic needs of the organism. _In liver_, these _hormonal controls are mediated through the action of the regulatory metabolite fructose-2,6-bisphosphate_, which is generated by phosphorylation of fructose-6-phosphate, in a reaction catalyzed by a kinase that transfers a phosphoryl group from ATP specifically to oxygen at C-2 of the substrate.

When the modifier fructose-2,6-bisphosphate binds to liver phosphofructokinase, the affinity of the enzyme for fructose at the catalytic site is strongly increased; as a result, the rate of glycolysis also increases. At the same time, the affinities of the enzyme's regulatory sites for ATP and citrate are decreased, which adds to the stimulatory effect. The modifier also inhibits the activity of fructose-bisphosphate phosphatase. Thus an increase in the concentration of fructose-2,6-bisphosphate stimulates glycolysis and inhibits gluconeogenesis, and a decrease in the concentration of the modifier favors gluconeogenesis over glycolysis.

The kinase that catalyzes the production of fructose-2,6-bisphosphate in liver is inactivated by phosphorylation. A cyclic-AMP-dependent protein kinase, apparently the same one that is responsible for the phosphorylation of phosphorylase kinase, catalyzes the phosphorylation and hence inactivation of the kinase that synthesizes fructose-2,6-bisphosphate. Quite remarkably, it appears that the phosphorylated kinase is itself the phosphatase that catalyzes the hydrolysis of fructose-1,6-bisphosphate to fructose-6-phosphate. A few other examples are known of enzymes that exist in two forms as a result of covalent modification, and that exert oppositely directed regulatory effects in the two forms.

Thus, in response to a low level of blood glucose, the level of glucagon increases. This causes an increase in the intracellular concentration of cyclic

α-D-Glucose-6-phosphate

Glucose-6-phosphate
dehydrogenase

NADP⁺ →
NADPH + H⁺ ←

6-Phospho-D-gluconolactone

H_2O
H^+
gluconeolactonase

6-Phospho-D-gluconate

NADD⁺
NADPH + H⁺
6-Phosphogluconate
dehydrogenase

3-Keto-6-phospho-D-gluconate

CO_2

D-Ribulose-5-phosphate

AMP. The cyclic-AMP-dependent protein kinase then phosphorylates the enzyme that generates fructose-2,6-bisphosphate, causing its inactivation. The resulting inhibition of glycolysis and stimulation of gluconeogenesis aids the liver in producing glucose to be supplied to the blood. Fructose-2,6-bisphosphate also affects muscle phosphofructokinase, although less strongly. The metabolic significance of this effect in muscle is not well understood.

In our consideration of the regulation of glycolysis and gluconeogenesis, we have seen that regulation occurs at reactions that link metabolic pools, rather than within the pools. Key enzymes at the linkage points are sensitive to regulatory factors that indicate metabolic needs. The resulting changes in enzyme activity and in affinity for substrate modulate reaction rates to meet those needs and to favor nearly unidirectional flow of intermediates, thus minimizing wasteful cycling. Regulation of the phosphoenolpyruvate-pyruvate pseudocycle will be discussed in the next chapter because of its close involvement in the TCA cycle.

THE PENTOSE PHOSPHATE PATHWAY

Many kinds of organisms and some mammalian organs, notably liver, contain an alternative pathway for the oxidation of hexoses. Rather than producing a cleavage into trioses, this *pentose phosphate pathway* consists of oxidation of glucose-6-phosphate to carbon dioxide and a five-carbon sugar phosphate. The carbon atoms of the pentose are then shuffled around in various ways by two enzymes that catalyze transfers of two-carbon and three-carbon pieces between molecules. The final products can contain from three to seven carbon atoms. They are triose phosphate, which can feed into the glycolytic pathway; erythrose-4-phosphate, which is used in the synthesis of several amino acids; ribose-5-phosphate, which is required in the synthesis of nucleic acids (Chapter 25); or regenerated hexose phosphate. The major functions of this pathway are generation of NADPH and supply of ribose-5-phosphate.

NADPH Is Generated by Oxidation of Glucose-6-phosphate

NADPH is required for many biosynthetic sequences. It is generated in different kinds of cells by a variety of reactions, including an $NADP^+$-linked oxidation of malate to pyruvate and CO_2 and transfer of hydride ion from NADH to $NADP^+$ in a mitochondrial reaction that is driven by metabolic energy. However, in some cases, including the mammalian liver, a major part of the NADPH requirement is met by oxidation of glucose-6-phosphate to ribulose-5-phosphate and CO_2. The four electrons that are released by the oxidation are transferred to $NADP^+$.

The reactions in this conversion are shown in Figure 14–22. The first oxidation, catalyzed by glucose-6-phosphate dehydrogenase, is at C-1, converting the hemiacetal derivative of the aldehyde group to the lactone of the corresponding acid, 6-phosphogluconic acid. After hydrolysis of the lactone, the second oxidation is at C-3, converting the secondary alcohol to a ketone. The expected product, 3-keto-6-phospho-gluconic acid, is decarboxylated to yield ribulose-5-phosphate, which is the product released by the enzyme.

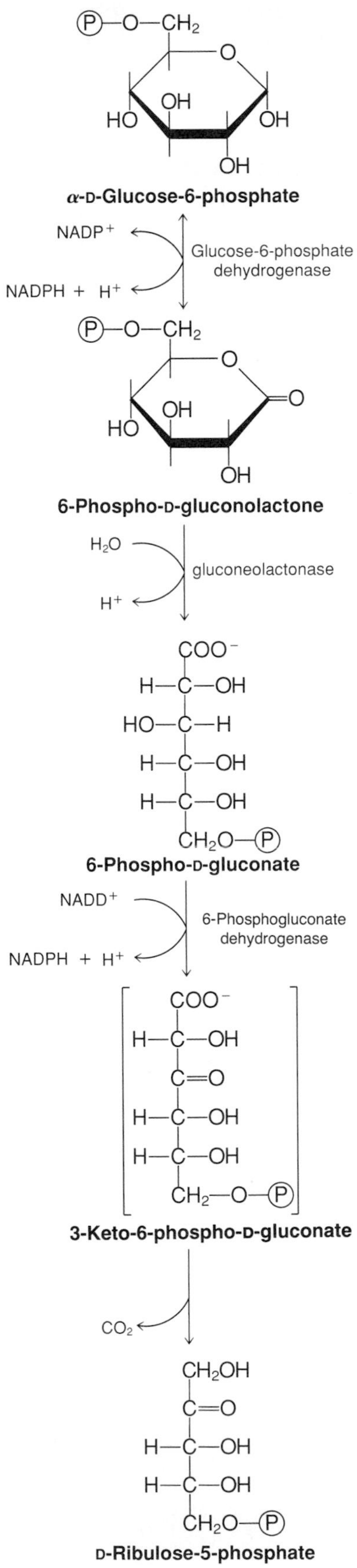

Figure 14–22
Stage 1 of the pentose phosphate pathway: the oxidation of glucose-6-phosphate to ribulose-5-phosphate and CO_2 and the reduction of $NADP^+$. Net Reaction: Glucose-6-phosphate + 2 $NADP^+$ ⟶ Ribulose-5-phosphate + CO_2 + 2 NADPH + 2 H^+.

At this point the metabolic function of the sequence, when it is serving to supply electrons for biosynthesis, has been fulfilled. Two molecules of NADPH have been generated for each molecule of glucose-6-phosphate that was oxidized. It is only necessary to convert ribulose-5-phosphate, the end product of the oxidative sequence, to compounds in the mainstream of metabolism.

Transaldolase and Transketolase Play a Variety of Roles

The key enzymes involved in these conversions—in addition to phosphofructokinase and aldolase, with which you are already familiar—are _transaldolase_ and _transketolase_. The two enzymes are similar in their substrate specificities. Both require a ketose as a donor and an aldose as an acceptor. The steric requirements for C-1 through C-4 are the same as the requirements of aldolase, except that aldolase requires phosphorylation at C-1, and both transaldolase and transketolase require a free hydroxyl group at C-1.

Transaldolase.   Transaldolase catalyzes a two-step conversion. The first step, an aldol cleavage of the bond between C-3 and C-4 of a ketose, is essentially identical to the reaction catalyzed by aldolase. However, the dihydroxyacetone that is produced in the transaldolase reaction from carbon atoms 1, 2, and 3 is not released. Rather, it is held at the catalytic site while the aldose product diffuses away and is replaced by another aldose molecule. An aldol condensation then generates the second product of the reaction, a ketose that contains the first three carbon atoms of the original ketose attached to C-1 of the acceptor aldose (Figure 14–23). The two steps, being an aldol cleavage and an aldol condensation, have no cofactor requirement.

Transketolase.   The reaction catalyzed by transketolase superficially resembles the transaldolase reaction. The substrate specificity is identical, but in

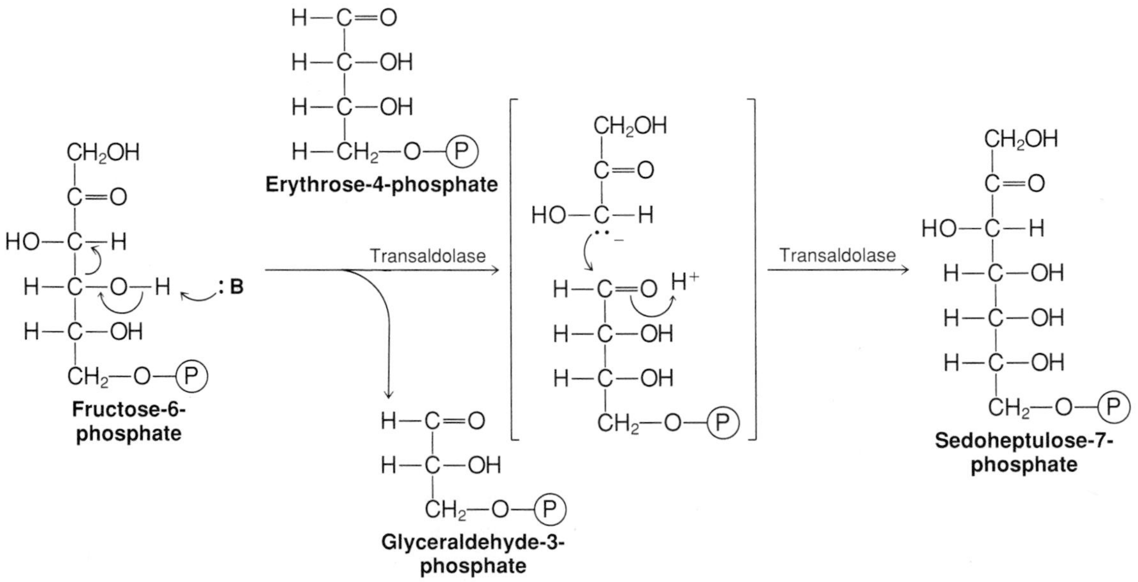

Figure 14–23
The transaldolase-catalyzed conversion of fructose-6-phosphate and erythrose-4-phosphate to glyceraldehyde-3-phosphate and sedoheptulose-7-phosphate. This is a two-step conversion. The first step is similar to the aldolase reaction except that the dihydroxyacetone produced is held at the catalytic site, while the aldose product diffuses away and is replaced by another aldose molecule. The second step involves an aldol condensation.

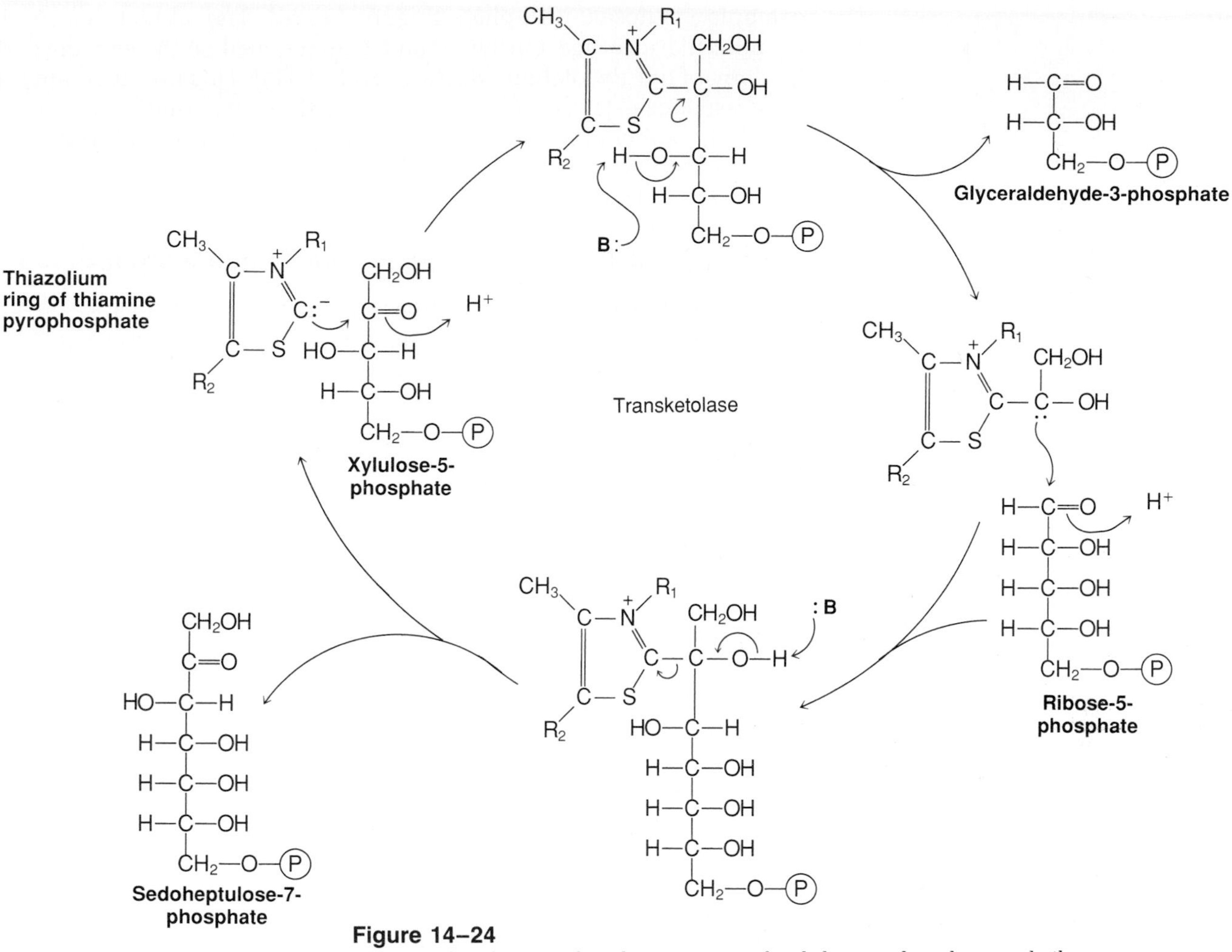

Figure 14–24

The transketolase-catalyzed conversion of xylulose-5-phosphate and ribose-5-phosphate to glyceraldehyde-3-phosphate and sedoheptulose-7-phosphate. Although the aldolase and ketolase reactions superficially resemble each other, they proceed by very different mechanisms. This is because in the aldolase reaction the carbon adjacent to a carbonyl acts as a nucleophilic agent, whereas the ketolase reaction involves an intermediate with a negative charge on what was originally a carbonyl carbon. The latter type of reaction is more complex and requires thiamine pyrophosphate.

this case the cleavage is between carbons 2 and 3 of the ketose. A two-carbon moiety is retained on the enzyme following cleavage and is subsequently transferred to an acceptor aldose. But in spite of the apparent similarity, this reaction is chemically quite different from the transaldolase reaction. Aldol condensations and cleavages, which occur readily in mildly alkaline solution, are feasible because protons on a carbon atom adjacent to a carbonyl carbon are moderately acidic. The carbanion formed on dissociation of such a proton can participate in a nucleophilic addition to the carbonyl carbon of another molecule of aldehyde or ketone, as in the transaldolase reaction. A ketol condensation, in contrast, would involve the carbonyl carbon as the nucleophile. That is not energetically feasible because the polarity of the carbonyl bond precludes a negative charge on carbon.

Transketolase is one of several enzymes that catalyze reactions that involve intermediates with a negative charge on what was initially a carbonyl carbon atom. All such enzymes require thiamin pyrophosphate (TPP) as a cofactor (Chapter 11). The transketolase reaction is initiated by addition of the thiamin pyrophosphate anion to the carbonyl of a ketose phosphate, for example xylulose-5-phosphate (Figure 14–24). The adduct next undergoes an aldol-like cleavage. Carbons 1 and 2 are retained on the enzyme surface in the form of the glycolaldehyde derivative of TPP. This intermediate can exist as a resonance-stabilized carbanion, and thus can condense with the carbonyl of another aldolase. If the reactants are xylulose-5-phosphate and ribose-5-phosphate, the products will be 3-phosphate-glyceraldehyde and the seven-carbon ketose sedoheptulose-7-phosphate (Figure 14–24).

Production of Ribose-5-phosphate and Xylulose-5-phosphate

Both transaldolase and transketolase require as substrates a ketose phosphate as the donor molecule and an aldose phosphate as the acceptor. Further, both enzymes require at carbons 3 and 4 the same steric configuration as is found in glucose and fructose. Ribulose-5-phosphate, the first pentose phosphate to be formed in the pentose phosphate pathway, does not have the correct configuration to serve as a substrate for transaldolase and transketolase. Both a suitable donor ketose and an acceptor aldose can be made by isomerizations of ribulose-5-phosphate, however, and enzymes that catalyze those isomerizations are found in cells that can carry out the pentose phosphate pathway (Figure 14–25).

Conversion of ribulose-5-phosphate to ribose-5-phosphate is a simple ketose-aldose isomerization similar to the interconversion of glucose-6-phosphate and fructose-6-phosphate. Conversion of ribulose-5-phosphate to xylulose-5-phosphate involves epimerization at C-3, and an enediol intermediate seems likely. Once a donor molecule, xylulose-5-phosphate, and an acceptor, ribose-5-phosphate, have been generated, conversion to triose phosphate and hexose phosphate is straightforward. Because of the alternative pathways that are possible, and that probably occur simultaneously, no single set of reactions will describe these conversions uniquely.

One possible set of pathways is shown in Figure 14–25. If the triose phosphate formed is converted to hexose phosphate, the overall pathway can be seen as regenerating five molecules of hexose phosphate for each six used initially.

$$6 \text{ glucose-6-phosphate} + 12 \text{ NADP}^+ \longrightarrow$$

$$6 \text{ CO}_2 + 5 \text{ fructose-6-phosphate} + 12 \text{ NADPH} \quad (30)$$

The pathway was first formulated in this form, hence the name pentose phosphate cycle.

Alternatively, it is possible to write a sequence of reactions, including action of phosphofructokinase and aldolase on seven-carbon intermediates, in which the carbon of ribulose-5-phosphate is converted mainly to glyceraldehyde-3-phosphate. Such a pathway, with the triose-phosphate entering the glycolytic sequence, would amount to a bypass or shunt around the first reactions of glycolysis, and the name *hexose monophosphate shunt* is sometimes used. It will be evident that any amount of ribose-5-phosphate or erythrose-4-phosphate that may be needed for biosynthetic sequences can also be obtained from this oxidative pentose phosphate pathway.

Since the transaldolase and transketolase reactions are symmetrical with respect to types of bonds cleaved and formed, their equilibrium constants are near 1. The pool of sugar phosphates is thus near equilibrium in cells that contain these enzymes. In a water-flow analogy, the sugar phosphates and the reactions that interconvert them resemble a large swamp, with ill-defined flows along many interconnecting channels. Water may be fed in from any direction, and may leave the swamp in any direction.

The swamp analogy is quite realistic for the reactions catalyzed by transaldolase and transketolase and the associated enzymes. The physiological function of these enzymes is to facilitate the conversion of one kind of sugar

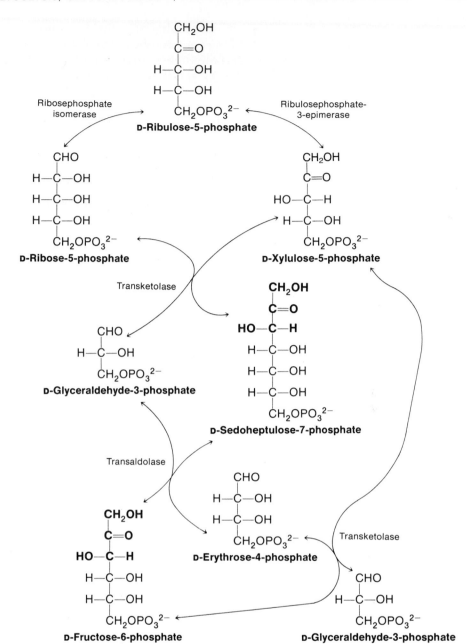

Figure 14–25

Stage 2 of the pentose phosphate pathway. The groups with color background are those transferred in transketolase-catalyzed reactions. The groups in bold type are transferred in the transaldose-catalyzed reactions. All of the reaction arrows in this figure are double-headed to indicate that the reactions can go in either direction with little change in free energy.

to another, and depending on the type of cell involved and on metabolic conditions, any of the component sugar phosphates may be added to the pool, and any others removed.

In addition to the regeneration of NADPH and the synthesis of ribose-5-phosphate, the reactions of the pentose phosphate pathway are used also in photosynthesis. For each molecule of hexose phosphate that is produced in photosynthesis there are ten molecules of glyceraldehyde-3-phosphate that must be converted to six molecules of ribulose-5-phosphate, which after phosphorylation serves as the acceptor for CO_2 in continuing photosynthesis (Chapter 17). In our water analogy, we may say that in the pentose phosphate pathway in liver, for example, pentose phosphate flows into the swamp at one point and triose phosphate or hexose phosphate flows out at another. In photosynthesis, triose phosphate is dumped into the swamp at one point and pentose phosphate flows out at another. The same enzymes are involved in both cases.

In keeping with their function in regeneration of NADPH, the enzymes of the pentose phosphate pathway are not found in muscle, a tissue in which carbohydrates are utilized almost exclusively for generation of mechanical energy and there is little biosynthetic activity. Liver, in which biosyntheses are important, contains considerable amounts of transaldolase and transketolase, as well as the enzymes that catalyze the oxidation of glucose-6-phosphate to ribulose-5-phosphate and the isomerization of the pentose phosphates.

Regulation of the Pentose Phosphate Pathway

Regulation of the pentose phosphate pathway has not been studied extensively. Both of the oxidative steps are regulated by the NADPH/NADP$^+$ ratio, and stimulation of glucose-6-phosphate dehydrogenase as the energy charge falls slightly has been observed. Clearly, in cells that contain the enzymes of the pentose phosphate pathway, the ratio of rates at which ATP and NADPH are regenerated will be determined largely by partitioning of hexose phosphate between the reactions catalyzed by phosphofructokinase and by glucose-6-phosphate dehydrogenase. Since both glycolysis and the pentose phosphate pathway lead to regeneration of ATP, it seems probable that the most important input into regulation at this branchpoint is the NADPH/NADP$^+$ ratio.

SUMMARY

The reactions that are discussed in this and the next chapter make up the central pathways of metabolism. They are closely related to all aspects of the biochemistry of the cell. They supply the starting materials for biosynthesis of the components of the cell and the reducing power needed for those biosyntheses, and also the starting materials and reducing power for long-term storage of energy as fat.

Quantitatively, the most important role of these pathways lies in the generation of energy to power the activities of the cell or organism, and in the short-term storage of energy as polysaccharides (starch and glycogen). The parts of the pathway that do not require oxygen and that are found in nearly all cells, aerobic or anaerobic, are discussed in this chapter. The discussion emphasizes the energetic aspects of glycolysis (the conversion of hexoses or storage polysaccharides to three-carbon compounds) and of gluconeogenesis (the conversion of three-carbon compounds to hexoses or storage polysaccharides).

The intermediates of glycolysis and gluconeogenesis are phosphorylated. The breakdown of storage polysaccharide yields glucose-1-phosphate, and the first step in the metabolism of free glucose is phosphorylation to glucose-6-phosphate. Those two compounds are freely interconvertible with fructose-6-phosphate. The three sugar phosphates make up the hexose phosphate pool, which is a major crossroads of energy metabolism. Glucose-1-phosphate is produced from storage polysaccharides and is used in their synthesis; glucose-6-phosphate is produced from glucose and is the precursor for blood glucose in mammals and the starting material for the pentose phosphate pathway; fructose-6-phosphate is produced by gluconeogenesis and photosynthesis and is the starting material for glycolysis.

Glycolysis and gluconeogenesis consist of pools of intermediates that are close to equilibrium. Most of the reactions that connect the pools are far from equilibrium in the metabolizing cell. The pools are the hexose phosphate pool (glucose-1-phosphate, glucose-6-phosphate, and fructose-6-phosphate), the aldolase pool (fructose-1,6-bisphosphate, glyceraldehyde-3-phosphate, and dihydroxyacetone phosphate), and the three-carbon pool (glycerate-3-phosphate, glycerate-2-phosphate, and phosphoenolpyruvate). Within each pool the direction of reaction is determined by simple mass-action considerations. The direction of overall conversion (glycolysis or gluconeogenesis) depends on the rates at which the connecting reactions remove materials from some pools and supply materials to others.

The connecting reactions (except for those catalyzed by glyceraldehyde-3-phosphate dehydrogenase and glycerate-3-phosphate kinase, which connect the aldolase pool and the three-carbon pool) have large free energies of reaction and are far from equilibrium in the cell. They are subject to regulation by signals that indicate the metabolic needs of the cell or organism. As a result of those regulatory interactions, polysaccharide is stored when nutrients are available in excess of current needs, and storage compounds are broken down to supply energy when necessary.

In mammals, the liver is a special case. It is a major site of storage of glycogen, and it regulates the level of glucose in blood. When blood glucose is low, the liver replenishes it by hydrolysis of glucose-6-phosphate, which is made simultaneously by breakdown of glycogen and by gluconeogenesis. When the liver's supply of glycogen is depleted, blood glucose is replenished by gluconeogenesis, and body proteins are broken down to obtain amino acids as starting materials.

SELECTED READINGS

Atkinson, D. E., *Cellular Energy Metabolism and Its Regulation*. New York: Academic Press, 1977. A general discussion, covering some topics in this and the preceding and following chapters in somewhat greater depth than the treatment in this book.

Beitner, R., *Regulation of Carbohydrate Metabolism*. Boca Raton, FL: CRC Press, 1985. Two-volume discussion, at a rather advanced level, of many aspects of the regulation of mammalian carbohydrate metabolism with frequent discussions of clinical conditions. Many references.

Dawes, E. A., *Microbial Energetics*. Glasgow and London: Blackie [New York: Chapman and Hall], 1986. Discussion of bacterial metabolism, with emphasis on the central pathways and the generation and use of ATP. A good source for learning something of the diversity of metabolic adaptations among different groups of bacteria.

Hochachka, P. W., and G. N. Somera, *Biochemical Adaptation*. Princeton: Princeton University Press, 1984. An excellent and extensive discussion of how biochemical processes, including many discussed in this book, are adapted by various types of organisms in fitting themselves for survival under specific and often difficult conditions.

Roehrig, K. L., *Carbohydrate Biochemistry and Metabolism*. Westport, CT: Avi Publishing Co., 1984. Carbohydrate metabolism discussed at a rather elementary level. Covers several topics that are not included in this book, including disorders of carbohydrate metabolism with brief discussions of many types of human genetic diseases in which carbohydrate metabolism is impaired.

PROBLEMS

1. An anaerobic bacterial culture is known to be accumulating lactate as fermentation progresses. Decide which of the following statements about the culture are true (T) and which are false (F).
   a. The culture is almost certainly growing on glucose, since few other compounds can be fermented by bacteria.
   b. The product(s) of fermentation are no more highly oxidized than the starting compounds, since no external electron acceptor is available.
   c. The culture cannot be producing any CO_2.
   d. If the culture is opened and air is bubbled through continuously, the lactate level in the culture medium will continue to increase.
   e. Addition of fluoride ions will result in a rapid increase in the ratio of glycerate-2-phosphate to phosphoenolpyruvate in the bacterial cells.

2. In the absence of oxygen, chemotrophic cells give rise to fermentation products that can be associated with bread, wine, and tired muscles. Explain each of these associations.

3. A cell-free extract of yeast known to contain all the enzymes required for alcoholic fermentation is incubated anaerobically in 100 ml of a medium initially known to contain 200 mM glucose, 20 mM ADP, 40 mM ATP, 2 mM NADH, 2 mM NAD$^+$, and 20 mM P$_i$.
   a. Assuming that the ethanol is removed continuously from the incubation medium as soon as it is formed, what is the maximum amount of ethanol (in millimoles) that can be formed? Explain your answer.
   b. Which of the following changes would be most likely to allow the production of the most additional ethanol, once the culture had reached the maximum of part (a)? Explain your answer. (i) Doubling the glucose concentration of the medium. (ii) Addition of 20 mM glyceraldehyde-3-phosphate. (iii) Addition of 20 mM pyruvate. (iv) Addition of ATPase, an enzyme known to degrade ATP to ADP and P$_i$.

   c. What is the maximum amount of ethanol that can be formed after making the change indicated in part (b)?

4. If a bacterial culture is carrying out alcoholic fermentation of ^{14}C-labeled glucose, in which position(s) of the glucose molecule would the radioactive ^{14}C atoms have to be located in order to ensure that the CO_2 produced during the fermentation process is labeled with ^{14}C?

5. Suppose that there exists a mutant form of glyceraldehyde-3-phosphate dehydrogenase that cleaves the oxidized product from the enzyme surface by hydrolysis rather than by phosphorolysis. What would be the effect, if any, on an obligately anaerobic organism that has the mutant form of the enzyme?

6. The disaccharide sucrose can be cleaved by either of two alternative means:

   $$\text{Sucrose} + \text{H O} \xrightarrow{\text{Invertase}} \text{glucose} + \text{fructose}$$

   $$\text{Sucrose} + \text{Pi} \xrightarrow[\text{phosphorylase}]{\text{Sucrose}}$$

   $$\text{glucose-1-phosphate} + \text{fructose}$$

   a. Assume that the $\Delta G^{\circ\prime}$ value for the invertase reaction is -7.0 kcal/mole. Estimate the $\Delta G^{\circ\prime}$ value for the sucrose phosphorylase reaction.
   b. What advantage would be gained by a cell in carrying out intracellular sucrose cleavage by the sucrose phosphorylase reaction rather than by the invertase route?

7. What concentration of glucose would be in equilibrium with 1.0 mM glucose-6-phosphate, assuming that hexokinase is present and that the ratio of ATP to ADP concentrations is 5? What can you say about the minimal concentration of glucose that would be necessary to maintain the concentration of glucose-6-phosphate at 1 mM in a cell that is actively metabolizing glucose?

8. When phosphoglucomutase is added to an equilibrium mixture of glucose-1-phosphate and glucose-6-phos-

phate, there will of course be no net reaction. If the glucose-6-phosphate is labeled with ^{32}P, would you expect the label to exchange into glucose-1-phosphate? If so, to what extent?

9. The standard free energy change for the hydrolysis of glycerate-2-phosphate is about -3 kcal/mole, while the value for the hydrolysis of phosphoenolpyruvate is about -14 kcal/mole. Since the standard free energy change for the enolase reaction is close to zero, it has sometimes been said that the free energies of formation of glycerate-2-phosphate and phosphoenolpyruvate are therefore nearly identical, and that it is paradoxical that the "bond energies" of the compounds (the free energies of hydrolysis) should be so different. It was thought that there must be a source of the extra energy that was believed to reside in the phosphoryl bond of phosphoenolpyruvate, and the explanation sometimes given was that "energy" that had been widely dispersed in the glycerate-2-phosphate molecule became concentrated in the phosphate bond of phosphoenolpyruvate, from which it could gush forth when the bond was broken. How many erroneous concepts can you spot in that view of the enolase reaction?

10. a. At each of the following concentrations of fructose-1,6-bisphosphate, what would be the equilibrium concentration of glyceraldehyde-3-phosphate in the presence of aldolase, assuming that the concentrations of glyceraldehyde-3-phosphate and dihydroxyacetone phosphate are equal? 1 M, 100 mM, 10 mM, 1 mM, 0.1 mM.
    b. How would the equilibrium ratios that you have just calculated be changed if triose phosphate isomerase were present as well as aldolase?

11. a. What would you expect to be the consequences in *Escherichia coli* of a mutation that caused total absence of phosphoglucokinase, which catalyzes the production of glucose-1,6-bisphosphate?
    b. What would you expect to be the consequences in a mammal of a mutation that caused total absence, in the liver, of the same enzyme? How might the metabolism of lactate and sucrose differ in such a mutant?

12. Define gluconeogenesis.

13. Why was it necessary for cells to evolve alternative pathways for synthesizing glucose, rather than merely reversing glycolysis?

14. What is the relative requirement of energy for synthesizing glucose from lactate, compared with the yield of energy in degrading glucose to lactate?

15. List some biologically important hexoses derived from glucose, and give their functions. What is the predominant nucleoside triphosphate utilized in their synthesis?

16. What are the key steps in the conversion of lactate to glucose?

17. In gluconeogenesis, glucose, which contains four chiral centers (asymmetric carbon atoms) is made from pyruvate, which is not chiral. Show which reaction in gluconeogenesis is responsible for the configuration at each of the four asymmetric carbon atoms of glucose.

18. An organism that possesses all the enzymes of the glycolytic and pentose phosphate pathways has remarkable metabolic flexibility. Consider, for example, the need that bacteria and plants have for erythrose-4-phosphate in the synthesis of aromatic amino acids (see Chapter 24).
    a. Devise a metabolic sequence that will accomplish net synthesis of erythrose-4-phosphate from glucose without the accumulation of any other carbon-containing compounds (except for coenzymes, if needful).
    b. Write an overall equation for the pathway.
    c. What assumptions have you made in devising the pathway?

19. Suppose that a culture of facultative bacteria is maintained under anaerobic conditions with the seven-carbon sugar sedoheptulose as its sole energy source. Assume that the organism possesses all the enzymes of the glycolytic and pentose phosphate pathways, as well as a heptokinase capable of phosphorylating sedoheptulose at the expense of ATP.
    a. Devise a pathway for the fermentation of sedoheptulose to lactate.
    b. Write an overall equation for the pathway.
    c. How does the ATP yield of sedoheptulose fermentation per carbon atom compare with that for glucose fermentation?

20. [3-^{14}C]Glucose is used for pentose phosphate synthesis by a cell extract in the presence of a concentration of fluoride sufficient to inhibit enolase. What positions of the pentose phosphates formed will be radiolabeled?
    a. If synthesis is only by the oxidative pathway, i.e., by means of 6-phosphogluconate.
    b. If synthesis is entirely nonoxidative, i.e., by means of glycolytic intermediates.

21. The metabolic flexibility of the pentose phosphate pathway is illustrated by the ability to carry out net oxidation of glucose-6-phosphate to CO_2. Devise a metabolic sequence that will accomplish the complete oxidation of glucose-6-phosphate to CO_2 by this route.

22. Arsenate (AsO_4^{3-}) is chemically similar to phosphate and can replace it in most, if not all, phosphorolytic reactions. However, arsenate esters are far less stable than phosphate esters and undergo spontaneous hydrolysis. For example, glyceraldehyde-3-phosphate dehydrogenase can use arsenate instead of phosphate to cleave the newly oxidized molecule from the enzyme surface (arsenolysis instead of phosphorolysis). The product, glycerate-1-arseno-3-phosphate, can then undergo nonenzymatic hydrolysis into glycerate-3-phosphate and free arsenate.
    a. In what sense might arsenate be called an uncoupler of substrate-level phosphorylation?
    b. Why is arsenate such a toxic substance for an organism that depends critically on glycolysis to meet its energy needs?
    c. Can you think of other reactions that are likely to be uncoupled by arsenate in the same way that the glyceraldehyde-3-phosphate dehydrogenase reaction is?

15

THE TRICARBOXYLIC ACID CYCLE

In Chapter 14 we saw that only a small fraction of the total free-energy content of glucose (or any other oxidizable substrate) is released under anaerobic conditions. This is because of the limited extent to which the oxidation of an organic substrate can occur in the absence of oxygen. Essentially, catabolism under anaerobic conditions means that every oxidative event in which electrons are removed from an organic compound must be accompanied by a reductive event in which electrons are returned to another organic compound, often closely related to the first compound. Since the difference in free energy in such coupled oxidation-reduction reactions is small, relatively little energy can be obtained from the starting substrate under these conditions. Thus a cell carrying out fermentation of glucose to lactate essentially transfers electrons from glyceraldehyde-3-phosphate by means of NADH to pyruvate and has access to only about 7 per cent of the total free-energy content of the glucose molecule. The cell must accordingly content itself with the generation of only two ATP molecules per molecule of glucose fermented. Most of the energy of the glucose molecule remains untapped in the lactate molecule.

Given access to oxygen, however, the cell can do much more with the oxidizable organic molecules available to it, and the energy yield increases dramatically. With oxygen available as the electron acceptor, the carbon atoms of glucose (or another substrate) can be oxidized fully to CO_2, and all the electrons that are removed during the multiple oxidation events are transferred ultimately to oxygen. In the process, the ATP yield per glucose is almost 20 times greater than that possible under anaerobic conditions. Therein lies the advantage of the aerobic way of life.

OVERVIEW OF AEROBIC CATABOLISM

It is not surprising, then, that aerobic processes capable of extracting further energy from pyruvate have come to play so prominent a role in energy metabolism. The overall process is _aerobic respiratory metabolism_, and its distinguishing characteristics are (1) the use of oxygen as the ultimate electron acceptor, (2) the complete oxidation of organic substrates to CO_2 and water, and (3) the conservation of much of the free energy as ATP.

The oxygen required for aerobic metabolism actually serves as the terminal electron acceptor only, providing for the continuous reoxidation of reduced coenzyme molecules (most prominent of which are NADH and $FADH_2$). It is these coenzyme molecules (in their oxidized form) that actually carry out the stepwise oxidation of organic intermediates derived from pyruvate. *Aerobic respiratory metabolism can therefore be thought of in terms of two separate but intimately linked processes: the actual oxidative metabolism, in which electrons are removed from organic substrates and transferred to coenzyme carriers, and the concomitant reoxidation of the reduced coenzymes by transfer of electrons to oxygen, accompanied indirectly by the generation of ATP.*

Under aerobic conditions, the glycolytic pathway becomes the initial phase of glucose catabolism. As shown in Figure 15–1, the other three components of respiratory metabolism are the *tricarboxylic acid (TCA) cycle*, which is responsible for further oxidation of pyruvate, the *electron-transport chain*, which is required for the reoxidation of coenzyme molecules at the expense of molecular oxygen, and the *oxidative phosphorylation* of ADP to ATP, which is driven by a "proton gradient" generated in the process of electron transport. Overall, this leads to the potential formation of 36 to 38 molecules of ATP per molecule of glucose in the typical eukaryotic cell.

In this chapter, discussion will focus on the TCA cycle and its central role in the aerobic catabolism of carbohydrates. The following chapter will then explain how the free energy present in the reduced coenzymes that are generated by glycolysis and the TCA cycle is conserved as ATP during the companion processes of electron transport and oxidative phosphorylation.

DISCOVERY OF THE TCA CYCLE

The discovery of the TCA cycle began with a series of biochemical experiments performed in the early 1900s on anaerobic suspensions of minced animal tissues. The experiments established that the suspensions contained enzymes that could transfer hydrogen atoms from various low-molecular-weight organic acids to other reducible compounds, such as the dye methylene blue. (Methylene blue was a convenient indicator in these experiments because it is converted from a blue to a colorless form by reduction.) Only a few organic acids were active in the reduction: succinate, fumarate, malate, and citrate. It was later observed that in the presence of oxygen the same suspensions oxidized these acids to CO_2 and water.

Albert Szent-Gyorgi found that when small amounts of these organic acids were added to a tissue suspension, considerably more oxygen was consumed than was required to oxidize the added acid. He concluded that in the presence of oxygen the organic acids had a catalytic effect on the oxidation of glucose or other carbohydrates in the tissue slices. Szent-Gyorgi also observed that a specific inhibitor of the succinate dehydrogenase, malonate, blocked the utilization of oxygen (respiration) by muscle suspensions. This finding suggested that oxidation of succinate is an indispensable reaction in muscle oxidative metabolism. When the same response was found in many other tissues, the phenomenon seemed to have fundamental and perhaps universal significance.

Next, Hans Krebs studied the interrelationships between the oxidative metabolism of different organic acids. For his experiments he used slices of pigeon flight muscle, which are particularly active in oxidative metabolism. He found a select group of organic acids that was oxidized very rapidly by extracts from the muscle. His list included the acids that Szent-Gyorgi had

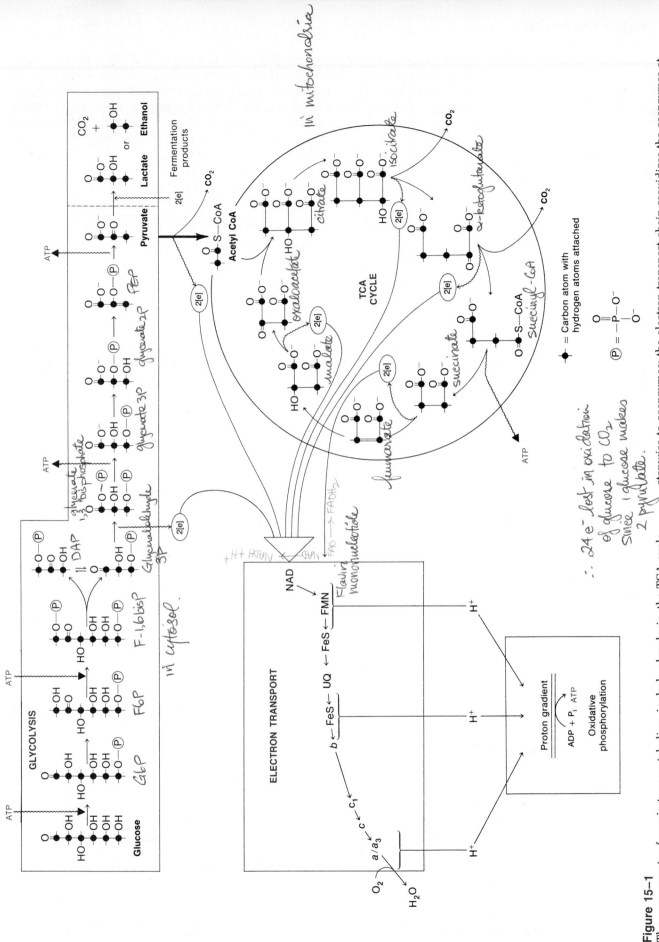

Figure 15–1

The components of respiratory metabolism include glycolysis, the TCA cycle, the electron-transport chain, and the oxidative phosphorylation of ADP to ATP. Glycolysis converts glucose to pyruvate; the TCA cycle fully oxidizes the pyruvate (by means of acetyl-coenzyme A) to CO_2 by transferring electrons stepwise to coenzymes; the electron-transport chain reoxidizes the coenzymes at the expense of molecular oxygen; and the energy of coenzyme oxidation is conserved in the form of a transmembrane proton gradient that is then used to generate ATP.

found plus oxaloacetate, α-ketoglutarate, isocitrate, and cis-aconitate. Like Szent-Gyorgi, Krebs also found that catalytic amounts of the same organic acids stimulated the oxidation of pyruvate or carbohydrate, and that their catalytic effect was blocked by malonate. Since malonate specifically inhibited the conversion of succinate to fumarate by succinic acid dehydrogenase, he concluded that this reaction was one of a series of steps essential for the complete oxidization of pyruvate or carbohydrate. Additional steps in the process were presumed to involve the other organic acids that had been found to catalyze the process. The organic acids with catalytic potential could be arranged in a chain related by biochemical conversions:

Citrate $\longrightarrow$ cis-aconitate $\longrightarrow$ isocitrate $\longrightarrow$ α-ketoglutarate$\longrightarrow$
succinate $\longrightarrow$ fumarate $\longrightarrow$ malate $\longrightarrow$ oxaloacetate

But if this chain of reactions were to act catalytically, then there must be some way of regenerating all of the intermediates of the chain from a single intermediate. A cyclical process would be the simplest way of doing this. For the chain to operate as a cycle, one or more additional reactions needed to be discovered. Krebs found that in the absence of oxygen, small amounts of citrate could be formed from added oxaloacetate and pyruvate. Could this reaction be the missing link in the cyclical process? Krebs thought so, and proposed that the first step in pyruvate oxidation involved its condensation with oxaloacetate to form citrate and CO_2. The oxidation of more complex carbohydrates then could be explained by their prior conversion to pyruvate through the glycolytic pathway.

Further support for the cyclical nature of the chain came from observations of malonate-inhibited muscle slices. In such preparations the addition of any of the intermediates of the chain led to accumulation of succinate. Even the addition of fumarate, the immediate product of succinate oxidation, led to accumulation of succinate.

When the oxidation of substantial amounts of pyruvate was blocked by malonate, stoichiometric amounts of pyruvate would react if either oxaloacetate or its precursors malate or fumarate was added. None of the precursors to succinate in the chain was effective in this regard. This strongly supported Krebs' hypothesis that oxidation of pyruvate involved its condensation with oxaloacetate and the subsequent series of conversions in the chain described by Krebs. The Krebs cycle in its originally proposed form is shown in Figure 15–2. It is also known as the tricarboxylic acid (TCA) cycle or the citrate cycle.

STEPS IN THE TRICARBOXYLIC ACID CYCLE

The tricarboxylic acid cycle (Figure 15–3) begins with acetyl-coenzyme A, which is obtained either by oxidative decarboxylation of pyruvate available from glycolysis or by oxidative cleavage of fatty acids, as described in Chapter 18. Regardless of source, acetyl-CoA transfers its acetyl group to a four-carbon acceptor (oxaloacetate), thereby generating citrate, the six-carbon compound for which the cycle was originally named. In a cyclic series of reactions, the citrate is subjected to two successive decarboxylations and several oxidative events, leaving a four-carbon compound from which the starting oxaloacetate is eventually regenerated. Each turn of the cycle involves the entry of two carbons from acetyl-CoA and the release of two carbons as CO_2, thus providing for regeneration of the oxaloacetate with which the cycle began. As a result, the cycle is balanced with respect to carbon flow and functions without net consumption or buildup of oxaloacetate (or any other intermediate), unless side reactions occur that either feed carbon into the cycle or drain it off into alternative pathways.

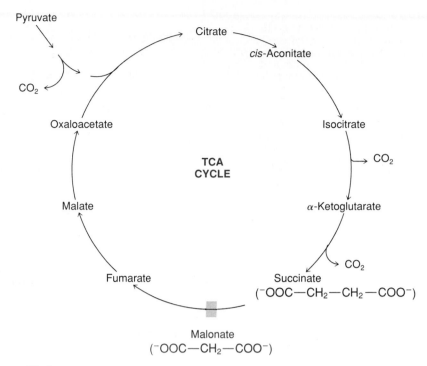

Figure 15–2

Original tricarboxylic acid (TCA) cycle proposed by Krebs. This cycle is also called the citrate cycle or the Krebs cycle. To start the cycle in operation, pyruvate loses one of its carbons and condenses with a four-carbon dicarboxylic acid, oxaloacetic acid, to form a six-carbon tricarboxylic acid, citrate. In one turning of the cycle, two carbons are lost as CO_2, thus returning the citrate to oxaloacetate. The conversion blocked by malonate is indicated by a colored bar.

The Oxidative Decarboxylation of Pyruvate Leads to Acetyl-CoA

Although the acetyl-CoA with which the cycle begins may be derived catabolically from fatty acids or amino acids, the major source in most cells is the pyruvate available from the glycolytic breakdown of carbohydrate. The gap from the glycolytic pathway to the TCA cycle is bridged by the oxidative decarboxylation of pyruvate (with which glycolysis ends) to yield acetate in the form of acetyl-CoA (with which the TCA cycle commences). The decarboxylation of pyruvate is achieved by a complicated sequence of events that is catalyzed by a cluster of three enzymes called the pyruvate dehydrogenase complex. In eukaryotic cells, this enzyme complex is located in the mitochondria, as are all the other reactions of aerobic energy metabolism beyond pyruvate. Since the glycolytic pathway occurs in the cytosol, it is as pyruvate that the carbon derived from glucose (or other carbohydrate substrates) enters the mitochondria.

Decarboxylation of an α-keto acid like pyruvate is a difficult reaction for the same reason as are the ketol condensations or cleavages that were discussed in the preceding chapter; both kinds of reactions require the participation of an intermediate in which the carbonyl carbon carries a negative charge. In all such reactions that occur in metabolism, the intermediate is stabilized by prior condensation of the carbonyl group with thiamin pyrophosphate. In Figure 15–4 thiamin pyrophosphate and its hydroxyethyl derivative are written in the ylid form (see Chapter 11) because those are the forms that actually participate in the reactions. The un-ionized forms will greatly predominate, however, and the reactive carbanions are shown only for clarity.

Figure 15–3
The tricarboxylic acid (TCA) cycle. In each turn of the cycle, acetyl-CoA from the glycolytic pathway or from β oxidation of fatty acids, enters and two fully oxidized carbon atoms leave (as CO_2). ATP is generated at one point in the cycle and coenzyme molecules are reduced. The two CO_2 molecules lost in each cycle originate from the oxaloacetate of the previous cycle rather than from incoming acetyl from acetyl-CoA. This point is emphasized by the use of color.

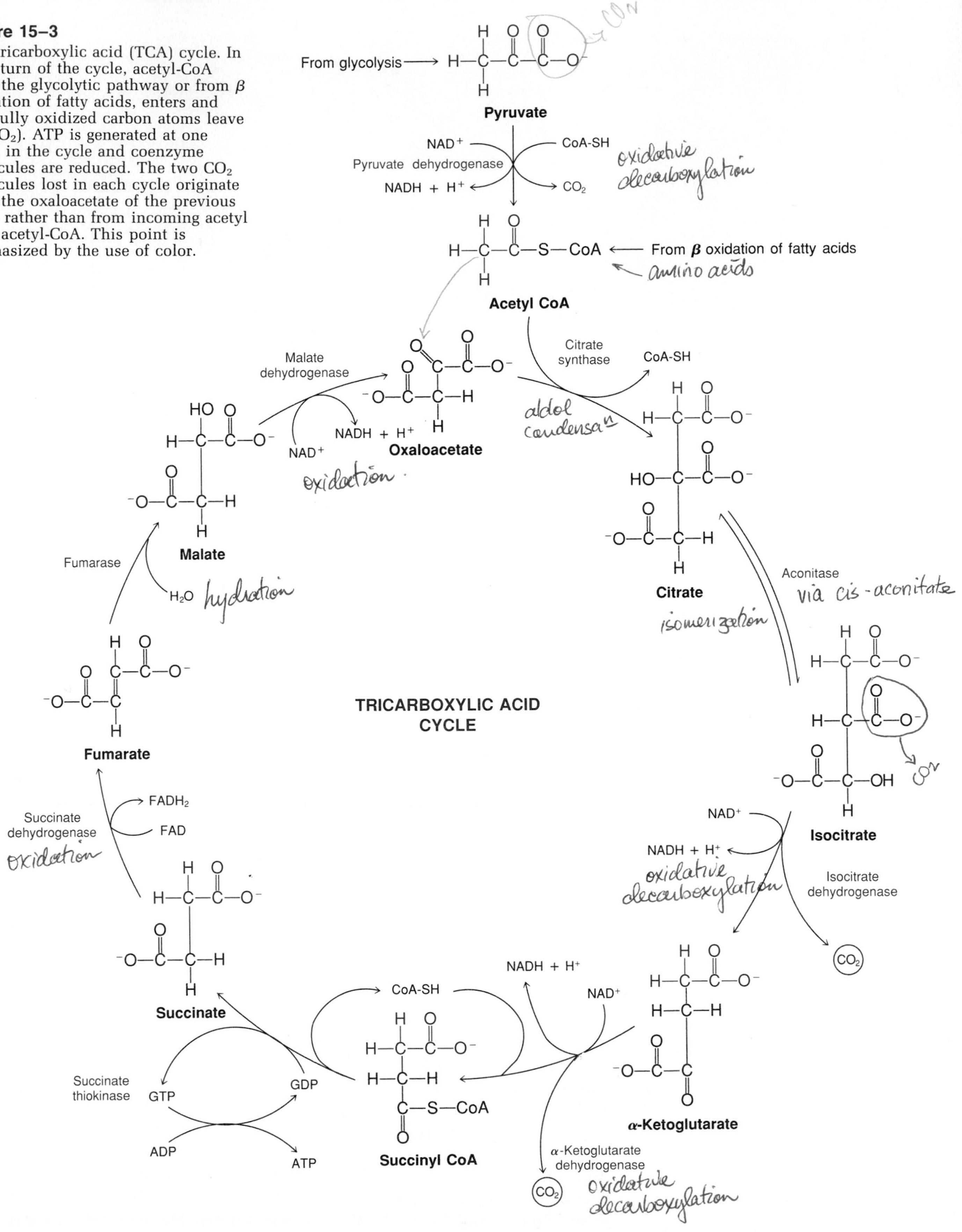

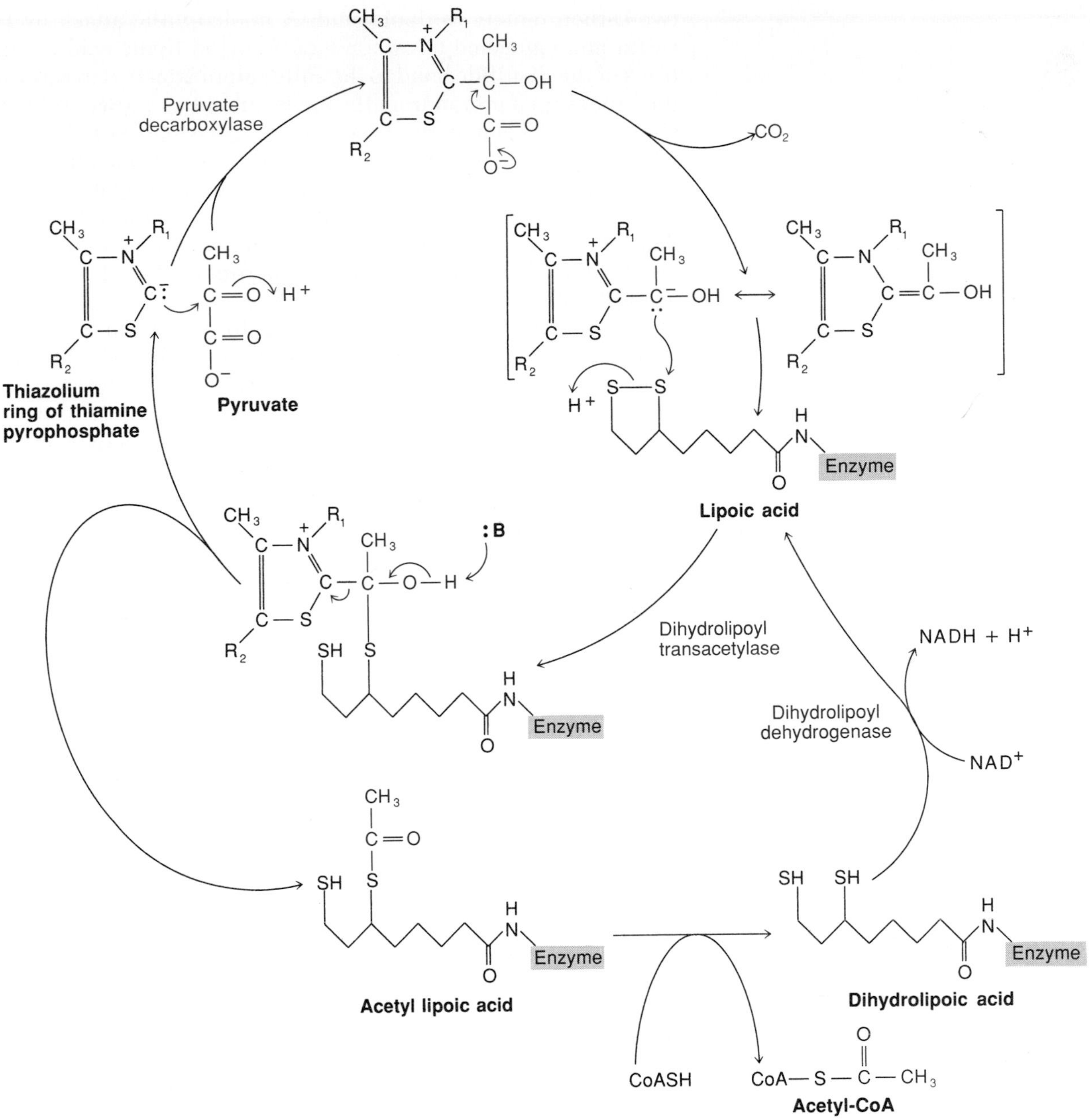

Figure 15–4
The conversion of pyruvate to acetyl-CoA. The reactions are catalyzed by the enzymes of the pyruvate dehydrogenase complex. This complex has three enzymes: pyruvate decarboxylase, dihydrolipoyl transacetylase, and dihydrolipoyl dehydrogenase. In addition, three coenzymes are required: thiamin pyrophosphate, lipoic acid, and NAD^+. Lipoic acid is covalently attached to the transacetylase component of the complex by an amide bond between the carboxyl group of lipoic acid and the terminal amino group of a lysine residue of the enzyme. Reactants and products are shown in color.

In the first step of the conversion, catalyzed by *pyruvate decarboxylase*, a carbon atom from thiamin pyrophosphate adds to the carbonyl carbon of pyruvate. Decarboxylation produces the key reactive intermediate, hydroxyethyl thiamin pyrophosphate (HETPP), which we saw as an intermediate in the conversion of pyruvate to acetaldehyde in the alcoholic fermentation (see Chapter 14). As shown in Figure 15–4, the ionized ylid form of HETPP is resonance-stabilized by the existence of a form without charge separation. The next enzyme, *dihydrolipoyl transacetylase*, catalyzes the transfer of the two-carbon moiety to lipoic acid. A nucleophilic attack by HETPP on the sulfur atom attached to carbon 6 of oxidized lipoic acid displaces the electrons of the disulfide bond to the sulfur atom attached to carbon 8. The sulfur then picks up a proton from the environment as shown in Figure 15–4. This simple displacement reaction is also an oxidation-reduction reaction, in which the attacking carbon atom is oxidized from the aldehyde level in HETPP to the carboxyl level in the lipoic acid derivative. The oxidized (disulfide) form of lipoic acid is converted to the reduced (mercapto) form. The fact that the two-carbon moiety has become an acyl group is shown more clearly after dissociation of thiamin pyrophosphate (TPP), which generates acetyl lipoic acid.

Further transfer of the acyl group to coenzyme A is catalyzed by the same enzyme. This displacement reaction produces reduced lipoic acid. A third enzyme, *dihydrolipoyl dehydrogenase,* catalyzes oxidation of this product back to the disulfide form. The electrons lost in that oxidation are transferred first to an enzyme-bound flavin (not shown in the figure) and then to NAD^+.

The overall equation for the conversion catalyzed by the pyruvate dehydrogenase complex is

$$\text{Pyruvate} + NAD^+ + \text{CoA} \longrightarrow \text{acetyl-CoA} + NADH + CO_2 + H^+$$

The standard free energy change for this conversion is about -8 kcal/mole.

The Nature of the Pyruvate Dehydrogenase Complex. Many molecules of each of the three enzymes that participate in the conversion of pyruvate to acetyl-coenzyme A are organized into a giant enzyme complex. In mammals, the complex has a molecular weight of about 9×10^6; it contains 60 molecules of the transacetylase and perhaps 20 to 30 each of the other two enzymes. The complex in *E. coli* is smaller and contains fewer molecules of each enzyme (see Figure 11–17). Each molecule of transacetylase within the multienzyme complex contains one molecule of lipoic acid covalently bound to the enzyme through an amide bond to the ϵ amino group of a lysine residue. The disulfide of lipoic acid is thus at the end of a long chain, and can sweep over a considerable area on the surface of the subunit to which it is attached and also reach the catalytic sites of neighboring molecules of pyruvate decarboxylase and of dihydrolipoyl dehydrogenase. Thus it appears that the catalytic activities of the complex depend on the ability of the sulfur atoms of the tethered lipoic acid to successively visit the three types of catalytic sites that are contained in the complex (see Figure 11–17). Pyruvate binds at the decarboxylase site and is converted to HETPP as discussed above; then the disulfide group of oxidized lipoic acid picks up the two-carbon moiety, oxidizing it to an acetyl group, and carries it to a site on the transacetylase subunit, where the acetyl group is transferred to coenzyme A. In the form of acetyl-CoA, the two-carbon unit is free from the enzymic tether and can diffuse away as the primary product of the sequence of reactions that are catalyzed by the complex. The reduced lipoic acid is oxidized at a site on the dehydrogenase, and is ready to pick up another acetyl group from the decarboxylase site. The NADH that is formed in the oxidation of dihydrolipoic acid dissociates from the enzyme and is available to the electron-transfer machinery of the mitochondrion.

At this point in the oxidation of glucose, four electrons per glucose molecule have been lost in the oxidation of glyceraldehyde-3-phosphate and four more in the conversion of pyruvate to acetyl-CoA. Thus of the total of 24 electrons that are lost in the oxidation of glucose to CO_2, 16 remain to be transferred to oxidizing agents in the course of the oxidation of two molecules of acetyl-CoA. A major function of the TCA cycle is to obtain these electrons for use in electron-transfer phosphorylation (Chapter 16).

Citrate Synthase Is the Gateway to the TCA Cycle

The first step of the TCA cycle is the reaction catalyzed by *citrate synthase,* in which acetyl-CoA enters the cycle and citrate is formed:

$$\text{Acetyl-CoA} + \text{oxaloacetate} \longrightarrow \text{citrate} + \text{CoA}$$

This reaction is an aldol condensation, in which a carbanion generated at C-2 of the acetyl group by loss of a proton to water (or to an acceptor group on the enzyme) adds to the carbonyl group of oxaloacetate (Figure 15–5). This is a common type of reaction in acetyl-CoA metabolism. In nearly all such cases, coenzyme A is released from the product while it is still bound to the enzyme, so that the products of the reaction are free coenzyme A and the condensation product.

The standard free energy change for this reaction is approximately -8 kcal/mole, so that the equilibrium constant is about 10^6. As we will see later, when we consider the energetics of the cycle, it is fortunate that this reaction is so highly favorable thermodynamically, because the concentration of oxaloacetate in cells is extremely low.

Aconitase Catalyzes the Isomerization of Citrate to Isocitrate

In the first reaction within the cycle, citrate is converted to its isomer isocitrate (Figure 15–6):

$$\text{Citrate} \rightleftharpoons \text{cis-aconitate} \rightleftharpoons \text{isocitrate}$$

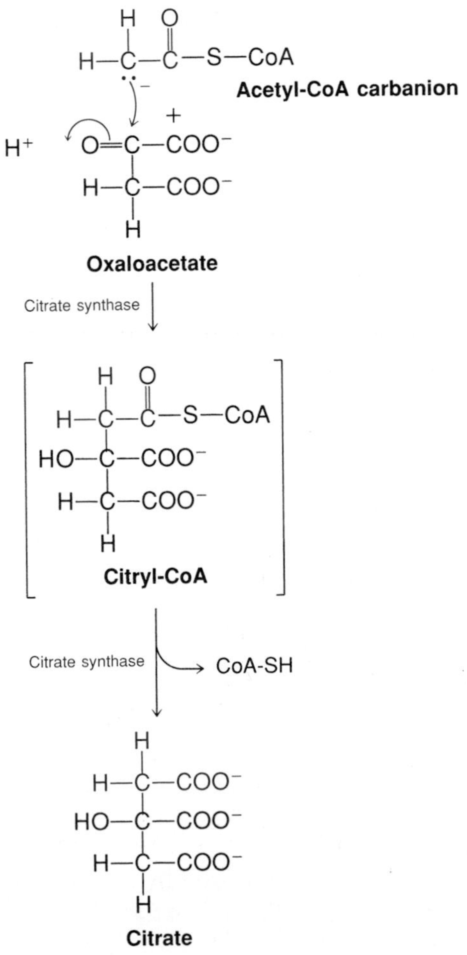

Figure 15–5
Formation of citric acid, catalyzed by citrate synthase. A carbanion of acetyl-CoA, generated by loss of a proton to water (or to an acceptor group on the enzyme), adds to the carbonyl group of oxaloacetate. The immediate product of the condensation is probably citryl-coenzyme A, but the products that dissociate from the catalytic site of the enzyme are citrate and free coenzyme A.

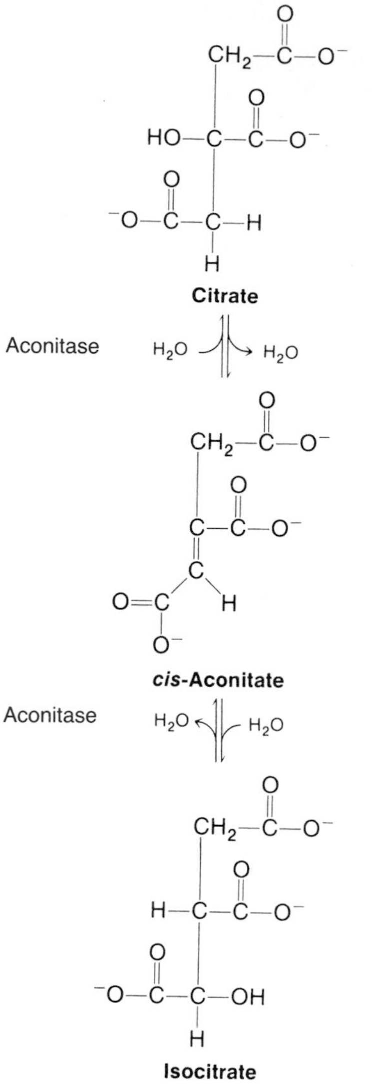

Citrate

Aconitase H_2O ⤵⤴ H_2O

cis-Aconitate

Aconitase H_2O ⤵⤴ H_2O

Isocitrate

Figure 15–6
Interconversion of citrate, cis-aconitate, and isocitrate, catalyzed by aconitase. At equilibrium the relative concentrations of these compounds are about 90, 6, and 4, respectively.

Aconitase, the enzyme that catalyzes this isomerization, is named for the fact that the unsaturated compound corresponding to citrate and isocitrate, cis-aconitate, can also serve as substrate or product. In any case, aconitase catalyzes the attainment of equilibrium between citrate, isocitrate, and cis-aconitate. At equilibrium the relative concentrations are about 90, 6, and 4 respectively. Aconitate is not known to have any function in metabolism, and it can be disregarded in the consideration of the TCA cycle.

Isocitrate Dehydrogenase Catalyzes the First Oxidation in the TCA Cycle

The first oxidative conversion in the TCA cycle is catalyzed by *isocitrate dehydrogenase*. Since the product is α-ketoglutarate, the conversion evidently involves two steps: oxidation of the secondary alcohol to a ketone, producing oxalosuccinate, followed by β decarboxylation (Figure 15–7).

$$\text{Isocitrate} + NAD^+ \longrightarrow \text{α-ketoglutarate} + NADH + H^+ + CO_2$$

NAD^+ is the electron acceptor for the oxidative step, and Mg^{2+} or Mn^{2+} is required for the decarboxylation. The oxalosuccinate that is presumably an intermediate does not dissociate from the enzyme. Isocitrate dehydrogenase is a large enzyme; the molecular weight of the mammalian enzyme is about 4×10^5. It has a range of regulatory properties, some of which will be discussed in the later section on regulation of the TCA cycle. The standard free energy change of the reaction is about -5 kcal/mole.

α-Ketoglutarate Dehydrogenase Catalyzes the Decarboxylation of α-Ketoglutarate to Succinyl-CoA

The second of two oxidative decarboxylation reactions of the TCA cycle is that catalyzed by *α-ketoglutarate dehydrogenase*. The reaction sequence is entirely analogous to that of pyruvate dehydrogenase, complete with the conservation of some of the energy of the oxidation in a coenzyme-containing derivative succinyl-CoA.

$$\text{α-ketoglutarate} + NAD^+ + CoA \longrightarrow \text{succinyl-CoA} + NADH + H^+ + CO_2$$

The enzyme complex involved in this reaction also is very similar to the pyruvate dehydrogenase complex. Indeed, the same dihydrolipoyl dehydrogenase subunit is used in both complexes. The product of such conversions is the coenzyme A ester of the acid containing one less carbon atom than the substrate, in this case succinyl-CoA (Figure 15–3). As might be expected, the standard free energy change for both reactions is similar also, about -8 kcal/mole.

Succinate Thiokinase Couples the Conversion of Succinyl-CoA to Succinate with the Synthesis of GTP

Succinyl-CoA is an activated intermediate. That activation is useful for the small amount of succinyl-CoA that is used in the synthesis of heme in animals. However, nearly all of the succinyl-CoA is retained in the TCA cycle, where it leads to the regeneration of the oxaloacetate that is needed for condensation with acetyl-CoA to keep the cycle operating. For that use, activation is not needed, and the thioester is merely converted to succinate.

$$\text{Succinyl-CoA} + GDP \longrightarrow \text{succinate} + GTP$$

The reaction is catalyzed by an enzyme that may have originally evolved for the reverse reaction (if succinyl-CoA was required in early anaerobic orga-

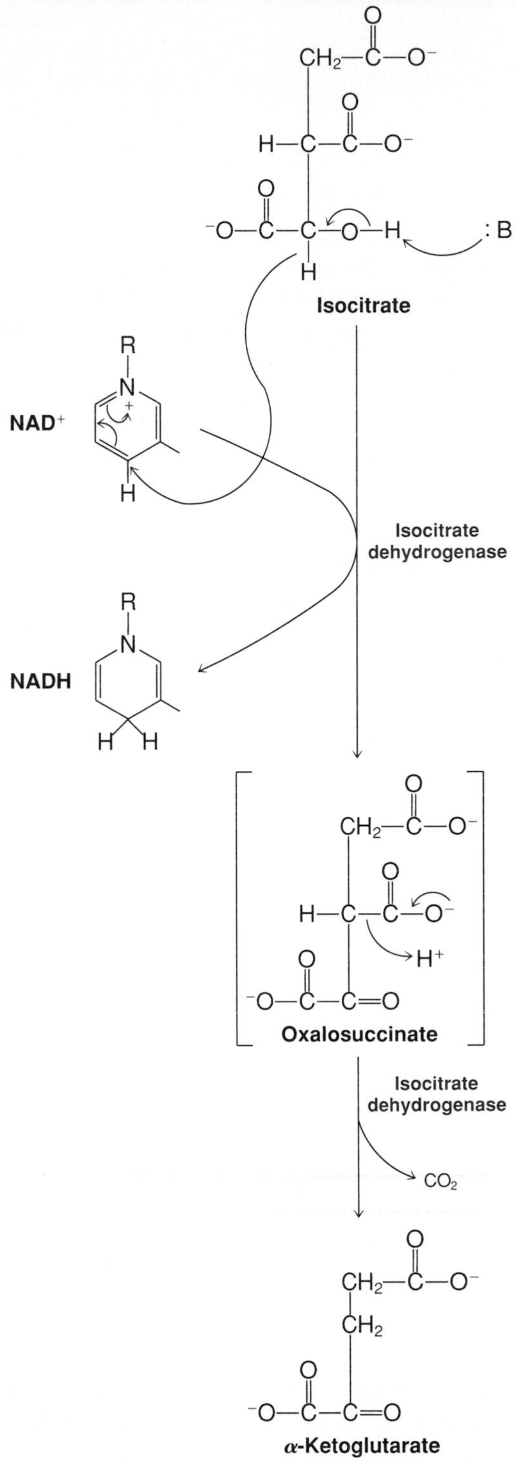

Figure 15–7

The oxidative decarboxylation of isocitrate to α-ketoglutarate, catalyzed by mitochondrial isocitrate dehydrogenase. The intermediate, oxalosuccinate, is not released from the enzyme. B: represents a catalytic side chain from the enzyme.

nisms). In any case, in the direction in which it occurs in the TCA cycle, the reaction produces one molecule of GTP in mammals (ATP in some other kinds of organisms) for each molecule of succinyl-CoA converted. The reaction pathway is complex, and involves an intermediate in which a phosphate group is attached to a histidine residue of the enzyme. Probably CoA is first displaced by inorganic phosphate, forming succinyl phosphate. A nitrogen atom of a specific histidine residue then attacks phosphorus, displacing succinate and forming an N-phosphoryl derivative. The final step is an attack by GDP (or ADP) on the phosphorus atom of that derivative, forming

GTP (or ATP). The standard free energies of hydrolysis of pyrophosphates, such as ATP and GTP, and of CoA esters are very similar; consequently the equilibrium constant for the overall conversion is small, about 3.

Succinate Dehydrogenase Catalyzes the Oxidation of Succinate to Fumarate

In terms of carbon atoms, we might say that since two carbons entered the cycle as the acetyl group and two have left as CO_2, the functional part of the cycle is finished at this stage, and that thus the four carbons of succinate must merely be converted to oxaloacetate to serve again as an acetyl acceptor. However, that assumption would be erroneous. When the TCA cycle serves as a means of oxidizing acetyl groups, that is, when it is not producing biosynthetic starting materials at a significant rate, its function is to supply electrons to the electron-transfer phosphorylation system. When we reach succinate, four electrons have been transferred to NAD^+. Thus, of the eight electrons that must be lost in the oxidation of acetate to CO_2, four remain in succinate. In terms of its oxidative function, the cycle is only half finished at this point.

The next step in the TCA cycle, the oxidation of succinate to fumarate, involves insertion of a double bond into a saturated hydrocarbon chain.

$$\text{Succinate} + \text{FAD} \longrightarrow \text{fumarate} + \text{FADH}_2$$

This is not an easy or common reaction in organic chemistry. It is, however, a very important type of reaction in metabolic chemistry, and is an integral step in the oxidation of carbohydrates, fats, and several amino acids.

Because of the nature of the reaction, a strong oxidizing agent is required. The electron acceptor that is used in most oxidative steps of catabolism, NAD^+, is not a strong enough oxidizing agent to allow a reasonable equilibrium constant. Flavoproteins are stronger oxidizing agents than NAD^+, and _succinate dehydrogenase_, which catalyzes the oxidation of succinate to fumarate, is a flavoprotein enzyme. The oxidation of succinate by the electron acceptor of succinate dehydrogenase is about 16 kcal (65 kJ) more favorable than it would be if the electron acceptor were NAD^+. This makes the equilibrium constant more favorable by a factor of about 10^{11}. In this example we see how important the choice of cofactors can be. In general FAD is a better oxidizing agent than NAD^+, and NADH is a better reducing agent than $FADH_2$.

The succinate dehydrogenase of mammalian mitochondria consists of two subunits with molecular weights of about 70,000 and 27,000. Each subunit contains iron-sulfide centers (Chapter 16) that appear to participate in electron transfer. The enzyme is integrally attached to the inner mitochondrial membrane (see Chapter 16). Because the $FADH_2$ that is produced in the reaction is covalently attached to the enzyme (in contrast to the NADH produced by other dehydrogenases of the cycle, which can diffuse from one catalytic site to another), succinate dehydrogenase presumably must be fixed in position relative to enzymes of the electron-transfer pathway, in order to facilitate direct transfer of electrons. Since the $FAD/FADH_2$ couple probably transfers electrons directly to other carriers, this reaction is probably ordinarily closely coupled to further electron-transfer reactions.

Fumarase Catalyzes the Addition of Water to Fumarate to Form Malate

Fumarate is converted to L-malate by stereospecific addition of water across the double bond.

$$\text{Fumarate} + H_2O \longrightarrow \text{L-malate}$$

The enzyme that catalyzes this reaction, named fumarase, is a tetramer of identical subunits, each with a molecular weight of about 48,000. The equilibrium constant for the reaction is about 4.

Malate Dehydrogenase Catalyzes the Oxidation of Malate to Oxaloacetate

The final oxidation, and final step, of the cycle is the conversion of malate to oxaloacetate.

$$\text{L-malate} + NAD^+ \longrightarrow \text{oxaloacetate} + NADH + H^+$$

Although oxidation of an alcohol to a carbonyl is not easy, this enzyme, rather surprisingly, uses NAD^+ as the oxidizing agent. The standard free energy change is about 7 kcal/mole, and the equilibrium constant is near 10^{-5}. Consequently the steady-state concentration of oxaloacetate is very low. This potential problem will be discussed further in the section on energetics of the cycle. The enzyme consists of two identical subunits, each with a molecular weight of about 66,000.

STEREOCHEMICAL ASPECTS OF TCA CYCLE REACTIONS

Some early applications of isotopic tracer techniques to the TCA cycle led to temporary confusion, but in the end they led to a new generalization concerning the stereochemistry of interaction between enzymes and a certain type of substrates.

In the early 1940s, before the discovery of ^{14}C and when acetyl-coenzyme A was unknown, two research groups used the stable isotope ^{13}C and home-made mass spectrometers (the technique was so new that instruments were not available commercially) to study carbon flow in the TCA cycle. Pyruvate and $^{13}CO_2$ were added to pigeon liver preparations to form carboxy-labeled oxaloacetate. Malonate was added to stop the TCA cycle at succinate. The expected result was that half of the ^{13}C would be found in succinate and the other half in CO_2 (Figure 15–8).

To the surprise of the investigators and the biochemical community, it was found that although pyruvate was consumed and succinate was produced, there was no ^{13}C in the succinate. Further experiments showed that the isotope had been incorporated, but was all lost in the oxidative decarboxylation of α-ketoglutarate to succinate. Since the citrate molecule has a plane of symmetry, it was taken for granted that the two $-CH_2-COO^-$ arms of the molecule must be chemically equivalent and that the $-OH$ group would have an equal chance of migrating into either arm in the aconitase reaction. In that case, the label would have an equal chance of being lost or retained in the decarboxylation of α-ketoglutarate. For a few years this experimental result was thought to prove that citrate as such could not be an intermediate of the cycle, and that some derivative or analog of citrate must be the immediate product of the condensation of oxaloacetate with the unknown active two-carbon intermediate.

Then, in 1948, Ogston suggested that citrate was not necessarily excluded by the isotopic evidence, because the two $-CH_2-COO^-$ arms might actually not be equivalent when citrate was the substrate for an enzymic reaction. He pointed out that if the substrate were attached to the enzyme at three points, its orientation would be fixed by those attachments and it would be impossible for the two identical arms to exchange positions. Thus

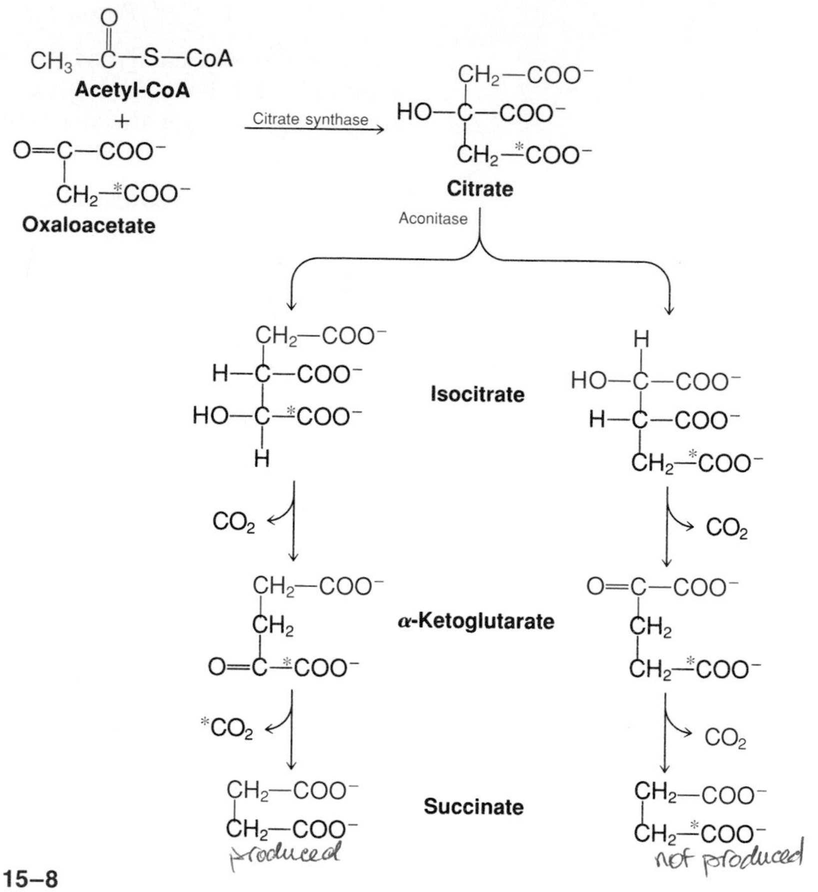

Figure 15–8

Stereochemical relationships in the synthesis and metabolism of citrate. When oxaloacetate labeled with ^{13}C in the carbonyl group β to the keto group (*) was used as substrate, it was expected that half of the label would be found in succinate and half in CO_2. That prediction was based on the assumption that the two $-CH_2-COO^-$ arms of citrate must be equivalent in every way. In fact, all of the label was found in CO_2. Thus only the intermediates shown on the left were produced. This result shows that both the condensation of acetyl-CoA with oxaloacetate and the isomerization of citrate are stereospecific reactions. The carbon atoms supplied by acetyl-CoA are shown in color. It is seen that neither of those atoms is lost in the first turn of the cycle after their entry.

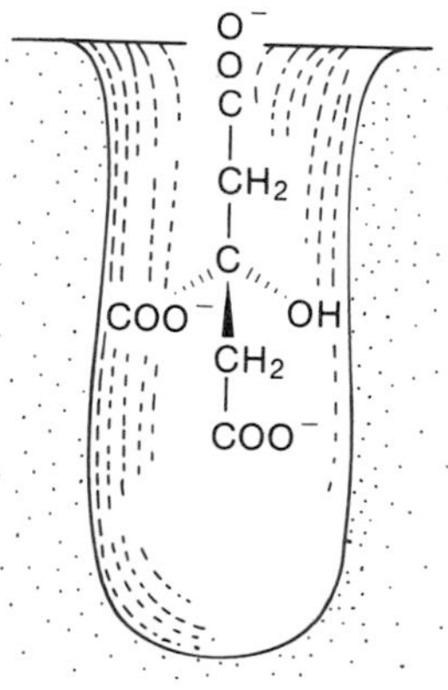

Figure 15–9

Citrate is shown with three of the substituents from the central carbon atom making contact with the enzyme surface. Binding in this way makes the two $-CH_2-COO^-$ groups nonequivalent, as they must be in the TCA cycle.

only one of them could occupy the position that allowed it to participate in the reaction (Figure 15–9; see also Box 15–A).

Ogston's concept is a valid and important generalization, but the suggestion of a three-point attachment is too restrictive. Three constraints are necessary to fix an object in three-dimensional space, but they need not all be points of attachment. In principle the two a groups of a molecule of the type Ca_2bd could be distinguished by an enzyme if the three constraints were one point of attachment, one pocket, into which b could fit but d could not, and the position of the reactive groups of the catalytic site. Westheimer and Vennesland showed a few years later that ethanol dehydrogenase distinguishes with 100% specificity between the two hydrogen atoms on C-1 of ethanol (see Chapter 11 and Figure 11–8). Three sterically distinct points of attachment between an enzyme and a substrate as small as ethanol seem unlikely.

Ogston's contribution led to an interesting extension of concepts concerning stereochemistry of enzyme action. Compounds of the type Ca_2bd are termed prochiral, and it is recognized that an enzyme that either synthesizes such a compound or uses it as a substrate will nearly always do so stereospecifically. Once the point has been made, it seems very reasonable. In the case

BOX 15–A

Potter's Demonstration of Prochirality

Ogston pointed out that an enzyme might discriminate between the two carboxymethyl groups of citrate, and thus that the evidence did not exclude citrate as an intermediate in the TCA cycle. Only further experimental evidence could show whether, in fact, citrate synthase and aconitase act with the kind of steric consequences that Ogston had suggested might be possible. What was necessary was to isolate citrate that had been produced from pyruvate and labeled oxaloacetate and then to use that citrate as the substrate for aconitase. V. R. Potter performed the critical experiment. An inhibitor of aconitase was used in the first incubation, so that citrate would accumulate and could be isolated. When this citrate was then used as substrate for a liver preparation poisoned with malonate, as in the previous experiment, the same result was obtained: all of the ^{13}C was in CO_2, and none was in succinate. This result proved that citrate synthase introduces the labeled carboxymethyl group into only one of the two possible steric positions in citrate, and that aconitase distinguishes between the two positions in

its action. Ogston's suggestion was confirmed, and citrate was shown to be an intermediate in the cycle.

This episode is an interesting specific illustration of the interplay between hypothesis (or reasoning) and experiment on which scientific advance is based. First, unexpected experimental results seemed to prove that citrate could not be an intermediate in the TCA cycle. Then it was pointed out that the accepted concepts were not necessarily valid, so that the results did not actually argue either for or against the participation of citrate as an intermediate. Finally, experiments that were performed because of the new hypothesis proved that citrate is actually formed by liver extracts and is converted to succinate by the same extracts, so that it is an intermediate. The results of the new experiments also showed that the novel suggestions of the new hypothesis (discrimination between identical groups in a nonchiral molecule) were valid. The experiments would not have been performed without the stimulus of the hypothesis; on the other hand, the hypothesis was only a suggestion until the experiments demonstrated its validity.

of citrate synthase, for example, it is inherently likely that the planar carbonyl carbon of oxaloacetate will lie flat on an enzyme surface, and that only one side of the atom will be available for attack by acetyl-coenzyme A.

ATP STOICHIOMETRY OF THE TCA CYCLE

By summing the component reactions of the TCA cycle, we arrive at the following overall summary reaction:

$$\text{Acetyl-CoA} + 2\,H_2O + 3\,NAD^+ + FAD + ADP + P_i \longrightarrow$$
$$2\,CO_2 + 3\,NADH + 3\,H^+ + FADH_2 + CoA + ATP$$

Looking at this summary reaction, you may wonder why it doesn't seem to reflect the substantially greater ATP yield that is supposed to be characteristic of aerobic metabolism. The answer is that ATP is stored in the reduced coenzyme molecules on the right-hand side of the reaction. Reoxidation of these compounds liberates a large amount of free energy. To see how that energy is released, we must wait until we examine the final phases of chemotrophic energy metabolism in Chapter 16. It is only as the electrons are transferred stepwise from the high-energy coenzyme carriers to molecular oxygen that the coupled generation of ATP occurs. We will see that the oxidative phosphorylation system regenerates three molecules of ATP for each pair of electrons from NADH, and two molecules of ATP for each pair of electrons from $FADH_2$. Thus in the normal aerobic mitochondrial metabolism, oxidation of one mole of acetyl groups leads to regeneration of 12 moles of ATP.

If we take pyruvate rather than acetyl-CoA as the starting point, the equation for the TCA cycle is

$$\text{Pyruvate} + 4\,NAD^+ + FAD + ADP \text{ (or GDP)} + P_i + 2\,H_2O \longrightarrow$$
$$3\,CO_2 + 4\,NADH + 4\,H^+ + FAD + ATP \text{ (or GTP)}$$

and we see that the mitochondrial part of carbohydrate metabolism yields 15 moles of ATP per mole of pyruvate, or 30 moles per mole of glucose.

ORGANIZATION OF THE REACTIONS OF THE TCA CYCLE

In the metabolic pathway for the oxidation of pyruvate there are four reactions (those catalyzed by pyruvate dehydrogenase, citrate synthase, isocitrate dehydrogenase, and α-ketoglutarate dehydrogenase) for which the equilibrium constants are large. For four others (those catalyzed by aconitase, succinate thiokinase, succinate dehydrogenase, and fumarase), the equilibrium constant is fairly close to 1, and for one reaction (catalyzed by malate dehydrogenase), it is very small.

Since the value of the equilibrium constant for the malate dehydrogenase reaction is about 10^{-5} and if the physiological ratio of $NAD^+/NADH$ is about 10, the ratio of oxaloacetate to malate at equilibrium should be about 10^{-4}. Thus if the concentration of malate is around 1 mM, the concentration of oxaloacetate would be about 0.1 μM, or 10^{-7} M, at equilibrium. Estimates of the concentration of oxaloacetate in the mitochondrial matrix range from around 10^{-8} M to slightly over 10^{-7} M, which is in good agreement with the thermodynamic expectations.

Clearly, the equilibrium constant for the malate dehydrogenase reaction is most unfavorable for operation of the TCA cycle. It seems strange that such an unfavorable reaction should occur in a sequence (the TCA cycle) that has one of the largest flux rates in metabolism. Probably this is possible only because the following reaction, catalyzed by citrate synthase, is so thermodynamically favorable. The equilibrium constant for this reaction is about 5×10^5. If the concentrations of acetyl-coenzyme A and of free coenzyme A are approximately equal, at equilibrium the ratio of citrate concentration to oxaloacetate concentration will be equal to the equilibrium constant. Then if [oxaloacetate]/[malate] is about 10^{-4} and [citrate]/[oxaloacetate] is about 5×10^5, the ratio [citrate]/[malate] could be as high as 50. That value is only approximate. The actual steady-state ratio would be expected to be significantly below the equilibrium ratio. Further, the values of the equilibrium constants are not precise; effects of ionic strength have not been taken into account, and the assumptions as to the [AcSCoA]/[HSCoA] and [NAD$^+$]/[NADH] ratios are tenuous. But crude as they must be, these calculations show that the very small equilibrium constant for the malate dehydrogenase reaction does not pose a serious thermodynamic problem for the operation of the TCA cycle.

THE AMPHIBOLIC NATURE OF THE TCA CYCLE

The sole purpose of the TCA cycle, when it is functioning as such, is the oxidation of acetate to CO_2, with concomitant conservation of the energy of oxidation as reduced coenzymes and eventually as ATP. Strictly speaking, then, the TCA cycle has but a single substrate, and that is acetyl-CoA. In most cells, however, there is considerable flux of four-, five-, and six-carbon intermediates into and out of the cycle, which occurs in addition to the primary function of the cycle. Such side reactions serve two main purposes: (1) to provide for the synthesis of compounds derived from any of several intermediates of the cycle, and (2) to replenish and augment the supply of intermediates in the cycle as needed. Because the TCA cycle can function both in a catabolic mode and as a source of precursors for anabolic pathways, it is often called an *amphibolic* pathway (from the Greek, *amphi* = both). Some of the main biosynthetic pathways that begin with intermediates in the

BOX 15–B

Generation of Biosynthetic Intermediates Under Anaerobic Conditions

As we saw in the preceding chapter, seven of the eleven metabolites that serve as starting materials for biosynthesis are intermediates in glycolysis or the pentose phosphate pathway (see Box 13–A for list of intermediates), and so are available when those sequences are active. The others, acetyl-coenzyme A, oxaloacetate, α-ketoglutarate, and succinyl-coenzyme A, can be produced from pyruvate, the end product of glycolysis. The individual reactions are considered in this chapter.

Acetyl-coenzyme A is produced from pyruvate by decarboxylation, followed by oxidation of the carbonyl carbon to the level of carboxylate and activation by formation of the thiol ester. Two electrons are lost in this conversion. Acetyl-coenzyme A is one of the most widely used biosynthetic starting materials. It supplies all of the carbon atoms of fats and several other classes of lipids, and also contributes some of the carbon atoms of several amino acids and other metabolites.

Oxaloacetate is produced by carboxylation of pyruvate. This reaction requires an ATP-to-ADP conversion, but no oxidation or reduction is involved. Oxaloacetate is required in the synthesis of aspartate, asparagine, threonine, isoleucine, methionine, and, in some species, lysine.

Oxaloacetate is converted to α-ketoglutarate by a sequence of reactions that is used frequently in metabolism for the conversion of an α-keto acid to the next higher homolog (the corresponding compound containing one more —CH_2— group). This sequence includes condensation with acetyl-CoA, isomerization by apparent migration of a hydroxyl group, oxidation, and β decarboxylation. α-Ketoglutarate is used in the syntheses of glutamate, glutamine, proline, arginine, and, in some species, lysine.

The production of acetyl-CoA and of α-ketoglutarate from pyruvate involves oxidation, and under anaerobic conditions this causes problems. Two electrons must be disposed of for each molecule of acetyl-CoA that is produced, and four electrons for each molecule of α-ketoglutarate. In the absence of oxygen or an alternative exogenous electron acceptor, those electrons must be fed back into metabolism. That is, they must be used in the reduction of some metabolite, just as the electrons from the 3-phosphoglyceraldehyde dehydrogenase are used in the reduction of pyruvate to lactate or ethanol in anaerobic glycolysis.

It would be possible to merely convert pyruvate to lactate in order to allow synthesis of α-ketoglutarate, but the common laboratory bacterium *E. coli*, and probably many other kinds of bacteria, when growing anaerobically use an alternative set of reactions that dispose of four electrons, rather than two, for each molecule of pyruvate. The pathway is the reduction of oxaloacetate to succinate. Two electrons are used in the first reduction, to malate. This product is dehydrated to fumarate, which is further reduced to succinate.

Thus the reduction of one molecule of oxaloacetate to succinate balances the oxidation that is involved in the production of two molecules of acetyl-CoA or one molecule of α-ketoglutarate. Since NAD^+ is the electron acceptor in the oxidation of pyruvate to acetyl-CoA and on to α-ketoglutarate, it is NADH that must be used in the reduction of oxaloacetate to malate and of fumarate to succinate.

Succinyl-coenzyme A is a starting material in the synthesis of heme. In bacteria, heme is used mainly in the production of cytochromes, which may not be needed in the absence of oxygen. If it is required, succinyl-CoA can be made under anaerobic conditions from succinate with the expenditure of one molecule of ATP or GTP.

Thus by use of the metabolic pathways considered in the preceding chapter and the conversions outlined here, all of the starting materials needed for the biosynthetic sequences of a cell can be made anaerobically from carbohydrates. These pathways are used by at least some bacterial species today in the absence of oxygen, and it is reasonable to assume that they were used in the necessarily anaerobic metabolism of organisms before oxygen was added to the atmosphere by the photosynthetic activity of cyanobacteria (Chapter 33) about three billion years ago. This example illustrates the most probable way that metabolic cycles may have evolved. The reactions of preexisting linear sequences are put to new use as a consequence of the develoment of one or a few reactions that convert the sequences into a cycle. In this case, only α-ketoglutarate dehydrogenase (and perhaps succinate thiokinase) were needed to convert the two branches of the biosynthetic pathway, which were balanced with respect to oxidation-reduction, to a cyclic oxidative pathway. Because of the difficulty of oxidizing succinate to fumarate (see text), the NAD-linked enzyme was not suitable, and a new FAD-linked enzyme (succinate dehydrogenase) was evolved. Even today, *E. coli* and probably many other organisms make the NAD-linked fumarate reductase under anaerobic conditions and FAD-linked succinate dehydrogenase in the presence of air. Under anaerobic conditions, the conversion of oxaloacetate to succinate consumes four electrons and thus permits necessary oxidative steps such as the conversion of pyruvate to α-ketoglutarate, whereas under aerobic conditions the conversion of succinate to oxaloacetate liberates four electrons for use in oxidative phosphorylation. The cyclic sequence still supplies all of the biosynthetic starting materials that were provided by the linear branched pathways.

TCA cycle, as well as the ways in which the supply of intermediates in the cycle is replenished, are indicated in Figure 15–10. We will discuss these pathways only briefly here, as they will be discussed in greater detail in subsequent chapters. (For fats, see Chapters 18, 21–23; for amino acids, see Chapters 19 and 24.)

draining of TCA cycle ✳

Oxaloacetate and α-ketoglutarate are used in the synthesis of several amino acids; succinyl-CoA is used in heme synthesis; and citrate is the source of the acetyl-CoA in the cytosol, which is used for the synthesis of fats and other lipids and some amino acids. These are some of the major drains on the TCA cycle. Now we will see how these drains are compensated for.

Reactions that replenish the intermediates in the TCA cycle are termed *anaplerotic*, from a Greek root that means "filling up." It is not necessary to replenish the intermediate that is used in a biosynthetic pathway directly, as the replenishment of any intermediate will occur by a feeding-in process at any point in the cyclical pathway.

Replenishing cycle

When carbohydrates are being metabolized, TCA cycle intermediates are replenished by production of oxaloacetate from pyruvate. In mammals, this reaction is catalyzed by *pyruvate carboxylase*, and one ATP-to-ADP conversion is associated with the carboxylation. Other properties of the reaction will be discussed later in this chapter, in connection with regulation of the TCA cycle and related metabolic sequences.

✳

In prokaryotic organisms and some eukaryotes, oxaloacetate is fed into the cycle by carboxylation of phosphoenolpyruvate. Energetically the carboxylation of phosphoenolpyruvate, catalyzed by *phosphoenolpyruvate carboxylase*, is equivalent to the sum of the pyruvate kinase and pyruvate carboxylase reactions, which are used by mammals:

$$\text{Phosphoenolpyruvate} + \text{ADP} \longrightarrow \text{pyruvate} + \text{ATP}$$

$$\text{Pyruvate} + \text{ATP} + CO_2 \longrightarrow \text{oxaloacetate} + \text{ADP}$$

Sum: $$\text{Phosphoenolpyruvate} + CO_2 \longrightarrow \text{oxaloacetate}$$

THE GLYOXYLATE CYCLE: AN ALTERNATIVE FATE FOR ACETYL-COENZYME A

In certain species of plants and microorganisms, an alternative fate for acetyl-CoA exists that is relevant to this chapter because it draws on some of the reactions of the TCA cycle, but in a way that is anabolic rather than catabolic. The result is a pathway called the *glyoxylate cycle*, named for glyoxylate, a two-carbon intermediate in the cycle. The glyoxylate cycle makes possible the synthesis of four-carbon (and, as we shall see shortly, six-carbon) compounds from the two-carbon level of acetyl-CoA.

The glyoxylate cycle uses four reactions in common with the TCA cycle and two reactions that are unique to the glyoxylate cycle. The conversions, not balanced with regard to oxidation, are

$$\text{Acetyl-CoA} + \text{oxaloacetate} \longrightarrow \text{citrate} \longrightarrow \text{isocitrate} \qquad \text{(a)}$$

$$\text{Isocitrate} \longrightarrow \text{succinate} + \text{glyoxylate} \qquad \text{(b)}$$

$$\text{Succinate} \longrightarrow \text{fumarate} \longrightarrow \text{oxaloacetate} \qquad \text{(c)}$$

$$\text{Acetyl-CoA} + \text{glyoxylate} \longrightarrow \text{malate} \qquad \text{(d)}$$

$$\text{Malate} \longrightarrow \text{oxaloacetate} \qquad \text{(e)}$$

Sum: $$2 \text{ Acetyl-CoA} \longrightarrow \text{oxaloacetate}$$

Reactions (b) and (d) are specific to the glyoxylate cycle. Reaction (b), catalyzed by *isocitrate lyase*, is an aldol-like cleavage. Reaction (d), catalyzed by

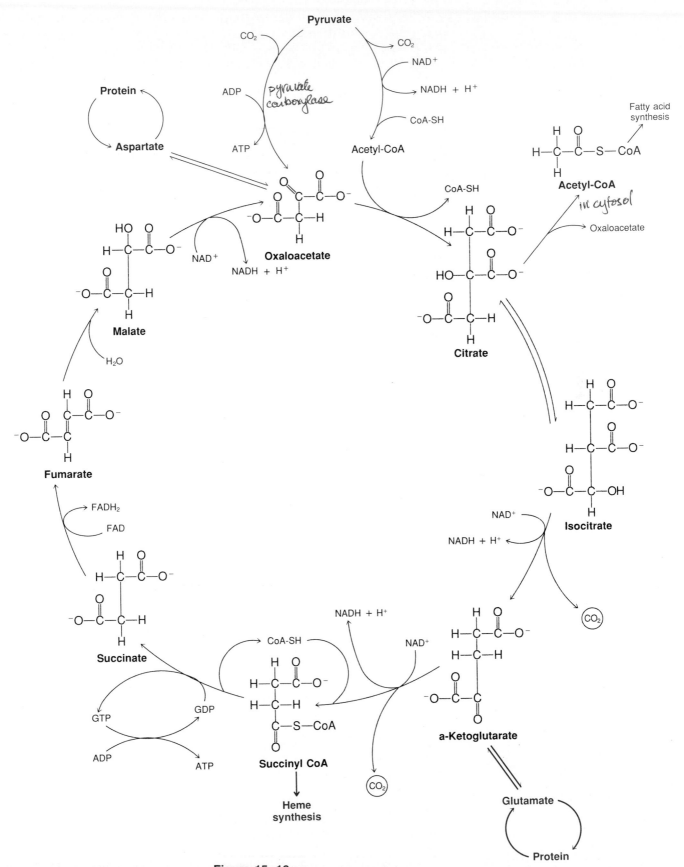

Figure 15–10
The TCA cycle, showing some of the branchpoint pathways (in color) that either drain or replenish the TCA intermediates.

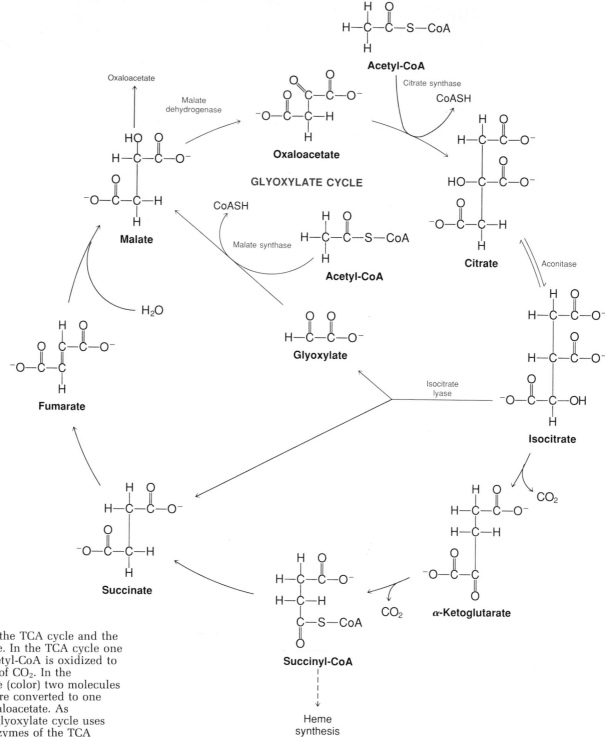

Figure 15–11
Comparison of the TCA cycle and the
glyoxylate cycle. In the TCA cycle one
molecule of acetyl-CoA is oxidized to
two molecules of CO_2. In the
glyoxylate cycle (color) two molecules
of acetyl-CoA are converted to one
molecule of oxaloacetate. As
indicated, the glyoxylate cycle uses
some of the enzymes of the TCA
cycle. Only enzymes operative in the
glyoxylate cycle are shown.

malate synthase, is an aldol condensation, closely analogous to that catalyzed by citrate synthase. Succinate and malate, the products of these two
reactions, are oxidized to oxaloacetate by the usual enzymes of the TCA
cycle.

The glyoxylate and TCA cycles are compared in Figure 15–11. The glyoxylate cycle bypasses both of the steps in which CO_2 is lost in the TCA
cycle, and also has a second point of entry for acetyl-CoA. Thus, in contrast

to the TCA cycle, which takes in two carbons as an acetyl group and loses two carbons as CO_2, the glyoxylate cycle takes in four carbons, loses no CO_2, and produces a molecule of oxaloacetate. Oxaloacetate can be converted to phosphoenolpyruvate and on to carbohydrates, as well as to all of the biosynthetic starting materials that are intermediates of glycolysis. Thus cells that possess the enzymes of the glyoxylate cycle can meet all of their carbohydrate needs from any substrate that can be converted to acetyl-CoA. For example, as mentioned in the preceding chapter, yeast is able to grow on ethanol because yeast can oxidize ethanol to acetyl-coenzyme A. When ethanol is the substrate, yeast cells produce isocitrate lyase and malate synthase, and obtain biosynthetic intermediates by way of the glyoxylate cycle.

The balanced overall equation for the glyoxylate cycle is

$$2 \text{ Acetyl-CoA} + \text{FAD} + 2 \text{ NAD}^+ + 3 \text{ H}_2\text{O} \longrightarrow$$
$$\text{oxaloacetate} + \text{FADH}_2 + 2 \text{ NADH} + 2 \text{ H}^+$$

Although no ATP is generated directly by the glyoxylate pathway, we will see that eight molecules of ATP are ultimately regenerated (three from each NADH and two from $FADH_2$) for each molecule of oxaloacetate produced (Chapter 16). This yield of four ATPs per acetyl-CoA is only one-third of the amount of ATP that is produced in the TCA cycle, but the main function of the glyoxylate cycle is production of oxaloacetate; the regeneration of ATP is only a side benefit.

OXIDATION OF OTHER SUBSTRATES BY THE TCA CYCLE

The TCA cycle, strictly speaking, has only one input fuel, acetyl-CoA. Catabolism of carbohydrates and fats leads to the production of acetyl-CoA, so the TCA cycle is ideally suited to serve as the major oxidative sequence in the catabolism of those types of compounds. However, degradation of the amino acids that are produced in the hydrolysis of protein produces a number of intermediates, among which are α-ketoglutarate, succinyl-CoA, fumarate, and oxaloacetate (Chapter 19). α-Ketoglutarate, succinyl-CoA, and fumarate can be oxidized to oxaloacetate, but the cycle as such cannot oxidize oxaloacetate further. In the presence of acetyl-CoA, each molecule of oxaloacetate used in the synthesis of citrate is regenerated in the cycle; thus there is no net oxidation of oxaloacetate in that case either.

The problem of how to oxidize oxaloacetate is solved by the action of phosphoenolpyruvate carboxykinase, which we discussed in connection with gluconeogenesis in the preceding chapter. This enzyme catalyzes the conversion of oxaloacetate to phosphoenolpyruvate with the help of ATP or GTP, and thus permits the total oxidation of oxaloacetate to CO_2 by the enzymes of the TCA cycle. The mechanism for the conversion of oxaloacetate to pyruvate is shown in Figure 15–12.

Oxaloacetate + ATP $\longrightarrow$ phosphoenolpyruvate + ADP + CO_2

Phosphoenolpyruvate + ADP $\longrightarrow$ pyruvate + ATP

CoA + pyruvate + NAD^+ $\longrightarrow$ acetyl-CoA + CO_2 + NADH + H^+

Oxaloacetate + acetyl-CoA $\longrightarrow$ citrate

Sum: 2 Oxaloacetate + NAD^+ $\longrightarrow$ citrate + 2 CO_2 + NADH + H^+

If we add the equation for the oxidation of citrate to oxaloacetate in one turn of the cycle,

Citrate + 3 NAD^+ + FAD + ADP (or GDP) $\longrightarrow$
$$\text{oxaloacetate} + 3 \text{ NADH} + \text{FADH}_2 + \text{ATP (or GTP)} + 2 \text{ CO}_2 + 3 \text{ H}^+$$

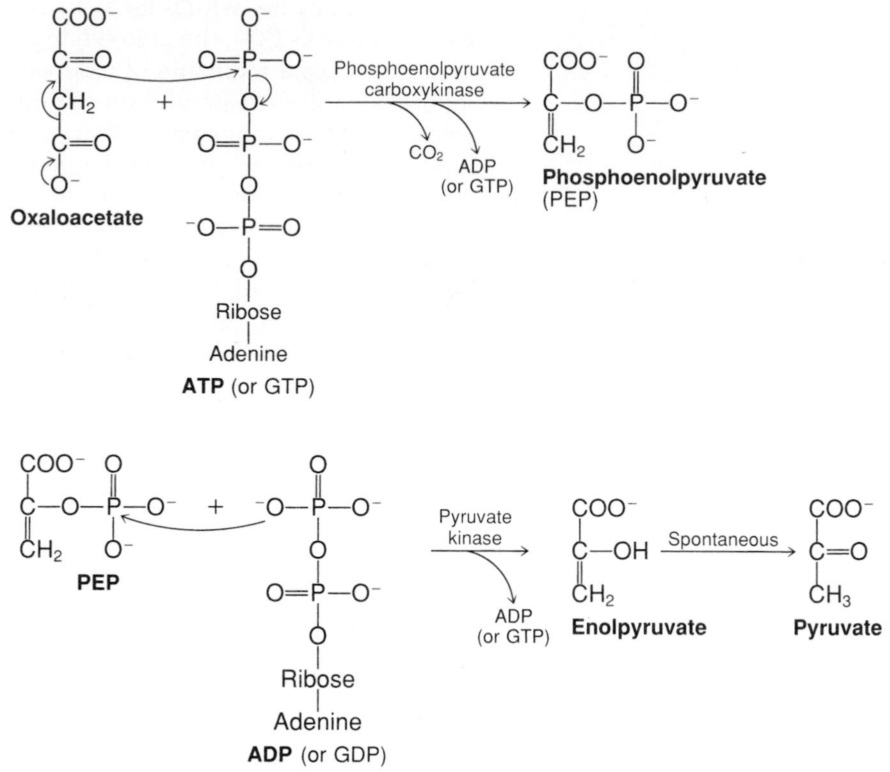

Figure 15–12
The conversion of oxaloacetate to pyruvate. The nucleotide triphosphate used in the first step is regenerated in the second step. Thus we have ATP operating as a true coenzyme in this reaction.

we obtain the equation for the total oxidation of oxaloacetate:

Oxaloacetate + 4 NAD$^+$ + FAD + ADP (or GDP) $\longrightarrow$

$$4\ CO_2 + 4\ NADH + FADH_2 + ATP\ (or\ GTP) + 4\ H^+$$

NADH and FADH$_2$ will feed electrons into the oxidative phosphorylation system, so the oxidation of a molecule of oxaloacetate will cause the regeneration of 15 molecules of ATP, which is the same as the yield from one molecule of pyruvate.

In addition to the amino acids that are converted to intermediates of the TCA cycle, others are converted to pyruvate or acetyl-CoA and thus enter the cycle in the usual way. In fact, all of the 20 protein amino acids are metabolized by way of the TCA cycle (see Chapter 19). Thus, although it is often thought of as part of the pathway for carbohydrate metabolism, the TCA cycle is actually the central oxidative sequence for all three of the major types of carbon and energy sources: carbohydrates, fats, and proteins.

REGULATION OF TCA CYCLE ACTIVITY

The TCA cycle is carefully regulated to ensure that its level of activity corresponds closely to cellular needs. The cycle serves two functions: (1) furnishing reducing equivalents (as NADH and to a lesser extent as FADH$_2$) to the electron-transport chain, and (2) by means of side reactions, providing substrates for biosynthetic reactions. Both of these functions are reflected in the regulation to which the cycle is subject.

In its primary role as a means of oxidizing acetyl groups to CO$_2$ and water, the TCA cycle is sensitive both to the availability of its substrate, acetyl-CoA, and to the accumulated levels of its principal end products, NADH and ATP. Actually the ratio NADH/NAD$^+$ and the energy charge or the ATP/ADP ratio are more important than the individual concentrations. Other regulatory parameters to which the TCA cycle is sensitive include the ratios of acetyl-CoA to free CoA, acetyl-CoA to succinyl-CoA, and citrate to oxaloacetate. The major known sites for regulation are shown in Figure 15–13. These include two enzymes outside the TCA cycle (pyruvate dehydrogenase and pyruvate carboxylase) and three enzymes inside the TCA cycle (citrate synthase, isocitrate dehydrogenase, and α-ketoglutarate dehydrogenase). As might be suspected *a priori*, each of these sites of regulation represents an important metabolic branchpoint.

Pyruvate Dehydrogenase Is Regulated by Small-molecule Effectors and Covalent Modification

In considering some of the regulatory events that appear to affect the flux through the cycle, we will begin with pyruvate dehydrogenase. The conversion of pyruvate to acetyl-CoA is not strictly speaking a part of the cycle.

In keeping with its strategic position in metabolism the pyruvate dehydrogenase complex is a site of multiple regulatory interactions. The affinity of the first enzyme of the complex (pyruvate decarboxylase) for pyruvate is reduced by a slight increase in the value of the energy charge (high ATP/ADP) and by a high concentration of the product of the overall conversion, acetyl-CoA. The rate of the overall reaction is also decreased by an increase in the NADH/NAD$^+$ ratio. All of these effects are in the correct direction to cause the rate of acetyl-CoA synthesis to vary appropriately with the need for electrons and for regeneration of ATP.

In addition to these typical noncovalent regulatory interactions, the pyruvate dehydrogenase complex, at least in mammals, is subject to regulation by covalent modification. Each of the giant complexes contains a few molecules of a protein kinase and of a protein phosphorylase. The kinase catalyzes phosphorylation of three specific serine —OH groups on the pyruvate carboxylase portion of the complex, and the phosphorylase catalyzes hydrolytic removal of those phosphoryl groups. The phosphorylated enzyme is relatively inactive. Thus the phosphorylation-dephosphorylation system also contributes to regulation of the rate of conversion of pyruvate to acetyl-CoA. The action of the kinase, and the resultant decrease in the activity of the pyruvate complex, is favored by high ATP/ADP and NADH/NAD$^+$ ratios and by high acetyl-CoA. These effects act in the same direction as the direct effects of the modifier metabolites on the enzymes of the complex, and the types of regulation must therefore reinforce each other.

As we will see below, these regulatory responses of pyruvate dehydrogenase are important also in the partitioning of pyruvate between conversion to acetyl-CoA for oxidation in the cycle and conversion to oxaloacetate to replenish cycle intermediates. For now we will continue down the oxidative pathway.

Citrate Synthase Is Negatively Regulated by NADH and the Energy Charge

Since citrate synthesis is the reaction by which carbon, in the form of acetyl-CoA derived from carbohydrate, fat, or protein, enters the cycle, regulation of the flux into the cycle is synonymous with regulation of the reaction catalyzed by citrate synthase. Regulation, by energy charge and other parameters, of the rate of glycolysis and of the pyruvate dehydrogenase reaction play important roles in controlling the rate of citrate synthesis. For example, yeast citrate synthase has been shown to respond sensitively to variation in the value of the energy charge. The enzymes from various bacteria and from mammals have been reported to be less sensitive to changes in the concentration of ATP. Synthesis of citrate is also favored by a low NADH/NAD$^+$

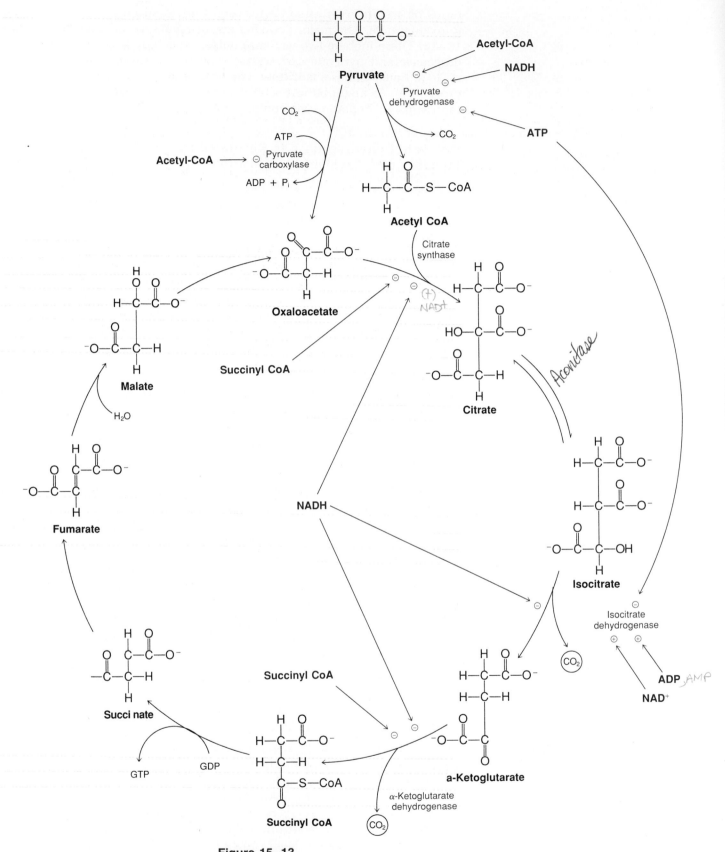

Figure 15–13

Major regulatory sites of the TCA cycle, with activators (⊕) and inhibitors (⊖) of specific reactions shown in color.

ratio. This is clearly metabolically desirable, since operation of the cycle should favor an increase in this ratio or the reduced form of the coenzyme.

Isocitrate Dehydrogenase Is Very Sensitive to the NADH/NAD$^+$ Ratio

The equilibrium constant for the conversion of citrate to isocitrate is small, as seen above, and the two intermediates thus make up a metabolic pool like those seen in the glycolysis-gluconeogenesis pathway. Regulation is not to be expected within a metabolic pool. Thus the next possible regulatory site is the conversion of isocitrate to α-ketoglutarate.

The mitochondrial NAD-specific isocitrate dehydrogenase is a highly cooperative enzyme. The activity of the enzyme from yeast varies as the fourth order with respect to isocitrate concentration and second order with respect to NAD$^+$ and the modifier AMP. Each of these ligands increases the affinity of the enzyme for the other ligands. In view of the fact that for this enzyme the affinity for substrate is a second-order function of the concentration of AMP, isocitrate dehydrogenase activity is especially sensitive to small changes in the concentration of AMP. The second-order response to changes in the concentration of NAD$^+$ is enhanced by a strong inhibitory effect of NADH, so the reaction is also very sensitive to the NADH/NAD$^+$ ratio, with the rate of reaction being significantly enhanced when that ratio falls even slightly. The properties of the mammalian enzyme are generally similar to those of the yeast enzyme, except that ADP replaces AMP as the positive modifier.

Why is isocitrate dehydrogenase subject to such intense regulation? We have emphasized that regulation is to be expected at metabolic branchpoints or at first reactions of sequences, which amounts to the same thing in the sense that the two reactions immediately following a branchpoint can be described as the first reactions in a sequence. Clearly, isocitrate dehydrogenase, being situated within the TCA cycle, cannot regulate the entry of carbon into the cycle. It would appear that regulation at this point could lead only to accumulation of isocitrate and citrate. Because of the large equilibrium constant of the citrate synthase reaction, accumulation of citrate could not significantly limit the entry of acetyl-CoA by mass action.

To understand the significance and the metabolic desirability of regulation at the isocitrate dehydrogenase step, we must expand our consideration beyond the TCA cycle, and indeed beyond the mitochondrion. The citrate-isocitrate metabolic pool is in fact a branchpoint of major importance, and the regulatory responses of isocitrate dehydrogenase appear to be an important factor in regulation of the partition ratio at this branch. That partitioning controls the rate of generation of acetyl-CoA in the cytosol of the cell.

Acetyl-coenzyme A is among the most important biosynthetic starting materials, in terms of the total amount used and of the number of sequences to which it contributes. It supplies most of the carbon atoms of lipids, some amino acids, and a wide range of vitamins, cofactors, and pigments. The synthesis of these compounds occurs in the cytosol, but there is no primary production of acetyl-CoA in the cytosol, since both the fatty acids derived from fat breakdown and the pyruvate that results from carbohydrate degradation enter the mitochondrion and are converted to acetyl-CoA there.

The acetyl-CoA needed for synthetic reactions in the cytosol is obtained secondarily by way of citrate exported from the mitochondria. In the cytosol this citrate is cleaved to yield acetyl-CoA and oxaloacetate. This step presents potential difficulties: you will remember that the equilibrium constant for the synthesis of citrate from acetyl-CoA and oxaloacetate is large. The lysis of citrate supplies another illustration of the generalization that any conversion can be made feasible by coupling to a sufficient number of ATP-to-ADP conversions. In this case, one ATP is enough to result in a favorable equilibrium constant, and the reaction catalyzed is

$$\text{Citrate} + \text{CoA} + \text{ATP} \longrightarrow \text{oxaloacetate} + \text{acetyl-CoA} + \text{ADP} + \text{P}_i.$$

This reaction, catalyzed by *citrate lyase,* is highly sensitive to the value of the energy charge, being favored by slight increases and slowed by slight decreases in the charge. That is the appropriate direction of regulation for a step that provides the starting material for a large number of biosynthetic sequences as well as for the major energy-storage process of fat synthesis.

The oxaloacetate that is produced in the cytosol as a byproduct of acetyl-CoA generation has several possible fates. Some cells contain transaminases (Chapter 19) in the cytosol that can convert oxaloacetate to aspartate, which is needed as such in protein synthesis and is also a precursor for the synthesis of asparagine, threonine, isoleucine, methionine, and lysine. (Of those amino acids, mammals can make only asparagine.) *Excess oxaloacetate beyond the amount required in biosyntheses is reduced to malate.* That is a standard pattern: oxaloacetate that is produced in the cytosol is usually immediately reduced to malate. Several reasons for this could be suggested. Oxaloacetate is somewhat unstable to decarboxylation; also it is reactive, and might participate in undesirable nonenzymic side reactions. But probably the most important reason is related to the feasibility of diffusion to the point of use. We saw earlier that at equilibrium the concentration of malate is several thousand times that of oxaloacetate. Since the net diffusion rate depends on the absolute concentration gradient rather than on a ratio of concentrations, malate therefore can diffuse (for example, to and into a mitochondrion) thousands of times faster than would be possible for the oxaloacetate that would be in equilibrium with it.

A similar relationship may be part of the reason why citrate, rather than isocitrate, is the compound that moves from the mitochondrion to the cytosol. The equilibrium concentration of citrate is about 15 times that of isocitrate so, at a given total concentration of the citrate-isocitrate pool, citrate can diffuse much faster than isocitrate.

There are two main paths by which the malate that is produced by reduction of cytosolic oxaloacetate is processed. The simpler is uptake by mitochondria and oxidation to oxaloacetate. In that case (Figure 15–14), the net overall reaction corresponding to the supply of acetyl-CoA to the cytosol is simply

$$\text{Acetyl-CoA}_{\text{mit}} + \text{ATP} \longrightarrow \text{acetyl-CoA}_{\text{cyt}} + \text{ADP} + \text{P}_i$$

and in effect the system is an ATP-driven pump for moving acetyl-CoA to the cytoplasm. Two electrons are also moved from the cytosol to the interior of the mitochondrion, where they can be used in electron-transfer phosphorylation.

The other path for malate is conversion to pyruvate in an oxidative decarboxylation catalyzed by an enzyme called the NADP-specific malate enzyme, after which the pyruvate formed is taken up by mitochondria.

$$\text{Malate} + \text{NADP}^+ \longrightarrow \text{pyruvate} + \text{CO}_2 + \text{NADPH}$$

In this case, part of the NADPH that is required in the synthesis of fatty acids (Chapter 21) is supplied by the oxidation of malate, since NADH had been used in the reduction of oxaloacetate to malate (see Figure 15–14). Continuing with this pathway, after entering the mitochondrion the pyruvate must be carboxylated to replace the oxaloacetate that was used in the initial production of citrate. The stoichiometry for this pathway is less simple than for the other.

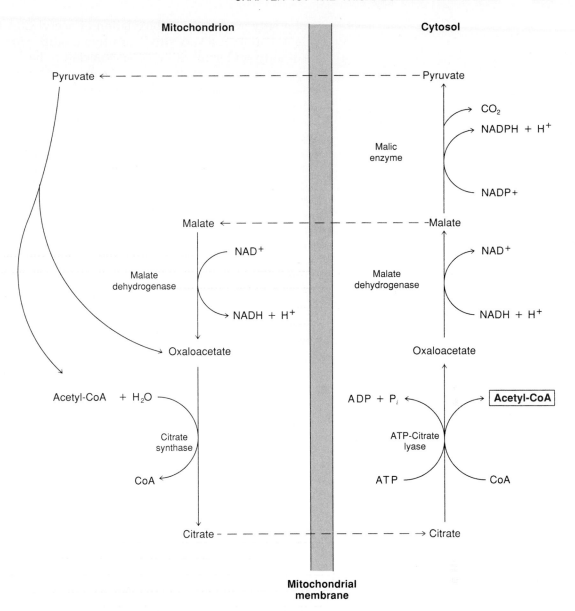

Figure 15–14

Cycling between mitochondria and cytosol in the supply of acetyl-CoA for use in biosynthetic sequences in the cytosol. Citrate moves from the interior of the mitochondria to the cytosol. In the cytosol it is cleaved to acetyl-CoA and oxaloacetate by citrate lyase. The equilibrium constant for this reaction is favorable because an ATP-to-ADP conversion is involved. Most of the oxaloacetate is reduced to malate. The malate may be taken up by mitochondria or oxidized to pyruvate and CO_2, generating NADPH for use in biosynthetic sequences in the cytosol. The pyruvate enters the mitochondria, where it may be converted to oxaloacetate or an acetyl-CoA by the usual routes (see Figure 15–13).

$$\text{Acetyl-CoA}_{\text{mit}} + 2\text{ ATP} + \text{NADH} + \text{NADP}^+ \longrightarrow$$
$$\text{acetyl-CoA}_{\text{cyt}} + 2\text{ ADP} + \text{NAD}^+ + \text{NADPH} + 2\text{ P}_i$$

Two molecules of ATP are used for each acetyl-CoA that is made available in the cytosol, but a pair of electrons are also promoted from the NAD pool to the NADP pool, where they can be used in reductive biosyntheses such as that of fatty acids.

In addition to supplying the carbon atoms of fatty acids, citrate is also a major regulator of the rate of fatty acid synthesis. The first step in the synthesis of a fatty acid is the carboxylation of acetyl-CoA to form malonyl-coenzyme A. Citrate is a very strong positive modifier for that reaction. Thus the rate of fatty acid synthesis can be expected to be a sensitive function of the concentration of cytosolic citrate. As we have seen in Chapter 14, citrate is also a negative modifier of phosphofructokinase. An increase in the concentration of citrate tends to decrease the rate of glycolysis, and a decrease in citrate concentration tends to increase that rate. Thus an increase in the concentration of citrate in the cytosol causes an increase in the production of storage fats and tends to limit the rate of carbohydrate breakdown. Conversely, a decrease in citrate concentration tends to increase the rate of carbohydrate breakdown and decreases the rate of production of storage fat. These effects are appropriate since the rate at which citrate moves from the mitochondria into the cytosol must be determined primarily by the effect that the energy charge in the mitochondrion has on the properties of isocitrate dehydrogenase. Thus an increase in citrate concentration is a sign of improving energy status and a decrease in that concentration indicates a deteriorating energy status. The concentration of citrate is also an indicator of the availability of the biosynthetic starting materials that are supplied by the catalytic sequences.

From this consideration of the roles of cytoplasmic citrate as acetyl-CoA precursor and metabolic signal, we can view the significance of the regulation of isocitrate dehydrogenase in a new light. The useful function of the regulatory properties of isocitrate dehydrogenase is probably control of the rate at which citrate is exported to the cytosol.

α-Ketoglutarate Dehydrogenase Is Negatively Regulated by NADH

α-Ketoglutarate is also a branchpoint metabolite, since it can be transaminated to form glutamate (Chapter 19). Glutamate is needed in protein synthesis directly and also is a precursor of a number of other amino acids. Thus it is consumed in the cytosol in rather large amounts in a cell that is synthesizing protein rapidly. Therefore we would expect that the reaction catalyzed by α-ketoglutarate dehydrogenase would be regulated so as to retain carbon in the cycle when energy is in short supply, and to allow the concentration of α-ketoglutarate to rise, facilitating its transamination to glutamate and exit from the cycle, when the energy supply is high. When NADH is low, the enzyme will compete more strongly for α-ketoglutarate and thus maximize the regeneration of ATP while tending to decrease biosynthetic activities that would use ATP. No adequate test of the effect of adenine nucleotides on this enzyme has been reported.

The Pyruvate Branchpoint Partitions Pyruvate Between Acetyl-CoA and Oxaloacetate

Acetyl-CoA is the only compound that can enter the TCA cycle when it is operating purely oxidatively, but one molecule of oxaloacetate must enter for each molecule of citrate, α-ketoglutarate, or succinyl-coenzyme A that is removed for use in biosynthesis. It follows that pyruvate is the most important branchpoint in the metabolism of a cell that is living on carbohydrate. The partitioning of pyruvate between decarboxylation to form acetyl-CoA and carboxylation to form oxaloacetate is in effect partitioning between the two major metabolic uses of pyruvate: oxidation of carbon for regeneration of ATP and conversion to suitable starting materials for biosynthesis.

Some of the regulatory interactions that affect partitioning at pyruvate are shown in Figure 15–13. We will focus first on the effects of acetyl-CoA, which is a negative modifier for pyruvate dehydrogenase and a very strong positive modifier for pyruvate carboxylase. To illustrate the regulatory roles of those effects, we may consider a mitochondrion in which the TCA cycle is functioning only oxidatively, so that no input of oxaloacetate is required. If now conditions change and α-ketoglutarate begins to be removed for biosynthesis, the rate at which oxaloacetate is regenerated by the cycle will be reduced by the rate of α-ketoglutarate removal. The citrate synthase reaction can proceed no more rapidly than the rate at which oxaloacetate, one of its substrates, is supplied, so the rate of citrate synthesis, too, will decrease. Consequently acetyl-CoA will be used more slowly, which will cause its concentration to increase slightly. The increase in acetyl-CoA concentration will lead to a decrease in the activity of the pyruvate dehydrogenase complex and an increase in the activity of pyruvate carboxylase. As a result of those effects, the partitioning of pyruvate is changed so as to produce more oxaloacetate, thus replenishing the cycle and allowing it to continue to function in spite of the removal of α-ketoglutarate.

Since the system responds to the concentration of acetyl-CoA and thus indirectly to the concentration of oxaloacetate, it will adjust automatically to maintain a functional concentration of oxaloacetate regardless of whether the intermediate removed from the cycle is citrate, α-ketoglutarate, succinyl-coenzyme A, oxaloacetate itself, or any combination of these. The same effects will cause the rate of conversion of pyruvate to oxaloacetate to decrease when the rate of removal of biosynthetic precursors decreases.

Green plants and many bacteria generate oxaloacetate from phosphoenolpyruvate rather than from pyruvate. This is a more direct route, and does not involve ATP:

$$\text{Phosphoenolpyruvate} + CO_2 \longrightarrow \text{oxaloacetate}$$

Although phosphoenolpyruvate carboxylase catalyzes a very different reaction, and does not require either ATP or biotin, it is modulated by acetyl-CoA in just the same way as is pyruvate carboxylase in eukaryotes where it is present. This fact illustrates once again the generalization that regulation mirrors function; regulatory interactions cannot be predicted from the equation for a reaction or from its mechanism.

SUMMARY

The major capabilities of the enzymes of the TCA cycle include the following:

1. Oxidation of acetyl-CoA, yielding electrons for regeneration of ATP by oxidative phosphorylation.
2. With the help of phosphoenolpyruvate carboxykinase, pyruvate kinase, and pyruvate decarboxylase, oxidation of oxaloacetate or any other intermediate of the cycle, yielding electrons for use in the oxidative phosphorylation system.
3. Conversion of oxaloacetate and acetyl-CoA to citrate, α-ketoglutarate, or succinyl-coenzyme A for use in biosynthesis.
4. When intermediates of the cycle are supplied—for example, from protein degradation—production of phosphoenolpyruvate as a starting material for gluconeogenesis.
5. With the help of isocitrate lyase and malate synthase, conversion of acetyl-CoA to oxaloacetate or any other in-

termediate of the cycle for use in biosynthesis or gluconeogenesis.

The rate of entry of carbon into the cycle is regulated by many factors, including the NADH/NAD$^+$ ratio, the energy charge, and the need for biosynthetic intermediates. Regulatory effects may be exerted on the behavior of pyruvate dehydrogenase and on the behavior of citrate synthase.

Partitioning between the oxidative and biosynthetic functions of the cycle occurs mainly at pyruvate in mammals. The amount of pyruvate that is carboxylated to form oxaloacetate is determined by the rate at which intermediates of the cycle are removed for use in biosyntheses. In many other kinds of organisms a similar partitioning occurs at phosphoenolpyruvate.

The TCA cycle is the main source of electrons for oxidative phosphorylation, and thus the major energetic sequence in the metabolism of aerobic cells or organisms. It serves as the main distribution center of metabolism, receiving carbon from the degradation of carbohydrates, fats, and proteins and, when it is appropriate, supplying carbon for the synthesis of carbohydrates, fats, or proteins. Every aspect of the metabolism of an aerobic organism is directly dependent on the TCA cycle.

The glyoxylate cycle provides an alternative fate for acetyl-CoA. Reactions of the glyoxylate cycle permit the synthesis of four- or six-carbon compounds from two-carbon compounds. This is accomplished with four enzymes from the TCA cycle and two that are unique to the glyoxylate cycle. The latter two enzymes are possessed by certain plants and microorganisms.

SELECTED READINGS

Atkinson, D. E., *Cellular Energy Metabolism and Its Regulation.* New York: Academic Press, 1977. A general discussion, covering some topics in this and the two preceding chapters in somewhat greater depth than the treatment in this book.

Beitner, R. *Regulation of Carbohydrate Metabolism.* Boca Raton, FL: CRC Press, 1985. Two-volume discussion, at a rather advanced level, of many aspects of the regulation of mammalian carbohydrate metabolism with frequent discussions of clinical conditions. Many referenes.

Estabrook, R. W., and P. Srere, (eds.), *Biological Cycles (Current Topics in Cellular Regulation, Vol. 18).* New York and London: Academic Press, 1981. A collection of papers written for a symposium commemorating Hans Kreb's eightieth birthday, dealing with many metabolic cycles, several of which are related to material presented in various chapters of this book.

Krebs, H. A., Excursion into the borderland of biochemistry and philosophy. *Bull. Johns Hopkins Hosp.* 95:45–51, 1954. An essay, by one of the greatest biochemists of this century, dealing with the importance of imagination in science and of a biological outlook in biochemistry. Although not directly related to the material of this chapter, the essay is valuable reading for any student of science and especially for anyone interested in biochemistry.

Krebs, H. A., The history of the tricarboxylic acid cycle. *Perspect. Biol. Med.* 14:154–170, 1970. Personal reminiscences of one of the most important advances in our understanding of how metabolic reactions are organized.

PROBLEMS

1. Mark each of the following statements with one of the following: T (for true), F (for false), or E (for eeny meeny, if you cannot tell whether the statement is true or false on the basis of the information given).
   a. All five cofactors or prosthetic groups of the pyruvate dehydrogenase complex are directly involved in electron transport.
   b. Thermodynamically, acetyl-CoA ought to be capable of driving the phosphorylation of ADP (or GDP) just as succinyl-CoA does.
   c. The methyl carbon of each acetyl-CoA entering the TCA cycle comes from carbon 3 of pyruvate.
   d. Even if aconitase were incapable of discriminating between the two ends of the citrate molecule, the CO_2 evolved by the isocitrate dehydrogenase reaction would still inevitably come from the oxaloacetate rather than the acetyl-CoA substrate of the citrate synthase reaction.
   e. Activation of pyruvate carboxylase means that one or more intermediates of the TCA cycle are being drawn off for biosynthetic purposes.
   f. Malate cannot be converted to fumarate in the mitochondria of an animal cell, because the TCA cycle operates only unidirectionally.

2. Assume that you have a solution containing the pyruvate dehydrogenase complex and all the enzymes of the TCA cycle but none of the metabolic intermediates.
   a. If you add 3 mM each of pyruvate, coenzyme A, NAD, FAD, ADP, GDP, and P_i, what will happen? How much CO_2 will be evolved?

   b. If in addition to the reagents specified in part (a), you add 3 mM each of citrate, isocitrate, α-ketoglutarate, succinate, fumarate, and malate, how much CO_2 will be evolved?
   c. If in addition to the reagents specified in part (a), you could add only one of the intermediates specified in part (b), which one would you choose? Why? Now how much CO_2 will be evolved?
   d. If in addition to the reagents specified in part (a), you add pyruvate carboxylase, what effect would you expect that to have on CO_2 evolution?

3. Suppose you feed [1-^{14}C] glucose to an aerobic culture of bacteria and then extract and separate intermediates of the glycolytic pathway and the TCA cycle. Which carbon atom(s) would you expect to find labeled first with ^{14}C in each of the following molecules?
   a. Glyceraldehyde-3-phosphate.
   b. Phosphoenolpyruvate.
   c. α-Ketoglutarate.
   d. Oxaloacetate.

4. In which carbon atom(s) of glucose fed to an aerobic bacterial culture would ^{14}C have to be located to ensure that labeled CO_2 was released first at the step catalyzed by the following?
   a. Pyruvate dehydrogenase.
   b. Isocitrate dehydrogenase.
   c. α-Ketoglutarate dehydrogenase.

5. In his landmark experiments that led to the elucidation of the TCA cycle, Hans Krebs observed that addition of

malonate to extracts made from pigeon flight muscle inhibited pyruvate utilization and resulted in the accumulation of succinate upon oxidative catabolism of a variety of added substrates.

a. Why do you suppose Krebs used minced flight muscle preparations in these studies?

b. Why does malonate inhibit pyruvate utilization by such preparations?

c. What would Krebs have been able to conclude when he found that succinate accumulated in malonate-treated muscle following addition of citrate, isocitrate, or α-ketoglutarate?

d. Why was the accumulation of succinate in malonate-treated muscle even more significant when the added substrate was fumarate, malate, or oxaloacetate instead of the substrates mentioned in part (c)?

e. Krebs also found that inhibition of pyruvate utilization could be overcome by adding oxaloacetate, malate, or fumarate along with the pyruvate. How would you explain these results?

6. The enthalpy of combustion (and hence the heat release during combustion) of many organic molecules is to a good approximation proportional to the number of electrons that are transferred to oxygen. Does this explain why the caloric values of foods, as determined by combustion, are nutritionally useful values?

7. Normal or wild-type *E. coli* can grow with glucose as the only source of carbon and energy. For each of the following mutant strains of *E. coli*, what compounds do you think you would need to add to the medium to obtain growth?

a. A strain lacking pyruvate carboxylase.

b. A strain lacking isocitrate dehydrogenase.

c. A strain lacking citrate lyase.

8. What would you expect to be the consequences of mutations in *E. coli* that led to the following conditions?

a. Inability to produce malate synthase.

b. A genetically altered pyruvate carboxylase that was not activated by acetyl-CoA.

c. An altered pyruvate dehydrogenase that was inhibited much more strongly than the normal enzyme by acetyl-CoA.

9. A principle of cellular energy metabolism is that alternative substrates are catabolized by bringing them as quickly as possible into the mainstream glycolytic/TCA cycle pathway. Hydroxypyruvate is such an alternative substrate. It is a three-carbon compound with the following structure:

$$HO-CH_2-\overset{\overset{\displaystyle O}{\|}}{C}-\overset{\overset{\displaystyle O}{\|}}{C}-O^-$$
Hydroxypyruvate

Despite the similarity between hydroxypyruvate and pyruvate, the only cellular route by which the former can be converted into the latter is a five-step pathway involving the four intermediates shown below. (The labels A, B, C, and D do not necessarily reflect the order in which the intermediates appear in the pathway.) The pathway begins with an NADH-mediated reaction and requires the presence of both ADP and ATP, although with no net consumption or generation of either.

a. Order the intermediates A, B, C, and D in the proper sequence to account for the stepwise conversion of hydroxypyruvate into pyruvate, and write a balanced equation for each of the five reactions.

b. Write an overall equation for the conversion of hydroxypyruvate to pyruvate.

10. Under anaerobic conditions, *E. coli* produces fumarate reductase, which uses NADH for the reduction of fumarate to succinate, rather than succinate dehydrogenase, which uses FAD for the oxidation of succinate to fumarate. Electrons from glycolysis (which are transferred to NAD^+ in the oxidation of 3-phosphate-glyceraldehyde) can then be fed into the reduction of pyruvate (by way of oxaloacetate) to succinate, rather than to lactate. In what way may this mode of fermentation be preferable to the production of lactate? If glucose were being converted to α-ketoglutarate (as a synthetic intermediate) and succinate, what fraction of the carbon atoms of glucose could be made available as α-ketoglutarate? Would this fraction be different if glucose were being converted to α-ketoglutarate and lactate?

11. For each of the enzymes listed below give the name of the compound that is most important in regulating the activity of the enzyme through allosteric activation or inhibition.

a. Phosphorylase *b*.

b. Glycogen synthetase.

c. Hexokinase.

d. Phosphofructokinase.

e. Pyruvate kinase.

f. Pyruvate carboxylase.

g. Citrate synthetase.

h. Isocitrate dehydrogenase.

i. α-Ketoglutarate dehydrogenase.

16

ELECTRON TRANSPORT AND THE PRODUCTION OF ATP

The catabolism of glucose or fatty acids to CO_2 and H_2O is an oxidative process. When glyceraldehyde-3-phosphate is oxidized to glycerate-1,3-bisphosphate in the glycolytic pathway, two electrons are transferred from the substrate to NAD^+. A similar transfer of electrons to NAD^+ occurs when pyruvate is oxidized to acetyl-CoA, and at three steps in the TCA cycle (isocitrate dehydrogenase, α-ketoglutarate dehydrogenase, and malate dehydrogenase). At another point in the TCA cycle, electrons are transferred from succinate to a flavin coenzyme, FAD. For each molecule of acetyl-CoA that is produced during the breakdown of a fatty acid, two molecules of NAD^+ are reduced to NADH and one molecule of FAD is reduced to $FADH_2$. Clearly, these reactions can continue to occur in a cell only if the reduced coenzymes, NADH and $FADH_2$, are reoxidized to NAD^+ and FAD.

Some types of cells that can live anaerobically and obtain energy by fermentation reoxidize NADH by transferring electrons back to an organic metabolite. Yeast reduce acetaldehyde to ethanol; some bacteria reduce pyruvate to lactate, and muscle cells do the same when they are deprived of oxygen. Several groups of bacteria survive by using inorganic oxidants such as nitrite, nitrate, or sulfate. On the primitive earth, before photosynthetic organisms began to release oxygen into the atmosphere, virtually all free-living organisms must have depended on processes similar to these. But the amount of ATP that can be obtained by fermentative reactions is relatively small. For each molecule of glucose that is converted to lactate, a muscle cell gleans only two molecules of ATP. Bacteria that transfer electrons to nitrite, nitrate, or sulfate do not appear to do any better. An additional molecule of ATP can be formed by the breakdown of succinyl-CoA in the TCA cycle, but this is hardly an impressive yield in view of the investment in enzymatic machinery that a cell must make in order to run the cycle.

Photosynthetic cyanobacteria developed the ability to produce oxygen approximately 2.4 billion years ago. The subsequent evolution of cells that could use molecular oxygen as an electron acceptor for the reoxidation of NADH and $FADH_2$ led to an enormous increase in the capacity to make ATP.

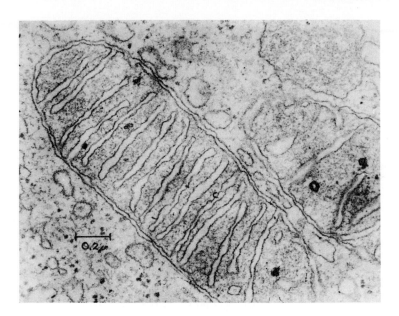

Figure 16–1
Thin-section electron micrograph of a mitochondrion in a frog kidney cell. (Courtesy Dr. J. Luft, University of Washington.)

Figure 16–2
Schematic cross-sectional view of a mitochondrion. The morphology illustrated here and in Figure 16–1 is typical of mitochondria in tissues that have an active aerobic metabolism, such as kidney. The cristae resemble flattened sacks extending most of the way across the mitochondrion. They are all extensions of the inner membrane, which separates the matrix from the intermembrane space. The space within a crista connects to the intermembrane space through five or six tubular channels around the edges. (In the micrograph shown in Figure 16–1, some of the cristae appear not to be connected to the inner membrane, but this is because the cross section does not pass through the connecting regions.) The intermembrane space communicates with the cytoplasm through pores in the outer membrane.

In this chapter, we shall explore the series of oxidation-reduction reactions by which electrons move from the reduced coenzymes to O_2, and discuss how cells couple these reactions to the synthesis of ATP.

MITOCHONDRIA: THE SITE OF AEROBIC ATP PRODUCTION

In eukaryotic cells, the TCA cycle and most of the associated reactions of aerobic energy metabolism occur in mitochondria, cylindrical organelles that typically are several μm long and about 0.5 μm in diameter. Figure 16–1 shows an electron micrograph of a mitochondrion in a kidney cell. The number of mitochondria per cell, their location in the cell, and their morphology vary considerably from tissue to tissue, depending on the tissue's demands for ATP. Mammalian erythrocytes have no mitochondria; human liver cells contain 500 to 1000 mitochondria, and kidney cells have even more.

Mitochondria generally can be recognized by their characteristic membrane structure (Figure 16–1). Close inspection shows that they have two separate membranes, a smooth outer membrane and an elaborately folded inner membrane (Figure 16–2). Extensive infoldings, known as *cristae*, give the inner membrane a substantial surface area. The inner membrane separates the mitochondrion into two distinct spaces, the internal or *matrix* space and the *intermembrane space*, between the inner and outer membranes.

The inner and outer membranes of mitochondria have very different properties. The outer membrane has only a few known enzymatic activities, and is permeable to molecules with molecular weights of up to about 5000. The inner membrane has an unusually high ratio of protein to phospholipid (about 3:1 by weight), and is impermeable to most ions and polar molecules. This permeability barrier prevents protons or cofactors such as NADH from moving freely between the mitochondrial matrix and the intermembrane space. As we shall see, the proteins associated with the inner membrane include enzyme systems that are responsible for oxygen consumption and the formation of ATP, and the barrier to movements of ions plays a key role in ATP formation. One of the enzymes of the TCA cycle, succinate dehydrogenase, also is associated with the inner membrane. The other enzymes of the TCA cycle are found in the matrix. The matrix also contains DNA and ribosomes, which are responsible for the synthesis of some of the proteins of the mitochondrion. (Other mitochondrial proteins are coded in nuclear DNA and are imported into the mitochondrion after synthesis on cytoplasmic ribosomes.) Several enzymes that participate in ATP utilization, including creatine kinase and adenylate kinase, reside in the intermembrane space.

The lipids found in the two membranes also differ. The phospholipids of the inner membrane have a high proportion of unsaturated fatty acids, and they include the unusual lipid diphosphatidylglycerol (cardiolipin). Cholesterol is found in the outer membrane, but not in the inner.

A role of mitochondria in oxygen uptake, or *respiration*, was suggested in histological studies by Michaelis in the early 1900s. In 1913, Otto Warburg found that oxygen consumption was associated with subcellular particles that he separated from soluble cellular components by filtration. In the late 1940s, Eugene Kennedy and Albert Lehninger showed that isolated mitochondria are able to carry out all of the reactions of the TCA cycle and the transport of electrons from substrates such as succinate to O_2, and that this flow of electrons provides energy for the formation of ATP.

ELECTRON TRANSPORT

Electron transport involves the reoxidation of coenzymes at the expense of molecular oxygen. The process can be summarized by the following overall reactions:

$$NADH + H^+ + \tfrac{1}{2}O_2 \longrightarrow NAD^+ + H_2O \qquad \Delta G^{\circ\prime} = -52.6 \text{ kcal/mole}$$

$$FADH_2 + \tfrac{1}{2}O_2 \longrightarrow FAD + H_2O \qquad \Delta G^{\circ\prime} = -43.4 \text{ kcal/mole}$$

The most important aspect of these reactions is the large amount of free energy released upon the oxidation of the two coenzymes. Both of the coenzymes are energy-rich molecules because each contains a pair of electrons with a high transfer potential. The transfer of electrons from these reduced coenzyme molecules to molecular oxygen releases the free energy that powers the synthesis of most of the ATP formed during aerobic metabolism.

The free energy changes that would accompany the direct oxidation of NADH and $FADH_2$ by oxygen are much larger than those usually encountered in biological reactions and are, in fact, sufficient to drive the synthesis of at least several ATPs. Hence it should not be surprising to find that electrons are not passed directly from reduced coenzymes to oxygen. Rather, the transfer is accomplished stepwise, by means of a series of reversibly oxidizable electron acceptors. In this way, the total free energy difference between reduced coenzymes and oxygen is parcelled out among a series of intermediates and is released in increments to maximize the opportunity for ATP generation.

A Chain of Cytochromes Transfers Electrons to O_2

Although many types of cells consume O_2 briskly, early biochemists noted that isolated metabolites such as sugars or organic acids do not react directly with O_2 at a significant rate. This fact suggested that an enzyme activates O_2 in some way to increase its reactivity. In the period around 1920, Warburg found that membranous fractions prepared from a variety of cells contained such an enzyme, and he concluded that iron in an organic form played a central role in its function. The evidence for his conclusion was indirect. Warburg had discovered a number of inhibitors of respiration, including CO, N_3^-, and CN^-. Of these, CO was particularly interesting, because the inhibition by CO could be reversed by light. CO was known to form complexes with Fe that had the property of dissociating when they were illuminated.

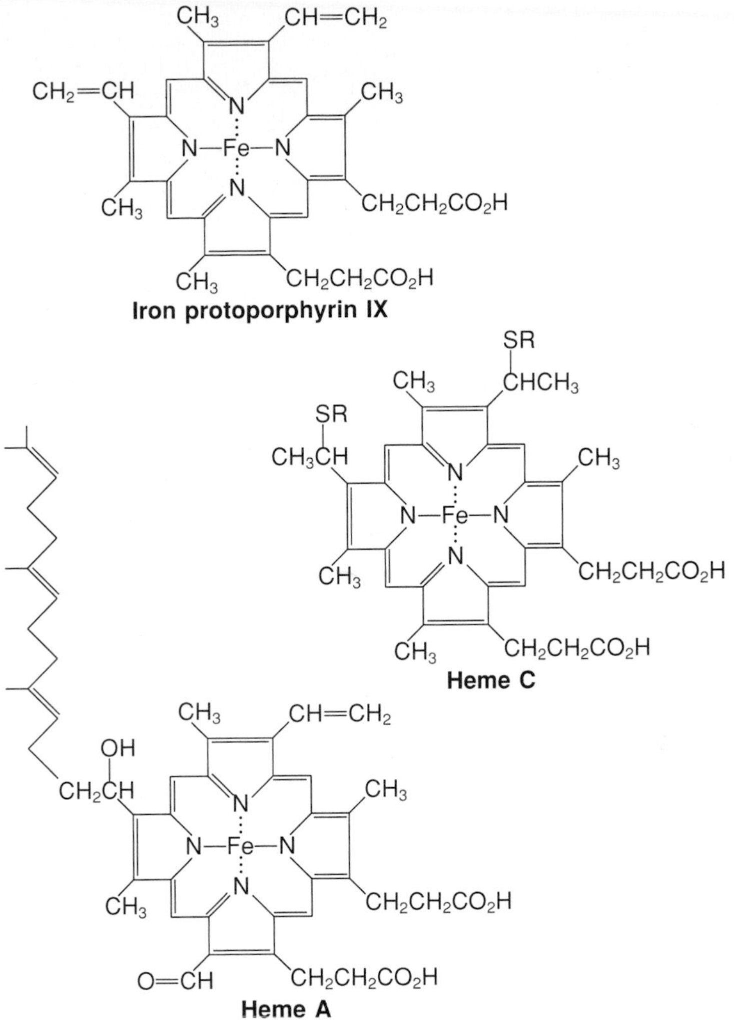

Figure 16–3

Iron protoporphyrin IX (heme) is found in the *b*-type cytochromes, and in hemoglobin and myoglobin. In heme C, cysteine residues of the protein (R) are attached covalently by thioether links to the two vinyl (—CH=CH₂) groups of protoporphyrin IX. Heme C is found in the *c* cytochromes. In heme A, which is found in the *a* cytochromes, a 15-carbon isoprenoid side chain is attached to one of the vinyls and a formyl group replaces one of the methyls.

The effectiveness of light in reversing the inhibition of respiration by CO depends on the wavelength of the light. The light, of course, must be absorbed in order to have any effect. By measuring the effectiveness of light of different wavelengths, Warburg was able to obtain an absorption spectrum of the enzyme-CO complex. From the spectrum he deduced that the iron is bound to the enzyme in the form of a heme. *Hemes are complexes of iron and a porphyrin* (Figure 16–3). The iron is attached to four pyrrole nitrogens that are part of the planar porphyrin ring. Hemes form the prosthetic groups of hemoglobin and myoglobin, the O_2 carriers in the blood and tissues (Chapter 2), and of a large group of proteins called *cytochromes*. Warburg's "respiratory enzyme" proved to be a cytochrome.

The wider participation of cytochromes in respiration was first recognized by David Keilin in the early 1930s. Keilin found that aerobic cells contain three main types of cytochromes, which he named *a*, *b*, and *c*. The three groups are distinguished by different substituents on the periphery of the porphyrin ring and by different modes of attachment of the porphyrin to the protein. The *b* cytochromes contain Fe-protoporphyrin IX, as do hemoglobin and myoglobin; the *a* cytochromes contain a modified prosthetic group, heme A (Figure 16–3). The *c* cytochromes contain heme C, in which the vinyl groups of protoporphyrin IX are attached covalently to the protein by thioether links to cysteine residues. In the *a* and *b* cytochromes and the O_2-carrying proteins, the heme binds noncovalently to the protein. The three

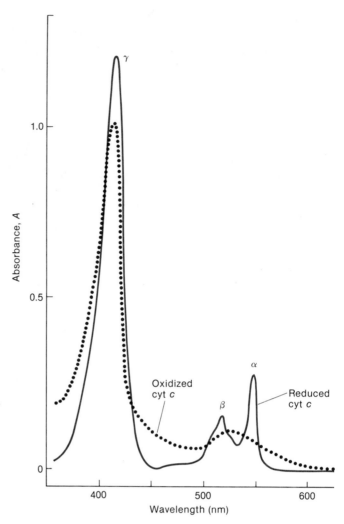

Figure 16–4
Optical absorption spectra of a 10 μM solution of cytochrome c in the reduced (──)
and oxidized (•••) states. The three characteristic bands in the spectrum of the reduced
cytochrome are called the α, β, and γ bands; the γ band is also referred to as the
Soret band.

types of cytochromes also have characteristic absorption spectra; their long-
wavelength absorption band (the α band) is near 600 nm in the a cytochromes, 560 nm in the b cytochromes, and 550 nm in the c cytochromes
(Figure 16–4).

Unlike hemoglobin and myoglobin, the b and a cytochromes are hydrophobic, membrane-bound proteins that are difficult to purify, and this is true
of some c cytochromes also. However, some c cytochromes are small, water-
soluble proteins. Keilin isolated the first member of the soluble c-type cytochromes from yeast. Since then, cytochromes homologous to this one have
been isolated from dozens of eukaryotic species, and their structures have
been studied intensively by a variety of techniques. Cytochromes of this
particular type are usually called simply cytochrome c; other members of the
c group are designated by a subscript (e.g., cytochrome c_1). With cytochromes that have been discovered more recently, the subscript often indicates the wavelength of the α absorption band (e.g., cytochrome b_{562}).

The structure of cytochrome c from horse heart has been determined by
x-ray crystallography (Figure 16–5). It is a compact, globular protein with a
molecular weight of 12,500. The planar heme group is packed in the center
of the molecule, surrounded mainly by hydrophobic amino acids. As is gen-

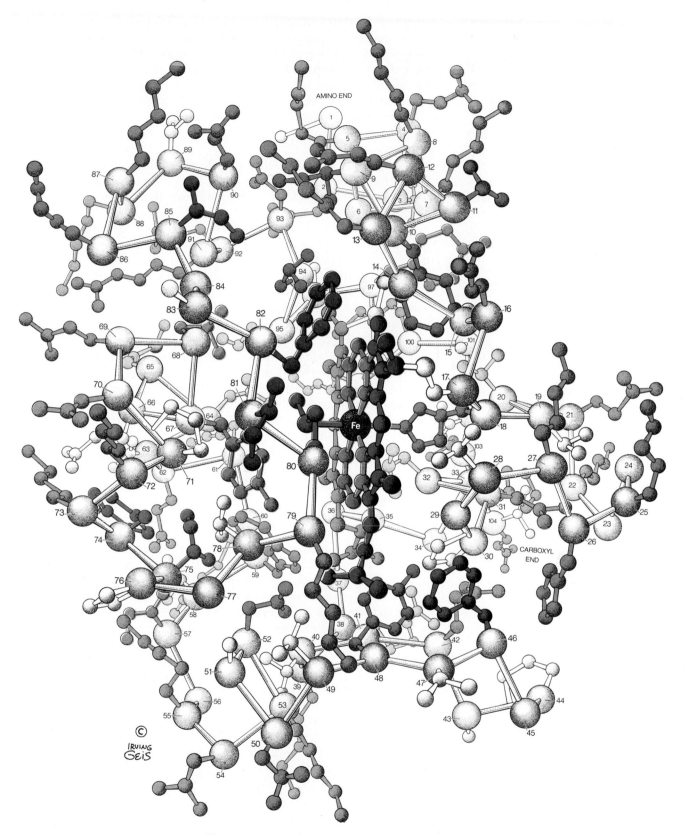

Figure 16–5
The three-dimensional structure of cytochrome *c* from horse heart. The lysines surrounding one edge of the heme are 8, 13, 25, 27, 72, 79, 86, and 87.

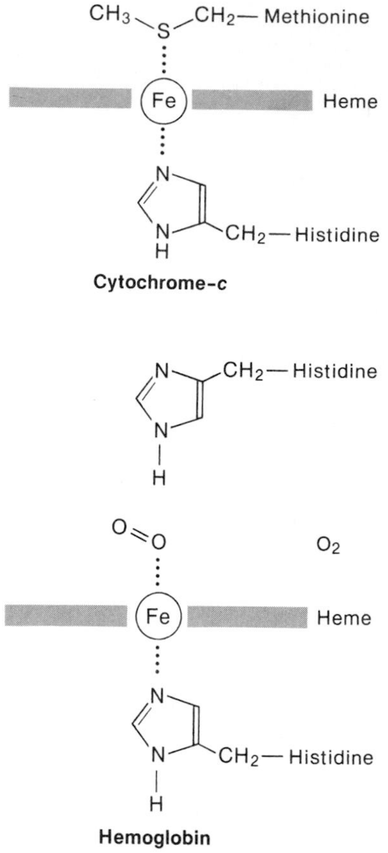

Figure 16–6

Edge-on views of the heme groups in a c cytochrome and in hemoglobin. The axial ligands in the c cytochrome are histidyl and methionyl residues of the protein. In hemoglobin, there are two histidyl residues at different distances from the Fe, leaving room for O_2 to bind to the Fe on one side.

erally true in water-soluble proteins, the amino acids on the outside of the protein are predominantly hydrophilic. Near one edge of the heme there is a strongly basic cluster of eight lysine residues. We shall see that this positively charged region probably plays a role in the activity of the cytochrome. The iron of the heme is bound on one side to the sulfur atom of a methionine residue and on the other to a histidyl nitrogen (Figure 16–6). By contrast, in hemoglobin and myoglobin, the methionine is replaced by a second, more distant histidine residue, leaving space for O_2, H_2O, or CO to bind to the Fe (Chapter 2).

Cytochrome c is associated loosely with the inner membrane of the mitochondrion. The b- and a-type cytochromes are firmly embedded in the same membrane, as parts of large complexes that are described in more detail below.

The iron atoms of the cytochromes undergo oxidation and reduction under physiological conditions, cycling between the ferrous (Fe^{2+}) and ferric (Fe^{3+}) oxidation states. The absorption spectra of the oxidized and reduced forms differ (Figure 16–4), and Keilin used this property to measure changes in the oxidation-reduction state of the cytochromes in living cells. Under anaerobic conditions, the cytochromes rapidly became reduced; in the presence of O_2, they became oxidized. If one of Warburg's inhibitors of respiration (CO, N_3^-, or CN^-) was added, it blocked the oxidation. Other types of inhibitors (amytal, rotenone, and malonate) blocked the reduction of cytochromes. Keilin found that purified cytochrome c was not oxidized directly by O_2, a fact suggesting that another component was required for the oxidation that occurred in intact cells. He therefore proposed a simple electron-transfer scheme, in which the transfer of electrons from a substrate such as succinate (AH_2) to cytochrome c was catalyzed by a dehydrogenase, and the transfer of electrons from cytochrome c to O_2 by a "cytochrome oxidase":

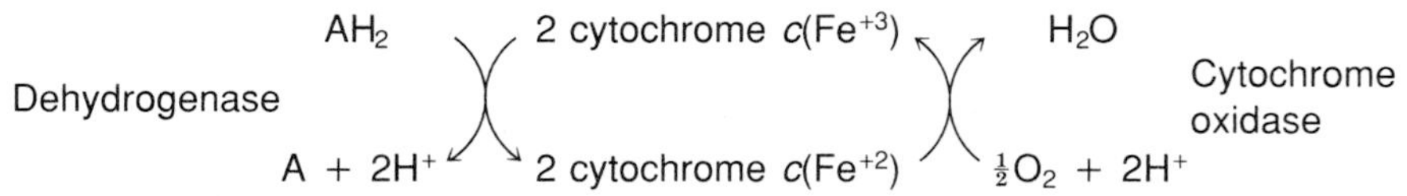

$$
\begin{array}{ccc}
& AH_2 \quad \diagdown \diagup \quad 2 \text{ cytochrome } c(Fe^{+3}) \quad \diagdown \diagup \quad H_2O & \\
\text{Dehydrogenase} & & \quad \text{Cytochrome} \\
& A + 2H^+ \diagup \diagdown \quad 2 \text{ cytochrome } c(Fe^{+2}) \quad \diagup \diagdown \quad \tfrac{1}{2}O_2 + 2H^+ & \text{oxidase}
\end{array}
$$

By 1940, it was clear that Keilin's cytochrome oxidase was identical with Warburg's respiratory enzyme, and that it involved two a-type cytochromes, a and a_3. Cytochrome c appeared to pass electrons to cytochrome a, which in turn reduced cytochrome a_3. Cytochrome a_3 evidently reacts with O_2 and is the site of inhibition by CO and CN^-:

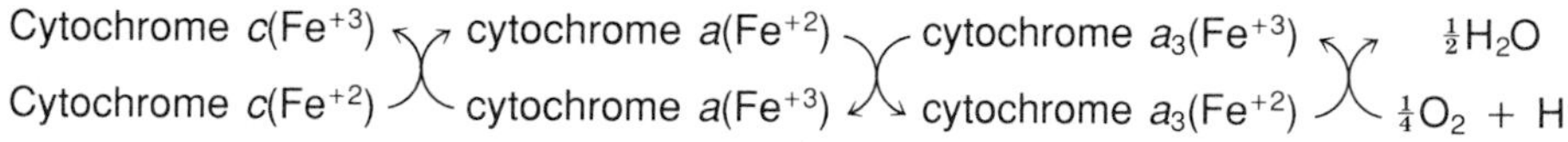

$$
\begin{array}{cccc}
\text{Cytochrome } c(Fe^{+3}) \diagdown\diagup & \text{cytochrome } a(Fe^{+2}) \diagdown\diagup & \text{cytochrome } a_3(Fe^{+3}) \diagdown\diagup & \tfrac{1}{2}H_2O \\
\text{Cytochrome } c(Fe^{+2}) \diagup\diagdown & \text{cytochrome } a(Fe^{+3}) & \text{cytochrome } a_3(Fe^{+2}) \diagup\diagdown & \tfrac{1}{4}O_2 + H^+
\end{array}
$$

The b cytochromes and another c cytochrome, cytochrome c_1, appeared to fit into the scheme somewhere between the reducing substrates and cytochrome c. The idea thus developed that *the respiratory apparatus includes a chain of electron carriers.*

Another way to write such a scheme is to use arrows to represent the path of electron flow:

$AH_2 \longrightarrow$ cytochromes b, $c_1 \longrightarrow$ cytochrome c $\longrightarrow$

cytochrome a $\longrightarrow$ cytochrome $a_3 \longrightarrow O_2$

Note, however, that we have not yet considered the stoichiometry of the reactions. The oxidation of a substrate AH_2 to A (e.g., the oxidation of succinate to fumarate, or of NADH to NAD^+) typically requires the removal of two electrons, whereas the reduction of a cytochrome from Fe^{3+} to Fe^{2+} involves the addition of only one; reduction of O_2 to 2 H_2O requires the addition of four electrons. If the cytochromes operate in a linear chain, cytochrome c must cycle between its oxidized and reduced states twice for each molecule of AH_2 that is oxidized, and four times for each molecule of O_2 that is reduced.

The next point that must be established is how the links in the respiratory chain are organized. Are all the electron carriers actually bound together physically like the links of a chain, or do they diffuse about independently and react only when they collide? If the individual carriers can diffuse independently, what prevents electrons from taking short-cuts such as jumping from a b cytochrome directly to a_3? Do electrons jump directly from the heme of one cytochrome to that of another, or do they have to work their way through the proteins? How does cytochrome a_3 reduce O_2 to H_2O, instead of simply binding O_2 and releasing it unmodified, as hemoglobin does? And how is the movement of electrons through the respiratory chain coupled to the formation of ATP? Before we can discuss these questions further we need to consider several other types of electron carriers that participate in the system along with the cytochromes.

Other Electron Carriers Work with the Cytochromes

In addition to the hemes of the cytochromes, *the mitochondrial electron-transport system contains at least four other types of electron carriers: flavins, quinones, iron-sulfur complexes, and copper atoms.*

Flavins. The dehydrogenases that remove electrons from succinate or NADH contain flavins as prosthetic groups. NADH dehydrogenase contains flavin mononucleotide (FMN); succinate dehydrogenase contains covalently bound flavin adenine dinucleotide (FAD). The structures of these coenzymes are shown in Figure 16–7. As discussed in Chapter 11, flavins can undergo one-electron reduction to semiquinone forms, or two-electron reduction to the dihydroflavins, $FMNH_2$ and $FADH_2$ (Figure 11–12). The reduction can be measured spectrophotometrically by the disappearance of a broad absorption band at 450 nm.

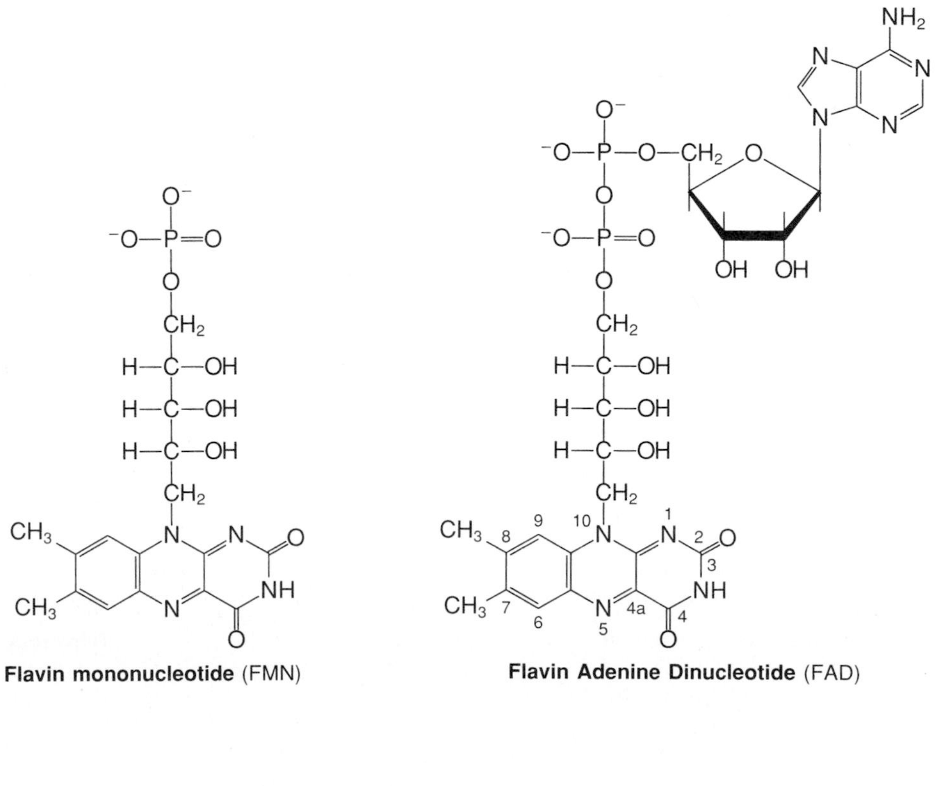

Figure 16–7
Structures of flavin mononucleotide (FMN) and flavin adenine dinucleotide (FAD).

Flavin mononucleotide (FMN)

Flavin Adenine Dinucleotide (FAD)

Protein

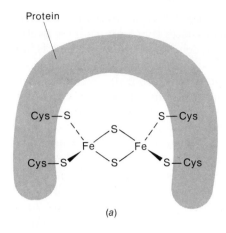

(a)

Protein

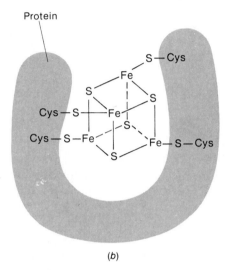

(b)

Figure 16–8
Structures of (a) 2Fe-2S and (b) 4Fe-4S iron-sulfur centers. In both structures, the arrangements of the sulfur ligands around the iron atoms are approximately tetrahedral. The cysteine residues are part of the polypeptide chain.

The inner membrane of the mitochondrion also has a flavoprotein, glycerol-3-phosphate dehydrogenase, that oxidizes glycerol-3-phosphate to dihydroxyacetone phosphate, reducing the bound FAD:

$$HOCH_2—CHOH—CH_2O℗ + FAD \longrightarrow HOCH_2—CO—CH_2O℗ + FADH_2$$

In the mitochondrial matrix, there are at least eight other flavoprotein dehydrogenases, including the acyl-CoA dehydrogenases that participate in the oxidation of fatty acids (Chapter 18). All of these dehydrogenases in the matrix transfer electrons to another membrane-bound flavoprotein, the *electron-transfer flavoprotein*, which reduces still another flavoprotein, the *electron-transfer flavoprotein ubiquinone oxidoreductase*. Like NADH dehydrogenase, succinate dehydrogenase, and glycerol-3-phosphate dehydrogenase, this enzyme passes electrons to the cytochromes by way of iron-sulfur complexes and a quinone.

Iron-sulfur Centers. NADH dehydrogenase, succinate dehydrogenase, and the electron-transfer flavoprotein ubiquinone oxidoreductase contain iron atoms that are bound by the sulfur atoms of cysteine residues of the protein, in association with additional, inorganic sulfide atoms. Proteins that contain such iron-sulfur complexes are termed *nonheme iron proteins*. The inorganic sulfur is frequently called "acid-labile sulfide," because it is released as H_2S if the protein is acidified. The *iron-sulfur centers* of most of the nonheme iron proteins that have been characterized contain either two Fe atoms and two labile sulfide atoms bound to four cysteines, or four Fe atoms and four sulfides, again bound to four cysteines. Structures of these complexes are shown in Figure 16–8. Succinate dehydrogenase appears to have three different iron-sulfur centers, one with a [2Fe-2S] structure, one with [4Fe-4S], and one with a cluster containing three Fe atoms and three (or possibly four) sulfides. Iron-sulfur centers undergo one-electron oxidation-reduction reactions, which can be viewed as involving a transition of one of the Fe atoms between its Fe^{2+} and Fe^{3+} states. However, the actual molecular orbitals of the iron-sulfur centers are more complex than this description suggests, and the changes in electron density are distributed over more than one Fe atom.

Ubiquinone. Ubiquinone (UQ) is a benzoquinone with a long side chain (Figure 16–9). It was discovered independently by two groups of investiga-

Figure 16–9
Structures of ubiquinone in its oxidized state (UQ) and in the fully reduced dihydroquinone or quinol state (UQH₂). The UQ found in eukaryotes has 50 carbons (10 prenyl units) in the side chain; shorter side chains are found in some bacteria. The side chain makes ubiquinone soluble in the phospholipid bilayer of the mitochondrial inner membrane.

Ubiquinone (UQ)

$2e^- + 2H^+$

Dihydroubiquinone (UQH₂)

tors about 1958, and it still goes under two different names, ubiquinone and *coenzyme Q*. The former name indicates the almost ubiquitous nature of the quinone in aerobic cells ranging from bacteria to plants and mammals. The length of the side chain varies from six *prenyl* (five-carbon) units in some bacteria to ten in mammals, but the different ubiquinones are functionally interchangeable. Some bacteria contain a naphthoquinone with a similar side chain (menaquinone, or vitamin K) in place of or in addition to UQ. With either UQ or the menaquinones, the hydrocarbon tail makes the molecule strongly hydrophobic. In bacteria, the UQ is found in the cytoplasmic membrane; in eukaryotes it is found mainly in the mitochondrial inner membrane. Compared with the concentrations of the cytochromes in these membranes, the concentration of UQ is relatively large. In heart mitochondria, for example, the concentration of UQ is about seven times that of cytochrome a_3. Although the mitochondrial inner membrane contains several different proteins that bind UQ, at any given time most of the UQ probably is not bound to proteins, but rather moves freely in the phospholipid bilayer of the membrane.

Ubiquinone undergoes a two-electron reduction to the dihydroquinone or quinol, UQH_2 (Figure 16–9). It also can accept a single electron and stop at the semiquinone, which can be either anionic ($UQ\cdot^-$) or neutral ($UQH\cdot$), depending on the pH and on the nature of the binding site when the semiquinone is bound to a protein. The $\cdot$ in $UQ\cdot^-$ or $UQH\cdot$ indicates that the semiquinone contains an unpaired electron.

The reduction of UQ can be measured by the disappearance of an absorption band at 275 nm. Using this technique, it was shown that the addition of a substrate such as succinate caused a rapid reduction of essentially all the UQ present in the mitochondrial membrane, and that the UQH_2 could be reoxidized by the cytochrome system in the presence of O_2. To determine whether UQ is a *necessary* participant in electron transport from succinate to O_2, the quinone was removed from the mitochondria by extraction with an organic solvent. The depleted mitochondria were incapable of respiration but recovered this activity when they were reconstituted with UQ. Similar results were obtained with electron donors other than succinate, including glycerol-3-phosphate and substrates that transfer their electrons initially to NADH, such as malate, α-ketoglutarate, and β-hydroxybutyrate (CH_3—$CHOH$—CH_2—CO_2^-). These results suggest that *the large pool of UQ in the mitochondrial inner membrane accepts electrons from several different dehydrogenases, and is responsible for passing these electrons on to the cytochrome system*. The kinetics of UQ reduction and oxidation are consistent with this view.

Redox Potentials Are a Measure of Electron Reduction Power

We have seen that the respiratory system includes a variety of molecules—cytochromes, flavins, ubiquinone, and iron-sulfur proteins—all of which can act as electron carriers. Our goal now is to understand how these carriers are organized to transport electrons from reduced substrates to O_2. To proceed, we first need to have a way of characterizing the relative tendencies of these different molecules to give off or accept electrons.

Consider the lactate dehydrogenase reaction, in which NADH reduces pyruvate to lactate:

$$NADH + H^+ + pyruvate \longrightarrow NAD^+ + lactate$$

Conceptually, we can imagine the reaction as occurring in two steps. In the first step, NADH undergoes oxidation to NAD^+, releasing two electrons and a proton; in the second, pyruvate undergoes reduction by accepting two electrons and two protons:

$$NADH \longrightarrow NAD^+ + 2e^- + H^+ \tag{1}$$

$$Pyruvate + 2e^- + 2H^+ \longrightarrow lactate \tag{2}$$

Actually, it is a hydride ion (H^-) that is transferred from NADH to pyruvate (see Chapter 11), but the details of the mechanism do not concern us for the moment. Each of the steps involves a *redox couple* of two molecules that are interconvertible by the transfer of electrons and protons.

The standard free energy change ($\Delta G^{\circ\prime}$) for the overall reaction between NADH and pyruvate is -6.22 kcal/mole. The negative sign implies that, under standard conditions (1 M NADH, NAD^+, lactate, and pyruvate, pH 7, and 25°C), the reaction proceeds spontaneously in the direction of lactate. This means that NADH must be a stronger electron donor than lactate. Now how does NADH compare with $FMNH_2$ or with cytochrome c? To deal quantitatively with questions like these, it is useful to define a parameter called *the standard redox potential, E°*, which *expresses the relative tendency of a pair of molecules like the NAD^+/NADH couple to release or accept electrons*. Like the electrical potentials that govern the direction of electron flow from one physical region to another, redox potentials are specified in units of volts. Because electron-transfer reactions frequently involve protons also, an additional symbol is used to indicate that the E° value applies to a particular pH; an $E^{\circ\prime}$ thus refers to an E° at pH 7. (People who work in bioenergetics commonly refer to E° as "the midpoint" potential and use the symbols E_m and $E_{m,7}$ in place of E° and $E^{\circ\prime}$.)

Suppose we have two redox couples D_{ox}/D_{red} and A_{ox}/A_{red}, where the subscripts ox and red indicate the oxidized and reduced forms of the molecules. Consider the reaction in which D_{red} acts as an electron donor (reductant) and A_{ox} as the electron acceptor (oxidant). We take the difference between the E° values of two couples (ΔE°) to be proportional to the standard free energy change for the reaction (see Box 12–C):

$$D_{red} + A_{ox} \longrightarrow D_{ox} + A_{red} \tag{3}$$

$$\Delta E^\circ = E_A^\circ - E_D^\circ = -\Delta G^\circ / nF \tag{4}$$

or

$$\Delta G^\circ = -nF \, \Delta E^\circ \tag{5}$$

Here n is the number of electrons transferred from the donor to the acceptor (two in the case of the lactate dehydrogenase reaction), and the proportionality constant F is the Faraday constant (23,060 cal per volt per mole of electrons). Note that these expressions specify a convention concerning the signs of E° values: the overall reaction is spontaneous (ΔG° negative) if $\Delta E^{\circ\prime}$ is positive, that is, if electrons are transferred from a couple with a more negative E° value to one with a more positive value. In other words, *relatively negative E° values are associated with strong reductants, and relatively positive values with strong oxidants*. With $n = 2$, a difference of 0.1 V between the E° values corresponds to a ΔG° of -4.61 kcal/mole.

The relationship between ΔG° and ΔE° is simply the expression that describes electron flow between two physical regions that are at different electrical potentials, such as the two terminals of a battery. Electrons tend to flow spontaneously in the direction of more positive potential, and the maximum amount of work that can be obtained from such flow is proportional to the number of electrons that move and to the potential difference. In fact, one can measure ΔE° values by separating the two redox couples physically into different compartments connected by a salt bridge and a voltmeter (Figure 16–10). Electrons can flow from one solution to the other through the meter. If the solutions in the two compartments are both under standard conditions, the meter will sense a voltage difference equal to the $\Delta E°$.

So far, we have defined only the difference between two $E°$ values. To set the individual values, it is merely necessary to choose a particular redox couple as a reference. The reference that is used most commonly by biochemists is the <u>standard hydrogen half-cell</u>, in which protons at 1 M [H^+] are reduced to H_2 at a pressure of 1 atm ($2H^+ + 2e^- \rightarrow H_2$). This half-cell is arbitrarily assigned an $E°$ value of zero. (Note that the standard condition for this half-cell is pH 0; this does not restrict one's choice of conditions for the cell in the other compartment of the apparatus.) Relative to the standard hydrogen half-cell, the NAD^+/NADH couple has an $E°'$ of -0.32 V, and the pyruvate/lactate couple has an $E°'$ of -0.19 V.

Table 16–1 gives the $E°'$ values of a number of biochemical redox couples, including some of the components of the respiratory chain. These are listed in order of increasing $E°'$, which means that under standard conditions a given couple will reduce any of the couples below it in the table.

As mentioned above, quinones and flavins can undergo one-electron reduction to semiquinone forms, or can be fully reduced by the addition of two electrons. The $E°$ values for the individual one-electron steps can be very different. In the case of UQ, the one-electron redox couples UQ/UQH· and UQH·/UQH$_2$ have $E°'$ values of approximately 0.03 and 0.19 V. Since the $E°'$ of the second couple is more positive than that of the first, the second electron is thermodynamically easier to add than the first. When the first electron is put on, a second usually follows, as long as two electrons are available from the donor. The $E°'$ for the overall two-electron reduction of UQ to

---

TABLE 16–1
$E°'$ Values of Biochemical Redox Couples

Redox Couple	$E°'$(V)	n^a
Succinate + CO_2 + $2H^+$ + $2e^-$ $\longrightarrow$ α-ketoglutarate + H_2O	-0.67	2
Glycerate-3-phosphate + $2H^+$ + $2e^-$ $\longrightarrow$ glyceraldehyde-3-phosphate + H_2O	-0.55	2
α-Ketoglutarate + CO_2 + $2H^+$ + $2e^-$ $\longrightarrow$ isocitrate	-0.38	2
NAD^+ + H^+ + $2e^-$ $\longrightarrow$ NADH	-0.32	2
Glycerate-1,3-bisphosphate + $2H^+$ + $2e^-$ $\longrightarrow$ glyceraldehyde-3-phosphate + P_i	-0.29	2
Lipoic acid + $2H^+$ + $2e^-$ $\longrightarrow$ dihydrolipoic acid	-0.29	2
FMN + $2H^+$ + $2e^-$ $\longrightarrow$ FMNH$_2$	-0.22^b	2
FAD + $2H^+$ + $2e^-$ $\longrightarrow$ FADH$_2$	-0.22^b	2
Acetaldehyde + $2H^+$ + $2e^-$ $\longrightarrow$ ethanol	-0.20	2
Pyruvate + $2H^+$ + $2e^-$ $\longrightarrow$ lactate	-0.19	2
Oxaloacetate + $2H^+$ + $2e^-$ $\longrightarrow$ malate	-0.17	2
Fumarate + $2H^+$ + $2e^-$ $\longrightarrow$ succinate	-0.03	2
Cytochrome-b_L(Fe^{3+}) + e^- $\longrightarrow$ cytochrome-b_L(Fe^{2+})	-0.03	1
UQ + H^+ + e^- $\longrightarrow$ UQH·	$+0.03^c$	1
Cytochrome-b_H(Fe^{3+}) + e^- $\longrightarrow$ cytochrome-b_H(Fe^{2+})	$+0.05$	1
UQ + $2H^+$ + $2e^-$ $\longrightarrow$ UQH$_2$	$+0.11^c$	2
UQH· + H^+ + e^- $\longrightarrow$ UQH$_2$	$+0.19^c$	1
Cytochrome-c_1(Fe^{3+}) + e^- $\longrightarrow$ cytochrome-c_1(Fe^{2+})	$+0.23$	1
Cytochrome-c(Fe^{3+}) + e^- $\longrightarrow$ cytochrome-c(Fe^{2+})	$+0.24$	1
Cytochrome-a(Fe^{3+}) + e^- $\longrightarrow$ cytochrome-a(Fe^{2+})	$+0.25$	1
Rieske Fe-S(Fe^{3+}) + e^- $\longrightarrow$ Fe-S(Fe^{2+})	$+0.28$	1
Cytochrome-a_3(Fe^{3+}) + e^- $\longrightarrow$ cytochrome-a_3(Fe^{2+})	$+0.35$	1
O_2 + $4H^+$ + $4e^-$ $\longrightarrow$ $2H_2O$	$+0.82$	4

^{a}n is the number of electrons transferred.

^{b}This value is for the free coenzyme. $E°'$ values for flavoproteins range from -0.3 to 0 V.

^{c}For UQ in aqueous ethanol.

UQH$_2$ is the mean of the $E°$ values for the two one-electron steps, 0.11 V. ($E°$ values for UQ depend strongly on the solvent; these values are for aqueous ethanol.) If only one electron is available, the reduction has to stop at the semiquinone.

$E°$ values are called standard potentials because they refer to conditions under which the oxidant and reductant of the couple are at standard concentrations. (Standard conditions are chosen usually to be 1 M concentrations of all reactants and products except for H$^+$, OH$^-$, and H$_2$O.) By the principle of mass action, a solution containing a redox couple D$_{ox}$/D$_{red}$ can be made more oxidizing by increasing the concentration ratio [D$_{ox}$]/[D$_{red}$], or more reducing by decreasing this ratio. The voltage measured with the apparatus shown in Figure 16–10 will reflect such changes. Suppose that the couple in one of the compartments is the standard hydrogen half-cell, and that the concentrations here are held constant while those of D$_{ox}$ and D$_{red}$ in the other compartment are varied. The voltage difference measured between the two compartments can be defined as the _redox potential_, E, in the solution containing D$_{ox}$ and D$_{red}$. Again, what actually is measured is the difference between the redox potentials in the two compartments (ΔE), and the redox potential of the standard half-cell is set equal to zero by convention. The relationship between E and $E°$ is the same as that between the free energy change in a reaction (ΔG) and the standard free energy change ($\Delta G°$). Considering how ΔG depends on $\Delta G°$ and on the concentrations of the reactants and products (Chapter 12), one can see that E must depend on the concentration ratio as follows:

$$E = E° + (2.303\ RT/nF)\ \log\{[D_{ox}]/[D_{red}]\} \qquad (6)$$

This is one of two expressions that go under the name of the _Nernst equation_. If [D$_{ox}$]/[D$_{red}$] is increased by a factor of 10, E increases by an amount

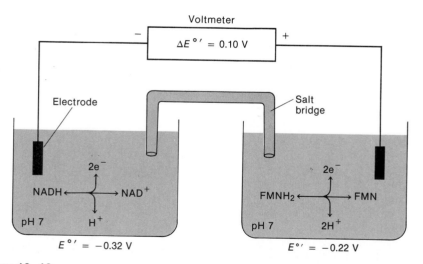

Figure 16–10
Apparatus for measuring the difference between the $E°$ values of two redox couples. The cell on the left contains equimolar concentrations of NADH and NAD$^+$; that on the right, equimolar concentrations of FMNH$_2$ and FMN. If both solutions are at pH 7, the voltmeter will sense the difference between the two $E°'$ values (0.10 V, negative on the left, in this example). To determine the $E°'$ value of one of the redox couples, the other couple is replaced by a standard redox couple such as the standard hydrogen half-cell (H$_2$/H$^+$ at pH 0 and 1 atm H$_2$). The dependence of one of the $E°$ values on pH can be measured by changing the pH in the experimental cell and holding that in the reference cell constant. Experimentally, the calomel redox couple (Hg/HgCl$_2$) is often used as a working reference instead of the hydrogen half-cell, and the results are corrected by subtracting the $E°$ value of the calomel couple (0.24 V). It may be necessary to add a catalytic amount of another organic redox couple as a mediator to facilitate the transfer of electrons to and from the metallic electrodes.

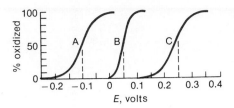

Figure 16–11
Theoretical titration curves for three different redox couples, showing how the portion of a couple that is in the oxidized form increases as the redox potential in the solution is made more positive. Curve A is for a redox couple that undergoes a one-electron reaction ($n = 1$) with an $E°$ value of -0.10 V. Curve B is for an $E°$ value of $+0.05$ V and $n = 2$. Curve C is for an $E°$ value of $+0.25$ V and $n = 1$. The ordinate shows $100 \times [D_{ox}]/([D_{ox}] + [D_{red}])$, as calculated from the Nernst equation [Equation (6)]. 50% of a redox couple is in the oxidized form when $E = E°$. The dotted lines indicate $E°'$ on the abscissa.

equal to $2.303\ RT/nF$. AT 25°C, $2.303\ RT/nF$ is equal to 0.060 V for $n = 1$, and 0.030 V for $n = 2$.

Another way to view the situation is that if we impose a redox potential of E on a solution containing D_{ox} and D_{red}—for example, by replacing the voltmeter in Figure 16–10 by a power supply, or by adding a large amount of some other reductant or oxidant to the solution—D_{ox} will accept electrons or D_{red} will release them, until the ratio $[D_{ox}]/[D_{red}]$ is in accord with the Nernst equation. The Nernst equation is analogous to the Henderson-Hasselbach equation (see Chapter 1), which describes how the ratio of base and acid concentrations depends on the pH and the pK_a of an acid. Figure 16–11 shows how the $[D_{ox}]/[D_{red}]$ ratio depends on E for several redox couples with different values of $E°$. Note that, because of the way n enters into Equation (6), the curves shown in Figure 16–11 are steeper if $n = 2$ than if $n = 1$. $[D_{ox}]/[D_{red}]$ increases by a factor of 10 for every 0.06 V increase in E when $n = 1$, and for every 0.03 V when $n = 2$.

Let us now return to the point that $E°$ values can depend on pH. When UQ, for example, is reduced to UQH_2, two protons are added to the molecule along with each pair of electrons. By mass action, increasing the proton concentration (lowering the pH) should favor the reduction. Lowering the pH thus must make UQ a stronger oxidant, which means that it must increase UQ's $E°$. This relation can be expressed quantitatively by including the proton concentration along with the concentrations of UQ and UQH_2 in the equation for the free energy change associated with the oxidation-reduction reaction. The result is that, if the pH is decreased by one pH unit, the $E°$ value becomes more positive by $(2.303\ RT/F)(n_{H^+}/n)$, where n_{H^+} is the number of protons that are taken up (two in the case of the UQ/UQH_2 couple) and n again is the number of electrons transferred (two in this case). At 25°C, $(2.303\ RT/F)(n_{H^+}/n)$ is 0.060 V per pH unit when n_{H^+} is equal to n:

$$\partial E°/\partial pH = -(2.303\ RT/F)(n_{H^+}/n) = -(0.060\ V)(n_{H^+}/n)^* \qquad (7)$$

Note that it is the *ratio* of n_{H^+} to n that is important, and not the absolute value of either of these quantities.

$E°$ Values of Electron Carriers. Given equal concentrations of reactants and products, electrons will tend to flow spontaneously from a carrier with a more negative $E°$ value to one with a more positive value. The sequence of carriers in the respiratory chain should, therefore, be consistent with the relative $E°$ values of the carriers. The linear sequence of b-, c-, and a-type cytochromes that was proposed above agrees with this expectation. This can be seen by plotting the $E°'$ values as a function of the carriers' suggested positions in the chain (Figure 16–12). There are, however, difficulties in using the $E°$ values to determine the exact organization of the respiratory chain, particularly in regions where there are numerous carriers with similar values of $E°$. First, the effective $E°$ values of the respiratory carriers are difficult to measure precisely, because they can depend on the local pH and electrical potential in the interior of the membrane. The $E°'$ of cytochrome c decreases by about 0.05 V when the soluble cytochrome binds to a phospholipid membrane, and the $E°'$ value of UQ in a membrane or on a UQ-binding protein could be very different from the value measured in ethanol. In addition, interactions between neighboring carriers can make it difficult to determine the $E°$ of one carrier in isolation from the other.

Keep in mind also that because the concentrations of reactants and products are rarely equal, it is not really $\Delta E°$ that determines the direction of electron flow, but the difference between the operating redox potentials, ΔE.

*In Equation (7), $\partial E°$ and ∂pH refer to small changes in $E°$ and pH.

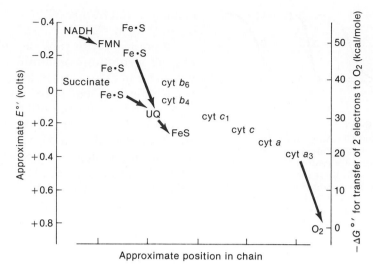

Figure 16–12
Approximate $E^{\circ\prime}$ values of some of the electron carriers in the respiratory chain, plotted as a function of the approximate positions of the carriers in the chain. The diagram includes components of the cytochrome bc_1 complex that are discussed later in the chapter. For simplicity, it shows a single $E^{\circ\prime}$ value for ubiquinone, although the $UQH_2/UQH\cdot$ and $UQH\cdot/UQ$ redox couples have different values. The $E^{\circ\prime}$ value for the FMN associated with NADH dehydrogenase is not known accurately; the value shown is for free FMN. The $E^{\circ\prime}$ values and positions for the iron-sulfur centers associated with NADH dehydrogenase also are uncertain; only two of the six to eight iron-sulfur centers in this region of the respiratory chain are shown. The right-hand scale gives the standard free energy change ($\Delta G^{\circ\prime}$) for transferring two equivalents of electrons from a carrier to O_2.

Electrons can move in the direction of more negative E° if the reactants (the reduced form of the electron donor and the oxidized form of the electron acceptor) are present at sufficiently high concentrations relative to the products (the oxidized donor and the acceptor). This is simply an application of the law of mass action.

Finally, even if electron transfer from a particular carrier to another is thermodynamically favorable, it may not be able to occur unless the two carriers are sufficiently close together. Contacts between carriers in the respiratory chain will depend on how the hemes, flavins, quinones, and iron-sulfur centers are positioned in the proteins that bind them, and on how the proteins are arranged in the membrane.

The Sequence of Carriers Can Be Determined with Spectral Analysis and Inhibitors

Because the sequence of electron carriers in the respiratory chain cannot be determined simply on the basis of the E° values, it is necessary to bring additional experimental techniques to bear on the problem. Spectrophotometric techniques have been particularly helpful, because they make it possible to examine the rates at which the different electron carriers accept and release electrons in intact, functioning mitochondria. Other approaches that have played key roles are the use of specific inhibitors of the electron-transfer reactions, and the isolation of protein complexes that represent small sections of the respiratory chain.

Difference Spectra. Extremely sensitive and rapid spectrophotometric techniques for examining the redox states of the respiratory carriers in intact mitochondria were developed by Britton Chance in the 1950s. Chance measured the differences between the optical absorption spectra of two samples

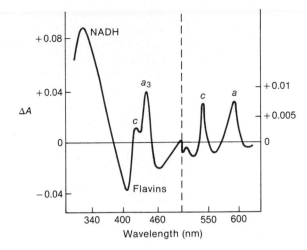

Figure 16–13
Reduced-minus-oxidized difference spectrum of the electron carriers in rat liver mitochondria. The plot shows the difference (ΔA) between the absorption spectrum of a sample of mitochrondria that had used up all of the O_2 in the solution and the spectrum of a reference sample that contained O_2. Both samples were provided with β-hydroxybutyrate, which can enter mitochondria and serve as a reductant for the internal NAD^+. The peak near 340 nm reflects reduction of NAD^+ to NADH; the sharp positive peaks at longer wavelengths reflect the reduction of cytochromes. The peaks near 420 and 440 nm, which are due mainly to the reduction of c and a cytochromes, are superimposed on a broad, negative band that results partly from the reduction of flavins. (For the individual absorption spectra of oxidized and reduced cytochrome c, see Figure 16–4.) Absorbance changes due to the b cytochromes were not well resolved from those due to the c and a cytochromes in this spectrum. A difference spectrum can be sensitive to small changes in one of the samples, even if both samples have a high absorbance or are cloudy. Mitochondria are relatively large particles and tend to scatter much of the light of the spectrophotometer, but the amount of scattering is similar in the two samples. (For details, see B. Chance and G. R. Williams, *J. Biol. Chem.* 217:395–407, 1955.)

of mitochondria that were under different conditions. Figure 16–13 shows such a *difference spectrum*. For this measurement, the electron carriers in the control or reference sample were predominantly in their oxidized states, because the solution contained excess O_2. The other sample was provided with an excess of a reducing substrate and was allowed to use up all its O_2, so that the electron carriers became largely reduced. The spectrum is therefore a reduced-minus-oxidized difference spectrum. The positive peak near 340 nm is due to the reduction of NAD^+ to NADH; the sharp positive peaks at longer wavelengths, to reduction of the cytochromes (compare the curves for reduced and oxidized cytochrome c in Figure 16–4). Reduction of flavins causes a broad, negative absorbance change in the region from 400 to 480 nm, underlying the sharp absorbance increases due to the cytochromes.

The reducing substrate used in the experiment shown in Figure 16–13 was β-hydroxybutyrate. A dehydrogenase in the mitochondrial matrix oxidizes β-hydroxybutyrate to β-ketobutyrate, transferring a pair of electrons to NADH:

$$CH_3\!-\!CHOH\!-\!CH_2\!-\!CO_2^- + NAD^+ \longrightarrow$$
$$CH_3\!-\!CO\!-\!CH_2\!-\!CO_2^- + NADH + H^+$$

The NADH dehydrogenase in the inner membrane then transfers electrons from NADH to the respiratory chain. Malate, α-ketoglutarate, or any other substrate that is capable of reducing mitochondrial NAD^+ to NADH gives a spectrum similar to that shown in the figure.

Succinate also gives a similar spectrum, except that the peak near 340 nm is not seen. Recall that succinate dehydrogenase is a flavin enzyme that does not involve NAD^+. This raises the question of whether succinate dehydrogenase reduces one set of cytochromes and NADH dehydrogenase reduces another independent set, or whether both enzymes reduce the same cytochromes. Chance found that if the strong chemical reductant $Na_2S_2O_4$ (sodium dithionite) was added to a mitochondrial sample that had been reduced with either β-hydroxybutyrate or succinate, there was little additional reduction of cytochromes. Any of the normal respiratory substrates evidently is capable of reducing virtually all of the cytochromes in the membrane. As mentioned above, either succinate or NADH-linked substrates also can reduce essentially all of the UQ. The flavins, however, are reduced only partially by succinate, but almost completely by a combination of succinate and malate. These observations fit the idea that _all of the cytochromes and UQ form a single chain or network, to which several different flavoprotein dehydrogenases can feed electrons._

Using similar techniques, Britton Chance, Quentin Gibson, and others also measured the kinetics with which the different electron carriers became oxidized, following the addition of O_2 to an anaerobic suspension of mitochondria, or became reduced after addition of substrate. When O_2 was added, cytochrome a_3 became oxidized first, followed by cytochrome a, the c cytochromes, the b cytochromes, and flavins, in that order. Later work by Klingenberg showed that UQ became oxidized at about the same time as the b cytochromes. These observations suggest that the electron carriers are arranged in the sequence:

$$\text{Succinate} \longrightarrow \text{FAD}$$
$$\searrow$$
$$\text{NADH} \longrightarrow \text{FMN} \longrightarrow [\text{UQ, cyt } b] \longrightarrow [\text{cyt } c_1, \text{ cyt } c] \longrightarrow$$
$$\text{cyt } a \longrightarrow \text{cyt } a_3 \longrightarrow O_2$$

This is consistent with the sequence shown in Figure 16–12. Kinetic measurements have to be interpreted cautiously, however. The observed rate of oxidation or reduction of a carrier depends on the difference between the rates of the reactions in which the carrier accepts electrons from components upstream and gives electrons to components downstream. In principle, the net rate might not accurately reflect the position of the carrier in the chain.

Inhibitors.   In addition to CO and CN^-, several _specific inhibitors have been useful experimentally for exploring the sequence of the electron carriers._ Amytal (a barbiturate), rotenone (a plant toxin that is used as a fish poison and an insecticide), and piericidin A specifically block NADH dehydrogenase. Malonate and thenoyltrifluoroacetone block succinate dehydrogenase. Antimycin, a toxin obtained from _Streptomyces griseus_, acts in the region of the b cytochromes. The structures of some of these inhibitors are shown in Figure 16–14.

The sites of action of the respiratory inhibitors were studied by measuring difference spectra between samples that were treated with the inhibitors and untreated controls. For these experiments, both of the mitochondrial suspensions contained a reducing substrate in addition to O_2, so that the electron carriers were in a steady state of partial reduction. If CO or CN^- was added to one of the samples, all of the electron carriers in the sample became more reduced. Such a result is consistent with the idea that these agents block the functioning of cytochrome a_3 at the downstream end of the respiratory chain, preventing the removal of electrons from the chain by O_2. If antimycin was added, NAD^+, flavins, and the b cytochromes became more reduced, but cytochromes c, c_1, a, and a_3 all became more oxidized. The situation here is analogous to the construction of a dam across a stream: when the gates are closed, the water level increases upstream from the dam, and decreases downstream. The results with antimycin suggest that this in-

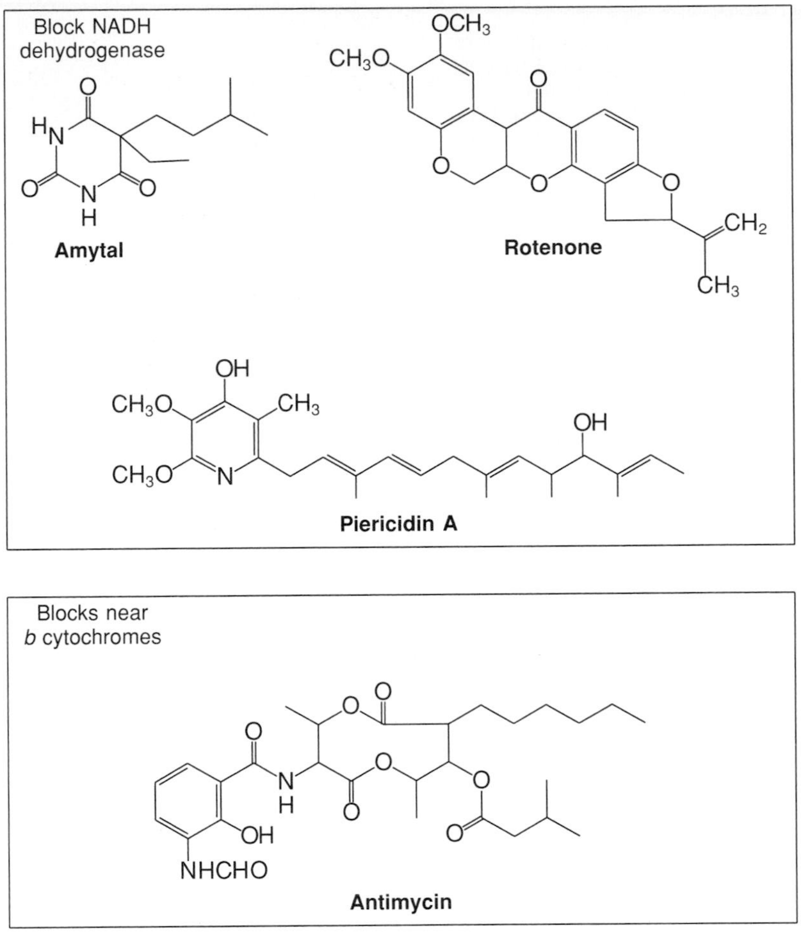

Figure 16–14

Structures of several inhibitors of electron transport. Amytal, rotenone, and piericidin A block NADH dehydrogenase; antimycin blocks in the region of the b cytochromes (see Figure 16–18).

hibitor blocks electron flow between the b cytochromes and cytochrome c_1. Although this conclusion has needed modification in the light of more recent work that is discussed below, antimycin was very helpful for establishing the general organization of the respiratory chain. The observation that antimycin did not inhibit the reduction of UQ, for example, showed that the quinone fits into the chain upstream of cytochromes c, c_1, a, and a_3.

Most of the Electron Carriers Occur in Large Complexes

Another way to explore the organization of the respiratory chain is to isolate fragments of the chain, after partial disruption of the mitochondrial inner membrane with detergents such as sodium deoxycholate. Following this approach, Yousef Hatefi, David Green, and others found that *all of the major electron carriers except for cytochrome c and UQ occur in the form of large complexes* (Figure 16–15). The isolated *NADH dehydrogenase complex* (NADH-ubiquinone oxidoreductase, or complex I) catalyzes the reduction of UQ by NADH; the *succinate dehydrogenase complex* (succinate-ubiquinone oxidoreductase, or complex II), catalyzes reduction of UQ by succinate. The *cytochrome bc_1 complex* (ubiquinone-cytochrome c oxidoreductase, or complex III) transfers electrons from reduced ubiquinone (UQH_2) to cytochrome c. And finally, *cytochrome oxidase* (complex IV) transfers electrons from cytochrome c to O_2.

Each of these complexes contains a large number of polypeptide subunits, including some that have no electron-carrying prosthetic groups

Figure 16–15

The multisubunit complexes of the respiratory chain. Complexes I (NADH dehydrogenase) and II (succinate dehydrogenase) transfer electrons from NADH and succinate to UQ. Complex III (the cytochrome bc_1 complex) transfers electrons from UQH_2 to cytochrome c, and complex IV (cytochrome oxidase), from cytochrome c to O_2. The arrows represent paths of electron flow. NADH and succinate provide electrons from the matrix side of the inner membrane, and O_2 removes electrons on this side. Cytochrome c is reduced and oxidized on the opposite side of the membrane, in the lumen of a crista or in the intermembrane space. The side of the inner membrane that faces the intermembrane space is generally referred to as the cytoplasmic side, because small molecules can move freely between the solution here and the cytosol.

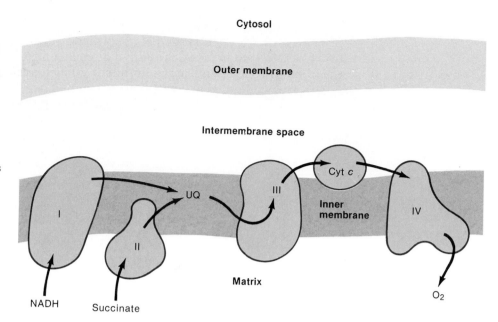

(Table 16–2). The presence of so many subunits is puzzling at first, in view of the seemingly simple nature of the electron-transfer reactions that the enzymes catalyze. _In addition to providing a path for the movement of electrons, however, these enzymes have the task of capturing some of the free energy that is released in the reactions._ Let us now consider the individual complexes in greater detail.

Complex I: The NADH Dehydrogenase Complex. The NADH dehydrogenase complex is the largest complex in the mitochondrial inner membrane. It appears to consist of at least 26 different polypeptides, with a total molecular weight of about 10^6. In addition to FMN, it contains approximately seven different iron-sulfur centers, which add up to about 23 Fe atoms per FMN. Although the structure of the complex is not known, a flavoprotein subunit and most of the iron-sulfur proteins appear to be clustered in the center, surrounded by a shell of hydrophobic proteins. The iron-sulfur centers include both 2-Fe and 4-Fe complexes. They fall in two groups, one with $E^{\circ\prime}$ values near that of $NAD^+/NADH$ (−0.32 V), and the other with $E^{\circ\prime}$ values

TABLE 16–2
Components of the Mitochondrial Electron-Transport Complexes

	Complex	Approx. mol. wt.[a]	Polypeptides[b]	Prosthetic Groups
I	NADH dehydrogenase complex	1×10^6	At least 26	FMN, Fe-S (23 Fe/FMN)
II	Succinate dehydrogenase complex	1×10^5	2 Fe-S proteins, 2 other polypeptides	FAD, Fe-S (9 Fe/FAD)
III	Cytochrome-bc_1 complex	2×10^5	2 cytochromes, 1 Fe-S protein, 6–8 others	b- and c-type hemes (cyt b_L, b_H, c_1), Fe-S (2 Fe/cyt c_1)
IV	Cytochrome oxidase	2×10^5	2 cytochromes, 4–11 others	a-type hemes (cyt a, a_3), Cu (2 Cu/cyt a)

[a]Complex III occurs as a dimer in the membrane. The molecular weight listed is for a monomer. The same is true of Complex IV.

[b]The contents of some of the smaller polypeptides in purified preparations of the complexes vary, depending on the sources of the enzymes. It is not yet clear whether all of these polypeptides are integral components of the complexes.

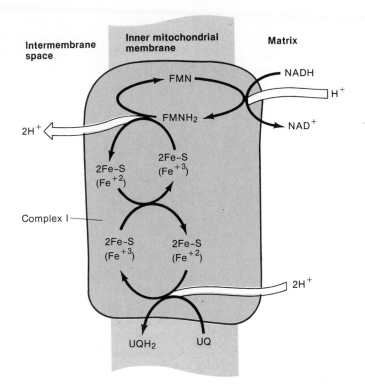

Figure 16–16

Electron transfer sequence in complex I. At the top, the reduction of FMN to $FMNH_2$ by NADH requires the uptake of one proton from the matrix. $FMNH_2$ subsequently transfers a pair of electrons to Fe-S centers, releasing protons to the solution on the cytoplasmic side of the membrane. Electrons then move back across the membrane to UQ on the matrix side via additional Fe-S centers. The detailed path of electrons through the six to eight different Fe-S centers in the complex is not known. When UQ is reduced to UQH_2, two more protons are taken up from the matrix.

near that of UQ/UQH_2 (approximately $+0.10$ V). NADH probably reacts with the FMN, reducing it to $FMNH_2$, and the reduced flavin then transfers electrons to an iron-sulfur center (Figure 16–16). Electrons then move from one iron-sulfur center to another, and eventually to UQ. In the presence of the inhibitors amytal, rotenone, or piericidin A, NADH is still able to reduce the flavin and most, if not all of the iron-sulfur centers, but the release of electrons to UQ is blocked.

NADH dehydrogenase is oriented <u>anisotropically</u> in the inner mitochondrial membrane, i.e. the protein has a definite orientation with respect to the two surfaces of the membrane. Its binding site for NADH faces inward toward the mitochondrial matrix space (Figure 16–15). This orientation is appropriate for oxidizing NADH that is generated in the matrix by the enzymes of the TCA cycle, isocitrate dehydrogenase, α-ketoglutarate dehydrogenase, and malate dehydrogenase. (Enzymatic shuttle systems that allow the oxidation of cytoplasmic NADH are discussed below.) The UQH_2 that is formed by complex I diffuses to complex III in the central, hydrophobic region of the membrane's phospholipid bilayer (see Figure 16–15), but the protons that are taken up when the quinone undergoes reduction come from the solution on the matrix side of the membrane (Figure 16–16). In the course of the electron-transfer reactions, protons also are released to the solution on the cytoplasmic side. We shall see the significance of these proton movements when we discuss the formation of ATP.

Complex II: The Succinate Dehydrogenase Complex. Succinate dehydrogenase is the only enzyme of the TCA cycle that is embedded in the mitochondrial inner membrane. It contains two iron-sulfur proteins, with molecular weights of about 70,000 and 27,000. The substrate oxidation site is on the larger of these subunits, and as with NADH dehydrogenase, is on the matrix side of the membrane. The large subunit has a 4-Fe center and a molecule of FAD that is bound covalently through the C-8a methyl group to a histidine residue of the protein. The role of the 4-Fe center is unclear, because the $E^{\circ\prime}$ of the center (-0.26 V) is substantially more negative than that of the succinate/fumarate couple (-0.03 V). Succinate could not reduce this iron-sulfur center to any appreciable extent, unless the [succinate]/[fumarate] concentration ratio was extremely high. The smaller iron-sulfur subunit has a 2-Fe center with an $E°$ near 0.0 V, and a 3-Fe center with a more positive $E°$. Complex II also has two other small polypeptides, which may participate in the interaction with UQ.

Complex III: The Cytochrome bc_1 Complex.   The cytochrome bc_1 complex contains two different b-type cytochromes. One of these, cytochrome b_L (or b_{566}), has an $E°'$ of about -0.03 V. The other, cytochrome b_H (or b_{562}), has an $E°'$ of about $+0.05$ V. (The subscripts L and H denote "low potential" and "high potential.") Both of the b cytochromes are on the same polypeptide, which has a molecular weight of about 30,000. The complex also contains cytochrome c_1 ($E°' \approx 0.23$ V), a 2-Fe iron-sulfur protein ($E°' \approx 0.28$ V), several subunits that have been implicated in interactions with UQ, and between four and six additional subunits without known prosthetic groups. The iron-sulfur protein is frequently called the _Rieske iron-sulfur protein_ after its discoverer, John Rieske. The entire cytochrome bc_1 complex can be isolated from heart mitochondria as a dimer with a total molecular weight of about 450,000. Similar complexes are found in the photosynthetic electron-transfer chains of chloroplasts and photosynthetic bacteria (Chapter 17). In mitochondria, the site at which cytochrome c undergoes reduction is on the side of the inner membrane that faces the intermembrane space (Figure 16–15).

Many of the subunits of complex III have been sequenced, and have been found to contain stretches of hydrophobic amino acids that could form transmembrane a helices stretching from one side of the inner membrane to the other. The cytochrome b subunit, for example, appears to have nine such transmembrane helices, connected by segments that contain predominantly hydrophilic amino acids. The two hemes probably are bound to pairs of histidyl residues in adjacent a helices, as shown in Figure 16–17. An electron jumping between the hemes would move from one side of the membrane to the other.

The Path of Electrons Through Complex III: The Q Cycle.   The path of electrons from UQ to cytochrome c is more circuitous than the observations that we have considered so far would suggest. The first indication of this was the remarkable finding that, under certain conditions, the addition of an _oxidant_, such as O_2 in the presence of cytochrome c and cytochrome oxidase, causes the b cytochromes to become _more reduced_, instead of more oxidized as would be expected. To see this effect, one first treats a suspension of mitochondria with antimycin, which blocks electron transfer in the region of the b cytochromes, as noted above. (The exact site at which antimycin acts will be clarified shortly.) Cytochromes c and c_1 and the Rieske iron-sulfur protein then are reduced by adding an electron donor that reacts directly with cytochrome c (usually sodium ascorbate in the presence of a mediator to speed up the reaction). The b cytochromes are not reduced under these conditions, because their $E°'$ values are more negative than the $E°'$ of ascorbate. If succinate is then added, the b cytochromes still do not become reduced. This is odd, because the $E°'$ of succinate (-0.03 V) is considerably more negative than that of cytochrome b_H, and about the same as that of cytochrome b_L. The succinate does reduce the UQ in the membrane. Finally, when O_2 is added, cytochromes c and c_1 become oxidized and the b cytochromes become reduced. Electrons evidently cannot be transferred from UQH_2 to the b cytochromes, unless electrons are simultaneously withdrawn from the cytochrome bc_1 complex.

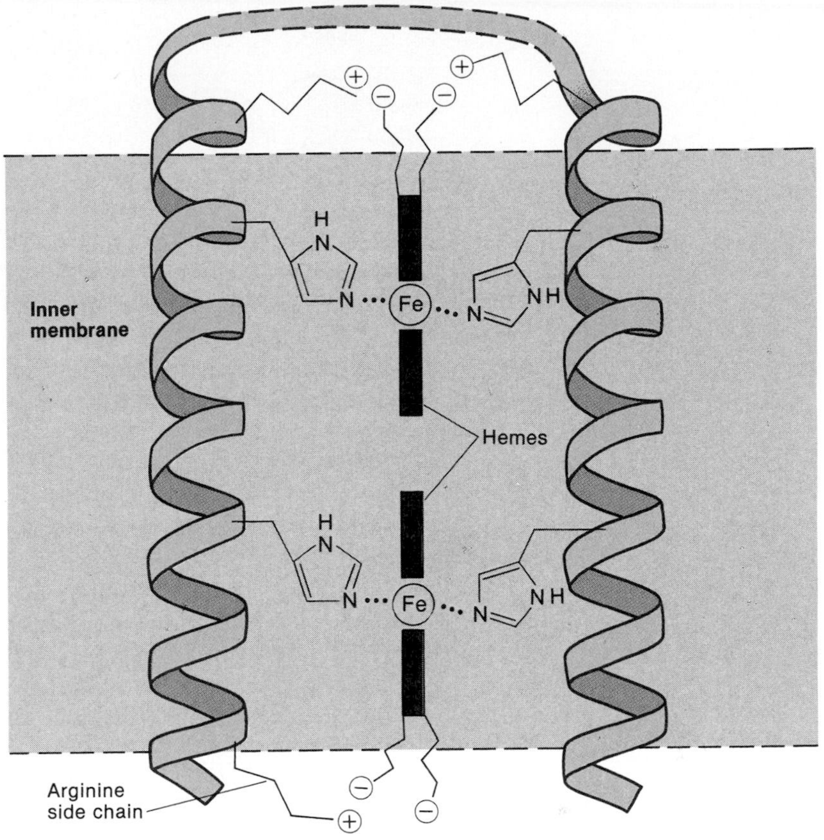

Figure 16–17
The hemes of cytochromes b_L and b_H probably are sandwiched between histidyl residues of two parallel α helices in the same subunit of the cytochrome bc_1 complex (complex III). Helices extending across the phospholipid bilayer are predicted to form in nine regions of the protein where there are stretches of approximately 20 predominantly hydrophobic amino acids. The positions of histidines in two of these regions are highly conserved in the amino acid sequences of the b cytochromes obtained from different species. The propionate side chains of the hemes (see Figure 16–3) probably form salt bridges to arginyl groups at the ends of the helices. (For details, see M. Saraste, *FEBS Lett.* 166:367–372.)

These puzzling observations can be explained best by the following reactions, in which UQH_2 *first transfers one of its two electrons to the iron-sulfur protein of the cytochrome bc_1 complex, and then gives the other electron to cytochrome b_L*:

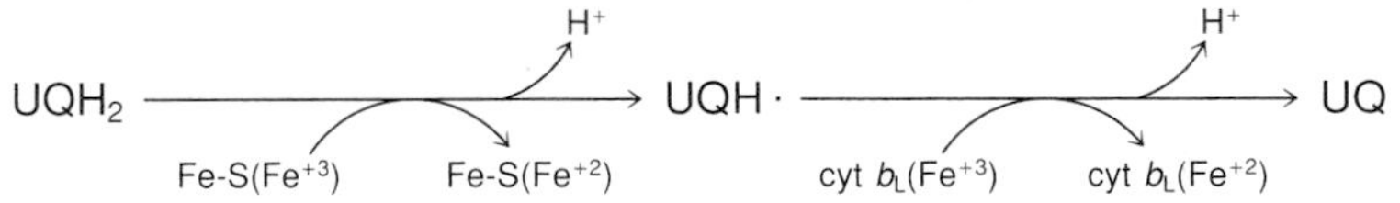

$$UQH_2 \xrightarrow[\text{Fe-S(Fe}^{+3})\quad\text{Fe-S(Fe}^{+2})]{\qquad\qquad H^+\qquad} UQH\cdot \xrightarrow[\text{cyt }b_L(\text{Fe}^{+3})\quad\text{cyt }b_L(\text{Fe}^{+2})]{\qquad\qquad H^+\qquad} UQ$$

In this scheme, the oxidation of UQH_2 by the iron-sulfur protein generates the semiquinone $UQH\cdot$, which then serves as the reductant for cytochrome b_L. Remember that the $UQ/UQH\cdot$ redox couple has a more negative $E°$ than the $UQH\cdot/UQH_2$ couple (see Table 16–1).

The reduced iron-sulfur protein subsequently transfers an electron to cytochrome c_1 and on to cytochrome c. Further experiments suggest that the reduced cytochrome b_L transfers its electron across the membrane to the other b cytochrome, cytochrome b_H, which then can contribute the electron for the reduction of another molecule of UQ at a second enzymatic site in the complex:

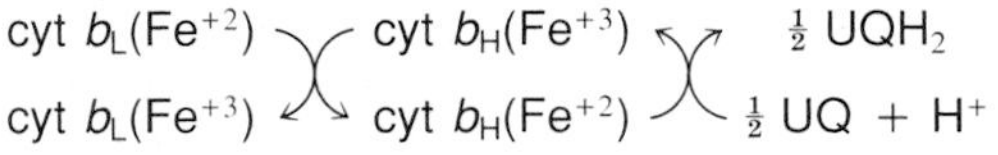

$$\text{cyt }b_L(\text{Fe}^{+2}) \underset{\text{cyt }b_L(\text{Fe}^{+3})}{\overset{}{\bigtimes}} \text{cyt }b_H(\text{Fe}^{+3}) \underset{\text{cyt }b_H(\text{Fe}^{+2})}{\overset{}{\bigtimes}} \begin{array}{l}\tfrac{1}{2}\,UQH_2\\[4pt]\tfrac{1}{2}\,UQ + H^+\end{array}$$

If each of these steps occurs twice, so that two electrons pass through the b cytochromes, a molecule of UQ could be reduced fully to UQH_2. At this point, we may seem to be back where we started. Note, however, that for each UQH_2 that is regenerated there has been a net oxidation of one UQH_2 to UQ, and two electrons will proceed to cytochrome c and on their way to O_2 (Figure 16–18).

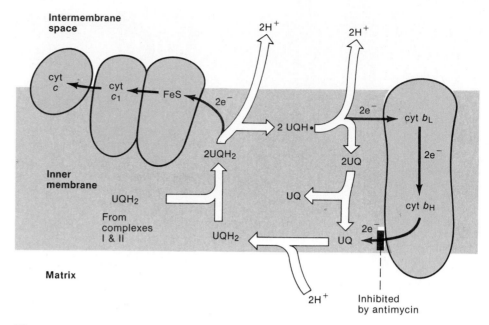

Figure 16–18

Electron transport through the cytochrome bc_1 complex: the Q cycle. In this figure, the black arrows represent the path of electron flow; the large arrows represent the paths of UQ in its various oxidation states and of protons. UQH_2 is oxidized to UQ in two steps at an enzyme site on the cytoplasmic side of the inner membrane (top part of the figure), transferring the first electron to the Fe-S protein and the second electron to cytochrome b_L. The stoichiometries indicated are for two molecules of UQH_2 undergoing these reactions. One of the two molecules of UQ that are produced diffuses to a site on the matrix side of the membrane (bottom), where it is reduced back to UQH_2. The two electrons required for the reduction come through the two b cytochromes. Antimycin inhibits this reaction. The UQH_2 that is generated at the reduction site diffuses back to the oxidation site, where it joins the pool of UQH_2 coming from complexes I and II. Electrons leaving the UQH_2 oxidation site reduce cytochrome c (top), which leaves the complex on the cytoplasmic side of the membrane to go to cytochrome oxidase. Protons also are released to the solution on this side of the membrane. Protons are taken up from the solution on the matrix side of the membrane at the UQ reduction site. The reduction of UQ to UQH_2 involves an intermediate semiquinone $(UQH\cdot)$ similar to that formed in the oxidation of UQH_2 to UQ, but the details of the reactions at the reduction site are not yet clear.

A scheme of the sort just discussed was proposed by Peter Mitchell in 1975. It has been elaborated in several versions by other investigators, and is generally called a Q *cycle*. As shown in Figure 16–18, the sites at which UQH_2 and $UQ\cdot$ undergo oxidation are on the cytoplasmic side of the membrane, facing the intermembrane space, and the UQ reduction site is on the matrix side. UQ and UQH_2 evidently diffuse through the membrane from one site to the other.

In addition to providing two distinct sites for the oxidation and reduction of UQ, the cytochrome bc_1 complex must impose constraints on the reactions of the quinone with the oxidant or reductant at each site. The Rieske iron-sulfur protein, for example, must be prevented from taking two electrons from UQH_2. If the iron-sulfur protein removed a second electron, oxidizing $UQH\cdot$ to UQ, reduction of the b cytochromes would not occur.

Antimycin blocks electron transfer between cytochrome b_H and UQ at the reduction site. Using antimycin to inhibit the reactions of cytochrome b_H at this site makes it possible to focus experimentally on the reduction of the b cytochromes at the UQH_2 oxidation site. Inhibitors that act at other points have been found, and also have been helpful for elucidating the scheme. The roles of the iron-sulfur protein and cytochrome c_1 have been explored by extracting the iron-sulfur protein from the cytochrome bc_1 complex. If the iron-sulfur protein is removed, electron transfer from UQH_2 to cytochrome c_1 is prevented, and oxidants cannot cause reduction of the b cytochromes. This shows that the iron-sulfur protein operates upstream of cytochrome c_1, even though its $E°$ is more positive than that of the cytochrome (see Figure 16–12).

As indicated in Figure 16–18, the _oxidation and reduction of UQ by the cytochrome bc_1 complex is coupled to the uptake of protons from the solution on the matrix side of the mitochondrial inner membrane and the release of protons on the cytoplasmic side._ Two protons are taken up on the matrix side and four protons are released on the cytoplasmic side for each pair of electrons that proceed to cytochrome c. We shall return to this point below.

Complex IV: Cytochrome Oxidase.   Cytochrome oxidase contains two atoms of copper, in addition to the hemes of cytochromes a and a_3. The Cu atoms undergo one-electron oxidation-reduction reactions between the cuprous (Cu^+) and cupric (Cu^{2+}) states. One of the Cu atoms (Cu_{a3}) is close to the Fe of cytochrome a_3 (Figure 16–19). The other (Cu_a) appears to be associated with cytochrome a, but not so intimately. Cytochrome a and Cu_a both have effective $E°'$ values of about 0.25 V, slightly more positive than that of cytochrome c (0.22 V); cytochrome a_3 and Cu_{a3} have $E°'$ values near 0.35 V. However, the interactions among the four electron carriers make it difficult to determine the $E°$ of any one of the components independently of the others.

The reduction of O_2 to 2 H_2O requires the addition of four electrons. Addition of a single electron to O_2 to produce superoxide (O_2^-) is thermodynamically unfavorable, and has to be pulled along by a further reduction to the level of peroxide (H_2O_2) or H_2O. (Although the $E°'$ value for the overall reduction of O_2 to H_2O is +0.82 V, the redox couple O_2/O_2^- has an $E°'$ value of −0.33 V). This is one of the reasons that O_2 is relatively unreactive in the absence of the enzyme. Cytochrome oxidase may get around this thermodynamic obstacle by binding the O_2 molecule to Cu_{a3} as well as to the Fe of cytochrome a_3, and then by transferring electrons to the O_2 in pairs (Figure 16–19).

Cytochrome oxidase isolated from mammalian mitochondria has between 6 and 13 different subunits, ranging in molecular weight from 5000 to 50,000. The total molecular weight of the complex is approximately 200,000. The three largest subunits are synthesized in the mitochondria; the smaller ones are made in the cytoplasm. The hemes and Cu atoms are all located in two of the large subunits. Several of the subunits span the phospholipid bilayer of the inner membrane, and a large domain of the protein protrudes from the membrane's cytoplasmic surface. The site at which cytochrome c oxidation occurs is on the cytoplasmic surface of the membrane, but the reduction of O_2 occurs on the matrix side (Figure 16–15). To get from one of these sites to the other, electrons must move across the membrane.

Electron Transfer from One Complex to the Next.   Having described the structures of the four multisubunit complexes that make up the respiratory chain, we now turn our attention to the question of how electrons move between the complexes. To investigate this point, the isolated electron-carrier complexes can be recombined so that they reconstitute longer stretches of the respiratory chain. For this, the complexes are mixed in the presence of phospholipids and a detergent such as sodium deoxycholate. When the detergent is removed by dilution or dialysis, the phospholipids assemble into vesicles (liposomes), and the proteins are incorporated into the phospholipid bilayers. If UQ is added, vesicles that contain the NADH dehydrogenase and cytochrome bc_1 complexes will catalyze electron transfer from NADH to cytochrome c. In the presence of cytochrome c, vesicles

Figure 16–19
Schematic drawing of the four electron carriers in cytochrome oxidase (complex IV). (a) The oxidized enzyme, in which cytochromes a (top) and a_3 (bottom) are both in the ferric (+3) states and the two Cu atoms are both cupric (+2). The hemes are viewed edge-on, as in Figure 16–6. The axial ligands in cytochrome a are both histidyl residues of the protein. In cytochrome a_3, one of the axial ligands probably is a histidine and the other is a sulfur atom that may also be a ligand of Cu_{a3}. When the electron carriers are reduced to the ferrous and cuprous states by electrons from four molecules of cytochrome c, the S moves away from the Fe of cytochrome a_3. This could make room for a molecule of O_2 to bind to the Fe and the Cu, giving the structure in (b). Four electrons now can be transferred to the O_2 (probably two at a time). Protons are picked up from the solution on the matrix side of the inner membrane, two H_2O molecules are released, and the enzyme returns to its original state. This scheme does not account for additional protons that move across the membrane in the course of the reaction. The evidence for the structures shown here comes mainly from spectroscopic techniques, including ESR and resonance Raman spectroscopy and x-ray absorption fine structure. The interpretation of some of this evidence is uncertain, particularly as to whether the Fe atom of cytochrome a_3 is connected to Cu_{a3} by a common ligand. (For discussion of possible intermediate electron transfer steps, see L. Powers et al., *Biophys. J.* 34:465, 1981.)

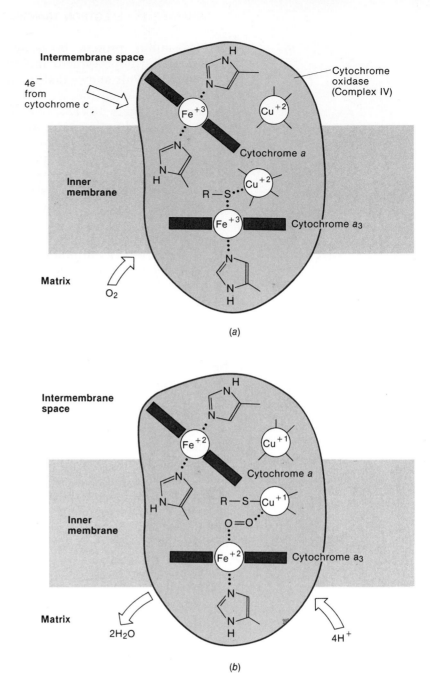

containing the cytochrome bc_1 and cytochrome oxidase complexes will transfer electrons from UQH_2 to O_2.

These observations raise the question of whether the different complexes all bind together to form a single, long chain in the mitochondrial inner membrane. Within each of the individual complexes, the electron carriers occur in a fixed molar ratio. In a reconstitution experiment, however, the complexes can be mixed in any ratio. The kinetics of electron transfer in the mixture suggest that the complexes do not bind to each other. Instead, *the movement of electrons from NADH dehydrogenase to the cytochrome bc_1 complex appears to be mediated by the diffusion of UQH_2 from one complex to the other within the phospholipid bilayer.* Similarly, *the movement of electrons from the cytochrome bc_1 complex to cytochrome oxidase appears to occur by the diffusion of reduced cytochrome c along the surface of the bilayer* (Figure 16–15). Remember that cytochrome c differs from the other cytochromes in being a water-soluble protein. It is attached loosely to the membrane surface by electrostatic interactions, which are relatively weak at physiological salt concentrations.

There is additional evidence that *the electron-transfer complexes are not connected in fixed, linear chains* in the mitochondrial membrane. First, the different complexes are not present in equimolar concentration ratios. In rat liver mitochondria, for example, the concentration of cytochrome oxidase is about twice that of the cytochrome bc_1 complex and more than six times that of NADH dehydrogenase. If the respiratory complexes were connected in discrete chains, each of which ended with a single cytochrome oxidase complex, only one-sixth of the chains would have NADH dehydrogenase. This would make it hard to explain how an NADH-linked substrate such as malate can rapidly reduce all of the cytochromes in the membrane. Second, when most of the cytochrome oxidase complexes in the membrane are inhibited with CO, the few molecules that remain uninhibited are still able to catalyze the oxidation of all the cytochrome c by O_2. This suggests that cytochrome c can diffuse rapidly from one cytochrome oxidase complex to another, rather than remaining bound to an individual complex. Direct measurements of the diffusion indicate that cytochrome c, UQ, and the complexes themselves diffuse laterally in the membrane at different rates, which means that they cannot all stay stuck together. The diffusion rates are such that random collisions between a complex and its reaction partner (reduced or oxidized UQ or cytochrome c) could occur at relatively high frequencies. Finally, the region of the cytochrome c molecule that interacts with the cytochrome bc_1 complex appears to be the same region that interacts with cytochrome oxidase. This indicates that cytochrome c could not stay fixed in position attached to one of these complexes and transfer an electron to the other; it must dissociate and at least turn around. The site on cytochrome c that interacts with both of the complexes is the cluster of lysine residues near one edge of the heme (Figure 16–5). It seems likely that the positively charged lysines also interact with the negatively charged phospholid head groups of the membrane, orienting the cytochrome c so that its reactive site faces the surface.

OXIDATIVE PHOSPHORYLATION

From the preceding discussion, it should be clear that the movement of a pair of electrons from NADH through the respiratory chain to O_2 involves a large, negative $\Delta G^{\circ\prime}$:

$$\Delta E^{\circ\prime} = E^{\circ\prime}(O_2/H_2O) - E^{\circ\prime}(NADH/NAD^+) = +0.82\ V - (-0.32\ V) = 1.14\ V$$

$$\Delta G^{\circ\prime} = -nF\ \Delta E^{\circ\prime}$$

$$= -(2\ \text{electrons/NADH molecule})(23{,}060\ \text{cal mole}^{-1}\ V^{-1})(1.14\ V)$$

$$= -52.6\ \text{kcal/mole}$$

The actual ΔG for the process under physiological conditions, though smaller than $\Delta G^{\circ\prime}$ because the concentration of O_2 is only about 200 μM, is probably still about -50 kcal/mole.

Evidence that the free energy released in electron-transfer reactions could be used to drive the formation of ATP was obtained about 1940 by Herman Kalckar, who showed that aerobic cells can make ATP from ADP and P_i by a process that depends on respiration. Belitser measured the stoichiometry of the process, and found that at least two molecules of ATP are formed per pair of electrons transferred to O_2. It was not clear, however, that this *oxidative phosphorylation* occurred in mitochondria, or that it involved NADH. When NADH was added to a crude homogenate of liver or muscle tissue, it was oxidized, but no ATP was made. We now know that the NADH dehydrogenase of animal mitochondria can only oxidize NADH that is in the mitochondrial matrix (Figures 16–15 and 16–16), and exogenous NADH does not get into the mitochondria. The oxidation that was observed was due to the endoplasmic reticulum. Resolution of the problem had to await the development of methods for preparing mitochondria free of other cellular constituents, and of presenting NADH on the appropriate side of the mitochondrial inner membrane. When these methods were devised in the early 1950s, Kennedy and Lehninger found that mitochondria do oxidize endogenous NADH, and that about three ATPs are generated per NADH oxidized. (Plant mitochondria, curiously, have a second NADH dehydrogenase that is distinct from complex I and can oxidize exogenous NADH.)

How is it possible to couple such different types of reactions as electron transfer and the formation of a phosphate anhydride bond? As we explore this question in the following sections, we shall be led to the conclusion that *the coupling mechanism hinges on the protons that are bound and released on either side of the inner membrane when electrons move down the respiratory chain*. The principal theory that has been developed to explain the role of proton movements in oxidative phosphorylation is known as the *chemiosmotic theory*. We shall develop the theory in some detail, partly because of its elegance and partly because the concepts that underlie it have broad ramifications in other areas of biochemistry. But first we must explore some additional phenomena associated with oxidative phosphorylation.

The P/O Ratio Measures the Efficiency of Coupling

The number of molecules of ATP formed per pair of electrons transferred down the respiratory chain is termed the $P/2e^-$ ratio. If the electrons travel all the way to O_2, the $P/2e^-$ ratio is usually called the *P/O ratio* because two electrons are used for each atom of oxygen that is reduced from the level of O_2 to H_2O. When NADH is used as the reducing substrate, the P/O ratio is about 3; with succinate it is about 2. (These values are somewhat idealized; the measured P/O ratios usually are closer to 2.5 and 1.5.) These observations suggest that *there are three coupling sites for ATP synthesis somewhere in the electron-transport chain between NADH dehydrogenase and O_2, and two coupling sites between succinate dehydrogenase and O_2.* Since the electron-transport chains for NADH and succinate converge at UQ, there would appear to be one coupling site between NADH and UQ, and two between UQH_2 and O_2 (Figure 16–20; also see Figure 16–15). The locations of these last two sites can be identified by using a substrate that feeds electrons directly to cytochrome c (such as ascorbic acid) in the presence of antimycin to block electron flow through complex III. This gives a P/O of about 1, indicating that there is one coupling site between cytochrome c and O_2 (Figure 16–20). Complementary experiments can be done by adding oxidants that remove electrons from the chain at various points. For example, the last coupling site can be bypassed by adding additional cytochrome c as an electron acceptor, along with CN^- to block cytochrome oxidase. Electron transfer from succinate to cytochrome c gives a $P/2e^-$ ratio of about 1, in agreement with the picture just described.

There thus appears to be one coupling site in each of complexes I, III, and IV, but none in complex II. Like succinate dehydrogenase (complex II), the glycerol-3-phosphate dehydrogenase and the electron-transfer flavoprotein and its ubiquinone-oxidoreductase do not include coupling sites. Efraim Racker has verified the capacity of the individual complexes I, III, and IV to support the formation of ATP, by incorporating purified preparations of the electron-transport complexes into phospholipid vesicles along

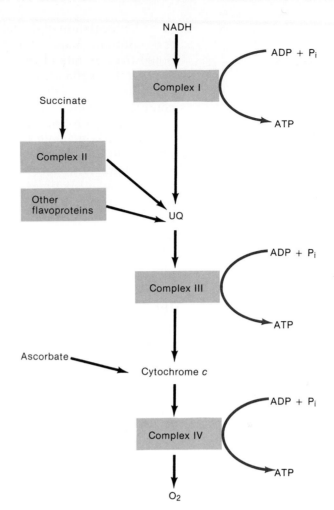

Figure 16–20

Approximately three molecules of ADP can be phosphorylated to ATP for each pair of electrons that traverse the electron-transport chain from NADH to O_2. Two ATPs are formed for a pair of electrons that enter the chain via succinate dehydrogenase or other flavoproteins such as glycerol-3-phosphate dehydrogenase, and one ATP is formed for a pair of electrons that enters via cytochrome c. It therefore appears that electron flow through each of complexes I, III, and IV is coupled to phosphorylation.

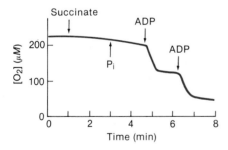

Figure 16–21

The rate of respiration by a suspension of mitochondria can be increased dramatically by the addition of a small amount of ADP. An oxidizable substrate (succinate) and P_i first are added in excess at the times indicated by the vertical arrows, but the respiratory rate remains low until ADP is added. The P/O ratio can be determined by dividing the amount of ADP added by the amount of oxygen taken up during the period of rapid respiration. The observation that another period of rapid respiration is obtained upon the addition of more ADP indicates that essentially all of the ADP gets used up.

with the mitochondrial ATP-synthase enzyme that is described below. By referring to Figure 16–12, you can see that the flow of two electrons through each of complexes I, III, and IV involves a sufficiently large negative $\Delta G^{\circ\prime}$ to support the phosphorylation of ADP to ATP. In the presence of 10 mM Mg^{2+}, $\Delta G^{\circ\prime}$ for the reaction ADP + $P_i \rightarrow$ ATP + H_2O is about 7.5 kcal/mole, which is equivalent to a $\Delta E^{\circ\prime}$ of about 0.16 V for $n = 2$ electrons/molecule. However, the actual ΔG for ATP formation in a suspension of respiring mitochondria can be as large as about 15 kcal/mole, because of a high [ATP]/[ADP][P_i] ratio.

The linkage between respiration and phosphorylation is so tight that the electron-transfer reactions are sharply inhibited if phosphorylation is blocked. For example, *respiration slows down greatly if mitochondria run out of ADP.* The experiment shown in Figure 16–21 illustrates this point. The trace shows a measurement of the concentration of O_2 in a suspension of intact mitochondria as a function of time. When succinate is added, O_2 is consumed at only a very low rate, indicating that something more than an electron donor is needed in order for respiration to occur. The addition of a small amount of ADP in the presence of excess P_i causes a brief period of brisk respiration, as reflected by a rapidly decreasing O_2 concentration. O_2 uptake evidently stops when all the ADP is converted to ATP, because it can be started again by the addition of more ADP.

Experiments like that shown in Figure 16–21 are one way of measuring P/O ratios. The P/O ratio is proportional to the amount of ADP that must be added in order to cause a certain amount of O_2 uptake.

The regulation of the rate of electron transport by ADP is called *respiratory control*. Chance and Williams, who first studied the phenomenon, noted that the rate of respiration can be limited by a variety of factors, including the availability of oxidizable substrates, ADP, and P_i, and the intrinsic cycling rates of the electron carriers. They used the term "state 4" for the condition in which ADP is limiting, and "state 3" for the condition in which ADP, P_i, and substrate are present in excess and the mitochondria are respiring at the maximal rate sustainable by the electron carriers. (States 1, 2, and 5 referred to other special conditions that are of less interest for the present discussion.) With succinate as substrate, the respiration rate in state 3 is typically about six times greater than that in state 4. Although it makes physiological sense for a cell to avoid oxidizing substrates wastefully when the concentration of ADP is low and the cell's energy charge is high, there is no evidence that respiratory control involves an allosteric regulatory mechanism. It simply reflects the fact that ADP and ATP are, in essence, substrates and products of the respiratory apparatus.

Inhibitors of Phosphorylation Act Directly or As Uncouplers

The coupling of respiration and phosphorylation can be lost if the integrity of the mitochondrial inner membrane is disrupted. For example, if mitochondria are placed in a hypotonic solution so that they swell, their rate of respiration can increase until it is similar to that in state 3, even though no ADP is added. The P/O ratio simultaneously decreases to zero. These effects can be related to an increase in the leakage of protons across the inner membrane.

There are also a number of different molecules that cause an uncoupling of electron transfer from the phosphorylation of ADP. Representative of these <u>uncouplers</u> of oxidative phosphorylation are 2,4-dinitrophenol and carbonylcyanide-p-trifluoromethoxyphenylhydrazone (Figure 16–22). Studies of the actions of uncouplers have played a pivotal role in attempts to understand the mechanism of oxidative phosphorylation. If an uncoupler is added to a suspension of mitochondria in the presence of an oxidizable substrate, O_2 uptake commences immediately and continues until essentially all of the O_2 in the solution is used up (Figure 16–23a). This happens even in the absence of added ADP or P_i. In damaged mitochondria, or in the presence of an uncoupler, the free energy that is released in the electron transport reactions is not captured in the form of ATP. Although the uncou-

2,4-Dinitrophenol

Carbonylcyanide-*p*-trifluoro-methoxyphenylhydrazone

Figure 16–22

Structures of two uncouplers of oxidative phosphorylation. 2,4-dinitrophenol (DNP) was one of the first uncouplers to be discovered; carbonylcyanide-p-trifluoromethoxyphenylhydrazone (FCCP) works at much lower concentrations than dinitrophenol (on the order of 1 μM). DNP has a weakly dissociable proton on the oxygen atom; FCCP has a similar proton on one of the nitrogens.

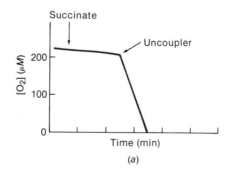
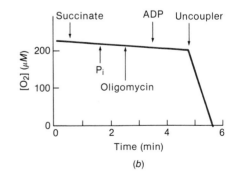

Figure 16–23

(a) Addition of an uncoupler to a suspension of mitochondria causes brisk O_2 consumption, which continues until all of the O_2 in the solution is used up. An oxidizable substrate (succinate) is added before the uncoupler, but P_i is not required. (b) Oligomycin, an inhibitor of the ATP-synthase, blocks the stimulation of respiration that is ordinarily caused by ADP. It does not block the stimulation caused by an uncoupler.

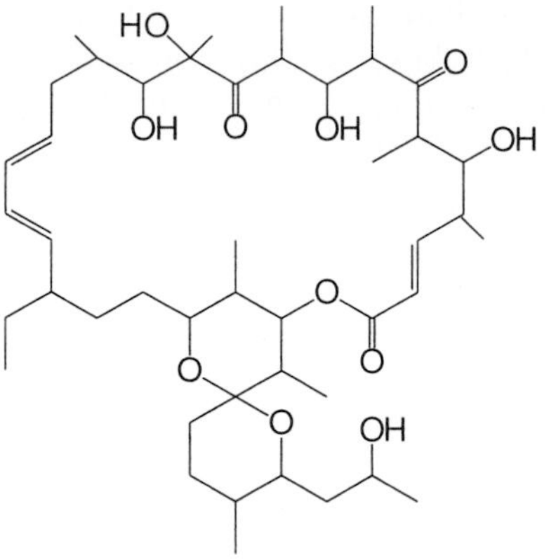

Oligomycin

$$C_6H_5-N=C=N-C_6H_5$$

Dicyclohexylcarbodiimide

Figure 16–24
Structure of oligomycin and
dicyclohexylcarbodiimide (DCCD), two
inhibitors of the ATP-synthase. Both
molecules also inhibit the ATPase
activity of the enzyme. Oligomycin,
like the electron-transport inhibitor
antimycin and many other antibiotics,
is obtained from species of
Streptomyces.

plers are a structurally diverse group of molecules, they are generally lipophilic weak acids; we shall return to this point below.

Other agents are found to interfere with the phosphorylation of ADP to ATP in a different manner. These true *inhibitors*, which include dicyclohexylcarbodiimide (DCCD) and oligomycin (Figure 16–24), prevent ADP from increasing the rate of respiration (Figure 16–23b). They do not block the stimulation of respiration that is caused by an uncoupler. This suggests that the inhibitors act directly on the ATP-synthase that converts ADP and P_i to ATP. (Another possibility would be an inhibition of the entry of exogenous ADP into the mitochondrion, or of the exit of ATP. However, this interpretation can be shown not to apply to oligomycin or DCCD.)

At the time experiments like those of Figures 16–22 and 16–23 were first done, it was reasonable to suggest that the mechanism of oxidative phosphorylation might resemble that of the glycolytic enzyme glyceraldehyde-3-phosphate dehydrogenase. As discussed in Chapter 14, the oxidation of glyceraldehyde-3-phosphate by NAD^+ results in the formation of an enzyme-bound thioester of 3-phosphoglyceric acid, which then reacts with P_i to form glycerate-1,3-bisphosphate (see Figure 14–11). The essence of the mechanism is that the substrate that undergoes oxidation is converted to a labile intermediate that later is broken down to yield ATP. If the intermediate is not broken down (for example, because no P_i is added), the oxidation-reduction reactions will come to a halt. One can generalize the mechanism as follows:

$$AH_2 + B \rightleftharpoons \sim A + BH_2$$

$$\sim A + ADP + P_i \rightleftharpoons A + ATP \text{ (in several steps)}$$

where AH_2 and B are the reductant and oxidant substrates, A and BH_2 are the oxidized and reduced products, and $\sim A$ is a labile intermediate form of one of the products. In the case of glyceraldehyde-3-phosphate dehydrogenase, the "squiggle" ($\sim$) represents a thioester bond to the enzyme; in other cases, it could represent a phosphoric anhydride bond, or any other structure that has a large, negative $\Delta G^{\circ\prime}$ of hydrolysis.

The effects of uncouplers of oxidative phosphorylation at first seem to fit in with this scheme. If the uncoupler somehow discharged $\sim A$, as, for example, by causing hydrolysis of a bond symbolized by the $\sim$, the electron carrier A would be set free and electron transport to O_2 could continue. However, there evidently is something different about oxidative phosphorylation, as compared to the phosphorylation catalyzed by glyceraldehyde-3-phosphate dehydrogenase or by the other soluble enzymes that make or use ATP, because uncouplers have no effects on most of the soluble enzymes. On the other hand, a molecule that acts as an uncoupler at any one of the three coupling sites of oxidative phosphorylation is found invariably to have a similar effect at the other two sites. This suggests that *uncouplers do not act on an intermediate that involves a particular electron carrier*, such as the species that we have written abstractly as $\sim A$, *but rather on some species or state that is formed at all three sites*. If we continue to denote this species or state as $\sim$, the reactions of oxidative phosphorylation would appear to need rewriting in the following form:

$$A_{red} + B_{ox} \rightleftharpoons A_{ox} + B_{red} + \sim \tag{8}$$

$$\sim + ADP + P_i \rightleftharpoons ATP + H_2O \quad \text{(blocked by oligomycin)} \tag{9}$$

$$\sim \xrightarrow{\text{Uncoupler}} \text{dissipated state} \tag{10}$$

where A_{red}, A_{ox}, B_{red}, and B_{ox} are the reduced and oxidized forms of electron carriers that participate at any one of the coupling sites. Although the nature of the $\sim$ has suddenly become rather vague, we can say at this point that the $\sim$ does not seem to involve phosphate or ADP, because the stimulation of respiration by uncouplers does not require the addition of either of these materials (Figure 16–23a). In the next section, we show that the $\sim$ that drives the formation of ATP can also be used to push electron transfer backwards.

Oxidative Phosphorylation Can Be Reversed

Note that the reactions that generate the hypothetical $\sim$ and use it to make ATP have been written as being reversible. Reversal of the reactions of oxidative phosphorylation can be demonstrated experimentally. Consider first a reversal of just the process that converts $\sim$ to ATP. Suppose that ATP is added to a suspension of mitochondrial inner membrane vesicles in the presence of a battery of inhibitors that block electron transport through the regions of all three coupling sites (amytal, antimycin, and CN^-, for example). A limited amount of ATP is hydrolyzed to ADP and P_i but then the hydrolysis stops. If an uncoupler is added, the hydrolysis of ATP continues. *Uncouplers thus can stimulate an ATPase activity in mitochondria*. These observations are consistent with reactions (9) and (10) above. As ATP is hydrolyzed, $\sim$ builds up, slowing down the hydrolysis. Removal of $\sim$ by an uncoupler, reaction (10), pulls reaction (9) to the left. The ATPase activity is blocked by oligomycin, just as the forward reaction that synthesizes ATP is, in agreement with the view that the ATPase and the ATP-synthase are the same enzyme.

Now suppose that the respiratory chain is inhibited in the cytochrome bc_1 region (complex III) by antimycin, but that electron transport through complexes I and II is not blocked (see Figure 16–25). Assuming that the reaction catalyzed by complex II is reversible, the following reactions can occur:

$$NADH + UQ \xrightleftharpoons{\text{Complex I}} NAD^+ + UQH_2 + \sim$$

$$Fumarate + UQH_2 \xrightleftharpoons{\text{Complex II}} UQ + succinate$$

If we neglect the $\sim$ that is formed in complex I, the equilibrium of the net reaction

$$NADH + fumarate \rightleftharpoons NAD^+ + succinate$$

would lie far to the right. (The $E^{\circ\prime}$ value of the NADH/NAD$^+$ couple is -0.32 V; that of the succinate/fumarate couple, -0.03 V.) But the actual net reaction includes the formation of $\sim$:

$$NADH + fumarate \rightleftharpoons NAD^+ + succinate + \sim$$

From this equation, we might expect that the electron-transfer reactions could be driven to the left if ATP was added to generate $\sim$. This is, in fact, observed (Figure 16–25). The reversal of electron transport is prevented by oligomycin, which blocks the formation of $\sim$ from ATP, or by an uncoupler, which dissipates the $\sim$ as rapidly as it is formed (Figures 16–25b and c).

A significant extension of these experiments is to generate the $\sim$, not by the addition of ATP, but by electron transport through one of the other coupling sites. For example, electron flow through complex IV will occur if ascorbate is added to reduce cytochrome c; antimycin can be added to block flow through complex III. We now can ask whether the $\sim$ that is generated in complex IV is capable of being used in complex I to drive electrons from succinate to NADH. The answer is yes (Figure 16–25, trace d). Again, this

Figure 16–25

Reversal of electron transport through complex I can be driven by ATP. The traces show the absorbance at 340 nm of suspensions of membrane vesicles made from the mitochondrial inner membrane. An increase in the absorbance reflects the reduction of NAD^+ to NADH (Figure 16–13). Antimycin has been added to block electron flow through complex III. Under these conditions, succinate alone cannot reduce NAD^+, but it can if ATP is added (trace a). The reduction of NAD^+ driven by ATP is blocked by oligomycin or uncouplers (traces b and c). In trace d, ascorbate is used to reduce cytochrome c, so that electrons can flow through complex IV, below the block imposed by antimycin. (N, N, N', N'-tetramethylphenylenediamine commonly would be added as a redox mediator to speed up the reduction of cytochrome c by ascorbate.) The free energy decrease associated with electron transfer in complex IV can be used to drive electrons from succinate to NAD^+ in the absence of ATP, ADP, and P_i. Oligomycin now has no effect, because the ATP-synthase does not participate in the reactions (trace e), but the process is still sensitive to uncouplers (trace f). Experiments like these are easiest to interpret if they are done in membrane vesicles ("submitochondrial particles"), because in intact mitochondria succinate can be converted to fumarate and then to malate, which can reduce NAD^+ directly.

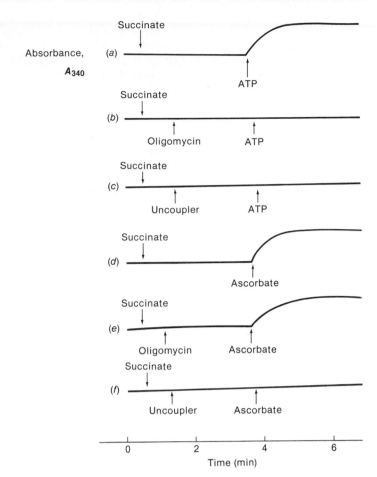

process is prevented by adding an uncoupler to dissipate the ~ as it is formed; but it is not blocked by oligomycin, because it does not involve the phosphorylation reactions directly. As expected, it also is independent of the presence of P_i and ADP.

Experiments of the type just described indicate that *the ~ that is generated at any of the three coupling sites can be used to drive electron transfer backwards in either complex I or complex III*. This conclusion agrees with the observation that uncouplers can dissipate the ~ that is formed at any of the sites. It also implies that, if the ~ is a compound, it must be able to diffuse relatively freely from one site to another. Alternatively, if the ~ is a state of some sort, the same state must be expressed at multiple sites. We shall see that this second interpretation appears to be the correct one.

The Chemiosmotic Theory Proposes That Phosphorylation Is Driven by a Proton Gradient

The effects of uncouplers suggest that the mechanism by which phosphorylation is coupled to electron transfer in the respiratory chain is fundamentally different from the known mechanisms of energy coupling in soluble enzymes. Although many investigators continued to favor the hypothesis that oxidative phosphorylation involved the generation and breakdown of "high-energy" intermediate forms of the electron carriers, such as the hypothetical species that we represented as ~A, determined searches for such intermediates were unsuccessful. Attempts to identify the ~ with various molecules other than the electron carriers also led to dead ends. Of course, negative results of this sort are not a very compelling argument. A labile

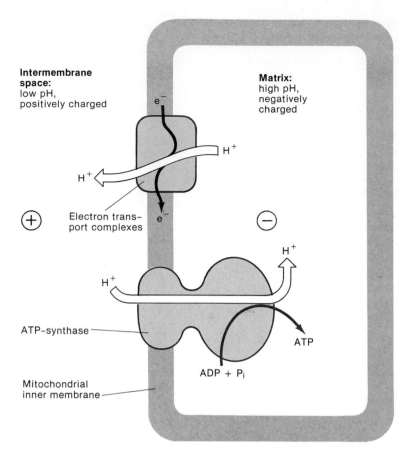

Figure 16–26
According to the chemiosmotic theory, the flow of electrons through the electron-transport complexes is coupled to the movement of protons across the inner membrane from the matrix to the intermembrane space. This raises the pH in the matrix, and leaves the matrix negatively charged with respect to the intermembrane space and the cytosol. Protons flow passively back into the matrix through a channel in the ATP-synthase, and this flow is coupled to the formation of ATP.

intermediate could escape detection if it decomposes when it is removed from an enzyme. However, the failure to find a molecular species with the properties of the $\sim$ does suggest the need to consider other possible ways in which electron transfer could be coupled to phosphorylation.

In 1961, Peter Mitchell suggested a radically different theory. He proposed that the $\sim$ is not a labile chemical bond, but rather _an electrochemical potential gradient for protons across the mitochondrial inner membrane._ Figure 16–26 illustrates the basic idea. Electron transport down the respiratory chain results in the movement of protons across the membrane, from the mitochondrial matrix to the intermembrane space and the cytoplasm. The removal of protons from the matrix causes the pH in this region to rise. Because protons are positively charged, the matrix space also becomes negatively charged electrically with respect to the cytoplasm. The differences in pH and electrical potential across the membrane provide a driving force that tends to pull protons from the cytoplasm back into the matrix. Part of the free energy decrease associated with the electron-transfer reactions thus is exchanged for an electrochemical potential difference for protons between the solutions on the two sides of the inner membrane. The membrane is relatively impermeable to protons, except for special channels that are part of an ATP-synthase enzyme. In principle, the enzyme could couple the inward movement of protons through these channels, down the electrochemical potential gradient, to the formation of ATP. This is called the _chemiosmotic_

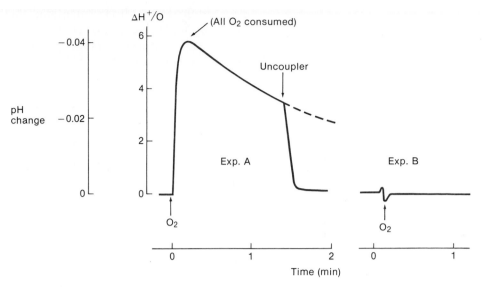

Figure 16–27
Proton extrusion by respiring mitochondria can be measured with a pH meter. A weakly buffered suspension of mitochondria is provided with an excess of an oxidizable substrate such as β-hydroxybutyrate, and is allowed to use up all the O_2 in the solution. The trace shows a measurement of the pH of the suspension, with pH decreases plotted upward. A small amount of O_2 is added at the upward arrow, allowing respiration to occur for a few seconds. The pH of the solution decreases suddenly at this point (experiment A). The pH change persists for a period of several minutes, long after the O_2 has been used up. If an uncoupler is added, the pH returns abruptly to nearly its original level. Experiment B was done in the presence of the detergent triton X-100, which disrupts the inner membrane's phospholipid bilayer and makes the membrane permeable to protons. Respiration still occurs under these conditions, but the pH change is not seen. In experiment A, the number of protons that move across the membrane can be estimated from the pH change, and can be related to the number of oxygen atoms consumed. In this experiment, about six protons appear to be translocated per oxygen atom consumed. (Since β-hydroxybutyrate reduces the mitochondrial NAD^+, a pair of electrons move through all three coupling sites for each oxygen atom that is consumed.) However, this is an underestimate of the amount of proton translocation, because the phosphate/OH^- transport system dissipates part of the pH gradient. Larger $\Delta H^+/O$ ratios are measured if the phosphate transporter is inhibited. (For more details see P. Mitchell and J. Moyle, *Nature* 208:147–151, 1965, and *Biochem. J.* 105:1147–1162, 1967; also A. Alexandre et al., *J. Biol. Chem.* 255:10721–30, 1980.)

theory, because it emphasizes that *chemical reactions can drive, or be driven by, the movements of molecules or ions between osmotically distinct spaces separated by membranes*. Let us examine some of the evidence that supports the theory.

Electron Transport Creates an Electrochemical Potential Gradient for Protons

Peter Mitchell and Jenifer Moyle showed that *protons are pumped out of mitochondria during respiration*. When O_2 is added to a weakly buffered suspension of mitochondria, the pH measured by a glass electrode immersed in the suspension drops, indicating that the proton concentration outside the mitochondria increases (Figure 16–27). The pH change is not seen if the integrity of the mitochondrial membrane has been broken by adding a detergent. This indicates that the protons probably emerge from the matrix space inside the mitochondria, rather than simply coming from a process that causes a net production of free protons, such as the hydrolysis of a phosphate ester. By making the membrane leaky to protons, the detergent allows any protons that are pumped out to go right back in.

Mitchell and Moyle calculated that mitochondria depleted of ADP can build up a pH gradient of about 0.05 pH units across the inner membrane, the value depending somewhat on the experimental conditions. Recent measurements indicate that between ten and twelve protons are pumped out of the mitochondrion for each pair of electrons that move down the full respiratory chain from NADH to O_2, and between six and eight protons for a pair of electrons coming from succinate. Between three and four protons thus appear to be extruded per pair of electrons that move through each of the three coupling sites.

The pH gradient that is created by respiration collapses abruptly if an uncoupler is added (Figure 16–27). This finding agrees with the proposal that the intermediate state ($\sim$) that couples electron transport to phosphorylation is an electrochemical potential gradient for protons across the membrane. In the absence of an uncoupler, the pH change caused by a brief period of respiration decays relatively slowly, in agreement with the view that the mitochondrial inner membrane is not very permeable to protons. Mitchell and Moyle reinforced this conclusion by experiments in which they added a small amount of HCl to a suspension of mitochondria and measured the rate at which protons leaked through the membrane. The leakage occurs on a time scale of minutes, which is much slower than the milliseconds-to-seconds scale on which oxidative phosphorylation occurs.

Although the idea that the membrane is largely impermeable to protons seemed hard to accept when Mitchell first proposed it, it fits well with what we now know about the structure of biological membranes (Chapters 6 and 34). A charged species such as a hydronium ion (H_3O^+) is unable to pass readily through the hydrocarbon region of the membrane's phospholipid bilayer. In the absence of an uncoupler, the effective conductance of the phospholipid bilayer to protons is approximately 10^6 times lower than that of the aqueous phases on either side of the membrane. Further experiments have shown that the phospholipid bilayer of the mitochondrial inner membrane also is largely impermeable to K^+, Cl^-, and most other ions, including ATP, ADP, NADH, and other ionic coenzymes. However, the membrane contains specific transport systems that catalyze the translocation of some of these species; we shall consider these systems below.

The ability of uncouplers to collapse a pH gradient across the membrane is understandable when we recall that most of the molecules that act as uncouplers are lipophilic weak acids. Because of its lipophilic character, the uncoupler can diffuse relatively freely through the phospholipid bilayer. Because it is a weak acid, it can pick up a proton from the solution on one side of the membrane, carry the proton across the bilayer, and release it on the opposite side. The unprotonated molecule then can diffuse back across the membrane to pick up another proton. Uncoupling also is caused by gramicidin A, a hydrophobic peptide antibiotic that can form a tubular channel extending across the membrane (Chapter 34). Protons and Na^+ or K^+ ions move rapidly in either direction through the gramicidin channel, short-circuiting the membrane.

The idea that *the pumping of protons across the membrane by the respiratory chain creates a trans-membrane electrical potential gradient, in addition to a pH gradient*, has been studied by examining the movements of lipophilic anions and cations. If the matrix space becomes negatively charged with respect to the region outside of the mitochondrion, there will be a driving force that tends to pull any positively charged ion into the matrix, and to push any negatively charged ion out. Whether or not a particular ion actually moves across the membrane will depend on whether the ion can pass through the phospholipid bilayer. As we have just discussed, a proton or a hydronium ion cannot traverse the bilayer readily in the absence

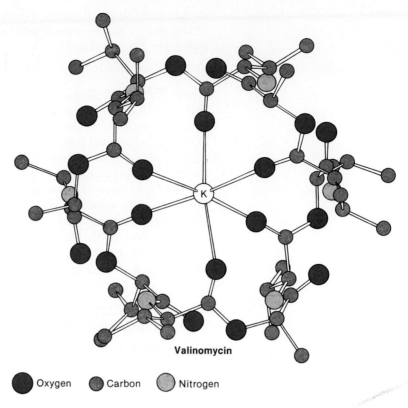

Valinomycin

● Oxygen   ● Carbon   ○ Nitrogen

Figure 16–28

Valinomycin, another antibiotic obtained from *Streptomyces* species, is one of a group of compounds, termed *ionophores*, that facilitate the movement of ions across phospholipid membranes. Valinomycin has the cyclic structure [-D-valine-lactate-L-valine-hydroxyvalerate-]$_3$. (Hydroxyvaleric acid is $(CH_3)_2CHOH—CO_2H$.) The links between D-valine and lactate and between L-valine and hydroxyvalerate are ester bonds; the links between lactate and L-valine and between hydroxyvalerate and D-valine are peptide bonds. Molecules with alternating ester and peptide linkages of this sort are called *depsipeptides*. This figure shows the crystal structure of the complex formed between valinomycin and K^+. The K^+ ion binds to the six ester oxygen atoms, which point in to the center of the ring. The amide oxygens are all hydrogen-bonded to nitrogens, and the methyl and isopropyl groups all point outward. The complex is soluble in organic solvents, and can pass through the phospholipid bilayer of the mitochondrial inner membrane. K^+ dissociates reversibly from the complex in the aqueous solution on either side of the membrane. The transport of K^+ across membranes by valinomycin is discussed further in Chapter 34.

of an uncoupler, but some lipophilic ions can. For example, in the triphenylmethylphosphonium and tetraphenylarsonium ions, $(C_6H_5)_3CH_3P^+$ and $(C_6H_5)_4As^+$, the positive charge is sufficiently buried by phenyl and methyl groups that the ion can move through the hydrocarbon center of the bilayer relatively freely. Other lipophilic cations that pass rapidly through phospholipid bilayers are the complexes of the ionophore valinomycin with K^+ or Rb^+ (Figure 16–28). Respiring mitochondria are found to take up lipophilic cations, such as the triphenylmethylphosphonium ion, or K^+ in the presence of valinomycin. In contrast, they extrude lipophilic anions such as thiocyanate (CNS^-). These observations are consistent with the idea that the efflux of protons driven by electron transport makes the interior of the mitochondrion negatively charged relative to the solution outside.

By measuring the concentrations of a lipophilic ion inside and outside, one can estimate the electrical potential difference across the mitochondrial membrane. The calculation proceeds as follows. The free energy change for the movement of one mole of a cation C^+ into the mitochondrion is the difference between the molar free energies for the ion on the two sides of the membrane. This difference depends both on the ratio of the concentrations of the ion on the two sides and on the difference between electrical potentials (See Box 12–C):

$$\Delta G_{C^+} = G_{C^+(in)} - G_{C^+(out)} = RT \ln \{[C^+]_{in}/[C^+]_{out}\} + zF \Delta\psi \tag{11}$$

where $[C^+]_{in}$ and $[C^+]_{out}$ are the concentrations of the cation inside and outside, z is the valency of the ion ($+1$ for a univalent cation), F is the Faraday constant, and $\Delta\psi$ is the difference in electrical potential across the membrane (the electrical potential inside minus that outside). Equation (11) is the second of the two Nernst equations. If the ion can pass freely across the membrane, the concentrations on the two sides will come to equilibrium, but the two concentrations generally will not be equal. Equilibrium means that $\Delta G_{C^+} = 0$, which occurs when $\ln \{[C^+]_{in}/[C^+]_{out}\} = -zF \Delta\psi/RT$, that is, when the difference in chemical potential exactly balances the difference in electrical potential. Measurements of ion concentration ratios have indicated that the electrical potential ($\Delta\psi$) across the inner membrane of respiring mitochondria is typically on the order of -0.15 V, inside negative.

How Do the Electron-transfer Reactions Drive Proton Translocation Across the Membrane?

We noted above that some of the components of the respiratory chain (NADH, flavins, and quinones) bind and release protons in addition to electrons when they undergo reduction and oxidation. Suppose that an electron-transfer reaction in which protons are taken up from the solution occurs on the matrix side of the inner membrane, and a reaction that releases protons to the solution occurs on the cytoplasmic side. If protons are transferred along with electrons between the two reaction sites, a flow of electrons through the system will result in a movement of protons across the membrane (Figure 16–29). The key point here is that, unlike reactions in free solution, enzymatic reactions in organized structures such as membranes can have a *vectorial*, or directional, character.

This basic idea accounts well for the proton translocation that occurs in complex III, the cytochrome bc_1 complex. As explained above (Figure 16–18), electron transport through complex III appears to involve the oxidation of UQH_2 to UQ on the cytoplasmic side of the inner membrane, and the reduction of UQ to UQH_2 on the matrix side. Hydrogen atoms are carried across the membrane by the diffusion of UQH_2 from one of these catalytic sites to the other. The net transfer of two electrons from UQH_2 through the complex to cytochrome c is coupled to the release of four protons to the solution on the cytoplasmic side, in agreement with the measured stoichiometry of proton translocation in this region of the respiratory chain.

The mechanism of proton translocation in complexes I and IV is not so clear. In the scheme that we presented for electron transfer through complex I (Figure 16–16), two protons are released to the intermembrane space for each pair of electrons that move from NADH to UQ. But this scheme fails to explain how FMN and $FMNH_2$, charged molecules that would not diffuse freely through the phospholipid bilayer, could have access to two enzymatic sites on opposite sides of the membrane. In addition, recent measurements indicate that four protons are transported per pair of electrons, two protons more than the scheme predicts. In complex IV, two protons are taken up from the solution on the matrix side for each $\frac{1}{2} O_2$ that is reduced to H_2O, or for each pair of electrons transferred. Since cytochrome c is oxidized on the cytoplasmic side of the membrane, while O_2 is reduced on the matrix side (Figure 16–15), there is a net transfer of electrons across the membrane from the cytoplasmic side to the matrix. This inward movement

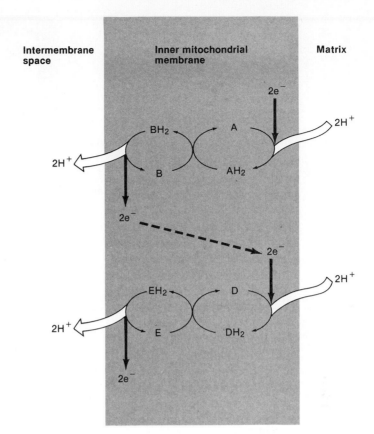

Figure 16–29
A general scheme showing how electron transport can result in proton translocation. A, B, D, and E are electron carriers that bind protons when they are reduced. A and D undergo reduction to AH_2 and DH_2 on the matrix side of the membrane, picking up protons from the solution on this side. They then transfer protons along with electrons to B and E. BH_2 and EH_2 are reoxidized on the cytoplasmic side, releasing protons here. Electrons move without protons from BH_2 to D. A set of reactions that move protons and electrons in one direction across a membrane, and move electrons alone back in the other direction, is frequently called a *loop*. The scheme shown here has one and a half loops. The scheme shown in Figure 16–16 has one loop.

of negative charge accounts for part of the contribution of complex IV to the electrical potential difference across the membrane. However, electron transfer from cytochrome c to O_2 also results in the inward movement of one to two additional protons per electron. How these protons are pumped is not known.

The Movement of Protons Back into the Matrix Drives the Formation of ATP

We have seen thus far that the flow of electrons from reducing substrates to O_2 causes protons to move out of the matrix space, setting up a pH gradient and an electrical potential difference across the mitochondrial inner membrane. The chemiosmotic theory postulates that _protons move back into the matrix via an ATP-synthase, driving the formation of ATP_. Evidence for this is that an electrochemical potential gradient for protons can support the formation of ATP in the absence of any electron-transfer reactions. A transient pH gradient that will tend to pull protons into the matrix can be set up by first incubating mitochondria at pH 9, so that the inside becomes alkaline, and then suddenly lowering the pH of the suspension medium to 7 (Figure 16–30). By using K^+ and valinomycin, it also is possible to set up an electrical potential gradient that adds to the force pulling protons into the matrix. The mitochondria first are loaded with K^+ in the absence of valinomycin, placed in a medium without K^+, and then suddenly treated with valinomycin. The valinomycin allows K^+ to flow rapidly out of the mitochondria, leaving a net negative charge inside. The generation of ATP in experiments of this sort was first achieved with chloroplast membranes by Andre Jagendorf (Chapter 17). Similar results subsequently were obtained by other investigators studying vesicles made from the mitochondrial inner membrane.

Additional evidence that proton movements are coupled to ATP synthe-

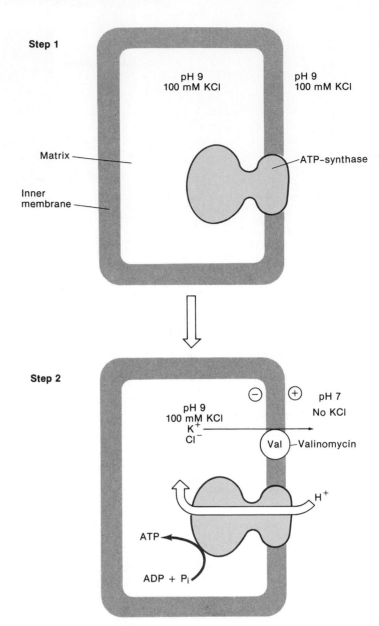

Step 1

Step 2

Figure 16–30

An electrochemical potential gradient for protons can be set up across the mitochondrial inner membrane in the absence of electron-transfer reactions. In step 1, mitochondria are incubated at high pH in the presence of KCl to reduce the proton concentration inside, and to load the inside with K^+ and Cl^-. In step 2, the proton concentration outside the mitochondria is increased (from pH 9 to pH 7), and the KCl concentration is decreased. The ionophore valinomycin (val) is added so that K^+ can come out. Because the counter ion Cl^- cannot move rapidly across the membrane, the efflux of K^+ down its concentration gradient leaves the inside of the mitochondrion with a net negative charge, relative to the outside. This difference in electrical potential, together with the higher pH inside, favors the influx of protons. Protons moving inward through the ATP-synthase can drive the formation of ATP.

sis comes from the observation that the hydrolysis of ATP that occurs when the ATP-synthase runs backwards results in the movement of protons *out* of the mitochondria. Remember that oxidative phosphorylation is reversible (the ATP-synthase also can act as an ATPase), and that the breakdown of ATP apparently generates the same $\sim$ that is created by respiration. In experiments analogous to those shown in Figure 16–27, the hydrolysis of ATP by mitochondria results in the extrusion of protons and the uptake of lipophilic cations. These processes are blocked by oligomycin, the inhibitor of the ATP-synthase/ATPase. (As one would predict, oligomycin does not block proton extrusion when this is driven by electron transport.)

Current estimates are that three protons move into the matrix through the ATP-synthase for each ATP that is synthesized. We shall see below that one additional proton moves into the mitochondrion in connection with the uptake of ADP and P_i and the export of ATP, giving a total of four protons per ATP. How large a free energy change is associated with moving these protons? Rewriting Equation (11) so that it applies to the movement of one mole of protons, we have

$$\Delta G_{H^+} = G_{H^+(in)} - G_{H^+(out)} = RT \ln \{[H^+]_{in}/[H^+]_{out}\} + F \Delta\psi$$
$$= -2.3 \; RT \; \Delta pH + F \; \Delta\psi$$

(12)

where $\Delta pH = pH_{in} - pH_{out}$. If the pH in the matrix space is 0.05 pH units above that in the cytoplasm, the term $-2.3 \; RT \; \Delta pH$ amounts to -0.07 kcal/mole. If the electrical potential difference across the membrane is -0.15 V, $F \Delta\psi$ is -3.46 kcal/mole. This gives a total ΔG_{H^+} of -3.53 kcal/mole, with the term $F \Delta\psi$ making the dominant contribution. If four protons move across the membrane for each molecule of ATP that is synthesized, the free energy change associated with the proton translocation would be $4 \times (-3.53)$, or -14.1 kcal per mole of ATP, which is about twice as large as the $\Delta G^{\circ\prime}$ of ATP hydrolysis $(-7.5$ kcal/mole). The free energy decrease associated with proton movement thus would be large enough to account for the high [ATP]/[ADP][P$_i$] ratio that is maintained under physiological conditions. The thermodynamic driving force for proton translocation is frequently termed the _proton motive force_ Δp, and is expressed quantitatively as $\Delta G_{H^+}/F$ in units of volts.

Deferring for the moment the question of exactly how the ATP-synthase is able to make ATP at the expense of the proton motive force, let us consider how the P/O ratios that we discussed above are related to the stoichiometry of proton movements. When mitochondria respire and form ATP at a constant rate, protons must be moving inward at a rate that just balances the rate at which protons are pumped out by the electron-transport reactions. Suppose that twelve protons are pumped outward for each pair of electrons that traverse the respiratory chain from NADH to O_2, and that four protons move back in for each ATP that is synthesized. Since the rates of proton efflux and influx must balance, three molecules of ATP (12/4) must be formed for each pair of electrons that go to O_2. The P/O ratio thus is given by the ratio of the proton stoichiometries. If the oxidation of succinate drives the extrusion of eight protons per pair of electrons, the P/O ratio for this substrate would be 8/4, or 2.0. These ratios agree with the maximum P/O ratios that have been associated with the two substrates. Note, however, that in the chemiosmotic theory the P/O ratio does not necessarily have to have an integer value. For example, if protons can leak across the membrane without passing through the ATP synthase, the P/O ratios will be lower than these maximal values.

The Proton-conducting ATP-synthase or ATPase: F_1 and F_o

Let us now turn to the ATP-synthase. What is its structure, and how is it able to use a proton motive force to drive the formation of ATP? Studies of the ATP-synthase began in the early 1960s, when Efraim Racker and his colleagues identified several different enzymatic components that appeared to participate in oxidative phosphorylation. The experimental strategy was to disrupt mitochondria and to fractionate the proteins that were released from the membranes. In some cases, the depleted membranes lost their ability to synthesize ATP, although the electron transfer reactions were undisturbed. Recombining the solubilized proteins, or _coupling factors_, with the membranes restored the capacity for oxidative phosphorylation.

The first coupling factor to be purified, F_1, was found to be an active ATPase when it was removed from the mitochondria. Unlike the membrane-bound ATPase of intact mitochondria, the solubilized F_1 ATPase was not inhibited by oligomycin. However, it recovered its sensitivity to oligomycin if it was reattached to the membranes in combination with another coupling factor, F_o. (The subscript o stood for oligomycin.)

F_o and F_1 both turned out to be parts of the proton-conducting ATPase or ATP-synthase. Together they form a multiprotein complex that, like the elec-

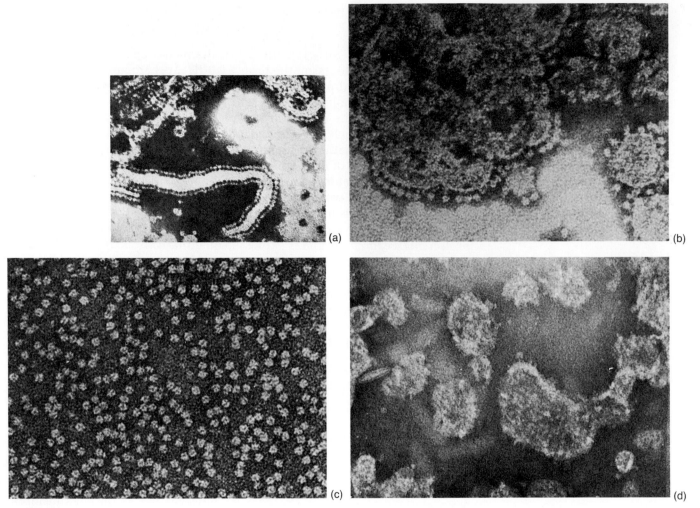

Figure 16–31

(a) Negatively stained electron micrograph of part of a mitochondrion, showing F$_1$ spheres lining the inner membrane. The worm-shaped white area is probably the lumen of a crista; the large dark areas are part of the matrix. The F$_1$ spheres show up as white spots, projecting from the membrane of the crista into the matrix. (Reproduced by permission from B. Chance and D. Parsons, *Science* 142:1176, 1963.) (b) Negatively stained electron micrograph of submitochondrial particles (membrane vesicles). These particles are capable of oxidative phosphorylation. F$_1$ spheres line the outer surfaces of the membranes; they can be seen best around the edges of the flattened vesicles. (c) Electron micrograph of the isolated F$_1$ coupling factor. The isolated F$_1$ is an active ATPase. (d) Submitochondrial particles that have been depleted of F$_1$ by treatment with urea and the proteolytic enzyme trypsin. These membranes contain a functional respiratory chain, but are incapable of making ATP. Note that the surfaces of the membranes appear smooth. (Micrographs B, C, and D courtesy Dr. E. Racker.)

tron-transfer complexes, is partially embedded in the mitochondrial inner membrane. F$_1$ includes five different polypeptide subunits (α, β, γ, δ, and ϵ) with the stoichiometry $\alpha_3\beta_3\gamma\delta\epsilon$. Its total molecular weight is about 360,000. In electron micrographs of mitochondria, the F$_1$ ATPase complexes can be seen as spheres with diameters of about 85 Å, protruding from the surface of the inner membrane (Figure 16–31). F$_o$, a complex of about five hydrophobic polypeptides, is an integral component of the inner membrane and acts as a base-piece and stalk that holds F$_1$ to the membrane (Figure 16–32). Complexes similar to the mitochondrial F$_o$ and F$_1$ have been isolated from the chloroplast thylakoid membrane (Chapter 17) and from several species of bacteria.

The F$_1$-F$_o$ ATPase complex is a useful marker for distinguishing the two surfaces of the mitochondrial inner membrane. In intact mitochondria, the spheres are on the side of the membrane that faces the matrix (Figure 16–

Figure 16–32
Components of the proton-conducting ATP-synthase. The F_1 head-piece includes three α and three β subunits and one copy each of three other subunits (γ, δ, and ϵ). F_o includes a cluster of about six copies of a small peptide, which appears to form a transmembrane channel for protons, and several additional subunits of unknown structure and function.

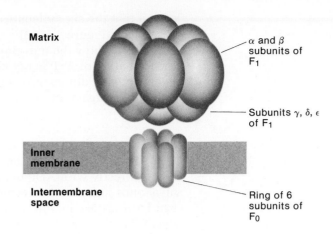

Matrix

α and β subunits of F_1

Subunits γ, δ, ϵ of F_1

Inner membrane

Intermembrane space

Ring of 6 subunits of F_0

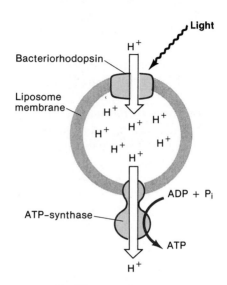

Figure 16–33
ATP synthesis can be obtained in an artificial system in which the mitochondrial ATP-synthase is incorporated into liposomes together with bacteriorhodopsin, a membrane protein obtained from *Halobacterium halobium*. When bacteriorhodopsin is excited with light, it pumps protons across the membrane (see Chapters 34 and 36). The solution inside the liposome thus becomes acidic and positively charged relative to the external solution. Protons can move back out through the ATP-synthase, and this movement is coupled to the formation of ATP. The formation of ATP is blocked by uncouplers or oligomycin. Note that the F_1 head-piece of the ATP-synthase, which is found on the inner surface of the mitochondrial inner membrane (Figures 16–26, 16–30, 16–31, and 16–32), is on the outside of the liposome in this artificial system. The ATP-synthase also faces outward in submitochondrial particles made by breaking the inner membrane with sonic oscillations. In all cases, ATP synthesis is associated with the movement of protons through F_o in the direction toward F_1. The hydrolysis of ATP results in proton movements in the opposite direction.

31a). In preparations of submitochondrial particles, made by breaking mitochondria by sonic oscillations, the ATPase is on the outside (Figure 16–31b). When the mitochondria are broken, the sacks formed by the cristae evidently pinch off at the edges and seal to form vesicles so that the side of the membrane that originally faced the matrix becomes the outer surface of the vesicle. You can see how this might occur by examining Figures 16–1 and 16–2. The idea that the matrix side of the membrane faces outward in the isolated vesicles is consistent with observations on the orientations of the electron carriers. NADH dehydrogenase, for example, faces the matrix in intact mitochondria but reacts with added NADH in submitochondrial particles; cytochrome c attaches to the cytoplasmic side of the membrane in mitochondria but to the inner surface in the particles. Further, the electron-transport chain pumps protons out of intact mitochondria, but *into* the isolated vesicles.

Mechanism of Action of the ATP-synthase

If the purified F_1-F_o complex is incorporated into a liposome phospholipid membrane and an electrochemical potential gradient for protons is set up across the membrane, the complex can synthesize ATP. To show this, Racker and Stoeckenius generated the proton gradient by incorporating a light-driven proton pump, bacteriorhodopsin from *Halobacterium halobium* (Chapters 6 and 34), into membrane vesicles (liposomes) along with the F_1-F_o complex. When the liposomes were illuminated, ATP was formed (Figure 16–33). This experiment provided a dramatic demonstration that *the ATP-synthase does not have to be attached directly to the electron-transport complexes* of the respiratory chain, for no electron carriers were included in the liposomes. It added strong support to the idea that the coupling of phosphorylation to electron transport is mediated by a proton motive force that can be generated at one point on the membrane and used at relatively distant points.

One of the subunits of F_o appears to constitute the proton-conducting channel of the ATPase. If this subunit is incorporated into liposomes, it makes the membranes leaky specifically to protons. The proton leaks can be blocked by the phosphorylation inhibitors oligomycin or dicyclohexylcarbodiimide (DCCD). Oligomycin reacts reversibly and noncovalently with the proton-conducting subunit; DCCD reacts irreversibly to form a covalent bond with a critical aspartate residue. The subunit that appears to form the channel is a small, hydrophobic polypeptide ($M_r \approx 8000$) that is present in approximately six copies in the F_1-F_o complex. Analysis of the amino acid sequence suggests that each of the subunit's polypeptides is folded into a hairpinlike structure with two transmembrane α helices. It has been suggested that the six molecules form a ring embedded in the membrane at the base of F_1 (Figure 16–32).

Exactly how the movement of protons through the ATP-synthase drives the formation of ATP is not known. However, studies by Paul Boyer, Harvey Penefsky, and others have shown that the enzyme binds ADP and ATP very tightly. The reaction

$$ADP + P_i \longrightarrow ATP + H_2O$$

occurs between the bound nucleotides on the enzyme, even in the absence of a proton motive force or any other external source of energy. P_i evidently reacts directly with ADP to give ATP in a single step, without the formation of a phosphorylated enzyme intermediate. (The proton-conducting ATP-synthase differs in this regard from the Na^+-K^+ and Ca^{2+} ATPases that move Na^+, K^+, and Ca^{2+} across the cellular plasma membrane. As discussed in Chapter 34, the catalytic mechanisms of the Na^+-K^+ and Ca^{2+} ATPases involve aspartyl-phosphate intermediates.) Remarkably, the equilibrium constant for the formation of ATP is close to 1 when the nucleotides are bound to the enzyme. When the nucleotides are free in solution, the equilibrium constant is about 10^{-5}.

Soluble enzymes sometimes display a substantial difference in the equilibrium constant of a reaction that forms ATP, depending on whether the reactants and products are bound to the enzyme or are free in solution. With both pyruvate kinase and 3-phosphoglycerate kinase, the equilibrium constant for the formation of bound ATP on the enzyme is much more favorable than the equilibrium constant for the reaction of the free materials. This difference could reflect the fact that the $\Delta G°'$ for ATP formation in solution includes an entropy decrease associated with the ordering of water molecules around the polyphosphate chain. Such an entropy decrease can be much smaller if the ADP, P_i, and ATP are bound on an enzyme.

The proton motive force apparently does not have much effect on the equilibrium constant for the formation of ATP on the ATP synthase. Instead, it alters the binding and release of ADP, P_i, and ATP from the enzyme. The nucleotides evidently are bound in an environment that favors the formation of ATP, and a movement of protons in response to the proton motive force can change the enzyme's conformation in such a way that the ATP is released.

Although there is much evidence that movement of protons through the ATP-synthase can drive the formation of ATP, changes in the [ATP]/[ADP][P_i] ratio have been measured in mitochondria under conditions when there is little change in the proton motive force. One way to explain such observations is to postulate that some of the protons that are pumped by the electron-transport reactions are not released into the aqueous solution at the surface of the membrane, but rather are conducted more directly to the ATP-synthase. The proton motive force that is measured between the solutions on the two sides of the membrane thus might reflect only a part of the driving force that is expressed at the ATP-synthase.

TRANSPORT OF SUBSTRATES, P_i, ADP, AND ATP INTO AND OUT OF MITOCHONDRIA

One of the principles underlying the chemiosmotic theory is that the mitochondrial inner membrane is basically impermeable to charged and highly polar molecules. As we have discussed, NADH, NAD^+, H_3O^+, and K^+ (in the absence of valinomycin) cannot pass freely through the phospholipid bilayer of the inner membrane, but protons do move through special channels associated with the ATP-synthase. This raises the question of how P_i, adenine

BOX 16–A

Natural Uncouplers: Thermogenesis in Brown Fat and Skunk Cabbage

Several unusual types of cells use electron transport that is uncoupled from phosphorylation. *In these cells the respiratory chain operates for thermogenesis, the generation of heat, rather than the formation of ATP.* Newborn mammals, and also the adults of some mammalian species that are adapted to live in cold climates, do this in a specialized adipose tissue called *brown fat.* The brown color is due largely to the cytochromes in the tissue's abundant mitochondria. Heat produced by brown fat mitochondria is important for maintaining body temperature in the newborn, and for the arousal of hibernating animals. Brown fat generates heat at a rate of about 4ᴐᴐ W/kg, far above the rate of about 1 W/kg that is typical of other resting mammalian tissues.

The inner membrane of brown fat mitochondria contains a protein called thermogenin, which acts as a channel for anions. The protein allows OH^- or Cl^- ions to pass rapidly across the membrane. Since a movement of OH^- from the mitochondrial matrix to the cytoplasm has the same consequences as a movement of H^+ in the opposite direction, thermogenin acts as an uncoupler. The free energy that is released in the electron-transfer reactions is stored transiently as an electrochemical potential gradient for protons, but then is degraded largely to heat.

The uncoupling activity of thermogenin is regulated by ATP, ADP, GTP, and GDP, which bind tightly to the protein and inhibit anion transport. Very low concentrations of fatty acids increase anion permeability. Fatty acids released from triacylglycerol stores in the tissue could be important in the hormonal control of thermogenesis. The amount of thermogenin in the brown fat mitochondria also changes in response to physiological conditions. In animals that live at low temperatures, it can represent as much as 15 per cent of the protein in the inner membrane.

The skunk cabbage and the arum lily flower use a different strategy for thermogenesis. They oxidize UQH_2 by an alternative pathway that bypasses the cytochrome bc_1 complex and cytochrome oxidase. Electron transport by this pathway does not result in proton translocation. Heat produced by respiration may help the plant to thrive early in Spring when the climate is cool.

nucleotides, and substrates such as pyruvate and citrate move into and out of mitochondria. The answer is that the inner membrane has a set of transport systems that specifically catalyze the movements of these materials.

Uptake of P_i and Oxidizable Substrates Is Coupled to Release of Other Compounds

The transport of materials across biological membranes is discussed in detail in Chapter 34, where we develop the idea that the movement of one substance across a membrane can be coupled to the movement of another substance in the same or the opposite direction. The uptake of β-galactosides by some bacteria, for example, is linked to the movement of protons into the cell. Like the proton-conducting ATP-synthase, transport systems in the cell membrane are able to harness the free energy decrease associated with moving protons down an electrochemical potential gradient. They use this free energy to move sugars and other nutrients thermodynamically uphill, against the concentration gradient of the nutrient. This concept was first advanced by Mitchell as an extension of the chemiosmotic theory, and it has turned out to apply to the movements of many materials into and out of mitochondria.

The uptake of P_i by mitochondria is coupled to an outward movement of OH^-. If the phosphate moves in the form of $H_2PO_4^-$, carrying one negative charge, an exchange of P_i for OH^- will be electrically neutral: there is no net movement of charge in either direction. The transport of P_i thus will be relatively insensitive to the electrical potential gradient $\Delta\psi$ that the respiratory chain sets up across the inner membrane. It will, however, be sensitive to the pH gradient. An outward movement of OH^- is essentially equivalent to an inward movement of H^+, and is thermodynamically downhill in respiring mitochondria because the pH is higher in the matrix than in the cyto-

Figure 16-34

(a) The phosphate/hydroxide exchange protein carries a phosphate ion $(H_2PO_4^-)$ in one direction across the mitochondrial inner membrane in exchange for a hydroxide ion moving in the opposite direction. These movements are reversible, but a net efflux of OH^- is thermodynamically downhill because the respiratory chain pumps protons out of the matrix and raises the pH there. Because OH^- efflux is linked to $H_2PO_4^-$ uptake, phosphate can be concentrated in the matrix. *(b)* The phosphate/dicarboxylic acid exchange protein exchanges succinate (or other dicarboxylic acids) for phosphate (HPO_4^{2-}). Phosphate efflux is thermodynamically favorable because the phosphate concentration in the matrix exceeds that in the cytoplasm, as a result of the action of the phosphate/hydroxide exchange protein. *(c)* The ADP/ATP exchange protein exchanges ADP^{3-} for ATP^{4-} An outward movement of ATP removes one negative charge from the matrix, and is favored because proton pumping by the respiratory chain gives the matrix a negative charge relative to the cytoplasm.

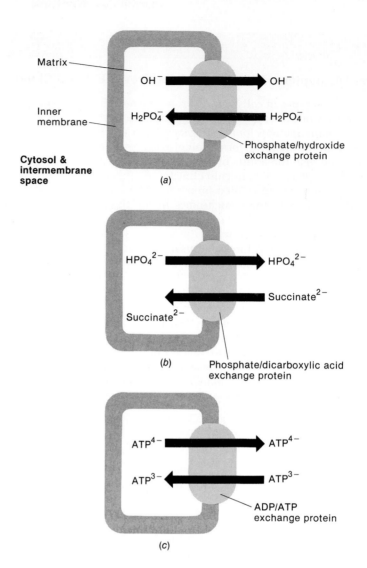

plasm (Figure 16-34a). Mitochondria thus are capable of taking up P_i from the cytoplasm even when the concentration of P_i inside exceeds that outside.

Pyruvate, which is generated in the cytoplasm by glycolysis but consumed in the mitochondria by the TCA cycle, also is transported into the mitochondria in exchange for OH^-. Succinate and malate are taken up by a transport system that can exchange either of these dicarboxylic acids for P_i (Figure 16-34b). The same transport system also can exchange one of the dicarboxylic acids for another. A separate system exchanges the two tricarboxylic acids citrate and isocitrate for each other or for a dicarboxylic acid. Mitochondria thus can achieve the uptake of any of these substrates by a series of exchanges of one carboxylic acid for another, of an acid for P_i, and ultimately of P_i for OH^-.

ATP Export Is Coupled to ADP Uptake

ATP, which is produced in the mitochondria largely for use elsewhere in the cell, is exported by a transport system that exchanges ATP for ADP. This exchange is not electrically neutral. At pH 7, the charge on ATP is approximately -4, whereas that on ADP is about -3. The outward movement of a molecule of ATP in exchange for the inward movement of an ADP removes one negative charge from the matrix, and is driven by the electrical gradient, $\Delta\psi$ (Figure 16-34c). The exchange is said to be *electrogenic*. The ATP/ADP

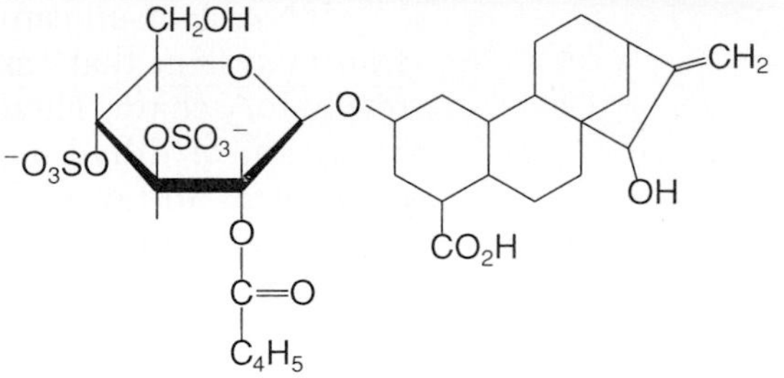

Atractyloside

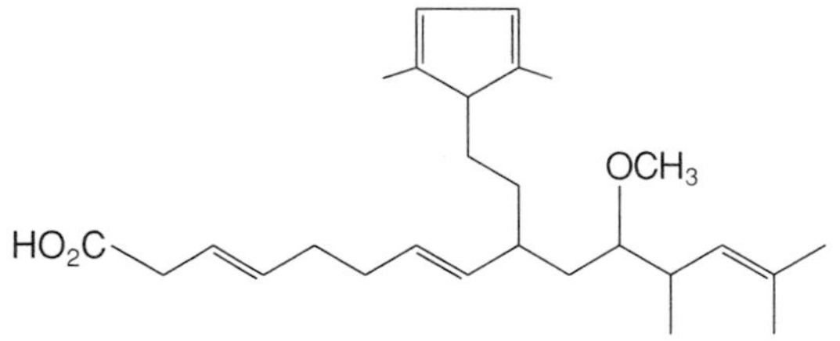

Bongkrekic acid

Figure 16–35
Structures of two inhibitors of the ATP/ADP exchange protein. Atractyloside is isolated from a species of thistle; bongkrekic acid, from a fungus found in decaying coconuts (*bongkrek* in Indonesian). As discussed in Chapter 34, the ATP/ADP exchange protein has two distinctly different conformational states. In one of these states it binds ADP and atractyloside; in the other, it binds ATP and bongkrekic acid.

exchange protein, or *adenine nucleotide translocator*, is actually the most abundant protein in the mitochondrial inner membrane. It is sensitive to several specific inhibitors, atractyloside, and bongkrekic acid (Figure 16–35), and can be isolated with bound derivatives of atractyloside. Its mechanism of operation is discussed in Chapter 34.

The combined effect of exchanging extramitochondrial ADP^{3-} and $H_2PO_4^-$ for mitochondrial ATP^{4-} and OH^- is to move one proton into the mitochondrial matrix for every molecule of ATP that the mitochondria release into the cytoplasm. This proton translocation must be considered, along with the movement of protons through the ATP synthase, in order to account for the P/O ratio of oxidative phosphorylation. If three protons pass through the ATP synthase, and the adenine nucleotide and P_i transport systems move one additional proton, then in total, four protons move into the matrix for each ATP that is provided to the cytoplasm. As explained above, the translocation of four protons per ATP is consistent with the relationship between the proton motive force and the measured ΔG for ATP formation under physiological conditions.

Reducing Equivalents from Cytoplasmic NADH Are Imported by Shuttle Systems

The mitochondrial inner membrane does not contain a transport system for NAD^+ or NADH. In animal cells, most of the NADH that must be oxidized by the respiratory chain is generated in the mitochondrial matrix by the TCA cycle or by the oxidation of fatty acids. However, NADH also can be generated by glycolysis in the cytoplasm, and this NADH must be reoxidized to NAD^+ in some manner. If O_2 is available, it clearly is advantageous to reoxidize the NADH by the respiratory chain, rather than by the formation of lactate or ethanol. As we have seen, approximately three molecules of ATP can be formed for each NADH that is oxidized in the mitochondria, whereas no ATP is made when NADH is oxidized by the cytoplasmic lactic dehydrogenase or alcohol dehydrogenase. That the NADH formed in the cytoplasm *is* oxidized by the respiratory chain can be seen from the observation that when muscle cells are stimulated to do work under anaerobic conditions they accumulate lactate, but under aerobic conditions they do not.

If NADH itself cannot enter the mitochondria, animal cells must have a *shuttle system* that can transfer electrons from cytoplasmic NADH to the respiratory chain. There are several such systems. The simplest involves the reduction of dihydroxyacetone phosphate to glycerol-3-phosphate in the cytoplasm, followed by the reoxidation of the glycerol-3-phosphate to dihydroxyacetone phosphate by the mitochondrial glycerol-3-phosphate dehydrogenase:

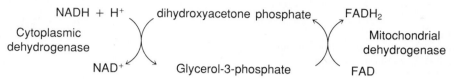

As we noted above, the catalytic site of the mitochondrial glycerol-3-phosphate dehydrogenase is on the cytoplasmic surface of the inner membrane, so the glycerol-3-phosphate does not have to pass through this membrane in order to be reoxidized. It does have to go through the mitochondrial outer membrane, but the outer membrane contains relatively large pores that allow molecules of this size to move through easily.

The dihydroxyacetone phosphate/glycerol-3-phosphate shuttle has the shortcoming that oxidation of glycerol-3-phosphate by the respiratory chain generates only two ATPs, instead of the three that can be generated by the oxidation of mitochondrial NADH (Figure 16–20). The mitochondrial glycerol-3-phosphate dehydrogenase feeds electrons to the respiratory chain at the level of UQ, below the coupling site in complex I. There is another, more complicated shuttle system that allows the electrons to pass through complex I. In this shuttle, cytoplasmic NADH reduces oxaloacetate to malate, which is carried across the inner membrane by a specific transporter. Inside the mitochondria, the malate is reoxidized to oxalacetate, reducing mitochondrial NAD^+ to NADH:

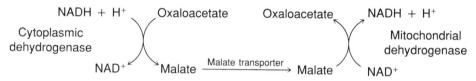

The internal NADH then can be oxidized by the respiratory chain, with the formation of three molecules of ATP.

Working in the opposite direction, these reactions play an important role in gluconeogenesis (Chapter 14). By themselves, they do not constitute a complete shuttle for the continuing oxidation of cytoplasmic NADH, because oxaloacetate is unable to move back out of the mitochondria; the inner membrane has no transporter for it. However, oxaloacetate can undergo transamination with the amino acid glutamate, to form aspartate and α-ketoglutarate, and the mitochondrial inner membrane is equipped to transport all three of these metabolites. The shuttle can be completed by coupling the inward movement of malate to the outward movement of α-ketoglutarate, and coupling the inward movement of glutamate to the outward movement of aspartate.

Complete Oxidation of Glucose Yields 36 to 38 ATPs

The shuttle systems that operate between the cytoplasm and mitochondria must be taken into account when we calculate the total yield of ATP from a molecule of glucose that is oxidized to CO_2 and H_2O. To make such a calculation, recall first that two molecules of ATP are formed in glycolysis, and two more in the TCA cycle. At the same time, two molecules of NADH are produced in the cytoplasm by glycolysis, and eight molecules of NADH are generated in the mitochondrial matrix by the pyruvate dehydrogenase complex and the TCA cycle. The two molecules of succinate proceeding through the succinate dehydrogenase step of the TCA cycle reduce two molecules of FAD to $FADH_2$. Reoxidation of the eight mitochondrial NADH and two $FADH_2$ by the respiratory chain can generate approximately 28 molecules of ATP (three for each NADH and two for each $FADH_2$). The two molecules of NADH that are formed in the cytoplasm can contribute their reducing equivalents to the mitochondria in the form of either malate or glycerol-3-phosphate. Operation of the glycerol-3-phosphate shuttle would provide two pairs of electrons at the level of $FADH_2$, allowing the formation of four molecules of ATP. The malate shuttle allows electrons to pass through complex I also, leading to the formation of six molecules of ATP. Thus, depending on which shuttle system is in operation, a total of either 32 or 34 molecules of ATP can be formed by oxidative phosphorylation in addition to the four molecules that are formed in glycolysis and the TCA cycle. This gives a grand total of either 36 or 38 ATP.

The actual yields of ATP under physiological conditions are less than these limiting values. As we noted above, the measured P/O ratio for NADH oxidation generally is closer to 2.5 than 3.0, and that for succinate is closer to 1.5 than 2.0. Some of the free energy that is stored transiently in the form of an electrochemical potential gradient for protons may be lost as a result of nonspecific leakage of protons or other ions through the membrane, or may be used to drive reactions other than the formation of ATP. In addition, NADH that is formed in the cytoplasm may be used there for reductive biosynthetic reactions such as the formation of fatty acids, rather than contributing electrons to the respiratory chain. Even so,.it is clear that respiration allows cells to make ATP in amounts that are far greater than the amounts provided by glycolysis.

ELECTRON TRANSPORT AND ATP SYNTHESIS IN BACTERIA

Electron-transport systems are found in the cytoplasmic membranes of numerous species of aerobic bacteria. Electron flow from reducing substrates to O_2 is coupled to transmembrane proton movements in these species, just as it is in the mitochondrial respiratory chain. Protons are pumped out of the bacterial cell, leaving the cytoplasm alkaline and with a net negative charge relative to the external solution. The cell membrane also contains an ATP-synthase/ATPase similar to the mitochondrial ATP-synthase, and protons moving back into the cell through the ATPase drive the formation of ATP.

The structure of the respiratory chain of *Escherichia coli* has been studied in some detail. Like the mitochondrial inner membrane, the cytoplasmic membrane of *E. coli* has a variety of flavoprotein dehydrogenases that transfer electrons to ubiquinone. However, the bacterial electron transport chain branches into two paths at UQH_2. In place of cytochrome c and complexes III and IV, there are two different multienzyme complexes, either of which can catalyze electron transfer from UQH_2 all the way to O_2. The *cytochrome d complex* consists of two polypeptides with four heme groups (cytochrome b_{558}, cytochrome b_{595}, and two copies of the unusual cytochrome d); it has no copper atoms or iron-sulfur centers. The alternative complex, the *cytochrome o complex*, also contains several hemes. Electron flow through either of these complexes is coupled to proton translocation.

A proton-translocating ATPase also is found in many species of anaerobic bacteria that do not have electron-transport systems. The role of the

ATPase in these species is to pump protons out of the cell, at the expense of ATP made by fermentative pathways such as glycolysis. The proton efflux creates an electrochemical potential gradient, which the bacteria use to drive the uptake of sugars and other nutrients. The enzymes that catalyze this transport of nutrients are discussed further in Chapter 34.

SUMMARY

In eukaryotes, the TCA cycle and most of the associated reactions of aerobic energy metabolism occur in mitochondria. The mitochondrion has two separate membranes. The inner membrane separates the mitochondrion into two distinct spaces, the internal matrix space and the intermembrane space. All but one of the enzymes of the TCA cycle are in the matrix space. (The exception, succinate dehydrogenase, is in the inner membrane.) In the catabolic reactions of the TCA cycle, electrons are passed from succinate, isocitrate, and α-ketoglutarate to NAD and FAD. An electron-transport system in the mitochondrial inner membrane reoxidizes the reduced coenzymes (NADH and $FADH_2$) at the expense of molecular oxygen. The flow of electrons from NADH and $FADH_2$ to O_2 releases the free energy that powers the synthesis of most of the ATP that is formed during aerobic metabolism.

Electron transfer to O_2 occurs in a stepwise manner, by means of a series of electron carriers that include flavoproteins, cytochromes (heme-proteins), iron-sulfur proteins, quinones, and Cu atoms. Most of the electron carriers are collected in four large complexes, which communicate via two mobile carriers, ubiquinone (UQ) and cytochrome c. Complex I transfers electrons from NADH to UQ, and Complex II transfers electrons from succinate to UQ. Both of these complexes contain flavins and numerous iron-sulfur centers. Complex III, which contains three cytochromes (cytochromes b_L, b_H, and c_1) and one iron-sulfur protein, passes electrons from reduced ubiquinone (UQH_2) to cytochrome c. Complex IV contains two cytochromes (a and a_3) and two Cu atoms, and passes electrons from cytochrome c to O_2. The transfer of electrons from UQH_2 to cyto-

chrome c by Complex III involves a cyclic series of reactions (a Q cycle), in which UQH_2 and UQ undergo oxidation and reduction at two distinct sites.

As electrons move through complexes I, III, and IV toward O_2, protons are taken up from the solution on the matrix side of the membrane and are released on the cytoplasmic side. This raises the pH of the matrix slightly above that of the cytoplasm and leaves the matrix negatively charged relative to the cytoplasm, creating an electrochemical potential difference or "proton motive force" that tends to pull protons from the cytoplasm back into the matrix. Proton movements through the F_o base-piece of the ATP-synthase in the inner membrane drive the formation of ATP by causing the release of bound ATP from the catalytic site on the F_1 head-piece of the enzyme.

Respiration is normally tightly coupled to the formation of ATP. Approximately three molecules of ATP are synthesized for each pair of electrons that pass down the electron transport chain from NADH to O_2. Uncouplers, which are lipophilic weak acids, can dissipate the proton motive force by carrying protons across the membrane. Respiration then occurs rapidly even in the absence of the phosphorylation reactions.

The proton motive force also drives the uptake of P_i and ADP into the mitochondrial matrix, and the export of ATP to the cytoplasm. By exchanging P_i for an organic acid, and exchanging one organic acid for another, mitochondria can concentrate pyruvate and other organic acids, thus obtaining the substrates that provide electrons to the electron transport chain.

SELECTED READINGS

Chance, B., and G. R. Williams, Respiratory enzymes in oxidative phosphorylation II. Difference spectra. *J. Biol Chem.* 217:395, 1955. One of a series of papers in which Chance developed kinetic techniques for elucidating the sequence of electron carriers in the respiratory chain and the sites of action of inhibitors.

Ernster, L. (ed.), *Bioenergetics (New Comprehensive Biochemistry, Vol. 9).* Amsterdam: Elsevier, 1984. A collection of authoritative reviews covering electron transport, the ATP-synthase, translocation of ions across the mitochondrial inner membrane, thermogenesis in brown fat, and other topics in bioenergetics.

Mitchell, P., and J. Moyle, Stoichiometry of proton translocation through the respiratory chain and adenosine triphosphatase systems of rat liver mitochondria. *Nature*

208:147, 1965. Evidence that electron transport drives the movement of protons outward across the mitochondrial inner membrane, and that ATP hydrolysis does the same. These findings led to the chemiosmotic theory.

Penefsky, H. S., M. E. Pullman, A. Datta, and E. Racker, Partial resolution of the enzymes catalyzing oxidative phosphorylation II. Participation of a soluble adenosine triphosphatase in oxidative phosphorylation. *J. Biol. Chem.* 235:3330, 1960. The discovery that the water-soluble ATPase (F_1) is an essential component of the ATP-synthase reaction.

Reynafarje, B., and A. L. Lehninger, The K^+/site and H^+/site stoichiometry of mitochondrial electron transport. *J. Biol. Chem.* 253:6331, 1978. Valinomycin is used to allow K^+ to move inward across the inner membrane, in response to the electrical potential difference that is created by H^+ efflux. The number of protons that are pumped out for each pair of electrons passing down the respiratory chain is found to be larger than previously estimated.

Sudha, B. P., S. N. Dixit, and J. Vanderkooi, Probing structure and motion of the mitochondrial cytochromes. *Curr. Top. Bioenerg.* 13:159, 1984. A review of the structures and interactions of cytochromes, and of the diffusion of electron carriers in the inner membrane.

Trumpower, B., Function of the iron-sulfur proteins of the cytochrome bc_1 segment in electron transfer and energy-conserving reactions of the mitochondrial respiratory chain. *Biochim. Biophys. Acta* 639:129, 1981. A review of the structure and operation of Complex III and the Q cycle.

PROBLEMS

1. The respiratory chain includes cytochromes, iron-sulfur proteins, flavoproteins, and quinones. Compare these different types of electron carriers with regard to the following.
   a. The nature of the prosthetic group or functional group that undergoes oxidation and reduction.
   b. The number of electrons taken up when the functional group is converted from the fully oxidized form to the fully reduced form.
   c. The number of protons taken up per electron.
   d. The formation of semiquinone intermediates.

2. Calculate the change in standard free energy ($\Delta G^{\circ\prime}$) for the following reactions at pH 7.0. (In this and the following two problems, you may refer to Table 16–1 for the standard reduction potentials.)
   a. cytochrome c (Fe^{2+}) + cytochrome a_3 (Fe^{3+}) $\longrightarrow$
          cytochrome c (Fe^{3+}) + cytochrome a_3 (Fe^{2+})
   b. 4 cytochrome c (Fe^{2+}) + O_2 + $4H^+$ $\longrightarrow$
          4 cytochrome c (Fe^{3+}) + $2H_2O$

3. Suppose a solution of cytochrome c is held at a redox potential of +0.30 V and pH 7.0. What per cent of the cytochrome will be in the oxidized form?

4. Given the standard reduction potentials of cytochrome c and ubiquinone at pH 7.0 (Table 16–1), what would the corresponding values be at pH 6.0 and 8.0?

5. Chemical modification of one or more of the lysine residues near the edge of the heme in cytochrome c decreases the rate at which the cytochrome is reduced by the cytochrome bc_1 complex (Complex III). Modification of the same lysines also decreases the rate at which the reduced cytochrome is oxidized by cytochrome oxidase (Complex IV). What does this suggest concerning the operation of the respiratory chain?

6. A preparation containing complexes II and III (succinate dehydrogenase and the cytochrome bc_1 reductase) was isolated from the mitochondrial inner membrane. The Rieske iron-sulfur protein was extracted from the preparation. Difference spectrum A shows the absorbance changes that resulted from adding succinate to the extracted preparation in the presence of antimycin. If the iron-sulfur protein was added back to the preparation, succinate then caused much larger absorbance changes, as shown in difference spectrum B. (Antimycin was added before the succinate in experiment B as well as in A.) In a parallel set of experiments, succi-

A

Fe–S protein removed, antimycin present

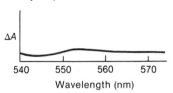

B

Fe–S protein restored, antimycin present

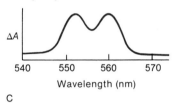

C

Fe–S protein removed, no antimycin

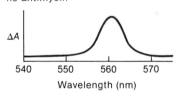

D

Fe–S protein restored, no antimycin

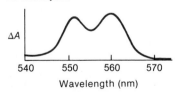

nate was added to the extracted and reconstituted preparations in the absence of antimycin. Spectrum C shows the absorbance changes caused by adding succinate to the extracted preparation in the absence of antimycin. Spectrum D shows the effect of succinate on the reconstituted preparation in the absence of antimycin. How would you interpret these results? (Hints: Remember that b- and c-type cytochromes can be distinguished by the positions of their α absorption bands, and that the electron-transfer reactions shown in Figure 16–18 are reversible.)

7. Rotenone is an extremely potent insecticide and fish poison. At the molecular level, its mode of action is to block electron transport from the FMN of NADH dehydrogenase to ubiquinone.
   a. Why do fish and insects die after digesting rotenone?
   b. Would you expect the use of rotenone as an insecticide to be a potential hazard to other forms of animal life as well (to people, for example)? Explain.
   c. Would you expect the use of rotenone as a fish poison to be a potential hazard to aquatic plants that might be exposed to the compound? Explain.

8. In his studies of alcoholic fermentation by yeast, Louis Pasteur noted that the sudden addition of oxygen (O_2) to a previously anaerobic culture of fermenting grape juice resulted in a dramatic decrease in the rate of glucose consumption. This Pasteur effect can be counteracted by the addition of 2,4-dinitrophenol, an uncoupler of oxidative phosphorylation.
   a. Why would the yeast cells consume less glucose in the presence of oxygen? Can you estimate how much less glucose they would use?
   b. Why would 2,4-dinitrophenol counteract or prevent the Pasteur effect?

9. Oxidative phosphorylation is normally tightly coupled with electron transport in the sense that no electrons will flow if ATP synthesis cannot occur concomitantly (owing, for example, to low levels of either ADP or P_i). Uncoupling agents dissociate ATP synthesis from electron transport, allowing the latter to occur in the absence of the former. The uncoupling agent described in Problem 8, 2,4-dinitrophenol (DNP), is highly toxic to humans, causing a marked increase in metabolism and temperature, profuse sweating, collapse, and death. For a brief period in the 1940s, however, sublethal doses of DNP were actually prescribed as a means of weight reduction in humans.
   a. Why would an uncoupling agent such as DNP be expected to cause an increase in metabolism, as evidenced by consumption of oxygen or catabolism of foodstuffs?
   b. Based on what you know about allosteric regulation of the glycolytic pathway and the TCA cycle, why would an uncoupling agent such as DNP be likely to have greater, more far-reaching effects on respiratory metabolism than might be predicted by the simple lack of control of the rate of electron transport?
   c. Why would consumption of DNP lead to an increase in temperature and to profuse sweating?
   d. DNP has been shown to carry protons across biological membranes. How might this observation be used to explain its uncoupling effect?
   e. Why would DNP have been considered as a drug for weight-reduction purposes? Can you guess why it was abandoned as a reducing aid?

10. A suspension of mitochondria is treated with rotenone and CO. When succinate is added subsequently, there is no uptake of O_2. However, if ferri(Fe^{3+})cytochrome c is added along with the succinate, the cytochrome undergoes reduction.
    a. How would you measure the rate of electron transfer to the cytochrome?
    b. If ADP and P_i are added along with the succinate and cytochrome c, approximately 0.5 mole of ATP is formed for each mole of cytochrome that is reduced. Explain this observation.
    c. How would the formation of ATP under these conditions be affected by the uncoupler carbonylcyanide-p-trifluoromethoxyphenylhydrazone? By antimycin? By oligomycin?
    d. The rotenone is omitted, and the succinate is replaced by a mixture of pyruvate and malate, which can reduce mitochondrial NAD^+ to NADH. How many moles of ATP would you expect to be formed for each mole of cytochrome that undergoes reduction under these conditions?

11. A suspension of mitochondria is incubated with pyruvate, malate, and ^{14}C-labeled triphenylmethylphosphonium (TPP) chloride under aerobic conditions. The mitochondria are collected by rapid centrifugation and the amount of ^{14}C that they contain is measured. The volume of the mitochondrial matrix space is measured in separate experiments using ^3H-labeled H_2O, so that the concentration of the TPP cation in the matrix can be calculated. The internal concentration is found to be 1000 times greater than the concentration in the external solution.
    a. What is the apparent electrical potential difference ($\Delta\psi$) across the inner membrane? (Be sure to give the units of your answer, and indicate which side of the membrane is positive.)
    b. Qualitatively, how might you expect $\Delta\psi$ to be affected by carbonylcyanide-p-trifluoromethoxyphenylhydrazone? By antimycin? By oligomycin? By an excess of ADP and P_i?

12. Consider a weak, organic acid HA with a pK_a of 7. Suppose that the unprotonated form of the molecule (A^-) is negatively charged, and cannot pass readily across the mitochondrial inner membrane. The protonated form (HA) is uncharged, and if it is sufficiently lipophilic, it may be able to pass freely across the membrane. When mitochondria are incubated with such an acid in the presence of pyruvate, malate, and O_2 at pH 7, they are found to take up the acid so that the total concentration ($[HA] + [A^-]$) in the matrix space becomes greater than the total concentration in the external solution. Show how a measurement of the concentration ratio could allow one to calculate the pH inside the matrix. (Hints: Use the Henderson-Hasselbach equation to relate $[H^+]$, $[HA]$, and $[A^-]$. Also, note that if HA moves freely and comes to equilibrium, there will be a simple relationship between the concentrations of HA inside and outside.)

13. Atractyloside, an inhibitor of the adenine nucleotide translocator, blocks oxidative phosphorylation in intact

mitochondria but not in submitochondrial particles (membrane vesicles) made by exposing mitochondria to supersonic oscillations. (Oligomycin and inhibitors of the electron transport chain inhibit oxidative phosphorylation in submitochondrial particles as well as in mitochondria, and uncouplers also have similar effects on both.) Explain this observation, and relate it to the direction of proton translocation in submitochondrial particles and in mitochondria. Would you expect submitochondrial particles to oxidize added NADH? Added cytochrome *c*?

17

PHOTOSYNTHESIS AND OTHER REACTIONS INVOLVING LIGHT

In Chapter 16, we saw that mitochondria convert the chemical free energy of electron-transfer reactions into a transmembrane electrochemical potential gradient for protons. The *vectorial* aspects of energy metabolism presented new ideas that biochemists initially found difficult to grasp. It was necessary to consider free energy changes that accompany movements of ions between different compartments of the cell, in addition to the free energy changes associated with the electron transfer reactions themselves. In this chapter, we come upon an even more esoteric transfer of energy, that between light and matter. We shall examine this phenomenon in three different biological situations: (1) the conversion of light into chemical energy in *photosynthesis*, (2) the effects of light on *phytochrome*, a pigment that regulates circadian and seasonal cycles of growth and flowering in plants, and (3) the production of light by chemical reactions in *bioluminescence*. We shall begin with photosynthesis, which will be the major concern of the chapter.

PHOTOSYNTHESIS

Heterotrophic organisms, which include animals, fungi, and most types of bacteria, live by degrading complex molecules that are provided by other organisms. Life on earth obviously could not continue indefinitely in this manner unless there were an independent mechanism for making complex molecules from simple ones. The energy needed to perform this task comes almost entirely from the sun, and is captured in the process of *photosynthesis*. Plants and other photosynthetic organisms convert, or "fix," about 10^{11} tons of carbon from CO_2 into organic compounds annually. In spite of the enormous magnitude of this conversion, the total amount of fixed carbon on earth appears to be decreasing as a result of consumption. As our reserves of energy and food diminish, it becomes increasingly important that we understand how photosynthesis works and how our activities affect it.

When most of us think of photosynthesis, we think of green trees or perhaps of fields of grain. Actually, about one-third of the carbon fixation that occurs on earth takes place in the oceans and is carried out by microorganisms. In addition to plants, several groups of bacteria are capable of photosynthesis. Bacteria have been extremely useful in the study of photosynthesis, because their photosynthetic apparatus is simpler than that of plants, and it has been easier to purify the pigment-protein complexes that are components of this apparatus in the bacteria. Although there are important differences between photosynthesis in bacteria and plants, the photochemical reactions that capture the energy of light are basically the same.

An overall equation for CO_2 fixation as it occurs in plants is

$$6\,CO_2 + 6\,H_2O + light \longrightarrow C_6H_{12}O_6 + 6\,O_2 \qquad (1)$$

or, more generally,

$$CO_2 + H_2O + light \longrightarrow (CH_2O) + O_2 \qquad (2)$$

where (CH_2O) represents part of a carbohydrate molecule. Electrons and protons are removed from H_2O, O_2 is evolved, and CO_2 is reduced to the level of a carbohydrate. One group of photosynthetic bacteria (the _cyanobacteria_, which are sometimes called "blue-green algae" although they are not true algae) carry out the same process. Other types of photosynthetic bacteria carry out similar overall processes except that they do not evolve O_2 because they use materials other than H_2O as a source of electrons.

In the absence of light, the equilibrium for the synthesis of glucose from CO_2 and H_2O lies vanishingly far to the left. The equilibrium constant at 27°C is 10^{-496}! Our goal is to explore how photosynthetic organisms use light to drive the reaction in the direction of carbohydrates, against this enormous thermodynamic gradient. How can light do chemistry?

The Photochemical Reactions of Photosynthesis Take Place in Membranes

In plants, the reactions of photosynthesis take place in specialized subcellular organelles, the _chloroplasts_. Figure 17–1 is an electron micrograph of a chloroplast from a lettuce leaf. Chloroplasts are bounded by an envelope of two membranes, and they have an extensive internal membrane called the _thylakoid membrane_. In electron micrographs of thin-sectioned chloroplasts, the thylakoid membrane gives the appearance of a large number of separate sheets or flattened vesicles. It actually is a single membrane that is highly folded and encloses a distinct compartment, the _thylakoid lumen_ (Figure 17–2). In places, the folded membrane is tightly stacked into structures called _grana_. The chlorophyll found in chloroplasts is bound to proteins that are integral constituents of the thylakoid membrane, and it is here that the initial conversion of light into chemical energy occurs. The thylakoid membrane also contains a collection of electron carriers and an ATP-synthase similar to the proton-translocating ATP-synthase of mitochondria. The enzymes responsible for the actual fixation of CO_2 and the synthesis of carbohydrates are soluble proteins and reside in the _stroma_ that surrounds the thylakoid membrane (Figures 17–1 and 17–2). The stroma also contains DNA and ribosomes, which are responsible for synthesizing some of the proteins found in the chloroplast.

Algae are members of the plant kingdom and contain chloroplasts similar to those of higher plants. Prokaryotic photosynthetic organisms, which include the cyanobacteria and several groups of purple or green bacteria, do not have chloroplasts. In prokaryotes, the photochemical reactions of photosynthesis take place in the membrane that encloses the cell. This membrane has extensive invaginations resembling the cristae of the mitochondrial

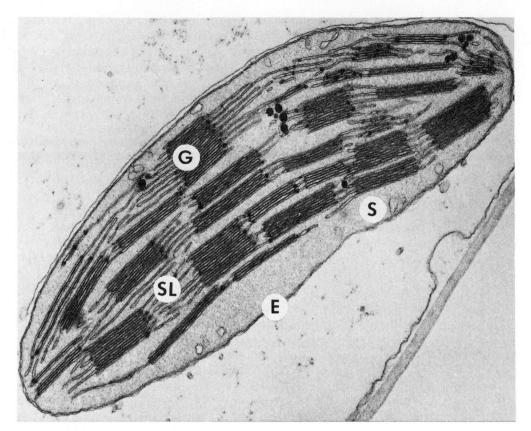

Figure 17–1

Cross-sectional view of a chloroplast in a lettuce leaf. The organelle is shaped like a flattened sausage with a width of about 10 μm and a thickness of about 3 μm. It is surrounded by a double outer membrane or envelope (E). The stroma (S) contains DNA, ribosomes, and the soluble enzymes of CO_2 fixation. Extending throughout the stroma is the *thylakoid membrane,* which is differentiated into stacked regions or *grana* (G) and unstacked *stromal lamella* (SL). (Courtesy Dr. Charles Arntzen.)

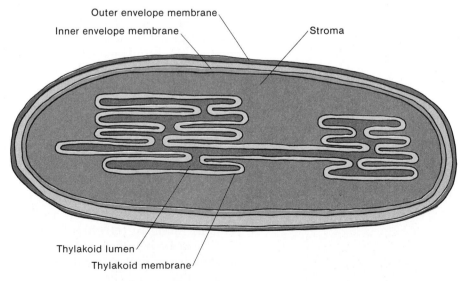

Figure 17–2

A schematic drawing of the chloroplast membrane systems. The highly folded thylakoid membrane separates the thylakoid lumen from the stroma. The initial conversion of light into chemical energy occurs in the thylakoid membrane.

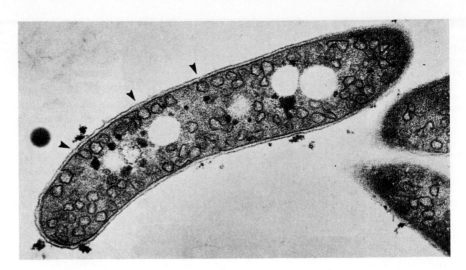

Figure 17–3

Cross-sectional view of a cell of *Rhodospirillum rubrum*, a purple photosynthetic bacterium. A double membrane system surrounds the cell. The inner membrane is extensively invaginated into tubules (arrows). These look circular when they are cut in cross section even though they represent an extension of the inner plasma membrane. (Courtesy Dr. Gerald Peters.)

inner membrane (Figure 17–3). Table 17–1 summarizes the distinctions between some of the major groups of photosynthetic organisms.

Photosynthesis Depends on the Photochemical Reactivity of Chlorophyll

With the exception of certain halophilic bacteria (see Chapter 36), all known *photosynthetic organisms take advantage of the photochemical reactivity of one or another type of chlorophyll*. Figure 17–4 shows the structures of several of the different types of chlorophyll that occur in nature. The chlorophylls are basically tetrapyrroles. They resemble hemes, the prosthetic groups of the cytochromes and hemoglobin, and are derived biosynthetically from protoporphyrin IX. However, they differ from hemes in four major respects:

1. Chlorophylls have an additional ring (ring V) with carbonyl and carboxylic ester substituents.
2. The central metal atom is magnesium rather than iron.
3. In the case of chlorophyll *a* and chlorophyll *b*, the two major chlorophylls in plants and cyanobacteria, one of the pyrrole rings (ring IV) is reduced by the addition of two hydrogens. In bacteriochlorophyll *a* and bacteriochlorophyll *b*, which occur in the purple and green bacteria, two of the rings are reduced (rings II and IV).

TABLE 17–1
Properties of Photosynthetic Organisms

Group	Evolve O_2?	Chloroplasts?	Type of Chlorophyll	Number of Photosystems
Higher plants and algae (eukaryotes)	Yes	Yes	Chlorophyll *a* and *b*	2
Cyanobacteria (prokaryotes)	Yes	No	Chlorophyll *a* and *b*	2
Purple bacteria (prokaryotes)	No	No	Bacteriochlorophyll *a* or *b*	1

Figure 17–4
Structures of protoporphyrin IX (the prosthetic group of hemoglobin, myoglobin, and the c-type cytochromes), and several types of chlorophyll and bacteriochlorophyll. Chlorophyll a and chlorophyll b are the main types of chlorophyll in plants and the cyanobacteria. Purple photosynthetic bacteria contain either bacteriochlorophyll a or bacteriochlorophyll b, depending on the bacterial species. (The green photosynthetic bacteria contain still another form of bacteriochlorophyll.) Chlorophyll b and bacteriochlorophyll b are the same as chlorophyll a and bacteriochlorophyll a except for the substituents on ring II. Pheophytins and bacteriopheophytins are the same as the corresponding chlorophylls or bacteriochlorophylls except that two hydrogen atoms replace Mg.

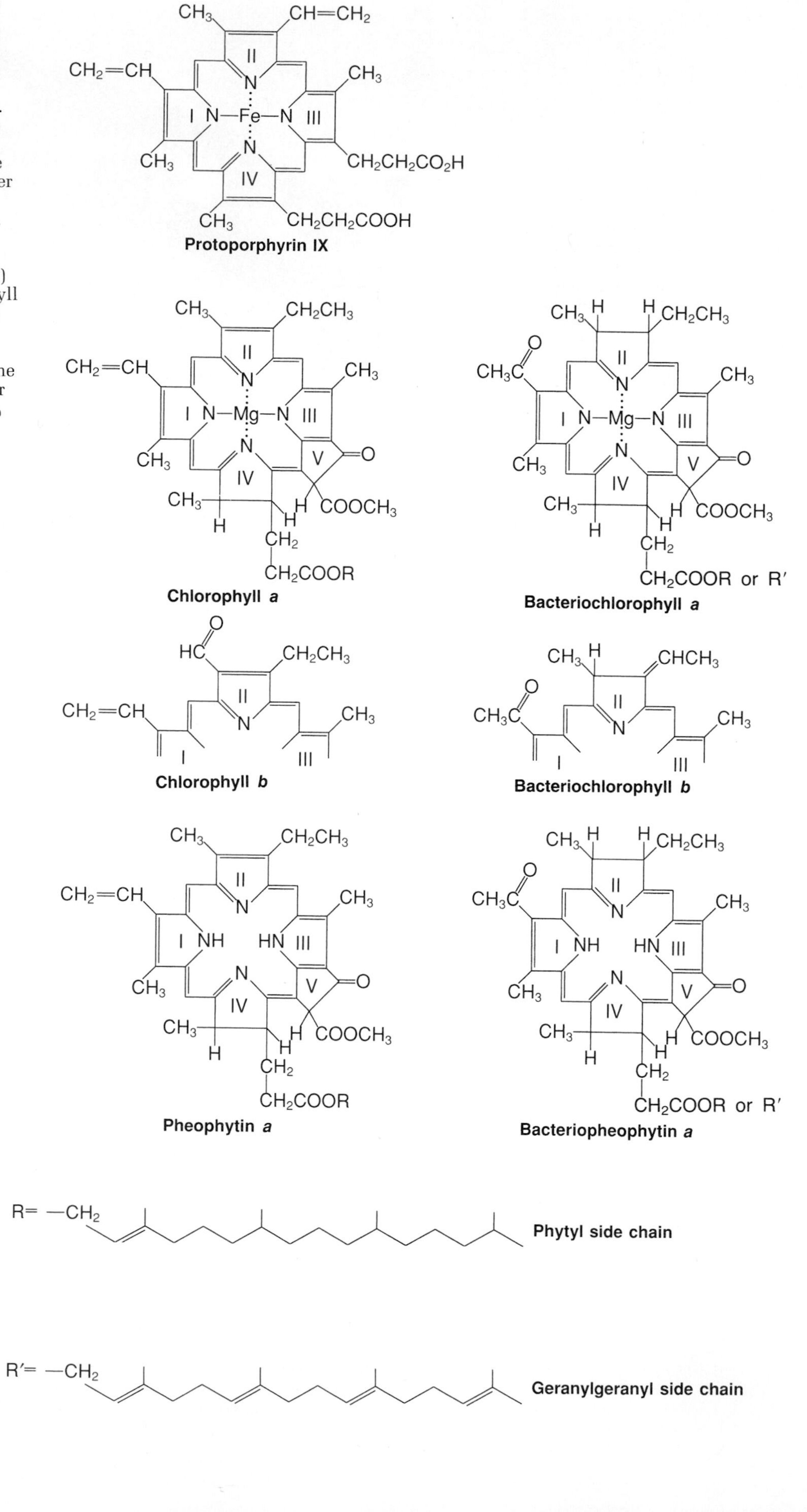

Protoporphyrin IX

Chlorophyll a

Bacteriochlorophyll a

Chlorophyll b

Bacteriochlorophyll b

Pheophytin a

Bacteriopheophytin a

R= —CH₂   Phytyl side chain

R′= —CH₂   Geranylgeranyl side chain

Figure 17–5
Absorption spectra of chlorophyll a and bacteriochlorophyll a in ether. Note that the long-wavelength absorption bands are about five times stronger than the α band of reduced cytochrome c (Figure 16–4). In the chlorophyll-protein complexes found in photosynthetic organisms, the long-wavelength absorption band generally is shifted to even longer wavelengths. This probably reflects interactions between neighboring chlorophylls, which are bound to the proteins as dimers or larger groups.

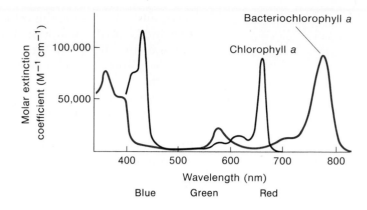

4. The propionyl side chain of ring IV is esterified with a long-chain isoprenoid alcohol. Chlorophylls a and b contain the alcohol _phytol_; bacteriochlorophylls a and b have either phytol or geranylgeraniol, depending on the species of bacteria.

Photosynthetic organisms also contain small amounts of _pheophytins_ or _bacteriopheophytins_, which are the same as the corresponding chlorophylls or bacteriochlorophylls except that Mg is replaced by two hydrogens (Figure 17–4). We shall see that the pheophytins and bacteriopheophytins play special roles as electron carriers in photosynthesis.

The reduction of ring IV in chlorophyll a or b makes the conjugated aromatic ring system decidedly asymmetrical. This changes the optical absorption spectrum of the molecule dramatically. Whereas the long-wavelength absorption band of a cytochrome (the α band) is relatively weak (see Figure 16–4), chlorophyll a has an intense absorption band at 676 nm (Figure 17–5). Chlorophyll b has a similar band at 642 nm. Bacteriochlorophylls a and b, in which the asymmetry of the conjugated system is even more pronounced, have extremely strong absorption bands in the region of 780 to 790 nm (Figure 17–5). All of the chlorophylls thus absorb light very well, particularly at relatively long wavelengths.

A Physical Definition of Light. Before we discuss how chlorophylls can transform light energy into chemical energy, we must review some of the basic properties of light. _Light is an electromagnetic field that oscillates sinusoidally in space and time_ (Figure 17–6). _It interacts with matter in packets, or quanta, called photons, each of which contains a definite amount of energy._ A mole of photons is called an _einstein_. The relationship between the energy ϵ of a photon and the frequency ν of the oscillating field is given by

$$\epsilon = h\nu \tag{3}$$

where h is Planck's constant (6.63×10^{-27} erg · sec or 4.12×10^{-15} eV · sec). The energy per einstein is

$$E = N\epsilon = Nh\nu \tag{4}$$

where N is Avogadro's number (6.02×10^{23} photons per einstein). The frequency ν is the number of oscillations per second at a given point in space. The wavelength λ of the oscillations (the distance between successive peaks in the amplitude of the field) depends on both ν and the velocity v at which the peaks move through space:

$$\lambda = v/\nu \tag{5}$$

Light travels with a velocity (v) of 3×10^{10} cm/sec in a vacuum. In a dense

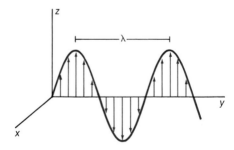

Figure 17–6
Light is an oscillating electromagnetic field. The lengths of the arrows in this diagram represent the strength of the electric field at a particular time, as a function of position in a ray of light proceeding along the y axis. A similar diagram could represent the field as a function of time at a particular point in space. The polarization of the light is defined by the orientation of the electric field; the beam shown here is polarized parallel to the z axis. The magnetic field, which is not shown here, is in phase with the electric field but is parallel to the x axis.

medium such as water the velocity is less, depending inversely on the refractive index of the medium. Blue light has a wavelength in the region of 450 nm ($\nu \approx 6.7 \times 10^{14}$ sec^{-1}) and an energy of about 2.8 eV (2.7×10^{-12} erg/einstein, or 64 kcal/einstein). Red light has a wavelength of about 650 nm.

Radiation with wavelengths much below 400 nm or above 750 nm is invisible to the human eye, and some authors prefer not to call this light. However, such radiation can be important biologically. Many photosynthetic bacteria are adapted to use radiation in the region between 800 and 900 nm, and some species do well with even longer wavelengths.

How Light Interacts with Molecules. The electrons in a molecule are held in a set of molecular orbitals, each of which is associated with a particular energy. In an isolated molecule, the energies of the orbitals depend mainly on the electrostatic interactions of the electrons with each other and with the nuclei, and are more or less independent of time. A molecule thus can have any of a series of different energies, depending on how its electrons are distributed among the available orbitals. For an organic molecule with n electrons, the lowest overall energy usually is obtained when there are two electrons with antiparallel spins in each of the first $n/2$ orbitals, leaving all the orbitals with higher energies empty (Figure 17–7). This is the <u>ground</u> <u>state</u> of the molecule. A molecule that is placed in this state will remain there indefinitely, as long as the electronic potential energies do not change.

When light interacts with a molecule, the oscillating electric field of the light makes the electronic potential energies strongly time-dependent. Sometimes, as a result, a photon is absorbed and <u>an electron moves from one</u>

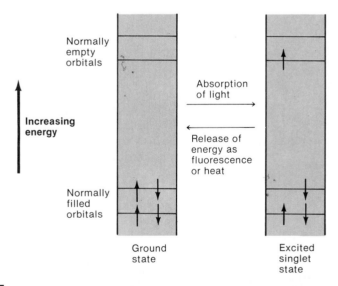

Figure 17–7

When a molecule absorbs light, an electron is excited to a molecular orbital with higher energy. The horizontal bars in this diagram represent molecular orbitals for electrons. Each orbital can hold two electrons with antiparallel spins (arrows pointing upward or downward). Only the top few of the many filled orbitals are shown here. The absorption of light can raise an electron from one of these orbitals to an orbital that is normally unoccupied. For this to occur, the energy of the photon must match the difference between the energies of the two orbitals. The choice of an upward or downward arrow is arbitrary, but there is no change of spin during the excitation. As long as the spins of the two unpaired electrons remain antiparallel, so that the molecule has no net electronic spin, the molecule is said to be in an excited singlet state. An excited molecule can return directly to the ground state by giving off energy as fluorescence or heat, or by transferring energy to another nearby molecule, or it can transfer an electron to another molecule (Figure 17–8).

of the occupied molecular orbitals to an unoccupied orbital with a higher energy (Figure 17–7). Two requirements must be met in order for this to occur. First, *the difference between the energies of the two orbitals must be the same as the photon's energy, hν.* This is why a given type of molecule absorbs light of some wavelengths and not of others. The second, more complex requirement has to do with the shapes of the two orbitals and the orientation (polarization) of the oscillating field relative to the disposition of the orbitals in space. *The two orbitals must have different geometrical symmetries and must be oriented in an appropriate way* with respect to the field. This requirement explains why some absorption bands are stronger than others.

If chlorophyll (or any other molecule) absorbs a photon, it is excited to a state that lies above the ground state in energy. This usually occurs with no change in electronic spin (Figure 17–7). As long as the spins of the two unpaired electrons remain antiparallel, so that the molecule has no net electronic spin, the molecule is said to be in an *excited singlet state*.

A molecule in an excited singlet state can decay back to the ground state by releasing energy in several different ways (Figure 17–7). One possibility is to *fluoresce*—that is, to emit a photon. The wavelength of the fluorescence is generally longer than that of the light that was originally absorbed, because readjustments of the molecular geometry decrease the energy of the excited molecule somewhat before the molecule fluoresces. The extra energy is given off to the environment as heat. If the molecule has several absorption bands, as the chlorophylls do, the wavelength of the fluorescence is usually slightly longer than that of the longest-wavelength absorption band. The molecule relaxes to the lowest, or "first," excited singlet state before the emission occurs. In some cases, the molecule can decay all the way to the ground state by radiationless processes, so that the excitation energy is converted entirely into heat. Another decay mechanism is to transfer the energy to a neighboring molecule by the process of *resonance energy transfer*. This phenomenon is extremely important in photosynthesis, and we shall return to it below. A fourth possibility is for the excited molecule to transfer an *electron* to a neighboring molecule.

Light Causes an Electron-transfer Reaction. *Electron transfer can be a favorable path for the decay of an excited molecule, because an electron in the upper, normally unoccupied orbital is bound less tightly than one in a lower, normally filled orbital.* The standard redox potential at pH 7, $E^{\circ\prime}$, for the removal of an electron from the excited molecule is more negative than the $E^{\circ\prime}$ for the molecule in the ground state, by approximately $h\nu/e$. Here $h\nu$ again is the excitation energy, which can be expressed in electron volts, and e is the charge of an electron. In the case of chlorophyll a, the $E^{\circ\prime}$ for oxidation in the ground state is approximately $+0.5$ V, and $h\nu$ for the long-wavelength absorption band is about 1.5 eV. The $E^{\circ\prime}$ of the excited molecule is approximately -1.0 V. This means that in the excited state chlorophyll a is a very strong reductant. For comparison, recall that $NAD^+/NADH$ has an $E^{\circ\prime}$ of only -0.32 V. The basic principle underlying the photochemistry of photosynthesis is that *excitation causes a molecule of chlorophyll or bacteriochlorophyll (or a complex of several such molecules) to release an electron.* The chlorophyll complex is oxidized, and another molecule becomes reduced (Figure 17–8). The oxidized chlorophyll species that is formed is a relatively strong oxidant and can extract an electron from a third molecule.

The idea that light drives the formation of oxidants and reductants was first advanced by C. B. van Niel in the 1920s. It was known at the time that the purple photosynthetic bacteria thrive only if they are provided with a reduced substrate such as an organic acid. Some species grow well on a

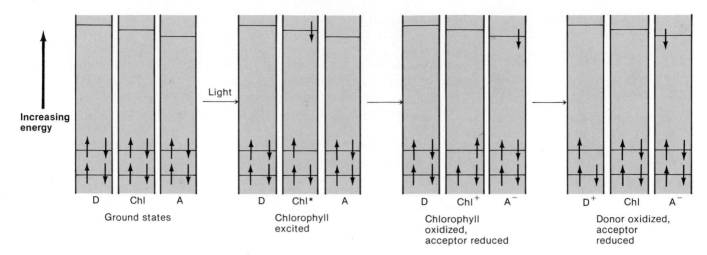

Increasing energy

D	Chl	A		D	Chl*	A		D	Chl⁺	A⁻

Ground states

Light →

Chlorophyll excited

Chlorophyll oxidized, acceptor reduced

Donor oxidized, acceptor reduced

Figure 17–8
The photochemical process that initiates photosynthesis is an electron-transfer reaction. The horizontal bars and vertical arrows represent molecular orbitals and electrons, as in Figure 17–7. Absorption of light increases the free energy of a chlorophyll complex (Chl) by $h\nu$, making electron transfer to an acceptor (A) thermodynamically favorable. The oxidized chlorophyll complex (Chl^+) extracts an electron from a donor (D).

reduced inorganic material such as H_2S. Van Niel noticed that although the bacteria do not evolve O_2, the reactions they carry out have a formal resemblance to the process that occurs in plants. If H_2B is used to represent a reduced substrate and B to represent an oxidized product, the process that occurs in the purple bacteria can be written

$$CO_2 + 2\ H_2B \xrightarrow{\text{Light}} (CH_2O) + H_2O + 2\ B \tag{6}$$

The equation for CO_2 fixation in plants and cyanobacteria, Equation (2), can be put in a similar form by replacing H_2B and 2B with H_2O and O_2:

$$CO_2 + 2\ H_2O \xrightarrow{\text{Light}} (CH_2O) + H_2O + O_2 \tag{7}$$

To van Niel, this suggested that *the essence of photosynthesis in both plants and bacteria is the photochemical separation of oxidizing and reducing power*. The substance that is reduced in the photochemical reaction could be used to reduce CO_2 to carbohydrates in enzyme-catalyzed reactions that do not require light. The material that is oxidized photochemically could be discharged by the oxidation of H_2O to O_2 in plants, or by the oxidation of some other material in bacteria (Figure 17–9).

Support for these ideas came from experiments done by Robin Hill in 1939. Hill discovered that isolated chloroplasts would evolve O_2 if they were illuminated in the presence of an added nonphysiological electron acceptor such as ferricyanide [$Fe(CN)_6^{3-}$]. The electron acceptor became reduced in the process. Since there was no fixation of CO_2 under these conditions, it was clear that the photochemical reactions of photosynthesis can be separated from the reactions that involve CO_2.

Photooxidation of Chlorophyll Generates a Cationic Free Radical

When chlorophyll or bacteriochlorophyll is oxidized, the product is a cationic free radical. The situation differs subtly from the oxidation of a cytochrome. When a cytochrome is oxidized, the electron that is removed comes from the iron, which changes from ferrous (Fe^{2+}) to ferric (Fe^{3+}). In chlorophyll, the electron is removed not from the magnesium, but rather from the

A   Light   A⁻
Chl → Chl   → Electron used to reduce CO_2
D   D⁺ → Electron replaced by one from H_2O or H_2B

Figure 17–9
Van Niel proposed that the reductant generated in a photochemical electron-transfer reaction (A⁻) is used to reduce CO_2 to carbohydrate. He suggested that the oxidant (D⁺) oxidizes H_2O to O_2 in plants, and oxidizes some other material (H_2B) in the purple bacteria. Only the initial charge separation requires light. (Chl = chlorophyll.)

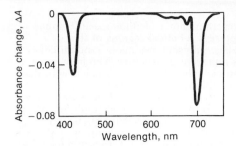

Figure 17–10

When a suspension of chloroplasts is illuminated, its optical absorbance decreases in the regions around 430 and 700 nm. The absorbance changes reflect the oxidation of a special chlorophyll complex (P700). Addition of a chemical oxidant such as potassium ferricyanide causes similar absorbance changes. (After B. Kok, Partial purification and determination of oxidation reduction potential of the photosynthetic chlorophyll complex absorbing at 700 mμ, *Biochim. Biophys. Acta* 48:527, 1961.)

Ubiquinone

Menaquinone

Plastoquinone

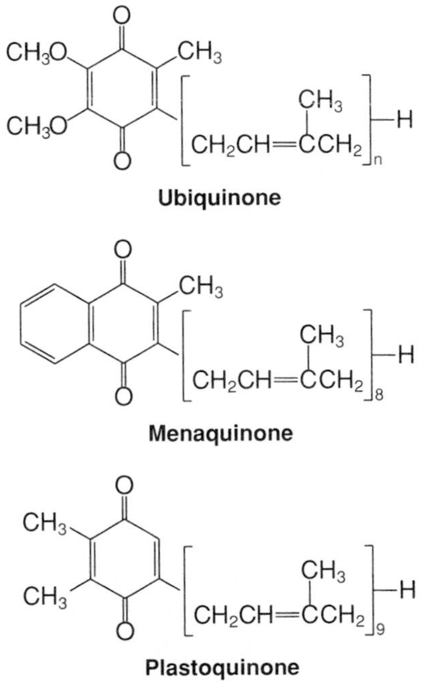

Figure 17–11

The structures of ubiquinone, menaquinone (vitamin K_2), and plastoquinone. Purple photosynthetic bacteria contain ubiquinone, menaquinone, or both, depending on the bacterial species; chloroplasts contain plastoquinone. The number of isoprenoid units in the ubiquinone side chain (*n*) varies from 8 to 10.

aromatic π-electron system of the molecule. The positive charge of the oxidized chlorophyll and the spin of the unpaired electron that remains behind are delocalized extensively over the π-electron system. This can be shown by studying the electron spin resonance (ESR) and electron nuclear double-resonance (ENDOR) spectra of the radical.

The photooxidation of chlorophyll can be detected by measuring changes in the optical absorption spectrum of the molecule. Oxidation results in the loss of the chlorophyll's characteristic absorption bands. The first measurements of this sort were made in the 1950s by Bessel Kok and Louis Duysens. Kok found that illumination of chloroplasts caused an absorbance decrease at 700 nm (Figure 17–10). He suggested that the decrease reflected the photooxidation of a reactive chlorophyll complex, which he called P700. P stood for "pigment," and 700 for the wavelength at which the unoxidized complex had its main absorption band. Duysens made similar observations on *Rhodospirillum rubrum*, a purple photosynthetic bacteria. Here the reactive bacteriochlorophyll complex absorbed at 870 nm, and Duysens called it P870. A second type of reactive complex in chloroplasts, P680, was discovered subsequently by H. Witt and his colleagues. We shall see below that P700 and P680 are parts of two distinct photochemical systems, <u>photosystem I</u> and <u>photosystem II</u>. P700 and P680 also are found in the cyanobacteria.

The photooxidation of P870, P700, or P680 generates a cationic radical (P870$^+$, P700$^+$, or P680$^+$) in which the charge and the unpaired electron are delocalized over the macrocyclic ring system. In fact, the ESR and ENDOR spectra of the P870$^+$ radical indicate that the unpaired electron is delocalized over *two* bacteriochlorophyll *a* molecules, which form a closely interacting pair. P700 and P680 may consist of similar dimers of chlorophyll *a*, but the evidence on this point is still uncertain. The strong interactions between the two bacteriochlorophyll molecules in P870 probably explain why the absorption band of the complex is at 870 nm, whereas the long-wavelength band of monomeric bacteriochlorophyll *a* in solution is at 780 nm (Figure 17–5). In bacterial species that contain bacteriochlorophyll *b* instead of bacteriochlorophyll *a*, the reactive complexes absorb at 960 nm instead of 870, and are often called P960. For simplicity, we shall neglect the differences in wavelength among the various species of purple bacteria, and shall use the term P870 in a general sense.

The Reactive Chlorophyll Is in Pigment-Protein Complexes Called Reaction Centers

The chlorophyll or bacteriochlorophyll that undergoes photooxidation is bound to a protein in a complex called a <u>reaction center</u>. Reaction centers have been purified from both plants and bacteria by disrupting chloroplasts or bacterial membranes with detergents. Purification was first achieved with purple photosynthetic bacteria, particularly by Roderick Clayton, and the structure of the bacterial complex is still much better understood than are the structures of the plant reaction centers. When they are excited with light, purified bacterial reaction centers are capable of carrying out the initial photochemical transfer of an electron from P870 to a series of electron acceptors.

Reaction centers isolated from the purple bacteria generally contain three polypeptides with a total molecular weight of about 100,000. Bound noncovalently to the protein are four molecules of bacteriochlorophyll and two molecules of bacteriopheophytin. The reaction center also contains two quinone molecules and one nonheme iron atom. In some bacterial species, both of the quinones are ubiquinone. In others, one of the quinones is menaquinone (vitamin K_2), a naphthoquinone that resembles ubiquinone in having a long isoprenoid side chain (Figure 17–11). Reaction centers from some

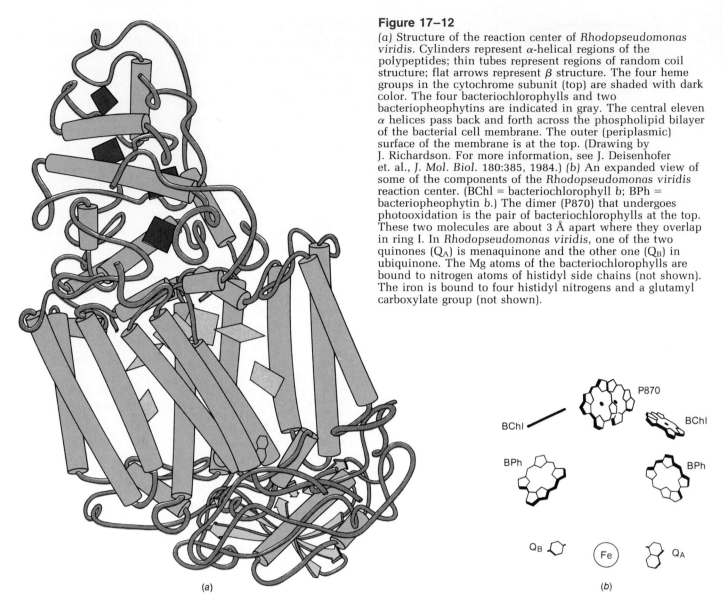

Figure 17–12

(a) Structure of the reaction center of *Rhodopseudomonas viridis*. Cylinders represent α-helical regions of the polypeptides; thin tubes represent regions of random coil structure; flat arrows represent β structure. The four heme groups in the cytochrome subunit (top) are shaded with dark color. The four bacteriochlorophylls and two bacteriopheophytins are indicated in gray. The central eleven α helices pass back and forth across the phospholipid bilayer of the bacterial cell membrane. The outer (periplasmic) surface of the membrane is at the top. (Drawing by J. Richardson. For more information, see J. Deisenhofer et. al., *J. Mol. Biol.* 180:385, 1984.) (b) An expanded view of some of the components of the *Rhodopseudomonas viridis* reaction center. (BChl = bacteriochlorophyll *b*; BPh = bacteriopheophytin *b*.) The dimer (P870) that undergoes photooxidation is the pair of bacteriochlorophylls at the top. These two molecules are about 3 Å apart where they overlap in ring I. In *Rhodopseudomonas viridis*, one of the two quinones (Q_A) is menaquinone and the other one (Q_B) in ubiquinone. The Mg atoms of the bacteriochlorophylls are bound to nitrogen atoms of histidyl side chains (not shown). The iron is bound to four histidyl nitrogens and a glutamyl carboxylate group (not shown).

purple bacteria, such as *Rhodopseudomonas viridis*, also have a cytochrome subunit with four *c*-type hemes.

The crystal structure of reaction centers from *Rhodopseudomonas viridis* was solved by Hartmut Michel, Johann Deisenhofer, Robert Huber, and their colleagues in 1984. This was a landmark achievement, because it was the first time anyone had determined the high-resolution crystal structure of a hydrophobic, integral membrane protein. Reaction centers from another species, *Rhodobacter sphaeroides*, subsequently proved to have essentially the same structure except that they lack the bound cytochrome. In both species, the bacteriochlorophyll and bacteriopheophytin molecules, the nonheme iron atom, and the quinones are all bound to two of the polypeptides, which are folded into a series of α helices that pass back and forth across the cell membrane (Figure 17–12a). The third polypeptide is located largely on the cytoplasmic side of the membrane, but it also has one transmembrane α helix. The cytochrome subunit in *Rhodopseudomonas viridis* is globular in shape, and sits on the external (periplasmic) surface of the membrane.

Figure 17–12b shows the arrangement of the pigments in greater detail. Two of the four bacteriochlorophylls are packed especially closely together.

Studies of the reaction center's optical absorption spectrum indicate that *this special pair of bacteriochlorophylls is P870, the reactive complex that releases an electron when the reaction center is excited with light.*

Preparations of reaction centers from photosystem I or photosystem II of plants typically contain from three to five different polypeptides with molecular weights ranging from about 10,000 to 65,000, and approximately 40 molecules of chlorophyll a. The reactive chlorophyll complex (P700 or P680) accounts for only a small part of this chlorophyll. Although the structures of the plant reaction centers are not yet known, photosystem II reaction centers resemble the reaction centers of the purple photosynthetic bacteria in a number of ways. The amino acid sequences of two of their polypeptides are homologous to those of the two polypeptides that hold P870 and the other pigments in the bacterial reaction center. Also, the reaction centers of photosystem II contain a nonheme iron atom and two molecules of plastoquinone, a quinone that is closely related to ubiquinone (Figure 17–11), and they contain one or more molecules of pheophytin a.

In Bacterial Reaction Centers, Electrons Move from P870 to Bacteriopheophytin and Then to Quinones

Let us now consider the acceptors that remove an electron from P870 in the bacterial reaction center. *The earliest detectable acceptor is a bacteriopheophytin.* Note that the reaction center contains two bacteriopheophytins and two additional bacteriochlorophylls, in addition to the special pair of bacteriochlorophylls that make up P870 (Figure 17–12b). When isolated reaction centers are excited with a very short flash of light, a transient state that appears to be an excited singlet state of the reactive bacteriochlorophylls (P870*) is formed essentially instantaneously. This state decays in about 3×10^{-12} sec, and as it does, a P870$^+$BPh$^-$ radical-pair is created. Here P870$^+$ is the cationic radical formed by removing an electron from P870, and BPh$^-$ is the anionic radical that is formed by adding an electron to a bacteriopheophytin (Figure 17–13). The formation of the radical-pair state can be detected spectrophotometrically by the disappearance of absorption bands of P870 and the bacteriopheophytin, and the formation of new absorption bands attributable to the two radicals. (The bacteriopheophytin that undergoes reduction is the one shown on the right in Figure 17–12a and b. The two bacteriopheophytins are bound to the polypeptides in slightly different ways, and are distinguishable by their different absorption spectra.) The P870$^+$BPh$^-$ state lasts for about 2×10^{-10} sec, decaying by the movement of an electron from BPh$^-$ to one of the quinones (Q_A). This leaves the reaction center in the state P870$^+Q_A^-$, where Q_A^- is the anionic semiquinone radical (Figure 17–13).

The electron carriers that participate in these first few steps are all fixed in position close together in the reaction center, and have little freedom of motion. One indication of this is that the electron-transfer reactions, in addition to being phenomenally fast, are almost independent of temperature. They actually increase slightly in speed with decreasing temperature. Therefore there is no need for the reactants to diffuse together in order for an electron to move from one molecule to another, and the reactants do not have to acquire any additional thermal energy from their surroundings. The initial electron-transfer steps are in a sense "solid state" processes. Further, the reactions are amazingly efficient. This is generally expressed in terms of the *quantum yield* of P870$^+Q_A^-$, which is the number of mols of P870$^+Q_A^-$ formed per einstein of light absorbed. The measured quantum yield in isolated reaction centers is 1.02 ± 0.04. *Essentially every time the reaction center is excited, an electron moves from P870 to Q_A.*

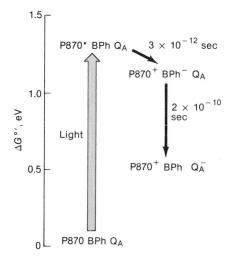

Figure 17–13

The initial electron-transfer sequence in reaction centers of purple photosynthetic bacteria. (P870* = the first excited singlet state of P870; BPh = bacteriopheophytin; Q_A = menaquinone or ubiquinone, depending on the bacterial species.) In this scheme, various states of the photosynthetic apparatus are positioned vertically according to their free energies, with the states that have the highest free energies at the top. Decay times are indicated. (For more details, see N. W. Woodbury, M. Becker, D. Middendorf, and W. W. Parson, Picosecond kinetics of the initial photochemical electron transfer reaction in bacterial photosynthetic reaction centers, *Biochem.* 24:7516, 1985.)

A Cyclic Electron-transport Chain Drives the Formation of ATP

From Q_A^-, an electron moves to the second quinone that is bound to the reaction center (Q_B in Figure 17–14). This step takes about 10^{-4} sec, considerably longer than the earlier steps, and it becomes even slower with decreasing temperature. In the meantime, *a c-type cytochrome replaces the electron that was removed from P870*, preparing the reaction center to operate again. Electron transfer between the cytochrome and $P870^+$ takes between 10^{-7} and 10^{-3} sec, depending on the species. In *Rhodopseudomonas viridis*, the electron donor is the bound cytochrome with four hemes that is shown at the top of Figure 17–12a. In other species, it often is a soluble c-type cytochrome with a single heme, more like the mitochondrial cytochrome c.

When the reaction center is excited a second time, a second electron is pumped from P870 to the bacteriopheophytin and on to Q_A and Q_B. This places Q_B in the fully (two-electron) reduced form, Q_B^{2-}. The uptake of two protons transforms the reduced quinone to the uncharged quinol, QH_2, which dissociates from the reaction center into the phospholipid bilayer of the cellular membrane. In intact bacteria, the protons that are taken up in the formation of QH_2 come from the cytosol of the cell (Figure 17–14). Like the mitochondrial inner membrane, the bacterial membrane contains a relatively high concentration of quinone, which again can be either ubiquinone or menaquinone, depending on the species of bacteria.

QH_2 is reoxidized by a cytochrome bc_1 complex that is very similar to complex III of the mitochondrial respiratory chain. Electrons move through the cytochrome bc_1 complex to a c-type cytochrome, which then diffuses to the reaction center and provides an electron for the reduction of $P870^+$ (Figure 17–14). As it does in mitochondria, the movement of electrons through the cytochrome bc_1 complex results in the release of protons on the extracellular surface of the membrane. The proton extrusion can be explained well by the same type of Q cycle that was discussed in connection with the mitochondrial complex (see Figure 16–18). *The flow of electrons from QH_2 back to P870 thus drives the movement of protons outward across the cell membrane, generating a transmembrane electrochemical potential gradient for protons.* The inside of the cell becomes negatively charged relative to the external medium, and the pH of the cytosol becomes higher than the external pH.

The flow of protons back into the bacterial cell, down the electrochemical potential gradient, is mediated by an ATP-synthase that resembles the proton-conducting ATP-synthase of the mitochondrial inner membrane (see Chapter 16). As in mitochondria, *the movement of protons through a channel in the F_o base-piece of the enzyme is linked to the formation of ATP* (Figure 17–14).

Note that the electron-transport system of purple photosynthetic bacteria is completely cyclic. Excitation of the reaction center with light creates a strong reductant (BPh^-) and a strong oxidant ($P870^+$), and electrons return from BPh^- to $P870^+$ via quinones, the cytochrome bc_1 complex, and c-type cytochromes. The cyclic flow of electrons results in the formation of ATP, but no net oxidation or reduction. How, then, can we explain van Niel's observation that the bacteria carry out a net transfer of electrons from organic acids to carbohydrates (Figure 17–9)? The biosynthesis of carbohydrates would seem to require a reductant such as NADH or NADPH, in addition to ATP. The answer to this puzzle is that the bacteria can use ATP to support the transfer of electrons from succinate or other substrates to $NADP^+$. These reactions are carried out by dehydrogenase complexes in the cell membrane. As was discussed in Chapter 16 (see Figure 16–25), a similar reduction of NAD^+ by succinate can be seen in mitochondria if ATP is added.

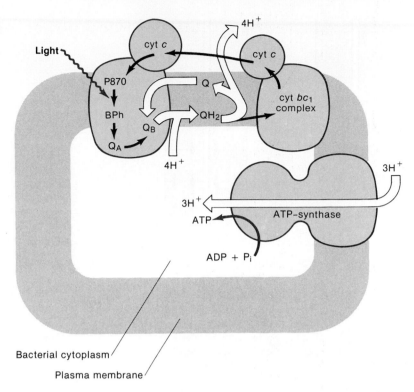

Figure 17–14

In purple photosynthetic bacteria, electrons return to P870$^+$ from the quinones Q_A and Q_B via a cyclic pathway. When Q_B is reduced with two electrons, it picks up protons from the cytosol and diffuses to the cytochrome bc_1 complex. Here it transfers one electron to an iron-sulfur protein an the other to a b type cytochrome, and releases protons to the extracellular medium. The electron transfer steps catalyzed by the cytochrome bc_1 complex probably include a Q cycle similar to that catalyzed by complex III of the mitochondrial respiratory chain (see Figure 16–18). The c-type cytochrome that is reduced by the iron-sulfur protein in the cytochrome bc_1 complex diffuses to the reaction center, where it either reduces P870$^+$ directly or provides an electron to a bound cytochrome that reacts with P870$^+$. In the Q cycle, four protons probably are pumped out of the cell for every two electrons that return to P870. The outward proton translocation creates an electrochemical potential gradient across the membrane. Protons move back into the cell through an ATP-synthase, driving the formation of ATP.

An Antenna System Transfers Energy to the Reaction Centers

The reactive chlorophyll or bacteriochlorophyll molecules of P700, P680, or P870 make up only a small fraction of the total pigment in photosynthetic membranes. Chloroplasts contain on the order of 300 chlorophyll molecules per molecule of P700 and P680, and the cell membranes of purple photosynthetic bacteria have from 25 to several hundred bacteriochlorophyll molecules per molecule of P870, depending on the species. Most of the chlorophyll or bacteriochlorophyll is not photochemically active. Instead, it serves as an antenna. *When one of the molecules in the antenna system is excited with light, it can transfer its energy to a neighboring molecule by resonance energy transfer. Energy absorbed anywhere in the antenna migrates rapidly from molecule to molecule until it is trapped by an electron-transfer reaction in one of the reaction centers* (Figure 17–15). Measurements of the lifetime of fluorescence from the antenna system indicate that, after the antenna absorbs a photon, the energy is trapped in a reaction center within about 10^{-10} sec in purple bacteria and within about 5×10^{-10} sec in chloroplasts.

The distinction between the antenna system and the reaction centers grew out of experiments by Robert Emerson and William Arnold in the

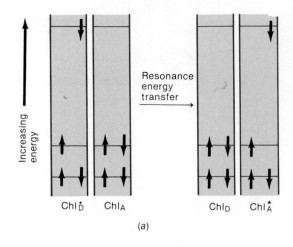

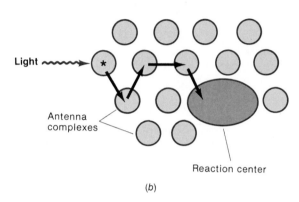

Figure 17–15

(a) A molecule in an excited state (Chl_D^*) can transfer its energy to another molecule (Chl_A). The donor molecule returns to its ground state (Chl_D) and the acceptor is elevated to an excited state (Chl_A^*). This requires that the difference in energy between Chl_D^* and Chl_D (the excitation energy of Chl_D) be the same as the difference between Chl_A^* and Chl_A (the excitation energy of Chl_A). When the energies match in this way, there is a resonance between the two states {Chl_D^* Chl_A} and {Chl_D Chl_A^*}. (b) The energy of light absorbed by pigment-protein complexes in the antenna system hops rapidly from complex to complex by resonance energy transfer, until it is trapped in an *electron*-transfer reaction in a reaction center.

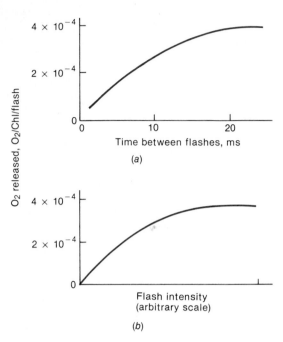

Figure 17–16

The amount of O_2 released when a suspension of algae (*Chlorella pyrenoidosa*) was excited with a train of short flashes of light, as a function of (a) the time between the flashes, and (b) the intensity of the flashes. Each flash lasted about 10 μs. The total amount of O_2 released was measured after several thousand flashes, and was divided by the number of flashes and by the amount of chlorophyll in the suspension. (What happens on the first few individual flashes will be discussed later.) The measurements in panel (a) were made with flashes that were comparable to the strongest flashes used in (b). For the measurements shown in (b), the flashes were spaced 20 ms apart. Note that the maximum amount of O_2 released per flash was only about one molecule of O_2 per 2500 molecules of clorophyll. With saturating flashes, the amount of O_2 evolution is limited by the concentration of reaction centers, whereas most of the chlorophyll in the algae is part of the antenna system.

1930s. Emerson and Arnold measured the amount of O_2 evolution that occurred when they excited suspensions of green algae with a train of short flashes of light. In order to obtain the maximum O_2 per flash, they found that they had to allow a period of about 2×10^{-2} sec of darkness between successive flashes. The amount of O_2 evolved per flash decreased if the flashes were spaced too close together (Figure 17–16a). Each flash evidently generated a product that had to be consumed before the photosynthetic apparatus was ready to work again. If the next flash arrived too soon, its energy was wasted, mainly as heat and fluorescence. We know now that *the chlorophyll complexes that undergo oxidation in the reaction centers (P700 and P680 in algae) have to be returned to their reduced states before they can react again. The acceptors that remove electrons from the chlorophyll complexes also have to be reoxidized.*

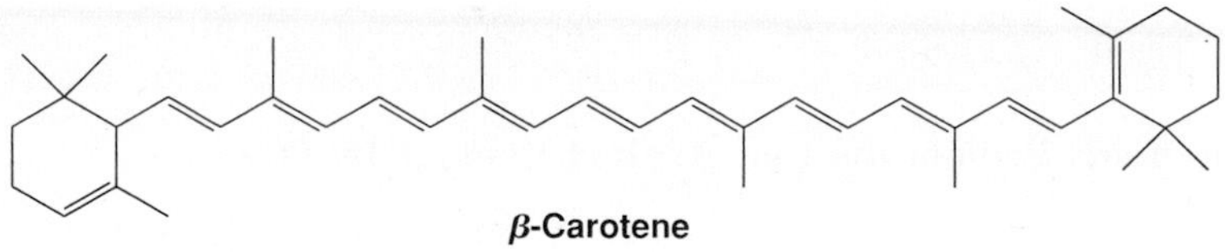

β-Carotene

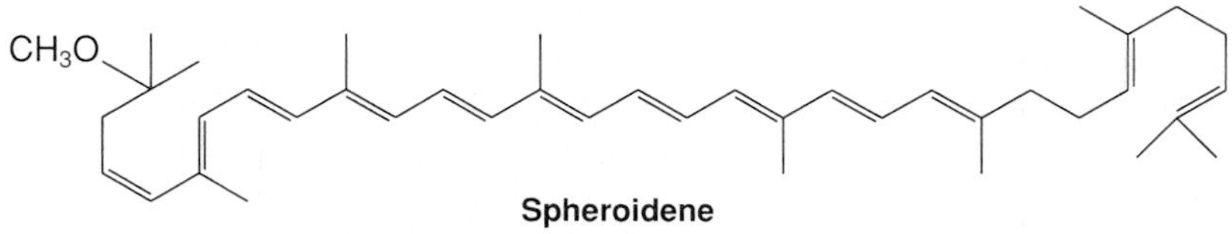

Spheroidene

Figure 17–17
Structures of β-carotene, a major carotenoid in many types of plants, and spheroidene, a common carotenoid of photosynthetic bacteria. There are about 350 different natural carotenoids. These vary widely in color, depending principally on the number of conjugated double bonds.

Emerson and Arnold found that the amount of O_2 that was evolved per flash increased as they raised the strength of the flashes (i.e., as they excited the algae with a larger and larger number of photons on each flash), but that it reached a plateau when the flashes were made sufficiently strong (Figure 17–16b). Even with flashes of saturating intensity and optimal timing, the amount of O_2 was very small relative to the chlorophyll content of the algae. The cells contained approximately 2500 molecules of chlorophyll for each molecule of O_2 that they evolved. This was extremely puzzling at the time, because it was generally accepted that chlorophyll was intimately involved in the photochemistry of CO_2 fixation and O_2 evolution. Now it is clear that *most of the chlorophyll is part of the antenna system*. When the flash intensity is high, the amount of O_2 evolution that can occur on each flash is limited by the concentration of reaction centers in the algae, and this is much smaller than the total concentration of chlorophyll. The apparent discrepancy between the number of chlorophyll molecules per molecule of P700 or P680 (about 300) and the number of chlorophyll molecules per molecule of O_2 (2500) will make more sense after we have discussed the fact that chloroplasts have to absorb about eight photons for each molecule of O_2 that is evolved.

The chlorophyll or bacteriochlorophyll molecules that make up the antenna are bound noncovalently to integral membrane proteins with molecular weights that vary from 6000 to 40,000, depending on the organism. Each polypeptide typically carries only two or three pigment molecules. These small complexes aggregate into larger arrays that allow the excitation to hop rapidly from complex to complex by resonance energy transfer.

Along with bacteriochlorophyll or chlorophyll, the antenna systems contain a variety of other pigments, such as carotenoids. These *accessory pigments fill in the antenna's absorption spectrum in regions where the chlorophylls do not absorb well*. As shown in Figure 17–5, chlorophyll a absorbs red or blue light well, but not green. Carotenoids, which are long, linear polyenes (Figure 17–17), have absorption bands in the green region. The energy that they absorb is transferred to the chlorophyll molecules of the antenna, and from there to the reaction centers. This transfer is thermodynamically downhill because the carotenoid's excited singlet state has a higher energy than the lowest excited singlet state of chlorophyll. For another role played by carotenoids, see Box 17–A.

Chloroplasts Have Two Photosystems Linked in Series

We mentioned earlier that chloroplasts contain two types of reactive chlorophyll complexes, P700 and P680. *Together with their antennae and their initial electron acceptors and donors, the reaction centers that contain P700 or P680 form two distinct assemblies called photosystem I and photosystem II, respectively, or more simply system I and system II.* Much

BOX 17–A

Carotenoids Protect the Cell Against Damage by O₂

In addition to transferring excitation energy to the chlorophylls, carotenoids play an important role in protecting the cell against damage by O_2 at high light intensities. If the antenna system is flooded with light too rapidly for the reaction centers to keep pace, the antenna chlorophylls discharge most of the extra energy as fluorescence and heat. However, the excited molecules also have an opportunity to evolve into an excited _triplet state_, in which the spins of the two unpaired electrons are parallel (Illustration 1a). Excited triplet states are relatively long-lived. They cannot decay to the ground (singlet) state unless the electronic spin changes again, and this does not happen readily. One way that they do decay is by reacting with molecular O_2, which has a triplet ground state. The reaction returns the chlorophyll to its ground state and promotes the O_2 to an excited _singlet_ state (Illustration 1b). This can have lethal consequences for the cell, because singlet O_2 is extremely toxic. It reacts irreversibly with a variety of groups in proteins, nucleic acids, and lipids.

Carotenoids intervene to prevent these destructive side reactions by quenching the excited triplet chlorophyll before it has a chance to react with O_2 (Illustration 1c). In this process, the chlorophyll returns immediately to its ground state, and the carotenoid is elevated to an excited triplet state. The carotenoid triplet state cannot generate singlet O_2, because it lies below singlet O_2 in energy. Instead, it decays harmlessly to the ground state. Carotenoids also can quench singlet O_2 itself.

Illustration 1

(a) An excited chlorophyll molecule can change its net electronic spin by a process called intersystem crossing, which converts the molecule from an excited singlet state ($^1Chl^*$) to an excited triplet state ($^3Chl^*$). In the triplet state, the spins of the two unpaired electrons are parallel. This gives the molecule a net electronic spin. There are three possible ways to align this spin in a magnetic field, and the three alignments have different energies. A superscript 1 is used in this figure to indicate that the ground state of chlorophyll (^1Chl) has no net spin and is therefore a singlet state. The lowest triplet states of chlorophylls lie well above the ground states, but below the lowest excited singlet states, in energy. Triplet states are not intermediates in the photosynthetic electron-transfer sequence; they are formed when the antenna is flooded with photons too rapidly for the reaction centers to handle. (b) Molecular O_2 is unusual in that its ground state is a triplet state ($^3O_2$) and its lowest excited state is a singlet state ($^1O_2^*$). Singlet O_2 can be formed by a simultaneous transfer of energy and spin from a chlorophyll molecule in an excited triplet state ($^3Chl^*$). (c) A chlorophyll in an excited triplet state also can transfer its energy and spin to a carotenoid. This raises the carotenoid from its ground state (^1car) to a triplet state ($^3car^*$). The carotenoid triplet state is too low in energy to form singlet O_2.

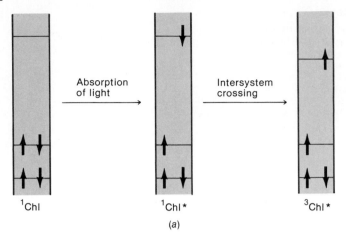

(a)

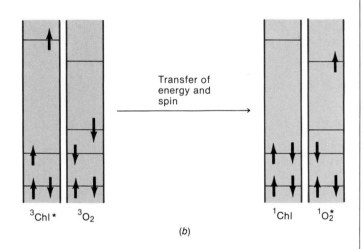

(b)

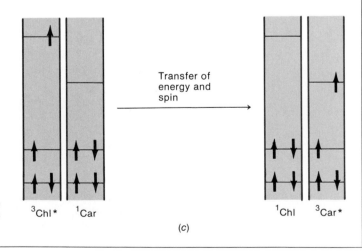

(c)

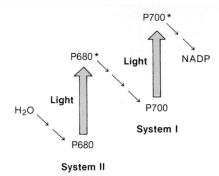

Figure 17-18

The Z scheme for the photosynthetic apparatus of plants. Two photochemical reactions are required to drive electrons from H_2O to NADP. In this scheme, the solid arrows represent paths of electron flow. The electron carriers are positioned vertically according to their $E^{\circ\prime}$ values, with the strongest reductants (most negative $E^{\circ\prime}$ values) at the top. Electron flow downward is thermodynamically spontaneous.

evidence indicates that *systems I and II are connected in series*, as shown in Figure 17–18. Excitation of P700 (system I) generates a strong reductant, which transfers electrons to NADP by way of several secondary electron carriers. Excitation of P680 (system II) generates a strong oxidant, which oxidizes H_2O to O_2. The reductant formed in system II injects electrons into a chain of carriers that connect the two photosystems. This scheme, which is called the *Z scheme*, was first suggested by R. Hill and F. Bendall in 1960. Note that the Z scheme differs from the bacterial electron-transport chain that we discussed above in being linear rather than cyclic.

Figure 17–19 gives a more detailed picture of the electron carriers that participate in the Z scheme. The electron acceptors on the reducing side of photosystem II resemble those of purple bacterial reaction centers. The acceptor that removes an electron from P680 appears to be a molecule of pheophytin *a*. The second and third acceptors are molecules of plastoquinone, whose structure was shown in Figure 17–11. As in the bacterial reaction center, electrons are transferred one at a time from the first plastoquinone to the second. When the second plastoquinone becomes doubly reduced, it picks up protons from the stromal side of the thylakoid membrane and dissociates from the reaction center to join a pool of plastoquinone molecules in the membrane.

The chain of carriers between the two photosystems includes the *cytochrome bf complex* and a copper protein called *plastocyanin*. The cytochrome *bf* complex is very similar to the mitochondrial and bacterial cytochrome bc_1 complexes. It contains a cytochrome with two *b*-type hemes, an iron-sulfur protein, and the *c*-type cytochrome *f*. As electrons move through the cytochrome *bf* complex from reduced plastoquinone to cytochrome *f*, the plastoquinone probably executes a Q cycle similar to the cycle that was presented for ubiquinone in the mitochrondrial complex III and in the bacte-

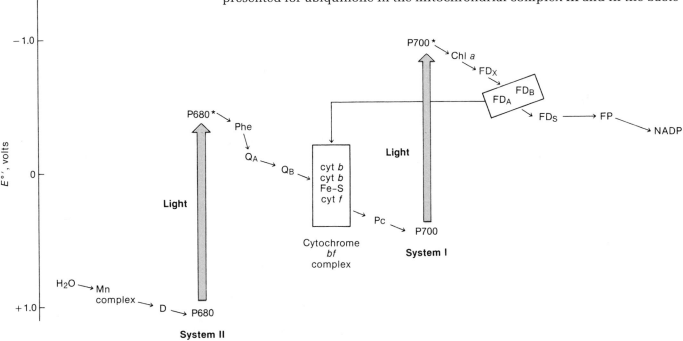

Figure 17-19

A more detailed version of the Z scheme. [Mn complex = a complex of four Mn atoms bound to a protein; D = an unidentified electron carrier; Phe = pheophytin *a*; Q_A and Q_B = two molecules of plastoquinone; Fe-S = an iron-sulfur protein similar to the Reiske iron-sulfur protein of the mitochondrial cytochrome bc_1 reductase; cyt *b* = *b*-type cytochromes; cyt *f* = cytochrome *f*; PC = plastocyanin;

Chl *a* = chlorophyll *a*; FD_X, FD_A, and FD_B = membrane-bound iron-sulfur proteins (bound ferredoxins); FD_S = soluble ferredoxin; FP = flavoprotein (ferredoxin-NADP oxidoreductase).] There may be an additional electron carrier in photosystem I between Chl *a* and FD_X. The sequence of electron transfer through FD_A and FD_B is not yet clear.

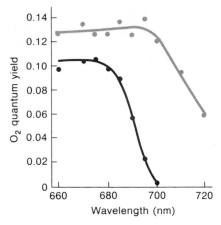

Figure 17–20

The quantum yield of O_2 evolution from a suspension of algae (*Chlorella pyrenoidosa*) as a function of the excitation wavelength. Measurements were made without any supplementary light (lower curve) and with supplementary blue-green light (upper curve). The quantum yield was calculated as mols of O_2 evolved per einstein of light incident on the sample; O_2 evolution caused by the supplementary light alone was subtracted. Without the supplementary light the quantum yield falls off precipitously in the far-red region, above 680 nm. The antenna of photosystem I absorbs light well in this region, but that of photosystem II does not. Supplementary blue-green light, which is absorbed well by system II, increases the quantum yield with which the far-red light can be used. (After R. Emerson et al., Some factors influencing the long-wavelength limit of photosynthesis, *Proc. Natl. Acad. Sci. USA* 43:133, 1957.)

rial photosynthetic apparatus (see Figure 16–18). The cytochrome *bf* complex provides electrons to plastocyanin, which transfers them to P700$^+$ in the reaction center of photosystem I. Additional details on the electron carriers between P700 and NADP and between H_2O and P680 will be presented later, after we discuss some of the evidence supporting the Z scheme.

Some of the earliest observations that led to the Z scheme came from measurements of how the quantum yield of O_2 evolution in algae depends on the excitation wavelength. The quantum yield of O_2 evolution is the molar ratio of O_2 evolved to photons absorbed. In green algae, the quantum yield is relatively independent of wavelength between 400 and 675 nm, but falls off drastically in the far-red region near 700 nm (Figure 17–20). This is odd, because most of the absorbance in the far-red region is due to chlorophyll *a*. Why would light absorbed by chlorophyll *a* be used less efficiently than light absorbed by carotenoids or other accessory pigments? A clue came from R. Emerson's finding in 1956 that 700-nm light could be used more efficiently if it was superimposed on a background of weak blue-green light (Figure 17–20). It was found subsequently that the blue-green light actually did not have to be presented *simultaneously* with the red light. Illumination with blue-green light improved the utilization of far-red light even if the blue-green light was turned off several seconds before the red light was turned on.

These observations can be explained by the Z scheme, if the absorption spectra of the antennas associated with systems I and II are somewhat different. Since the two systems must operate in series, light will be used most efficiently when the flux of electrons through system II is equal to that through system I. If light of some wavelengths excites one of the systems more frequently than it does the other, some of the light will be wasted.

The effectiveness with which light of different wavelengths excites the two photosystems can be explored by blocking one of the systems. For example, herbicides such as 3-(3,4-dichlorophenyl)-1,1-dimethylurea (DCMU) block electron flow between the two molecules of plastoquinone in the reaction center of system II. System I continues to function well in the presence of DCMU if a reductant is added to provide electrons to one of the carriers in the chain connecting the two systems. If the excitation wavelength is varied, it becomes clear that system I absorbs and uses virtually all wavelengths of light up to about 740 nm. Similar studies of system II show that it is not excited well by far-red light. The pool of electron carriers between the two systems will, therefore, be drained of electrons if algae are illuminated with far-red light in the absence of added reductants. P700 will remain oxidized and will be unable to respond to the light. System II does absorb blue-green light well, so illumination with green light will replenish the pool. Electrons will remain in the pool when the blue-green light is turned off, hence far-red light can be used effectively for a time even after a period of darkness.

Subsequent experiments by Louis Duysens provided strong support for the idea that *system I withdraws electrons from carriers situated between the two photosystems, while system II feeds electrons to these components.* Duysens and his colleagues measured the redox states of several of the carriers directly. The easiest component to measure is cytochrome *f*, because its absorption spectrum changes markedly when the cytochrome undergoes oxidation. (Refer to Figure 16–4 to see the absorption spectra of a *c*-type cytochrome in the oxidized and reduced forms, and to Figure 17–19 for the position of cytochrome *f* in the Z scheme.) Illumination of algae with far-red light causes essentially all of the cytochrome *f* in the cells to become oxidized, as would be expected if far-red light drives system I but not system II (Fig. 17–21). When a supplementary green light is turned on, some of the cytochrome returns quickly to the reduced state. This makes sense if green

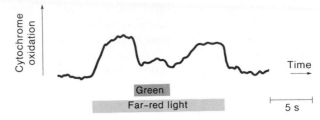

Figure 17–21

Illumination of a suspension of red algae *(Porphyridium cruentum)* with far-red light (680 nm) causes an oxidation of cytochrome *f*, as measured by an absorbance change at 422 nm. (The box below the trace indicates the period when the far-red light was on.) Green (562-nm) light superimposed on top of the far-red light causes the cytochrome to become more reduced. Additional measurements showed that green light also could cause cytochrome oxidation, if it was turned on in the absence of the far-red light, when the cytochrome was initially reduced. This result indicates that the green light can drive either system I or system II. Far-red light can only cause cytochrome oxidation, because it is absorbed only by system I.

light can drive system II. During the illumination, individual cytochrome molecules will cycle repeatedly between the oxidized and reduced states, so that the cytochrome population as a whole will be in a steady state of partial oxidation. When the green light is turned off, the cytochrome returns to the fully oxidized state. Green light actually can cause either a net oxidation or a net reduction of cytochrome *f*, depending on the initial conditions, so it must be able to excite either system II or system I. Far-red light, however, can only cause oxidation of the cytochrome.

The electron carriers in system II are located mainly in the stacked, granal regions of the thylakoid membrane, whereas those of system I occur mainly in the unstacked stromal lamellae (Figure 17–22). The cytochrome *bf* complexes that participate in the transport of electrons between the two photosystems are found in both regions of the membrane. The distribution of photons between the two photosystems depends partly on the relative numbers of antenna complexes in the two regions of the membrane, and can change as a result of movements of antenna complexes from one region to another. Movements of one type of antenna complex appear to be regulated by phosphorylation of the protein. When chloroplasts are illuminated with

Figure 17–22

The photosystem II complexes are located mainly in the stacked (granal) regions of the thylakoid membrane, whereas photosystem I complexes are most abundant in the unstacked stromal lamellae. The cytochrome *bf* complex is located in both regions of the membrane. Some types of antenna complexes also are found in both regions, and can shift from one region to the other in response to a change in conditions. The ATP-synthase occurs mainly in the unstacked regions. These conclusions came from studies in which thylakoid membranes were fragmented gently by ultrasound and then separated by density-gradient centrifugation into fractions • representing stacked and unstacked membranes.

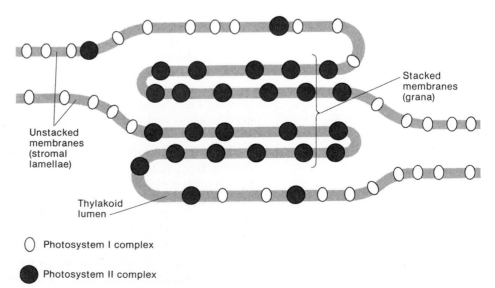

Stacked membranes (grana)

Unstacked membranes (stromal lamellae)

Thylakoid lumen

○ Photosystem I complex

● Photosystem II complex

light that is absorbed better by photosystem I than by photosystem II, the mobile antenna complex moves into the stacked regions, and a larger fraction of the excitations then are passed to system II.

If the reaction centers of system I and system II are segregated into separate regions of the thylakoid membrane, how can electrons move readily from system I to system II? The answer appears to be that the plastoquinone that is reduced in the reaction center of system II can diffuse rapidly in the plane of the membrane, just as ubiquinone does in the mitochondrial inner membrane. Plastoquinone thus carries electrons from system II to the cytochrome *bf* complex. Similarly, plastocyanin is a small protein that can diffuse in the thylakoid lumen. Plastocyanin thus can act as a mobile electron carrier from the cytochrome *bf* complex to a reaction center of system I, just as cytochrome *c* carries electrons from the mitochondrial cytochrome bc_1 complex to cytochrome oxidase and as a *c*-type cytochrome provides electrons to the reaction centers of purple bacteria (Figure 17–14).

System I Reduces NADP by Way of Iron-sulfur Proteins.   On the reducing side of photosystem I, the earliest electron acceptor appears to be a molecule of chlorophyll *a* (Figure 17–19). In this respect, the reaction center of system I resembles the reaction centers of system II and the purple photosynthetic bacteria, which use pheophytin *a* and bacteriopheophytin *a* as initial electron acceptors. From this point on, however, system I is different. Its next set of electron carriers consists of *iron-sulfur proteins* instead of quinones.

The first of the iron-sulfur proteins to be characterized was a soluble protein called *ferredoxin*. This protein is found in the chloroplast stroma. It is designated as FD_S in Figure 17–19. FD_S contains two iron atoms and two atoms of inorganic sulfide, which are bound to the sulfur atoms of four cysteine residues in the type of complex that was illustrated in Figure 16–8*a*. Photosystem I also contains three additional iron-sulfur proteins that are often called "bound" ferredoxins because they cannot be solubilized readily from the thylakoid membrane. These are designated FD_X, FD_A, and FD_B in Figure 17–19. FD_A and FD_B probably contain four iron atoms and four inorganic sulfides each, held by four cysteines in the cubic structure shown in Figure 16–8*b*. FD_X also is likely to contain a cluster of four iron and four sulfur atoms, but it is possible that it has only two irons. As was discussed in Chapter 16, iron-sulfur proteins undergo one-electron oxidation-reduction reactions. The $E^{\circ\prime}$ values for these transitions are about -0.7 V, -0.59, -0.54, and -0.40 V for FD_X, FD_B, FD_A, and FD_S, respectively. (Some iron-sulfur proteins have much more positive $E^{\circ\prime}$ values. The $E^{\circ\prime}$ value for the iron-sulfur protein in the cytochrome *bf* complex is about $+0.30$ V.)

The chlorophyll molecule that is reduced by P700 in photosystem I transfers an electron to FD_X by way of another electron carrier whose nature is not yet clear. FD_X reduces FD_A and FD_B, which in turn reduce FD_S (Figure 17–19). From FD_S, electrons move to a flavoprotein (*ferredoxin-NADP oxidoreductase*) and then to NADP.

O_2 Evolution Requires the Accumulation of Four Oxidizing Equivalents in System II.   The oxidation of H_2O to O_2 requires the removal of four electrons for each O_2 produced. According to the Z scheme each electron must traverse the photochemical reactions of both system I and system II, so at least 8 photons (2×4) have to be absorbed for each O_2 that is released. Currently accepted experimental measurements of the number of photons that are needed, the *quantum requirement*, are indeed on the order of 8 to 12. (The quantum requirement is the reciprocal of the quantum yield of O_2 evolution.) The quantum requirement will exceed 8 if some of the photons that are

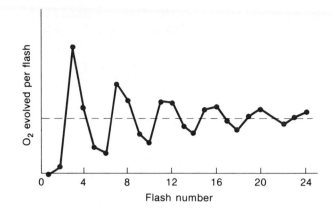

Figure 17–23

O_2 evolution from a suspension of isolated chloroplasts that was excited with a series of 24 flashes after having been kept in the dark for several minutes. Little or no O_2 is evolved on the first two flashes. O_2 evolution peaks on the third flash and on every fourth flash thereafter. The oscillations in O_2 evolution are damped, and after many flashes the yields converge on the level indicated by the horizontal line. This level is the same as the average yield measured in experiments like those illustrated in Figure 17–16.

absorbed are lost as heat or fluorescence from the antenna, or if some of the electrons pumped through the photosystems are not removed to NADP, but instead cycle back into the electron-transport chain between the two photosystems. As we shall discuss below, some cyclic electron transport of this sort does occur in chloroplasts.

If the photosystem II reaction center can transfer only one electron at a time, how does the photosynthetic apparatus assemble the four oxidizing equivalents that are needed for the oxidation of H_2O to O_2? One possibility would be that several different system II reaction centers cooperate, but this seems not to happen. Instead, each reaction center progresses through a series of oxidation states, advancing to the next state each time it absorbs a photon. O_2 evolution occurs only when the reaction center has accumulated four oxidizing equivalents. This conclusion comes principally from measurements of the amount of O_2 that is evolved on each flash when algae or chloroplasts are excited with a series of short flashes after a period of darkness. Pierre Joliot found that essentially no O_2 is released on the first or second flashes (Figure 17–23). On the third flash, however, there is a burst of O_2. After this, the amount of O_2 released on each flash oscillates, going through a maximum every fourth flash.

Bessel Kok pointed out that this pattern can be explained if the reaction center cycles through five different oxidation states, S_0 through S_4, as shown in Figure 17–24. When the system reaches state S_4, O_2 is given off, and the reaction center returns to state S_0. The fact that the first burst of O_2 comes on the third flash instead of the fourth can be explained if, during the dark period before the flashes, the reaction centers relax mainly into S_1 rather than into S_0. By adding reductants, it is possible, in fact, to convert the reaction centers from S_1 to S_0, so that the first peak of O_2 occurs on the fourth flash. The gradual damping of the oscillations in Figure 17–23 is due to reaction centers that get out of phase, either because they miss being excited on one of the flashes or because they are excited twice and advance two steps.

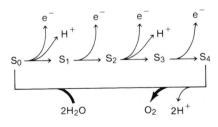

Figure 17–24

The five different oxidation states of the O_2-evolving apparatus, S. One electron (e^-) is removed photochemically in each of the transitions between states S_0 and S_4. S_4 decays spontaneously, releasing O_2. Protons are released at several steps of the cycle.

The component that undergoes oxidation, as the photosystem II reaction center progresses from one of its oxidation states to the next, appears to be a complex of four atoms of manganese bound to a protein. The manganese atoms probably are oxidized sequentially from the Mn(III) level to Mn(IV). The manganese complex transfers electrons to $P680^+$ via an additional electron carrier (D in Figure 17–19), whose nature is still uncertain.

The $E^{\circ\prime}$ of the $P680^+/P680$ redox couple must be on the order of $+1$ V, because the secondary oxidants that $P680^+$ generates are powerful enough to oxidize water. The O_2/H_2O couple has an $E^{\circ\prime}$ of $+0.82$ V. In contrast, $P700^+/P700$ has an $E^{\circ\prime}$ of approximately $+0.50$ V. It is puzzling that $P680^+$ is so much stronger an oxidant than $P700^+$, if they both consist of chlorophyll a.

However, we would expect the redox potential of the chlorophyll radical to depend strongly on the polarity of the environment.

The Flow of Electrons from H₂O to NADP Is Linked to Proton Transport

As in mitochondrial inner membrane and in the cell membrane of the purple photosynthetic bacteria, the flow of electrons through the chloroplast's cytochrome *bf* complex results in the translocation of protons across the thylakoid membrane. When plastoquinone undergoes reduction in photosystem II and in the cytochrome *bf* complex, protons are taken up from the stromal side of the membrane. When the plastoquinone is reoxidized, protons are released into the thylakoid lumen. This lowers the pH in the lumen, and makes the inside positively charged with respect to the stroma. In current schemes of the Q cycle catalyzed by the cytochrome *bf* complex, four protons move across the membrane for each pair of electrons that proceed to photosystem I. Two more protons are released in the lumen for each molecule of H_2O that is oxidized to $\frac{1}{2} O_2$, and one proton is taken up from the stroma when NADP is reduced to NADPH (Figure 17–25).

Proton movement back out through the thylakoid membrane is conducted by an ATP-synthase, and this movement drives the formation of ATP. The chloroplast ATP-synthase is structurally very similar to the mitochon-

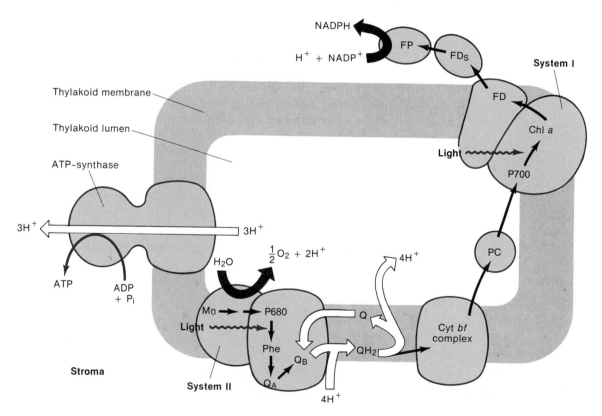

Figure 17–25

The transport of two electrons from photosystem II through the cytochrome *bf* complex to photosystem I results in the movement of four protons from the chloroplast stroma to the thylakoid lumen. The proton translocation probably occurs as a result of a Q cycle resembling that illustrated in Figure 16–18. Two more protons are released in the lumen for each molecule of H_2O that is oxidized to O_2, and one additional proton is taken up from the stroma for each molecule of $NADP^+$ that is reduced to NADPH. The movements of protons from the thylakoid lumen back to the stroma through an ATP-synthase (CF_o-CF_1) drives the formation of ATP. The abbreviations used in this figure are the same as in Figure 17–19.

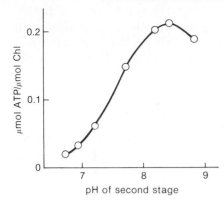

Figure 17–26

Chloroplasts can use an electrochemical potential gradient for protons to form ATP in the dark. In a two-step experiment, spinach chloroplasts first were equilibrated at an acidic pH (pH 4.0). The pH of the solution then was raised quickly to a value between 6.7 and 8.8, as indicated on the abscissa of the graph, and ADP and P_i were added. Because the pH in the thylakoid lumen was buffered at pH 4, the sudden increase in the external pH created a pH gradient of between 2.7 and 4.8 pH units across the thylakoid membrane. Proton efflux through the ATP-synthase can drive the formation of ATP. Under optimal conditions, one molecule of ATP was formed for approximately every five chlorophyll molecules in the sample (see the ordinate scale on the graph). For comparison, the concentration of reaction centers is about one molecule of P700 and one of P680 per 300 chlorophyll molecules. (After A. Jagendorf and E. Uribe, ATP formation caused by acid-base transition of spinach chloroplasts, *Proc. Natl. Acad. Sci. USA* 55:197, 1966.)

drial ATP-synthase (see Chapter 16). Its head-piece, which contains the catalytic sites, is generally called CF_1. Its hydrophobic base-piece, CF_o, includes the proton-conducting channel. As in the case of the mitochondrial enzyme, CF_1 acts as an ATPase if it is removed from CF_o.

Observations on chloroplasts played a key role in the development of the chemiosmotic theory of oxidative phosphorylation, which was discussed in detail in Chapter 16. Andre Jagendorf and his colleagues discovered that if they illuminated chloroplasts in the absence of ADP, the chloroplasts developed the capacity to form ATP when ADP was added later, after the light was turned off. The amount of ATP that could be synthesized was much greater than the number of electron-transport assemblies in the thylakoid membranes, so the energy to drive the phosphorylation could not have been stored in an energized form of one of the electron carriers. Protons were found to be taken up from the solution outside the thylakoid membranes during the illumination, and the ability to form ATP was correlated with a pH gradient across the membrane. If a pH gradient in the right direction was set up across the thylakoid membrane by suddenly raising the pH on the outside, the chloroplasts were able to make ATP without any illumination at all (Figure 17–26).

The formation of ATP by photosynthetic systems is often called <u>*photophosphorylation*</u> to distinguish it from the process that is coupled to respiration in mitochondria. Although the two processes are very similar, they do differ in a few details. The respiratory chain pumps protons outward across the inner membrane, from the mitochondrial matrix to the intermembrane space (Chapter 16). The chloroplast electron-transport chain pumps protons into the thylakoid lumen. The electrochemical potential gradient that is set up for protons across the mitochondrial membrane consists mainly of an electrical potential; in chloroplasts it consists mainly of a pH gradient, which can be greater than three pH units during strong illumination. In both cases, however, the F_1 head-piece of the proton-conducting ATP-synthase is on the side of the membrane that becomes more alkaline. The relative importance of the pH gradient and the electrical potential probably depends on the permeability of the membrane to other ions such as Cl^-. Movement of Cl^- into the thylakoid lumen tends to collapse the electrical potential difference across the membrane, and allows the buildup of a larger pH gradient.

Cyclic Electron Transport Also Occurs in Chloroplasts

In the Z scheme, we have seen how photosystem II, the cytochrome *bf* complex, and photosystem I operate in series to move electrons from H_2O to NADP and to create an electrochemical potential gradient for protons across the thylakoid membrane. In addition to this linear pathway, chloroplasts also have a cyclic electron-transfer pathway that includes photosystem I and the cytochrome *bf* complex, but not photosystem II. When the concentration of NADP is low, electrons ejected by the reaction center of photosystem I can be passed from one of the iron-sulfur proteins FD_A and FD_B to a *b*-type cytochrome (cytochrome b_{563}), and from there to the plastoquinone pool between the two photosystems (Figure 17–19). From plastoquinone, electrons can return to P700 via the cytochrome *bf* complex and plastocyanin. The Q cycle that is included within this cyclic electron flow results in proton translocation across the thylakoid membrane, contributing to the transmembrane electrochemical potential gradient that drives the ATP-synthase.

Cyclic electron transport probably occurs to some extent in chloroplasts even when NADP is not limiting. Its role presumably is to bolster the supply of ATP for CO_2 fixation and for other energy-requiring processes. As we shall discuss in the following section, the conversion of CO_2 into fructose-6-

phosphate requires two equivalents of NADPH and three equivalents of ATP for each CO_2 that is fixed. During continuous illumination, the NADPH and approximately two of the three equivalents of ATP probably are obtained from noncyclic electron transport, and the third equivalent of ATP is obtained from cyclic electron transport.

Carbon Fixation Results from the Reductive Pentose Cycle

The ATP and NADPH that are generated by the photosynthetic electron-transfer reactions are used to drive the fixation of CO_2. Reactions in which CO_2 is incorporated into carbohydrates were discovered in the early 1950s by Melvin Calvin and his coworkers in a series of studies that were among the first to use radioactive tracers. A key experiment was to expose algae to a brief period of illumination in the presence of $^{14}CO_2$ and then to disrupt the cells quickly and search for organic molecules that had become labeled with ^{14}C. The material that turned out to be labeled most rapidly was glycerate-3-phosphate, and almost all of its ^{14}C was found to be in the carboxyl group. To identify the precursor of the glycerate-3-phosphate, Calvin's group looked for changes in the steady-state concentrations of other compounds when they turned a continuous light on or off or when they suddenly raised or lowered the concentration of CO_2. These experiments showed that glycerate-3-phosphate is formed from ribulose-1,5-bisphosphate. A molecule of CO_2 is incorporated in the process, so that one molecule of the five-carbon sugar ribulose-1,5-bisphosphate gives rise to two molecules of the three-carbon glycerate-3-phosphate (Figure 17–27). The reaction is catalyzed by *ribulose bisphosphate carboxylase*, which is found in the chloroplast stroma.

Further studies revealed that other enzymes in the stroma can convert glycerate-3-phosphate back to ribulose-1,5-bisphosphate. The reactions of this *reductive pentose cycle*, or *Calvin cycle*, are shown in Figure 17–28. The glycerate-3-phosphate is first phosphorylated to glycerate-1,3-bisphosphate at the expense of ATP and then reduced to glyceraldehyde-3-phosphate by NADPH. (Note that the nucleotide specificity in the reductive step differs from that of the cytoplasmic glyceraldehyde-3-phosphate dehydrogenase, which uses NAD and NADH.) Glyceraldehyde-3-phosphate and its isomerization product dihydroxyacetone phosphate can combine to form fructose-1,6-bisphosphate under the influence of aldolase. Fructose-6-phosphate, formed by hydrolysis of the fructose-1,6-bisphosphate, combines with another molecule of glyceraldehyde-3-phosphate, generating xylulose-5-phosphate and erythrose-4-phosphate. This reaction involves the coen-

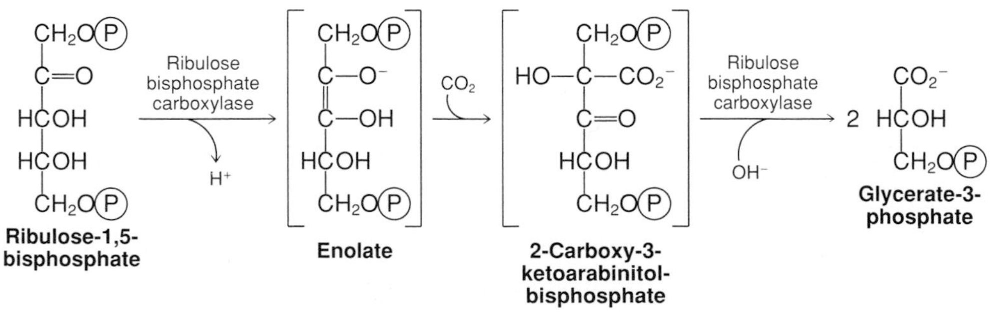

Figure 17–27

The reaction catalyzed by ribulose biphosphate carboxylase involves 2-carboxy-3-ketoarabinitol-1,5-bisphosphate as an enzyme-bound intermediate. The intermediate probably forms by the addition of CO_2 to the enolate of ribulose-1,5-bisphosphate. The substrate is known to be CO_2 rather than bicarbonate.

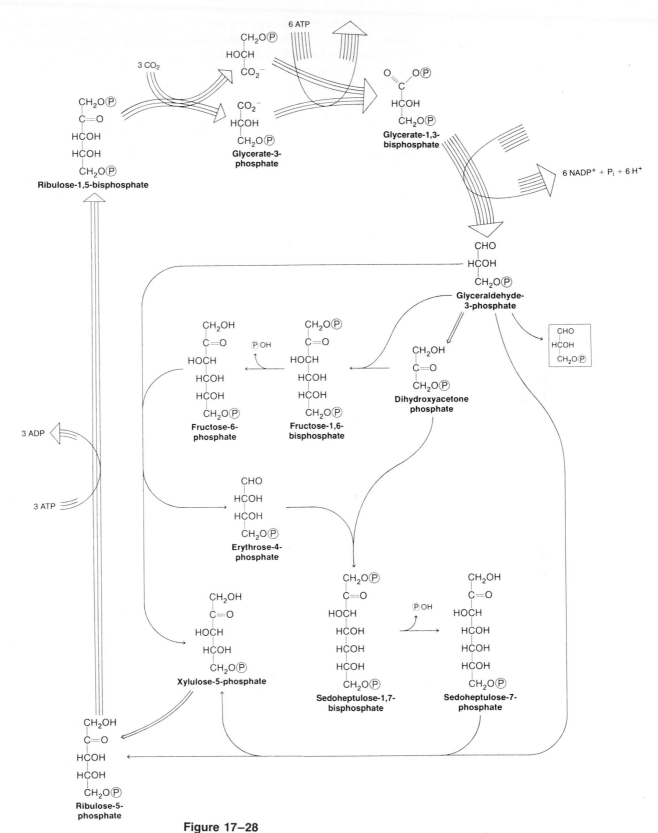

Figure 17–28

The reductive pentose cycle, or Calvin cycle. The number of arrows drawn at each step in the diagram indicates the number of molecules proceeding through that step for every three molecules of CO_2 that enter the cycle. The entry of three molecules of CO_2 results in the formation of one molecule of glyceraldehyde-3-phosphate *(box on right)*, and requires the oxidation of six molecules of NADPH to NADP$^+$ and the breakdown of nine molecules of ATP to ADP.

zyme thiamine pyrophosphate and is similar to the reactions catalyzed by transketolase in the pentose phosphate pathway (Chapter 14). Erythrose-4-phosphate combines with dihydroxyacetone phosphate to form sedoheptulose-1,7-bisphosphate in a second reaction catalyzed by aldolase. After hydrolysis to sedoheptulose-7-phosphate, the seven-carbon sugar reacts with glyceraldehyde-3-phosphate in another transketolase reaction, forming ribulose-5-phosphate and a second molecule of xylulose-5-phosphate. Xylulose-5-phosphate can be isomerized to ribulose-5-phosphate. Finally, ribulose-5-phosphate is phosphorylated to ribulose-1,5-bisphosphate at the expense of an additional ATP, completing the cycle.

In Figure 17–28, note that three molecules of ribulose-1,5-bisphosphate are regenerated for every three that are carboxylated. In the process, three molecules of CO_2 are taken up, and there is a gain of one molecule of glyceraldehyde-3-phosphate. The expenses are nine molecules of ATP converted to ADP and six molecules of NADPH oxidized to NADP, or three molecules of ATP and two of NADPH consumed per CO_2 fixed. The glyceraldehyde-3-phosphate that is saved is exported from the chloroplast to the cytoplasm or converted to hexoses for storage in the chloroplast as starch.

PHOTORESPIRATION AND THE C-4 CYCLE

Ribulose bisphosphate carboxylase typically accounts for more than half the soluble protein in a leaf. It is surely the world's most abundant enzyme, and probably the most abundant protein. The enzyme found in plants contains eight copies each of two types of subunits. The larger subunit, which has a molecular weight of 56,000 and contains the catalytic site, is synthesized within the chloroplast under the direction of chloroplast DNA and ribosomes. The smaller subunit, with a molecular weight of 14,000, is synthesized on cytoplasmic ribosomes under the direction of nuclear DNA and has to be brought into the chloroplast for assembly of the complete enzyme. The smaller subunit is lacking in the enzyme isolated from some species of bacteria, and what role it plays in plants is still unknown.

In addition to serving as a substrate, CO_2 activates ribulose bisphosphate carboxylase by binding to the ϵ amino group of a lysyl residue in the large subunit to form a carbamate (lysine-NH-CO_2^-). This process requires Mg^{2+}, which probably binds to the carboxyl group of the carbamate. The Mg^{2+} in turn forms part of the binding site for a second molecule of CO_2, which acts as the substrate in the carboxylase reaction.

Studies of ribulose bisphosphate carboxylase took on additional complexity with the discovery that the same active site on the enzyme also catalyzes a competing reaction in which O_2 takes the place of CO_2 as substrate. The products of the _oxygenase_ reaction are glycerate-3-phosphate and a two-carbon acid, 2-phosphoglycolate (Figure 17–29). Phosphoglycolate is oxidized to CO_2 by O_2 in a series of additional reactions involving enzymes in the cytosol, mitochondria, and another organelle, the peroxisome. This _photorespiration_ is not coupled to oxidative phosphorylation, and it appears to constitute a severe drain on chloroplast metabolism. In some plants, on the order of one-third of the CO_2 that is fixed is released again by photorespiration. If photorespiration has any benefit to the plant, it is not obvious.

The outcome of the competition between the carboxylase and oxygenase reactions depends on the concentrations of CO_2 and O_2. The K_m of the activated enzyme for the substrate molecule of CO_2 is about 20 μM and that for O_2 is about 200 μM, so CO_2 is the preferred substrate. In illuminated chloroplasts, however, the concentration of O_2 is elevated as a result of the photosynthetic splitting of H_2O. The concentrations of the two gases are frequently

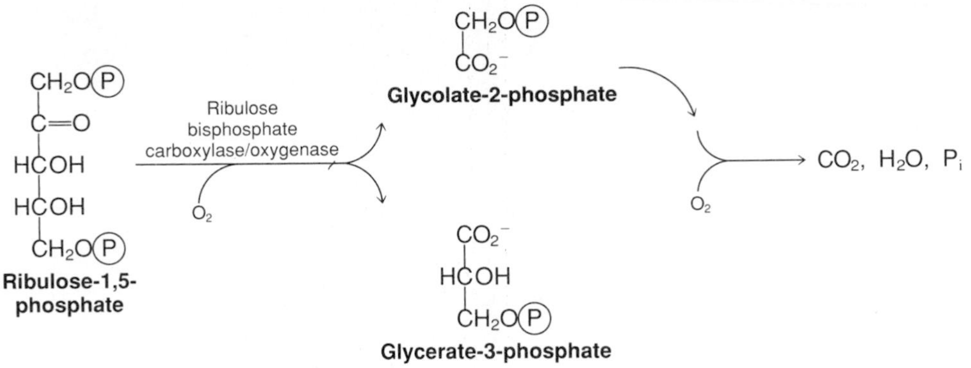

Figure 17–29

Photorespiration results from the oxygenation reaction catalyzed by ribulose bisphosphate carboxylase/oxygenase. Glycolate-2-phosphate generated by the reaction moves from the chloroplast to the cytosol, where other enzymes carry out its further oxidation to CO_2, H_2O, and P_i. The oxygenation reaction, like carboxylation, does not require light directly. It occurs mainly during illumination, however, because the formation of the substrate, ribulose-1,5-bisphosphate, requires ATP and NADPH (see Figure 17–28).

on the order of the K_m values, making O_2 a serious competitor. Plants that live in dry climates face the additional problem that opening the stomata in the leaves to improve the exchange of CO_2 and O_2 with the atmosphere can result in dehydration.

If photorespiration is only a liability, you might think that plants would have evolved a ribulose bisphosphate carboxylase that had minimal activity as an oxygenase, but there is no evidence that evolution has moved in this direction. The oxygenase activity of the enzyme isolated from higher plants is similar to that of the enzymes from algae or photosynthetic bacteria. _Some species of plants have, however, evolved an alternative strategem for favoring the carboxylase reaction over the oxygenase: they increase the concentration of CO_2 in the region of the enzyme._ These plants, which include corn, sugar cane, and numerous tropical species, have a layer of specialized _mesophyll_ cells at the outer surface of the leaf. The mesophyll cells take up CO_2 from the air by the carboxylation of phosphoenolpyruvate to obtain oxalacetate (Figure 17–30). Oxalacetate is reduced to malate, which then moves from the mesophyll cells to the _bundle sheath cells_ that surround the vascular structures in the interior of the leaf. Here the malate is decarboxylated to pyruvate in an oxidative reaction that reduces NADP to NADPH. The pyruvate returns to the mesophyll cell, where it is phosphorylated to phosphoenolpyruvate. The phosphorylation is driven by the splitting of ATP to AMP and pyrophosphate and by the subsequent hydrolysis of the pyrophosphate to phosphate.

The result of this cyclic series of reactions is the delivery of CO_2 and reducing power (NADPH) to the bundle sheath cells at a cost of the breakdown of ATP to AMP (Figure 17–30). The bundle sheath cells fix the CO_2 by the ribulose bisphosphate carboxylase reaction and the reductive pentose cycle. Plants that have this mechanism for concentrating CO_2 can grow considerably more rapidly than species that do not, particularly in hot, dry climates. The auxiliary cycle is called the _C-4 cycle_ because it involves the four-carbon acids malate and oxalacetate. In some species, oxalacetate is transaminated to aspartate rather than being reduced to malate, and it is aspartate that moves to the bundle sheath cells. There is currently interest in the possibility of increasing agricultural productivity by using genetic engineering techniques to introduce the C-4 pathway into plant species that lack it.

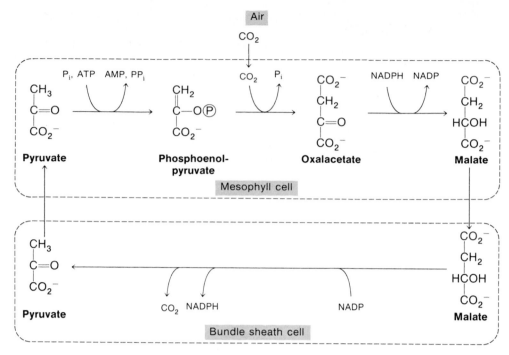

Figure 17–30

The C-4 pathway involves the cooperation of two types of cells. Mesophyll cells *(top)* take up CO_2 from the air, and export malate to the bundle sheath cells *(bottom)*. The bundle sheath cells return pyruvate to the mesophyll cells, and fix the CO_2 using ribulose bisphosphate carboxylase and the reductive pentose cycle.

OTHER BIOCHEMICAL PROCESSES INVOLVING LIGHT

Living things depend on light as a source of energy. They also have found a number of other uses for light. Vision is one of these and is taken up separately in Chapter 36, where it can be treated in the necessary detail. In the remainder of this chapter, we shall consider bioluminescence and the role of light in the synchronization of circadian and seasonal rhythms.

The Synchronization of Circadian and Seasonal Rhythms in Plants Utilizes Phytochrome

Most eukaryotic organisms have endogenous rhythms of metabolic activity with periods of about 24 hours. These *circadian*, or *diurnal*, rhythms are triggered by light, and under natural conditions they are locked in phase with the 24-hr cycle of day and night. Many organisms also have seasonal cycles that are set by the length of the day. In mammals, the synchronization of circadian and seasonal rhythms depends mainly on visual responses to light, but organisms that lack nervous systems depend on other types of photoreceptors. In some unicellular organisms, such as paramecia, the photoreceptive pigment appears to be a flavin or flavoprotein. In higher plants, the receptor is a pigment-protein complex called *phytochrome*.

In addition to the synchronization of diurnal and seasonal rhythms, phytochrome has been implicated in literally hundreds of other responses of plants to light. These include the germination of seeds, changes in ion transport, and synthesis of numerous enzymes. Some of its effects involve changes in gene expression, including an increase in the rate of transcription of the gene for the small subunit of ribulose bisphosphate carboxylase. Other effects, such as changes in the movement of Ca^{2+} across the cytoplasmic

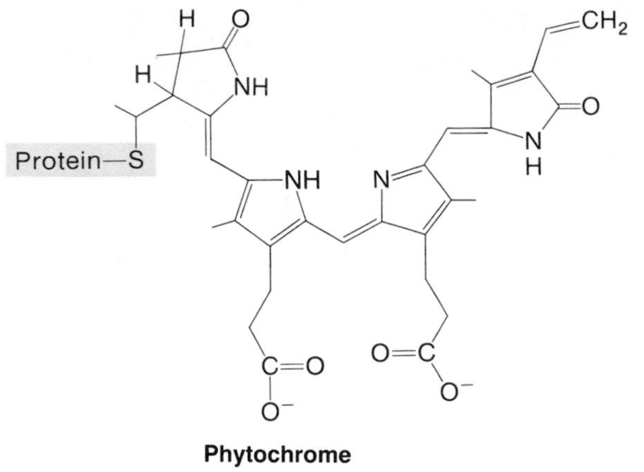

Phytochrome

Figure 17–31
The prosthetic group of phytochrome is an open-chain tetrapyrrole, which is bound to the protein as a cysteine thioether.

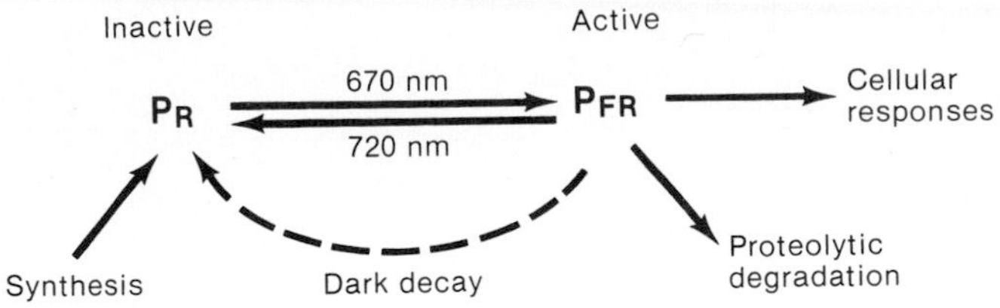

Figure 17–32
Phytochrome exists in two interconvertible forms, P_R and P_{FR}, which have different absorption spectra. The form that is synthesized initially, P_R, has an absorption band at 670 nm. When it is excited with light, P_R is converted to the biologically active species, P_{FR}, which absorbs at 720 nm. P_{FR} decays back to P_R in the dark, and it can be converted to P_R by excitation with light. P_{FR} also is degraded relatively quickly by proteolysis. P_R and P_{FR} may be conformational isomers that share a common excited state.

membrane, seem too rapid for this, and may reflect a more immediate site of action. Some of the effects appear to be mediated by the Ca^{2+}-binding protein calmodulin.

The phytochrome protein is a dimer of subunits with molecular weights of about 120,000. The pigment is an open-chain tetrapyrrole (Figure 17–31). Phytochrome has two interconvertible forms (Figure 17–32). When the molecule is in one form, called P_R, phytochrome has an optical absorption maximum near 670 nm. The absorption of light converts phytochrome from P_R to the other form (P_{FR}), which absorbs maximally at longer wavelengths (720 nm). (The subscripts R and FR stand for "red" and "far red.") P_{FR} is evidently the biologically active form of phytochrome, but its structure is still unknown, and exactly what it does is not clear.

Phytochrome decays slowly from P_{FR} back to P_R in the dark. It also can be returned to P_R immediately by exciting it with light (Figure 17–32). Because P_R absorbs maximally at 670 nm, and P_{FR} at 720 nm, phytochrome can be cycled back and forth between the active and inactive forms by excitation flashes that alternate between the two wavelengths. Continuous illumination sets up a steady-state mixture, in which the amount of P_{FR} depends on the color of the light. Since chlorophyll absorbs red light better than it does far-red light, the color of the light reaching the leaf depends on whether the leaf is shaded by other leaves. The information collected by phytochrome can therefore be useful in controlling the direction and rate of growth of the plant.

Bioluminescence Is Biochemically Produced Light

Conceptually, *bioluminescence is just the reverse of photosynthesis. A reduced substrate reacts with O_2 and is converted to an oxidized product in an excited electronic state. The excited molecule then decays to the ground state, emitting light.* The process occurs in several groups of bacteria and fungi; in marine invertebrates such as sponges, shrimp, and jellyfish; and in a variety of terrestrial creatures, including earthworms, centipedes, and insects. The bacteria that emit light generally live symbiotically with fish in special luminous organs. In some cases, the evolutionary benefits of bioluminescence seem clear: fireflies use it for communication; fish, to attract or locate prey, or to confuse predators. In other cases, such as in fungi, the benefits are not obvious, and one is struck by the expense of the process. The

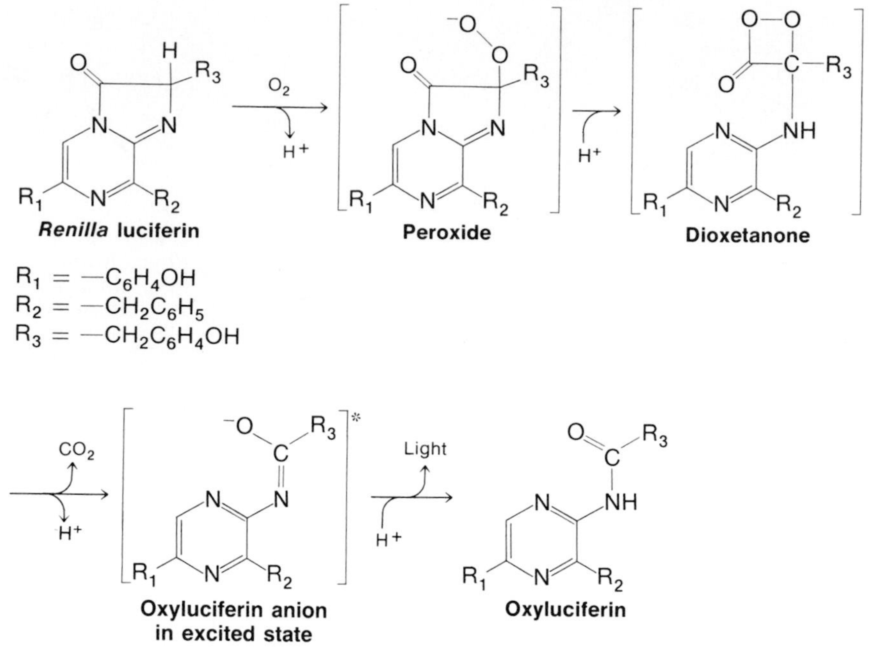

$R_1 = -C_6H_4OH$
$R_2 = -CH_2C_6H_5$
$R_3 = -CH_2C_6H_4OH$

Figure 17–33
Bioluminescent reactions in the sea pansy *Renilla reniformis*. The structures in brackets are plausible enzyme-bound intermediates, but have not been identified conclusively. They are based partly on analogous nonenzymatic reactions of dioxetanones and on the observation that one of the O atoms of the CO_2 that is released comes from the O_2.

energy that is released in the oxidation-reduction reaction could, in principle, be directed into ATP production instead of being emitted as light.

The substrates that undergo oxidation with the emission of light are all called *luciferins*, although their structures vary among different bioluminescent species. The oxidized products are termed *oxyluciferins*, and the enzymes that catalyze the oxidations are called *luciferases*. Figures 17–33 and 17–34 show the structures of the luciferins and oxyluciferins used by two species, the sea pansy (*Renilla reiformis*) and the firefly (*Photinus*). The

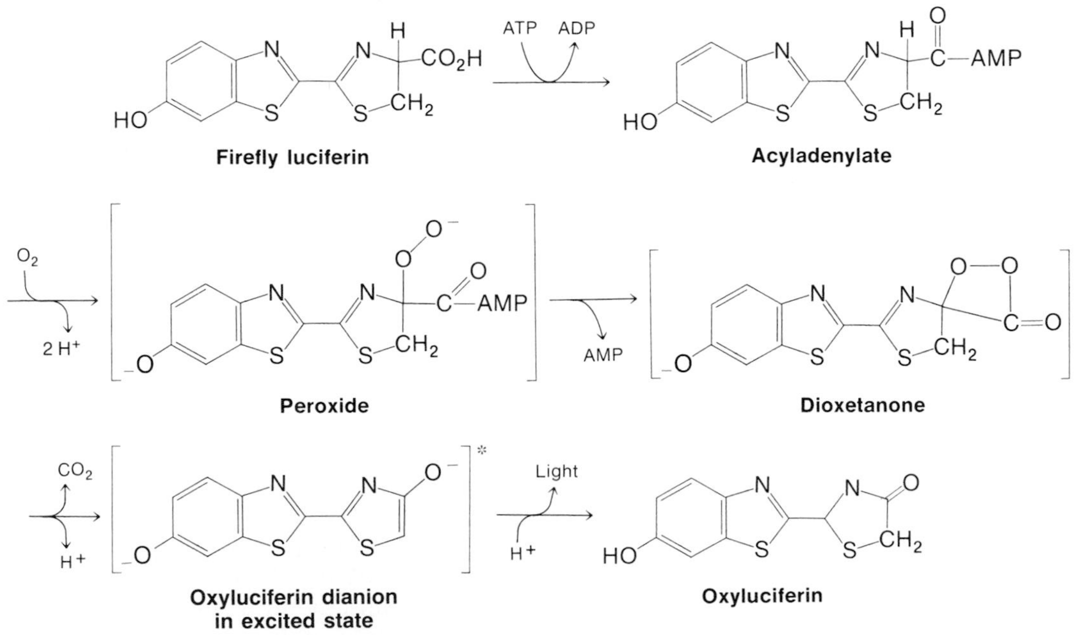

Figure 17–34
Bioluminescent reactions of fireflies. The excited oxyluciferin that emits light is believed to be a dianion, in which the phenolic and enolic oxygens are both ionized. The ionization state can be deduced by comparing the emission spectrum of the luminescence with those of model compounds.

figures also indicate likely intermediates in the oxidation reactions. Although the luciferins have rather different structures in the two species, the chemistry appears to be much the same. O_2 reacts with a heterocyclic ring, forming a peroxide at a carbon that is bound to a nitrogen and to a carboxylic amide or ester. The peroxide cyclizes to give a cyclic peroxide, or dioxetanone, with the release of the amine or alcoholic group that was bound to the carboxyl. The cyclic peroxide then decomposes with the release of CO_2, forming the oxyluciferin in an excited electronic state.

The decomposition of the dioxetanone intermediate may involve an internal oxidation-reduction reaction in which an electron is transferred to the peroxide from another part of the molecule. Such a reaction could generate a linked pair of free radicals $(D^+ \cdot A^- \cdot)$ analogous to the species that are formed in the photochemical reactions of photosynthesis. After CO_2 splits off, an electron could return from the reduced component to the oxidized, putting the product in an excited state.

One difference between fireflies and *Renilla* is that the luciferase of fireflies activates the carboxyl group of the luciferin by forming an acyladenylate (Figure 17–34). This step resembles the activation of amino acids for protein synthesis (Chapter 28), but it is unique to fireflies among the bioluminescent systems that have been studied. Since the activation requires ATP, measurements of the light that is emitted by firefly oxyluciferin provide a sensitive assay for ATP.

Luminescent bacteria use a long-chain aliphatic aldehyde such as decanal as a luciferin, oxidizing it to the carboxylic acid. $FMNH_2$ is oxidized simultaneously to FMN. Again, a peroxide appears to be an intermediate in the oxidation. The excited product is probably a hydroxyderivative of FMN, which subsequently loses H_2O to form FMN.

In bioluminescent organisms, the molecule that actually emits the light is sometimes not the oxyluciferin itself, but a pigment on another protein, to which the excited oxyluciferin transfers its energy. This is conceptually like a reverse operation of the light-harvesting antenna systems of photosynthesis.

SUMMARY

The ultimate source of energy to carry out biosynthesis is sunlight. Photosynthesis, the process of capturing sunlight and converting it into chemical energy, occurs in numerous species of bacteria and algae, in addition to higher plants. In plants, the photochemical reactions of photosynthesis occur in the chloroplast thylakoid membrane. In bacteria, they occur in the cytoplasmic membrane. Almost all photosynthetic organisms take advantage of the fact that chlorophyll or bacteriochlorophyll becomes a strong reductant when it is excited with light. Photooxidation of chlorophyll or bacteriochlorophyll occurs in pigment-protein complexes, the reaction centers. Chloroplasts contain two types of reaction centers: photosystem I, which contains a reactive chlorophyll complex called P700, and photosystem II, which contains P680. The reactive complex in purple photosynthetic bacteria (P870) is a bacteriochlorophyll dimer.

The thylakoid membrane and the bacterial cell membrane also contain molecules that serve as antennas. The antenna systems consist of small pigment-protein complexes assembled into large arrays that can contain hundreds of pigment molecules per reaction center. When the antenna absorbs a photon, the excitation energy moves rapidly from complex to complex by resonance energy transfer until the energy is trapped in a reaction center.

The bacterial reaction center contains two molecules of bacteriopheophytin and two of bacteriochlorophyll in addition to the two bacteriochlorophylls of P870. When P870 is excited, it transfers an electron to one of the bacteriopheophytins. The bacteriopheophytin then reduces a quinone, Q_A, which in turn reduces another quinone, Q_B. In some bacteria, Q_A and Q_B are both ubiquinone; in other species one or both of the quinones is menaquinone. When P870 is excited a second time, a second electron is sent through the same carriers to Q_B. The electron released by P870 is replaced by one from a c-type cytochrome. The doubly reduced Q_B picks up protons from the solution on the inside of the cell. Electrons move from the quinol (QH_2) to the c-type cytochrome via a cytochrome bc_1 complex, and protons are released to the solution outside the cell. The movement of protons across the membrane creates an electrochemical potential gradient across the membrane. Proton movement back into the cell is mediated by an ATP-synthase and drives the formation of ATP.

Photosystems I and II of chloroplasts operate in series. The photooxidation of P680 in photosystem II generates a strong oxidant that can oxidize H_2O to O_2. In this process, a manganese complex bound to a protein evidently progresses through four different oxidation states (S_0 to S_4), advancing to the next higher state each time P680 undergoes photooxidation. When state S_4 is reached, O_2 is given off. The electron released by P680 appears to go to a pheophytin, then to a molecule of plastoquinone, and from there to a second plastoquinone. When the second plastoquinone has been doubly reduced, it dissociates from the reaction center and picks up protons from the chloroplast stromal space. Electrons move from the reduced plastoquinone to P700 of photosystem I by means of a cytochrome *bf* complex and a copper protein (plastocyanin).

Photooxidation of P700 in system I appears to reduce a chlorophyll, which transfers electrons to a series of membrane-bound iron-sulfur proteins (ferredoxins). From the bound iron-sulfur proteins, electrons move to a soluble ferredoxin, and then to a flavoprotein that reduces NADP. There also is some cyclic electron flow, in which electrons go from the bound iron-sulfur proteins to the carriers between the two photosystems.

Electron movement through the cytochrome *bf* complex between the chloroplast photosystems results in proton translocation from the stroma to the thylakoid lumen. In addition, protons are released in the lumen when H_2O is oxidized, and are taken up in the stromal space when NADP is reduced. Protons move from the thylakoid lumen back to the stroma through an ATP-synthase (CF_o-CF_1), driving the formation of ATP.

ATP and NADPH are used for the incorporation of CO_2 into carbohydrates. CO_2 reacts with ribulose-1,5-bisphosphate to give two molecules of 3-phosphoglycerate, under the influence of ribulose bisphosphate carboxylase/oxygenase. 3-Phosphoglycerate can then be converted back to ribulose-1,5-bisphosphate at the expense of ATP and NADPH in the reductive pentose cycle. For every three molecules of CO_2 that enter the cycle, there is a gain of one molecule of glyceraldehyde-3-phosphate.

Ribulose bisphosphate carboxylase/oxygenase also catalyzes a reaction in which ribulose-1,5-bisphosphate reacts with O_2, generating glycolate-2-phosphate and glycerate-3-phosphate. Glycolate-2-phosphate is oxidized to CO_2 and H_2O by additional enzymes. This photorespiration appears to be a wasteful process. Some species of plants use an auxiliary cycle of reactions (the C-4 cycle) to concentrate CO_2 in the cells that carry out the reductive pentose cycle. This allows the plants to fix CO_2 at elevated rates.

Biological systems have found other uses for light. In plants, light regulates a broad range of physiological processes through phytochrome, a complex of a protein and an open-chain tetrapyrrole. Phytochrome has two interconvertible forms, P_R and P_{FR}. P_R is converted to P_{FR} when it is excited with red light. P_{FR} reverts to P_R upon excitation with far-red light. The formation of P_{FR} initiates numerous responses, but the mechanism of its action is unknown.

Bioluminescence involves the oxidation of a small molecule (a luciferin) by O_2, creating a product in an excited electronic state. The product decays to its ground state by emitting light. Luciferins vary from species to species, but many are heterocyclic compounds that react with O_2 to form a cyclic peroxide that decomposes by releasing CO_2.

SELECTED READINGS

Clayton R. K., *Photosynthesis: Physical Mechanisms and Chemical Patterns*. London: Cambridge University Press, 1980. An excellent general treatment of photosynthesis.

Duysens, L. N. M., and J. Amesz, Function and identification of two functional systems in photosynthesis. *Biochim. Biophys. Acta* 64:243, 1962. Evidence supporting the Z scheme is obtained by measuring oxidation and reduction of cytochrome *f* in algae illuminated with light of various colors.

Forbush, B., B. Kok, and M. McGloin, Cooperation of charges in photosynthetic O_2 evolution: II. Damping of flash yield, oscillation, deactivation. *Photochem. Photobiol.* 14:307, 1971. Oscillations in O_2 yield during a train of flashes provide evidence for the four S states.

Govindjee, X. (ed.), *Photosynthesis: Energy Conversion by Plants and Bacteria*. New York: Academic Press, 1982. A collection of review articles.

Jagendorf, A. T., and E. Uribe, ATP formation caused by acid-base transition of spinach chloroplasts. *Proc. Natl. Acad. Sci. USA* 55:197, 1966. Chloroplasts can form ATP in the dark if an electrochemical potential gradient for protons is set up across the thylakoid membrane.

Lorimer, G. H., The carboxylation and oxygenation of ribulose-1,5-bisphosphate: The primary events in photosynthesis and photorespiration. *Ann. Rev. Plant Physiol.* 32:349, 1981. Mechanisms and biochemical significance of the carboxylase and oxygenase reactions.

Michel, H., O. Epp, and J. Deisenhofer, Pigment-protein interactions in the photosynthetic reaction centre from *Rhodopseudomonas viridis*. *EMBO Journal* 5:2445, 1986. The structure of a bacterial reaction center is revealed by x-ray crystallography.

Staehelin, L. A., and C. J. Arntzen, (eds.), Photosynthesis III. Photosynthetic membranes and light harvesting systems. *Encyclopedia of Plant Physiology*, Vol. 19. Berlin: Springer-Verlag, 1986. Current reviews of many aspects of photosynthesis.

Woodbury, N. W., M. Becker, D. Middendorf, and W. W. Parson, Picosecond kinetics of the initial photochemical electron transfer reaction in bacterial photosynthetic reaction centers. *Biochem.* 24:7516, 1985. Fast spectrophotometric techniques are used to follow the initial steps in reaction centers purified from photosynthetic bacteria.

PROBLEMS

1. The lowest-energy absorption band of P870 is at 870 nm.
   a. Calculate the energy of an einstein of light with this wavelength.
   b. Estimate the effective standard redox potential $E°$ of P870 in its first excited singlet state if the $E°$ for oxidation in the ground state is +0.45.

2. The traces below show measurements of optical absorbance changes at 870 and 550 nm when a suspension of chromatophores was excited with a short flash. Downward deflections of the traces represent absorbance decreases. Explain the observations. (Absorption spectra of a c-type cytochrome in its reduced and oxidized forms are shown in Chapter 16.)

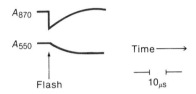

3. Ubiquinone has an absorption band at 275 nm. This band bleaches when the quinone is reduced to either semiquinone or dihydroquinone. The anionic semiquinone has an absorption band at 450 nm, but neither the quinone nor the dihydroquinone has a band at this wavelength. If a suspension of purified bacterial reaction centers is supplemented with extra ubiquinone and reduced cytochrome c and is then excited with a series of short flashes, the absorbance at 275 nm decreases on the odd-numbered flashes as shown in the first curve below. The absorbance at 450 nm increases on the odd-numbered flashes, but returns to its original level on the even-numbered flashes, as shown in the second curve.

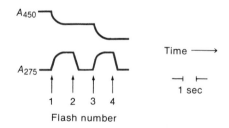

   a. Explain the patterns of absorbance changes at the two wavelengths.
   b. Why is it necessary to have reduced cytochrome c present in order to see these effects?

4. A suspension of algae is excited with a train of 1000 short flashes spaced 20 ms apart.
   a. Draw a graph showing how the total amount of O_2 that the algae evolve during the flash train will depend on the intensity of the flashes. Use meaningful units for the scales on the abscissa and ordinate of the graph.
   b. How could you use the data from such a graph to calculate the average quantum yield of O_2 evolution?
   c. How could you use the data to calculate the effective size of the antenna system?

   d. Why is it necessary to use a large number of flashes for these determinations?

5. You add a nonphysiological electron donor to a suspension of chloroplasts. When you illuminate the chloroplasts, the donor becomes oxidized. How could you determine whether this process involves both photosystem I and photosystem II? (In principle, the donor could transfer electrons either to some component on the O_2 side of photosystem II or to a component between the two photosystems.)

6. The photoreduction of the early electron carrier FD_X in photosystem I can be detected by the optical absorption spectrum of the reduced species. The reduced carrier is not ordinarily seen in chloroplasts that are illuminated under physiological conditions, but it can be seen when chloroplasts are illuminated at very low temperatures after the ambient redox potential has been poised at a strongly negative level.
   a. Why might it be necessary to lower the temperature and the redox potential?
   b. How negative do you think the redox potential would have to be in order to make the photoreduction of FD_X readily observable?
   c. Does the fact that the photoreduction of FD_X can be seen under these conditions prove that FD_X participates in the Z scheme under physiological conditions? Why or why not?

7. When chloroplasts are illuminated with weak continuous light, the intensity of fluorescence from the antenna chlorophylls is relatively low. (About 2 per cent of the photons absorbed are reemitted as fluorescence.) The intensity of the fluorescence from the antenna chlorophyll of photosystem II increases severalfold if the ambient redox potential is lowered below about 0 mV. Explain these observations.

8. When chloroplasts are illuminated with a strong continuous light, the intensity of fluorescence from the chlorophyll of the photosystem II antenna increases with time, as shown in the accompanying graph (curve A).
   a. Explain the increase in fluorescence.
   b. In the presence of 3-(3,4-dichlorophenyl)-1,1-dimethylurea (DCMU), the fluorescence increases more abruptly (curve B). Explain why there is a lag in the increase in the absence of DCMU (curve A) but not in the presence of DCMU (curve B).

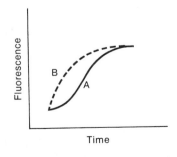

18

CATABOLISM OF FATTY ACIDS

As we learned in Chapter 5, fatty acids are the major components of complex lipids (e.g., triacylglycerols and phospholipids). The fatty acids in triacylglycerols are the most important form of energy storage in animals. An organism gains two main advantages by storing energy as fatty acid rather than as polysaccharides. First, the carbon in fatty acids, when oxidized, yields more ATP than any other form of carbon. Second, fatty acids in lipids are less hydrated than polysaccharides. Hence the lipids pack very close to each other and occupy less space. In addition, the storage of triacylglycerols in fat (adipose) tissue provides insulation from cold temperatures. This function is most obvious in cold-water mammals such as arctic whales (beluga whales), which have vast stores of fat (blubber). This chapter will focus on the mechanism by which fatty acids are mobilized from adipose tissue and the biochemical steps that lead to the oxidation of fatty acids to CO_2 and H_2O, yielding energy as ATP.

MAJOR SOURCES OF FATTY ACIDS FOR OXIDATION

Fatty acids are mainly derived from the diet, but they are also made from carbohydrates and, to a lesser extent, amino acids. Thirty to forty per cent of the calories ingested each day in the human diet are provided by the fatty acid components of complex lipids (e.g., triacylglycerols, phospholipids). These lipids are degraded by enzymes known as *lipases* and *phospholipases*, which are secreted into the intestinal lumen. The fatty acids released by this process are absorbed in the intestine, and triacylglycerols are resynthesized in cells (Figure 18–1). The triacylglycerols are packaged into *lipoproteins*. Lipoproteins are spherical particles that have surfaces composed of phospholipids, cholesterol, and proteins; the core of each particle contains triacylglycerols and cholesterol esters. The lipoproteins will be discussed in detail in Chapter 23.

Chylomicron is a large lipoprotein that is secreted by the intestine into lymphatic vessels (Figure 18–1) that connect with the circulatory system at

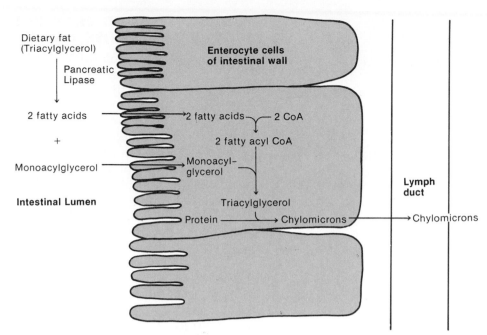

Figure 18–1
Triacylglycerols are digested in the intestine by pancreatic lipase. The fatty acids and monoacylglycerol are absorbed into the intestinal wall. The cells in the wall, enterocytes, resynthesize triacylglycerol and package the lipid with proteins to give lipoprotein particles called chylomicrons, which are secreted from the enterocytes into the lymph.

the jugular vein. In the circulatory system the triacylglycerol components of chylomicrons are degraded to fatty acids and glycerol by an enzyme, _lipoprotein lipase_, that is bound to the exterior of the endothelial cells lining the lumen of the capillary vessels of adipose, heart, muscle, and other tissues (Figure 18–2). The released fatty acids are transported into these tissues for generation of energy (the main topic of this chapter), synthesis of phospholipids (Chapter 22), or storage of energy in triacylglycerols (Chapter 22).

When there is an abundance of energy, dietary fatty acids and fatty acids made from carbohydrate are stored as components of triacylglycerols, primarily in adipose tissue. When energy requirements are high and food is scarce, the animal mobilizes this stored energy. We shall now examine the biochemical processes for degradation of triacylglycerols in adipose tissue and transport of the fatty acids to other tissues for β oxidation.

Figure 18–2
The chylomicrons are delivered to other tissues in the body via the blood stream. In the tissues the chylomicrons bind to the endothelial cells of the capillaries and are degraded to fatty acids and glycerol by lipoprotein lipase. This enzyme is secreted by cells in the tissue and is bound to the endothelial cells.

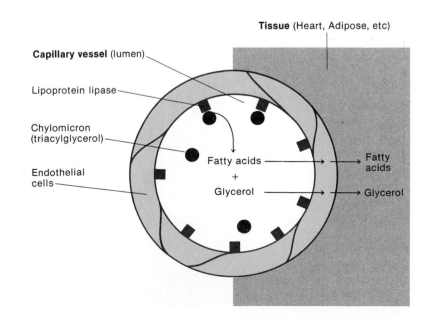

MOBILIZATION AND TRANSFER OF FATTY ACIDS FROM ADIPOSE TISSUE

Although triacylglycerols are found in the liver and intestine, they are primarily found in adipose (fat) tissue, which functions as a storage depot for this lipid. The specialized cell in this tissue is called the *adipocyte*. The cytoplasm of the cell is full of lipid vesicles that are almost exclusively triacylglycerols (Figure 18–3) and serve as an energy reserve for mammals. *When the energy supply from the diet becomes limited, the animal responds to the deficiency with a hormonal signal* that is transmitted to the adipose tissue by the release of *epinephrine, glucagon* and other hormones. The hormones bind to the plasma membrane of the adipocyte and stimulate the synthesis of cyclic AMP (cAMP), as previously discussed for the mobilization of glycogen (Chapter 14). As shown in Figure 18–4, cAMP activates a protein kinase that phosphorylates and activates *triacylglycerol lipase*. The latter enzyme hydrolyzes the triacylglycerol to diacylglycerol with release of a fatty acid from carbon 1 or 3 of the glycerol backbone. This reaction is thought to be the rate-limiting step in the complete hydrolysis of the triacylglycerols. The diacylglycerols and monoacylglycerols are rapidly hydrolyzed to yield fatty acids and glycerol. While it is not certain that the diacylglycerol lipase is a separate enzyme or the same enzyme as triacylglycerol lipase, it is clear that monoacylglycerol lipase is a separate enzyme.

The unesterified fatty acids move through the plasma membranes of the

Figure 18–3
Scanning electron micrograph of white adipocytes from rat adipose tissue (500×). (Courtesy Dr. A. Angel and Dr. M. J. Hollenberg of the University of Toronto.)

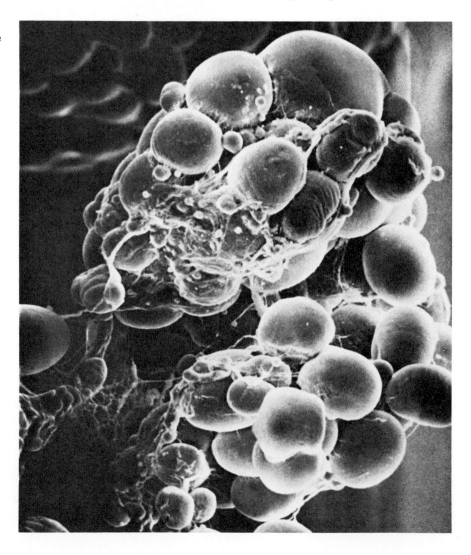

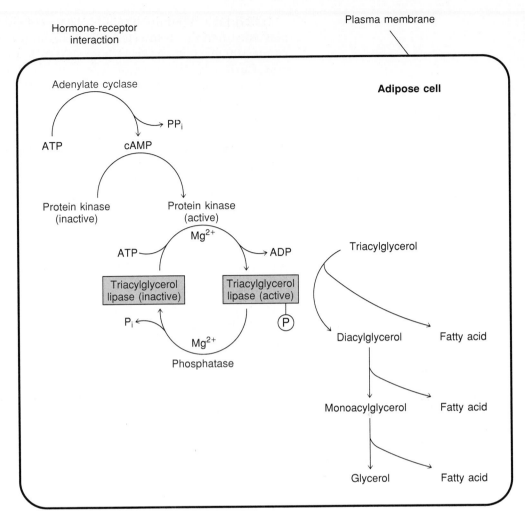

Figure 18–4
When certain hormones (eg., epinephrine) bind to their receptors in adipose tissue, adenylate cyclase is activated. The cAMP that is formed activates protein kinase A, which phosphorylates triacylglycerol lipase. The phosphorylated form of this enzyme is the active species, and triacylglycerols are degraded to fatty acids. The fatty acids are released into the bloodstream, bound by albumin, and delivered to energy-deprived tissues.

adipocytes and endothelial cells of the blood capillaries and are bound by *albumin*, the major protein in plasma. Passive diffusion appears to account for the movement of fatty acids across the adipocyte membrane into the blood plasma. Hence the rate of transfer depends on the concentrations of fatty acids, both in the adipocytes and in the plasma. Albumin carries the fatty acids to other tissues in the body. The glycerol also can be released into the plasma and removed by the liver for glucose production.

Albumin is the most abundant protein in human plasma, constituting about 50 per cent (4 g/100 ml) of the plasma proteins. The protein exists as a monomer with a molecular weight of 66,200 and has the capacity to bind ten molecules of fatty acid, although on the average less than two are bound. Longer-chain fatty acids bind more readily to albumin than do shorter-chain fatty acids (e.g., stearate binds more readily than palmitate). Monounsaturated fatty acids bind with higher affinity to albumin than saturated fatty acids (e.g., oleate binds more readily than stearate). However, linoleate binds less readily than stearate. The time (half-life) it takes for one-half the fatty acids bound to albumin in plasma to be taken up by various tissues is 1 to 2 min.

Although albumin has long been considered an essential protein of plasma, there are a few people with depressed levels of albumin (4.6 to 24 mg/100 ml, compared with 4 g/100 ml), a condition called *analbuminemia*. Curiously, these people are virtually asymptomatic, and *analbuminemia* is usually diagnosed as a result of routine blood analyses. Clearly, large quantities of albumin are not essential for life. The transport function of albumin in people suffering from analbuminemia is probably assumed by the lipoproteins.

Albumin carries the fatty acids to energy-deficient tissues, where the fatty acids move from the plasma to the tissues by a process of diffusion. The amount of fatty acid removed by a tissue depends on the relative concentrations, both in the plasma and in the cells of the tissues. Cardiac muscle utilizes fatty acids as the major oxidative source of energy for ATP synthesis and therefore removes large amounts from the circulation.

ACYL-CoA LIGASES: THE ENZYMES THAT ACTIVATE FATTY ACIDS

The fatty acids taken into the cells are activated in the cytoplasm by reaction with coenzyme A (CoA) and ATP to yield fatty acyl-CoA in a reaction catalyzed by *acyl-CoA ligase* (also known as thiokinase):

$$\text{RCOO}^- + \text{ATP} + \text{CoA} \underset{\text{Mg}^{2+}}{\rightleftharpoons} \text{RCO—CoA} + \text{PP}_i + \text{AMP}$$

As is the case with other reactions in which PP_i is a product, PP_i is rapidly hydrolyzed to $2\,\text{P}_i$. The hydrolysis of ATP to AMP has a $\Delta G^{\circ\prime}$ of about -10.0 kcal/mole, whereas the $\Delta G^{\circ\prime}$ for acyl-CoA hydrolysis is about -7.5 kcal/mole. Hence the synthesis of acyl-CoA is favored. Moreover, the concentration of PP_i in the cells (0.01 mM in rat liver) is very low; thus the pyrophosphorolysis of the acyl-CoA by acyl-CoA lipase is prevented.

There are several acyl-CoA ligases, which differ in subcellular location and specificity with respect to chain length of the fatty acid. Short-chain and medium-chain acyl-CoA ligases are found in the mitochondria. Long-chain acyl-CoA ligase is bound to the endoplasmic reticulum and the outer mitochondrial membrane. This enzyme utilizes both saturated (10 to 18 carbons) and unsaturated (16 to 20 carbons) fatty acids.

TRANSPORT OF FATTY ACIDS INTO THE MITOCHONDRIA

Fatty acids cannot be utilized for the energy requirements of cells until they have been transported into the mitochondrion, the major site of fatty acid oxidation. *Fatty acyl-CoAs cannot cross the inner membranes of mitochondria, but are converted to their acyl carnitine derivatives, which can cross this membrane:*

$$\text{RCO—CoA} + (\text{CH}_3)_3\text{N}^+\text{—CH}_2\text{CHCH}_2\text{COO}^- \rightleftharpoons (\text{CH}_3)_3\text{N}^+\text{—CH}_2\text{CHCH}_2\text{COO}^- + \text{CoA}$$

OH	O
	RC=O
Carnitine	**Acyl carnitine**

This reaction is catalyzed by *carnitine acyltransferase*. There are at least three acyltransferases associated with mitochondria: one specific for short-chain fatty acids, *carnitine acetyltransferase*, and two specific for the longer-chain fatty acids, *carnitine acyltransferases I and II*. There also is evidence for an acyltransferase with intermediate-chain-length specificity.

There is a protein carrier in the inner mitochondrial membrane that can transport carnitine, acetylcarnitine, and short- and long-chain acyl carnitine

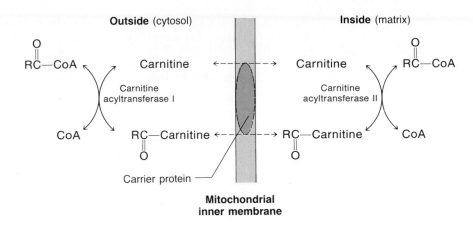

Figure 18–5

Acyl-CoA is not transported across the inner membrane of the mitochondrion. Instead, the acyl-CoA reacts with carnitine to yield the acyl carnitine derivative, which is carried across the membrane by a protein carrier. Once inside the mitochondrion, the acyl carnitine is converted back to its CoA derivative, the substrate for β oxidation.

derivatives across the membrane. The transfer of fatty acyl carnitine into the mitochondria appears to involve an exchange with free carnitine, as illustrated in Figure 18–5. Once inside the mitochondria, the reaction is reversed by _carnitine acyltransferase II_ to yield a fatty acyl-CoA inside the mitochondria. Thus there are at least two distinct pools of acyl-CoA in the cell, one in the cytosol and the other in the mitochondria. Acyltransferases I and II are different enzymes located on opposite sides of the mitochondrial inner membrane.

OXIDATION OF FATTY ACIDS

Our current understanding of fatty acid oxidation did not start to develop in detail until the early 1950s. However, in 1905, Fritz Knoop, a German biochemist, reported a series of experiments that indicated that fatty acids were oxidized by removal of two carbons at a time. He fed rabbits fatty acids in which the methyl group had been replaced with a phenyl group. If the altered fatty acid contained an even number of carbons [for example, C_6H_5—$CH_2(CH_2)_2COOH$], the primary metabolite was phenylacetic acid (C_6H_5—CH_2COOH), which was excreted as a glycine conjugate, phenylaceturic acid (C_6H_5—CH_2CO—$NHCH_2COOH$). When he fed the rabbits a phenyl derivative of a fatty acid with an odd number of carbons [for example, C_6H_5—$CH_2(CH_2)_3COOH$], he found benzoic acid (C_6H_5COOH), which was excreted in the urine as its glycine conjugate, hippuric acid (C_6H_5CO—NH—CH_2COOH). Knoop postulated that _fatty acids are oxidized at the β carbon (hence β oxidation) and degraded to acetic acid and a fatty acid with two fewer carbons_:

$$RCH_2CH_2COOH \longrightarrow R\overset{O}{\overset{\|}{C}}CH_2COOH \longrightarrow R\overset{O}{\overset{\|}{C}}OH + CH_3COOH$$

The next major experimental step was the demonstration in 1943 by J. M. Munoz and Luis Leloir that fatty acids could be oxidized in a cell-free system. This was followed by Albert Lehninger's demonstration that the process of fatty acid oxidation occurred in liver mitochondria and, apparently, involved an "active acetate." Experiments by Fritz Lipmann proved that CoA was involved in the formation of "active acetate":

$$\text{Acetate} + \text{ATP} + \text{CoA} \longrightarrow \text{"active acetate"}$$

Subsequently, in 1951, Feodor Lynen, working with yeast, demonstrated that "active acetate" was acetyl-CoA. At this stage, several laboratories conceived of the idea that CoA might play a role in the activation of fatty acids for β oxidation, and by 1954, the basic outline of β oxidation as we know it today was developed.

The Oxidation of Fatty Acids Occurs in Mitochondria

The outline for β oxidation of a saturated fatty acid is shown in Figure 18–6. In the first reaction, the acyl-CoA is dehydrogenated to yield the α-β (or 2-3)-*trans*-enoyl-CoA and FADH$_2$. This enoyl-CoA is subsequently hydrated stereospecifically to yield the 3-L-hydroxyacyl-CoA. The hydroxyl group is oxidized by NAD$^+$ and a dehydrogenase to yield β-ketoacyl-CoA and NADH. The final step in the sequence involves a thiolytic cleavage to form acetyl-CoA and an acyl-CoA that is two carbons shorter than the initial substrate for β oxidation (Figure 18–6). This acyl-CoA can undergo another round of β oxidation to yield FADH$_2$, NADH, acetyl-CoA, and acyl-CoA. The enzymatic steps are repeated until, in the last sequence of reactions, butyryl-CoA (CH$_3$CH$_2$CH$_2$CO—CoA) is degraded to two acetyl-CoAs (Figure 18–6b).

The first reaction in β oxidation (Figure 18–6) is catalyzed by *acyl-CoA dehydrogenases*, which have FAD tightly, but noncovalently, bound. There are three different soluble matrix enzymes with similar molecular weights (180,000) that show short-, medium- or long-chain acyl-CoA specificity. In each case the products are the *trans* α,β-enoyl-CoA and enzyme-bound FADH$_2$. The electrons from FADH$_2$ are channeled into the electron-transport chain via ubiquinone (see Chapter 16).

The second reaction (Figure 18–6) is catalyzed by *enoyl-CoA hydrase*. Two different mitochondrial enzymes have hydrase activity. One is *crotonase*, a soluble matrix enzyme ($M_r = 165,000$) that prefers short-chain acyl-CoAs. The other prefers long-chain acyl-CoAs and is probably membrane-associated. *3-L-Hydroxyacyl-CoA dehydrogenase* is a soluble matrix enzyme that catalyzes the oxidation of the β-hydroxyl group with NAD$^+$ as a specific cofactor (Figure 18–6).

The last reaction in Figure 18–6 is catalyzed by *thiolase*. Mitochondria contain two types of thiolase in the matrix. One is specific for acetoacetyl-CoA and is involved in ketone body metabolism. The second thiolase acts on 3-ketoacyl-CoAs of various chain lengths and is involved in β oxidation.

Microorganisms also have the ability to oxidize fatty acids. When *E. coli* are grown on fatty acids instead of glucose, the enzymes of β oxidation are produced in much larger quantities. The enzyme activities (except acyl-CoA dehydrogenase, which is a separate enzyme) are associated with a multienzyme complex with a molecular weight of approximately 250,000 and an $\alpha_2\beta_2$ structure. The α subunit ($M_r = 78,000$) has the enoyl-CoA hydratase and the 3-hydroxyacyl-CoA dehydrogenase activities, whereas the β subunit ($M_r = 42,000$) has the thiolase activity. Two other activities associated with the α subunit are *enoyl-CoA isomerase*, which is involved in β oxidation of unsaturated fatty acids (discussed in next section), and *hydroxyacyl-CoA epimerase*, which is involved in β oxidation of D-β-hydroxy fatty acids.

The equations for the complete oxidation of palmitoyl-CoA are shown in Table 18–1. Equation (1) in the table shows the oxidation of palmitoyl-CoA by the enzymes of β oxidation. Each of the products of Equation (1) is further oxidized by the respiratory chain, Equations (2) and (3), or by the tricarboxylic acid cycle and the respiratory chain, Equation (4). When the reactions of Equations (1) to (4) are added together, the result is Equation (5). Hence one molecule of palmitoyl-CoA can be oxidized to yield 131 ATP + 16 CO$_2$ + 146 H$_2$O and CoA. If the starting material were palmitic acid, its complete oxidation would yield 129 ATP + 16 CO$_2$ + 145 H$_2$O. The formation of the CoA derivative of the fatty acid requires two ATP equivalents, and one molecule of H$_2$O is consumed in the hydrolysis of the PP$_i$ produced by

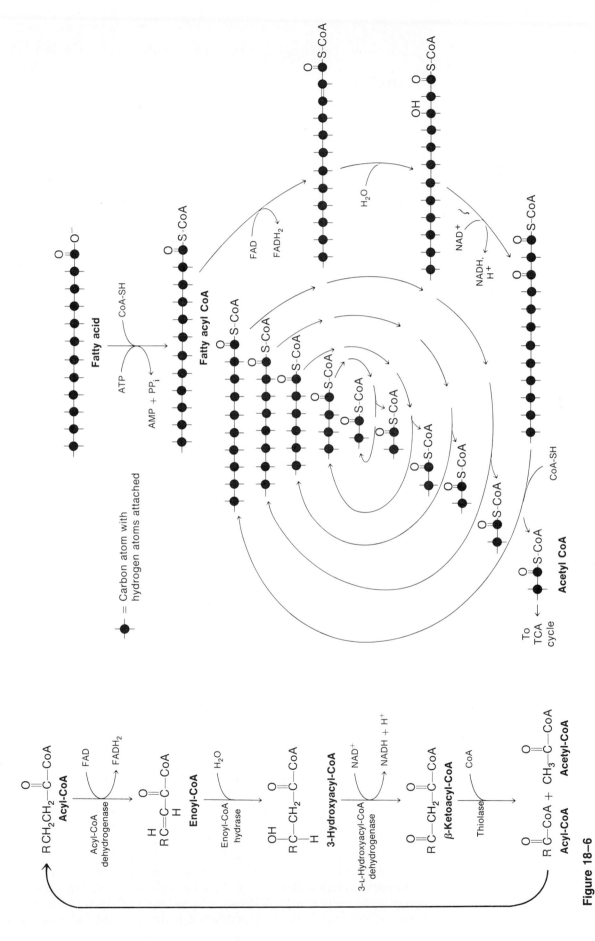

Figure 18–6

(a) The β oxidation of fatty acids consists of four reactions in the matrix of the mitochondrion. Each cycle of reactions results in the formation of acetyl-CoA and an acyl-CoA with two fewer carbons. (b) The cyclic nature of the β oxidation cycle. In some illustrations the —SH group of CoA is indicated for emphasis. Likewise in some illustrations the —S— group in an acyl-CoA structure is indicated.

605

TABLE 18–1
Equations for the Complete Oxidation of Palmitoyl-CoA to CO₂ and H₂O

$$CH_3(CH_2)_{14}CO—CoA + 7\ FAD + 7\ H_2O + 7\ NAD^+ + 7\ CoA \longrightarrow$$
$$8\ CH_3COSCoA + 7\ FADH_2 + 7\ NADH + 7\ H^+ \quad (1)$$

$$7\ FADH_2 + 14\ P_i + 14\ ADP + 3\tfrac{1}{2}\ O_2 \longrightarrow 7\ FAD + 21\ H_2O + 14\ ATP \quad (2)$$

$$7\ NADH + 7\ H^+ + 21\ P_i + 21\ ADP + 3\tfrac{1}{2}\ O_2 \longrightarrow$$
$$7\ NAD^+ + 28\ H_2O + 21\ ATP \quad (3)$$

$$8\ Acetyl\text{-}CoA + 16\ O_2 + 96\ P_i + 96\ ADP \longrightarrow$$
$$8\ CoA + 104\ H_2O + 16\ CO_2 + 96\ ATP \quad (4)$$

$$CH_3(CH_2)_{14}CO—CoA + 131\ P_i + 131\ ADP + 23\ O_2 \longrightarrow$$
$$131\ ATP + 16\ CO_2 + 146\ H_2O + CoA \quad (5)$$

the thiokinase reaction. Interestingly, oxidation of fatty acids can be used as a major source of H₂O, e.g., in the killer whale (which is actually a dolphin). This animal lives in the sea, but does not drink the seawater. Instead, the whale obtains a significant amount of its water from dietary fatty acids.

The oxidation of palmitate to CO₂ and H₂O yields a $\Delta G^{\circ\prime}$ of -2340 kcal/mole. The formation of 129 ATPs has a $\Delta G^{\circ\prime}$ of $+942$ kcal/mole. Thus, approximately 40 per cent (942/2340) of the standard free energy for the oxidation of palmitate is conserved in the formation of ATP.

The yield of ATP from the oxidation of palmitic acid can be compared with that from glucose. Both are major sources of energy in our bodies. Since palmitate has 16 carbons and glucose has 6 carbons, the comparison should be made between 1 palmitate and $2\tfrac{2}{3}$ glucose molecules. As you learned in Chapter 16, 36 ATPs are produced from the complete oxidation of one glucose molecule to CO₂ and H₂O. The yield from $2\tfrac{2}{3}$ glucose molecules is therefore 96 ATPs. Thus oxidation of palmitate yields an additional 33 molecules of ATP, making palmitate a more efficient molecule than glucose for storage of energy. The chemical reason for the difference in oxidative energy yield between glucose and palmitate is that palmitate is almost completely in the reduced state, whereas glucose is partially oxidized, with six oxygens in the molecule.

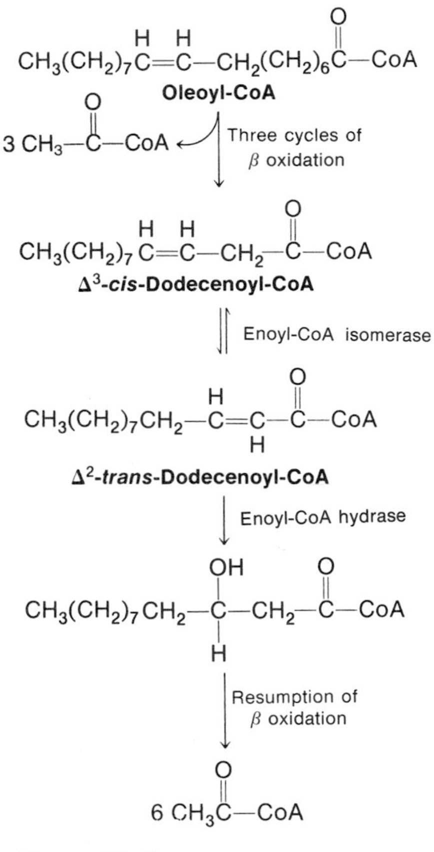

Unsaturated Fatty Acids Are Also Oxidized in Mitochondria

Unsaturated fatty acids are degraded by β oxidation, but additional enzymes are required. As seen in Figure 18–7, oleoyl-CoA can be degraded in the same manner as stearoyl-CoA through the first three cycles of β oxidation. The result, Δ^3-cis-dodecenoyl-CoA ($12:1\ \Delta^3$), however, is not a substrate for the acyl-CoA dehydrogenase. This step is bypassed by an isomerization of the double bond by _enoyl-CoA isomerase_ to the Δ^2-trans-dodecenoyl-CoA, which is a normal substrate for enoyl-CoA hydrase, and the normal route for β oxidation resumes. The enzyme has been purified from liver mitochondria ($M_r = 90,000$). The isomerase is active with Δ^3-cis or Δ^3-trans acyl-CoAs that contain from 6 to 16 carbons.

Polyunsaturated fatty acids also are degraded by β oxidation, but the process requires _enoyl-CoA isomerase_ and an additional enzyme, _2,4-dienoyl-CoA reductase_ (Figure 18–8). For example, the degradation of linoleoyl-CoA begins, as with oleoyl-CoA, with three rounds of β oxidation and results in a Δ^3-cis unsaturated fatty acid that is not a substrate for the acyl-CoA dehydrogenase. Isomerization of the double bond to the Δ^2-trans position by enoyl-CoA isomerase allows the resumption of one cycle of β oxidation, and generation of a Δ^4-cis-enoyl-CoA. Subsequently, acyl-CoA dehydrogenase produces Δ^2-trans, Δ^4-cis-dienoyl-CoA (Figure 18–8), which is converted by an NADPH-dependent 2,4-dienoyl-CoA reductase to Δ^3-trans

Figure 18–7
Unsaturated fatty acids such as oleoyl-CoA can also be degraded to acetyl-CoA in mitochondria. Three cycles of β oxidation result in an acyl-CoA (Δ^3-cis-dodecenoyl-CoA) that is not a substrate for acyl-CoA dehydrogenase. This problem is circumvented by isomerization of the double bond to the Δ^2-trans position by enoyl-CoA isomerase. Complete oxidation of the remainder of the molecule by the enzymes of β oxidation is now possible.

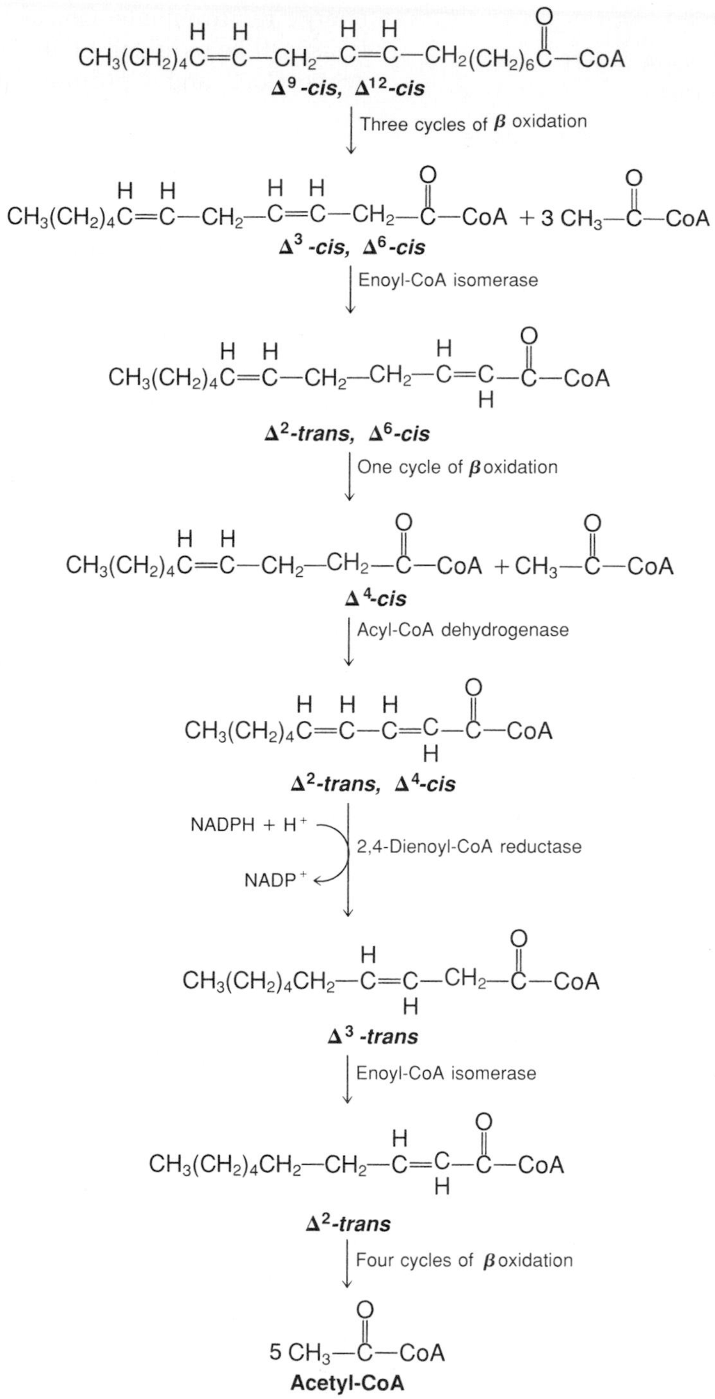

$$CH_3(CH_2)_4\overset{\displaystyle H}{C}=\overset{\displaystyle H}{C}-CH_2-\overset{\displaystyle H}{C}=\overset{\displaystyle H}{C}-CH_2(CH_2)_6\overset{\displaystyle O}{\overset{\|}{C}}-CoA$$

Δ^9-*cis*, Δ^{12}-*cis*

Three cycles of β oxidation

$$CH_3(CH_2)_4\overset{\displaystyle H}{C}=\overset{\displaystyle H}{C}-CH_2-\overset{\displaystyle H}{C}=\overset{\displaystyle H}{C}-CH_2-\overset{\displaystyle O}{\overset{\|}{C}}-CoA \;+\; 3\, CH_3-\overset{\displaystyle O}{\overset{\|}{C}}-CoA$$

Δ^3-*cis*, Δ^6-*cis*

Enoyl-CoA isomerase

$$CH_3(CH_2)_4\overset{\displaystyle H}{C}=\overset{\displaystyle H}{C}-CH_2-CH_2-\overset{\displaystyle H}{C}=\underset{\displaystyle H}{C}-\overset{\displaystyle O}{\overset{\|}{C}}-CoA$$

Δ^2-*trans*, Δ^6-*cis*

One cycle of β oxidation

$$CH_3(CH_2)_4\overset{\displaystyle H}{C}=\overset{\displaystyle H}{C}-CH_2-CH_2-\overset{\displaystyle O}{\overset{\|}{C}}-CoA \;+\; CH_3-\overset{\displaystyle O}{\overset{\|}{C}}-CoA$$

Δ^4-*cis*

Acyl-CoA dehydrogenase

$$CH_3(CH_2)_4\overset{\displaystyle H}{C}=\overset{\displaystyle H}{C}-\overset{\displaystyle H}{C}=\underset{\displaystyle H}{C}-\overset{\displaystyle O}{\overset{\|}{C}}-CoA$$

Δ^2-*trans*, Δ^4-*cis*

NADPH + H$^+$ ⟶ 2,4-Dienoyl-CoA reductase
NADP$^+$ ⟵

$$CH_3(CH_2)_4CH_2-\overset{\displaystyle H}{C}=\underset{\displaystyle H}{C}-CH_2-\overset{\displaystyle O}{\overset{\|}{C}}-CoA$$

Δ^3-*trans*

Enoyl-CoA isomerase

$$CH_3(CH_2)_4CH_2-CH_2-\overset{\displaystyle H}{C}=\underset{\displaystyle H}{C}-\overset{\displaystyle O}{\overset{\|}{C}}-CoA$$

Δ^2-*trans*

Four cycles of β oxidation

$$5\, CH_3-\overset{\displaystyle O}{\overset{\|}{C}}-CoA$$

Acetyl-CoA

Figure 18–8
Polyunsaturated fatty acids such as linoleoyl-CoA are also oxidized in the mitochondria. In addition to the enzymes of β oxidation and enoyl-CoA isomerase, the process requires 2,4-dienoyl-CoA reductase.

enoyl-CoA, then to the Δ^2-*trans* isomer by enoyl-CoA isomerase. With the double bond in the Δ^2-*trans* configuration, β oxidation can resume. The 2,4-dienoyl-CoA reductase has also been purified from liver ($M_r = 124,000$) and is specific for NADPH. NADH is not a substrate nor an inhibitor of the reductase.

Fatty Acids with an Odd Number of Carbons Are Oxidized to Propionyl-CoA

Fatty acids with an odd number of carbons are rare in many mammalian tissues. However, in ruminant mammals such as cows and sheep, the oxidation of odd-chain fatty acids can account for as much as 25 per cent of their

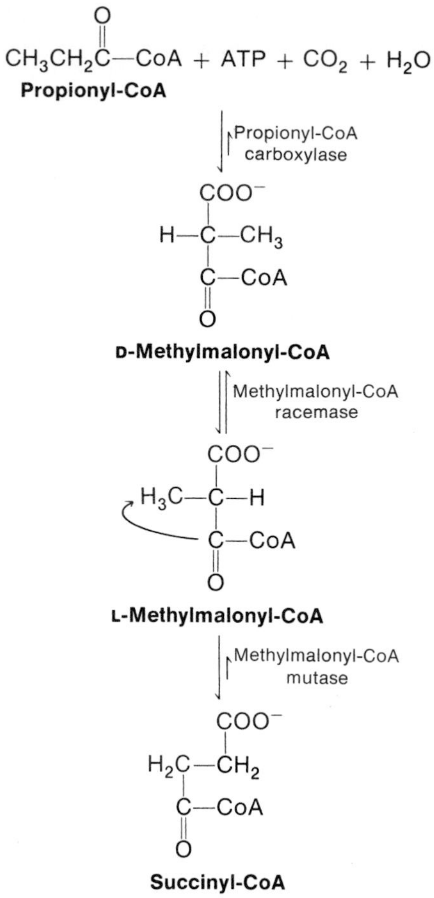

$$CH_3CH_2\overset{O}{\underset{}{C}}—CoA + ATP + CO_2 + H_2O$$
Propionyl-CoA

Propionyl-CoA
carboxylase

$$\begin{array}{c} COO^- \\ | \\ H—C—CH_3 \\ | \\ C—CoA \\ \| \\ O \end{array}$$

D-Methylmalonyl-CoA

Methylmalonyl-CoA
racemase

$$\begin{array}{c} COO^- \\ | \\ H_3C—C—H \\ | \\ C—CoA \\ \| \\ O \end{array}$$

L-Methylmalonyl-CoA

Methylmalonyl-CoA
mutase

$$\begin{array}{c} COO^- \\ | \\ H_2C—CH_2 \\ | \\ C—CoA \\ \| \\ O \end{array}$$

Succinyl-CoA

Figure 18–9
Propionyl-CoA generated by the
β oxidation of odd-chain fatty acids is
converted to succinyl-CoA, an
intermediate of the tricarboxylic acid
cycle.

energy requirements. Consequently, straight-chain fatty acids with 17 carbons will be oxidized by the normal β oxidation sequence and give rise to 7 acetyl-CoAs and 1 propionyl-CoA:

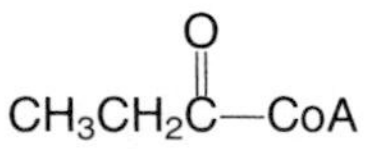

$$CH_3CH_2\overset{O}{\underset{}{C}}—CoA$$
Propionyl-CoA

This three-carbon acyl-CoA also is a product of degradation of the amino acids valine and isoleucine (see Chapter 19). The propionyl-CoA is converted to succinyl-CoA by three enzymatic steps, as indicated in Figure 18–9. The initial carboxylation is catalyzed by _propionyl-CoA carboxylase_, which utilizes biotin as a cofactor. In the second reaction, D-methylmalonyl-CoA is converted to its optical isomer, L-methylmalonyl-CoA, by _methylmalonyl-CoA racemase_. The last step in the sequence, catalyzed by _methylmalonyl-CoA mutase_, involves an unusual migration of the carbonyl-CoA group to the methyl group in an exchange for hydrogen. The product, succinyl-CoA, can be metabolized in the tricarboxylic acid cycle.

Methylmalonyl-CoA mutase is a mammalian enzyme that requires cobalamin (see Chapter 11) for activity. The enzyme has been purified from sheep liver and human placenta. The human enzyme has a molecular weight of 145,000. The absence of this enzymatic activity in children with _congenital methylmalonicaciduria_ results in death during childhood.

A Defect in α Oxidation Causes Refsum's Disease

Although β oxidation is quantitatively the most significant pathway for catabolism of fatty acids, α oxidation of some fatty acids is essential to our well-being. In a normal diet, small amounts of _phytol_, a component of chlorophyll, are ingested. As shown in Figure 18–10, this long-chain alcohol is oxidized to _phytanic acid_, which is a more important dietary component, present in ruminant fat and dairy products. The estimated daily intake of phytanic acid is somewhere between 50 and 100 mg. Because of the methyl substitution on carbon 3, phytanic acid is not a substrate for acyl-CoA dehydrogenase, the first enzyme in β oxidation. This step is circumvented by another mitochondrial enzyme, fatty acid α-hydroxylase, that hydroxylates the α carbon of phytanic acid. The hydroxyl intermediate is decarboxylated to yield _pristanic acid_ and CO_2 (Figure 18–10). A thiokinase reaction activates the acid to its CoA derivative. Pristanyl-CoA is unsubstituted at carbon 3 and can be oxidized by acyl-CoA dehydrogenase and the normal enzymes of β oxidation to produce propionyl-CoA and an acyl-CoA. The latter can be degraded by four cycles of β oxidation, which yield, alternately, acetyl-CoA and propionyl-CoA. The final sequence of reactions produces acetyl-CoA and isobutyryl-CoA. The latter acyl-CoA can be converted into succinyl-CoA and subsequently metabolized by the tricarboxylic acid cycle.

Our current understanding of how humans metabolize phytol and phytanic acid is largely a result of the studies of Daniel Steinberg. His experiments were prompted by interest in _Refsum's disease_, an inherited and extremely rare disorder characterized by numerous neurologic malfunctions—tremors, unsteady walking, constricted visual field, and poor night vision. The symptoms are probably due to an accumulation of phytanic acid throughout the nervous system. In patients with the disease, 5 to 30 per cent of the plasma fatty acids (20–100 mg/100 ml) and approximately 50 per cent of the liver fatty acids is phytanic acid. In contrast, the normal amount of phytanic acid in plasma is below 1 per cent (0.3 mg/100 ml). The disease is now known as _phytanic acid storage syndrome_. A series of biochemical studies has demonstrated that this fatty acid accumulates because of a defi-

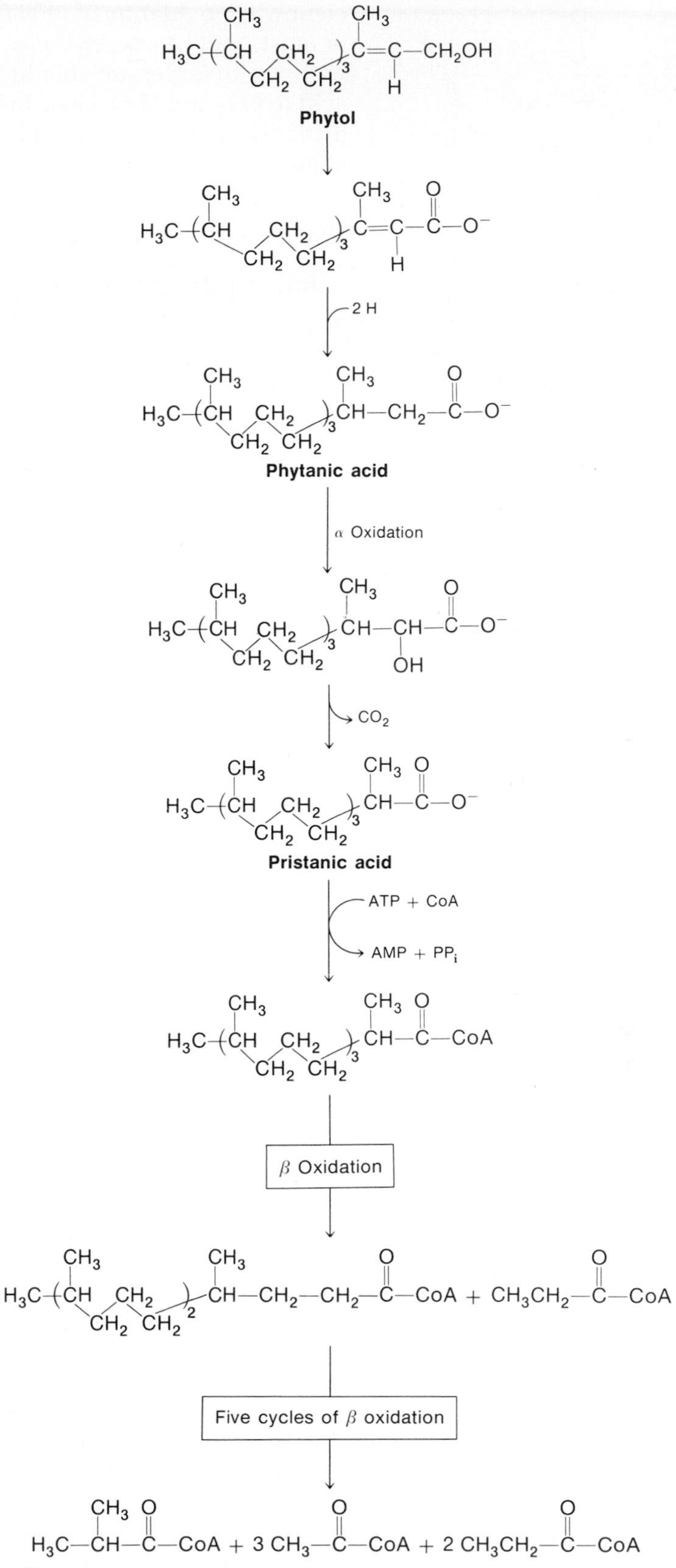

Figure 18–10
The oxidation of phytol and phytanic acid involves oxidation of the α carbon and decarboxylation. The product, pristanic acid, can be degraded by β oxidation after conversion to the CoA derivative. For the most part, enzymes involved in this series of reactions are not well characterized.

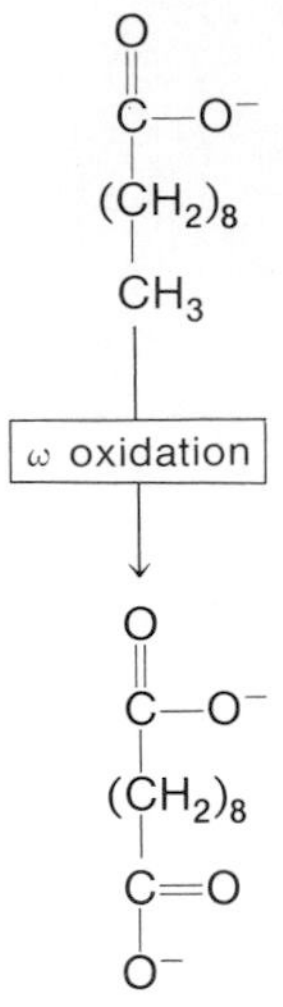

Figure 18–11
The oxidation of the methyl group (ω oxidation) of certain fatty acids is known to occur. However, the enzymes involved have not been well described.

ciency in α oxidation of phytanic acid to pristanic acid. The metabolic defect is most likely to be in the α hydroxylation of phytanic acid, since people with the disorder are able to oxidize phytol to phytanic acid and pristanic acid to CO_2 and H_2O. Once these patients are identified, the symptoms of the disease can be diminished by a strict dietary regimen in which foods that contain phytanic acid are restricted.

Fatty Acids Can Also Be Oxidized at the Methyl Group

A minor pathway for the oxidation of fatty acids has been observed in rat liver microsomes. This pathway involves oxidation of the terminal methyl, called the ω (omega) carbon, or the adjacent methylene carbon of fatty acids by NADPH and molecular oxygen (Figure 18–11). The pathway is probably not quantitatively significant for the oxidation of long-chain fatty acids, and the enzymes involved have not been characterized. However, ω oxidation may be important for the metabolism of fatty acids with 6 to 10 carbons.

Ketone Bodies Are Formed in the Liver and Used for Energy in Other Tissues

Once fatty acids are degraded in the mitochondria of liver, the acetyl-CoA can undergo a number of metabolic fates. Of central importance, as we learned in Chapter 15, is utilization of acetyl-CoA by the tricarboxylic acid cycle. An alternative fate is the synthesis of *ketone bodies* (acetoacetate, β-hydroxybutyrate, and acetone), which takes place only in the mitochondrial matrix, as depicted in Figure 18–12. In the first reaction, catalyzed by *acetoacetyl-CoA thiolase*, two acetyl-CoAs condense to form *acetoacetyl-CoA*. A third molecule of acetyl-CoA reacts with acetoacetyl-CoA to yield β-*hydroxy-*β-*methylglutaryl-CoA* (HMG-CoA) in a reaction catalyzed by *HMG-CoA synthase*. As we shall see in Chapter 23, the same two reactions are the first steps in cholesterol biosynthesis, which, however, take place only in the cytosol. In the formation of ketone bodies, the next reaction is catalyzed by *HMG-CoA lyase* and yields acetoacetate and acetyl-CoA. The acetoacetate can be reduced to β-hydroxybutyrate by an enzyme on the inner membrane of the mitochondrion, β-*hydroxybutyrate dehydrogenase*. Although the acetoacetate also can be decarboxylated to form acetone, this is normally of minor importance. However, patients with uncontrolled diabetes have high levels of ketone bodies in their plasma, and their breath smells of acetone.

Ketone body synthesis is primarily a liver function, since HMG-CoA synthase is present in large quantities only in this tissue. Acetoacetate and β-hydroxybutyrate diffuse into blood, where they are carried to other tissues and converted into acetyl-CoA, as described in Figure 18–13. The reactions catalyzed by β-*hydroxybutyrate dehydrogenase* and *thiolase* are common to both the synthesis and degradation of the ketone bodies. However, the second enzyme in the sequence (Figure 18–13), β-*oxoacid-CoA-transferase*, is present in all tissues but liver. Hence the ketone bodies are largely made in the liver and are metabolized to CO_2 and energy in nonhepatic tissues. Ketone bodies are an important source of energy for the brain during prolonged starvation.

β Oxidation Can Also Occur in Peroxisomes

Peroxisomes are subcellular organelles that have flavin-requiring oxidases and that regenerate oxidized flavin by reaction with oxygen, which is reduced to H_2O_2. Catalase then rapidly reduces the H_2O_2 to H_2O. In 1976, Paul

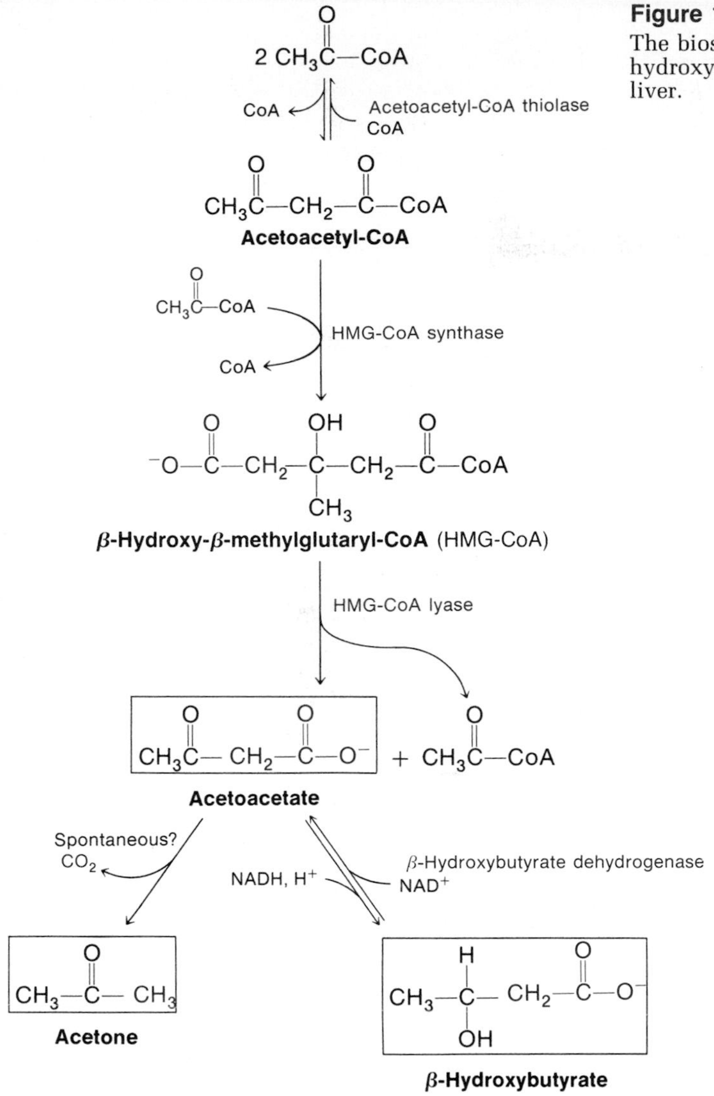

Figure 18–12

The biosynthesis of ketone bodies (acetoacetate, hydroxybutyrate, and acetone) occurs in the mitochondria of liver.

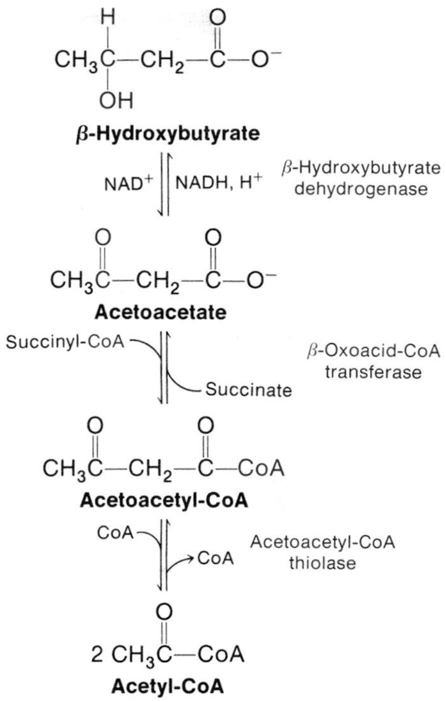

$$2\ CH_3\overset{O}{\overset{\|}{C}}\!-\!CoA$$

Acetoacetyl-CoA thiolase
CoA ⟵   CoA

$$CH_3\overset{O}{\overset{\|}{C}}\!-\!CH_2\!-\!\overset{O}{\overset{\|}{C}}\!-\!CoA$$
Acetoacetyl-CoA

$CH_3\overset{O}{\overset{\|}{C}}\!-\!CoA$
HMG-CoA synthase
CoA ⟵

$$^-O\!-\!\overset{O}{\overset{\|}{C}}\!-\!CH_2\!-\!\overset{OH}{\underset{CH_3}{\overset{|}{\underset{|}{C}}}}\!-\!CH_2\!-\!\overset{O}{\overset{\|}{C}}\!-\!CoA$$
β-Hydroxy-β-methylglutaryl-CoA (HMG-CoA)

HMG-CoA lyase

$$CH_3\overset{O}{\overset{\|}{C}}\!-\!CH_2\!-\!\overset{O}{\overset{\|}{C}}\!-\!O^-\ +\ CH_3\overset{O}{\overset{\|}{C}}\!-\!CoA$$
Acetoacetate

Spontaneous?
CO_2

NADH, H$^+$ β-Hydroxybutyrate dehydrogenase   NAD$^+$

$$CH_3\!-\!\overset{O}{\overset{\|}{C}}\!-\!CH_3$$
Acetone

$$CH_3\!-\!\overset{H}{\underset{OH}{\overset{|}{\underset{|}{C}}}}\!-\!CH_2\!-\!\overset{O}{\overset{\|}{C}}\!-\!O^-$$
β-Hydroxybutyrate

$$CH_3\overset{H}{\underset{OH}{\overset{|}{\underset{|}{C}}}}\!-\!CH_2\!-\!\overset{O}{\overset{\|}{C}}\!-\!O^-$$
β-Hydroxybutyrate

NAD$^+$ ‖ NADH, H$^+$ β-Hydroxybutyrate dehydrogenase

$$CH_3\overset{O}{\overset{\|}{C}}\!-\!CH_2\!-\!\overset{O}{\overset{\|}{C}}\!-\!O^-$$
Acetoacetate

Succinyl-CoA   β-Oxoacid-CoA transferase
Succinate

$$CH_3\overset{O}{\overset{\|}{C}}\!-\!CH_2\!-\!\overset{O}{\overset{\|}{C}}\!-\!CoA$$
Acetoacetyl-CoA

CoA   Acetoacetyl-CoA thiolase
CoA

$$2\ CH_3\overset{O}{\overset{\|}{C}}\!-\!CoA$$
Acetyl-CoA

Figure 18–13

Ketone bodies are converted back to acetyl-CoA in the mitochondria of nonhepatic tissues and used as a source of energy in these tissues.

Lazarow and Christian de Duve showed that peroxisomes from rat liver contained a β-oxidation system. The reactions of β oxidation in peroxisomes are similar to those in mitochondria (Figure 18–6), with the notable exception of the initial dehydrogenation. This reaction is catalyzed by the FAD-containing enzyme *acyl-CoA oxidase*:

1) $E\!-\!FAD + RCH_2CH_2\overset{O}{\overset{\|}{C}}\!-\!CoA \longrightarrow R\overset{H}{\underset{H}{C}}\!\!=\!\!C\overset{O}{\overset{\|}{C}}\!-\!CoA + E\!-\!FADH_2$

2) $E\!-\!FADH_2 + O_2 \longrightarrow H_2O_2 + E\!-\!FAD$

Acyl-CoA oxidase is inactive with hexanoyl-CoA or butyryl-CoA. It is suspected that these acyl-CoAs are further catabolized in the mitochondria. Hence the peroxisome will not completely catabolize fatty acids to acetyl-CoA. The relative importance of β oxidation in peroxisomes and mitochondria is a question currently under investigation.

Glyoxysomes are peroxisomes that also contain the enzymes of the glyoxylate pathway. The glyoxysomes of germinating seeds contain β oxidation enzymes similar to those of rat liver peroxisomes.

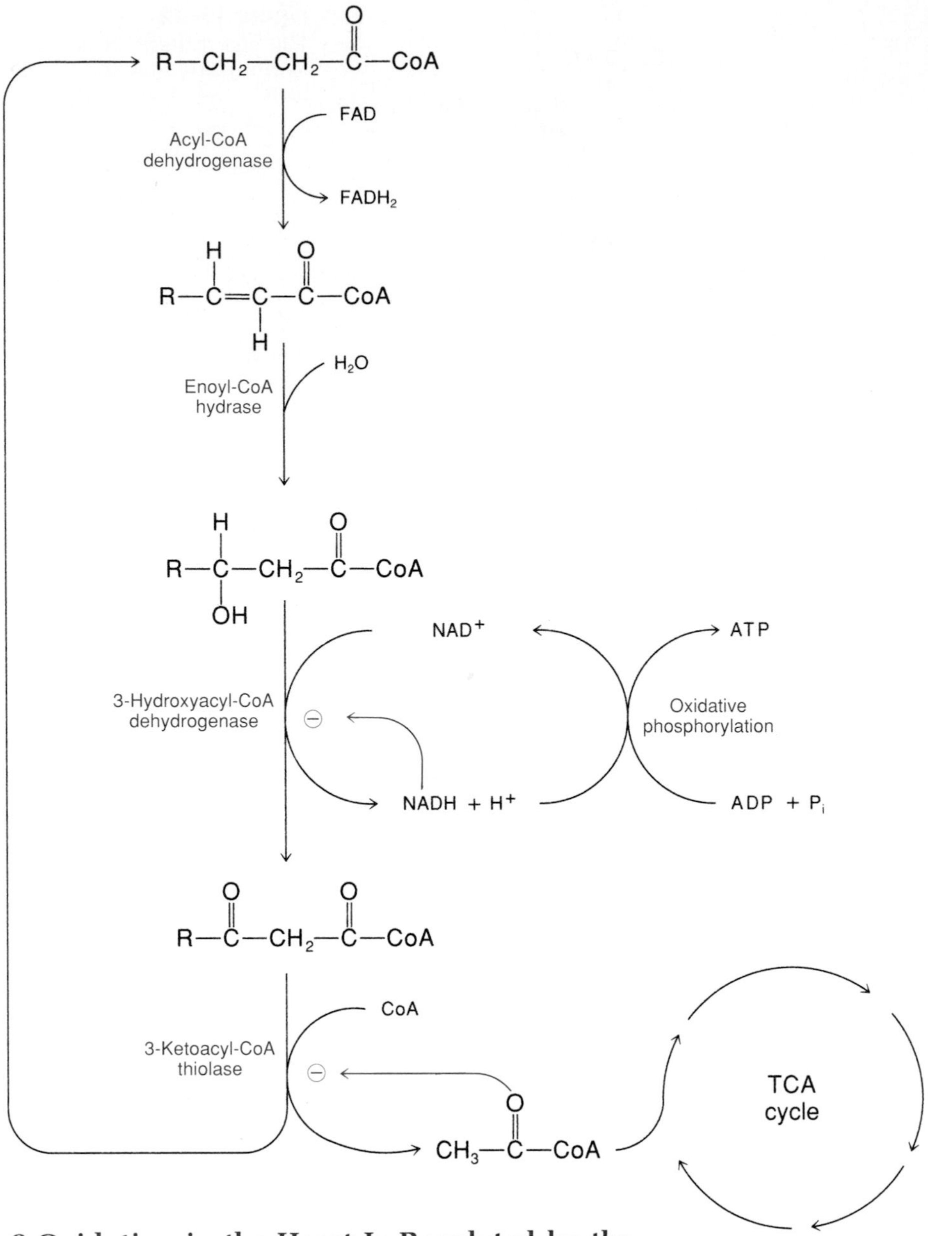

Figure 18–14
The rate of β oxidation in the heart appears to be regulated by the supply of fatty acids to the heart and by the feedback inhibition of thiolase by acetyl-CoA. The colored arrows indicate inhibition. The product inhibition of 3-hydroxyacyl-CoA dehydrogenase by NADH may also be an important regulatory mechanism.

β Oxidation in the Heart Is Regulated by the Supply of Fatty Acids and Acetyl-CoA

Regulation of β oxidation is a complex topic that relates to lipid biosynthesis and carbohydrate metabolism. For this reason the regulation of β oxidation in liver will be discussed in connection with fatty acid biosynthesis in Chapter 21. The regulation of β oxidation in heart and skeletal muscle is less complicated, since these tissues utilize fatty acids primarily for generation of energy. The rate of β oxidation in the heart is regulated by the supply of fatty acids from diet, biosynthesis, or adipose tissue. If energy is not required, the reduced activity of the tricarboxylic acid cycle and oxidative phosphorylation will cause an accumulation of acetyl-CoA and NADH. An increase in acetyl-CoA in the mitochondria inhibits thiolase and thus inhibits β oxidation (Figure 18–14). The increase in NADH and the lack of NAD$^+$ for the 3-hydroxylacyl-CoA dehydrogenase might also be important in retarding β oxidation. It is noteworthy that the regulation of β oxidation in heart tissue seems to occur at the later enzymes in the β oxidation cycle and not at the initial reactions, as is usually the case in regulation of a metabolic pathway.

SUMMARY

The fatty acids in triacylglycerols are the predominant form of energy storage in animals. The liberation of energy stored in fatty acids begins with a cAMP-mediated lipolysis. The fatty acids are carried to energy-deficient tissues by albumin, the major protein in human plasma. Inside the cells the fatty acids are activated to their acyl-CoA derivatives and transported into the mitochondria by carnitine acyltransferase reactions.

The fatty acids are degraded to acetyl-CoA by the enzymes of β oxidation. The acetyl-CoA can be oxidized to CO_2, H_2O, and energy by the tricarboxylic acid cycle and the electron-transport chain. The β oxidation of oleoyl-CoA requires an additional enzyme, enoyl-CoA isomerase. Com-

plete oxidation of polyunsaturated fatty acids requires the enoyl-CoA isomerase and 2,4-dienoyl-CoA reductase. Odd-chain fatty acids are oxidized to yield acetyl-CoA and one propionyl-CoA, which is converted to succinyl-CoA.

Phytanic acid is a dietary component found in ruminant fat and dairy products. Initial metabolism of this acid occurs by α oxidation to pristanic acid, which can be degraded by β oxidation.

Acetoacetate and β-hydroxybutyrate (ketone bodies) are formed from acetyl-CoA in liver mitochondria and used for energy in other tissues. β oxidation in the heart is regulated by the supply of fatty acids and acetyl-CoA.

SELECTED READINGS

Deuel, H. J., *The Lipids: Biochemistry*, Vol. 3, New York: Interscience, 1957. This is a comprehensive and classical treatise on the biochemistry of lipids until the mid-1950s.

Najjar, V. A., *Fat Metabolism*, Baltimore, Md: The Johns Hopkins Press, 1954. A good summary of the early work on fatty acid metabolism.

Stanbury, J. B., J. B. Wyngaarden, D. S. Fredrickson, J. L. Goldstein, and M. S. Brown, *The Metabolic Basis of*

Inherited Disease, 5th ed. New York: McGraw-Hill, 1983. An excellent compendium of important metabolic diseases including a chapter on phytanic acid storage disease.

Schulz, H., Oxidation of fatty acids. In D. E. Vance and J. E. Vance, eds., *Biochemistry of Lipids and Membranes*. Menlo Park, Calif.: Benjamin/Cummings, 1985. This is a comprehensive and authoritative discussion of fatty acid oxidation in an advanced textbook.

PROBLEMS

1. Palmitic acid is a 16-carbon saturated fatty acid ($C_{16}H_{32}O_2$) that can be oxidized completely to CO_2 and H_2O by a combination of β oxidation and the TCA cycle (see also Chapter 15).
   a. Calculate the net number of ATP molecules generated by the complete catabolism of palmitic acid to the two-carbon (acetyl-CoA) level, assuming that the ATP yield upon coenzyme oxidation is 3 for NADH and 2 for $FADH_2$.
   b. Next, calculate the number of ATP molecules generated by the further oxidation of the eight resulting acetyl-CoA molecules to CO_2 and H_2O, making the same assumptions as in part (a).
   c. Finally, calculate the total number of ATP molecules generated by the complete oxidation of one molecule of palmitic acid to CO_2 and H_2O, and write a balanced equation for this process.

2. Higher plants and animals store energy reserves preferentially as fat rather than as carbohydrate because fat has a higher energy content per unit weight. The calculations specified here are designed to quantify this difference (refer to Chapters 14 and 15).
   a. Consider first the utilization of glucose ($C_6H_{12}O_6$; $M_r = 180$) as an energy source and calculate the moles of ATP generated during the complete oxidative metabolism of 1 g of glucose. Assuming that the ΔG value for the hydrolysis of ATP under physiological conditions is -12 kcal/mole, how much free en-

ergy is conserved as ATP upon aerobic catabolism of 1 g of glucose?
   b. Repeat the calculations of part (a), but for the complete oxidation of 1 g of palmitic acid ($C_{16}H_{32}O_2$; $M_r = 256$). [See part (c) of Problem 1.]
   c. How much more efficient on a per-gram basis is fat as a form of energy storage compared with carbohydrate, assuming the values for glucose and palmitic acid to be representative of carbohydrates and fats in general? Why do organisms as diverse as castor beans and humans prefer fat as a means of storing energy reserves?
   d. If a friend laments to you that she is "getting fat" and is already 5 kg overweight, you can console her that she would be even more overweight if she were "getting carbohydrate" instead. Why is this so? How much overweight would she in fact be if the human body stored energy reserves as carbohydrate instead of fat?
   e. Bearing in mind that respiratory metabolism is essentially an oxidative process, how might you explain the difference in energy content of carbohydrate and fat on a per-gram basis?

3. What would be the net yield of ATP from the complete oxidation of oleic acid?

4. Draw the pathway for the complete β oxidation of linolenic acid.

5. There once was a student whose only vice in life was eating odd-chain fatty acids. The student was discovered to be deficient in vitamin B_{12}. After consulting a physician, the student was told to stop this unusual eating habit until the vitamin deficiency was improved. What is the biochemical reasoning for the doctor's advice?

6. What effect would a deficiency of carnitine have on β oxidation?

7. Compare the organization of the β oxidation enzymes in mitochondria and E. coli.

8. Why do patients with phytanic acid storage syndrome not excrete the excess quantities of phytanic acid in the urine?

9. Explain why the liver is the tissue that makes ketone bodies, whereas other tissues use ketone bodies for energy.

10. In patients with diabetes that is untreated, fatty acids are mobilized as a major source of energy. Why? These patients also have high levels of plasma ketone bodies and sometimes acetone breath. Why would this happen?

11. What effect would elevated levels of plasma glucose have on fatty acid oxidation in the heart? Explain the probable mechanism.

19

CATABOLISM OF AMINO ACIDS

Proteins are usually thought of primarily in their roles as enzymes, structural proteins, and molecules serving other specific functions. However, proteins or, more specifically, the amino acids of which they are composed, also can be used as a source of energy or as substrates for other biosynthetic pathways.

Amino acids that are catabolized come from three different sources: dietary proteins, storage proteins, and metabolic turnover of endogenous proteins. Catabolism of dietary proteins and amino acids is a characteristic of higher animals, while the catabolism of storage protein is best illustrated during the germination of protein-storing seeds such as beans or peas. In addition, all cells undergo metabolic turnover of most proteins; in this process protein-containing structures, and the amino acids to which they are degraded, can be recycled into either proteins or other derivatives that involve amino acids as precursors. Amino acids in excess are partially degraded to yield carbon skeletons for biosynthesis, or are completely degraded for energy production. When the acids are not needed for biosynthesis or for energy, degradation usually serves simply to eliminate excess amino acids, which can have toxic effects, especially in mammals.

There are numerous examples, however, where protein degradation serves more selective functions, not relating to material or energy needs. These will be discussed in Chapter 28. In this chapter we will focus on the main routes of degradation of the amino acids commonly found in proteins.

PROTEIN DEGRADATION

Protein catabolism begins with hydrolysis of the covalent peptide bonds that link successive amino acid residues together in a polypeptide chain (Figure 19–1). This process is termed _proteolysis_, and the enzymes responsible for the action are called _proteases_. For ingested proteins, proteolysis occurs in the gastrointestinal tract and depends on proteases secreted by the stomach,

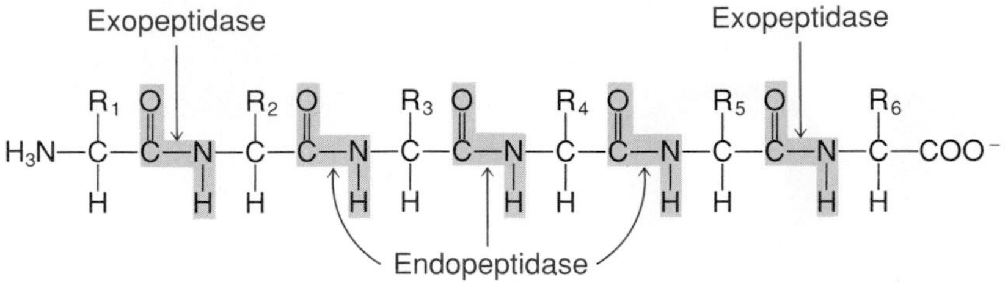

Figure 19-1
A protease hydrolyzes a peptide bond. Proteases have varying degrees of specificity, depending on the chemical nature of the R group and the location of the peptide linkage. Exopeptidases attack one or both ends of a polypeptide chain, while endopeptidases attack interior linkages.

pancreas, and small intestine. For endogenous protein, proteolytic digestion occurs within the cell.

The products of proteolytic digestion are free amino acids and small peptides. Further digestion of peptides then depends on *peptidases*, which are characteristic of the intestinal mucosa. Peptidases act on their substrate either by hydrolyzing internal peptide bonds (in the case of *endopeptidases*) or by removing successive amino acids from the end of the peptide (*exopeptidases*). Exopeptidases are referred to as *aminopeptidases* or *carboxypeptidases*, depending on the end of the peptide from which digestion proceeds.

Even single-cell microorganisms make proteases and peptidases, which they secrete into the surrounding medium so as to break down potential nutrient proteins to a size suitable for absorption.

NITROGEN REMOVAL FROM AMINO ACIDS

Degradative pathways for most amino acids begin by removal of the α-amino nitrogen. There are two major routes of deamination: *transamination* and *oxidative deamination*. Both of those processes are of major importance.

Transamination Is the Most Widespread Form of Nitrogen Removal

The process of transamination is illustrated in Figure 19-2 for an undesignated amino acid donating its amino group to the TCA cycle intermediate α-ketoglutarate. The reaction leads to an α-keto acid and glutamate.

Nearly all *transaminases contain pyridoxal-5'-phosphate as the coenzyme*. The mechanism for transamination involving pyridoxal phosphate was discussed in Chapter 11 (see Figure 11-6a). First the amino acid forms a Schiff base adduct with the coenzyme. This becomes protonated and then hydrolyzed, releasing the α-keto acid derivative of the amino acid while retaining the amino group. Another α-keto acid then combines with the coenzyme-bound amino group, ultimately releasing an amino acid and regenerating the pyridoxal-5'-phosphate.

Most transaminases are specific for α-ketoglutarate but show considerably less specificity for the amino acid.

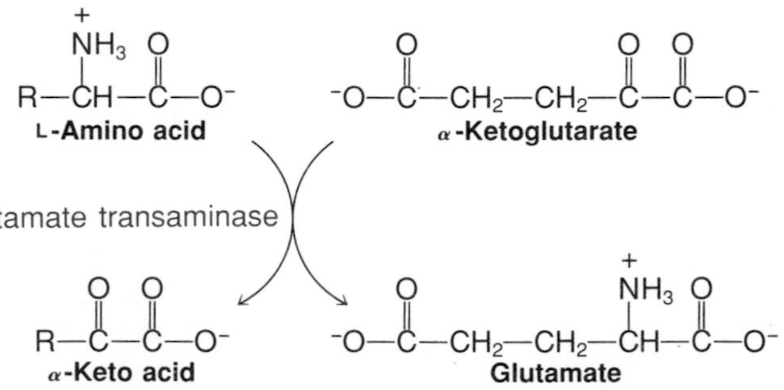

Figure 19-2
Glutamate transaminase catalyzes the transfer of the α amino group of an amino acid to α-ketoglutarate. The reaction is highly reversible, since the reacting functional groupings of the products are identical to those of the reactants.

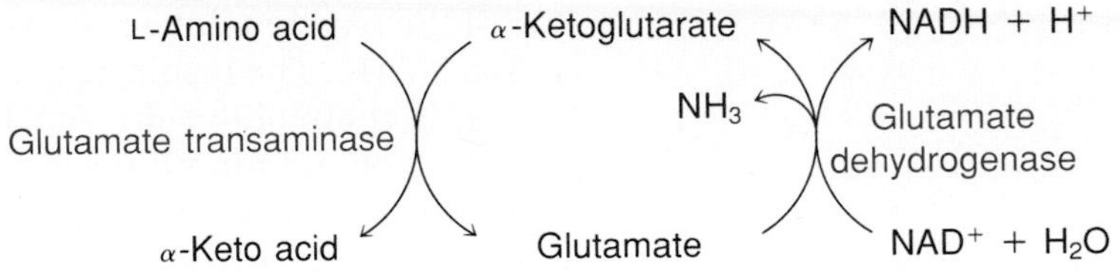

Figure 19–3
Transamination is not deamination. Transamination yields ammonia only if it is linked to another type of deamination process. Here net deamination results from the combined action of glutamate transaminase and glutamate dehydrogenase. In this process the α-ketoglutarate is recycled.

Oxidative Deamination Is Required for Net Deamination

Transamination does not result in any net deamination, since one amino acid is replaced by another amino acid. *The main function of transamination is to funnel the amino nitrogen into one or a few amino acids.* For glutamate to play a role in the net conversion of amino groups to ammonia, a mechanism for glutamate deamination is needed so that α-ketoglutarate can be regenerated for further transamination. The regeneration is accomplished by the oxidative deamination of glutamate, a reaction catalyzed by an NAD-linked enzyme, *glutamate dehydrogenase*. This broadly distributed enzyme is located in the mitochondria of eukaryotic cells. It catalyzes release of the α amino group of glutamate, leading to the regeneration of α-ketoglutarate.

$$\text{Glutamate} + \text{NAD}^+ + \text{H}_2\text{O} \longrightarrow \alpha\text{-ketoglutarate} + \text{NH}_4^+ + \text{NADH}$$

The overall process of transamination of α-ketoglutarate and regeneration of the α-ketoglutarate is shown in Figure 19–3.

The catalysis by glutamate dehydrogenase involves covalent bond formation between an intermediate and the enzyme (Figure 19–4). In the first

Figure 19–4
Mechanism of the glutamate-dehydrogenase-catalyzed reaction. The reaction involves hydride transfer from glutamate to NAD$^+$, followed by transimidation to an α amino group of a lysyl side chain of the enzyme, and finally by hydrolysis to α-ketoglutarate.

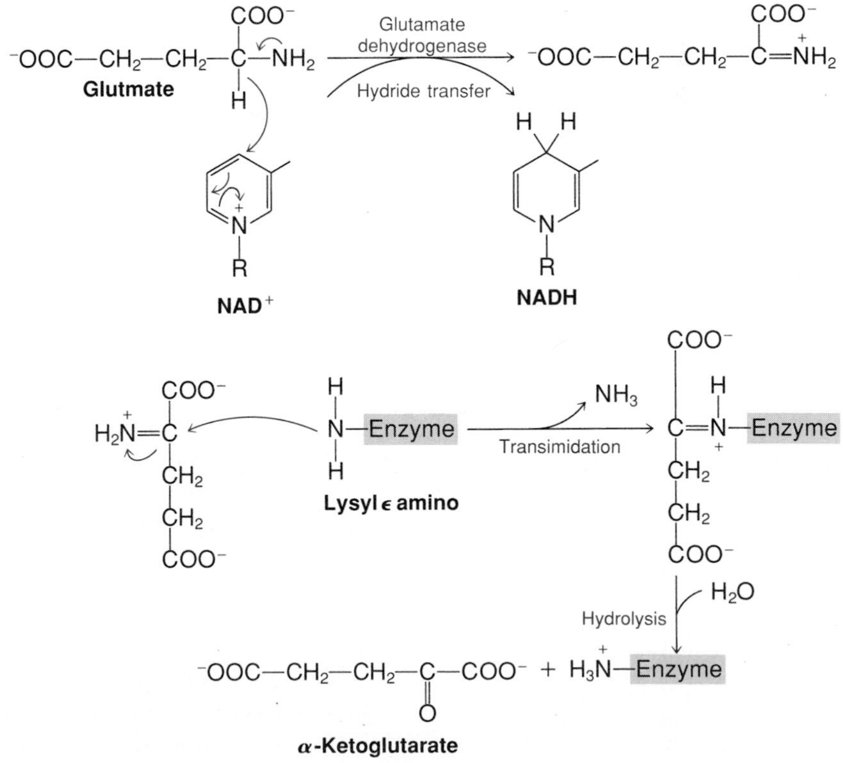

step of the reaction there is a hydride transfer from the α carbon of the amino acid to NAD^+. The resulting electron-deficient imino carbon is attacked by a lysyl side chain of the enzyme, leading to the displacement of ammonia and formation of an imino linkage with the enzyme. In the last step a hydrolysis restores the enzyme to its original state and α-ketoglutarate is released. Like glutamate, some amino acids that can undergo transamination can be deaminated more directly by oxidative reactions, either by a flavoprotein or by an NAD-linked enzyme.

The broadly specific flavin-linked D amino acid oxidases of animals as well as of microorganisms are of special interest, since D amino acids are so seldom encountered by animals except as breakdown products of bacterial cell walls. Thus the physiological importance of the widely distributed and highly active D amino acid oxidases remains obscure. The mammalian enzymes also exhibit activity against glycine.

Whether transamination or direct deamination is more important as an initial step in amino acid breakdown probably depends on the organism or tissue under investigation. However, where two mechanisms are available for one amino acid in a given cell type, it may well be that both mechanisms are employed.

THE FATE OF NITROGEN DERIVED FROM AMINO ACID BREAKDOWN

The NH_3 resulting from deamination of amino acids is converted to ammonia either directly or indirectly (e.g., by means of a transamination to yield a readily deaminated product such as glutamate). In microorganisms using a single amino acid as a nitrogen source, the ammonia so liberated is assimilated and used to form other nitrogen-containing cell components. When the amino acid is a carbon source, much more ammonia is liberated than is needed for biosynthesis, and it is disposed of by excretion to the medium. This simple disposal mechanism is adequate for free-living microorganisms, since the ammonia is carried away in the surrounding medium or escapes into the atmosphere.

Ammonia is also the major nitrogenous end product in some of the simpler aquatic and marine animal forms, such as protozoa, nematodes, and even bony fishes, aquatic amphibia, and amphibian larvae. Such animals are called ammonotelic. But in many animals, NH_3 is toxic, and its removal by simple diffusion would be difficult. Thus, in terrestrial snails and amphibia, as well as in other animals with environments in which water is limited, urea is the principal end product (Figure 19–5). Urea formation also helps in maintaining osmotic balance with seawater in the cartilagenous fishes. In such animals, most of the urea secreted by the kidney glomerulus is reabsorbed by the tubules. Indeed, the amount of nitrogen excreted by the kidneys of fishes is small compared with that excreted by the gills, and in most fishes, ammonia is the major form of excreted nitrogen.

Another form of "detoxified" ammonia that is used in nitrogen excretion is uric acid. Uric acid is the predominant nitrogen excretory product in birds and terrestrial reptiles (turtles excrete urea, whereas alligators excrete ammonia unless they are dehydrated, when they, too, excrete uric acid). Uric acid formed as a product of amino acid catabolism involves the de novo pathway of purine biosynthesis; therefore, its formation from NH_3 liberated upon amino acid catabolism will not be described here. In mammals, uric acid is exclusively an intermediate in purine catabolism, and in most mammals (primates excluded), it is further converted by uricase to allantoin.

Ammonia NH_3

Urea $H_2N{-}\underset{\underset{O}{\|}}{C}{-}NH_2$

Uric acid

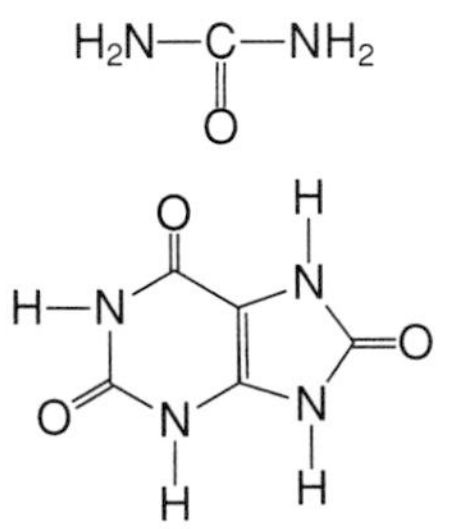

Figure 19–5

Excretory forms of nitrogen in different organisms. NH_3 is the most common end product of nitrogen metabolism. In many organisms NH_3 is toxic. To prevent the harmful excess of ammonia it is converted to urea or uric acid before excretion.

Urea Formation in the Animal Liver Detoxifies NH₃

The formation of urea in the liver involves the multistep conversion of ornithine to arginine. Urea itself is formed from arginine by the action of _arginase_, which regenerates ornithine. The overall cyclic pathway was deduced by Krebs and Henseleit in 1932 from their biochemical investigations on rat liver slices. The details, shown in Figure 19–6, have been developed through the efforts of many workers, particularly P. P. Cohen, S. Grisolia, and S. Ratner. We will discuss in detail the urea cycle as it occurs in the mammalian liver.

The complete urea cycle uses five enzymes, _argininosuccinate synthase,_ _arginase_ and _argininosuccinate lyase,_ which function in the cytosol, and

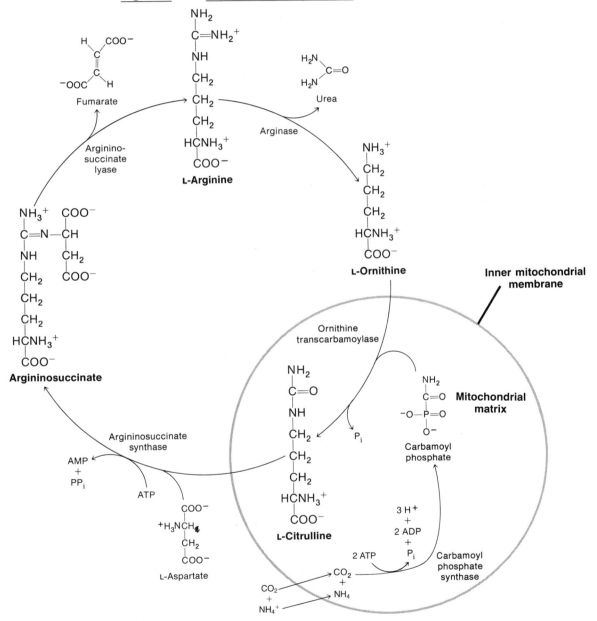

Figure 19–6
The urea cycle is a mechanism for removing unwanted nitrogen. Sources of nitrogens involved in urea formation are shown in color. Five enzymes are used in the urea cycle. Three of these function in the cytosol and two, as shown, function in the mitochondrial matrix. Specific carriers in the inner mitochondrial membrane transport ornithine, citrulline, ammonium ion, and HCO₃⁻ (CO₂) into and out of the mitochondrial matrix.

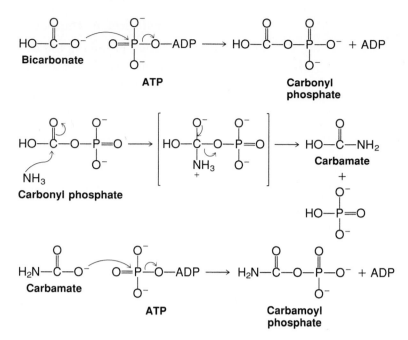

Figure 19–7

The mechanism of formation of carbamoyl phosphate. The reaction involves three steps, all of which take place on the same enzyme, carbamoyl phosphate synthase.

ornithine transcarbamoylase and *carbamoyl phosphate synthase*, which function in the mitochondria. Additional specific transport proteins are required for the mitochondrial uptake of L-ornithine, NH_4^+, and HCO_3^- and for the release of L-citrulline.

The free ammonia formed by oxidative deamination of glutamate is converted into <u>carbamoyl phosphate</u> in a reaction requiring two ATP molecules:

$$NH_4^+ + HCO_3^+ + H_2O + 2\ ATP \longrightarrow$$
$$\text{carbamoyl phosphate} + HPO_4^{2-} + 2\ ADP$$

The reaction involves three steps, all of which take place on the same enzyme (Figure 19–7). In the first step the bicarbonate ion is activated and prepares the carbon for a nucleophilic attack by ammonia, which leads to the intermediate carbamate (step 2). In a reaction closely related to step 1 a second phosphoryl group is transferred to carbamate to form carbamoyl phosphate (step 3).

The carbamoyl group has a high group transfer potential, which is displayed by its transfer to the terminal amino group of ornithine to form L-citrulline (see Figure 19–6). In the process inorganic phosphate is released. Before a further reaction can occur the citrulline must be transported across the mitochondrial membrane to the cytosol, where the remaining reactions leading to urea formation occur. Citrulline reacts with L-aspartate in an ATP-dependent reaction to form argininosuccinate, AMP, and PP_i. The PP_i is subsequently hydrolyzed to inorganic phosphate, so in effect the cost of this step is two ATP molecules. Argininosuccinate is cleaved to fumarate and L-arginine. The fumarate returns to the pool of TCA cycle intermediates, whereas the arginine becomes hydrolyzed to urea and ornithine. The ornithine is reutilized in further rounds of the urea cycle. Urea diffuses through the bloodstream and is ultimately eliminated through the kidney in the urine. The stoichiometry for the urea cycle is

$$CO_2 + NH_4^+ + 3\ ATP + \text{aspartate} + 2\ H_2O \longrightarrow$$
$$\text{urea} + 2\ ADP + 2\ P_i + AMP + PP_i + \text{fumarate}$$

In each turning of the urea cycle two nitrogens are eliminated, one originating from the oxidative deamination of glutamate and the other coming from

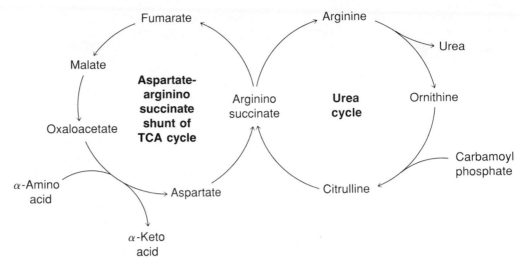

Figure 19–8
The "Krebs bicycle" involves interaction between components of the TCA cycle (on the left) and the urea cycle (on the right). This interaction explains the origin of the amino group contributed by aspartate to urea formation. The amino group originates from a transamination reaction involving oxaloacetate. The resulting aspartate is deaminated to fumarate, which can be recycled to oxaloacetate.

the α amino group of aspartate. Since the PP_i is subsequently hydrolyzed, it takes four high-energy phosphates to form a single molecule of urea. Thus the cost of this form of detoxification of ammonia is surprisingly high.

The Urea Cycle and the TCA Cycle Are Linked

In the urea cycle the carbon skeleton of the aspartate is released as fumarate. This product links the urea cycle with the TCA cycle. Fumarate is hydrated to malate, which is oxidized to oxaloacetate. The carbons of oxaloacetate can stay in the citrate cycle by condensation with acetyl-CoA to form citrate, or they can leave the TCA cycle either by gluconeogenesis to form glucose or by transamination to form aspartate as shown in Figure 19–8. Since Krebs was involved in the discoveries of both the urea cycle and the TCA cycle, the interaction between the two cycles shown in Figure 19–8 is sometimes referred to as the Krebs bicycle.

Different Carriers Transport Ammonia to the Liver

The urea cycle is a unique function of the liver. Ammonia formed in other tissues must be carried in a nontoxic form to the liver. In many tissues glutamine serves as the carrier of excess nitrogen. The glutamine is formed in the tissues in question in a reaction, catalyzed by *glutamine synthase*, that combines NH_3 with glutamate:

$$ATP + NH_4^+ + glutamate \xrightleftharpoons{\text{Glutamine synthase}} ADP + P_i + glutamine + H^+$$

This reaction involves activation of the γ carboxyl group of glutamate to yield a γ-glutamyl enzyme complex, together with the cleavage of ATP to ADP and P_i (Figure 19–9). In a second step, the γ-glutamyl group is transferred to NH_4^+.

After the ammonia reaches the liver, the enzyme *glutaminase* releases the ammonia from the glutamine by the reaction

$$Glutamine + H_2O \longrightarrow glutamate + NH_4^+$$

CH$_2$COO$^-$
|
CH$_2$
|
HCNH$_3^+$
|
COO$^-$
L-Glutamate

Glutamine synthase ⟍ ATP + NH$_4^+$

$$
\left[
\begin{array}{c}
NH_3 \\
CH_2\!-\!CO\!-\!P_i \\
CH_2 \\
HCNH_3^+ \\
COO^-
\end{array}
\quad \boxed{\text{Enzyme}} \quad ADP
\right]
$$

Enzyme-bound γ-L-glutamyl phosphate

⟍ P_i + ADP + H$^+$

CH$_2$—CONH$_2$
|
CH$_2$
|
HCNH$_3^+$
|
COO$^-$
L-Glutamine

Figure 19–9
Glutamine synthase catalyzes the synthesis of glutamine in an ATP-dependent reaction. Free ammonia is used as the amino group donor, and the active intermediate is an enzyme-bound γ-L-glutamyl phosphate.

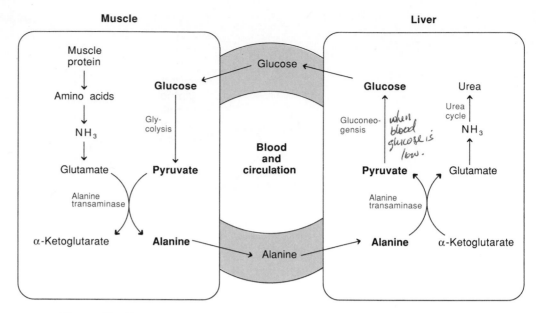

Figure 19-10

The glucose-alanine cycle. Active muscle functions anaerobically and synthesizes alanine by a transamination reaction between glutamate and pyruvate. The alanine is transported to the liver, where the pyruvate is regenerated and converted to glucose by gluconeogenesis. The glucose then is transported back to the muscle tissue, where it is used for energy production in glycolysis.

Skeletal muscle transports NH_3 to the liver in the form of the amino acid alanine. The alanine is formed in the muscle tissue by a transamination reaction between pyruvate and glutamate. Then the alanine is transported through the bloodstream to the liver, where it reacts with α-ketoglutarate to re-form pyruvate and glutamate. This reaction is catalyzed by _alanine transaminase_. The nitrogen originating from the glutamate is processed by the urea cycle. When the blood glucose concentration is low, the pyruvate resulting from alanine transamination is used to make glucose via the gluconeogenesis pathway. The glucose can be returned to the skeletal muscle to supply quick energy. Thus the transport of alanine from muscle to liver results in a reciprocal transfer of glucose to muscle. The entire cyclical process is referred to as the _glucose-alanine cycle_ (Figure 19-10). Its importance is proportional to the muscular activity of the organism. Recall that active muscle tissue operates anaerobically, producing large quantities of pyruvate and consuming large quantities of glucose.

AMINO ACIDS AS A SOURCE OF CARBON AND ENERGY

Thus far we have been considering the deamination of amino acids and the fate of the resulting ammonium ion. The carbon skeleton remaining after deamination can be used in various biosynthetic pathways, or it can be degraded to produce energy.

Catabolism of amino acids usually entails their conversion to intermediates in the central metabolic pathways. All amino acids can be degraded to carbon dioxide and water by appropriate enzyme systems. In every case, _the pathways involve the formation, directly or indirectly, of a dicarboxylic acid intermediate of the tricarboxylic acid cycle, of pyruvate, or of acetyl-CoA_ (Figure 19-11).

Acetyl-CoA so formed can be oxidized to carbon dioxide by means of the tricarboxylic acid cycle or, when cycle function is restricted, can be converted to acetoacetate and lipid. Amino acids metabolized to acetoac-

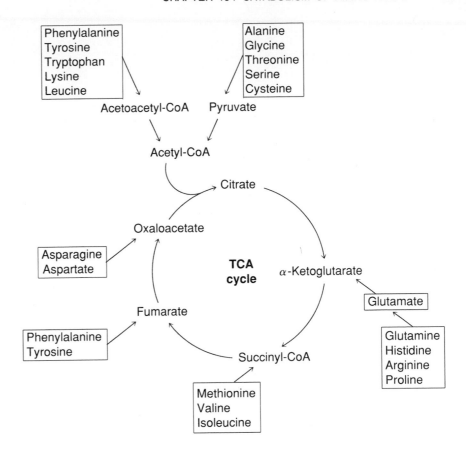

Figure 19–11
Breakdown products resulting from amino acid catabolism funnel into intermediates of the central metabolic pathways. These intermediates may be further degraded for energy production or utilized for biosynthesis.

etate and acetate are considered _ketogenic_. This is because in animals there cannot be any net conversion of two-carbon units into glucose. In contrast, α-ketoglutarate or the four-carbon dicarboxylic acids derived from amino acid breakdown can stimulate tricarboxylic acid function, since they play a catalytic role in the cycle. For their further metabolism they must leave the cycle by one of two routes (see Chapter 15). By one route, the conversion of oxaloacetate to phosphoenolpyruvate results in gluconeogenesis when carbohydrate utilization is restricted. For this reason, such amino acids are considered _glycogenic_. By the other route, pyruvate is formed and, after conversion of the latter to acetyl-CoA, can be oxidized completely to carbon dioxide and water.

Our discussion of the catabolic pathways is organized into groups of amino acids that give rise to the same main pathway intermediates.

Five Amino Acids Degrade to Acetyl-CoA by Way of Pyruvate

The carbon skeletons of ten amino acids yield acetyl-CoA. Five of these, _alanine, glycine, threonine, serine and cysteine, are degraded to acetyl-CoA by way of pyruvate_ (Figure 19–12). Another five, phenylalanine, tyrosine, tryptophan, lysine, and leucine, go by way of acetoacetyl-CoA. We will discuss first the amino acids that are converted to pyruvate.

Alanine. Alanine undergoes a reversible transamination directly to pyruvate. Recall that this reaction is part of the glucose-alanine cycle described above.

Threonine. Threonine is degraded in more than one way. In the pathway shown in Figure 19–12, _threonine dehydrogenase_ converts threonine via

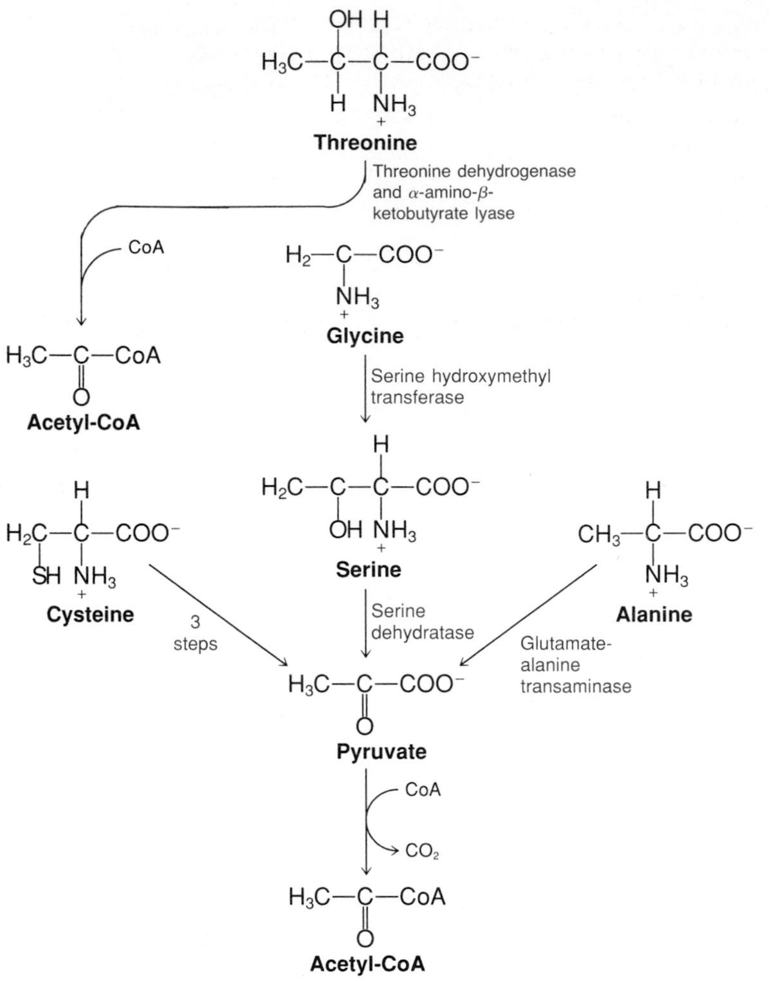

Figure 19–12
Outline of the catabolism of threonine, serine, cysteine, and glycine to acetyl-CoA by way of pyruvate. When a single enzyme is involved in the transition, the enzyme name is indicated next to the reaction arrow. Otherwise the number of steps is indicated.

two enzymes, _threonine dehydrogenase_ and the acetyl-CoA-dependent _α-amino-β-ketobutyrate lyase_.

Glycine and Serine. Glycine itself is degraded in two ways, only one of which leads to pyruvate. The pathway to pyruvate involves the conversion of glycine to serine by addition of a hydroxymethyl group carried by 5,10-methenyltetrahydrofolate (see Figure 11–19). Subsequently the serine is converted to pyruvate by a specific _serine dehydratase_, unless, of course, there is a shortage of serine for biosynthesis.

The major pathway for the catabolism of glycine involves the oxidative cleavage of glycine to CO_2, NH_4^+, and a methylene group ($-CH_2-$), which is accepted by tetrahydrofolate in a reversible reaction catalyzed by glycine synthase:

$$\text{Glycine} + FH_4 + NAD^+ \rightleftharpoons N^5,N^{10}\text{-methylene } FH_4 + CO_2 + NADH + NH_4^+$$

Thus even though glycine does not enter the TCA cycle by this mode of degradation, its degradation products are not wasted: the methyl group is donated to the coenzyme for use in one-carbon metabolism, and the NADH also produced in this process can be used directly or indirectly in biosynthesis. Recall that biosynthesis has a general requirement for reducing power.

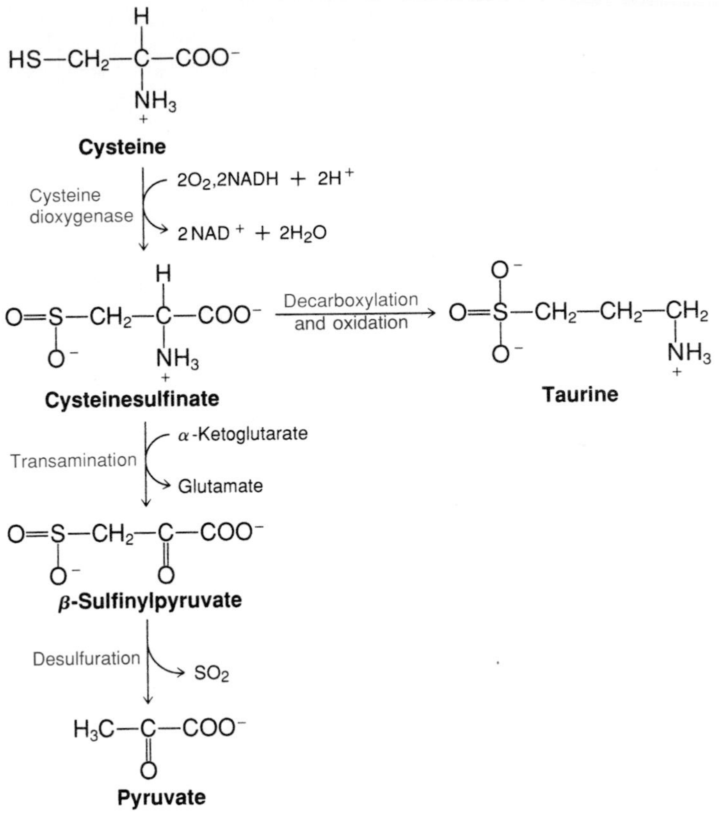

Figure 19–13

The main pathway for the conversion of cysteine to pyruvate in animals takes place in three steps. The first intermediate in this pathway, cysteinesulfinate, is a branchpoint that can also lead to taurine. Taurine is a component of certain bile acids (see Chapter 23).

Cysteine. There are several pathways for the catabolism of cysteine. All of these ultimately lead to the formation of pyruvate. The main pathway in animal cells occurs in the three steps, shown in Figure 19–13. Cysteine sulfinate, an intermediate in this pathway, is also a biosynthetic intermediate; upon decarboxylation and oxidation it produces *taurine* (2-amino-ethanesulfonate), a component of certain bile acids (see Chapter 23). Frequently two cysteines are disulfide-linked into a cystine. On such occasions the cystine is first reduced by an NADH-linked *cystine reductase*.

Five Amino Acids Degrade to Acetyl-CoA by Way of Acetoacetyl-CoA

The pathways for the degradation of *phenylalanine, tyrosine, tryptophan, lysine*, and *leucine* also *lead to acetyl-CoA* but they go *by way of acetoacetyl-CoA* rather than pyruvate (Figure 19–14). All of the pathways contain many steps, as indicated by the numbers next to the reaction arrows.

Lysine and Leucine. The pathways for leucine, lysine, and tryptophan are similar in the final steps and resemble the steps in the β oxidation of fatty acids (see Chapter 18). No more will be said here about the catabolism of lysine and leucine.

Tyrosine. The pathways for the aromatic amino acids phenylalanine, tyrosine, and tryptophan are noteworthy not only because their side chains create special complications for degradation, but because they provide numerous intermediates for the biosynthesis of useful compounds.

The oxidation of tyrosine (and phenylalanine) by the liver proceeds by way of acetoacetate and the dicarboxylic acid fumarate. Thus tyrosine and phenylalanine are both ketogenic and glycogenic.

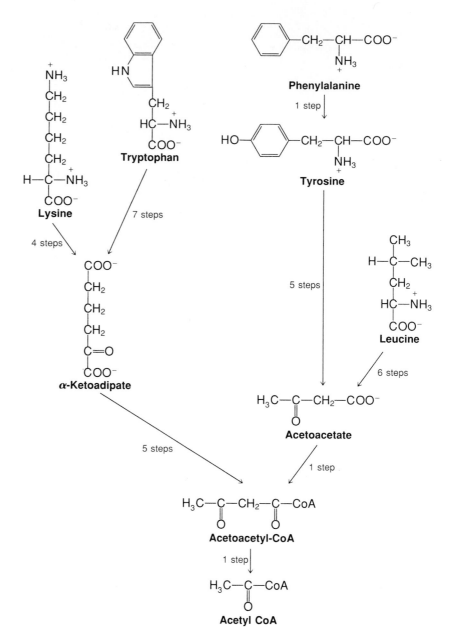

Figure 19–14
Outline of the catabolism of lysine, tryptophan, phenylalanine, tyrosine, and leucine. The numbers of enzyme-catalyzed steps are indicated next to the reaction arrows.

The first step in tyrosine catabolism involves its conversion to 4-hydroxyphenylpyruvate by a _tyrosine-glutamate transaminase_ of rather broad specificity (Figure 19–15). The next step is catalyzed by _4-hydroxyphenylpyruvate dioxygenase_, a copper-containing enzyme that is stimulated by ascorbate. The enzyme is called a dioxygenase because both atoms of the oxygen become incorporated into the product. The product, homogentisate, results from oxidation of the aromatic ring and an oxidative decarboxylation and migration of the side chain. The aromatic ring is further oxidized and cleaved by _homogentisate-1,2-dioxygenase_ to 4-maleylacetoacetate. As is apparent from the name, this enzyme is also a dioxygenase. Nearly all cleavages of aromatic rings in biological systems are catalyzed by dioxygenases. The enzyme requires ferrous iron and is also stimulated by ascorbate. An isomerase, maleylacetoacetate isomerase, yields the _trans_ compound 4-fumarylacetoacetate, which is hydrolytically cleaved to fumarate and acetoacetate by fumarylacetoacetate hydrolase.

Another route of tyrosine metabolism is that leading to melanin, which results from a two-stage attack on tyrosine by _tyrosinase_ (Figure 19–16),

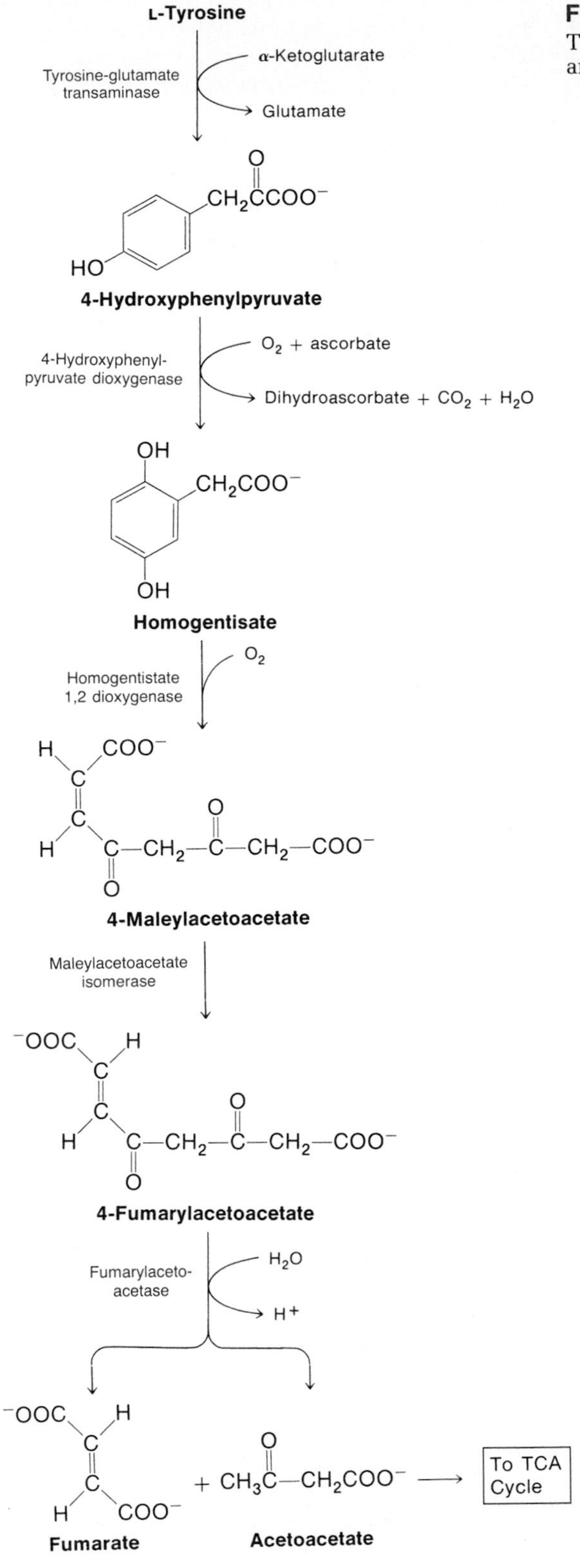

L-Tyrosine

Tyrosine-glutamate transaminase

α-Ketoglutarate

Glutamate

4-Hydroxyphenylpyruvate

4-Hydroxyphenyl-pyruvate dioxygenase

O_2 + ascorbate

Dihydroascorbate + CO_2 + H_2O

Homogentisate

Homogentisate 1,2 dioxygenase

O_2

4-Maleylacetoacetate

Maleylacetoacetate isomerase

4-Fumarylacetoacetate

Fumarylaceto-acetase

H_2O

H^+

Fumarate   **Acetoacetate**

+ $CH_3\overset{O}{\overset{\|}{C}}-CH_2COO^-$ ⟶ To TCA Cycle

Figure 19–15
The conversion of tyrosine to fumarate and acetoacetate.

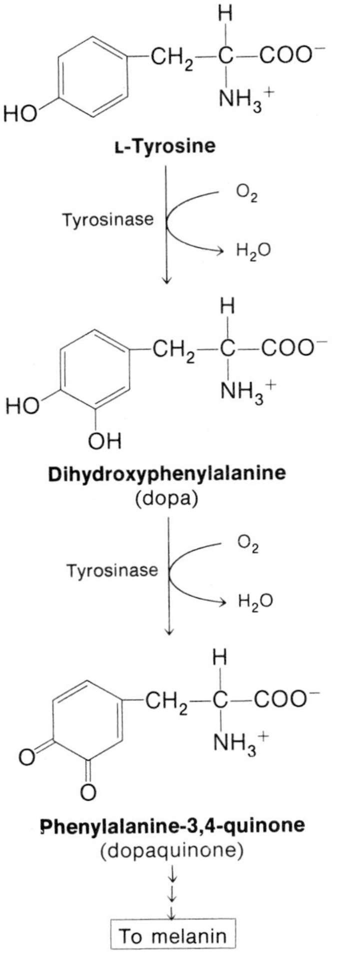

L-Tyrosine

Tyrosinase

O_2

H_2O

Dihydroxyphenylalanine
(dopa)

Tyrosinase

O_2

H_2O

Phenylalanine-3,4-quinone
(dopaquinone)

To melanin

Figure 19–16
Melanin formation from tyrosine. Melanin is the black pigment found in hair and skin.

yielding first dihydroxyphenylalanine (dopa). The latter is oxidized as a cosubstrate by tyrosinase to yield the 3,4-quinone. The quinone is unstable and undergoes a series of spontaneous reactions that ultimately lead to melanin, a black pigment. In animals tyrosinase is only found in the organelles known as melanosomes, which are present in specialized pigment-producing melanocytes found in the epidermis and certain other tissues.

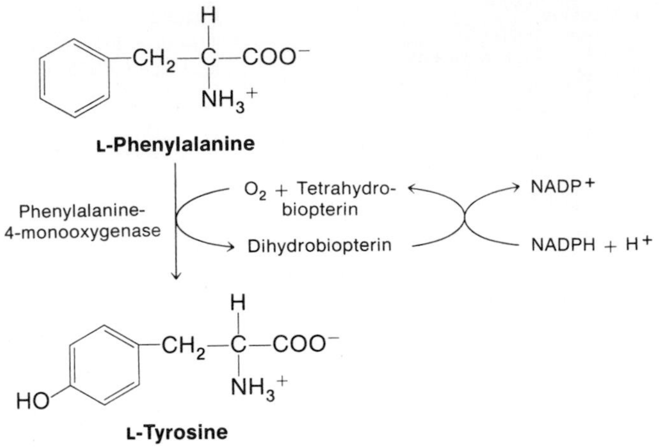

Figure 19–17
The formation of tyrosine from phenylalanine. This reaction occurs in one step. The enzyme requires tetrahydrobiopterin, a folic-acid-like compound, as a cosubstrate. Tetrahydrobiopterin is occasionally used as an electron carrier coenzyme.

Dopa is also synthesized from tyrosine by *tyrosine hydroxylase* as a precursor to certain neurohormones (*norephinephrine* and *epinephrine*). This reaction occurs exclusively in the adrenal glands (see Chapter 35). Between tyrosinase and tyrosine hydroxylase we have an example of two enzymes that carry out the same reaction but for totally different purposes.

Phenylalanine. Phenylalanine is broken down normally by way of tyrosine through the action of *phenylalanine-4-monooxygenase*, as indicated in Figure 19–17. The enzyme requires *tetrahydrobiopterin*, a folic-acid-like compound, as a cosubstrate. Tetrahydrobiopterin is an infrequently used electron carrier coenzyme (Figure 19–18). Dihydrobiopterin is its oxidized form. The biopterin is kept in the reduced form by NADPH, the ultimate hydrogen donor in the hydroxylation reaction. The presence of phenylalanine-4-monooxygenase accounts for the fact that tyrosine is not an essential amino acid in mammals, provided the dietary supply of phenylalanine is sufficient.

Minor pathways for phenylalanine breakdown in animals involve transamination to yield phenylpyruvate. Although phenylpyruvate can be reduced to phenyllactate, and metabolized to other phenyl derivatives, the disposal of dietary phenylalanine by these routes is insufficient, so that in the inherited absence of the hydroxylation to tyrosine, high blood levels of phenylalanine and phenylpyruvate result. The condition is known as *phenylketonuria*. The precise causes of the mental retardation accompanying phenylketonuria are unknown, owing in part to the fact that there are several metabolic effects of the high levels of phenylalanine metabolites. Heritable disorders in phenylalanine and tyrosine metabolism, such as phenylketonuria, are among the most studied of the inborn metabolic errors in humans. These and other inborn errors will be considered in a later section of this chapter.

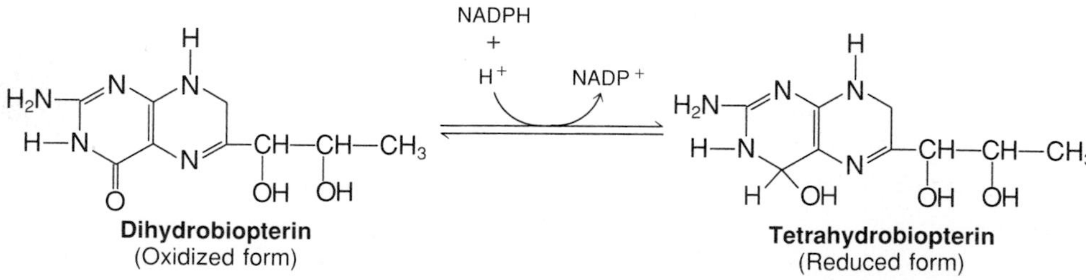

Figure 19–18
The structures of dihydrobiopterin and tetrahydrobiopterin. Dihydrobiopterin is the oxidized form of the coenzyme.

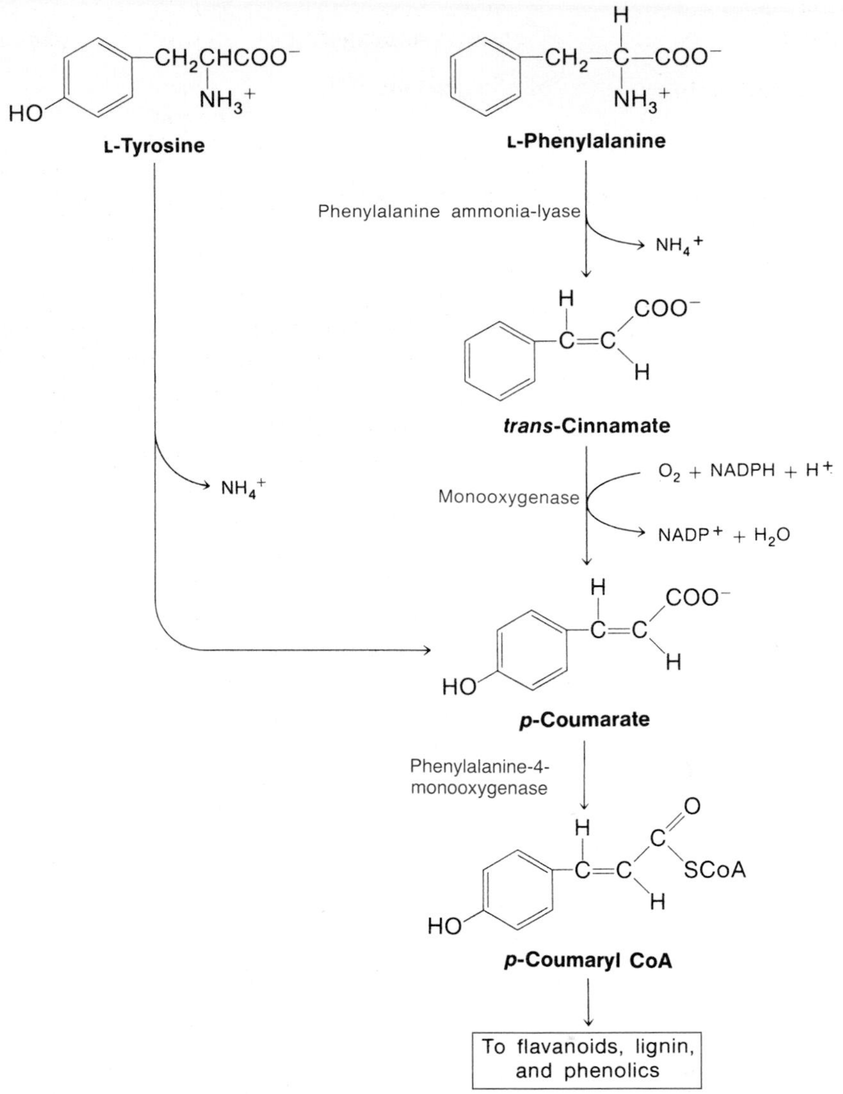

Figure 19–19

Conversion of phenylalanine and tyrosine to p-coumaryl-CoA, the precursor to flavonoids, lignin, and other compounds in plants. The phenylalaninine ammonia-lyase enzyme uses phenylalanine or tyrosine as a substrate.

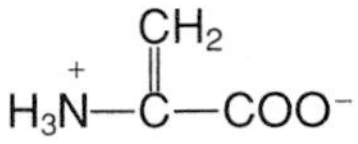

Dehydroalanine

Figure 19–20

The structure of dehydroalanine. The enzyme phenylalanine ammonia-lyase contains at its N-terminal end a dehydroalanine residue that directly participates in the catalysis.

Another important route for phenylalanine utilization is found in plants. There, the formation of flavonoids, lignin, and other derivatives of phenolic compounds plays an important role by means of the intermediate formation of p-coumarate-coenzyme A (Figure 19–19). A key reaction in the conversion of phenylalanine to p-coumarate-CoA is catalyzed by _phenylalanine ammonia-lyase_. The enzyme, which catalyzes the removal of the hydrogen from the α carbon that is _cis_ to the amino group to yield _trans_-cinnamate, contains a dehydroalanine residue at the N-terminal end of the peptide chain (Figure 19–20). The amino group of the dehydroalanine is thought to be in an imine linkage with some other group on the protein that provides the electron sink required for elimination of the amino group from the α carbon. The _trans_-cinnimate is, in turn, oxidized to 4-hydroxycinnamate (p-coumarate) by _trans-cinamate-4-monooxygenase_. The enzyme requires FAD and NADPH, which is oxidized in the hydroxylation process. p-Coumarate can also be formed directly from tyrosine through the action of phenylalanine ammonia-lyase on tyrosine. p-Coumarate is then converted to its coenzyme A derivative by _p-coumaryl-CoA synthase_. It is the coenzyme A derivative that is the branchpoint for a variety of biosynthetic routes found in plants.

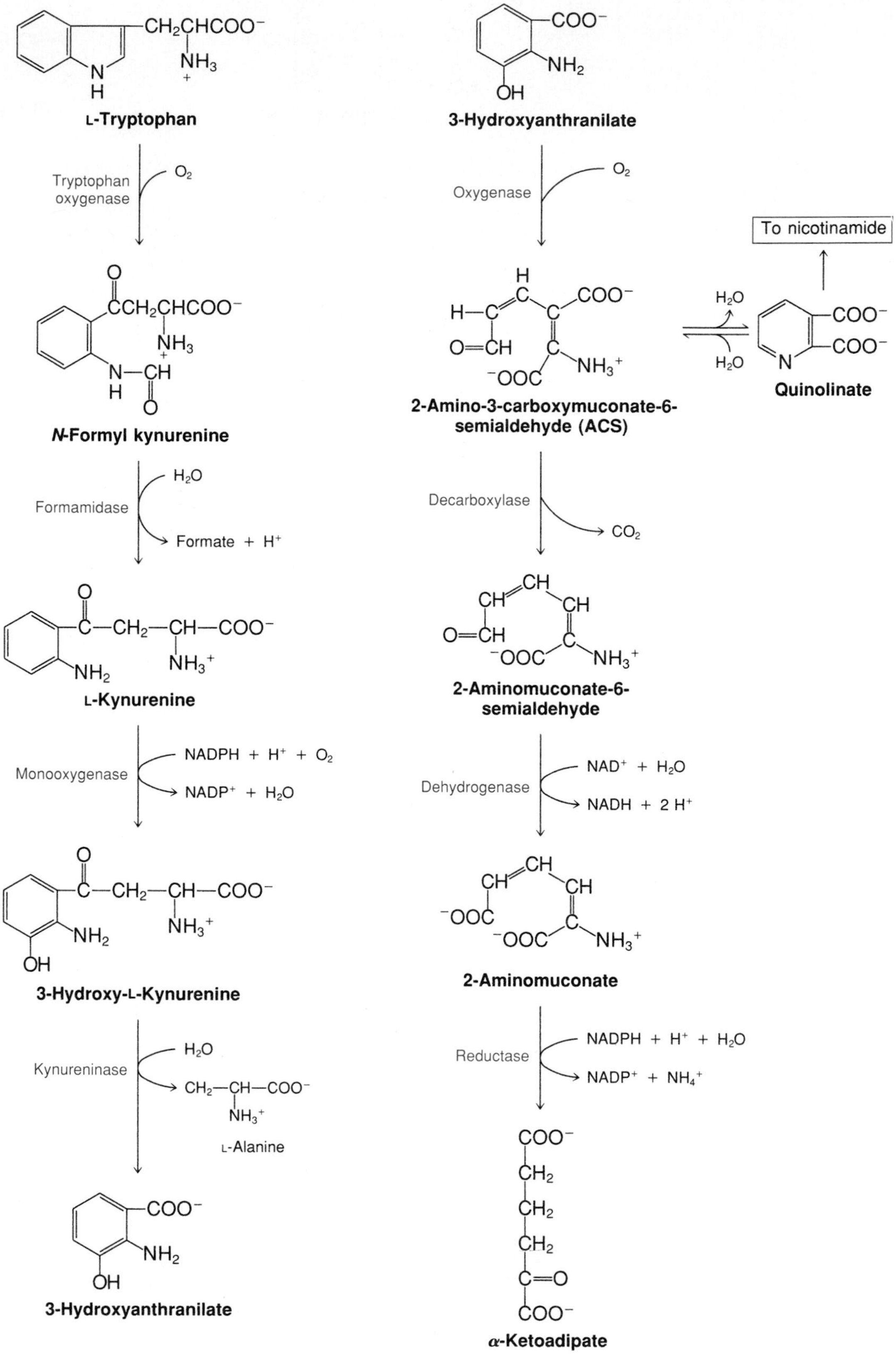

Figure 19–21
The main route of tryptophan degradation in mammals leads to α-ketoadipate. The intermediate 2-amino-3-carboxymuconate-6-semialdehyde (ACS) is a branchpoint metabolite that can also lead to nicotinamide via quinolate as indicated.

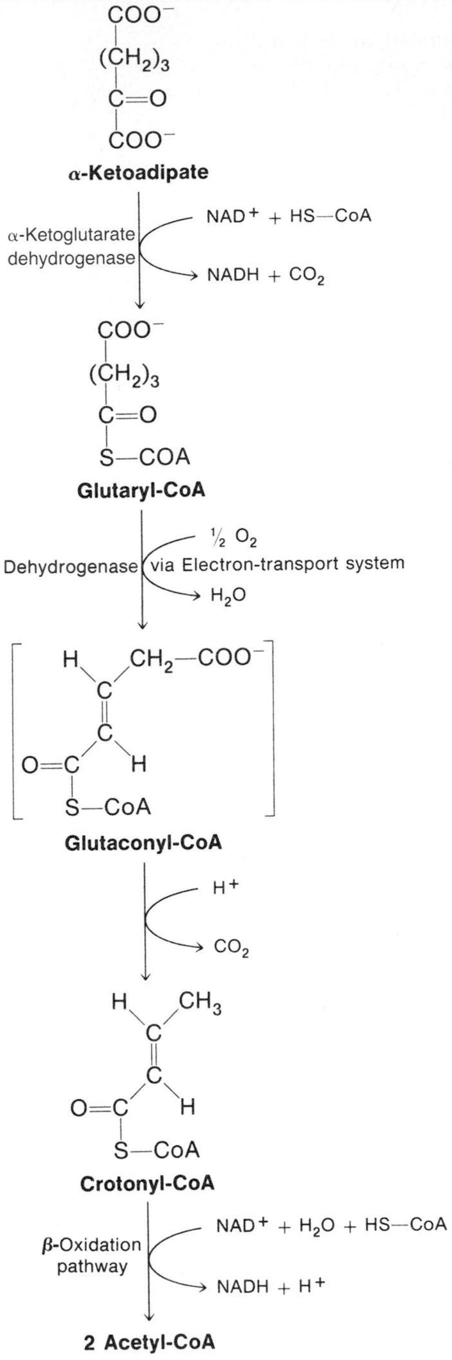

COO⁻
|
(CH₂)₃
|
C=O
|
COO⁻

α-Ketoadipate

α-Ketoglutarate ⟶ NAD⁺ + HS—CoA
dehydrogenase
⟶ NADH + CO₂

COO⁻
|
(CH₂)₃
|
C=O
|
S—COA

Glutaryl-CoA

Dehydrogenase ⟶ ½ O₂ (via Electron-transport system)
⟶ H₂O

$$\begin{bmatrix} \text{H} \quad \text{CH}_2\text{—COO}^- \\ \text{C} \\ \| \\ \text{C} \\ \text{O=C} \quad \text{H} \\ \text{S—CoA} \end{bmatrix}$$

Glutaconyl-CoA

⟶ H⁺
⟶ CO₂

H   CH₃
 \ /
  C
  ‖
  C
 / \
O=C   H
|
S—CoA

Crotonyl-CoA

β-Oxidation ⟶ NAD⁺ + H₂O + HS—CoA
pathway
⟶ NADH + H⁺

2 Acetyl-CoA

Figure 19–22
The conversion of α-ketoadipate to acetyl-CoA. The sulfur in the CoA-containing compounds is indicated. α-ketoadipate is also formed in the liver by the breakdown of lysine (see Figure 19–14). Thus the degradative pathways for tryptophan and lysine converge at the level of α-ketoadipate. Two molecules of CO₂ are released and two molecules of NADH are produced in the process of converting α-ketoadipate into two molecules of acetyl-CoA.

Tryptophan. The major pathways for tryptophan catabolism in the mammalian liver and for many microorganisms proceed by way of kynurenine (Figure 19–21). Kynurenine itself can be metabolized in liver by way of α-ketoadipate, which is also an intermediate in lysine degradation. An interesting variant of the kynurenine pathway allows for the synthesis of the coenzyme nicotinamide.

The first step in the breakdown of tryptophan is catalyzed by _tryptophan oxygenase_, which yields N-formylkynurenine (see Figure 19–21). The enzyme is a dioxygenase and cleaves the indole ring by incorporating an oxygen atom on both C-2 and C-3 of the indole ring. Kynurenine itself is formed by the liberation of formate by _kynurenine formamidase_.

Kynurenine is converted to 3-hydroxykynurenine by the NADPH-dependent _kynurenine-3-monooxygenase_ (Figure 19–21). _Kynureninase_, a pyridoxal phosphate enzyme, catalyzes a hydrolytic cleavage of the alanine side chain to yield 3-hydroxyanthranilate. The aromatic ring is cleaved to 2-amino-3-carboxymuconate-6-semialdehyde (ACS) by _3-hydroxyanthranilate oxygenase_. Again, this enzyme is a dioxygenase, and oxygen atoms are incorporated on both the carbons at the site of ring cleavage. Ferrous ions are required by the enzyme. ACS can be spontaneously cyclized to quinolinate with the liberation of a molecule of H₂O. Quinolinate is an intermediate in the biosynthesis of nicotinamide, a synthesis that many animals have a limited capacity to perform. The ACS is decarboxylated by a specific decarboxylase to yield 2-aminomuconate-6-semialdehyde. 2-aminomuconate-6-semialdehyde is oxidized by an NAD-dependent _aminomuconate semialdehyde dehydrogenase_. The resulting 2-amino muconate is reduced to α-ketoadipate by an NAD(P)H-dependent reductase.

The further catabolism of α-ketoadipate results in the liberation of two molecules of CO₂ and two of acetyl-CoA (Figure 19–22). The first step is a coenzyme-A-dependent oxidative decarboxylation by an enzyme probably identical to α-ketoglutarate dehydrogenase. The product, glutaryl-CoA, is oxidized by a flavin-linked dehydrogenase to an intermediate common to the oxidation of fatty acids, crotonyl-CoA. (The α-keto derivative glutaconyl coenzyme is probably an enzyme-bound intermediate in this reaction.) Finally, two molecules of acetyl coenzyme are formed by the action of the fatty acid oxidizing enzymes (see Chapter 18). It should be remembered that α-ketoadipate is also formed in the liver by the breakdown of lysine. Thus the degradative pathways for lysine and tryptophan converge at the level of α-ketoadipate.

Five Amino Acids Degrade to α-Ketoglutarate

α-Ketoglutarate is the endpoint for degradation of five amino acids, _arginine, histidine, proline, glutamic acid_, and _glutamine_ (Figure 19–23). As we will see (Chapter 24), it is also the starting point for the synthesis of these five amino acids. Hence these are reversible conversions, like those we saw in the glycolytic-gluconeogenic interconversions (see Chapter 14), but fewer enzymes are used in common.

Glutamine and Glutamate. Reactions involving the interconversions of glutamine, glutamate, and α-ketoglutarate were discussed earlier in this chapter when we were considering deamination.

Proline. Proline is converted into Δ¹-pyrolline-5-carboxylate in a reaction catalyzed by _proline oxidase_. This compound is in equilibrium with glutamate-γ-semialdehyde, which is also an intermediate in arginine catabolism (Figure 19–24).

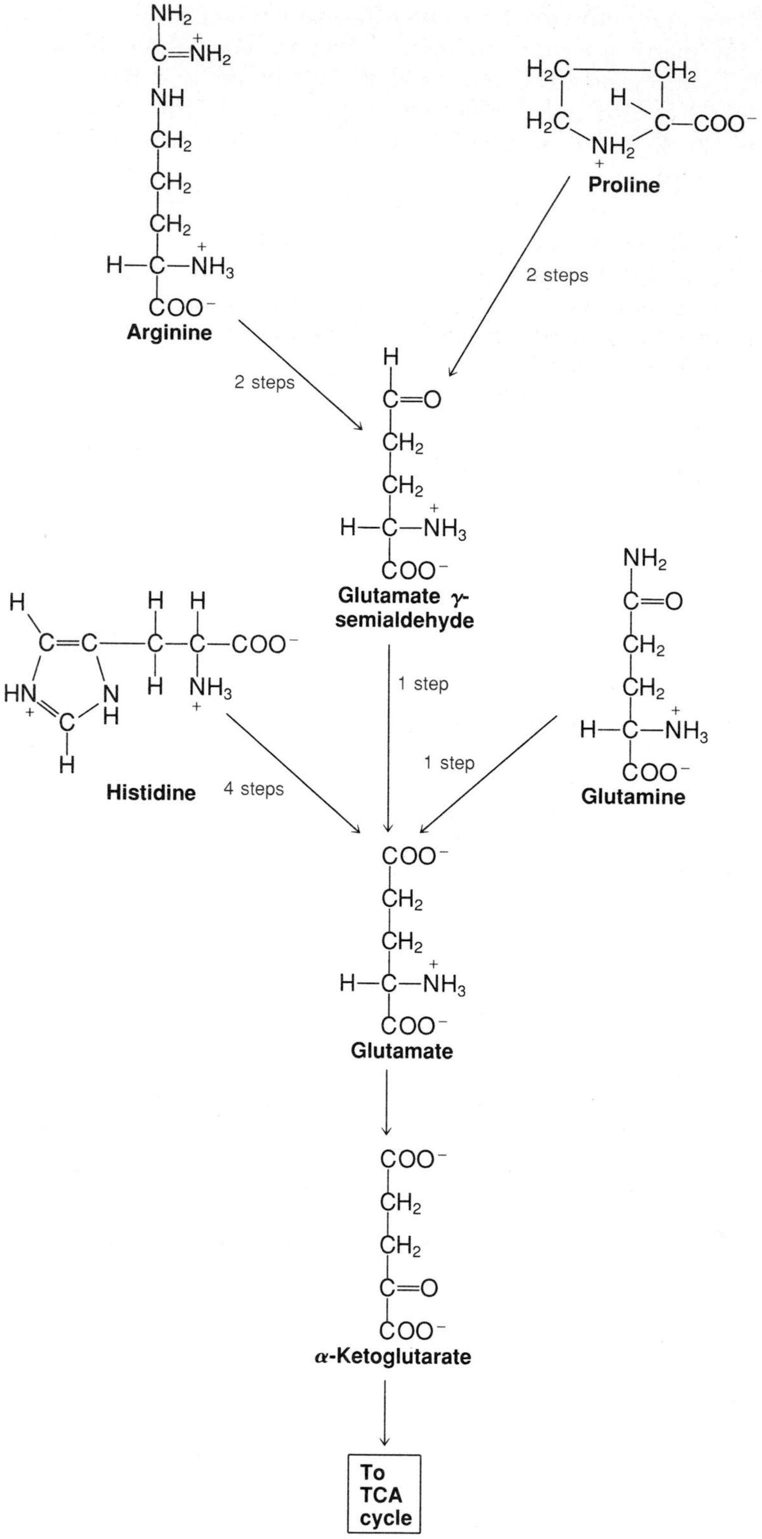

Figure 19–23
Outline of the catabolism of histidine, arginine, proline, and glutamine to glutamate and then to the TCA component α-ketoglutarate.

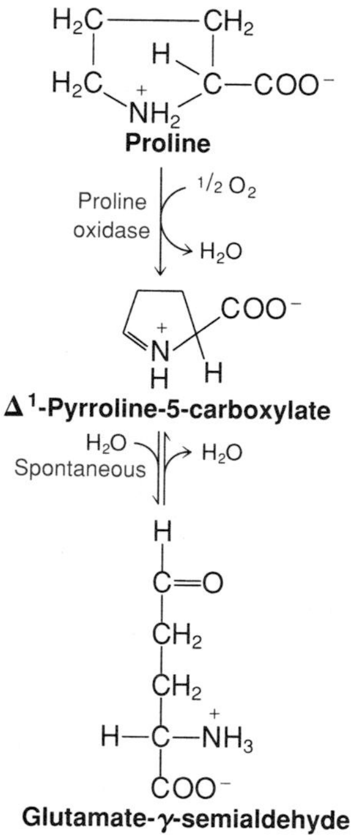

Figure 19–24
Degradation of proline to glutamate-γ-semialdehyde. This can be thought of as a two-step reaction. The first step is enzyme-catalyzed. The second step is spontaneous and reversible.

Arginine. Arginine is one of the few amino acids for which there is no transaminase reaction. In addition to serving as a source of carbon and energy, arginine is a precursor of various essential polyamines and, as we have seen, is required in the urea cycle.

Arginine is converted to ornithine by two different routes (Figure 19–

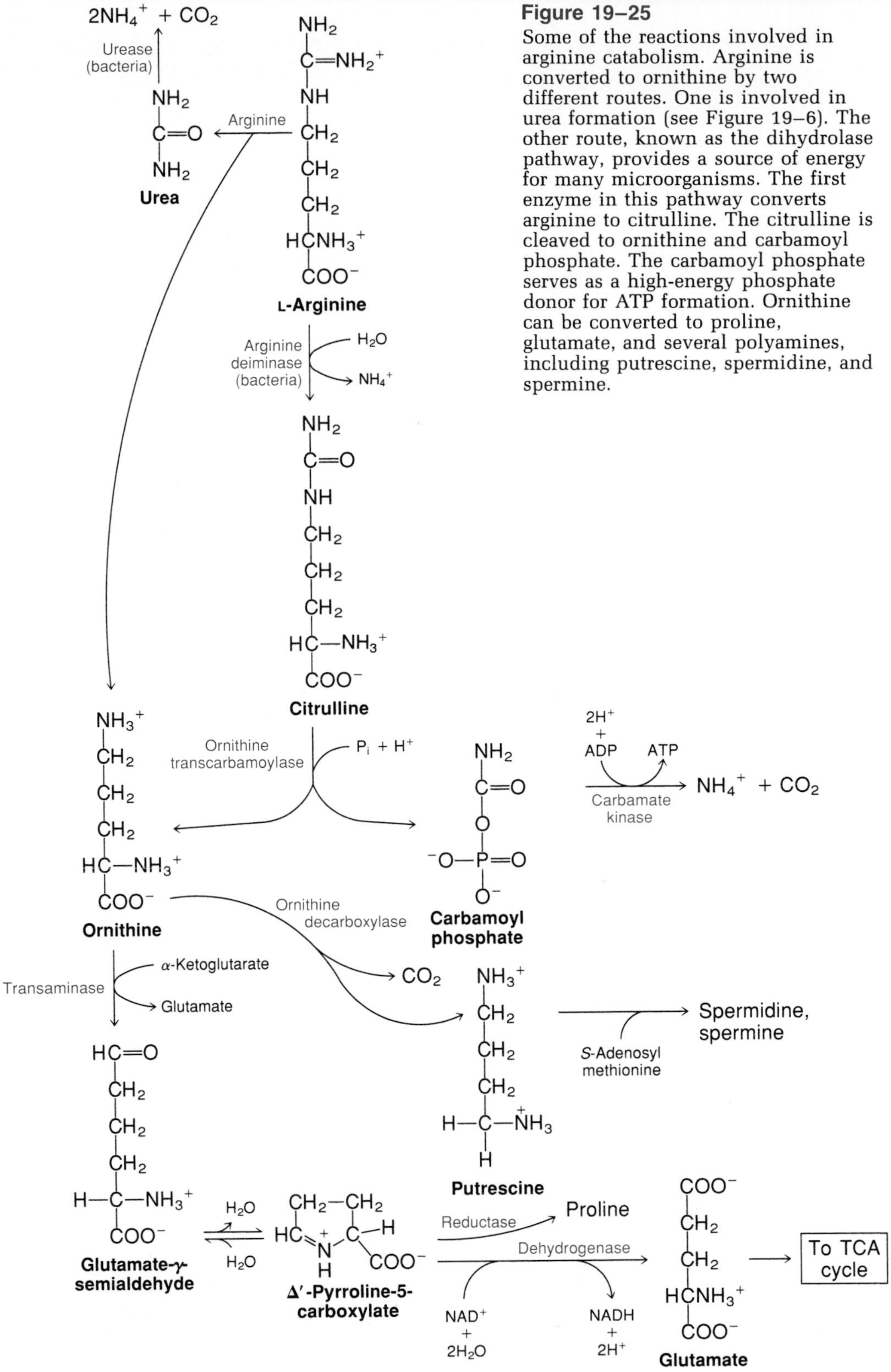

$2NH_4^+ + CO_2$

Urease
(bacteria)

Arginine

Urea

L-Arginine

Arginine
deiminase
(bacteria)

H_2O

NH_4^+

Citrulline

Ornithine
transcarbamoylase

$P_i + H^+$

Carbamoyl phosphate

$2H^+ + ADP$ ATP

Carbamate
kinase

$NH_4^+ + CO_2$

Ornithine

Ornithine
decarboxylase

CO_2

Transaminase

α-Ketoglutarate

Glutamate

Spermidine,
spermine

S-Adenosyl
methionine

Putrescine

Glutamate-γ-semialdehyde

H_2O H_2O

Δ'-Pyrroline-5-carboxylate

Reductase   Proline

Dehydrogenase

$NAD^+ + 2H_2O$ $NADH + 2H^+$

Glutamate

To TCA
cycle

Figure 19–25
Some of the reactions involved in arginine catabolism. Arginine is converted to ornithine by two different routes. One is involved in urea formation (see Figure 19–6). The other route, known as the dihydrolase pathway, provides a source of energy for many microorganisms. The first enzyme in this pathway converts arginine to citrulline. The citrulline is cleaved to ornithine and carbamoyl phosphate. The carbamoyl phosphate serves as a high-energy phosphate donor for ATP formation. Ornithine can be converted to proline, glutamate, and several polyamines, including putrescine, spermidine, and spermine.

25). We have discussed one, which provides a mechanism for urea formation and is important in the nitrogen metabolism in many animal species. The other route is one that provides a source of energy for many microorganisms. It is called the *arginine dihydrolase* pathway. The first enzyme, *arginine deiminase*, converts arginine to citrulline, with the liberation of NH_3 (Figure

19–25). Citrulline can be cleaved by a degradative ornithine transcarbamoylase to yield ornithine and carbamoyl phosphate. The carbamoyl phosphate so formed serves as a high-energy phosphate donor for ATP formation in a reaction catalyzed by *carbamate kinase*. In some animal tissues it appears that the same reaction can be catalyzed by an acetate kinase.

Ornithine, whether formed by the arginase of the urea cycle or arginine dihydrolase pathway, is broken down in most organisms by a transaminase to yield glutamate-γ-semialdehyde or its cyclized derivative, Δ^1-pyrroline-5-carboxylate. This compound can be further oxidized to glutamate by Δ^1-*pyrroline-5-carboxylate dehydrogenase*, an NAD$^+$-linked enzyme, or it can be reduced to proline by the normal proline biosynthetic enzyme Δ^1-*pyrroline-5-carboxylate reductase* in an NADPH-requiring reaction. This route to proline from ornithine accounts for the interconvertibility of ornithine and proline that is seen in many cells and tissues.

Amines Produced from Arginine.   *Putrescine*, the decarboxylated product of ornithine, is important as an intermediate in *spermidine* and *spermine* formation. These two polyamines of unknown function are generally found in association with nucleic acids. The decarboxylation is catalyzed by *ornithine decarboxylase*, and in most bacteria this route is the sole or primary route of putrescine formation (see Figure 19–25).

The synthesis of the omnipresent polyamines spermidine and spermine requires not only the formation of putrescine, but the generation of a propylamine group. The propylamine donor is a product derived from S-adenosylmethionine by a specific decarboxylase (Figure 19–26). The same enzyme transfers an aminopropyl group to spermidine to yield spermine.

Histidine.   The major route for histidine catabolism in mammals involves the conversion of histidine to glutamate. We shall see that in the process of

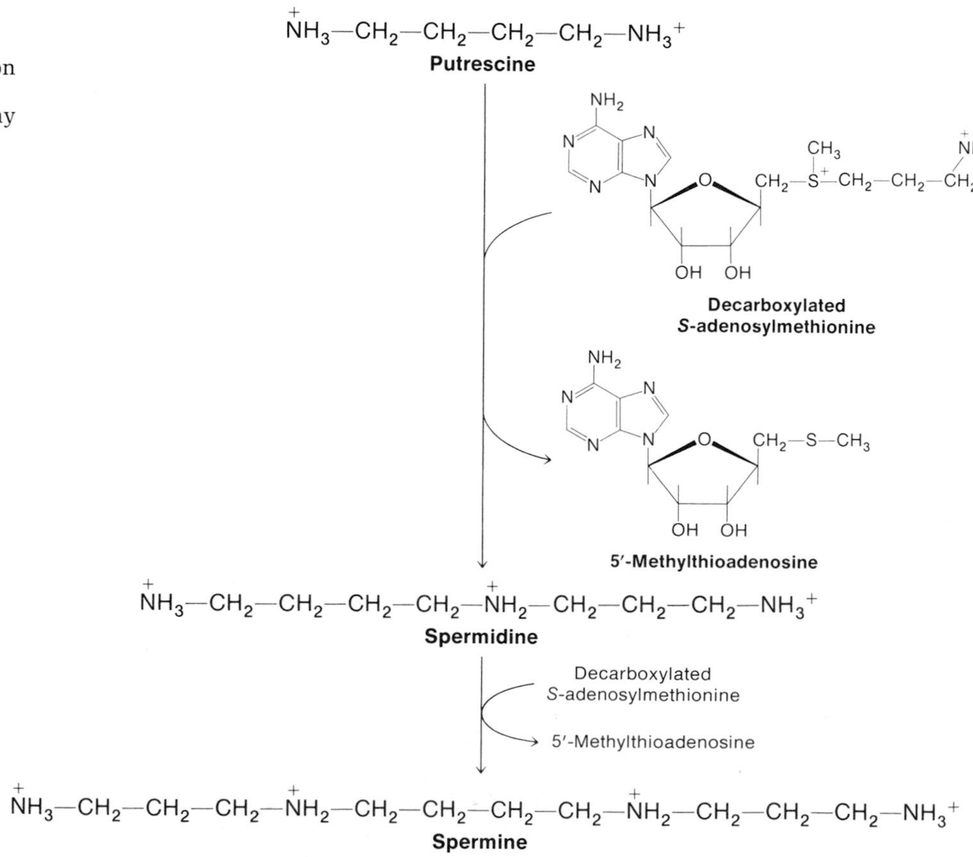

Figure 19–26

The utilization of putrescine and S-adenosylmethionine for the formation of spermidine and spermine. Polyamines are thought to have many functions; they are invariably found complexed with nucleic acids, both DNAs and RNAs.

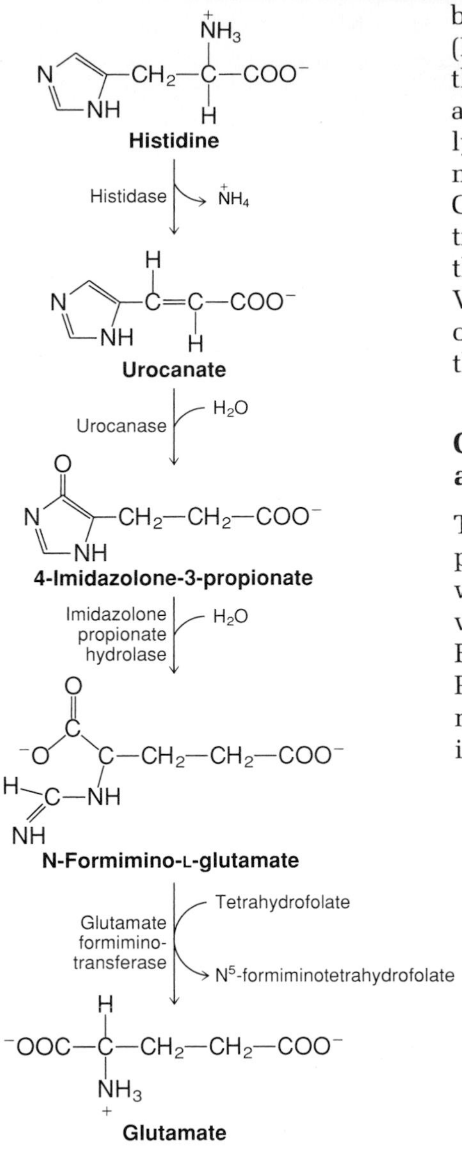

breakdown, histidine contributes a carbon atom to one-carbon metabolism (Figure 19–27). In the first step *histidase* catalyzes the removal of NH_3 with the formation of urocanate. This step is followed by an internal oxidation and reduction involving addition of the elements of water in a reaction catalyzed by *urocanase*. The resulting intermediate, 4-imidazolone-3-propionate, contains an imidazolone ring which is opened by a hydrolytic reaction. Cleavage of the product N-formino-L-glutamate leads to formimino group transfer to the N^5 position of tetrahydrofolate and free glutamate. Animals that are deficient in folic acid excrete large amounts of formiminoglutamate. Vitamin B_{12} deficiency leads to the same syndrome, a fact suggesting that the one-carbon unit on N^5-formiminotetrahydrofolate is normally transferred to this vitamin (see Chapter 11 for further discussion on vitamin B_{12}).

Catabolism of Methionine, Isoleucine, and Valine Leads to Succinyl-CoA

The carbon skeletons of *methionine, valine, and isoleucine are degraded* by pathways that lead *to succinyl-CoA* (Figure 19–28). Although these pathways are rather long, there are some simplifying features: Isoleucine and valine undergo identical reactions in the first four steps of degradation (see Figure 19–29); methionine and isoleucine are reduced to propionyl CoA. Propionyl-CoA is converted into methylmalonyl-CoA in two steps. Methylmalonyl-CoA, a common intermediate in all three pathways, is converted into succinyl-CoA in one step. The reactions between propionyl-CoA and

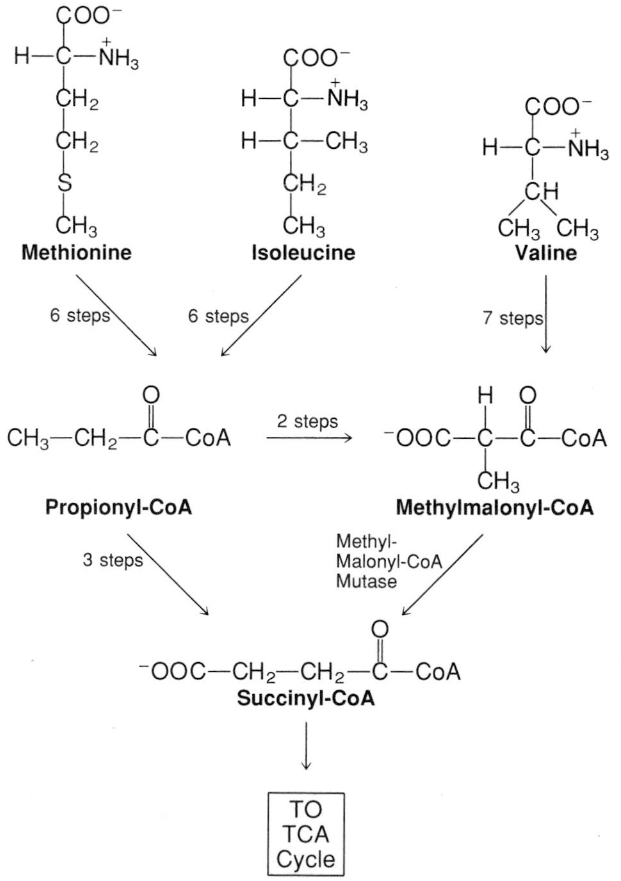

Figure 19–27
The catabolism of histidine. In the final step of histidine breakdown to glutamate, tetrahydrofolate is converted to N^5-formiminotetrahydrofolate. In this way histidine breakdown contributes a carbon atom to C-1 metabolism.

Figure 19–28
Outline of the catabolism of methionine, isoleucine, and valine to succinyl-CoA.

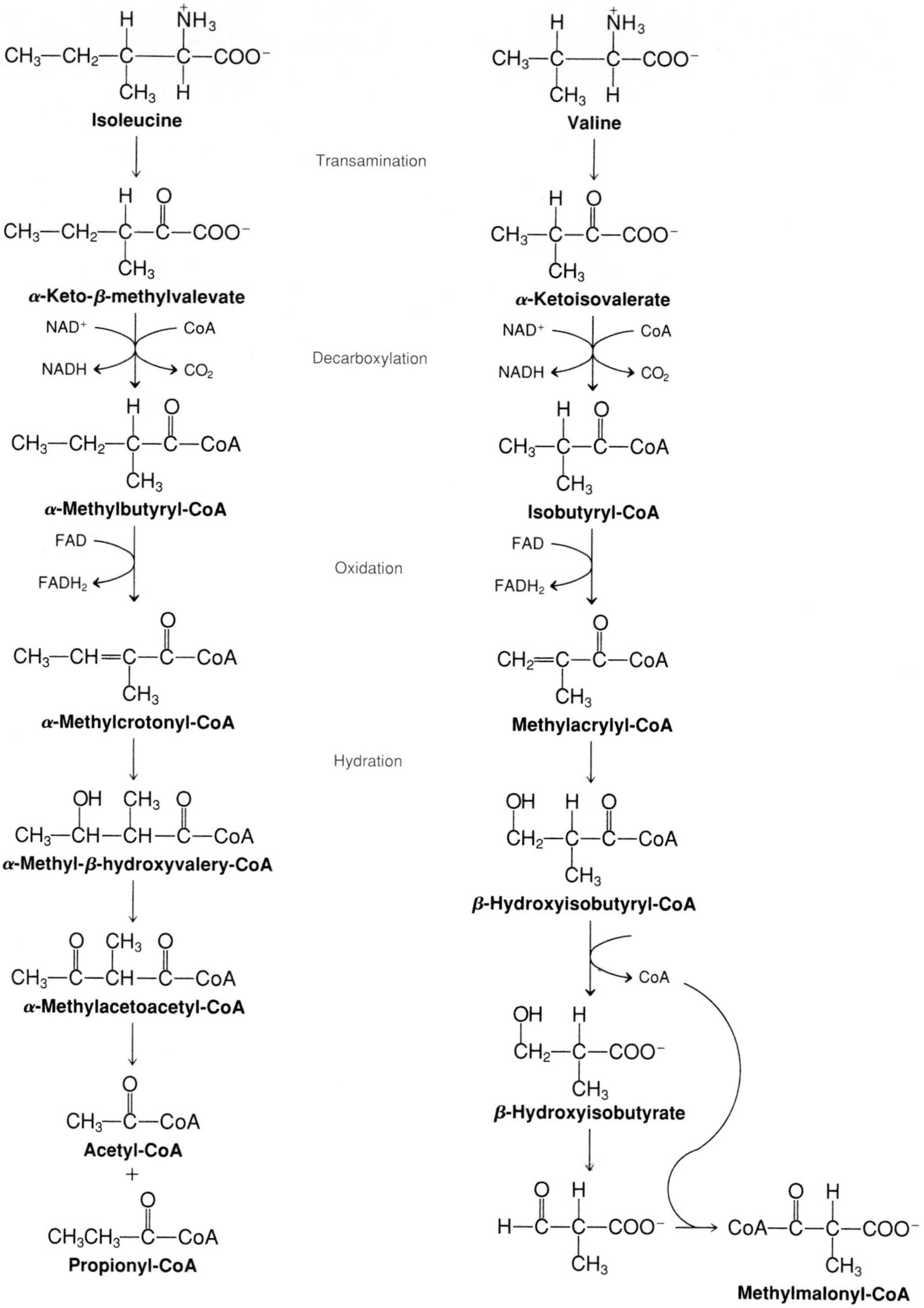

Figure 19–29
Isoleucine and valine undergo identical reactions in the first four steps of degradation.

succinyl-CoA have already been discussed in Chapter 18. Once again we see the strong overlap of reactions in lipid and amino acid catabolism, where the carbon skeletons contain several saturated carbon–carbon linkages.

The three keto acids derived by deamination of valine, isoleucine, and leucine are decarboxylated by the same enzyme complex. This enzyme com-

TABLE 19-1
Some Inborn Errors of Amino Acid Metabolism in Humans

Amino Acid Catabolic Pathway Involved	Condition	Distinctive Clinical Manifestation	Enzymatic Block or Deficiency
Arginine and the urea cycle	Argininemia and hyperammonemia	Mental retardation	Arginase
	Hyperammonemia	Neonatal death, lethargy, convulsions	Carbamoyl phosphate synthase
	Ornithinemia	Mental retardation	Ornithine decarboxylase
Glycine	Hyperglycinemia	Severe mental retardation	Glycine-cleavage system
Histidine	Histidinemia	Speech defects, mental retardation in some; in others, none	Histidase
Isoleucine, leucine, and valine	Branched-chain ketoaciduria ("maple syrup urine disease")	Neonatal vomiting, convulsions, and death; mental retardation in survivors	Branched-chain keto acid dehydrogenase complex
Isoleucine, methionine, threonine, and valine	Methylmalonic acidemia	Similar to preceding except that methylmalonate accumulates	Methylmalonyl-CoA mutase (some patients respond to vitamin B_{12} therapy)
Leucine	Isovaleric acidemia	Neonatal vomiting, acidosis, lethargy, and coma; survivors mentally retarded	Isovaleryl-CoA dehydrogenase
Lysine	Hyperlysinemia	Mental retardation and some noncentral nervous system abnormalities	Lysine-ketoglutarate reductase
Methionine	Homocystinuria	Mental retardation common; several eye diseases and thromboembolism common; osteoporosis and faulty bone structures	Cystathionine-β-synthase
Phenylalanine	Phenylketonuria and hyperphenyl-alaninemia	Vomiting is an early neonatal symptom, but mental retardation and other neurologic disorders develop in the absence of dietary treatment	Phenylalanine hydroxylase
Proline	Hyperprolinemia, type I	Probably not etiologically associated with any disease; proline excreted	Proline oxidase
Tyrosine	Alkaptonuria	Homogentisic acid in urine darkens on standing; in adult years, pigment deposits cause darkening of skin, cartilage; arthritis develops	Homogentisic acid oxidase
	Albinism	The most common type, oculocutaneous albinism, results in white hair, pink skin, and an extreme photophobia owing to lack of pigment in the eye	Tyrosinase of the melanocyte is absent

plex also acts on pyruvate and α-ketobutyrate, which are products of both threonine and methionine metabolism, respectively. Some persons are genetically defective for the presence of this enzyme complex. They accumulate substantial amounts of branched chain α-keto acids in the urine and suffer from a variety of disorders (see Table 19–1). Strict control of the diet, allowing only low amounts of these three amino acids, alleviates the immediate symptoms of this disease, but unless begun early after birth would not prevent or reverse the mental retardation.

Methionine. The catabolism of methionine involves nine steps leading to succinyl-CoA. Figure 19–30 illustrates the first six steps of this pathway; the last three steps, from propionyl-CoA, have already been discussed in Chap-

Figure 19–30
Degradation of methionine to propionyl-CoA. The first step in methionine breakdown leads to the formation of S-adenosylmethionine (SAM), which is used in many transmethylating reactions. Another useful product formed during methionine breakdown is cysteine.

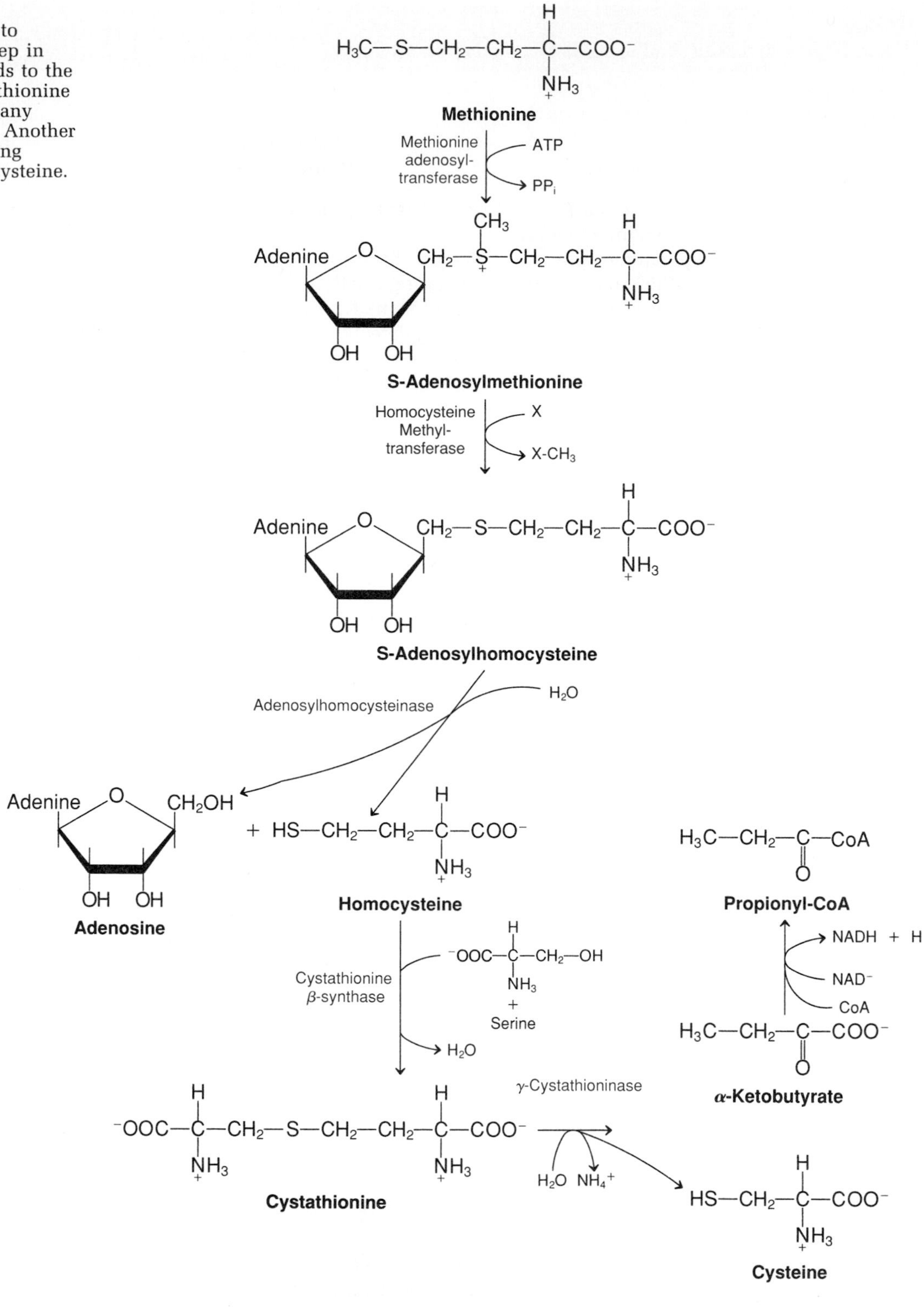

ter 18. We will focus on the first six steps, which are unique to methionine catabolism. These reactions illustrate some interesting one-carbon and sulfur metabolism processes.

In the first step methionine is adenylated to _S-adenosylmethionine_ (SAM). SAM is probably the most used transmethylating agent in the cell. We will see some further examples of the use of SAM in Part IV, which deals with biosynthesis. Transfer of the methyl group from SAM to an appropriate receptor leads to S-adenosylhomocysteine. This is hydrolyzed to adenosine and homocysteine. The homocysteine is condensed with serine to yield cystathionine, which in one more step is converted to cysteine and α-ketobutyrate. Cysteine, if it is present in excess, can be catabolized by the three-step pathway described in Figure 19–13. The α-ketobutyrate is converted in one step into propionyl-CoA.

Aspartate and Asparagine Are Deaminated to Oxaloacetate

The last two amino acids to be considered, to complete our discussion of amino acid catabolic pathways, are *aspartate* and *asparagine*. The entry of these amino acids into the TCA pool via oxaloacetate involves only two enzymes. *Asparaginase* converts asparagine to aspartate, and aspartate is reversibly converted into oxaloacetate in a typical transamination reaction with glutamate:

$$\text{Asparagine} + H_2O \longrightarrow \text{aspartate}^- + NH_4^+$$

$$\text{Aspartate}^- + \alpha\text{-ketoglutarate} \rightleftharpoons \text{oxaloacetate} + \text{glutamate}^-$$

Recall that aspartate is also converted into fumarate in the urea cycle.

INBORN ERRORS IN CATABOLISM OF AMINO ACIDS IN THE HUMAN

The concept that a gene might specify the formation of a specific enzyme was introduced by Garrod in 1902, after he had analyzed the occurrence of homogentisic acid excretion (alkaptonuria) in some of his patients and their families. He recognized the inheritable and therefore genetic nature of the condition in families of several alkaptonuric patients and postulated that a genetically controlled enzymatic deficiency underlay this metabolic error. We now know that the accumulation of homogentisate indicates a block in the third step in tyrosine breakdown. The condition results from a defective gene for the dioxygenase enzyme that catalyzes this reaction (see Figure 19–17 and Table 19–1).

Since 1902, researchers have described many metabolic diseases that are due to the inability of the affected individuals to dispose of certain dietary components. The diseases may be difficult to treat in the cases of errors in amino acid catabolism, since the culprit amino acid is one of the normal constituents of protein and is required for growth and development as well as for replacement of those body proteins that undergo rapid turnover. Because the affected fetus is usually carried by a mother heterozygous for the deficiency (carrying one normal and one defective gene) and whose own metabolism is essentially normal, the development of the fetus is essentially normal. Thus management of these diseases is possible, but it is dependent on a diagnosis soon after birth. Treatment consists in the use of a low-protein diet, carefully selected to supply enough of the culprit amino acid for protein formation but not enough to allow high plasma levels of it or of the offending metabolites. Supplements of nonoffending amino acids, prepared by synthesis or by fermentation processes, could be employed to compensate in part for the low-protein diet. Indeed, the chemical industries have made such preparations available.

Some of the diseases of amino acid catabolism are listed in Table 19–1. To appreciate the nature of the disease listed, you should refer to the text describing the metabolism of the individual amino acids. These naturally occurring defects have been invaluable in demonstrating the obligatory nature of some of the steps in amino acid breakdown. A mutation that causes a defective enzyme usually leads to (1) a substantial accumulation of the intermediate that is a substrate for that enzyme, and (2) a drastic lowering of all the intermediates below that step in the pathway.

REGULATION OF AMINO ACID BREAKDOWN

In animal cells the amino acid catabolic enzymes are subject most frequently to a hormonal control, although diet also influences the level of certain catabolic enzymes. Some catabolic enzymes that attack amino acids appear to be developmentally controlled and are formed only in certain tissues or at certain times during development. The developmentally programmed appearance of certain catabolic enzymes could, in fact, be mediated by hormonal signals that are themselves developmentally programmed.

A few enzymes that degrade amino acids have been measured at various stages during the development of animals, from late fetal periods to the adult stage. Thus tryptophan oxygenase is very low in the newborn rat, but it increases after about 12 days, concomitantly with a rise in adrenal activity. That at least part of the developmentally controlled formation of the enzyme is mediated by adrenal activity is indicated by the fact that glucocorticoids induce the de novo formation of the enzyme in young rats and stimulate its formation in adults. Liver serine deaminase also increases around this time, but it exhibits a transient increase in activity at birth. Ornithine transcarbamoylase and the other urea-forming enzymes appear shortly after birth and, like serine and threonine deaminases and ornithine transaminase, are further induced by high-protein diets. Table 19–2 lists several amino-acid-degrading enzymes that have been shown to be induced or repressed in animal cells and tissues, and some of the factors affecting their formation.

Our knowledge of the regulation of formation of amino acid degradation enzymes is far more extensive in bacteria and fungi. Toxic affects of amino acid excesses are far less a problem in unicellular organisms. As a result, regulation in microorganisms is dominated by considerations of nutritional and energy needs.

Some bacteria and fungi can use certain amino acids as a sole source of carbon and energy. In so doing, there is considerable nitrogen liberated, so that the amino acid used as a carbon source provides an excess of nitrogen, which is then excreted into the medium, usually as ammonia. However, most forms will utilize a carbohydrate or an organic acid as a carbon and energy source in preference to an amino acid. In such forms, the induced formation of enzymes metabolizing the amino acid is prevented as long as

TABLE 19–2
Regulation of Some Typical Amino Acid Degradative Enzymes in Animal Cells and Tissue

Amino Acid	Enzyme	Cell or Tissue	Factors Affecting Formation
Arginine and its precursor, ornithine	Ornithine-glutamate transaminase	Liver	High-protein diet; glucagon or cyclic AMP stimulate formation in vivo and in cell culture; corticosteroids and glucose repress in vivo
	Urea cycle enzymes	Liver	High-protein diet stimulates
Serine	Serine deaminase	Liver	High-protein diet stimulates
Tryptophan	Tryptophan oxygenase	Liver	Glucocorticoids stimulate formation
Tyrosine	Tyrosine-glutamate transaminase	Liver	Glucocorticoids stimulate formation
Threonine	Threonine deaminase	Liver	High-protein diet stimulates formation

TABLE 19–3
The Regulation of Some Typical Amino Acid Degradative Pathways in Bacteria and Fungi

Amino Acid	Organism	Inducer	Key Enzymes in Degradation
Alanine	E. coli	D- or L-alanine	Alanine racemase, D-alanine dehydrogenase
Arginine	S. cerevisiae	Arginine	Arginase, ornithine transaminase
Glutamate	Fungi	Glutamate	NAD-linked glutamate dehydrogenase
Histidine	K. aerogenes	Urocanate	Histidase, urocanase
Proline	E. coli	Proline	Proline oxidase
D-Serine	E. coli	D-Serine	D-Serine deaminase
Tryptophan	E. coli	Tryptophan	Tryptophanase

the preferred carbon and energy source is present. Such an antagonism of the induction of enzymes catabolizing one compound by a preferred or more readily used carbon source is an example of *catabolite repression* (see Chapter 29).

Microorganisms also can occur in environments in which a good (readily metabolizable) carbon source is available, but the only source of nitrogen is an amino acid. For some microorganisms, this nitrogen source would be unavailable, since its breakdown is prevented by catabolite repression. A fairly common finding, however, is that catabolite repression can be bypassed by an induced formation of the particular catabolic pathway as a result of a control signal that is conditioned through nitrogen starvation in the cell.

When nitrogen is in excess, the predominant mode of amino-group formation in many bacteria is by means of the reductive amination of α-ketoglutarate by free NH_3. Under these conditions less glutamine is required and glutamine synthase formation is repressed. When nitrogen is limited, NH_3 is more efficiently utilized by the energy-dependent glutamine synthase reaction, and the amide group of glutamine is used in the reductive amination of α-ketoglutarate catalyzed by glutamate synthase. Under these conditions, formation of glutamine synthase is induced. Some examples of microorganism enzymes that are induced when carbon and energy limit growth are given in Table 19–3.

In addition to regulation of its synthesis, the activity of glutamine synthase is directly subject to an elaborate set of controls that sense the need of the cell for nitrogen. Although the enzyme is regulated in most animals and plants in a similar manner, our most detailed understanding of how it is regulated comes from studies of the E. coli enzyme. The regulatory properties of this enzyme will be discussed in Chapter 24.

SUMMARY

Amino acids can serve as a source of energy, carbon, or nitrogen. In addition to these uses amino acids are frequently catabolized simply because they are present in potentially harmful excess.

For most amino acids the α amino group is removed at an early stage in catabolism, usually in the first step. Transaminases are specific for different amino acids. Frequently

α-ketoglutarate is the acceptor for the amino group, in which case it is converted into glutamate. The α-ketoglutarate can be regenerated from the glutamate by oxidative deamination.

A great deal of excess NH_3 frequently results from amino acid catabolism. This excess ammonia must be eliminated. In bacteria and lower eukaryotes the ammonia can usually be removed by simple diffusion, but in higher eukaryotes this is not feasible. Since the ammonia is frequently quite toxic, before removal it is detoxified by conversion to uric acid or urea. An intricate pathway resulting in the conversion of ammonia into urea involves five enzymes, three located in the cytoplasm and the remaining two in the mitochondrial matrix.

All amino acids can be degraded to CO_2 and water via the TCA cycle by the appropriate enzymes; the pathways are often complex and contain branchpoints to useful biosynthetic products. In every case, the pathways involve the formation of a dicarboxylic acid intermediate of the TCA cycle, of pyruvate, or of acetyl-CoA.

The discussion of amino acid catabolism is organized according to the common intermediates formed during degradation. Alanine, glycine, threonine, serine, and cysteine are degraded to acetyl-CoA by way of pyruvate. Phenylala-nine, tyrosine, tryptophan, lysine, and leucine also lead to acetyl-CoA, but they go by way of acetoacetyl-CoA rather than pyruvate. Arginine, histidine, proline, glutamic acid, and glutamine all degrade to α-ketoglutarate. Catabolism of methionine, valine, and isoleucine leads to succinyl-CoA. Aspartate and asparagine are converted to oxaloacetate on degradation.

The importance of catabolic pathways is underscored by a broad spectrum of human metabolic diseases in each of which one enzyme for normal amino acid catabolism is either missing or defective.

The formation of amino acid degradative enzymes is frequently regulated by hormones or nutritional factors. In microorganisms the formation of degradative enzymes is most often dependent on whether or not the amino acid is needed to supply energy, carbon, or nitrogen for growth.

SELECTED READINGS

Barker, H. A., Amino acid degradation by anaerobic bacteria. *Ann. Rev. Biochem.* 50:23, 1981. A review of an important group of fermentation pathways of amino acid breakdown that occur in nature and could not be covered in this chapter.

Christen, P., and D. E. Metzler, eds., *Transaminases.* New York: John Wiley and Sons, 1985. A series of review chapters describing in detail the scope and mechanisms of transamination reactions.

Ledley, F. D., H. E. Grenett, M. McGinnis-Shelnutt, and S. L. C. Woo, Retroviral-mediated gene transfer of human phenylalanine hydroxylase into NIH 3T3 and hepatoma cells. *Proc. Natl. Acad. Sci. USA* 83:409, 1986.

Mazelis, M., Amino Acid Catabolism. In B. J. Mifflin ed., *The Biochemistry of Plants*, Vol. 5. New York: Academic Press, 1980. Pp. 541–567. A survey of some of the amino acid catabolic pathways that have been found in plants.

Meister, A., *Biochemistry of the Amino Acids*, Vol. 2, 2d ed. New York: Academic Press, 1965. Pp. 593–1084. A very complete survey of amino acid catabolic pathways as they were known up to that time.

Wellner, D., and A. Meister, A survey of inborn errors of metabolism and transport in man. *Ann. Rev. Biochem.* 50:911, 1981. This review documents the importance of the pathways that break down amino acids in humans.

PROBLEMS

1. What consequences would you expect a severe deficiency in pyridoxal phosphate to have for amino acid metabolism?

2. Illustrate schematically the fate of nitrogen derived from the deamination of an amino acid in skeletal muscle, ending with the production of urea. Indicate where in the body each of these reactions occurs.

3. Given the scheme derived in Problem 2, calculate the energy cost, in moles of ATP hydrolyzed, that is involved in the deamination of a single mole of aspartate.

4. Draw the Kreb's bicycle, indicating the subcellular locations of each of the reactions and the points of the reaction scheme at which amino groups enter and exit (see also Chapter 15).

5. What is the function of the glucose-alanine cycle?

6. Define the terms glycogenic and ketogenic as applied to amino acids. Group the amino acids according to these definitions.

7. Compare the major fates of carbon skeletons derived from histidine and glycine.

8. α-Ketoadipate is formed as an intermediate in the catabolism of lysine by the liver. In light of the way ketoadipate is converted to lysine in fungi (see Chapter 24), propose a pathway for its formation from lysine in liver. Consider carefully the choice of pyridine nucleotide.

9. The pathways used in the breakdown of several amino acids include reactions common to β oxidation of fatty acids. For which amino acids is this the case?

10. Isoleucine, valine, and methionine are all glycogenic amino acids. Inborn errors that affect catabolism of these amino acids include defects in propionyl-CoA carboxylase and methylmalonyl-CoA mutase. Given this information, can you infer the common pathway to glucose-6-phosphate used by these three amino acids?

11. Some patients with the deficiencies described in Problem 10 who accumulate propionate, and others who

accumulate β-methylcrotonate, can be treated success-fully using biotin therapy. Explain why this might be the case. Why would vitamin B_{12} therapy remedy some patients' methylmalonic acidemia (see also Chapter 24)?

12. Inborn errors in leucine catabolism involve deficiencies in β-methylcrotonyl-CoA carboxylase and β-hydroxy-β-methylglutaryl-CoA lyase, in addition to those listed in Table 19–1. On the basis of these deficiencies, propose a pathway for leucine breakdown.

13. A patient is found to have a defect in the enzyme 4-hydroxyphenylpyruvate dioxygenase. What diet would you advise as treatment for this condition?

IV

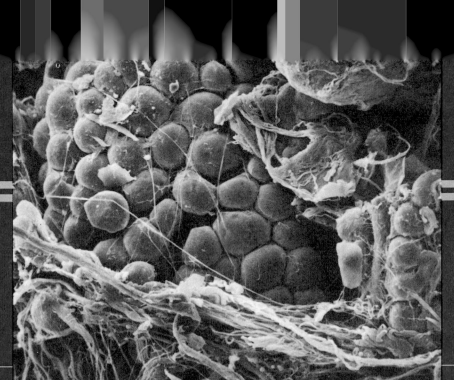

Scanning electron micrograph of fat cells (adipocytes) from the Rhesus monkey. Magnification 320×. (B. F. King, Univ. of California School of Medicine/BPS.)

ANABOLISM

PART IV

In Part IV the focus is on biosynthesis of carbohydrates, lipids, and the intermediates involved in the synthesis of proteins and nucleic acids, meaning amino acids and nucleotides, respectively.

The biosynthesis of simple sugars and simple sugar polymers has already been discussed in Chapter 14. In Chapter 20 we discuss the synthesis of sugars other than glucose—the disaccharides, as well as polysaccharides of different degrees of complexity. We also discuss the synthesis of oligosaccharides and polysaccharides that are covalently linked to proteins in glycoproteins. The chapter concludes with a discussion of bacterial cell wall synthesis.

Chapters 21, 22, and 23 are concerned mainly with the biosynthesis of fatty acids and lipids of various sorts. In Chapter 21 the biosynthesis of fatty acids and the regulation of fatty acid synthesis and breakdown is discussed.

Chapter 22 is entitled "Complex Lipids." This chapter examines the metabolism of glycerolipids, sphingolipids, and eicosanoids. Both anabolism and catabolism are discussed, and the regulation of these processes is also considered at select points. Most of the lipids described in this chapter serve structural roles, which have been considered in Chapter 6. By contrast, the eicosanoids act like hormones, and their functions are briefly considered in this chapter.

Cholesterol and cholesterol-related molecules, such as steroid hormones and bile acids, are the focus of Chapter 23. The discussion is concentrated on the metabolic processes as they take place in mammals. To appreciate the complexities of the metabolism it is essential to consider the events taking place in various parts of the organism.

In Chapter 24 we describe the biosynthesis of amino acids and a select group of products of which amino acids are the precursors. We consider first the nutritional aspects of amino acid metabolism. Then we discuss the fixation of atmospheric nitrogen into liquid or solid nitrogen-containing compounds. After that, we describe the biosynthesis of all twenty amino acids as it occurs in microorganisms, fungi, and plants. Also included is a discussion of some nonprotein amino acids found in microbes and a description of natural amino acid analogs. The chapter ends with a look at certain products derived from amino acids, including porphyrin, biologically active amines, glutathionine, gramicidin, and phosphocreatine.

Chapter 25 is concerned with the synthesis and breakdown of nucleotides. We devote the first part of the chapter to the biosynthesis of all nucleotides that serve as precursors for DNA and RNA. We then discuss the antimetabolite properties of nucleotide analogs and the regulation of nucleotide metabolism. The concluding topic in this chapter is the biosynthesis of nucleotide coenzymes.

20

CARBOHYDRATES

In earlier chapters we considered the structure and functions of sugars (Chapter 4) and the metabolism of hexoses that are involved in the central metabolic pathways (Chapter 14). The most important hexoses in this regard are phosphorylated derivatives of glucose and fructose. The variety of hexoses involved in structural polysaccharides is far greater than this.

In this chapter we will first describe the routes of synthesis of some of the more important hexoses found in oligo- and polysaccharides. Then we will consider some of the main pathways for polysaccharide synthesis.

BIOSYNTHESIS OF HEXOSES

The biosynthesis of glucose and fructose from simpler starting materials was discussed in Chapters 14 and 17. The biosynthesis of other hexoses can be linked to glucose by a complex network of single-step reactions (Figure 20–1). Glucose can serve as the precursor for the synthesis of any other hexose without any rearrangement of the central carbon atoms.

Although the list of hexoses is large, there are certain generalizations that can be made about the pathways. One of them is that neutral hexoses are never interconverted. This statement should not be surprising because neutral hexoses are rarely found intracellularly, for the reasons discussed in Chapter 14. Several hexose interconversions occur at the level of the monophosphorylated hexose. Many interconversions also occur at the level of nucleotidyl sugars that are suitably activated for polymerization. Hexose modification usually takes place before polymerization.

The Hexose Monophosphate Pool Includes
Glucose, Fructose, and Mannose

Derivatives of glucose, mannose, fructose, and galactose are the most common sugars found in oligo- and polysaccharides. Three of these hexoses—glucose, mannose, and fructose—belong to the same hexose monophosphate

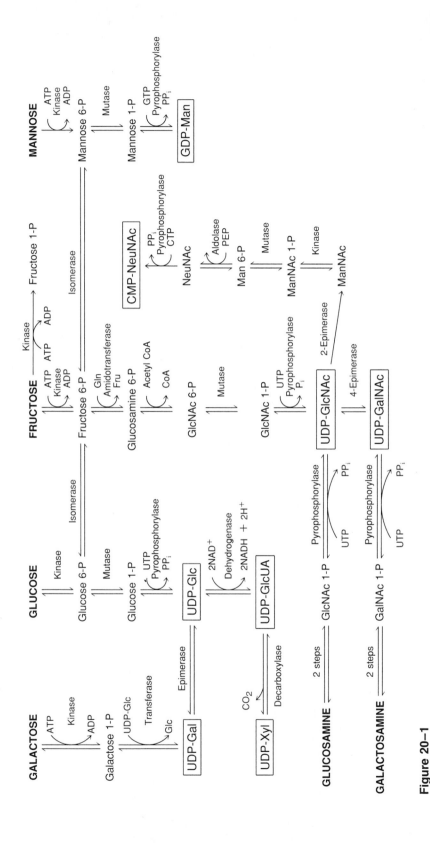

Figure 20-1

Hexose interconversions. The following abbreviations are used here and throughout the chapter: Gal = galactose, Xyl = xylose, Glc = glucose, GlcUA = glucuronic acid, GlcNAc = N-acetylglucosamine, GalNAc = N-acetylgalactosamine, Man = mannose, ManNAc = N-acetylmannosamine, NeuNAc = N-acetylneuraminic acid (equivalent to Sia = sialic acid). Actually, sialic acid is a more general term including N- and O-substituted derivatives of neuramic acid. Compounds in capital boldface type are the unmodified sugars or their amine derivatives. The nucleotide sugars are boxed.

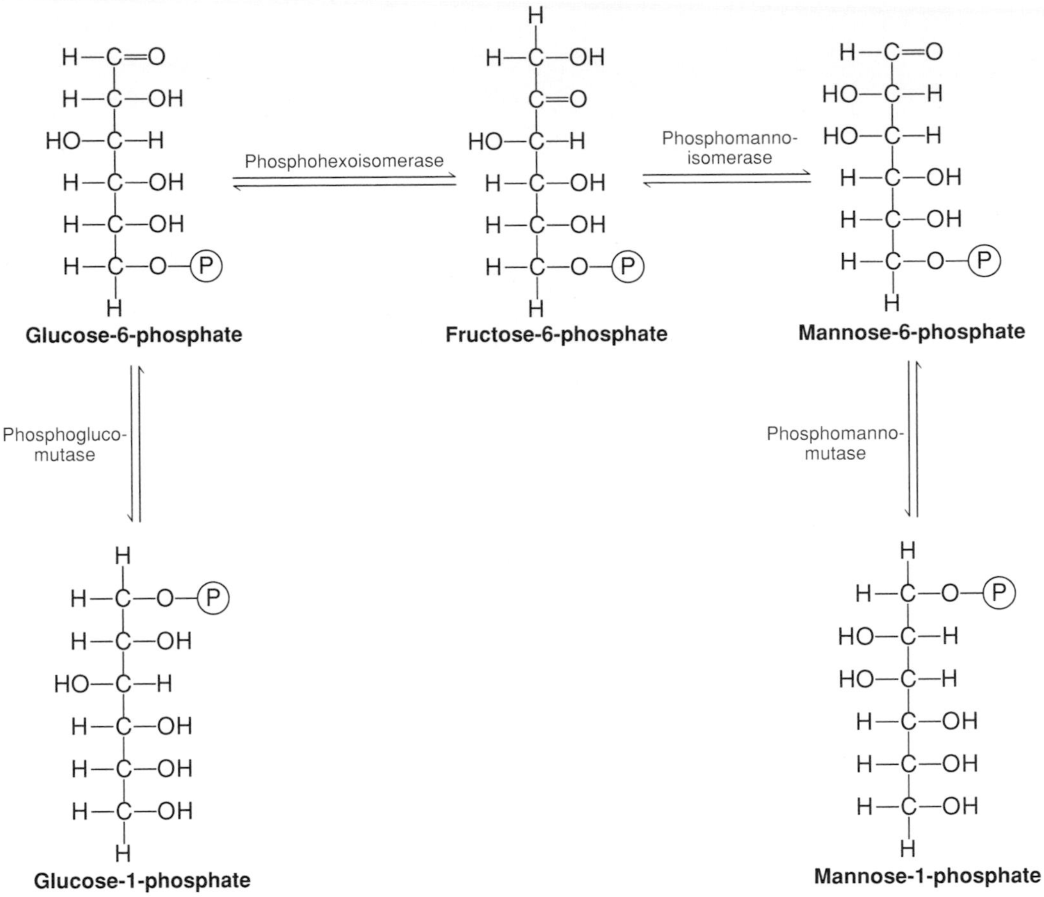

Figure 20–2
Members of the hexose monophosphate pool. The sugars shown are all freely interconvertible, with little change in free energy involved in the conversions.

pool (Figure 20–2). They are readily interconverted, with little or no difference in the free energy of the different compounds involved.

The interconversion of glucose and fructose phosphates was discussed in Chapter 14, in conjunction with glycolysis and gluconeogenesis. Mannose was left out of that discussion because it is not centrally involved in either process. Mannose-6-phosphate, the 2-epimer of glucose-6-phosphate, can be made directly from fructose-6-phosphate in a reaction analogous to the interconversion of fructose-6-phosphate and glucose-6-phosphate (Figure 20–3). *Phosphomannoisomerase* is the enzyme that catalyzes this isomerization. Evidently it holds the substrate so that a proton adds to the planar C-2 of the intermediate enediol on the side opposite that on which addition occurs in the phosphoglucoisomerase reaction (see Figure 14–6). Mannose-6-phosphate can then be converted to mannose-1-phosphate by a specific *phosphomannomutase*. Mannose and mannose derivatives are incorporated into complex macromolecules by way of GDP-mannose, which is made from mannose-1-phosphate and GTP in reactions analogous to the activation reactions that we have seen (see Figure 14–16).

Galactose Is Not a Member of the Pool

Galactose, the other main hexose found in structural polysaccharides, is not a member of the central hexose monophosphate pool. The only route from the main pool and galactose goes through UDP-activated derivatives of glu-

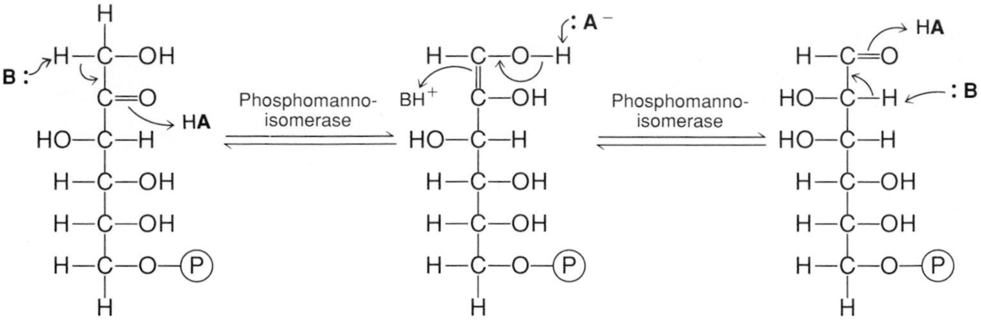

Figure 20–3

Mechanism of the interconversion of fructose-6-phosphate and mannose-6-phosphate. This interconversion is very similar to that observed for glucose-6-phosphate and fructose-6-phosphate, involving the same enediol intermediate (see Figure 14–6). The A and B groups refer to catalytic sites on the enzyme.

cose (see Figure 14–1). By the simplest route galactose, the 4-epimer of glucose (see Figure 4–11), is produced by way of UDP-glucose. The reactions are

$$\text{Glucose-1-phosphate} + \text{UTP} \longrightarrow \text{UDP-glucose} + \text{PP}_i$$

$$\text{UDP-glucose} \rightleftharpoons \text{UDP-galactose}$$

UDP-glucose 4-epimerase, which catalyzes the isomerization of UDP-glucose to UDP-galactose, contains a tightly bound molecule of NAD^+ or $NADP^+$, even though no net oxidation is involved. It seems likely that the —OH group at C-4 is transiently oxidized to a carbonyl, which can then be reduced by addition of a hydride ion to either side, thus producing either UDP-glucose or UDP-galactose.

Hexose Modifications Involve Alterations or Additions

Thus far we have considered only the isomerizations of hexoses. Many modifications of hexoses involve the alteration of existing groups or the addition of small groups to the hexose moiety.

Oxidation of carbon 6 to a carboxylate group produces a uronic acid. The dehydrogenases that catalyze those oxidations use NAD^+ as the oxidizing agent. For example, UDP-glucose (UDP-Glc) is converted to UDP-glucuronic acid (UDP-GlcUA) in this way.

Replacement of the —OH group at carbon 2 by —NH_2 yields an amino sugar, which is invariably modified further. For example, fructose-6-phosphate is the monosaccharide precursor of several important derivatives (Figure 20–4). The process begins with glucosamine formation, a one-step reaction in which the amide nitrogen of glutamine is transferred to the C-2 carbon of the fructose. This is the first step in a biosynthetic pathway leading to several different sugar monomers. The amine is acetylated to N-acetylglucosamine-6-phosphate, which is activated by reaction with UTP to form UDP-N-acetylglucosamine. This derivative may be directly incorporated into polymer or may be converted to other polymer precursors. In one step UDP-N-acetylglucosamine can be epimerized to UDP-N-acetylgalactosamine, or in six steps it can be converted to CMP-N-acetylneuraminic acid, also known as a CMP-sialic acid (see Figure 4–11 for the structure of this sialic acid). The CMP-sialic acids are the only nucleotide sugars that occur as nucleoside monophosphate derivatives.

Little is known about the regulation of synthesis of the various nucleotide sugar derivatives. If that regulation follows the general scheme sug-

Figure 20–4
Some derivatives formed from fructose-6-phosphate. This figure elaborates on one branch of Figure 20–1.

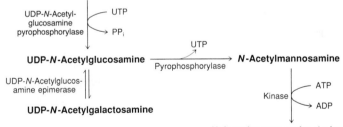

$$H-\underset{\underset{H}{|}}{\overset{|}{C}}-OH$$
$$H-C=O$$
$$HO-C-H$$
$$H-C-OH$$
$$H-C-OH$$
$$H-\underset{\underset{H}{|}}{\overset{|}{C}}-O-\text{(P)}$$

Fructose-6-phosphate

Glutamine : fructose-6-phosphate amido-transferase — Glutamine → Glutamate

Glucosamine-6-phosphate

Acetyl-CoA → CoA

$$H-\underset{\underset{H}{|}}{\overset{|}{C}}-OH$$
$$C-NH-\overset{O}{\overset{\|}{C}}-CH_3$$
$$HO-C-H$$
$$H-C-OH$$
$$H-C-OH$$
$$H-\underset{\underset{H}{|}}{\overset{|}{C}}-O-\text{(P)}$$

N-Acetylglucosamine-6-phosphate

Mutase

N-Acetylglucosamine-1-phosphate

UDP-N-Acetyl-glucosamine pyrophosphorylase — UTP → PP$_i$

UDP-N-Acetylglucosamine   UTP → **N-Acetylmannosamine**
Pyrophosphorylase

UDP-N-Acetylglucos-amine epimerase

UDP-N-Acetylgalactosamine

Kinase — ATP → ADP

N-Acetylmannosamine-1-phosphate

Mutase

N-Acetylmannosamine-6-phosphate

N-Acetylneura-minate aldolase — Phosphoenolpyruvate → P$_i$

N-Acetylneuraminic acid-9-phosphate

Phosphatase → P$_i$

N-Acetylneuraminic acid

Sialic acid pyrophosphorylase — GTP → PP$_i$

CMP-N-Acetylneuraminic acid

gested in Chapter 12 for a biosynthetic pathway, then the first enzyme in a sequence should be negatively regulated by the end product of the sequence. In agreement with this principle, it has been found in rat liver that UDP-N-acetylglucosamine regulates its own synthesis by inhibiting the amidotransferase that catalyzes the conversion of fructose-6-phosphate to glucosamine-6-phosphate, the first step that is specific to this pathway (see Figure 20–4).

DISACCHARIDE SYNTHESIS

In Chapter 14 we considered some of the problems associated with the catabolism of two well-known disaccharides, lactose (galactose-β-1,4-glucose) and sucrose (glucose-α-1,2-fructose). At this juncture we will describe the biosynthetic pathways of these two disaccharides.

In disaccharide synthesis only one of the participating hexoses is activated. The disaccharide lactose is formed in the mammary gland from D-glucose and UDP-galactose by the action of *lactose synthase* (Figure 20–5). The reaction involves nucleophilic displacement of UDP from UDP-galactose by the C-4 hydroxyl group of a free glucose. This type of nucleophilic displacement reaction, whether it involves a C-4 hydroxyl or another hydroxyl from the acceptor molecule, is the standard mechanism for forming glycosidic bonds (also see Figure 14–16).

Formation of lactose involves an unusual mechanism for controlling enzyme specificity. Lactose synthase is actually a complex of two proteins. Protein A, otherwise known as galactosyltransferase, is found not only in the mammary gland, but also in the liver and small intestine. It catalyzes the reaction

UDP-galactose + N-acetyl-D-glucosamine $\longrightarrow$ UDP + N-acetyllactosamine

Protein B is known as the α-lactalbumin of milk. It has no catalytic activity of its own. Rather, it alters the specificity of protein A so that the latter will

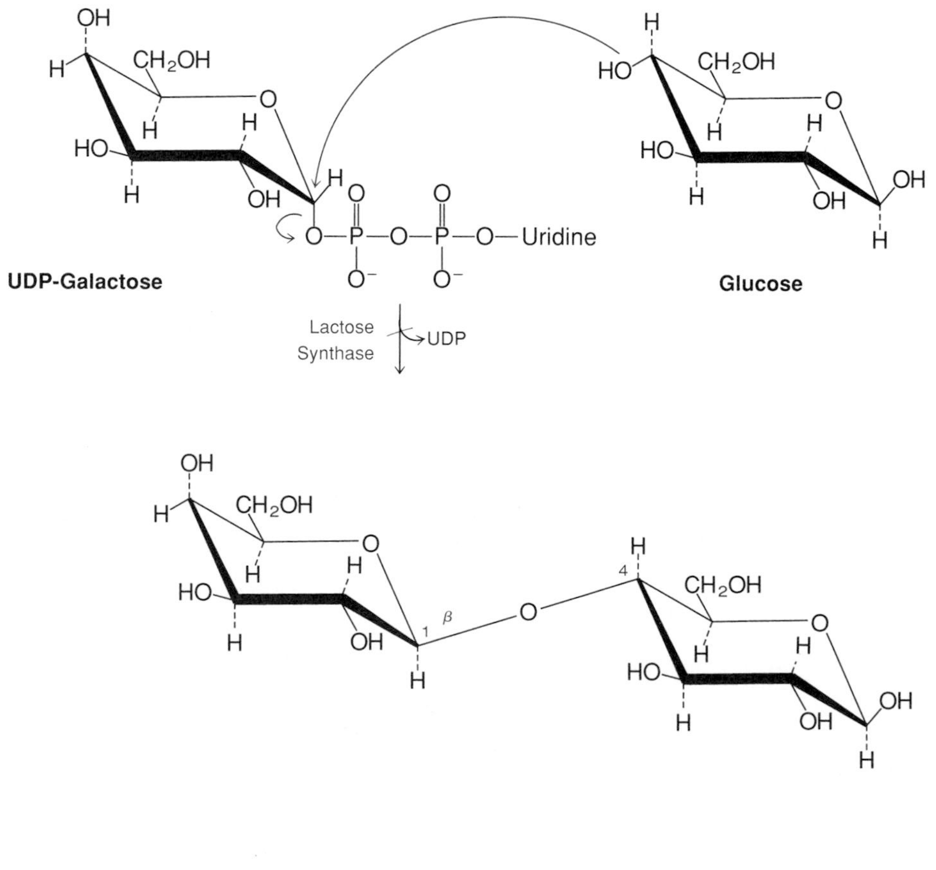

Figure 20–5
The mechanism of lactose formation. The more realistic chair forms are shown for the hexoses to make it easier to appreciate the stereochemistry of the reaction. Note how the configuration at C-1 becomes inverted in this reaction.

utilize D-glucose instead of N-acetyl-D-glucosamine as the galactose acceptor. As a result, protein A makes lactose instead of N-acetyllactosamine:

$$\text{UDP-Galactose} + \text{D-glucose} \longrightarrow \text{UDP} + \text{lactose}$$

Sucrose is synthesized from glucose and fructose. First, fructose-6-phosphate is produced from glucose-6-phosphate. The latter is also converted via glucose-1-phosphate to uridine diphosphate glucose (UDP-glucose), which then reacts with fructose-6-phosphate to give UDP and sucrose-6-phosphate. The phosphate is removed by a single enzymatically catalyzed hydrolysis to yield sucrose and inorganic phosphate. In some plants, sucrose is formed simply by the reaction of UDP-glucose with fructose.

POLYSACCHARIDE SYNTHESIS

The topic of polysaccharide synthesis has been partially covered in Chapter 14, where the focus was on energy-generating metabolism and its regulation. In this chapter we will focus more on the different types of polysaccharides that are made, and the pathways that are used for their synthesis.

Energy-storage Polysaccharides Are Simple Homopolymers

Most polysaccharides that are used for energy storage are simple homopolymers of glucose linked by $\alpha(1,4)$-glycosidic bonds. The basic mode of synthesis of these polyglucose molecules was described in Chapter 14. In all cases the C-1 of the monomer is activated as in disaccharide synthesis (Figure 20–6). Sometimes this activation is supplied by a UDP-derivative, sometimes by an ADP-derivative (see Table 20–1). The $\Delta G^{\circ\prime}$ of this reaction is about -3.2 kcal/mol.

The enzyme _glycogen synthase_ requires as a primer an $\alpha(1,4)$-polyglucose chain having at least four glucose residues, to which it adds successive glucosyl groups. It works best with long-chain glucose polymers as primers. In addition to $\alpha(1,4)$ bonds, glycogen contains $\alpha(1,6)$ bonds. The

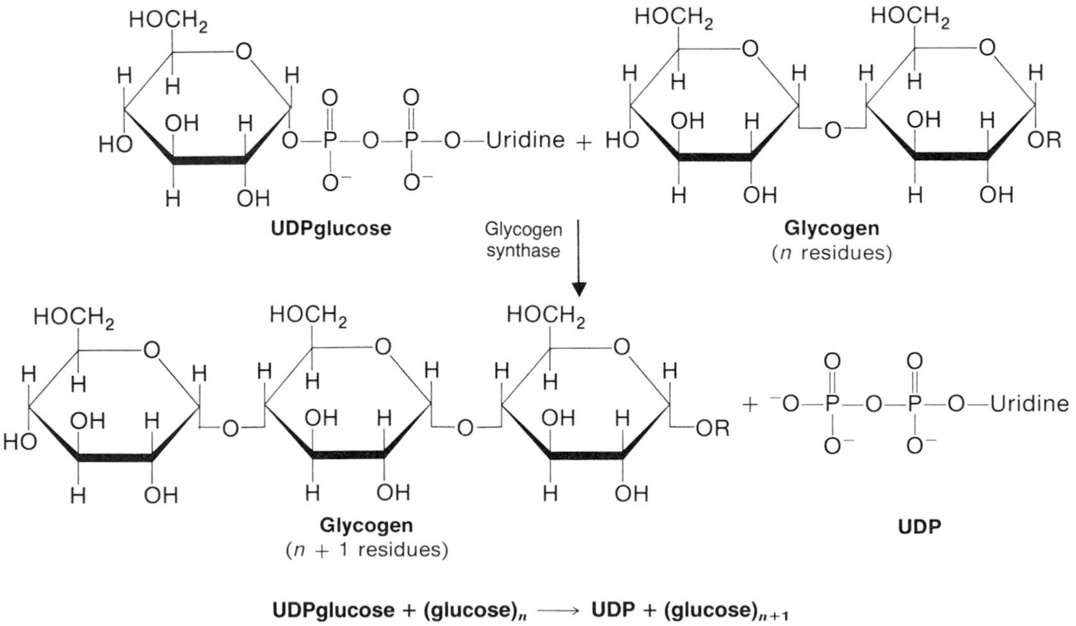

$$\text{UDPglucose} + (\text{glucose})_n \longrightarrow \text{UDP} + (\text{glucose})_{n+1}$$

Figure 20–6
Elongation step in glycogen synthesis. The chemistry of this reaction is quite similar to that of lactose synthesis except that there is no inversion about C-1.

TABLE 20–1
Some Storage Polysaccharides

Source	Polysaccharide	Monosaccharide Component(s)	Glycosyl Donor	Polymer Structure
Primarily muscle and liver cells of animals	Glycogen	Glucose	UDPglucose	$\alpha(1,4)$ with $\alpha(1,6)$ branch points
Bacterial glycogen	Glycogen	Glucose	ADPglucose	$\alpha(1,4)$ with $\alpha(1,6)$ branch points
Green algae	Amylose	Glucose	ADPglucose	Linear $\alpha(1,4)$
Leaves, stem, roots, and seeds of higher plants	Amylopectin	Glucose	ADPglucose	Linear $\alpha(1,4)$ with $\alpha(1,6)$ branch points
Some bacteria	Dextran	Glucose	Sucrose	Linear $\alpha(1,6)$ with $\alpha(1,2)$, $\alpha(1,3)$ or $\alpha(1,4)$ branch points

latter bonds are made by the branching enzyme <u>amylol(1,4-1,6)-trans-glyco-sylase</u>. This enzyme transfers a terminal oligosaccharide fragment of six or seven glucosyl residues from the end of the main glycogen chain to the 6-hydroxyl group of a glucose residue somewhere in a glycogen chain. The reaction produces a branched-chain polymer from a straight-chain polymer, as shown in Figure 20–7. As might be expected, the free energy change in this reaction is very small, since very similar chemical linkages are involved.

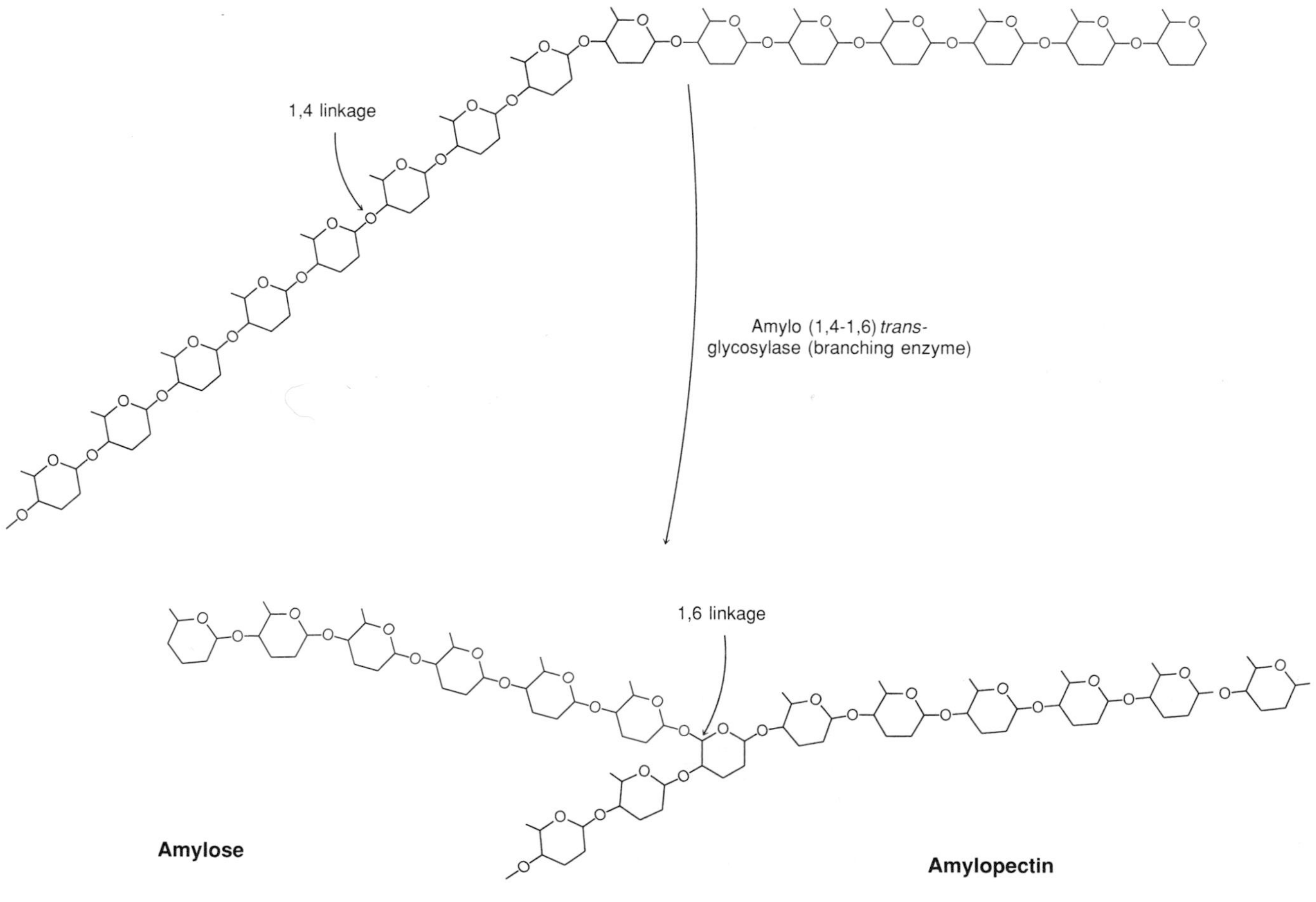

Figure 20–7
Schematic diagram showing the action of the "branching enzyme" in glycogen formation. A terminal hexasaccharide fragment is shown as being transferred from a 1,4 straight-chain linkage to a 1,6 branchpoint. No activation is involved because the energy change on reaction is very small.

In plant tissues, starch synthesis occurs by an analogous pathway that is catalyzed by _amylose synthase_. ADP-glucose is the preferred glucose donor.

Since glycogen is just a convenient way of storing sugar until it is required for energy purposes, its breakdown is as important as its synthesis. Glycogen breakdown proceeds by a different route, involving the action of inorganic phosphate and the enzyme glycogen phosphorylase on the polymer (see Chapter 14). One of the most interesting aspects of glycogen metabolism has to do with the intricate mechanism that controls synthesis and breakdown, which relates to the organism's energy requirements. This, too, is discussed in Chapter 14.

Structural Polysaccharides Include Homo-polymers and Heteropolymers

The most abundant structural polysaccharide is plant cellulose, a straight-chain homopolymer of glucose with a $\beta(1,4)$ linkage. Cellulose is formed by the same general mechanism as glycogen, using nucleoside diphosphate sugars. In addition, chitin in insects, which is a $\beta(1,4)$ homopolymer of N-acetylglucosamine (GlcNAc), is formed in a similar reaction from UDP-N-acetylglucosamine (UDP-GlcNAc).

The animal polysaccharide hyaluronic acid presents a variation characteristic of glycosaminoglycans (see Table 4–1). Like many linear polysaccharides found as protein or lipid conjugates, hyaluronic acid consists of a strict alternating sequence of two different hexoses (see Table 4–1). The monomers of hyaluronic acid are N-acetylglucosamine and glucuronic acid (GlcUA). Hyaluronic acid is formed by successive alternating reactions of UDP-glucuronic acid and UDP-N-acetylglucosamine with the end of the growing chain. Two enzymes, one specific for UDP-glucuronic acid and one specific for UDP-N-acetylglucosamine, are used in the polymerization stage.

In most polysaccharide syntheses, nucleoside diphosphate sugars are used as activated substrates. A striking exception is seen in the case of certain bacteria that synthesize dextran, a predominantly $\alpha(1,6)$ polymer of glucose. The substrate for dextran synthesis is sucrose, and the energy of the glycosidic bond between glucose and fructose in this disaccharide drives the reaction:

$$\text{n Sucrose} \xrightarrow[\text{sucrase}]{\text{Dextran}} \text{dextran} + \text{n fructose}$$

Dextrans formed by bacteria growing on the surface of teeth are an important component of dental plaque.

OLIGOSACCHARIDE SYNTHESIS IN HIGHER ANIMALS

One of the most complex and potentially exciting areas in contemporary carbohydrate research has to do with the role of oligosaccharides attached to proteins (or lipids) that are free or membrane-bound. In only a few cases have specific functions been clearly established. Many potential functions are, however, being considered.

Since glycoproteins that become membrane-bound always have the oligosaccharide oriented toward the noncytoplasmic side of the membrane, it has been suggested that the oligosaccharide may be essential for proper orientation of the glycoprotein in the membrane. In specific instances where inhibitors of oligosaccharide synthesis have been introduced, the same proteins, though lacking their oligosaccharide surface structure, orient normally in the target membrane. Nevertheless, in many cases the oligosaccharide portion of the glycoprotein may play an important role in anchoring and orienting the glycoprotein in the membrane. The idea is at least worthy of further exploration.

Perhaps the most significant role that has been suggested for the oligosaccharides on glycoproteins is that they serve as a recognition feature. Since there are so many possible structures for oligosaccharides, it seems highly likely that an oligosaccharide bound to the surface of a protein or membrane could play an important role in specific cell–cell, cell–protein, or protein–protein interactions. The clearest indications that oligosaccharides dictate specific interactions come from immunology. Protein-bound oligosaccharides or membrane-bound oligosaccharides are highly antigenic, that is, they interact strongly with specific antibodies. Recall (Chapter 2) that antibodies are proteins with specific interaction sites for antigens. The sites consist of highly specific cavities on the antibody surface. The oligosaccharide, by contrast, has a protruding structure that could insert into a properly designed cavity like a hand in a glove. A little later we will discuss an example in which oligosaccharides play an important antigenic role. The immunologic literature abounds with further examples.

Another function that has been suggested for the protein-bound oligosaccharide is that it guides newly synthesized protein to its correct final destination. We will mention one such example for lysosomal enzymes, which are made in one cell organelle and guided to another by their bound oligosaccharide.

Carbohydrate-binding proteins called _lectins_ are widely distributed in all living cells. The existence of these proteins provides added support for the specificity function of protein-linked oligosaccharides. Plant lectins are the best characterized. The jackbean lectin, _concanavalin A_, binds oligosaccharides containing terminal mannose. Peanut lectin is specific for galactose, and _wheat germ agglutinin_ binds to N-acetylglucosamine. In many cases there is also some specificity for the penultimate hexose as well. As might be expected, lectins are usually found in membrane structures. They have been implicated in specific protein absorption and cellular differentiation in different organisms.

All Glycoproteins Are Synthesized in the Endoplasmic Reticulum

The synthesis of glycoproteins is complicated by the fact that oligosaccharide synthesis is usually initiated before protein synthesis is completed. We will not discuss the mechanism of protein synthesis in any detail until Chapter 28, but there are certain facts that must be presented here if we are to appreciate the significance of the oligosaccharide synthesis associated with glycoproteins.

In a eukaryotic cell, protein synthesis takes place on the ribosomes (see Introduction). All the ribosomes in a cell are identical except that they become transiently associated with different messenger RNA molecules. The messenger RNA determines the specific protein that will be synthesized. Ribosomes in the process of protein synthesis exist as independent bodies in the cell cytosol or as membrane-complexed bodies on the _endoplasmic reticulum_. The endoplasmic reticulum is a membrane-bounded organelle that is found in the cytoplasm (see Figure I–2). Ribosomes bind to the endoplasmic reticulum when the partly synthesized polypeptide chain contains a signal sequence with a high affinity for the endoplasmic reticulum. This sequence guides the ribosome synthesizing such a polypeptide to the endoplasmic reticulum. The polypeptide eventually passes through the membrane into the _lumen_ (noncytoplasmic side) of the endoplasmic reticulum (Figure 20–8). All glycoproteins are synthesized in the endoplasmic reticulum. Thus the endoplasmic reticulum seems to be an area for segregating proteins to which oligosaccharide addition is desirable.

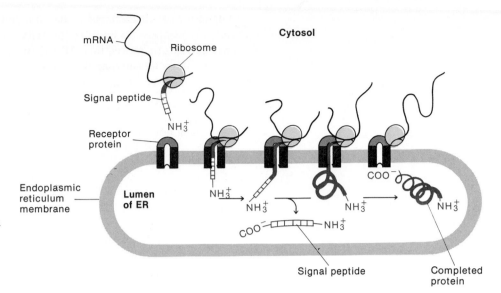

Figure 20–8
Schematic view of the effect of the "signal peptide" on determining the destination of the newly synthesized polypeptide. Certain polypeptide chains carry a peptide at their NH₂-terminals called the signal peptide. This draws the peptide-ribosome complex to the endoplasmic reticulum (ER). The polypeptide penetrates the membrane to the ER lumen. The protein usually loses the signal peptide by a specific hydrolysis reaction during or after synthesis.

As we mentioned earlier, the synthesis of the oligosaccharide part of the glycoprotein begins before polypeptide synthesis is complete. The partially completed glycoprotein, containing the entire protein but only part of the oligosaccharide, is transferred by a budding process to another class of membrane-bounded organelles, collectively known as the *Golgi apparatus* (Figure 20–9). In the Golgi apparatus, further additions and processing of the oligosaccharide occur. An additional budding process directs the glycoproteins to their final destination. We can now discuss specific steps in the synthesis of the oligosaccharide portion of the glycoprotein in greater detail.

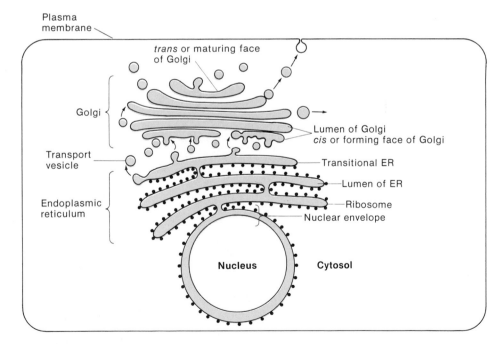

Figure 20–9
Schematic diagram showing the relative locations of nucleus, endoplasmic reticulum, Golgi apparatus, and plasma membrane. Glycoproteins synthesized in the lumen of the endoplasmic reticulum (ER) pass from the ER to the Golgi apparatus by a sequential process of budding and fusion to membranes. The Golgi membranes are classified as *cis*, medial, and *trans* in the order of increasing distance from the nucleus. The mature proteins exit from the *trans* Golgi in membrane-bounded vesicles formed from the Golgi. Depending on the nature of these membranes, such vesicles may have one of several different fates, distributing their contents to different sites inside or outside the cell.

Different oligosaccharides are distinguished by the number and type of sugars they contain as well as the way in which these sugars are interconnected. There are more than 100 different kinds of glycosidic linkages in glycoprotein oligosaccharides and a comparable number of specific glycosyltransferases that catalyze reactions of the following general type:

$$\text{Glycosyl}_1\text{-phosphonucleotide} + \text{glucose}_2 \longrightarrow$$

Nucleotide sugar   Acceptor
donor substrate   substrate

$$\text{glycosyl}_1\text{-}O\text{-glycose}_2 + \text{nucleotide}$$

Glycoside

The oligosaccharides in glycoproteins are most frequently attached to the constituent protein through the oxygen of a serine or a threonine side chain, or through the amide nitrogen of an asparagine side chain (see Figure 4–20). The former oligosaccharides are referred to as oxygen-linked or O-linked, and the latter as nitrogen-linked or N-linked.

Biosynthesis of N-Linked Oligosaccharides.   More is known about the synthesis of N-linked oligosaccharides, so we will discuss these first and in greater detail. The N-linked oligosaccharides have been classified into three major types: complex, high mannose, and hybrids of complex and high mannose types. One member of each type is illustrated in Figure 20–10. Clearly, the variety obtainable through the use of different residues and different branch points is virtually unlimited. Whereas specific glycosyltransferases are available for making different types of linkages, there is still very little understanding of what constellation of factors determines the final oligosaccharide structure associated with specific proteins. This is a most challenging and most important problem that must be answered by future research.

Dolichol phosphate, a membrane-associated lipid, is a key component

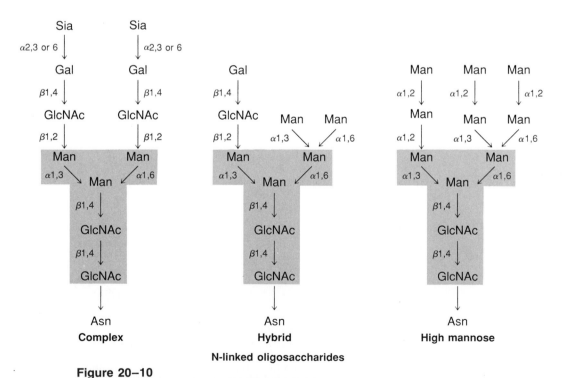

N-linked oligosaccharides

Figure 20–10

Three types of N-linked oligosaccharides. All N-linked oligosaccharides are linked to protein through an amide nitrogen of an asparagine side chain. N-linked oligosaccharides are classified into three types: complex, hybrid, and high mannose. Note that the different classes have a common base of five hexoses.

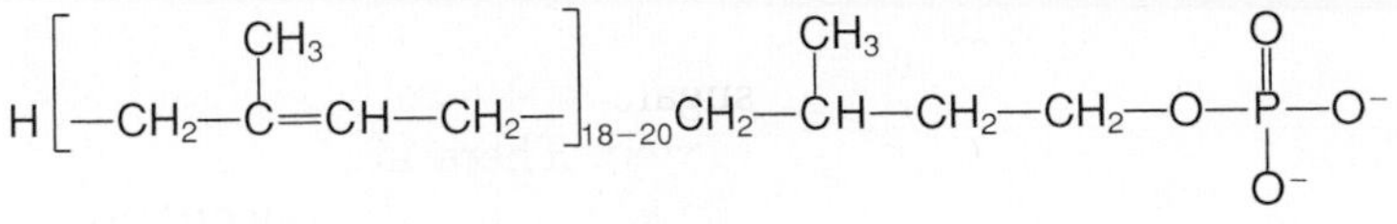

Figure 20–11
The structure of dolichol phosphate. The hydrocarbon portion of the molecule has a high affinity for membrane structures. The phosphate end forms an activated complex with oligosaccharide intermediates.

$$H\left[-CH_2-\underset{\underset{CH_3}{|}}{C}=CH-CH_2-\right]_{18-20}CH_2-\underset{\underset{CH_3}{|}}{CH}-CH_2-CH_2-O-\underset{\underset{O^-}{|}}{\overset{\overset{O}{\|}}{P}}-O^-$$

Dolichol

in the synthesis of all glycoproteins (Figure 20–11). This lipid serves as an anchor for the synthesis of the oligosaccharide and also provides the activation essential for transglycosylation and transfer to protein. In the early stages, nucleoside-diphosphate-activated sugar monomers are condensed in stepwise fashion from the phosphate end of the dolichol phosphate. At some intermediate point in the synthesis of the oligosaccharide, the oligosaccharide-lipid complex is translocated to the lumen of the endoplasmic reticulum. Further sugar additions to the initial sugar cluster are made in stepwise fashion by transglycosylations from hexose-lipid-activated complexes formed on the cytoplasmic side of the endoplasmic membrane. The new sugar units are then translocated, like the initial cluster, to the lumen. Ultimately the oligosaccharide-lipid complex transfers the oligosaccharide to the amide side chain of an asparagine residue. This complex process of oligosaccharide translocation serves the function of preventing the glycosylation of polypeptide chains that do not carry the signal sequence for penetration of the endoplasmic reticulum.

Let us now consider a typical pattern for the synthesis of an N-linked glycoprotein. The structure of the oligosaccharide-lipid ultimately transferred to a protein asparagine side chain is shown in Figure 20–12. The first seven residues are condensed onto a dolichol phosphate through a series of seven reactions, outlined in Figure 20–13. In these first seven reactions of

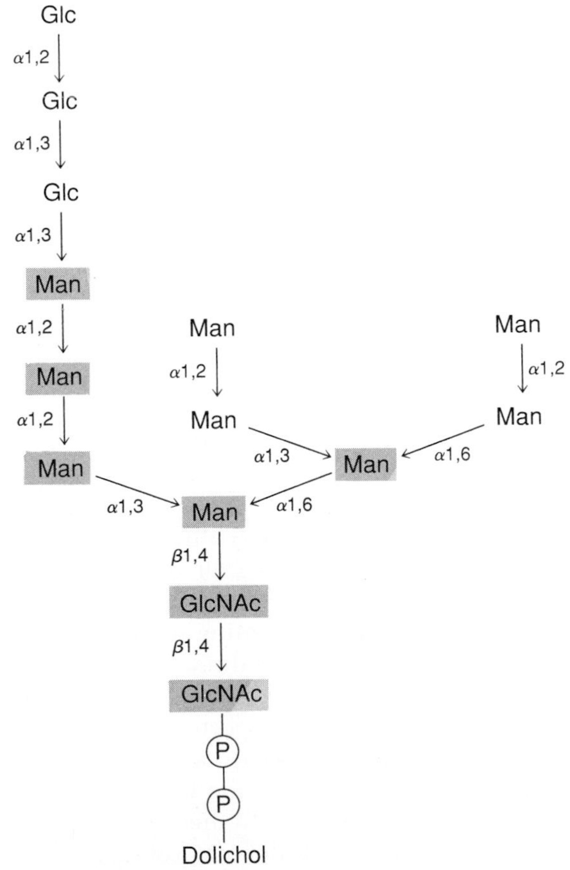

Figure 20–12
A complex N-linked oligosaccharide. Seven residues, highlighted in color, are the first ones to be attached to the dolichol phosphate (—P—P—Dol). These first seven reactions occur on the cytoplasmic side of the endoplasmic reticulum. The remaining sugars are attached in the lumen, each sugar being transferred from a hexose—P—P—Dol complex.

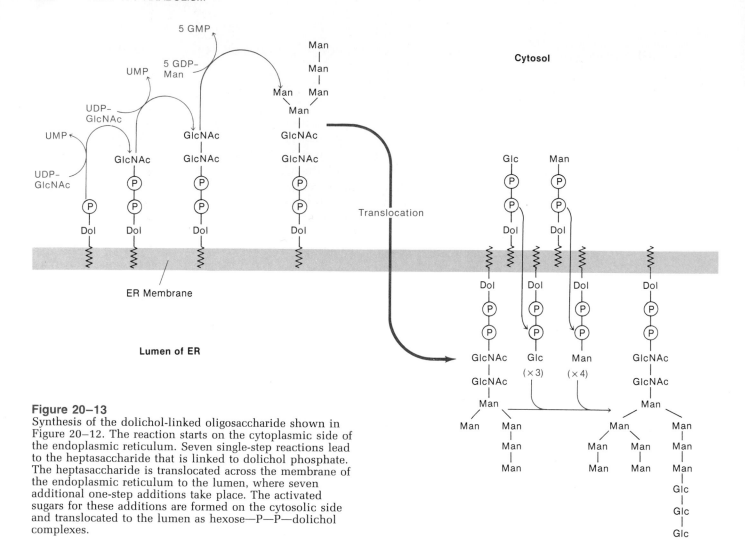

Figure 20–13
Synthesis of the dolichol-linked oligosaccharide shown in
Figure 20–12. The reaction starts on the cytoplasmic side of
the endoplasmic reticulum. Seven single-step reactions lead
to the heptasaccharide that is linked to dolichol phosphate.
The heptasaccharide is translocated across the membrane of
the endoplasmic reticulum to the lumen, where seven
additional one-step additions take place. The activated
sugars for these additions are formed on the cytosolic side
and translocated to the lumen as hexose—P—P—dolichol
complexes.

the synthesis, the individual sugar residues are all activated as nucleoside
diphospho sugars as in standard polymer formation. The resulting heptasaccharide complex is translocated to the lumen.

In the lumen, further sugar additions are made from monomer sugar
complexed in pyrophosphate fashion to a dolichol phosphate. These complexes were initially formed on the cytoplasmic side of the endoplasmic
reticulum from dolichol phosphate and the appropriate UDP-hexose. Subsequently they were translocated to the lumen like the original complex. Once
the complex shown in Figure 20–12 is fully formed, it is ready for addition
to a growing polypeptide chain. This and subsequent steps in the glycoprotein synthesis are shown in Figure 20–14.

After transfer, the polypeptide synthesis continues to completion. Simultaneously, certain hexoses are removed by specific glycosidases (called
trimming enzymes) that are located in the membrane on the endoplasmic
reticulum. Except for glycoproteins that become integrated into the endoplasmic reticulum membrane, the newly synthesized glycoproteins are
transported to the Golgi by means of vesicles that bud from the endoplasmic
reticulum membrane. The newly synthesized glycoproteins are then fused
with the Golgi membrane (Figure 20–14).

In the Golgi, considerable additional processing occurs, which varies for
different proteins and different cell types. It can involve phosphorylations,

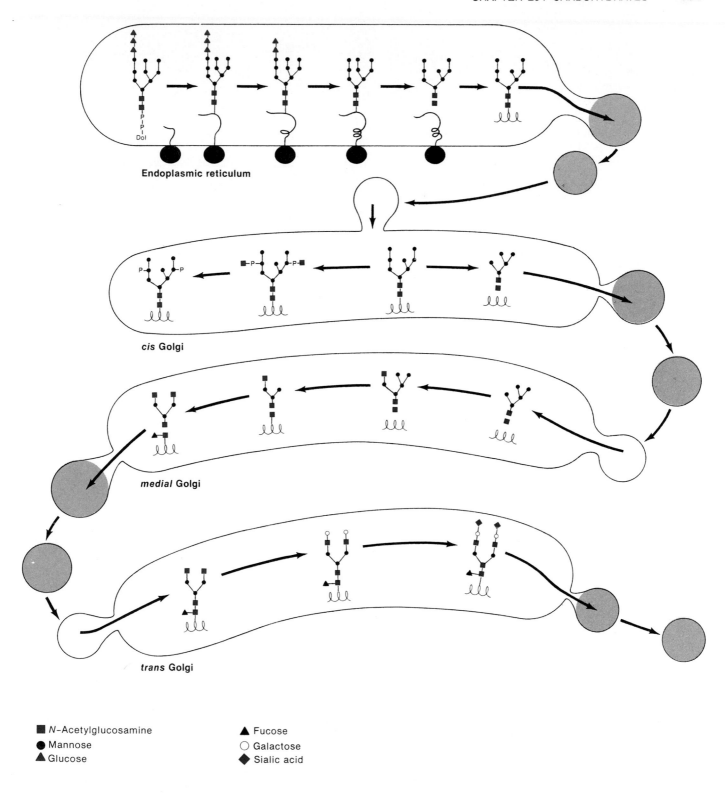

Endoplasmic reticulum

cis Golgi

medial Golgi

trans Golgi

■ *N*–Acetylglucosamine   ▲ Fucose
● Mannose   ○ Galactose
▲ Glucose   ◆ Sialic acid

Figure 20–14

Schematic diagram illustrating the formation, processing, and transfer of a glycoprotein. In the endoplasmic reticulum the newly synthesized oligosaccharide attaches to the amino acid side chain of a growing polypeptide chain. As the synthesis of the polypeptide chain continues, the oligosaccharide undergoes processing reactions. These include the removal and addition of hexoses. After polypeptide synthesis is complete, the resulting glycoprotein is transferred by budding to the Golgi apparatus. Processing of the oligosaccharide continues in the Golgi. The budding mechanism is used to transfer the glycoprotein successively through the *cis*, medial, and *trans* Golgi and then to its final destination. The final structure of the protein-attached oligosaccharides is believed to be an important recognition factor in determining the final destination of the different proteins.

Figure 20-15
Schematic diagram of the sorting and distribution of newly synthesized glycoproteins. By a mechanism that is unclear, proteins with a similar destination and containing similar oligosaccharides cluster together in the *trans* Golgi. Buds form vesicles carrying the glycoproteins to their final destination. Depending on their particular properties, the glycoproteins may become integral parts of a membrane or free proteins within a membranous organelle such as the lysozyme, or they may be secreted outside the cell when the vesicle membrane fuses with the plasma membrane.

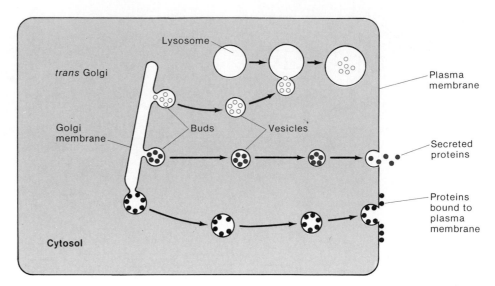

additional glycosylations, or deglycosylations. The final steps of complex oligosaccharide synthesis that occur in the Golgi (Figure 20-14) consist of addition of outer-chain galactose and sialic acid residues. The newly synthesized glycoproteins then exit the Golgi and are transported to their final destination.

The specific localization of the processing enzymes in the rough endoplasmic reticulum or the Golgi is a major factor in determining the final glycoprotein structure. It is believed that functionally related proteins exit from the Golgi together by budding (Figure 20-15). The resulting protein-packed vesicles make their way to the appropriate final destination (Figure 20-15). Oligosaccharides are almost certainly an important part of the recognition signal that guides the vesicles so that they become bound to specific structures or liberated as secretory protein.

As suggested in Figure 20-15, certain glycoproteins bud off and become part of the lysosome organelles. These glycoproteins are destined to become _lysosomal enzymes_. Lysosomes contain a variety of hydrolyzing enzymes, glycosidases, lipases, and proteases. The lysosomal enzymes serve a scavenging function; that is, they break down unneeded polymers of different types. All lysosomal enzymes contain N-linked oligosaccharides rich in mannose and mannose-6-phosphate. If glycosylation is blocked, the uptake of the specific lysosomal enzymes by the lysosomes does not occur. This finding suggests that the carbohydrate moiety of the lysosomal enzyme is an important part of the recognition signal assuring that the enzyme will become part of the appropriate organelle.

Biosynthesis of O-Linked Oligosaccharides.   Humans and most other vertebrates carry complex oligosaccharides attached to serine side chains of certain membrane proteins. The synthesis of the O-linked oligosaccharides is believed to follow the same general scheme as that observed for the N-linked oligosaccharides, except that there is no intermediate lipid involved. O-linked polysaccharide complexes frequently are exposed on the outer surface of human cells and excite a specific immune response when cells carrying them are injected into individuals that do not contain the same cell surface antigens (see Box 4-B). In the case of humans, this response has led to the classification of individuals according to different blood group types. Individuals of similar blood group types can accept transfusions from one another, whereas individuals of different types frequently cannot, because of immunological re ection.

TABLE 20-2
The Human ABO Blood-Group Scheme

Gene Type	Antigen
AA	A
AB	A,B
BB	B
AO	A
BO	B
OO	—

One of the better-known blood-grouping schemes is the ABO scheme. Blood is considered to be of type A, B, or O. All individuals contain, in most of their cells, two genes for blood type. The possible combinations give rise to six different gene types and four different combinations of blood group antigens in different individuals (Table 20-2). Cells of these individuals carry A antigens (in gene types AA or AO), B antigens (in gene types BB or BO), a mixture of A and B antigens (in gene type AB), or neither of these antigens (in gene type OO). The explanation for these correlations between gene types and blood group antigens is that the A gene and the B gene encode different glycosyltransferases. The A gene encodes a glycosyltransferase that catalyzes the addition of a terminal N-acetylgalactosamine (GalNAc) residue onto a core polysaccharide, and the B gene encodes a similar enzyme that adds a galactose (Gal) residue to the same site (Figure 20-16). When A and B genes are present, both structures are found, but when only O genes are present, the site on the oligosaccharide is left naked. The presence or absence of these glycosyltransferases is readily detectable in the milk of the lactating female.

BACTERIAL CELL WALL SYNTHESIS

Earlier in the text we described the structure of the peptidoglycan that comprises the bacterial cell wall (see Chapter 4 and Figure 4-26). Briefly, a segment of the cell wall has the structure of a two-dimensional network containing a parallel array of linear heteropolysaccharide strands extended in one direction, which are cross-linked with strands of an oligopeptide in the perpendicular direction. Biosynthesis of the cell wall is unusual in two respects: (1) it is the first example we have seen (and the only one we will consider) that involves the synthesis of a regularly cross-linked polymer; (2) part of the synthesis takes place inside the plasma membrane and part takes place outside the plasma membrane.

For descriptive purposes, the synthesis of peptidoglycan can be conveniently broken into three stages, which occur at different locations in the cell: (1) synthesis of UDP-N-acetylmuramylpentapeptide, (2) polymerization of N-acetylglucosamine and N-acetylmuramyl-pentapeptide to form the linear peptidoglycan strands, and (3) cross-linking of the peptidoglycan strands.

Figure 20-16
The structure and reactions at the oligosaccharide termini of the ABO human blood group antigens. Two enzymes add different hexoses to the termini of plasma membrane O-linked glycoproteins. Individuals may carry both or neither or only one of these enzymes. Individual differences are reflected in the structures of their blood group antigens. These differences are genetically inherited. Fuc stands for fucose, a 6-deoxy sugar.

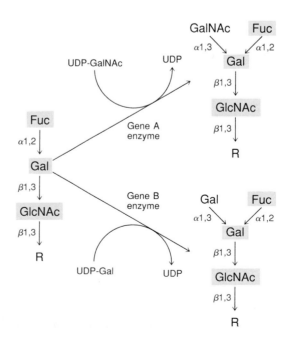

Synthesis of the UDP-*N*-acetylmuramyl-pentapeptide Monomer, Occurs in the Cytoplasm

The first stage in cell wall synthesis (Figure 20–17) involves the synthesis of UDP-*N*-acetylmuramyl-pentapeptide. First, the condensation of *N*-acetylglucosamine-1-phosphate with UTP leads to the formation of UDP-*N*-acetylglucosamine. A specific transferase catalyzes a reaction with phosphoenolpyruvate to give the 3-enolpyruvylether of UDP-*N*-acetylglucosamine. The pyruvyl group is then reduced to lactyl by an NADPH-linked reductase, thus forming the 3-*O*-D-lactylether of *N*-acetylglucosamine. This compound is known as UDP-*N*-acetylmuramic acid (see Figure 4–11 for its detailed structure).

Conversion of UDP-*N*-acetylmuramic acid to its pentapeptide form occurs by the sequential addition of the necessary amino acids. Each step requires ATP and a specific enzyme that ensures the addition of amino acids in the proper sequence. L-alanine is added first, followed by D-glutamic acid, L-lysine (attached by its α amino group to the α carboxyl group of the glutamic acid), and finally the dipeptide D-alanyl-D-alanine as a unit. The latter dipeptide is formed by two enzymatic reactions: conversion of L-alanine to D-alanine by a racemase, followed by the linking of the two alanine residues in an ATP-requiring reaction to form D-alanyl-D-alanine. All of these reactions occur in the cytoplasm of the bacterial cell.

Figure 20–17
The first stage of cell wall synthesis: formation of UDP-*N*-acetylmuramyl-pentapeptide (full structure shown at bottom). Points of inhibition by the antibiotic penicillin and phosphonomycin are indicated.

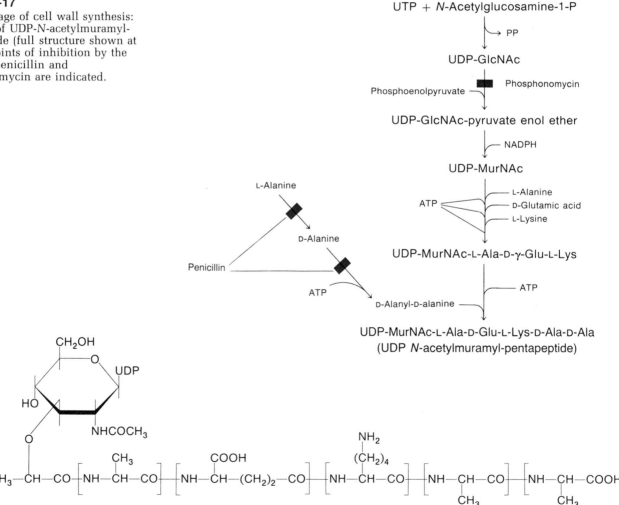

UDP *N*-Acetylmuramyl-pentapeptide

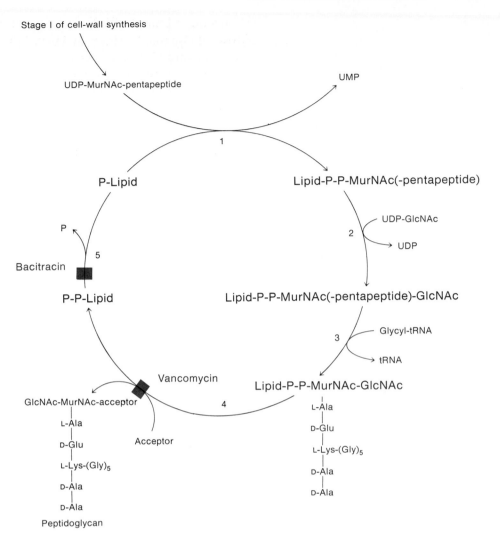

Figure 20–18
The second stage of cell wall synthesis. An ATP-requiring amidation of glutamic acid that occurs between steps 2 and 3 has been omitted. Points of action of the antibiotic inhibitors bacitracin and vancomycin are indicated.

Formation of Linear Polymers of the Peptidoglycan is Membrane-associated

The most complex stage in peptidoglycan synthesis takes place on the plasma membrane. It may be divided into five steps, which are illustrated in Figure 20–18. This stage involves the polymerization of N-acetylglucosamine and N-acetylmuramyl-pentapeptide-containing residues into linear peptidoglycan strands.

In step 1 (Figure 20–18), UDP-acetylmuramyl-pentapeptide reacts with a 55-carbon (C_{55}) isoprenyl alcohol known as undecaprenol phosphate. The structure of this membrane-associated lipid is shown in Figure 20–19. The lipid is similar in structure and function to the dolichol phosphates involved in glycoprotein synthesis in eukaryotes. A pyrophosphate linkage is formed with the lipid, and UMP is released.

Undecaprenol phosphate

Figure 20–19
The structure of undecaprenol phosphate. Isoprene phosphates such as undecaprenol phosphate are important carriers and activators in the synthesis of oligosaccharides. The synthesis of undecaprenol phosphate is described in Chapter 22.

In step 2 (Figure 20–18), N-acetylglucosamine is added to the lipid intermediate by means of a typical transglycosylation from UDP-N-acetylglucosamine, and UDP is released. In step 3, five glycine residues are sequentially added to the ϵ amino group of lysine. Curiously, these glycines are activated by ester formation to a transfer RNA molecule. This form of amino acid activation is otherwise seen only in protein synthesis (see Chapter 28).

In step 4, the disaccharide-oligopeptide unit is transferred from the lipid intermediate to the growing peptidoglycan, and lipid pyrophosphate is generated. It is in this step that we see the dual function of the undecaprenol lipid. The lipid not only serves to activate the monomer for addition to polymer, but it transports the monomer from the cytoplasmic side of the membrane to the extracellular side of the membrane, where cell wall assembly must take place. Little is known about the details of this transport process.

In the fifth and final step, one phosphate is hydrolyzed to regenerate the phospholipid, which then can react once again with UDP-acetylmuramyl-pentapeptide and participate in another cycle, resulting in the addition of a new unit to the growing peptidoglycan strand. The antibiotic bacitracin is a specific inhibitor of the dephosphorylation of the pyrophosphate form of the lipid.

Cross-linking of the Peptidoglycan Strands Occurs on the Noncytoplasmic Side of the Plasma Membrane

The cross-linking of peptidoglycan strands takes place outside the cell membrane, at the site of the preexisting wall. Since there is no ATP or other obvious energy source available there, a mechanism independent of any external energy source has evolved for this reaction. The reaction is a transpeptidation in which the terminal amino end of an open cross-bridge attacks the terminal peptide bond in an adjacent strand to form a cross-link. The terminal alanine residue from the strand that becomes cross-linked is thus eliminated (Figure 20–20; also see Figure 4–26c).

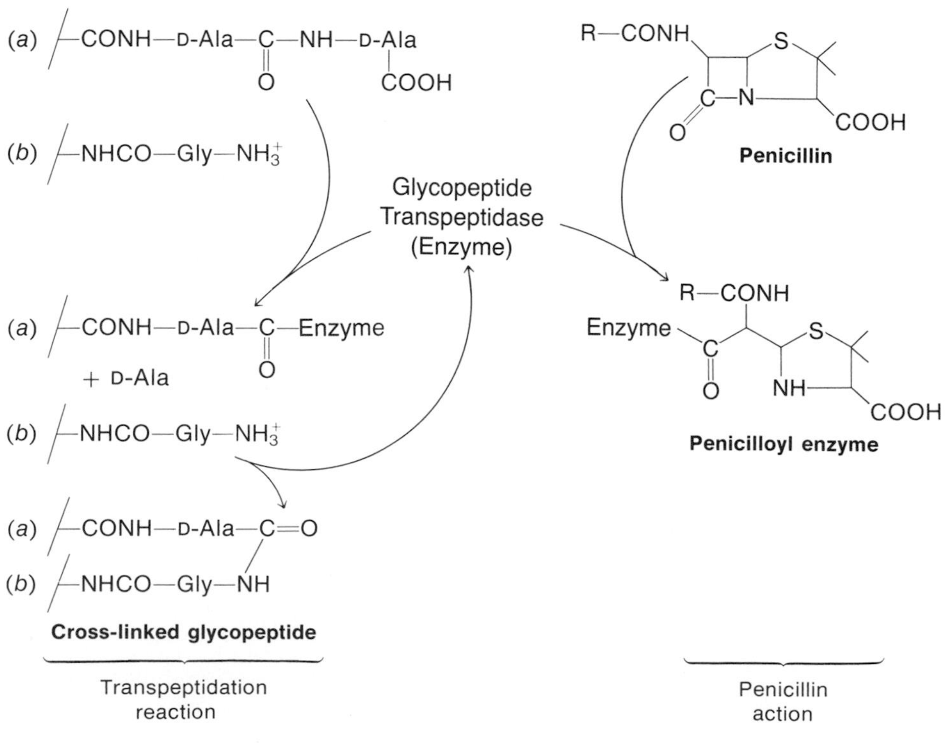

Figure 20–20
Mechanism of the third stage of cell wall synthesis. This diagram shows the cross-linking reaction and the mechanism of inhibition by penicillin in the bacterium *Staphylococcus aureus.* (a) The end of the peptide side chain of a glycan strand; (b) the end of the pentaglycine substituent from an adjacent strand. You may wish to refer to Figure 4–26 for details of the structure of the peptidoglycan.

Penicillin Inhibits the Transpeptidation Reaction.   Penicillin has been unequaled for usefulness in combatting bacterial diseases and infections. During the 50 years since Fleming brought penicillin to the attention of microbiologists, many biochemists and pharmacologists have been interested in the mechanism by which this potent antibiotic kills bacteria. It is now known that penicillin inhibits the cross-linking reaction by acting as a structural analog of the terminal D-alanyl-D-alanine residue of the peptidoglycan strand. The similarity between the conformation of the penicillin molecule and one of the conformations of the dipeptide D-alanyl-D-alanine is shown in Figure 4–27.

It is thought that the transpeptidase (TPase) first reacts with the substrate to form an acyl enzyme intermediate, with the elimination of D-alanine, and that this active intermediate then reacts with another strand to form the cross-link and regenerate the enzyme. Because penicillin is an analog of alanylalanine, it should fit the substrate binding site, with the highly reactive —CO—N— bond in the β-lactam ring in the same position as the bond involved in the transpeptidation. It thus has the potential to acylate the enzyme, forming a penicilloyl enzyme, and thereby inactivate it (Figure 20–20). In support of this view is the fact that penicilloyl is the piece of the antibiotic found in inhibited enzymes. An acyl group derived from the substrate is used to form an acyl enzyme intermediate. Most important for verifying the proposed mechanism of penicillin action, the antibiotic-derived penicilloyl moiety and the substrate-derived acyl moiety are substituted on the same site in the penicillin-sensitive enzymes from a variety of genera of bacteria.

A number of other antibiotics in addition to penicillin (phosphonomycin, bacitracin, and vancomycin) block other stages in cell wall synthesis. Their points of action are indicated in Figures 20–17 and 20–18. In addition to their biological and medical importance, these antibiotics have been extremely useful in elucidating the biosynthetic pathway. This is because they permit accumulation of the product before the blocked step; the product can then frequently be isolated and confirmed as a genuine intermediate in the pathway.

SUMMARY

Hexoses, which are the primary building blocks of oligosaccharides and polysaccharides, come in a large variety of types. All hexoses can be thought of as derivatives of glucose through a series of conversions. These conversions usually occur at the level of the monophosphorylated sugar or the nucleoside diphosphate sugar. The nucleoside sugar is also the activated substrate for formation of disaccharides, oligosaccharides, and polysaccharides.

In higher animals a great variety of branched-chain oligosaccharides are found as conjugates in glycolipids and glycoproteins. A large number of sugars and specific glycosyltransferases are involved in oligosaccharide synthesis. Two types of oligosaccharides are distinguished, according to their mode of attachment to the protein in the glycoprotein. The O-linked oligosaccharides are synthesized directly on the amino acid side chain hydroxyl group of a serine or a threonine. The N-linked oligosaccharides are synthesized first on a long-chain dolichol phosphate and then transferred to the asparagine side chain of a receptor protein. The protein-attached oligosaccharide is processed by removal of certain sugars and addition of others.

The synthesis of glycoproteins takes place in two cytoplasmic organelles, the endoplasmic reticulum and the Golgi apparatus. The mature glycoproteins leave the Golgi apparatus in the form of microvesicles formed by budding. The oligosaccharide portion of the glycoprotein is believed to be instrumental in guiding the glycoprotein to its final destination. The oligosaccharide can serve other important recognition functions in addition to assuring the location of the glycoprotein.

The bacterial cell wall contains a heteropolymeric polysaccharide chain that is cross-linked by peptide linkages. The complexity of the resulting peptidoglycan results in part from the complex repeating units and in part from the fact that a cross-linked polymer is being made outside the cell. The partially completed polysaccharide structures are transferred from the cytoplasm to extracellular space by attachment to a long-chain bactoprenol lipid that can traverse the cell membrane. This lipid is similar in structure and function to the dolichol phosphate used in oligosaccharide synthesis in animals. Various antibiotics that block specific steps in cell wall synthesis are of great medical importance. They have also been very helpful in elucidating the biochemical pathway.

SELECTED READINGS

Ashwell, G., and Joe Harford, Carbohydrate-specific receptors of the liver. *Ann. Rev. Biochem.* 51:531–54, 1982.

Edelman, G. M., Cell adhesion and the molecular processes of morphogenesis. *Ann. Rev. Biochem.* 54:135–136, 1985.

Hanson, R. W., and M. A. Mehlman (eds.), *Gluconeogenesis, Its Regulation in Mammalian Species.* New York: Wiley, 1976.

Hubbard, S. C., and R. J. Ivatt, Synthesis and processing of asparagine-linked oligosaccharides. *Ann. Rev. Biochem.* 50:555, 1981. A recent review of glycoprotein synthesis in eukaryotes.

Kochetkov, N. K., and V. N. Shibaev, Glycosyl esters of nucleoside pyrophosphates. *Adv. Carbohydr. Chem. Biochem.* 28:307, 1973. A concise review of the chemistry

and biochemistry of nucleoside pyrophosphate sugars and derivatives.

Kornfield, R., and S. Kornfield, Assembly of asparagine-linked oligosaccharides. *Ann. Rev. Biochem.* 54:631–664, 1985.

Tipper, D. J., and A. Wright, The Structure and Biosynthesis of Bacterial Cell Walls. In J. R. Sokatch and L. N. Ornston (eds.), *The Bacteria,* Vol. VII, *Mechanisms of Adaptation.* New York: Academic Press, 1979. Pp. 291–426.

Van Schaftingen, C., and H.-G. Hers, Synthesis of a stimulation of phosphofructokinase, most likely fructose 2,6-bisphosphate from phosphoric acid and fructose 6-phosphoric acid. *Biochem. Biophys. Res. Commun.* 96:1524, 1980.

PROBLEMS

1. What are the three basic stages of oligosaccharide biosynthesis?

2. For each of the antibiotics listed below, what intermediate of cell wall synthesis would accumulate as a result of its action?
   a. Phosphonomycin   b. Bacitracin   c. Vancomycin

3. Penicillin kills only growing cells. Explain.

4. Penicillin has been helpful in elucidating cell wall synthesis but not in obtaining mutants in enzymes involved in cell wall synthesis. Explain.

7. The peptide hormones made in the pituitary (see Chapter 31), which include LH, FSH, and TSH, are composed of two peptides, α and β, held together noncovalently. The α chain in all these peptides comes from the same gene, but the β chain is different in each case. Even though the sequences of the α chains are identical they are glycosylated differently. Why do you think this is so?

8. Sialic acid residues can be labelled with ^{3}H (tritium) by oxidation followed by reductive tritiation using a labeled borohydride:

5. Cell wall and O-antigen synthesis are both unusual in that part of the synthesis is carried on outside the cell. What complications does this introduce in making polymeric linkages in the two cases, and how are these complications overcome?

6. a. The sequence of a protein is determined by the sequence of the mRNA from which it is translated (see Chapter 28). What determines the sequence of oligosaccharides that are attached to proteins?
   b. Which are capable of more sequence diversity, proteins or oligosaccharides? Why?
   c. Virtually all hydroxyl groups of the hexoses can and do serve as reactive sites *in vivo.* Given that this is so, illustrate (using an example from the text) how specific linkages are actually made with such high fidelity.

All of the serine proteases involved in blood clotting can be labeled in this manner.
   a. What type of carbohydrate do they probably contain? (Refer to Figure 20–10).
   b. Describe in brief the likely sequence of events in glycosylation of these proteins, with particular emphasis on subcellular localization.
   c. Prothrombin, one of the clotting enzymes, contains four carbohydrate moieties per protein molecule. When prothrombin is proteolytically cleaved to thrombin, the enzyme that clots fibrin, about three-fourths of the carbohydrate becomes soluble in trichloracetic acid (TCA) and the remainder is attached to thrombin. (TCA precipitates large proteins, but smaller peptides remain soluble.) How might this fact be used to study the kinetics of thrombin production from prothrombin?

21

FATTY ACID BIOSYNTHESIS

In the next three chapters we will describe the biosynthesis of the broad range of lipid compounds whose structures and functions were described in Chapters 5 and 6. We will begin with the fatty acids.

Fatty acids are synthesized from acetyl-CoA generated by the catabolism of other molecules, mainly carbohydrates or amino acids. Cells use the fatty acids they make for fabricating membrane lipids (see Chapters 5 and 6) or for long-term energy storage in the form of triacylglycerols. In animals, liver is the primary tissue in which fatty acids are synthesized, and their usual fate is to be used in biosynthesis of triacylglycerols. Some triacylglycerols are stored in the liver, but the vast majority of them are secreted into the bloodstream as components of lipoproteins (see Chapter 23). Catabolism of triacylglycerols results in the release of fatty acids that are absorbed by various tissues. There they can be degraded to satisfy immediate energy needs or reesterified into phospholipids for membrane assembly. When the concentration of circulating fatty acids exceeds that required to satisfy immediate needs, they are delivered to adipose tissues, where they become stored as triacylglycerols.

Both the synthesis and the catabolism of fatty acids involve units of two-carbon atoms, but there the similarity between the two processes ends. Completely different enzymes and intermediates are used in the catabolic and the anabolic reactions, and the two processes are located in different cellular compartments. Regulatory factors insure that the two processes do not occur simultaneously; thus the operation of a pseudocycle is prevented.

In this chapter we will describe the synthesis of saturated and unsaturated fatty acids and the regulation of fatty acid synthesis in the liver.

BIOSYNTHESIS OF SATURATED FATTY ACIDS

The finding that most fatty acids contain an even number of carbon atoms, between four and twenty, led Rapier in 1907 to postulate that they were produced by condensation of an activated two-carbon compound. After the introduction of isotopes in the late 1930s and early 1940s as a fundamental tool for the biochemist, experiments were performed that clearly implicated an acetate derivative as the two-carbon compound. In a series of experiments done in 1944 and 1945, David Rittenberg and Konrad Bloch fed mice acetate that had been labeled with both deuterium and ^{13}C ($C^2H_3{}^{13}COOH$) and found that both isotopes were incorporated into fatty acids. Subsequently, Fyodor Lynen discovered that "active acetate" was *acetyl-CoA*, and a central role for this compound in fatty acid biosynthesis was demonstrated. Precisely how acetyl-CoA was converted into fatty acids eluded workers until the late 1950s, when the involvement of *malonyl-CoA* was discovered. From this point on, progress was rapid and the scheme for fatty acid biosynthesis as we know it today was elucidated.

The synthesis of saturated fatty acids is very similar in all organisms. In eukaryotes, synthesis occurs in the cytosol. The overall reaction for the formation of the commonly synthesized 16-carbon *palmitic acid* is

$$CH_3\overset{O}{\overset{\|}{C}}-CoA + 7\ {}^-O-\overset{O}{\overset{\|}{C}}-CH_2-\overset{O}{\overset{\|}{C}}-CoA + 14\ NADPH + 14\ H^+ \longrightarrow$$

Acetyl-CoA   **Malonyl-CoA**

$$CH_3(CH_2)_{14}COO^- + 7\ CO_2 + 8\ CoA + 14\ NADP^+ + 7\ H_2O$$

Palmitic acid

If this equation is compared with the equation for the complete oxidation of palmitoyl-CoA (see Table 18–1, equation 1), we find two major differences in carriers and intermediates. The principal electron carrier system in the anabolic pathway is the $NADPH$-$NADP^+$ system; the principal electron carriers in the catabolic pathway are FAD-$FADH_2$ and NAD^+-$NADH$. The second striking difference between the two pathways is that malonyl-CoA is the principal substrate in the anabolic pathway, but plays no role in the catabolic pathway. These differences reflect the fact that the two pathways do not use any enzymes in common. Indeed, in animal cells the reactions even occur in separate cell compartments; biosynthesis takes place in the cytosol, whereas degradation occurs in the mitochondria.

Acetyl-CoA and NADPH Are the Main Substrates for Fatty Acid Synthesis

The first prerequisite for any pathway to function is that the essential substrates be present. The overall equation for any fatty acid synthesis, such as the one just presented for palmitic acid, indicates a need for acetyl-CoA, malonyl-CoA, and NADPH. As we shall see, malonyl-CoA is derived from acetyl-CoA, so we can think of acetyl-CoA as the main substrate for fatty acid synthesis. The supply of acetyl-CoA and NADPH necessary for fatty acid synthesis is generated from substrates that originate in the glycolytic pathway and the TCA cycle. Thus glycolysis generates pyruvate, and in the mitochondria pyruvate dehydrogenase converts the pyruvate to acetyl-CoA. In animals, excess acetyl-CoA cannot be directly utilized for fatty acid synthesis. This is because acetyl-CoA cannot cross the mitochondrial membrane to the site of fatty acid synthesis in the cytosol. The mitochondrial membrane is generally quite selective with respect to the molecules that can pass through it and frequently requires special carriers for that purpose.

The problem of acetyl-CoA transport is overcome with the help of *citrate synthase* (an enzyme of the TCA cycle), which converts acetyl-CoA and oxaloacetate to citrate. The citrate can cross the mitochondrial membrane into the cytosol, where *ATP-citrate lyase* converts it back into acetyl-CoA and oxaloacetate. In this indirect way the acetyl-CoA generated in the mito-

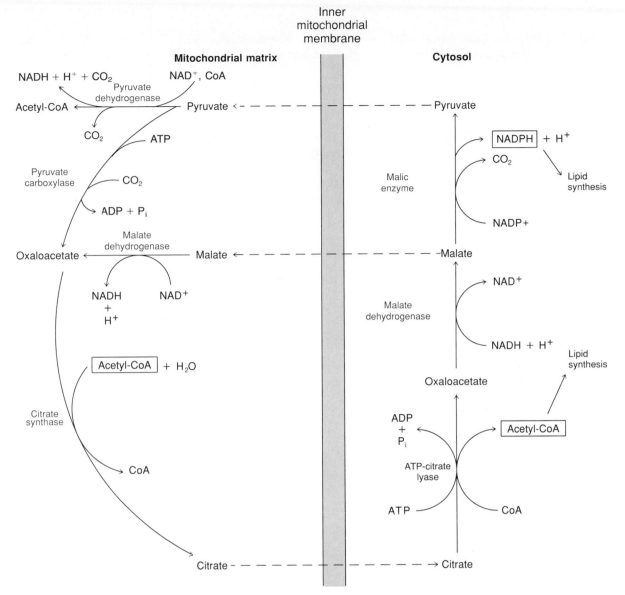

Figure 21–1

Sources of acetyl-CoA and NADPH for fatty acid synthesis. Because of the permeability barrier of the inner mitochondrial membrane, the excess acetyl-CoA produced in the mitochondrion cannot be used directly for fatty acid synthesis. However, the citrate generated from the condensation of acetyl-CoA with oxaloacetate can. The cytosolic citrate is converted back to oxaloacetate to regenerate cytosolic acetyl-CoA. Further reactions involving oxaloacetate lead to the production of NADPH.

chondria can be made available for fatty acid synthesis in the cytosol (Figure 21–1).

The oxaloacetate generated in the cytosol must be returned to the mitochondria to preserve the oxaloacetate needed for the TCA cycle. It cannot be directly returned to the mitochondria because oxaloacetate, like acetyl-CoA, cannot pass across the mitochondrial membrane. However, oxaloacetate can be returned to the mitochondria after conversion to malate (Figure 21–1). Alternatively, the malate can be oxidatively decarboxylated to pyruvate by *malic enzyme* in the cytosol (see Figures 21–1 and 15–14). The pyruvate can be transported back to the mitochondria and converted to acetyl-CoA by pyruvate dehydrogenase or to oxaloacetate by pyruvate carboxylase. The

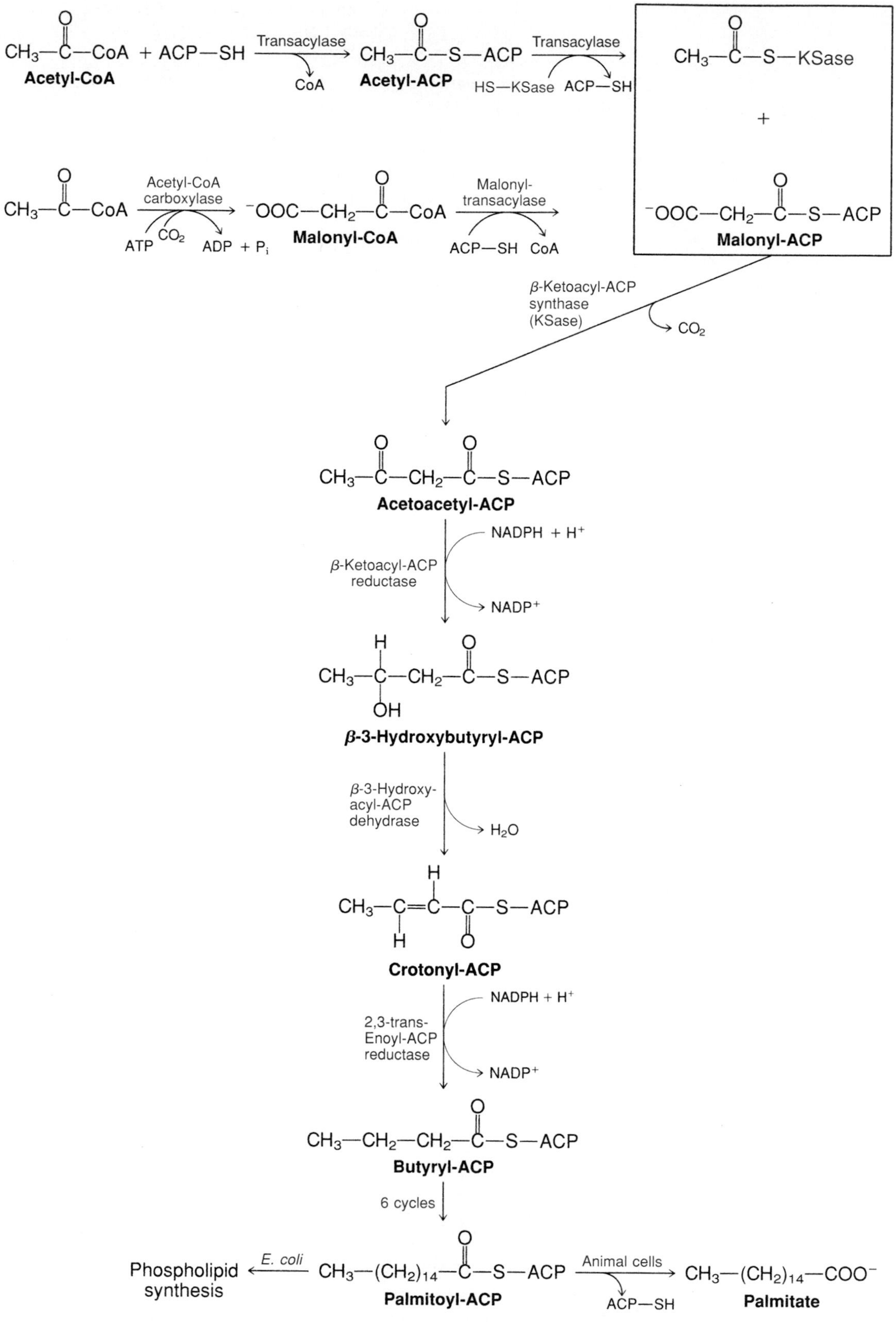

Figure 21–2

Outline of the reactions for fatty acid synthesis. Fatty acids grow in steps of two-carbon units. To make a 16-carbon fatty acid requires many repeats of the initial cycle of reactions. In animal cells the palmitate is released as the free acid. In *E. coli* the ACP derivative is used directly for the biosynthesis of phospholipids (discussed in Chapter 22). Most of the reactions of fatty acid synthesis are believed to take place on a multiprotein complex.

malic enzyme reaction in the cytosol is important in another respect: it generates NADPH needed for fatty acid synthesis. Two other reactions, one catalyzed by glucose-6-phosphate dehydrogenase and one catalyzed by 6-phosphogluconate dehydrogenase, also are important contributors of NADPH for fatty acid synthesis (see Chapter 14).

Fatty Acid Synthesis in *E. coli* and Plants Is Catalyzed by Separate Enzymes

There are eight enzyme-catalyzed reactions involved in the conversion of acetyl-CoA into fatty acid (Figure 21–2). In *E. coli* and plants, each of the enzyme activities exists on a separate protein, which can be isolated from the others by conventional methods of purification. Table 21–1 summarizes some of the main properties of the *E. coli* enzymes. Fatty acid synthesis in *E. coli* also requires an additional protein, the *acyl carrier protein (ACP)*. The main function of the ACP is served by a covalently linked *4'-phosphopantetheine coenzyme*. The coenzyme is bound to the ACP by an ester linkage between the phosphate of the coenzyme and a serine hydroxyl side chain of the protein (Figure 21–3). The sulfhydryl group of the phosphopantetheine is the attachment site for the intermediates during fatty acid synthesis. Apparently, the long flexible arm of phosphopantetheine permits the attached substrate to visit successively all of the active sites during the course

TABLE 21–1
The Enzymes of Fatty Acid Synthase from *E. coli*

Enzyme	M_r	Subunits	Specificity	Miscellaneous
Acetyl-CoA-ACP transacylase	—	—	Acetyl-CoA 100% Butyryl-CoA 10% Hexanoyl-CoA 4.5%	Acetyl-CoA + enz. $\rightleftharpoons$ acetyl-enz. + CoA Acetyl-enz. + ACP $\rightleftharpoons$ acetyl-ACP + enz.
Malonyl-CoA-ACP transacylase	36,700	Single	Acetyl-CoA is a competitive inhibitor; $K_i = 115\ \mu M$	Malonyl-CoA + enz. $\rightleftharpoons$ malonyl-enz. + CoA Malonyl-enz. + ACP $\rightleftharpoons$ malonyl-ACP + enz. Malonate is esterified to serine on the enzyme
β-Ketoacyl-ACP synthase I	80,000	$2 \times 40,000$, apparently identical	Active with C_2–C_{14} ACP, but inactive with C_{16} ACP; inactive with CoA derivatives	Acetyl-ACP + enz. $\rightleftharpoons$ acetyl-enz. + ACP Acetyl-enz. + malonyl-ACP $\rightleftharpoons$ acetoacetyl-ACP + CO_2 + enz.
β-Ketoacyl-ACP synthase II	88,000	$2 \times 44,000$	Active with $16:1^{\Delta 9}$-ACP as substrate and C_2– to C_4–ACP	
β-Ketoacyl-ACP reductase	—	—	Specific for NADPH; active with C_4–C_{16} β-ketoacyl-ACP	The product has the D configuration
β-Hydroxyacyl-ACP dehydrase	—	—	Specific for D isomer; inactive with L-β-hydroxyacyl-ACP; active with C_4–C_{16} derivatives; lowest activity with C_{10} substrate; inactive with CoA derivatives	The product is *trans*
Enoyl-ACP reductase	—	—	There may be two enzymes, one specific for NADH and one for NADPH	

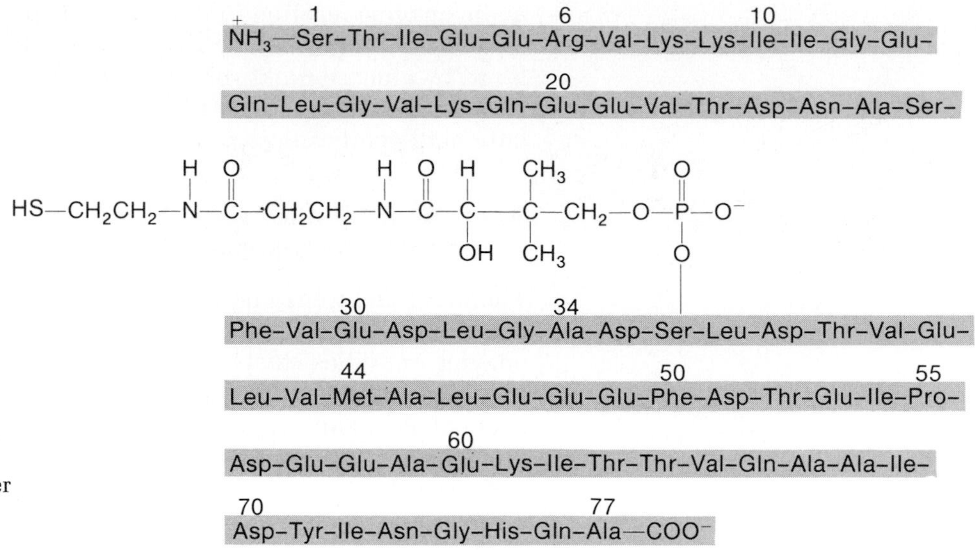

Figure 21–3
The amino acid sequence of acyl carrier protein (ACP) from *E. coli*, showing the phosphopantetheine coenzyme covalently linked to the side chain hydroxyl of serine 36. The sulfhydryl group of the phosphopanthetheine forms a thioester with the acetyl group of the growing fatty acid.

of the reactions. As postulated in Chapter 11 (Figure 11–15), these enzymes may exist inside the cell as a complex that disaggregates when the cells are homogenized. However, such an arrangement remains to be demonstrated.

The Enzymes for Fatty Acid Synthesis in Animals Are Organized in a Complex

Except for the first reaction, involving the synthesis of malonyl-CoA from acetyl-CoA, in animals all of the reactions of fatty acid synthesis (Figure 21–2) occur on a multienzyme complex known as *fatty acid synthase*. In addition to enzymes, the complex includes ACP. The use of a multienzyme complex, together with the coenzyme attachment of intermediates, greatly increases the efficiency of fatty acid synthesis. The next enzyme in the pathway is always near at hand, and the possibility of loss of intermediates is minimized. Thus it is not surprising that similar multienzyme complexes have been found in species ranging from more advanced bacteria to higher eukaryotes (see Box 21–A).

The subunit structure of fatty acid synthases differs according to the sources from which they are isolated. For example, the fatty acid synthase from yeast contains two different polypeptide chains, each with multiple enzyme activities. In liver all of the enzyme activities are located on a single polypeptide chain. Despite these differences in subunit structure, the synthases from different eukaryotic sources exist as multienzyme complexes that carry out identical reactions, a fact testifying to the efficiency of the organization of this pathway.

We now turn to a discussion of the individual reaction steps.

Acetyl-CoA Carboxylase Consists of Three Protein Components

The formation of malonyl-CoA from acetyl-CoA is catalyzed by *acetyl-CoA carboxylase* and is the only reaction in *animal* fatty acid synthesis that does not take place on the multienzyme synthase. The properties of this carboxylase are quite similar to those of pyruvate carboxylase, which is important in the gluconeogenesis pathway [see Chapter 14, Equation (13)]. We shall see that the activity of acetyl-CoA carboxylase plays an important role in the control of fatty acid biosynthesis in most organisms. This is not surprising, since the carboxylase represents the first step in fatty acid synthesis. The enzyme contains the biotin coenzyme, covalently linked by means of the ϵ group of a lysine in the protein (Figure 21–4).

BOX 21–A

Diversity in the Structure of Fatty Acid Synthases

The organization of the enzymes that catalyze fatty acid synthesis has evolved in diverse ways. The fatty acid synthase from *E. coli* (Table 21–1) exists as a group of enzymatic activities that can be separated from one another by conventional methods of purification. It is possible that within the bacterium they are organized as a multienzyme complex, but this has not been proven.

Most bacteria have a fatty acid synthase that resembles the *E. coli* enzyme. However, phylogenetically more advanced bacteria such as mycobacteria have high-molecular-weight fatty acid synthases with multifunctional polypeptides. Studies on the enzyme from *Mycobacterium smegmatis* have revealed an enzyme with many unusual features. The enzyme has a molecular weight of 2.0×10^6 and is composed of six to eight copies of apparently identical subunits, each with a molecular weight of 290,000. In contrast to the enzyme from *E. coli*, liver, and yeast, the major products of the mycobacterial synthase are 16-, 18-, 22-, and 24-carbon CoAs. Another unusual feature is that the synthase is markedly stimulated by the addition of *methylated polysaccharides*, either a polymer with units of 3-O-methylmannose ($M_r = 2100$) or a polymer that contains 6-O-methylglucose as the major unique sugar ($M_r = 4100$). These polysaccharides, which can be isolated from the mycobacteria and other closely related bacteria, have hydrophobic sites that bind equimolar amounts of long-chain acyl-CoA. Without the polysaccharides, the highly hydrophobic acyl-CoA products (22- or 24-carbon CoA) do not readily diffuse from the enzyme to feedback inhibit fatty acid synthesis. Thus the rate-limiting step for the synthesis of fatty acids in this organism is the diffusion of the very-long-chain fatty acyl-CoAs from the active site of the enzyme. The hydrophobic polysaccharides apparently "catalyze" this diffusion and relieve feedback inhibition of the fatty acid synthase by 22- and 24-carbon CoAs.

Similar to the mycobacterial enzyme, the fatty acid synthase of most types of eukaryotic cells exists as a multienzyme complex. (Plant cells have an ACP-dependent synthase that is reminiscent of the *E. coli* enzyme.) One of the best-characterized is the fatty acid synthase from yeast. In 1961, Fyodor Lynen published a classic article that described many of the important features of the purified enzyme from yeast. Lynen and others originally viewed the synthase as an aggregate of seven distinct enzymes, each present in several copies, held together by strong noncovalent interactions. In the mid-1970s the yeast fatty acid synthase was shown to arise from two genetically unlinked loci that code for two different multifunctional polypeptides. Subunit A has a molecular weight of 185,000 and contains the ACP region, β-ketoacyl-ACP synthase, and β-ketoacyl-ACP reductase activities. Subunit B has a molecular weight of 175,000 and contains the remaining activities of the synthase. Among these activities is an enzyme that transfers palmitate from ACP to CoA to give the major product, palmitoyl-CoA. The subunit composition of the yeast enzyme is A_6B_6 and has the theoretical capacity to synthesize six fatty acids at one time.

The fatty acid synthases from many different animal tissues (e.g., insects, brain, mammary gland) have also been isolated. The organization of these enzymes resembles that of liver as described in the preceding text and Figure 21–8.

The apparently rapid evolution of the multienzyme polypeptides in bacteria and their retention in eukaryotes other than plants suggests that this arrangement of the fatty acid synthase may have conferred a selective advantage. Clearly, the multifunctional polypeptides avoid the accumulation of intermediates and provide equivalent stoichiometry for each of the component enzyme activities.

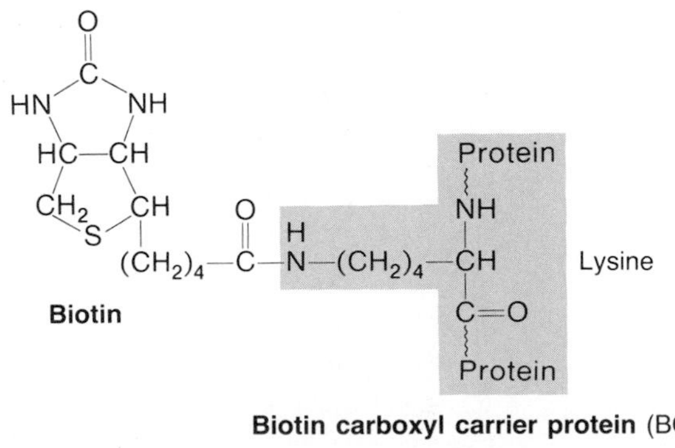

Figure 21–4
Structure of biotin linked to biotin carboxyl carrier protein (BCCP). BCCP is one of the components of acetyl-CoA carboxylase isolated from *E. coli*.

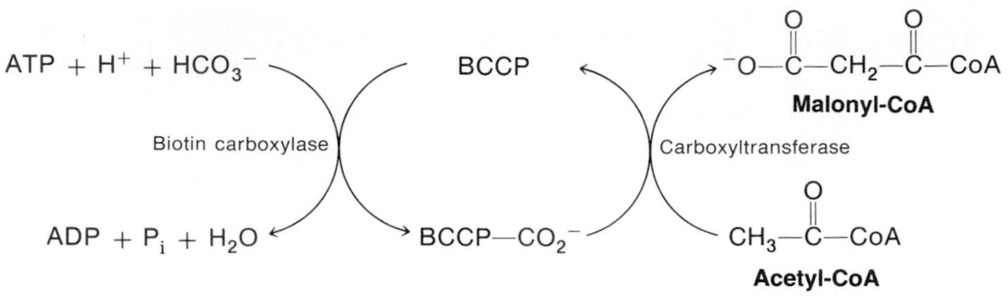

Figure 21–5
Reactions catalyzed by acetyl-CoA carboxylase. In *E. coli*, BCCP and the two enzymatic activities (biotin carboxylase and carboxyltransferase) can be separated from each other. In contrast, in liver all three components exist on a single multifunctional polypeptide.

The acetyl-CoA carboxylase of *E. coli* consists of three protein components: *biotin carboxyl carrier protein (BCCP)* ($M_r = 22{,}500$), *biotin carboxylase* ($M_r = 98{,}000$; two subunits of 49,000), and *carboxyltransferase* ($M_r = 130{,}000$). The carboxyltransferase component has an A_2B_2 structure, and the molecular weights of the two types of subunits are 35,000 and 30,000. The reaction sequence (Figure 21–5) involves an initial carboxylation of BCCP, catalyzed by biotin carboxylase. Subsequently, the carboxyltransferase transfers the CO_2 from BCCP to acetyl-CoA. The probable mechanism for the carboxyltransferase reaction involves a carbanion attack on the biotin-linked carboxyl group, as shown in Figure 21–6.

A distinctly different form of acetyl-CoA carboxylase is found in the cytosol of animal tissues. The rat liver enzyme has a molecular weight of 520,000 and is composed of two subunits ($M_r = 260{,}000$) with one biotin per subunit. In contrast to the situation in *E. coli*, the three functional parts of acetyl-CoA carboxylase in rat liver occur as a single multifunctional polypeptide. The dimeric form of the enzyme has very low activity, but incubation with citrate results in activation of the enzyme (Figure 21–7) and subsequent aggregation of the dimer to a polymer with a molecular weight between 4 and 8 million. The enzyme is deactivated and depolymerized

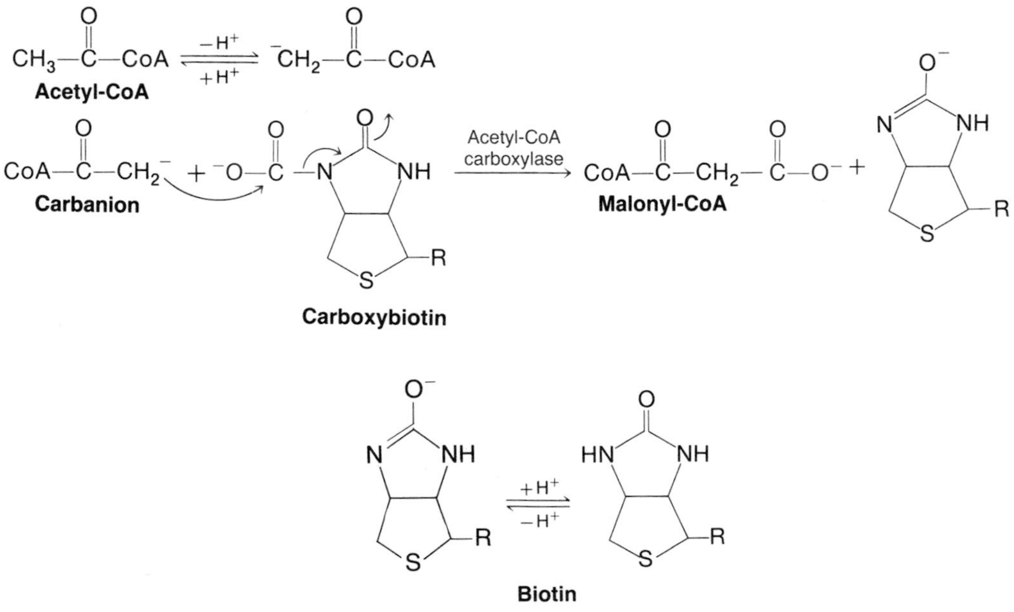

Figure 21–6
Mechanism for the carboxylation of acetyl-CoA. Acetyl-CoA forms a carbanion by proton loss. This carbanion attacks the carbon in carboxybiotin to give malonyl-CoA and biotinate. The biotinate anion is returned to the neutral form by addition of a proton. The R group attached to carboxybiotin is the carboxyl carrier protein.

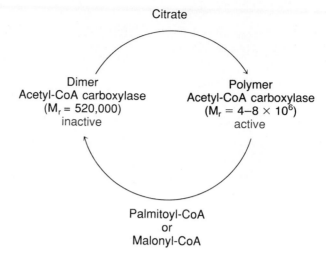

Figure 21–7
Activation of acetyl-CoA carboxylase from rat liver. Citrate activates the enzyme and converts it to a polymer. Either palmitoyl-CoA or malonyl-CoA can reverse this process. Citrate does not activate or polymerize the *E. coli* enzyme.

when incubated with malonyl-CoA or palmitoyl-CoA. The significance of these small-molecule modulators on the rate of fatty acid biosynthesis will be discussed when we look at the regulation of fatty acid metabolism, later in the chapter. Incidentally, citrate does not activate or polymerize the *E. coli* enzyme.

Seven Reactions Are Catalyzed by the Fatty Acid Synthase

Once malonyl-CoA has been synthesized, the remaining steps in fatty acid synthesis in animals (see Figure 21–2) take place on the fatty acid synthase multienzyme complex. This complex contains seven distinct enzyme activities. The organization of the seven reactions in liver is schematically represented in Figure 21–8.

Binding of Acetyl-CoA and Malonyl-CoA.   The first reaction to occur on the synthase involves the attachment of acetyl-CoA, first to the —SH of the ACP and then to a cysteine —SH of the β-ketoacyl-ACP synthase (KSase in Figure 21–8). The *acetyl transacylase* enzyme catalyzes the transfer of the acetyl group from acetyl-CoA to the —SH group of the ACP. To make way for the malonyl derivative, the acetyl group is further transferred to a cysteine —SH group on the β-ketoacyl-ACP synthase. With the acetyl group in place on the condensing enzyme, the ACP is now ready to receive a malonyl group from malonyl-CoA in a reaction catalyzed by *malonyl transacylase*. This reaction is very similar to the formation of acetyl-ACP except that the malonyl group remains linked to the pantetheine —SH group of ACP.

The Condensation Reaction.   In the condensation reaction the acetyl group is transferred from the cysteine —SH group to the malonyl group so that the acetyl group becomes the methyl terminal two-carbon unit of the new aceto-acetyl group. The reaction is catalyzed by *β-ketoacyl-ACP synthase*. In this reaction we see a major reason why malonyl-CoA is used instead of a second molecule of acetyl-CoA. The release of CO_2 in this condensation reaction provides the extra thermodynamic push to make the reaction highly favorable. It also makes the central carbon a better nucleophilic agent for attacking the carbonyl carbon of the acetyl group (Figure 21–9).

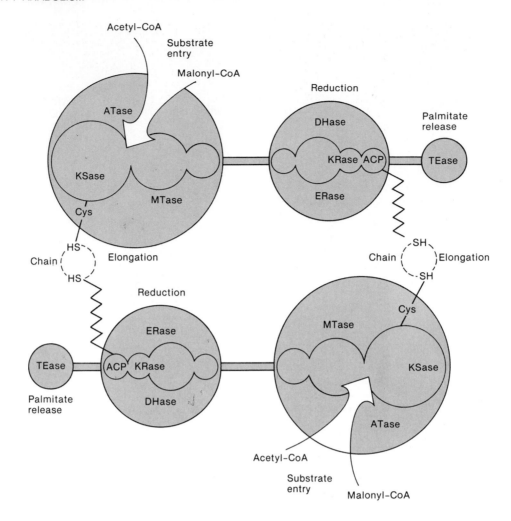

Figure 21–8

A proposed organization of the enzymatic activities of fatty acid synthase from animal liver. The multienzyme complex exists as a dimer of two giant identical polypeptides (M_r = 250,000). The two peptides are organized in a head-to-tail configuration in such a way that it is theoretically possible for the complex to make two fatty acids at the same time. The substrates, acetyl-CoA and malonyl-CoA, are transferred onto the enzyme by acetyl-CoA-ACP transacylase (ATase) and malonyl-CoA-ACP transacylase (MTase). Subsequently, β-ketobutyryl-ACP and CO_2 are formed in a condensation reaction catalyzed by β-ketoacyl-ACP synthase (KSase). The polypeptide with this condensing activity has a cysteine residue (Cys—SH) at the active site. The β-ketobutyryl-ACP is reduced to the β-hydroxy derivative by β-ketoacyl-ACP reductase (KRase), dehydrated to enoyl-ACP by β-hydroxyacyl-ACP dehydrase (DHase) and finally reduced to butyryl-ACP by enoyl-ACP reductase (ERase). The butyryl group is then transferred to the cysteine residue of KSase and another malonyl group is transferred to ACP by MTase. Another condensation occurs and the β-ketohexanoyl group is converted to hexanoyl-ACP by the enzymes KRase, DHase, and ERase. This enzymatic process continues to recycle until palmitoyl-ACP is made. The palmitate is released from the complex by a thioesterase (TEase). This model is similar to the one proposed by Dr. S. Wakil and coworkers (*J. Biol. Chem.* 258:15312, 1983).

Figure 21–9

Formation of acetoacetyl-ACP, catalyzed by 3-ketoacyl-ACP synthase (KSase). The acetyl group is bound to the cysteine residue of KSase (labeled enzyme in the figure). The carbonyl group of the enzyme-bound acetyl is attacked by the central carbon on the malonyl bound to ACP. Finally the C—S linkage is broken, resulting in acetoacetyl-ACP.

BOX 21–B

Specificity in Electron Transfer

The electrons that are liberated in the oxidation of carbohydrates, fats, and amino acids serve two main functions in metabolism. Most of them are transferred to the oxidized form of the biological electron carrier nicotinamide adenine dinucleotide (NAD^+) and then on through a series of other electron carriers to the ultimate acceptor, oxygen, in a series of oxidation-reduction reactions that are obligately coupled to the phosphorylation of ADP to ATP. Those electrons thus drive the energy metabolism of the cell.

Other electrons are used in the reductive steps of biosynthetic sequences. These electrons are transferred to nicotinamide adenine dinucleotide phosphate, converting its oxidized form, $NADP^+$, to the reduced form, NADPH. NADPH, in turn, is the reducing agent in nearly all biosynthetic reductions.

The extra phosphoryl group in NADP does not affect the redox properties of the molecule and would be of no significance in any nonbiological reactions that involved the nicotinamide end of the molecule. But in metabolism, that phosphoryl group determines which enzymes will be able to bind the cofactor, and thus what reactions it can participate in. This is a relatively simple but very important example of the symbolic use of chemical structure in biology—that is, the use of atomic groupings or other structural features to convey information that is totally unrelated to the inherent chemistry of the group, but that guides and controls metabolic transformations.

The Reduction Reactions. The object of the next three reactions in the fatty acid synthesis cycle is to reduce the β carbonyl group. This is accomplished in three steps. The carbonyl is first reduced to a hydroxyl. This is dehydrated to produce a *trans* double bond, which is further reduced to give a fully saturated derivative. Chemically, this is exactly the reverse of three steps in the fatty acid oxidation pathway except that different cofactors and different enzymes are used. See Box 21–B for a discussion of specificity in electron transfer.

Why are different enzymes used, if the three reactions involved in the reduction are merely the reverse of those in fatty acid catabolism? The primary reason is that the enzymes are organized in different parts of the cell. The physical separation of degradation and synthesis is most clearly observed in animals, where the degradative pathway is localized in the mitochondria and the biosynthetic pathway is localized in the cytosol. By contrast, in the glycolysis-gluconeogenesis system, where several enzymes are shared, both processes take place in the cytosol.

To return to the pathway, we find that *β-ketoacyl-ACP reductase* catalyses the first reduction of acetoacetyl-ACP to β-hydroxybutyryl-ACP. Mechanistically this reduction is similar to the reduction of pyruvate to lactate, discussed in Chapter 9. Following this reduction, the elements of water are removed by *β-hydroxyacyl-ACP dehydrase* to yield crotonyl-ACP. The double bond resulting from the dehydration is further reduced by *2,3-trans-enoyl-ACP reductase*, again using NADPH.

Continuation Reactions. At this point we have seen one full round of reactions on the fatty acid synthase. Each enzyme activity of the complex has been used precisely once. The resulting butyryl group, formed in the last step, is transferred from the pantetheine —SH group to a cysteine —SH group on the β-ketoacyl-ACP synthase, as was the ACP-bound acetyl group in the first cycle. The process is now ready to repeat itself. It continues, usually for six more cycles, until the palmitoyl group is formed; this group is released from the enzyme by a *thioesterase* (Figures 21–2 and 21–8).

Further modifications of the palmitoyl group are described later in this chapter; its incorporation into more complex lipids is described in the next chapter. Box 21–C describes an assay for the fatty acid synthase from liver.

BOX 21–C

Assaying the Activity of Fatty Acid Synthase

The activity of fatty acid synthase can be assayed by the incorporation of [2-^{14}C]malonyl-CoA or [^{3}H]acetyl-CoA into fatty acid. This scheme illustrates a common principle often utilized in the assay of lipid biosynthetic enzymes. The radioactive substrate is a water-soluble molecule that can easily be separated from the lipid product by extraction of the reaction mixture with an organic solvent such as petroleum ether.

Step 1. Prepare the incubation mixture, which contains:

0.3 mM NADPH
0.1 M phosphate buffer, pH 6.8
5 μM mercaptoethanol
3 μM EDTA
50 μM malonyl-CoA
12 μM[^{3}H]acetyl-CoA   (specific   radioactivity = 2.0 × 10^6 dpm/μmol)
Enough distilled H$_2$O to bring the final volume to 1 ml

Mercaptoethanol is used to keep the SH residues of the enzyme in the reduced state. EDTA chelates divalent cations such as Mg^{2+}, which might inhibit the reaction. All of the materials can be purchased from companies that sell chemical compounds. Radioactive compounds (e.g., [^{3}H]acetyl-CoA) are sold by companies that specialize in the manufacture of radioisotopes.

Step 2. Equilibrate the mixture for 5 min at 37°C in a shaking water bath.

Step 3. Add enzyme (e.g., 2 mg protein), mix thoroughly, and incubate at 37°C for 5 min.

Step 4. Stop the reaction by the addition of 0.1 ml 18% perchloric acid. The perchloric acid lowers the pH to ~1, and thus denatures and inactivates the enzyme.

Step 5. Add 1 ml ethanol and 2 ml petroleum ether, mix thoroughly, and allow the phases to separate. Transfer the upper ether layer to a tube and extract the lower aqueous phase two more times with petroleum ether. The ethanol and water separate from the ether layer, which floats on the aqueous layer. Because the pH is approximately 1, the fatty acid product is protonated (RCOOH) and can therefore be easily extracted into the ether phase.

Step 6. Evaporate the combined petroleum ether extracts, add liquid scintillation fluid, and determine the radioactivity by liquid scintillation spectrophotometry. Liquid scintillation counters are sophisticated instruments that are obtainable from companies that specialize in their manufacture and sale. Their cost ranges between $15,000 and $30,000.

Step 7. Calculate the specific activity of the enzyme. To do this, divide the dpm incorporated into the fatty acid by the specific radioactivity of the acetyl-CoA, the time of the incubation, and the milligrams of protein added to the assay. For example:

Specific activity =

$$\frac{\text{dpm in fatty acid}}{\text{specific radioactivity of acetyl-CoA} \cdot \text{min} \cdot \text{mg protein}}$$

$$= \frac{50,000 \text{ dpm}}{2 \times 10^6 \text{ dpm}/\mu\text{mol} \cdot 5 \text{ min} \cdot 2 \text{ mg protein}}$$

$$= 2.5 \times 10^{-3} \text{ } \mu\text{mol fatty acid formed/min} \cdot \text{mg protein}$$

The specific activity is a measure of the activity of an enzyme as a function of time and amount of protein. When fatty acid synthesis is reduced, for example during a fast, the specific activity will be much lower by about tenfold than after a carbohydrate-rich meal. The specific activity can also be a measure of the purity of an enzyme preparation. Thus fatty acid synthase in rat liver cytosol might have an activity of 1×10^{-3} μmole fatty acid formed per minute per milligram of protein. By contrast, a pure enzyme, free of all other cytosolic proteins, might have 1000-fold higher specific activity (1 μmole fatty acid formed per minute per milligram of protein).

When the fatty acid synthase is highly purified, it also can be assayed by a spectrophotometric method in which the oxidation of NADPH is followed. As NADP$^+$ is formed, there is a decrease in the absorbance at 340 nm.

BIOSYNTHESIS OF MONOUNSATURATED FATTY ACIDS

Two chemically distinct mechanisms exist for the introduction of a *cis* double bond into saturated fatty acids—the <u>anerobic pathway</u>, as typified in *E. coli*, and the <u>aerobic pathway</u>, found in many eukaryotes and studied mainly in mammalian liver.

As the name "anaerobic" implies, the double bond of the fatty acid is inserted in the absence of oxygen. Biosynthesis of monounsaturated fatty acids follows the pathway described previously for saturated fatty acids until the intermediate β-hydroxydecanoyl-ACP is reached (Figure 21–10). At this point, there is an apparent competition between <u>β-hydroxy-</u>

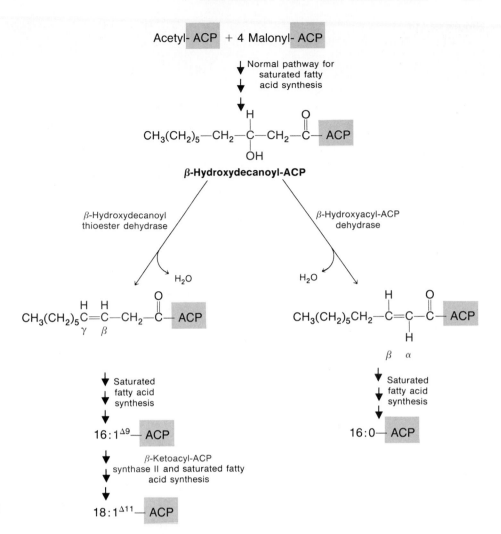

Figure 21–10
Anaerobic pathway for biosynthesis
of monounsaturated fatty acids in
E. coli. Synthesis of monounsaturated
fatty acids follows the pathway
described previously for saturated
fatty acids until the intermediate
β-hydroxydecanoyl-ACP is reached.
At this point there is an apparent
competition between the enzymes
involved in saturated and unsaturated
fatty acid synthesis.

decanoylthioester dehydrase, which forms a β,γ-*cis* double bond, and *β-hydroxyacyl-ACP dehydrase*, which forms an α,β *trans* double bond. Both enzymes are highly specific for the C_{10}-β-hydroxyacyl-ACP. The β,γ unsaturated acyl-ACP is subsequently elongated by the *E. coli* fatty acid synthase to yield palmitoleoyl-ACP ($16:1^{\Delta 9}$).* The conversion of this compound to the major unsaturated fatty acid of *E. coli*, *cis*-vaccenic acid ($18:1^{\Delta 11}$), appears to involve β-ketoacyl-ACP synthase II, which shows a preference for palmitoleoyl-ACP as a substrate (see Table 21–1). The subsequent conversion of β-keto-*cis*-vaccenyl ACP to *cis*-vaccenyl-ACP is catalyzed by the usual enzymes of fatty acid biosynthesis.

In contrast to the anaerobic pathway found in *E. coli*, the aerobic pathway in eukaryotic cells introduces double bonds after the C_{16} or C_{18} saturated fatty acid has been synthesized. In rat liver and other eukaryotic cells, an enzyme complex associated with the endoplasmic reticulum desaturates stearoyl-CoA ($18:0$) to oleoyl-CoA ($18:1^{\Delta 9}$). This reaction requires NADH and O_2 and results in the remarkable formation of a double bond in the middle of an acyl chain with no activating groups nearby. Although many elegant experiments have been performed, the chemical mechanism for desaturation of long-chain acyl-CoAs remains unclear.

Desaturation requires the cooperative action of two enzymes, *cytochrome b_5 reductase and stearoyl-CoA desaturase*, and the action of *cytochrome b_5*. A scheme for this set of reactions is shown in Figure 21–11.

*Terminology for lipid structure is explained in Chapter 5.

Figure 21–11
The aerobic pathway for formation of oleoyl-CoA in eukaryotes. In eukaryotes the double bonds are introduced after the C_{16} and C_{18} saturated fatty acid has been synthesized.

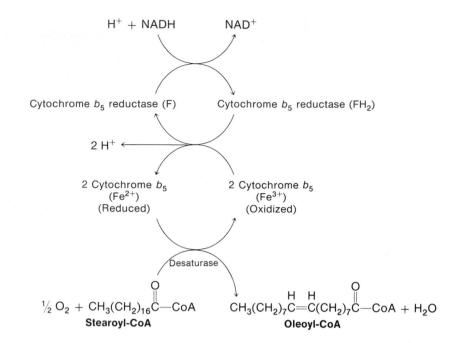

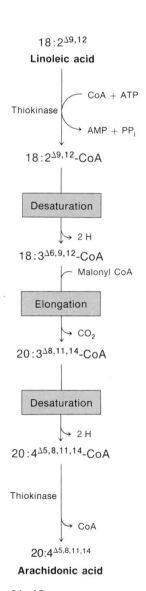

$$18:2^{\Delta 9,12}$$
Linoleic acid

Thiokinase — CoA + ATP

→ AMP + PP_i

$$18:2^{\Delta 9,12}\text{-CoA}$$

Desaturation

→ 2 H

$$18:3^{\Delta 6,9,12}\text{-CoA}$$

— Malonyl CoA

Elongation

→ CO_2

$$20:3^{\Delta 8,11,14}\text{-CoA}$$

Desaturation

→ 2 H

$$20:4^{\Delta 5,8,11,14}\text{-CoA}$$

Thiokinase

→ CoA

$$20:4^{\Delta 5,8,11,14}$$

Arachidonic acid

Figure 21–12
Synthesis by mammalian tissues of arachidonic acid from linoleic acid. The Δ^5 and Δ^6 desaturases are separate enzymes and are also different from the Δ^9 desaturase (Figure 21–11). The mechanisms, however, seem to be the same, involving cytochrome b_5 and cytochrome b_5 reductase. The enzymes for elongation of unsaturated fatty such as 18:3 to 20:3 occur on the endoplasmic reticulum.

Cytochrome b_5 reductase ($M_r = 43,000$) is a flavoprotein that transfers electrons from NADH by means of flavin (F) to cytochrome b_5, a heme-containing protein ($M_r = 16,700$) in which Fe^{3+} is reduced to Fe^{2+}. Both cytochrome b_5 and the reductase are amphipathic proteins; that is, each has a hydrophobic peptide tail that anchors the protein into the membrane of the endoplasmic reticulum and a hydrophilic portion that is outside the membrane surface (see Chapter 6 for details about membrane-bound proteins). *Stearoyl-CoA desaturase* utilizes two electrons from cytochrome b_5 coupled with an atom of oxygen to form a *cis* double bond in the Δ^9 position of stearoyl-CoA. The desaturase ($M_r = 53,000$) has 62 per cent nonpolar amino acids, which is probably the main reason it is tightly embedded in the membrane. There is also one atom of nonheme iron per molecule of enzyme.

BIOSYNTHESIS OF POLYUNSATURATED FATTY ACIDS

E. coli does not have polyunsaturated fatty acids, whereas eukaryotes produce a large variety of polyunsaturated fatty acids. Mammals cannot desaturate between the Δ^9 position and the methyl end of an acyl chain, whereas plants have the enzymes to desaturate at positions Δ^{12} and Δ^{15}. Thus mammals have a dietary requirement for linoleic acid ($18:2^{\Delta 9,12}$) and linolenic acid ($18:3^{\Delta 9,12,15}$), which are essential components of phospholipids in membranes. However, enzyme complexes occur in the endoplasmic reticulum of animal cells that desaturate at Δ^5 if there is a double bond at the Δ^8 position, or at Δ^6 if there is a double bond at the Δ^9 position. These enzymes are different from the Δ^9-desaturase, but they do appear to utilize cytochrome b_5 reductase and cytochrome b_5.

The major polyunsaturates of mammals are derived either from diet or from desaturation and elongation of $18:2^{\Delta 9,12}$ or $18:3^{\Delta 9,12,15}$. A scheme for the synthesis of arachidonic acid ($20:4^{\Delta 5,8,11,14}$) from linoleic acid is shown in Figure 21–12. This example illustrates the principle by which polyunsaturated fatty acids are made in animals. The elongation step is catalyzed by a series of membrane-bound enzymes that are present in the endoplasmic reticulum. These enzymes use malonyl-CoA as the donor for the two-carbon unit, and the chemical mechanism seems to be similar to that described earlier for fatty acid synthesis (Figure 21–2). The liver enzymes also will elongate other polyunsaturated fatty acyl-CoAs. In addition, endoplasmic reticulum enzymes will elongate C_{16}- and C_{18}-CoAs to produce the C_{22}- and C_{24}-CoAs characteristic of sphingolipids (see Chapter 22). The latter elongation enzymes are most active in brain tissue during the synthesis of myelin.

REGULATION OF FATTY ACID METABOLISM

Our discussion of the control of fatty acid metabolism will focus on the mammalian liver, about which much is known. Fatty acids are useful for energy storage and as precursors for a variety of membrane lipids. It seems likely that the structural needs for fatty acids will be met by the organism even when the nutritional supply is skimpy. We will not comment on the regulatory mechanisms that guarantee meeting these needs; those mechanisms are discussed for E. coli in the next chapter. Instead, we will describe the regulatory mechanisms that control fatty acid synthesis in mammals under conditions of dietary excess.

Fatty acid synthesis for energy storage is something that occurs only when the energy charge is high. In this sense fatty acid synthesis is like glycogen synthesis. However, the mammalian body is designed in such a way that it can store only a few hundred grams of glycogen in the liver and muscle. This is just enough to supply the body's energy needs for one day or so. By contrast, the body stores lipids for future energy needs in special fat cells that are distributed in many parts of the body. The extra storage space permits the average human to bank sufficient fat to satisfy basic energy needs for as long as two months. Clearly, energy storage for long-term needs is extremely important in cases where nutrients are not available on a daily basis, and we should know something about the mechanisms that regulate the storage process.

With this function in mind, we can apply the principles of regulation discussed in Chapter 13 to anticipate the types of controls that ought to be present if the system is to operate efficiently and effectively. First, we should expect there to be a regulatory switch that turns when the organism has satisfied its immediate energy needs and the small storage space for glycogen has been filled. The turning of this switch should direct nutrients to favor fatty acid synthesis. There should be one or more additional switches that prevent the simultaneous synthesis and breakdown of fatty acids. Taken together, the synthesis and breakdown of fatty acids constitute a giant pseudocycle. To have both halves of this pseudocycle operating simultaneously would simply cause a net loss of ATP, an undesirable result. Finally, we would expect the regulatory switches to be embedded in regulatory proteins and enzymes that function at the first committed step in any given pathway. The activity of these proteins should respond primarily to pathway end products and the energy status of the cell.

In addition to these mechanisms, which operate at the level of the existing protein machinery, we should be alert for mechanisms that vary the amount of enzymes available. Such mechanisms usually operate by regulating enzyme synthesis, but they can also operate by regulating enzyme degradation.

Now let us see how well the regulation of fatty acid metabolism measures up to our expectations.

Acetyl-CoA Carboxylase Activity and the Availability of Acetyl-CoA Are Controlling Factors

One major mechanism for control of fatty acid synthesis is a change in the concentration of active enzyme(s) available for catalysis. Genetic factors, stage of development, hormones, and energy supply are important elements that dictate the amount of acetyl-CoA carboxylase, fatty acid synthase, malic enzyme, and other enzymes related to fatty acid synthesis. For example, the concentrations of fatty acid synthase and acetyl-CoA carboxylase in rat liver are reduced by four- to fivefold after fasting. When the rat is allowed to eat again, the concentrations of fatty acid synthase and acetyl-CoA carboxylase rise dramatically. When the rat is fed a fat-free diet, the concentration of fatty acid synthase is 14-fold higher than in a rat maintained on a normal rat chow. Current evidence indicates that the levels of these enzymes are governed by the rate of enzyme synthesis, not degradation. It appears that the change in the rate of enzyme synthesis is due to a fluctuation in the supply of mRNA, which in turn is controlled by the rate of transcription of DNA. A question of current interest is how this transcription of DNA is regulated.

The alteration in enzyme levels is a long-term change, since the response occurs over a period of hours or days. Faster responses, which operate at the level of direct control of enzyme activity, must exist if immediate needs are to be met. Major points of control of enzyme activity are indicated in Figure 21–13.

Figure 21–13
Overview of the conversion of carbohydrate to lipid in rat liver cells and its regulation. Colored arrows with pluses and minuses indicate points of activation and inhibition, respectively. Glucagon and epinephrine are the main hormones involved in regulation. Citrate and palmitoyl-CoA are the main substrates involved in regulation.

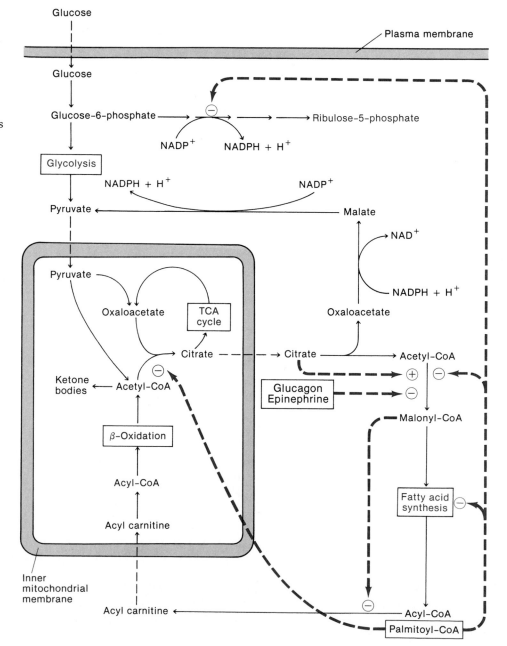

As might be suspected, acetyl-CoA carboxylase, the first enzyme in the pathway for fatty acid synthesis, is a primary target for regulation. This enzyme is activated by citrate and inhibited by palmitoyl-CoA. We have already seen that citrate activates the liver acetyl-CoA carboxylase and converts it to a high-molecular-weight polymer (see Figure 21–7). The activation by citrate is appropriate because excess citrate is a strong indication that the energy needs of the cell are satisfied (see Chapter 15). Furthermore, citrate is the most important source of cytoplasmic acetyl-CoA, which is supplied by a mechanism previously described (Figure 21–1). In this connection, recall from Chapter 15 that isocitrate dehydrogenase is strongly inhibited by a high $NADH/NAD^+$ ratio, another good indication that the energy status of the cell is high. The inhibition of isocitrate dehydrogenase blocks the TCA cycle and leads to a buildup of citrate, which is a precursor of the cytosolic acetyl-CoA.

The inhibition of acetyl-CoA carboxylase by palmitoyl-CoA is also an appropriate regulatory response, since the latter is the end product of fatty acid synthesis. Palmitoyl-CoA also inhibits glucose-6-phosphate degradation by the pentose pathway. This pathway is a major supplier of the NADPH needed for fatty acid synthesis. Hence the response is entirely appropriate; the need for NADPH should be less if there is a high level of fatty acid already present. Palmitoyl-CoA likewise inhibits the fatty acid synthase, but to a lesser extent than it inhibits the carboxylase.

Another way in which acetyl-CoA carboxylase is inhibited is as a result of the action of the hormones glucagon and epinephrine. High levels of these hormones are a signal that energy is in short supply or soon may be needed in large amounts (see Chapter 14). At such a time it would be inappropriate to divert energy to the synthesis of fatty acids. The hormonal effects, as usual, proceed through a chain of events (Figure 21–14). The hormones bind to receptors on the plasma membrane and stimulate adenylate cyclase, the enzyme that catalyzes cAMP synthesis. This in turn stimulates a cAMP-dependent kinase, which phosphorylates acetyl-CoA carboxylase. The phosphorylation results in the inactivation of the carboxylase. Recall that hormones are active only on cells that carry the specific hormone receptors. Both liver and adipose (fat) tissue respond to glucagon and epinephrine.

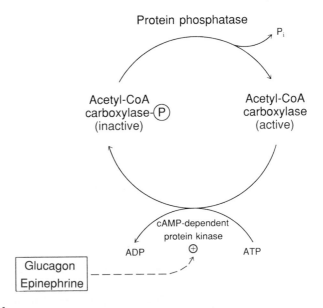

Figure 21–14
Regulation of acetyl-CoA carboxylase by phosphorylation and dephosphorylation. The cAMP-dependent phosphorylation results in inactivation of the carboxylase. This inactivation is reversed by protein phosphatase-1.

Incidentally, insulin, which often counteracts the action of glucagon, has the opposite effect on the carboxylase; the mechanism of insulin action is under study.

Malonyl-CoA and Hormones Inhibit Simultaneous Synthesis and Breakdown of Fatty Acids

So far we have considered signals that directly inhibit or stimulate the synthesis of fatty acids. Now let us consider signals that prevent the simultaneous synthesis and degradation of fatty acids. These signals are easiest to understand in the eukaryote, where there is a physical separation of the anabolic and catabolic processes in the cytosol and the mitochondria, respectively. Malonyl-CoA, the immediate product of acetyl-CoA carboxylase, is an important signal in this regard. Malonyl-CoA is a potent inhibitor of carnitine acyltransferase I. We discussed this transferase in Chapter 18 (see Figure 18–5), where we saw that it is the enzyme responsible for transferring fatty acids from CoA to carnitine. Only in this form can fatty acids be transported from the cytosol to the mitochondrion for degradation. Thus high levels of malonyl-CoA, which indicate that fatty acid synthesis is in progress, also prevent fatty acid catabolism. As a result, the biosynthetic part of the fatty acid pseudocycle is strongly favored when the malonyl-CoA level is high.

The hormonal effects of glucagon and epinephrine also prevent simultaneous synthesis and degradation. These hormones both stimulate catabolism of stored lipids (see Figure 18–4), while they inhibit the synthesis of fatty acids through their inhibitory effects on acetyl-CoA carboxylase. In the absence of these hormones, synthesis is favored over degradation.

Chain Length and Ratio of Unsaturated to Saturated Fatty Acids Are Regulated in *E. coli*

The chain length of fatty acids in *E. coli* is regulated by the activities and specificities of several enzymes. First, neither of the β-ketoacyl-ACP synthases of *E. coli* (see Table 21–1) uses 18-carbon substrates effectively. Second, β-ketoacyl-ACP synthase II is required for elongation of palmitoleate ($16:1^{\Delta 9}$) to *cis*-vaccenate ($18:1^{\Delta 11}$) and is, therefore, responsible for *E. coli* having unsaturated fatty acids with 18 carbons. Third, the lengths of fatty acids are also determined by the rate of utilization of acyl-ACPs for phospholipid synthesis. Thus there appears to be a competition between the rate of elongation of acyl-ACPs and that of utilization of acyl-ACPs for phospholipid synthesis. This competition appears to be the main reason why palmitate, and not stearate, is the most plentiful saturated fatty acid in *E. coli*.

The proportion of unsaturated and saturated fatty acids synthesized is regulated also. The ratio of monounsaturated to saturated fatty acids is a function of temperature—the higher the temperature, the lower the content of unsaturated fatty acids. In this connection we should remember that unsaturated fatty acids do not pack as well in membrane structures as saturated fatty acids (Chapter 6). The major fatty acid made at lower temperatures is *cis*-vaccenate, and its concentration decreases when *E. coli* is grown at higher temperatures. In contrast, the level of palmitoleate does not change with temperature. This observation, coupled with studies on an *E. coli* mutant defective in β-ketoacyl-ACP synthase II, has demonstrated that this enzyme is solely responsible for thermal regulation of the fatty acid composition of *E. coli*. How the lack of synthase II inhibits the initial introduction of a *cis* double bond by β-hydroxyldecanoyl thioester dehydrase (Figure 21–10) is not understood.

SUMMARY

When there is an abundance of nutrients, the production of excess acetyl-CoA and NADPH occurs. These are the two main substrates required for fatty acid synthesis, which is therefore stimulated. In animals the excess acetyl-CoA produced in the mitochondria cannot be directly transported to the cytosol, the site of fatty acid synthesis. It must first be condensed with oxaloacetate to form citrate, which can pass through the mitochondrial membrane and be converted back into oxaloacetate and acetyl-CoA. One of the reactions (catalyzed by malic enzyme) involved in the return of oxaloacetate to the mitochondria also makes NADPH. This is needed in large quantities to supply the reducing power for fatty acid synthesis.

Fatty acid synthesis begins with the carboxylation in the cytosol of acetyl-CoA to malonyl-CoA, a reaction catalyzed by acetyl-CoA carboxylase. Malonyl-CoA and acetyl-CoA are subsequently converted into fatty acids, especially the 16-carbon fatty acid palmitate, by a series of reactions that add two carbons at a time to a growing chain. These reactions take place in animals on a multienzyme complex known as fatty acid synthase. Fatty acid synthase in liver is a multifunctional polypeptide that contains seven enzyme activities; the intermediates in fatty acid synthesis are not released from the synthase until palmitate is made. In *E. coli* and plants the enzymes of fatty acid synthesis occur separately.

Monounsaturation can be introduced into fatty acids in bacterial cells during the process of chain elongation. In contrast, in eukaryotic cells, double bonds are introduced after the 16- or 18-carbon saturated fatty acids have been made.

Fatty acid metabolism is controlled by complex interactions similar to those in other metabolic pathways. These involve regulation of the rates of synthesis and degradation of enzymes, modulation of enzyme activities, cellular compartmentalization, alterations in the concentrations of substrates, products, coenzymes, and hormonal control. In liver the rate of fatty acid synthesis seems to be regulated primarily by the supply of malonyl-CoA, which is a product of the reaction catalyzed by acetyl-CoA carboxylase. The activity of this enzyme is controlled by the rate of enzyme synthesis, covalent phosphorylation, citrate activation, and palmitoyl-CoA inhibition. The rate of β oxidation is inhibited by the action of malonyl-CoA on carnitine acyltransferase I; this prevents the transport of fatty acids to the site of β oxidation in the mitochondrion. Hence the simultaneous synthesis and catabolism of fatty acids is minimal in liver.

SELECTED READINGS

Cook, H. W., Fatty acid desaturation and chain elongation in eukaryotes. In D. E. Vance and J. E. Vance (eds.), *Biochemistry of Lipids and Membranes*. Menlo Park, Calif: Benjamin/Cummings, 1985. This chapter provides an advanced treatment of fatty acid desaturation.

Deuel, H. J., *The Lipids: Biochemistry*, Vol. 3. New York: Interscience, 1957. This is a comprehensive and classical treatise on the biochemistry of lipids until the mid-1950s.

Goodridge, A. G., Fatty acid synthesis in eucaryotes. In D. E. Vance and J. E. Vance (eds.), *Biochemistry of Lipids and Membranes*. Menlo Park, Calif: Benjamin/Cummings, 1985. This chapter reviews at an advanced level fatty acid synthesis and its control in eukaryotes.

McGarry, J. E., and D. W. Foster, Regulation of hepatic fatty acid oxidation and ketone body production. *Ann. Rev. Biochem.* 49:395, 1980. A review in which the role of malonyl-CoA is discussed.

Najjar, V. A., *Fat Metabolism*. Baltimore: The Johns Hopkins Press, 1954. A good summary of the early work on fatty acid metabolism.

Rock, C. O., and J. E. Cronan, Lipid metabolism in procaryotes. In D. E. Vance and J. E. Vance (eds.), *Biochemistry of Lipids and Membranes*. Menlo Park, Calif.: Benjamin/Cummings, 1985. This chapter reviews at an advanced level fatty acid synthesis in *E. coli*.

PROBLEMS

1. A defect in ATP-citrate lyase might be expected to have severe consequences for fatty acid biosynthesis. Explain.

2. Fatty acid biosynthesis proceeds by the successive addition of two carbon units to the growing chain, and yet malonyl-CoA rather than acetyl-CoA is the substrate for each addition. Explain.

3. For an *in vitro* synthesis of fatty acids with purified fatty acid synthase, the acetyl-CoA was supplied as the tritium derivative:

The other reactants, including malonyl-CoA, were not radioactive.
   a. Where would the tritium label be found in the product, palmitic acid?

b. If malonyl-CoA were supplied as the only labeled compound in which both hydrogens on carbon 2 were replaced by tritium, i.e.,

$$^-O\overset{O}{\underset{\|}{C}}-C^3H_2-\overset{O}{\underset{\|}{C}}-CoA$$

how many tritium atoms would be incorporated per molecule of palmitate?

c. If [3-^{14}C]malonyl-CoA, i.e.,

$$^-O-^{14}\overset{O}{\underset{\|}{C}}-CH_2\overset{O}{\underset{\|}{C}}-CoA$$

were used in the reaction, which carbons in palmitate would be labeled?

4. How many molecules of ATP, NADPH, and acetyl-CoA would be required for the synthesis of a molecule of palmitic acid from acetyl-CoA by each of the following?
   a. An *E. coli* cell.
   b. A liver cell.

5. Describe the difference in the organization of enzymes involved in fatty acid biosynthesis in *E. coli* and mammalian liver.

6. There is a rare disease called "thin-person syndrome." (It is so rare we are still looking for the first case.) The problem with these people is that they are unable to make fatty acids. A new drug called "gainabit" was tested in animal models and caused an induction of fatty acid synthase. Treatment of people with "thin-person syndrome" with this new drug resulted in a level of fatty acid synthase that was fivefold higher than normal. However, the drug had no effect on fatty acid synthesis in these patients. Offer an explanation for why this drug was ineffective. What other enzyme should have been studied prior to the human drug trial?

7. How many molecules of ATP, NADPH, and acetyl-CoA would be required for a liver cell to synthesize a molecule of:
   a. Oleic acid?
   b. Linoleic acid?

8. Draw a scheme that illustrates how mammals would synthesize 22:6$^{\Delta 4,7,10,13,16,19}$ from linolenic acid.

9. Compare the mechanisms used in the synthesis of unsaturated fatty acids in *E. coli* and mammalian liver.

10. Explain why variation in the concentration of oxaloacetic acid might affect the rate of fatty acid synthesis.

11. The term Sherringtonian metaphor is sometimes used to describe regulatory systems that involve inhibitory and stimulatory components operating synergistically. [The term derives from Sherrington's classic studies of muscle action.] Describe an example of a Sherringtonian metaphor in the control of fatty acid metabolism in mammals.

22

COMPLEX LIPIDS

Thus far we have been concerned with the metabolism of fatty acids as it relates to the storage and release of energy (see Chapters 18 and 21). In this chapter we will focus primarily on the metabolism of lipids that serve other roles. In particular, we will consider the metabolism of fats that are integral components of membrane structures and the metabolism of lipophilic compounds with potent physiological activity.

Fatty acids are a major component of _glycerolipids_, of which there are two types: _triacylglycerols_ and _phospholipids_. Triacylglycerols are important for storage of energy (see Chapters 5 and 18). Phospholipids have a central function in membrane structure (see Chapter 6); they are now recognized as important precursors of hormones such as _eicosanoids_, and hormone-related substances such as _inositol-1,4,5-P$_3$_. In eukaryotic cells, fatty acids are also an important component of another class of membrane lipids, the _sphingolipids_.

In this chapter we consider various aspects of the metabolism of these complex lipids in both prokaryotes and eukaryotes.

GLYCEROLIPIDS

Glycerolipids are those lipids that contain glycerol as a backbone. The chief glycerolipids are triacylglycerol and phospholipids. The importance of phospholipids to living organisms is underscored by the nearly complete lack of genetic defects in the metabolism of these lipids in humans. Presumably, any such defects are lethal at any early stage of development and, therefore, are never observed.

Phospholipids Are Amphipathic Substances

By definition, phospholipids are those lipids that contain phosphate. The phosphoglycerides are quantitatively the most important structural group in this lipid class. In the first part of this chapter we will discuss phosphoglycerides. Sphingomyelin is also an important phospholipid, and it will be discussed with the sphingolipids.

The structures of phospholipids and their roles in membrane structures have been discussed in detail in Chapters 5 and 6. This is but a brief summary of those properties. All the phosphoglycerides have a glycerol-3-phosphate backbone. Carbons 1 and 2 are usually acylated with fatty acids in most phospholipids; the fatty acid substituent at carbon 1 is saturated and that at carbon 2 is unsaturated. In some instances, the substituent on carbon 1 is an alkyl ether or an alkenyl ether (see Figure 5–6).

Phosphoglycerides are classified according to the substituent on the phosphate group. Major phospholipids are listed in Table 5–3.

Although phospholipids are essential components of all biological membranes, there is great variation in the composition of lipids that make up the membranes in different cells. The major lipids of animal cell plasma membranes are phosphatidylcholine, phosphatidylethanolamine, sphingomyelin, and cholesterol. The first three of these are discussed in this chapter; the metabolism of cholesterol is the main topic of Chapter 23.

In contrast to eukaryotic membranes, the membranes of E. coli have a simple lipid composition and lack phosphatidylcholine, phosphatidylinositol, cholesterol, and sphingolipids. The major phospholipid of E. coli is phosphatidylethanolamine.

Phospholipids are amphipathic (dual sympathy) molecules, because they have both polar and nonpolar portions. The polar head groups prefer an aqueous environment (are hydrophilic), whereas the nonpolar (hydrophobic) acyl substituents do not. It is this amphipathic property that causes most phospholipids to form spontaneously a bilayer (see Figure 6–11). The bilayer is the dominant structural feature of most membranes in which proteins are embedded (see Chapter 6). The amphipathic nature of phospholipids has a great influence on the mode of biosynthesis of the phospholipids. Most of the reactions involved in lipid synthesis occur on the surface of membrane structures with enzymes that themselves are amphipathic.

Phospholipid Synthesis in *E. coli* Leads to Phosphatidylethanolamine, Phosphatidylglycerol, and Cardiolipin

E. coli contains three important classes of phospholipids: phosphatidylethanolamine (75–85%), phosphatidylglycerol (10–20%), and cardiolipin (5–15%). All three phospholipids share the same biosynthetic pathway up to the formation of CDP-diacylglycerol (Figure 22–1), after which the pathways branch.

The enzymes for phospholipid synthesis are located on the inner plasma membrane of E. coli, except for phosphatidylserine synthase, which is mostly found associated with the ribosomes.

Glycerol-3-phosphate acyltransferase, the first enzyme in the pathway, preferentially utilizes saturated fatty acyl derivatives (palmitoyl-CoA or palmitoyl-ACP) for the initial acylation of glycerol-3-phosphate. The general mechanism for acyltransferase-catalyzed reactions is described in Box 22–A. The second enzyme (Figure 22–1), 1-acylglycerol-3-phosphate acyltransferase, is membrane-bound; it will use either acyl-CoA or acyl-ACP as substrate and prefers acyl residues with a double bond. Phosphatidic acid is the product. The substrate specificity of both of these acyltransferases can largely account for the finding of mostly saturated fatty acids in the sn-1* position and unsaturated fatty acids preferentially in the sn-2 position of the E. coli phospholipids. The next enzyme in the sequence (Figure 22–1), phos-

*The sn terminology is explained in Chapter 5.

BOX 22–A

Mechanism of Action of Acyltransferases

Nucleophilic addition to the neutral activated acyl group is a favored process and coenzyme A is a good leaving group from the tetrahedral intermediate. This point was discussed in Chapter 11 in connection with the formation of acetoacetyl-CoA, and we see it again here in the acyltransferase reactions involving glycerol or glycerol derivatives and acyl-CoA compounds. The reaction is initiated by a nucleophilic attack of acyl-CoA by a hydroxyl, giving rise to a tetrahedral intermediate. This is followed by expulsion of the CoA.

phatidate cytidylyltransferase, will utilize either cytidine triphosphate (CTP) or deoxy CTP (dCTP) as a substrate, and both CDP-diacylglycerol and dCDP-diacylglycerol are products in _E. coli_. No functional difference between dCDP-diacylglycerol and CDP-diacylglycerol has been demonstrated.

CDP-diacylglycerol is a branchpoint intermediate and has two fates. One fate for this compound (Figure 22–1) is reaction with serine, catalyzed by _phosphatidylserine synthase_, to form phosphatidylserine. In the final reaction of this sequence, phosphatidylserine is decarboxylated by _phosphatidylserine decarboxylase_ to yield phosphatidylethanolamine. In an alternative sequence the CDP-diacylglycerol is converted to phosphatidylglycerol phosphate by _phosphatidylglycerol phosphate synthase_. Then the phosphatidylglycerol phosphate is dephosphorylated by _phosphatidylglycerol phosphate phosphatase_, yielding phosphatidylglycerol. Subsequently, cardiolipin can be made by the reaction of two phosphatidylglycerol molecules in a reaction catalyzed by _cardiolipin synthase_.

Phospholipid Synthesis in _E. coli_ Is Regulated at an Early Stage in Fatty Acid Synthesis

The rate of phospholipid synthesis in _E. coli_ appears to be regulated primarily at an early stage of fatty acid biosynthesis. This was shown with the help of a mutant of _E. coli_ that requires an abnormally high concentration of glycerol-3-phosphate for the synthesis of phospholipids. When the concentration of glycerol-3-phosphate in the medium was limited, phospholipid synthesis in these mutants was curtailed and acyl-ACP accumulated. Conversely, when phospholipid synthesis was active (because of high glycerol-3-phosphate in the medium), acyl-ACP levels in the cell were low. Thus it

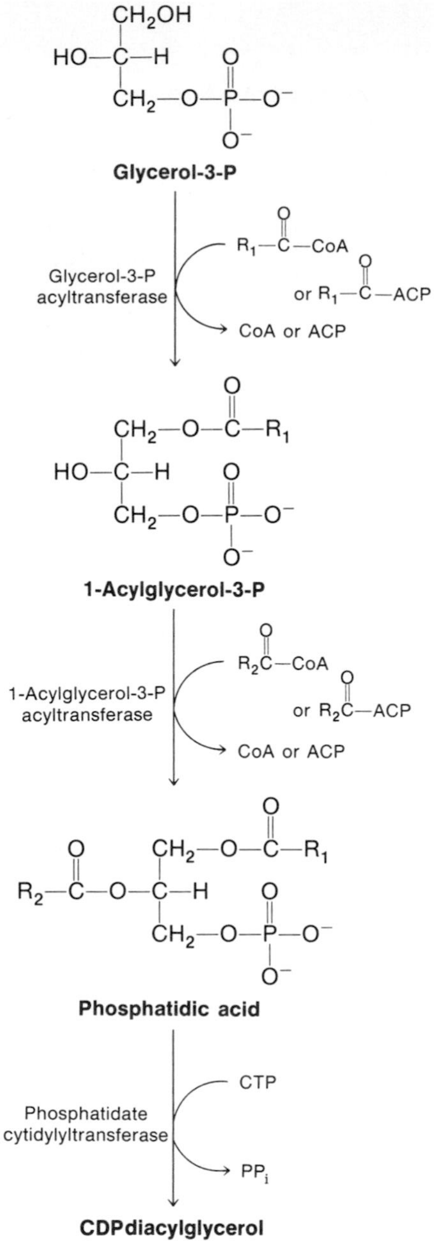

Figure 22–1 *(Facing pages)*
Scheme for the biosynthesis of phospholipids in *E. coli*.

appears that acyl-ACP synthesis is rate-limiting for phospholipid synthesis. However, the precise step of fatty acyl-ACP synthesis that is regulated has not yet been identified.

Control of phospholipid synthesis at the level of fatty acid supply makes good sense because fatty acids in *E. coli* (in contrast to mammals) are primarily used for phospholipid synthesis. Furthermore, most of the energy required for phospholipid synthesis (94%) is expended in the manufacture of fatty acids.

A summary scheme for the regulation of fatty acid and phospholipid synthesis in *E. coli* is shown in Figure 22–2. The following comments relate to the numbered points in the figure.

1. The rate of fatty acid and phospholipid biosynthesis is controlled at an early, but unspecified, step in fatty acid synthesis.

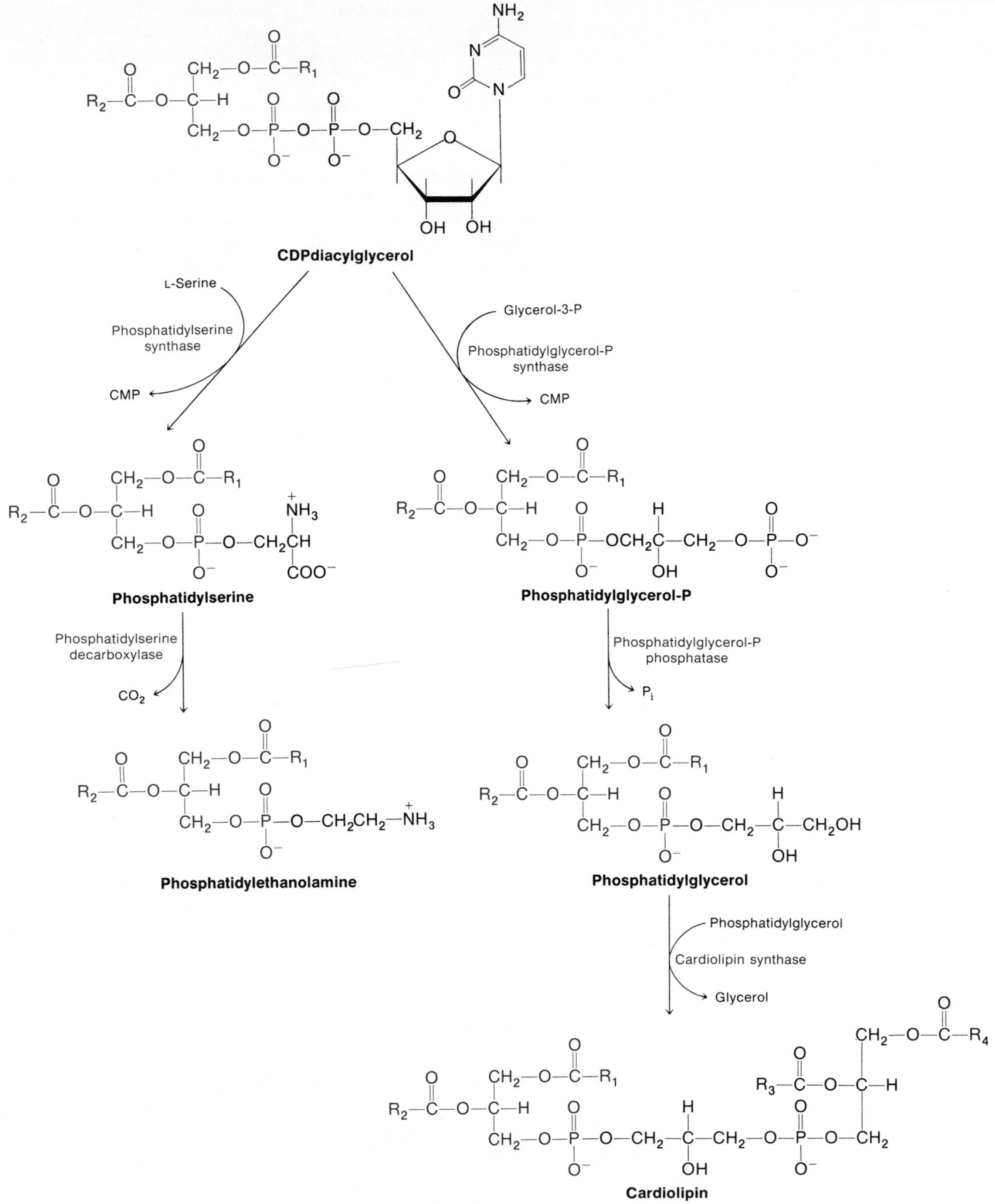

Figure 22–1 *(Continued)*

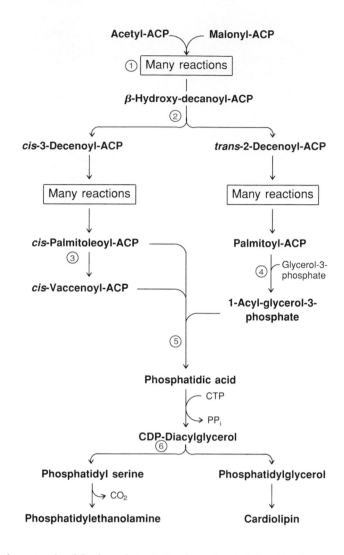

Figure 22–2
Scheme for regulation of phospholipid synthesis in *E. coli*. Control points are numbered and explained in the text.

2. Either the *cis* double bond is introduced at this site during fatty acid synthesis or the ACP derivative is committed to synthesis of a saturated fatty acid.

3. The formation of *cis*-vaccenoyl-ACP by β-ketoacyl-ACP synthase II is thermally regulated. Low temperatures promote and high temperatures inhibit this reaction. Reduced activity for this synthase II is somehow also reflected in reduced synthesis of the initial *cis* double bond.

4. The preferential introduction of a saturated fatty acid at the sn-1 position of glycerol-3-phosphate is regulated by the substrate specificity of glycerol-3-phosphate acyltransferase.

5. 1-Acyl-glycerol-3-phosphate acyltransferase is selective for unsaturated acyl-ACPs.

6. The regulation of head group selection for the *E. coli* phospholipids is not well understood.

Phospholipid Synthesis in Eukaryotes Is More Complex

In eukaryotes there are more types of phospholipids. Just as in bacteria, phosphatidic acid is a central biosynthetic intermediate in eukaryotic cells and the nucleotide CTP is required for phospholipid biosynthesis. In contrast to bacteria and other prokaryotes, eukaryotic cells also synthesize triacylglycerol from phosphatidic acid. Triacylglycerol is the major energy storage compound in many organisms.

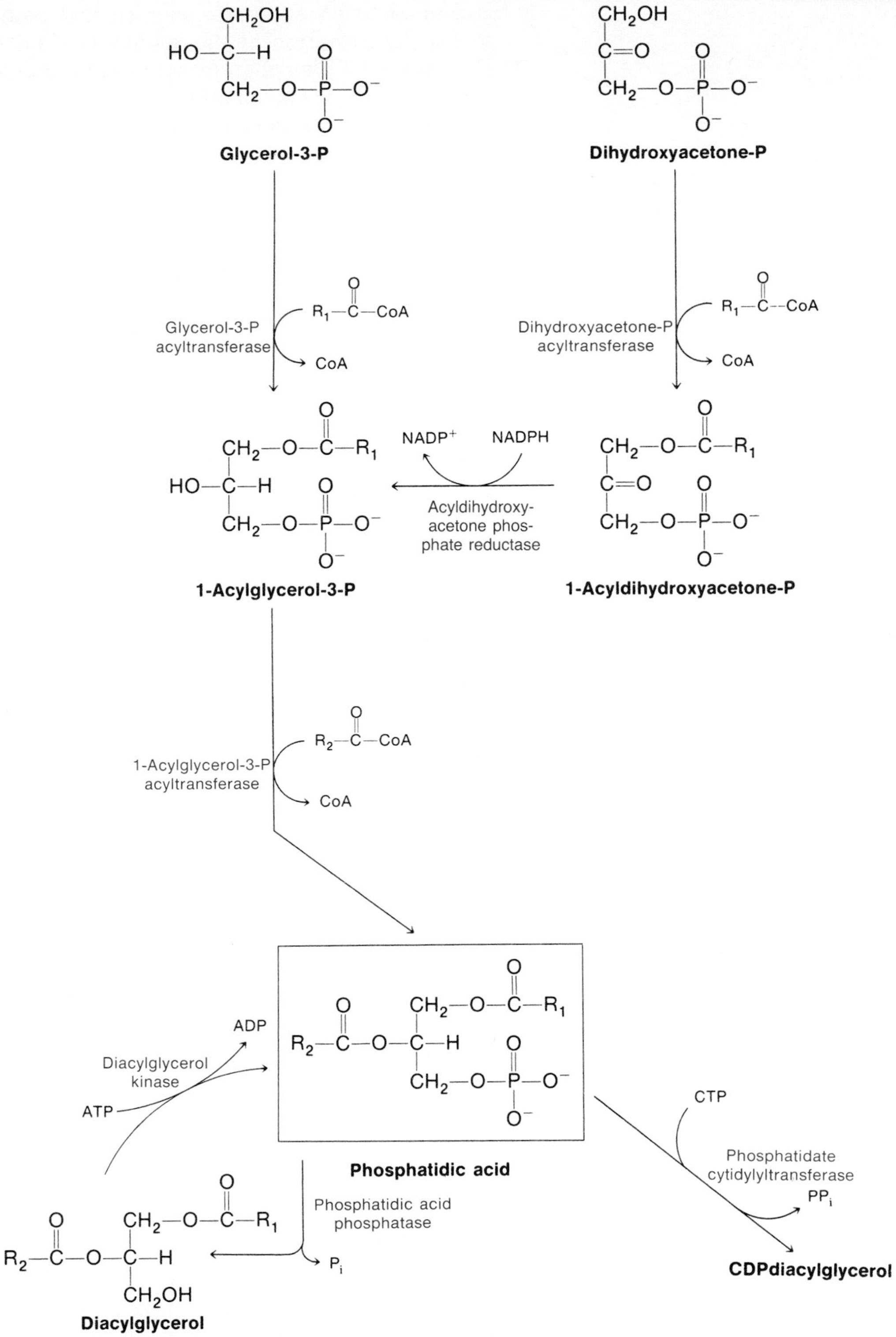

Figure 22–3
Metabolism of phosphatidic acid in eukaryotes.

In eukaryotes there are three pathways for the synthesis of phosphatidic acid (Figure 22–3). One major biosynthetic route is identical to that observed in *E. coli*, where glycerol-3-phosphate is acylated by the successive actions of *glycerol-3-phosphate acyltransferase* and *1-acylglycerol-3-phosphate acyltransferase*. Alternatively, dihydroxyacetone phosphate can serve as a precursor of phosphatidic acid. In this pathway, an initial acylation precedes the reduction of the ketone to yield *1-acylglycerol-3-phosphate*. A third route for phosphatidic acid synthesis is by the phosphorylation of diacylglycerol. Once phosphatidic acid is made, it is rapidly converted to *diacylglycerol* or *CDP-diacylglycerol*, which are metabolized as described in the following discussion.

Biosynthesis of Phosphatidylcholine and Phosphatidylethanolamine. Phosphatidylcholine and phosphatidylethanolamine, which are quantitatively the most important phospholipids in eukaryotic cells (Table 6–5), are derived from diacylglycerol as shown in Figure 22–4. The biosynthesis of phosphatidylcholine begins with the transport of *choline* into the cell. Choline is an essential ingredient in the human diet and cannot be made by animals except indirectly, as noted below. Once inside the cell, the choline is rapidly phosphorylated to *phosphocholine* by the action of *choline kinase*, a cytosolic enzyme. Phosphocholine and CTP form CDP-choline in a reaction catalyzed by *CTP:phosphocholine cytidylyltransferase* (Figure 21–4). The product, CDP-choline, immediately reacts with diacylglycerol in a reaction catalyzed by *CDP-choline:1,2-diacylglycerol phosphocholinetransferase*, an enzyme localized on the endoplasmic reticulum.

The biosynthesis of phosphatidylethanolamine (Figure 22–4) proceeds from ethanolamine in a comparable series of reactions catalyzed sequentially by *ethanolamine kinase* (cytosol), *CTP:phosphoethanolamine cytidylyltransferase* (cytosol), and *CDP-ethanolamine:1,2-diacylglycerol phosphoethanolaminetransferase* (endoplasmic reticulum). It appears that choline and ethanolamine phosphorylations are catalyzed by the same kinase. However, the cytidylyltransferases involved in the biosynthesis of CDP-choline and CDP-ethanolamine are distinctly different enzymes, as are the phosphocholine and phosphoethanolamine transferases.

In liver, yeast, and the bacteria *Pseudomonas*, there is a pathway for the conversion of phosphatidylethanolamine to phosphatidylcholine in which methyl groups are transferred from *S-adenosylmethionine* (SAM) to phosphatidylethanolamine in three consecutive reactions (Figure 22–5). The enzyme for all three reactions, *phosphatidylethanolamine-N-methyltransferase*, is a small protein ($M_r \cong 20,000$) located in the endoplasmic reticulum. This methylation of phosphatidylethanolamine, together with the subsequent degradation of phosphatidylcholine, is the only known mechanism by which liver can produce choline. The mechanism of methylation by SAM is discussed in Box 22–B.

Lung tissue manufactures a specialized species of phosphatidylcholine, *dipalmitoylphosphatidylcholine*, in which palmitic acid is the fatty acyl substituent on both the *sn*-1 and *sn*-2 positions of the glycerol backbone. This species of phosphatidylcholine is the major component (approximately 50–60%) of *lung surfactant*. This substance is mostly lipid, with less than 20% protein, and maintains surface tension in the lung alveoli so that they do not collapse when air is expelled. The secretion of lung surfactant is defective in newborns with *respiratory distress syndrome*.

The probable pathway for the synthesis of dipalmitoylphosphatidylcholine is illustrated in Figure 22–6. The starting species of phosphatidylcholine is made by the CDP-choline pathway (see Figure 22–4). The fatty acid at the *sn*-2 position is hydrolyzed by *phospholipase A₂*, and the lysophosphatidylcholine is reacylated with palmitoyl-CoA. This modification allows alteration of the properties of the phospholipid without resynthesis of the entire molecule. The synthesis of dipalmitoylphospha-

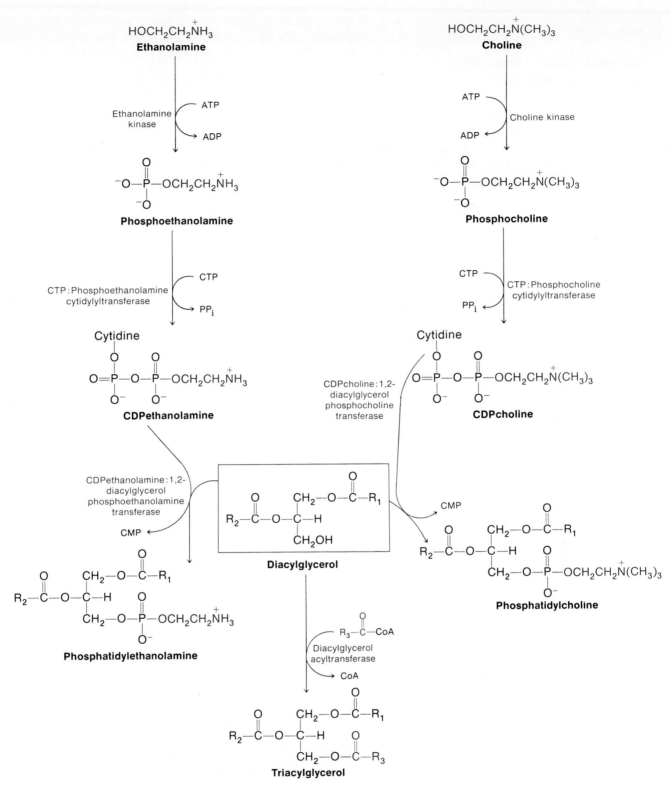

Figure 22–4

Metabolism of diacylglycerol in eukaryotes.

tidylcholine in lung tissue is a specific example of the modulation of the fatty acid composition of a phospholipid. The deacylation-reacylation reaction also occurs in other tissues and provides an important route for the introduction of polyunsaturated fatty acids into the sn-2 position of phospholipids (Figure 22–7).

BOX 22-B

The Mechanism of Methylation by *S*-Adenosylmethionine

S-Adenosylmethionine (SAM) is the primary methylating agent in the cell. Despite the importance of SAM, we have had only a single encounter with this cofactor so far. It is not considered a coenzyme, so we did not discuss it in Chapter 11. Only in Chapter 19 did we see it being formed and functioning in the degradative pathway from methionine to propionyl-CoA (see Figure 19–30). SAM also methylates many other compounds, including the ϵ amino group of lysine, the guanidino group of arginine, the nitrogens of the imidazole side chain of histidine, and the bases of DNA and RNA. SAM is discussed here because of its role in the remarkable trimethylation of phosphatidylethanolamine to form phosphatidylcholine.

The positive charge on the sulfur atom of *S*-adenosylmethionine makes SAM a powerful alkylating agent. This is because the plus charge on the sulfur converts the *S*-adenosylhomocysteine moiety into an excellent leaving group. Alkylation occurs as a bimolecular substitution reaction.

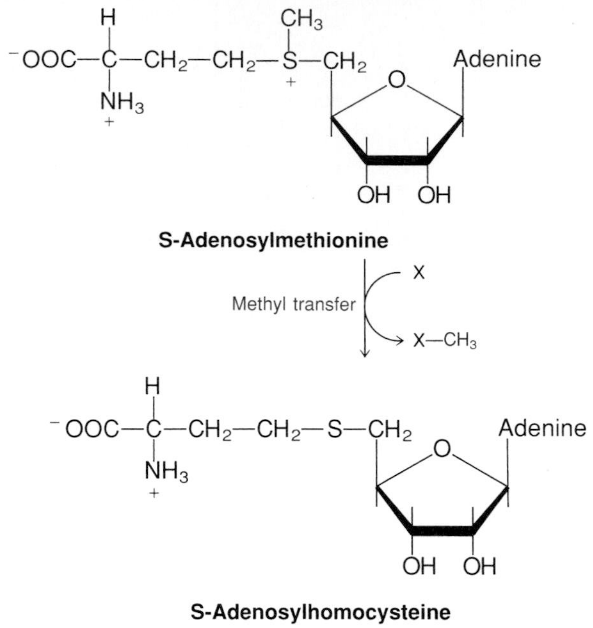

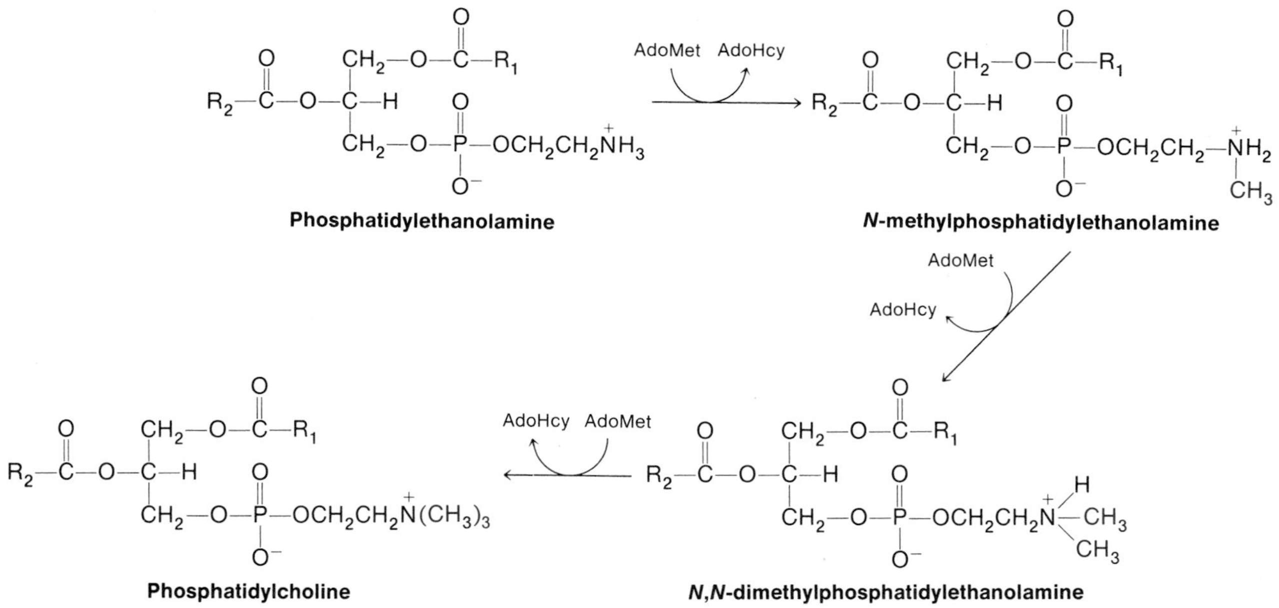

S-Adenosylmethionine

Methyl transfer

S-Adenosylhomocysteine

Figure 22–5

Conversion of phosphatidylethanolamine to phosphatidylcholine by phosphatidylethanolamine-N-methyltransferase. (AdoMet is a standard abbreviation for S-adenosyl-L-methionine. Recall that SAM may also be used. AdoHcy is a standard abbreviation for S-adenosyl-L-homocysteine.) The structures of AdoMet and AdoHcy and the general mechanism of the methylation reaction are presented in Box 22–B.

Biosynthesis of Phosphatidylserine, Phosphatidylglycerol, and the Inositol Phospholipids. Phosphatidylserine is made in mammalian cells by a Ca^{2+}-mediated base exchange enzyme found on the endoplasmic reticulum (Figure 22–8). The phospholipid substrate for the base exchange enzyme can be phosphatidylethanolamine or other phospholipids. Phosphatidylserine can

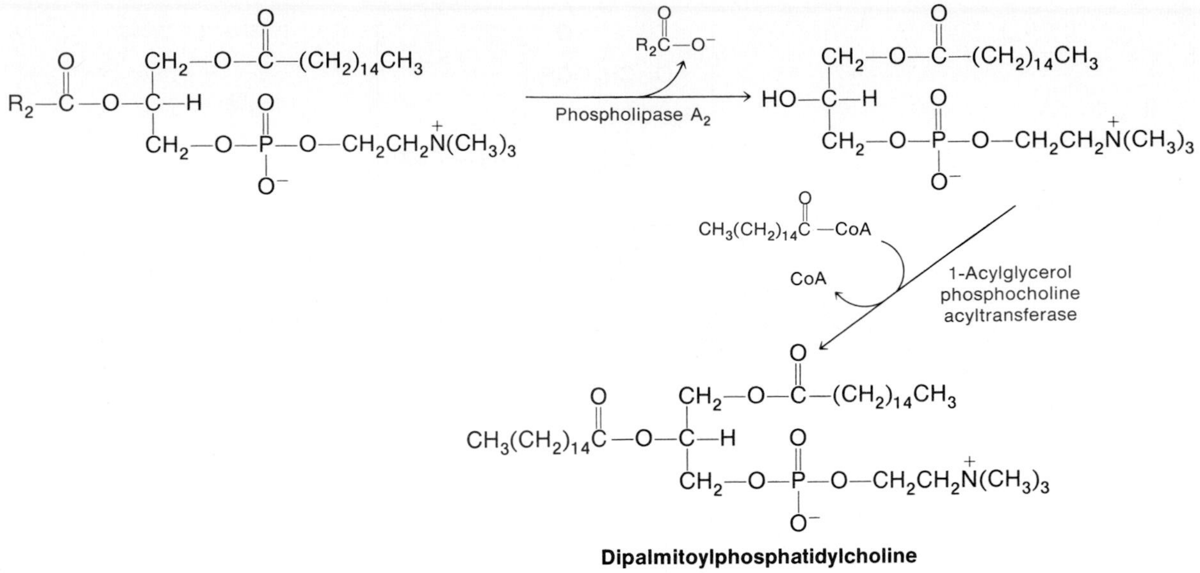

Dipalmitoylphosphatidylcholine

Figure 22–6
Biosynthesis of dipalmitoylphosphatidylcholine. R_2 is usually an unsaturated fatty acid. Thus this two-step reaction usually results in the replacement of an unsaturated by a saturated fatty acid at the C-2 position on the glycerol backbone.

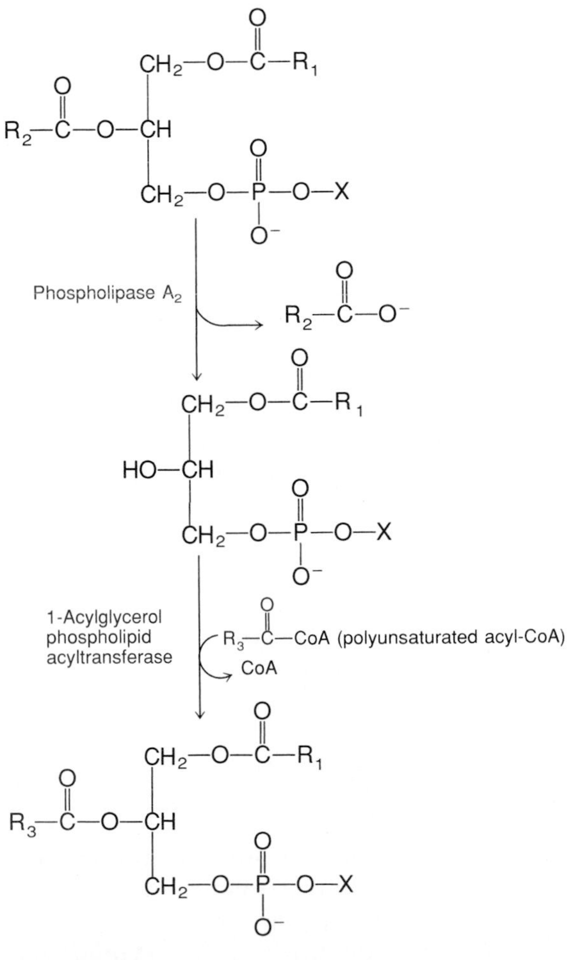

Figure 22–7
A mechanism for the enrichment of phospholipids with arachidonic acid ($20:4^{\Delta5,8,11,14}$) or other polyunsaturated fatty acids.

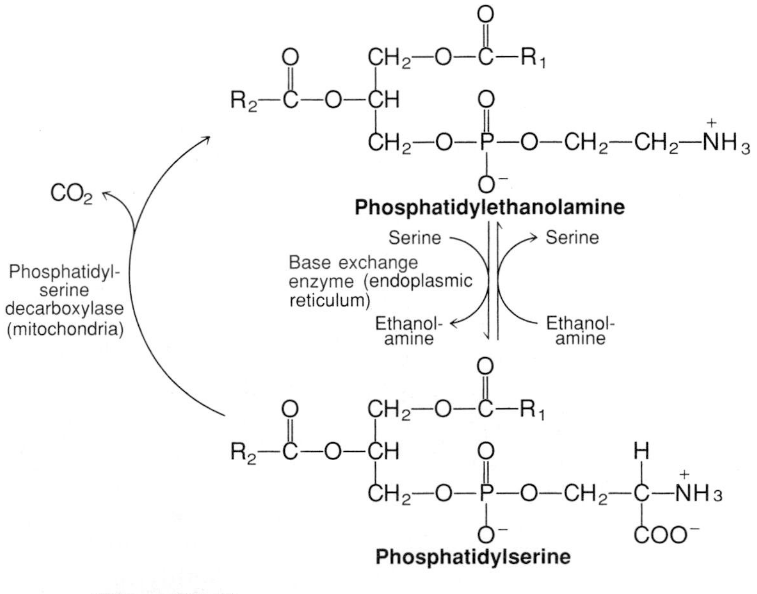

Phosphatidylethanolamine

Phosphatidylserine

Figure 22–8
Phosphatidylserine biosynthesis is catalyzed by a base exchange enzyme on the endoplasmic reticulum. Decarboxylation of phosphatidylserine occurs in mitochondria.

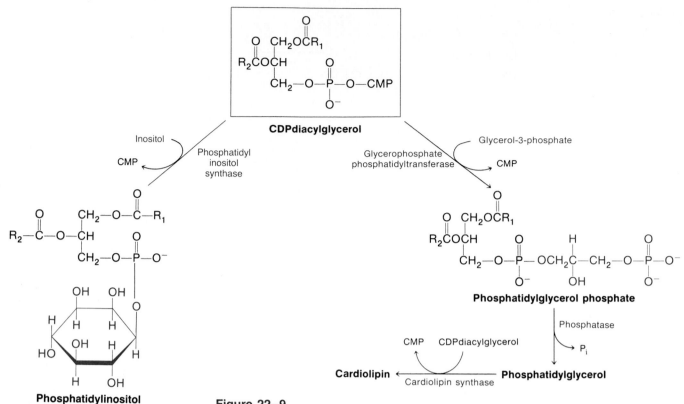

Figure 22–9
Biosynthesis of phosphatidylglycerol, phosphatidylinositol, and cardiolipin in eukaryotes. Glycerophosphate phosphatidyltransferase is found in endoplasmic reticulum and mitochondria. Cardiolipin synthase and the lipid cardiolipin are found exclusively in the mitochondria.

subsequently be decarboxylated to phosphatidylethanolamine in the mitochondria by *phosphatidylserine decarboxylase*. Hence, a cycle can occur which has the net effect of converting serine to ethanolamine, and this cycle appears to be a major mechanism for the biosynthesis of ethanolamine (Figure 22–8) in eukaryotic cells. There is no known enzyme capable of the direct decarboxylation of serine to ethanolamine.

In yeast, phosphatidylserine is made by a reaction of serine with CDP-diacylglycerol as in *E. coli* (Figure 22–1). The base exchange enzyme has not been detected in this organism.

CDP-diacylglycerol is also an intermediate for the biosynthesis of phosphatidylinositol and phosphatidylglycerol in both mammalian cells and yeast (Figure 22–9). Inositol is a hexahydroxycyclohexane that is biosynthesized from D-glucose-6-phosphate. Phosphatidylglycerol synthesis occurs in mitochondria and the endoplasmic reticulum. In mitochondria, CDP-diacylglycerol condenses with phosphatidylglycerol to produce cardiolipin (Figure 22–9), whereas in *E. coli*, two phosphatidylglycerols condense to form cardiolipin (see Figure 22–1). Cardiolipin is found primarily in the mitochondria in mammalian cells (see Table 6–5).

Phosphatidylinositol accounts for about 5% of the lipids present in membranes (see Table 6–5). Also present, at much lower concentrations (1–3% of phosphatidylinositol), are phosphatidylinositol-4-phosphate and phosphatidylinositol-4,5-P_2 (Figure 22–10). It has been known for a long time that the inositol-containing phospholipids turn over (are synthesized and then degraded) much more rapidly than other phospholipids after the addition of certain hormones or neurotransmitters to cells or tissue homogenates. However, until relatively recently, the function of the rapid turnover

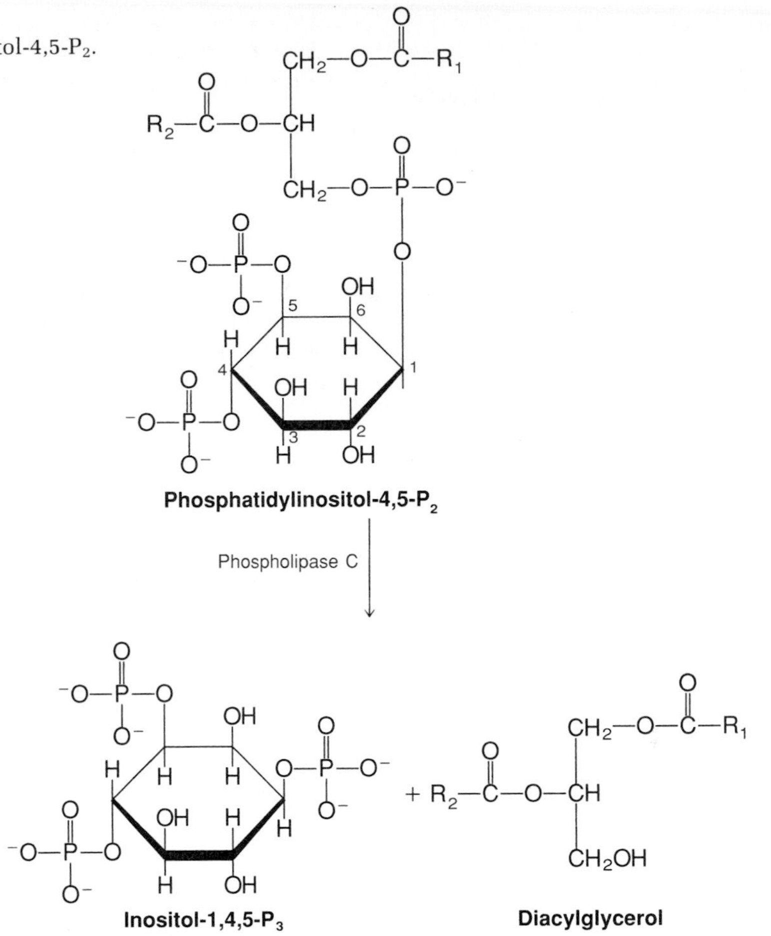

Figure 22–10

Phospholipase C degradation of phosphatidylinositol-4,5-P$_2$.

Phosphatidylinositol-4,5-P$_2$

Phospholipase C

Inositol-1,4,5-P$_3$ **Diacylglycerol**

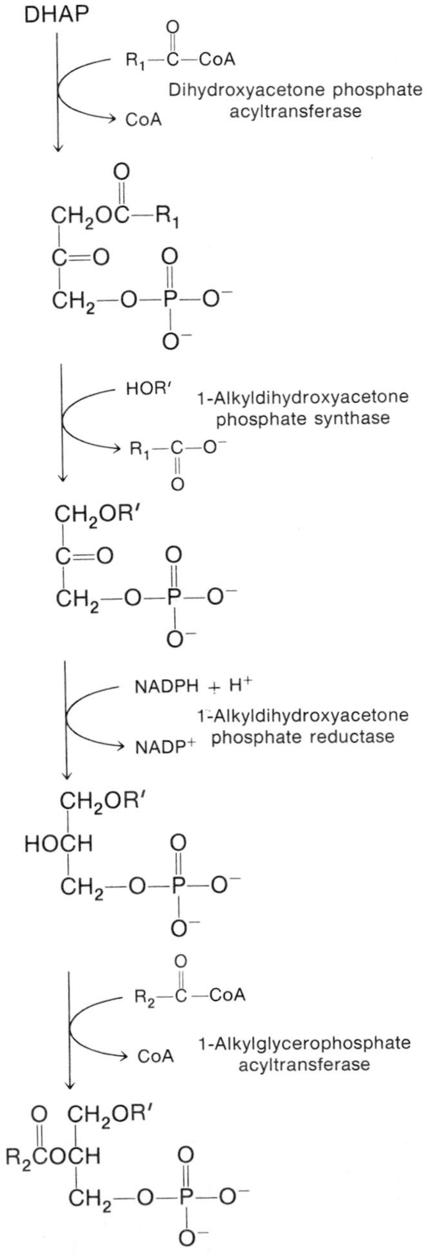

Figure 22–11

Biosynthesis of the alkyl ether species of phosphatidic acid.

of the inositol lipids was unknown. Now it is known that phosphatidylinositol-4,5-P$_2$ is degraded to inositol-1,4,5-P$_3$ and diacylglycerol by *phospholipase C* (Figure 22–10). Each of these degradation products has important regulatory functions (see Chapter 31). *The inositol-1,4,5-P$_3$ is involved in mobilization of calcium from intracellular stores* (endoplasmic reticulum) *and the rise in cytosolic calcium activates various enzymes*. For example, treatment of liver cells with the hormone vasopressin causes the breakdown of phosphatidylinositol-4,5-P$_2$ (see Figure 31–16). The inositol-1,4,5-P$_3$ causes an increase in cytosolic calcium, which in turn activates glycogen phosphorylase. The other product of the phospholipase C reaction, diacylglycerol, activates an enzyme called *protein kinase C*. This enzyme also requires calcium and phosphatidylserine for activity. Protein kinase C phosphorylates an extremely diverse group of proteins. The role of inositol-1,4,5-P$_3$ as a mediator of hormone action is discussed further in Chapter 31. The phosphatidylinositol cycle is described in Box 22–C.

Biosynthesis of Alkyl and Alkenyl Ethers. As discussed in Chapter 5, some phospholipids contain either an *O*-alkyl or an *O*-alkenyl ether species at the sn-1 position, instead of the more common acyl ester linkage. The biosynthetic pathway for the alkyl ether species of phosphatidic acid in eukaryotes is summarized in Figure 22–11. The initial step is the acylation of dihydroxyacetone phosphate, a reaction already discussed in connection with phosphatidic acid biosynthesis (see Figure 22–3). Subsequently, an exchange reaction replaces the 1-acyl group with an alkyl group derived from an alcohol (Figure 22–11). The long-chain alcohol used in this reaction is formed by

BOX 22–C

The Phosphatidylinositol Cycle

A great deal of attention has been focused in the past few years on the metabolism of the inositol-containing lipids because of the second messenger role of the hydrolysis products of phosphatidylinositol-4,5-P_2—diacylglycerol and inositol-1,4,5-P_3. Both of these compounds can be recycled to phosphatidylinositol-4,5-P_2 as shown in the illustration.

The diacylglycerol is phosphorylated by ATP with the enzyme diacylglycerol kinase to give phosphatidic acid. This reacts with CTP to yield CDP-diacylglycerol as described in the text.

The inositol-1,4,5-P_3 has two known fates. In one, it can be degraded to inositol-P_2, inositol-phosphate, and inositol through the sequential action of cellular phosphatase activities. This inositol can then react with CDP-diacylglycerol to yield phosphatidylinositol. The cycle is completed by the subsequent phosphorylations of phos-

phatidylinositol to give phosphatidylinositol-4-phosphate and phosphatidylinositol-4,5-P_2.

A second fate of inositol-1,4,5-P_3 has recently been discovered. A soluble kinase can phosphorylate it to inositol-1,3,4,5-P_4. There is evidence that this inositol tetraphosphate is involved in the stimulation of calcium entry into cells from the extracellular environment, whereas inositol-1,4,5-P_3 appears to function in the mobilization of the calcium stores from inside the cell. The inositol-1,3,4,5-P_4 is converted to inositol-1,3,4-P_3 by a 5-phosphomonoesterase. Cellular phosphatases are believed to degrade the inositol-1,3,4-P_3 to inositol. Thus the metabolism of the inositol lipids and the inositol phosphates is much more extensive than was anticipated. It is likely that there are still additional inositol derivatives and functions to be discovered.

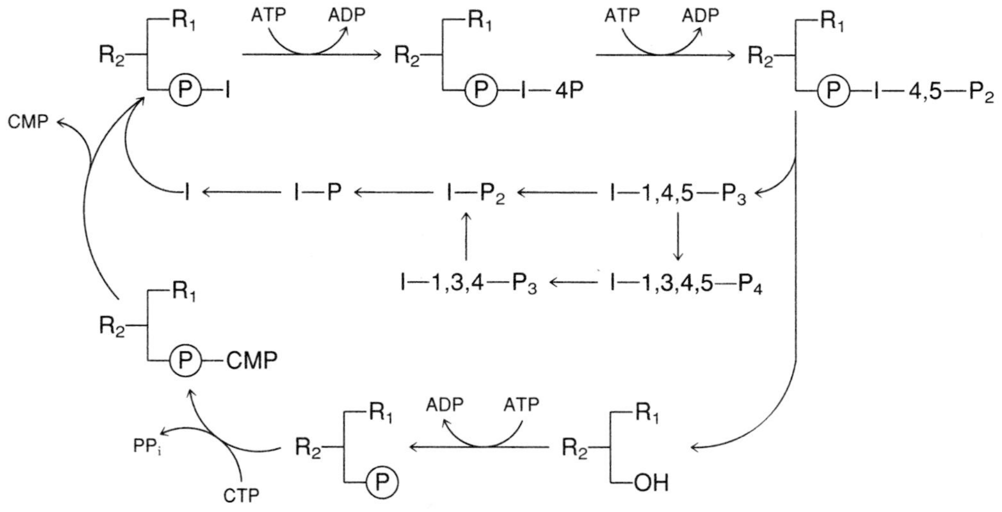

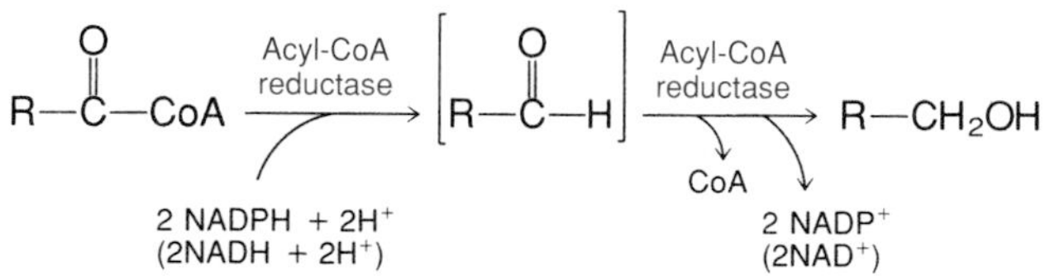

Figure 22–12
Acyl-CoA reductase catalyzes the formation of long-chain alcohols from fatty acyl-CoA.

reduction of an acyl-CoA with two molecules of reduced pyridine nucleotide (NADPH or NADH) (Figure 22–12). The subsequent reduction of the ketone and acylation of the two-hydroxyl group are similar reactions to those previously noted (see Figure 22–3). Once the 1-alkyl ether derivative of phosphatidic acid is formed, it is used for the synthesis of other phospholipids, as already discussed.

In many tissues, 1-alkyl-1-acylphosphatidylethanolamine can be desaturated by an endoplasmic reticulum enzyme, _1-alkyl-1-acylglycerophosphoethanolamine desaturase_, to yield the corresponding unsaturated derivative called a _plasmalogen_ (Figure 22–13). This enzyme requires O_2, NADH, and cytochrome b_5, the same cofactors required for the desaturation of ste-

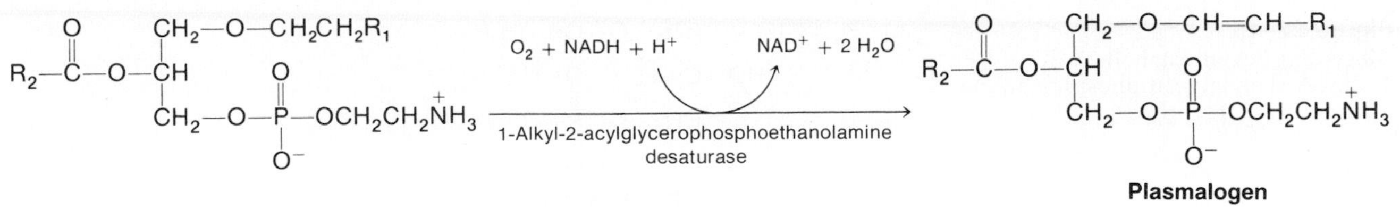

Figure 22–13
Formation of plasmalogens. A specific enzyme found in the endoplasmic reticulum desaturates the alkyl group and is specific for the ethanolamine phospholipid. The product of this reaction is called plasmalogen and is abundant in heart tissue.

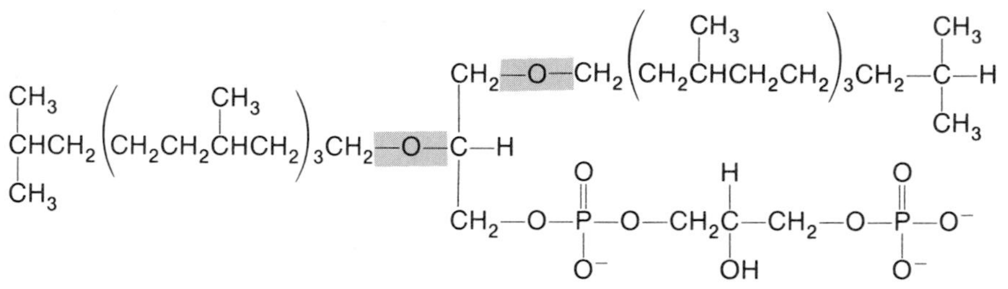

Figure 22–14
Structure of a dialkyl phosphatidylglycerol phosphate from *Halobacterium cutirubrum*. This is a very unusual phospholipid and may be related to this bacterium's capacity to grow in high (4 M) salt concentrations.

aroyl-CoA (see Figure 21–11). However, stearoyl-CoA desaturase appears to be different from the alkyl-acylglycerophosphoethanolamine desaturase. Although the plasmalogens are minor constituents in many tissues, nearly 50 per cent of the phospholipids in heart tissue contain the alkenyl ether at position sn-1. The biological function of the ether phospholipids remains an enigma.

The ether phospholipids also are found in microorganisms, notably the eukaryotic protozoa and the halophilic bacteria. *Halobacterium cutirubrum*, which requires 4 M NaCl for growth, produces the unusual phosphatidylglycerol phosphate shown in Figure 22–14. The alkyl ether bonds are stable to alkaline hydrolysis and relatively stable to acid hydrolysis when compared with esters (alkenyl ethers are more labile than alkyl ethers to acid hydrolysis). Perhaps the ability of *H. cutirubrum* to survive in such high concentrations of salt is related to the stability of these phospholipids.

Medical research has led to the discovery of an unusual species of phosphatidylcholine that possesses extremely potent biological activity. This compound, *1-alkyl-2-acetylglycerophosphocholine* (Figure 22–15), also known as *platelet activating factor*, has a striking ability to decrease blood pressure in hypertensive rats and, at concentrations of 10^{-10} M, will cause blood platelets to aggregate. The biosynthesis of this lipid from 1-alkyl-2-lysophosphatidylcholine is catalyzed by *acetyl-CoA:1-alkyl-2-lyso-*

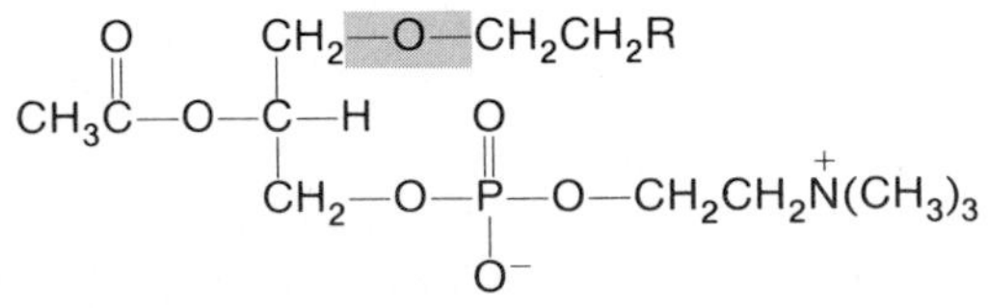

Figure 22–15
Structure of 1-alkyl-2-acetylglycerophosphocholine.

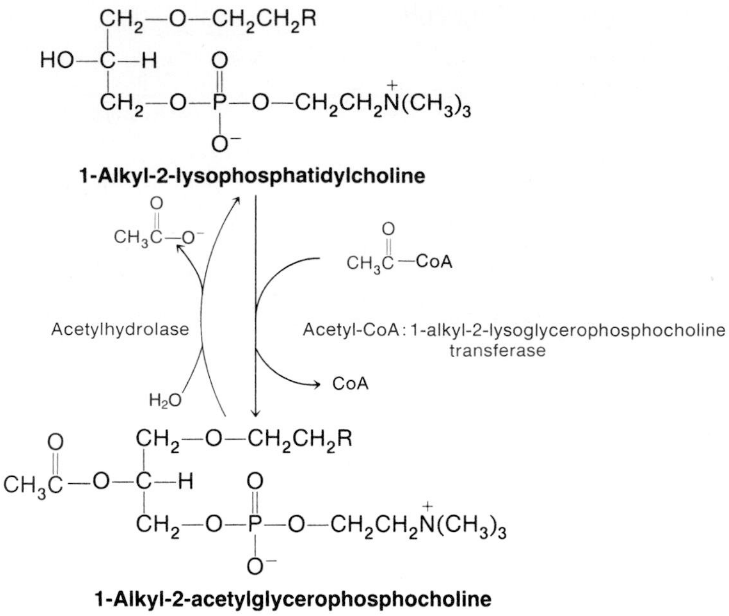

Figure 22–16
Biosynthesis and catabolism of
1-alkyl-2-acetylglycerophosphocholine.

1-Alkyl-2-lysophosphatidylcholine

Acetylhydrolase

Acetyl-CoA:1-alkyl-2-lysoglycerophosphocholine transferase

1-Alkyl-2-acetylglycerophosphocholine

2-Monoacylglycerol

Monoacylglycerol acyltransferase

Diacylglycerol

Diacylglycerol acyltransferase

Triacylglycerol

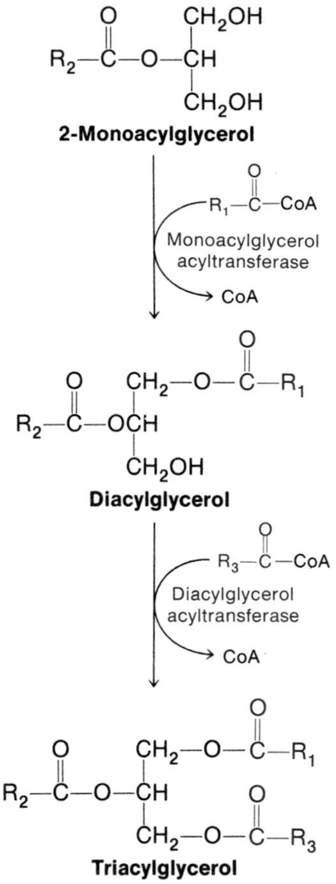

Figure 22–17
Biosynthesis of triacylglycerol in
intestinal epithelial cells of
nonruminant animals.
2-Monoacylglycerol is produced in the
lumen of the intestine from ingested
triacylglycerols, and is absorbed by
the intestinal epithelium.
Triacylglycerol is re-formed by two
transacylations of fatty acids from acyl-
CoAs.

glycerophosphocholine transferase (Figure 22–16). Significant activity of this enzyme has been found in the endoplasmic reticulum from rat spleen, kidney, and other tissues. An enzyme that cleaves the acetyl group, *acetylhydrolase* (Figure 22–16), has been found in cytosol from kidney and some other tissues. The discovery of this unusual species of phosphatidylcholine opens a new and unexpected chapter in the story of phospholipid metabolism.

Biosynthesis of Triacylglycerols. Triacylglycerol is an important glycerolipid made from diacylglycerol in liver, intestine, and adipose tissues. It is in the triacylglycerols that fatty acids are stored as an energy source. The biosynthetic pathway in liver and adipose tissues involves the acylation of diacylglycerol by *diacylglycerol acyltransferase* (Figure 22–4). This enzyme is tightly bound to the cytosolic side of the endoplasmic reticulum.

A different biosynthetic pathway for triacylglycerol synthesis is found in the intestine of nonruminant animals (Figure 22–17). Dietary triacylglycerol is degraded by *pancreatic lipase* in the intestinal lumen to 2-monoacylglycerol and fatty acids. These products enter the epithelial cells of the small intestine, where triacylglycerols are resynthesized by direct acylation of monoacylglycerol and diacylglycerol, as shown in Figure 22–17. This pathway accounts for 80 per cent of the triacylglycerol made in this tissue. The remainder is made from diacylglycerol synthesized from glycerol phosphate and dihydroxyacetone phosphate, as described previously (Figure 22–4). In ruminant animals, the bacteria in the rumen degrade the monoacylglycerol to glycerol and fatty acid. Hence the pathway for acylation of monoacylglycerol (Figure 22–17) is not important in these animals and triacylglycerol is made from fatty acids and glycerol phosphate or dihydroxyacetone phosphate.

Regulation of Glycerolipid Synthesis in the Liver Favors Structural Lipids over Energy-storage Lipids

Our knowledge about the regulation of glycerolipid synthesis in eukaryotic cells is in a rudimentary stage. The present discussion will be restricted to

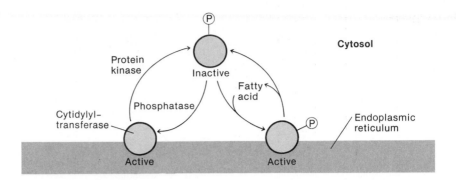

Figure 22–18
Proposed mechanism for regulation of CTP-phosphocholine cytidylyltransferase in rat liver. Phosphate removal or fatty acid binding encourages membrane binding. Specific phosphorylation discourages membrane binding.

the mechanisms in rat liver, about which some definite information is available.

It is well documented that a plentiful supply of fatty acids promotes glycerolipid synthesis. In situations of energy excess (e.g., after a meal), the supply of fatty acids tends to increase as a result of both diet and biosynthesis from carbohydrate. Increased insulin levels after a meal stimulate the synthesis of malonyl-CoA, which promotes fatty acid synthesis and inhibits fatty acid oxidation. Consequently, acyl-CoA tends to be available for both the synthesis of phosphatidic acid (Figure 22–3) and the acylation of diacylglycerol (Figure 22–4).

The rate of phosphatidylcholine biosynthesis appears to be regulated principally by the activity of CTP:phosphocholine cytidylyltransferase (Figure 22–4). This enzyme is a tetramer whose subunits each have a molecular weight of 45,000. The enzyme is recovered in both the cytosolic and endoplasmic reticulum fractions of rat liver homogenates. The enzyme in the cytosol is inactive. Translocation to the endoplasmic reticulum results in activation of the cytidylyltransferase (Figure 22–18). Two mechanisms appear to be important for regulation of the translocation of the cytidylyltransferase. Cyclic-AMP-dependent protein kinase is believed to phosphorylate the enzyme and cause its release from the membrane (Figure 22–18), a process that can be reversed by a phosphatase (Figure 22–18). In addition, fatty acid promotes binding of the cytidylyltransferase to the endoplasmic reticulum (in spite of the phosphate group), while removal of the fatty acid releases the enzyme into the cytosol (Figure 22–18). This novel mechanism for control of enzyme activity allows the liver to modulate phosphatidylcholine biosynthesis in a rapid and reversible fashion.

The conversion of phosphatidic acid to diacylglycerol by *phosphatidic acid phosphatase* is also regulated. The activity of this phosphatase increases and decreases with the rate of synthesis of triacylglycerol in a number of different metabolic situations. Like the cytidylyltransferase of phosphatidylcholine biosynthesis, phosphatidic acid phosphatase appears to be active when associated with membranes and inactive when free in the cytosol. And, as is true with cytidylyltransferase, an increase in the supply of fatty acids causes binding of cytosolic enzyme to endoplasmic reticulum where the phosphatase is activated. The process is reversed upon removal of the fatty acids. However, unlike the supply of cytidylyltransferase, the amount of phosphatidic acid phosphatase is also controlled by regulating the rate of enzyme biosynthesis. A class of steroid hormones (glucocorticoids) is involved (see Chapter 31).

We do not understand very well what regulates the flux of diacylglycerol to phosphatidylcholine, phosphatidylethanolamine, or triacylglycerol. Apparently, at least in liver, the requirements for the synthesis of the essential membrane components, phosphatidylcholine and phosphatidylethanolamine, are met before an appreciable amount of energy storage lipid

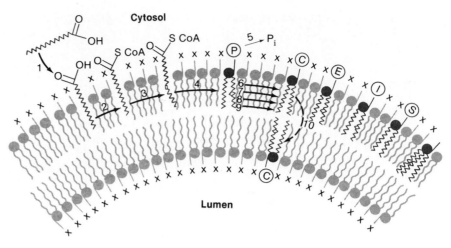

Figure 22–19

Glycerolipid synthesis on the endoplasmic reticulum from rat liver. Fatty acids are inserted into the cytoplasmic surface of the endoplasmic reticulum (1), and activated to form acyl-CoA thioesters (2). The acyl chains may be elongated and/or desaturated (3). Glycerol-phosphate undergoes acyl-CoA-dependent esterification to form phosphatidic acid (4). The action of phosphatidic acid phosphatase (5) forms diacylglycerols that are converted to phosphatidylcholine and phosphatidylethanolamine by acquisition of phosphocholine and phosphoethanolamine polar head groups (6). Phosphatidylserine synthesis occurs by base exchange (7). Triacylglycerol synthesis occurs by esterification of diacylglycerol (8). CDP-diacylglycerol is an intermediate in the synthesis of phosphatidylinositol (9). Once formed, the glycerolipids may move to the lumenal surface of the endoplasmic reticulum (10). (C = choline; E = ethanolamine; I = inositol; S = serine; X = polar head group C, E, I, or S; P = PO$_4$.) (From R. M. Bell, L. M. Ballas, and R. A. Coleman, Lipid topogenesis. *J. Lipid Res.* 22:391, 1981.)

(triacylglycerol) is made. The regulation of the synthesis of these lipids, as well as the synthesis of phosphatidylserine, phosphatidylglycerol, cardiolipin, and the inositol phospholipids, is intimately related to the maintenance, proliferation, and functions of the plasma membrane and internal cellular membranes such as those of the endoplasmic reticulum and the Golgi.

Mechanisms for Final Distribution of Lipids Between Membranes Are Still Unclear

The final reactions for the biosynthesis of phosphatidylcholine, phosphatidylethanolamine, phosphatidylserine, and phosphatidylinositol occur on the cytosolic surface of the endoplasmic reticulum in rat liver (Figure 22–19). Recently it has been shown that the Golgi membranes also contain some of the enzymes for phospholipid biosynthesis. Phosphatidylglycerol and cardiolipin are synthesized in and remain largely in the mitochondria. Two presently unsolved questions arise: (1) Since phospholipid synthesis occurs on the cytosolic side of the endoplasmic reticulum (Figure 22–19), and phospholipids are found on both leaflets of the bilayer, how do these lipids reach the inner leaflet of the bilayer? (2) How does the cell sort and transport phospholipids from the site of synthesis to other membranes in the cell?

The first question was addressed in Chapter 6. In answer to the second, one possibility is that phospholipids are translocated within the cell as part of membranes. In this view, membrane vesicles with specific membrane proteins bud from the endoplasmic reticulum and move through the cytoplasm, eventually fusing with the membrane of a particular organelle. Specific membrane proteins present on the surface of the vesicle may direct the vesicle to a specific organelle or permit fusing upon collision. Another possible mechanism for intracellular lipid transfer is provided by phospholipid ex-

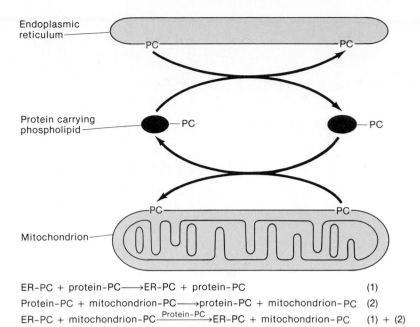

Figure 22–20
An example of phospholipid exchange between two membranes. (ER = endoplasmic reticulum; PC = phosphatidylcholine.)

ER-PC + protein-PC $\longrightarrow$ ER-PC + protein-PC   (1)

Protein-PC + mitochondrion-PC $\longrightarrow$ protein-PC + mitochondrion-PC   (2)

ER-PC + mitochondrion-PC $\xrightarrow{\text{Protein-PC}}$ ER-PC + mitochondrion-PC   (1) + (2)

change proteins found in the cytosol of eukaryotic cells. These proteins catalyze the exchange of phospholipid molecules between two membranes. For example, a phosphatidylcholine molecule on the endoplasmic reticulum exchanges with a molecule of phosphatidylcholine bound to the exchange protein. The protein moves to a mitochondrion, where another exchange occurs (Figure 22–20). Thus a phosphatidylcholine molecule can be moved from the endoplasmic reticulum to a mitochondrion. Some isolated exchange proteins show no specificity for the polar head group, while other proteins are highly specific.

The phosphatidylcholine exchange protein from beef liver has been purified and partially sequenced. The protein is present in very small amounts in the cells, has a molecular weight of 28,000, and is highly specific for phosphatidylcholine. There appears to be one hydrophobic, noncovalent binding site for phosphatidylcholine per molecule of protein. The amino acid sequence at this site is

$$R_1-\overset{\overset{\textstyle H}{|}}{N}-Val-Phe-Met-Tyr-Tyr-Phe-\overset{\overset{\textstyle O}{\|}}{C}-R_2.$$

Since most phospholipids are synthesized on the endoplasmic reticulum, a mechanism is required for transfer of these lipids to other parts of the cell, such as the nucleus or cytoplasmic membrane. The exchange proteins are candidates for this job. However, for the membranes to grow, a *net* transfer of lipid is required, and this has been difficult to demonstrate with these proteins. At a minimum, the proteins could function in renewal of the phospholipids in various membranes by exchange, since the phospholipids in eukaryotic cells do turn over.

Phospholipases Degrade Phospholipids

Enzymes that degrade phospholipids are called *phospholipases*. They are classified according to the bond cleaved in a phospholipid, as shown in Figure 22–21. Phospholipases A_1 and A_2 selectively remove fatty acids from the sn-1 and sn-2 positions, respectively. Phospholipase C cleaves between glycerol and the phosphate moieties; phospholipase D hydrolyzes the head group moiety, X, from the phospholipid. The lysophospholipids are de-

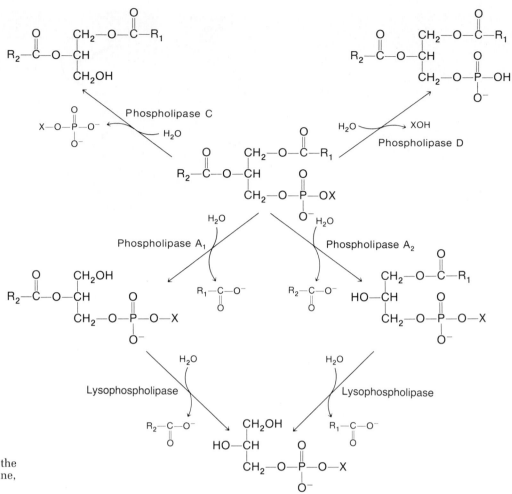

Figure 22–21
Reactions catalyzed by phospholipases. X can be any of the head groups: choline, ethanolamine, serine, glycerol, or inositol.

graded by _lysophospholipases_. Phospholipases are found in all types of cells and in various subcellular locations within eukaryotic cells. Some of these enzymes show specificity for particular polar head groups; others are non-specific. The enzymes from pancreatic tissue and snake venom have been studied extensively because of the relatively high concentrations of phospholipases in these sources.

Figure 22–22 shows the structure of phospholipase A_2 from rattlesnake venom determined by Paul Sigler and coworkers. The structure of the pancreatic enzyme has also been determined, and the subunit structures of the two phospholipases are homologous. However, the pancreatic enzyme functions as a monomer, whereas the venom enzyme functions as a dimer. The pancreatic enzyme is made as its zymogen, which has very low activity and can be converted into its active form by the removal of a heptapeptide from the N terminal by _trypsin_. The phospholipase shows great stability, since it is not denatured by 8 M urea, and retains 85% of its activity in a solution of 2% sodium dodecylsulfate. Such stability is an attractive attribute for an enzyme that is involved in phospholipid degradation in the intestine.

The regulation of phospholipase activity has become an area of active investigation as a result of the discovery of the protein _lipocortin_ (M_r = 37,000), an inhibitor of phospholipase A_2. The synthesis of lipocortin is induced by certain steroids called _glucocorticoids_. Lipocortin appears to inhibit the phospholipase by sequestering the phospholipid substrate. The physiological significance of this protein is currently uncertain.

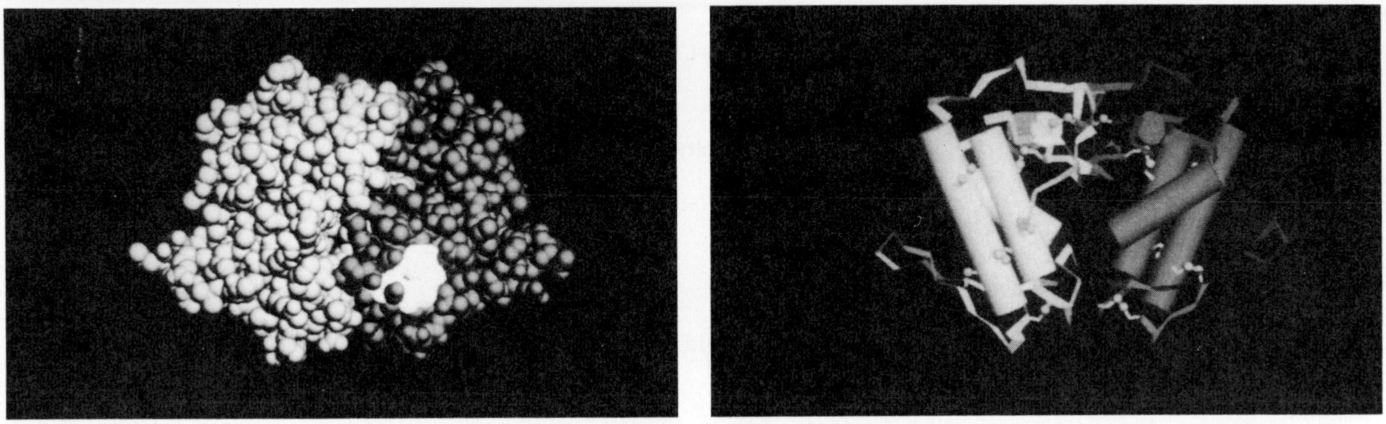

Figure 22–22

The 2.5-Å crystal structure of phospholipase A_2 from the rattlesnake *Crotalus atrox*. The enzyme is a dimer composed of identical subunits of 122 residues. Both views of the enzyme are from the same angle. The two shades denote the two subunits. The white patch of the space-filling model (left) is where the enzyme is believed to bind to phospholipid micelles or membranes. (Photographs kindly supplied by Dr. Paul B. Sigler.) Also see S. Brunie et al., 1985. *J. Biol. Chem.* 260:9742–49.

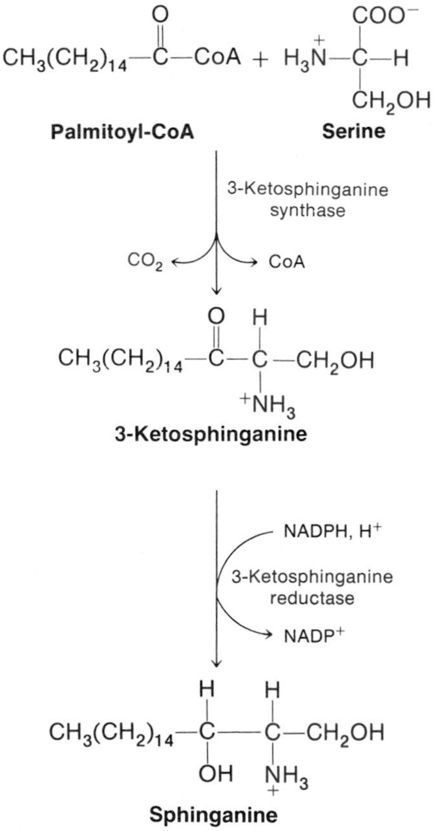

SPHINGOLIPIDS

The common structural feature of sphingolipids is a long-chain, hydroxylated secondary amine (see Table 5–4). These lipids usually occur as components of more complex lipids. Sphingolipids are major components of myelin sheath, a multilayered membranous structure that protects and insulates nerve fibers. They are also found as components of plasma lipoproteins (see Chapter 23). Even eukaryotic microorganisms such as yeast contain sphingolipids, but most prokaryotes do not. The structure of sphingolipids and their location in membranes has been discussed in Chapters 5 and 6, respectively.

Sphingolipids Are Synthesized on Membranes

Sphingosine (sphingenine) and sphinganine are the chief long-chain bases present in sphingolipids. The biosynthesis of sphinganine (Figure 22–23) occurs on the endoplasmic reticulum and involves a condensation of palmitoyl-CoA with serine, catalyzed by *3-ketosphinganine synthase*, and a subsequent reduction of the 3-ketone by *3-ketosphinganine reductase*. The 3-ketosphinganine synthase contains pyridoxal phosphate as an essential cofactor. Once the long-chain base is formed, it rapidly reacts with the acyl-CoA to form ceramide (Figure 22–24).

Although sphingosine is the most abundant long-chain base, the origin of the Δ^4-*trans* double bond has been an enigma for many years. Evidence now favors two alternative biosynthetic pathways in different organisms. In yeast, palmitoyl-CoA can be desaturated to yield *trans*-2-palmitoleoyl-CoA, which condenses with serine to yield sphingosine (Figure 22–25). Once the sphingosine is formed, it is quickly acylated to form ceramide (Figure 22–25). In mouse cells grown in cell culture, the *trans* double bond of sphingosine is introduced after sphinganine has been N-acylated.

Sphingomyelin is synthesized by the transfer of the phosphocholine moiety of phosphatidylcholine to ceramide (Figure 22–26). The enzyme is membrane-bound and is located on either the Golgi membrane or the plasma membrane.

The synthesis of glycosphingolipids shows many similarities to the synthesis of glycoproteins (see Chapter 20). The biosynthesis of some of the

Figure 22–23

Biosynthesis of sphinganine. In the first reaction we see another example of CoA serving as an excellent leaving group when the carbonyl carbon of an acyl-CoA is attacked by a nucleophilic agent. In this case the nucleophilic agent is a carbanion formed from decarboxylation of serine. See Box 22–A for the general mechanism by which CoA fuctions as a leaving group.

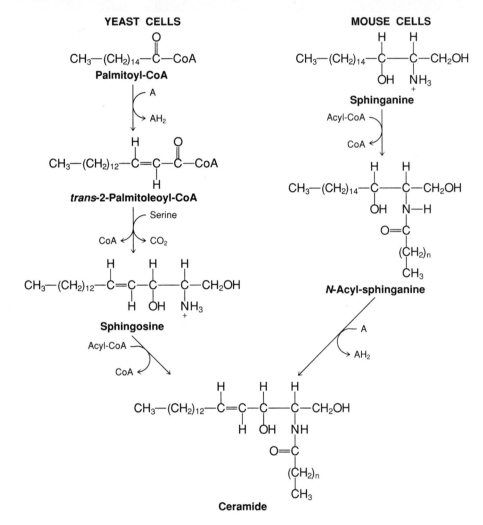

Figure 22–24
Biosynthesis of ceramide. Sphinganine, once formed, reacts rapidly with acyl-CoA to form ceramide.

Figure 22–25
Formation of ceramide containing sphingosine. Introduction of the 4-*trans* double bond in the sphingosine base occurs at the level of acyl-CoA in yeast, but is introduced in the N-acyl-sphingosine of mouse cells. The nature of the acceptor (A) for the two hydrogens is unknown.

glycosphingolipids is shown in Figure 22–27. The principle governing the synthesis of these lipids is that the carbohydrates are added to the acceptor lipid by transfer from a sugar nucleotide. For example, *UDP-glucose*, which is the glucosyl donor for the synthesis of glycogen (Chapter 14), is also the donor of glucose for the synthesis of the glycosphingolipids. The enzymes involved in these reactions are called *glycosyltransferases* and are thought to be specific for each reaction. Most of the glycosyltransferases involved in the

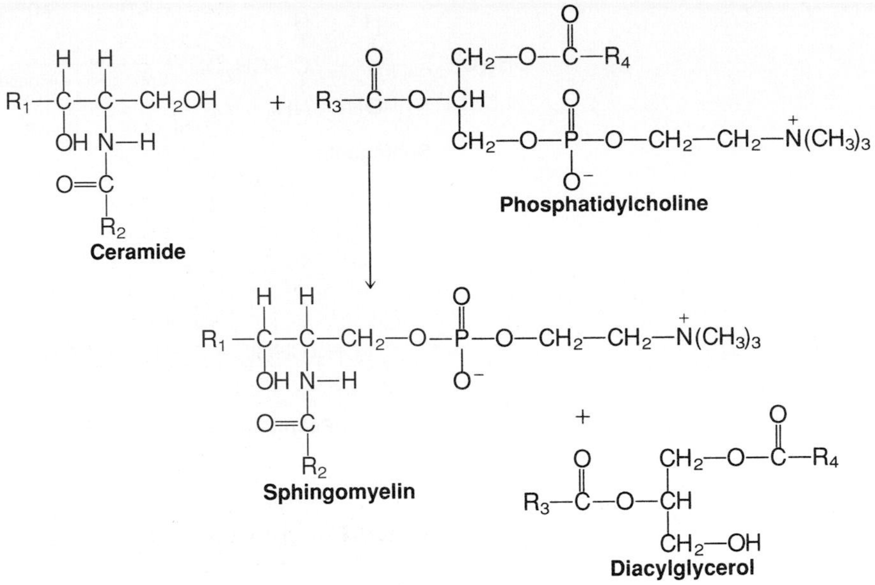

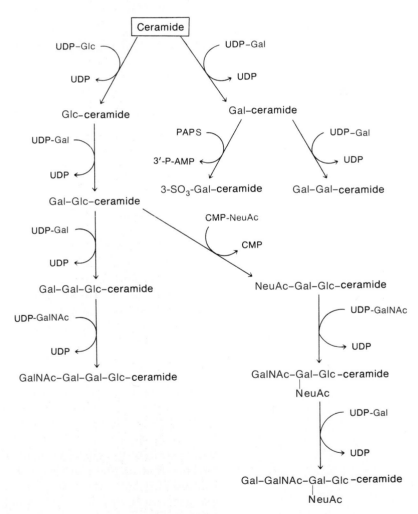

Figure 22-26

The pathway of sphingomyelin biosynthesis.

Figure 22-27

Outline of biosynthesis of some glycosphingolipids. (Glc = glucose, Gal = galactose, GalNAc = N-acetylgalactosamine, NeuAc = N-acetylneuraminic acid, PAPS = 3'-phosphoadenosine, 5'-phosphosulfate.) PAPS is a sulfate analog of 3'-phospho ADP. It has a high transfer potential for sulfate. The formation of PAPS is discussed in Chapter 24.

biosynthesis of the glycosphingolipids are located on the lumen side of the Golgi apparatus. The nucleotide sugars used in biosynthesis (Figure 22-27) are transferred across the membrane into the lumen of the Golgi by a transporter located on the Golgi membrane.

There is still a great deal to be learned about the enzymes involved in

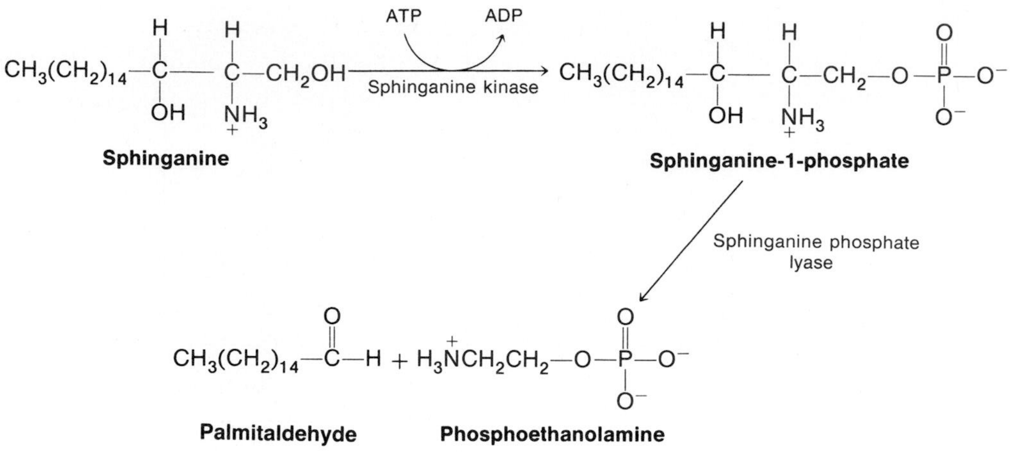

Figure 22–28
Catabolism of sphinganine.

sphingolipid biosynthesis and the regulation of the corresponding pathways. The enzymes are hard to study because they are present in small quantities, are membrane-bound, and use amphipathic substrates.

Defects in Sphingolipid Catabolism Are Associated with Metabolic Diseases

The degradation of long-chain sphingosine bases begins with a phosphorylation by _sphinganine kinase_ to yield _sphinganine-1-phosphate_. This is subsequently cleaved by _sphinganine phosphate lyase_ to form palmitaldehyde and phosphoethanolamine (Figure 22–28). The palmitaldehyde can either be reduced to the C_{16} alcohol or oxidized to palmitic acid. The phosphoethanolamine appears to enter the major metabolic pool of this compound and is therefore converted to phosphatidylethanolamine (see Figure 22–4).

Investigations on the structure and biochemistry of sphingolipids have been stimulated by the occurrence of a number of human genetic diseases, the _sphingolipidoses_, each of which results from the accumulation of one of these lipids. In all but a few cases, the diseases are caused by a deficiency in the activity of one of the enzymes involved in the catabolism of the sphingolipids. The degradation of glycosphingolipids proceeds in a stepwise fashion, as shown in Figure 22–29. The catabolic enzymes are localized in the lysosomes. Hence the sphingolipidoses are part of a larger group of metabolic diseases that result from impaired function of lysosomal enzymes.

The assay of these _glycosyl hydrolases_ in vitro is complicated because in aqueous systems the glycolipids aggregate to form micelles. As a result, the glycolipids are not readily degraded by the glycosyl hydrolases. The problem is alleviated by the addition of detergents (such as _bile acids_; see Chapter 23), which make the lipids more soluble in aqueous environments. Lysosomes do not contain bile acids; yet these organelles are the sites for degradation of the glycosphingolipids. In the past few years it has been discovered that in place of detergents, certain activator proteins occur in the soluble portion of the lysosomes. The lipid binds to one of these proteins, and as a result, degradation by the hydrolase can more easily proceed. Separate activator proteins have been identified which aid the hydrolysis of gangliosides,* glucosylceramide, and 3′-sulfate galactosylceramide.

The principal diseases in which a defect in the catabolism of a sphingo-

---

*A ganglioside is a glycosphingolipid containing one or more molecules of N-acetylneuraminic acid.

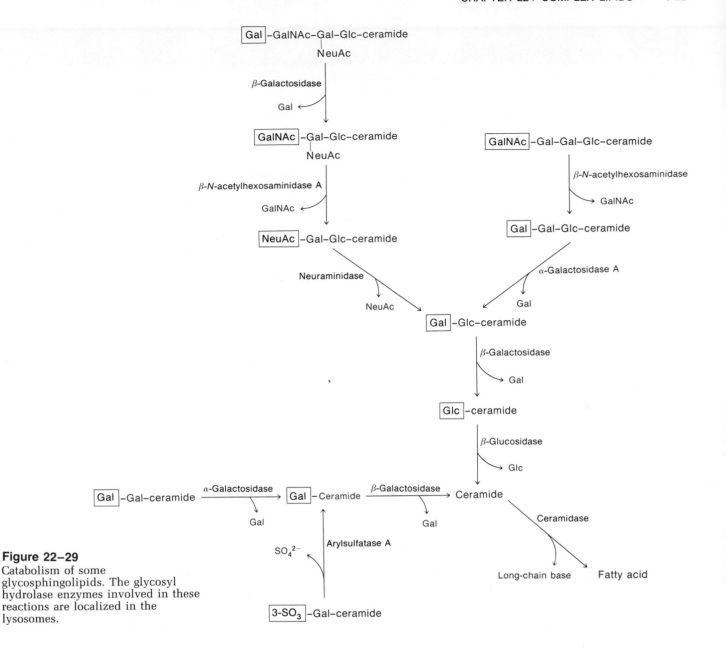

Figure 22–29
Catabolism of some glycosphingolipids. The glycosyl hydrolase enzymes involved in these reactions are localized in the lysosomes.

lipid has been detected are shown in Table 22–1. Many of them have variations that are not included in the table. For more information about these and other metabolic diseases, read *The Metabolic Basis of Inherited Disease* (see Selected Readings). Here we will describe briefly just two of the disorders listed in Table 22–1.

Fabry's disease was first described in 1898 by J. Fabry and independently by W. Anderson. Some characteristic symptoms are skin rash, pain in the extremities, and renal impairment, which leads to hypertension. Patients lead a reasonably normal life until approximately their fourth decade, when the kidneys usually fail. Little progress was made in understanding the cause of this disease until Charles Sweeley and Bernard Klionsky described, in 1963, the structure of *Gal-Gal-Glc-ceramide* as the major lipid that accumulates in the Fabry kidneys. Subsequently, in 1967, Roscoe Brady demonstrated that the enzymatic defect was a deficiency of the enzyme that degrades trihexosylceramide, and this was later shown by J. A. Kint to be *α-galactosidase A* (Figure 22–29). It is now possible to diagnose this X-

TABLE 22–1
Inherited Diseases of Sphingolipid Catabolism

Disease	Enzyme Activity that Is Deficient	Reaction
1. Ceramidase deficiency: Farber's lipogranulomatosis	Ceramidase	Ceramide $\longrightarrow$ fatty acid + long-chain base
2. Sphingomyelin lipidosis: Niemann-Pick disease	Sphingomyelinase	Sphingomyelin $\longrightarrow$ ceramide + phosphocholine
3. Glucosylceramide lipidosis: Gaucher's disease	β-glucosidase	Glc-ceramide $\longrightarrow$ Glc + ceramide
4. Galactosylceramide lipidosis: globoid cell leukodystrophy	β-galactosidase	Gal-ceramide $\longrightarrow$ Gal + ceramide
5. Sulfatide lipidosis: metachromatic leukodystrophy	Arylsulfatase A	$3'\text{-}SO_3^-$-Gal-ceramide $\longrightarrow$ Gal-ceramide + SO_4^{2-}
6. Fabry's disease	α-Galactosidase A	Gal-Gal-Glc-ceramide $\longrightarrow$ Gal + Gal-Glc-ceramide
7. GM_1† gangliosidosis	GM_1-β-galactosidase	$GM_1 \longrightarrow$ Gal + GM_2
8. Tay-Sachs disease (GM_2 gangliosidosis)	Hexosaminidase A	$GM_2 \longrightarrow GM_3$ + GalNAc
9. Sandhoff's disease	Hexosaminidases A + B	$GM_2 \longrightarrow GM_3$ + GalNAc

†See Table 5–5 for abbreviations.

chromosome-linked disease with biochemical tests based on these discoveries. For example, in prenatal diagnosis, cells are obtained from the amniotic fluid by amniocentesis and cultured; then the activity of α-galactosidase A is determined.

Unfortunately, there is presently no effective treatment for Fabry's disease. Preliminary studies have suggested that patients might respond to infusions of purified α-galactosidase A (enzyme replacement therapy). Such studies have had limited success because of the short life time of the infused enzyme in the blood. Fortunately, Fabry's disease is rare, with only a few hundred cases reported throughout the world.

The disease described by Warren Tay in 1881, now known as Tay-Sachs disease, is quantitatively more important. It is estimated that 30 to 50 children with Tay-Sachs disease are conceived each year in the United States, largely by Jewish parents. This disease is devastating; the children usually do not survive beyond the age of 3. As is clear from Table 22–1, in Tay-Sachs disease there is an accumulation of ganglioside GM_2 (see Table 5–5), especially in the brain, as a result of a deficiency of hexosaminidase A.

There is no effective treatment for Tay-Sachs disease. Moreover, enzyme replacement is not considered a likely remedy because infused enzyme would not penetrate the blood-brain barrier. However, the incidence of the disease has been dramatically reduced by prenatal diagnosis. Tay-Sachs disease can arise only if both parents are carriers, i.e., if each parent carries a single defective gene for the hexosaminidase A enzyme. In this case there is a 25 per cent chance that an offspring will have the disease.

Glycosphingolipids Function As Structural Components and As Specific Cell Receptors

It is clear that a major function of glycosphingolipids in some membranes is structural. Galactosylceramide represents 15 per cent of the myelin membrane, and globoside (see Table 5–5 for structure) represents approximately 7 per cent of the human red blood cell membrane. Moreover, in all instances examined, the glycolipids are oriented asymmetrically in the bilayer of the plasma membrane, facing the outside of the cell. Recall that with glycoproteins a similar asymmetry is observed (see Chapter 20).

Many intriguing observations indicate that the glycosphingolipids are more than just structural lipids and have specific cellular functions that are just becoming understood. In this respect they are similar to the glycoproteins discussed in Chapters 6 and 20. For example, cells growing in culture dishes in the presence of protein growth factors (for growth factors, see Chapter 31) are inhibited when the ganglioside GM_1 or GM_3 is added. Apparently the gangliosides inhibit a protein kinase activity of the receptor that is stimulated by the growth factors. This inhibition of cellular growth by gangliosides is interesting in light of another observation. When cells are transformed by tumor viruses, the newly formed tumor cells have a greatly reduced content of GM_3 in some kinds of tumors and a lower content of GM_1 in other types of tumors. Possibly the reduced amount of the gangliosides is related to the abnormal growth of the tumors.

In addition to regulatory functions, the glycosphingolipids also seem to serve as cell receptors. For example, the ganglioside GM_1 can act as a receptor for cholera toxin. (The binding of this toxin provides no obvious advantage to the cell.) Some glycosphingolipids are also blood-group antigens. This means that persons with, for example, the B blood type have a different glycosphingolipid on their cell surfaces than persons with the A blood type. These glycosphingolipids have six sugar residues attached to ceramide and occur at very low concentrations in the plasma membrane. The only structural difference is that the A antigen has an N-acetylgalactosamine in place of a galactose residue that is found on the B antigen. (These antigens are also carried by certain glycoproteins; see Chapter 20.) It has also been shown that the glycosphingolipids expressed on a cell surface differ according to the stage of development of the organism. It is thought that this variation in expression of glycosphingolipids may be involved in cell–cell recognition and growth of tissues.

Finally, the relatively high concentrations of gangliosides in neurons suggest a function for these lipids in nerve transmission, but what that function might be is unknown.

THE EICOSANOIDS: PROSTAGLANDINS, THROMBOXANES, LEUKOTRIENES, AND HYDROXY-EICOSAENOIC ACIDS

Up to this point we have focused on the metabolism of lipids as it relates to the energy and structural requirements of the cell. Switching gears, we shall now explore a heterogeneous class of lipids called eicosanoids, which are structurally and functionally very different. The term _eicosanoid_ refers to any twenty-carbon (C_{20}) fatty acid. In this chapter we will be discussing oxygenated eicosanoids, which are hormonelike substances that act near their site of synthesis; they are effective at very low concentrations.

Oxygenated eicosanoids appear to be unique to animal cells. The _prostaglandins_ were the first oxygenated eicosanoids to be discovered. They were extracted from human semen in the 1930s by Ulf Von Euler in Sweden. He found that when components in this fluid were injected into animals, the uterus contracted and blood pressure was lowered. He believed the potent compounds originated from the prostate, hence the name prostaglandin. The prostaglandins in semen were later shown to be derived from the seminal gland. It is now known that prostaglandins are present in most tissues of both male and female animals.

Von Euler was able to establish that the compounds were hydroxy fatty acids. However, the techniques available at that time did not permit complete purification and structural analysis. With the advent of gas chromatog-

BOX 22-D

Analysis of Prostaglandins and Thromboxanes

The prostaglandins and thromboxanes can be extracted from acidified (low pH) aqueous solution by organic solvents such as ethyl acetate or ether. These eicosanoids are separated from other lipids by silicic acid chromatography and fractionated into individual components by high-pressure liquid chromatography or thin-layer chromatography.

There are, however, several problems that arise in doing research on prostaglandins. One difficulty is that many of the compounds are unstable. For example, the half-life ($t_{1/2}$) for PGG_2 in H_2O at 37°C is 5 min. TXA_2 is even more unstable ($t_{1/2} = 30-40$ sec) and decomposes to TXB_2.

A second problem is quantitative analysis. The normal *in vivo* concentrations are hard to estimate, since the amounts of prostaglandins in a tissue vary according to the procedure used for preparation of the tissue. For example, less than 10 nanograms (ng) of PGE_2 could be detected per gram of rat tissue homogenized in ethanol, whereas 200 to 300 ng of PGE_2 per gram of tissue was obtained after homogenization in 0.9% saline at 0°C for 5 min (1 ng = 10^{-9} g). These differences in the amounts of prostaglandins are due to the rapid synthesis of prostaglandins in recently excised tissue.

A third difficulty results from the low concentrations of prostaglandins found in tissues. Accurate measurements can be made by gas chromatography with an electron capture detector, by gas chromatography combined with mass spectrometry, or by radioimmunoassay. With these procedures it has been possible to determine that the concentration of $PGE_{2\alpha}$ in human serum is only 2 picograms (pg) per milliliter (1 pg = 10^{-12} g).

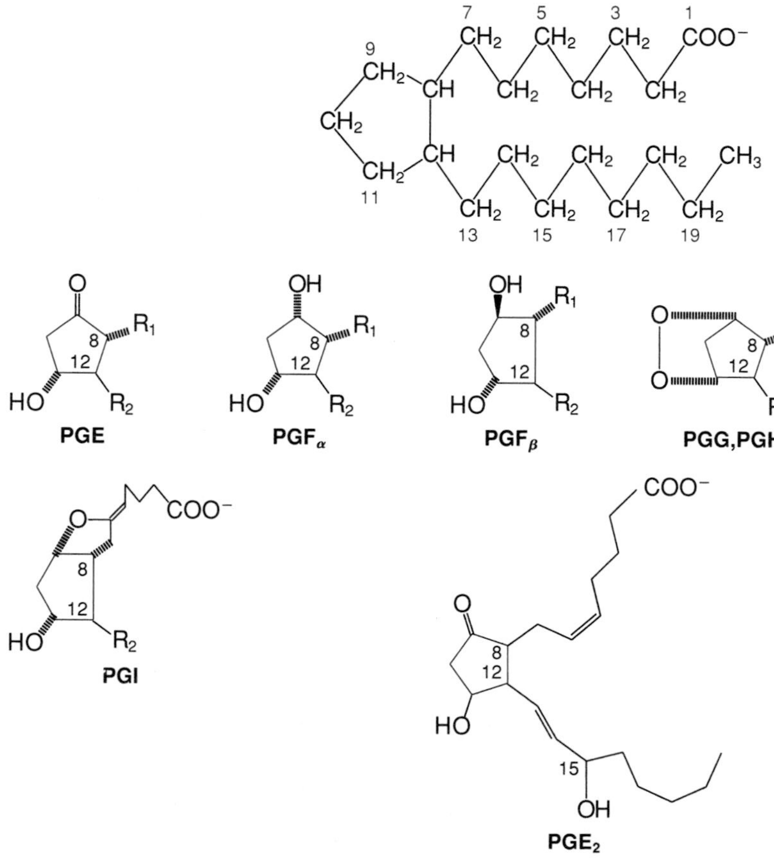

Figure 22-30

Structures of some prostaglandins. All prostaglandins are derived from C_{20} fatty acids. The names and numbering system are based on the hypothetical parent compound, prostanoic acid, shown at the top. The full structure is shown for PGE_2. The PG stands for prostaglandin. The E specifies the type, and the subscript refers to the number of double bonds. From the structure it can be seen that the double bonds of PGE_2 are *trans*Δ^{13}, *cis*Δ^5.

raphy and mass spectrometry, the first structures were determined in the early 1960s by Sune Bergstrom and Jon Sjovall, also from Sweden. At that time two compounds were isolated. One was more soluble in ether and was referred to as _prostaglandin E_. The other was more soluble in phosphate buffer and was named _prostaglandin F_. An explosion of activity in prostaglandin research identified other prostaglandins and led to the elucidation of metabolic pathways of many of these compounds.

Pharmacological interest in prostaglandins was awakened in 1971 by John Vane's discovery that aspirin blocked the synthesis of prostaglandins. Many of aspirin's pharmacologic actions, such as its anti-inflammatory effects, are believed to be due to the inhibition of the synthesis of prostaglandins and related compounds. The discovery of many related compounds (_thromboxanes_, _prostacyclin_, and _leukotrienes_) has continued to make eicosanoid research one of the most rapidly developing fields in biochemistry (see Box 22–D).

All Eicosanoids Are Related to C_{20} Polyunsaturated Acids

Prostaglandins are cyclopentanoic acids that are biosynthetically derived from C_{20} polyunsaturated fatty acids. The compounds are named from the hypothetical compound prostanoic acid (Figure 22–30). Classes of prostaglandins differ from each other in the structure of the substituted cyclopentane ring. There are at least nine different cyclopentane ring structures, six of which are shown in Figure 22–30. The names of the prostaglandins are abbreviated as PG, followed by a letter of the alphabet.

The R_1 and R_2 groups attached to the cyclopentane ring differ in the various types of prostaglandins. All prostaglandins except PGG contain a hydroxyl substituent at C-15 (as shown for PGE_2 in Figure 22–30); in PGG there is a peroxide (—O—OH) at C-15. The prostaglandins are named according to the number of carbon–carbon double bonds in groups R_1 and R_2, as shown in Figure 22–30 for PGE_2. Subscript numerals 1, 2, or 3 refer to the number of these double bonds in the molecule.

Thromboxanes (TX) are oxygenated eicosanoids closely related to prostaglandins. They were first isolated from thrombocytes (blood platelets), hence the name. They differ from prostaglandins in the ring structure, which for thromboxanes is a cyclic ether (oxane ring). The structures of the rings for TXA and TXB and the complete structure of TXA_2 are shown in Figure 22–31. The rules for naming of thromboxanes parallel those for naming of prostaglandins. All thromboxanes have the hydroxyl substitution at C-15, as shown for TXA_2 in Figure 22–31.

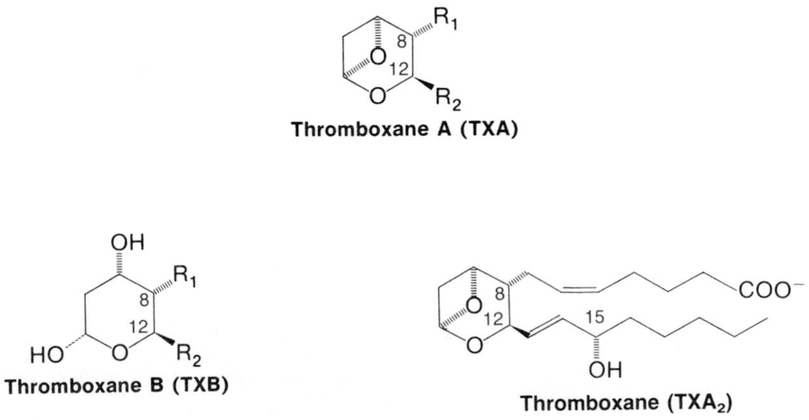

Figure 22–31
The structures of some thromboxanes. Thromboxanes differ from prostaglandins in having a cyclic ether (oxane) ring structure.

Thromboxane A (TXA)

Thromboxane B (TXB)

Thromboxane (TXA$_2$)

Precursor fatty acid	Product: prostaglandin or thromboxane
$C_{20:3}\Delta^{8,11,14}$ $\longrightarrow$	PG_1 or TX_1
$C_{20:4}\Delta^{5,8,11,14}$ $\longrightarrow$	PG_2 or TX_2
$C_{20:5}\Delta^{5,8,11,14,17}$ $\longrightarrow$	PG_3 or TX_3

Figure 22–32
Fatty acid precursors and the related prostaglandins and thromboxanes. All prostaglandins and thromboxanes contain two fewer double bonds than their fatty acid precursors.

Arachidonic acid

$2 O_2$

Fatty acid cyclooxygenase

PGG₂

Peroxidase

PGH₂

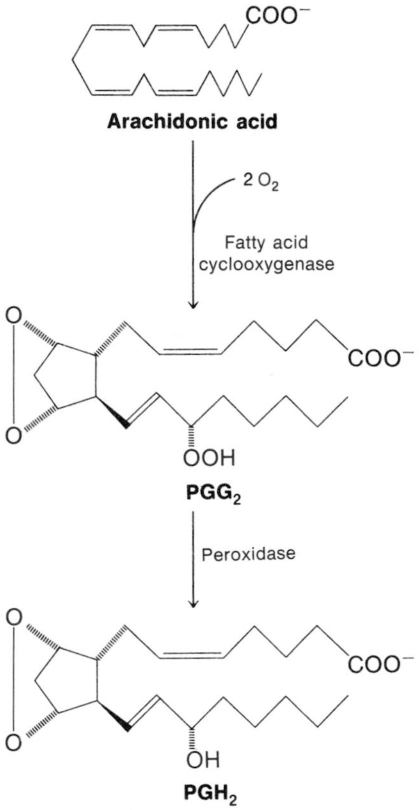

Figure 22–33
Reaction catalyzed by prostaglandin endoperoxide synthase. This enzyme has two catalytic activities, as indicated next to the reaction arrows.

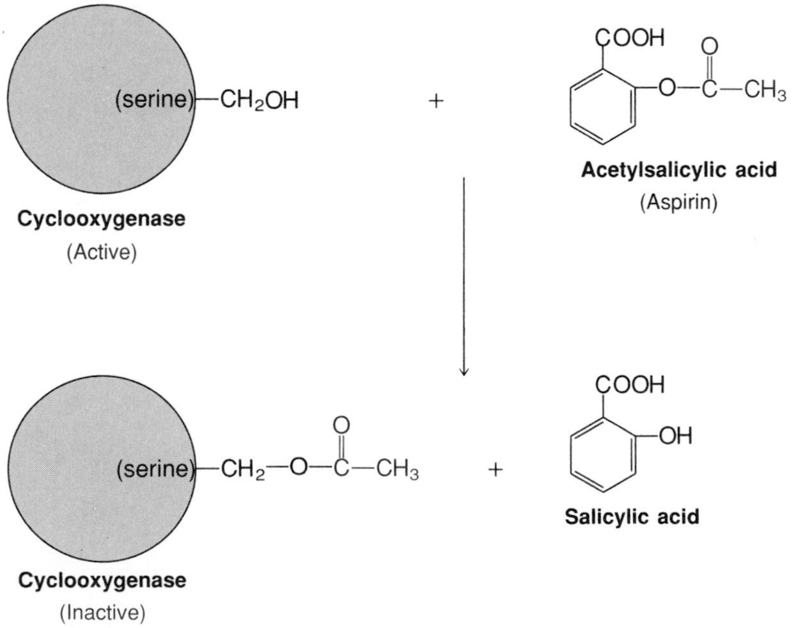

Most Eicosanoids Are Synthesized from Arachidonic Acid

Prostaglandins and thromboxanes arise from C_{20} polyunsaturated fatty acids and contain two fewer double bonds than the parent fatty acids (Figure 22–32). Because *arachidonic* acid (Figure 22–33) is the most plentiful C_{20} polyunsaturated fatty acid in most mammals, it is not surprising that the PG_2 and TX_2 series are the predominant classes of these compounds. The precursor fatty acids are derived from phospholipids by the release of the fatty acid by a specific phospholipase A_2. Alternatively, the phospholipid may be degraded by a phosphatidylinositol-specific phospholipase C to yield diacylglycerol, which is subsequently catabolized by diacylglycerol lipase to yield arachidonic acid. At the present time, both possible routes should be considered. The release of arachidonic acid appears to be the rate-limiting step for the synthesis of prostaglandins and thromboxanes.

The initial steps for the synthesis of all prostaglandins and thromboxanes are the oxidation and cyclization of arachidonic acid to yield PGG_2 and then PGH_2 in reactions catalyzed by *prostaglandin endoperoxide synthase,* an endoplasmic reticulum enzyme (Figure 22–33). This enzyme (subunit $M_r = 70,000$) has two catalytic activities and has been purified from bovine and sheep vesicular glands. The formation of PGG_2 is catalyzed by a *fatty acid cyclooxygenase* component, and the subsequent formation of PGH_2 is catalyzed by a *peroxidase* component. The cyclooxygenase activity has the unusual property of catalyzing its own destruction. Approximately once in every 400 substrate turnovers, the cyclooxygenase activity is irreversibly inactivated. This "suicide reaction" occurs both *in vivo* and *in vitro*, but the chemical mechanism has not been described.

The anti-inflammatory effect of aspirin appears to be due to the acetylation of a serine at the active site of the cyclooxygenase, which irreversibly inactivates the cyclooxygenase activity of the enzyme (Figure 22–34). Aspirin has no effect on the peroxidase activity of the enzyme.

PGH_2 is the precursor of various prostaglandins and thromboxanes. Each step is catalyzed by a separate tissue-specific enzyme as indicated in Figure 22–35. For example, blood platelets contain *TXA₂ synthase,* which

(serine)—CH₂OH

Cyclooxygenase
(Active)

+

Acetylsalicylic acid
(Aspirin)

(serine)—CH₂—O—C—CH₃

Cyclooxygenase
(Inactive)

+

Salicylic acid

Figure 22–34
Aspirin inactivates cyclooxygenase by acetylation of a serine, probably at or near the active site.

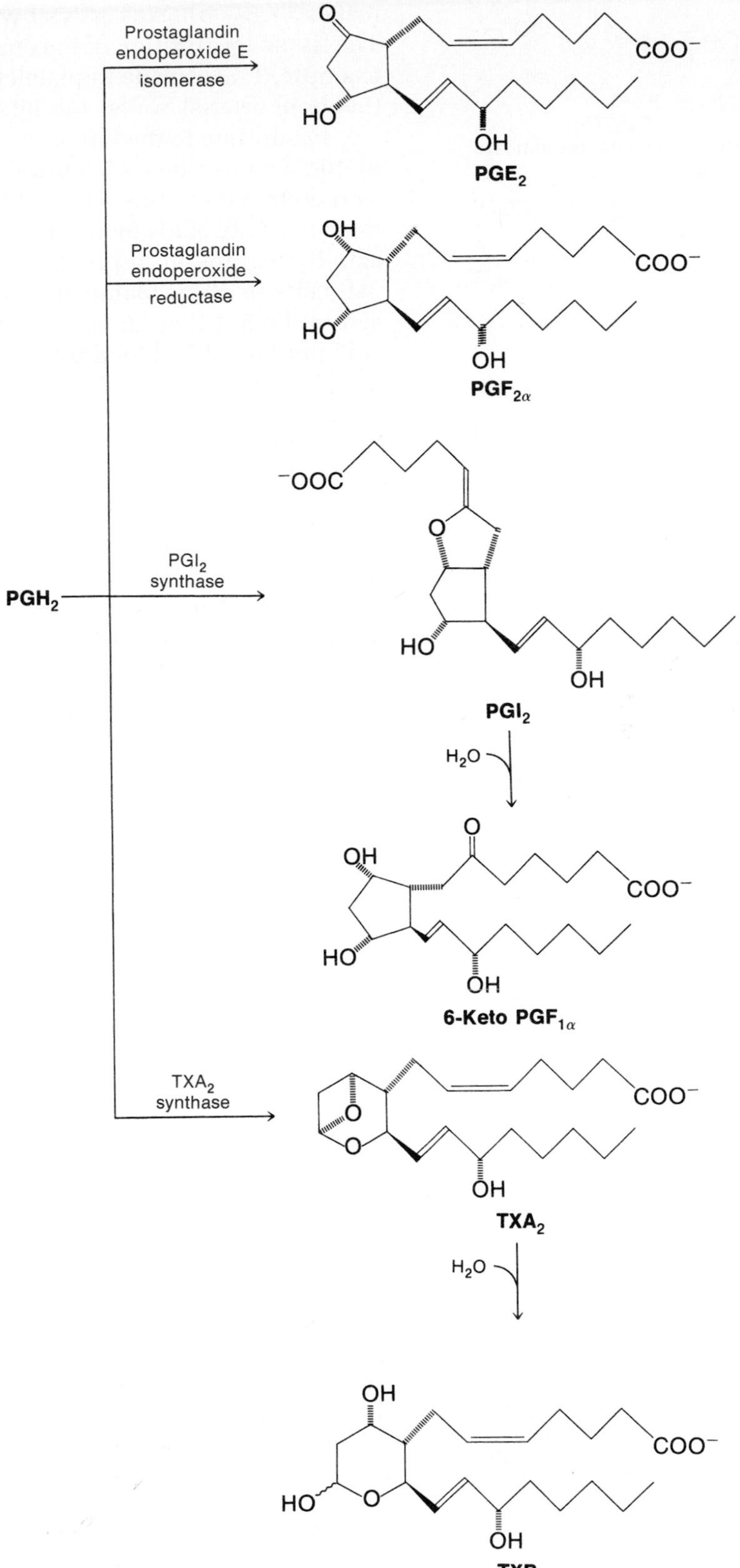

Figure 22-35
Formation of prostaglandins and thromboxanes.

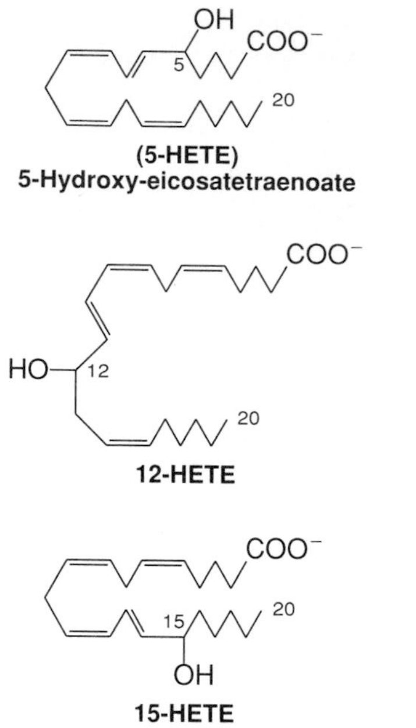

(5-HETE)
5-Hydroxy-eicosatetraenoate

12-HETE

15-HETE

Figure 22–36
Structures of three common hydroxy-eicosatetraenoic acids (HETEs). These compounds are all formed directly from arachidonic acid by the action of a specific lipoxygenase, which inserts a hydroxyl group at the indicated location.

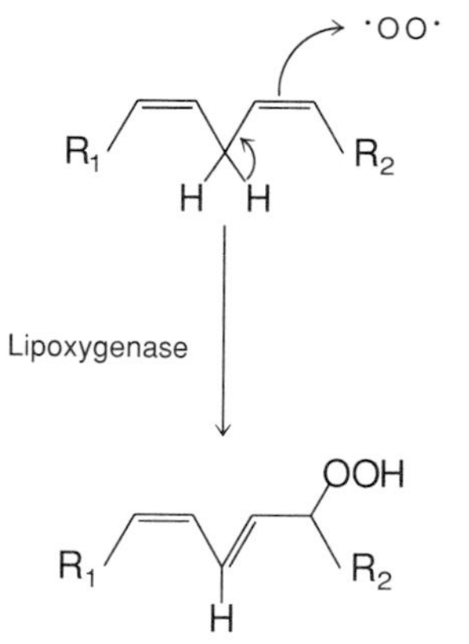

Lipoxygenase

Figure 22–37
The mechanism of the reaction catalyzed by lipoxygenases. This is a simple dioxygenase reaction; there is no net oxidation-reduction of either the fatty acid or the oxygen. A *cis–trans* conjugated diene is formed in the reaction.

makes TXA$_2$, whereas arterial walls have *PGI$_2$ synthase*, which makes PGI$_2$. The tissue localization of the enzyme relates to the role of the eicosanoid; for example, TXA$_2$ (made in platelets) causes platelet aggregation, whereas PGI$_2$ (made in arterial walls) inhibits this process.

In addition to the formation of prostaglandins and thromboxanes, arachidonic acid can be metabolized to several other compounds. The hydroxy-eicosatetraenoic acids (HETEs) (Figure 22–36) are formed from polyunsaturated C$_{20}$ fatty acids by the action of the enzyme *lipoxygenase*. This enzyme activity was found in plants four decades ago, but was first reported in animal cells in 1973. Mammalian lipoxygenases catalyze the insertion of oxygen in the 5, 12, or 15 position of various eicosanoic acids. These activities are found in white blood cells (neutrophils, macrophages, platelets, and others). The mechanism of the reaction involves the addition of oxygen to a double bond with the formation of a conjugated *cis–trans* diene as shown in Figure 22–37. The hydroperoxy group is subsequently reduced to an alcohol to form the corresponding hydroxy-eicosanoic acid. The physiological functions of hydroxy-eicosanoic acids are not well understood; their formation is not inhibited by aspirin as are the cyclooxygenases involved in prostaglandin and thromboxane formation.

Another family of products derived from arachidonic acid is that of the *leukotrienes*. The leukotrienes are so named because they were first shown to be formed in leukocytes and contain a conjugated triene (Figure 22–38). The biosynthesis of leukotrienes begins with a 5-lipoxygenase reaction. Subsequently a dehydration reaction leads to the formation of a 5,6-epoxide (leukotriene A). Addition of glutathione to C-6 results in the formation of leukotriene C, which can be metabolized by peptidases to yield leukotriene D and leukotriene E. Leukotrienes, formerly known as *slow-reacting substance*, are extremely potent muscle contractants that can severely constrict the small airways of the lung, producing asthma.

The rate of synthesis of lipoxygenase-derived compounds is controlled by both the release of arachidonic acid and the activation of the lipoxygenase. Certain peptides, which promote chemically stimulated movement (chemotaxis) of cells, cause the release of arachidonic acid and the activation of the 5-lipoxygenase.

If the eicosanoids are to serve useful regulatory functions, there must be a rapid means for their removal as well as their synthesis. Indeed, it has been found that eicosanoids have short half-lives *in vivo* and are rapidly metabolized to physiologically inactive compounds. The PGE and PGF compounds do not normally survive a single pass through the circulatory system. They are rapidly catabolized to 15-keto-13,14-dihydro derivatives by the lungs (Figure 22–39). The number of final metabolites of these prostaglandins is very large. For example, in the human female, at least 15 metabolites of PGF$_{2\alpha}$ have been identified. It would serve no useful purpose to elaborate on all these metabolic transformations. However, as one example, the pathway for the metabolism of PGF$_{2\alpha}$ to its major human urinary metabolite, 5α,7α-dihydroxy-11-ketotetranorprostane-1,16-dioic acid, is shown in Figure 22–39.

Eicosanoids Are Hormones That Exert Their Action Locally

Eicosanoids have a wide range of biological effects, some of which have already been mentioned. They are generally considered to be hormones that exert their effect locally without altering functions throughout the body. Hence a comprehensive scheme that depicts the mode of action of these compounds is not yet available. However, a few general points can be made.

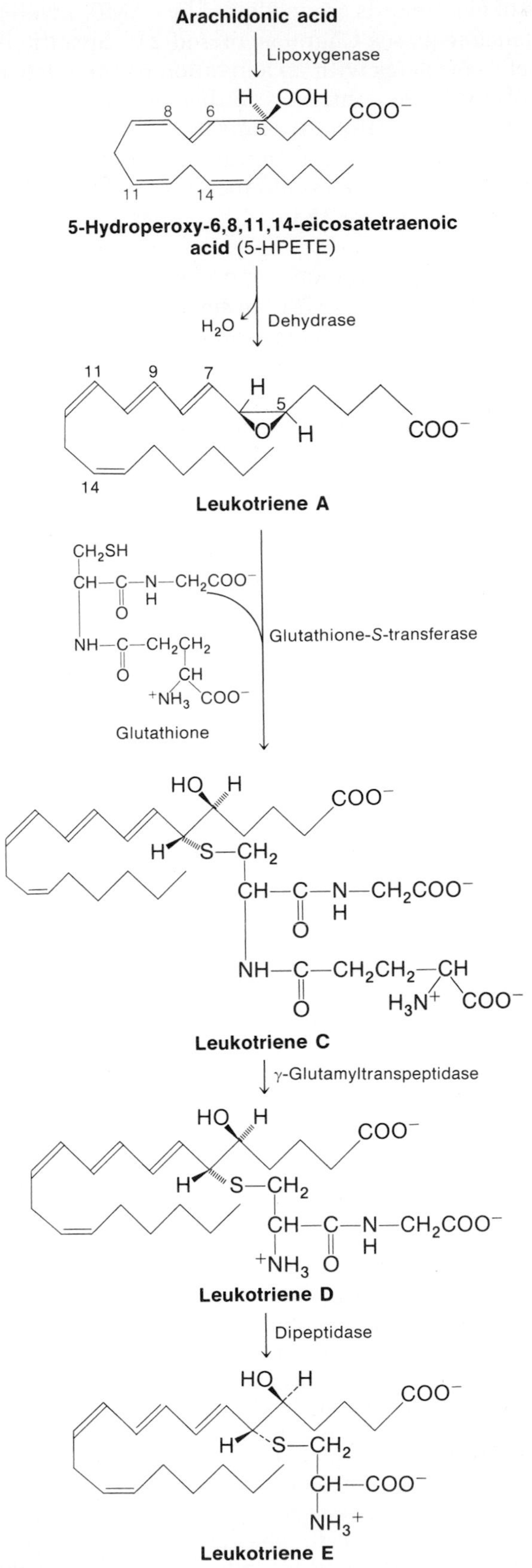

Arachidonic acid

Figure 22–38
Biosynthesis of the leukotrienes.

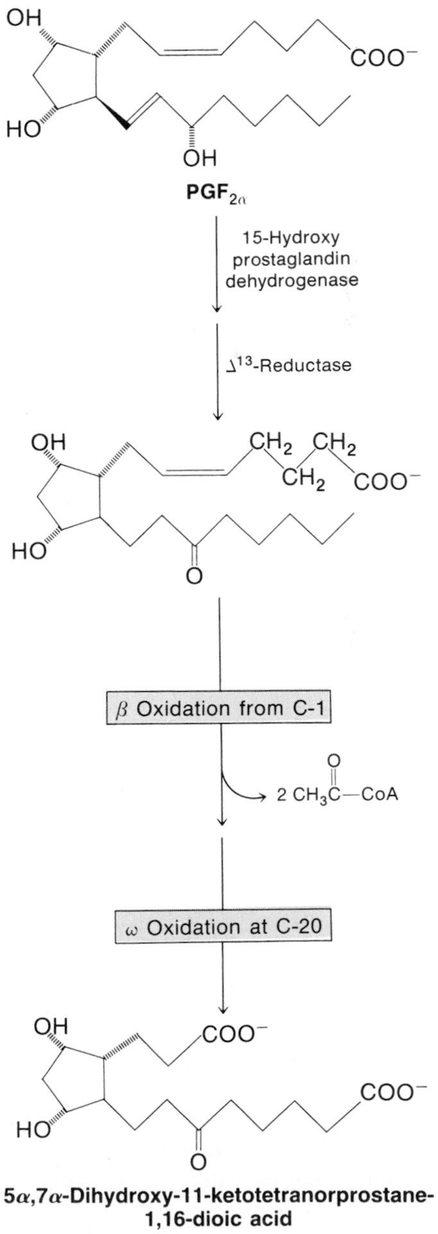

Figure 22–39
Major catabolic pathway of PGF$_{2\alpha}$ in humans.

Many of the effects of eicosanoids are mediated by cAMP, a well-known hormone effector substance (e.g., see Chapters 14 and 21). Specific binding of PGE to a variety of cells correlates with an activation of adenylate cyclase and the accumulation of cAMP. In contrast, PGF has no such effect. Moreover, different cell types respond differently to PGE. In human adipocytes, PGE_1 causes a 15-fold increase in the concentration of cAMP. Fibroblasts isolated from human adipose showed a 95-fold increase in cAMP after exposure to PGE_1. In platelets, PGI_2 rather than PGE_2 appears to mediate the increase in cAMP.

PGE_1 relaxes, whereas $PGF_{2\alpha}$ contracts venous smooth muscle. $PGF_{2\alpha}$ has been shown to increase the levels of cGMP in several types of tissues and cells. Whether or not the antagonistic effects of PGE and PGF are mediated by cAMP and cGMP, respectively, remains to be established.

There is evidence that specific, high-affinity receptors exist for certain prostaglandins. $PGF_{2\alpha}$ appears to inhibit progesterone secretion and regression of the corpus luteum, and it has been shown that a specific protein receptor for $PGF_{2\alpha}$ is localized in the plasma membrane of bovine corpus luteum.

The eicosanoids have been implicated as mediators of tissue inflammation since aspirin's anti-inflammatory effect was shown to be the result of its inactivation of cyclooxygenase. How the eicosanoids cause tissue inflammation is the focus of much current research.

The possible function of prostaglandins in blood clotting is also generating a great deal of research and speculation. PGI_2 lowers blood pressure, relaxes coronary arteries, and inhibits platelet aggregation. TXA_2 has the opposite effects. PGI_2 is made in the endothelial lining of blood vessel walls. It appears to inhibit platelet aggregation by binding to a receptor on the plasma membrane, and this subsequently results in an increase of cAMP in platelets. TXA_2 appears to suppress the increase in cAMP caused by PGI_2. It has been speculated that the synthesis of PGI_2 may prevent platelets from binding to arterial walls. In damaged areas of arteries, the synthesis of PGI_2 may be decreased and the presence of TXA_2 would cause the platelets to aggregate, leading to the formation of a thrombus (blood clot). The physiological relevance of these observations is unclear, and further studies are needed to provide a clear picture of the function of these two compounds in blood clotting.

SUMMARY

Fatty acids are components of a rich variety of more complex lipid molecules that play structural roles in membranes. Fatty acids are also the precursors of compounds with hormonelike activity. In this chapter we have focused on the biosynthesis of these compounds, with some mention of their functions.

Lipid synthesis is unique in that it is almost exclusively localized to the surface of membrane structures. The reason for this is the amphipathic nature of the lipid molecules. In bacteria, where there are no organelles, lipid synthesis is localized to the inner plasma membrane. In eukaryotes, various organelles are involved in lipid synthesis, most notably the endoplasmic reticulum and the Golgi apparatus.

Phospholipids are biosynthesized by acylation of either glycerol-3-phosphate or dihydroxyacetone-phosphate to form phosphatidic acid. This central intermediate can be converted into phospholipids by two different pathways. In one of these, phosphatidic acid reacts with CTP to yield CDP-diacylglycerol, which is converted to phosphatidylserine, phosphatidylinositol, phosphatidylglycerol, or cardiolipin. In *E. coli*, the major phospholipid, phosphatidylethanolamine, is synthesized by means of this route by the decarboxylation of phosphatidylserine. In the second pathway, found in eukaryotes, phosphatidic acid is hydrolyzed to diacylglycerol, which reacts with CDP-ethanolamine or CDP-choline to yield phosphatidylethanolamine or phosphatidylcholine, respectively. Alternatively, the diacylglycerol may react with acyl-CoA to form triacylglycerol.

There are numerous reactions by which the acyl groups or polar head groups of phospholipids might be modified or exchanged. Phospholipids are degraded by specific phospholipases. The transfer of phospholipids among membranes is facilitated by phospholipid exchange proteins. Budding and fusion also play a role in the distribution of lipids between the different membranes.

Regulation of lipid synthesis has been studied most thoroughly in *E. coli* and the rat liver. In *E. coli*, lipids are used almost exclusively for phospholipid synthesis, and regulation is known to occur at an early stage in fatty acid synthesis. In rat liver, fatty acids are important precursors of both structural phospholipids and also the energy storage lipid, triacylglycerol. The need for structural lipids is satisfied before fatty acids are shunted into energy storage. The rate of triacylglycerol lipid synthesis is regulated by the availability of fatty acid, from both diet and biosynthesis. That availability is primarily a function of nutrient excess, as described in the previous chapter. The regulation of phosphatidylcholine biosynthesis occurs primarily at the reaction catalyzed by CTP:phosphocholine cytidylyltransferase. This enzyme is activated by translocation from the cytosol, where it is inactive, to the endoplasmic reticulum, where it is activated. The regulation of the biosynthesis of other phospholipids in normal cells is not well understood.

The sphingolipids are important structural lipids found in eukaryotic membranes. The acylation of sphingosine produces ceramide, which reacts with phosphatidylcholine to give sphingomyelin. Ceramide also can react with activated carbohydrates (e.g., UDP-glucose) to form the glycosphingolipids. Studies on the catabolism of the sphingolipids have revolved around many inherited diseases that result from a defect in the enzymatic degradation of single sphingolipids.

Prostaglandins, thromboxanes, and leukotrienes are related compounds known as eicosanoids, which have a large variety of biological activities. Prostaglandins and thromboxanes are biosynthesized from C_{20} polyunsaturated fatty acids, primarily arachidonic acid. In the initial reaction, arachidonic acid is converted to PGH_2 by prostaglandin endoperoxide synthase. PGH_2 is then converted to PGE_2, $PGF_{2\alpha}$, TXA_2, or PGI_2. Alternatively, arachidonic acid can be converted to a 5-hydroperoxy derivative, which is dehydrated to form a 5,6-epoxide (leukotriene A). Leukotriene A reacts with glutathione to yield leukotriene C, which can be converted into leukotriene D by γ-glutamyltranspeptidase.

SELECTED READINGS

Brindley, D. N., Metabolism of triacylglycerols. In D. E. Vance and J. E. Vance (eds.), *Biochemistry of Lipids and Membranes.* Menlo Park, Calif.: Benjamin/Cummings, 1985. This chapter provides an advanced and authoritative discussion of phosphatidic acid and triacylglycerol metabolism.

Dennis, E. A., Phospholipases. In *Enzymes XVI.* New York: Academic Press, 1983. This article provides a thorough review of phospholipases.

Rock, C. O., and J. E. Cronan, Lipid metabolism in procaryotes. In D. E. Vance and J. E. Vance (eds.), *Biochemistry of Lipids and Membranes.* Menlo Park, Calif.: Benjamin/Cummings, 1985. This chapter provides an advanced treatment of phospholipid metabolism in *E. coli.*

Smith, W. L., and P. Borgeat, The eicosanoids: prostaglandins, thromboxanes, leukotrienes, and hydroxyeicosaenoic acids. In D. E. Vance and J. E. Vance (eds.), *Biochemistry of Lipids and Membranes.* Menlo Park, Calif.: Benjamin/Cummings, 1985. This chapter provides advanced information on the biochemistry of the eicosanoids.

Snyder, F., Metabolism, regulation, and function of ether-linked glycerolipids. In D. E. Vance and J. E. Vance (eds.), *Biochemistry of Lipids and Membranes.* Menlo Park, Calif.: Benjamin/Cummings, 1985. This chapter contains advanced information on the biochemistry of the ether-containing lipids.

Stanbury, J. B., J. B. Wyngaarden, D. S. Fredrickson, J. L. Goldstein, and M. S. Brown, *The Metabolic Basis of Inherited Disease,* fifth ed. New York: McGraw-Hill, 1983. This book contains in-depth chapters on many of the sphingolipidoses.

Sweeley, C. C., Sphingolipids. In D. E. Vance and J. E. Vance (eds.), *Biochemistry of Lipids and Membranes.* Menlo Park, Calif.: Benjamin/Cummings, 1985. This is an advanced chapter on the chemistry, metabolism, and function of the sphingolipids.

Vance, D. E., Phospholipid metabolism in eucaryotes. In D. E. Vance and J. E. Vance (eds.), *Biochemistry of Lipids and Membranes.* Menlo Park, Calif.: Benjamin/Cummings, 1985. This chapter provides advanced information on phospholipid biochemistry.

Waite, M., Phospholipases. In D. E. Vance and J. E. Vance (eds.), *Biochemistry of Lipids and Membranes.* Menlo Park, Calif.: Benjamin/Cummings, 1985. This chapter provides advanced knowledge about phospholipases and the hydrolysis of phospholipids.

PROBLEMS

1. Would an enzymatic defect in the biosynthesis of CDP-ethanolamine probably be lethal to a liver cell? Would such a defect be lethal to *E. coli*?

2. cAMP-dependent protein kinases figure in the regulation of both fatty acid and phospholipid biosynthesis in the mammalian liver. Describe something of their role in each case.

3. The majority of phospholipids contain an unsaturated fatty acid at the *sn*-2 position. Give an example of a phospholipid in which this is not the case, and describe the pathway used to synthesize it.

4. A graduate student in biochemistry was asked to devise an experiment to prove that the synthesis of phosphatidylcholine *in vitro* occurs on the outer surface of microsomes. Can you help?

5. What distinguishes the pathways used in the synthesis of triacylglycerols in the liver and the intestinal epithelia of mammals?

6. If [1-^{14}C]sphingosine were injected into a rat, would you expect to find any radioactivity in phosphatidylcholine?

7. Until recently, the biosynthesis of sphingomyelin was thought to occur by the transfer of phosphocholine from CDP-choline to ceramide. Phosphatidylcholine is now thought to be the immediate donor of the phosphocholine moiety. Suggest an experiment that would help prove that phosphatidylcholine is the immediate precursor of sphingomyelin.

8. If you discovered a disease in which a lipid with the structure Gal(β1-4)Glc(β1-1)ceramide accumulated abnormally because of an enzymatic defect, what enzymatic reaction would you expect to be defective?

9. An investigator was presented with a liver biopsy from a patient with Gaucher's disease. Much to his surprise, when he assayed the β-glucosidase, he found normal enzyme activity. As always, his assay included a bile acid to aid in the solubilization of the substrate, glucosylceramide. Suggest an explanation for the results observed.

10. Arachidonic acid is the major precursor of prostaglandins and thromboxanes. If a person were unable to absorb arachidonic acid from the diet but could absorb linoleic acid, would that person still be able to make PGE_2?

11. After the discovery of TXA_2, it was postulated that regular treatment of people with aspirin might prevent platelet aggregation and thus help prevent thromboses from forming. Subsequently, PGI_2 and its biosynthetic pathway was discovered. In light of this new knowledge, would continuation of aspirin treatment for TXA_2-mediated thrombosis still make sense?

12. A person was admitted to the emergency room of a hospital with extremely low levels of fatty acids in plasma. The attending physician was going to attempt to increase fatty acid levels by injecting the person with either PGE_1 or PGF_1. Which prostaglandin would the physician choose and how might it help?

13. In what way may the control of inflammation be regarded as linked to phospholipid degradation?

23

CHOLESTEROL AND RELATED DERIVATIVES

Steroids are tetracyclic hydrocarbons that can be considered derivatives of perhydrocyclopentanophenanthrene (see Figure 5–16). They differ from one another in the degree of saturation of each of the four hydrocarbon rings and in the side chain substituents attached to these rings.

Steroids are much more common in eukaryotes than in prokaryotes. Cholesterol, the most prominent member of the steroid family, is an important component of many eukaryotic membranes (see Chapter 6). In addition, it is the precursor of the other two major classes of steroids: the steroid hormones and the bile acids.

Steroid hormones play a key role in the regulation of metabolism. These hormones come in a rich variety, each interacting in a highly specific manner with a receptor protein to effect gene expression in the appropriate target tissue (see Chapter 31). Bile acids are the primary degradation product of cholesterol. The bile acids are made in the liver, stored in the gall bladder, and secreted into the small intestine. There they aid in the solubilization of lipids, facilitating their digestion by intestinal lipases.

All of these biologic roles of the steroids figure prominently in human well-being. Defects in cholesterol metabolism are a major cause of cardiovascular disease. It is no wonder that steroids are a central concern in medical biochemistry. In this chapter we will discuss the metabolism of these complex lipids and the plasma lipoproteins in which they and other complex lipids are transported to various tissues.

BIOSYNTHESIS OF CHOLESTEROL

Early in the 1930s, the structure of cholesterol was finally determined, an achievement that concluded a brilliant chapter in structural organic chemistry. However, the solution to that problem led to the formulation of many new ones. In particular, it was not clear how such a complex structure could be assembled from small molecules.

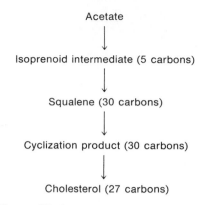

Figure 23–1

Basic scheme for cholesterol biosynthesis proposed by Bloch in 1952.

Work on the biosynthesis of cholesterol began in earnest after Rudolf Schoenheimer and David Rittenberg, at Columbia University, developed isotopic tracer techniques for the analysis of biochemical pathways. In 1941, Rittenberg and Konrad Bloch were able to show that deuterium-labeled acetate ($C^2H_3COO^-$) was a precursor of cholesterol in rats and mice. Subsequently, in collaboration with Edward Tatum and others, Bloch proved that the carbon skeleton of the sterol of <u>Neurospora crassa</u> (ergosterol) was entirely derived from acetate. In 1949, James Bonner and Barbarin Arreguin postulated that three acetates could combine to form a single five-carbon unit called <u>isoprene</u>:

$$CH_2{=}\underset{\underset{H}{|}}{\overset{\overset{CH_3}{|}}{C}}{-}C{=}CH_2$$

This proposal agreed with an earlier prediction of Sir Robert Robinson, that cholesterol was a cyclization product of squalene, a 30-carbon polymer of isoprene units. Thus Bloch postulated a scheme for the biosynthesis of cholesterol as shown in Figure 23–1. And in 1952, Bloch and Robert Langdon demonstrated that squalene could readily be converted into cholesterol.

Two difficult problems remained. What was the structure of the isoprenoid intermediate, and how did squalene cyclize to form cholesterol? In 1953, R. B. Woodward and Bloch postulated a cyclization scheme for squalene (Figure 23–2) that was later shown to be correct. In 1956, the unknown isoprenoid precursor was identified as <u>mevalonic acid</u> by Karl Folkers and others at Merck, Sharpe and Dohme Laboratories. The discovery of mevalonate provided the missing link in the basic outline of cholesterol biosynthesis. Since that time, the sequence and the stereochemical course for the biosynthesis of cholesterol have been defined in detail.

Mevalonate Is a Key Intermediate in Cholesterol Biosynthesis

The sequence of cholesterol biosynthesis begins with a condensation in the cytosol of two molecules of acetyl-CoA, a reaction catalyzed by *thiolase* (Figure 23–3). The next step requires the enzyme *β-hydroxy-β-methylglutaryl-CoA (HMG-CoA) synthase*. This enzyme catalyzes the condensation of a third acetyl-CoA with β-ketobutyryl-CoA to yield <u>HMG-CoA</u>. HMG-CoA is then reduced to mevalonate by <u>HMG-CoA reductase</u>. The activity of this reductase is primarily responsible for control of the rate of cholesterol biosynthesis.

<u>HMG-CoA is an important intermediate for the biosynthesis of both cholesterol and ketone bodies</u> (see Chapter 18). The biosynthesis of cholesterol is catalyzed by enzymes in the cytosol and enzymes bound to the endoplasmic reticulum. The synthesis of ketone bodies, however, is restricted to the mitochondrial matrix. Thus thiolase and HMG-CoA synthase are found in both mitochondria and cytosol of rat liver. In contrast, <u>HMG-CoA lyase</u>, which cleaves HMG-CoA to ketone bodies (see Chapter 18), is located only in mitochondria. HMG-CoA reductase is bound to the endoplasmic reticulum.

HMG-CoA synthase from chicken liver mitochondria has a molecular weight of 105,000 and is composed of two subunits. In chicken liver cytosol there are four isozymes of HMG-CoA synthase, all of which are different from the mitochondrial enzyme. The molecular weights of the cytosolic enzymes are approximately 100,000, and each enzyme is composed of two subunits. Unlike avian liver, rat liver has only a single cytosolic species of HMG-CoA synthase.

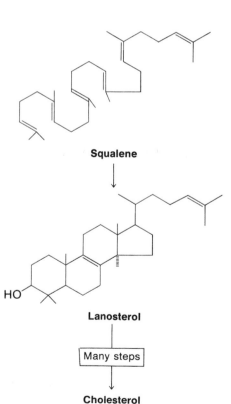

Squalene

Lanosterol

Many steps

Cholesterol

Figure 23–2

Scheme proposed for the cyclization of squalene by Woodward and Bloch in 1953.

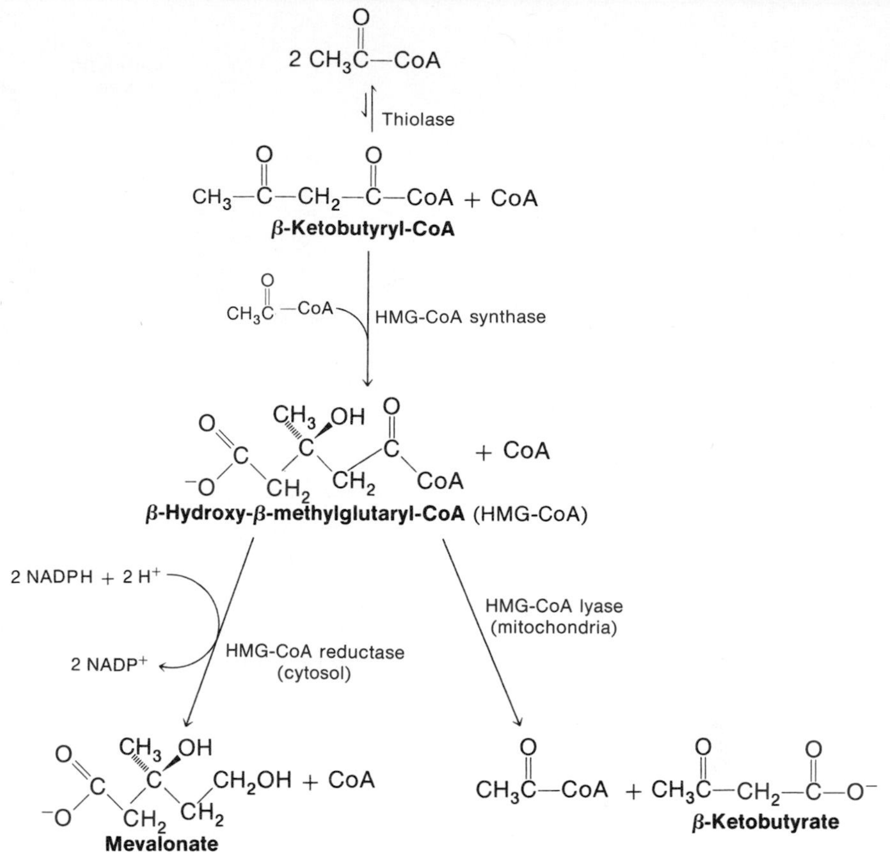

Figure 23–3
Formation of mevalonate. The first two enzymes, thiolase and synthase, are found in both cytosol and mitochondria. The lyase that catalyzes ketone body formation is found only in the mitochondria. The reductase that catalyzes mevalonate formation is found only in the cytosol.

The Rate of Mevalonate Synthesis Determines the Rate of Cholesterol Biosynthesis

The thiolase and HMG-CoA synthase exhibit some regulatory properties in rat liver (cholesterol feeding causes a decrease in these enzyme activities in the cytosol, but not in the mitochondria). However, *primary regulation of cholesterol biosynthesis appears to be centered on the HMG-CoA reductase reaction.* HMG-CoA reductase is found on the endoplasmic reticulum, has a molecular weight of 97,092, and consists of 887 amino acids in a single polypeptide chain. The structure of the enzyme was deduced by Michael Brown and Joseph Goldstein from the sequence of a piece of DNA (cDNA) derived from mRNA that codes for the reductase. The enzyme has two domains (Figure 23–4). The amino-terminal domain has a molecular weight of 35,000, with seven hydrophobic segments that are thought to cross the membrane as shown in Figure 23–4. The carboxyl-terminal domain has a molecular weight of 62,000, contains the catalytic site of the enzyme, and is thought to protrude into the cytosol.

The activity of the reductase is regulated by three distinct mechanisms. The first control point is at the level of gene expression. The amount of mRNA produced is modulated by the supply of cholesterol. When cholesterol is in excess, the amount of mRNA for HMG-CoA reductase is reduced, hence less enzyme is made. Depletion of cholesterol enhances the synthesis of reductase mRNA.

The second regulatory mechanism involves the rate of degradation of HMG-CoA reductase. As stated in Chapter 13, the amount of an enzyme in a cell is determined by both its rate of synthesis and its rate of degradation. It has been known for some time that the half-life for HMG-CoA reductase is between 2 and 4 h, about 10-fold lower than that of many other proteins on

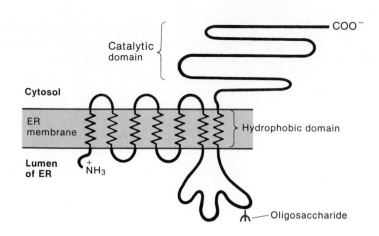

Figure 23–4
Proposed structure of HMG-CoA reductase, derived from studies of recombinant DNA that codes for the enzyme. The enzyme is attached to the endoplasmic reticulum membrane and consists of two domains: the hydrophobic domain, embedded in the membrane, and the catalytic domain, which protrudes into the cytosol. (Adapted from Liscum et al. *J. Biol. Chem.* 260:522–530, 1985.)

the endoplasmic reticulum. In other words, HMG-CoA reductase is rapidly degraded within the cell. The rate of degradation of the reductase appears to be modulated by the supply of cholesterol. Thus when cholesterol is abundant, the rate of enzyme degradation is twice as fast as when there is a limited supply of cholesterol. The effect of cholesterol on enzyme degradation is mediated by the membrane domain of the enzyme. In support of this point, a mutant enzyme lacking the membrane domain was found to be active in the synthesis of mevalonic acid even though free in the cytosol, and the mutant enzyme had an extended half-life of five times the normal. Moreover, the supply of cholesterol did not affect the half-life. More work needs to be done to determine the exact mechanism of regulating the half-life of this enzyme.

The third regulatory mechanism is phosphorylation/dephosphorylation of the reductase, which causes inactivation and activation as outlined in Figure 23–5. *HMG-CoA reductase kinase kinase* catalyzes the phosphorylation and activation of *HMG-CoA reductase kinase*. This kinase phosphorylates HMG-CoA reductase, which results in an inactivation of the reductase. Both kinases are cAMP-independent. Their effects are reversed by the action of *HMG-CoA reductase kinase phosphatase* and *HMG-CoA reductase phosphatase*. It is thought that these kinases and phosphatases mediate short-term changes in the activity of HMG-CoA reductase, but the precise mechanism is unclear.

Figure 23–5
Scheme for modulation of the activity of HMG-CoA reductase by phosphorylation/dephosphorylation. The kinases and phosphatases shown in this figure are believed to mediate short-term changes in the activity of HMG-CoA reductase.

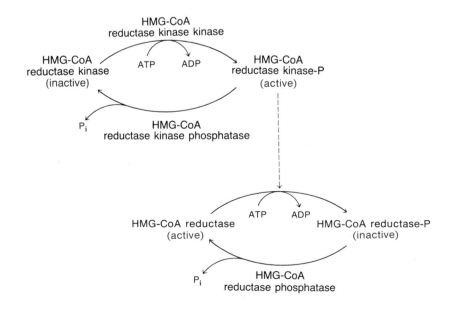

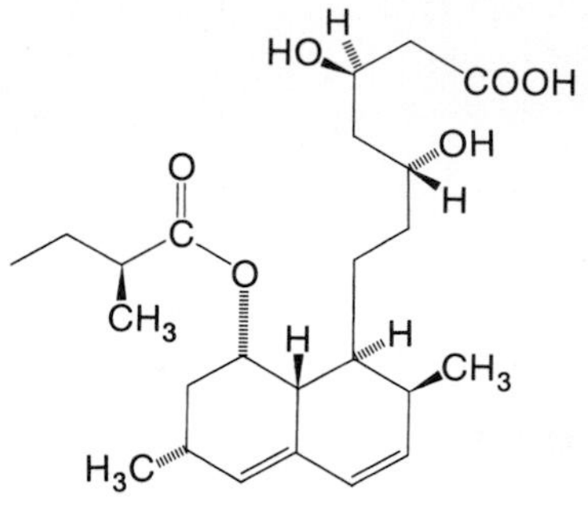

Figure 23–6
Structure of lovastatin acid, a potent competitive inhibitor of HMG-CoA reductase.

Cholesterol biosynthesis is also controlled by plasma low-density lipoproteins, which will be discussed in the context of lipoprotein metabolism later in this chapter.

It has long been recognized that a correlation exists between a high level of serum cholesterol and cardiovascular disease (e.g., most heart diseases, stroke). Most serum cholesterol originates from the liver; thus a drug that would specifically reduce cholesterol biosynthesis has been sought. It would be logical for such a drug to inactivate HMG-CoA reductase, since this enzyme catalyzes the key regulatory step in the pathway. Several fungal metabolites have been isolated that are competitive inhibitors of HMG-CoA reductase. One of the most active compounds is *lovastatin* (Figure 23–6), which competes favorably ($K_I = 0.6$ nM) with HMG-CoA for the reductase. Small doses of this drug (8 mg/kg of body weight) lower the levels of plasma cholesterol in dogs by 30 per cent. This drug has been recently approved in the United States for the treatment of patients with hypercholeterolemia.

It Takes Six Mevalonates and Ten Steps to Make Lanosterol, the First Tetracyclic Intermediate

In the next segment of the pathway, *mevalonate is converted to squalene, which is cyclized to form lanosterol.* The first stage in this sequence of reactions is the synthesis of the five-carbon isoprenoid intermediates *isopentenyl pyrophosphate* and *dimethylallyl pyrophosphate*. The synthesis involves four enzymes (Figure 23–7). The action of *mevalonate kinase* and *phosphomevalonate kinase* produces 5-pyrophosphomevalonate. The third enzyme (Figure 23–7), *pyrophosphomevalonate decarboxylase*, catalyzes a decarboxylation and elimination of the 3-hydroxyl group. The decarboxylase probably acts by an initial phosphorylation of the 3-hydroxyl group with ATP, followed by the *trans* elimination of the carboxyl and phosphate to give *isopentenyl pyrophosphate*. This intermediate can be enzymatically isomerized to *3,3-dimethylallyl pyrophosphate*. These two C_5 isoprenoid pyrophosphates react to produce the C_{10} intermediate *geranyl pyrophosphate* (Figure 23–8). Subsequently, a C_{15} intermediate, *farnesyl pyrophosphate*, is formed by the reaction of another molecule of isopentenyl pyrophosphate

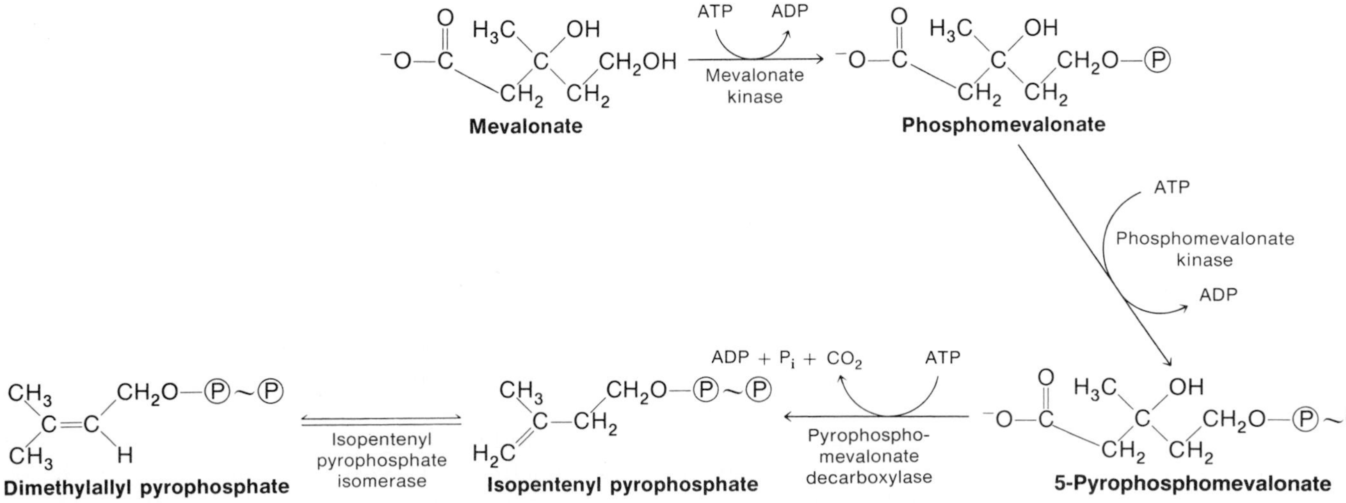

Figure 23–7
The conversion of mevalonate to isopentenyl pyrophosphate and dimethylallyl pyrophosphate. Mevalonate is converted to isopentenyl pyrophosphate in three steps. Each of these steps requires one ATP cleavage. The conversion of isopentenyl pyrophosphate to dimethylallyl pyrophosphate is readily reversible and does not require any further expenditure of ATP.

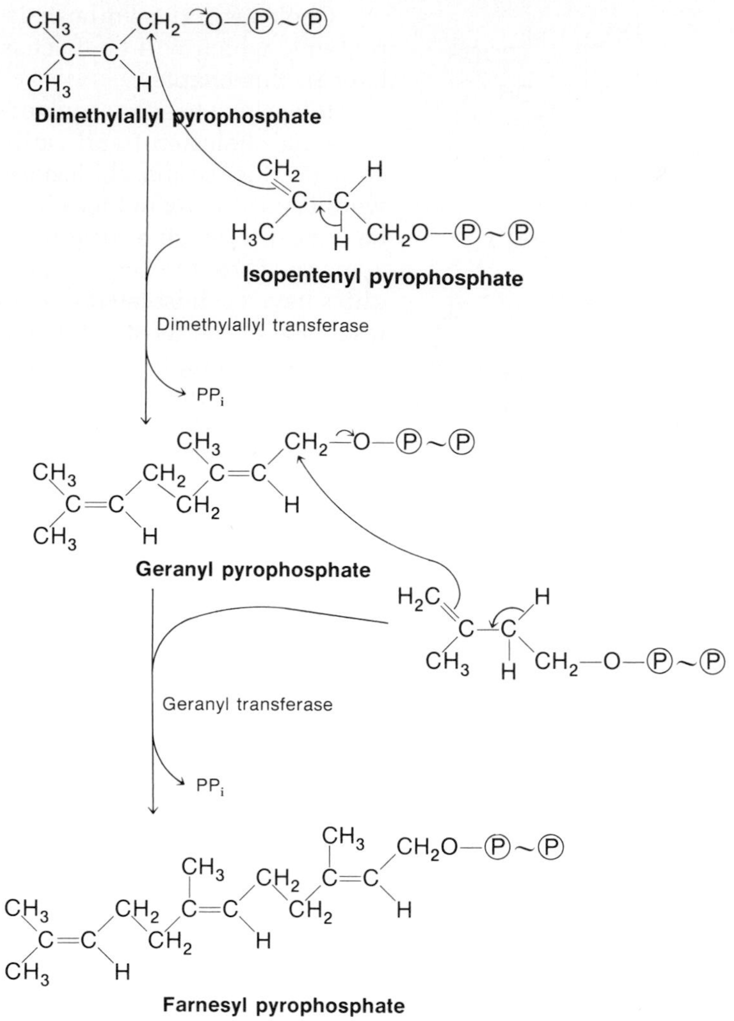

Figure 23–8

Biosynthesis of farnesyl pyrophosphate. Farnesyl pyrophosphate is a C_{15} intermediate containing three C_5 isoprenoid subunits. The two transferase reactions involved in the formation of farnesyl pyrophosphate occur by virtually identical mechanisms as shown.

with geranyl pyrophosphate. Two molecules of farnesyl pyrophosphate react to form _presqualene pyrophosphate_, which rearranges with the elimination of PP$_i$ to yield the C_{30} intermediate _squalene_ (Figure 23–9).

The enzymes that convert mevalonate to farnesyl pyrophosphate are probably cytosolic, whereas _farnesyl transferase_ (squalene synthase) is tightly associated with the endoplasmic reticulum (Figure 23–9). Even though such membrane-bound enzymes are difficult to solubilize, this enzyme has been isolated from yeast. In a polymeric form and in the presence of NADPH, farnesyl transferase will convert two farnesyl pyrophosphates to squalene. When dissociated into its depolymerized form, the farnesyl transferase catalyzes only the formation of presqualene pyrophosphate.

The two remaining reactions in the biosynthesis of lanosterol are shown in Figure 23–10. In the first of these reactions, _squalene-2,3-oxide_ is formed from squalene by _squalene monooxygenase_, an endoplasmic-reticulum-bound enzyme, in a reaction that requires O_2, NADPH, FAD, phospholipid, and a cytosolic protein. As can be seen in Figure 23–9, squalene is a symmetrical molecule, hence the formation of squalene oxide can be initiated from

BOX 23–A

Stereochemistry of the Conversion of Mevalonate to Squalene

As we have seen, the biosynthesis of cholesterol from acetyl-CoA is accomplished by means of a complicated route that involves more than 30 different enzymes, numerous cofactors, and at least two cytosolic proteins. Although this certainly represents enough complexity for most people, George Popjak and John Cornforth, in Britain, recognized that it was possible to define the conversion of mevalonate to squalene in a more precise and elegant manner. These two scientists observed in the 1960s that there were 14 "stereochemical ambiguities" in the conversion of pyrophosphomevalonate to squalene. In other words, there were 2^{14}, or 16,384, theoretically possible stereochemical pathways by which mevalonate could be transformed into squalene. This in itself was a remarkable observation. More remarkably, these two men and their collaborators subsequently were able to define precisely which one of these 16,384 possible stereochemical pathways actually occurred.

It is beyond the scope of this introductory text for us to examine each of the 14 stereochemical ambiguities and their resolution. Two examples should demonstrate the principles involved. The reaction catalyzed by *pyrophosphomevalonate decarboxylase* could involve either a *cis* or *trans* elimination of the carboxyl and hy-droxyl groups to produce isopentenyl pyrophosphate (Figure 23–7). Cornforth et al. solved this stereochemical ambiguity by the synthesis of a stereospecifically deuterium-labeled pyrophosphomevalonate (Illustration 1) that was incubated with the decarboxylase. The product of the reaction was isolated, and after several chemical transformations, Cornforth and co-workers were able to distinguish which of the two possible isomers of isopentenyl pyrophosphate was formed. The product was solely the result of a *trans* elimination (Illustration 1). Thus the first stereochemical ambiguity was resolved.

Another stereochemical problem was to determine which of the two hydrogens from C-2 of isopentenyl pyrophosphate was lost in its isomerization to dimethylallyl pyrophosphate (Illustration 2). Isopentenyl pyrophosphate was chemically labeled with deuterium on the 2-R or 2-S position (the use of the RS terminology is explained in Figure 11–8). Incubation of this substrate with the isomerase and subsequent characterization of the product demonstrated that the proR hydrogen (H_R) was specifically removed during the isomerization reaction (Illustration 2). Further information on the 14 stereochemical ambiguities and their resolution can be found in the book by Ronald Bentley (see Selected Readings).

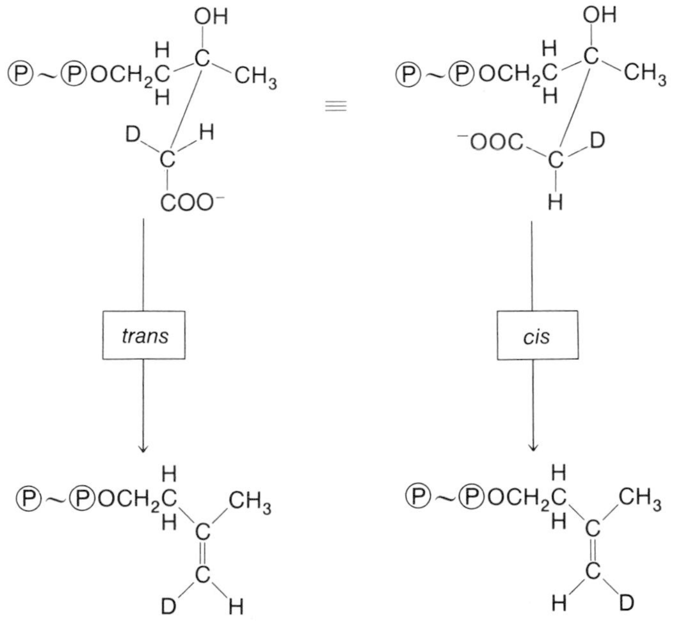

Illustration 1

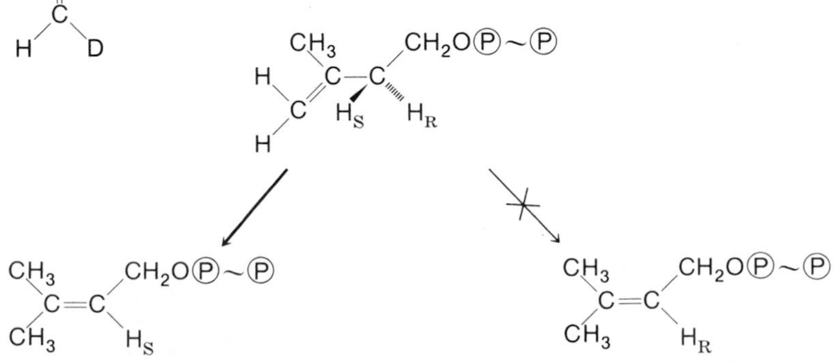

Illustration 2

Figure 23–9

Formation of squalene from farnesyl pyrophosphate. Farnesyl transferase is tightly complexed to the endoplasmic reticulum. The enzyme is active in both protomeric and polymeric forms. In the protomeric form the enzyme catalyzes only to the intermediate presqualene pyrophosphate. Only in the polymeric form does the reaction proceed beyond the first step to squalene.

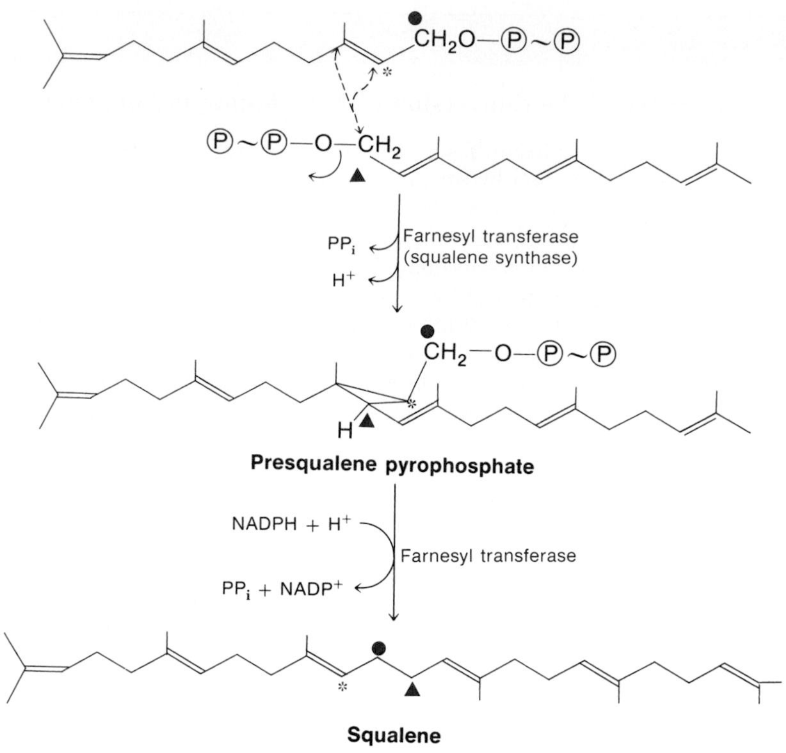

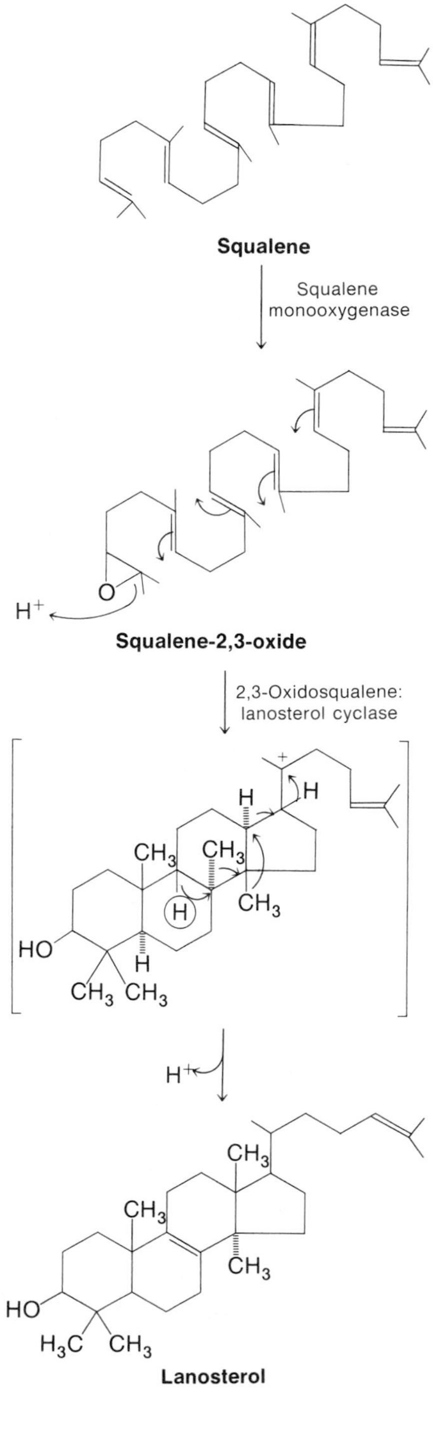

Squalene

Squalene-2,3-oxide

Lanosterol

Farnesyl transferase (squalene synthase)

Presqualene pyrophosphate

Farnesyl transferase

Squalene

either end of the molecule. The oxide is converted into lanosterol by another endoplasmic-reticulum-bound enzyme, *2,3-oxidosqualene lanosterol cyclase*. The reaction can be formulated as proceeding by means of a protonated intermediate that undergoes a concerted series of *trans*-1,2 shifts of methyl groups and hydride ions to produce lanosterol (Figure 23–10).

From Lanosterol to Cholesterol Takes Another Twenty Steps

The last sequence of reactions in the biosynthesis of cholesterol involves approximately 20 enzymatic steps, starting with lanosterol. In mammals the major route involves a series of double-bond reductions and demethylations (Figure 23–11). The exact position in the scheme for the reduction of the Δ^{24} double bond is not established. Otherwise, the sequence of reactions involves the oxidation and removal of the 14α methyl group followed by the oxidation and removal of the two methyl groups at position 4 in the sterol. The final reaction is a reduction of the Δ^7 double bond in *7-dehydrocholesterol*.

An alternative pathway from lanosterol to cholesterol (Figure 23–11) initially involves three demethylations to give *zymosterol* and then isomerization of the Δ^8 double bond to the Δ^5 position to produce *desmosterol* (Figure 23–11). The final reaction in this pathway is the reduction of the Δ^{24} double bond.

The enzymes involved in the transformation of lanosterol to cholesterol are all located on the endoplasmic reticulum. In addition to these enzymes, two cytosolic proteins have been found that stimulate several of the mem-

Figure 23–10

The transformation of squalene into lanosterol. The squalene monooxygenase reaction requires O$_2$, NADPH, FAD, phospholipid, and a cytosolic protein. The cyclase reaction has no cofactor requirements. The reaction proceeds by means of a protonated intermediate that undergoes a concerted series of *trans*-1,2 shifts of methyl groups and hydride ions to produce lanosterol.

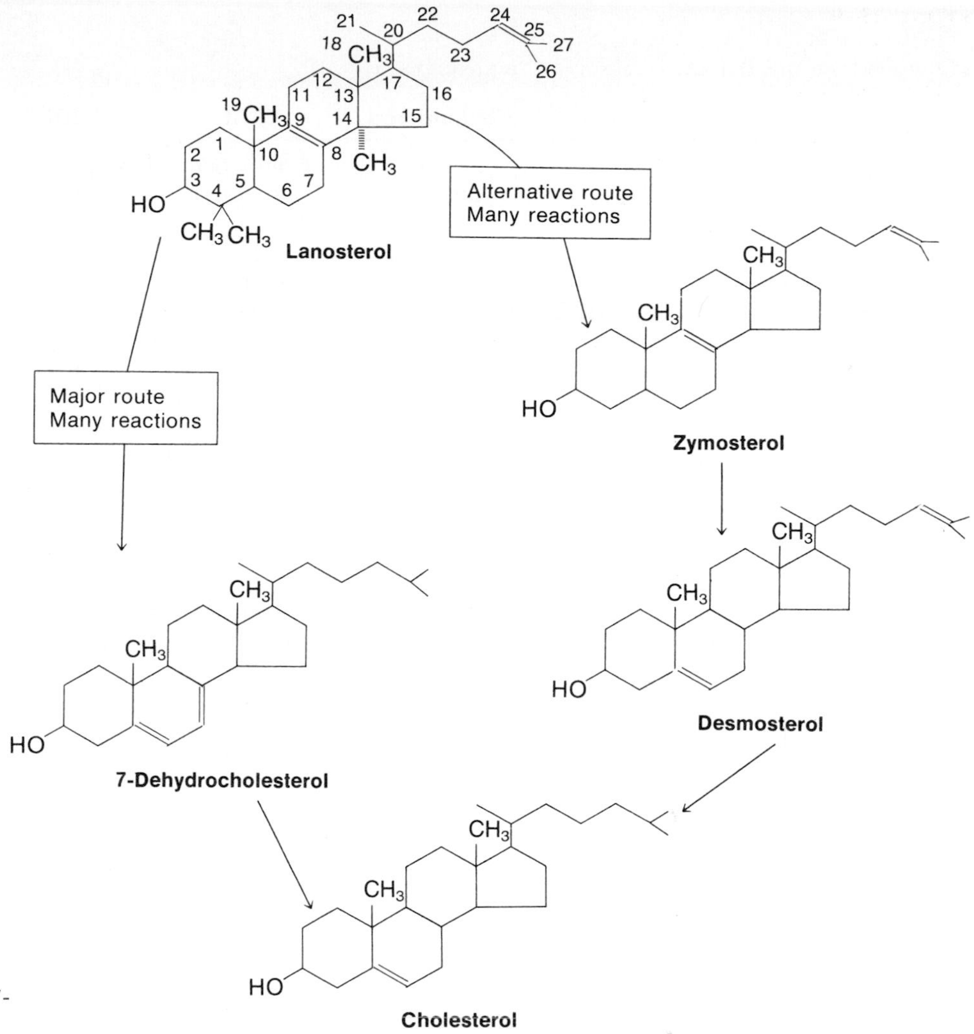

Figure 23–11
Two pathways for the conversion of lanosterol to cholesterol. The major route in mammals proceeds through 7-dehydrocholesterol.

brane-associated reactions that convert squalene to cholesterol. How these soluble proteins actually function in cholesterol biosynthesis is a problem of current interest.

LIPOPROTEIN METABOLISM

Unesterified fatty acids are carried in plasma by albumin (Chapter 18). The plasma also transports more complex lipids among the various tissues as components of _lipoproteins_ (particles composed of lipids and proteins). In this section we will be concerned with the structure and metabolism of these lipoproteins.

There Are Five Types of Lipoproteins in the Human Plasma

The amounts and types of lipids found in human plasma fluctuate accord-
ing to the dietary habits and metabolic states of the individual. The normal ranges for the lipid levels in plasma are shown in Table 23–1. In plasma these lipids are associated with proteins in the form of lipoproteins, which are classified into five major types on the basis of their density (Table 23–2). The lipoproteins of lowest density, the _chylomicrons,_ are the largest in size and contain the most lipid and the smallest percentage of protein. At the other extreme are the _high-density lipoproteins_ (HDL), which are the small-

TABLE 23–1
Normal Concentrations of the Major Lipid Classes in Plasma in Humans

Lipid	Concentration (g/l)
Total lipid	3.6–6.8
Cholesterol and cholesterol ester	1.3–2.6
Triacylglycerol	0.8–2.4
Phospholipid	1.5–2.5

TABLE 23–2
Composition and Density of Human Lipoproteins

	Chylomicron	VLDL	IDL	LDL	HDL
Density(g/ml)	<0.95	0.95–1.006	1.006–1.019	1.019–1.063	1.063–1.210
Diameter (nm)	500–1000	30–70	25–50	20–25	10–15
Components (% dry weight)					
Protein	1–2	10	18	25	33
Triacylglycerol	83	50	31	10	8
Cholesterol and cholesterol esters	8	22	29	46	30
Phospholipids	7	18	22	22	29
Apoprotein composition	A-I, A-II B_{48}, C-I, C-II, C-III	B_{100} C-I, C-II, C-III E	B_{100} C-I, C-II, C-III E	B_{100}	A-I, A-II C-I, C-II, C-III D E
Classification by electrophoresis	Omega	Pre-beta	Between beta and pre-beta	Beta	Alpha

est particles and contain the highest percentage by weight of protein and lowest percentage of lipid. Between these two classes, in both size and composition, are the _low-density lipoproteins_ (LDL), the intermediate-density lipoproteins (IDL), and the _very-low-density lipoproteins_ (VLDL).

The structures of the various lipoproteins appear to be similar (Figure 23–12). Each of the lipoprotein classes contains a neutral lipid core composed of triacylglycerol and/or cholesterol ester. Around this core is a layer of protein, phospholipid, and cholesterol oriented with the polar portions exposed to the surface of the lipoprotein.

There are at least eight apoproteins* (Table 23–2) associated with the lipoproteins, as well as several enzymes and a _cholesterol ester transfer protein_. The structure and function of these apoproteins has been intensely studied in the past decade, and some of the properties of these apoproteins are summarized in Table 23–3. Most of the apoproteins have been sequenced

*The apoprotein is the protein part of the lipoprotein.

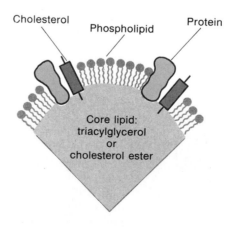

Figure 23–12
Generalized structure of human lipoproteins. The lipoproteins are spherical and vary in diameter from 10 nm to as much as 1000 nm, depending on the particular proteins and lipids. Each of the lipoprotein classes contains a neutral lipid core composed of triacylglycerol and/or cholesterol ester. Around the core is a layer of protein, phospholipid, and cholesterol that is oriented with the polar portions exposed to the surface of the lipoprotein.

TABLE 23-3
Properties of the Apoproteins of the Major Human Lipoprotein Classes

Apoprotein	M_r	Plasma Concentration (mg/100 ml)	Miscellaneous
A-I	28,300	90–120	Major protein in HDL (64%); contains 245 amino acids and no carbohydrate; activates LCAT*
A-II	17,400	30–50	Mainly in HDL (20% of dry mass); two identical chains with 77 amino acids each, joined by a disulfide at residue 6.
B-100	513,000	80–100	Major protein in LDL; very difficult to solubilize in detergents
B-48	241,000	<5	A protein found exclusively in chylomicrons.
C-1	7,000	4–7	Contains 57 amino acids
C-II	10,000	3–8	Contains 80–85 amino acids; activates lipoprotein lipase
C-III	9,300	8–15	Contains 79 amino acids and Gal, GalNAc, and NeuNAc
D	35,000	8–10	Also known as cholesterol ester transfer protein; associated with HDL and LCAT
E	33,000	3–6	Also known as arginine-rich lipoprotein

*LCAT = Lecithin-cholesterol acyltransferase.

and contain regions that are rich in hydrophobic amino acids, which facilitate binding of phospholipid.

Lipoproteins Are Made in the Endoplasmic Reticulum of the Liver and Intestine

Of the various lipid components of the lipoproteins, only the biosynthesis of *cholesterol esters* has not been mentioned. Cholesterol ester is the storage form of cholesterol in cells. It is synthesized from cholesterol and acyl-CoA by *acyl-CoA:cholesterol acyltransferase* (ACAT) (Figure 23–13), which is located on the cytosolic surface of liver endoplasmic reticulum. The synthesis of the apoproteins takes place on ribosomes that are bound to the endoplasmic reticulum. As mentioned previously, the biosynthesis of cholesterol, triacylglycerols, and phospholipids also occurs on the endoplasmic reticulum.

How the various components of the lipoproteins are assembled and secreted into the plasma is not known. Current ideas for this process suggest the transfer of the components from the endoplasmic reticulum to the Golgi apparatus, where secretory (membrane-encapsulated) vesicles are formed. These vesicles would subsequently fuse with the plasma membrane and release their lipoprotein contents into plasma (Figure 23–14).

The plasma lipoproteins appear to be made mainly in the liver and intestine. In the rat, approximately 80 per cent of the plasma apoproteins originate from the liver; the rest come from the intestine. Most of the components of chylomicrons, including apoprotein A, apoprotein B, phospholipid, cholesterol, cholesterol ester, and triacylglycerols, are products of the intes-

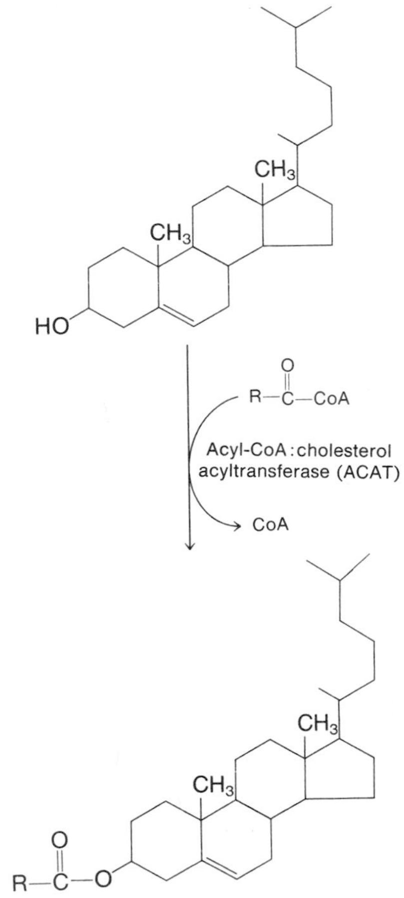

Figure 23–13
Biosynthesis of cholesterol esters. The acyl-CoA : cholesterol acyltransferase involved in cholesterol ester synthesis is located on the cytosolic surface of liver endoplasmic reticulum.

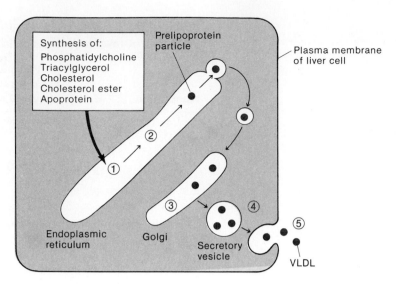

Figure 23–14
Postulated scheme for the synthesis, assembly, and secretion of VLDL by a hepatocyte (liver cell). (1) Synthesis: the apoproteins, phosphatidylcholine, triacylglycerol, cholesterol, and cholesterol esters are synthesized in the endoplasmic reticulum. (2) Assembly: these components are assembled into a prelipoprotein particle in the lumen of the endoplasmic reticulum. (3) Processing: the particle moves to the Golgi, where additional phospholipids and perhaps also cholesterol and cholesterol ester are added. (4) Vesicle formation: a secretory vesicle containing the lipoprotein particles is formed and fuses with the plasma membrane. (5) Secretion: the VLDL is released into the circulation.

tinal cells. The chylomicrons are secreted into lymphatic capillaries, which eventually enter the bloodstream at the large subclavian vein and therefore bypass the liver. The liver appears to be the major source of VLDL and HDL, which include apo A-I, A-II, B, C-I, C-II, C-III, and E and the lipid components of these lipoproteins. Low-density lipoprotein is produced from VLDL, as will be discussed below.

Chylomicrons and Very-low-density Lipoproteins (VLDL) Transport Cholesterol and Triacylglycerol to Other Tissues

Chylomicrons serve as the mode of transport of triacylglycerol and cholesterol ester from the intestine to other tissues in the body. Very-low-density lipoprotein functions in a similar manner for the transport of lipid from the liver to other tissues. These two types of triacylglycerol-rich particles are initially degraded by the action of _lipoprotein lipase_, an extracellular enzyme that is most active within the capillaries of adipose tissue, cardiac and skeletal muscle, and lactating mammary gland. Lipoprotein lipase catalyzes the hydrolysis of triacylglycerols (see Figure 18–2). The enzyme is specifically activated by apoprotein C-II, which is associated with chylomicrons and VLDL (Table 23–2). As a result, this lipase supplies the heart and adipose tissue with fatty acids, derived from these lipoproteins in plasma. In both the heart and adipose tissue, the fatty acids produced by lipoprotein lipase can be used for energy or stored as a component of triacylglycerols. Alternatively, the fatty acids can be bound by albumin and transported to other tissues.

As the lipoproteins are depleted of triacylglycerol, the particles become smaller. Some of the surface molecules (apoproteins, phospholipids) are transferred to HDL. In the rat, "remnants" that result from chylomicrons and VLDL catabolism are taken up by the liver. In humans, the uptake of remnant VLDL also occurs, but much of the VLDL appears to be gradually degraded,

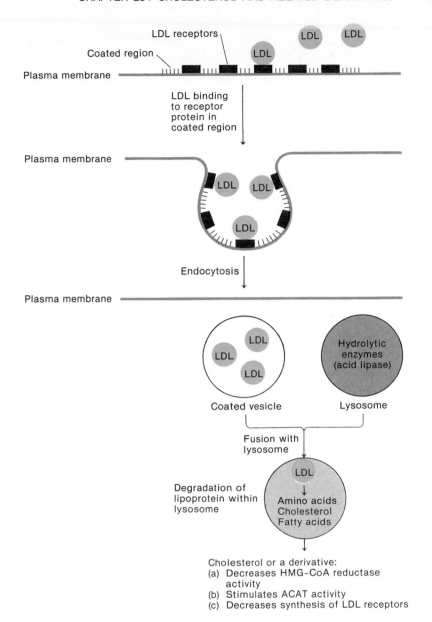

Figure 23–15

Receptor-mediated uptake of LDL by human skin fibroblasts. Specific LDL receptors are located in coated regions of the plasma membrane. LDL binding results in uptake by endocytosis and formation of a coated vesicle. This vesicle fuses with a lysosome containing many hydrolytic enzymes that degrade the lipoprotein, releasing cholesterol.

so that LDL particles are produced. The half-life for clearance of chylomicrons and remnants from plasma of man is 4 to 5 min. The clearance value for VLDL is 1 to 3 hr.

Low-density Lipoproteins (LDL) are Removed from the Plasma by the Liver, Adrenals, and Adipose Tissue

Each day approximately 45 per cent of the plasma pool of low-density lipoprotein is removed from human plasma by both the liver and extrahepatic tissues (particularly the adrenals and adipose tissue). The mechanism for the uptake of LDL in extrahepatic tissue has been extensively described by Michael Brown and Joseph Goldstein. They studied the uptake of LDL by human skin fibroblasts grown in cultures in Petri dishes. LDL particles bind to the cell surface by specific receptors that congregate in areas of the plasma membrane called *coated regions* (Figures 23–15 and 23–16). These areas of plasma membranes engulf the LDL particles (in a process called *endocytosis*) to form "coated vesicles," which are somehow directed toward and fuse

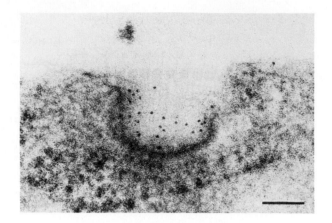

Figure 23–16
Electron micrograph of LDL particles (made electron-dense with covalently bound ferritin) bound to coated regions of a human skin fibroblast (97,000×). (From R. G. W. Anderson, M. S. Brown, and J. L. Goldstein, Role of the coated endocytic vesicle in the uptake of receptor-bound low density lipoprotein in human fibroblasts. *Cell* 10:351, 1977. Reprinted with permission.)

with lysosomes. The LDL particles are degraded within the lysosomes by the action of proteases and *lysosomal acid lipases* (lipid degradative enzymes). The cholesterol, or a derivative of cholesterol, diffuses from the lysosomes, suppresses the activity of *HMG-CoA reductase*, and stimulates the activity of *acyl-CoA:cholesterol acyltransferase* (ACAT). ACAT catalyzes the synthesis of cholesterol esters (Figure 23–13), which are then stored within the cell. The cholesterol (or its derivative) also suppresses the synthesis of the LDL receptors and thereby limits the uptake of LDL.

The LDL receptor is a glycoprotein with 839 amino acids and consists of five domains (Figure 23–17). Domain 1 is the binding site for lipoproteins that contain apo B_{100} and apo E. Domain 2 is 35 per cent homologous to part of the extracellular domain of the *precursor* to epidermal growth factor (growth factors are discussed in Chapter 31). The function of this domain is not clear. The cytosolic portion (domain 5) is required for congregation of the LDL receptors in the coated regions of the plasma membrane (Figure 23–15). The number of LDL receptors per cell can vary between 15,000 and 70,000, depending on the cell's requirement for cholesterol. Once an LDL molecule is bound to the receptor, both are rapidly internalized by endocytosis ($t_{1/2}$ = 3 min).

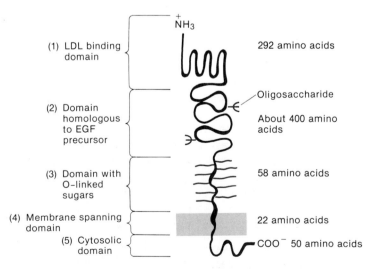

Figure 23–17
The LDL receptor: a single protein with five domains. The cytosolic portion (domain 5) is required for congregation of the LDL receptors in the coated regions of the plasma membrane (see Figure 23–15). Once an LDL molecule is bound to the receptor, both are rapidly internalized by endocytosis.

Serious Diseases Result from Cholesterol Deposits

Brown and Goldstein were able to deduce the pathway for the uptake and catabolism of LDL largely as a result of their studies on the inherited disease *familial hypercholesterolemia*. Patients with the homozygous (two defective genes) form of this disease have grossly elevated levels of plasma cholesterol (650–1000 mg/100 ml) (see Table 23–1 for normal values), which is largely carried by an elevated concentration of LDL. One result is the formation of cholesterol deposits in the skin (xanthomas) in various areas of the body. Of greater consequence is the deposit of cholesterol in arteries, which results in *atherosclerosis*, a condition that is the underlying cause of most cardiovascular diseases. In fact, patients with homozygous familial hypercholesterolemia have symptoms of heart disease by the early teens and usually die from cardiovascular disease before the age of 20. The heterozygotes (individuals with one normal and one defective gene) manifest similar but less severe symptoms. Their plasma cholesterol is in the range of 250–550 mg/100 ml of plasma, and they generally do not have a heart attack before the age of 40. The frequency of the heterozygous form of familial hypercholesterolemia has been estimated at 1 in 500, and that of the homozygous form is, in all likelihood, about 1 in 1 million.

Four different classes of biochemical mutations have been shown to cause familial hypercholesterolemia (Figure 23–18). The most common defect (class 1) is in the synthesis of the receptor. The other classes of mutations are defects in the transport of the receptor to the Golgi (class 2), defects in the binding of LDL (class 3), and inability of the receptors to cluster in coated pits (class 4).

A related disorder, Wolman's disease, has provided further evidence for the receptor-mediated pathway of LDL uptake (Figure 23–15). Wolman's disease is a very rare inborn error of metabolism (approximately 20 cases diagnosed since 1956) that is characterized by the accumulation of cholesterol esters and triacylglycerols in various tissues. The disease can be diagnosed within several weeks of birth but is fatal, usually within 6 months. It is caused by a complete lack of a lysosomal acid lipase, which is responsible for the normal catabolism of cholesterol esters and triacylglycerols in lysosomes. *Cholesterol ester storage disease* is a related disorder, caused by a substantial reduction in the activity of lysosomal acid lipase (1 to 20 per cent of normal). The symptoms are far less severe than in Wolman's disease, and patients have survived to the age of 40.

Figure 23–18

Four classes of mutations that disrupt the structure and function of the LDL receptor. Each class of mutation interferes with a different step in the process by which the receptor is synthesized. Class 1 mutations result in defective receptor synthesis in the endoplasmic reticulum. Class 2 mutations lead to defective receptor processing in the Golgi. Class 3 mutations result in defective receptor binding sites for LDL particles. Finally, class 4 mutations result in the inability of a receptor to cluster in coated pits.

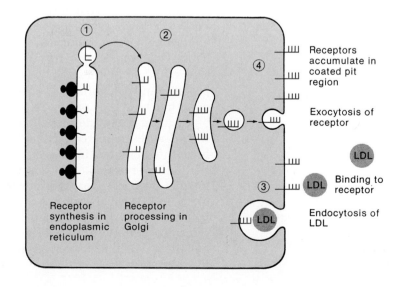

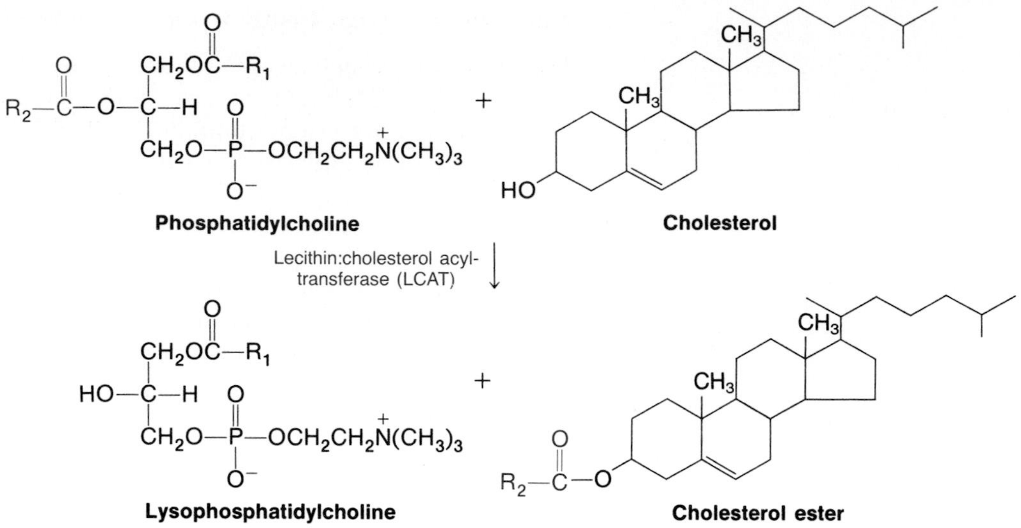

Figure 23–19
Reaction catalyzed by lecithin: cholesterol acyltransferase (LCAT). The resulting cholesterol ester is transferred to VLDL and LDL particles by a lipid transfer protein.

The importance of the work done on the LDL receptor and familial hypercholesterolemia was recognized in 1985 when Joseph Goldstein and Michael Brown were awarded the Nobel Prize in Physiology or Medicine.

High-density Lipoproteins (HDL) May Reduce Cholesterol Deposits

The catabolism of high-density lipoproteins is a complex process that is currently under investigation. The half-life of HDL in human plasma (5 to 6 days) is much longer than for the other lipoproteins. When HDL is secreted into plasma from liver, it has a discoid shape and is almost devoid of cholesterol ester. These newly formed HDL particles are converted into spherical particles by the accumulation of cholesterol ester. The cholesterol ester is derived from cholesterol and phosphatidylcholine on the surface of the HDL particle in a reaction catalyzed by *lecithin:cholesterol acyltransferase* (LCAT) (Figure 23–19). LCAT is a glycoprotein (24 per cent carbohydrate by weight) with a molecular weight of 59,000. This enzyme is associated with HDL in plasma and is activated by apoprotein A-I, a component of HDL (Table 23–3). Associated with the LCAT-HDL complex is *apoprotein D*, also known as *cholesterol ester transfer protein*. It is currently thought that after cholesterol esters are formed in HDL, apoprotein D catalyzes the transfer of cholesterol esters from HDL to VLDL or LDL. In the steady state, cholesterol esters that are synthesized by LCAT would be transferred to these other lipoproteins and catabolized as noted earlier. The HDL particles themselves turn over, but how they are degraded is not firmly established.

Although elevated levels of cholesterol and LDL in human plasma are linked with an increased incidence of cardiovascular disease, recent data have shown that an increase in concentration of HDL in plasma is correlated with a lowered risk of coronary artery disease. Why does an elevated HDL level in plasma appear to protect against cardiovascular disease, whereas an elevated LDL level seems to cause this disease? The answer to this question is not known. An explanation currently favored is that HDL functions in the return of cholesterol to the liver, where it is metabolized and secreted. The net effect would be a decrease in the amount of plasma cholesterol available for deposit in arteries.

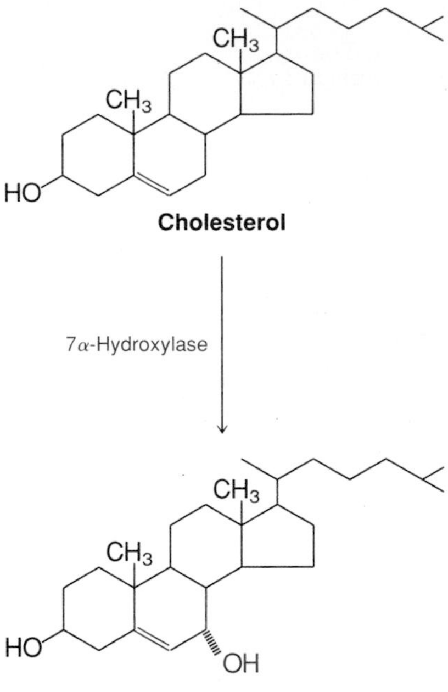

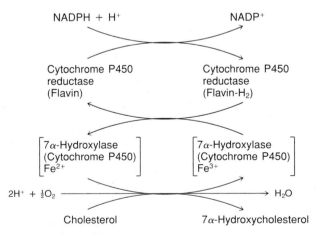

Cholesterol

7α-Hydroxylase

7α-Hydroxycholesterol

Figure 23–20

Formation of 7-hydroxycholesterol. The committed and rate-limiting reaction for bile acid synthesis is catalyzed by the endoplasmic reticulum enzyme 7α-hydroxylase.

BILE ACID METABOLISM

The conversion of cholesterol to bile acids is quantitatively the most important mechanism for degradation of cholesterol. In a normal human adult approximately 0.5 g of cholesterol is converted to bile acids each day. The regulation of this process operates at the initial biosynthetic step catalyzed by the endoplasmic reticulum enzyme *7α-hydroxylase* (Figure 23–20). The 7α-hydroxylase is one of a group of enzymes called *mixed function oxidases*, which are involved in the hydroxylation of the sterol molecule at numerous specific sites. A mixed function oxidase is an enzyme complex that catalyzes hydroxylation of a substrate and production of H_2O from a single molecule of O_2. The 7α-hydroxylase is one of several enzymes referred to as cytochrome P450. The hydroxylation of cholesterol also requires *NADPH:cytochrome P450 reductase* (Figure 23–21).

The subsequent conversion of 7-hydroxycholesterol to cholic acid is outlined in Figure 23–22. These reactions involve oxidation of the 3β-hydroxyl group, isomerization of the double bond, 12α-hydroxylation, reduction of the double bond, and reduction of the 3-keto group to a 3α-hydroxyl group. Additional hydroxylations and oxidation reactions on the side chain lead to cholic acid, one of the two major human bile acids (see Chapter 5).

The bile acids are mostly converted to the corresponding bile salts, as shown for the formation of glycocholate in Figure 23–23. The bile salts are critically important for the solubilization of lipids in the intestine, as we saw in Chapter 5.

METABOLISM OF STEROID HORMONES

The structures of the major steroid hormones in mammals and their function were given in Chapter 5 (Table 5–7). More information on the hormonal functions of steroids and the mechanism by which steroids work is presented in the chapter on hormone action (Chapter 31). Here we will focus on the enzymatic processes by which cholesterol is converted to the major steroid hormones.

The initial reaction in steroid hormone biosynthesis is catalyzed by *desmolase* (side-chain cleavage complex), which is found in the mitochondria of steroid-producing tissues (e.g., adrenals, gonads). The reaction is

NADPH + H⁺   NADP⁺

Cytochrome P450 reductase (Flavin)   Cytochrome P450 reductase (Flavin-H₂)

[7α-Hydroxylase (Cytochrome P450) Fe²⁺]   [7α-Hydroxylase (Cytochrome P450) Fe³⁺]

2H⁺ + ½O₂ ⟶   ⟶ H₂O

Cholesterol   7α-Hydroxycholesterol

Figure 23–21

The reaction for the mixed function oxidase activity of 7α-hydroxylase.

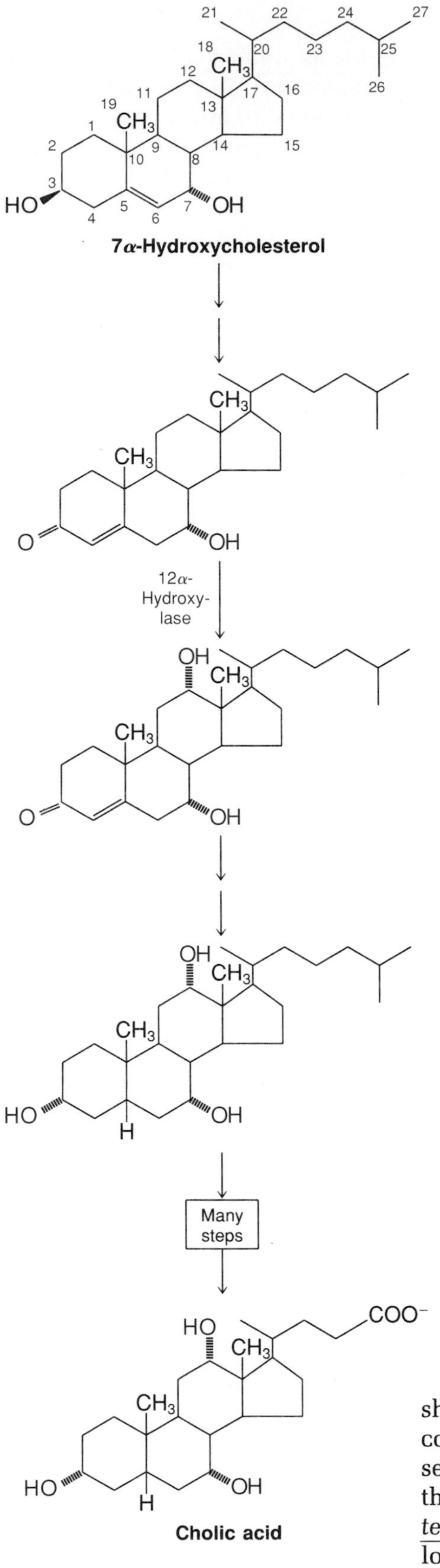

Figure 23–22
Conversion of 7α-hydroxycholesterol to cholic acid. The increase in the number of polar groups in the converison to cholic acid improves the water solubility.

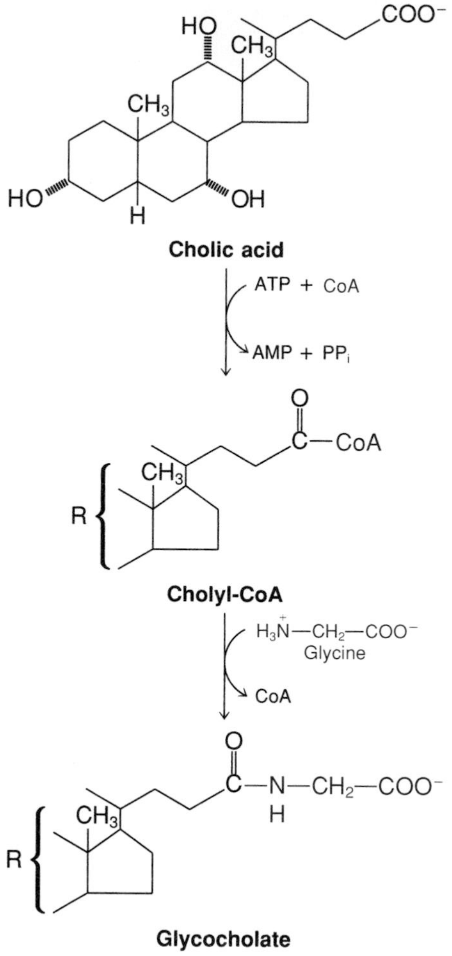

7α-Hydroxycholesterol

12α-Hydroxy-lase

Many steps

Cholic acid

Cholic acid

ATP + CoA

AMP + PP$_i$

Cholyl-CoA

$H_3\overset{+}{N}$—CH_2—COO^-
Glycine

CoA

Glycocholate

Figure 23–23
Conversion of a bile acid into a bile salt in the formation of glycocholate. Glycocholate is an example of a bile salt. The bile salts solubilize lipids in the small intestine so that they can be degraded by lipases.

shown in Figure 23–24. Desmolase appears to consist of two hydroxylases containing cytochome P450, and a lyase. The product, *pregnenolone*, is subsequently transferred to the endoplasmic reticulum, where an oxidation of the hydroxyl group and isomerization of the double bond produces *progesterone* (Figure 23–24). Progesterone is a steroid hormone, and it or pregnenolone is the biosynthetic precursor of all other steroid hormones.

The initial step in the conversion of progesterone to aldosterone (Figure 23–25) is catalyzed by a *21-hydroxylase*, present on endoplasmic reticulum from the adrenal cortex but absent in gonads and placenta. This enzyme

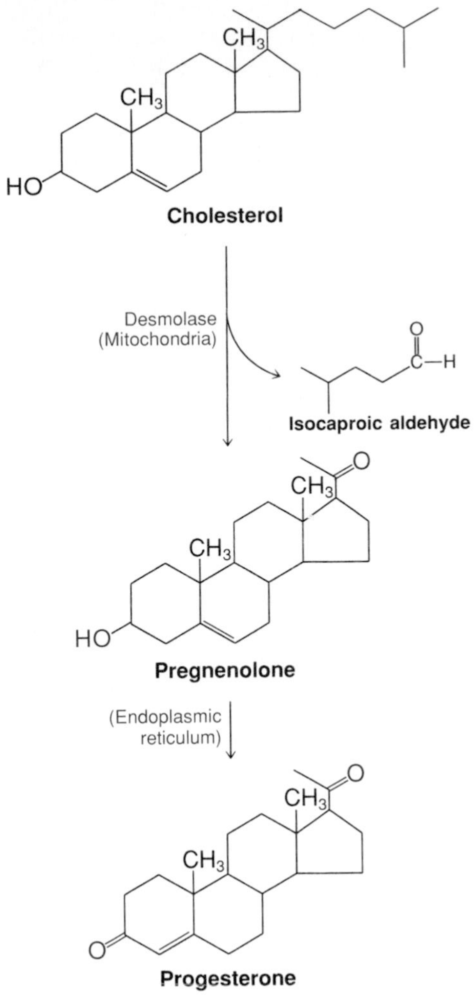

Figure 23–24

Biosynthesis of progesterone. Desmolase is found in the mitochondria of steroid-producing tissues. It is converted to pregnenolone in the mitochondria and then transferred to the endoplasmic reticulum, where it is converted to progesterone. Progesterone is a hormone as well as the precursor of several other hormones.

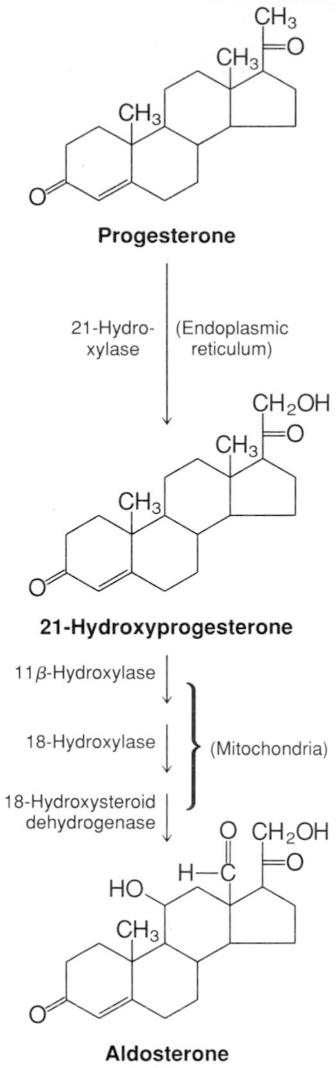

Figure 23–25

Conversion of progesterone to aldosterone. The initial reaction, involving the 21-hydroxylase enzyme, occurs on the endoplasmic reticulum of the adrenal cortex. Two mitochondrial enzymes are involved in the next step, leading to aldosterone.

(M_r = 47,000) is also a cytochrome P450 protein and has been purified from adrenocortical endoplasmic reticulum. It is distinct from the _11-β-hydroxylase_ and _18-hydroxylase_, two mitochondrial enzymes also involved in aldosterone synthesis (Figure 23–25). It is curious that the biosynthesis of aldosterone from cholesterol begins in mitochondria with desmolase. Then the next reactions are catalyzed by enzymes on the endoplasmic reticulum, and finally the 21-hydroxy-progesterone is carried back to the mitochondria for the last enzymatic steps. The reason for this subcellular compartmentation is not obvious, and the mechanism by which the cell directs the intermediates from one subcellular site to another is not known.

The _17α-hydroxylase_ (Figure 23–26) is a mixed function oxidase found on the endoplasmic reticulum in all steroid-secreting organs. _It is the key enzyme for directing steroids into the synthesis of glucocorticoids, androgens, and estrogens._ Hydroxylations at the 11β-position and the 21-position direct the 17-OH progesterone into cortisol, a glucocorticoid made in the adrenals (Figure 23–26). Alternatively, the gonads will direct 17-OH proges-

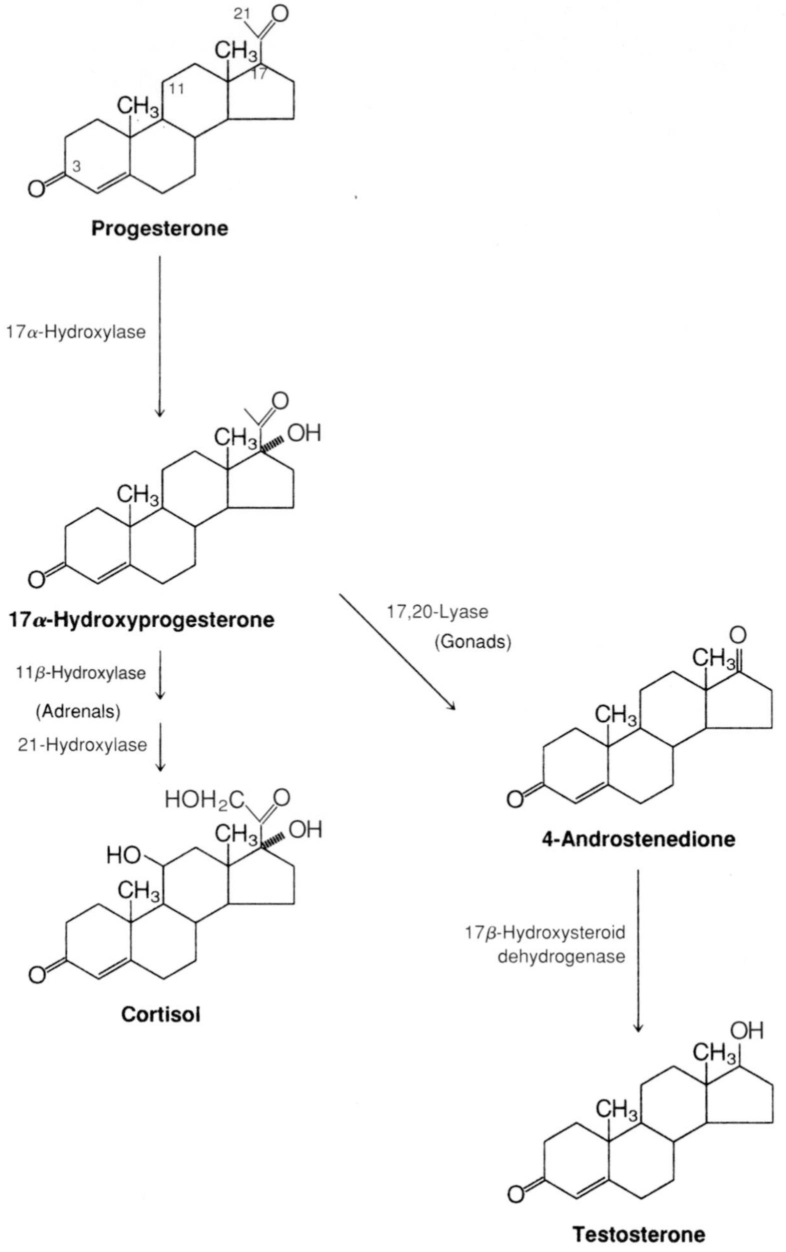

Figure 23–26

Two pathways for the conversion of progesterone to other steroid hormones. The first reaction, catalyzed by 17α-hydroxylase, is found in all steroid-secreting organs. Subsequent conversions to cortisol and testosterone are organ-specific as shown.

terone to the synthesis of *testosterone* as the result of the action of *17,20-lyase* and a reduction of the 17-keto group (Figure 23–26). Testosterone can be converted to *dihydrotestosterone* as discussed in Chapter 31.

The female sex hormones arise from testosterone via the 19-hydroxylated intermediate (Figure 23–27). The enzyme complex responsible for this conversion is called the *aromatase system* and is one of the few reactions by which mammals can make an aromatic ring. This complex is associated with the endoplasmic reticulum of cells in the ovary and placenta.

Thus the second major degradative fate of cholesterol in mammals is the formation of steroid hormones. The enzymes involved are found in mitochondria and endoplasmic reticulum of steroid-producing tissues. The

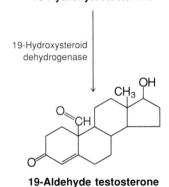

Testosterone

↓ 19-Hydroxylase

19-Hydroxytestosterone

↓ 19-Hydroxysteroid
dehydrogenase

19-Aldehyde testosterone

↓ 10,19-Lyase

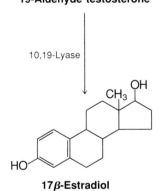

17β-Estradiol

Figure 23–27
The conversion of testosterone into estradiol. These reactions are catalyzed by a complex of enzymes called the "aromatase system," which consists of three enzyme activities located on the endoplasmic reticulum. The aromatase system is found in the ovaries and the placenta.

UDP—Glucuronic acid
+
"Inactive steroid"—OH

↓ Glucuronyl
transferase

Figure 23–28
Inactivated steroid hormones are conjugated to glucuronic acid prior to excretion in the urine.

hydroxylases are in each case mixed function oxidases that have cytochrome P450 at the active site. The remarkable effects of these potent hormones are discussed in Chapter 31.

There is no known pathway in mammals by which the steroid ring nucleus can be degraded to smaller molecules such as acetate. The carbon atoms of the ring system cannot, therefore, be used as a source of metabolic energy. However, many reactions occur, particularly in the liver, for the partial catabolism and inactivation of the steroid hormones. Frequently, these reactions involve reduction of ketone groups or double bonds. These inactive steroids are conjugated to *glucuronic acid* (Figure 23–28) or *sulfate*. The excretion of such derivatives results in a very large number of steroid metabolites in the urine. For example, 20 different metabolites of estrogen, conjugated to sulfate or glucuronic acid, have been identified in human urine.

OVERVIEW OF MAMMALIAN CHOLESTEROL METABOLISM

The metabolism of cholesterol in mammals is extremely complex. A summary sketch (Figure 23–29) helps to draw the major metabolic interrelationships together. Cholesterol is taken in through the diet (Figure 23–29b) or biosynthesized largely in the liver (Figure 23–29a). From the intestine, cholesterol is secreted into the plasma mainly as a component of chylomicrons. These particles are quickly degraded by lipoprotein lipase and the remnants are removed by the liver. Apoproteins and lipid components of the chylomicrons and remnants appear to exchange with HDL. Cholesterol made in the liver (Figure 23–29a) has several alternative fates. It can be (1) secreted into plasma as a component of HDL and VLDL, (2) stored in droplets as cholesterol ester, (3) used as a structural component of cell membranes, or (4) converted into bile salts. In plasma, VLDL is degraded to IDL and LDL by the action of lipoprotein lipase and through exchange reactions with HDL. The LDL serves as a major carrier of cholesterol to extrahepatic cells, which include the adrenals and gonads. LDL and HDL are also returned to the liver. The steroid hormones made in the adrenals and gonads are delivered to various target tissues and promote a wide range of metabolic effects. The steroid hormones are eventually excreted as glycosyl conjugates in the urine. The bile salts made in the liver are delivered to the upper intestine, where they aid in the solubilization of dietary lipid. Most of the bile salts are resorbed in the lower intestine and returned to the liver by the portal vein.

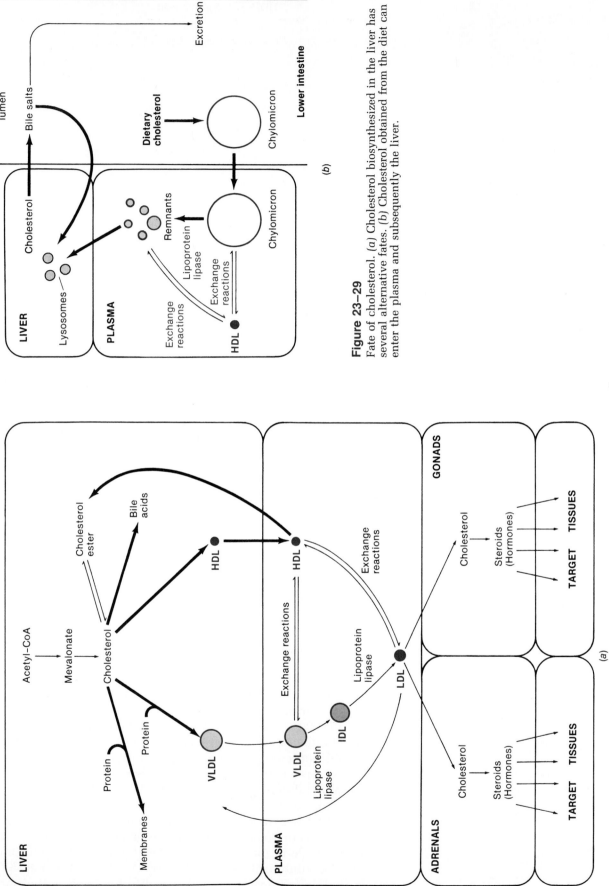

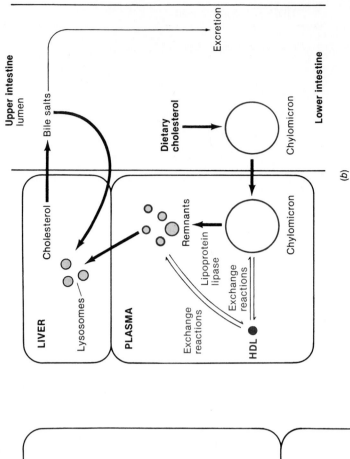

Figure 23–29

Fate of cholesterol. (a) Cholesterol biosynthesized in the liver has several alternative fates. (b) Cholesterol obtained from the diet can enter the plasma and subsequently the liver.

SUMMARY

Steroids are derivatives of the tetracyclic hydrocarbon perhydrocyclopentanophenanthrene. The biosynthesis of steroids begins with the conversion of three molecules of acetyl-CoA into mevalonate, the decarboxylation of mevalonate, and its conversion to isopentenyl pyrophosphate. Six molecules of isopentenyl pyrophosphate are polymerized into squalene, which is cyclized to yield lanosterol. Lanosterol is converted to cholesterol, which is the precursor of bile acids and steroid hormones.

The rate of cholesterol biosynthesis appears to be regulated primarily by the activity of HMG-CoA reductase. This key enzyme is controlled by the rate of enzyme synthesis and degradation and by phosphorylation/dephosphorylation reactions, and synthesis of the reductase is inhibited by cholesterol delivered to cells by means of low-density lipoproteins (LDL).

Complex lipids such as cholesterol and phospholipids are carried in plasma by lipoproteins, which are synthesized and secreted by the intestine and liver. The major lipoproteins are chylomicrons, very-low-density lipoproteins (VLDL), low-density lipoproteins (LDL), and high-density lipoproteins (HDL). The triacylglycerols in chylomicrons and VLDLs are degraded in plasma by lipoprotein lipase, and the fatty acids and monoacylglycerols are absorbed primarily by heart, skeletal, and adipose tissue. LDLs are removed from plasma by an endocytotic process after binding to specific LDL receptors on the plasma membrane. The LDLs are enzymatically degraded in the lysosomes. In familial hypercholesterolemia, the specific receptors for LDL uptake are inactive. High levels of LDL are associated with an increased risk of cardiovascular disease, whereas high levels of HDL seem to protect against this disease.

Bile acids are C_{24} carboxylic acids that are biosynthetically derived from cholesterol. The 7α-hydroxylation of cholesterol is the committed and rate-limiting reaction in the synthesis of bile acids. Salts formed from the bile acids are secreted into the small intestine and aid the solubilization and digestion of lipids.

Steroid hormones are biosynthesized from cholesterol in the adrenal cortex, gonads, and placenta. These steroids are important hormones for many specific physiological processes.

SELECTED READINGS

Bentley, R., *Molecular Asymmetry in Biology*, Vol. 2. New York: Academic Press, 1970. This book contains a very lucid explanation of the stereochemistry of cholesterol biosynthesis.

Brown, M. S., and J. L. Goldstein, A receptor-mediated pathway for cholesterol homeostasis. *Science* 232:34–47, 1986. An article describing their research on the LDL receptor and familial hypercholesterolemia.

Fielding, C. J., and P. E. Fielding, Metabolism of cholesterol and lipoproteins. In D. E. Vance and J. E. Vance (eds.), *Biochemistry of Lipids and Membranes*. Menlo Park, Calif.: Benjamin/Cummings, 1985. This chapter provides an advanced treatment of the complexities of cholesterol and lipoprotein metabolism.

Makin, H. L. J., *Biochemistry of Steroid Hormones*, second ed. Oxford: Blackwell Scientific Publication, 1984. An advanced and comprehensive treatment of steroid hormones.

Stanbury, J. B., J. B. Wyngaarden, D. S. Fredrickson, J. L. Goldstein, and M. S. Brown, *The Metabolic Basis of Inherited Disease*, fifth ed. New York: McGraw-Hill, 1983. This book has many chapters on inherited disorders of lipid and lipoprotein metabolism.

PROBLEMS

1. Which enzyme catalyzes the rate-limiting reaction in cholesterol biosynthesis? Draw the structures of the substrates and products for this reaction.

2. During routine investigations, the plasma from a family of rats was found to have very low concentrations of cholesterol. When the microsomal HMG-CoA reductase from liver was assayed, there was extremely low activity. When the cytosol from normal rats was added to the microsomes that had low HMG-CoA reductase activity, the enzyme activity was gradually restored to normal values. What enzyme activity (or activities) might be deficient in these unusual rats?

3. One way of reducing the high serum cholesterol associated with cardiovascular disease is by inhibiting cholesterol biosynthesis. Give an example of a treatment designed to achieve this aim and explain its underlying rationale.

4. A person with diabetes is found to show no signs of ketone bodies in plasma even when in diabetic shock. Which enzymes in ketone body synthesis might be deficient? If cholesterol synthesis were normal, would this be a clue as to which enzyme might be deficient?

5. Compare the pathways by which fatty acids and cholesterol are synthesized.

6. Sketch schematically the relationship between the high-density lipoproteins (HDL) and the low-density lipoproteins (LDL), indicating the approximate half-

lives of each class of particle. How do these half-lives reflect the function of each class of lipoproteins?

7. A deficiency of apoprotein C-II results in the disease hyperlipoproteinemia type I, in which there is a massive increase in the concentration of plasma triacylglycerol. Provide an explanation for this clinical finding.

8. What is the hereditary defect in Wolman's disease? Would you expect HMG-CoA reductase activity to be high or low in skin fibroblasts cultured from patients with this disorder? Would the number of LDL receptors be high or low in these fibroblasts?

9. There is an inherited disease in which lecithin-cholesterol acyltransferase (LCAT) is deficient. What effect would you expect this deficiency to have on the composition of HDL and other lipoproteins in plasma?

10. Describe briefly the role of mixed function oxidases in cholesterol metabolism. What is the principal cofactor involved in cholesterol degradation?

24

AMINO ACIDS AND AMINO-ACID-DERIVED PRODUCTS

Amino acids are best known as the building blocks of proteins, and indeed, that is the main function of the twenty L-amino acids most commonly found in proteins (see Chapter 1). These and additional amino acids also serve as precursors for a number of other important metabolites. Another role of amino acids is to serve as vehicles for nitrogen and sulfur fixation into biomolecules. Finally, all amino acids can be catabolized, and the resulting carbon skeletons and energy can contribute to other biosynthetic processes. Thus far, we have discussed only the catabolism of amino acids (see Chapter 19), and not the other aspects of their metabolism.

In view of the central importance of amino acids in proteins, we might expect that all organisms would possess the necessary enzymes to synthesize the protein amino acids. Quite surprisingly, we find that this is not the case. In mammals, for example, fewer than half of the protein amino acids can be synthesized by *de novo* pathways. The remainder must be supplied by nutrients. Frequently this leads to nutritional problems, since many diets that might contain sufficient carbohydrates and other essential components are deficient in one or more of the required amino acids.

Why did the evolutionary process not favor the preservation of pathways for synthesizing all the amino acids needed in higher animals? Can this be a case of too great a genetic load? That seems unlikely, since only 84 genes are required to encode all of the enzymes needed to make the 20 amino acids found in proteins. Mammals are believed to have in excess of 50,000 genes. Another possible reason for the loss of competency is that some of the intermediates in amino acid biosynthesis may have toxic effects on other sophisticated biochemical processes occurring in higher eukaryotes. There is no proof for this explanation, but it is at least a tenable hypothesis. We do know that amino acid excess and certain degradative intermediates can lead to metabolic problems (see Chapter 19).

Because of the biosynthetic insufficiencies that are found in many higher eukaryotes, we will devote a high percentage of this chapter to amino acid metabolism in microorganisms. We begin, however, with a discussion of the amino acid needs of rats and humans. Next we consider the broader question of nitrogen fixation in biomolecules. Then we describe the metabolism associated with amino acid biosynthesis, and finally we discuss the synthesis of some nonprotein amino acids and the utilization of amino acids for nonprotein products.

AMINO ACIDS AND NUTRITIONAL NEEDS

The inability of mammals to synthesize all of the amino acids they require has led to the classification of amino acids as *essential* and *nonessential*. An "essential" amino acid, in this classification, means one that must be supplied in the diet if the organism is to maintain a positive nitrogen balance. As we will see, the absence of a *de novo* pathway for the biosynthesis of an amino acid does not automatically mean that the amino acid must be obtained from the diet. For one thing, it is frequently possible for the organism to make one amino acid from another amino acid. Furthermore, an alternative pathway for synthesis may supply sufficient amino acid for maintenance, if not for growth.

The classical differentiation between essential and nonessential amino acids was made for rats by W. C. Rose. He based his identification on the weight gain of growing white rats that were fed diets containing 19 of the 20 amino acids found in proteins. His results appear in the first two columns of Table 24–1. For humans, there have been only a few studies. These were based not on weight gain or loss, but on short-term maintenance of a positive nitrogen balance. If there was a negative nitrogen balance (total nitrogen excretion exceeding total nitrogen intake) during the period when a single amino acid was excluded from the diet, that was taken to mean that tissue protein was being degraded and used to supply the missing amino acid. Table 24–1 lists the findings. For comparison, the amino acids found to be essential and nonessential for a strain (L) of mouse fibroblasts grown in cell culture are also included in Table 24–1.

TABLE 24–1
The Essential Amino Acids

For Weight Gain in Protein-Starved Adult Rats		For Positive Nitrogen Balance in Adult Humans		For Mouse L Cells in Culture	
Essential	**Nonessential**	**Essential**	**Nonessential**	**Essential**[a]	**Nonessential**
Histidine	Alanine	Isoleucine	Alanine	Arginine	Alanine
Isoleucine	Arginine[b]	Leucine	Arginine	Cysteine	Asparagine
Leucine	Asparagine	Lysine	Asparagine	Glutamine	Aspartate
Lysine	Aspartate	Methionine	Aspartate	Histidine	Glutamate
Methionine	Cysteine	Phenylalanine	Cysteine	Isoleucine	Glycine
Phenylalanine	Glutamate	Threonine	Glutamate	Leucine	Proline
Threonine	Glutamine	Tryptophan	Glutamine	Lysine	Serine
Tryptophan	Glycine	Valine	Glycine	Methionine	
Valine	Proline		Histidine[c]	Phenylalanine	
	Serine		Proline	Threonine	
	Tyrosine		Serine	Tryptophan	
			Tyrosine	Tyrosine	
				Valine	

[a]The medium also contained 0.25 to 1 per cent dialyzed horse serum.

[b]Arginine is required in the diet of young rats.

[c]Histidine is required in infant humans.

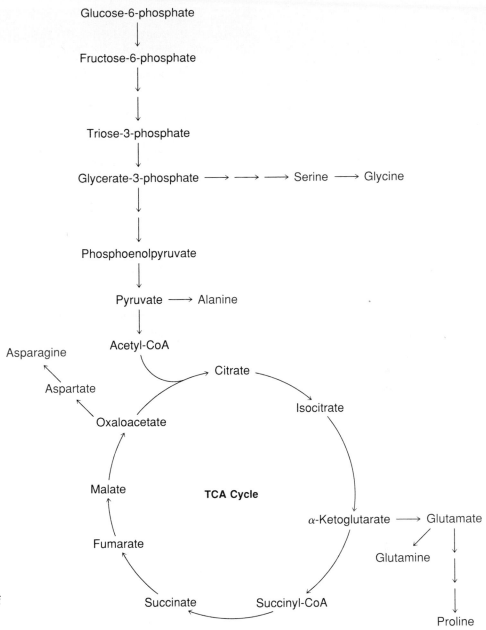

Figure 24–1
Amino acids that are synthesized *de novo* in mammals. All such amino acids are related by a small number of steps to glycolysis or TCA cycle intermediates.

The results in Table 24–1 do not mean, for example, that rats have all the enzymes they need for converting the several central metabolites to amino acids listed in column 2, or that the human has the enzymes needed to form those in column 4. Rather, they indicate that given 19 other amino acids, the particular amino acid can be formed either by the *de novo pathway* used by plants or by a *"salvage" pathway* at the expense of another amino acid.

It is true that many animals, including humans, do contain the enzymes of several of the "short" biosynthetic pathways—those to alanine, glutamate, glutamine, proline, aspartate, asparagine, serine, and glycine (Figure 24–1). The enzymes needed for proline biosynthesis (a kinase and two reductases) are present in animal tissues. In addition, animals contain a proline oxidase that yields Δ^1-pyrroline-5-carboxylic acid, which because of the equilibrium with glutamic-γ-semialdehyde (see Figure 19–24) and a transaminase, allows for a nonacetylated route to ornithine. Ornithine, in turn,

TABLE 24–2
"Salvage" Pathways Allowing the Formation of Certain Nonessential Amino Acids from Other Amino Acids

Amino Acid Formed	Formed From	Enzymes Required
Arginine	Proline	Proline oxidase Ornithine-glutamate transaminase Ornithine transcarbamoylase Argininosuccinate synthase Argininosuccinate lyase
Cysteine	Methionine	S-adenosylmethionine synthase α-methyltransferase S-adenosylhomocysteinase Cystathionine-β-synthase Cystathionine-γ-lyase
Tyrosine	Phenylalanine	Phenylalanine-4-monooxygenase

can be converted to arginine by means of the enzymes of liver that are essential for urea formation (see Figure 19–6). Thus, although the pathway by which bacteria and plants form arginine is not present in animal tissue, arginine can be formed, but, in some organisms (e.g., rats), not at a rate sufficient for growth. Two other amino acids normally present in the diet also can be formed from other amino acids by unidirectional routes. For example, cysteine can be formed from dietary methionine (see Figure 19–30), and tyrosine can be formed by hydroxylation of phenylalanine by a pteridine-dependent monooxygenase (see Figure 19–17). These routes and the corresponding enzymes are listed in Table 24–2.

Our discussion of amino acid biosynthesis in mammals will be combined with our discussion of the corresponding pathways in microorganisms. But before considering this topic, we must deal with a much more basic consideration, the way in which nitrogen is fixed in biomolecules.

SOURCES OF NITROGEN FOR AMINO ACID BIOSYNTHESIS: THE NITROGEN CYCLE

The reason we discuss nitrogen fixation in connection with amino acid biosynthesis is that nitrogen, in the form of ammonia, enters organic molecules primarily through the biosynthesis of the amino acid glutamate. Nitrogen itself actually passes through various forms and valence states as a result of its interactions with different living forms. Collectively the transitions constitute a cycle that is illustrated in Figure 24–2. The valence states range from +5 in nitrates to −3 in ammonia or organic materials. In the 0 valence state, nitrogen is a gas. In all other valence states, it usually exists as a solid. The passage of nitrogen from one form to another involves a chain of widely distributed organisms, as indicated in Figure 24–2.

NH_3 is the form in which nitrogen is incorporated into organic materials, but it is less often available to plants or bacteria for biosynthesis than other forms of nitrogen. When present for any length of time in nature, NH_3 will either be assimilated into organic materials or be oxidized by nitrifying bacteria (such as *Nitrosomonas* and *Nitrobacter*) to nitrite and nitrate. Reduction of nitrate by plants or by bacteria seldom yields NH_3 in excess. Nitrogen fixation, by which nitrogen of the atmosphere is reduced to NH_3, occurs in a very limited number of microorganisms and plants. NH_3 is rarely produced in excess by this process either. Thus most plants must assimilate

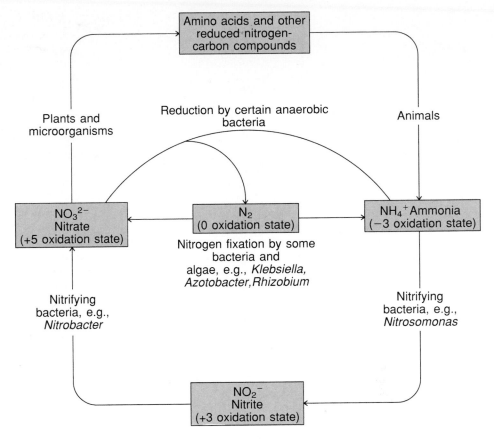

Figure 24–2

The nitrogen cycle depicts the flow of nitrogen in the biological world. The proteins of animals and plants are cleaved by many microorganisms to free amino acids from which ammonia (or ammonium ion) is released by deamination. Urea, the main nitrogen excretion product of animals, is hydrolyzed to NH_3 and CO_2. *Nitrosomonas* soil bacteria obtain their energy by oxidizing NH_3 to nitrite, NO_2^-. *Nitrobacter* obtain their energy by oxidizing nitrite, NO_2^-, to nitrate, NO_3^-. Plants and many microorganisms reduce nitrate for incorporation into amino acids, completing the cycle. Other microorganisms reduce nitrate partly to NH_3 and partly to N_2, which is lost to the atmosphere. Atmospheric nitrogen, N_2, can be recaptured, reduced, and converted into organic substances by a limited number of nitrogen-fixing bacteria and algae.

NH_3 at the low levels that are generated by their own nitrate and nitrite reductases or from the reduction of nitrogen by microorganisms.

Nitrogen Fixation Involves an Enzyme Complex Called Nitrogenase

The biological fixation of nitrogen by both free-living and symbiotic nitrogen-fixing bacteria involves an enzyme complex called <u>nitrogenase</u> (Figure 24–3). Nitrogenases consist of two proteins. One is an iron-containing protein called component II, and the other, containing molybdenum and iron, is called component I (Figure 24–3). The Fe protein from *Clostridium pasteurianum* is a dimer of two identical subunits, each with a molecular weight of 29,000, surrounding an iron-sulfur center that contains four iron atoms and four acid-labile sulfur atoms. The Mo-Fe protein is considerably more complex, and in *C. pasteurianum* is a tetramer of molecular weight 220,000, with two molybdenum atoms, about 30 iron atoms, and about as many acid-labile sulfur atoms. The four subunits are of two kinds, two with a molecular weight of 50,000 and two with a molecular weight of 60,000. There are probably four iron-sulfur centers, each containing four iron atoms

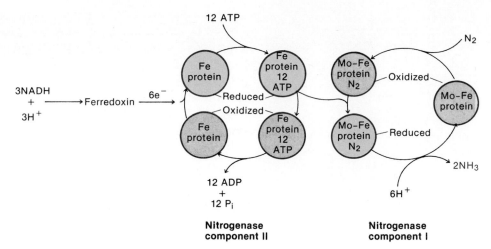

Figure 24–3

Nitrogenase reduction of N_2 to NH_3. Nitrogenase consists of two multiprotein complexes known as components I and II. The reduction starts with the reduction of the Fe in component II by ferredoxin. Electrons from the reduced Fe are transferred from component II to the molybdenum-iron protein of component I. Four ATPs are hydrolyzed for each electron pair that is transferred. The reduced Mo-Fe protein then reduces the nitrogen to NH_3.

and four labile sulfur atoms, and two extractable iron-molybdenum-sulfur "cofactors" that contain eight iron atoms, six labile sulfur atoms, and one molybdenum atom.

Nitrogen fixation is a metabolically costly process, requiring 12 high-energy phosphate bonds and 6 reducing equivalents for each mole of N_2 reduced. Thus *C. pasteurianum* consumes 70 per cent more carbon source when grown under nitrogen-fixing conditions than when grown in excess NH_3.

For nitrogen fixation to occur, the Fe protein (component II) is reduced by one or both of two low-potential reductants, ferredoxin or flavodoxin, which are generated during carbohydrate fermentation in *C. pasteurianum* or in the presence of a high $NADPH/NADP^+$ ratio. Electrons are transferred from the Fe protein to the Mo-Fe protein. This transfer is coupled to the hydrolysis of four ATPs for each electron pair. The reduced Mo-Fe protein then reduces nitrogen to NH_3. The reaction occurs in three discrete steps, each involving two electrons. First, nitrogen is reduced to the diimine

$$N_2 + 2\,H^+ + 2e^- \longrightarrow HN{=}NH$$

then to hydrazine

$$H_2N{=}NH_2 + 2\,H^+ + 2e^- \longrightarrow H_2N{-}NH_2$$

and finally to ammonia

$$H_2N{-}NH_2 + 2\,H^+ + 2e^- \longrightarrow 2\,NH_3$$

Nitrogenase also can reduce acetylene to ethylene, CN^- to CH_4 and NH_4^+, N_2O to N_2 and H_2O, and H^+ to H_2. This last reaction, the conversion of H^+ to H_2, can make nitrogen fixation even more costly. Thus, when the ATP supply is low or the reduction of the Fe protein is retarded, the six-electron reduction of the Mo-Fe protein is sluggish and the two-electron reduction of protons is favored. Furthermore, in the presence of H_2, the first intermediate, $HN{=}NH$ is converted to N_2 and $2\,H_2$:

$$HN{=}NH + H_2 \longrightarrow N_2 + H_2$$

Such a futile cycle apparently always occurs to some extent, so that for each dinitrogen reduced, a pair of protons is also reduced.

Nitrogenases appear to be highly conserved proteins. One of their characteristics is the extreme lability of the enzyme from all sources in the presence of oxygen. For strict anaerobes, this feature is not a special problem, since the organism can grow only anaerobically. For facultative anaerobes, such as *Klebsiella pneumoniae*, nitrogen fixation occurs only under anaerobic conditions. Some nitrogen fixers, however, are strict aerobes. One genus, *Azotobacter*, protects its nitrogenase by a very active electron-transport system that removes oxygen at a rapid rate. Another genus, *Rhizobium*, which fixes nitrogen symbiotically in the root nodules of legumes, depends on the leghemoglobin, a special hemoglobin made by plants that absorbs O_2 and transports it to the respiring bacteria at a partial pressure of oxygen low enough to allow nitrogenase to function.

Nitrate and Nitrite Reduction Play Two Physiological Roles

Nitrate and nitrite reduction play two physiological roles. One, exhibited primarily by bacteria, is a dissimilatory role in which nitrate and nitrite serve as terminal electron acceptors. In other words, the reduction serves to oxidize reducing equivalents, e.g., NADH, generated during oxidation of substrates. The other, exhibited by bacteria, fungi, and plants, is an assimilatory role in which nitrite is formed from nitrate and is reduced to NH_3 at a rate no greater than that required for synthesis of nitrogenous compounds during growth.

In general, the assimilatory nitrate and nitrite reductases are soluble enzymes that utilize reduced pyridine nucleotides or reduced ferredoxin. In contrast, the dissimilatory nitrate reductases are membrane-bound terminal electron acceptors that are tightly linked to cytochrome b_1 pigments. Such complexes allow one or more sites of energy conservation (ATP generation) coupled with electron transport.

The NH_4^+ produced by fermentative bacteria that utilize nitrite as an oxidant, or produced during the decomposition of organic materials, can be utilized by plants and bacteria for cell material. However, under suitably aerobic conditions, NH_3 is rapidly converted again to nitrite and nitrate by the nitrifying bacteria, such as *Nitrosomonas europaea* and *Nitrobacter agilis*, which are chemolithotrophs that gain energy from the oxidation of NH_3. The ubiquity of such organisms has permitted the assimilation of nitrate and nitrite so important to green plants and, ultimately, to animals.

THE ADVANTAGES OF MICROORGANISMS FOR THE STUDY OF AMINO ACID BIOSYNTHESIS

E. coli has been extensively investigated for amino acid biosynthesis. The reason is that with this organism researchers can combine genetic and biochemical techniques to analyze the pathways for all 20 amino acids commonly found in proteins.

Typically, research begins by isolating *E. coli* mutants that are deficient in only one amino acid needed for growth (auxotrophs). Such a deficiency usually results from a mutation in a single gene that encodes an enzyme required for a single step in the biosynthesis of the amino acid. The mutation produces a defective enzyme. There are two consequences (Figure 24–4): (1) no amino acid, the final product of the pathway, is produced; and (2) the substrate for the defective enzyme is produced in large excess. Frequently this substrate intermediate, or a closely related derivative, can be readily

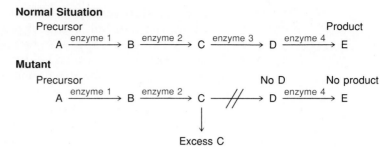

Figure 24–4

Immediate consequences of a point mutation in a biosynthetic pathway. When the mutation leads to an inactive enzyme, the chain of reactions leading to the end product in the pathway is broken, and frequently large amounts of intermediate are produced, accumulate in the cell, and may leak to the environment.

isolated because it is excreted in abundance into the surrounding medium. As you can see from Figure 24–4, several different mutations could disrupt the synthesis of a single end product. Through genetic analysis and accumulation of different intermediates, the researchers can begin to trace the outlines of the pathway. Eventually, all of the intermediates in the pathway can be identified and a coherent series of chemical transformations can be worked out.

The next step is to identify and in some cases to isolate the enzymes responsible for the individual conversions. For this purpose, partially fractionated protein extracts made from normal ("wild type") cells or from mutant cells are tested in conjunction with different intermediates to see whether the extracts possess the enzyme activity to carry out a particular conversion. In favorable cases, by progressive subfractionation of the extract it is ultimately possible to purify the enzyme that carries out a specific step in the biosynthetic pathway.

This sounds like a lot of work, and it is. Nevertheless, this general procedure, coupled with the use of isotopes and all the other tools of modern biochemistry, has led to a tremendous amount of information about amino acid biosynthesis—far more than we can present in a single chapter.

The _pathways to amino acids_ do not arise uniquely from the carbon source, but rather _arise as branchpoints from a few key intermediates in the central metabolic pathways_ to all cells, regardless of the carbon source. By the central metabolic pathways, we mean the glycolytic pathway, the pentose phosphate pathway, and the tricarboxylic acid cycle (Chapters 14 and 15). The relationships between these carbohydrate pathways and the branchpoints that lead from them to specific amino acids are indicated in Figure 24–5. Most of the carbon flow from the central metabolic routes is irreversible. However, in nearly all cases the flow is an orderly one that provides to the cell just those amounts of amino acids needed for cellular synthesis. The orderly flow is achieved by relatively simple, yet almost faultless regulatory mechanisms.

THE BIOSYNTHESIS OF AMINO ACIDS OF THE GLUTAMATE FAMILY: L-GLUTAMATE, L-GLUTAMINE, L-PROLINE, L-LYSINE, AND L-ARGININE

A common element in all amino acids is the α amino group. Directly or indirectly, the α amino groups are all derived from ammonia by way of the amino groups of L-glutamate. From glutamate the other amino acids of the glutamate family are synthesized. These include glutamine, proline, and arginine. In fungi they also include lysine.

Source of starting metabolite	Family	Pathway

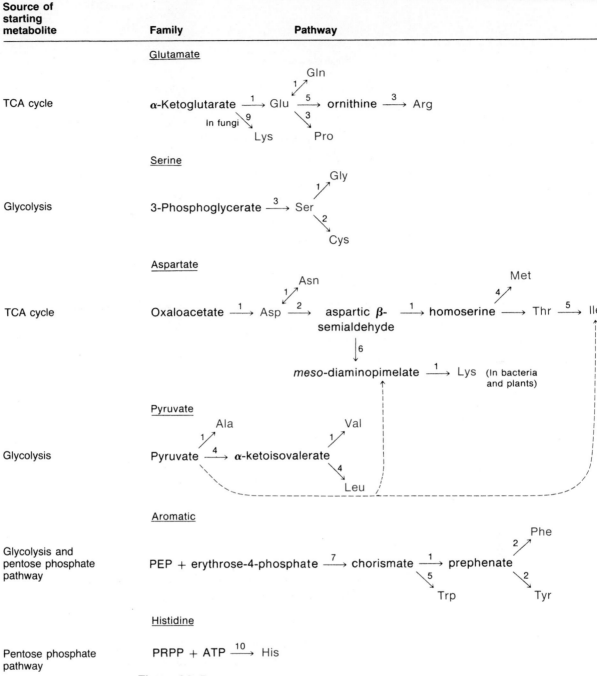

Figure 24–5

Amino acids are ordered into families according to the specific metabolite or amino acid that serves as the starting point for their synthesis. Solid arrows indicate one or more discrete biochemical steps. Dashed arrows from pyruvate to both lysine and isoleucine indicate that pyruvate contributes some of the carbon atoms for each of these amino acids. Only amino acids and key intermediates are specified. The numbers of steps between key intermediates or end points are indicated over the arrows. Notice that lysine is unique in that two completely different pathways exist for its biosynthesis.

The Direct Amination of α-Ketoglutarate Leads to Glutamate

The simplest route to glutamate (and therefore, amino group formation) is that exhibited by many bacteria when grown in a medium containing an ammonium salt as the sole nitrogen source. The reaction is a reductive amination reaction catalyzed by *glutamate dehydrogenase* (Figure 24–6). In or-

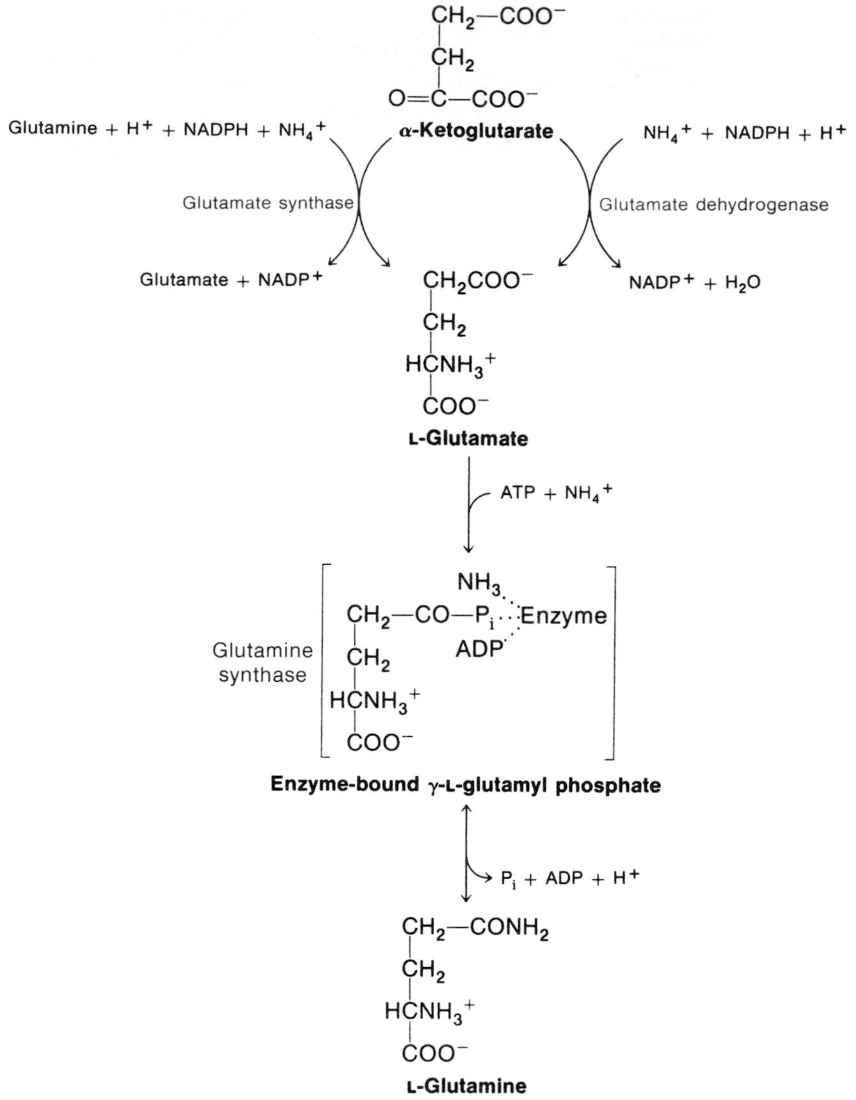

Figure 24-6
The converison of ammonia into the α-amino group of glutamate and into the amide group of glutamine. The direct amination of α-ketoglutarate by NH_4^+ occurs only under conditions of high NH_4^+ concentrations, which are rarely found in nature. Glutamine synthase is regulated to control the assimilation of nitrogen (see Figure 24-7).

ganisms such as *E. coli*, the enzyme is specific for NADPH as the hydrogen donor, as might be expected of a biosynthetic reaction involving a reductive step.

Some organisms have both an NAD^+/NADH- and an $NADP^+$/NADPH-dependent glutamate dehydrogenase. In such cases, the NAD^+-dependent enzyme is considered to play a catabolic role that converts glutamate back to α-ketoglutarate. Both enzymes, however, catalyze reversible reactions, and assignment to anabolic or catabolic functions is largely inferential. However, in some organisms (e.g., certain water molds), the presumed catabolic enzyme is inhibited by ATP, CTP, and fructose bisphosphate, indicators of a high-energy charge, and is stimulated by AMP.

Studies on the glutamate dehydrogenases from green plants revealed that the plant enzymes can use either NADH or NADPH as the hydrogen donor in the amination reaction. However, for reasons given below, it is unlikely that the glutamate dehydrogenase of plants plays a significant role in glutamate biosynthesis. It is more likely that in green plants, as in many bacteria and fungi, and in probably all bacteria when they are not being grown under NH_4^+ excess, the formation of the amino group of glutamate occurs not from NH_4^+, but from the amide group of glutamine (see the following discussion). In such cases, the primary conversion from NH_4^+ to organic nitrogen is catalyzed by glutamine synthase.

Amidation of Glutamate to Glutamine Is a Highly Regulated Process

The reaction catalyzed by *glutamine synthase* is one that involves activation of the γ carboxyl group of glutamate to yield a γ-glutamylenzyme complex and the cleavage of ATP to ADP and P_i (Figure 24–6). In a second step, γ-glutamyl transfer to NH_4^+ occurs.

Glutamine synthase is a key enzyme in the flow of ammonia nitrogen to organic compounds, and its activity is subject to elaborate controls that sense the cell's need for nitrogen-containing compounds.

The enzyme from *E. coli* has been studied extensively, and much is known of its structure and its regulatory behavior. The studies of Stadtman and Ginsberg have shown that glutamine synthase of *E. coli* is composed of 12 identical 50,000-molecular-weight subunits arranged symmetrically in two hexameric rings. *The activity of the enzyme is regulated in a complex pattern of feedback inhibition in which a partial inhibition is effected by each of eight different nitrogenous compounds:* carbamylphosphate, glucosamine-6-phosphate, tryptophan, alanine, glycine, histidine, cytidine triphosphate, and AMP. All these compounds except glycine and alanine receive the amide nitrogen of glutamine directly during their biosynthesis and are thus end products of reaction sequences leading from glutamine. Glycine and alanine, although not directly end products, could be looked upon as "indicators" of the sufficiency of the nitrogen supply of the cell.

Even more important in the regulation of *E. coli* glutamine synthase activity is the reversible ATP-dependent adenylylation of a specific tyrosyl residue on each subunit. As the enzyme becomes progressively more adenylylated (up to the fully adenylylated form of 12 AMP groups per enzyme molecule), the enzyme becomes progressively less active.

The adenylylation reaction and its reversal by a phosphorolytic deadenylylation are regulated by the nitrogen supply in the cell. The immediate small-molecule effectors of this regulatory system are glutamine and α-ketoglutarate. A high glutamine concentration or a high glutamine/α-ketoglutarate ratio signals nitrogen excess. A high α-ketoglutarate concentration or a high α-ketoglutarate/glutamine ratio signals nitrogen limitation. Nitrogen excess leads to inactivation of glutamine synthase, just as nitrogen limitation leads to activation (Figure 24–7). The activation of glutamine synthesis favors fixation of ammonia in a condensation with glutamate to form glutamine. Although the enzyme catalyzes the reverse reaction (like any enzyme), the equilibrium is very much in favor of glutamine formation. Both the activation (deadenylation) and the inactivation (adenylation) are controlled by a cascade of regulatory interactions illustrated in Figure 24–7. As in any cascade, a metabolic signal is greatly amplified by the cascade arrangement (e.g., see Figure 14–20). The cascades involved here are unusual in that the same two regulatory proteins are involved in both cascades. Both regulatory proteins have two enzymatically distinct sites arranged in such a way that only one site is active. Which site is active in the two regulatory proteins depends on the relative concentrations of glutamine and α-ketoglutarate, as already indicated.

Let us first consider the situation under conditions of nitrogen excess (see Figure 24–7). The first regulatory protein in the cascade at high glutamine is converted into a uridylyl-removing enzyme. This enzyme hydrolyzes UMP from a PII·4UMP protein that acts in concert with adenylyltransferase, at a high glutamine/α-ketoglutarate ratio, to adenylate glutamine synthase. The resulting adenylated enzyme is inactive.

Under conditions of nitrogen limitation, the first regulatory enzyme in the cascade is converted into a uridylyltransferase at high α-ketoglutarate concentrations. This enzyme uridylates the PII protein. The PII·4UMP pro-

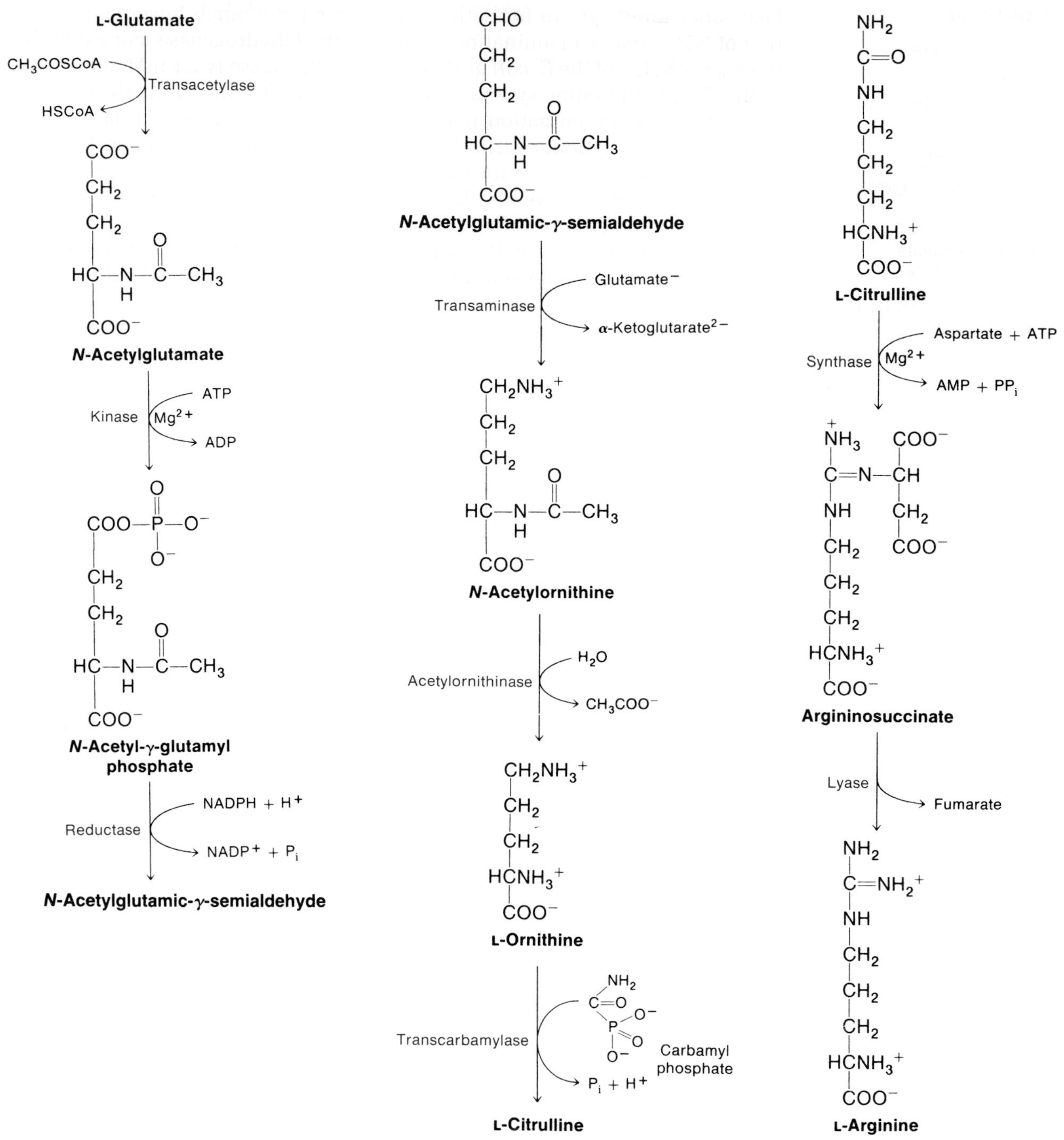

Figure 24–9 *(above)*

The biosynthesis of arginine. This synthesis begins by the N-acetylation of the α amino group. The γ carboxyl group is phosphorylated before reduction to an aldehyde that is subsequently transaminated. In some organisms the acetyl group, whose purpose is to protect the α amino group from reacting during intermediate steps, is removed as shown. In others, it is preserved by transfer to another glutamate as the initial step in arginine biosynthesis. In the latter, the acetyl-CoA-dependent formation of N-acetylglutamate serves only an anaplerotic role. The ornithine formed after the deacetylation is converted to arginine by means of a series of reactions that were encountered in urea formation (see Figure 19–6).

Figure 24–10 *(facing page)*

The biosynthesis of lysine in fungi and the eukaryotic green alga *Euglena*. The initial steps in the pathway result in the lengthening of the carbon chain of α-ketoglutarate by one CH_2 unit to α-ketoadipate. The α-ketoadipate is aminated in a transamination with glutamate. The δ carboxyl group is activated by phosphoadenylation and subsequently reduced to an aldehyde. The aldehyde is then aminated by the amino group of glutamate in a NADPH-specific reductive condensation reaction in which saccharopine is formed. The saccharopine is oxidatively cleaved to yield lysine and α-ketoglutarate. The completely different pathway for lysine biosynthesis found in bacteria and plants is illustrated in Figure 24–17.

Figure 24–10 *(Legend appears on facing page.)*

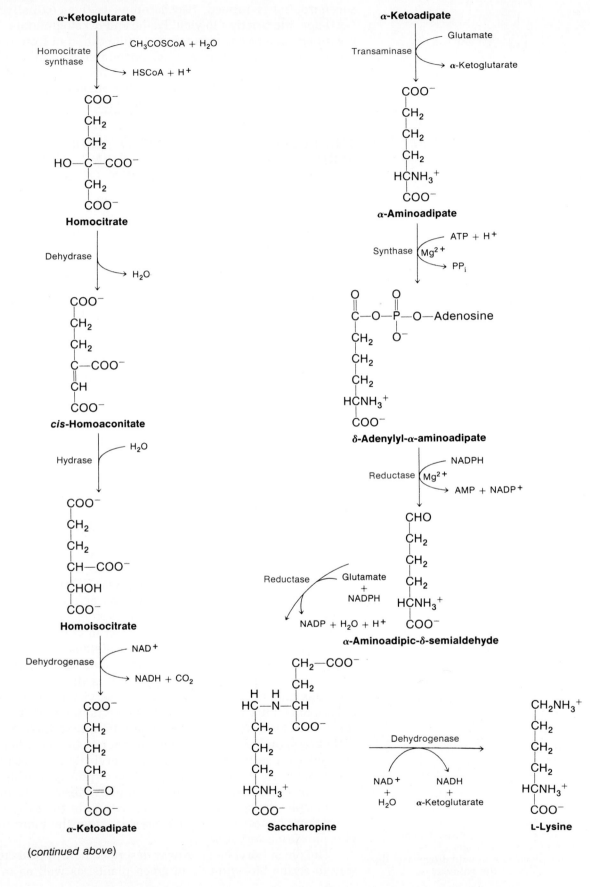

(continued above)

mate in a unique NADPH-specific reductive condensation reaction in which *saccharopine* is formed. Saccharopine is then oxidatively cleaved in an NAD-specific reaction to yield lysine and α-ketoglutarate. The two reactions are, in effect, a transamination, but quite different from the pyridoxal-phosphate-dependent transaminations encountered in other pathways. The enzyme forming saccharopine has been termed *α-aminoadipic-δ-semialdehyde-glutamate reductase*; the enzyme cleaving saccharopine has been termed *saccharopine dehydrogenase*.

THE BIOSYNTHESIS OF AMINO ACIDS OF THE SERINE FAMILY (L-SERINE, GLYCINE, AND L-CYSTEINE) AND THE FIXATION OF SULFUR

The diversion of 3-phosphoglycerate from the glycolytic pathway into the serine biosynthetic pathway is important, not only for the formation of L-serine, L-cysteine, and glycine needed for incorporation into protein, but also for the many other functions these amino acids serve. For example, the conversion of serine to glycine serves to generate one-carbon units that can be used in purine, thymine, and methionine biosynthesis as well as to replenish the methyl group transferred from methionine in many methylation reactions. The carbons of glycine contribute to purine and heme-containing compounds, to glutathione (as do those of cysteine), and in animals, to certain detoxification products. Oxidation of glycine provides an additional source of one-carbon units from the α carbon. In many plants and microorganisms, sulfur in the form of sulfide is incorporated first into cysteine and then later is transferred to methionine and other sulfur-containing compounds. In so doing, the carbons of cysteine are returned to the glycolytic pathway in the form of pyruvate. Serine itself is incorporated directly into phospholipids and into tryptophan, which, therefore, also might be considered a member of the serine family of amino acids. However, it will be more convenient to consider tryptophan biosynthesis along with the formation of the other aromatic amino acids.

Three Enzymes Convert 3-Phospho-D-glycerate (Glycerate-3-phosphate) to Serine

The initial reaction in L-serine biosynthesis is the oxidation of 3-phosphoglycerate by an NAD-linked dehydrogenase (Figure 24–11). Even though the reaction is freely reversible, the enzyme activity is regulated by the end product of the biosynthetic sequence, serine. The product of the reaction, 3-phosphohydroxypyruvate, is converted to *O*-phosphoserine by a specific *phosphoserine-glutamate transaminase*. Removal of the phosphate group by a specific *phosphoserine phosphatase* yields serine. Phosphoserine itself is also found as a protein constituent, but it is almost always derived from the posttranslational phosphorylation of a seryl residue by a specific protein kinase. Such kinases serve an important function in regulating the activity of some enzymes. However, there are minor serine-accepting suppressor tRNAs (responding to UGA codons) that can be phosphorylated after being charged with serine to yield phosphoseryl tRNA. These tRNAs have been shown, *in vitro*, to permit incorporation of phosphoserine directly into protein. Thus there may be two routes to phosphoserine in proteins. However, the phosphorylation of serine which occurs in the fully formed protein by the action of specific protein kinases is by far the more important mode of phosphoserine formation.

The phosphorylated pathway described here appears to be the sole pathway to serine biosynthesis in green plants, as well as in most microorga-

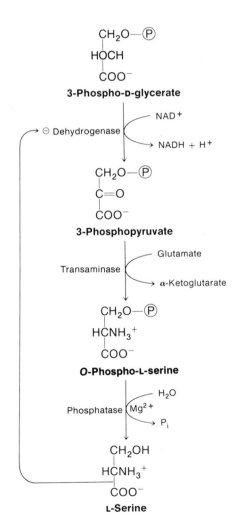

Figure 24–11
The biosynthesis of serine. The end product, serine, inhibits 3-phosphoglycerate dehydrogenase, the first enzyme in the pathway.

nisms. A series of reversible nonphosphorylated reactions and a route from glycine to serine that can be demonstrated in many microorganisms as well as mammals are probably catabolic reactions.

Two More Enzymes Convert L-Serine to Glycine

The pathway from serine to glycine consists of a single but complex step catalyzed by serine hydroxymethyltransferase (Figure 24–12). The reaction is a pyridoxal-phosphate-dependent aldol cleavage to yield glycine and an "active" formaldehyde that is transferred to a tetrahydrofolate cofactor. This reaction is an important one in supplying one-carbon units for other biosynthetic reactions. It has been calculated, however, that the amount of _methylene tetrahydrofolate_ generated during the biosynthesis of the glycine in bacterial protein is not quite sufficient to account for the synthesis of those cellular constituents that require one-transfers of carbon for their formation (you may wish to review the reactions involving the folic acid coenzymes in Chapter 11). The most likely source of these required extra one-carbon units is the α carbon of glycine, which is converted to NH_3, CO_2, and a one-carbon unit as 5,10-methylene tetrahydrofolate by a four-protein complex, the glycine oxidase system. This system has been found in bacteria, plants, and animals.

Cysteine Biosynthesis Involves Sulfhydryl Transfer to Activated Serine

The biosynthesis of L-cysteine consists essentially of a sulfhydryl transfer to an activated form of serine. The form of the sulfhydryl group used in cells is unclear. For most plants and microorganisms, the sulfur source is sulfate, which must be reduced to the level of sulfide. This reduction is poorly understood and will not be considered in this section. For some organisms, it appears that the initial sulfhydryl transfer is made to an activated form of homoserine, a methionine precursor. Transfer of the sulfur to cysteine would then occur by a transsulfuration pathway.

The direct sulfhydrylation pathway to L-cysteine has been most thoroughly studied in _E. coli_ and the related _Salmonella typhimurium_ and in the eukaryotic green alga _Chlorella_. The initial step, which is inhibited by cysteine, is a transfer of the acetyl group of acetyl-CoA to serine, to yield _O-_

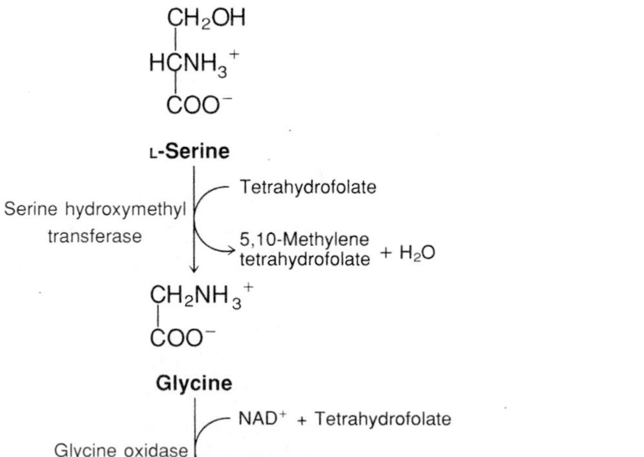

Figure 24–12
Biosynthesis and oxidation of glycine. Both of these reactions are important for supplying one-carbon units for metabolism by way of the tetrahydrofolate coenzyme.

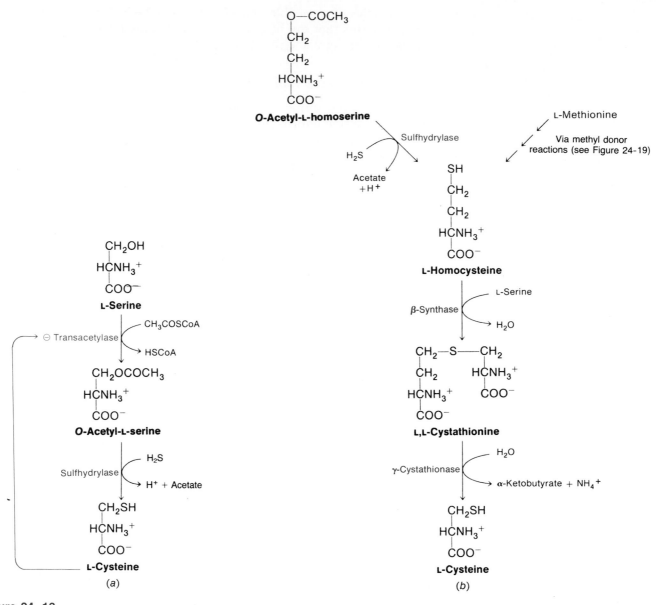

Figure 24–13

The biosynthesis of cysteine by direct sulfhydrylation and by a transsulfuration route in which the sulfur is derived from homocysteine. The direct sulfhydrylation pathway (a) is indicated as occurring with H_2S as the source of sulfur. The transsulfuration pathway (b) passes through homocysteine.

Two routes leading to homocysteine are shown. One starts with L-methionine and proceeds through reactions described in Figure 24–19. This is the sole route in animals. The second route involves the homocysteine-synthase-catalyzed conversion of O-acetyl-L-homoserine as shown.

acetylserine (Figure 24–13a). The reaction is catalyzed by <u>serine transacetylase</u>. The formation of cysteine itself is catalyzed by <u>O-acetylserine sulfhydrylase</u>, a reaction in which O-acetylserine serves as a β-alanyl donor (alanine with the amino group on the β carbon instead of the α carbon). In vitro, the enzyme will react with H_2S as the β-alanyl acceptor.

The direct sulfhydrylation of O-acetylserine to yield cysteine is the most common mechanism for sulfide incorporation. Nevertheless, in some forms the major, if not the sole, mechanism for sulfur incorporation is by means of <u>homocysteine synthase</u>, an enzyme considered later in the chapter, when we discuss L-methionine biosynthesis. Under such conditions, cysteine formation occurs by transsulfuration, with the intermediate formation of L,L-cystathionine (Figure 24–13b).

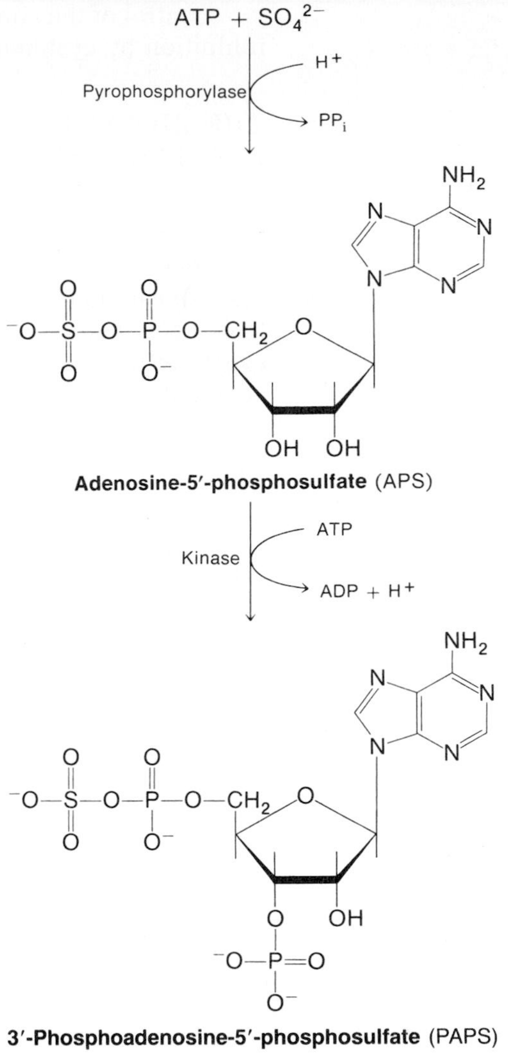

$$ATP + SO_4^{2-}$$

Pyrophosphorylase — H^+

→ PP_i

Adenosine-5'-phosphosulfate (APS)

Kinase — ATP

→ ADP + H^+

3'-Phosphoadenosine-5'-phosphosulfate (PAPS)

Figure 24–14

Formation of 3'-phosphoadenosine-5'-phosphosulfate, an active intermediate involved in sulfate reduction. The eight-electron reduction of SO_4^{2-} to H_2S is poorly understood except for the initial steps involved in the activation of sulfate. Reduction in yeast and plants involves the APS derivative shown. In *E. coli* it involves the PAPS derivative.

Cystathionine biosynthesis from serine and homocysteine is presumably catalyzed as a simple condensation by *cystathionine-β-synthase*. The cleavage of cystathionine to yield cysteine, α-ketobutyrate, and NH_4^+ is catalyzed by *γ-cystathionase*, a pyridoxal-phosphate-containing enzyme. This transsulfuration pathway from methionine is one route for methionine catabolism (see Figure 19–30).

The eight-electron reduction of SO_4^{2-} to H_2S is poorly understood. Initially, an activation of sulfate (Figure 24–14) is necessary. The first step, catalyzed by *adenylylsulfate pyrophosphorylase*, yields an "active" form of sulfate, *adenosine-5'-phosphosulfate* (APS). Further activation is required by *E. coli* and certain other bacteria and is achieved by phosphorylation of the 3'-OH by APS kinase to yield 3'-phosphoadenosine-5'phosphosulfate (PAPS) with ATP as the phosphate donor. The reduction of the sulfonyl moiety of APS (in yeast and plants) or of PAPS (in *E. coli*) to sulfite occurs by transfer of the sulfonyl group to a thiol acceptor, such as thioredoxin (a vicinal dithiol protein), to yield an $—S—SO_3^-$ derivative. Reaction with a second thiol group (on thioredoxin or on a second molecule of glutathione) yields sulfite and oxidized acceptor.

The six-electron reduction of sulfite to sulfide is catalyzed by *sulfite reductase* without the release of any free intermediates. Sulfite reductase is a multisubunit complex composed of a flavoprotein and a protein containing both heme and iron-sulfur centers.

Control of the conversion of sulfate to sulfide may occur in *E. coli* by an inhibition by cysteine of the active transport of sulfate into the cell.

THE BIOSYNTHESIS OF AMINO ACIDS OF THE ASPARTATE FAMILY: L-ASPARTATE, L-ASPARAGINE, L-METHIONINE, L-THREONINE, AND L-ISOLEUCINE

The formation of the aspartate family of amino acids and the conversion of aspartate to the pyrimidine nucleotides account for an even more significant drain of carbon from the tricarboxylic acid cycle in organisms such as *E. coli* than does that of the α-ketoglutarate family. In addition, the nitrogen of aspartate is used in both the formation of inosinate and its conversion to adenylate and in the conversion of citrulline to arginine.

Aspartate Is Formed from Oxaloacetate in a Transamination Reaction

L-Aspartate is formed from oxaloacetate in a transamination reaction with glutamate as the amino donor (Figure 24–15). In *E. coli*, the major protein exhibiting aspartate-glutamate transaminase activity is *transaminase A*, an enzyme that also exhibits activity with the aromatic amino acids but not with leucine.

Asparagine Biosynthesis Requires ATP and Glutamine

In most organisms it is likely that the formation of L-asparagine occurs by an ATP-dependent transfer of the amide group of glutamine to the β carboxyl of aspartate by *asparagine synthase* (Figure 24–15). Thus the basic mechanism of amidation is different from that catalyzed by glutamine synthase in that it involves β-aspartyladenylate as an enzyme-bound intermediate.

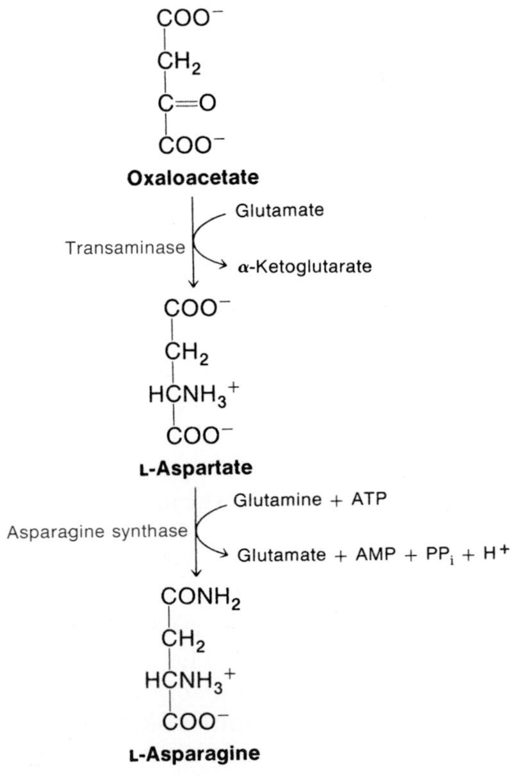

Figure 24–15
The biosynthesis of aspartate and asparagine. The more prevalent transamination reaction is shown. In a limited number of microorganisms an ammonia-dependent asparagine synthase has been found. The asparagine synthase involved in the typical transamination reaction shown in the figure also catalyzes a pyrophosphate cleavage of ATP.

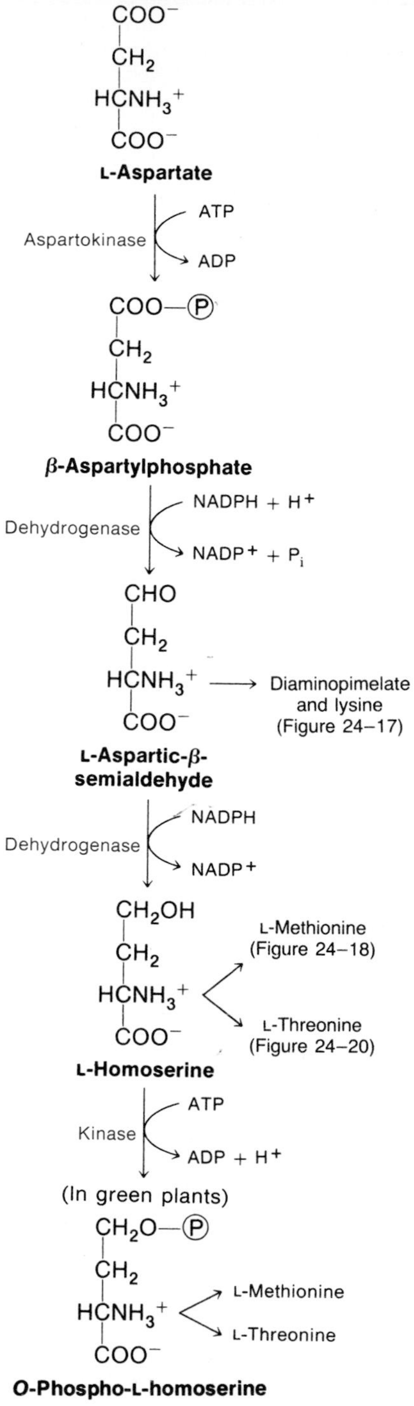

L-Aspartic-β-semialdehyde Is a Common Intermediate in L-Lysine, L-Methionine, and L-Threonine Synthesis

In the conversion of L-aspartate to L-lysine, L-methionine, and L-threonine, a reduction of the β carboxyl group is necessary. As in the case of the reduction of the glutamate β carboxyl group (see Figure 24–8) or in the conversion of glycerate-3-phosphate to glyceraldehyde-3-phosphate (see Figure 14–1), an activation by ATP is needed (Figure 24–16). This reaction is catalyzed by *aspartokinase*.

The NADPH-dependent reduction of β-aspartyl phosphate by *aspartic-β-semialdehyde dehydrogenase* yields aspartic-β-semialdehyde, which is a branchpoint compound from which lysine biosynthesis in plants and bacteria proceeds (see Figure 24–16). The common pathway is longer for methionine and threonine biosynthesis. In most living forms, *homoserine* is a branchpoint compound. Homoserine itself is formed by the reduction of aspartic-β-semialdehyde by *homoserine dehydrogenase*. Whereas some homoserine dehydrogenases use either NADPH or NADH in vitro, it would be anticipated that NADPH would more likely serve as the hydrogen donor in bacteria and in those eukaryotic cells in which the enzyme is not sequestered from the mitochondrial electron-transport system.

The intricacies of the control of carbon flow over the common aspartate family pathway will be considered after the specific branches have been described.

The Pathway to L-Lysine in Bacteria and Plants Is Different from That in Fungi

The pathway for synthesis of lysine from α-ketoglutarate in fungi has already been described. As already indicated, there is another pathway for lysine biosynthesis characteristic of bacteria and plants. We cannot say for certain why there should be a distinct route found for different organisms. However, we note that the pathway observed in bacteria has two branchpoints, which provide compounds used in bacterial spores (*dipicolinate*) and bacterial cell walls (*diaminopimelate*). The need for these compounds would certainly favor retention of this pathway, at least in bacteria.

As mentioned, the plant and bacterial pathway to L-lysine branches from the common aspartate family pathway with the appearance of aspartic-β-semialdehyde (as shown in Figure 24–17). The first specific step is a condensation of aspartic-β-semialdehyde and pyruvate with the elimination of water to yield a cyclic compound, 2,3-dihydrodipicolinate. (Thus lysine could be looked upon as a member of both the aspartate family and the pyruvate family of amino acids.) The enzyme *dihydrodipicolinate synthase* is inhibited by lysine. 2,3-Dihydrodipicolinate is itself a branchpoint compound in the spore-forming bacilli and clostridia, since it is oxidized to dipicolinate, an essential constituent of bacterial spores. In at least some bacilli, the condensing enzyme is desensitized to lysine inhibition at the onset of sporulation.

2,3-Dihydrodipicolinate is reduced by an NADPH-requiring reductase to Δ^1-piperideine-2,6-dicarboxylate (2,3,4,5-tetrahydrodipicolinate). In some organisms, such as *E. coli*, this cyclic compound is then hydrolytically opened and maintained in a straight-chain form by succinylation to yield N-succinyl-ϵ-keto-L-α-aminopimelate. In other organisms, an acetyl group serves the same function. The ϵ-keto group is aminated by a specific transaminase to yield N-α-succinyl-L,L-α,ϵ-diaminopimelate. Cleavage by succinyl-diaminopimelate succinylase yields L,L-α,ϵ-diaminopimelate.

In bacteria, *diaminopimelate* is also a branchpoint compound, since in

Figure 24–16
The common aspartate family pathway. In some green plants, O-phosphohomoserine rather than homoserine itself is the point at which methionine biosynthesis and threonine biosynthesis diverge.

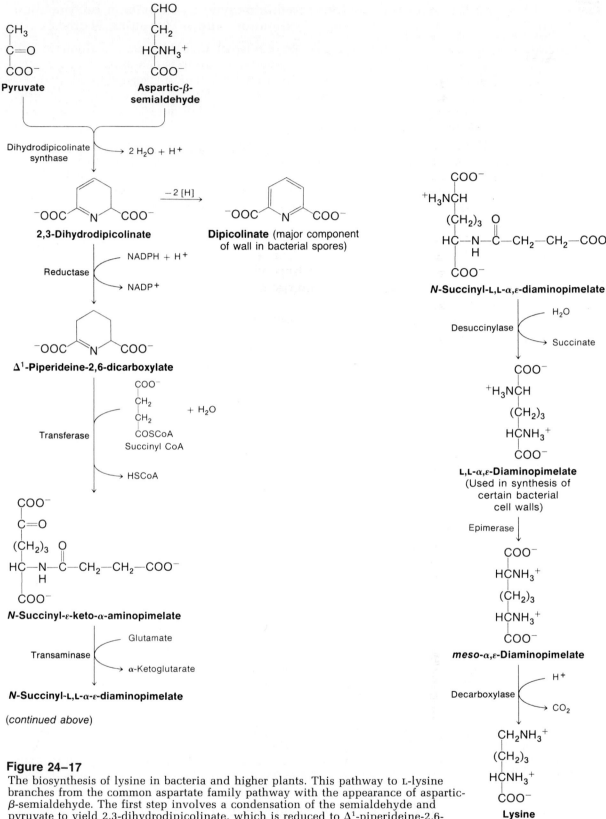

Figure 24–17

The biosynthesis of lysine in bacteria and higher plants. This pathway to L-lysine branches from the common aspartate family pathway with the appearance of aspartic-β-semialdehyde. The first step involves a condensation of the semialdehyde and pyruvate to yield 2,3-dihydrodipicolinate, which is reduced to Δ¹-piperideine-2,6-dicarboxylate. In *E. coli* this compound is hydrolytically opened by succinylation. Four more enzymes are required for lysine formation by this pathway. Note that two of the intermediates in this pathway are useful in certain bacteria for other purposes. The 2,3-dihydrodipicolinate is a precursor of dipicolinate, a major component of wall in bacterial spores. The L,L-α,ε-diaminopimelate is used in synthesis of certain bacterial cell walls.

many bacteria it is incorporated into the peptidoglycan, a highly crosslinked polymer of several amino acids (including some D-amino acids) and two amino sugars that form the rigid sacculus of the bacterial cell wall. In some bacteria, lysine is used instead of diaminopimelate (see Chapter 20).

The meso form of diaminopimelate is the substrate for _diaminopimelate decarboxylase,_ the enzyme that forms lysine. In some bacteria, the decarboxylase is inhibited by lysine. Since lysine is formed by way of an L,L intermediate, we would expect that one-half the carboxyl carbon of lysine is derived from the carboxyl carbon of pyruvate and one-half from the α carboxyl of aspartate.

Methionine Is Both an End Product and a Precursor of *S*-Adenosylmethionine

The conversion of L-homoserine to L-methionine occurs in more than one way. One route, used in some bacteria, consists of acylation of the hydroxyl group of homoserine, a condensation with cysteine to yield cystathionine, cleavage to homocysteine, and methylation to yield methionine (see Figure 24–18). There are several variations to be found in the acylation reactions.

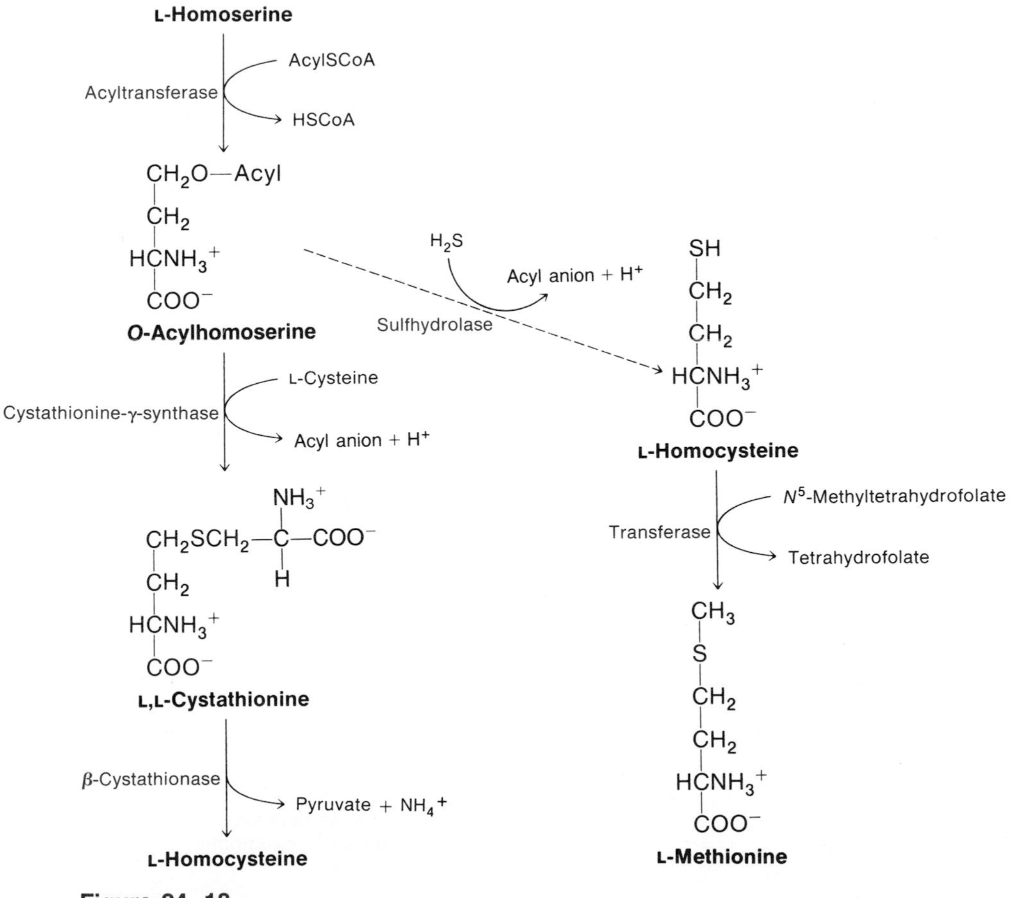

Figure 24–18

The biosynthesis of methionine. The acyl group employed to activate homoserine varies among different organisms. When a phosphoryl group is used as in green plants, the intermediate, O-phosphohomoserine, is a branchpoint compound that is converted to either threonine or methionine (see Figures 24–20 and 24–13). The direct sulfhydrylation of activated homoserine (by either H_2S or carrier-bound sulfide) that is found in some forms is indicated by a broken line. Finally, the methylation of homocysteine may occur by either a tetrahydrofolate-dependent route (shown) or a cobalamin-dependent route (not shown).

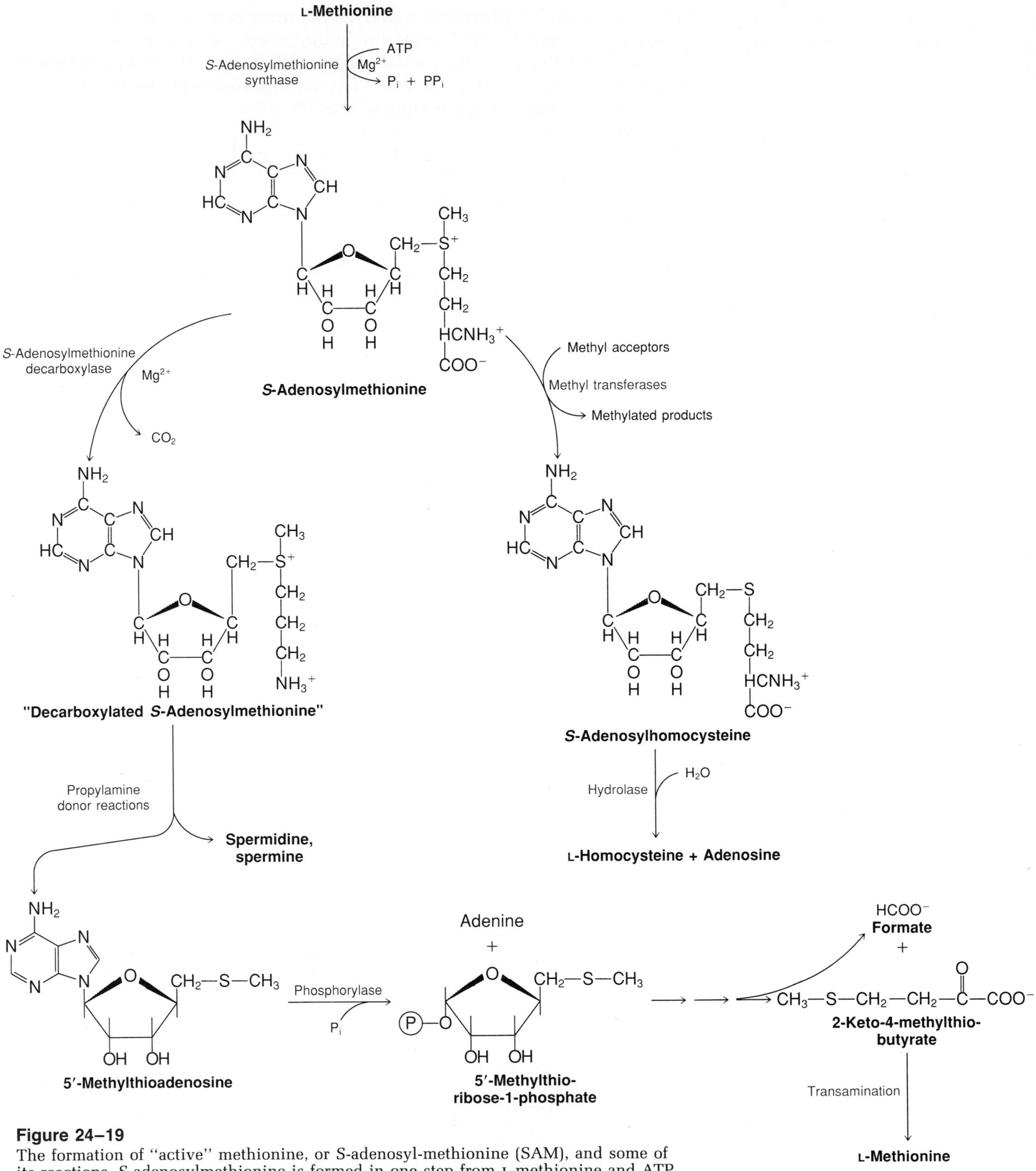

Figure 24–19
The formation of "active" methionine, or S-adenosyl-methionine (SAM), and some of
its reactions. S-adenosylmethionine is formed in one step from L-methionine and ATP.
The SAM so formed can serve as a propylamine donor or a methyl donor as shown.
In the methyl donor reaction, the S-adenosyshomocysteine formed is cleaved to
homocysteine and adenosine, which can be used to regenerate methionine and ATP,
respectively. The methylthioadenosine liberated in propylamine donor reactions is
converted to 5′methylthioribose-1-phosphate, which is converted to methionine and
formate by a series of reactions.

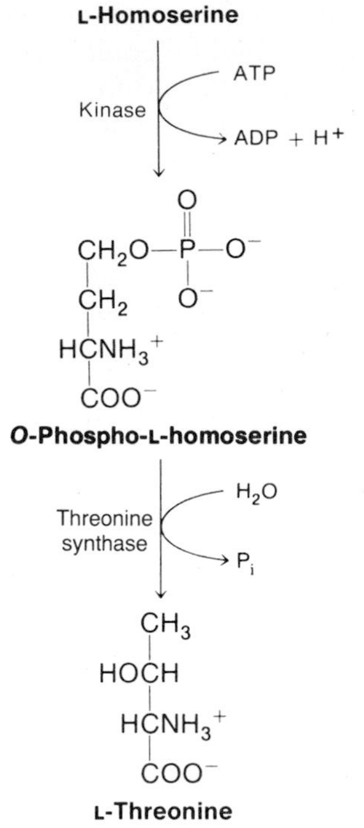

L-Homoserine

CH₂OH
|
CH₂
|
HCNH₃⁺
|
COO⁻

Kinase — ATP

→ ADP + H⁺

$$\begin{array}{c} O \\ \| \\ CH_2O{-}P{-}O^- \\ | \\ CH_2 \quad\ \ O^- \\ | \\ HCNH_3^+ \\ | \\ COO^- \end{array}$$

***O*-Phospho-L-homoserine**

Threonine synthase — H₂O

→ Pᵢ

$$\begin{array}{c} CH_3 \\ | \\ HOCH \\ | \\ HCNH_3^+ \\ | \\ COO^- \end{array}$$

L-Threonine

Figure 24–20

The biosynthesis of threonine. The first step in this reaction is a simple phosphorylation. The second step is complex. It involves formation of an adduct between the α amino group of O-phospho-L-homoserine and pyridoxal phosphate. This is followed by β,γ elimination of the phosphate, hydration of the resulting β,γ double bond, and a final hydrolysis to release threonine and regenerate the pyridoxal phosphate.

Although methionine is itself an end product of the pathway and is incorporated into protein, another important biosynthetic intermediate is *S-adenosylmethionine (SAM)*. This intermediate serves as a methyl donor in many reactions (see Figure 19–26) and as a precursor of the propylamine groups in spermidine and spermine. In some organisms SAM participates in the control of the pathway leading from homoserine to methionine.

An adenosylation of methionine by <u>SAM synthase</u> serves to activate methionine for either methyl group or, with a decarboxylase, propylamine transfers. As Figure 24–19 shows, the activation is unusual in that all three phosphate groups of ATP are cleaved in the adenosylation reaction. In the methyl donor reaction, *S*-adenosylhomocysteine is liberated by one of several methyltransferases and then cleaved by <u>*S*-adenosylhomocysteine hydrolase</u> to homocysteine and adenosine, which can be used to regenerate methionine and ATP, respectively. In the reactions in which a propylamine group is transferred to putrescine or to spermidine, methylthioadenosine is liberated (see Figure 19–26). A phosphorylase converts the latter to 5-methylthioribose-1-phosphate and free adenine. The latter can be returned to the adenylate pool by one of the pyrophosphorylases described in Chapter 25, and 5-methylthioribose-1-phosphate is converted to methionine and formate by a series of reactions that are being studied currently in animal, bacterial, and plant extracts.

Two Enzymes Convert 2-Homoserine into Threonine

For the formation of L-threonine from L-homoserine, the —OH group must be transferred from the γ to the β carbon, as shown in Figure 24–20. This transfer is done indirectly, starting by phosphorylation of the hydroxyl group of homoserine by <u>homoserine kinase</u>. For most organisms that form threonine, the kinase reaction is the first specific step in the pathway, and the reaction is inhibited by threonine. The next step is catalyzed by a pyridoxal-dependent enzyme.

As indicated in Figure 24–16, in some green plants, *O*-phosphohomoserine is a branchpoint compound, being a common intermediate for both methionine and threonine. In some cases, at least, the one-step route to threonine is regulated not by the end product, i.e., threonine, but by the "need" for the competing reaction (transsulfuration to cystathionine), for which *O*-phosphohomoserine is also a substrate. Thus threonine synthase is inhibited by L-cysteine (the cosubstrate for transsulfuration) and stimulated by *S*-adenosylmethionine (a product derived ultimately by means of transsulfuration). This general pattern may exist in other systems when the more "typical" pattern of end product inhibition is not found.

Isoleucine Biosynthesis and Valine Biosynthesis Are Related

L-Isoleucine is usually considered a member of the aspartate family, since in most bacteria and plants four of its six carbons are derived from aspartate and only two from pyruvate. However, since four of the five enzymes in isoleucine biosynthesis are also involved in valine biosynthesis, isoleucine biosynthesis is more appropriately considered with the pyruvate family of amino acids.

The Carbon Flow in the Aspartate Family Is Regulated at the Aspartokinase Step

The flow of carbon into the common aspartate family pathway must be regulated in a way that provides ample amounts of the branchpoint compounds aspartic-β-semialdehyde (except in fungi) and homoserine (or *O*-phosphohomoserine in most green plants) but does not lead to oversynthesis of either. In most organisms there appear to be negligible pools of either branchpoint compound—perhaps only those amounts bound to product or substrate sites of the respective enzymes. Therefore feedback control cannot be exerted by these intermediates. Rather, in one way or another, control must be exerted by the end products (lysine, methionine, and threonine), which do accumulate in measurable pools.

Although an essentially common pathway has evolved for the formation of the aspartate family precursors, the patterns of feedback control have been found to vary considerably. They are of two general kinds, with considerable specific variation from the two basic patterns. In one, there are single aspartokinases, but the control of the enzyme is _multivalent_. By multivalent, we mean that more than one of the amino acid end products is required to inhibit the enzyme. The other basic pattern is one of multiple aspartokinases, each of which is controlled differently.

The most extensively studied pattern is that of the enteric bacteria, as exemplified by _E. coli_ (Figure 24–21). In _E. coli_, there are three different proteins with aspartokinase activity. One enzyme is inhibited by threonine. The enzyme is part of a protein that also exhibits homoserine dehydrogenase activity, which is also sensitive to inhibition by threonine. The enzyme is called _aspartokinase I–homoserine dehydrogenase I_. Another distinctly different protein, _aspartokinase II–homoserine dehydrogenase II_, is not inhibited by any of the multiple end products of the aspartate family pathway, but its synthesis is repressed by methionine (as are other methionine biosynthetic enzymes). The synthesis of aspartokinase I–homoserine dehydrogenase I is repressed by a combination of two end products, threonine and isoleucine (multivalent repression).

Figure 24–21

Regulation of the aspartokinases of _E. coli_. The activity of these enzymes is regulated at the level of enzyme synthesis and at the level of enzyme activity. In _E. coli_ there are three aspartokinases that catalyze the conversion of L-aspartate to β-aspartylphosphate. The formation of these aspartokinases is repressed by different amino acids that are end products of this highly branched pathway. In addition, the activities of aspartokinase I and homoserine dehydrogenase I are inhibited by threonine and that of aspartokinase III is inhibited by lysine (arrows for inhibited reactions are not shown).

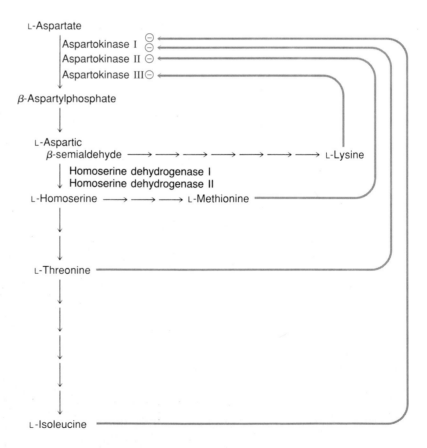

The third aspartokinase in *E. coli*, *aspartokinase III*, is inhibited by lysine. The inhibition is enhanced synergistically by phenylalanine, leucine, or, to a lesser extent, methionine. Any physiological advantage that results from this synergistic inhibition has not been explained. Unlike the other two aspartokinases, aspartokinase III has no other activity associated with it. Its synthesis is also repressed by lysine.

It should be emphasized that the intermediates of the common aspartate family pathway are not being channeled into the branches leading to lysine, methionine, and threonine, but rather, there is probably a common pool of intermediates (albeit small) from which materials needed for the three specific pathways are drawn. In some strains of *E. coli*, lysine-sensitive aspartokinase III is the predominant aspartokinase, and there is very little of the methionine-repressible aspartokinase II–homoserine dehydrogenase II in an amino-acid-free medium. Under conditions of strong inhibition and repression (i.e., repressed synthesis of the enzyme) of aspartokinase I–homoserine dehydrogenase I and of aspartokinase III, a starvation for methionine would be prevented by a derepression (i.e., increased synthesis of the enzyme) of the aspartokinase II–homoserine dehydrogenase II.

In those organisms in which there are single aspartokinases, it is common to find that the aspartokinases are inhibited by lysine and threonine. Usually lysine or threonine alone is weakly inhibitory, so that the pattern is actually a strongly synergistic one. In such organisms there is only a single homoserine dehydrogenase, and it is usually inhibited by threonine alone. In some cases, the addition of lysine and threonine is strongly inhibitory to growth, and the growth inhibition is reversed by methionine. It may be that aspartokinase II–homoserine dehydrogenase II of *E. coli* provides one means of avoiding this complication. In other organisms the inhibition is less severe, and the proteins are relatively insensitive to the feedback inhibitors after growth in the presence of the inhibitory amino acids. The physical basis of the desensitization is unknown.

The precise patterns of regulation of synthesis of amino acids of the aspartate family vary widely, and in other microorganisms and in plants the studies have not been as thorough as they have in *E. coli* and related organisms, which appear to have a rather unique pattern of regulation.

THE BIOSYNTHESIS OF AMINO ACIDS OF THE PYRUVATE FAMILY: L-ALANINE, L-VALINE, AND L-LEUCINE

The pyruvate family of amino acids consists of L-alanine, L-valine, and L-leucine. In addition, pyruvate contributes two carbons to isoleucine and, on the average, two and one-half carbons to lysine. As mentioned earlier, the biosynthesis of isoleucine, a member of the aspartate family, is carried on by a pathway that parallels that of valine. For this reason, the biosynthesis of isoleucine is considered here. There are two additional ways that have been evolved to form isoleucine, but these will not be considered in this section.

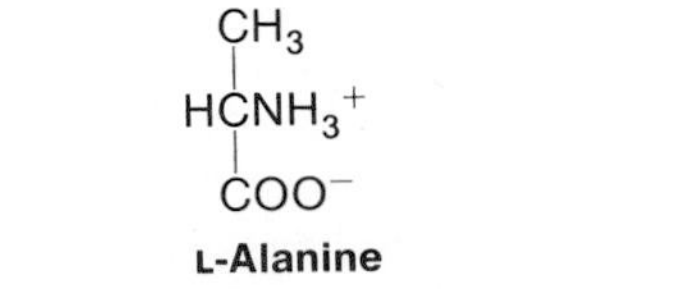

L-Alanine Is Formed from Pyruvate in a Transamination Reaction

The formation of L-alanine occurs by a transamination reaction, with glutamate as the amino donor and pyruvate as the acceptor (Figure 24–22). There is no feedback control over alanine formation, and in many forms of bacteria large intracellular pools of alanine are present unless the nitrogen supply is restricted. However, since the transaminases catalyze completely reversible reactions, this accumulation of alanine does not effect a drain on the supply of pyruvate.

Figure 24–22
The biosynthesis of alanine. This reaction is not regulated. It is readily reversible and therefore constitutes no major drain on the pyruvate supply.

Isoleucine Biosynthesis and Valine Biosynthesis Share Four Enzymes

In all forms, the four enzymes required for valine biosynthesis are also required for the last four, parallel steps in isoleucine biosynthesis (Figure 24–23). The first step in valine biosynthesis is a condensation between pyruvate and "active" acetaldehyde (probably hydroxyethyl thiamine pyrophosphate) to yield α-acetolactate. The enzyme usually has a requirement for FAD, which, in contrast to most flavoproteins, is rather loosely bound to the protein. The same enzyme transfers the acetaldehyde group to α-ketobutyrate, yielding α-aceto-α-hydroxybutyrate, the isoleucine precursor. The α-ketobutyrate, unlike pyruvate, is not one of the key intermediates in any of the central metabolic routes; therefore a specific pathway to ketobutyrate must be present.

For nearly all plants, fungi, and bacteria, the normal route to α-ketobutyrate is that from aspartate by way of threonine, which in turn is deaminated to α-ketobutyrate. The enzyme threonine deaminase contains pyridoxal phosphate and functions as a dehydratase (see Chapter 11), presumably liberating α-aminocrotonate, which upon rearrangement to α-iminobutyrate spontaneously yields α-ketobutyrate and ammonia.

Conversion of the acetohydroxy acids to the β-dihydroxy acid precursors of valine and isoleucine is a complex reaction catalyzed by _acetohydroxy acid isomeroreductase_. The α,β-dihydroxy acids are both converted to the α-keto acid precursors of valine and isoleucine by a _dihydroxy acid dehydrase_. Finally, the two amino acids are formed in transamination reactions in which glutamate is the amino donor _(branched-chain amino acid-glutamate transaminase)_.

The pathways to isoleucine and valine illustrate well the way studies with nutritionally deficient mutants (auxotrophs), isotope incorporation experiments, and enzymatic analysis have been used to decipher the biosynthetic pathways to the amino acids.

Early studies with mutants of both Neurospora and E. coli revealed that certain mutants, presumably altered in but a single gene, required not one, but two amino acids, isoleucine and valine. This finding was an apparent contradiction to the one gene, one enzyme concept. Examination of the mutants genetically and nutritionally revealed several classes of these doubly auxotrophic mutants. One class was found to accumulate material in the culture fluids that fed mutants of several other classes. Analysis revealed that the active material consisted of the α-keto acid precursors of valine and isoleucine. One class of the mutants that responded to the keto acids accumulated material that fed other isoleucine and valine auxotrophs. This accumulated material was identified as α,β-dihydroxyisovalerate and α,β-dihydroxy-β-methylvalerate. Both these findings were followed by the demonstration of a lack of the branched-chain amino acid-glutamate transaminase in the class accumulating the α-keto acids and the lack of what is now called dihydroxy acid dehydrase in the dihydroxy acid accumulators. _This loss of a single enzyme that catalyzes the corresponding step in both pathways thus accounted for the unexpected double auxotrophy._

The valine and isoleucine auxotrophs that responded to the dihydroxy acids did not feed any other class except one, which appeared to be blocked only in isoleucine biosynthesis and which responded as well to α-ketobutyrate as to α-aminobutyrate. These singly blocked auxotrophs were later shown to lack threonine deaminase, the enzyme required only for isoleucine biosynthesis.

While these nutritional analyses were in progress, isotopic studies with _Neurospora_ and yeast indicated that pyruvate carbons were being incorpo-

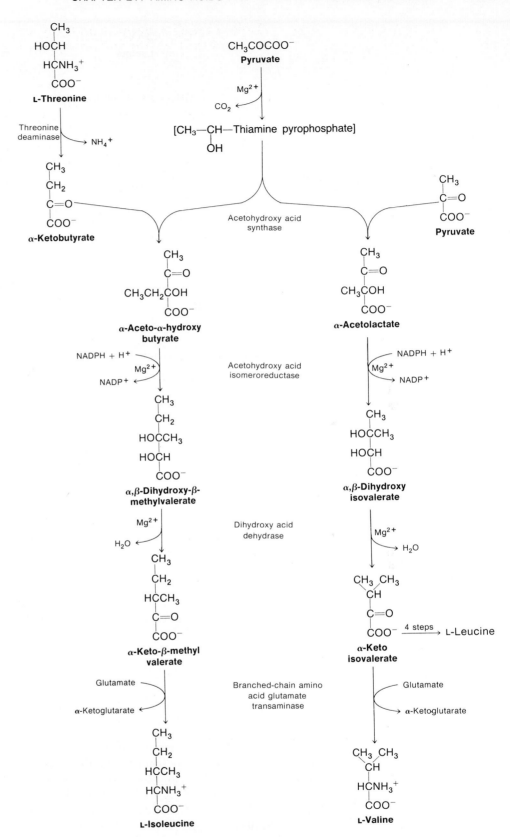

Figure 24–23

The biosynthesis of isoleucine and valine. The reactions leading to valine are catalyzed by the same enzymes that catalyze the corresponding reactions in isoleucine biosynthesis.

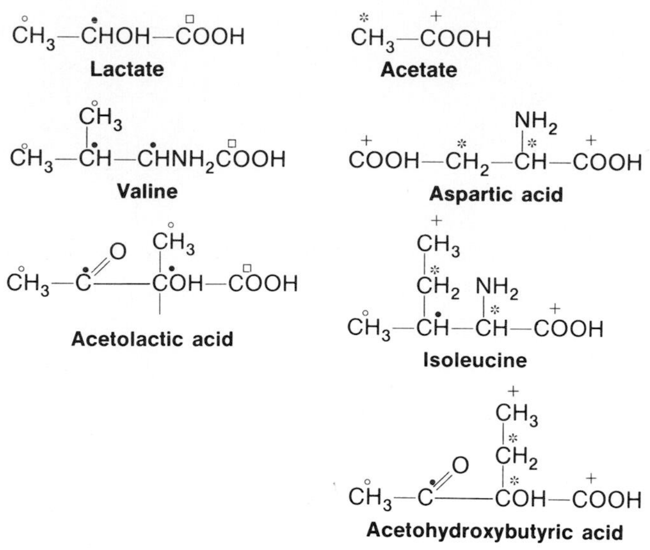

CH₃—CHOH—COOH
Lactate

CH₃—COOH
Acetate

CH₃—CH—CHNH₂COOH
            |
           CH₃
Valine

COOH—CH₂—CH—COOH
            |
           NH₂
Aspartic acid

CH₃—C—COH—COOH
        O |
            CH₃
Acetolactic acid

CH₃—CH—CH—COOH
        | |
       CH₂ NH₂
        |
       CH₃
Isoleucine

CH₃—C—COH—COOH
        O |
            CH₂
             |
            CH₃
Acetohydroxybutyric acid

Figure 24–24

Incorporation of lactate and acetate carbon into valine and isoleucine. The colored symbols next to the carbon atoms in lactate, acetate, and the other compounds indicate where the carbons are situated in the precursors and products. Radioactive labeling patterns of this type provide important information for pathway analysis. The isotope studies indicate that no sequence of three carbons in valine could have arisen from lactate (or pyruvate) directly, nor is there any four-carbon sequence in isoleucine that is labeled with acetate carbon. Thus the incorporation of these carbons into the amino acids must be complex.

rated into valine and that exogenous nonradioactive aspartate, homoserine, threonine, and α-ketobutyrate competed effectively with radioactive glucose for incorporation into isoleucine. The paradox was that when valine and isoleucine were degraded carbon by carbon after growth with either labeled lactate (metabolically equivalent to pyruvate) or labeled acetate, there was no sequence of three carbons in valine (see Figure 24–24) that could have arisen from pyruvate directly, nor was there any four-carbon sequence in isoleucine that was labeled with acetate carbon, as aspartate and threonine were labeled. It was proposed by Strassman and Weinhouse that a condensation of a two-carbon fragment derived from pyruvate with either pyruvate (to yield acetolactate) or α-ketobutyrate (to yield acetohydroxybutyrate) followed by an intramolecular migration could account for the isotopic distribution in valine and isoleucine. This postulated rearrangement was shortly followed by the demonstration of acetolactate accumulating in the culture fluids of mutants unable to reduce and rearrange the molecule and by the demonstration of the condensing enzyme that formed both acetohydroxy acids and the isomeroreductase that led to the formation of the dihydroxy acids.

Valine and isoleucine biosynthesis in plants has become of considerable interest with the finding that three classes of herbicides are potent inhibitors of most of the plant acetohydroxy acid synthases examined and of some of the acetohydroxy acid synthase isozymes of bacteria.

The control of metabolite flow over the pathways to valine and isoleucine is subject to one complication. In many organisms, acetohydroxy acid synthase is inhibited by valine. The complete inhibition of this enzyme not only would prevent oversynthesis of valine, but also would prevent isoleucine biosynthesis, for which acetohydroxy acid synthase is the second enzyme in the pathway. It is perhaps for this reason that many organisms also contain a valine-insensitive acetohydroxy acid synthase. Such an enzyme

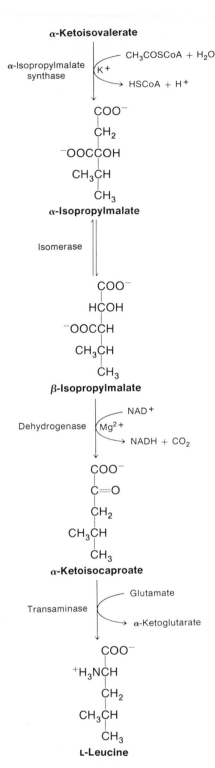

Figure 24–25
The biosynthesis of leucine from α-ketoisovalerate, the branchpoint intermediate also used in L-valine synthesis. In the first step the carbon chain of α-ketoisovalerate is lengthened by one carbon atom (this is comparable to the reaction in Figure 24–10). The steps to leucine are completed by specific isomerase, dehydrogenase, and transaminase reactions.

allows isoleucine synthesis to occur in the presence of valine, but also potentially allows an oversynthesis of valine under some conditions. Control of the synthesis of isoleucine itself is achieved by an inhibition of threonine deaminase, the first enzyme in the pathway to isoleucine.

L-Leucine Is Formed from α-Keto-Isovalerate in Four Steps

The biosynthesis of L-leucine involves the lengthening of the carbon chain of α-ketoisovalerate by one carbon to yield α-ketoisocaproate (Figure 24–25). This lengthening is comparable to that by which α-ketoglutarate is converted to α-ketoadipate in the pathway to lysine in fungi (see Figure 24–10).

THE BIOSYNTHESIS OF THE AROMATIC FAMILY OF AMINO ACIDS: L-TRYPTOPHAN, L-PHENYLALANINE, AND L-TYROSINE

The aromatic amino acids phenylalanine, tyrosine, and tryptophan are all formed by means of a common pathway to build the benzene ring. This pathway, often referred to as the *shikimate pathway*, is also important for the formation of the aromatic nuclei in or the aromatic precursor of vitamins E and K, folic acid, ubiquinone, and plastoquinone and certain metal chelators, such as enterochelin. The branchpoint compound for all these diverse products is *chorismate*, which has a prearomatic cyclohexadiene nucleus (Figure 24–26).

Chorismate Is a Key Intermediate

The overall route of chorismate synthesis, beginning with the condensation of two intermediates from the central metabolic routes, phosphoenolpyruvate and erythrose-4-phosphate, is illustrated in Figure 24–26. Chorismate is the final product of the common pathway to all aromatic compounds, including phenylalanine, tyrosine, and tryptophan.

Prephenate Is a Common Intermediate in L-Phenylalanine and L-Tyrosine Synthesis

In some organisms, the pathways from chorismate to L-phenylalanine and L-tyrosine shown in Figure 24–27 are truly separate pathways, even though the first step from chorismate is the same for the two pathways. For example, in *E. coli* and related organisms, one protein, *chorismate mutase P-prephenate dehydratase*, converts chorismate to phenylpyruvate, with the intermediate, *prephenate*, being enzyme-bound. A second protein, the NAD-dependent *chorismate mutase T-prephenate dehydrogenase*, converts chorismate to 4-hydroxyphenylpyruvate, again with the formation of prephenate occurring as an enzyme-bound intermediate. Both enzymes utilize prephenate, the phenylalanine biosynthetic enzyme yielding phenylpyruvate, and the tyrosine biosynthetic enzyme yielding 4-hydroxyphenylpyruvate. The final aromatization step, the removal of water from prephenate in phenylalanine biosynthesis, or the removal of hydrogen in tyrosine biosynthesis, is accompanied by the loss of the ring carboxyl as CO_2.

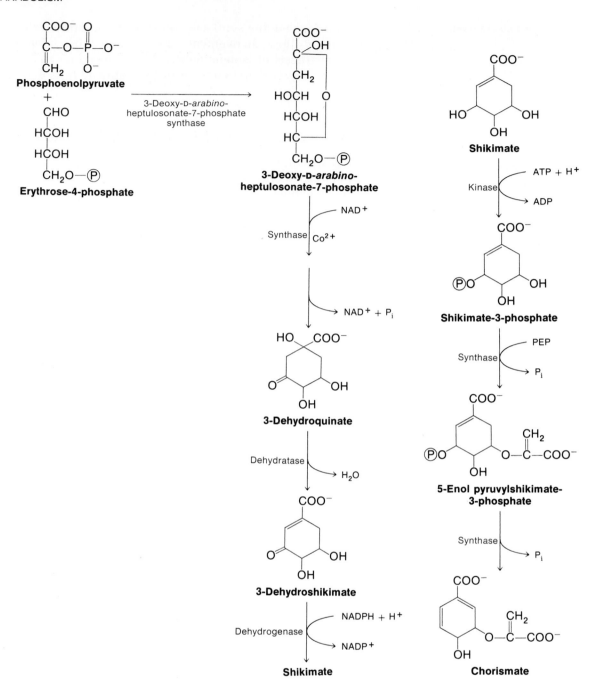

Figure 24–26
The common aromatic (shikimate) pathway leading to chorismate biosynthesis. Note that NAD$^+$ is required for the conversion of 3-deoxy-*arabino*-heptulosonate-7-phosphate to 3-dehydroquinate, but there is no net change in the redox state during the conversion of substrate to product. This was also seen in the epimerization of fructose (see Chapters 14 and 20) and is indicative of an oxidized intermediate. Chorismate formed by this series of reactions is a common intermediate for phenylalanine tyrosine and tryptophan biosynthesis.

Tryptophan Is Synthesized in Five Steps from Chorismate

The pathway leading from chorismate to L-tryptophan (Figure 24–28) is the most thoroughly studied of any biosynthetic pathway. In *E. coli*, the details of the enzymatic steps, the correlation between DNA sequence and the protein products, and the factors controlling the transcription of the structural

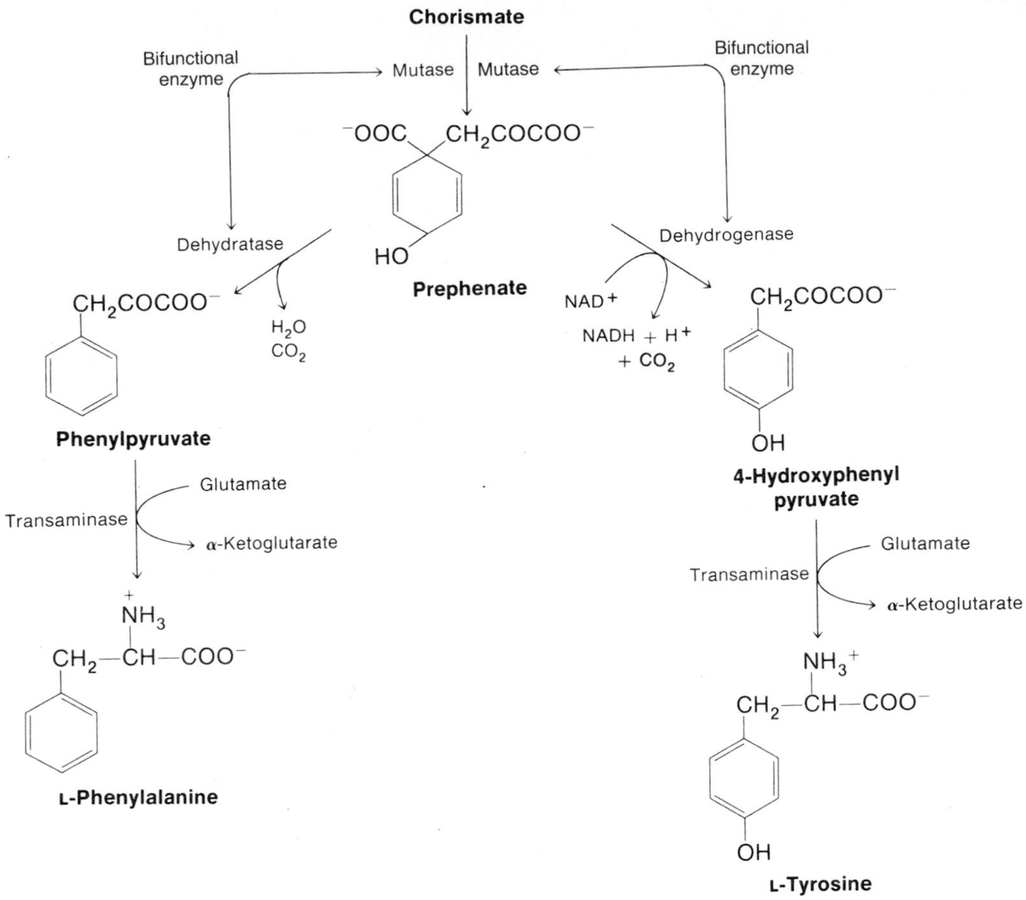

Figure 24–27

The biosynthesis of phenylalanine and tyrosine from the branchpoint compound chorismate. The lines between the mutase and the dehydratase and between the mutase and the dehydrogenase indicate that in both pathways we are dealing with bifunctional enzymes that are exclusive to the indicated pathways. Thus, although prephenate is the first intermediate in both pathways, it is not a branchpoint intermediate in the usual sense.

genes far exceed those known for any other set of related genes. Although this had not been the work of any one group, the extensive gene-enzyme analysis of Charles Yanofsky laid the foundation for others to explore details of some of the enzymatic steps by physical and kinetic approaches. Comparative studies in other bacteria and in fungi have revealed a variation upon the themes found in *E. coli*, particularly with respect to the distribution on one protein or another of the sequence of enzyme activities, which are identical in all forms. In addition, these studies have also revealed differences in the way the genes are arranged in the DNA and in the way expression of those genes is controlled.

The first specific step in tryptophan biosynthesis is the glutamine-dependent conversion of chorismate to the simple aromatic compound *anthranilate*. Like most other glutamine-dependent reactions, the reaction can also occur with ammonia as the source of the amino group. However, high concentrations of ammonia are required. Thus far, almost all the anthranilate synthases examined have the glutamine amidotransferase activity (component II) and the chorismate-to-anthranilate activity (component I) on separate proteins.

Anthranilate is transferred to a ribose phosphate chain in a phosphoribosyl-pyrophosphate-dependent reaction catalyzed by *anthranilate*

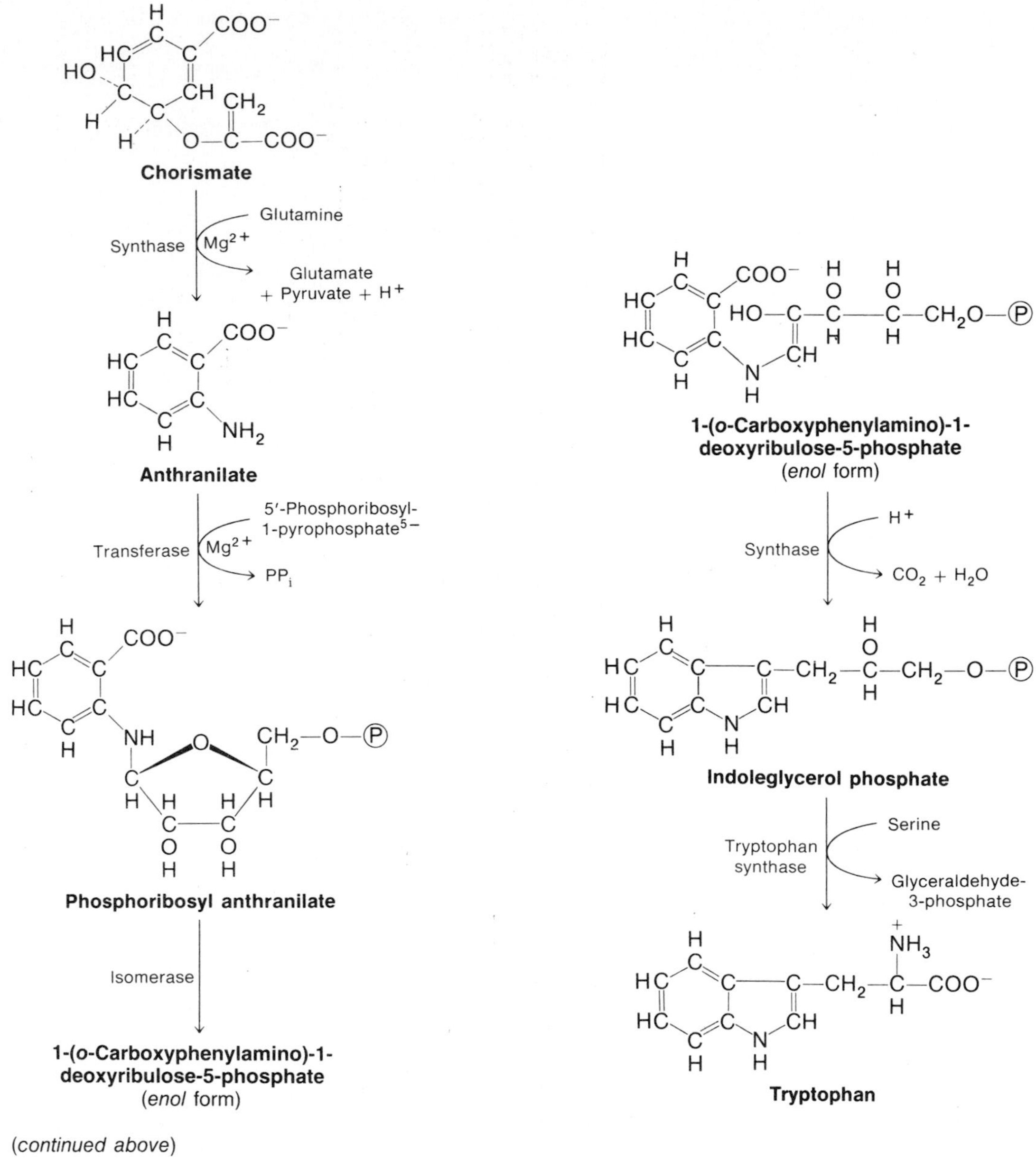

Figure 24–28

The biosynthesis of tryptophan from the branchpoint compound, chorismate in *E. coli.* The first step involves the conversion of chorismate to the aromatic compound anthranilate. The anthranilate is transferred to a ribose phosphate chain. The product is cyclized to indoleglycerol phosphate by the removal of water and loss of the ring carboxyl by indoleglycerol phosphate synthase. Finally, in a replacement reaction catalyzed by tryptophan synthase, glyceraldehyde-3-phosphate is removed from indoleglycerol phosphate and the enzyme-bound indole is condensed with serine.

phosphoribosyltransferase. *Phosphoribosylanthranilate* undergoes an *Amadori rearrangement*, in which the ribosyl moiety becomes a ribulosyl moiety. The product, 1-(*o*-carboxyphenylamino)-1-deoxyribulose-5'-phosphate, is cyclized to *indoleglycerol phosphate* by the removal of water and loss of the ring carboxyl by *indoleglycerol phosphate synthase.* The final step in tryptophan biosynthesis is a replacement reaction, catalyzed by *tryptophan synthase,* in which glyceraldehyde-3-phosphate is removed from indoleglycerol phosphate and the enzyme-bound indole is condensed with serine.

TABLE 24–3
Distribution of Tryptophan Biosynthetic Enzyme Activities on Different Proteins in Bacteria and Fungi

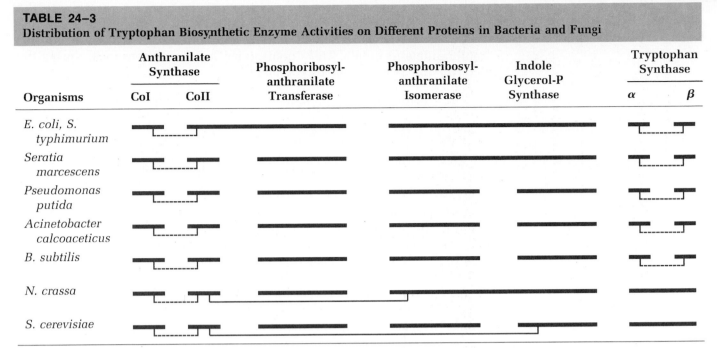

Organisms	Anthranilate Synthase		Phosphoribosyl-anthranilate Transferase	Phosphoribosyl-anthranilate Isomerase	Indole Glycerol-P Synthase	Tryptophan Synthase	
	CoI	CoII				α	β
E. coli, S. typhimurium							
Seratia marcescens							
Pseudomonas putida							
Acinetobacter calcoaceticus							
B. subtilis							
N. crassa							
S. cerevisiae							

Note: ⌐⌐⌐ = covalent linkage; ⌐----⌐ = obligatory association required for full activity. A single band covering two activities indicates a single polypeptide.

Among different organisms these five enzyme activities are distributed on different proteins (Table 24–3). For example, in *E. coli*, indoleglycerol phosphate synthase catalyzes both the isomerization of phosphoribosylanthranilate and the cyclization step. Of particular interest is the occurrence on a single protein of the catalytic activities for nonconsecutive reactions in some cases. If in such cases the proteins were separate from each other in the cell, this arrangement, for example, in *Neurospora*, would necessitate the product of one reaction leaving the product site of one enzyme to be acted upon by another enzyme and then returning to the substrate site of a third enzyme on the same protein that exhibited the first enzyme activity. *The persistence of this arrangement during evolution makes attractive the idea that all the tryptophan biosynthetic enzymes exist in the cell as a single multienzyme (and multiprotein) complex.* However, if so, the complex must be quite labile, since individual gene products are so readily separated.

The tryptophan pathway provides another example in which nutritional studies with mutants of *Neurospora* and *E. coli*, isotope incorporation studies, and enzymatic analyses have been exploited to reveal the steps in a biosynthetic pathway. For example, early studies with tryptophan-requiring organisms found in nature revealed that some could use indole (a compound known to be formed by the microbial degradation of tryptophan; see Chapter 19), while others could use anthranilate. Later, after Beadle and Tatum introduced the approach of studying metabolism with mutants of the bread mold *Neurospora*, tryptophan-requiring mutants of this organism were found that could use indole or either anthranilate or indole. Still later, similar mutants of *E. coli* were found, and mutants of both organisms were described that accumulated one or the other of these compounds. Clearly, *those mutants that grew on anthranilate or indole were blocked in some step before these compounds, and those that accumulated them were blocked in the step after them.*

Incorporation studies with isotopes showed that when anthranilate was converted to tryptophan, the carboxyl group of anthranilate was lost as carbon dioxide, but the nitrogen was retained. Because the enzymes in the tryptophan biosynthetic pathway have only a limited specificity, it was possible to substitute 4-methylanthranilate in *E. coli* extracts that could convert anthranilate to indole. This "nonisotope" label was conserved during the conversion to yield 6-methyl indole:

4-Methylanthranilate   6-Methyl indole

It was thus clear that some two-carbon unit replaced the carboxyl carbon of anthranilate. Further studies with such *E. coli* extracts indicated that phosphoribosyl pyrophosphate was a good cosubstrate for the formation of indole from anthranilate. Fractionation of these extracts, as well as examination of mutants blocked between anthranilate and indole, revealed that an intermediate in this conversion was indole-3-glycerol phosphate. Extracts from one group of such mutants could not form indole-3-glycerol phosphate, while the other group could not convert it to indole and glyceraldehyde-3-phosphate. The latter group was found to accumulate the dephosphorylated derivative, indole-3-glycerol, in culture fluids.

The two intermediates in the conversion of anthranilate to indole-3-glycerol phosphate, phosphoribosylanthranilate and 1-(o-carboxyphenylamino)-1-deoxyribulose-5-phosphate, were originally postulated to account for the involvement of phosphoribosyl pyrophosphate in indole-3-glycerol phosphate formation. Support for the postulate was obtained when the dephosphorylated derivative of the second of these intermediates was found in the culture fluids of certain bacterial tryptophan-requiring mutants. The corresponding derivative of the first intermediate has not been found, probably because of its instability. Indeed, this compound, when formed in extracts, is rapidly broken down to anthranilate and ribose-5-phosphate.

For several years, indole, which was accumulated by some mutants and used to satisfy the tryptophan requirement by others, was considered an intermediate in tryptophan biosynthesis. Such a role for indole would have been of interest, since it appeared to be an exception to the generalization that biosynthetic intermediates had to bear a charge. It was found that extracts of cells that utilized indole did indeed catalyze the condensation of indole with serine, and extracts of cells that accumulated indole catalyzed the cleavage of indole-3-glycerol phosphate to indole and glyceraldehyde-3-phosphate. Furthermore, *E. coli* mutants of these two classes clearly were affected in separate genes. However, the two products of these genes catalyzed their corresponding reactions faster when they were associated in a complex of the form $\alpha_1\beta_2$, where α and β stand for different protein subunits. The complex itself catalyzes the overall reaction

Indole-3-glycerol phosphate + serine $\longrightarrow$

tryptophan + glyceraldehyde-3-phosphate

faster than either of the separate reactions. The same was found with extracts of *Neurospora* in which the two partial reactions were catalyzed by the same protein and with which no evidence for indole as a free intermediate could be found. Thus it became clear that indole, although historically important in deciphering the pathway to tryptophan, occurs only as a bound intermediate.

Carbon Flow in the Biosynthesis of Aromatic Amino Acids Is Regulated at Branchpoints

Metabolite flow to tryptophan is controlled by inhibition of anthranilate synthase by tryptophan. Regulation of metabolite flow in phenylalanine and tyrosine biosynthesis varies from organism to organism, owing to the variety of enzyme patterns in the conversion of chorismate to the two amino acids. In *E. coli* and related organisms, phenylalanine inhibits both activities of chorismate mutase P-prephenate dehydratase, whereas tyrosine inhibits only the mutase activity of chorismate mutase T-prephenate dehydrogenase.

There are two general patterns of control over the common aromatic pathway. One is that in *E. coli* and related organisms. The pattern is similar to that of the common aspartate family pathway of the same organism in that there are three isozymic deoxy-*arabino*-heptulosonate-7-phosphate synthases. Each is inhibited by one of the three aromatic amino acids. (There is, in addition, a tryptophan-specific repression of the tryptophan-sensitive enzyme, a tyrosine-specific repression of the tyrosine-sensitive enzyme, and a tryptophan plus phenylalanine-specific multivalent repression of the phenylalanine-sensitive enzyme.) As in the synthesis of the intermediates in the aspartate family common pathway, the three enzymes contribute to a common pool of deoxy-*arabino*-heptulosonate-7-phosphate that is drawn upon for all the compounds formed from chorismate. Indeed, in some strains, the phenylalanine-sensitive enzyme is predominant, whereas in others, the tyrosine-sensitive enzyme is predominant.

Another pattern is found in *Bacillus subtilis*. The single deoxy-*arabino*-heptulosonate-7-phosphate synthase is carried on the same protein that exhibits chorismate mutase activity. The protein is complexed with another protein that exhibits shikimate kinase activity. Both the deoxy-*arabino*-heptulosonate-7-phosphate synthase activity and the shikimate kinase activity are inhibited by chorismate and prephenate, which may inhibit by virtue of binding to the substrate and product sites of the chorismate mutase. The chorismate mutase activity is inhibited by prephenate. Prephenate dehydratase is inhibited by phenylalanine, whereas prephenate dehydrogenase is inhibited by tyrosine.

THE BIOSYNTHESIS OF HISTIDINE

The pathway to histidine in all plants and bacteria involves the transfer of N-1 and C-2 of the adenine moiety of ATP to the ribose phosphate moiety of phosphoribosyl pyrophosphate (PRPP) as shown in Figure 24–29. This transfer is initiated by a condensation reaction between the two parent compounds. After a series of additional reactions involving ring opening, isomerization, and an amido-group transfer, the residue from the ATP molecule is released as 5-aminoimidazole-4-carboxamide ribotide. The latter, an intermediate in the purine nucleotide biosynthetic pathway (see Chapter 25) is then "recycled." In this cyclic process, the 6-amino group of the adenine ring which had been derived from aspartate becomes the amide group of the 5-aminoimidazole-4-carboxamide ribotide. During purine biosynthesis, this amide group is also derived from aspartate and becomes N-1 of the purine ring. Thus there is no way to distinguish an adenosine derivative that has been formed directly by the *de novo* pathway from one that has been recycled from the histidine pathway.

Another interesting feature of the histidine pathway is the conversion of imidazoleglycerol phosphate to a carbonyl derivative that can undergo transamination. Following the incorporation of the α amino group, which pro-

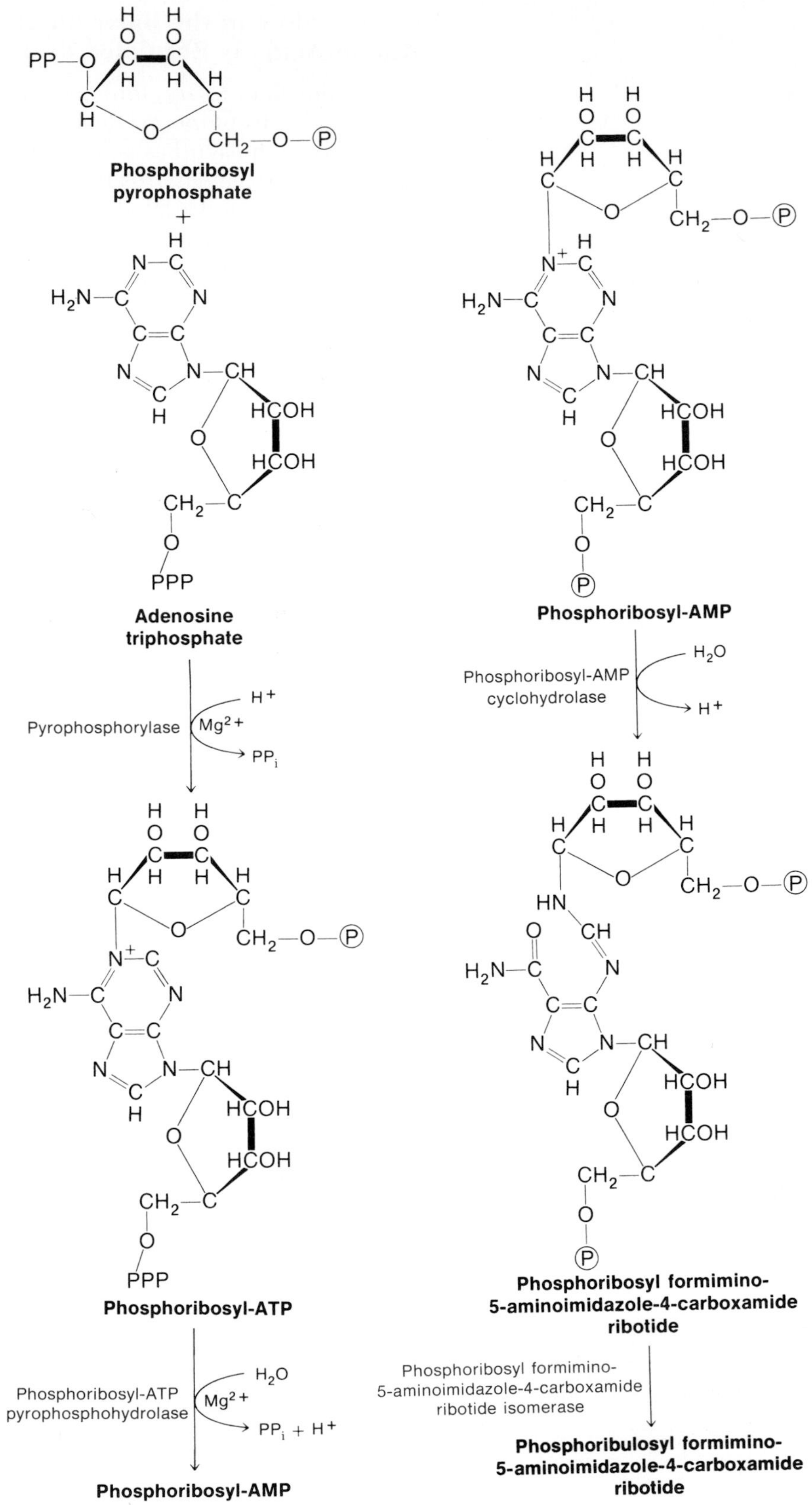

**Phosphoribosyl
pyrophosphate**

+

**Adenosine
triphosphate**

Pyrophosphorylase $\Big\{$ Mg^{2+}

H^+

$\rightarrow PP_i$

Phosphoribosyl-ATP

Phosphoribosyl-ATP
pyrophosphohydrolase $\Big\{$ Mg^{2+}

H_2O

$\rightarrow PP_i + H^+$

Phosphoribosyl-AMP

Phosphoribosyl-AMP

Phosphoribosyl-AMP
cyclohydrolase

H_2O

$\rightarrow H^+$

**Phosphoribosyl formimino-
5-aminoimidazole-4-carboxamide
ribotide**

Phosphoribosyl formimino-
5-aminoimidazole-4-carboxamide
ribotide isomerase

**Phosphoribulosyl formimino-
5-aminoimidazole-4-carboxamide
ribotide**

(continued on next page)

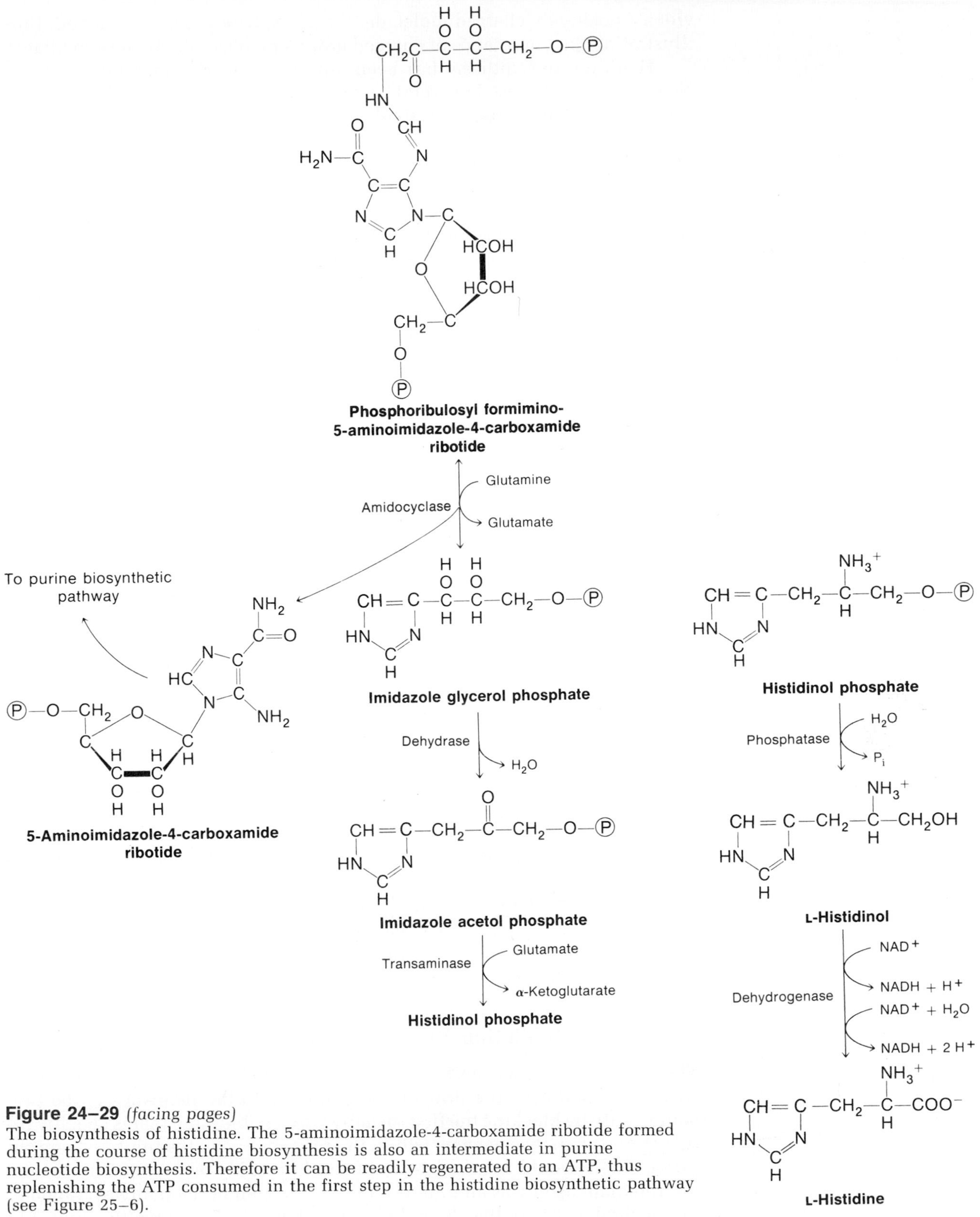

Figure 24-29 *(facing pages)*
The biosynthesis of histidine. The 5-aminoimidazole-4-carboxamide ribotide formed during the course of histidine biosynthesis is also an intermediate in purine nucleotide biosynthesis. Therefore it can be readily regenerated to an ATP, thus replenishing the ATP consumed in the first step in the histidine biosynthetic pathway (see Figure 25-6).

vides a positively charged molecule, the phosphate group is removed, thus illustrating the importance of charged groups on biosynthetic intermediates.

Histidine biosynthesis has been studied primarily in a few bacteria, *Neurospora*, and yeast. In several bacteria, the phosphatase and the dehydratase activities are carried on a single bifunctional protein. In yeast there is a single protein that exhibits the pyrophosphatase, cyclohydrolase, and dehydrogenase activities (the second, third, and tenth steps). These findings, along with the fact that significant "pools" of the intermediates are not found, raise the question of whether all the enzymes of histidine biosynthesis might not be arranged in a single complex, as has been suggested for the enzymes involved in tryptophan biosynthesis.

ADDITIONAL ASPECTS OF THE REGULATION OF AMINO ACID BIOSYNTHESIS

The extent to which amino acid biosynthesis is controlled and the mechanisms that have evolved to effect the controls vary widely in different pathways. They also vary widely for the same pathway in different plant and microbial systems. This variation in patterns of regulation results probably because the selection of control mechanisms during evolution was superimposed on and became secondary to the selection of the capacity to form a given amino acid. Thus the emergence of a variant route to an amino acid may have made a preexisting control mechanism inefficient or even inoperative, and the emergence of a variant control mechanism may have soon followed.

It would therefore not be possible in a short space to describe or even to catalog all the ways in which the biosynthesis of amino acids is controlled. In only the enterobacteria (and specifically in *E. coli* and the closely related *S. typhimurium*) has the regulation of all 20 amino acids been studied sufficiently to describe some details of the regulation. However, there have been enough studies with other bacteria, fungi, and plants to show that some of the *E. coli*-type control systems are far from being universally distributed. It is clear that no matter how well studied some regulatory interaction has been in *E. coli* or how elegant and potentially universal it seems to be, one should not conclude *a priori* that the same pattern is to be found in the pathway to any other amino acid in *E. coli*, or in the same pathway in any other organism.

Regulation of amino acid biosynthesis may occur at two levels— regulation of enzyme activity or metabolite flow over a pathway, and regulation of enzyme amount. The former will be considered here. The latter is discussed in Chapter 29.

End-product Inhibition Takes on Various Forms

The production of an amino acid can be quenched by blocking any step within the pathway; the flow of metabolites into the pathway can be prevented only by blocking the first (usually irreversible) step that is specific for that amino acid. If an enzyme within the sequence were the inhibited one, an intermediate would accumulate and would probably be excreted unless the intermediate itself was an inhibitor of the first step in the pathway. Some of the general patterns that have been found to block metabolite flow into amino acid biosynthetic pathways are illustrated in Figure 24–30.

The simplest kind of feedback loop in the living cell is the direct inhibition of the initial enzyme in an amino acid biosynthetic sequence by the amino acid itself (Figure 24–30). That such interactions are significant has been demonstrated in several cases by the isolation of microbial mutants in

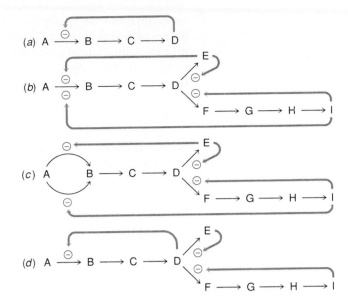

Figure 24-30
Patterns of end-product inhibition in amino acid biosynthetic pathways. (a) Simple end-product inhibition: End product D inhibits the first enzyme in the pathway. (b) Concerted end-product inhibition of a common pathway: End products E and I act in concerted fashion to inhibit the first enzyme in a common pathway, and each inhibits the first enzyme in the branch pathway leading specifically to it. (A special case, not shown, would be synergistic inhibition, in which E and I singly are only weak inhibitors, but when both are present, a strong synergistic inhibition occurs.) (c) Multiple enzymes specifically controlled by end products of different branch pathways: Two enzymes convert A to B. One enzyme is inhibited by E, the end product of one branch pathway. The other enzyme is inhibited by I, the end product of the other branch pathway. The two branch pathways are controlled by simple end-product inhibition. (d) Sequential end-product inhibition: The ultimate end products E and I inhibit only the first enzyme in their own branch pathways. Intermediate D inhibits the first enzyme in the common pathway.

which the loss of sensitivity of an initial enzyme to its respective end product is accompanied by an overproduction and excretion of that end product.

Those reactions known to be sensitive to end-product inhibition in at least one organism have been so noted in the text. Alanine, aspartate, and glutamate are three amino acids for which no form of end-product inhibition is known. These amino acids, however, are in an essential equilibrium by means of reversible reactions with their corresponding α-keto acids, which are key intermediates in the central metabolic routes. In most bacteria, the pools of these three amino acids are fairly high, at least under conditions of excess nitrogen supply, and actually are "leaked" to the external medium in small amounts.

Glycine is another amino acid formed by a single reversible reaction. Glycine pools are usually small, but the enzyme is not product-inhibited. It may be controlled in part by the availability of the acceptor for one-carbon units, tetrahydrofolate.

Metabolite flow into the biosynthetic pathways of all the remaining 16 amino acids is controlled by one means or another. The specific controls for these 16 amino acids, as found in E. coli, have been cited earlier. In addition, some of the examples in which there are different controls in other forms also have been cited.

Most of the departures from the patterns exhibited by E. coli are related to the different ways the common aspartate family and common aromatic pathways have evolved. It is of interest that in some forms, a means has evolved for sequential feedback inhibition to occur (see Figure 24–30). The basic requirement for such a pattern of regulation to be effective is that the early enzyme be inhibited when the inhibitory intermediate is no longer being utilized. However, pool levels of biosynthetic intermediates are usually very small, so that fluctuations in pool size alone might not be sufficient to provide the range of control required. In *B. subtilis* this control may have been achieved by the single 3-deoxy-D-*arabino*-heptulosonate-7-phosphate synthase being part of the same protein that exhibits chorismate mutase activity and by both the substrate and product of the mutase being inhibitors of the synthase. Whether this is the general mechanism by which intermediates exert feedback inhibition has not been widely examined.

The multiple-enzyme pattern found in the *E. coli* common aspartate pathway might be considered costly, and in some organisms, a different pattern of regulation has been selected: the concerted (or highly synergistic) inhibition by lysine plus threonine. In some bacteria this pattern results in a methionine deficiency in a medium containing threonine plus lysine. The fact that such organisms have survived in nature, and that no organisms containing a threonine-plus-lysine–inhibited aspartokinase along with a methionine-repressed aspartokinase have been found, may be evidence that the threonine-plus-lysine combination does not often occur in the absence of methionine. In one organism that has a threonine-plus-lysine–sensitive aspartokinase, *B. subtilis*, the need for diaminopimelate as a cell wall constituent may have provided a more important selective pressure, since a second aspartokinase inhibited by diaminopimelate is found. The pattern found in some plants of two aspartokinases, one inhibited by threonine, the second by a synergistic effect of lysine plus S-adenosylmethionine, represents still another mechanism to achieve control in this highly branched pathway.

Sometimes Enzymes Are Activated

Just as end-product inhibition (or in the few cases, inhibition by biosynthetic intermediates) *has been invoked as an important mechanism for the control of amino acid biosynthesis, so has the less frequently encountered activation of enzyme activity by precursors or by intermediates in other pathways that depend on the function of a converging pathway for their further utilization.* The most striking example of the latter encountered in this chapter has been found in the regulation of those carbamoyl phosphate synthases that function for both pyrimidine and arginine biosynthesis (Figure 24–31). The inhibitor is UMP. Alone, this interaction could lead to a quenching of arginine biosynthesis. The inhibition, however, is reversed by ornithine. This is very important, because ornithine is the intermediate that would accumulate if (1) arginine were no longer in high enough concentration to block the first enzyme in the pathway and (2) there were no carbamoyl phosphate available. The antagonism between ornithine (the positive effector) and UMP (the negative effector) is thus a compensatory mechanism controlling metabolite flow.

The activity of cystathionine-γ-synthase of *Neurospora crassa*, an essential enzyme for methionine formation, is stimulated by 5-methyltetrahydrofolate (Figure 24–32). 5-Methyltetrahydrofolate might accumulate if the flow from cystathionine to homocysteine were retarded, and thus it might provide an appropriate "signal" for a methionine deficiency. In green plants, S-adenosylmethionine, the ultimate end product in one biosynthetic branch from phosphohomoserine, stimulates threonine synthase, which constitutes another (competing) branch leading from phosphohomoserine (Figure 24–33).

As we mentioned in Chapter 13, many (but certainly not all) end-product-sensitive enzymes exhibit sigmoid substrate saturation curves. In

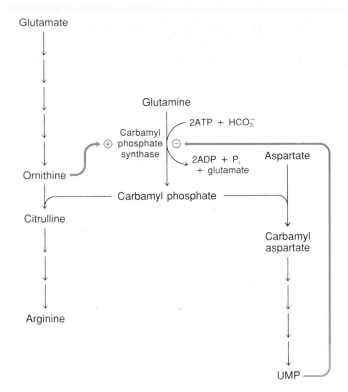

Figure 24–31

Activation of carbamoyl phosphate synthase by ornithine. Carbamoyl phosphate is required for both arginine and UMP synthesis. UMP inhibits the enzyme, and ornithine overcomes this inhibition and activates the enzyme. The activation by ornithine assures sufficient carbamoyl phosphate for arginine biosynthesis even when the UMP supply is sufficient. For more details on these two pathways, see Figures 24–9 and 25–10.

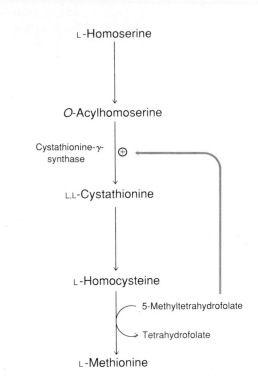

Figure 24–32

Activation of cystathionine-γ-synthase by N-methyltetrahydrofolate. An excess of 5-methyltetrahydrofolate might accumulate if the formation of methionine were for any reason retarded. Therefore, stimulation of the first reaction in the L-methionine pathway by 5-methyltryptophan would be an appropriate signal to overcome a methionine deficiency, at least under some circumstances. See Figure 24–18 for more details on this pathway.

some cases, the sigmoid curves (showing cooperative binding) are dependent on the presence of some end-product inhibitor. Where cooperative binding does occur, the substrate itself can be looked upon as an activator.

Protein–Protein Interactions Can Regulate Enzyme Activity

Both activation and inhibition of enzyme activity have been found to accompany binding to a second protein. A striking example is the inhibition of yeast ornithine transcarbamylase upon binding of arginase, an enzyme induced upon addition of arginine to the growth medium. Thus arginase,

Figure 24–33

Activation of threonine synthase by S-adenosylmethionine. In some green plants O-phospho-L-homoserine is a branchpoint intermediate for L-threonine, L-methionine, and S-adenosylmethionine synthesis. A sufficient supply of S-adenosylmethionine leads to activation of threonine synthase, the first enzyme of the other branch. See Figures 24–16, 24–19, and 24–20 for more details on these pathways.

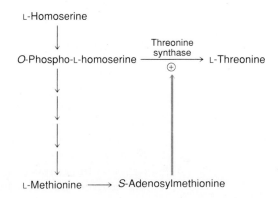

which converts arginine to ornithine, prevents the futile cycling of ornithine back to arginine.

There also are examples of stimulation of activity of enzymes upon binding to other proteins. The best examples might be those in which one component of a multienzyme complex exhibits a low activity alone but an enhanced activity in the complex, such as the stimulation of the ammonia-dependent anthranilate synthase activity of *E. coli* component I by component II. Many other examples of such an activation have been found, and it may be that, within the cell, many less tightly bound protein–protein associations can occur that do, in fact, greatly influence rates of enzyme-catalyzed reactions.

NONPROTEIN AMINO ACIDS

Not all amino acids in plants, bacteria, or animals are found as protein constituents. Some are found in peptide linkages in compounds that are important as cell wall or capsular structures in bacteria or as antibiotic substances produced by bacteria and fungi. Others are found as free amino acids in seeds and other plant products. Among the amino acids that occur in these forms are not only the 20 normal amino acids, but also some amino acids never found in proteins. These *nonprotein amino acids*, numbering into the hundreds, *include precursors of normal amino acids*, such as homoserine and diaminopimelate, *intermediates in catabolic pathways*, such as pipecolic acid, D-enantiomers of "normal" amino acids, *and amino acid analogs*, such as azetidine-2-carboxylic acid and canavanine, *that might be formed by unique pathways or by modification of normal amino acid biosynthetic pathways*.

A Wide Variety of D-Amino Acids Are Found in Microbes

Certain D-amino acids along with some L-enantiomers are commonly found both in microbial cell walls (see Chapter 19) and in many peptide antibiotics. For example, the peptidoglycans of bacteria contain both D-alanine and D-glutamate. The latter is present in a γ-glutamyl linkage. In some forms, the α carboxyl of the D-glutamyl residue is either amidated or in peptide linkage with glycine. D-Lysine or D-ornithine is found in the glycopeptide of some Gram positive organisms. The capsule of the anthrax bacillus is composed of a nearly pure homopolymer of D-glutamate in γ linkage. Other bacilli also produce γ-linked polyglutamates, some of which form separate D-glutamate and L-glutamate chains, while others form a copolymer of D- and L-glutamate. A wide variety of D-amino acids have been found in antibiotics, and some of these are listed in Table 24–4.

Racemases Are Often Involved in D-Amino Acid Formation

At least some of the D-amino acids are formed by racemases as free intermediates and are subsequently incorporated into peptide bonds. An example is D-alanine found in the bacterial cell wall peptidoglycan. L-Alanine is converted to D-alanine by a racemase that contains pyridoxal phosphate as a cofactor. The racemization is followed by the formation of a D-alanyl-D-alanine dipeptide, which is accompanied by the conversion of ATP to ADP. The dipeptide is subsequently incorporated into the glycopeptide.

In most cases of formation of D-amino acid–containing peptides studied thus far, the L form of the amino acid is the substrate for the incorporating enzyme. In contrast, the free D-amino acid is ordinarily a poor substrate for the incorporation reaction. Whether the racemization occurs on the enzyme

TABLE 24–4
Some D-Amino Acids Found in Peptide Antibiotics

Antibiotic	D-Amino Acids Present	Produced by
Actinomycin C_1 (D)	D-Valine	*Streptomyces parralus* and others
Bacitracin A	D-Asparagine, D-glutamate, D-ornithine, D-phenylalanine	*Bacillus subtilis*
Circulin A	D-Leucine	*Bacillus circulans*
Fungisporin	D-Phenylalanine, D-valine	*Penicillium* species
Gramicidin S	D-Phenylalanine	*Bacillus brevis*
Malformin A_1, C	D-Cysteine, D-leucine	*Aspergillus niger*
Mycobacillin	D-Aspartate, D-glutamate	*Bacillus subtilis*
Polymixin B_1	D-Phenylalanine	*Bacillus polymyxa*
Tyrocidine A, B	D-Phenylalanine	*Bacillus brevis*
Valinomycin	D-Valine	*Streptomyces fulrissimus*

or inversion occurs afterwards remains to be determined in most cases. In the case of the D-valyl residue formed in penicillin, a tripeptide derivative containing L-valine is an intermediate, and conversion is thought to occur by way of an α-β-dehydro form of the valyl residue.

There Are Hundreds of Naturally Occurring Amino Acid Analogs

Among the hundreds of nonprotein amino acids found in nature are many that might be considered naturally occurring amino acid analogs and many that are toxic and antagonistic to the usual 20 amino acids found in proteins. Some are found as components of antibiotics, but others have been identified as antibiotic substances themselves. The frequent occurrence of toxic amino acids in the seeds of plants suggests that they might play a role in the protection of the seeds from insects or other predators.

Some typical examples of these naturally occurring analogs are given in Table 24–5 along with their sources and the antagonistic L-amino acid, where such an antagonism is known. It should be pointed out, however, that not all the "analogs" are toxic. For example, pipecolic acid, the next higher homolog of proline and an intermediate in lysine degradation, does not interfere in any demonstrable way with proline metabolism.

THE CONVERSION OF AMINO ACIDS TO OTHER AMINO ACIDS AND TO OTHER METABOLITES

As stated at the outset of this chapter, the primary fate of amino acids is their incorporation into protein. However, amino acids also serve a number of other important functions. These include processes leading to the formation of the porphyrin nucleus found in many oxygen- and electron-carrying proteins, of biologically active amines, of glutathione, of peptide antibiotics, and of several other important metabolites. These topics are considered in the remainder of this chapter.

Porphyrin Biosynthesis Starts with the Condensation of Glycine and Succinyl-CoA

The early isotope tracer experiments of David Shemin permitted the elucidation of the formation of the immediate precursor of the porphyrin needed for the cytochromes and for hemoglobin. These studies indicated that the gly-

TABLE 24–5
Some Naturally Occurring Nonprotein Amino Acids

Compound	Occurrence	Remarks
Branched-chain and cyclopropane amino acids		
$CH_3-CH_2-CH(CH_3)-CH_2-CHNH_2-COOH$ 2-Amino-4-methylcaproic acid (homoisoleucine)	California buckeye	Leucine antagonizes toxicity

$(CH_3)_2-NCH_2-CHNH_2-COOH$
    2-Amino-3-dimethylaminopropionic acid (azaleucine)
	*Streptomyces
	    neocaliberis*
		Leucine antagonizes toxicity

$CH_2=C-CHCHNH_2-COOH$ (with CH$_2$ cyclopropane ring) 2-(Methylenecyclopropyl)glycine	Lychee seeds	Leucine antagonizes toxicity
Sulfur-containing amino acids		

$CH_3-SCH_2-CHNH_2-COOH$
    S-Methylcysteine
	Broad bean
	    (*Phaseolus vulgaris*)
		—

$CH_3-CH=CH-S-CH_2-CHNH_2-COOH$ S-(Prop-1-enyl)cysteine	Garlic	—
Aromatic and heterocyclic amino acids		

CH_2-CHNH_2-COOH (phenyl ring with COOH)
    3-(3-Carboxyphenyl)alanine
	Iris
	    (*Iris pseudacoras*)
		—

CH_2-CHNH_2-COOH (3-hydroxy-4-pyridone ring) β-N-(3-Hydroxy-4-pyridone)alanine (mimosine)	Mimosa tree	Tyrosine; toxic to nonruminants

(piperidine ring CHCOOH)
Pipecolic acid
	Widely distributed
	    in plants
		Probably not toxic; an inter-
		    mediate in lysine catabolism

Acidic amino acids		
$HOOC-CH(CH_3)-CH_2-CHNH_2-COOH$ 4-Methylglutamic acid	Sweet pea	—
Basic amino acids		

$H_2N-C(=NH)-NH-O-CH_2-CH_2-CHNH_2-COOH$
    Canavanine
	Jack bean and
	    other legumes
		Arginine antagonizes toxicity

$NH_2-CH_2-CHNH_2-COOH$
    2,3-Diaminopropionic acid
	Seeds of acacia
	    and mimosa
		As the oxalyl derivative, acts
		    as a neurotoxin

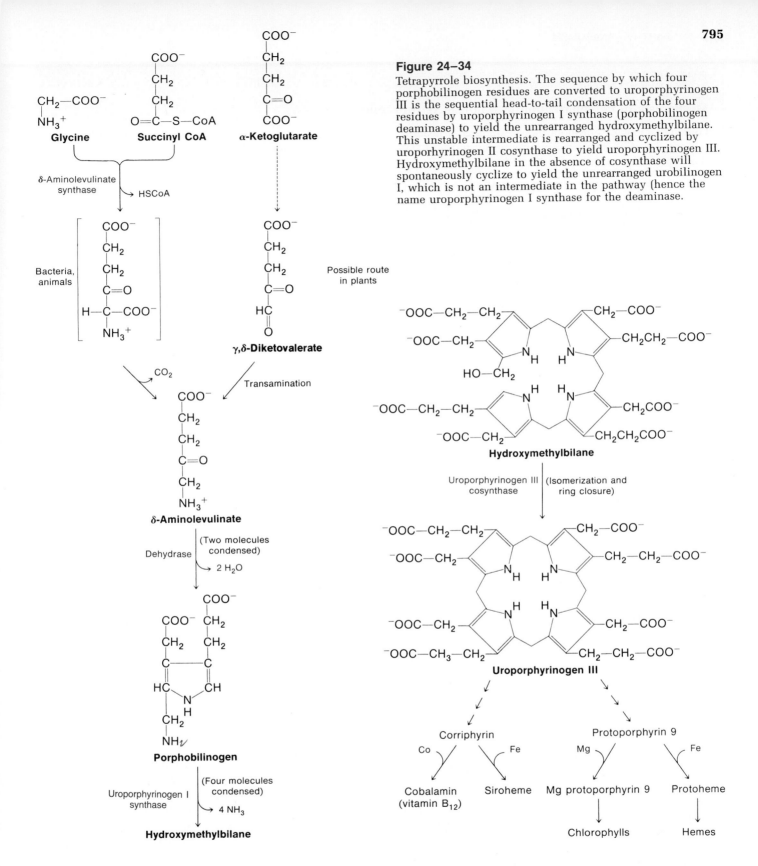

Figure 24–34

Tetrapyrrole biosynthesis. The sequence by which four porphobilinogen residues are converted to uroporphyrinogen III is the sequential head-to-tail condensation of the four residues by uroporphyrinogen I synthase (porphobilinogen deaminase) to yield the unrearranged hydroxymethylbilane. This unstable intermediate is rearranged and cyclized by uroporhyrinogen II cosynthase to yield uroporphyrinogen III. Hydroxymethylbilane in the absence of cosynthase will spontaneously cyclize to yield the unrearranged urobilinogen I, which is not an intermediate in the pathway (hence the name uroporphyrinogen I synthase for the deaminase.

cine methylene carbon and nitrogen were incorporated along with both carbons of acetate. Subsequent enzymic studies in both bacteria and animals revealed a condensation reaction between succinyl-CoA and glycine to yield δ-_aminolevulinate_ and CO_2 (presumably by way of an enzyme-bound β-keto acid, α-amino-β-ketoadipate) (Figure 24–34).

BOX 24–A

Mechanism of Polymerization in a Linear Tetrapyrrole

Four molecules of porphobilinogen undergo a head-to-tail condensation catalyzed by uroporphyrinogen I synthase to yield a tetrapyrrole. Asterisks indicate nitrogen and carbon atoms derived from glycine; the others are derived from succinyl-CoA.

Aminolevulinate is also the precursor to porphobilinogen in plants and in the blue-green algae (cyanobacteria), but it is not formed by a condensation of glycine and succinyl-CoA. Rather, it is probably formed by the reduction of the α carboxyl group of α-ketoglutarate to yield γ,β-diketovalerate followed by an ω amination to yield δ-aminolevulinate.

The pyrrole monomer porphobilinogen arises from the condensation of two molecules of δ-aminolevulinate with the loss of two water molecules. The reaction is catalyzed by *δ-aminolevulinate dehydrase*. The condensation of four porphobilinogen molecules to yield the branchpoint compound in tetrapyrrole synthesis, uroporphyrinogen III, is a complex reaction requiring two enzymes, *uroporphyrinogen I synthase*, which catalyzes a head-to-tail condensation of four porphobilinogen molecules (see Box 24–A), and *uroporphyrinogen III cosynthase*, which inverts one of the units and closes the ring.

The Conversion of Amino Acids to Biologically Active Amines Entails Decarboxylation

Among the more important metabolites derived from amino acids are those derived directly or indirectly by decarboxylation. *Amino acid decarboxyl-*

ases often use *pyridoxal phosphate* as the coenzyme. The amines or their derivatives play a variety of roles in both bacteria and animals.

One of the simplest amines, ethanolamine, is usually formed by decarboxylation of the serine moiety of phosphatidylserine (Chapter 22). Upon the transfer of three methyl groups, phosphatidylcholine is formed. Choline can be liberated from the phosphatide to give rise to the important neurotransmitter *acetylcholine*, or it can be recycled into phospholipids. Ethanolamine is also a constituent of certain lipopolysaccharides of Gram negative bacteria.

Two polyamines found complexed with DNA are spermidine and spermine. The formation of *spermidine* and *spermine* is dependent on the decarboxylation of both ornithine (or arginine) and S-adenosylmethionine. Ornithine is decarboxylated to *putrescine* in both animals and bacteria (see Figures 19–25 and 19–26). A second route to putrescine is found in some bacteria and involves the decarboxylation of arginine.

Several physiologically important metabolites are derived indirectly by amino acid decarboxylation. Among these are *epinephrine* and *serotonin*, which are formed by the decarboxylation of hydroxylated derivatives of tyrosine and tryptophan, respectively. Their biosynthetic pathways are shown in Figure 24–35. The direct decarboxylation products of the aromatic amino acids phenylethylamine, tyramine, and tryptamine are, like epinephrine and serotonin, neurologically active compounds that serve as vasoconstrictors (i.e., constrictors of blood vessels) but with much less activity than either epinephrine or serotonin. Histamine, the direct decarboxylation product of the basic amino acid histidine, acts as a vasodilator.

Glutathione Is γ-Glutamylcysteinylglycine

The tripeptide γ-glutamylcysteinylglycine, or glutathione, is found in nearly all cells and plays a variety of roles. The tripeptide is formed in two steps

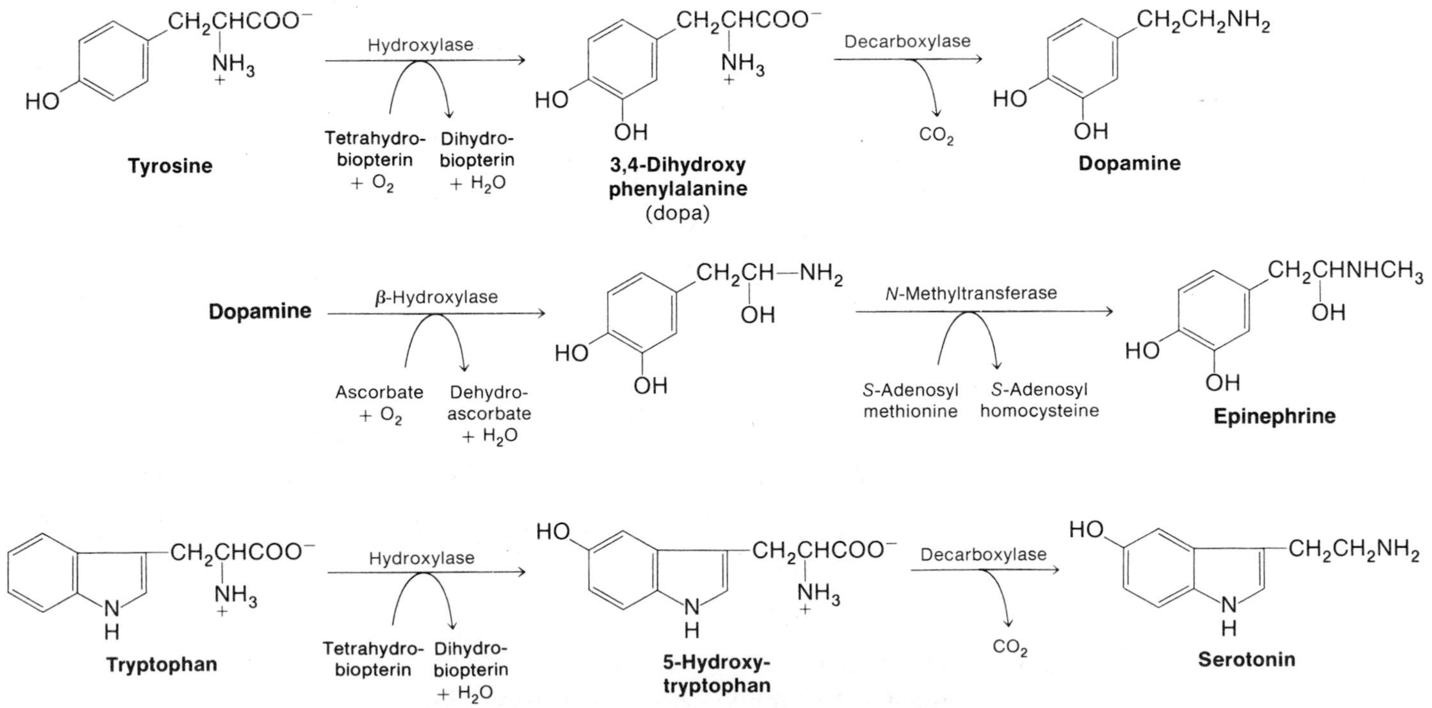

Figure 24–35
The conversions of tyrosine and tryptophan to epinephrine and serotonin, respectively. These hormones are made in specialized tissues (see Chapter 31).

catalyzed by ATP-requiring reactions. The first step is the condensation of glutamate with cysteine:

$$\text{Glutamate} + \text{cysteine} + \text{ATP} \xrightarrow{\substack{\gamma\text{-Glutamylcysteine} \\ \text{synthase}}}$$

$$\gamma\text{-glutamylcysteine} + \text{ADP} + \text{P}_i$$

The second step is the condensation of the dipeptide with glycine:

$$\gamma\text{-Glutamylcysteine} + \text{glycine} + \text{ATP} \xrightarrow{\substack{\text{Glutathione} \\ \text{synthase}}}$$

$$\text{glutathione} + \text{ADP} + \text{P}_i$$

Glutathione has often been considered important in maintaining the sulfhydryl groups of proteins in the cell in a reduced state, presumably by a nonenzymatic reaction. In contrast, an enzyme, protein-disulfide reductase, does catalyze sulfhydryl-disulfide interchanges between glutathione and proteins. The enzyme is important in insulin breakdown and is probably important in the reassortment of disulfide bonds during polypeptide chain folding.

Mutants of *E. coli* have been isolated that are essentially devoid of glutathione owing to the loss of one or the other of the two synthases. Such cells are viable and have normal growth rates, but they are more sensitive to sulfhydryl reagents, such as mercurials.

Glutathione also plays a role as a reduced carrier for the reduction of glutaredoxin, which, like thioredoxin, is a hydrogen donor for nucleotide reductase (see Chapter 25) and for the reduction of activated sulfate to sulfite (see above). Glutathione is also important in maintaining the iron of hemoglobin in the ferrous state. In all these roles, the reduction of oxidized glutathione by the NADPH-dependent glutathione reductase provides for the regeneration of reduced glutathione.

A role for glutathione that is independent of its reducing property is one as a γ-glutamyl donor in *the γ-glutamyl cycle* (Figure 24–36). Of particular significance is the fact that cells exhibiting the activities of the cycle contain substantial amounts of γ-glutamyl transpeptidase activity on the outer surface of their cell membranes. This enzyme is thought to transfer the γ-glutamyl group of extracellular glutathione to an extracellular amino acid. The γ-glutamylamino acid is transported into the cell, where, as a substrate for γ-glutamyl cyclotransferase, the amino acid and 5-oxoproline are released. 5-Oxoprolinase is converted to glutamate in an ATP-dependent cleavage catalyzed by 5-oxoprolinase. The glutathione is then regenerated from glutamate and the cysteine and glycine that were released by cysteinylglycine dipeptidase.

The γ-glutamyl cycle enzymes are found in those tissues for which the transport of glutathione into cells is an important function. Whereas glutathione is exported by most cells, it is efficiently transported only into cells that contain the membrane-bound γ-glutamyl transpeptidase. Thus the *γ-glutamyl transpeptidase appears to facilitate salvage of glutathione secreted by some tissues into the blood stream as well as to permit an energy-driven transport system for amino acids.* Either oxidized or reduced glutathione is a substrate for the transpeptidase. The enzyme also catalyzes the hydrolysis of glutathione to glutamate and cysteinylglycine and glutamine to glutamate and ammonia. Many amino acids can serve as acceptors for the γ-glutamyl group, including γ-glutamylamino acids (to yield γ-glutamyl-γ-glutamylamino acids) and even glutathione itself. *Such a transport across cell membranes is probably especially important for cysteine and methionine, as well as for glutathione.*

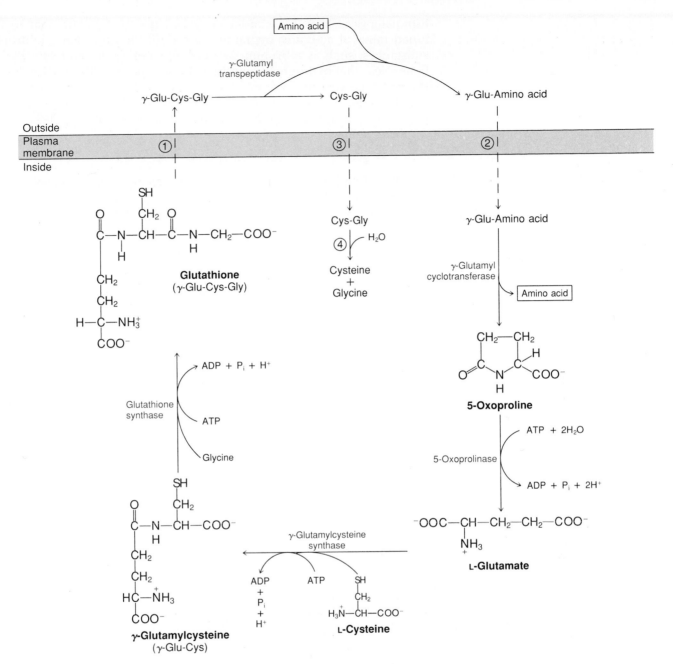

Figure 24–36

The γ-glutamyl cycle proposed by Alton Meister. The cycle involves enzymes forming glutathione, the excretion of glutathione (1) and, in cells that import glutathione, the γ-glutamyltranspeptidase-dependent transport (2) of another amino acid (or peptide), the cleavage of the intracellular γ-glutamyl amino acid, and the ATP-dependent conversion of 5-oxoproline to glutamate. The cysteinylglycine (Cys-Gly) formed in the reaction is transported by an uncharacterized transport system (3) and cleaved by an intracellular protease (4) or cleaved by a membrane-bound protease and the free amino acids transported. The γ-glutamyl cycle enzymes are found in those tissues for which the transport of glutathione into cells is an important function. This includes liver and kidney cells in animals.

That this cycle does play a role in transport is shown by the fact that incorporating glutamyl transpeptidase isolated from kidney into erythrocyte membranes stimulates the uptake of glutamate and alanine supplied along with glutathione to such preparations.

There is another class of enzymatic conversions involving glutathione that occurs in the cytosol of the liver. The enzymes are called glutathione

S-transferases. They are similar enzymes, few in number, that exhibit a broad range of activities owing to their ability to bind many hydrophobic substances, such as bilirubin, steroids, and polycyclic aromatic hydrocarbons. Near this hydrophobic binding site is a second site specific for glutathione. Whether the hydrophobic ligand is also a substrate depends on whether it has an electrophilic atom that can undergo nucleophilic attack by glutathione resulting in conjugation with glutathione and release of the electrophilic atom:

$$RX + GSH \longrightarrow RSG + HX$$

The glutathione conjugation product may be further metabolized or excreted as such. The glutathione S-transferases thus serve to solubilize, detoxify, and initiate the catabolism of a wide variety of hydrophobic substances.

Gramicidin Is a Cyclic Decapeptide Synthesized on a Protein Template

The antibiotic gramicidin S produced by the bacterium *Bacillus brevis* is a cyclic decapeptide (Figure 24–37) composed of a repeated sequence of five amino acids (–D-Phe–L-Pro–L-Val–L-Orn–L-Leu–)$_2$. Whereas in glutathione biosynthesis the order of amino acids is determined by a specific enzyme for each peptide linkage made, in gramicidin the ordering is mainly a function of the attachment points of amino acids on the enzyme surface. Thus one of the two enzymes involved in gramicidin synthesis appears to be serving a dual role of enzyme and template. The synthesis of more complex polypeptides requires the intricate biochemical machinery used in protein synthesis. Gramicidin synthesis is divided into five phases.

1. *Activation, Thioesterification, and Racemization of L-Phenylalanine.* The light enzyme of the gramicidin-forming system activates L-phenylalanine as the aminoacyl adenylate and transfers the phenylalanyl residue to a thiol group on the enzyme. Racemization of L-phenylalanine occurs at this thioester stage:

$$ESH + ATP + \text{L-Phe} \longrightarrow (\text{L-Phe} \sim AMP)ESH + PP_i \longrightarrow$$
$$ES \sim \text{L-Phe} + AMP \longrightarrow ES \sim \text{D-Phe}$$

2. *Activation and Thioesterification of L-Proline, L-Valine, L-Ornithine, and L-Leucine.* The heavy enzyme of the gramicidin-forming system activates the

Figure 24–37
The structure of the antibiotic gramicidin S. Gramicidin S is a cyclic decapeptide composed of a repeated sequence of five amino acids.

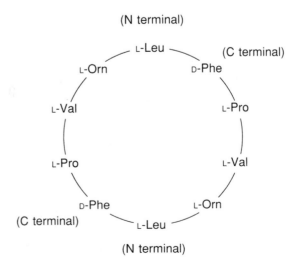

other four amino acids found in gramicidin S and transfers each to a specific thiol-containing site on the protein:

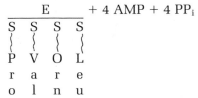

$$\frac{E}{\underset{H\ H\ H\ H}{S\ S\ S\ S}} + Pro + Val + Orn + Leu + 4\ ATP \longrightarrow$$

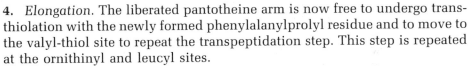

$$\frac{E}{\underset{\substack{P\ V\ O\ L\\ r\ a\ r\ e\\ o\ l\ n\ u}}{S\ S\ S\ S}} + 4\ AMP + 4\ PP_i$$

3. *Transfer of D-Phenylalanine to the Heavy Enzyme and Initiation of Peptide Formation.* The heavy enzyme contains a covalently linked 2-nm-long pantotheine arm that is thought to serve as a carrier of the growing peptide chain. It is to the —SH group of this pantotheine arm on the heavy enzyme that the D-phenylalanyl group is probably transferred from the light enzyme (transthiolation). The first peptide bond would then be formed by a transpeptidation reaction that liberates the pantotheine thiol group.

4. *Elongation.* The liberated pantotheine arm is now free to undergo transthiolation with the newly formed phenylalanylprolyl residue and to move to the valyl-thiol site to repeat the transpeptidation step. This step is repeated at the ornithinyl and leucyl sites.

5. *Cyclization.* After the pentapeptide is formed, it is cyclized with an identical peptide in a head-to-tail fashion. One possible mechanism, implied in Figure 24–38, involves the transfer of the first pentapeptide to a thiol "waiting" site, and when a second pentapeptide has been completed, cyclization occurs by two additional transpeptidation reactions (Figure 24–38). Another possibility is that the cyclization occurs by an *intermolecular* transpeptidation involving two heavy enzymes, each containing one completed pentapeptidyl residue at the terminal thiol site (not shown).

Phosphocreatine Is an Important Energy Reservoir in Skeletal Muscle

In skeletal muscle of vertebrates and, to a lesser extent, in other tissues, *creatine* (α-methylguanidoacetate) is an important reservoir of high-energy phosphate groups. In many invertebrates, arginine plays a similar role. Creatine is synthesized by means of a pair of reactions in which the amidino group of arginine is transferred to glycine to yield ornithine and guanidinoacetate, and a methyl group is transferred to the α nitrogen. These reactions are shown in Figure 24–39. The reversible phosphorylation of the

(a)

(b)

(c)

(d)

(e)

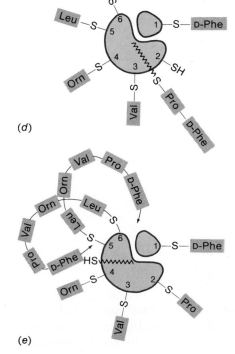

Figure 24–38
The formation of gramicidin S on a protein template. (a) The activated amino acids are held in thioester linkage on the light and heavy enzymes. The phenylalanyl residue has undergone racemization and is being transferred to the pantotheine arm on the heavy enzyme. (b) The first peptide bond is about to be formed by transfer of the phenylalanyl group from the pantotheine group to the prolyl residue. (c) The phenylalanylprolyl residue is about to be transferred to the free thiol group of the pantotheine arm. The light enzyme has accepted another phenylalanyl residue that has already undergone racemization. (d) The phenylalanylprolyl residue is about to be transferred to the valyl residue. (e) The first pentapeptidyl group after being made was transferred to the waiting site 6. The second pentapeptidyl group has just been completed and is about to be condensed with the first to yield the decapeptide gramicidin S. The pantotheine arm is now free to repeat the process.

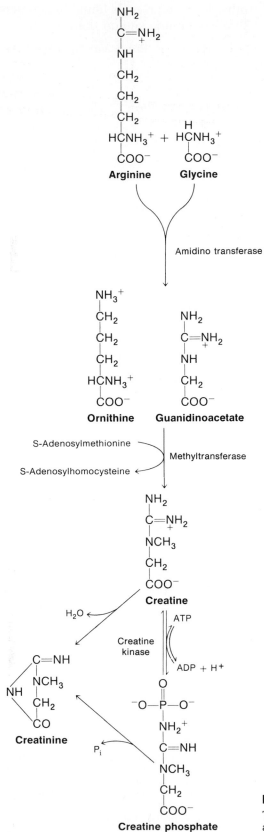

Figure 24–39

The formation of creatine phosphate and creatinine.

terminal amino group results in the high-energy storage compound phos-phocreatine. Creatine is lost from the metabolic pool by the spontaneous cyclization of either phosphocreatine or creatine itself to yield creatinine, which is excreted in the urine.

SUMMARY

Essential amino acids are those that must be supplied in the diet. Bacteria and most plants can make all of the amino acids found in proteins. Many animals, including humans, cannot. Only eight of the *de novo* pathways for amino acid biosynthesis can be found in humans. In order to study all of the amino acid pathways we must turn to the microorganisms.

For their carbon skeletons, the pathways leading to the 20 amino acids found in proteins draw upon a few intermediates from the central metabolic routes. The resulting pathways are highly branched, and the amino acids can be grouped into families based on key intermediates that serve as progenitors of the families.

Inorganic nitrogen for amino acid biosynthesis must be derived from nitrate, nitrite, or ammonia in the environment. However, it is incorporated into organic form only as ammonia. For most cells and under most conditions, this conversion occurs by means of the amidation of glutamate to yield glutamine. This amide group is then used as a donor of "active" ammonia in numerous reactions, including the reductive amination of α-ketoglutarate. The amino group of glutamate thus formed then serves as the source of amino groups in the biosynthetic pathways of all the amino acids found in proteins. In only some cell types, growing in the presence of a high concentration of ammonia, does the amination of α-ketoglutarate and the amidation of aspartate (to yield asparagine) occur directly by free ammonia.

The pathway for synthesizing the α-ketoglutarate family of amino acids has two main branches: one involves a chain-lengthening process that yields lysine, only known to occur in fungi and *Euglena*, and the other occurs by way of glutamate. Glutamate is also a precursor of proline and arginine. Both routes require activation and reduction of a carboxyl group. In the route to arginine, protection of the α amino group is required to prevent the cyclization reaction essential for proline biosynthesis. The formation of the guanidine group of arginine requires carbamoyl phosphate, which is also a precursor of the pyrimidines.

The biosynthetic route for the aspartate family is a highly branched pathway leading to its amide, asparagine, to lysine following condensation with pyruvate, to methionine following sulfur and methyl group transfer, and to threonine. In these pathway branches, the α amino group of aspartate is preserved. The amino group is lost, however, in the route to isoleucine, by which threonine contributes four carbons to isoleucine.

The pyruvate family consists of its α-amino analogue, alanine, valine (derived from the condensation of two pyruvate molecules), and leucine (made by lengthening the carbon skeleton of valine, much like the chain-lengthening reaction in the fungal lysine pathway). The steps to valine are paralleled by those to isoleucine and, indeed, are catalyzed by the same enzymes—the difference being that α-ketobutyrate rather than pyruvate is one of the precursors of isoleucine.

The aromatic amino acids tyrosine, phenylalanine, and tryptophan derive their aromatic rings from the shikimate pathway, with final aromatization of the ring occurring only in the specific branch pathways leading to the final products. The ring itself arises from a condensation between erythrose-4-phosphate and phosphoenolpyruvate, followed by cyclization. The α amino group of tryptophan arises only indirectly by transamination with glutamate, since there is an exchange of serine for three carbons that had originated from ribose-5-phosphate.

The histidine pathway is a complex one in which a —C—N— unit of adenine serves as a nucleus for condensation with a ribosylphosphate moiety and another nitrogen derived from glutamine. The residue from adenine is, in fact, an intermediate in the purine nucleotide biosynthetic pathway and can thus be recycled by replenishing the lost —C—N— unit. The phosphate group is retained until after the α amino group is incorporated, illustrating the principle that metabolic intermediates bear charged groups.

The control of carbon flow through the amino acid biosynthetic pathways can occur at various levels and by a variety of mechanisms. The most common pattern is simple inhibition of enzyme activity at the initial committal step in the pathway by the amino acid end product. The control over some of the branching routes is more complicated, involving regulation of the steady-state levels of enzymes in the pathways.

Many biologically important routes of amino acid utilization—other than those leading to incorporation into proteins—are known. Some of these routes are distinctly anabolic pathways in which the amino acids, whether supplied from exogenous sources or synthesized *de novo* by the organism, serve as an initial substrate in an independent biosynthetic pathway. Among the simplest of these pathways are those leading to some of the physiologically important amines, such as epinephrine and serotonin. Other simple pathways involve the conversion of one amino acid to another, such as the formation of tyrosine from phenylalanine or ornithine from proline. These pathways are not usually found in organisms that form the amino acids *de novo*. On the other hand, the utilization of glycine in the formation of porphyrin derivatives occurs by very complex and highly branched pathways. Some of the other biologically important pathways lead to the biosynthesis of small peptides, including glutathione and the antibiotic gramicidin.

SELECTED READINGS

Battersby, A. R., C. J. R. Fookes, G. W. J. Matcham, and E. McDonald, Biosynthesis of the pigments of life: Formation of the macrocycle. *Nature* 285:17, 1980. This paper discusses the steps in tetrapyrrole biosynthesis and the pathways diverting this nucleus to chlorophylls, hemes, cytochromes, and other macrocyclic pigments.

Cohen, G. N., *Biosynthesis of Small Molecules.* New York: Harper and Row, 1967. A small book reviewing some of the early literature and describing the way isotope incorporation and mutant and enzymatic analyses were used to unravel metabolic pathways.

Fowden, L., P. J. Lea, and E. A. Bell, The nonprotein amino acids of plants. *Adv. Enzymol.* 50:117, 1979. A discussion of the occurrence and biosynthesis of naturally occurring amino acid analogs in plants.

Griffith, O., R. J. Bridges, and A. Meister, Formation of γ-glutamylcyst(e)ine *in vivo* is catalyzed by γ-glutamyltranspeptidase. *Proc. Natl. Acad. Sci. USA* 78:2778, 1981. This and earlier references cited therein provide a discussion of the physiological role of the γ-glutamyl cycle.

Herrmann, K. M., and R. L. Somerville (eds.), *Amino Acids: Biosynthesis and Genetic Regulation.* Reading, Mass.: Addison-Wesley, 1983. Twenty-two detailed review chapters describing the regulation of amino acid biosynthesis.

Katz, E., and A. L. Demain, The peptide antibiotics of *Bacillus:* Chemistry, biogenesis and possible functions. *Bacteriol. Rev.* 41:449, 1977. A description of several peptide antibiotics showing the distribution of D-amino acid in these compounds.

Meister, A., *Biochemistry of the Amino Acids*, Vols. 1 and 2. New York: Academic Press, 1965. The two-volume classic provides a thorough discussion of amino acid literature, occurrence, properties, and metabolism of amino acids up to that time.

Mifflin, B. J., and P. J. Lea, Amino acid metabolism. *Ann. Rev. Plant Physiol.* 28:299, 1977. A review of amino acid biosynthesis in the higher plants with emphasis on primary amino group formation.

Umbarger, H. E., Amino acid biosynthesis and its regulation. *Ann. Rev. Biochem.* 47:533, 1978. A review of amino acid biosynthesis in bacteria, fungi, and plants and a consideration of factors controlling metabolite flow and enzyme amount in these pathways.

Yamada, K., S. Kinoshita, T. Tsunoda, and K. Aida (eds.), *The Microbial Production of Amino Acids.* New York: John Wiley and Sons, 1972. A collection of essays describing microbial processes used in Japanese industry for the production of amino acids. Includes examples in which the regulatory mechanisms functioning in most cells have been modified or bypassed.

PROBLEMS

1. Taken at face value, it would seem paradoxical that the reaction catalyzed by glutamine synthase, which consumes L-glutamate, constitutes in most instances the first step in L-glutamate biosynthesis. What is the explanation for this apparent paradox?

2. A mutant of *E. coli* is discovered with a defect in serine hydroxymethylase. What would you expect the nutritional requirement of this mutant to be? Although the mutant does not require L-methionine, its growth is stimulated when this amino acid is added to the medium. Can you suggest an explanation?

3. Illustrate schematically the feedback control system governing biosynthesis of amino acids of the aspartate family in *E. coli*. Give an example of another pathway that is regulated in a similar fashion.

4. Draw a scheme illustrating the relationship between pathways involved in the biosynthesis of L-cysteine and L-methionine, indicating the steps at which sulfhydrylation occurs. Suggest an explanation for the observation that L-cysteine but not L-methionine is a nonessential amino acid in humans.

5. A number of years ago it was discovered that a strain of *E. coli* was unable to grow in a medium containing L-valine but lacking L-isoleucine and L-leucine, even though the same strain could grow in a medium lacking all three amino acids. Suggest an explanation.

6. Recent examination of some lunar samples revealed the presence of bacterialike organisms that contained lipid, carbohydrate, protein, and nucleic acids remarkably similar to those found in prokaryotic cells on Earth. However, the proteins contained 21 amino acids but no isoleucine. The two unusual amino acids were the L-enantiomers of

$$CH_3CH_2\underset{\underset{NH_3^+}{|}}{C}HCOO^- \quad \text{and} \quad CH_3CH_2CH_2\underset{\underset{NH_3^+}{|}}{C}HCOO^-$$

On the basis of known amino acid biosynthetic pathways, would you expect these amino acids to belong to the pyruvate, the aspartate, or the glutamate family? By what route might they be formed? Indicate the labeling you would expect to find in the new compounds if you were able to use $1\text{-}^{14}C$-labeled aspartate, glutamate, pyruvate, and acetate as tracers in a growth experiment.

7. In some plants, threonine synthase is inhibited by L-cysteine and stimulated by S-adenosylmethionine. What is the underlying rationale for this pattern of regulation? How does it differ from that underlying the activation in *N. crassa* of cystathionine-γ-synthase by N^5-methyltetrahydrofolate?

8. In what sense may indole be viewed as an intermediate in L-tryptophan biosynthesis?

9. The biosynthesis of L-proline from L-glutamate may be thought of as involving a "*cis*"-amination reaction. Describe what you think is meant by this term, and give an example of another amino acid biosynthetic pathway that includes a variation on the theme of this reaction.

10. The accumulation of biosynthetic intermediates, or of metabolites derived from those intermediates (as in the case of indole-3-glycerol accumulation by certain tryptophan auxotrophs), has proved to be a valuable tool in

the analysis of biosynthetic pathways. It was found, however, that these accumulations occurred only after the required amino acid had been consumed and growth had stopped. How do you account for this fact?

11. At the beginning of the chapter a diagram was provided showing the flow of carbon into the various amino acids from key intermediates in the central metabolic routes. Prepare an analogous diagram showing the flow of nitrogen from ammonia into all 20 amino acids. (Assume the organism is one that contains an NADPH-dependent glutamate dehydrogenase and is growing in a high ammonia environment. Remember, too, that some nitrogen is discarded during amino acid biosynthesis.)

12. Contrast the pathways involved in the biosynthesis of L-glutathione and gramicidin S. How do these differ in principle from the mechanism of protein synthesis catalyzed by the ribosome? In what way does the mechanism of gramicidin S synthesis resemble that of fatty acid synthesis?

25

NUCLEOTIDES

This chapter deals with the biosynthesis of ribonucleotides and deoxyribonucleotides, their role in metabolic processes, and the pathways for their degradation. The biosynthesis of nucleotides is a vital process, since these compounds are indispensable precursors for the synthesis of both RNA and DNA. Without RNA synthesis, protein synthesis is halted; and unless cells can synthesize DNA, they cannot divide.

It is not surprising that inhibitors of nucleotide biosynthesis are very toxic to cells. As we will see, their toxicity has been used to advantage in the treatment of cancer as well as in the treatment of certain diseases resulting from infections by viruses, bacteria, or protozoa.

Nucleotides play important roles in all major aspects of metabolism. ATP, an adenine nucleotide, is the major substance used by all organisms for the transfer of chemical energy from energy-yielding reactions to energy-requiring reactions such as biosynthesis. Other nucleotides are activated intermediates in the synthesis of carbohydrates, lipids, proteins, and nucleic acids. Adenine nucleotides are components of many major coenzymes, such as NAD^+, $NADP^+$, FAD, and CoA. The critical role played by nucleotides as regulators of metabolism in both prokaryotic and eukaryotic organisms is described in other chapters.

Pathways for the metabolic degradation of nucleotides are also very important to the organism, as demonstrated by the fact that several genetic defects causing blocks in these pathways have serious consequences for the health of the organism.

The physiological importance of nucleotides is reflected in the careful regulation of their intracellular levels as well as intra- and extracellular levels of nucleosides and nucleobases. Levels of ATP, ADP, and AMP are tightly controlled in a variety of cells, and intracellular levels of dATP, dCTP, dGTP, and dTTP are also carefully regulated under normal conditions. This regulation is determined by the concentration and location of the enzymes of nucleotide metabolism and by levels of substrates, products, and effectors.

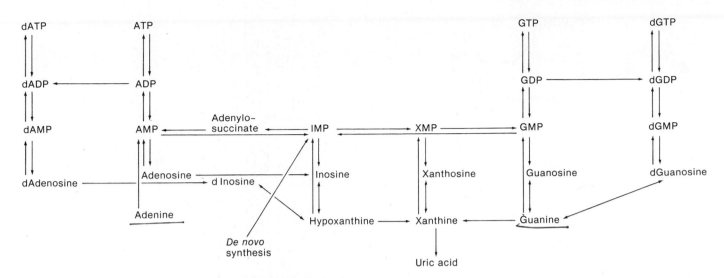

Figure 25–1

Pathways of purine metabolism. Double-headed arrows indicate reversible enzymatic reactions. Separate arrows in opposite directions between metabolites indicate a different enzyme participating in each direction. The diagram is arranged in tiers: purines at the bottom, nucleosides at the next level, then nucleoside mono-, di-, and triphosphates (in ascending order).

The pathways by which cells synthesize, interconvert, and catabolize various purine and pyrimidine nucleotides are summarized schematically in Figures 25–1 and 25–2. Cells of different types, or even the same cells in different stages of development, differ greatly in their ability to carry out some of the reactions involved, with some cells favoring one set of reactions and others another. The rest of the chapter deals with the details of these pathways.

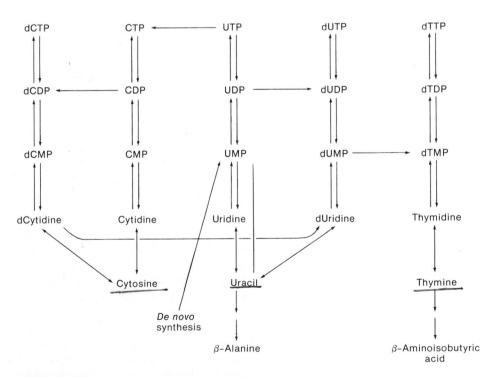

Figure 25–2

Pathways of pyrimidine metabolism. Arrows and layout have a similar significance as in Figure 25–1.

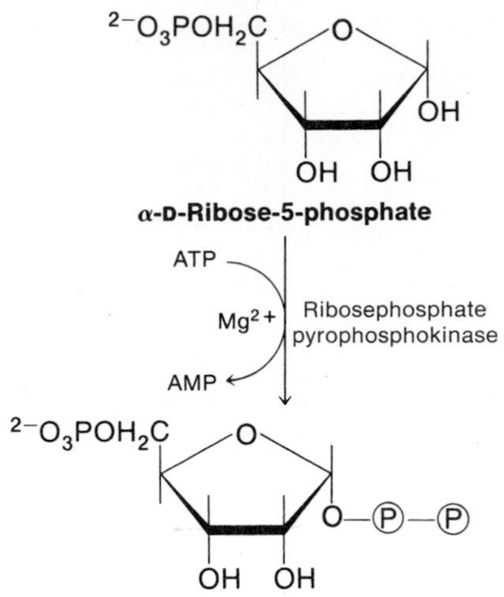

Figure 25-3
Synthesis of phosphoribosyl pyrophosphate (PRPP). This is an unusual kinase-catalyzed reaction because the group transferred is the pyrophosphate group rather than the phosphate group.

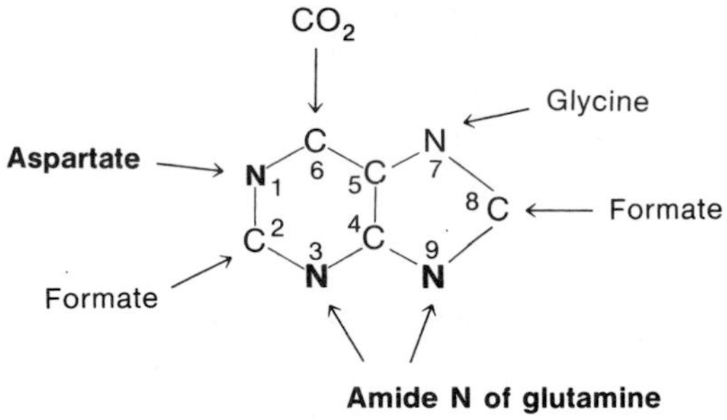

Figure 25-4
Precursors of the purine ring of uric acid in pigeons as determined by isotope labeling experiments. The indicated precursor substances were administered one at a time to pigeons. Each precursor was labeled with isotopic nitrogen or carbon, and in each case the excreted uric acid was purified and degraded chemically. The isotope content of the various degradation products indicated that precursors contributed the specific atoms indicated.

SYNTHESIS OF PURINE RIBONUCLEOTIDES *DE NOVO*

Purine nucleotides can be synthesized in three ways: by *de novo* synthesis, by reconstruction from purine bases by the addition of a ribose phosphate moiety, or by phosphorylation of nucleosides. The first two pathways are the more important quantitatively, and in both of them *phosphoribosylpyrophosphate (PRPP)* is an essential precursor. It has a similar important role in pathways for pyrimidine nucleotide biosynthesis. PRPP is synthesized from ribose-5-phosphate, which cells synthesize either from glucose-6-phosphate by an oxidative pathway or from intermediates of glycolysis by a nonoxidative pathway (Chapter 14).

The formation of PRPP from ribose-5-phosphate and ATP is catalyzed by *ribose-5-phosphate pyrophosphokinase* (Figure 25-3). This is an unusual kinase because the pyrophosphoryl group is transferred rather than the phosphoryl group. As might be anticipated for an enzyme with a key position in several biosynthetic pathways, its activity is regulated by a number of metabolites. Inorganic phosphate is an activator, and the Mg^{2+} ion is both cofactor and activator. Inhibitors include ADP and glycerate-2,3-bisphosphate, which are competitive with respect to ribose-5-phosphate. By contrast, AMP and GDP are noncompetitive inhibitors. At any specific time, the concentration of these metabolites as well as those of the substrates will determine the activity of the kinase. In particular, the nucleotide inhibitors will curtail synthesis of PRPP when the energy stores of the cell are low.

The ultimate precursors of the purine ring were established by administering isotopically labeled compounds to pigeons and tracing the incorporation of labeled atoms into the purine ring of uric acid. Birds were used in these experiments because they excrete waste nitrogen largely as uric acid, a purine derivative that is easily isolated in pure form. Chemical degradation of the uric acid revealed the origins of the atoms as depicted in Figure 25-4. A flow scheme showing the successive incorporation of atoms from these precursors is shown in Figure 25-5.

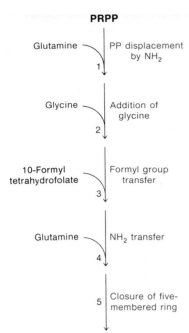

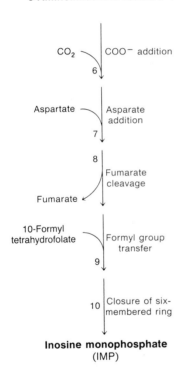

Figure 25–5

Summary of incorporation of precursors into the purine ring of IMP. Formate as well as other indirect donors of one carbon, such as serine, donate their atoms by means of the formyl group of a folate derivative in steps 3 and 9.

Inosine Monophosphate (IMP) Is the First Purine Nucleotide Formed

The pathway from PRPP to the first complete purine nucleotide, inosine monophosphate, involves ten steps and is shown in Figure 25–6. It would seem logical that the purine should be built up first, followed by addition of ribose-5-phosphate, but this is not the case. The starting point is PRPP, to which the imidazole ring is added; the six-membered ring is built up afterward.

The first step, in which phosphoribosylamine is formed, is catalyzed by *glutamine phosphoribosylpyrophosphate amidotransferase,* an enzyme containing nonheme iron. The reaction involves inversion of the configuration at C-1 of the ribose and leads to the β configuration that is characteristic of naturally occurring nucleotides. This step involves commitment of PRPP to the purine biosynthetic pathway and, as might be expected, is subject to important feedback inhibitory effects by purine nucleotides. We will examine these effects a little later.

In step 2, an amide bond is formed by the synthase between the carboxyl group of glycine and the amino group of phosphoribosylamine, with ATP supplying energy and being hydrolyzed to ADP and inorganic phosphate. After these two steps have introduced atoms 4, 5, 7, and 9 of the purine ring, the remaining atoms are introduced one by one (steps 3, 4, 6, 7, and 9).

In step 3, carbon 8 is introduced as a formyl group that is transferred from 10-formyltetrahydrofolate (10-formyl-H$_4$folate). (To review the way tetrahydrofolate derivatives accept a one-carbon unit from donors such as serine, glycine, or formate and transfer it to a suitable acceptor in biosynthetic reactions, see Chapters 11 and 24.) The five-membered (imidazole) ring is now ready for closure, but before this occurs, N-3 of the purine ring is introduced (step 4) by transfer of another amino group from glutamine to phosphoribosylformylglycinamide. ATP provides energy for the amido group transfer, being itself hydrolyzed to ADP and phosphate.

The imidazole ring is closed in an essentially irreversible cyclization requiring the presence of Mg^{2+} and K^+ (step 5). Then C-6 of the purine ring is introduced by addition of bicarbonate in the presence of a specific carboxylase (step 6). This carboxylation is unusual in that it does not seem to involve biotin and is not coupled with any energy-yielding process such as ATP hydrolysis. The equilibrium is unfavorable for the formation of the carboxylate. *In vivo* the reaction proceeds at physiological concentrations of bicarbonate because of coupling with subsequent steps that are thermodynamically favorable.

Next, in steps 7 and 8, N-7 of the purine ring is contributed by aspartate. Aspartate forms an amide with the 4-carboxyl group, and the succinocarboxamide so formed is then cleaved with release of fumarate. Energy for carboxamide formation is provided by ATP hydrolysis to ADP and phosphate. These reactions resemble the conversion of citrulline to arginine in the urea cycle (Chapter 19) and the conversion of IMP to AMP (see Figure 25–7). It is worth noting that fumarate released in all these synthetic pathways can be converted to oxaloacetate by fumarase and malate dehydrogenase, and in the process, NAD^+ is reduced to NADH (by malate dehydrogenase). Reoxidation of the NADH by the electron-transport chain then generates three ATPs from ADP and P$_i$. The ATPs supply some of the energy required for purine synthesis, and the oxaloacetate formed can be used to replenish the supply of aspartate by transamination with glutamate (Figure 25–7).

The final atom of the purine ring is provided in step 9 by donation of a formyl group from 10-formyltetrahydrofolate (10-formyltetrahydropteroyl-

5-Phospho-α-D-ribosyl-
1-pyrophosphate

Glutamine + H₂O ⟶ ① Glutamine
       Mg²⁺   PRPP
Glutamate + PPᵢ ⟵   amidotransferase

$^{2-}O_3POH_2C$ O NH₂

OH OH

5-Phospho-β-D-ribosylamine

Glycine + ATP ⟶ ② Synthase

ADP + Pᵢ ⟵

CH₂—NH₂
O=C
     NH
$^{2-}O_3POH_2C$ O

OH OH

5'-Phosphoribosylglycinamide

10-Formyl
tetrahydrofolate ⟶ ③ Formyltransferase

Tetrahydrofolate ⟵

CH₂—NH
O=C   CHO
     NH
Ribose-5-phosphate

5'-Phosphoribosyl-N-formylglycinamide

ATP + Gln + H₂O ⟶ ④ Synthase

ADP + Glu + Pᵢ ⟵

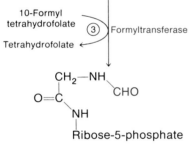

CH₂—NH
HN=C   CHO
     NH
Ribose-5-phosphate

5'-Phosphoribosyl-N-formylglycinamidine

5'-Phosphoribosyl-
N-formylglycinamidine

ATP ⟶ ⑤ Mg²⁺,K⁺ Synthase

ADP + Pᵢ ⟵

N
H₂N   N

Ribose-5-phosphate

5'-Phosphoribosyl-5-aminoimidazole

CO₂ ⟶ ⑥ Carboxylase

⁻OOC   N
H₂N   N

Ribose-5-phosphate

5'-Phosphoribosyl-5-aminoimidazole-
4-carboxylate

ATP + Asp ⟶ ⑦ Mn²⁺ Synthase

ADP + Pᵢ ⟵

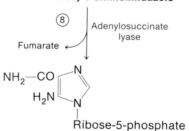

COO⁻
HC—NH—CO   N
CH₂   H₂N   N
COO⁻
Ribose-5-phosphate

5'-Phosphoribosyl-4-(N-succino-
carboxamide)-5-aminoimidazole

⑧ Adenylosuccinate
         lyase

Fumarate ⟵

NH₂—CO   N
H₂N   N

Ribose-5-phosphate

5'-Phosphoribosyl-4-carboxamide
5-aminoimidazole

5'-Phosphoribosyl-4-carboxamide-
5-aminoimidazole

10-Formyl-
tetrahydrofolate ⟶ ⑨ Formyltransferase

Tetrahydrofolate ⟵

O
H₂N—C   N
OCH—HN   N

Ribose-5-phosphate

5'-Phosphoribosyl-4-carboxamide-
5-formamidoimidazole

⑩ Cyclohydrolase

H₂O ⟵

O
HN   N
N   N

Ribose-5-phosphate

Inosine-5'-monophosphate (IMP)

Figure 25–6
Biosynthetic pathway to inosine monophosphate. Color indicates the group or atom introduced at each step. The first step, in which phosphoribosylamine is formed, involves inversion of the configuration at C-1 of the ribose. In step 2 an amide bond is formed by the synthase between the carboxyl group of glycine and the amino group of phosphoribosylamine. An additional formyl group is added to the glycine amino group in the next step. An additional amide group is donated by a glutamine, and next the five-membered imidazole ring is formed. Further groups are introduced through successive attachments to the imidazole, and finally the purine ring is completed in inosine-5'-monophosphate.

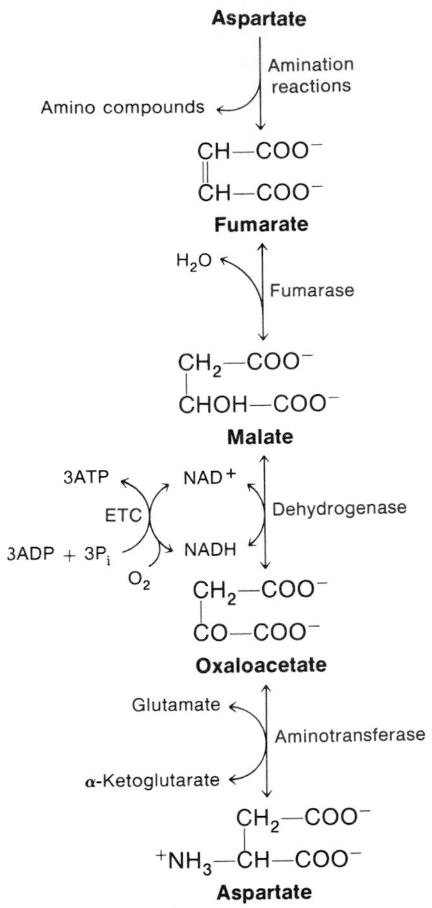

Aspartate

Amination
reactions

Amino compounds

CH—COO⁻
‖
CH—COO⁻

Fumarate

H_2O ← Fumarase

CH_2—COO⁻
|
CHOH—COO⁻

Malate

3ATP   NAD⁺

ETC   Dehydrogenase

3ADP + 3P$_i$ NADH

O_2

CH_2—COO⁻
|
CO—COO⁻

Oxaloacetate

Glutamate

Aminotransferase

α-Ketoglutarate

CH_2—COO⁻
|
⁺NH₃—CH—COO⁻

Aspartate

Figure 25–7
Utilization of fumarate formed in reactions in which aspartate acts as donor of an amino group to generate ATP and regenerate aspartate (ETC = electron-transport chain).

glutamate, or 10-formyl-H$_4$PteGlu) to the 5-amino group of the almost completed ribonucleotide. In the final step, ring closure is effected by elimination of water to form IMP (inosine monophosphate), the first product with a complete purine ring. Although this final ring closure does not require energy from ATP, closure of the imidazole ring does, and the synthesis of IMP from ribose-5-phosphate requires a total of six high-energy phosphate groups from ATP (assuming hydrolysis of pyrophosphate released during phosphoribosylamine synthesis, step 1 of Figure 25–6).

IMP Is Converted into AMP and GMP

IMP does not accumulate in the cell, but is converted to AMP, GMP, and the corresponding diphosphates and triphosphates. The two steps of the pathway from IMP to AMP (Figure 25–8) are typical reactions by which the amino group from aspartate is introduced into a product. The 6-hydroxyl group of IMP (tautomeric with the 6-keto group) is first displaced by the amino of aspartate to give adenylosuccinate, and the latter is then cleaved nonhydrolytically by _adenylosuccinate lyase_ to yield fumarate and AMP. In the condensation of aspartate with IMP, cleavage of GTP to GDP and phosphate provides energy to drive the reaction.

Conversion of IMP to GMP also proceeds by a two-step pathway (Figure 25–8): first, dehydrogenation of IMP to xanthosine-5′-phosphate (XMP), and second, transfer of an amino group from glutamine to C-2 of the xanthine ring to yield GMP. The second reaction also involves the cleavage of ATP to AMP and inorganic pyrophosphate. The latter is in turn hydrolyzed to inorganic phosphate by the ubiquitous inorganic pyrophosphatase in a reaction with a very favorable equilibrium. This hydrolysis is coupled with the GMP synthase reaction because pyrophosphate is a product of the latter and a substrate of the pyrophosphatase. The net result is that the release of two high-energy phosphate groups is used to drive the GMP synthase reaction to completion.

SYNTHESIS OF PYRIMIDINE RIBONUCLEOTIDES _DE NOVO_

Like purine nucleotides, pyrimidine nucleotides can be synthesized either _de novo_ or by the "salvage pathways" from nucleobases or nucleosides (Figure 25–2). However, salvage is less efficient because, except in the case of uracil, nucleobases are not converted to nucleotides directly but only via nucleosides. Uridine monophosphate (UMP) plays a central role in pyrimidine nucleotide synthesis, a role analogous to that of IMP in purine nucleotide synthesis.

The biosynthetic pathway to pyrimidine nucleotides is simpler than that for purine nucleotides, reflecting the simpler structure of the base. _In contrast to the biosynthetic pathway for purine nucleotides, in the pyrimidine pathway the pyrimidine ring is constructed before ribose-5-phosphate is incorporated into the nucleotide_. The first pyrimidine mononucleotide to be synthesized is orotidine-5′-monophosphate (OMP), and from this compound, pathways lead to nucleotides of uracil, cytosine, and thymine. OMP thus occupies a central role in pyrimidine nucleotide biosynthesis, somewhat analogous to the position of IMP in purine nucleotide biosynthesis. Like IMP, OMP is found only in low concentrations in cells and is not a constituent of RNA.

Early clues to the nature of the pyrimidine pathway were provided by

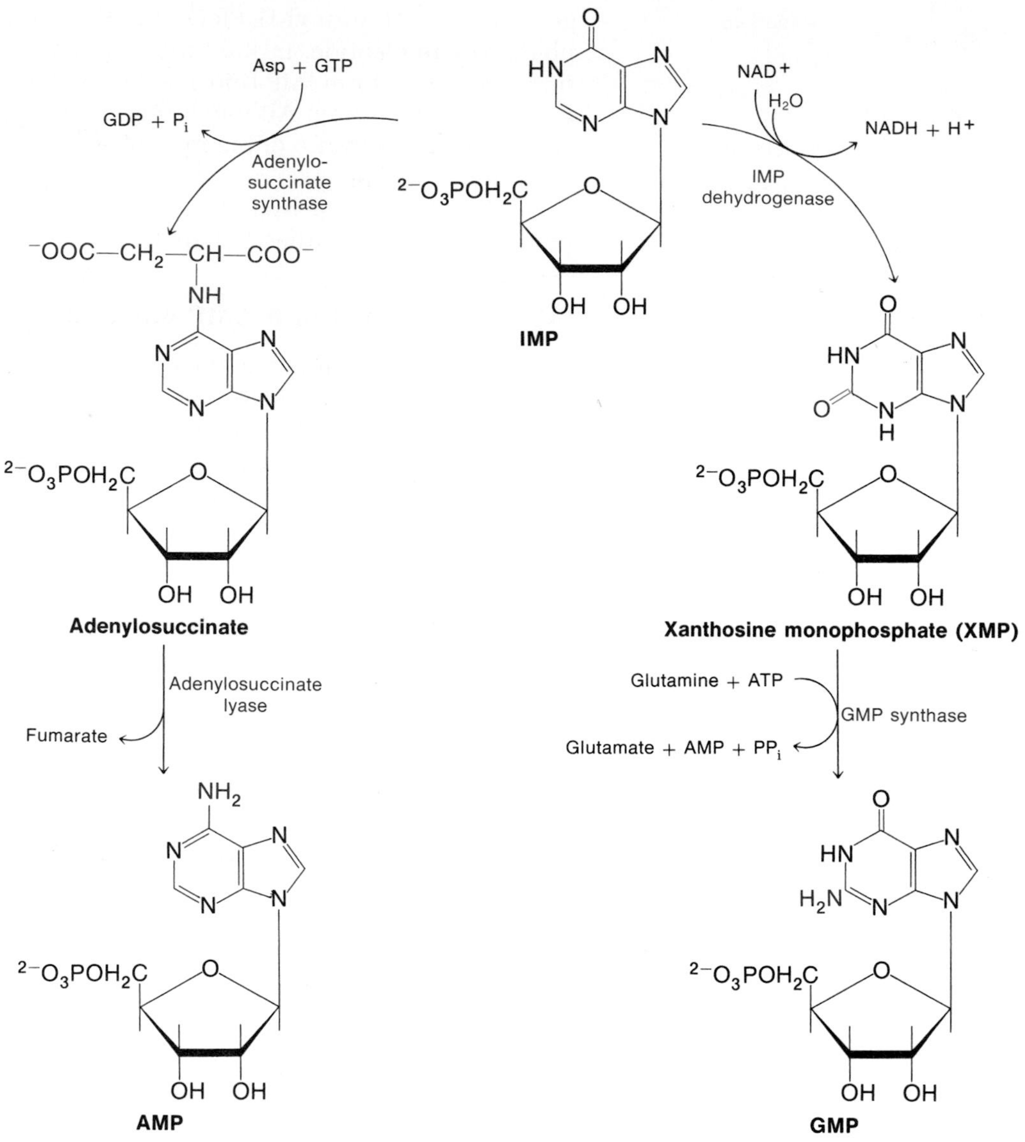

Figure 25–8
Conversion of IMP to AMP and GMP. In both cases two steps are required. It is of interest that the formation of AMP requires GTP and the formation of GMP requires ATP. This tends to balance the flow of the IMP down the two pathways.

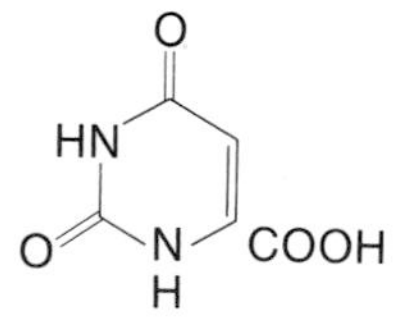

Figure 25–9
The structure of orotic acid. Loss of proton leads to orotate. Deposits of sodium orotate cause a painful condition.

the observations that orotic acid (6-carboxyuracil, Figure 25–9) can satisfy the growth requirement of mutants of *Neurospora* that are unable to make pyrimidines, and that isotopically labeled orotate is an immediate precursor of pyrimidines in *Neurospora* and a number of bacteria.

UMP Is a Precursor of All Pyrimidine Mononucleotides

The pathway for UMP synthesis is shown in Figure 25–10. It starts with the synthesis of carbamoyl phosphate, catalyzed by *carbamoyl phosphate synthase* (glutamine). This enzyme is present in microorganisms and in the cytosol of all eukaryotic cells that are capable of forming pyrimidine nucleotides. Eukaryotes also have a carbamoyl phosphate synthase, a distinct enzyme that uses ammonia as a substrate instead of glutamine. It is associated with citrulline formation in the pathway for arginine biosynthesis, and in

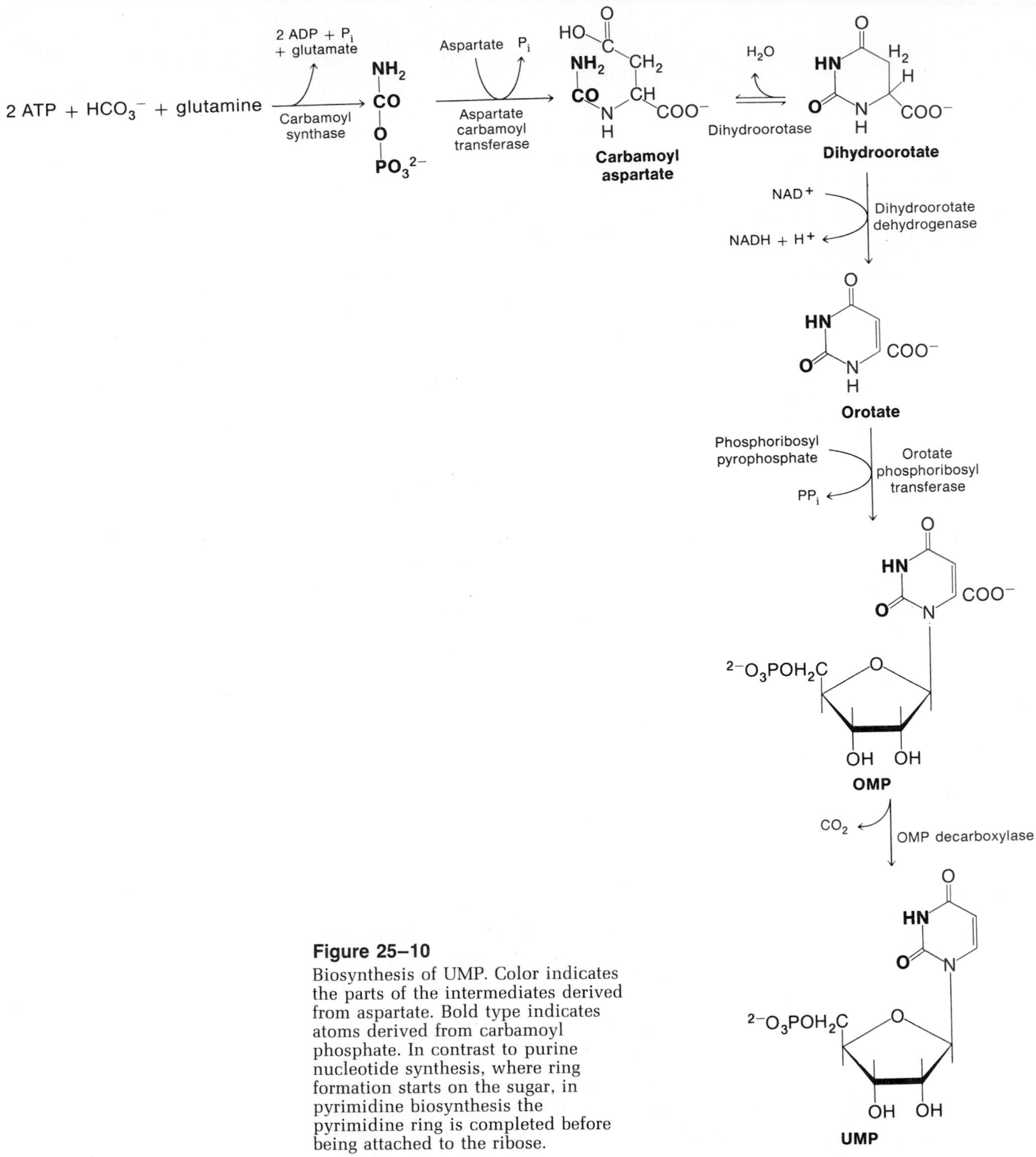

Figure 25-10

Biosynthesis of UMP. Color indicates the parts of the intermediates derived from aspartate. Bold type indicates atoms derived from carbamoyl phosphate. In contrast to purine nucleotide synthesis, where ring formation starts on the sugar, in pyrimidine biosynthesis the pyrimidine ring is completed before being attached to the ribose.

mammals it is a mitochondrial enzyme present predominantly in the liver, where it catalyzes a step in the urea cycle (Chapter 19).

Carbamoyl phosphate synthase does not contain biotin and is not activated by it. The enzyme product, carbamoyl phosphate, next reacts with aspartate to form carbamoyl aspartate. The reaction is catalyzed by aspartate carbamoyltransferase (aspartate transcarbamoylase or aspartate transcarbamylase), and the equilibrium greatly favors carbamoyl aspartate synthesis.

In the third step, the pyrimidine ring is closed by dihydroorotase to form L-dihydroorotate. Dihydroorotate is then oxidized to orotate by dihydrooro-

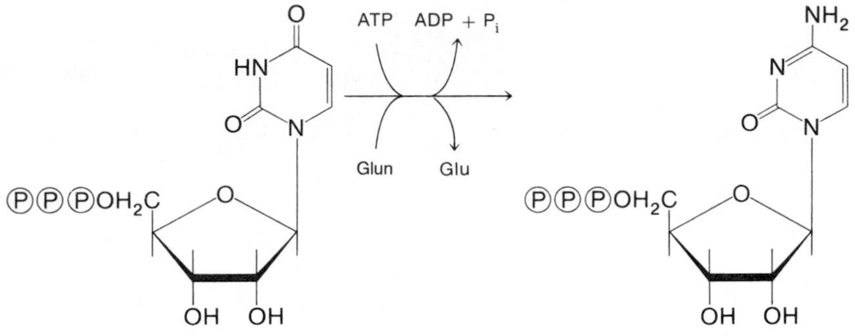

Figure 25–11
The production of CTP by amination of UTP. The conversion of the pyrimidine ring of uracil to cytosine occurs at the level of the nucleotide triphosphate. The enzyme responsible for this conversion is known as cytidine triphosphate synthase.

tate dehydrogenase. This flavoprotein in some organisms contains FMN and in others both FMN and FAD. It also contains nonheme iron and sulfur. In eukaryotes it is a lipoprotein associated with the inner membrane of the mitochondria. In the final two steps of the pathway, orotate phosphoribosyltransferase yields orotidine-5′-phosphate (OMP), and a specific decarboxylase then produces UMP.

Low activities of orotidine phosphate decarboxylase and (usually) orotate phosphoribosyltransferase are associated with a genetic disease in children that is characterized by abnormal growth, megaloblastic anemia, and the excretion of large amounts of orotate. When affected children are fed a pyrimidine nucleoside, usually uridine, the anemia decreases and the excretion of orotate diminishes. A likely explanation for the improvement is that the ingested uridine is phosphorylated to UMP, which is then converted to other pyrimidine nucleotides so that nucleic acid and protein synthesis can resume. In addition, the increased intracellular concentrations of pyrimidine nucleotides would inhibit carbamoyl phosphate synthase, which is required for orotate synthesis.

CTP Is Formed from UTP

After UMP is synthesized, it is phosphorylated to UTP. Then CTP is formed by reaction of UTP with glutamine in a reaction driven by the concomitant hydrolysis of ATP to ADP and inorganic phosphate (Figure 25–11). Both the mammalian and bacterial cytidine triphosphate synthases can use ammonia as a donor in place of glutamine, but this reaction is of no physiological significance because the K_m for ammonia is very high and the reaction rate low. With glutamine as the amino donor, GTP is an allosteric activator for CTP synthase from *E. coli* and probably for the mammalian enzyme as well (Figure 25–12). However, there is no stimulation of CTP synthesis from ammonia, a result interpreted to mean that the allosteric effect of GTP is specifically on the release of ammonia from glutamine. This ammonia is then channeled into reaction with UTP in the enzyme active site.

BIOSYNTHESIS OF DEOXYRIBONUCLEOTIDES

It was once supposed that 2′-deoxyribonucleotides were built up from 2-deoxyribose-5-phosphate by biosynthetic pathways analogous to those used for ribonucleotides. The current view is very different. Tracer studies with isotopically labeled precursors have shown that _in both mammalian tissues and in microorganisms, deoxyribonucleotides are formed from corresponding ribonucleotides_ by replacement of the 2′-OH group with hydrogen.

There are two types of _ribonucleotide reductase_ that catalyze this reduction of the ribose ring. The type that is widely distributed in nature, and that

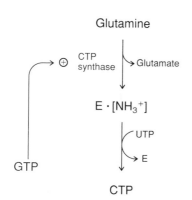

Figure 25–12
Cytidine triphosphate synthase is activated by GTP.

is present in mammalian cells, contains nonheme iron. It is composed of more than one type of polypeptide chain, and is specific for the reduction of diphosphates (ADP, GDP, CDP, and UDP). The second type appears to be restricted to certain microorganisms, including species of the bacteria *Clostridium*, *Lactobacillus*, and *Rhizobium*, a number of species of algae, such as *Euglena* and *Phormidium*, and the fungus *Pithomyces chartarum*. Ribonucleotide reductase of this type requires adenosylcobalamin (coenzyme B_{12}) as an obligatory coenzyme. It does not contain nonheme iron; it uses either nucleoside diphosphates or triphosphates, depending on the source of the enzyme; and it consists of only one type of polypeptide chain (although this may form oligomers in some cases).

The most completely studied ribonucleotide reductase of the first type is that from *E. coli*. It consists of two distinct proteins, B1 and B2, each consisting of two identical or very similar polypeptide chains and both apparently contributing to the catalytic site. The best-studied example of the second type is that from *Lactobacillus leichmannii*. This reductase is a monomeric protein and reduces ribonucleoside triphosphates. Both types of reductase are very unusual in that they appear to catalyze reactions involving radical mechanisms. The *E. coli* reductase B2 subunit contains a stable free organic radical, and the unpaired electron appears to be localized on the benzene ring of a tyrosine side chain. In the case of the *L. leichmannii* enzyme, a radical pair is generated by homolytic cleavage of the C—Co bond of the adenosylcobalamin coenzyme (coenzyme B_{12}).

For both types of reductase, the physiological reducing substrate is a small-molecular-weight (13,000) electron-transport protein, *thioredoxin*. Thioredoxin has two half-cystine residues that are separated in the polypeptide chain by two other residues. The oxidized form of thioredoxin, with a disulfide bridge between the half-cystines, is reduced by NADPH in the presence of a flavoprotein, thioredoxin reductase. The reduced form of thioredoxin, with two cysteine residues present, is the reducing substrate for ribonucleotide reduction. The flow of electrons from NADPH to ribose is shown in Figure 25-13. The mechanism of the reaction in *L. leichmannii* is discussed in Chapter 11 [see Equations (49)–(51)].

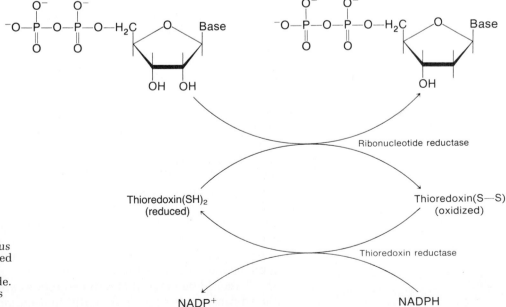

Figure 25–13
Ribonucleotide reductase and the thioredoxin system. In *Lactobacillus leichmannii*, vitamin B_{12} is involved in the reduction of ribonucleotide diphosphate to deoxyribonucleotide. The mechanism for this reaction is discussed in Chapter 11.

An alternative electron-transport system for ribonucleotide reduction has been discovered in *E. coli*. In this case, the ultimate source of electrons is again NADPH, but they are passed to glutathione (in a reaction catalyzed by glutathione reductase), and the reduced glutathione, in turn, reduces a small protein called glutaredoxin. It is the reduced glutaredoxin that acts as the reducing substrate in ribonucleotide reduction. The distribution of glutaredoxin in nature and the relative importance of the glutaredoxin and thioredoxin systems in *E. coli* remain to be determined.

Thymidylate Is Formed from dUMP

Since DNA contains thymine (5-methyluracil) as a major base instead of uracil, *the synthesis of thymidine monophosphate (dTMP or thymidylate) is essential to provide dTTP (thymidine triphosphate), needed for DNA replication* together with dATP, dGTP, and dCTP.

Thymidylate is synthesized from dUMP, and there are several pathways by which the latter may be formed in cells. A possible pathway is the deamination of dCMP according to a reaction catalyzed by deoxycytidylate deaminase:

$$dCMP + H_2O \longrightarrow dUMP + NH_3$$

Deamination of 5-methyldeoxycytidylate and 5-hydroxymethyldeoxycytidylate is also catalyzed by deoxycytidylate deaminase. This enzyme is widely distributed in animal tissues, but several lines of evidence suggest that the reaction shown is not the major source of dUMP.

A route to dUMP that is probably of greater importance is the reduction of UDP to dUDP, followed by phosphorylation of dUDP to dUTP (or direct reduction of UTP to dUTP in some microorganisms). The dUTP is then hydrolyzed to dUMP. This circuitous route to dUMP is dictated by two considerations. First, the ribonucleotide reductase in most cells acts only on ribonucleoside diphosphates, probably because this permits better regulation of its activity. Second, cells contain a powerful deoxyuridine triphosphate diphosphohydrolase (dUTPase). It prevents the incorporation of dUTP into DNA by keeping intracellular levels of dUTP low by means of the reaction

$$dUTP + H_2O \longrightarrow dUMP + PP_i$$

Methylation of dUMP to give thymidylate is catalyzed by *thymidylate synthase* and utilizes 5,10-methylenetetrahydrofolate as the source of the methyl group. This reaction is unique in the metabolism of folate derivatives in that the folate derivative acts both as a donor of the one-carbon group and also as its reductant, using the reduced pteridine ring as the source of reducing potential. Consequently, in this reaction, unlike any other in folate metabolism, dihydrofolate is a product (Figure 25–14). Since folate derivatives are present in cells at very low concentrations, continued synthesis of thymidylate requires regeneration of 5,10-methylenetetrahydrofolate from dihydrofolate. As shown in Figure 25–14, this occurs in two steps catalyzed by the enzymes dihydrofolate reductase, an $NADP^+$-linked dehydrogenase, and serine hydroxymethyltransferase. Thymidylate synthesis can be interrupted, and consequently the synthesis of DNA arrested, by the inhibition of either thymidylate synthase or dihydrofolate reductase. Many potent inhibitors are known for each of these enzymes, the best known being 5-fluoro-dUMP for thymidylate synthase and methotrexate, trimethoprim, and related compounds for dihydrofolate reductase. These and other inhibitors of nucleotide synthesis are discussed in a later section.

It should be noted that reduced folates are present in cells as polyglutamate forms, with up to five additional glutamate residues attached to the

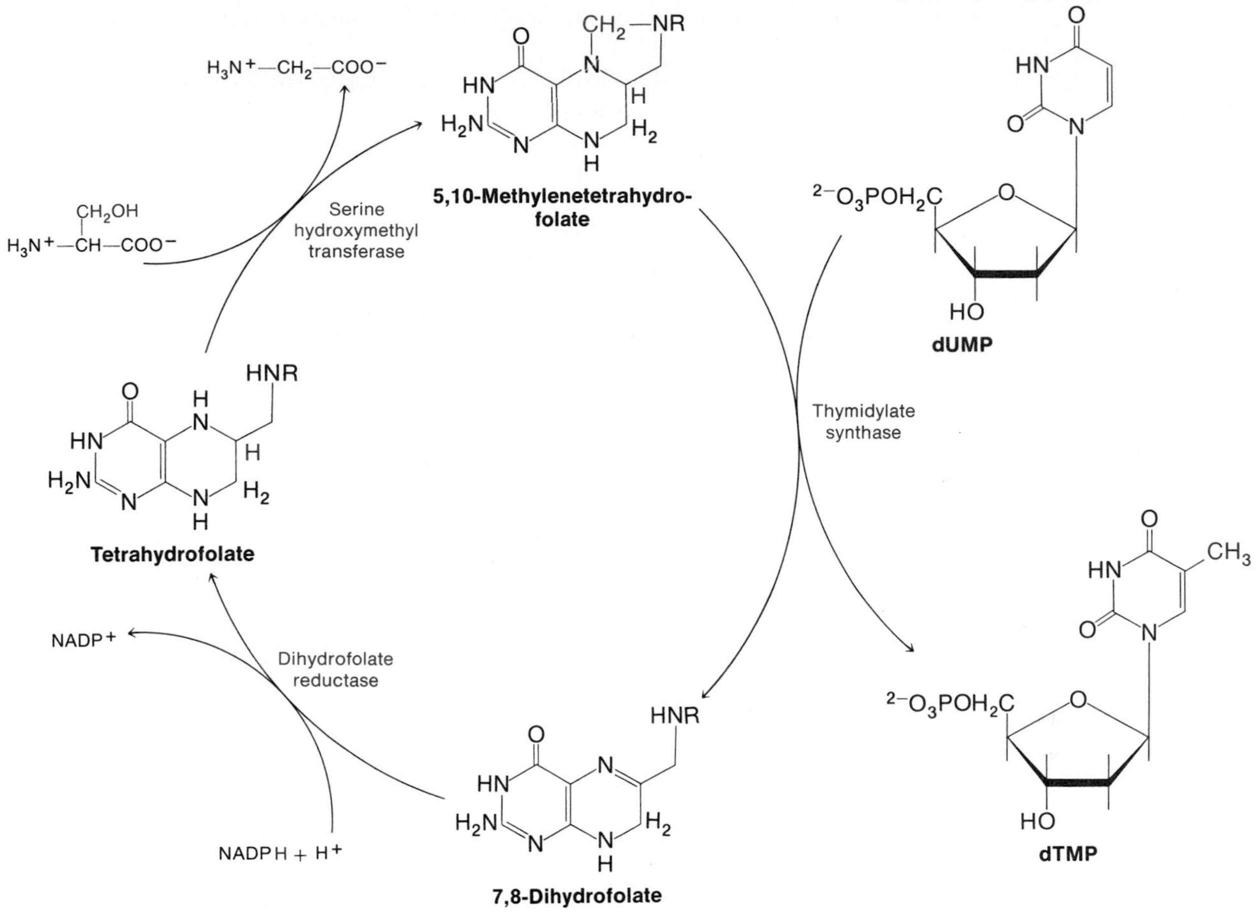

Figure 25-14
Thymidylate biosynthesis. R = p-aminobenzoyl-L-glutamate. Colored symbols indicate atoms that are precursors of the methyl group of thymidylate. The mechanism of this reaction is discussed in Chapter 11 (see Figure 11-21). The dTMP is formed from dUMP. 5,10-methylenetetrahydrofolate donates the methyl group in this reaction. This methyl group donor is regenerated by a two-step process involving first the reduction of 7,8-dihydrofolate to tetrahydrofolate and thence to 5,10-methylenetetrahydrofolate, in which serine supplies the methylene group.

terminal carboxyl of the folates. The glutamate residues are attached to each other in γ-peptide linkage. These polyglutamate forms of methylenetetrahydrofolate are the true substrates for thymidylate synthase; they have much lower K_m values and give higher maximum velocity than the monoglutamate. Analogous polyglutamate forms of 10-formyltetrahydrofolate are the true cofactors for purine synthesis.

FORMATION OF NUCLEOSIDE MONOPHOSPHATES FROM BASES (SALVAGE PATHWAYS)

In addition to the pathways for synthesis *de novo, certain enzymes* that are widely distributed in both mammalian tissues and microorganisms *can catalyze the synthesis of mononucleotides from purine and pyrimidine bases.* *Phosphoribosyltransferases* catalyze reactions of the following type:

$$\text{Base} + \text{PRPP} \rightleftharpoons \text{ribonucleoside-5'-phosphate} + \text{PP}_i$$

The equilibrium of this reaction is in favor of nucleotide synthesis, and since the inorganic pyrophosphate released is rapidly hydrolyzed by inorganic pyrophosphatase, the coupling of these reactions makes the synthesis of nucleotide irreversible.

Purine Phosphoribosyltransferases Convert Purines to Nucleotides

In mammals, two phosphoribosyltransferases that convert purine bases to nucleotides are present in many organs. The first, _adenine phosphoribosyltransferase_, catalyzes the formation of AMP from adenine. The second, _hypoxanthine-guanine phosphoribosyltransferase_, catalyzes the conversion of hypoxanthine to IMP and of guanine to GMP, and it is probably the enzyme that converts xanthine to XMP. Some microorganisms contain a separate xanthine phosphoribosyltransferase in addition to the adenine and hypoxanthine-guanine phosphoribosyltransferases.

The exact role of these enzymes in metabolism remains controversial. Originally they were considered to function solely for the recovery of bases released during the degradation of nucleic acids. That enzymes are involved in the salvage of bases seems likely, since turnover of some types of mRNA is rapid. To test this hypothesis, researchers have compared the excretion of purines (as uric acid) in humans who lack hypoxanthine-guanine phosphoribosyltransferase with purine excretion in normal humans. They have also studied purine metabolism in cultures of cells from both groups. The results indicate that most of the purines formed by degradation of nucleic acids are utilized for the resynthesis of nucleotides and nucleic acids. Another type of study has focused on humans who lack the enzyme xanthine oxidase (to be discussed later) and who therefore excrete purines as xanthine. Using the isotope-dilution technique, researchers measured the amounts of xanthine and hypoxanthine formed daily by such subjects, and then compared these amounts with the amounts of xanthine and hypoxanthine excreted. The results indicated that about 90 per cent of the purines formed were reutilized. This high recovery rate is presumably possible because of the restricted distribution, in higher animal tissues, of catabolic enzymes acting on free purines.

Hypoxanthine and xanthine appear to be the major purine bases produced and released by cells. Since there is little or no release of adenine (discussed below), adenine phosphoribosyltransferase does not seem to have a significant role in salvage of nucleotides. This enzyme perhaps serves mainly to utilize small amounts of adenine that are produced during intestinal digestion of nucleic acids or in the metabolism of 5'-deoxy-5'-methylthioadenosine, a byproduct of polyamine synthesis. Hypoxanthine-guanine phosphoribosyltransferase probably also functions to convert to nucleotides the small amounts of hypoxanthine and guanine released into the circulation by intestinal degradation of dietary nucleic acids.

Besides this salvage role, however, hypoxanthine-guanine phosphoribosyltransferase is probably important also for the transfer of purines from liver to other tissues. Purine biosynthesis _de novo_ is especially active in the liver, and there is evidence to suggest that extrahepatic cells that have a low capacity for the synthesis of purines _de novo_, such as erythrocytes and bone marrow cells, depend on uptake of hypoxanthine and xanthine from the blood to fulfill their needs for purine nucleotides. It seems likely that blood levels of xanthine and hypoxanthine, which are normally about 0.04 mM, are maintained by release of these bases from the liver. Some evidence suggests that the bases released by the liver are largely taken up by red blood cells and converted to purine nucleotides that are later broken down again with release of bases to tissues. If this is the case, the factors regulating their release and breakdown are unknown. The uptake of purine bases by extrahepatic tissues, as well as by bacteria, appears to be closely linked to the activity of the purine phosphoribosyltransferases. At least in some cases, this is so because the transferases are membrane proteins.

The neurologic disorder of children called the _Lesch-Nyhan syndrome_

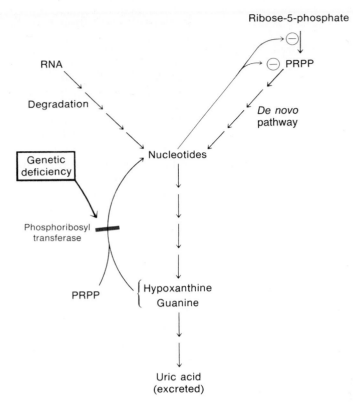

Figure 25–15

Mechanism of overproduction of purine nucleotides in the congenital deficiency of hypoxanthine-guanine phosphoribosyltransferase. The loss of the transferase prevents the recycling of hypoxanthine and guanine. This increases uric acid production as well as the *de novo* synthesis of purine nucleotides.

is due to a congenital lack of hypoxanthine-guanine phosphoribosyltransferase. The disorder is characterized by aggressive behavior, mental retardation, spastic cerebral palsy, and self-mutilation. Purine metabolism is profoundly disturbed, with greatly increased *de novo* biosynthesis of purines (200 times normal), overproduction of uric acid (6 times normal), and elevated blood levels of uric acid. Severe gout is caused by the latter in some individuals. The increased rate of purine biosynthesis is probably due to several factors (Figure 25–15). Because nucleotide production is depressed by the lack of phosphoribosyltransferase, the normal feedback inhibition that controls the production of PRPP amidotransferase is lifted. At the same time, decreased utilization of PRPP by phosphoribosyltransferase makes a higher intracellular concentration of PRPP available for PRPP amidotransferase. Finally, decreased nucleotide pools also would increase the PRPP level by deregulating the activity of ribose-5-phosphate pyrophosphokinase.

Although purine nucleosides are intermediates in the catabolism of nucleotides and nucleic acids in higher animals and humans, these nucleosides do not accumulate and are normally present in blood and tissues only in trace amounts. Nevertheless, cells of many vertebrate tissues contain kinases capable of converting purine nucleosides to nucleotides. Typical of these is *adenosine kinase*, which catalyzes the reaction:

$$\text{Adenosine} + \text{ATP} \longrightarrow \text{AMP} + \text{ADP}$$

2'-Deoxyadenosine is also a substrate, though a relatively poor one. In some mammalian cells deoxycytidine kinase also phosphorylates deoxyadenosine and deoxyguanosine (see next section). These kinases have not been studied extensively, and their specificity and role in various tissues and organs is uncertain.

Conversion of Pyrimidines to Mononucleotides Goes Through Nucleoside Intermediates

Many bacteria take up pyrimidine bases efficiently for nucleotide synthesis, and a phosphoribosyltransferase for uracil has been identified in a few species of bacteria. By contrast, pyrimidines are utilized poorly by mammalian cells. An alternative route for conversion of uracil to UMP, both in bacteria and in higher animals, is by successive reactions catalyzed by *uridine phosphorylase* and *uridine kinase*, respectively:

$$\text{Uracil + ribose-1-phosphate} \rightleftharpoons \text{uridine} + P_i$$

$$\text{Uridine + ATP} \xrightarrow{\text{Mg}^{2+}} \text{UMP + ADP}$$

Mammalian uridine phosphorylase will also accept deoxyuridine as a substrate, and thymidine is slowly attacked, but other nucleosides are not substrates. Uridine kinase activity has been demonstrated in a variety of bacteria and animal cells, including tumors, and is especially high in cells of high growth rate. Cytidine is the only other physiological nucleoside to act as a substrate for uridine kinase, but a variety of nucleoside triphosphates can act as phosphoryl donors.

Thymine is similarly converted to dTMP according to the following reactions:

$$\text{Thymine + deoxyribose-1-phosphate} \rightleftharpoons \text{thymidine} + P_i$$

$$\text{Thymidine + ATP} \longrightarrow \text{dTMP + ADP}$$

Thymine phosphorylase has been purified from bacteria and mammalian tissues. Thymidine is the preferred substrate, but deoxyuridine and 5-substituted deoxyuridines are also cleaved. The *thymidine kinase* is widely distributed. Its activity in cells increases dramatically during rapid growth and DNA synthesis. Infection of cells with any of several viruses also induces thymidine kinase activity, together with certain other enzymes concerned with deoxyribonucleotide synthesis. The viral thymidine kinase is quite different from the host enzyme. Significantly, the viral enzymes are subject to none of the allosteric controls of the host enzymes.

A *deoxycytidine kinase* has been purified from certain bacteria, from thymus, and from tumor cells. It catalyzes the reaction

$$\text{Deoxycytidine + ATP} \longrightarrow \text{dCMP + ADP}$$

Deoxycytidine kinase has several isoenzymes that differ in Michaelis constants and substrate specificity. Thus mitochondria contain an isoenzyme that is different from the cytosolic kinase. Deoxyadenosine and deoxyguanosine are also substrates for deoxycytidine kinase, but the purine deoxyribonucleosides have much higher K_m values than that of deoxycytidine for the cytosol enzyme.

Conversion of Nucleoside Monophosphates to Triphosphates Goes Through Diphosphates

The products of the biosynthetic pathways discussed in the preceding sections are, in most cases, mononucleotides. In cells, *a series of kinases (phosphotransferases) converts these mononucleotides to their metabolically active diphosphate and triphosphate forms*.

Nucleoside Monophosphate Kinases.   Bacteria and other microorganisms, as well as animal cells, contain a variety of kinases that catalyze reactions of the general type

$$(d)NMP + ATP \rightleftharpoons (d)NDP + ADP$$

Four types of nucleoside monophosphate kinases are known. These catalyze, respectively, the phosphorylation of (1) GMP and dGMP, (2) AMP and dAMP, (3) dCMP, CMP, and UMP, and (4) dTMP.

The second of the kinases mentioned, which uses AMP as substrate, is referred to as _adenylate kinase_. Its activity is high in tissues where the turnover of energy from adenine nucleotides is great, e.g., in liver and muscle, and its activity is also high in mitochondria. In these tissues its function is to make more energy available as ATP bond energy. When ADP is formed from ATP in energy-consuming reactions, more ATP can be formed from ADP according to the reaction

$$2\ ADP \rightleftharpoons AMP + ATP$$

Under conditions where energy-generating reactions convert intracellular ADP to ATP, AMP will be phosphorylated by running the preceding reaction from right to left. Adenylate kinase is therefore important in biological systems for maintaining equilibrium among adenine nucleotides as they are depleted or formed by energy transfers.

Nucleoside Diphosphate Kinases.   Enzymes of this type have been found in many tissues of animals, plants, and microorganisms. They catalyze reactions of the following general type:

$$N_1TP + N_2DP \rightleftharpoons N_1DP + N_2TP$$

where N_1 and N_2 are purine or pyrimidine ribonucleosides or deoxyribonucleosides. The activity of NDP kinases is relatively high, usually 10- to 100-fold greater than the activity of the monophosphate kinases. As a result, intracellular concentrations of triphosphates are normally much higher than those of diphosphates, which in turn are often higher than those of monophosphates. Unlike the monophosphate kinases, which are substrate-specific, NDP kinases from all sources are active with a wide range of nucleoside diphosphates and triphosphates. They require a divalent cation for activity, and although many metal ions can satisfy this requirement, Mg^{2+} is the physiological cofactor.

Nucleoside diphosphate kinases from many sources have been shown to function by forming a phosphoryl-enzyme intermediate:

$$E + N_1TP \rightleftharpoons E \sim P + N_1DP$$

$$E \sim P + N_2DP \rightleftharpoons E + N_2TP$$

In several cases it has been shown that the phosphoryl group is attached to N-1 of a histidine side chain.

INHIBITORS OF NUCLEOTIDE SYNTHESIS

There are several distinct types of inhibitors of nucleotide biosynthesis, each type acting at different points in the pathways to purine or pyrimidine nucleotides. All these inhibitors are very toxic to cells, especially rapidly growing cells such as those of tumors or bacteria, because interruption of the supply of nucleotides of any class seriously limits the cell's capacity to synthesize the nucleic acids necessary for protein synthesis and cell replication. In some cases, the toxic effect of such inhibitors makes them useful in cancer chemotherapy or in the treatment of bacterial infections. However, these agents can also damage the replicating cells of the intestinal tract and bone marrow. This danger imposes limits on the doses that can be used safely.

6-Mercaptopurine (Table 25–1) and related thiopurines are potent inhibitors of purine nucleotide biosynthesis, but they are inactive until they are converted to the corresponding ribonucleoside 5′-phosphates. 6-Mercaptopurine is converted by the action of hypoxanthine-guanine phosphoribosyltransferase to the nucleotide 6-thioinosine-5′-monophosphate (T-IMP). The latter inhibits several enzymes of purine biosynthesis, and it is uncertain which effect is primarily responsible for the toxicity. It blocks conversion of IMP to adenylosuccinate and to XMP, key reactions in the formation of AMP and GMP (Figure 25–8). T-IMP is also capable of "pseudofeedback" inhibition of glutamine PRPP amidotransferase, the first committed step in the purine nucleotide pathway. This enzyme is highly responsive to intracellular concentrations of both normal ribonucleoside 5′-monophosphates and analogs. 6-Mercaptopurine is used clinically in the treatment of leukemia.

Tiazofurin is a thiazole-C-nucleoside with anticancer activity. It is metabolized to tiazofurin-5′-monophosphate and then to thiazole-4-carboxamide adenine dinucleotide (TAD), an analog of NAD. TAD specifically inhibits IMP dehydrogenase (Figure 25–8), with consequent depletion of the pools of GMP, GDP, and GTP. The corresponding analog with selenium in place of sulfur is also under trial as an anticancer agent.

Three inhibitors specifically interfere with steps in pyrimidine nucleotide biosynthesis. N-(Phosphonacetyl)-L-aspartate (PALA) was synthesized as an analog of the transition state intermediate postulated for the aspartate carbamoyl transferase reaction (Figure 25–10), and is a powerful inhibitor of this enzyme. DUP-785 is a new synthetic inhibitor that appears to inhibit dihydroorotate dehydrogenase specifically. Pyrazofurin is a fermentation product. It is converted by kinase action to the 5′-phosphate, which appears to mimic the substrate of orotidylate decarboxylase, for which it is a powerful inhibitor. All three compounds are under trial as anticancer agents.

5-Fluorouracil, which is an agent used in treating solid tumors, interferes with thymidylate synthesis. Its major inhibitory effect occurs after its conversion in the cell to 5-fluoro-2′-deoxyuridine-5′-monophosphate (Table 25–1). The latter acts as an analog of dUMP and binds very tightly to thymidylate synthase in the presence of methylenetetrahydrofolate, forming a covalent complex with the enzyme that is unable to undergo the normal catalytic reaction. 5-Fluorouracil may also exert a cytotoxic effect through incorporation of 5-fluorouridine phosphate into RNA.

Among inhibitors that interfere with the synthesis of both pyrimidine and purine nucleotides are hydroxyurea and α-(N)-heterocyclic carboxaldehyde thiosemicarbazone (Table 25–1). These compounds interfere with the synthesis of deoxyribonucleotides by inhibiting ribonucleotide reductase of mammalian cells, an enzyme that is crucial and probably rate-limiting in the biosynthesis of DNA. They probably act by disrupting the iron-tyrosyl radical structure at the active site of the reductase, and hydroxyurea is in clinical use as an anticancer agent.

Reactions involving glutamine as a substrate are inhibited by the glutamine analogs azaserine and 6-diaza-5-oxo-L-2-aminohexanoic acid (Table 25–1). Inhibition is irreversible, with formation of a covalent bond between the inhibitor and an amino acid side chain at the catalytic site. The specific reactions inhibited are those catalyzed by glutamine PRPP amidotransferase and phosphoribosyl-N-formylglycinamidine synthase in the de novo purine pathway and carbamoyl phosphate synthase and CTP synthase in the pyrimidine pathway.

TABLE 25–1
Some Inhibitors of Nucleotide Metabolism Together with Corresponding Physiological Metabolites, Active Inhibitor Products, and Enzymes Inhibited

Inhibitor	Corresponding Metabolite	Active Form of Inhibitor	Enzymes Inhibited
Inhibitors of purine nucleotide biosynthesis			
6-Mercaptopurine	Hypoxanthine	6-Thioinosine-5'-phosphate	Figure 25–8, adenylosuccinate synthase, IMP dehydrogenase; Figure 25–6, reaction 1
Tiazofurin	Nicotinamide ribonucleoside-5'-phosphate	Thiazole-4-carboxamide adenine dinucleotide	Figure 25–8, IMP dehydrogenase

TABLE 25–1 (continued)

Inhibitor	Corresponding Metabolite	Active Form of Inhibitor	Enzymes Inhibited
Inhibitors of pyrimidine nucleotide biosynthesis			
COO⁻ CH—NH—CO—CH₂—PO₃²⁻ CH₂ COO⁻ **N-(Phosphonacetyl)-L-aspartate**	COO⁻ CH—NH··CO—NH₂··O—PO₃²⁻ CH₂ COO⁻ **Transition state intermediate**	—	Figure 25–10, aspartate carbamoyl transferase
DUP-785	Unknown	—	Figure 25–10 dihydroorotate dehydrogenase

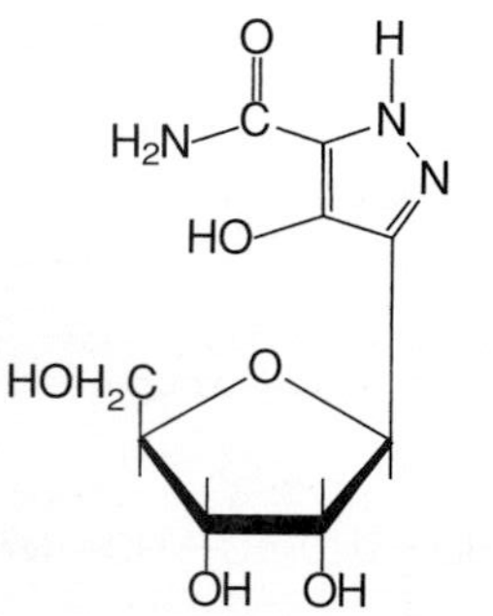

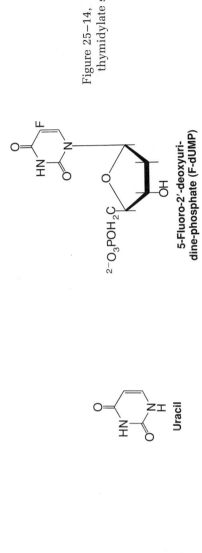

Figure 25–10, orotidylate decarboxylase

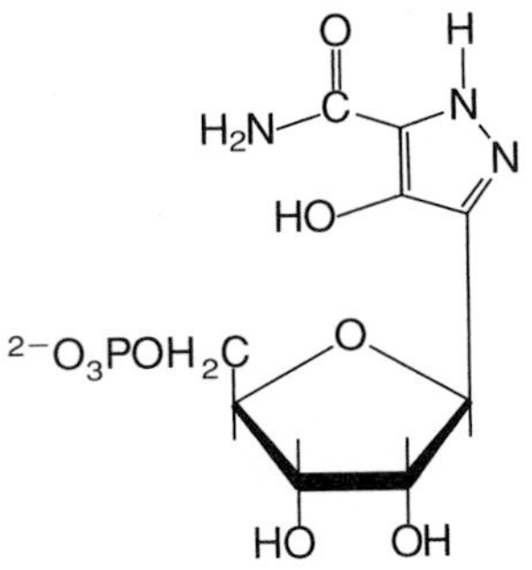

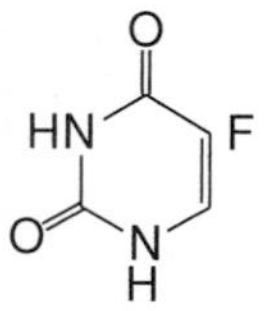

Pyrazofurin-5'-phosphate

OMP

Pyrazofurin (pyrazomycin)

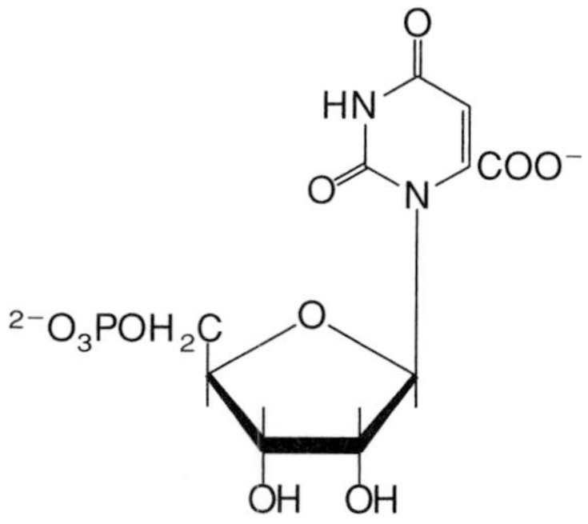

Figure 25–14, thymidylate synthase

5-Fluoro-2'-deoxyuridine-phosphate (F-dUMP)

Uracil

5-Fluorouracil

TABLE 25–1 (continued)

Inhibitor	Corresponding Metabolite	Active Form of Inhibitor	Enzymes Inhibited
Inhibitors of both purine and pyrimidine nucleotide biosynthesis			
H$_2$N—C(=O)—NHOH **Hydroxyurea**	—	—	Figure 25–13, ribonucleotide reductase
(ring)—CH=N—NH—C(=S)—NH$_2$ **α-(N)-Heterocyclic carboxaldehyde thiosemicarbazone**	—	—	
$^-$N≡N$^+$=CH—C(=O)—O—CH$_2$—CH(NH$_3^+$)—COO$^-$ **Azaserine (O-diazoacetyl-L-serine)**	NH$_2$—C(=O)—CH$_2$—CH$_2$—CH(NH$_3^+$)—COO$^-$ **L-Glutamine**	—	Figure 25–6, reactions 1 and 4; Figure 25–10, reaction 1; Figure 25–12
$^-$N≡N$^+$=CH—C(=O)—CH$_2$—CH$_2$—CH(NH$_3^+$)—COO$^-$ **6-Diazo-5-oxo-L-2-aminohexanoic acid (6-diazo-5-oxo-norleucine, DON)**	NH$_2$—C(=O)—CH$_2$—CH$_2$—CH(NH$_3^+$)—COO$^-$ **L-Glutamine**	—	Figure 25–6, reactions 1 and 4; Figure 25–8, GMP synthase; Figure 25–10, reaction 1
(isoxazoline ring with Cl)—CH(H$^+$NH$_3$)—COO$^-$ **Acivicin**	—	—	

Indirect inhibitors of nucleotide biosynthesis

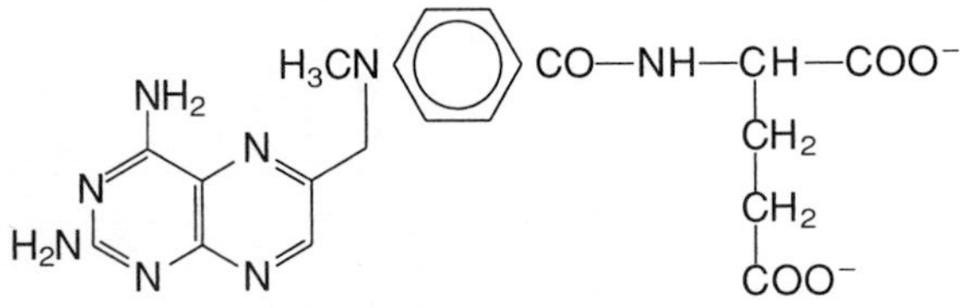

Methotrexate (amethopterin)

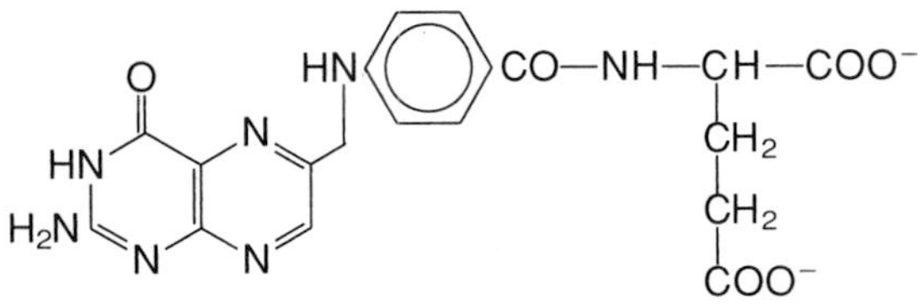

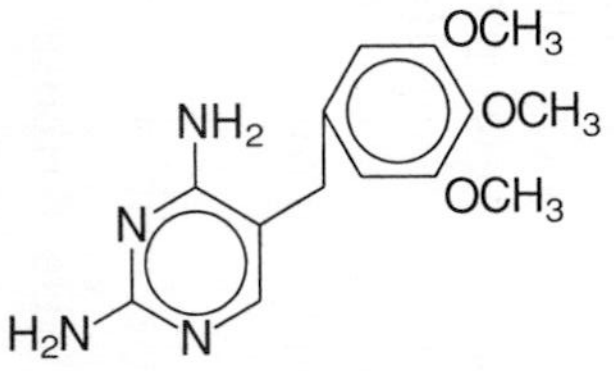

Trimethoprim

Sulfonamide

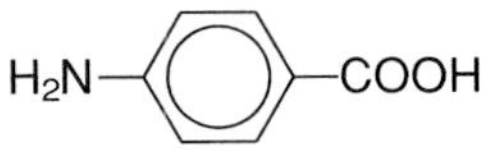

Folic acid (pteroylglutamic acid)

p-Aminobenzoic acid

Figure 25–14, dihydrofolate reductase

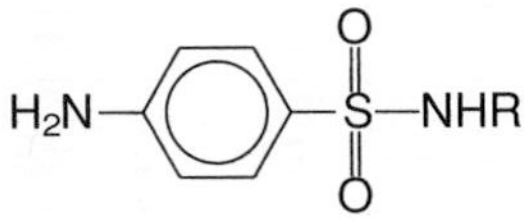

Dihydropteroate synthase

Acivicin is an analog with a more distant resemblance to glutamine. It is, nevertheless, a potent inhibitor of several steps in purine nucleotide biosynthesis that utilize glutamine. The enzymes it inhibits are glutamine PRPP amidotransferase (step 1, Figure 25–6), phosphoribosyl-N-formylglycinamidine synthase (step 4, Figure 25–6), and GMP synthase (Figure 25–8). In pyrimidine nucleotide biosynthesis the enzymes inhibited are carbamoyl synthase II (step 1, Figure 25–10) and CTP synthase (Figure 25–11). Acivicin is under trial for the treatment of some forms of cancer.

Another group of inhibitors prevents nucleotide biosynthesis indirectly by depleting the level of intracellular tetrahydrofolate derivatives, including 10-formyltetrahydrofolate (required for purine synthesis). As seen from Table 25–1, *sulfonamides* are structural analogs of p-aminobenzoic acid, and they competitively inhibit the bacterial biosynthesis of folic acid at a step at which p-aminobenzoic acid is incorporated into folic acid. Sulfonamides are widely used in medicine because they inhibit growth of many bacteria. When cultures of susceptible bacteria are treated with sulfonamides, they accumulate 4-carboxamide-5-aminoimidazole in the medium, because of a lack of 10-formyltetrahydrofolate for the penultimate step in the pathway to IMP (Figure 25–6). *Methotrexate, trimethoprim,* and a number of related compounds inhibit the reduction of dihydrofolate to tetrahydrofolate, a reaction catalyzed by dihydrofolate reductase. These inhibitors are structural analogs of folic acid (Table 25–1) and bind at the catalytic site of dihydrofolate reductase, an enzyme catalyzing one of the steps in the cycle of reactions involved in thymidylate synthesis (Figure 25–14). These inhibitors therefore prevent synthesis of thymidylate and purine nucleotides in replicating cells. Methotrexate is used as an anticancer drug, and trimethoprim, which specifically inhibits bacterial dihydrofolate reductase, is used in combination with sulfonamides for treating certain bacterial infections.

In addition to the inhibitors just described, there are a number of other nucleoside analogs that are useful in treating cancer or infectious diseases. These act by inhibiting DNA polymerase or by being incorporated into DNA and interfering with subsequent DNA chain elongation, DNA replication, or transcription. Among them are *cytosine arabinoside* (araC, anticancer drug), *acyclovir* (antiviral agent), and *azidothymidine* (in trials for use against AIDS).

CATABOLISM OF NUCLEOTIDES

Dietary nucleic acids are unaffected by gastric enzymes, but in the small intestine, ribonuclease and deoxyribonuclease I, which are secreted in the pancreatic juice, hydrolyze nucleic acids mainly to oligonucleotides. The oligonucleotides are further hydrolyzed by phosphodiesterases, also secreted by the pancreas, to yield 5′- and 3′-mononucleotides. Most of the mononucleotides are then hydrolyzed to nucleosides by various group-specific nucleotidases or by a variety of nonspecific phosphatases. The resulting nucleosides may be absorbed intact by the intestinal mucosa, or they may undergo phosphorolysis by nucleoside phosphorylases and by nucleosidases to free bases. Little is known about these enzymes and their specificity. The small intestinal mucosa are rich in nucleoside phosphorylase, and most of the remaining nucleoside is probably hydrolyzed to bases in this tissue.

$$\text{Nucleoside} + \text{P}_i \rightleftharpoons \text{base} + \text{ribose-1-phosphate}$$

$$\text{Nucleoside} + \text{H}_2\text{O} \longrightarrow \text{base} + \text{ribose}$$

Experiments with labeled nucleic acid indicate that both purines and pyrimidines of ingested nucleic acids are used only to a small extent for synthesis of tissue nucleic acids, and in the case of purines, most of the bases were shown to be catabolized. This finding is consistent with the presence in intestinal mucosa of a high level of xanthine oxidase, a catabolic enzyme.

Intracellular Catabolism of Nucleotides Is Highly Regulated

Nucleotides are also catabolized within cells by several types of intracellular nucleotidase. One of those present in mammalian cells is probably located in lysosomes. It has a pH optimum of 5.0 and a broad substrate specificity. In lymphocytes there are two cytoplasmic purine nucleotidases, one acting on deoxyribonucleotides, the other on ribonucleotides. In these cells three types of cytoplasmic nucleotidase act on pyrimidine nucleotides; one acts on ribonucleotides, another on deoxyribonucleotides, and a third is specific for thymidylate.

The intracellular nucleotidases are under strict regulatory control, but nevertheless appear to be involved in cycling of nucleotides along the pathways shown in Figures 25–1 and 25–2. This is evident from results of experiments with inhibitors of some of the enzymes involved in these pathways, such as adenosine deaminase and purine nucleoside phosphorylase. Among their functions is the disposal of nucleotides formed from the breakdown of mRNA. In certain tissues they have a specific role. For example, one of the acid phosphatases present in bone, where its role is to provide inorganic phosphate, is an iron-stimulated nucleotidase acting on di- and triphosphates.

One ecto-5′-nucleotidase, which is attached to the outer surface of the plasma membrane of many types of mammalian cell, dephosphorylates purine and pyrimidine ribo- and deoxyribonucleoside monophosphates to the corresponding nucleosides. An important function of the ecto-nucleotidase may be the assimilation of nucleotides arising from the dissolution of dying cells. The nucleosides formed by the ecto-5′-nucleotidase are then transported into the cell and are reconverted to nucleotides via reactions shown in Figures 25–1 and 25–2. There is evidence to indicate that transport of nucleosides into mammalian cells is achieved by at least three independent systems. They are specific, saturable, rapid, reversible, and temperature-dependent.

Purines Are Catabolized to Uric Acid and Then to Other Products

After purine nucleotides have been converted to the corresponding nucleosides by 5′-nucleotidases and by phosphatases, inosine and guanosine are readily cleaved to the nucleobase and ribose-1-phosphate by the widely distributed *purine nucleoside phosphorylase*. The corresponding deoxynucleosides yield deoxyribose-1-phosphate and base with the phosphorylase from most sources. Adenosine and deoxyadenosine are not attacked by the phosphorylase of mammalian tissue, but much AMP is converted to IMP by an aminohydrolase (deaminase), which is very active in muscle and other tissues (Figure 25–16). It has recently been discovered that inherited deficiency of purine nucleoside phosphorylase is associated with a deficiency in the cellular type of immunity, but not in humoral immunity.

An adenosine aminohydrolase (deaminase) is also present in many mammalian tissues. This enzyme is of interest because hereditary deficiency of the enzyme is linked to a severe (usually fatal) defect in the immune system, marked by a serious deficiency in lymphocytes and consequent inability to combat infections.

Inosine formed by either route is then phosphorolyzed to yield hypoxanthine. Although we have previously seen that much of the hypoxanthine and guanine produced in the mammalian body is converted to IMP and GMP

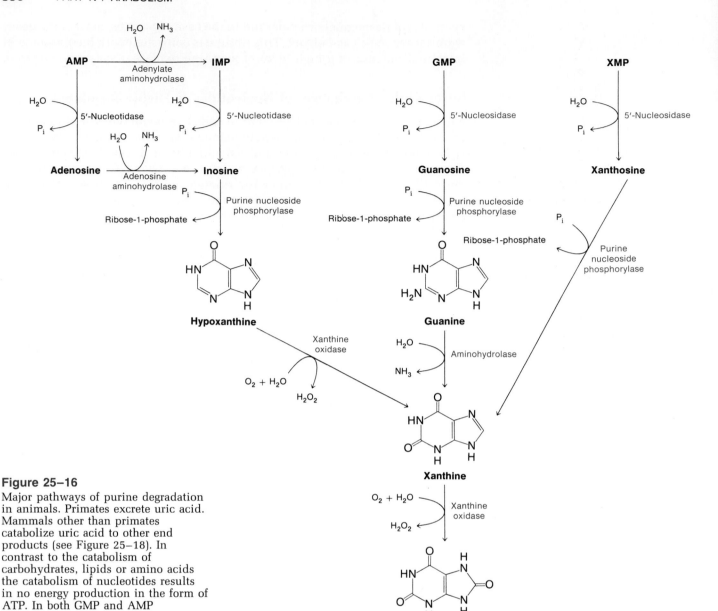

Figure 25–16

Major pathways of purine degradation in animals. Primates excrete uric acid. Mammals other than primates catabolize uric acid to other end products (see Figure 25–18). In contrast to the catabolism of carbohydrates, lipids or amino acids the catabolism of nucleotides results in no energy production in the form of ATP. In both GMP and AMP catabolism ribose-1-phosphate is released.

by a phosphoribosyltransferase, about 10 per cent is catabolized. Xanthine oxidase, an enzyme present in large amounts in liver and intestinal mucosa and in traces in other tissues, oxidizes hypoxanthine to xanthine and xanthine to uric acid (Figure 25–16). Xanthine oxidase contains FAD, molybdenum, iron, and acid-labile sulfur in the ratio $1:1:4:4$, and in addition to forming hydrogen peroxide, it is also a strong producer of the superoxide anion $\cdot O_2$, a very reactive species. The enzyme oxidizes a wide variety of purines, aldehydes, and pteridines.

Guanine aminohydrolase (guanine deaminase or guanase), present in liver, brain, and other mammalian tissues, provides another pathway to xanthine, this time from guanine. Subsequent oxidation of xanthine to uric acid then occurs.

Gout is a relatively common (≈ 3 per 1000 persons) derangement of purine metabolism that is associated with elevated plasma levels of uric acid. The excessive uric acid leads to painful deposits of monosodium urate in the cartilage of joints, especially of the big toe. Uric acid deposits also may

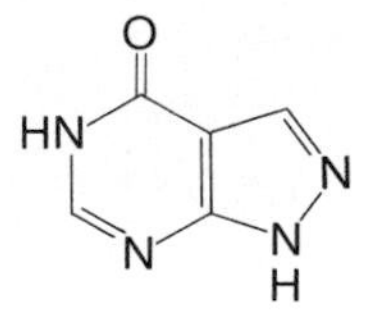

Figure 25-17
The structure of allopurinal, an analog of hypoxanthine. Allopurinol inhibits xanthine oxidase and is used in the treatment of gout.

occur as calculi in the kidney, with resultant renal damage. The genetics are complex and incompletely understood. Individuals suffering from gout may be treated with the xanthine oxidase inhibitor allopurinol (Figure 25-17), an analog of hypoxanthine. This treatment produces a gradual decrease in urate levels in blood and urine and increased excretion of xanthine.

Mammals other than primates further oxidize urate by a liver enzyme, urate oxidase, which is a copper protein. The product, allantoin, is excreted. Humans and other primates, as well as birds, lack urate oxidase and hence excrete uric acid as the final product of purine catabolism. In many animals other than mammals, allantoin is metabolized further to other products that are excreted: allantoic acid (some teleost fish), urea (most fishes, amphibia, some mollusks), and ammonia (some marine invertebrates, crustaceans, etc.). This pathway of further purine breakdown is shown in Figure 25-18.

Pyrimidines Are Catabolized to β-Alanine, NH_3, and CO_2

A number of deaminases present in many cells are able to deaminate cytosine or its nucleosides or nucleotides to the corresponding uracil derivatives. Cytosine aminohydrolase (deaminase) appears to occur only in microorganisms (yeast and bacteria), but cytidine aminohydrolase is widely distributed in bacteria, plants, and mammalian tissues. A distinct deoxycytidine aminohydrolase is present in various mammalian tissues and tumors, in plants, and in bacteria. A deoxycytidylate aminohydrolase that is similarly distributed produces dUMP, which is susceptible to attack by 5'-nucleotidase to give deoxyuridine. Although the physiological function of these aminohydrolases is not completely understood, the uridine and deoxyuridine formed can be further degraded by uridine phosphorylase (see above) to uracil, so that these reactions provide a pathway for converting nucleotides of uracil and cytosine to uracil and ribose-1-phosphate or deoxyribose-1-phosphate (Figure 25-19). Similarly, thymine nucleosides and nucleotides can be converted by 5'-nucleotidase and phosphorylase to thymine.

Enzymes present in mammalian liver are capable of the catabolism of both uracil and thymine. The first reduces uracil and thymine to the corresponding 5,6-dihydro derivatives. This hepatic enzyme uses NADPH as the reductant, whereas a similar bacterial enzyme is specific for NADH. Similar enzymes are apparently present in yeast and plants. Hydropyrimidine

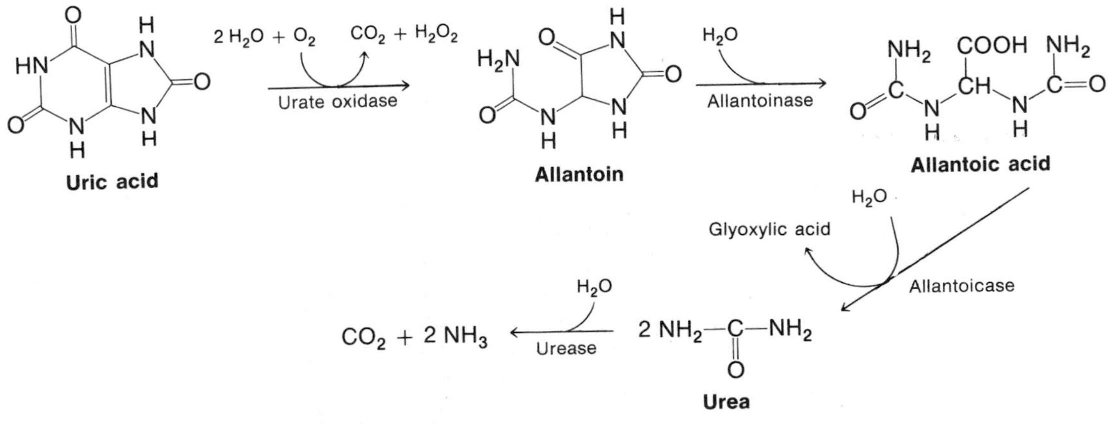

Figure 25-18
Degradation of uric acid to excretory products. Mammals other than primates oxidize uric acid further to allantoin. Humans and other primates as well as birds lack urate oxidase and hence excrete uric acid as the final product of purine catabolism. In many animals other than mammals, allantoin is metabolized further to urea or ammonia and CO_2 as shown.

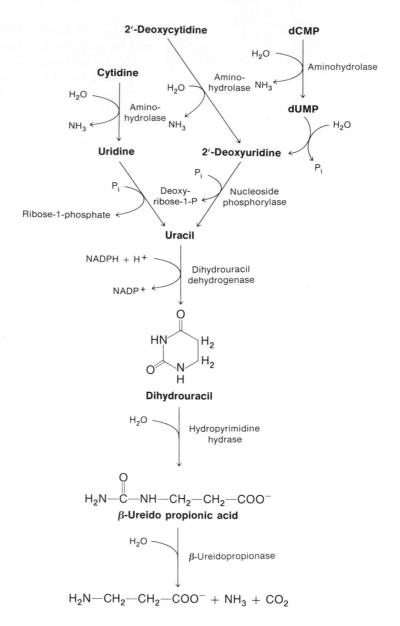

Figure 25–19
Degradation of pyrimidine bases. Parts of this pathway are widely distributed in nature. The entire pathway is found in mammalian liver. As in purine nucleotide catabolism, no ATP results from catabolism, and the ribose-1-phosphate is released during catabolism before destruction of the base.

hydrase then opens the reduced pyrimidine ring, and finally the carbamoyl group is hydrolyzed off from the product to yield β-alanine or β-aminoisobutyric acid, respectively, from uracil and thymine (Figure 25–19).

REGULATION OF NUCLEOTIDE METABOLISM

Among the reaction pathways described earlier are many possibilities for futile cycles, in which nucleotides built up in the biosynthetic pathways are broken down in catabolic pathways to products closely related to the starting materials. As an example, AMP synthesized from IMP by adenylosuccinate synthase and adenylosuccinate lyase may be hydrolyzed back to IMP by adenylate aminohydrolase (Figure 25–16). The net result is the conversion of aspartate to fumarate and ammonia and the hydrolysis of GTP to GDP and P_i. To avoid such futile cycles, both biosynthetic and catabolic processes are under tight regulatory controls. The efficiency of these controls is demonstrated by the increased activity of many enzymes involved in nucleotide biosynthesis when cells are proliferating. Evidently, regulatory mechanisms increase nucleotide biosynthesis as intracellular nucleotides are used for the

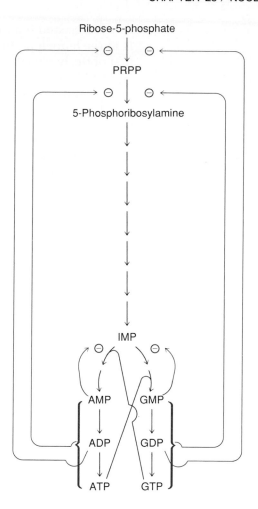

Figure 25–20
Regulation of purine biosynthesis. Red arrows show points of feedback inhibition. In addition to the feedback inhibition, GTP stimulates ATP synthesis and ATP stimulates GTP synthesis. This helps to assure a balance between the pools of the two nucleoside triphosphates. The full biosynthetic pathways are shown in Figures 25–6 and 25–8.

synthesis of RNA and DNA. As we have seen, drastic consequences can attend impairment of the control machinery, as in the Lesch-Nyhan syndrome or intervention with drugs such as 6-mercaptopurine.

Although much remains to be discovered about the details of the regulation of nucleotide metabolism, a number of important control points are rather well understood and will be discussed together with their known effects on intracellular nucleotide pools.

Purine Biosynthesis Is Regulated at Two Levels

Many lines of evidence indicate that *the first committed step in de novo purine nucleotide biosynthesis, production of glutamine PRPP amidotransferase, is rate-limiting for the entire sequence.* Consequently, regulation of this enzyme is probably the most important factor in control of purine synthesis *de novo* (Figure 25–20). The enzyme is inhibited by purine-5'-nucleotides, but the nucleotides that are most inhibitory vary with the source of the enzyme. Inhibition constants (K_I) are usually in the range 10^{-3} to 10^{-5} M. The maximum effect of this end-product inhibition is produced by certain combinations of nucleotides (e.g., AMP and GMP) in optimum concentrations and ratios, indicating two kinds of inhibitor binding sites. This is an example of a concerted feedback inhibition.

The rate of the amidotransferase reaction is also governed by intracellular concentrations of the substrates L-glutamine and PRPP. Competing metabolic reactions or drugs that alter the supply of these substrates also affect the rate of IMP synthesis.

The second important level of regulation of purine nucleotide synthesis is in the branch pathways from IMP to AMP and to GMP (Figure 25–8). The first of the two reactions leading from AMP to IMP is the irreversible synthesis of adenylosuccinate. This requires GTP as a source of energy and is inhibited by AMP. Of the two reactions required to convert IMP to GMP, the first is irreversible and is inhibited by GMP, while the second requires ATP as a source of energy. Thus there are two types of regulation at this level of purine nucleotide synthesis: (1) a "forward" control, by which increased GTP accelerates AMP synthesis and increased ATP accelerates GMP synthesis, and (2) feedback inhibition, by which AMP and GMP each regulate their own synthesis. Excess AMP also may be converted to IMP by adenylate aminohydrolase and thus can serve as a source of GMP. Adenylate aminohydrolase is activated by ATP and inhibited by GTP, which may serve to control this potential conversion of adenine nucleotides to guanine nucleotides. Finally, when the energy reserves of the cell are low, feedback inhibition of ribose-5-phosphate pyrophosphokinase by ADP and GDP restricts the synthesis of PRPP.

Pyrimidine Biosynthesis Is Regulated at the Level of Carbamoyl Aspartate Formation

In bacteria, the first committed step in pyrimidine nucleotide biosynthesis is the formation of carbamoyl aspartate from carbamoyl phosphate and aspartate. In *E. coli*, the enzyme catalyzing this step, aspartate carbamoyltransferase, is powerfully inhibited by CTP, which acts chiefly by decreasing the affinity of the enzyme for aspartate (see Chapter 10). ATP has the opposite effect, activating the enzyme by increasing its affinity for aspartate. Concentrations of ATP and CTP in *E. coli* are high enough for these nucleotides to influence the intracellular activity of aspartate carbamoyltransferase. However, this is not a regulatory enzyme in all bacterial species and is not involved in regulation of pyrimidine nucleotide synthesis in animal cells.

In eukaryotes, carbamoyl phosphate synthase is inhibited by pyrimidine nucleotides and stimulated by purine nucleotides; it appears to be the most important site of feedback inhibition of pyrimidine nucleotide biosynthesis in mammalian tissues. However, it has been suggested that under some conditions, orotate phosphoribosyltransferase may be a regulatory site as well.

Ribonucleotide Reduction Is Regulated by Both Activators and Inhibitors

The manner in which the reduction of ribonucleotides to deoxyribonucleotides is regulated has been studied with reductases from relatively few species. The enzymes from *E. coli* and from Novikoff rat liver tumor have a complex pattern of inhibition and activation (Figure 25–21). In both cases, dATP inhibits the reduction of all substrates, dTTP inhibits the reduction of CDP and UDP but activates the reduction of ADP and GDP, and ATP activates the reduction of UDP and CDP. Since there is evidence that ribonucleotide reductase may be the rate-limiting step in deoxyribonucleotide synthesis in a least some animal cells, these allosteric effects may be important in controlling deoxyribonucleotide synthesis.

The adenosylcobalamin-requiring ribonucleotide reductases from lactobacilli and certain other microorganisms have a different pattern of allosteric effects, the principal one being specific activation effects. For example, in the case of the *Lactobacillus leichmannii* enzyme, dGTP activates ATP reduction, dATP activates CTP reduction, and dCTP activates UTP reduction. These effects may serve to adjust the relative rates of reduction of the various

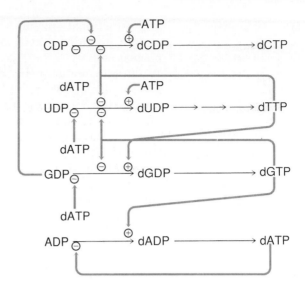

Figure 25–21
Proposed scheme for the regulation of deoxyribonucleotide synthesis in *E. coli* and mammalian cells. Red arrows indicate points of activation and inhibition, respectively. (Adapted from L. Thelander and P. Reichard, Reduction of ribonucleotides, *Ann. Rev. Biochem.* 48: 133, 1979.)

substrates to more equal values. In addition, the synthesis of the *L. leichmannii* enzyme is repressed by the presence in the growth medium of an excess of vitamin B_{12} (cyanocobalamin) or of a deoxyribonucleoside such as thymidine. The repressor for enzyme synthesis is probably dTTP or a closely related nucleotide, which accumulates in the cell when rapid ribonucleotide reduction occurs as a result of an ample cobalamin supply or when deoxynucleoside is supplied. Further deoxyribonucleotide synthesis is then slowed by the decreased rate of reductase synthesis.

Metabolites Are Channeled Along the Nucleotide Biosynthesis Pathways

In addition to the regulatory controls described in the preceding sections, which are mainly allosteric feedback mechanisms, evidence is accumulating that nucleotide biosynthetic pathways are closely controlled through the phenomenon of <u>channeling</u>. This involves an assembly of enzymes catalyzing successive steps in the biosynthetic pathway so that metabolic intermediates pass directly from one enzyme to the next. Thus the <u>metabolites are</u> <u>channeled along the metabolic pathway, with restricted opportunity for</u> <u>their diffusion into the medium or entry into the general metabolic pool in</u> <u>the cell</u>.

There are two major ways in which channeling of nucleotide precursors is achieved. The first involves multifunctional enzymes, in which several catalytic activities occur on a single polypeptide chain. The second involves noncovalent association of many pathway enzymes in complexes that may sometimes be concentrated in a particular intracellular location. In addition, channeling may be assisted by complete or partial confinement of metabolites within a specific intracellular compartment, e.g., within the nucleus.

Examples of multifunctional enzymes are provided by the pyrimidine biosynthetic pathway. Although in most prokaryotes, six structural genes code for the six enzymes involved in the *de novo* synthesis of UMP, in eukaryotes the number of genes is reduced because of the production of multifunctional proteins. In *Neurospora fungi*, a single protein has both carbamoyl phosphate synthase activity and aspartate carbamoyltransferase activity, but in mammalian cells a single protein not only has both these activities, but also has dihydroorotase activity. The latter protein is an oligomer (probably a trimer) of a large polypeptide ($M_r = 200,000$), and there is evidence to indicate that the multifunctional polypeptide is a single gene product. This

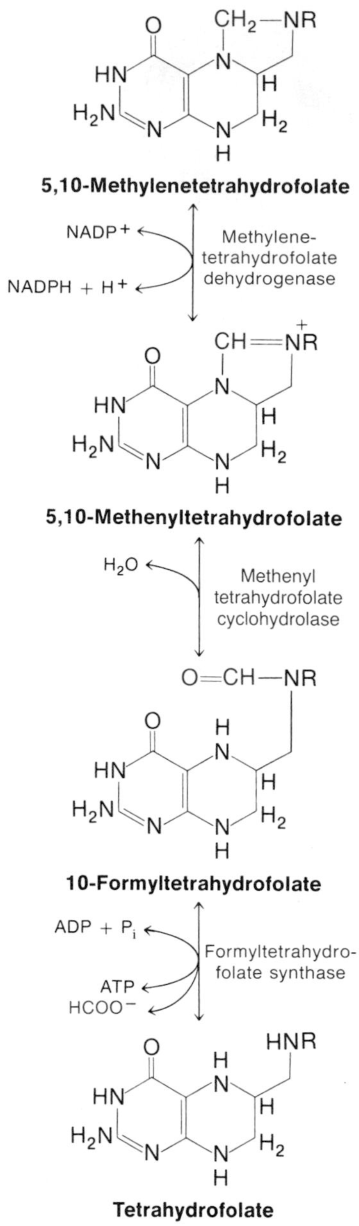

5,10-Methylenetetrahydrofolate

NADP+

NADPH + H+

Methylene-
tetrahydrofolate
dehydrogenase

5,10-Methenyltetrahydrofolate

H_2O

Methenyl
tetrahydrofolate
cyclohydrolase

10-Formyltetrahydrofolate

ADP + P_i

ATP
HCOO⁻

Formyltetrahydro-
folate synthase

Tetrahydrofolate

Figure 25–22
Formation of formyl group donors for purine biosynthesis (R = p-aminobenzoyl-L-glutamate). A trifunctional enzyme catalyzes the three reactions shown, which generate the formyl donors for steps in purine nucleotide synthesis (see steps 3 and 9 in Figure 25–6).

multifunctional enzyme channels carbamoyl phosphate and carbamoyl aspartate, provided dihydroorotate is rapidly removed, which is the case in the normal cell.

In mammalian cells, the last two steps of the pathway to UMP (Figure 25–10) also appear to be catalyzed by a multifunctional protein that is the product of a single gene. This protein therefore has both orotate phosphoribosyltransferase and OMP decarboxylase activities. Although added OMP is accepted as substrate for the decarboxylase, OMP formed from orotate and PRPP is not released, but is preferentially utilized at the decarboxylase site for UMP formation.

In the purine pathway, a trifunctional enzyme catalyzes three reactions concerned with generation of the formyl donors for steps 3 and 9 (Figure 25–6). The enzymatic reactions catalyzed by this protein are shown in Figure 25–22. Since 10-formyltetrahydrofolate inhibits the cyclohydrolase, this may serve to regulate the relative amounts of the two formyl donors that are available. Channeling is further enhanced by noncovalent association of other enzymes of the pathway with the dehydrogenase-cyclohydrolase-synthase enzyme. Evidence has been obtained that this loose complex contains serine hydroxymethyltransferase (which generates methylenetetrahydrofolate), the transformylases catalyzing steps 3 and 9 of the purine pathway and probably all the other enzymes of the pathway. These enzyme activities largely remain associated through certain mild purification procedures. This association of pathway enzymes greatly increases channeling and permits the efficient generation and use of the hydrolytically and oxidatively unstable formyl donors.

Research with fibroblast cells in culture suggests that during the DNA-synthesizing phase of the cell cycle (S phase), a complex of at least six enzymes associated with DNA synthesis is present in the cell nucleus. Possibly its assembly signals the initiation of the S phase. The complex apparently involves DNA polymerase and many enzymes involved in the synthesis of deoxyribonucleoside triphosphates. These include thymidine kinase, dCMP kinase, nucleoside diphosphate kinase, thymidylate synthase, and dihydrofolate reductase. When the cells are quiescent, or in the G_1 phase, these enzymes are largely present in the cytoplasm and are no longer associated in a complex. Channeling by the nuclear complex is indicated by the observation that ribonucleoside diphosphates are incorporated more efficiently into DNA than deoxyribonucleoside triphosphates. This observation also suggests that ribonucleotide reductase is part of the complex.

Intracellular Nucleotide Pools Vary with the Stages in the Cell Cycle

Methods are now available for analysis of nucleotides in eukaryotic cells and tissues in various phases of the cell cycle (Figure 25–23). Concentrations are frequently expressed in terms of picomoles per 10^6 cells or picomoles per microgram of DNA, since it is easier to express analyses on this basis than as intracellular molar concentration. However, some estimates in molar terms are available, and an example is given in Table 25–2. Deoxyribonucleotide concentrations in nondividing cells, such as erythrocytes, unstimulated lymphocytes, or cultured cells blocked in G_1, are low but significant. dTTP is frequently present at the highest concentration (although not in Table 25–2), and dGTP is usually lowest in hamster or mouse cells, whereas dCTP is as low as dGTP, or lower, in human cell lines. *As cells enter the S phase, there is an increase in the amount of each deoxyribonucleoside triphosphate in the cell*, the increase being greater for dCTP and dGTP. *There is also a marked change in the relative distribution of nucleotides between the nu-*

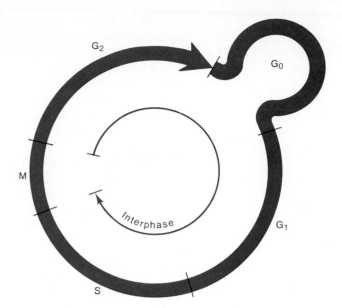

Figure 25–23

Phases in the life cycle of a typical eukaryotic cell. The cell cycle is divided into the resting stage (G_0), the prereplication stage (G_1), the synthesis or replication stage (S), the mitotic stage (M), and the postreplication stage (G_2).

cleus and the cytoplasm (Table 25–2), with a much greater increase in concentrations in the nucleus. This difference is particularly marked in the case of dTTP, which moves almost entirely into the nucleus by the end of the S phase.

In prokaryotes, the levels of deoxyribonucleotides vary from undetectable to 200 picomoles per 10^6 cells (compared with 3 to 30 picomoles of dATP, dCTP, and dGTP per 10^6 eukaryotic cells). As in eukaryotic cells, dTTP is usually present at higher concentrations than the other deoxyribonucleoside triphosphates.

The arrest of cell growth by many agents is associated with depletion of one or more of the deoxyribonucleotide pools. Thus thymidine at millimolar concentrations arrests cell growth and decreases the concentration of dCTP dramatically, whereas the concentrations of dATP, dGTP, and especially dTTP increase. This effect is considered to be mediated by the allosteric inhibition of ribonucleotide reductase, referred to earlier. Hydroxyurea, another agent that arrests cell growth by blocking DNA synthesis, depletes the pools of dATP and dGTP, and it is the effect on the latter, also brought about by ribonucleotide reductase inhibition, that is probably critical.

TABLE 25–2

Nuclear and Cytoplasmic Concentrations of Deoxyribonucleoside Triphosphates in Chinese Hamster Ovary Cells in the G_1 and S (DNA-Synthesizing) Phases of the Cell Cycle

	In G_1 Phase			In S Phase		
	Nuclear Concentration (μM)	Cytoplasmic Concentration (μM)	Total Amount (pmol/ μg DNA)	Nuclear Concentration (μM)	Cytoplasmic Concentration (μM)	Total Amount (pmol/ μg DNA)
dATP	6	12	3.2	40	19	3.4
dCTP	14	13	4.0	140	55	10.4
dGTP	1	0.7	0.16	10	2.5	0.76
dTTP	38	7	3.2	100	<1	3.1

Source: From B. Bjursell and L. Skoog, Control of nucleotide pools in mammalian cells, *Antibiotics Chemother.* 28:78, 1980.

Ribonucleotides are present in cells in much higher concentrations than deoxyribonucleotides, the general range of the concentration of the triphosphates in mammalian cells being 600 to 4000 picomoles per 10^6 cells. This concentration is equivalent to about 0.2 to 10 mM. The relative concentrations of the four triphosphates vary with conditions and the cell type, but ATP is normally highest in concentration. Levels of ribonucleotides are also perturbed by inhibitors. For example, they are increased by inhibitors of ribonucleotide reductase or decreased by inhibitors of the biosynthetic pathways.

There is some evidence that equilibration of ribonucleotides in the nuclei of mammalian cells with those in the cytoplasm occurs at different rates in different cell types. In most cultured cell lines it is very rapid, but in others (Novikoff hepatoma) it is quite slow, and in still others (HeLa) equilibration occurs at an intermediate rate. The rates were determined by examining the extent to which the incorporation into RNA of intracellular radiolabeled uridine is affected by putting the cells in a medium containing a high concentration of unlabeled uridine. The rapid influx of unlabeled uridine into the cytoplasm can decrease incorporation of labeled uridine in the nucleus only as cytoplasmic and nuclear nucleotides equilibrate.

T4 Bacteriophage Infection Stimulates the Nucleotide Metabolism

Infection of *E. coli* by T4 phage results in the induction of nearly 30 proteins. Many of these are enzymes that ensure an abundant supply of deoxynucleotides above those produced by the host. As a result, deoxynucleotide concentrations are increased many times. Phage-coded enzymes and related proteins include thioredoxin, ribonucleotide reductase, dihydrofolate reductase, dCMP deaminase, thymidylate synthase, and deoxyribonucleotide kinase. However, in some instances, the phage relies completely on host enzymes, which are normally present at high levels. Examples of such enzymes are adenylate kinase and nucleoside diphosphate kinase. Completely novel enzymes coded by the phage are endonucleases II and IV, which supply nucleotides directly by degrading host DNA.

The phage DNA contains no cytosine; instead hydroxymethylcytosine is incorporated. To accomplish this the phage induces enzymes that hydrolyze dCTP and dCDP, synthesize 5-hydroxymethyl dCMP from dCMP and methylenetetrahydrofolate, and phosphorylate hydroxymethyl dCMP. All these enzymes help to ensure rapid and specific synthesis of phage DNA while preventing synthesis of host DNA.

BIOSYNTHESIS OF NUCLEOTIDE COENZYMES

Many of the nucleotides considered thus far play important roles in metabolism and are discussed in other chapters. However, nucleotide coenzymes such as flavin nucleotides, NAD^+, $NADP^+$, and coenzyme A are also extremely important in metabolism. In the following, the pathway for the completion of each coenzyme is discussed.

Riboflavin, i.e., 7,8-dimethyl-10-(1'-D-ribityl)isoalloxazine, is synthesized by microorganisms such as the fungus *Eremothecium* and mutants of the yeast *Saccharomyces* in a pathway that starts from GTP. Riboflavin is an essential dietary constituent for mammals and is converted in the body to the mononucleotide or dinucleotide forms that function as the prosthetic groups of many enzymes. Riboflavin is converted to riboflavin-5'-phosphate, more commonly called *flavin mononucleotide* (FMN), by flavokinase (ATP:riboflavin phosphotransferase), as shown in Figure 25–24. The en-

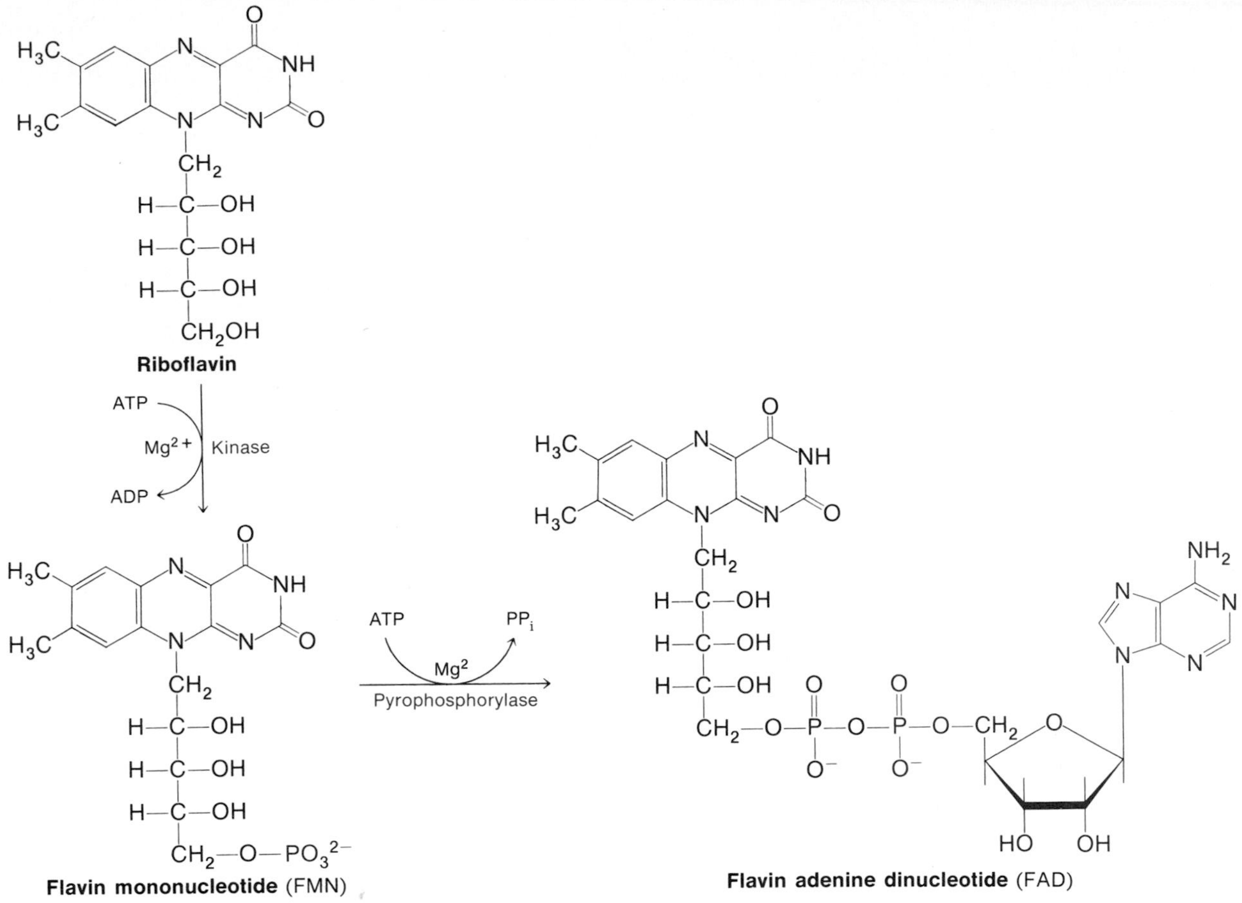

Figure 25–24

Biosynthesis of flavin mononucleotide (FMN) and flavin adenine dinucleotide (FAD) from riboflavin. In the first reaction a kinase transfers a single phosphate to the terminal hydroxyl of the ribose. In the second reaction the AMP moiety is transferred to the phosphate.

zyme has been purified from yeast, plants, and liver. It is also present in a variety of other animal tissues (kidney, brain, spleen, and heart).

The other nucleotide form of riboflavin, flavin adenine dinucleotide (FAD), is formed from FMN in a reversible reaction catalyzed by *flavin nucleotide pyrophosphorylase* (Figure 25–24). This enzyme is also widely distributed in nature and has been observed in plants, yeast, lactobacilli, and many animal tissues.

The nicotinamide moiety of the coenzymes nicotinamide adenine dinucleotide (NAD^+) and nicotinamide adenine dinucleotide phosphate ($NADP^+$) is synthesized by several routes. In liver and other animal tissues, tryptophan degradation forms, among other products, quinolinic acid (Chapter 19), which is converted to nicotinate mononucleotide (deamidonicotinamide mononucleotide, deamido-NMN) by quinolinate phosphoribosyltransferase (Figure 25–25). In the cytosol of cells of many mammalian tissues, and in yeast and other microorganisms, there is present a nicotinate phosphoribosyltransferase that also forms deamido-NMN (Figure 25–25). A very similar phosphoribosyltransferase present in the cytosol of all animal tissues investigated acts on nicotinamide. These transferases are responsible for utilization of nicotinate and nicotinamide in the diet. The role of ATP in these reactions is unclear. Some transferases do not require it,

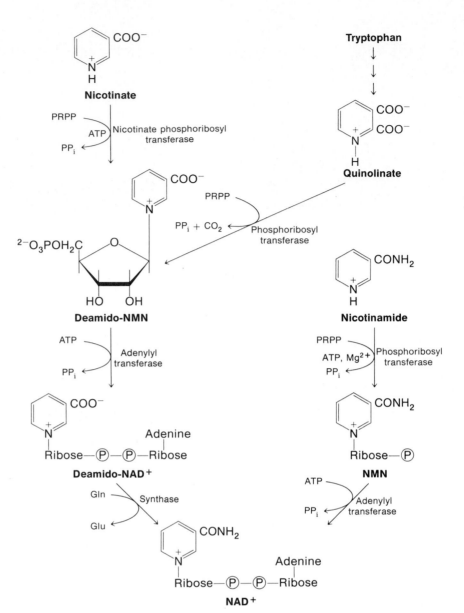

Figure 25–25

Biosynthesis of NAD$^+$. In animal tissues tryptophan degradation leads to quinolinate, which is converted to deamido-nicotinamide mononucleotide (deamido-NMN). Deamido-NMN can also be formed from nicotinate in some mammalian tissues and in many microorganisms. The deamido-NMN is subsequently converted into NAD$^+$. Another route to NAD$^+$ starts from nicotinamide. The phosphoribosyltransferase required for this pathway is found in the cytosol of animal tissues. These transferases are responsible for utilization of nicotinate and nicotinamide in the diet.

for others it seems to be an allosteric regulator, and in yet other cases ATP seems to be hydrolyzed to yield ADP and P$_i$ in equimolar amounts with deamido-NMN formation.

The mononucleotides so formed are converted to the corresponding dinucleotides by NMN adenyltransferase (Figure 25–25). In mammalian cells it appears to be a single enzyme that catalyzes both reactions, but an adenylyltransferase acting only on NMN has been isolated from some bacteria (*Lactobacillus fructosus*). A cytoplasmic NAD$^+$ synthase present in yeast, liver, and other tissues transfers the amino group from glutamine at the expense of ATP hydrolysis (Figure 25–25).

A cytoplasmic kinase present in liver, mammary gland, and brain is responsible for the formation of NADP$^+$ from NAD$^+$:

$$NAD^+ + ATP \longrightarrow NADP^+ + ADP$$

NADH is not a substrate and inhibits competitively with respect to NAD$^+$.

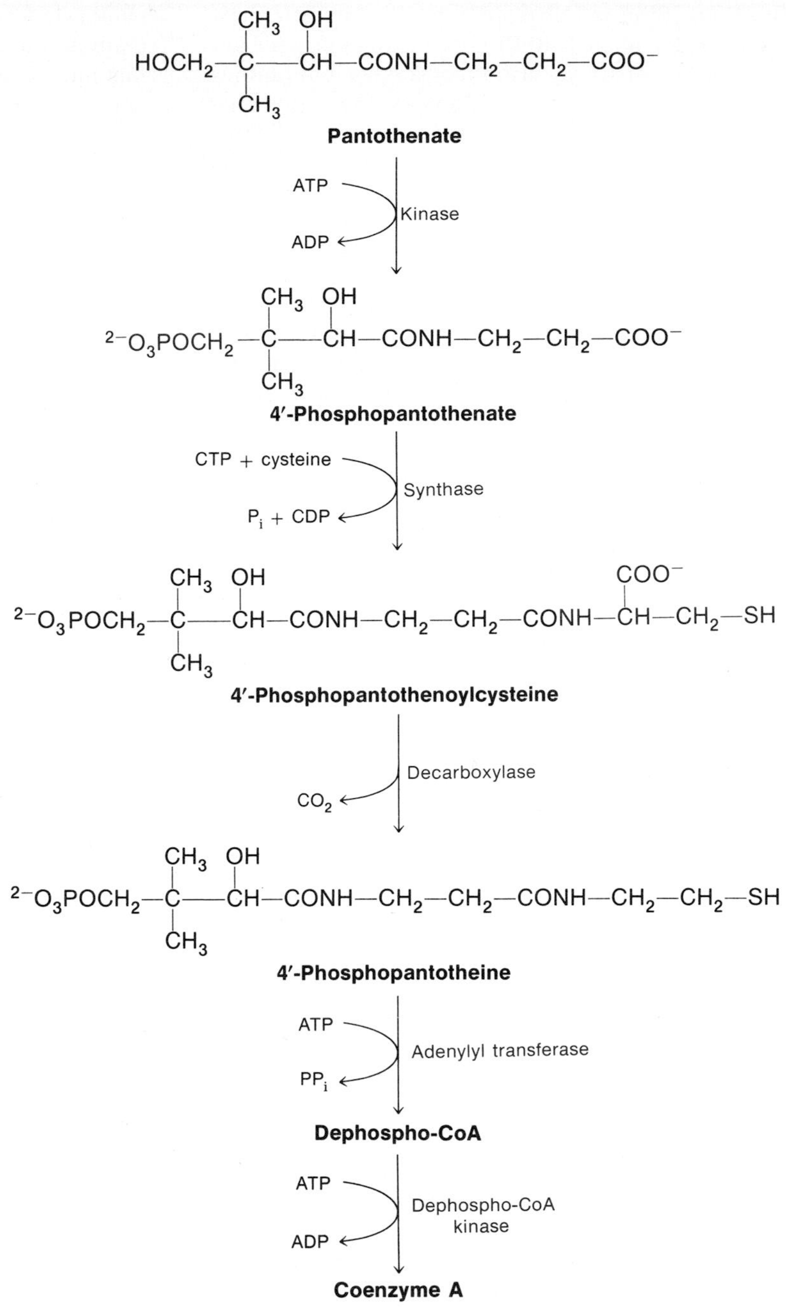

$$\text{HOCH}_2-\overset{\overset{\displaystyle CH_3}{|}}{\underset{\underset{\displaystyle CH_3}{|}}{C}}-\overset{\overset{\displaystyle OH}{|}}{CH}-\text{CONH}-\text{CH}_2-\text{CH}_2-\text{COO}^-$$

Pantothenate

ATP → Kinase
ADP ←

$$^{2-}\text{O}_3\text{POCH}_2-\overset{\overset{\displaystyle CH_3}{|}}{\underset{\underset{\displaystyle CH_3}{|}}{C}}-\overset{\overset{\displaystyle OH}{|}}{CH}-\text{CONH}-\text{CH}_2-\text{CH}_2-\text{COO}^-$$

4'-Phosphopantothenate

CTP + cysteine → Synthase
P_i + CDP ←

$$^{2-}\text{O}_3\text{POCH}_2-\overset{\overset{\displaystyle CH_3}{|}}{\underset{\underset{\displaystyle CH_3}{|}}{C}}-\overset{\overset{\displaystyle OH}{|}}{CH}-\text{CONH}-\text{CH}_2-\text{CH}_2-\text{CONH}-\overset{\overset{\displaystyle COO^-}{|}}{CH}-\text{CH}_2-\text{SH}$$

4'-Phosphopantothenoylcysteine

Decarboxylase
CO_2 ←

$$^{2-}\text{O}_3\text{POCH}_2-\overset{\overset{\displaystyle CH_3}{|}}{\underset{\underset{\displaystyle CH_3}{|}}{C}}-\overset{\overset{\displaystyle OH}{|}}{CH}-\text{CONH}-\text{CH}_2-\text{CH}_2-\text{CONH}-\text{CH}_2-\text{CH}_2-\text{SH}$$

4'-Phosphopantotheine

ATP → Adenylyl transferase
PP_i ←

Dephospho-CoA

ATP → Dephospho-CoA kinase
ADP ←

Coenzyme A

Figure 25–26
Biosynthesis of coenzyme A from pantothenate. This synthesis occurs in the mammalian liver. Pantothenate must be supplied in the diet. Color indicates the groups introduced at the kinase and synthase steps.

Coenzyme A is synthesized in the mammalian liver from pantothenic acid (pantoyl-β-alanine), which is required in the mammalian diet. The five steps in the synthesis are shown in Figure 25–26. In the last step, a specific kinase transfers a phosphoryl group to the 3'-hydroxyl of the adenylate portion of the molecule.

SUMMARY

Nucleotides are the building blocks for nucleic acids; they are also involved in a wide variety of metabolic processes. They serve as the carriers of high-energy phosphate and as the precursors of several coenzymes and regulatory small molecules. Nucleotides can be synthesized de novo from small-molecule precursors or, through salvage pathways, from the partial breakdown products of nucleic acids.

The ribose for nucleotide synthesis comes from glucose, either by means of the pentose phosphate pathway or from glycolytic intermediates through transketolase-transaldolase reactions. Ribose-5-phosphate is converted to phosphoribosylpyrophosphate (PRPP), the starting point for purine synthesis. This pathway also incorporates into purines atoms from glycine, aspartate, glutamate, CO_2, and one-carbon fragments carried by folates. IMP synthesized by this route is converted by two-step pathways to AMP and GMP, respectively.

The biosynthetic pathway to UMP starts from carbamoyl phosphate and results in the synthesis of the pyrimidine orotate, to which ribose phosphate is subsequently attached. CTP is subsequently formed from UTP. Deoxyribonucleotides are formed by reduction of ribonucleotides (diphosphates in most cells). Thymidylate is formed from dUMP.

All biosynthetic pathways are under regulatory control by key allosteric enzymes that are influenced by the end products of the pathways. For example, the first step in the pathway for purine biosynthesis is inhibited in a concerted fashion by the adenine and guanine nucleotides. Either of these nucleotides by itself inhibits the conversion of inosine monophosphate (IMP) into nucleotide end product. On the other hand, adenine nucleotide by itself stimulates the conversion of IMP into GMP.

There are several inhibitors known for nucleotide biosynthesis. Each of these inhibitors is extremely toxic, especially to rapidly growing cells, where the need for continuous nucleic acid synthesis is most strongly felt. In limited amounts, some of the inhibitors have chemotherapeutic value in the treatment of cancer and other illnesses.

Nucleic acids and nucleotides are degraded to nucleosides or free bases before they are ingested. Nucleotides or their partial degradation products may be reutilized for nucleic acid synthesis, or they may be further catabolized for excretion or for use in the synthesis of other products. Purine nucleotides are degraded via guanine, hypoxanthine, and xanthine to uric acid, which in some species is degraded further before excretion. Inherited deficiencies in some of the enzymes involved in nucleotide degradation and salvage cause severe impairment of health, a fact testifying to the importance of the degradative pathways. Pyrimidine nucleotides are degraded via uridine and uracil to simpler substances such as β-alanine.

In addition to the nucleotides used as substrates in nucleic acid synthesis, there are a number of other nucleotide-containing molecules that serve various purposes in the cell. These include the coenzymes NAD^+, $NADP^+$, FAD, and CoA.

SELECTED READINGS

Anderson, E. P., Enzyme pattern-targeted chemotherapy. *Adv. Enzyme Regul.* 24:118, 1985. This volume contains several chapters about inhibitors of nucleotide metabolism and their mechanism of actions.

Becker, M. A., K. D. Raivio, and J. E. Seegmiller, Synthesis of phosphoribosylpyrophosphate in mammalian cells. *Adv. Enzymol.* 49:281, 1979. Review of information about PRPP synthetase and its regulation—a key element in nucleotide regulation.

Foster, J. W., and A. G. Moat, Nicotinamide adenine dinucleotide biosynthesis and pyrimidine nucleotide cycle metabolism in microbial systems. *Microbiol. Rev.* 44:83–105, 1980. Comprehensive review.

Hoffee, P. A., and M. E. Jones (eds.), *Methods in Enzymology*, Vol. 51. *Purine and Pyrimidine Nucleotide Metabolism*. New York: Academic Press, Inc. 1978. Contains short summaries of information about most enzymes of the pathways, with references and details of preparation and assay.

Jones, M. E., Pyrimidine nucleotide biosynthesis in animals: Genes, enzymes and regulation of UMP biosynthesis. *Ann. Rev. Biochem.* 49:253, 1980. Authoritative outline of the regulatory properties of the two multifunctional proteins responsible for pyrimidine nucleotide synthesis in animals.

Reichard, P., and A. Ehrenberg, Ribonucleotide reductase—a radical enzyme. *Science* 221:514, 1983. Short review on the radical mechanism of reaction.

Stadel, J. M., A. D. Lean, and R. J. Lefkowitz, Molecular mechanisms of coupling in hormone receptor-adenylate cyclase systems. *Adv. Enzymol.* 53:1–43, 1982. A current account by a major contributing group, emphasizing the more biochemical aspects of this important system.

Strauss, P. R., J. F. Henderson, and M. G. Goodman, Nucleosides and lymphocytes—an overview. *Proc. Soc. Exp. Biol. Med.* 179:413–418, 1985. Excellent short summary introducing a series of papers on related topics in this volume.

Thelander, L., and P. Reichard, Reduction of ribonucleotides. *Ann. Rev. Biochem.* 48:133, 1979. The most recent review on this topic by authors who have been the major contributors.

PROBLEMS

1. Compare the pathways by which purine and pyrimidine nucleotides are synthesized *de novo*. How are the differences between these pathways reflected in the choice of metabolites regulating each?

2. Which atoms of nucleotide bases isolated from a hydrolysate of DNA would be labeled by the following precursors?
   a. [3-^{14}C]Serine.

b. [^{15}N]Serine.

c. [^{15}N]Aspartic acid.

d. [2-^{14}C]Glucose.

e. [^{14}C]CO_2.

3. Compare the number of high-energy phosphate bonds required for the synthesis of GTP and the synthesis of CTP. Assume that PRPP and folate one-carbon derivatives are available.

4. The pathway for *de novo* synthesis of dTTP is more complex than that for the other deoxyribonucleotides. Illustrate this difference by reference to known inhibitors of DNA synthesis.

5. Some of the inhibitors referred to in Problem 4 have been used as anticancer drugs. Would you predict that an inhibitor of thymidine kinase would be effective as an anticancer treatment? Would you expect the same to be true of an inhibitor of deoxycytidine kinase?

6. From what you know of the genetic defect responsible for Lesch-Nyhan syndrome, would you expect the symptoms of individuals suffering from this disorder to include anemia? If so, why?

7. An analog of 2'-deoxyadenosine is extremely cytotoxic to human cells in culture.

a. Suggest a possible mechanism involving nucleotide metabolism that might be responsible for the cytotoxicity.

b. How would you obtain data to support the proposed mechanism?

c. Deoxycoformycin, an inhibitor of adenosine deaminase, increases the cytotoxicity of deoxyadenosine but not of the analog. Explain.

8. Allopurinol administered together with 6-mercaptopurine under certain conditions enhances the anticancer effectiveness of the latter. How can the known site of action of allopurinol explain this?

9. A pig liver multienzyme protein preparation was incubated with 22 μM 5,10-methylenetetrahydrofolate and 36 μM NADP$^+$. After 3 min, the amount of 5,10-methenyltetrahydrofolate formed was 1.0 nmol and the amount of 10-formyltetrahydrofolate was 2.5 nmol. Over this interval, the formation of both compounds proceeded at a constant rate. What does this indicate about the kinetic behavior of the multienzyme protein? What result would be expected in the absence of this phenomenon?

10. A rat was injected subcutaneously with 13.5 μmol of cytidine labeled in all carbons with ^{14}C. The specific radioactivity of the base moiety of this cytidine was 144 cpm/μmol, while the pentose moiety had 125 cpm/μmol. Twenty hours after the injection the animal was killed, and the DNA and RNA from the tissues were converted to nucleosides. Specific activities of the base (B) and pentose (P) moieties of nucleosides were as follows: cytidine: B,55; P,40; deoxycytidine: B,30; P,22. What conclusion can be drawn about deoxyribonucleotide synthesis in the rat? Do the results exclude the possibility of a pathway not involving ribose compounds?

V

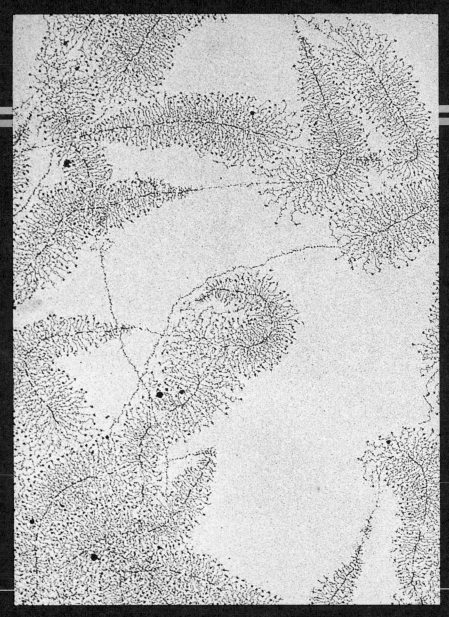

Electron micrograph of nucleolar DNA showing a cluster of ribosomal RNA genes and growing RNA chains. (From O. L. Mller, Jr., and B. R. Beatty, Portrait of a Gene, *J. Cell. Physiol.* 74(Suppl. 1):225, 1969.)

NUCLEIC ACID AND PROTEIN METABOLISM

PART V

Growth and reproduction are two of the main characteristics associated with all biological systems. The cellular architecture involves tens of thousands of molecules with precise structures. These molecules interact to form larger aggregates, which also have precise structures. Overall, then, an enormous amount of information is required to replicate cellular structures. In Part V we are concerned with the storage, transmission, and utilization of this biochemical information.

Biochemical information is stored in the genes. Each gene consists of a unique linear sequence of the four bases that are found in a single strand of duplex DNA. (From an informational point of view, the two complementary strands in a DNA duplex are identical.) Growth resulting from cell division is preceded by DNA replication and a segregation process that ensures that each daughter cell will receive an identical complement of the information-bearing DNA.

The biochemistry of DNA replication is discussed in Chapter 26. Other metabolic processes involving DNA degradation and repair of DNA damage are also discussed in Chapter 26, as are the chemical synthesis and the sequence analysis of DNA.

To make use of the sequence information stored in the DNA, the information is first transcribed into RNA chains, and polypeptide chains are then synthesized from these templates. Transcription and other metabolism involving RNA are discussed in Chapter 26. The second process, polypeptide synthesis, is called translation. This is because nucleic acids and proteins use two different languages. The nucleic acid language has four "letters," namely, the four commonly occurring nucleotides. The protein language has twenty "letters," which are the twenty commonly occurring amino acids found in most proteins. In the synthesis of a polypeptide chain, the nucleic acid language must be translated into the language of the polypeptide chain. Translation and other aspects of protein metabolism are discussed in Chapter 27.

Beyond the gross need for RNA and proteins, there is the need to regulate production so that the amounts of specific gene products made in any particular cell are consistent with the metabolic needs of the cell. Mechanisms designed to regulate transcription and translation are described for prokaryotes in Chapter 29 and for eukaryotes in Chapter 30.

In multicellular organisms, cells of different types must produce substances for the entire organism, not just for themselves. To regulate the types and amounts of substances produced, some form of intercellular communication is needed. For most multicellular organisms this need is fulfilled by hormones that pass between cells as intercellular messengers. These messengers serve to coordinate the metabolic activities of various tissues. The behavior of hormones, which has been mentioned at many points in this text, is the focal point of Chapter 31.

In Chapter 32 we describe some of the ways in which biochemists can now manipulate genes. Such techniques have greatly increased the possible gene combinations available for fundamental studies and commercial purposes.

The final chapter of Part V, Chapter 33, deals with the rapidly developing subject of the origin of life. Our ever-increasing knowledge of biological systems makes it meaningful to ask how life began. As we will see, there is general agreement that nucleic acids in some form must have preceded the first self-replicating systems.

26

DNA METABOLISM

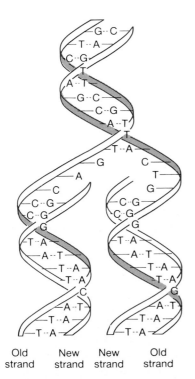

Figure 26–1
Watson-Crick model for DNA replication. The double helix unwinds at one end. New strand synthesis begins by absorption of mononucleotides to complementary bases on the old strands. These ordered nucleotides are then covalently linked into a polynucleotide chain, a process resulting ultimately in two daughter DNA duplexes.

Old strand New strand New strand Old strand

DNA is the central topic of discussion in three chapters of this text. In Chapter 7 we introduced DNA by briefly discussing its biological significance as the information-bearing component of the chromosome. Chapter 7 then focused on the structure and chemistry of DNA. In this chapter we describe the metabolic behavior of DNA, and in Chapter 32 we will discuss how DNA can be manipulated so that genes or parts of genes may be analyzed and used in various ways.

DNA metabolism is a multifaceted subject. Our emphasis in this chapter will be on DNA replication. Other topics that we will consider are DNA degradation, postreplicative modification, repair, and recombination.

DNA REPLICATES SEMICONSERVATIVELY

The transfer of genetic information from one generation to the next is one of the foremost concerns of geneticists. The key to understanding the transfer process was provided by the discovery of the complementary duplex nature of DNA. From the structure of DNA alone, the overall mechanism of genetic information transfer became clear. Watson and Crick articulated this mechanism in their model of DNA replication (Figure 26–1). First, the double helix unwinds, then mononucleotides pair to complementary sites on each polynucleotide strand, and finally these mononucleotides become linked to yield two identical daughter DNA duplexes. This simple yet elegant explanation has been supported by subsequent research.

DNA replication as described by the Watson-Crick model is called _semiconservative_ because each of the daughter duplexes arising from replication contains one old (conserved) strand and one new strand. Historically, two other models for replication were considered. These are known as the _conservative_ and the _dispersive_ models for replication (Figure 26–2). In the conservative model, the original parental duplex chains remain together, while the other duplex contains two newly synthesized strands. In the dis-

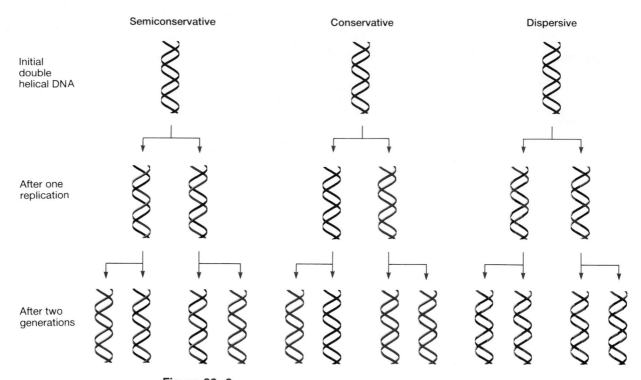

Figure 26–2
Three models for DNA replication: semiconservative, conservative, and dispersive. These three models predict different distributions of new and old DNA following replication.

persive model for replication, a mixture of new and old segments is found in both strands of the daughter duplexes.

Matthew Meselson and Franklin Stahl conceived of a way to determine which of these three modes of replication is actually correct. Their approach made use of isotopes that would result in DNAs of altered densities after replication. For this purpose *E. coli* cells were grown for several generations on a defined medium in which all the nitrogen was the heavy ^{15}N isotope (normal nitrogen is ^{14}N). As a result, the DNA in the descendant cells had a greater than normal density, since the ^{15}N became incorporated into the bases of the DNA during replication. Next, the bacteria were transferred to growth medium containing normal ^{14}N-nitrogen, and the cells were allowed to go through one or more doublings. The DNA from these cells was isolated and analyzed by CsCl density-gradient centrifugation (see Box 7–B). Pure ^{15}N-DNA produces a single band of DNA (Figure 26–3, frame 1). The same is true for ^{14}N-DNA (Figure 26–3, frame 2). The only difference is that denser DNA produces a band farther down the centrifuge tube. Thus the location of the DNA in the tube made it possible to monitor the density of the DNA. When cells containing pure ^{15}N-DNA were allowed to grow in ^{14}N medium for precisely one generation time, the only band visible in the isolated DNA was that corresponding to ^{15}N-^{14}N hybrid DNA (Figure 26–3, frame 4).

Those data argue strongly against a conservative mode of replication, but they do not discriminate between the semiconservative and the dispersive modes of replication. For this purpose we must consider the results when cells continue to replicate for additional generations. Only in the semiconservative mode would we expect to find equal amounts of DNA with the hybrid density and the light density after two generations (see Figure 26–2). This was the observed result, as shown in Figure 26–3, frame 5. In the third generation we still see a band with the hybrid density, but the amount of

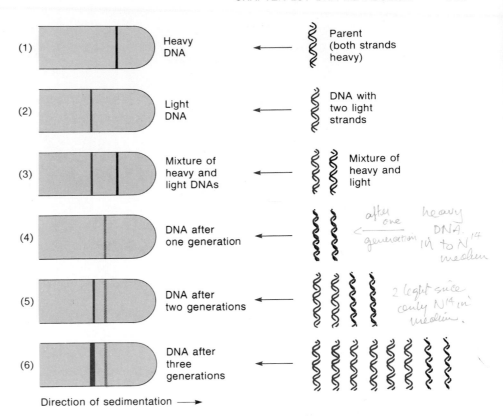

Figure 26–3
The Meselson-Stahl experiment demonstrating semiconservative replication for *E. coli* chromosomal DNA. CsCl density-gradient centrifugation is used to discriminate between DNAs of different densities. *E. coli* DNA has different densities when cells are grown in ^{14}N or ^{15}N medium (frames 1 and 2). When cells containing pure heavy DNA (^{15}N-^{15}N DNA) are grown in ^{14}N medium for one generation, all of the DNA is of intermediate density (^{14}N-^{15}N). After two generations of growth in ^{14}N medium, the cells contain equal amounts of light (^{14}N-^{14}N) DNA and intermediate density DNA (frame 5). In subsequent generations the hybrid DNA reappears in constant amounts, but the amount of light DNA increases. These results support the model of a semiconservative mode of DNA replication.

(1) Heavy DNA ← Parent (both strands heavy)

(2) Light DNA ← DNA with two light strands

(3) Mixture of heavy and light DNAs ← Mixture of heavy and light

(4) DNA after one generation ←

(5) DNA after two generations ←

(6) DNA after three generations ←

Direction of sedimentation ⟶

^{14}N-^{14}N pure light DNA has increased (Figure 26–3, frame 6). Those results strongly support the semiconservative mode for DNA replication.

Similar experiments have been performed on mammalian cells grown in tissue culture, using bromouracil as a density label. Bromouracil contains a bromine atom instead of a methyl group on the 5 position of thymine. Bromouracil, when incorporated into DNA, can substitute for most of the thymidine, leading to DNA with a substantially higher than normal density. The results with eukaryotic DNA were found to parallel those in *E. coli*. Thus it appears that semiconservative replication of cellular DNA is a general phenomenon.

Taylor, Woods, and Hughes devised another method for labeling DNA to follow its replication. Chromosomal DNA was selectively labeled by the incorporation of tritiated (^3H-labeled) thymidine, and the distribution of the radioactive label was visualized directly in autoradiographs of chromosomes (Figure 26–4*a*); the pattern of labeling was also observed after the cells subsequently replicated in the absence of tritiated thymidine (Figure 26–4*b*). Bean seedlings were used for those experiments because cell division is very rapid in their growth tips. A sufficient period of time was allowed for some of the cells in the seedlings to undergo one round of DNA duplication (incorporating the radioactive label) and cell division. After this, the seedlings were transferred to a fresh solution containing colchicine with the ^3H label. Colchicine is a plant alkaloid that inhibits normal mitosis by interacting with microtubule proteins necessary for mitotic spindle formation; it does not inhibit DNA synthesis or the replication of chromosomes but delays the formation of daughter cells by inhibiting chromatid segregation. The advantage of colchicine treatment in this experiment is that in tissue so treated many cells become arrested in the state where the sister chromatids are paired. Cells that were examined shortly after the transfer from the [^3H] thymidine medium were examined at mitosis, and all the chromosomes ap-

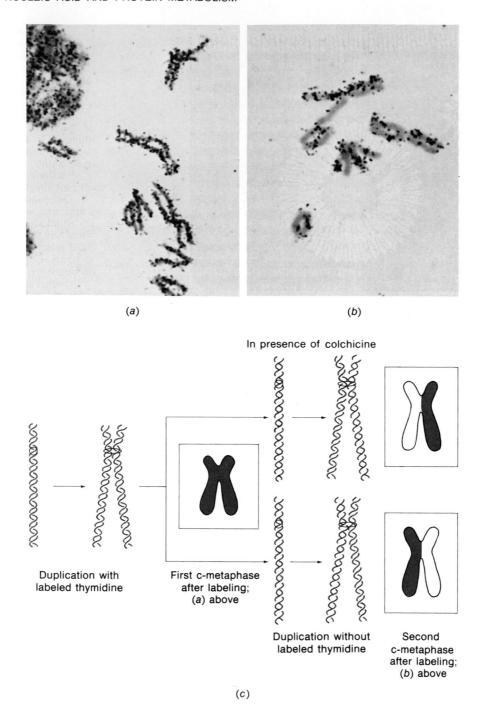

In presence of colchicine

Duplication with
labeled thymidine

First c-metaphase
after labeling;
(a) above

Duplication without
labeled thymidine

Second
c-metaphase
after labeling;
(b) above

(a)   *(b)*

(c)

Figure 26–4

Autoradiographs of *Vicia faba* chromosomes labeled with [³H]thymidine. The labeled thymidine becomes incorporated into the chromosomal DNA. A suitably labeled preparation is flattened and subject to film exposure. Small dots indicate radioactive disintegration in the exposed film. *(a)* The first metaphase after replication in the presence of [³H]thymidine. *(b)* The second metaphase after an additional replication in nonradioactive medium. *(c)* A diagrammatic interpretation of the results shown in *(a)* and *(b)*. Radioactive single strands of DNA are shown in color. Radioactive chromatids at metaphase are also indicated in color. Colchicine has been used to inhibit spindle fiber formation and thus the anaphase separation of sister chromatids. Under these "C-metaphase" conditions, separation of sister chromatids is delayed. In *(a)* both sister chromatids are labeled uniformly. In *(b)* the sister chromatids are not labeled uniformly. The large chromosome at the top has one chromatid labeled and one virtually unlabeled. The homolog to its right has two exchanges (a labeled segment moved into the lower chromatid). The two small chromosomes to the lower left of it are lying one on top of the other. Both have one sister chromatid exchange. The small chromosome to the upper left has one sister chromatid exchange and the one at the lower left is lightly labeled but probably has two exchanges. (Autoradiographs generously donated by J. H. Taylor.)

peared to be labeled uniformly (Figure 26–4a). When the cells were allowed to duplicate their chromosomes in unlabeled medium, only one of the two chromatids in each chromosome pair was labeled (Figure 26–4b). That is exactly the result that would be expected if each chromatid is composed of a single linear duplex of DNA that replicates semiconservatively (Figure 26–4c). Similar results have been obtained for other eukaryotic cell types.

The universality of the semiconservative mode of DNA replication follows from the complementary nature of the DNA duplex. As the duplex unwinds, it presents two templates for the binding of complementary nucleotides. Subsequently, polymerization results in two duplexes, each with one old strand and one new strand. Thus the genetic information contained in the base sequence is directly transferred from one generation of DNA to the next by the capacity of DNA single strands to serve as templates for the assembly of complementary mononucleotides.

DNA SYNTHESIS IN PROKARYOTES

More is known about the synthesis of DNA in *E. coli* than in any other cell. The *E. coli* bacterium contains a single circular chromosome with about 3×10^6 base pairs. Most other bacteria have chromosomes of about the same size and, in some cases, have been found to be circular as well. The *E. coli* chromosome encompasses about 2000 genes, each one represented by a unique sequence of bases and encoding the genetic information to direct the synthesis of a specific RNA molecule. Each gene is located at a defined point on the chromosome that can be determined by genetic mapping. The number of genes required for DNA synthesis is probably between 20 and 50. In Figure 26–5 the locations of some of these genes are indicated, together with the locations of unique initiation (*oriC*) and termination points for DNA replication. Under optimal growth conditions the *E. coli* chromosome is replicated in 20–30 min.

Bacterial DNA Replication Is Bidirectional

Replication of the *E. coli* chromosome can be visualized by autoradiography of intact ³H-labeled chromosomes, using a gentle isolation technique developed by Cairns and Davern. After completing one round of replication in

Figure 26–5
A diagram of *E. coli* chromosome. The origin and approximate region of termination of replication are indicated. Locations of some genes involved in DNA replication are also indicated. The functions of many of these genes are described in Tables 26–1 and 26–2.

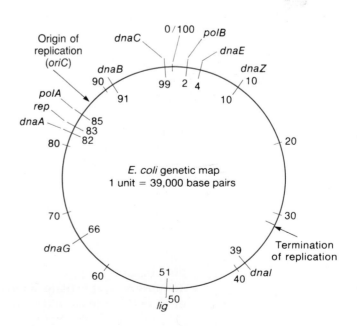

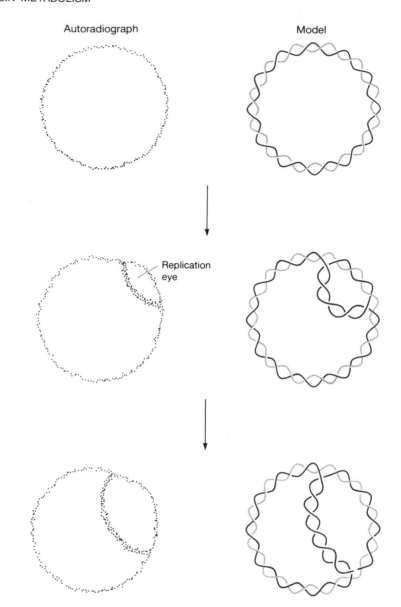

Autoradiograph   Model

Replication eye

Figure 26–6
Simulated autoradiographs of the
E. coli chromosome after one or more
replications in the presence of
[³H]thymidine. After one round of
replication the autoradiograph shows a
circular structure that is uniformly
labeled. The second round of
replication begins with the formation
of a replication eye. One branch in the
replication eye is twice as strongly
labeled as the remainder of the
chromosome, indicating that this
branch contains two labeled strands.
This structure is consistent with
semiconservative replication for the
E. coli chromosome.

labeled medium, chromosomes show up as being circular and uniformly
labeled (Figure 26–6, top). Initiation of a second round of replication leads
to the formation of a replication eye (Figure 26–6, middle). As synthesis
proceeds in labeled medium, the size of the replication eye increases; the
replicating chromosome at this stage is referred to as a *theta structure* because it has the appearance of the Greek letter θ (Figure 26–6, bottom). Semiconservative replication is consistent with the density of the autoradio-
graphic tracks observed on parts of the chromosome after one and two
rounds of replication in [³H]thymidine.

It is reasonable to conclude that the replication eye contains two partially separated parental DNA strands that are base-paired with strands of
newly synthesized DNA. Not resolved by this type of cytologic observation
is the question of whether replication occurs in one direction or both directions about the origin of replication. If growth is unidirectional we would
expect *one* growth point (Figure 26–7*a*), called a growth fork (or replication
fork); if growth is bidirectional we would expect *two* growth points or
growth forks (Figure 26–7*b*). Although examples of both types of replication
are observed, it is believed that most bacterial chromosomes, including that
of *E. coli*, replicate bidirectionally.

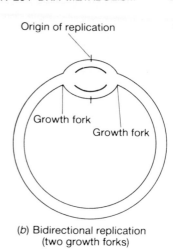

Figure 26–7

Schematic diagrams of two different modes of DNA synthesis at the growth fork(s). In unidirectional replication (a) there is one growth fork; in bidirectional replication (b) there are two. Color indicates regions containing newly synthesized DNA.

(a) Unidirectional replication (one growth fork)

(b) Bidirectional replication (two growth forks)

Convincing evidence of bidirectional replication in *E. coli* was obtained by measuring gene frequency during replication. Shortly after initiation, those genes near the origin of replication that have been duplicated must be present at twice the number of copies as those genes far removed from the origin. The numbers of genes can be assayed biologically by transformation (Chapter 7) or transduction. If replication is unidirectional, a steady increase in gene number should occur in one direction along the circular chromosome (Figure 26–8a). On the other hand, if replication is bidirectional, the increase in gene number should proceed in both directions from a common point, the origin (Figure 26–8b). Detailed measurements indicated that the latter situation is true for the chromosomes of *E. coli* and *Bacillus subtilis*. These measurements also made it possible to locate the initiation and termination points on the chromosomes (Figure 26–5).

The genetic evidence for bidirectional replication of the *E. coli* chromosome was supported by direct cytologic evidence. Bacteria were grown for a very short time in the presence of radioactive thymidine, and replicating chromosomes were examined by autoradiography; both of the forks in the replicating structures were intensely labeled (as in the colored portions of Figure 26–7b). This result shows that both forks must be active during replication, a finding consistent with bidirectional growth. Replication was also shown to be bidirectional in bacteriophage λ DNA, through electron microscopy. By analyzing circular λ molecules at different stages of their replication it was found that both ends of the replication eye moved relative to a single, unique cut made by a restriction endonuclease (the properties of restriction endonucleases are considered later in this chapter and more fully in Chapter 32).

Figure 26–8

Measurements of gene frequency make it possible to distinguish between unidirectional and bidirectional synthesis. (a) In unidirectional replication, the gene frequencies are highest for genes located on one side of the origin. (b) In bidirectional replication, the gene frequencies are equally high for regions symmetrically disposed about the origin.

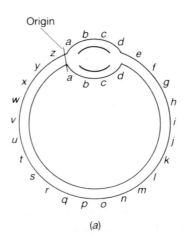

(a)

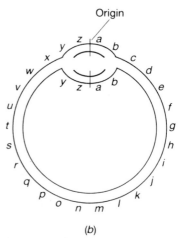

(b)

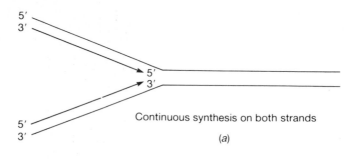

Continuous synthesis on both strands

(a)

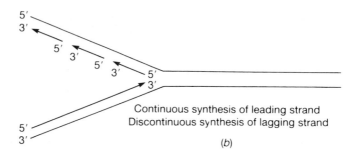

Continuous synthesis of leading strand
Discontinuous synthesis of lagging strand

(b)

Figure 26–9

Models for synthesis at the replication fork. *(a)* Continuous synthesis on both strands. Note that both growth arrows are pointing in the same direction, which would require growth in the 5′→3′ direction on one strand and in the 3′→5′ direction on the other strand. If growth occurs only in the 5′→3′ direction, synthesis would have to be discontinuous on one strand, as in *(b)*. Alternatively, it could be discontinuous on both strands *(c)*.

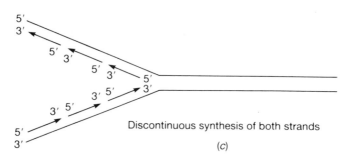

Discontinuous synthesis of both strands

(c)

Growth Is Discontinuous on at Least One Strand

Continuous synthesis on both strands of a replication fork would require synthesis in the 5′→3′ direction on one strand and in the 3′→5′ direction on the other strand because of the antiparallel nature of the DNA duplex (Figure 26–9a). That possibility seems unlikely because, as we will see, the only known enzymes that catalyze DNA synthesis do so in the 5′→3′ direction. For this reason replication was postulated to be discontinuous on one of the branches at the replication fork (Figure 26–9b). Careful electron microscopic examination of replication forks in bacterial viruses has in fact shown that transient gaps sometimes are apparent on one of the daughter DNA strands, close to the replication fork. Observations such as this led to the notion of a *leading strand* and a *lagging strand* (Figure 26–9b). In this model, 5′→3′ synthesis of the leading strand could occur continuously in the same direction as the unwinding of the replication fork. Synthesis of the lagging strand in the 5′→3′ direction could occur in discontinuous spurts in a direction *opposite* to DNA unwinding.

This concept of discontinuous synthesis was supported by Okazaki, who found that at least half the newly synthesized DNA is first made as small pieces (*Okazaki fragments*) that later become incorporated into large segments of DNA. Small replication fragments were detected by exposing growing cells to tritiated thymidine for a very short time (2 to 10 sec) (a

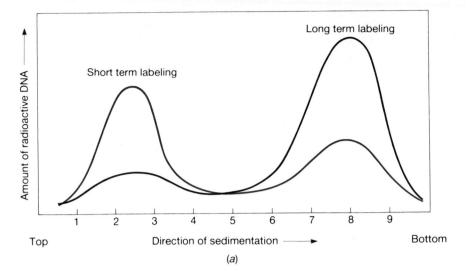

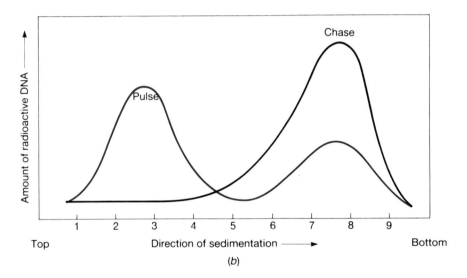

Figure 26–10

Sedimentation analysis of *E. coli* DNA from cells labeled with [³H]thymidine for different lengths of time. Sedimentation analysis is done on an alkaline sucrose gradient. The alkali denatures the DNA so that it becomes single-stranded. (*a*) Short-term labeling (2 to 10 sec) preferentially labels the most slowly sedimenting DNA, in the size range of 1000–2000 base pairs. In long-term labeling (1 to 2 min) most of the labeled DNA is of much higher molecular weight. (*b*) In a pulse labeling experiment, the short-term labeling is done in the same way as in (*a*). The long-term labeling includes the addition of a large amount of nonradioactive substrate after a short time. This added substrate is called a chase, as it lowers the specific radioactivity of the labeled substrate so that no further incorporation of radioactivity is detectable. Long-term labeling is done in order to follow the fate of the short-term-labeled DNA.

"pulse"), followed by rapid isolation of the radioactively labeled DNA. If replication was allowed to continue in unlabeled medium for several minutes (the "chase" period) most of the labeled DNA was found in much larger pieces of DNA (Figure 26–10). Sedimentation analysis in alkali (the alkali denatures the DNA into single strands) provided an estimated length of 1000–2000 bases for the bacterial Okazaki fragments. It was concluded that the replicating polymerase must operate by synthesizing short fragments on the lagging strand as new points for the initiation of replication are presented by the progressive unwinding of the double helix at the growth fork.

A closer examination of the Okazaki fragments revealed short stretches of ribonucleotides at their 5′ ends. From this and many other observations made with different *in vitro* systems, the researchers concluded that a new DNA chain can only be initiated by attaching the first deoxynucleotide through its 5′ phosphate to the 3′-OH of a short RNA chain. An RNA oligonucleotide that functions in this capacity is called a primer. Primers are synthesized *de novo* at various secondary priming sites along the chromosome, and they base-pair with the single-stranded template DNA in the regions where they are found. Whether these secondary initiation sites are specific or are chosen at random is not known.

A summary of the currently accepted mechanism for discontinuous synthesis is given in Figure 26–11. First, RNA primers are made on the single-

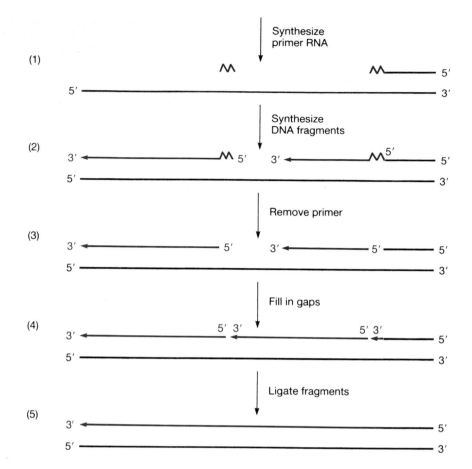

Figure 26–11
A model for discontinuous DNA synthesis. Synthesis occurs in a region that has been partially single-stranded. First, RNA primers are formed at various points on the single-stranded region (1). The DNA synthesis starts at the 3' ends of the primers (2). The primers are removed (3). Gaps between DNA fragments are filled in by further DNA synthesis (4). The fragments are ligated to make one long continuous piece of DNA (5). Newly synthesized RNA and DNA is indicated in color.

strand region of the template; then DNA is synthesized. Finally, the RNA is removed from the fragments and the gaps are filled and ligated. In *E. coli* the *dnaG* gene encodes a *primase* that generates primers from either ribonucleotide or deoxyribonucleotide precursors. The role of RNA polymerase in DNA synthesis is primarily in transcriptional activation of the DNA template (by exposing single-stranded regions). The primers that have been characterized in bacteria and virus SV40 replication *in vivo* are consistent with their being products of primases.

Many Proteins Are Required for DNA Replication

Many proteins are required for DNA replication in addition to the polymerizing enzyme. Table 26–1 lists subunits of *E. coli* polymerase III and Table 26–2 the other proteins implicated in DNA replication and their likely function. Figure 26–12a shows in a highly diagrammatic form how *E. coli* DNA polymerase III might be organized. But how is it possible to analyze such a complex situation? Historically, three general methods have been used for identifying and characterizing the proteins involved in DNA replication. They are purification, reconstitution, and mutation. Ideally, all three methods are used together in suitable organisms.

The first method, purification, involves isolating proteins with enzymatic activities that are logically related to the replication process, such as DNA polymerases and ligases. This is the classical biochemical approach and can be applied to any biological system. After an enzyme is isolated and characterized, several approaches may be used to demonstrate that the purified enzyme is active in the replication process *in vivo*. Sometimes this can be done by using inhibitors that act on both the purified protein and the

TABLE 26–1
Structural Genes, Subunits, and Molecular Weights of *E. coli* Polymerase III

Gene	Subunit	Size
dnaE	α	130,000
dnaZ	γ	52,000
dnaX	δ	32,000
dnaN	β	40,600
dnaQ (mutD)	ϵ	27,500
dnaZX	τ	71,000
—	θ	10,000

The *dnaZ* and *dnaX* genes are adjacent and together (*dnaZX*) have a sequence that can code for a protein with a molecular weight of 72,000. Subunit τ appears to be encoded by *dnaZX*. It is currently believed that *dnaX* encodes subunit δ. The gene encoding subunit θ has not yet been established.

TABLE 26–2
Other Proteins Involved in DNA Synthesis in _E. coli_

Gene	Function of Encoded Protein	Size (Daltons × 10³)
polA	In DNA polymerase I, extending Okazaki fragments and removing primers	109
dnaB	Interacts with primase (functions as hexamer)	50 monomer
dnaC	Stimulates priming	25 monomer
dnaG	Primase (acts as dimer)	60 monomer
dnaY	—	—
lig	Ligase	75
rep	Helicase, ATP-dependent	66
ssb	Single-strand DNA-binding protein	74
gyrA	Subunit of topoismerase II encoding GyrA (sensitive to naladixic acid)	100 monomer
gyrB	Subunit of topoisomerase II encoding GyrB (sensitive to coumermycin and novobiocin)	95 monomer

The collection of the _dnaB_-, _dnaC_-, and _dnaG_-encoded proteins, together with i, n, n″ and Y, constitute the primosome. Factor Y first recognizes a primosome assembly site, stimulating its DNA-dependent ATPase. The n protein then binds to the factor Y:DNA complex, with the subsequent addition of protein i, n″, and _dnaB⁻_ and _dnaC_-encoded proteins, which together constitute the _prepriming complex_. It can move along DNA in the 5′→3′ direction when primase is bound (forming the _primosome_), primers are synthesized at various locations on the DNA template.

Figure 26–12
The intact (holoenzyme) DNA polymerase III contains many subunits and is involved in the replication of the _E. coli_ chromosome. There are at least seven subunits. The catalytic core consists of α, ϵ, and θ; α is encoded by _dnaE_, ϵ by _dnaQ_ (carries out 3′→5′ proofreading); the function of θ is unknown. The other subunits and the genes encoding them are: β (_dnaN_), γ _dnaZ_) and δ (_dnaX_). The subunit γ dimerizes the polymerase III subunits; the dimer structure may be important to coordinate the synthesis of both the leading and lagging DNA strands at the replication fork. (Adapted from C. S. McHenry, _Mol. Cell. Biochem._ 66:71–85, 1985).

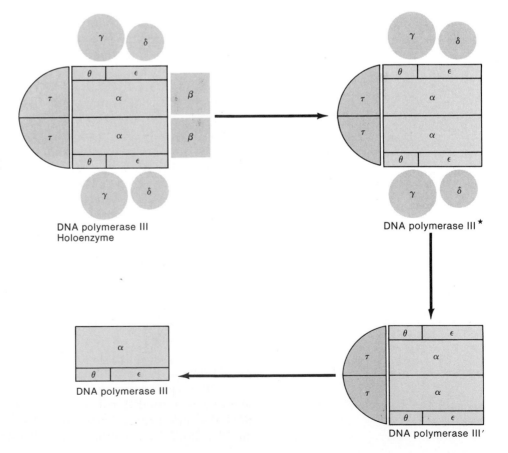

DNA polymerase III Holoenzyme

DNA polymerase III*

DNA polymerase III′

DNA polymerase III

cellular process. The concentration of inhibitor required to inhibit the purified enzyme *in vivo* should be approximately the same as that required to inhibit the purified enzyme *in vitro*. In prokaryotes, mutations in specific genes have been very useful for confirming the functions of isolated proteins in the replication process. The induction of a new enzyme activity associated with a biological process, such as virus infection or cell proliferation, also provides useful evidence.

A second method used to identify proteins needed for replication is reconstitution. Whole cell lysates containing all of the components necessary for replication are fractionated and the DNA replication system is then reconstituted with various combinations of the purified or partially purified proteins. Components of the replication system are recognized on the basis of their ability to restore overall activity *in vitro*. This procedure can be applied to any organism, even when relevant genetic mutants are not available. However, mutant studies are usually required for final confirmation that a protein carries out a particular function *in vivo*. We will see an example of such confirmation a little later, when we discuss the role of the DNA polymerase I.

The third method of analysis uses mutants as the primary tool. This approach is most effective with organisms having well-characterized means of carrying out genetic analysis and exchange. It requires the isolation of *conditionally lethal mutants*, i.e., mutants that grow normally under one set of conditions but fail to do so under another set of conditions. Most commonly, *temperature-sensitive* DNA replication mutants are used. Such mutants have been isolated in *E. coli* and they grow normally at a low temperature (25–30°C), called the permissive temperature, but poorly or not at all at a high temperature (~41°C), referred to as the nonpermissive temperature. Preliminary analysis of the temperature-sensitive step provides clues to the stage of replication affected.

For example, the length of time required for DNA synthesis to stop, after cells have been shifted from permissive to nonpermissive temperatures, can indicate whether the mutation occurs in a protein involved in initiation or elongation reactions of DNA synthesis. If the mutation is in the gene for a protein required for elongation, DNA synthesis will stop almost immediately at the nonpermissive temperature. If the mutation affects a protein required only for initiation of replication, DNA synthesis will continue for some time and stop when the round of replication in progress is completed.

In vitro assays are then used to aid in purifying the corresponding proteins from wild-type cells. Extracts from cells with the temperature-sensitive defect will not synthesize DNA at the elevated temperature, but activity can be restored by adding the corresponding protein from normal wild-type cells. This *complementation* test can be used as an assay for the purification of particular replication proteins. To prove that the correct protein has been purified from wild-type cells, the proteins from the temperature-sensitive mutant also must be purified and shown to be abnormal. Frequently they exhibit unusual thermolability.

An extension of the mutant approach is to identify genes that code for second-site suppressors of DNA synthesis conditional mutants. In one such class of revertants (from mutants selected at the nonpermissive temperature), the mutant genes suppress by protein–protein interaction—the process in which a defect in the initial mutant protein can be corrected by altering a protein with which the first protein interacts (both may be components of the DNA replication complex). If the mutant bearing a second-site suppressor has a conditional phenotype (e.g., cold sensitivity) of its own, so much the better, since that allows the suppressor gene to be mapped, cloned, and identified. By combining two different conditional mutants, each with a distinctive phenotype in the same cells, the dominance of one over the other in temperature shift experiments can sometimes tell us the order of function of the two genes.

A Polymerase with the Expected Characteristics Is Discovered. The first DNA-polymerizing enzyme to be discovered was found in cell-free extracts of *E. coli*. We shall examine this polymerase in some detail as an example of how biochemical and genetic studies can be used to determine the properties of an enzyme and its normal *in vivo* role. We shall then consider the other major classes of proteins involved in *E. coli* DNA replication.

The Watson-Crick proposal of a complementary duplex structure for DNA stimulated a search for a DNA polymerase enzyme with certain implied properties. The enzyme should require an intact DNA chain to serve as a template for the binding of complementary bases, and the *de novo* synthesized DNA should be complementary to a template DNA chain. Kornberg and his coworkers isolated from cells of *E. coli* a DNA-synthesizing enzyme that satisfied these requirements. They gave it the name DNA polymerase; it is now known as <u>DNA polymerase I</u> (Pol I) or the <u>Kornberg enzyme</u>.

In its purified state the Kornberg enzyme requires a DNA template, the four commonly occurring deoxynucleoside triphosphates, and Mg^{2+} ions for synthesizing DNA. The enzyme catalyzes the addition of mononucleotides to the 3'-OH end of a growing chain (Figure 26–13a). At the same time, the linkage between the α and β phosphates is broken, releasing an inorganic

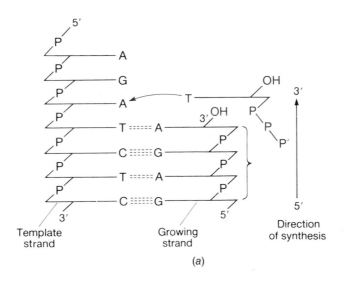

(a)

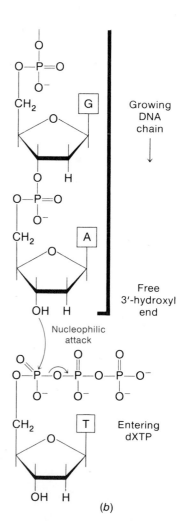

(b)

Figure 26–13

Template and growing strands of DNA. (a) Nucleotides are added one at a time to the 3'-OH end of the growing chain. Only residues that form Watson-Crick H-bonded base pairs with the template strand are added. (b) Covalent bond formation between the 3'-OH end of the growing chain and the 5' phosphate of the mononucleotide is accompanied by removal of the two terminal phosphates from the substrate nucleoside triphosphate.

pyrophosphate group (P—P). In fact, the energy provided by the cleavage drives the linking of the mononucleotide to the growing DNA chain. The bases added to the growing chain are determined by the sequence of bases in the DNA template. As nucleotides complementary to those on the template are added, the single-stranded DNA template is progressively converted to a double helix.

Physiochemical studies have shown that the enzyme has a complex surface, with specific attachment sites for the template chain, the growing chain, and monomer nucleoside triphosphate. The enzyme is highly selective, since it links to the growing chain only those nucleotides that form Watson-Crick base pairs with the template strand. As the new chain becomes lengthened by synthesis, the enzyme moves along the template one base at a time. That the newly synthesized DNA is a replica of the template strand has been shown in a number of ways—by gross base composition analysis, by melting curve profile of the hybrid formed between the template strand and the newly synthesized strand (see Chapter 7 for an explanation of nucleic acid melting curves), and finally, through a series of manipulations, by the *de novo* synthesis of genetically active DNA. A unique property of the enzymes that take part in the synthesis of DNA (and RNA) is that they require a template to function.

Phosphodiester bond formation probably occurs as a nucleophilic attack, by the 3′ hydroxyl group of the terminal mononucleotide residue at the growing end of the chain, on the α phosphorus atom of the entering nucleoside-5′-triphosphate. The attack causes displacement of the pyrophosphate group and formation of the internucleotide linkage (Figure 26–13b).

In addition to catalyzing the characteristic template-directed polymerization, DNA polymerase I is associated with an activity that can remove mononucleotides from the 3′ end of a polynucleotide strand. This second function is an _exonuclease_ activity; it leads to hydrolytic cleavage of the bond between the 5′ phosphate of the terminal residue and the 3′-OH group of the penultimate residue. Since this is the same linkage made during synthesis, the degradation reaction may be thought of as a reversal of polymerization. For net chain propagation to occur beyond the 3′-OH end of the DNA strand, the polymerization rate must exceed the depolymerization rate. Polymerization is much faster than depolymerization if the correct base-paired nucleotide is inserted in the growing chain. If a mismatched base is accidentally inserted, the opposite is true and the mismatched base is usually removed. The combination of polymerization-depolymerization reactions has been viewed as an "editing" or "proofreading" error-reducing mechanism in DNA synthesis. If the editing function is defective because of a mutation, a generalized increase in mutations ("mutator") occurs in cells harboring the altered nuclease.

Pol I has another associated activity, catalyzing the 5′→3′ degradation of DNA. Whereas the 3′→5′ nuclease activity of the enzyme is much more effective on unpaired or mispaired bases, the 5′→3′ activity preferentially cleaves base-paired regions. The ability of DNA Pol I to degrade DNA (or RNA) in the 5′→3′ direction, as well as to carry out a polymerization reaction, suggests a possible function in removing the RNA primer prior to gap filling (see Figure 26–11). In fact, RNA primer removal and gap filling may be catalyzed in rapid succession by the same DNA polymerase I molecule.

The question arises whether the three activities of Pol I just described all originate from a single active site or more than one active site. Current evidence favors the notion that the polymerization activity and the 3′→5′ nuclease activity are associated with one distinct site. Thus Pol I can be cleaved by the protease subtilisin to produce a small fragment ($M_r = 30,000$) with 5′→3′ nuclease activity and a large fragment ($M_r = 70,000$), called the

Klenow fragment, exhibiting the polymerization and $3' \rightarrow 5'$ depolymerization activities.

E. coli Has Three DNA Polymerases.   All of Kornberg's early work was done on DNA polymerase in cell-free systems. The fact that the enzyme had so many of the properties expected for a DNA-replicating enzyme led most observers to believe that it was indeed the DNA-replicating enzyme of _E. coli_. But to obtain final proof that an enzyme functions in a given capacity _in vivo_, it is essential to isolate mutants in which the enzyme's behavior has been altered. Our current understanding of the role of DNA polymerase I has come about by correlating _in vitro_ biochemical behavior with the behavior of mutants in which DNA replication and repair are affected.

From the time of the discovery of DNA polymerase I, about 20 years were needed to clarify its physiological function. A major step in that direction was taken by Cairns and De Lucia, who laboriously scanned several thousand colonies of heavily mutagenized _E. coli_ and found some that contained almost no DNA polymerase I polymerizing activity (1 to 2 per cent of normal). That mutant grew well under normal conditions, a fact suggesting that the DNA-polymerizing activity of the Kornberg enzyme was _not_ needed for replication. However, the mutant was more sensitive to ultraviolet irradiation than its wild-type parent. Ultraviolet irradiation was known to damage DNA. Therefore the increased sensitivity of the mutant suggested that, although DNA polymerase I was not the chromosome-replicating polymerase, it might be involved in repairing chromosome damage caused by ultraviolet light.

The discovery made by Cairns and De Lucia strongly influenced experimental design in subsequent studies on DNA biosynthesis. Their unexpected finding underscored the importance of the principle we referred to earlier—that a cellular function should not be assigned to an enzyme on the basis of its _in vitro_ properties alone. Only by isolating mutants in which the activity of the enzyme is affected can we make meaningful _in vivo_ correlations.

In cell-free extracts from a mutant _E. coli_ strain that did not contain DNA polymerase I polymerizing activity, two additional DNA-polymerizing enzymes were subsequently detected. These were named DNA polymerases II and III. The behavior of conditional lethal mutants of DNA polymerase III led to the conclusion that DNA polymerase III is the main replication enzyme. Some of the properties of the three enzymes are summarized in Table 26–3 and Figure 26–12.

For a few years following the observations of Cairns and De Lucia, it was assumed that DNA polymerase I was not important in replication but only in repair. However, further genetic and biochemical studies proved otherwise. The original mutant DNA polymerase I was defective only in its polymerizing function; its two degradation functions were intact. Subsequently, workers isolated conditional lethal mutants in which the $5' \rightarrow 3'$ degradation function of Pol I was affected; the mutants cannot elongate DNA under nonpermissive conditions. Thus it was demonstrated that the degradation activity of DNA polymerase I is indispensable for chromosome replication.

DNA polymerase III, which functions as the main replicating enzyme, is believed to be composed of a complex of at least seven different proteins (Figure 26–12). The complete enzyme, or _holoenzyme_, is assayed by testing its ability to convert primed single-stranded circular DNA of phage ϕX174 to the duplex form (see the discussion of viral chromosome replication, below). Genetic loci for some of the protein subunits found in the polymerase III enzyme have been identified (see Table 26–1 and Fig. 26–12). Mutants in

TABLE 26–3
Properties of Polymerases I, II, and III of *E. coli*

	Pol I	Pol II	Pol III
Molecules per cell	400	—	15
Turnover number[a]	600	30	9000
Structural gene[b]	*polA*	*polB*	*polC* (dnaE)
Conditional lethal mutant	+	−	+
5′→3′ Polymerizing activity	+	+	+
3′→5′ Exonuclease activity	+	+	+
5′→3′ Exonuclease activity	+	−	+
Mutant loci	*polA*	*polB*	*polC, dnaN, dnaX, dnaZ, dnaQ*
Lethality	Viability reduced only when 5′→3′ exonuclease affected	No effect	Conditional lethality

[a]Nucleotides polymerized/min/molecule of enzyme at 37°C.
[b]Only the structural gene for the largest protein subunit in the enzyme is recorded.

which θ is affected have not yet been isolated, so the physiological significance of θ remains unknown.

Polynucleotide Ligase Links Chains Together. Discontinuous DNA synthesis requires the existence of an enzyme for joining the newly synthesized segments (see Figure 26–11). Initially, two enzymes of this type were discovered, and they have been given the general name *polynucleotide ligase*. The ligase-catalyzed reaction involves the formation of a single phosphodiester bond between long runs of discontinuous chains held in proper juxtaposition by a template chain (Figure 26–14). Since the joining reaction is vital to DNA replication, it is not surprising to find that ligases are ubiquitous in living cells. They differ according to the source of activation energy used for phosphodiester bond formation. The *E. coli* bacterial ligase (encoded by the

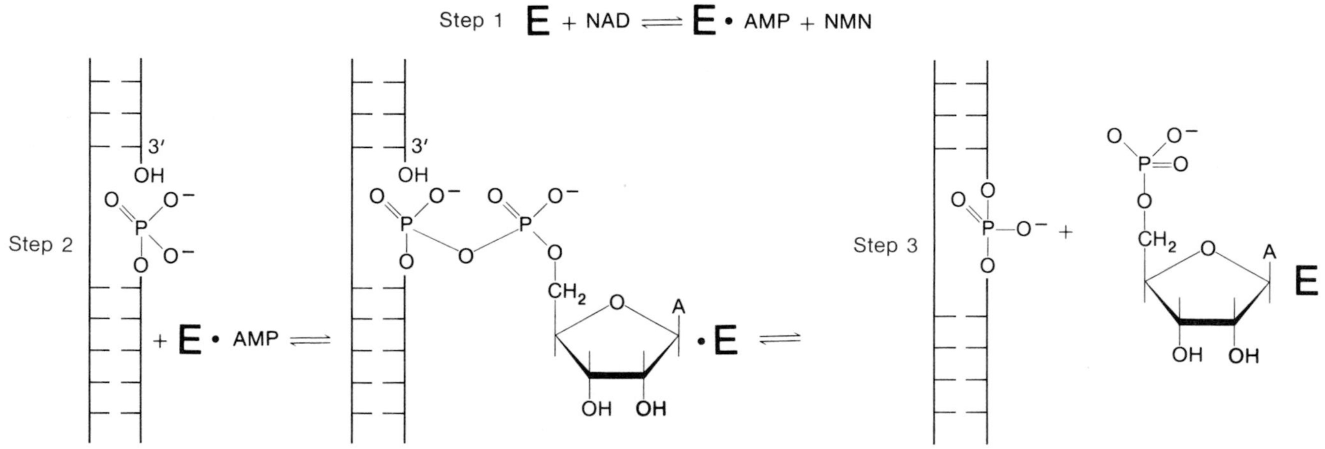

Step 1   E + NAD ⇌ E • AMP + NMN

Figure 26–14
Steps in the sealing of a DNA nick, catalyzed by DNA ligase. The bacterial ligase uses NAD⁺ to make an enzyme-AMP intermediate. Mammalian DNA ligases and bacteriophage T4 ligase use ATP for the same purpose.

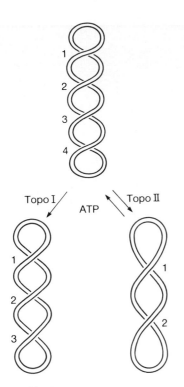

Figure 26–15

Type I and type II topoisomerases relax negatively supercoiled DNA in steps of one and steps of two, respectively. Type II topoisomerases can also add additional negative supercoils, as indicated by the double reaction arrow. The latter reaction requires energy input, which is encoded by ATP cleavage.

lig gene, Table 26–1) uses nicotinamide adenine dinucleotide (NAD) as the source of energy: the T4 bacteriophage and mammalian DNA ligases use ATP. Ligase is important in recombinant DNA techniques since it can seal fragments that have single-stranded complementary termini. Such fragments can be generated by restriction endonucleases. If the concentration of DNA fragments is high enough, ligase will also covalently link blunt-ended fragments that lack single-stranded complementary termini.

Topoisomerases Catalyze Supercoiling.   Supercoiled helices are frequently found in closed circular DNA molecules or in otherwise constrained DNA segments (see Chapter 7). With the help of supercoiled DNA and advances in the technology of DNA analysis by agarose gel electrophoresis (Figure 7–27), researchers have discovered several kinds of enzymes that can either reduce or increase the linking number (*L*) of a supercoiled helix. Those enzymes, called *topoisomerases*, were first identified as activities capable of relaxing negatively supercoiled DNA in *E. coli* and in mouse embryo cells.

Topoisomerases are believed to be essential for DNA metabolism and they are widely distributed in living organisms. The enzymes can break and rejoin DNA repeatedly without the need for any added cofactor to supply the energy for rejoining. Thus it appears that the reaction intermediate conserves the energy of the DNA phosphodiester bond. The proposed mechanism for supercoiling involves covalent-bond formation between the protein and the broken end of the DNA, followed by resealing of the DNA and dissociation of the enzyme, after the linking number has been changed by one or more units. A 3′-phosphotyrosine linkage has been detected as an intermediate.

Topoisomerases are classified as type I or type II, according to whether they change the linking number in steps of one or steps of two, respectively. Type I enzymes produce transient single-strand breaks in the double helix, whereas type II enzymes produce transient double-strand breaks (Figure 26–15). Type I topoisomerase of *E. coli*, also known as ω protein, is the best-understood enzyme of this class. Its molecular weight is 110,000 and the native protein exists as a monomer in solution. The enzyme shows a preference for highly negatively supercoiled DNA and is inactive on positively supercoiled DNA. This enzyme can also catalyze the *catenation* (interlocking) of double-stranded circular DNA, or the separation of catenanes into simple circles provided at least one circle contains a single-stranded break (Figure 26–16). Type II topoisomerases can carry out the catenation reaction as well; in this case, no single-strand breaks are required. Type I topoisomerases that have been isolated from eukaryotic cells differ in two important respects from the *E. coli* protein: they do not require Mg^{2+} ions and they can relax positively as well as negatively supercoiled DNA.

DNA gyrase, a type II topoisomerase found only in prokaryotes, differs from other topoisomerases in being able to catalyze the conversion of relaxed duplex DNA into a high-energy negatively superhelical form (Figure 26–15). That reaction is ATP-dependent. In the absence of ATP, gyrase can still catalyze relaxation of superhelical DNA or the catenation reactions described earlier. The gyrase enzyme contains two different subunits, encoded by the *gyrA* and the *gyrB* genes (Table 26–2). Normally it exists as a tetramer with two subunits of each type.

Topoisomerases are believed to serve vital functions, since mutants carrying defective topoisomerases grow very poorly. We do not know all the functions of topoisomerases, but probably they include the elimination of tangled DNA by decatenation reactions, as well as the elimination of the extreme supercoiling that would otherwise result from the unwinding of the duplex structure during DNA replication (see below). Topoisomerase is likely to act in the final stages of the separation of replicating circular DNA.

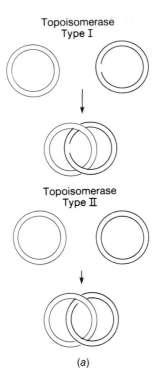

Topoisomerase
Type I

Topoisomerase
Type II

(a)

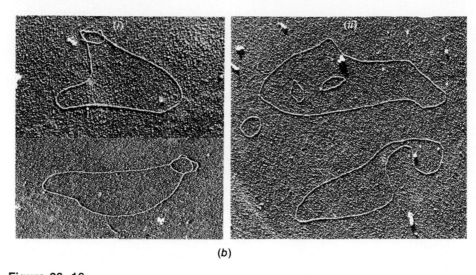

(b)

Figure 26–16
Catenation by topoisomerases. *(a)* Two circular DNAs can be catenated by type I topoisomerase only if one of the DNAs is nicked. This is not necessary when using a type II topoisomerase. *(b)* Electron micrographs of catenated DNA before (i) and after (ii) incubation with DNA gyrase. The catenane contains one large circular DNA and one small circular pBNP66 plasmid DNA. (Adapted from M. Gellert, L. M. Fisher, H. Ohmori, M. H. O'Dea, and K. Mizuchi, DNA gyrase: Site-specific interactions and transient double-strand breakage of DNA, *Cold Spring Harbor Symp. Quant. Biol.* 45:301, 1981.)

Helicases Catalyze DNA Unwinding.   The parental double helix must be unwound at the growth fork so that it can present additional single-stranded regions to serve as templates for continued replication. The enzymes that catalyze unwinding are called *helicases*. The rep protein of *E. coli* is believed to be a helicase, since DNA elongation is slower when the rep protein is not present. During the unwinding of duplex DNA by helicase, ATP is consumed, showing that the reaction requires energy. Helicase is one of the components of the prepriming *E. coli* replication complex, probably facilitating both priming and replication fork movement.

The unwinding of a circular duplex should tend to generate positively supercoiled DNA. As unwinding progresses, the increasing degree of positive supercoiling would lead to considerable strain in the duplex structure. It seems likely that those tensions are relieved by the action of topoisomerases. For example, the tension could be relieved in bacterial systems by the counteracting tendency of gyrase to cause negative supercoiling. Type I bacterial topoisomerases cannot relax positively supercoiled DNA and therefore would be of little value in this regard. By contrast, in eukaryotes containing no gyrase-like enzyme, the type I topoisomerases are competent in relaxing positively supercoiled DNA.

Single-strand Binding Protein Stabilizes Single Strands.   Single-strand DNA-binding protein (SSB) is found in abundance in *E. coli*. It has several functions, including the presentation of the template to the polymerase and productive interactions with several replication proteins. This and other evidence described below indicate that it plays a role in DNA replication (and probably in repair and recombination as well). It is believed to act in concert with the rep protein to facilitate unwinding at the growth fork. Since it binds preferentially to single-stranded DNA, it probably facilitates unwinding by inhibiting rewinding. Possibly, also, it protects the single-stranded DNA from attack by endogenous nucleases.

REPLICATION OF A VIRAL CHROMOSOME

Our understanding of prokaryotic DNA replication has proceeded most rapidly through investigations on viruses because of their relative simplicity. Bacteriophages that infect *E. coli* vary considerably in size and structure. Each type replicates its chromosome in a unique manner and relies on the host enzymatic machinery in a way that suits its needs. In general, there is an inverse relationship between the size of the virus chromosome and the degree of dependency on host enzymes. Large viruses, such as bacteriophage T4, encode several genome-replicating functions that have properties analogous to those of closely related host proteins, but that for various reasons are preferred by the virus.

The bacteriophage ϕX174 has a small, single-stranded circular DNA chromosome (5386 bases) that depends on many *E. coli* proteins for replication. ϕX174 DNA passes from a single-stranded to a double-stranded form and then back to a single-stranded form during replication. The replication process is particularly well understood, and findings with ϕX174 not only reveal the replication process of the viral DNA but also give insights concerning certain host proteins used in host chromosome replication. The ease with which ϕX174 DNA can be isolated intact and assayed biologically, through its ability to infect *E. coli* spheroplasts, makes it an ideal template for the study of the fidelity of its replication *in vitro*.

When ϕX174 infects *E. coli* it loses its protein coat, so that only the circular viral DNA enters the cytoplasm. During the first phase of the replication process, the single-stranded viral DNA must be converted to a double-stranded form. This double-stranded form serves two purposes: as an intermediate in replication (replicative form) and as a template for synthesis of viral RNA. Since no viral RNA or proteins can be made until after the viral DNA is converted to the double-stranded form, it follows that the single-to-double-strand transition must be carried out by pre-existing host cellular proteins. This point has been verified by experiments showing that the double-stranded form of the virus is made in the presence of chloramphenicol, which inhibits protein synthesis, whereas further steps in viral DNA replication cannot occur in the presence of this antibiotic.

Kornberg, his coworkers, and Hurwitz and his colleagues have constructed a cell-free system containing ϕX174 DNA and purified proteins that can carry out many of the steps believed to be involved in ϕX174 DNA replication *in vivo*. Before DNA synthesis can be initiated on the single-stranded circular DNA, a primer must be synthesized by the host *primase*. Two types of complexes have been made in which primase is active. The first complex is formed by adding primase to a preformed complex of DNA and *dnaB*-encoded protein. Primer formation occurs in this system at random sites on the DNA. In the second complex, which is sequence-specific, several additional proteins are required before dnaB protein or primase can bind (Figure 26–17). These include n, n′, n″, dnaC, and single-strand binding (SSB) protein. When the DNA is uniformly coated with SSB, random primer formation is prevented because the dnaB protein cannot displace SSB from the DNA to form a complex. Under these conditions n′ makes an initial complex at a specific site on the DNA (step I in Figure 26–17). This site is adjacent to a 44-base region on the single-stranded DNA that forms a hairpin double helix to which SSB does not bind. Other proteins—n, n″—join the initial complex, resulting in the displacement of some SSB. Finally, the dnaB, dnaC, and i proteins become part of the complex, as more SSB is displaced; the resulting multiprotein complex is known as the *preprimosome*. [Addition of primase to this complex completes assembly of

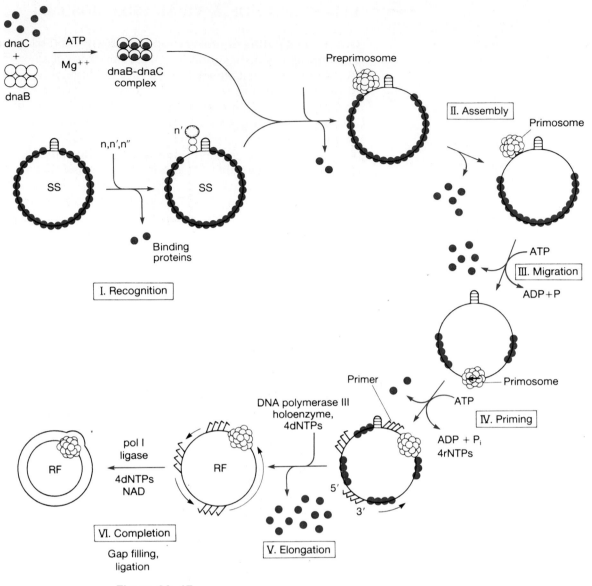

Figure 26–17
Proposed mechanism for the conversion of single-stranded ϕX174 DNA to the duplex replicative form (SS→RF reaction). (Adapted from K. Arai, R. Low, J. Kobori, J. Shiomai, and A. Kornberg, *J. Biol. Chem.* 256:5280, 1981).

the <u>primosome</u> (step II, Figure 26–17).] The primosome can migrate on the single-stranded DNA in the 5'→3' direction. In the process, still more SSB is removed from the single-stranded DNA (step III, Figure 26–17).

This movement of the primosome is an energy-dependent reaction requiring ATP. At various points during its journey the primosome pauses to synthesize a short stretch of primer (step IV, Figure 26–17). DNA polymerase III holoenzyme uses these primers as initiation sites for DNA synthesis (step V, Figure 26–17). Although primase initiates limited stretches of RNA synthesis at several locations on the template, the necessary primosome can be assembled only at a unique site around base 2300 (step I, Figure 26–17). The primosome complex is very stable and can survive many rounds of replication. The Okazaki fragments formed by primase and DNA polymerase III (step V, Figure 26–17) require DNA polymerase I and ligase for completion (step VI, Figure 26–17). DNA polymerase I closes the gaps as it removes

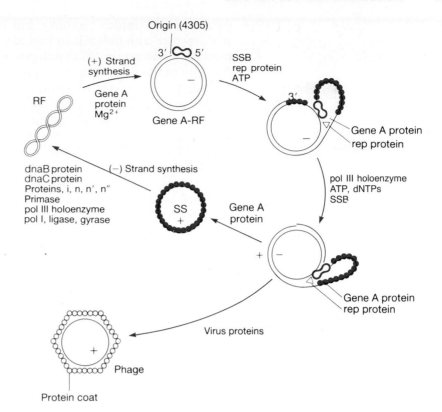

Figure 26-18

Scheme for φX174 RF replication in two stages. Continuous replication initiated by gene A protein cleavage generates viral (+) circles, and discontinuous replication of the viral circles by the SS→RF system produces RF. In the presence of phage-encoded maturation and capsid proteins, viral circles (RF→SS) are encapsulated rather than replicated. (Adapted from A. Kornberg, *DNA Replication*, W. H. Freeman, San Francisco, 1980, p. 510.)

the RNA present on the 5′ ends of the fragments. Ligase joins the nicked fragments of DNA together. The circular duplex so formed is referred to as *replicative form* DNA (RF). If gyrase is present, this molecule will become negatively supercoiled. The RF is used as a template to synthesize more RF and viral RNA. This RNA serves in turn as the template for the synthesis of viral proteins that are essential in subsequent steps in viral DNA replication.

After single-stranded viral DNA (SS DNA) has been converted to the replicative form duplex (SS→RF), the parental RF is duplicated to produce multiple copies of RF (RF→RF). In this reaction the viral DNA, known as the *plus strand*, and the complementary DNA, known as the *minus strand*, are replicated by distinct mechanisms. Synthesis of the viral plus strand begins when the viral gene A protein induces the cleavage of the viral plus strand in the RF at position 4305 (see Figure 26-18). The bifunctional A protein not only cleaves the plus strand at this point, but also becomes covalently attached to the 5′ end of the interrupted strand (see Figure 26-18). More copies of the replicative duplex form are initiated from this point. Nucleotides are added continuously to the 3′-OH end of the plus strand. The new plus strand displaces the old plus strand during this phase of the reaction. After one viral single-stranded plus strand has been displaced, it becomes covalently closed through the action of the viral gene A protein (acting as a specific ligase) and can either serve as the template for synthesis of more RF or become a component (in the genome) of the mature virus. During the synthesis of plus strand it can be seen that the minus strand is being used as the template in a continuous fashion; its action could be likened to that of a rolling circle. The *rolling circle mode of replication* is used by a number of bacterial viruses (e.g., bacteriophage λ) at some stage during their replication cycle.

Host proteins required during plus-strand synthesis include the host *rep protein*, or helicase, as well as DNA polymerase III and single-strand binding protein (SSB). As we have noted, the function of helicase is to facilitate unwinding of the duplex. Helicase is not absolutely required for DNA synthesis using a duplex template, but it greatly accelerates the process. SSB acts together with helicase by complexing with single-stranded regions, thus inhibiting rewinding. The viral gene A protein is crucial in maintaining the integrity of the displaced virus plus strand during synthesis. It does this by holding the otherwise free end and ensuring that the strand closes at the proper point without releasing itself from the template.

What we know about additional steps in the replication process comes primarily from an analysis of the events occurring *in vivo*. After about 60 copies of the replicative form have been made (20 min after infection), the infected cell switches to making single-stranded circles exclusively. Presumably some regulatory device is involved to ensure that further single-stranded circles do not get converted into duplex forms at this stage. The protein encoded by the phage genes *C* and *J* are thought to play a key role in diverting the single-stranded DNA into mature viral particles. In that process, the plus-strand circles get packaged into phage protein shells as they are being synthesized, in preparation for making completed viruses. The packaging step concludes the replication cycle for ϕX174 DNA.

REPLICATION OF THE *E. COLI* CHROMOSOME

Replication of the *E. coli* chromosome has been much more difficult to study than replication of phage DNA, in part because of its large size. As stated earlier, genetic evidence has indicated that host proteins involved in ϕX174 replication might also be involved in similar ways in *E. coli* DNA replication. In particular, *the movement of the primosome complex suggests a mechanism for discontinuous DNA synthesis on the lagging strand of the replication fork*. The primosome could migrate on the lagging strand as the template unwinds, stopping periodically to allow primase to form primer (Figure 26–19). The elongation steps, so clearly elucidated in ϕX174 minus-strand synthesis, could account for lagging-strand synthesis at the *E. coli* replication fork.

That leaves two aspects of *E. coli* chromosome replication to be accounted for: initiation at the origin of replication (location indicated in Fig-

Figure 26–19
Role of the primosome at the bacterial replication fork. The primosome, assembled at or near the DNA replication origin, migrates continuously on the lagging strand as the replication fork moves. (Adapted from K. Arai, R. L. Low, and A. Kornberg, Movement and site selection for priming by the primsome in phage ϕX174 DNA replication, *Proc. Natl. Acad. Sci. USA* 78:711, 1981.)

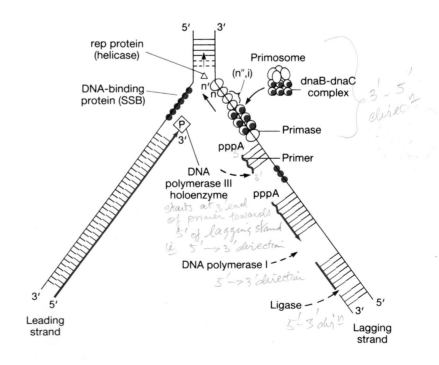

ure 26–5) and termination. Study of termination awaits the reconstitution of an *in vitro* system containing all the proteins necessary for replication. To explore the mechanism of initiation, Kornberg and his coworkers have developed a cell-free system for <u>bidirectional</u> replication using small circular DNA molecules, known as hybrid plasmids, that contain the origin of the *E. coli* chromosome (*oriC*) together with purified proteins from *E. coli*. These ongoing studies are beginning to show how bacterial chromosome synthesis is initiated.

The use of the plasmid is crucial in such studies because it greatly facilitates detection of initiation on a much smaller and more tractable piece of DNA. The plasmid contains about 245 base pairs (bp) of the minimal origin sequence from the *oriC* region of the *E. coli* chromosome. The 245-bp *oriC* region has a considerable number of GATC sites in which the cytosine residues are specifically methylated by an *E. coli* enzyme encoded by the gene *dam* (discussed further below). Methylation may be important in the control of initiation of replication at *oriC*.

The initiation of replication at *oriC* can be divided into three steps:

1. Generation of a template (requiring DNA gyrase, topoisomerase I, and a histone-like DNA-binding protein called HU).
2. Transcription and priming (requiring RNA polymerase and the *dnaA*-encoded protein).
3. Primer processing (requiring RNAse H, an enzyme that degrades RNA when it is base-paired to DNA or RNA, and some yet unidentified factor).

In addition to the primosomal proteins and the Pol III holoenzyme, other, as yet unidentified proteins are required for *oriC*-dependent DNA replication. The requirements for RNA polymerase and DNA gyrase have been shown by their specific sensitivities to the inhibitors rifampicin and nalidixic acid (and novobiocin), respectively. Rifampicin is known specifically to inhibit RNA polymerase (see Chapter 27), and nalidixic acid and novobiocin specifically inhibit the action of the subunits of DNA gyrase (see Chapter 7). The characterization of the proteins required in this system is under intensive investigation. It is noteworthy that RNA polymerase catalyzes primer synthesis at *oriC* but that primase is used at other sites along the *E. coli* chromosome.

DNA SYNTHESIS AND CHROMOSOMAL REPLICATION IN EUKARYOTES

The most productive approach to understanding DNA replication in eukaryotes has been to isolate proteins that exhibit activities logically related to DNA replication and to isolate replicative intermediates. Different forms of DNA polymerase, DNA ligases that require ATP, topoisomerases, single-stranded DNA-binding proteins, unwinding enzymes, and degradation enzymes have all been isolated from eukaryotic cells. It is presumed that many of these have vital functions *in vivo*, but the functions have not been verified because relevant mutants have not been found.

Three major systems are currently being exploited in studying eukaryotic DNA synthesis. They are adenovirus, simian virus 40 (SV40), and yeast. The yeast cell system and its extensively known genetics make it possible to isolate mutants by conventional means. The use of yeast also makes it possible to exploit reverse genetics—that is, purified proteins can be used to clone the genes encoding them, then the cloned genes can be used to disrupt, by integrative transformation, their genomically located counterparts.

Eukaryotic Cells Have Several DNA Polymerases

Four species of DNA polymerases have been characterized in eukaryotes. These have been named α, β, γ, and δ in the order of their discovery. In some respects the eukaryotic DNA polymerases behave like prokaryotic enzymes. For example, they use a template to direct the polymerization of mononucleoside triphosphates and cannot initiate new chains without a primer. There are, however, many important differences. One outstanding difference is that DNA polymerases α, β, and γ do not contain either of the associated exonuclease activities present in the bacterial enzymes. Presumably, when required, these additional activities are contributed by other enzymes, not yet described.

The amount of DNA polymerase α varies greatly with cell type and the physiological state of the cell. Whereas all the eukaryotic polymerases require template and primer, the α enzyme is the only one that can use an RNA primer. Several lines of evidence indicate that DNA polymerase α is the main enzyme involved in the replication of cellular DNA. It is also thought to be important in the replication of some mammalian viruses, such as SV40 and polyoma. That evidence includes the following.

1. The level of DNA polymerase α activity is correlated with cellular proliferation and is absent in certain terminally differentiated cells, such as nerve cells, that do not undergo further duplication.
2. The extent to which certain compounds inhibit cellular DNA replication is directly proportional to their inhibitory activity on DNA polymerase α (see below).
3. DNA polymerase α copurifies with replicating chromosomes of the SV40 animal virus when these chromosomes are separated from other cellular components in a crude lysate.

DNA polymerases β and γ are present in eukaryotic cells in resting and differentiated states. The levels of β and γ enzymes do not change markedly in proliferating cells. The function of the β enzyme is unclear, although its localization in the nucleus has led to the speculation that it may be involved in repair or replication of nuclear DNA. DNA polymerase γ is usually localized in mitochondria, and is believed to be involved in replication of the organelle's DNA.

Eukaryotic Chromosomes Contain Multiple Origins of Replication

Although the general features of DNA replication in eukaryotes are thought to be similar to those in prokaryotes, there are some interesting differences. The chromosomes of higher eukaryotic organisms are quite large, in some cases a thousand times larger than their bacterial counterparts. In order for such large DNA molecules to replicate in a reasonable length of time, they must have _multiple origins of replication_. The simultaneous synthesis of DNA at several points along the chromosome has been demonstrated by incorporating radioactive nucleotides into the replicating chromosomes for a short period and then observing the distribution of radioactive DNA by autoradiography (Figure 26–20). Multiple regions (or replicons) of incorporated label are observed, with replication proceeding _bidirectionally_ from each of these regions. Termination of replication occurs at the point where the growth forks from two adjacent replication units collide (Figure 26–20). DNA on at least the lagging strand of a fork is made discontinuously. The Okazaki fragments are much shorter than those found in prokaryotes, averaging between 100 and 200 nucleotides in length. Synthesis of these DNA fragments is initiated on RNA primers that are found covalently attached to

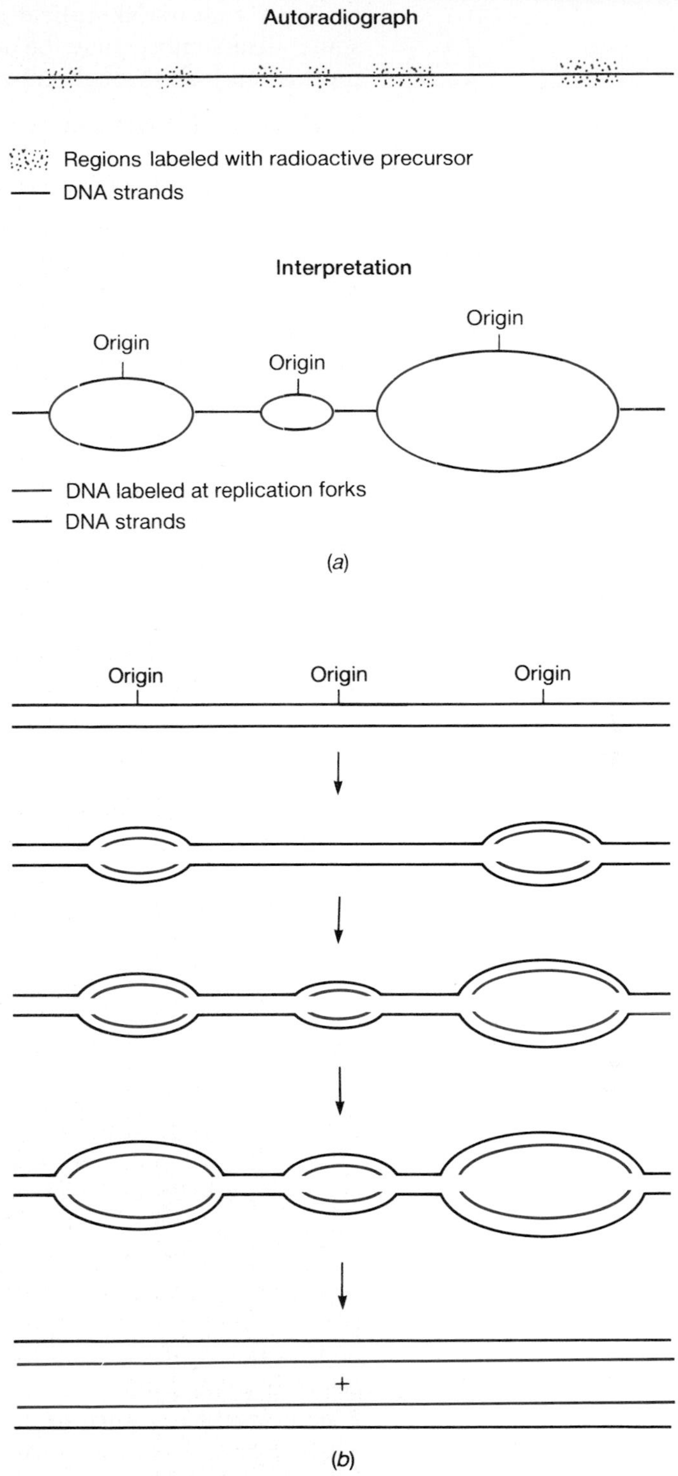

Autoradiograph

░░░ Regions labeled with radioactive precursor
—— DNA strands

Interpretation

—— DNA labeled at replication forks
—— DNA strands

(a)

(b)

Figure 26–20
Multiple-origin model for eukaryotic chromosomal DNA replication. (a) Autoradiograph of short-term labeling of a eukaryotic chromosome during replication and its interpretation. (b) Overall replication scheme for a eukaryotic chromosome. Only a short region of the chromosome is shown. It is believed that replication origins are relatively free of proteins.

the 5′ ends of newly synthesized fragments. Chromosomal replication occurs only during the S phase of the cell cycle.

DNA in eukaryotic chromosomes is associated with histones in complexes called _nucleosomes_ (the nucleosome structure is described in Chapter 30). These nucleoprotein complexes serve to condense exceedingly long DNA molecules into much more compact structures, but during replication the nucleosomes must be disassembled so that the DNA strands can be separated. The disassembly of the DNA-histone complex presumably occurs directly in front of the replication fork. Two observations suggest that nucleosome disassembly may be a rate-limiting step in the migration of the replication fork through chromatin:

1. The rate of migration of replication forks is slower in eukaryotes (with nucleosomes) than in prokaryotes (lacking nucleosomes).
2. The length of the Okazaki replication fragments is similar to the length of DNA between adjacent nucleosomes (about 200 base pairs).

Before the newly replicated DNA is reassembled into nucleosomes, the RNA primers must be removed by an as yet unknown enzyme, the gaps must be filled in by DNA polymerase, and the replication fragments must be linked together by DNA ligase. Whereas the origin of DNA replication in prokaryotes has been characterized, the structure of the multiple cellular origins of replication in higher eukaryotes remains unclear. Replication origins are activated in a specific temporal order, the overall number of origins depending on the time required to replicate the cell's genome. In yeast, autonomously replicating sequences *(ARS)* have been characterized structurally and functionally and are thought to be origins of replication.

Summaries of two eukaryotic DNA replication systems, those of SV40 and *S. cerevisiae*, are given below.

In SV40, the entire duplex, circular genome (5243 bp) has been sequenced. The DNA is associated with nucleosomes and its replication is bidirectional, starting at a unique origin (about 65 bp) that has been identified. The viral T antigen is involved in the initiation of viral DNA replication, bypassing the chromosomal initiation machinery. It binds to DNA at the origin of replication and has a helicase activity that results in DNA unwinding. Since synthesis is sensitive to the inhibitor amphidicolin, it is assumed that DNA polymerase α is the DNA elongation enzyme. Replication *in vitro* generates relaxed viral DNA monomers.

The yeast genome has 17 chromosomes with a combined size of 1.35×10^4 kb. There are about 400 *ARS* sequences, where DNA replication is believed to be initiated. *ARS* sequences allow plasmids to replicate autonomously in yeast by providing origins of replication. DNA-binding proteins that bind specifically to *ARS* sequences have been identified; they could be analogous to the SV40 viral T antigen. Only a few mutant genes that affect replication in yeast have been isolated; those identified include thymidine kinase *(cdc21)*, DNA ligase *(cdc9)*, repair deficient *(cdc40)*, and mutants in topoisomerase I and II. Reverse genetics (isolation of a gene by making use of the purified protein it encodes) has been used to clone and integratively disrupt the genes encoding yeast DNA polymerase I (equivalent to DNA polymerase α) and the single-strand DNA-binding protein SSB-1. The gene for polymerase I is a single gene and, from disruption experiments, is thought to be essential. Polymerase I contains a primase that synthesizes a 5'-terminal oligoribonucleotide, about 10 nucleotides in length and of random sequence. Ribonuclease H copurifies with polymerase I, but its function is dispensable. The gene encoding SSB-1 is also dispensable, since its disruption shows it to be unessential for yeast viability. *In vivo* experiments have demonstrated that topoisomerase II is essential and that its function is likely to be in the segregation of replicated chromosomes and plasmids.

Mitochondrial DNA Replicates Continuously on Both Strands

Mitochondria in mammalian cells contain circular, supercoiled DNA (mtDNA) about 5 μ in size ($M_r = 10 \times 10^6$; about 15,000 bp). It appears to be replicated by DNA polymerase γ, since this is the only DNA polymerase present in mammalian mitochondria. Furthermore, mtDNA replicates even

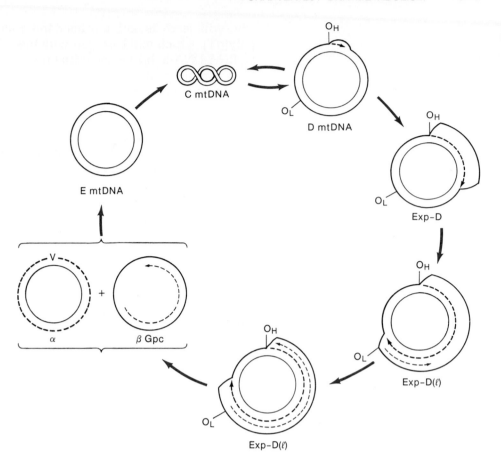

Figure 26-21

Replication model for mouse mitochondrial DNA. (Thick solid line; parental heavy (H) strands. Thin solid lines: parental light (L) strands. Thick dashed lines: daughter H strands. Thin dashed lines: daughter L strands.) The order of replication is clockwise, starting at D mtDNA. O_H and O_L are the origins of H- and L-strand synthesis, respectively. The double arrows reflect the metabolic instability of D-loop strands and consequent equilibrium between D mtDNA and C mtDNA. Expanded D-loop replicative intermediates are termed Exp-D prior to initiation of L-strand synthesis and Exp-D(l) after initiation of L-strand synthesis. The carat marks the interruption of at least one phosphodiester bond in the H strand of the daughter molecule. (β Gpc = gapped circular daughter molecule) Each replicative form is discussed in order in the text.

in cells where DNA polymerase α is absent or inhibited by aphidicolin, a highly specific inhibitor of this polymerase.

The replication of animal mtDNA has been studied most extensively in mouse L cells (Figure 26-21). There are two origins of replication, one for each strand (referred to as H and L). DNA replication is initiated at the first site on the H strand (O_H) to form a displacement loop (D loop). It is a triple-stranded structure that includes the newly synthesized short daughter H strand. Replication continues unidirectionally until completion. When H strand synthesis is two-thirds complete, L strand synthesis is initiated (O_L) and elongated in the direction opposite to that of H strand synthesis. The daughter molecules segregate and synthesis proceeds to completion prior to closure with the introduction of about 100 negative superhelical turns. The overall rate of mtDNA synthesis is about 270 nucleotides/min, about 0.5% the rate at which *E. coli* DNA replicates.

An interesting observation is that mammalian mtDNA contains a small percentage of ribonucleotides, which have been monitored by measuring susceptibility to alkali and to ribonuclease H. It has been found that mouse L-cell mtDNA has ribosubstitutions in the two replication origin regions. Their function is unknown.

Some Polymerases Function Without a DNA Template

RNA viruses that cause tumors (oncogenic RNA viruses) contain an unusual DNA polymerase called *reverse transcriptase*. This enzyme carries out template-directed polymerization and shares many features with the cellular DNA polymerization reaction. The main difference is that RNA is the preferred template for reverse transcriptase, which catalyzes the polymerization of deoxynucleoside triphosphates only. (RNA with appropriate initiation sites will serve as both a primer and a template.) If a primer is added [e.g., poly(dT), which will base-pair with the poly(A) at the 3' end of mRNA], then an RNA-DNA hybrid is the initial product of the reaction. A second round of DNA synthesis on the hybrid produces a double-stranded DNA (see Chapter 32). In this manner, viral RNA sequences can be transcribed into double-stranded DNA sequences (called cDNA). The duplex DNA copy (cDNA) of the viral RNA is subsequently integrated into the host chromosome, where it functions in the synthesis of further viral RNA, often causing malignant transformation of the cells. In the laboratory, reverse transcriptase is used extensively for conversion of mRNA species into DNA sequences for cloning and sequencing studies (see Chapter 32).

The only deoxynucleoside triphosphate polymerizing enzyme that does not use any template for polymerization is called _terminal deoxynucleotidyl transferase_. The enzyme is usually isolated from calf thymus. This enzyme will extend the 3'-OH end of a polydeoxyribonucleotide with any deoxynucleoside triphosphate, singly or in combinations. The biological function of this enzyme has not been established; its occurrence in undifferentiated lymphocyte populations from primary lymphoid organs (thymus and marrow) suggests a special role in differentiating lymphocytes. In acute lymphoblastic leukemia, the population of cells containing terminal transferase is greatly expanded. Measurement of this enzyme in blood and marrow cells is frequently used in the diagnosis of leukemia. In the laboratory, this enzyme has been used to generate single-stranded, complementary strands on DNA fragments that are to be cloned (Chapter 32).

CONTROL OF DNA REPLICATION

Little is known about the biochemical mechanisms that are used to regulate DNA synthesis. At this point, all we can do is mention certain problems in control that are known to exist in various systems.

Once a round of bacterial DNA synthesis has been initiated, it continues to completion. Only extreme conditions, such as substrate starvation, will interfere with elongation. All indications point to initiation as being the primary site for regulating DNA replication, even though very little is known about the factors involved. Protein synthesis is required to initiate a new round of E. coli chromosomal DNA replication. Addition of protein synthesis inhibitors such as chloramphenicol will, however, allow a round of DNA synthesis to continue to completion. Some bacterial plasmids, such as ColEl and pBR322, do not require protein synthesis to initiate a new round of replication. This fact is exploited in the amplification of such plasmids; the addition of chloramphenicol to E. coli harboring pBR322 will inhibit chromosomal DNA replication but allow the plasmid to accumulate to a level of several hundred to a thousand copies.

In prokaryotes such as E. coli, the rate of reinitiation is a sensitive function of the cell division time (40 min at 37°C) and actually can exceed the cell division time (as low as 20 min). In such situations, initiation of replication for subsequent cell divisions occurs before the division of the current generation. Each of the daughter cells then receives a chromosome that is already in the process of replication.

In eukaryotes, the replication of DNA occurs only during one phase of the cell cycle in actively dividing cells (Figure 26-22). Following mitosis (M) and cell division, there is a gap in time (G_1 phase) before DNA replication begins (S phase). Upon completion of DNA synthesis, there is another pause (G_2 phase) before mitosis can occur. Actively dividing cells grown under the same conditions spend comparable times in each phase of the cell cycle. Therefore, the initiation of replication is not a random event in the life

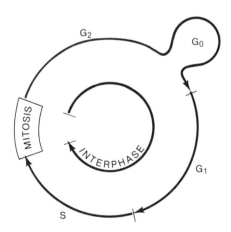

Figure 26-22

Life cycle of a eukaryotic cell. The cell cycle is divided into the resting stage (G_0), the prereplication stage (G_1), the synthesis or replication stage (S), the mitotic stage (M), and the postreplication stage (G_2). Except for G_0, which is extremely variable in length, the duration of each stage of the cell cycle is measured in fractions of hours to several hours. These times are similar for similar cell types. The term interphase applies to all phases other than mitosis.

of a cell and presumably occurs in response to some intracellular signal. RNA and protein synthesis must occur before DNA synthesis can begin, but the exact nature of the signals for initiation of DNA replication is still a mystery. Furthermore, in eukaryotes, protein synthesis is required during S phase. Chromosomal proteins, especially histones, are usually synthesized at the same time as the DNA.

In cultured mammalian cells, there are approximately 20,000 origins of replication for each haploid equivalent of DNA in the genome. If replication were to proceed evenly throughout S phase, all the replication units would be active simultaneously. In fact, this is not the case. Considerable variation occurs with different types of cells and culture conditions. During the S phase of the cell cycle, all of the DNA in the genome is replicated once, and (with the exception of certain special cases noted below) each segment of DNA is replicated only once. Some replication units complete DNA synthesis before other units initiate; therefore there is probably a mechanism that distinguishes between the origins of replication on DNA that have already been replicated during the current cell cycle.

Some specialized cells require extensive specific gene expression. In order to achieve a high rate of RNA synthesis, these genes are selectively amplified so that many more copies exist at this stage of development than at other times. For example, in the South African frog *Xenopus laevis*, the genes for the ribosomal RNAs are amplified during the maturation of oocytes. At this stage there are about 4000 extra ribosomal RNA genes in the genome, which are not propagated when the oocyte divides. A mechanism that normally ensures a single replication cycle must be bypassed to allow for specific gene amplification of the type described here.

SPECIFIC INHIBITORS OF DNA REPLICATION

The study of replication has been greatly aided by the availability of specific chemical inhibitors. The most useful inhibitors are those that block the action of specific proteins involved in the replication process. Through combined *in vitro* and *in vivo* studies with such inhibitors it is frequently possible to determine the mechanism of action of the target protein. Inhibitors have also been useful in isolating mutant microorganisms that contain altered target proteins. Some inhibitors of DNA replication have therapeutic value in the treatment of cancer. This is particularly true in the case of fast-growing tumors, since the inhibitors do not show much selectivity for cell type except to attack cells that are in S phase.

Inhibitors can be divided into three general categories:

1. Inhibitors that interact with DNA.
2. Inhibitors that affect the synthesis of deoxyribonucleotides.
3. Inhibitors of enzymes involved in DNA synthesis.

The second category has already been discussed in Chapter 25.

Some Inhibitors Interact Directly with the DNA

Compounds such as acridine, actinomycin D, and ethidium bind between the stacked bases of the DNA duplex. Hydrophobic planar portions of these molecules become inserted between the hydrophobic faces of the adjacent base pairs in the DNA. This intercalation disrupts the normal structure of the DNA, so that some unwinding of the DNA helix must occur (see Chapter 7). Netropsin and a related antibiotic, distamycin A, through hydrogen bonding and van der Waals forces, bind tightly to the exterior of the DNA duplex and along the minor groove to regions rich in A and T residues. The binding of

TABLE 26–4
DNA Replication Inhibitors that Act on DNA Polymerase

Inhibitor	Polymerase Affected	Action
Aphidicolin	Mammalian DNA polymerase α	Competitive with dCTP
Arylhydrazinopyrimidines	*B. subtilis* DNA polymerase III	Competitive with dGTP or dATP
Arabinosylcytosine[a]; arabinosyladenine[a]	DNA polymerases; several other enzymes in nucleic acid metabolism	Chain terminator
2',3'-Dideoxythymidine triphosphates[b]	Mammalian DNA polymerases ($\gamma > \beta \gg \alpha$); prokaryotic DNA polymerases	Competitive with dTTP (or other dNTP); chain terminator
N-Ethylmaleimide	Mammalian DNA polymerase ($\alpha > \gamma \gg \beta$)	Reacts with sulfhydryl groups

[a]Probably converted to triphosphate in cells.

[b]Works only *in vitro* because triphosphates cannot pass through the cell membrane.

the compounds inhibits the use of the DNA as a template for both replication and transcription by a simple blocking reaction. At low concentrations, acridine is a mutagen that produces additions or deletions of single bases during the replication process. This effect of acridine (addition or deletion of bases) is believed to be due to the way in which acridine stacks between the base pairs of the DNA.

Bleomycin and the antibiotic protein neocarzinostatin introduce breaks in DNA. The reaction of bleomycin is mediated by an Fe(II) chelate that generates an active oxygen species at susceptible sites on DNA. The breaks caused by neocarzinostatin and bleomycin cannot be sealed by DNA ligase. Bleomycin splits the $C_{3'}$-$C_{4'}$ bond in deoxyribose and thereby releases a base (usually cytosine) attached to a three-carbon sugar. These breaks disrupt the continuity of the DNA template and result in nonproductive binding of DNA polymerase.

Some Inhibitors Interact with the Enzymes Involved in DNA Synthesis

Several compounds that act directly on the DNA polymerases show some selectivity toward the individual polymerases (Table 26–4). The differences in sensitivity to some of these compounds have been most useful in assigning the functions of DNA polymerases *in vivo*, especially if mutants are not available.

The structures of three compounds that show selective inhibition of the eukaryotic polymerases are shown in Figure 26–23. The 2',3'-dideoxy derivative of thymidine lacks the 3' hydroxyl group of the sugar moiety. When a dideoxynucleotide is incorporated into a growing chain, polymerization is terminated (see Box 26–A for use in DNA sequencing). DNA polymerase γ is much more sensitive to ddTTP than DNA polymerase α. Its sensitivity is directly related to the extent to which the polymerase recognizes the dideoxytriphosphate as a substrate for incorporation into a growing DNA chain. ddTTP also inhibits adenovirus and mitochondrial DNA replication

2',3'-Dideoxythymine

β-Arabinofuranosylcytosine

Aphidicolin

Figure 26–23
Selective inhibitors of eukaryotic DNA polymerases. The 2',3'-dideoxythymine lacks the 3' hydroxyl group of the sugar. When it is incorporated into a growing chain, polymerization is terminated. The β-arabinofuranosylcytosine has a hydroxyl group in the *trans* configuration at the 2' position of the sugar. When incorporated into a growing chain, this alteration also blocks elongation. The first compound is recognized by DNA polymerase γ much more than the DNA polymerase α. The second compound is recognized more by polymerase α than β. Aphidicolin is a highly specific inhibitor that acts on DNA polymerase α directly.

BOX 26–A

The Sanger Method for Sequencing DNA

In Chapter 7 we discussed the Maxam-Gilbert chemical method for sequencing DNA. Sanger's laboratory developed a quite different method, which involves the use of chain-terminating dideoxynucleoside triphosphates to produce a continuous series of fragments in reactions catalyzed by _polymerase_. Dideoxynucleoside triphosphates lack a 3'-OH group; otherwise they are normal in structure. They may add to a growing chain during enzyme-catalyzed polymerization, but may not be added onto. Either DNA or RNA may be sequenced by the Sanger method. For RNA to be sequenced, _reverse transcriptase_ is used to make a DNA copy of the RNA. When DNA is being sequenced, DNA polymerase is used to make a complementary copy of a primed single-stranded DNA fragment. By choosing an appropriate primer, the region of the nucleic acid that is copied can be predetermined—see the illustration, part (a).

Synthetic reaction mixtures are then set up, each containing all four radioactive deoxyribonucleoside triphosphates to label the fragments for detection by autora-

diography. Each mixture contains a single, _limiting_ amount of dideoxynucleoside triphosphate to randomly terminate the synthesized fragments at one of the four nucleotides. The products of four separate reaction mixtures, each containing a different dideoxynucleoside triphosphate, are analyzed—see the illustration, part (b). Reaction 1, using dideoxyadenosine triphosphate (ddATP), contains all fragments with an A terminus; reaction 2, using ddCTP, contains all C terminations; and so on. Following synthesis, the reaction products are separated from the template by denaturation and displayed by electrophoresis on polyacrylamide gels. After electrophoresis the positions of the fragments on the gel are detected by autoradiography. The sequence is read directly from the autoradiogram, starting with the fastest-moving (smallest) fragment at the bottom, and moving up the gel—see the illustration, part (c). If the first band is in reaction 3, it is a G residue; the next highest band, appearing from reaction 4, would be T; and so on. Up to 200 residues can be read from a single gel.

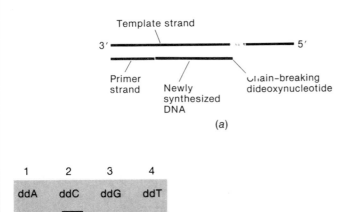

(a)

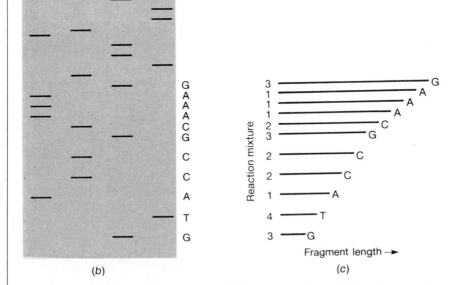

(b)

(c)

The Sanger dideoxynucleoside method of sequencing DNA. (a) A suitable template is chosen, and the primer is chosen so that DNA synthesis begins at the point of interest. In addition to the template-primer complex the reaction mixture contains all four radioactive deoxyribonucleoside triphosphates and small amounts of a single dideoxynucleoside triphosphate. The dideoxy compound serves as a chain terminator. (b) After synthesis in the presence of DNA polymerase I, the products of the reaction mixture are separated by gel electrophoresis and analyzed by autoradiography. For a given dideoxy compound all fragments terminating with that particular base should give rise to bands on the gel. The interpretation of the gel pattern is given in (c). The smallest labeled fragment moves the fastest and appears at the bottom of the gel.

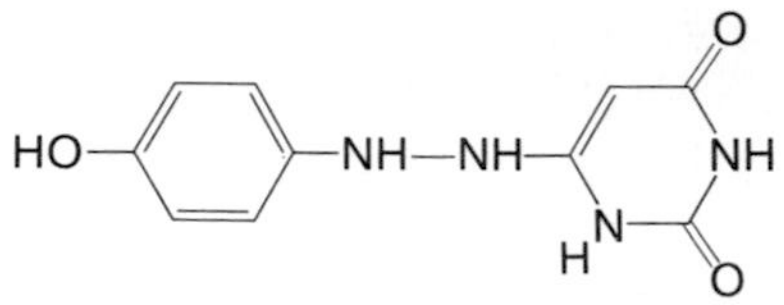

6-(p-Hydroxyphenylhydrazino)-uracil
[OHPh(NH)₂Ura]

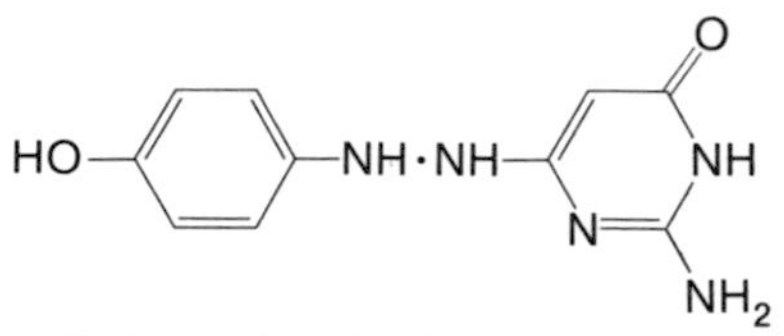

6-(p-Hydroxyphenylhydrazino)-isocytosine
[OHPh(NH)₂Iso]

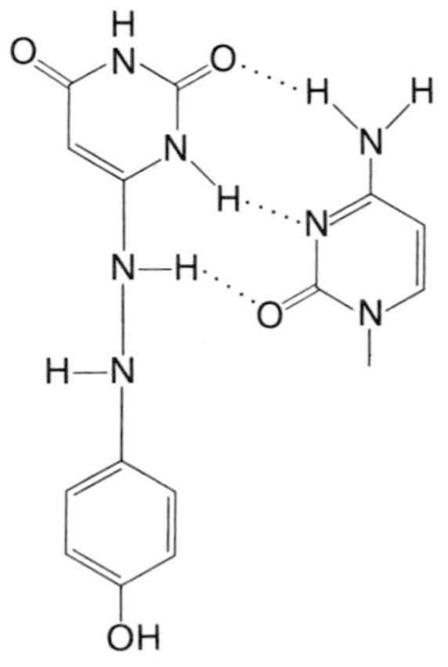

Hydrogen bonding between OHPh(NH)₂Ura
and cytidine

Figure 26–24
Arylhydrazinopyrimidines are
inhibitors of *Bacillus subtilis* DNA,
polymerase III. These compounds have
been used to isolate mutants of
B. subtilis polymerase III and to
establish the role of this enzyme in
DNA replication. The inhibitory action
is believed to be due to the formation
of ternary 1:1:1 complexes with DNA
polymerase III and the DNA template
primer.

at the same concentration that it inhibits DNA polymerase γ *in vitro*, but it has no effect on SV40 virus or cellular DNA replication. This has been taken as evidence that DNA polymerase γ participates in mitochondrial DNA replication.

The arabinose analogs of CTP and ATP have a hydroxyl group in the *trans* configuration at the 2' position of the sugar. This altered configuration affects the ability of the arabinosyl derivatives to act as nucleotide acceptors when they are incorporated into a growing DNA chain. DNA polymerase α is more sensitive than DNA polymerase β to inhibition by β-arabinosyl CTP and α-arabinosyl ATP. The arabinosyl nucleoside analogs, most likely as triphosphate derivatives, also specifically inhibit cellular DNA replication.

N-Ethylmaleimide (NEM) reacts generally with sulfhydryl groups of proteins, so it is not a specific DNA polymerase inhibitor. It is useful for comparing enzyme sensitivity. SV40 viral and cellular DNA synthesis *in vitro* are as sensitive to NEM as DNA polymerase α; DNA polymerase γ is less sensitive, and polymerase β is completely resistant. Aphidicolin is a highly specific inhibitor of DNA polymerase α. This compound blocks chain elongation in SV40 viral and cellular DNA replication.

The action of the arylhydrazinopyrimidines on *B. subtilis* DNA polymerase III has been studied by enzymatic, genetic, and physical methods (Figure 26–24). These compounds have been used to isolate mutants of *B. subtilis* polymerase III and to establish the role of this enzyme in DNA replication. The inhibitory action of the arylhydrazinopyrimidines is believed to be due to the formation of ternary 1:1:1 complexes with DNA polymerase III and the DNA template primer. The presence of primer is required for formation of this ternary complex. Inhibitor binding to the template is a base-specific interaction, with hydroxyphenylhydrazinouracil [OHPhe(NH)₂Ura] hydrogen bonding to C residues and the isocytosine [OH-Phe(NH)₂Iso] derivative hydrogen bonding to T residues.

DNA DEGRADATION

Chemical degradation of DNA was considered in Chapter 7. In this chapter we will consider enzymatic degradation. Certain enzymes involved in DNA synthesis also have a capacity for cleaving phosphodiester linkages. The bacterial polymerases I and III both have exonucleolytic activity associated with them, removing nucleotides one at a time from either end of a polymer. All topoisomerases have endonucleolytic activity, making transient breaks which they invariably mend. There are also a large number of nucleases and other enzymes that are exclusively degradative in their action. It is presumed that these enzymes serve functions useful to the cell; many of them also have been useful in manipulating DNA *in vitro*.

Nucleases Attack Phosphodiester Bonds

Some nucleases attack phosphodiester bonds in a polynucleotide chain from one end (or the other); others carry out a random attack on internal bonds. Nucleases attack phosphodiester bonds (Figure 26–25). Hydrolysis of an internal linkage is classified as endonucleolytic and is carried out by enzymes called *endonucleases*. The site of phosphodiester bond cleavage may be on the 3'-phosphate side, leading to the formation of 5'-phosphates, or on the 5'-phosphate side, leading to the formation of 3'-phosphates. The attack from the ends of chains is an exonucleolytic process, and the enzymes are grouped as *exonucleases*. Exonucleases may be further distinguished according to whether the direction of cleavage is from the 3' or the 5' end or, in some cases, from both ends of the polynucleotide chain. The initial products

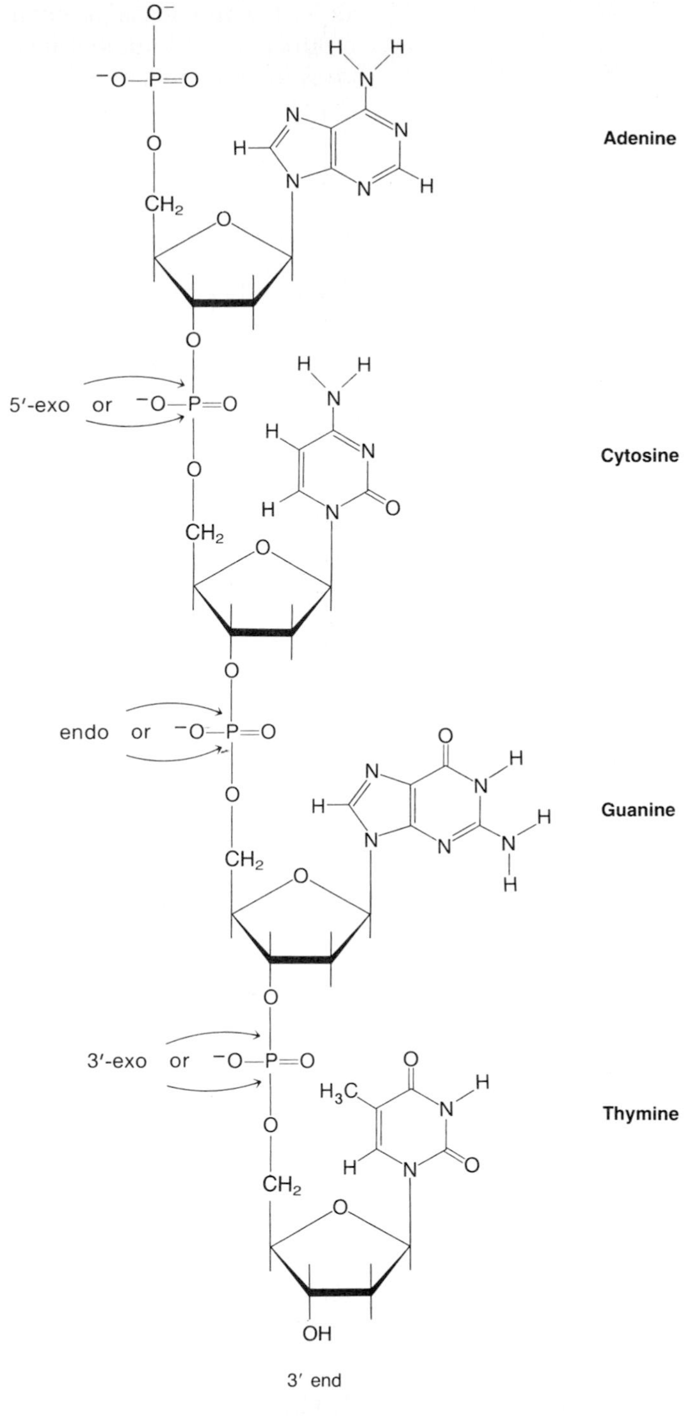

Figure 26–25
A polydeoxyribonucleic acid chain, indicating points of attack by exonucleases and endonucleases. Enzymes that attack the 5′ end are referred to as 5′ exonucleases and those that attack the 3′ end are referred to as 3′ exonucleases. Depending on the P-O linkage attacked, these produce 3′- or 5′- phosphomononucleotides.

of endonucleases are mixtures of oligonucleotides of varying chain lengths. The products of exonucleases are usually either 3′- or 5′-phosphomononucleotides. Exonucleases have proven useful in characterizing circular DNAs (on which they do not act) and in generating single-stranded termini (for structural or sequence studies or deleting sequences).

A phosphodiesterase isolated from rattlesnake venom (*Crotalus adamanteus*) is a 3′→5′ exonuclease (3′-exonuclease) that acts on single polydeoxynucleotide or polyribonucleotide chains to produce only 5′-nucleotides. The enzyme has an alkaline pH optimum and requires no added divalent metal ions for activity. The opposite specificity is exhibited by a phosphodiesterase isolated from bovine spleen. Spleen phosphodiesterase is a 5'→3 exonuclease (5'-exonuclease) that acts on DNA or RNA substrates to produce only 3'-nucleotides. Most rapid action is observed if the 5'-terminal phosphate of the polynucleotide is first removed by a monophosphatase. *E. coli* exonuclease I exhibits 3'→5' exonuclease activity, much like venom phosphodiesterase. In contrast to the venom enzyme, exonuclease I requires Mg²⁺, will not split the terminal 5'-dinucleotide of a DNA chain, and will hydrolyze glycosylated DNAs (encountered in the T-even series of *E. coli* bacteriophages). *E. coli* exonuclease III and bacteriophage T4 exonuclease are only capable of degrading double-stranded DNA. Both are 5'-nucleotide formers, but T4 exonuclease attacks at the 5' end of a polynucleotide, whereas exonuclease III starts at the 3' end. If the DNA chain has a 3'-phosphate, exonuclease III can hydrolyze that phosphomonoester bond before beginning 3'→5' degradation to generate 5'-nucleotides. Since exonuclease III will act on DNA fragments with a 5' (but not a 3') overlap, it can be very useful in digesting DNA from one end only by first cutting duplex DNA with appropriate *restriction enzymes*. Some of these well-characterized exonucleases, which cleave duplex DNAs at specific sequences, are listed in Table 26–5.

Deoxyribonuclease I, a digestive enzyme secreted by the pancreas, acts on single- or double-stranded DNA. Initially, only one chain of the double helix is cleaved, generating a 5'-phosphate and 3'-OH termini at the point of cleavage. Continued digestion eventually leads to mixtures of di-, tri-, tetra-, and some higher oligonucleotides. Deoxyribonuclease II, a lysosomal enzyme isolated from the spleen, splits both chains of DNA simultaneously. S1 endonuclease (from *Aspergillus oryzae*) is a single-strand-specific enzyme that attacks both DNA and RNA. It is often used to remove single-stranded (nonhybridized) regions after solution-hybridization experiments or single-stranded regions generated by restriction enzymes or enzymes such as exonuclease III (in either case generating blunt ends) (also see Chapters 27 and 32). Ribonuclease H degrades RNA but only when it is base-paired with RNA or DNA. Another enzyme acting on nucleic acids is alkaline phosphatase (referred to as BAP when isolated from bacteria and as CAP when isolated from calf intestine). This enzyme catalyzes the removal of 5' phosphate from DNA, RNA, oligonucleotides, and deoxyribonucleotide triphosphates. The highest degree of sequence specificity on duplex DNA is exhibited by site-specific endonucleases known as restriction enzymes (see below and Chapter 32).

AP endonucleases (apurinic or apyrimidinic) exhibit specific cleavage at DNA sites lacking a base. Sites of this kind are produced when DNA glycohydrolases (described below) remove damaged or incorrect bases from DNA. These enzymes are widely distributed and apparently participate in certain DNA repair processes.

TABLE 26–5
Exonucleases

Name	Source	Substrate	Product	Direction	Divalent Cation Requirement
Venom phosphodiesterase	Snake venom	Oligo or poly	5'-dXMP	3'→5'	−
Spleen phosphodiesterase	Beef spleen	Oligo or poly	3'-dXMP	5'→3'	−
E. coli Exo I	Bacteria	Oligo or poly	5'-dXMP	3'→5'	+
E. coli Exo III[a]	Bacteria	Double-stranded DNA	5'-dXMP	3'→5'	+
λ-Exonuclease	λ-Infected *E. coli*	Double-stranded DNA	5'-dXMP	5'→3'	+

[a]Has an associated 3'-phosphatase activity.

Glycohydrolases Attack *N*-glycosidic Bonds

The hydrolysis of the N-glycosidic bonds of purine and pyrimidine mononucleosides is a well-established enzymatic reaction in the degradation pathway for nucleic acids. Enzymes that degrade nucleosides do not act on intact nucleic acids, but other special nucleic acid-base glycohydrolases that can cleave N-glycosidic bonds in DNA have been described. They appear to act only on DNA with abnormal bases or abnormal base pairing and are thought to be part of the cellular mechanism for repair of DNA damage.

When uracil appears in DNA, as a result of misincorporation of dUTP or deamination of cytosine, it can be removed by a uracil-DNA glycohydrolase. Similarly, hypoxanthine—resulting from dITP incorporation or from deamination of adenine—can be removed by a hypoxanthine-DNA glycohydrolase. Separate enzymes for removing alkylated bases also have been discovered. All these enzymes participate in the repair process by removing inappropriate bases in a process known as base excision repair (see below).

Postreplicative Modification Protects DNA from Certain Nucleases

Prokaryotes contain special types of DNA modification enzymes that provide protection from specific endonucleases. The discovery of this phenomenon, known as *restriction-modification*, dates back to the studies of Salvator Luria on the T-even *E. coli* bacteriophages. The DNA of bacteriophages T2, T4, and T6 contains 5-hydroxymethylcytosine (HMC) instead of cytosine. The HMC residues of the DNA of the T-even bacteriophages are uniquely glycosylated postreplicatively. Luria found that sometimes a phage preparation was unable to grow (was restricted) on an alternate *E. coli* host. Studies showed that when growth did occur, glycosylation (modification) protected the infecting phage DNA from certain degradative (restriction) enzymes produced by the host.

The studies on the T-even phages were extended to phage λ and the small, single-stranded *E. coli* phages. Phage that was propagated (and its DNA modified) on one strain (e.g., *E. coli* strain B) would show a high degree of plating efficiency on the same strain, but plated very inefficiently (growth was restricted) on an alternate host strain (e.g., *E. coli* strain K-12). Chemical analyses of the small phage DNAs revealed that modification resulted in a small increase in methylated bases. More precise biochemical observations by Meselson and Yuan revealed that a *modification enzyme* methylated (using S-adenosylmethionine as donor) duplex unmethylated DNA at specific sites and that a *restriction enzyme* degraded the unmethylated DNA.

These pioneering studies on restriction-modification led to the discovery of a class of restriction endonucleases now known as type I enzymes. Even though the type I restriction enzymes recognize specific unmethylated sites, the DNA restriction sites are located randomly and are usually far removed from these specific recognition sites. The type I restriction endonucleases, found for example in strains *E. coli* B and *E. coli* K-12, are multifunctional proteins that can carry out both restriction and modification. They have three subunits, one of which recognizes a particular DNA sequence that can be either methylated or restricted, in a mutually exclusive manner, by either of the other two subunits. Type I enzymes modify by methylating the 6-amino group of adenine in a specific DNA sequence.

Shortly after the type I restriction enzymes were characterized, specific endonucleases that cleaved duplex DNA at specific sites were detected and characterized. Thus Hamilton Smith in 1970 found that *Haemophilus influenzae* cell-free extracts cut heterologous DNA at specific sequences, but failed to cut its own (modified) DNA. Since then about 200 different type II restriction enzymes (found only in bacteria), recognizing different specific cleavage sites, have been characterized. In the type II systems, different enzymes methylate either the 6-amino group of adenine or carbon 5 of cytosine. The type II restriction enzymes will be discussed in greater detail in Chapter 32, as they have been very important in DNA manipulations *in vitro*. In a third class of restriction-modification enzymes, the type III enzymes, the specific site recognized is usually 24–26 bases away from the cleavage site.

E. coli also possesses DNA adenine and cytosine methylases that are different from those involved in restriction-modification. The *dam* gene encodes an adenine methylase that recognizes the sequence pGATC in double-stranded DNA. The *dcm* gene encodes a cytosine methylase. These methylations may serve as signals for other metabolic processes; for example, they may signal the initiation of DNA or RNA synthesis.

In *E. coli* the *dam* methylase plays an important role in *mismatch repair*. Mismatches are generated during DNA replication (those not removed by the proofreading function of polymerase III) or recombination. It is estimated that base misincorporation occurs in *E. coli* at a frequency of 10^{-5} to 10^{-4} per base pair per generation. The ϵ subunit of DNA polymerase III supplies a proofreading function that increases the fidelity to 10^{-8} to 10^{-7}. The remaining mismatches are reduced to 10^{-10} to 10^{-9} by mismatch repair. Prior to DNA replication the parental DNA is already methylated at GATC sequences, but the newly synthesized strand (with its potential mismatch) remains unmethylated for a while. This provides a signal for a mismatch to be repaired, the excised base being primarily on the newly synthesized strand.

There are no restriction enzymes in eukaryotes. Nevertheless, some eukaryotes show postreplicative methylation. The principal methylated bases in mammalian DNA contain only 5-methylcytosine, found chiefly in the sequence C*pG. Analysis of specific sequences for methylation in different cells of a multicellular organism shows strikingly different patterns of methylation. Methylation at the 5′ upstream region of some genes has been correlated with gene expression and differentiation. In certain cases undermethylation (or hypomethylation) has been shown to correlate with gene activation and excess methylation with decreased gene activity. Less frequently, genes appear to be *activated* by methylation. Many eukaryotes, for example, the fruit fly *Drosophila* and the yeast *Saccharomyces cerevisiae*, contain no methylated DNA.

DNA REPAIR

Abnormalities in DNA in the form of a mispaired base or biochemical alteration of the preformed structure would obviously cause high mutation rates or lethal effects if there were no mechanism for correcting them. One mechanism, the proposed "proofreading" 3′-exonuclease function associated with DNA polymerases I and III, has already been described as a means of correcting some errors made during synthesis. For correcting damage to preformed DNA, more elaborate systems exist. The field of DNA repair has received increasing attention as it has become clear that most types of repairable damage in DNA are both mutagenic and carcinogenic.

Damage to DNA is caused by a variety of agents, including ultraviolet light, ionizing radiation, and reactive chemicals. Damage can be in the form of a missing, incorrect, or modified base or an alteration in the structural integrity of the DNA strands by breaks, cross-links, or base dimers.

Some chemicals that modify the bases in nucleic acids are very specific for one base. Hydroxylamine (NH_2OH) forms a specific adduct with cytosine

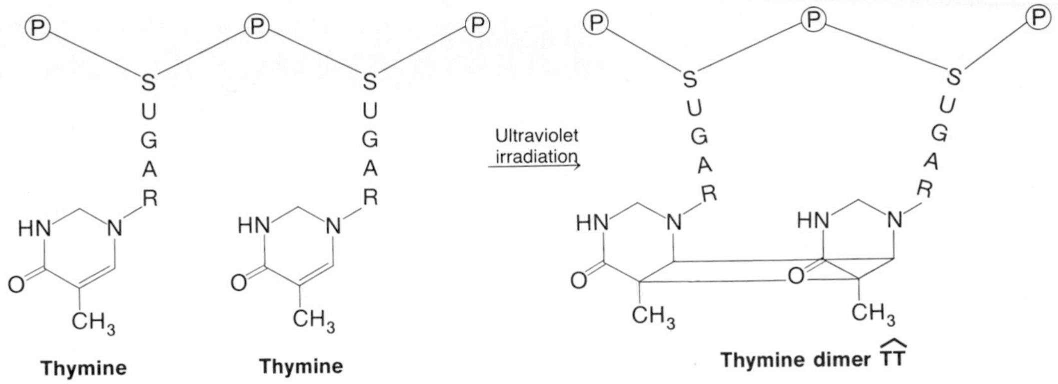

Figure 26–26

Structure of thymine dimer formed by adjacent thymines when DNA is irradiated with ultraviolet light.

that can then base-pair with adenine. Nitrous acid, however, is a general reagent used to deaminate bases, converting adenine to hypoxanthine, guanine to xanthine, and cytosine to uracil. Other compounds that cause DNA damage alkylate the bases or form other covalent adducts. Nitrosamine derivatives usually must be converted to more reactive species by biological oxidation before becoming active as base-alkylating reagents. The formation of alkyl adducts may alter the pairing characteristics of the bases. N-methyl-N-nitrosoguanidine is a particularly potent nitrosamine that has found widespread use in the laboratory for generating mutations in bacteria. It acts with greater efficiency at the growing point during DNA synthesis.

Aromatic polycyclic hydrocarbons, such as 2-acetamidofluorene, benzo(a)pyrene, and the mycotoxin (known as aflatoxin B), form adducts with nucleic acids. Most of these compounds are inactive by themselves but can be converted to highly reactive derivatives within organisms. The reactive derivatives are nonselective and can modify most nucleic acid bases. The bulky groups that are introduced by these adducts apparently act by preventing base pairing rather than by causing errors of base pairing, such as those seen with the simpler alkylating agents.

Adjacent pyrimidine bases in a DNA strand form dimers with high efficiency after absorption of ultraviolet light (Figure 26–26). By contrast, purines are quite resistant to damage by ultraviolet. Pyrimidine dimers formed within an otherwise intact DNA duplex have provided a useful substrate to assay for DNA repair. These dimers can be repaired directly by enzymatic photoreactivation (Figure 26–27). The photoreactivation enzyme binds to the DNA containing the pyrimidine dimer and uses visible light to cleave the dimer without breaking any phosphodiester bonds.

Pyrimidine dimers and other forms of DNA damage can also be removed by a general *excision repair* mechanism. In this form of repair the lesion in the DNA is removed and repaired. The well-defined genetic system of *E. coli* has promoted an understanding of the enzymatic steps involved in excision repair, particularly in response to UV-induced damage. The genes involved in DNA excision repair fall into three categories: those that are exclusively involved in repair processes, those that are involved in both repair and recombination processes, and those that are also involved in other aspects of DNA replication (Table 26–6).

The removal of ultraviolet-induced pyrimidine dimers in bacterial cells is catalyzed by two mechanisms (Figure 26–28). Special glycosylases that recognize abnormal or incorrectly paired bases can cleave N-glycosidic bonds to generate an apurinic or apyrimidinic (AP) site (Figure 26–28a). Alternatively, a double incision is made in the strand that carries the lesion

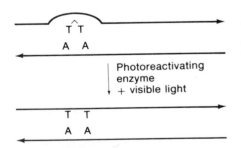

Figure 26–27

Thymine dimers may be monomerized from DNA by enzymatic photoreactivation. In this case no nucleotides are removed in the repair reaction.

TABLE 26–6
Some *E. coli* Genes that Affect Responses to DNA Damage

Gene	Map Location	Function
***uvr* Genes**		
uvrA	91	Gene products work together to make initial incision at or near the site of DNA damage.
uvrB	17	
uvrC	42	
uvrD		Helicase
***rec* Genes**		
recA	58	Structural gene for recA protein. *RecA* carries a highly specific protease that is activated by DNA damage.
recB	60	Encodes two subunits of exonuclease V. Believed to make the initial nick required to initiate DNA recombination.
recC	60	
Other genes		
lexA	90	Controlling gene for *recA* and other SOS functions
polA	85	Structural gene for DNA polymerase I
polC	4	Structural gene for DNA polymerase III (same as *dnaE*)
lig	51	Structural gene for DNA ligase

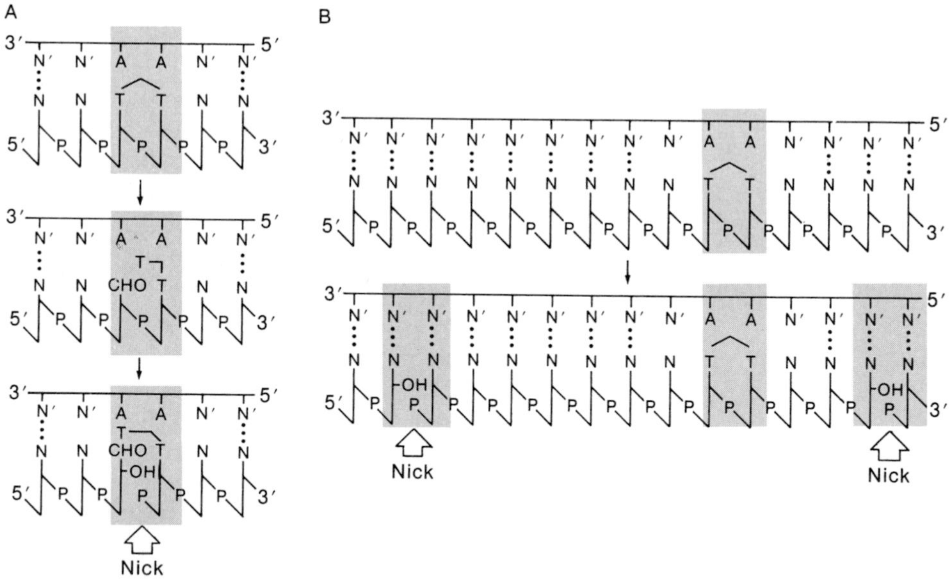

Figure 26–28
Pyrimidine dimers and other forms of DNA damage can be removed by a general excision repair mechanism. The first reaction in this form of repair involves forming nicks about the damaged region of the DNA. In A we see the mode of incision of UV-irradiated DNA by the pyrimidine-dimer-specific glycosylase and AP endonuclease activities of *M. luteus* and bacteriophage T4. In B we see the mode of incision of the UvrABC endonuclease of *E. coli*. (Adapted from Graham Walker, *Ann. Rev. Biochem.* 54:425, 1985.)

by the enzymes of the *uvrA*, *uvrB*, and *uvrC* system (Figure 26–28b). In the case of pyrimidine dimers, one incision is made seven nucleotides 5' to the pyrimidine dimer, and then a second cut is made three or four nucleotides 3' to the same dimer. The *uvrD* gene product (shown to be helicase II), together with DNA polymerase I and possibly a single-strand DNA-binding protein, appears to act to release the 12–13-nucleotide oligomer generated by the incision of the uvrABC complex. After the release of the oligomer, DNA polymerase I and ligase are required to resynthesize the excised region and

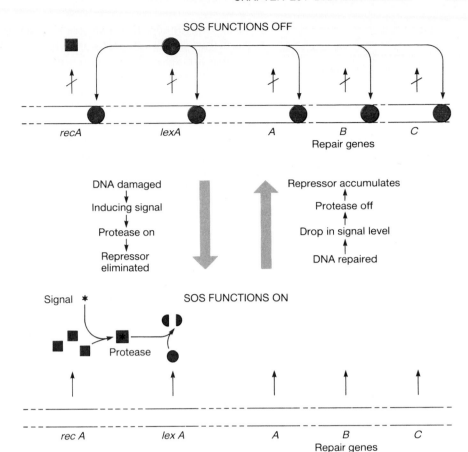

Figure 26–29

Model for the SOS regulatory system. In normally growing cells the SOS functions associated with DNA repair are not expressed. This is because lexA repressor inhibits their transcription. LexA repressor also inhibits its own expression and that of recA. SOS functions are turned on by a series of reactions that starts with DNA damage. DNA damage results in an inducing signal that activates the protease function of recA. This protease cleaves lexA protein, so that all genes that were formerly inhibited by lexA can be expressed. Once the damage is repaired, the level of lexA repressor builds up again and the SOS genes return to their usual repressed state.

ligate the nicks. Recall that certain Pol I mutants have an increased sensitivity to ultraviolet light. There is very little expansion of the gap (called a _short patch_, about 20 nucleotides long). When larger sequences are involved, the repair is called a _long patch_.

A very interesting aspect of bacterial repair is the regulation of the uvr genes, together with other genes involved in a variety of repair and genetic processes. The activities of these genes are regulated by the so-called SOS system, responding to the distress signal of DNA damage. At the heart of the SOS network are two genes, _lexA_ and _recA_ (Figure 26–29). Under normal conditions the lexA protein acts as a repressor binding to about 17 SOS genes, whose encoded proteins are involved in excision repair, gap repair, double-strand break repair, and methyl-directed mismatch repair as well as SOS processing. These genes are collectively referred to as _din_ (damage-inducible) genes.

The recA protein has two roles. It catalyzes recombination between homologous DNA molecules in _E. coli_ (discussed in the following section) and regulates, in an activated state, the induction of the SOS responses. It does this by mediating the cleavage of the lexA protein at an -Ala-Gly- bond as a result of the interaction of the two proteins. Thus an insult to DNA that leads to DNA damage somehow activates the protease function of recA. The protease cleaves the lexA protein, and the result is a greatly augmented synthesis of proteins associated with DNA repair. Once the DNA damage is removed, the activated _recA_ function is turned off and newly synthesized lexA protein again represses the DNA repair genes.

In mammalian cells there is no evidence for inducible long-patch excision repair. If the lesions are not excised, perhaps they are bypassed rather than repaired, and are subsequently eliminated by recombinational processes.

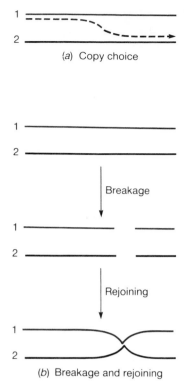

(a) Copy choice

Breakage

Rejoining

(b) Breakage and rejoining

Figure 26–30
Two basically different ways of
visualizing the recombination process.
The figures represent two
chromosomes, 1 and 2, undergoing
recombination. (a) In copy choice, the
enzyme jumps during synthesis from
copying one chromosome to copying
the other chromosome. (b) In breakage
and rejoining, the two chromosomes
break at homologous points and rejoin
to give recombinants. The process
could be reciprocal as shown, or
nonreciprocal, in which case only one
chromosome would survive.

In humans, the inability to repair DNA damage is associated with several rare genetic syndromes. The best known is *Xeroderma pigmentosum;* there are several variations of this condition. People with the disease are unable to repair the DNA damage caused by exposure to ultraviolet light and some chemicals. Individuals with different syndromes associated with defective DNA repair mechanisms show different sensitivities to damaging agents. This fact suggests that there are several enzyme systems for DNA repair in humans. Hypersensitivity to DNA-damaging agents may be caused by a defect in a single enzyme in any one of the different pathways used for DNA repair. The multiplicity of defects in repair in different patients has been confirmed by complementation studies carried out with cells grown in culture; there are about six to eight complementation groups, suggesting at least an equal number of loci that encode enzymes involved in repair of DNA damage in human cells.

DNA RECOMBINATION

Recombination involves the rearrangement of DNA on the same or on different chromosomes. There are three basically different types of recombination. The first and most common is called *general recombination*. This form of recombination takes place between identical or nearly identical segments of chromosomes such as homologously paired eukaryotic chromosomes during meiosis, or bacteriophage chromosomes during multiplication. The second type is *site-specific recombination*. This kind of recombination is limited to highly select regions of the genome. The best-understood example of site-specific recombination involves integration of bacteriophage DNA into the *E. coli* host chromosome. The third type, *nonspecific recombination*, occurs between nonhomologous regions of chromosomes; little or no site specificity is involved.

In the 1960s Meselson proposed *a three-step model for general recombination: (1) formation of staggered breaks in two parental DNAs; (2) base pairing between single-stranded regions of the two parental types; and (3) repair synthesis* (Figure 26–30). Meselson's model explained many experimental results. In fact, it went beyond that. The proposal that base pairing takes place between single-stranded regions of the different parental DNAs provided an explanation for why chromosomes tend to recombine in homologous regions. *The main approach to the enzymology of recombination has been to correlate mutations that affect recombination with actual enzyme activities.* The ultimate goal of these studies is to reconstitute a cell-free system with purified enzymes that can carry out recombination. Most of this work has taken place on *E. coli*.

Before discussing some of the actual results, we might speculate on some of the enzyme activities that would be needed, given the activities required by the model for recombination. First, an endonuclease activity is required to make the initial incision or incisions in the DNA. Second, an activity is required to catalyze single-strand invasion of a duplex structure and subsequent unwinding and rewinding of base-paired structures. Third, enzymes are required for repair, synthesis, and ligation. Fourth, an enzyme is required that removes (but not always) mismatched bases. Fifth and finally, another endonuclease is required that makes scissions that resolve the complex into separate chromosomes. The third and fourth requirements are probably common to DNA synthesis and damage-repair processes that have already been discussed. Nothing is known about the fifth item. Consequently, the discussion here will be focused on the first two activities, about which a great deal is known in *E. coli*.

The search for enzyme activity associated with recombination has been based on analysis of organisms carrying mutations that affect general recombination. Such mutants, known as *rec* mutants, were first discovered more than 15 years ago, and new ones are still being found. We will discuss only a few of the most important results. The first *E. coli rec* mutations, *recA*, *recB*, and *recC*, reduce recombination efficiency to different extents. The *recA* mutants are totally deficient in homologous recombination, while *recB* or *recC* mutants can recombine only about 10^{-4} times as well as wild-type cells. In addition, mutations in two other genes will reverse the defect of *recB* or *recC* mutants; no comparable second-site suppressors of *recA* have been found. This lack suggests that *recA* is absolutely required for homologous recombination but that there may be alternate pathways to recombination that do not require *recB* and *recC*.

Since the recA protein is indispensable for general recombination, we shall discuss it first. Quite surprisingly, the *recA* gene product ($M_r = 40,000$) carries out two types of reactions that have little biochemical similarity. First, it acts as a protease to cleave certain specific polypeptides, such as the lexA protein (see Fig. 26–29). The protease function of recA will not be discussed further here, since it appears to have nothing directly to do with recombination.

The second class of reactions catalyzed by recA protein suggests that the enzyme is involved in the initiation of homologous pairing during genetic recombination. Thus the purified recA protein catalyzes various forms of complex formation and exchange between duplex and single-stranded DNAs in reactions that entail the cleavage of ATP (Figure 26–31). D-loop formation can result when the recipient is a circular duplex and the donor is a single strand (Figure 26–31a). Partial circular duplex formation can result from the exchange between a circular single strand and a linear duplex (Figure 26–31b). Strand exchange can occur where a single-stranded fragment is removed in the presence of a double-stranded fragment (Figure 26–31c). Finally, recA protein will also cause a complex to form between two circular helices, provided that at least one input helix is gapped on one strand (Figure 26–31d). Interestingly, while the two helices must contain homologous sequences in order to form a four-stranded structure, the gap may lie in a nonhomologous region. The fact that mutations in *recA* completely eliminate recombination, as well as the finding that the recA-related protein catalyzes strand exchange reactions like those required in all three of the recombination models we have been discussing, makes it highly likely that *recA* plays a key role in these processes *in vivo*. If this is so, then recA protein is clearly situated at the hub of activities in the recombination complex.

The recA enzyme will not catalyze any recombination event unless at least one free end of a single strand or a DNA duplex is available. Thus it seems unlikely that the recA protein could be involved in making the initial incision(s) in the duplex structure that are required to initiate recombination. There are reasons for believing that the *recB* and *recC* genes are involved in this capacity. First recall that a mutation in either of these genes reduces recombination by a factor of about 10^{-4}. Second, it has been shown that these two genes together encode the subunits of the enzyme exonuclease V (exoV). This enzyme is both a helicase and a nuclease. The helicase activity is best demonstrated *in vitro* under conditions where the nuclease activity is blocked; this can be done by adding Ca^{2+}. When linear duplex DNA is incubated with exoV in the presence of ATP and Ca^{2+}, it begins to unwind at one end. As the unwinding progresses, the single-stranded regions collapse back on each other to re-form a duplex. A double-loop structure is maintained in the vicinity of the migrating enzyme (Figure 26–32).

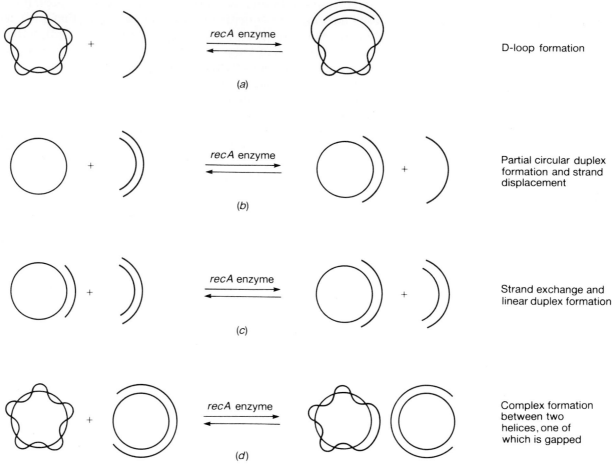

Figure 26–31

Reactions catalyzed by purified recA protein *in vitro*. RecA catalyzes a number of different reactions between DNA strands, all of them involving the unwinding and winding of base-paired structures. *(a)* D-loop formation by interaction between supercoiled circular duplex DNA and single-stranded DNA. *(b)* Partial circular duplex formation by displacement of one strand from a duplex structure. *(c)* Strand exchange between a gapped circular duplex structure and a linear duplex structure. *(d)* Complex formation between two helices, one of which is gapped.

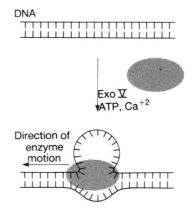

Figure 26–32

DNA unwinding by exoV enzymes. This is best observed *in vitro* in the presence of ATP and Ca^{2+}. The Ca^{2+} inhibits the exonucleolytic activity of the enzyme. The ATP provides the energy that drives the enzyme in a concerted manner into the duplex structure. The duplex unwinds in front of the path of the enzyme and rewinds in back of the enzyme.

What is the point of this migrating behavior of *recB*? There are recent indications that the recBC enzyme may be scanning the duplex structure to find a site where it would be appropriate to make the strand scission required to initiate recombination. These indications come from detection of an octanucleotide sequence known as *chi*, 5'-G-C-T-G-G-T-G-G-3', which is associated with regions highly active for recombination. The *chi* sequence does not stimulate recombination in special *recBC* mutants where exoV retains some recombination proficiency but lacks detectable nuclease activity. This observation has led to the suggestion that *recBC* normally scans the duplex and makes strand incisions in or near regions containing the *chi* sequence.

SUMMARY

This chapter deals with reactions involved in DNA synthesis, degradation, and repair. DNA replication was shown to take place by the synthesis of one new strand on each of the parental strands. This mode of replication is called semiconservative and it appears to be universal. DNA synthesis initiates from a primer at a unique point on a prokaryotic template such as the *E. coli* chromosome. From the initiation point, DNA synthesis proceeds bidirectionally on the circular bacterial chromosome. This bidirectional mode of synthesis is not followed by all chromosomes. For some chromosomes, usually small in size, replication is unidirectional.

In eukaryotic systems replication can start at several points (not well defined) along the chromosome. Replication is usually bidirectional about each initiation site. The termination points of replication are interspersed between initiation sites. In most cases of unidirectional or bidirectional replication, synthesis occurs nearly (but not exactly) simultaneously on both strands of the parent DNA template. Since synthesis can occur only in the 5'→3' direction on the growing chain, and since the two strands in the parent duplex are oriented in opposite directions, this means that synthesis can occur continuously on only the leading strand. On the other (lagging) strand it must pause for the template to unwind. Synthesis on the lagging strand does not occur continuously but rather in small discontinuous spurts, generating Okazaki fragments.

Many proteins are required for DNA synthesis and chromosomal replication. In addition to the main polymerizing enzymes, a number of other enzymes and proteins are required for optimal DNA synthesis: (1) helicases, that unwind the parental duplex; (2) enzymes that fill in the gaps

and join the ends in the case of lagging-strand synthesis; (3) enzymes that synthesize RNA primers at various points along the DNA template; (4) topoisomerases, which permit rotation and supercoiling, and (5) single-strand DNA-binding proteins, which stabilize single-stranded regions that are transiently formed during replication. Most of these proteins have been isolated from whole cells and studied in cell-free systems.

In *E. coli*, mutations have been obtained in the genes encoding a number of these enzymes. Many of these mutations are conditional, since the functional enzymes involved are required for DNA synthesis and cell viability. Mutants carrying mutationally altered proteins have been important in confirming their roles as predicted from cell-free studies.

In eukaryotes most of the work on enzymes involved in DNA synthesis has been done without mutants. Considerable progress has been made in studying the *in vitro* replication of animal viruses such as SV40 and adenovirus. The importance ascribed to the enzymes that have been characterized is largely based on a comparison of their properties with similar prokaryotic enzymes whose functions are better understood.

Some enzymes that act on DNA are involved in processes other than DNA synthesis. They include DNA repair enzymes, DNA degradation enzymes, and DNA recombination enzymes. The most useful class of degradation enzymes, for recombinant DNA research purposes, have been the restriction enzymes, which permit the cleavage of high-molecular-weight DNAs at discrete target sites. Their use is discussed extensively in Chapter 32.

SELECTED READINGS

Bollum, F. J., Mammalian DNA Polymerases in Progresses. In W. E. Cohn (ed.), *Nucleic Acid Research and Molecular Biology*, Vol. 15. New York: Academic Press, 1975. Pp. 109–144. A review of research on eukaryotic DNA polymerases.

Campbell, J. L., Eukaryotic DNA Replication. *Ann. Rev. Biochem.* 55:733, 1986.

Grossman, L., and K. Moldave (eds.), *Methods in Enzymology*, Vol. 65, Part I: *Nucleic Acids*. New York: Academic Press, 1980. Discusses the use of restriction endonucleases and RNA and DNA sequencing techniques.

Kornberg, A., *DNA Replication*. San Francisco: W. H. Freeman, 1980. An enlarged version of "DNA Synthesis," treating prokaryotic DNA replication in outstanding detail. See also the 1982 supplement to this book.

Low, R. L., K.-I. Arai, and A. Kornberg, Conservation of the primosome in successive stages of ϕX174 DNA replication. *Proc. Natl. Acad. Sci. USA* 78:1436, 1981.

Marians, K., Enzymology of DNA replication in prokaryotes. *Crit. Revs. Biochem.* 17:153, 1984.

McHenry, C. A., DNA Polymerase III holoenzyme of *Escherichia coli*: components and function of a true replicative complex. *Mol. Cell. Biochem.* 66:71, 1985.

Strauss, B. S., Cellular aspects of DNA repair. *Adv. Cancer Res.* 45:45–62, 1985. A thorough review of DNA repair in eukaryotic cells.

Walker, G. C., Inducible DNA repair systems. *Ann. Rev. Biochem.* 54:425–452, 1985. A very thorough review of DNA repair in prokaryotes.

Wu, R., *Methods in Enzymology*, Vols. 68 and 101. New York: Academic Press, 1979. State-of-the-art treatment of the concepts of DNA sequencing and recombinant DNA technology.

PROBLEMS

1. The circular map below represents the distribution of eight genes (a–h) as they appear on a bacterial chromosome. The plot is a graphic representation of the average number of gene copies per cell of each of the eight genes for a cell growing with a doubling time equal to the time required for a complete round of DNA replication.
   a. Estimate the location of the origin of replication.
   b. Infer whether replication is bidirectional or unidirectional.
   c. Draw a gene copy plot for the same cell growing with a doubling time equal to half the time required for a complete round of DNA replication.

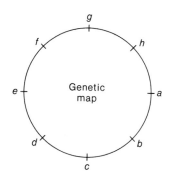

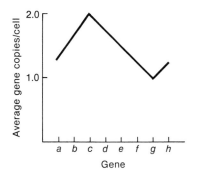

2. a. Diagram in detail the steps involved in the synthesis of the ϕX174 phage minus strand (ssDNA → RF) and briefly describe the function of each protein required at every step. What aspects of the replication of the *E. coli* chromosome are these steps likely to reflect?
   b. Would you expect the replication of either ϕX174 DNA (ssDNA → RF → ssDNA) or the *E. coli* chromosome to be sensitive to inhibition by the antibiotic rifampicin, and if so at what stage?

3. λ phage DNA is linear and double-stranded with overlapping complementary ends ("sticky ends"). DNA ligase converts linear λ DNA into closed circular DNA. Pretreatment of linear λ DNA with either bacterial alkaline phosphatase or spleen phosphodiesterase prevents circularization by DNA ligase. Explain. (For more about recombinant DNA methodology, see Chapter 32.)

4. a. You have isolated a conditionally lethal temperature-sensitive mutant of *E. coli* that lacks the 3'→5' exonuclease activity of DNA polymerase III. What consequences would you expect this mutation to have for DNA replication at the nonpermissive temperature?
   b. You find that a secondary mutation in the gene for DNA polymerase I suppresses the lethality of the first. Suggest a possible basis for this suppression.

5. T7 bacteriophage DNA is a linear duplex that replicates bidirectionally, starting at a point near the end of the duplex. There are special problems associated with the synthesis of DNA at the 3' ends of the template. What are these and how might they be resolved? (Hint: T7 DNA is terminally redundant, in that it contains a direct repeat of 160 bp at either end of the DNA.)

6. Describe briefly the roles of rep protein and topoisomerases I and II in *E. coli* DNA replication. In what sense may these roles be viewed as complementary?

7. Devise a protocol for the preparation of pBR322 plasmid DNA with five negative supercoils, starting from a preparation of relaxed duplex DNA (refer also to Chapter 7).

8. What roles do DNA methylases play in bacteria, and why is it necessary for methylation to occur in most instances near the replication fork?

9. The ends of eukaryotic chromosomes, called telomeres, have been found to consist predominantly of tandemly repeated units of simple (G + C)-rich sequences of variable length. Although yeast and *Tetrahymena* telomeric sequences differ, it has been observed that both will serve as templates for the addition of repeat units to their ends in *Tetrahymena* cell extracts. In both instances the repeat unit added is the same and is that characteristic of *Tetrahymena* telomeres.

a. To what problem in eukaryotic DNA replication do you think the activity responsible for this addition represents a solution?

b. Given the stated properties of the activity, would you expect the generation of telomeric ends to be sensitive to inhibition by aphidicolin? Explain.

10. Describe an enzymatic test for distinguishing between the following.

a. Single-stranded and double-stranded DNA.

b. Single-stranded linear and single-stranded circular DNA.

c. Duplex linear and duplex circular DNA.

11. Describe the steps involved in sequencing DNA by the Maxam-Gilbert and the Sanger techniques. In what principal ways do these techniques differ, and in what ways are they similar? (Refer to Chapter 7 for the Maxam-Gilbert method.)

27

RNA METABOLISM

In the early 1950s, Gamov suggested that DNA might serve as a template for protein synthesis. However, biochemical investigations showed that in eukaryotes the processes of RNA and protein synthesis are neatly compartmentalized in the nucleus and cytoplasm, respectively. For this reason, it seemed unlikely that DNA could serve directly as a template for protein synthesis.

Base composition studies by Erwin Chargaff, which had provided information that was vital in deducing DNA structure and in suggesting a mechanism for DNA replication, were less helpful in regard to the structure and synthesis of RNA. Total cellular RNA did not reflect the base composition of the DNA. However, Volkin and Astrachan, in a monumental experiment, were able to show that when a T2 bacteriophage infects an *E. coli* cell, the newly synthesized RNA (which is exclusively transcribed from the phage DNA template) reflects the base composition of the DNA. From experiments such as this, and from an understanding of the DNA duplex structure, there emerged the hypothesis that DNA, by unwinding, serves as a template for its own synthesis and for the synthesis of RNA.

Crick elegantly rephrased these ideas into the "*central dogma*," which states that genetic information flows from DNA to RNA and thence to protein:

$$\text{DNA} \longrightarrow \text{RNA} \longrightarrow \text{protein}$$

This dogma has provided the theoretical framework that experimentalists used to design ways of demonstrating the existence of a transcribing enzyme that copies the DNA sequence into an RNA molecule.

The first enzyme discovered that could catalyze polynucleotide synthesis was polynucleotide phosphorylase. This enzyme, isolated by Severo Ochoa and Marianne Grunberg-Manago in 1955, could make long chains of $5' \rightarrow 3'$ linked polyribonucleotides starting from nucleoside diphosphates. However, no template dependence could be found for this synthesis, and the sequence was uncontrollable except in a crude way, namely, by adjusting the relative concentrations of different nucleotides in the starting materials.

Sam Weiss was the first investigator to obtain evidence for a true transcribing activity in cell-free extracts of rat liver (1959). His experimental design was influenced by the theoretical framework described earlier and by Arthur Kornberg's discovery that *in vitro* DNA synthesis required nucleoside triphosphates for substrates. With crude liver extracts, Weiss was able to demonstrate a capacity for RNA synthesis that was severely inhibited by DNAse. This was the beginning of systematic investigations of the biochemistry of RNA metabolism. A continuous expansion of research effort in related areas over the past quarter century has led us in many directions and has provided us with a wealth of understanding about the transcription process and other aspects of RNA metabolism.

Various aspects of nucleotide metabolism (Chapter 25) and RNA structure (Chapter 7) have already been considered. This chapter will focus on the basic aspects of RNA metabolism. In subsequent chapters, consideration will be given to various related topics: (1) regulation of gene expression in prokaryotes (Chapter 29), (2) regulation of gene expression in eukaryotes (Chapter 30), and finally, (3) regulation of transcription by hormones in multicellular organisms (Chapter 31).

DIFFERENT CLASSES OF RNA

Although most RNAs are synthesized as a result of transcription from a DNA template, different strategies are used in their synthesis and in their posttranscriptional modification and processing. The strategy used in a given case is strongly related to the function of the RNA. For this reason, it is important that we discuss the different classes of RNA found in the cell before we consider their mode of synthesis.

There are three major types of cellular RNA involved in protein synthesis: *messenger RNA* (mRNA), *ribosomal RNA* (rRNA) and *transfer RNA* (tRNA). In addition, there are other RNAs, some of which function as primers for DNA synthesis, as component parts of ribonucleases, and in the posttranscriptional modification of eukaryotic mRNA precursors. The properties of the RNAs found in *E. coli* are summarized in Table 27–1. This table should be referred to during the course of the discussion below.

Messenger RNA Carries the Information for Polypeptide Synthesis

Messenger RNA is transcribed from DNA by RNA polymerase, and its sequence contains the information for the sequence of amino acids in the protein product. Specific mRNAs are synthesized by the cell in response to

TABLE 27–1
Types of RNA in *E. coli*

Type	Function	Number of Different Kinds	Number of Nucleotides	Per Cent of Synthesis	Per Cent of Total RNA in Cell	Stability
mRNA	Messenger	Thousands	500–6000	40–50	3	Unstable ($t_{1/2}$ = 1 to 3 min)
rRNA	Structure and function of ribosomes	3 {23S / 16S / 5S}	2800 / 1540 / 120	50	90	Stable
tRNA	Adapter	50–60	75–90	3	7	Stable
RNA primers	DNA replication	?	<50	<1	<1	Unstable
RNA component of RNAse P	?	1 or 2	250–350	<1	<1	?

the conditions under which it finds itself, and the cell can thereby control the kinds and amounts of proteins it produces. Prokaryotic cells have unstable mRNAs that turn over rapidly, with an average half-life of 1 to 3 min, whereas eukaryotic cells, which do not have to be able to respond as rapidly to changing conditions, often have more stable mRNAs.

The mRNA fraction is heterogeneous in size, ranging from 500 to 6000 nucleotides in *E. coli*. This range reflects not only the heterogeneity in size of the proteins of the cell, but also the fact that some prokaryotic mRNAs contain the information to encode more than one protein. Such mRNAs are referred to as *polycistronic mRNAs*, each cistron containing the information for the synthesis of a single polypeptide chain.

In prokaryotes, ribosomes begin binding to mRNA and synthesizing protein even before the entire mRNA molecule has been synthesized. This rapid utilization of nascent transcript minimizes the opportunity for processing of the transcript. In contrast, in eukaryotes, where the processes of transcription and translation are sharply divided, the nascent transcript undergoes an elaborate regimen of processing, usually at both ends and frequently internally, before it is transported to the cytoplasm for use as mRNA.

Ribosomal RNA Is an Integral Part of the Ribosome

Ribosomal RNA is also transcribed from DNA and is a structural and functional component of ribosomes, the cellular organelles responsible for protein synthesis. Ribosomes in prokaryotes are referred to as 70S ribosomes, a measure of their rate of sedimentation in a centrifuge and hence their size (S refers to Svedberg units, which are defined in Chapter 3). A 70S ribosome consists of two subunits, the 50S subunit and the 30S subunit, each of which is made up of RNA and protein. The 50S subunit contains 23S and 5S rRNAs and 33 different ribosomal proteins. The 30S subunit contains 16S rRNA and 21 different ribosomal proteins (see Chapter 28 for further information on ribosomes). Eukaryotic ribosomes are similar in structure, although they are somewhat larger (80S) and contain mostly larger RNAs (25–28S, 18S, 5S, and an additional 5.5 to 5.8S; this additional rRNA corresponds in sequence to the first 150 nucleotides of the prokaryotic large rRNA subunit). Chloroplasts have ribosomes and rRNAs that are distinctly different from those present in the cytoplasm and strongly resemble those of prokaryotes. These RNAs are encoded in the organelle DNA and are transcribed by organelle-specific RNA polymerases.

Transfer RNA Carries Amino Acids to the Ribosome

The third major type of RNA is transfer RNA. It, too, is transcribed from a DNA template. tRNAs contain both a site for the attachment of an amino acid and a site, the *anticodon*, that recognizes the corresponding three-base codon on the mRNA (see Chapters 7 and 28). There are about 50 different kinds of tRNAs in a bacterial cell. Each amino acid is enzymatically attached to the 3′ end of one or more tRNAs by a specific aminoacyl-tRNA synthase that recognizes both the amino acid and the tRNA. This two-step process is discussed in Chapter 28.

A great deal is known about the structure of tRNA because the entire nucleotide sequences of a large number of tRNAs have been determined, and the three-dimensional structures of some of them have been obtained by x-ray crystallography (Chapter 7). Four hydrogen-bonded hairpin stems are present, forming a structure that looks somewhat like a cloverleaf. In the three-dimensional structure of yeast phenylalanine tRNA, the loops at the

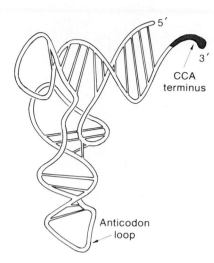

Figure 27-1
Schematic diagram of the three-dimensional structure of yeast phenylalanine tRNA. (Redrawn from S. H. Kim, J. L. Sussman, F. L. Suddath, G. L. Quigley, A. McPherson, A. H. J. Wang, N. C. Seeman, and A. Rich, The general structure of transfer RNA molecules. *Proc. Natl. Acad. Sci. USA* 71:4970, 1974. Used with permission.)

ends of the stems are folded, giving an L-shaped structure. The amino acid accepting site is at one end of the L and the anticodon is at the other (Figure 27-1).

Up to 5 per cent of the ribonucleotide bases in tRNA differ from the usual four. tRNA is initially synthesized from the four normal bases, which are then enzymatically modified by a large variety of enzymes. Some of the more than 30 such unusual bases formed are shown in Figure 7-44. Their function is not understood, but it may involve interacting with the mRNA, the ribosomes, or the aminoacyl-tRNA synthases.

Other Cellular RNAs Are Involved in DNA Synthesis and Processing

DNA synthesis requires a primer—an oligonucleotide that is base-paired with the template strand and provides a 3′ hydroxyl group upon which to start polymerizing a new DNA strand (see Chapter 26). The cell uses small RNAs as primers for DNA replication. In some cases in prokaryotic cells these primers are synthesized by RNA polymerase (the main transcribing enzyme), but most primers in *E. coli* are made by a special enzyme, called *DNA primase*, that is encoded by the *dnaG* gene.

One of the processing ribonucleases in *E. coli*, RNAse P, contains a small RNA complexed to the protein. This RNA is about 350 nucleotides long and is essential for enzyme activity. Under certain conditions the RNA alone is catalytically active. The protein by itself has no catalytic activity.

Eukaryotic cells contain in their nuclei a unique collection of small RNAs, called *small nuclear RNAs* (snRNAs), which are complexed with certain proteins to form ribonucleoprotein particles. These RNAs have been named U1, U2, U3, U4, U5, and U6 RNA and range in size from about 110 to 220 bases. One, U3, is found in the nucleolus, the site of rRNA synthesis. All are very abundant, and there may be as many as 1 million copies per nucleus. Their base sequences are highly conserved between organisms, and all seem to contain unusual trimethylguanosine structures at their 5′ ends. The role of these RNAs is not fully known, but U1 and U2 appear to be involved in processing RNA. This point is discussed below. It seems likely that additional RNAs will be found that play regulatory or structural roles in the cell.

DNA-DEPENDENT SYNTHESIS OF RNA

The enzymatic activity that Weiss discovered in rat liver extracts is capable of joining ribonucleoside triphosphates by 3′-5′ phosphodiester bonds under the direction of a DNA template, releasing pyrophosphate. This enzyme is now referred to as DNA-dependent RNA polymerase and carries out the following reaction:

$$NTP + (NMP)_n \xrightarrow[\text{DNA}]{\text{Mg}^{2+}} (NMP)_{n+1} + PP_i$$

The DNA template strand determines which base will be added to the growing RNA molecule, by base pairing similar to that used to direct the semiconservative replication of DNA. For example, a cytosine in the template strand of DNA means that a complementary guanine will be incorporated into the corresponding position of the RNA. Synthesis proceeds in a 5′→3′ direction, with each new nucleotide being added on to the 3′-OH end of the growing RNA chain. This reaction will be described in detail later in this chapter. Figure 27-2 summarizes a method for assaying RNA polymerase.

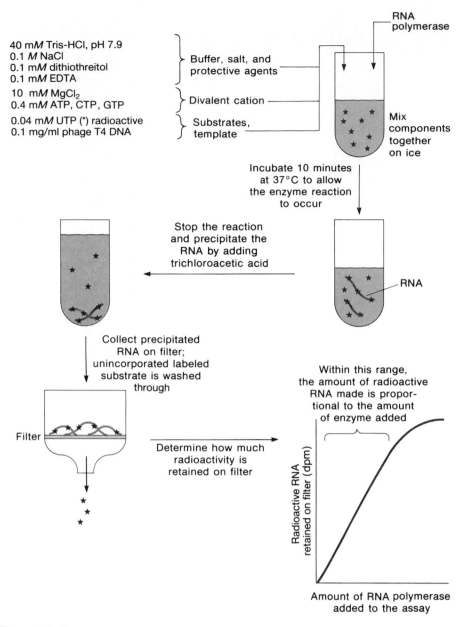

Figure 27–2
A method for assaying the activity of RNA polymerase.

RNA Polymerase Is Composed of Several Protein Subunits

In this section we will focus on the process of transcription in the bacterium *E. coli,* since it is in this organism that it is best understood. The process is quite similar in other prokaryotes.

Subunit Structure and Function.   The purification of substantial amounts of RNA polymerase from *E. coli* has made it possible to study the structure of the enzyme. The active enzyme molecule is a pentamer containing four different polypeptide chains with a total molecular weight of about 500,000. The subunits of the enzyme can be separated by electrophoresis on polyacrylamide gels, as illustrated in Figure 27–3 (this technique is explained in Chapter 3). The four different polypeptide chains, termed β', β, σ^{70}, and α, have molecular weights of 155,000, 151,000, 70,000, and 36,500, respectively. An additional subunit called ω ($M_r = 10,000$) usually copurifies with

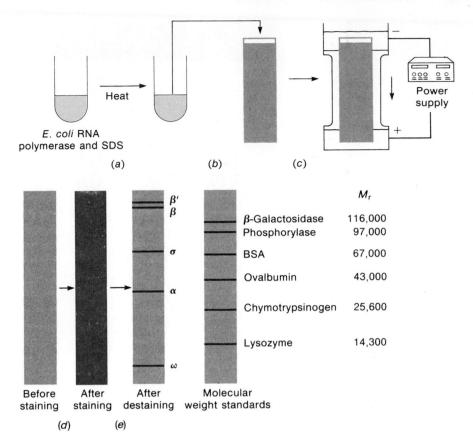

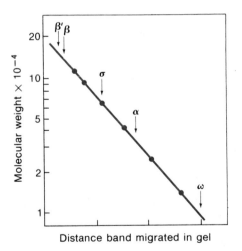

Figure 27–3
Analysis of the subunits of RNA polymerase by polyacrylamide gel electrophoresis. *(a)* The protein is denatured by treatment with the detergent sodium dodecyl sulfate (SDS). This causes the subunits to bind the negatively charged SDS and dissociate from each other. *(b)* The denatured enzyme is layered on top of a gel of polymerized acrylamide in a cylindrical tube. *(c)* A voltage is applied across the gel, causing the polypeptides, which are negatively charged owing to the bound SDS, to move through the gel toward the positively charged anode. The smaller the protein, the faster it is able to migrate through the gel. *(d)* After electrophoresis, the gel is removed from the tube and placed in dye that stains proteins. *(e)* Excess dye is removed and stained bands, containing as little as 1 μg protein, are visible. The molecular weights of the stained subunit bands can be estimated by comparing their positions with those of proteins of known molecular weights electrophoresed under identical conditions. The latter are indicated by the (unlabeled) points on the graph. An additional subunit called ω (M_r = 10,000) usually copurifies with the rest of the enzyme, but has no known function in RNA synthesis.

the rest of the enzyme, but no function for ω has been established.* Recently, additional sigmalike proteins, σ^{32} and σ^{54}, have been identified; these will be discussed later. Some of the properties of the polymerase subunits are summarized in Table 27–2.

A complex with the subunit structure $\alpha_2\beta\beta'\sigma^{70}$ can carry out the functions necessary for correct and efficient synthesis of RNA and is referred to as *holoenzyme*. Holoenzyme can be reversibly separated into two components by chromatography on a phosphocellulose column:

$$\underset{\text{Holoenzyme}}{\alpha_2\beta\beta'\sigma^{70}} \;\rightleftharpoons\; \underset{\substack{\text{Core} \\ \text{polymerase}}}{\alpha_2\beta\beta'} \;+\; \underset{\text{Sigma-70}}{\sigma^{70}}$$

---

*This ω should not be confused with the ω that is a DNA topoisomerase in *E. coli*.

TABLE 27–2
E. coli RNA Polymerase Subunits and Regulatory Factors

Subunit or Protein	Gene Name	Map Position (min)	Polypeptide MW	No. in Enzyme	Function	Properties
β' (beta')	rpoC	90	155,000	1	DNA binding?	Basic
β (beta)	rpoB	90	151,000	1	Active site	Acidic
σ^{70} (sigma-70)	rpoD	67	70,000	1	Promoter recognition, initiation	Acidic
σ^{54}	ntrA	70	54,000	1	Initiation	Acidic
σ^{32} (sigma-32)	htpR	76	32,000	1	Initiation	Acidic
α (alpha)	rpoA	73	36,500	2	?	Acidic
ω (omega)	rpoZ	?	10,000	1	?	Acidic
CAP	crp	74	23,000	2	Activation	Basic
ρ (rho)	rho	85	46,000	6	Termination	Basic
nusA	nusA	69	55,000	2	Termination	Acidic
tau	?	?	30,000?	?	Termination	—

One component, called <u>core polymerase</u> ($\alpha_2\beta\beta'$), <u>retains the capability to synthesize RNA, but it is defective in the ability to bind and initiate transcription at the appropriate initiation sites</u> (promoters) on the DNA. The other component, called <u>sigma-70 (σ^{70}), has no RNA synthetic activity, but when added back to core polymerase, it re-forms holoenzyme with its ability to bind tightly and selectively at promoters and initiate RNA chains efficiently</u>. Because of its role in binding and initiation, sigma-70 is often referred to as an <u>initiation factor</u>.

The precise functions of the various subunits of core are not known. β' is a basic (positively charged) polypeptide thought to be involved in DNA binding. The β subunit is the site of binding of several inhibitors of transcription and is thought to contain most or all of the active sites for phosphodiester bond formation. The α subunit is necessary in order to reconstitute active enzyme from separated subunits.

Genetics of the Subunits.   Since RNA polymerase is an essential cellular enzyme, most mutations in the polymerase genes are lethal. Those that can be studied are the ones that affect the enzyme activity only under certain conditions, e.g., at higher than normal temperatures. As we saw in Chapter 26, mutants that can grow normally at one temperature but not at another are called <u>temperature-sensitive mutants</u>. Another class of mutants is composed of those in which the enzyme is resistant to an inhibitor of RNA synthesis, which acts by binding to the enzyme. Examples of such inhibitors are discussed later in this chapter.

Strains of *E. coli* have now been obtained with mutations in the α, β, β', and σ^{70} and σ^{32} subunits of RNA polymerase. These mutants have made it possible to map the genes for the corresponding subunits. The circular map of the *E. coli* genome is shown in Figure 27–4; on it we show the positions of the genes known to be involved in RNA synthesis, modification, processing, and degradation. It is evident that these genes are scattered throughout the genome. At first it may seem surprising that the genes encoding the subunits of RNA polymerase are not all clustered together. Their dispersion is probably due to the fact that only β and β' are required in equimolar amounts. While the genes for β and β' are in the same region at 90 min, this region also

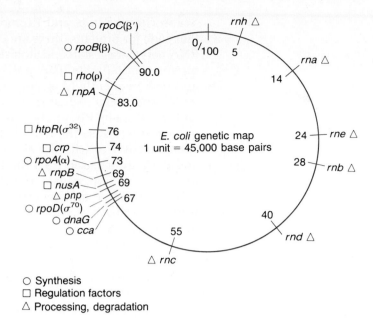

Figure 27–4

Genetic map of *E. coli* showing the location of some genes involved in RNA metabolism. The map is divided into 100 minutes. Three types of genes are indicated: (1) genes involved in RNA synthesis, (2) genes involved in regulation, and (3) genes involved in processing and degradation. (From B. J. Bachmann, Linkage Map of *E. coli* K12, edition 7, *Microbiol. Rev.* 47:180, 1983.)

contains the genes for two ribosomal proteins. The gene for α is found far from those for β and β', at 73 min on the map, and also is in a region containing genes for several ribosomal proteins. The gene for the sigma-70 subunit is found at 67 min, in a region containing the genes for DNA primase and a ribosomal protein.

Transcription Involves Initiation, Elongation, and Termination

The overall transcription cycle, involving binding, initiation, elongation, and termination, is shown in Figure 27–5 and is discussed in detail below.

Figure 27–5

Schematic of the overall transcription cycle. (Adapted from J. D. Watson, *Molecular Biology of the Gene*, 3rd ed., Benjamin, Menlo Park, Calif., 1976. Used with permission.)

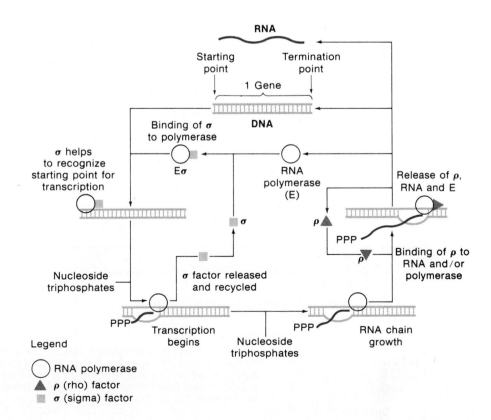

Transcription Units and Promoter Signals. In addition to the sequences that code for proteins, there are precise sequences along the DNA that signal RNA polymerase start and stop sites. *The promoter region contains the information that tells the RNA polymerase where to bind, how tightly to bind, and how frequently to initiate an RNA chain.* It also often contains sites at which additional regulatory proteins bind; these proteins influence binding and initiation by RNA polymerase. The terminator region contains DNA sequences that cause the RNA polymerase to stop transcribing. Then, either spontaneously or with the aid of a termination factor, the RNA polymerase and RNA are released from the template. The transcribed region, including the start and stop signals, is called a *transcription unit* or *operon*. It may include the structural genes for several proteins whose syntheses are controlled coordinately. An example of a transcription unit in *E. coli* is shown in Figure 27–6.

A large number of transcription units have been studied, and many promoter and terminator regions have been sequenced. Common features have been identified in the sequence of over 100 different promoters, and in the case of several promoters, further information about DNA bases important for binding has been obtained. Promoter features are summarized in Figure 27–6.

RNA polymerase binding at the promoter protects about 60 base pairs of DNA from digestion by DNAse. The protected region runs from −40 (40 bp before the RNA chain starting site) to +20 (20 bp after the starting site). *Two sites within this 60-bp region, one centered at −35 and one centered at −10, contain specific recognition sequences that are common to most but not all bacterial promoters.* Mutations in these two sites affect the ability of RNA polymerase to bind and initiate.

The DNA base sequence of each promoter differs somewhat from the "average" promoter sequence. This is to be expected because promoters differ tremendously in their "strength" or frequency of RNA initiation. Some

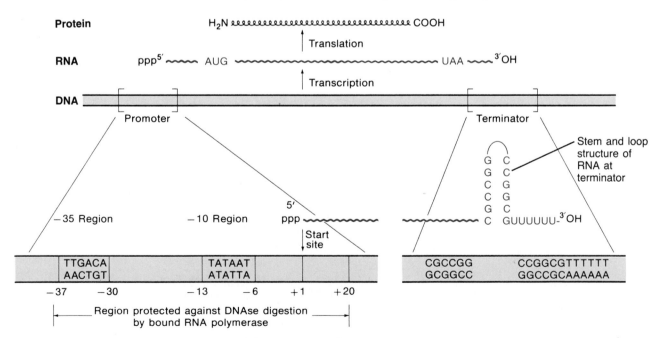

Figure 27–6
Important features of a typical transcription unit. DNA is shown with promoter and terminator regions expanded below. RNA is transcribed starting in the promoter region at +1 and ending after the stem and loop of the terminator. The protein resulting from translation of this RNA is shown above with its N and C termini indicated.

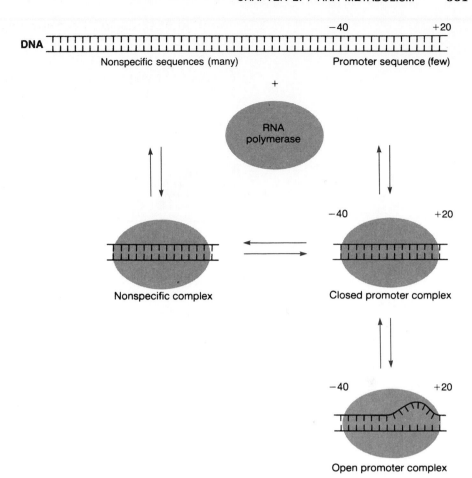

Figure 27–7
Various types of RNA polymerase-DNA binding complexes. A nonspecific complex is formed at any point along the DNA. The closed promoter complex is formed at a polymerase binding site. Following the formation of the closed promoter complex, an open promoter complex is formed at the same site.

promoters are very "weak"; for example, the promoter for the *lac* repressor* produces an mRNA only once in 20 to 40 min. This frequency is estimated as follows: Each *E. coli* contains about 12 molecules of *lac* repressor. Since each repressor is composed of four identical subunits, this represents about 48 molecules of repressor polypeptide. If each mRNA is translated by about 50 ribosomes before it decays, then only one mRNA is needed per generation of 20 to 40 min. In contrast, the promoters for rRNA genes must be utilized once every second or two in order to produce over 10,000 rRNAs per generation from the seven copies of the rRNA gene. This 2000-fold difference in promoter strength is encoded in the base sequence of the promoter.

Binding at Promoters. In the cell, RNA polymerase transcribes only selected regions of the DNA and makes RNA complementary to only one of the DNA strands in any particular region. This selectivity is possible because holoenzyme is able to recognize and form a stable complex with DNA at specific promoter regions. RNA polymerase is able to form unstable nonspecific complexes at any place on the template, mainly by an interaction with the DNA phosphates, but it either rapidly dissociates and rebinds, or slides along the DNA until it reaches a promoter region. It then forms a moderately stable complex with the promoter DNA, most likely interacting stereospecifically with particular nucleotides in the −35 region of the promoter. These complexes can form at 0°C, where the DNA remains double helical or unmelted, and are referred to as *closed promoter complexes* (Figure 27–7). The next step involves a conformational change in the polymerase and a temper-

---

*The *lac* repressor will be discussed in Chapter 29. It is not important at this point that we know anything about the *lac* repressor except what is stated here.

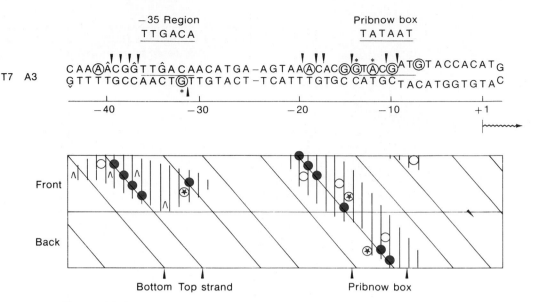

Figure 27–8

(Upper) Sequence of the T7 A3 promoter, showing the stronger contacts to the RNA polymerase. (❙ = phosphate contacts; ○ and ∧ = purines that the polymerase protects from methylation or whose susceptibility to this methylation is enhanced, respectively; ✳ = methylated purines that interfere with polymerase binding.) The most probable bases for the Pribnow box and the −35 region are shown above the corresponding regions in the A3 promoter sequence: +1 represents the start of transcription. The minimal region unwound by the polymerase is represented by a separation of the strands. *(Lower)* Planar representation of the cylindrical projection of the DNA molecule [10.5 base pairs per turn], with contacts to the polymerase marked. (● = phosphate contacts, ⊛ = methylated purines that interfere with polymerase binding and other symbols are as in the upper drawing.) Contact regions, strands, front view, back view, and initiation site are indicated. Regions likely to interact with polymerase are shaded with vertical lines. (From U. Siebenlist and W. Gilbert, Contacts between *E. coli* RNA polmerase and an early promoter of phage T7, *Proc. Natl. Acad. Sci. USA* 77:122, 1980.)

ature-dependent melting of about 10 base pairs of DNA from positions −9 to +2. The resulting very stable complex is termed the <u>open promoter complex</u>.

The existence of consensus sequences in both the −35 and −10 regions of the promoter carries with it the implication that these regions are important in forming the open promoter complex with RNA polymerase. One possibility is that both regions are complexed with polymerase in the open promoter complex. Another possibility is that one region serves for initial recognition of the polymerase, but that in the final complex the polymerase binds only to the other region.

Several approaches used to establish more precisely the regions of contact (or close approach) between polymerase and DNA were developed in Walter Gilbert's laboratory. One strategy used was to make the polymerase-DNA complex and then determine which regions showed altered susceptibility to specific chemical probes. Another strategy was to carry out a partial reaction with a specific chemical probe and see which sites interferred with polymerase complex formation. For these experiments, two chemical probes have been most useful. The first, dimethylsulfate, methylates double-stranded DNA at the N-7 position of guanine in the major groove and the N-3 position of adenine in the minor groove. The methylated purine will depurinate on heating; alkali then can cause a β-elimination reaction and create a series of breaks in the DNA chain. The second probe, ethylnitrosourea, preferentially ethylates the phosphates in DNA. The resulting phosphotriesters serve as cleavage sites when the DNA is exposed to alkali.

The results of these experiments are depicted in Figure 27–8. Domains of interaction are focused on the −35 region and the −10 region. *This result*

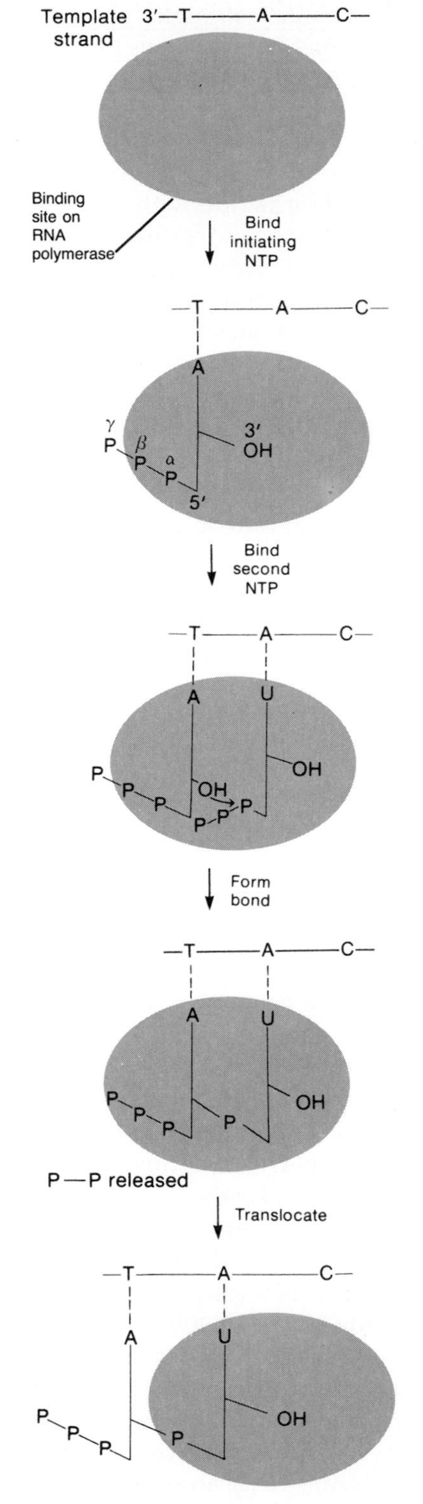

Figure 27–9
Details of phosphodiester bond formation. The α, β, and γ phosphates are indicated on the initiating NTP, which in this case is ATP. The colored ovals represent NTP binding sites on the RNA polymerase. The biochemistry of bond formation in RNA synthesis is very similar to that in DNA synthesis.

provides strong support for the conclusion that polymerase binds simultaneously to both of these regions.

From the experiments with dimethylsulfate, evidence emerged that *the polymerase produces an unwinding of the double helix in the −9 to +2 region.* Dimethylsulfate, which reacts with the N-1 of adenine and the N-3 of cytosine only in single-stranded DNA, reacted with this region in the presence of polymerase. In the double helix these nitrogens are normally unavailable for reaction.

Another type of experiment was used to establish the thymine contacts with polymerase. For this purpose, thymine was replaced in the DNA by 5-bromouracil. Ultraviolet light normally cleaves DNA at the bromouracils. A nearby bound protein can either protect the DNA from such cleavage or become cross-linked to the DNA at position 5′ of the bromouracil. This approach was particularly valuable in suggesting which of the polymerase subunits are in direct contact with the DNA. Observations for the *lac*UV5 promoter show that the sigma-70 subunit cross-links at position −3 of the sense strand (the sense strand carries the same sequence as the transcribed RNA) and that the β subunit cross-links at position +3 of the sense strand.

The chemical-probe approach used to establish points of close contact between polymerase and DNA also has been exploited to obtain useful information on regulatory protein interaction with DNA (see the discussion on CAP and *lac* repressor in Chapter 29).

Initiation at Promoters. Once the polymerase binds to the promoter and strand separation occurs, initiation usually proceeds rapidly (1 to 2 sec). The first, or initiating, nucleoside-5′-triphosphate, which is usually ATP or GTP, binds to the enzyme. The binding is directed by the complementary base in the DNA template strand at position +1, the start site. A second nucleoside-5′-triphosphate (NTP) binds, and initiation occurs upon formation of the first phosphodiester bond by a nucleophilic attack of the 3′ hydroxyl group of the initiating NTP on the α phosphorus atom of the second NTP. Inorganic pyrophosphate derived from the second NTP is a product of the reaction. This process is illustrated in Figure 27–9.

The initiation process can be followed during an *in vitro* transcription reaction in several ways. NTP radioactively labeled with ^{32}P in the β or γ phosphate positions can be used in transcription reactions. Since the initiating NTP retains its 5′ triphosphate end, the beginning of the RNA chain, or the 5′ triphosphate end, will be exclusively labeled. Another way is to add ^{32}P-labeled pyrophosphate ($\overset{*\,*}{p}p_i$) to an otherwise unlabeled reaction and measure initiation by a reaction called *pyrophosphate exchange*. The labeled pyrophosphate participates in the reverse of the phosphodiester-bond formation reaction, allowing the 3′-terminal nucleotide to be removed and the label to be incorporated into its β and γ positions. Pyrophosphate exchange also can occur during chain elongation:

$$\text{pppApU} + {}^{32}\overset{*\,*}{p}p_i \underset{\text{Bond formation}}{\overset{\text{Pyrophosphorolysis}}{\rightleftharpoons}} \text{pppA} + \overset{*\,*}{p}p\text{pU}$$

A third method for measuring initiation depends on the finding that in the absence of the nucleotide needed for the third position on the RNA chain, the dinucleotide formed, pppXpY, dissociates from the active site and the initiation process must start again. This phenomenon is called *abortive initiation.* For example, if only ATP and [α-^{32}P]UTP are added to a reaction mixture where an RNA chain that begins with the sequences pppApUpCp is being synthesized, the reaction produces the dinucleotide pppA^{32}pU. The production of this dinucleotide can be quantified to measure initiation of this chain. Some promoters *in vitro* undergo considerable abortive initiation, even in the presence of all four nucleoside triphosphates, releasing oligonucleotides of two to eight residues several times before finally succeeding in producing a long RNA chain. Therefore, _promoter strength should be equated not strictly with the frequency of initiation at a promoter, but rather with the frequency with which long RNA chains are produced_.

Elongation of RNA. After initiation has occurred, chain elongation proceeds by the successive binding of the nucleoside triphosphate complementary to the next base in the template strand, bond formation with pyrophosphate release, and translocation of the polymerase one base farther along the template strand. Transcription proceeds in the $5' \rightarrow 3'$ direction, antiparallel to the $3' \rightarrow 5'$ strand of the template DNA. Once elongation has produced an RNA chain about 10 bases long, the σ^{70} subunit dissociates from the holoenzyme to yield core polymerase, which continues the elongation reaction until a termination signal is reached. The released σ^{70} is available to bind to a free core polymerase and re-form holoenzyme capable of binding at a promoter and initiating a new RNA chain (see Figure 27–5).

In vivo, mRNA chains grow at a rate of about 45 nucleotides per second, nicely matched with the rate of translation of 15 amino acids per second. rRNA is synthesized about twice as fast as mRNA.

As the polymerase travels along the DNA, it must continually cause an opening or strand separation of the DNA so that a single DNA template strand is available at the active site of the enzyme. For the transcription reaction to be energetically feasible, one base pair must re-form behind the active site for every base pair opened in front of it. It is likely that a short transient RNA-DNA hybrid duplex forms between the newly synthesized RNA and the 10-base-long unpaired region of the DNA and helps hold the RNA to the elongating complex.

Termination of Transcription. Termination of transcription involves stopping the elongation process at a region on the DNA template that signals termination and releases the RNA product and RNA polymerase. _The majority of terminators that have been studied are similar in that they code for a double-stranded RNA stem-and-loop structure just preceding the 3' end of the transcript_ (see Figure 27–6). It is thought that such a structure causes the RNA polymerase to pause or stop elongating.

Two main types of terminators have been distinguished. The first is capable of termination with no accessory factors and contains about six uridine residues following the stem and loop. Apparently, a particularly weak interaction between the 3' terminal oligo(U) and the oligo(dA) region in the template DNA strand allows the release of the RNA from the transcription complex. The second type of terminator lacks the oligo(U) region and requires _termination factor rho_ for RNA chain release. Rho, discovered by Jeffery Roberts in 1969, is a hexamer of subunits (M_r = 46,000 per subunit). Rho does not bind tightly to DNA or to RNA polymerase. It does bind tightly to RNA, especially cytosine-rich RNA, and in its presence rho hydrolyzes ribonucleoside triphosphates to nucleoside diphosphates. A current model is that rho binds to sites on RNA and, by hydrolyzing nucleoside triphosphates, either moves along the RNA or winds the RNA around it until it reaches an RNA polymerase that has stopped at a terminator, where rho causes release.

Several other termination factors, including nusA and tau, have been identified recently. NusA is an acidic protein with a molecular weight of 55,000 that is able to bind to core polymerase and aid in the release of RNA and RNA polymerase at some terminators. The nusA protein, also known as L factor, is further discussed in Chapter 29. Tau factor has an apparent molecular weight of about 30,000 and stimulates termination at several terminators previously thought to terminate independently of factors *in vitro*.

Other Factors Regulating Transcription.   As more transcription units are studied in detail, it becomes apparent that the basic processes just described can be modulated in a great number of ways, both by including additional signals in the DNA sequence and by supplying additional regulatory factors. Some of these regulatory mechanisms are discussed in more detail in Chapters 29 and 30.

Many transcription units have terminator signals called *attenuators* preceding the structural gene or between two structural genes. Attenuators sometimes cause termination and sometimes allow read-through to the structural part of the gene. The efficiency of termination at these sites is variable and can be regulated in response to changes in the growth conditions of the cell. Attenuators are discussed in Chapter 29.

Protein factors can act to inhibit transcription (negative control) or to stimulate it (positive control). The binding of a repressor molecule to a specific site (operator) overlapping the promoter region prevents transcription by sterically interfering with RNA polymerase binding. This is an example of *negative control*. An example of *positive control* is the stimulation of transcription caused by the binding of the catabolite activator protein (CAP) in the presence of cyclic AMP (cAMP) to sites just adjacent to the promoters of a number of genes coding for enzymes involved in the catabolism of a variety of sugars. In *in vitro* transcription reactions, these promoters bind RNA polymerase holoenzyme only weakly and are greatly activated by the addition of CAP and cAMP.

Although most transcription in *E. coli* appears to utilize σ^{70} as an initiation factor, recently several additional initiation factors have been discovered that bind to core polymerase to form hololike enzymes that recognize different promoters. One such factor is the product of the *htpR* gene, which is involved in heat shock regulation. This factor is called sigma-32 (σ^{32}) because of its molecular weight of 32,000. It becomes bound to core polymerase in *E. coli* cells that have been subjected to heat shock and directs the polymerase to bind to and initiate at a class of promoters (the heat shock promoters) responsible for high-level expression of a class of a dozen or so heat shock proteins. Another such factor σ^{54} is a protein ($M_r = 54,000$) encoded by the *ntrA* gene that is needed for expression of certain genes involved in nitrogen metabolism. It seems likely that additional members of the sigma family will be found in *E. coli*.

Multiple sigmas have also been shown to be present in the bacterium *Bacillus subtilis*. Several different sigmalike subunits are present in growing cells, and additional ones appear during sporulation to allow expression of sporulation-specific genes. Furthermore, the infection of *B. subtilis* with certain bacteriophages, such as SP01 or SP82, results in virus-coded sigmalike factors that bind to core polymerase in place of σ^{70} and direct RNA polymerase to viral promoters. All of these bacterial and bacteriophage sigmalike factors exhibit significant regions of amino acid sequence homology, a feature implying that they evolved from a common ancestor.

EUKARYOTIC TRANSCRIPTION

Unlike prokaryotes, in which all major types of RNA are synthesized by one RNA polymerase holoenzyme, eukaryotic cells have become more specialized in their transcription capabilities. They contain at least four different DNA-dependent RNA polymerases, three of which are nuclear, each responsible for synthesizing a different class of RNA. The basic mechanism of RNA

TABLE 27–3
Comparison of Eukaryotic DNA-Dependent RNA Polymerases

Type	Location	RNAs Synthesized	Sensitivity to α-Amanitin
RNA polymerase I	Nucleolus	Pre-rRNA	Resistant
RNA polymerase II	Nucleoplasm	hnRNA, mRNA	Sensitive
RNA polymerase III	Nucleoplasm	Pre-tRNA, 5S RNA	Sensitive to very high levels
Mitochondrial	Mitochondria	Mitochondrial	Resistant
Chloroplast	Chloroplasts	Chloroplast	Resistant

synthesis for all these enzymes is very similar to that of prokaryotic RNA polymerase.

There Are Three Nuclear RNA Polymerases

Nuclear extracts can be fractionated by chromatography on DEAE-cellulose to give three peaks of RNA polymerase activity (the use of DEAE-cellulose is explained in Chapter 3). These three peaks correspond to three different RNA polymerases (I, II, and III), which differ in relative amount, cellular location, type of RNA synthesized, subunit structure, response to salt and divalent cation concentration, and sensitivity to the mushroom-derived toxin α-amanitin. Three polymerases and their properties are summarized in Table 27–3.

RNA polymerase I is located in the nucleolus and synthesizes a large precursor that is later processed to form rRNA. It is resistant to inhibition by α-amanitin at concentrations even greater than 1000 μg/ml. *RNA polymerase II is located in the nucleoplasm and synthesizes large precursor RNAs (sometimes called heterogeneous nuclear RNA or hnRNA) that are processed to form cytoplasmic mRNAs.* It is also responsible for synthesis of most viral RNA in virus-infected cells. It is very sensitive to α-amanitin, being inhibited 50 per cent by about 0.05 μg/ml. *RNA polymerase III is also located in the nucleoplasm and synthesizes small RNAs such as 5S RNA and the precursors to tRNAs.* This enzyme is somewhat resistant to α-amanitin, requiring about 5 μg/ml to reach 50 per cent inhibition.

All three of these enzymes have been purified extensively and have complex subunit structures. All contain two subunits with molecular weights larger than 120,000 and six to ten smaller subunits. Although each has some unique subunits, some subunits are common to two or three. At present, very little is known about the functions of the individual subunits. The subunit structure for RNA polymerase II is quite similar for enzymes purified from a variety of eukaryotes, including human, calf, mouse, wheat, cauliflower, acanthamoeba, yeast, and slime molds.

Recently the genes for the largest subunits of RNA polymerases I and II from yeast and of RNA polymerase II from *Drosophila* have been cloned and sequenced. Similar amino acid sequences are present in several regions of these three subunits, and, surprisingly, are also found in the β' subunit of *E. coli*. Thus there seems to be a remarkable relatedness between enzymes with similar functions in very diverse organisms.

Although purified RNA polymerases have not been shown to be capable of selective transcription of DNA *in vitro*, crude enzyme preparations of RNA polymerases I, II, and III recently have been shown to be capable of selective transcription of defined DNA templates, initiating at sites known in some cases to be utilized *in vivo*. Fractionation of these extracts has revealed additional proteins that are necessary for *in vitro* selectivity. It is likely that the purified enzymes are more like core polymerases than holoenzymes and that these additional components are needed to obtain selective binding and initiation *in vitro*.

Because many eukaryotic genes have been cloned during the last five years, it has become possible to compare the DNA sequences preceding the genes that may act as promoterlike signals for RNA polymerase II. One feature that stands out is a common sequence, TATAAATA, often called a *TATA box*, found about 25 to 30 bases before the transcription start site in many but not all genes. Another fairly common sequence, located 60–80 bases upstream from the start site, is GGCCAATCT or GGTCAATCT, often called the *CAAT box*. Certain proteins selectively bind to the TATA and CAAT boxes. Although these proteins have not been fully characterized, they appear to be essential for the initiation of transcription.

Several additional promoter regions have been identified that are needed for enhanced levels of transcription. These have been called *enhancers* and they have the unusual property of being effective at enhancing transcription relatively independently of their orientation and distance from the start site. Enhancers and enhancerlike elements have been found to be very widespread in eukaryotes. Their properties and possible modes of action will be discussed in greater detail in Chapter 30.

In contrast to the complex for RNA polymerase II promoters, the region necessary for selective transcription of 5S RNA by RNA polymerase III is located in a region 40 to 80 bases after the start site, well into the 5S RNA transcript. One of the additional proteins needed for selective 5S RNA transcription *in vitro* binds to this site and somehow directs RNA polymerase III to bind and start in the correct place. The binding site for this additional regulatory protein was determined by the DNA "footprinting" technique (see Figure 30–16). The promoter regions for RNA polymerases II and III are shown in Figure 27–10.

Some Organelles Have Their Own RNA Polymerases

In addition to nuclear RNA polymerases, eukaryotic cells also contain mitochondrial and, in plants, chloroplast RNA polymerases that are responsible for transcription of the mitochondrial and chloroplast DNAs. These enzymes are present in very small amounts and have proved quite difficult to purify. Despite these difficulties, evidence has been obtained for two types of transcription in chloroplasts. An enzyme tightly bound to DNA is primarily involved in rRNA synthesis, while an enzyme that appears related to prokaryotic RNA polymerase is able to transcribe certain chloroplast genes and tRNA *in vitro*. Genes for chloroplast RNA polymerase subunits, closely re-

Figure 27–10
Promoter regions for eukaryotic RNA polymerases II and III. The polymerases are depicted as large colored ovals. Polymerase II has several protein transcription factors associated with it at the promoter. The precise situation is unclear. The regulatory protein associated with the 5S RNA promoter is depicted by a smaller colored oval. The wavy line indicates the RNA starting at position +1. The starting position for RNA polymerase III is determined exclusively by the regulatory protein, which interacts with a specific sequence located (approximately) between nucleotides +50 and +87 on the DNA.

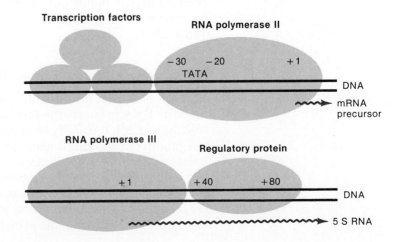

lated to *E. coli* α, β, and β' subunit genes, are encoded in chloroplast DNA. Mitochondrial RNA polymerases from both yeast and humans are multicomponent enzymes. They consist of catalytic and selectivity factors and may resemble prokaryotic enzymes.

Important Differences Exist Between Eukaryotic and Prokaryotic Transcription

Although the basic mechanism by which RNA is synthesized is quite similar in prokaryotes and eukaryotes, there are several important differences. The DNA in eukaryotic cells is complexed with proteins, primarily small basic proteins called histones, to form chromatin (see Chapter 7). The chromatin, in turn, is condensed to form chromosomes. Only a small fraction of the chromosome is actively being transcribed. In transcriptionally active regions, the DNA must be partially exposed and be more accessible to RNA polymerase. The DNA in these regions is found to be more sensitive to cleavage by mild treatment with bovine pancreatic DNAse I and appears to contain bound RNA polymerase, additional nonhistone proteins, and perhaps modified histones. In addition, DNA in active regions is undermethylated when compared with DNA from regions not active in transcription. Although prokaryotic DNA is not complexed with histones, it seems to be coated with small basic proteins; however, as far as we know, most parts of the bacterial DNA are accessible to RNA polymerase binding and transcription.

As stated earlier, transcription and translation occur simultaneously in prokaryotes. RNA polymerase moves along a gene generating an mRNA chain, and as soon as a bit of mRNA is synthesized, ribosomes bind to it and start translating it. The situation is very different in eukaryotes, where transcription and translation occur in separate compartments of the cell. The nucleus, where the chromosomes are located, is the site of DNA-dependent RNA synthesis. Large RNA precursors to mRNA are synthesized in the nucleus, become complexed with proteins to form ribonucleoprotein particles (RNPs), and then are modified, processed to form the smaller mRNAs, and transported across the nuclear membrane to the cytoplasm, where they bind to the ribosomes and are translated into protein. The processing steps that eukaryotic mRNAs undergo are described in a later section.

OTHER RNA SYNTHESIS

Although most RNA in cells is synthesized by cellular DNA-dependent RNA polymerases, there are several other enzymes that are capable of forming phosphodiester bonds and of synthesizing additional RNA in cells or cells infected with viruses. Some properties of these enzymes are summarized in Table 27–4.

DNA Primase Makes Primer RNA for DNA Synthesis

RNA primers used for the initiation of DNA synthesis during replication are made by DNA primase, the protein product of the *dnaG* gene of *E. coli* (also see Chapter 26). It usually binds to DNA in association with another protein, the product of the *dnaB* gene, although it can initiate primer synthesis by itself at certain hairpin structures in single-stranded DNA. The primase can use either NTPs or dNTPs as substrates *in vitro* and synthesizes primers 10 to 50 nucleotides long that are complementary to the DNA template. Synthesis is in the 5'→3' direction. As was mentioned earlier, RNA polymerase is also capable of synthesizing primers. The use of primase or RNA polymerase

TABLE 27–4
RNA-Synthesizing Enzymes

	Template	Primer	Molecular Weight of Subunit(s)	Gene Name	Substrate	Inhibition by Rifampicin
Template-dependent						
Enzymes from bacteria						
Holoenzyme (*E. coli*)	DNA	—	155,000 151,000 70,000 36,500	rpoC rpoB rpoD rpoA	4 NTPs	Yes
DNA primase	DNA	—	65,000	dnaG	4 NTP, 4 dNTP	No
Enzymes from phage or phage-infected bacteria						
T7 RNA polymerase	T7 DNA	—	99,000	T7 gene1	4 NTPs	No
N4 RNA polymerase	N4 DNA	?	350,000	Viral	4 NTPs	No
Qβ replicase	Qβ RNA	—	65,000 55,000 43,000 35,000	rpsA Viral tuf tsf	4 NTPs	No
Template-independent						
CCA enzyme	—	3′ end tRNA	45,000	cca	CTP ATP	No
Poly(A) polymerase (eukaryotic)	—	3′ end mRNA			ATP	No
Polynucleotide phosphorylase	—	3′ end RNA	86,000 48,000	pnp	4 NDPs	No

to make primer is controlled by the DNA. One or the other is used exclusively in specific instances.

Many Viruses Make Their Own RNA Polymerases

There seem to be three types of DNA viruses, each having a different strategy to accomplish transcription of viral DNA. The first type utilizes the host RNA polymerase, in some cases modifying it or synthesizing new promoter-specificity factors to direct it to read the viral promoters. Examples of such viruses are bacteriophage φX174, λ, and T4 (of *E. coli*), and SP01 and SP82 (of *B. subtilis*), as well as animal viruses SV40 and adenovirus.

The second type of virus utilizes the host RNA polymerase to transcribe some early expressed viral genes. One of these "early" genes codes for a new RNA polymerase that transcribes exclusively the remaining "late" viral genes. *E. coli* bacteriophages T7, T3, and SP6 are the best-known examples of this type. T7 RNA polymerase is a single polypeptide with a molecular weight of 99,000. It is not inhibited by the antibiotics rifampicin and streptolydigin, which inhibit host RNA polymerase. It recognizes specifically the T7 late promoters, all of which contain a nearly identical sequence of 18 to 22 nucleotides just before the 5′ triphosphate terminal GTP start site. T7 RNA polymerase is also capable of termination at specific points on the template. Thus it seems to be able to carry out the basic polymerization reaction and specific initiation and termination with a much simpler sub-unit structure than the host holoenzyme. However, with simplicity it loses versatility. Unlike the host polymerase, it is not able to recognize a wide variety of related but nonidentical promoter sequences, nor is it able to be regulated by positive and negative control factors that act at sites of initiation and termination.

A third type of virus, exemplified by bacteriophage N4, carries a viral RNA polymerase within its virion. This enzyme enters the cell along with the viral DNA and transcribes some early viral genes. Some of these genes code for specificity factors that direct the host RNA polymerase to transcribe late genes. Vaccinia virus is another example of a virus that contains a virion-encapsulated RNA polymerase. It shares some of the properties of the N4 enzyme.

RNA-dependent RNA Polymerases of RNA Viruses.   The RNA genomes of single-stranded RNA bacterial viruses (such as $Q\beta$, MS2, R17, and f2) are themselves mRNAs. Bacteriophage $Q\beta$ codes for a polypeptide ($M_r \approx$ 55,000) that combines with three host proteins to form an RNA-dependent RNA polymerase (replicase). The three host proteins are ribosomal protein S1 (*rpsA* gene) and two elongation factors for protein synthesis, EF-Tu (*tuf* gene) and EF-Ts (*tsf* gene) (see Table 27–4). The $Q\beta$ replicase can use only the $Q\beta$ RNA plus strand as a template. It first makes a complementary RNA transcript (minus strand) and ultimately uses it to make more viral RNA plus strands. Like the DNA-dependent RNA polymerases, the replicase utilizes ribonucleoside-5'-triphosphates and transcribes in a 5'→3' direction. The phage RNA must first act as an mRNA to direct the synthesis of a component of the replicase, since uninfected cells do not have an RNA-dependent RNA polymerase or replicase.

RNA tumor viruses (retroviruses) that infect animal cells exhibit a different replication strategy. They carry in their virion an enzyme that can use the RNA viral genome as a template to synthesize a DNA copy. Since this process is the reverse of transcription, the enzyme is called *reverse transcriptase*. The result is a DNA-RNA hybrid. A second DNA strand is then synthesized, displacing the RNA strand. Once a double-stranded DNA copy is made, it is integrated into the host genome, and additional virus RNA is synthesized by the host RNA polymerase II in a normal DNA-dependent fashion. The fascinating story of how the retrovirus converts its genome into a duplex DNA molecule is described in Box 27–A.

Reverse transcriptase activity has recently been found in virus-free cultures of yeast and fruit flies, suggesting that this enzyme may be involved in normal cellular metabolism. There is evidence that certain segments of the genomic DNA can jump from one location to another with the combined assistance of RNA polymerase and reverse transcriptase. This capability could provide a major mechanism for evolutionary change.

3'-End Addition Enzymes Add Nucleotides to tRNA

Three enzymes are known that add ribonucleotides posttranscriptionally to the 3' hydroxyl end of RNA. None of them are DNA-dependent. One adds the CCA sequence that all tRNAs have at their 3' ends. The 3' terminal adenine is the base to which the amino acid is covalently attached by the aminoacyl-tRNA synthase. The 3'-CCA is relatively unstable and is continually being added when needed by this enzyme, called the CCA enzyme, or tRNA nucleotidyltransferase.

In eukaryotes, 100 to 200 adenosine residues are added to the 3' end of most mRNAs by a poly(A) polymerase. This addition occurs in the nucleus before the mRNA is processed and transported to the cytoplasm, as will be described later. The third enzyme capable of adding nucleotides posttran-

BOX 27–A

Synthesis of Viral DNA from the Viral RNA Genome

Integration of a retrovirus genome into a host genome is mediated by the activity of a virus-encoded enzyme, RNA-dependent DNA polymerase (reverse transcriptase). This enzyme was first detected in 1970 in virions of avian and murine retroviruses. It is now known to be a universal constituent of retroviruses, whether or not they are oncogenic. Reverse transcriptase is encoded by the *pol* gene and carries two major enzymatic activities: a DNA polymerase capable of copying RNA or DNA templates, and a ribonuclease (RNAse H) that specifically removes the RNA from DNA-RNA hybrids.

Synthesis of the first (minus) strand of DNA begins near the 5′ end of the viral genome, at the place where a tRNA primer is bound—see parts *(a)* and *(b)* of the illustration. The growing DNA strand, still covalently attached to the primer, is extended to the 5′ end of the

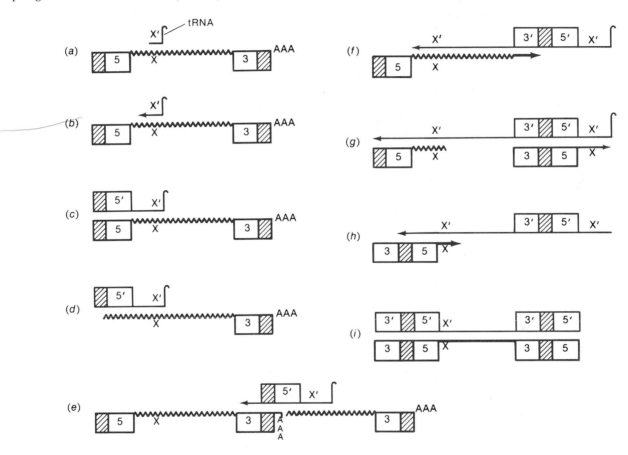

A model of the generation of double-stranded DNA carrying two copies of a long terminal repeat (LTR = [3′ 5′]). The sequence X is a marker for the plus strand; the complementary sequence X′ occurs on the minus strand. *(a–c)* Synthesis of minus-strand DNA from the genomic RNA template using a tRNA primer (represented by ⌐ in step *(a)*. This initial DNA synthesis of minus strand reaches a pause site at the end of the RNA template as shown in *(c)*. *(d)* Degradation of the RNA portion of the resulting DNA-RNA hybrid by RNAse H partially exposes the newly synthesized DNA. *(e)* Bridge formation between the newly synthesized segment of minus strand and repeated sequences at the 3′ end of genomic RNA permits continued synthesis of the minus-strand DNA. This synthesis of minus-strand DNA continues until it reaches the end of the RNA template. *(f)* At the same time, DNA synthesis of the plus-strand DNA begins in the region of the LTR on the minus-strand DNA using the plus-strand RNA as a primer. *(g)* This synthesis pauses after 300 bases, that is, when it reaches the end of the template. The tRNA primer is degraded by RNAse H *(h)*. Continued synthesis of plus-strand DNA requires bridge formation between sequences repeated in a completed minus-strand DNA and in the 300-nucleotide fragment. *(i)* Completion of synthesis and tidying up results in a duplex with copies of the LTR at both ends. The DNA duplex is now ready for the next series of reactions, which results in the integration of the viral DNA sequences into the host genome.

template (c). Part or all of the RNA in the resulting DNA:RNA hybrid is then degraded by RNAse H (d). The redundant sequence in the newly synthesized DNA is thus exposed and is able to form a bridge to the complementary sequence at the same or another RNA molecule (e). Synthesis of minus-strand DNA is now free to proceed along the RNA molecule (f), displacing the primer tRNA if necessary.

Long before synthesis of the minus strand is completed, however, the plus strand of DNA begins to be made. Much less is known about its synthesis, although it is generally assumed that the plus strand is copied from a minus-strand template, perhaps using as a primer oligonucleotides generated by the action of RNAse H on the minus-strand RNA hybrid. One of the first plus-strand segments to be synthesized is a 300-nucleotide sequence complementary to the part of the minus strand where initiation of DNA synthesis took place (g). The 300-nucleotide DNA therefore is complementary to both the 5′ and 3′ ends of the RNA, to the primer binding site, and perhaps to the tRNA itself. When the minus strand reaches the end of the viral RNA, a second "bridge" can be formed to the primer binding site within the 300-nucleotide plus-strand segment (h). The minus strand would then be extended further using the 300-nucleotide segment as template. In turn, the plus strand is extended;

it ultimately consists of a full-length copy of the minus strand (i).

Reverse transcription of retroviral RNA therefore produces linear duplex DNA, several hundred nucleotides longer than the RNA itself, in which both ends contain an identical sequence called the long terminal repeat (LTR). The LTR contains 250 to 1200 nucleotide pairs (the exact size varies from retrovirus to retrovirus) of the region unique to the 3′ terminus of viral RNA, a single copy of the short sequence present at both ends of the RNA, and about 100 to 150 nucleotides from a region unique to the 5′ terminus.

The synthesis of linear duplex DNA occurs in the cytoplasm of the infected cell, presumably in partially degraded virions. A portion of the linear duplex DNA enters the nucleus and some of the molecules are then converted into circular forms. This process seems chiefly to occur by either of two mechanisms. Direct joining of the ends would produce circles carrying two copies of the LTR; homologous recombination between the redundant ends would produce circles carrying one copy of the LTR. Both sorts of molecules have been isolated, cloned, and sequenced. Although direct evidence is lacking, it seems likely that these circular forms of retrovirus DNA are the substrates for integration.

scriptionally to the end of RNA is polynucleotide phosphorylase, which is described in the next section.

Polynucleotide Phosphorylase Makes Polynucleotides from Ribodiphosphates

Polynucleotide phosphorylase was the first enzyme found that could synthesize long polynucleotide chains *in vitro*. Two forms of the enzyme have been isolated, a form with three identical subunits ($M_r = 86,000$) and a form with these subunits and two additional subunits ($M_r = 48,000$). Unlike all the enzymes we have discussed, polynucleotide phosphorylase utilizes ribonucleoside-5′-diphosphates instead of triphosphates as substrates for RNA synthesis. It catalyzes the reaction

$$NDP + (NMP)_n \xrightleftharpoons{Mg^{2+}} (NMP)_{n+1} + P_i$$

Polynucleotide phosphorylase does not require a template and randomly incorporates bases into RNA, depending on the relative concentration of the four NDPs in the reaction medium. The enzyme takes advantage of a primer if one is available to provide a 3′ hydroxyl end and synthesizes RNA in the 5′→3′ direction. The reaction is readily reversible, and it is not known whether this enzyme plays a role primarily in degradation or in synthesis of RNA in the cell.

RNA Ligase of Bacteriophage T4 Links RNAs Together

Perhaps the most unusual enzyme capable of synthesizing RNA is the bacteriophage T4 RNA ligase. It can link together, in an ATP-dependent reaction, a 3'-OH terminus on an "acceptor" to a 5'-PO$_4$ terminus on a "donor," as shown below:

$$ATP + \cdots + XpY\text{-}3'\text{-}OH + PO_4\text{-}5'Zp \cdots \longrightarrow$$

$$\qquad\qquad \underset{\text{Acceptor}}{\qquad} \underset{\text{Donor}}{\qquad}$$

$$\cdots XpYpZp + \cdots + ADP + PP_i$$

$$\underset{\text{Ligated product}}{\qquad}$$

A template strand is not required to align the ends of the reactants. The donor and acceptor end can be on the same RNA molecule, in which case a circular RNA product is produced. The enzyme has been used to end-label the 3'-OH end of RNA molecules by the addition of the short donor [5'-^{32}P]Cp.

This enzyme is now widely used to synthesize defined sequences of RNA. The ligase also can accept deoxyribose polymers as donors. It is not known whether this enzyme carries out the ligation reaction *in vivo* after T4 infection of *E. coli*.

POSTTRANSCRIPTIONAL MODIFICATION AND PROCESSING OF RNA

Almost all the major types of RNA synthesized by cellular DNA-dependent RNA polymerases undergo changes before they can carry out their functions. Two types of changes are usually distinguished. *Modification* involves additions to or alterations of existing bases or sugars. In some cases it involves addition of one or more nucleotides. *Processing* involves phosphodiester bond cleavage and loss of certain nucleotides. These changes are summarized in Table 27–5.

Processing and Modification of tRNA Requires Several Enzymes

Transfer RNAs are processed from larger precursors in prokaryotic and eukaryotic cells. This processing involves two types of enzymes that can cleave phosphodiester bonds in RNA. Endoribonucleases cleave at internal sites in the RNA, resulting in two smaller RNAs. Exonucleases sequentially remove single nucleotides from one end of the RNA. Some of the processing ribonucleases of *E. coli* are shown in Table 27–6.

As an example, the processing of the *E. coli* tyrosine tRNA$_1$ is dia-

TABLE 27–5
Summary of RNA Modification and Processing

RNA	Precursor	Modification	Processing	Products
mRNA				
Prokaryotic	None	None? Polyadenylation?	In some cases specific cleavage by endoribonucleases	mRNAs
Eukaryotic	hnRNA	Capping, methylation, polyadenylation	In most cases splicing out introns	mRNAs
rRNA	Pre-rRNA	Methylation	Specific cleavage	16S, 23S, 5S, spacer tRNA (18S, 28S and 5.8S in eukaryotes)
tRNA	Pre-tRNA	Many modified bases	Specific cleavage by endonucleases, trimming by exonucleases, CCA addition, removal of intervening sequences in eukaryotes	Mature tRNAs

TABLE 27–6
Representative Enzymes Involved in Processing and Degradation

Enzyme	For *E. coli* Genes		Type	Product	Specificity
	Gene Name	Map Position			
Processing					
RNAse III	rnc	55'	endo	3'-OH, 5'-PO$_4$	Specific, long, double-stranded RNA
RNAse D	rnd	40'	3'→5' exo	5'-NMPs	Nonspecific, but stops at CCA
RNAse E	rne	24'	endo		Specific
RNAse F	rnf	?	endo		Specifically cuts 3' to tRNA-like structures
RNAse P	{ rnpA	83'	endo	3'-OH, 5'-PO$_4$	Specifically cuts 5' to tRNA-like structures
	{ rnpB	70'			
RNAse M16			endo(s)		Specific, cuts pre-16S to 16S RNA
RNAse M23			endo(s)		Specific, cuts pre-23S to 23S RNA
RNAse M5			endo		Specific, cuts pre-5S to 5S RNA
Degradation					
RNAse I	rna	14'	endo	3'-PO$_4$ oligos	Nonspecific
RNAse II	rnb	28'	3'→5' exo	5'-NMP	Nonspecific
Polynucleoside phosphorylase	pnp	69'	3'→5' exo	5'-NDP	Nonspecific
RNAse H	rnh	5'	endo		Nonspecific, digests RNA out of RNA-DNA duplex
Bovine pancreatic RNAse A			endo	Py-3'-PO$_4$	Specific, cuts 3' to pyrimidines
Aspergillus RNAse T1			endo	G-3'-PO$_4$	Specific, cuts 3' to guanine
Aspergillus S1 nuclease			endo	5'-NMP	Nonspecific, cuts single-stranded RNA or DNA
Bovine spleen phosphodiesterase			5'→3' exo	3'-NMP	Nonspecific
Snake venom phosphodiesterase			3'→5' exo	5'-NMP	Nonspecific

grammed in Figure 27–11. The initial transcript has, in addition to the 85 nucleotides of the final product, 41 nucleotide residues at the 5' end and 225 residues at the 3' end. The initial transcript is able to fold up to form the typical cloverleaf tRNA structure. Processing begins when a specific endonuclease, called RNAse F, cleaves the precursor at a site three nucleotides

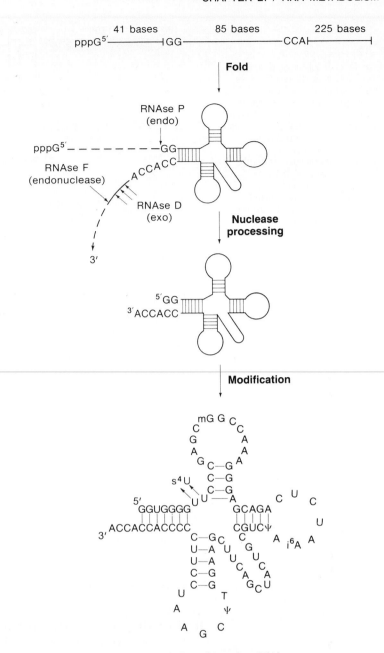

Figure 27–11
Processing and modification of E. coli tyrosine tRNA. (T = ribothymidine; ψ = pseudouridine; i⁶A = isopentyl adenosine; mG = methylguanosine; s⁴U = thiouridine.)

beyond what will be the 3′ end of the mature tRNA. Another endonuclease, RNAse P, then cleaves the remaining RNA to produce the mature 5′ end. At the 3′ end, exonuclease RNAse D sequentially removes the additional nucleotides and stops, leaving the 3′ terminal CCA sequence.* Individual bases on the tRNA molecule are then modified by a variety of enzymes, including methylases, deaminases, thiolases, pseudouridylating enzymes, and transglycosylases. Some of the modified bases are presented in Figure 7–44. In addition, some eukaryotic tRNAs are processed to remove internal RNA sequences, as will be discussed later.

* Some tRNA genes encode the CCA terminal sequence; some do not.

Figure 27–12

Processing of *E. coli* ribosomal RNA. The ribosomal RNA is transcribed as one long RNA molecule, which contains the sequences for the three ribosomal RNAs and one or two tRNA molecules. There are many processing sites and many different enzymes involved in the processing, as indicated by the vertical arrows and the symbols associated with these arrows. The various nucleases are described in Table 27–6. (Adapted from D. Apirion and P. Gegenheimer, Processing of bacterial RNA, *FEBS Lett.* 125:1, 1981.)

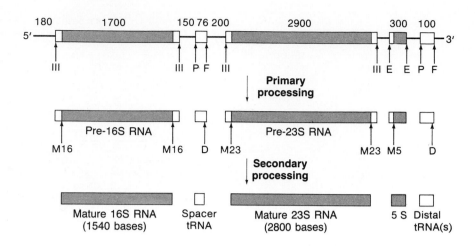

Processing of Ribosomal RNA Precursor Leads to Three RNAs

Both eukaryotic and prokaryotic cells synthesize large precursors to rRNA that are processed to produce the mature rRNAs. The most detailed studies of the numerous enzymatic steps involved in this process have been carried out in *E. coli*. This processing scheme is summarized in Figure 27–12.

The initial transcript is over 5500 nucleotides long and includes, reading from the 5′ end of the RNA: the 16S rRNA, a spacer region that includes one or two tRNAs, the 23S rRNA, the 5S rRNA, and in some cases one or two more tRNAs. Extra bases are found preceding and following each of these RNAs. Primary processing events include the endonucleolytic action by RNAse III to produce pre-16S and pre-23S rRNAs and then the action of specific ribonucleases to produce the tRNAs and pre-5S rRNA. Secondary processing by endonucleases M16, M23, and M5 results in mature 16S, 23S, and 5S RNAs, respectively. These three endonuclease activities, as well as that of RNAse F, have not been completely purified and characterized, and each may represent more than one enzyme. Extra bases on the 3′ end of the tRNAs are removed by exonuclease RNAse D. This processing scheme has been deduced by observing the accumulation of intermediates in mutant strains defective in one or more of the nucleases and by cleaving the intermediates *in vitro* with purified or partially purified nucleases.

Modification and Processing of Messenger RNA Is of Major Importance in Eukaryotes

In prokaryotes, many mRNAs function in translation with no prior modification or processing. There is evidence, however, that some are processed by specific endonucleolytic cleavage, often cutting polycistronic messengers into smaller units. In eukaryotes, a much more complex process occurs to produce functional mature mRNA. The discoveries of these modification and processing steps during the last decade have provided some surprising results and provocative new concepts.

Capping the 5′ End. In 1975, Aaron Shatkin and coworkers found that most viral and cellular mRNAs contain an unusual methylated nucleotide at the 5′ terminus. This entire methylated terminal oligonucleotide is called a *cap structure* and is shown in Figure 27–13. The cap structure is formed in the nucleus by a series of enzymatic reactions outlined in Figure 27–14. Its formation frequently takes place before the nascent transcript is completed. First, a triphosphatase converts a 5′-triphosphate to a 5′-diphosphate terminus. Second, a guanylyltransferase adds a GMP to the 5′ end to form an

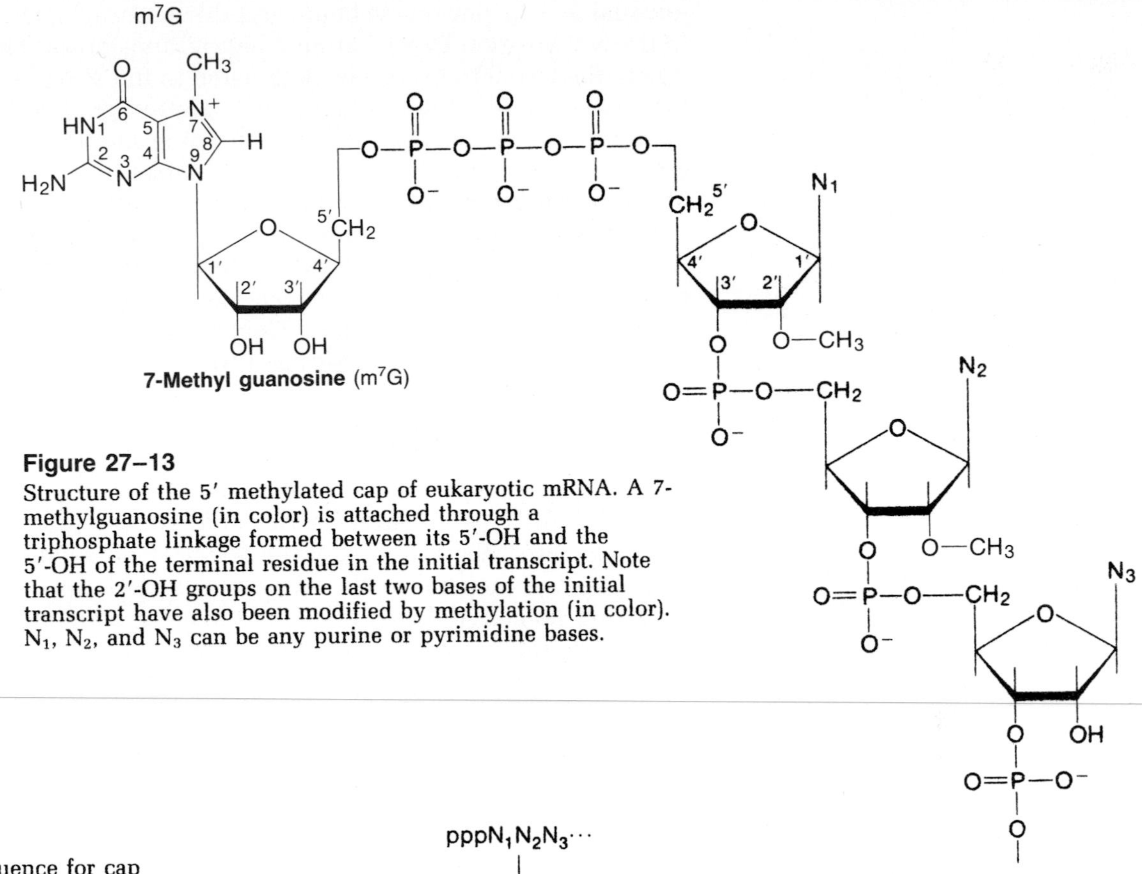

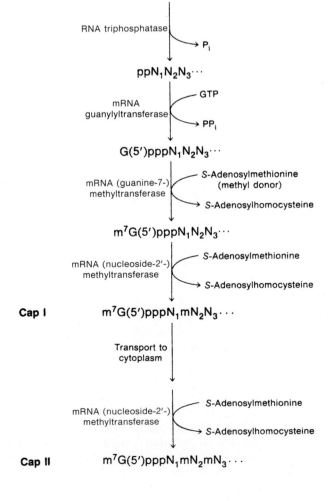

m⁷G

7-Methyl guanosine (m⁷G)

Figure 27–13
Structure of the 5′ methylated cap of eukaryotic mRNA. A 7-methylguanosine (in color) is attached through a triphosphate linkage formed between its 5′-OH and the 5′-OH of the terminal residue in the initial transcript. Note that the 2′-OH groups on the last two bases of the initial transcript have also been modified by methylation (in color). N_1, N_2, and N_3 can be any purine or pyrimidine bases.

Figure 27–14
Proposed reaction sequence for cap formation in HeLa cells. The enzyme catalyzing each reaction is shown on the left. (Adapted from S. Venkatesan and B. Moss, Donor and acceptor specificities of HeLa cell mRNA guanylyl transferase, *J. Biol. Chem.* 255:2835, 1980.)

$$pppN_1N_2N_3\cdots$$

RNA triphosphatase → P_i

$$ppN_1N_2N_3\cdots$$

mRNA guanylyltransferase ⟨ GTP → PP_i

$$G(5')pppN_1N_2N_3\cdots$$

mRNA (guanine-7-)methyltransferase ⟨ *S*-Adenosylmethionine (methyl donor) → *S*-Adenosylhomocysteine

$$m^7G(5')pppN_1N_2N_3\cdots$$

mRNA (nucleoside-2′-)methyltransferase ⟨ *S*-Adenosylmethionine → *S*-Adenosylhomocysteine

Cap I $m^7G(5')pppN_1mN_2N_3\cdots$

Transport to cytoplasm

mRNA (nucleoside-2′-)methyltransferase ⟨ *S*-Adenosylmethionine → *S*-Adenosylhomocysteine

Cap II $m^7G(5')pppN_1mN_2mN_3\cdots$

unusual 5'-5' triphosphate bond, and this terminal guanosine is methylated in the N-7 position by a guanine-7-methyltransferase. Then the first nucleotide in the initial transcript is methylated in the 2'-0 position of the ribose to form what is called a *cap I structure*. Methyl groups are derived from the methyl donor S-adenosylmethionine. Some mRNAs, after transport to the cytoplasm, become methylated in the 2'-0 position of the second nucleotide of the initial transcript to form a *cap II structure*.

The cap structure facilitates binding of ribosomes prior to initiation of translation of eukaryotic mRNAs; it may also function to stabilize the mRNA, since uncapped messengers have considerably reduced half-lives. rRNA and tRNA are not capped. Most small nuclear RNAs (snRNAs) have a different cap structure, containing a trimethylguanosine.

Polyadenylation of the 3' End. Most eukaryotic mRNAs found associated with ribosomes contain 50 to 150 adenine nucleotides on their 3' ends. This poly(A) is not coded by the DNA template but is added to the mRNA before it leaves the nucleus. In several cases studied, the process occurs in at least two steps. First, the RNA is cleaved about 12 nucleotides past an AAUAAA sequence, and then 200 to 250 adenylate residues are added, one residue at a time, by a poly(A) polymerase. After the mature mRNA is transported to the cytoplasm, the poly(A) is shortened somewhat as the mRNA ages. Poly(A) addition does not appear to be essential for transport or translation of all mRNAs, since some eukaryotic mRNAs, including histone mRNAs, do not contain poly(A).

Removal of Intervening Sequences. It had long been known that some of the mRNA precursors in the nucleus (hnRNAs) are much larger than the mRNAs found in the cytoplasm associated with ribosomes. Therefore, it came as no surprise to learn that processing occurs to remove parts of the precursor RNAs. What was surprising was a finding made simultaneously by Tom Broker's laboratory and Phil Sharp's laboratory in 1977, which was that the parts that are removed are not at the ends of the molecules but are interspersed with the coding regions. The phenomenon was first observed for adenovirus transcripts.

One of the early demonstrations of sequence removal resulted from finding that mouse β-globin precursor mRNA did not form a perfect hybrid with DNA complementary to mature mRNA. When the hybrid was observed with an electron microscope, a loop in the DNA of the heteroduplex appeared, which suggested that the precursor contained internal sequences not present in the mature mRNA (Figure 27–15). These noncoding intervening sequences (also called *introns*) are interspersed with the coding sequences (*exons*) in a large number of eukaryotic mRNAs that have been studied.

As an example, the primary transcript of the chicken ovalbumin gene (Figure 27–16) is 7700 bases long and has 7 introns and 8 exons. The intervening sequences, or introns, are removed and the exons are spliced together, giving a final product that is only 1872 nucleotides long. It includes 1158 nucleotides that code for the 386 amino acids of ovalbumin and 714 nucleotides from untranslated regions at the 5' and 3' ends.

The function of intervening sequences is not yet understood. Some genes (e.g., histones) do not contain them, yet function well. When a particular intervening sequence was removed from SV40 virus DNA before using the latter to infect cells, the mRNA that formed was unstable and was not transported from nucleus to cytoplasm. However, similar experiments with other mRNAs have not caused abnormal mRNA function. Frequently, splice points are correlated with "domains" that define protein structural regions, and similar domains are often seen in different proteins. For example, the

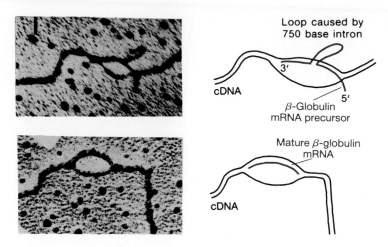

Figure 27–15
Electron micrograph showing mouse β-globin precursor mRNA (nascent transcript) hybridized with DNA (cDNA) complementary to mature mRNA *(upper photo)*. A control experiment *(lower photo)* shows that mature mRNA forms a perfect hybrid with DNA complementary to mature mRNA (cDNA) as expected. The intron region in the nascent transcript is indicated by a loop in the upper figure. Schematics indicating the RNA and cDNA are shown on the right. A second small intron is present in the precursor mRNA near the 5' end and is the reason that the 5' end of the RNA is not hybridized to the cDNA in the upper photo. (From A. Kinniburgh, J. Mertz, and J. Ross, The precursor of mouse β-globin contains two intervening sequences, *Cell* 14:681, 1978. Used with permission.)

Figure 27–16
Maturation of ovalbumin mRNA. First, the entire ovalbumin gene is transcribed into a precursor RNA, the primary transcript. The transcript is capped at the 5' end and the poly(A) tail is added at the 3' end. Then the transcripts of the introns are excised and the adjacent exon transcripts are ligated in a series of splicing steps; an intermediate, from which five of the seven intron transcripts have been eliminated, is illustrated. These steps are accomplished in the cell nucleus. After splicing, mature messenger is transferred to cytoplasm. *L* indicates the 5' leader region, which is part of the mature RNA that is not translated. (From P. Chambon, Split genes, *Sci. Am.* 244:60, 1980. Used with permission.)

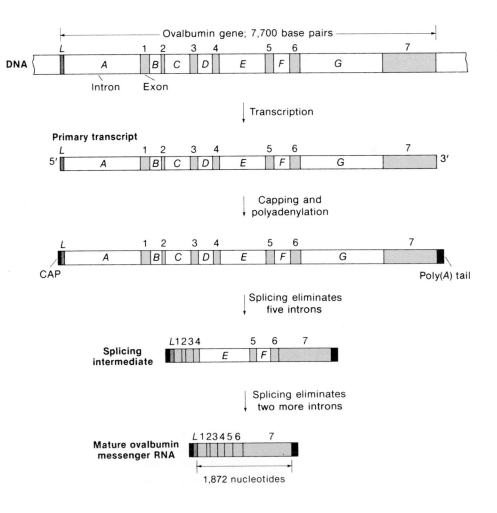

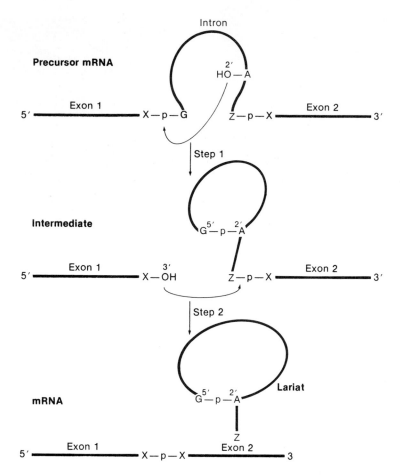

Figure 27–17
Splicing scheme for pre-mRNA. In step 1 the 2'-OH on an adenosine attacks a phosphate that is 5' linked to a guanine residue. This leads to a lariat configuration connected to the distal exon. The lariat is formed via a 2'-5' phosphodiester bond, which joins the 5' terminal guanosine of the intron to an adenosine residue within the intron, 18–40 nucleotides upstream of the 3' splice site. In the next step there is a cleavage at the 3' splice site to generate the free intron RNA and a ligation of the two exons via a 3'-5' phosphodiester bond.

exons of hemoglobin encode three structural domains of different types, while the heavy-chain immunoglobulin exons encode domains that are quite similar in structure.

By attaching the promoter for the bacteriophage SP6 RNA polymerase to the β-globin gene, it has become possible to transcribe the gene *in vitro* with SP6 RNA polymerase to produce abundant amounts of β-globin pre-mRNA. This source of precursor mRNA has been used to determine the steps and factors involved in splicing. The process (see Figure 27–17) involves cleavage of the pre-mRNA at the 5' splice site to generate the proximal exon and an RNA species containing the intron in a "lariat" configuration connected to the distal exon (Figure 27–17, step 1). The lariat is formed via a 2'-5' phosphodiester bond, which joins the 5' terminal guanosine of the intron to an adenosine residue within the intron at a spot 18–40 nucleotides upstream of the 3' splice site. These two RNA species are probably held together in a noncovalent complex until the next step (Figure 27–17, step 2) in the reaction, which is cleavage at the 3' splice site to generate the free intron RNA and ligation of the two exons via a 3'-5' phosphodiester bond. U1 and U2 snRNAs as well as several additional protein factors appear to be necessary for the reactions to occur *in vitro*. These various factors are sometimes seen as a large 40S–60S complex, termed a *spliceosome*.

Abnormal mRNA splicing appears to be one cause of the human disease β⁺-thalassemia. The inefficient splicing of β-globin mRNA precursors in affected individuals seems to be due to mutations in the intron and leads to very low levels of mature mRNA and thus to a β-globin deficiency.

Removal of intervening sequences in eukaryotes is not restricted to mRNA processing. It also occurs in the processing of rRNA and some tRNAs. A temperature-sensitive mutant of yeast has been isolated that accumulates

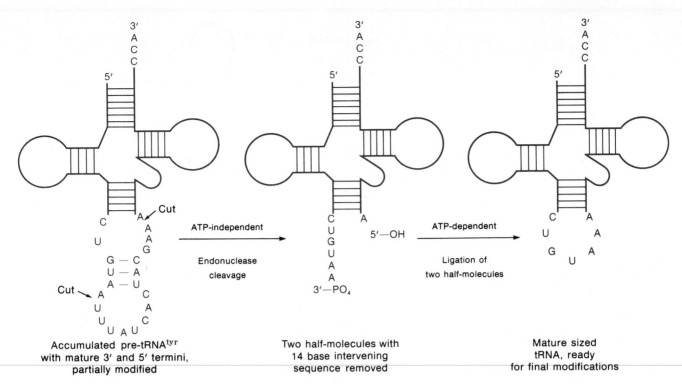

Figure 27–18
Processing of yeast tyrosine tRNA to remove intervening sequence. First, a 14-base sequence is removed by an ATP-independent endonucleolytic cleavage. The two newly produced termini are then ligated in a second reaction, which requires ATP.

certain tRNA precursors at the nonpermissive temperature. One of these is a tyrosine tRNA precursor that has a 14-base intervening sequence that can be removed by cell extracts *in vitro*. The reaction occurs in two steps and is shown in Figure 27–18.

First, a 14-base sequence is removed by an ATP-independent endonucleolytic cleavage to produce two half tRNA molecules. These cleavages produce 3'-PO$_4$ and 5'-OH termini, unlike the 3'-OH and 5'-PO$_4$ termini produced by most RNA processing enzymes, such as RNAse III or RNAse P. The two termini are then ligated in a second reaction that requires ATP.

Some yeast tRNAs have intervening sequences that are up to 60 bases long, but they seem to be processed by the same endonuclease, which appears to recognize a structural feature in the precursor rather than a particular sequence. The function of the yeast intervening sequences is not known, nor is it known whether their splicing is at all similar to the splicing that occurs in mRNA.

Some RNAs Are Self-splicing

In 1982, the splicing of rRNA from the protozoan *Tetrahymena* was shown by Tom Cech to involve a startling mechanism. When the pre-rRNA was incubated under the appropriate conditions of salt, Mg^{2+}, and guanosine nucleotide, splicing of the RNA occurred without the help of a protein enzyme! This ability of RNA to act as a catalyst for its own splicing has given rise to the term "ribozyme" and has broadened our definition of enzymes.

The mechanism of the self-splicing reaction is shown in Figure 27–19. The first step is a transesterification reaction in which the 3' hydroxyl group of the guanosine cofactor attacks the phosphodiester bond at the 5' splice site. The second step is another transesterification reaction in which the 3'

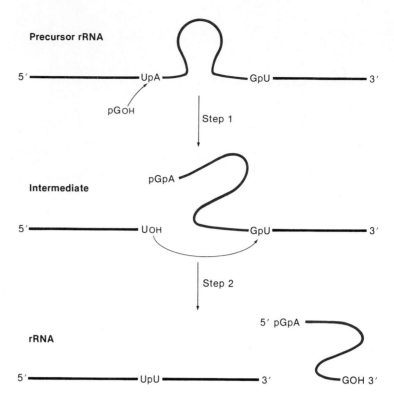

Figure 27–19

Self-splicing of pre-rRNA from the protozoan *Tetrahymena*. The first step is a transesterification reaction in which the 3′ hydroxyl group of a guanosine attacks the phosphodiester bond at the 5′ splice site. The second step involves another transesterification reaction in which the 3′ hydroxyl group of the upstream exon attacks the phosphodiester bond at the 3′ splice site and displaces the 3′ hydroxyl group of the intron.

hydroxyl group of the upstream exon attacks the phosphodiester bond at the 3′ splice site and displaces the 3′ hydroxyl group of the intron. The final reaction products are the joined exons and the excised intron.

Since its initial discovery, self-splicing has been found to occur for RNAs from a wide variety of organisms. However, the fraction of unprocessed RNAs that can be shown to undergo self-splicing *in vitro* is quite small in all cases. Recently it has been discovered that certain precursor RNAs that exhibit self-splicing produce intron lariats, just like those seen in the commonly observed splicing reactions that take place in most precursor RNAs with introns. In this case the guanosine nucleotide is not required. This fact suggests that originally RNAs may have been self-splicing, and that the spliceosome components have evolved to improve catalytic efficiency and perhaps to regulate the process.

Self-splicing of RNA shows that in certain cases RNA has enzymelike activity. However, it does not by itself demonstrate true enzyme behavior. Recall that an enzyme is a substance that acts as a catalyst—it accelerates a reaction without itself being consumed in the reaction. Recently Zaug and Cech have shown that a fragment of the self-splicing ribosomal RNA intervening sequence of *Tetrahymena thermophila* can act as an enzyme *in vitro*. This RNA fragment has been shown to catalyze the breakage and rejoining of oligonucleotide substrates in a sequence-dependent manner, with a $K_m = 42 \mu M$ and a $k_{cat} = 2 \text{ min}^{-1}$. With pentacytidylic acid as the substrate, successive cleavage and rejoining reactions lead to the synthesis of higher polymers of polycytidylic acid (Figure 27–20).

The discovery of Cech and his colleagues that RNA can function as an enzyme raises many possibilities both for biochemical reactions as they are currently understood and for reactions that may have played a major role in prebiotic evolution. Most molecular evolutionists are of the opinion that nucleic acids make good templates but poor enzymes (e.g., see Chapter 33). The results of Cech suggest that polynucleotide chains may have replicated by autocatalysis before the advent of polypeptide enzymes.

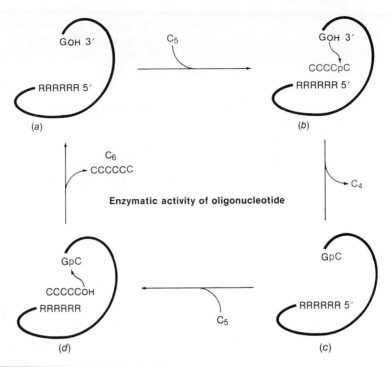

Figure 27–20
Model for the enzymatic mechanism of the 19-base oligonucleotide intervening sequence isolated by Cech from a *Tetrahymena* precursor rRNA. (*a*) The oligonucleotide enzyme (ribozyme) is shown with the oligopyrimidine binding site (RRRRRR), a sequence of six purines near its 5′ end, and a guanine with a 3′ hydroxyl group at its 3′ end. (*b*) The enzyme binds its substrate, a pentacytidylic acid. (*c*) Nucleophilic attack by the terminal 3′-OH of the guanine residue leads to formation of the covalent intermediate and displacement of the tetracytidylic acid. (*d*) A second pentacytidylic acid binds at the enzyme site and, by transesterification, a hexanucleotide is formed.

DEGRADATION OF RNA BY RIBONUCLEASES

Although a large number of nonspecific exonucleases and endonucleases have been identified in many organisms, the role of most of them is not understood. Some are extracellular or secreted enzymes that presumably function in breaking down RNA to recycle the purine and pyrimidine bases. Intracellular ribonucleases also may be involved in recycling, as well as in some aspects of processing, as described earlier. Some of the better-known enzymes are listed in Table 27–6.

E. coli RNAse I is an endonuclease that is located in the periplasmic space between the cell membrane and the cell wall. RNAse II is an intracellular 3′ exonuclease that rapidly degrades RNA fragments. Mutants defective in these enzymes were very important in the development of *in vitro* translation systems, since extracts from cells defective in these enzymes allowed mRNA to be translated without being rapidly degraded.

RNAse H has the unusual property of degrading the RNA strand from a RNA-DNA hybrid molecule and may be involved in the removal of RNA primers during DNA replication. Some endonucleases have marked preferences for cleavage after certain bases. Pancreatic RNAse A cleaves at the 3′ side of pyrimidines, while RNAse T1 cleaves only after guanosine residues. Both leave 3′ phosphate termini. Because of their specificity, these enzymes have proven very useful in the analysis of RNA sequences. S1 nuclease digests single-stranded DNA or RNA.

Finally, there are exonucleases that digest in the 3′→5′ direction and others that digest in the 5′→3′ direction. It is likely that degradation of mRNA proceeds by numerous endonucleolytic cleavages and then exonuclease digestion of the resulting fragments. All known intracellular exonucleases in *E. coli* produce 5'-NMPs, which can readily be recycled for use in RNA synthesis.

INHIBITORS OF RNA METABOLISM

A large variety of inhibitors of RNA synthesis have been identified. These inhibitors have proved useful in elucidating transcription mechanisms, and some have allowed the isolation of mutant strains with enzymes that are resistant to their inhibition. The inhibitors fall into several classes, as described below.

Some Inhibitors Act by Binding to DNA

The best-known example of inhibitors that bind to DNA is actinomycin D, an antibiotic produced by *Streptomyces antibioticus*. The inhibition of RNA synthesis is caused by the insertion (intercalation) of its phenoxazone ring between two G-C base pairs, with the side chains projecting into the minor groove of the double helix, hydrogen-bonded to guanine residues. RNA polymerase binding to DNA that contains actinomycin D is only slightly impaired, but RNA chain elongation in both eukaryotes and prokaryotes is blocked. Ethidium bromide also intercalates into DNA and at low concentrations preferentially binds to negatively supercoiled DNA (see Chapter 7). It has been used to selectively inhibit transcription in mitochondria, which contain supercoiled DNA.

Some Inhibitors of Transcription Bind to DNA Gyrase

DNA gyrase is an enzyme in *E. coli* that is needed to maintain the proper degree of supercoiling in DNA. Transcription of some genes seems to be affected by the supercoiled state of the template, and transcription of these genes can be selectively inhibited or stimulated by compounds that interfere with the action of DNA gyrase. Examples of compounds that inhibit DNA gyrase are naladixic acid, which binds to one of the two gyrase subunits, and coumermycin and novobiocin, which bind to the other (see Chapters 7 and 26).

Some Inhibitors Bind to RNA Polymerase

Rifampicin is a synthetic derivative of a naturally occurring antibiotic, rifamycin, that inhibits bacterial DNA-dependent RNA polymerase but not T7 RNA polymerase or eukaryotic RNA polymerases. It binds tightly to the β subunit. While it does not prevent promoter binding or formation of the first phosphodiester bond, it effectively prevents synthesis of longer RNA chains. It does not inhibit elongation when added after initiation has occurred. Another antibiotic, streptolydigin, also binds to the β subunit, but it is able to prevent all bond formation, whether involved in initiation or elongation.

The most useful inhibitor of eukaryotic transcription has been α-amanitin, a major toxic substance in the poisonous mushroom *Amanita phalloides*. The toxin preferentially binds to and inhibits RNA polymerase II. At high concentrations it also can inhibit RNA polymerase III, but not RNA polymerase I or bacterial, mitochondrial, or chloroplast RNA polymerase.

Two inhibitors, cordycepin and DRB, in their 5'-triphosphorylated forms are substrate analogs and can be incorporated into growing RNA chains by most RNA polymerases. Cordycepin is a 3'-deoxyadenosine. It causes chain termination, since it does not contain the 3' hydroxyl group necessary for the formation of the next phosphodiester bond. DRB is a 5,6-dichloro-1-β-D-ribofuranosylbenzimidazole. The major effect of DRB *in vitro* and *in vivo* appears to be the inhibition of a protein kinase somehow involved in initiation of RNA synthesis by RNA polymerase II.

SUMMARY

The basic aspects of RNA metabolism include synthesis, modification, processing, function, and degradation. RNA is synthesized in the 5'→3' direction by the formation of 3', 5'-phosphodiester linkages between the four ribonucleoside triphosphate substrate precursors, analogous to the process of DNA synthesis. The sequence of RNA transcripts catalyzed by DNA-dependent RNA polymerases is specified by the complementary DNA template strand. Some newly synthesized RNA transcripts are the functional species, while other newly synthesized RNAs are precursor transcripts that must be modified and/or processed into the mature functional species. Modifying enzymes add nucleotides to the 5' or 3' terminals or alter bases within the RNA, such as by methylation of specific residues. Specific processing enzymes cleave RNA internally, splice together noncontiguous regions of a transcript, or remove nucleotides from the 5' or 3' terminals.

The major classes of RNA in both prokaryotes and eukaryotes are messenger RNA, ribosomal RNA, and transfer RNA. These distinct classes of RNA play specific functional or structural roles involved in the translation of genetic information into proteins. Other classes of RNAs transcribed by DNA-dependent RNA polymerases include primers for DNA synthesis, small RNAs that may be directly involved in processing of larger RNA precursors, and still others that may play as yet unidentified regulatory or structural roles in cells.

DNA-dependent synthesis of RNA in *E. coli* is catalyzed by one enzyme, consisting of five polypeptide subunits. The complete holoenzyme is composed of four polypeptides (the core enzyme) and an additional polypeptide that confers specificity for initiation at promotor sequences in the DNA template. The steps involved in transcription—binding of polymerase, initiation, elongation, and termination—have been studied in great detail. Elucidation of these steps, as well as the identification and function of the proteins and DNA sequences involved at each step, has been greatly facilitated by the use of specific inhibitors of RNA synthesis, mutants, and a variety of *in vitro* techniques.

In eukaryotes, transcription is catalyzed by at least four distinct polymerases, each specializing in the synthesis of a different class of RNA. RNA polymerases I, II, and III are responsible for the synthesis of rRNA, mRNA, and small RNA (such as tRNA) transcripts, respectively. Specific RNA polymerases are responsible for transcription in mitochondria and chloroplasts. Although the basic mechanism of RNA synthesis by these enzymes is similar to that of prokaryotic RNA polymerase, many details of the steps, proteins, and DNA regulatory sequences involved in eukaryotic transcription are less well understood than in prokaryotes.

Various other RNA polymerases have been identified and studied. These include DNA primase, specific viral-induced or viral-encoded polymerases, and polymerases that are not dependent on DNA templates for RNA synthesis. Other enzymes involved in RNA metabolism include processing nucleases, degradation nucleases, ligases, and modifying enzymes. Many of these enzymes have been at least partially purified and characterized, and some have proved useful for the *in vitro* manipulation and analysis of RNAs.

One of the more interesting processing reactions of nascent transcripts involves the removal of intervening sequences. This type of processing is generically referred to as splicing. Most splicing reactions appear to require host proteins. However, it is clear that certain splicing reactions require only RNA, a fact demonstrating that the RNA is capable of functioning like an enzyme in the making and breaking of phosphodiester linkages in polyribonucleotides.

SELECTED READINGS

Apirion, D., and P. Gegenheimer, Processing of bacterial RNA. *FEBS Lett.* 125:1, 1981. A summary of RNA processing in bacteria.

Chamberlin, M., Bacterial RNA Polymerase and Bacteriophage RNA Polymerase. In P. D. Boyer (ed.), *The Enzymes*, 3rd Ed., Vol. 15B. New York: Academic Press, 1982. Pp. 61–108. A review of prokaryotic and bacteriophage RNA polymerases.

Chambon, P., Split genes. *Sci. Am.* 244:60, 1981. A description of the processing of eukaryotic mRNA, in particular the mRNA for chicken ovalbumin.

Lewis, M. K., and R. R. Burgess, Eukaryotic RNA Polymerases. In P. D. Boyer (ed.), *The Enzymes*, 3rd Ed., Vol. 15B. New York: Academic Press, 1982. Pp. 109–153. A review of eukaryotic RNA polymerases.

Losick, R., and J. Pero, Cascades of sigma factors. *Cell* 25:582, 1980. A mini review of the multiple sigmas found in *B. subtilis* during sporulation and after bacteriophage infection.

Reznikoff, W. S., D. A. Siegele, D. W. Cowing and C. A. Gross, The regulation of transcription initiation in bacteria. *Ann. Rev. Genet.* 19:355–87, 1985.

Sharp, P. A., Splicing of messenger RNA precursors. *Science* 235:766–771, 1987.

Siebenlist, U., R. Simpson, and W. Gilbert, *E. coli* RNA polymerase interacts homologously with two different

promoters. *Cell* 20:269, 1980. A summary of the structure of promoters.

Zaug, A. J., and T. R. Cech, The intervening sequence RNA tetrahymena is an enzyme. *Science* 231:470–475, 1986.

PROBLEMS

1. One strand of a DNA molecule is completely transcribed into mRNA by RNA polymerase. The base composition of the DNA template strand is: G = 20%, C = 25%, A = 15%, T = 40%. What would you expect to be the base composition of the newly synthesized RNA molecule?

2. a. Calculate the moles of UTP incorporated in the RNA polymerase assay depicted in Figure 27–2, given a reaction volume of 1 ml, a specific activity of the radioactive UTP of 100 Cu/mole (1 Cu equals 2.2×10^{12} disintegrations per minute), and 220,000 dpm of radioactivity incorporated into RNA in a 10-min reaction.

   b. Assuming that the RNA is 40 per cent U, calculate the moles of total nucleotide incorporated into RNA.

   c. Why is the concentration of UTP lower than that of the other three NTPs?

   d. The molecular weight of RNA polymerase is 460,000. If 1 μg of polymerase is added to the preceding reaction, and if each polymerase molecule initiates at once and catalyzes chain elongation at a constant rate during the course of the reaction, how long is the average RNA chain and what is the average chain growth rate?

3. The illustration is a schematic drawing representing a portion of Miller's electron microscope picture of transcription and translation in progress in *E. coli*.

   a. Identify 1, 2, 3, and 4.

   b. Indicate the 3′ and 5′ ends of the sense, or template, strand of DNA.

   c. Indicate the 3′ and 5′ ends of the mRNA.

   d. Draw four peptides in the process of being synthesized, indicating relative lengths.

   e. Indicate the N- and C-terminal ends of the longest peptide.

   f. Indicate with an arrow the direction in which RNA polymerase is moving.

   g. What parts of this diagram would be different in eukaryotes?

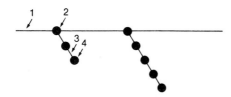

4. Although 40 to 50 per cent of the RNA being synthesized in *E. coli* at any given time is mRNA, only about 3 per cent of the total RNA in the cell is mRNA. Explain.

5. In *E. coli* the precise spacing between the −35 and −10 conserved promoter elements has been found to be a critical determinant of promoter strength *in vivo* and *in vitro*. What do you think this means in terms of the interaction between RNA polymerase and these elements? What other experimental evidence described in this chapter has immediate bearing on this question?

6. The conserved −10 element of *E. coli* promoters is notably (A + T)-rich, as is the TATA box found 25–30 bases upstream of the initiation site of many eukaryotic genes transcribed by RNA polymerase II. Can you suggest why this might be so?

7. What is the maximum theoretical rate of initiation at a promoter, assuming that the diameter of RNA polymerase is about 200 Å and the rate of RNA chain growth *in vivo* is 45 nucleotides per second?

8. Cordycepin-5′-triphosphate, or 3′-deoxy ATP, is a ribonucleoside triphosphate analog that can be bound as a substrate by RNA polymerase. When it is added to an *in vitro* transcription reaction, it is incorporated into the RNA, but RNA synthesis terminates. This fact was used to argue that the direction of transcription is 5′→3′. Explain why the argument is a reasonable one.

9. The *E. coli* transcription termination factor rho has recently been found to possess RNA-DNA helicase activity. From what you know of the process of transcription termination in *E. coli* suggest a mechanism for rho-mediated termination that incorporates this finding. Compare this mechanism with that thought to operate in rho-independent transcription termination.

10. Given two oligonucleotides, 5′PO$_4$-GUUC-PO$_4$3′ and 5′OH-ACCG-OH3′, describe how to synthesize 5′OH-ACCGGUUCAAAA---3′-OH. You may use enzymes described in this chapter and the enzyme bacterial alkaline phosphatase, which is capable of removing a nucleotide 5′-PO$_4$ or 3′-PO$_4$ to produce a nucleotide 5′-OH or 3′-OH.

11. The following hypothetical RNA is a primary precursor transcript made in a eukaryotic nucleus and its synthesis is inhibited by low levels of α-amanitin:

    5′-pppAUUAUGCCGAUAAGGUAAGUA (100 bases)
        AUCUCCCUGCAGGGCGUAACCAAUAAA
            CGACGACGACGUCACC---3′-OH

    Indicate the final processed RNA found in the cytoplasm and point out important features.

12. A particular eukaryotic DNA virus was found to code for two mRNA transcripts, one shorter than the other,

from the same region of its genome. Analysis of the translation products revealed that the two polypeptides shared the same amino acid sequence at their amino terminals, but differed at their carboxy terminals. Surprisingly, the longer of the two polypeptides was coded for by the shorter mRNA. Suggest an explanation.

13. In a recent study, deproteinized RNA was isolated from *E. coli* that had been infected with bacteriophage T4. When the RNA was incubated in the presence of [α-^{32}P]GTP under selected conditions, it gave rise to multiple labeled species. Can you suggest an explanation for this result?

14. A short RNA chain synthesized *in vitro* by *E. coli* RNA polymerase has the sequence 5'-AUGUACCGAAGUG-GUUU-3'-OH.
    a. Draw the phosphate groups and star those that would be radioactive when the transcription is carried out in the presence of [γ-^{32}P]ATP.
    b. Do the same for [α-^{32}P]UTP.
    c. Show where RNAse T1 and pancreatic RNAse A would cleave this RNA.
    d. Show the oligonucleotides produced by RNAse T1 digestion of the [α-^{32}P]UTP-labeled RNA and indicate which are labeled. Suggest how this result might be exploited to analyze RNA sequences.

28

PROTEIN METABOLISM

Proteins are informational macromolecules, the ultimate heirs of the genetic information encoded in the sequence of nucleotide bases within the chromosomes. Each protein is composed of one or more polypeptide chains and each polypeptide chain is a linear polymer of amino acids. There are twenty amino acids commonly found in polypeptide chains (see Table 1–1). The order of these amino acids in the polypeptide polymer is determined by the order of nucleotides in the corresponding messenger RNA. In this chapter we will be concerned primarily with the complex process whereby these amino acids are ordered and polymerized into polypeptide chains. We will also deal with questions of posttranslational modification and processing, and protein degradation.

THE CELLULAR MACHINERY OF PROTEIN SYNTHESIS

In Chapter 27 we described three types of RNA molecules that play crucial roles in protein synthesis. These are ribosomal RNA (rRNA), transfer RNA (tRNA), and messenger RNA (mRNA). The tRNAs and the mRNAs exist as independent entities or in complexes with other molecules; ribosomal RNAs exist solely as part of a ribonucleoprotein complex in the ribosome.

Much of the early work pinpointing the importance of *ribosomes* as the site of peptide synthesis, and of *transfer RNA* as the agent that transfers the amino acid to the ribosome, was done in Paul Zamecnick's laboratory in the early 1950s. The biochemical pathway followed by an amino acid was traced by injecting the acid in a radioactive form into a whole animal and subsequently fractionating the subcellular components for their content of the radioactive amino acid. In this way, it was shown that the amino acids first become linked to a low-molecular-weight RNA fraction (the transfer RNA), which then migrates to a large ribonucleoprotein particle (the ribosome), where the peptide linkage is formed. During protein synthesis the *messenger RNA* (mRNA), which carries the genetic message for ordering the amino acids into a polypeptide chain, also becomes complexed to the ribosome.

In the following discussion, we will first describe the structures of the transfer RNA, the ribosomes, and the mRNA, which constitute the main cellular machinery of protein synthesis. Then we will see how these structures interact with various protein enzymes during protein synthesis.

Transfer RNA Transports Amino Acids to the Template

Transfer RNAs serve a dual function in protein synthesis; they contain a site for attachment of the amino acid, and they contain a site that interacts with the mRNA. In Chapter 7 we discussed the structure of the transfer RNA for phenylalanine in some detail (see Figure 7–42). Each amino acid is associated with a specific transfer RNA, and some with more than one. In writing, the different classes of tRNAs are distinguished by a superscript. For example, tRNA^{Phe} and tRNA^{Ser} refer to the family of tRNAs that form complexes with the amino acids phenylalanine and serine, respectively.

The transfer RNAs specific for all the amino acids found in proteins and derived from all biological sources except mammalian mitochondria are very similar in secondary structure. A single figure can describe the generalities of their structure (Figure 28–1). These molecules fold back on themselves to form a cloverleaf with either four or five double-stranded base-paired stems and either three or four single-stranded loops. They fold in such a way as to bring the 5′ and 3′ termini together in what is known as the *acceptor stem*. The amino acid attaches to the 3′ end of the acceptor stem. Nearly all acceptor stems are composed of seven regular Watson-Crick base pairs, formed between regions at the 3′ and 5′ ends of the tRNA. The most consistent feature of the amino acid attachment site is the invariant sequence of three bases at the 3′ terminus: this sequence is CCA.

The unpaired regions (loops) of tRNA are named according to their unique structural features. Loop I varies in size from 7 to 11 unpaired bases and frequently contains dihydrouracil; it is designated the D loop. The tRNA brings its specific amino acid to the correct site on the mRNA template by pairing three of its bases, the *anticodon*, with three bases, the *codon*, in the

Figure 28–1
The primary and secondary structure of tRNA. The solid line is the phosphodiester backbone. Base pairing between nucleotides is indicated by dashed lines, and nucleotides are represented by open circles. A is the 3′ end. The solid circles indicate nucleotides that may or may not be present. (Adapted from H. Weissbach and S. Pestka, *Molecular Mechanisms of Protein Biosynthesis*, Academic Press, New York, 1977.) See Figure 7–42 for a more realistic three-dimensional structure of tRNA.

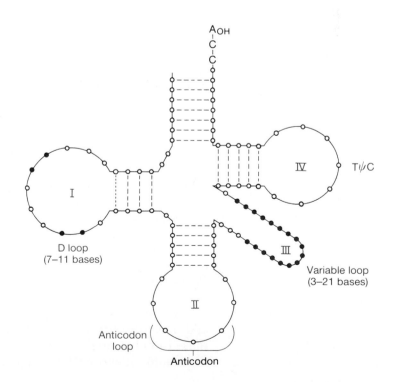

mRNA. Loop II, containing the anticodon sequence, is known as the anticodon loop. Loop III, the variable loop, may contain as few as 3 or as many as 21 bases, making it the major site of size variability in tRNA. Loop IV usually contains ribothymidine and pseudo-uridine as part of an invariant sequence; for this reason it is known as the TψC loop. Beyond the anticodon loop, the functional significance of the structural features common to the other loops has not been established. As noted in Chapter 27, tRNAs vary in nucleotide sequence and content of rare nucleotides.

Ribosomes Are the Site of Protein Synthesis

The assembly of amino acids into polypeptide chains in all living systems occurs on ribosomes. The ribosomes in bacteria have a particle mass of about 2.5 million daltons, a sedimentation coefficient of 70S, and are about one-third by weight protein and two-thirds by weight RNA. They are somewhat smaller than the ribosomes found in the cytoplasm of eukaryotes, which have particle masses of approximately 4 million daltons and sedimentation coefficients of about 80S, and which contain about equal weights of RNA and protein.

All ribosomes are composed of two ribonucleoprotein subunits of unequal size. In bacteria, these subunits have sedimentation coefficients of 30S and 50S, while in the cytoplasm of eukaryotes, the subunits have sedimentation coefficients of approximately 40S and 60S. The E. coli ribosome contains three ribosomal RNAs. A 16S RNA of 1542 bases forms part of the small ribosomal subunit; the larger ribosomal subunit contains both a 23S RNA of 2904 bases and a 5S RNA of 120 bases. The ribosomes in eukaryotic cytoplasm contain four ribosomal RNAs; two of these correspond to the larger RNAs found in bacterial ribosomes, and there are two small ribosomal RNAs, a 5S and a 5.8S species. The current view, with some specific exceptions, is that ribosomal RNA plays a structural role, providing a scaffold for the correct assembly and positioning of the ribosomal proteins.

With respect to protein composition, there appears to be only one type of ribosome within a given E. coli cell. The ribosome contains approximately 900,000 daltons of protein divided among 56 molecules. All of these proteins have been sequenced; one of them appears in four copies and 52 of them appear only once. A simplified diagram of the E. coli ribosome and its components is presented in Figure 28–2. Further aspects of E. coli ribosome structure and assembly are discussed in Box 28–A. In general, eukaryotic ribosomes appear to contain more proteins (between 70 and 90) than bacterial ribosomes, and these proteins appear to have greater molecular weights on the average than those found in bacteria.

The catalytic center responsible for the formation of the peptide bond during protein synthesis resides entirely within the large ribosomal subunit. Similarly, the specific binding of messenger RNA during protein synthesis occurs entirely on the small ribosomal subunit. There are two adjacent binding sites for tRNAs; these binding sites are mainly on the large subunit (see Figure 28–3). Beyond these generalities, the assignment of specific functions to the various ribosomal components has been difficult.

Whereas most of the ribosomes are located in the cytoplasm in eukaryotic cells, some ribosomes are localized in certain intracellular organelles. These ribosomes, which are used to make proteins that function exclusively within their own organelles, are somewhat different in structure. The ribosomes from chloroplasts resemble bacterial ribosomes in both composition and size. Ribosomes from the mitochondria of mammals are similar in size to those from bacteria, but they have substantially smaller sedimentation coefficients (55S) and contain a much higher percentage of protein.

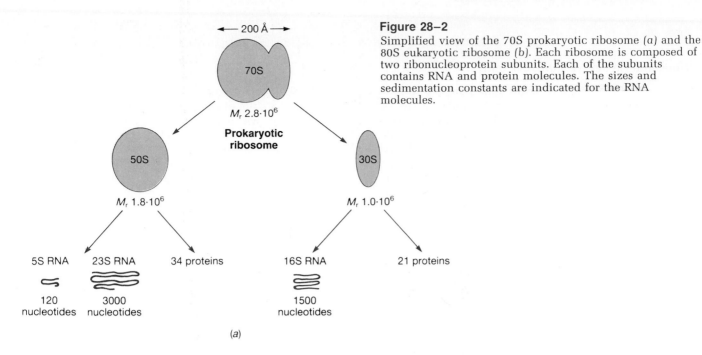

Figure 28-2

Simplified view of the 70S prokaryotic ribosome (a) and the 80S eukaryotic ribosome (b). Each ribosome is composed of two ribonucleoprotein subunits. Each of the subunits contains RNA and protein molecules. The sizes and sedimentation constants are indicated for the RNA molecules.

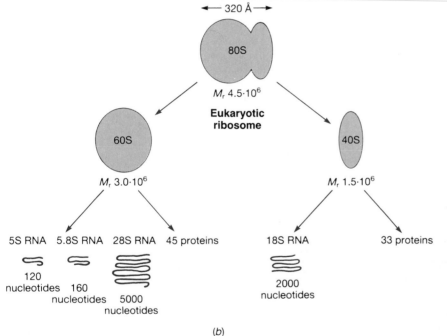

Figure 28-3 (Below)

Diagram of the ribosome showing the binding sites for mRNA and tRNA. (a) The mRNA binds to the 30S ribosomal subunit. The tRNAs bind to either of two sites on the 50S ribosomal subunit, the P site or the A site. Three bases in each tRNA are hydrogen-bonded to three bases in the mRNA. (b) Correct relative sizes and shapes of ribosome and tRNA.

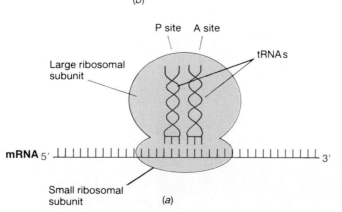

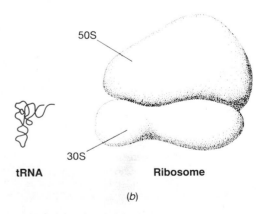

BOX 28–A

Ribosome Structure and Assembly

Each ribosome in *E. coli* contains approximately 56 protein molecules. By convention, an L designates those proteins from the large ribosomal subunit and an S, those from the smaller ribosomal subunit. These letters are followed by a number that designates the protein mobility in a two-dimensional electrophoresis system (see Chapter 3). Thus, for example, the slowest-moving protein from the 30S subunit is designated S1 (Illustration 1).

Of the 56 proteins found on the ribosome, 53 are unique. One protein partitions between the ribosomal subunits upon their dissociation and hence occurs partially in the small subunit (S20) and partially in the large subunit (L26). Another protein is present in four copies per large subunit, and its amino terminus is partly acetylated so that it gives rise to two electrophoretic spots (L7/L12). Beyond these two exceptions, the small subunit contains one equivalent each of 20 different proteins (S1 to S19, and S21), and the large subunit contains one equivalent each of 31 different proteins (L1 to L6, L8 to L11, L13 to L25, and L27 to L34).

The protein composition of eukaryotic cytoplasmic ribosomes is not nearly so clear. In general, these ribosomes appear to contain more proteins (between 70 and 90) than bacterial ribosomes, and the proteins appear to have greater molecular weights on the average than those found in bacteria. We don't yet know whether ribosomes in the cytoplasm of eukaryotes are heterogeneous with respect to their protein composition.

Electron microscopy indicates that the individual ribosomal subunits of *E. coli* are both irregular and asymmetric in shape (Illustration 2). Since the complete structure of individual ribosomal components is not known, the relative location of components within the ribosome can only be known in a general way. What, then, do we know about the three-dimensional arrangement of the 56 proteins and the three RNAs that constitute the *E. coli* ribosome?

First of all, it is clear that the ribosome is held together by interactions both among and between proteins and RNA. Some individual r-proteins, for example, bind to specific regions of rRNA and protect these regions from digestion by nucleases, even in the absence of other ribosomal components. Other r-proteins depend on the presence of both rRNA and r-protein in order to remain within the ribosome. Finally, conservation of base pairing (see Chapter 7) makes it clear that double-stranded regions and higher-order structures exist within rRNA.

Neighbor relationships, distances between, and locations of ribosomal components are all being investigated. First, it has been possible to chemically cross-link two components within the ribosome with a bifunctional reagent; it therefore appears that the two components, or chemically reactive portions of them, are adjacent to each other in the ribosome. Second, distances between components are being mapped by measuring the degree of energy transfer between fluorophores on individual components in the ribosome. The locations of components are being investigated in two ways. Neutron scattering studies of ribosomes containing individual deuterated components are beginning to give information about the location of r-protein (Illustration 3). The locations of individual antigenic components on the ribosome surface

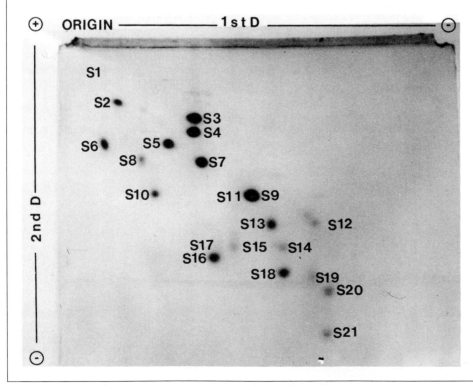

Illustration 1
A two-dimensional gel electropherogram of *E. coli* 30S proteins. (Courtesy of Bishwajit Nag and Robert R. Traut.)

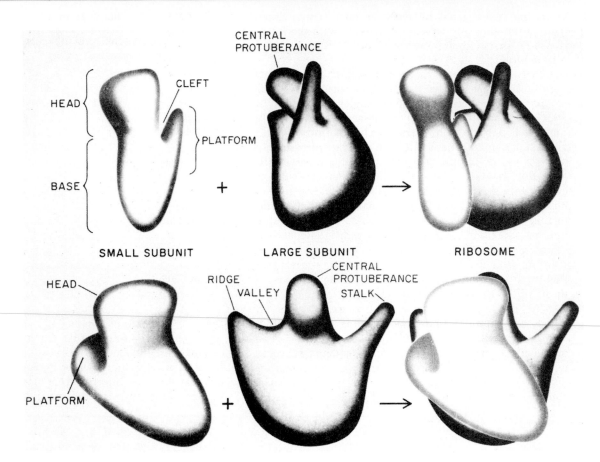

Illustration 2

The morphological view of the *E. coli* ribosome that has emerged from electron micrographic studies. (From J. A. Lake, Evolving ribosome structure: domains in archaebacteria, eubacteria, eocytes and eukaryotes, *Ann. Rev. Biochem.* 54:507, 1985.)

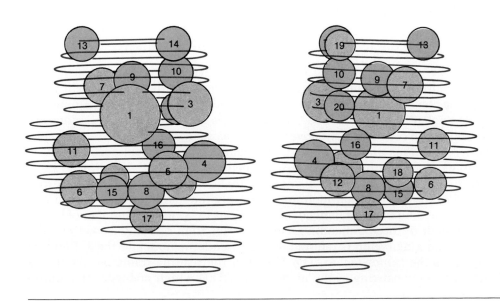

Illustration 3

Two computer graphic views of the three-dimensional distribution of *E. coli* 30S ribosomal proteins, deduced from neutron scattering studies superimposed on an electron microscopic model of the 30S subunit. The "bottom" of the 30S subunit appears to be composed almost entirely of rRNA. (Courtesy Peter B. Moore.)

BOX 28–A *(Continued)*

are being investigated by electron microscopy. Since many antibodies are capable of reacting with the intact ribosome, this method promises to provide a great deal of information.

Nomura has shown that it is possible, after separating the ribosome into its 59 individual macromolecular components, to recombine these under conditions in which they will spontaneously reassemble into a fully functional particle. With various combinations of individual components, he was able to analyze the assembly process and define the sequence of events involved (Illustration 4). Assembly of the small ribosomal subunit begins with the independent binding of several ribosomal proteins to 16S RNA. Additional proteins then add to this structure, ultimately giving rise to an assembly intermediate that requires a temperature-dependent structural rearrangement prior to the addition of the remaining proteins. A number of experiments suggest that this same assembly pathway occurs during the biosynthesis of the ribosome within the cell.

By conducting partial-assembly experiments in which one or more ribosomal proteins were omitted, Nomura was able to construct an "assembly map." The map defines not only the approximate sequence of events in assembly, but also those proteins whose binding requires the prior assembly of other proteins. It is reasonable to conclude that these obligatory relationships result

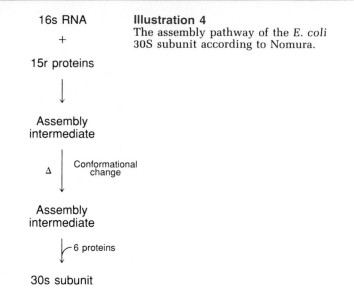

Illustration 4
The assembly pathway of the *E. coli* 30S subunit according to Nomura.

from the interactions among proteins within the ribosome, and that proteins that depend on each other for assembly are in fact close to one another within the ribosome. In general, there is good agreement between neighbor relationships inferred from the assembly map and those established by the means described earlier.

When actively involved in protein synthesis, the ribosome moves along the mRNA chain in the 5′→3′ direction. Most commonly, a single molecule of mRNA is read simultaneously by a number of ribosomes, each engaged in the synthesis of a single polypeptide chain; the resulting structure resembles a string of beads and is known as a *polysome* (Figure 28–4). Polysomes producing proteins that are destined for transport across membrane-bounded structures are in turn bound to lipoprotein membrane structures. In eukaryotes, the complex of the ribosomes with the lipoprotein membranes is known as the *rough endoplasmic reticulum* (Figure 28–5).

Messenger RNA Is the Template for Protein Synthesis

The mRNA molecule carries the genetic message, in the form of a sequence of nucleotides that determines the order of amino acids in the protein polypeptide chain. Each amino acid is represented in the mRNA by a sequence of three nucleotides. These three-base sequences are called *codons*. Codons are arranged in a contiguous, nonoverlapping manner. The detailed relationship between codons and amino acids is discussed in the next section of this chapter. Translation of the mRNA proceeds in the 5′→3′ direction as indicated above. That region of the message that contains the codons is referred to as the *translation reading frame* or simply the reading frame (Figure 28–6). The reading frame is flanked on both sides by a variable number of bases. The length of these noncoding regions varies from one mRNA to the next. The 5′ flanking region is referred to as the *leader region*, and the 3′ flanking region as the *trailer region*. The 5′ end of the reading frame begins with a *start codon* consisting of the nucleotides AUG, which codes for the amino

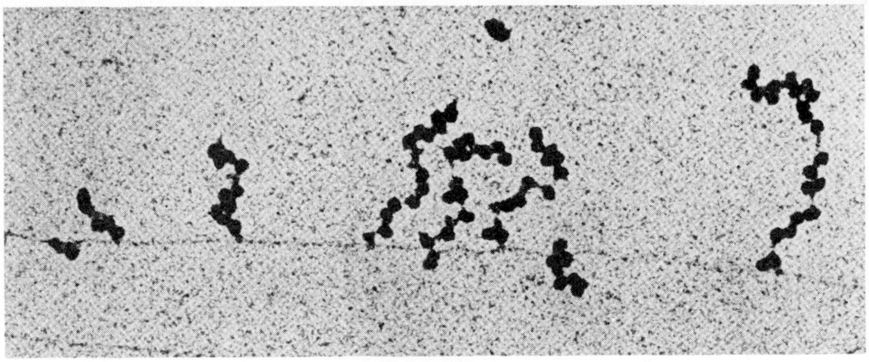

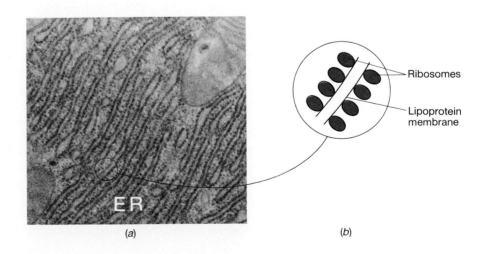

(a)

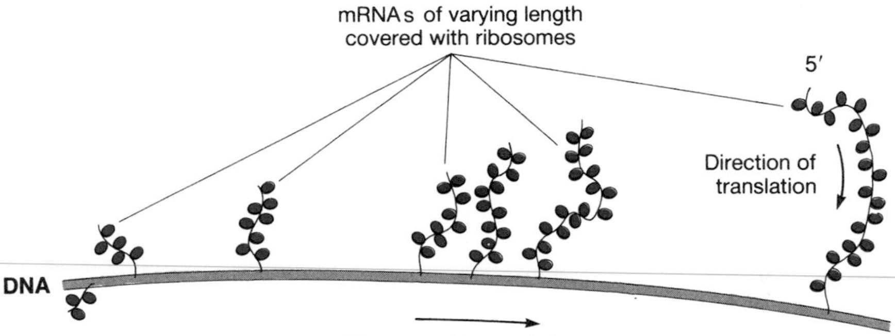

Figure 28–4

(a) Electron micrograph of *E. coli* polysomes. Ribosomes are the dark structures connected by the faintly visible mRNA strand. A DNA strand connecting the polysomes from which the mRNA is being transcribed is visible as a horizontal line. (From O. L. Miller et al., Visualization of bacterial genes in action, *Science* 169:392, 1970. Used with permission.) (b) Line drawing for clarification. In bacteria, translation usually begins before transcription is completed. This results in a DNA-mRNA-ribosome complex as shown. As the mRNA grows, the number of ribosomes associated with it increases. Here the mRNA appears to be growing from left to right. It is not possible to see the growing polypeptide chains on the ribosome because of their small size.

mRNAs of varying length covered with ribosomes

5′

Direction of translation

DNA

Direction of transcription

(b)

acid methionine. (Occasionally, GUG is also read as an initiating methionine codon, but in internal positions it is always read as valine.) The 3′ end of the reading frame invariably contains one of the three *stop codons*, UAA, UAG, or UGA. Stop codons do not represent any amino acid but rather serve as signals for termination of the polypeptide chain.

Although the AUG codon is usually the start codon of a reading frame, not every AUG represents a start codon. Some AUG codons occur within the reading frame, in which case they code for internal methionines in the polypeptide chain. One of the major demands of protein synthesis is to select the appropriate AUG start codon. Recognition of an AUG as a start codon is done in different ways in prokaryotes and eukaryotes. In eukaryotes the AUG sequence closest to the 5′ end of the mRNA usually serves as the start codon

Figure 28–5

(a) Electron micrograph of mammalian rough endoplasmic reticulum. Continuous sheets of membrane create a compartment distinct from the surrounding cytosol. (b) Clarifying line drawing shows expanded region of endoplasmic reticulum with ribosomes attached to the surface of the membrane facing the cytosol. The term rough endoplasmic reticulum arose because at low magnification the attached ribosomes give the endoplasmic reticulum a rough appearance.

ER

(a)

Ribosomes

Lipoprotein membrane

(b)

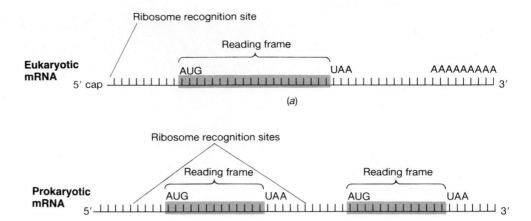

Figure 28–6
Simplified diagram of mRNA structure. (a) Typical eukaryotic mRNA. An AUG start codon is located near the 5′ end of the mRNA. The single reading frame ends with one of the three trinucleotide sequences that represents a stop codon. Frequently, but not always, the 5′ end of the mRNA is capped and the 3′ end contains a poly(A) tail. The cap structure is described in Chapter 27. (b) Typical bacterial mRNA. Some bacterial mRNAs contain more than one reading frame; some contain only one. Each reading frame contains a start and a stop codon. Recognition of the start codons is facilitated by the presence of a ribosome recognition sequence (Shine-Dalgarno). The space between any two reading frames in the same transcript varies from one transcript to the next.

for the single protein encoded by each mRNA molecule (Figure 28–6). In some cases the second or even the third AUG from the 5′ end is preferentially recognized. The optimal sequence for initiation by eukaryotic ribosomes includes a few bases in addition to the start codon (ACCAUGG).

In bacterial systems the AUG start codon may occur at any point within the mRNA and more than once within the same mRNA (Figure 28–6). How do bacterial ribosomes select the appropriate initiating code word for translation? The answer was suggested when Shine and Dalgarno noticed that 16S ribosomal RNA contains a seven-base pyrimidine-rich sequence near its 3′ terminus, and bacterial mRNA contains a complementary purine-rich region (which has become known as the _Shine-Dalgarno sequence_) centered approximately 10 bases toward the 5′ side of the AUG sequence. They proposed that base pairing between these rRNA sequences and the mRNA sequences could serve to align the initiating AUG for decoding. Subsequent sequence analysis of mRNAs has lent further support to the functional importance of the Shine-Dalgarno sequence. Some representative examples are shown in Table 28–1. The sequences are somewhat variable in both their length and position relative to the AUG codon and appear to involve $G \cdot U$ as well as conventional Watson-Crick base pairs. The degree of pairing probably affects the affinity of the ribosome for the message and hence the efficiency with which it is translated; thus those messages whose ribosome recognition sequences form the best match with 16S rRNA are read most often.

The difference between prokaryotes and eukaryotes in the recognition of initiator AUG codons has a fundamental effect on messenger utilization in the two systems. In bacteria, mRNAs frequently code for more than one polypeptide chain. Each eukaryotic mRNA, in contrast, appears to specify only a single polypeptide chain, since only the 5′ proximal AUG is recognized as a translation start site.

Eukaryotic mRNAs also differ from prokaryotic mRNAs as a result of posttranscriptional modification; the former mRNAs frequently contain a

TABLE 28–1
The Sequence of the 3′ End of *E. coli* 16S RNA and Some Shine-Dalgarno Sequences at the 5′ End of Bacterial mRNAs

	The pyrimidine-rich complement to the Shine-Dalgarno sequence
16S rRNA	3′ ··· HO̅A̅U̅U̅C̅C̅U̅C̅C̅A̅C̅U̅A ··· 5′
lacZ mRNA	5′ ··· ACAC̲A̲G̲G̲A̲AACAGCUA̲U̲G̲ ··· 3′
trpA mRNA	5′ ··· ACG̲A̲G̲G̲G̲G̲AAAUCUGA̲U̲G̲ ··· 3′
RNA polymerase β mRNA	5′ ··· GAGCUG̲A̲G̲G̲AACCCUA̲U̲G̲ ··· 3′
r-Protein L10 mRNA	5′ ··· CCA̲G̲G̲A̲G̲CAAAGCUAA̲U̲G̲ ··· 3′

The purine-rich Shine-Dalgarno sequence   The initiation codon

cap structure at the 5′ end (see Figure 27–14) and a poly(A) tail at the 3′ end (see "Posttranscriptional Modification and Processing of RNA" in Chapter 27).

THE STEPS IN TRANSLATION

Protein synthesis is the most complex biochemical process known, involving more than one hundred different proteins and more than thirty kinds of RNA molecules. The process begins by the attachment of amino acids to specific tRNA molecules. Subsequent steps take place on the ribosome; amino acids are transported to the ribosome on their tRNA carriers and they do not leave the ribosome until they have become an integral part of a polypeptide chain.

Amino Acids Are Activated Before Being Linked to Transfer RNA

The incorporation of free amino acids into proteins requires their intermediate covalent attachment to tRNAs. This attachment serves two functions: (1) The tRNA directs the amino acid to a designated place on the mRNA template so that the amino acid will be incorporated at the designated location in the polypeptide chain. (2) The linkage between the amino acid and the tRNA activates the amino acid, making the subsequent formation of a peptide linkage energetically favorable.

Formation of peptide linkages from free amino acids is not energetically favorable.

The covalent attachment between amino acid and tRNA is catalyzed by a class of enzymes known as *aminoacyl-tRNA synthases*, one for each of the twenty amino acids commonly found in proteins. These synthases vary widely in size, and, of course, they have different specificities, but they all catalyze the same general reaction by means of a similar mechanism.

The attachment of an amino acid to a tRNA involves the formation of an ester linkage between the carboxyl group of an amino acid and the 3′-terminal hydroxyl group of a tRNA (Figure 28–7). This energetically unfavorable reaction is driven by the hydrolysis of ATP to AMP and pyrophosphate. In almost every case, the interaction between the synthase, the amino acid, ATP, and tRNA proceeds in two separate steps (see Figure 28–7). The first step is formation of an aminoacyl-AMP complex from amino acid and ATP.

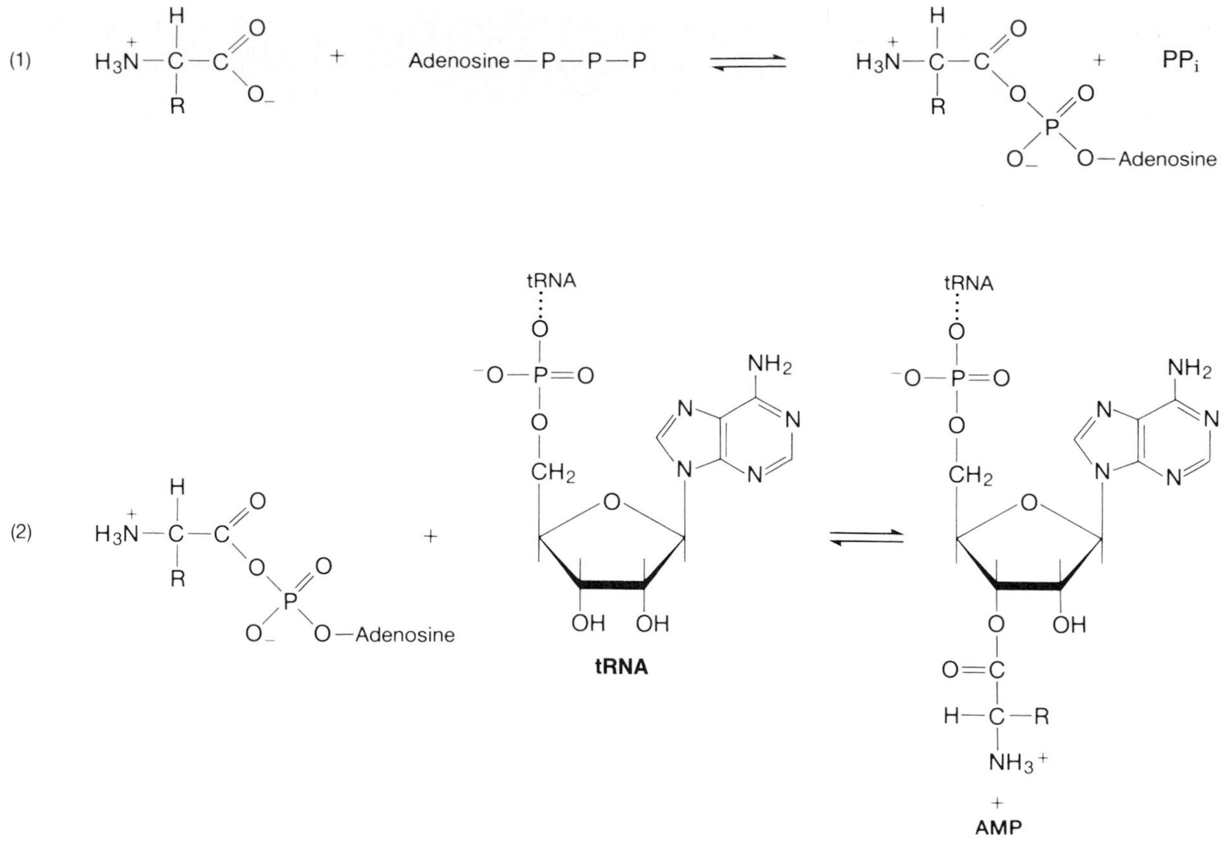

Figure 28–7
Formation of aminoacyl-tRNA. This is a two-step process involving a single enzyme that links a specific amino acid to a specific tRNA molecule. In the first step (1) the amino acid is activated by the formation of an aminoacyl-AMP complex. This complex then reacts with a tRNA molecule to form an aminoacyl-tRNA complex (2).

This step activates the amino acid so that it is energetically capable of forming a complex with the tRNA. In the next step the amino acid exchanges its AMP for an ester linkage with tRNA. Both of these steps are carried out by the same enzyme. In the second step the enzyme shows exquisite specificity. It must attach the correct amino acid to the correct tRNA. Otherwise the amino acid would appear at an incorrect location in the polypeptide chain. This is because once an amino acid is attached to a tRNA it is the tRNA which determines its location in the polypeptide chain. The specificity of the amino acid-tRNA reaction is discussed further in Box 28–B.

The decisive role of the tRNA in directing the location of the amino acid in the polypeptide chain was demonstrated by a classic experiment performed in 1962 in both Fritz Lipmann's and Seymour Benzer's laboratories. The experiment itself was very simple. First cysteinyl-tRNA^{Cys} was formed by incubating cysteine, ATP, and tRNA^{Cys} with the cysteine-specific aminoacyl-tRNA synthase. Next cysteinyl-tRNA^{Cys} was reacted with Raney nickel, which reduced the cysteine sulfhydryl to hydrogen sulfide and replaced it with a proton. This chemical treatment converted cysteine to alanine without removing it from the tRNA and thus yielded alanyl-tRNA^{Cys}. Finally, the alanyl-tRNA^{Cys} was used in cell-free protein synthesis. It was found that the alanine so complexed was incorporated in the protein polypeptide chain at sites normally occupied by cysteine.

Among the amino acids found in proteins, methionine plays a unique role, as it is found at the beginning (i.e., the amino terminus) *of all polypep-*

BOX 28–B

Specificity Considerations in Aminoacyl-tRNA Formation

In general, cells contain a single aminoacyl-tRNA synthase but multiple tRNAs for each amino acid. Since incorrect attachment of amino acids to tRNA inevitably leads to the production of faulty proteins, it is crucially important that this reaction proceed with high fidelity. Fidelity is achieved in two ways, both of which depend on the recognition by synthases of subtle structural differences between the various amino acids and the various tRNAs. Despite extensive structural studies, we are not yet in a position to define the molecular basis of this recognition. Nonetheless, it is clear that the enzymatic specificity of the aminoacyl-tRNA synthases is governed by the same rules of active-site geometry that apply to all enzymes.

Not only do the synthases employ specificity in the attachment of the amino acid to tRNA, but they also are capable of recognizing and hydrolyzing inappropriate reaction products. These *proofreading hydrolysis reactions*, which are not a simple reversal of synthesis, are best understood in the case of valine and isoleucine synthases. Discrimination between these amino acids is the most demanding because they differ by only a single methylene group. If isoleucyl-tRNA synthase either erroneously forms or encounters valyl-tRNA^{Ile} in free solution, it rapidly hydrolyzes it in the absence of ATP:

$$\text{Valyl-tRNA}^{Ile} + H_2O \longrightarrow \text{Val} + \text{tRNA}^{Ile}$$

Thus not only does the isoleucine tRNA synthase prefer isoleucine over valine in the synthetic reaction, but it also preferentially hydrolyzes incorrectly acylated tRNAs in an entirely separate reaction. This two-step discrimination is common to many aminoacyl-tRNA synthases and overall produces an error frequency of less than 1 in 10,000.

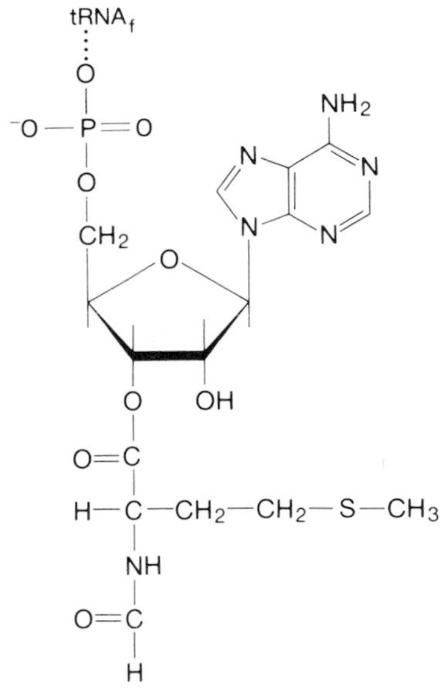

Figure 28–8
The 3′ terminal structure of bacterial initiator tRNA. The aminoacyl-tRNA for the initiator tRNA is special in bacterial systems, as the esterified amino acid becomes N-formylated.

tide chains*; it is also found internally like other amino acids. Only one codon, AUG, represents methionine in all protein-synthesizing systems. Despite this fact, two tRNAs can accept methionine in response to a single synthase. One of these tRNAs, tRNA$_m^{Met}$, reads internal methionine codons on the mRNA, while the other, tRNA$_f^{Met}$, plays a unique and indispensable role in initiation. Bacteria contain an enzyme that formylates the terminal amino group of the initiator methionine tRNA to yield the product shown in Figure 28–8.

$$\text{tRNA}_f^{Met} \xrightarrow{\text{Synthase}} \text{methionyl-tRNA}_f^{Met} \xrightarrow{\text{Transformylase}} \text{fMet-tRNA}_f^{Met}$$

$$\text{tRNA}_m^{Met} \xrightarrow{\text{Synthase}} \text{methionyl-tRNA}_m^{Met} \xrightarrow{\text{Transformylase}} \text{No reaction}$$

Polypeptide Synthesis Is Initiated on the Ribosome

Although specific differences distinguish the initiation reactions in eukaryotes and prokaryotes, three steps are required to initiate synthesis in all protein-synthesizing systems. In the first two steps, the small ribosomal subunit binds to the initiator tRNA carrying a methionine, and the mRNA binds at an appropriate AUG start codon. In the third step in initiation, the large ribosomal subunit joins the complex, and protein initiation factors (IF) dissociate from it (Figure 28–9).

In *E. coli* there are three protein initiation factors. These factors are bound to a small pool of 30S ribosomal subunits that is waiting to initiate the events of protein synthesis. One of these factors, IF-3, serves to hold the 30S and 50S subunits apart after termination of a previous round of protein synthesis. The other two factors, IF-1 and IF-2, function to promote the binding of fMet-tRNA$_f^{Met}$ and mRNA to the 30S subunit. Binding between the two RNAs occurs in such a way that the anticodon on the tRNA is specifically

---

*Methionine is not found at the amino terminus in all polypeptide chains in mature bacterial proteins because posttranslational processing reactions result in the cleavage of one or more residues from the amino terminal end. Eukaryotes don't formylate but they still contain two types of methionine tRNA.

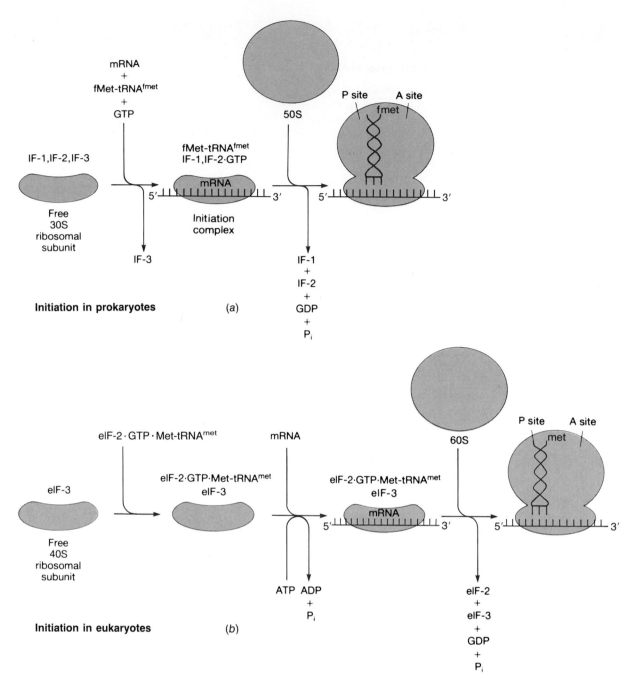

Figure 28–9
Formation of initiation complex for protein synthesis. (a) The initiation events in
E. coli. (b) The initiation events in eukaryotes. The main events are very similar in
the two systems. Note that the methionyl-tRNA complexes directly to the P site. All
other aminoacyl-tRNAs bind first to the A site. Once the initiation complex is formed,
the initiation factors are displaced in both systems.

complexed to the start AUG codon on the mRNA. At this point, IF-3 dissociates from the 30S subunit, permitting the 50S subunit to join the complex.
This in turn results in the release of the other two initiation factors, IF-1 and
IF-2. The overall binding reaction requires the participation of GTP, which,
upon addition of the 50S subunit, becomes hydrolyzed to GDP and P_i.

Eukaryotic cells contain a more complex spectrum of initiation factors.
These factors are also found primarily associated with the small ribosomal
subunit and are even more difficult to distinguish from ribosomal proteins

than their bacterial counterparts. Protein synthesis is initiated on the small ribosomal subunit by the binding of the ternary complex composed of an initiation factor protein (eIF-2), GTP, and Met-tRNA^{Met}. (The prefix e in eIF-2 designates its eukaryotic origin.) In contrast to binding in prokaryotes, this binding appears to be a prerequisite to the interaction of the small subunit with an mRNA (Figure 28–9b). The small eukaryotic ribosomal subunit binds at or near the 5′ end of mRNA and moves along it until it encounters the first AUG sequence. Initiation in eukaryotes requires the participation of ATP and its hydrolysis to ADP and P$_i$. This nucleotide is not required for the early ribosomal events in prokaryotes. A possible role for ATP hydrolysis is that it provides the motive force to propel the ribosome from the site of its initial binding, at the capped message end, to the start codon in eukaryotes.

Once the small subunit is properly positioned at the initiation site, the large ribosomal subunit joins the complex in a factor-assisted reaction, and GTP, originally bound in complex with eIF-2, is hydrolyzed to GDP and P$_i$. At this point, both eIF-3 and eIF-2 dissociate and the ribosome is primed to begin elongation.

Elongation Reactions Involve Peptide Bond Formation and Translocation

Elongation involves all of the reactions related to peptide synthesis, from the first peptide linkage to the last. A new group of dissociable protein factors, called *elongation factors*, comes into play at the time of elongation.

Elongation begins with the binding of a second aminoacyl-tRNA at a site adjacent to the methionyl-tRNA. Recall that there are two sites on the ribosome for tRNA binding (see Figure 28–3), the P site and the A site. *The initiator tRNA is the only aminoacyl-tRNA that binds directly to the P site. All of the remaining aminoacyl-tRNAs bind first to the A site* (later they become translocated to the P sites as described below). Binding of an aminoacyl-tRNA to the A site is mediated by the protein factor EF-Tu (the EF stands for elongation factor). The EF-Tu factor forms a complex with GTP and an aminoacyl-tRNA. This complex binds to the A site on the ribosome. A productive complex is formed if, and only if, the anticodon of the tRNA in the GTP-EF-Tu-aminoacyl-tRNA complex interacts specifically with the codon next to the start codon. Following this, the GTP hydrolyzes to GDP and P$_i$ and the resulting EF-Tu-GDP complex dissociates from the ribosome.

A second elongation factor, EF-Ts, catalyzes the regeneration of the EF-Tu-GTP to prepare it for interaction with other aminoacyl-tRNAs. The series of reactions involved in this regenerative process is depicted in Figure 28–10. In eukaryotes a single multisubunit protein, EF-1, seems to combine the functions of the prokaryotic factors EF-Tu and EF-Ts.

With two aminoacyl-tRNAs complexed to the ribosome, the system is now ready for peptide bond formation. The actual formation of the peptide bond, called transpeptidation, is the only subreaction of protein synthesis that does not require the participation of either a nonribosomal protein or GTP. Given appropriately bound substrates, this reaction is catalyzed by peptidyl transferase, an enzyme contained entirely within the large ribosomal subunit. The reaction does not involve the participation of a nucleotide, and its energetic requirements are entirely satisfied by the cleavage of the high-energy bond through which the amino acid is attached to tRNA (see Box 28–C).

The free amino group of the newly bound aminoacyl-tRNA attacks the carbonyl group of the adjacently bound methionyl-tRNA in a nucleophilic displacement reaction entailing the replacement of the ester bond with a peptide bond. The net result of this reaction is the transfer of the methionine

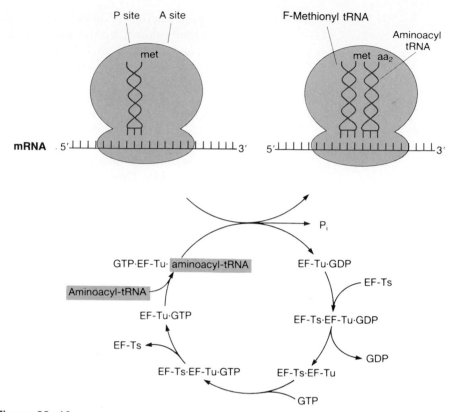

Figure 28–10

Addition of the second aminoacyl-tRNA to the ribosome complex and the accompanying EF-Tu, EF-Ts cycle in *E. coli*. The purpose of the cycle is to regenerate another protein aminoacyl-tRNA complex suitable for transferring further aminoacyl-tRNAs to the A site on the ribosome.

from one tRNA to the next, simultaneous with peptide bond formation (Figure 28–11).

Following *transpeptidation*, the ribosomal P site is occupied by a deacylated tRNA, and a peptidyl-tRNA occupies the A site. Further peptide synthesis cannot occur until the peptidyl-tRNA complex is moved from the A site to the P site. At the same time the mRNA must move precisely three bases along the ribosome so that the interaction between the peptidyl-tRNA and the mRNA is preserved. This precision is essential to insure correct translation of the mRNA. The combined movement of the peptidyl-tRNA and the mRNA is known as the translocation reaction. *Translocation* can be viewed as a reordering of the binding so as to expel the vacant tRNA, reposition the peptidyl-tRNA at the P site, and place the next adjacent codon at the A site for its subsequent decoding. These collective reactions require the hydrolysis of GTP to GDP and P, and the participation of EF-G in bacteria and EF-2 in eukaryotes (Figure 28–12).

The three reactions just described—adsorption of the aminoacyl-tRNA to the ribosome A site, transpeptidation, and translocation—are successively repeated until each codon on the mRNA has been translated to produce a fully formed polypeptide and the ribosome reaches a stop codon on the mRNA.

Termination of Translation Requires Special Termination Codons

The translation of natural messages yields protein products that are detached from tRNA and released from the ribosome. This overall reaction, the release reaction, comes about through the special reading of termination

BOX 28–C

The Puromycin Reaction

The antibiotic puromycin has played a key role in defining the mechanism of ribosomal reactions. It was observed in the early 1960s that _puromycin bears a structural resemblance to aminoacyl-tRNA and that in inhibiting protein synthesis it becomes covalently incorporated into peptide chains_ (Illustration 1). The nascent proteins with puromycin attached to their C-terminal residue are released from the ribosome. Puromycin competes with aminoacyl-tRNA as a substrate for peptidyl transferase.

This substrate role of puromycin can be clearly seen in its reaction with ribosome-bound fMet-tRNA^{fMet}. Incubation of puromycin with ribosomes bearing bound fMet-tRNA^{fMet} from the initiation reaction yields fMet-puromycin without further additions. Thus, fMet-tRNA^{fMet} is bound by initiation factors so that it can participate in peptide-bond formation, and this reaction is catalyzed entirely by the ribosome. If the ribosome, after initiation, is reacted with an appropriate aminoacyl-tRNA, a dipeptide is formed, and this dipeptide cannot react with puromycin. Thus puromycin can react with peptidyl-tRNA when it is bound in the P site, but not when it is bound in the A site.

A variety of experimental findings has served to localize peptidyl transferase to the large ribosomal subunit. Normally, of course, both subunits are required for this reaction, because of the contribution the small subunit makes to the binding of peptidyl-tRNA. However, it has been empirically observed that peptidyl-puromycin is formed when the large subunit alone is incubated with peptidyl-tRNA and puromycin in the presence of an organic solvent. Apparently the organic solvent alters the structure of the subunit so that sufficient substrate binding occurs to allow slow catalysis.

Exactly why puromycin should be such an effective competitor with aminoacyl-tRNA for the ribosomal A site is not clear. Thus far, chemical modifications of puromycin have either altered its activity only slightly or significantly reduced it. Although the A site must have broad specificity so that it can accept any amino acid, puromycin must combine a number of structural features that uniquely facilitate its interaction in the absence of the remainder of the usual tRNA structure.

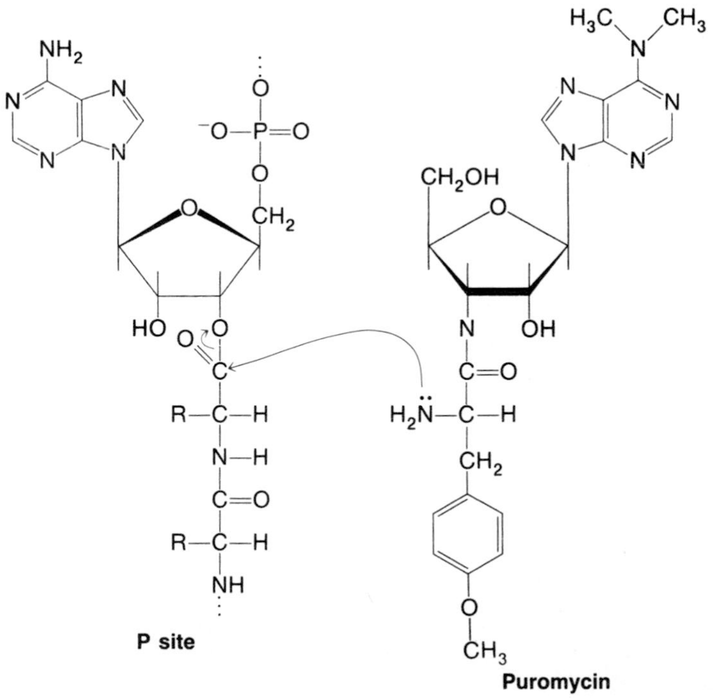

Illustration 1
The puromycin reaction. Puromycin is mistaken for an aminoacyl-tRNA.

---

codons, a process that requires the participation of _release factors_ (Figure 28–13).

The normal termination codons (UAA, UAG, and UGA)—also called opal, amber, and ochre codons, respectively—are the only 3 of 64 codons that do not specify amino acids. Of the three release factors in _E. coli_, RF-1 responds specifically to the triplets UAA and UAG, while RF-2 responds

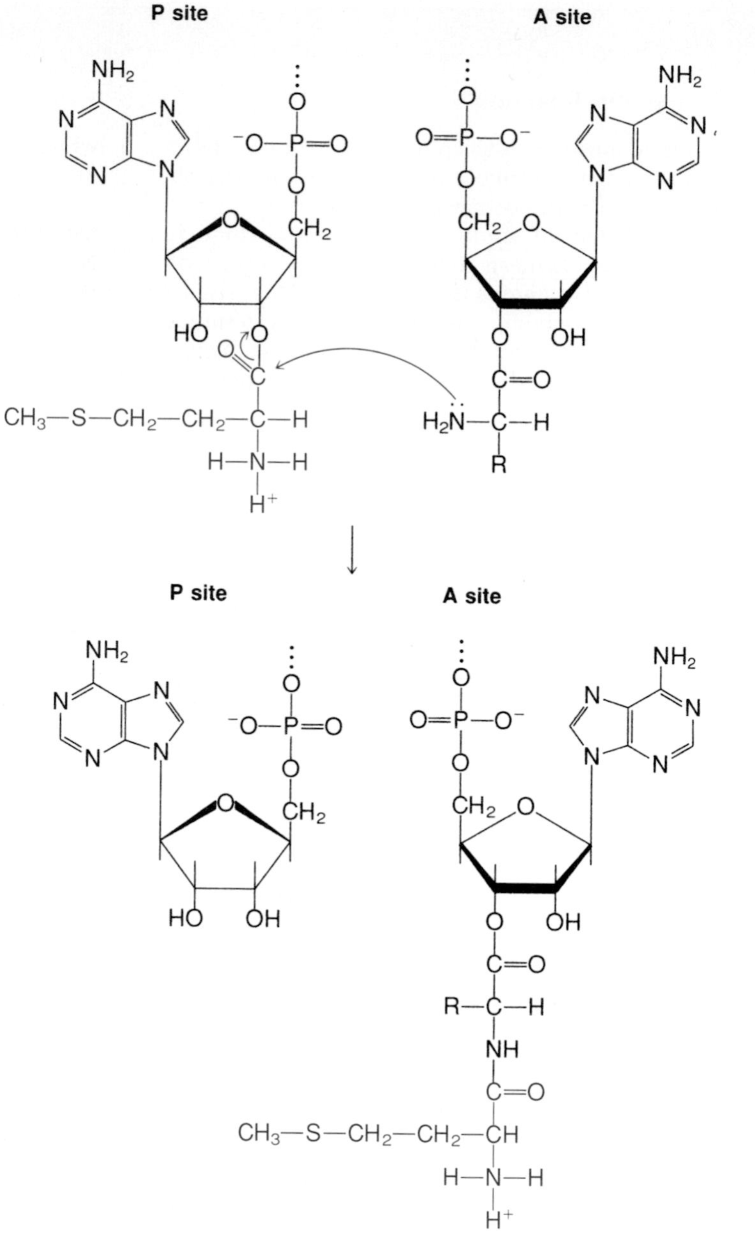

P site   A site

Figure 28–11

Formation of the first peptide linkage.
The formylmethionine group is
transferred from its tRNA at the P site
to the amino group of the second
aminoacyl-tRNA at the A site on the
ribosome. This involves nucleophilic
attack by the amino group of the
second amino acid on the carboxyl
carbon of the methionine. The
resulting bond formation attaches both
amino acids to the tRNA at the A site.

Figure 28–12

The translocation reaction in *E. coli*.
The translocation reaction occurs
immediately after peptide synthesis. It
involves displacement of the
discharged tRNA from the P site and
concerted movement of the peptidyl-
tRNA and mRNA so that the peptidyl-
tRNA is bound to the P site and the
same three nucleotides in the mRNA.
The A site is vacated and ready for
the addition of another aminoacyl-
tRNA. Translocation in eukaryotes is
similar except that the EF-2 factor is
involved instead of the EF-G factor.

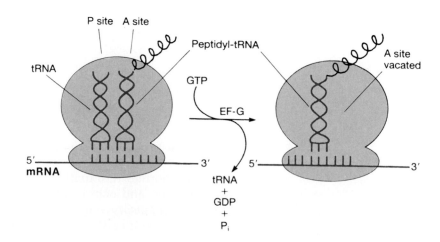

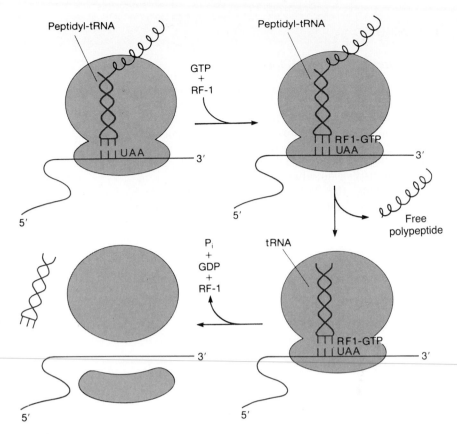

Figure 28–13

The release reaction in *E. coli*. The release reaction occurs when the codon adjacent to the anticodon-codon complex is one of the stop codons, e.g., UAA. The stop is recognized by release factor proteins that cause the peptidyl transferase to transfer the nascent polypeptide to water, forming a free polypeptide. Following the release of the polypeptide, the final tRNA and the mRNA dissociate from the ribosome, and the ribosome dissociates into its constituent subunits.

specifically to UAA and UGA. RF-3 does not have release activity by itself, but stimulates the reaction catalyzed by either RF-1 or RF-2. The release factors themselves recognize and bind weakly to termination codons in the absence of ribosomes. Most of this information about release factors was obtained with the help of a simple *in vitro* assay for termination (see Box 28–D).

When the termination codons are bound to the decoding site on the ribosome, then the release factors, in turn, bind. This codon-directed binding of either RF-1 or RF-2 causes some as yet unknown change in the peptidyl transferase so that it transfers the completed polypeptide chain to water, rather than to an amino group of an aminoacyl-tRNA. Said another way, when a release factor occupies the ribosomal A site, peptidyl-tRNA is hydrolyzed. Subsequently the final tRNA is released, the mRNA dissociates from its complex with the ribosome, and the two ribosomal subunits separate.

In the eukaryotic cytoplasm, researchers have observed only a single release factor, which recognizes all three termination codons. This release factor requires the presence of GTP, which, as in other factor reactions, is hydrolyzed to GDP to P_i. The function served by GTP hydrolysis in this reaction may be related to the recycling of the termination factor. The role of GTP in this and other steps in protein synthesis was clarified with the help of the nonhydrolyzable analog GMPPCP (see Box 28–E).

The sequence of steps involved in protein synthesis is long and complex (Figure 28–14). Most of the complexity arises from the heavy demands

BOX 28–D

An *In Vitro* Assay for Release Factors

The study of the release reaction was facilitated by the discovery of a much simpler reaction. When fMet-tRNA^{fMet}, an AUG triplet, and the ribosome are incubated with a termination codon, formylmethionine is released. Two features of this reaction indicate that it provides a model for the normal release reaction. First, it requires a termination triplet; no other codon will promote the reaction. Second, the factors identified and obtained through the use of this assay promote the normal release reaction.

BOX 28–E

The Role of GTP in Ribosomal Reactions

GTP and its hydrolysis to GDP and P_i play a conspicuous role in ribosomal reactions. As noted before, *each major step in protein synthesis, except peptide-bond formation itself, involves the hydrolysis of GTP to GDP.* For these reasons, a great deal of effort has been focused on explaining the enzymology of GTP in ribosomal reactions. A general picture of the role played by GTP has begun to emerge. As far as the chemistry of hydrolysis is concerned, none of these reactions has been found to involve a covalent intermediate. GTP hydrolysis is not used to phosphorylate a ribosomal protein, for example. What alteration, then, is brought about by GTP hydrolysis?

Some answers to this question have come from experiments with the nonhydrolyzable analog of GTP, GMPPCP. This analog can satisfy the GTP requirements in several but not all ribosomal reactions. It will, for example, replace GTP in the binding of aminoacyl-tRNA catalyzed by EF-Tu. Most significantly, however, after tRNA binding, the EF-Tu·GMPPCP complex does not dissociate from the ribosome. From this observation the view has developed that the purpose of GTP hydrolysis is to change the conformation of the factor so that it will dissociate from the ribosome. Another way of stating this in the case of EF-Tu is that when GTP is bound to the protein, it has high affinity for both the ribosome and aminoacyl-tRNA. When GDP is bound to the protein, its conformation is such that it will bind neither the ribosome nor aminoacyl-tRNA. Thus *GTP hydrolysis in general appears to change the conformation of the protein-synthesis factors so that they will cycle on and off the ribosome.*

The structure of guanylyl methylene diphosphonate (GMPPCP).

Why do factors exist separate from the ribosome? Why must they cycle on and off, and why are their functions not performed by ribosomal proteins? Answers to these questions appear to lie in a certain economy of structure that results from the cycling of proteins on and off the ribosome. As the factors come and go from the ribosome, they each appear to occupy the same or a nearby site. For example, no two factors, at least those which employ GTP, appear to be able to interact simultaneously with the ribosome. In addition, efficient interaction with each of the GTP-requiring factors specifically requires the full complement of four copies of the ribosomal protein L7/L12. Thus *the ribosome seems to have essentially a single site through which the various factors cycle.*

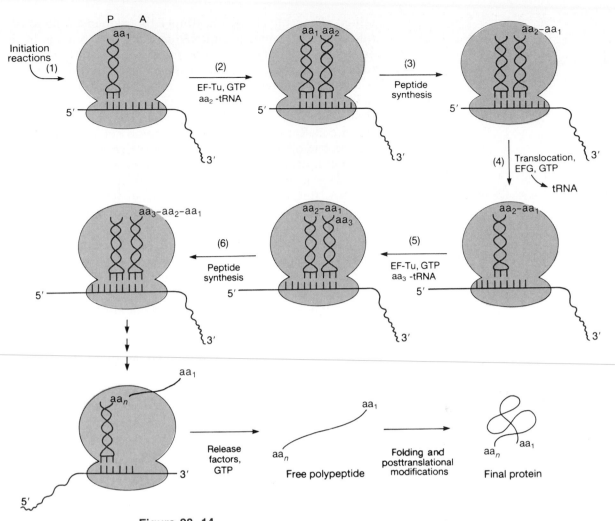

Figure 28–14

Overview of reactions in protein synthesis. Formation of the initiation complex (reaction 1) is followed by a series of reactions (2, 3, and 4) collectively known as the elongation reactions. These reactions are precisely repeated for each codon in the mRNA until a stop signal is reached. At this point, release factors hydrolyze the nascent polypeptide from the final tRNA, and the rest of the complex breaks up into its components—the ribosomal subunits, the tRNA, and the mRNA. The free polypeptide folds into a three-dimensional structure, and is often accompanied by certain post-translational modifications.

placed on the system to make a polypeptide chain with a strictly ordered sequence of amino acids.

Throughout this discussion on protein synthesis, we have stressed how important it is that the anticodon in the tRNA should interact with a specific codon in the mRNA, in order to achieve the correct arrangement of amino acids in the protein polypeptide chain. We have also indicated that each codon consists of a sequence of nucleotides, and that the codons are arranged in side-by-side fashion in the mRNA. The codons for initiation (AUG) and termination (UAA, UAG, and UGA) have also been indicated. In the next section we will consider some of the biochemical experiments that helped to clarify the relationships between codons, anticodons, and amino acids.

DECIPHERING THE GENETIC CODE

The concept of a genetic code grew out of the realization that both nucleic acids and proteins are linear polymers made of a limited number of building blocks, plus the knowledge that the structure of proteins is genetically determined. The *genetic code* is simply the sequence relationship between the nucleotides in genes (or mRNA) and the amino acids in the proteins they encode. The *coding ratio* is the number of nucleotides in an mRNA required to represent an amino acid. Since there are four different nucleotides in RNA and twenty different amino acids in proteins, it is clear from simple arithmetic that the minimum acceptable coding ratio is 3. We can rule out a singlet code, in which one nucleotide represents one amino acid; such a system could code for only four different amino acids, because there are only four different nucleotides in DNA or RNA. Even a doublet code, in which two nucleotides code for a single amino acid, would not work. It could code for only sixteen amino acids, since there are sixteen possible doublet sequences in RNA. But a triplet code—made from four nucleotides taken three at a time—generates a total of 64 possible triplet sequences. This would be more than enough to represent the twenty amino acids formed in proteins. In fact, a triplet code creates a problem of a different sort. What function can be served by the extra trinucleotide sequences? Allowing for the three triplets that represent stop signals (as indicated above), we still have 61 possible triplets to code for just twenty amino acids. If the code is *nondegenerate*— that is, if a unique triplet codes for each amino—then 41 triplet sequences will never be found in translation reading frames. On the other hand, if the code is *degenerate*—that is, if all 61 triplets are used—then each amino acid, on the average, must be represented by more than one triplet sequence.

Another concern relates to the location of codons in the translation reading frame. Codons could be arranged in a close-packed, side-by-side arrangement. In this event each nucleotide within the translation reading frame would represent a code letter within one, and only one, codon. Alternatively, codons could be separated by one or more "spacer nucleotides." In this event some nucleotides within the reading frame would represent code letters within codons, while others would serve as spacer nucleotides. Finally, we could imagine an overlapping codon arrangement, in which all nucleotides represent code letters and some nucleotides represent code letters in more than one codon. The biochemical experiments presented in this chapter demonstrated a nonoverlapping triplet code without spacers.

Synthetic Polynucleotides Can Serve As Templates for Polypeptide Synthesis

Our understanding of the mechanism of protein synthesis has come mainly from reconstructed cell-free systems. Such systems have been prepared from crude extracts of bacterial, animal, or plant cells; they contain all of the components necessary for protein synthesis. Thus when essential factors are added or withheld, the effects on protein synthesis can be observed. In this way the significance of each essential factor can be directly determined.

In 1961, Nirenberg and Matthei were attempting to synthesize proteins in a cell-free system. They were using crude extracts of *E. coli* cells that contained ribosomes, tRNA, and all of the proteins necessary for protein synthesis. Their system was fortified with the necessary salts and substrates for protein synthesis, in particular ATP, GTP, magnesium ions, and amino acids. Radioactive amino acids were used so that small amounts of peptide or protein synthesis could be detected by incorporation of radioactive isotope. For their source of mRNA, they chose to use viral RNA from tobacco mosaic virus (TMV). Their goal was to make some of the viral proteins. Indeed, the TMV RNA did stimulate polypeptide synthesis when added to the Nirenberg cell-free system, but it was never demonstrated that complete viral proteins were made. Thus Nirenberg and Matthei did not achieve their primary goal, but ironically, they achieved something much more important.

For, during the course of their cell-free synthesis studies, they found that a synthetic polynucleotide, containing only uridine residues—poly(U)—specifically stimulated the synthesis of polyphenylalanine. The logical inference was that the amino acid phenylalanine was encoded by a sequence of U's. A wave of excitement was produced by this result, because it appeared to mark the beginning of a solution to the coding problem.

The result with poly(U) set in motion a series of similar experiments in which various synthetic polynucleotides were used to stimulate polypeptide synthesis. For instance, it was established that poly(A) specifically promotes polylysine synthesis and that poly(C) specifically promotes polyproline synthesis. These results suggested that the amino acid lysine is coded by a sequence of A's and the amino acid proline by a sequence of C's.

At the time (1961–1963), it was difficult to extend this approach to the use of synthetic messengers containing more than one nucleotide, because it was not then possible to make mixed polynucleotides of defined sequence. Mixed copolymers with a random sequence were used with limited success. For example, a mixed copolymer of A and C was found to stimulate the incorporation of six amino acids—lysine, proline, threonine, asparagine, glutamine, and histidine. This result suggested that these six amino acids required A or C or both A and C in their codons. The earlier results with homopolymers had showed that only lysine and proline are incorporated by poly(A) or poly(C), respectively. Therefore the four remaining amino acids in this list—threonine, asparagine, glutamine, and histidine—were presumed to be coded by a mixture of A and C residues. By using copolymers with different ratios of A and C residues, some tentative conclusions could also be drawn about the relative numbers of A and C residues in the codons for these four amino acids. Many conclusions drawn from data of this type have turned out to be correct. The main failing of the mixed-copolymer approach was that its conclusions were statistical in nature, and nothing could be learned about the sequence of nucleotides within the individual codons.

Nucleotide Triplets Stimulate Ribosome Binding of Aminoacyl-tRNA

In 1964 Leder and Nirenberg devised an ingenious scheme for determining amino acid codons without the need for long mRNAs with a known sequence of nucleotides. Their scheme was based on the fact that aminoacyl tRNAs bind to ribosomes in a message-specific way and that this binding does not require all of the components necessary for protein synthesis. Beyond ribosomes, template, and a suitable buffer, the only requirement for binding is a relatively high magnesium ion concentration. Under these conditions, binding of aminoacyl-tRNA occurred without polymerization. Leder and Nirenberg also found that nucleotide triplets were effective in stimulating the binding reaction. Thus no binding of phenylalanyl-tRNA occurred with mono(U) or di(U), but tri(U) was nearly as effective as higher oligomers of U. Incidentally, this fact is consistent with a triplet code, but it does not prove it.

Once triplets of the desired sequence could be synthesized, it became possible to test the effects of different triplets on the binding of different aminoacyl-tRNAs. For this purpose there was devised a rapid means of measuring binding. The method makes use of nitrocellulose membrane filters, which have a general affinity for protein but not nucleic acid. As a result, they will retain ribosomes and their bound aminoacyl-tRNA, but not unbound aminoacyl-tRNA. In a typical series of experiments, a single radioactively labeled amino acid was used to synthesize aminoacyl-tRNA. The product was then mixed with ribosomes in the presence of different trinucle-

TABLE 28–2
Trinucleotide-Stimulated Binding of Aminoacyl-tRNAs to Ribosomes

Trinucleotide	Aminoacyl-tRNA Bound
UUU, UUC	Phenylalanine
UUA, UUG, CUU, CUC, CUA, CUG	Leucine
AUG	Methionine
UCU, UCC, UCA, UCG	Serine
CCC, CCU, CCA, CCG	Proline
AAA, AAG	Lysine

otides, and the mixture was passed through a nitrocellulose filter. Any trinucleotide that caused selective retention of a particular aminoacyl-tRNA was assumed to be a codon for the corresponding amino acid. This technique allowed the rapid assignment of amino acids to particular triplet code words. Some typical results are presented in Table 28–2.

There were two limitations to this approach. First, some triplets were very inefficient in promoting aminoacyl-tRNA binding to the ribosome, while others appeared to promote nonspecific binding. More troubling was the possibility of misinterpretation, owing to the fact that not all the specificity of protein synthesis was involved in the assay. Nevertheless, these results significantly advanced our general understanding of the sequence relations of the code, even though they could not provide a complete picture.

Repeating-sequence Synthetic Polynucleotides Are mRNAs for Simple Repeating Polypeptides

The key set of experiments that completed the definition of the genetic code was performed by Gobind Khorana and his coworkers; as a result, Khorana shared a 1968 Nobel Prize with Marshall Nirenberg and Robert Holley, the latter for having been first to sequence an RNA molecule. Khorana's experiments resulted from years of painstaking work on the chemical synthesis of polynucleotides of defined sequence. A careful analysis of the coding specificity of these synthetic polynucleotides and the resulting polypeptides unambiguously defined the properties of the genetic code.

The nature of these results is illustrated by one particularly clever experiment that characterized the translation products of a perfectly alternating copolymer of U and C. If we assume a triplet code, such a polymer could contain only two codons, UCU and CUC. It was found that this template stimulated the incorporation of only two amino acids, serine and leucine, and these were incorporated in equal amounts. Peptide bonds involving the amino group of serine are unusually sensitive to acid hydrolysis. Khorana exploited this property to characterize the translation products. Upon partial acid hydrolysis, the *in vitro* translation products of poly(UC) were quantitatively converted to the dipeptide Ser-Leu. Hence the original polypeptide must have been a perfectly alternating copolymer of the two amino acids (Figure 28–15). Previous triplet binding studies (see Table 28–2) had suggested that UCU codes for serine and that CUC codes for leucine. Khorana's results were consistent with this suggestion, but they went beyond it. The combined sequence data on polynucleotide and polypeptide suggested that the synthetic mRNA represented a continuous sequence of nonoverlapping codons, that is, a nonoverlapping triplet code without spacer sequences.

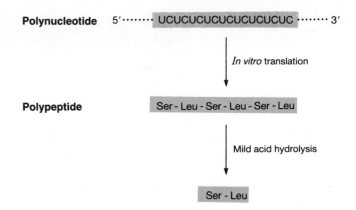

Figure 28–15
Demonstration that poly(UC) codes for a Ser-Leu repeating peptide. When poly(UC) containing a strict alternating sequence of U and C is used as an mRNA, only serine and leucine are incorporated. Analysis of the polypeptide product indicates that it contains an alternating sequence of serine and leucine.

Start and Stop Triplets Were Identified by *In Vitro* Studies

A polynucleotide containing a repeat of the AUG triplet should contain three triplets, AUG, UGA, and GAU. Hence one might expect such a polynucleotide to stimulate incorporation of three different amino acids, depending on which nucleotide was recognized as the first letter in the start codon. However, when Khorana used the poly(AUG) he found that it specified the incorporation of only two amino acids, methionine and aspartic acid. Studies on termination indicated that the UGA triplet is one of the stop signals. Hence only the remaining two triplets in the poly(AUG) messenger, AUG and GUA, would be expected to function as codons. Genetic experiments relating to stop codon assignments have confirmed the assignments of stop codons made by purely biochemical means.

The AUG triplet was known to code for methionine, leaving GAU to code for aspartic acid. As expected, methionine and aspartic acid were incorporated into separate homopolymers. This is strong support for the notion that mRNAs are read as a continuous sequence of triplets. If AUG is the first codon recognized, then contiguous AUGs will be read as codons and polymethionine will be synthesized. On the other hand, if GAU is the first triplet to be recognized, then translation will continue in this reading frame and polyaspartic acid will be the product.

Methionine was much more effectively incorporated with the poly(AUG) template than was aspartic acid, especially at low magnesium ion concentrations (Figure 28–16). Perhaps most significant was the observation that polymethionine, particularly when formed at low magnesium ion concentrations, was always blocked at its amino terminus with a formyl group. This was consistent with the initiator tRNA being used at the amino terminus, since methionine becomes formylated only if it is attached to the initiator tRNA. The fact that aspartic acid was incorporated only at high magnesium ion concentrations was consistent with a number of other results, for example, that incorporation in the Nirenberg cell-free system has a

Figure 28–16
Translation of poly(AUG). The poly(AUG) stimulates the synthesis of polymethionine and polyaspartic acid. The amino terminal methionine is formylated. Polymethionine is made at low and high magnesium concentrations. Polyaspartic acid is made only when the concentration of magnesium is high. The low magnesium conditions are believed to simulate the *in vivo* situation more closely.

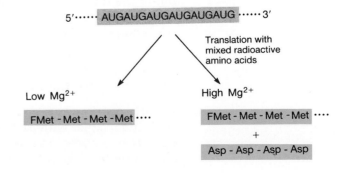

TABLE 28–3
The Genetic Code

5'-OH Terminal Base	Middle Base				3'-OH Terminal Base
	U	C	A	G	
U	Phe	Ser	Tyr	Cys	U
	Phe	Ser	Tyr	Cys	C
	Leu	Ser	Term[b]	Term[b]	A
	Leu	Ser	Term[b]	Trp	G
C	Leu	Pro	His	Arg	U
	Leu	Pro	His	Arg	C
	Leu	Pro	Gln	Arg	A
	Leu	Pro	Gln	Arg	G
A	Ile	Thr	Asn	Ser	U
	Ile	Thr	Asn	Ser	C
	Ile	Thr	Lys	Arg	A
	Met[a]	Thr	Lys	Arg	G
G	Val	Ala	Asp	Gly	U
	Val	Ala	Asp	Gly	C
	Val	Ala	Glu	Gly	A
	Val[a]	Ala	Glu	Gly	G

[a]Sometimes used as initiator codons. When GUG is used as an initiation codon, it codes for Met.

[b]Stands for termination or stop codon.

Amino acid abbreviations: Phe (phenylalanine); Leu (leucine); Ile (isoleucine); Met (methionine); Val (valine); Ser (serine); Pro (proline); Thr (threonine); Ala (alanine); Tyr (tyrosine); His (histidine); Gln (glutamine); Asn (asparagine); Lys (lysine); Asp (aspartic acid); Glu (glutamic acid); Cys (cysteine); Trp (tryptophan); Arg (arginine); Gly (glycine).

general requirement for high magnesium. In retrospect, the meaning of the requirement is that normal initiation requires a formylmethionyl-tRNA at the amino terminal end, but this requirement can be overcome *in vitro* by the use of high magnesium.

CODE WORD ASSIGNMENTS

The culmination of these biochemical experiments on coding—the statement of the relationship between base sequences in mRNAs and amino acid sequences in proteins—is presented in Table 28–3. Three of the triplets represent stop codons; the remaining 61 triplets and 20 amino acids can be neatly summarized by grouping codons with similar first and second bases into a grid. The four major horizontal rows are composed of codons with the same first base. The four major vertical columns are composed of codons with the same second base. The individual boxes generated by the intersection of the rows and columns are code word families, which differ only in their 3'-terminal base. For example, the code words UCU, UCC, UCA, and UCG comprise a family encoding serine.

The 5' Base in the Anticodon Can Frequently Pair with More than One Base in the Codon

If all 61 triplets act as codons, then there must be tRNAs that can interact in a specific manner with each of these codons. Strict Watson-Crick base pairing between codon and anticodon would require 61 different anticodons and, correspondingly, 61 different tRNAs. As the characterization of tRNAs

TABLE 28–4
The Wobble Rules of Codon-Anticodon Pairing

5′ Base of Anticodon	3′ Base of Codon
C	G
A	U
U	A or G
G	C or U
I	U, C, or A

progressed it became clear that in many cases individual tRNAs could recognize more than one codon. In all cases the different codons recognized by the same tRNA have been found to contain identical nucleotides in the first two positions and a different nucleotide in the third position (i.e., the 3′ position of the codon). Codons recognized by the same tRNA are always represented by the same amino acid.

The 3′-terminal redundancy of the genetic code and its mechanistic basis were first appreciated by Crick in 1966. Crick proposed that codons in mRNA and anticodons in tRNA interact in an antiparallel manner on the ribosome in such a way as to require strict Watson-Crick pairing (that is, A-U and G-C base pairing) in the first two positions of the codon, but to allow certain other pairings as well in the 3′-terminal position. This formation of nonstandard base pairs between the 3′-terminal position of the codon and the 5′-terminal position of the anticodon demands an altered geometry between the paired bases; Crick's proposal was appropriately labeled the _wobble hypothesis_.

At the time of Crick's proposal, it was recognized that some tRNA molecules can pair specifically with two or three different codons, whereas other tRNAs could interact with only a single codon. Using molecular models, Crick examined the geometric relationships between all possible hydrogen-bonded base pairs and found some combinations that could be made with the minimum geometric distortions from normal Watson-Crick base pairs. Considerations of pairing were complicated by the fact that an adenosine group in the 5′ position of the anticodon is usually converted to an inosine group by a cellular enzyme (the structure of inosine is shown in Figure 7–44). Taking this additional base into account, Crick suggested that the most acceptable base pairs for the wobble position are those given in Table 28–4. According to this table, a C or an A in the 5′ position of an anticodon can pair only with a G or a U, respectively, in the 3′ position of a codon (codons and anticodons pair in an antiparallel manner), but U, G, or I in this position would permit pairing with two or three codons differing in their 3′-terminal base. The structures resulting from these abnormal pairings are described in Box 28–F.

A careful comparison of these wobble rules (Table 28–4) with the genetic code (Table 28–3) indicates that the minimum number of tRNAs required to translate all 61 codons is 31. If an additional methionyl-tRNA is added to this list for the initiator tRNA, the total comes to 32. Nevertheless, a typical cell usually contains many more (50 to 60) different tRNA molecules. Certain tRNAs with different anticodons can pair with the same codon, within the constraints of the wobble hypothesis, and some tRNAs with different overall sequences have the same anticodons. In every case, multiple tRNAs that correspond to a single amino acid are all recognized by a single aminocyl-tRNA synthase. It is not clear why cells have so many more tRNAs than the minimum number required. In many multicellular organisms, the pattern of physically different tRNAs frequently changes with the developmental state and is usually different in cells of different organs. It has been suggested—but not yet proved—that this changing pattern of tRNAs is in some way related to a changing pattern of gene expression.

The Genetic Code Is Not Completely Universal

The very earliest experiments concerning the genetic code led, after some uncertainty, to the belief that the genetic code was universal. The essential conclusion was that while the relative numbers of the different codons might differ between organisms, each of the 61 codons specified the same amino acid in all living systems. For example, beyond special sequences that are

BOX 28–F

Dimensions of Wobble Pairs

The illustration shows the dimensions of wobble base pairs. In part (a) are shown two wobble pairs: inosine-adenine and guanine-uracil. Part (b) shows the overall dimensions of standard Watson-Crick H-bonded base pairs and a variety of abnormal base pairs. Only the positions of the glycosidic bonds are indicated. The point X, at far left, represents the position of the C-1′ atom of the glycosidic bond in the anticodon. All glycosidic bonds are indicated in color. The other label points show where the C-1′ atom and glycosidic bond would fall in various base pairs. The wobble code uses the four codon positions to the right in the diagram, but not the three close positions (U-U, U-C, or C-U).

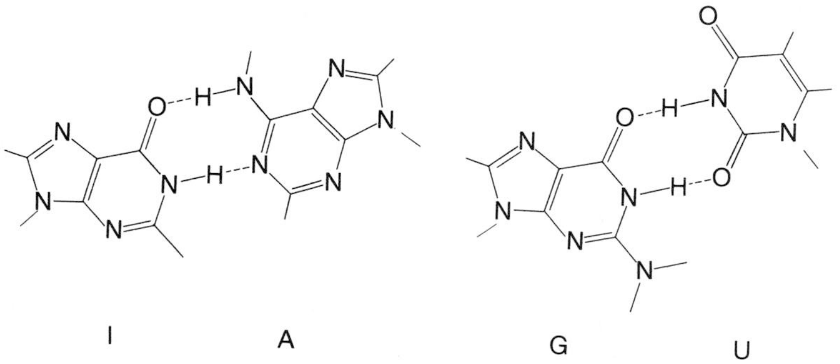

(a)

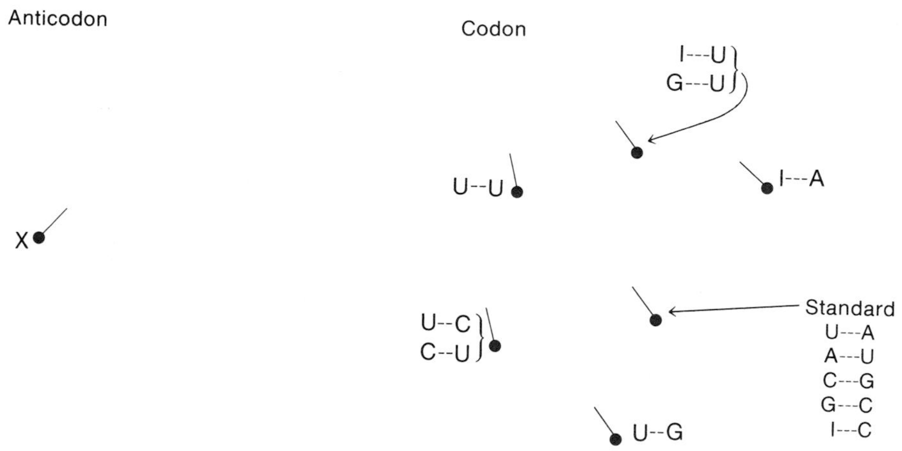

(b)

involved in initiating synthesis, the messenger RNA for rabbit hemoglobin will elicit the production of globin in a bacterial extract. Thus the code was thought to have a universal meaning in all genetic systems. More recently, that conclusion has been shown not to apply completely to all genetic systems.

The results on mitochondria and chloroplasts of certain eukaryotes are particularly well documented. Both mitochondria and chloroplasts contain a genetic system that is semi-independent of that contained in the nucleus of eukaryotic cells. Thus, for example, mitochondrial DNA codes for a limited number of intramitochondrial polypeptides, plus ribosomal RNA and mitochondrial tRNAs. The ribosomes, protein-synthesis factors, tRNA, and aminacyl-tRNA synthases contained within the mitochondrion are different from those contained within the cytosol. Surprisingly, the genes for the aminoacyl-tRNA synthases and the majority of the ribosomal proteins and factors used in protein synthesis within the mitochondrion are contained within the cellular nucleus, and their mRNAs are translated in the cytosol, after which the proteins are transported into the mitochondria.

The first clue that something was peculiar about mitochondrial genetics derived from the realization that some mitochondria simply do not code for enough tRNAs to permit genetic translation following the "universal" rules previously described. Detailed analyses of both mitochondrial tRNAs and their genes have shown, in fact, that they contain fewer than the 32 tRNAs minimally required for the translation of all 61 codons by Crick's wobble rules. One possible solution to the conundrum is that mitochondrial genes do not contain sequences for all 61 codons. Another possibility is that _the rules that govern the interaction between codons and anticodons on mitochondrial ribosomes, i.e., wobble, are different from those governing other protein-synthesizing systems_. Recent information indicates that the latter possibility is true.

Six amino acids (Val, Ser, Pro, Ala, Arg, Gly) are specified in the mitochondrial genetic code by four-codon families with the same two bases in the first two positions (Table 28–5). Threonine (Thr) is specified by two-codon families with the same two bases in the first two positions. The wobble rules require that at least two different tRNAs are needed to translate each of these families. Mitochondria appear to have developed a means of reading an entire four-codon family with a single tRNA and in this way have reduced the minimum number of tRNAs needed to translate the code words from 32 to 24. In all cases except for the Arg codons, these tRNAs contain U in the anticodon "wobble" position. These tRNAs on mitochondrial ribosomes appear to follow a different set of wobble rules that allow all four bases to pair with the unmodified U in the wobble position of the anticodon. The other six mitochondrial tRNAs that have a U in the wobble position show normal behavior; these are all chemically modified as indicated in Table 28–5.

TABLE 28–5
The Genetic Code of Yeast Mitochondria

UUU	Phe	AAG		UCU				UAU	Tyr	AUG		UGU	Cys	ACG
UUC				UCC	Ser	AGU		UAC				UGC		
UUA	Leu	AAU*		UCA				UAA	Term			UGA	Trp	ACU*
UUG				UCG				UAG				UGG		
CUU				CCU				CAU	His	GUG		CGU		
CUC	Thr	GAU		CCC	Pro	GGU		CAC				CGC	Arg	GCA[b]
CUA				CCA				CAA	Gln	GUU*		CGA		
CUG				CCG				CAG				CGG		
AUU	Ile	UAG		ACU				AAU	Asn	UUG		AGU	Ser	UCG
AUC				ACC	Thr	UGU		AAC				AGC		
AUA	Met	UAC[a]		ACA				AAA	Lys	UUU*		AGA	Arg	UCU*
AUG				ACG				AAG				AGG		
GUU	Val	CAU		GCU				GAU	Asp	CUG		GGU		
GUC				GCC	Ala	CGU		GAC				GGC	Gly	CCU
GUA				GCA				GAA	Glu	CUU*		GGA		
GUG				GCG				GAG				GGG		

Source: Adapted from S. G. Bonitz et al., Codon recognition rules in yeast mitochondria, _Proc. Natl. Acad. Sci. USA_ 77:3167, 1980.

The codons (5'→3') are at the left and the anticodons (3'→5') are at the right in each box (* designates U in the 5' position of the anticodon that carries the —CH₂NH₂CH₂COOH grouping on the 5 position of the pyrimidine).

[a]Two tRNAs for methionine have been found. One is used in initiation and one is used for internal methionines.

[b]Although an Arg tRNA has been found in yeast mitochondria, the extent to which the CGN codons are used is not clear.

In some mitochondria, the genetic code itself appears to have different meanings, but at the moment the generality of this picture is not clear. For example, yeast mitochondria appear to read the codon family CUN (where N is any base) as threonine rather than as leucine, while human and *Neurospora crassa* mitochondria appear to read the same codon family as leucine.

INHIBITORS OF PROTEIN SYNTHESIS

The study of various inhibitors of protein synthesis has played a conspicuous role in defining the mechanistic basis of the individual reactions involved in translation. A few inhibitors, such as the nonhydrolyzable analog of GTP, GPPCP, have been intentionally synthesized for this purpose (see Box 28–E), but most of the compounds are of natural origin (Figure 28–17).

Given the essential role of translation in overall metabolism and the complexity of the process, it is not surprising that many antibiotics function by inhibiting translation. Because bacterial ribosomes are structurally different from eukaryotic cytoplasmic ribosomes, antibiotics that inhibit bacterial protein synthesis frequently do not operate against eukaryotic protein synthesis. On the other hand, several proteins have been discovered that inhibit protein synthesis; these generally act against eukaryotic systems and are inactive against bacterial protein-synthetic components.

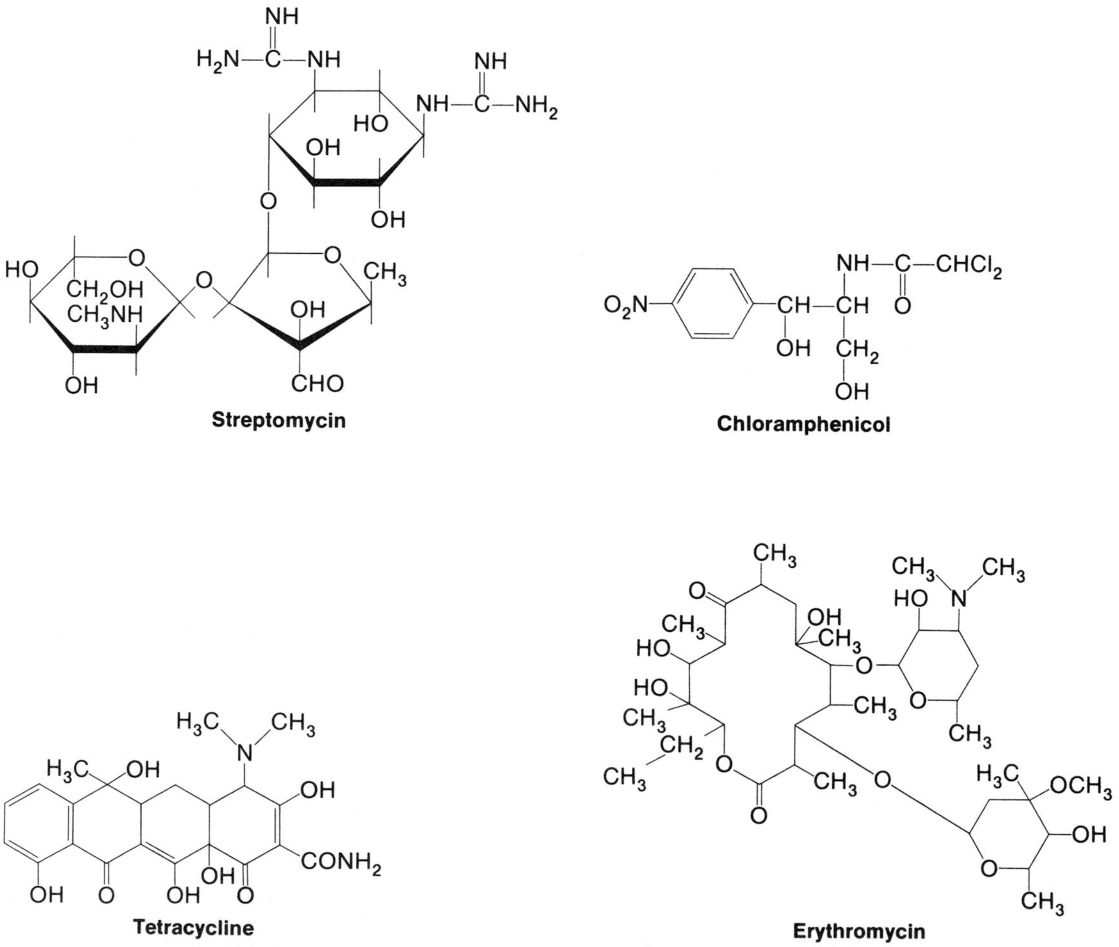

Streptomycin

Chloramphenicol

Tetracycline

Erythromycin

Figure 28–17
The structures of some antibiotic inhibitors of protein synthesis. All of the inhibitors shown function by binding to specific sites on the ribosome.

The Tetracyclines Prevent Anticodon Binding

The tetracyclines are a family of chemically related antibiotics. Tetracyclines interact primarily with the small ribosomal subunit so as to prevent the anticodon binding of aminoacyl-tRNA. Two features of the action of these antibiotics have, however, limited a definitive assessment of their mode of action. First, they bind to many macromolecules in a nonspecific way, so it has been difficult to obtain ribosomes bearing tetracycline-resistant mutations. Most bacterial resistance to this drug results from altered membrane permeability or antibiotic-inactivating enzymes. It is interesting to note that eukaryotic ribosomes themselves are also sensitive to tetracycline, but as it cannot pass through the eukaryotic membranes, it does not produce *in vivo* inhibition of eukaryotic protein synthesis.

Streptomycin Leads to Mistranslation

From the point of view of both structure and function, a great many antibiotic inhibitors of protein synthesis belong to the class known as aminoglycosides. Of these, _streptomycin_ is the best known and most extensively investigated in relation to its effect on protein synthesis.

Streptomycin is bacteriocidal in that it causes cell death, and although it affects a variety of cellular functions, its primary lethal action appears to be on protein synthesis at the level of the ribosome. Thus resistance to the effects of streptomycin can result from altered ribosomes. Ribosomes from streptomycin-sensitive but not streptomycin-resistant cells have a high affinity for the antibiotic, and binding occurs to the small ribosomal subunit.

When bound to the ribosome, streptomycin produces a variety of functional alterations. One of the first alterations to be recognized was the loss of translational fidelity of the ribosome. When bound to the small ribosomal subunit, streptomycin, in some unknown way, appears to distort the interaction between codons and anticodons so as to allow the incorporation of incorrect amino acids. Indeed, mutations to streptomycin resistance frequently prevent antibiotic binding to the ribosome and involve alterations in proteins that are involved in maintaining ribosomal fidelity in translation. The synthesis of erroneous proteins in the presence of streptomycin, however, is clearly not a sufficient explanation of its bacteriocidal action. Rather, it appears that streptomycin binding distorts ribosome structure in such a way as to cause the ribosome to dissociate from the message and prevent its normal reinitiation. Thus streptomycin sensitivity is dominant over streptomycin resistance, because the sensitive ribosome in the presence of the antibiotic reinitiates protein synthesis in an abortive manner.

Chloramphenicol Inhibits Peptidyl Transferase

Chloramphenicol inhibits the growth of a wide range of both Gram positive and Gram negative bacteria, and it was the first broad-spectrum antibiotic to be used clinically. Although eukaryotic cells are generally resistant to the effects of the drug, its clinical utility has been severely curtailed as a result of toxic side effects commonly associated with its use. The toxic effects are at least partly due to the fact that mitochondrial ribosomes are sensitive to the antibiotic.

The relative structural simplicity of chloramphenicol and the fact that only one of its four diastereoisomers is active in inhibiting protein synthesis might suggest that its mode of action by now should be well understood. The facts are otherwise. At inhibitory levels, one molecule of the antibiotic binds to the 50S ribosomal subunit. A variety of tests indicate that chloramphenicol specifically inhibits the peptidyl transferase reaction. Because of these circumstances, it would seem that chloramphenicol provides the ideal means of dissecting the nature and location of this reaction. Unfortunately, however, this has not proved to be the case, and this failure typifies the frustration associated with the dissection of ribosomal reactions.

For example, it has not been possible to show unambiguously that substrate binding is altered by chloramphenicol. Puromycin inhibits the binding of chloramphenicol, and under some conditions, chloramphenicol also inhibits the binding of aminoacyl-tRNA to the ribosome. In the same vein, the chemical attachment of chloramphenicol to proteins in the vicinity of its binding has led to conflicting results. At best, *these studies lead to the view that one molecule of chloramphenicol interacts with a variety of proteins in the vicinity of the peptidyl transferase, and this binding produces multiple effects on their functions.*

Diphtheria Toxin Catalyzes the ADP-Ribosylation of Elongation Factor EF-2

Diphtheria is among the best understood of infectious diseases. The pathogenesis of the disease results entirely from the elaboration by *Corynebacterium diphtheria* of a single exotoxin that kills cells by inhibiting protein synthesis. The pathogenic consequences of the disease are entirely prevented by immunization with toxoid, an inactivated form of the purified toxin. Curiously, the structural gene for the protein is carried by a bacterial virus, called β_1, that must infect the bacterium to induce toxin production.

The toxin, as released into the circulation of the infected host, is a single polypeptide chain ($M_r = 63,000$) with two intramolecular disulfide bonds. Its toxic action requires that a portion of the molecule, the A fragment ($M_r = 21,000$), enter the cytoplasm, where it acts catalytically and specifically to inactivate EF-2. A single molecule is enough to kill the cell.

The entry of the A fragment into the cytoplasm proceeds by a poorly understood mechanism. Essentially stated, the intact toxin interacts with an unidentified receptor on the surface of sensitive cells and is both proteolytically cleaved and disulfide-reduced to yield the A and a larger B fragment ($M_r = 42,000$). The B fragment then facilitates the penetration of the A fragment through the cell membrane. This penetration appears to involve formation by the B fragment of a pore in the membrane through which the A fragment passes.

Within the cell, the A fragment acts as a very specific protein-modifying enzyme. It catalyzes the ADP-ribosylation and consequent inactivation of EF-2 by the following reaction:

$$\text{EF-2} + \text{NAD}^+ \rightleftharpoons \text{ADP-ribosyl-EF-2} + \text{nicotinamide} + \text{H}^+$$

The reaction as shown is reversible when conducted *in vitro*, but under the intracellular conditions of pH and nicotinamide concentration, it is irreversible. Thus diphtheria toxin kills by irreversibly destroying the ability of EF-2 to participate in the translocation step of protein chain elongation.

The enzymatic specificity of diphtheria toxin deserves special comment. The catalytic A fragment *in vitro* will ADP-ribosylate EF-2 in the cytoplasm of apparently all eukaryotic cells, whether sensitive to toxin *in vivo* or not, but it will not modify any other protein, including the bacterial counterpart of EF-2. This narrow enzymatic specificity of the toxin has called attention to an unusual, posttranslational histidine derivative, diphthamide (Figure 28–18), which is present in EF-2 at the site of its ADP-ribosylation. While the unique occurrence of diphthamide in EF-2 explains the specificity of the toxin, it raises questions about the function of the residue in the trans-

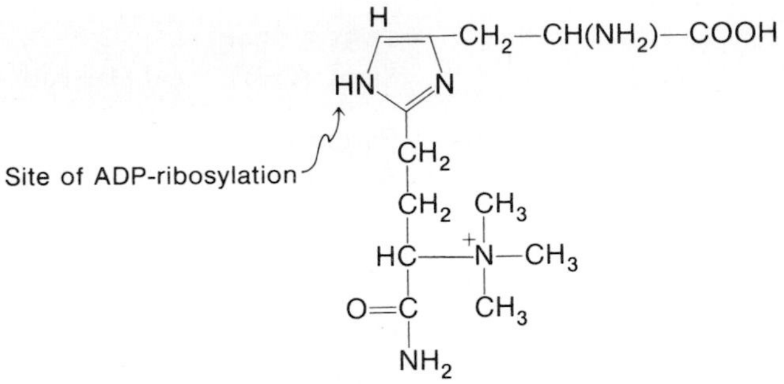

$$H$$

(structure diagram)

CH₂—CH(NH₂)—COOH

HN   N

Site of ADP-ribosylation

CH₂

CH₂   CH₃

HC——N⁺—CH₃

O=C   CH₃

NH₂

Figure 28–18
The structure of diphthamide.

location. Rather interestingly, some mutants of eukaryotic cells selected for resistance to the toxin lack one of several enzymes that carry out the post-translational modification of EF-2 necessary for toxin action, but these cells seem perfectly capable of protein synthesis. Thus the *raison d'être* of diphthamide, as well as the biological origin of the toxin that modifies it, remains a mystery.

Specific Nucleases Can Inactivate Ribosomes

Two protein inhibitors of protein synthesis are known to inactivate the ribosome by introducing specific cleavages into rRNA. One of these proteins, colicin E3, cleaves rRNA in the prokaryotic small subunit near its 3′ terminus and produces a short fragment that is released from the ribosome. This fragment contains the pyrimidine-rich region that pairs with the Shine-Dalgarno sequence in mRNA during the initiation step. Prokaryotic ribosomes inactivated by colicin E3 are unable to participate in initiation. Eukaryotic ribosomes are not sensitive to colicin E3.

Alpha-sarcin is a cytotoxin, produced by a mold, that inactivates the large subunit of both the prokaryotic and eukaryotic ribosome. In this case cleavage occurs in a highly conserved sequence about 500 bases from the 3′ terminus of the large rRNA, releasing what is known as the alpha-fragment. Ribosomes inactivated by alpha-sarcin appear to be deficient in the binding of aminoacyl-tRNA during elongation.

POSTTRANSLATIONAL MODIFICATION OF PROTEINS

Despite the fact that only 20 amino acids (plus formylmethionine in prokaryotic systems) are directly specified by the genetic code, analysis of proteins has revealed that they contain well over 100 different amino acids. These amino acids are all structural variants on the original 20. In addition, proteins infrequently function in their final form with the full complement of amino acids assembled during their translation. Most often their primary translation products are proteolytically cleaved in producing their final three-dimensional structure, in their transport, or in the activation of proenzymes or prohormones. This collection of structural alterations of proteins is considered together as posttranslational modification.

Amino Acids in Proteins Sometimes Undergo Covalent Modification

A partial list of covalently modified amino acids known to occur in proteins is shown in Table 28–6. A number of simple modifications are observed on different amino acids, particularly as they occur at the termini of proteins. Acetylated and amidated amino acids are examples. In addition, a number of interesting single amino acid transfer reactions have been observed at the N

TABLE 28–6
Some Amino Acid Derivatives Found During Posttranslational Modification

Parent Amino Acid	Derivative
Ala	N-acetylalanine, N-methylalanine
Arg	N^ω-methylarginine, ADP-ribosyl-arginine, citrulline, ornithine
Asn	Aspartic acid
Asp	Aspartic acid α-amide, N-acetylaspartic acid
Cys	Cystine, S-galactosylcysteine
Glu	Glutamic acid α-amide, γ-methylglutamic acid, γ-carboxyglutamic acid
Gly	Glycinamide, N-formylglycine
His	π-Methylhistidine, diphthamide
Lys	N^Σ-trimethyllysine, N^Σ-phosphopyridoxyllysine, desmosine-δ-hydroxylysine
Met	Methioninamide, N-acetylmethionine
Phe	Phenylalanine amide, β-hydroxyphenylalanine
Pro	4-hydroxyproline, O^4-arabinosylhydroxyproline
Ser	Pyruvate, O^β-phosphonoserine, O^β-mannosylserine
Thr	α-ketobutyrate, O^β-mannosylthreonine
Tyr	Tyrosine-O^4-sulfate, 3,5-diiodotyrosine, O-adenosyltyrosine

Source: From F. Wold, Posttranslational modifications of proteins, *Ann. Rev. Biochem.* 50:783, 1981. Used with permission.

and C termini of proteins. These one-step reactions occur by nonribosomal means and result in amino acids that are joined at the ends of proteins by peptide bonds not specified by the message. In general, the functional significance of these terminal modifications is obscure, but it likely reflects some regulatory modification. In this context, it is interesting to consider the *E. coli* ribosomal proteins L7 and L12. These two proteins are the product of the same gene and differ only by an N-terminal acetyl group. The fraction of total L7/L12 that is acetylated varies with bacterial growth rate, but thus far no functional alteration has been found to result from this modification.

Most readily detected are modifications that result in altered protein functions or are essential to protein function. Regulation of protein function has been observed to result from methylation, phosphorylation, nucleotidylylation, and ADP-ribosylation. In these cases, usually quite specific enzymes modify regulatory targets in key enzymes. In this category are the coenzymes that are essential to enzyme function and are covalently attached by amino acid side chains.

Carbohydrates of a wide variety become attached to proteins in many different ways (see Chapter 20). In some cases, the carbohydrate addition is essential to the participation of the protein in membrane or cell wall structure. In higher organisms, the carbohydrate plays a key role in the circulation of blood proteins. These proteins may be targeted to certain tissues or cleared from the circulation by the liver on the basis of their carbohydrate content.

Some of the most exotic protein alterations are employed to form cross-links between proteins, particularly in the connective-tissue protein collagen. These cross-links are important to the mechanical stability of connective tissue. In this context it should be noted that cystine formed by the posttranslational cross-linking of two cysteine residues probably serves a similar stabilizing role, particularly in extracellular enzymes. The disulfide cross-linking of the two chains of insulin is a clear example (see below).

In general, the techniques of protein chemistry are tuned to the analysis of standard amino acids, and it is relatively easy to overlook unstable or infrequent derivatives. It is noteworthy that two of the more interesting modifications of proteins, γ-carboxyglutamic acid in coagulation proteins and diphthamide in EF-2, were discovered as a result of the study of the biological inhibitors dicumarol and diphtheria toxin, respectively. Dicumarol inhibits the formation of γ-carboxyglutamic acid in the vitamin K-dependent blood proteins, and diphtheria toxin inactivates EF-2 by modifying its single diphthamide residue, as described earlier. It is possible that the actual list of functionally important amino acid derivatives is much longer.

Proteolytic Processing of Proteins Is a Maturation Reaction for Some Proteins

Proteolytic activation of zymogens or proproteins to functional proteins has been recognized for many years. First seen in the activation of proteolytic enzymes (e.g., trypsinogen → trypsin; prothrombin → thrombin), a similar phenomenon has since been observed with a variety of other proteins and hormones (e.g., proalbumin → albumin, proinsulin → insulin). The proteolytic cleavage or cleavages involved in proprotein → protein conversion can take a variety of forms, most of which lead to the removal of a peptide fragment or fragments. The propeptide of proproteins may be either terminal or internal to the protein chain. Both internal and terminal peptides are removed from prothrombin; an N-terminal 6-residue peptide is removed from proalbumin, and a 29-residue peptide (the C peptide) is removed from the interior of proinsulin. Many of the N-terminal proprotein cleavages occur on the C-terminal side of Arg–Arg sequences and appear to involve a common enzyme system.

The in vitro translation of messages for proteins destined for extracellular transport (e.g., proalbumin, proinsulin) has revealed yet another class of posttranslational proteolytic cleavages. Translation of these messages in protease-deficient in vitro translation systems (e.g., reticulocyte and wheat germ extracts) invariably yields translation products larger than the proproteins observed in vivo. These in vitro translation products, known as preproteins or preproproteins, in general, begin with Met and contain 16 to 30 additional amino acids at their amino terminals.

These presequences, or "signal" sequences, as they have been designated by Gunter Blobel, are rich in hydrophobic amino acids, especially leucine. The ribosomes that produce proteins containing signal sequences are generally attached to intracellular membranes, and the sequences are not normally seen on completed proteins because they are removed cotranslationally, i.e., while the nascent protein is still being assembled on the ribosome. When wheat germ or reticulocyte translation systems, which accumulate preproteins, are supplemented with endoplasmic membrane vesicles, typically from dog pancreas, the signal sequences are removed and the translation products are found within the vesicles.

Observations of this type led Blobel to propose the "signal hypothesis" (Figure 28–19). According to this hypothesis, the presequence serves to guide the polysome producing the protein that is destined for extracellular transport to the endoplasmic reticulum, where the peptide inserts into the membrane because of its hydrophobic character. In some as yet undefined way, the nascent protein threads through the membrane as it is being assembled and the signal peptide is removed by a protease on the distal side of the membrane, where the completed protein is ultimately sequestered.

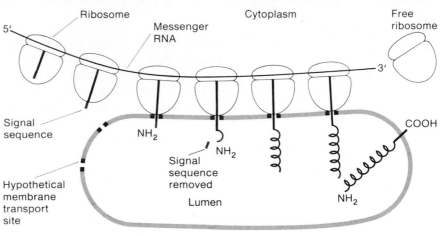

Figure 28–19

The signal hypothesis as proposed by Blobel and Dobberstein. Polyribosomes are bound to the membrane by means of an N-terminal signal sequence on the secreted protein that is hypothesized to interact with a transmembrane pore. In the example shown, this sequence is cleaved by a membrane-bound protease before translation is complete.

BOX 28–G

Observations Relating to the Signal Hypothesis (Jacobson and Saier)

The signal hypothesis describes a means whereby proteins destined to become membrane proteins or to be exported can be localized outside of the cytoplasm. The hypothesis is that the N-terminal sequence of the polypeptide chain serves as a signal for membrane attachment; as a result, the nascent polypeptide chain is extruded through a pore in the membrane, the signal sequence being cleaved off before chain completion (see Figure 28–19).

A considerable amount of evidence has accumulated in favor of certain elements of the signal hypothesis in both eukaryotes and prokaryotes. A number of mitochondrial and chloroplast membrane proteins, as well as proteins of the outer membrane of *E. coli*, have been shown to be synthesized as precursors that are then cleaved into mature proteins during or after translation. Several periplasmic proteins of *E. coli*, which must be secreted across the inner membrane, are also translated with signal sequences by membrane-bound ribosomes, and these sequences are rich in hydrophobic amino acids, which might be expected to "lead" the nascent polypeptide along with the ribosome to the membrane.

Genetic studies by Jon Beckwith and Tom Silhavy with *E. coli* cells have demonstrated the importance of the signal sequence in the membrane insertion and export of some proteins. In these experiments, most of the gene for the soluble enzyme β-galactosidase (*lacZ*) was fused to portions of each of several genes coding for proteins involved in the transport of the disaccharide maltose (see illustration). These maltose-specific proteins, products of the genes *malF*, *malE*, and *malB*, are normally located in the inner membrane, periplasmic space, and outer membrane, respectively (Chapter 6). Cells containing these hybrid genes synthesize hybrid proteins that include (1) portions of the membrane or exported protein at the N terminus and (2) a covalently linked, functional β-galactosidase molecule that is easy to assay spectrophotometrically. With these cells, the localization of the three hybrid proteins could easily be determined

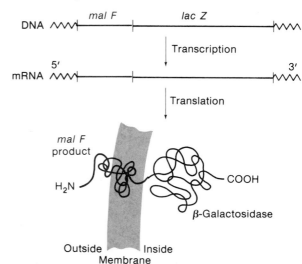

The use of gene fusion to study membrane protein localization and assembly in bacteria. Genetic techniques have been developed for the fusion of specific genes in *E. coli*. In this example, part of the gene for an inner membrane protein involved in maltose transport (*malF*) is fused to most of the β-galactosidase gene (*lacZ*). β-Galactosidase, normally a soluble protein, becomes membrane-bound in this case, showing that the information for membrane localization of the malF product is carried by the amino acid sequence of this portion of the malF protein itself. Fusions of various parts of the *malF* gene to *lacZ* can therefore be used to determine which portions of the protein are important for insertion simply by determining the location of β-galactosidase, which is easy to assay.

by cell fractionation and β-galactosidase activity measurements. As long as a sufficient amount of the N terminus of the maltose transport proteins was present, β-galactosidase activity was found in the inner membrane for the first two fusion products and in the outer membrane for the last, even though β-galactosidase is nor-

Thus, for example, insulin mRNA is translated in wheat germ extracts as a single polypeptide chain, preproinsulin, that contains the 84 residues of proinsulin plus a 23-residue signal peptide. Within the islets of Langerhans, the N-terminal sequence is removed cotranslationally in targeting proinsulin to the Golgi apparatus, and the two disulfides joining the ends of the molecule are formed. Following this, the C-peptide region of proinsulin is removed to yield the circulating form of insulin with 51 residues in two disulfide-linked peptides (Figure 28–20). Further evidence in support of the signal hypothesis is presented in Box 28–G.

INTRACELLULAR PROTEIN DEGRADATION

The proteins of which living cells are made are much more labile than the cells themselves and are subject to continual renewal; they are degraded back to amino acids and replaced by new synthesis. For example, the aver-

mally a soluble protein (see Table). Thus in accordance with the signal hypothesis, the signal sequence was able to direct membrane or exported proteins to the membrane, even though a large hydrophilic molecule was attached to it. Furthermore, mutations were isolated in which the lamB-lacZ hybrid protein was entirely cytoplasmic. These mutations were mapped at the extreme N terminus of the lamB portion, showing that defects in the signal sequence can lead to defective insertion of membrane proteins.

Several membrane proteins have been studied that do not conform to the signal mechanism of assembly. For example, the lactose transport protein of E. coli (lactose permease; Chapter 34) is an integral cytoplasmic membrane protein that appears not to be synthesized in precursor form and therefore lacks a signal sequence. Similar observations have been made for several other bacterial membrane proteins, as well as for some proteins found in mitochondrial, chloroplast, and endoplasmic reticular membranes. Furthermore, a major lipoprotein found in the E. coli outer membrane, which is synthesized with a removable signal sequence, has been found

in its correct location even when a mutation was introduced in the signal sequence that blocked removal of the N-terminal extension. These results show that *neither a signal sequence nor precursor processing is necessary for the proper insertion of some membrane proteins*.

Genetic studies of some bacterial membrane and periplasmic proteins have led to the conclusion that even in the case of those polypeptides synthesized with an N-terminal extension, *the signal sequence is not by itself sufficient to ensure proper localization*. When the N-terminal half (including the signal sequence) of the periplasmic protein necessary for maltose transport in E. coli was fused to β-galactosidase, the hybrid protein was found in the inner membrane, not in the periplasm (see Table). This and other observations make it likely that internal amino acid sequences are also important in the proper localization of some membrane and exported proteins.

In conclusion, it seems likely that the complex problems of membrane protein assembly and the secretion of extracellular polypeptides may be solved by the cell in more than one way.

Localization of Hybrid Proteins Produced by Various Maltose Gene-*lacZ* Fusion Strains

Fusion	Approximate Amount of Maltose Gene DNA in Hybrid Gene	Cellular Localization of Hybrid Protein (β-Galactosidase Activity)
malF (inner membrane protein)-*lacZ*	1/11	Cytoplasm (92%)
	9/11	Inner membrane (80%)
malE (periplasmic protein)-*lacZ*	2/13	Cytoplasm (88%)
	4/13	Inner membrane (65%)
lamB (outer membrane protein)-*lacZ*	1/9	Cytoplasm (90%)
	6/9	Outer membrane (80%)

Source: Data from T. J. Silhavy, P. J. Bassford, Jr., and J. R. Beckwith, A genetic approach to the study of protein localization in *Escherichia coli*. In M. Inouye (ed.), *Bacterial Outer Membranes: Biogenesis and Functions*, Wiley, New York, 1979.

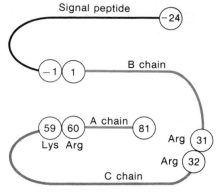

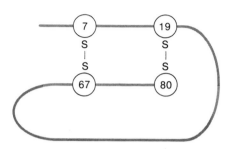

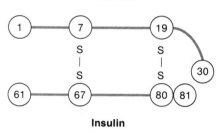

Figure 28–20
Biosynthesis of insulin. Insulin is synthesized by membrane-bound polysomes in the β cells of the pancreas. The primary translation product is preproinsulin, which contains a 24-residue signal peptide preceding the 81-residue proinsulin molecule. The signal peptide is removed by signal peptidase, cutting between Ala (-1) and Phe $(+1)$, as the nascent chain is transported into the lumen of the endoplasmic reticulum. Proinsulin folds and two disulfide bonds cross-link the ends of the molecule as shown. Before secretion, a trypsin-like enzyme cleaves after a pair of basic residues 31, 32 and 59, 60; then a carboxypeptidase B-like enzyme removes these basic residues to generate the mature form of insulin.

age protein in rat liver has a half-life of approximately 1 day, and in brain or muscle, 3 or 6 days. By contrast, liver cells divide about once per year, while muscle and liver cells are non-dividing. Individual enzymes in these cells may turn over with half-lives as short as 30 min to 1 hr, and in bacteria certain proteins involved in the regulation of gene transcription are completely hydrolyzed within a few minutes after their synthesis. At first glance this continuous degradation of cell proteins appears to be a highly wasteful process. In fact, *this process is of major importance in the regulation of enzyme levels, in protecting the organism against the accumulation of highly abnormal proteins, in the control of cell growth and in the organism's ability to adapt to poor nutritional conditions.*

The Rate of Breakdown of a Protein Helps Determine Its Intracellular Level

The level of any protein is determined by the balance between its rate of synthesis and degradation, and therefore the inherent differences in degradative rates of proteins can have important implications for the regulation of enzyme levels. A rapid rate of degradation insures that the concentration of an enzyme falls very quickly when its synthesis is reduced. On the other hand, the concentration of a short-lived protein rises to a new steady-state level, especially rapidly when its rate of synthesis is enhanced.

The half-lives of individual cytoplasmic or nuclear proteins are not uniform. Even the polypeptides comprising an organelle, such as the mitochondria, are synthesized and degraded at distinct rates. The degradation of specific proteins in cells has always been found to obey first-order kinetics (like the decay of radioactive nuclei). Therefore, the rate of degradation is defined normally by the half-life of the protein, the period during which 50% of it is degraded. This value is independent of its rate of synthesis, indicating that cell proteins do not undergo an aging process (unlike erythrocytes or human beings).

The rates of degradation of a large number of enzymes have been measured in rat liver (Table 28–7). From these values, it is clear that the *enzymes*

TABLE 28–7
Half-Lives of Some Proteins in Mammalian Cells

Enzyme	Half-life (hr)
Rapidly degraded	
1. c-myc, c-fos, p53 oncogenes	0.5
2. Ornithine decarboxylase	0.5
3. α-Amino levulinate synthase	1.1
4. RNA polymerase 1	1.3
5. Tyrosine aminotransferase	2.0
6. Tryptophan oxygenase	2.0
7. β-Hydroxylβ-methylglutaryl coenzyme A reductase	2.0
8. Deoxythymidine kinase	2.6
9. Phosphoenol pyruvate carboxykinase	5.0
Slowly degraded	
1. Arginase	96
2. Aldolase	118
3. Cytochrome b_5	122
4. Glyceraldehyde-3-phosphate dehydrogenase	130
5. Cytochrome B	130
6. Lactic dehydrogenase (isoenzyme 5)	144
7. Cytochrome C	150

degraded most rapidly in liver all catalyze reactions that are particularly important metabolic control points. Included in this group (Table 28–7), for example, are the rate-limiting enzymes in cholesterol and heme synthesis, whose levels change rapidly under different physiological conditions. Among the most labile proteins are several products of oncogenes; these proteins help regulate cell division, and if they accumulate can induce cancerous growth. In general, these labile proteins also tend to have very short-lived mRNAs. By contrast, the liver proteins with especially long half-lives are rarely, if ever, the sites of metabolic control.

Whether a protein has a short or long half-life must be determined by its primary or secondary structure. Surveys by Rechsteiner indicate the existence of certain domains (called PEST sequences) common to short-lived proteins in eukaryotic cells. These domains are particularly rich in proline, serine, and acid residues. Either the proteolytic machinery recognizes such regions or these sequences promote the unfolding of the proteins and thus cause rapid degradation.

Alterations in the half-lives of specific enzymes have also been documented under various physiological conditions. In a number of instances, decreased breakdown of a degradative enzyme, such as tryptophan oxygenase, occurs in the presence of its substrate (tryptophan). By itself, *such a reduction in degradation can lead to an accumulation of the enzyme even in the absence of any change in its synthesis.* Physiologically, this property also insures that levels of these enzymes fall when the substrates are present in low amounts. A related phenomenon has been demonstrated in which the product of the enzyme glutamine synthase actually promotes the enzyme's degradation. In this way, excess glutamine retards its own production by promoting the degradation of the biosynthetic enzyme. These types of regulation would appear to be very useful mechanisms for ensuring that enzyme levels are appropriate for metabolic needs. *Thus, degradative rates of proteins, like the catalytic or allosteric properties of proteins, have evolved to optimize metabolic regulation.*

Abnormal Proteins Are Selectively Degraded

One very important function of protein breakdown is to protect the organism against the intracellular accumulation of abnormal polypeptides. This process thus serves as a sort of cellular "sanitation system."

Both bacterial and animal cells rapidly hydrolyze proteins with abnormal structures as may arise from mutations, errors in RNA or protein synthesis, or by intracellular denaturation. Even mitochondria and chloroplasts contain a complete pathway for rapid breakdown of abnormal organellar proteins. Such mutant proteins are synthesized at the same rates as normal gene products but fail to accumulate to the usual extent because of their rapid hydrolysis. For example, in *E. coli*, incomplete chains of β-galactosidase are degraded with half-lives as short as a few minutes, even though the normal polypeptide is completely stable. Incomplete short-lived proteins may result from nonsense mutations, from biosynthetic errors, or through incorporation of puromycin (which causes the premature termination of polypeptides). Certain drugs (e.g., streptomycin) or mutations that reduce the fidelity of the protein synthetic apparatus promote the production of complete proteins that can have abnormal conformations. Finally, many unusual polypeptides generated by the techniques of genetic engineering (such as protein fusions) also tend to be rapidly degraded. In fact, a large fraction of the proteins cloned and expressed in bacteria fail to assume their proper conformations due to the lack of disulfide bridges, to the lack of glycosylation in bacteria, or to the absence of stabilizing cofactors. Many of these labile polypeptides are of industrial and medical importance, such as insulin or viral antigens, and their commercial production has been facilitated in some cases by the use of bacterial mutants with defects in their proteolytic machinery.

Many studies in this area have followed the fate of proteins that have incorporated synthetic analogs of the natural amino acids. The incorporation of such analogs interferes with the normal tertiary folding of proteins and the resulting polypeptides are rapidly hydrolyzed *in vivo*. For example, hemoglobin is probably the most stable intracellular protein and normally lasts the life span of the red cell (110 days in humans). However, globin containing the valine analog aminochlorobutyrate fails to form tetramers or bind heme and has a half-life of about 10 min.

The very rapid degradation of abnormal hemoglobins is also an important factor in many human genetic diseases. Several human hemoglobin variants (e.g., the "unstable" hemoglobinopathies) are rapidly degraded within reticulocytes, as are the free α-chains produced in excess in β-thalassemia. Their selective hydrolysis appears highly advantageous, since it helps reduce the deleterious accumulation of these partially denatured molecules.

Highly abnormal proteins must be continually generated in all cells through spontaneous denaturation or chemical modification of normal cell constituents. There is now strong evidence that oxidative damage to hemoglobin or other proteins through free radicals can trigger their rapid degradation. Many common enzymes, such as xanthine oxidase or other mixed function oxidases, generate highly reactive products, such as peroxide, superoxide, or hydroxyl radicals, that can rapidly inactivate cell proteins by modifying susceptible amino acids (methionine, histidine, tyrosine). Although many enzyme mechanisms exist to protect cell proteins from such damage (e.g., met hemoglobin reductase, superoxide dismustase, catalase), cells also have very efficient mechanisms to recognize and quickly degrade the damaged proteins. In fact, the rate-limiting step in the degradation of many cell proteins may be their irreversible unfolding and their assuming an "abnormal" conformation.

Upon exposure of cells to high temperatures or to various toxic agents (e.g., heavy metals), cell proteins may become partially denatured. *Prokaryotic and eukaryotic cells adapt to such stressful conditions through programmed changes in gene transcription and translation, called the "heat-shock response."* They reduce overall protein synthesis and, for a brief time, express a small repertoire of genes, called "heat-shock genes" (or *Hsps*). This response has a clear adaptive value in promoting the organism's viability at high temperatures, and it also enhances the cell's capacity to degrade the abnormal polypeptides. Various conditions that generate large amounts of abnormal proteins induce the heat-shock response, including exposure of *E. coli* to streptomycin or puromycin, incorporation of amino acid analogs, high level expression of a single cloned abnormal protein, or simply microinjection of a denatured protein into cells. In bacteria, the heat-shock response requires a specific sigma factor of RNA polymerase that serves as a positive regulator of the heat-shock genes, and mutants with defects in this protein have a reduced capacity to degrade abnormal proteins.

The Soluble Pathway for Protein Breakdown Requires ATP

In eukaryotic cells, there exist at least two major proteolytic systems serving distinct physiological functions: the lysosomal system and soluble ATP-requiring pathway (Figure 28–21). It had long been assumed that the lysosome is the only, or at least the primary, site for intracellular protein degradation, since this membrane-enclosed organelle contains many proteases

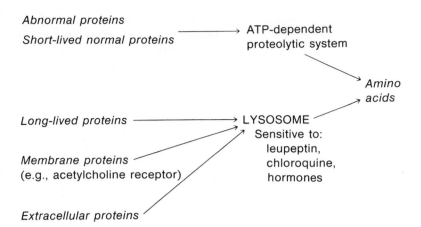

Figure 28–21
Pathways for protein degradation in mammalian cells. There are at least two major systems for protein breakdown in the mammalian cytoplasm: (1) the organelle-localized lysosomal system, and (2) the soluble alkaline system. Surprisingly, the soluble system is ATP-dependent.

able to hydrolyze proteins rapidly (see page 000). However, prokaryotic cells and certain eukaryotic cells (e.g., red blood cells) lack such organelles but have the capacity to selectively degrade abnormal proteins. Knowledge about the nonlysosomal pathways has developed rapidly since the development by Alfred Goldberg of cell-free preparations from animal and bacterial cells that selectively degrade abnormal proteins in a fashion similar to intact cells. This proteolytic pathway is derived from the soluble fraction of the cytoplasm. It is not enclosed within membrane residues and is most active at pH 7.8 (unlike the lysosomal proteases). Therefore, precise mechanisms must exist to prevent the autodigestion of the cell by these proteases.

The most interesting biochemical feature of these soluble systems is that they require ATP. *In vivo*, the degradation of proteins can be prevented with inhibitors of oxidative phosphorylation and glycolysis that prevent ATP generation, and ATP was shown to be essential for the initial cleavage reactions on the protein. An energy requirement for protein hydrolysis would not be anticipated on thermodynamic grounds, since the hydrolysis of peptide bonds is an exergonic reaction. This ATP requirement was therefore an important clue to the eventual discovery of the responsible degradative systems and the elucidation of novel biochemical mechanisms. In *E. coli*, the ATP appears to be essential for the function of a new type of protease isolated by Goldberg, called protease La, the product of the *lon* gene. Most well-known proteases, such as the pancreatic proteins, are relatively small enzymes (M_r 20,000 to 30,000). By contrast, the ATP-dependent proteases are unusually large multimeric enzymes (with M_r of 3.6×10^5 to 10^6), and have appreciable structural complexity. The most interesting property of protease La is that it *is an ATPase as well as a protease, and these two functions are tightly coupled*. Protein substrates stimulate ATP hydrolysis, but nondegraded protein (ones in their native conformations) do not. For every peptide bond hydrolyzed in proteins, protease La consumes 2 ATPs, which is almost as much ATP as the cell uses in peptide bond formation. In the absence of ATP, this enzyme lacks activity, but after ATP binding its proteolytic site is allosterically activated. Hydrolysis of ATP then takes place, and enables the enzyme to attack quickly the next proteolytic site. Thus, proteolysis involves repeated cycles of ATP hydrolysis, until the polypeptide substrate is digested to small inactive oligopeptides, which are subsequently degraded by various peptidases that function independently of ATP.

Two unusual features of protease La seem particularly important in preventing nonspecific proteolysis *in vivo*: (1) Its ability to cleave peptide bonds is quite low until protease La binds to a protein substrate. Thus, the protease has 2 sites for recognizing substrates—the proteolytic site and a regulatory site, whose occupancy by an unfolded protein activates the enzyme. And (2), this enzyme has a very high affinity for ADP, which is a potent inhibitor of proteolysis. Consequently, ADP serves to limit its activity *in vivo*. However, following interaction with an unfolded protein, protease La loses all its affinity for ADP, quickly binds ATP, and becomes active. This activation mechanism, which appears similar to the GTP-GDP exchange mechanisms in Elongation Factor Tu and in the G-protein of adenylate cyclase, probably evolved to prevent excessive proteolysis *in vivo*.

The synthesis of protease La, like its activity, is precisely regulated. *It is a heat-shock protein, whose synthesis rises when large amounts of abnormal proteins are generated.* Lon mutants have a reduced capacity to degrade such proteins, but also have other physiological defects due to their inability to degrade certain critical regulatory proteins. Susan Gottesman has shown that this protease also rapidly degrades short-lived normal proteins that regulate cell division following DNA damage *(sulA)* or that control transcription of specific operons *(cpsA, N)*. Bacteria also contain additional ATP-dependent and -independent proteases that also must contribute to degradation of abnormal proteins and normal proteins during starvation.

The ATP-Dependent Pathway in Eukaryotic Cells Requires Ubiquitin

In eukaryotic cells, the soluble ATP-dependent pathway appears more complex than in bacteria. *This process involves an additional reaction in which protein substrates undergo postsynthetic modification prior to their hydrolysis.* Hershko, Ciechanover and Rose showed that the *ATP-dependent pathway in reticulocytes consists of multiple components, one of which is the small heat-stable polypeptide, "ubiquitin."* This polypeptide, although lacking in bacteria, is abundant in all eukaryotic cells from yeast to man, and is one of the most conserved proteins throughout evolution. In the presence of ATP, ubiquitin is covalently linked through an isopeptide bond to protein substrates by a system involving at least three distinct enzymes. One or more ubiquitin molecules become conjugated through their carboxyl terminal residues to the epsilon-amino groups on lysines on the proteins. In this process, the carboxyl group on the ubiquitin is first activated in an ATP-dependent reaction process involving formation of an ubiquityl-AMP intermediate. This reaction resembles those involved in the activation of amino acids for charging to tRNA. The activated ubiquitin is then transferred via a carrier protein to lysine residues on proteins.

This conjugation process marks proteins for rapid digestion. Ubiquitination occurs more rapidly with denatured proteins, although it is uncertain what structural features of proteins are being recognized. After the conjugation step, the proteins are degraded by a large enzyme complex ($M_r \sim 10^6$) called the Ubiquitin-Conjugate Degrading Enzyme (UCDEN). This protease requires ATP and is active only against proteins linked to ubiquitin.

Ubiquitin is a heat-shock protein whose level rises dramatically under various stressful conditions. The increased production of ubiquitin is essential for cell viability at high temperatures, presumably because proteolysis is particularly important at such times. This pathway does not simply eliminate highly abnormal proteins; it is also very active in maturing reticulocytes or in atrophying muscles, where normal proteins are subject to rapid degradation. In addition, ubiquitin exists attached to certain stable proteins, including histone (UH2A) and on cell surface receptors. Thus, this modification probably serves other functions aside from protein breakdown.

Many Proteases and Acid Hydrolases Are Localized in the Lysosome

The classic studies of DeDuve in the 1960s established that mammalian cells contain a degradative organelle, the *lysosome*, in which are concentrated a large number of proteases and other acid hydrolases, including acid phosphatases, various polysaccharidases, RNAse, DNAse, cholesterol esterase, etc. Together these degradative enzymes are capable of completely hydrolyzing various macromolecules found within cells as well as shorter chain lipids, nucleotides and oligosaccharides. In yeast and in higher plants, there exists a similar degradative organelle called the "vacuole," which also is enriched in acid hydrolases. The small products of these degradative enzymes easily exit from this organelle for use by the cell. In a variety of human genetic diseases, called "Lysosomal Storage Diseases," one or more of these acid hydrolases is missing. As a consequence, there is a slow accumulation within the lysosome of undegraded molecules. The failure to degrade such molecules can have devastating consequences. For example, Tay-Sachs is a recessive neurological disorder leading to mental retardation and death before the age of five. It is caused by the absence of a specific lysosomal hexosaminidase and a failure to degrade gangliosides, a lipid normally found in brain. The various lysosomal storage diseases illustrate the importance of these acid hydrolases in the continual turnover of various cell constituents.

The lysosomal proteases, called cathepsins, and the other lysosomal hydrolases all function optimally in the acid pH range, and cells expend ATP to maintain the pH within this organelle below 4.8. This low pH is generated by an ATP-linked proton pump on the lysosomal membrane. The lysosomal hydrolases themselves are quite resistant to other lysosomal enzymes, probably due to their extensive glycosylation. A useful experimental tool to probe the function of lysosomes has been exposing cells to weak organic bases, such as the antimalarial drug, chloroquine. These agents easily penetrate lipid membranes and accumulate to high levels in the acidic milieu of the lysosome. In doing so, they raise intralysosomal pH, block function of the organelle, and cause a build-up of undegraded materials.

Several mechanisms function to prevent damage to the cells by these potent enzymes. A complex biosynthetic pathway exists to target the newly synthesized hydrolases to this organelle. It involves proteolytic processing, addition of polysaccharides, and specific sugar receptors on the lysosome. If these enzymes ever leak out of the organelle, they tend to have little activity because cytosolic pH is 7.0 or higher. Also, there exist in the cytosol inhibitory proteins which inactivate lysosomal proteases and thus help protect the cell from autodigestion.

Lysosomes Function in Degradation of Both Endocytosed and Cellular Proteins

Probably the best established function of the lysosomal hydrolases is in the breakdown of proteins and larger particles taken into animal cells from their surrounding medium by *endocytosis*. In receptor-mediated endocytosis, a single extracellular protein (e.g., a polypeptide hormone) or small complexes (e.g., lipoprotein particles) bind initially to specific high-affinity receptors on the outside of the plasma membrane. The occupied receptors become localized in distinct regions where there occurs an invagination of the plasma membrane. These invaginations are then pinched off to form small vesicles containing within them the ingested materials. Most cells can also engulf larger materials by the less specific processes of pinocytosis and phagocytosis. In phagocytosis, mammalian cells, especially macrophages and tissue monocytes, take up whole bacteria, viruses, antigen-antibody complexes, parasites, and even dead or damaged cells, and form membrane-enclosed vesicles about them. Whether they are small or large, the endocytic vesicles fuse with lysosomes to form "secondary lysosomes," where the ingested materials are quickly hydrolyzed.

In addition, lysosomes play an important role in the breakdown of intracellular proteins, primarily in nutritionally poor environments. For example, in the liver or in skeletal muscles of fasting animals or in cultured cells deprived of serum, there is a large increase in the rate of degradation of proteins that are otherwise quite stable. In such catabolic conditions, the lysosomes are enlarged and often contain whole bits of the cytoplasm, including other organelles. Such membrane-enclosed structures are called "autophagic vacuoles," and they form by a specific mechanism in which certain expendable proteins containing specific consensus sequences somehow become enclosed in vesicles derived from the endoplasmic reticulum. These vesicles then fuse with lysosomes to form the "autophagic vacuoles."

This enhancement of overall proteolysis can be blocked with inhibitors of lysosomal function such as weak bases, or with certain inhibitors, which inactivate lysosomal proteases, such as the antibiotic leupeptin. However, these same inhibitors do not reduce rapid degradation of highly abnormal proteins or of short-lived enzymes, which are catalyzed by the nonlysosomal pathway. The quantitative importance of the lysosomal system in proteolysis thus depends on the nutritional and endocrine status of the organism. In rapidly growing cultures, inhibitors of lysosomal function have little effect on overall proteolysis. In contrast, in nongrowing, contact-inhibited cells or in hepatocytes exposed to the hormone glucagon, overall protein breakdown is accelerated through this autophagic mechanism. Similarly, in eukaryotic microorganisms, such as yeast, during starvation, the vacuolar proteases assume particular importance.

Because mobilization of amino acids from tissue proteins is important in energy homeostasis, especially in fasting, the breakdown of proteins is precisely regulated by several hormones. For example, in the liver and in skeletal muscle, insulin not only stimulates protein synthesis but also inhibits protein breakdown by suppressing the rate of autophagic vacuole formation. In fasting or diabetes, when insulin falls, intralysosomal proteolysis rises. The resulting amino acids are released by this tissue for oxidation as an energy substrate or for use in gluconeogenesis in the liver. It is also noteworthy that lysosomal protein breakdown is also enhanced in various pathological states, where there is a failure of normal growth, and tissue atrophy develops. For example, in muscular dystrophy, in hyperthyroidism, or in patients suffering from burns or febrile illness, there is excessive breakdown of tissue proteins. Under these conditions, large increases in lysosomal enzyme content have been reported.

Overall Protein Breakdown in Microorganisms Is Regulated

In bacteria, as in mammalian cells, there also exist mechanisms that regulate the overall rate of protein breakdown, even though these cells lack lysosomes. During exponential growth of bacteria, most cell proteins are very stable. However, starvation of bacteria for required amino acids, for nitrogen, or a carbon source prevents further growth and quickly leads to a 2- to 4-fold stimulation of overall protein degradation. This process also requires ATP but does not involve protease La. This response is triggered by the lack of a full complement of amino acids and rises when a specific aminoacyl tRNA limits ribosomal function. The amino acids generated by the breakdown of cell proteins can provide the needed residues. Thus, one important role of protein breakdown in nutritionally poor environments is to provide precursors for further protein synthesis. In addition, this process can provide metabolizable amino acids to cells lacking a carbon source. Thus, in *E. coli*, overall protein breakdown is coordinately regulated with a number of growth-related processes, including the synthesis of ribosomal RNA and phospholipid. When these synthetic processes fall, proteolysis rises, and both responses are signalled by the unusual nucleotide guanosine tetraphosphate.

A dramatic acceleration of proteolysis also occurs in many microorganisms under poor nutritional conditions when they undergo the process of differentiation, called sporulation. Sporulation involves extensive breakdown of preexistent cell proteins, and an essential step in this process is the induction of new proteases. The amino acids generated are then used for production of spore-specific proteins. Similarly, when environmental conditions improve, the spores undergo the process of germination, in which new endoproteases are synthesized to digest the spore and to provide amino acids for synthesis of proteins characteristic of vegetative cells. In complex eukaryotes, there are many analogous adaptive responses in which the breakdown of preexistent proteins is required for the synthesis of new cell enzymes (e.g., the degradation of egg albumin as the chick develops). Thus, when cells undergo adaptations in shape or in composition, proteolysis can be especially important both to remove preexistent cellular structures and to obtain the essential building blocks for new proteins.

SUMMARY

The assembly of amino acids into polypeptides is the first step in the metabolism of proteins. Small RNA molecules called tRNAs play a crucial role in the process. The structures of tRNA molecules as a class are strikingly similar. Each tRNA contains between 73 and 93 bases in a single polynucleotide chain that folds back on itself to generate a secondary structure comprising a three- or four-leaf clover. The leaves and stems of the cloverleaf then fold on each other to generate an L-shaped tertiary structure. At one end of the L of each tRNA is the 3' terminal of the polynucleotide chain, to which amino acids are attached. At the other end of the L is a stretch of three unpaired bases, called an anticodon, which serves to specify the incorporation of an attached amino acid. Once attached, the specificity of incorporation of the amino acid into protein is entirely determined by the structure of the tRNA to which it is joined. The anticodon of each aminoacyl-tRNA is matched against the codons read sequentially, and its attached amino acid is incorporated into the polypeptide being synthesized. The correct matching of codon and anticodon requires their interaction in an antiparallel manner in such a way as to result in normal Watson-Crick pairing in the first two positions of the codon. A limited ambiguity at the third (3'-OH) base of the codon, termed wobble, allows recognition by anticodons of codons imperfectly matched at this position.

The mRNA templates are read from 5' to 3', and the encoded polypeptides are polymerized from amino to carboxyl terminal. The process of translation, which is catalyzed by the ribosome, comprises three stages: initiation, elongation, and termination. Each of these stages requires the participation of nonribosomal protein factors together with the hydrolysis of GTP.

The relationship between mRNA nucleotide and protein amino acid sequence is termed the genetic code. The fundamental features of the genetic code were deduced from experiments using synthetic polynucleotides as mRNAs. These experiments demonstrated that each amino acid is specified in translation by trinucleotide or triplet code words (codons). They also defined precisely the relation between code word and amino acid for the 64 possible triplets and 20 amino acids encoded. The genetic code revealed by these studies has several notable features. First, it is degenerate, in that some amino acids are coded for by more than one codon. Second, only 61 triplets of the total 64 correspond to amino acids. One codon (AUG or GUG) specifies the translation start point. Three triplets without coding function (UAA, UAG, and UGA) specify translation stop points. Although the genetic code was believed to be universal, a variant code has been found to operate in mitochondria.

A number of antibiotics and protein toxins exert their biological effect by interfering with discrete steps of protein synthesis. Some of those agents affect only prokaryotes or eukaryotes, owing to differences in the protein synthetic apparatus of the two classes of organisms.

The majority of polypeptides undergo posttranslational modification of one kind or another before assuming their functions. These modifications range from proteolytic cleavages, such as those associated with zymogen activation or protein transport, to covalent modification of specific amino acid residues.

The ultimate fate of all polypeptides synthesized is to be broken down to their component amino acids. However, individual proteins vary considerably in their rate of degradation. Rapidly degraded proteins tend to be those with particular significance for regulation, for example, enzymes catalyzing reactions at important metabolic control points. Both bacterial and animal cells rapidly break down proteins with damaged or defective structures. The pathway respon-

sible for this degradation, which requires ATP in the initial stage of breakdown, may function also in the selective degradation of normal proteins.

Overall rates of protein breakdown are determined by the activity of a pathway distinct from that responsible for selective degradation. In mammalian cells, an organelle called the lysosome is the site of nonselective breakdown. A similar system operates in bacteria, even though these organisms lack lysosomes. The rates of degradation catalyzed by nonselective pathways are responsive to nutritional demand.

SELECTED READINGS

Beckwith, J., and C. Lee, Cotranslational and posttranslational translocation in prokaryotic systems. *Ann. Rev. Cell Biology* 2:581, 1986. Review article with an emphasis on protein secretion and membrane protein biogenesis in bacteria.

Gale, E. F., E. Cundliffe, P. E. Reynolds, H. H. Richmond, and M. H. Waring, *The Molecular Basis of Antibiotic Action*, 2nd ed. New York: Wiley, 1981, pp. 402–549. A description of the ways in which antibiotics and toxins inhibit protein synthesis.

Glaumann, H. and F. J. Ballard, *Lysosomes: Their Role in Protein Breakdown*. New York: Academic Press, 1987.

Hardesty, B., and G. Kramer, *Structure, Function and Genetics of Ribosomes*. Springer Series in Molecular Biology. Springer Verlag, 1986. The third in a series of books describing current research on protein synthesis.

Knauf, P. A., and J. S. Cook. (eds.), *Current Topics in Membranes and Transport*, Vol. 24. *Membrane Protein Biosynthesis and Turnover*. New York: Academic Press, 1985. A collection of review articles on the biochemistry and genetics of membrane protein biogenesis.

Lake, J. A., Evolving ribosome structure: domains in archae-

bacteria, eubacteria, eocytes and eukaryotes. *Ann. Rev. Biochem.* 54:507, 1985. The picture of the ribosome that is emerging from electronmicroscopic studies.

Noller, H. F., Structure of ribosomal RNA. *Ann. Rev. Biochem.* 54:507, 1985. A description of the structures of ribosomal RNAs and their evolutionary conservation.

Weissbach, H., and S. Pestka. *Molecular Mechanisms of Protein Biosynthesis*. New York: Academic Press, 1977. A comprehensive view of protein synthesis.

Wickner, W. T., and H. F. Lodish, Multiple mechanisms of protein insertion into and across membranes. *Science* 230:400, 1985. Review article summarizing some of the mechanisms and problems of membrane and secreted protein biogenesis.

Witmann, H. G., Architecture of prokaryotic ribosomes. *Ann. Rev. Biochem.* 52:35, 1983. Describes the methods that are being used to deduce the three-dimensional structure of the ribosome.

The Genetic Code: Cold Spring Harbor Symposium On Quantitative Biology, Vol. 31. New York: Cold Spring Harbor Press, 1965. A description of the original experiments that defined the genetic code.

PROBLEMS

1. Compare the template translation initiation signals in prokaryotic and eukaryotic systems, and describe the features of mRNAs in each case that determine the frequency with which a particular message is translated. What consequences do these differences have for the mechanism of translation initiation and gene organization in the two systems?

2. Both prokaryotes and eukaryotes contain two types of methionyl-tRNA, one of which is used exclusively for translation initiation. Suggest a possible explanation for the requirement for these different methionyl-tRNAs.

3. The antibiotic viomycin, an inhibitor of protein synthesis, specifically affects translocation. When viomycin is added to endogenous bacterial polysomes engaged in polypeptide synthesis *in vitro*, chain elongation is rapidly curtailed. What effect would you expect to result

from addition of puromycin to viomycin-treated polysomes?

4. Protein factors play an important role in translation, cycling on and off the ribosome during initiation, elongation, and termination. Suggest a possible advantage afforded the system by the use of such factors.

5. Transfer RNA molecules are rather large considering the fact that the anticodon is a trinucleotide. Suggest probable reasons why this is the case.

6. Approximately how much energy is required to synthesize a single peptide bond in protein synthesis? How does this compare with the free energy of formation of the peptide linkage, which is about 5 kcal/mole? Explain qualitatively in thermodynamic terms why there is such a huge discrepancy.

7. Bromouracil is a base replacement mutagen, while acridine causes single base additions or deletions. Mutations caused by acridine frequently can be compensated for by secondary mutations several nucleotides distant from the first mutation, while compensation for those caused by bromouracil usually requires changes within the same codon. The difference in the requirement for recovery from mutations induced by the two agents derives from a fundamental feature of the genetic code. Explain.

8. Assume you have a random copolymer of A and U containing equimolar amounts of the two bases. When the copolymer is used as an mRNA, which amino acids would be incorporated and in what ratios?

9. a. Assuming that translation begins at the first codon, deduce the amino acid sequence of the polypeptide encoded by the following mRNA template:

    AUGGUCGAAAUUCGGGACACCCAUUU

    GAAGAAACAGAUAGCUUUCUAGUAA

   b. Although this polypeptide does not contain serine, a mutation involving a single base substitution alters the composition such that it contains a single serine residue. How else might the composition of the mutant protein be expected to differ from the wild type?

10. The following abstract appeared in a paper by K. Itakura et al., *Science* 198:1056, 1977:

    Abstract. A gene for somatostatin, a mammalian peptide (14 amino acid residues) hormone, was synthesized by chemical methods. This gene was fused to the *E. coli* β-galactosidase gene on the plasmid pBR322. Transformation of *E. coli* with the chimeric plasmid DNA led to the synthesis of a polypeptide including the sequence of amino acids corresponding to somatostatin. *In vitro*, active somatostatin was specifically cleaved from the large chimeric protein by treatment with cyanogen bromide. This represents the first synthesis of a functional polypeptide product from a gene of chemically synthesized origin.

    The synthetic gene in question has the following sequence:

    5'-AATTCATGGCTGGTTGTAAGAACTTCTTTTGG

    AAGACTTTCACTTCGTGTTGATAG-'3

    3'-GTACCGACCAACATTCTTGAAGAAAACCTTC

    TGAAAGTGAAGCACAACTATCCTAG-'5

    a. Given this information, deduce the C- and N-terminal amino acid residues of active somatostatin.
    b. What advantages are afforded in terms of increased efficiency of somatostatin synthesis by the construction of such a synthetic gene fusion?

11. The effect of single-point mutations on the amino acid sequence of a protein can provide precise identification of the codon used to specify a particular residue. Assuming a single base change for each step, deduce the wild-type codon in each of the following instances.
    a. Gln ⟶ Arg ⟶ Trp
    b. Glu ⟶ Lys ⟶ Ile
    c.   Leu
          ↙   ↓   ↘
        Ser   Val   Met
    d.   Thr
          ↙   ↓   ↘
        Ile   Pro   Lys

12. In the chapter it is stated that 32 anticodons are the minimum required to decode all 61 codons. Explain.

13. Bearing in mind the rules known to govern the specificity of codon-anticodon interaction and the need for fidelity in translation, predict the anticodons used to specify the following amino acids.
    a. Serine
    b. Tryptophan
    c. Isoleucine
    Indicate the corresponding codons in each case.

14. a. The amino acid composition of a functional protein may in many instances differ from that prescribed by its mRNA. Explain.
    b. Describe an example of a protein in which such differences have been shown to be of functional significance.

15. In a hypothetical test of expression it is discovered that *E. coli* transformed with the construction described in Problem 10 yield far less of the somatostatin product than expected. Suggest a reason for the discrepancy, and outline a possible strategy for increasing the yield of this peptide.

16. a. In the context of protein metabolism, what are PEST sequences? Would you expect a regulatory protein to be more or less likely to contain such sequences than a structural protein? Explain.
    b. Given what is known of the pathways involved in protein degradation in eukaryotes, suggest a possible mechanistic basis for the observation that, while normal hemoglobin is very stable, the same protein containing the valine analog aminochlorobutyrate is very unstable.

17. E-64 is a mold product that selectively inactivates several lysosomal proteases by covalent modification of their active sites. How would you expect this product to affect the various proteolytic processes occurring in the liver cells of normal and fasted animals?

29

REGULATION OF GENE EXPRESSION IN PROKARYOTES

We turn our attention now from the basic biochemical mechanisms for expression of genes to the way in which gene expression is regulated. We have seen that the DNA sequence carries signals for the starting and stopping of transcription (see Chapter 27). Similarly, the mRNA sequence carries signals for the binding of ribosomes as well as the initiation and termination of polypeptide chains. In addition to these signal sequences, the DNA and sometimes the mRNA carry signal sequences for the binding of regulatory proteins that modulate the level of synthesis of RNA or protein. Such nucleic acid-protein complexes are a major means for regulating gene expression. Another way in which gene expression is regulated involves small-molecule effectors that interact with the regulatory proteins or the RNA polymerase.

Our discussion of these regulatory mechanisms is divided into two chapters, one on prokaryotes and one on eukaryotes. This most fundamental dividing point between life forms is also a useful dividing point for the discussion of gene regulatory mechanisms. In this chapter our discussion will center on the bacterium *E. coli* and the bacterial virus λ. These two systems are particularly well understood and exemplify a wide variety of regulatory mechanisms.

REGULATION OF GENE EXPRESSION IN *E. COLI*

E. coli contains a single chromosome with about 4.4×10^6 base pairs; this is sufficient to code for about 4400 genes. The system is regulated so that under conditions of active growth, only about 5 per cent of the genome is being actively transcribed at any given time. The remainder of the genome is either silent or being transcribed at a very low rate. When growth conditions change, some active genes are turned off and some inactive genes are turned on. The cell always retains its totipotency, so that within a short time (seconds to minutes in most cases), and given appropriate circumstances, any gene can be fully turned on.

The level of transcription for any particular gene usually results from a complex series of control elements organized into a hierarchy that coordinates all the metabolic activities of the cell. For example, when the rRNA genes are highly active, so are the genes for ribosomal proteins, and the latter are regulated in such a way that stoichiometric amounts of most of the ribosomal proteins are produced. When glucose is abundant, most genes involved in processing more complex carbon sources are turned off in a process called catabolite repression. If the glucose supply is depleted and lactose is present, then the genes involved in lactose catabolism are expressed. In *E. coli*, the production of most RNAs and proteins is regulated exclusively at the transcriptional level, although there are notable exceptions. Rapid response to changing conditions is ensured partly by a short mRNA lifetime— on the order to 1 to 3 min for most mRNAs. Some mRNAs have appreciably longer lifetimes (10 min or longer) and the consequent potential for much higher levels of protein synthesis per mRNA molecule. These atypical mRNAs, at least in some instances, also may be subject to translational control.

Initiation Point Is a Major Site for Regulating Gene Expression

The most commonly known way of regulating gene expression in bacteria involves controlling the rate of initiation of transcription. The basic mechanics of the transcription process were described in detail in Chapter 27. For purposes of this discussion, it is sufficient to remember that prior to initiation of transcription, the RNA polymerase holoenzyme becomes attached to a 35- to 40-nucleotide segment of the DNA called the promoter. *The affinity between DNA and polymerase is controlled by a sequence of bases in the DNA.* Two main areas of contact have been recognized, one centered in the −35 region of the promoter, with a favored sequence of TTGACA, and one centered in the −10 region, with the favored sequence TATAAT. *These two hexanucleotide regions are referred to as the polymerase binding site 1 (PBS1) and the polymerase binding site 2 (PBS2), respectively.* The favored sequences, or *consensus sequences*, as they are more commonly called, were identified by comparing the sequences found in about 300 different promoters. Each base in the consensus sequence represents the base most commonly found in that position. Naturally occurring promoters have never been found with the consensus sequences at both sites. However, artificially constructed promoters bearing the consensus sequence have been shown to be highly active in transcription. Furthermore, the activity of naturally occurring promoters is frequently proportional to the concordance between their promoter sequences and the consensus sequence. Future research may show that other regions are also important in promoter-RNA polymerase interaction.

From the time it first makes contact with the promoter to the time that it has achieved the proper orientation for initiation, the promoter-polymerase complex may go through several metastable states. The final conformation of this complex, adopted immediately before initiation, is referred to as the *rapid-start complex* or *open-promoter complex*. In this state, the polymerase is in contact with PBS1 and PBS2, and about 11 base pairs from the −9 to the +1 positions at the origin of transcription are unpaired (Figure 29–1). Once the rapid-start complex has been formed, initiation of RNA synthesis is rapid, taking only a fraction of a second in the presence of ribonucleotide triphosphates.

The rate of initiation of transcription can be regulated in several ways, all of which influence the rate of formation of the rapid-start complex. The

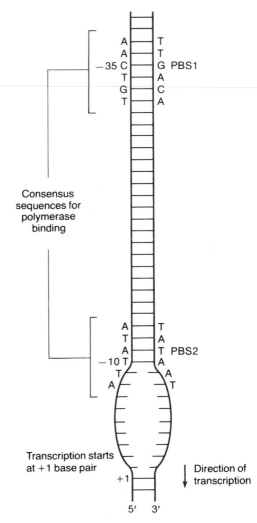

Figure 29–1
Schematic diagram of DNA conformation in the rapid-start complex. Two regions, PBS1 and PBS2, most important in polymerase binding, are lettered with most favored sequences. Transcription starts at the +1 base pair.

primary sequence of nucleotides in the promoter region is the first factor that should be considered. The closer this sequence is to the consensus sequence, the greater will be the affinity of the polymerase for the promoter. For some promoters, *negative supercoiling of the DNA* serves as an appreciable stimulus to transcription. This probably results from the fact that negative supercoiling facilitates unwinding and unpairing of the double helix (see Chapter 7), such as is observed in the -9 to $+1$ region of the promoter in the rapid-start complex.*

The rate of initiation of transcription also can be altered by changes in the RNA polymerase structure (see Chapter 27). This structure can be altered by *subunit replacement, subunit covalent modification, or small-molecule-induced allosteric transition*. During a temperature upshift ($30 \rightarrow 42°C$), the usual σ subunit (σ^{70}) is partially replaced by an alternative σ factor (σ^{32}), producing a change in the types of promoters recognized by the polymerase. In bacteriophage T4 infection, the subunits of the polymerase become ribose-adenylated, lowering the affinity of polymerase for bacterial promoters and raising the affinity for phage promoters. Although not finally proved, it is believed that binding of guanosine tetraphosphate (ppGpp) to RNA polymerase changes the structure of the polymerase so that it has a greatly lowered affinity for rRNA, tRNA, and ribosomal protein promoters and at the same time a somewhat greater affinity for some other promoters.

Finally, the rate of initiation of RNA synthesis can be controlled by auxiliary regulatory proteins that affect the rate of formation of the rapid-start complex in either a positive or a negative way; such regulatory proteins are known as *activators* or *repressors*, respectively. Each promoter has a natural affinity for RNA polymerase that is determined by its promoter sequence. Activators increase this affinity, thereby increasing the rate of transcription. Repressors have the opposite effect. The *lac* repressor inhibits polymerase binding to the *lac* operon promoter. The CAP protein is an activator that stimulates polymerase binding to the same promoter. Strictly speaking, CAP should be referred to as an *apoactivator* rather than an activator because it promotes transcription only when it is complexed to the small molecule *coactivator* 3′5′-cyclic AMP (cAMP). The ways in which CAP and the *lac* repressor function are described in detail below.

The *lac* Operon Is a Cluster of Three Genes Regulated As a Unit

The stretch of chromosome known as the *lac* operon contains three genes associated with the metabolism of the disaccharide lactose. The story of its characterization reveals one of the clearest and best-understood pictures of the molecular mechanism of gene regulation. The historical presentation used here is intended to give a sense of the close interplay between genetics and biochemistry and of how important various techniques have been to progress in this field.

Significant events in elucidation of the concept and role of the *lac* operon have spanned a period of almost 40 years. The principal reason for the relatively slow progress was the unavailability of genetic and biochemical skills that were needed to solve the problem. Improvement in technical skills and advances in our understanding of the *lac* operon have gone hand in hand. The development of various genetic techniques, coupled with elucidation of the basic mechanisms of DNA, RNA, and protein synthesis, has

---

*Negative supercoiling does not always stimulate transcription. Some genes are unaffected by the extent of supercoiling. The gene for DNA gyrase is actually much more active when the DNA is relaxed. This is a very interesting finding, since DNA gyrase is the enzyme that catalyzes negative supercoiling. This result indicates that relaxed DNA serves as a signal for the synthesis of DNA gyrase.

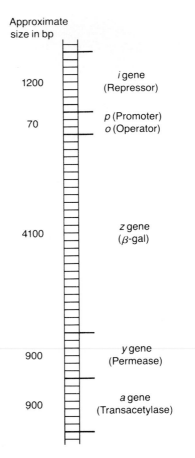

Approximate
size in bp

1200 — *i* gene
(Repressor)

70 — *p* (Promoter)
o (Operator)

4100 — *z* gene
(β-gal)

900 — *y* gene
(Permease)

900 — *a* gene
(Transacetylase)

Figure 29–2
Different genetic elements of the *lac* operon. The operon contains a control region, the promoter-operator region, and three structural genes, z, y, and a. The *i* gene, a repressor, is also shown. It is not part of the operon, but it is located at an adjacent site on the genome.

been essential to progress. More recently, cell-free synthesis techniques, use of restriction enzymes for isolation and cloning of small discrete segments of DNA, and methods for determining nucleotide sequences have played important roles in advancing our knowledge of the *lac* operon.

The *lac* DNA is a region of the *E. coli* chromosome with a molecular weight of about 4×10^6, constituting about 0.1 per cent of the chromosome. The DNA is separated into two functional portions: the controlling elements of the operon, and the structural genes, which code for the three proteins specified by the *lac* operon: β-galactosidase, lactose permease, and thiogalactoside transacetylase (Figure 29–2).

β-Galactosidase (z gene) hydrolyzes β-galactosides, in particular lactose, to produce the monosaccharides glucose and galactose. The permease protein (y gene) is associated with the β-galactoside active transport system. High concentrations of permease lead to the concentration of galactosides by as much as 100-fold over their concentration in the external medium. The *a* gene codes for thiogalactoside transacetylase. Unlike the situation with the other structural genes of the operon, defective *a* genes do not affect the ability of the bacteria to grow on lactose. Transacetylase is known to catalyze the transfer of an acetyl group from acetyl-CoA to a thiogalactoside to form an acetyl-thiogalactoside. Surprisingly, the physiological role of this enzyme is not clear despite all the work that has been done on the *lac* operon.

The elements controlling the operon consist of a promoter locus p, an operator locus o, and a repressor gene i. These will be introduced here and discussed in greater detail later. The operator is the site on the chromosome where the *lac* repressor binds. The promoter contains a site for RNA polymerase binding and an adjacent site for binding the apoactivator protein CAP. The repressor gene i is located near the operon and encodes the lac repressor protein.

Enzyme Induction. Wild-type *E. coli* cells grown in the absence of a galactoside contain an average of 0.5 to 5.0 molecules of β-galactosidase per cell, whereas bacteria grown in the presence of an excess of lactose or certain lactose analogs contain 1000 to 10,000 molecules per cell. Radioactive amino acid has been used as a tracer to show that the increase in enzyme activity observed on induction results from *de novo* protein synthesis. When excess β-galactoside inducer is added, enzyme activity increases at a rate proportional to the increase in total protein within the culture (Figure 29–3). At 37°C, enzyme formation reaches its maximum rate within 3 min after inducer is added. After the inducer is removed, enzyme synthesis ceases in about the same amount of time.

A large number of compounds have been tested for their capacity to induce β-galactosidase. All inducers contain an intact, unsubstituted galac-

Figure 29–3
Effect of inducer on β-galactosidase synthesis. Differential plot expressing accumulation of β-galactosidase as a function of increase in mass of cells in a growing culture of *E. coli*. Since the abscissa and ordinate are expressed in the same units (micrograms of protein), the slope of the straight line gives galactosidase as the fraction *(P)* of total protein synthesized in the presence of inducer. (After Melvin Cohn, 1957.)

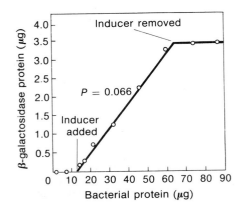

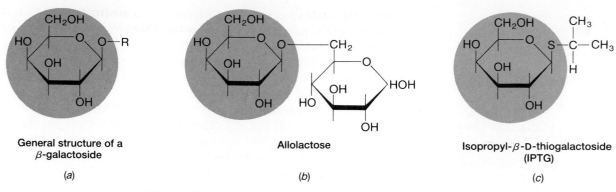

General structure of a
β-galactoside

(a)

Allolactose

(b)

Isopropyl-β-D-thiogalactoside
(IPTG)

(c)

Figure 29–4
Inducers of the *lac* operon. (a) All inducers have the β-galactoside structure shown, in which R can be a variety of substituents. (b) Allolactose is the natural inducer when cells are grown on lactose. (c) Isopropyl-β-D-thiogalactoside (IPTG) is a synthetic inducer useful in the laboratory; the β oxygen is replaced by a β sulfur atom. This prevents hydrolysis by β-galactosidase. For this reason IPTG is called a gratuitous inducer.

tosidic residue (Figure 29–4). Many compounds that are not themselves substrates for β-galactosidase, such as thiogalactosides, are good inducers (called gratuitous inducers). No correlation exists between affinity for β-galactosidase and the capacity to induce. Lactose, the natural substrate of the operon, is not the inducer *in vivo*. Rather, *allolactose*, which is formed as an intermediate in lactose metabolism in the presence of the very limited amount of β-galactosidase that occurs in the uninduced cells, is believed to be the natural inducer (Figure 29–5). The three proteins of the *lac* operon are *coordinately induced*, i.e., they are induced to the same extent by the same inducer. These results suggest that *the receptor molecule for inducer is distinct from the structural components of the operon* and that *there is one site where inducer acts*.

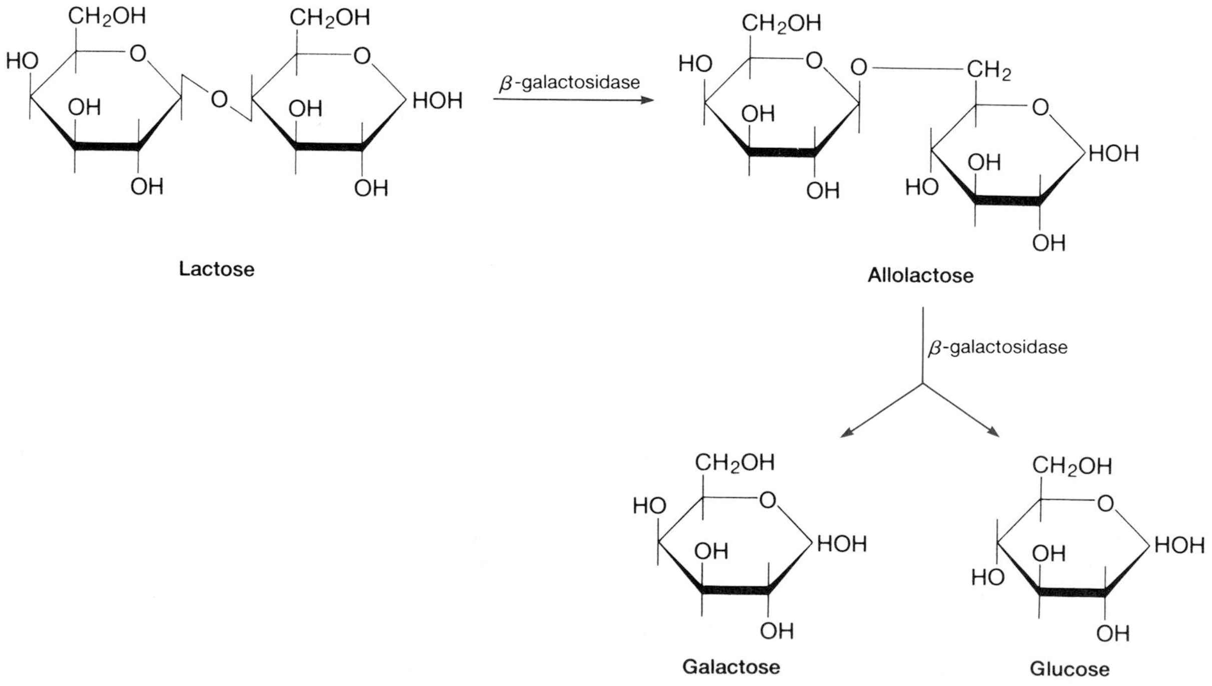

Lactose

β-galactosidase →

Allolactose

β-galactosidase

Galactose

Glucose

Figure 29–5
Conversion of lactose to allolactose, the natural inducer of the *lac* operon. Ultimately lactose is broken down to its constituent monosaccharides, galactose and glucose.

Discovery of the Repressor Gene.   Two distinct types of mutations have been observed in the lactose system. One class of mutations includes _structural gene mutations_: (1) β-galactosidase mutations ($z^+ \rightarrow z^-$), expressed as the loss of the capacity to synthesize active β-galactosidase; (2) permease mutations ($y^+ \rightarrow y^-$), expressed as the loss of the capacity to concentrate lactose; and (3) transacetylase mutations ($a^+ \rightarrow a^-$), expressed as the loss of the capacity to form thiogalactoside transacetylase. The other class of mutations involves _controlling elements_ of the operon, such as i gene mutations ($i^+ \rightarrow i^-$), expressed as the capacity to synthesize large amounts of β-galactosidase in the absence of inducer. This type of i^- mutation is called a _constitutive mutation_. Structural gene mutations usually affect only the enzyme in whose gene the alteration occurs.* In contrast, _constitutive mutations in the i gene invariably affect the amounts of β-galactosidase, permease, and transacetylase synthesized, but not their structures_ (at this point you may wish to refer to the review of genetic notation and concepts in Box 29–A).

Some of the most informative early studies were performed with partial diploids (_merodiploids_) that contain the relevant genes on both the cellular chromosome and an F-factor plasmid. Merodiploids of the type $z^+y^-a^-/$ $Fz^-y^+a^+$ or $z^-y^+a^+/Fz^+y^-a^-$ are wild-type, that is, they behave normally. They metabolize lactose and form normal amounts of both β-galactosidase and transacetylase. Furthermore, it does not matter which _lac_ operon is on

---

* This is strictly true only when synthetic inducer is used. As explained earlier, a small amount of β-galactosidase is required to convert lactose to allolactose, the natural inducer.

BOX 29–A

Genetic Concepts and Genetic Notation

Much of the early work on the _lac_ operon was purely genetic. It is essential that certain aspects of genetics be understood. The information in this box should be adequate for those with no prior exposure to genetics except for what they have already encountered in this text.

Genes are specified by one or more small letters in italic. Thus z indicates the gene for β-galactosidase and _lac_ indicates the operon. Frequently a superscript is appended to the genetic symbol. The two most common superscripts are $+$, indicating a normal (wild-type) gene, and $-$, indicating a nonfunctioning (mutant) gene. Different representations of the same gene are referred to as _alleles_. Thus z^+ and z^- are both alleles of the z gene.

Cells that carry a single copy of each gene are referred to as _haploids_. Cells that carry two copies of each gene are referred to as _diploids_. Bacteria are haploid cells, as they carry a single chromosome with a unique representation for each gene. Bacterial cells that are partial diploids (_merodiploids_) may occur naturally or they may be selected for by genetic techniques.

A favored method for constructing merodiploids is to infect the bacterial cell with a virus or a plasmid DNA that carries the extra genes of interest. The F plasmid is commonly used for this purpose. The genetic representation for a cell carrying the _lac_ operon on the chromosome and the F plasmid would be $z^+y^+a^+//Fz^+y^+a^+$, where the

diagonal lines represent the host chromosome to the left and the plasmid chromosome to the right. As a rule, only one diagonal line is used to separate the genetic symbols. Also, for convenience, if all the alleles for a given gene are wild type they may not be shown. Thus $z^+y^+a^+/$ $Fz^-y^+a^+$ and z^+/Fz^- may be taken as representations of the same genetic state in cases where it is known that the _lac_ operon is present on both the host and the plasmid chromosome.

Two genetic elements located on the same chromosome are said to be in the _cis_ orientation. Two genetic elements located on different chromosomes in the same cell are said to be in the _trans_ orientation. In the merodiploid $z^-y^-a^+/Fz^+y^+a^+$, the two mutant genes are in the _cis_ orientation. In the merodiploid $z^-y^+a^+/$ $Fz^+y^-a^+$, they are in the _trans_ orientation.

A major reason for using merodiploids is to study the interaction between different alleles of the same gene. This often tells us a great deal about how a gene or the gene product functions. The two simplest types of interactions which may be described are _dominant and recessive_. A cell that is z^+/Fz^- behaves like a z^+ cell as far as the metabolism of β-galactosidase is concerned. Therefore the z^+ allele is dominant to the z^- allele, or conversely, the z^- allele is recessive to the z^+ allele.

TABLE 29–1
Synthesis of β-Galactosidase and Galactoside Transacetylase by Haploids and Partial Diploids of *E. coli* Regulator Mutants

Strain No.	Genotype	β-Galactosidase		Galactoside-transacetylase	
		Noninduced	Induced	Noninduced	Induced
1	$i^+z^+y^+$	<0·1	100	<1	100
2	$i_6^-z^+y^+$	100	100	90	90
3	$i_3^-z^+y^+$	140	130	130	120
4	$i^+z_1^-y^+/Fi_3^-z^+y^+$	<1	240	1	270
5	$i_3^-z_1^-y^+/Fi^+z^+y_U^-$	<1	280	<1	120
6	$i_3z_1^-y^+/Fi^-z^+y^+$	195	190	200	180
7	$\Delta_{izy}/Fi^-z^+y^+$	130	150	150	170
8	$i^sz^+y^+$	<0·1	<1	<1	<1
9	$i^sz^+y^+/Fi^+z^+y^+$	<0·1	2	<1	3

Bacteria are grown in glycerol as carbon source and induced, when stated, by isopropyl-thiogalactoside (IPTG), 10^{-4} M. Values are given as a percentage of those observed with induced wild-type. Δ_{izy} refers to a deletion of the whole *lac* region. It should be noted that organisms carrying the wild allele of one of the structural genes (z or y) on the F factor form more of the corresponding enzyme than the haploid. This is presumably due to the fact that several copies of the F*lac* plasmid are present per cell. In i^+/i^- heterozygotes, values observed with uninduced cells are sometimes higher than in the haploid control. This is due to the presence of a significant fraction of i^-/i^- homozygous recombinants in the population. Subscripts in column labeled genotype refer to different mutants. (Data obtained from J. Monod.)

the *E. coli* chromosome and which is on the F-factor chromosome. This efficient complementation between structural gene mutations indicates that they belong to independent genes or cistrons. The most significant feature of i^- mutations is that they simultaneously affect all three gene-product proteins, each independently determined by its different structural genes, z, y, or a.

The study of merodiploids of the types i^+z^-/Fi^-z^+ or i^-a^+/Fi^+a^- demonstrated that the i^+ inducible allele is dominant to the i^- constitutive allele, and that it is active on the *same chromosome (cis)*, or on a *different chromosome (trans)* with respect to both a^+ and z^+ (Table 29–1). The fact that it is effective in the *trans* position as well as the *cis* position shows that *i* gene mutations belong to an independent gene, governing the expression of the z, y, and a genes through production of a diffusible cytoplasmic component. *The dominance of the inducible to the constitutive allele suggests that the former corresponds to the active form of the i gene.*

Further understanding of *i* gene function has come from study of a mutant gene designated i^s. This mutant has lost its capacity to synthesize all structural gene products of the *lac* operon. In merodiploids of the constitution i^s/i^+, the i^s is dominant, that is, the merodiploids cannot synthesize either β-galactosidase or transacetylase even when inducer is present (Table 29–1). The most reasonable explanation for the i^s mutant is that it is an allele of *i* in which the structure of the repressor is changed so that it can no longer be antagonized by the inducer.

Discovery of Operator Mutations. In the *lac* system, rare dominant constitutive mutants (o^c) were isolated by selecting for constitutivity in cells diploid for the *lac* region, including the *i* gene, thus virtually eliminating the much more frequently occurring recessive (i^-) constitutive mutants. (Since two copies of *i* gene are present in such cells, both *i* genes would have to mutate simultaneously to i^- to give constitutive behavior.) If the probability of an $i^+ \to i^-$ mutation is 10^{-6}, the probability of two such simultaneous events in the same cell would be 10^{-12}. By recombination studies, the o^c mutations were mapped in the *lac* region between the i and z loci, generating the gene order shown in Figure 29–2. Genetic and biochemical evidence has shown that the o locus is adjacent to but distinct from the z gene. Thus constitutive o mutations affect the quantity of β-galactosidase synthesized but not its structure.

In merodiploids of the type o^c/o^+, β-galactosidase and transacetylase are constitutively synthesized, showing that the o^c mutation is dominant to o^+. In merodiploids of the type $o^c z^+/o^+ z^-$, the o^c mutation is dominant; but in $o^c z^-/o^+ z^+$, it is recessive for galactosidase expression. Thus the o^c mutation is dominant only in the *cis* position, i.e., when it is adjacent to a wild-type z^+ gene. From this evidence, Jacob and Monod inferred that the $o^+ \to o^c$ mutations correspond to a modification of the specific repressor-accepting structure of the operator. This identifies the operator locus, i.e., the genetic segment responsible for the structure of the operator, but not necessarily the operator itself.

The Operon Hypothesis.   The behavior of the various mutations we have just discussed led François Jacob and Jacques Monod to propose a model for the regulation of protein synthesis. The genetic elements of this model consist of a structural gene or genes, a regulator gene, and an operator locus (Figure 29–6).

1. The *structural gene* produces a messenger RNA molecule that serves as the template for protein synthesis.
2. The *regulator gene* (not itself part of the operon) produces a *repressor* that can interact with the operator locus.

Figure 29–6

Schematic model illustrating the operon hypothesis. This diagram is modified from the original proposed by Jacob and Monod, who thought i gene repressor was an RNA rather than a protein. (*a*) The i gene encodes a repressor that binds tightly to the operator o locus, thereby preventing transcription of the mRNA from the z, y, and a structural genes. (*b*) When inducer is present, it combines with repressor, changing its structure so it can no longer bind to the operator locus. Inducer also can remove repressor already complexed with the o locus.

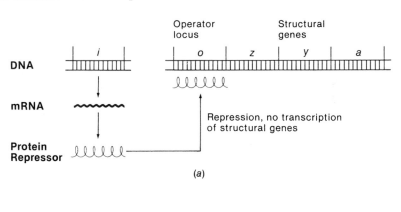

(a)

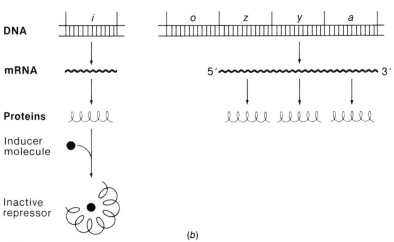

(b)

3. The _operator_ is always adjacent to the structural genes it controls.
4. The _operator_ and its associated structural genes are referred to as the _operon_.

The repressor molecule combines with the operator locus to prevent the structural gene(s) from synthesizing messenger RNA. In induction, the _inducer_ (or _antirepressor_, a more descriptive term) combines with the repressor to prevent its interaction with the operator; this region of the genome is then free to combine with RNA polymerase. The Jacob-Monod operon hypothesis has provided a tremendous stimulus for investigations directed toward understanding not only the _lac_ system, but other genetic regulatory systems as well.

Isolation and Action of the _i_ Gene Repressor.   Genetic studies were most important in arriving at a hypothesis to explain how the repressor worked, but biochemical studies were necessary to demonstrate that the hypothesis was correct. To do the biochemical experiments, it was first necessary to isolate repressor protein.

According to the operon hypothesis, antirepressor is supposed to combine with repressor. In order to isolate repressor, Walter Gilbert and Benno Muller-Hill used ^{14}C-labeled isopropylthiogalactopyranoside (IPTG), one of the strongest known anti-repressors (see Figure 29–4), to monitor repressor purification from a crude cell extract. Unlike the natural inducer allolactose, IPTG not only binds strongly to repressor, but it is completely stable in _E. coli_ cells or crude extracts. This makes it most useful for experimental purposes. A crude cell-free extract was fractionated by standard protein purification procedures, and the fraction containing the repressor that binds IPTG was found (see Box 29–B for details of measuring binding).

With purified repressor available, the Jacob-Monod hypothesis about binding of repressor to operator could be directly tested. The binding of purified repressor to DNA was studied in two ways. First, ^{35}S-labeled repressor was mixed with λ_lac_ DNA (i.e., λ bacteriophage DNA containing the _lac_ operon) and centrifuged through a sucrose gradient for a limited time. In the absence of binding, the repressor, detected by its radioactivity, sediments much more slowly than the DNA. In the absence of inducer, some repressor

---

BOX 29–B

Measuring the Binding of Inducer to Repressor

The theory for equilibrium binding is discussed in Chapter 10. Here we describe a popular technique for measuring binding of a small molecule to a protein.

Binding of inducer to repressor in any particular repressor-containing extract can be measured by equilibrium dialysis. In this procedure, the extract is placed inside a semipermeable dialysis bag that is impermeable to protein but permeable to IPTG. The extract is allowed to equilibrate with the external solution by gentle agitation in the presence of a buffer containing radioactive inducer. At the end of the dialysis, a certain amount of free inducer (I_f) should be detectable outside the bag. The inducer inside the bag should exceed that outside by the amount that is bound to the protein (I_b).

The strength of binding of repressor to IPTG can also be determined by equilibrium dialysis. When the ratio I_b:I_f is plotted versus I_b, the slope should equal the negative of the formation constant for binding, K_f, where K_f is defined by the equation

$$K_f = \frac{\text{repressor} - \text{IPTG}}{(\text{repressor})(\text{IPTG})}$$

The formation constant for this complex, using wild-type repressor, is approximately $10^6\ M^{-1}$, or 10^6 moles^{-1} liters; the K_f for allolactose binding to repressor is about 100 times higher. Variants of wild-type repressor purified to the same degree have also been examined. The mutant repressor i^t, which was isolated from strains that are easier to induce, has about double this affinity. The repressor encoded by i^s, isolated from noninducible strains, shows no affinity for IPTG, as predicted (see main text).

comigrates with the DNA. Because the DNA is so much larger than the repressor, the position of the DNA in the sedimentation pattern is only slightly affected by the binding of repressor. Proof that this binding actually represents specific binding of repressor to the *lac* operator was provided by several experiments: (1) the binding was eliminated in the presence of $1.2 \times 10^{-4} M$ IPTG; (2) the binding of repressor was greatly reduced if DNA containing an o^c mutant was used; and (3) binding was not observed for wild-type λ DNA.

A faster and more versatile method, known as the membrane-filter binding technique, was also used to demonstrate the binding of repressor to DNA. This technique is based on the fact that most proteins bind to nitrocellulose filter membranes, but duplex DNA does not. However, DNA complexed to protein does bind to the membrane filter. With this technique, radioactively labeled DNA is used in conjunction with unlabeled purified repressor. A solution containing the labeled DNA, with or without added repressor, is passed through the membrane filter with the help of mild suction. Only when repressor is present is the label retained by the filter. Other variables were tested, and the same conclusions were reached as when the centrifugation technique was used.

The operator binding site for repressor was further characterized by sequence analysis. This analysis revealed a region of about 36 bases, extending from about −8 to +28, that shows extensive symmetry (Figure 29–7). Twenty-eight of the 36 bases in this region are arranged so that if the double helix is rotated 180°, the same sequence is observed. This type of symmetry in DNA is referred to as *dyad symmetry*. The *lac* repressor is composed of an even number of identical subunits (four) and most proteins so composed also contain axes of dyad symmetry. As a result, the repressor binding site probably consists of two symmetrically disposed binding sites on the DNA surface that simultaneously bind to complementary sites on the protein. This symmetrical property is common to many but not all DNA-regulatory protein interactions.

Several pieces of evidence add support to the notion that this region includes the repressor binding site. (1) All eight o^c mutations thus far located involve base replacements near the center of this region. (2) Repressor binding substantially decreases the reactivity toward dimethylsulfate of several purine bases in this region (Figure 29–7). Dimethylsulfate reacts with the N-7 position on guanine and the N-3 position on adenine in the double helix (see Chapter 7 for a discussion of these chemical reactions). Finally, if all the thymines in the DNA are replaced by bromouracil, a number of the bromouracil bases become cross-linked to the repressor in the presence of ultraviolet light. Taking into account the normal twist of the double helix, all the groups shown by chemical methods to be in the vicinity of the repressor

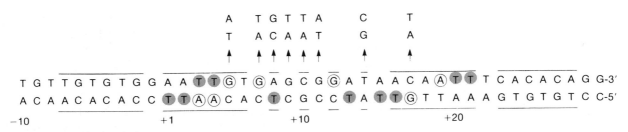

Figure 29–7

The operator locus (the presumptive repressor binding site). Bases are numbered +1 for the first base transcribed and −1 for the base before that. Regions showing dyad symmetry are underlined and overlined. Arrows indicate point mutations leading to the constitutive phenotype (o^c). Circled bases are those groups that are strongly protected against reaction with dimethylsulfoxide when lac repressor is bound. Shaded circles indicate those groups that become cross-linked to repressor in the presence of ultraviolet light when thymine in the DNA is replaced by 5-bromouracil.

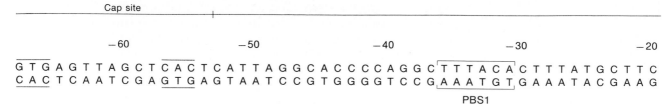

Cap site

```
        −60             −50             −40             −30             −20
G T G A G T T A G C T C A C T C A T T A G G C A C C C C A G G C T T T A C A C T T T A T G C T T C
C A C T C A A T C G A G T G A G T A A T C C G T G G G G T C C G A A A T G T G A A A T A C G A A G
                                                              PBS1
```

Figure 29–8

DNA sequence in the *lac* promoter-operator region. Bases are numbered +1 for the first base transcribed and −1 for the base before that. CAP and repressor-binding sites are indicated. Regions within these two sites showing dyad symmetry are underlined and overlined. Two regions, PBS1 and PBS2, where polymerase binds most strongly in a number of promoters, are bracketed in color. Note that PBS1 has five of six bases in the idealized sequence (TT-ACA), whereas PBS2 has four out of six (TAT--T). Arrows indicate replacements in the mutant *UV5* promoter that lead to high-level expression in the absence of cAMP-CAP.

are situated on one side of the double helix. This strongly supports the notion that *repressor binds on one side of the double helix, over the 36-base region covered by the symmetry axis*.

The RNA polymerase and repressor binding sites (promoter and operator) overlap (Figure 29–8), so that the binding of one protein would preclude the binding of the other. Whereas there are other DNA sequences where the repressor might bind to prevent transcription, *most repressors bind so they overlap the promoter region, thereby preventing the initial formation of the RNA polymerase-DNA complex*.

Catabolite Repression and the Catabolite Activator Protein CAP. In the absence of active repressor or when the repressor is combined with inducer, RNA polymerase can bind to the *lac* operon and transcribe *lac* mRNA. The extent to which it does this depends on other factors in the growth medium, particularly on the presence of sugars other than lactose. For example, when *E. coli* cells are grown in the presence of glucose and lactose as the sole carbon sources, the cells will first utilize glucose and afterwards utilize the lactose. This phenomenon is referred to as *diauxic growth*. Careful analysis of cells growing under such conditions has shown that during the glucose phase of growth, the *lac* operon is poorly expressed and induction of the operon must occur before the lactose can be utilized. Repression of *lac* operon expression by glucose or its closely related derivatives, such as glucose-6-phosphate or fructose, is known by the general term *catabolite repression* (or glucose repression). Enzymes of the glycolytic pathway that directly utilize glucose are always present in the cell. The enzymes of the *lac* operon are not needed as long as this more directly utilizable carbon source is available. Many genes in *E. coli* associated with catabolism are subject to this type of repression for the same reason.

A turning point in our understanding of catabolite repression was provided by the finding of Sutherland that the level of intracellular cAMP (Figure 29–9) in *E. coli* is drastically lowered in the presence of glucose, from about 10^{-4} to 10^{-7} M. Subsequently it was shown by I. Pastan and R. Perlman that large quantities of cAMP in the growth medium could partially reverse the glucose catabolite repression effect on the *lac* operon. *The fact that cAMP has a stimulatory effect on the expression of genes subject to catabolite repression* was demonstrated directly in a cell-free system in Zubay's laboratory by Donald Chambers. The system contained λ*lac* DNA, a cytoplasmic extract of *E. coli*, and the substrates and other low-molecular-weight components necessary for transcription of the DNA and translation of the resulting mRNA. In this system it was shown that 10^{-3} M cAMP stimulated β-galactosidase synthesis as much as 30-fold.

Subsequently, mutants of *E. coli* that were permanently catabolite-

cAMP

Figure 29–9

The structure of 3′,5′-cyclic AMP (cAMP).

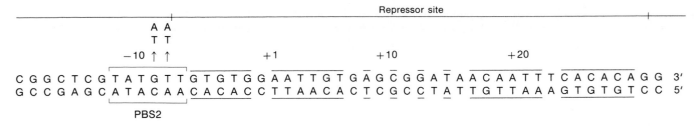

Figure 29–8 (Continued)

repressed were isolated by Beckwith and his colleagues. One of the observed properties (phenotype) of such mutants was that they were incapable of expressing a number of genes that required cAMP for activation. These mutants fell into two categories, those that could be phenotypically corrected by growing in the presence of 5×10^{-3} M cAMP and those that could not. The first class of mutants was defective in the synthesis of cAMP, and the latter class of mutants was defective in the protein(s) with which cAMP interacts to bring about stimulation of β-galactosidase synthesis (as well as that of other catabolite-repressed proteins). Further work showed that these two types of mutants originate in the genes now referred to as *cya* and *crp*, respectively. When the cell-free system was prepared from extracts of *crp*⁻ cells, it produced a low level of β-galactosidase that was not stimulated by adding cAMP. The defect could be corrected by adding soluble protein from *crp*⁺ cells. By making use of this *in vitro* complementation assay, Zubay and Beckwith and their coworkers were able to fractionate the *crp*⁺ cells and isolate a single protein that was responsible for the activity.

The protein encoded by the *crp* gene, the apoactivator CAP, is a dimer composed of identical subunits with a molecular weight of 22,000. CAP binds to DNA, and this binding is greatly stimulated in the presence of cAMP. The structural alterations produced in CAP by cAMP binding apparently alter its conformation so that it can form a strong complex with DNA. CAP binds preferentially to the *lac* promoter region, although the selectivity for preferential binding to the specific site is substantially less than is the case for *lac* repressor.

It is known from a series of genetic deletions that the site necessary for CAP stimulation of the *lac* operon is in the −50 to −80 base-pair region. Inspection of the sequence in this region reveals a 14-base segment between −53 and −68 that shows dyad symmetry for 12 of the 14 base pairs (Figure 29–10).

As in the case with *lac* repressor, chemical-probe experiments have

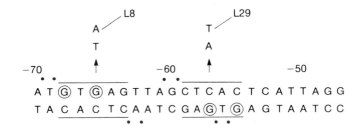

Figure 29–10

CAP-binding region in the *lac* promoter. Regions showing dyad symmetry that are believed to interact strongly with CAP are overlined and underlined. Circled bases are those that are strongly protected from reaction with dimethylsulfate when CAP is bound. Point mutations L8 and L29, which produce a promoter that is not stimulated in transcription by cAMP, are indicated. Colored dots indicate those phosphate positions which, if ethylated by ethylnitrosourea, block CAP binding. If the area of interest exists as a normal DNA double helix when binding CAP, then most of the groups implicated in CAP binding would appear on one side of the DNA.

been carried out to characterize the region binding to CAP. In these experiments, a small DNA segment containing the CAP binding site was used. Dimethylsulfate (DMS), which reacts with the adenine N-3 and the guanine N-7 positions, and ethylnitrosourea (EtNu), which reacts with the phosphates, were used. Experiments with the DMS probe showed that CAP binding protects the two outer guanines on each side of the presumptive binding site from methylation. Conversely, if the DNA is first premethylated, binding of CAP is strongly inhibited. The pattern of protection demonstrates symmetry and interaction in two adjacent, large grooves of the DNA. Experiments with EtNu confirmed the importance of the symmetry and also showed that, like *lac* repressor, *CAP interacts with only one side of the DNA helix, over a region covered by the symmetry axis.*

In the absence of bound repressor, the full sequence of reactions involved in initiation of transcription is summarized by the following set of equations. First, the coactivator cAMP combines with the apoactivator CAP, which then binds in the −60 region of the promoter. This complex stimulates the binding of RNA polymerase to an adjacent site on the promoter. CAP stimulates binding of RNA polymerase by an affinity between CAP and polymerase.

$$\text{cAMP} + \text{CAP} \rightleftharpoons \text{cAMP-CAP}$$

$$\text{cAMP-CAP} + \text{DNA} \rightleftharpoons \text{cAMP-CAP-DNA}$$

$$\text{cAMP-CAP-DNA} + \text{polymerase} \rightleftharpoons \text{cAMP-CAP-DNA-polymerase}$$

The transcription rate of the *lac* operon in the presence of cAMP-CAP is enhanced 20- to 50-fold, and this is directly due to the increased affinity of RNA polymerase holoenzyme for the promoter in the presence of CAP and cAMP.

The overall strategy in creating a promoter sequence responsive to cAMP-CAP activation could be summarized as follows. The nucleotide sequence in the RNA polymerase binding site is adjusted so that polymerase by itself produces a low level of transcription. An adjacent site for binding CAP is created so that when CAP is bound, the additional affinity contributed by favorable contacts between the CAP and polymerase converts this sequence into a high-level promoter. A mutant *(UV5)* containing two base replacements in the *lac* promoter produces high-level transcription in the absence of cAMP-CAP. These base replacements in the PBS2 site (see Figure 29–8) bring the promoter sequence closer to the consensus sequence and therefore should lead to a stronger polymerase binding site in the absence of the cAMP-CAP complex, as is in fact observed.

Gene Regulatory Proteins Usually Bind First to the DNA

Gene regulatory proteins, like CAP and the *lac* repressor, are produced in limited quantities and they bind very selectively to specific sites on the DNA. Only a few regulatory proteins have been characterized well enough to comment on their structures in detail. These seem to have certain common structural features that permit them to interact with specific sequences along the DNA duplex.

Two of the most thoroughly examined proteins are CAP and cro. CAP has already been discussed. Cro is a regulatory protein of λ bacteriophage that we will have more to say about shortly. Each of these regulatory proteins is a dimer with identical subunits. Each contains two protruding segments of α helix, which are believed to bind to adjacent large grooves on one side of the DNA. A more detailed description of the structure of regulatory proteins that bind to DNA and the way in which they bind to DNA is presented in Box 29–C.

Another common feature shared by many but not all regulatory proteins is their sensitivity to small-molecule effectors. For example, the *lac* repressor adopts two conformations, one that is favored in the absence of small-molecule effectors like allolactose, and one that is favored when binding to allolactose. The former has a high affinity for the *lac* operator, the latter does not. A similar situation exists for CAP, except that in this case the role of the small-molecule effector is reversed. That is, the binding of cAMP converts the protein from a structure with a low affinity to one with a high affinity for a specific site on the DNA. The precise way in which the regulatory protein structure is altered by the small-molecule effector is not known, but the analogy to allosteric enzymes is very compelling (see Chapter 10).

Many Genes Important in Catabolism Are Stimulated by CAP

The cAMP-CAP activator is an essential component for regulating a wide variety of genes involved in catabolism. This complex usually functions in series with a gene-specific regulator so that any particular gene subject to *cAMP-CAP activation will be fully expressed only if both control switches are in the "on" position*. The galactose (*gal*) operon, which encodes enzymes for galactose catabolism, has a promoter regulated by cAMP-CAP as well as by a specific repressor that is antagonized by galactose. The *gal* operon appears to have a second start signal close by (5 bases before the first start signal) that is not subject to cAMP-CAP control; the function of this second promoter is not understood at the present time. It may be related to the fact that galactose can serve two roles in *E. coli*: one as a carbon source (catabolite) and one as a substrate for cell wall synthesis (anabolite). The enzymes of the *gal* operon convert galactose to an activated complex, uridine diphosphogalactose (UDPG), that is an intermediate in both the catabolic and the anabolic pathways. Thus the *gal* operon should be responsive to two types of controls that reflect different needs.

The *arabinose (ara) operon* also uses the pleiotropic cAMP-CAP activator and a specific regulator, but here the similarity with *lac* and *gal* ends. The specific regulatory protein, encoded by the *araC* gene, regulates its own synthesis by functioning as a repressor. The sugar L-arabinose serves as both antirepressor and coactivator. When it complexes with the araC protein, the complex dissociates from the repressor binding site and becomes converted into an activator complex that binds at a different site adjacent to the polymerase binding site for the operon. In this case the binding of cAMP-CAP stimulates the binding of the L-arabinose-araC complex to an adjacent site, which in turn stimulates the nearby binding of RNA polymerase (Figure 29–11).

The studies on *lac* and *ara* have led to the impression that *activators stimulate expression by an energetically favorable interaction between the*

Figure 29–11

The control regions of the arabinose operon and the arabinose operon regulatory protein. As shown, the gene for the regulatory protein (*araC*) is transcribed to the left and three structural genes of the operon (*araB*, *araA*, and *araD*) are transcribed to the right. In the absence of the sugar L-arabinose, the genes are expressed at a low level. Under these conditions the araC protein acts as a repressor, binding to the *araC* gene promoter. When L-arabinose is present, it complexes with the araC protein, which changes its structure so that it no longer binds to the *araC* promoter. This permits expression of the *araC* gene. Optimal expression of the *araBAD* operon involves the sequential binding of cAMP-CAP, L-arabinose-araC protein, and RNA polymerase at adjacent sites as shown. Regulation of the arabinose operon may be even more complex than depicted. R. Schleif has found another binding site for the araC protein to the left of the binding site shown in the bottom part of this figure that appears to enhance repression of the *araBAD* operon.

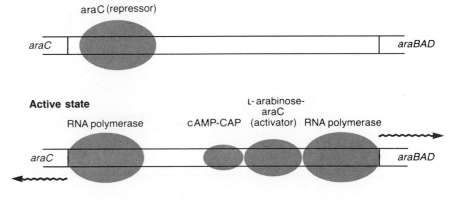

BOX 29–C

Interaction Between DNA and Specific Regulatory Binding Proteins

The precise ways in which proteins bind to double-stranded DNA require cocrystallization of the proteins with fragments of appropriate DNA. Very few observations of this type have been made (see reference 1). However, some features of the interaction have been inferred from crystallographic studies on the regulatory proteins alone. Such studies have been carried out on several regulatory proteins including the cro repressor of phage λ and the catabolite gene activator protein (CAP) of *E. coli.*

The cro repressor binds to a specific site on λ bacteriophage DNA, thereby preventing expression of certain λ genes. Cro repressor is a small protein containing 66 amino acid residues. In a 2.8 = Å resolution x-ray study, the polypeptide chain is seen to be organized in a three-stranded segment of antiparallel β-sheet structure, together with three α helices and a C-terminal tail (Illustration 1). The cro repressor is known from DNA methylation studies to bind to a region of DNA that is 17 base pairs in length. The protein binds in such a way that it prevents purine alkylation in the large groove (to guanine N-7 at key locations), but permits purine alkylation in the narrow groove (to all adenine N-3 sites in the general region of the binding site). The suggestion has been made that the cro helix extending from amino acid residue 27 to residue 36 (α_3 in Illustration 1) binds to the wide groove of B DNA (Illustration 2). The cro molecule exists as a dimer, and the symmetry-related α_3 helices from the two subunits could bind to adjacent major grooves along the DNA molecule.

Tom Steitz has found that the CAP protein has a similar structure to cro repressor protein in that it exists as a dimer with two protruding segments of α helix. However, these helices are oriented so that they would have to bind to the DNA in a different way. Other regulatory proteins that bind to specific sites on the DNA appear to share the structural motif of two protruding α helices.

A detailed analysis of the model for cro binding to DNA predicts several specific contacts between each cro monomer and the edges of base pairs in the major groove. It seems likely that these types of contacts will be found quite commonly in similar specific DNA-protein complexes, so we shall consider them in some detail. These contacts all involve side chains in or near the α helix. Three of the proposed H-bonding complexes are shown in Illustration 3. The side-chain amide of a glutamine donates a hydrogen bond to N7 of an adenine and accepts a hydrogen bond from N6 of the same adenine (Illustration 3, top). The hydroxyl group of a serine forms two hydrogen bonds with N6 and N7 of an adenine (Illustration 3, middle). Finally, the guanidinium group of an arginine donates two hydrogen bonds to O6 and N7 of guanine (Illustration 3, bottom). Some direct experimental

Illustration 1
Model for *Cro* binding to DNA suggested by Brian Mathews and his coworkers.

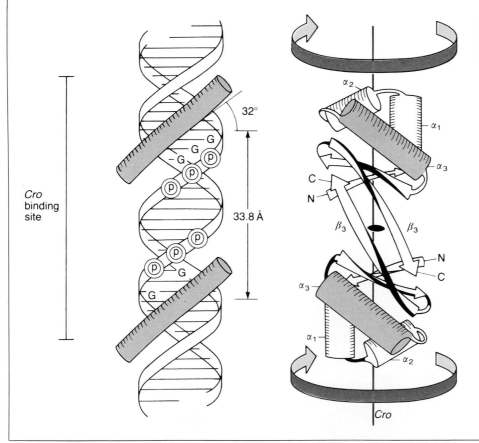

evidence has been raised in support of this cro-DNA model that involves determining the effect of amino acid changes or of modification by genetic or biochemical techniques, respectively. Change in side chains that are believed to be involved in the DNA-protein complex in all cases lowers the affinity between the two macromolecules. Final proof that the structure is correct will require a crystallographic analysis of the DNA-protein complex itself.

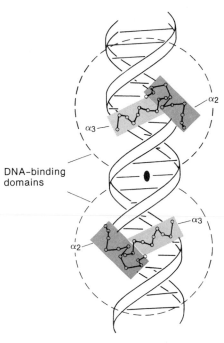

Illustration 2
Model for location of α_3 and α_2 segments when cro binds to DNA.

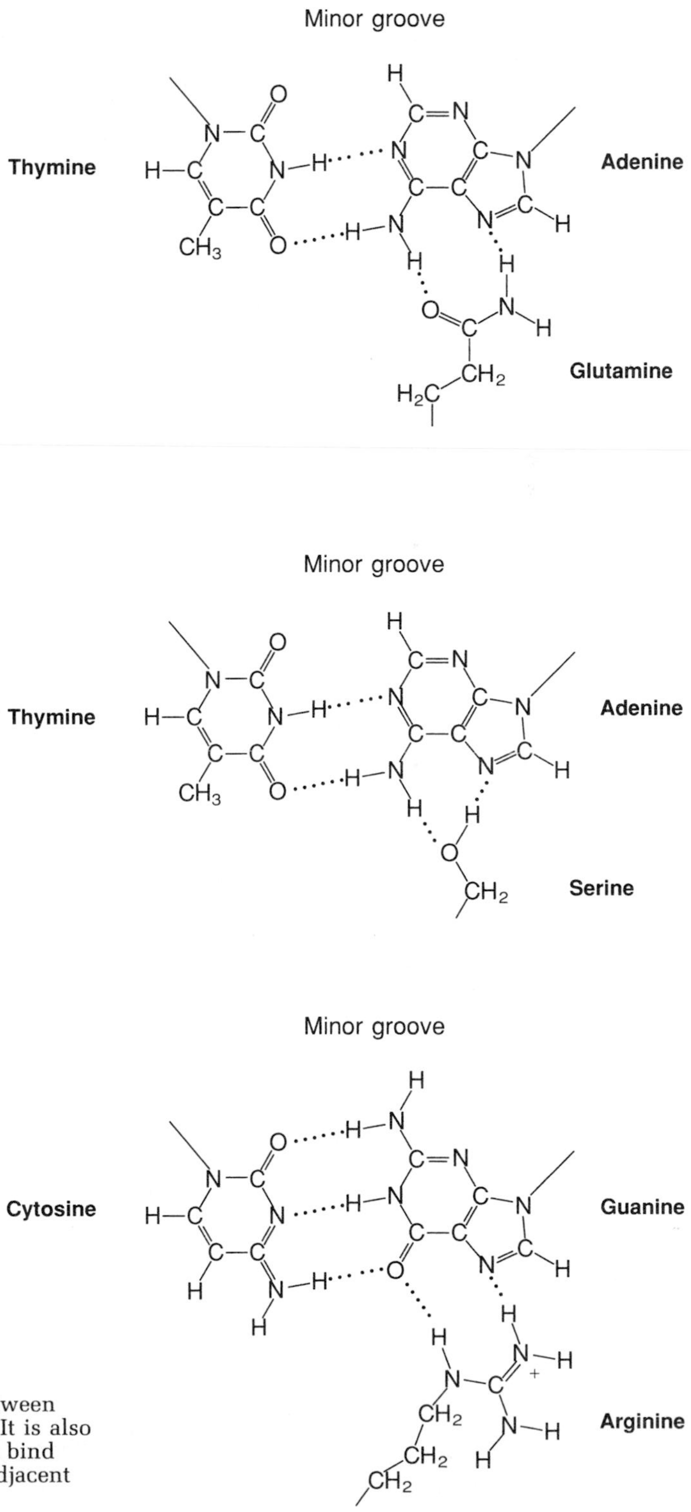

DNA-binding domains

Minor groove

Thymine   Adenine
Glutamine

Minor groove

Thymine   Adenine
Serine

Minor groove

Cytosine   Guanine
Arginine

Illustration 3
Possible hydrogen bonds between amino acids and base pairs. It is also known that amino acids can bind simultaneously to sites on adjacent base pairs.

protein surface of the activator and the protein surface of another activator or the polymerase bound to an adjacent site on the DNA.

It is clear that the repressor proteins discussed thus far interfere with the transcription process in one of two ways: (1) through binding at a site that overlaps the polymerase binding site at the promoter (as in the *lac* operon) or (2) through binding at a site that interferes indirectly with the polymerase binding at the promoter by interfering with CAP binding (as in the *ara* operon). By contrast, the activator complex cAMP-CAP acts by binding at a site adjacent to the polymerase binding site, thereby augmenting polymerase-promoter interaction. *It may be that the most fundamental difference between activators and repressors is not how they bind but where they bind to the DNA.* If this is correct, then we might expect the same regulatory protein to act as a repressor or an activator in different situations. A possible example is araC protein, which serves as an activator for the *araBAD* genes and a repressor for transcription of its own mRNA.

Amino Acid Biosynthesis Is Regulated at the Level of Transcription

Of all the genes concerned with anabolic processes, the genes uniquely involved in synthesis of the amino acid tryptophan are the best understood. Our understanding is mostly a result of the efforts of Charles Yanofsky and his colleagues, who have used a wide variety of genetic and biochemical techniques to probe the complexities of this system. Expression of the genes involved in tryptophan biosynthesis is regulated at the level of transcription. The way in which transcription is regulated in a biosynthetic operon is quite different from the way in which it is regulated for the catabolic genes. We will see that these differences reflect the different strategies involved in assuring that the cell has sufficient amounts of a particular substrate required for protein biosynthesis.

Wild-type *E. coli* has the capacity to synthesize all 20 amino acids from simpler substrates, but for most of them, it does so only when they are not available in adequate amounts from the external growth medium. For example, the production of enzymes for tryptophan synthesis is sharply reduced when the external tryptophan supply is high. A lowering of the available L-tryptophan selectively stimulates synthesis of the messenger RNA and the five polypeptide chains that form the enzymes associated with the biosynthesis of tryptophan from chorismate (Figure 29–12). The five contiguous

Figure 29–12
The tryptophan operon, indicating the size of the different genes, the polypeptide chains, the resulting enzyme complexes, and the reactions catalyzed by the enzyme complexes.

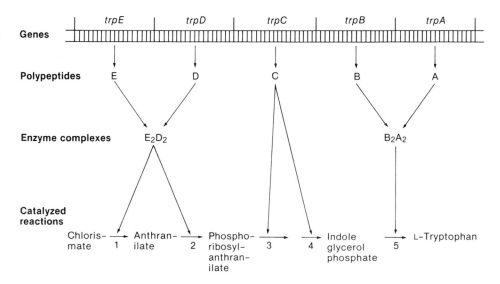

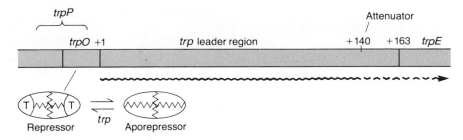

Figure 29–13

Schematic diagram of the repressor control of *trp* operon expression. The *trp* promoter *(P)* and *trp* operator *(O)* regions overlap. The *trp* aporepressor is encoded by a distantly located *trpR* gene. L-tryptophan binding converts the aporepressor to the repressor that binds at the operator locus. This complex prevents the formation of the polymerase-promoter complex and transcription of the operon that begins in the leader region *(trpL)*. Only a fraction of the transcripts extend beyond the attenuator locus in the leader region. The regulation of this fraction is discussed in the text.

structural genes are transcribed as a single polygenic messenger RNA approximately seven kilobases in length. Initiation of transcription is regulated in part by the interaction of the tryptophan aporepressor, the protein product of the *trpR* gene, with its target site on the DNA, the *trp* operator, *trpO* (Figure 29–13). Binding of L-tryptophan to the aporepressor causes a structural alteration essential for strong specific binding to the *trpO* locus. Like *lac* repressor, the *trp* repressor binds at a site that overlaps the RNA polymerase binding site, and the region where the repressor binds contains a number of bases arranged with dyad symmetry (Figure 29–14). It seems likely that the bases on the dyad-symmetry axis are strongly involved in the binding site for repressor and probably reflect similar symmetry in the repressor itself. The repressor is composed of four identical subunits, each with a molecular weight of 12,400.

The most significant difference between the action of the *trp* and *lac* repressors relates to the function of the small-molecule effector. *In the case of lac, the effector molecule allolactose acts as an antirepressor, causing release of repressor from the operator; in the case of trp, the effector molecule L-tryptophan acts as a corepressor, stimulating the binding of repressor to the operator.* It should be obvious that the difference in action of these small-molecule effectors, whose concentrations dictate the level of operon activity, is well suited to the different metabolic needs of the cell satisfied by the two operons.

The Attenuator: A Means for Controlling Transcription After Initiation. From this point on, the close parallel between the regulation of the *trp* operon and the *lac* operon ends. The *trp* operon has no positive control system like cAMP-CAP, but it does use another means of regulating transcription.

Figure 29–14

The promoter-operator region of the tryptophan operon. Two regions, PBS1 and PBS2, where polymerase binds are bracketed. Regions within the repressor-binding site showing dyadic symmetry are underlined and overlined. Single-base changes that lead to operator constitutive (o^c) mutants are indicated below the duplex.

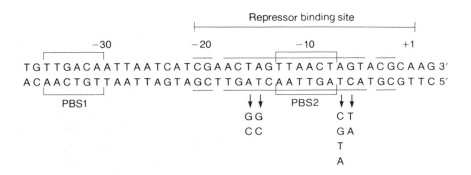

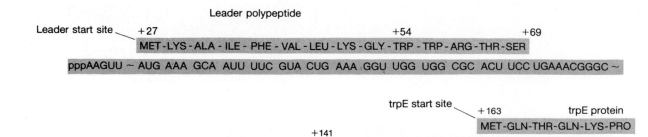

Figure 29–15
The leader region for the tryptophan operon. The region of the leader RNA containing the hypothesized leader polypeptide is shown. The translation start of the trpE protein is also shown.

This is a provisional stop signal called an *attenuator*, located 141 bases downstream from the initiation site for transcription. Only a fraction of the messenger RNAs (0.1 to 0.9 depending on metabolic conditions) that are initiated transcribe through this provisional stop signal to the end of the operon. The existence of the stop signal was first suspected when it was discovered that a genetic deletion of some of the bases between the initiation site for transcription and the first structural gene *(trpE)* raised the level of expression of the operon 8- to 10-fold. This was true even in strains with a defective repressor gene *(trpR⁻ strains)*. How could this be, and what could be the mechanism of action? An important clue was provided by sequence analysis of the 162 bases in the *trp leader region*, that is, the region between the initiation site for transcription and the initiation site for translation of the first structural gene (Figure 29–15). This leader region contains a potential initiation codon (bases 27 to 29), two tandem *trp* codons (bases 54 to 59), and a terminator codon (bases 69 to 71). A so-called *leader peptide* of 14 amino acids would result from translation of this region.

There are numerous reasons for believing that translation of the leader peptide up to or through the *trp* codons regulates attenuation of mRNA transcription. First of all, selective starvation of cells for tryptophan relieves attenuation and permits most RNA polymerase molecules to read through the leader region. The only other amino acid that relieves attenuation of the *trp* operon when it is lacking is arginine; arginine starvation is about 80 per cent as effective as tryptophan starvation. It should be noticed that an *arg* codon is located adjacent to the two *trp* codons in the leader region. Most telling of all is the finding that a leader mutation resulting in the replacement of the AUG start codon by AUA, which should eliminate translation of the leader peptide, also prevents transcription beyond the attenuator.

Other experiments indicated that the fraction of tRNA^Trp that is charged with an amino acid is a crucial factor in the attenuation response. This has been examined *in vivo* by comparing the *trp* operon enzyme levels in *trpR⁻* strains that are otherwise normal with strains that are defective in some respect in charged tRNA^Trp (Table 29–2). This work shows that such structural defects in tRNA^Trp or in the charging enzyme elevates expression, probably by permitting polymerase to transcribe through the attenuator.

The preceding results support the hypothesis that *transcription read-through requires partial translation of the leader sequence. However, only if the translation pauses or stops in the region where the trp or arg codons occur is read-through favored.* A careful examination of the secondary-structure possibilities in the attenuator region suggests why this is so. The leader region RNA between bases 50 and 141 possesses the potential for forming a variety of base-paired conformations. Figure 29–16 illustrates the most

TABLE 29–2
Attenuation of the *trp* Operon Is Affected by *trpT*, *trpX*, and *trpS* Mutations in *trpR⁻* Strains

Strain (*trpR* Derivative)	*trp* Enzyme Levels (Normalized)	
	Wild type *(trpa⁺)*	*trpΔLD102 (trpa⁻)*
Wild type	1.0	7.1
trpT$_{ts}$	7.1	—
trpX	4.5	6.7
trpS9969	2.7	7.5

All strains were *trpR⁻* (derepressed) and grown in the presence of excess tryptophan; *trpΔLD102* deletes the attenuator. *trpT$_{ts}$⁻* is a temperature-sensitive lesion in the tRNA^{Trp} gene. *trpX⁻* leads to a defect in the modification of the adenine adjacent to the anticodon sequence of the tRNA. *trpS9969* is a mutant in the tryptophenyl-tRNA synthase gene; the resulting enzyme has one-hundredth the affinity for tryptophan shown by the wild-type enzyme, resulting in much lower levels of charged tRNA^{Trp} than normal.

likely secondary structures that form in terminated *trp* leader RNA. These are based on analysis of regions of the transcript that show resistance to RNAse T1 digestion under mild conditions and the base pairing established by studies of defined oligonucleotides. Four regions of base pairing that are capable of forming three stem-and-loop structures have been proposed. Region 1, which includes the tandem *trp* codons and the leader peptide translation stop codon (bases 54 to 68), can base-pair with region 2 (bases 76 to 91). Although region 2 (bases 74 to 85) also should be capable of base pairing with region 3 (bases 108 to 119), stem-and-loop 2 · 3 has not been observed *in vitro*, presumably because stem-and-loop 3 · 4 and stem-and-loop 1 · 2 form preferentially. Region 3 (bases 114 to 121) can base-pair with region 4 (bases 126 to 134). The existence of this stem-and-loop is inferred from the resistance of the GC-rich region from residue 107 to the 3′ end of the transcript to RNAse T1 digestion. The stem-and-loop structure formed between

Figure 29–16
Proposed secondary structures in the leader RNA. Four regions, labeled 1, 2, 3, and 4 at the left, can base-pair to form three stem-and-loop structures (1.2, 2.3, and 3.4). The arrows in the main figure mark the RNAse T1 cleavage sites.

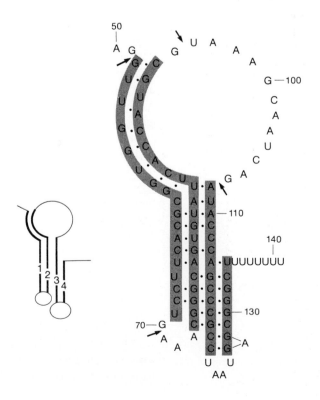

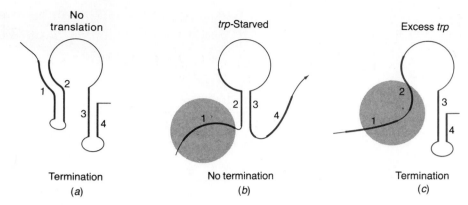

Figure 29–17

Model for attenuation in the *trp* operon, showing ribosome and leader RNA. *(a)* Where no translation occurs, as when the leader AUG codon is replaced by an AUA codon, stem-and-loop 3.4 is intact and termination in the leader is favored. *(b)* Cells are selectively starved for tryptophan so that the ribosome stops prematurely at the tandem *trp* codons. Under these conditions, stem-and-loop 2.3 can form, and this is believed to lead to the disruption of stem-and-loop 3.4, permitting attenuation. *(c)* All amino acids, including excess tryptophan, are present so that stem-and-loop 3.4 is present.

regions 3 and 4, followed by a sequence of U residues, is a common structure in transcription termination (see Chapter 27). Hence it is expected that conditions under which this structure is preserved would favor transcription termination. In support of this, a number of single-base replacement mutations have been isolated that lead to mispairing in the 3 · 4 stem, and all of these lower the level of transcription termination in the leader to some extent.

The absence of translation of the leader would do nothing to perturb this 1 · 2 and 3 · 4 structure (Figure 29–17a). Consistent with this, changing the initiator codon of the leader peptide by a single base prevents read-through, as discussed earlier. On the other hand, selective starvation resulting from either tryptophan or arginine deprivation stimulates transcription read-through. Most likely this is because the ribosome stalls in the region of the *trp* or *arg* codons (bases 54 to 62). The resulting rupture of the base-pairing 1 · 2 structure would make region 2 available for pairing with region 3. This would encourage disruption of the stem-and-loop 3 · 4 structure, resulting in transcription read-through (Figure 29–17b). In the presence of an adequate supply of all amino acids, translation would proceed beyond this critical region, so that region 2 would not be available for base pairing and the 3 · 4 loop would be maintained, favoring transcription termination at the attenuator (Figure 29–17c).

This attenuator mechanism of control is amazingly simple, since it involves no proteins other than those normally used for transcription and translation. One might expect such a simple and effective mechanism to be used repeatedly for other operons involved in amino acid biosynthesis. An indication that the attenuator mechanism might be functioning would be the finding that the biosynthetic enzymes for a particular amino acid were increased when charging of the cognate tRNA was reduced. Indeed, for the several other amino acid biosynthetic pathways in *E. coli* for which tRNA charging is involved in regulation, attenuator mechanisms have been found. Analysis of the leader region preceding the structural genes for the biosynthesis of three such amino acids in *E. coli*, phenylalanine, histidine, and leucine, reveals structures with the features of this type of control mechanism: (1) leader sequences contain several consecutive codons for that amino acid followed by (2) sequences that can be organized into stem-and-loop structures that could signal termination. Even the multivalently controlled* threonine operon (*thr*) and isoleucine and valine (*ilv*) operon of *E. coli* exhibit this mechanism of expression. Thus the leader sequence of the *thr* operon specifies a 21-amino-acid peptide containing 8 threonine codons and 4 isoleucine codons, and that of the *ilv* operon specifies a 32-amino-acid peptide, of which 14 are leucine, isoleucine, or valine. Comparable studies of one of the genes involved with arginine biosynthesis (*argF*), however, failed to show attenuation. This gene, which is required for arginine biosynthesis, appears to be exclusively regulated by a repressor mechanism. By contrast, the *his* operon has no repressor and appears to be regulated exclusively with an attenuator.

These contrasting findings provoke the question as to why both types of transcription controls are used in the *trp* and certain other operons. Two conceivable advantages have been suggested for the tryptophan operon having preserved two control switches in series. First, for a system with two such metabolic switches, the amplification factor between fully on and fully off should be the product of what could be achieved with either switch alone. Second, under some conditions, for example, during very rapid protein synthesis, it is possible that a higher level of amino acid is required to keep the tRNA at an adequately charged level. Finally, it should be mentioned that the *trpR* protein functions in other roles; it is also a repressor for the *trpR* gene and for the *aroH* gene, which encodes one of three isozymes that catalyze the first reaction in aromatic biosynthesis.

Genes for RNA Polymerase and Ribosomes Are Coordinately Regulated

Over 100 genes in *E. coli* are involved in the synthesis of the RNAs and proteins that comprise the enzymatic machinery for transcription and translation. The relevant gene products make up between 20 and 40 per cent of the dry cell mass. In rapidly growing cells, about 85 per cent of the RNA is ribosomal, 10 per cent is tRNA, and most of the remainder is mRNA. The various RNAs and proteins are produced according to their need. For ribosomes and tRNAs, this results in a synthesis rate that is roughly proportional to the cell growth rate; the relative amounts of the three rRNAs (16S, 23S, and 5S), the 60 or so tRNAs, and the 50 ribosomal proteins is consistent with the stoichiometric needs for making ribosomes. For RNA polymerase, which contains four different subunits (see Chapter 27), this results in equimolar amounts of β and β', approximately equal in sum to the number of α subunits. The dissociable σ subunit of RNA polymerase is produced in somewhat less than stoichiometric amounts. The synthesis of RNA polymerase subunits is loosely coupled to the synthesis of ribosomal protein subunits. Demands for large amounts of all these RNAs and proteins and precise regulation of their relative amounts have led to complex gene arrangements and controls at both the transcription and translation levels.

Control of rRNA and tRNA Synthesis by the *rel* Gene. Under conditions of rapid growth, *E. coli* cells contain about 10^4 ribosomes. The maximum rate of reinitiation at the ribosomal gene promoter is about one per second. In rapid growth, *E. coli* can duplicate once every 20 min, which would allow for the synthesis of only about 1200 molecules of rRNA if there were only one gene for ribosomal RNA. In fact, there are seven copies for ribosomal RNA operons in the bacterial chromosome, which makes it possible for

---

*The *thr* operon is repressed in the presence of excess threonine and isoleucine. Repression of the *ilv* operon requires leucine, isoleucine, and valine in excess.

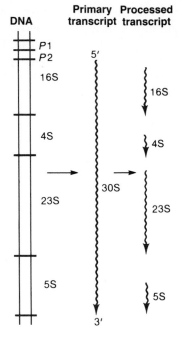

Figure 29-18

A typical rRNA *(rrn)* operon contains two promoters and genes for 16S, 23S, and 5S rRNA and a single 4S tRNA gene. The four fully processed RNAs are derived from a single intact 30S primary transcript.

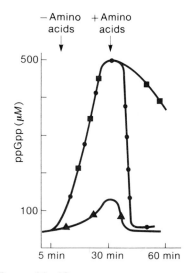

Figure 29-19

Guanosine tetraphosphate (ppGpp) concentration under normal conditions, after amino acid starvation, and after readdition of amino acids (● = wild-type cells; ▲ = relA cells; and ■ = spoT cells).

rRNA synthesis to maintain the necessary pace under conditions of rapid growth. These operons are dispersed at seven locations around the circular *E. coli* chromosome. Each operon is transcribed into one long transcript that is processed into one each of 16S, 23S, 5S, and usually at least one 4S transfer RNA (Figure 29–18). The order of the genes in the operon, starting from the initiation site for transcription, is 16S, 4S, 23S, and 5S. In some of the operons, additional 4S genes for tRNA are located downstream from the 5S gene. All the known ribosomal RNA operons appear to have two strong promoters located in tandem; this probably permits a more rapid rate of reinitiation for the genes than would otherwise be possible.

As stated earlier, rRNA synthesis is usually maintained at a rate that is proportional to the gross rate of protein synthesis. In a normal wild-type cell, when protein synthesis is limited (e.g., by amino acid availability), M. Cashel and T. Gallant showed that there is a rapid rise in ppGpp concentration from about 50 to 500 μM (Figure 29–19). Concomitantly, there is an abrupt cessation of rRNA synthesis. This is part of the phenotype known as the *stringent response*: First, *amino acid deprivation or other factors that slow down protein synthesis provoke an increased rate of accumulation of ppGpp; this in turn leads to an inhibition of rRNA synthesis.*

Observations on different mutants indicate that the concentration of ppGpp is regulated by a careful balance between its *rate of synthesis* (controlled by the *rel* gene) *and rate of breakdown* (controlled by the *spoT* gene). Correlated observations indicate that the synthesis of rRNA is inversely proportional to the ppGpp concentration. In a *relA* mutant cell, neither the rapid rise in ppGpp concentration nor the cessation of rRNA synthesis is seen when amino acids are removed. The *relA* gene encodes a protein that is required for ppGpp synthesis. If amino acids are reintroduced into the growth medium of a wild-type culture, the ppGpp concentration falls rapidly (half-life about 20 sec), and the rate of rRNA synthesis rises rapidly. In a mutant called *spoT*, the normal rise in ppGpp level is observed on amino acid starvation, but the level of ppGpp falls much more slowly on readdition of amino acids to the growth medium. Correlated with this, the rate of rRNA synthesis in an *spoT* mutant also increases very slowly on readdition of amino acids.

The synthesis of ppGpp has been studied both in crude cell-free extracts of *E. coli* and in a partially purified system to determine what factors influence its rate of synthesis. It was found that ppGpp is synthesized on the ribosome from GTP in the presence of the protein encoded by the wild-type *relA* gene. Maximum ppGpp synthesis occurs in the presence of ribosomes associated with mRNA and uncharged tRNA with anticodons specified by the mRNA. If the uncharged tRNA bound to the ribosome acceptor site (A site) is replaced by charged tRNA, the rate of ppGpp synthesis is greatly lowered. If uncharged tRNA anticodons are not complementary to mRNA codons exposed on the ribosome for protein synthesis, ppGpp synthesis does not occur.

Cell-free synthesis studies have also provided strong support for the notion that *ppGpp directly inhibits RNA synthesis.* Thus in a cell-free system the DNA-directed synthesis of rRNA with *E. coli* RNA polymerase has been shown to be strongly and selectively inhibited by 100 to 200 μM ppGpp. Such experiments have led to the hypothesis that ppGpp, by binding to RNA polymerase, alters its structure so that it has a lowered affinity for rRNA promoters.

The *in vivo* and *in vitro* studies on ppGpp and rRNA have resulted in a model for how the level of amino acid charging of tRNA controls the rate of rRNA synthesis (Figure 29–20). First, uncharged tRNA that is codon-specific for the exposed codons on the mRNA becomes bound to the ribosome accep-

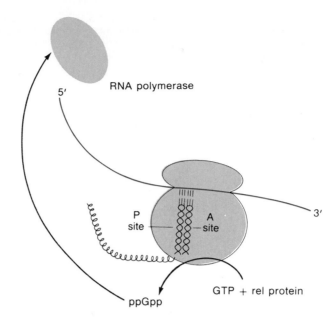

Figure 29-20

Schematic diagram of ppGpp synthesis and the hypothesized mechanism for its action. ppGpp is synthesized on the ribosome when there is a peptidyl-tRNA on the P site of the ribosome and an uncharged tRNA on the A site. The ppGpp probably inhibits rRNA synthesis by complexing with the RNA polymerase.

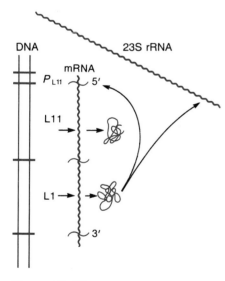

Figure 29-21

Schematic diagram explaining autogenous inhibition by L1. L1 can form a complex either with the 5′ end of its own mRNA or with 23S rRNA. If there is an excess of L1 over the available 23S rRNA, then L1 will bind to the 5′ end of the mRNA and inhibit both L1 and L11 synthesis. The inhibition of L1 translation by this binding depends upon the fact that the translation of L1 and L11 is somehow coupled.

tor site, creating a situation unfavorable for protein synthesis but favorable for ppGpp formation. Second, the ppGpp diffuses and binds to RNA polymerase, thereby lowering its affinity for rRNA promoters.

This alteration of the RNA polymerase structure by ppGpp affects the ability of RNA polymerase to interact with promoters in a differential way. For the promoters of the *lac* and *trp* operons discussed above, the interaction is stimulatory; this is evidenced by an elevated expression of these operons at high concentrations of ppGpp. For the promoters of the rRNA operons, the polymerase-promoter interaction is strongly inhibited by ppGpp. Final proof that the proposed mechanism is correct will require the isolation of RNA polymerase mutants that are not affected by ppGpp; in fact, polymerase mutants with an altered response to ppGpp (*in vivo*) have been described.

The observations on ppGpp involvement in rRNA synthesis show that this small molecule is an important control factor regulating rRNA synthesis, but it does not eliminate the possibility that there may be other factors that also affect the level of rRNA. *In vivo* and *in vitro* evidence exists that the inhibitory effect of ppGpp on transcription extends to most tRNA and ribosomal protein genes. Ribosomal protein gene expression also appears to be regulated at the translation level.

Translational Control of Ribosomal Protein Synthesis. In exponentially growing *E. coli* cells, synthesis rates of all ribosomal proteins (except L7, L12, and S6) are identical and coordinately regulated. Nomura and his coworkers have suggested that *free ribosomal proteins inhibit the translation of their own mRNA and that as long as the assembly of ribosomes removes ribosomal proteins, the corresponding mRNA escapes this feedback inhibition.* This hypothesis has been tested *in vitro* using a protein-synthesizing system with various template DNA molecules carrying ribosomal protein genes; it has been tested *in vivo* by examining the effect of overproduction of certain ribosomal proteins on the synthesis of other ribosomal proteins using various recombinant plasmids. By these means it was found that certain ribosomal proteins selectively inhibit the synthesis of other ribosomal proteins whose genes are part of the same operon as their own, and that this autogenous or self-imposed inhibition occurs at the level of the translation of the mRNA rather than at the level of transcription of mRNA. Figure 29-21 illustrates how this scheme works for the regulatory protein L1. L1 and L11

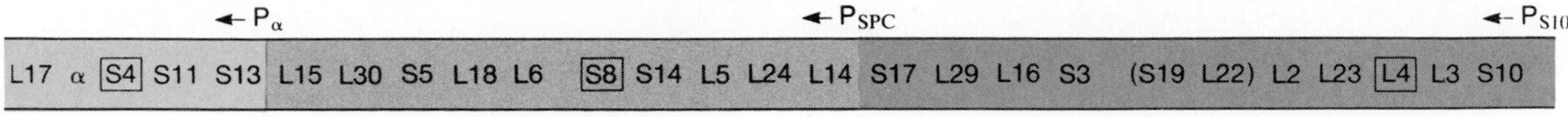

← P$_\alpha$ ← P$_{SPC}$ ← P$_{S10}$

| L17 | α | S4 | S11 | S13 | L15 | L30 | S5 | L18 | L6 | S8 | S14 | L5 | L24 | L14 | S17 | L29 | L16 | S3 | (S19 L22) | L2 | L23 | L4 | L3 | S10 |

Figure 29–22

A giant cluster containing 25 r-protein genes and the gene for RNA polymerase. This cluster is divided into three operons with promoters P$_\alpha$, P$_{spc}$, and P$_{S10}$. The corresponding messages are subject to autogenous inhibition at the translation level by S4, S8, and L4 protein, respectively. S8 regulates only genes downstream from L5; these include S14, S8, L6, L18, S5, L30, and L24. Translation is also autogenously regulated by one or both of these proteins. Parentheses around S19 and L22 indicate that the order of genes is uncertain.

are encoded by the P$_{L11}$ operon. L1 can form a complex with either the 5′ end of its own mRNA or with 23S rRNA. It binds more strongly to the 23S rRNA. However, if there is an excess of L1 over the available 23S rRNA, then L1 will bind to the 5′ end of the mRNA, thereby inhibiting the synthesis of both L1 and L11. In this way the amounts of L1 and L11 protein synthesized are kept in register with the amounts of rRNA synthesized. Other operons carrying ribosomal protein genes are believed to be regulated in a similar manner.

A meaningful overall pattern of control appears to be emerging that is consistent with the requirement to balance the relative amounts of rRNA and ribosomal proteins synthesized. Most of the ribosomal protein genes exist in clusters or operons with promoters at one end (e.g., Figure 29–22). In some cases, individual operons are regulated by one of the encoded ribosomal proteins; in other cases, operons appear to be subdivided into "units of regulation," and individual units of regulation are regulated by one of the gene products (e.g., the P$_{SPC}$ operon on Figure 29–22).

Regulation of RNA Polymerase Synthesis Is Coordinated with Ribosomal Protein Synthesis. The genes for the RNA polymerase subunits β and β' (rpoB and rpoC, respectively) and L7/L12 and L10 (rplJ and rplL, respectively) belong to a single transcription unit (Figure 29–23). The gene for the α subunit (rpoA) is cotranscribed with four ribosomal protein genes coding

Figure 29–23

DNA and transcripts in the β operon. Only a small number of transcripts read through the attenuator (ATT) located 69 nucleotides beyond the 3′ end of the rplL gene. The read-through transcripts are cleaved by RNAse III about 200 nucleotides beyond the end of the rplL gene. L10 ribosomal protein and RNA polymerase are believed to inhibit translation of the promoter-proximal and promoter-distal messages, respectively. L10 may only inhibit its own synthesis. Normally about five times as much L7/L12 is synthesized at L10. Ribosomes can initiate independently at the initiation site for L7/L12, and it is possible that L7/L12 may regulate this initiation. Bruckner and Matzura have shown that another promoter, two genes upstream from the P$_{L10}$ promoter, may be the true initiation point for this operon in vivo.

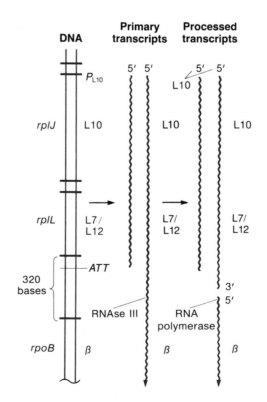

for the proteins S13, S11, S4, and L17 (see Figure 29–22). This suggests that the _regulatory system for RNA polymerase subunit synthesis and that for ribosomal protein synthesis are shared_. Even more fascinating is the finding that the gene for the sigma subunit of RNA polymerase (_rpoD_) is located on the same operon as the gene for DNA primase (_dnaG_) and the small ribosomal subunit protein S21 (_rpsU_). This operon, known as the sigma operon, clearly encodes proteins essential to DNA, RNA, and protein synthesis, which suggests that regulation of the operon may be important to balance all three of these biosynthetic processes.

Despite the use of a common promoter, the synthesis of the RNA polymerase proteins and certain nearby ribosomal proteins is not always coordinately regulated. Only the β operon will be discussed because it is better understood. For brief times, synthesis of L7/L12 and L10 is inhibited _in vivo_ under conditions of amino acid deprivation, while the synthesis of β and β' is not. Furthermore, the RNA polymerase inhibitor rifampicin causes a transient stimulation of β and β' synthesis, but not of L10 and L7/L12. This evidence for lack of coordinate regulation implies the existence of a separate regulatory device for controlling L10 and L7/L12 and for controlling β and β' subunit synthesis. There are 320 untranslated bases between the _rplL_ gene and the downstream _rpoB_ gene (Figure 29–23); this is a suspiciously large segment for an intercistronic region, suggesting that it functions as more than just a spacer between cistrons. Under normal growth conditions, only 20 per cent of the transcripts that initiate at the L10 promoter P_{L10} read through the entire operon. The remainder terminate 69 nucleotides beyond the 3' end of the _rplL_ gene. The nonattenuated transcript is normally cleaved by RNAse III in the intercistronic region at a point about 200 nucleotides beyond the end of the _rplL_ gene. This divides the mRNA for the operon into two segments, one encoding the ribosomal proteins and the other the two RNA polymerase proteins. The L10 ribosomal protein is an autogenous regulator, inhibiting translation of its own synthesis. Clearly this translational control would have no direct effect on the translation of the message for β and β'. Recently it has been discovered that RNA polymerase holoenzyme inhibits the translation of β and β'. Thus both segments of the operon appear to be regulated at the translation level by different specific autogenous regulatory proteins.

The finding that RNA polymerase inhibits the synthesis of its own subunits immediately suggests an explanation for the differential _in vivo_ effect of amino acid deprivation (increased ppGpp) and rifampicin on expression of the ribosomal protein genes and the polymerase protein genes. Whereas ppGpp or rifampicin should inhibit initiation of transcription that occurs at the P_{L10} promoter, inhibiting both ribosomal protein and polymerase subunit synthesis, the binding of either rifampicin or ppGpp to RNA polymerase could interfere with its ability to act as an autogenous inhibitor. This should have a stimulatory effect on β and β' synthesis that involves all the messages for β and β' already synthesized, as well as any that are in the process of being made.

OVERVIEW OF GENE REGULATION IN _E. COLI_

We have seen that in most cases gene expression in _E. coli_ is regulated at the level of transcription. However, there are notable exceptions, for example, in the case of the ribosomal protein genes, where autogenous regulation plays an important controlling role at the translational level. Controls for the expression of any particular gene are frequently multiple and organized into a hierarchy. In the case of the _lac_ operon, for example, transcription is positively regulated by the cAMP-CAP complex that also regulates a large number of catabolic genes in a similar manner; the *lac* operon is also negatively regulated in a highly specific way by the *lac* repressor and lactose.

Bacterial regulatory systems are geared toward a rapid response to changing environmental conditions. Since environmental conditions may change in either direction, the regulatory responses must also be of a highly reversible type. Bacterial systems manage this by maintaining low levels of regulatory proteins, which are controlled by the concentration of small-molecule effectors. The concentration of the effector molecules can vary over wide limits.

For the remainder of this chapter we turn our attention to bacteriophage control systems. The patterned responses of phage systems have led to very different strategies for control of gene expression. By and large, phage systems are regulated by the concentrations of the regulatory proteins themselves; these regulatory proteins are not subject to the influence of small-molecule effectors.

REGULATION OF GENE EXPRESSION IN BACTERIAL VIRUSES

Virus metabolism differs from normal cellular metabolism in several notable respects. First, viruses are obligate parasites and borrow heavily on the host cell to supply them with the activated substrates and machinery for nucleic acid and protein synthesis. Second, events in the infectious cycle whereby viruses replicate proceed rapidly and usually end with the destruction of the infected cell. These differences between virus metabolism and cellular metabolism are reflected by differences in the patterns of gene regulation: Bacterial genes are regulated in a highly reversible manner according to metabolic needs; viral genes are usually turned on only once and then in a predesignated sequence.

Mature viruses are metabolically inactive. Their genome can be either DNA or RNA (see Chapter 7). The size of the genome is a good measure of the number of genes carried by the virus. For bacterial viruses (bacteriophages) that infect *E. coli* this varies from 3000 bases (about three genes) for the single-stranded R17 RNA virus to 166,000 base pairs (about 166 genes) for the double-stranded T4 DNA virus (Table 29–3). Each bacteriophage (or phage) has a characteristic shape, which is determined by its protein coat. In addition to protecting the virus genome, the protein coat frequently carries specific receptors enabling the virus to attach to a host cell in preparation for infection.

The λ phage is a medium-size DNA virus with 48.6×10^3 base pairs. It has a head-and-tail structure characteristic of many *E. coli* phages (Figure 29–24). The phage infectious cycle begins by attachment of the tail structure to the cell surface. Once this attachment is secure, λ injects its entire linear genome into the host cell, leaving the virus protein on the cell surface (Figure 29–25). The λ DNA carries complementary 12-base 5′ overhangs, which readily anneal to form a circular structure. Immediately after circularization this structure is sealed and supercoiled by host enzymes. From this point on, the fate of the phage DNA depends on a delicate balance between the genes of the virus that express immediately after infection and the physiologic state of the cell. In the remainder of this chapter, we will focus on the events that occur after infection by the λ DNA, with an emphasis on regulatory events.

Bacteriophage λ Can Develop Along Two Different Pathways

All bacteriophages have a _permissive cycle_, which involves (a) the synthesis of viral enzymes, nucleic acids, and structural proteins, (b) the assembly of mature virus particles, and, finally, (c) the escape of the virus from the cell.

TABLE 29-3
Sizes of Some *E. coli* Phages and Genomes

Phage	Particle Shape	DNA $M_r(\times 10^6)$	DNA Length (bp $\times 10^3$)	Structure of Chromosome	Special Properties of Genome
ϕX174	Icosahedron	1.8	5.4[a]	Single strand, circular	Duplex replicative form
M13	Filament	2.1	6.4[a]	Single strand, circular	Duplex replicative form
T7	Head, short tail	26.5	40.0	Duplex, linear	Terminal redundancy
λ	Head, tail	32.0	48.6	Duplex, linear	Cohesive ends form replicative circles; lysogenic
P1	Head, tail	58.6	88.0	Duplex, linear	Terminal redundancy; circularly permuted
T4	Head, tail	110.0	166.0	Duplex, linear	Terminal redundancy; circularly permuted
R17	Icosahedron	1.0	3.0[a]	Single strand, linear	RNA (not DNA)
Mu		25.00	38.00	Duplex, linear	Replicates in the integrated state
T5	Head, tail	75.0	113.0	Duplex, linear	One strand interrupted

[a] For single-stranded phages, length is for bases (b) rather than base pairs (bp).

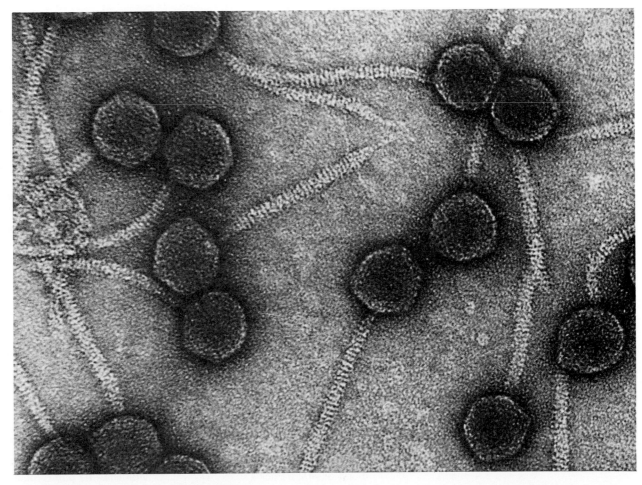

Figure 29-24
Electron micrograph of bacteriophage lambda. (Courtesy Roger Hendrix.)

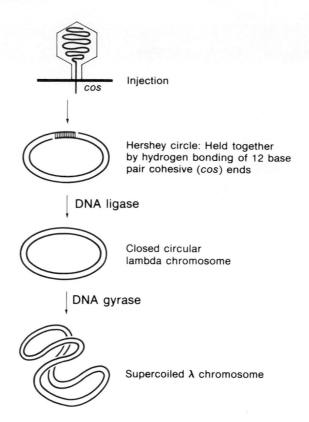

Figure 29–25

Early steps in lambda infection result in circularization and supercoiling of the infecting lambda DNA.

The escape phase of the permissive cycle usually involves cell lysis; thus the term _lytic cycle_ is often used to describe the entire permissive cycle. The term _lytic virus_ is used to describe a virus that invariably enters the lytic cycle on infecting a suitable host cell. From Table 29–3, note that the T phages T7, T4, and T5 are all lytic viruses.

The term _temperate virus_ is used to describe a virus like λ (or P1 or Mu, see Table 29–3), which can enter either an active lytic cycle or a dormant phase. In the dormant phase only a small portion of the phage genome is expressed, and replication occurs at the same rate as for the host genome. This latter pathway of development takes on different forms for different viruses. For λ it is described as a _lysogenic_ pathway. The phage chromosome inserts into the bacterial chromosome and remains as an inert _prophage_, which is repressed by a phage-encoded repressor until alterations in the bacterial physiology result in its activation. At this point the prophage excises from the host chromosome and enters the lytic pathway, replicates and packages its DNA, and destroys its host. The two pathways of λ development are depicted in Figure 29–26.

Figure 29–26

Lambda development: lytic or lysogenic growth. The major events in lytic or lysogenic growth are outlined.

λ infection

Lytic growth

1. DNA replication
2. Formation of virus coats, and packaging of DNA
3. Cell lysis and release of burst of viral particles (about 100 particles per cell)

Lysogenic growth

1. Repression of lytic viral functions
2. Integration of lambda genome into host chromosome
3. Passive replication of viral DNA as part of host DNA. No host killing.

TABLE 29–4
Regulatory Proteins of λ

Gene	Function
cI	At low concentrations a repressor of P_R and P_L and an activator of P_{RM}; at high concentrations also represses P_{RM}
cII	An activator of P_{RE} and P_{INT}
cIII	Inhibits cII breakdown
cro	At low concentrations a repressor of P_{RM}; at high concentrations also a repressor of P_L and P_R
N	An antiterminator at t_{L1}, t_{R1}, t_{R2}, and other terminators
Q	An antiterminator at t_{6S}

Implementation of the course followed by λ DNA is governed by the action of at least six regulatory proteins encoded by the phage genome (Table 29–4), cI, cII, cIII, cro, N, and Q. The cI, cII, and cro proteins are most like the bacterial regulatory proteins we have discussed already. They bind to the DNA at specific points, where they act as activators or repressors. The exclusive function of the cIII protein is to stabilize the cII protein; we will not discuss it further. The N and Q proteins bind directly to the RNA polymerase and influence its ability to recognize termination signals. We shall see that the temporal pattern of expression of the regulatory genes is the key to λ development.

Early transcription of λ is catalyzed by the unmodified host RNA polymerase. The phage promoters active at this time are P_R, P_L, and P_R' (Figure 29–27); these promoters and the remaining promoters are described in Table 29–5. Terminators t_{R1}, t_{R2}, t_{L1}, and t_{6S} prevent more than limited transcription

Figure 29–27
Bacteriophage lambda DNA. Above: Circular map indicating locations of main control genes and early and late functions. Below: Expanded region containing control genes and main promoters and terminators. Arrows indicate main transcripts and conditions under which they are active. In the absence of N protein, about half the transcription beginning at P_R reads through t_{R1} to t_{R2} (indicated by dashed portion of arrow). More information on regulatory proteins and promoters is given in Tables 29–4 and 29–5.

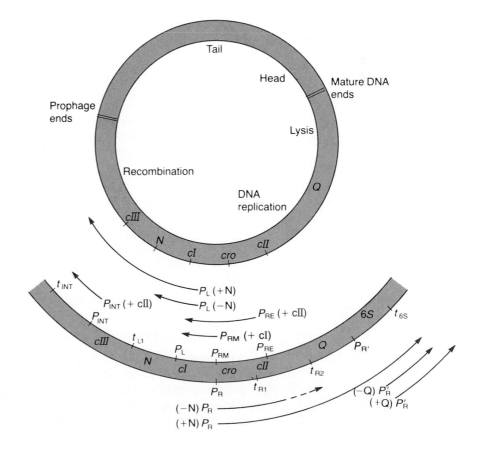

TABLE 29–5
Promoters of λ

Promoter	Function
P_R	Main rightward transcription
P_L	Main leftward transcription
P_{RM}	Transcription for repressor maintenance
P_{RE}	Transcription for repressor establishment
P_R'	Transcription of 6S RNA; also late rightward transcription
P_{INT}	Transcription of genes for integration and excision

from these promoters (see Figure 29–27). Whether the lytic or lysogenic pathway is followed depends on the relative rates of synthesis of products from these early transcripts. This sounds like a chancy process, and it is. The system is designed so that about half the time one pathway is followed and about half the time the other pathway is followed. Some of the factors that influence which pathway is followed are described in the following sections.

The Establishment of Lysogeny Requires Rapid Synthesis of cI Protein

Early transcription from P_R leads to the expression of the cII gene, which encodes a regulatory protein, cII. High concentrations of cII favor the establishment of lysogeny. This is because cII stimulates transcription from P_{RE} and P_{INT}. Transcription from P_{RE} leads to the synthesis of cI protein, which is a repressor that inhibits transcription from both P_R and P_L. Transcription from P_{INT} leads to synthesis of the Int protein, which is required for integration of the λ chromosome into the host chromosome. If sufficient cII protein is made at very early times after infection, lysogeny will result.

The Maintenance of the Lysogenic State Is Favored by cI Synthesis

During the lysogenic response, the P_R promoter gradually becomes inhibited as the level of cI repressor rises. With the shutoff of P_R activity, the synthesis of cII product comes to a halt, and the levels of the inherently unstable cII protein rapidly fall. Without the cII activator, the P_{INT} and P_{RE} promoters become inactive. The high rate of cI gene expression characteristic of the P_{RE} transcript is replaced by the much slower rate of cI product synthesis that results from the P_{RM} transcript. The lower rate of cI product synthesis from P_{RM} is not due to a difference in the strengths of the P_{RM} and P_{RE} promoters. Rather, it is because the shorter P_{RM} transcript lacks the Shine-Dalgarno sequence for ribosome recognition of the cI reading frame. As a result of this deficiency, the cI sequences of the P_{RM} transcript are translated only about 10 per cent as efficiently as the same sequences in the P_{RE} transcript.

The Lytic Cycle Is Favored by Destruction of the cI Repressor

The general strategy of events occurring during the lytic cycle is similar for most bacteriophages. Phage-encoded products are made as they are needed for virus replication and cease to be made when they are present in sufficient concentrations or no longer needed. Enzymes required for phage recombination and replication are made early in the lytic cycle. Structural proteins are made later. Finally, lytic proteins are made as the mature phage particles become fully assembled.

The expression of phage gene products in this sequential manner is controlled by regulatory proteins encoded by the phage. These regulatory proteins themselves are expressed in a sequential manner.

In λ-infected cells, the synthesis of specific transcripts is modulated by the regulatory proteins described in Table 29–4. Induction of the λ prophage begins with the destruction of the cI repressor. This is the result of a cleavage reaction that is catalyzed by the highly specific protease function of the host recA protein. It should be remembered that the protease function of recA is activated under SOS conditions (see Chapter 26). Once the cI protein has been eliminated, the derepressed P_R and P_L promoters become active. This triggers a cascade of transcriptional events associated with the lytic cycle. Early during activation, the prophage becomes excised and circularized. From this point on, events in the lytic cycle are similar for the activated prophage of λ DNA infecting the cell from without.

The first gene to be transcribed from rightward synthesis starting from the P_R promoter is *cro*; *cro* encodes a repressor that inhibits cI synthesis from P_{RM}. Cro protein prevents cI protein from reappearing and aborting the lytic cycle. Later during the lytic cycle, the concentration of Cro increases, and additional Cro proteins bind between P_{RM} and P_R. These extra molecules repress P_R activity. This is entirely appropriate at this stage in the lytic cycle, since synthesis of genes controlled by the P_R promoter is no longer required. Although cI protein can serve as a repressor or activator depending on its concentration, Cro protein acts only as repressor. The complex actions of these two regulatory proteins are explained in Box 29–D.

Early leftward transcription from P_L leads to the synthesis of the N protein. In the absence of N, early transcription stops at the t_{L1}, t_{R1}, and t_{R2} terminators (see Figure 29–27). When N protein is present, this situation changes. N binds to polymerase, and this binding of N alters the properties of the transcribing polymerase so that it no longer recognizes t_{R1}, t_{L1}, and t_{R2} as terminators. This leads to longer transcripts and the synthesis of additional proteins (see Figure 29–27 for how the P_L and P_R transcripts will be extended when N is bound to the polymerase).

The action of the N protein has been explored in great depth. This protein, in combination with RNA polymerase, suppresses transcription termination at both phage and bacterial terminators. Despite this fact and the fact that N protein is freely diffusible, the action of N in λ-infected cells is confined to elongation of the P_L and P_R transcripts. The specificity of N is due to the *nut* (N utilization) sequences that lie in the phage DNA. One of these, *nut*$_L$, lies some 50 base pairs to the left of P_L; the other, *nut*$_R$, is approximately 250 base pairs to the right of P_R, between *cro* and t_{R1} (Figure 29–28).

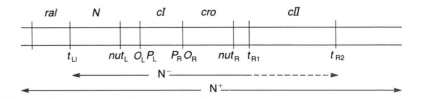

Figure 29–28
Early transcripts. The early left transcript is initiated from the P_L promoter. In the absence of N protein, this transcript terminates at t_{L1}. In the presence of N protein, the polymerase picks up an N protein at the N utilization site, *nut*$_L$. This makes it possible for the polymerase to transcribe through t_{L1}. The early right transcript tends to terminate at t_{R1} unless N protein is present. In this case the N protein becomes bound to the polymerase at the *nut*$_R$ site.

BOX 29–D

Control of the λ Immunity Regions by cI and cro Repressor Proteins

The cI and cro repressor proteins bind to the same regions but their effects are quite different. A very extensive investigation by Mark Ptashne has revealed that the different actions of these two repressors are due to the complex nature of the binding sites. The cI repressor binds to the λ operators that regulate the activity of the P_L and P_R promoters. The two operators, O_L and O_R, are each composed of three binding sites with strong sequence homology. The binding sites are 17 base pairs in length and they are separated by spacers of 6 to 7 base pairs (see illustration).

The affinity of the different sites for cI repressor is not identical, nor is the effect of cI repressor binding to the sites. This has been examined in detail for the right operator, P_R. The formation constant K_f for sites O_{R1} and O_{R2} is about $3 \times 10^8 M$; that for O_{R3} is some 25 times less. This difference means that under physiological conditions O_{R1} and O_{R2} are usually occupied and O_{R3} is usually not occupied. The apparent high affinity of O_{R2} is the consequence of cooperative binding interactions between the cI repressor dimers at their carboxyl terminals. Thus, if O_{R1} is eliminated by a mutation, the affinity of O_{R2} for cI repressor falls to that of O_{R3}.

The binding of cI repressor to O_{R1} and O_{R2} inhibits the activity of P_R by providing steric hindrance to the access of RNA polymerase to this promoter. By contrast, the cI repressor bound at O_{R2} stimulates the activity of P_{RM}, located within the operator cluster, presumably by a favorable interaction between the bound cI repressor and the polymerase. Repressor cI is therefore an activator for its own gene. Binding of cI repressor at O_{R3} occurs only at very high levels of repressor; this blocks the activity of P_{RM}. The autogenous regulation ensures that the concentration of repressor is low enough (100 cI repressor monomers per cell) to allow for rapid induction under the appropriate circumstances, but is high enough to repress the prophage completely.

The cro repressor binds to the λ operators at a subset of contact points utilized by the cI repressor. But in contrast to cI repressor, it fills O_{R3} first and then O_{R2} and O_{R1} at higher concentrations. The effect of either repressor at O_{R3} is the same; they block the activity of P_{RM}. When the O_{R2} and O_{R1} sites are filled by cro, the activity of P_R is inhibited; thus the expression of the cro gene is autoregulated. However, the concentration of cro during the early stages of induction is just sufficient to inactivate P_{RM}, thereby inhibiting cI-repressor synthesis; this effect makes the lytic induction process irreversible.

Repression of both the P_L and P_R promoters by cro occurs at later stages during the lytic cycle and is required for phage production, although the reason for this is not completely clear.

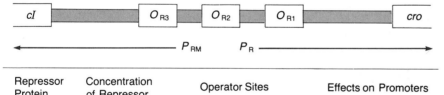

Structure of the control region between the *cI* and *cro* genes.

Repressor Protein	Concentration of Repressor	Operator Sites			Effects on Promoters	
		O_{R1}	O_{R2}	O_{R3}	P_R	P_{RM}
cI	Low	+	+	−	↓	↑
	High	+	+	+	↓	↓
cro	Low	−	−	+		↓
	High	+	+	+	↓	↓

+ = Site occupied − = Site unoccupied
↑ = Promoter stimulation ↓ = Promoter repression

Without the *nut* sequences, N cannot suppress transcription termination; the absence of *nut* regions in bacterial operons is the reason why bacterial transcripts are not acted upon by N product. Current indications are that polymerase loses its sigma factor shortly after the initiation of transcription (see Chapter 27). This permits another bacterial protein, known as the NusA protein, to bind to the elongating core polymerase. At the *nut* sites the polymerase pauses, and if N protein is present it will bind to the NusA protein on the polymerase and remain bound until the polymerase completes the elongation process. This explains the specificity of N protein even though it does not explain why N will bind to polymerase only at a *nut* site, nor how the binding of N protein suppresses termination.

Efficient synthesis of late gene products required for head and tail structural proteins and cell lysis requires the viral Q protein (see Figure 29–27). This regulatory protein, like the N protein, is a transcription antiterminator that functions by binding to the polymerase. Q protein binding by the transcribing polymerase eliminates termination at the t_{6S} terminator so that transcription initiating at P_R' (see Figure 29–27) continues through the genes required for late functions. This completes the transcription that is necessary for the lytic cycle.

The contrast between the way in which λ genes are regulated during the lytic cycle and the way in which bacterial genes are normally regulated is quite striking. In this sense λ is prototypic of most phages that enter a lytic cycle. Bacterial regulatory proteins are present most of the time. Whether or not they are active depends on the concentration of small-molecule effectors like cAMP or allolactose. This probably results from the desirability of maintaining most bacterial genes in an "always prepared" state, so that they can be readily expressed when needed or rapidly turned off when they are not needed. By contrast, phage genes are called upon to express only at fixed times during the lytic cycle by an irreversible process, which permits the concentrations of the regulatory proteins to vary over wide ranges. There is no need for small-molecule effectors to modulate the activity; rather, the activity of the genes is controlled by the concentrations of the regulatory proteins themselves.

SUMMARY

In this chapter the regulatory systems of the *E. coli* bacterium and the lambda bacteriophage are discussed. To judge from the size of its genome, *E. coli* carries about 4400 genes. Only a small fraction of the genome is actively transcribed at any given time. But all of the genes are in a state where they can be readily turned on or turned off in a reversible fashion. The level of transcription is regulated by a complex hierarchy of control elements.

In the most common form of control, expression is regulated at the initiation site of transcription. There are several ways of doing this, all of them revolving around protein or small-molecule factors that influence the binding of RNA polymerase at the transcription start site.

The *lac* operon, a cluster of three genes involved in the catabolism of lactose, exemplifies both positive and negative forms of control that influence the rate of initiation of transcription. The foundation for understanding the action of regulatory proteins was laid by the combined genetic and biochemical studies of Jacob and Monod on the *lac* operon. The repressor was identified as a negative control element that is *trans* dominant. The operator was identified as a *cis* dominant site for binding the repressor. Transcription is initiated by RNA polymerase at the promoter, which overlaps the operator site on one side of the three structural genes of the *lac* operon. The tight complex between repressor and operator prevents initiation. This complex is broken when lactose is present. The lactose is readily converted to allolactose, which binds to the *lac* repressor. This changes the structure of the repressor so that it dissociates from the DNA.

Initiation of transcription proceeds at a greatly increased rate when cyclic AMP is present. This is because cyclic AMP forms a complex with the CAP activator protein, which then binds at a site adjacent to the polymerase binding site. The CAP protein enhances polymerase binding at the adjacent site by cooperative binding.

This system of control is very enonomic. Lactose is the substrate of the enzymes of the *lac* operon. In the absence of lactose, there is no use for enzymes of the *lac* operon. Understanding the cAMP effect is a little more complicated. CAP and cAMP activate a large number of genes in *E. coli* that are concerned with catabolism. When glucose is present, the cAMP is greatly lowered and the *lac* operon is expressed at a very low level, even when lactose is present. This is because glucose is a more readily metabolizable carbon source than lactose.

The second gene cluster examined is the *trp* operon. This contains a cluster of five structural genes associated with tryptophan biosynthesis. Initiation of transcription of the *trp* operon is regulated by a repressor protein. This repressor functions similarly to the lac repressor. The main difference is that the trp repressor action is subject to control by the small-molecule effector tryptophan. When tryptophan binds the repressor, the repressor binds to the *trp* operator. Thus the effect of the small-molecule effector here is opposite to its effect on the *lac* operon. Again, this is logical because of the functions involved. When tryptophan is present, there is no need for the enzymes that synthesize tryptophan.

The *trp* operon also has a control locus about 150 bases after the transcription initiation site. The control site, called an attenuator, is regulated by the level of charged tryptophan tRNA, so that between 10 and 90 per cent of the elongating RNA polymerases transcribe through this site to the end of the operon. Low levels of *trp* tRNA encourage transcription through the attenuator.

Ribosomal RNA and protein synthesis are both controlled at the level of initiation of transcription. This is a result of the direct binding of guanosine tetraphosphate, ppGpp, to the RNA polymerase. This binding decreases the affinity of RNA polymerase for the initiation sites of transcription. Guanosine tetraphosphate is synthesized when the general level of amino-acid-charged tRNA is low.

The synthesis of ribosomal proteins is also regulated at the level of translation. Certain ribosomal proteins bind to specific sites on the ribosomal RNAs or their own mRNAs. In the absence of the ribosomal RNAs, they will bind to their own mRNAs, which inhibits their translation. This form of translational control seems to be geared to regulating the synthesis of ribosomal proteins so that it does not exceed the rate of ribosomal RNA synthesis.

Viruses are obligate parasites; they borrow heavily on the host machinery to obtain energy for synthesis, as well as for replication, transcription, and translation. The virus life cycle is strongly irreversible. Virus infection is followed by the gradual turning on of viral genes. In the early phases of virus infection, viral enzymes are the most frequent viral gene products; in late virus infection the virus structural proteins are favored. The irreversible lytic cycle of the virus is directed by a cascade of controls.

In λ the host RNA polymerase is used throughout. Regulation is achieved through a series of repressors and activators, as well as two viral proteins that bind directly to the RNA polymerase. The viral proteins that bind to the polymerase modify it so that it can transcribe through provisional stop signals.

When λ phage infects an *E. coli* cell, it does not always produce viral progeny. Sometimes it integrates its genome into the host genome and replicates only as the host genome replicates. This so-called lysogenic state can be disrupted by SOS conditions. Under these conditions the dormant viral genome enters the active replication cycle.

Whereas bacterial genes and viral genes use many similar types of regulatory elements, there are two outstanding differences: Bacterial regulatory proteins are controlled by small-molecule effectors; viral regulatory proteins are not. Bacterial genes are regulated in a highly reversible manner; viral genes are usually turned on only once.

SELECTED READINGS

Anderson, J. E., M. Ptashne, and S. C. Harrison, Structure of the repressor-operator complex of bacteriophage 434. *Nature* 326:846–852, 1987.

Botstein, D., and R. Maurer, Genetic approaches to the analysis of microbial development. *Ann. Rev. Genetics* 16:61–84, 1982.

Ebright, R. H., P. Cossart, B. Gicquel-Sanzey, and J. Beckwith, Mutations that alter the DNA sequence specificity of the catabolite gene activator protein of *E. coli*. *Nature* 311:232–235, 1984. Genetic approach to studying the interaction between protein activator and DNA.

Freifelder, D., Bacteriophages, Lytic Phages, in *Molecular Biology*, Chapter 15. New York: Science Books International, 1983.

Gilbert, W., and B. Muller-Hill, Isolation of the *lac* repressor. *Proc. Natl. Acad. Sci.* 56:1891–1898, 1966. First isolation of a repressor protein.

Gottesman, S., Bacterial regulation: Global regulatory networks. *Ann. Rev. Genetics* 18, 415–442, 1984.

Miller, J. H., and W. S. Reznikoff, *The Operon*. Cold Spring Harbor Laboratory, 1978. A collection of papers.

Pabo, C. O., and R. T. Sauer, Protein-DNA recognition. *Ann. Rev. Biochem.* 53:293–322, 1984. How regulatory proteins interact with DNA.

Raibaud, O., and M. Schwartz, Positive control of transcription initiation in bacteria. *Ann. Rev. Genetics* 18:173–206, 1984.

Takeda, Y., D. H. Ohlendorf, W. F. Anderson, and B. W. Matthews, DNA-binding proteins. *Science* 221:1020–1026, 1983. How regulatory proteins interact with DNA.

Weber, I. T., and T. A. Steitz, Model of specific complex between catabolite gene activator protein and B-DNA suggested by electrostatic complementarity. *Proc. Natl. Acad. Sci.* 81:3973–3977, 1984. How CAP activator protein interacts with DNA.

Yanofsky, C., Attenuation in the control of expression of bacterial operons. *Nature* 289:751–758, 1981.

Zhang, R.-g., C. L. Lawson, R. W. Schevitz, Z. Otwinowski, and P. B. Sigler, The crystal structure of *trp* aporepressor at 1.8 Å shows how binding tryptophan enhances DNA affinity. *Nature* 18:591–597, 1987.

Zubay, G., M. Lederman, and J. DeVries, DNA-directed peptide synthesis III. Repression of β-galactosidase synthesis and inhibition of repressor by inducer in a cell-free system. *Proc. Natl. Acad. Sci.* 58:1669–1675, 1967. Showing that repressor works in a cell-free system.

Zubay, G., D. Schwartz, and J. Beckwith, Mechanism of activation of catabolite-sensitive genes: A positive control system. *Proc. Natl. Acad. Sci.* 66:104–110, 1970. First isolation of an activator protein.

PROBLEMS

1. Which of the following originally led Jacob and Monod to suggest the existence of a repressor in *lac* operon regulation? Explain.
   a. *Cis* dominance of the *i* gene.
   b. *Trans* dominance of the *i* gene.
   c. *Cis* dominance of the *o* locus.
   d. Isolation of *i* repressor.

2. In a cell that is z^-, what would be the relative thiogalactoside transacetylase concentration, compared with wild type, under the following conditions?
   a. After no special treatment.
   b. After addition of lactose.
   c. After addition of IPTG.

3. Consider a negatively controlled operon with two structural genes (*A* and *B*, for enzymes A and B), an operator gene (*O*), and a regulatory gene (*R*). The first line of data below gives the enzyme levels in the wild-type strain after growth in the absence or presence of the inducer. Complete the table for the other cultures.

	Uninduced		Induced	
Strains	**Enz A**	**Enz B**	**Enz A**	**Enz B**
Haploid strains:				
(1) $R^+O^+A^+B^+$	1	1	100	100
(2) $R^+O^cC^+B^+$				
(3) $R^-O^+A^+B^+$				
Diploid strains				
(4) $R^+O^+A^+B^+/R^+O^+A^+B^+$				
(5) $R^+O^cA^+B^+/R^+O^+A^+B^+$				
(6) $R^+O^+A^-B^+/R^+O^+A^+B^+$				
(7) $R^-O^+A^+B^+/R^+O^+A^+B^+$				

4. Why do most mutations in the *lac* operon that give rise to the o^c phenotype involve more than a single base change? How would you go about isolating such mutants?

5. Choose the best answer:
   When lactose is added to a culture of *E. coli* growing on glucose, the enzymes involved in lactose metabolism:
   a. Will be synthesized because lactose is the natural inducer for the *lac* operon.
   b. Will not be synthesized because of catabolite repression in the presence of glucose.
   c. Will be synthesized partially and then turned off at the translational level.
   d. Will not be affected whether there is glucose present or not.

6. In a diploid situation, would you expect a crp^+ to be dominant to a crp^-? Referring to *lac* expression, describe the phenotypes of crp^-, crp^+, and crp^+/crp^-.

7. The effect of cAMP and CAP on DNA-directed synthesis of β-galactosidase *in vitro* was studied using dialyzed cell-free extracts derived from the strains (all i^-) listed in the following table. Fill in the last column with values for the β-galactosidase activity that might be expected in each case.

Source of Bacterial Extract	cAMP	CAP	β-Galactosidase
crp^-	−	−	5
crp^-	+	−	
crp^-	−	+	
crp^-	+	+	
crp^+	−	−	
crp^+	+	−	
crp^+	−	+	
crp^+	+	+	100

8. Three mutants with alterations in the machinery of catabolite repression are isolated and found to have the following characteristics:

Mutant	A	B	C
lac operon	Not inducible	Not inducible	Not inducible
ara operon	Not inducible	Not inducible	Inducible
Effect of added cAMP	Both inducible	No effect	No effect
Dominance	Recessive	Recessive	*Cis* dominant

   What is the most likely alteration in each mutant?

9. Although *E. coli* promoters generally conform to a rather well-defined consensus sequence, no perfect match to this consensus has ever been observed in a naturally occurring promoter. Suggest a possible explanation.

10. The crystal structures of *trp* aporepressor and the tryptophan-aporepressor complex have recently been determined. A principal effect of tryptophan binding has been shown to be the movement in the aporepressor dimer of two symmetrically disposed flexible bihelical motifs. Given this observation, suggest a possible molecular basis for the influence of tryptophan on regulation by the *trp* repressor. Would you expect binding of allolactose to the *lac* repressor to involve a similar conformational change? Explain.

11. A mutation in the *trp* leader region is found to result in a reduction in the level of *trp* operon expression when the mutant is grown in rich medium. However, when the mutant is grown in a medium lacking glycine specifically, a stimulation in the level of trp enzymes is observed. Can you suggest an explanation for these observations? What would you anticipate would be the effect of growing the mutant in a medium lacking both glycine and tryptophan?

12. The isolàtion of a *cis* dominant o^c mutation for the histidine operon was reported several years ago. Explain why this claim must have been wrong. What is the most likely explanation for the mutation?

13. Certain antibiotics inhibit protein synthesis by binding to the ribosome but do not result in a rise in ppGpp concentration. Suggest an explanation.

14. Describe the principal differences between patterns of control of gene expression used by host bacterial and bacteriophage systems.

15. In bacteriophage λ some regulatory proteins act as both activators and repressors. How does the mechanism underlying this dual influence compare to that underlying the regulation of the arabinose operon by araC protein in the host?

16. In the infection of a host cell by λ, cI repressor favors lysogeny, while cro repressor favors the lytic cycle, even though both repressors bind to the same sites. Explain the molecular basis of their different effects.

30

REGULATION OF GENE EXPRESSION IN EUKARYOTES

The fundamental difference between eukaryotes and prokaryotes is that the DNA in eukaryotes is carried in a membrane-bounded organelle, the nucleus. Eukaryotes tend to have larger cells and almost always have more than one chromosome. Eukaryotes frequently occur as multicellular organisms, with cells within the same organism playing different roles. All of these differences between prokaryotes and eukaryotes are reflected by increasingly complex mechanisms for gene regulation. We cannot present a comprehensive treatment of this subject because it is too vast and, at the same time, not that well understood. We have chosen to emphasize modes of gene regulation that are different and present new and interesting possibilities.

We will begin the discussion by considering a relatively simple eukaryote, the budding yeast *Saccharomyces cerevisiae*. Then we will take a peek at some of the complex problems associated with higher eukaryotes that show extensive cellular differentiation.

GENE REGULATION IN UNICELLULAR EUKARYOTES

A cell of the yeast *Saccharomyces* occupies more than 10 times the volume of an *E. coli* cell. Its genome is about three to four times the size of that of *E. coli*, and is distributed among 17 chromosomes. Yeast cells are nearly spherical in shape, divide by budding, and each bounded by a cytoplasmic membrane that is surrounded by a thick polysaccharide cell wall (Figure 30–1). Structures visible in the electron microscope include a nucleus, mitochondria, microsomes as well as ribosomes, a Golgi apparatus with secretory vesicles, and several types of granular and vesicular inclusions. The number of mitochondria, microsomes, and ribosomes fluctuates widely with growth conditions, reflecting the presence of regulatory devices that control their numbers.

As in all eukaryotes, the presence of a nuclear membrane leads to separation of the biochemical activities of transcription and translation. Attenuator-type control mechanisms that necessitate the close coupling of tran-

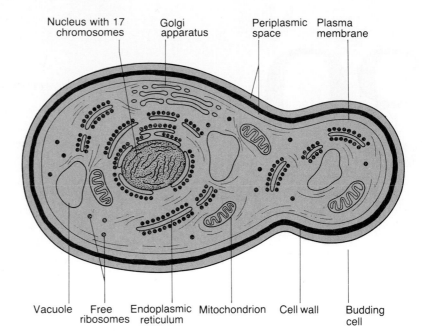

Figure 30–1

Diagram of a haploid yeast cell. A yeast cell contains many of the organelles characteristic of a typical eukaryotic cell. Duplication occurs by a budding process. The bud gradually grows until, just before pinching off, it contains a nuclear equivalent of chromosomes, as well as some mitochondria and other elements present in the cytoplasm. Mechanical strength of the yeast cell is guaranteed by a thick cell wall.

scription and translation clearly could not exist in eukaryotes. However, the separation of transcription from translation permits more elaborate schemes for the processing of nascent transcripts prior to their translation. Nascent transcripts destined to become mRNA are modified by capping the 5′ end and are tailed by 3′-terminal polyadenylic acid, just as they are in higher eukaryotes (see Chapter 27). Splicing of transcripts occurs in some cases during mRNA synthesis (encountered predominantly in ribosomal protein genes in yeast), as well as in tRNA synthesis. Once a processed transcript reaches the cytoplasm, only the 5′ proximal initiator codon (AUG) is usually recognized as a translation start, although important exceptions exist. This restriction eliminates the value of polygenic mRNAs and probably accounts for the greater dispersal of genes serving common pathways that are found in yeast.

If we compare the gene arrangements and messages for histidine biosynthesis in *E. coli* and *S. cerevisiae*, we can see the contrast in gene arrangement and expression. In *E. coli*, all ten histidine biosynthetic enzymes are encoded by a single polygenic mRNA, which is used for the synthesis of the corresponding individual polypeptide chains. In yeast, genes for three of these activities (steps 2, 3, and 10 in the biosynthesis) are clustered at the *HIS4* locus, yielding a single transcript that specifies a single polypeptide chain of molecular weight 9.5×10^4. This polypeptide chain folds into a multifunctional enzyme that carries out three enzymatic functions (Figure 30–2) that in *E. coli* are executed by separate proteins.

Galactose Metabolism Exemplifies Gene Regulation in Yeast

The genes for galactose utilization present another example of the clustering of some common functions. Four enzyme activities are required for galactose utilization. Three of these activities are specified by three tightly linked genes (*GAL7*, *GAL10*, and *GAL1*)* on chromosome II, whereas the fourth, for galactose transport, is specified by a gene (*GAL2*) located on chromosome XII. Expression of genes associated with galactose metabolism is regulated by positive and negative controls that are of a highly specific type. Coordi-

*In yeast, the normal (wild-type) genotype is capitalized; the mutant genotype is written in lower case. Both designations are italicized.

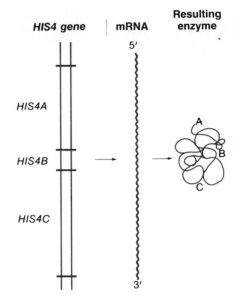

Figure 30–2
The *HIS4* gene encodes three enzyme activities. It produces a single transcript that results in a single enzyme. The enzyme is multifunctional, carrying out steps 2 (B), 3 (A), and 10 (C) in histidine biosynthesis.

nate transcription of the *GAL1*, *GAL7*, and *GAL10* genes is induced several thousandfold by galactose. Regulation appears to occur at the transcriptional level.

Specific Controls Affect Expression of the *GAL* Genes. Despite their clustering, the *GAL7*, *GAL10*, and *GAL1* genes are not transcribed into a single mRNA molecule (Figure 30–3a). The *GAL7* and *GAL10* genes are transcribed from the same DNA strand. However, the *GAL1* gene, approximately 600 base pairs from *GAL10*, is transcribed from the complementary DNA strand. Discrete transcripts are generated for the synthesis of three distinct proteins that contain the different enzyme activities: transferase (*GAL7*), epimerase (*GAL10*), and kinase (*GAL1*). The *MEL1* gene, encoding the enzyme melibiase, is coordinately controlled together with the *GAL* genes.

A genetic analysis of the *GAL* system has revealed two *trans*-acting gene products that regulate expression. *GAL4* (located on chromosome XVI) is required for induction of the *GAL* genes, and *GAL80* (on chromosome VIII) inhibits *GAL4* function in the absence of inducer. *GAL80*^{s} and *gal4* mutants are uninducible, whereas *GAL4*^{c} and *gal80* mutants result in constitutive

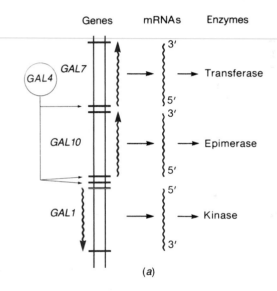

(a)

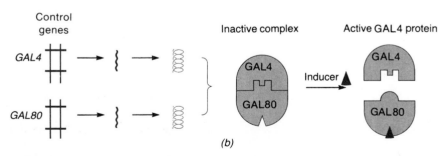

(b)

Figure 30–3
Model for the negative and positive regulation of enzymes involved in galactose metabolism. (a) Three structural genes synthesize distinct mRNAs and enzymes (transferase, epimerase, and kinase). Arrows next to the genes indicate direction of transcription. Synthesis requires the *GAL4* transcriptional activator. (b) GAL4 protein is synthesized constitutively from the *GAL4* gene. However, it is inactive in the absence of inducer because of complex formation with the GAL80 protein. The GAL4 protein can be reactivated in the presence of inducer, which forms a complex with the GAL80 protein. The inducer is either galactose or, more likely, a metabolic derivative of galactose. It is not clear whether this reactivation results in the release of GAL4 from the complex as shown or in a structural alteration without release.

expression of the *GAL* genes. *GAL80*ˢ mutants are dominant to wild type. The *GAL4* gene has been cloned and encodes a protein of 881 amino acids with a molecular weight of 99,000 that is located in the cell nucleus. The GAL4 protein is transcribed constitutively; it promotes RNA polymerase II-dependent transcription of the target genes, but the GAL80 product represses transcription by forming an inactive complex with the GAL4 protein (Figure 30–3b). Induction is assumed to involve the dissociation of the GAL4-GAL80 encoded complex, allowing the GAL4 protein to function. Gene dosage of GAL4 (cloned on a multicopy vector) has shown that its amplification over *GAL80* results in an escape expression* of the cluster of *GAL* genes. Glucose strongly inhibits *GAL4*-directed transcription, regardless of the presence of any *GAL80* allele.

An intense effort is currently being made to study the *cis*-acting sequences in the intergenic region between *GAL1* and *GAL10* that are the site of action of the positively acting GAL4 protein. These GAL4 protein-binding sequences, being part of the upstream promoter elements, have been referred to as UAS, or upstream activating sequences. In the case of *GAL* gene regulation, the recognition site for the regulatory protein is designated UAS$_G$. The GAL4 protein binds to four related 17-bp dyad symmetric sequences upstream from *GAL1* and *GAL10*, activating their (divergent) transcription.

That the GAL4 protein binds to specific DNA sequences could be shown *in vivo* by methylating the DNA with dimethylsulfate and then following the methylation pattern by nucleotide sequencing to determine which guanine bases have been protected by the regulatory protein. The sequences that showed *in vivo* protection in a wild-type *GAL4* strain were unprotected in a *gal4* mutant derivative. When the 17-bp recognition site was synthesized chemically and introduced in various positions upstream of the *GAL1* gene devoid of such sequences, a single synthetic site (that is nearly a consensus of the four natural sites) was found sufficient to allow nearly wild-type levels of regulated, *GAL4*-mediated inducibility of the engineered yeast cells. When yeast is grown on glucose as a carbon source, the GAL4 protein-induced protection of its target DNA sequences from dimethyl sulfate is not observed. Thus glucose repression (discussed below) may involve, in part, inhibition of the binding of the GAL4 protein to DNA.

Recent studies from Mark Ptashne's laboratory have used deletion and gene fusion experiments to characterize the functional domains of the GAL4 protein. There are two distinct and separable domains; first, an amino terminal domain of about 100 amino acids that represents the *GAL4* DNA binding site and second, two regions that specifically activate *GAL* transcription (when fused to the DNA-binding region). Interestingly, as much as 80 per cent of the GAL4 protein can be deleted without drastic loss of the transcription activation function.

Catabolite Gene Products Are Regulated by Repression and Inactivation. Induction of the *GAL* genes in yeast not only requires the presence of galactose but the absence of glucose in the growth medium. Glucose can cause carbon catabolite repression, similar to the effect encountered in *E. coli* (Chapter 24). The mechanism, however, is quite different. *In yeast, carbon catabolite repression can result from enzyme inactivation and/or inhibition of transcription or translation.* Since the inactivation mechanism does not occur in *E. coli*, the term *catabolite inactivation* is used to distinguish this mechanism from catabolite repression. Both effects are general, involving a variety of enzymes that are capable of converting their substrates to metabo-

---

*Called escape expression because there is not enough GAL80 protein to complex all of the GAL4 protein.

lites that the cell can obtain independently and more readily from the metabolism of glucose.

Surprisingly little is known about the precise mechanisms leading to catabolite repression and inactivation in yeast. cAMP does not play a key role, since yeast mutants defective in cAMP synthesis are unaffected in carbon catabolite repression. In recent studies, mutations in *cis*-acting regions upstream from various different genes (*ADHII*, alcohol dehydrogenase, and *GAL10*) have been used to locate the sites where carbon catabolite repression is exerted. Such sites have been identified; however, it is not clear whether the glucose effect is exerted through a repressor that affects downstream transcription. Studies on the catabolite repression of invertase have provided some clues as to a possible mechanism. Carlson has found that a key gene *(SNF1)* essential for the expression of various glucose-repressible genes encodes a protein kinase. It is possible, then, that glucose exerts its effect by phosphorylating proteins that regulate the expression of specific genes.

Positive and Negative Controls of Transcription Are Common in Yeast

Cloning, *in vitro* mutagenesis, the use of genetically defined mutants of regulatory genes, gene fusions, and biochemical studies all have helped to define the *cis*-acting sites, or elements, that modulate transcription of the *GAL* structural genes. These promoter (or controlling) elements may be found up to several hundred nucleotides from the transcription initiation sites and have been referred to as *upstream activation sites* or *UAS*. They include components of the promoter where RNA polymerase II may enter to move downstream toward the TATA box.

In addition to the *GAL* system, the *cis*-acting regulatory regions of other yeast genes have been investigated (Figure 30–4). The gene encoding iso-1-cytochrome *c (CYC1)* is also under positive control; the gene *HAP1* encodes a CYC1-like activator whose activity or synthesis is dependent on heme. The transcription of *HIS3* and *HIS4* (involved in histidine biosynthesis) is derepressed by starvation for histidine or other amino acids (general amino acid control). Starvation for a single amino acid brings about a derepression of amino acid biosynthetic genes. These coregulated genes have two to four

Figure 30–4
Location of control elements for a select group of yeast genes. The genes have all been found to contain upstream activation sites (UASs) or upstream sites of repression. Locations of TATA boxes and initiation sites are also indicated. In some cases there is more than one initiation site.

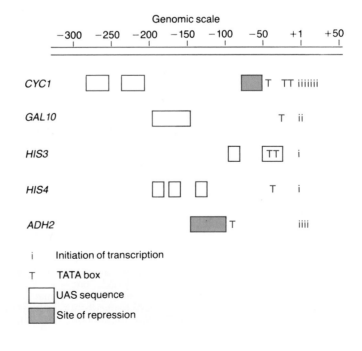

copies of a *cis*-active DNA sequence, 5′-TGACTC-3′, in far upstream regions. Several *trans*-acting genes act in a regulatory hierarchy in the starvation response. The gene *GCN4* functions most directly as a positive regulator of transcription; it does so by an interaction of the GCN4 protein with the TGACTC sequence in the promoter regions of the genes regulated by the general control.

Figure 30–4 summarizes the *cis*-acting sites of some of these genes. In all cases these include TATA boxes (T), another promoter element that is also recognized by RNA polymerase II. The open boxes indicate the sites that mediate positive control of transcription. For example, the site that interacts with the GAL4 protein that regulates *GAL10* gene expression is located more than 100 base pairs upstream of its TATA box. The *CYC1* gene has two homologous sites, at about −275 and −225, that mediate activation by heme. UAS control sites can be located at considerable distances from the genes they influence. Furthermore, they can be oriented in either direction relative to the genes they influence. In these respects they appear to be quite different from controls found in prokaryotes, but quite similar to enhancer control elements (discussed later in this chapter).

Specific negative regulators that act on an operator locus are rare in yeast. An example is the protein encoded by one of the genes of the MAT_α locus (see later discussion).

How different are the controlling regions in yeast from bacteria? In yeast there can be several controlling sites that are variable in their spacing and that are located large distances from the sites of mRNA initiation. The multiple UASs for a given gene permit a variety of regulatory transcriptional responses, yet the TATA boxes still determine the sites for transcription initiation.

Mating Type Is Determined by Transposable Elements

Yeast is a typical eukaryote in that it has two haploid cell mating types, MAT_a (or simply *a*) and MAT_α (or *α*) that can fuse to make a single diploid (*a/α*) cell. The mating type of *S. cerevisiae* is determined by the *MAT* locus. Haploid cells with the <u>homothallism (HO)</u> gene are able, at almost every cell division, to switch between the *a* and *α* mating types in a direct fashion. This efficient switching generates progeny of opposite mating types from a single cell, giving rise to *a/α* diploids that do not switch or mate, but can sporulate when starved. Haploid <u>heterothallic (ho)</u> strains interchange their mating types rarely, on the order of 10^{-6} events/generation. During interconversion of the mating type, the DNA sequences of *MAT* are replaced by either the *a* or *α* sequences copied from one of the two silent storage copies of mating-type information, or cassettes, HML_α or HMR_a (Figure 30–5). When the se-

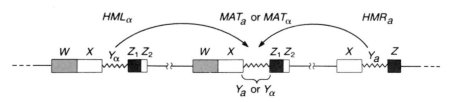

Figure 30–5

Yeast chromosome III, showing the *HML*, *MAT*, and *HMR* loci. *W* is a region common to *MAT* and *HML*, but not *HMR* (~750 bp). *X* is a region found at *MAT*, HML_α, and HMR_a (~700 bp). Y_a is a specific substitution found at MAT_a and HMR_a (~600 bp). Y_α is a specific substitution found at HML_α and MAT_α (750 bp). Z_1 is a region found at *MAT*, HML_α, and HMR_a (~250 bp). Z_2 is a region found at *MAT* and HML_α (~70 bp).

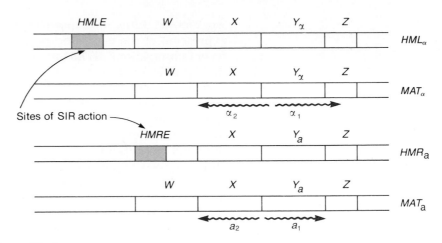

Figure 30–6
Structure of mating loci determinants in yeast. The genetic regions HML_α and HMR_a are normally silent, whereas MAT_a or MAT_α are active. MAT_a which contains the Y_a segment, expresses two transcripts, a_1 and a_2. Similarly MAT_α, which contains the Y_α segment, expresses two transcripts, α_1 and α_2. The structures in and around the Y_a and Y_α segments are the same at the storage locus and the expression locus. The inactivity at the storage loci results from far upstream *cis*-acting elements not present at MAT. These elements, in conjunction with the four SIR gene products, negatively regulate expression of mating type genes at the storage loci.

quences are transposed from HML_α to MAT, the α mating type is expressed; transposition from HMR_a to MAT leads to expression of the opposite, or a, mating type.

Two immediately interesting questions arise about regulation associated with mating type switching of haploid yeast strains: (1) What regulates mating type switching? And (2) why do the a and α sequences express only when they are present at the MAT locus? Partial answers are available to both of these questions. In both cases it appears that *trans*-acting control elements are involved.

The a and the α sequences that are stored at the HML and HMR loci are normally expressed only when they are present at the MAT locus. It is possible to imagine several reasons for this, but the explanation was revealed by the finding of at least four unlinked genes, $SIR1–SIR4$, that appear to be regulatory in nature. In all *sir* mutant strains the a and α sequences at HML and HMR that are normally silent express and provide the same functions as they provide when present at MAT. Repression is mediated by one or more *trans*-acting regulatory factors, encoded by the SIR genes, and acting on "silencer" sites ($HMLE$ and $HMRE$) adjacent to HMR and HML that are more than a kilobase away from the regions of transcription initiation (Figure 30–6). These *cis* sites are not found at the mating type locus (MAT) so that the a and α coding sequences can be transcribed when transposed to MAT. The *cis*-acting $HMRE$ sequence is able to turn off function in either orientation and, when placed in front of promoters of other genes, can act to repress at distances 2600 bp away.

The important question remains as to how a single genetic locus determines the haploid yeast cell mating type. Cells that carry MAT_a behave as a cells; cells that carry MAT_α behave as α cells. It was proposed in 1974 that the MAT locus encodes regulatory proteins that control unlinked genes that are involved in cell type. An example of an α-specific gene is $MF_{\alpha 1}$, encoding the α-factor mating peptide. An example of an a-specific gene is STE6, involved in a-factor production. It was later proposed that the MAT_α locus contains two genes, $MAT_{\alpha 1}$ and $MAT_{\alpha 2}$ that encode two regulators, $\alpha 1$ and $\alpha 2$, respectively. According to the model proposed by Herskowitz and coworkers, α1 _turns on_ expression of the α-specific genes and α2 _turns off_ expression of the a-specific genes. Supporting the model are a number of recent observations. STE6 (an a-specific gene) mRNA is synthesized in cells that lack the α2 product. But more significantly, it has been shown that α2 is a sequence-specific DNA-binding protein. It recognizes a 32-bp operator sequence upstream of STE6. The sequence is also found in four other a-specific genes. When this operator is placed between the CYC1 UAS and its transcription initiation site, α2 brings about repression, or negative control, _in vivo_. Even when the operator is placed upstream of the UAS sequence it still represses expression of CYC1, but to a lesser degree. These studies clearly show that the operator need not overlap with essential promoter sequences (as in prokaryotes) to permit repression by α2.

How then does a eukaryotic repressor like α2 act? Herskowitz considers three models: (1) direct steric hindrance, where α2 blocks transcription by binding to a DNA region required by RNA polymerase or factor; (2) indirect steric hindrance, which would involve α2 interacting specifically with some proteins, such as histone, that would in turn block the availability of promoter sequences; and (3) diffusion blockage, in which α2 bound to DNA blocks the movement of RNA polymerase or factor.

The α2 protein resembles to some extent a bacterial repressor protein such as lexA, which acts directly upstream of several dispersed genes involved in bacterial recombination and repair to repress their transcription. The α2 protein, on the other hand, controls regulated genes concerned with cell specialization. The main difference between the two repressors is that the operator for α2 does not overlap the promoter that it regulates.

The control of mating type in yeast illustrates one use of movable genetic elements. Movable genetic elements are commonly found in both prokaryotes and eukaryotes, but they rarely insert at specific sites as is required in this case. In particular instances they are called _insertion elements_ or _transposons_, but the more general term is _movable genetic elements_. Because of their random behavior, most movable genetic elements are more apt to be vehicles for evolutionary change than mechanisms for regulating specific genes as in yeast mating type switching. In higher organisms, mating type (sex) is controlled in part by gene dosage. For example, in humans, males carry one X chromosome whereas females carry two X chromosomes. The mechanism used by yeast and other fungi is more susceptible to factors in the external environment and may be more suitable for a unicellular organism, which is more exposed to the environment.

This concludes our discussion of the regulatory behavior of yeast. We have stressed cases where there are differences between the regulatory mechanisms used by yeast and _E. coli_. But there is one way in which the two organisms are very similar. All of the regulatory processes are highly reversible, so that each cell retains its _totipotency_—that is, its potential to express any of its genes. This situation does not hold true in more complex eukaryotes.

GENE REGULATION IN MULTICELLULAR EUKARYOTES

In terms of the types of regulatory mechanisms used, yeast is probably typical of the simple, relatively undifferentiated eukaryote. Even in yeast we see the beginnings of _cell differentiation processes_ and _hormonal systems for intracellular communication_ that typify complex eukaryotes. Thus yeast shows differentiation into three cell types, the a and α haploid types and a/α diploids. Furthermore, haploid yeast cells secrete peptides known as _pheromones_ that impinge on cells of the opposite mating type and change their pattern of gene expression so as to favor cell-to-cell contact and mating. Such similarities between yeast and more complex eukaryotes have made yeast valuable as a model system for studying cellular differentiation and hormonal control, but they do not eliminate the need to study these processes more directly in higher forms.

There are a number of new regulatory problems that must be confronted in higher eukaryotes:

1. The arrangement of DNA in the nucleus must be efficiently disposed so that the small number of genes (probably less than 1 per cent) that get expressed in any particular type of differentiated cell are accessible.
2. As the embryo develops, new genes must be expressed and already expressed genes in some cases must be turned off. Similar problems were confronted by the developing λ bacteriophage (see Chapter 29) but there are obviously additional complications. The process must be carefully timed and spatially organized so that different cell types are produced in the proper numbers and orientation relative to one another.
3. Even in the adult organism there are developmental processes underway. This problem is particularly well understood for the immune system.
4. Finally, in the developing organism and the adult organism alike, there is a need for intercellular communication to regulate biochemical activities occurring in different cells. This need is met by a complex array of diffusible biochemical signals known as hormones, which trigger responses at the cell membrane in some cases and at the nucleus in others.

We will touch on all these subjects except hormones in the following sections. Hormones are dealt with in Chapter 31.

Nuclear Differentiation Occurs During Early Development

In the mature adult state, a higher eukaryote contains many cells, each with a potential for the constitutive or inducible expression of only a small select fraction of the total genome. Whereas most differentiated cells in the mature adult organism contain the normal amount of genomic DNA, the expression of only a fraction of the genome in any particular cell type suggests that the average nucleus may have lost its totipotency.

The first direct evidence that nuclei undergo a process of irreversible differentiation during development came from the work of Briggs and King in 1952. Working with frog eggs, they showed that the original nucleus from an unfertilized egg could be replaced by the diploid nucleus from a developing animal (Figure 30–7). If the new nucleus came from a frog very early in development (blastula), a mature frog developed. If, however, the nucleus came from a later stage of development, no growth or only limited growth ensued. These nuclear transplantation experiments demonstrated that _during development the nucleus assumes a pattern of expression that is difficult or impossible to reverse_ even when the nucleus is returned to the environment of the egg cytoplasm.

In 1962 Gurdon took this experiment one step further by showing that adult frogs could be developed by injecting enucleated eggs with nuclei from the intestinal epithelium of tadpoles. The success frequency was much lower than when nuclei from earlier stages of development were used. Nevertheless, such results show the influence of the cytoplasm on nuclear expression; they also demonstrate that in certain cases nuclei already tentatively committed to a specific pathway of development can be reprogrammed by placing them in a different environment. The most important lesson to be gained from such experiments is that _the nucleus tends to assume a stable pattern of expression that is ultimately dictated by its surroundings_.

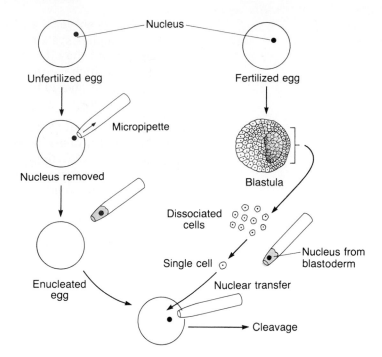

Figure 30–7
Nuclear transplantation technique. The nucleus of an unfertilized egg is mechanically removed with a micropipette *(left)*, and a nucleus from a blastula cell is removed and microinjected into the enucleated egg.

These pioneering studies of Briggs and King and of Gurdon, demonstrating nuclear differentiation and dedifferentiation, are supported by a broad range of genetic and biochemical findings that indicate that chromosomal structural differences often can be correlated with changes in the potential for gene expression.

Chromosome Structure Varies with Gene Activity

In interphase cells, chromosomes exist in a dispersed form called *chromatin*. Chromatin can exist in either a compact form known as *heterochromatin* or a swollen form known as *euchromatin*. Most nuclei contain both types of chromatin. The heterochromatin is generally concentrated around the nucleolus and the inside of the nuclear envelope. Some chromosomes or parts of chromosomes are always heterochromatic *(constitutive heterochromatin)*; others are heterochromatic only during certain times of the cell cycle or in certain cell types *(facultative heterochromatin)*. The ratio of euchromatin to heterochromatin increases with increasing protein synthetic activity. Many studies indicate that euchromatin is relatively active in RNA synthesis and heterochromatin relative inactive. Whereas *the euchromatic state seems to be necessary for a high state of transcription, for most genes it is not sufficient*. This has been demonstrated by autoradiography of cells grown briefly in the presence of ^{3}H-labeled RNA precursors. Those regions of the genome that are transcriptionally active show labeling, and the extent of labeling is proportional to the amount of RNA synthetic activity. Some euchromatic regions appear active by this test, whereas others do not.

Giant Chromosomes Permit Direct Visualization of Active Genes. *Polytene chromosomes* are unusually large chromosomes found in the salivary glands of certain insects. Because of their size, they are convenient vehicles for studying the relationship between chromosome structure and activity. Polytene chromosomes are produced by the repeated replication of interphase chromosomes without separation, resulting in a large number of chromosomes that remain laterally aligned. Thus, for example, the DNA content in the giant salivary gland cells of *Drosophila melanogaster* may be as much as

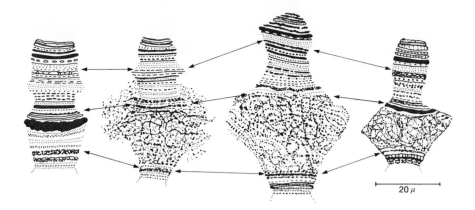

Figure 30–8
A segment of giant chromosome from the midge larva of *Rhyncosciara* at different stages during development. The arrows and connecting lines indicate comparable bands. Changes in the extent of swelling of different regions reflect the activity of those regions in transcription. (Drawing based on the results of M. F. Breuer and C. Pavin, *Chromosoma* 7:275–280, 1946.)

1000 times that of other cells in the fruit fly. Microscopically, the stained chromosomes appear as cross-banded extended bodies (Figure 30–8). It has been estimated that there are about 5000 bands in *Drosophila*, and it is tempting to associate single bands with single genes. However, the average DNA content found in each band is 30,000 base pairs, which is considerably more DNA than necessary to encode the average protein (about 1000 base pairs).

Transcription in polytene chromosomes usually is associated with local swellings of the bands, called puffs (see Figure 30–8). Sometimes puffing results in a broadening and lengthening, and sometimes the extended DNA projects laterally into loops that combine to form a large ringlike structure. As a rule, a puff originates from a single band, but it can result from a swelling of one or more bands (Figure 30–9). *The extent of incorporation of precursors into a puff is approximately proportional to the size of the puff.* A detailed investigation of the puffing patterns as a function of the physiological state of the organism has shown that different bands become activated, swollen, and transcriptionally active in a sequentially related, tissue-specific fashion.

The correlation of puffs with genetic functions has focused on two types of gene products, secretory proteins and proteins formed in response to heat

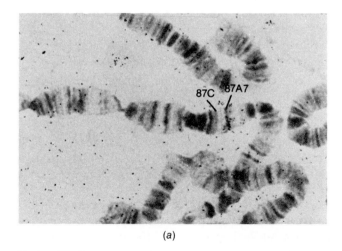

(a)

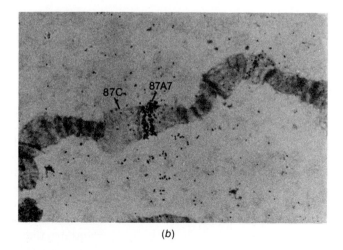

(b)

Figure 30–9
In situ hybridization to salivary gland polytene chromosomes. Heat shock locus 87A7 is cytogenetically localized by hybridization to labeled DNA probes containing 87A7 DNA sequences. The segment is labelled before heat shock (a) and after heat shock (b). It can be seen that the region in and around the 87A7 locus swells dramatically in response to heat shock. An adjacent locus, 87C, is also part of the region that swells on heating, but does not label because DNA in this region does not contain 87A7 specific sequences. (Autoradiographs of A. Udvardy, E. Maine, and P. Schedl, Department of Biology, Princeton University.)

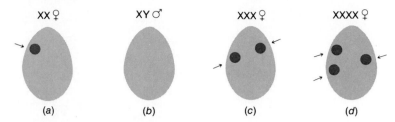

Figure 30–10
Diagram of nuclei obtained from cells in the mucous membrane of the human mouth. *(a)* Nucleus from a female, showing one Barr body (arrow). *(b)* Nucleus from a male, with no Barr body. *(c)* Nucleus from an XXX female, showing two Barr bodies. *(d)* Nucleus from an XXXX female, showing three Barr bodies.

shock. Certain puffs in *Drosophila* can be artificially induced by the insect hormone *ecdysone*, which is instrumental in regulating development. Some puffs are induced directly by exposure to the hormone, whereas other puffs are induced indirectly, being dependent for puffing on the gene products of the more directly induced puffs. This dependence has been shown by adding drugs that inhibit protein synthesis, with the result that secondary puffs do not form.

In Some Cases Entire Chromosomes Are Inactive.   In some cases entire chromosomes are heterochromatized and stay that way through successive cell divisions. Male and female mammals are distinguished by the fact that females carry two X chromosomes, whereas males carry only one. Invariably, one of the X chromosomes in the somatic tissue of females exists in the condensed heterochromatic state known as a *Barr body* (Figure 30–10). The consequences of X-chromosome inactivation are readily observed in females who are heterozygous for an X-linked mutation.

For example, the enzyme glucose-6-phosphate dehydrogenase (G-6-PD) is encoded by an X-linked gene. A female that is heterozygous for this gene may carry two alleles that produce electrophoretically distinct forms of the same enzyme (A and B). When isolated cells from a skin biopsy are cloned, each clone contains either the A or the B form of the G-6-PD enzyme, but never both. If every skin cell of a single organism were to be analyzed for the enzyme, large homogenous patches expressing one or the other of the G-6-PD alleles would be found. This pattern has been interpreted as indicating that *the decision as to which X chromosome of the female is to be inactivated is made at the multicellular stage in the embryo. Once the decision has been made, that X chromosome remains inactive through successive cell divisions*. There is an equal chance that the X chromosome from the maternal or the paternal parent will be inactivated. In genetically abnormal cells that contain three or four X chromosomes, two or three Barr bodies respectively can be found (see Figure 30–10).

X-chromosome inactivation provides a simple means of maintaining equal amounts of active X-linked genes in both males and females. This form of so-called *dosage compensation* is not found in all species. For example, in *Drosophila melanogaster*, neither of the X chromosomes in the female cells is condensed into a Barr body. In this species, dosage compensation, which regulates the activity of specific alleles, operates by a different mechanism so that many alleles in the female X are about 50 per cent as active as the corresponding alleles in the male. However, not all alleles in the *Drosophila* X chromosome are regulated in this way, so that for certain genes twice as much gene product is produced in females as in males.

Chromosomal inactivation takes a different form in the ciliated proto-zoan *Tetrahymena*. In the vegetative phase each cell contains two nuclei, a *micronucleus* and a *macronucleus*. The micronucleus is transcriptionally inactive; it divides mitotically during normal cellular duplication and it undergoes meiosis during the sexual phase of the life cycle. The macronucleus contains much more DNA than the micronucleus, but only about 10 per cent of the sequences present in the micronucleus are represented in the macronucleus. Clearly, the average sequence in the macronucleus is represented many times. The macronucleus divides in an imprecise fashion during normal cell divisions. It is completely lost during meiosis and therefore must be generated anew following each sexual phase.

All nuclear transcripts are made in the macronucleus. From its size and limited sequence complexity, it is clear that the macronucleus must contain many copies of selected regions of the genetic nucleus. How the number of copies made from different genes is regulated is unknown. Selective hybridization studies on rDNA have shown that the micronucleus contains about 20 copies of the rDNA genes, whereas the macronucleus contains about 200. These observations on *Tetrahymena* demonstrate that even in a relatively simple eukaryote the percentage of the nuclear DNA that is transcriptionally active is quite small (about 10 per cent). This intricate compartmentalization and amplification of transcriptionally active regions of the genome in the vegetative nucleus provides a mechanism for the organism to regulate the number of genes according to its needs.

A good deal of the preceding discussion could be summarized by the statement that *chromatin active in transcription tends to exist in a more swollen state than inactive chromatin*. Several biochemical changes accompany these gross morphological differences. These changes include a redistribution of proteins along the DNA duplex, their chemical modification, and an alteration in the pattern of chromosomal proteins binding to the DNA. At the present time, most of these changes can be discussed only in a descriptive manner, since we do not know their causes or their consequences.

Histones Are the Main Proteins Found in Chromatin.   In order to explore the question of chromatin structure further, we must consider the detailed way in which DNA is complexed with protein in the chromatin.

In the interphase state of a eukaryotic cell, DNA exists as a nucleoprotein complex. The proteins present are of two types: approximately equal weights of five basic proteins, which are collectively known as histone, and variable amounts of less well characterized proteins, which are collectively referred to as non-histone proteins. Certain cells, such as chicken erythrocytes or calf thymocytes, that are especially low in transcription activity provide a source of almost pure nucleohistone.

A good deal of effort has gone into the characterization of histones and the nucleohistone complex. It seems likely that histones are largely responsible for the degree of compaction seen in the interphase nucleus. The net positive charge of the histones is sufficient to neutralize about 80 per cent of the nucleic acid phosphates. The electrostatic interaction between DNA and histone is undoubtedly important in stabilizing the DNA-histone complex, but it cannot be the whole story. Since the same kind of structure is formed with almost any DNA, it is clear that the specific sequence of bases is not a major factor in determining the stability of the nucleohistone complex. The uniformly grooved shape of the DNA duplex appears to be ideally suited to binding to the multiprotein histone complex (see page 1025).

Electron microscopic and x-ray diffraction studies on chromatin have led to the proposal that chromatin exists as a coiled structure. Because of the complexity of the structure, precise information has been difficult to obtain. A breakthrough came when Olins and Olins observed that chromatin viewed

Figure 30–11
Swollen fibers of chromatin from the nucleus of the chicken red blood cell. The electron micrograph is enlarged about 325,000 × and negatively stained with uranyl acetate. The chromatin shows a beaded structure with a spacing between beads of about 140 Å and a bead diameter of about 100 Å. (Micrograph generously donated by Ada L. Olins and Donald E. Olins.)

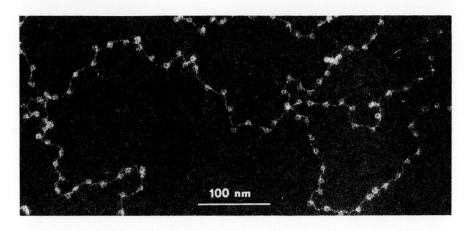

after sudden swelling in water shows a beaded structure (Figure 30–11). The beads, called _nucleosomes_, contain most of the histone; they are about 100 Å in diameter, and the spacing between the beads is about 140 Å. Brief enzymatic digestion of chromatin (using micrococcal nuclease or some other endonucleases) fragments this structure. The DNA from such partial digestion gives rise to a banded pattern on agarose gel electrophoresis that indicates nucleoprotein structures containing 200 base pairs of DNA or multiples thereof (400, 600, 800 bp, etc.). Examination of the fractions in the electron microscope shows a direct correlation between the size of the DNA estimated on gels and the number of nucleosomes. Thus the most rapidly moving DNA band seen on gels is derived from a structure containing one nucleosome, and the second fastest migrating species contains nucleosome dimers, etc. Evidently, the brief treatment with endonuclease preferentially cleaves DNA in the internucleosomal region where the DNA is least likely to be protected from enzyme attack.

The histones present in chromatin have been extensively characterized, and their sequences are known (Table 30–1). The lysine-rich histone H1 is not present in the nucleosome core particle. Its precise binding site is unknown. There are eight other histones—two each of the four histones H2A, H2B, H3, and H4. These form a tightly complexed core particle that is conserved when chromosomes duplicate. The structure of this _histone octamer_ (H2A-H2B-H3-H4)$_2$ has been determined by E. N. Moudrianakis and his colleagues to a resolution of 3.3 Å. The octamer has the approximate shape of a prolate ellipsoid 110 Å long and 65 to 75 Å in diameter (Figure 30–12a). The most striking feature of the histone octamer is its tripartite organization into a central (H3-H4)$_2$ tetramer flanked by two H2A-H2B dimers. Spectroscopic measurements indicate that nucleosomal DNA has a secondary structure quite similar to B DNA, with 10.4 base pairs per turn. The DNA duplex probably wraps around the octamer in a left-handed superhelical fashion,

TABLE 30–1
Characteristics of Histones

Histone	Ratio of Lysine to Arginine	M_r	Copies per Nucleosome
H1	20	21,000	1 (not in bead)
H2A	1.2	14,500	2 (in bead)
H2B	2.5	13,700	2 (in bead)
H3	0.7	15,300	2 (in bead)
H4	0.8	11,300	2 (in bead)

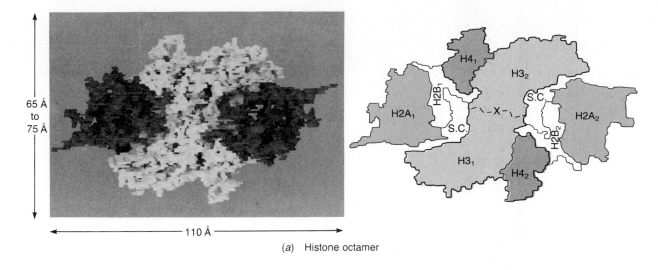

(a) Histone octamer

Figure 30–12
Structure of the histone octamer (a) and proposed structure of the complex between the histone octamer and DNA (b). The histone octamer structure was determined by x-ray crystallography. The DNA is believed to wind around the octamer to form a left-handed superhelix. Two full turns of DNA superhelix require 168 base pairs; the ends of the DNA in two full turns are separated by 75 Å. In (a), $H2B_1$ and $H2B_2$ are barely visible from the angle shown. The initials S.C. stand for solvent channels. (After R. W. Burlingame, W. E. Lover, Bi-Chen Wang, R. Hamlin, N.-H. Xuong, and E. N. Moudrianakis, *Science* 228:546– 553, 1985. Copyright 1985 by the AAAS. Used with permission.)

(b) Histone octamer with DNA

following the contour dictated by the grooves and ridges on the octamer surface (Figure 30–12b). A single nucleosome unit contains one octamer and approximately 168 base pairs in two full superhelical turns of DNA. The DNA appears as a spring holding the H2A-H2B dimers at either end of the $(H3-H4)_2$ tetramer.

The nucleosome structure *in vivo* exists in a variety of states, with different spacings between individual nucleosomes and different degrees of compaction of the overall structure. It seems likely that the H1 histone and other chromosomal proteins play a role in determining the higher-order structures and the different degrees of compaction found in chromosomes.

Histones Are Often Modified in Active Chromatin. *Histones undergo transient modifications that may have a general effect on transcription.* The amino group in the side chain of lysine is subject to acetylation and methylation. Methylation also results in modification of certain arginine and histidine residues. Histones H3 and H4 are the main targets of these reactions, but H2A and H2B are also affected. Modification reaches a maximum during

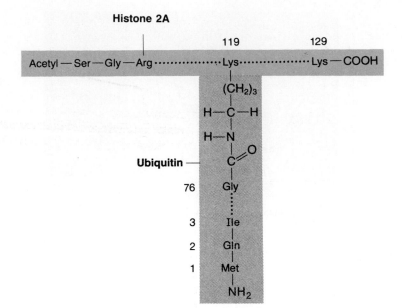

Figure 30–13

Histone H2A can become linked to ubiquitin to form UH2A. The linkage involves the α-COOH on the C-terminal glycine of ubiquitin and the side-chain amino group on a lysine residue of the histone.

S phase when the DNA is duplicated in synchronized cultures. Certain serine groups in histone H1 are subject to phosphorylation. The maximum in phosphorylation occurs at a later time during mitosis. _These covalent modifications all have the effect of reducing the net positive charge of the histones, which could lower the affinity between the histones and the DNA;_ however, no correlations between the covalent alterations and gene activities have been established.

Some of histone H2A is modified by combination with a 76-residue protein called _ubiquitin_. The complex formed between these two proteins involves the C-terminus glycine residue in ubiquitin and the side-chain amine group of the lysine at position 119 of H2A (Figure 30–13). Between 5 and 15 per cent of the H2 exists in the form of this complex. Usually only one of the two H2A molecules in a histone octamer carries ubiquitin. The heterochromatic satellite DNA from _Drosophila_ does not contain ubiquitin. _Although ubiquitin has a greater tendency to be associated with euchromatin, it is released from all chromatin at some time during mitosis._

Active Chromatin Is Most Susceptible to DNAse Degradation.   _To the extent that transcriptionally active chromatin is more swollen, it seems likely that it might also be more accessible._ This possibility has been elegantly demonstrated for the hemoglobin and ovalbumin genes in the chicken. Chromatin was isolated from chicken erythrocytes, in which the hemoglobin genes were recently very active, and from the oviduct, in which the ovalbumin gene is very active. These two chromatin preparations were subjected to degradation by DNAse I using an assay that permitted the measurement of gross DNA degradation as well as the DNA degradation associated with the specific genes. In both cases the rate of degradation of DNA within the active gene was much greater than that of the average DNA. Thus in erythrocyte chromatin the globin gene DNA and in oviduct chromatin the ovalbumin gene DNA were rapidly digested. Additional observations indicated that substantial regions around the active genes were also quite susceptible to DNAse I attack.

Active Chromatin Is Associated with High-Mobility Group (HMG) Proteins.   _In addition to the histones and ubiquitin, chromatin is associated with a group of proteins known as the high-mobility group (HMG) proteins,_ so named because of their high electrophoretic mobility in low pH polyacrylamide gels. There are four major HMG proteins in calf thymus—HMG1, 2, 14, and 17—all with molecular weights of less than 30,000. The HMG proteins can be selectively removed from the chromatin with 0.35 M NaCl, evidence that they are bound less firmly than the histones. The major HMG proteins, HMG1 and HMG2, contain approximately 25 per cent basic and acidic amino acid residues. Amino acid sequence studies show that the charge distribution within the HMG molecules is nonuniform; major clusters of acidic amino acids can be found. The HMG1 protein contains an extraordinary sequence of 41 continuous aspartic and glutamic acid residues.

The total nuclear concentration of HMGs is low relative to that of the histones (about 3 per cent). Their distribution within the chromatin suggests they are located in the transcriptionally active regions of the chromatin. Two of the nonhistone proteins, HMG14 and HMG17, are preferentially released from chromatin on partial digestion with DNAse I. The notion that HMG proteins are associated with active chromatin regions is supported by reconstitution experiments. Selective removal of HMG from chicken erythrocyte chromatin by 0.35 M NaCl extraction eliminated the selective sensitivity of the globin gene to DNAse I. The globin gene in such a preparation regained its DNAse I sensitivity on reconstitution with HMG14 and HMG17. This result suggests that there are factors present in the active regions of the genome that favor HMG protein binding; it also indicates that HMG protein binding enhances the activity of already active genes.

Enhancer Sequences Are General Stimulators of Transcription

Some of the primary sequence signals in eukaryotic DNA can be understood in conventional terms. This is not true of _enhancer_ signals. What makes enhancer signals unique is their ability to stimulate transcription at appreciable distances from where they are located in the genome. Enhancers may be located a thousand base pairs or more upstream or downstream from the promoters they stimulate. Even the orientation of the enhancer element relative to the promoter it affects is not critical to its action.

Enhancers have been found in many systems; some enhancers are active in most cells, others are quite tissue-specific. They undoubtedly constitute a major mechanism for regulating gene expression and a major mystery to those who would like to know how they function. It is possible that enhancers account for the difference between euchromatic and heterochromatic regions of the chromatin.

The first evidence for the existence of enhancer sequences came from a study of a virus called simian virus 40 (SV40), which grows on monkey cells. The SV40 viral genome is a circular duplex molecule with about 5200 base pairs. In SV40-infected cells, viral RNA synthesis is divided into early and late phases. Early transcription originates from a unique site illustrated in Figure 30–14. This early transcription start site has a TATA box, typical of eukaryotic gene promoters (see Chapter 27), about 27 base pairs upstream from the transcription initiation point. It also has sites for binding a repressor and an activator protein near the polymerase binding site. Further upstream from the normal transcription start site for early SV40 transcription there is a tandemly repeated 72-bp sequence (−116 to −188 and −189 to −261 from the 5′ end of the messenger). Removal of one of these sequences has no effect on transcription _in vivo_. But removal of both of the 72-bp sequences results in a drastic lowering of early transcription.

Additional observations indicate, most surprisingly, that the precise position or orientation of the 72-bp segment is not critical. Genetic manipulations indicate that the 72-bp segment may be oriented in either direction or

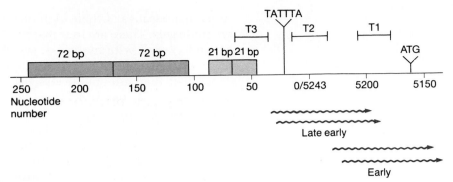

Figure 30–14

Region in and around the early transcription start sites of the SV40 genome. The base-pair number on the circular genome is indicated. There are several transcription start sites. One cluster of start sites (early), used initially after infection, is located about 27 base pairs downstream from the TATA box. The other cluster of start sites (late early) is used at later times. This cluster of start sites is actually located upstream of the TATA box. One of the products of early transcription is the protein known as large T antigen. This protein, which binds at three sites (T1, T2, and T3), represses early transcription. The upstream positive control *cis* elements include the tandem 72-bp enhancer sequences and the tandem 21-bp sites. The latter have been shown to function in conjunction with a host-encoded protein known as Sp1. (This drawing was generously contributed by D. Lewis and J. Manley.)

located much further upstream or downstream from the transcription start site without losing its ability to stimulate transcription. Some foreign genes inserted into DNA containing the enhancer segment are similarly stimulated (e.g., β-globin). This shows that there is no specific relationship between the enhancer and the gene it stimulates.

Other animal viruses carry segments with a similar activating function, even though the sequences are very different. Gruss and his colleagues have shown that the SV40 72-bp repeat may be functionally replaced by a sequentially unrelated 72-bp repeat from either end of the cloned Moloney murine sarcoma provirus (MSV) DNA. Enhancers have also been found in the host genome. For example, a tissue-specific enhancer has been found in the intron region of the immunoglobulin gene; it is tissue-specific because it stimulates expression of nearby genes only in antibody-forming lymphocyte cells.

The finding of enhancer elements constitutes one of the most exciting discoveries in the field of gene regulation. Upstream activating sequences and upstream repressing sequences, subsequently discovered in yeast, appear to be closely related in structure and function to enhancer sequences except that they seem to be functional only when located upstream (in either orientation) from the transcription start of the gene.

Thus far we do not know how these *cis*-acting elements manage to be effective at regulating transcription over such long distances. Several models have been proposed: (1) The element may function as an attachment point to a structural component of the nucleus to stimulate transcription; (2) it may serve as an initial binding site for some factor required for transcription, which must subsequently move along the duplex to the initiation point for transcription; (3) it may induce folding of the chromosome, bringing the *cis* element into close contact with the gene it stimulates.

Of these three models, there are reasons for discounting the first and the third. Model (1) is disfavored by the observation that enhancer elements are effective in solubilized *in vitro* systems for transcription, where there is no nucleus. Model (3) tends to be excluded by the observation that when cross-linking agents react with the DNA between the enhancer site and the affected gene, the enhancer effect is eliminated. Although these observations, by default, lend support to model (2), they leave open the question of what factor(s) travel from the enhancer site to the stimulated gene. Is it the whole RNA polymerase that makes this journey, some of the polymerase subunits, or ancillary factors? Or is there another model for the enhancer that we have failed to consider?

REGULATORY PHENOMENA ASSOCIATED WITH DEVELOPMENT

As cells proliferate within the embryo, they assume different properties; this change is evidenced by the genes that they express and the proteins that they synthesize. Developmental differences can be maintained or altered during subsequent cell duplication. During growth of the embryo, cells that are initially capable of following any pathway of development become _committed_ to a particular pathway. As a rule, pathways have many branchpoints, so that when progeny cells of partially committed cells reach a branchpoint they may differentiate further, down a more specialized pathway. This process of gradual commitment, repeating itself many times, gives rise to a complex pattern of cell lineages, all arising from the initial fertilized nucleus.

It seems likely that the pivotal events in the evolution of a differentiated cell involve changes in gene expression that result from a complex hierarchy of controls. The key to understanding differentiation, therefore, is to identify the regulatory genes that are responsible for the controls involved in differentiation and to explain how they act. If this is true, then our main task is to determine how different regulatory mechanisms interact to make a developmental program.

In this section we will focus on genetic regulatory events that occur in development. We will first discuss some of the classical studies of gene expression in the embryogenesis of sea urchins and amphibians, and then turn to some of the developmental events in fruit flies and mice, two of the best-characterized developmental systems.

During Embryonic Development in the Amphibian, Specific Gene Products Are Required in Large Amounts

During early development, when cell divisions are occurring with great rapidity, there is an increased demand for certain products associated with the genome (histones) and the translation apparatus (ribosomes). Special regulatory devices exist to insure that these gene products are produced in the required amounts and that they are not overproduced in the more mature organism.

Ribosomal RNA in Frog Eggs Is Elevated by DNA Amplification. In eukaryotic organisms the main ribosomal RNA genes are present in clusters of hundreds to thousands of copies. The nascent transcript, a 45S molecule, undergoes an elaborate series of processing reactions to produce three kinds of ribosomal RNAs—28S (25–28S), 18S (17–18S), and 5.8S (5.5–5.8S). Multiple copies of ribosomal RNA genes are organized into tandem arrays separated by nontranscribed spacers, which vary in length in different species from about 2000 to 30,000 bp. _In situ_ hybridization studies, using radioactive rRNA and autoradiography, have demonstrated that the ribosomal RNA genes are localized around the nucleolus, where rRNA processing and ribosome assembly take place. Germ cells and, in particular, immature eggs or _oocytes_ often raise the level of rDNA per nucleus by amplification to satisfy the high demand for rRNA during the very rapid early cleavage stages. This type of amplification has been most thoroughly investigated in the frog _Xenopus laevis_ (Figure 30–15). In this organism _amplification_

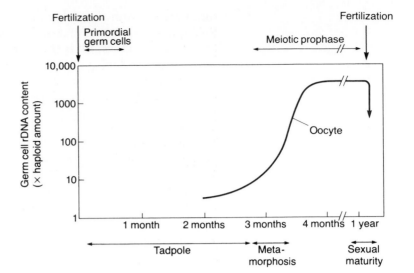

Figure 30–15

Time course of rDNA amplification in the *Xenopus laevis* oocyte. The amount of rDNA per cell is plotted against time. Note that the ordinate is a logarithmic scale. The sharp drop in rDNA at fertilization is thought to be due to dilution of the amplified rDNA copies by cell division, rather than to their destruction. (After Prescott.)

raises the number of rRNA genes about 1000-fold. These extrachromosomal genes are transcribed during oogenesis and subsequently discarded. The mechanism for this amplification is not understood. An entirely different type of mechanism is used to elevate the level of 5S rRNA synthesis during oocyte maturation.

5S RNA Synthesis in Frogs Requires Regulatory Protein. Eukaryotic ribosomes contain two small rRNA components (5S and 5.8S), while prokaryotic ribosomes contain only one (5S rRNA). Although the eukaryotic 5S molecule is the same size (about 120 nucleotides) as the prokaryotic 5S species, they differ in both their sequences and mode of synthesis.

The 5S rRNA genes of most eukaryotes are present in multiple gene copies and are organized in simple tandem multigene families. In amphibians the 5S rRNA genes are organized into different multigene families that are under developmental control. For instance, in *Xenopus laevis* the normal haploid genome contains about 20,000 copies of the 5S rRNA genes, organized into three different multigenic families. The genes of two major families, comprising about 98 per cent of the total 5S rRNA genes, are expressed only in growing oocytes. The genes of the third family, containing about 400 copies, are active in most (somatic) cells and growing oocytes. *This developmental control enables the oocyte to accumulate 5S rRNA at rates 1000-fold higher than is possible in somatic cells.*

The developmental control for 5S rRNA synthesis resides in a regulatory protein whose properties have been characterized by the studies of Brown and Gurdon. They have shown that purified 5S DNA is transcribed accurately when injected into the nuclei of living oocytes. Extracts of these nuclei also can transcribe either high-molecular-weight genomic 5S DNA or individual repeating units of 5S DNA joined to bacterial plasmids and cloned in *Escherichia coli*. The 5S genes are transcribed by RNA polymerase III. One cloned 5S DNA of *Xenopus borealis* somatic 5S DNA was subject to a more detailed analysis of the sequences that influence its accurate transcription. For this purpose a collection of plasmids containing deletions of the 5S gene and its flanking regions was constructed. Two deletion series were made, one removing increasing amounts of DNA from the 5′ end and another removing increasing amounts of DNA from the 3′ end of the gene.

Individual repeating units of 5S DNA were accurately transcribed into 5S RNA, demonstrating that *the signals for initiation and termination are present in each repeating unit,* rather than at one end of a tandem array of genes. Studies of deletions from the 5′ end showed that the entire flanking sequence could be removed without interfering with proper initiation. As much as the first third of the gene could be removed without interfering appreciably with accurate initiation. In this event, initiation started at a comparable location in the plasmid DNA. Deletions proceeding downstream beyond the +50 residue showed a loss of proper initiation. When deletions were made from the 3′ end, initiation was not affected until the deletion proceeded upstream beyond the +87 residue. The surprising conclusion from these observations is inescapable. *The control region for initiation is within the gene and extends approximately from the 50th to the 87th residue.* Internal promoters have thus far not been found in prokaryotes. Likewise, genes transcribed by the eukaryotic polymerase II polymerase invariably have their promoters located upstream from the transcription start site.

Further studies have shown that *it is not the RNA polymerase III that binds in the +50 to +87 region, but rather a control protein that influences the binding of polymerase III at an upstream location.* A protein that seems to fulfill this role has been purified. When oocyte nuclear extracts are pretreated with antibody to this protein, they are no longer competent in initiating *in vitro* transcription from the 5S gene. In order to confirm the site of action of this protein (known as TFIIIA; $M_r \cong 40,000$), the approximate location of its binding site on the 5S gene was determined by the so-called *footprinting* technique (Figure 30–16). DNA fragments containing the 5S gene were 5′-end labeled with ^{32}P, mixed with the protein, and digested with DNAse I. The resulting DNA fragments were electrophoresed on a polyacrylamide gel and autoradiographed. Regions of the DNA that were protected from DNAse attack by TFIIIA binding appear as a blank spot ("footprint") on the autoradiogram. The footprint shows that the protected region is situated between the 45th and the 90th base pair. This result is in reasonable agreement with the hypothesis that TFIIIA protein binds specifically to the control region (+50 to +87) of the 5S gene.

A further analysis of the binding was undertaken to determine which sites in the 5S gene were instrumental in the binding. For this purpose, DNA

Figure 30–16

DNAse I protection (footprinting) experiment on 5S DNA of *Xenopus*. The diagram on the left indicates the region on the gel that corresponds to the 5S RNA gene. Arrow points in the direction of transcription. Cross-hatched area indicates the region that binds transcription factor protein. Column labeled Xbs refers to an intact gene containing 160 bp of the 5′ flanking sequence, the 5S RNA gene (120 bp), and the 3′ flanking sequence (138 bp). The various deleted 5S DNAs are preceded by 74 bp of the plasmid pBR322 sequence. Numbers in other columns refer to portions of the 5S gene that have been deleted. All samples were subjected to partial digestion with DNAse I, then were electrophoresed and autoradiographed. In + columns, transcription factor protein was added before DNAse I treatment. (Courtesy Donald D. Brown of the Carnegie Institute of Washington.)

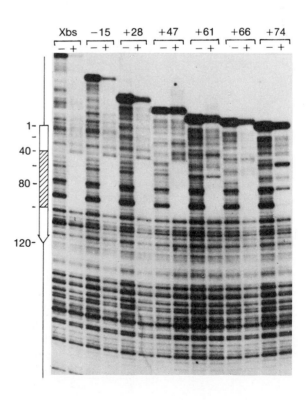

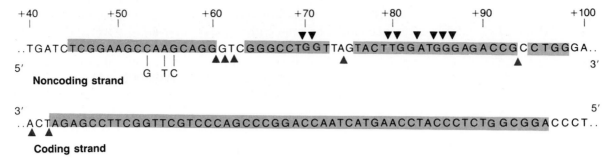

Figure 30–17

Summary of the interaction of the positive transcription factor and the internal control region of the *Xenopus* 5S RNA gene. Xbs 5S DNA sequences of the noncoding and coding strands between residues 40 and 100 are indicated on the top and bottom, respectively. Guanine residues that interfere, when methylated, with binding of the transcription factor are indicated by color letters. On the sequence, DNA phosphate residues that make contact with the transcription factor are marked by black triangles. Sequences protected from DNAse I digestion by the factor are shaded in color, while the residues at which cleavages are enhanced are designated by colored triangles. Bases that differ in both the major and trace *X. laevis* oocyte-type genes from the somatic 5S DNA sequence are indicated as changes underneath the noncoding strand. (After Sakonju and Brown, Contact points between a positive transcription factor and the *Xenopus* 5S RNA, *Cell* 31:395–405, 1982.)

was subjected to chromosomal modification such that the guanine residues were specifically methylated. In a parallel series of experiments the phosphates were specifically ethylated. Those residues whose modification interfered with the binding reaction are indicated in Figure 30–17. Quite remarkably, almost all of these residues are located on the noncoding DNA strand. This result led Brown to suggest that the TFIIIA protein may not dissociate from the gene as the polymerase migrates downstream during elongation. Brown's suggestion would help account for the ability of the regulatory protein to serve as an activator protein even though it is located downstream from the polymerase binding site. The protein may function as a more efficient activator protein by facilitating the unwinding of the DNA, but that remains to be seen. The presence of transcriptional regulatory sequences within a gene allows such a gene to be amplified and remain functional without having to amplify upstream regions as well.

Sequence analysis of TFIIIA has revealed an imperfectly repeated sequence of about 30 amino acids, including two cysteines and two histidines in conserved positions, with a consensus sequence (Tyr-Phe)-X-Cys-X_4-Cys-X_3-Phe-X_3-Leu-X_2-His-X_3-His-X_{2-6}, where X is any amino acid. The combination of two cysteine and two histidine residues could make an ideal binding site for Zn^{+2}, which is found in complexes of TFIIIA with nucleic acid. These metal-binding complexes (fingers) within each of the 30mers are believed to be the functional units in DNA binding. The detailed three-dimensional structure of TFIIIA remains a mystery, but it seems clear that it represents a different type of structure from that observed for CAP and cro (see Chapter 29). Researchers have found other regulatory proteins with this repeating sequence in eukaryotes, a fact suggesting that TFIIIA may represent a second major type of DNA-binding regulatory protein.

Competition binding experiments in which both somatic and oocyte 5S DNAs are exposed to limited amounts of the TFIIIA protein show that the regulatory protein binds considerably more firmly to the somatic 5S genes. This probably explains why the latter are uniquely active in somatic tissues when the amounts of TFIIIA protein are much lower. Under any conditions, it seems clear that the large amounts of 5S rRNA that are synthesized in the oocyte, but not in somatic cells, are strongly related to the concentration of TFIIIA regulatory protein in the two situations.

The TFIIIA protein has another function. It interacts with 5S RNA to form a stable 7S nucleoprotein complex that accumulates in the cytoplasm

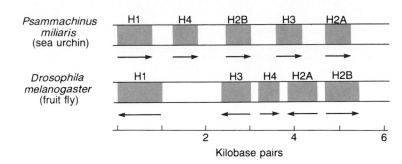

Figure 30–18

Molecular maps of histone gene repeat units of a sea urchin and a fruit fly. The histone gene organizational maps are calibrated in base pairs. The mRNA coding regions, including their leader and trailer sequences, are depicted by the colored boxes. The direction of transcription is indicated by the arrows.

of young oocytes. In growing oocytes, 5S RNA is not incorporated immediately into ribosomes; rather, it is synthesized actively for several weeks before 18S and 28S rRNA synthesis begins. In addition to its storage function, it is conceivable that TFIIIA protein also serves a transfer function, facilitating the transport of 5S rRNA to the nucleolus, where ribosome assembly takes place. This function has not been verified.

Demands for Histones Are Met by Gene Amplification and Storage. Only a few gene families that encode mRNAs exist consistently in multiple copies. One such family is the histone family of genes. The existence of multiple copies of the histone genes is easy to rationalize in terms of the needs most cells have for a large increment in the quantity of histones over a short period of time, especially during early embryogenesis. They are usually but not always synthesized during S phase, when the DNA is being replicated in approximately equal weight proportions to the histone. The five histone genes invariably occur in clusters containing transcribed and nontranscribed regions (Figure 30–18). Each cluster is usually repeated many times in a tandem array.

Most of the detailed structural analysis of histone genes has been carried out on sea urchin histone genes. Sea urchins contain several hundred copies of the genes for histones, whereas in the human, mouse, or *Drosophila* there are only 10–50 copies. *It seems likely that histone genes are reiterated in a very high frequency in sea urchin because of the requirement for very large amounts of histones during early embryogenesis, when cell cleavages occur once every 30 minutes.* Indeed, those histone genes that are active during early embryogenesis constitute a distinct class. The set of histone genes that are active late in embryogenesis of the sea urchin are present in only 5–12 copies per genome; unlike the clustered, tandemly arranged early histone genes, they are dispersed and irregularly arranged. It is not clear what mechanism controls this class switching between early and late histone genes.

In *Xenopus laevis* it takes, on average, 14 minutes for the number of cells to increase from 50 to 5000 during early embryogenesis. Since the genome size in the frog is larger than in the sea urchin, it is clear that the demand for histones in this species is even greater during early embryogenesis. In fact, the reiteration of histone genes is less in the frog than the sea urchin. *The problems of histone demand in the frog have been solved by accumulation of a large store of histone proteins and histone mRNA in the egg.* Histone genes, like other structural genes, are transcribed by RNA polymerase II. Other than this, little is known about those factors that regulate histone gene expression in either the frog or the sea urchin.

Specific Genes Regulate Development in Fruit Flies

Whereas sea urchins and frogs have proved valuable in the study of early embryonic events and the structure and composition of the embryo, they have not been very useful in genetic studies. Thus it has been necessary to

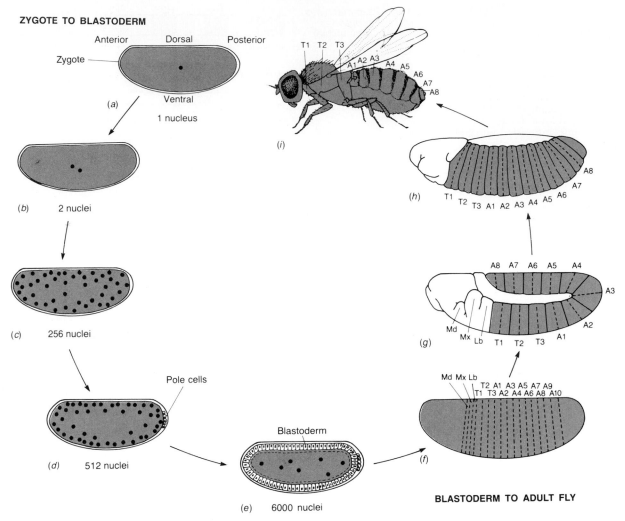

Figure 30–19

Steps in the development of *D. melanogaster*. The fertilized egg contains a single zygotic nucleus (*a*). This divides (*b*) every ten minutes. After eight divisions, when there are 256 nuclei, nuclear migration toward the outer cortex structure begins (*c*). Eventually all the nuclei form a monolayer on the cortex surface. The first nuclei to become enclosed are the pole cells, which will become germ cells in the adult organism (*d*). The fully formed blastoderm contains about 6000 cells, which form a monolayer around the cortex (*e*). Even before a cell membrane has formed around the nuclei, the embryo has become functionally divided into a segmented structure (*f*). During gastrulation (*g*), there is continued cell duplication, a folding of sheets of cells, and mass migration of segments. Eventually this structure hatches into the first larval stage (*h*). The remaining stages between the larva and the adult fly (*i*) are not illustrated. In (*f*) through (*i*), various segments are labeled. Three segments, Md, Mx, and Lb, fuse to make the head structure. Thoracic segments T1–T3 and abdominal segments A1–A8 retain their segmental appearance in the adult.

turn to other organisms to identify the genes or the gene products involved in regulating development. Many fundamental aspects of developmental biology have been studied in *Drosophila melanogaster*, because for this organism researchers have learned to use an extensive body of genetic and molecular biology techniques.

The steps in the development of *Drosophila* are shown in Figure 30–19. Beginning shortly after fertilization, the new diploid nucleus undergoes a rapid series of divisions (one every ten minutes) with no segregation of nuclei into separate cells. After the eighth division, when there are 256 nuclei (Figure 30–19*c*), the nuclei begin to migrate to the periphery of the egg cytoplasm. This migration of nuclei is carried out in a very orderly fashion so that each nucleus maintains its position relative to the other nuclei. After the next nuclear doubling, the first cell membranes are formed around a group of cells at the posterior end of the egg. These cells are progenitors of germ cells for the subsequent generation. The remaining nuclei continue to divide until there are about 6000 nuclei at the periphery (Figure 30–19e). At this point, about two hours after fertilization, membranes are formed separating the nuclei into a monolayer of cells. The resulting structure, known as the _blastoderm_, is essentially a cell monolayer enclosing the yolk.

Even before the blastoderm forms, regions of it become committed to forming specialized body structures (Figure 30–19f). The blastoderm is divided into 14 different segments: Md, Mx, Lb, T1–T3, and A1–A8. Three of these segments, Md, Mx, and Lb, become part of the head structure. The remaining segments become subdivided into two compartments, with anterior and posterior parts. During gastrulation there is a continued cell duplication, folding of sheets of cells, and mass migration of segments (Figure 30–19g). The embryo eventually hatches into the first larval stage (Figure 30–19h). Cells within the larva are of two types. About 80 per cent of them are fully functional in the larva. The remaining 20 per cent are embryonic precursors to adult tissues rather than larval tissues. These latter cells form packets within the larva, called _imaginal disks_, that are arrested in development until pupation, when the hormone ecdysone signals their differentiation into specific adult structures.

At some time during the preblastoderm stage the embryo takes on a segmented structure, which becomes cytologically visible soon after blastoderm formation. This segmented appearance is preserved continuously to adulthood. Mutagenesis screens have detected three types of genes that affect the segmentation pattern; these are called _maternal effect genes_, _segmentation genes_, and _homeotic genes_.

Maternal Effect Genes.   Maternal effect genes are involved in specifying positional information along the anterior–posterior and the dorsal–ventral axes in the young embryo. Mutations that disrupt these axes include _dorsal_ and _bicaudal_. _Bicaudal_ is a strict maternal gene mutation that leads to the development of two posterior ends arranged in mirror-image symmetry and lacking anterior (head and thoracic) segments. _Dorsal_ is a maternal effect mutation that results in the transformation of dorsal into ventral cells. Proof that dorsal is a strict maternal mutation is provided by the fact that the abnormal dorsal phenotype can be prevented in the mutant by injecting RNA from wild-type unfertilized eggs into mutant fertilized eggs. Maternal effect mutations are interesting because they indicate that some early steps in segmentation development result exclusively from factors originating from the maternal genome.

Segmentation Genes.   Segmentation genes affect the number and polarity of the body segments, but not the properties that ultimately distinguish one segment from another. One of the most studied segmentation mutations is called _fushi terazu (ftz)_. Mutations in the _ftz_ gene lead to embryos bearing half the normal number of segments, the anterior portion of one segment being fused with the posterior part of the next segment throughout the segmented structure of the embryo. Such embryos invariably die before hatching.

In situ hybridization studies on normal embryos, in which the cloned _ftz_ gene is used as a probe, indicate that transcription is quite active 150 minutes after fertilization. The nuclei are lined up with cortical cytoplasm at this time, but the membranes have not yet formed around them. The periodic hybridization pattern takes on a banded appearance because transcription is most active in the posterior half of one segment and the adjacent anterior half of the next segment (Figure 30–20). Seven bands are clearly visible. It is

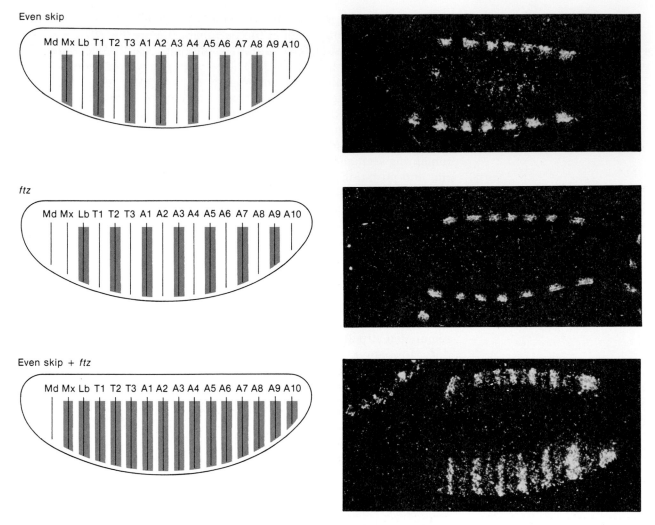

Figure 30–20

Localization of *even skip* and *ftz* transcripts in the wild-type developing embryo. Localization of the transcripts is determined by *in situ* labeling with a ^{32}P-labeled specific DNA probe. All embryos are oriented so that anterior is to the left and dorsal is up. In the left half of the figure the labeling pattern is diagrammatically indicated relative to the location of the different segments. It can be seen that *even skip* transcripts are located in adjacent anterior and posterior regions of every other segment (top). The same is true of *ftz* except that complementary regions are labeled (middle). This point is made most clear by the approximately uniform labeling pattern observed in a double labeled section where the embryo is simultaneously labeled with both probes (bottom). On the right, dark-field photomicrographs of tissue autoradiograms are shown. The labeling pattern is seen only near the dorsal and ventral edges of the embryos because the labeling is done on sections cut through the middle of the embryos. It should be remembered that cells are concentrated near the surface in the early embryo. (Photomicrographs made by K. Harding and M. Levine.)

just those regions that show labeling in the normal embryo that are missing in the mutant *ftz* embryo.

A second segmentation gene, known as *engrailed*, becomes transcriptionally active shortly after the *ftz* gene. *In situ* hybridization studies show that the *engrailed* gene is expressed in the posterior parts of all the segments. Weak *engrailed* mutants have the correct number of segments, but the posterior compartments of the segments are not fully differentiated. Strong *engrailed* mutants show no segmental compartmentation whatsoever. For example, in flies homozygous for *engrailed*, posterior wing compartments are converted into mirror images of anterior compartments; there is no effect on anterior wing structure. Similarly, posterior compartments of leg disks develop like anterior compartments. Observations such as this indicate that the *engrailed* gene plays an essential role in compartment formation within each segment. These experiments strongly indicate a connection between the reg-

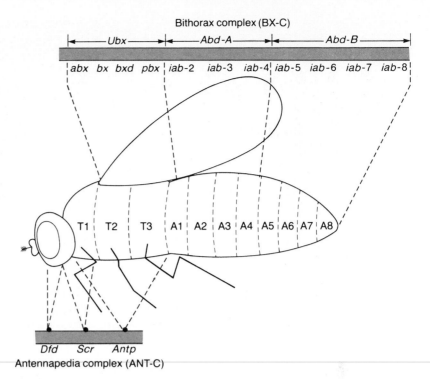

Figure 30–21

Genetic maps of the antennapedia complex (ANT-C) and the bithorax complex (BX-C). The colored dashed lines specify the regions affected by mutations in a given locus. The bithorax complex is shown in greater detail. By lethal complementation assay, Sanchez-Herrero has shown that the BX-C is divided into three complementation groups: *Ubx*, *Abd-A*, and *Abd-B*. Within each of these complementation groups there are nonlethal point mutations (such as *abx*, *bx*, *bxd*, and *pbx* in *Ubx*) which cover specific phenotypes in the regions encompassed by the dashed lines. *Ubx* is required for the development of the posterior compartment of the mesothorax (T2p) through the anterior compartment of the first abdominal segment (A1a). The *Abd-A* function is required for morphogenesis of A1p through A4, and *Abd-B* is required for A5 through A8. At least three essential homeotic functions have been assigned to the antennapedia complex: *Dfd*, *Scr*, and *Antp*. These are shown on the genetic map below the schematic of the fly. The primary domains of ANT-C function are indicated. *Antp* function is required for proper segment morphogenesis of the thorax. Analysis of *Scr⁻* and *Dfd⁻* mutant embryos suggest that these genes are required for the differentiation of the prothorax and posterior head regions.

ulation of expression of *ftz*, *engrailed*, and other segmentation genes and the establishment of positional identities in different blastoderm cells. Once we understand the factors that regulate the spatial and temporal expansion of these genes, we will better understand the mechanism of segmentation.

Homeotic Genes. *Hemeotic mutations* result in changes in the specific type of structures that are associated with individual segments or subsegments. There are homeotic genes specific for the structures associated with each type of segment. Most of the homeotic genes in *Drosophila* occur in two regions of the genome (Figure 30–21): the *antennapedia complex (ANT-C)*, which is largely involved in the diversification of segment identity of anterior structures, and the *bithorax complex (BX-C)*, whose products affect the development of segments posterior to the first thoracic segment. *Homeotic mutations* never affect the total number, size, or polarity of the segments. Rather, they *cause intersegment transformations*. Thus a homeotic mutation can cause wing structures to develop from the eye imaginal disks, or wings to develop in the segment where halteres normally develop (Figure 30–22). Homeotic mutations that give rise to a temperature-sensitive phenotype can

Figure 30–22
A four-winged adult fruit fly, showing complete transformation of the third thoracic into a second thoracic segment. The fly is actually a hemizygote, with one chromsome having the bithorax complex deleted and the other carrying bithorax and post-bithorax mutations. (From E. Lewis, California Institute of Technology.)

be used, in temperature shift experiments, to determine the time during development when the activity of the homeotic gene product is required. The temperature-sensitive periods occur during the second and third instar larval stage. After that, the fate of each imaginal disk is determined such that it cannot be changed by any homeotic locus.

The genetic regions containing the homeotic genes have been cloned, and probes have been made that are suitable for analyzing transcription within different segments by in situ hybridization. In this way Michael Levine and Walter Gehring and others have shown that there is a general correlation between the region that is affected by a mutation and the region in which the gene is strongly expressed. Thus *Ubx* mutants affect the T2 and T3 thoracic segments and the anterior portion of the A1 segment; indeed, *Ubx* transcripts are most active in A1. Similarly, an *Abd-A* mutation usually affects the posterior part of A1 and A2–A4; this genetic region is normally transcriptionally active in A2–A7. Finally, *Abd-B* mutations usually affect the abdominal segments in the A5–A8 region, and the *Abd-B* gene is transcriptionally most active in the anterior half of A7 and the posterior part of A8/A9. It seems most reasonable that a gene should be most active transcriptionally in the region of the organism that is most affected by mutations of that gene. Why the correlation is not perfect is impossible to say at this point; some of the mutations may be in a regulatory rather than a structural gene.

Various explanations have been offered for how homeotic genes regulate development. This is an area of intense investigation. One universal feature of homeotic genes is that they all possess a common 180-base segment in the 3′ exon. This base sequence homology ranges between 60 and 80 per cent, depending on the specific genes being compared; the amino acid homology is even higher (up to 87 per cent). The high basicity of the amino acid sequence in this region suggests that the run of 60 amino acids, called a *homeo box*, would cause the protein to have a high affinity for DNA. This likelihood has led to the notion that most homeotic genes encode regulatory proteins that function by binding to DNA. The notion that the homeo box proteins are DNA-binding proteins is strengthened by the observation that some homeotic gene-related proteins have been found to be localized in the nucleus. The relevant measurements were made with specific immunofluorescent probes that bind to single homeotic proteins.

One of the most exciting findings about the homeo box is that homologous sequences have been found in other animals such as the frog and the mouse (Figure 30–23). This suggests that there may be homeo-box-encoded proteins in a wide range of organisms, possibly playing similar roles in regulating developmental processes.

```
                  Thr        5                        10                        15                        20
Arg  Lys  Arg  Gly  Arg  Gln  Thr  Tyr  Thr  Arg  Tyr  Gln  Thr  Leu  Glu  Leu  Glu  Lys  Glu  Phe
Ser            Gly


                       25   Ile              30                  Asp       35   Asn              40   Ser
His  Phe  Asn  Arg  Tyr  Leu  Thr  Arg  Arg  Arg  Arg  Ile  Glu  Ile  Ala  His  Ala  Leu  Cys  Leu
     Thr       His                                                 Met       Tyr


Ser            45                        50                        55   Ser              Asp  Arg  60
Thr  Glu  Arg  Gln  Ile  Lys  Ile  Trp  Phe  Gln  Asn  Arg  Arg  Met  Lys  Trp  Lys  Lys  Glu  Asn
                                                                       Leu                      Ile
```

Figure 30–23

The homeo box sequence found in three *Drosophila* regulatory proteins. The amino acid sequence of the *Antp* homeo box is given in the center. The sequence for the ftz protein is given above where it differs, and the sequence for the ubx protein is given below where it differs. Certain proteins isolated from the human and mouse embryos have segments with similar sequences. The high basicity of the sequence is notable. This should favor electrostatic binding to nucleic acid.

Gene Changes and Cell Amplification Lead to Specific Antibody Synthesis in Vertebrates

The immune system provides us with a most fascinating example of cell development in the mature vertebrate organism. A feature of the immune system is the constant generation of new cells that represent variations in gene organization of the immune system. The object of this variability is to provide the organism with immune cells that have the widest possible range of specificities. When the organism is invaded by foreign agents, usually viruses or bacteria, the intrusion excites the immune system. Those immune cells that carry the specific immune receptors for interacting with the foreign agent are stimulated to proliferate. In a short while, usually a matter of days, clones of immune cells with the appropriate specificity have been produced and the organism fends off the invader with the specific immunologic tools provided by these cells. The clones of cells tend to persist for some time, which accounts for the fact that the second time the same foreign invader makes its presence known, it is rejected more promptly and without crisis. Our concern here is with how the cells that make specific immunoglobulins arise.

B Cells Cooperate with T Cells to Make Antibodies. There are two types of white cells or lymphocytes associated with the immune response; these are called *B cells* and *T cells*. B cells originate from precursor cells in the marrow and become implanted in peripheral lymphoid tissue, the spleen and lymph nodes (Figure 30–24). Each B cell has undergone genetic changes so that it has the potential for making a unique type of antibody protein. Its surface contains a limited number of highly specific receptors. The receptors are a special class of membrane-bound antibodies. When the organism is exposed to a specific foreign agent or antigen, the antigen reacts with B cells that carry complementary receptors. This contact stimulates proliferation of the original B cells (stem cells) as well as the further differentiation of some of them to make large quantities of antibodies (plasma cells). The plasma cells are considerably larger and contain a great deal of endoplasmic reticulum for the synthesis of specific antibody.

T cells also originate from stem cells in the marrow, but they mature in the thymus gland. They can also be stimulated to proliferate by exposure to an appropriate antigen but, unlike the B cells, their specific effector molecules remain firmly bound to the cellular membrane, as opposed to being secreted. Several types of T cells with different functions have been recog-

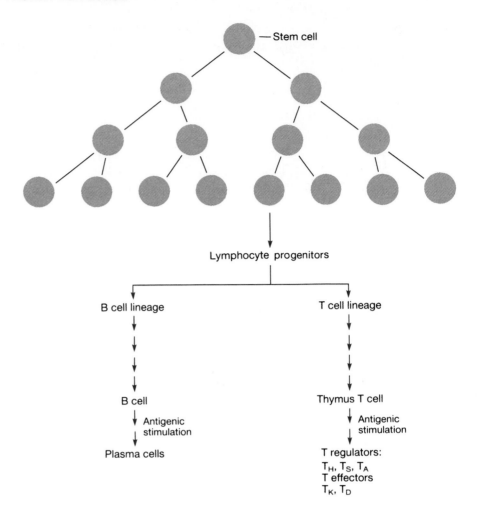

Figure 30–24
Pluripotent stem cells in the marrow give rise to more stem cells and to various progenitors. The lymphocyte progenitors give rise to B and T cell lineages. The final differentiation step in both B cell and T cell development requires antigenic stimulation.

nized. One class of T cells, known as *T helper cells*, promotes the maturation of antigen-stimulated B cells.

DNA Splicing Has Been Demonstrated in Antibody-forming Cells. Antibodies have been studied most extensively in the mouse, an ideal vertebrate for both genetic and biochemical manipulations. It is believed that mice can synthesize more than 10^6 antibodies with different antigenic specificities. This enormous diversity of proteins is generated from a limited amount of genetic information with the help of two mechanisms: DNA splicing and point mutations.

The involvement of somatic recombination has been demonstrated by examining the primary sequence of specific antibody genes in embryonic and adult immunoglobulin-forming tissue. The adult tissues favored for such studies are myelomas. These are tumors that produce a homogeneous population of polypeptide chains. They have served as a convenient source of pure immunoglobulin polypeptide chains as well as their mRNAs.

The generalized structure of the predominant serum antibody, IgG, consists of two identical heavy chains and two identical light chains (see Figure 2–46). These chains are held together by disulfide linkages and other noncovalent forces. Each heavy chain has an N-terminal region of about 110 amino acids, a variable (V) region that exhibits extensive sequence diversity, and a C-terminal portion, the constant (C) region, that exhibits negligible diversity. In the mouse there are two classes of light chains, which differ appreciably in the constant regions of the polypeptide chains. These are referred to as the κ and the λ light chains. Using highly inbred genetically identical (isogenic)

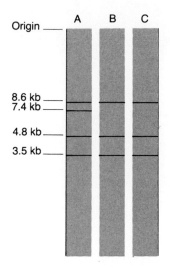

Figure 30–25

Analysis of DNA fragments containing λ_1 gene sequences from mouse embryo and myeloma cells. High-molecular-weight DNAs extracted from myeloma H2020, a λ chain producer (A), from a 13-day-old BALB/d embryo (B) and from myeloma MOPC321, a κ chain producer (C) were digested to completion with EcoR1, electrophoresed on agarose gel, and transferred to nitrocellulose membrane filters and hybridized with a nick-translated Hhα1 fragment of the plasmid B1 DNA. (After Brack, Hirama, Lemhard-Schuller, and Tonegawa, A complete immunoglobulin gene is created by somatic recombination, Cell 15:1–14, 1978.)

strains of mice, Tonegawa and his coworkers isolated the DNA from the embryo and two different myeloma tumor cells that produce homogeneous λ light chains (strain H2020) and κ light chains (strain MOPC321). These DNAs were exhaustively digested with the EcoR1 restriction enzyme and electrophoresed on an agarose gel. The gels contained an enormous variety of restriction fragments representing the entire nuclear DNA, and consequently no discrete pattern of bands could be seen with a stain for nucleic acid. However, if the electrophoresed gels containing the DNA were denatured and hybridized with ^{32}P-labeled DNA containing the sequences found in the RNA for the λ chain, a specific pattern of bands could be seen after autoradiography (Figure 30–25). The R1 digest of the λ-containing myeloma (H2020) DNA showed four bands; the DNAs of the embryo and myeloma (MOPC321) showed three bands in common with the first DNA but were missing the fourth band.

These results indicate that at some point during development from the embryonic state, a rearrangement took place in the H2020 myeloma cells on one of the homologous pairs of chromosomes carrying the λ chain gene. Further analysis with more specific radioactive probes containing sequences from either C_λ (the constant region of the λ chain) or V_λ (the variable region of the λ chain) were made. Only the 7.4-kb fragment originating from the λ myeloma cell hybridized to both probes. Thus the EcoR1 fragment must contain both V_λ and C_λ DNA. The 8.6-kb fragment hybridized to the C_λ but not the V_λ probe. Therefore this fragment must contain C_λ sequences but no V_λ sequences. Both the 3.5-kb fragment and the 4.8-kb fragment hybridized exclusively to the V_λ probe, indicating that they contain only V_λ sequences. Further analyses indicated that the 7.4-kb fragments arose from a recombinational event between the 3.5-kb fragment and the 8.6-kb fragment (Figure 30–26). The 4.8-kb fragment originates from another V_λ sequence located in the same chromosome. The mouse has only two V_λ sequences in the embryo. As a rule, only one of the pairs of homologous chromosomes recombines to yield a productive antibody gene. This explains why the 3.5-kb and the 8.6-kb bands are still visible in the H2020 myeloma DNA preparation. The κ myeloma cell producer shows the same pattern as the embryonic cell. Had it been probed with a DNA carrying the V_κ antibody sequence, it would have shown differences from the embryonic pattern.

The success achieved by Tonegawa and others in this sequence rearrangement analysis of the immunoglobulin genes has led to a massive research effort and a general picture of antibody gene organization. Many V gene segments are arranged in tandem on the same chromosomes as one or more C gene segments. A specific somatic recombination event creates the potential for synthesis of one V-C related antibody (Figure 30–27).

Whereas the discovery of somatic recombination has helped to account for the variety of immunoglobulins that can be produced in somatic cells, it

Figure 30–26

Arrangement of mouse λ_1 gene sequences in embryos and λ_1 chain-producing plasma cells. The vertical arrows point to EcoR1 restriction sites. The colored dashed diagonal lines point to hypothesized splice points that will explain the difference in structure of the region in the two cell types. The boxed regions represent coding regions, and I_1 and I_2 indicate first and second introns in the spliced gene.

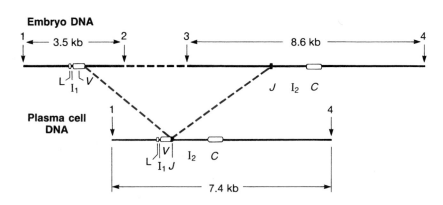

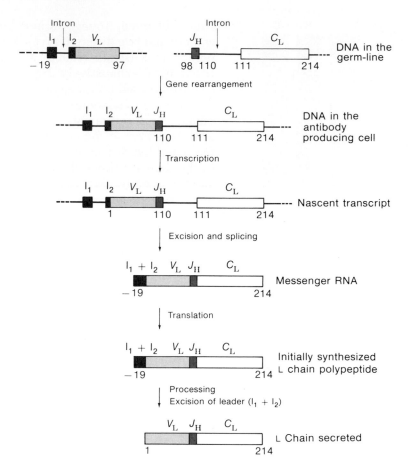

Figure 30–27

Origin of the antibody light chain (mouse L chain). Steps include rearrangement of the germ-line DNA, transcription of the antibody gene, splicing of the nascent transcript to give messenger RNA, translation of the mRNA, and finally processing of the polypeptide chain.

is insufficient to explain the vast numbers of antibodies that are potentially available to the organism. Within the variable gene segments there are so-called _hypervariable regions_, which show discrete nucleotide sequence changes in the somatic cells destined to become antibody producers. Such alterations probably result from a high level of _somatic mutations_ by some as yet to be determined mechanism.

The mechanism used to generate antibodies is quite unusual but also most appropriate for the situation. The alternative to generating the necessary genetic diversity in relevant tissues (bone marrow and lymphoid tissues) would be to carry a staggering number of genes for all possible antibodies in the germ line tissue.

Another method for generating diversity from a limited amount of genetic information is alternative mRNA splicing. If a gene, such as myosin, has many introns, then alternative splicing programs generate a variety of proteins from the translation of the mature processed mRNA.

SUMMARY

Gene regulation in eukaryotes differs appreciably from that in prokaryotes because of the contrasting modes of transcription and translation and the absence of polygenic mRNAs in eukaryotes. These differences are reflected in the greater dispersion of functionally related genes and in the relatively greater emphasis given to posttranscriptional regu-

lation in eukaryotes. Eukaryotes also differ in that they show cellular differentiation, i.e., cells that are basically different in their gene expression patterns, even though they result from a common ancestor cell during development.

In this chapter we first examined some of the better-understood regulatory patterns found in the relatively simple eukaryotic protist S. cerevisiae. We then went on to consider some of the regulatory problems that exist in higher eukaryotes.

One of the most intriguing mechanisms of gene regulation in yeast is the one that controls mating type. There are two mating types in yeast, MAT_a and MAT_α. Haploid cells of opposite mating types can fuse to make a diploid cell. A haploid (HO) cell of one mating type can efficiently switch to the other mating type. The mechanism for controlling mating type involves three loci. Two of these loci store mating-type genetic information. These are repressed (silent) loci. The expressed mating type is determined by which of these genetic loci is present at an expression locus located between the information storage loci on chromosome III. In mating type switching, the genetic information present at one of the silent loci is replaced by that present at the other silent locus. Regulatory genes that catalyze this conversion and regulatory genes that prevent expression of the sequences stored at the silent loci have been identified.

Higher eukaryotes differ from lower eukaryotes in several respects that influence the way in which genes are regulated. First, they contain much more DNA than lower eukaryotes. The genome is organized so that in regions that are expressed the chromatin is swollen, whereas in regions that are not expressed the chromatin is condensed. Several biochemical differences are also associated with transcriptionally active chromatin; these include histone modification (e.g., acetylation) and the presence of nonhistone proteins. There are no operons in eukaryotes. Eukaryotic mRNAs, while they can be quite large, encode only a single protein. Most eukaryotic mRNAs are processed by the removal of introns.

The most surprising element to be found in the DNA primary sequences that affect transcription is the enhancer. This element has a cis-stimulating effect on genes in its proximity. The enhancer may be located upstream or downstream from the gene it affects and may be oriented in either direction. Current evidence favors the view that specific proteins bind to the enhancer initially and then move along the DNA to the site of transcription.

Gene regulatory mechanisms in higher eukaryotes affect gene expression in the developing organism as well as the adult organism. In the developing embryo, special mechanisms exist for the expression of ribosomal RNA genes and histone genes that are not expressed in the adult. This difference is due to the increased demands for certain gene products during early, rapid-cleavage stages.

Studies on the fruit fly Drosophila melanogaster have led to the identification of a family of regulatory genes that affect the development of the normal segmentation pattern of the fly. A number of the genes have been isolated, and some of the gene products have been characterized. A common, highly basic region of 60 amino acids has been found; this region is called the homeo box. The homeo box may favor binding to DNA; such a suggestion is supported by cytological observations. Genes that contain homeo boxes have been found in a variety of organisms, including the frog and the mouse. These results are exciting, since they suggest that regulatory proteins in different species may share common features. If so, then it will be much easier to study development in organisms that are not as genetically amenable to work with as are fruit flies.

The vertebrate immune system exemplifies a developmental regulatory process that is fully active in the adult organism. Cells originating in the bone marrow are constantly differentiating by a process that includes DNA splicing and somatic mutation into cells that form specific antibodies.

In addition to the regulatory mechanisms considered in this chapter, to regulate the metabolism of the entire multicellular organism there are complex hormonal signals that serve as a means of communication between cells. This subject is taken up in Chapter 31.

SELECTED READINGS

Baker, B. S., and K. Ridge, Sex and the single cell. I. On the action of major loci affecting sex determination in Drosophila melanogaster. Genetics 94:383–423, 1980.

Brown, D. D., How a simple animal gene works. In The Harvey Lectures, Series 76, pp. 27–44. New York: Academic Press, 1982.

Chandler, V. L., B. A. Maler, and K. R. Yamamoto, DNA sequences bound specifically by glucocorticoid receptor in vitro render a heterologous promoter responsive in vivo: A steroid specific enhancer. Cell 33:489–499, 1983.

Duncan, I., The bithorax complex. Ann. Rev. Genetics 21, 1987.

Edstrom, J.-E., Structural manifestation of nonribosomal gene activity. In L. Goldstein and D. M. Prescott (eds.), Cell Biology, pp. 215–264. New York: Academic Press, 1980.

Gartler, S. M., and A. D. Riggs, Mammalian X-chromosome inactivation. Ann. Rev. Genetics 17:155–190, 1983. One of the most puzzling and challenging problems in gene regulation.

Groudine, M., and H. Weintraub, Propagation of globin DNAse I-hypersensitive sites in absence of factors required for induction: A possible mechanism for determination. Cell 30:131–139, 1982.

Guarente, L. P., Regulatory proteins in yeast. Ann. Rev. Genetics 21, 1987.

Gurdon, J., Egg cytoplasm and gene control in development. The Croonian Lecture, 1976. Proc. R. Lond. B. 198:211–247, 1977.

King, T., and R. Briggs, Serial transplantation of embryonic nuclei. Cold Spring Harbor Symp. Quart. Biol. 21:271–290, 1956.

McGinnis, W., R. L. Garber, J. Wing, A. Kuroiwa, and W. J. Gehring, A homologous protein-coding sequence in

Drosophila homeotic genes and its conservation in other metazoans. *Cell* 37:403–408, 1984.

Struhl, K., Genetic properties and chromatin structure of the yeast *gal* regulatory element: An enhancer-like sequence. *Proc. Natl. Acad. Sci.* 81:7865–7869, 1984.

Tonegawa, S., Somatic generation of antibody diversity. *Nature* 302:575–581, 1983.

Topol, J., D. M. Ruden, and C. S. Parker, Sequences required for in vitro transcriptional activation of a *Drosophila* hsp 70 gene. *Cell* 42:537–537, 1985.

PROBLEMS

1. Why is clustering of functionally related genes more common in bacteria than in yeast? How is the lack of such clustering related to the frequent use, in yeast, of multifunctional enzymes?

2. A mutation in GAL^{C_1} of yeast leads to constitutivity for galactose fermentation in haploid cells. Describe the nature of this mutation at the molecular level. Would you expect expression of the *GAL* genes to be responsive to glucose repression in this mutant?

3. In a recent experiment the DNA-binding domain of the lexA protein of *E. coli* was fused to a segment of *GAL4* lacking DNA-binding activity. When the ability of this fusion protein to activate *GAL1* transcription was tested, using a plasmid lacking the UAS but containing instead the *lexA* operator, inserted either upstream or in the midst of the *GAL1* gene, it was found to mirror the effect observed for the wild-type GAL4 protein and UAS_G. In contrast, lexA protein itself did not activate transcription of *GAL1*. What do these results suggest about the mechanism of transcription activation by GAL4? In a situation in which the lexA-GAL4 protein was overproduced, *GAL1* expression was found to be constitutive. Suggest an explanation for this observation.

4. How would you expect a deletion of *HMLE* to affect the expression of mating-type genes in yeast? Compare with the effect of deleting the $\alpha2$ gene from MAT_α. Consider both homothallic and heterothallic backgrounds.

5. Compare the roles and suggested mechanisms of action of UAS and *HMLE/HMRE* elements in the control of gene expression in yeast. In what ways do these elements resemble the enhancers of higher eukaryotes?

6. From what we know about enhancers, is there any reason to believe that they could not exist in bacteria?

7. Briggs and King were able to grow a differentiating embryo from a *Rana pipiens* egg whose chromosomes were replaced with a single diploid nucleus from another embryo. What does this tell us about the state of the nucleus in the developing embryo?

8. Given that hemophilia is an X-chromosome-linked disorder, and bearing in mind the phenomenon of X-chromosome inactivation, suggest an explanation for the observation that females heterozygous for the defective gene are essentially asymptomatic.

9. How would you expect the molecular basis of histone interaction with DNA to differ from that of regulatory protein-DNA interaction?

10. List four characteristics that distinguish active from inactive chromatin.

11. During some stages of development the cell division rate may exceed the ability of one or a few gene copies to supply sufficient amounts of a particular essential protein. What alternative mechanisms are used in different systems to remedy this situation?

12. In eukaryotes there appear to be at least two distinct types of site-specific DNA-protein interaction. The first involves binding of regulatory proteins to DNA sequences with rotational symmetry, and is essentially similar to that described for cro (see Chapter 29). The second, exemplified by TFIIIA of *Xenopus*, involves the recognition of binding sites that lack rotational symmetry. How are the differences between these types of binding sites reflected in the structures of the proteins that recognize them?

13. Suggest a possible mechanistic basis for the influence of maternal genes in specifying positional information in the *Drosophila* embryo.

14. Given that specific subsets of homeotic genes are required for the development of specific segments of the *Drosophila* embryo, suggest a possible mechanism whereby the necessary spatially restricted pattern of homeotic gene expression might be achieved. Incorporate the observed effect of homeotic mutations upon segment morphology into your model.

15. In the germ-line mouse cell, the promoter(s) for the heavy-chain immunoglobulin gene are located a considerable distance upstream of the enhancer for this gene. In what way is this arrangement of regulatory elements appropriate to the need for control of antibody synthesis during development of the B cell? Which feature of this system determines the tissue-specific expression of the heavy-chain gene?

31

HORMONE ACTION

Communication among cells of multicellular organisms is achieved by chemical messengers that pass between them. These messengers serve to coordinate the metabolic activity of various tissues, allow the organism to adapt to a changing environment, and prepare it for reproduction. Some chemical messengers travel only between neighboring cells, while other messengers enter the bloodstream and travel throughout the body.

Neurotransmitters, such as acetylcholine, dopamine, and norepinephrine, are examples of chemical messengers that are used to communicate between adjacent cells. They are liberated from one nerve cell that impinges upon other nerve or muscle cells. There are potent mechanisms for reuptake and degradation of these neurotransmitters to ensure that they do not travel very far or act for long (microseconds). We will have more to say on neurotransmitters in Chapter 35. In contrast, the classical _hormones_, such as insulin, growth hormone, and testosterone, are liberated into the bloodstream from ductless glands, and generally circulate for longer times (minutes). The chemical messengers employed by the local, short-acting and the dispersed, longer-acting systems are similar in nature. In fact, in some cases the same chemicals are used by both systems; for example, norepinephrine and cholecystokinin act as both neurotransmitters and circulating hormones.

In addition to neurotransmitters and hormones, there are many other messengers that travel between cells and act in an analogous manner. For example, many _peptide growth factors_ that have recently been identified act like hormones. In some cases, growth factors act on the cells that synthesize them as well as on peripheral cells. These factors generally stimulate replication of specific cell types. Other hormonelike factors serve an alarm function. For example, most cells make _interferons_ in response to viral infection. Interferons stimulate other cells to make proteins that will help protect them against viruses. Finally, there is only a slight distinction between the mechanism of action of the various messengers that travel within the body and _pheromones_, which travel between organisms.

All of these chemical messengers act by binding to _receptors_—proteins that act as antennae to detect the presence of the messenger. For each messenger there is a very specific receptor that recognizes that molecule (and possibly related molecules) but not other messengers. Different cells carry different sets of receptors. Hence they are capable of responding only to those messengers for which they have cognate receptors. Binding of the messenger to the receptor elicits an intracellular response that depends on the type of cell to which the receptor is attached. Thus the same chemical message may give rise to different responses in different cells. Some messengers cannot penetrate the cell membrane; hence their receptors must lie on the cell surface, embedded within the membrane. In these cases, there must be a mechanism to transduce the receptor-messenger interaction into an intracellular response. When the messenger can enter the cell, the receptor may mediate intracellular events more directly.

In this chapter, we will discuss how hormones are synthesized, transported, and degraded. Then we will explore the basics of hormone-receptor interactions and the more direct biochemical consequences of those interactions. We will also indicate some of the ways in which hormonal circuits are regulated. Most of the examples will relate to mammalian hormones, but the action of growth factors, interferons, and plant hormones will be covered to show how the principles apply to all chemical messengers used by any organism.

HORMONE SYNTHESIS

Table 31–1 lists many of the better-known hormones of vertebrates. Although many different classes of compounds are used as hormones by animals and plants, most vertebrate hormones fall into three classes: polypeptides, amino acid derivatives, and steroids.

Polypeptide Hormones Are Synthesized from Larger Precursors

Insulin, the first polypeptide hormone to be identified, was discovered by Banting and Best in 1922. They found that this substance, which they isolated from the pancreas, would restore normal glucose utilization in experimental animals lacking a pancreas. Insulin was also the first protein to be sequenced, a landmark accomplishment achieved by Fred Sanger in 1955. About twenty years later, Steiner discovered that the two polypeptide chains of insulin are synthesized as a single polypeptide, _proinsulin_, which folds and is cross-linked by disulfide bonds. An internal peptide is then removed by the concerted action of specific proteases. All of these events occur within the pancreatic β cells that synthesize insulin.

With the advent of techniques to isolate and translate mRNA in cell-free systems, it was discovered that the primary translation product of insulin mRNA—called _preproinsulin_—is even larger than proinsulin. Like all secreted polypeptide hormones, proinsulin is synthesized with a hydrophobic signal sequence at the amino terminus that directs the nascent polypeptide into the endoplasmic reticulum and is then removed by a signal peptidase. As a rule, in fact, the primary translation product of the mRNA is much larger than the functional hormone; the hormone is liberated from the precursor by specific proteases that frequently cleave at pairs of basic residues (arginine and lysine).

Surprisingly, in some cases several different peptide hormones can be liberated from the same precursor by proteolytic processing. A striking ex-

TABLE 31-1
Vertebrate Hormones[a]

Endocrine Organ	Hormone	Structure	Function
Pineal	Melatonin	N-Acetyl-5-methoxytryptamine	Regulates circadian rhythms
Hypothalamus[b]	Corticotropin-releasing factor (CRF or CRH)	Polypeptide (41 residues)	Stimulates ACTH and β-endorphin secretion
	Gonadotropin-releasing factor, (GnRF) or (GnRH)	Polypeptide (10 residues)	Stimulates LH and FSH secretion
	Prolactin-releasing factor (PRF)	(May be TRH)	Stimulates prolactin secretion
	Prolactin-release inhibiting factor (PIF)	(May be 56-residue peptide from GnRH precursor)	Inhibits prolactin secretion
	Growth hormone-releasing factor, (GRF or GRH)	Polypeptide (40 and 44 residues)	Stimulates GH secretion
	Somatostatin (Growth hormone-release inhibiting factor, SIF)	Polypeptides (14 and 28 residues)	Inhibits GH and TSH secretion
	Thyrotropin-releasing factor, (TRF or TRH)	Polypeptide (3 residues)	Stimulates TSH and prolactin secretion
Pituitary			
Neurohypophysis	Oxytocin (ocytocin)	Polypeptide (9 residues)	Uterine contraction, milk ejection
	Vasopressin (antidiuretic hormone, ADH)	Polypeptide (9 residues)	Blood pressure, water balance
	Melanocyte-stimulating hormones (MSH)	α Polypeptide (13 residues) β Polypeptide (18 residues) γ Polypeptide (12 residues)	Pigmentation
Adenohypophysis	Lipotropin (LPH)	β Polypeptide (93 residues) γ Polypeptide (60 residues)	Fatty acid release from adipocytes
	Corticotropin (adrenocorticotropic hormone, ACTH)	Polypeptide (39 residues)	Stimulates adrenal steroid synthesis
	Thyrotropin (thyroid-stimulating hormone, TSH)	2 Polypeptides (α, 96 residues; β, 112 residues	Stimulates thyroid hormone synthesis
	Growth hormone (GH)	Polypeptide (191 residues)	General anabolic effects; stimulates release of insulinlike growth factor-I
	Prolactin	Polypeptide (197 residues)	Stimulates milk synthesis
	Luteinizing hormone (LH)	2 Polypeptides (α, 96 residues; β, 121 residues)	Ovary: luteinization, progesterone synthesis; testis: interstitial cell development, androgen synthesis
	Follicle-stimulating hormone, (FSH)	2 Polypeptides (α, 96 residues; β, 120 residues)	Ovary: follicle development, ovulation, estrogen synthesis; testis: spermatogenesis

TABLE 31–1 (Continued)

Endocrine Organ	Hormone	Structure	Function
Thyroid	Thyroxine and triiodothyronine	Iodinated dityrosine derivatives (see Figure 31–3)	General stimulation of many cellular reactions
	Calcitonin	Polypeptide (32 residues)	Ca^{2+} and P_i metabolism
	Calcitonin gene related peptide (CGRP)	Polypeptide (37 residues)	Vasodilator
Parathyroid	Parathyroid hormone (PTH)	Polypeptide (84 residues)	Ca^{2+} and P_i metabolism
Alimentary tract[c]	Gastrin	Polypeptide (17 residues)	Stimulates acid secretion from stomach and pancreatic secretion
	Secretin	Polypeptide (27 residues)	Regulates pancreas secretion of water and bicarbonate
	Cholecystokinin	Polypeptide (33 residues)	Secretion of digestive enzymes
	Motilin	Polypeptide (22 residues)	Controls gastrointestinal muscles
	Vasoactive intestinal peptide (VIP)	Polypeptide (28 residues)	Gastrointestinal relaxation; inhibits acid and pepsin secretion
	Gastric inhibitory peptide (GIP)	Polypeptide (43 residues)	Inhibits gastrin secretion
	Somatostatin	Polypeptide (14 residues)	Inhibits gastrin secretion; inhibits glucagon secretion
Heart	Atrial natriuretic peptide (ANP)	Several active peptides cleaved from precursor polypeptide of 126 residues	Smooth muscle relaxation; diuretic activity
Pancreas	Insulin	2 Polypeptides (21 and 30 residues)	Glucose uptake, lipogenesis, general anabolic effects
	Glucagon	Polypeptide (29 residues)	Glycogenolysis, release of lipid
	Pancreatic polypeptide	Polypeptide (36 residues)	Glycogenolysis, gastrointestinal regulation
	Somatostatin	Polypeptide (14 residues)	Inhibition of somatotropin and glucagon release

Source	Hormone	Structure	Function
Adrenal cortex	Glucocorticoids	Steroids (cortisol, corticosterone)	Many diverse effects on protein synthesis and inflammation
	Mineralocorticoids	Steroids (aldosterone)	Maintains salt balance
Adrenal medulla	Epinephrine	Tyrosine derivative (see Figure 31-4)	Smooth-muscle contraction, heart function, glycogenolysis, lipid release
	Norepinephrine	Tyrosine derivative (see Figure 31-4)	Arteriole contraction, lipid release
Gonads	Estrogens (ovary)	Steroids (estradiol, estrone)	Maturation and function of secondary sex organs
	Progestins (ovary)	Steroids (progesterone)	Ovum implantation, maintenance of pregnancy
	Androgens (testes)	Steroids (testosterone)	Maturation and function of secondary sex organs
	Inhibins A and B	1 polypeptide (α, 134 residues; β, 115 and 116 residues)	Inhibit FSH secretion
Placenta	Estrogens	Steroids	Maintenance of pregnancy
	Progestins		
	Choriogonadotropin	2 Polypeptides (α, 96 residues; β, 147 residues)	Similar to LH
	Placental lactogen	Polypeptide (191 residues)	Similar to prolactin
	Relaxin	2 Polypeptides (22 and 32 residues)	Muscle tone
Liver	Angiotensin[d]	Polypeptide (8 residues)	Responsible for essential hypertension
Kidney	1,25-dihydroxyvitamin D_3	Steroid	Calcium uptake, bone formation

[a] Only the more common hormones of known structure are listed.

[b] Most of the hypothalamic releasing factors are also called hypothalamic regulatory hormones.

[c] Many of these peptides are also bound in the brain, where they may modulate neural activity.

[d] The liver secretes α_2-globulin, which is cleaved by renin, a kidney enzyme, to give a decapeptide, proangiotensin, from which the carboxy terminal dipeptide is removed to give angiotensin.

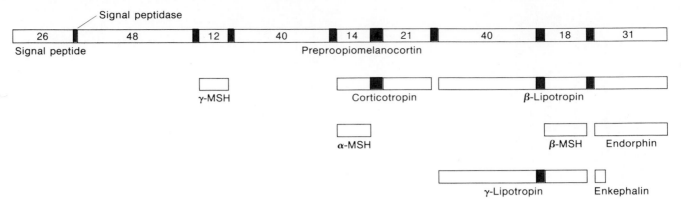

Figure 31–1

Processing pathway of preproopiomelanocortin. This precursor polypeptide is cleaved into a variety of active peptides. With the exception of the signal peptidase cleavage site, the cleavage sites are generally pairs of basic amino acids, although one site contains four. Which active peptides are produced depends on the processing pathway, which varies in different cell types.

ample is _proopiomelanocortin_ (Figure 31–1), which is a precursor for corticotropin (ACTH), β-lipotropin, three melanocyte-stimulating hormones (MSH), endorphin, and an enkephalin. When several different polypeptide hormones are cleaved from a common precursor, the cleavage pattern can vary to yield a different spectrum of peptides, depending on the cell type.

In some cases, researchers have been able to recognize in the precursor certain novel peptides that may turn out to be predicted hormones. For example, the precursor of gonadotropin regulatory hormone (GnRH) contains another peptide of 56 amino acids that may be the long sought-after prolactin-inhibiting factor (PIF). When this peptide was chemically synthesized, on the basis of the sequence predicted from the cDNA, it turned out to inhibit prolactin secretion in cultured pituitary cells.

To generalize, _all polypeptide hormones are synthesized from mRNA as precursors_, which contain signal peptides that direct them into the lumen of the endoplasmic reticulum. In a few cases (e.g., growth hormone and prolactin), no further processing is required to make the mature hormone; however, in most cases further processing is required. The peptides synthesized by neurosecretory cells of the hypothalamus, for example, are often much larger than the final hormones. Oxytocin and vasopressin, each 9 residues long, represent the amino terminus of precursors called proneurophysins, which are 160 and 215 amino acids long, respectively. After oxytocin or vasopressin is cleaved from proneurophysin in the hypothalamus, where it is synthesized, the hormone remains associated with the neurophysin as it passes down the axons to the posterior pituitary, where it is secreted. The neurophysin may serve to protect the hormone from degradation prior to secretion. Somatostatin, another hypothalamic hormone, is the 14-amino-acid product cleaved from the carboxy terminal of a precursor that contains 121 amino acids. _Thyrotropin regulatory hormone (TRH)_ is the smallest known polypeptide hormone (pyroGlu-His-ProNH$_2$). In the synthesis of TRH, the glutamyl residue is cyclized and the carboxy-terminal amide originates from an adjacent glycine. The precursor polypeptide (255 amino acids) contains five copies of the sequence Lys-Arg-Gln-His-Pro-Gly-Arg-Arg within it (Figure 31–2).

Some peptide hormones—for example, luteinizing hormone, follicle-stimulating hormone, and thyroid-stimulating hormone—are composed of two polypeptides, α and β, held together by noncovalent interactions. The α chains of all of these hormones are derived from the same gene, whereas the

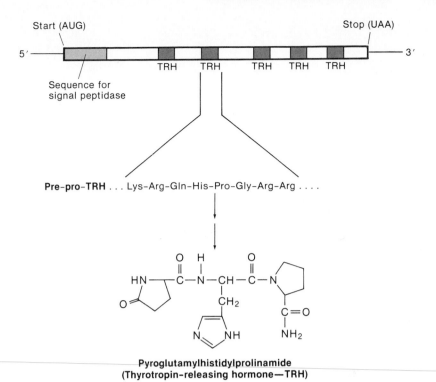

Figure 31–2

Biosynthesis of thyrotropin releasing hormone (TRH). Five copies of TRH are contained within a 255-amino-acid precursor polypeptide (pre-pro-TRH). The precursor has a hydrophobic signal peptide (shaded) and five copies of the sequence Lys-Arg-Gln-His-Pro-Gly-Arg-Arg (color). The dibasic amino acids are recognized by specific proteases to liberate Gln-His-Pro-Gly, which is subsequently converted into TRH. The amino-terminal glutamine is converted into pyroglutamine, and the carboxy-terminal amide is derived from the neighboring glycine, which is removed by a specific enzyme.

β chains come from different genes. The association of the two chains occurs after they pass into the lumen of the endoplasmic reticulum. Although the sequences of the α chains are identical, the chains are nevertheless glycosylated differently as they pass through the Golgi, apparently because of their association with different β chains. Inhibins, which originate in the gonads, are also composed of two peptide chains derived from different mRNAs, but they are linked by disulfide bonds.

Peptide hormones are usually stored in secretory granules after their passage through the endoplasmic reticulum and Golgi. Release of the hormones into the bloodstream is accomplished by fusing the secretory granule membranes with the plasma membrane. As we will see, this event is often regulated by other hormones.

Thyroid Hormones and Epinephrine Are Amino Acid Derivatives

Thyroxine (T_4) and the more potent triiodothyronine (T_3) are cleaved from a large precursor protein called thyroglobulin. Thyroglobulin exists as a dimer of two identical polypeptides ($M_r = 330,000$). It is a storage protein for iodine and can be considered a prohormone of the circulating thyroid hormones. Thyroglobulin is secreted into the lumen of the thyroid gland, where several of the tyrosine residues are iodinated in one or two positions by a special peroxidase; then two iodinated residues condense as shown in Figure 31–3.

The secretion of thyroid hormones starts with endocytosis of the modified thyroglobulin, followed by fusion of the endocytotic vesicles with lysosomes. The lysosomal enzymes then degrade the thyroglobulin, liberating triiodothyronine and thyroxine into the circulation. Only about five molecules of T_3 and T_4 are generated from each molecule of thyroglobulin. Thyroid hormone secretion is stimulated by thyrotropin (TSH), a pituitary hormone that activates adenylate cyclase in its target cells.

Epinephrine, originally called adrenalin, was the first hormone to be isolated, characterized, and synthesized. These achievements occurred at

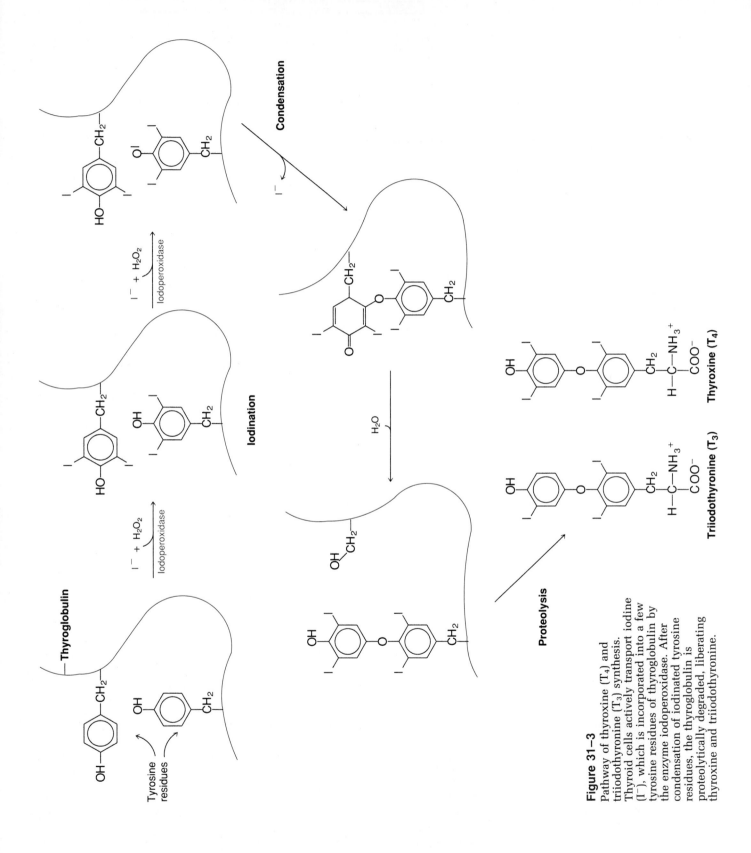

Figure 31-3
Pathway of thyroxine (T₄) and
triiodothyronine (T₃) synthesis.
Thyroid cells actively transport iodine
(I⁻), which is incorporated into a few
tyrosine residues of thyroglobulin by
the enzyme iodoperoxidase. After
condensation of iodinated tyrosine
residues, the thyroglobulin is
proteolytically degraded, liberating
thyroxine and triiodothyronine.

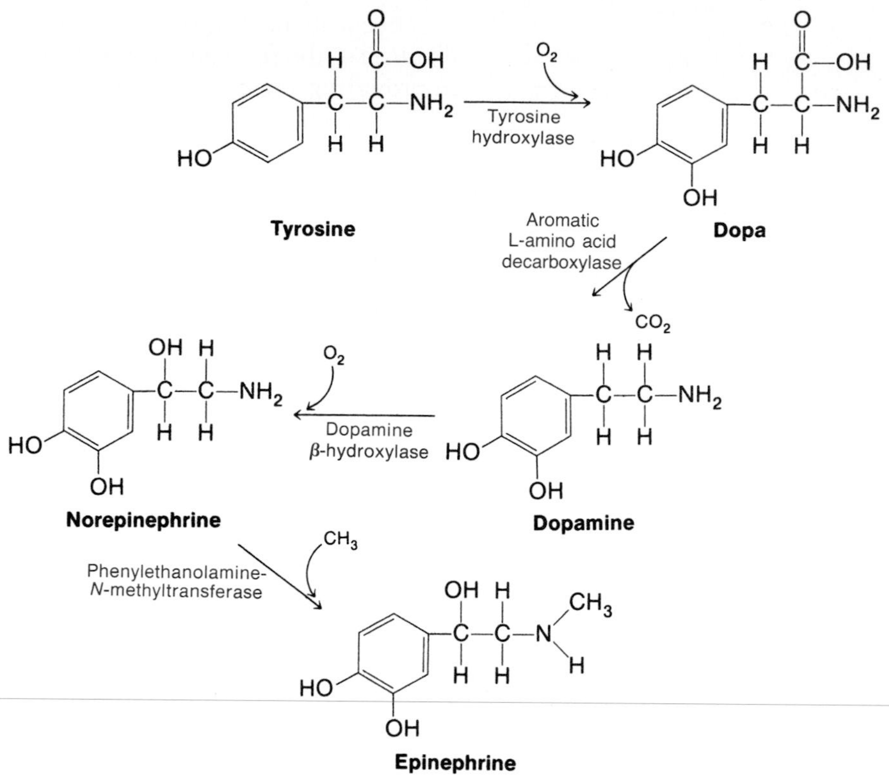

Figure 31–4
Pathway of epinephrine synthesis. Epinephrine and its precursor, norepinephrine, are synthesized from tyrosine. The synthesis occurs in the chromaffin cells of the adrenal medula and in neurons of the central and peripheral nervous system. The first step, which is catalyzed by tyrosine hydroxylase, is the rate-limiting step in the pathway.

the end of the nineteenth century. *Epinephrine and its precursor, norepinephrine, are synthesized from tyrosine* in the chromaffin cells of the adrenal medulla. They are also synthesized by neurons of the central and peripheral nervous system. The biosynthetic pathway is shown in Figure 31–4.

The first step, which involves oxidation of tyrosine to 3,4-dihydroxyphenylalanine (dopa), is catalyzed by tyrosine hydroxylase. This is the rate-limiting enzyme in the pathway. The amount and activity of tyrosine hydroxylase are regulated by cAMP-dependent mechanisms that are responsive to the neurotransmitter acetylcholine. The latter is liberated by special neurons that impinge upon the chromaffin cells. Second, dopa is decarboxylated to dopamine, which is then β-hydroxylated to produce norepinephrine. Finally, the N-methylation of norepinephrine (S-adenosylmethionine is the methyl donor) produces epinephrine. Both of these catecholamines (norepinephrine and epinephrine) are stored in chromaffin granules, where they are complexed with ATP and proteins called chromogranins. Neural stimulation of the medulla is mediated by acetylcholine, which binds to receptors on the membranes of medullary cells. This event leads to a local depolarization and influx of calcium. As a result, the chromaffin granules fuse with the cell membrane, and a packet of catecholamines, ATP, and protein is extruded into the extracellular fluid.

Steroid Hormones Are Derivatives of Cholesterol

Steroid hormones are derived from cholesterol by a stepwise removal of carbon atoms and hydroxylation. The steroid hormones are synthesized by cells of the adrenal cortex (in the case of glucocorticoids and mineralocorticoids) and the gonads (in the case of estrogens, progestins, and androgens). The hormone names just mentioned are generic names for entire classes of compounds that interact with specific receptors; for example, estrogen is the generic name for a family that includes 17β-estradiol, estrone, and diethylstilbestrol (DES), a nonsteroidal estrogen.

Because most of the cholesterol of cells exists as cholesterol esters, hydrolysis of the fatty acid esters by _cholesterol ester hydrolase_ is the first step in providing substrates for steroid biosynthesis. Cleavage of the side chain in the mitochondria is performed by an enzyme complex called _cholesterol desmolase_, which hydroxylates the side chain in two positions and then cleaves it to yield pregnenolone, as discussed in Chapter 23. These are the rate-limiting steps in steroid biosynthesis in both adrenals and gonads; consequently they are tightly regulated. Corticotropin (ACTH) regulates the enzymes in the adrenals, and follicle-stimulating hormone (FSH) and luteinizing hormone (LH) regulate them in the gonads. These hormones bind to membrane receptors that activate adenylate cyclase. Cholesterol ester hydrolase is known to be activated by a cAMP-dependent protein kinase, but the regulation of desmolase is less well understood. Steroid hormones, unlike most hormones, are not stored in granules for subsequent secretion. Consequently, the circulating level of steroids is controlled largely by the rate of their biosynthesis. Nevertheless, if an animal is stressed, for example, the serum concentration of corticosterone can rise severalfold within a few minutes.

The step-by-step synthesis of steroid hormones from cholesterol (C_{27}) was presented in Chapter 23 and is summarized in Figure 31–5, with emphasis on the number of carbon atoms. Note that pregnenolone (C_{21}) and progesterone (C_{21}) are intermediates in the biosynthesis of all of the major adrenal steroids, including cortisol (C_{21}), corticosterone (C_{21}), and aldosterone (C_{21}). The same two compounds are intermediates in the synthesis of the gonadal steroid hormones, testosterone (C_{19}) and 17β-estradiol (C_{18}). Because the synthesis of all these hormones follows a common pathway, a defect in the activity or amount of an enzyme along that pathway can lead to both a deficiency in the hormones beyond the affected step and an excess of the hormones, or metabolites, prior to that step.

Deficiencies in each of the six enzymes involved in the conversion of cholesterol to aldosterone have been observed in humans. Each deficiency gives rise to a characteristic steroid hormone imbalance, with telling clinical consequences. For example, a deficiency in _17-hydroxylase_ gives rise to inadequate levels of cortisol as well as inadequate levels of androgens and estrogens, with severe effects on sexual maturation. A deficiency in the next enzyme along the pathway, _21-hydroxylase_, blocks the synthesis of adrenal glucocorticoids and mineralocorticoids and leads to an overproduction of testosterone by the adrenals. This overproduction of testosterone is due to metabolic shunting of progesterone into the sex steroid pathway. The synthesis of androgens is exacerbated by the lack of feedback inhibition of cortisol on the hypothalamus; the result is chronic production of CRF and ACTH and perpetual activation of adrenal steroid biosynthesis. In females, excessive androgen production, due to a defect in 21-hydroxylase, leads to masculinization or, to use the clinical term, _female pseudohermaphrodism_—that is, a genetic female with male appearance. This reversal of sexual phenotype is explained by the fact that during embryonic development of mammals, the external genitalia develop from common precursor cells. In the absence of hormonal stimulation, they will develop into female structures, but androgens will direct their development into male structures.

Testosterone is both a hormone and a prohormone. The high levels of testosterone normally produced in the male by the testes play a major role in the growth and function of many tissues in addition to reproductive organs. Essentially all the sexual differences in nonreproductive tissues, such as muscle, liver, and brain, are a consequence of androgen action. Although testosterone is the major circulating androgen, many target cells reduce this steroid to 5α-dihydrotestosterone, a steroid that binds to the androgen recep-

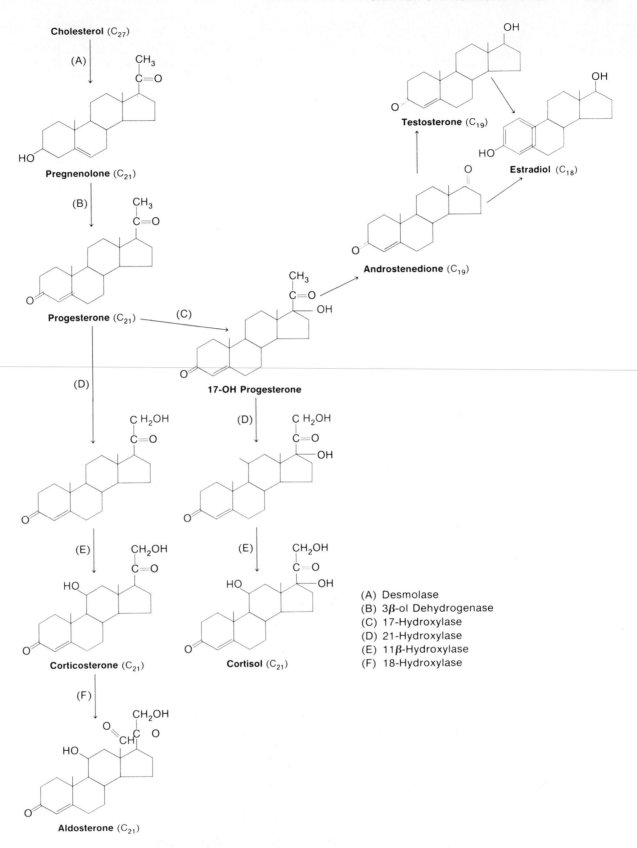

(A) Desmolase
(B) 3β-ol Dehydrogenase
(C) 17-Hydroxylase
(D) 21-Hydroxylase
(E) 11β-Hydroxylase
(F) 18-Hydroxylase

Figure 31–5

Schematic pathway of steroid hormone biosynthesis. Note that pregnenolone and progesterone are intermediates in the synthesis of all of the major adrenal steroids, including cortisol, corticosterone, and aldosterone. They are also intermediates in the synthesis of the gonadal steroid hormones, testosterone and 17β-estradiol.

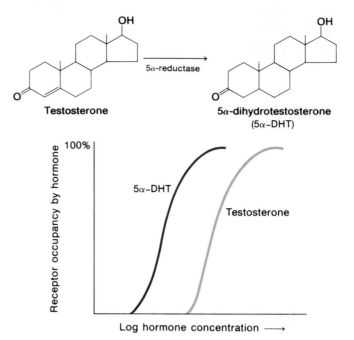

Figure 31–6

Conversion of testosterone to 5α-dihydrotestosterone (5α-DHT). Receptor-binding studies indicate that 5α-DHT has a higher affinity for the androgen receptor than testosterone.

tor with higher affinity than testosterone (Figure 31–6). When 5α-reductase is defective, the androgen receptors are only partially activated and a full androgen response is not obtained. A deficiency in this enzyme therefore leads to abnormal development of male genitalia, i.e., they are of female phenotype (a clinical condition referred to as *male pseudohermaphrodism, type 2*). In this example, testosterone can be considered a prohormone of a more active androgen.

Testosterone is also a prohormone of metabolites that bind to different receptors. For example, some of the effects of androgens on the production of red blood cells (erythropoiesis) are due to the reduction of testosterone to 5β-dihydrotestosterone within precursor cells. 5β-Dihydrotestosterone binds to a receptor that is distinct from the one that binds testosterone and 5α-dihydrotestosterone (Figure 31–7). Testosterone also influences erythropoiesis by stimulating the kidney to produce a specific growth factor, *erythropoietin*. Another striking example of testosterone as a prohormone occurs in the brain, where testosterone influences neural development and activity (e.g., male-specific mating behavior and bird songs) by being converted into 17β-estradiol, which interacts with estrogen receptors. Testosterone is metabolized to 17β-estradiol by *aromatase*, the same enzyme involved in 17β-estradiol synthesis in the ovary (Figure 31–7).

Vitamin D_3 is a precursor of the hormone 1,25-dihydroxyvitamin D_3. Vitamin D_3 is essential for normal calcium and phosphorus metabolism. It is formed from 7-dehydrocholesterol by ultraviolet photolysis in the skin. Insufficient exposure to sunlight and absence of vitamin D_3 in the diet leads to *rickets*, a condition characterized by weak, malformed bones. Vitamin D_3 is inactive, but it is converted into an active compound by two hydroxylation reactions that occur in different organs. The first hydroxylation occurs in the liver, which produces 25-hydroxyvitamin D_3, abbreviated $25(OH)D_3$; the second hydroxylation occurs in the kidney and gives rise to the active product, 1,25-dihydroxyvitamin D_3 $(1,25(OH)_2D_3)$ (Figure 31–8). The hydroxylation at the 1 position that occurs in the kidney is stimulated by parathyroid hormone (PTH), which is secreted from the parathyroid gland in response to low circulating levels of calcium. In the presence of adequate calcium, $25(OH)D_3$ is converted into an inactive metabolite, $24,25(OH)_2D_3$. The active derivative of vitamin D_3 is considered a hormone because it is transported

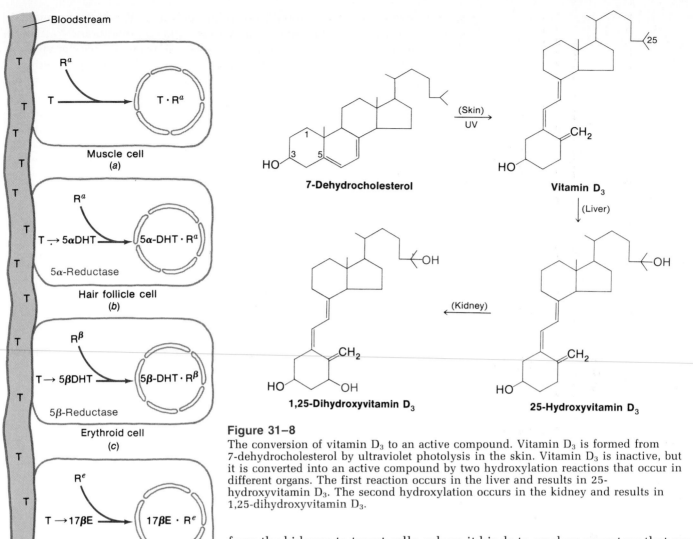

Figure 31-7
Metabolic conversion of testosterone by target cells. Testosterone (T) is the predominant androgen in the bloodstream. When testosterone enters target cells, it can be metabolized in a variety of different ways. It can either (a) bind directly to androgen receptors (R^a) or (b) be reduced to 5α-dihydrotestosterone (5α-DHT), which then binds to R^a with higher affinity. (c) Other target cells reduce testosterone to 5β-dihydrotestosterone (and other 5β metabolites), which bind to a distinct receptor (R^β). (d) Yet other cells convert testosterone into an estrogen, 17β-estradiol, which binds to estrogen receptors (R^e).

Figure 31-8
The conversion of vitamin D_3 to an active compound. Vitamin D_3 is formed from 7-dehydrocholesterol by ultraviolet photolysis in the skin. Vitamin D_3 is inactive, but it is converted into an active compound by two hydroxylation reactions that occur in different organs. The first reaction occurs in the liver and results in 25-hydroxyvitamin D_3. The second hydroxylation occurs in the kidney and results in 1,25-dihydroxyvitamin D_3.

from the kidneys to target cells, where it binds to nuclear receptors that are analogous to those of typical steroid hormones. $1,25(OH)_2D_3$ stimulates calcium transport by intestinal cells and increases calcium uptake by osteoblasts (precursors of bone cells).

REGULATION OF CIRCULATING HORMONE CONCENTRATION

The serum concentrations of many hormones fall in the range between 10^{-7} to 10^{-12} M. The biological assays that were originally used to measure these low concentrations have now been supplemented by, and to a large extent supplanted by, extremely sensitive _radioimmunoassays_ (RIA). The essential ingredients of an RIA are an antibody that specifically binds the hormone and a radioactively labeled hormone. The principle of an RIA is competition of the unlabeled hormone (standard or unknown) with radioactive hormone for a limiting amount of antibody. The sensitivity of these assays is limited by the specific radioactivity of the hormones. It is not unusual to be able to measure 10^{-12} to 10^{-15} moles of hormones with these assays.

The occupancy of hormone receptors can fluctuate greatly and is ultimately determined by the concentration of "free" hormone in the blood. The major determinants of hormone concentration are (1) the rate of hormone secretion from endocrine cells and (2) the rate of hormone removal by clearance or metabolic inactivation. As we have seen, most hormones (with the

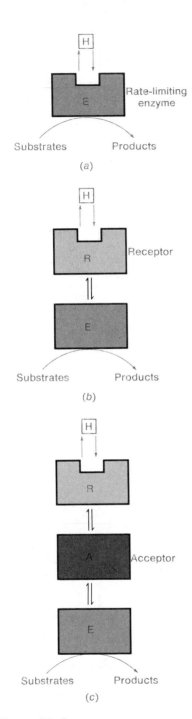

Figure 31–9

Possible mechanisms of hormone action. (a) The hormone (H) could theoretically activate an enzyme (E) directly as an allosteric effector. (b) Alternatively, there might be a separate binding protein for the hormone, called a receptor (R), which then activates an enzyme. (c) Another possibility interposes an acceptor protein (A) between the receptor and the enzyme. Each interaction is reversible.

exception of steroids) are stored in secretory granules. When the hormone is needed, the granule membranes fuse with the plasma membrane to liberate their contents into the bloodstream. This event is triggered by signals from other hormones or by neural signals. Stimulation of hormonal secretion is usually coupled with an increase of hormone synthesis, so that hormonal stores are replenished.

Most hormones have a half-life in the blood of only a few minutes because they are cleared or metabolized very rapidly. The rapid degradation of hormones allows target cells to respond transiently. Polypeptide hormones are removed from circulation by serum and cell-surface proteases, by endocytosis followed by lysosomal degradation, and by glomerular filtration in the kidney. Steroid hormones are taken up by the liver and metabolized by hydroxylation, sulfation, or conjugation with glucuronic acid to inactive forms; then the inactive forms are often excreted into the bile duct or back into the blood for removal by the kidneys. Catecholamines are metabolically inactivated by O-methylation, by deamination, and by conjugation with sulfate or glucuronic acid.

Thyroid hormones and most steroid hormones are associated with carrier proteins in the serum. The carrier proteins are called, appropriately, thyroxine-binding globulin, transcortin (for cortisol), and sex-steroid binding protein. These proteins have high affinity ($K_d \sim 10^{-8}$ to 10^{-9} M) for their respective hormones. They presumably function to buffer the concentration of "free" hormone and to retard hormone degradation and excretion. The carrier proteins are distinguishable from the intracellular receptors for these hormones.

HORMONE ACTION IS MEDIATED BY RECEPTORS

Binding of any hormone induces a conformational change in its receptor, and this change is detected by other macromolecules. Hence the hormone-receptor interaction can be transduced from one molecule to another. In the simplest scheme, a hormone receptor might be a rate-limiting enzyme or be coupled to a rate-limiting enzyme. The insulin receptor, for example, appears to be a rate-limiting enzyme. In cases where the receptors regulate adenylate cyclase, there is another protein interposed between the receptor and the adenylate cyclase. The protein directly activated by hormone-receptor interaction is sometimes referred to as an acceptor protein. GTP-binding proteins (G_s and G_i) are the acceptors for all receptors that activate or inhibit adenylate cyclase (described in more detail below). For many membrane receptors, the molecular intermediates are still unknown. Generalized schemes for hormone-receptor activation of physiological responses are shown in Figure 31–9.

Receptors for steroid and thyroid hormones are located in the nucleus and bind to specific DNA sequences when they are activated by the appropriate hormones. The binding of these receptors to DNA in the promoter region of a gene activates transcription, presumably by helping to assemble an efficient initiation complex. The acceptor proteins in these cases may be other transcription factors.

The important point is that all hormones act by binding to receptors that are located either on the cell membrane or inside the cell. Binding of the hormone induces a conformational change in the receptor that is transmitted to its active site (if it is an enzyme) or to other macromolecules. In either case, a chain of events is elicited that ultimately affects a vast array of metabolic processes, ranging from alterations in enzyme activities to changes in gene expression. With time, these changes may lead to profound alterations in cell growth, morphology, and function.

Hormone Binding to Receptors Changes Their Functional Activity

Receptors transduce chemical signals; hence they must have at least two binding sites: one for the hormone and one for the macromolecular acceptor. The interaction of receptors with hormones is analogous to the interaction of allosteric enzymes with effectors (see Chapter 10) in that binding of the hormone to one site on the receptor affects the functional activity (conformation) of another site on the same molecule. Since receptors are essential for hormone action, it follows that (1) target cells are only those cells that have receptors and (2) the magnitude of the response in target cells should be related to the number of functional receptors. However, not all target cells respond to hormone-receptor binding in the same way. Their response depends on what genes have been activated (committed) during prior developmental stages.

The initial binding of hormones to their receptors, be they polypeptide hormones interacting with membrane receptors or steroid hormones interacting with nuclear receptors, can be thought of in terms of the Michaelis-Menten enzyme kinetics discussed in Chapter 8. That is, we can measure and assign a dissociation constant (K_d) to the hormone-receptor complex ($H \cdot R$), just as we can for an enzyme-substrate complex ($E \cdot S$). The formation of $H \cdot R$ is reversible, as is that of $E \cdot S$; but unlike $E \cdot S$, the hormone-receptor complex is not converted to products. Because the affinity of receptors for hormones is generally very high (K_d values range from 10^{-7} to 10^{-12} M), it is usually possible to measure $H \cdot R$ directly by using radioactive hormones and rapidly separating $H \cdot R$ from unbound hormone.

Hormones that bind to receptors in a productive manner are often called _agonists_, a pharmacological term used to describe a productive interaction of a drug with a protein. (Many effective drugs interact with hormone receptors.) Chemical modification of various constituents of natural hormones usually leads to a family of agonists with either increased or decreased affinity relative to the natural hormone. These modifications not only help define the stereochemistry of the active site on the receptor, but they also have practical implications for effective drug design.

Many polypeptide hormones fall into homologous families (Table 31–2). The similar chemical structures of different members of a family suggest that they probably all evolved from a single gene by gene duplication and divergence. This process, along with parallel duplication and divergence of receptor genes, could lead to increasingly fine regulation of metabolic processes during evolution. Because of their homology, peptides within a family often show some affinity for all the receptors within that family. For example, most of the actions of prolactin are mimicked by placental lactogen. Likewise, many of the mitogenic actions attributed to high doses of insulin are probably due to insulin binding to receptors for a homologous hormone, insulinlike growth factor-I (IGF-I).

Compounds that bind to receptors in a nonproductive manner are called _antagonists_ (or antihormones). These compounds compete with agonists for binding but they do not elicit a productive conformational change when they bind. Although most members of an agonist/antagonist family are usually chemically related, there are numerous examples of totally unrelated compounds that have similar biological activity. In many cases these compounds are stereochemically similar, but in other cases even the stereochemical relationship is not obvious.

For example, the three most abundant natural estrogens are 17β-estradiol, estrone, and estriol; these compounds are obviously related, but they differ by a factor of 10 in their affinity for estrogen receptors (Figure 31–10). On the other hand, diethylstilbestrol (DES) is a synthetic estrogen chemically unrelated to steroids, yet it has an affinity for estrogen receptors that is

TABLE 31–2
Homologous Polypeptide Hormones

Prolactin
Placental lactogen
Growth hormone

Thyrotropin
Luteinizing hormone
Follicle-stimulating hormone
Choriogonadotropin

Insulin
Relaxin
Insulinlike growth factors (IGF-I, IGF-II)

Nerve growth factor

Glucagon
Vasoactive intestinal peptide
Gastric inhibitory peptide
Secretin

Gastrin
Pancreozymin

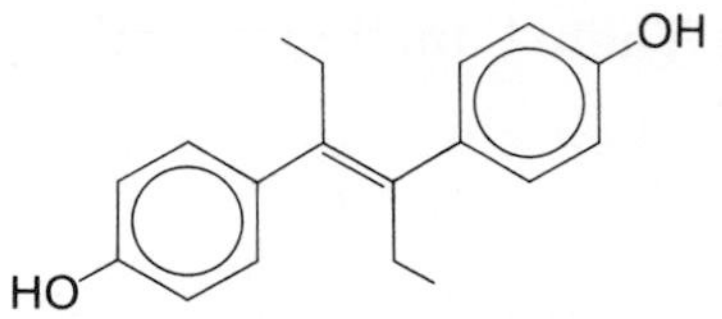

Diethylstilbestrol (DES) (300%)

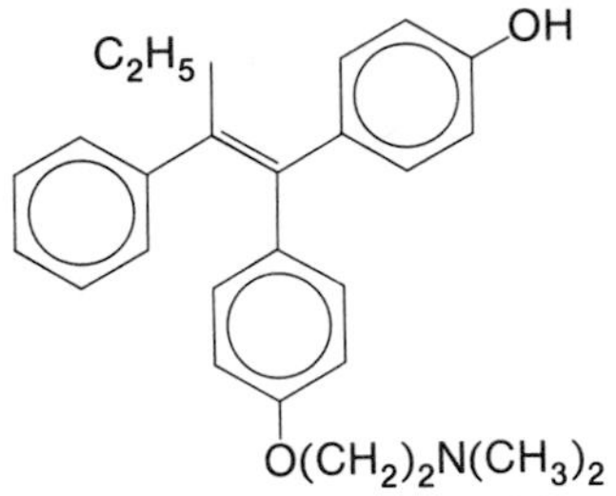

Hydroxytamoxifen (200%)

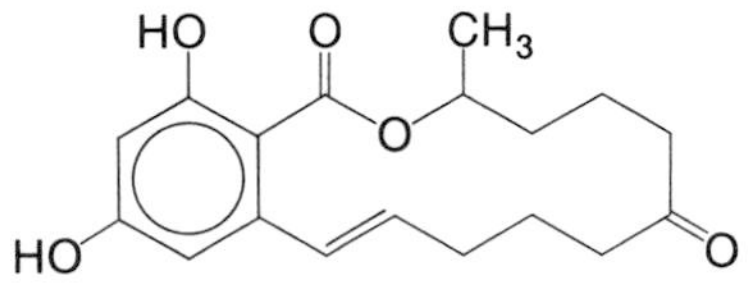

Zearalenone (2%)

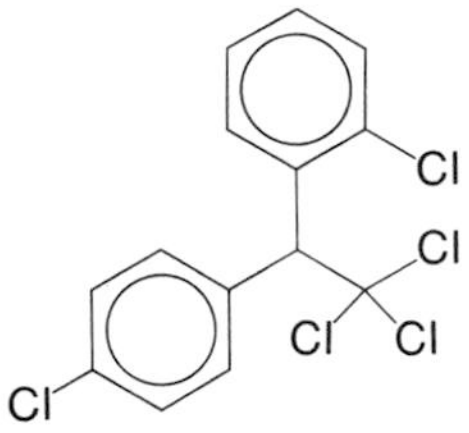

o,p'-DDT (0.004%)

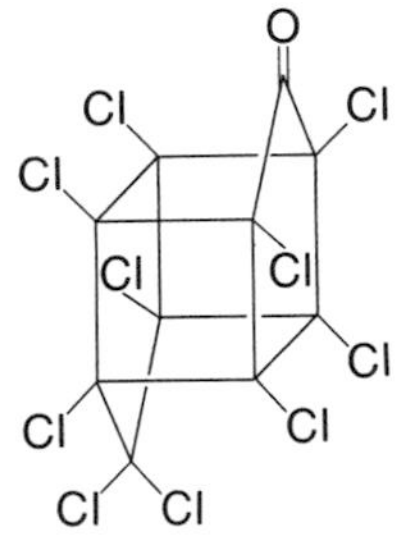

Kepone (0.04%)

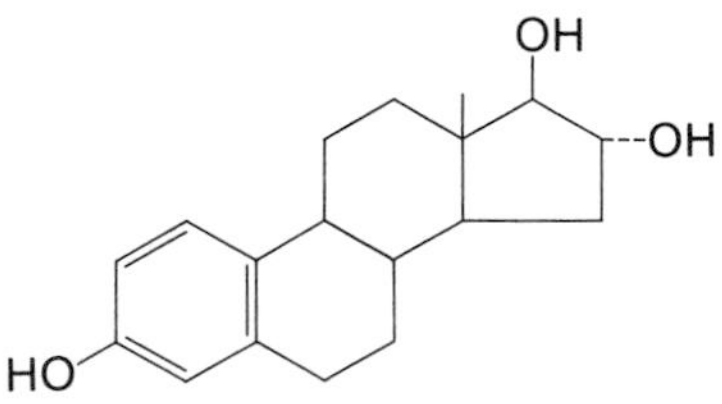

Estriol (10%)

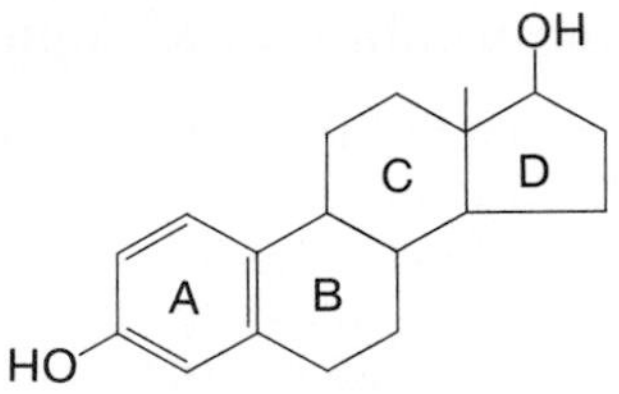

17β-Estradiol (100%)

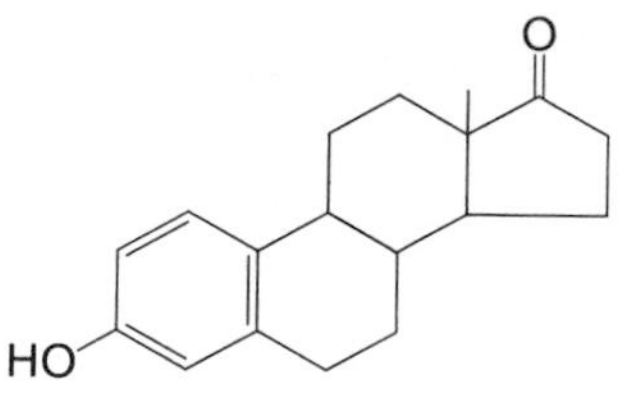

Estrone (10%)

Figure 31–10

Structures of some natural and some synthetic estrogens. The relative affinity for estrogen receptors is shown in parentheses, with that of 17β-estradiol being 100 per cent.

comparable to that of 17β-estradiol, and it promotes all of the biological effects of natural estrogens. (It is still not clear whether the harmful side effects of DES are associated with its estrogenic activity or with nonestrogenic metabolites. Perhaps the latter were responsible for the high rates of malformation and cancer found among the children of those millions of women who were treated with DES during pregnancy in the 1950s.) Hydroxytamoxifen is related to DES, but it is an estrogen antagonist used for the treatment of estrogen-dependent tumors. Although DES and hydroxytamoxifen are not chemically related to steroids, three-dimensional models reveal that the two phenolic rings of these compounds lie in approximately the same position as the A and D rings of 17β-estradiol. Likewise, the binding of zearalenone, an estrogen produced by the fungus *Fusarium*, can be envisaged with stereochemical models. (The estrogen activity of zearalenone can be a serious problem when this fungus infests corn or other livestock foods.) Several insecticides, including o,p'-DDT and Kepone, have weak estrogenic activity; however, the structural relationship of these compounds to steroids is not apparent.

The interaction of catecholamine receptors (usually called α- or β-adrenergic receptors) with agonists and antagonists illustrates some of the common principles. These receptors, which are found on smooth muscle, heart, liver, and nerve cells, normally respond to epinephrine and norepinephrine. However, the ways in which agonists and antagonists bind to the receptors from different tissue sources are not identical, a fact suggesting that the receptors are different. It is now clear that there are at least four distinct catecholamine receptors, designated α_1, α_2, β_1, and β_2, that can be distinguished on the basis of their affinities for different agonists and antagonists. Many agonists have some affinity for each receptor; in contrast, there are antagonists that are quite specific in blocking one receptor but not another. Although both agonists and antagonists must bind to the active site on the receptor, the greater specificity of antagonists may be due to their binding to peripheral domains as well. These peripheral domains may be quite different on α and β receptors and hence allow discrimination by antagonists. Thus, by using an appropriate combination of agonists and antagonists, it is possible to activate a particular catecholamine receptor. This approach has many clinical applications.

Hormone Receptors Are Detected by Their Ability to Bind Hormone

Hormone receptors can be identified by their binding to radioactive hormones. A particularly effective approach involves *affinity labeling*. The principle is to covalently cross-link a radioactively labeled hormone with its receptor, either by using a bifunctional reagent that reacts with both the hormone and the receptor or by modifying the hormone so that it is chemically reactive and can form a covalent bond when it binds in the active site of the receptor. One approach involves incorporation into the hormone of a latent chemical group that can be activated by light after the hormone binds to its receptor; the photoactivation step minimizes spurious chemical reactions with other proteins. Once the receptor is covalently labeled, it can be purified by chromatographic and electrophoretic techniques.

Because hormone receptors are present in low abundance, usually only a few thousand molecules per cell, obtaining homogeneous samples is an arduous task, requiring 10,000- to 100,000-fold purification. *Affinity chromatography* is frequently used for this purpose. (See Chapter 3 for a description of the basic technique.) An agonist or antagonist is covalently attached to an insoluble support (such as agarose) and then a solubilized receptor preparation is passed through a column containing this material. Because of the high affinity of the receptor for the immobilized hormone, the receptor binds tightly while the contaminating proteins pass through the column. The receptor is then eluted with an agonist or antagonist with higher affinity. Receptors for many steroid hormones and catecholamines have been purified by affinity chromatography. Purified receptors not only allow more detailed biochemical characterization (e.g., determination of amino acid sequence), but also provide a means for generating specific antibodies. These antibodies can be used to localize and quantify receptors in the absence of hormones.

The complete amino acid sequence of the *insulin receptor* has been deduced from a full-length cDNA clone. In this case, the cDNA clone was detected with synthetic oligonucleotides that were based on knowledge of a partial amino acid sequence of purified receptor. Although the mature protein is composed of four chains (two α and two β chains) linked together by disulfide bonds (Figure 31–11), both chains are derived from the same mRNA by proteolytic processing of a precursor polypeptide. The α chain is 719 residues long and represents the NH_2 terminus of the precursor, while the β chain is 619 residues long. The two polypeptides are separated by four basic residues, which are presumably recognized by a specific protease. The precursor also has a signal peptide that directs the α chains to the outside of the cell. The α chains form the insulin binding site, whereas the β chains have a hydrophobic patch that spans the plasma membrane. The cytoplasmic domain of the β chain contains a protein kinase activity that phosphorylates tyrosine residues. Thus binding of insulin to the external domain of this receptor is thought to transduce the signal through the membrane to activate the internal kinase domain. Although the natural substrate(s) for this protein kinase are not yet defined, their phosphorylation is presumed to be a key intermediate step in insulin activation of cellular response. The tyrosine kinase domain of the insulin receptor is homologous to serine kinase domains of other proteins, as well as several cellular proto-oncogenes (e.g., *src* and *abl*) and several growth factors (discussed below).

cDNA clones coding for *glucocorticoid and estrogen receptors* have also been isolated. In these cases, the strategy involved cloning the cDNAs into an expression vector that would allow the synthesis of the receptor protein in bacteria. Bacteria containing the sequences of interest were then identified by means of antibodies directed against the receptor. These receptors and those for progesterone and 1,25-dihydroxy vitamin D have a common molecular organization: The steroid-binding domain lies at the carboxy terminus and the DNA-binding domain lies in the middle; the amino terminus presumably interacts with other transcription factors.

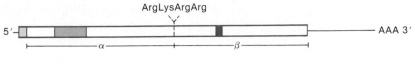

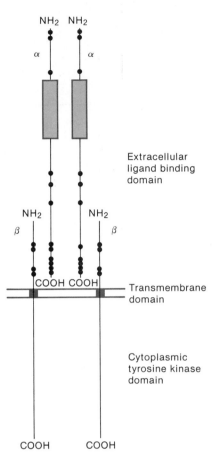

TABLE 31–3
Hormones that Do Not Affect Adenylate Cyclase

Catecholamines (acting on α_1 receptors)
Placental lactogen
Growth hormone
Insulin
Oxytocin
Prolactin
Somatostatin

TABLE 31–4
Hormones that Activate or Inhibit Adenylate Cyclase

Activators

Corticotropin (ACTH)
Calcitonin
Catecholamines (acting on β_1 and β_2 receptors)
Choriogonadotropin
Follicle-stimulating hormone (FSH)
Glucagon
Gonadotropin-releasing hormone (GnRH)
Luteinizing hormone (LH)
Lipotropin (LPH)
Melanocyte-stimulating hormones (MSH)
Parathormone (PTH)
Secretin
Thyrotropin regulatory hormone (TRH)
Thyrotropin (TSH)
Vasoactive intestinal peptide (VIP)
Vasopressin

Inhibitors

Angiotensin
Catecholamines (acting on α_2 receptors)

Figure 31–11
Biosynthesis of the insulin receptor. The mRNA for the insulin receptor codes for a 1370-amino-acid precursor with a signal peptide (shaded 5′ end). The precursor is cleaved at a set of four dibasic acids into an external α subunit and a transmembrane β subunit. The α subunit has a cysteine-rich domain (shaded) that is also found in other receptors (e.g., the EGF receptor); the membrane-spanning region is very hydrophobic (full color). The α and β subunits are cross-linked by disulfide bonds; the locations of cysteines that may be involved are indicated by solid circles.

Most Membrane Receptors Generate a Diffusible Intracellular Signal

Activation of most membrane-associated hormone receptors generates a diffusible intracellular signal, often called a *second messenger*. Many different receptors generate the same second messenger. *Five intracellular messengers are currently known: cyclic AMP, cyclic GMP, inositol triphosphate, diacylglycerol, and calcium.* There are a few membrane receptors for which second messengers do not exist or have not yet been identified (Table 31–3).

The Adenylate Cyclase Pathway.   Many hormones bind to receptors that act through an intermediary protein to either activate or inhibit *adenylate cyclase* (Table 31–4). Cyclic AMP (cAMP) is formed from ATP by the action of adenylate cyclase, which is bound to the inside of the cell membrane (Figure 31–12). Although this reaction is only slightly exergonic, the synthesis of

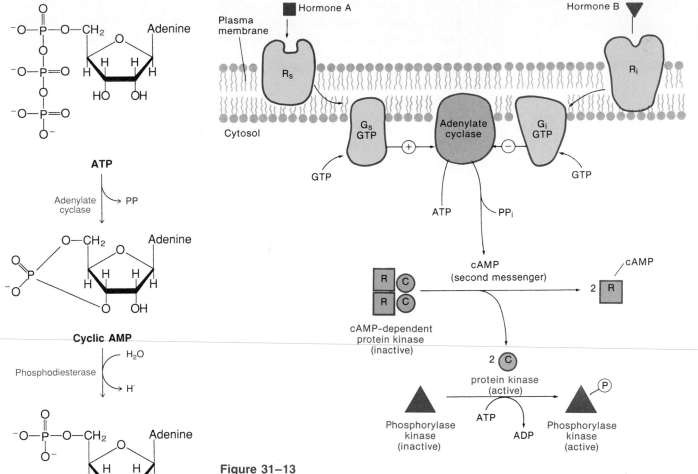

Figure 31–12

Synthesis and degradation of cyclic AMP. The adenylate cyclase is bound to the inside of the cell membrane. The phosphodiesterase, which degrades the cAMP formed by the cyclase, is a soluble enzyme.

Figure 31–13

The adenylate cyclase pathway of hormone receptor action. When hormone receptors (R_s and R_i) bind their appropriate hormones they activate GTP-binding proteins (G_s or G_i), which then either stimulate or inhibit adenylate cyclase (AC). Adenylate cyclase catalyzes the synthesis of cAMP (the second messenger), which binds to the regulatory subunits (R) of cAMP-dependent protein kinase, liberating the catalytic subunits (C), which are then active. The C subunit catalyzes the phosphorylation of numerous protein substrates; for example, phosphorylation of phosphorylase kinase converts it into a more active form. The activity of other enzymes (e.g., glycogen synthase) is inhibited by phosphorylation.

cAMP is driven by the subsequent hydrolysis of inorganic pyrophosphate. cAMP is a stable compound in the absence of *phosphodiesterase,* which degrades it into adenosine monophosphate. cAMP serves as a second messenger by activating a protein kinase that phosphorylates a variety of proteins to alter their function (Figure 31–13). The elucidation of this pathway provided the first molecular explanation of enzyme activation by covalent modification, as well as giving insight into the molecular mechanism of hormone action (also see Chapter 14). Remarkably, the entire pathway, up to phosphorylation of specific enzymes, can be reproduced in preparations of lysed cells; this feature has greatly simplified the elucidation of the individual steps.

Three membrane proteins transduce extracellular hormone binding into intracellular cAMP production. Some tissue culture cells are killed by prolonged exposure to hormones that stimulate cAMP production or by incubation with exogenous cAMP. However, certain mutant cells are resistant to the killing effect of cAMP. These mutants fall into several classes. One class has defective regulatory (R) subunits of protein kinase that do not respond to cAMP. Another class lacks functional catalytic (C) subunits of protein kinase. A third class lacks adenylate cyclase, and a fourth class lacks hormone receptors. A fifth, unexpected class lacks a GTP-binding protein that mediates the signal generated by receptor occupancy into activation of adenylate cyclase. The GTP-binding protein is called G protein. It is noteworthy that a mutation in G, adenylate cyclase, or protein kinase eliminates responses to all hormones acting via cAMP. Clearly, there is a single, unbranched pathway leading from G proteins to cAMP-dependent protein kinase. The fact that cells containing any of these mutations can grow normally is consistent with the notion that hormones stimulate specialized functions rather than maintain basal metabolic activities.

Biochemical experiments indicate that the hormone receptor is embedded in the membrane with the hormone-binding site facing the outside of the cell. When the receptor binds an agonist, it undergoes a conformational change that is detected by a G protein, which is also embedded in the membrane but has its GTP-binding site facing the inside of the cell. When a G protein is activated by a hormone-receptor complex it binds GTP, dissociates from the receptor, and can activate adenylate cyclase, which also has its active catalytic site facing the inside of the cell. These three proteins—the receptor, the GTP-binding protein, and adenylate cyclase—are not likely to be coupled prior to hormonal activation; rather, they probably find each other by lateral diffusion within the membrane (Figure 31–13). G proteins have a GTPase activity that hydrolyzes GTP to GDP and thereby terminates the activation of adenylate cyclase, preventing permanent activation. In cell extracts, maximal G-protein activity can be obtained with the nonhydrolyzable GTP analog guanosine-5' (β,γ-imido) triphosphate (GppNHp).

Evidence for the lateral diffusion of adenylate cyclase within the membrane came from experiments in which mutant cells that lacked adenylate cyclase, but contained functional receptors, were fused to cells that lacked receptors, but contained functional adenylate cyclase. Shortly after cell fusion, the hybrid cells became responsive to the hormone by producing cAMP. It seems clear that the receptors from one cell had diffused through the membrane into the other to activate adenylate cyclase.

G protein can be isolated in a detergent extract and that extract can be added to mutant cells lacking a functional G protein to restore hormonal activity in the mutant cells. This technique has been extended to study the interaction of receptors and G protein by mixing two detergent extracts—one containing only functional β-adrenergic receptors and the other only functional G protein—and then assaying for activated G protein by inserting these components into mutant cells that lack G protein and assaying for adenylate cyclase activity. Only when both membrane extracts and a potent agonist are used is significant activation of adenylate cyclase achieved. Moreover, the activity of G protein is maintained even after the receptor is inactivated with a potent antagonist. This observation and the kinetics of the reaction indicate that the receptors act catalytically to activate many molecules of G protein. The hormonal response is further amplified because the G protein activates many molecules of adenylate cyclase.

Some hormones inhibit adenylate cyclase. Angiotensin and epinephrine (acting on α_2 receptors, labeled R_i in Figure 31–13) inhibit adenylate cyclase by a mechanism that requires GTP and is potentiated by sodium ions. These hormones, along with a much larger group of compounds including opiates, adenosine, and some neurotransmitters, are thought to act through receptors that interact with a different G protein than the one we have just discussed. Thus, there are at least two different classes of G proteins—G_s, which stimulates adenylate cyclase when it is activated, and G_i, which inhibits cAMP production when it is activated (Figure 31–13). This

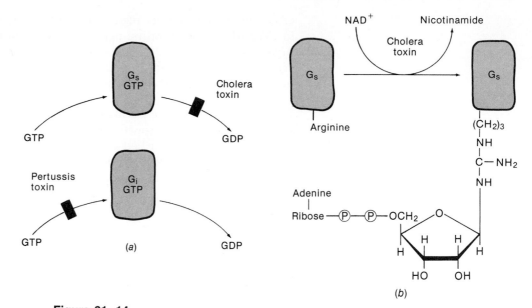

Figure 31–14
(a) Cholera toxin, pertussis toxin, and the ADP-ribosylation of GTP-binding proteins. Cholera toxin enzymatically cleaves NAD^+ (b) and attaches the ADP-ribose moiety to an arginine residue of G_s, thereby preventing GTP hydrolysis and permanently activating this stimulator of adenylate cyclase. In a similar reaction, pertussis toxin ADP-ribosylates G_i, which prevents GTP binding and hence prevents the normal inhibition of adenylate cyclase by G_i.

dual system for regulating cAMP synthesis provides another means of integrating the different environmental stimuli that impinge upon cells.

The two G proteins can be distinguished biochemically by the action of bacterial toxins. One of the subunits of _cholera toxin_ can permanently activate the G_s protein, leading to continuous adenylate cyclase activity, whereas _pertussis toxin_ can permanently inactivate the G_i protein, thereby preventing inhibition of adenylate cyclase. Cholera toxin activates G_s protein by enzymatically cleaving NAD^+ and attaching the ADP-ribose moiety to an arginine residue of the G protein (Figure 31–14). This process, called _ADP-ribosylation_, leads to permanent activation by inactivating the GTPase activity. The extreme diarrhea and dehydration characteristic of cholera are caused by the chronic activation of adenylate cyclase in mucosa cells of the intestine. The resulting increase in cAMP stimulates a large efflux of sodium ions and water into the gut. Pertussis toxin and diphtheria toxin (which inhibits protein synthesis) also act by ADP-ribosylation.

Activation of cAMP-dependent Protein Kinase. Cyclic AMP binds to the regulatory subunits of a cAMP-dependent protein kinase that is composed of four subunits: two catalytic subunits, C; and two regulatory subunits, R. Binding of cAMP to the regulatory subunits promotes the dissociation of the two catalytic subunits, which are then active (Figure 31–14). _The catalytic subunits phosphorylate a wide variety of proteins on serine or threonine residues_, inducing conformational changes that alter their function. For example, many substrates of the kinase are rate-limiting enzymes in key metabolic pathways. The activity of cAMP-dependent protein kinase is determined by the intracellular cAMP concentration, which is a function of its rate of synthesis by adenylate cyclase and its rate of degradation by _cAMP phosphodiesterases_. There are several phosphodiesterases and their activities can be hormonally regulated. Thus there are many levels at which the activity of cAMP-dependent protein kinases can be regulated.

Some hormones that utilize cAMP as a second messenger ultimately induce transcription of specific genes. The mechanism is uncertain. The cAMP may bind to other regulatory proteins, which act in a manner similar to CAP (the cAMP-binding protein of bacteria that mediates catabolite repression; see Chapter 29), or the regulatory subunits may serve this function directly, or some of the proteins phosphorylated by the catalytic subunit may be regulatory transcription factors.

The Guanylate Cyclase Pathway.   Many cells have membrane-bound and soluble guanylate cyclases that convert GTP into cyclic GMP in a reaction analogous to that described for production of cAMP. Likewise, there is a *cGMP-dependent protein kinase*, but it differs from the cAMP-dependent kinase in that the catalytic and regulatory activities are part of the same polypeptide, rather than existing on two subunits.

In many systems, the levels of cAMP and cGMP tend to move in opposite directions in response to hormonal stimuli. However, the signals that couple hormone-receptor occupancy to guanylate cyclase activity remain obscure. In silkmoths, a polypeptide hormone called eclosion hormone is liberated from an endocrine organ in the brain when the moth is ready to emerge from its cocoon. This hormone elicits a characteristic behavioral pattern composed of wriggling movements that help the moth escape from its confinement. In response to the hormone, the level of cGMP rises dramatically in nerve cells; indeed, the sterotyped neural behavior can be generated by exogenous cGMP. The newly discovered natriuretic hormone, which is made by the heart and relaxes the smooth muscle surrounding blood vessels, thereby reducing blood pressure, also stimulates a rise in cGMP levels. cGMP levels usually rise in concert with an increase in intracellular calcium; this finding suggests that the second messengers calcium or inositol triphosphate may activate guanylate cyclase.

Calcium and the Inositol Triphosphate Pathway.   The outline of another important second messenger system has been elucidated during the last few years. Chemical messengers that act via this system include a variety of hormones (e.g., vasopressin and thyrotropin regulatory hormone) as well as some growth factors (e.g., platelet-derived growth factor, PDGF), neurotransmitters (e.g., acetylcholine acting on pancreatic acinar cells to stimulate secretion of digestive enzymes, or acting on pancreatic β cells to stimulate insulin secretion) and antigens (acting on mast cells to stimulate histamine synthesis). The receptors for these hormones and hormonelike signals are membrane-bound proteins with their binding sites facing the outside of the cell. Their activation by an appropriate signal is transmitted, at least in some cases, by a G protein (yet another GTP-binding protein, distinct from the ones described earlier), which then activates a phosphodiesterase (*phospholipase C*) that cleaves the polar *inositol triphosphate (IP$_3$)* from *phosphatidylinositol-4,5-bisphosphate (PIP$_2$)*. As a result, IP$_3$ can enter the cytoplasm, while the diacylglycerol moiety remains in the membrane (Figure 31–15). Both breakdown products of PIP$_2$ play important second messenger roles. IP$_3$ stimulates the release of calcium from intracellular stores (residing in the endoplasmic reticulum) into the cytoplasm. The calcium is bound by *calmodulin*, which then *activates one or more calcium-dependent protein kinases*. Meanwhile, the *diacylglycerol* (along with phosphatidylserine, which is an essential cofactor) *activates a membrane-associated protein kinase (C-kinase)* that phosphorylates serine and threonine residues (Figure 31–16). Thus several kinases are activated by this system and they, like the cAMP-dependent protein kinases, modify the function of rate-limiting enzymes and regulatory proteins involved in a variety of metabolic pathways. In addition to intracellular messengers, the breakdown of PIP$_2$ also stimulates the production of extracellular modulators of hormone activity because, in the

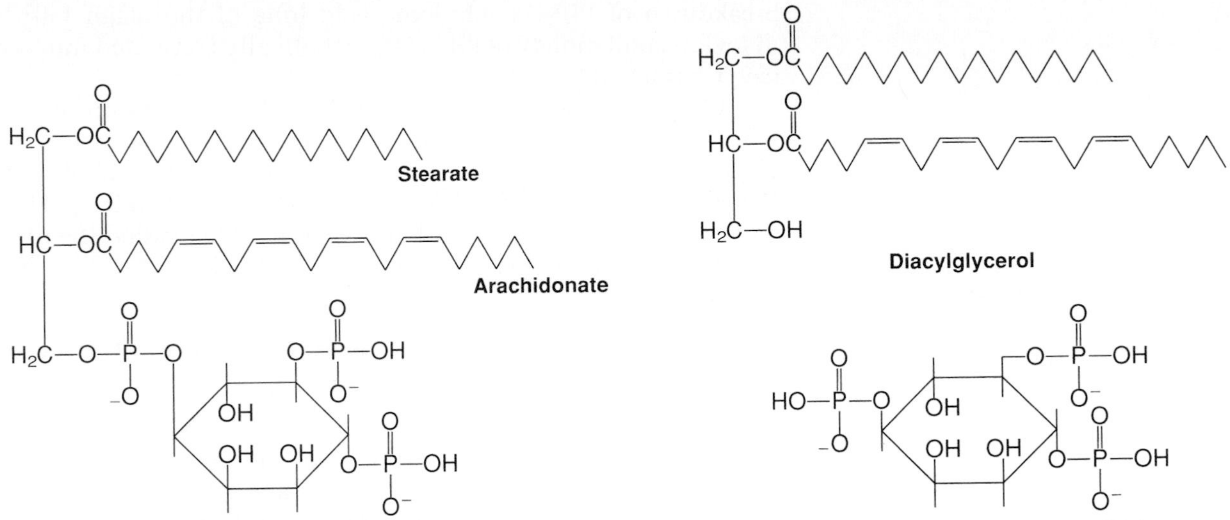

Stearate

Arachidonate

Phosphatidylinositol-4,5-bisphosphate (PIP$_2$)

Diacylglycerol

D-Inositol-1,4,5-triphosphate (IP$_3$)

Figure 31–15
Phosphatidylinositol-4,5-bisphosphate (PIP$_2$) and the two second messengers, diacylglycerol and inositol triphosphate, that are derived from it.

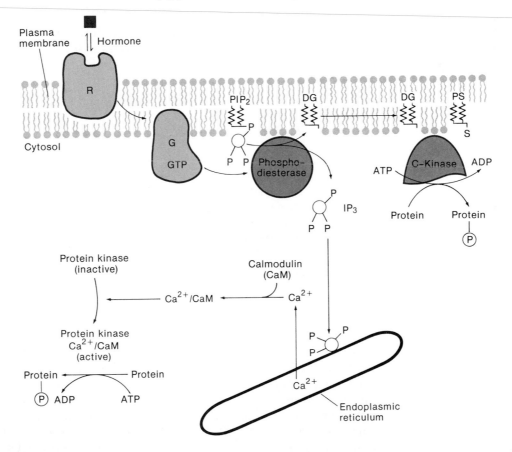

Figure 31–16
The phosphatidylinositol (PIP$_2$) pathway. Binding of certain hormones and some other ligands to a variety of membrane receptors (R) leads to activation of GTP-binding proteins, which then activate a phosphodiesterase that cleaves the inositol triphosphate (IP$_3$) moiety from PIP$_2$ and leaves diacylglycerol (DG) in the membrane. Diacylglycerol, in conjunction with phosphoserine (PS), activates a protein kinase (C-kinase) that phosphorylates enzymes associated with key metabolic pathways, thereby activating or inactivating them. Meanwhile, the polar IP$_3$ moiety binds to intracellular receptors on the endoplasmic reticulum, resulting in the liberation of calcium into the cytosol. The calcium binds to calmodulin, which then activates another group of protein kinases (the Ca^{2+}/CaM-dependent protein kinases). Hormonal stimulation can be short-lived because IP$_3$ and DG are rapidly degraded to inactive forms that are ultimately recycled to PIP$_2$; Ca^{2+} is pumped back into the endoplasmic reticulum, where it is sequestered.

breakdown of PIP_2, arachidonic acid (one of the major fatty acids in the diacylglycerol moiety of PIP_2) is metabolically converted into prostaglandins (see Chapter 22).

IP_3 is rapidly degraded to inactive IP_2 and then on to inositol. Meanwhile, diacylglycerol is phosphorylated and then converted to CDP-diacylglycerol, which combines with inositol to form phosphatidylinositol. The latter is subsequently phosphorylated in two steps to PIP_2. The degradation and resynthesis of PIP_2 completes what has been called the phosphatidylinositol, or PI, cycle (see Chapter 22, Box 22–C).

Steroid Receptors Modulate the Rate of Gene Transcription

The receptors for all classes of steroid hormones, including the receptors for 1,25-dihydroxyvitamin D, are soluble proteins. These proteins are not very abundant, usually only 10^2 to 10^5 molecules per cell. Initially they were thought to be cytoplasmic proteins, because during cell fractionation they were recovered in the supernatant after all membranous structures were removed by centrifugation. However, more recent studies, based in part on immunological localization, suggest that the proteins are normally confined to the nucleus, but may leak into the cytoplasmic fraction when the cell is disrupted.

Formation of a hormone-receptor complex leads to activation of the receptor, which most probably represents a conformational change. Activation proceeds readily at 37°C, but may take hours at 0°C; it may involve dephosphorylation of the receptor and dissociation of subunits. The activated receptor has an increased affinity for DNA and becomes tightly associated with chromatin. Thus, after hormone treatment, the receptors are retained in the nuclear fraction during cell fractionation.

Cells that contain _steroid receptors can be visualized by autoradiography_. In this technique the tissues are exposed to a radioactive steroid, sectioned, fixed onto slides, and coated with photographic emulsion. After the slides are developed, the presence of silver grains over the nucleus identifies those cells with receptors (Figure 31–17).

When it became clear that activated steroid receptors are localized in the nucleus, the suggestion arose that these receptors might regulate gene expression. But in the 1960s, when these observations were made, there were no direct assays of gene expression. It was known that steroid treatment stimulated the growth of some organs (e.g., the mammalian uterus and avian oviduct) and that hormone treatment induced an accumulation of certain proteins, but the mechanisms involved were uncertain. The first indication that steroid hormones affect mRNA levels came from the development of cell-free translation systems that would allow initiation of protein synthesis as well as elongation. With these assays it was shown that the amount of functional mRNA for specific proteins (e.g., ovalbumin, which is the major eggwhite protein) increased dramatically following hormonal treatment. The discovery of reverse transcriptase and improved methods of mRNA purification then led to the development of hybridization assays with radioactive complementary DNA (cDNA). These assays showed that the total amount of mRNA sequence increased in parallel with functional mRNA; thus hormone treatment involved the accumulation of mRNA rather than activation of previously existing mRNA. Table 31–5 lists many examples of mRNA induction by different classes of steroid hormones. But mRNA accumulation could be due to either an increase in the rate of mRNA synthesis or a decrease in the rate of mRNA degradation, or possibly to both.

To demonstrate the effect of steroid hormones on mRNA synthesis, a transcriptional assay was needed. One approach was to isolate nuclei from

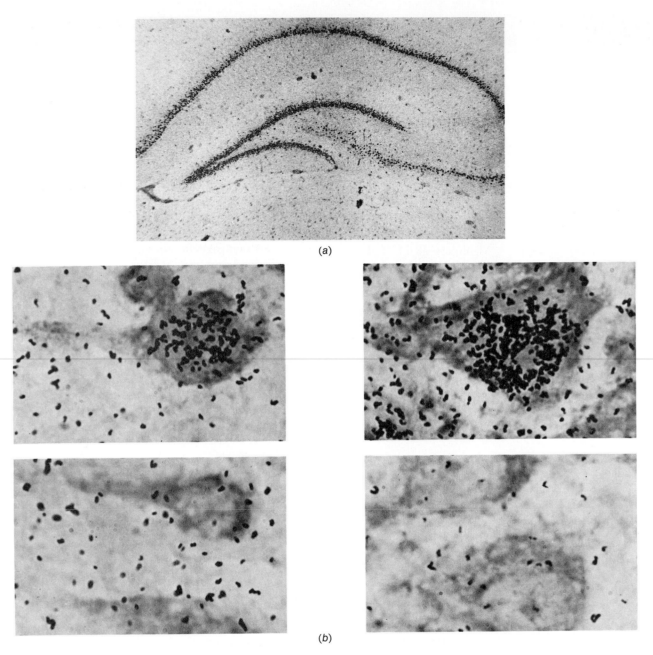

Figure 31–17
Localization of steroid receptors by autoradiography. Rats were injected with radioactive estradiol. Then frozen sections of the brain were prepared, mounted on slides, and dipped in photographic emulsion. After several weeks the slides were developed. The location of the silver grains reveals which cells took up the hormone. *(a)* A low magnification of the hippocampal region. *(b)* A higher-power magnification showing the location of the receptors in the nuclei of specific cells. (Reprinted with permission of Bruce McEwen of the Rockefeller Institute.)

control and hormone-stimulated cells and then allow the endogenous RNA polymerases to elongate RNA chains that were initiated *in vivo* in the presence of radioactive nucleotides. The labeled RNA chains were hybridized to plasmids containing the gene sequences of interest. The percentage of RNA that specifically hybridized presumably reflects the number of RNA polymerases on the gene at the time the nuclei were isolated. *In most cases, steroid hormones do, indeed, increase the number of RNA polymerases transcribing the gene,* which is interpreted as an increase in the rate of transcription (Figure 31–18). However, in some cases, the increase in the rate of transcrip-

TABLE 31–5
Regulation of Specific Genes by Steroid and Thyroid Hormones

Glucocorticoids

Tyrosine aminotransferase	Liver
Tryptophan oxygenase	Liver
Glutamine synthase	Liver, retina
Phosphoenolpyruvate carboxykinase	Kidney
Ovalbumin	Oviduct
Conalbumin	Oviduct (liver)[a]
α-Fetoprotein ($\downarrow$)	Liver
α_{2_μ}-Globulin	Liver
Metallothionein	Liver
Proopiomelanocortin ($\downarrow$)	Pituitary
Mammary tumor virus	Mammary gland

Estrogens

Ovalbumin	Oviduct
Conalbumin	Oviduct (liver)
Ovomucoid	Oviduct
Lysozyme	Oviduct
Vitellogenin	Liver
apo-VLDL	Liver
Glucose-6-P-dehydrogenase	Uterus

Progestins

Avidin	Oviduct
Ovalbumin	Oviduct
Conalbumin	Oviduct
Uteroglobin	Uterus

Androgens

β-Glucuronidase	Kidney
Aldolase	Prostate
Prostate-binding proteins	Prostate
Ovomucoid	Oviduct
Ovalbumin	Oviduct

1,25-Dihydroxyvitamin D_3

Calcium-binding protein	Intestine

Ecdysone

Dopa-decarboxylase	Epidermis
Vitellogenin	Fat body
Larval serum protein I	Fat body

Thyroid hormones

Carbamyl phosphate synthase	Liver
Growth hormone	Pituitary
Prolactin ($\downarrow$)	Pituitary
α-Glycerophosphate dehydrogenase	Liver (mitochondria)
Malic enzyme	Liver

[a]In the liver, the product of the conalbumin gene is called *transferrin*.
Note: ($\downarrow$) means that mRNA levels are decreased by hormone.

tion does not fully account for the increase in mRNA levels; this result suggests that the steroids increase mRNA stability as well. In an extreme case, the induction of hepatic α_1-glycoprotein mRNA by glucocorticoids occurs without any change in the rate of transcription.

Ecdysone, a steroid hormone found in insects, induces characteristic puffs in the polytene chromosomes from the salivary gland of certain insects (e.g., *Drosophila*, see Chapter 30). These puffs represent sites of active transcription, as indicated by autoradiography following pulse-labeling with [3H]uridine. Furthermore, antibodies directed against ecdysone localized the hormone to these same puffs (Figure 31–19). In sum, the cytological studies in insects suggested that steroid hormones activate transcription of specific genes, perhaps as a consequence of receptors binding near the genes that are being transcribed. It was also observed that some puffs regressed in response to β-ecdysone; perhaps, then, steroids can also inhibit transcription of some genes. All of these observations confirmed the hypothesis that steroid hormone receptors regulate gene transcription and suggested that they probably bind near the genes that they regulate (Figure 31–20).

The next step involved precise identification of the binding sites for steroid hormones. Two approaches have been used. In one, genes that are regulated by steroids are cloned, and then the wild-type gene, and various mutants derived from it, are introduced into test cells that contain the appropriate steroid receptors. Mutations that abolish steroid regulation are presumed to impinge upon the receptor-binding site(s). In the other approach,

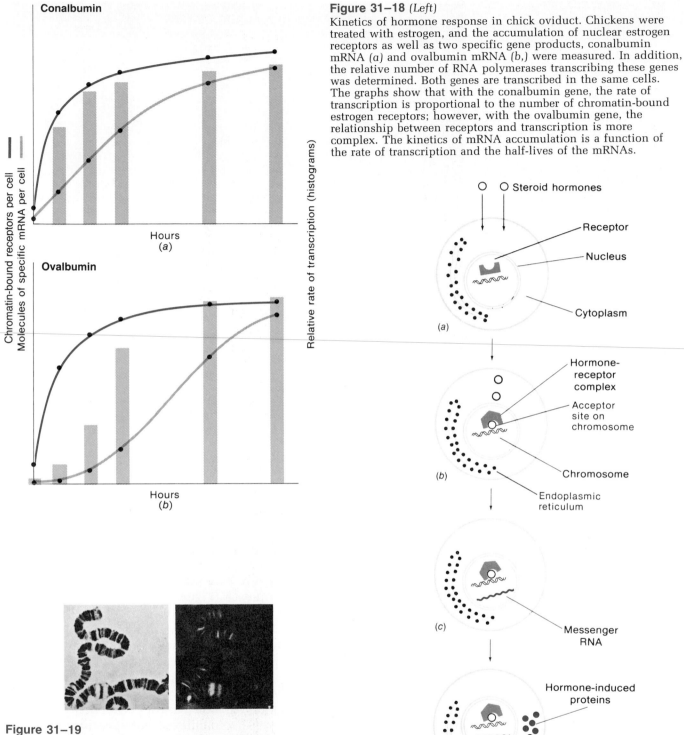

Figure 31–18 *(Left)*
Kinetics of hormone response in chick oviduct. Chickens were treated with estrogen, and the accumulation of nuclear estrogen receptors as well as two specific gene products, conalbumin mRNA *(a)* and ovalbumin mRNA *(b,)* were measured. In addition, the relative number of RNA polymerases transcribing these genes was determined. Both genes are transcribed in the same cells. The graphs show that with the conalbumin gene, the rate of transcription is proportional to the number of chromatin-bound estrogen receptors; however, with the ovalbumin gene, the relationship between receptors and transcription is more complex. The kinetics of mRNA accumulation is a function of the rate of transcription and the half-lives of the mRNAs.

Figure 31–19
Association of ecdysone receptors with specific chromosomal puffs. The salivary chromosomes of *Drosophila* are naturally amplified about 1000-fold, which makes them easily visible. When specific genes are activated, the chromosomal region corresponding to that gene expands or puffs. *Drosophila* salivary glands were treated with ecdysone. A few minutes later, the ecdysone was covalently linked to the receptor by ultraviolet light. Then the nuclear hormone-receptor complexes were visualized with fluorescent antibodies prepared against ecdysone. The ecdysone receptors are localized over those puffs which normally appear in response to this hormone. (Photographs kindly supplied by O. Pongs.)

Figure 31–20
Receptors for steroid hormones activate transcription. Steroid hormones diffuse through the membrane *(a)* and bind to receptors, which become activated and bind to specific DNA sequences *(b)*. This interaction leads to the activation (or sometimes inhibition) of transcription of adjacent genes *(c)*. The messenger RNA that is synthesized is translated into proteins *(d)*.

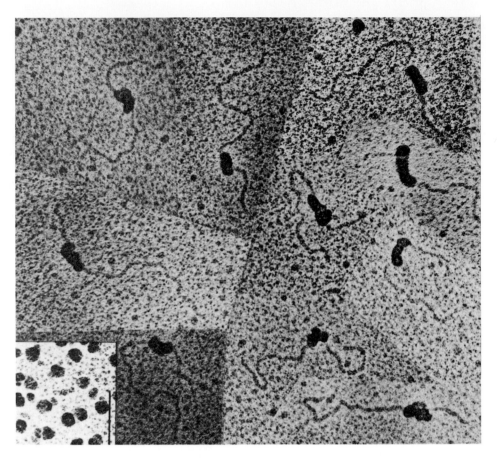

Figure 31–21
Electron micrographs of glucocorticoid receptors bound to the promoter region of
mouse mammary tumor virus (MMTV) DNA. Purified receptors were incubated with
MMTV DNA, then spread on special grids and shadowed for visualization by electron
microscopy. Note that several receptors appear to bind close together, judging by the
extended size of the dark blobs. The inset shows receptors in the absence of DNA.
(Photos kindly provided by K. Yamamoto.)

the binding of purified receptors to cloned genes has been studied by a variety of techniques.

Direct visualization of receptor-DNA interaction can be achieved by
electron microscopy of the complexes (Figure 31–21). Alternatively, the
gene can be cut into a number of fragments with restriction enzymes, and
then the fragments can be radioactively labeled, incubated with receptors,
and passed through a nitrocellulose membrane under conditions that allow
DNA to pass through the membrane unless it is complexed with protein. The
DNA that is retained is then identified by electrophoresis. This filter-binding
assay can be repeated with smaller and smaller DNA fragments to localize
the binding site to within a few hundred base pairs of DNA. For more precise
localization, end-labeled DNA can be complexed with receptor and then
mildly treated with DNase to generate a footprint of the binding site (Figure
31–22). The footprint in this case represents the protection of DNA from
DNase digestion. (Other applications of footprinting have been described in
Chapters 29 and 30.)

_These techniques have revealed that there are specific binding sites,
located in the promoter region of genes, that are regulated by steroid hor-
mones._ The chicken lysozyme gene and a human metallothionein gene have
two and one binding sites, respectively, for glucocorticoids; the binding sites
occur within a few hundred base pairs of the transcription start sites. The
mouse mammary tumor virus, which is regulated by glucocorticoids, has

many functional receptor binding sites within the promoter region as well as in the coding region. All of these binding sites share a short, conserved DNA sequence.

Other experiments have provided formal proof that this sequence is sufficient to confer glucocorticoid regulation. For example, small DNA fragments containing the receptor binding site were inserted into a thymidine kinase gene that is not normally regulated by steroids. When this hybrid gene was introduced into cells, the receptor-binding region conferred glucocorticoid inducibility to the thymidine kinase gene. Furthermore, the exact position and orientation of the receptor-binding region relative to the promoter of the thymidine kinase gene were relatively unimportant.

Despite the identification of the receptor binding sites, _it is not yet clear how steroid receptors activate transcription_. Presumably they help establish an efficient initiation complex for RNA polymerase II, but development of a cell-free transcription system that responds to the presence of functional receptors will be necessary before the molecular details can be unraveled.

In many cases, steroid hormones can activate gene transcription within a few minutes; thus the activated receptors can find their binding sites relatively rapidly. In other cases, however, transcriptional activation takes many hours. The binding of receptors to DNA does not require protein synthesis, but sometimes transcriptional activation by steroids can be prevented with inhibitors of protein synthesis. These observations suggest that for some genes, regulation by steroid hormones involves other rate-limiting factors (some of which may depend on continued synthesis, because they have short half-lives). In addition, the presence of functional receptors is not sufficient to activate transcription of a particular gene. For example, the ovalbumin gene is not transcriptionally responsive to estrogen in the liver, where the vitellogenin gene responds well. Such differences indicate that developmental events are important in committing a particular gene to be responsive to steroid receptors.

Thyroid hormones are thought to act in a manner analogous to that of steroid hormones. The receptors for T_3 and T_4 appear to be tightly associated with chromatin, regardless of whether they are activated with hormone or not. The thyroid hormone receptor gene has been cloned and it is homologous to steroid receptor genes and the viral oncogene, _erbA_. Like the steroids, _thyroid hormones stimulate transcription of a number of genes_, and their receptor binding sites are being characterized.

REGULATION OF HORMONAL RESPONSES

The fertilized egg is not known to respond to hormones, yet subsequent development produces cells that are capable of responding to a wide variety of hormones. Certainly one important developmental event is the activation of genes coding for specific receptors. Other developmental events activate

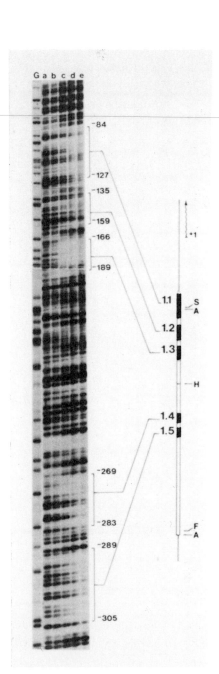

Figure 31–22
A DNA footprint of glucocorticoid receptors bound near the promoter of mouse mammary tumor virus (MMTV). A DNA restriction fragment several hundred base pairs long was labeled at one end with ^{32}P, then incubated with increasing amounts of glucocorticoid receptor and subjected to partial digestion with DNase. The products were then electrophoresed on an acrylamide gel. Lane G, markers; lane a, no receptor added; lanes b through e, increasing amounts of receptor. When a receptor binds, it protects the DNA from degradation by DNase at that site, thus creating a hole or footprint in the pattern. Five such footprints are revealed in this piece of DNA. Their positions, relative to the mRNA start site (+1), are shown to the right. They are located within a 220-base-pair region upstream of the mRNA start site and are labeled 1.1, 1.2, etc. These closely spaced binding sites account for the clustering of receptors shown in Figure 31–21. (Photograph kindly supplied by K. Yamamoto.)

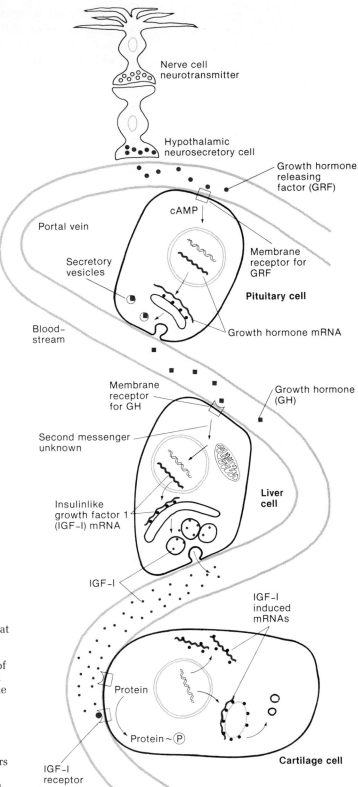

Figure 31–23

Growth hormone cascade. Neurosecretory cells in the hypothalamic region of the brain are activated by neurotransmitters from other neurons to secrete growth hormone releasing factor (GRF), a 44-amino-acid peptide that travels through the portal circulation to the anterior pituitary, where it binds to membrane receptors on somatotroph cells. GRF binding stimulates the production of cAMP, which activates growth hormone (GH) synthesis and secretion. GH, a 191-amino-acid polypeptide, passes into the bloodstream and travels to the liver, where it binds to membrane receptors that probably produce a second messenger (as yet unknown) that activates transcription of insulin-like growth factor I (IGF-I) gene, which codes for a 71-amino-acid peptide. IGF-I is secreted from the liver into the bloodstream and ultimately binds to membrane receptors (which are tyrosine kinases similar to the insulin receptor Figure 31–11), located on many peripheral cells. Activation of these receptors leads to cellular proliferation under appropriate conditions.

(commit) genes that will be either directly or indirectly responsive to these receptors. For example, the metamorphosis of a tadpole into a frog is triggered by thyroid hormones. Tadpole liver cells do not respond to estrogen, whereas shortly after exposure to thyroid hormones, estrogen can induce vitellogenin, a liver-specific protein. In this case, thyroid hormones may induce the synthesis of estrogen receptors.

Another level of regulation involves _feedback control_. The synthesis of many hormones is regulated by a cascade of hormones (Figure 31–23). Frequently, a hypothalamic hormone impinges on the pituitary to stimulate synthesis of a hormone that activates hormone synthesis in yet another organ. The end products of these cascades generally feedback-inhibit the production of hormones at the beginning of the cascade (Figure 31–24). Feedback inhibition can be revealed by injection of hormones, by perfusion of hormones using osmotic minipumps, or by producing transgenic animals that synthesize copious amounts of particular hormones from an ectopic (inappropriate) tissue.

For example, transgenic mice were produced that synthesized large amounts of growth hormone (GH) in the liver (and other tissues) because the GH structural gene was fused to a metallothionein gene promoter, which functions well in the liver. The fusion gene was microinjected into fertilized eggs so that it could be incorporated into the mouse genome prior to cell division; as a result, the mice carried the foreign gene in all of their cells,

Figure 31–24
Control of hormone synthesis and secretion in the anterior pituitary. Neurosecretory neurons in the hypothalamus liberate polypeptides that either stimulate ⊕ or inhibit ⊖ hormone synthesis by specialized pituitary cells containing the appropriate receptors. For example, GnRH stimulates gonadotroph cells in the pituitary to synthesize and secrete LH and FSH. These pituitary hormones then impinge upon target cells, typically stimulating them to make other low-molecular-weight hormones. The end products of these cascades feedback-inhibit hormone production at either or both hypothalamic and pituitary levels.

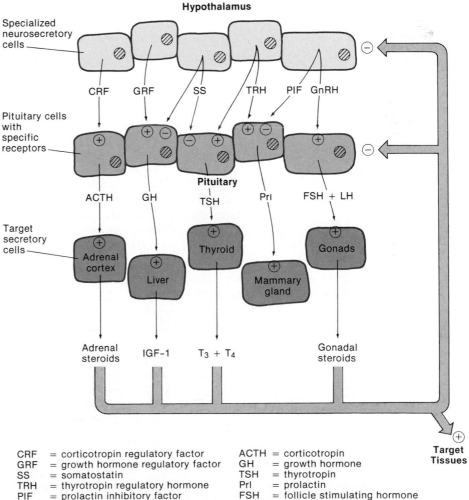

CRF = corticotropin regulatory factor	ACTH = corticotropin
GRF = growth hormone regulatory factor	GH = growth hormone
SS = somatostatin	TSH = thyrotropin
TRH = thyrotropin regulatory hormone	Prl = prolactin
PIF = prolactin inhibitory factor	FSH = follicle stimulating hormone
GnRH = gonadotropin regulatory hormone	LH = luteinizing hormone
IGF-1 = insulinlike growth factor 1	T_3, T_4 = thyroid hormones

Figure 31–25
A transgenic mouse that synthesizes rat growth hormone ectopically, and a normal littermate. A chimeric gene with the mouse metallothionein promoter fused to the structural gene of rat growth hormone (rGH) was introduced into the germline of mice by microinjection of cloned DNA into fertilized eggs. Because of the metallothionein promoter, these genes were expressed in many large organs (such as liver, kidney, heart, intestine) that do not normally make GH. As a consequence, the circulating GH level was elevated several hundredfold, an effect that led to increased growth of those mice that inherited the gene. For further information, see R. D. Palmiter, R. L. Brinster, R. E. Hammer, and R. M. Evans in *Nature* 300:611–615, 1982. (Picture courtesy Ralph Brinster.)

including the germline. When the pituitaries of these mice were examined, the somatotroph cells of the pituitary, which normally produce GH, were not detectable. Presumably GH or insulinlike growth factor-I (which is also elevated in response to GH) caused feedback inhibition of GH production. In a similar experiment, a gene for metallothionein-growth hormone releasing factor was introduced. The result was massive production of growth hormone releasing factor (GRF) in the liver, which then stimulated the somatotroph cells to proliferate and expand their production of GH. In both cases, offspring of those mice that inherited the fusion genes grew almost twice as large as their normal littermates (Figure 31–25).

Another common mechanism for modulating hormonal response involves two (or more) hormonal inputs with both positive and negative effects. The hypothalamic peptides somatostatin and GRF have opposite effects on GH synthesis and secretion. Similarly, glucagon and insulin have opposite effects on gluconeogenesis in the liver, and some of the effects of ecdysone on gene expression in insects are blocked by juvenile hormone (a terpene derivative; Figure 31–26).

There is also considerable cross-talk between different hormones at the level of receptor function. For instance, the diacylglycerol-activated protein kinase C phosphorylates and thereby inhibits the activity of insulin and EGF receptor kinases. Likewise, when cAMP-dependent protein kinases are active they can inhibit β-adrenergic receptors by phosphorylating them.

Cells also have mechanisms that tend to prevent chronic stimulation. Exposure of cells to epinephrine leads to an initial sharp rise in cAMP levels; however, cAMP levels fall nearly to basal levels within an hour or so, despite the continuous presence of saturating amounts of epinephrine. Furthermore, if the agonist is removed and the cells are challenged within a few hours, the secondary response is submaximal. This phenomenon, whereby prior exposure to an agonist leads to an acute decrease in responsiveness, is called _desensitization_. Desensitization is associated with both an uncoupling of receptors from adenylate cyclase activation and a decrease in the number of receptors accessible to hormone binding. It occurs only after productive receptor function; hence it is not achieved with antagonists. The consequence

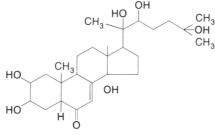

β-Ecdysone (steroid)

Juvenile hormone (JH-1)
(terpene derivative)

Figure 31–26
Structure of β-ecdysone and juvenile hormone. These hormones play major roles in the growth and maturation of insects by controlling the timing for molting of the insect exoskeleton.

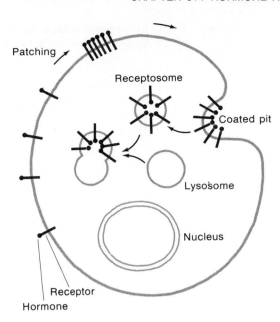

Figure 31–27
The down-regulation of receptors by endocytosis. The hormone-mediated loss of receptors is often referred to as down-regulation. Continuous activation of receptors by hormone often leads to patching, a clustering of receptors as if they were cross-linked. Endocytosis of the patches removes them from the cell surface. The endocytotic vesicles, sometimes called receptosomes, fuse with lysosomes, where the contents are degraded by the lysosomal enzymes. Receptosomes also may allow entry of receptors into other cell compartments, such as the nucleus, by fusing with these organelles.

of desensitization is that exposure to hormone results in a transient activation of cellular events, rather than chronic activation.

The response to many polypeptide hormones also diminishes with chronic stimulation, but the time course is much longer (many hours to days) and the mechanism is different from that involved in desensitization. Binding of insulin, calcitonin, LH, TRH, and EGF to their respective receptors promotes a physiological response but ultimately leads to the clearance of these receptors from the surface and a blunting of that response. The hormone-mediated loss of receptors is often referred to as *down-regulation*. In this case, the receptors are internalized and degraded. Hormone binding leads to a clustering of receptors, as if they were cross-linked. These clusters, or patches, aggregate in membrane structures called *coated pits* (the intracellular side of the pits is coated with a scaffolding protein called *clathrin*). Small vesicles that engulf the receptors and their ligands (sometimes referred to as *receptosomes*) bud off from the coated pits, migrate within the cell, and associate with other membranous structures known collectively as GERL (Golgi-endoplasmic-reticulum-lysosomes). After a few hours they fuse with lysosomes, at which point the lysosomal enzymes degrade both the receptor and the hormone (Figure 31–27). Replacement of the hormone receptors requires protein synthesis. Hormone action is generally assumed to occur while the receptors are on the cell surface, but it is not clear whether receptor function ceases during endocytosis or whether endocytosis may be essential for the action of some hormones.

Most hormones are released in an episodic manner (examples include insulin, glucagon, growth hormone, and many of the hypothalamic releasing hormones). In some cases the cycles appear to be autonomously regulated, but in others they are clearly entrained by neural and hormonal signals that may vary, depending on the developmental stage, season, or physiological condition. The periodic release of peptide hormones avoids the down-regulation of receptors described above.

ENDOCRINOPATHIES

Human diseases related to endocrine dysfunction can be broadly grouped into: (1) overproduction of a particular hormone, (2) underproduction of a hormone, and (3) target cell insensitivity to a hormone.

Overproduction of Hormones Is Commonly Caused by Tumor Formation

Most cases of hormonal overproduction are associated with hypertrophy of the normal endocrine organ, frequently owing to a tumor. Pituitary neoplasms usually affect the production of only one pituitary hormone as a result of the cancerous proliferation of the cell type that normally synthesizes that hormone. Examples include giantism (acromegaly), which is associated with proliferation of the somatotroph cells that synthesize growth hormone, and Cushing's syndrome, which is usually due to overproduction of ACTH. Adrenal and parathyroid tumors that lead to the overproduction of various adrenal steroids and parathyroid hormones have also been described.

Occasionally, tumors of organs that do not usually produce a given hormone (ectopic tumors) synthesize and secrete peptide hormones, as in ACTH synthesis by certain lung tumors and GRF production from a pancreatic islet tumor. Expression of hormones by these ectopic tumors represents a curious activation of gene expression in an inappropriate tissue. The same result can be obtained experimentally by producing animals in which a structural gene codes for a hormone that is under the transcriptional control of a promoter/enhancer from a gene expressed in another tissue. This type of overproduction was illustrated in the experiments with metallothionein-growth hormone and metallothionein-GRF (see Figure 31–25).

The excessive production of thyroid hormone in Grave's disease is associated with an enlarged thyroid gland, but in this case a circulating immunoglobulin that mimics the activity of thyroid stimulating hormone (TSH) is implicated. Finally, we saw in an earlier section how an inappropriate steroid hormone may be secreted in excessive amounts when specific enzymes in the adrenal steroid biosynthetic pathway are present at inadequate levels.

Underproduction of Hormones Has Multiple Causes

A wide variety of defects could lead to inadequate production of hormones. In some conditions, there are not enough cells producing a given hormone. For example, in juvenile-onset diabetes there is a decrease in the number of pancreatic islet cells that synthesize insulin. In other, more extreme cases, the gene coding for the hormone may be missing; for example, some forms of dwarfism are due to a lack of the growth hormone gene. In most cases, however, the gene coding for the hormone is present but inadequate amounts of active hormone are produced. Such defects can have many possible explanations. Some of them are genetic. For example, a mutation may affect the rate of synthesis of mRNA coding for the hormone precursor, or a mutation may affect the processing of the mRNA, or an amino acid substitution may decrease the activity or processing of a polypeptide hormone. The techniques of molecular biology are being used to discover the causes of many of these genetic disorders. Other defects in hormone synthesis are the consequence of an inadequate supply of precursors. For example, iodine is essential for thyroid hormone synthesis, and its absence leads to goiter. This condition, involving hypertrophy of the thyroid, is caused by high concentrations of TSH that result from lack of normal feedback by T_3 and T_4 on the hypothalamus. Likewise, synthesis of vitamin D requires ultraviolet irradiation of 7-dehydrocholesterol, without which the formation of $1,25(OH)_2D_3$ cannot occur, so that rickets ensues. A rare cause of rickets involves a defect in the enzyme that converts $25(OH)D_3$ to $1,25(OH)_2D_3$, the active form of vitamin D. Similarly, defects in enzymes involved in adrenal steroid biosynthesis lead to Addison's disease, and a defect in 5α-reductase leads to one form of testicular feminization, the result of inadequate production of dihydrotestoster-one.

Target-cell Insensitivity Results from a Lack of Functional Receptors

The most dramatic examples of target-cell insensitivity are those in which the correct receptors are lacking. In *complete testicular feminization*, androgen receptors are missing from all cells. The consequence is phenotypic expression of female characteristics in genotypic males. A rare form of dwarfism (Laron dwarfs) is associated with high plasma levels of growth hormone (GH) but low levels of insulinlike growth factor-I (IGF-I); consequently, a defect in GH receptors is suspected. Occasionally, antibodies are directed against receptors and interfere with normal hormone binding, e.g., in one form of *adult-onset diabetes*. Remarkably, the antibody itself promotes insulin effects.

GROWTH FACTORS

In addition to the many hormones that have been discussed, there exist a large number of growth factors that have hormonelike activities. As their name implies, *growth factors stimulate the proliferation of particular cells*. Unlike hormones, most growth factors are synthesized by a variety of cell types rather than in a specialized endocrine gland. However, the mechanisms of action of growth factors and hormones are likely to be very similar.

All of the growth factors that have been characterized so far are proteins (Table 31–6). Many of them are cleaved from larger precursors. For example, the 53-amino-acid epidermal growth factor (EGF) is cleaved from a precursor of 1168 amino acids. This precursor appears to be a membrane-spanning protein, with the EGF moiety and nine related sequences in the extracellular domain. EGF is homologous to several other growth factors, including α-tumor growth factor and a protein secreted by vaccinia virus. These factors bind to membrane receptors and are thought to trigger intracellular events in much the same way as hormones.

The EGF receptor, for example, shows striking similarity to the insulin receptor in overall organization and amino acid sequence; however, instead of separate α and β chains, the EGF receptor is composed of a single polypeptide that appears to be a composite of the α and β chains of the insulin receptor. As is true with the insulin receptor, the intracellular domain of the EGF receptor has tyrosine kinase activity.

The receptor for insulinlike growth factor-I (IGF-I) is also similar to the insulin receptor in structure and tyrosine kinase activity. Thus, in this case it is likely that the hormones (insulin and IGF-I) and their receptors have each evolved from common ancestral genes by duplication and divergence. However, the receptors for insulinlike growth factor-II appear to be quite different in that they do not have tyrosine kinase activity, even though IGF-II is also homologous to insulin.

The receptors for one of the colony-stimulating factors (M-CSF) and for platelet-derived growth factor (PDGF) are also membrane proteins with tyrosine kinase activity. Although the mechanisms by which these receptors regulate cellular activities are unknown, a major hypothesis is that they phosphorylate (and thereby either activate or inhibit) regulatory enzymes, including some serine/threonine protein kinases. PDGF also appears to activate the inositol triphosphate/diacylglycerol pathway.

Normal cells require growth factors (mitogens) for proliferation. In the absence of these factors they reversibly withdraw from the cell cycle and

TABLE 31–6
Vertebrate Growth Factors

Factor[a]	M_r	Cell Types Affected
Epidermal growth factor (EGF)[b]	6400	Epithelial and meso-dermal cells
Tumor growth factor-α (TGF-α)	7000	Same
Insulinlike growth factor-I (IGF-I)	7000	Same
Insulinlike growth factor-II (IGF-II)	7000	Same
Fibroblast growth factor (FGF)	13,000	Same
Platelet-derived growth factor (PDGF)[b]	31,000	Same
Nerve growth factor (NGF)	13,000	Sensory and sympathetic neurons
Erythropoietin	23,000	Erythroid cell precursors
Macrophage colony stimulating factor (M-CSF or CSF-I)[b]	70,000 dimer	Macrophage precursors
Granulocyte colony stimulating factor (G-CSF)	25,000	Granulocyte precursors
Granulocyte-macrophage colony stimulating factor (GM-CSF)	23,000	Granulocytes and macro-phage precursors
Multicolony stimulating factor (Multi-CSF) or Interleukin-3 (IL-3)	25,000	Precursors of most hematopoietic cells
Interleukin-2 (IL-2)	13,000	T lymphocytes

[a] The genes for most of these growth factors and many of their receptors have been cloned.

[b] The EGF receptor is homologous to the *erbB* and *neu* oncogenes, and the M-CSF receptor is homologous to the *fms* oncogene; PDGF is homologous to the *sis* oncogene.

become arrested in the G_1/G_0 phase. Transformed cells (tumor cells) have a relaxed cell cycle control and can traverse the cell cycle in the absence of added growth factors. Cell transformation can be achieved by activation or inappropriate expression of a variety of cellular or viral genes. The isolation and characterization of many of these transforming genes (oncogenes) has revealed striking homology with growth factors or their receptors. For example, the oncogene product *sis* resembles PDGF in its sequence and its affinity for specific receptors. The EGF receptor is homologous to the *erbB* oncogene, and the M-CSF receptor is clearly related to the *fms* oncogene. Thus a prevalent idea is that many or all oncogenes lead to abnormal cell proliferation by directly or indirectly providing a rate-limiting factor, be it a growth factor, its receptor, or an intracellular intermediate.

INTERFERON

Although *interferons* are not usually considered hormones, they resemble hormones in their action. Interferons are proteins that are liberated by cells in response to viral infection; these proteins circulate to neighboring cells and stimulate them to make antiviral proteins, as shown schematically in Figure 31–28. Three types of interferons have been described: leukocyte (or α), fibroblast (or β), and immune (or γ). The names refer to the types of cells that synthesize them. Although all three interferons induce antiviral proteins, γ-interferon has additional effects on cells of the immune system.

Upon entering an animal cell, viruses lose their protective coat, synthesize specialized proteins, and replicate their nucleic acids, which are eventually packaged into mature progeny viruses. In the process, they gener-

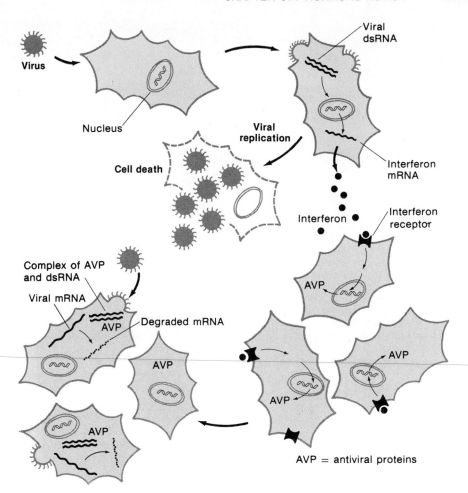

Figure 31–28
Induction and action of interferon.
Interferon mRNA is induced by
double-stranded RNA (dsRNA) that is
generated by viruses during
replication. The mRNA is translated
into interferon, which binds to
membrane receptors of other cells,
stimulating them to synthesize
antiviral proteins (AVP). When these
AVPs are activated by dsRNA they
inactivate viral mRNA and prevent its
translation (see Figure 31–29).

AVP = antiviral proteins

ally monopolize cellular biosynthetic apparatus and kill the host cell. But
while replicating, viruses also induce the host cell to synthesize and secrete
interferon. The inducer of interferon gene expression is thought to be
double-stranded RNA, because these molecules are the most potent synthetic inducers. A polymer of inosine and cytosine, poly(I · C), is one of the
most frequently used synthetic inducers because it is relatively stable. Perhaps all viruses, be they single-stranded or double-stranded RNA or DNA
viruses, generate small amounts of double-stranded RNA during their replication, and these molecules then interact with a regulatory protein (conceptually similar to a steroid receptor) that binds in the promoter region of the
interferon gene to activate transcription.

Interferons are secreted proteins. They exist as dimers in solution, with
subunit molecular weights of about 20,000. There is one gene for
β-interferon, but there are 10 or so for α-interferon. Their amino acid sequences are homologous but they have remarkably different properties in
different cells, the significance of which is just beginning to be explored. The
number of distinct receptors for interferons is not yet clear; however, they
are known to be membrane proteins with very high affinity ($K_d \sim 10^{-11}$ M).
Interaction of interferon with these receptors leads to the induction of antiviral proteins, but the intracellular train of events is unknown.

Two of the best-characterized antiviral proteins are oligoadenylate synthase and eIF-2 protein kinase (Figure 31–29). Both of these enzymes are
activated by double-stranded RNA; thus they remain dormant until virus
infection. Oligoadenylate synthase catalyzes the synthesis of short oligomers
of adenosine linked by 2′-5′ phosphodiester bonds instead of the 3′-5′ bonds
usually found. These oligomers, often referred to as 2-5A, serve as a second

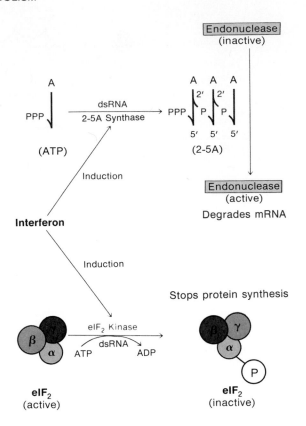

Figure 31–29

Interferon activation of antiviral proteins. Interferon binds to membrane receptors and stimulates target cells to synthesize the 2-5A synthase and eIF-2 kinase. When double-stranded RNA is present (as a consequence of viral infection), these enzymes are activated. The 2-5A synthase converts ATP into unusual oligoadenylate molecules with 2′-5′ phosphodiester bonds. These 2-5A molecules activate an endonuclease that participates in degradation of viral RNA. Meanwhile, activation of eIF-2 kinase results in phosphorylation (and inactivation) of the α subunit of an important initiation factor for protein synthesis.

messenger to activate an endonuclease that degrades RNA. The eIF-2 protein kinase phosphorylates the smallest (α) subunit of initiation factor 2 (eIF-2), thereby inhibiting protein synthesis. These activities presumably protect the cell by preventing translation of the viral mRNA and degrading it. However, neither enzyme is specific for viral mRNA; therefore, it remains to be determined how the specificity is generated or whether in some cases the infected cell may kill itself along with the unprotected virus. Much of the excitement about interferons stems from the observation that they protect cells from nearly all viruses. They also appear to inhibit the proliferation of certain cancer cells, but the mechanisms involved are not understood.

PLANT HORMONES

Plant hormones (also called growth factors) coordinate the growth and development of plants. The major hormones discovered to date fall into six classes: _auxins, cytokinins, gibberellins, abscisic acid, oligosaccharins,_ and _ethylene._ There are many different natural compounds within some classes that have hormonal activity; e.g., over 60 gibberellins are known. The structures of representative members of each class are shown in Figure 31–30. All of these hormones are low-molecular-weight compounds; indeed, one hormone, ethylene, is a gas. Polypeptide and steroid hormones have not been described for higher plants, although some yeasts and fungi use these compounds as mating factors. Each class of compounds elicits many diverse responses, and there is considerable interaction among different plant hormones in the control of physiological processes. Furthermore, the same process is controlled by different hormones (or combinations of hormones) in different species. These considerations make it difficult to analyze, and generalize about, the mechanism of plant hormone action.

Auxins are synthesized from tryptophan in the apical buds of growing shoots. They stimulate growth of the main shoot, but inhibit the develop-

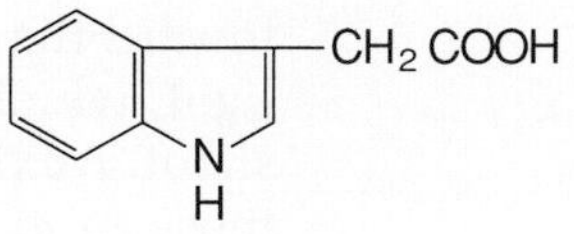

Figure 31-30

Some common plant hormones. All of these hormones are low-molecular-weight compounds. One hormone, ethylene, is a gas.

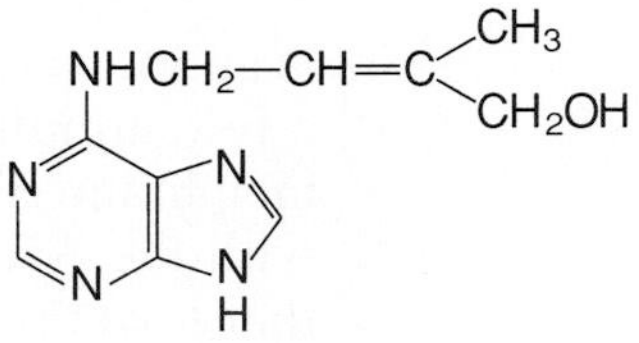

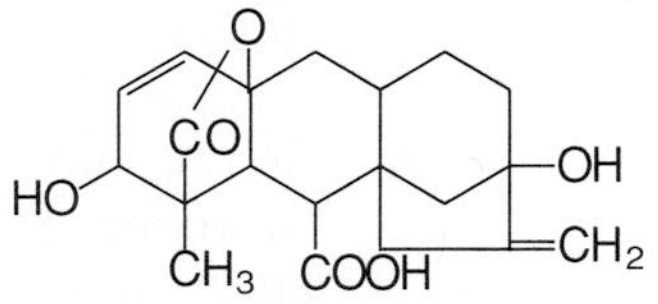

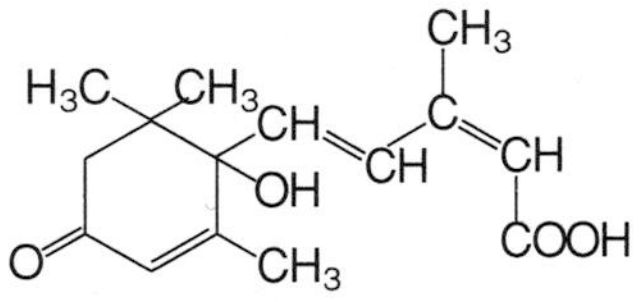

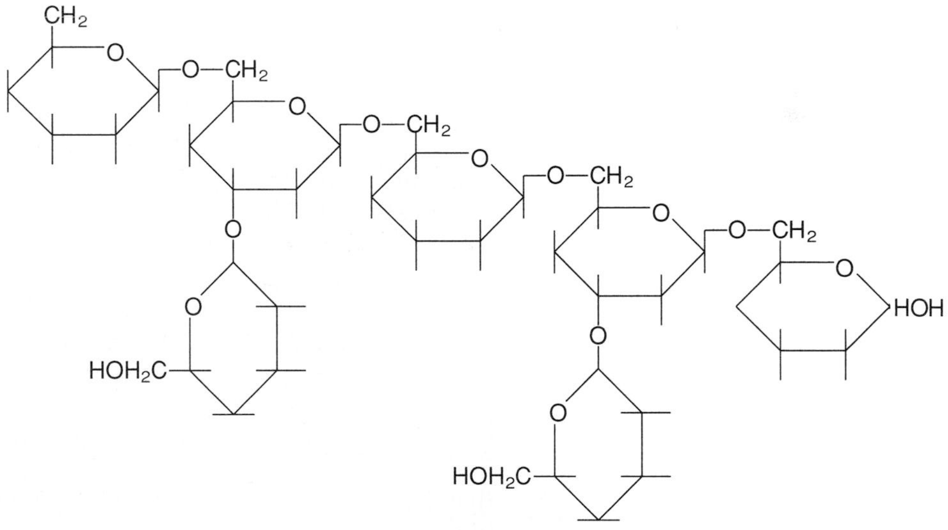

ment of lateral shoots; this effect has led to the horticultural practice of pinching off the apical buds to stimulate the formation of bushy plants. Auxins are transported from cell to cell in a unidirectional manner, as well as through the plant circulatory system. Directional growth is a consequence of the resulting gradients in auxin concentration. The curvature of plants toward the light (phototropism) is due to transport of auxins away from the light, which stimulates more rapid growth of cells on the darker side of the shoot. Auxins bind to specific membrane proteins. The affinity of auxins for these proteins, coupled with their location and abundance on target cells, supports the view that the membrane proteins may be the receptors that mediate auxin action.

Auxin stimulation of growth occurs in two phases. The earliest response is an increase in proton transport out of the cell, which occurs after a lag of a few minutes. This hydrogen ion pump is thought to be coupled with a membrane ATPase, but it is not clear whether the receptor interacts directly with the ATPase or whether other intermediates are involved. It is thought that lowering the extracellular pH activates enzymes that partially degrade the cell wall, thereby loosening it and allowing for cell expansion. A subsequent effect of auxin is to increase the synthesis of proteins and nucleic acids, resulting in sustained growth. Auxin has been shown, for example, to increase the amount of cellulase mRNA during pea cell expansion.

Cytokinins are adenine derivatives that are produced in the roots and that promote growth and differentiation of numerous tissues. Cytokinins can overcome the auxin-mediated inhibition of lateral shoots; in other tissues both hormones act synergistically. Plant cells can be grown in culture if auxins and cytokinins are provided. With relatively balanced concentrations of both hormones, cells will proliferate but remain undifferentiated. If the ratio of auxins to cytokinins is high, roots will develop; if the ratio is low, shoot development will be favored. Intermediate concentrations of both hormones will promote undifferentiated growth, with neither roots nor shoots. As these results show, plant cells can display a wide range of physiological responses. They are much more plastic in their developmental potential than animal cells. Indeed, normal plants can be grown from a single tissue culture cell. While this diversity of responses is fascinating, it has also made it very difficult to define the mechanism of action of plant hormones.

Gibberellins also promote shoot elongation, and frequently they act synergistically with auxins. Gibberellins stimulate the accumulation of specific mRNAs, e.g., amylase mRNA in germinating seeds. This fact suggests that their receptors or a second messenger act at the genetic level.

Abscisic acid is antagonistic to many other plant hormones. It inhibits germination, growth, bud formation, and promotes leaf senescence. Wilting stimulates the synthesis of abscisic acid in the chloroplasts within mesophyll cells of the leaves. The abscisic acid in this case is a stress signal that stimulates the guard cells to close and thus minimize water loss through the stomata. Abscisic acid has a rapid inhibitory effect on potassium uptake by guard cells; thus it may act by means of membrane receptors to modulate ion pumps, as was suggested for auxins.

Oligosaccharins are a diverse set of complex oligosaccharides that function in defense against disease, in control of plant growth, and in differentiation. They are released from the cell wall by specific degradative enzymes. For example, when bacteria or fungi infect plants, they produce enzymes that degrade the cell wall, liberating oligosaccharins that diffuse to neighboring cells. There they stimulate the plant cells to produce antibiotics that inhibit the growth of the pathogen. Note the similarity to interferon action, discussed earlier. When certain plant cells are damaged they can produce the enzymes required to liberate oligosaccharins; this allows a hormonal response even when the pathogen does not produce the appropriate enzymes. Auxins also stimulate the production of oligosaccharins, which then counteract the auxin-stimulated growth; thus these oligosaccharins serve as feedback inhibitors. The inhibition of lateral shoot growth that has been attributed to auxin may actually be due to the production of oligosaccharins.

CH_3—S—CH_2—CH_2—CH—COO^−
Methionine

| ATP

PP_i + P_i

CH_3—S—CH_2—CH_2—CH—COO^−
Adenosine
S-Adenosylmethionine

CH_3—S
|
Adenosine
+
1-Aminocyclopropane-1-
carboxylic acid (ACC)

O_2

HCOOH + NH_3 + CO_2

H_2C=CH_2
Ethylene

Figure 31–31
Synthesis of ethylene, a gaseous plant hormone involved in fruit ripening and flower senescence.

Addition of oligosaccharins to combinations of auxin plus cytokinins also influences the differentiation of shoots and roots in tissue culture. Indeed, many of the pleiotropic effects originally attributed to auxins and other plant hormones may actually be mediated by oligosaccharins.

Ethylene, although gaseous, has effects that are comparable to those of other hormones. It plays an important role in transverse rather than longitudinal growth of plants, by redirecting auxin transport. It also stimulates fruit ripening and flower senescence, and it inhibits seedling growth. Ethylene is synthesized from S-adenosylmethionine (SAM), as shown in Figure 31–31. The conversion of SAM to 1-aminocyclopropane-1-carboxylic acid, the immediate precursor of ethylene, is stimulated by auxins, cytokinins, wounding, and anaerobiosis. Once again we see the interplay of hormones in the regulation of cell activity.

Plant tumors result from uncontrolled hormone production. Crown gall tumors, for example, are due to the infection of plant wounds by certain strains of *Agrobacterium*. These bacteria carry a large plasmid, the tumor-inducing or T1 plasmid, part of which is incorporated into the plant genome (see Chapter 32). This DNA encodes several genes that stimulate rapid cell proliferation. One gene product is a rate-limiting enzyme involved in auxin biosynthesis, and another controls cytokinin biosynthesis. Together they promote the rapid, but undifferentiated, growth that is the crown gall. Interestingly, inactivation of one of the genes promotes the growth of shoots at the site of infection, while inactivation of the other promotes the growth of roots (Figure 31–32). These effects are reminiscent of the action of auxins and cytokinins in tissue culture.

Figure 31–32
Crown gall tumors of plants are caused by certain strains of the bacterium *Agrobacter* that can infect plant cells and introduce new genetic information coding for plant hormones. The transforming DNA of these bacterial strains is carried on a large plasmid; it contains genes coding for rate-limiting enzymes in hormone biosynthesis. (*a*) The transforming DNA of wild-type bacterial plasmid codes for rate-limiting enzymes involved in both auxin and cytokinin biosynthesis, resulting in undifferentiated growth (called callus). (*b*) Mutations that disrupt the gene involved in auxin biosynthesis give rise to tumors that produce only cytokinins, which leads to a proliferation of leaves and shoots. (*c*) Alternatively, mutations that disrupt the gene involved in cytokinin synthesis give rise to tumors that produce only auxin. In these tumors, only roots differentiate.

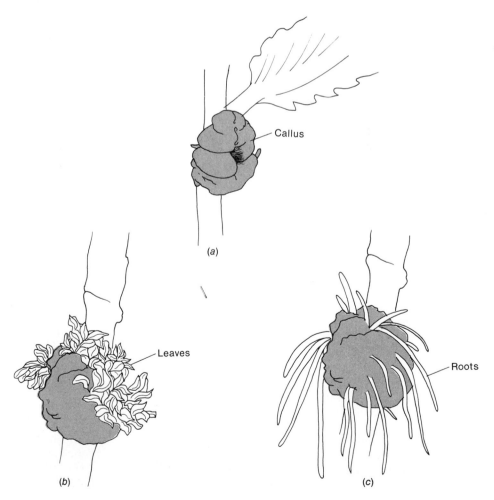

SUMMARY

Hormones are chemical messengers that circulate between tissues of multicellular organisms. They serve to coordinate metabolic activities, maintain homeostasis of essential nutrients and minerals, and prepare the organism for reproduction. Hormones act by reversibly binding to proteins called receptors, an event that results in a conformational change that is detected by other macromolecules (acceptors) and that eventually leads to activation of rate-limiting enzymes. Each class of hormones binds to specific receptors that activate other membrane proteins.

The best-understood hormonal system involves activation of adenylate cyclase. Three proteins are involved—the receptor, a GTP-binding protein, and adenylate cyclase. Activation of adenylate cyclase promotes the accumulation of cyclic AMP, which acts as a second messenger by activating a protein kinase that phosphorylates many other rate-limiting enzymes. About half the polypeptide hormones act by this mechanism. Other examples of second messengers that are synthesized in response to hormone action include cyclic GMP and inositol triphosphate. Steroid hormones penetrate the cell and bind to receptors in the nucleus, and activate (or sometimes repress) transcription of specific genes. Thyroid hormones act similarly.

In addition to classical hormones, there are many other chemical messengers. Some serve to coordinate growth of tissues during development and in response to injury (growth factors) or to protect against viral infection (interferons). A large number of diseases are due to either overproduction or underproduction of hormones, or to insensitivity of target tissues to circulating hormones. Knowledge of hormone biosynthesis, secretion, and interaction with target cells is essential to an understanding of the biochemical basis of these disorders.

SELECTED READINGS

Annual Reviews of Biochemistry and *Annual Reviews of Physiology*. Over 40 volumes in each series with many relevant reviews in the area of hormone receptors and hormone action. Provides good access to primary literature.

Baxter, J. D., and K. M. MacLoed, Molecular Basis for Hormone Action. In P. K. Bondy and L. E. Rosenberg (eds.), *Metabolic Control and Disease*. Philadelphia: Saunders, 1980. pp. 104–160. A useful summary.

deGroot, L. J., et al. (eds.). *Endocrinology*, 3 Vols. New York: Grune and Stratton, 1979. Over 2000 pages of comprehensive treatment; primarily from a medical point of view.

Guillemin, R. Peptides in the brain: The new endocrinology of the neuron. *Science* 202:390, 1978. Nobel laureate speech.

O'Malley, B. W., and L. Birnbaumer, *Receptors and Hormone Action*, 3 Vols. New York: Academic Press, 1977. Review articles by many authors cover most aspects of hormone action.

Recent Progress in Hormone Research. New York: Academic Press. An annual publication with nearly 40 volumes. A good place to find a recent summary.

Riddiford, L. M., and J. W. Truman, Biochemistry of Insect Hormones and Insect Growth Regulators. In *Biochemistry of Insects*. New York: Academic Press, 1978. Pp. 307–357.

Sutherland, E. W., Studies on the mechanism of hormone action. *Science* 177:401, 1972. Nobel laureate speech related to discovery of cAMP as second messenger.

PROBLEMS

1. Why are all known hormone receptors proteins? Could other macromolecules serve as receptors? Or as acceptors?

2. What advantages or disadvantages accrue to each of the hypothetical receptor-mediated pathways depicted in Figure 31–9?

3. Sex determination in mammals can be divided into two stages: (1) differentiation of the uncommitted gonad (ovotestis) into either testis or ovary, and (2) differentiation of identical internal structures (Wolffian and Müllerian ducts) and identical external structures (urogenital sinus, tubercle, etc.) into appropriate male or female organs. Normally there is no confusion between genetic and phenotypic sex because the Y chromosome of males (XY) codes for hormonelike proteins (called Y antigens because they have only been identified immunologically). These antigens are postulated to direct the differentiation of the ovotestis into a testis, which then synthesizes androgens that direct the differentiation of the remaining sex organ primordia in the male direction. Thus, in the absence of the Y antigen or androgens, the embryo will develop into a female. With this highly simplified background and your knowledge of steroid biosynthesis and action, describe the various defects that could lead to an ambiguity between genetic and phenotypic sex, i.e., pseudohermaphrodism.

4. Compare the defects that result from 21-hydroxylase deficiency with those resulting from 5α-reductase deficiency. What might happen if there were a deficiency in both of these enzymes?

5. The contraceptive pill contains synthetic progestin or progestin plus estrogen. Given what you have learned from the text, how do you suppose the pill works?

6. Describe how you might go about determining which region of a gene confers upon it androgen regulation. You may draw upon analogous examples from the text.

7. List several reasons why polypeptide hormones are synthesized as precursors.

8. A patient has a hypothyroid condition. He has low serum T_3 and T_4 levels and elevated serum TSH. Upon injection of TRH his serum TSH goes even higher. Is his defect primary (thyroid), secondary (pituitary), or tertiary (hypothalamus)?

9. Imagine that you want to establish a radioimmunoassay for a hormone that you have discovered. How would you proceed? Draw your standard curve. Describe the sensitivity of your assay and how you calculated it. What artifacts might you encounter?

10. The steady-state concentration of a protein or an mRNA is determined by its rate of synthesis (S) and its rate of degradation, usually expressed as a half-life ($t_{1/2}$). If S and $t_{1/2}$ are constant, then [protein] or [mRNA] = (S × $t_{1/2}$)/ln 2. Assuming that the half-life of insulin mRNA is 24 hr, what rate of synthesis is required to accumulate 1000 molecules per cell? How often does RNA polymerase II initiate transcription on the insulin genes to accomplish this rate of synthesis? What assumptions did you have to make? If each insulin mRNA is translated 10 times a minute and insulin has a half-life of 1 hr, what would be the steady-state concentration of insulin produced by a single β cell? How many β cells would it take to maintain a serum concentration of 10^{-10} M, assuming a serum volume of 5 liters?

11. Draw hypothetical dose-response and receptor-binding curves for steroid hormones. What would you conclude if the response curve were displaced to the left or right of the binding curve? Suppose it were superimposed on the binding curve?

12. Differential hormone response depends ultimately on which genes are expressible in a particular cell. The process by which a gene becomes expressible is sometimes referred to as gene commitment. What mechanisms may be involved in gene commitment?

13. Make a table listing each of the mutations involved in the pathway leading from receptor to cyclic AMP-dependent protein kinase. Describe biochemical assays that would allow you to identify which gene is mutated.

14. A particular culture cell line is killed by prolonged growth in X, requires Y for survival, and is induced to proliferate in Z. X, Y, and Z are all polypeptides of varying sizes. Three mutants (A, B, and C) are isolated by selection for growth in X. The data are shown in the table.

	Growth in X	Y Requirement	Z Induction	Growth in Cholera Toxin	Growth in Pertussis Toxin
Mutant A	+	+	+	−	−
Mutant B	+	+	+	+	−
Mutant C	+	−	+	+	+
Normal	−	+	+	−	−

a. Can you identify possible target genes for the mutants that might explain the data?

b. How might the effect of Z be explained and what experiments could be done to prove it?

15. Interferon is not usually considered a hormone, yet it shares some properties with peptide hormones. What are they?

16. How might interferon-induced antiviral proteins selectively prevent viral replication without killing the cell?

17. Inhibitors of protein synthesis have been shown to block both the rapid and the slow auxin-mediated growth responses. How would you explain these observations?

32

RECOMBINANT DNA METHODOLOGY

From its inception, genetics has been a manipulative science in which major advances reflect our increasing control over the genetic destiny of living things. Mendel was not content to observe the consequences of natural biological processes. He guided the mating activities of his plants by selective breeding and simple surgical techniques. Nevertheless, Mendel and his followers for many years to come were limited to the genetic variability that arose naturally, through spontaneous mutation and recombination.

In the 1930s the first mutagenic agents were discovered, and it became possible to increase mutation rates artificially. This was the beginning of our efforts to directly influence the structure of the gene, albeit in an undirected manner. Still, treating whole organisms with mutagenic agents does not permit much control over the types of genetic changes we can achieve. To obtain a particular change in a particular gene by such an approach, we are limited to the simplest microorganisms, ones that allow us to scan enormous populations for the organism with the desired genotype.

In the 1940s and 1950s the search for greater control over genetic processes resulted in the exploitation of natural and artificial procedures for exchanging DNAs between microorganisms. It became possible to isolate gene-sized segments of cellular DNA attached to much smaller viral chromosomes. Concurrent progress in biochemistry and molecular biology permitted detailed investigations of the genes of prokaryotic organisms at both the structural and functional level. This dramatic progress in our understanding of microorganisms was not shared by parallel advances for higher organisms. The much greater complexity of the chromosomes of higher organisms and our inability to isolate genes of particular interest from these organisms precluded the detailed molecular analysis of most eukaryotic genes.

The early 1970s saw a bringing together of the biochemical procedures for manipulating DNA *in vitro* with the genetic techniques for transferring DNA from cell to cell. It became possible to propagate individual segments of DNA by a process known as _DNA cloning_. Presently, genes of any organism can be taken from their natural surroundings, sequence-analyzed or altered, and reinserted into the same type of organism or into a different one. The potential usefulness of this new methodology is enormous. It is possible to produce large quantities of precious human hormones by implanting the gene in microorganisms that can be grown on an industrial scale. Projects are in progress to cure certain human molecular diseases by simple DNA transfection and to radically alter the nutritional requirements of agriculturally important plants by suitable gene transplantation. The ultimate control over the gene has been achieved and multitudinous applications lie before us.

The central feature of DNA cloning is the propagation of a particular fragment of DNA in a line of dividing cells. Ordinarily, the DNA fragment to be cloned lacks the ability for self-propagation in the host cell. Thus it must be linked to a carrier molecule, or *vector*, that is able to replicate itself in the host, and the composite DNA molecule must be introduced into a cell that is capable of reproduction. Also important is the ability to identify or select, from a large population of cells, those individuals that contain the DNA fragment of interest. Collectively, DNA cloning procedures have become known as *recombinant DNA methodology*; the term "*recombinant DNA*" is also used in a narrower sense to refer to *composite DNA molecules that result from the physical joining of DNA segments derived from different sources*.

In this chapter we will consider the conceptual and methodological foundations of cloning. Examples of cloning in bacteria, yeast, mammalian cells, and plants will be discussed.

HISTORICAL PERSPECTIVES AND BASIC METHODOLOGY

Work carried out in a number of laboratories in the late 1960s and early 1970s contributed to the conceptual and methodological foundations of DNA cloning. This work provided the four essential ingredients:

1. *A method for physically joining two DNA segments together.*
2. *A self-replicating segment of DNA* (a cloning vehicle or vector) *that is able to propagate in the host organism, and that can be linked to the DNA segment to be cloned.*
3. *A procedure for introducing the composite DNA molecules into a biologically functional recipient cell.*
4. *A means of selecting those organisms that have acquired the desired composite molecule.*

Separate DNA Fragments Can Be Joined *In Vitro*

Before cloning could be seriously attempted, it was necessary to have ways of joining segments of DNA that would lead to predictable results. The procedures that have been developed rely heavily on the use of enzymes discovered by biochemists during the course of investigating DNA metabolism. Some of these enzymes and their functions are described in Table 32–1.

The first experiments aimed at joining DNA segments depended on the ability of complementary nucleotides on single-stranded DNA to form duplex regions by base pairing. The usefulness of such base pairing in linking DNA termini had been known since the early 1960s, when it was discovered that the termini of bacteriophage λ DNA have single-stranded 5' projections that are complementary to each other (Figure 32–1). Cohesion of the ends of λ DNA by hydrogen bonding produces circular molecules, which open upon heating to yield linear λ DNA molecules. λ DNA ends held together by hydrogen bonding provided the substrate for one of the assays that led to the discovery of DNA ligase (see Chapter 26), which converts hydrogen-bonded

TABLE 32–1
Some Enzymes Used in DNA Recombinant Methodology

Enzyme	Function
Polynucleotide ligase	Joining of two DNA molecules
Polynucleotide kinase	Adding a phosphate to the 5'-OH on a polynucleotide
Terminal transferase	Adding homopolymer tails to the 3'-OH ends of a linear duplex
Exonuclease III	Removing phosphates from the 3' ends of a DNA
DNA polymerase I	Repair synthesis, filling in gaps in duplexes by stepwise addition of nucleotides to 3' ends
λ exonuclease	Stepwise removal of nucleotides from 5' ends of a duplex to expose the 3' ends
Type II restriction endonucleases	Specific cleavage of DNAs to produce staggered ends or blunt ends
Alkaline phosphatase	Removal of terminal phosphates from either the 5' or 3' end or both
Reverse transcriptase	Making a DNA copy of an RNA molecule
Type II restriction enzymes	Cleaving duplex DNAs at recognition sites to give 5' phosphate and 3'-OH termini (Some restriction enzymes produce blunt-ended fragments; others produce ends with 3' or 5' overhangs.)

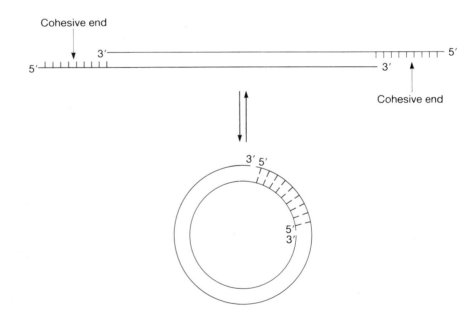

Figure 32–1

Structure of the circular and linear forms of λ DNA. λ DNA possesses complementary 5' projections. Cohesion of the ends by hydrogen bonding occurs spontaneously. The circular form can be covalently closed with DNA ligase.

λ DNA circles to covalently closed circles. *Since complementary sequences on single-stranded DNA projections could hold together ends that were part of the same DNA molecule, potentially they could be used for linking the ends of different molecules.*

Experiments of this type were first carried out by Khorana and his collaborators, who created duplex DNA molecules by serially linking single-stranded DNA segments. By the late 1960s, methods had been developed for the chemical synthesis of polydeoxyribonucleotides of defined sequence. However, there was a severe practical limit on the length of the polynucleotide chains that could be assembled unambiguously by such methods. The

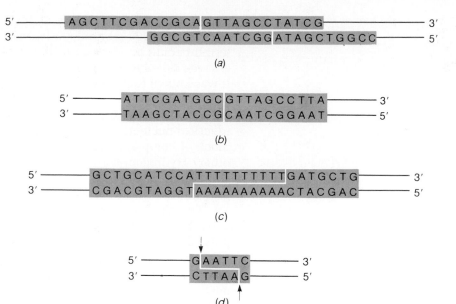

Figure 32–2

Different types of DNA complexes that can be covalently linked by DNA ligase. (a) Complementary duplexes with staggered nicks. (b) Duplexes with blunt ends. (c) Complementary duplexes with staggered nicks. (d) Complementary duplexes with short staggered nicks. (a), (c), and (d) may be ligated with *E. coli* DNA ligase or T4 DNA ligase. (b) can be ligated only with T4 DNA ligase.

procedure employed by Khorana and coworkers circumvented this limitation. First, they synthesized a series of single-stranded polydeoxyribonucleotide chains, each corresponding to a segment of the DNA sequence that was known to encode alanine tRNA. Two separate single-stranded polynucleotide chains containing sequences adjacent to each other in the native tRNA gene were hydrogen bonded to a third single-stranded segment that contained a nucleotide sequence complementary to a part of the sequence on each of the other two chains. This yielded duplex DNA in the region of overlap (Figure 32–2a). The 5′ ends of the duplex segment were phosphorylated using *polynucleotide kinase*, and the adjacent polynucleotide chains, which were being held in position by hydrogen bonding between nucleotides on the overlapping complementary sequence, were joined by DNA ligase. Because ligase introduces phosphodiester bonds at the appropriate sites and accomplishes covalent linkage of separate polynucleotide chains, the chemically synthesized segments could be joined together to form the full-length structural gene. The procedure was laborious, but it worked.

In 1970, Sgaramella, van de Sande, and Khorana reported the discovery of a novel property of the bacteriophage *T4 polynucleotide ligase*, which they were using for their gene synthesis work. Unlike the ligase encoded by the *E. coli* chromosome, the T4 ligase was able to bring about *end-to-end joining of blunt-ended termini on duplex DNA segments* when the nucleoside at a 5′ terminus had a phosphate group and the nucleoside opposite to it had a free 3′ hydroxyl group. By means of the T4 ligase, separate short pieces of synthetic duplex DNA having *blunt ends* could be joined together covalently to form dimers or oligomers; thus DNA joining could be accomplished without the synthesis of complementary termini (Figure 32–2b). Although the importance of these experiments was not generally appreciated for several years, they nevertheless provided the first example of covalent joining of two duplex DNA molecules, albeit synthetic DNA, *in vitro*.

Addition of Complementary Homopolymeric Termini to DNA.   Notwithstanding the discovery of the remarkable DNA-joining properties of the T4 ligase by Sgaramella and his coworkers, most approaches to linking molecules of natural DNA nevertheless were focused on the use of "cohesive" termini having single-strand projections of complementary nucleotides. As already noted, the synthesis of such projections by the stepwise chemical addition of different nucleotides was laborious; potentially, complementary termini could be created on DNA molecules much more simply by using an enzyme called *terminal deoxynucleotidyl transferase* (terminal transferase). It was known that this enzyme can catalyze the stepwise addition of nucleotide residues to the 3′ termini of single-stranded polynucleotide chains. If the enzyme could also accept double-stranded DNA as a primer, a block of identical nucleotides could be added to the 3′ ends of one DNA duplex and a block of complementary nucleotides could be added to the 3′ ends of another. Hydrogen bonding between the complementary nucleotide bases could hold together the two DNAs, which could then be linked covalently by the action of the DNA ligase (Figure 32–2c).

The first attempts at such experiments were reported by Jensen and coworkers in 1971. Homopolymeric blocks of polydeoxyadenosine—poly(dA)—residues were added at the 3′ ends of one population of duplex DNA molecules isolated from the bacterial virus T7, and a block of poly(dT) residues were added to another. When the populations were mixed and annealed, hydrogen bonding between the dA and dT bases resulted in end-to-end joining. Fast-sedimenting molecules that were shown by electron microscopy to be two to three times larger than normal T7 DNA were formed. However, attempts to covalently link the segments of these concatamers using the DNA ligase were unsuccessful; their components came apart when the molecules were treated with alkali.

The problem of *in vitro* ligation of DNA fragments held together by homopolymeric extensions added to their 3′ ends by terminal transferase was solved by Lobban and Kaiser. The entire procedure that they used is summarized in Figure 32–3. The 3′ ends on the linear DNA segments are extended by cutting back on the 5′ ends with λ exonuclease. This step facilitates the action of terminal transferase, which is used to produce the homopolymeric tails of A and T on the two DNAs to be linked. Next, the two DNAs to be linked are mixed together. They spontaneously anneal to form a gapped circular structure. Then exonuclease III and DNA polymerase I (*polI*) are added. The exonuclease III removes 3′ terminal phosphate that resulted from contaminating nuclease in the terminal transferase enzyme preparation. The 5′-OH termini created by the same contaminant are removed by the 5′→3′ exonuclease activity of *polI*. The polI enzyme also fills in the gaps with the appropriate A and T residues. Finally, the two DNAs are covalently attached by DNA ligase.

Jackson, Symons, and Berg used the dA-dT "tailing" method to covalently join molecules of simian virus 40 (SV40) to molecules of the bacteriophage λ derivative, λ olv*gal*. In the Jackson experiment, the duplex DNA segments that were linked together were taken from *different biological species*, thus yielding the first '*recombinant DNA' molecules*. Such λ-SV40 hybrid molecules cannot be cloned, because the site on λ at which the SV40 segment was joined is located within a gene required for replication of the bacterial virus; interruption of the continuity of the essential gene by an inserted DNA fragment prevents λ from functioning as a vector. In any case, concern about possible biohazards related to the SV40 component of the hybrid λ-SV40 molecules led Jackson and his colleagues to decide not to try cloning the molecules.

Restriction Enzymes Cleave DNA to Fragments Ideal for Cloning

Restriction enzymes have been important in DNA recombinant methodology in two ways: (1) they provide a means of cutting the genome into well-defined fragments suitable for cloning, and (2) in some cases the termini

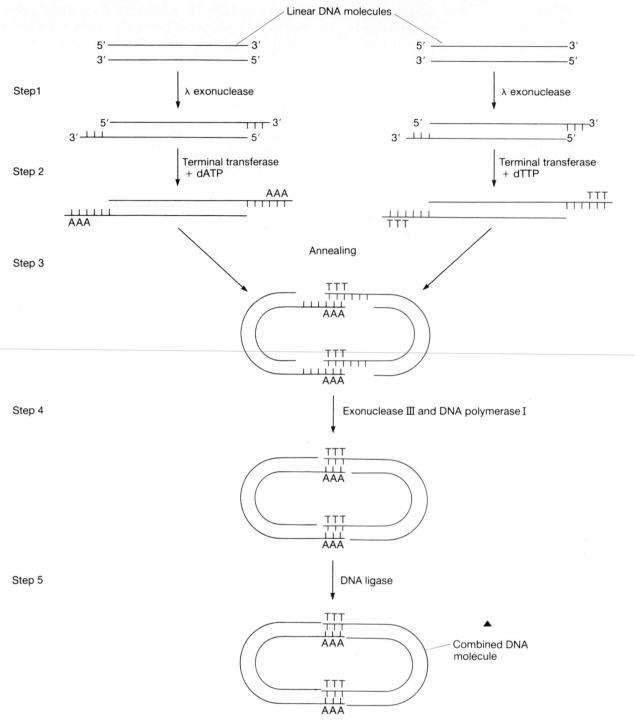

Figure 32–3
Terminal transferase procedure for joining two different DNA molecules. Step 1: Linear duplexes to be joined are briefly treated with λ exonuclease to expose their 3' ends. This facilitates the action of terminal transferase in the next step. Step 2: The 3' ends of the DNAs are extended with A's and T's, respectively, in the presence of terminal transferase.

Step 3: The two DNAs are mixed. They anneal spontaneously either to linear concatemers or to circles (as shown). Step 4: The circles are repaired by adding exonuclease III and DNA polymerase I. Step 5: The repaired circles are covalently closed with DNA ligase.

generated during cleavage have single-strand extensions complementary to each other. Such cohesive termini facilitate the joining process.

Restriction enzymes of type II (see Chapter 26) catalyze double-strand cleavage at recognition sequences that are usually four or six bases in length. In each case the recognition sites were found to consist of a nucleotide se-

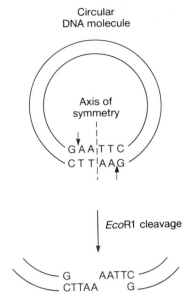

Figure 32–4
A circular DNA molecule with a single recognition site for *EcoRI* restriction enzyme is cleaved to a linear duplex with complementary ends. Note symmetry in recognition site and cleavage pattern.

quence that reads the same on both DNA strands. Sometimes restriction endonucleases cleave both DNA strands at precisely opposite points on the two strands, yielding blunt-ended fragments. On other occasions the enzyme introduces staggered breaks, resulting in termini with either 3′ or 5′ overhangs. In all cases the restriction enzyme hydrolyzes the phosphodiester linkages on the 3′ side, leaving termini with free 3′-OH groups and 5′ phosphate groups. Termini of this type make ideal substrates for DNA ligase.

The *EcoRI* enzyme, discovered in the laboratories of Dussoix and Boyer, was one of the first restriction enzymes to be exploited by DNA recombinant technologists. This enzyme is now known to make staggered cuts at a six-base recognition sequence, thereby generating overlapping cohesive ends (Figure 32–4). The ability of the *EcoRI* enzyme to generate fragments that have complementary projecting single-strand termini was simultaneously discovered by Sgaramella and by Mertz and Davis, in 1972. Sgaramella found that the *E. coli* ligase, which (unlike the T4 ligase) was not able to accomplish end-to-end joining of blunt-ended molecules, could convert molecules of *EcoRI*-cleaved P22 viral DNA to oligomeric products. Mertz and Davis observed that covalently closed molecules of SV40 DNA that had been cleaved by the *EcoRI* endonuclease could re-form themselves by hydrogen bonding into circular molecules, which could then be sealed covalently by the *E. coli* DNA ligase. The reconstituted molecules of SV40 DNA were fully infectious in animal cells growing in tissue culture. The conclusion by both groups—that the *EcoRI* enzyme can generate complementary DNA termini—was confirmed by Boyer and his coworkers by direct analysis of the terminal DNA sequences.

Thus, by late 1972, *separate molecules of duplex DNA could be joined: (1) by the action of the T4 ligase on blunt-ended DNA duplexes, or (2) by the action of either the E. coli ligase or the T4 ligase on the cohesive-ended duplex DNA molecules. Cohesive-ended duplexes could be generated (1) by restriction endonuclease or (2) by the enzymatic addition of complementary homopolymeric single-strand "tails."* The availability of such methods for physically joining DNA segments *in vitro* was of great importance in the development of procedures for the actual cloning of DNA.

Small Plasmids Were the First Vectors to Be Used for Cloning

The initial DNA cloning experiments grew out of studies aimed at elucidating the molecular nature of a class of plasmids responsible for antibiotic resistance in bacteria. *Plasmids,* which *are covalently closed circular genetic elements that replicate separately from chromosomal DNA*, had been known for more than a decade to be present in many bacterial species. They can encode a variety of genetic traits that are not essential to the bacterial host cell but are biologically advantageous under certain circumstances (Table 32–2). Antibiotic resistance is one of these properties.

Plasmids often are present in multiple copies in a bacterial cell. For this reason, a recessive mutation in a gene of one of the plasmid copies is usually masked by the coexistence of nonmutated copies of the same plasmid. To employ classical genetic methods for the study of plasmid mutations, and to investigate the organization of genetic information on plasmid DNA, it was necessary to study the progeny of individual molecules of plasmid DNA. To accomplish this, a procedure for introducing single molecules of plasmid DNA into bacteria (that is, a transformation procedure) was required. Methods for transforming bacteria for chromosomally encoded traits had been developed for *Pneumococcus*, *Hemophilis*, *Bacillus*, and certain other organisms, but at the time of the first DNA cloning experiments, generalized transformation had not been shown for *E. coli* or other closely related bacte-

TABLE 32–2
Some Properties Encoded by Naturally Occurring Plasmids

Antibiotic resistance
Fertility (ability to transfer genetic material by conjugation)
Production of bacteriocins
Antibiotic production
Heavy-metal resistance (Cd^{2+}, Hg^{2+})
Ultraviolet resistance
Enterotoxin production
Virulence factors, hemolysin, K88 antigen
Metabolism of camphor, octane, and other polycyclic hydrocarbons
Tumorigenicity in plants
Restriction/modification enzymes

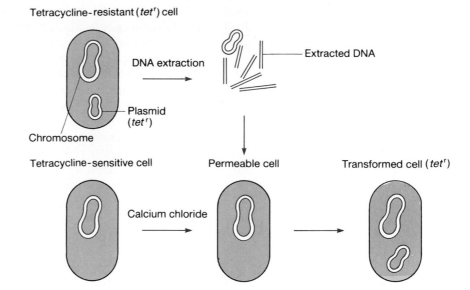

Figure 32–5

Transformation of a normal tetracycline-sensitive *E. coli* cell to a plasmid-containing tetracycline-resistant cell. Cells carrying a plasmid with a tetracycline resistance gene are used to make DNA. Normal plasmidless cells are permeabilized by calcium chloride treatment and transfected with plasmid-containing DNA. Tetracycline-resistant transformed cells may be isolated by plating cells on tetracycline-containing agar plates.

ria that were the natural hosts for most of the plasmids that were being studied.

The successful transformation of *E. coli* with plasmid DNA was a direct outgrowth of earlier work carried out with the DNA of bacteriophage λ. Since 1960, it had been possible to introduce purified λ DNA into *E. coli* cells. Viable phage particles could be obtained by such transfection methods, provided that genetically related "helper" phages were introduced into the cell concurrently. The helper phages function at least in part by altering cell wall permeability, and high concentrations of calcium salt were found to improve the infectivity of helper-aided λ DNA. In 1970, Mandel and Higa reported that DNA from the bacteriophages P2 and λ could infect a sensitive host in the _absence_ of helper phage when *E. coli* cells were incubated with the DNA and calcium chloride at 0°C, and then transferred to 37°C. This result showed that calcium-chloride treated bacteria were not only able to take up purified phage DNA molecules, but also were able to serve as vessels for the production of phage particles.

In 1972, Cohen, Chang, and Hsu, using a modification of the procedure worked out by Mandel and Higa, found that *E. coli* could take up circular plasmid DNA molecules. Transformants in the bacterial population could readily be identified and selected by utilizing the antibiotic resistance genes carried by the plasmid; when the cells were spread on Petri plates containing the appropriate antibiotic, only those bacteria that had acquired the plasmid could grow. Cells transformed with plasmid DNA reproduced themselves normally, and produced a clone of bacteria (Figure 32–5). Since each cell in the clone contained a DNA species having the same genetic and molecular properties as the plasmid DNA molecule that was taken up by a single bacterial cell, the procedure made possible the cloning (and thus, the biological purification) of individual molecules even if they were originally present in a heterogeneous population of plasmids.

The Initial DNA Cloning Experiments Involved Antibiotic Resistance Genes

The cloning of individual plasmid DNA molecules provided a possible way to clone other DNA fragments that might be linked to the plasmid's replication system. In nature, plasmids had evolved by the linking of genes encoding a variety of different traits to DNA segments that encode replication functions. Potentially, plasmids might also be able to propagate DNA segments linked to them by *in vitro* ligation.

The earliest cloning experiments were reported in 1973 by Cohen, Chang, Boyer, and Helling. *EcoRI* endonuclease was used to generate fragments of the large antibiotic resistance plasmid R6-5. Fragments carrying antibiotic resistance genes were ligated to fragments that carried the replication machinery. Such experiments showed that *fragments of a plasmid that were themselves not capable of self-propagation could be cloned by linking them individually to the plasmid's replication region,* provided that the cloned fragment contained a gene useful in the selection of transformed bacteria. While the *EcoRI*-generated fragment containing the replication region of the R6-5 plasmid could thus serve as a carrier molecule or vector, to be generally suitable for DNA cloning, a vector would have to contain both replication machinery and a selectable gene on the *same* restriction-endonuclease-generated fragment.

The first plasmid found to meet these requirements was pSC101. It is 9000 bp in length and contains only a single cleavage site for the *EcoRI* endonuclease. It carries a gene conferring resistance to the antibiotic tetracycline (tet^r). Since insertion of a fragment of foreign DNA at its single *EcoRI* cleavage site does not interfere with either the replication functions of the plasmid or expression of its tetracycline resistance gene, pSC101 can be used as a *directly selectable cloning vector*.

pSC101 plasmid DNA was mixed with DNA of a plasmid that carries a kanamycin resistance gene (kan^r) bracketed by two *EcoRI* cleavage sites, and the DNA fragments in the mixture were treated with the endonuclease, ligated, and introduced into *E. coli* by transfection (Figure 32–6). Bacteria that expressed resistance to both kanamycin and tetracycline were selected on Petri dishes containing the two antibiotics.

The structures of newly constructed hybrid plasmids, such as the pSC101 derivative just described, are usually confirmed by reisolation of the plasmid from a large population of the cloned cells. The plasmid is treated with a characteristic enzyme and then gel-electrophoresed to size the resulting restriction fragments. Remember that fragments migrate in reciprocal relation to their size. In Figure 32–7 (lane A) we see the electrophoretic pattern of the *EcoRI*-treated DNA isolated from a kan^r, tet^r cell. Also shown are the *EcoRI* digest patterns from pSC101 (lane D), from pSC102, which is the source of the kan^r fragment (lane C), and from a mixture of pSC101 and pSC102 (lane B). The pSC101 derivative can be seen to carry two separate *EcoRI*-generated fragments, one of them identical in length to the pSC101 plasmid. The other is indistinguishable from the *EcoRI* endonuclease-generated fragment carrying the kanamycin resistance gene (middle fragment, lane C). The electrophoretic patterns therefore show that the hybrid plasmid must contain the pSC101 plasmid with the kan^r fragment of pSC101 inserted at the *EcoRI* site.

This result demonstrated that the pSC101 vector could be used to propagate a nonreplicating segment of another *E. coli* plasmid. The next step was to determine whether the same procedure could be used to propagate and genetically express DNA from an unrelated bacterial species. It was possible that the arrangement of genetic information on totally foreign DNA, or still another unknown factor, might prevent the survival of interspecies hybrid molecules in *E. coli*. Furthermore, even if DNA from a different bacterial species could be propagated in *E. coli* by joining it to the pSC101 vector, there was no assurance that genes carried by the foreign DNA would be expressed in the heterospecific host.

To determine whether the DNA cloning methodology developed for

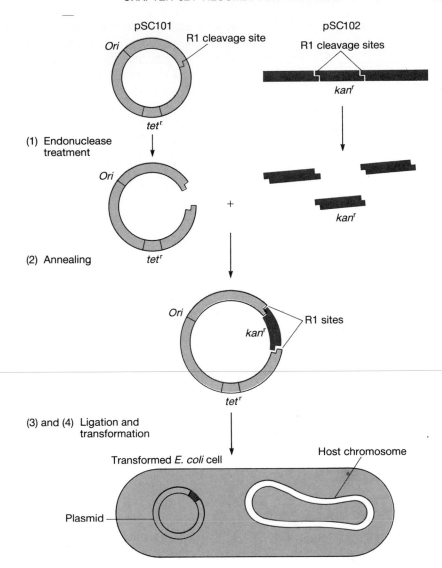

Figure 32-6

Preparation of hybrid plasmid containing pSC101 and the kanamycin resistance gene (kan^r) from pSC102. Step 1: Both plasmids are digested with the EcoRI endonuclease. Step 2: The digestion products are mixed. Some spontaneous annealing takes place because of the unpaired complementary ends. After ligation (step 3) cells are transfected (step 4) and plasmid-containing cells are selected by plating on agar medium with both kanamycin and tetracycline. Only cells carrying the newly formed hybrid plasmid can survive exposure to both drugs. Note that the newly formed hybrid plasmid contains two EcoRI sites, whereas the original pSC101 plasmid contained only one.

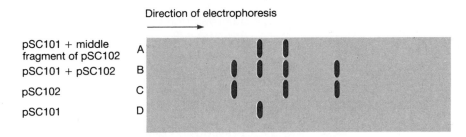

Figure 32-7

Electrophoretogram of EcoRI digests of pSC101 (D), pSC102 (C), a mixture of pSC101 and pSC102 (B), and a pSC101 derivative carrying a kanamycin resistance gene (A). This type of experiment is done to confirm the structure of a newly constructed plasmid. In this case the newly constructed plasmid is the one in lane A. This plasmid originates from a cell possessing both tetracycline and kanamycin resistance. When pSC101 is digested with EcoRI, it yields a single linear fragment (lane D). When pSC102 is digested, it yields three fragments (lane C). The pSC101 derivative, when digested, yields two fragments (lane A). One of these has the same mobility as pSC101. The other has the same mobility as the middle fragment from pSC102. This indicates that the pSC101 derivative is composed of pSC101 and the middle fragment of pSC102 inserted at the EcoRI site. Also see Figure 32-6. The electrophoresis is run from left to right; fragments are stained with ethidium bromide and photographed with UV light.

DNA fragments derived from *E. coli* would work also for foreign DNA, the researchers mixed *Eco*RI-cleaved preparations of the pSC101 plasmid with DNA from the *Staphylococcus aureus* plasmid PI258, which carries a gene conferring resistance to ampicillin (*amp*ʳ). The mixture was then treated with DNA ligase and introduced into *E. coli* by transfection. Cells that expressed the ampicillin resistance gene carried by the *S. aureus* plasmid as well as the tetracycline resistance gene encoded by pSC101 were isolated; gel electrophoresis demonstrated that these cells contained a plasmid that included the entire pSC101 plasmid DNA segment, plus an additional fragment produced by cleavage of the PI258 plasmid by *Eco*RI endonuclease. The latter fragment contained the *amp*ʳ resistance gene, originally carried by PI258.

The replication and expression in *E. coli* of a gene derived from an organism not known to ordinarily exchange DNA with *E. coli* suggested that interspecies genetic combinations might be generally obtainable by means of DNA cloning methods. These methods could therefore be used to introduce into *E. coli,* and thus isolate and study, genes specifying metabolic and synthetic functions indigenous to a variety of other prokaryotes and possibly also eukaryotes.

In the foregoing we have described the series of discoveries that led to cloning procedures and shown that certain foreign DNAs can be cloned in *E. coli.* We turn now from this historical presentation to an account of current procedures used for cloning in *E. coli* and other systems, and of some of the uses to which these procedures have been put.

DNA CLONING MATERIALS AND PROCEDURES IN *E. COLI*

Although cloning procedures have become very widespread and are carried out in a number of organisms, most cloning is still done in *E. coli.* It is partly for historical reasons that *E. coli* has been the most popular organism used by molecular biologists. However, there are also practical reasons for this popularity. *E. coli* is very easy to manipulate, both genetically and biochemically, and it is the natural host of many plasmids and viruses that make ideal cloning vectors.

The Choice of Cloning Vector Depends on the Research Objective

In order for a segment of DNA to serve as a cloning vector, (1) it must encode functions that enable it to replicate independently of the chromosome of the host cell, (2) it should carry a gene that enables the selection of cells that acquire it, (3) it should be easily isolatable from the host as a discrete molecule, and (4) it should contain an assortment of unique restriction endonuclease cleavage sites.

Plasmids.   The vectors initially used for DNA cloning were bacterial plasmids. These extrachromosomal genetic elements range in size from 5 to 400 kilobase (kb) pairs and commonly are present in two to up to several hundred copies per cell. Plasmids present in a large number of copies per cell are most often used as cloning vectors, since they enable the DNA segments linked to them to be amplified many times and thus increase the yield of the cloned DNA fragment and the gene products encoded by it.

The pSC101 plasmid carries a tetracycline resistance gene and contains a single cleavage site for the *Eco*RI endonuclease, as already discussed; however, it is present in only four to six copies per chromosome. Shortly after the initial DNA cloning experiments, the usefulness of plasmid vectors such as ColEI for obtaining larger amounts of cloned DNA fragments was demon-

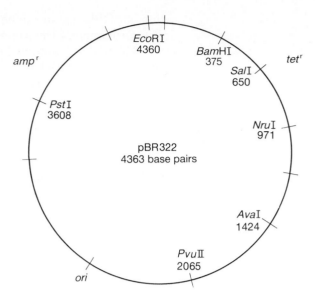

Figure 32–8

Structure of the pBR322 plasmid. Unique sites for various restriction enzymes are indicated. Also indicated are the locations of the tetracycline (tet^r) and the ampicillin (amp^r) resistance genes and the origin for DNA replication.

strated. ColEI is normally present in 10 to 20 copies per chromosome, but the number of copies can be amplified many times by treatment of plasmid-containing cells with the antibiotic chloramphenicol. This antibiotic inhibits protein synthesis and host chromosomal synthesis, but it does not inhibit plasmid DNA synthesis, which can increase disproportionately in its presence. While ColEI and other naturally occurring multicopy vectors do not normally carry antibiotic resistance genes, such genes can be added by the use of *in vitro* joining procedures, thus facilitating the selection of bacteria that take up the plasmids. One of the most widely employed cloning vectors, pBR322, was made in this way.

The plasmid pBR322 is a 4363-bp covalently closed circular DNA molecule that includes component segments derived originally from pMB8, a small ColEI-like plasmid, plus segments of the pSC101 plasmid and the β-lactamase gene of transposon Tn3 (Figure 32–8). pBR322 contains single cleavage sites for a number of commonly used restriction endonucleases, and also carries genes that encode resistance for ampicillin (amp^r) and tetracycline (tet^r). Foreign DNA segments can be introduced into a unique cleavage site in the ampicillin resistance gene, and bacteria that acquire the plasmids can be selected using tetracycline resistance. Those cells that contain plasmids having foreign DNA inserts can be identified because the ampicillin resistance gene is inactivated by the inserted DNA, with the result that the host cell becomes sensitive to the action of ampicillin. Methods have also been developed to enable the direct selection of clones showing sensitivity to tetracycline, ampicillin, or other antibiotics, thus facilitating the isolation of plasmids that carry an insert in either of the resistance genes.

Bacteriophage λ Vectors.   Bacteriophage λ possesses a number of advantages that make it particularly attractive as a cloning vector. DNA fragments as large as 24 kb can be propagated using this vector. The primary pool of clones can be amplified by limited phage growth as plaques, and the entire collection of phage clones (recognized as plaques) can be stored for long periods in a small volume. Screening of phage plaques can also be done by rapid and sensitive methods.

Since λ phage will not accommodate molecules of DNA that are much longer than the viral genome, the use of phage λ as a vector requires the removal of a significant portion of the viral DNA. Fortunately, the central third of the λ genome contains genes that are not essential for plaque production, and this region can be replaced by an externally derived DNA fragment of approximately the same size without influencing phage growth.

The native λ genome (approximately 50 kb) is large in size relative to the size of the bacterial plasmids used as cloning vectors (approximately 5–10 kb); thus phage λ DNA normally contains multiple cleavage sites for even those endonucleases that require a 6-bp recognition sequence. To use bacteriophage λ as a vector, it ordinarily is necessary to remove all except one cleavage site for a particular endonuclease, and to ensure that the remaining site is located in a nonessential region of the phage genome. During 1974, three groups working concurrently but independently utilized the wealth of genetic and biochemical information available for bacteriophage λ to construct phage vectors that contain only a single target site for the EcoR1 endonuclease. The λ vectors constructed by Camaron and Davis lack all EcoR1 cleavage sites except those bracketing the nonessential region in the middle of the λ genome. These workers then ingeniously used previous knowledge that there is a minimum and maximum amount of DNA that must be packaged to form a viable phage particle. By removing enough nonessential DNA, they constructed a phage variant having a genome too small to be packaged; insertion of any DNA fragment of sufficient length to yield a molecule longer than the minimal packaging size enabled the phage to grow and to form a plaque. The resulting λ phage ensures that *all of the plaques produced are derived from phage genomes that contain an inserted DNA fragment.*

Bacteriophage λ vectors that accommodate foreign DNA fragments generated by a variety of restriction endonucleases have since been constructed. The recombinant DNA molecules that incorporate some of these vectors can be introduced directly into E. coli by means of a transfection procedure developed earlier for bacteriophage λ, using calcium-chloride-treated cells. Alternatively, recombinant DNA molecules can be packaged into phage particles *in vitro*. The availability of systems for the *in vitro* packaging of DNA into phage particles set the stage for the development of still another type of vector, the cosmid, which incorporates certain desirable features of both phage and plasmid vectors.

Cosmids.   While plasmids and bacteriophage λ are both highly useful as vectors, their use is limited in two ways. First, the size of the DNA fragments that can be cloned in such vectors is limited. With plasmids, the larger the inserted fragment of foreign DNA, the lower the efficiency of ligation and transfection; thus the cloning of DNA fragments larger than 15 kb is experimentally difficult. In the case of λ vectors, the length of the nonessential region of λ DNA limits the fragment size to 24 kb or less. Second, the original λ vectors do not allow propagation of viable bacterial cells that carry the inserted DNA fragment; the insert is propagated as part of a virus that lyses the cell.

Cosmids were developed as vectors specifically designed for the cloning of large fragments of DNA. The first part of their name, *cos,* comes from the fact that they contain the cohesive ends or *cos* sites of normal λ (Figure 32–1). These ends are essential for the packaging of any DNA in λ phage heads. The last part of their name, *mid,* comes from the fact that cosmids carry a plas*mid* origin of replication. The first cosmids consisted of a ColE1-derived plasmid carrying the *cos* sites of phage λ. Such cosmids can be used for cloning in the same way as any other plasmid vector. However, because cosmids also contain the *cos* sites, cosmid DNA along with an inserted DNA fragment can be packaged as a λ phage. The result after packaging is a defective but nevertheless infectious phage particle. Once the cosmid and the inserted DNA fragment have been introduced by infection into a λ-sensitive cell, the plasmid replication system allows the vector to replicate as a plasmid. Since in cosmids virtually the entire bacteriophage λ genome has been deleted, except for the region adjacent to the *cos* sites, these vectors can propagate exogenously derived DNA fragments up to 40–50 kb in length.

Several cosmid vectors have been developed. These differ in size (and therefore cloning capacity), useful cloning sites, and selectable properties (markers). The pJB8 vector is a 5.1-kb plasmid containing several possible cloning sites and the *amp*^{r} selectable marker. Its cloning capacity is limited to DNA fragments between 30 and 47 kb. A feasible protocol for using this vector is described in Box 32–A.

Bacteriophage M13 Vectors.   A series of highly useful cloning vectors also has been developed from the single-stranded DNA bacteriophage M13. When this phage infects bacteria, the strand that is packaged in the phage's filamentous protein capsid (the plus strand) replicates to form a double-stranded intermediate, known as the replicative form (RF). The RF of M13 is structurally and functionally similar to a plasmid and can be isolated from bacterial cells by the methods used to obtain plasmid DNA. Foreign DNA fragments can be inserted at any one of a number of unique restriction sites in the 7200-bp M13 genome. Depending on the orientation, one or the other strand of the inserted fragment will be packaged into progeny virus particles along with the plus strand of M13. Since the entire sequence of M13 is known, a short polynucleotide complementary to the region of M13 adjacent to the cloning site can be synthesized chemically. This polynucleotide can serve as a primer for the sequencing of any DNA fragment that is cloned at the same site in M13, using the dideoxy method of Sanger (see Box 26–A). A single primer complementary to the segment adjacent to a particular cleavage site can thus be used to sequence a variety of different DNA fragments inserted into the site, greatly facilitating analysis of the cloned DNA.

Shuttle Vectors.   Vectors that include replication systems derived from more than one host species are known as *shuttle vectors*. Such vectors commonly include a replication system able to function in *E. coli*, as well as a replication system able to function in a second host—which may be bacterial or eukaryotic. Initial cloning and amplification of the DNA segment to be studied is often carried out in *E. coli* because it is easier to make large quantities in a culture of *E. coli*. The recombinant DNA molecule—consisting of the "bifunctional vector" plus the cloned segment of DNA—is then introduced into the second host, where the purpose is usually to measure the expression of the genes carried by the vector. Shuttle vectors that can replicate in both *E. coli* and yeast are quite common.

The Ideal Clone Bank Contains All the Sequences with a Minimum of Repeats

A *DNA library* is a collection of independently isolated vector-linked DNA fragments derived from a single organism. A library serves as the source of well-defined sequences from a given organism. Within each clone of the library there is a purified sequence from the desired organism. Within the entire library there may be several repeats of a particular sequence, but other sequences may be missing. The ideal library is one that represents all of the sequences with the smallest possible number of clones. This ideal is never achieved; it can only be approached.

There are two basically different ways of preparing a library from the same cell or organism. In the first, the genome may be fragmented and ligated to the appropriate vector to produce a *genomic DNA library*. The second type of library is called a *cDNA library*; the DNA fragments for the cDNA

BOX 32–A

Construction of a Cosmid

The pJB8 vector is a 5.1-kb plasmid containing a λ *cos* site, an *amp*[r] selectable marker, an origin of replication, and several possible cloning sites. To construct it, first one of the staggered-end ligation sites of the plasmid was opened with the appropriate restriction enzyme (step 1). In this case the *Bam*HI site was chosen. The linearized vector was then treated with alkaline phosphatase (step 2) to prevent recircularization of the vector in the subsequent ligation step. The remaining procedures were dictated by the goal of this protocol, which was to pro-

duce very large cloned segments of DNA. Thus ligation had to yield large DNA fragments that were produced by random cutting (and thus were representative of the entire genome) and that could be inserted into the *Bam*HI site of the cosmid vector.

On a random basis, any very large DNA fragment is likely to be cut several times by even a 6-cutter restriction enzyme under conditions of complete digestion. In order to avoid this problem and to obtain maximum randomness of fragments of the appropriate size, the *Mbo*I re-

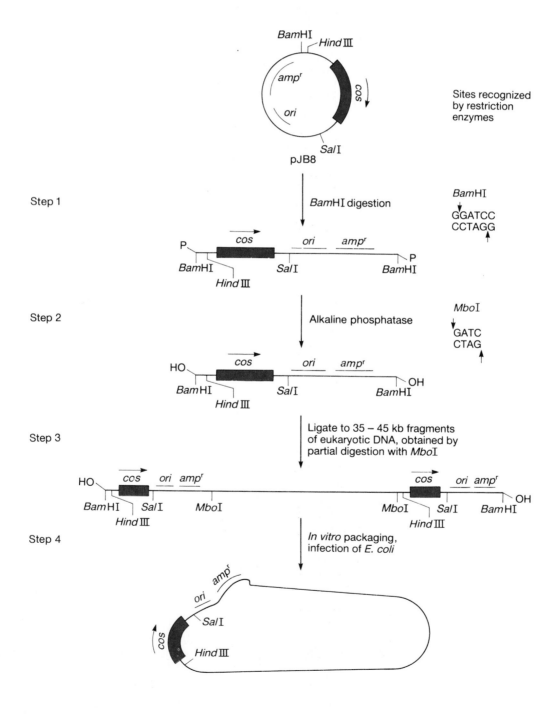

striction enzyme was used under conditions of very limited digestion. This enzyme recognizes the four-base sequence GATC, whereas BamHI recognizes a six-base sequence, GGATCC. The two enzymes produce identical overlapping fragments (see upper right of illustration). Thus the eukaryotic fragments should ligate efficiently with the BamHI restricted cosmid vector (step 3). Because limited digestion conditions were used, only some

of the MboI (and BamHI) sites were cleaved, and these varied from molecule to molecule on a random basis. Thus some molecules in the population were likely to contain an uncleaved enzyme recognition site at any given position. After ligation, the resulting concatamers were packaged in vitro into λ particles and introduced into a suitable E. coli strain (step 4). Transformants were selected with ampicillin.

library are obtained by reverse transcription from the cellular RNA. As we shall see, each of these libraries has advantages and disadvantages. For a specific purpose, one library is usually preferred to the other.

The wealth of information available in today's DNA libraries is incalculable, but the vast majority of DNAs within the libraries are uncharacterized. It is like having a library full of books with words but no titles. The task of finding the desired genes or sequences within a library is usually much greater than that of constructing the library.

Preparation of a Genomic DNA Library. In making a genomic DNA library, the aim is usually to maximize the probability that all genes are represented at least once. If the genomic DNA is prepared by cutting with a restriction enzyme, there is always the possibility that the restriction enzyme chosen will cleave the gene of interest at one or more sites. To increase the likelihood of isolating the desired gene in one piece, different restriction enzymes can be used on different parallel preparations. But even if the gene of interest is not cut by the enzyme(s) chosen, the DNA fragments produced may be uncomfortably small to work with. An enzyme that recognizes a sequence of six bases (a 6-cutter) gives an average fragment size of 4096 bp,* which is a reasonable size for making a plasmid library but much smaller than the size desirable for cloning in λ or a cosmid vector. Therefore, when large randomly generated fragments are desired, the method of choice is usually to make an incomplete digest using a 4-cutter restriction enzyme, which produces overlapping ends that can be readily cloned into the chosen vector as described earlier. The extent of digestion is controlled so that cleavage occurs at only some of the restriction enzyme recognition sites and the average size of the fragments produced is in the desired range. The conditions chosen thus depend on whether the product is going to be cloned in a plasmid, a λ phage, or a cosmid vector.

Table 32–3 gives the minimum number of clones (that is, the size of the library) required to fully represent the entire genome in a genomic DNA library, as a function of the average size of the cloned fragments and the size of the genome. Since DNA fragments in a population are cloned on a random basis, there is a 50 per cent chance of finding a given single-copy gene in a library of the indicated size. A clone bank should be three to ten times the minimum size to insure a high probability that a particular segment will be represented.

Preparation of a cDNA Library. A cDNA library consists of a collection of clones that contain DNA copies of the cellular or organismic RNA. If the RNA is obtained from a differentiated multicellular organism, then the library will vary in composition with the type of cell used as the RNA source

---

*This size is calculated from the probability of finding a sequence of six bases. The probability of any given base at a particular site is $\frac{1}{4}$. The probability of a particular sequence of six bases is the product probability $\frac{1}{4} \times \frac{1}{4} \times \frac{1}{4} \times \frac{1}{4} \times \frac{1}{4} \times \frac{1}{4} = (\frac{1}{4})^6 = \frac{1}{4096}$.

TABLE 32-3
Theoretical Number of Clones Required to Fully Represent the Entire Genome of Various Organisms

Size of Cloned DNA Fragment (bp)	Genome Size (bp)		
	2×10^6 (e.g., bacteria)	2×10^7 (e.g., fungi)	3×10^9 (e.g., mammals)
5×10^3	400	4000	600,000
10×10^3	200	2000	300,000
20×10^3	100	1000	150,000
40×10^3	50	500	75,000

and also the physiological state of the cell. This variation is a reflection of the varying abundances of particular messengers RNAs made by different cell types and under different experimental conditions. If a cDNA species corresponding to a particular gene product is desired, it is often possible to select a cell type known to synthesize a large amount of the corresponding mRNA. Thus pituitary cells can be used if cDNA encoding for growth hormone is desired, and liver cells can be used if a serum albumin cDNA is the goal. mRNAs present in low amounts will clearly require the screening of a larger library than mRNAs present in medium or high abundance (Table 32–4).

Sometimes prepurification of crude RNA is possible to increase the chances of finding a particular cDNA in the library. A technique commonly used with RNA from eukaryotic cells is to purify the crude RNA by chromatography over a cellulose column containing short poly(dT) chains. This will lead to an enrichment for RNA species that contain a 3′-poly(A) tail (i.e., most of the mRNAs). The poly(A) tail provides an important feature for the subsequent synthesis of the cDNA (Figure 32–9); the crude poly(A)-mRNA fraction serves as a template for the synthesis of the first strand of the cDNA, using the enzyme _reverse transcriptase_ (step 1, Figure 32–9). When poly(A) mRNA is the template, a short poly(dT) chain makes an ideal primer as it complexes with the poly(A)-containing portion at the 3′ end of the RNA. Synthesis of the polynucleotide chain then proceeds along the mRNA template in the 5′ direction. The product of this reaction is an RNA-DNA hybrid duplex, which in favorable cases extends to the full length of the mRNA. The RNA strand is removed from the duplex by alkali treatment (step 2, Figure 32–9) and the reisolated single strand of enzymatically synthesized DNA is used as a template for the synthesis of a complementary second strand of DNA by DNA polymerase I (step 3, Figure 32–9). This enzyme also has a requirement for primer; commonly the 3′ end of the first strand of DNA seems to be able to loop around and serve as a primer and template for its

TABLE 32-4
Distribution of Population of mRNA Molecules of Typical Eukaryotic Cells

Abundance Class	Fraction of Total mRNA Population in Class	Number of Different mRNA Sequences in Class	Number of Copies of a Unique mRNA Sequence Per Cell
High	22%	30	3500
Medium	49%	1090	230
Low	29%	10670	14

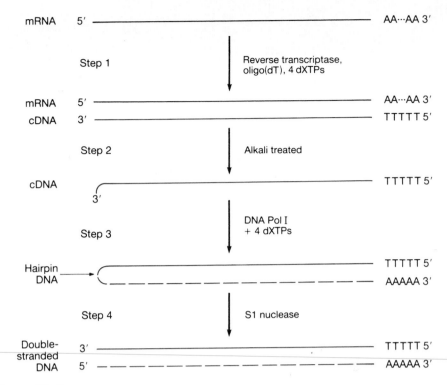

Figure 32–9
Formation of a cDNA duplex from a poly(A) mRNA. Step 1: To the poly(A) mRNA are added reverse transcriptase, oligo(dT), and the four deoxynucleotide triphosphates. In favorable circumstances this results in a full-length DNA-RNA hybrid duplex. Step 2: The RNA is removed and degraded by alkali treatment. Step 3: The remaining single-stranded DNA serves as both template and primer for the synthesis of a complementary DNA strand. The newly synthesized DNA is linked to the primer template DNA by a hairpin turn. Step 4: The hairpin is cleaved with S1 nuclease, giving a duplex molecule.

own synthesis, leading to the formation of a duplex DNA molecule having its strands joined together at one end like a hairpin. The next step (step 4, Figure 32–9) in the preparation of a cDNA duplex involves scission of the hairpin. This is usually done by brief treatment with S1 nuclease, which acts as an endonuclease on single-stranded regions of the DNA.

After S1 cleavage, appropriate ends are attached to the newly synthesized duplex to make it suitable for insertion into a plasmid or other vector. There are two general ways of doing this. One involves homopolymer tailing. The basic strategy of homopolymer tailing has already been described in Figure 32–3. In this case the double-stranded cDNA would be tailed with one base, and a suitable plasmid vector such as pBR322 would, after opening, be tailed with the complementary base. A detailed protocol for this is shown in Figure 32–10. The double-stranded cDNA is 3' tailed with C residues. The plasmid is opened by treatment with *Pst*I restriction endonuclease at a unique site in the *amp*ʳ gene. This treatment results in a linear duplex with a four-base 3'-OH overhang, TGCA. This is 3' tailed with G residues. The two-tailed duplexes are mixed together, whereupon they spontaneously anneal. The annealed complex can be directly transfected, leaving the final repairs to take place in the bacterial cytoplasm.

Another way of inserting the cDNA into a vector is to tail it with _linkers_. Linkers are synthetic single-stranded oligonucleotide segments (6, 8, 10, or 12 bases in length) that self-associate to form symmetrical blunt-ended double-stranded molecules containing the recognition sequence for a particular restriction enzyme. An example is illustrated in Figure 32–11 for an eight-base linker (CCTGCAGG) containing a *Pst*I recognition site. This linker

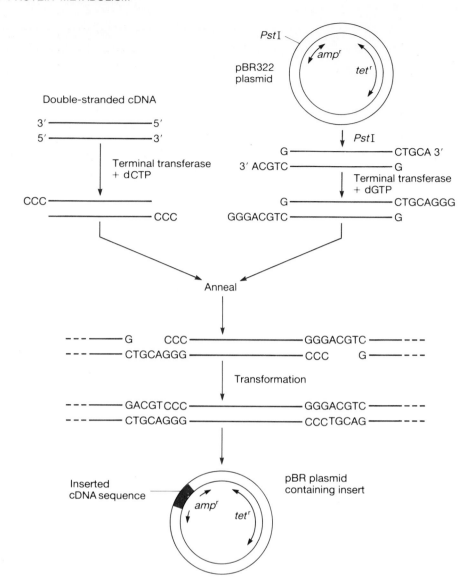

Figure 32–10
Insertion of cDNA into pBR322 plasmid by the homopolymer tailing method. The cDNA duplex is tailed with terminal transferase and dCTP. pBR322 is opened with the *Pst*I enzyme and tailed with G residues. The two DNAs spontaneously anneal when mixed. The complex is transfected directly, allowing for the repair synthesis and ligation to take place *in vivo*.

self-associates to produce an eight-base blunt-ended duplex structure that adds to the double-stranded cDNA in the presence of T4 ligase. The resulting product is treated with *Pst*I to produce the characteristic 3′ overhang. The plasmid, linearized with *Pst*I, and the two DNAs are mixed and reacted with ligase to produce plasmid with the insert. Considerable plasmid will reclose without incorporating the insert, but this material can be distinguished from plasmid with the insert because the latter has lost its ampicillin resistance. Alternatively, the linearized plasmid DNA can be treated with alkaline phosphatase to remove the 5′ phosphates from its termini. Phosphate removal prevents the plasmid from recircularizing. When the terminally dephosphorylated duplex is reacted with the cDNA insert in the presence of ligase, a circular structure with two nicks is formed. Upon introduction into a bacterial cell these nicks are mended.

Although a cDNA library requires more work to prepare than a genomic DNA library, it has major advantages for certain purposes: (1) Since only a small fraction of the genomic DNA of eukaryotes is expressed at any one time and in any one type of cell, the variety of clones is greatly reduced. If the desired genetic sequence is among those cloned, screening is greatly facilitated. (2) cDNA contains a minimum number of repetitive sequences,

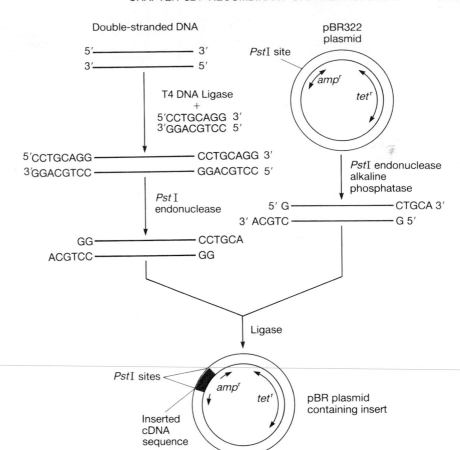

Figure 32–11

Insertion of cDNA into pBR322 plasmid by the linker method. The strategy here is to open up the plasmid with a restriction enzyme that makes staggered cuts and to attach linkers that contain the same recognition site to the cDNA. After the linkers are attached to the cDNA, the duplex is treated with the same restriction enzyme (PstI) to expose the overhangs. The two DNAs are mixed together and ligated. After transfection, cells containing the hybrid plasmids are recognized by tetracycline resistance and ampicillin sensitivity. Identification of the insert is discussed in the text.

which can complicate the screening procedure when the cloned DNA is used as a probe. A probe with repetitive sequences is likely to anneal to any structure that has similar repetitive sequences, whereas the goal is usually to use a probe that will anneal to a unique sequence in the genome. (3) Because introns present in eukaryotic DNA are not present in the RNA (and consequently in the cDNA), the cloned cDNA frequently can be used in E. coli or other microorganisms to obtain the proteins encoded by the eukaryotic gene. A DNA containing introns cannot be properly expressed in E. coli because the bacterium lacks the splicing apparatus. (4) A single cDNA clone is pure in the sense that it contains only the sequences corresponding to a single mRNA. For this reason the results obtained when using the cDNA probe are often easier to interpret than data from probing experiments that employ cloned genomic DNA segments, which sometimes may contain several different genetic sequences.

There Are Several Approaches to Picking the Correct Clone from a Library

A library can contain thousands of different clones (Table 32–3) and it can be a time-consuming task to isolate a clone with the DNA of interest. In the vast majority of cases, cloned DNAs do not include easily selectable genetic markers, such as a gene conferring resistance to a specific antibiotic.

Most of the currently used procedures for screening large numbers of colonies for plasmids or phage that contain specific DNA inserts are variants of the _colony hybridization method_ developed by Grunstein and Hogness. This procedure makes use of a highly specific radioactive probe that contains some sequences complementary to those in the DNA of interest. The

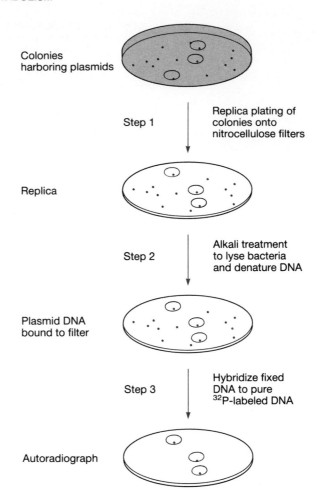

Figure 32–12
Colony hybridization procedure used to identify bacterial clones harboring a plasmid containing a specific DNA. Step 1: Replica-plate the colonies containing plasmids onto nitrocellulose paper. Step 2: Lyse cells with NaOH and fix denatured DNA to paper. Step 3: Hybridize to ^{32}P-labeled DNA carrying desired sequence and autoradiograph the product. Locations of desired DNA are observed in autoradiographs. Clones carrying desired plasmids (circled) may be isolated by referring back to the agar replica plates.

colonies to be screened are first grown on agar Petri plates (Figure 32–12). A replica of each plate is made on another agar plate and stored for reference. A replica is also made on a nitrocellulose filter. The colonies formed on the filter are lysed and the contents denatured simultaneously by treatment with sodium hydroxide. After heating, the denatured DNA is fixed on the filter at each site where a colony was located. The DNA on the filter is then hybridized with a radioactively labeled nucleic acid probe complementary to the specific DNA sequence to be selected. The presence of hybridized probe at sites occupied by DNA derived from colonies that include the DNA fragment of interest is detected by autoradiography on film. The colony whose DNA hybridizes with the nucleic acid probe can then be picked from the reference plate, which contains a viable bacterial colony at a corresponding location.

An analogous procedure developed by Benton and Davis has made it possible to screen bacteriophage λ recombinant DNA clones by hybridization to single plaques *in situ*. The number of phage plaques that can be placed on a single Petri plate is much larger than the number of individual bacterial colonies that can be placed on the plate; therefore the method is especially useful for the screening of large libraries of eukaryotic DNA for genes that may be present at a low frequency (i.e., single-copy genes).

Sometimes no probe containing specific sequences is available but the protein product is available. Sherman Weissman and his coworkers devised a procedure called *reverse translation* for construction of suitable probes. This procedure requires some knowledge of the amino acid sequence near the carboxy-terminal end of the protein. For example, a portion of the amino

Amino acid sequence of a tetrapeptide from the carboxyl end of an HLA antigen	– – –Met—Trp—Arg—Arg – – –

Probable nucleotide sequence of corresponding region in the DNA	5′ – – –AUG UGG AG(Pu) AG(Pu)

Complementary oligonucleotide primer	3′ TAC ACC TCT TCT

Figure 32–13

A procedure known as reverse translation has been used to construct probes suitable for selection of clones from a cDNA library. This procedure requires some knowledge of the amino acid sequence of the protein encoded by the desired gene. From this sequence the corresponding mRNA sequence is deduced. Then a complementary oligonucleotide sequence is synthesized and used as a primer for the synthesis of the complement to the mRNA. The sequence for the peptide is shown at the top. The most likely sequence for the RNA is shown below this (Pu refers to purine, either A or G) and the most likely sequence for a complementary DNA is shown at the bottom.

acid sequence from the carboxyl end of a particular HLA-antigen is known to contain the tetrapeptide Met-Trp-Arg-Arg. The first step in reverse translation is to guess at the sequence in the RNA, given the amino acid sequence. The codons for Met and Trp are unique, and this uniqueness gives an unambiguous sequence for six adjacent nucleotides. Furthermore, all arginines have G as the middle base in the codon; the 5′ base in the codons can be either C or A, whereas the 3′ base can be any of the four bases. The sequence AG(Pu) is more likely than CG(N) because of the known bias against the occurrence of the dinucleotide CG in most animal cell mRNAs. The possible nucleotide sequence of the corresponding region in the mRNA of the tetrapeptide is given in Figure 32–13. The deoxyoligonucleotide complement of this sequence is chemically synthesized. Since AG(Pu) has been chosen for arginine in the mRNA, T is chosen for the position corresponding to the Pu in the DNA complement, because T can form a normal base pair with A or a wobble pair with G (see Chapter 28). Next the 12mer of DNA is used as a primer, in conjunction with reverse transcriptase and a crude mRNA preparation isolated from cells known to contain the mRNA for the HLA antigen. The DNA is made with highly radioactive deoxytriphosphates so that it can be used as a probe. The theory behind this approach is that even though the RNA preparation is very crude, there will be very little reverse transcription from the RNAs except for those where primers are available. In fact, Weissman was able to isolate from the synthetic mixture a labeled 30mer that acted as an efficient probe in clone selection for the HLA antigen gene. Although this method seems very complex and indirect, it has proven to be extremely valuable in instances where the protein of a particular gene, but not the gene or mRNA sequences, is available.

CLONING IN OTHER SYSTEMS

Despite the success and broad applications of *E. coli* cloning systems, there are instances where gene products cannot be made in *E. coli*. Either they are not synthesized in their entirety, or they are rapidly broken down after synthesis. In addition, the study of certain processes indigenous to other species (e.g., photosynthesis, antibiotic production) often requires the use of a host bacterial species that naturally carries out the process; no selection system may exist for the gene product in *E. coli*.

Effective gene cloning systems have now been developed for a wide variety of bacterial hosts. The non-*E. coli* bacterial hosts used most extensively in DNA cloning experiments include *Bacillus subtilis*, *Streptomyces* species, and *Agrobacter tumefaciens*. Cloning systems have also been developed for eukaryotic hosts. In this section we will consider examples of cloning in three eukaryotic systems: the yeast *Saccharomyces cerevisiae*; mammalian cells in tissue culture; and plant cells.

Yeast Is the Most Popular Eukaryotic Cell for Cloning

Simple and generally useful techniques exist for isolating and amplifying virtually any yeast gene in *E. coli*. Moreover, by transformation these genes can be returned to the yeast cell. The availability of such procedures has greatly facilitated the molecular dissection of yeast genes; it has also provided a variety of new approaches to gene mapping. Yeast gene isolations usually begin by taking fragments of total yeast DNA and cloning them into an *E. coli* vector, either λ bacteriophage or a plasmid. After transfection or transformation, respectively, *E. coli* cells containing specific regions of the yeast DNA are selected. Cells containing the desired fragments are then amplified, purified, and sometimes modified, after which the fragments may be returned to yeast by transformation.

The first isolation of a gene encoding a protein in yeast took advantage of the fact that many yeast genes function in *E. coli*. For example, a clone containing an origin for replication in *E. coli* and the yeast *LEU2** gene was selected from a yeast plasmid library by its ability to restore to a *leu⁻* mutant of *E. coli* the ability to grow in the absence of added leucine. The purified clone containing the *LEU2* gene could then be returned by transfection to a yeast cell defective in *LEU2*. Subsequent analysis showed that stable transformants contained the *LEU2* gene, integrated at the homologous site in the yeast genome. Thus the yeast recombination system, like the *E. coli* recombination system, appears to have the capacity to guide a gene introduced by transfection to a site of homology. Indeed *yeast is one of the few eukaryotes in which it has been possible to demonstrate homologous integration of introduced DNA*.

The transformation frequency in yeast, which is usually quite low, may be greatly increased by incorporating special yeast DNA sequences into the plasmid used for transfection. The best-known of these special sequences derives from the so-called *2μ plasmid* found in most *Saccharomyces* strains. Addition of a fragment of the 2μ plasmid into a plasmid containing a selectable gene, such as *LEU2*, increases the transformation frequency a thousandfold or more. High-frequency transfection using all or part of the 2μ plasmid is usually due to autonomous replication of the plasmid, which is maintained in multiple copies. Other types of high-frequency transformation have been observed when certain yeast sequences, referred to as autonomously replicating sequences, are an integral part of the transforming plasmid. These probably contain a yeast chromosomal origin for DNA synthesis.

Integrative transformation with specially constructed plasmids has opened the door to many forms of gene mapping that would be much more difficult to do by conventional mapping techniques. In addition, it has permitted the construction of unusual gene arrangements. For example, if a yeast-integrating plasmid contains a gene to be transposed and a segment of yeast DNA homologous to another location on the yeast gene, then it should be possible to transpose the selectable gene to another position on the genome.

---

*In yeast, wild-type genes are usually indicated by three capital letters, mutants by three small letters. + and − have their usual meanings.

Cloning in Mammalian Cells Employs Tissue Culture Cells

Mammalian cells from various sources can be adapted for growth as single cells in liquid culture or on plates. Such cells can be formally treated like bacteria or yeast; single cells give rise to genetically homogeneous colonies. Cultured cells have proved to be effective recipients of cloned DNAs.

Introducing DNA into Mammalian Cells.   Several methods are available for introducing DNA into mammalian cells grown in tissue culture. The method depends principally on the type of cell involved. Two of the most widely used methods will be mentioned.

Method Using Aggregated DNA.   The most popular procedure for transfection of mammalian cells utilizes purified DNA. The DNA is mixed with calcium chloride and sodium phosphate to give a finely divided calcium phosphate-DNA precipitate. Treated cells appear to take up the DNA by endocytosis. Transfection efficiencies, measured as the fraction of treated cells that become transfected, vary from 10^{-4} to several per cent by this method.

Microinjection.   DNA can be injected directly into the nuclei of many cell types by means of a fine needle. Use of this technique is confined to experiments in which we want to be certain that all of the cells in the recipient population have taken up the DNA to be cloned. A method of transferring cloned DNA to whole mice, initiated by Rudy Jaenisch, is to inject the DNA directly into the embryo. Mice to which cloned DNA has been transferred are called transgenic mice.

Electroporation.   Exposure of animal and plant cells to high voltage renders the cell membrane permeable to the uptake of DNA. Such cells have been shown to express genes carried by the introduced DNA molecules.

Lipofection.   DNA encapsulated into lipid vesicles fuses with cell membranes.

Selection of Genetic Markers Introduced into Mammalian Cells.   Cultured mammalian cells permit many types of studies that would be impractical to carry out on whole animals. There are only a few mutants that provide readily selectable genetic markers; this creates a problem in selecting for transformed colonies. One of the best systems available for selection involves the thymidine kinase (tk) gene of herpes simplex virus. Mutants that are tk^- have been isolated for a few cell types, including mouse L cells and rat fibroblast cells. Such cells cannot grow on the selective medium known as HAT, which contains hypoxanthine (H), aminopterin (A), and thymidine (T). The tk^- cells can be transformed to tk^+ cells by plasmid DNA that contains the herpes simplex virus tk gene. Other genes of interest can be inserted into the same plasmid vector so that selection of tk^+ cells after transformation leads to a high probability of coselection of the adjacent genes. Moreover, since certain cells in the population appear to be particularly prone to the uptake of DNA, the cells identified as having taken up a tk gene are likely to also acquire the nonselected gene, even if the two genes are not covalently linked.

The effectiveness of this strategy has been demonstrated in the cloning of a number of different genes. However, a significant disadvantage of this selection procedure is that tk^- mutants have been isolated for only a few cell types.

The search for dominant selectable markers that can be used with any cell type has met with limited success. The E. coli xgprt gene encodes the enzyme xanthine-guanine phosphoribosyl transferase (XGPRT), which is the bacterial analog of the mammalian enzyme hypoxanthine-guanine phosphoribosyl transferase (HGPRT). The bacterial enzyme will convert xanthine

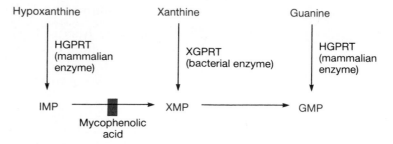

Figure 32–14
Purine metabolism relevant to the XGPRT positive selection system in mammalian cells. In the *de novo* synthesis of GMP the penultimate step involves the conversion of IMP to XMP. This step is inhibited by mycophenolic acid. The growth inhibition by mycophenolic acid can be overcome by adding xanthine to the growth medium, provided the cells have been transformed to contain the XGPRT bacterial enzyme that converts xanthine to XMP. Other genes of interest can be linked to the XGPRT gene for cotransformation.

into XMP, which is a precursor to GMP, whereas the mammalian enzyme will utilize only hypoxanthine or guanine (Figure 32–14). Berg and his co-workers have shown that the *E. coli* gene *xgprt* expresses functional enzyme in mammalian cells. This finding has permitted the development of a positive selection method that involves the *xgprt* gene and the specific inhibitor mycophenolic acid. Mycophenolic acid blocks the conversion of IMP into XMP, which is required for the *de novo* synthesis of GMP. Normal mammalian cells can grow in the presence of mycophenolic acid only if the medium is supplemented with guanine. This growth ability is due to the fact that the phosphoribosyl transferase enzyme (HGPRT) can convert guanine to GMP. Xanthine cannot be substituted for guanine because the mammalian HGPRT enzyme cannot convert xanthine to XMP and thus GMP cannot be produced. If, however, the mammalian cells contain the *E. coli* xgprt enzyme, growth of the cells on xanthine is possible. The *xgprt* marker has been used as a positive selection tool to incorporate other markers in much the same way as the *tk* gene. It has the distinct advantage over the *tk* marker that it does not require any mutation of the recipient cell; in the presence of mycophenolic acid, even wild-type cells serve as selectable recipients.

With the selection procedures described thus far, a gene other than the selectable marker has been the gene of interest. The purpose of the selection procedure was to identify those cells that have taken up DNA. It has also been possible to directly identify a particular gene when the gene can be recognized by the phenotype that it confers on a mammalian cell. However, even in such cases indirect selection procedures often must be used because of the limitations in the cloning vehicles currently available for mammalian cells. A useful general strategy has been to make a genomic library of mammalian DNA in *E. coli* and then to test clones from the library in subgroups in mammalian cells until a clone that expresses the desired properties is identified. The DNA of interest is then propagated in *E. coli*.

This selection method is illustrated in Figure 32–15 for the isolation of an _oncogene_* from chicken lymphoma DNA. A genomic library was prepared from DNA of a chicken lymphoma; it contained 200,000 phage clones. This library was divided into *sublibraries*, and the DNA of the clones in each sublibrary was introduced by transfection into mouse fibroblast cells growing in tissue culture. Cellular transformation was assayed by the appearance of a clone or clones or rapidly dividing cells (*foci*) superimposed on a background of slowly dividing, nontransformed cells. The sublibrary that as-

---

*An oncogene is a gene of cellular or viral origin that causes rapid, unruly growth, usually associated with cancer cells.

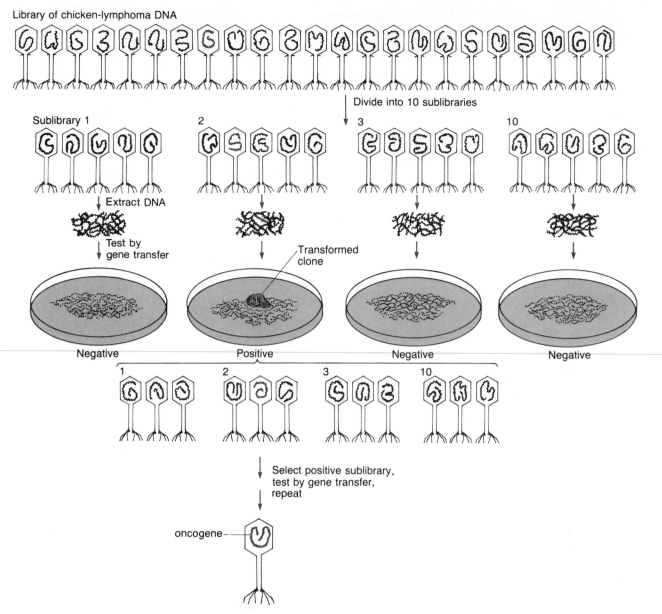

Figure 32–15

Subculture cloning procedure devised by Jeffrey Cooper and his coworkers. Subculture cloning is used in successive steps to isolate the clone that contains the gene of interest. In this case it was an oncogene originating from a chicken lymphoma, identifiable by the rapid growth characteristics it confers on transformed cells. A λ library was made with the chicken lymphoma nuclear DNA. The library was subdivided into ten approximately equal lots. From each lot the DNA was isolated and used in a transformation assay. Lot 2 scored positively. This lot was further subdivided and the procedure was repeated again and again until a pure plaque carrying the oncogene was isolated.

sayed positively was further subdivided and retested. This procedure was repeated until a clone that conferred the rapid growth properties associated with oncogene transformation was isolated.

The procedure described here, called _subculture cloning_, is dependent on the sensitivity of the assay used for detecting a small number of positive clones in a heterogeneous population. More rapid procedures have been devised, but they still employ the basic strategy of carrying out the cloning of such genes in bacteria and then detecting them in mammalian cells. However, it is possible that the availability of papilloma viruses (described

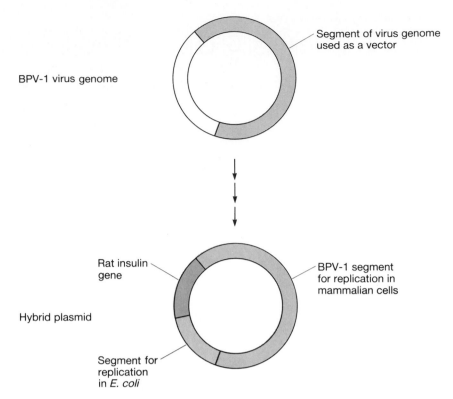

Figure 32–16

The bovine papilloma virus (BPV-1) genome and associated shuttle vector. The region of the genome used as a vector in transformation is indicated by color shading. Hybrid plasmid constructed from BPV-1 contains the rat insulin gene and the origin for replication in *E. coli*.

below), which can function as true plasmid vectors in mammalian cells, will make possible the cloning and selection of mammalian genes in one step.

Vectors Used for Cloning in Mammalian Cells. Because virtually any DNA introduced into mammalian cells by transfection integrates itself into the host chromosome, it is a challenging problem to find a suitable vector for propagation of cloned genes in mammalian cells. Considerable excitement has been generated by the finding that the genomes of the papilloma viruses replicate extrachromosomally as circular plasmids in virus-transformed mouse cells. It is unclear why these viruses are exceptional in this respect (all other known transforming viruses integrate into the host genome) but this phenomenon makes papilloma viruses prime candidates as vectors for introducing foreign DNA fragments into cells that are susceptible to papilloma virus-mediated transformation. The genome of bovine papilloma virus (BPV), which is the most extensively studied of these viruses, contains 7945 base pairs, and the complete sequence is known. Only 69 per cent of the genome is required for transformation and autonomous extrachromosomal replication (Figure 32–16), leaving considerable space for the insertion of foreign DNA for various purposes. A composite DNA molecule, containing the rat gene for insulin, a vector permitting replication in *E. coli*, and the 69 per cent fragment of BPV that contains the transformation and replication functions, was constructed by Hawley. This shuttle vector carrying the gene for insulin could be introduced into *E. coli*, using ampicillin as a selectable marker, or into mammalian cells, using transformation to rapid growth as the selectable phenotype. This prototypical experiment suggests that BPV and related papilloma viruses may be valuable vectors for the _direct_ cloning of mammalian DNA in mammalian cells.

Cloning in Plants Has Been Accomplished with a Bacterial Plasmid

Tissue culture systems for plants are difficult to manipulate and grow for extended periods of time. The difficulties of cloning DNA in plants are compounded by the fact that there is no plasmid that can be introduced directly into plant cells. A good deal of current work in plant DNA cloning is focused on a naturally occurring host-vector system involving _Agrobacter tumefaciens_, a pathogenic soil bacterium that causes so-called crown gall tumors (Chapter 31) in a wide variety of dicotyledonous plants.* Such tumor production has been shown to be due to the presence of tumor-inducing (T1) plasmids in some agrobacteria; related bacteria that lack the plasmid are not tumorigenic. The T1 plasmids are very large circular DNA duplexes ranging in size from 160 to 240 kb. Bacterial infection of the plant leads to development of a tumor, usually near the junction of root and stem, as a consequence of a plasmid segment called T-DNA. T-DNA is integrated into the nuclear genome of the plant cells, where it alters their metabolism, causing them to become tumorous. It also alters the metabolism of the tumor cells so that small molecules known as opines are synthesized in large quantities. Different T1 plasmids encode genes that lead to the synthesis of different types of opines. While these compounds are of no apparent use to the plants' cells, they provide a growth substance for the agrobacteria that contain the T1 plasmid. Other genes carried by the T1 plasmid confer upon the host bacteria the ability to metabolize the T-DNA-encoded opine and use it as a substrate to promote bacterial growth.

It is impractical to use the entire T1 plasmid as a cloning vector; because of its size, T1 DNA has multiple cleavage sites for restriction enzymes, and it is too large to be introduced into cells efficiently by transfection. However, foreign genes have been transferred into plants by introducing them into the T-DNA segment of T1, reinserting the T-DNA into T1, and then letting the natural biological process of T-DNA transfer to the plant cells to do the rest (see Box 32–B).

USES OF CLONED DNA

Even prior to the discovery of DNA cloning techniques, it was possible to "capture" discrete regions of bacterial chromosomes onto the genomes of certain prokaryotic plasmids and transducing bacteriophages. However, there was no generally applicable way of obtaining defined segments of DNA from any species. Even in instances where gene capture by recombination _in vivo_ can be accomplished, there are many things that can be done more effectively with the help of cloning techniques. These include gene mapping, controlled mutagenesis, and increased production of specific bacterial gene products. However, the major revolution that has resulted from the development of DNA cloning methods has occurred in eukaryotic biology. Because such methods allow defined segments of any DNA to be isolated and characterized, molecular approaches that have dominated research on microorganisms for almost two generations are being extended to even the most complex eukaryotes. The following are some applications that have been either made possible by, or greatly facilitated by, DNA cloning.

---

*It is noteworthy that this is the only known case of naturally occurring nucleic acid transfer between a prokaryote and a eukaryote.

BOX 32–B

Construction of a T1 Plasmid Derivative Suitable for Cloning

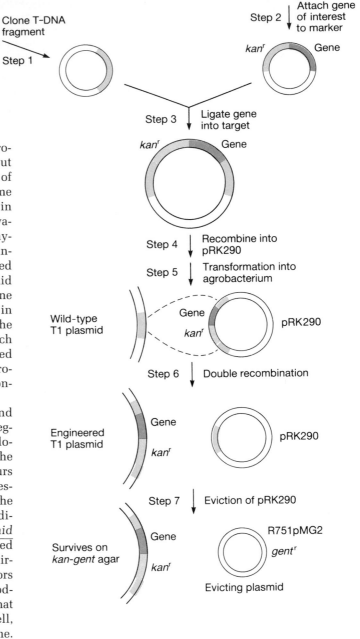

The following procedure is quite complex, but it introduces some new and very worthwhile principles about cloning for the sophisticated reader. First, a segment of T-DNA that included a preselected restriction enzyme cleavage site was cloned into pBR322 for replication in *E. coli* (step 1). The plant gene to be cloned was covalently linked to a selectable genetic marker, the kanamycin resistance (*kan*[r]) gene (step 2). The segment containing both the *kan*[r] gene and the plant gene was spliced into a T-DNA segment cloned in the pBR322 plasmid (step 3); this was used for amplification of the plant gene in *E. coli*. However, since pBR322 does not replicate in agrobacteria, the segment containing the T-DNA and the *kan*[r] gene was inserted into the plasmid pRK290, which replicates in either bacterial host (step 4). The modified T-DNA fragment, now attached to pRK290, was introduced by transformation into an agrobacterium that contained a T1 plasmid (step 5).

Transformed cells retained both the T1 plasmid and several copies of the pRK290 plasmid. The T-DNA segment in pRK290 recombines *in vivo* with the homologous T-DNA region in the T1 plasmid, thus inserting the modified T-DNA segment (step 6). This event occurs spontaneously at a very low frequency, so it was necessary to devise a way of screening for those cells in the population that contained the T1 plasmids with a modified T-DNA segment. A strategy referred to as *plasmid eviction* was used to accomplish this. It was reasoned that if pRK290 could be evicted under conditions requiring kanamycin resistance, the most likely survivors should be T1 plasmid recombinants containing the modified T-DNA segment with the *kan*[r] gene. It is known that when two closely related plasmids cohabit the same cell, *incompatibility* between them results in the loss of one. R751pMG2, a plasmid closely related to pRK290 but carrying a gentamycin resistance gene (*gent*[r]) and lacking a T-DNA segment, was introduced by conjugation with *E. coli* containing pRK290 and the T1 plasmid. To direct the process so that pRK290 was evicted instead of R751pMG2, the conjugation was done in the presence of both gentamycin and kanamycin (step 7). Only cells that carried both of the drug resistance genes could survive;

these cells had lost the pRK290 plasmid but retained the *kan*[r] gene in T1 plasmids that had acquired the T-DNA/ *kan*[r] gene segment by recombination. The R751pMG2 plasmid was also present. The T1 plasmid in the agrobacterium could then be used for transfer of the desired gene to a plant cell system.

DNA Sequence Analysis. DNA sequence analysis requires homogeneous duplex segments. Two methods for sequencing DNA are described in Chapters 7 and 26. Both of these methods depend on the availability of purified DNA segments that are produced by cloning.

Site-directed Mutagenesis. Small regions of cloned DNA can be specifically modified and reintroduced into the organism. This process is known as site-directed mutagenesis. After modification, the altered DNA can be amplified in bacteria and the effect of the alteration can be studied either *in vitro* or *in vivo*. In some organisms, most notably *E. coli* and yeast, it has been possible to replace the normal gene with the altered gene by homologous recombination following introduction of the modified DNA. In more complex mammalian organisms, the altered gene can be integrated into the host genome at a nonhomologous location.

Gene Mapping. Gene-mapping studies using cloned DNA have been mentioned for *Saccharomyces cerevisiae*. Even in hosts where homologous recombination between plasmid and host DNA does not occur, valuable advances in gene mapping have resulted from the use of cloned DNA. One important approach involves the development of the technique known as *chromosome walking*. This is a process of isolating a set of cloned DNAs so that a contiguous region of the genome is represented in the population of clones. Chromosome walking is very useful for investigating the organization of genes in complex eukaryotes, for which detailed genetic maps are almost impossible to obtain by other means. The technique involves obtaining a series of clones that have overlapping segments of a particular region of the genome. The procedure has been successfully applied to selected regions of the *Drosophila* and mammalian genome, leading to maps extending for distances of 200 kb or more.

Studies of Gene Transcription *In Vivo* and *In Vitro*. Genetically engineered DNA inserted into small plasmids provides an ideal concentrated source of a gene, as noted above. Such DNA can be used directly as a template to study transcription *in vitro* or can be expressed *in vivo* after introduction into a suitable host cell. Alternatively, the DNA can be used as a hybridization probe to assess transcription from the genome, either *in vivo* or *in vitro*.

Expression Vectors for Making Large Amounts of Proteins. Certain proteins of medical and biological importance are desired in much larger amounts than can be obtained from their natural sources. In such cases, the genes for the proteins can be implanted in plasmid or bacteriophage vectors and preceded by efficient promoters and ribosome recognition sites. Commonly, such synthesis is done in *E. coli* using an appropriate prokaryotic plasmid vector. As much as 20 per cent of the protein mass of *E. coli* can be expressed as a single protein in certain instances.

Possible Medical Applications. The knowledge gained through DNA cloning is already providing important information for use in medical problems involving gene abnormalities. Significant medical applications can be anticipated in at least three areas. The first is in the manufacture of certain proteins, such as human insulin, interferon, and growth hormone, for medical use. As was previously indicated, such mammalian hormones can be produced in large quantities by cloning their genetic sequences into prokaryotic plasmids adjacent to strong promoters and ribosome recognition sites, so that they will express at a high level in the bacterial cells. A second important area is in diagnosis. Many genetic diseases that yield developmental abnormalities can be detected by characteristic patterns in the DNA primary structure. Such mutational changes in DNA sequence are identified by restriction fragment analysis and Southern blotting, using appropriate DNA probes. Analyses of this type could be performed either pre- or postnatally. Similarly, cloned genes are being used as hybridization probes for the clinical diagnosis of infectious microorganisms that contain homologs of such genes. Finally, potential medical applications are envisioned in the direct treatment of diseases by the introduction of functional genes into individuals that carry gene mutations interfering with normal gene expression.

SUMMARY

Recombinant DNA methodology involves all of those procedures that are used to clone specific fragments of DNA. Four things are required for cloning: (1) a way of joining two DNAs together; (2) a self-replicating segment of DNA (a cloning vehicle or vector) that can propagate in the host organism, and that can be linked to the DNA segment to be cloned; (3) a procedure for introducing the composite DNA molecule into the host organism, and (4) a way of selecting organisms that have acquired the desired composite molecule.

A number of enzymes are used in DNA recombinant methodology. Type II restriction endonucleases make cleavages at specific sites. These specific sites usually consist of a sequence of four or six bases that read the same on both strands. The specific DNA fragments arising from restriction enzyme cutting can be integrated into a cloning vehicle. This procedure requires at least one additional enzyme for ligating the two DNAs together. Cloning vehicles of various types exist. Two of the most popular are small plasmids that carry drug resistance markers, and specially rebuilt λ DNAs that have nonessential DNA removed so they can accommodate foreign DNA.

Most cloning has been done using E. coli as the host cell. When a plasmid cloning vector is used, the cells are transformed with DNA from a hybrid plasmid that contains the desired foreign DNA integrated into it. Successfully transfected cells are usually selected with the drug corresponding to the drug resistance marker carried by the cloning vector.

When transformation starts with a mixture of hybrid plasmids containing different DNA fragments, an additional procedure must be used to identify the transformed cells carrying the desired DNA. One of the most popular procedures utilizes a radioactive DNA probe containing sequences homologous to the one desired. A clone of transformed E. coli cells carrying these sequences is identified by replica plating and autoradiography. When λ phage is used as a cloning vector, specific clones are similarly identified, except that in this case the clone consists of a phage plaque.

Among the nonbacterial systems used for cloning, yeast has been used in the greatest number of applications. A large number of genes have been exchanged between E. coli and yeast with the help of shuttle vectors that carry origins for replication in both organisms.

Although it has been possible to introduce various genes into mammalian cells, actual cloning in such cells has been difficult because the would-be cloning vehicles usually integrate into the cellular genome and cease to function as independent plasmids. Recently it has been discovered that bovine papilloma virus may be an exception to this rule; this finding has raised a great deal of excitement about the possibility of direct cloning in mammalian cells.

The bacterium known as Agrobacter tumefaciens carries a plasmid T1 which can naturally transfect dicotyledonous plant cells in plants carrying Agrobacter infections. These plasmids cause crown gall tumors in the plants. The T1 plasmid provides a cloning vehicle that has been used to introduce various genes into plant cells.

SELECTED READINGS

Bevan, M. W., and M. D. Chilton, T-DNA of the agrobacterium T1 and R1 plasmids. *Ann. Rev. Genetics* 16:357–384, 1982.

Cohen, S. N., A. Chang, H. Boyer, and R. Helling, Construction of biologically functional bacterial plasmids *in vitro. Proc. Natl. Acad. Sci.* 70:3240–3244, 1973.

Jackson, D. A., R. H. Symons, and P. Berg, Biochemical method for inserting new genetic information into DNA of Simian Virus 40: Circular SV40 DNA molecules containing lambda phage genes and the galactose operon of *Escherichia coli. Proc. Natl. Acad. Sci.* 69:2904–2909, 1972.

Watson, J. D., J. Tooze, and D. T. Kurtz, *Recombinant DNA: A Short Course.* New York: W.H. Freeman and Co., 1983. Excellent description of experimental procedures and good lists of classified references.

PROBLEMS

1. You have just isolated a novel recombinant clone and purified the desired insert from the vector as a 10.0-kilobase-pair (kbp) linear duplex DNA. Now you wish to map the recognition sequences for restriction endonucleases A and B. You cleave the DNA with these enzymes and fractionate the digestion products according to size by gel electrophoresis. You observe the following:
   a. Digestion with A alone gives two fragments, of lengths 3.0 and 7.0 kbp.
   b. Digestion with B alone gives three fragments, of lengths 0.5, 1.0, and 8.5 kbp.
   c. Digestion with A plus B gives four fragments, of lengths 0.5, 1.0, 2.0, and 6.5 kbp.
   Draw a restriction map for the insert, showing the relative positions of the cleavage sites with respect to one another.

2. Describe the procedure you would use to clone a DNA insert into the *Bam*Hl site of pBR322.

3. A small circular duplex viral DNA with about 5000 base pair (bp) contains a 72-bp direct repeat in tandem. This 72-bp segment contains a cleavage site for the *Eco*R1 restriction enzyme that is not present in the rest of the virus. In order to explore the function of the 72-bp repeat, it would be desirable to remove one of the 72-bp segments. Describe how you would construct such a modified viral chromosome.

4. Mapping restriction sites for a given enzyme in a DNA molecule (as in Problem 1) is simple enough if such sites are relatively few, but becomes increasingly difficult as the number of sites increases. Eventually it becomes necessary to map all sites from a fixed point, rather than mapping them with respect to one another. Suggest how this might be done.

5. Describe a protocol for constructing and isolating a hybrid plasmid containing the gene for *E. coli* DNA polymerase I.

6. How large a genomic library should you construct in order to detect and isolate a 15-kbp gene out of a genome containing 3×10^9 bp?

7. Describe the steps you would use to isolate from a mouse genomic library a DNA clone containing an intact *HGPRT* gene. Note that *HGPRT*⁻ mouse cell lines are available, and can be used in your protocol.

8. Table 32–3 indicates the theoretical number of clones that would be required to represent the entire genome of various organisms. As can be seen, 75,000 clones would be required to represent a genome containing 3×10^9 bp using cloned fragments whose size is approximately 40,000 bp. The probability that one such cloned fragment will contain an intact, full-length gene is so small that many more clones must in fact be screened in order to isolate the desired fragment, especially if the gene of interest is large.

The restriction enzyme *Bgl*I cuts within the following nonspecific sequence that interrupts the recognition site:

$$5'\text{-GCCNNNN}^{\downarrow}\text{NGGC-3'}$$
$$3'\text{-CGGN}_{\uparrow}\text{NNNNCCG-5'}$$

(The arrows represent the cleavage sites, and N represents any base.) The enzyme thus leaves a 3-bp 3′ overhang that is derived from the nonspecific region; that is, the "sticky ends" generated by *Bgl*I digestion are characteristic of the locus being cut, rather than of the enzyme.

It is sometimes desirable to routinely clone a given gene from populations of cells (for instance, in order to analyze the sequence of mutants selected for *in vivo*). But, as we have suggested, the labor involved in the screening of very large genomic libraries is formidable. Under such circumstances, the ideal answer would be a procedure allowing for the screening of much smaller libraries that are somehow enriched for the gene of interest.

The gene encoding the enzyme dihydrofolate reductase (DHFR) in Chinese hamster ovary cells (CHO) is spread over 26 kbp of DNA, and is contained on a 41-kbp *Bgl*I genomic fragment. Suggest a procedure for cloning the intact gene for DHFR into a specialized cosmid vector that takes advantage of the unusual characteristics of the *Bgl*I restriction enzyme.

9. Explain how reverse translation might be used to detect specific DNA clones in a genomic library. From what you have learned about the genetic code in previous chapters, can you suggest what the potential problems inherent to this method are? How are these problems minimized?

10. It is sometimes difficult to obtain full-length cDNA clones (e.g., extending to the 5′ end of a gene) from a cDNA library made with reverse transcriptase. Why do you think this is?

11. Why isn't a primer required to promote second-strand synthesis using the single-stranded DNA produced by reverse transcriptase from an RNA template?

12. Describe a protocol for isolating a cDNA library representing only liver-specific functions, and excluding functions expressed by other tissue types.

13. The analysis of DNA regulatory sequences can be undertaken via the technique of DNA-mediated gene transfer, using a gene that provides a functional assay for the effect of such putative regulatory sequences on gene expression. Can you suggest how this might be done?

14. What are the advantages of using the BPV vector to introduce DNA sequences into mammalian cells?

33

ORIGINS OF LIFE

Biochemistry is largely a study of complex organic molecules that are self-organizing and self-replicating. These molecules are synthesized and interconverted by enzyme-catalyzed processes in which the enzyme interacts at specific locations of these structures. The ultimate driving force for these reactions is the solar energy absorbed by specific chromophores in photosynthetic organisms. What we have not emphasized thus far is how complex biomolecules initially formed and how life processes spontaneously resulted from their interactions.

Life on Earth today is the product of nearly four billion years of evolution and, consequently, may have only a superficial resemblance to the earliest living systems. The original forms of life no longer are present because they were unable to compete with the higher organisms that evolved from them. In addition, it appears unlikely that life as we know it will be found elsewhere in our solar system. No life was discovered in the vicinity of the two Viking landers on Mars, and there is only a remote possibility that life will be found on the satellites of Jupiter and Saturn. Mercury and Venus are too hot for life to exist, while Uranus, Neptune, and Pluto are too cold. There are levels of the atmospheres of Jupiter and Saturn warm enough for life, but the turbulent winds and convection currents would most probably carry any living systems to upper or lower levels, where they would be destroyed. It appears likely that some of the chemical reactions that took place prior to the origins of life have taken place on Titan, the largest moon of Saturn. Compounds generally believed to be essential for the origins of life, including HCN, HC≡CCN, $(CN)_2$, HC≡CH, and HC≡CC≡CH, were detected in its atmosphere by the Voyager I probe in 1980. But the low surface temperature of Titan (95 K) probably precludes the origin of life on this moon as well.

Even though we are unable to test theories of the origin of life by studying life on other planets in our solar system, those theories represent an increasingly active research area. *The goal of many experimental studies is to establish the reaction conditions on the primitive Earth and to determine whether biological molecules will form under these conditions.* Since there is little hope of exactly duplicating the process of the origin of life on Earth, the objective of these experiments is simply to determine whether there exist reasonable pathways by which life processes may have originated. If convincing processes can be demonstrated for the origin of life on Earth, then it would be reasonable to conclude that life is a natural outcome of the chemistry of the universe. Such a finding would suggest (1) that there must be abundant life in the universe and (2) that humans may not be among the more highly evolved forms of life in the universe.

In this chapter we describe the reasoning and experiments that relate to the study of the origins of life. Of course, in this limited space we cannot cover all aspects of prebiotic chemistry. Therefore, we have chosen to focus our attention on polynucleotides and polypeptides and on their building blocks, the nucleotides and amino acids. The books and reviews listed at the end of the chapter contain discussions of other important topics, such as prebiotic paths to lipids and some coenzymes, studies on the origins of chirality, and the possible role of metal ions and hydrothermal vents in prebiotic synthesis.

WHAT IS LIFE?

The object of any search must be defined before the search can begin. The question "What is life?" is not a trivial one, because if the first living forms evolved spontaneously from their environment, they must have differed very little from it. Consequently, sophisticated tests would be needed to distinguish the living from the nonliving. One definition of life is that it is a process capable of *self-duplication* and *mutation.* Self-duplication provides for the formation of new generations of the new life form, while mutation permits it to evolve, under the pressure of natural selection, to forms that can compete more readily for energy and nutrients. This definition of life was derived directly from observation of the life processes on Earth.

Other general definitions of life have been proposed, some of them encompassing systems organized around minerals or microcomputers. We recognize that a definition such as ours, based on life as it is observed on Earth, may be limited in scope; however, it is better suited to the discussion in this chapter.

UNDERLYING ASSUMPTIONS ABOUT THE ORIGINS OF LIFE

Since little is known about the chemistry of the Earth at the time when life originated, and since there is essentially no knowledge of the biochemistry of the first living systems, a model must be devised before experiments can be performed. A few of the assumptions that provide the basis for this model include the following (Figure 33–1).

1. Life originated on Earth.
2. The biochemistry of the initial living system was based on some of the biochemical subunits that are of importance to contemporary biochemistry. It is assumed that amino acids and nucleotides had a central role in the initial life forms, but it is recognized that the biochemistry was much simpler and also less efficient than that of contemporary life.
3. Life originated as a consequence of the spontaneous formation of complex molecules on the primitive Earth. These complex structures were formed from simple molecules present in the Earth's atmosphere or were transported to the Earth by comets and meteorites. The chemical processes leading to life's origins took place in the presence of water and minerals,

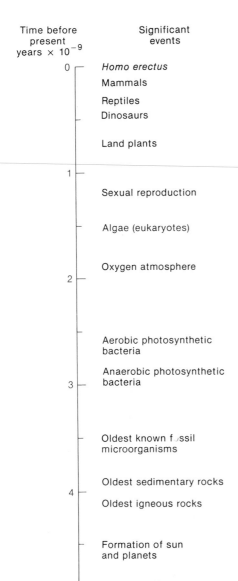

Time before present years × 10^{-9}	Significant events
0	*Homo erectus*
	Mammals
	Reptiles
	Dinosaurs
	Land plants
1	
	Sexual reproduction
	Algae (eukaryotes)
	Oxygen atmosphere
2	
	Aerobic photosynthetic bacteria
	Anaerobic photosynthetic bacteria
3	
	Oldest known fossil microorganisms
	Oldest sedimentary rocks
4	Oldest igneous rocks
	Formation of sun and planets
5	

Figure 33–1

A time line showing major chemical and biological events that occurred on the Earth since its formation 4.6×10^9 years ago. The origin of life occurred between 4.5 and 3.5×10^9 years ago, possibly about $4.0 \pm 0.2 \times 10^9$ years ago.

TABLE 33–1
Energy Sources on the Primitive Earth[a]

Source	Energy (cal/cm²/yr)
Total solar radiation	260,000
<300 nm	3,400
<250 nm	563
<200 nm	41
<150 nm	2
Electric discharges	4
Shock waves	1
Radioactivity	1
Volcanoes	0.1

Source: Adapted from S. L. Miller and L. E. Orgel, *The Origins of Life on the Earth*, Prentice-Hall, Englewood Cliffs, N.J., 1974, p. 55. Used with permission.

[a]The solar luminosity in the ultraviolet region (<300 nm) was much greater when the sun first ignited than it is today. For example, the output at wavelengths below 200 nm was 5000 cal/cm²/yr, or 100 times greater than now.

where the pH of the aqueous phase was approximately 7. The pH of the ocean today is controlled at 8 by the buffering action of clay minerals. It appears likely that the same buffers maintained the nearly neutral pH of the oceans four billion years ago, when life originated.

4. Chemical evolution was driven by solar energy in the form of direct irradiation by ultraviolet and visible light or indirectly in the form of lightning (electric discharges and shock waves). Some thermal energy was supplied by volcanic action and the heat resulting from radioactive decay. An estimate of all the available energy sources on the primitive Earth is given in Table 33–1.

POSSIBLE STAGES IN CHEMICAL EVOLUTION

The synthesis of the biological molecules essential for the origin of life proceeded in the 0.5 billion year period after the formation of the Earth. Since it is difficult to model this process in a laboratory experiment, various stages in the process are investigated separately. The stages may be described as follows.

1. The conversion of simple inorganic and organic compounds into the building blocks for biopolymers. *Electric discharges, ultraviolet light*, and other energy sources are used to initiate the formation of amino acids, purine and pyrimidine bases, and sugars or their precursors.
2. The formation of ordered biopolymers from the monomer units produced in stage 1. The search for *templates* to direct the formation of the ordered polymers in an important aspect of this research.
3. The self-replication of the biopolymers formed in stage 2 in a process that may have taken place in the first living system. At the present time, the only experiments in which stage 3 has been investigated have made use of an intact biological system (e.g., a virus) that is separated into its molecular components. The regeneration of the intact biological system is then investigated using only those biopolymers that direct the synthesis of other biopolymers. For example, a viable RNA virus has been regenerated from its RNA, the RNA polymerase, nucleotides, and helper enzymes of the host bacterium of the virus.

It should be emphasized that the experimental studies in this area are fragmentary, and often there is no well-defined connection between one stage and the next. For example, we will discuss the formation of nucleotides from nucleosides, even though there has been limited success in devising a prebiotic synthesis of nucleosides. Consequently, it will not be possible for us to provide a neat, well-defined picture of the origins of life in this chapter.

FORMATION OF OUR GALAXY AND SOLAR SYSTEM

Insight into the origin of life on Earth resulted from studies of the constituents of the galaxy to which our Earth belongs. In addition to billions of stars, the galaxy is also made up of *planets, planetesimals* (solid objects much smaller than planets), *dust, simple molecules*, and *atoms*. These constituents are arrayed in the shape of a disk, with the bulk of the mass present in a central bulge in the form of massive stars and interstellar dust. Tenuous spiral arms extend out from the disk, and our solar system is located in the periphery of one of the arms.

Our sun and solar system formed in the outer regions of the galaxy as the result of the collapse of a dust cloud. This collapse, which may have been triggered by the explosion of a nearby star to a supernova, led to the formation of a spinning disc of dust and molecules from which the sun and planets eventually condensed. The inner planets, i.e., those near the sun, are more dense because only the heavier elements condense at the higher temperatures in the vicinity of the sun. The much lower temperatures ($< -150°C$) at the orbit of Jupiter favor the condensation of the light gases, including hydrogen, helium, methane, and ammonia.

The Earth continued to accrete significant amounts of dust and planetesimals for about 0.1 billion years after it initially condensed. The impact craters on the moon are direct evidence for a continuing bombardment of the Earth and moon with this material (Figure 33–2). *There was a preferential accretion of the heavier elements in the early stages of the formation of the Earth. Consequently, the lighter elements, including the main ones essential for the origins of life (C, H, N, O) were swept up by the orbiting Earth during the later stages of the accretion process. The partial segregation of these essential ingredients in the crust of the Earth set the stage for the origins of life.*

COSMOCHEMISTRY

The chemistry of our solar nebula can be deduced from a spectroscopic observation of the molecular species present in interstellar gases in our galaxy and in the other galaxies in the universe. These include an abundance of carbon compounds, such as CO, CH_4, CH_2O, HCN, and $HC≡CCN$, together with H_2O and NH_3 (Table 33–2). Many of these same compounds are considered to have initiated the abiotic chemical processes leading to the origins of life. Most of these carbon and nitrogen compounds were probably not transferred intact to the atmosphere of the primitive Earth, but their presence in the solar nebula provides strong support for the hypothesis that compounds of this type can be readily formed and they were likely starting materials for the formation of biological molecules on the primitive Earth. *The widespread occurrence of these organic compounds in our galaxy and other galaxies suggests strongly that carbon chemistry is not only the chemistry of the cosmos, but probably the basis for the origin of life elsewhere in the universe, as well as on the Earth.*

Meteorites, relics of the condensation of the solar dust cloud to the solar system, also provide clues concerning the chemistry of the primitive Earth. There was a large influx of meteorites during the first 0.5 billion years after

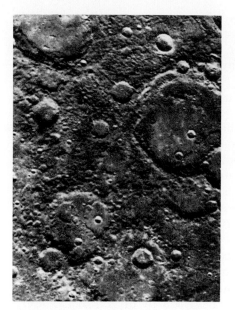

Figure 33–2
The existence of many impact craters on the moon, and on other planets and moons in our solar system, supports the theory that the Earth has been subject to extensive bombardment by meteorites.

TABLE 33–2	
Some Interstellar Molecules	
HCN	CH_3OH
H_2O	$HCONH_2$
H_2S	CH_3CN
SO_2	CH_3CHO
$H_2C=O$	$CH_2=CHCN$
$HN=C=O$	CH_3NH_2
$H_2C=S$	$CH_3C≡CH$
NH_3	$HC≡CC≡CCN$
$CH_2=NH$	HCO_2CH_3
$HC≡CCN$	CH_3CH_2OH
HCO_2H	CH_3OCH_3
NH_2CN	CH_3CH_2CN
CH_4	$HC≡CC≡CC≡CCN$
	$HC≡CC≡CC≡CC≡CCN$

the Earth formed. These 4.6-billion-year-old meteorites are still striking the Earth today, and _those few which contain carbon possess organic molecules that have some of the same monomeric units present in biological molecules._ Some meteorites also contain clays, which are formed only in the presence of water; this fact suggests that the specimens may be fragments of much larger bodies, such as asteroids, that had sufficient mass to attract and maintain water on their surfaces.

The carbon compounds present in the carbonaceous meteorites include alcohols, aldehydes, amines, amino acids, hydrocarbons, ketones, purines, pyrimidines, and other heterocyclic compounds. These meteoritic carbon compounds are probably the result of photochemical and other reactions of simple precursors on the surfaces of interstellar dust particles and the subsequent thermal reaction of these organics at the high temperatures and pressures of the contracting solar nebula. Further transformations took place on the colder planetesimals from which the meteorites were derived.

It is not clear whether the organic compounds in meteorites survived after reaching the primitive Earth. The Earth's crust was probably very hot after its initial formation, having high levels of radioactivity. The carbon compounds present in meteorites may have been pyrolyzed to CO, CO_2, COS, CH_4, $CH_3CH{=}CH_2$, CH_3COCH_3, CH_3CN, C_6H_6, and higher-molecular-weight hydrocarbons and heterocycles. As we will see later, many of these compounds have been postulated as starting materials for the synthesis of biomolecules on the primitive Earth.

Comets also served as a source of carbon, hydrogen, and nitrogen compounds on the primitive Earth. They probably formed in the colder regions of the solar nebula at the time that the solar system formed. Most comets have orbits that extend beyond Pluto and Neptune. Comets visible from the Earth were probably perturbed into eccentric orbits as a result of gravitational interaction with a passing star so that they pass near the sun. These comets cross the Earth's orbit and could collide with the Earth.

The "_dirty ice_" model is the one most widely accepted for comets. That is, the comet nucleus consists mainly of ice with dust, rocks, and organic molecules embedded in the ice matrix. In addition, there may be a core consisting of more dense, stony material. When a comet approaches the sun, the solar radiation vaporizes the icy coating to release dust particles and molecular species, which are photolyzed to ionic species and radicals (Figure 33–3). The dust, neutral molecules, ions, and radicals constitute the cometary tail, which often extends some 10 million km from the head. Peri-

Figure 33–3

A model for the chemical processes that occur when a comet approaches the sun.

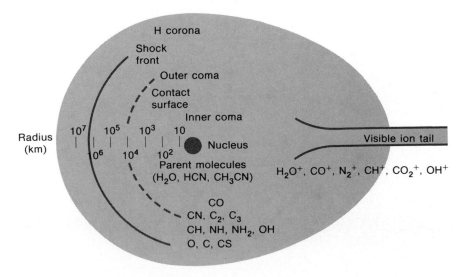

odic meteorite showers occur when the Earth crosses the trail of dust and larger particles sloughed off a comet. Some of the stony meteorites are believed to have been a comet's core from which all the volatiles have been vaporized as a result of its orbiting close to the sun.

This theoretical model is consistent with the data obtained from the 1986 flybys of Comet Halley, but with one modification; Halley's is covered with a surface of black debris. Jets of dust and gases are emitted from Halley's 7×15 km, potato-shaped body. The black, sooty-appearing surface appears to be enriched in high-molecular-weight organic matter. This conclusion is based on preliminary data that indicate an unexpectedly high level of C^+ emissions, which cannot originate from simple carbon compounds such as CH_4, CO, and CO_2.

The known molecular constituents of comets (Figure 33–3) have all been detected in the interstellar medium. This observation is consistent with the postulate that comets are composed mainly of the less dense molecular species and that they condensed in the colder regions of the solar nebula, distant from the sun. Comets may have brought significant amounts of water and cyano compounds when they collided with the primitive Earth. The icy portion of comets vaporized completely as they passed through the Earth's atmosphere, and only the less volatile material would have impacted with the Earth's crust.

THE ATMOSPHERE OF THE PRIMITIVE EARTH

As we have seen, comets and meteorites may have been important sources of the constituent compounds of the atmosphere of the primitive Earth. Comets brought *water, carbon dioxide, hydrogen cyanide, and cyano compounds.* Carbonaceous meteorites may have delivered complex organics intact to the surface of the Earth or may have brought compounds similar to those present in comets if the meteorite pyrolyzed on impact. The principal atmospheric constituents from these two sources were probably carbon dioxide and water vapor.

The Earth's crust was probably the most important source of atmospheric gases. There would have been extensive volcanic activity on the young Earth owing to the heating of the crust by radioactivity. In addition, holes were punched in the crust by the impacting planetesimals, which caused lava flows and degassing of the crust. *Carbon dioxide and water vapor were probably the predominant gases resulting from outgassing the Earth's crust,* since these are the main compounds emitted from contemporary volcanoes. Smaller amounts of the reduced forms of carbon and nitrogen also were formed, because the crust had some reducing equivalents in the form of ferrous iron and sulfides.

The observation that carbon dioxide and water are major constituents of the atmospheres of Venus and Mars (Figure 33–4), planets that flank the Earth, is consistent with the conclusion that these compounds were major constituents of the atmosphere of the primitive Earth. The low concentration of carbon dioxide in the atmosphere of the Earth today reflects the deposition of most of the atmospheric CO_2 as limestone.

From these data we can derive a model for the primitive Earth at the time when the prebiological formation of biological molecules took place. *There was an atmosphere, consisting of water vapor and carbon dioxide, that contained lesser amounts of carbon monoxide, methane, and nitrogen. Large bodies of water started to accumulate and these had a pH of 7.5 ± 1.5,* depending on how effectively the clay minerals buffered and calcium ions precipitated the dissolved carbon dioxide. *Smaller amounts of ammonia, hydrogen sulfide, hydrogen cyanide, and simple organic compounds also*

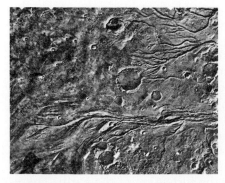

Figure 33–4
Channels on Mars are believed to have been caused by ancient rivers that have since evaporated.

were dissolved in the primitive oceans and lakes. Most of the radioactivity of the heavy elements was probably dissipated, so that the temperature of the Earth was probably only slightly warmer than it is today. The presence of water vapor would have allowed for charge separation between the clouds and the Earth's surface, so thunderstorms, with the resulting lightning and shock waves, would have been prevalent. In addition, ultraviolet light of wavelengths between 220 and 400 nm would have reached the surface of the Earth, and even shorter-wavelength light would have been absorbed by the molecules of the atmosphere.

The challenge to the chemist is to determine what chemical processes took place under these conditions and how these processes led to the formation of biopolymers and the eventual origin of life.

ANCIENT FORMS OF LIFE

Plausible reaction conditions on the primitive Earth are derived from our understanding of the origin of our galaxy and solar system. The conditions place some constraints on the laboratory experiments that can be used to simulate the chemical events on the primitive Earth leading to the origins of life. It would be possible to constrain the chemical system further if we knew the biochemistry of the first life on Earth. This information is not available; however, there is strong evidence for the presence of life on Earth 3.5 billion years ago. It is helpful to our understanding of the processes leading to the origin of life to know that *life was present on the Earth only 1 billion years after it formed.*

Evidence for this ancient life, together with the chemical processes on the primitive Earth, has been found in studies of ancient sedimentary rocks. The oldest known sedimentary rocks are located at Isua in Greenland. These rocks are 3.8 billion years old and in many respects resemble recent sedimentary rocks. The presence of sedimentary rock 3.8 billion years old indicates that there was weathering of the Earth's crust at that time. This suggests that a hydrosphere similar to the contemporary hydrosphere, minus molecular oxygen, was present on the Earth over 3.8 billion years ago.

Graphitized carbon is present in the Isua sediments, and its distribution and isotopic constitution suggest that it may have been originally present in microorganisms. However, the Isua sediments have been heated to such high temperatures (500°C) since they were deposited that it appears unlikely that anyone can prove that the carbon is biological in origin.

Fossilized remains of microorganisms would be expected to survive longer than the organic constituents of these microorganisms. A 3.5 million year old *stromatolite*, the fossilized remains of microbial colonies, is the oldest evidence of life. Some stromatolites are still forming in the brackish water of Shark Bay, Australia (Figure 33–5). These dome-shaped structures form because limestone, sand, and bits of rock stick to the slimy coating of each successive generation of bacterial growth. The fossilized stromatolite, which was found by radioactive dating techniques to be 3.5 billion years old, was located in western Australia. Morphologic studies on this stromatolite revealed its similarity to contemporary stromatolites. Since the algae prevalent in contemporary stromatolites are photosynthetic, it is assumed that the microorganisms responsible for the formation of this ancient stromatolite also were photosynthetic. This indicates that these microorganisms were quite highly evolved 1 billion years after the formation of the Earth. Therefore, the origins of life probably took place much earlier, about 4 ± 0.2 billion years ago (Figure 33–1). Thus the chemical processes resulting in the origins of life may have been completed in less than 0.5 billion years.

Figure 33–5
Contemporary stromatolites in Shark Bay, Australia. Stromatolites still grow in this brackish water because microbial predators cannot survive in the salty environment.

SYNTHESIS OF BIOMONOMERS

The simple biomonomers essential for the origins of life may have been formed from the components of the primitive atmosphere in reactions driven by the energy sources listed in Table 33–1. The precursors of these biomonomers also may have been brought to Earth by comets (e.g., HCN and nitriles) or by the pyrolysis of carbonaceous meteorites (HCN, nitriles, aliphatic and aromatic hydrocarbons) when they impacted on the Earth's hot crust. Alternatively, some of the biomolecules present in carbonaceous meteorites may have reached the Earth intact if they accreted after the crust cooled.

Lightning and the shock waves resulting from lightning were probably the most effective energy sources for the conversion of the components of the primitive atmosphere into the precursors for biomonomers. The energy is released close to the surface of the ocean, and the products formed can be easily dissolved and shielded from ultraviolet light. Ultraviolet light caused the syntheses of some biomonomers; however, it also effected the photodecomposition of organics to simpler carbon and nitrogen derivatives. Thermal energy was important in the initial stages of chemical evolution when the Earth's crust was strongly heated by the rapid decay of radioactive elements, but most of the radioactive isotopes were short-lived and the crust cooled rapidly.

Amino Acids Are Formed in Experiments Using Electric Discharges

The first experiment that dramatically demonstrated that it is possible to perform laboratory investigations of the origins of life was performed by Stanley Miller while working as a graduate student with Harold Urey. Miller arced an electric discharge for 1 week through a mixture of methane, ammonia, and water and discovered that amino acids formed. Hydroxy acids, aldehydes, and hydrogen cyanide are produced efficiently by the discharge, which suggests that the observed products are formed by their condensation in the aqueous reaction mixture (Figure 33–6). The relative yields of hydroxy acids and amino acids depended on the concentration of ammonia in the system. The levels of ammonia present in the primitive lakes and oceans are a subject of considerable disagreement. Ammonia may have been photolyzed to nitrogen and hydrogen by solar ultraviolet light; however, there may have been a low steady-state concentration of ammonia resulting from the hydrolysis of hydrogen cyanide and other nitriles.

About half the 20 primary amino acids are produced directly in electric-discharge experiments with CH_4, N_2, NH_3, and H_2O as the starting materials. Methionine is formed when hydrogen sulfide is added to the reaction mixture. Some of the aromatic amino acids may be produced when aromatic

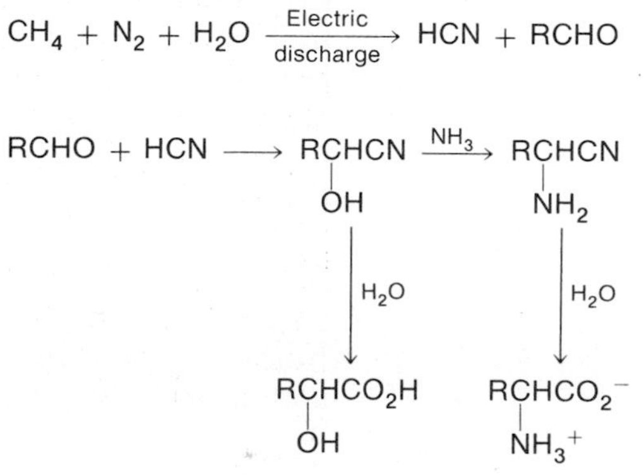

Figure 33–6

Synthesis of amino acids from the aldehydes, ammonia, and HCN produced when an electric discharge is passed through a mixture of methane, nitrogen, and water.

$$CH_4 + N_2 + H_2O \xrightarrow[\text{discharge}]{\text{Electric}} HCN + RCHO$$

$$RCHO + HCN \longrightarrow \underset{\underset{OH}{|}}{RCHCN} \xrightarrow{NH_3} \underset{\underset{NH_2}{|}}{RCHCN}$$

hydrocarbons are added to the reaction mixture. Aromatic hydrocarbons are formed by the pyrolysis of methane and carbonaceous meteorites.

Primitive-Earth simulation experiments in which methane and ammonia are the major constituents have been criticized on the grounds that most geochemical models predict that the amounts of these highly reduced forms of carbon and nitrogen will be much lower than the amounts of carbon dioxide, carbon monoxide, and nitrogen. The primitive atmosphere must have contained reducing equivalents in some form to yield amino acids, since no biomolecules or their precursors are formed when a mixture of carbon dioxide, water, and nitrogen is sparked. Amino acids are formed when hydrogen is added to the mixture of carbon dioxide, water, and nitrogen, or when a mixture of carbon monoxide, water, nitrogen, and hydrogen is sparked. The amino acid yields are lower using these starting mixtures, while formaldehyde and HCN are major reaction products.

Amino acids also have been produced by the action of heat (900°C), gamma rays, shock waves, and short-wavelength ultraviolet light on postulated primitive atmospheres. From these findings it may be concluded that *amino acids were produced as a consequence of any form of energy interacting with the constituents of the primitive atmosphere*. Consequently, it appears very likely that amino acids were present in significant amounts on the primitive Earth.

Amino Acids Are Formed by Hydrolysis of HCN Polymers

Amino acids are released by hydrolysis of the oligomers that form by the self-condensation of hydrogen cyanide in aqueous solution (Figure 33–7). Hydrogen cyanide is produced in appreciable amounts by the action of electric discharges or shock waves on simulated primitive atmospheres. Pyrolysis of organic nitrogen compounds is another potential source of hydrogen cyanide on the primitive Earth. Glycine, alanine, and aspartic acid, together with five other amino acids that are not among the 20 proteinaceous amino acids, have been identified. Numerous other ninhydrin-positive compounds have been detected using an amino acid analyzer. These findings indicate the HCN alone may have been a direct source of some of the amino acids essential for the origins of life.

Amino Acids Are Present in Carbonaceous Meteorites

Carbonaceous meteorites may have contributed to the supply of amino acids present on the primitive earth. Many of the principal amino acids extracted from meteorites are identical with those obtained in primitive-Earth simulation experiments. The meteoritic amino acids include glycine, glutamic acid, alanine, aspartic acid, valine, β-alanine, and α-aminoisobutyric acid. These amino acids are present as racemic mixtures in meteorites, a result that proves that the amino acids were not present as a consequence of terrestrial contamination. This finding does not prove that the amino acids were formed abiotically, since the half-life for the racemization of amino acids is about 10^6 years at 0°C and the meteorites are about 10^9 years old. However, the presence in the meteorites of β-alanine and α-aminoisobutyric acid, amino acids that are not utilized in contemporary life, provides support for the contention that these organic compounds were formed in abiological processes. The *abiotic formation of a similar ensemble of amino acids, both on a planetesimal that fragmented into meteorites and in primitive-Earth simulation experiments, provides strong support for the validity of the simulation experiments*.

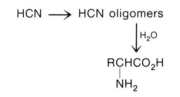

Figure 33–7
The formation of amino acids by hydrolysis of the oligomers formed by the self-condensation of HCN.

Nucleic Acid Bases Can Be Made from Hydrogen Cyanide and Cyanoacetylene

Hydrogen cyanide and cyanoacetylene were the most likely starting materials for the synthesis of the nucleic acid bases on the primitive Earth. These bases also may have been brought intact to the primitive Earth by meteorites.

Purines. When hydrogen cyanide condenses under the conditions described previously for the synthesis of amino acids, it also yields purines and pyrimidines. Hydrolysis of HCN oligomers yields *adenine* and *5-aminoimidazole-4-carboxamide* (Figure 33–8). There is tentative evidence for the formation of xanthine and hypoxanthine as well. Purines also may have been produced from *5-aminoimidazole-4-carbonitrile* (Figure 33–8), a compound produced by the photolysis of diaminomaleonitrile (Figure 33–8).

Hydrogen cyanide may have been produced in a variety of ways on the primitive Earth, but it seems unlikely that a high concentration ($>10^{-2}$ M) of cyanide was present if the oceans had the same volume as they do today. A high concentration of cyanide may have existed shortly after the oceans first formed, when their volume was low and the amount of hydrogen cyanide was high, owing to a high rate of impact of comets and meteorites. The hydrogen cyanide may have been concentrated in certain areas as a result of a high frequency of electrical storms. Since hydrogen cyanide is more volatile than water, a concentration mechanism based on the evaporation to near dryness of a lake that contains small amounts of cyanide would not give concentrated cyanide solutions. But, since the eutectic concentration of hydrogen cyanide is 74.5 mol% at $-23.4°C$, the cyanide would have been concentrated sufficiently for diaminomaleonitrile formation if water containing cyanide were cooled to $-20°C$. We can imagine the oligomerization of dilute cyanide solutions each winter as water droplets or lakes that contain hydrogen cyanide freeze. In the summer, the further reaction of diaminomaleonitrile proceeds, along with the continued synthesis of hydrogen cyanide.

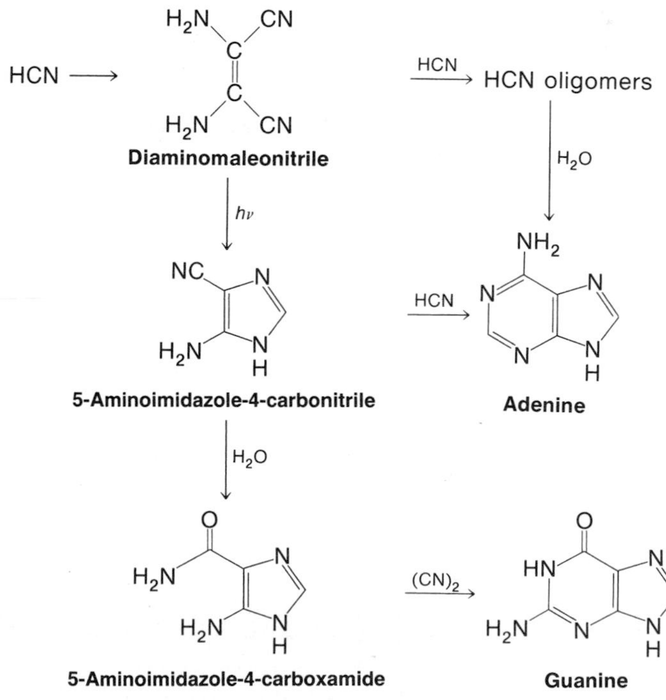

Figure 33–8

Purine bases may have been formed from HCN on the primitive Earth. (CN)$_2$ stands for cyanogen.

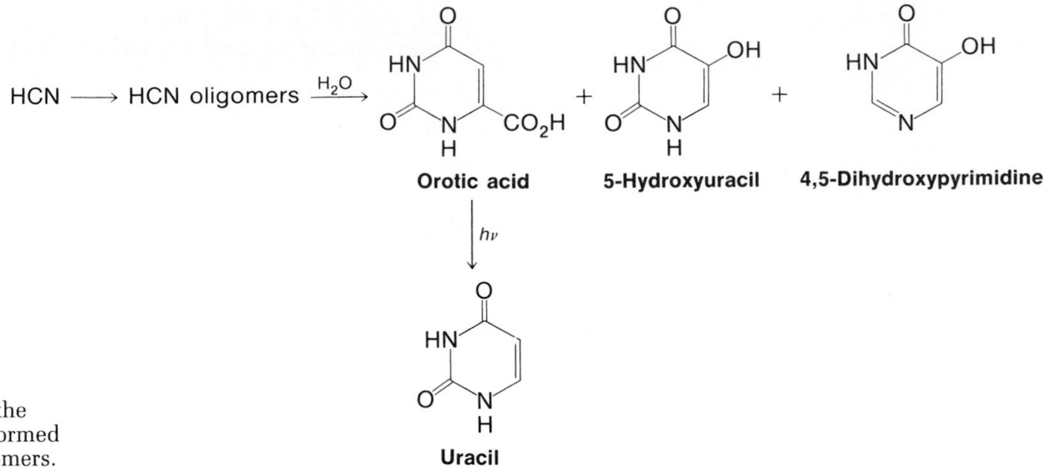

Figure 33–9
Pyrimidine bases essential for the origins of life may have been formed by the hydrolysis of HCN oligomers.

Another possibility is catalysis of diaminomaleonitrile formation by the cyanohydrin of formaldehyde, a compound formed by the reaction of formaldehyde and cyanide. The yield of adenine from frozen mixtures of HCN and the cyanohydrin of formaldehyde is greater than that from a frozen solution of HCN alone. This result suggests that formaldehyde may have facilitated the oligomerization of HCN in glaciers on the primitive Earth.

Pyrimidines. Pyrimidines also are released by the hydrolysis of HCN oligomers (Figure 33–9). Orotic acid is photochemically decarboxylated to uracil, one of the pyrimidines present in RNA. The nucleotide derivative of orotic acid (orotidine-5′-phosphate, OMP) is photochemically decarboxylated to uridine-5′-phosphate (UMP) via a reaction pathway that is similar to the enzymic conversion of OMP to UMP.

Cyanoacetylene, a major product formed by the action of an electric discharge on methane-nitrogen mixtures, is the starting material for an alternative pyrimidine synthesis (Figure 33–10). *Cytosine*, a pyrimidine base present in both RNA and DNA, is formed directly by the reaction of cyanoacetylene with cyanate. *Uracil* is formed by the hydrolysis of cytosine. The validity of this reaction sequence as prebiotic synthesis of pyrimidines has been questioned because both cyanoacetylene and cyanate are rapidly hydrolyzed in mildly basic solutions. These hydrolytic reactions would limit the possibility of the bimolecular reaction of cyanate and cyanoacetylene on the primitive Earth.

An alternative pyrimidine synthesis includes the hydrolysis of cyanoacetylene to *cyanoacetaldehyde* as the first step (Figure 33–10). The

Figure 33–10
Cyanoacetylene, a compound formed in electric discharge experiments, may also have been a starting point for the synthesis of pyrimidines.

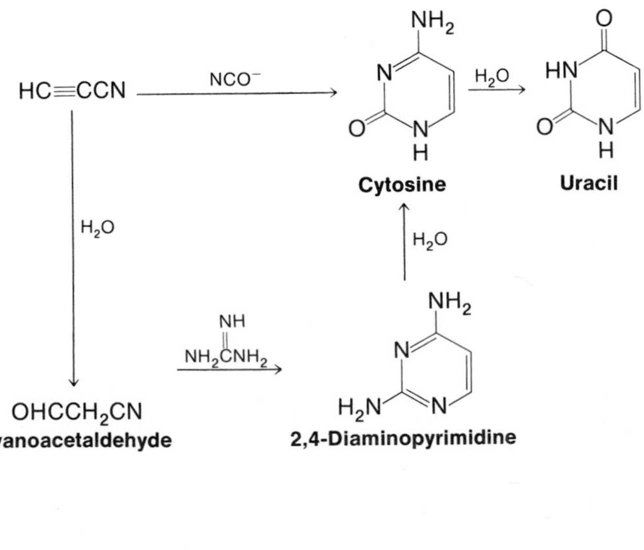

cyanoacetaldehyde reacts with guanidine in aqueous solution to give 2,4-diaminopyrimidine, a compound that is hydrolyzed to cytosine and uracil.

The efficiency of the synthesis of pyrimidines from cyanoacetylene is much greater than from hydrogen cyanide. However, hydrogen cyanide is formed in greater yield in primitive-Earth simulation experiments. As a consequence, it is not possible to assess whether hydrogen cyanide or cyanoacetylene was the more important starting material for the synthesis of pyrimidines.

Ribose May Have Been Formed from Formaldehyde

Ribose is an essential structural element in the backbone of RNA but it is not clear whether it or other sugars had a similar role in the primitive RNA used in the first life forms. *Formaldehyde was probably the starting material for the synthesis of ribose and other sugars* on the primitive Earth. Formaldehyde is formed readily in primitive-Earth simulation experiments by the photolysis of CH_4-H_2O, CO-H_2O, or CO_2-H_2O-Fe^{2+} mixtures and by the passage of an electric discharge through CH_4-H_2O or CO-H_2O-N_2 mixtures.

The condensation of formaldehyde to sugars is catalyzed by divalent cations, alumina, and clays under mildly basic conditions. The reaction proceeds by the stepwise condensation of formaldehyde to a dimer (glycolaldehyde), a trimer, and a tetramer. The tetramer then condenses with the smaller units to generate pentoses, hexoses, heptoses, and so forth. If ribose were used in the first RNA, a selection process must have been operative to segregate it from the other sugars that were formed.

Nucleosides and Nucleotides Can Be Made by Dry-Phase Heating of Precursors

The first life forms probably contained nucleic acids for the storage of genetic information. This information-storage function may have been performed by RNA-like molecules, with the DNA in contemporary cells evolving from RNA after life originated. Consequently, there must have been a pathway by which the nucleotide building blocks of RNA analogs were synthesized.

There has been limited success in elucidating pathways for the prebiotic synthesis of nucleosides. Purine nucleosides have been prepared by *dry-phase heating* of the heterocyclic bases with ribose in the presence of divalent ions such as magnesium or calcium. In this synthesis, an aqueous solution of the reactants is evaporated to dryness and heated in the dry state in a synthetic procedure that models the evaporation to dryness of a prebiotic lake or pond. Yields range from 2 to 15 per cent of a mixture of α- and β-anomers, depending on the purine base used in the reaction. Pyrimidine nucleosides are not formed in this process. Further studies are required to develop a more convincing prebiotic synthesis of nucleoside analogs.

Solid-state reactions that model the evaporation of a lake or pond to dryness, and the subsequent heating of the precipitated salts in the dry state, provide an efficient prebiotic nucleotide synthesis. Dry heating of the nucleoside at 60°C to 100°C with inorganic phosphate and urea gives initially the *5'-phosphate* as the major reaction product, along with traces of the 2'- and 3'-phosphates. The *2',3'-cyclic phosphate* is the major product on further heating of the reaction mixture (Figure 33–11). Addition of Mg^{2+} to the mixture accelerates the rate of formation of pyrophosphates so that the major product is the nucleoside 5'-diphosphate, while low yields of the 5'-triphosphate are observed (Figure 33–11). The total yield of phosphorylated derivatives exceeds 90 per cent, a result that suggests that this is a plausible prebi-

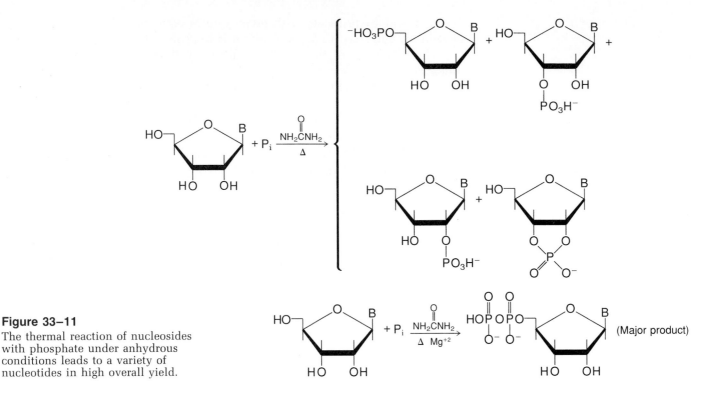

Figure 33–11
The thermal reaction of nucleosides with phosphate under anhydrous conditions leads to a variety of nucleotides in high overall yield.

otic pathway for nucleotide synthesis. Since both the 3'- and 5'-phosphates may be used directly in polynucleotide synthesis and the 2'-phosphate can be converted to the 3'-phosphate by means of the 2',3'-cyclic diester, all the simple phosphorylated products may be utilized for the synthesis of oligonucleotides. The polyphosphates and 2',3'-cyclic phosphate may have served as a prebiotic energy source for the formation of polynucleotides on the primitive Earth.

Condensing Agents May Have Driven Prebiological Reactions

ATP is the energy source that powers the conversion of monomers to polymers in contemporary biochemical systems. Since it is unlikely that much ATP was available on the primitive Earth, other high-energy compounds have been suggested as the condensing agents that drove prebiological reactions. One class of compounds comprises the cyano compounds derived from hydrogen cyanide (equation 1 in Figure 33–12, and diiminosuccinonitrile, Figure 33–13). A second class of compounds comprises phosphoric acid anhydride derivatives (trimetaphosphate and linear polyphosphates) formed by heating phosphate minerals. Both types of condensing agents effect the stepwise conversion of monomers to polymers by the elimination of a molecule of water from two monomeric units, with the free energy of hydration of the condensing agent driving the reaction.

Cyanamide and Related Structures. Cyanamide has been evaluated extensively as a condensing agent for the formation of peptides and nucleotides. It has been detected in the interstellar medium (Table 33–2) and is formed under prebiotic conditions by the irradiation of _ferrocyanide_ in the presence of ammonia and halide ions. This facile synthesis suggests that cyanamide may have been present on the primitive Earth.

Cyanamide effects the condensation of amino acids to dipeptides at pH 5 or less. This is a much more acidic environment than is assumed to have been likely for the primitive ocean (pH 7.5 ± 1.5). The absence of polypep-

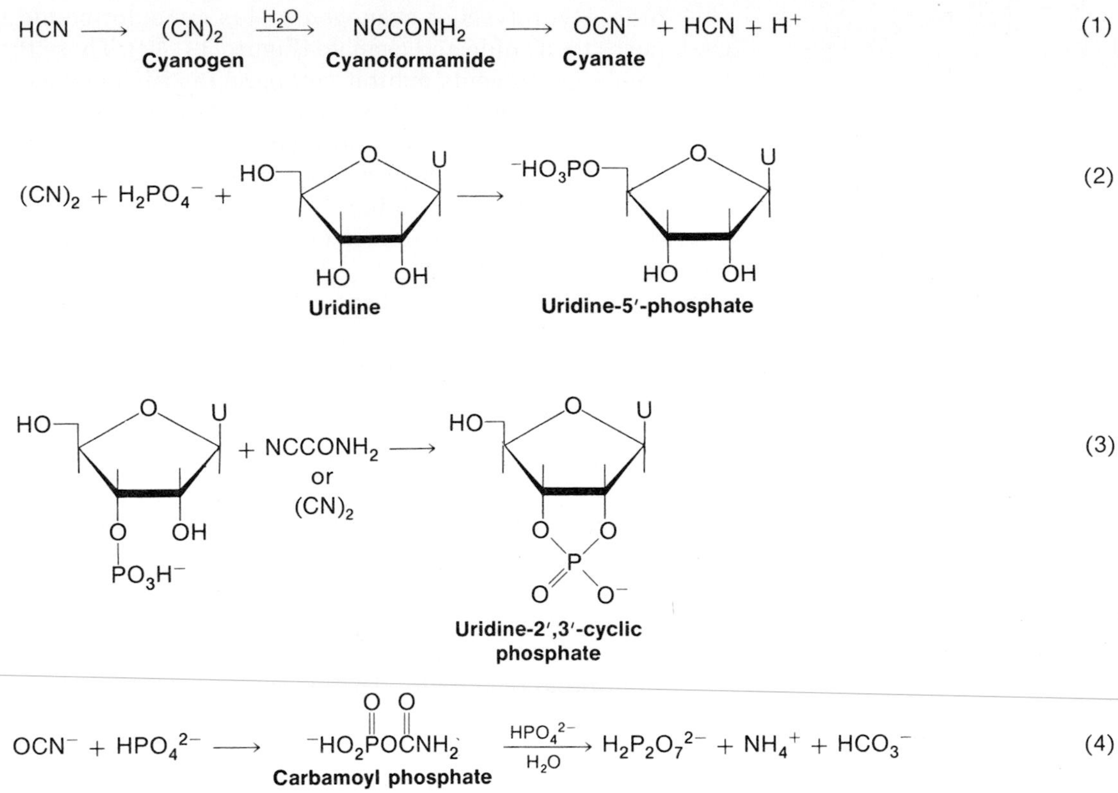

HCN $\longrightarrow$ (CN)$_2$ $\xrightarrow{H_2O}$ NCCONH$_2$ $\longrightarrow$ OCN$^-$ + HCN + H$^+$ (1)

Cyanogen   **Cyanoformamide**   **Cyanate**

(CN)$_2$ + H$_2$PO$_4^-$ + [structure: **Uridine**] $\longrightarrow$ [structure: **Uridine-5′-phosphate**] (2)

[structure] + NCCONH$_2$ $\longrightarrow$ [structure] (3)
or
(CN)$_2$

Uridine-2′,3′-cyclic phosphate

OCN$^-$ + HPO$_4^{2-}$ $\longrightarrow$ [structure] $\xrightarrow[H_2O]{HPO_4^{2-}}$ H$_2$P$_2$O$_7^{2-}$ + NH$_4^+$ + HCO$_3^-$ (4)

Carbamoyl phosphate

Figure 33–12
The possible roles of cyanogen, cyanoformamide, and cyanate as prebiotic condensing agents.

tide formation and the high acidity required for reaction detracts from its plausibility as a peptide-forming reagent on the primitive Earth.

The oligomerization of deoxythymidine-5′-phosphate is brought about in low yield by aqueous solutions of cyanamide at pH 7. Cyanamide is more effective when it is used in the presence of a nucleotide triphosphate under the dehydrating conditions that might result from a pond drying up on the primitive Earth. The synthesis of pyrophosphates, deoxyoligonucleotides, polypeptides, and glycerides has been demonstrated using solid-phase synthesis conditions. The solid-phase synthesis of polypeptides is not a plausible prebiotic reaction because it requires an initial pH that is less than or equal to 4. Both the solution-phase synthesis at pH 7 and the other solid-phase reactions are plausible prebiotic syntheses.

Cyanogen, Cyanoformamide, and Cyanate. Cyanogen is formed readily by the photolysis of (or passing an electric discharge through) hydrogen cya-

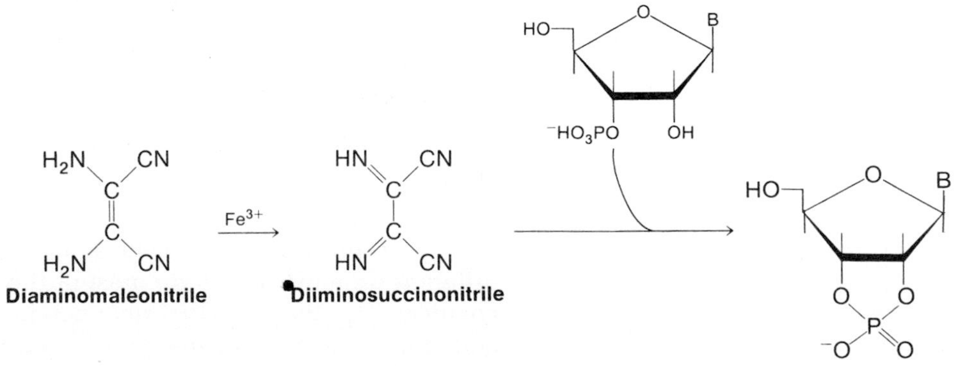

Figure 33–13
The use of diiminosuccinonitrile, a compound formed from HCN, to form the phosphodiester bond present in oligonucleotides.

H$_2$N—C(CN)—C(CN)—NH$_2$ [**Diaminomaleonitrile**] $\xrightarrow{Fe^{3+}}$ HN=C(CN)—C(CN)=NH [**Diiminosuccinonitrile**] $\longrightarrow$ [structure]

nide. Hydrolysis of cyanogen yields cyanoformamide, which then decomposes to cyanide and cyanate (Figure 33–12). These three chemically related condensing agents exhibit comparable efficiency as condensing agents.

Cyanogen effects the phosphorylation of uridine, cytidine, and adenosine in low yields. Cyanoformamide, cyanate, and cyanogen all bring about the cyclization of uridine-3'-phosphate to the corresponding cyclic phosphate, a simple phosphorylation reaction, in 10 per cent yields. None of these reagents has been effective in condensing simple nucleotides to oligonucleotides.

Pyrophosphate is formed by the action of cyanate on the mineral *hydroxyapatite*. Presumably this reaction proceeds by the formation of carbamoyl phosphate, a phosphorylating agent that reacts with a second phosphate anion to give pyrophosphate (Figure 33–12). Cyanate also will effect the condensation of glycine to diglycine in about 10 per cent yields when heated with apatite in the dry state.

Since cyanogen, cyanoformamide, and cyanate do not effect the direct formation of oligonucleotides, they could be considered as important prebiotic condensing agents only to the extent that the polymerization of the nucleoside cyclic 2',3'-phosphate derivatives was a major route for the formation of nucleic acids.

Diaminomaleonitrile. Diaminomaleonitrile has been shown to be a plausible prebiotic condensing agent. Since diaminomaleonitrile is an intermediate in the formation of purines, pyrimidines, and amino acids from hydrogen cyanide, it is reasonable to assume that it also may have served as a condensing agent on the primitive Earth. It brings about the condensation of glycine to diglycine in 5 per cent yields, but it does not effect the phosphorylation of uridine. *Diiminosuccinonitrile*, an oxidation product of diaminomaleonitrile, is a more likely condensing agent, since it would be expected to give the nitrogen analog of a mixed anhydride on reaction with carboxylate or phosphate. These mixed anhydride analogs react to form peptide or phosphoric acid ester linkages, respectively. An example of the formation of the phosphoric ester linkage is the cyclization of adenosine-3'-phosphate to the corresponding 2',3'-cyclic phosphate (Figure 33–13).

Trimetaphosphate and Polyphosphates. *Linear inorganic polyphosphates* and the cyclic phosphate trimer *trimetaphosphate* are effective prebiotic condensing agents. Their prebiotic syntheses proceed by the thermolysis of phosphate to yield linear polyphosphates, which in turn cyclize to trimetaphosphate in aqueous medium.

Nucleoside-5'-polyphosphates are formed by the dry-phase reaction of nucleoside-5'-phosphates with trimetaphosphate in the presence of Mg^{2+} (Figure 33–14). These *nucleoside polyphosphates may have served as an energy source for primitive life forms, just as ATP does in contemporary biochemical systems*. These polyphosphates degrade rapidly to the corresponding triphosphates, which in turn decompose more slowly to diphosphates and monophosphates.

POLYMERIZATION OF BIOMONOMERS

In contemporary life the flow of information is generally from nucleic acids to nucleic acids and from nucleic acids to proteins. The nucleic acids represent the repository of the genetic message, and specify the production of different proteins, which do most of the work—catalytic, structural, or mechanical. The discovery that ribonucleic acid can act as a catalyst in cleavage and joining reactions of another RNA molecule (Chapter 27) suggests that the

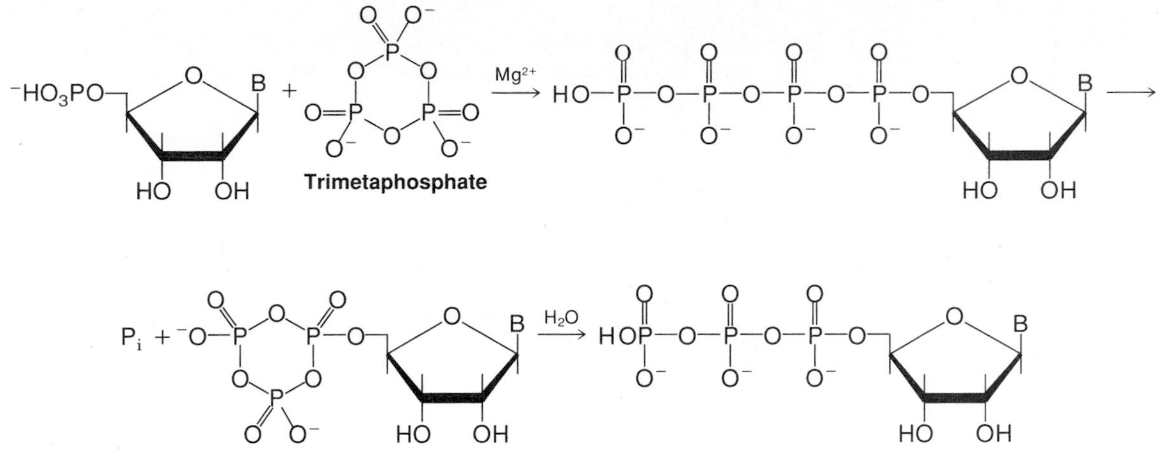

Figure 33-14
A possible route to nucleoside triphosphates starting from trimetaphosphate.

earliest living things could have been based on nucleic acids alone, but at the moment the range of reactions that RNA is known to catalyze is limited. RNA once may have carried a greater diversity of functional groups than it does today, and could therefore have been a versatile catalyst. However, there is not complete agreement that the earliest living forms were based on molecules that at all resembled today's nucleic acids and proteins. It has even been suggested that something like a clay may have been the earliest genetic material, where "information" was passed from generation to generation in the form of *lattice defects* in the clay structure. An additional feature of this system is that the genetic material could be its own phenotype and thus could itself interact directly with the environment—a *direct-acting gene*. Eventually, of course, the present system must have taken over. A genetic role for clay is an ingenious idea, but there is an understandable tendency among experimentalists to simplify the overall problem by reducing the number of possible genetic systems under consideration.

We can now consider the joining together of monomer units to make polymeric strands. We will assume that there is a sufficient supply of condensing agents and of monomeric amino acids and nucleotides. *It is important to distinguish between template-directed polymerization and the nontemplate variety.* Note that the latter term is not necessarily synonymous with "random" polymerization, since there may be some self-ordering property, however slight, in the monomers themselves. However, the nontemplate type of ordered synthesis is unlikely to give rise to a self-reproducing system of molecules—one in which a daughter strand preserves information originally present in a parent strand, and one in which, over several generations, that information gradually changes by the processes of *mutation* and *selection*. A distinction also can be made between "dry-state" synthesis and synthesis that occurs in aqueous solution, but since many dry-state reactions appear to proceed best under slightly moist or alternately moist and dry conditions, this distinction is not always very meaningful.

We will first discuss the polymerization of nucleotides, both nontemplate and template-directed, and then turn to the formation of peptides from amino acids in the absence of nucleotides. (This order of presentation is not intended to endorse either a "nucleic acid first" or a "protein first" school of chemical evolution. Indeed, a complete separation of the chemical reactions and functions of amino acids and of nucleotides in the early days of this planet seems unlikely.) Finally, the coupling of these two

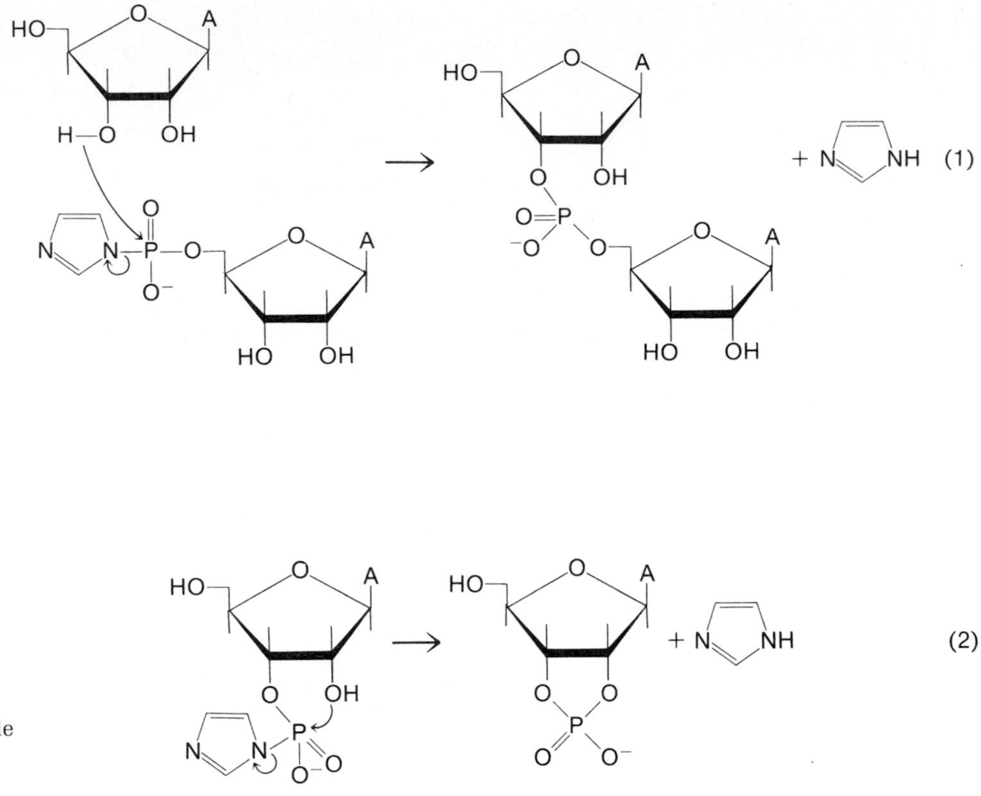

Figure 33–15
The formation of an internucleotide bond from a nucleoside and a nucleoside-5′-phosphorimidazolide (reaction 1). A 3′-phosphorimidazolide would rapidly give a 2′,3′-cyclic phosphate (reaction 2).

classes of biomolecules will be discussed in light of recent suggestions for the origin of the genetic code.

Formation of Polynucleotides by Nontemplate Reactions Requires Dry Heat or Condensing Agents

The polymerization of monomeric nucleotides requires the removal of water. This can be accomplished either by using conditions of low humidity (usually combined with elevated temperatures) or by adding a condensing agent such as a polyphosphate or cyanamide. In some experiments, an activated nucleotide has been prepared ahead of time and has then been added to the polymerization mixture. A 5′-phosphorimidazolide (Figure 33–15, reaction 1) is a typical example of such an activated nucleotide, and these imidazolides can be formed under simulated prebiotic conditions (see Figure 33–16). Other 5′ activating groups include carboxylates or polyphosphates. The activating group is merely a good leaving group and can be displaced by a ribose hydroxyl group of another nucleoside. This will result in the formation of a phosphodiester bond.

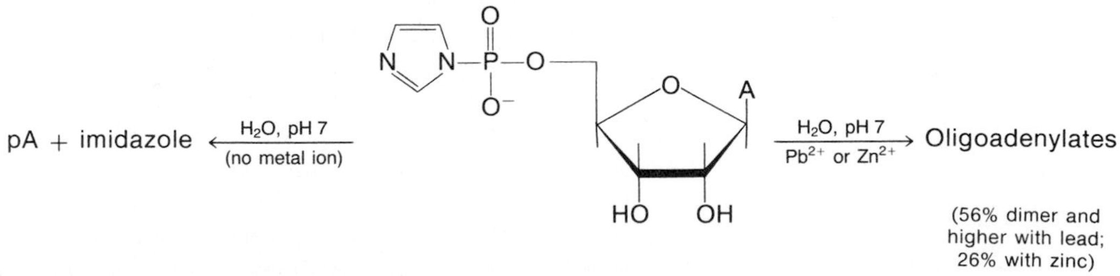

Figure 33–16
The effect of a metal ion on the polymerization of adenosine-5′-phosphorimidazolide.

If a similar attempt is made to activate a ribonucleoside-2′- or 3′-phosphate, the adjacent hydroxyl group on the ribose will displace the imidazole in a rapid intramolecular reaction, and a 2′,3′-cyclic phosphate will result (Figure 33–15, reaction 2). The same product will be obtained whatever the means of activation employed. This is of some importance, for nonenzymatic hydrolytic degradation of RNA eventually yields a mixture of nucleoside-2′- and 3′-phosphate esters. It therefore seems probable that if RNA-like molecules were undergoing alternate degradation and repolymerization (in the absence of enzymes), nucleoside-2′(3′)-phosphates and the related 2′,3′-cyclic phosphates would have been the most abundant of all monomer nucleotides. This is not to say that the 5′-phosphate monoesters would have been totally absent. They could have been made, for example, from the 2′(3′)-esters by hydrolysis and rephosphorylation. But they would probably have been present in relatively low concentration compared with the cyclic phosphates. These nucleoside-2′,3′-cyclic phosphates are still "activated" (they have a large negative standard free energy of hydrolysis compared with an acyclic phosphate diester) and can be polymerized, but they are not as active as, for example, a 5′-phosphorimidazolide. _The situation presents an interesting choice to the experimentalist: 5′ activated esters can give longer oligomers in polymerization experiments, but they may not have been present in great abundance on the primitive Earth. The 2′,3′-cyclic phosphates may have been more abundant, but they tend to give shorter oligomers._

Evaporation of a solution of a 2′,3′-cyclic phosphate, followed by mild heating of the residue, results in the formation of oligonucleotides. This dry-state synthesis has been investigated in some detail. Adenosine-2′,3′-cyclic phosphate gives oligomers up to at least the 13mer, with about 5 per cent being heptamer and higher, although polymerization appears to be inhibited if sodium salts are present. The ratio of 3′,5′ to 2′,5′ internucleotide bonds formed is about 1.8:1.

The possibility of this type of reaction occurring on the primitive Earth was underscored by an experiment performed in the Anza-Borrego Desert in California in July of 1972. A solution of adenosine-2′,3′-cyclic phosphate in aqueous ethylenediamine buffer at pH 9.5 was poured (at night) over a number of different surfaces, such as rock, sand, and glass, and left out for 41 hrs. About 4.5 per cent of material longer than the dimer was found among the products.

In what may seem at first to be an unrelated reaction, oligouridylates (27 per cent dimer, 7 per cent trimer) can be made by holding a mixture of uridine, urea, and ammonium dihydrogen phosphate at 100°C for 11 days. However, it is likely that this reaction goes by way of the intermediate formation of uridine-2′,3′-cyclic phosphate, which has been isolated from the reaction products in 45 per cent yield. Cyanamide has been used as a condensing agent to give oligothymidylates (70 to 90 per cent dimer and higher oligomers) from thymidine-5′-phosphate. This reaction proceeds best under drying conditions at elevated temperatures in the presence of an acid salt such as ammonium chloride. Circular oligomers that have no 3′ or 5′ ends appear to be the major products of this reaction.

The polymerization of nucleotides in aqueous solution in the absence of a template has proved to be difficult. This is understandable when we remember that in order to make an internucleotide bond, a ribose hydroxyl at a concentration that may be only a few hundredths molar must compete successfully with 55 M water for an activated phosphate group. Nevertheless, addition of Pb^{2+} or Zn^{2+} ions to an aqueous solution of a nucleoside-5′-phosphorimidazolide does allow the formation of a modest yield of small oligomers (Figure 33–16). The bonds formed are 80 to 90 per cent the 2′,5′-isomer. Presumably the metal ions complex two or more nucleotides in such

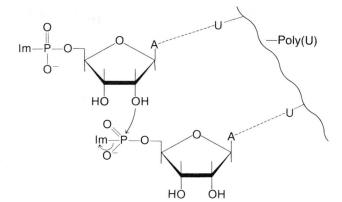

Figure 33–17
The use of a poly(U) strand as a template to line up adenosine-5′-phosphorimidazolide monomers prior to reaction (ImH = imidazole).

a way that the ribose hydroxyls of one molecule are held close to the phosphorus of another so that reaction can occur more readily.

Polynucleotide Templates Favor Polynucleotide Synthesis

Another way of trying to bring the potential partners of the reaction into proximity with each other is to use a *template* that consists of a complementary polynucleotide (Figure 33–17). If this could be done with precision and in such a way that the Watson-Crick rules of base pairing were always preserved (A with U, C with G), then we would have a nonenzymatic mechanism for copying the information present in a polynucleotide strand.

The first template-directed joining reaction of nucleotides was recorded by Naylor and Gilham in 1966. They showed that *hexathymidylate* molecules [(pdT)$_6$] could be lined up on a poly(A) template, and when a water-soluble carbodiimide was added, 5 per cent of the dodecamer [(pdT)$_{12}$] was produced. This reaction was not in itself considered prebiotic, largely on account of the nature of the condensing agent that was used, but it did serve to show that reactions of this type could succeed.

It was soon found that whereas poly(U) was able to enhance the coupling of monomer derivatives of adenosine, poly(A) was quite without effect on the coupling of monomeric derivatives of uridine. A similar situation was found for G and C; that is, poly(C) enhanced monomer G coupling, but poly(G) had no effect on monomer C coupling. These observations can be understood on the basis of the greater tendency of purines compared with pyrimidines to self-stack in aqueous solution. Even when they are not joined by a ribose phosphate diester backbone, the flat purine bases tend to stack together, just as they do in the double helix. By contrast, the stacking energy for pyrimidines is much less, and it has not been possible to demonstrate stable complex formation in any poly(purine):monomer(pyrimidine) system, even at 0°C.

There is another complication in that the complex formed from poly(U) and a monomeric or oligomeric adenosine derivative is actually a *triple* rather than a double helix. The stoichiometry is always A:2U, and each triplet of bases is held together by hydrogen bonds, as shown in Figure 33–18. The poly(U) strand (denoted U1) is antiparallel to both the A strand and the other poly(U) strand. Thus the U2 strand is parallel to the A strand. When monomer adenosine units are used, it would perhaps be more accurate to refer to a "stack" of A units, but even here some polarity (5′ end to 3′ end) is evident from the direction in which the separate ribose groups all line up.

Molecules of adenosine-2′,3′-cyclic phosphate or oligoadenylates that end in a 2′,3′-cyclic phosphate will join together when they are allowed to

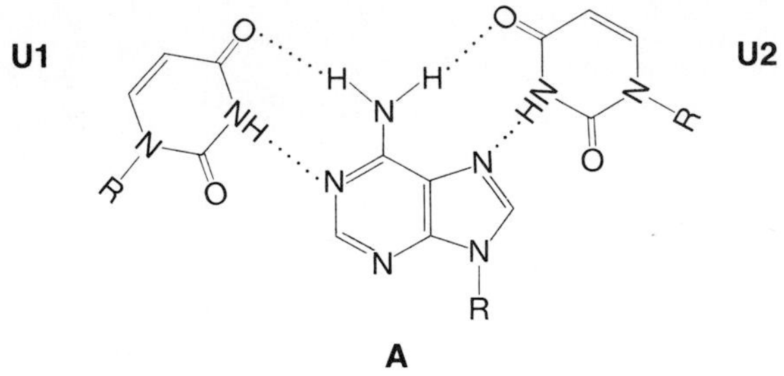

Figure 33–18
The geometry of the bases in the triple helix of stoichiometry A:2U.

complex with poly(U) in an amine buffer at pH 8. Coupling between two adjacent units occurs with a typical yield of about 25 per cent after 5 days at 2°C. In contrast to the related nontemplate "solid-state" reaction, which gave a preponderance of the 3′,5′ internucleotide bond, here the new bond formed is almost entirely the "unnatural" 2′,5′-isomer (96 per cent). (An explanation for this is given below.) Although monomeric uridine derivatives will not couple on a poly(A) template, _oligomers_ of uridylate, from the hexamer on up, do form a complex with poly(A), just as hexathymidylate will complex with poly(A); coupling will therefore occur if there is an activated phosphate group at one end of the template-bound oligomer. If this activated group is a 2′,3′-cyclic phosphate, the new internucleotide bond formed is again about 95 per cent the 2′,5′-isomer.

Orgel and coworkers have examined the base specificity of a number of related reactions. In a coupling reaction that was driven by the addition of carbodiimide, poly(U) was found to enhance the coupling of pA with A, but not with U, C, or G. Similarly, poly(C) enhanced the coupling of pG with G, but not with U, C, or A.

A remarkable discrimination has been found for the polymerization of the 2-methylimidazolide of guanosine-5′-phosphate on a poly(C) template. Noncomplementary bases (A, C, or U) were excluded from the product with an error rate as low as 0.2 per cent. This reaction gave predominantly 3′,5′-linked oligomers of G, in over 80 per cent yield, with a chain length up to the 50mer.

In earlier work with the imidazolides, rather than the 2-methylimidazolides, the error rate was 0.5 per cent when Zn^{+2} was present, and about 10 per cent with Pb^{2+} in place of the Zn^{2+}. The lead- and zinc-ion-catalyzed reactions of ImpG on poly(C) both gave oligomers of G up to thirty to forty units in length, with the lead reaction giving 85 per cent of material that had a mean chain length of seventeen. When the poly(C) template was present, but the metal ion was absent, the total yield of oligomer was about 50 per cent, and the product had an average chain length of only about five. The uncatalyzed and lead-ion-catalyzed reactions both gave a preponderance of 2′,5′-linked oligomers, while the zinc-ion-catalyzed reaction gave very largely 3′,5′-linked material. The zinc-ion-catalyzed ImpG/poly(C) system is significant in that it was the first template-directed synthesis discovered that gave almost entirely a 3′,5′-linked oligomer as a product.

A mixture of 2-MeImpG and 2-MeImpC can use the pentamer CpCpGpCpC as a template; the major pentameric product, obtained in 17 per cent yield, has the expected complementary sequence pGpGpCpGpG, and is predominantly 3′,5′ linked. Mixed dimers that are activated at the 5′-position can be prepared under mild conditions and have been shown to couple on an appropriate alternating-sequence polymer template (Figure 33–19). The yield appears to be strongly dependent on the sequence of the dinucleotide.

There still remains the question of the origin of the template strand, but this does not present much difficulty. It could have been formed in one of the nontemplate reactions discussed above, and there is even a demon-

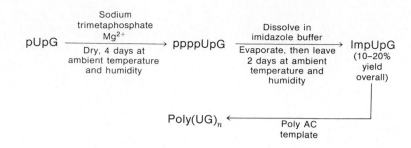

Figure 33–19
Formation of the 5'-imidazolide of a dinucleotide under simulated prebiotic conditions and its subsequent coupling on a complementary template.

strated mechanism for concentrating longer oligonucleotides from a mixture of chain lengths by selective adsorption onto hydroxyapatite. The longer molecules stick more effectively, and therefore better resist being washed away. A similar process is often used for the separation of a mixture of oligonucleotides by column chromatography.

Stability Favors the 3',5' Linkage Over the 2',5' Linkage

We have seen that in the products of some template-directed reactions there is a preponderance of the "unnatural" 2',5'-internucleotide linkage. This is particularly true for template-directed polymerization of 2',3'-cyclic phosphates, and in this case it is clearly the geometry of the reaction that is responsible for the result. When a 5' hydroxyl group attacks a 2',3'-cyclic phosphate of an adjacent nucleotide unit, the geometry of the helix ensures that the incoming 5' oxygen is in a direct line with the phosphorus atom and the 3' oxygen of the cyclic phosphate. Under mildly basic conditions, only the 3' oxygen can be displaced (by an in-line attack in which the attacking nucleophile enters opposite to the leaving group), and the result is a 2',5'-linked product (Figure 33–20). However, the reverse of this process also can occur, i.e., attack on phosphorus by the 3' hydroxyl to displace the 5' oxygen, and this happens much more rapidly when the 2',5'-linked nucleotides are bound in a helix. In contrast, in a helical 3',5'-linked oligomer, the internucleotide bonds are strongly stabilized against cleavage.

These experimental observations suggest how the daily cycles in temperature and humidity could have given rise to the gradual formation of complementary sequences of largely 3',5'-linked oligonucleotides. We start with a mixture of nucleoside-2',3'-cyclic phosphates. During the day, this undergoes (nontemplate) solid-state polymerization to give random copolymers that contain 3',5'- and 2',5'-internucleotide bonds in the ratio of about 2:1. This mixture dissolves in evening dew or rain, and at night the temperature may drop sufficiently so that short helices can form between oligonucleotides that happen to have complementary sequences. The next day the tem-

Figure 33–20
The geometry of attack on a 2',3'-cyclic phosphate by the 5' hydroxyl group of an adjacent nucleotide unit. Both nucleotides are bound in a helix, and the in-line attack results in the formation of the 2'-5' internucleotide bond.

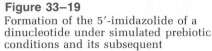

perature increases slowly, and helical 2',5' bonds will degrade most readily, nonhelical 2',5' and 3',5' bonds less so, while the helical 3',5' bonds will be most stable. As the temperature rises further, most of the short helices will melt into separate strands, and eventually the solution will dry out. At this stage, the oligomers that are left have a higher percentage of 3',5' bonds, but are shorter in length than previously. Another round of dry-state synthesis can now take place, followed that night by another pairing of complementary sequences. Next day there will be another round of degradation and further synthesis. It must be emphasized that some aspects of this cyclical scheme have not yet been demonstrated experimentally, but in theory it could provide longish double-stranded oligomers that are mostly 3',5'-linked.

We have seen that in the template-directed coupling of 2',3'-cyclic phosphates, the formation of the 2',5' bond is inevitable, but that it is not necessarily a problem. We also have seen that this geometric restriction does not apply to template-directed reactions of activated nucleoside-5'-phosphates, and that under some circumstances the 3',5' bond can be formed directly. It is interesting to remember that contemporary RNA polymerases all give 3',5'-linked products, and all require 5'-activated nucleotides as substrates.

RNA Probably Preceded DNA

Circumstantial evidence suggests that RNA-like polymers probably preceded DNA. First, ribose forms more readily than 2'-deoxyribose in prebiotic simulation experiments. Second, the polymerization of ribonucleotides is easier than that of deoxyribonucleotides. Third, in modern biochemistry, deoxyribonucleotides are made from ribonucleotides. Fourth, nonenzymatic hydrolytic degradation of unprotected RNA to monomer units occurs relatively easily. The monomer units can then be reassembled into new and potentially more useful (and stable) sequences. Chemical degradation of DNA into monomer nucleotide units is difficult (although cleavage of the purine and pyrimidine bases is readily achieved). Fifth, there is a reasonable evolutionary sequence of events that starts with the use of ribonucleoside-2',3'-cyclic phosphates (which give much 2',5'-linked product) followed by the use of activated ribonucleoside-5'-phosphates (which can more easily give a high yield of the potentially more stable 3',5'-linked oligomers). Finally, activated 2'-deoxyribonucleoside-5'-phosphates were used; these give as product a polymer (DNA) that is still more stable toward hydrolysis than is double-helical 3',5'-linked RNA (Figure 33–21). Any crude polymerases that aided the second process would have required only minor changes in structure to be able to aid the third. By this time there may well have evolved crude DNAses that could degrade the DNA for possible recycling of the monomer units.

Figure 33–21
A hypothetical scheme for the evolution of the use of different types of monomers in polynucleotide formation.

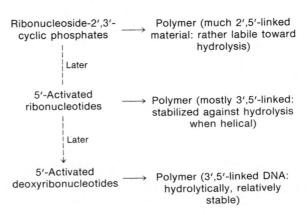

Although many sugars can be formed under the conditions that give rise to ribose, there is experimental evidence to suggest that the incorporation of beta-ribonucleosides into a polynucleotide strand is a favored process. We mentioned earlier that ImpA on a poly(U) template will react with adenosine; however, it will not couple nearly as well with 2′-deoxyadenosine, alpha-adenosine, 3′-deoxyadenosine, or arabinosyl adenine. Some workers have pointed out that ribose (and deoxyribose) are not indefinitely stable in the presence of basic amines. This is true, but conversion to the nucleosides would have eliminated the immediate cause of this instability by masking the aldehyde carbonyl group.

Formation of Polypeptides

Most suggestions for nontemplate polypeptide synthesis envision the polymerization of amino acids, either through the application of heat under drying conditions, or by the addition of condensing agents. Fox and coworkers have synthesized thermal copolymers of amino acids from mixtures of the monomers. In the early work, rather high temperatures were used (150–180°C), but in some cases temperatures below 100°C were found to be sufficient if the reactions were allowed to proceed for extended periods of time. These "thermal proteinoids" appear to contain linkages that are not found in proteins, as well as the normal peptide bond, and since several polyfunctional amino acids were present in the original mixture of monomers, this is probably not very surprising. The use of high temperatures in the early experiments was criticized on the grounds that such temperatures probably would not have existed over extended regions of the Earth's surface, and in addition, prolonged exposure to temperatures in this range would tend to cause decomposition of any peptides that may have been produced. The mixtures of pure amino acids that were employed are not very realistic, but this is a type of criticism that can be leveled against many prebiotic simulation experiments.

The presence of clay can aid peptide bond formation in these dry-state reactions, and workers at NASA's Ames Research Center have found that when cycles of temperature and humidity are used, the yield is further enhanced. _Homoionic bentonite clays_ were investigated and a Cu^{2+}-substituted clay was found to be somewhat better than Ni^{2+} or Zn^{2+} at promoting the synthesis of oligoglycine and oligoalanine from the monomeric amino acids. The maximum yields and chain lengths were not very high, however, reaching 6 per cent of total oligoglycines with a maximum chain length of five units, and 2 per cent of alanylalanine. The highest temperature employed during each cycle was 94°C. When a polyribonucleotide or a small amount of histidylhistidine was added to a similar kaolinite clay system, the yield of oligoglycine was found to increase about two to four times. In contrast to these successful "dry-state" reactions, it has been shown, at least for the formation of lysyllysine from the monomer, that clays alone cannot help at all if one attempts to carry out peptide bond formation in aqueous solution at temperatures up to 90°C.

Various condensing agents have been used to make peptides from amino acids. As one example, cyanamide together with ATP and aminoimidazolecarboxamide in a dry-state reaction at 90°C for 24 hours gave 5 per cent Gly_n (up to the tetramer), 17 per cent Ile_n (up to the dimer), or 66 per cent Phe_n (up the tetramer). Somewhat better results came from the use of polyphosphates as a condensing agent. In a "solid-state" reaction that required cyclic trimetaphosphate, Mg^{2+}, and imidazole, glycine gave 25 per cent of total oligomers, with 1.1 per cent higher than the heptamer, after 10 days at 65°C. After only four hours at 100°C, oligomers up to the decamer were produced.

By contrast, a reaction between glycine and trimetaphosphate that was run in aqueous solution at mildly alkaline pH gave only low yields of the dimer and trimer. A reaction of a very different type occurs when the hexadecyl ester of glycine spreads in a monolayer on a water–air interface. Hydrophobic interactions between the hydrocarbon tails cause the ester molecules to come into close enough proximity that a fairly efficient synthesis of oligoglycine can occur.

In the peptide syntheses that we have considered so far, there is no obvious copying or translation of information. Except that there may be some slight self-ordering property in the growth of a polypeptide that affects the sequence of amino acids, the products of these reactions will tend to be random copolymers. There have been suggestions that a polypeptide could act as a template for the production of a complementary polypeptide, and thus by a repetition of this process a more or less faithful copy of the original could be produced. Such schemes are usually based on interactions between amino acid side chains. Thus a positively charged side chain (e.g., lysine) may select a carboxylate anion (aspartate or glutamate) in the side chain of its "complement." A hydrophobic group or neutral group may select another hydrophobic or neutral group directly. It could be argued that a highly accurate copying mechanism is unnecessary; the function of the resulting polypeptide will perhaps not be much changed if aspartate is substituted for glutamate, or valine for leucine. Some parts of these suggestions seem plausible, and the idea as a whole would have obvious importance to the "protein first" school of chemical evolution, but there is as yet little experimental evidence that really suggests that such a scheme should work.

Some production of polypeptides on the primitive Earth seems unavoidable, granted present assumptions concerning the conditions that existed then. If an orderly evolution of those polypeptides is out of the question, then what part did they play in the origin of life, and what is the relevance of the polymerization reactions? We are now four billion years removed in time and cannot really know, but perhaps some of them acted as weak nonspecific catalysts that helped to establish the operations of replication and translation. These operations soon would have evolved the ability to make their own polypeptides, which were better catalysts for the same reactions, and the original makeshift peptides would no longer have been of any use, except perhaps as a reservoir of amino acids.

Coevolution of Polynucleotides and Polypeptides—The Origin of Translation

Proteins can be marvelous catalysts, but they do not self-reproduce; nucleic acids are well suited to self-replication, but at least in their present form they are not very versatile catalysts. For instance, we do not know if they would be able to catalyze the formation of their own building blocks from simpler precursors if the reservoir of endogenous nucleotides became depleted. If the ability of a polynucleotide to replicate could be coupled in the right way with the ability of a polypeptide to act as a catalyst, both parties to this cooperation would benefit. The polynucleotide could specify the formation of the polypeptide, which in turn could catalyze a variety of reactions, including replication of the polynucleotide and possibly also the translation process. The importance of a working translation system is so great that for a number of people, the origin of translation is almost synonymous with the origin of life.

Not all the experimental work that links amino acids with nucleotides has been concerned with translation. Aminoacyl adenylates (Figure 33–22) are part amino acid and part nucleotide, and are highly reactive by virtue of

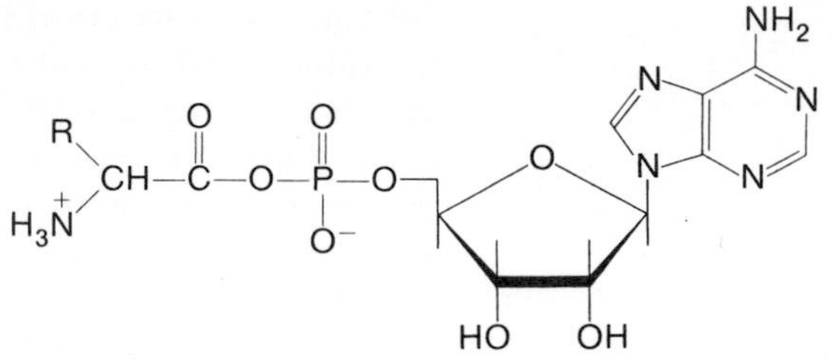

Figure 33–22
An aminoacyl adenylate.

the mixed carboxylic-phosphoric anhydride linkage. They are also intermediates in contemporary protein biosynthesis (Chapter 28), and are therefore of intrinsic interest. When such an aminoacyl adenylate is dissolved in aqueous buffer, it hydrolyzes rapidly to give the amino acid and pA, together with about 30 per cent of di- and tripeptide. However, in the presence of the clay mineral montmorillonite, or when absorbed into *micelles* or reversed micelles, these aminoacyl adenylates can give polypeptides up to 40–50 amino acid units long. The total yield of polypeptide in some cases is as high as 95 per cent. Although this is an impressive result, it has not yet been shown to be directly relevant to the process of translation. It is true that the adenylate group is being made use of as a handle, with which the amino acids can be made to line up so that they can react, but it is not really being made use of as a specific handle. Indeed, in one related case where an attempt was made to utilize the specific base-pairing capability of the adenylate group, the results were disappointing: phenylalanyl adenylate gave no more peptide in the presence of poly(U) than in its absence.

Origin of the Genetic Code.   Many theories for the origin of translation and the genetic code have been proposed. Some are more philosophical and some more chemical, but by and large they can be put in one or the other of two categories according to their author's view of how particular codons came to be assigned to particular amino acids. There is the "frozen accident" school, and there is the "specific interaction" school.

According to the former view, the assignment was purely a random event, but once that event occurred, the system that used it and was able to carry out translation would have had a large selective advantage over any nontranslating system; the assignment would have become fixed. As an example of how this could occur, suppose the sequence of nucleotides in a small RNA molecule caused it to fold up in such a way as to generate a pocket that exactly matched a particular amino acid, say alanine, which therefore bound to this site. In a different place on that RNA molecule there may have been an exposed triplet of nucleotides, say AGC, that allowed the RNA to bind to a message strand at the "codon" GCU, and thus incorporate alanine at the corresponding position on a growing peptide chain. One codon for alanine could thus have become fixed as GCU even though there was no special interaction between alanine and its codon or anticodon.

The "frozen accident" theory was dominant until the late 1970s, but lost ground when it was demonstrated that *there is a small but significant correlation between the polarity and hydrophobicity of an amino acid and the same properties of its anticodon nucleotides.* The effect is most clearly seen with the homocodonic amino acids—those amino acids that have a codon that is a set of three identical bases, e.g., CCC or UUU.

The comparison of properties between amino acids and their anticodons has been explored in several ways. In one set of experiments, Weber and Lacey measured the R_F values of amino acids and dinucleotides on paper chromatography in a solvent that contained a high salt concentration (Figure 33–23). The salt was present to mask the extreme polarity of the charged

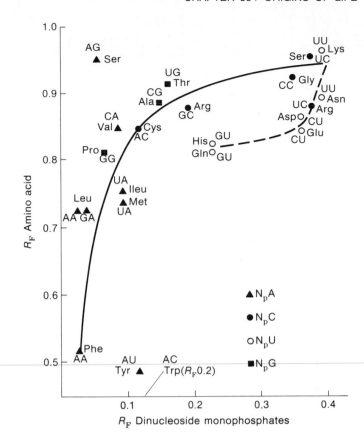

Figure 33–23

R_F values (the solvent was 10 volume per cent saturated ammonium sulfate, pH 7.0) of amino acids plotted versus the R_F values of the dinucleoside monophosphates representing the first two letters (3′→5′) of their anticodons. (From A. L. Weber and J. C. Lacey, Jr., Genetic code correlations: Amino acids and their anticodon nucleotides, *J. Mol. Evol.* 11:199, 1978. Reprinted with permission.)

phosphate residue in each dinucleotide. A relationship was found between the R_F values for a particular amino acid and the dinucleotide that was *complementary* to the first two bases of its codon. Thus one of the most hydrophobic amino acids is phenylalanine, and the most hydrophobic dinucleotide is ApA; both of the codons for Phe start with UU. The most polar amino acid is lysine, and the most polar dinucleotide is UpU; both of the codons for Lys start with AA. There are some anomalies in the results, but plausible explanations can be found for most of them.

It is one thing to show that an amino acid side chain and its modern anticodon have some properties in common, but it is quite another to incorporate this information into a working translation model of protein synthesis. The ideal sort of reaction in which to examine the effects of specific interactions between nucleotides and amino acids would be an efficient oligonucleotide-directed peptide synthesis reaction that conceivably could have occurred around the time of the origin of life. Unfortunately, we have not yet discovered any such reaction.

Translation. *There are two basic requirements for a working translation system: (1) the selection of a particular amino acid, and (2) the efficient formation of the peptide bond.* A possible mechanism for the first of these has been suggested by Hopfield. Consideration of the nucleotide sequences of different tRNA molecules from *E. coli* led him to the tentative conclusion that these molecules may once have been able to fold up in a different way. The important feature of his alternative conformation is that the nucleotides of the anticodon would sit quite close to the amino acid that is esterified at the 3′ terminus (Figure 33–24a). In the usual structure of a modern tRNA, the anticodon is as much as 70 Å away from the amino acid (Figure 33–24b). We have seen that a hydrophobic amino acid has a hydrophobic anticodon, and thus it appears possible that a mutual interaction of the esterified amino

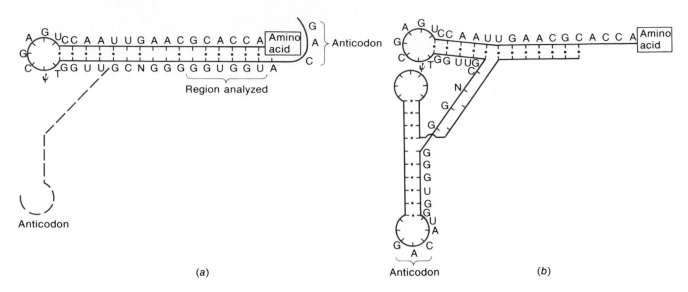

Figure 33–24
(a) A possible alternative conformation for an early tRNA. The normal conformation is shown in (b). (From J. J. Hopfield, Origin of the genetic code: A testable hypothesis based on tRNA structure, sequence, and kinetic proofreading, *Proc. Natl. Acad. Sci. USA* 75:4334, 1979. Reprinted with permission.)

acid with its anticodon ("like attracts like") could, for example, stabilize the amino acid ester bond against hydrolysis. As a result, there would be a small preference for a hydrophobic amino acid to remain in the site adjacent to a hydrophobic anticodon. There are ways in which this effect could be amplified, and some additional specificity could come from more subtle interactions between the shape or polarity of the amino acid side chain and the anticodon nucleotides.

Peptide Bond Formation. Turning now to the peptide bond formation step, we find that although many theories have been published that are concerned with the origin of the genetic code, remarkably few of them indicate precisely how this key reaction is supposed to take place. It is therefore hard to put them to an experimental test. What should we look for in such a theory? Actually, the requirements can be spelled out simply enough. We should remember that whatever the mechanism we choose, it should be possible for it to have evolved, by way of a series of reasonable steps, to the present system of protein biosynthesis. For example, it has been pointed out by Crick that a primordial doublet code is unacceptable, as all the information present in a message would be rendered useless when the transition to a triplet code was made. Ways around this type of objection can be found, but they tend to be cumbersome.

As far as peptide bond formation itself is concerned, we need to consider both the thermodynamic and the kinetic factors. Is there enough driving force to make it go, and if so, will it go fast enough to be useful? An amino acid ester is sufficiently activated for our purposes. The equilibrium constant for the formation of a dipeptidyl ester from two aminoacyl esters is probably about 3000 at 25°C and pH 7. This is perfectly adequate to drive a peptide synthesis reaction. Aminoacyl adenylates could be even better, but there is often a price to be paid for higher reactivity, namely, that the material is hydrolyzed more rapidly when it is dissolved in water.

So esters have enough driving force, but can they be made to react at a reasonable rate in a prebiotic reaction, without the aid of modern enzymes? All we can say for certain is that 2′(3′)-glycyl adenosine does show a modest

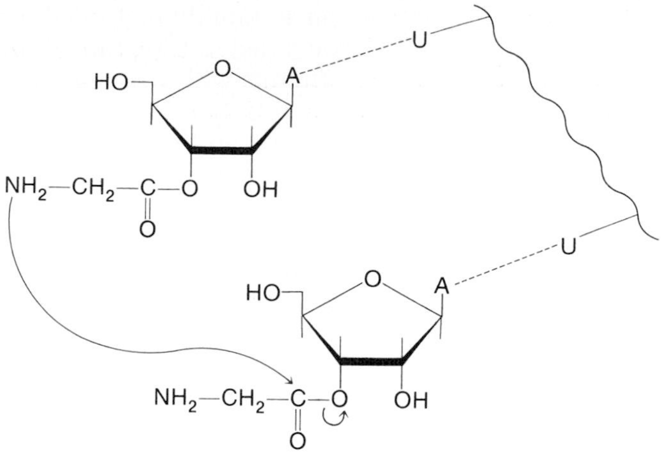

Figure 33–25

The formation of glycylglycine from two adjacent molecules of 2'(3')-glycyl adenosine that are complexed with poly(U). The complex is a triple helix, but in the diagram the second poly(U) strand has been omitted.

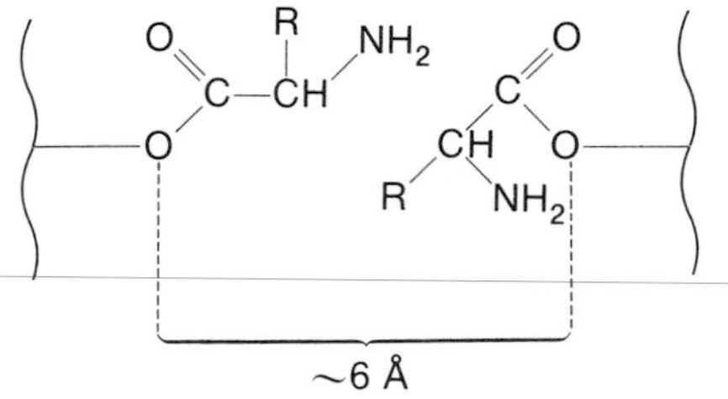

Figure 33–26

In order for two amino acid esters to be able to react and form a peptide bond, the ester oxygens should be no farther than about 6 Å apart.

increase in the rate of peptide bond formation (Figure 33–25) when it is complexed to poly(U). The templating effect of the poly(U) is more evident at low concentrations of ester, where the ordinary bimolecular reaction of one amino acid ester with another is very slow. The template should serve to concentrate the adenosine units out of dilute solution, as well as to align them for reaction. Perhaps this reaction is not very efficient because the reactants are too "floppy," meaning they are not held rigidly enough in the right position for reaction to occur. The study of intramolecular reactions as models of enzyme action has shown that what may appear to be a small increase in the rigidity of a system can give a large increase in rate.

Consideration of the stereochemical requirements of the attack of an amino group on an ester bond leads to the conclusion that for reaction to be possible, the distance between the two esterified oxygens (Figure 33–26) should not be greater than about 6 Å. For example, this requirement rules out efficient peptide bond formation between amino acids that are esterified to 2' hydroxyls on opposite strands of a double-helical RNA in the A-form. The closest approach of any two 2'-OH groups in this structure is about 10 Å. An interesting possibility, however, is reaction between two amino acids, one of which is esterified to a 3' terminus of an oligonucleotide and the other to the 5' terminus of an adjacent oligomer. The ester oxygens in this case are only 2.5 Å apart.

The reaction of 2'(3')-glycyl adenosine, mentioned above, actually gives more *diketopiperazine* (DKP) than diglycine. A diketopiperazine is a cyclic dimer that forms when the free amino group of a dipeptidyl ester attacks its own ester bond in an intramolecular reaction (Figure 33–27). This can hap-

Figure 33–27

(a) The formation of diketopiperazine by an intramolecular cyclization.
(b) The reaction is prevented by the presence of the N-acetyl group.

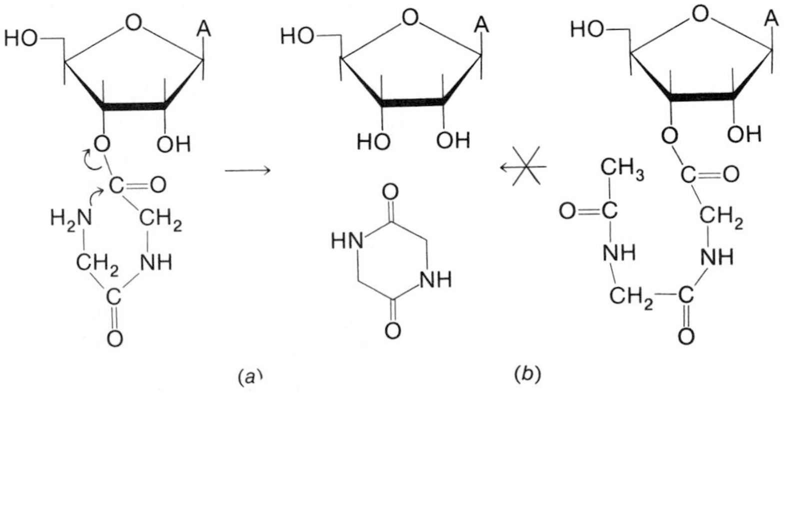

(a)   (b)

pen so rapidly that the dipeptidyl ester may not survive long enough to be converted to tripeptide. If the dipeptidyl ester stage is safely passed, the N terminus will then tend to be sufficiently distant from the C-terminal ester that this type of cyclization reaction becomes unimportant. It has been pointed out that the unwanted reaction to make DKP is blocked if the first amino acid to enter peptide synthesis (i. e., the N-terminal amino acid) carries a group such as acetate or formate. Perhaps the use of N-formylmethionine to initiate protein synthesis in prokaryotes is a remnant from a time when there had to be a relatively simple chemical method for preventing DKP formation.

Selection of α-Amino Acids in a Template Reaction. The process of bringing amino acids into reaction by means of oligonucleotide carriers that bind to a template has other advantages in a prebiotic setting than just the formation of peptides by translation. The formation of α-amino acids in prebiotic simulation experiments is accompanied by the production of a variety of other compounds such as β-amino acids, simple carboxylic acids, and amines. Why should these compounds not interfere with the formation of a peptide chain? After all, the primitive Earth environment was unlikely to have provided pools that contained pure amino acids and pure nucleotides just waiting to get acquainted.

Fortunately, we can invoke the special characteristics of a template synthesis to provide a partial answer to this problem. First, a simple amine that cannot become esterified to an oligonucleotide could attack the peptidyl ester bond of a growing chain, and so interrupt the synthesis, but it would have to do so in an ordinary bimolecular reaction with no advantage from the templating effect. The effective concentration of the amino group of an amino acid in an efficient template-directed reaction could be several hundred molar, and this would easily swamp any unwanted interference from a simple amine at a concentration of a few hundredths molar. A simple carboxylic acid could become esterified to an oligonucleotide carrier, but it would have no amino group, and so as long as the direction of translation is from N terminus to C terminus, it could not participate or interfere with the formation of peptide, except at the N-terminal end. As we have just seen, an N-terminal formate or acetate could be a distinct advantage. Discrimination against β- or γ-amino acids could occur because of the stereochemical requirements of the template-directed reaction itself, but until this reaction has been identified, we cannot test this point. Indeed, it is hard to avoid the feeling that when an efficient template-directed peptide bond formation reaction is discovered, it will reveal a great deal more about related phenomena (such as the origin of the genetic code), than we had previously expected to find out from this source alone.

MEMBRANES AND COMPARTMENTATION

If we could go back in time about four billion years and look around for some evidence of life, we would probably start by trying to find objects that resembled cells. We would be unlikely to award the title of "living" to a puddle or small pond, even though it may at that moment possess a system of self-reproducing molecules, and be the site of synthesis of any number of interesting compounds. This is only partly semantic prejudice, for such a puddle or small pond has some features that make it ill-adapted to any further development. A sudden downpour of rain could easily disperse the contents of a special puddle to an extent that the component molecules would no longer be able to interact. A pond is not a unit that can compete for limited resources, nor can it evolve by the processes of replication, mutation, and selection. The formation in such a place of a crude replicase enzyme (by translation of the sequence of bases in an oligonucleotide) would result in an increased rate of replication of *all* the oligonucleotides in that pond, unless the replicase were able to recognize and therefore duplicate only its own gene. This latter feat would be asking quite a lot of a primitive enzyme. Similarly, the formation of a superior "enzyme" for catalyzing translation could result in an enhanced translation of all the oligonucleotides in the pool, and the accumulation of useless polypeptides.

If a self-reproducing system were enclosed instead in a cell-like structure, some of these problems would disappear. Dispersion of the contents would be unlikely to result from a shower of rain, and although recognition of its own gene by a replicase would still be advantageous, it would no longer be absolutely essential—particularly if the cell were able to grow and divide, so that there would be a finite chance that one of the progeny would *not* contain any nonsense RNA, which was being unnecessarily copied and translated.

A cell is a compartment, and even an early cell must therefore have had a membrane that separated it from everything "outside." What material could have served as a membrane to encapsulate the first "living" set of molecules? Contenders include some form of lipid, the thermal polypeptides of Fox, or coacervate droplets. Analogy with the modern cell would favor the lipid vesicle as a container for this purpose, but each of these materials can be induced to show some morphological similarity to a cell. For example, when thermally produced polymers of amino acids are boiled in water and then cooled, microspheres of about 2 μm diameter are formed in abundance. But it does not seem very likely that these could have served as the first cells, for they are too leaky; small molecules can pass in or out far too readily. Coacervate droplets were investigated by Oparin, but they were usually formed from materials such as gum arabic and gelatin, and these would have required the prior existence of living organisms for their formation. However, it is possible that similar structures could be formed from abiotic polymers.

At present it seems most likely that the first membrane was made largely of lipid. Lipids can form bilayer vesicles relatively easily, and these vesicles not only are self-repairing, but they can trap small polar molecules.

Experiments by Deamer and Hargreaves have given support to this idea. Mixtures of glycerol, fatty acids, and orthophosphate (sometimes with silica or clay) were subjected to cycles of temperature and humidity in a similar way to that described for the formation of oligonucleotides and oligopeptides. Among the products were mono- and diglycerides, and a small amount of phosphatidic acid. Under the same conditions, these lipid products appeared to form vesicles, and when the external water was allowed to evaporate, the vesicles opened and were able to exchange material with the outside world. When the system was rehydrated, the vesicles again formed their separate units. If the pH of the outside solution became different from what it was when the vesicles had reclosed, the resulting pH gradient could be used to drive the uptake of various compounds through the membrane and into the vesicle. For example, if the interior were at a lower pH than the exterior, a weak base amine in its neutral unprotonated form could cross the membrane, become protonated, and thus be trapped inside as the lipid-insoluble cation.

Other schemes for facilitating the transport of ionic molecules have been suggested by Stillwell. Amino acids could be rendered lipid-soluble by formation of a Schiff base with a long-chain aldehyde; nucleotides could be introduced as complexes with metal ions and a sugar as a Schiff base with a long-chain amine. Once inside, the amino acids could be liberated by hydrolysis of the Schiff base, and retained inside by polymerization. This latter step would require the introduction of a condensing agent and the removal of the products of its hydrolysis, as well as the presence of a polymerase catalyst.

The prebiotic formation of lipids has not been extensively investigated, but the limited experiments that have been performed suggest that lipidlike materials probably formed spontaneously on the primitive Earth. For example, when an electric discharge is passed through a mixture of methane and water, a mixture of carboxylic acids is obtained that contains two to twelve carbon atoms. However, all the acids with more than six carbon atoms have branched chains instead of the linear chains found in the lipids present in the membranes of contemporary cells. Linear fatty acids and hydrocarbons with up to 18 carbon atoms are produced when carbon monoxide and hydrogen are heated at high pressures to 200–400°C in the presence of an iron catalyst (*Fischer-Tropsch synthesis*). The Fischer-Tropsch reaction has been suggested as a possible source of the complex carbon compounds in the solar nebula, but it is unlikely that the high temperatures and pressures required for this synthesis were prevalent on the primitive Earth. Since carboxylic acids have been detected in carbonaceous meteorites, they may have been synthesized by a Fischer-Tropsch process in the solar nebula and then brought to the Earth in the later stages of meteoritic infall.

In spite of what has been accomplished so far, we seem to be still a long way away from showing in detail how a working cell could have arisen. We will need to demonstrate not only that a functioning genetic system can be constructed, but that it can be incorporated into a compartment such as a lipid vesicle.

SUMMARY

The Earth formed about 4.5 billion years ago and the first life originated in less than a billion years. The original living systems were assembled from complex organic molecules that had formed spontaneously under the reaction conditions prevalent on Earth during those first billion years. The original life forms were much simpler and less efficient than contemporary living systems, but evolution to more complex systems occurred rapidly. The oldest evidence for life is a 3.5-billion-year-old fossilized algal stromatolite. It shows some striking similarities to contemporary stromatolites.

The starting materials for the formation of the original biopolymers may have been brought to the Earth by meteorites and comets, or they may have been synthesized here by the action of electric discharges, ultraviolet light, and other forms of energy from the components of the primitive atmosphere (CO_2, CO, CH_4, N_2, and H_2O). When different forms of energy interact with a reducing atmosphere, HCN, nitriles, aldehydes, ketones, alcohols, acids, and amines are formed. Amino acids may have formed on the primitive Earth from amino nitriles that were produced either directly in an electric discharge or else by the reaction of HCN, carbonyl compounds, and ammonia. Alternatively, they may have formed directly from HCN. Cyanoacetylene and HCN were likely starting materials for the synthesis of the purine and pyrimidine bases present in nucleic acids. Sugars may have formed directly from formaldehyde, and nucleosides were formed by the reaction of the purine and pyrimidine bases with these sugars.

Phosphorylation reactions, to give nucleotides, polynucleotides, and phospholipids, were brought about either by dry heating or by high-energy condensing agents such as cyanamide and trimetaphosphate. Similar reagents converted amino acids into peptides.

The earliest self-reproducing system was likely to have been based on both nucleic acids and proteins, although a mineral such as clay has been suggested as an alternative genetic material for the first living forms. The earliest polynucleotides probably resembled RNA more than DNA. Simulation experiments have shown that the condensation of activated nucleotides on a complementary polynucleotide template can be facilitated by metal ions, especially divalent zinc or lead. Weak but significant correlations between some properties of amino acids and the same properties of their anticodon nucleotides have been found. This argues against the "frozen accident" theory for the origin of the genetic code. However, an efficient peptide-synthesis scheme that is based on oligonucleotide templates has not yet been found.

SELECTED READINGS

Cairns-Smith, A. G., *Genetic Takeover*. Cambridge, England: Cambridge University Press, 1982.

Ferris, J. P., The chemistry of life's origins, *Chem. Eng. News*, Aug. 27, 1984, pp. 22–35. A general overview of most aspects of the problem of life's origins.

Goldsmith, D., and T. Owen, *The Search for Life in the Universe*, Menlo Park, Calif.: Benjamin/Cummings, 1980. An authoritative and readable book.

Horowitz, N. H., On the evolution of biochemical synthesis. *Proc. Natl. Acad. Sci. USA* 31:153, 1945. A discussion of biochemical pathways as evidence for prebiological synthesis; a theory proposed prior to the initial Miller experimental synthesis of amino acids.

Miller, S. L., and L. E. Orgel, *The Origins of Life on Earth*, Englewood Cliffs, N.J.: Prentice-Hall, 1974. An excellent critical discussion of chemical evolution.

Orgel, L. E., RNA catalysis and the origins of life. *J. Theor. Biol.* 123:127–149, 1986.

Schopf, J. W. (ed.), *Earth's Earliest Biosphere, Its Origin and Evolution*. Princeton, N.J.: Princeton University Press, 1983. A comprehensive overview of the geology and biology of the early Precambrian (primitive) Earth.

Walker, J. C. G., *Earth History*, Boston: Jones and Bartlett Publishers Inc., 1986.

PROBLEMS

1. What functions are essential to the maintenance of a simple living system? Are the processes that you defined limited to the organic molecules found in contemporary life forms? Could these processes be performed by inorganic structures? Explain.

2. a. Metal ions play an important catalytic role in contemporary life forms, and it is likely that they also served as catalysts in primitive life. Can one conclude that the metals found in contemporary life were the only ones utilized in the first living systems? Explain on the basis of information presented in the text (refer to various metalloenzymes in Chapters 9 and 11).

   b. An alternative approach to deducing which metal ions may have had a role in primitive life is to consider those metal ions that are found in the present oceans, since it is very likely that the first life originated and maintained itself in the presence of water. What basic assumption concerning the primitive ocean is made in this argument? How would you adjust the metal ion content of the contemporary ocean to derive a list of metal ions that may have had a role in primitive life?

3. At first glance it appears to be a contradiction that HCN, an extremely toxic compound, is believed to have had a central role in the origins of life. Review the role of cyanide as an inhibitor of electron transport (Chapter 16) and explain why there is no inconsistency between the toxicity of HCN and its proposed role in life's origins.

4. Given the structures of ATP, NAD, CoA, FAD, etc., it is generally concluded that coenzymes were formed on the primitive Earth at the same time as the nucleic acids. Explain.

5. The proponents of the idea that nucleic-acid-like molecules were the most important prebiotic polymers in primitive life suggest that coenzymes may have been covalently incorporated into these structures. What structural units common to nucleic acids and coenzymes suggest the possible incorporation of coenzymes in nucleic acids? What would have been the advantage to having coenzymes incorporated in nucleic-acid-like molecules?

6. a. The biosynthesis of histidine is cited as evidence for the hypothesis that the imidazole ring was derived from and was initially incorporated into a primitive nucleic acid. Explain.

   b. Those who do not agree with this conclusion point out that the biosynthesis of histidine is a very complicated multistep process. Explain why the complexity of histidine biosynthesis can be used as an argument against its origin from a nucleic acid precursor.

7. Suppose that a living organism has just been found on a planet in another solar system. It looks very unlike any organism found on Earth, and yet it contains poly(alpha-amino acids) that perform various catalytic and structural functions. Would you expect these foreign "proteins" to contain the same amino acids as our proteins? Explain.

VI

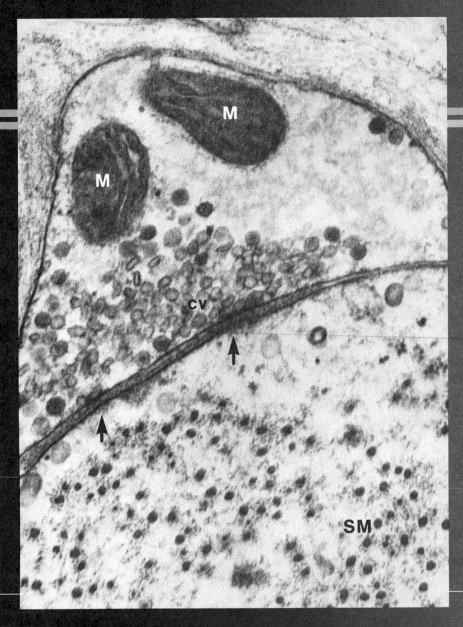

Neuromuscular junction on vascular muscle in the mollusc *Aplysia* (a sea slug). This electron micrograph shows details of the specialized membranes and organelles associated with a nerve terminal and its target smooth muscle fiber. The small clear vesicles (CV) are involved in the release of neurotransmitter serotonin (5-hydroxytryptamine). Serotonin is released at the active zones (arrows), diffuses across the intercellular cleft, binds to receptors on the muscle membrane, and causes contraction of the fiber. Thick contractile myofilaments are distributed throughout the muscle cytoplasm, each surrounded by a number of much thinner filaments (M = mitochondrion, and SM = smooth muscle fiber cut in cross section). (Magnification approximately 90,000×.) (Micrograph courtesy Dr. C. Price.)

MEMBRANE-ASSOCIATED REACTIONS

PART VI

In Chapter 6 we considered the structures and some of the functions of biological membranes. There and in subsequent chapters we showed that membranes are involved directly or indirectly in such diverse phenomena as transmembrane transport, lipid biosynthesis, respiration, photosynthesis, hormone secretion and reception, protein synthesis, and cell division. The detailed mechanisms by which membrane constituents participate in these various processes are, for the most part, still poorly understood. But recent advances in the technology necessary for studying biological membranes and their components hold great promise. Perhaps soon we will understand the details of membrane-related phenomena to the degree, for example, that we understand the molecular mechanisms of such enzymes as chymotrypsin (Chapter 9).

In the following chapters we will examine three membrane-related processes that are, from a biochemical standpoint, among the most extensively studied. The first of these, transmembrane transport (Chapter 34), is probably the most ubiquitous of membrane functions because of its role in nutrient acquisition and energy generation, and because the activities of membrane transport proteins are often necessary for other membrane-related functions. Two such related processes are considered in Chapters 35 and 36— neurotransmission and vision. In these chapters, we shall see how the activities of certain transport proteins, in this case ion channels, are necessary for the conduction of nerve impulses and for the translation of light energy, impinging on the retinal cells of the eye, into an image of our surroundings by the brain. As we shall also see, many of the techniques and principles introduced in these chapters will have general applicability to the study and understanding of membrane-related processes in a wide variety of different cell types and organisms.

34

MEMBRANE TRANSPORT

In Chapter 6 we considered the current state of knowledge about the structure and assembly of biological membranes. Two of the most important conclusions to be drawn are that membranes are both _dynamic_ and _asymmetric_ structures. As we shall see, these properties allow _transmembrane_ processes to be carried out, often, but not always, in a unidirectional manner. The most prominent of these processes are the transport of nutrients and inorganic electrolytes into the cell and the export of toxic substances and waste materials out of the cytoplasm. In both instances, the problem is usually the same: how do small, hydrophilic metabolites traverse the hydrophobic and generally impermeable lipid bilayer at rates sufficient to satisfy the maintenance and growth needs of the cell? Often this problem is solved by integral membrane proteins, which, by analogy to enzymes, specifically interact with their substrates, but, unlike enzymes, have as their primary purpose the movement of a molecular species across a membrane rather than the catalysis of a chemical reaction.

Only recently have sufficiently general techniques been developed for the purification of integral membrane proteins, not the least of which is the use of detergents to allow their study and separation in a soluble form (Chapter 6). Accordingly, our understanding of the molecular mechanisms of transport proteins, or _permeases_, has lagged considerably behind that of soluble enzymes. Nevertheless, in several instances, preparations of sufficient purity have been obtained to conduct physical, chemical, and mechanistic studies that should eventually reveal the molecular mechanisms of the "catalysis" of transport.

In this chapter, we shall consider the theory, energetics, and mechanisms of transmembrane transport in cells. We shall further explore a number of biological processes that rely on transport, directly or indirectly, for their proper functioning (Chapters 35 and 36). Examples of these seemingly diverse phenomena include nerve impulse propagation, cellular behavior and differentiation, and sensory reception and transduction. Indeed, it should become apparent that nearly all cellular and organismal functions are dependent on biological transport and its efficient functioning. As in Chapter 6, examples from both prokaryotic and eukaryotic worlds will show us that many fundamental principles concerning the biochemistry of transport and related functions are probably valid throughout the living kingdom.

THE THEORY AND THERMODYNAMICS OF BIOLOGICAL TRANSPORT

Before we consider specific examples of transport across biological membranes, we shall discuss the thermodynamics of this process, as well as specific ways in which such a transfer can occur. With this foundation, it becomes possible to assign certain predicted characteristics to each kind of transport mechanism. And, on the basis of these predictions, it becomes possible to design experiments that will give useful information regarding the events responsible for the transmembrane movement of a solute.

Most Solutes Are Transported by Specific Carriers

In principle, a solute could move across a barrier such as a membrane by either of two fundamentally different mechanisms: *simple* (or *passive*) *diffusion* or *carrier-mediated transport*. In the first mechanism, the molecule in question must be somewhat soluble in the hydrophobic phase of the bilayer, while in the second mechanism, the solute's journey across the membrane is facilitated by another molecule, in most cases a protein. For these reasons, simple diffusion is usually of importance in biological transport systems only for molecules having a largely hydrophobic character or in membrane systems possessing nonspecific aqueous pores, as will be discussed in a later section. An exception to this rule, however, is water, which traverses many biological membranes by simple diffusion. Most other hydrophilic molecules, which comprise the majority of species transported into and out of cells, are usually recognized by transport systems that include one or more protein carriers.

In order to illustrate the characteristics of simple diffusion, let us consider two dilute solutions of a solute So, one twice as concentrated as the other, which are separated by a barrier that is permeable to So, as in Figure 34–1. As a consequence of random thermal motion, a certain proportion of molecules originally present on one side of the barrier (one-quarter in Figure 34–1) will move to the other side in a specified length of time t. This *flux* is bidirectional, so that after time t, the ratio of the concentration of So on the left side of the barrier to that on the right will have been reduced from 2:1 to 7:5. After many such time intervals, this ratio will approach 1:1, or equilibrium. The number of molecules of So that move from left to right in time interval t is therefore directly proportional to the initial concentration of So on the left side:

$$v \propto [So] \qquad (1)$$

while a similar relationship holds true for movement in the opposite direction:

$$v' \propto [So'] \qquad (2)$$

where v and v' refer to rates of transfer from left to right and right to left, respectively, and $[So]$ and $[So']$ are the initial concentrations of the solute on the left and right sides of the barrier. The net rate of transfer V from the more

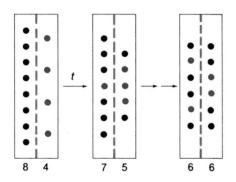

8 4   7 5   6 6

Figure 34–1
Illustration of simple diffusion in a two-compartment system divided by a semipermeable membrane. After time t, one-fourth of the molecules present originally on one side of the membrane have diffused by random thermal motion to the other side. After many such intervals, equilibrium is approached. The molecules in this illustration are identical, but different colors are used to indicate their initial locations. In simple diffusion, the *number* of molecules diffusing from one side to the other in any given time interval is directly proportional to their concentration on the first side. (Adapted from K. D. Neame and T. G. Richards, *Elementary Kinetics of Membrane Carrier Transport*, Blackwell Scientific Publications, Oxford, 1972.)

concentrated to the more dilute solution is therefore the difference between these two component rates:

$$V = v - v' \quad \text{or} \quad V \propto [So] - [So'] \tag{3}$$

Hence, at equilibrium, $V = 0$.

If the barrier in Figure 34–1 is, for example, a biological membrane of finite thickness l, we also must take this parameter into account in the expression for overall transfer rate:

$$V \propto \frac{[So] - [So']}{l} \tag{4}$$

Finally, the rate of transfer of So will depend on such factors as temperature, viscosity of the solvent, and the solubilities of So in the solvent and in the membrane. In a given membrane system, these factors often can be held relatively constant, so that we can express relationship (4) as an equation:

$$V = \frac{D([So] - [So'])}{l} \tag{5}$$

where D is a constant, the *diffusion coefficient*, which has the dimensions of area per unit of time (e.g., cm^2/sec). Equation (5) is essentially *Fick's law*, and it shows that the net rate of diffusion is directly proportional to the concentration gradient of So across the membrane barrier.

It should be apparent from Equation (5) that in a system in which So traverses the membrane only by simple diffusion, a plot of V versus $[So] - [So']$ should give a straight line that passes through the origin with a slope of D/l (Figure 34–2a). In practice, it is often possible to measure V as a function of the concentration gradient of So across, for example, the plasma membrane of a cell by methods to be considered later in this chapter. Provided that a wide range of solute concentration gradients is studied, *a linear plot of V versus [So] − [So'] is evidence that So is taken up by simple diffusion*.

The uptake kinetics shown in Figure 34–2a for simple diffusion are, in fact, seen only in a minority of cases for transport of molecules across cytoplasmic membranes of cells. More often, the curve approximates that of a *rectangular hyperbola*, as illustrated in Figure 34–2b. In this case, the transport rate approaches a limiting value at high $[So] - [So']$ that is termed V_{max} by analogy to steady-state enzyme kinetics (Chapter 8). This *saturation* phenomenon implies that uptake of So is dependent on a discrete number of *carriers* in the membrane, all of which become filled at high concentration gradients of So. This situation is formally analogous to saturation of an enzyme by its substrate, and the rate equation for this process, *carrier-mediated transport*, is essentially the same as the Henri-Michaelis-Menten equation derived in Chapter 8:

$$V = \frac{V_{max}[So]}{K_m + [So]} \tag{6}$$

In the derivation of Equation (6), $[So']$ is assumed to be negligible (just as is $[P]$ in enzyme kinetics) and is taken to be the initial concentration of So inside the cell or vesicle being studied. Experimentally, this situation is usually easy to attain, so that $[So]$ equals the concentration initially present outside, $[So'] = 0$, and V is an *initial rate of transport* of So into the system enclosed by the membrane. As in enzyme kinetics, K_m is the solute concentration at which $V = \frac{1}{2}V_{max}$ (Figure 34–2b). It should be noted that the value of K_m, although often assumed to be a measure of the affinity of a carrier for its "substrate," is really a ratio of rate constants and is not equivalent to the dissociation constant of the solute from its carrier.

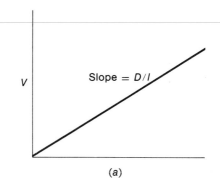

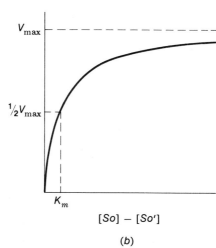

Figure 34–2
Initial rate kinetics of simple diffusion (a) and carrier-mediated transport (b). Plots of velocity versus the difference in substrate concentrations on the two sides of the membrane give a straight line with a slope of D/l in (a) and a rectangular hyperbola approaching a maximum velocity (V_{max}) at high $[So] - [So']$ in (b).

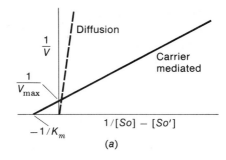

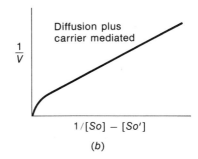

Figure 34–3
Double-reciprocal (Lineweaver-Burk) plots of carrier-mediated transport and simple diffusion. For diffusion, the line passes through the origin, while for carrier-mediated processes the x intercept is equal to $-1/K_m$ and the y intercept is equal to $1/V_{max}$, as shown in (a). In (b), the transport kinetics for a cell transporting a given solute by *both* simple diffusion and a carrier-mediated system are illustrated. The line shown is a combination of the processes plotted in (a) and passes through the origin. In the case of two carrier-mediated systems, a similar line would be obtained that would intersect the ordinate above the origin. (Adapted from K. D. Neame and T. G. Richards, *Elementary Kinetics of Membrane Carrier Transport,* Blackwell Scientific Publications, Oxford, 1972.)

From the foregoing discussion it is clear that in many cases, by studying the rate of transport, V, as a function of [So], simple diffusion can be distinguished from carrier-mediated transport. Often, however, the data so obtained are plotted in a double-reciprocal fashion (Lineweaver-Burk plot; Chapter 8). Plots of 1/V against 1/[So] yield straight lines in both instances, but the line for simple diffusion passes through the origin, while that for carrier-mediated transport intersects the ordinate at $1/V_{max}$ and the abscissa at $-1/K_m$ (Figure 34–3a). This, of course, reflects the fact that carrier systems are saturable (V_{max} is finite), while simple diffusion systems are not ($V_{max} = \infty$; $1/V_{max} = 0$).

Unfortunately for biochemists interested in studying a single biological transport system, double-reciprocal plots for the uptake of a specific solute into a cell, or vesicles derived from cells, are sometimes nonlinear. This usually indicates that *more than one process is involved in the transport of this substance into the cell*. These different processes may include both simple diffusion and carrier mediation, or both may be carrier-mediated (Figure 34–3b). In either case, however, it is first necessary to find experimental conditions under which only the system of interest is functional before conclusions concerning this transport mechanism can be drawn. This may be accomplished by working at solute levels recognized only by one carrier, by the use of mutants lacking one system, by using specific inhibitors, or even by purification and reconstitution of the carrier of interest (see the following discussion).

Finally, carrier-mediated transport can in many cases be distinguished from simple diffusion by its stereospecificity. For example, in a membrane system that is permeable to glucose only by simple diffusion, both D- and L-glucose are found to traverse the membrane at equal rates. In contrast, many biological membranes contain carriers specific for glucose that recognize only the D-stereoisomer. In these cells, the initial transport rate for D-glucose is much higher than that for the L-isomer and shows saturation kinetics. In fact, the rate of transport of L-glucose is often used in such instances to correct the total rate observed for diffusion in order to obtain the "true" carrier-mediated rate for the D-isomer. *Stereospecificity in biological transport, as in enzyme-catalyzed reactions, is a consequence of the highly selective configuration of the "active site" of the transport protein for its "substrate."*

Transport Against a Concentration Gradient Requires Energy

In the preceding section we saw that a solute will tend to diffuse "down" its concentration gradient (from a region of higher to a region of lower concentration) spontaneously, until equilibrium is reached. Such a process requires no outside input of energy into the system and thus occurs with a negative change in free energy ΔG. This is a consequence of the fact that the equilibrium situation ([So] = [So'] in Figure 34–1) is more "random" or "disordered" than the starting situation (in which [So] > [So']) and thus has relatively more entropy S. Because $\Delta G = \Delta H - T\,\Delta S$, an increase in entropy results in a negative free energy change (Chapter 12). The magnitude of ΔG can be calculated from the expression

$$\Delta G = 2.3RT \log \frac{[So']}{[So]} \tag{7}$$

for the situation illustrated in Figure 34–1. *Net transport from left to right will occur spontaneously in this system only if [So']/[So] < 1, while at equilibrium, ΔG = 0 ([So'] = [So]).*

In biological systems, both simple diffusion and *facilitated diffusion* are

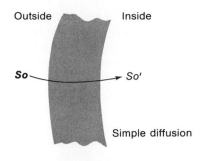

Outside   Inside

$So \longrightarrow So'$

Simple diffusion

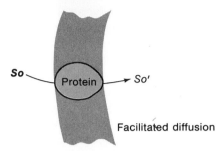

$So \longrightarrow$ Protein $\longrightarrow So'$

Facilitated diffusion

Figure 34–4

Schematic illustration of diffusion processes in biological transport. In both simple and facilitated diffusion, the concentrations of the solute on both sides of the membrane are the same at equilibrium, and no metabolic energy is expended.

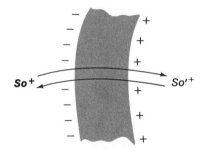

$So^+ \longrightarrow So'^+$

Figure 34–5

At equilibrium, a *charged* solute, So^+, may be unequally distributed across a biological membrane if an independently maintained potential exists across it. This situation can occur even if So^+ is transported by simple or facilitated diffusion. Bold type indicates higher concentration.

transport mechanisms that result in net transport of a solute only in the direction of negative ΔG, i.e., usually from higher to lower concentrations. In contrast to simple diffusion, facilitated diffusion is a carrier-mediated process and hence is saturable (Figure 34–4). Both it and simple diffusion generally result in *equilibration* of a solute across a membrane, unless the solubility properties of the solute are significantly different on both sides of the membrane or the solute has a net charge. *If the solute is charged, both types of diffusion can lead to an equilibrium situation in which [So] ≠ [So'] if there is an independently maintained electrical potential, $\Delta\psi$, across the membrane* (see Chapter 12):

$$\Delta G = 2.3RT \log \frac{[So']}{[So]} + ZF\,\Delta\psi \qquad (8)$$

In this equation, Z is the net charge on So, and F is the Faraday constant. Thus, at equilibrium ($\Delta G = 0$), unequal concentrations of So can exist on either side of a membrane as long as the term $ZF\,\Delta\psi$ is not equal to zero (Figure 34–5). The relationship between free energy and membrane potential is very similar to the relationship between free energy and electromotive potential discussed in Chapter 16.

Both simple and facilitated diffusion are examples of *energy-independent* biological transport mechanisms. Although they occur in fundamentally different ways (one is carrier-mediated and one is not), neither requires the input of energy, nor can either lead to concentration of the solute on one side of the membrane relative to the other in the absence of a charged solute and an electrical potential across the membrane. Most cells, however, can maintain relatively high intracellular concentrations of some metabolites and ions relative to the extracellular environment. For example, many cells have an intracellular level of K^+ that is at least 30 times the concentration of this ion in the surrounding medium. It therefore follows that if carriers for accumulated molecules, such as K^+, exist in the cell membrane, they must be able to transport their substrates in the direction of a positive ΔG, i.e., up a concentration gradient of the solute. In order to accomplish this thermodynamically unfavorable movement, *such processes must be coupled to events that have negative ΔG values*, i.e., that can yield energy to "push" the solute "uphill." *Concentrative accumulation of a solute inside a cell, organelle, or vesicle with the consequent expenditure of metabolic energy is termed active transport.* In a related process, *group translocation*, energy for the transmembrane accumulation of a solute is provided by chemical modification of the molecule as it passes through the membrane.

Active Transport and Group Translocation Are Used to Concentrate Solutes

The use of metabolic energy to drive active transport implies that the transport system itself must somehow be able to harness the energy in a particular form and use it to do work, i.e., to transport a solute against its concentration gradient. The molecular mechanisms by which these interconversions take place are still, in most cases, obscure, although a few clues are emerging that we will consider later in this chapter. Because systems that carry out active transport are invariably carrier-mediated (almost by definition), it seems likely that in most cases integral membrane permeases themselves will directly participate in this energy interconversion process. A major direction of research in membrane-transport biochemistry in recent years has therefore been to study the energetics of active transport and, whenever possible, to isolate and characterize the proteins that comprise a particular system.

Two basic types of active transport have been recognized. In *primary*

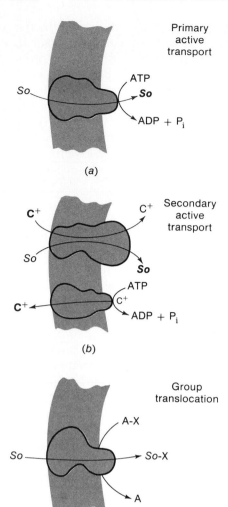

Figure 34–6
Schematic illustrations of primary active transport (a), secondary active transport (b), and group translocation (c). In (a), the energy source can be ATP, as shown, or electron transport or light can be used. In (b), active accumulation of So is energized by cotransport of an ion that flows down its concentration gradient. In (c), So is accumulated as a derivative, So-X, which is formed by a chemical reaction catalyzed by the carrier during transport through the membrane.

active transport, energy is provided directly by the hydrolysis of ATP, by electrons flowing down an electron transport chain, or by light. Examples of these include the Na$^+$-K$^+$ ATPase of animal cells, which pumps K$^+$ into the cell and Na$^+$ out of the cell at the expense of ATP; the respiratory electron-transport chain, which pumps H$^+$ out of the mitochondrion (Chapter 16), and the light-driven H$^+$ pump of *Halobacterium*, bacteriorhodopsin. In secondary active transport, ion gradients across the membrane, themselves created by active transport systems, are used to drive the concentrative uptake of other ions or metabolites. Because an ion gradient is a form of potential energy, a process involving its collapse releases this energy, which can be used to perform work. In mitochondria and bacteria, the potential energy stored in gradients of H$^+$ is used during respiratory metabolism to synthesize ATP (Chapter 16). As we shall see, gradients of H$^+$ as well as other ions also are an important source of energy for active transport in both prokaryotic and eukaryotic cells.

Related to primary active transport is the process of group translocation, a term coined by Peter Mitchell over 25 years ago. While molecules transported by active systems are translocated across the membrane without chemical changes occurring, in group translocation the transported solute is chemically modified on its journey through the membrane. The best-documented example of group translocation occurs in many bacteria, which convert sugars taken up from the surrounding medium into sugar monophosphates simultaneously with their transport through the membrane. Energy is expended in the process, since the phosphoryl donor is the high-energy phosphate glycolytic intermediate, phosphoenolpyruvate. In this case, the permease for the sugar is also an enzyme that catalyzes both the transport and the phosphorylation of its substrate. Figure 34–6 illustrates schematically the differences between primary and secondary active transport and group translocation, while Table 34–1 summarizes modes of energy coupling for selected transport systems of both animal and bacterial cells.

ENERGY-COUPLING MECHANISMS IN BIOLOGICAL TRANSPORT

It is the ultimate goal of the biochemist interested in transmembrane transport mechanisms to understand at the molecular level the physicochemical properties of a transport protein that allow it to carry out its function. In this section, we shall briefly describe experimental systems commonly used to study biological transport and show how these systems can be used to investigate energy-coupling mechanisms at the molecular level. In so doing, we shall place emphasis on some of the best-characterized biological transport systems. With this as a background, we will then venture one step further in the next section and attempt to explain how the structures and properties of integral membrane permeases permit them to carry out the myriad of transport processes so vital to the cell.

Isotopes, Analogs, and Specially Prepared Vesicles Are Used to Study Transport

In order to measure the rate of transport of a solute into a cell, some method must be available for easily identifying it and distinguishing molecules that have indeed crossed the membrane. Most commonly, radioactively labeled solutes are employed for this purpose, and time-dependent uptake of radio-activity into a cell is evidence that the compound is being transported through the cell membrane. With many animal cells grown in culture, it is often sufficient to separate intracellular label from that remaining in the

TABLE 34–1
Energy-coupling Modes of Well-studied Transport Systems

Solute	Polarity of Pump	Energy Coupling	Organism/Tissue
Na^+-K^+	Out-in	Primary active (ATP) (Na^+-K^+ ATPase)	Animal cells
Ca^{2+}	Out	Primary active (ATP) (Ca^{2+} ATPase)	Sarcoplasmic reticulum
H^+	Out	Primary active (ATP) (H^+ ATPase)	Plasma membrane of stomach epithelial cells
H^+	Out	Primary active (light) (bacteriorhodopsin)	*Halobacterium halobium*
H^+	Out	Primary active (electron transport)	Mitochondria, bacteria
H^+	Out	Primary active (ATP) (F_1/F_o ATPase)	Bacteria (anaerobic)
Various sugars and amino acids	In	Primary active (chemical energy; binding protein systems)	Gram negative bacteria
Lactose	In	Secondary active (H^+ cotransport)	*E. coli* and some other bacteria
Glucose	None	Facilitated diffusion	Most animal cells
Glucose	In	Secondary active (Na^+ cotransport)	Certain animal cells
Glucose	In	Group translocation (phosphotransferase system)	*E. coli* and some other bacteria
HCO_3^--Cl^-	None	Facilitated diffusion (anion channel)	Erythrocytes
ATP-ADP	Out-in	Facilitated diffusion, but sensitive to membrane potential (ATP/ADP exchanger)	Mitochondria

medium by quickly washing the cells (which adhere to the vessel in which they are grown) with buffer or medium that is free of the solute being tested. With bacteria and small, free-living eukaryotes such as yeast and protozoa, separation of cells from the medium for this purpose can be accomplished quickly by centrifugation or filtration.

In a few special instances, rapid and simple spectrophotometric assays have been developed to measure transport rates into cells. The most prominent example of this has been the use of o-nitrophenyl-β-galactoside (ONPG) to study lactose transport in E. coli. ONPG is transported into the cell by means of the lactose permease (lacY gene product) and is subsequently cleaved by β-galactosidase inside the cell to galactose and o-nitrophenol (Figure 34–7), which is yellow and can be quantitated spectrophotometrically. As long as transport is the rate-limiting step in this process, the appearance of o-nitrophenol in cell suspensions can be used to measure the transport rate of ONPG by the lactose carrier. The convenience of this assay has allowed detailed studies of the mechanism of lactose transport in E. coli.

One of the problems associated with measuring transport rates, especially in whole cells, is the fact that subsequent metabolism of the transport substrate may affect the rate observed. In order to uncouple transport from metabolism, nonmetabolizable analogs of some solutes have been used. Ex-

Figure 34–7

o-Nitrophenyl-β-galactoside (ONPG) is a convenient substrate for measuring transport by the lactose permease in *E. coli*. Upon entering the cell, it is cleaved by β-galactosidase, yielding galactose and o-nitrophenol, which can be quantitated spectrophotometrically.

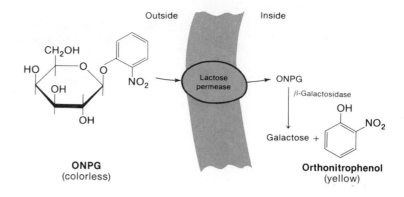

ONPG
(colorless)

Orthonitrophenol
(yellow)

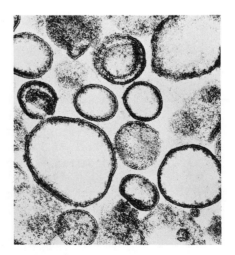

Methyl-α-glucoside

2-Deoxyglucose

Figure 34–8

Methyl-α-glucoside and 2-deoxyglucose are two analogs that are useful in the study of glucose transport in many cells. Usually they are not metabolized beyond the hexose-phosphate stage, which avoids the complications of subsequent reactions that might affect the measurement of transport rates.

amples of compounds that have been employed to study the transport of glucose, a common carbon and energy source for many cells, are *methyl-α-glucoside* and *2-deoxyglucose* (Figure 34–8). Both compounds cannot be metabolized beyond the hexose-phosphate stage by most cells. Another technique that has been useful in this regard, especially in bacteria, is the isolation of mutants that are able to transport a given solute, but that lack one or more enzymes for the further metabolism of the compound. By these procedures, it is often possible to determine many characteristics of the transport system itself, without any complications owing to other reactions that take place after the solute enters the cell.

Finally, in recent years, vesicles derived from the membrane system of interest have become favorite tools of the biochemist interested in membrane transport phenomena. Cytoplasmic membrane vesicles often can be isolated from eukaryotic cells after mild homogenization or N_2 cavitation (Chapter 6), which leave many of the intracellular organelles intact. In bacteria, the use of membrane vesicles was pioneered in the 1960s by H. R. Kaback, who developed a method, similar to that shown in Figure 6–6, for preparing membrane vesicles from *E. coli*. Most of these vesicles (Figure 34–9) appear to be "right side out"; i.e., the orientations of integral membrane proteins appear to be the same as those in the intact cell membrane. Kaback and others have shown that many substances transported by *E. coli* also are taken up by these membrane vesicles as long as the required energy sources are provided. These vesicles have the advantage of lacking most cytoplasmic enzymes, so that metabolism of the transport substrate is severely limited. Furthermore, they are especially suited for the study of energy-coupling mechanisms in active transport because the investigator can easily introduce various energy

Figure 34–9

Electron micrograph of membrane vesicles obtained from *E. coli* spheroplasts subjected to osmotic lysis. These closed, spherical structures have been shown to transport a variety of compounds normally accumulated by whole cells as long as any required energy source is provided. They also have the advantage of lacking cytoplasmic components that could affect the transport rate (77,000×). (From H. R. Kaback, Transport studies in bacterial membrane vesicles, *Science* 186:882, 1974. Reprinted with permission.)

sources into the vesicle preparations and test their effects on the transport system of interest. Such experiments are sometimes difficult to interpret with whole cells, again because of the complications of multiple metabolic events occurring in the cytoplasm.

Examples of Facilitated Diffusion

Transport systems that appear to operate only by facilitated diffusion have been identified both in bacterial and animal cells. Glycerol, which can act as the sole carbon and energy source for many bacterial cells, has been shown to be transported in several bacteria by nonconcentrative systems. Saturation studies as well as genetic analyses using *E. coli* by E. Lin and his collaborators have provided evidence for a "facilitator protein" that allows glycerol permeation in this organism. The facilitator increases the rate at which glycerol crosses the phospholipid bilayer but cannot accumulate it within the cell.

Recent experimental work on the glycerol facilitator has shown that this permease exhibits characteristics of a fairly nonspecific pore. The transport system allows a variety of straight-chain polyols (tetritols, pentitols, and hexitols, in addition to glycerol) to enter the bacterial cell. The analogous sugars, present in a ring form, cannot be transported, presumably owing to steric hindrance. Other small, neutral, or zwitterionic molecules including urea, glycine, and D,L-glyceraldehyde are transported by the system. Moreover, glycerol and xylitol do not effectively compete with each other as transport substrates, and these compounds do not exhibit saturation kinetics at physiological concentrations. Transport is relatively insensitive to temperature, which is suggestive of a diffusion process. Taken together, these observations suggest that the glycerol facilitator is a nonspecific aqueous channel with an inner diameter of about 0.4 nm. No other nonspecific pore in the inner cytoplasmic membrane of *E. coli* is known.

In animal cells, glucose, a common energy source, appears to be transported by a variety of mechanisms depending on cell type. Both active (see the following discussion) and passive glucose transport systems have been characterized, sometimes both in the same type of cell. In the human erythrocyte, glucose is apparently transported only by facilitated diffusion. The carrier for this sugar in red blood cells has been isolated in Triton X-100 by P. Hinkle and coworkers and reintroduced into phospholipid vesicles. The reconstituted carrier was shown to rapidly equilibrate glucose across the bilayer, but no evidence was found for its ability to carry out active transport, as shown in Figure 34–10.

A third example of facilitated diffusion in biological membranes is the process carried out by the anion-transport (band 3) protein of erythrocytes, whose structure we considered in Chapter 6. The anion transporter allows CO_2 accumulated in the erythrocyte from respiring tissues to rapidly diffuse out of the red blood cell in the lung capillaries in the form of the bicarbonate anion HCO_3^-. This process has been shown to be accompanied by influx of Cl^- into the erythrocyte. Much work has demonstrated that *the anion transporter catalyzes the one-for-one exchange of HCO_3^- for Cl^-*. These two molecules always flow in a fashion determined by the sums of their concentration gradients, and thus this protein catalyzes a type of facilitated diffusion. Transport of HCO_3^- out of the cell is *obligatorily coupled* to the influx of a monovalent anion such as Cl^-, as shown in Figure 34–11. This process, called *exchange diffusion*, is an example of a transport process termed *antiport*. *Antiport involves the coupled transport of two different molecules that move in opposite directions across a membrane.* "Antiporters" may carry

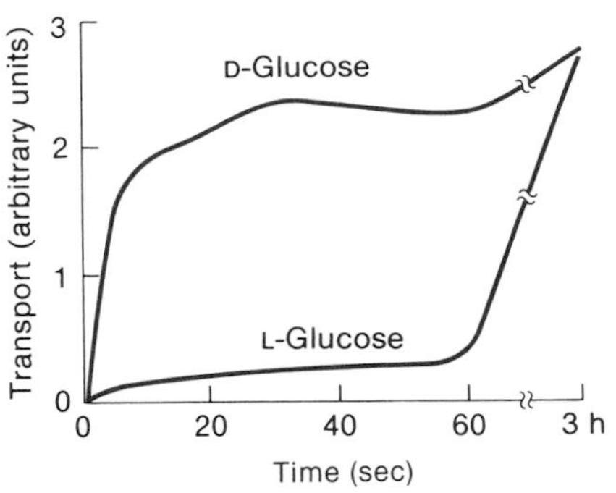

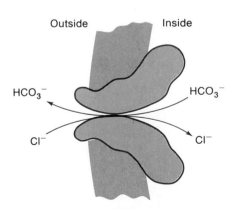

Figure 34–10

Facilitated diffusion carried out by the glucose carrier from human erythrocytes. The solubilized carrier was introduced into phospholipid vesicles, and transport was measured with D-glucose and L-glucose as substrates. The initial transport rate was much greater for D-glucose, but L-glucose, which accumulates in these vesicles by simple diffusion, eventually reaches the same interior concentration as D-glucose. (Adapted from M. Kasahara and P. C. Hinkle, Reconstitution and purification of the D-glucose transporter from human erythrocytes, *J. Biol. Chem.* 252:7384, 1977.)

Figure 34–11

Exchange diffusion of anions carried out by the human erythrocyte anion-channel (band 3) protein. This exchange is important in lung capillaries for the removal of CO_2 generated by cellular metabolism from the red cell cytoplasm.

out facilitated diffusion, as does the anion-transport protein, or they can catalyze active transport, as discussed below.

The Na⁺-K⁺ ATPase Is an Active Transport System

One of the first biological transport systems to be studied in detail was that responsible for maintaining the normally high concentrations of K^+ and low levels of Na^+ inside animal cells. In order for this situation to occur, both cations must be transported essentially unidirectionally against their concentration gradients. A breakthrough in explaining how this could be accomplished came in the 1950s when J. C. Skou identified an enzyme from crab nerves that *hydrolyzes ATP to ADP only in the presence of both Na^+ and K^+.* This *sodium-potassium ATPase* has subsequently been shown to be a *transmembrane protein* that pumps Na^+ out of the cell and K^+ into the cell at the expense of the energy contained in the β-γ-phosphate anhydride bond of ATP. This coupling of chemical energy to concentrative transport is an example of primary active transport, as well as of an active antiport system. The same protein catalyzes both the ion translocations and the hydrolysis of ATP, as shown in Figure 34–12. By a mechanism we will explore later in this chapter, the energy released by ATP hydrolysis is "translated" by the Na^+-K^+ ATPase into potential energy in the form of Na^+ and K^+ electrochemical gradients across the membrane. These gradients may then be used to drive secondary active transport processes, as we shall discuss shortly.

An extremely important tool in the study of transport mechanisms has been the use of *specific inhibitors*. In the case of the Na^+-K^+ ATPase, *ouabain* (Figure 34–13), a compound isolated from certain plants, specifically binds to the outer surface of the enzyme and blocks both Na^+ and K^+ translocation events as well as ATP hydrolysis. This observation independently confirms the obligatory coupling of these three processes as catalyzed by a single protein. Furthermore, it has been demonstrated that K^+ has a much higher affinity for the outside surface of Na^+-K^+ ATPase, while Na^+ binds much more tightly to the cytoplasmic domain. Finally, the site of ATP hydrolysis has been localized to the inner surface of the cytoplasmic mem-

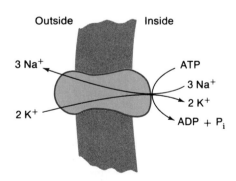

Figure 34–12
Primary active transport of Na^+ and K^+ by the sodium-potassium ATPase found in many animal cells. This enzyme maintains the normally high K^+ and low Na^+ concentrations within the cell at the expense of chemical energy stored in ATP. Each cycle of the enzyme results in the extrusion of three molecules of Na^+ out of the cell, the transport of two molecules of K^+ into the cell, and the hydrolysis of one molecule of ATP. The subunit structure and stoichiometry of this enzyme have been established. One $\alpha\beta$ heterodimer is the active catalytic unit. The α subunit ($M_r = 112,200$) is associated with all known catalytic properties of the enzyme complex, while the β subunit ($M_r = 57,000$) is a glycoprotein of unknown function.

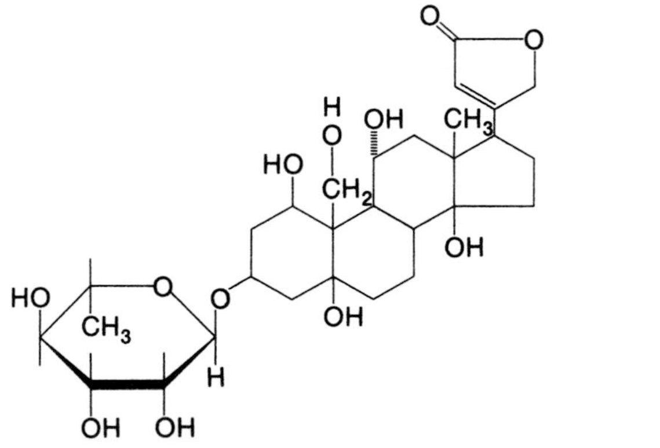

Figure 34–13
The structure of ouabain, a specific inhibitor of the Na^+-K^+ ATPase. Ouabain is a member of a class of compounds called *cardiotonic steroids*. *Digitalis*, the general name for a family of cardiotonic steroids that inhibit the Na^+-K^+ ATPase, indirectly causes an increase in intracellular Ca^{2+}, which is due to increased cytoplasmic Na^+ that arises because the cardiac sarcolemma contains a Na^+/Ca^{2+} antiport system. The increase in the intracellular Ca^{2+} concentration appears to directly stimulate contraction in cardiac muscle cells. This explains the use of digitalis in treating such conditions as congestive heart failure.

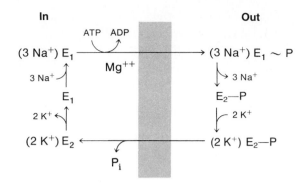

Figure 34–14

Postulated mechanism for the functioning of the Na^+-K^+ ATPase. The grey boundary separates events in which cation association or dissociation occurs on the inside and outside surfaces of the membrane, while the arrows relating E_1 and E_2 denote conformational changes in the enzyme, but are not meant to imply its physical translocation across the membrane. The enzyme is hypothesized to have at least two conformations, E_1 and E_2, both of which can exist in a phosphorylated state. $E_1{\sim}P$ is a high-energy phosphoryl bond, in equilibrium with the β-γ pyrophosphoryl bond of ATP. By contrast, E_2—P is of low energy, in equilibrium with inorganic phosphate. The changes in bond energies are presumed to be due to the free energies of Na^+, K^+, and Mg^{2+} binding. It should be noted that because ion binding may induce secondary conformational changes, both E_1 and E_2 may exist in both "empty" and "full" conformations. The "full" conformations, $E_2(K^+)$ and $E_1{\sim}P(Na^+)$, contain occluded (2 K^+) and (3 Na^+) portions, respectively. E_1 and E_2—P have their transport sites facing in and out, respectively. Only E_1 possesses high affinity for Na^+, while only E_2 possesses high affinity for K^+. Phosphorylation of E_1 drives the cyclical process in a unidirectional fashion, causing Na^+ to be translocated from the inside to the outside. Conversion of $E_1{\sim}P$ to E_2—P decreases the affinity of the enzyme for Na^+ and increases its affinity for K^+. Dephosphorylation of E_2 drives translocation of K^+ from the outside to the inside. Conversion of E_2 to E_1 decreases the affinity of the enzyme for K^+ and enhances its affinity for Na^+. Release of K^+ from E_2 is greatly accelerated by ATP binding. Translocation of Na^+ out of the cell and K^+ into the cell is the net consequence of this series of conformational changes in the protein. Ouabain inhibits by binding to E_2—P, preventing dephosphorylation in the presence of K^+.

brane. These observations support the schematic illustration shown in Figure 34–12 and demonstrate that the Na^+-K^+ ATPase must be *asymmetrically oriented* in the cytoplasmic membrane.

Because ouabain binds on the outside of the Na^+-K^+ ATPase and yet inhibits ATPase activity on the cytoplasmic surface, information must be conducted through the membrane by means of *conformational changes* in the enzyme. Studies with purified Na^+-K^+ ATPase have shown that the protein is *covalently phosphorylated* during its catalytic cycle, the site of phosphorylation being an aspartic acid residue. Phosphoryl Na^+-K^+ ATPase is formed in the presence of Na^+ and ATP (plus Mg^{2+}), while the dephosphorylation step is K^+-dependent. Ouabain binds tightly to the enzyme-phosphate intermediate and prevents dephosphorylation in the presence of K^+.

These observations, and others, have led to the formulation of the model depicted in Figure 34–14 for the functioning of the Na^+-K^+ ATPase. Transport of Na^+ out of the cell and of K^+ into the cytoplasm is thought to be mediated by conformational changes in the enzyme triggered by phosphorylation, dephosphorylation, and the binding of ligands at both surfaces. While Figure 34–14 illustrates a reasonable mechanism for Na^+-K^+ ATPase, it says little about how conformational changes in the protein can lead to unidirectional transport of ions through the membrane. Although this question has yet to be satisfactorily answered in any system, a number of models have emerged for the molecular mechanisms of such translocation events. These will be considered later in this chapter, after we describe a number of other well-characterized transport systems.

Other Ion-translocating ATPases Exist for H^+, Na^+, K^+, and Ca^{2+}

Muscle cells contain a specialized endoplasmic reticular membrane system called the *sarcoplasmic reticulum*. This organelle acts as a repository for Ca^{2+} in resting muscle, and electric stimulation of muscle cells causes rapid

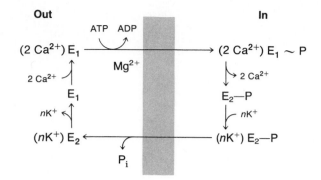

Figure 34–15

Proposed mechanism for the Ca^{2+} ATPase of sarcoplasmic reticulum. Phosphorylation of the enzyme by ATP and the binding of Ca^{2+} and of ATP plus Mg^{2+} are all believed to play roles in causing the enzyme to cycle in an essentially irreversible fashion from its conformation with the cation binding site on the inside of the membrane to that with its cation binding site exposed to the outside, and back again. All interconversions are believed to occur by mechanisms that are analogous to those described in Figure 34–14 for the Na^+-K^+ ATPase. The reactions in this sequence have been shown to be reversible, but are driven in the direction shown by the hydrolysis of ATP. An obligatory involvement of K^+ has yet to be established. The grey boundary and arrows relating E_1 and E_2 have the same significance as in Figure 34–14.

release of this ion from the sarcoplasmic reticulum into the sarcoplasm, where it triggers contraction of the myofibrils. Reuptake of Ca^{2+} into the reticular lumen is effected by a *calcium ATPase*, which catalyzes the following reaction:

$$2\ Ca^{2+}(out) + ATP(out) \xrightarrow{Mg^{2+}} 2\ Ca^{2+}(in) + ADP(out) + P_i(out)$$

In this reaction sequence, (out) refers to the outside of the sarcoplasmic reticulum (sarcoplasm), and (in) corresponds to the lumen of the system.

The Ca^{2+} ATPase has been extensively characterized, and like the Na^+-K^+ enzyme, it spans the membrane in an asymmetric fashion. In the presence of Ca^{2+} and ATP, formation of a covalent phosphoryl enzyme also can be demonstrated in this system. Thus many of the aspects of Ca^{2+} transport in the sarcoplasmic reticulum parallel those of Na^+ and K^+ translocation in the cytoplasmic membranes of animal cells. Again, conformational changes in the protein, triggered by bound ligands and the state of phosphorylation of the enzyme, are hypothesized to be involved in the translocation event, as shown in Figure 34–15.

A third ion-translocating ATPase is the H^+-translocating enzyme found in the cytoplasmic membrane of epithelial cells lining the stomach. This enzyme is responsible for acidification of the lumen and appears to be structurally related to the Na^+-K^+ and Ca^{2+} ATPases. For example, each of the three enzymes has a large, catalytic polypeptide chain with a molecular weight of about 100,000 that is transiently phosphorylated during transport. H^+-translocating ATPases of similar structure have also been identified in fungi and in yeast cells. Other H^+-translocating ATPases have been found in secretory granules (e.g., chromaffin cells), clathrin-coated vesicles, vacuoles of *Neurospora crassa*, lysosomes, and possibly Golgi vesicles as well. All of the ion pumps that form a phospho-enzyme intermediate are inhibited by pentavalent vanadate (VO_4^-), which, because of its trigonal bipyramidal structure, seems to be a transition-state analog of phosphate and "locks" the enzyme in the transition state.

All of these ion-translocating ATPases, in which a mixed anhydride phosphoryl enzyme intermediate drives the cycle of ion translocation, have been shown to be structurally, and presumably evolutionarily, related. Extensive sequence homology has been demonstrated between the Na^+-K^+ ATPase and the Ca^{2+} ATPase of animal cells, as well as a K^+ ATPase of *E. coli*. These enzymes are quite distinct in structure and in mechanism of action from the H^+-translocating F_1/F_o ATPase found in bacteria, mitochondria, and chloroplasts. The latter enzymes do not form a phosphoryl enzyme intermediate and are not inhibited by vanadate.

A fourth well-characterized ion-translocating ATPase was discussed in Chapter 16. The F_1/F_o *ATPases* (or *ATP synthases*) of mitochondria, chloroplasts, and bacteria appear to have a similar function in all three systems. Proton gradients formed as a result of electron transport (see the following discussion) are used to synthesize ATP from ADP and P_i as a consequence of H^+ flow *down* its electrochemical gradient through the F_1/F_o ATPase (Figure 34–16*a*).

It is also possible, however, for the F_1/F_o ATPase to work in the reverse direction, i.e., to hydrolyze ATP and translocate protons *against* a concentration gradient (Figure 34–16*b*). This fact has been established in some bacteria under conditions where respiratory electron transport does not occur (e.g., anaerobiosis or the absence of a functional electron-transport chain as in some strict anaerobes). Thus the F_1/F_o ATPase also can be classified as a protein complex carrying out primary active transport by pumping H^+ against a concentration gradient using the chemical energy stored in ATP. Unlike Na^+-K^+, Ca^{2+}, and H^+ ATPases of animal cell plasma membranes, however, there is no evidence that F_1/F_o ATPases are covalently phosphorylated during proton translocation. As we shall see in a later section, proton gradients generated by the F_1/F_o ATPase can be used to provide energy for the uptake of nutrients into bacteria under anaerobic conditions.

The structure of the mitochondrial F_1/F_o ATPase was described in Chapter 16, and the chloroplast and bacterial enzymes appear to have very similar structures. The F_1 portion, which possesses the ATPase activity, consists of five different subunits (named α through ϵ), each of which has a similar molecular weight in each of these three sources, as summarized in Table 34–2. The F_o portion, which presumably spans the membrane, has been shown to be the proton channel. It consists of three polypeptides (named a, b, c) with molecular weights of 28,000, 19,000, and 8,500 respectively, in a subunit ratio 1:2:6. While subunits a and b are believed to function in binding F_1 to F_o, the c subunits, which contain bound lipid, function as the proton channel. Removal of F_1 from membranes results in an increased permeability of the bilayer to protons, suggesting that F_1 acts as a "cap" that couples proton translocation to ATP synthesis or degradation. The F_o portion, which is rich in hydrophobic amino acids, has been shown to cause proton conductivity when incorporated into artificial membrane vesicles.

Elegant reconstitution studies by Y. Kagawa and coworkers have led to a model of how the various protein subunits of the F_1/F_o ATPase may be arranged on the membrane. As a result of information gained by purifying each

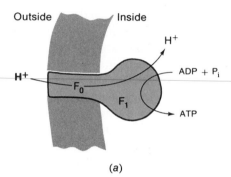

(a)

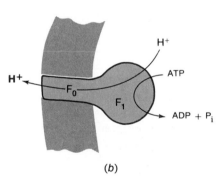

(b)

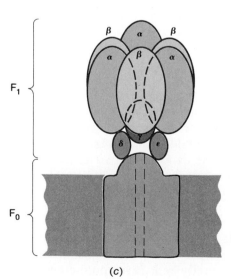

(c)

Figure 34–16
The F_1/F_o ATP synthase or ATPase of bacterial, mitochondrial, and chloroplast membranes. (*a*) A proton gradient generated by electron transport can be used for ATP synthesis by means of chemiosmotic coupling. (*b*) Alternatively, in bacteria growing anaerobically, ATP hydrolysis can be used to *generate* a proton gradient in the reverse reaction. This gradient can then be used to energize secondary active transport systems as well as for other processes in bacteria that use chemiosmotic energy, such as flagellar motility. (*c*) Reconstitution studies of the F_1/F_o ATPase from the thermophilic bacterium PS3 have led to a model for the arrangement of the subunits in the F_1 portion of this enzyme. (Adapted from M. Yoshida, H. Okamoto, N. Sone, H. Hirata, and Y. Kagawa, Reconstitution of thermostable ATPase capable of energy coupling from its purified subunits, *Proc. Natl. Acad. Sci. USA* 74:936, 1977.)

TABLE 34–2
Subunit Compositions of F$_1$ ATPases

Source	Molecular Weight ($\times 10^3$)	Subunit Sizes ($\times 10^3$)[a]				
		α	β	γ	δ	ϵ
E. coli	380	56	52	32	22	11
Thermophilic bacterium (PS3)	380	56	53	32	15	11
Beef heart mitochondrion	360	54	50	33	17	11
Chloroplast	325	59	56	37	17	13

Source: Adapted from D. B. Wilson and J. B. Smith, Bacterial transport proteins, in B. P. Rosen (ed.), *Bacterial Transport*, Dekker, New York, 1978.

[a] The stoichiometry of subunits in the F$_1$ ATPase is still controversial; it is probably $\alpha_3\beta_3\gamma\delta\epsilon$ for the bacterial enzymes.

subunit from the thermophilic bacterium PS3 and recombining them in various ways, the structure depicted in Figure 34–16c was proposed. In this model, the complex of α and β subunits ($\alpha_3\beta_3$), which possesses ATPase activity by itself, is bound to the F$_o$ portion through subunits δ and ϵ. Subunit γ is believed to act as the "cap" or gate through which protons flow. This conclusion was reached through experiments that showed that membranes containing F$_o$ and the δ plus ϵ subunits of F$_1$ were permeable to protons, but addition of the γ subunit blocked proton conductance. Because the subunit structures and functions of F$_1$/F$_o$ ATPases from a number of sources are very similar, it seems reasonable to expect that the functions and topographies of the individual protein subunits will be related as well.

Some Transport Systems Are Driven by Electron Transport or Light

Membrane-bound electron-transport chains have as a primary function the extrusion of protons out of mitochondria and bacteria and into chloroplasts of plants and algae (Chapter 16). These gradients can then be used for ATP biosynthesis, as described earlier, or for secondary active transport processes linked to ion translocation (see the following discussion). The exact mechanisms of proton translocation during electron transport remain a matter of controversy (see Chapter 16), and both unidirectional protein channels and coenzyme Q may be involved. In any case, energy derived from the oxidation-reduction reactions that take place within the membrane is more or less directly transformed into a chemical gradient of protons. For this reason, electron transport is generally thought of as a primary active transport event.

Light energy can likewise be used to drive active transport in a number of systems. For example, in photosynthesis, light energy absorbed by chlorophyll molecules is converted into a proton gradient at least partly by means of the electron-transport reactions that subsequently take place. In halophilic bacteria, such as *Halobacterium halobium*, a remarkable light-driven proton pump, bacteriorhodopsin, has evolved in part to allow this organism to synthesize ATP under anaerobic conditions. The structure of bacteriorhodopsin, which we considered in Chapter 6, in some manner allows it to actively transport protons from inside the cell to the outside in response to light absorption.

In 1971, Walter Stoeckenius and his coworkers showed that if O$_2$ was limited in cultures of *Halobacterium*, a purple membrane appeared on the surface of the cell in addition to its normally red, carotenoid-containing

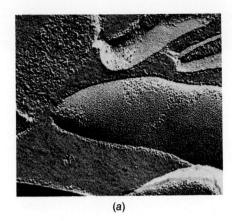

(a)

(b)

(c)

Figure 34–17

Bacteriorhodopsin (BR). (a) Freeze-fracture electron micrograph of the membrane of *H. halobium*. Fine-grain regions are patches of purple membrane containing bacteriorhodopsin. (Magnification 48,000×.) (Courtesy Walter Stoeckenius.) (b) Bacteriorhodopsin contains a covalently bound molecule of retinal, a derivative of vitamin A, attached by means of a Schiff base linkage to a lysine residue of the protein. Both the all-*trans* isomer and the 13-*cis* form can be isolated from the protein. Retinal is also the light-gathering chromophore of the vertebrate retina

where it exists as the 11-*cis* isomer attached by means of a Schiff base to the protein opsin (Chapter 31). Light absorption by the retinal moiety of bacteriorhodopsin results in a transient conversion of the form, absorbing maximally at 570 nm (BR_{570}), to one having a maximum absorption at 412 nm (BR_{412}). This transition is accompanied by a deprotonation of the Schiff base, as shown in (c), in a process believed to be involved in proton translocation by bacteriorhodopsin.

membrane. The purple membrane is 75 per cent protein by weight, and bacteriorhodopsin is the sole protein component. The purple color (λ_{max} = 570 nm) is due to the covalent association of one molecule of *retinal*, a derivative of vitamin A, to a lysine residue of bacteriorhodopsin by means of a Schiff base linkage (Figure 34–17a). It is the retinal moiety that absorbs photons and is responsible, at least partly, for the conversion of this form of energy into a proton gradient that can be used for ATP synthesis.

Although the photochemical reactions of bacteriorhodopsin are extremely complex, it is believed that *light absorption by retinal leads to deprotonation of the Schiff base* to a form that absorbs light maximally at 412 nm. Reassociation of a proton with the retinal moiety then takes place slowly, and the bacteriorhodopsin protein cycles back to the 570-nm form through several intermediate conformations that can be detected spectrophotometrically (Figure 34–17b). It is highly probable that this protonation-deprotonation cycle involving multiple protein conformational states is a key to the mechanism of proton translocation and thus energy transformation carried out by bacteriorhodopsin. A molecular model for how this process may occur will be discussed later in this chapter.

Some Bacterial Systems Use Specific Binding Proteins

The last primary active transport mechanisms we shall consider are the so-called *binding protein transport systems* of the Gram negative bacteria. As discussed in Chapter 6, proteins that bind specific sugars, amino acids, and

TABLE 34-3
Binding Protein Transport Systems in Gram Negative Bacteria

Organism	Substrate	Molecular Weight of Binding Protein ($\times 10^3$)
E. coli	Phosphate	41
	Leucine/isoleucine	36
	Glutamine	25
	Lysine/arginine/ornithine	27
	Arabinose	38
	Ribose	30
	Maltose	37
	Galactose/glucose	35
Salmonella typhimurium	Sulfate	31
	Histidine	26
	Ribose	31
Salmonella enteritidis	Galactose/glucose	35

Source: Adapted from D. B. Wilson and J. B. Smith, Bacterial transport proteins, in B. P. Rosen (ed.), *Bacterial Transport*, Dekker, New York, 1978.

inorganic ions have been shown to be localized to the periplasmic spaces of these organisms. Some of the compounds for which such binding proteins have been demonstrated are listed in Table 34–3. Cold osmotic shock of cells or the transformation of the culture to spheroplasts releases these proteins, and a concomitant decrease in the ability of these cells to transport binding protein-linked substrates is observed. Thus these periplasmic proteins apparently are important either in concentrating the substrate at the cell surface or in interacting with the transmembrane permeases (or both). In all cases that have been studied, however, the binding proteins are not themselves responsible for the transmembrane transport of the molecules they recognize. Rather, specific integral membrane proteins exist for this purpose (Figure 34–18).

The integral membrane proteins that, in part, comprise the binding protein-dependent systems have been characterized in several instances. One such system is the maltose transport system in *E. coli*. It is known that four distinct proteins are required for the translocation of maltose across the cytoplasmic membrane. These are illustrated in Figure 34–18. In addition to the maltose-binding protein in the periplasm (the E protein), two integral membrane proteins (F and G) probably span the membrane. While the E protein associates with the transmembrane transport components on the external surface, the K protein is localized to the inner surface, probably owing to a specific association with the G protein (Figure 34–18). The K protein functions not only as an essential transport component, but also in the regulation of transport activity.

Energy for the active accumulation of binding protein substrates is apparently provided by ATP or a high-energy metabolic derivative of it. The evidence for this is that arsenate, which greatly reduces the intracellular ATP pool, is generally a strong inhibitor of binding protein-dependent transport. Furthermore, compounds such as fructose, which increase the ATP concentration in the cell by means of substrate-level phosphorylation, are stimulatory to the uptake of some binding protein substrates. However, an electron donor such as D-lactate is unable to stimulate these systems in *E. coli* cells that lack a functional F_1/F_o ATPase. Since D-lactate is metabolized mainly by oxidative phosphorylation in *E. coli*, the proton gradient

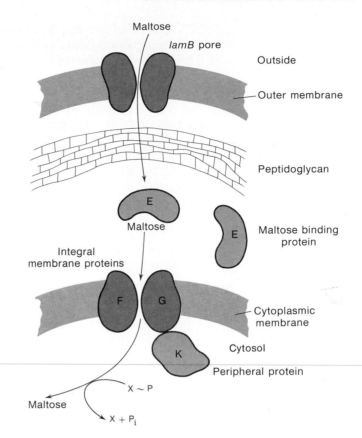

Figure 34–18

Schematic illustration of solute transport by the binding protein systems of Gram negative bacteria using the maltose system as an example. Maltose passes through a pore in the outer membrane (the *lamB* gene product, which is also the receptor for bacteriophage λ), interacts with its periplasmic binding protein (E) and is then transported by inner membrane permeases (F and G) in a process dependent on a high-energy phosphate compound. The K protein is a peripheral membrane protein that is necessary for maltose transport, but its exact role in this process is unknown (see text). The localizations and interactions of these proteins have been determined by biochemical and genetic analyses. The *malE* binding protein has been shown to interact both with the *lamB* protein pore in the outer membrane and with the maltose permease complex in the cytoplasmic membrane. It may act to facilitate and provide stereospecificity to the transport processes across both membranes.

resulting from its metabolism is unable to form ATP in such mutants and thus cannot promote uptake of the binding protein substrate.

The evidence just discussed appears to rule out any mechanism in which these binding protein systems are energized solely by a secondary mechanism, such as a proton gradient (see the following discussion). At least one exception appears to exist to this rule, however. The transport system for dicarboxylic acids such as succinate in *E. coli* has been shown to have a periplasmic binding protein and yet to probably be energized by a proton gradient. The method by which ATP or another high-energy phosphate compound donates its energy to support active transport in the majority of binding protein systems remains obscure. This is mainly due to the unavailability of purified integral membrane permeases from many of these systems. Such purified preparations will be necessary to determine, for example, if transient phosphorylation of the transport protein is a key step in the translocation event, as it is for the Na^+-K^+ and Ca^{2+} ATPases.

Secondary Active Transport Can Result When the Uptake of One Solute Is Coupled to the Uptake of Another Solute

In Chapter 16 we learned that a gradient of protons across a biological membrane is a form of potential energy that can be used to drive the synthesis of ATP from ADP and P_i. Expulsion of protons out of mitochondria or bacterial cells during electron transport results in both a difference in H^+ concentration [ΔpH] and in electric charge (ΔΨ; interior negative) across the membrane. Recall that both these components contribute to the *proton motive force* (pmf or Δp; see Chapter 16), which can be expressed as

$$\Delta p = \Delta\Psi - \frac{2.3RT}{F}\Delta pH \qquad (9)$$

The pmf, or Δp, is related to the free energy change experienced by protons, which tends to drive them across the cellular or organellar membrane (see Chapter 12). Since lipid bilayers are relatively impermeable to protons, such a flow must be mediated by membrane proteins. In the example of the F_1/F_o ATPase, the exergonic flow of H^+ through the enzyme, driven by Δp, is used to drive the endergonic synthesis of ATP. It also should be apparent that Δp *could be used to drive active transport if the inward flow of H^+ were coupled to the uptake of a particular solute*. Examples of such proton *cotransport* or *symport* mechanisms have now been established in mitochondria, bacteria, and lower eukaryotic cells. Because the primary active transport of H^+ is responsible for Δp, active transport systems dependent on Δp for energization are said to carry out *secondary active transport* by means of *chemiosmotic coupling*.

Undoubtedly the most well-characterized H^+ symport system is the lactose transport system of *E. coli*. This system has been extensively studied by P. Mitchell, T. H. Wilson, and H. R. Kaback, among many others. Evidence that the active accumulation of this sugar by means of the *lacY* gene product involves the cotransport of H^+ can be summarized as follows:

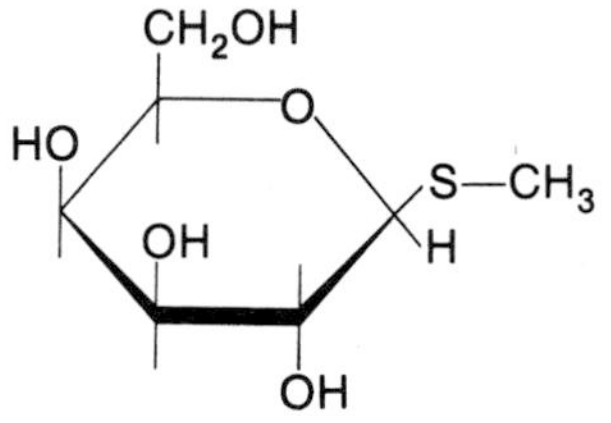

1. Agents such as 2,4-dinitrophenol (Figure 16–22), which collapse proton gradients across membranes, inhibit lactose transport in whole cells and membrane vesicles.
2. Uptake of lactose, or the nonmetabolizable analog *thiomethyl-β-D-galactopyranoside* (TMG; Figure 34–19), is stimulated by reducing the extracellular pH of energy-depleted cells.
3. A 1:1 ratio of entry of TMG and H^+ has been demonstrated in energy-depleted cells.
4. Lactose transport in *E. coli* membrane vesicles, which cannot synthesize ATP in the absence of a source of ADP, is greatly stimulated by compounds such as D-lactate, which are oxidized and donate electrons to the electron-transport chain in this system.
5. A mutation in the *lacY* gene has been isolated in which lactose uptake is not tightly coupled to H^+ entry (i.e., is relatively insensitive to Δp). This mutant protein is severely impaired in the active transport of lactose, but has an increased ability to facilitate its diffusion across the membrane compared with the normal permease.
6. The lactose permease has been purified and the transport function of the pure protein has been reconstituted in a vesicular phospholipid membrane. The active accumulation of lactose in response to an artificially imposed Δp has been demonstrated. In fact, the highly purified, reconstituted permease was found to exhibit essentially the same kinetic properties as the transport system embedded in *E. coli* membrane vesicles. This observation shows that a single polypeptide chain, the protein encoded by the *lacY* gene, is both necessary and sufficient for catalysis of lactose:proton symport in a process that is responsive to the transmembrane proton electrochemical gradient.

The preceding observations establish that the active accumulation of lactose in *E. coli* is obligatorily coupled to the simultaneous entry of H^+ flowing down its concentration and/or electrical gradients (Figure 34–20a). If this is true, then *a gradient of lactose (or TMG) should produce a Δp in cells in which this value is initially zero*. Indeed, it has been demonstrated that addition of TMG to the outside of energy-depleted cells causes acidification of the cytoplasm as a result of H^+-TMG symport, as mentioned earlier (Figure 34–20b).

The structure of the lactose permease is now fairly well understood. The most significant advance leading to the elucidation of its structure resulted

Figure 34–19
Thiomethyl-β-D-galactopyranoside (TMG), a nonmetabolizable substrate of bacterial lactose permeases. TMG has been useful in elucidating the H^+ symport mechanism of these proteins in the active transport of lactose.

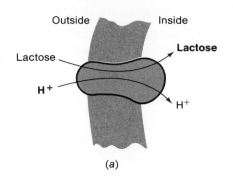
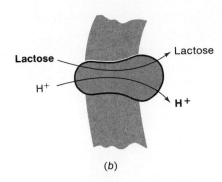
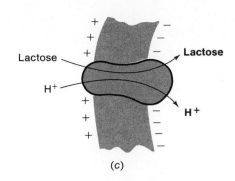

Figure 34–20
Transport events carried out by the bacterial lactose permeases. (a) In response to a proton gradient, lactose is actively accumulated in the cell. (b) In the presence of a lactose gradient, protons can be concentrated in the cytoplasm. (c) In the presence of a $\Delta\psi$, both lactose and protons can be actively accumulated. The physiological situation in respiring cells corresponds to (a), in which a $\Delta\psi$ is usually also present.

from the cloning and sequencing of the *lacY* gene. From the nucleotide sequence of the gene, the amino acid sequence of the protein was deduced. The amino acid sequence was then subjected to analysis by a computer method that revealed the distribution of hydrophobic versus hydrophilic amino acid residues along the linear sequence of the protein. These and other analyses led to the suggestion that the protein consists of twelve extended hydrophobic segments, with a mean length of 24 ± 4 residues per segment. Both the C- and N-termini appear to be localized to the cytoplasmic surface of the membrane. On the basis of circular dichroic measurements conducted with the purified lactose permease protein, approximately 85 per cent of the amino acid residues are probably arranged in helical secondary structure. Since about 70 per cent of the 417 amino acid residues were found to be included within the twelve hydrophobic segments, it was predicted that the embedded segments are largely α-helical, with each helix extending the entire thickness of the membrane, perpendicular to the plane of the membrane. This structure is reminiscent of the seven-helix structure described for bacteriorhodopsin (see Figure 6–22).

As a result of the application of biochemical and genetic approaches, the structure of the active site of the permease is now coming to light. The galactoside binding site resides within a segment of the protein that is embedded within the phospholipid bilayer. Replacement of either of two aminoacyl residues in the native protein with alternative residues changes the substrate specificity of the permease. Moreover, a specific histidyl residue has been implicated in H^+-binding. Current attempts to correlate altered structure with function should reveal the details of this sugar:H^+ symport system.

Transport systems such as the *E. coli* lactose permease, in which electrical charges are carried across the membrane without simultaneous compensation of the electrical potential so generated (for example, by the movement of other charged species across the membrane) are termed *electrogenic. Electrogenic transport systems are always affected by the electric potential $\Delta\Psi$ across a biological membrane, even if the transported ion is not H^+.* In the lactose transport system, a $\Delta\Psi$ (interior negative) is sufficient to drive active accumulation of this sugar in the absence of ΔpH because Δp has both $\Delta\Psi$ and ΔpH components [Equation (9)], as shown in Figure 34–20c. In animal cells and in some bacteria, such as *Halobacterium*, that live in high concentrations of Na^+, Na^+-solute symport systems are common means for the accumulation of salts, carbohydrates, and amino acids. These systems are also electrogenic and thus are sensitive to both $\Delta\Psi$ and $\Delta[Na^+]$, but not to ΔpH.

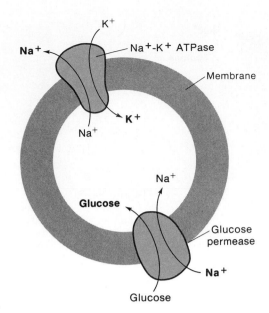

Figure 34–21
Na$^+$ symport is a common means of actively transporting sugars and amino acids in animal cells. For example, glucose is actively accumulated by this mechanism in intestinal and renal epithelial cells. The Na$^+$-K$^+$ ATPase maintains the Na$^+$ gradient essential to these secondary active transport systems.

The Na$^+$-K$^+$ ATPase in animal cells ensures that Na$^+$ flowing in through these Na$^+$-solute "symporters" is rapidly pumped back out to maintain a high [Na$^+$]out/[Na$^+$]in ratio (Figure 34–21).

These results establish the essential mechanistic features of cation-solute symport. Thus in the case of the lactose permease, a single protein alone catalyzes the tightly coupled, unidirectional transport of sugar and a proton. The stoichiometry of this process is 1:1. While these facts appear well-established, further work will be required to ascertain the detailed mechanistic features of the translocation process. For example, it is not yet known if the protein functions by a carrier-type or a channel-type mechanism, or by a process that incorporates mechanistic features of both models. These problems will be considered in greater detail later in this chapter. A list of some of the better-characterized secondary active transport systems is given in Table 34–4.

The Mitochondrial ATP/ADP Exchanger Expels ATP Because of a Membrane Potential

Because ATP is synthesized within the mitochondria of eukaryotic cells, while metabolic processes that utilize ATP are largely confined to the cytoplasm, a mechanism must exist for transport of this molecule across the

TABLE 34–4
Some Secondary Active Transport Systems

Substrate	Cotransported Ion	Organism/Tissue
Neutral amino acids	Na$^+$	Eukaryotic cells
Glucose	Na$^+$	Some animal cells (intestine, kidney, choroid plexus)
Lactose	H$^+$	E. coli and some other bacteria
Dicarboxylic acids	H$^+$	E. coli
Proline	H$^+$	E. coli
Glutamate	Na$^+$	E. coli and H. halobium
Melibiose	Na$^+$	E. coli and S. typhimurium
Alanine	H$^+$	Thermophilic bacterium PS3

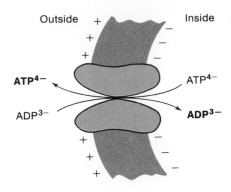

Figure 34–22
The ATP/ADP exchanger of the mitochondrial inner membrane. This dimeric protein carries out the exchange of intramitochondrial ATP for ADP formed in the cytoplasm by metabolic reactions. A membrane potential (interior negative) favors this exchange because of its electrogenic nature at pH values near neutrality.

inner mitochondrial membrane. The protein responsible for this function has been shown to be an *ATP/ADP exchanger* that carries out exchange of intramitochondrial ATP for ADP formed from metabolic reactions in the cytoplasm. This antiport is electrogenic because, at pH 7, ATP molecules average about one more negative charge than do ADP molecules (Figure 34–22).

Despite the fact that the [ATP]/[ADP] ratio is usually higher in the cytoplasm than inside mitochondria in actively respiring cells, the ATP/ADP exchanger still preferentially expels ATP from the organelle with the concomitant inward movement of ADP. This can be explained by the presence of a membrane potential $\Delta\Psi$ (inside negative), resulting, in part, from the Δp formed across the mitochondrial membrane during respiration. This potential favors outward transport of ATP and inward transport of ADP because this process results in the net translocation of one negative charge from inside to outside (Figure 34–22).

The ATP/ADP exchanger is an example of a transport system that does not fit neatly into either the category of facilitated diffusion or of active transport as we have defined them. On the one hand, the transport can be thought of as active because concentrative accumulation can occur and metabolic energy is sacrificed ($\Delta\Psi$ is partially collapsed as a result of ATP/ADP exchange). On the other hand, the ATP/ADP exchanger essentially corresponds to our definition of facilitated diffusion, in which the solute is charged, and an independent potential is maintained across the membrane [Equation (7)]. This situation reflects more the somewhat arbitrary definitions we have made by convention rather than any real confusion as to the overall transport reaction carried out by this protein.

The ATP/ADP exchanger has been purified by M. Klingenberg and his associates and has been shown to be a dimer of identical polypeptides (M_r = 30,000) in detergent solution and possibly in the membrane as well. Its dimeric structure may help explain the molecular mechanism by which the exchange reaction takes place, as we shall discuss in a later section.

The Bacterial PEP:Sugar Phosphotransferase System Exemplifies Group Translocation

The final mechanism of energy coupling in transport processes that we shall consider is group translocation. In this mechanism, chemical modification of the transported solute is part of the translocation step itself (Figure 34–6). The most famous example of group translocation is the *phosphoenolpyruvate(PEP)-dependent sugar phosphotransferase system* (PTS) in bacteria, which was discovered by Saul Roseman in 1964. The PTS both transports and phosphorylates sugars as they pass through the cytoplasmic membrane, with PEP as the phosphoryl donor. It consists of several soluble and membrane-bound proteins in *E. coli* that catalyze the following set of reactions:

$$\text{PEP} + \text{enzyme I} \rightleftharpoons \text{enzyme I}{\sim}\text{P} + \text{pyruvate} \tag{10}$$

$$\text{enzyme I}{\sim}\text{P} + \text{HPr} \rightleftharpoons \text{HPr}{\sim}\text{P} + \text{enzyme I} \tag{11}$$

$$\text{HPr} + \text{sugar}^a{}_{(in)} = \text{P}$$

$$\text{HPr}{\sim}\text{P} + \text{sugar}^a{}_{(out)} \xrightarrow[\text{(Enzyme III}^a)]{\text{Enzyme II}^a} \text{HPr} + \text{sugar}^a\,\text{P}_{(in)} \tag{12}$$

In this reaction scheme, enzyme I and HPr are soluble, cytoplasmic proteins that participate in phosphoryl transfer reactions common to all sugars transported by the PTS in *E. coli*. In contrast, enzyme II is an integral membrane protein that is usually specific for only one sugar (as indicated by the

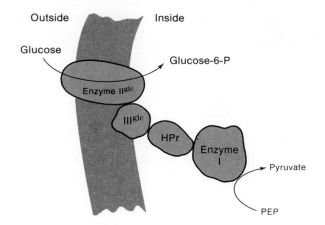

Figure 34–23

Schematic representation of the reactions carried out by the bacterial sugar phosphotransferase system (PTS). A phosphoryl moiety from PEP is sequentially transferred to enzyme I, HPr, and enzyme III^{glc} before its ultimate transfer to glucose by an integral membrane protein specific for this sugar (enzyme II^{glc}). Although the general PTS enzymes, enzyme I and HPr, are soluble proteins, there is some evidence that they are associated peripherally with the membrane as shown. In *E. coli* there are at least seven different enzymes II, each specific for the sugars recognized by the PTS in this organism: glucose, mannose, fructose, N-acetyglucosamine, mannitol, glucitol, and galactitol. While HPr has been shown to be a monomer ($M_r = 9,500$) and enzyme I a dimer (subunit $M_r = 70,000$), the subunit structures of the enzymes II are as yet unknown. The mannitol-specific enzyme II (or mannitol permease) has been shown to consist of a single kind of polypeptide chain ($M_r = 68,000$) (see Table 6–7).

superscript) and acts as the permease as well as the phosphoryl transfer enzyme. Enzyme III, which is required for the transport of some but not all sugars, also is sugar-specific and may or may not be membrane-bound. The spatial arrangement of the reactions that occur during PTS sugar transport in enteric bacteria is depicted in Figure 34–23. The PTS allows for unidirectional sugar transport because the sugar-phosphate product is very impermeable to the lipid bilayer and is "trapped" once it is transported into the cell. The cost to the cell is one ATP equivalent (in the form of PEP) for each molecule of sugar accumulated by means of the PTS. The importance of this fact will be discussed in the next section.

The phosphotransferase systems in *E. coli* and *Salmonella typhimurium* have been among the most extensively studied. All the general proteins have been purified and shown to be transiently phosphorylated, as shown in reactions (10) through (12). The phosphate is attached to a histidine residue in both enzyme I and HPr, and it is also known that the enzymes II and III are similarly transiently modified before the phosphate is transferred to the incoming sugar. Such enzyme phosphorylation reactions may provide the driving force for transport by means of conformational changes in the enzymes II, as it apparently does for the ion-translocating ATPases.

One of the enzymes II, that specific for the sugar mannitol, has been solubilized from the *E. coli* membrane, purified to homogeneity, and reconstituted in an artificial membrane. Additionally, the gene encoding the protein has been cloned and sequenced, so that like the lactose permease, the mannitol enzyme II is well understood both structurally and functionally. The N-terminal half of this protein ($M_r = 68,000$) is strongly hydrophobic, probably traversing the membrane seven times. The C-terminal half of the protein exhibits the hydrophilic properties of a typical water-soluble protein and is localized to the cytoplasmic (inner) surface of the membrane. This particular protein, which functions without the aid of an enzyme III, is believed to be phosphorylated twice during catalytic turnover. A phosphoryl group is first transferred from phospho HPr to the N-3 position of a histidyl residue in the C-terminal part of the protein; next, the phosphoryl group is transferred to another residue, possibly in the N-terminal half of the protein; and finally, dephosphorylation of this second residue is accompanied by the coupled phosphorylation and transport of the sugar across the membrane.

Group translocation mechanisms also have been proposed for the uptake of fatty acids and purine and pyrimidine bases in bacteria. The evidence for these is much weaker than that for the PTS. In the *E. coli* PTS, the one permease that has been purified, the mannitol-specific enzyme II, for example, has been shown to possess coupled sugar-phosphorylating and transport activities in a reconstituted phospholipid-enzyme II vesicular proteoliposome system. This establishes the tight and obligatory coupling of transport and phosphorylation by a single membrane protein. In order to demonstrate true group translocation in other systems, similar properties (i.e., tightly coupled translocation and enzymatic activities) must be observed for the permeases involved.

Energy Interconversion and Active Transport in Bacteria Are Regulated by Growth Conditions

In Chapter 6, the structural similarities between mitochondria and bacteria were pointed out. In this chapter we have seen that both mitochondria and many bacteria possess similar F_1/F_o ATPases and electron-transport chains in their membranes. In energy generation, however, *facultatively anaerobic bacteria* (able to live either aerobically or anaerobically) such as *E. coli* are considerably more versatile than mitochondria. They can generate energy either by aerobic respiration, as do mitochondria, or by anaerobic glycolysis, which does not take place in the eukaryotic organelle. This situation can best be illustrated by considering the pathways of energy interconversion that exist in many facultative anaerobes, as illustrated in Figure 34–24.

During aerobic respiration, a Δp generated by electron transport is used to drive ATP synthesis by means of the F_1/F_o ATPase and to energize H^+ symport and other electrogenic transport systems. Under anaerobic conditions, bacteria obtain energy in the form of ATP by substrate-level phosphorylation during glycolytic fermentation of sugars. *In order for Δp-linked processes to occur in the absence of O_2, protons must be expelled from the cell by reversal of the F_1/F_o ATP synthase reaction.* This has been shown to occur in *E. coli*, as well as in anaerobic bacteria, such as *Streptococcus lactis*, which

Figure 34–24
Energy interconversion pathways in anaerobic and facultatively anaerobic bacteria growing on glucose as the sole carbon and energy source. Note that ATP can be used to create a proton motive force by means of the F_1/F_o ATPase under anaerobic conditions. Energy flow that occurs in the absence of O_2 is indicated by the colored arrows, while all these reactions except formation of a Δp by ATP hydrolysis occur in the presence of oxygen.

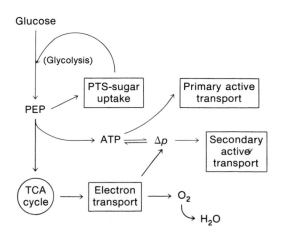

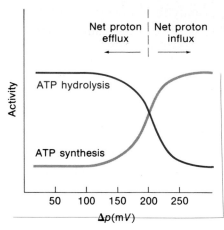

Figure 34–25

Illustration of the functioning of the bacterial F_1/F_o ATPase as a function of Δp. The enzyme has a "threshold" of around 200 mV, above which net proton influx and net ATP synthesis occur. Below this value, the enzyme hydrolyzes ATP to pump protons out of the cell. The latter situation generally occurs under anaerobic conditions. This is a highly schematic diagram, since the threshold Δp can vary depending on the organism and the cell growth conditions.

completely lacks a respiratory electron-transport chain and has as its only source of Δp the F_1/F_o ATPase. The energy-conversion reactions that take place under anaerobic conditions in these bacteria are indicated by the colored arrows in Figure 34–24.

Because respiration of a sugar such as glucose results in many more ATP equivalents than its fermentation, *E. coli* cells grow much more slowly in the absence of O_2. *In order to grow anaerobically, bacteria must make judicious use of the ATP formed.* An illustration of this is the fact that the *sugar-phosphotransferase transport system (PTS) is almost exclusively found in bacteria that can live anaerobically.* It is generally absent in strictly aerobic bacteria. The probable reason for this can be demonstrated by considering the energetics of these situations. Bacteria growing anaerobically on glucose use one ATP equivalent (as PEP) to both transport and phosphorylate this sugar by means of the PTS. Bacteria lacking a PTS must use a partial ATP equivalent to actively transport the sugar and another ATP in the hexokinase reaction to phosphorylate it in preparation for glycolysis. Thus the PTS is likely to have evolved as an efficient means of conserving energy during anaerobic growth of bacteria on a fermentable sugar.

Finally, it is reasonable to ask what dictates the direction of operation of the F_1/F_o ATPase. The answer appears to be that *there is a threshold value of Δp below which the enzyme does not work in the direction of ATP synthesis but only works in the direction of ATP hydrolysis.* This threshold value has been shown to be about 200 mV (negative inside) for *Streptococcus lactis* (Figure 34–25). Measurements by E. Kashket and her coworkers of Δp values in several actively growing facultative anaerobes are consistent with this threshold value. Under aerobic conditions, values of 200 mV or greater were measured, while Δp dropped considerably below this value when these bacteria were grown in the absence of O_2. Thus the pathways of energy flow and transport are finely regulated in these organisms to allow adaptation to a variety of growth conditions.

Lactose Permease Function Is Regulated by Glucose

In virtually all cells that have been studied, transport processes have been found to be subject to regulation. In animal cells, hormones can influence both the activities and the rates of synthesis of the permease proteins. Numerous examples of transport regulation have been documented in both prokaryotic and eukaryotic cells, and in a few cases, the mechanisms have been established.

One of the best-studied examples of carbohydrate transport regulation in *E. coli* involves the inhibition of lactose uptake by extracellular glucose. While lactose enters the cell by proton symport, glucose is transported by the phosphotransferase-catalyzed group translocation process. It is now clear that glucose influences the phosphorylation state of a *cytoplasmic regulatory protein*, RPr. When RPr is fully phosphorylated, the lactose permease is not inhibited and exhibits maximal activity. When the regulatory protein is dephosphorylated, however, RPr binds to an allosteric regulatory site on the cytoplasmic surface of the lactose permease, thereby inducing a less active conformation of the protein that transports lactose at low rates. Extracellular glucose, during transport into the cell, is thought to promote the dephosphorylation of RPr, thereby inhibiting lactose uptake. The allosteric regulatory protein RPr has recently been shown by direct biochemical means to be the glucose enzyme III of the PTS. This example illustrates that in bacteria, as well as in animal cells, complex regulatory phenomena have evolved to allow integration of the multiple facets of the biochemical machinery.

MOLECULAR MECHANISMS OF BIOLOGICAL TRANSPORT

In the last section we examined the wide variety of energy-coupling mechanisms used by cells to accumulate and expel nutrients and ions. In order to fully understand transport at the molecular level, however, it is necessary to determine how the structures of transport permeases allow the often unidirectional translocation of hydrophilic molecules through the hydrophobic lipid bilayer. It is reasonable to assume that such proteins will contain specific binding sites for their transport substrates by analogy to enzymes, but that unlike most soluble enzymes, permeases also must be able to carry out a *vectorial* process: "picking up" a substrate on one side of a membrane and depositing it on the other. It will become clear in our discussion that in no instance do we yet know the molecular details of how this process is accomplished. However, as more transport proteins are purified and their structures determined, clues are emerging that will undoubtedly allow this question to be answered for at least a few systems in the near future. In this section we shall examine these clues as well as models for the function of transport permeases.

Mobile Carriers or Pores Could Explain Transport

Two conceptual models have been put forth to explain the transport of a substance through a biological membrane. In the *mobile-carrier* mechanism, the transporter or a solute-binding moiety of the permease is assumed to physically change its orientation in the membrane during translocation, either by shuttling back and forth across the bilayer or by rotation through the plane of the membrane (Figure 34–26a). On the other hand, the *pore* model assumes that the protein is more or less fixed in its intramembrane orientation and forms a hydrophilic pore that is stereospecific for its transport substrate. This pore is usually thought of as being *gated* in the sense that it opens only transiently in response to proper solute recognition (Figure 34–26b).

As model systems for studying biological transport, the *ion-translocating antibiotics* have provided evidence for both of the mechanisms illustrated in Figure 34–26. *Valinomycin*, an antibiotic isolated from a species of the bacterium *Streptomyces*, is a cyclic depsipeptide containing the sequence D-valine, L-lactate, L-valine, and D-hydroxyisovalerate repeated three times (Figure 34–27a). It is extremely selective for binding K^+ and makes both biological membranes and artificial bilayers permeable to this cation. This can be explained by the fact that the valinomycin-K^+ complex effectively shields the hydrophilic groups of valinomycin on the interior of the molecule, while the hydrophobic side chains are exposed to solvent, as shown in Figure 34–27b. For this reason, the valinomycin-K^+ complex is soluble in lipid bilayers, while free K^+ is highly water-soluble.

A second well-studied antibiotic is *gramicidin A*, also isolated from a bacterium (*Bacillus brevis*). It is a 15-amino-acid linear polypeptide that is a cation-specific *ionophore* with much less specificity than valinomycin for any particular cation (Figure 34–28). The mechanisms of ion translocation carried out by both valinomycin and gramicidin A have been extensively studied. In one experiment, the conductance of an artificial phospholipid bilayer ($T_m = 41°C$) to K^+ was measured as a function of temperature in the presence and absence of these antibiotics. The results showed that K^+ conductance was high in this model system throughout the temperature range studied with gramicidin A. In contrast, the K^+ permeability of the bilayer in the presence of valinomycin was low (below 41°C) but sharply increased when the temperature was increased above this value. Hence ion transloca-

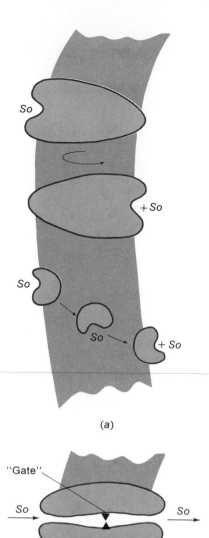

(a)

(b)

Figure 34–26
In theory, a solute could be transported across a membrane by a mobile carrier that either rotates through or traverses the plane of the membrane *(a)*, or by a more or less fixed channel or pore that may be controlled by a gate *(b)*. Examples of both have been found among the ion-translocating antibiotics.

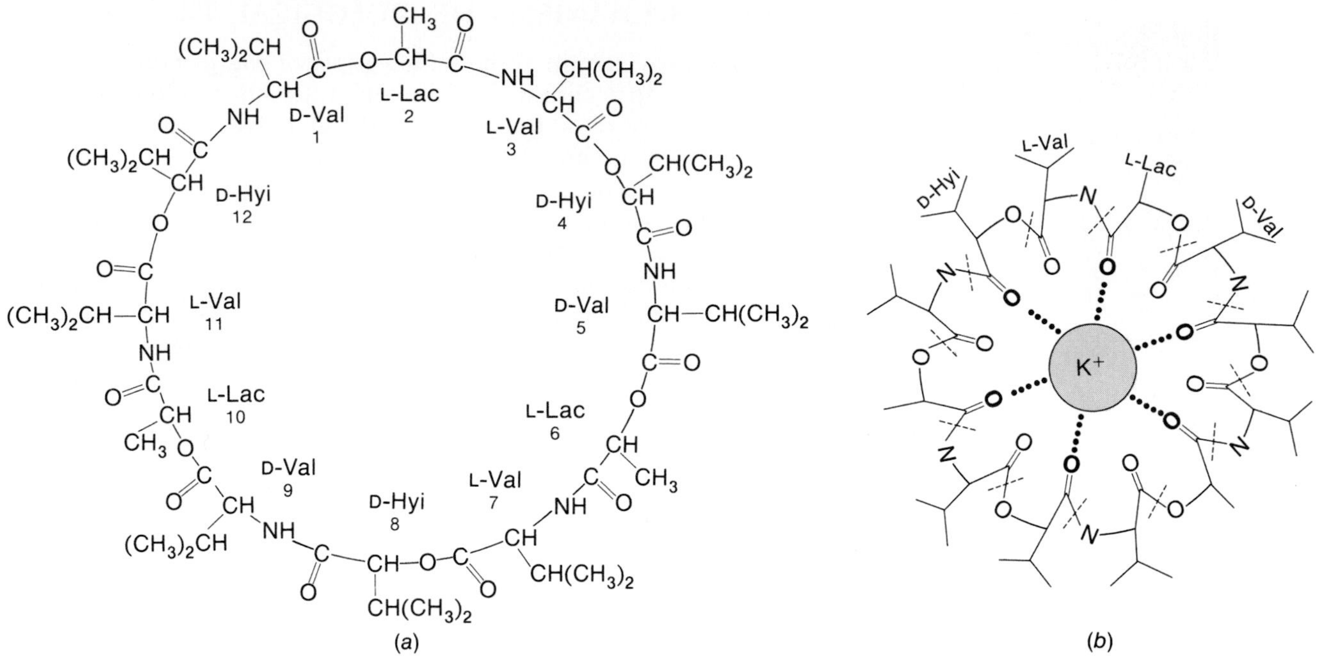

Figure 34–27

Valinomycin, an ionophore specific for K⁺. (a) Its chemical structure is a cyclic peptide containing the sequence D-valine (D-Val), L-lactate (L-Lac), L-valine (L-Val), and D-hydroxyisovalerate (D-Hyi) repeated three times. (b) Its complex with K⁺ is illustrated, showing that the hydrophilic groups complexing the ion are "buried" inside the "donut-shaped" molecule, while the hydrophobic side chains are exposed around the perimeter of the molecule. This makes the K⁺-valinomycin complex highly soluble in lipid bilayers. (The structure in b is adapted from B. C. Pressman, Biological applications of ionophores, *Ann. Rev. Biochem.* 45:501, 1976.)

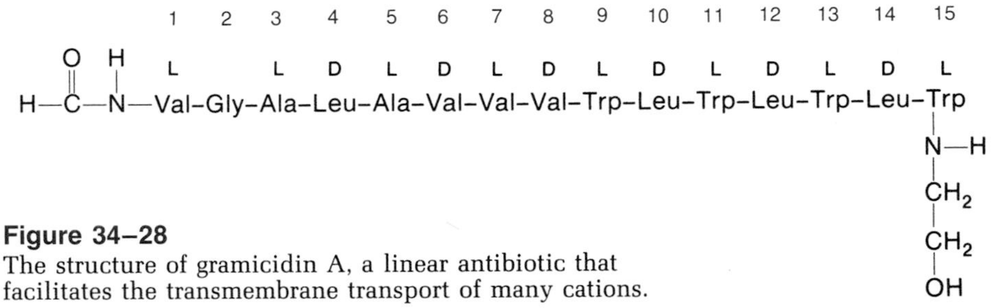

Figure 34–28

The structure of gramicidin A, a linear antibiotic that facilitates the transmembrane transport of many cations.

tion by valinomycin apparently depends on the physical state of the phospholipid bilayer, while the permeability induced by gramicidin A is insensitive to this parameter. This and many other experiments, including kinetic and physicochemical measurements, have led to the conclusion that *valinomycin is a mobile carrier that can diffuse rapidly through the bilayer only above the T_m, while gramicidin A forms a static pore through the bilayer and does not require an environment of low viscosity for its function.*

The mechanisms of ion translocation by valinomycin and gramicidin A as they are believed to occur are illustrated in Figures 34–29 and 34–30. A single valinomycin-K⁺ complex traverses the bilayer by simple diffusion. In contrast, physical measurements show that gramicidin A forms a transmembrane pore by means of head-to-head dimerization of helical monomers, each of which has a hydrophilic aqueous channel through the axis of the helix. Thus, at least in these model systems, both mobile-carrier and pore mechanisms of solute translocation can be readily demonstrated.

A vast array of both natural and synthetic ionophores of varying specificities has now been studied. In Chapter 16, proton ionophores such as 2,4-dinitrophenol were shown to be uncouplers of oxidative phosphoryla-

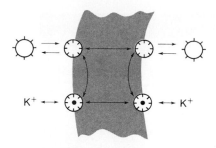

Figure 34–29

Mechanism of ion translocation by valinomycin. The carbonyl oxygens chelating K^+ are shown as "spokes" on the circular antibiotic molecule. In the aqueous phase, these oxygens are more freely available to the solvent; i.e., valinomycin undergoes a conformational change upon going from a lipid to an aqueous environment, as illustrated in the top of the figure. The lipid-soluble form is the one that complexes K^+ and traverses the membrane by diffusion, as shown in the lower part of the figure. (Adapted from Y. A. Ovchinnikov, Physico-chemical basis of ion transport through biological membranes: Ionophores and ion channels, *Eur. J. Biochem.* 94:321, 1979.)

← 2.5–3 nm →

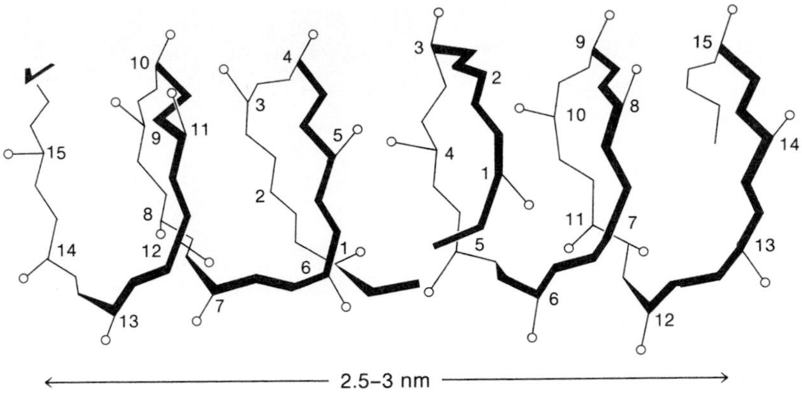

Figure 34–30

Head-to-head dimer channel formed by gramicidin A in lipid bilayers. Each monomer is a so-called $\pi_{L,D}$ helix formed by intramolecular hydrogen bonds between amino acid residues of the gramicidin molecule (numbered 1–15 from the L-valyl terminus). The length of the dimer, about 3 nm, is consistent with the length of the hydrophobic region of a typical phospholipid bilayer. (Adapted from Y. A. Ovchinnikov, Physico-chemical basis of ion transport through biological membranes: Ionophores and ion channels, *Eur. J. Biochem.* 94:321, 1979.)

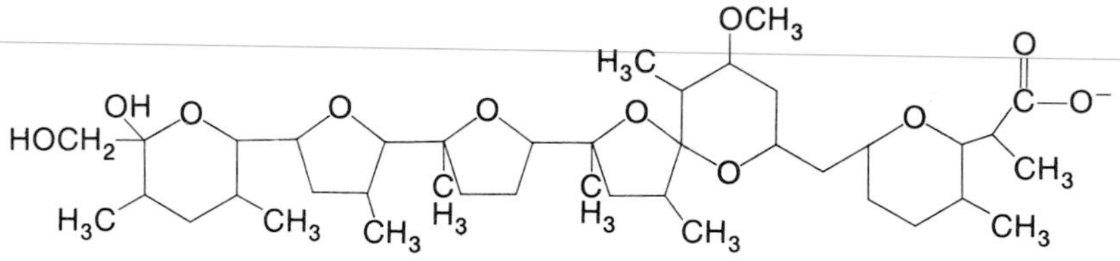

Figure 34–31

The structure of nigericin, a polycyclic ether carboxylic acid that exchanges H^+ for K^+ across membranes.

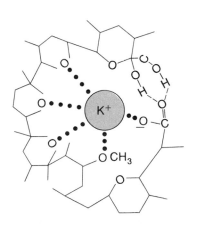

Figure 34–32

Circular complex formed between nigericin and K^+. Note that the negatively charged carboxylate oxygen is involved in chelating the cation. In order for nigericin, which has deposited K^+ on one side of the membrane, to return to the other side, this carboxylate must be protonated (neutral), resulting in a 1:1 exchange of H^+ for K^+. (Adapted from B. C. Pressman, Biological applications of ionophores, *Ann. Rev. Biochem.* 45: 501, 1976.)

tion. Because of its size and lipid solubility, 2,4-dinitrophenol is undoubtedly a mobile H^+ carrier. Another ionophore that also transports its substrates by diffusion is <u>nigericin</u>, a polycyclic ether carboxylic acid (Figure 34–31). Nigericin has the interesting property of catalyzing the one-for-one exchange of H^+ and K^+ across biological and artificial membranes. The reason for this is indicated in Figure 34–32. The nigericin-K^+ complex is circular, much like that formed with valinomycin, and the negatively charged carboxylate interacts with the cation in this state. This complex is freely soluble in the bilayer, so that K^+ is conducted to the other side. In order to return to the first side, this carboxylate anion must be neutralized. A nigericin-H^+ complex is one form in which it can diffuse back. The net result is the exchange of K^+ for H^+, and transport by nigericin, in contrast to that by valinomycin or gramicidin, is by necessity electroneutral.

Most Biological Transport Systems Use Pores

Studies on the topographies of integral membrane proteins that we considered in Chapter 6 suggest a more or less static and asymmetric disposition of these polypeptides across biological membranes. Furthermore, any large movements, such as rotations through the membrane plane (Figure 34–26a), are unfavorable, especially for charged membrane proteins with a considerable proportion of their mass in contact with the aqueous phase. In fact, considerable evidence leads to the conclusion that several classes of proteins that carry out transport do so by means of transmembrane pores or channels. These structures may or may not be static. Nonspecific transport, where

Figure 32–33
Arrangement of porin "pores" in the outer membrane of *E. coli*. Computer processing of many electron micrographs gives the image shown in (*a*), while a technique involving optical diffraction and filtering of micrographs gives the image shown in (*b*). A schematic interpretation of these images, shown in (*c*), suggests "triplet indentations" that are penetrated by the negative stain (*dark circles*) and which are the presumptive pores. A, B, and C are centers of local threefold symmetry and L, the lattice constant, is 7.7 nm. The center-to-center spacing D of the pores is 3 nm, and each is partially surrounded by a kidney-shaped region that excludes the negative stain and is probably part of a single molecule of porin. The view is perpendicular to the plane of the outer membrane. (From A. C. Steven, B. ten Heggeler, R. Müller, J. Kistler, and J. P. Rosenbusch, Ultrastructure of a periodic protein layer in the outer membrane of *Escherichia coli*, *J. Cell Biol.* 72:292, 1977. Reprinted with permission.)

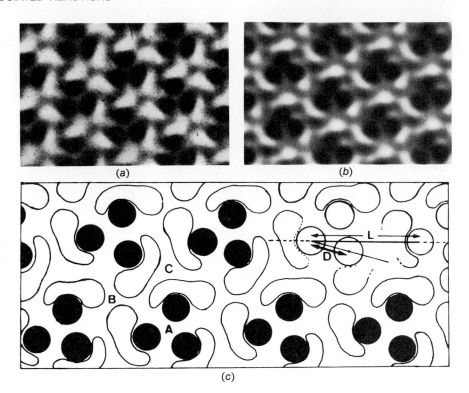

solute recognition is minimal, is presumably mediated by relatively static pore structures. By contrast, stereospecific solute permeation may involve conformational changes in the permease that transiently "open" the channel only in response to binding of the specific solute recognized by the transport system.

In all cases for which detailed structure-function information is available, the evidence is in favor of a pore-type mechanism for biological transport proteins. In several instances, presumptive transmembrane pores formed by an integral membrane protein have been visualized by electron microscopy. The outer membrane of *E. coli* has been shown to be permeable to hydrophilic molecules with molecular weights below about 600. This permeability is conferred by a class of proteins called porins, which are integral constituents of the outer membrane. Porin molecules are arranged in a hexagonal lattice of trimers in the outer membrane, and optical filtration and computer processing of many negatively stained preparations viewed in the electron microscope give images that have been interpreted as representing transmembrane pores (Figure 34–33). The diameter of these pores is about 1 nm, which is consistent with the molecular-weight exclusion limit given earlier. *When incorporated into artificial phospholipid vesicles, porin molecules allow rapid diffusion of most small hydrophilic molecules through the bilayer, and thus solute recognition by this protein is minimal.* Channels similar to those formed by bacterial porins are also present in the outer membranes of chloroplasts and mitochondria. *Thus, for many hydrophilic molecules, specific transport systems reside only in the inner membranes of Gram negative bacteria, chloroplasts, and mitochondria.*

Nonspecific transmembrane pores also have been demonstrated in the plasma membranes of eukaryotic cells. Well-studied examples of these pores are the so-called gap junctions, which connect neighboring cells of similar types in tissues such as liver, intestine, kidney, brain, and cardiac muscle. In the region of the gap junction, the plasma membranes of two different cells are closely apposed and appear to have channels connecting them (Figure

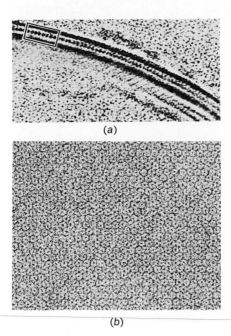

(a)

(b)

Figure 34–34

Isolated gap-junction sheets viewed in the electron microscope (a) parallel and (b) perpendicular to the planes of the two apposed membranes. The view in (a), part of which is outlined within a rectangle, arose from a curling up of a sheet on its edge and shows links (gaps or pores) penetrated by negative stain bridging the space between the two cell membranes. The perpendicular view in (b) shows annuli (rings), each composed of a protein hexamer surrounding a pore that has been filled with negative stain. The specimens were stained with uranyl acetate (300,000×). (From P. N. T. Unwin and G. Zampighi, Structure of the junction between communicating cells, Nature 283: 545, 1980. Reprinted with permission.)

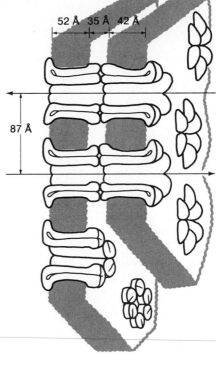

Figure 34–35

Molecular model of gap-junction pores as inferred from chemical, electron microscopic, and x-ray diffraction studies. A cross section, nearly perpendicular to the two membrane planes, is shown. The arrows, representing aqueous channels, are drawn through cross sections of two pores, formed by apposition of two hexamers (or connexons), each spanning its own lipid bilayer membrane. The end-on views of each connexon at the right and bottom of the illustration show that they protrude beyond the lipid bilayer on both sides of the plasma membrane. (Adapted from L. Makowski, D. L. D. Caspar, W. C. Phillips, and D. A. Goodenough, J. Cell Biol. 74:629, 1977.)

34–34a). In the surface view of a gap junction, the region appears as a hexagonal lattice of protein hexamers, each of which seems to form a hydrophilic pore (Figure 34–34b). Ultrastructural studies have produced detailed models of the structure of the gap junction, one of which is shown in Figure 34–35. _The protein hexamer of each membrane is thought to span the bilayer, and apposition of two such hexamers forms an aqueous channel between the cells._

Gap junction pores have been shown to be permeable to hydrophilic molecules with masses as large as 1000 to 2000 daltons, depending on the cell type. They are believed to play roles in conducting electric impulses through particular cell types (nerve, muscle, or epithelia), as well as in providing avenues for the free flow of small metabolites between cells. The permeability of gap junction pores is regulated by the cytoplasmic concentration of Ca^{2+}; low concentrations ($<10^{-7}$ M) lead to open channels, while higher concentrations tend to close the channels in a graded manner. This sensitivity to $[Ca^{2+}]$ may play a role in the regulation of intercellular communication between cells connected by gap junctions.

Although porins and gap junctions are relatively _nonspecific_ aqueous channels in biological membranes, it is reasonable to imagine that _solute-specific_ pores can be formed by permease proteins as a result of their binding

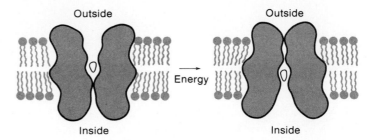

Figure 34–36

Generalized molecular model for transport involving a conformational rearrangement of subunits of an oligomeric carrier protein. In active transport, energy (such as that from the hydrolysis of ATP) could be used to effect this conformational change as shown. Alternatively, binding of the transport substrate could be enough to trigger this rearrangement in systems carrying out facilitated diffusion. (Adapted from S. J. Singer, The molecular organization of membranes, *Ann. Rev. Biochem.* 43:805, 1974.)

specificities. It further seems reasonable that such pores might be formed easily _between_ subunits of oligomeric transmembrane proteins. Given these considerations, a number of workers have suggested similar models for how specific permeases recognize and translocate their substrates. Aqueous channels present between transmembrane subunits may respond to solute binding by undergoing _conformational changes that alter the relative positions of the subunits to each other and thereby open the channels_ (Figure 34–36). This conformational change might be triggered solely by the substrate itself, in which case facilitated diffusion would be the consequence. Alternatively, metabolic energy might expedite this process, in which case active transport could result. Although conformational changes resulting in solute translocation are most easily envisioned between subunits of an oligomeric protein, there is also evidence that at least some permeases may accomplish this process in a monomeric form.

Pore-type transport mechanisms appear well established for nonspecific transport systems (porins and gap junctions) as well as for certain stereospecific transport systems (such as those in bacteria, which utilize periplasmic binding proteins). However, other permeases, such as those which catalyze solute-cation cotransport, may function by mechanisms that incorporate features of the carrier model. Thus, while the *E. coli* glycerol channel, bacterial porins, and periplasmic binding-protein-dependent transport systems are relatively insensitive to phospholipid temperature transitions, solute-cation symport systems are generally more sensitive to these physical changes in the membrane. Moreover, detailed kinetic analyses of lactose transport in *E. coli* are most easily interpreted in terms of a carrier-type mechanism. Finally, it should be noted that in addition to naturally occurring ionophore antibiotics, the coenzymes Q and plastoquinones are examples of nonprotein carriers that operate in a mobile fashion. These molecules play a role in H^+ translocation during electron transport in respiration and photosynthesis, respectively. They are believed to be able to diffuse freely in the hydrophobic phase of the phospholipid bilayer, although the exact mechanism by which they function in unidirectional H^+ transport is still unclear.

Examples of Molecular Mechanisms in Membrane Transport

Several lines of evidence have been used to lend support to pore-type transport mechanisms for specific solutes and to the importance of protein conformational changes in the overall process. E. Racker and P. Hinkle obtained support for a channel-type mechanism of H^+ translocation by bacteriorho-

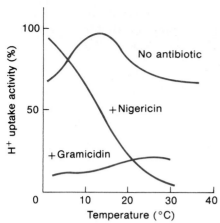

Figure 34–37

The effect of temperature on H^+ translocation by bacteriorhodopsin inserted into vesicles of dimyristoylphosphatidylcholine (T_m = 24°C). The top curve shows that proton transport is relatively insensitive to temperature, suggesting a pore-type mechanism. Nigericin, a mobile carrier that would collapse any proton gradient formed, does so only at higher temperatures, demonstrating that these membranes do, indeed, undergo a phase transition in the range studied. Gramicidin, a pore-forming antibiotic, inhibits proton translocation at all temperatures as expected. The curves with antibiotic present are plotted as percentages of the activity with no antibiotic, while the upper curve is in arbitrary units. (Adapted from E. Racker and P. C. Hinkle, Effect of temperature on the function of a proton pump, J. Memb. Biol. 17:181, 1974.)

dopsin in 1974. They incorporated the purified protein into vesicles of dimyristoylphosphatidylcholine (T_m = 24°C). In these vesicles, bacteriorhodopsin molecules were shown to have the *opposite* orientation when compared with *Halobacterium* cells, so that H^+ was accumulated *inside* the vesicles in response to light. By measuring H^+ uptake as a function of temperature, they showed that *the H^+ pumping ability of bacteriorhodopsin was insensitive to the phase of the bilayer* (gel or liquid crystalline), as shown in Figure 34–37. A mobile-carrier mechanism in which the protein undergoes large changes in orientation in the membrane thus appeared to be ruled out by this experiment.

It should be pointed out, however, that this type of experiment does not always lead to unequivocal results concerning transport mechanism. For example, many transport proteins do undergo abrupt changes in their activation energies for transmembrane transport at a temperature that does not correspond to the T_m of the membrane system in which they are being studied. Nevertheless, these and other experiments strongly suggest that bacteriorhodopsin contains a proton channel, with the protonated Schiff base (formed between retinal and the protein) acting as a gate. One attractive mechanism is that the Schiff base may change its physical position in the membrane in response to light absorption as a consequence of a conformational change in the protein. This might "deliver" the proton to hydrogen-bonded groups near the outer surface of the membrane. Loss of the proton might then be accomplished by a return of the chromophore to its original position, where it would pick up a proton from hydrogen-bonded groups nearer the inner membrane surface. This light-driven "proton shuttle" is a plausible mechanistic model for the functioning of bacteriorhodopsin, but it has not yet been proven experimentally.

In primary active transport systems driven by the hydrolysis of ATP, ample evidence has accumulated that the translocation steps take place by means of a gated-channel mechanism. The structure of the F_1/F_o ATPase (Figure 34–16c), with its large, hydrophilic F_1 portion projecting into the cytoplasm, makes any mobile-carrier mechanism extremely unlikely for the translocation of protons by this protein. Similarly, experiments with both the Na^+-K^+ and Ca^{2+} ATPases have shown that under certain conditions, *antibody molecules that are bound to these proteins at the membrane surface do not interfere with the functions of these proteins*. Again, because of the large amount of activation energy that would be required to translocate hydrophilic antibodies across the bilayer, pore mechanisms can be postulated for these ATPases as well. This is consistent with their asymmetric orientations in the membrane, as was discussed earlier.

The translocation mechanisms of both the Na^+-K^+ and Ca^{2+} ATPases have been studied by substrate-binding experiments, as well as by physical and kinetic measurements. In both instances, *the translocation of ions is believed to occur by means of a conformational change in the protein that moves the ion-binding site from one surface of the membrane to the other.* This is similar to the model presented in Figure 34–36, in which elements of the classical channel and carrier models can be recognized. Both the Na^+-K^+ and Ca^{2+} ATPases may possibly be oligomeric, so a slight shift in the orientations of subunits with respect to each other could be the conformational change associated with translocation. Alternatively, the channel could be formed within a single protein subunit that spans the membrane and translocates ions as a result of a conformational change. A mechanism consistent with these possibilities was first proposed by O. Jardetzky in 1966 for the functioning of the Na^+-K^+ ATPase and is shown in Figure 34–38. A conformational change in the *phosphorylated* Na^+-enzyme complex is thought to expel Na^+ from the cell, and interaction of K^+ with this form of

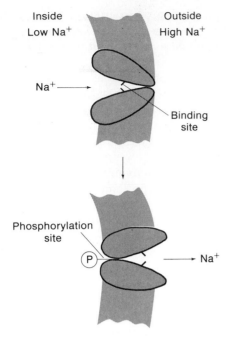

Figure 34–38

Model proposed by O. Jardetzky for the functioning of the Na$^+$-K$^+$ ATPase. The binding of Na$^+$ to a high-affinity site on the inside triggers phosphorylation by ATP, which in turn causes a conformational change exposing the cation binding site to the outside. The latter conformation has a low affinity for Na$^+$, but a high affinity for K$^+$. The binding of K$^+$ to this form reverses the conformational change and is accompanied by dephosphorylation of the enzyme and pumping of K$^+$ into the cytoplasm (not shown). (Adapted from O. Jardetzky, Simple allosteric model for membrane pumps, *Nature* 211:969, 1966.)

the enzyme on the outside of the cell causes dephosphorylation and translocation of K$^+$ into the cytoplasm. Essential to this model is that *the affinity of the enzyme is higher for Na$^+$ than for K$^+$ on the inside, while the opposite is true when the binding site is exposed to the outside.* This ensures unidirectional transport of both ions.

Although Jardetzky's model was based primarily on theoretical considerations, it remains consistent with most of the experimental evidence obtained on this system. The energy for active transport is thus obtained by phosphorylation of the enzyme by ATP. Phosphorylation is stimulated by Na$^+$ and subsequently drives the conformational change that is responsible for translocation against a concentration gradient. The return cycle is triggered by dephosphorylation, which is stimulated by K$^+$ and allows the enzyme to return to its original state. Such a mechanism is analogous to a *Ping-Pong* mechanism in an enzyme-catalyzed reaction (see Chapter 9) except that vectorial transport rather than chemical transformation of the substrates is involved. A similar series of events has been proposed for Ca^{2+} translocation by the Ca^{2+} ATPase, and it is likely that other primary transport systems that use ATP also may operate by this type of conformational coupling mechanism.

As a final example of a molecular model for membrane transport, let us consider the ATP/ADP exchange protein of mitochondria. As mentioned earlier, this protein is a dimer of identical subunits. Because the protein loses its binding affinity for transport substrates if it is dissociated into monomers, a dimer may be necessary for carrying out the exchange process. Biochemical and kinetic studies on the ATP/ADP exchanger support a two-state gated-pore mechanism for this protein. The evidence for this suggestion can be summarized as follows:

1. There appears to be only one nucleotide binding site per dimeric ATP/ADP exchanger.
2. The nucleotide binding site has a higher affinity for ADP on the outer surface of the membrane than on the inner surface, while the opposite is true for ATP.
3. This site was shown never to be on both sides of the membrane simultaneously.
4. The respiratory poison *atractylic acid* binds to the form preferring ADP, while the antibiotic *bongkrekic acid* (see Chapter 16) binds to the form that shows specificity toward ATP.

These results strongly support the transport model presented in Figure 34–39. A single nucleotide binding site per dimer is involved in transport by this protein. The configuration of this site, and thus its binding specificity, depends on the side of the membrane that it faces, and these two states are interconvertible through protein conformational changes that also result in translocation of any nucleotide bound to the protein. A conformational

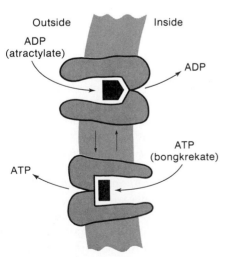

Figure 34–39

Molecular model for the functioning of the ATP/ADP exchanger of the mitochondrial inner membrane. The dimeric protein appears to have only one binding site for adenine nucleotides, perhaps involving amino acid residues from both subunits. When facing the outer surface of the membrane, this site has a high affinity for ADP (*pentagon*), and a low affinity for ATP (*rectangle*). The opposite specificities are observed when the site faces the lumen of the organelle. These states are interconvertible, even in the purified carrier, demonstrating that they arise from different conformations of the same protein. This model is analogous to those shown in Figures 34–36 and 34–38 except that energy is apparently not required for the conformational interconversion. (Adapted from M. Klingenberg, Membrane protein oligomeric structure and transport function, *Nature* 290:449, 1981.)

change can be triggered by binding of either ADP at the outside or ATP at the inside of the membrane surface. Once transported through the membrane, the nucleotide is now in a binding site that has little affinity for it, and the complex dissociates. Accordingly, the two forms of the adenine nucleotide exchanger appear to have quite different conformations. Antibodies prepared against the atractylic acid-protein complex do not cross-react with the form that binds bongkrekic acid, and vice versa. These two forms are, however, interconvertible by adding the appropriate inhibitor or adenine nucleotide. This conformational change is possibly the cause of structural alterations seen in the inner mitochondrial membrane when exposed to adenine nucleotides.

All the transport permeases we have discussed in this section have been purified. They are among the transport systems which are best understood at the molecular level. Secondary active transport and group translocation systems also should be amenable to detailed mechanistic studies, now that some of these transport proteins are available in purified form. Such studies will be aided by reconstitution of isolated transport proteins into artificial membrane systems, a powerful investigative tool developed in recent years by biochemists interested in transport. This technique will be discussed in the following section.

RECONSTITUTION OF PURIFIED TRANSPORT PROTEINS

In Chapter 6, the importance of phospholipids in maintaining the structural and functional integrity of integral membrane proteins was stressed. This can be best illustrated by considering the lipid requirements of those transport permeases that still exhibit enzymatic activity after dissociation from the membrane, such as the ion-translocating ATPases. For example, the Na^+-K^+ ATPase isolated from rabbit kidney loses ATPase activity if residual bound phospholipid molecules are removed from the purified protein. About 90 moles of phospholipid per mole of enzyme are required for maximal activity. Similarly, a molar ratio of 30:1 (phospholipid to protein) is necessary for optimal functioning of purified Ca^{2+} ATPase from sarcoplasmic reticulum. By replacing bound phospholipid in these purified preparations with lipids of known structure, it has further been possible to define which of these promote optimal activity of ATPase function in the isolated state, and presumably in the membrane as well.

These types of studies have set the stage for a fairly recent development in experimental membrane transport biochemistry, the reconstitution of isolated permeases into artificial membrane systems. By the introduction of such purified proteins into pure phospholipid bilayers, it has become possible to study the functional properties of a single transport system without interference by other membrane proteins or associated metabolic processes. Accordingly, _reconstitution of membrane transport systems offers one of the more powerful tools available to membrane biochemists_ who are interested purely in the mechanism of the transport event itself. The results of several of these types of experiments have been referred to elsewhere in this chapter. We shall briefly consider the types of information that reconstitution has contributed to our understanding of transport mechanisms.

Racker and coworkers have been successful in incorporating many of the enzymes involved in oxidative phosphorylation and electron transport, as well as other transport proteins, into artificial phospholipid vesicles. A number of techniques, including sonication of purified membrane proteins with phospholipid and dilution or dialysis of protein-detergent complexes in solutions containing phospholipid, have led to the preparation of _proteoliposomes_ (protein-phospholipid vesicles) capable of transporting the

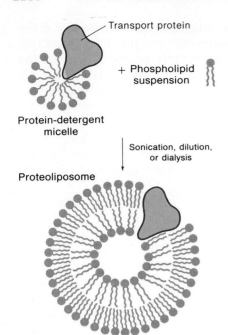

Figure 34–40
Schematic illustration of the reconstitution of a purified transport protein (in a detergent micelle) into a phospholipid vesicle to form a proteoliposome. Such proteoliposomes are very useful in studying transport mechanisms in a well-defined system.

appropriate solutes (Figure 34–40). In a particularly important experiment, Racker and Stoeckenius reconstituted both purified bacteriorhodopsin and the mitochondrial F_1/F_o ATPase into proteoliposomes. The orientations of both these molecules were shown to be opposite to those in the bacterial and mitochondrial membranes, respectively. _In the presence of light, these reconstituted vesicles catalyzed the synthesis of ATP from ADP and P_i, and this reaction required both proteins to be present simultaneously_ (Figure 34–41). Thus, in one reconstitution experiment, both the role of bacteriorhodopsin as a proton pump and the chemiosmotic hypothesis of Mitchell were confirmed (see Chapter 16).

A second useful reconstitution method in membrane biochemistry is the introduction of purified permeases into _planar phospholipid bilayers_. An apparatus for accomplishing this is schematically illustrated in Figure 34–42. Typically, a solution of phospholipid in an organic solvent is painted over an aperture in a thin wall separating two compartments of the apparatus that contain aqueous solutions. After several minutes, a bilayer of the phospholipid spontaneously forms between the two halves of the apparatus.

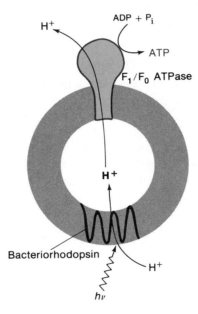

Figure 34–41
Reconstitution experiment of Racker and Stoeckenius that demonstrated that a proton gradient formed by illumination of bacteriorhodopsin in a proteoliposome could drive the synthesis of ATP by the mitochondrial F_1/F_o ATPase incorporated into the same membrane. ATP was synthesized in the extravesicular space only if both proteins were included in the reconstitution and only in the presence of light. Both proteins in the reconstituted vesicles assumed orientations opposite to those they normally have in their native membranes. This was one of several experiments which were instrumental in proving one of the essential concepts of Mitchell's chemiosmotic hypothesis of oxidative phosphorylation.

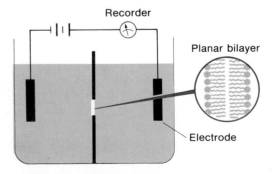

Figure 34–42
Cross section of an apparatus used to form planar lipid bilayers. The partition separating the two aqueous compartments has a circular aperture over which the bilayer film is formed. Proteins can be introduced into such bilayers either during or after their formation. This apparatus has the advantages of convenience in controlling and measuring the compositions of the solutions on both sides of the membrane, and of the ability of the investigator to measure electrical potentials and current flux through the reconstituted membrane.

Membrane proteins often can be introduced into the bilayer as protein-detergent or protein-lipid complexes added to either aqueous compartment. Alternatively, the phospholipids can be mixed with the protein of interest before the bilayer is formed. The potential advantages of this system over proteoliposomes include the following.

1. Ability to easily control and determine the composition of the solutions on both sides of the reconstituted membrane.
2. Ease of applying voltages across the reconstituted bilayer in the study of transport proteins sensitive to the membrane potential.
3. Possibility of measuring transport continuously by monitoring current flow if the transported species has a net charge.
4. The ability, in some cases, to control the orientation of the reconstituted protein in the bilayer.

By the use of such planar bilayer systems, it has been possible to study the functional properties of a number of solute pumps and carriers. For example, planar membranes into which bacteriorhodopsin is incorporated in predominantly one orientation induce a potential between the two aqueous compartments in response to light energy applied to the aperture supporting the membrane. This potential has been shown to be due to the net pumping of protons into one of the compartments of an apparatus similar to that shown in Figure 34–42. In another experiment, a purified *E. coli* porin preparation (Figure 34–33) was incorporated into planar bilayers. *When an electric potential was externally applied across these membranes, an ionic current was measured.* The current was not constant, but *fluctuated in a stepwise manner* about the steady state level (Figure 34–43). The fluctuation has been interpreted as the opening and closing of single porin channels in the reconstituted membranes. This interpretation is supported by the fact that similar results are obtained when pore-forming antibiotics such as gramicidin A are introduced into a planar bilayer system. In fact, such incremental changes in the permeability of a membrane for a particular solute are often taken as evidence that the transport protein acts as a gated pore.

Several additional purified transport proteins have been incorporated into proteoliposomes and/or planar bilayers in functionally active states. These techniques therefore hold the promise of solving many unanswered questions about the mechanisms of biological transport. It should be pointed out, however, that reconstituted membranes are indeed artificial systems, so that what is measured in a particular experiment may not occur under "real"

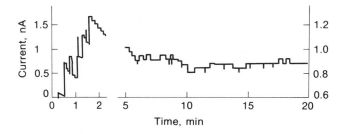

Figure 34–43

Current measurements on planar phospholipid bilayers containing *E. coli* porin, using an apparatus similar to that depicted in Figure 34–42. Porin channels are initially closed, but are induced to open by application of a potential across the membrane (240 mV at time zero in this experiment). After about 5 min, the current fluctuates in a stepwise manner about a steady-state value. (Adapted from H. Schindler and J. P. Rosenbusch, Matrix protein from *Escherichia coli* outer membranes forms voltage-controlled channels in lipid bilayers, *Proc. Natl. Acad. Sci. USA* 75:3751, 1978.)

conditions in a biological membrane. Reconstitution, therefore, has the potential to tell a great deal about transport *per se*, but it must be combined with observations on unfractionated membranes and even whole cells in order to ascertain the role of a particular transport system in the overall physiology of an organism.

SUMMARY

Cells and organelles must be able to rapidly transport hydrophilic solutes across the hydrophobic membrane barrier that bounds them. This process can occur by simple diffusion or can be carrier-mediated. In simple diffusion, the solute must be appreciably soluble in the domain of the bilayer. The kinetics of its movement across the membrane follows Fick's law, and the net transport rate in either direction is directly proportional to the concentration gradient of the solute. In carrier-mediated transport, the transport rate is saturable with respect to solute concentration because only a discrete number of permeases exist in the membrane. The kinetics of this process is described by the Henri-Michaelis-Menten equation when initial transport rates are measured. Thus kinetic measurements frequently distinguish between these two mechanisms, although in many cases cells may use more than a single system to transport a given solute.

Net transport of a molecule across a biological membrane in a given direction will occur spontaneously only if ΔG for the process is negative. For an uncharged molecule, this means that in the absence of energy input, the compound will reach equal concentrations on both sides of the membrane barrier at equilibrium. Simple and facilitated diffusion, therefore, can lead only to equilibration of an uncharged solute across a membrane. However, if the molecule being transported has a net charge, unequal concentrations of it can occur at equilibrium without the input of additional energy if there is an independently maintained electric potential across the membrane. Both simple and facilitated diffusion are therefore examples of energy-independent biological transport mechanisms.

Many cellular transport systems, however, can carry out the active accumulation of metabolites and ions. Because ΔG for solute accumulation is positive, the process must be coupled to an exergonic event in which ΔG is negative. Primary active transport systems may use the chemical energy of ATP, light energy, or electron flow to drive the uphill concentration of the solute. In secondary active transport, ion gradients, which themselves are maintained by primary active transport processes, are used to drive the active accumulation of a second solute. Finally, in group translocation, the transported substrate is chemically modified by the membrane permease during the transport event. This modification may result in the expenditure of metabolic energy, as in the case of many bacterial sugar transport systems, and may provide the driving force for solute accumulation.

Primary active transport systems that use ATP as a source of energy have been identified in both eukaryotic and prokaryotic cells. The Na^+-K^+ ATPase of animal cells is an asymmetrically oriented transmembrane protein that maintains high intracellular levels of K^+ by carrying out the active accumulation of this ion and the concomitant extrusion of Na^+. A second, well-studied ion-translocating ATPase, the Ca^{2+} ATPase of sarcoplasmic reticulum, is believed to catalyze the active accumulation of Ca^{2+} into the lumen of this organelle by a similar mechanism. The F_1/F_o proton-translocating ATPase found in bacteria and eukaryotic organelles couples H^+ translocation with ATP synthesis, or hydrolysis, thereby allowing the interconversion of chemiosmotic and chemical energy. Primary active transport of H^+ is also carried out by carriers of the electron-transport chains of bacteria, chloroplasts, and mitochondria, as well as by bacteriorhodopsin in *Halobacterium halobium*.

Na^+ and proton gradients across cellular and organellar membranes are forms of potential energy that can be used to drive secondary active transport. In this type of transport, the flow of one of these cations down its electrochemical gradient can be coupled by cotransport, or symport, with the transport of another molecule against its concentration gradient. An example of this type of mechanism is the H^+-lactose symport system of *E. coli* and other bacteria.

The ATP/ADP exchanger found in the inner mitochondrial membrane was considered as a case where membrane potential can drive the transport of negatively charged molecules. The phosphoenolpyruvate(PEP)-dependent sugar phosphotransferase system (PTS) in bacteria is the best-studied group translocation system.

A membrane permease could function mechanistically as a mobile carrier or a relatively static pore. Examples of both types of transport processes are found among the ion-translocating antibiotics. Examples of nonspecific pores in cells include the glycerol permease in the cytoplasmic membrane of *E. coli*, the porins of Gram negative bacteria, and gap junctions of animal cells. Examples of molecular transport mechanisms that have been studied in detail include bacteriorhodopsin, the Na^+-K^+ and Ca^{2+} ATPases, and the mitochondrial ATP/ADP exchanger.

Finally, reconstitution of purified membrane permeases into artificial membranes holds the promise of greatly increasing our knowledge of biological transport mechanisms. Proteoliposomes and reconstituted planar bilayers have been used to establish the transport characteristics of purified carriers. The future appears bright for an understanding of biological transport mechanisms at the molecular level.

SELECTED READINGS

Bronner, F., and A. Kleinzeller (eds.), *Current Topics in Membranes and Transport.* New York: Academic Press. Continuing series reviewing some of the current problems in biological transport.

Ghosh, B. K. (ed.), *Organization of Prokaryotic Cell Membranes,* Vols. I and II. Boca Raton, Florida: CRC Press, 1981. Reviews in these volumes deal with transport of small molecules across the inner and outer membranes of *E. coli* as well as the structures of these membranes.

Hobbs, A. S., and R. W. Albers, The structure of proteins involved in active membrane transport. *Ann. Rev. Biophys. Bioeng.* 9:259, 1980. Includes information on the Na^+-K^+ and Ca^{2+} ATPases, the F_1/F_o ATPase, and bacteriorhodopsin.

Oxender, D., A. Blume, I. Diamond, and C. F. Fox, *Membrane Transport and Neuroreceptors.* New York: Alan R. Liss, Inc., 1981. Summarizes diverse experimental approaches concerned with specific aspects of the structures and modes of action of a wide variety of well-characterized transport systems and neuroreceptors.

Racker, E., *Reconstitutions of Transporters, Receptors, and Pathological States.* Orlando, Florida: Academic Press, 1985. Reviews the methodology of transport reconstitution as well as the molecular details of some of the better-characterized ion-transporting and solute:cation symporting permeases.

Saier, M. H., Jr., *Mechanisms and Regulation of Carbohydrate Transport in Bacteria.* Orlando, Florida: Academic Press, 1985. Reviews the current literature (up to 1985) dealing with the transport of carbohydrates across bacterial cell membranes and the mechanisms by which these transport processes are regulated.

PROBLEMS

1. a. Using Fick's law, show that the diffusion coefficient D has the dimensions of area per unit time.
   b. The diameter of a porin channel is about 10^{-9} m and its length is 4×10^{-9} m. In planar bilayers, glucose traverses this channel at the rate of about 50 molecules per channel per second at room temperature when the concentration of glucose is 3×10^{-6} M on one side of the membrane. Calculate the diffusion coefficient for glucose through porin channels under these conditions.

2. The Nernst equation relates the electric potential $\Delta\Psi$ resulting from an unequal distribution of a charged solute across a membrane permeable to that solute to the ratio between the concentrations of solute on one side and on the other:

$$m \, \Delta\Psi = \frac{-2.3RT}{F} \log \frac{[So]_1}{[So]_2}$$

where m is the charge on the solute, $2.3RT/F$ has a value of about 60 mV at 37°C, and $[So]_1$ and $[So]_2$ refer to the concentrations of solute on either side of the membrane. (For more about the Nernst equation, see Chapter 35.) Consider a planar phospholipid bilayer separating two compartments of equal volume. Side 1 contains 50 mM KCl and 50 mM NaCl, while side 2 contains 100 mM KCl.
   a. If the membrane is made permeable only to K^+, e.g., by addition of valinomycin, what will be the magnitude of $\Delta\Psi$?
   b. If the membrane is made permeable to H^+ and K^+, in which direction will H^+ initially flow?
   c. If the membrane could be made selectively permeable to both K^+ and Cl^-, what would be the value of $\Delta\Psi$ and the ion concentrations on both sides of the membrane at equilibrium? (Hint: Initially, K^+ would diffuse down its concentration gradient accompanied by an equivalent amount of Cl^-. Equilibrium would be established when the potentials due to K^+ and Cl^- were equal to each other and to the overall membrane potential.)

3. Membrane vesicles of *E. coli* that possess the lactose permease are preloaded with KCl and are suspended in an equal concentration of NaCl. It is observed that these vesicles actively, although transiently, accumulate lactose if valinomycin is added to the vesicle suspension. No such active uptake is observed if KCl replaces NaCl in the suspending medium. Explain these results in light of what you know about the mechanism of lactose transport and the properties of valinomycin.

4. Intracellular vacuoles in the yeast *Saccharomyces cerevisiae* are membrane-bounded organelles that are known to concentrate within them a variety of basic amino acids, including arginine (net charge = +1). Vesicles prepared from these vacuoles lack an electron-transport chain, and arginine uptake into them is dependent on extravesicular ATP. A membrane potential $\Delta\Psi$ has no effect on ATP-dependent arginine uptake in the absence of a proton gradient, while proton ionophores and dicyclohexylcarbodiimide (a known inhibitor of the F_1/F_o ATPase) greatly inhibit accumulation of arginine by this system. Upon addition of ATP in the absence of arginine, the intravesicular pH of these vesicles drops. Describe a mechanism for the energization of arginine transport in this system, taking into account all these observations.

5. In *E. coli*, lactose is taken up by means of proton symport, maltose by means of a binding protein system, melibiose by means of Na^+ symport, and glucose by means of a phosphotransferase system (PTS). Although this bacterium normally does not transport sucrose, suppose you have isolated a strain that does. How would you determine whether one of the four mechanisms just listed is responsible for sucrose transport in this mutant strain?

6. In some instances, the efflux of a radioactively labeled transport substrate out of preloaded cells or vesicles is transiently stimulated by addition of the same nonradioactive transport substrate to the outside. This phenomenon is known as trans-stimulation and occurs with transport systems that are reversible (i.e., can operate in either direction). Can you think of an explanation for trans-stimulation in view of what is known about the molecular mechanisms of transmembrane transport?

7. Outline a molecular mechanism by which, and the conditions under which, an H^+ symport system (such as the *E. coli* lactose permease system) might operate to actively accumulate a metabolite such as lactose.

8. Predict the effects of the following on the initial rate of glucose transport into vesicles derived from animal cells that accumulate this sugar by means of Na^+ symport. Assume that initially $\Delta\psi = 0$, $\Delta pH = 0$ (pH = 7), and the outside medium contains 0.2 M Na^+, while the vesicle interior an equivalent amount of K^+.

   a. Valinomycin.
   b. Gramicidin A.
   c. Nigericin.
   d. Preparing the membrane vesicles at pH 5 (in 0.2 M KCl), resuspending them at pH 7 (in 0.2 M NaCl), and adding 2,4-dinitrophenol.

35

THE NERVOUS SYSTEM: NEUROTRANSMISSION

The structures of biological membranes and the mechanisms by which ions and metabolites are transported across them were considered in Chapters 6 and 34. Clearly, however, cells must do much more than simply assimilate nutrients and expel waste materials. Unicellular, free-living organisms such as bacteria and protozoa must be able to navigate and successfully compete in environments that often contain many perils to their livelihood. Multicellular organisms, however, have the additional problem of coordination and signaling among many diverse, differentiated cell types to ensure the efficient functioning of the individual as a whole. Interactions of a cell with its environment and other cells that are necessary for these types of communication must obviously involve processes associated with the cytoplasmic membrane. In many cases, transmembrane transport is integrally involved in these signaling processes between cells and their environment.

One type of communication mechanism in animals was considered in Chapter 31. Hormones released in one part of the body interact with receptors, which are often membrane-bound, to affect processes occurring in other cells that are often quite distant in space from the cells releasing the signal. This mechanism allows for efficient coordination of the wide variety of metabolic events continually taking place in higher organisms. Another, quite different sensory phenomenon takes place in the retina cells of the eye of vertebrate animals. Light energy is converted by these cells into nerve impulses, which are translated by the brain into an image of our surroundings. This interconversion involves transmembrane movements of ions and will be considered in Chapter 36.

In this chapter, we shall examine the mechanism by which nerve impulses are propagated, to illustrate how excitable tissues function and interact in higher organisms. As will be seen, the fundamental mechanisms revealed by studies of membrane transport in both prokaryotic and eukaryotic cells (Chapter 34) can account in large part for the seemingly more complex phenomena of nerve-impulse propagation and transmission.

NERVE-IMPULSE PROPAGATION

As early as the late eighteenth century, from experiments by Galvani and Volta, it was suspected that the transmission of nerve impulses and muscular contraction involved electric signals. In 1902, J. Bernstein first proposed that the unequal distribution of K^+ across the nerve-cell membrane and the selective permeability of this membrane for this ion were responsible for a _resting potential_ known to exist in nerve and muscle fibers. He further believed that excitation of a nerve cell involved a transient collapse in this selective permeability such that other ions were able to penetrate the nerve-cell membrane and abolish the resting potential. If these changes in ion permeability could move down the axon of a nerve cell and be transmitted to other cells, they could provide the basis for the propagation of nerve impulses.

In the 1930s, the isolated _squid giant nerve axon_ became available for experimentation, and its size was especially amenable to electrophysiologic measurements. Experiments pioneered by A. L. Hodgkin and A. F. Huxley soon established the essential ionic movements associated with impulse propagation, and the resultant local changes in the membrane potential could be measured. It became clear that the transmembrane transport of ions was important in nerve cells for their signaling function and for the actual signal conductance itself. Relatively little is still known at the molecular level about how these changes in ion permeability in nerve cells are accomplished. Somewhat more is known about how these signals are chemically transmitted between nerve cells and from nerves to muscle; these mechanisms will be considered as well. An examination of such mechanisms is likely to provide insights into ways in which all different types of cells can communicate with each other in response to changes in their environments.

An Unequal Distribution of Ionic Species Results in a Resting Transmembrane Potential

To understand how nerve impulses are generated along the axon of a nerve cell (Figure 35–1), the basis for electric potentials that exist across the neuronal membrane must first be considered. An unequal distribution of ionic species across a biological membrane that is permeable to these molecules can result in a transmembrane electric potential, $\Delta\psi$ (Chapter 34). For a membrane system permeable to several ionic species, the numerical value of $\Delta\psi$ can be approximated by the _Goldman equation_, derived by D. E. Goldman in 1943:

$$\Delta\psi \text{ (``in'' relative to ``out'')} = \frac{2.3RT}{F} \log_{10} \left(\frac{\Sigma P_c[\text{C}]_{\text{out}} + \Sigma P_a[\text{A}]_{\text{in}}}{\Sigma P_c[\text{C}]_{\text{in}} + \Sigma P_a[\text{A}]_{\text{out}}} \right) \qquad (1)$$

where C and A are univalent cations and anions, respectively, and P_c and P_a refer to their _permeability coefficients_* across the membrane of interest. Since multivalent ions are generally not quantitatively significant in contributing to $\Delta\psi$ in resting neuronal membranes, they are usually ignored in calculating $\Delta\psi$. If the membrane is selectively permeable to one ion only, for example C, Equation (1) reduces to the familiar _Nernst equation_:

$$\Delta\psi = E_c = \frac{2.3RT}{F} \log_{10} \frac{[\text{C}]_{\text{out}}}{[\text{C}]_{\text{in}}} \qquad (2)$$

where E_c refers to the equilibrium electric potential of C.

---

*The permeability coefficient is equal to the diffusion coefficient D divided by the width of the membrane l.

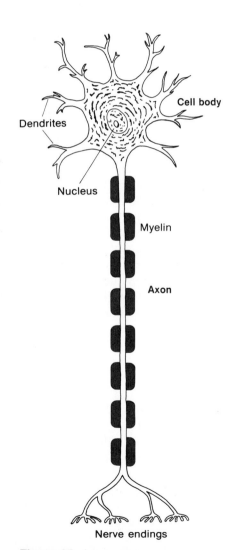

Figure 35–1

Schematic diagram of a typical motor neuron (a nerve cell conducting impulses to muscle cells).

Cell body

Dendrites

Nucleus

Myelin

Axon

Nerve endings

TABLE 35–1
Ionic Concentrations Inside (Axoplasm) and Outside (Blood) the Squid Giant Axon

Ion	Inside (mM)	Outside (mM)
Na⁺	50	440
K⁺	400	20
Cl⁻	40–150	560

Source: Adapted from S. W. Kuffler and J. G. Nicholls, *From Neuron to Brain.* Sunderland, Mass.: Sinauer Associates, 1976.

It is the unequal distribution of protons and other cations that gives rise to transmembrane potentials that can drive ATP synthesis and secondary active transport in many cells (Chapter 34). In nerve cells, a similar situation exists as summarized in Table 35–1. Thus the external environment of the nerve cell contains a high concentration of sodium ions and a low concentration of K⁺, while the reverse is true for the cytoplasm (*axoplasm*). Furthermore, the extracellular concentration of Cl⁻ is 5 to 10 times that of the axoplasm. A number of other impermeant anions, largely organic molecules, proteins, and nucleic acids, maintain an approximate charge neutrality in the axoplasm. *Resting nerve cells are highly permeable to K⁺ but not Na⁺, and it is this selective permeability that allows an electric potential to develop across the membrane in the presence of a K⁺ concentration gradient.* The unequal distributions of K⁺, Na⁺, and Cl⁻ are partially determined by passive processes leading to what is called a *Donnan equilibrium* and partly by the activity of the Na⁺-K⁺ ATPase (Chapter 34).

The passive distribution of ions across a membrane was predicted by F. Donnan. He theorized that if a hypothetical cell having membrane-impermeable anions inside (such as proteins and nucleic acids) were placed in a KCl solution, Cl⁻ would diffuse into the cell, down its concentration gradient, accompanied by an equivalent amount of K⁺ to maintain electroneutrality, until a state of equilibrium was achieved. If K⁺ were the counterion present initially inside the cell, Donnan predicted that the equilibrium concentration of K⁺ inside would be much higher than that of Cl⁻ (and higher than that of K⁺ outside the cell if a sufficient concentration of nondiffusible anions were present inside). More specifically, Donnan showed that the final equilibrium concentrations of K⁺ and Cl⁻ inside and outside the cell would be related by the following equation:

$$\frac{[K^+]_{in}}{[K^+]_{out}} = \frac{[Cl^-]_{out}}{[Cl^-]_{in}} \tag{3}$$

This relationship describes a Donnan equilibrium (Figure 35–2). Because both K⁺ and Cl⁻ are permeable to the resting axonal membrane, their unequal distributions (Table 35–1) can be partly explained by this passive process. *The Na⁺ gradient, however, is due to the relative impermeability of the membrane to this ion and to the Na⁺-K⁺ ATPase that actively pumps Na⁺ out of the cell.* The activity of this enzyme also alters somewhat the value of the K⁺ gradient from that predicted by a Donnan equilibrium.

The predicted resting membrane potential $\Delta\psi$ across the axonal membrane can be calculated from Equation (1) using the values in Table 35–1, the permeabilities of Na⁺ and Cl⁻ relative to K⁺ (0.04 and 0.45, respectively), and assuming an intracellular Cl⁻ concentration of 50 mM:

$$\Delta\psi = 60 \log\left(\frac{20 + 0.04(440) + 0.45(50)}{400 + 0.04(50) + 0.45(560)}\right) = -62 \text{ mV}$$

This value is close to the experimentally measured value of the resting membrane potential across a squid axonal membrane.

An Action Potential Is the Transient Change in Membrane Potential Occurring During Nerve Stimulation

The use of giant axons from squid nerve cells in electrophysiologic experiments has greatly aided our understanding of the electrical events that take place during nerve stimulation. An experimental apparatus for measuring changes in the potential across the membrane of such an axon, which has a diameter of approximately 0.5 mm, is schematically illustrated in Figure

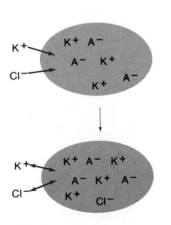

Figure 35–2
If a cell containing the potassium salt of a nondiffusible anion (A⁻) is placed in a KCl solution, K⁺ and Cl⁻ will diffuse into the cell (*above*) until a Donnan equilibrium is reached [also see Equation (3)].

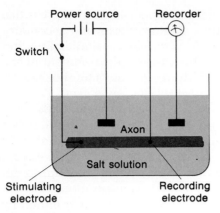

Figure 35–3

A device for eliciting and recording action potentials along the squid giant nerve axon. Brief closure of the switch connected to the stimulating electrode causes a current pulse into the axoplasm. If an impulse is generated, resultant potential changes can be detected by the recording electrode, which is connected to an oscilloscope or other recording device.

35–3. It consists of a pair of stimulating electrodes connected to a current source and a second pair of recording electrodes located slightly farther down the axonal segment. The latter electrodes are connected to a sensitive recording device, such as an oscilloscope, and can be used to measure time-dependent changes in the membrane potential.

Initially, the potential measured in the system shown in Figure 35–3 is about −60 mV; i.e., the resting membrane potential (Figure 35–4). If a brief current is applied at the stimulating electrodes, a time-dependent change in the membrane potential may be recorded on the oscilloscope, as shown in Figure 35–4a. *This so-called action potential only occurs if the stimulus is sufficient to depolarize the membrane by about 20 mV* (i.e., to about −40 mV). Weaker stimuli give small local potential changes, while current pulses greater than this *threshold* value give a curve similar in shape and height to that shown in Figure 35–4a, independent of the magnitude of the stimulus. During the development of the action potential, the value of $\Delta\psi$ across the axonal membrane rises in about 1 ms to nearly +40 mV. This is followed by a somewhat slower return to the resting potential, during which time the membrane potential drops transiently below the resting value, to about −75 mV. This is close to the value predicted by the Nernst equation if the membrane were permeable only to K^+ (the potassium equilibrium potential) and is referred to as *hyperpolarization*.

Classic experiments by A. L. Hodgkin and A. F. Huxley have established that the *changes in membrane potential occurring during nerve stimulation are due to transient changes in the permeability of the membrane to Na^+ and K^+ ions.* As illustrated in Figure 35–4b, the rapid rise in $\Delta\psi$ to a positive value is accompanied by a large increase in the relative permeability of Na^+, while the return of the membrane to the resting potential is correlated with inactivation of Na^+ permeability and a transient increase in K^+ permeability. An important conclusion from these observations is that *the permeabilities*

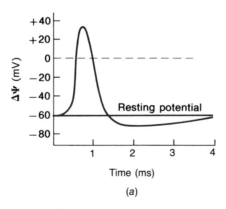

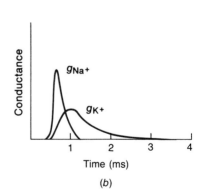

Figure 35–4

(a) A typical action potential that might be recorded by the instrument in Figure 35–3 if the stimulating current is sufficient to depolarize the membrane by at least 20 mV. The membrane potential eventually returns to its resting level of about −60 mV within about 5 ms. (b) Changes in relative Na^+ permeability (g_{Na^+}) and K^+ permeability (g_{K^+}) as a function of time during passage of the action potential in (a) along a point on the axonal membrane. Depolarization is correlated with an increase in g_{Na^+}, while repolarization is accompanied by a decrease in g_{Na^+} and a transient increase in g_{K^+}. Note that the membrane potential transiently becomes more negative than the resting value (hyperpolarizes) until g_{K^+} returns to its normal value. The time of appearance of the action potential after stimulation at $T = 0$ depends on the distance between the recording and stimulating electrodes. (Adapted from A. L. Hodgkin and A. F. Huxley, A quantitative description of membrane current and its application to conduction and excitation in nerve, *J. Physiol.* 117:500, 1952.)

of the axonal membrane of Na$^+$ and K$^+$ depend on the membrane potential. Thus depolarization above the threshold, leading to a more positive $\Delta\psi$, first leads to an increased permeability of Na$^+$ followed by inactivation of this phenomenon and an increase in the membrane permeability of K$^+$. The latter events tend to hyperpolarize the membrane, increasing the negative $\Delta\psi$ and decreasing the Na$^+$ permeability.

As mentioned in the preceding section, the membrane potential depends on the relative permeabilities and concentration gradients of electrolytes across the membrane [Equation (1)]. The resting potential is largely dependent on the K$^+$ gradient, since unstimulated nerve membranes have a high permeability only for this cation. At the height of depolarization, however, the membrane is much more permeable to Na$^+$ than to K$^+$. Because the Na$^+$ gradient is opposite to that of K$^+$, Na$^+$ influx causes the membrane potential to become positive rapidly, approaching but never reaching the value it would have if the membrane were permeable only to Na$^+$. This can be seen if one substitutes the values of [Na$^+$]$_{in}$ and [Na$^+$]$_{out}$ into Equation (2). A value of +57 mV is thereby obtained for $\Delta\psi$ if the membrane were permeable only to Na$^+$. This value is never attained because an increased permeability of the membrane to K$^+$ follows the change in Na$^+$ permeability, and the opening of the Na$^+$ channel is only transient (Figure 35–4*b*).

The ionic movements leading to an action potential can therefore be summarized as follows:

1. Stimulation leads to an influx of Na$^+$ into the axoplasm, down its electrochemical gradient, owing to an increased permeability of the membrane to this cation.
2. The change in $\Delta\psi$ resulting from this flow increases the membrane permeability of K$^+$, which flows out of the cell, down its electrochemical gradient; this reestablishes a negative $\Delta\psi$ and is accompanied by inactivation of the influx of Na$^+$.
3. The membrane potential eventually becomes sufficiently negative to return both K$^+$ and Na$^+$ permeabilities to their normal values, and $\Delta\psi$ reassumes its resting level.

It should be pointed out that *the ion fluxes that accompany these events and lead to changes in $\Delta\psi$ are actually very small compared with the concentrations of Na$^+$ and K$^+$ inside and outside the cell.* From the capacitance of a squid giant axon membrane, it can be calculated that for a 100-mV change in $\Delta\psi$ (from -60 to $+40$ mV), only about 10^{-12} mol Na$^+$ per cm^2 of cell surface need enter the cell, while an equal amount of K$^+$ must leave the axoplasm to return the potential to the resting value. This corresponds, for example, to about one in 10^6 molecules of K$^+$ leaving the cell per action-potential spike, or to a change in the intracellular K$^+$ concentration of only 0.0001 per cent.

The events shown in Figure 35–4 record fluctuations in $\Delta\psi$ and ionic currents at one point on the axonal membrane as a function of time during passage of an action-potential wave through this point. *The action potential, however, is conducted down the axon as a wave of depolarization-repolarization events* through the following mechanism: depolarization of a given area of the membrane causes current to flow in the axoplasm from the more positive (depolarized) region to neighboring regions. This, in turn, triggers an action potential across the neighboring section of membrane, and so forth (Figure 35–5). Thus the action potential provides a mechanism whereby the transmitted signal is constantly amplified to maintain a constant amplitude. In a regular cable without amplification, the propagated pulse decreases with distance due to resistance and leakage. Without the action potential, a current pulse would therefore be reduced to an insignificant level after traveling a very short length along the axon. In the nerve cells of invertebrate

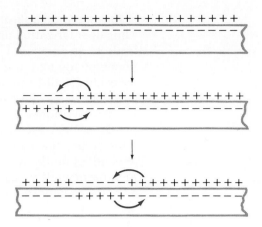

Figure 35–5

Nerve impulses in unmyelinated nerves are conducted through local current movements that propagate the action potential. A portion of the resting axonal membrane is shown at the top. Arrival of an action potential causes local depolarization of the membrane (*middle*), which is propagated from left to right by local currents shown by the arrows. (Adapted from A. L. Hodgkin, *Proc. R. Soc. Lond.* [*Biol.*] 148:1, 1957.)

animals, as well as in many cells in vertebrates, nerve impulses are therefore conducted along axons and dendrites by these local currents and action potentials.

The axons of many nerve cells of higher animals, however, are also surrounded by a multilayered *myelin sheath*, each layer consisting of a typical lipid bilayer membrane (Figure 35–6). At intervals, the spacing of which depends on the fiber diameter, this myelin insulation is interrupted by the so-called *nodes of Ranvier* (Figure 35–7). Nerve impulses in these types of nerve fibers are conducted in a *saltatory* manner, with action potentials "jumping" from node to node where the axonal membranes are in direct contact with the extracellular fluid (Figure 35–7). The insulation provided by the myelin sheath allows for efficient current conduction within the axoplasm by preventing signal loss, and as a result, this type of impulse propagation can be much more rapid than that observed in unmyelinated fibers of similar diameter. Indeed, impulse propagation velocities of *over 100 m/sec* have been recorded in myelinated nerve fibers.

Nerve Cell Membranes Have Separate Channels for K^+ and Na^+

By a slight modification of the experimental setup shown in Figure 35–4 it is possible to hold the membrane potential across an axonal membrane constant at any predetermined value. This is accomplished by connecting the recording electrodes to a device called a *feedback amplifier*, which compensates for any potential change sensed by these electrodes by applying a current to keep the voltage constant. In such a *voltage-clamped* situation, ionic movements can be inferred from the amount of current necessary to hold $\Delta \psi$ at its predetermined value.

It was the use of such a voltage clamp by Hodgkin and Huxley that allowed them to deduce the movements of Na^+ and K^+ that accompany the appearance of the action potential. Because the changes in membrane permeability to Na^+ and K^+ were not superimposable in time, it was tentatively concluded that two different "channels" were involved in the transmembrane movements of these ions. This conclusion has since been confirmed by the discovery of compounds that specifically block the conductance of the nerve membrane to either Na^+ or K^+. *Tetrodotoxin* and *saxitoxin* (Figure

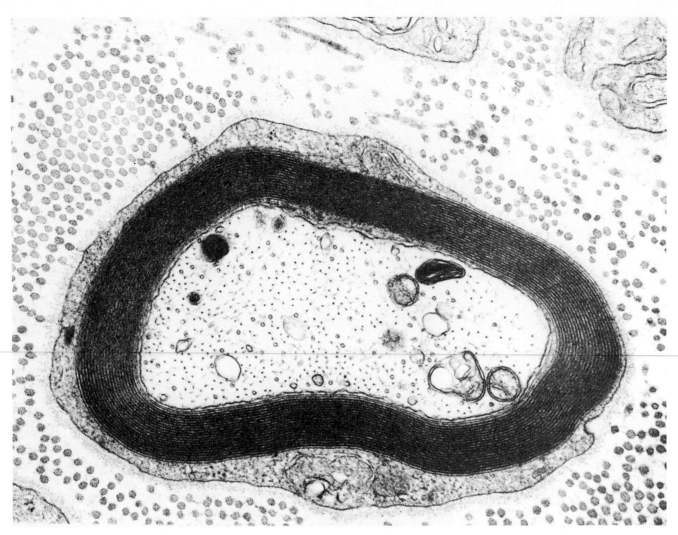

Figure 35–6
Cross section of a myelinated nerve axon from the superior cervical ganglion of the rabbit. Note the multilayered membrane of the myelin sheath that serves as an electric insulator. (Micrograph courtesy Dr. T. Lentz.)

Figure 35–7
Schematic illustration of a longitudinal cross section of a myelinated axon. The sheath is interrupted at various intervals by the nodes of Ranvier, where the cytoplasmic membrane is exposed to the surrounding fluid. Nerve impulses are conducted in a saltatory manner in myelinated axons, as illustrated in this schematic diagram. Impulse propagation is from left to right and the arrows show the accompanying local current movements.

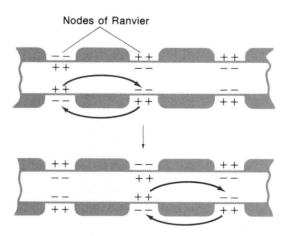

35–8) are both nerve poisons that specifically inhibit the transmembrane movement of Na^+ by binding to the outside of nerve membranes without affecting K^+ permeability, as measured in voltage-clamp experiments. Similarly, *tetraethylammonium ions* (Figure 35–9) have been shown to bind to

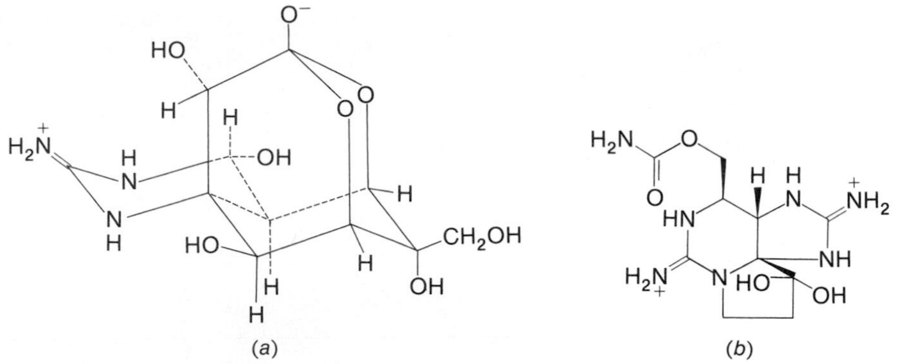

Figure 35–8

(a) The structure of tetrodotoxin, a compound that specifically blocks Na^+ channels in nerve cell membranes. This extremely toxic compound is found in the liver and ovaries of the Japanese puffer fish (*Spheroides rubripes*). (b) The structure of saxitoxin, a compound found in certain marine dinoflagellates ("plankton"), which are constituents of the so-called red tide. Mussels and clams that have fed upon these organisms are therefore extremely poisonous, and commercial shell fishing is banned in areas where these dinoflagellates appear. Saxitoxin is also a Na^+-channel blocking agent, and both compounds most probably interact with the channel through their positively charged guanidino groups.

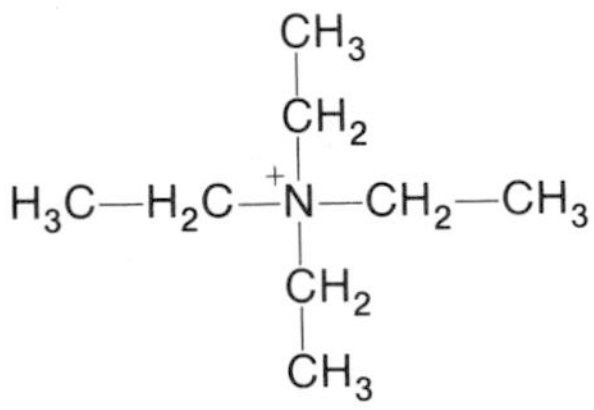

Figure 35–9

Tetraethylammonium ion, a compound that specifically inhibits K^+ channels in nerve cell membranes. The ethyl groups presumably sterically hinder passage of this cation through the channel, "plugging" the channel and therefore blocking the transport of K^+.

the axoplasmic membrane surface and to specifically block the outward flow of K^+.

A large body of evidence has now accumulated that supports the suggestion that K^+ and Na^+ movements through axonal membranes are mediated by separate channels that act as gated pores (Chapter 34). During a typical action potential lasting about 5 ms, it has been calculated that *at least 60 Na^+ ions pass through a single Na^+-specific channel* (i.e., about 12,000 per second). Since the neuronal Na^+-K^+ ATPase normally pumps less than 200 Na^+ ions per second out of the cell, this pump does not seem to be directly involved in the development of the action potential. Instead, a process involving passive Na^+ diffusion through a transmembrane pore appears to best explain these rapid Na^+ fluxes. Furthermore, voltage-clamp studies by B. Hille have established that *molecules such as the K^+-monohydrate complex with dimensions larger than about 0.3 to 0.5 nm^2 in cross section* (the size of a monohydrated Na^+ ion) *do not pass readily through the Na^+ channel*. However, smaller molecules, such as monohydrated Li^+ ions, and cations approximately the same size as a hydrated Na^+ ion, such as hydroxylamine and hydrazine, do. This observation, that is, exclusion based on size rather than on chemical structure, is more characteristic of a pore than a specific membrane permease (Chapter 34).

Additional properties of the Na^+ channel have been inferred by extensions of the specificity studies just outlined. Thus the pH dependence of Na^+ permeation as well as other experimental observations suggest that a *negatively charged carboxylate group* ($—COO^-$) *is present in the pore* and is involved in interacting with the cationic molecules that can traverse the Na^+ channel. It is presumably this interaction that allows the positively charged guanidinium groups of tetrodotoxin and saxitoxin (Figure 35–8) to bind to the channel, but the size of these poisons prevents their passage through the membrane and results in blockage of Na^+ conduction. An additional observation is that methylamine, a cation with dimensions that should allow its more or less free passage through the pore, is nevertheless much less permeable than Na^+. This observation has been interpreted as evidence that the sodium channel has one or more oxygen atoms at its narrowest point, including the carboxylate group. Monohydrated Na^+, hydroxylamine, and hydrazine are thought to form hydrogen bonds with these oxygen atoms

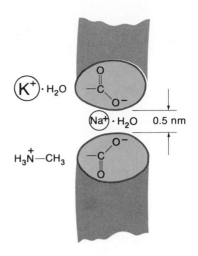

Figure 35–10
Schematic representation of a sodium channel in a nerve cell membrane. The protein(s) comprising the channel are believed to form a pore with dimensions such that only molecules with a diameter ≤0.5 nm will pass through (excluding ions such as monohydrated K^+). The impermeability of methylamine, which is about the size of monohydrated Na^+, suggests that other interactions, such as hydrogen bonding to one or more oxygen atoms within the channel, are also necessary for penetration of an ion through the pore to occur.

through their H_2O, —OH, and —NH_2 groups, respectively, thereby facilitating passage of these cations through the channel. The relative impermeability of methylamine can thus be explained by the inability of its —CH_3 moiety to participate in this interaction (Figure 35–10).

It has been possible to estimate the number of Na^+ channels in a variety of nerve types by the use of radioactively labeled tetrodotoxin or saxitoxin molecules. From these studies it has become apparent that in unmyelinated nerve fibers, the density of these channels in the membrane is exceedingly low. For example, the olfactory nerve of the garfish has only about 30 to 40 Na^+ channels per square micrometer of membrane, which corresponds to only about 0.2 per cent of the total surface area of the phospholipid bilayer. In the squid giant axon, which has a fiber diameter some 2500 times that of the garfish olfactory nerve, this value is only increased to several hundred Na^+ channels per square micrometer, or about 2 per cent of the surface area. The situation is quite different, however, in myelinated nerves of vertebrate animals. In this case, Na^+ channels are found in significant numbers only at the nodes of Ranvier, as might be anticipated from the mechanism of impulse propagation in these types of nerve cells (Figure 35–7). Here, the channel density is on the order of 10^4 channels per square micrometer, corresponding to approximately 60 per cent of the total membrane surface area. Indeed, Na^+ fluxes in these regions of myelinated nerves have been estimated to be 10 to 100 times larger than those in their unmyelinated counterparts during propagation of the action potential.

In contrast to the Na^+ channel, much less is known about the properties of the channel responsible for the outward flow of K^+ from the axon during nerve excitation. This is due partly to the fact that there are even fewer K^+ channels than Na^+ channels in the axonal membrane (about one-tenth the number in the squid giant axon). Furthermore, tightly binding specific inhibitors of the K^+ channel, analogous to the poisons of the Na^+ channel, have not been found. Nevertheless, it has been demonstrated that the K^+ pore is quite specific, barring the passage of cations both smaller (Na^+) and larger (Cs^+; tetramethylammonium ions) than potassium. It seems likely, therefore, that both the diameter of the channel and specific interactions of channel components with the $K^+ \cdot H_2O$ complex confer cation selectivity on this channel as well.

Gated Pores Regulate Na^+ Ion Transmembrane Flux

How do the channels selective for Na^+ and K^+ ions "sense" the membrane potential and react by opening or closing during various stages of action-potential propagation? Although this question is still far from being answered in molecular detail, a number of clues are emerging from recent work on the gating properties of these channels and their purification and reconstitution into artificial membrane systems. The results of these investigations suggest that *ion channels in nerve and muscle cell membranes can assume more than one conformation, one of which is more permeable to the ion in question than the others.*

A fairly recent observation that bears on this question is the discovery of *gating currents*, first predicted by Hodgkin and Huxley. Under appropriate conditions, a small current opposite in direction to that carried by Na^+ during the opening of the Na^+ channels can be detected during the development of an action potential. This current is short in duration (0.1 ms) and precedes the opening of the sodium channels (as detected by the inward movement of Na^+) (see Figure 35–11). One attractive model is that the Na^+ channel has a "built in" or closely associated "voltage sensor" that has a large dipole moment. Depolarization of the membrane, which elicits the action potential,

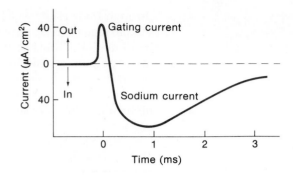

Figure 35–11

Illustration of the gating current that precedes the inward-directed Na$^+$ current associated with depolarization during the action potential. The gating current is believed to be related to the voltage-dependent opening of the Na$^+$ channels and perhaps may reflect rearrangement of a dipolar "voltage sensor" associated with the channel gate. (Adapted from C. M. Armstrong and F. Bezanilla, Charge movement associated with the opening and closing of the activation gates of the Na channels, *J. Gen. Physiol.* 63:533, 1974.)

would then result in the displacement or rearrangement of this dipole in response to the electric field. This process would be detected as the gating current. The change in orientation of this dipole could then be the "trigger" by which the Na$^+$ channel is opened, presumably by means of a conformational change in this protein.

Additional evidence for voltage-dependent conformational changes that allow gating of the Na$^+$ channel has been obtained using nerve poisons isolated from North African scorpions and the sea anemone. These toxins, which are basic polypeptides, bind to sites on the Na$^+$ channel that are distinct from those which bind the channel-blocking agents tetrodotoxin and saxitoxin. They appear to exert their physiological effects by *slowing inactivation* of the Na$^+$ channel after its initial activation (opening) during the action potential. Binding of scorpion toxin to Na$^+$ channels in nerve and muscle-cell membranes has been shown to be *voltage-dependent*, suggesting that this toxin recognizes a conformation of the channel that also depends on the membrane potential. This observation has led W. A. Catterall to hypothesize that toxins of the scorpion type may interact with the voltage-sensor component of the channel, perhaps the same moiety responsible for the gating current described earlier. Toxin binding would thus reflect the conformational state of the channel, which, in turn, depends on $\Delta \psi$. These and other experiments have led to the proposal that the Na$^+$ channel can exist in at least three conformations, depending on $\Delta \psi$: *closed* (resting $\Delta \psi$), *open*, and *inactive*. The latter two states are responsible for the transient increase and then decrease in Na$^+$ permeability observed during the action potential. Hyperpolarization as a result of increased K$^+$ permeability then recycles the channel to its closed conformation.

Perhaps the most promising approach to understanding structure-function relationships in the ion channels of nerve-cell membranes will be to purify these proteins and to study their properties in artificial membrane systems, as we described for transport permeases in Chapters 6 and 34. Significant progress has already been made in this direction. For example, M. A. Raftery and coworkers have purified a tetrodotoxin-binding protein from the *electric organ* of the electric eel (*Electrophorus electricus*) that is rich in Na$^+$ channels. A major component of this preparation was a protein with a polypeptide-chain molecular weight of about 260,000. Similarly, W. A. Catterall and collaborators have purified from rat brain a saxitoxin receptor that consists of three polypeptides, α, β_1, β_2, with molecular weights of 260,000, 39,000, and 37,000 respectively. They also have demonstrated that similar-sized polypeptides can be labeled in cultured neuroblastoma cells with a photoactivatable derivative of scorpion toxin. Thus Na^+ channels from quite different sources appear to be proteins of similar structure.

Reconstitution of several highly purified preparations of the Na^+ channel has been achieved. In accordance with the behavior of native Na^+ channels, the purified and reconstituted rat brain Na^+ channel was activated by channel activators such as veratridine and blocked by channel inhibitors such as saxitoxin and tetrodotoxin. Local anaesthetics such as tetracaine and lidocaine also blocked Na^+ fluxes through the channel, just as in native membranes. The cation specificity of the reconstituted complex was the same as that of the native Na^+ channel: $Li^+ = Na^+ > K^+ > Rb^+ > Cs^+$. Further, it retained the same voltage dependency in its purified form as was observed for the complex in its native environment. These results establish that the purified $\alpha\beta_1\beta_2$ complex is sufficient for most of the biological functions attributed to the Na^+ channel in nerve and muscle cells. The recent demonstration that the Na^+ channel can be phosphorylated by a cyclic AMP-dependent protein kinase adds a new degree of complexity to an understanding of the regulation of its function. Recently, S. Numa and collaborators have cloned and sequenced the structural gene coding for the large subunit of the Na^+ channel. Analysis of the deduced amino acid sequence of the protein suggests that it may consist of four similar intramembrane domains, which could possibly form the ion pore in the membrane. Correlation of this emerging structural information with Na^+ channel function, as well as assignment of specific functions to the smaller subunits, remains as work for the future.

The Synapse Is a Chemical Connection for Communication Between Nerve Cell and Target Cell

Nerve impulses, propagated along the axon by the mechanisms outlined in the preceding sections, must be transmitted between nerve cells or from nerve cells to muscle or glandular tissues in order for their effects (e.g., contraction or secretion) to take place. This type of *intercellular communication can occur either by an electrical mechanism of coupling or by chemical transmission across specific connections between nerve cells and target cells called synapses.* An example of electrical transmission was given in Chapter 34. Direct cell-cell contact by means of *gap junctions* can allow action potentials to be transmitted by ionic mechanisms between cells connected in this manner. The more common mechanism, however, involves the release of specific chemicals, called *neurotransmitters*, at the *presynaptic membrane* of one cell in response to membrane depolarization and their diffusion to *receptors* in the *postsynaptic membrane* of the recipient cell, where a new action potential may be generated.

The process of synaptic transmission is depicted schematically in Figure 35–12 for *acetylcholine*, the excitatory neurotransmitter at vertebrate neuromuscular junctions (motor end plates). Acetylcholine is synthesized in nerve cells by the enzyme *cholineacetyltransferase* (Figure 35–13) and is packaged in units ("quanta") of 10^3 to 10^4 molecules within *synaptic vesicles*, which are abundant near the cytoplasmic membrane of the presynaptic axon. The arrival of an action potential triggers a large increase in the permeability of the presynaptic membrane to Ca^{2+}, and this ion flows into the axoplasm down its chemical gradient. Fusion of the synaptic vesicles with the plasma membrane and the concomitant release of acetylcholine into the *synaptic cleft* are promoted by this increase in intracellular Ca^{2+}. Several hundred synaptic vesicles empty their contents into the synaptic cleft in a typical neuromuscular synaptic junction by this mechanism in response to a

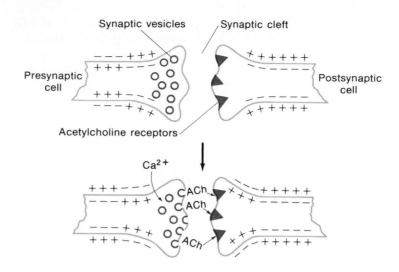

Figure 35–12
Schematic diagram of a synaptic junction in which acetylcholine is the chemical transmitter. Arrival of an action potential at the terminus of the presynaptic cell *(top)* stimulates Ca^{2+} uptake, which triggers release of acetylcholine (ACh) from vesicles near the terminus of the presynaptic cell. Release is accomplished by vesicle fusion with the plasma membrane, and the interaction of acetylcholine with its receptors in the postsynaptic membrane triggers depolarization, thus propagating the action potential in the postsynaptic cell *(bottom)*.

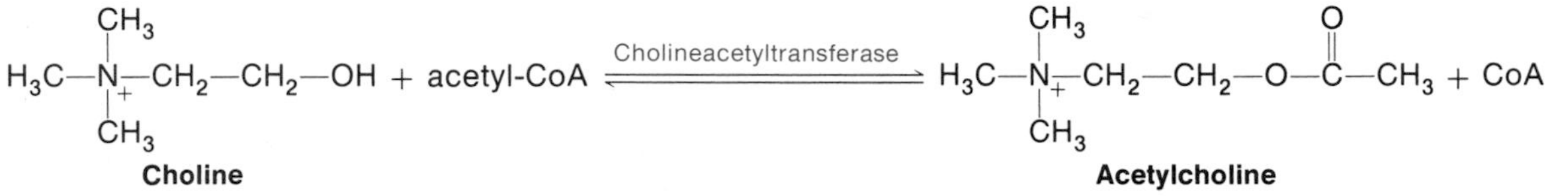

Figure 35–13
Acetylcholine is synthesized by cholineacetyltransferase, which esterifies choline with an acetyl group from acetyl-CoA.

single action potential. The resultant large increase in the local concentration of acetylcholine is "sensed" by a protein, the *acetylcholine receptor*, located in the cytoplasmic membrane of the postsynaptic cell. Binding of the neurotransmitter to many receptor molecules triggers an action potential in the recipient cell. The acetylcholine is subsequently rapidly degraded into acetate and choline by an enzyme in the synaptic cleft called *acetylcholinesterase*, and the resting potential of the postsynaptic membrane is rapidly restored.

How does the binding of a neurotransmitter to its receptor promote depolarization of the postsynaptic membrane? One of the first clues came from studies by B. Katz and collaborators who showed that *acetylcholine increases the ionic permeability of the postsynaptic membrane*. Subsequent studies with radioactive tracers and by the voltage-clamp technique established that *the membrane permeabilities to both Na^+ and K^+ were increased simultaneously by the action of acetylcholine*. Because the electrochemical Na^+ gradient is somewhat larger than that of K^+ across postsynaptic membranes, depolarization results from the inward flow of Na^+, tending to collapse $\Delta\psi$. This local perturbation in $\Delta\psi$ is enough to initiate a new action potential in the recipient neural or muscular membrane *as long as a sufficient number of receptor molecules bind the neurotransmitter*. In fact, as we shall see in a later section, *the receptor itself contains the ion channel* through which both K^+ and Na^+ can flow. The number of occupied receptors at a given moment thus dictates the magnitude of the inward flow of Na^+ and the resulting magnitude of the change in the membrane potential. Rapid hydrolysis of acetylcholine by acetylcholinesterase bound to the postsynaptic membrane then quickly reduces the number of transmitter-receptor complexes and repolarizes the membrane until a new action potential triggers the release of more acetylcholine quanta from the presynaptic membrane.

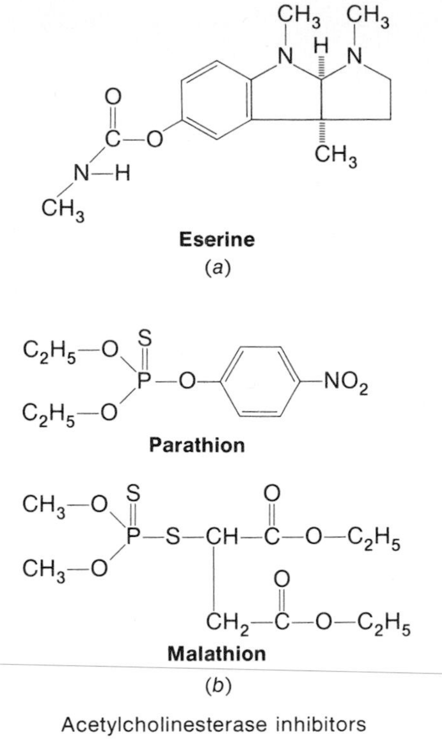

Eserine

(a)

Parathion

Malathion

(b)

Acetylcholinesterase inhibitors

Figure 35–14
Acetylcholinesterase inhibitors.
(a) Eserine, or *physostigmine*, is an alkaloid that forms a relatively stable covalent carbamoyl intermediate with an active-site serine on the enzyme that is hydrolyzed only very slowly. *(b)* Parathion and malathion are organophosphorus compounds that also form stable covalent complexes with the active-site serine of acetylcholinesterase. They are widely used agriculturally as insecticides.

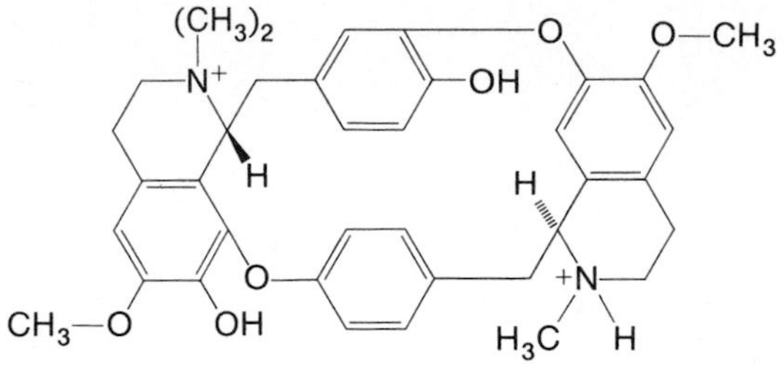

Figure 35–15
The structure of d(+)-tubocurarine, an active component of the neurotoxin curare. This compound binds to the acetylcholine-binding site on its receptor, preventing synaptic transmission and subsequent depolarization of the postsynaptic cell membrane.

The elucidation of the steps just outlined has been aided greatly by the use of specific inhibitors of both acetylcholinesterase and the acetylcholine receptor. Compounds such as *eserine* (Figure 35–14a) block the hydrolytic enzyme and thus can be used to study the effects of acetylcholine in *cholinergic* systems (those using acetylcholine as a transmitter) under conditions where this neurotransmitter cannot be hydrolyzed. Likewise, certain organophosphorus compounds efficiently inhibit acetylcholinesterase by forming stable covalent intermediates with an active-site serine in the enzyme. The widely used insecticides *parathion* and *malathion* are examples of this class of nerve poisons (Figure 35–14b). Neuromuscular junctions exposed to acetylcholinesterase inhibitors are paralyzed because the persistent presence of acetylcholine prevents repolarization of the postsynaptic membrane to restore its excitability. In fact, such a situation eventually results in the acetylcholine receptor becoming *desensitized*, i.e., remaining closed to ion flow for long intervals even in the presence of the neurotransmitter.

Specific blocking agents of the acetylcholine receptor include *d-tubocurarine*, an active component of the neurotoxin curare (Figure 35–15), and the snake venom poisons *α-bungarotoxin* (from snakes of the genus *Bungarus*) and *cobratoxin*. The latter are small basic proteins with masses around 7000 daltons. All three of these substances interact noncovalently with the receptor and interfere with acetylcholine binding, thus blocking depolarization of the postsynaptic membrane. They are referred to as *antagonists* of the cholinergic systems. Another type of acetylcholine receptor-inhibitor is exemplified by the divalent cation *decamethonium* (Figure 35–16), which "locks" the ion channel of the receptor in the open state and thus leads to a constant depolarization of the recipient cell membrane. Such compounds, referred to as *agonists*, mimic the effect of acetylcholine but cannot be rapidly inactivated, thus blocking resensitization of the postsynaptic membrane. These substances have allowed workers to investigate properties of the acetylcholine receptor in the "open" and "closed" states and their effects on the permeability of the postsynaptic membrane. They also have been useful in the purification of this protein, as we shall describe shortly.

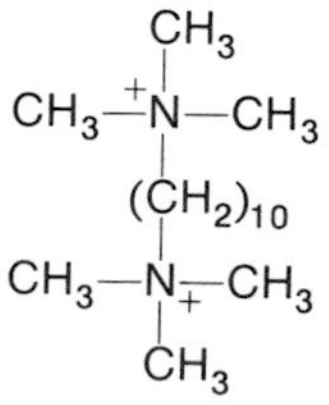

Figure 35–16
Decamethonium ion, an agonist of cholinergic systems. This compound binds to the acetylcholine receptor, but because it cannot be degraded, it causes persistent depolarization of the postsynaptic membrane.

A Number of Compounds Serve As Neurotransmitters in Addition to Acetylcholine

A number of other compounds have been implicated as neurotransmitters in addition to acetylcholine. The best-documented examples of these are the *catecholamines*, certain *amino acids* (and derivatives), and a variety of *peptides*. The catecholamines are all derived biosynthetically from L-tyrosine (Figure 35–17) and include the hormones *norepinephrine* and *epinephrine* (adrenaline). Because these compounds are also synthesized in the adrenal

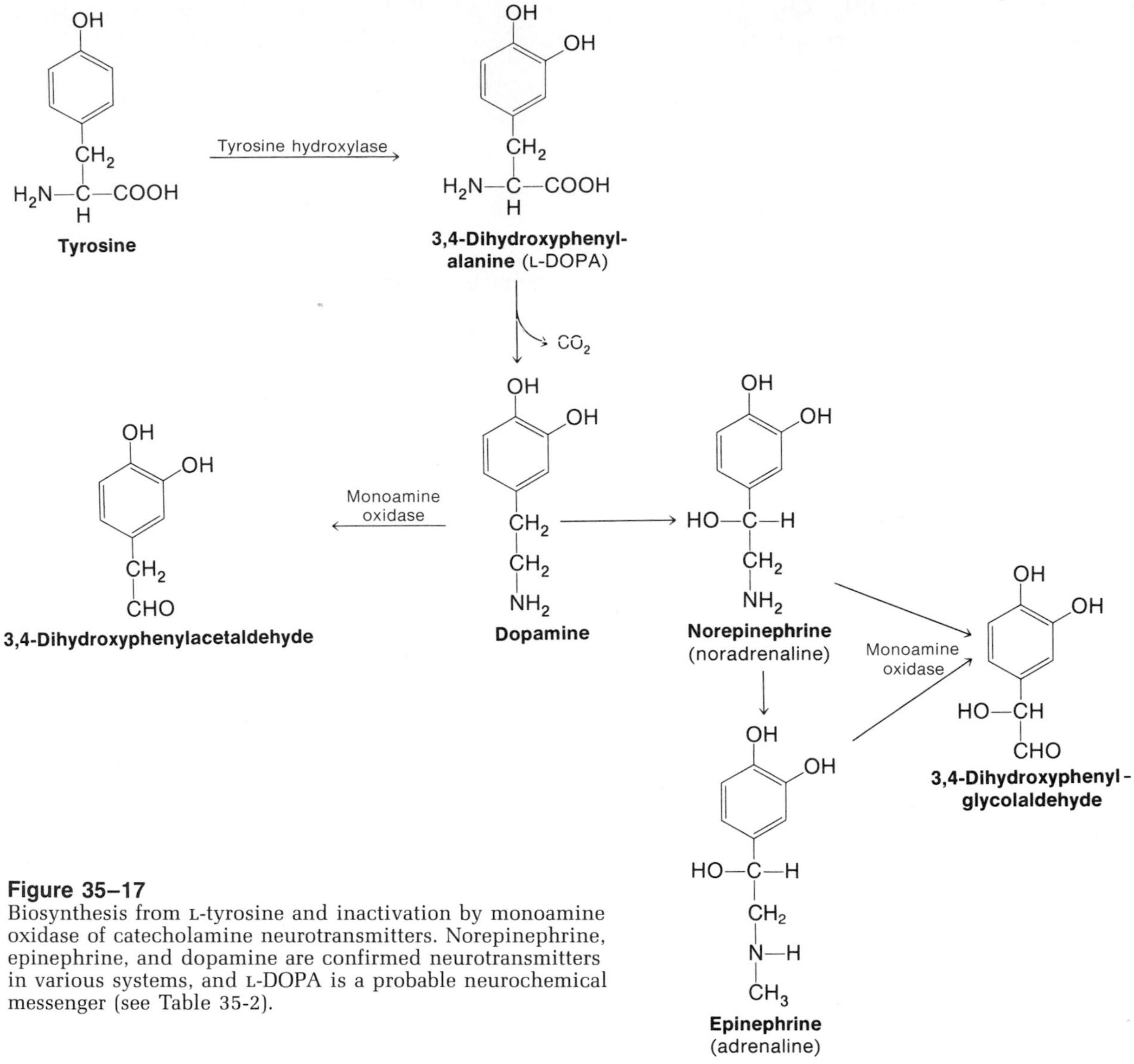

Figure 35–17
Biosynthesis from L-tyrosine and inactivation by monoamine oxidase of catecholamine neurotransmitters. Norepinephrine, epinephrine, and dopamine are confirmed neurotransmitters in various systems, and L-DOPA is a probable neurochemical messenger (see Table 35-2).

gland (see Table 31–1), neurons that use these substances as chemical transmitters are said to be *adrenergic*. Sympathetic nerve fibers that innervate smooth-muscle cells in internal organs such as the heart, spleen, and gut have been shown to release norepinephrine at their terminals by mechanisms similar to those used by cholinergic neurons. Norepinephrine and related amines also have been shown to serve as neurotransmitters in a number of nerve pathways in the brain. Like acetylcholine, catecholamines may be inactivated by chemical modifications. Inactivation may be effected by a methylation reaction or by an oxidation reaction catalyzed by the enzyme *monoamine oxidase* (Figure 35–17). In some cases, they also can be resorbed through the presynaptic membrane after their release, providing an additional mechanism for removal from the synaptic cleft.

The hydroxylation of tyrosine, which is the first unique step in the biosynthesis of catecholamines, is catalyzed by *tyrosine hydroxylase* and yields the compound *3,4-dihydroxyphenylalanine* (L-DOPA). The neurological disorder *Parkinson's disease* is associated with an underproduction in the human brain of the catecholamine transmitter *dopamine,* which is derived from L-DOPA by a decarboxylation reaction (Figure 35–17). L-DOPA has

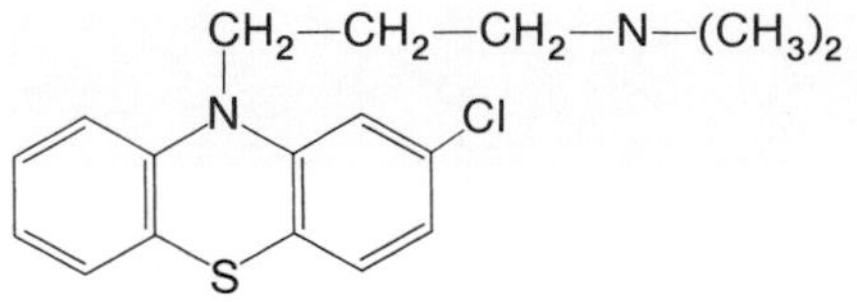

Figure 35–18

The structure of chlorpromazine, a drug that blocks dopamine receptors and which has been used in the treatment of psychological disorders such as schizophrenia.

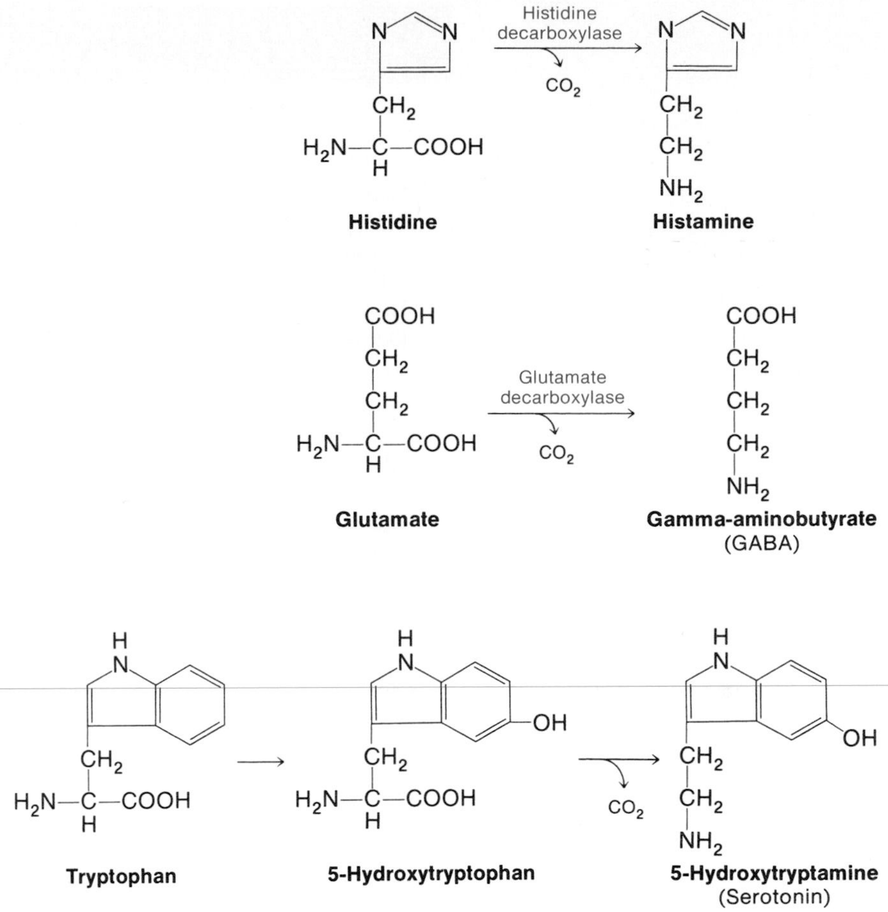

Figure 35–19

Biosynthesis of the neurotransmitters histamine, gamma-aminobutyrate (GABA), and serotonin (5-hydroxytryptamine).

therefore been found to be an effective drug in many instances in the treatment of Parkinson's disease. Interestingly, overproduction of dopamine in the brain also occasionally occurs and appears to be associated with psychological disorders such as schizophrenia. In this case, _dopamine-receptor blocking drugs_, such as _chlorpromazine_ (Figure 35–18), have been found to be useful therapeutic agents.

Amino acids that are believed to have roles as neurotransmitters include _glutamic acid_ and _glycine_. The amino acid derivatives _histamine_ (synthesized by the decarboxylation of histidine), _5-hydroxytryptamine_ (or _serotonin_; derived from tryptophan), and _gamma-aminobutyric acid_ (or _GABA_; a decarboxylation product of glutamic acid) have all been shown to be transmitters in various systems as well. The reactions involved in the biosynthesis of these compounds are diagrammed in Figure 35–19. Gamma-aminobutyrate is used most often as an _inhibitory transmitter_. Interaction of this compound with its receptor on many postsynaptic membranes results in a large _increase in the membrane permeability to Cl^- and/or K^+ ions_. This inhibits the postsynaptic cell, often by hyperpolarizing its membrane (recall, for example, that the equilibrium potential of K^+ is more negative than the resting potential). _Most target tissues, in fact, are innervated by more than one type of nerve fiber, each using a different neurotransmitter. This allows for different signals to be relayed to the recipient cells, some stimulatory and others inhibitory._ Given the complexity of the nervous systems of higher animals, additional neurotransmitters are likely to be discovered. Some compounds currently thought to be chemical transmitters are listed in Table 35–2.

TABLE 35–2 Neurochemical Messengers	
Compounds	**Status**[a]
1. Acetylcholine	C
2. Catecholamines	
Norepinephrine (noradrenaline)	C
Epinephrine (adrenaline)	C
L-DOPA	P
Dopamine	C
Octopamine	C
3. Amino acids (and derivatives)	
Glutamate	C
Aspartate	P
Glycine	C
Proline	Pos
Gamma-aminobutyrate (GABA)	C
Tyrosine	Pos
Taurine	P
Alanine	Pos
Cystathione	Pos
Histamine	C
Serotonin (5-hydroxytryptamine)	C
4. Peptides	
Substance P	P
Cholecystokinin	P
Neurotensin	P
Enkephalins	P
Somatostatin	P

[a]C = confirmed neurotransmitter; P = probable; Pos = possible.

The Acetylcholine Receptor Is the Best-understood Neurotransmitter Receptor

The structure and properties of the acetylcholine receptor are by far the best understood among all neurotransmitter receptors. One reason for this is their abundance in postsynaptic membranes found in the electric organs of the electric eel (*Electrophorus*) and the electric ray (*Torpedo*). These organs contain stacks of cells (*electroplaxes*), each cell of which receives nerve endings on one side, but not on the other. Release of acetylcholine by the presynaptic nerve endings thus depolarizes the postsynaptic membrane by virtue of the binding of neurotransmitter to the membrane-bound receptors, while the other face of the cell remains at its resting potential. In this way, each cell, when stimulated, can attain a potential difference of over 100 mV between its two faces, and thousands of such stacked cells, present in the electric organ, can consequently emit an electric discharge of several hundred volts.

Acetylcholine-receptor-rich membranes can easily be isolated from the electric organs of *Electrophorus* and *Torpedo*. These membranes contain from 10 to 50×10^4 receptor sites per square micrometer, as measured by the binding of radioactively labeled antagonists such as α-bungarotoxin. Electron microscopy and x-ray diffraction have further revealed that such membrane fragments have a similar density of particles, each about 8 nm in diameter, arranged in a regular hexagonal lattice (Figure 35–20). Each particle is actually a rosette, apparently consisting of five subunits arranged around a central axis, as shown in Figure 35–20. The density of these oligomeric protein molecules on the membrane, their size (see below), and their interac-

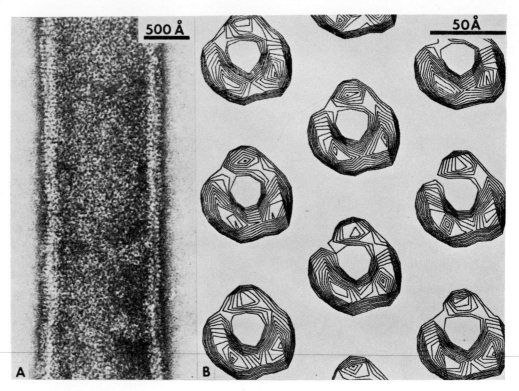

Figure 35–20
A view of the arrangement of acetylcholine receptors on electroplax membranes from
Torpedo californica. (a) Membrane tubes formed spontaneously from membrane
vesicles showing a crystalline surface lattice of acetylcholine receptors. Stained with
uranyl acetate. (b) Computer filtration of micrographs similar to that in (a), showing
the arrangement of protein density in a single receptor around the central depression
(presumed to be the ion channel). (From J. Kistler, R. M. Stroud, M. W. Klymkowsky,
R. A. Lalancette, and R. H. Fairclough, *Biophys. J.* 37:371, 1982.) (Micrograph courtesy
Dr. R. Stroud. Used with permission.)

tion with antibodies leave little doubt that they are the acetylcholine receptors themselves. Furthermore, closed membrane vesicles derived from such
membrane fragments ("microsacs") can be made permeable to Na^+ by adding agonists of the acetylcholine receptor, an effect that is blocked by antagonists such as the snake venom toxins.

Purification of the acetylcholine receptor to apparent homogeneity has
been achieved by solubilization of electroplax membranes of *Torpedo* by
nonionic detergents and affinity chromatography on columns containing
covalently bound cobratoxin. Alternatively, preparations of similar purity
can be obtained by fractionation of membrane fragments using sucrose density-gradient centrifugation and subsequent removal of peripherally associated proteins by treatment at high pH values. Both these procedures yield a
glycoprotein consisting of four subunits ($M_r = 40$, 50, 60, and 65×10^3,
named α, β, γ, and δ) in a molar ratio of $2:1:1:1$. The simplest structure for
the acetylcholine receptor is therefore a pentamer $(\alpha_2\beta\gamma\delta)$ with a mass of
about 255,000 daltons. This value agrees with hydrodynamic measurements
of the molecular weight of the purified receptor in detergent, as well as with
the dimensions and apparent subunit composition of the particles seen by
electron microscopy of acetylcholine-receptor-rich membranes (Figure 35–
20). Interestingly, the four different subunits of this protein, all of which
have been fully sequenced, exhibit extensive amino acid sequence homolo-
gies, indicating that they all arose by duplication and divergence of a single
ancestral gene. Computer analyses of the amino acid sequences of the four

subunits suggest a common motif of secondary structure for proteins of the subunits inside the lipid bilayer.

Reactive affinity labels that covalently bind to the acetylcholine receptor have been prepared by chemically modifying known agonists and antagonists of cholinergic systems. In most cases examined, the α subunit was labeled by these treatments. It is now generally accepted that this subunit contains the binding site (or sites) for acetylcholine in the receptor complex. The purified receptor from *Torpedo* has been functionally reconstituted into vesicles consisting of lipids isolated from electroplax membranes. These proteoliposomes became permeable to Na^+ in the presence of an acetylcholine analog, while α-bungarotoxin blocked Na^+ permeability. The availability of this experimental system should allow detailed studies of structure-function relationships in the receptor protein, e.g., the determination of which subunits are necessary to form the ion channel and whether agonist/antagonist binding sites can be separated from the channel-forming parts of the molecule.

Because all five subunits of the acetylcholine receptor span the postsynaptic membrane, as demonstrated by their susceptibility to tryptic hydrolysis at both membrane surfaces, it is possible that the channel corresponds to the central "hole" in the rosette seen by electron microscopy of electroplax postsynaptic membranes (Figure 35–20). In fact, the five membrane-spanning subunits of the acetylcholine receptor have recently been resolved in electron microscope images. They consist of five rod-shaped structures, of about equal cross section, which lie largely perpendicular to the membrane plane. The five subunits are contained within a pentagonally symmetrical cylindrical shell, 140 Å long and 80 Å in diameter, which delineates a water-filled channel along the central axis of the cylinder. The cylinder extends about 40 Å into the cytoplasm and nearly 70 Å into the synaptic cleft. The channel opening to the synaptic cleft is wide but narrows as it extends through the membrane into the interior of the cell. This information provides a clear conceptual picture of the acetylcholine receptor and allows us to visualize possible conformational alterations that accompany ligand binding.

Considerable evidence points to a gated-pore-type mechanism for ion permeability conferred by the acetylcholine receptor, as is the case for the Na^+ channel. At least two conformations of the protein, corresponding to "open" and "closed" states, have been recognized in the presence of agonists and antagonists, respectively. Furthermore, voltage-clamp studies of muscle fibers containing cholinergic receptors in the presence of agonists have revealed current pulses having a square shape and a constant amplitude similar to those seen with bacterial porins and certain ionophore antibiotics in artificial membrane systems (Chapter 34). If these events correspond to the opening and closing of individual channels, as seems likely, then it can be calculated that *about 10^4 molecules of Na^+ flow through the receptor in vivo in the millisecond or so that it is open*. A pore model allowing more or less free diffusion of ions through the channel in its open state, rather than a slower, carrier-mediated mechanism, is therefore favored for the acetylcholine receptor.

SUMMARY

Nerve cells are highly specialized for the reception of external signals and transmission of those signals to other cells in the organism. The mechanism of signal transmission within a neuron involves waves of membrane depolarization-repolarization events that travel the length of the axon to the nerve endings. The arrival of such a wave at a single region of the axon can be recorded as an action potential, which is generated by a transient influx of Na^+, followed by a transient efflux of K^+ until the membrane potential returns to its resting state. Transmission of this signal from a neuron to another cell occurs most commonly by the release of chemical neurotransmitters at the synapse, triggered by the arrival of the action potential at the terminus of the axon. Neurotransmitters bind to receptors in the postsynaptic membrane, initiating a response in the target cell, again triggered by the influx of cations. Recent applications of molecular biological techniques to the study of the proteins comprising the ion channels and the neurotransmitter receptors hold the promise of providing detailed information on the molecular mechanisms involved in these processes.

It should be pointed out that, although nerve cells are among the most extensively studied excitable cells, excita-bility has been demonstrated not only in other cell types in higher animals but also in organisms as simple as unicellular protozoa and in algae. In ciliated protozoa, for example, membrane depolarization events have been shown to regulate swimming behavior, while in certain algae localized transmembrane movements of ions appear to play a role in early development. It is therefore likely that membrane excitability is a common phenomenon in many different types of cells, and plays a role in a variety of physiological processes. In the following chapter, one such process, vision, is examined in detail. In this case, photons—impinging upon specialized nerve cells containing photoreceptors—initiate nerve impulses that are sent to the brain, which translates these signals into an image of our environment.

SELECTED READINGS

Brisson, A., and P. N. T. Unwin, Quaternary structure of the acetylcholine receptor. *Nature* 315:474, 1985. This article describes the three-dimensional electron image analysis of tubular crystals of the acetylcholine receptor grown from native membrane vesicles.

Catterall, W. A., The molecular basis of neuronal excitability. *Science* 223:653, 1984. This short review summarizes the recent biochemical work defining the Na^+ channel using purified, reconstituted preparations. It also reviews briefly the earlier physiological work.

Guy, H. R., and P. Seetharamulu, Molecular model of the action potential sodium channel. *Proc. Natl. Acad. Sci. USA* 83:508–512, 1986.

Hille, B., *Ionic Channels of Excitable Membranes.* Sunderland, Mass.: Sinauer Associates, 1984. This book summarizes recent research on a variety of nerve and muscle channels, including those specific for Na^+, K^+, Cl^-, and Ca^{2+} and a few responsive to neurotransmitters. Theory of ion channel function and techniques of analysis are also reviewed.

Kao, C. Y., and S. R. Levinson (eds.), Tetrodotoxin, saxitoxin and the molecular biology of the sodium channel. *Ann. N.Y. Acad. Sci.* vol. 479, 1986.

Katz, B., *Nerve, Muscle and Synapse.* New York: McGraw-Hill, 1966. Although over 20 years old, this classical book offers an excellent overview of the subject.

Keynes, R. D., Ion channels in the nerve-cell membrane. *Sci. Am.* 240:126, 1979. Reviews properties of the Na^+ and K^+ channels in nerve cell membranes.

Kistler, J., R. M. Stroud, M. W. Klymkowsky, R. A. Lalancette, and R. H. Fairclough, Structure and function of an acetylcholine receptor. *Biophys. J.* 37:371, 1982. Review of the structure and function of this receptor in *Torpedo californica* electroplax membranes.

Kuffler, S. W., and J. G. Nicholls, *From Neuron to Brain.* Sunderland, Mass.: Sinauer Associates, 1976. This book covers aspects of neural transmission and neurophysiology in an easily readable form.

PROBLEMS

1. a. Calculate the membrane potential $\Delta\psi$ across a resting nerve-cell membrane in the presence of tetrodotoxin. (Assume that the resting permeability to Na^+ is due to a small, steady-state level of "open" Na^+ channels.)

b. What would the value of $\Delta\psi$ be if the resting axonal membrane were permeable only to Cl^-? (Assume an axoplasmic Cl^- concentration of 50 mM.)

2. Explain why a nerve-cell membrane exhibits an "all or none" response (i.e., action potential) independent of the magnitude of an electric or chemical stimulus (above a threshold value).

3. List the criteria for demonstrating that a particular compound acts as a neurotransmitter in a given system, assuming that the mechanism of synaptic transmission in most instances is analogous to that found in cholinergic systems.

4. Solubilization of Na^+ channels from nerve membranes can be achieved by treatment with sodium cholate. These channels, as well as other proteins solubilized by this procedure, can be incorporated into proteoliposomes such that each artificial vesicle receives only one, or at most a few, solubilized proteins. "Open" Na^+ channels are known to allow the diffusion of Cs^+, a dense monovalent cation, through the membrane, although the rate of this diffusion is much slower than that of Na^+. Veratridine, an alkaloid that binds to the Na^+ channel (see structure on the following page), causes persistent activation of these channels in the reconstituted state. On the basis of these observations, devise a procedure for purifying proteoliposomes containing the Na$^+$ channel from a heterogeneous population of reconstituted vesicles.

Veratridine

5. In reconstituted transport systems, it is often important to demonstrate that the rate of transport is similar to that observed *in vivo*, i.e., that the transport protein is fully functional in the reconstituted state. For systems in which *in vivo* fluxes are very rapid (e.g., cation flux through the acetylcholine receptor), it is often difficult to measure these rates directly in reconstituted vesicles. Thallous ion (Tl$^+$) is known to pass readily through the "open" state of the acetylcholine receptor. It also very efficiently quenches the fluorescence emission of the fluorophore 8-aminonaphthalene-1,3,6-trisulfonate (ANTS), which is relatively impermeable to phospholipid bilayers.

a. Using this information, outline a series of experiments to measure Tl$^+$ fluxes in reconstituted proteoliposomes containing purified acetylcholine receptors.

b. Actual Tl$^+$ fluxes into proteoliposomes containing an average of two acetylcholine receptor channels per vesicle in the presence of agonists have been measured to be 200 moles/(liter · sec). If the average inner diameter of such vesicles is 400 Å, what is the number of Tl$^+$ ions transported per second by each activated acetylcholine receptor channel? How does this value compare with the rate of Na$^+$ flux measured *in vivo*?

6. Excitable cells, such as neurons, are not limited to multicellular higher organisms. For example, ciliated unicellular protozoa such as *Paramecium* are known to undergo a behavioral response called the "avoidance reaction." When such cells collide with an obstacle, the direction of beating of their cilia (which normally propel the cell in a forward direction) reverses, and the protozoa "back up" to avoid repeated collisions with the obstacle. After a short time, the cells resume their normal forward motion in a new direction. *Paramecium* is known to have a "resting potential" across its cytoplasmic membrane of about -30 mV when swimming normally. The avoidance reaction and the direction of ciliary beating in this organism are critically dependent on the extracellular concentration of Ca^{2+} ions (e.g., no Ca^{2+}, no response). Using this information, as well as your knowledge of the mechanism of action potential production in nerve cells, postulate a reasonable mechanism for the avoidance reaction in *Paramecium*.

36

VISION

The photochemistry of vision is different from that of photosynthesis. Perhaps this is not surprising. Animals that can see use light to obtain information; photosynthetic organisms use light as a source of energy. However, *vision and photosynthesis both start with the excitation of an electron from one molecular orbital to another orbital of higher energy*. The excited molecule must then undergo a transformation to a metastable product. Since chlorophyll can undergo such a transformation with a quantum yield near 1.0, it is not hard to imagine an eye that uses electron-transfer reactions like those that work so well in photosynthesis. In fact, it is possible to make a synthetic eye that works very much that way—a silicon diode array television camera tube. But the eyes of multicellular animals are different. Instead of chlorophyll, their light-sensitive cells contain a complex called *rhodopsin*, which consists of a protein, *opsin*, and a linear polyene, *11-cis-retinal*. Instead of undergoing oxidation when it is excited with light, the retinal isomerizes.

THE VISUAL PIGMENTS ARE FOUND IN MEMBRANES IN ROD AND CONE CELLS

The eyes of vertebrates are marvelously complex organs, with many different types of specialized cells (Figure 36–1). Light rays entering the eye are refracted by the *cornea*, the clear tissue at the front of the eye. The light traverses an aqueous chamber and reaches the lens, which is densely packed with proteins called *crystallins*. Adjustments in the shape of the lens focus a sharp optical image onto the *retina*, a thin layer of tissue that lines the back of the eye. The retina is a neural tissue with several different layers of cells. Some of these cells, *the rod and cone cells, contain the visual pigments. Other cells make synaptic connections to the rods or cones and to additional neural cells that carry impulses to the brain.*

In both rods and cones, the light-sensitive molecules are collected in a layered system of membranes at one end of the cell (Figures 36–2 and 36–3).

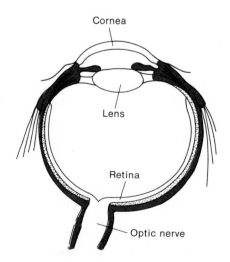

Figure 36–1
Structure of the human eye.

Figure 36–2
Thin-section electron micrograph (59,200×) of a portion of a rod cell in a rabbit retina. Part of the outer segment is shown at the top, and part of the inner segment at the bottom. (D = disks; M = mitochondrion; C = cilium.) (Courtesy Dr. Ann Bunt-Milam.)

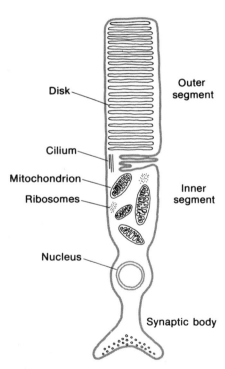

Figure 36–3
Schematic diagram of a rod cell. The orientation of the cell is the same as that in Figure 36–2. In the retina, many rod cells are stacked side by side, with the outer segments all pointing out to the periphery of the eye. Light enters the cells end-on through the inner segment, after passing through several layers of other neural cells. In cone cells, the outer segments are shorter, and are conical rather than cylindrical in shape.

The membranes form by invaginations of the cytoplasmic (plasma) membrane near the middle of the cell. In cone cells, the membranes remain contiguous with the cytoplasmic membrane. In rods, they pinch off to form a stack of autonomous flattened vesicles, or _disks_, in the _outer segment_ of the cell. The outer segment of each rod contains from 500 to 2000 of these disks. This region of the cell is connected by a thin cilium to the _inner segment_, which is packed with mitochondria and ribosomes. The basal part of the cell ends in a synaptic junction with another neural cell called a _bipolar cell_. In a living rod cell, disks are constantly forming at the base of the outer segment and moving in a file to the tip of the outer segment, where they are sloughed off and phagocytosed by the underlying epithelial cells. It takes about 10 days for a disk to make this journey. Exactly why the disks need to be replaced so rapidly is unclear.

Rod cells are specially adapted for vision in dim light. Cones provide visual acuity in bright light, and also serve for perception of color. Animals such as owls, which have very high visual sensitivity in dim light but cannot distinguish colors, have only rod cells. Some animals, such as pigeons, have only cones and are skilled at distinguishing colors in bright light but inept at night vision. Primates have both rods and cones, with rods considerably outnumbering cones in all but the central portion of the retina.

Rhodopsin Consists of 11-_cis_-Retinal Bound to a Protein, Opsin

Figure 36–4 shows the structures of 11-_cis_-retinal and its more stable isomer all-_trans_-retinal. The retinals are related to the alcohol _retinol_, or _vitamin A₁_. These compounds cannot be synthesized _de novo_ by mammals, but they can be formed from carotenoids, such as β-carotene, which are abundant in carrots and some other vegetables. A deficiency of vitamin A causes night blindness, along with a serious deterioration of the eyes and a number of other tissues.

There are animals with eyes in several different phyla, including Mollusca, Arthropoda, and Annelida, in addition to Chordata. The eyes of arthropods show substantial anatomical differences from those of mollusks and chordates, and they apparently originated independently, after the

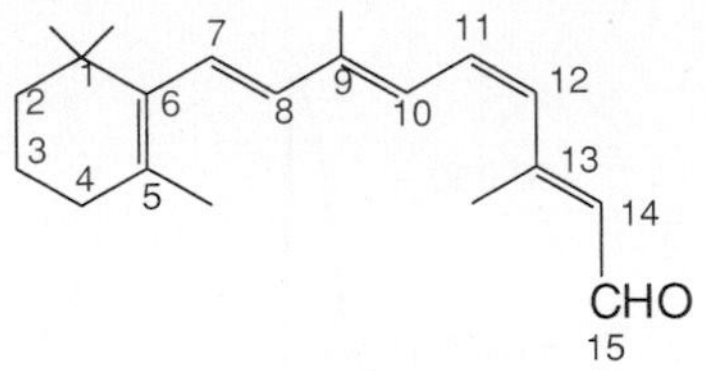

11-*cis*-Retinal

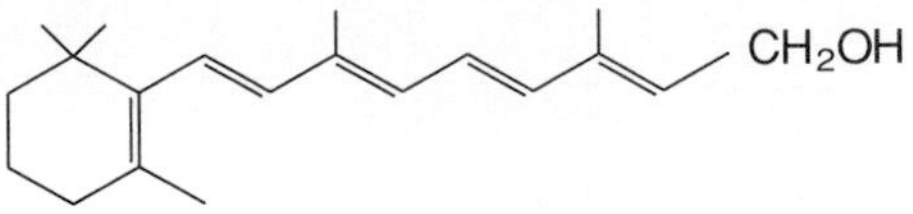

All-*trans*-retinol
(Vitamin A₁)

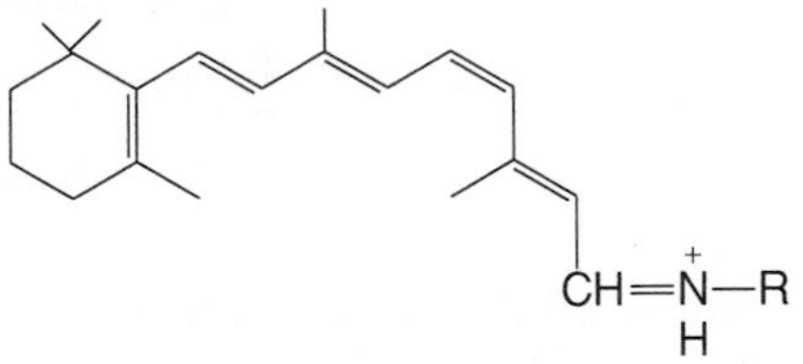

**Protonated Schiff base
of 11-*cis*-retinal**

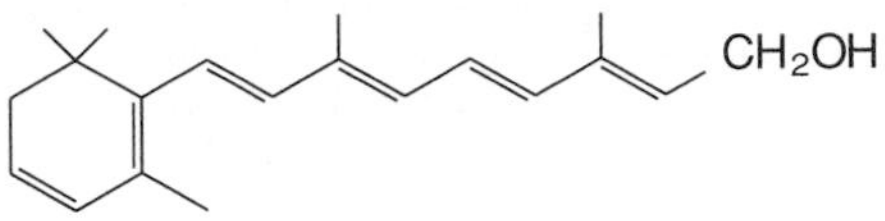

Vitamin A₂

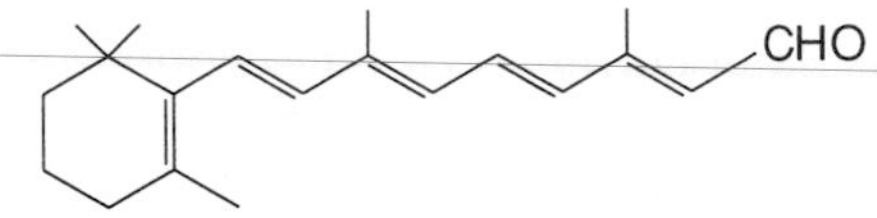

All-*trans*-retinal

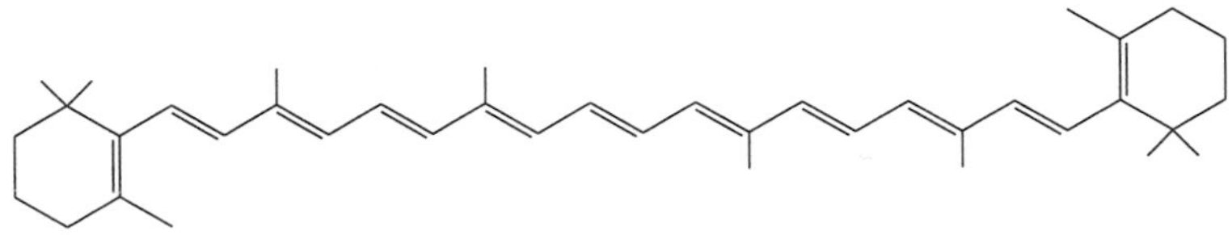

β-Carotene

Figure 36–4

Structures of retinals, retinols, and β-carotene. The structure of 11-*cis*-retinal *(top)* indicates the numbering system used for the carbons. For simplicity, the molecules are shown here as being planar, although the most stable conformer of 11-*cis*-retinal actually is twisted about the C-6—C-7 and C-12—C-13 single bonds to reduce steric crowding of the methyl groups. In rhodopsin, 11-*cis*-retinal is bound by a protonated Schiff's base linkage to a lysine of opsin.

phyla had separated in evolution. Some animals, such as sea turtles and amphibians in certain stages of development, use a retinal that has an additional double bond in the ring. The parent molecule in this case is vitamin A₂, or 3,4-dehydroretinal (Figure 36–4). *In all cases, however, the photochemically active protein complex is made from the 11-cis isomer of the aldehyde.* 11-*cis*-Retinals must somehow be particularly fit for the task of responding to light. (Some unicellular organisms such as *Euglena* have light-sensitive organelles that contain a carotenoid instead of a retinal, but little is known of the biochemistry of these primitive receptors.)

Opsin is an intrinsic membrane protein with a molecular weight of about 38,000. It accounts for about 95 per cent of the protein in the disk membranes. Labeling studies of isolated disks indicate that the carboxyl-terminal end of rhodopsin is exposed to the cytoplasmic space outside the disk, and the amino-terminal end to the space inside the disk. Seven hydrophobic α-helical stretches of the protein probably lace back and forth across

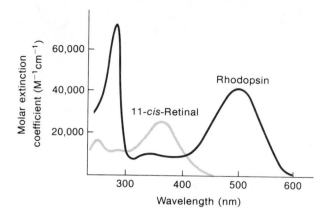

Figure 36–5

Absorption spectra of rhodopsin and of 11-*cis*-retinal in hexane solution. The absorption of rhodopsin in the 280-nm region is due mainly to the opsin.

the phospholipid bilayer. The retinal is bound to a lysyl-NH$_2$ group by a *Schiff's base*, or aldimine linkage (Figure 36–4) in a region that is buried in the bilayer. Rhodopsin also has several carbohydrate groups bound near the amino-terminal end.

In solution, 11-*cis*-retinal absorbs maximally near 380 nm, but in rhodopsin the peak is at 500 nm (Figure 36–5). *The absorption spectrum of rhodopsin is essentially identical to the spectrum of the sensitivity of the rod cells* (after correction for absorption in the cornea and lens). *The perception of color by cones depends on the fact that there are three different types of cones with different absorption spectra.* One of these absorbs blue light (440 nm) maximally, the second absorbs green (530 nm), and the third absorbs yellow (570 nm). In some species of birds, the third component absorbs maximally at 630 nm. However, all cones contain 11-*cis*-retinal bound to proteins that are similar to opsin. (Some fish use another strategy altogether for color vision. Their eyes have carotenoid-containing oil droplets of three colors, and these act as filters in front of the receptor cells.)

Light Isomerizes the Retinal of Rhodopsin to All-*trans*

The discovery that rhodopsin contains the 11-*cis*-isomer of retinal came as a surprise. In the early 1950s, Ruth Hubbard, George Wald, and their coworkers found that rhodopsin decomposes into retinal and the apoprotein opsin when the retina is exposed to light. The retinal that was released was the all-*trans* isomer. When opsin was mixed with a crude preparation of retinal, rhodopsin was regenerated. Crystalline all-*trans*-retinal, however, did not bind to opsin. The regeneration achieved with the crude preparation proved to be due to a small amount of the 11-*cis* isomer in the mixture. Since rhodopsin binds 11-*cis*-retinal in the dark, but releases all-*trans*-retinal after illumination, Hubbard and Kropf concluded that *the action of light in vision is to isomerize the chromophore about its C-11≡C-12 double bond.* This change in structure must somehow be translated into an electrophysiological signal that can be transmitted to the brain.

Transformations of Rhodopsin Can Be Detected by Changes in Its Absorption Spectrum

The release of all-*trans*-retinal from opsin takes several minutes and is too slow to be an obligatory step in visual perception. It appears to be a step in the regeneration of the active form of rhodopsin. The all-*trans*-retinal probably leaves the rod cell and is isomerized back to the 11-*cis* form before it returns. An enzyme that catalyzes the isomerization of all-*trans*-retinol to 11-*cis*-retinol is found in the epithelial tissue that covers the outside of the

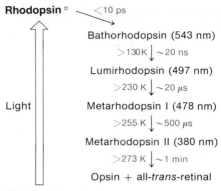

Rhodopsin *   <10 ps

Bathorhodopsin (543 nm)

>130K ↓ ~20 ns

Lumirhodopsin (497 nm)

>230 K ↓ ~20 μs

Light   Metarhodopsin I (478 nm)

>255 K ↓ ~500 μs

Metarhodopsin II (380 nm)

>273 K ↓ ~1 min

Opsin + all-*trans*-retinal

Rhodopsin

Figure 36–6
Photochemical transformations of rhodopsin. The numbers in parentheses indicate the optical absorption maxima of the intermediates. The numbers in color on the right are approximate half-times for the conversions in rod outer segment membranes near 37°C. With isolated rhodopsin, some of the reactions are slower and are multiphasic. The steps following the formation of bathorhodopsin have progressively higher thermal activation energies. They can be blocked by lowering the temperature below the temperatures indicated on the left.

Ground state
+
N
H
11
12

Excited state
+
11
12
N
H
R

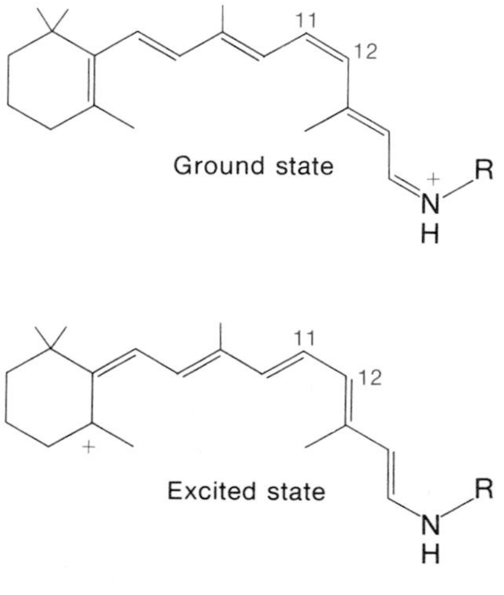

Figure 36–7
Excitation of the protonated Schiff's base causes a movement of positive charge from the nitrogen toward the ring. The C-11═C-12 bond loses much of its double-bond character. The valence bond diagrams shown here should not be taken too literally, but they give a good qualitative picture of the redistribution of electrons that occurs when the molecule is excited.

retina, but the energy source for this endothermic process is unknown. In the eyes of invertebrates, all-*trans*-retinal does not come off opsin at all but is isomerized back to the 11-*cis* form on the protein photochemically and probably also enzymatically.

To explore the changes in rhodopsin that precede the release of all-*trans*-retinal from the protein, Toru Yoshizawa and Wald measured the optical absorbance changes that occurred when they illuminated rhodopsin at low temperatures. Illumination at liquid N_2 temperature (77 K) caused the absorption band of the rhodopsin to shift from 500 to 543 nm. The product of this transformation is now called *bathorhodopsin*. Bathorhodopsin is stable indefinitely in the dark at 77 K, but if it is warmed above about 130 K it decays spontaneously to a species that absorbs maximally at 497 nm. This is called *lumirhodopsin*. If the sample is warmed further, to about 230 K, lumirhodopsin decays to *metarhodopsin I*, which absorbs at 478 nm. Above about 255 K, metarhodopsin I decays to *metarhodopsin II*, which absorbs at 380 nm. These transformations are outlined in Figure 36–6.

Subsequent kinetic measurements by other investigators have shown that rhodopsin passes through the same series of states if it is excited with a short flash at physiological temperatures. The numbers on the right side of Figure 36–6 give approximate half-times for the transformations at 37°C. Agreement between the kinetic studies and low-temperature experiments may seem routine, but it might not have turned out this way at all. The trapping of a metastable state at low temperature can depend on the thermodynamic properties of the state, as well as on its position in the kinetic sequence. For example, side products that normally are unimportant could accumulate when the normal pathway is blocked by lowering the temperature.

How can the absorption spectrum of rhodopsin go through such wild changes from 500 to 543 nm, and eventually to 380 nm? In thinking about this, we are reminded that rhodopsin's initial absorption spectrum is already shifted by 120 nm compared with the spectrum of free 11-*cis*-retinal (Figure 36–5). Further, the absorption maxima of the cone pigments vary from 450 to 630 nm, in spite of the fact that these pigments all contain 11-*cis*-retinal. The explanation for these spectral differences depends partly on the fact that *the nitrogen atom of the Schiff's base linkage in rhodopsin is protonated and is therefore positively charged* (Figure 36–7). The protonation state of the nitrogen has been demonstrated clearly by nuclear magnetic resonance and resonance Raman spectroscopy. When the retinal absorbs light and is raised to an excited state, the electron density on the nitrogen increases, and positive charge moves to the opposite end of the molecule, as represented roughly in Figure 36–7. Because of the redistribution of charge, the relative energies of the excited and ground states are extremely sensitive to the positions of other charged, dipolar, or polarizable groups nearby. An arrangement of charged groups that stabilizes the excited state relative to the ground state will shift the absorption spectrum to longer wavelengths.

There must be at least one charged group near the Schiff's base in rhodopsin, the anionic counterion that is needed to balance the positive charge on the nitrogen. The anion is probably an amino acid residue of the protein, rather than a free ion, because the Schiff's base linkage is buried in the protein and is not accessible to the solution. *The structure of the protein thus can determine the distance between the counterion and the nitrogen, and this will affect the absorption spectrum of the pigment profoundly.* Moving the counterion farther away from the nitrogen will destabilize the ground state and shift the absorption spectrum to longer wavelengths. However, this may not be the whole story. Studies of the absorption spectra of the complexes of opsin with a variety of synthetic derivatives of retinal suggest that the binding site contains an additional negatively charged or dipolar group near C-12 of the retinal.

Isomerization of the Retinal Causes Other Groups in the Protein to Move

Let us now consider bathorhodopsin, which appears to be the first metastable product of the photochemical reaction. If bathorhodopsin is excited with long-wavelength light at 77 K, it can be converted back to rhodopsin. Resonance Raman measurements support the view that _the retinal in batho-rhodopsin has isomerized to the all-trans form, but that it continues to be held as a protonated Schiff's base_. Additional evidence that bathorhodopsin contains all-_trans_-retinal has been obtained by studying a modified form of rhodopsin, _isorhodopsin_, which contains the 9-cis isomer of retinal. When isorhodopsin is illuminated, it gives rise to bathorhodopsin within 10^{-11} sec, just as rhodopsin (11-_cis_) does. The only common product that could form directly from both the 9-_cis_ and 11-_cis_ isomers is the all-_trans_ isomer.

Isomerization of the retinal Schiff's base can occur when the molecule is excited with light, because the C-11=C-12 bond loses much of its double-bond character in the excited state. The valence bond diagrams of Figure 36–7 illustrate this point qualitatively. In the ground state of rhodopsin, the potential energy barrier to rotation about the C-11=C-12 bond is probably on the order of 30 kcal/mole. This barrier essentially vanishes in the excited state. In fact, molecular orbital calculations suggest that the energy of the excited molecule is minimal when the C-11=C-12 bond is twisted by about 90° (Figure 36–8). The 11-_cis_ and all-_trans_ molecules thus have a common excited state. When the molecule decays from the excited state to a ground state, it can end up in either of these isomeric forms. It turns out that the excited state decays to bathorhodopsin (all-_trans_) about 67 per cent of the time and to rhodopsin (11-_cis_) about 33 per cent of the time. Isorhodopsin (9-_cis_) and other isomers are formed only in small amounts (about 1 per cent). Similar ratios of products are obtained no matter whether excitation starts with rhodopsin or bathorhodopsin.

The high energy barrier to the isomerization of rhodopsin in the dark is physiologically important because it limits the "noise" in our perception of light. If the barrier were low enough to be overcome thermally, the discrimination of light from dark would be more difficult.

Because opsin binds 11-_cis_-retinal, but not all-_trans_-retinal, the all-_trans_ molecule that is created in bathorhodopsin must find itself initially in a binding state that is tailored for the 11-_cis_ structure. In solution, 11-_cis_-retinal is less stable than all-_trans_-retinal, because of steric repulsion between the methyl group on C-13 and the hydrogen atom on C-10 (see Figure 36–4). On the protein, there must be compensating interactions that decrease the free energy of the 11-_cis_ complex relative to that of the all-_trans_ complex. These interactions will be disrupted when the pigment is isomerized. The pronounced shift of the absorption spectrum to longer wavelengths in bathorhodopsin could be explained if the isomerization results in a movement of the Schiff's base nitrogen, increasing the distance between the positively charged nitrogen and its counterion (Figure 36–9). Such a movement would destabilize the ground state of bathorhodopsin relative to the excited state, as discussed earlier for rhodopsin. A repositioning of the Schiff's base also is likely to lead to changes in the structure of the protein surrounding the binding site. If the positively charged nitrogen moves closer to a second anionic group, as illustrated in Figure 36–9, a new hydrogen bond could

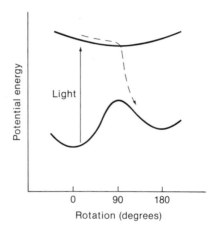

Figure 36–8

Potential energy surfaces of ground and excited states in rhodopsin, as functions of the angle of rotation of the bond between carbons 11 and 12. A rotational angle of 0° means that the retinyl group is 11-_cis_ (rhodopsin); 180° means that it is all-_trans_ (bathorhodopsin). In the ground state (_lower curve_), rhodopsin is stabilized with respect to bathorhodopsin, possibly because the positive charge of the protonated Schiff's base interacts more favorably with a negatively charged group of the protein (Figure 36–9). The energy of the excited state (_upper curve_) is less sensitive to the position of the Schiff's base than the energy of the ground state is, because the positive charge has moved to the opposite end of the retinal molecule (Figure 36–7). The energy of the excited state is minimal when the rotation angle is about 90°. Rhodopsin is raised from the ground state to the excited state by light (_vertical arrow_). It relaxes back to the ground state along a path like that indicated by the dashed arrow, ending up as bathorhodopsin.

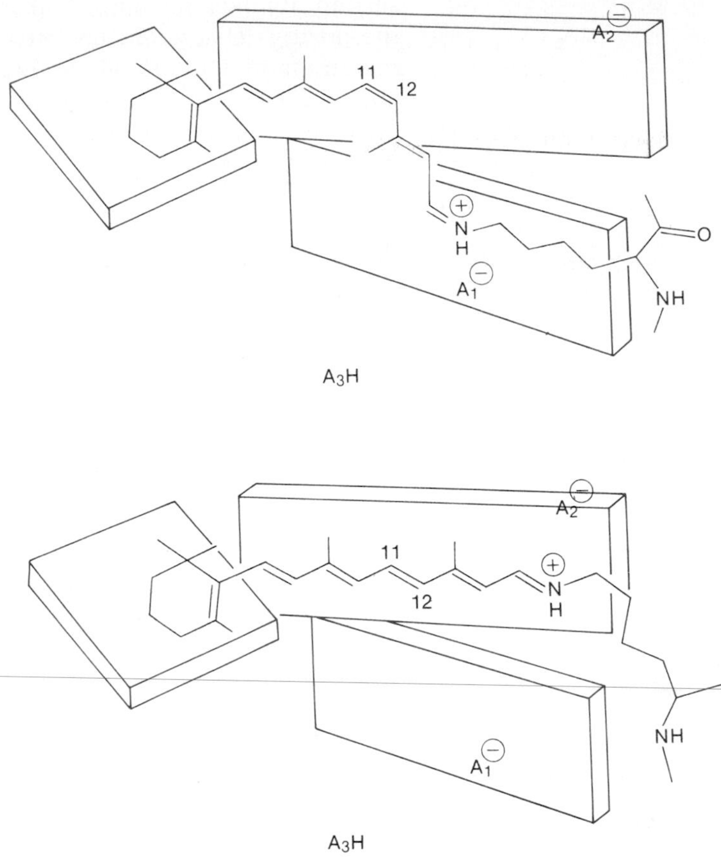

Figure 36–9
A model for the photochemical conversion of rhodopsin *(top)* to bathorhodopsin *(bottom)*. In rhodopsin, the proton of the Schiff's base nitrogen is hydrogen bonded to a counterion, A_1^-. The retinal is twisted about the C-12—C-13 and C-6—C-7 single bonds. (As noted in Figure 36–4, free retinal is twisted in this way, but the amount of twisting in rhodopsin is still uncertain.) When rhodopsin is converted to bathorhodopsin *(bottom)*, the ring end of the retinal is presumed to be locked in position. Isomerization about the C-11=C-12 double bond flips the protonated nitrogen and its positive charge away from the counterion A_1^-. The separation of electrical charge could account for the shift of the absorption spectrum to longer wavelengths. The isomerization also can lead to proton movements. The proton of the Schiff's base could be passed to another basic group (A_2^-), and A_1^- might pick up a proton from another acidic group (A_3H). Other groups of the protein, including the lysine attached to the retinal, also must adjust to the new geometry of the retinyl Schiff's base. (For additional discussion, see B. Honig et al., An external point-charge model for wavelength regulation in visual pigments, *J. Am. Chem. Soc.* 101:7084, 1979, and Photoisomerization, energy storage, and charge separation: A model for light-energy transduction in visual pigments and bacteriorhodopsin, *Proc. Natl. Acad. Sci. USA* 76:2503, 1979.)

form between this group and the Schiff's base nitrogen, and we would expect the protein to respond by adjusting the positions of other nuclei nearby.

The reorganization of rhodopsin that is set in motion by isomerization of the retinyl Schiff's base continues as the system relaxes through the states lumirhodopsin, metarhodopsin I, and metarhodopsin II. Physical measurements indicate that isolated rhodopsin undergoes particularly substantial changes in protein conformation during the transition from metarhodopsin I to metarhodopsin II. The extent of these changes and the kinetics of formation of metarhodopsin II are sensitive to the type and number of phospholipids attached to the protein. Conformation changes also undoubtedly occur in this step when the rhodopsin is in place in the disk membrane, but they are subtler and may be confined to small regions of the protein. The nitrogen of the retinyl Schiff's base loses its proton as metarhodopsin II is formed, and a second, unidentified group takes up a proton from the solution. The loss of the positive charge on the nitrogen accounts for the shift of the absorption maximum to 380 nm, where unprotonated Schiff's bases of all-*trans*-retinal absorb. The Schiff's base linkage also becomes accessible to reagents in the aqueous solution during or shortly after the formation of metarhodopsin II.

The formation of metarhodopsin II is fast enough to be an obligatory step in visual transduction. It clearly is associated with changes in the interactions between rhodopsin and its surroundings. A reasonable hypothesis, therefore, is that the changes in protein structure allow metarhodopsin II to initiate an interaction with some other component of the disk membrane. We shall explore the nature of this component in the following sections.

Absorption of a Photon Causes a Change in the Cation Conductivity of the Cytoplasmic Membrane

The human eye is amazingly sensitive. After being in the dark for a time, we can perceive continuous light that is so weak that an individual rod cell absorbs a photon, on the average, only once every 38 minutes. A flash of light is detectable if approximately six rods each absorb one photon. This means that *the absorption of a single photon by any one of the approximately* 3×10^7 *molecules of rhodopsin in a rod must be sufficient to excite the cell* and trigger a neuronal response. The combined responses of six rods can elicit the sensation of seeing.

The electrophysiological response of a rod or cone to light involves a change in the permeability of the cytoplasmic membrane to cations. In the inner segment and basal parts of the cell, the cytoplasmic membrane contains a Na^+-K^+ pump, which uses ATP to move Na^+ out of the cell and K^+ in. (Ion pumps have been discussed in detail in Chapter 34.) K^+ can diffuse back out of the cell relatively freely, and its efflux causes the membrane to become negatively charged by about 20 mV on the inside relative to the outside. Na^+, on the other hand, cannot pass readily back across the membrane into the inner segment. Electrophysiological measurements by William Hagins showed that Na^+ flows outside the cell from the inner segment to the outer, where it reenters the cell through channels that are selectively permeable to cations (Figure 36–10). Once back inside, the Na^+ flows to the inner segment to be pumped out again. The round trip takes on the order of a minute. *When a rod cell of a vertebrate's retina is excited with light, some of the cation channels in the outer segment suddenly close,* decreasing the inward movement of Na^+. The interruption of the Na^+ current causes an increase in the electrical potential across the cytoplasmic membrane. This *hyperpolarization* causes a change in the movement of an unidentified chemical transmitter to the bipolar cell that makes a synaptic junction with the rod. The bipolar cell then sends a signal to a *ganglion cell*, the third in the hierarchy of retinal neurons.

The hyperpolarization of the rod and the response of the bipolar cell are graded effects. Absorption of a single photon causes the membrane potential to increase by about 1 mV, with the effect peaking about 1 sec after the excitation. Up to about 100 photons per rod, the more light the cell absorbs, the larger the amplitude of the hyperpolarization and the faster the hyperpolarization occurs. The reaction of the ganglion cell, however, is all-or-none. When the ganglion cell is triggered, it responds with an action potential that proceeds to the brain. Sensitivity to weak light is enhanced (with a sacrifice in spatial resolution) by having multiple rod cells connected to each bipolar cell and multiple bipolar cells connected to each ganglion cell. The output of cone cells is not summed in this way. This partly explains why the cones provide better visual acuity but lower sensitivity than the rods.

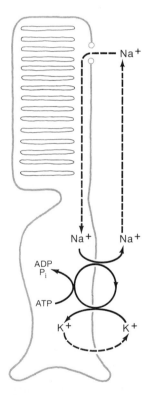

Figure 36–10
Na^+ is pumped out of the inner segment *(solid arrows)* and diffuses back into the cell through channels in the outer segment *(dashed arrows)*. In rods of vertebrates, the Na^+ channels are held open in the dark, and they close in the light. In invertebrates, the channels open in the light.

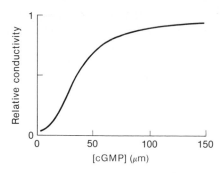

Figure 36-11
c-GMP increases the electrical conductivity of the rod-cell cytoplasmic membrane. The conductivity was measured with a patch of membrane held at the tip of a micropipet. By varying the ions in the solution, it is possible to show that cGMP affects the movement of Na^+ and other cations across the membrane. Note the sigmoidal concentration dependence: the Hill coefficient is 1.6. (For more details, see E. E. Fesenko et al., *Nature* 313:310, 1985.)

Analogous events occur in the eyes of invertebrates, except that light causes an increase in the cation permeability of the receptor cell rather than a decrease. In the rods of vertebrates, the absorption of a single photon decreases the current of Na^+ into the outer segment by about 3 per cent. The inflow of approximately 10^6 Na^+ ions is transiently prevented. Exactly how many Na^+ channels have to close in order to achieve this effect is uncertain, but estimates have ranged from 25 to 1000. Since only one molecule of rhodopsin is excited, *the cell must have a mechanism for amplifying the effect of light* by a substantial factor. Note also that the cytoplasmic membrane of the rod outer segment is not contiguous with the disk membranes that hold the rhodopsin (Figure 36-3). This suggests that the excitation of rhodopsin causes a change in the concentration of a *diffusible transmitter*, which moves from the disk to the cytoplasmic membrane.

The Effect of Light Is Mediated by Guanine Nucleotides

There is strong evidence that *the diffusible component that moves between the disk and the cytoplasmic membrane is 3′,5′-cyclic-GMP* (cGMP). Electrophysiological measurements have shown that cGMP causes an increase in the permeability of the cytoplasmic membrane to Na^+ (Figure 36-11). This appears to be a direct effect of cGMP on the Na^+ channels, rather than an indirect effect mediated by a kinase, because it can be seen in the absence of ATP.

The disk membranes of the rod outer segment contain a *phosphodiesterase*, which hydrolyzes 3′,5′-cyclic-GMP (cGMP) to 5′-GMP. If the disk membranes are kept in the dark, the phosphodiesterase remains in a relatively inactive state. When the disks are illuminated and rhodopsin is converted to metarhodopsin II, the activity of the phosphodiesterase increases. For each molecule of rhodopsin that is excited, on the order of 500 molecules of the phosphodiesterase are activated. The outer segment also contains a *guanylate cyclase*, which forms cGMP from GTP, but this enzyme is not greatly affected by light. *Illumination thus results in a decrease in the cGMP content of the cell.* A drop in the cGMP concentration is in the right direction to cause a decrease in Na^+ permeability of the cell membrane (Figure 36-11), and thus to cause a hyperpolarization of the membrane. In agreement with this scheme, injecting extra cGMP into rod outer segments temporarily inhibits the hyperpolarization caused by light.

Unlike rhodopsin, the phosphodiesterase is a peripheral membrane protein that can be readily solubilized. Its activation by light requires the presence of GTP and is associated with the binding of GTP to a second peripheral membrane protein that has been called the *G protein*, or *transducin*. Like the G proteins that participate in the activation or inhibition of adenylate cyclase by hormones in other tissues (see Chapter 31), transducin consists of three subunits, α, β and γ (Figure 36-12). In the resting state, the α subunit contains a molecule of bound GDP. *When rhodopsin is transformed to metarhodopsin II by light, it interacts with transducin, causing GTP to displace the bound GDP. Once GTP is attached, the α subunit separates from the β and γ subunits and binds to an inhibitory subunit of the phosphodiesterase. The removal of the inhibitory component activates the phosphodiesterase.*

Each molecule of metarhodopsin II that is generated by light appears to be able to trigger about 500 molecules of transducin to bind GTP in place of GDP. This agrees with the number of phosphodiesterase molecules that are activated, and accounts for much of the amplification in the response to light. Additional amplification results from the enzymatic action of each phosphodiesterase on many molecules of cGMP. These effects can be dem-

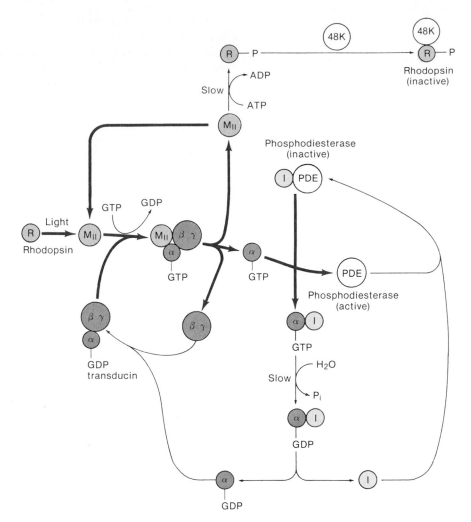

Figure 36–12

The large arrows show the main steps in the activation of the cGMP phosphodiesterase by light. Light converts rhodopsin (R) to a metastable form (M_{II}, probably metarhodopsin II). M_{II} reacts catalytically with transducin, which contains GDP bound to one of its three subunits (α). The interaction with M_{II} changes the specificity of the nucleotide binding site so that GDP is released and GTP is bound. This causes the α subunit to dissociate from the other two subunits of transducin (β and γ). In the dark, the phosphodiesterase (PDE) is inactive, because of the presence of an inhibitory polypeptide (I). The I polypeptide is removed by binding to the complex of the transducin α subunit and GTP. The active phosphodiesterase then can hydrolyze cGMP to 5'-GMP. M_{II} can go on to activate several hundred additional molecules of transducin, before slower enzymatic reactions (*thin arrows*) gradually convert it to an inactive form. The inactivation involves conversion of the rhodopsin to a phosphorylated form (R-P) and an interaction with another protein (48K). The GTP bound to transducin's α subunit eventually is hydrolyzed to GDP. When this happens, I dissociates and returns to the phosphodiesterase, the three subunits of transducin reassemble, and the system returns to its resting state. (For details, see M. Chabre, *Ann. Rev. Biophys. Biophys. Chem. 14,* 331, 1985.)

onstrated with purified preparations of rhodopsin, transducin, and the phosphodiesterase. Purified, illuminated rhodopsin that has been incorporated into phospholipid vesicles is capable of activating transducin, and the complex of transducin's α subunit with GTP is capable of activating the phosphodiesterase in the absence of rhodopsin.

In order for the eye to respond rapidly to changing light intensities, it is necessary that the activation of the phosphodiesterase be a transient response that quickly switches off again. The activation of transducin's α subunit is reversed by hydrolysis of the bound GTP to form bound GDP and free inorganic phosphate (Figure 36–12). When this happens, the α subunit separates from the inhibitory polypeptide of the phosphodiesterase and recombines with the β and γ subunits of transducin. The inhibitory polypeptide then recombines with the phosphodiesterase, returning the phosphodiesterase to its resting, inactive state. Activation of the phosphodiesterase can be prolonged indefinitely if, instead of adding GTP, one adds a nonhydrolyzable analog of GTP.

To switch off the response to light, metarhodopsin II also must decay to a form that is incapable of activating additional molecules of transducin (Figure 36–12). Although the details of this step are not yet clear, they appear to involve phosphorylation of the rhodopsin, as well as an interaction with another protein (the "48K protein," or "arrestin").

Ca^{2+} ions appear to participate in the responses of rods and cones to light, but exactly what role they play is still uncertain. Ca^{2+} ions enter the rod outer segment through the same channels that admit Na^+, and they are exported by a transport protein that exchanges Ca^{2+} for Na^+. The Ca^{2+} concentration in the cell thus decreases upon illumination, when the cation channels close. The activities of the cGMP phosphodiesterase and the guanylate cyclase, and the effect of cGMP on the Na^+ permeability of the membrane are all sensitive to the concentration of Ca^{2+}.

Rhodopsin Can Move Around in the Disk Membrane

Each molecule of rhodopsin that is converted to metarhodopsin II is capable of causing some 500 molecules of transducin to bind GTP. Since this happens within about 0.5 sec after the absorption of a photon, either rhodopsin or transducin, or both, must move about rapidly enough to encounter thousands of reaction partners per second. Rhodopsin is firmly embedded in the disk membrane and transducin is attached to the surface of the membrane, so they need to diffuse only in two dimensions in the plane of the membrane. But is it likely that they could move this rapidly? Actually, evidence that rhodopsin can diffuse rapidly within the membrane was obtained prior to the discovery of transducin and the phosphodiesterase.

The initial indications that rhodopsin moves around in the membrane came from studies of _linear dichroism_ of rod outer segment disks. A material is said to exhibit linear dichroism if the strength of its optical absorption, when measured with polarized light, depends on the orientation of the polarizer. The polarization of a light beam is defined by the orientation of the light's electric field, as indicated in Figure 17–6. Molecules in solution usually do not exhibit linear dichroism because they are free to tumble about. At any given time, the solution contains molecules with all possible orientations relative to the polarizer, so the absorbance of the solution will not change if we rotate the polarizer. But if all the molecules are fixed in position with their molecular axes aligned, the system generally will exhibit linear dichroism. In the case of retinal or its Schiff's base, light is absorbed best when the polarization makes the electric vector of the light parallel to the long axis of the molecule. You can get a feeling for this by examining the diagrams in Figure 36–7. In order to convert the molecule from the ground state to the excited state, the electric field of the light must move electron density away from the ring end of the molecule and in the direction of the nitrogen.

The eyes of vertebrates are well suited for measurements of linear dichroism, because the rod cells are neatly aligned with their long axes perpendicular to the plane of the tissue. Suppose that we send a weak beam of

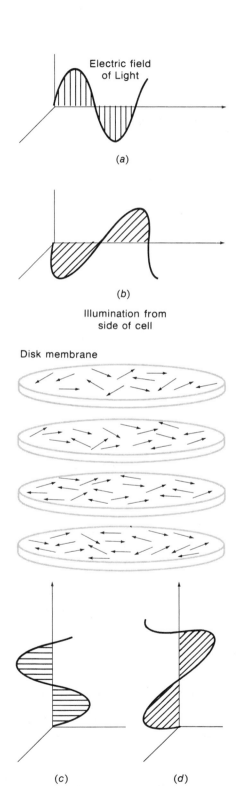

Electric field
of Light

(a)

(b)

Illumination from
side of cell

Disk membrane

(c)   (d)

Illumination of cell end-on

polarized light through the rods from the side (Figure 36–13a and b). The polarization of the beam then could be made either parallel or perpendicular to the long axis of the cells. The planes of the disk membranes in the rod outer segments are aligned perpendicular to the cell axis, so the polarization of the light will be either perpendicular or parallel to the membrane surface. It turns out that the rhodopsin absorbs much more strongly when the light is polarized parallel to the membranes than it does when the polarization is perpendicular. This means that _the long axes of the retinyl groups must be held more or less parallel to the plane of the membrane_. Now suppose that we send the light beam through the cells end-on (Figure 36–13c and d). The polarization then must be parallel to the plane of the disk membranes, no matter how the polarizer is turned. With light coming through the cells end-on, the absorbance of the rhodopsin is found to be independent of the direction of the polarization. This means that the _retinyl groups are not aligned in any particular direction within the plane of the membrane, but instead can take on all possible orientations in this plane_.

It is interesting to note in passing that the orientation of the retinyl groups in the retinas of vertebrates optimizes our visual sensitivity for unpolarized light. Most of the light reaching the retinas passes through the rods end-on and is absorbed well, whatever its polarization (Figure 36–13c and d). Insects, however, can distinguish between horizontally and vertically polarized light. This can be advantageous, because light that is reflected by smooth surfaces is partially polarized. Insects may make this distinction by having separate sets of cells with differently aligned retinyl groups.

How can we use measurements of linear dichroism to see whether rhodopsin molecules move about within the membrane? Suppose that the retina is exposed to a flash of polarized light that enters the rods end-on. Some of the retinyl groups in the disk membranes will be oriented parallel to the polarization, and some will not. The light will _selectively_ excite those molecules that are parallel to the polarization, because these have the highest absorbance. Most of the molecules that are excited will be converted to bathorhodopsin and the series of states that descend from it, and their absorption spectrum will change. If we measure the absorbance changes at a short time after the excitation flash, using polarized measuring light that passes through the cells end-on, the absorbance changes will depend on how the polarization of the measuring light is oriented with respect to the excitation polarization. The polarized measuring light selectively measures molecules that have a particular orientation. Before the excitation flash, there is no linear dichroism for measurements made end-on; after the flash, there is. However, _if the rhodopsin molecules can rotate in the plane of the membrane, the orientations of the molecules that were excited will in time become randomized, and the linear dichroism that is induced by the excitation will disappear_. By measuring the kinetics of this disappearance, it is possible to determine how rapidly rhodopsin rotates. Experiments of this sort were done by Richard Cone. Rhodopsin proved to rotate surprisingly rapidly. The time required for a rotation is about 2×10^{-5} sec. This means that _the environment of rhodopsin in the membrane must be highly fluid_.

To determine whether rhodopsin can diffuse laterally in the membrane,

Figure 36–13

When rod cells are illuminated from the side, the light can be polarized either perpendicular (a) or parallel (b) to the planes of the disk membranes. The absorbance of the rhodopsin is much greater for parallel polarization than it is for perpendicular. When the cells are illuminated end-on, the light's electric field has to be parallel to the membranes, no matter how the light is polarized (c and d). With end-on illumination, the absorbance of the rhodopsin is independent of the polarization. These observations show that the 11-_cis_-retinyl groups _(small arrows in disks)_ are held parallel to the plane of the membrane but can point in any direction in this plane.

Cone and Paul Leibman independently used a microspectrophotometric technique. They excited individual rod outer segments end-on near one edge of the disks through a microscope. Initially, the excitation caused absorbance changes only in the region that was illuminated, but within a few seconds after the excitation, the absorbance changes became uniformly distributed across the disk. This means that *rhodopsin must be able to diffuse rapidly in the plane of the membrane*. The diffusion coefficient is calculated to be about 5×10^{-9} cm^2/sec. At this speed, rhodopsin molecules would collide with each other 10^5 to 10^6 times a second, and it would take only about a second for a single rhodopsin to collide with most of the molecules of transducin on the surface of the disk.

BACTERIORHODOPSIN: A BACTERIAL PIGMENT-PROTEIN COMPLEX THAT RESEMBLES RHODOPSIN

Halobacterium halobium is a red-colored, halophilic (salt-loving) bacterium that thrives in tidal salt flats. Its red color comes from two components of its cytoplasmic membrane: carotenoids and a purple pigment-protein complex that bears a striking resemblance to rhodopsin. The purple complex, *bacteriorhodopsin*, was first described by Walther Stoeckenius and his colleagues. Bacteriorhodopsin is not distributed randomly throughout the cell membrane, but rather is collected in patches that are held together by strong interactions among the bacteriorhodopsins. The patches, or *purple membranes*, can be isolated relatively simply, if one disrupts the cells by putting them in distilled water. Bacteriorhodopsin is the only protein in the purple membrane. It forms a highly ordered two-dimensional array that is almost crystalline in nature. Electron diffraction studies have shown that the individual proteins are folded into seven α-helical stretches extending from one side of the membrane to the other (see Chapter 2).

Like rhodopsin, bacteriorhodopsin contains retinal bound to a lysine as a protonated Schiff's base. The retinal isomerizes when it is excited with light. However, bacteriorhodopsin differs from rhodopsin in that the retinal starts out in the all-*trans* form rather than as the 11-*cis* isomer. The main product of the isomerization is the 13-*cis* isomer. Like rhodopsin, the excited bacteriorhodopsin progresses through a series of metastable states with different absorption spectra. The first of these states resembles bathorhodopsin in having an absorption spectrum that is shifted markedly to longer wavelengths. A later state resembles metarhodopsin II in that the retinyl Schiff's base has lost its proton. A major difference, however, is that the transformations of bacteriorhodopsin are cyclic. Rather than dissociating from the protein, the retinal returns to the all-*trans* form, and the complex relaxes back to its original state and is ready to operate again.

The role of bacteriorhodopsin in *H. halobium* is to pump protons across the cytoplasmic membrane. Some aspects of this pumping have been discussed in Chapter 34. As the excited bacteriorhodopsin undergoes its cyclic series of transformations, protons are taken up from the solution inside the cell and are released on the extracellular side of the membrane. Two protons appear to be translocated during each cycle. One of these could be the proton that is initially on the nitrogen atom of the Schiff's base, because this proton is lost and then replaced again in the course of the cycle. The isomerization of the retinal could move this proton with the nitrogen from a position where it equilibrates with the cytosol to a position where it equilibrates with the extracellular solution. Figure 36–9 will serve to illustrate this idea. The figure was drawn to show the isomerization of 11-*cis*-retinal to all-*trans*-retinal that occurs in rhodopsin, but the principles here are conceptually the same. The isomerization of the retinyl chain pushes the proton on the Schiff's base nitrogen from the region of one functional group (A_1^- in the figure) to the region of another (A_2^-). From A_2H, a proton could be conducted to the solution on the extracellular side of the membrane. When the retinal then returns to its original state, the proton lost by the nitrogen could be replaced by one from a third group (A_3H) on the cytoplasmic side of the membrane. It is not yet clear, however, how a second proton could be pumped simultaneously.

As in other types of photosynthetic bacteria, the protons that are pumped across the membrane by bacteriorhodopsin in the light can reenter the cell by way of a proton-conducting ATP-synthase, generating ATP. *H. halobium* also contains a transport system that moves Na^+ ions out of the cell in exchange for protons coming in. Removal of Na^+ is a major chore for an organism that lives in water containing 4 M NaCl. The cytoplasmic membrane also has two other pigment-protein complexes that execute photochemical cycles similar to the transformations of bacteriorhodopsin. One of these, _halorhodopsin_, acts as a light-driven pump for Cl^-. The other, _sensory rhodopsin_, is involved in phototaxis, the tendency of the cells to swim in the direction of a source of light.

SUMMARY

Light rays entering the eye of a vertebrate are refracted by the cornea and focused to form an image on the retina. The retina contains several layers of cells. Some of these cells, the rods and cones, contain the visual pigments that are responsible for the initial response to light; other cells make synaptic connections to rods or cones and to additional neural cells that carry impulses to the brain. The light-sensitive protein complex in the rods, rhodopsin, consists of the compound 11-*cis*-retinal bound as a Schiff's base to a protein, opsin. Rhodopsin is an integral constituent of membranes that form a stack of disks at one end of the cell. Cone cells, which are responsible for the perception of color, contain similar complexes in infoldings of the plasma membrane. When rhodopsin absorbs light, the retinal isomerizes to the all-*trans* isomer. This initiates a series of transformations of the pigment-protein complex that result in an interaction with another protein, transducin. The interaction with rhodopsin causes one of transducin's subunits to take up a molecule of GTP in exchange for bound GDP, to dissociate from the other subunits, and to react with a phosphodiesterase that hydrolyzes cGMP. This activates the phosphodiesterase. The resulting drop in cGMP concentration leads to a decrease in the Na^+ permeability of the cytoplasmic membrane. A hyperpolarization of the membrane then triggers the transmission of a signal at the rod's synaptic junction to the adjacent neural cell.

Bacteriorhodopsin, a bacterial pigment-protein complex, resembles rhodopsin (it contains all-*trans*-retinal), but serves a different function. Like rhodopsin, bacteriorhodopsin progresses through a series of metastable states when it is excited with light. In the course of these transformations, protons are taken up from the solution inside the cell and are released on the extracellular side of the membrane, generating an electrochemical potential gradient for protons across the membrane. Protons reenter the cell by way of an ATP-synthase, energizing the formation of ATP.

SELECTED READINGS

Chabre, M., Trigger and amplification mechanisms in visual phototransduction. *Ann. Rev. Biophysics Biophysical Chem.* 14:331, 1985. A review covering the structure of rhodopsin and the roles of cGMP, GTP, the phosphodiesterase, and the G-protein ("transducin") in the response of the rod cell to light.

Cone, R., Rotational diffusion of rhodopsin in visual photoreceptor membrane. *Nature* 236:39, 1972. Excitation of the retina with polarized light creates a transient linear dichroism. The decay of the dichroism is explained by rotation of rhodopsin.

Fesenko, E. E., S. S. Kolesnikov, and A. L. Lyubarsky, Induction by cyclic GMP of cation conductance in plasma membrane of retinal rod outer segment. *Nature* 313:310, 1985. cGMP increases the Na^+ conductance of the cytoplasmic membrane.

Honig, B., T. Ebrey, R. Callender, V. Dinur, and M. Ottolenghi, Photoisomerization, energy storage, and charge separation: A model for light energy transduction in visual pigments and bacteriorhodopsin. *Proc. Natl. Acad. Sci. USA* 76:2503, 1979. Isomerization of the retinal could cause movement of the protonated N of the Schiff's base. This could explain the capture of energy and the movement of protons in rhodopsin and bacteriorhodopsin.

Stoeckenius, W., Purple membrane of halobacteria: A new light-energy converter. *Acc. Chem. Res.* 13:337, 1980. A review of the structure and function of bacteriorhodopsin and the purple membrane.

Wald, G., The molecular basis of visual excitation. *Nature* 219:800, 1968. Wald's Nobel Prize address, describing early work on the isomerization of retinal in rhodopsin.

PROBLEMS

1. Draw traces showing how the optical absorbance of a fresh suspension of rod outer segment disks might change as a function of time when the suspension is excited with a short flash of light at 37°C. Show the absorbance at (a) 545 nm and (b) 480 nm. Select the time scale for each trace judiciously, so that the traces illustrate the kinetics of the major absorbance changes that occur at the two wavelengths. (You may need to use two traces with different time scales at each wavelength to show both the initial absorbance change and its decay.)

2. If rhodopsin is illuminated at 500 nm at 77 K, the absorbance of the sample at 500 nm decreases. If the sample is then illuminated at 550 nm (still at 77 K), the absorbance at 500 nm increases again. Explain.

3. Why is the absorption spectrum of metarhodopsin II so different from that of metarhodopsin I?

4. Draw traces showing how the membrane potential of a rod cell changes with time when the cell is excited with (a) one photon, and (b) two photons. Assume that each photon converts one molecule of rhodopsin to batho-rhodopsin and on to metarhodopsin II. (The probability of conversion actually is only about 0.67.)

5. Provide evidence and a model for the involvement of cAMP as a mediator in visual transduction.

6. A frog retina is excited with a flash of polarized light that passes through the rods end-on. The flash causes optical absorbance changes at 500 nm, which are measured with polarized light that also passes through the rods end-on.
   a. Draw traces showing the kinetics of the absorbance changes measured with light polarized parallel to the excitation polarization and with light polarized perpendicular to the excitation polarization. The traces should cover the time period up to 100 μs after the excitation.
   b. Similar experiments are done with a retina that has been treated with glutaraldehyde, which causes cross-linking of the rhodopsins in the membrane. Draw traces showing the kinetics of the absorbance changes expected in this case. Assume that the cross-linking immobilizes the rhodopsin molecules but does not otherwise affect their photochemical transformations.

GLOSSARY

A form. A duplex DNA structure with right-handed twisting in which the planes of the base pairs are tilted about 70° with respect to the helix axis.

Acetal. The product formed by the successive condensation of two alcohols with a single aldehyde. It contains two ether-linked oxygens attached to a central carbon atom.

Acetyl-CoA. Acetyl-coenzyme A, a high-energy ester of acetic acid that is important both in the tricarboxylic acid cycle and in fatty acid biosynthesis.

Actin. A protein found in combination with myosin in muscle and also found as filaments constituting an important part of the cytoskeleton in many eukaryotic cells.

Actinomycin D. An antibiotic that binds to DNA and inhibits RNA chain elongation.

Activated complex. The highest free energy state of a complex in going from reactants to products.

Active site. The region of an enzyme molecule that contains the substrate binding site and the catalytic site for converting the substrate(s) into product(s).

Active transport. The energy-dependent transport of a substance across a membrane.

Adenine. A purine base found in DNA or RNA.

Adenosine. A purine nucleoside found in DNA, RNA, and many cofactors.

Adenosine diphosphate (ADP). The nucleotide formed by adding a pyrophosphate group to the 5'-OH group of adenosine.

Adenosine triphosphate (ATP). The nucleotide formed by adding yet another phosphate group to the pyrophosphate group on ADP.

Adenylate cyclase. The enzyme that catalyzes the formation of cyclic 3',5' adenosine monophosphate (cAMP) from ATP.

Adipocyte. A specialized cell that functions as a storage depot for lipid.

Aerobe. An organism that utilizes oxygen for growth.

Affinity chromatography. A column chromatographic technique that employs attached functional groups that have a specific affinity for sites on particular proteins.

Alcohol. A molecule with a hydroxyl group attached to a carbon atom.

Aldehyde. A molecule containing a doubly bonded oxygen and a hydrogen attached to the same carbon atom.

Alleles. Alternative forms of a gene.

Allosteric enzyme. An enzyme whose active site can be altered by the binding of a small molecule at a nonoverlapping site.

Angstrom (Å). A unit of length equal to 10^{-8} cm.

Anomers. The sugar isomers that differ in configuration about the carbonyl carbon atom. This carbon atom is called the anomeric carbon atom of the sugar.

Antibiotic. A natural product that inhibits bacterial growth (is bacteriostatic) and sometimes results in bacterial death (is bacteriocidal).

Antibody. A specific protein that interacts with a foreign substance (antigen) in a specific way.

Anticodon. A sequence of three bases on the transfer RNA that pair with the bases in the corresponding codon on the messenger RNA.

Antigen. A foreign substance that triggers antibody formation and is bound by the corresponding antibody.

Antiparallel β-pleated sheet (β sheet). A hydrogen-bonded secondary structure formed between two or more extended polypeptide chains.

Apoactivator. A regulatory protein that stimulates transcription from one or more genes in the presence of a coactivator molecule.

Asexual reproduction. Growth and cell duplication that does not involve the union of nuclei from cells of opposite mating types.

Asymmetric carbon. A carbon that is covalently bonded to four different groups.

Attenuator. A provisional transcription stop signal.

Autoradiography. The technique of exposing film in the presence of disintegrating radioactive particles. Used to obtain information on the distribution of radioactivity in a gel or a thin cell section.

Autoregulation. The process in which a gene regulates its own expression.

Autotroph. An organism that can form its organic constituents from CO_2.

Auxin. A plant growth hormone usually concentrated in the apical bud.

Auxotroph. A mutant that cannot grow on the minimal medium on which a wild-type member of the same species can grow.

Avogadro's number. The number of molecules in a gram molecular weight of any compound (6.023×10^{23}).

B cell. One of the major types of cells in the immune system. B cells can differentiate to form memory cells or antibody-forming cells.

B form. The most common form of duplex DNA, containing a right-handed helix and about 10 (10.5 exactly) base pairs per turn of the helix axis.

β bend. A characteristic way of turning an extended polypeptide chain in a different direction, involving the minimum number of residues.

β sheet. A sheetlike structure formed by the interaction between two or more extended polypeptide chains.

β oxidation. Oxidative degradation of fatty acids that occurs by the successive oxidation of the β carbon atom.

Base analog. A compound, usually a purine or a pyrimidine, that differs somewhat from a normal nucleic acid base.

Base stacking. The close packing of the planes of base pairs, commonly found in DNA and RNA structures.

Bidirectional replication. Replication in both directions away from the origin, as opposed to replication in one direction only (unidirectional replication).

Bilayer. A double layer of lipid molecules with the hydrophilic ends oriented outward, in contact with water, and the hydrophobic parts oriented inward.

Bile salts. Derivatives of cholesterol with detergent properties that aid in the solubilization of lipid molecules in the digestive tract.

Biochemical pathway. A series of enzyme-catalyzed reactions that results in the conversion of a precursor molecule into a product molecule.

Bioluminescence. The production of light by a biochemical system.

Blastoderm. The stage in embryogenesis when a unicellular layer at the surface surrounds the yolk mass.

Bond energy. The energy required to break a bond.

Branchpoint. An intermediate in a biochemical pathway that can follow more than one route in subsequent steps.

Buffer. A conjugate acid-base pair that is capable of resisting changes in pH when acid or base is added to the system. This tendency will be maximal when the conjugate forms are present in equal amounts.

cAMP. 3′,5′ cyclic adenosine monophosphate. The cAMP molecule plays a key role in metabolic regulation.

CAP. The catabolite gene activator protein, sometimes incorrectly referred to as the CRP protein. The latter term, in small letters (crp), should be used to refer to the gene but not to the protein.

Capping. Covalent modification involving the addition of a modified guanidine group in a 5′-5″ linkage. It occurs only in eukaryotes, primarily on mRNA molecules.

Carbohydrate. A polyhydroxy aldehyde or ketone.

Carboxylic acid. A molecule containing a carbon atom attached to a hydroxyl group and to an oxygen atom by a double bond.

Carcinogen. A chemical that can cause cancer.

Carotenoids. Lipid-soluble pigments that are made from isoprene units.

Catabolism. That part of metabolism that is concerned with degradation reactions.

Catabolite repression. The general repression of transcription of genes associated with catabolism that is seen in the presence of glucose.

Catalyst. A compound that lowers the activation energy of a reaction without itself being consumed.

Catalytic site. The site of the enzyme involved in the catalytic process.

Catenane. An interlocked pair of circular structures, such as covalently closed DNA molecules.

Catenation. The linking of molecules without any direct covalent bonding between them, as when two circular DNA molecules interlock like the links in a chain.

cDNA. Complementary DNA, made *in vitro* from the mRNA by the enzyme reverse transcriptase and deoxyribonucleotide triphosphates.

Cell commitment. That stage in a cell's life when it becomes committed to a certain line of development.

Cell cycle. All of those stages that a cell passes through from one cell generation to the next.

Cell line. An established clone originally derived from a whole organism through a long process of cultivation.

Cell lineage. The pedigree of cells resulting from binary fission.

Cell wall. A tough outer coating found in many plant, fungal, and bacterial cells that accounts for their ability to withstand mechanical stress or abrupt changes in osmotic pressure. Cell walls always contain a carbohydrate component and frequently also a peptide and a lipid component.

Chelate. A molecule that contains more than one binding site and frequently binds to another molecule through more than one binding site at the same time.

Chemiosmotic coupling. The coupling of ATP synthesis to an electrochemical potential gradient across a membrane.

Chimeric DNA. Recombinant DNA whose components originate from two or more different sources.

Chiral compound. A compound that can exist in two forms that are nonsuperimposable images of one another.

Chlorophyll. A green photosynthetic pigment that is made of a magnesium dihydroprophyrin complex.

Chloroplast. A chlorophyll-containing photosynthetic organelle, found in eukaryotic cells, that can harness light energy.

Chromatin. The nucleoprotein fibers of eukaryotic chromosomes.

Chromatography. A procedure for separating chemically similar molecules. Segregation is usually carried out on paper or in glass or metal columns with the help of different solvents. The paper or glass columns contain porous solids with functional groups that have limited affinities for the molecules being separated.

Chromosome. A thread-like structure, visible in the cell nucleus during metaphase, that carries the hereditary information.

Chromosome puff. A swollen region of a giant chromosome; the swelling reflects a high degree of transcription activity.

Cis dominance. Property of a sequence or a gene that exerts a dominant effect on a gene to which it is linked.

Cistron. A genetic unit that encodes a single polypeptide chain.

Citric acid cycle. See tricarboxylic acid (TCA) cycle.

Clone. One of a group of genetically identical cells or organisms derived from a common ancestor.

Cloning vector. A self-replicating entity to which foreign DNA can be covalently attached for purposes of amplification in host cells.

Coactivator. A molecule that functions in conjunction with a protein apoactivator. For example, cAMP is a coactivator of the CAP protein.

Codon. In a messenger RNA molecule, a sequence of three bases that represents a particular amino acid.

Coenzyme. An organic molecule that associates with enzymes and affects their activity.

Cofactor. A small molecule required for enzyme activity. It could be organic in nature, like a coenzyme, or inorganic in nature, like a metallic cation.

Complementary base sequence. For a given sequence of nucleic acids, the nucleic acids that are related to them by the rules of base pairing.

Configuration. The spatial arrangement in which atoms are covalently linked in a molecule.

Conformation. The three-dimensional arrangement adopted by a molecule, usually a complex macromolecule. Molecules with the same configuration can have more than one conformation.

Consensus sequence. In nucleic acids, the "average" sequence that signals a certain type of action by a specific protein. The sequences actually observed usually vary around this average.

Constitutive enzymes. Enzymes synthesized in fixed amounts, regardless of growth conditions.

Cooperative binding. A situation in which the binding of one substituent to a macromolecule favors the binding of another. For example, DNA cooperatively binds histone molecules, and hemoglobin cooperatively binds oxygen molecules.

Coordinate induction. The simultaneous expression of two or more genes.

Cosmid. A DNA molecule with *cos* ends from λ bacteriophage that can be packaged *in vitro* into a virus for infection purposes.

Cot curve. A curve that indicates the rate of DNA-DNA annealing as a function of DNA concentration and time.

Cytidine. A pyrimidine nucleoside found in DNA and RNA.

Cytochromes. Heme-containing proteins that function as electron carriers in oxidative phosphorylation and photosynthesis.

Cytokinin. A plant hormone produced in root tissue.

Cytoplasm. The contents enclosed by the plasma membrane, excluding the nucleus.

Cytosine. A pyrimidine base found in DNA and RNA.

Cytoskeleton. The filamentous skeleton, formed in the cytoplasm, that is largely responsible for controlling cell shape.

Cytosol. The liquid portion of the cytoplasm, including the macromolecules but not including the larger structures, such as subcellular organelles or cytoskeleton.

D loop. An extended loop of single-stranded DNA displaced from a duplex structure by an oligonucleotide.

Dalton. A unit of mass equivalent to the mass of a hydrogen atom (1.66×10^{-24} g).

Dark reactions. Reactions that can occur in the dark, in a process that is usually associated with light, such as the dark reactions of photosynthesis.

De novo pathway. A biochemical pathway that starts from elementary substrates and ends in the synthesis of a biochemical.

Deamination. The enzymatic removal of an amine group, as in the deamination of an amino acid to an alpha keto acid.

Dehydrogenase. An enzyme that catalyzes the removal of a pair of electrons (and usually one or two protons) from a substrate molecule.

Denaturation. The disruption of the native folded structure of a nucleic acid or protein molecule; may be due to heat, chemical treatment, or change in pH.

Density-gradient centrifugation. The separation, by centrifugation, of molecules according to their density, in a gradient varying in solute concentration.

Dialysis. Removal of small molecules from a macromolecule preparation by allowing them to pass across a semipermeable membrane.

Diauxic growth. Biphasic growth on a mixture of two carbon sources in which one carbon source is used up before the other one is mobilized. For example, in the presence of glucose and lactose, *E. coli* will utilize the glucose before the lactose.

Difference spectra. Display comparing the absorption spectra of a molecule or an assembly of molecules in different states, for example, those of mitochondria under oxidizing or reducing conditions.

Differential centrifugation. Separation of molecules and/or organelles by sedimentation rate.

Differentiation. A change in the form and pattern of a cell and the genes it expresses as a result of growth and replication, usually during development of a multicellular organism. Also occurs in microorganisms (e.g., in sporulation).

Diploid cell. A cell that contains two chromosomes (2N) of each type.

Dipole. A separation of charge within a single molecule.

Directed mutagenesis. In a DNA sequence, an intentional alteration that can be genetically inherited.

Dissociation constant. An equilibrium constant for the dissociation of a molecule into two parts (e.g., dissociation of acetic acid into acetate anion and proton).

Disulfide bridge. A covalent linkage formed by oxidation between two SH groups either in the same polypeptide chain or in different polypeptide chains.

DNA. Deoxyribonucleic acid. A polydeoxyribonucleotide in which the sugar is deoxyribose; the main repository of genetic information in all cells and most viruses.

DNA cloning. The propagation of individual segments of DNA as clones.

DNA library. A mixture of clones, each containing a cloning vector and a segment of DNA from a source of interest.

DNA polymerase. An enzyme that catalyzes the formation of 3′-5′ phosphodiester bonds from deoxyribonucleotide triphosphates.

Domain. A segment of a folded protein structure showing conformational integrity. A domain could comprise the entire protein or just a fraction of the protein. Some proteins, such as antibodies, contain many structural domains.

Dominant. Describing an allele whose phenotype is expressed regardless of whether the organism is homozygous or heterozygous for that allele.

Double helix. A structure in which two helically twisted polynucleotide strands are held together by hydrogen bonding and base stacking.

Duplex. Synonymous with double helix.

Dyad symmetry. Property of a structure that can be rotated by 180° to produce the same structure.

Ecdysone. A hormone that stimulates the molting process in insects.

Edman degradation. A systematic method of sequencing proteins, proceeding by stepwise removal of single amino acids from the amino terminal of a polypeptide chain.

Eicosanoid. Any fatty acid with 20 carbons.

Electrophoresis. The movement of particles in an electrical field. A commonly used technique for analysis of mixtures of molecules in solution according to their electrophoretic mobilities.

Elongation factors. Protein factors uniquely required during the elongation phase of protein synthesis. Elongation factor G (EF-G) brings about the movement of the peptidyl-tRNA from the A site to the P site of the ribosome.

Eluate. The effluent from a chromatographic column.

Embryo. Plant or animal at an early stage of development.

Enantiomorphs. Isomers that are mirror images of one another.

Endergonic reaction. A reaction with a positive standard free energy change.

End-product (feedback) inhibition. The inhibition of the first enzyme in a pathway by the end product of that pathway.

Endocrine glands. Specialized tissues whose function is to synthesize and secrete hormones.

Endonuclease. An enzyme that breaks a phosphodiester linkage at some point within a polynucleotide chain.

Endopeptidase. An enzyme that breaks a polypeptide chain at an internal peptide linkage.

Endoplasmic reticulum. A system of double membranes in the cytoplasm that is involved in the synthesis of transported proteins. The rough endoplasmic reticulum has ribosomes associated with it. The smooth endoplasmic reticulum does not.

Energy charge. The fractional degree to which the AMP-ADP-ATP system is filled with high-energy phosphates (phosphoryl groups).

Enhancer. A DNA sequence that can stimulate transcription at an appreciable distance from the site where it is located. It acts in either orientation and either upstream or downstream from the promoter.

Entropy. The randomness of a system.

Enzyme. A protein that contains a catalytic site for a biochemical reaction.

Epimers. Two stereoisomers with more than one chiral center that differ in configuration at one of their chiral centers.

Equilibrium. The point at which the concentrations of two compounds are such that the interconversion of one compound into the other compound does not result in any change in free energy.

Escherichia coli (E. coli). A Gram negative bacterium commonly found in the vertebrate intestine. It is the bacterium most frequently used in the study of biochemistry and genetics.

Established cell line. A group of cultured cells derived from a single origin and capable of stable growth for many generations.

Ether. A molecule containing two carbons linked by an oxygen atom.

Eukaryote. A cell or organism that has a membrane-bound nucleus.

Excision repair. DNA repair in which a damaged region is replaced.

Excited state. An energy-rich state of an atom or a molecule, produced by the absorption of radiant energy.

Exergonic reaction. A chemical reaction that takes place with a negative change in standard free energy.

Exon. A segment within a gene that carries part of the coding information for a protein.

Exonuclease. An enzyme that breaks a phosphodiester linkage at one or the other end of a polynucleotide chain so as to release single or small nucleotide residues.

F factor. A large bacterial plasmid, known as the sex-factor plasmid because it permits mating between F^+ and F^- bacteria.

Facultative aerobe. An organism that can use molecular oxygen in its metabolism but that also can live anaerobically.

Fatty acid. A long-chain hydrocarbon containing a carboxyl group at one end. Saturated fatty acids have completely saturated hydrocarbon chains. Unsaturated fatty acids have one or more carbon–carbon double bonds in their hydrocarbon chains.

Feedback inhibition. *See* end-product inhibition.

Fermentation. The energy-generating breakdown of glucose or related molecules by a process that does not require molecular oxygen.

Fingerprinting. The characteristic two-dimensional paper chromatogram obtained from the partial hydrolysis of a protein or a nucleic acid.

Fluorescence. The emission of light by an excited molecule in the process of making the transition from the excited state to the ground state.

Frameshift mutations. Insertions or deletions of genetic material that lead to a shift in the translation of the reading frame. The mutation usually leads to nonfunctional proteins.

Free energy. That part of the energy of a system that is available to do useful work.

Furanose. A sugar that contains a five-membered ring as a result of intramolecular hemiacetal formation.

Futile cycle. *See* pseudocycle.

G_1 phase. That period of the cell cycle in which preparations are being made for chromosome duplication, which takes place in the S phase.

G_2 phase. That period of the cell cycle between S phase and mitosis (M phase).

Gametes. The ova and the sperm, haploid cells that unite during fertilization to generate a diploid zygote.

Gel-exclusion chromatography. A technique that makes use of certain polymers that can form porous beads with varying pore sizes. In columns made from such beads, it is possible to separate molecules, which cannot penetrate beads of a given pore size, from small molecules that can.

Gene. A segment of the genome that codes for a functional product.

Gene amplification. The duplication of a particular gene within a chromosome two or more times.

Gene splicing. The cutting and rejoining of DNA sequences.

General recombination. Recombination that occurs between homologous chromosomes at homologous sites.

Generation time. The time it takes for a cell to double its mass under specified conditions.

Genetic map. The arrangement of genes or other identifiable sequences on a chromosome.

Genome. The total genetic content of a cell or a virus.

Genotype. The genetic characteristics of an organism (distinguished from its observable characteristics, or phenotype).

Globular protein. A folded protein that adopts an approximately globular shape.

Goldman equation. An equation expressing the quantitative relationship between the concentrations of charged species on either side of a membrane and the resting transmembrane potential.

Golgi apparatus. A complex series of double-membrane structures that interact with the endoplasmic reticulum and that serve as a transfer point for proteins destined for other organelles, the plasma membrane, or extracellular transport.

Gluconeogenesis. The production of sugars from nonsugar precursors such as lactate or amino acids. Applies more specifically to the production of free glucose by vertebrate livers.

Glycogen. A polymer of glucose residues in 1,4 linkage and 1,6 linkage at branchpoints.

Glycogenic. Describing amino acids whose metabolism may lead to gluconeogenesis.

Glycolipid. A lipid containing a carbohydrate group.

Glycolysis. The catabolic conversion of glucose to pyruvate with the production of ATP.

Glycoprotein. A protein linked to an oligosaccharide or a polysaccharide.

Glycosaminoglycans. Long, unbranched polysaccharide chains composed of repeating dissacharide subunits in which one of the two sugars is either N-acetylglucosamine or N-acetylgalactosamine.

Glycosidic bond. The bond between a sugar and an alcohol. Also the bond that links two sugars in disaccharides, oligosaccharides, and polysaccharides.

Glyoxylate cycle. A pathway that uses some of the enzymes of the TCA cycle and some enzymes whereby acetate can be converted into succinate and carbohydrates.

Glyoxysome. An organelle containing some enzymes of the glyoxylate cycle.

Gram molecular weight. For a given compound, the weight in grams that is numerically equal to its molecular weight.

Ground state. The lowest electronic energy state of an atom or a molecule.

Growth factor. A substance that must be present in the growth medium to permit cell proliferation.

Growth fork. The region on a DNA duplex molecule where synthesis is taking place. It resembles a fork in shape, since it consists of a region of duplex DNA connected to a region of unwound single strands.

Guanine. A purine base found in DNA or RNA.

Guanosine. A purine nucleoside found in DNA and RNA.

Hairpin loop. A single-stranded complementary region that folds back on itself and base-pairs into a double helix.

Half-life. The time required for the disappearance of one half of a substance.

Haploid cell. A cell containing only one chromosome of each type.

Heavy isotopes. Forms of atoms that contain greater numbers of neutrons (e.g., ^{15}N, ^{13}C).

Helix. A spiral structure with a repeating pattern.

Heme. An iron-porphyrin complex found in hemoglobin and cytochromes.

Hemiacetal. The product formed by the condensation of an aldehyde with an alcohol; it contains one oxygen linked to a central carbon in a hydroxyl fashion and one oxygen linked to the same central carbon by an ether linkage.

Henderson-Hasselbalch equation. An equation that relates the pK_a to the pH and the ratio of the proton acceptor (A^-) and the proton donor (HA) species of a conjugate acid-base pair.

Heterochromatin. Highly condensed regions of chromosomes that are not usually transcriptionally active.

Heteroduplex. An annealed duplex structure between two DNA strands that do not show perfect complementarity. Can arise by mutation, recombination, or the annealing of complementary single-stranded DNAs.

Heteropolymer. A polymer containing more than one type of monomeric unit.

Heterotroph. An organism that requires preformed organic compounds for growth.

Heterozygous. Describing an organism (a heterozygote) that carries two different alleles for a given gene.

Hexose. A sugar with a six-carbon backbone.

High-energy compound. A compound that undergoes hydrolysis with a high negative standard free energy change.

Histones. The family of basic proteins that is normally associated with DNA in most cells of eukaryotic organisms.

Holoenzyme. An intact enzyme containing all of its subunits with full enzymatic activity.

Homologous chromosomes. Chromosomes that carry the same pattern of genes, but not necessarily the same alleles.

Homopolymer. A polymer composed of only one type of monomeric building block.

Homozygous. Describing an organism (a homozygote) that carries two identical alleles for a given gene.

Hormone. A chemical substance made in one cell and secreted so as to influence the metabolic activity of a select group of cells located at other sites in the organism.

Hormone receptor. A protein that is located on the cell membrane or inside the responsive cell and that interacts specifically with the hormone.

Host cell. A cell used for growth and reproduction of a virus.

Hybrid (or chimeric) plasmid. A plasmid that contains DNA from two different organisms.

Hydrogen bond. A weak attractive force between one electronegative atom and a hydrogen atom that is covalently linked to a second electronegative atom.

Hydrolysis. The cleavage of a molecule by the addition of water.

Hydrophilic. Preferring to be in contact with water.

Hydrophobic. Preferring not to be in contact with water, as is the case with the hydrocarbon portion of a fatty acid or phospholipid chain.

Hydrophobic bonding. The association of nonpolar groups with each other in aqueous solution.

Hydroxyapatite. A calcium phosphate gel used, in the case of nucleic acids, to selectively absorb duplex DNA-RNA from a mixture of single-stranded and duplex nucleic acids.

Icosahedral symmetry. The symmetry displayed by a regular polyhedron that is composed of 20 equilateral triangular faces with 12 corners.

Imine. A molecule containing a nitrogen atom attached to a carbon atom by a double bond. The nitrogen is also covalently linked to a hydrogen.

Immunofluorescence. A cytological technique in which a

specific fluorescent antibody is used to label an antigen. Frequently used to determine the location of an antigen in a tissue or a cell.

Immunoglobulin. A protein made in a B plasma cell and usually secreted; it interacts specifically with a foreign agent. Synonymous with antibody. It is composed of two heavy and two light chains linked by disulfide bonds. Immunoglobulins can be divided into five classes (IgG, IgM, IgA, IgD, and IgE) based on their heavy-chain component.

In vitro. Literally, "in glass," describing whatever happens in a test tube or other receptacle, as opposed to what happens in whole cells of the whole organism (*in vivo*).

Induced fit. A change in the shape of an enzyme that results from the binding of substrate.

Inducers. Molecules that cause an increase in a protein when added to cells.

Initiation factors. Those protein factors that are specifically required during the initiation phase of protein synthesis.

Intercalating agent. A chemical, usually containing aromatic rings, that can sandwich in between adjacent base pairs in a DNA duplex. The intercalation leads to an adjustment in the DNA secondary structure, as adjacent base pairs are usually close-packed.

Interferon. One of a family of proteins that are liberated by special host cells in the mammal in response to viral infection. The interferons attach to an infected cell, where they stimulate antiviral protein synthesis.

Intervening sequence. *See* intron.

Intron. A segment of the nascent transcript that is removed by splicing. Also refers to the corresponding region in the DNA. Synonymous with intervening sequence.

Inverted repeat. A chromosome segment that is identical to another segment on the same chromosome except that it is oriented in the opposite direction.

Ion-exchange resin. A polymeric resinous substance, usually in bead form, that contains fixed groups with positive or negative charge. A cation exchange resin has negatively charged groups and is therefore useful in exchanging the cationic groups in a test sample. The resin is usually used in the form of a column, as in other column chromatographic systems.

Isoelectric pH. The pH at which a protein has no net charge.

Isomerase. An enzyme that catalyzes an intramolecular rearrangement.

Isomerization. Rearrangement of atomic groups within the same molecule without any loss or gain of atoms.

Isozymes. Multiple forms of an enzyme that differ from one another in one or more of the properties.

K_m. *See* Michaelis constant.

Ketogenic. Describing amino acids that are metabolized to acetoacetate and acetate.

Ketone. A functional group of an organic compound in which a carbon atom is double-bonded to an oxygen. Neither of the other substituents attached to the carbon is a hydrogen. Otherwise the group would be called an aldehyde.

Ketone bodies. Refers to acetoacetate, acetone, and β-hydroxybutyrate made from acetyl-CoA in the liver and used for energy in nonhepatic tissue.

Ketosis. A condition in which the concentration of ketone bodies in the blood or urine is unusually high.

Kilobase. One thousand bases in a DNA molecule.

Kinase. An enzyme catalyzing phosphorylation of an acceptor molecule, usually with ATP serving as the phosphate (phosphoryl) donor.

Kinetochore. A structure that attaches laterally to the centromere of a chromosome; it is the site of chromosome tubule attachment.

Krebs cycle. *See* tricarboxylic acid (TCA) cycle.

Lampbrush chromosome. Giant diplotene chromosome found in the oocyte nucleus. The loops that are observed are the sites of extensive gene expression.

Law of mass action. The finding that the rate of a chemical reaction is a function of the product of the concentrations of the reacting species.

Leader region. The region of an mRNA between the 5' end and the initiation codon for translation of the first polypeptide chain.

Lectins. Agglutinating proteins usually extracted from plants.

Ligase. An enzyme that catalyzes the joining of two molecules together. In DNA it joins 5'-OH to 3' phosphates.

Linkers. Short oligonucleotides that can be ligated to larger DNA fragments, then cleaved to yield overlapping cohesive ends, suitable for ligation to other DNAs that contain comparable cohesive ends.

Linking number. The net number of times one polynucleotide chain crosses over another polynucleotide chain. By convention, right-handed crossovers are given a plus designation.

Lipid. A biological molecule that is soluble in organic solvents. Lipids include steroids, fatty acids, prostaglandins, terpenes, and waxes.

Lipid bilayer. Model for the structure of the cell membrane based on the hydrophobic interaction between phospholipids.

Lipopolysaccharide. Usually refers to a unique glycolipid found in Gram negative bacteria.

Lyase. An enzyme that catalyzes the removal of a group to form a double bond, or the reverse reaction.

Lysogenic virus. A virus that can adopt an inactive (lysogenic) state, in which it maintains its genome within a cell instead of entering the lytic cycle. The circumstances that determine whether a lysogenic (temperate) virus will adopt an inactive state or an active lytic state are often subtle and depend upon the physiologic state of the infected cell.

Lysosome. An organelle that contains hydrolytic enzymes

designed to break down proteins that are targeted to that organelle.

Lytic infection. A virus infection that leads to the lysis of the host cell, yielding progeny virus particles.

M phase. That period of the cell cycle when mitosis takes place.

Meiosis. Process in which diploid cells undergo division to form haploid sex cells.

Membrane transport. The facilitated transport of a molecule across a membrane.

Merodiploid. An organism that is diploid for some but not all of its genes.

Mesosome. An invagination of the bacterial cell membrane.

Messenger RNA (mRNA). The template RNA carrying the message for protein synthesis.

Metabolic turnover. A measure of the rate at which already existing molecules of the given species are replaced by newly synthesized molecules of the same type. Usually isotopic labeling is required to measure turnover.

Metabolism. The sum total of the enzyme-catalyzed reactions that occur in a living organism.

Metamorphosis. A change of form, especially the conversion of a larval form to an adult form.

Metaphase. That stage in mitosis or meiosis when all of the chromosomes are lined up on the equator (i.e., an imaginary line that bisects the cell).

Micelle. An aggregate of lipids in which the polar head groups face outward and the hydrophobic tails face inward; no solvent is trapped in the center.

Michaelis constant (K_m). The substrate concentration at which an enzyme-catalyzed reaction proceeds at one-half maximum velocity.

Michaelis-Menten equation (also known as the Henri-Michaelis-Menten equation). An equation relating the reaction velocity to the substrate concentration of an enzyme.

Microtubules. Thin tubules, made from globular proteins, that serve multiple purposes in eukaryotic cells.

Mismatch repair. The replacement of a base in a heteroduplex structure by one that forms a Watson-Crick base pair.

Missense mutation. A change in which a codon for one amino acid is replaced by a codon for another amino acid.

Mitochondrion. An organelle, found in eukaryotic cells, in which oxidative phosphorylation takes place. It contains its own genome and unique ribosomes to carry out protein synthesis of only a fraction of the proteins located in this organelle.

Mitosis. The process whereby replicated chromosomes segregate equally toward opposite poles prior to cell division.

Mobile genetic element. A segment of the genome that can move as a unit from one location on the genome to another, without any requirement for sequence homology.

Molecularity of a reaction. The number of molecules involved in a specific reaction step.

Monolayer. A single layer of oriented lipid molecules.

Mutagen. An agent that can bring about a heritable change (mutation) in an organism.

Mutagenesis. A process that leads to a change in the genetic material that is inherited in subsequent generations.

Mutant. An organism that carries an altered gene or change in its genome.

Mutarotation. The change in optical rotation of a sugar that is observed immediately after it is dissolved in aqueous solution, as the result of the slow approach of equilibrium of a pyranose or a furanose in its α and β forms.

Mutation. The genetically inheritable alteration of a gene or group of genes.

Myofibril. A unit of thick and thin filaments in a muscle fiber.

Myosin. The main protein of the thick filaments in a muscle myofibril. It is composed of two coiled subunits (M_r about 220,000) that can aggregate to form a thick filament, which is globular at each end.

Nascent RNA. The initial transcripts of RNA, before any modification or processing.

Negative control. Repression of biological activity by the presence of a specific molecule.

Nernst equation. An equation that relates the redox potential to the standard redox potential and the concentrations of the oxidized and reduced form of the couple.

Nitrogen cycle. The passage of nitrogen through various valence states, as the result of reactions carried out by a wide variety of different organisms.

Nitrogen fixation. Conversion of atmospheric nitrogen into a form that can be converted by biochemical reactions to an organic form. This reaction is carried out by a very limited number of microorganisms.

Nitrogenous base. An aromatic nitrogen-containing molecule with basic properties. Such bases include purines and pyrimidines.

Noncompetitive inhibitor. An inhibitor of enzyme activity whose effect is not reversed by increasing the concentration of substrate molecule.

Nonsense mutation. A change in the base sequence that converts a sense codon (one that specifies an amino acid) to one that specifies a stop (a nonsense codon). There are three nonsense codons.

Northern blotting. *See* Southern blotting.

Nuclease. An enzyme that cleaves phosphodiester bonds of nucleic acids.

Nucleic acids. Polymers of the ribonucleotides or deoxyribonucleotides.

Nucleohistone. A complex of DNA and histone.

Nucleolus. A spherical structure visible in the nucleus during interphase. The nucleolus is associated with a site on the chromosome that is involved in ribosomal RNA synthesis.

Nucleophilic group.   An electron-rich group that tends to attack an electron-deficient nucleus.

Nucleosome.   A complex of DNA and an octamer of histone proteins in which a small stretch of the duplex is wrapped around a molecular bead of histone.

Nucleotide.   An organic molecule containing a purine or pyrimidine base, a five-carbon sugar (ribose or deoxyribose), and one or more phosphate groups.

Nucleus.   In eukaryotic cells, the centrally located organelle that encloses most of the chromosomes. Minor amounts of chromosomal substance are found in some other organelles, most notably the mitochondria and the chloroplasts.

Okazaki fragment.   A short segment of single-stranded DNA that is an intermediate in DNA synthesis. In bacteria, Okazaki fragments are 1000–2000 bases in length; in eukaryotes, 100–200 bases in length.

Oligonucleotide.   A polynucleotide containing a small number of nucleotides. The linkages are the same as in a polynucleotide; the only distinguishing feature is the small size.

Oligosaccharide.   A molecule containing a small number of sugar residues joined in a linear or a branched structure by glycosidic bonds.

Oncogene.   A gene of cellular or viral origin that is responsible for rapid, unruly growth of animal cells.

Operon.   A group of contiguous genes that are coordinately regulated by two cis-acting elements, a promoter and an operator. Found only in prokaryotic cells.

Optical activity.   The property of a molecule that leads to rotation of the plane of polarization of plane-polarized light when the latter is transmitted through the substance. Chirality is a necessary and sufficient property for optical activity.

Organelle.   A subcellular membrane-bounded body with a well-defined function.

Osmotic pressure.   The pressure generated by the mass flow of water to that side of a membrane-bounded structure that contains the higher concentration of solute molecules. A stable osmotic pressure is seen in systems in which the membrane is not permeable to some of the solute molecules.

Oxidation.   The loss of electrons from a compound.

Oxidative phosphorylation.   The formation of ATP as the result of the transfer of electrons to oxygen.

Oxido-reductase.   An enzyme that catalyzes oxidation-reduction reactions.

Palindrome.   A sequence of bases that reads the same in both directions on opposite strands of the DNA duplex (e.g., GAATTC).

Pentose.   A sugar with five carbon atoms.

Pentose phosphate pathway.   The pathway involving the oxidation of glucose-6-phosphate to pentose phosphates and further reactions of pentose phosphates.

Peptide.   An organic molecule in which a covalent amide bond is formed between the α amino group of one amino acid and the α carboxyl group of another amino acid, with the elimination of a water molecule.

Peptide mapping.   Same as fingerprinting.

Peptidoglycan.   The main component of the bacterial cell wall, consisting of a two-dimensional network of heteropolysaccharides running in one direction, cross-linked with polypeptides running in the perpendicular direction.

Periplasm.   The region between the inner cytoplasmic membrane and the cell wall.

Permease.   A protein that catalyzes the transport of a specific small molecule across a membrane.

Peroxisomes.   Subcellular organelles that contain flavin-requiring oxidases and that regenerate oxidized flavin by reaction with oxygen.

Phenotype.   The observable trait(s) that result from the genotype in cooperation with the environment.

Phenylketonuria.   A human disease caused by a genetic deficiency in the enzyme that converts phenylalanine to tyrosine. The immediate cause of the disease is an excess of phenylalanine, which can be alleviated by a diet low in phenylalanine.

Pheromone.   A hormonelike substance that acts as an attractant.

Phosphodiester.   A molecule containing two alcohols esterified to a single molecule of phosphate. For example, the backbone of nucleic acids is connected by 5′-3′ phosphodiester linkages between the adjacent individual nucleotide residues.

Phosphogluconate pathway.   Another name for the pentose phosphate pathway. This name derives from the fact that 6-phosphogluconate is an intermediate in the formation of pentoses from glucose.

Phospholipid.   A lipid containing charged hydrophilic phosphate groups; a component of cell membranes.

Phosphorylation.   The formation of a phosphate derivative of a biomolecule.

Photoreactivation.   DNA repair in which the damaged region is repaired with the help of light and an enzyme. The lesion is repaired without excision from the DNA.

Photosynthesis.   The biosynthesis that directly harnesses the chemical energy resulting from the absorption of light. Frequently used to refer to the formation of carbohydrates from CO_2 that occurs in the chloroplasts of plants or the plastids of photosynthetic microorganisms.

Pitch length (or pitch).   The number of base pairs per turn of a duplex helix.

Plaque.   A circular clearing on a lawn of bacterial or culture cells, resulting from cell lysis and production of phage or animal virus progeny.

Plasma membrane.   The membrane that surrounds the cytoplasm.

Plasmid. A circular DNA duplex that replicates autonomously in bacteria. Plasmids that integrate into the host genome are called episomes. Plasmids differ from viruses in that they never form infectious nucleoprotein particles.

Polar group. A hydrophilic (water-loving) group.

Polar mutation. A mutation in one gene that reduces the expression of a gene or genes distal to the promoter in the same operon.

Polarimeter. An instrument for determining the rotation of polarization of light as the light passes through a solution containing an optically active substance.

Polyamine. A hydrocarbon containing more than two amino groups.

Polycistronic messenger RNA. In prokaryotes, an RNA that contains two or more cistrons; note that only in prokaryotic mRNAs can more than one cistron be utilized by the translation system to generate individual proteins.

Polymerase. An enzyme that catalyzes the synthesis of a polymer from monomers.

Polynucleotide. A chain structure containing nucleotides linked together by phosphodiester (5'-3') bonds. The polynucleotide chain has a directional sense with a 5' and a 3' end.

Polynucleotide phosphorylase. An enzyme that polymerizes ribonucleotide diphosphates. No template is required.

Polypeptide. A linear polymer of amino acids held together by peptide linkages. The polypeptide has a directional sense, with an amino and a carboxy terminal end.

Polyribosome (polysome). A complex of an mRNA and two or more ribosomes actively engaged in protein synthesis.

Polysaccharide. A linear or branched chain structure containing many sugar molecules linked by glycosidic bonds.

Porphyrin. A complex planar structure containing four substituted pyrroles covalently joined in a ring and frequently containing a central metal atom. For example, heme is a porphyrin with a central iron atom.

Positive control. A system that is turned on by the presence of a regulatory protein.

Posttranslational modification. The covalent bond changes that occur in a polypeptide chain after it leaves the ribosome and before it becomes a mature protein.

Primary structure. In a polymer, the sequence of monomers and the covalent bonds.

Primer. A structure that serves as a growing point for polymerization.

Primosome. A multiprotein complex that catalyzes synthesis of RNA primer at various points along the DNA template.

Prochiral molecule. A nonchiral molecule that may react with an enzyme so that two groups that have a mirror-image relationship to each other are treated differently.

Prokaryote. A unicellular organism that contains a single chromosome, no nucleus, no membrane-bound organelles, and has characteristic ribosomes and biochemistry.

Promoter. That region of the gene that signals RNA polymerase binding and the initiation of transcription.

Prophage. The silent phage genome. Some prophages integrate into the host genome; others replicate autonomously. The prophage state is maintained by a phage-encoded repressor.

Prophase. The stage in meiosis or mitosis when chromosomes condense and become visible as refractile bodies.

Proprotein. A protein that is made in an active form, so that it requires processing to become functional.

Prostaglandin. An oxygenated eicosanoid that has a hormonal function. Prostaglandins are unusual hormones in that they usually have effects only in that region of the organism where they are synthesized.

Prosthetic group. Synonymous with coenzyme except that a prosthetic group is usually more firmly attached to the enzyme it serves.

Protamines. Highly basic, arginine-rich proteins found complexed to DNA in the sperm of many invertebrates and fish.

Protein subunit. One of the components of a complex multicomponent protein.

Proteoglycan. A protein-linked heteropolysaccharide in which the heteropolysaccharide is usually the major component.

Protist. A relatively undifferentiated organism that can survive as a single cell.

Proton acceptor. A functional group capable of accepting a proton from a proton donor molecule.

Proton motive force (Δp). The thermodynamic driving force for proton translocation. Expressed quantitatively as $\Delta G_{H^+}/F$ in units of volts.

Proto-oncogene. A cellular gene that can undergo modification to a cancer-causing gene (oncogene).

Pseudocycle. A sequence of reactions that can be arranged in a cycle but that usually do not function simultaneously in both directions. Also called a futile cycle, since the net result of simultaneous functioning in both directions would be the expenditure of energy without accomplishing any useful work.

Pulse-chase. An experiment in which a short labeling period is followed by the addition of an excess of the same, unlabeled compound to dilute out the labeled material.

Purine. A heterocyclic ring structure with varying functional groups. The purines adenine and guanine are found in both DNA and RNA.

Puromycin. An antibiotic that inhibits polypeptide synthesis by competing with aminoacyl-tRNA for the ribosomal binding site A.

Pyranose. A simple sugar containing the six-membered pyran ring.

Pyrimidine. A heterocyclic six-membered ring structure. Cytosine and uracil are the main pyrimidines found in RNA, and cytosine and thymine are the main pyrimidines found in DNA.

Pyrophosphate. A molecule formed by two phosphates in anhydride linkage.

Quaternary structure. In a protein, the way in which the different folded subunits interact to form the multisubunit protein.

R group. The distinctive side chain of an amino acid.

R loop. A triple-stranded structure in which RNA displaces a DNA strand by DNA-RNA hybrid formation in a region of the DNA.

Rapid-start complex. The complex that RNA polymerase forms at the promoter site just before initiation.

Recombination. The transfer to offspring of genes not found together in either of the parents.

Redox couple. An electron donor and its corresponding oxidized form.

Redox potential (E). The relative tendency of a pair of molecules to release or accept an electron. The standard redox potential ($E°$) is the redox potential of a solution containing the oxidant and reductant of the couple at standard concentrations.

Regulatory enzyme. An enzyme in which the active site is subject to regulation by factors other than the enzyme substrate. The enzyme frequently contains a nonoverlapping site for binding the regulatory factor that affects the activity of the active site.

Regulatory gene. A gene whose principal product is a protein designed to regulate the synthesis of other genes.

Renaturation. The process of returning a denatured structure to its original native structure, as when two single strands of DNA are reunited to form a regular duplex, or an unfolded polypeptide chain is returned to its normal folded three-dimensional structure.

Repair synthesis. DNA synthesis following excision of damaged DNA.

Repetitive DNA. A DNA sequence that is present in many copies per genome.

Replica plating. A technique in which an impression of a culture is taken from a master plate and transferred to a fresh plate. The impression can be of bacterial clones or phage plaques.

Replication fork. The Y-shaped region of DNA at the site of DNA synthesis; also called a growth fork.

Replicon. A genetic element that behaves as an autonomous replicating unit. It can be a plasmid, phage, or bacterial chromosome.

Repressor. A regulatory protein that inhibits transcription from one or more genes. It can combine with an inducer (resulting in specific enzyme induction) or with an operator element (resulting in repression).

Resonance hybrid. A molecular structure that is a hybrid of two structures that differ in the locations of some of the electrons. For example, the benzene ring can be drawn in two ways, with double bonds in different positions. The actual structure of benzene is in between these two equivalent structures.

Restriction-modification system. A pair of enzymes found in most bacteria (but not eukaryotic cells). The restriction enzyme recognizes a certain sequence in duplex DNA and makes one cut in each unmodified DNA strand at or near the recognition sequence. The modification enzyme methylates (or modifies) the same sequence, thus protecting it from the action of the restriction enzyme.

Reverse transcriptase. An enzyme that synthesizes DNA from an RNA template, using deoxyribonucleotide triphosphates.

Rho factor. A protein involved in the termination of transcription of some messenger RNAs.

Ribose. The five-carbon sugar found in RNA.

Ribosomal RNA (rRNA). The RNA parts of the ribosome.

Ribosomes. Small cellular particles made up of ribosomal RNA and protein. They are the site, together with mRNA, of protein synthesis.

RNA (Ribonucleic acid). A polynucleotide in which the sugar is ribose.

RNA polymerase. An enzyme that catalyzes the formation of RNA from ribonucleotide triphosphates, using DNA as a template.

RNA splicing. The excision of a segment of RNA, followed by a rejoining of the remaining fragments.

Rolling circle replication. A mechanism for the replication of circular DNA. A nick in one strand allows the 3′ end to be extended, displacing the strand with the 5′ end, which is also replicated, to generate a double-stranded tail that can become larger than the unit size of the circular DNA.

S phase. The period during the cell cycle when the chromosome is replicated.

Salting in. The increase in solubility that is displayed by typical globular proteins upon the addition of small amounts of certain salts, such as ammonium sulfate.

Salting out. The decrease in protein solubility that occurs when salts such as ammonium sulfate are present at high concentrations.

Salvage pathway. A family of reactions that permits, for instance, nucleosides as well as purine and pyrimidine bases resulting from the partial breakdown of nucleic acids to be reutilized in nucleic acid synthesis.

Satellite DNA. A DNA fraction whose base composition differs from that of the main component of DNA, as revealed by the fact that it bands at a different density in a CsCl gradient. Usually repetitive DNA or organelle DNA.

Second messenger. A diffusible small molecule, such as cAMP, that is formed at the inner surface of the plasma membrane in response to a hormonal signal.

Secondary structure. In a protein or a nucleic acid, any repetitive folded pattern that results from the interaction of the corresponding polymeric chains.

Semiconservative replication. Duplication of DNA in which the daughter duplex carries one old strand and one new strand.

Sigma factor. A subunit of RNA polymerase that recognizes specific sites on DNA for initiation of RNA synthesis.

Single-copy DNA. A region of the genome whose sequence is present only once per haploid complement.

Somatic cell. Any cell of an organism that cannot contribute its genes to a subsequent generation.

SOS system. A set of DNA repair enzymes and regulatory proteins that regulate their synthesis so that maximum synthesis occurs when the DNA is damaged.

Southern blotting. A method for detecting a specific DNA restriction fragment, developed by Edward Southern. DNA from a gel electrophoresis pattern is blotted onto nitrocellulose paper; then the DNA is denatured and fixed on the paper. Subsequently the pattern of specific sequences in the Southern blot can be determined by hybridization to a suitable probe and autoradiography. A Northern blot is similar, except that RNA is blotted instead onto the nitrocellulose paper.

Splicing. *See* RNA splicing.

Sporulation. Formation from vegetative cells of metabolically inactive cells that can resist extreme environmental conditions.

Stacking energy. The energy of interaction that favors the face-to-face packing of purine and pyrimidine base pairs.

Steady state. In enzyme-kinetic analysis, the time interval when the rate of reaction is approximately constant with time. The term is also used to describe the state of a living cell where the concentrations of many molecules are approximately constant because of a balancing between their rate of synthesis and breakdown.

Stem cell. A cell from which other cells stem or arise by differentiation.

Stereoisomers. Isomers that are nonsuperimposable mirror images of each other.

Steroids. Compounds that are derivatives of a tetracyclic structure composed of a cyclopentane ring fused to a substituted phenanthrene nucleus.

Structural domain. An element of protein tertiary structure that recurs in many structures.

Structural gene. A gene encoding the amino acid sequence of a polypeptide chain.

Structural protein. A protein that serves a structural function.

Subunit. Individual polypeptide chains in a protein.

Supercoiled DNA. Supertwisted, covalently closed duplex DNA.

Suppressor gene. A gene that can reverse the phenotype of a mutation in another gene.

Suppressor mutation. A mutation that restores a function lost by an initial mutation and that is located at a site different from the initial mutation.

Svedberg unit (S). The unit used to express the sedimentation constant S: $1\ S = 10^{-13}$ sec. The sedimentation constant S is proportional to the rate of sedimentation of a molecule in a given centrifugal field and is related to the size and shape of the molecule.

Synapse. The chemical connection for communication between two nerve cells or between a nerve cell and a target cell such as a muscle cell.

Synapsis. The pairing of homologous chromosomes, seen during the first meiotic prophase.

Tandem duplication. A duplication in which the repeated regions are immediately adjacent to one another.

TCA cycle. See tricarboxylic acid cycle.

Template. A polynucleotide chain that serves as a surface for the absorption of monomers of a growing polymer and thereby dictates the sequence of the monomers in the growing chain.

Termination factors. Proteins that are exclusively involved in the termination reactions of protein synthesis on the ribosome.

Terpenes. A diverse group of lipids made from isoprene precursors.

Tertiary structure. In a protein or nucleic acid, the final folded form of the polymer chain.

Tetramer. Structure resulting from the association of four subunits.

Thioester. An ester of a carboxylic acid with a thiol or mercaptan.

Thymidine. One of the four nucleosides found in DNA.

Thymine. A pyrimidine base found in DNA.

Topoisomerase. An enzyme that changes the extent of supercoiling of a DNA duplex.

Transamination. Enzymatic transfer of an amino group from an α-amino acid to an α-keto acid.

Transcription. RNA synthesis that occurs on a DNA template.

Transduction. Genetic exchange in bacteria that is mediated via phage.

Transfection. An artificial process of infecting cells with naked viral DNA.

Transfer RNA (tRNA). Any of a family of low-molecular-weight RNAs that transfer amino acids from the cytoplasm to the template for protein synthesis on the ribosome.

Transferase. An enzyme that catalyzes the transfer of a molecular group from one molecule to another.

Transformation. Genetic exchange in bacteria that is mediated via purified DNA. In somatic cell genetics the term is also used to indicate the conversion of a normal cell to one that grows like a cancer cell.

Transgenic. Describing an organism that contains transfected DNA in the germ line.

Transition state. The activated state in which a molecule is best suited to undergoing a chemical reaction.

Translation.   The process of reading a messenger RNA sequence for the specified amino acid sequence it contains.

Transport protein.   A protein whose primary function is to transport a substance from one part of the cell to another, from one cell to another, or from one tissue to another.

Tricarboxylic acid (TCA) cycle.   The cyclical process whereby acetate is completely oxidized to CO_2 and water, and electrons are transferred to NAD^+ and flavine. The TCA cycle is localized to the mitochondria in eukaryotic cells and to the plasma membrane in prokaryotic cells. Also called the Krebs cycle.

Trypsin.   A proteolytic enzyme that cleaves peptide chains next to the basic amino acids arginine and lysine.

Tryptic peptide mapping.   The technique of generating a chromatographic profile characteristic of the fragments resulting from trypsin enzyme cleavage of the protein.

Tumorigenesis.   The mechanism of tumor formation.

Turnover number.   The maximum number of molecules of substrate that can be converted to product per active site per unit time.

Ultracentrifuge.   A high-speed centrifuge that can attain speeds up to 60,000 rpm and centrifugal fields of 500,000 times gravity. Useful for characterizing and/or separating macromolecules.

Unidirectional replication.   See bidirectional replication.

Unwinding proteins.   Proteins that help to unwind double-stranded DNA during DNA replication.

Urea cycle.   A metabolic pathway in the liver that leads to the synthesis of urea from amino groups and CO_2. The function of the pathway is to convert the ammonia resulting from catabolism to a nontoxic form, which is subsequently secreted.

UV irradiation.   Electromagnetic radiation with a wavelength shorter than that of visible light (200–390 nm). Causes damage to DNA (mainly pyrimidine dimers).

van der Waals forces.   Refers to two types of interactions, one attractive and one repulsive. The attractive forces are due to favorable interactions among the induced instantaneous dipole moments that arise from fluctuations in the electron charge densities of neighboring nonbonded atoms. Repulsive forces arise when noncovalently bonded atoms or when molecules come too close together.

Viroids.   Pathogenic agents, mostly of plants, that consist of short (usually circular) RNA molecules.

Virus.   A nucleic-acid-protein complex that can infect and replicate inside a specific host cell to make more virus particles.

Vitamin.   A trace organic substance required in the diet of some species. Many vitamins are precursors of coenzymes.

Watson-Crick base pairs.   The type of hydrogen-bonded base pairs found in DNA, or comparable base pairs found in RNA. The base pairs are A-T, G-C, and A-U.

Wild-type gene.   The form of a gene (allele) normally found in nature.

Wobble.   A proposed explanation for base pairing that is not of the Watson-Crick type and that often occurs between the 3' base in the codon and the 5' base in the anticodon.

X-ray crystallography.   A technique for determining the structure of molecules from the X-ray diffraction patterns that are produced by crystalline arrays of the molecules.

Ylid.   A compound in which adjacent, covalently bonded atoms, both having an electronic octet, have opposite charges.

Z form.   A duplex DNA structure in which there is the usual type of hydrogen bonding between the base pairs but in which the helix formed by the two polynucleotide chains is left-handed rather than right-handed.

Zwitterion.   A dipolar ion with spatially separated positive and negative charges. For example, most amino acids are zwitterions, having a positive charge on the α amino group and a negative charge on the α carboxyl group but no net charge on the overall molecule.

Zygote.   A cell that results from the union of haploid male and female sex cells. Zygotes are diploid.

Zymogen.   An inactive precursor of an enzyme. For example, trypsin exists in the inactive form trypsinogen before it is converted to its active form, trypsin.

ANSWERS TO
SELECTED PROBLEMS

(Please see accompanying Student's Solution Guide for fully worked solutions to all problems)

Chapter 1

1. a. The pK_a's are approximately 2, 4, and 10.
   b. Aspartic acid or glutamic acid.
   c. A: $\alpha - NH_3^+$, $\alpha - COOH$, $\gamma - COOH$
      B: $\alpha - NH_3^+$, $\alpha - COO^-$, $\gamma - COOH$
      C: $\alpha - NH_3^+$, $\alpha - COO^-$, $\gamma - COO^-$
      D: $\alpha - NH_2$, $\alpha - COO^-$, $\gamma - COO^-$
      pI = 2.95.
   d. 0.2 moles.
4. pH = 7: Asp^-, Glu^-, Lys^+, Arg^+, His^+.
   pH = 2: Arg^+, Lys^+, His^+.
   pH = 13: Glu^-, Asp^-, Cys^-, Tyr^-.
6. Donors: Arg, Lys, Trp.
   Donors/acceptors: Ser, Cys, Tyr, Asn, Gln, Asp, Glu, His, and Thr.

Chapter 2

1. Hydrophobic interactions.
2. Primary, secondary, tertiary, quaternary.
4. Subtilisin.
5. b, d, e, and f.
8. Decreased entropy of surrounding medium.
9. Two A subunits and two B subunits.
10. Steric constraints due to R groups near peptide bond.

Chapter 3

1. $(NH_4)_2SO_4$ step gives best yield, but DEAE step gives best purification.
3. M_r = 37,450 daltons.
5. The presence of Ca^{+2} causes hexamer formation.
7. a. Proteins 1, 3, 4, 2.
   b. Proteins 2, 3, 1, 2, 3, 4. (but see Student's Solution Guide).
9. It has 2 different polypeptide chains linked by disulfide bonds.

Chapter 4

1. c. 16 for straight chain, 32 for pyranose ring form.
2. 36.4% α-D-glucose and 63.6% β-D-glucose.
3. a. 13%.
   b. 7.7.
   c. 22.9 mmol.
   d. 17,283.
4. a. Cellulose is used for structure and glycogen for energy storage.
   b. Cellulose has $\beta(1,4)$ linkages; glycogen has α (1,4) linkages.
   c. The linkages are catalyzed by separate enzymes.
5. Only growing cells are in the process of making a cell wall.

Chapter 5

3. PC variants differ in the fatty acyl substituents at positions $sn = 1$ and $sn = 2$ of the glycerol backbone.
5. Both classes of molecule are amphipathic, containing polar head group and hydrophobic portions.
7. Cholesterol: nonpolar, integral membrane component.
   Bile acids: amphipathic, lipid-solubilizing agents.
   Steroid hormones: nonpolar, intercellular chemical messengers (cytoplasmic site of action).

Chapter 6

2. a. More mobility.
   b. More mobility; decrease in T_m.
   c. More mobility; decrease in T_m.
   d. Less mobility; phase transition abolished.
3. a. Bulkiness of polar head group.
   b. PC has a bulkier head group than PE.
5. a. Sodium deoxycholate.
   b. Triton x-100; dialysis or gel filtration.
   c. Sodium deoxycholate.
7. a. Partitioning of PE between the inner and outer leaflet.
   b. 37.5% of newly synthesized PE in outer leaflet.

Chapter 7

2. T4 gene 32-encoded protein: lowers T_m, binds to single-stranded DNA.
   NaCl: raises T_m, forms salt bridges between phosphate groups.
   Formamide: lowers T_m, disrupts hydrogen bonding between strands.
   Alkali: lowers T_m, disrupts hydrogen bonding.
4. a. Salt bridges reduce repulsion between phosphate groups of the phosphodiester backbone, which are greater in Z than in B form DNA.
   b. Sequences within plasmid A undergo a transition from B to Z form, resulting in a reduction in the number of right-handed supercoils in this plasmid.
   c. Plasmid A must contain an uninterrupted stretch of alternating purines and pyrimidines, while plasmid B probably lacks such a sequence.
8. A is an uninterrupted RNA: DNA heteroduplex; B is the DNA strand of A; C is the RNA strand of A.
11. Hydroxyapatite chromatography overestimates the extent of renaturation because of the retention of partially duplexed molecules. Hypochromic shift and S1 nuclease digestion both provide an accurate measure of renaturation.

Chapter 8

7. When $[S] = 10^{-2}$ M or 10^{-3} M, v_i is unchanged.
   When $[S] = 10^{-6}$ M, $v_i = 5 \times 10^{-3}$ mol/min.
8. When $[S] = 0.1$ M, $[I] = 0.09$ M.
   When $[S] = 0.01$ M, $[I] = 0.009$ M.

9. a. $K_m = 1 \times 10^{-4}$ M.
  b. $V_{max} = 10^{-8}$ M·sec^{-1}
  c. $k_3 = 10^2$ sec^{-1}
  d. $v_i = 1.67 \times 10^{-9}$ M·sec
10. a. Competitive inhibitor contaminant.
  c. $K_m = 2 \times 10^{-5}$ M.

Chapter 9

2. pH = 6.0.
4. Mass action.
5. Posttranslational processing.
7. a. Geometric strain.
  b. Transition state stabilization.
  c. Binding energy of substrate.
  d. Dielectric constant surrounding Zn^{2+} is decreased.

Chapter 10

5. Amino acid substitutions at interfaces between subunits will effect cooperativity.
6. Separated Hb subunits would bind more O_2. DPG would have no effect.
7. a. Sensitive control of metabolic reactions.
  b. CTP inhibits ATCase by augmenting cooperativity when pyrimidine synthesis is not necessary.
  c. When ATP is high, pyrimidine synthesis is favored due to high energy charge (see Chapter 13).

Chapter 11

1. a. Step 1. Formation of Schiff base.
    Step 2. Formation of quinonoid intermediate.
    Step 3. H^+ addition to cause racemization.
  b. α decarboxylation.
  c. β aspartate decarboxylation.
3. Each can bind an acyl group in stable form and transport it through space.
6. The esterification with pantetheine coenzymes eliminates the carboxylate anion and lowers the barrier to enolization.
8. a. B_{12}.
  c. FAD.
  e. Biotin.

Chapter 12

2. Extensive: volume, mass, and energy. Intensive: density, temperature, and pressure.
5. $dH = d(E + PV) = dE + P\,dV + V\,dP$. V and P are constant in most biochemical reactions; therefore $dH = dE$.
7. ΔH and ΔS for solvation are negative and $\Delta G = \Delta H - T \Delta S$.
12. Mary.
13. 9.534×10^3 cal/mol.
15. Formation of citrate from oxaloacetate and acetyl coA ($\Delta G^{\circ\prime} = -7.9$ kcal/mol) pulls formation of oxaloacetate.
17. a. -1398 kcal/mol.
  b. 40 %.
20. a. pH = 3.5.
  b. pH = 4.6.

Chapter 13

1. Both are equally fundamental.
2. a. False.
  b. False.
  c. True.
3. a. Sensitive control of metabolic reactions.
  b. CTP inhibits ATCase by augmenting cooperativity when pyrimidine synthesis is not necessary.

c. When ATP is high, pyrimidine synthesis is favored due to high energy charge.

Chapter 14

1. a. False.
  b. True.
  c. False.
  d. False.
  e. True.
3. a. 2 mmol.
  b. iv.
  c. 40 mmol.
5. No ATP could be formed, so the organism could not survive.
7. 50 nM glucose.
8. There would be exchange only if bisphosphate were added or if the enzyme were phosphorylated. If these conditions are met, we would expect to find labeled and unlabeled glucose-1-phosphate molecules in a ratio of 19:1.
10. a.

Concentration of fructose-1, 6-bisphosphate	Concentration of glyceraldehyde-3-phosphate and dihydroxyacetone phosphate
1 M	10^{-2}
100 mM	$\sqrt{10^{-5}}$
10 mM	10^{-3}
1 mM	$\sqrt{10^{-7}}$
0.1 mM	10^{-4}

  b. The concentration ratio between the two phosphates and the biphosphate would remain the same, but the concentration of dihydroxyacetone phosphate would rise to 22 times that of glyceraldehyde-3-phosphate.
13. Because there are three essentially irreversible steps in glycolysis.
14. There is a net difference of 8 high-energy phosphate bonds.

Chapter 15

1. a. F
  c. E
  e. E
2. a. 3 mmol of CO_2.
  b. 9 mmol of CO_2.
  c. Any one of the intermediates; 9 mmol of CO_2.
  d. 9 mmol of CO_2 minus the number of mmoles of pyruvate used to maintain cycle intermediates.
4. a. C_3 and/or C_4.
  b. C_2 and/or C_5.
  c. C_1 and/or C_6.
9. a. B→D→C→A.
  b. Hydroxypyruvate + NADH + H^+→pyruvate + NAD^+ + H_2O.
11. a. AMP.
  c. glucose-6-phosphate ($-$).
  e. ATP, acetyl-CoA ($-$), fructose-1, 6-diphosphate ($+$).
  g. NADH, succinyl-CoA ($-$).
  i. ATP, NADH, succinyl-CoA ($-$).

Chapter 16

2. a. -2.54 kcal/mole.
  b. 12.9 kcal/mole.
  For cytochrome c, no change since there is no proton exchange. For ubiquinone, the values would be 0.17 V at pH 6.0 and 0.05 V at pH 8.0.
5. Electron transfer complexes are not connected in fixed linear chains.

8. a. Energy utilization is much more efficient; 94% less glucose would be used.

b. DNP eliminates this added efficiency by uncoupling.

10. a. Measure rate of appearance of peak at 550 nm in a difference spectrum.

b. $FADH_2$ (Complex II) $\xrightarrow[]{2e^- \ (Fe^{3+} \longrightarrow Fe^{2+})} 2$ cytochrome $- c \xrightarrow[]{2e^-} 1$ ATP (Complex III)

Result: 1 mole of ATP to 2 moles of cytochrome c.

c. CTMPH would eliminate ATP production; antimycin would reduce the rate of ATP production; oligomycin would eliminate ATP production.

d. 1 mole of ATP per mole of cytochrome c.

Chapter 17

1. a. 33.0 kcal/einstein.

b. -0.97 V.

3. b. Cytochrome c is required to reduce $P870^+$ after each flash.

4. b. Slope of the curve at low flash intensity.

c. Asymptote of the curve at high flash intensity.

5. Excite with far-red light or treat with DCMU.

6. a. To prevent A_2^- oxidation.

b. -0.54 V to -0.59 V or more negative.

Chapter 18

2. a. 2.5 kcal/g glucose.

b. 6.0 kcal/g palmitate.

c. Fat is 2.4-fold better as a source of energy than carbohydrate on a per gram basis.

d. 12 kg.

e. Fat is more highly reduced than carbohydrate.

5. a. The student would accumulate L-methylmalonyl-CoA which, as an inhibitor of acetyl-CoA carboxylase, would impair his ability to synthesize fatty acids.

9. HMG-CoA synthase is present only in the liver, while β-oxoacid-CoA transferase is present only in extrahepatic tissues.

Chapter 19

1. Transamination would be impaired because pyridoxal phosphate is needed as a cofactor of transaminases, and as a result amino acids that are not substrates for oxidative deamination would accumulate.

5. The glucose-alanine cycle transports the nitrogen removed from amino acids in skeletal muscle to the liver, in exchange for the transfer of glucose from the liver to muscle.

6. The carbon skeletons of glycogenic amino acids feed the TCA cycle either directly or via pyruvate, while ketogenic amino acids are degraded to acetoacetyl-CoA. The glycogenic amino acids are Ala, Gly, Thr, Ser, Cys, Asp, Asn, Val, Met, Ile, Glu, Gln, His, Arg, Pro, (Phe), (Tyr). The ketogenic amino acids are Trp, Lys, Leu, (Phe), (Tyr).

11. a. Some defects in propionyl-CoA carboxylase or β-methyl-crotonyl-CoA carboxylase activity may be compensated for by the increased concentration of their cofactor, biotin.

b. The same applies for methylamalonyl-CoA mutase, whose cofactor is the vitamin B_{12}-derivative cobalamin.

Chapter 20

1. Hexose conversion, activation, and polymerization.

2. a. UDP-GlcNAc

b. P-P-lipid and UDP-MurNAc-pentapeptide

c. Lipid-P-P-MurNAc-pentapeptide-(Gly)$_5$-GlcNAc

4. Such mutants would be lethal.

6. a. Specific enzymes.

b. Oligosaccharides.

c. Two enzymes involved in hyaluronic acid synthesis are described in the text.

Chapter 21

1. Such a defect would hinder the transport of acetyl-CoA from the mitochondrial matrix into the cytosol, and thus prevent its use in fatty acid biosynthesis.

4. The requirements would be the same in each case: 7 ATP, 14 NADPH, and 8 acetyl-CoA.

6. Either the amount of acetyl-CoA carboxylase or the activity of acetyl-CoA carboxylase or fatty acid synthase might be responsible for the defect.

9. In E. coli, double bonds are introduced during fatty acid synthesis without the requirement of O_2, whereas in the liver, double bonds are introduced after synthesis of the fatty acid chain, in reactions requiring O_2.

Chapter 22

1. In the liver, the major pathway used in the synthesis of phosphatidylethanolamine (PE) involves the condensation of CDP-ethanolamine with diacylglycerol, whereas in E. coli, PE is produced by the reaction of CDP-diacylglycerol and serine.

6. Yes. Sphingosine may first be degraded to phosphoethanolamine, then converted to phosphatidylethanolamine and finally used to generate phosphatidylcholine, all the while retaining the ^{14}C label at the C-1 position of the phosphate substituent.

9. The defect probably lies in the activity of the activator protein required to solubilize Glc(β1-1) ceramide.

10. Yes. PGE_2, and prostaglandins generally, may be synthesized by desaturation and elongation of linoleic acid.

Chapter 23

3. Mevinolinic acid lowers plasma cholesterol. It is a competitive inhibitor of HMG-CoA reductase, the enzyme that catalyzes the rate-limiting step in cholesterol biosynthesis.

7. A defect in apoprotein C-II would impair the degradation of chylomicrons and VLDLs catalyzed by lipoprotein lipase, and result in the accumulation of these lipoprotein particles in the bloodstream.

10. Bile acids and steroid hormones are both produced by the degradation of cholesterol, and the synthesis of both classes of compound involves hydroxylation reactions catalyzed by mixed function oxidases.

Chapter 24

1. In most organisms, the synthesis of every amino acid, including glutamate, can be regarded as beginning with the amination of glutamate, catalyzed by glutamine synthase, in which NH_4^+ is converted to organic nitrogen.

2. a. Glycine.

b. The reaction catalyzed by serine hydroxymethyltransferase generates 1-C units that may be used for methionine biosynthesis.

5. The strain probably contains only one form of the enzyme acetohydroxy acid synthase that is sensitive to feedback inhibition by valine.

9. a. Transamination reactions involve the exchange of an amine group of an amino acid for the carbonyl group of a α-keto acid. In proline biosynthesis the carbonyl and amine groups are present on the same molecule; L-glutamate-β-semialdehyde.

b. Histidine.

Chapter 25

3. The synthesis of 1 molecule of GTP requires the hydrolysis of 8 high-energy phosphate bonds, while the synthesis of the same amount of CTP requires hydrolysis of only 5 such bonds.

5. a. Yes. Thymidine kinase catalyzes the conversion of thymidine to dTMP in the salvage pathway for TTP synthesis; its inhibition would retard the rapid growth characteristic of neoplastic cells. The existence of the alternative route to dTMP involving methylation of dUMP would lessen its overall cytotoxic effect.

   b. Yes, and for the same reason. In this case, the alternative route is the conversion of CDP to dCDP catalyzed by ribonucleotide reductase.

7. a. The analog would be converted to one of dATP, which would inhibit ribonucleotide reductase and result in the depletion of dNTP pools.

   b. Demonstrate the reduction in dNTP levels.

   c. The inhibitor's effect on the cytotoxicity of deoxyadenosine is due to the accumulation of dATP and the resultant inhibition of ribonucleotide reductase. The lack of effect on toxicity of the analog indicates that the latter is not degraded by a pathway involving adenosine deaminase.

Chapter 26

1. a. Gene c.

   b. Bidirectional: the number of gene copies declines gradually to 1 on both sides of the origin.

   c. The overall shape of the plot is the same, but the gradient is steeper, from 4 copies (gene c) to 1 copy (gene g).

4. a. An increased error/mutation rate, due to the lack of a proofreading step during synthesis.

   b. An increase in the level of 5' to 3' exonuclease activity of DNA polymerase I, resulting in the substitution of its proofreading function for that of polymerase III.

9. a. The need to replicate each chromosome to its end.

   b. Yes. Only one strand of each telomere would be synthesized by the activity described. Synthesis of the complementary strand would require the activity of DNA polymerase α, which is inhibited by aphidicolin.

10. a. S1 nuclease degrades preferentially single-stranded DNA, while DNAase II degrades specifically double-stranded DNA.

    b. Exonucleases, which require free 5' or 3' ends, will degrade linear but not circular single-stranded DNA.

    c. Treatment with exonuclease III, followed by DNA polymerase I in the presence of ^{32}P-labeled dNTPs will label linear but not circular double-stranded DNA.

Chapter 27

4. In E. coli, mRNA is degraded rapidly, whereas rRNA and tRNA are turned over very slowly.

5. a. The importance of spacing implies that RNA polymerase interacts with -35 and -10 regions simultaneously.

   b. Experiments in which regions of the template protected by RNA polymerase were mapped using dimethylsulfate and ethylnitrosourea.

8. Cordycepin-5'-triphosphate would be incorporated into the growing chain only if chain growth were in the 5'→3' direction, and once incorporated would result in termination of transcription because it lacks the 3'-OH necessary for further addition to the chain.

9. Transcription termination in E. coli requires melting of the RNA:DNA heteroduplex immediately preceding the transcription complex. The helicase activity of rho provides for this melting in rho-mediated termination, whereas in factor-independent termination, melting is facilitated by the U-rich composition of the RNA strand of the RNA:DNA duplex.

12. The mRNAs encoding the two polypeptides are spliced from the same pre-mRNA but using different splice sites, their common amino terminals deriving from a segment of the pre-mRNA retained in both spliced mRNAs. The larger mRNA codes for the shorter polypeptide because it contains a translation termination codon spliced out of the shorter mRNA.

Chapter 28

1. In eukaryotes, translation begins at the AUG triplet that is nearest to the 5' end, and translation efficiency is conditioned by the sequence context of the initiation codon. In prokaryotes, initiation requires both an AUG triplet and an upstream Shine-Dalgarno sequence, and efficiency is determined primarily by the sequence and spacing of the upstream element. In prokaryotes, coordinating the expression of a set of genes is generally accomplished by the synthesis of polycistronic mRNAs. In eukaryotes, other strategies have had to be devised to achieve the same coordination.

3. None. After viomycin treatment, peptidyl tRNA is positioned in the A site. Consequently, puromycin cannot bind to the ribosome to elicit release of the nascent polypeptide.

7. The genetic code is nonoverlapping. Translation involves the sequential reading of adjacent triplets in frame. Mutations involving additions or deletions that change the frame may therefore be compensated for by distal mutations restoring the frame.

10. a. N terminal: alanine; C terminal: cysteine.

11. a. Gln (CAG); Arg (CGG); Trp (UGG)

    b. Glu (GAA); Lys (AAA); Ile (AUA)

    c. Leu (UUG); Ser (UCG); Val (GUG); Met (AUG)

    d. Thr (ACA); Ile (AUA); Pro (CCA); Lys (AAA)

15. a. The protein is degraded because of its abnormal structure.

    b. Make the protein in a lon^- strain.

Chapter 29

1. b. This result indicated that the i gene belongs to a different cistron than z, y, and a, and that its effect is mediated through the action of a diffusible factor.

6. a. Yes.

   b. crp^+: permanent catabolite repression even in the absence of glucose, inducible by lactose.
   crp^+ and crp^+/crp^-: normal catabolite repression, inducible by lactose.

10. a. Binding of tryptophan to the aporepressor changes its conformation in such a way that part of the molecule possesses a shape characteristic of a DNA-binding protein.

    b. No. Allolactose is an inducer. Its binding to the lac repressor would be more likely therefore to have the reverse effect, changing its conformation in a way that would prevent its binding to the lac operator.

12. The his operon in E. coli is not subject to control by a repressor-operator system. The mutation most likely affects attenuation.

14. Bacteriophage gene expression is regulated according to a set program, and is in this sense committed. Bacterial gene expression is subject to continually changing demands dictated by the environment, and thus requires considerably greater regulatory flexibility.

Chapter 30

2. a. The mutant GAL4 protein is unable to complex with its negative regulator, the product of the GAL80 gene.

   b. Yes. Glucose inhibits GAL4-induced expression regardless of the existence of a functional GAL80 allele.

7. The nuclei of cells of the early embryo are totipotent, in

the sense that they are not committed to a program of gene expression characteristic of a particular cell type.

10. Active chromatin is more susceptible to DNAase I digestion, is associated with modified histones and high mobility group (HMG) proteins, and is under-methylated when compared with inactive chromatin.

13. Two possibilities are generally considered. The mosaic model invokes a nonrandom distribution of qualitatively different regulatory molecules in the zygote. The gradient model hypothesizes a molecular gradient in which quantitative differences are the determining factor.

15. DNA rearrangement, which occurs during development of the B cell, brings the promoter for the heavy-chain variable region within proximity of the enhancer.

Chapter 31

3. Defects in 21-hydroxylase, 5α-reductase, aromatase, or androgen receptors.

4. Speculation: Females would probably be normal, but males would probably be pseudohermaphrodites.

7. Reasons include signal sequence, peptide chain folding, control of hormone activation, coordinate production of multiple hormones, and storage.

8. Primary.

10. The required rate of synthesis is $S = 29$ molecules/cell per hour. Transcription is initiated once every 2 minutes. The steady-state concentration is 8.6×10^5 molecules per β cell. A serum concentration of $10^{-10}M$ would require 3.5×10^8 β cells.

12. Mechanisms postulated are DNA methylation, altered composition of nuclear proteins in chromatin, and, nucleosome position with respect to gene.

Chapter 32

3. Digest the viral DNA with *Eco*R1, then either isolate the desired DNA fragment by gel electrophoresis and religate using DNA ligase, or simply religate the mixture of fragments after diluting severalfold.

5. Make a DNA library consisting of random fragments of *E. coli* DNA using a plasmid vector, then screen the hybrid plasmids for the ability to complement a conditionally lethal mutation in the gene for DNA polymerase I at the nonpermissive temperature.

6. 2×10^5 is the minimum: between 6×10^5 and 2×10^6 is preferred.

12. Prepare cDNA using poly-A$^+$ RNA isolated from liver, and pass through a column to which poly-A$^+$ RNA from other tissues has been coupled. Clone the cDNAs that do not bind to the column.

13. The analysis generally involves the construction of hybrid genes, in which the regulatory region under investigation is fused to the body of a gene whose product may be easily assayed.

Chapter 33

1. Essential functions are self-duplication and mutation. Some inorganic systems (crystals or clays) could possibly perform these processes.

3. Aerobic metabolism evolved long after the origins of life, so HCN could not have been toxic to the first life forms.

4. The coenzymes are very similar in structure to the nucleotides that make RNA.

Chapter 34

1. b. $D = 5.7 \times 10^{-6}$ cm^2/sec.

2. a. $\Delta\Psi = +18$ mV.
   b. From side 1 to side 2.
   c. Side 1: 64.3 mM K$^+$, 50 mM Na$^+$, 144.3 mM Cl$^-$.
      Side 2: 85.7 mM K$^+$, 85.7 mM Cl$^-$.
      $\Delta\Psi = 7.5$ mV.

8. a. Stimulation.
   b. Inhibition.
   c. No effect.
   d. Stimulation.

Chapter 35

1. a. $\Delta\Psi = -71$ mV.
   b. $\Delta\Psi = -63$ mV.

3. Criteria are (i) excitation of proper postsynaptic cell, (ii) presence of the compound in presynaptic terminal, (iii) release upon stimulation, and (iv) receptors on postsynaptic membrane.

6. Step 1: Collision triggers opening of Ca^{2+} channels, causing local depolarization.
   Step 2: Nearby voltage-sensitive Ca^{2+} channels open and "action potential" spreads over entire cell.
   Step 3: Cell swims backwards.
   Step 4: Resting potential is regained by action of Ca^{2+} pump and K$^+$ channels.

Chapter 36

2. Rhodopsin $\xrightarrow{500 \text{ nm}}$ bathorhodopsin $\xrightarrow{550 \text{ nm}}$ rhodopsin.

3. The Schiff's base linkage between the retinal and opsin is protonated in metarhodopsin I but not in metarhodopsin II.

INDEX

Page numbers in *italics* refer to illustrations.

A band of sarcomere, 89, *89*
A form of DNA, 228, *229*, 1228
a genes, 977
Abietic acid, 172
    structure of, *173*
ABO blood group, 663
Abscisic acid, *1083*, 1084
Absorption spectra in different states, 1231
Acceptor protein, 1058
Acetals, 134, 1228
    formation of, 15
Acetate
    activated, in metabolic pathways, 418, *420*
    incorporation of carbon units into valine and
        isoleucine, *778*
Acetoacetate
    formation from tyrosine, 627
    synthesis in liver, 610, *611*
Acetoacetyl-ACP, formation of, *678*
Acetoacetyl-CoA, 366–367, 610
    from amino acid catabolism, 625–631
Acetoacetyl-CoA thiolase, 610
Acetogenins, 174
Acetohydroxy acid isomeroreductase, 776
Acetone synthesis in liver, 610, *611*
Acetylcarnitine, 602
Acetylcholine, 797, 1203–1205, *1204*
    and adrenal medulla stimulation, 1053
    receptor for, 1204, 1208–1210
        inhibitors of, 1205
Acetylcholinesterase, 266, 1204–1205
    inhibitors of, 1205
Acetyl-CoA, 366–367, 1228
    from amino acid catabolism
        by way of acetoacetyl-CoA, 625–631
        by way of pyruvate, 623–625
    binding of, 677
    carboxylation of, *676*
    condensation to citric acid formation, 407, *407*
    in fatty acid biosynthesis, 670–673, *671*
    in glyoxylate cycle, 486–501
    production by citrate cleavage, 505–506
    and pyruvate partitioning, 508–509
    pyruvate as source of, 485–488, *487*, 497
    in tricarboxylic acid cycle, 484
Acetyl-CoA-ACP transacylase, 673
Acetyl-CoA:L-alkyl-2-lysoglycerophosphocoline
    transferase, 703–704
Acetyl-CoA carboxylase, 266, 674–677
    activation of, *677*, 685
    inhibition of, 685
    regulation of, 683–686
N-Acetylgalactosamine
    linkage with serine, *147*
    structure of, *139*
N-Acetylglucosamine
    conformation of, *311*
    linkage with asparagine, *147*
    structure of, *139*
Acetyl-hydrolase, 704
N-Acetylmuramic acid
    conformation of, *311*
    structure of, *139*

N-Acetylneuraminic acid, 164
    structure of, *139*, *165*
Acetylphenylalanine methyl ester, *294*
O-Acetylserine sulfhydrylase, 766
Acetylthiamine pyrophosphate, 349, *350*
Acetyl transacylase, 677
N-Acetyl-L-tryptophan amide, hydrolysis of,
    297, *298*, 297–298
Acid dissociation constant, 34
Acid hydrolases in lysosomes, 969
Acid hydrolysis, 40
Acivicin, 826, 828
Aconitase, 489–490
ACP. *See* Acyl carrier protein
Acromegaly, 1078
ACTH. *See* Corticotropin
Actin, 89–90, 201, 1228
Actinomycin D, 1228
    amino acids in, 793
    inhibiting RNA synthesis, 924
Action potentials, 1195–1198
Activated complex, 1228
Acyclovir, 828
Acyladenylate, in bioluminescent reaction, *594*
Acyl carnitine, structure of, *602*
Acyl carrier protein (ACP), 366–369, 673
    amino acid sequence of, *674*
Acyl-CoA:cholesterol acyltransferase (ACAT),
    735, 738
Acyl-CoA dehydrogenases, 604
    reactions catalyzed by, 363
Acyl-CoA ligases, 602
Acyl-CoA ocidase, 611
Acyl-CoA reductase, *702*
1-Acylglycerol-3-phosphate, 696
1-Acylglycerol-3-phosphate acyltransferase, 690,
    695
Acyltransferases, mechanism of action, 691
Addison's disease, 1078
Adenine, 1228
    in DNA and RNA, 218, 223
    formation on primitive Earth, 1129
    hydrogen bonding to thymine, *223*
    structure of, *219*
Adenine nucleotide translocation, 557
Adenine phosphoribosyltransferase, 818
Adenohypophysis, hormones of, 1047
Adenosine, 1228
    base found in, 219
    in tRNA, *253*
    uncharged and protonated forms of, *220*
Adenosine diphosphate (ADP), 23, 1228
    and electron-transport rate, 539–540
    mitochondrial ATP/ADP exchanger, 556–557,
        1174–1175, 1186
    phosphorylation of, 423–424
    ratio to ATP, 432–433
    structure of, *23*
Adenosine kinase, 819
Adenosine monophosphate (AMP)
    cyclic (cAMP), 1229
        coactivator, 976
        and eicosanoid actions, 722
        glucose affecting intracellular levels of, 984
        and protein kinase activation, 468, *468*

    as second messenger, 1062–1063
    structure of, *984*
    synthesis of, *468*, *1063*
    formation from inosine monophosphate, 811,
        812
    and fructose-bisphosphate phosphatase
        regulation, 472
    ionization constants of, 219
    molar absorbance of, *220*
Adenosine-5′-phosphate. *See* Adenosine
    monophosphate
Adenosine-5′-phosphosulfate (APS), 767
Adenosine triphosphatase. *See* ATPase
Adenosine triphosphate (ATP), 23, 1228
    biosynthesis of, 25
        in bacteria, 559–560
        in catabolic pathways, 418, *420*
        and electron-transport system in photosyn-
            thetic bacteria, 576, *577*
        in glycolysis, 449–452
        in photosynthesis, 587
        proton movements affecting, 549–551
    consumption in anabolic pathways, 418, *420*
    consumption in gluconeogenesis, 453–455
    exchange for ADP in mitochondria, 556–557,
        1174–1175, 1186
        inhibitors of, 557
    hydrolysis of, 408–411
        alternative routes of, *408*
        equilibrium constant for, factors affecting,
            410–411
    in phosphoenolpyruvate production, 455–457
    in protein degradation, 966–968
    ratio to ADP, 432–433
    structure of, *23*
    yields of
        in glucose catabolism, 449–450
        in glucose oxidation, 558–559
        in glyoxylate cycle, 501
        in tricarboxylic acid cycle, 495–496
S-Adenosylhomocysteinase, 359, *360*
S-Adenosylhomocysteine hydrolase, 773
S-Adenosylmethionine (SAM), 638–639, 773
    formation of, *772*
    methylation by, 698
S-Adenosylmethionine synthase, 773
Adenylate aminohydrolase, 834
Adenylate cyclase, 1228
    activation by hormones, 1062
    inhibition by hormones, 1062, 1064–1065
    pathway of hormone receptor action, *1062*,
        1062–1066
Adenylate kinase, 821
Adenylic acid, base found in, 219
Adenylosuccinate lyase, 811
Adenylosuccinate synthase, inhibitors of, 823
δ-Adenylyl-α-aminoadipate reductase, 761
Adipocyte, 159, 600, *600*, 1228
Adipose tissue
    fatty acid transfer from, 600–602
    thermogenesis in brown fat, 555
ADP. *See* Adenosine diphosphate
ADP-ribosylation, 1065, *1065*
Adrenal hormones, 1049
Adrenalin. *See* Epinephrine

Adrenergic neurons, 1206
Adrenergic receptors, 1060
Adrenocorticotropic hormone. *See* Corticotropin
Adrenodoxin, 383
Aerobes, 1228
    facultative, 1232
Aerobic catabolism, 481–482
Aerobic pathway for monounsaturated fatty
        acid biosynthesis, 681–682, *682*
Affinity chromatography, 122, 190, 1061, 1228
Affinity labeling of hormone receptors,
        1060–1061
Agglutinin, wheat germ, 656
Agonists
    of cholinergic systems, 1205
    hormone, 1059
Alanine
    biosynthesis of, 775, *775*
    catabolism of, 623
        in bacteria and fungi, 641
    cotransport with hydrogen, 1174
    pK values for, 36
    from pyrimidine catabolism, 831
    structure of, 33
    titration curve of, *35*
Alanine transaminase, 622
Albinism, 637
Albumin, 601–602
    physical constants of, 106
Alcohol, 1228
    dehydration reactions, 14–15
Alcohol dehydrogenase, 265, 451
    binding domains for, *317*
    nicotinamide side-specificities of, 361
Alcoholic fermentation, 451–452
Aldehyde, 15–16, 1228
Aldehyde dehydrogenase, 624
Aldehyde hydrate, 134, *135*
Aldimine formation, and pyridoxal phosphate
        activity, 353
Aldolase, 266
    and cleavage of fructose-1,6-bisphosphate,
        445–446, *446*
Aldoses, 131, 133
Aldosterone, 169
    biosynthesis of, *743*, *1055*
    function of, 1049
Algae, 565
Alimentary tract, hormones of, 1048
Alkaline phosphatase, action on nucleic acids,
        880
Alkaptonuria, 637, 639
Alkenyl ether phospholipids, 160, *160*
    biosynthesis of, 701
1-Alkyl-2-acetylglycerophosphocholine, 703,
        703, *704*
1-Alkyl-α-Acylglycerophosphoethenolamine
        desaturase, 702
Alkyl ether phospholipids, 160, *160*
    biosynthesis of, 701
Allantoin, 831
Alleles, 979, 1228
Allolactose, 978, 991
Allopurinol, structure of, *831*
p-Allose, structure of, *133*
Allosteric enzymes, 325, 336–343, 1228
Allosteric proteins, 325–336
Alpha helix (α helix), 71, 71–72, 84–85
    in keratins, 76
    in membranes, 193–194
    in trypsin family of enzymes, 288
Alpha oxidation of fatty acids (α oxidation),
        608–610
Alpha-sarcin, 959
p-Altrose, structure of, *133*
α-Amanitin, 924
Amide formation, 15
Amide plane, 38
Amines, 15
Amino acid(s), 9, 32–38
    acid and base properties of, 34–36
    acidic, 794
    amino terminal identification, 44, 45

analyzer of, 42, *42*
    in antibiotics, 792, 793
    aromatic, 794
    attachment to tRNA, 937–939
    in bacterial cell walls, 792
    basic, 794
    biosynthesis of, 26, *26*, 752–793
        in aromatic family, 779–785
            carbon flow regulation in, 785
        in aspartate family, 768–775
            and regulation of aspartokinases, 773–775
        branchpoints in, 756
        de novo pathway in, 751, *751*
        from electric discharges, *1127*, 1127–1128
        end-product inhibition in, 788–790, *789*
        enzyme activation in, 790–791, *791*
        in *Escherichia coli*, 755–756
        in glutamate family, 756–764
        histidine, 785–788
        by hydrolysis of hydrogen cyanide poly-
            mers, 1126, *1126*
        nitrogen sources in, 752–755
        in nonprotein amino acids, 792–793
        protein–protein interactions in, 791–792
        in pyruvate family, 775–779
        racemases in, 792–793
        regulation of, 788–792, 990–995
        salvage pathway in, 751, 752
        in serine family, 764–768
    branched-chain, 794
    in carbonaceous meteorites, 1128
    carboxy terminal identification, 45, 46
    catabolism of, 615–643
        to acetyl-CoA
            by way of acetoacetyl-CoA, 625–631
            by way of pyruvate, 623–625
        in bacteria and fungi, 641
        inborn errors of, 637, 639–640
        to α-ketoglutarate, 631–635
        to oxaloacetate, 639
        regulation of, 640–641
        to succinyl-CoA, 635–639
    charge–charge interactions between, 63
    chirality of, 36–37
        and conformation of proteins, 79–85
    composition of, quantitative analysis of, 40–42
    cotransport with sodium, 1174
    covalent modification in proteins, 959–960
    conversion to other amino acids and metabo-
        lites, 793–802
    cyclopropane, 794
    p form of, 37, *37*
    enzymes affecting, 22, *22*
    essential, 750
    families of, *757*
    glycogenic, 623, 1232
    heterocyclic, 794
    hydrophilicity of side chains, 56
    hydrophobicity of side chains, 56
    ketogenic, 623, 1234
    L form of, 37, *37*
    metabolism of, and pyridoxal-5′-phosphate
        activity, 352–356
    naturally occurring analogs of, 793
    as neurotransmitters, 1207
    nitrogen removal from, 616–618
        fate of, 618–622
        net deamination in, 617–618
        by transamination, 616, *617*
    nonessential, 750
    nonprotein, 792–793, 794
    and nutritional needs, 750–752
    peptide synthesis from, 1142–1143
    pK values for, 36
    R groups of, 9, *9*, 33, 36, 1238
    sequence determination in, 42–49
        and protein conformation, 54–56
        and secondary structures of proteins, 74–75,
            75
    as source of carbon and energy, 622–639
    structure of, 9, *32*, 32–33
    sulfur-containing, 794
    titration curves for, 35–36, *35*, *36*

p-Amino acid, 37, *37*
L-Amino acid, 37, *37*
Amino acid decarboxylases, 796–797
Amino acid oxidase, *365*
    reactions catalyzed by, 363
Aminoacyl adenylate, *1144*
Aminoacyl-tRNA formation, *938*, 939
Aminoacyl-tRNA synthases, 937–939
    proofreading hydrolysis reactions of, 939
α-Aminoadipic-δ-semialdehyde-glutamate
        reductase, 764
γ-Aminobutyric acid
    biosynthesis of, *1207*
    as neurotransmitter, 1207
5-Aminoimidazole-4-carbonitrile, 1129
5-Aminoimidazole-4-carboxamide, 1129
δ-Aminolevulinate, 795
δ-Aminolevulinate dehydrase, 796
Aminomuconate semialdehyde dehydrogenase,
        631
Aminopeptidases, 616
Ammonia
    assimilation into organic materials, 752–753
    as end product in nitrogen metabolism, 618
    hyperammonemia, 637
    from pyrimidine catabolism, 831
    transport to liver, 621–622
AMP. *See* Adenosine monophosphate
cAMP phosphodiesterase, 470
Amphibolic pathways in tricarboxylic acid
        cycle, 496–498
Amphipathic properties
    of membranes, 185
    of phospholipids, 162, 689–690
Ampholytes, 109
Amplification, gene, 1029–1030, 1232
Amylol (1,4–1,6)-*trans*-glycosylase, 654
Amylopectin, 142
Amylose, 142
    helical structure of, 144, *144*
Amylose synthase, 655
Amytal
    as electron-transport inhibitor, 528
    structure of, *529*
Anabolism, 23, 418
    and end-product inhibition, 428–429
Anaerobes, facultative, 450, 1177
Anaerobic pathway for monounsaturated fatty
        acid biosynthesis, 680–681, *681*
Analbuminemia, 602
Anaplerotic reactions, 498
Androgens
    and gene regulation, 1070
    structure and function of, 1049
4-Androstenedione, *744*
Anemia, sickle-cell, 335
Angiotensin, structure and function of, 1049
Ångstrom unit, 1228
Animals
    cell structure in, *3*
    evolution of, *2*
Anion transport protein, *192*, 192–193
    domains of, 193
    properties of, 191
Anisotropic orientation of NADH
        dehydrogenase, 531
Anomers, 135, 1228
Antagonists
    of cholinergic systems, 1205
    hormone, 1059
Antenna system in photosynthesis, 577–579
Antennapedia complex (ANT-C), 1037, *1037*
Anthranilate, 781
Anthranilate phosphoribosyltransferase, 781–782
Antibiotics, 1228
    amino acids in, 792, 793
    ion-translocating, 1179–1181
    macrolide, 174
Antibodies, 90, 91, 96–98, 1228, 1234
    complex with antigens, 90, *91*
    constant region in, 1040
    heavy chains in, 1040
    light chains in, 1040

Antibodies (Continued)
  synthesis of, 1039–1042
  variable region in, 1040
Anticodon, 894, 929, 1228
  pairing with codon, wobble rules in, 953, 954, 1240
Antidiuretic hormone. See Vasopressin
Antigens, 90, 1228
  complex with antibodies, 90, 91
  O antigen, 151–152
Antimycin
  as electron-transport inhibitor, 528
  structure of, 539
Antiport, 1163
Aphidicolin
  as DNA replication inhibitor, 876
  structure of, 876
Apoactivator, 976, 1229
Apoenzyme, 264
Apoprotein, 734
  properties of, 735
Apoprotein D, 740
Aquocobalamin, 377, 379
ara operon, 987, 987
β-Arabinofuranosylcytosine, structure of, 876
D-Arabinose, 133
Arabinose operon, 987, 987
Arabinosyladenine, as DNA replication inhibitor, 876
Arabinosylcytosine, as DNA replication inhibitor, 876
Arachidic acid, structure of, 155
Arachidonic acid, 156
  biosynthesis of, 682, 682–683
  and eicosanoid synthesis, 718–720
  phospholipids enriched with, 699
  structure of, 155
Arginase, 619
Arginine
  amines produced from, 634
  biosynthesis of, 761, 762
  catabolism of, 632, 632–634
    in bacteria and fungi, 641
  ionization reactions for side chains of, 37
  pK values for, 36
  structure of, 33
Arginine deiminase, 633
Arginine dihydrolase pathway, 633
Argininemia, 637
Argininosuccinate lyase, 619
Argininosuccinate synthase, 619
Aromatase, 744, 1056
Arylhydrazinopyrimidines
  as DNA replication inhibitors, 876, 878
  structure of, 878
Arylsulfatase A deficiency, 714
Asexual reproduction, 1229
Asparaginase, 639
Asparagine
  biosynthesis of, 768, 768
  catabolism of, 639
  linkage with N-acetylglycosamine, 147
  pK values for, 36
  structure of, 33
Asparagine synthase, 768
Aspartate
  biosynthesis of, 768, 768
  catabolism of, 639
  and chymotrypsin-catalyzed hydrolysis, 293
  decarboxylation of, 355
  ionization reactions for side chains of, 37
  pK values for, 36
  structure of, 33
Aspartate aminotransferase, 265
  Michaelis constant for, 270
Aspartate carbamoyltransferase, 336–343
  allosteric behavior of, 339–340
  compression mechanism for, 338
  and cytidine triphosphate synthesis, 428
  inhibitors of, 824, 834
  molecular weight and subunit composition of, 88
  pathway for substrate binding, 337–339

subunit and domain structure of, 340
x-ray diffraction studies of, 340–343
Aspartic-β-semialdehyde-dehydrogenase, 769
Aspartokinase, 769
  regulation of, 773–775, 774
Aspirin, and cyclooxygenase inactivation, 718
Asymmetric carbon, 1229
Atherosclerosis, 739
Atmosphere of primitive Earth, 1125–1126
Atoms, 1122
ATP. See Adenosine triphosphate
ATP-citrate lyase, 670
ATP/ADP exchanger, mitochondrial, 556–557, 1174–1175, 1186
ATP synthase, 551–554, 1167, 1177–1178
  mechanism of action, 553–554
  proton-conducting, 551–553
ATPase
  calcium, 1166, 1186
  F_1/F_0, 551–553, 1167–1168, 1177–1178
  factors affecting activity of, 542
  hydrogen, 1166
  sodium-potassium, 1164–1165, 1185–1186
Atractylic acid, 1186
Atractyloside, structure of, 557
Atrial natriuretic peptide, structure and function of, 1048
Attenuator, 1229
Autophagic vacuoles, 970
Autoradiography, 1068, 1069, 1229
Autoregulation, 1229
Autotroph, 417, 1229
Auxins, 1082–1084, 1083, 1229
Auxotroph, 1229
Avogadro's number, 105, 395, 569, 1229
Axon of squid, 1194
Axoplasm, 1195
Azaleucine, 794
Azaserine, 822, 826
Azidothymidine, 828

B cells, 1039, 1040, 1229
B form of DNA, 228, 229, 1229
Bacitracin
  amino acids in, 793
  inhibiting bacterial cell wall synthesis, 665, 667
Bacteria
  amino acid catabolism regulation in, 640–641
  anaerobic, facultative, 450, 1177
  cell envelopes of, 180–182
  cell structure in, 3
  cell wall of, 149–151
    amino acids in, 792
    synthesis of, 663–667
      inhibition by penicillin, 664, 667
  chemical composition of cells in, 6
  electron transport and ATP synthesis in, 559–560
  lysine biosynthesis in, 769–771, 770
Bacteriochlorophyll, 567, 568
  electron release by, 571
  molecular weight and subunit composition of, 88
  in reaction centers, 573–575
  reactive complexes of, 573
Bacteriophage
  DNA in, 213–215, 217
    replication of, 865–868
  gene expression regulation in, 1000–1007
  lambda (λ), 1000–1004
    cI and cro repressor proteins, 1006
    development of, 1002
    promotors of, 1004
    regulatory proteins of, 1003
    lysogenic pathway for, 1002
    lytic cycle in, 1002
    permissive cycle in, 1000–1002
    properties of, 1001
  RNA polymerases in, 909–910
  T4
    affecting nucleotide metabolism, 838
    polynucleotide ligase of, 1091

RNA ligase of, 912–913
  as vectors in DNA cloning, 1099–1100, 1101
Bacteriopheophytin, 569, 575
  structure of, 568
Bacteriorhodopsin, 193–194, 1168–1169, 1169, 1185, 1225–1226
  amino acid sequence of, 195
  properties of, 191
  structure of, 194
Bactoprenol, 172, 184, 184
Band 3 protein, 192–193
Barr bodies, 1022, 1022
Base analog, 1229
Base pairs, Watson-Crick, 1240
Base sequence, complementary, 1230
Base stacking, 1229
Bathorhodopsin, 1217, 1218
Beer-Lambert law, 112
Behenic acid, structure of, 155
Bentonite clays, homionic, 1142
Benzoylalanine methyl ester, 294
Beta bends (β bends), 73–74, 74, 1229
Beta oxidation of fatty acids (β oxidation), 603–608, 1229
Beta-pleated sheets (β sheets), 73, 1229
  antiparallel, 73, 73, 83, 1229
  barrel shape, 81, 81, 82
  connections between strands of, 79–80, 80
  crossover connections in, 79–80, 82
  in globular proteins, 79–83
  parallel, 73, 73, 83
  saddle shape, 81, 82, 82
  in trypsin family of enzymes, 288
Bicarbonate, and carboxylation of biotin, 372–373
Bidirectional replication, 1229
Bilayer, lipid, 162, 186–187, 195–196, 1229, 1234
Bile, 168
  components in, 170
Bile acids, 168–170, 712
  metabolism of, 741
Bile duct, common, 168
Bile salts, 168, 1229
  enterohepatic circulation of, 170
  formation of, 742
Bilirubin in bile, 170
Biochemical pathways, 25, 25–26, 1229
Biogenesis, 205
Bioluminescence, 593–595, 1229
Biomonomers
  polymerization of, 1134–1148
  synthesis of, 1127–1134
Biotin, 371–373
  structure of, 371
Biotin carboxyl carrier protein (BCCP), 676
  structure of, 675
Biotin carboxylase, 676
Bithorax complex (BC-C), 1037, 1037
Blastoderm, 1035, 1229
Blood group, ABO, 663
Boat forms
  of pyranoses, 136, 136
  of steroids, 155
Boltzmann constant, 395, 396
Bonds
  disulfide, 40, 40
  energy values for, 62, 1229
  glycosidic, 8, 137, 1233
    in nucleotides, 218
  hydrogen, 17, 1233
  hydrophobic, 1233
  peptide, 9, 38
Bongkrekic acid, 1186
  structure of, 557
Branched-chain amino acid, 794
Branched-chain amino acid-glutamate transaminase, 776
Branched-chain pathway of metabolism, 429
Branching enzyme in glycogen formation, 654, 654
Branchpoints, 26, 1229
  in amino acid biosynthesis, 756

and enzyme activity, 429, 431
  in metabolic pathways, 419, 421, *421*
Briggs-Haldane equation, 268–269, 294–295
Brown fat, thermogenesis in, 535
Buffer, 1229
Buffered solutions, 35
Bundle sheath cells, in C-4 pathway, 591, *592*
α-Bungarotoxin, 1205

cI repressor protein, bacteriophage, 1003, 1006
Calcitonin, 1048
Calcitonine gene related peptide, structure and
  function of, 1048
Calcium
  cytosolic, inositol-triphosphate affecting, 701
  and inositol triphosphate pathway, 1066–1068
  and membrane fluidity, 200
Calcium ATPase, 1166, *1166*, 1186
Calmodulin, 1066
Calories, 395
Calorimetry, differential scanning, *199*, 199–200
Calvin cycle, 588, *589*
Canavanine, 794
CAP protein, 976, 985–986, 1229
  binding to DNA, 988
  and stimulation of gene expression, 987–990
Cap structure of mRNA, 916–918, *917*, 1119
Carbamate kinase, 634
Carbamoyl aspartate, and pyrimidine
  biosynthesis, 834
Carbamoyl phosphate, 336
  formation of, 620, *620*
Carbamoyl phosphate synthase, 620, 812
  inhibition of, 834
Carbohydrates, 7–8, 131–153, 1229
  bacterial cell wall synthesis of, 663–667
  disaccharides, 138, *138*
    synthesis of, 652–653
  heteropolysaccharides, 144–152
  hexose biosynthesis, 647–652
  in membranes, 148–149, 178
  monosaccharides, 131–137
  oligosaccharide synthesis, 655–663
  polysaccharides, 138–144
    synthesis of, 653–655
3-Carbomethoxy-3,4-dihydro-isocarbostyril, *294*
Carbon
  anomeric, 135
  asymmetric, 1229
  C-4 cycle and photorespiration, 590–592
  compounds in solar system, 1123–1124
  valence of, 14
Carbon dioxide
  in atmosphere of primitive Earth, 1125
  fixation of
    in C-4 cycle, 591, 592
    in photosynthetic electron-transfer reactions,
      588–590
  and oxygen release from hemoglobin, 328
  production and utilization of, 24
  from pyrimidine catabolism, 831
  utilization by methanogens, 383
Carbonic anhydrase, molecular weight and
  subunit composition of, 88
Carbonylcyanide-*p*-trifluoromethoxyphenylhy-
  drazone, structure of, *540*
N¹-Carboxybiotinyl enzymes, 373
γ-Carboxyglutamic acid in proteins, 171, *171*
Carboxylation, biotin in, 371–373
Carboxylic acid, 1229
  enolization of, 366–368
  and ester formation, 14
  formation of, 16
Carboxyltransferase, 676
Carboxypeptidases, 298–303, 616
  β-sheet structures in, 80, *81*
  catalytic mechanism in, 301–303
  competitive inhibitor of, 303, *303*
  structure of enzyme-substrate complex, 299–
    301
  substrate specificity, 298–299
  transition-state analogs of, 303
3-(3-Carboxyphenyl)alanine, 794

Carcinogen, 1229
Cardiolipin, 161
  biosynthesis of, *700*
    in *Escherichia coli*, 690
  in membrane preparations, 185
Cardiotonic steriods, *1164*
Carnitine, 602
Carnitine acetyltransferase, 602
Carnitine acyltransferases, 602–603
  inhibition of, 686
β-Carotene, 172
  structure of, *579*, *1215*
Carotenoids, 580, 1229
Carriers in transmembrane transport, 1156–1158
Catabolism, 23, 418, 1229
  aerobic, 481–482
  end products of, 418
Catabolite activator protein. *See* CAP protein
Catabolite repression, 641, 1229
  of *lac* operon, 984–986
Catalase, 381
  Michaelis constant for, 270
  physical constants of, 106
Catalysis, concerted, 286, 303–308
Catalysts, 264, 284–286, 1229
  enzyme, properties of, 264–266
Catalytic site, 1229
Catecholamines, as neurotransmitters,
  1205–1207. *See also* Epinephrine;
    Norepinephrine
Catenane, 1229
Catenation, 863, *864*, 1119
Cavitation, nitrogen, 178
CCA enzyme, 910
cca gene, 909
CCA sequence, 929
CDP-choline:1,2-diacylglycerol phosphocholine
  transferase, 696
CDP-diacylglycerol, 696
CDP-ethanolamine:1,2-diacylglycerol phosphoethanolamine transferase, 696
Cedrol, 172
  structure of, *173*
Cell(s), 3–12
  animal, *3*
  bacterial, *3*
    chemical composition of, 6
  biochemical reactions in, 25
    regulation of, 26–27
  commitment of, 1029, 1229
  diauxic growth of, 984, 1231
  division of, *11*
  DNA content in, 216–217
  in eukaryotes, 176
  host, 1233
  membrane of. *See* Membranes
  organelles in, 176, *177*
  plant, *3*
  in prokaryotes, 176
  rod and cone, 1213–1214
  second messengers in, 1062
  somatic, 1239
  specialized types of, 5
  structure of, *3*
Cell cycle, 1229
  phases in, 836–837, *837*, 874, *874*, 1232, 1235
Cell line, 1229
  established, 1232
Cell lineage, 1230
Cell wall, 3, 149–151, 180, 1230
  amino acids in, 792
  synthesis of, 663–667
    inhibition by penicillin, *664*, 667
Cellobiose, structure of, *138*
Cellulose, 140
  configuration and role of, 142–144
  methylation of, 140
  structure of, *140*
  synthesis of, 655
Central dogma, 892
Centrifugation
  density-gradient, 235, 1231
  differential, 120, 178, *178*, 1231

isopycnic, 178, *178*
Ceramidase deficiency, 714
Ceramide, *163*, 163–164
  biosynthesis of, *710*
Cerotic acid, structure of, 155
Ceruloplasmin, molecular weight and subunit
    composition of, 88
Cetyl trimethylammonium bromide, 188, *188*
Chair forms
  of pyranoses, 136, *136*
  of steroids, 166
Chelate, 1230
Chelation effect, 399–400
Chemical reactions
  first-order, 260
  molecularity of, 261
  rates of, 259–260
    thermodynamics and kinetics affecting,
      262–264
  reverse, 260–261
  second-order, 260
Chemiosmotic coupling, 1230
  in membrane transport, 1172
  in oxidative phosphorylation, 543–545, 587
Chemoautotrophs, 417
Chenodeoxycholic acid, 168
  structure of, *169*
Chimeric DNA, 1230
Chimeric plasmid, 1233
Chiral centers of molecules, 131–132, *359*
  in solutions, 113
Chiral compounds, 1230
Chiral properties
  of amino acids, 36–37
  of polypeptide chains, 78–83
Chitin, 144, *144*
    Chloramphenicol
  inhibiting protein synthesis, 957–958
  structure of, *956*
Chlorophyll, 567, 1230
  absorption spectra of, *569*
  in antenna system, 579
  electron release by, 571
  photooxidation of, 572–573
  in reaction centers, 573–575
  reactive complexes of, 573
  structure of, *568*
Chloroplasts, 565, *566*, 1230
  photosystems in, 579–586
Chlorpromazine, 1207, *1207*
Cholecalciferol, 171, 1056–1057
Cholecystokinin, structure and function of, 1048
Cholera toxin, 1065, *1065*
Cholestanol, structure of, *167*
Cholesterol, 166–170
  in bile, 170
  bile acids as products of, 168–170
  biosynthesis of, 725–733
    lanosterol in, 729–733
    mevalonic acid in, 726–729
  concentrations in plasma, 733
  conversion to bile acids, 741
  hypercholesterolemia, familial, 739
    in heterozygotes, 739
    in homozygotes, 739
  in membranes, 100, 184
  metabolism of, 745, *746*
  and steroid hormone synthesis, 168,
    1053–1057
  structure of, *167*
  transport of, 736–737
Cholesterol desmolase, 1054
Cholesterol ester
  biosynthesis of, 735, *735*
  storage disease, 739
Cholesterol ester hydrolase, 1054
Cholesterol ester transfer protein, 734, 740
Cholic acid, 168
  formation from 7α-hydroxycholesterol,
    742
  structure of, *169*
Choline acetyltransferase, 1203
Choline kinase, 696

Cholinergic systems, 1205
  agonists of, 1205
  antagonists of, 1205
Chondroitin sulfate, structure of, *145*
Choriogonadotropin, structure and function of, 1049
Chorismate
  biosynthesis of, 779, *780*
  end products from, 421, *421*
Chorismate mutase P-prephenate dehydratase, 779
Chorismate mutase T-prephenate dehydrogenase, 779
Chromatin, 102, 1230
  active, 1023, 1025–1027
  histones in, 1023–1026
  inactive, 1023
Chromatography, 1230
  affinity, 122, 190, 1061, 1228
  column, *121*, 121–122
  gel-exclusion, 107–108, *107–108*, 121, 1232
  hydrophobic interaction, 189, *190*
  ion-exchange, 41, *41*, 121–122
  liquid, high-performance, 122–123
Chromosome(s), 11, 1230
  analysis with gel electrophoresis, 245
  general recombination of, 1232
  giant, 1020–1021, *1021*
  homologous, 1233
  inactivation of, 1022–1023
  lampbrush, 1234
  polytene, 1020–1021, *1021*
  puffs in, 1021, 1070, *1071*, 1230
  structural differences in, 1020–1027
Chromosome walking, 1117
Chylomicron, 598–599, *599*
  transporting cholesterol and triacylglycerol, 736–737
Chymotrypsin, 47, 287, *287*
  amino acid sequence of, *289*
  catalytic mechanism in, 288–298
    kinetic parameters in, 295–297
    pH affecting, 297–298
    steady-state analysis of, 294–295
  Michaelis constant for, 270
  molecular weight and subunit composition of, 88
  peptide and ester substrate kinetics, 292–294
  specificity pocket of, 291–292, *292*
  structure of, *291*
Chymotrypsinogen
  activation of, 297–298
  physical constants of, 106
trans-Cinamate-4-monooxygenase, 629
Circadian rhythms in plants, 592–593
Circulin A, amino acids in, 793
cis orientation of genetic elements, 979, 1230
Cistron, 1230
Citrate
  activating acetyl-CoA carboxylase, 685
  cycling between mitochondria and cytosol, 506–508, *507*
  metabolism of, *494*
  and phosphofructokinase regulation, 471
  synthesis of, 407, *407*, 489, *494*
Citrate lyase, 506
Citrate synthase, 670
  regulation of, 503–505
  in tricarboxylic acid cycle, 489
Citric acid cycle. *See* Tricarboxylic acid cycle
Citrinin, 174
  structure of, *174*
Clathrate structures, 58, *58*
Clathrin, 1077
Clay
  genetic role for, 1135
  homoionic bentonite, 1142
Clone, 1230
Cloning of DNA. *See* DNA, recombination of
Cloning vector, 1230
CMP. *See* Cytidine monophosphate
Coactivator cAMP, 976, 1230
Coated pits of membranes, 1077

Coated regions of membranes, 737
Codon, 11, 929, 934, 1230
  pairing with anticodons, wobble rules in, 953, 954, 1240
  start, 934
  stop, 935
  termination, 942–947
Coenzyme(s), 12, 264, 347–391, 1230
  biotin, 371–373
  flavin, 361–365
  folate, 373–377
  iron-containing, 381–383
  α-lipoic acid, 369–371
  in methanogenesis, 383–386
  nicotinamide, 356–361
  nucleotide, biosynthesis of, 838–841
  phosphopantetheine, 366–369
  pyridoxal-5'-phosphate, 351–356
  thiamine pyrophosphate, 348–351
  vitamin B_{12}, 377–381
Coenzyme A, 366
  biosynthesis of, 841, *841*
Coenzyme M, 385
Coenzyme Q. *See* Ubiquinone
Cofactor, 1230
Colicin E3, 959
Collagen, structure of, 77, 77–78
Colony hybridization for selection of clones from DNA library, 1107–1108, *1108*
Colony-stimulating factors, 1080
Comets, *1124*, 1124–1125
Commitment, cellular, 1029, 1229
Complementary base sequence, 1230
Complementation test of DNA replication proteins, 858
Concanavalin A, 656
  molecular weight and subunit composition of, 88
Concentration work, 405
Condensing agents, prebiotic, 1132–1134
Cone and rod cells, 1213–1214
  response to light, 1220–1221
Configuration, 1230
  compared to conformation, 136*n*
Conformation, 53, 1230
  compared to configuration, 136*n*
Consensus sequence, 1230
Conservation
  of energy, 394
  of mass, 423
Constitutive enzymes, 1230
Contraction of muscle, sliding-filament model of, 89, *90*
Conversions, metabolic, 424–426
Cooperative binding, 1230
  negative, 327, 344
  positive, 327, 340
    advantages of, 343–344
Cooperativity
  in folding of protein, 72
  of regulatory enzymes, 429–430, 431
Coordinate induction, 1230
Coprostanol, 170
  structure of, *167*
Cordycepin, 924–925
Cornea, 1213
Corticosterone
  biosynthesis of, *1055*
  function of, 1049
Corticotropin, 1050
  structure and function of, 1047
  releasing factor, 1047
Cortisol, 169, *744*
  biosynthesis of, *1055*
  function of, 1049
Cosmids, 1230
  construction of, 1102–1103
  as vectors in DNA cloning, 1100–1101
Cosmochemistry, 1123–1125
Cot curve, 1230
Coulomb, 395
p-Coumaryl-CoA synthase, 629, *629*
Coupling reactions, 407

Covalent linkages, 14
Creatine, 801
Creatine phosphate formation, *802*
Creatinine formation, *802*
Cro repressor protein, bacteriophage, 988, 1003, 1006
Crotonase, 604
crp gene, 985
Cruciforms
  in DNA and RNA synthesis, 232
  and renaturation of DNA, 241
Crystallins, 1213
Crystallography, X-ray, 54, 115–119, *116–118*, 1240
CTP. *See* Cytidine triphosphate
CTP:phosphocholine cytidylyltransferase, 696
  regulation of, *705*
CTP:phosphoethanolamine cytidylyltransferase, 696
Cushing's syndrome, 1078
Cyanamide as prebiotic condensing agent, 1132–1133
Cyanate as prebiotic condensing agent, *1133*, 1133–1134
Cyanide as enzyme inhibitor, 275
Cyanoacetylene in synthesis of nucleic acid bases, 1130–1131
Cyanobacteria, 565
Cyanocobalamin, 377
Cyanoformamide as prebiotic condensing agent, *1133*, 1133–1134
Cyanogen as prebiotic condensing agent, *1133*, 1133–1134
Cyanogen bromide, for polypeptide-chain cleavage, 46–47, *47*
Cyclic AMP. *See* Adenosine monophosphate, cyclic
Cyclic GMP. *See* Guanosine monophosphate, cyclic
2',3'-Cyclic phosphates, 1131–1132, 1137
Cyclooxygenase inactivation by aspirin, *718*
Cyclopropane amino acids, 794
Cyclopropane ring, fatty acid, 156
γ-Cystathionase, 767
Cystathionine-β-synthase, 767
Cysteine
  biosynthesis of, 765–768, *766*
  catabolism of, 625, *625*
  ionization reactions for side chains of, 37
  pK values for, 36
  structure of, 33
Cystine reductase, 625
Cytidine, 1230
  base found in, 219
  in tRNA, 253
Cytidine monophosphate (CMP)
  ionization constants of, 219
  molar absorbance of, *220*
Cytidine triphosphate (CTP)
  biosynthesis of, 428
  and inhibition of aspartate carbamoyl-transferase, 337
Cytidine triphosphate synthase, biosynthesis of, 814, *814*
Cytidylic acid, base found in, 219
Cytochrome(s), 515, 1230
  amino acid content of, 43
Cytochrome *a*, 515, 516
Cytochrome *a₃*, 518
Cytochrome *b*, 515, 516
Cytochrome *b₅*, 191
Cytochrome *b₅* reductase, 191, 681–682
Cytochrome *b₅₆₂*, 516
Cytochrome *bc₁* complex, in electron transport, 532–535
Cytochrome *bf* complex, 581
Cytochrome *c*, 92, 94, 515–519
  heme groups in, *518*
  molecular weight and subunit composition of, 88
  physical constants of, 106
  side-chain model of, 58, *59*
  structure of, 19, *517*
    diversification of, *95*

Cytochrome c_1, 516
Cytochrome oxidase, 202, 518, 535
Cytochrome P-450, 383
Cytokinins, *1083*, *1084*, 1230
Cytoplasm, 4, 1230
Cytoplasmic membrane, lipid composition of, 185
Cytoplasmic regulatory protein (RPr), 1178
Cytosine, 1230
  in DNA and RNA, 218
  formation of, 1130
  hydrogen bonding to guanine, *223*
Cytosine arabinoside, 828
Cytoskeleton, 4, 201, 1230
Cytosol, 4, 1230
  citrate cycling from mitochondria, 506–508, *507*

D loop of DNA, 1230
Dalton unit, 1230
dam gene, 882
Dansyl chloride method for N-terminal amino acid determination, 46
Dark reactions, 1230
dcm gene, 869, 882
o,p'-DDT, structure of, *1060*
Deamination, 1230
Decamethonium, 1205, *1205*
α-Decarboxylases, and pyridoxal phosphate activity, 355
Decarboxylation
  of aspartate, 355
  of pyruvate, 485–488, *487*
Degradation, Edman, 1231
Dehydroalanine, structure of, *629*
7-Dehydrocholesterol, 732
Dehydrogenase, 1231
  reactions catalyzed by, 363
Dehydrogenation reaction, alcohol, 15
Denaturation, 1231
  of DNA, 236–238
  of protein, 54–55
De novo pathways, 751, *751*, 808–814, 1230
Density-gradient centrifugation, 235, 1231
Deoxyadenosine, structure of, *218*
5'-Deoxyadenosylcobalamin, *377*, 378
Deoxycholic acid, 168
  structure of, *169*
Deoxycytidine, structure of, *218*
Deoxycytidine kinase, 820
Deoxycytidylate aminohydrolase, 831
2-Deoxyglucose, 1162, *1162*
Deoxyguanosine, structure of, *218*
Deoxyhemoglobin, structure of, *330*
Deoxynucleotidyl transferase, terminal, 874
Deoxyribonucleotide, structure of, *222*
Deoxyribonuclease I, 880
Deoxyribonuclease II, 880
Deoxyribonucleic acid, 10–12. *See also* DNA
Deoxyribonucleoside, structure of, *217*
Deoxyribonucleotides
  biosynthesis of, 814–817
    regulation of, 834–835, *835*
  structure of, *217*
Deoxyribose, in DNA, 217
Deoxythimidine, structure of, *218*
Deoxyuridine monophosphate, and thymidylate biosynthesis, 816–817
Deoxyuridine triphosphate diphosphohydrolase, 816
Depolarization, and action potential, 1197
Desensitization
  of acetylcholine receptor, 1205
  and hormone responses, 1076–1077
Desmolase, 741
Desmosterol, 732
Detergents, 188–189
  critical micellar concentration of, 189
  hydrophilic-lipophilic balance in, 189
  properties of, 187
Development, regulatory mechanisms, 1029–1043
  in frogs, 1029–1033

  in fruit flies, 1033–1039
Dextran, synthesis of, 655
Dextrorotatory rotation, 114, 132
Diabetes mellitus
  adult-onset, 1079
  juvenile-onset, 1078
Diacylglycerol, 696
  metabolism in eukaryotes, *697*
  structure of, *1067*
Diacylglycerol acyltransferase, 704
Dialkyl phosphatidylglycerol phosphate, 703, *703*
Diaminomaleonitrile, as prebiotic condensing agent, 1134
Diaminopimelate, 769–771
Diamonopamelate decarboxylase, 771
2,3-Diaminopropionic acid, 794
Diauxic growth of cells, 984, 1231
6-Diaza-5-oxo-L-2-aminohexanoic acid (DON), 822, 826
Dicarboxylic acid, cotransport with hydrogen, 1174
Dichroism, circular, 114, *114*, *115*
Dicyclohexylcarbodiimide, structure of, *541*
2',3'-Dideoxythymidine, as DNA replication inhibitor, 876
2',3'-Dideoxythymine, structure of, *876*
2,4-Dienoyl-CoA reductase, 606
Diethylstilbestrol, structure of, *1060*
Difference spectra, 1231
Differential centrifugation, 178, *178*, 1231
Differentiation, 1231
Diffraction, X-ray, 54, 115–119, *116–118*
  of DNA, 224
Diffusion, 1156, *1156*, *1159*
  exchange, 1163, *1163*
  facilitated, 1159
    examples of, 1163–1164
Diffusion coefficient, 1157
  for rhodopsin, 1225
Diffusion constant, protein, 105–106
Diglyceride, 158
Dihydrobiopterin, 628, *628*
Dihydrodipicolinate synthase, 769
Dihydrofolate reductase, inhibitors of, 827
Dihydrolipoyl dehydrogenase, 488
Dihydrolipoyl transacetylase, 370, 487–488
Dihydroorotate dehydrogenase, inhibitors of, 824
Dihydropteroate synthase, inhibitors of, 827
Dihydrosphingosine, 164
Dihydrotestosterone, 744, *1056*
Dihydroubiquinone, structure of, *520*
Dihydrouracil dehydrogenase, 265
Dihydrouridine, in tRNA, *253*
Dihydroxyacetone
  formation of, *132*, *133*
  structure of, *134*
Dihydroxyacetone phosphate
  and glycolysis, 445
  interconversion with glyceraldehyde-3-phosphate, 446–447
Dihydroxy acid dehydrase, 776
Dihydroxyphenylalanine (dopa), 1053, 1206
1,24-Dihydroxyvitamin D₃, 171, 1056–1057
  structure and function of, 1049
Diiminosuccinonitrile, as prebiotic condensing agent, *1133*, 1134
Diisopropylfluorophosphate (DFP), as enzyme inhibitor, 275, *276*
Diketopiperazine, *1147*, 1147–1148
3,3-Dimethylallyl pyrophosphate, 729
N²,N²-Dimethylguanosine, in tRNA, *253*
2,4-Dinitrophenol, structure of, *540*
Dinucleotides, 221
  structure of, *10*
Dioxetanone, in bioluminescent reactions, 594
Dioxygenases, 364
Dipalmitoylphosphatidylcholine, 696
  biosynthesis of, *699*
Diphosphates, nucleoside, 221, *221*
Diphosphatidylglycerol, 161. *See also* Cardiolipin
Diphosphopyridine nucleotide (DPN), 356

Diphthamide, 958–959
  structure of, *959*
Diphtheria toxin, inhibiting protein synthesis, 958–959
Dipicolinate, 769
Diploid cells, 979, 1231
Dipolar properties of water, 57, *57*
Dipole, 1231
Disaccharides, 138, *138*
  catabolism of, 462–464
  synthesis of, 652–653
Disks in rod outer segment, 1214
Dissociation constant, 34, 1231
  relationship to Michaelis constant, 270
Disulfide bonds, 40, *40*
  cleavage of, 44, *45*
  location in proteins, determination of, 49
Disulfide bridge, 1231
Diterpenes, 172, *173*
Diurnal rhythms, 592
DNA, 10–12, 1231
  A form of, 228, *229*, 1228
  actinomycin D binding to, 924
  anti conformation in, 228, *231*
  B form of, 228, *229*, 1229
  in bacteriophage, 213–215, 217
  base composition of, 223
  catenation of, 863, *864*
  chemical synthesis of, 247–249
  chimeric, 1230
  complementary (cDNA), 1103–1107, 1229
    coding of clones for hormone receptors, 1061
  content in cells, 216–217
  D loop of, 1230
  degradation of, 878–882
    glycohydrolases in, 881
    nucleases in, 878–880
  denaturation of, 236–238
    cooperative binding in, 238
    hyperchromic shift in, 236
  double helix structure of, 20, 222–223, *225*
    conformational variants of, 226–232, *227–231*
    pitch length of, 224, 1235
    stability of, 226
    unwinding of, 903
  dyad symmetry in, 983
  evolution of, 1141–1142
  footprinting technique for, 1031, *1031*
  hybridization of, 239
  information transfer to protein, 12, *13*
  interaction with histone, 1023–1024
  interaction with regulatory binding proteins, 988–989
  intercalation with ethidium bromide, 233, *233*
  library of, 1101–1107, 1231
    cDNA, 1103–1107
    genomic, 1103
    selection of clones from, 1107–1109
      colony hybridization in, 1107–1108, *1108*
      reverse translation in, 1108–1109, *1109*
    sublibraries, 1112
  polymerase binding sites, 975
  postreplicative modification of, 881–882
  recombination of, 886–889, 1088–1119
    antibiotic resistance genes in, 1095–1098
    in antibody-forming cells, 1040–1041
    bacteriophage vectors in, 1099–1100, 1101
    cosmids in, 1100–1101
    and DNA library, 1101–1109
    enzymes used in, 1090
    in *Escherichia coli*, 1098–1109
    general, 886
    interspecies combinations in, 1098
    and ligation of fragments *in vitro*, 1089–1092
    in mammalian cells, 1111–1115
      vectors used in, 1114–1115
    nonspecific, 886
    in plants, 1115
    plasmids as vectors in, 1094–1095, 1098–1099

DNA (*Continued*)
  restriction enzymes in, 1092–1094
  shuttle vectors in, 1101, 1114
  site-specific, 886
  subculture cloning in, *1113*, 1113–1114
  uses of, 1115–1118
  vectors in, 1114–1115, 1230
    in yeast, 1110
relaxed, 233, *234*
renaturation of, 238–242, *241*
  nucleation in, 240
  zippering in, 240
repair of, 882–886, 1238
  excision, 883, 1232
  long patch in, 885
  mismatch, 882
  removal of pyrimidine dimers in, 883–885,
    884
  removal of thymine dimers in, 883, *883*
  short patch in, 885
  SOS system in, 885
repetitive, 241, 1238
replication of, 26, 847–751
  autonomously replicating sequences in, 872
  in bacteriophage, 865–868
  bidirectional, 851–853, 869, 870
  conservative, 847, *848*
  control of, 874–875
  dcm gene in, 869
  and discontinuous growth on one strand,
    854–856
  dispersive, 847, *848*
  DNA polymerase I in, 859–861
  in *Escherichia coli*, 851–856, 868–869
    polymerases in, 856, *857*, 861–862, *862*
  in eukaryotes, 869–974
    multiple origins of, 870, *871*
    polymerases, 870
  helicases in, 864, 867–868
  inhibitors of, 875–878
  leading strand and lagging strands in, 854
  in mitochondria, 872–873, *873*
  nucleosomes in, 871–872
  Okazaki fragments in, 854–855
  at *oriC* region, 869
  plus strand and minus strands in, 768
  polynucleotide ligase in, 862–863
  preprimosome in, 865
  primase in, 856
  primer in, 855
  primosome in, 866
  in prokaryotes, 851–864
  proteins required for, 856–864
    complementation test of, 858
  rolling circle model of, 867, 1238
  semiconservative, 847–851, *848*
  single-strand binding protein in, 864
  temperature-sensitive mutants in, 858
  theta structure in, 852
  topoisomerases in, 863–864, *863*, *864*
  Watson-Crick model for, *947*
replicative form of, 867
restrictive-modification of, 881–882
and RNA synthesis, 895–905
satellite, 237, 242, 1238
selfish, 242
sequencing of, 49, 242–247, *246–247*
  autoradiogram of, 244, *244*
  Maxam-Gilbert method of, 242–247
  Sanger method of, 877
single-copy or unique, 241, 1239
structural properties of, 216–247
supercoiling of, *232*, 232–236, 1239
  biological importance of, 236
  electrophoresis of, 235, *235*
  linking number in, 233, *234*, 1234
  negative, 232, 234, 863, 976
  positive, 232, 234
syn conformation in, 228, *231*
temperature affecting, 236–237, *237*,
  238–239
template and growing strands of, *859*
and transformation in bacteria, 211–213, *212*

viral, 909
  synthesis from viral RNA genome, 911–912
  X-ray diffraction of, 224
  Z form of, 228, *230*, 1240
DNA gyrase, 234, 863
  inhibitors of, 924
DNA ligase, 1089–1090
DNA polymerase, 1231
  RNA-dependent. *See* Reverse transcriptase
DNA polymerase I, 859–861, 877
  of *Escherichia coli*, 861, 862
  exonuclease activitiy of, 860
DNA polymerase II, of *Escherichia coli*, 861,
  862
DNA polymerase III, 856, *857*
  of *Escherichia coli*, 856, *857*, 761, 862
  holoenzyme, *957*, 861
DNA polymerase α, 870
DNA polymerase β, 870
DNA polymerase δ, 870
DNA polymerase γ, 870
DNA primase, 856, 865, 895, 908–909
dnaB gene, 857, 908
dnaC gene, 857
dnaE gene, 856
dnaG gene, 856, 857, 895, 909
dnaN gene, 856
dnaQ gene, 856
dnaX gene, 856
dnaY gene, 857
dnaZ gene, 856
dnaZX gene, 856
Dolichol, 172, 208
Dolichol phosphate, 658–659
  structure of, *659*
Dolichyl phosphate, 172
Domains, structural, 86–87, *86–87*, 1231, 1239
Dominance, 1231
Donnan equilibrium, 1195
Dopa, 1053, 1206
Dopamine, 1053
  biosynthesis of, *1206*
  overproduction of, 1207
  underproduction of, 1206
Double helix, 1231
DRB, inhibiting RNA synthesis, 925
Drosophila melanogaster, development of,
  1034, 1034–1039
DUP–785, 822
Duplex, 1231
Duplication, tandem, 1239
Dust in galaxy, 1122
Dwarfs, Laron, 1079
Dyad symmetry in DNA, 983, 1231

Eadie-Hofstee plot, 269, *269*
Earth primitive
  atmosphere of, 1125–1126
  energy sources on, 1122
  polymerization of biomonomers on,
    1134–1148
  synthesis of biomonomers on, 1127–1134
Ecdysone, 1022, 1070, *1071*, 1231
  structure of, 1076
Edman degradation procedure, 47–48, *48*, 1231
EDTA, treatment of cells with, 180, *181*
Eicosanoids, 170, 689, 715–722, 1231
  derived from C_{20} polyunsaturated fatty acids,
    717
  local actions of, 720–722
  synthesis from arachidonic acid, 718–720
Einstein unit, 569
Elastase, 95, 95–96, 287, *287*
  amino acid sequence of, *289*
  catalytic mechanisms in, 288
  domains, 86, *86*
  specificity pocket of, 292, *292*
Electric discharges
  and amino acid formation, *1127*, 1127–1128
  as energy source, 1122
Electric organ of electric eel, 1202, 1208
Electrical potentials in electron transport,
  546–548

Electrical work, and free energy, 405, 406
Electrogenic transport, 556–557, 1173
Electromotive force or potential, 406
Electron microscopy, freeze-fracture, 202, *203*
Electron paramagnetic resonance spectroscopy,
  198, *198*
Electron transport, 482, 512–563
  ADP affecting, 539–540
  in bacteria, 559–560
  chemiosmotic theory of, 543–545
  cyclic chain in
    chloroplasts, 587–588
    photosynthetic bacteria, 576, *577*
  cytochrome activity in, 514–519, 532–535
  cytochrome oxidase in, 535
  electrical potentials in, 545–548
  electrochemical potential gradient in
    chloroplasts, 587
    photosynthetic bacteria, 576
  and electromotive force, 406
  flavin-mediated, 361–365, 519–520
  hemes in, 515–519
  inhibitors of, 528–529
  iron-sulfur clusters in, 382, 520
  light affecting, 571–572
  lipoic acid in, 369–371
  mechanisms in, 535–537
  NADH dehydrogenase complex in, 530–531
  nicotinamide coenzymes in, 357
  and oxidative phosphorylation, 537–554
  in photosystems, 575–579
  P/O ratio in, 538–540
  proton translocation in, 548–549, *549*
    and ATP formation, 549–551
    motive force in, 551
    photosynthesis, 586–587
  Q cycle in, 532–535
  redox potentials in, 521–525
  reversal of, 542–543
  specificity in, 679
  spectral analysis of carriers in, 526–528
  succinate dehydrogenase complex in, 531–532
  and thermogenesis, 555
  ubiquinone in, 520–521
Electrophoresis, 108–111
  in protein purification procedures, 123
  slab-gel, 109, *109*
    in chromosome analysis, 245
    sodium dodecyl sulfate, 109–111, *110–111*
  of supercoiled DNA, 235, *235*
Electroplaxes, 1208
Electrostatic forces in protein structure, 62–63
Elongation factors, 1231
Eluate, 1231
Embden-Meyerhof pathway, 437
Embryo, 1231
  development of, regulatory mechanisms in,
    1029–1043
Enantiomers, 36
End-product inhibition, 428–429, 1231
Endergonic pathways in metabolism, 418
Endergonic reaction, 405, 1231
Endocrine glands, 1231
Endocrinopathies, 1077–1079
Endocytosis, 737, 969
Endonucleases, 878, *879*, 880, 923, 1231
Endopeptidases, 46, *47*, 616, *616*, 1231
Endoplasmic reticulum, 4, 1231
  glycoprotein synthesis in, 656–658
  lipid composition of membrane, 180, 185
  lipoprotein synthesis in, 735–736
  rough, 934, *935*
  vesicles in, properties of, 179
Endotoxins, 184
Energy
  in biochemical reactions, 22–24
  bond, 1229
  conservation of, 394
  free, 262, 400–402, 1232. *See also* Free energy
  internal, 393–395
  sources on primitive Earth, 1122
  stacking, 1239
  storage in polysaccharides, 653–655

for transmembrane transport, 1158–1159
Energy charge, 1231
    and phosphofructokinase regulation, 471
    and regulation of metabolic pathways,
        430–431
Engrailed gene, 1036–1037
Enhancer sequences, 1027–1029, 1231
Enolase, 449
Enolization, 16
    of carboxylic acids, 366–368
Enoyl-ACP reductase, 673, 679
Enoyl-CoA hydrase, 604
Enoyl-CoA isomerase, 604, 606
Enterohepatic circulation of bile salts, 170
Enthalpy, 395–396
Entropy, 396–400, 1231
Enzyme(s), 9, 1231
    active site of, 264
        catalytic efficiency of, 271
    allosteric, 325, 336–343, 1228
    as catalysts, 284–286
        unique features of, 286
    catalytic efficiency of, 271
    catalytic site on, 264, 286
    classification of, 265–266
    cofactors for, 264
    constitutive, 1230
    in DNA recombinant methodology, 1090
    feedback inhibition of, 324–325, 1231
    half-of-the-sites reactivity in, 344
    induced fit to substrates, 300, 302
    induction of, 977–978
    inhibitors of, 272–277
        competitive, 272–275, 273
        irreversible, 275
        noncompetitive, 273, 275, 1235
        reversible, 272–275
        uncompetitive, 273, 275
    kinetic analysis of, 266–272
        and steady-state concept, 268
    lysosomal, 662
    Michaelis constant for, 270
    modification, 881
    multisubstrate, 278–280
        pathways for, 278
        Ping-Pong mechanism in, 280, 280, 1186
    organization according to function, 423
    pH affecting, 277, 277
    R-side-specific, 360
    redox, 381–383
    regulation of, 427–428
    regulatory, 26, 324–346, 438–430, 1238
        cooperative behavior of, 429–430, 431
        in metabolic pathways, 426–427
    replacement therapy, 714
    restriction, 880, 881–882, 1092–1094
    S-side specific, 360
    specificity constant for, 271
    substrates of, 21, 264
        binding site for, 286
    temperature affecting, 277, 277
    trypsin family of, 287–298
    turnover numbers of, 271
Enzyme-enzyme complexes, metabolite transfer
    by, 460
Epidermal growth factor, 1079, 1080
Epimers, 1231
Epinephrine, 467, 469–470, 600, 628, 797,
    1051–1053
    biosynthesis of, 1052, 1053
    and inhibition of acetyl-CoA carboxylase, 685
    as neurotransmitter, 1205–1206
    structure and function of, 1049
Equilibrium, 1232
    Donnan, 1195
Equilibrium constant, 262
    for ATP hydrolysis, factors affecting, 410–411
    calculation of, 403
    and standard free energy, 402–404
Equilibrium-density-gradient centrifugation, 235
Equilibrium sedimentation, 106
Erythrocyte membranes, lipid composition of,
    185

Erythromycin, structure of, 956
Erythropoietin, 1056
D-Erythrose, structure of, 133
Erythrose-4-phosphate, 473
    in transaldolase reaction, 474
D-Erythrulose, structure of, 134
Escherichia coli, 1232
    amino acid biosynthesis in, 24, 26, 27,
        755–756
    amino acid catabolism in, 641
    chain length of fatty acids in, 686
    DNA replication in, 851–856, 868–869,
        1098–1109
        polymerases in, 856, 857, 861–862, 862
    fatty acid synthesis in, 672, 673–674
    gene expression in, 974–1000
    lactose carrier protein, purification of,
        126–128, 127
    membranes of, 177
    phages and genomes of, 1001
    phospholipid synthesis in, 690–694, 692–693
        regulation of, 691–694, 694
    respiratory chain in, 559
Eserine, 1205, 1205
Established cell line, 1232
Esters
    cholesterol, 735
    formation of, 14–15
Estradiol, 169, 745
    biosynthesis of, 1055
    function of, 1049
    structure of, 1060
Estriol, structure of, 1060
Estrogens
    and gene regulation, 1070
    receptors for, cDNA clones coding for, 1061
    structure and function of, 1049
Estrone
    function of, 1049
    structure of, 1060
Ethanol. See Alcohol
Ethanolamine kinase, 696
Ether, 1232
    formation of, 15
Ethylene, 1083, 1085
    synthesis of, 1085
N-Ethylmaleimide, as DNA replication inhibitor,
    876, 878
Euchromatin, 1020
Eukaryotes, 4, 1232
    cell cycle in, 874, 874
    cell structure in, 176
    chromosomes in, 11
    DNA replication in, 869–874
    evolution of, 2
    gene expression in, 1011–1044
    membranes of, analysis of, 178–180
    phospholipid biosynthesis in, 694–704
    RNA transcription in, 905–908
Evolution
    chemical, stages in, 1122
    and diversification of proteins, 93–96
Evolutionary tree, 1–2, 2
Excision repair, 883, 1232
Excited state, 1232
Exergonic pathways in metabolism, 418
Exergonic reaction, 405, 1232
Exons, 918, 1232
Exonuclease, 878, 879, 880, 923, 1232
Exonuclease activity of DNA polymerase I, 860
Exopeptidases, 616, 616
Extinction coefficient, 112
Eye, vision mechanisms in, 1213–1227

F_1/F_0 ATPases, 551–553, 1167–1168, 1177–1178
F_{420} coenzyme, 385, 385–386
F_{430} coenzyme, 385, 385–386
F factor, 1232
F plasmid, 979
Fabry's disease, 713–714
Facultative organisms
    aerobes, 1232
    anaerobes, 450, 1177

Faraday constant, 395, 406
Farber's lipogranulomatosis, 714
Farnesyl pyrophosphate, 729
    biosynthesis of, 730
Farnesyl transferase, 730
Fatty acid(s), 154–159, 1232
    activation by enzymes, 602
    biosynthesis of, 669–683
        condensation reaction in, 677
        continuation reactions in, 679
        in Escherichia coli and plants, 672,
            673–674
        monounsaturated fatty acids, 680–682
            aerobic pathway for, 681–682, 682
            anaerobic pathway for, 680–681, 681
        polyunsaturated fatty acids, 682–683
        reactions involved in, 672, 674–679
        reduction reactions in, 679
        regulation of, 683–686
        saturated fatty acids, 669–680
        substrates for, 670–673, 671
    catabolism of, 598–614
    chain length regulation in Escherichia coli,
        686
    cyclopropane ring of, 156
    and eicosanoid synthesis, 717, 718
    gas chromatography of methyl esters,
        157, 158
    metabolism regulation in, 683–686
    monounsaturated, 154
        biosynthesis of, 680–682
    oxidation of, 603–613
        alpha, 608–610
        beta, 603–608
        defect in, and Refsum's disease, 608
        in heart, 612
        and ketone body formation, 610
        at methyl group, 610
        in mitochondria, 604–607
        omega, 610
        in peroxisomes, 610–611
        to propionyl-CoA, 607–608
        saturated fatty acids, 604–606
        unsaturated fatty acids, 606–607
    in phosphoglycerides, 182–183
    in phospholipids, 183–184
    polyunsaturated, 154
        biosynthesis of, 669–680, 682–683
        saturated, 154
        oxidation of, 604–606
    sources of, 598–599
    transfer from adipose tissue, 600–602
    transport into mitochondria, 602–603
    unsaturated, 154
        biosynthesis of, 680–683
        oxidation of, 606–607
Fatty acid cyclooxygenase, 718
Fatty acid synthase, 368–369, 369, 674
    activities of, 678
        assay of, 680
    diversity in structure of, 675
    reactions catalyzed by, 677–679
Fatty acyl-CoA, 602
Feedback inhibition
    in amino acid biosynthesis, 788
    of enzymes, 324–325, 1231
    in hormone synthesis, 1075
Fehling's solution, 139
Feminization, testicular, 1078–1079
Fermentation, 1232
    alcohol, 451–452
Ferredoxin, 584
Ferredoxin-NADP oxidoreductase, 584
Ferritin, molecular weight and subunit
    composition of, 88
Ferrocyanide, 1132
Fibrinogen, physical constants of, 106
Fibroblast growth factor, 1080
Fibronectin, 201
Fick's first law of diffusion, 105
Fingerprinting, 1232
Fischer projection, 132, 135, 136, 159
Fischer-Tropsch reaction, 1150

Flavin adenine dinucleotide (FAD), 361–365
  biosynthesis of, 839, *839*
  structure of, *519*
Flavin coenzymes, 361–365
  oxidation states of, *363*
Flavin mononucleotide (FMN), 361–365
  biosynthesis of, 838, *839*
  structure of, *519*
Flavins in electron transport, 514, 519–520
Flavodoxin, β-sheet structures in, 80, *81*
Flavone, 174
  structure of, *174*
Flavonoids, 174
Flavoprotein
  electron-transfer, 520
  reactions catalyzed by, *363*
Fluorescence, 1232
  in photobleach recovery technique, 197–198
5-Fluorouracil, 376, *376*, 822, 825
Folding pattern for proteins, 53
Folic acid, 348, 373
Follicle-stimulating hormone, 1050
  structure and function of, 1047
Footprinting technique in DNA studies, 1031,
  1031
Formaldehyde, 377
  in ribose formation, 1131
Formyl-methanofuran, structure of, *384*
Formlyphenylalanine methyl ester, *294*
10-Formyltetrahydrofolate, 373–374
  structure of, *374*
Fossils of microbial colonies, 1126
Frameshift mutations, 1232
Free energy, 262, 400–401, 1232
  of activation, 263
  and electrical work, 406
  of formation, 401–404
  and hydrolysis of ATP, 408–411
  standard, 401–404
    and equilibrium constant, 402–404
  in useful work, 404–405
Freeze-fracture electron microscopy, 202, *203*
Frictional coefficient of protein, 105
Friedelin, 172
  structure of, *173*
Frogs, RNA synthesis in, 1029–1033
Fructose
  excess of, complications from, 462–463
  structure of, *134*
Fructose-1,6-bisphosphate, 443–445
  cleavage by aldolase, 445–446, *446*
  conversion to fructose-6-phosphate, 459
  and glycolysis, 445
  interconversion with fructose-6-phosphate,
    regulation of, 470, 472–473
  production from phosphoenopyruvate,
    457–459
Fructose-2,6-bisphosphate, 472–473
Fructose-bisphosphate phosphatase, 459
  regulation of, 472
Fructose-6-phosphate, 438, 439, 649
  derivatives formed from, *651*
  interconversion with fructose-1,6-bisphos-
    phate, regulation of, 470, 472–473
  interconversion with glucose, 649
  interconversion with glucose-6-phosphate, 443
  interconversion with mannose-6-phosphate,
    650
  production from fructose-1,6-bisphophate, 459
  in transaldolase reaction, *474*
Fruit flies, development regulation in,
  1033–1039
ftz gene, 1035–1036, *1036*
Fumarase, 492–493
  Michaelis constant for, 270
Fumarate
  converstion to malate, 492–493
  formation from succinate, 492
  formation from tyrosine, *627*
Fungi
  amino acid catabolism regulation in, 640–641
  evolution of, *2*
  lysine biosynthesis in, 761, *763*

Fungisporin, amino acids in, 793
Furanoses, 135, 137, 1232
Fushi terazu gene, 1035–1036, *1036*
Futile cycles. See Pseudocycles

G proteins, 1064, 1066
  classes of, 1064–1065
  transducin, 1221
G_1 and G_2 phases in cell cycle, 1232. See also
  Cell cycle
gal genes in yeast, 1012–1015
gal operon, 987
Galactocerebroside, 165
Galactokinase, 463
Galactose
  linkage with hydroxylysine, *147*
  metabolism of, 1012–1015
    abnormal, 463–464
  structure of, *133, 139*
  synthesis of, 649–650
Galactose operon, 987
Galactose phosphate uridyltransferase, 463
Galactosemia, 463
α-Galactosidase A, deficiency of, 713–714
β-Galactosidase, 977
  deficiency of, 714
Galactosylceramide, 165
  structure of, *166*
Galaxy formation, 1122–1123
Gallbladder, 168
Gametes, 1232
Ganglion cells, retinal, 1220
Ganglioside, 164, 712n
Gangliosidoses, 714
Gap junctions, membrane, 1182–1183, *1183*,
  1203
Gas constant, in thermodynamics, 395
Gastrin, structure and function of, 1048
Gastrointestinal tract, hormones of, 1048
Gating currents in nerve cell membrane
  channels, 1201–1203
Gaucher's disease, 714
Gel electrophoresis, 109, *109*
  in chromosome analysis, 245
Gel-exclusion chromatography, 107–108,
  107–108, 121, 1232
Gene(s), 1232
  amplification of, 1029–1030, 1232
  and development regulation
    in frogs, 1029–1033
    in fruit flies, 1033–1039
  direct-acting, 1135
  engrailed, 1036–1037
  expression of
    in amino acid biosynthesis, 990–995
    in bacteriophages, 1000–1007
    CAP protein affecting, 987–990
    enhancer sequences in, 1027–1029, 1231
    in *Escherichia coli*, 974–1000
    in eukaryotes, 1011–1044
      regulatory mechanisms in, 1029–1043
    hormones affecting, 1070
    internal promoters in, 1031
    lac operon in, 976–986
    in prokaryotes, 974–1010
    proteins regulating, 986–987
    regulation of transcription initiation in,
      975–976
    for RNA polymerase and ribosomes,
      995–999
    steroid receptors affecting, 1068–1073
    TFIIIA protein in, 1031–1033
    in yeast, 1011–1018
  homeotic, 1037–1039
  hypervariable regions in, 1042
  mapping studies, 1117, 1232
  maternal effect, 1035
    bicaudal mutation, 1035
    dorsal mutation, 1035
  mutations in
    constitutive, 979
    structural, 979
  notations for, 979

  regulatory, 1238
  repressor, in *lac* operon, 979–980
  segmentation, 1035–1037
  splicing of, 1232. See also DNA,
    recombination of
  and protein evolution, 96
  structural, 1239
  suppressor, 1239
  wild-type, 1240
Generation time, 1232
Genetic code, 11, 947–952
  code word assignments in, 952
  coding ratio in, 948
  degenerate, 948
  experiments for definition of, 948–952
  of mitochondria, 954–955
  nondegenerate, 948
  origin of, 1144–1145
  wobble rules of codon-anticodon pairing in,
    953, 954, 1240
Genetic concepts, 979
Genetic elements, movable, 1018, 1235
Genome, 1232
Genotype, 1232
Geranyl pyrophosphate, 729
Gibberellins, 172, *1083*, 1084
Gilbert-Maxam sequencing of DNA, 242–247
Globoside, 165
  structure of, *165*
Globular protein, 1232
  folding pattern in, 78–83
Glucagon, 467, 469–470, 600
  and inhibition of acetyl-CoA carboxylase, 685
  molecular weight and subunit composition
    of, 88
  structure and function of, 1048
Glucocorticoids, 708
  and gene regulation, 1070
  receptors for, cDNA clones coding for, 1061
  structure and function of, 1049
Glucokinase, 441
Gluconeogenesis, 439, 452–473, 1232
  consumption of ATP in, 453–455
  energetics and regulation of, 464–473, *465*
  pathway in, *454*
Glucose
  and cAMP levels in cells, 984
  biosynthesis of. See Gluconeogenesis
  catabolism of. See Glycolysis
  conversion of pyruvate, 418, *419*, 438
  cotransport with sodium, 1174
  and formation of glucose-6-phosphate, 23–24
  homopolymers of, 141–142
  interconversion with fructose phosphates, 649
  and lactose permease function, 1178
  oxidation of, and yields of ATP, 558–559
  phosphorylation to ATP, 407
  structure of, *133, 139*
Glucose-alanine cycle, 622, *622*
Glucose-1,6-bisphosphate, 442
Glucose oxidase, 265
  reactions catalyzed by, *363*
Glucose-6-phosphatase, 266
Glucose-1-phosphate, 438, 649
  conversion to starch or glycogen, 459–462
  interconversion with glucose-6-phosphate,
    441–442
Glucose-6-phosphate, 24, 438, 649
  interconversion with fructose-6-phosphate, 443
  interconversion with glucose-1-phosphate,
    441–442
  oxidation of, and NADPH generation, 473–474
Glucose-6-phosphate dehydrogenase,
    nicotinamide side-specificities of, 361
β-Glucosidase deficiency, 714
Glucosides, 137, *137*
Glucosylceramide, 165
Glucuronic acid, 745
  structure of, *139*
Glutamate
  biosynthesis of, 757–761
  catabolism of, 631, *632*
    in bacteria and fungi, 641

cotransport with sodium, 1174
ionization reactions for side chains of, 37
as neurotransmitter, 1207
pK values for, 36
structure of, 33
titration curves of, 36
Glutamate dehydrogenase, 617, 757–758
nicotinamide side-specificities of, 361
Glutamate synthase, 760–761
Glutamate transaminase, 616
Glutaminase, 621
Glutamine
biosynthesis of, 759–760
catabolism of, 631, 632
pK values for, 36
structure of, 33
Glutamine phosphoribosylpyrophosphate amidotransferase, 809, 833
Glutamine synthase, 621, 759–760
molecular weight and subunit composition
of, 88
γ-Glutamyl cycle, 798, 799
γ-Glutamylcysteinylglycine, 797–800
Glutathione, 797–800
Glutathione reductase, 365
reactions catalyzed by, 363
Glyceraldehyde
formation of, 131–132, 132
structure of, 133
Glyceraldehyde-3-phosphate
and glycolysis, 445
interconversion with dihydroxyacetone
phosphate, 446–447
oxidation to glycerate-3-phosphate, 447–448
in transaldolase reaction, 474
in transketolase reaction, 475
Glycerate-2,3-bisphosphate, structure of, 327
Glycerate-2-phosphate, interconversion with
glycerate-3-phosphate, 448–449, 449
Glycerate-3-phosphate
conversion to serine, 764
interconversion with glycerate-2-phosphate,
448–449, 449
production of, 447–448
Glycerol-3-phosphate, 160, 690
Glycerol-3-phosphate acyltransferase, 690, 695
Glycerolipids, biosynthesis of, 689–709
regulation of, 704–706
Glycine, 168
biosynthesis of, 765, 765
catabolism of, 624, 624
hyperglycinemia, 637
as neurotransmitter, 1207
pK values for, 36
and porphyrin biosynthesis, 793–796
structure of, 33, 170
Glycocholate, 168
formation of, 742
structure of, 170
Glycogen, 141, 141–142, 1232
biosynthesis of, 653
configuration and role of, 142–144
formation from glucose-1-phosphate, 459–462
Glycogen phosphorylase, 467
molecular weight and subunit composition
of, 88
Glycogen synthase, 470, 653
Glycogenic amino acids, 623, 1232
Glycohydrolases, 881
Glycolipids, 144, 1232
in membranes, 178, 185
Glycolysis, 422, 437–452, 1232
and amino acid biosynthesis, 757
and ATP generation, 449–452
energetics and regulation of, 464–473, 465
first metabolic pool in, 438–445
hexoses in, 438
organization of, 452
pathway in, 454
second metabolic pool in, 445–448
third metabolic pool in, 448–452
Glycophorin, 191–192, 192
properties of, 191

Glycoproteins, 144, 172, 1233
in membranes, 178
oligosaccharides in, synthesis of, 655–663
synthesis in endoplasmic reticulum, 656–658
Glycosaminoglycans, 145–149, 1233
Glycosides, 137
Glycosidic bonds, 8, 137, 1233
in nucleotides, 218
Glycosphingolipids, 164–165
biosynthesis of, 709–715, 711
catabolism of, 713
function of, 714–715
in membranes, 182
Glycosylhydrolases, 712
Glycosyltransferases, 710
Gluoxylate cycle, 498–501, 1233
compared to tricarboxylic acid cycle, 500
Glyoxysomes, 611, 1233
GMP. See Guanosine monophosphate
GMP synthase inhibitors, 826
Goiter, 1078
Goldman equation, 1194, 1232
Golgi apparatus, 4, 657, 1232
lipid content in membrane of, 180, 185
properties of, 179
Gonadal hormones, 1049
Gonadotropin-releasing hormone, 1050
structure and function of, 1047
Gout, 830–831
Gram molecular weight, 1233
Gram stain reaction, 180
negative, 180
positive, 180
Gramicidin, 800–801
amino acids in, 793
formation of, 801
ion translocation by, 1179–1180
structure of, 800, 1180
Granulocyte colony stimulating factor, 1080
Granulocyte-macrophage colony stimulating
factor, 1080
Grave's disease, 1078
Ground state of molecules, 570, 1233
Group translocation, 1160, 1175–1177
Growth, diauxic, 984, 1231
Growth factors, 1045, 1079–1080, 1233
plant, 1082–1085
Growth fork, 1233
Growth hormone, 1074
release inhibiting factor. See Somatostatin
releasing factor, 1047
structure and function of, 1047
GTP. See Guanosine triphosphate
GTP-binding protein, 1058, 1064
Guanase, 830
Guanidinoacetate methyltransferase, 265
Guanine, 1233
in DNA and RNA, 218, 223
hydrogen bonding to cytosine, 223
structure of, 219
Guanine aminohydrolase, 830
Guanine deaminase, 830
Guanosine, 1233
base found in, 219
in tRNA, 253
uncharged and protonated forms of, 220
Guanosine monophosphate (GMP)
cyclic, 1066
and electrophysiological responses to light,
1221–1223
formation from inosine monophosphate, 811, 812
ionization constants of, 219
molar absorbance of, 220
Guanosine tetraphosphate (ppGpp), in rRNA
synthesis, 996–997
Guanosine triphosphate (GTP)
biosynthesis of, 490–492
in phosphoenolpyruvate production, 455–457
in ribosomal reactions, 945, 946
Guanylate, base found in, 219
Guanylate cyclase
pathway of hormone receptor action, 1066
in rod outer segment, 1221

D-Gulose, structure of, 133
gyrA gene, 857, 863
Gyrases, and production of supercoiled DNA,
234, 235
gyrB gene, 857, 863

H zone in sarcomere, 89, 89
Hairpin loop, 1233
Half-life, 1233
of hormones in blood, 1058
or proteins in cells, 964
Halophilic bacteria, purple membrane in, 193,
1168–1169, 1225
Halorhodopsin, 1226
Haploid cells, 979, 1233
in yeast, mating types of, 1016–1018
Haworth structures, 135–136, 136
Heart
atrial natriuretic peptide, 1048
regulation of β oxidation in, 612
Heat
and enthalpy charges, 395
equivalent to work, 394
thermogenesis in brown fat and skunk
cabbage, 555
Heat-shock genes, 966
Heavy isotopes, 1233
Helicases, 864, 867–868
Helix, 1233
alpha, 71, 71–72, 84–85
in keratins, 76
in membranes, 193–194
in trypsin family of enzymes, 288
double helix structure of DNA, 20, 222–232,
1231
pitch length of, 224, 1235
unwinding of, 903
triple helix in adenosine derivatives, 1136
Hematin, 381
Hematoside, 165
Heme, 92–93, 381–383, 515, 1233
in electron transport, 515–519
Hemiacetals, 15, 1233
intramolecular, 134–137
Hemoglobin, 88–89, 326–336, 515
abnormal, degradation of, 966
heme groups in, 518
invariant residues in, 334, 334–335
molecular weight and subunit composition
of, 88
mutations in, 335
oxygen binding by, 326–327
carbon dioxide affecting, 328
and changes in conformation, 332–335
pH affecting, 327–328
sequential model of, 336, 336
symmetry model of, 336, 336
physical constants of, 106
two conformations for, 329–331
Henderson-Hasselbach equation, 35, 1233
Henri-Michaelis-Menten equation, 267
Heparan sulfate, structure of, 145
Heparin, 147
structure of, 145
Herpes simplex virus, 1111
Heterochromatin, 1020, 1233
constitutive, 1020
facultative, 1020
α-(N)-Heterocyclic carboxaldehyde
thiosemicarbazone, 822, 826
Heteroduplex, 1233
Heterokaryon, from cell fusions, 198
Heteropolymers, 140, 1233
Heteropolysaccharides, 144–152
Heterothallic strains of yeast, 1016
Heterotrophs, 417, 1233
Heterozygosity, 1233
Hexathymidylate, 1138
Hexokinase, 440–441
domains in, 87, 87
Michaelis constant for, 270
molecular weight and subunit composition
of, 88

Hexosaminidases, deficiency of, 714
Hexose, 1233
    biosynthesis of, 647–652
    interconversions of, 648
Hexose monophosphate shunt, 476
Hexose phosphates
    conversion to polysaccharides, 459–462
    in glycolysis, 438–439
High-energy compound, 1233
High-mobility group (HMG) proteins, 1026–1027
Hill plot, and multiple binding, 328
HIS4 gene, 1012, 1013
Histamine
    biosynthesis of, 1207
    as neurotransmitter, 1207
Histidase, 635
    biosynthesis of, 785–788, 786–787
    catabolism of, 634–635, 635
        in bacteria and fungi, 641
    and chymotrypsin-catalyzed hydrolysis, 293
    ionization reactions for side chains of, 37
    pK values for, 36
    structure of, 33
Histidine ammonia-lyase, 266
Histidinemia, 637
Histone(s), 1023–1026, 1233
    characteristics of, 1024
    linkage with ubiquitin, 1026, 1026
    modifications of, 1025–1026
    multiple copies and storage of, 1033
Histone octamer, 1024, 1025
HMG-CoA. See β-Hydroxy-β-methylglutaryl-CoA
HMG proteins, 1026–1027
HML locus of yeast chromosome, 1016, 1016–1017
Holoenzyme, 264, 1233
    DNA polymerase III, 857, 861
    RNA polymerase, 897–898
Homeo box, 1038, 1039
Homeotic genes, 1037–1039
Homocysteine synthase, 766
Homocystinuria, 637
Homogenization, mild, for membrane analysis, 178
Homogentisate-1,2-dioxygenase, 626
Homoionic bentonite clays, 1142
Homoisoleucine, 794
Homologous chromosomes, 1233
Homopolymers, 140–142, 1233
Homoserine, 769
Homoserine dehydrogenase, 769
Homoserine kinase, 773
Homothallism, 1016
Hormones, 168, 1045–1087, 1233
    activating adenylate cyclase, 1062
    affecting glycogen metabolism, 467–470
    biosynthesis of, 1048–1057
    and endocrinopathies, 1077–1079
    and gene regulation, 1070
    half-life in blood, 1058
    inhibiting adenylate cyclase, 1062, 1064–1065
    mechanisms of action, 1058
    overproduction of, 1078
    plant, 1082–1085
    receptors for, 1046, 1058–1073, 1233
        absence of, 1079
        acceptors for, 1058
        adenylate cyclase pathway for, 1062, 1062–1066
        binding to, 1059–1060
        detection of, 1060–1061
        down–regulation of, 1077, 1077
        guanylate cyclase pathway for, 1066
        inositol triphosphate pathway for, 1066–1068
    regulation of responses to, 1073–1077
        feedback control in, 1075
    regulation of serum concentrations of, 1057–1058
    release of, 1051, 1077
    storage in granules, 1051
    underproduction of, 1078
Host cell, 1233

htpR gene, 905
Humidity, and formation of nucleotides, 1140–1141
Hyaluronic acid, 146–147
    structure of, 145, 146
    synthesis of, 655
Hybrid
    nucleic acid, 239
    plasmid, 1233
    resonance, 39, 1238
Hydration, entropy in, 398
Hydrocarbons, hydroxylation of, 383
Hydrogen
    contrasport with substrates, 1174
    standard half-cell, 523
    transfer in nicotinamide coenzymes, 358–360, 359
    valence of, 14
Hydrogen ATPase, 1166
Hydrogen bonds, 17, 1233
    as forces in protein structure, 63–64
    of water, 57
Hydrogen cyanide
    in atmosphere of primitive Earth, 1125
    hydrolysis of polymers in amino acid formation, 1128, 1128
    in synthesis of nucleic acid bases, 1129–1131
Hydrogen sulfide, in atmosphere of primitive Earth, 1125
Hydrogenation reaction, 15
Hydrolases, 266
Hydrolysis, 1233
    acid, 40
    of ATP, 408–411
Hydrophilicity, 7, 1233
    of amino acid side chains, 56
Hydrophobic bonding, 1233
Hydrophobic interaction chromatography, 189, 190
Hydrophobicity, 7, 1233
    of amino acid side chains, 56
    in protein structure, 64–65
    of resins, 189–190
β-Hydroxyacyl-ACP dehydrase, 681
3-L-Hydroxyacyl-CoA dehydrogenase, 604
Hydroxyacyl-CoA epimerase, 604
3-Hydroxyanthranilate oxygenase, 631
Hydroxyapatite, 1134, 1233
β-Hydroxybutyrate, synthesis in liver, 610, 611
β-Hydroxybutyrate dehydrogenase, 610
7α-Hydroxycholesterol
    conversion to cholic acid, 742
    formation of, 741
Hydroxycobalamin, 377, 379
β-Hydroxydecanoylthioester dehydrase, 680–681
Hydroxyeicosatetraenoic acids (HETEs), structures of, 720
β-Hydroxyacyl-ACP dehydrase, 673, 679
7α-Hydroxylase, 741
11-β-Hydroxylase, 743
17α-Hydroxylase, 743
    deficiency of, 1054
18-Hydroxylase, 743
21-Hydroxylase, 742
    deficiency of, 1054
Hydroxylation of hydrocarbons, 383
Hydroxylysine, linkage with galactose, 147
β-Hydroxy-β-methylglutaryl-CoA (HMG-CoA), 610
β-Hydroxy-β-methylglutaryl-CoA lyase, 610, 726
β-Hydroxy-β-methylglutaryl-CoA reductase, 726, 727–728, 728, 738
β-Hydroxy-β-methylglutaryl-CoA reductase kinase, 728
β-Hydroxy-β-methylglutaryl-CoA reductase kinase kinase, 728
β-Hydroxy-β-methylglutaryl-CoA reductase kinase phosphatase, 728
β-Hydroxy-β-methylglutaryl-CoA reductase phosphatase, 728
β-Hydroxy-β-methylglutaryl-CoA synthase, 610, 726
Hydroxyphenylhydrazinoisocytosine, structure of, 878

Hydroxyphenylhydrazinouracil, structure of, 878
4-Hydroxyphenylpyruvate dioxygenase, 626
17α-Hydroxyprogesterone, 744
21-Hydroxyprogesterone, 743
β-N-(3-Hydroxy-4-pyridone) alanine, 794
Hydroxytamoxifen, structure of, 1060
19-Hydroxytestosterone, 745
5-Hydroxytryptamine, 797
    biosynthesis of, 1207
    as neurotransmitter, 1207
Hydroxyurea, as enzyme inhibitor, 826
25-Hydroxyvitamin D₃, 171
Hyperpolarization, 1196
    of rod cells, 1220
Hypervariable regions in gene segments, 1042
Hypothalamus, hormones of, 1047
Hypotonic solutions, 178n
Hypoxanthine-guanine phosphoribosyl-transferase, 818, 1111–1112
    deficiency of, 819, 819

I band of sarcomere, 89, 89
i gene, 977, 979–980
Ice, arrangement of molecules in, 57, 57
Icosahedral symmetry, 1233
D-Idose, structure of, 133
L-Iduronic acid, structure of, 139
ilv operon, 995
Imaginal disks, 1035
Imines, 15, 1233
Immune system, 90
    antibody synthesis in, 1039–1042
Immunofluorescence, 1233–1234
Immunoglobulins, 90, 91, 1234. See also Antibodies
IMP. See Inosine monophosphate
Indolegylcerol phosphate, 782
Indoleglycerol phosphate synthase, 782
Induced fit, 300, 302, 1234
Inducers, 1234
Induction of enzymes, 977–978
Inhibins A and B, structure and function of, 1049
Inhibitors of enzymes, 272–277
    noncompetitive, 273, 275, 1235
Initiation factors, 1234
Inosine, in tRNA, 253
Inosine monophosphate (IMP)
    biosynthesis of, 809–810
    conversion into AMP and GMP, 811, 812
Inosine monophosphate dehydrogenase, inhibitors of, 823
Inositol phospholipids, biosynthesis of, 700–701
Inositol triphosphate, 689, 701, 702
    pathway of hormone receptor action, 1066–1068
    structure of, 1067
Insertion elements or transposons, 1018
Insulin, 469
    amino acid content of, 43
    biosynthesis of, 963, 964, 1046
    molecular weight and subunit composition of, 88
    receptor for, biosynthesis of, 1061, 1062
    structure and function of, 1048
Insulin-like growth factor, 1079, 1080
Intercalation, 233, 233, 1234
Interferons, 1045, 1080–1082, 1234
    activation of antiviral proteins, 1081–1082, 1082
    induction and action of, 1081
Interleukins, 1080
Intervening sequence, 1234
Intestines, lipoprotein synthesis in, 735–736
Introns, 918, 1234
Inverted repeat, 1234
In vitro, 1234
Iodoacetate, as enzyme inhibitor, 275
Ion-exchange chromatography, 41, 41, 121–122
Ion-exchange resin, 1234
Ionic strength, and protein solubility, 101, 102
Ionization reactions, 16
    entropy in, 398

Ionophores, 547, 1179–1181
Iron, and oxygen binding by hemoglobin, 332–334
Iron-protoporphyrin IX, 515, *515*
Iron-sulfur clusters, 382
   in electron transfer, 520
   structures of, *382, 520*
Iron-sulfur proteins
   in photosystem I, 584
   of Rieske, 532
Isocitrate dehydrogenase, 490, 508
   regulation of, 505–508
Isocitrate lyase, 498
Isoelectric focusing, 109
Isoelectric pH, 35, 101, 1234
Isoenzymes of lactate dehydrogenase, 321
Isoleucine
   biosynthesis of, 773, 776–779, *777*
   catabolism of, *635*, 635–637, *636*
   pK values for, 36
   structure of, 33
Isomerases, 266, 1234
Isomerization, 16, 1234
Isoosmotic solutions, 180n
N^6-Isopentenyladenosine, in tRNA, 253
Isopentenyl pyrophosphate, 729
Isoprene, 171–172, 726
Isoprenoids in membranes, 184
Isopycnic centrifugation, 178, *178*
Isorhodopsin, 1218
Isotopes, heavy, 1233
Isovaleric acidemia, 637
Isozymes, 1234

Joule, 395
Juvenile hormones, 172, *173*
   structure of, *1076*

K_m. *See* Michaelis constant
Kepone, structure of, *1060*
Keratan sulfate, structure of, *145*
Keratins, 75–77
Ketoaciduria, branched-chain, 637
β-Ketoacyl-ACP reductase, 673, 679
β-Ketoacyl-ACP synthases, 673, 677
α-Ketoadipate, catabolism of, 631, *631*
Ketogenic amino acids, 623, 1234
α-Ketoglutarate
   amination of, 757–758, 760–761
   from amino acid degradation, 631–635
   and lysine biosynthesis, 761
   produced from pyruvate, 497
α-Ketoglutarate dehydrogenase, 490
   regulation of, 508
Ketone, 15–16, 1234
   formation of, 15
Ketone bodies, 610, *611*, 1234
   biosynthesis of, 726
Ketone monooxygenase, reactions catalyzed by, 363
Ketoses, 131, 133
Ketosis, 1234
3-Ketosphinganine reductase, 709
3-Ketosphinganine synthase, 709
Kilobase, 1234
Kinase, 1234
Kinetic barrier, 262
Kinetics of enzyme-catalyzed reactions, 266–272
Kinetochore, 1234
Kornberg enzyme, 859–861
Krebs bicycle, 621, *621*
Krebs cycle. *See* Tricarboxylic acid cycle
Kynureninase, 631
Kynurenine formamidase, 631
Kynurenine-3-monooxygenase, 631

lac operon, 976–986
   catabolite repression of, 984–986
   genetic elements of, *977*
   inducers of, 977–978, *978*
   operator of, 977
      mutations in, 980–981
   promoter of, 977

promoter-operator region of, *984*
   repressor of, 901, 979–980
      binding of inducer to, 982
      isolation and action of, 982–984
Lactase, 464
Lactate
   formation from pyruvate, 521–522
   incorporation of carbon into valine and
      isoleucine, *778*
Lactate dehydrogenase, 315–321, 450
   binding sites of, 316–319
   as bisubstrate enzyme, 315
   catalysis of, 319–320, *320*
   isoenzymes of, 321
   kinetic studies of, 319
   molecular weight and subunit composition of, 88
   nicotinamide side-specificities of, 361
   reactions catalyzed by, 363
Lactate oxidase, *365*
   reactions catalyzed by, 363
Lactobacillic acid, structure of, 155
Lactogen, placental, 1049
β-Lactoglobulin
   physical constants of, 106
   titration curve of, 101, *101*
Lactose
   carrier protein of *Escherichia coli*, purification
      of, 126–128, *127*
   catabolism of, 463–464
   conversion to allolactose, 978
   cotransport with hydrogen, 1174
   structure of, *138*
Lactose operon, 976–986. *See also lac* operon
Lactose permease, 963, 977, 1161, 1172–1173,
      1173
   glucose affecting, 1178
   properties of, 191
Lactose synthase, 652
Lactosylceramide, 165
lacZ gene, 962
Lampbrush chromosome, 1234
Lanosterol, 172, 729
   and cholesterol biosynthesis, 729–733
Lattice defects, 1135
Law of mass action, 1234
Leader region, 992, 1234
Lecithin, 160, 161. *See also* Phosphatidylcholine
Lecithin: cholesterol acyltransferase (LCAT), 740
Lectins, 656, 1234
Lens of eye, 1213
Lesch-Nyhan syndrome, 818–819
Leucine
   biosynthesis of, 779, *779*
   catabolism of, 625, *626*
   in gramicidin synthesis, 800
   pK values for, 36
   structure of, 33
Leukodystrophy, globoid cell, 714
Leukotrienes, 717
   biosynthesis of, 720, *721*
Levorotatory rotation, 114, 132
lexA gene, 884, 885
Library, DNA, 1101–1107, 1231
Life
   ancient forms of, 1126
   definitions of, 1121
   origins of, 1120–1151
      assumptions in, 1121–1122
      and atmosphere of primitive Earth, 1125–
         1126
      and coevolution of nucleotides and pep-
         tides, 1143–1148
      condensing agents in, 1132–1134
      and cosmochemistry, 1123–1125
      galaxy and solar system formation in,
         1122–1123
      membranes and compartmentation in,
         1148–1150
      nucleotide polymerization in, 1136–1141
      peptide formation in, 1142–1143, 1146–
         1148
      and stages in chemical evolution, 1122
      time line for, *1121*

lig gene, 857, 863, 884
Ligases, 266, 1234
Light
   in bioluminescence, 593–595
   and electron-transfer reaction, 571–572
   and electrophysiological responses of rods or
      cones, 1220–1221
      guanine nucleotides affecting, 1221–1223
   interaction with molecules, 570–571
   and isomerization of retinal, 1216
   and membrane transport, 1168–1169
   and photosynthesis, 564–590
   physical definition of, *569*, 569–570
   polarized, 113
   and rhythms in plants, 592–593
   scattering procedure for molecular weight
      determination, 103, *103*, 104
   ultraviolet. *See* Ultraviolet radiation
Lightning, as energy source, 1127
Lignoceric acid, structure of, 155
Limonene, 172
   structure of, *173*
Lineweaver-Burk equation, 269, *269*
   and plot of carrier-mediated transport, 1158
Linkers, 1234
Linking number, 233, *234*, 1234
Linoleic acid, 156
   and arachidonic acid synthesis, 682, *682*
   structure of, 155
Linolenic acid, 156
   conformational model of, *157*
   structure of, 155
Lipases, 598
   intestinal, 168
Lipid(s), 7, 154–175, 1234
   in bilayers, 162, 186–187, 195–196, 1229,
      1234
      planar, 1188–1189
   fatty acids, 154–159
   in membranes, 178, 182–185, 1149
      asymmetry of, 203–204
      in organellar membranes, 180
      synthesis and insertion of, 207
   monolayer of, 185, 1235
   in myelin, 165, 185
   phospholipids, 160–166
   plasma concentrations of, 733
   prebiotic formation of, 1150
   solubilization by bile salts, 168
   structure of, 7
   types found in plasma, 733–735
Lipid A, 152
Lipid-soluble vitamins, 170–171
Lipocortin, 708, 722
α-Lipoic acid, 369–371
Lipopolysaccharides, 144, 151–152, 1234
   in bacterial cell envelope, 180, *181*, 184
      antigenic behavior of, 152
   in membranes, 178
Lipoprotein, 168, 598, 733
   apoproteins of, 734, 735
   biosynthesis in liver and intestine, 735–736
   composition of, 734
   high-density, 733–734
      and cholesterol levels in plasma, 740
   intermediate-density, 734
   low-density, 165, 734
      removal from plasma, 737–738
   structure of, *734*
   very-low-density, 734
      synthesis in liver, *736*
      transporting cholesterol and triacylglycerol,
         736–737
Lipoprotein lipase, 599, 736
Lipotropin, 1050
   structure and function of, 1047
Lipoxygenase, 720
Liver
   ammonia transport to, 621–622
   gluconeogenesis in, 466
   ketone bodies formed in, 610, *611*
   lipoprotein synthesis in, 735–736
   urea formation in, 619–621

Lovastatin, 729
Luciferases, 594
Luciferins, 594, *594*
Lumirhodopsin, 1217, 1219
Lung surfactant, 696
Luteinizing hormone, 1050
  structure and function of, 1047
Lyases, 266, 744, 1234
Lycopene, 172
  structure of, *173*
Lymphocyte B and T cells, 1039–1040
Lysine
  biosynthesis of
    in bacteria and plants, 769–771, *770*
    in fungi, 761–764, *763*
  catabolism of, 625, *626*
  hyperlysinemia, 637
  ionization reactions for side chains of, *37*
  pK values for, 36
  structure of, 33
Lysogenic pathway for bacteriophage develop-
    ment, 1002
Lysogenic virus, 1234
Lysolecithin, 188, *188*
Lysophosphatidylcholine, 162
Lysophospholipases, 708
Lysophospholipids, 161–162, *162*
  in membranes, 186
Lysosomal acid lipases, 738
Lysosomes, 4, 662, 1234–1235
  lipid composition of membranes, 180, 185
  properties of, 179
  in protein degradation, 966–967, *967*, 969–
    970
Lysozyme, 308–315
  active site in, 309
  amino acid residues in, *310*
  binding of substrates, 21, 311–313
  catalytic mechanism in, 313–315
  interaction with hexamer, 311, *313*
  molecular weight and subunit composition
    of, 88
  structure of, *78*
  treatment of cells with, 180, *181*
D-Lyxose, structure of, *133*

M line of sarcomere, 89, *89*
M phase of cell cycle, 1235. *See also* Cell cycle
Macrolide antibiotics, 174
Macrophage colony stimulating factor, 1080
Magnesium
  and ATP hydrolysis, 410–411
  and membrane fluidity, 200
mal genes, 962
Malate
  formation from fumarate, 492–493
  formation from oxaloacetate, 506
  oxidation to oxaloacetate, 493
  transport into mitochondria, 556
Malate dehydrogenase, 493
  binding domains for, *317*
  nicotinamide side-specificities of, 361
Malathion, 1205, *1205*
Malformin, amino acids in, 793
Malic enzyme, 671
Malonyl-CoA
  binding of, 677
  and fatty acid synthesis and regulation, 670,
    686
Malonyl-CoA-ACP transacylase, 673
Malonyltransacylase, 677
Maltose, structure of, *138*
Mannitol permease, properties of, 191
D-Mannose, structure of, *133*, *139*
Mannose-1-phosphate, 649
Mannose-6-phosphate, 649
  interconversion with fructose-6-phosphate, *650*
Maple syrup urine disease, 637
Mapping
  of genes, 1117, 1232
  peptide, 1236
    tryptic peptide, 1240
Mars, atmosphere of, 1125

Mass, conservation of, 423
Mass action, law of, 1234
MAT locus on yeast chromosome, *1016*, 1016–
    1017
Maternal effect genes, 1035
  bicaudal mutation, 1035
  dorsal mutation, 1035
Maxam-Gilbert sequencing of DNA, 242–247
Mechanical work, 405
Meiosis, 1235
Melanin formation from tyrosine, *627*
Melanocyte-stimulating hormone, structure and
    function of, 1047
Melanosomes, 627
Melatonin, structure and function of, 1047
Melibiose cotransport with sodium, 1174
Melting, entropy in, 397
Membranes, 176–209
  amphipathic properties of, 185
  asymmetry in, 197, 202–204, 1155
  bilayer structures of, 186–187, 195–196,
    1229, 1234
    leaflets of, 202–203
    planar, 1188–1189
  biosynthesis and assembly of, 205–208
  carbohydrates in, 148–149
  coated pits in, 1077
  coated regions of, 737
  constituents of, 178–195
  diffusion in, 198–199
    factors affecting, 199–202
  dynamic, 1155
  electron paramagnetic resonance spectroscopy
    of, 198, *198*
  in eukaryotes, analysis of, 178–180
  flip-flop of molecules in, 199, 206
  fluid mosaic model of, 196, *196*
  fluorescence photobleach recovery studies of,
    197–198
  gap junctions in, 1182–1183, *1183*, 1203
  lipids in, 178, 182–185, 1149
    asymmetry of, 203–204
    in bilayer, 162, 186–187, 195–196, 1229, 1234
    in organellar membranes, 180
    synthesis and insertion of, 207
  micelles of, 186
  mitochondrial, 513–514
  monolayers of, 185
  of organelles, properties of, 179
  permeability barrier in, 196–197
  photosynthesis in, 565–567, *566*
  postsynaptic, 1203
  prebiotic formation of, 1148–1150
  presynaptic, 1203
  proteins in, 178, 179
    arrangement of, 191–195
    asymmetry of, 202
    mobility of, 197–198
    in organellar membranes, 180
    properties and isolation of, 187–191
    synthesis and insertion of, 207
  structure of, 195–204
  thylakoid, 565
  transport across, 1155–1192, 1235. *See also*
    Transport across membranes
  turnover process in, 207
  vectorial character of, 548
  visual pigments in, 1213–1225
Menaquinone, 171, 521
  structure of, *171*, 573
6-Mercaptopurine, 822, 823
p-Mercuribenzoate, as enzyme inhibitor, 275
Merodiploids, 979–980, 1235
Merrifield process for peptide synthesis, 50, *51*
Mesophyll cells, 591, *592*
Mesosomes, 1235
Messenger RNA, 12, 893–894, 1235. *See also*
    RNA, messenger
Metabolic turnover, 1235
Metabolism, 1235
  aerobic respiratory, 481–482
  and direct transfer of intermediates by en-
    zyme-enzyme complexes, 460

  futile cycles in, 426. *See also* Pseudocycles
  intermediate, 416
  kinetic and thermodynamic factors in, 432–
    433
  pathways in, 418–423
    anabolic, 23, 418
    branched-chain, 429
    branchpoints of, 419, 421, *421*
    catabolic, 23, 418, 1229
    convergence of, 419, *420*
    conversions in, 424–427
    coupled to ATP-ADP system, 423–424
    divergence of, *420*, 421
    energy charge affecting, 430–431
    functional interrelationships in, 422
    regulation of, 427–433
    regulatory enzymes in, 428–430
    pseudocycles in, 426. *See also* Pseudocycles
  sequences in, 425, *425*
Metabolites, secondary, 172
Metalloenzymes, zinc, 298–303
Metalloflavoproteins, 364
Metamorphosis, 1235
Metaphase, 1235
Metarhodopsins, 1217, 1219–1220, 1221
Meteorites, 1123–1124
  amino acids in, 1128
Methane, in atmosphere of primitive Earth,
    1125
Methanofuran, structure of, *384*
Methanogenesis, 383–386
5,10-Methenyltetrahydrofolate, 374–375
  structure of, *374*
Methenyl-tetrahydromethanopterin, structure of,
    384
Methionine
  biosynthesis of, *771*, 771–773
  catabolism of, 635, 637–639
  and initiation of protein synthesis, 938–939
  pK values for, 36
  structure of, 33
Methotrexate, 376, *376*, 827, 828
1-Methyladenosine, in tRNA, *253*
Methylcobalamin, 381
Methyl-CoM, 385
S-Methylcysteine, 794
5-Methylcytidine, in tRNA, *253*
5-Methylcytosine, in DNA, 223
2-(Methylenecyclopropyl) glycine, 794
Methylene tetrahydrofolate, 764
2,3,6-tri-O-Methylglucose, 140, *140*
Methyl-α-glucoside, 1162, *1162*
4-Methylglutamic acid, 794
1-Methylguanosine, in tRNA, *253*
1-Methylinosine, in tRNA, *253*
Methylmalonic acidemia, 637
Methylmalonic aciduria, congenital, 608
Methylmalonyl-CoA, 635
Methylmalonyl-CoA mutase, 608
Methylmalonyl-CoA racemase, 608
5-Methyltetrahydrofolate, *374*, 375
Methymycin, 174
  structure of, *174*
Mevalonic acid, 726
  biosynthesis of, *727*
  and cholesterol biosynthesis, 726–729
  conversion to squalene, 731
Mevinolinic acid, structure of, *729*
Micelles, 186, 1235
  aminoacyl adenylates in, 1144
  mixed, 188–189, *189*
Michaelis constant (K_m), 268, 1235
  for ester hydrolysis, 295
  relationship with dissociation constant, 270
  significance of, 269–270
Michaelis-Menten equation, 267, 1235
Microfilaments, 201
Microtubules, 201, 1235
Mimosine, 794
Mineralocorticoids, structure and function of,
    1049
Mismatch repair, 882, 1235
Missense mutation, 1235

Mitochondria, 4, *513*, 513–514, 1235
  ATP/ADP exchanger in, 556–557, 1174–1175, 1186
  cycling of citrate to cytol, 506–508, *507*
  DNA replication in, 872–873, *873*
  fatty acid oxidation in, 604–607
  fatty acid transport into, 602–603
  genetic code of, 954–955
  membranes of, 513–514
    lipid composition of, 180, 185
  properties of, 179
Mitogens, 1079
Mitosis, 11, *11*, 1235
Mixed function oxidases, 741
Mixed micelles, 188–189, *189*
Modification enzymes, 881
Molecular weight, gram, 1233
Molecularity of reactions, 261, 1235
Molecules
  interstellar, 1123
  simple, 1122
Monoamine oxidase, 1206
  reactions catalyzed by, 363
Monoglyceride, 158
Monolayer, 185, 1235
Mononucleotide formation by salvage pathways, 817–821
Monooxygenases, reactions catalyzed by, 363
Monophosphates, nucleoside, 221, *221*
Monosaccharides, 8, 131–137
  configurational relationships of, 133, *133–134*
  formation of intramolecular hemiacetals, 134–137
  and glycosidic bond, 137
Monoterpenes, 172, *173*
Moon, craters on, *1123*
Mosaics, protein, 198
Motilin, structure and function of, 1048
Motive force, proton, 551, 1171–1172, 1237
Movable genetic elements, 1018, 1235
Multicolony stimulating factor, 1080
Muscle contraction, sliding-filament model of, 89, *90*
Mutagen, 1235
Mutagenesis, 1235
  directed, 1231
Mutant, 1235
  in DNA replication, 858
Mutarotation, 135, 1235
Mutation, 1235
  constitutive, 979
  and definition of life, 1121
  frameshift, 1232
  missense, 1235
  nonsense, 1235
  polar, 1237
  in polymerization of biomonomers, 1135
  somatic, 1042
  structural, 979
  suppressor, 1239
Mycobacillin, amino acids in, 793
Mycolic acids, 156
  structure of, 155
Myelin, lipids in, 165, 185
Myelin sheath, 1198
Myeloma cells, 1040
Myofibril, 1235
Myoglobin, 92, 325–326, 515
  molecular weight and subunit composition of, 88
  physical constants of, 106
Myosin, 89–90, 1235
  physical constants of, 106
Myristic acid, structure of, 155

N protein, bacteriophage, 1003, 1007
NAD. See Nicotinamide adenine dinucleotide
NADH. See Nicotinamide adenine dinucleotide, reduced form of
NADH dehydrogenase complex, in electron transport, 530–531
NADP. See Nictonamide adenine dinucleotide phosphate

NADPH
  in fatty acid synthesis, 670–673, *671*
  generation by glucose-6-phosphate oxidation, 473–474
NADPH: cytochrome P_{450} reductase, 741
Naringenin, 174
  structure of, *174*
Nascent RNA, 1235
Negative control, 1235
Nernst equation, 406, 524–525, 1235
Nerve growth factor, 1080
Nerve-impulse propagation and transmission, 1194–1210
Neurohypophysis, hormones of, 1047
Neurotransmission, 1193–1212
  action potentials in, 1195–1198
  gated pores in, 1201–1203
  and membrane channels for sodium and potassium, 1198–1201
  and resting membrane potential, 1194–1195
  synapse in, 1203–1205
Neurotransmitters, 1045
  acetylcholine, 1203–1205, *1204*
  amino acid, 1207
  catecholamine, 1205–1207
  inhibitory, 1207
  receptors for, 1203, 1204, 1208–1210
Nicotinamide, 348
  coenzymatic forms of, 356–361
    hydrogen transfer in, 358–360, *359*
  mechanisms of action, *360–361*
  structure of, *357*
Nicotinamide adenine dinucleotide (NAD), 356–361
  biosynthesis of, 839–840, *840*
  in oxidation reactions, 315
  ratio to NADH, and isocitrate dehydrogenase activity, 505–508
  reduced form of (NADH), 315–357
    in electron transport, 514
    and α-ketoglutarate dehydrogenase regulation, 508
    and pyruvate reduction to lactate, 521–522
    ratio to NAD, and isocitrate dehydrogenase activity, 505–508
    shuttle system for movement of, 557–558
    structure of, *357*
  structure of, *357*
Nicotinamide adenine dinucleotide phosphate (NADP), 356–357
  biosynthesis of, 839–840
  structure of, *357*
Niemann-Pick disease, 714
Nigericin
  ion translocation by, 1181, *1181*
  structure of, *1181*
Ninhydrin reaction, 41, *41*
Nitrate, reduction of, 755
Nitrogen
  in atmosphere of primitive Earth, 1125
  cavitation for membrane analysis, 178
  fixation of, 753–755, 1235
  production and utilization of, 24
  removal from amino acids, 616–618
  valence of, 14
Nitrogen cycle, 752–753, *753*, 1235
Nitrogen-linked oligosaccharides, 658–662
Nitrogenase, 382, 753–755
Nitrogenous base, 1235
p-Nitrophenol, formation of, 295, *295*
o-Nitrophenyl-β-galactoside (ONPG), 1161, *1162*
Nitroxide groups in spin labels, 198
Nonsense mutation, 1235
Norepinephrine, 628
  biosynthesis of, 1053, *1053*, 1206
  as neurotransmitter, 1205–1206
  structure and function of, 1049
Northern blotting, 239, 1235
Novobiocin, 236
ntrA gene, 905
Nuclease, 1235
  in DNA degradation, 878–880

Nucleic acids, 10–12, 211–256, 1235. *See also* DNA; RNA
  genetic significance of, 211–216
  interaction with proteins, 252–254
  structure of, *10*
  synthesis of bases in primitive Earth, 1129–1131
Nucleohistone, 1235
Nucleolus, 1235
Nucleophilic group, 1236
Nucleoproteins, 252–254
Nucleoside(s), 217
  formation by dry-phase heating of precursors, 1131–1132
Nucleoside diphosphate, 221, *221*
Nucleoside diphosphate kinase, 455, 821
Nucleoside monophosphate, 221, *221*
  formation by salvage pathways, 817–821
Nucleoside monophosphate kinases, 821
Nucleoside triphosphates, 221, *221*
Nucleosomes, 871–872, 1024, 1236
Nucleotide(s), 10, 217–221, 1236
  biosynthesis of, 806–843
    in bacteriophage T4 infection, 838
    in cell cycle phases, 836–838
    channeling of metabolites in, 835–836
    by de novo pathways, 808–814
    deoxyribonucleotides, 814–817
    inhibitors of, 821–828
    purine ribonucleotides, 808–811
    pyrimidine ribonucleotides, 811–814
    regulation of, 832–838
    by salvage pathways, 817–821
  catabolism of, 828–832
    regulation of, 829
  coevolution with peptides, 1143–1148
  formation by dry-phase heating of precursors, 1131–1132
  forming diphosphates and triphosphates, 221
  polymerization of
    linkages in, 1140–1141
    nontemplate reactions in, 1136–1138
    templates in, 1138–1140
  polynucleotide chain, 221
  structure of, *10*, 217, *217*
Nucleotide coenzymes, biosynthesis of, 838–841
Nucleotide triplets, 949–952
Nucleus, cellular, 4, 1236
  differentiation during early development, 1019–1020
  lipid composition of membrane, 180, 185
  macronucleus, 1023
  micronucleus, 1023
  properties of, 179
  transplantation technique, 1019, *1020*
nusA factor for RNA transcription termination, 904
Nutrition, amino acids in, 750–752

O antigen, 151–152
Octamer, histone, 1024, *1025*
Octylglucoside, 188, *188*
Okazaki fragments, 854–855, 1236
Oleic acid, 156
  structure of, 155
Oligomers of uridylate, 1139
Oligomycin, structure of, *541*
Oligonucleotides, 221, 1236
Oligopeptides, 38
Oligosaccharides, 148, *148*, 1236
  nitrogen-linked, 658–662
  oxygen-linked, 662–663
  synthesis of, 655–663
Oligosaccharins, *1083*, 1084–1085
Omega oxidation of fatty acids (ω oxidation), 610
Oncogenes, 1080, 1112, 1236
Oocytes, 1029
Operon, 900, 981–982, 1236
  arabinose, 987, *987*
  galactose, 987
  lactose, 976–986. *See also lac* operon
  threonine, 995
  tryptophan, 990–991

Operon (*Continued*)
  valine, 995
Opsin, 172, 1213, 1215–1216
Optical rotatory dispersion, 114, *114*, *115*, 1236
Organelles, cellular, 4, 176, *177*, 1236
  in eukaryotes, RNA polymerases in, 907–908
  membranes of, 180
    protein and lipid in, 180
  properties of, 179
Organophosphoric acid, 16
oriC region of *Escherichia coli* chromosome, 869
Ornithine
  catabolism of, 634
  in gramicidin synthesis, 800
Ornithine decarboxylase, 634
Ornithine-glutamate transaminase, regulation of, 640
Ornithine transcarbamoylase, 620
Ornithinemia, 637
Orotate 5′-monophosphate decarboxylase, and biosynthesis of UMP, 123–126
Orotate phosphoribosyltransferase, and biosynthesis of UMP, 123–126
Orotic acid, structure of, *812*
Orotidine-5′-monophosphate (OMP), in pyrimidine nucleotide biosynthesis, 811
Orotidylate decarboxylase, inhibitors of, 825
Osmotic pressure, 102–103, 1236
Osmotic shock, 179
Osmotic work, 405
Ouabain, as inhibitor of sodium-potassium ATPase, *1164*, 1164–1165
Ovalbumin mRNA, 918, *919*
Ovary, hormones of, 1049
Oxaloacetate
  from amino acid catabolism, 639
  conversion to pyruvate, 501–502, *502*
  formation from malate, 493
  formation from pyruvate, 497
  malate produced from, 506
  production by citrate cleavage, 505–506
  regulation of, 509
Oxidases
  mixed function, 741
  reactions catalyzed by, 363
Oxidation, 1236
  alpha, 608–610
  beta, 603–608, 1229
  of fatty acids, 603–613
  omega, 610
  of oxaloacetate, 501–502
  photooxidation of chlorophyll, 572–573
Oxidative phosphorylation, 537–554, 1236
Oxidoreductases, 265, 1236
2,3-Oxidosqualene lanosterol cyclase, 732
β-Oxoacid-CoA-transferase, 610
Oxygen
  binding by hemoglobin, 325–336
  electron transport to. *See* Electron transport
  production and utilization of, 24
  valence of, 14
Oxygen-linked oligosaccharides, 662–663
Oxygenase reaction, 590, *591*
Oxyhemoglobin, structure of, *330*
Oxyluciferins, 594, *594*
Oxytocin, 1050
  structure and function of, 1047

Packing between secondary structures of proteins, 83–86
Palindromes, DNA, 241, 1236
Palmitic acid, 154
  structure of, 155
Palmitoleic acid, structure of, 155
Palmitoyl-CoA, inhibiting acetyl-CoA carboxylase, 685
Pancreas, hormones of, 1048
Pancreatic lipase, 704
Pancreatic polypeptide, structure and function of, 1048
Pancreatic RNAse, 303–308
Pantetheine coenzymes, sulfhydryl groups of, 366–369

Panthothenic acid, structure of, *366*
Papain, *47*
  domains in, 85, *86*
Papilloma virus, as vector in DNA cloning, 1114
Parathion, 1205, *1205*
Parathyroid hormone
  structure and function of, 1048
  and vitamin D₃ formation, 1056
Parkinson's disease, 1206–1207
Pathways, biochemical, 25, 25–26
  branchpoints in, 26
Penicillin, 151, *151*
  inhibiting bacterial cell wall synthesis, 664, 667
Pentose, 1236
Pentose cycle, reductive, 588–590, *589*
Pentose phosphate pathway, 473–478, 1236
  and amino acid biosynthesis, *757*
  regulation of, 478
  stage 1 of, *473*
  stage 2 of, *477*
Pepsin, *47*
Peptidases, 616
Peptide(s), 38–40, 1236
  analysis of, 47–49
  chemical synthesis of, 49–51, *49–51*
  coevolution with nucleotides, 1143–1148
  formation on primitive Earth, 1142–1143, 1146–1148
  mapping of, 1236
    tryptic peptide, 1240
Peptide bonds, 9
  structure of, 39, *39*
Peptide growth factors, 1045
Peptidoglycans, 144, 1236
  of bacterial cell wall, 149–151, 180
  synthesis of, 663–667
Perhydrocyclopentanophenanthrene, structure of, *166*
Periplasm, 1236
Periplasmic proteins, 180
Periplasmic space, 180
  constituents of, 182
Permeability barrier in membranes, 196–197
Permeability coefficients, 1194
Permeases, 1155, 1236
Peroxidase, 381–382
  molecular weight and subunit composition of, 88
  and prostaglandin synthesis, 718
Peroxide, in bioluminescent reactions, *594*
Peroxisomes, 4, 1236
  fatty acid oxidation in, 610–611
Pertussis toxin, 1065, *1065*
pH, 34
  and chymotrypsin activity, 297–298
  and enzyme activity, 277, *277*
  isoelectric, 35, 101, 1234
  and oxygen binding by hemoglobin, 327–328
  and redox potential, 522, 525
  and RNAse activity, 306
Phage. *See* Bacteriophage
Phenanthrene, structure of, *166*
Phenotype, 1236
Phenylalanine
  biosynthesis of, 779, *781*
  catabolism of, *626*, 628–629
  in gramicidin synthesis, 800
  hyperphenylalaninemia, 637
  p*K* values for, 36
  structure of, 33
Phenylalanine ammonia-lyase, 629
Phenylalanine-4-monooxygenase, 628
Phenylketonuria, 628, 637, 1236
Pheophytin, 569
  structure of, *568*
Pheromones, 1018, 1045, 1236
Phosphates
  in ATP hydrolysis, 409–410
  uptake by mitochondria, 555–556
Phosphatidic acid, 161
  alkyl ether species of, biosynthesis of, *701*
  in membrane preparations, 185

Phosphatidic acid cytidylytransferase, 690–691
Phosphatidic acid phosphatase, 705
Phosphatidylcholine, 160, 161
  in bile, 170
  biosynthesis of, 696–698
  constituents of, 182–183, *183*
  in membrane preparations, 185
  structure of, *161*
  thin-layer chromatography of, *163*
Phosphatidylethanolamine, 161
  biosynthesis of, 696–698
  in *Escherichia coli*, 690
  in membrane preparations, 185
  structure of, *183*
  thin-layer chromatography of, *163*
Phosphatidylethanolamine-N-methyltransferase, 696
Phosphatidylglycerol, 161
  biosynthesis of, 698–700, *700*
  in *Escherichia coli*, 690
  in membrane preparations, 185
  thin-layer chromatography of, *163*
Phosphatidylglycerol phosphate, dialkyl, from halophilic bacteria, 703, *703*
Phosphatidylglycerol phosphate phosphatase, 691
Phosphatidylglycerol phosphate synthase, 691
Phosphatidylinositol, 161
  biosynthesis of, *700*, 700–701, *702*
  in membrane preparations, 185
  thin-layer chromatography of, *163*
Phosphatidylinositol-4, 5-bisphosphate (PIP₂), 1066, *1067*
Phosphatidylserine, 161
  biosynthesis of, 698–701, *699*
  in membrane preparations, 185
  thin-layer chromatography of, *163*
Phosphatidylserine decarboxylase, 691, 700
Phosphatidylserine synthase, 691
3′-Phosphoadenosine-5′phosphosulfate, formation of, *767*
Phosphocholine, 696
Phosphocreatine, 801–802
Phosphodiester, 1236
Phosphodiesterase, 266, 879–880, 1063
  in rod outer segment, 1221
Phosphoenolpyruvate
  conversion to fructose-1,6-bisphosphate, 457–459
  conversion to pyruvate, 449–452
  formation from pyruvate, 455–457
Phosphoenolpyruvate carboxykinase, 455, 498, 501
Phosphoenolpyruvate carboxylase, regulation of, 509
Phosphoenolpyruvate-dependent sugar phosphotransferase system, 1175
Phosphoenolpyruvate synthase, 457
Phosphofructokinase, 444–445
  molecular weight and subunit composition of, 88
  regulation of, 471
Phosphoglucomutase, 441–443, 449
Phosphogluconate pathway, 1236
Phosphoglucose kinase, 442
3-Phosphoglyceraldehyde dehydrogenase, 447, 450
3-Phospho-D-glycerate. *See* Glycerate-3-phosphate
3-Phosphoglycerate kinase, 447
  domain organization of, 86, *87*
Phosphoglycerides, 160
  fatty acids in, 182–183
  in membranes, 182
Phosphohexoisomerase, 443
Phospholipase, 598, 707–708
Phospholipase A₂, 696, *709*
Phospholipase C, 701, 1066
Phospholipids, 18, *18*, 160–166, 689, 1236
  amphipathic property of, 162–163, 689–690
  biosynthesis of, 689–709
    alkyl and alkenyl ethers in, 701–704
    in *Escherichia coli*, 690–694, *692–693*
      regulation of, 691–694, *694*

in eukaryotes, 694–704
 regulation of, 704–706
degradation by phospholipases, 707–708
distribution between membranes, 706–707
fatty acids of, 183–184
in membranes, 182–185
midtransition temperatures for aqueous
 suspensions of, 200, *200*
plasma concentrations of, 733
Phosphomannoisomerase, 649
Phosphomannomutase, 649
Phosphomevalonate kinase, 729
N-(Phosphonacetyl)-L-aspartate (PALA), 822, 824
 structure of, *337*
Phosphopantetheine coenzymes, 366–369, 673
5'-Phosphorimidazolide, 1136, *1136*
Phosphoribosylanthranilate, 782
Phosphoribosylpyrophosphate (PRPP), 808
Phosphoribosyltransferases, 817–820
Phosphoric acid, 16
 and ester formation, 15
Phosphorus, valence of, 14
Phosphoryl group, in nucleotides, 217
Phosphorylase, 439–440
 polynucleotide, 1237
Phosphorylase *a*, 467
Phosphorylase *b*, 467
Phosphorylase kinase, 468
Phosphorylation, 1236
 of ADP, 423–424
 oxidative, 537–554, 1236
  chemiosmotic theory of, 543–545, 587
  coupling factors in, 551
  inhibitors of, 541
  reversal of, 542–543
  uncouplers of, *540*, 540–542
Phosphoserine-glutamate transaminase, 764
Phosphoserine phosphatase, 764
Phosphotransferase systems, bacterial,
 1175–1177
Photoautotrophs, 417
Photoheterotrophs, 417
Photons, 569
Photophosphorylation, 587
Photoreactivation, 1236
Photorespiration, 590–592
Photosynthesis, 564–590, 1236
 antenna system in, 577–579
 chlorophyll in, 567
 in chloroplasts, 565, *566*
 electron-transfer reactions in, 575–590
 photochemical process in, 571–572, *572*
 in purple bacteria, 567, *567*
 Z scheme for, 581, *581*
Photosystems, 573
 in bacterial reaction centers, 575–579
 in chloroplasts, 579–590
Phototaxis, 1226
Phylloquinone, 171
 structure of, *171*
Physostigmine, *1205*
Phytanic acid, 608
 oxidation of, *609*
 storage syndrome, 608–610
Phytochrome, 592–593, *593*
Phytol, 172, 184, 569, 608
 oxidation of, *609*
 structure of, *173*
Phytosphingosine, 163, 164
Piercidin A
 as electron-transport inhibitor, 528
 structure of, *529*
Pigment, visual, 1213–1225
Pigment-protein complexes, 573–575
Pineal hormones, 1047
Pinene, 172
 structure of, *173*
Ping-Pong mechanism in enzyme kinetics, 280,
 280, 1186
Pipecolic acid, 794
Pitch length of duplex helix, 224, 1235
Pituitary hormones, 1047
 synthesis and secretion of, *1075*

pK values
 for amino acids, 36
 for ribonucleotides, 219
Placenta, hormones of, 1049
Placental lactogen, structure and function of, 1049
Planck's constant, 569
Planet(s), 1122
Planetesmials, 1122
Plants
 cell structure in, 3
 chlorophyll in. *See* Chlorophyll
 chloroplasts in, 565, *566*, 579–586, 1230
 DNA cloning in, 1115
 evolution of, *2*
 fatty acid synthesis in, 673–674
 hormones of, 1082–1085
 lysine biosynthesis in, 769–771, *770*
 photosynthesis in. *See* Photosynthesis
 rhythms in, light affecting, 592–593
 tumors in, 1085, *1085*, 1115
Plaque, 1236
Plasma cells, 1039, *1040*
Plasma membrane, 176, 1236. *See also*
 Membranes
Plasmalogens, 160, 702
 formation of, *703*
 in membranes, 182
Plasmids, 1237
 construction of T1 derivative, 1116
 DNA in, 1116
 eviction of, 1116
 F plasmid, 979
 hybrid or chimeric, 1233
 incompatibility of, 1116
 as vectors in DNA cloning, 1094–1095,
  1098–1099
Plastocyanin, 581
Plastoquinone, structure of, *573*
Platelet activating factor, 703
Platelet-derived growth factor, 1079, 1080
pnp gene, 909
P/O ratio in electron transport, 538–540
polA gene, 857, 884
Polar group, 1237
Polar mutation, 1237
Polarimetry, 113, 1237
polC gene, 884
Polyamine, 1237
Polycistronic mRNA, 894, 1237
Polyketides, 174, *174*
Polymer, 418
Polymerase, 1237
Polymerization of biomonomers on primitive
 Earth, 1134–1148
Polymyxin B, amino acids, in, 793
Polynucleotide, 10, 221, 1237
Polynucleotide kinase, 1091
Polynucleotide ligase, 862–863
 bacteriophage, 1091
Polynucleotide phosphorylase, 912, 1237
Polypeptides, 9, *9*, 38, *39*, 1237
 chemical synthesis of, 49–51
 chiral properties of, 78–83
 cleavage of, 46–47, *47*
Edman degradation for sequence determination,
 47–48, *48*
Polyphenols, 174
Polyphosphates as prebiotic condensing agents,
 1134
Polyprenols, 172
Polyribosome, 1237
Polysaccharides, 8, *8*, 138–144, 1237
 energy storage in, 653–655
 methylated, 675
 production from hexose phosphates, 459–462
 synthesis of, 653–655
Polysomes, 934, *935*, 1237
Pores, transmembrane, 1179, *1179*, 1181–1184
 gap junction, 1182–1183, *1183*, 1203
 gated, 1201–1203
 negatively charged carboxylate group in, 1200
Porins, 1182
 properties of, 191

Porphyrin, 1237
 biosynthesis of, 793–796
 iron complex with, 515
Positive control, 1237
Postsynaptic membrane, 1203
Posttranslational modification, 1237
Potassium
 channels in nerve cell membranes, 1198–1201
 membrane permeability to, 1196–1197
 sodium-potassium ATPase, 1164–1165,
  1185–1186
Potential
 action, 1195–1198
 electromotive, 406
 redox, 521–525, 1238
 resting, 1194–1195
 transmembrane, 406
  in electron transport, 546–548
ppGpp, 996–997
Precipitation, differential, 120
Precursors in hormone synthesis, 1046, 1050
Pregnenolone, 742, 1054
Prephenate, 779
Preprimosome, 865
Preproinsulin, 1046
Presqualene pyrophosphate, 730
Pressure, osmotic, 102–103, 1236
Presynaptic membrane, 1203
Primary structure, 1237
Primer, 1237
Primosome, 866, 1237
Pristanic acid, 608
Prochiral centers, *359*
Prochirality, 494–495, 1237
Progesterone, 169, 742, 1054
 biosynthesis of, *743*
 function of, 1049
Progestins
 and gene regulation, 1070
 structure and function of, 1049
Proinsulin, 1046
Prokaryotes, 4, 1237
 cell envelopes of, 180–182
 cell structure in, 176
 chromosomes in, 11
 DNA replication in, 851–864
 evolution of, *2*
 gene expression in, 974–1010
 photosynthesis in, 565
Prolactin
 inhibiting factor, 1050
  structure and function of, 1047
 releasing factor, structure and function of,
  1047
 structure and function of, 1047
Proline
 biosynthesis of, 761, *761*
 catabolism of, 631, *632*
  in bacteria and fungi, 641
 cotransport with hydrogen, 1174
 in gramicidin synthesis, 800
 hyperprolinemia, 637
 pK values for, 36
 structure of, 33
Proline oxidase, 631
Promoter, 1237
Proofreading hydrolysis reactions, 939
Proopiomelanocortin, 1050, *1050*
Propane, entropy of, 396
S-(Prop-l-enyl) cysteine, 794
Prophages, 1002, 1237
Prophase, 1237
Propionyl-CoA, generation by fatty acid
 oxidation, 607–608
Propionyl-CoA carboxylase, 608
Proprotein, 1237
Prostacyclin, 717
Prostaglandin(s), 715–722, 1237
 analysis of, 716
 biosynthesis of, 719
 catabolism of, 721
 structures of, 716
Prostaglandin endoperoxide synthase, 718

Prostaglandin synthase, 720
Prosthetic groups, 92, 92–93, 347n, 1237
Protamines, 1237
Protease, 615, 616, 969
Protease La, 967–968
Protein, 9, 53–98
　acid hydrolysis of, 40, 40
　allosteric, 325–336
　amino acids in, 9, 32–38
　　covalently modified, 959–960
　in bile, 170
　binding-protein transport systems, 1169–1171,
　　1171
　catabolism of, 615–616
　characterization of, 100–119
　conformation of, 53
　degradation of, 615, 963–971
　　in abnormal proteins, 965–966
　　ATP in, 966–968
　　lysosomes in, 966–967, 967, 969–970
　　rate affecting intracellular levels, 964–965
　　regulation of, 970–971
　　ubiquitin in, 968
　denaturation of, 54–55
　in DNA replication, 856–864
　fibrous, 75–78
　folding pattern for, 53
　　amino acid sequence affecting, 55–56
　　chiral properties affecting, 79–83
　　cooperativity in, 72
　　electrostatic forces in, 62–63
　　forces determining, 56–68
　　in globular proteins, 78–83
　　hydrogen bond forces in, 63–64
　　hydrophobic forces in, 64–65
　　intermediates in, 69, 69
　　van der Waals forces in, 63
　　water properties affecting, 56–58
　functional diversification of, 90–98
　gene regulatory, 986–987
　globular, 75, 78, 1232
　　compared to fibrous proteins, 75, 78
　　structural organization of, 78–83
　in Gram-negative bacteria, 182
　half-lives in cells, 964
　heme-containing, 92–93
　high-mobility group (HMG), 1026–1027
　interaction with nucleic acids, 252–253
　isoelectric point of, 101
　in membranes, 178, 179
　　arrangement of, 191–195
　　asymmetry of, 202
　　mobility of, 197–198
　　in organellar membranes, 180
　　properties and isolation of, 187–191
　　synthesis and insertion of, 207
　mosaics of, 198
　nonheme iron, 520
　overlapping sequences in, 520
　periplasmic, 180
　physical constants of, 106
　pigment-protein complexes, 573–575
　posttranslational modification of, 959–963
　primary structure of, 42, 70
　prosthetic groups in, 92, 92–93
　proteolytic processing of, 961–963
　purification methods, 119–127
　quaternary structure, 70, 71, 87–88
　renaturation of, 55–56
　salting in and salting out of, 101–102
　secondary structure, 70, 71–87
　　alpha helical, 84–85, 85
　　and amino acid sequence, 74–75, 75
　　common elements in, 71–74
　　and fibrous proteins, 75–78
　　packing between, 83–86
　signal hypothesis for, 961–963, 962
　solubility of, 101
　　and differential precipitation, 120–121
　structural, 1239
　structural domains of, 71, 1239
　structure of, 53–98
　　Ramachandran plot of, 67–68, 67–68

visualization of, 60–61
subunits of, 9, 1237, 1239
synthesis of, 928–937
　aminoacyl-tRNA synthases in, 937–939
　elongation reactions in, 941–942
　in heat-shock response, 966
　inhibitors of, 956–959
　initiation reactions in, 939–941
　messenger RNA in, 934–937
　operon hypothesis of, 981–982
　ribosomes as site of, 930–934
　steps in, 937–947
　termination of, 942–947
　　and GTP hydrolysis, 945, 946
　　release factors in, 943, 945, 946
　transfer RNA in, 929–930
　translocation reaction in, 942, 944
　transpeptidation in, 942
tertiary structure, 70, 71
　βαβ loops in, 83
　domains of, 86–87, 86–87
　X-ray diffraction and crystallography of,
　　115–119
transport, 1240
unwinding, 1240
Protein kinase
　activation by cAMP, 468, 468, 1065–1066
　calcium-dependent, 1066
　catalytic subunits of, 1063, 1064, 1065
　cGMP-dependent, 1066
　regulatory subunits of, 1063, 1063, 1065
Protein kinase C, 701
Protein phosphatases, 470
Proteoglycans, 144, 1237
　structure of, 146
Proteoliposomes, 1187–1188
Proteolysis, 615, 963–971. See also Protein,
　degradation of
Protist, 1237
Proton acceptor, 1237
Proton motive force, 551, 1171–1172, 1237
Proton pump, light-driven, 1168
Proton translocation in electron transport,
　548–549, 549
　and ATP formation, 549–551
Proton transport in photosynthesis, 586–587
Proto-oncogene, 1237
Protoporphyrin IX, 92, 515, 515, 568
　structure of, 382
Pseudocycles, 426, 1237
　in fatty acid metabolism, 683
　glycolysis and gluconeogenesis in, 453, 456
Pseudohermaphrodism
　female, 1054
　male, 1056
Pseudouridine in tRNA, 253
D-Psicose, structure of, 134
Puffs, chromosomal, 1070, 1071, 1230
　and genetic functions, 1021–1022
Pulse-chase, 1237
Purification of proteins, 119–128
Purine, 1237
　biosynthesis of
　　in primitive Earth, 1129–1130
　　regulation of, 833, 833–834
　catabolism of, 829–831
　conversion to nucleotides, 818–820
　metabolism of, 807
　in nucleotides, 217
Purine nucleoside phosphorylase, 829
Purine phosphoribosyltransferases, 818–820
Purine ribonucleotides, biosynthesis of, 808–811
Puromycin, 943, 1237
Purple bacteria, photosynthesis in, 567, 567
Purple membrane, 193, 1168–1169, 1225
Putidaredoxin, 383
Putrescine, 634, 634, 797
Pyranoses, 135, 1237
　chair and boat forms of, 136, 136
Pyrazofurin, 822, 825
Pyridoxal, 348, 351
Pyridoxal-5′-phosphate, 351–356, 616
　mechanism of action, 354

Pyridoxamine, 351
Pyridoxine, 351
Pyrimidine, 1237
　biosynthesis of
　　in primitive Earth, 1130–1131
　　regulation of, 834
　catabolism of, 831–832
　conversion to mononucleotides, 820
　metabolism of, 807
　in nucleotides, 217
Pyrimidine ribonucleotides, biosynthesis of,
　811–814
Pyrophosphatase, 460
Pyrophosphate, 16, 1238
　exchange in RNA transcription, 903
5-Pyrophosphomevalonate, 729
Pyrophosphomevalonate decarboxylase, 729, 731
Δ′-Pyrroline-5-carboxylate dehydrogenase, 634
Δ′-Pyrroline-5-carboxylate reductase, 634
Pyruvate
　from amino acid catabolism, 623–625
　conversion to phosphoenolpyruvate, 455–457
　formation from glucose, 418, 419, 438
　formation from oxaloacetate, 501–502, 502
　oxidative decarboxylation of, 485–488, 487
　partitioning between acetyl-CoA and
　　oxaloacetate, 508–509
　reduction to lactate, 521–522
　transport into mitochondria, 556
Pyruvate carboxylase, 455, 498
Pyruvate decarboxylase, 266, 451, 487
Pyruvate dehydrogenase, 488
　molecular weight and subunit composition
　　of, 88
　regulation of, 503
Pyruvate kinase, 449–450
　β-sheet structures in, 81
　domains of, 96, 96
Pyruvate phosphate dikinase, 457

Q cycle in electron transport, 532–535
　in photosynthetic bacteria, 576, 577
Q protein, bacteriophage, 1003, 1007
Quantum requirement in photosynthesis,
　584–585
Quaternary structure, 70, 71, 87–88, 1238

R groups of amino acids, 9, 9, 33, 36, 1328
R loop, 1238
Racemases, in amino acid formation, 792–793
Racemic mixtures, 132
Radiation, for protein separations, 111–118
Radioimmunoassays, 1057
Ramachandran plot of protein structures, 67–68,
　67–68
Ranvier nodes, 1198
Rapid-start complex, 1238
Reaction rates, chemical, 259–262
rec genes, 884, 885, 887
Receptors
　for hormones, 1046, 1058–1073, 1233
　for neurotransmitters, 1203, 1204, 1208–1210
Receptosomes, 1077
Recombination, 1238
　general, 1232
Redox couple, 1238
Redox enzymes, 381–383
Redox potentials, 521–525, 1238
5α-Reductase, deficiency of, 1056
Refsum's disease, 608
Regulatory enzyme, 1238
Regulatory gene, 1238
rel gene, 995–997
Relaxin, structure and function of, 1049
Renaturation, 1238
　of DNA, 238–242, 241
rep gene, 857, 867
Repair synthesis, 1238
Repeat, inverted, 1234
Repetitive DNA, 1238
Replica plating, 1238
Replication
　bidirectional, 1229

rolling circle, 867, 1238
semiconservative, 1239
unidirectional, 1240
Replication fork, 1238
Replicon, 1238
Repolarization, and action potential, 1197
Repressor gene in *lac* operon, 979
Repressor protein, 1238
Reproduction, asexual, 1229
Resins
hydrophobic, 189–190
ion-exchange, 41, 1234
Resonance hybrid, 39, 1238
Respiration, mitochondrial, 514
Respiratory chain, electron transport in. *See* Electron transport
Respiratory distress syndrome, 696
Respiratory metabolism, aerobic, 481–482
Resting potentials, 1194–1195
Restriction enzymes, 880, 881–882, 1092–1094
Restriction-modification system, 1238
Reticulum
endoplasmic, 4, 1231. *See also* Endoplasmic reticulum
sarcoplasmic, 1165–1166
Retina, 1213
11-*cis*-Retinal, 172, 184, 1213
light absorption by, 1169
in rhodopsin, isomerization of, 1216, 1218
structure of, 1215
Retinols
structure of, *1215*
trans-retinol, 171
Reverse transcriptase, 245, 873–874, 877, 910, 911, 1104, 1238
Reverse translation, for selection of clones from DNA library, 1108–1109, *1109*
Rho factor for RNA transcription termination, 904, 1238
Rhodanese
domains in, 86, *86*
molecular weight and subunit composition of, 88
Rhodopsin, 172, 1213, 1214–1218
absorption spectra of, *1216*, 1216–1218
diffusion coefficient for, 1225
movements in disk membrane, 1223–1225
photochemical transformations of, *1217*
retinal in, isomerization of, 1216, 1218
sensory, 1226
Riboflavin, 348, 361
biosynthesis of, 838–839
structure of, *362*
Ribonuclease, 303–308, 923
amino acid content of, 43
molecular weight and subunit composition of, 88
Ribonuclease I, 923
Ribonuclease II, 923
Ribonuclease A, pancreatic, 303–308
amino acid sequence of, *305*
catalytic mechanism in, 307–308
chemical studies on active site, 304–306
crystal structure studies of, 306
derivatives of, 305, *305*
pH affecting, 306
Ribonuclease H, 880, 923
Ribonucleic acid (RNA), 12, 1238. *See also* RNA
Ribonucleoside, structure of, *217*
Ribonucleotide
ionization constants of, 219
purine, biosynthesis of, 808–811
pyrimidine, biosynthesis of, 811–814
reduction, regulation of, 834–835
structure of, *217*
Ribonucleotide reductase, 380, 814–815, *815*
inhibitors of, 826
Ribose, 1238
formation from formaldehyde, 1131
in RNA, 217
structure of, *133*
Ribose-5-phosphate
in transketolase reaction, *475*

production of, 476–478
Ribose-5-phosphate pyrophosphokinase, 808
Ribosomal RNA, 12, 894, 1238. *See also* RNA, ribosomal
Ribosomes, 12, 1238
genes for, regulation of, 995–999
in protein synthesis, 930–934
structure and assembly of, 252, 932–934
Ribothymidine, in tRNA, *253*
Ribozyme, 921
D-Ribulose, structure of, *134*
Ribulose bisphosphate carboxylase, 588, 590
Ribulose-5-phosphate, *473*, 473–474
Ribulose phosphate epimerase, 266
Rickets, 1056, 1078
Rieske iron-sulfur protein, 532
RNA, 12, 1238
classes of, 893–895
conformational behavior of, 249–252
degradation by ribonucleases, 923–924
DNA-dependent synthesis of, 895–905
evolution of, 1141–1142
hybridization of, 239
messenger, 12, 893–894, 1235
cap structure of, 916–918, *917*
eukaryotic, *936*
leader region of, 934
modification and processing of, 913, 916–921
ovalbumin, 918, *919*
polyadenylation of 3' end, 918
polycistronic, 894, 1237
prokaryotic, *936*
in protein synthesis, 934–937
reading frames for, 934, *936*
removal of intervening sequences in, 918–921
Shine-Dalgarno sequence in, 936, 937
splicing of, 920
abnormal, 920
structure of, *936*
trailer region of, 934
nascent, 1235
posttranscriptional modification and processing of, 913–923
enzymes involved in, 914
ribosomal, 12, 894, 1238
control of synthesis by *rel* gene, 995–997
DNA amplification affecting, 1029–1030
modification and processing of, 913, 916, *916*
in protein synthesis, 930–934
structure of, 252
type 5S, 916, 930
synthesis in frogs, 1030–1033
type 16S, 916, 930
type 23S, 916, 930
self-splicing of, 921–923
small nuclear, 895
splicing of, 1238
synthesis in cells, 25
enzymes involved in, 908–913
inhibitors of, 924–925
in tobacco mosaic virus, 215, 215–216
transcription cycle for, 899–905
attenuators of, 905
binding at promotors in, 901–903
closed promoter complexes in, 901, *901*
elongation of chains in, 904
in eukaryotes, 905–908
compared to prokaryotes, 908
initiation at promoters, 903–904
abortive, 903
negative control of, 905
open promoter complex in, *901*, 902
positive control of, 905
in prokaryotes, 895–905
compared to eukaryotes, 908
promoter signals in, 900–901
regulation of, 904–905
termination of, 904
transfer, 12, 894–895, 1239
acceptor stem in, 929

control of synthesis by *rel* gene, 995–997
modification and processing of, 913–915
nucleotides added to, posttranscription, 910
in protein synthesis, 929–930
removal of intervening sequences in yeast, 921, *921*
structure of, 249–252, *929*
viral
reverse transcriptase in, 873–874
RNA-dependent RNA polymerases of, 910
RNA ligase, of bacteriophage T4, 912–913
RNA polymerase, 895, 896–899, 1238
assay of activity, 896
core polymerase component, 898
eukaryotic, 906–907
in organelles, 907–908
genes for, regulation of, 995–999
genetics of subunits, 898–899
holoenzyme, 897
inhibitors of, 924–925
initiation factor in, 898
sigma-70 component of, 898, 905
subunit structure and function of, 896–898, *897*
temperature-sensitive mutants, 898
in viruses, 909–910
RNA polymerase I, eukaryotic, 906
RNA polymerase II, eukaryotic, 906
RNA polymerase III, eukaryotic, 906
RNAse. *See* Ribonuclease
Rocks, sedimentary, 1126
Rod and cone cells, 1213–1214
response to light, 1220–1221
Rolling circle replication, 867, 1238
Rotation
dextrorotatory, 132
levorotatory, 132
Rotenone
as electron-transport inhibitor, 528
structure of, *529*
rpo genes, 909, 998–999
RPr protein, 1178
rpsA gene, 909
rrn operon, 996
RS system for naming of compounds, 159

S phase of cell cycle, 1238. *See also* Cell cycle
Saccharomyces cerevisiae, gene expression in, 1011–1018
Saccharopine, 764
Saccharopine dehydrogenase, 764
Salicylate hydroxylase, reactions catalyzed by, 363
Saltatory conduction, 1198
Salting in or salting out, 1238
and protein solubility, 101–102
Salvage pathways, 751, 752, 817–821, 1238
Sandhoff's disease, 714
Sanger method for DNA sequencing, 877
Santonin, 172
structure of, *173*
α-Sarcin, 969
Sarcomere, 89, *89*
Sarcoplasmic reticulum, 1165–1166
Satellite DNA, 237, 242, 1238
Saturation, 1157
Saxitoxin, 1198–1199, *1200*
Schiff base
deprotonation of, 1169
protonated
in bacteriorhodopsin, 1225
in rhodopsin, 1217
and pyridoxal phosphate activity, 353
Schizophrenia, 1207
Second messenger, intracellular, 1062, 1238
Secondary structure, 70, 71–87, 1238
Secretin, structure and function of, 1048
Sedimentation constant or coefficient, 104–107
Sedoheptulose-7-phosphate, in transaldolase reaction, 474
Segmentation genes, 1035–1037
Selection process, in polymerization of biomonomers, 1135

Self-duplication, and definition of life, 1121
Self-splicing of RNA, 921–923
Semiconservative replication, 1239
Sequenators for polypeptide sequence determination, 48
Serine
  biosynthesis of, *764*, 764–765
  catabolism of, 624, *624*
    in bacteria and fungi, 641
  and chymotrypsin-catalyzed hydrolysis, *293*
  ionization reactions for side chains of, *37*
  linkage with N-acetylgalactosamine, *147*
  pK values for, 36
  structure of, 33
Serine deaminase, regulation of, 640
Serine dehydratase, 624
Serine hydroxymethyltransferase, 265
Serine transacetylase, 766
Serotonin, 797
  biosynthesis of, *1207*
  as neurotransmitter, 1207
Sesquiterpenes, 172, *173*
Shine-Dalgarno sequence of mRNA, 936, 937
Shock
  heat-shock genes, 966
  osmotic, 179
Shock waves as energy source, 1127
Shuttle vectors in DNA cloning, 1101, 1114
Sialic acid, 164
  structure of, *139, 165*
Sickle cell anemia, 335
Sigma factor, 898, 905, 1239
Signal hypothesis for protein assembly, 961–963, *962*
Silk, structure of, *76, 77*
Single-copy DNA, 1239
Single-strand DNA-binding protein, 864
SIR genes, 1017
Skunk cabbage, thermogenesis in, 555
Slow-reacting substance, 720
Sodium
  channels in nerve cell membranes, 1198–1201, *1201*
    gating currents in, 1201–1203
  channels in rod cells, 1220
  cotransport with substrates, 1174
  membrane permeability to, 1196–1197
Sodium dodecyl sulfate (SDS), 188, *188*
  electrophoresis, 109–111, *110–111*
  properties of, 189
Sodium-potassium ATPase, 1164–1165, 1185–1186
Solar system formation, 1122–1123
Solvation, and entropy, 397–398
Somatic cell, 1239
Somatic mutations, 1042
Somatostatin, 1050
  structure and function of, 1047, *1048*
Sonication of spheroplasts, 180, *181*
D-Sorbose, structure of, *134*
SOS system in DNA repair, 885
Southern blotting, 239, 1239
Specificity constant, enzyme, 271
Specificity pockets of trypsin family of enzymes, 291–292, *292*
Spectrophotometry of electron carriers, 526–528
Spectroscopy
  electron paramagnetic resonance, 198, *198*
  ultraviolet absorption, 111–113, *112–113*
Spermidine, 634, *634*, 797
Spermine, 634, *634*, 797
Spheroidene, structure of, *579*
Spheroplasts, 180, *181*
Sphinganine
  biosynthesis of, *709*, 709
  catabolism of, *712*
Sphinganine kinase, 712
Sphinganine-1-phosphate, 712
Sphinganine phosphate lyase, 712
Sphingolipid(s), 154, 163–165, 689
  biosynthesis of, 709–715
  catabolism of, defects in, 712–714
  long-chain bases of, 163, 164

in membranes, 182, 185
Sphingolipidoses, 712–714
Sphingomyelin, 164
  biosynthesis of, 709, *711*
  in membranes, 182, 185
  structure of, *164*
  thin-layer chromatography of, *163*
Sphingomyelinase deficiency, 714
Sphingosine, 163, 164
  biosynthesis of, 709, *710*
Spin labels, nitroxide groups in, 198
Spliceosome, 920
Sporulation, 1239
Squalene, 729
  biosynthesis of, 730, 731, *732*
  cyclization of, 726
  transformation to lanosterol, *732*
Squalene monooxygenase, 730
Squalene-2,3-oxide, 730
Squid giant nerve axon, 1194
ssb gene, 857
Stacking energy, 1239
Starch, *141*, 141–142
Starch synthase, 460
Steady-state condition, 268, 416, 1239
  active, 433
  passive, 433
Stearic acid, 154
  conformational model of, *157*
  structure of, 155
Stearoyl-CoA desaturase, 682
Stem cells, 1039, *1040*, 1239
Stereoisomers, 1239
Steroids, 169, 725, 1239
  biosynthesis of, 168, 1053–1057, *1055*
  cardiotonic, 1164
  function of, 1049
  inactivated, conjugated to glucuronic acid, 745, *745*
  in membranes, 184
  metabolism of, 741–745
  receptors for, and gene transcription rate, 1068–1073
Stoichiometry, evolved coupling, 423
Streptomycin
  inhibiting protein synthesis, 957
  structure of, *956*
Stromatolites, 1126, *1126*
Structural domain, 1239
Structural gene, 1239
Structural protein, 1239
Substrates, enzyme, 21, 264
  and multisubstrate enzymes, 278–280
Subtilisin, 95, 96, 304
Subunit, 1239
Succinate
  formation from succinyl-CoA, 490–492
  oxidation to fumarate, 492
  transport into mitochondria, 556
Succinate dehydrogenase, 492
  in electron transport, 531–532
  reactions catalyzed by, 363
Succinate thiokinase, 490–492
Succinyl-CoA, 490
  from amino acid catabolism, 635–639
  conversion to succinate, 490–492
  formation from propionyl-CoA, 608, *608*
  formation from pyruvate, 497
  and porphyrin biosynthesis, 793–796
Sucrose
  catabolism of, 462–463
  structure of, *138*
Sugars. See Carbohydrates
3′-Sulfate-galactosylceramide, 165
Sulfhydryl groups of pantetheine coenzymes, 366–369
Sulfhydrylation, in cysteine biosynthesis, 765–767
Sulfite reductase, 767
Sulfonamide, 827, 828
Sulfur
  amino acids containing, 794
  valence of, 14

Sun, formation of, 1122–1123
Supercoiled DNA, *232*, 232–236, 1239
Suppressor gene, 1239
Suppressor mutation, 1239
Surfactant, pulmonary, 696
SV40 virus
  DNA replication in, 872
  transcription in, 1027–1028, *1028*
Svedberg units, 894, 1239
  for sedimentation constant, 105
Symmetry
  dyad, 1231
  icosahedral, 1233
Symport, 1172
  sodium, *1174*
Synapse, 1203–1205, 1239
  to rods or cones, 1213
Synapsis, 1239
Synaptic cleft, 1203
Synaptic vesicles, 1203
Synaptosome membrane, lipid composition of, 185
Synthetic work, 405

T cells, 1039–1040, *1040*
  helper, 1040
D-Tagatose, structure of, *134*
D-Talose, structure of, *133*
Tandem duplication, 1239
Tau factor for RNA transcription termination, 904
Taurine, 168, 625
  structure of, *170*
Taurocholate, 168
  structure of, *170*
Tay-Sachs disease, 714
Tay-Sachs ganglioside, 165
  structure of, *166*
TCA cycle. See Tricarboxylic acid cycle
Teichoic acid in bacterial cell envelope, 180, *181*
Temperature
  affecting DNA, 236–237, *237*, 238–239
  annealing, 238–239
  DNA replication mutants and, 858
  enzyme activity and, 277, *277*
  hydrogen translocation by bacteriorhodopsin and, 1185
  membrane function and, 199–201
  nucleotide formation and, 1140–1141
  peptide formation and, 1142
  protein denaturation and, 966
  RNA polymerase subunit mutations and, 898
  units of, 395
Template, 1239
  in DNA synthesis, *859*
  in polymerization of nucleotides, 1138–1140
Terminal deoxynucleotidyl transferase, 874, 1092
Termination factors, 1239
Terpenes, 172–173, 1239
  in membranes, 184
Terramycin, 174
  structure of, *174*
Tertiary structure, *70, 71*, 1239
Testes, hormones of, 1049
Testosterone, 169, 744, 1054–1056
  biosynthesis of, *1055*
  conversion to 5α-dihydrotestosterone, *1056*
  conversion to estradiol, 745
  function of, 1049
  metabolic conversion by target cells, *1057*
Tetracyclines, 174
  inhibiting protein synthesis, 957
  structure of, *956*
Tetraethylammonium ion, 1199–1200, *1200*
Tetrahydrobiopterin, 628, *628*
Tetrahydrofolate, 373–377
  structure of, *374*
Tetrahydromethanopterin, structure of, *384*
Tetramer, 1239
Tetrapyrrole biosynthesis, *795*
Tetraterpenes, 172, *173*

Tetrodotoxin, 1198–1199, *1200*
TFIIIA protein, 1031–1033
β-Thalassemia, 920
Thermodynamics, 392–414
  constants used in calculations of, 395
  and enthalpy, 395–396
  and entropy, 396–400
  first law of, 394
  and free energy, 400–408
  nonequilibrium or irreversible, 412
  in open systems, 411–412
  and reaction rates, 262
  and regulation of metabolic pathways,
    432–433
  second law of, 396
  in transmembrane transport, 1156–1160
Thermogenesis in brown fat and skunk cabbage,
  555
Thermogenin, 555
Theta structure in DNA replication, 852
Thiamine, 348
Thiamine pyrophosphate, 348–351
  mechanisms, of action, *350*
  structure of, *348*
  and transketolase reaction, 475–476
2-Thiocytidine, in tRNA, *253*
Thioester, 1239
Thioesterase, 679
Thiogalactoside transacetylase, 977
Thiokinase, 602
Thiol esters, 15
  enolization of, 367
Thiolase, 610
Thiomethyl-β-D-galactopyranoside, 1172, *1172*
Thioredoxin, 380, 815, *815*
4-Thiouridine, in tRNA, *253*
thr operon, 995
Threonine
  biosynthesis of, 772, *773*
  catabolism of, 623–624, *624*
  ionization reactions for side chains of, *37*
  pK values for, 36
  structure of, *33*
Threonine deaminase, regulation of, 640
Threonine operon, 995
D-Threose, structure of, *133*
Thromboxane
  analysis of, 716
  biosynthesis of, *719*
  structure of, *717*
Thromboxane A₂ synthase, 718
Thylakoid membrane, 565
Thymidine, 1239
  base found in, 219
Thymidine kinase, 820
Thymidine kinase gene of herpes simplex
  virus, 1111
Thymidine monophosphate. *See* Thymidylate
Thymidine triphosphate (dTTP), 816
Thymidylate
  base found in, 219
  biosynthesis of, 816–817, *817*
Thymidylate synthase, 375–376, 816
  inhibitors of, 825
  mechanism of action, *376*
Thymine, 1239
  catabolism of, 831–832
  in DNA, 218, 223
  hydrogen bonding to adenine, *223*
Thymine phosphorylase, 820
Thyroglobulin, 1051
Thyroid hormones, 1048
  and gene regulation, 1070, 1073
Thyroid-stimulating hormone. *See* Thyrotropin
Thyrotropin, 1050
  releasing factor
    biosynthesis of, 1050, *1051*
    structure and function of, 1047
  structure and function of, 1047
Thyroxine (T₄), 1051
  biosynthesis of, *1052*
  structure and function of, 1048
Tiazofurin, 822, 823

Titration curves of amino acids, 35–36, *35*, *36*
tk gene, 1111
dTMP. *See* Thymidylate
Tobacco mosaic virus
  molecular weight and subunit composition
    of, 88
  physical constants of, 106
  RNA in, *215*, 215–216
Tobacco necrosis virus, satellite, molecular
  weight and subunit composition of, 88
α-Tocopherol, 171
  structure of, *171*
Tollen's reagent, 139
Topoisomerases, 235, 1239
  in DNA replication, 863, 864, *863*, *864*
Totipotency of cells, 1018
trans orientation of genetic elements, 979
Transaldolase, 474
Transaminases, 616, 768
Transamination, 616, *617*, 1239
Transcarboxylase, 373
Transcriptase, reverse, 245, 873–874, 877, 910,
  911, 1104, 1238
Transcription, 1239
Transducin, 1221
Transduction, 1239
Transfection, 215, 1239
Transfer RNA, 12, 894–895, 1239. *See also*
  RNA, transfer
Transferases, 265, 1239
Transformation, 1239
  bacterial, DNA role in, 211–213, *212*
Transgenic organisms, 1239
  mice, 1075–1076, *1076*, 1111
Transition state, 1239
Transketolase, 474–476
Translation, 1240
  origin of, 1145–1146
Translocation, group, 1160, 1175–1177
Transport across membranes, 1155–1192, 1235
  active, 1159–1160
    primary, 1159–1160, 1164–1169
    secondary, 1160, 1171–1174
  analogs in studies of, 1161–1162
  antiport in, 1163
  ATP/ADP exchanger in, 1174–1175, 1186
  binding-protein systems in, 1169–1171, *1171*
  calcium ATPase in, 1166, *1166*, 1186
  carrier-mediated, 1156–1158
  chemiosmotic coupling in, 1172–1174
  cotransport or symport mechanisms in, 1172
  electrogenic, 1173
  electron transport in, 1168
  energy for, 1158–1159
  energy-coupling mechanisms in, 1160–1178
  energy-independent, 1159
  energy-interconversion pathways in, 1177–
    1178
  F₁/F₀ ATPases in, 1167–1168
  and group translocation, 1160, 1175–1177
  hydrogen ATPase in, 1166
  inhibitors of, 1164–1165
  initial rate of, 1157
  ion-translocating antibiotics in, 1179–1181
  isotope studies of, 1160–1161
  light as energy in, 1168–1169
  metabolism of substrates in, 1161–1162
  mobile-carrier mechanism in, 1179, *1179*
  molecular mechanisms in, 1179–1187
    examples of, 1184–1187
  phosphotransferase systems in, 1175–1177
  pores in, 1179, *1179*, 1181–1184
  and reconstitution of transport proteins,
    1187–1190
  sodium-potassium ATPase in, 1164–1165,
    1185–1186
  theory and thermodynamics of, 1156–1160
  vesicles in, *1162*, 1162–1163
Transport protein, 1240
Transposons, 1018
Triacylglycerol, 154, 158–159, 689
  biosynthesis of, 704, *704*
  conformational model of, *159*

plasma concentrations of, 733
  transport of, 736–737
Triacylglycerol lipase, 600
Tricarboxylic acid, ATP stoichiometry of, 495–
  496
Tricarboxylic acid cycle, 422, 481–511, *485*,
  486, 1240
  and amino acid biosynthesis, *757*
  amphibolic nature of, 496–498
  branchpoint pathways in, *499*
  compared to glyoxylate cycle, *500*
  discovery of, 482–484
  linked to urea cycle, 621, *621*
  organization of reactions in, 496
  oxidation of oxaloacetate in, 501–502
  regulation of, 502–509, *504*
  steps in, 484–493
  stereochemical aspects of reactions in,
    493–495
Triglyceride, 158
Trihexosylceramide, 165
Triiodothyronine (T₃), 1051
  biosynthesis of, *1052*
  structure and function of, 1048
Trimetaphosphate, as prebiotic condensing
  agent, 1134
Trimethoprim, 827, 828
2,4,6-Trinitrobenzene sulfonic acid, 204, *204*
Triose phosphate, 473
Triose phosphate dehydrogenase, 447
  nicotinamide side-specificities of, 361
Triose phosphate isomerase, 446–447
  β-sheet structures in, 80, *81*
Triphosphates, nucleoside, 221, *221*
Triple helix in adenosine derivatives, 1138
Triplets
  carotenoid, 580
  nucleotide, 949–952
Triterpenes, 172, *173*
Triton X-100, 188, *188*
trp operon, 990–991
  attenuation of, 991–995
  leader region in, *992*
  promoter-operator region in, *991*
  repressor control of, 991, *991*
Trypsin, 47, 287–288, 1240
  amino acid sequence of, *289*
  catalytic mechanism in, 288
  inhibition of, *276*, 276–277
  specificity pocket of, 291, *292*
  structure of, *290*
Trypsinogen, activation of, *290*
Tryptic peptide mapping, 1240
Tryptophan
  biosynthesis of, 418, *418*, 780–784, *782*
  catabolism of, *630*, 631
    in bacteria and fungi, 641
  conversion to serotonin, 797
  pK values for, 36
  structure of, 33
Tryptophan operon, *990*, 990–995. *See also trp*
  operon.
Tryptophan oxygenase, 631
  regulation of, 640
Tryptophan synthase, 782
tsf gene, 909
dTTP, 816
D-Tubocurarine, 1205, *1205*
Tubulin, 201
tuf gene, 909
Tumor formation
  and hormone overproduction, 1078
  in plants, 1085, *1085*, 1115
Tumorigenesis, 1240
Turnover, metabolic, 1235
Turnover number, 271, 1240
Tyrocidine, amino acids in, 793
Tyrosinase, 626–627
Tyrosine
  biosynthesis of, 779, *781*
  catabolism of, 625–628, *626*
  in catecholamine biosynthesis, 797, 1053,
    1053, 1206, *1206*

Tyrosine (*Continued*)
formation from phenylalanine, 628
ionization reactions for side chains of, 37
pK values for, 36
structure of, 33
Tyrosine-glutamate transaminase, 626
regulation of, 640
Tyrosine hydroxylase, 628, 1053, 1206
Tyrosyl-tRNA synthase, 266

Ubiquinone
in electron transfer, 520–521
structure of, 520, 573
Ubiquinone oxidoreductase
complex with NADH, 530–531
complex with succinate, 531–532
Ubiquitin
histone linkage with, 1026, 1026
in protein degradation, 968
UDP. *See* Uridine diphosphate
Ultracentrifuge, 1240
Ultraviolet radiation, 1240
absorption spectroscopy, 111–113, 112–113
as energy source, 1122
and vitamin D$_3$ formation, 1056
UMP. *See* Uridine monophosphate
Undecaprenol, 172, 184, 184
Unidirectional replication, 1240
Unwinding proteins, 1240
Uracil
catabolism of, 831–832
formation of, 1130
in RNA, 218
Urea, 618
formation in liver, 619–621
structure of, 618
Urea cycle, 619, 619–621, 1240
and arginine biosynthesis, 761
linked to TCA cycle, 621, 621
regulation of enzymes in, 640
Urease, physical constants of, 106
Uric acid, 618
biosynthesis of, 830, 830–831
degradation of, 831
structure of, 618
Uridine
base found in, 219
in tRNA, 253
Uridine diphosphate galactose-4-epimerase, 359, 360
Uridine diphosphate glucose, 710
Uridine kinase, 820
Uridine monophosphate (UMP)
biosynthesis of, 812–814, 813
ionization constants of, 219
molar absorbance of, 220
Uridine 5'-monophosphate synthase, purification of, 123–126, 124, 125
Uridine phosphorylase, 820
Uridine triphosphate (UTP), and cytidine triphosphate synthase formation, 814, 814
Uridylate
base found in, 219
oligomers of, 1139
Urocanase, 635
Uroporphyrinogen I synthase, 796
Uroporphyrinogen III, biosynthesis of, 795
Uroporphyrinogen III cosynthase, 796

uvr genes, 884

Vaccenic acid, 156
structure of, 155
Vacuoles, autophagic, 970
Valence forces, secondary, 17–18
Valence states, 14
Valine
biosynthesis of, 773, 776–779, 777
catabolism of, 635, 635–637, 636
in gramicidin synthesis, 800
pK values for, 36
structure of, 33
Valine operon, 995
Valinomycin
amino acids in, 793
ion translocation by, 1179, 1180, 1181
structure of, 547, 1180
Vancomycin, inhibiting bacterial cell wall synthesis, 665, 667
Van der Waals forces, 63, 1240
attractive, 63
repulsive, 63
Van der Waals separation, 63
Van't Hoff's law, 103
Vaporization, entropy in, 397
Vasoactive intestinal peptide, structure and function of, 1048
Vasopressin, 1050
structure and function of, 1047
Vectorial character of membranes, 548
Venus, atmosphere of, 1125
Vesicles
derived from bacterial membranes, 180–181, 181
in membrane transport, 1162, 1162–1163
synaptic, 1203
Vesicular bilayers of membranes, 186, 187
Viroids, 216, 1240
Viruses, 1240
bacterial. *See* Bacteriophage
DNA in, 217
transcription of viral DNA, 909
genetic information in, 213–216
lysogenic, 1234
lytic, 1002
RNA
reverse transcriptase in, 873–874
RNA-dependent RNA polymerases of, 910
SV40
DNA replication in, 872
transcription in, 1027–1028, 1028
temperate, 1002
Vision, 1213–1227
Vitamin, 1240
lipid-soluble, 170–171
water-soluble, 348
Vitamin A, 171–172, 1214
structure of, 171
Vitamin B$_1$, 348
Vitamin B$_2$, 348, 361
Vitamin B$_6$, 348, 351
Vitamin B$_{12}$, 348, 377–381
Vitamin D$_3$, 171, 1056–1057
Vitamin E, 171
structure of, 171
Vitamin K$_1$, 171
structure of, 171

Vitamin K$_2$, 171
structure of, 171, 573
Voltage clamp studies, 1198–1200

Water
in atmosphere of primitive Earth, 1125
dipolar properties of, 57, 57
interaction with other molecules, 6
interaction with polypeptides, 56–58
structure of, 6
Watson-Crick base pairs, 1240
Watson-Crick model of DNA replication, 847
Wheat germ agglutinin, 656
Wild-type gene, 1240
Wobble rules of codon-anticodon pairing, 953, 954, 1240
Wolman's disease, 739
Work
electrical, 405
equivalent to heat, 394
mechanical, 405
osmotic, 405
synthetic, 405
useful, energy available for, 404–405

X chromosomes, 1022
Xanthine-guanine phosphoribosyl transferase, 1111
Xanthine oxidase, 265, 830
Xeroderma pigmentosum, 886
xgprt gene, 1112
X-ray diffraction and crystallography, 54, 115–119, 116–118, 1240
of DNA, 224
D-Xylose, structure of, 133
D-Xylulose, structure of, 134
Xylulose-5-phosphate, in transketolase reaction, 475

y gene, 977
Yeast
in alcohol fermentation, 452
DNA cloning in, 1110
DNA replication in, 872
gene expression, 1011–1018
catabolite repression and inactivation in, 1014–1015
controlling elements in, 1015–1016
mating type determination in, 1016–1018
turned off, 1018
turned on, 1018
upstream activation sites in, 1015
mitochondrial genetics in, 955
tRNA intervening sequences in, removal of, 921, 921
Ylid, 349n, 1240

Z form of DNA, 228, 230, 1240
z gene, 977, 979
Z line of sarcomere, 89, 89
Z scheme for photosynthesis, 581, 581
Zearelenone, structure of, 1060
Zinc metalloenzyme carboxypeptidase A, 298–303
Zwitterions, 17, 32, 1240
Zygote, 1240
Zymogen, 1240
Zymosterol, 732